Other Books by the Same Author

1. Electrical Measurements and Measuring Instruments
2. Electronic Measurements and Instrumentation
3. Questions and Answers in "Electrical Engineering"
4. Questions and Answers in "Electronics and Communication Engineering"
5. Mechatronics

PREFACE TO THE FOURTH EDITION

I am pleased to present the "**Fourth Edition**" of this book on "**Electrical and Electronic Measurements and Instrumentation**". The warm reception which the previous editions have received is a matter of much satisfaction to me.

In this edition, the book has been thoroughly *revised and updated by adding* ***new topics*** wherever necessary, as per the requirements of syllabi of various *Indian Universities*. In addition to this, *"PART-III"* with the following *'Contents'* has been added to make this treatise a comprehensive book on this subject :

PART-III : GATE AND UPSC EXAMINATIONS' QUESTIONS :

(Latest-selected with Answers/Solutions)

A. Conventional Questions with Answers/Solutions.

B. Multiple-choice Questions with Answers and "Explanations".

I'm thankful to the Management Team and the Editorial Department of S. Chand & Company Pvt. Ltd., New Delhi for all help and support in the publication of this book.

Er. R.K. RAJPUT
(Author)

ELECTRICAL AND ELECTRONIC MEASUREMENTS AND INSTRUMENTATION

For Engineering Students of "B.E./B.Tech.", AMIE-Section B (India) and "Competitive Examinations"

Er. R.K. RAJPUT

M.E. (Hons.) Gold Medallist; Grad.-Mech. Engg. & Elect. Engg.; MIE (India); MSESI; MISTE; CE (India)

Recipient of :

"Best Teacher (Academic) Award"

"Distinguished Author Award"

"Jawahar Lal Nehru Memorial Gold Medal" for an Outstanding Research Paper (Institution of Engineers—India)

Principal (Formerly)

- Thapar Polytechnic College;
- Punjab College of Information Technology;

PATIALA

S. CHAND
PUBLISHING

S Chand And Company Limited

(ISO 9001 Certified Company)

S Chand And Company Limited

(ISO 9001 Certified Company)

Head Office: D-92, Sector–2, Noida – 201301, U.P. (India), Ph. 91-120-4682700

Registered Office: A-27, 2nd Floor, Mohan Co-operative Industrial Estate, New Delhi – 110 044, Phone: 011-49731800

www.**schandpublishing.com;** e-mail: **info@schandpublishing.com**

Marketing Offices:

Chennai	:	Ph: 23632120; chennai@schandpublishing.com
Guwahati	:	Ph: 2738811, 2735640; guwahati@schandpublishing.com
Hyderabad	:	Ph: 40186018; hyderabad@schandpublishing.com
Jalandhar	:	Ph: 4645630; jalandhar@schandpublishing.com
Kolkata	:	Ph: 23357458, 23353914; kolkata@schandpublishing.com
Lucknow	:	Ph: 4003633; lucknow@schandpublishing.com
Mumbai	:	Ph: 25000297; mumbai@schandpublishing.com
Patna	:	Ph: 4011400; patna@schandpublishing.com

First Edition 2008

Subsequent Editions and Reprints 2010, 2011, 2012, 2013, 2015, 2018 (Twice), 2019, 2020, 2022 (Twice), 2023, 2025

Reprint 2026

ISBN : 978-93-856-7601-7 **Product Code :** H3EEI52ETRN10ENAD0XO

PRINTED IN INDIA

By Shri Shyamlal Printing Press Private Limited, Plot 20/4, Site-IV, Industrial Area Sahibabad, Ghaziabad – 201 010 and Published by S Chand And Company Limited, A-27, 2nd Floor, Mohan Co-operative Industrial Estate, New Delhi – 110 044.

CONTENTS

PART–I

ELECTRICAL AND ELECTRONIC MEASUREMENTS AND MEASURING INSTRUMENTS

PART–II

ELECTRICAL AND ELECTRONIC INSTRUMENTATION

PART – I

ELECTRICAL AND ELECTRONIC MEASUREMENTS AND MEASURING INSTRUMENTS

1. Units, Dimensions and Standards
2. Basic and Digital Electronics
3. Concepts of Measurements and Measurement Systems
4. Analog Electromechanical Instruments
5. Instrument Transformers
6. Electronic Instruments
7. Cathode Ray Oscilloscope (CRO)
8. Digital Instruments
9. Potentiometers
10. Measurement of Resistance, Inductance and Capacitance
11. Magnetic Measurements
12. High Voltage Measurements
13. Illumination
14. Signal Generators
15. Signal Analysis

CHAPTER

1

Units, Dimensions and Standards

1.1. Units – Definition of unit – Absolute units – Fundamental and derived units. **1.2. Dimensions; 1.3. Systems of electric units; 1.4. S.I. units** – Introduction to S.I. units – Definitions – Salient features of S.I. units – Conversion factors – Dimensions of quantities. **1.5. Dimensions in electrostatic and electromagnetic systems; 1.6. Determination of absolute units** – Absolute measurement of current – Absolute measurement of resistance. **1.7. Standards of measurement; 1.8. Standard of mass and length; 1.9. Atomic frequency and time standards; 1.10. E.m.f. standards** – Primary standard of e.m.f. – Saturated weston cell – Secondary standard of e.m.f. – Unsaturated weston cell – Zener diode reference standard – Josephon–Effect standard. **1.11. Resistance standards; 1.12. Inductance standards** – Standard and sub-standard of mutual inductance (M) – Standard and sub-standard of self inductance (L). **1.13. Capacitance standards** – Primary standard of capacitance – Secondary capacitance standard – *Highlights – Objective Type Questions – Theoretical Questions – Unsolved Examples.*

1.1 UNITS

1.1.1. Definition of Unit

The standard measure of each kind of physical quantity is called a **unit.**

Measurement implies *comparison with a standard value*. The magnitude of a physical quantity can be written as :

Magnitude of a physical quantity = (Numerical ratio) × (Unit)

The ***numerical ratio*** (or *numerical multiplier*) is the number of times the unit occurs in any given amount of the same quantity and, therefore, is called the *number of measures*. For example, when it is said that length is 8 metres, it indicates that the metre is the unit of length and that the number of units of length (*i.e.,* numerical ratio) is 8. So the physical quantity, length is defined by the unit metre. The number of measures has no physical meaning without unit.

1.1.2. Absolute Units

A system in which the various units are all expressed in terms of a small number of fundamental units is called an **absolute system of units.**

- The word "absolute" does not imply any special standard of accuracy; it has been associated with relating units to units of *length, mass* and *time*.
- Absolute measurements do not compare the measured quantity with arbitrary units of the same type but are made in terms of *fundamental units*.

1.1.3. Fundamental and Derived Units

In science and engineering, two kinds of units are used : *fundamental units* and *derived units*.

- Experience has shown that *three* basic ideas or units are adequate to describe quantitatively all the phenomena encountered in mechanical science. These are *"length"*, *"mass"* and *"time"* and are called ***fundamental units***. Since these units are fundamental to most other physical quantities besides those in mechanics, they are called the *"primary fundamental units"*.

Measures of certain physical quantities in the thermal, electrical, and illumination disciplines are also represented by fundamental units. These units are used only when these particular disciplines are involved and they may therefore be called as *"auxiliary fundamental units"*.

- *All other units which can be expressed in terms of the fundamental units are called **derived units**.*

— Every derived unit *originates from some physical law or equation defining that unit.*

— A derived unit is always recognised by its *dimension*, which can be defined as the complete algebraic formula for the derived unit.

1.2 DIMENSIONS

The unique quality of every quantity which distinguishes it from all other quantities is called **dimension.**

The dimensional symbols for fundamental units of length, mass and time are L, M and T respectively. The various powers of the fundamental units represent the dimensions of any derived unit. For example, the dimensional symbols for the *derived units* of volume, speed and acceleration are $[L]^3$, $[LT^-]$ and $[LT^{-2}]$ respectively. These dimensional formulae of the derived units are very *useful for converting units from one system to another.*

Some derived units, for convenience, have been given new names such as the derived unit of force in the S.I. system is called the newton (N) instead of dimensionally correct name kg m/s^2.

All other mechanical quantities can be expressed in terms of three fundamental quantities, *i.e.*, length (L), mass (M) and time (T); a few of them are given below :

Force : mass × acceleration *i.e.* $[F] = [M][LT^{-2}] = [MLT^{-2}]$

Work : force × distance *i.e.* $[W] = [MLT^{-2}][L] = [ML^2T^{-2}]$

Power : $\dfrac{\text{Work}}{\text{Time}}$ *i.e.* $[P] = \dfrac{[ML^2T^{-2}]}{[T]} = [ML^2T^{-3}]$

Such expressions indicate the nature of derived quantities in terms of fundamental dimensions $[L]$, $[M]$ and $[T]$ and are known as ***dimensional equations***.

The dimensional equations help in serving the following *purposes* :

(*i*) Conversion from one system of units to another one.

(*ii*) Derivation of equations for physical quantities.

(*iii*) Checking the accuracy of an equation.

1.3 SYSTEMS OF ELECTRICAL UNITS

The British Association Committee on Practical Standards for Electrical Measurements adopted the centimetre, gram, and second as the fundamental units of length, mass and time respectively. It was called **CGS system** *of units*.

- *Two* systems of CGS units existed ; these are :
  - (*i*) One having only the *permittivity* K of the medium as well as the units of length, mass and time. This system was called *electrostatic* CGS system of units, normally abbreviated as ***ESU system** of units.*
  - (*ii*) The other system involves *permeability* μ as well as units of length, mass and time. This system was called *electromagnetic* CGS system of units, abbreviated as ***EMU system** of units.*
- The system which uses metre, kilogram and second as the units of length, mass and time is called **MKS system** *of units.*
- The concept of unit flux issuing from unit magnetic pole or from unit charge instead of a flux of 4π is called *rationalisation*. The system using this concept is called **rationalised system** *of units.*

— Rationalisation means the removal of quantity π from the places, where its appearance is unnatural, irrational and geometrically unjustified to places where it properly belongs.

1.4 S.I. UNITS

1.4.1. Introduction to S.I. units

S.I., the *International System of units,* are divided into three classes :

1. Base units
2. Derived units
3. Supplementary units.

From the scientific point of view, division of S.I. units into these classes is to a certain extent arbitrary, because it is not essential to the physics of the subject. Nevertheless, the General Conference, considering the advantages of a single, practical, world-wide system for international relations, for teaching and for scientific work, decided to base the international system on a choice of six well-defined units given in Table 1.1 below :

Table 1.1: S.I. Base Units

Quantity	*Name*	*Symbol*
length	metre	m
mass	kilogram	kg
time	second	s
electric current	ampere	A
thermodynamic temperature	kelvin	K
luminous intensity	candela	cd
amount of substance	mole	mol

The second class of S.I. units contains derived units, *i.e.,* units which can be formed by combining base units according to the algebraic relations linking the corresponding quantities. Several of these algebraic expressions in terms of base units can be replaced by special names and symbols can themselves be used to form other derived units.

Derived units may, therefore, be classified under three headings. Some of them are given in Tables 1.2, 1.3 and 1.4.

Table 1.2: Examples of S.I. Derived Units Expressed in Terms of Base Units

Quantity	*S.I. Units*	
	Name	*Symbol*
area	square metre	m^2
volume	cubic metre	m^3
speed, velocity	metre per second	m/s
acceleration	metre per second squared	m/s^2
density, mass density	kilogram per cubic metre	kg/m^3
concentration (of amount of substances)	mole per cubic metre	mol/m^3
activity (radioactive)	1 per second	s^{-1}
specific volume	cubic metre per kilogram	m^3/kg
luminance	candela per square metre	cd/m^2

Table 1.3: S.I. Derived Units with Special Names

Quantity	*S.I. Units*			
	Name	*Symbol*	*Expression in terms of other units*	*Expression in terms of S.I. base units*
frequency	hertz	Hz	–	s^{-1}
force	newton	N	–	$m.kg.s^{-2}$
pressure	pascal	Pa	N/m^2	$m^{-1}.kg.s^{-2}$
energy, work, quantity of heat power	joule	J	N.m	$m^2.kg.s^{-2}$
radiant flux, quantity of electricity	watt	W	J/s	$m^2.kg.s^{-3}$
electric charge	coulomb	C	A.s	s.A
electric tension, electric potential	volt	V	W/A	$m^2.kg.s^{-3}.A^{-1}$
capacitance	farad	F	C/V	$m^{-2}.kg^{-1}.s^4$
electric resistance	ohm	Ω	V/A	$m^{-2}.kg.s^{-3}.A^{-2}$
conductance	siemens	S	A/V	$m^{-2}.kg^{-1}.s^3.A^{-2}$
magnetic flux	weber	Wb	V.S.	$m^2.kg.s^{-2}.A^{-1}$
magnetic flux density	tesla	T	Wb/m^2	$kg.s^{-2}.A^{-1}$
inductance	henry	H	Wb/A	$m^2.kg.s^{-2}.A^{-2}$
luminous flux	lumen	lm	–	cd.sr
illuminance	lux	lx	–	$m^{-2}.cd.sr$

Table 1.4: Examples of S.I. Derived Units Expressed by Means of Special Names

Quantity	*S.I. Units*		
	Name	*Symbol*	*Expression in terms of S.I. base units*
dynamic viscosity	pascal second	Pa.s	$m^{-1}.kg.s^{-1}$
moment of force	metre newton	N.m	$m^2.kg.s^{-2}$
surface tension	newton per metre	N/m	$kg.s^{-2}$
heat flux density, irradiance	watt per square metre	W/m^2	$kg.s^{-3}$
heat capacity, entropy	joule per kelvin	J/K	$m^2.kg.s^{-2}.K^{-1}$
specific heat capacity, specific entropy	joule per kilogram kelvin	J/(kg.K)	$m^2.s^{-2}.K^{-1}$
specific energy	joule per kilogram	J/kg	$m^2.s^{-2}$
thermal conductivity	watt per metre kelvin	W/(m.K)	$m.kg.s^{-3}.K^{-1}$
energy density	joule per cubic metre	J/m^3	$m^{-1}.kg.s^{-2}$
electric field strength	volt per metre	V/m	$m.kg.s^{-3}.A^{-1}$
electric charge density	coulomb per cubic metre	C/m^3	$m^{-3}.s.A$
electric flux density	coulomb per square metre	C/m^2	$m^{-2}.s.A$
permittivity	farad per metre	F/m	$m^{-3}.kg^{-1}.s^4$
current density	ampere per square metre	A/m^2	–
magnetic field strength	ampere per metre	A/m	–
permeability	henry per metre	H/m	$m.kg.s^{-2}.A^{-2}$
molar energy	joule per mole	J/mol	$m^2.kg.s^{-2}mol^{-1}$
molar heat capacity	joule per mole kelvin	J/(mol.K)	$m^2.kg.s^{-2}.K^{-1}.mol^{-1}$

The S.I. units assigned to third class called "Supplementary units" may be regarded either as base units or as derived units. Refer to Tables 1.5 and 1.6.

Table 1.5: S.I. Supplementary Units

Quantity	*S.I. Units*	
	Name	*Symbol*
plane angle	radian	rad
solid angle	steradian	sr

Table 1.6: Examples of S.I. Derived Units Formed by Using Supplementary Units

Quantity	*S.I. Units*	
	Name	*Symbol*
angular velocity	radian per second	rad/s
angular acceleration	radian per second squared	1 rad/s^2
radiant intensity	watt per steradian	W/sr
radiance	watt per square metre steradian	$W\text{-}m^{-2}.sr^{-1}$

Table 1.7: S.I. Prefixes

Factor	*Prefix*	*Symbol*	*Factor*	*Prefix*	*Symbol*
10^{12}	tera	T	10^{-1}	deci	d
10^{9}	giga	G	10^{-2}	centi	c
10^{6}	mega	M	10^{-3}	milli	m
10^{3}	kilo	k	10^{-6}	micro	μ
10^{2}	hecto	h	10^{-9}	nano	n
10^{1}	deca	da	10^{-12}	pico	p
			10^{-15}	fasnto	f
			10^{-18}	atto	a

1.4.2. Definitions

The S.I. *seven* base units and two supplementary units are defined below :

(*i*) **Metre.** The metre is the length equal to 1,650, 763.73 wavelengths in vacuum of the radiation corresponding to the transition between the levels $2p_{10}$ and $5d_5$ of the krypton-86 atom.

(*ii*) **Kilogram.** One kilogram is equal to the mass of the international prototype of the kilogram.

(*iii*) **Second.** The second is the duration of 9,192,631,770 periods of the radiation corresponding to the transition between the two hyperfine levels of the ground state of the cesium-133 atom.

(*iv*) **Ampere.** One ampere is that constant current which, if maintained in two straight parallel conductors of infinite length, of negligible circular cross-section and placed one metre apart in vacuum, would produce between these conductors a force equal to 2×10^{-7} newton per metre of length.

(*v*) **Kelvin.** The kelvin is the fraction $\frac{1}{273.16}$ of thermodynamic temperature of the triple point of water.

(*vi*) **Candela.** The candela is the luminous intensity, in the perpendicular direction, of a surface of a $\frac{1}{600,000}$ square metre of a black body at a temperature of freezing platinum under a pressure of 101, 325 newton per square metre.

(*vii*) **Mole.** The mole is the amount of substance of a system which contains as many elementary entities as there are atoms in 0.012 kg, carbon 12. When the mole is used, the elementary entities must be specified and may be atoms, molecules, ions, electrons, other particles or specified groups of such particles.

(*viii*) **Radian.** The radian is the plane angle between two radii of a circle that cut off on the circle an arc equal in length to the radius.

(*ix*) **Steradian.** The steradian is the solid angle which, having its vertex in the centre of a sphere, cuts off an area of the surface of the sphere equal to that of a square with sides of length equal to the radius of the sphere.

Some Other Definitions :

Newton. The newton (N) is a derived unit of force and is defined as the unit of force per second, which when acting on a mass of 1 kilogram gives it an acceleration of one metre per second per second. Since acceleration due to gravity equals 9.81 m/s^2, one kilogram force equals 9.81 newtons.

Joule. The joule (J) is a derived unit of energy, work or quantity of heat and is defined as the work done when a force of one newton acts so as to cause a displacement of one metre. Energy is defined as the capacity to do work. A unit of energy in nuclear physics is the electron volt (eV) which is defined as the energy gained by an electron in rising through a potential difference of one volt.

$$1 \text{ eV} = 1.6021 \times 10^{-19} \text{ J}.$$

Watt. The watt (W) is a unit of power (*i.e.*, rate of doing work)

$$\text{Power in watts} = \frac{\text{work (or energy) in joules}}{\text{time in seconds}}$$

Thus 1 watt equals 1 joule/sec.

1 kilowatt-hour (kWh) = 1000 watt-hours = 3600000 joules.

Coulomb. The coulomb (C) is the derived unit of charge. It is defined as *the quantity of electricity passing a given point in a circuit when a current of I A is maintained for 1 second.*

$$Q = I.t$$

where, Q = Charge in coulombs,

I = Current in amperes, and

t = Time in seconds.

1 coulomb represents 6.24×10^{18} electrons.

Ohm. The ohm (Ω) is the unit of electric resistance and is defined as *the resistance in which a constant current of I A generates heat at the rate of 1 watt.*

Siemen. The siemen is a unit of electric conductance (*i.e.*, reciprocal of resistance). If a circuit has a resistance of 5 ohms, its conductance is 0.2 siemen. A more commonly used name for siemen is *mho* (℧).

Volt. The volt is a unit of potential difference and electromotive force. It is defined as *the difference of potential across a resistance of 1 ohm carrying a current of 1 ampere.*

Hertz. The hertz (Hz) is a unit of frequency. 1 Hz = 1 cycle per second.

Horse-power. It is a practical unit of mechanical output. BHP (British horse power or brake horse power) equals 746 watts. The metric horse power equals 735.5 watts. To avoid confusion between BHP and metric horse power, the mechanical output of machines in S.I. units is expressed in watts or kilowatts.

1.4.3. Salient Features of S.I. Units

The salient features of S.I. units are as follows :

1. It is a coherent system of units, *i.e.*, product or quotient of any two base quantities results in a unit resultant quantity. For example, unit length divided by unit time gives unit velocity.
2. It is a rationalised system of units, applicable to both magnetism and electricity.
3. It is a non-gravitational system of units. It clearly distinguishes between the units of mass and weight (force) which are kilogram and newton respectively.
4. All the units of the system can be derived from the base and supplementary units.
5. The decimal relationship between units of same quantity makes possible to express any small or large quantity as a power of 10.
6. For any quantity there is one and only one S.I. unit. For example, joule is the unit of energy of all forms such as mechanical, heat, chemical, electrical and nuclear. However, kWh will also continue to be used as unit of electrical energy.

Advantages of S.I. Units :

1. Units for many different quantities are related through a series of simple and basic relationship.
2. Being an absolute system, it avoids the use of factor '*g*', *i.e.*, acceleration due to gravity in several expressions in physics and engineering which had been a nuisance in all numericals in physics and engineering.
3. Being a rationalised system, it ensures all the advantages of rationalised MKSA system in the fields of electricity, magnetism, electrical engineering and electronics.
4. Joule is the only sole unit of energy of all forms and watt is the sole unit of power hence a lot of labour is saved in calculations.
5. It is a coherent system of units and involves only decimal coefficients. Hence it is a very convenient and quick system for calculations.
6. In electricity, all the practical units like volt, ohm, ampere, henry, farad, coulomb, joule and watt accepted in industry and laboratories all over the world for well over a century have become absolute in their own right in the SI system, without the need for any more practical units.

Disadvantages :

1. The non-S.I. time units 'minute' and 'hour' will still continue to be used until the clocks and watches are all changed to kilo seconds and mega seconds etc.
2. The base unit kilogram (kg) includes a prefix, which creates an ambiguity in the use of multipliers with gram.
3. S.I. units for energy, power and pressure (*i.e.*, joule, watt and pascal) are too small to be expressed in science and technology, and, therefore, in such cases the use of larger units, such as MJ, kW, kPa, will have to be made.
4. There are difficulties with regard to developing new S.I. units for apparent and reactive energy while joule is the accepted unit for active energy in S.I. systems.

1.4.4. Conversion Factors

1. Force :

$$1 \text{ newton} = \text{kg.m/sec}^2 = 0.012 \text{ kgf}$$
$$1 \text{ kgf} = 9.81 \text{ N}$$

2. Pressure :

$$1 \text{ bar} = 750.05 \text{ mm Hg} = 0.9869 \text{ atm} = 10^5 \text{ N/m}^2$$
$$= 103 \text{ kg/m-sec}^2$$
$$1 \text{ N/m}^2 = 1 \text{ pascal} = 10^{-5} \text{ bar} = 10^{-2} \text{ kg/m-sec}^2$$
$$1 \text{ atm} = 760 \text{ mm Hg} = 1.03 \text{ kgf/cm}^2 = 1.01325 \text{ bar}$$
$$= 1.01325 \times 10^5 \text{ N/m}^2$$

3. Work, Energy or Heat :

$$1 \text{ joule} = 1 \text{ newton metre} = 1 \text{ watt-sec}$$
$$= 2.7778 \times 10^{-7} \text{ kWh} = 0.239 \text{ cal}$$
$$= 0.239 \times 10^{-3} \text{ kcal}$$
$$1 \text{ cal} = 4.184 \text{ joule} = 1.1622 \times 10^{-6} \text{ kWh}$$
$$1 \text{ kcal} = 4.184 \times 10^3 \text{ joule} = 427 \text{ kgf m} = 1.1622 \times 10^{-3} \text{ kWh}$$
$$1 \text{ kWh} = 8.6042 \times 10^5 \text{ cal} = 860.42 \text{ kcal} = 3.6 \times 10^6 \text{ joules}$$
$$1 \text{ kgf-m} = \left(\frac{1}{427}\right) \text{ kcal} = 9.81 \text{ joules}$$

4. Power :

1 watt = 1 joule/sec = 0.86 kcal/h

1 h.p. = 75 m kgf/sec = 0.1757 kcal/sec = 735.3 watts

1 kW = 1000 watts = 860 kcal/h

5. Specific heat :

1 kcal/kg-°K = 4184 joules/kg-K

6. Thermal conductivity :

1 watt/m-K = 0.8598 kcal/h-m-°C

1 kcal/h-m-°C = 1.16123 watt/m-K = 1.16123 joules/s-m-K.

7. Heat transfer coefficient :

1 watt/m^2-K = 0.86 kcal/m^2-h-°C

1 kcal/m^2-h-°C = 1.163 watt/m^2-K.

The following conversion factors may be used to convert the quantities in non-S.I. units into S.I. units.

To convert	*To*	*Multiply by*
angstroms	m	10^{-10}
atmospheres	kg/m^2	10332
bars	kg/m^2	1.02×10^4
Btu	joules	1054.8
Btu	kWh	2.928×10^{-4}
circular mils	m^2	5.067×10^{-10}
cubic feet	m^3	0.02831
dynes	newtons	10^{-5}
ergs	joules	10^{-7}
ergs	kWh	0.2778×10^{-13}
feet	m	0.3048
foot-pounds	joules	1.356
foot-pounds	kg-m	0.1383
gauss	tesla	10^{-4}
grams (force)	newton	9.807×10^{-3}
horse power (metric)	watts	735.5
lines/sq. inch	tesla	1.55×10^{-5}
Maxwell	webers	10^{-8}
mho	siemens	1
micron	metre	10^{-6}
miles	km	1.609
mils	cm	2.54×10^{-3}
poundals	newton	0.1383
pounds	kilogram	0.454
pounds (force)	newtons	0.448
pounds/sq. ft.	N/m^2	47.878
pounds/sq. inch	N/m^2	6894.43

1.4.5. Dimensions of Quantities

Different units can be represented dimensionally in terms of units of length L, mass M, time T and current I. The dimensions can be derived as under :

1. Velocity = length/time = $L/T = LT^{-1}$
2. Acceleration = velocity/time = $LT^{-1}/T = LT^{-2}$
3. Force = mass × acceleration = MLT^{-2}
4. Charge (coulomb) = current × time = IT
5. Work or energy = force × distance = ML^2T^{-2}
6. EMF or potential = work/charge
   $= ML^2T^{-2}/IT = ML^2I^{-1}T^{-3}$
7. Power = work/time = $ML^2T^{-2}/T = ML^2T^{-3}$
8. Current density = current/area = $I/L^2 = IL^{-2}$
9. Resistance = emf/current = $ML^2I^{-1}T^{-3}/I = ML^2I^{-2}T^{-3}$
10. Electric flux density = electric flux or charge/area = $IT/L^2 = ITL^{-2}$
11. MMF = current × number of turns = I
12. Conductance = 1/resistance = $1/ML^2I^{-2}T^{-3} = I^2T^3M^{-1}L^{-2}$
13. Electric field intensity = volt/metre = $ML^2I^{-1}T^{-3}/L = MLI^{-1}T^{-3}$
14. Resistivity = $\dfrac{\text{resistant} \times \text{area}}{\text{length}}$
    $= (ML^2I^{-2}T^{-3})(L^2)/L$
    $= ML^3I^{-2}T^{-3}$
15. Magnetic field intensity (H) = MMF/length
    $= I/L = IL^{-1}$
16. Magnetic flux = emf × time = $(ML^2I^{-1}T^{-3})(T) = ML^2I^{-1}T^{-2}$
17. Magnetic flux intensity= magnetic flux/area
    $= (ML^2I^{-1}T^{-2})/L^2 = MI^{-1}T^{-2}$
18. Impedance = emf/current = $ML^2I^{-2}T^{-3}$
19. Admittance = I/impedance = $I^2T^3M^{-1}L^{-2}$
20. Inductance = magnetic flux/current
    $= ML^2T^{-2}I^{-1}/I = ML^2T^{-2}I^{-2}$
21. Capacitance = electric charge/potential
    $= IT/ML^2T^{-3}I^{-1} = M^{-1}L^{-2}T^4I^2$

1.5 DIMENSIONS IN ELECTROSTATIC AND ELECTROMAGNETIC SYSTEMS

1.5.1. Dimensions in Electrostatic System

In an electrostatic system the dimensions of permittivity $\in$ is taken as the *fourth* fundamental dimension.

1. **Charge.** The force exerted between two charges Q_1 and Q_2 according to Coulomb's law is given as:

$$F = \frac{Q_1Q_2}{\in d^2}$$

where, d = Distance between charges Q_1 and Q_2.
Inserting the dimensions of force and distance, we have

$$[MLT^{-2}] = \frac{[Q^2]}{[\in][L^2]}$$

$\therefore$ Dimensions of charge $[Q] = [\in^{1/2}\ M^{1/2}\ L^{3/2}\ T^{-1}]$ ...(1.1)

2. **Current.** Current is *charge per unit time.* Let (I) represent the dimensions of current.

$$[I] = \frac{[Q]}{[T]} = \frac{[\in^{1/2}\ M^{1/2}L^{3/2}T^{-1}]}{[T]} = [\in^{1/2}\ M^{1/2}L^{3/2}T^{-2}] \quad ...(1.2)$$

3. **Potential difference or e.m.f.** Potential difference is *work done per unit charge*

$$[E] = \frac{[W]}{[Q]} = \frac{[ML^2T^{-2}]}{[\in^{1/2}\ M^{1/2}L^{3/2}T^{-1}]} = [\in^{-1/2}\ M^{1/2}L^{1/2}T^{-1}] \quad ...(1.3)$$

4. **Capacitance.** Capacitance is the *ratio of charge to e.m.f. i.e.* $C = \frac{Q}{E}$

$\therefore$ The dimensions of capacitance are :

$$[C] = \frac{[Q]}{[E]} = \frac{[\in^{1/2}\ M^{1/2}L^{3/2}T^{-1}]}{[\in^{-1/2}\ M^{1/2}L^{1/2}T^{-1}]} = [\in L] \quad ...(1.4)$$

5. **Resistance.** Resistance is the *ratio of e.m.f. to current, i.e.,* $R = \frac{E}{I}$

$\therefore$ Dimensions of resistance are :

$$[R] = \frac{[E]}{[I]} = \frac{[\in^{-1/2}\ M^{1/2}L^{1/2}T^{-1}]}{[\in^{1/2}\ M^{1/2}L^{3/2}T^{-2}]} = [\in^{-1}\ L^{-1}T] \quad ...(1.5)$$

6. **Inductance.** Inductance, $L = \frac{\text{emf}}{\text{rate of change of current}} = \frac{E}{dI/dt}$

$\therefore$ Dimensions of inductance are :

$$[L] = \frac{[E]}{[I]/[T]} = \frac{[E][T]}{[I]} = \frac{[\in^{-1/2}\ M^{1/2}L^{1/2}T^{-1}][T]}{[\in^{1/2}\ M^{1/2}L^{3/2}T^{-2}]}$$

$$= [\in^{-1}\ L^{-1}T^2] \quad ...(1.6)$$

1.5.2. Dimensions in Electromagnetic System

In an electromagnetic system the *fourth* dimension is the dimension of permeability μ.

1. **Pole strength.** Force, $F = \frac{m_1 m_2}{\mu d^2}$

where, d = Distance between poles of strengths m_1 and m_2.
Writing the dimensions, we have :

$$[MLT^{-2}] = \frac{[m^2]}{[\mu][L^2]}$$

Thus, dimensions of pole strength are, $[m] = [\mu^{1/2} M^{1/2} L^{3/2} T^{-1}]$ ...(1.7)

2. **Magnetising force.** This force H is measured by force exerted on a unit pole.

∴ The dimensions of magnetising force are :

$$[H] = \frac{[F]}{[m]} = \frac{[MLT^{-2}]}{[\mu^{1/2} M^{1/2} L^{3/2} T^{-1}]} = [\mu^{-1/2} M^{1/2} L^{-1/2} T^{-1}] \quad ...(1.8)$$

3. **Current.** The magnetising force at the centre of a loop of radius r is given as :

$$H = \frac{2\pi I}{r} \text{ if } I \text{ is in e.m. units}$$

or,

$$[H] = \frac{[I]}{[L]}$$

∴ Dimensions of current are, $[I] = [H][L] = [\mu^{-1/2} M^{1/2} L^{1/2} T^{-1}]$ ...(1.9)

4. **Charge.** Charge is the *product of current and time, i.e.,* $Q = I \times t$

∴ The dimensions of charge are :

$$[Q] = [I][T] = [\mu^{-1/2} M^{1/2} L^{1/2} T^{-1}][T] = [\mu^{-1/2} M^{1/2} L^{1/2}] \quad ...(1.10)$$

5. **Potential difference.** It is the *work done per unit charge.*

The dimensions of potential difference are :

$$[E] = \frac{[W]}{[Q]} = \frac{[ML^2T^{-2}]}{[\mu^{-1} M^{1/2} L^{1/2}]} = [\mu^{1/2} M^{1/2} L^{3/2} T^{-2}] \quad ...(1.11)$$

6. **Capacitance.** The dimensions of capacitance are :

$$[C] = \frac{[Q]}{[E]} = \frac{[\mu^{-1/2} M^{1/2} L^{1/2}]}{[\mu^{1/2} M^{1/2} L^{3/2} T^{-2}]} = [\mu^{-1} L^{-1} T^2] \quad ...(1.12)$$

7. **Resistance.** The dimensions of resistance are :

$$[R] = \frac{[E]}{[I]} = \frac{[\mu^{1/2} M^{1/2} L^{3/2} T^{-2}]}{[\mu^{-1/2} M^{1/2} L^{1/2} T^{-1}]} = [\mu L T^{-1}] \quad ...(1.13)$$

8. **Inductance.** Inductance $L = \dfrac{E}{dI/dt}$

∴ The dimensions of inductance are :

$$[L] = \frac{[E]}{[I]/[T]} = \frac{[E][T]}{[I]} = \frac{[\mu^{1/2} M^{1/2} L^{3/2} T^{-2}][T]}{[\mu^{-1/2} M^{1/2} L^{1/2} T^{-1}]} = [\mu L] \quad ...(1.14)$$

Similarly the dimensions of other quantities can be found out.

Relationship Between Electrostatic and Electromagnetic Systems of Units :

The dimensions of charge in the above systems of units is given as :

$$[Q] = [\in^{1/2} M^{1/2} L^{3/2} T^{-1}] \quad \text{e.s.u. system}$$

$$[Q] = [\mu^{-1/2} M^{1/2} L^{1/2}] \quad \text{e.m.u. system}$$

Equating the two dimensions, we get

$$\in^{1/2} M^{1/2}L^{3/2}T^{-1} = \mu^{-1/2}M^{1/2}L^{1/2}$$

or,
$$[LT^{-1}] = [\mu^{-1/2} \in^{-1/2}]$$

But LT^{-1} are the dimensions of velocity,

$$\therefore \quad \frac{1}{\sqrt{\mu \in}} = \text{a velocity} \qquad \text{...(1.15)}$$

If we equate the dimensional equations of other quantities, we shall again get the same result, *i.e.*, some power of $\sqrt{\mu \in}$ is always involved. It should be understood that it is not possible to obtain separate expressions for μ and $\in$.

The permeability of free space μ_0 and permittivity of free space $\in_0$, in any system of units, are related by relation :

$$\mu_0 \in_0 = \frac{1}{v^2} \qquad \text{...(1.16)}$$

Where v is the velocity of light in system of units considered.

Eqn. (1.16) can be used to convert the dimensions of any electrical quantity from those of e.s.u. system to e.m.u. system, and vice-versa.

Example 1.1. *Derive the dimensions of the following quantities in L, M, T, I :*

(i) E.m.f. ; (ii) Magnetic flux density ; (iii) Electric flux density ; (iv) Current density ; (v) Permeability ; (vi) Permittivity ; (vii) Resistivity ; (viii) Conductivity.

Solution. (*i*) **E.m.f. (E) :**

$$\text{E.m.f.} = \frac{\text{Work done}}{\text{Charge}}$$

$$[E] = \frac{[ML^2T^{-2}]}{[IT]} = [ML^2T^{-3}I^{-1}] \quad \textbf{(Ans.)}$$

(*ii*) **Magnetic flux density (B) :**

$$\text{Magnetic flux density, } B = \frac{\phi\,[\text{Flux}]}{A\,(\text{Area})}$$

$$\text{E.m.f.} = \text{Rate of change of flux linkages} = N\frac{d\phi}{dt}$$

or,
$$\text{Flux } (\phi) = [\text{e.m.f.}][\text{time}]$$

$$\therefore \quad [B] = \frac{[\phi]}{[A]} = \frac{[\text{e.m.f.}][\text{time}]}{[\text{area}]} = \frac{[ML^2T^{-3}I^{-1}][T]}{L^2} = [MT^{-2}I^{-1}] \quad \textbf{(Ans.)}$$

(*iii*) **Electric flux density (D) :**

$$\text{Electric flux density, } D = \frac{\text{Electric flux}}{\text{Area}} = \frac{\text{Charge}}{\text{Area}}$$

$$\therefore \quad [D] = \frac{[IT]}{[L^2]} = [L^{-2}TI] \quad \textbf{(Ans.)}$$

(*iv*) **Current density (J) :**

Current density, $J = \dfrac{\text{Current}}{\text{Area}}$

$\therefore \quad [J] = \dfrac{[I]}{[L^2]} = [L^{-2}I]$ **(Ans.)**

(*v*) **Permeability (μ) :**

$$\text{Flux} = \text{Flux density} \times \text{area}$$

or, $[\phi] = [MT^{-2}I^{-1}]\,[L^2] = [ML^2T^{-2}I^{-1}]$

$$\text{M.m.f.} = \text{Turns} \times \text{current} = [I]$$

$$\text{Reluctance (S)} = \frac{\text{M.m.f.}}{\text{Flux}} \quad \therefore \quad [S] = \frac{[I]}{ML^2T^{-2}I^{-1}} = M^{-1}L^{-2}T^2I^2$$

Also, $\text{Reluctance} = \dfrac{\text{Length}}{\text{Permeability} \times \text{area}}$

$\therefore$ Permeability, $\mu = \dfrac{\text{Length}}{\text{Reluctance} \times \text{area}}$

or, $[\mu] = \dfrac{[L]}{[M^{-1}L^{-2}T^2I^2][L^2]} = [MLT^{-2}I^{-2}]$ **(Ans.)**

(*vi*) **Permittivity (∈) :**

$$\text{Force} = \frac{Q_1Q_2}{\in d^2}, \text{ where } \in = \text{permittivity}$$

$$\therefore \quad \in = \frac{[\text{Charge}]^2}{[\text{Force}][\text{Distance}]^2}$$

$$= \frac{[\text{Current}]^2[\text{Time}]^2}{[\text{Force}][\text{Distance}]^2} = \frac{I^2T^2}{[MLT^{-2}][L^2]} = [M^{-1}L^{-3}T^4I^2] \text{ **(Ans.)**}$$

(*vii*) **Resistivity (ρ) :**

Resistance $R = \dfrac{\text{E.m.f.}}{\text{Current}}$

or, $[R] = \dfrac{[ML^2T^{-3}I^{-1}]}{[I]} = [ML^2T^{-3}I^{-2}]$

Also, $\text{Resistance} = \dfrac{\text{Resistivity} \times \text{length}}{\text{Area}}$

$\therefore \quad [\text{Resistivity}] = \dfrac{[\text{Resistivity}][\text{area}]}{[\text{Length}]}$

or, $[\rho] = \dfrac{[ML^2T^{-3}I^{-2}]}{[L]} = [ML^3T^{-3}I^{-2}]$ **(Ans.)**

(*viii*) **Conductivity (σ) :**

$$\text{Conductivity} = \frac{I}{\text{Resistivity}}$$

or, $[\sigma] = [\rho^{-1}] = [M^{-1}L^{-3}T^3I^2]$ **(Ans.)**

Example 1.2. *Show that ω^2LC is non-dimensional, ω being the angular frequency of the applied voltage.*

Solution. We know that dimensions of ω, L (inductance) and C (capacitance) are as under :

$$[\omega] = [T^{-1}];$$
$$[L] = [ML^2T^{-2}I^{-2}]$$
$$(C) = [M^{-1}L^{-2}T^4I^2]$$

Now, $$[\omega^2LC] = [\omega^2][L][C]$$
$$= [T^{-2}][ML^2T^{-2}I^{-2}][M^{-1}L^{-2}T^4I^2] = M^0L^0T^0I^0$$

Hence, ω^2LC *is a non-dimensional quantity* - - - - -**Proved.**

Example 1.3. *The energy stored in a parallel plate capacitor per unit volume (energy density) is given by :*

$$w = K \in^a V^b d^c$$

where, K = *A constant,*

$\in$ = *Permittivity of medium,*

V = *Voltage between plates, and*

d = *Distance between plates.*

Determine the values of a, b, and c. **(Roorkee University)**

Solution. Dimensional equation $w = K\in^aV^bd^c$ *(Given)* ...(*i*)

Writing the dimensions of various quantities in LMTI system, we have :

Energy density, $$w = \frac{\text{Energy}}{\text{Volume}}$$

or, $$[w] = \frac{[ML^2T^{-2}]}{[L^3]} = [ML^{-1}T^{-2}]$$

Also, Force, $$F = \frac{Q_1Q_2}{\in d^2}$$

$\therefore$ Permittivity, $$\in = \frac{Q_1Q_2}{Fd^2}$$

or, $$[\in] = \frac{[I^2T^2]}{[MLT^{-2}][L^2]} = [M^{-1}L^{-3}T^4I^2]$$

and $$[\text{Voltage}] = [V] = [ML^2T^{-3}I^{-1}], \quad d = [L]$$

Substituting the dimensions of various quantities in expression (*i*), we get :

$$[ML^{-1}T^{-2}] = [M^{-1}L^{-3}T^4I^{-2}]^a[ML^2T^{-3}I^{-1}]^b[L]^c$$

or $$[ML^{-1}T^{-2}] = [M]^{-a+b}\ [L]^{-3a+2b+c}\ [T]^{4a-3b}\ [I]^{2a-b}$$

Equating the components of M, L, T and I respectively, we have :

For M : $1 = -a + b$ ...(*ii*)

For L : $-1 = -3a + 2b + c$ ...(*iii*)

For T : $-2 = 4a - 3b$ ...(*iv*)

For I : $0 = 2a - b$ ...(*v*)

Solving expressions (*ii*), (*iii*), (*iv*) and (*v*), we get

$$a = 1, b = 2, c = -2$$

Hence, energy stored per unit volume,

$$w = K\in V^2 d^{-2} = \frac{K\in V^2}{d^2} \quad \textbf{(Ans.)}$$

Example 1.4. *The expression for mean torque T of an electrodynamic wattmeter is written in the form*

$$T \propto M^a E^b Z^c$$

where M is mutual inductance between fixed and moving coils, E-applied voltage, Z-impedance of coil.

From first principles determine the values of constants a, b and c.

(Punjab University)

Solution. $T \propto M^a E^b Z^c$ ...(*Given*)

or, $T = KM^a E^b Z^c$ ...(*i*)

where K is a dimensionless number.

Writing the dimensions of various quantities in *LMTI* system, we have :

$$[T] = [ML^2T^{-2}]$$
$$[M] = [ML^2T^{-2}I^{-2}]$$
$$[E] = [ML^2T^{-3}I^{-1}]$$
$$[Z] = [ML^2T^{-3}I^{-2}]$$

Substituting the dimensions of quantities involved in expression (*i*), we get :

$$[ML^2T^{-2}] = [ML^2T^{-2}I^{-2}]^a[ML^2T^{-3}I^{-1}]^b[ML^2T^{-3}I^{-2}]^c$$

or, $$[ML^2T^{-2}] = [M]^{a+b+c}\,[L]^{2a+2b+2c}\,[T]^{-2a-3b-3c}\,[I]^{-2a-b-2c}$$

Equating the components of *M, L, T* and *I* respectively, we get :

For M : $1 = a + b + c$ ...(*ii*)

For L : $2 = 2a + 2b + 2c$ ...(*iii*)

For T : $-2 = -2a - 3b - 3c$ ...(*iv*)

For I : $0 = -2a - b - 2c$ ...(*v*)

Solving above expressions, we get :

$$\mathbf{a = 1,\ b = 2,\ \text{and}\ c = -2} \quad \textbf{(Ans.)}$$

Example 1.5. *The following expression was found for eddy currents produced in a metallic former moving in the field of a permanent magnet :*

$$I_e = \frac{KBlbA}{(2b + l)\rho}$$

where, K = *A constant,*

B = *Flux density,*

l = *Length of former,*

b = *Width of former,*

A = *Area of former, and*

ρ = *Resistivity of conducting former.*

It is suspected that the term angular frequency is missing. Use LMTI system for dimensional analysis to detect the error.

Solution. The dimensions of L.H.S. = [I]

The dimensions of various quantities of RHS are :

$$[B] = [MT^{-2}I^{-1}]$$
$$[l] = [L]$$
$$[b] = [L]$$
$$[A] = [L^2]$$
$$[\rho] = [ML^3T^{-3}I^{-2}]$$

After substituting the dimensions of various quantities, we have :

$$\text{R.H.S.} = \frac{[MT^{-2}I^{-1}][L][L][L^2]}{[L][ML^3T^{-3}I^{-2}]} = [TI]$$

For making the two sides *dimensionally correct, the R.H.S. must be multiplied by a term having a dimension T^{-1}, i.e.,* ω.

∴ The correct expression is,

$$I_e = \frac{BlbA\omega}{(2b+l)\rho} \quad \textbf{(Ans.)}$$

Example 1.6. *Determine the indices in the following expression for eddy current loss W per cm of wire having circular cross-section :*

$$W = f^a B_{max}^b d^c \rho^g$$

where, f = *Frequency of the supply in Hz,*

B_{max} = *Maximum value of flux density,*

d = *Diameter of the wire,*

ρ = *Resistivity of wire material.* **(AMIE)**

Solution.

Dimensionally we can write : $W = f^a B_{\max}^b d^c \rho^g$ ...(*Given*)

$$[V] = \frac{[F][d]}{[Q]} = \frac{[MLT^{-2}][L]}{[IT]} = [ML^2T^{-3}I^{-1}]$$

$$[W] = \frac{[V][I]}{[L]} = \frac{[ML^2T^{-3}I^{-1}][I]}{[L]} = [MLT^{-3}]$$

$$[f] = [T^{-1}]$$

$$[B_{\max}] = \frac{[\phi]}{[A]} = \frac{[V][T]}{[A]} = \frac{[ML^2T^{-3}I^{-1}][T]}{[L^2]} = [MT^{-2}I^{-1}]$$

$$[d] = [L]$$

$$[\rho] = \frac{[R][A]}{[L]} = \frac{[V/I][A]}{[L]} = \frac{[ML^2T^{-3}I^{-1}][L^2]}{[I][L]} = [ML^3T^{-3}I^{-2}]$$

Substituting the dimensions of the quantities involved in the given expression, we get :

$$[MLT^{-3}] = [T^{-1}]^a\ [MT^{-2}I^{-1}]^b\ [L]^c\ [ML^3T^{-3}I^{-2}]^g$$

or, $$[MLT^{-3}] = [M]^{b+g}\ [L]^{c+3g}\ [T]^{-a-2b-3g}\ [I]^{-b-2g}$$

Equating the indices of *M, L, T* and *I* respectively, we get :

For M : $1 = b + g$ ...(*i*)

For L : $1 = c + 3g$ ...(*ii*)

For T : $-3 = -a - 2b - 3g$ ...(*iii*)

For I : $0 = -b - 2g$ ...(iv)

Solving the above expressions, we have :

$$a = 2;\ b = 2;\ c = 4;\ g = -1 \quad \textbf{(Ans.)}$$

Hence the expression is : $W = f^2 B_{max}^2 d^4 \rho^{-1}$ **(Ans.)**

Example 1.7. *The following expression is obtained in deriving expression for resistance in Hay's bridge :*

$$R = \frac{\omega^2 R_1 R_2 R_3 C^2}{1 + \omega^2 C R_2^2}$$

where R, R_1, R_2, R_3 are resistances; C, a capacitance and $\omega = 2\pi f$, where f is frequency, Hz

Determine whether the expression is dimensionally correct or incorrect. In case there is an error, point out the term which needs correction. **[Delhi University]**

Solution.
$$R = \frac{\omega^2 R_1 R_2 R_3 C^2}{1 + \omega^2 C R_2^2} \quad \text{...(Given)}$$

The dimensions of various terms in e.m. system are :

$$[R] = [\mu L T^{-1}]$$
$$[C] = [\mu^{-1} L^{-1} T^2]$$
$$[\omega] = [T^{-1}]$$

The dimensions of *L.H.S.* are :

$$[R] = [\mu L T^{-1}]$$

The dimensions of *R.H.S.* are :

$$\frac{[T^{-1}]^2 [\mu L T^{-1}][\mu L T^{-1}][\mu L T^{-1}][\mu^{-1} L^{-1} T^2]^2}{1 + [T^{-1}]^2 [\mu^{-1} L^{-1} T^2][\mu L T^{-1}]^2} = \frac{[\mu L T^{-1}]}{1 + [\mu L T^{-2}]}$$

From the above expressions we find that the dimensions of left hand and right hand terms do not balance and, therefore, the equation is dimensionally *incorrect*.

The dimensions of left hand terms are $[\mu L T^{-1}]$. The dimensions of the numerator of right hand terms are $[\mu L T^{-1}]$ and in the denominator we have two terms, *i.e.*, 1 and $\omega^2 C R_2^2$. Now 1 is a dimensionless quantity and, therefore, in order that the dimensions of left hand term balance those of the right hand term, the denominator of right hand term should also be *dimensionless* quantity. The dimensions of second term in the denominator are $[\mu L T^{-2}]$ and in order that it may be a dimensionless quantity it should be multiplied with a term having dimensions $[\mu^{-1} L^{-1} T^2]$ which are the *dimensions of capacitance*. Therefore, the second term in the denominator should contain another capacitance term or this term should be $(\omega^2 C R_2^2 \times C)$ or $\omega^2 C^2 R_2^2$. Therefore, the correct equation is :

$$R = \frac{\omega^2 R_1 R_2 R_3 C^2}{1 + \omega^2 C^2 R_2^2} \quad \textbf{(Ans.)}$$

Example 1.8. *Check the following equation dimensionally :*

$$E = \frac{2Ir\omega^2 C (L + C^2 r^2)}{(1 - \omega^2 LC)^2 - 4\omega^2 C^2 r^2}$$

where ω is the angular frequency of resonance. **[Punjab University]**

Solution. $$E = \frac{2Ir\omega^2 C\,(L + C^2 r^2)}{(1 - \omega^2 LC)^2 - 4\omega^2 C^2 r^2} \quad ...(Given)$$

The dimensions of various terms in *M, L, T, Q* system are :

$$[E] = [ML^2T^{-2}Q^{-1}]$$
$$[r] = [ML^2T^{-1}Q^{-2}]$$
$$[\omega] = [T^{-1}]$$
$$[L] = [ML^2Q^{-2}]$$
$$[C] = [M^{-1}L^{-2}T^2Q^2]$$
$$[I] = [T^{-1}Q]$$

The dimensions of L.H.S. are :

$$[E] = [ML^2T^{-2}Q^{-1}]$$

The dimensions of R.H.S. are :

$$\left[\frac{2Ir\omega^2 C(L + C^2r^2)}{(1-\omega^2LC)^2 - 4\omega^2C^2r^2}\right] =$$

$$\frac{\{[T^{-1}Q][ML^2T^{-1}Q^{-2}][T^{-2}][M^{-1}L^{-2}T^2Q^2]\} \times \{[ML^2Q^{-2}] + [M^{-2}L^{-4}T^4Q^4][M^2L^4T^{-2}Q^{-4}]\}}{\{[L^0M^0T^0Q^0] - [T^{-2}][ML^2Q^{-2}][M^{-1}L^{-2}T^2Q^2]\} - [T^{-1}]^2[M^{-2}L^{-4}T^4Q^4][M^2L^4T^{-2}Q^{-4}]}$$

$$= [T^{-2}Q][ML^2Q^{-2} + T^2]$$

This clearly shows that if the dimensions of left hand and right hand terms are to be dimensionally equal, then the terms in the second bracket of R.H.S. should be of dimensions ML^2Q^{-2}, which is possible if the expression is :

$$E = \frac{(2Ir\omega^2C)(L + Cr^2)}{(1 - \omega^2LC)^2 - 4\omega^2C^2r^2}$$

In this the dimensions of R.H.S. terms will be

$$= [T^{-2}Q]\ \{[ML^2Q^{-2}] + [M^{-1}L^{-2}T^2Q^2][M^2L^4T^{-2}Q^{-4}]$$
$$= [T^{-2}Q]\ \{[ML^2Q^{-2}] + [ML^2Q^{-2}]\}$$
$$= [T^{-2}Q][ML^2Q^{-2}]$$
$$= [ML^2T^{-2}Q^{-1}] = \text{Dimensions of L.H.S.}$$

Hence, *correct expression* must be,

$$E = \frac{2Ir\omega^2C\,(L + Cr^2)}{(1 - \omega^2LC)^2 - 4\omega^2C^2r^2} \quad \textbf{(Ans.)}$$

Example 1.9. *An expression of the following form was arrived at while working for a derivation:*

$$I = E\left\{\frac{1}{Z_1} + \frac{j\omega M}{Z_2}\left(\frac{1}{R} + \frac{C}{L}\right)\right\}$$

Show that there must be an algebraic error, and point out the term or terms which require correction.

Solution. $$I = E\left\{\frac{1}{Z_1} + \frac{j\omega M}{Z_2}\left(\frac{1}{R} + \frac{C}{L}\right)\right\} \quad ...(Given)$$

The dimensions of various quantities in e.m. system are :

$$[I] = [\mu^{-1/2}\ L^{1/2}\ M^{1/2}\ T^{-1}]$$

$$[E] = [\mu^{1/2}\, L^{3/2}\, M^{1/2}\, T^{-2}]$$
$$[Z] = [\mu L T^{-1}]$$
$$[M] = [\mu L]$$
$$[C] = [\mu^{-1}\, L^{-1} T^{2}]$$
$$[R] = [\mu L T^{-1}]$$
$$[L] = [\mu L]$$
$$[\omega] = [T^{-1}]$$

The dimensions of L.H.S. are :

$$[I] = [\mu^{-1/2}\, L^{1/2}\, M^{1/2}\, T^{-1}]$$

The dimensions of R.H.S. are :

$$\left[E\left\{\frac{1}{Z_1}+\frac{j\omega M}{Z_2}\left(\frac{1}{R}+\frac{C}{L}\right)\right\}\right] = [\mu^{1/2}L^{3/2}M^{1/2}T^{-2}]\times\left\{\frac{1}{[\mu LT^{-1}]}+\frac{[T^{-1}][\mu L]}{[\mu LT^{-1}]}\left(\frac{1}{[\mu LT^{-1}]}+\frac{[\mu^{-1}L^{-1}T^{2}]}{[\mu L]}\right)\right\}$$

$$= [\mu^{1/2}\, L^{3/2}M^{1/2}T^{-2}] \times \{[\mu^{-1}L^{-1}T] + [\mu^{-1}L^{-1}T] + [\mu^{-2}L^{-2}T^{2}]\}$$

$$= \{[\mu^{-1/2}L^{1/2}M^{1/2}\, T^{-1}] + [\mu^{-1/2}L^{1/2}M^{1/2}T^{-1}] + [\mu^{-3/2}L^{-1/2}M^{1/2}T^{0}]\}$$

For the two sides to balance dimensionally, the dimensions of R.H.S. should be $[\mu^{-1/2}L^{1/2}M^{1/2}T^{-1}]$. We find that the dimensions of all the terms on the R.H.S. are not $[\mu^{-1/2}L^{1/2}M^{1/2}T^{-1}]$ and *hence there is an algebraic error.* This error is due to the last term having dimensions of $[\mu^{-3/2}L^{-1/2}M^{1/2}T^{0}]$. The error can be eliminated if it has the dimensions $[\mu^{-1/2}L^{-1/2}M^{1/2}T^{-1}]$. In order to achieve this we must multiply the last term by a quantity which has the dimensions

$$\frac{[\mu^{-1/2}L^{1/2}M^{1/2}T^{-1}]}{[\mu^{-3/2}L^{-1/2}M^{1/2}T^{0}]} = \mathbf{[\mu LT^{-1}]}$$

But $[\mu LT^{-1}]$ are the dimensions of resistance or impedance. Hence, the *last term of the R.H.S.,* $\frac{C}{L}$, *should be multiplied by a term having the dimensions of either resistance or impedance.* **(Ans.)**

Example 1.10. *The following expression was obtained while deriving the theory of Biff-circuit for testing of current transformer:*

$$\frac{I_p}{I_s} = \frac{R_2[(1+\omega^2C^2r_1^2)\{R_1^2\,(1+\omega^2C^2r_1^2)+2R_1r_1+r_1^2\}]^{1/2}}{R_1\,(R_2+r)\,(1+\omega^2C^2r_1^2)}$$

Examine the term dimensionally and suggest the correction, if any necessary.

[Bangalore University]

Solution. Writing the terms dimensionally in the S.I. units, we get :

$$[I_p] = [I_s] = [I]$$
$$[\omega] = [T^{-1}]$$
$$[R_1] = [R_2] = [r_1] = \frac{[V]}{[I]} = \frac{[ML^2T^{-3}I^{-1}]}{[I]} = [ML^2T^{-3}I^{-2}]$$
$$[C] = \frac{[I]}{[\omega][V]} = \frac{[I]}{[T^{-1}][ML^2T^{-3}I^{-1}]} = [M^{-1}L^{-2}T^4I^2]$$

The dimensions of L.H.S. are :

$$\frac{[I]}{[I]} = \text{Dimensionless quantity}$$

The dimensions of R.H.S. are :

$$\frac{R_2\,[(1+\omega^2C^2r_1^2)\,\{R_1^2\,(1+\omega^2C^2r_1^2)+2R_1r_1+r_1^2\}]^{1/2}}{R_1\,(R_2+r)\,(1+\omega^2C^2r_1^2)}$$

$$=\frac{\left[[ML^2T^{-3}I^{-2}][(M^0L^0T^0I^0+T^{-2}M^{-2}L^{-4}T^8I^4M^2L^4T^{-6}I^{-4})\left\{[M^2L^4T^{-6}I^{-4}]\,(M^0L^0T^0I^0+T^{-2}M^{-2}L^{-4}T^8I^4M^2L^4T^{-1}I^{-4})+(ML^2T^{-3}I^{-2})(ML^2T^{-3}I^{-2})+M^2L^4T^{-6}I^{-4}\right\}\right]^{1/2}}{(ML^2T^{-3}I^{-2})(ML^2T^{-3}I^{-2}+ML^2T^{-3}I^{-2})(M^0L^0T^0I^0+T^{-2}M^{-2}L^{-4}T^8I^4M^2L^4T^{-6}I^{-4})}$$

Since the sum of terms which have the same dimensions as one another must have the same dimensions as its constituent terms, therefore, R.H.S. terms can be written dimensionally as :

$$\frac{[ML^2T^{-3}I^{-2}][M^2L^4T^{-6}I^{-4}+M^2L^4T^{-6}I^{-4}+M^2L^4T^{-6}I^{-4}]^{1/2}}{[ML^2T^{-3}I^{-2}][ML^2T^{-3}I^{-2}]}$$

$$=\frac{[ML^2T^{-3}I^{-2}][M^2L^4T^{-6}I^{-4}]^{1/2}}{[M^2L^4T^{-6}I^{-4}]}=\frac{M^2L^4T^{-6}I^{-4}}{M^2L^4T^{-6}I^{-4}}$$

= Dimensionless quantity.

Since the quantities on both sides are dimensionless, therefore, **no correction** *is required.* **(Ans.)**

Example 1.11. *Find the dimensions of each term in the following equation and make a dimensional check of equation :*

$$E = RI + j\omega LI + \frac{\mu_0 NAM}{\delta^2}v$$

N = Being a constant, it is dimensionless; $A = [L^2]$; $\delta = [L]$

where, E = e.m.f.; I = current; R = resistance; ωL = inductive reactance; μ_0 *= permeability of free space; v = the velocity; M = magneto-motive force; N = number of turns; A = area of pole pieces; δ = length of the air.*

Solution.

$$E = RI + j\omega LI + \frac{\mu_0 NAM}{\delta^2}v \quad \text{...(Given)}$$

The dimensional equations of the various terms are :

$$E = \frac{\text{Work done}}{\text{Charge}} = \frac{\left[ML^2T^{-2}\right]}{[IT]} = \left[ML^2T^{-3}I^{-1}\right]$$

$$R = \frac{\text{E.m.f.}}{\text{Current}} = \left[\frac{[ML^2T^{-3}I^{-1}]}{[I]}\right] = [ML^2T^{-3}I^{-2}]$$

$$\omega L = 2\pi fL = [T^{-1}\mu L]$$

μ_0 = Being a constant, it is dimensionless

$$v = [LT^{-1}]$$

Substituting these value in the given equation, we have :

$$[ML^2T^{-3}I^{-1}] = [ML^2T^{-3}I^{-2}][I]+[\mu LT^{-1}][I]+\frac{[L^2][I]}{[L^2]}[LT^{-1}]$$

$$= [ML^2T^{-3}I^{-1}] + [\mu LT^{-1}][I] + [LT^{-1}I]$$

But, $$[I] = [\mu^{-1/2} M^{1/2} L^{1/2} T^{-1}]$$

or, $$[\mu] = \frac{[MLT^{-2}]}{[I^2]}$$

$$\therefore \quad [ML^2T^{-3}I^{-1}] = [ML^2T^{-3}I^{-1}] + \frac{[MLT^{-2}][LT^{-1}I]}{[I^2]} + [LT^{-1}I]$$

$$= [ML^2T^{-3}I^{-1}] + [ML^2T^{-3}I^{-1}] + [LT^{-1}I]$$

Thus we observe that the *third term on the R.H.S. does not check dimensionally, hence the equation is inaccurate.* **(Ans.)**

1.6 DETERMINATION OF ABSOLUTE UNITS

The determination of the absolute values of electrical quantities is very essential for establishing a satisfactory system of measurements. *It involves the measurement of a unit in terms of physical quantities which are measured practically.* Thus if we can determine the units of, say, current and e.m.f. by practical measurements, standards for all other electrical quantities like resistance etc. can be set up in terms of current and e.m.f.

- There is not much difficulty in making absolute measurements provided that an accuracy of a few parts in a hundred is all that is required and in that case absolute measurements of current and resistance can be made in any reasonably well equipped laboratory by a number of methods.
- The determination of absolute unit of e.m.f. involves the measurement of energy which is difficult to measure and, therefore, the determination of absolute unit of e.m.f. is *not* usually attempted. Normal practice is to determine the units of *current* and *resistance* and to express the units of all other quantities *in terms of these quantities.*
- The unit of power is determined in terms of the units of current and potential difference.
- Mutual or self inductors are set up in forms for which the inductance can be calculated in terms of its dimensions. Similarly, the capacitance is also measured in terms of mutual inductance.

1.6.1. Absolute Measurement of Current

The dimensions of current are $[\mu^{-1/2} M^{1/2} L^{1/2} T^{-1}]$ or $[M^{1/2} L^{1/2} T^{-1}]$ if permeability μ is regarded as non-dimensional quantity. The dimensions of $(\text{current})^2$ are $[MLT^{-2}]$ which are the dimensions of the force; thus the absolute measurement of current involves the measurement of force. This force may be exerted in the following *two ways* :

1. By the current in a solenoid upon a suspended magnetic needle ... as in a *tangent or sine galvanometer.*
2. By the current in one part of the circuit upon other part of the circuit in series with it and carrying the same current ... as in an *electro-dynamometer* or a *current balance.*

- *"Galvanometer methods"* of determination of current suffer from the following *disadvantages* :

(*i*) There is always some uncertainty about the exact position of the poles of the magnetic needle used.

(*ii*) Horizontal component of the magnetic field of earth is required to be determined separately with great accuracy.

(*iii*) The following assumptions made in deriving the expression for deflection of galvanometer are not justifiable in practice :

— The needle is suspended exactly at the centre of coil.

— The axis of needle is horizontal.

— The needle is infinitesimally small.

— The plane of galvanometer coil lies exactly in the magnetic meridian, and is exactly vertical.

- In *"electro-dynamometer method"*, the current is measured in terms of the torsion of suspension wire or of a bifilar suspension, therefore this method is also *not satisfactory.*
- The methods which make use of some form of current balance are the most satisfactory and, therefore, commonly used. The absolute measurement of current by *Rayleigh's current balance method* is described below.

Rayleigh current balance :

Fig. 1.1 shows a Rayleigh's current balance :

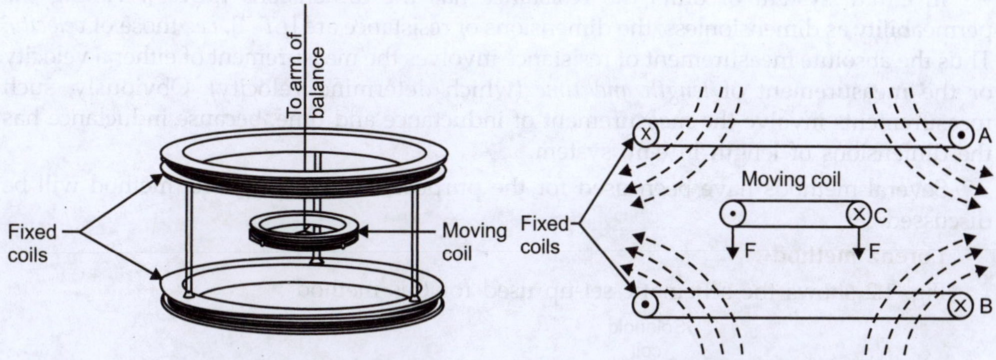

(a) Arrangement of coils in the Rayleigh current balance

(b) Arrangement of coils–working principle

Fig. 1.1. Rayleigh current balance — For absolute measurement of current.

— It consists of two fixed coils (A and B) and a small moving coil (C) which is suspended from the arm of a sensitive balance. The three coils are connected in *series.*

— If I is the value of current through them, then the force F exerted on the suspended coil is given as :

$$F = I^2 \frac{dM}{dx} \text{ newton} \qquad ...(1.17)$$

where, $\frac{dM}{dx}$ = The rate of change of mutual inductance between the fixed and moving coils. This is expressed in henry/metre.

— The value of $\frac{dM}{dx}$ can be obtained from the physical dimensions of the coils and F is purely a mechanical quantity which can be accurately measured by the balance. *Thus the value of I is dependent upon units of purely physical quantities and is independent of other electrical units.*

— For measuring the value F, it is usual to counterbalance the suspended coil when the force on it is upwards. The current through the suspended coil is then reversed thus reversing the force. The additional counterbalance weight gives the value $2F$.

It should be noted that the expression for the force is in newton and since the weights used in counterbalancing the force are in kg, the value of 'g' acceleration due to gravity must be known. The accuracy of the current measurement will depend on the accuracy with which the value of 'g' is known.

Precautions necessary for high accuracy :

1. Precision balance of special form should be used.
2. Much care is necessary in construction to ensure that the flexible leads, used for leading current into the moving coil or coils, must not exert any appreciable torque upon the moving system.
3. The cooling of the coils is very important. Water jackets are used for the fixed coils and water cooled chambers are provided for the moving coils.
4. In order to check the error due to distortion of magnetic field, bobbins made of non-magnetic materials should be used.

1.6.2. Absolute Measurement of Resistance

In e.m.u. system of units the resistance has the dimensions [μLT^{-1}]. Taking the permeability as dimensionless, the dimensions of resistance are [LT^{-1}], *i.e.*, those of *velocity*. Thus the absolute measurement of resistance involves the measurement of either a velocity or the measurement of *length and time* (which determine velocity). Obviously, such measurements involve the measurement of inductance and time, because inductance has the dimensions of length in e.m. system.

Several methods have been used for the purpose, here only Lorenz method will be discussed :

Lorenz method :

Fig. 1.2 shows the schematic set-up used for this method :

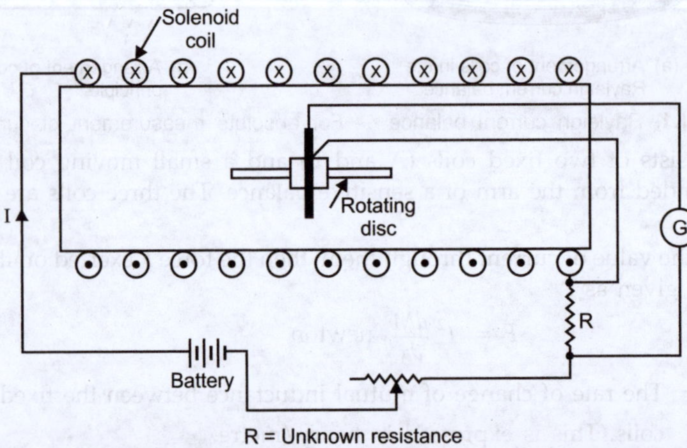

Fig. 1.2. Lorenz method – For absolute measurement of resistance.

— In this method/arrangement there is a circular metallic disc which is mounted concentrically inside a solenoid. The disc is rotated at a *uniform speed.*

— A steady current (I) is passed through the solenoid.

— R is a *low resistance whose absolute value is to be determined.* It is connected across the disc with the help of leads and two brushes. One brush presses the edge of the disc and the other brush makes a contact with the disc near its centre. A sensitive galvanometer G is included in one of the leads.

— The magnetising current (I) for the solenoid is supplied by a battery. This current passes the resistance R also.

— When current is passed through the solenoid, a magnetic field is produced. As the disc rotates an e.m.f. is induced in it. The current flowing through R produces a voltage drop across it. *The e.m.f. generated in the disc is balanced against the voltage drop* in resistance R, the balance being indicated by the galvanometer G. The *disc speed is adjusted till a balance is obtained.*

Let M be the mutual inductance between disc and solenoid in henry. Since inductance is defined as flux linkages per unit current, therefore, the value of *magnetic flux linkages* produced by the disc when a curent of I ampere passes through the solenoid is MI.

But, flux linkages of the disc = number of turns × flux linking the disc = flux linking the disc (since number of terms in a disc is one). Thus we conclude that the value of flux passing perpendicularly through the disc surface is MI weber when a current I ampere passes through the solenoid.

Let n be the speed of the disc in r.p.s. when the deflection of galvanometer is *zero*.

$$\therefore \quad \text{E.m.f. induced in the disc} = \text{Rate of change of flux linkages}$$

$$= MI \times n = MIn \text{ volt} \qquad ...(1.18)$$

Also, voltage drop in resistance $R = IR$ volt

where, I and R are in ampere and ohm respectively.

Now, when the deflection of galvanometer is zero,

$$MIn = IR$$

or,

$$R = Mn \qquad ...(1.19)$$

Knowing the value of mutual inductance M, which is calculated from the dimensions of the solenoid and disc, and from their relative positions and the value of rotational speed of disc n in r.p.s. which can be obtained by stroboscopic method, the value of unknown resistance (R) can be determined.

Precautions necessary for accuracy :

1. In order to determine the value of M accurately, both the disc and solenoid must be carefully constructed and their dimensions are accurately measured with the help of precision measuring instruments.

2. The dimensions of the solenoid relative to the disc diameter must be chosen such that the value of M may depend, as little as possible, upon the effective dimensions of the disc, which cannot be determined, when rotating, with great accuracy.

3. The effect of earth's magnetic field upon the e.m.f. induced should be made small. This is accomplished by arranging the plane of the disc to be in the magnetic meridian and making two measurements, one with current flowing in one direction and the other with the direction of current reversed.

4. The thermo-electric e.m.f. of the brush contacts should be reduced. This is done by using two phosphor-bronze discs of special construction and two solenoids.

5. The speed n should be as accurate as possible. It is determined by stroboscopic methods, or by a driven chronograph. It is advisable to incorporate flywheel to ensure uniform speed.

Example 1.12. *The following data relate to a simple form of Lorenz apparatus for absolute measurement of resistance :*

Diameter of solenoid = 60 mm

Length of solenoid = 600 mm

Number of turns of wire on the solenoid = 7500
Diameter of the rotating disc = 50 mm
Speed of the disc when the galvanometer indicates balance condition = 5845 r.p.m.
Calculate :
(i) The mutual inductance between the disc and the coil in henry.
(ii) The value of resistance in ohm.

Solution. Given: Radius of solenoid $r_s = \frac{d}{2} = \frac{60}{2}$, 30 mm or 0.03 m ; l_s = 600 mm or 0.6 m ; N = 7500 ; Radius of disc, $r_d = \frac{50}{2}$ = 25 mm = 0.025 m ; $n = \frac{5845}{60}$ = 97.417 r.p.s.

(i) **Mutual inductance, M :**

Intensity of field at the centre, for a long solenoid,

$$H = \frac{NI}{2\sqrt{(l/2)^2 + r_s^2}} \qquad ...(i)$$

Field intensity at the centre, $B = \mu_0 H = \frac{\mu_0 NI}{2\sqrt{(l_s/2)^2 + r_s^2}}$ ...(ii)

Flux linked by the disc, $\phi = B \times$ area of the disc

$$= \frac{\mu_0 NI}{2\sqrt{(l_s/2)^2 + r_s^2}} \times \pi r_d^2$$

$$= \frac{\pi\mu_0 r_d^2 NI}{2\sqrt{(l_s/2)^2 + r_s^2}} \qquad ...(iii)$$

The disc is treated as a single turn winding.

∴ Flux linkages of disc, $\psi = \frac{\pi\mu_0 r_d^2 NI}{2\sqrt{(l_s/2)^2 + r_s^2}}$ ...(iv)

Mutual inductance between disc and solenoid,

$$M = \frac{\psi}{I} = \frac{\pi\mu_0 r_d^2 N}{2\sqrt{(l_s/2)^2 + r_s^2}}$$

$$= \frac{\pi \times 4\pi \times 10^{-7} \times (0.025)^2 \times 7500}{2\sqrt{(0.6/2)^2 + (0.03)^2}} \text{ H} = 3.07 \times 10^{-5} \text{ H}$$

= **0.0307 mH** **(Ans.)**

(ii) **Value of resistance R :**

Resistance $R = Mn = 0.0307 \times 10^{-3} \times 97.417 =$ **2.99 mΩ** **(Ans.)**

1.7 STANDARDS OF MEASUREMENT

The ***measurement standard*** *is the physical embodiment of the unit of measurement as well as that of its submultiple value.*

As the measurement is a process of comparison, therefore, regardless of our measurement method, we must employ a basis of comparison – *standardised units.* The standard must be precisely defined, and because different systems of units exist, the method of conversion from system to system must be mutually agreed upon.

— It is of significant importance that a relatonship between the standards and the read out scale of each measuring system must be established through a process known as *calibration.*

Classification of Standards :

The standards of measurement are often *classified* as follows :

1. International Standards.
2. Primary Standards.
3. Secondary Standards.
4. Working Standards.

1. **International standards :**

— These standards are defined on the basis of international agreement.

— They represent the units of measurements which are closest to the possible accuracy attainable with present-day technological and scientific methods.

— These are checked and evaluated regularly against *absolute measurements* in terms of the fundamental units.

— These standards are maintained at the International Bureau of Weights and Measures and are *not* available to the ordinary user of measuring instruments for the purposes of calibration and comparison.

- The improvements in the accuracy of absolute measurements have made the International units superfluous and they have been *replaced by absolute units.*

2. **Primary standards :**

The highest standard of either a base unit or a derived unit is called a **primary standard.**

- These standards essentially are copies of international prototypes and are kept throughout the world in national standard laboratories and institutions of similar standing.
- These standardized units are :

— *Quite stable;*

— *Independent;*

— *Invariant;*

— *Not relative but finite.*

While setting up the primary standards, the following points are given due considerations :

(*i*) Long time stability of the material.

(*ii*) Low temperature coefficient.

(*iii*) No deterioration in its characteristics with changes in humidity, temperature and other environmental conditions.

(*iv*) Rigidity of construction.

(*v*) Accuracy of machining.

(*vi*) Most sophisticated techniques to be used for measuring physical dimensions which determine the accuracy of the standard.

- The primary standards *constitute the ultimate basis of reference* and are used for the purpose of *"verification and calibration of secondary standards".*

— These standards have the *"highest possible accuracy"* but are *very expensive* to own and maintain.

— The primary standards are *not available for use outside the national laboratories.*

3. **Secondary Standards :**

The secondary standards are the *reference calibrated standards designed and calibrated from the primary standards.*

- These are sent periodically to the national standard laboratories for their calibration (the interval of calibration depends upon the accuracy and the type of standard being maintained).
- These standards are kept by the measurement laboratories and the industrial organisations to check and calibrate the general tools for their accuracy and precision.

4. **Working Standards :**

- These standards have an accuracy of *one order lower than that of the secondary standards.*
- These are the normal standards which are *used by the workers and technicians who actually carry out the measurements.*

1.8 STANDARDS OF MASS AND LENGTH

Standards of Mass :

Primary standards :

- *It is a* prototype kilogram kept at national physical laboratory of every country.
- These standards have an accuracy of 1 part in 10^8 and occasionally verified against the standard at the International Bureau.

Secondary standards :

- These are kept by the industrial laboratories, have an accuracy of 1 ppm (part per million) and may be checked against the primary standard kept at the national physical laboratories.

Working standards :

- These standards, which are used commercially, are available in a wide range of values to suit almost every application.
- These have an accuracy of the order of 5 ppm.
- These standards are verified against the secondary standards.

Standards of Length :

- The metre (the metric unit of length) was initially defined as the ten-millionth part of the meridianal quadrant through Paris. This was materially represented by the distance between two lines engraved on a platinum-irridium bar preserved at the International Bureau of Weights and Measures near Paris.
  - — Later on, metre was redefined more accurately in terms of an *optical standard*, namely, the orange-red radiation of *krypton atom*. The internationally specified krypton-86 discharge lamp, excited and observed under well-defined conditions, emits orange light whose wavelength now constitutes the basic standards of length to 1 ppm.
  - — The metre, the unit of length in S.I. system, is now defined as equal to 1,650,763.73 wavelengths in vacuum of the orange-red light radiation of the krypton-86 atom in its transition between level ${}^2p_{10}$ and ${}^5d_5$.

— In, 1983, the metre was redefined in terms of time which can be specified accurately using atomic standards. The metre is now the length travelled by light in vacuum in a time interval of $\frac{1}{299792458}$ second. The speed of the light therefore is now 299792458 m/s.

- The *working standards* of length, which are most widely used in industries, are precision gauge blocks, made of steel. These steel blocks have two parallel surfaces, at specified distance apart, with accuracy tolerances in the 0.5–0.25 micron range. Such precision gauges are cheaper in cost and of high accuracy.

1.9 ATOMIC FREQUENCY AND TIME STANDARDS

These days, due to technological advancements, it is very easy to construct clocks of great precision and accuracy by frequency conversion (because frequency $\propto \frac{1}{\text{time}}$).

The transition between two energy levels E_1 and E_2 of an atom is accompanied by the emission (or absorption) by radiation given by :

$$E_2 - E_1 = h\nu \quad \text{...(1.20)}$$

where, h = Planck's constant, and

ν = A physical constant (which is a frequency) provided that the energy states are not affected by external conditions such as magnetic fields. This constant depends only on the internal structure of the atom.

Since frequency is inversely proportional to time, atomic clocks can be constructed on the above mentioned principles.

— In a *passive* atomic resonator, transitions are induced by driving the atomic device with a signal derived from a quartz oscillator, which in turn is stabilised to produce the proper driving frequency.

— The *active* resonators are micro-wave masers such as atomic hydrogen masers, which produce stimulated emission in the microwave range.

The *main* "**atomic standards**" used are :

1. *Primary or absolute standards :*
   (*i*) Caesium beam standard.
   (*ii*) Hydrogen maser standard.
   - These standards do not need to be referenced to any other standard for checking or calibration.
2. *Secondary standards :*
   (*i*) Rubidium vapour standard.
   - This standard needs to be initially set up against a primary standard, and then periodically recalibrated.
   - Quartz oscillator is often used to provide feedback frequency to caesium or rubidium atomic standards, which are passive.

Advantages of atomic standards :

1. Results obtained with these standards are more uniform.
2. They form a better and convenient time base.
3. Their accuracy is much higher than that of standards based upon astronomical measurements.

4. They form the *primary frequency standard.*
   - Some of the atomic clocks are in operation which give a precision of 1 μs per day.

1.10 E.M.F. STANDARDS

The primary, secondary and working standards of e.m.f. are described below :

- Several types of voltaic cells were used as a voltage reference in the past. These cells were developed over a long period and to a high degree of excellence. Such cells *have a long life and serve as a standard in most satisfactory way.*

1.10.1. Primary Standard of E.m.f. – Saturated Weston Cell

The primary voltage standard is the *"Saturated normal Weston cell"*. It was developed by Edward Weston in 1892.

- This is a *voltaic cell.*

Construction :

Fig. 1.3 shows the construction of a saturated Weston cell :

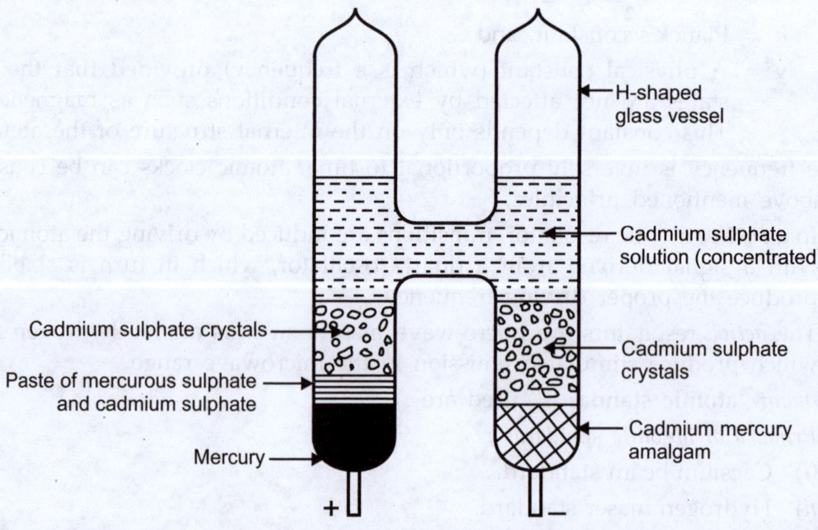

Fig. 1.3. Saturated Weston cell.

— It consists of a H-shaped glass vessel, each leg being 25 mm in diameter containing saturated solution of cadmium sulphate ($CdSO_4$) in the upper part, then crystals of cadmium sulphate, the paste of mercurous sulphate (Hg_2SO_4) and cadmium sulphate over the mercury at the bottom of one leg and crystals of cadmium sulphate and cadmium mercury amalgum at the bottom of second leg. The two legs are hermetically sealed.

— *"Cadmium sulphate solution"* acts as an *electrolyte, "mercury"* acts as a *+ve electrode,* cadmium mercury amalgam (a *"solution"* of 1 part of cadmium in 7 parts of mercury) acts as a *–ve electrode* and *"paste of mercurous sulphate and cadmium sulphate"* acts as a *depolariser.*

A simplified representation of the cell reaction is :

$$Cd + Hg_2SO_4 = CdSO_4 + 2Hg \quad \text{...(1.21)}$$

The e.m.f. of the cell remains absolutely constant so long as the amalgam has both a solid and liquid phase present.

— The connections of the cell to the external circuit are made by platinum wires sealed into the glass.

- Saturated cells are normally allowed to age for a year before they are considered to be of constant e.m.f. Following this period, their life span appears indefinite excluding the possibility of accidents.

Characteristics :

(*i*) The e.m.f. of the cell is *1.01864 volt.*

The e.m.f. of the cell changes with temperature according to the following equation :

$$E_t = E_{20} - 46 \times 10^{-6} (t - 20) - 0.95 \times 10^{-6} (t - 20)^2 + 10 \times 10^{-9} (t - 20)^3 \quad ...(1.22)$$

where, E_t = E.m.f. of cell at any temperature t°C, and

E_{20} = E.m.f. of cell at 20°C.

(*ii*) The cell voltage drops by 40 μV/°C. The cell is kept in an oil bath to control its temperature within 0.01°C.

(*iii*) It has a life span of about 10 to 20 years. The drift in the value of voltage is about 1 μV per year.

(*iv*) Its temperature coefficient is –40 μV/°C.

Uses :

- Since this cell is a standard for maintenance of the volt, therefore, it is *only used in standardising laboratories.*
- These cells are *not* used for ordinary work because of temperature requirements and also because they are not portable.

Example 1.13. *A standard cell has a voltage rating of 1.018500 V and internal resistance of 500 Ω. The insulation resistance between its terminals is 5.5 MΩ.*

Calculate :

(i) The current drain due to insulation resistance.

(ii) The terminal voltage.

Solution. *Given :* Voltage rating of the cell = 1.018500 V; Internal resistance = 500 Ω; Insulation resistance between the terminals = 5.5 MΩ.

(*i*) **The current drain :**

The current flowing through the insulation resistance

$$= \frac{1.018500}{500 + 5.5 \times 10^6} = 0.185\ \mu A$$

Thus there is a *constant drain of current of* **0.185 μA.** **(Ans.)**

(*ii*) **Terminal voltage :**

Voltage drop in internal resistance = $0.185 \times 10^{-6} \times 500$ V = 92.5 μV

∴ Terminal voltage = Voltage rating of the cell (e.m.f.) – voltage drop in internal resistance

$= 1.018500 - 92.5 \times 10^{-6}$ = **1.0184075 V** **(Ans.)**

1.10.2. Secondary Standard of e.m.f. – Unsaturated Weston Cell

The secondary standard of e.m.f. is unsaturated Weston cell (Fig. 1.4).

Construction :

— This cell is similar to a saturated cell except that it *does not have the cadmium sulphate crystals.*

— It has the same electrodes but the solution is at 4°C, and therefore, *unsaturated at room temperature.*

— In this cell, in addition, a 'sputum' or retaining member (porous plugs in this case) is used over each electrode to *hold the material in place* and hence these cells are considered *portable.*

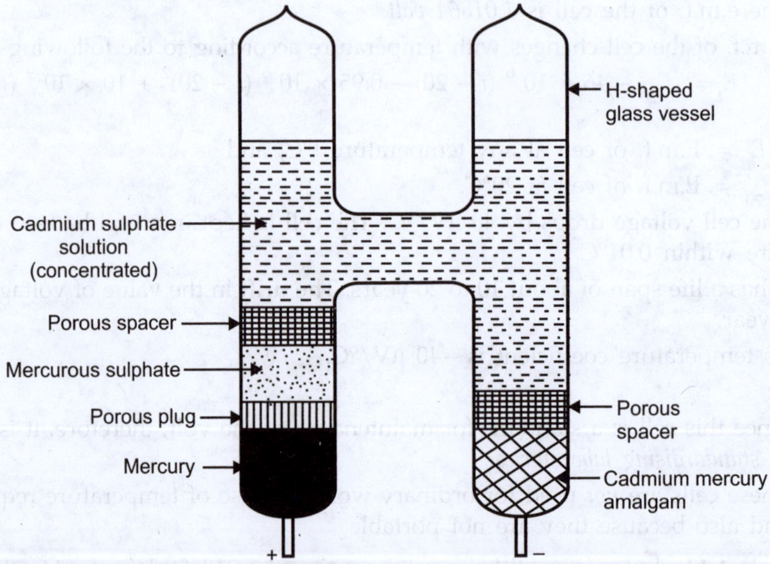

Fig. 1.4. Unsaturated Weston cell.

— The cells are enclosed in a bakelite case which is lined with 1.5 mm thick copper to help keep all parts of the cell at a *uniform temperature.*

— After being moved from one place to another, the cell should be allowed a few days to settle so that the results obtained from measurements involving cells are accurate.

Characteristics :

(*i*) The e.m.f. of new unsaturated cells varies; it lies between 1.0180 and 1.0194 volt. Therefore, each cell must be calibrated against the primary standard in order to know its exact e.m.f. The temperature at which this e.m.f. is measured should be specified.

(*ii*) Even though this type of cell becomes less stable with age, yet it is remarkably good.

(*iii*) The decrease in e.m.f. is about 30 to 50 μV per year.

(*iv*) Its life is shorter than a saturated type. These cells, if handled by qualified personnel in proper way, can last up to 20 years or even more. These cells should be checked every year or at an interval of two years against primary standard.

(*v*) Its temperature coefficient is –10 μV/°C. It has an accuracy of 0.005 per cent.

(*vi*) This cell has a positive temperature effect at one electrode and negative at the other, therefore, it is important to keep all parts of the cell at the same temperature.

Uses : Such cells are used as *secondary standard of volt.*

Precautions in the use of Standard Cells :

1. Mercurous sulphate is sensitive to light and, therefore, it should not be exposed to light.
2. Saturated cells should be kept in an upright position because if they are tilted there is always the danger of mixing of electrode materials. Therefore these cells cannot be shifted and are sent through special messengers.
3. Laboratory conditions, which would result in the condensation of moisture on the cell case, should be avoided.
4. The standard cells should be kept at dry places having uniform temperatures of about 15°C to 20°C for the purpose of storage, so that the hysteresis effect due to change in temperature and possibility of leakage currents owing to moisture on the insulating materials between the terminals of the cell are avoided.
5. The main limitation of the use of standard cells is the current drawn from the cell at any time; this should be negligibly small. Standard cell being a potential device gets injured if an appreciable current is drawn from it; a very minute current (not exceeding 10 μA) should be drawn from the cell and that also only for a short duration. Hence such cells should be used in measurements in *either with potentiometer or some other open-circuited device.* Protective resistance of high value is inserted in series with the cell to keep the current drawn to a minimum value while balancing the potentiometer.
6. Such a cell should never be short-circuited accidentally as voltage of the cell gets changed permanently on short-circuit. Of course with time, it recovers its voltage but not completely.

- A voltmeter should never be used for measuring the voltage of a standard cell due to the following reasons :

(*i*) Due to the current there is a possibility of damage to the cell;

(*ii*) This measurement is meaningless since the cells have a high internal resistance (about 600 to 800 ohms).

The voltage should be measured with a potentiometer only because at the balance the current drawn from the cell is zero. When using the cell in potentiometric work, a resistance of 20 kΩ should be inserted in series with the cell in order to avoid excessive drain of current when there is unbalance.

1.10.3. Zener Diode Reference Standard

Zener diode (a special junction diode) provides an alternative method of a reference standard. It is a *reverse-biased diode operating in the breakdown region.* It is *quite accurate for many measurement requirements.*

Fig. 1.5 shows the Zener diode in a typical application as a laboratory standard :

— The supply voltage *E* is usually much higher than the Zener voltage of diode assuring breakdown of the diode.

— Resistor *R* usually has a high value and is placed in series with the diode and battery. It serves to limit the

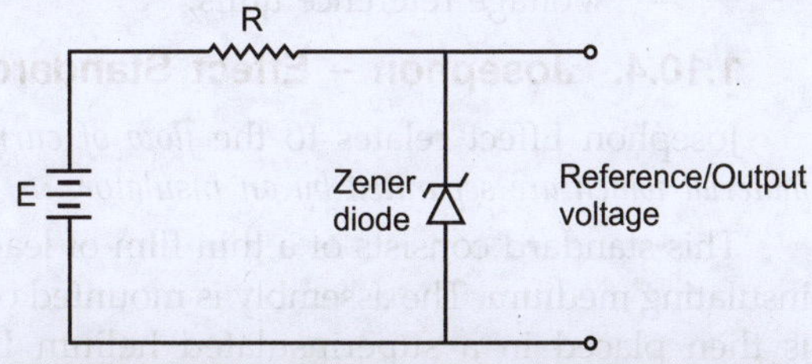

Fig. 1.5: Zener diode laboratory standard.

current through the diode during breakdown to a safe value. Also, when the supply voltage varies for any reason, most of this variation is taken up in the voltage drop across R with a very small change in diode current and as a result the *reference voltage* (*output voltage*) *remains practically the same.*

- Commercial voltage standards using zener diode essentially consist of :
  (*i*) A zener-controlled voltage source placed in a temperature-controlled environment to improve its long-term stability;
  (*ii*) A precision output voltage divider.
- If a standard voltage considerably greater than 5 V is to be provided, there is an advantage in using several low-voltage diodes in *series*, rather than one high-voltage diode as the temperature coefficient strongly depends upon the zener voltage for which the diode is made, being low for diodes of 4 to 5 volt zener voltage and higher for higher-voltage diodes.
- Versatile *laboratory working standards* have been developed with accuracies of the order of standard cell accuracy, in which zener-controlled voltage source is used in a temperature-controlled environment with a precision output voltage divider.

> **Note.** The breakdown of Zener voltage depends upon the nature of semiconductor and the amount of doping and ranges from about *3V to several hundred volts.* If the diode is *heavily doped,* depletion layer will be *thin* and consequently the junction will breakdown at a *low reverse voltage.* On the other hand, a *lightly doped diode* will break down at *higher reverse voltage.* The breakdown phenomenon is reversible and harmless so long as the safe operating temperature is maintained.

Advantages :

1. The advantage of these standards over standard cell is that, unlike the standard cell, the *semiconductor device is immune to the effects of short circuit and overload conditions.*
2. For chemical cells, spring board or other applications where vibration is constantly present do not present ideal conditions. The Zener sources which are rugged may furnish the accuracy required under such conditions.
3. The Zener sources can be substituted for standard chemical cells in recording potentiometers which may be subjected to freezing temperatures. This is due to the fact that the standard cells cannot work below 4°C while Zener sources give good accuracies at all temperatures if properly compensated.
4. Whereas chemical cells can be built to voltage level of about 1 volt, the Zener sources can be built to any voltage level.
5. The constant voltage drop of Zener diode may be utilised in :
   — Voltage regulators; — Compensators;
   — Limiters; — Clippers;
   — Voltage reference units; — Voltage standards.

1.10.4. Josephon – Effect Standard

Josephon Effect relates to the *flow of current between two pieces of a superconducting material which are separated by an insulator.*

This standard consists of a thin film of lead separated by lead oxide which acts as an insulating medium. The assembly is mounted on a glase substrate. The probe, thus formed, is then placed in a superinsulated helium Dewar system. The tunnelling junction is excited by a microwave source having a frequency of nearly 9 GHz, and is biased by a supply current source. Under these conditions the Josephon junction produces a voltage as per the following equation :

$$E = \frac{Nhf}{2e} \qquad \text{...(1.23)}$$

where, N = An integer (its value can be controlled by the bias current, thereby varying the voltage produced in discrete steps),

h = Planck's constant,

f = Frequency of microwave source, and

e = Electron charge.

- These voltage standards have an accuracy of about 0.05 ppm and, therefore, can be employed for calibration of other e.m.f. standards using potentiometers.

1.11 RESISTANCE STANDARDS

In S.I. system of units, the absolute value of the *ohm* is defined in terms of fundamental units of *length* and *mass.*

Absolute measurement of the ohm is done against a group of resistance standards, which are maintained at the International Bureau of Weights and Measures in Severes. This group is periodically checked against each other, and with similar groups of standards kept at other "standards laboratories" in the world.

Standard resistor. Fig. 1.6 shows the construction of a typical standard resistor :

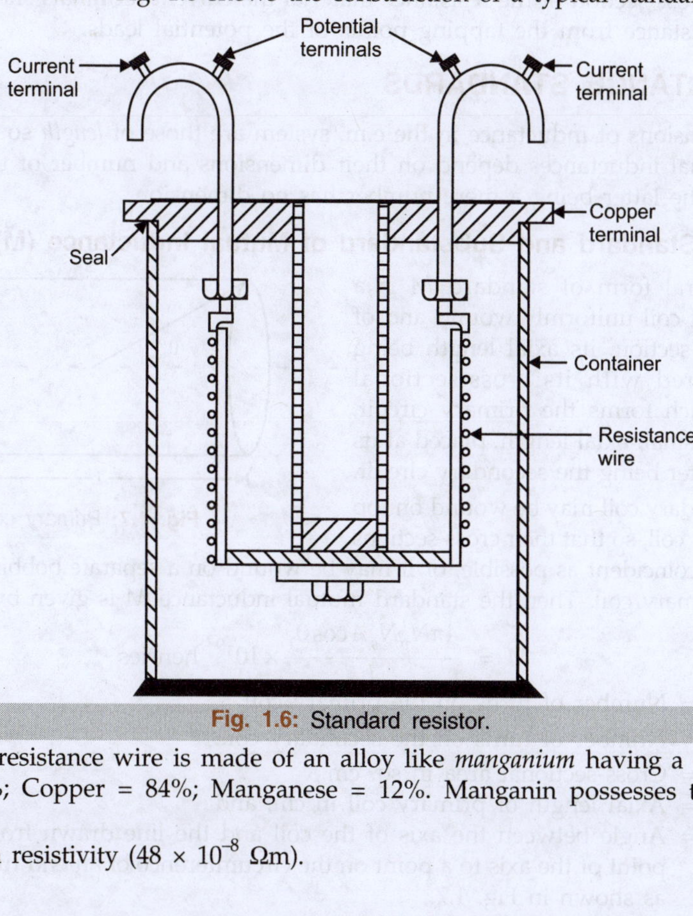

Fig. 1.6: Standard resistor.

— The resistance wire is made of an alloy like *manganium* having a composition : Nickel = 4%; Copper = 84%; Manganese = 12%. Manganin possesses the following properties :

(*i*) High resistivity (48×10^{-8} Ωm).

(*ii*) Low thermal e.m.f. with copper (Below 3 μV/°C).

(*iii*) Low temperature coefficient of resistance (40×10^{-6}/°C at 20°C).

— The resistance coil is so wound that it reduces stress.

— The coil is immersed in oil for temperature stability.

— The resistor is housed in a sealed moisture-free container.

— The connections to the resistance coil are silver soldered and the terminal hooks are made of nickel plated copper. The resistance of the standard at any temperature *t*°C is given by :

$$R_t = R_{25} + \alpha (t - 25) + \beta (t - 25)^2 \quad ...(1.24)$$

where, R_{25} = Normal value of resistance at 25°C in Ω, and

α, β = Thermal coefficients whose values are 10×10^{-6} and -0.5×10^{-6} respectively so that change in resistance for a change in temperature of 10°C is less than 50 ppm.

- The major difference between the *low resistance* standards and those used for *reference purpose* only (which are not required to carry large currents) is that the *former ones are fitted with current terminals as well as potential terminals* (See Fig. 1.6.)

— The *potential terminals* are the points, between which the normal resistance of the standard is measured. The *current terminals* are the points by means of which standard resistor is connected to supply circuit. For obtaining uniform distribution of current through the cross-section of the resistance material the current terminals should be at an appreciable distance from the tapping points of the potential leads.

1.12 INDUCTANCE STANDARDS

The dimensions of inductance in the e.m. system are those of *length* so standards of self and mutual inductances depend on their dimensions and number of turns of wire in them but the latter being a mere number has no dimension.

1.12.1. Standard and Substandard of Mutual Inductance (M)

The general form of standard *M* is a single layer of coil uniformly wound and of circular cross-section, its axial length being large compared with its cross-sectional diameter, which forms the primary circuit, with a coil of small axial length, placed at its centre, the latter being the secondary circuit.

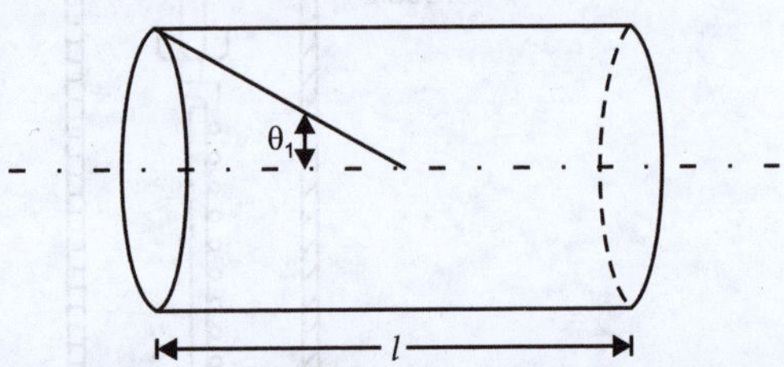

Fig. 1.7: Primary coil.

The secondary coil may be wound on top of the primary coil, so that their cross-sections are as nearly coincident as possible, or it may be wound on a separate bobbin and placed inside the primary coil. Then the standard mutual inductance *M* is given by :

$$M = \frac{4\pi N_p N_s A \cos\theta_1}{l} \times 10^{-9} \text{ henries}$$

where, N_p = Number of turns on the primary coil,

N_s = Number of turns on the secondary coil,

A = Cross-sectional area in sq. cm.,

l = Axial length of primary coil in cm, and

θ_1 = Angle between the axis of the coil and the line drawn from the centre point of the axis to a point on the circumference of an end turn of the coil as shown in Fig. 1.7.

Substandards of M are used as standard of mutual inductances for general laboratory purposes. Since the standards of *M* are not absolute standard, it is not essential that their dimensions shall be determined with great accuracy. The essential thing is that the substandard of *M* is as nearly as possible to the nominal value for which it is designed. When constructed they are compared with standard ones and their mutual inductance is adjusted.

Substandard *M* normally consists of two coils wound on a bobbin of *marble,* the end being separated by a flange. The wire is standard copper, with double silk coverings. After winding, the coils and bobbin are immersed in hot paraffin wax. When withdrawn and allowed to cool, the wax firmly fixes the wire in the coil in position.

Adjustment to the value of mutual inductance required is carried out by carrying the end of the coils through a further arc of circle in order to give the effect of a fraction of turn.

1.12.2. Standard and Substandard of Self Inductance (L)

The magnitudes of standard of *L* are calculated from formulae and performance is ensured by winding the coil of bare hard drawn copper wire under tension inside screw thread cut in a marble cylinder which should be very carefully ground so as to be as true cylindrical as possible.

As in the case of substandard of *M,* substandards *L* are constructed to have a nominal value which is usually a simple fraction of one henry. Such substandards are compared with standards L that are used as reference standards for general laboratory work.

1.13 CAPACITANCE STANDARDS

In capacitance standards, standard capacitors of fixed or variable values are used; the capacitance of such capacitors can be determined accurately by means of a formula and from dimensions.

Capacitance in electrostatic system of units has the dimensions of length and, therefore, the capacitance of absolute capacitors can be expressed in terms of their dimensions. Hence these dimensions should be known *very accurately* and should *not* change with the passage of time, once the capacitor has been made.

- The effect of time, frequency, temperature and other relevant factors on the capacitance of standard capacitors should be *as small as possible* and they should have *low dielectric losses and a high insulation resistance.*

1.13.1. Primary Standards of Capacitance

Primary standards of capacitance are of the following three types :

1. The concentric sphere type.
2. The concentric cylinder with guard ring.
3. The parallel plate type with guard plates.

The air is always preferred as dielectric because its *permittivity is definitely known and it is also free from absorption and dielectric loss.*

"Parallel plate capacitor" is used *rarely* as it requires very careful adjustments of plates to get the calculated value of capacitance accurately.

Standard air capacitors :

Usually these are *concentric cylinder type* capacitors in which *guard rings* of same diameter are used at both ends in order to shield the capacitor from external electrostatic influences and to define the effective area of electrodes accurately so that area to be used in calculation does not have any uncertainty.

— Air gap between the electrodes is kept large in order to withstand the applied voltage.

— Ends of the *inner cylinder* are made *bell shaped* and project beyond both ends of outer cylinder by a considerable length.

— Guard rings and *outer cylinder* are kept at same potential but are insulated from each other. Thus these capacitors are free from brush discharge and with efficient screening.

Fig. 1.8 shows a general sketch of an air capacitor.

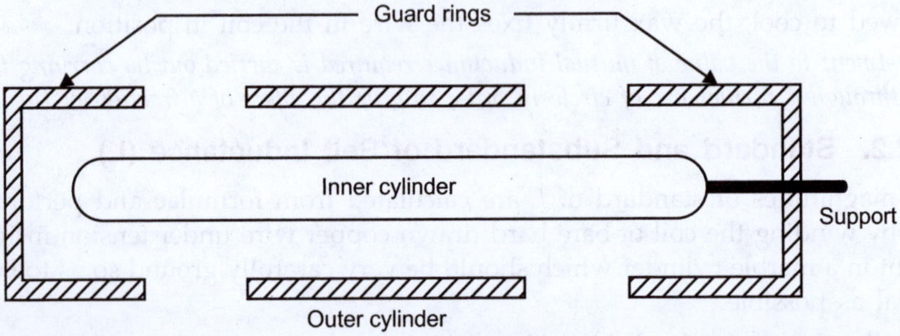

Fig. 1.8: Air capacitor.

- *These capacitors are used in measurement of dielectric loss and power factor of capacitors at high voltages.*

Compressed gas capacitors :

— In a compressed gas capacitor the gas used may be either air or nitrogen, but it should be *clean* and *dry*.

— In these capacitors gas filled at 10 bar increases the operating voltage as breakdown voltage of gas increases with the increase in pressure while the dielectric constant remains unchanged.

— These capacitors can be made for voltages up to 1500 kV.

— The capacitance of such capacitors is in the range of 10–200 pF and its loss angle is less than 0.00001 degree.

Precautions. Since air as dielectric has some disadvantages, therefore, following precautions need to be taken when using it as dielectric material in standard capacitors :

1. Owing to very low permittivity of air and long gap between electrodes, the air capacitor, even of small value, is *very bulky*.
2. Due to low dielectric strength of air long gap is required, so that air does not break down on the application of high voltage.
3. In air capacitors, dry air should be used otherwise dust particles settling in the gap between the plates may cause leakage troubles.
4. The electrodes of capacitors should be rigidly fixed in position by supports of some solid dielectrics such as fused quartz and amberite (which have no tendency to wrap and cause displacement of the plates from their original positions).

1.13.2. Secondary Capacitance Standards

The dimensions of the secondary capacitors need not to be known accurately since these are *calibrated capacitors*.

— In secondary standard capacitors a number of plates are used instead of only two as in the primary standards, so *greater value of capacitance can be obtained in small size.*

These have value up to about 0.02 μF as compared to 200 pF in case of primary standard capacitors.

— These capacitors have long period permeance of 1 part in 20,000 and temperature coefficient less than 1 part/10^5/°C.

— Their power factors are found to be less than 0.00001 at 1 kHz up to radio frequencies.

— These capacitors are properly screened to prevent the effect of stray electrostatic fields.

HIGHLIGHTS

1. The standard measure of each kind of physical quantity is called a *unit.*
2. A system in which the various units are all expressed in terms of a small number of fundamental units is called an *absolute system of units.*
3. The unique quality of every quantity which distinguishes it from all other quantities is called *dimension.*
4. S.I., the International system of units, has the following six base units : length; mass; time; electric current; thermodynamic temperature; luminous intensity.
5. *Rayleigh's current balance* method is employed for the absolute measurement of current.
6. *Lorenz method* is used for the absolute measurement of resistance.
7. The *measurement standard* is the physical embodiment of the unit of measurement as well as that of its submultiple value. The standards of measurement are often classified as :

   (*i*) International standards; (*ii*) Primary standards;
   (*iii*) Secondary standards; (*iv*) Working standards.
8. The primary voltage standard is "*Saturated Weston cell*", while the secondary voltage standard is "*Unsaturated Weston cell*".
9. Josephon-Effect relates to the flow of current between two pieces of a superconducting material which are separated by an insulator.
10. Standard air capacitors are used in measurement of dielectric loss and power factor of capacitors at high voltages.

OBJECTIVE TYPE QUESTIONS

Choose the Correct Answer :

1. The maximum current that can be drawn from a saturated Weston cell without damaging it, is

   (*a*) 100 mA momentarily (*b*) 100 μA continuously
   (*c*) 100 μA momentarily (*d*) 1 mA momentarily.
2. The internal resistance of a saturated Weston cell is of the order of :

   (*a*) a few ohm (*b*) a few kilo ohm
   (*c*) a few hundreds of ohm (*d*) a fraction of ohm.
3. The decrease in the value of emf of an unsaturated Weston cell is :

   (*a*) about 1 μV per year (*b*) about 1 mV per year
   (*c*) about 5 – 10 μV per year (*d*) about 30-50 μV per year.

4. The symbol *nm* stands for :
   (*a*) newton metre (*b*) nano metre
   (*c*) nano milli (*d*) number of moles.
5. In measuring the absolute value of ampere using Rayleigh's balance, the value of gravitational constant g must be accurately known because
   (*a*) the expression for the force is in newton and weights used in counterbalancing the force are in kg.
   (*b*) value of g is different at different places.
   (*c*) corrections have to be made for the weight of moving coil.
   (*d*) none of the above.
6. The material of wires used for making resistance standards is usually
   (*a*) Manganin (*b*) Nichrome
   (*c*) Copper (*d*) Phosphor bronze.
7. In order to reduce skin effect in resistance standards when they are used on high frequency A.C. :
   (*a*) the diameter of the resistance wire should be small.
   (*b*) adjacent turns of the resistance coil should carry currents in the opposite direction.
   (*c*) they should be provided with four terminals.
   (*d*) all of the above.
8. The units whose sizes cannot be chosen independently are called
   (*a*) Derived units (*b*) Fundamental units
   (*c*) Absolute units (*d*) Auxiliary fundamental units.
9. The dimensions of inductance in CGS e.s. system are :
   (*a*) $[\in L^{-1} T^2]$ (*b*) $[\in^{-1} ML^{-1} T^2]$
   (*c*) $[\in^{-1} LT^{-2}]$ (*d*) $[\in^{-1} L^{-1} T^2]$
10. In Lorenz method for absolute measurement of ohm, two measurements are made one with current in one direction and the other with current in the opposite direction. This is done to
    (*a*) eliminate the effect of thermal emfs.
    (*b*) eliminate the effect of earth's magnetic field.
    (*c*) eliminate the effects of both thermal emfs and earth's magnetic field.
    (*d*) eliminate the effect of temperature on coil and disc.
11. Zener sources when used as voltage standards
    (*a*) are reverse biased.
    (*b*) can be designed to cover a wide range of voltages.
    (*c*) are immune to the effects of short circuits.
    (*d*) all of the above.
12. The output voltage Josephon effect standard
    (*a*) can be changed in a continuous manner.
    (*b*) can be changed only in discrete steps.
    (*c*) can be changed continuously as well as in discrete steps.
    (*d*) cannot be changed.
13. Caesium beam standard is normally used as primary atomic standard for frequency because
    (*a*) caesium beam standard has an accuracy 3 parts in 10^{12}, has short-term stability, low drift and small size.

(b) hydrogen maser standard, though it is the most stable frequency source, has a large size.

(c) quartz crystal's operating frequency changes with change in load capacitance and also the drift in frequency is asymptotic with time.

(d) all of the above.

14. The disadvantages of atomic standards are :

(a) Their accuracy is lower than that of standards based upon astronomical measurements.

(b) The time taken for making measurements is long.

(c) The results obtained with atomic standards are not uniform.

(d) None of the above.

15. The dimensions of resistance CGS e.m. system are :

(a) $[\mu^{-1} LT^{-1}]$ (b) $[\mu^{-1} MLT^{-1}]$

(c) $[\mu LT^{-1}]$ (d) $[\mu LT^{-2}]$.

16. The base units in SI system are :

(a) metre, kilogramme, second

(b) metre, kilogramme, second, ampere

(c) metre, kilogramme, second, ampere, kelvin, candela, mole

(d) metre, kilogramme, second, ampere, kelvin, candela.

17. The international standard of length is defined in terms of

(a) length of earth's meridian passing through Paris.

(b) distance between two lines engraved on a platinum-iridium bar.

(c) wavelength in vacuum, of radiation of krypton – 86 atom in its two specified transitions.

(d) none of the above.

18. The most stable primary atomic standard for frequency is :

(a) Caesium beam standard (b) Hydrogen maser standard

(c) Quartz standard (d) Rubidium vapour standard.

ANSWERS

1. (c)	**2.** (c)	**3.** (d)	**4.** (b)	**5.** (a)	**6.** (a)	**7.** (a)
8. (a)	**9.** (d)	**10.** (c)	**11.** (d)	**12.** (b)	**13.** (d)	**14.** (d)
15. (c)	**16.** (c)	**17.** (c)	**18.** (b).			

THEORETICAL QUESTIONS

1. Define the following terms :

(i) Units (ii) Absolute units

(iii) Fundamental units (iv) Derived units.

2. What is the difference between rationalised and unrationalised systems of units?

3. Discuss S.I. system of units.

4. How many base and supplementary units exist in S.I. system? Give their definitions.

5. Define base units of S.I. system.

6. What are the advantages of S.I. units?

7. Prove that $\frac{1}{\sqrt{\mu_0 \epsilon_0}}$ = velocity of light, where μ_0 and ϵ_0 are respectively the permeability and permittivity of free space.

8. What is meant by dimension of an electrical quantity? Obtain the dimensions of resistance and current.

9. Derive dimensional equations for current and e.m.f. in both e.s. and e.m. systems.
10. Derive the dimensions of the following quantities in *L, M, T, I* system :
(*i*) Magnetic flux; (*ii*) Inductance; (*iii*) Capacitance; (*iv*) Magnetic flux density; (*v*) MMF; (*vi*) Reluctance; (*vii*) Magnetising force; (*viii*) Electric flux; (*ix*) Permittivity; (*x*) Resistance; (*xi*) Permeability.
11. Explain briefly an experiment for the absolute measurement of resistance. Give the precautions necessary to ensure accuracy of measurement.
12. Describe the method of absolute measurement of ohm.
13. Explain a method for the absolute measurement of current. Give the necessary precautions.
14. Discuss with the help of a diagram the principle and working of one form of current balance.
15. Explain the constructional details of a Weston standard cell and give the precautions in using standard cells.
16. Describe Josephon-effect standard for e.m.f. Explain how the output voltage can only be varied in discrete steps.
17. Explain briefly primary standard for mutual inductance.

UNSOLVED EXAMPLES

1. Prove that the following equation is dimensionally correct :

$$e = Blv$$

where, e = E.m.f. generated in a conductor,
B = Flux density,
l = Length of conductor, and
v = Peripheral velocity of conductor.

2. The pull of an electromagnet is given by the expression :

$$F \propto B^a A^b \mu^c$$

where, B = The flux density in the air gap,
A = Cross-sectional area of the gap, and
μ = Permeability of the air in the gap.

Find the values of *a*, *b* and *c* by analysing the equation dimensionally.

[**Ans.** $a = 2$; $b = 1$; $c = -1$]

3. Use dimensional equations to find out whether the following expression is dimensionally correct or not :

$$R_4 = \frac{R_2(1+\omega^2 R_3^2 C_3^2)}{\omega^2 R_1 R_3 C_3}$$

where R_1, R_2, R_3, R_4 are resistances, C_3 is a capacitance, and $\omega = 2\pi f$ (f being frequency in Hz).

$\left(\textbf{Ans.} \text{ Incorrect, correct eqn.} : R_4 = \dfrac{R_2(1+\omega^2 R_3^2 C_3^2)}{\omega^2 R_1 R_3 C_3^2}\right)$

4. Using dimensional analysis check whether the following expression is correct or not.

$$R = \frac{1+\omega^2 C^2 R^2}{\omega^2 CR}$$

$\left(\textbf{Ans.} \text{ No, correct expression} : R = \dfrac{1+\omega^2 C^2 R^2)}{\omega^2 C^2 R}\right)$

5. Prove that the following equations are dimensionally correct :

(*i*) $r = \dfrac{R\omega^2 M_1 M_2}{R^2 + \omega^2 L^2}$; (*ii*) $C = \dfrac{1}{\omega^2 M}$

where, r, R = Resistances,
M_1, M_2, M = Mutual inductances,
L = Self inductance,
C = Capacitance, and
ω = Angular frequency.

6. In one of the bridges, an expression for self inductance was derived as under :

$$L = \frac{R_2 R_3 C_4^2}{1 + \omega^2 C_4^2 R_4^2}$$

Find out if this expression is correct or not. If not correct, suggest a suitable correction.

$\left(\textbf{Ans.}\ \text{No, correct expression} : L = \dfrac{R_2 R_3 C_4}{1 + \omega^2 C_4^2 R_4^2}\right)$

7. The following expression was obtained in the course of a calculation :

$$I = \frac{V\omega M}{[(\omega M + R_1 R_2)^2 + \omega^2 L_1 L_2 R_1^2]^{1/2}}$$

Check whether the equation is dimensionally correct. If not, suggest any corrections.

$\left(\textbf{Ans.}\ \text{Incorrect}; I = \dfrac{V\omega M}{[(\omega^2 M^2 + R_1 R_2)^2 + \omega^2 L_1 L_2 R_1^2]^{1/2}}\right)$

8. In the course of calculation the following expression for current in a circuit was arrived at and it is suspected that an erorr has been made in it :

$$I = \frac{\omega M E}{[(\omega^2 M^2 + R_1 R_2)^2 + \omega^2 L_1 R_1^2]^{1/2}}$$

where, E = Voltage,
w = Angular velocity,
M = Mutual inductance,
L_1 = Self inductance, and
R_1, R_2 = Resistances.

Examine if the expression is dimensionally correct and if necessary, make a correction.

$\left(\textbf{Ans.}\ \text{No; Correct expression} : I = \dfrac{\omega M E}{[(\omega^2 M^2 + R_1 R_2)^2 + \omega^2 L_1 L_2 R_1^2]^{1/2}}\right)$

9. In a simple form of Lorenz apparatus for absolute measurement of resistance, the solenoid, is 50 mm in diameter and 500 mm long. It is uniformly wound with 8000 turns of wire. The rotating copper disc is 40 mm in diameter and the galvanometer indicates balance condition when the speed of the disc is 5952 r.p.m.

Calculate :

(*i*) The mutual inductance between the disc and the coil in henry.

(*ii*) The value of resistance in ohm. [**Ans.** (*i*) 0.0251 mH; (*ii*) 2.49 mΩ]

10. A standard cell has a voltage rating of 1.018500 V and an internal resistance of 500 Ω. The insulation resistance between its terminals is 5 MΩ. Find the current drain due to insulation resistance and the terminal voltage. [**Ans.** 0.204 μA; 1.018398 V]

CHAPTER

2 Basic and Digital Electronics

2.1 Electronic Components : Introduction – Active components – Passive components. **2.2. Electronic Devices :** General aspects - Semiconductors - Intrinsic semiconductor - Extrinsic semiconductor - PN junction diode - Zener diode - Tunnel diode - Bipolar junction transistor (BJT) - Field-effect transistor (FET) - Unijunction transistor (UJT) - Thyristor - Optoelectronic devices - Rectifiers. **2.3. Digital Electronics :** Introduction - Advantages and disadvantages of digital electronics - Digital circuit - Logic gates - Universal gates - Half adder - Full adder - Boolean algebra - Boolean laws - De Morgan's theorems - Operator precedence - Duals - Logic system - Flip - flop circuits - Counters - Register - Logic families - Integrated circuits - Operational amplifiers. *Highlights – Objective Type Questions – Theoretical Questions.*

2.1 ELECTRONIC COMPONENTS

2.1.1. Introduction

In order to obtain a particular function electronics circuits are designed with a number of electronic components suitably connected. A few *basic components* used in all the electronic circuits are :

Tube devices ; Semiconductor devices ..... called ***Active components.***

Resistors ; Capacitors ; Inductors ; ..... called ***Passive components.***

2.1.2. Active Components

*The electronic components which are capable of amplifying or processing an electrical signal are called **active components.***

Examples :

(*i*) *Tube devices :*

— *Vacuum tubes* (*e.g.,* vacuum diode, vacuum triode, vacuum pentode, etc.)

— *Gas tubes* (*e.g.,* gas diode, thyratron etc.)

(*ii*) *Semiconductor* (*solid state*) *devices* (e.g., junction diode, zener diode, transistor, FET. UJT, SCR, etc.)

2.1.2.1. Tube devices

The various types of tube devices are discussed below :

(*a*) **Vacuum tubes :**

(*i*) *Vacuum diode.* Its symbol is shown in Fig. 2.1 (*i*).

- It is used as a rectifier and detector.

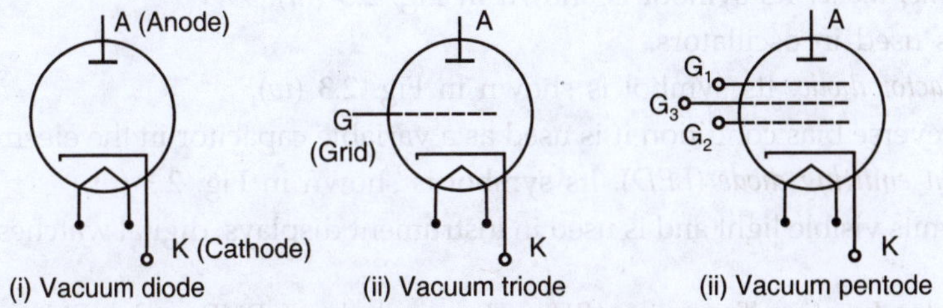

Fig. 2.1. Vacuum tubes.

(*ii*) *Vacuum triode.* Its symbol is shown in Fig. 2.1 (*ii*).

- It is used as amplifier and oscillator.

(*iii*) *Vacuum pentode.* Its symbol is shown in Fig. 2.1 (*iii*).

- It is used as amplifier and oscillator.

(*b*) **Gas tubes :**

(*i*) *Gas diode.* Its symbol is shown in Fig. 2.2 (*i*).

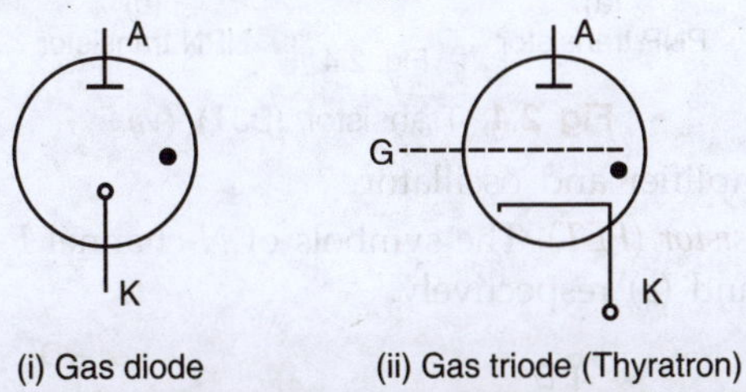

Fig. 2.2. Gas tubes.

- It is used as voltage regulator and in neon signs.

(*ii*) *Gas triode. (thyratron).* Its symbol is shown in Fig. 2.2 (*ii*).

- It is used as controlled rectifier.

2.1.2.2. Semiconductor devices

The various semiconductor devices are discussed as follows :

(*i*) *Junction diode.* Its symbol is shown in Fig. 2.3 (*i*).

- It is used as rectifier, detector and in switching circuits.

(*ii*) *Zener diode.* Its symbol is shown in Fig. 2.3. (*ii*).

- It is used as voltage regulator.

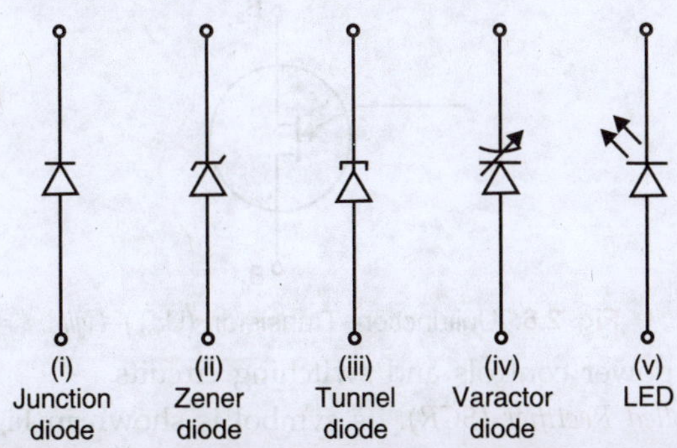

Fig. 2.3. Diodes.

(iii) *Tunnel diode.* Its symbol is shown in Fig. 2.3 (*iii*).

- It is used in oscillators.

(iv) *Varactor diode.* Its symbol is shown in Fig. 2.3 (*iv*).

- In reverse bias condition it is used as a variable capacitor in the electronic circuits.

(v) *Light emitting diode (LED).* Its symbol is shown in Fig. 2.3 (*v*).

- It emits visible light and is used in instrument displays, digital watches, calculators, etc.

(vi) *Bipolar Junction Transistor (BJT).* The symbols of PNP and NPN transistors are shown in Fig. 2.4 (*a*) and (*b*) respectively.

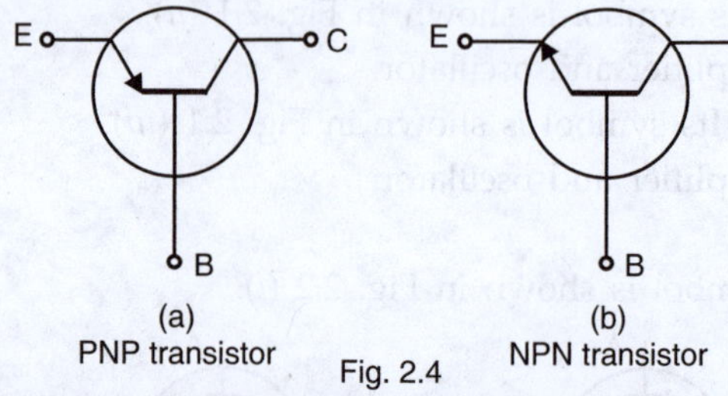

Fig 2.4. Transistor (BJT)–(*vi*)

- It is used as amplifier and oscillator.

(vii) *Field Effect Transistor (FET).* The symbols of *N*-channel *P*-channel FET are shown in Fig. 2.5. (*a*) and (*b*) respectively.

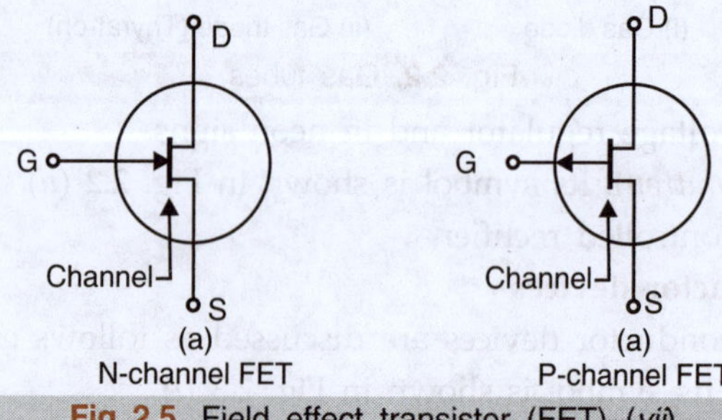

Fig 2.5. Field effect transistor (FET)–(*vii*)

- It is used as amplifier and oscillator.

(viii) *Unijunction Transistor (UJT).* Its symbol is shown in Fig. 2.6.

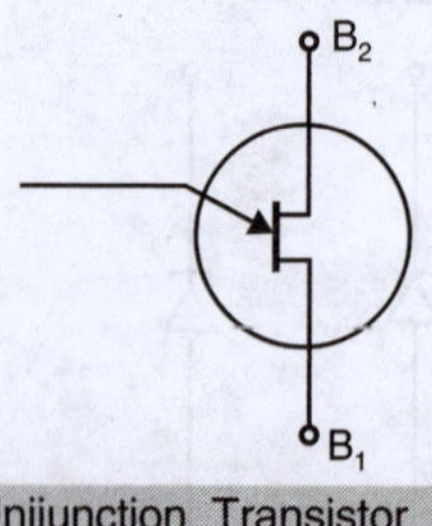

Fig 2.6. Unijunction Transistor (UJT)–(*viii*)

- It is used in power controls and switching circuits.

(ix) *Silicon Controlled Rectifier (SCR).* Its symbol is shown in Fig. 2.7 (*a*).

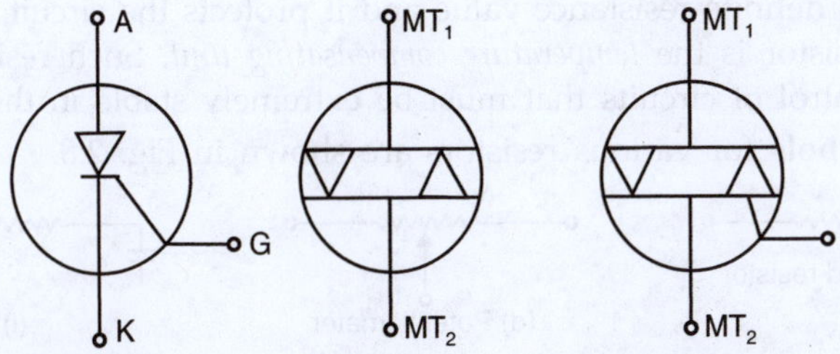

(a) SCR - (ix) (b) Diac - (x) (c) Triac - (xi)

Fig. 2.7. SCR, Diac, Triac.

- It is used for speed control of motors and power controls.

(x) *Diac.* Its symbol is shown in Fig. 2.7 (*b*).

- Generally it is used to give a pulse to the gate of triac.

(xi) *Triac.* Its symbol is shown in Fig. 2.7 (*c*).

- It is a bidirectional device and is used to obtain regulated A.C. at the output.

(xii) *Visual display devices.* Cathode ray tube (*CRT*) is the major visual display device.

— Digital watches and electronic calculators use LEDs (Light emitting diodes) or LCDs (Liquid crystal diodes) for the digital display.

2.1.3. Passive Components

The electronic components which are not capable of amplifying or processing an electrical signal are called **passive components.**

Examples. *Resistors ; inductors ; capacitors.*

These components are as important as active ones, since the active devices cannot process the electrical signals without their assistance.

2.1.3.1. Resistors

A resistor entails the following *two main characteristics* :

(i) Its *resistance* (*R*) in ohms. ..... The resistors are available from a fraction of an ohm to many mega ohms.

(ii) The *wattage rating.* ..... The power rating may be as high as several hundred watts or as low as $\frac{1}{10}$ watt. *Power rating indicates the maximum wattage the resistor can dissipate without excessive heat* (Too much heat can make the resistor burn open).

Classification of resistors :

The resistors are *classified* as follows :

1. *Fixed resistors.* The fixed resistor is the simplest type of resistor. Fixed means that the unit is so constructed that its resistance value is *constant* and *unchangeable.* These are made of a *carbon composition* and have a cover of black or brown hard plastics.

2. *Tapped resistors.* A tapped resistor is a resistor which has a tap, or connection, somewhere along the resistance material. These resistors are usually wire wound type. If they have more than one tap, they will have a separate terminal for each.

3. *Variable resistors.* A variable resistor has a movable contact that is used to adjust or select the resistance value between two or more terminals. A variable resistor is commonly called a *control.*

4. *Special resistors.* The most common type of special resistor is the fusible type. A

fusible resistor has a definite resistance value and it protects the circuit much like a fuse. Another special resistor is the *temperature compensating unit.* Such resistors are used to provide special control of circuits that must be extremely stable in their operation.

Schematic symbols for various resistors are shown in Fig. 2.8.

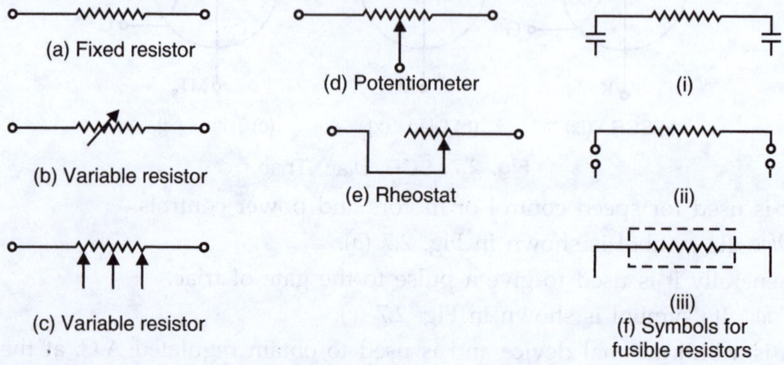

Fig. 2.8. Schematic symbols for various resistors.

The following types of resistors are used in electrical circuits :

(*i*) Carbon resistors.

(*ii*) Wire-wound resistors on ceramic or plastic forms (as in case of rheostats etc.).

(*iii*) Deposited carbon resistors on ceramic base.

(*iv*) Deposited metal resistors on ceramic base.

(*v*) Printed, painted or etched circuit resistors.

Resistor colour coding :

— Resistance is measured in units called *ohms.*

— *Wire wound resistors normally have their values in ohms and tolerance in percent stamped on them.*

— *For carbon or composition resistors a colour code is used.*

The resistance values, for several years have been coded by *three coloured bands* painted around the body of the resistors. If the tolerance is either 5 or 10 percent, a *fourth colour* band is added. Position of the bands is shown in Fig. 2.9.

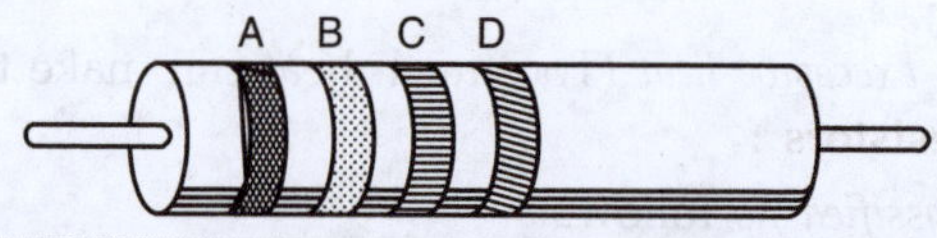

Fig. 2.9. The colour code system : colour bands indicate resistance value.

Colours and numbers :

Each of the colours represents one of the ten digits—0 through 9—as follows :

Colour	*Number*	*Colour*	*Number*
Black	0	Green	5
Brown	1	Blue	6
Red	2	Violet	7
Orange	3	Grey	8
Yellow	4	White	9

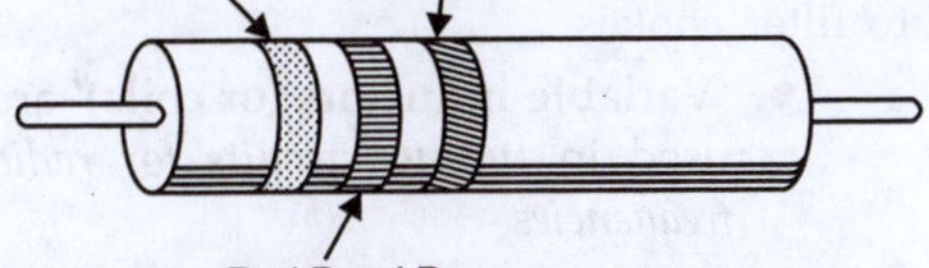

Fig 2.10. Colour code used on a 62000-ohm resistor.

The *bands are read from the end of the resistor toward the middle.*

— The first fwo colours (*A* and *B* in Fig. 2.9.) tells the first two digits in the resistance value.

— The third band (C) tells how many *zeros follow the first two digits.*

— Sometimes a fourth band (*D*) is present. This band tells the tolerance and will be either gold or silver. A gold band means 5% tolerance, silver 10% and no fourth band, 20%. The tolerance band tells how close the resistance should be to the value shown by the other three bands.

The procedure of reading the bands is given below. Refer to Fig. 2.10 :

Band	*A*	*B*	*C*	*D*
Colour	Blue	Red	Orange	No band
Numbers	6	2	3 zeros	20% tolerance

The blue-red-orange bands signify 62 followed by three zeros and would be read as 62000 ohms ± 20%.

2.1.3.2. Inductors

An ***inductor*** *is an electronic component (usually a coil) which opposes the change of current in the circuit.*

— *The property of the coil due to which it opposes any increase or decrease of current or flux through it, is known* as ***Self-inductance.***

— Self induction is sometimes analogously called electromagnetic or electrical inertia.

— The unit of inductance (L) is *henry* (H).

- An inductor offers high impedance (opposition) to A.C. but very low impedance to D.C.
- In an electronic circuit the usual function of an inductor is to *block* A.C. signal *but to pass D.C. signal or voltage.*

Classification of inductors :

The inductors can be broadly *classified* as follows :

1. Fixed inductors.
2. Variable inductors.

The schematic symbols of fixed and variable inductors are shown in Fig. 2.11. (*a*) and (*b*) respectively.

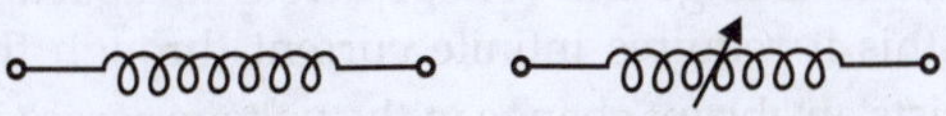

Fig 2.11. Inductors.

Filter chokes and Radio-frequency (RF) chokes :

— A *filter choke* [See Fig. 2.12 (*a*)] is an inductor used in the filter section of a D.C. power supply. Most of the power supplies use filter chokes having inductane ranging from about 1 H to 50 H, capable of carrying current upto 0.5 A.

— *Radio-frequency* chokes [See Fig. 2.12 (*b*)] are smaller in size as compared to filter chokes.

- **'Variable inductors (or coils)'** are used in *tuning circuits for radio frequencies.*
- A **transformer** is just similar in appearance to an inductor (filter choke). In electronic circuits, the transformers which are generally used are known as : *power transformers, output transformers* and *intermediate -frequency transformers.*

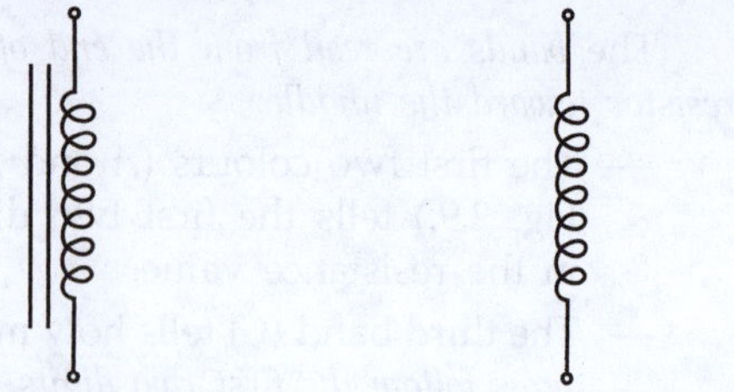

(a) Schematic symbol of filter choke. (b) Schematic symbol of RF choke.

Fig. 2.12

2.1.3.3. Capacitors

A **capacitor** *is a device capable of storing electric charge.*

- It consists of two conducting surfaces (may be in the form of either circular or rectangular plates or of spherical or cylindrical shape) separated by an insulating material called a *dielectric.*
- ***Capacitance*** *is a measure of ability of a capacitor to store an electric charge.* It is the ratio of the charge (Q) that can be stored to the voltage applied (V) across the plates. Mathematically, $C = \frac{Q}{V}$. Energy stored in a capacitor $= \frac{1}{2}CV^2$.

— The capacitance may be expressed in *F* (*Farads*) or *μF* or *pF*.

- This component (*i.e.,* capacitor) offers low impedance to *A.C.* but very high impedance (resistance) to D.C. The usual function of a capacitor is to *block D.C. voltage but pass the A.C. signal voltage,* by means of charging and discharging. These applications include *coupling, by passing* and *filtering for an A.C. signal.*

Fig. 2.13 (*a*) and (*b*) shows the fixed and variable capacitors respectively.

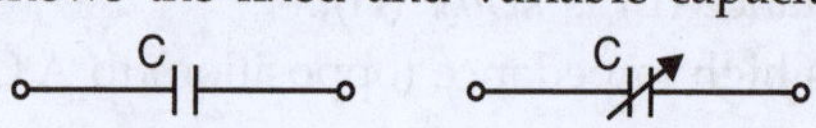

(a) Fixed capacitor (b) Variable capacitor

Fig. 2.13 Fixed and variable capacitors.

- The *variable capacitors are mostly air-gang capacitors.*

Some important properties of capacitors :

1. The capacitor never dissipates energy, but only stores it.
2. A capacitor is sort of open circuit to D.C.
3. It the voltage across a capacitor is not changing with time, the current through it is zero.
4. It is not possible to change the voltage across a capacitor by a finite amount in zero time, for this it requires infinite current through the capacitor.
5. A capacitor resists an abrupt change in the voltage across it in a manner analogous to the way a spring resists an abrupt change in its displacement.

Types of capacitors :

The various types of capacitors are enumerated and discussed below :

1. Paper capacitors
2. Mica capacitors
3. Plastic film capacitors
4. Electrolytic capacitors
5. Ceramic capacitors
6. Air capacitors.

1. ***Paper capacitors :***

- Dry paper is good insulator and has high dielectric strength. It can withstand high potential difference without breaking down. It is commonly used in the manufacture of capacitors.
- There are *two* basic forms of capacitors :

— In one form it consists of two rolls of aluminium foils or tin foils sandwiching at tissue paper rolled by a machine so that the final shape is that of a small cylindrical tube. The entire cylinder is generally placed in a cardboard coated with wax or encased in a plastic paper. These capacitors are available in a wide range of capacitance values and voltage ratings. The physical size for 0.05 μF is typically 2.5 cm long with 1 cm diameter.

Fig. 2.14. Paper capacitor.

— In another form a "metallised" paper is used. A long strip of paper is metallised with aluminium by a special process. The strip is rolled to form a small cylinder. The capacitor is inserted into waxed cardboard case or plastic case.

- These capacitor should *not* be used in radio-frequency tuned circuits because they are not electrically stable enough.

2. ***Mica capacitors :***

Mica capacitors or *paraffined capacitors* are widely used in radio circuit where *fixed volue capacitors are required.* Both these have metal foil sheets forming the coating and separated by a flat mica sheet or paraffin paper ; the dielectric paraffin paper capacitor of fairly large value is made by placing alternatively sheets or paraffin paper and the foil one above the other. Alternated tin foil sheets are connected together to form the two coatings.

- These capacitors are very small in size having 10 mm length and 3 mm thickness. These are often used for small capacitance values ranging from 50 to 500 pF, with voltage ratings ranging from 200 V to 1000 V.
- Silver mica capacitors are more stable electrically than foil-type capacitors and are used in high stability frequency determining circuits.

3. ***Plastic film capacitors :***

- Polyester is a thermoplastic material. It has better performance at high frequencies.
- The method of manufacture is same as in the case of paper capacitors, *i.e,* two strips of aluminium foils are separated by a thin film of polystyrene, then rolled and placed in aluminium container.

4. ***Electrolytic capacitors :*** Refer to Fig. 2.15

Electrolytic capacitors are used in the power supply circuits of "*Radio and TV circuits.* They have higher losses than paper capacitors.

The *working principle* of an electrolytic capacitor is as follows :

"When current is passed through a solution of aluminium borate or sodium phosphate with aluminium electrode, a layer of aluminium oxide forms at the positive electrode. This film acts as dielectric between the plates. As the film is very thin, a very high capacitance may be obtained. In the wet type oxide layer is reformed after being broken by a large potential difference applied. As this type has to be mounted vertically, the solution gets evaporised as such it has been replaced by dry type of electrolytic capacitor".

Being polarized they are *suitable only on D.C. supply.*

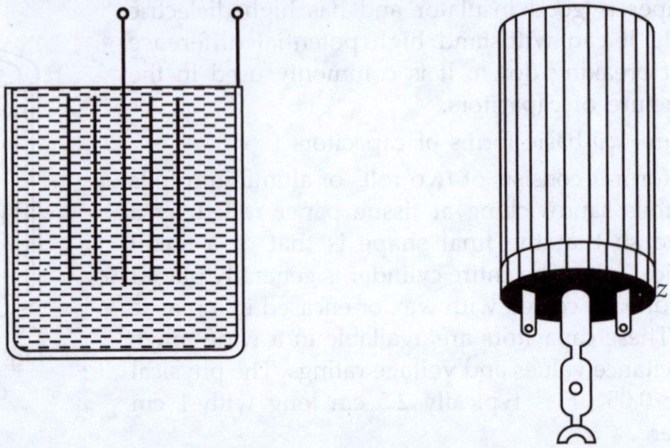

Fig. 2.15. Electrolytic capacitor.

- *Tantalum-electrolytic capacitor* consists of two tantalum foils with a tissue paper integrated with a non-corrosive electrolyte. The dielectric is pentaoxide layer which is electrochemically formed on the anode. The solid tantalum capacitors are available only in polarised form.

5. ***Ceramic capacitors :*** Refer to Fig. 2.16.

Deelectric constant of ceramic is high so that large capacitors can be obtained in a comparatively small space. It, however, suffers from the disadvantage of having higher losses than mica.

Ceramic capacitors are available in the following forms and shapes :

(i) Disc ceramics
(ii) Tubular ceramics
(iii) Moulded ceramics
(iv) Button ceramics.

The general construction of disc type consists of application of silver coatings on both sides of ceramic plates, in tubular type silver coating is applied on the inside and outside of hollow ceramic tube.

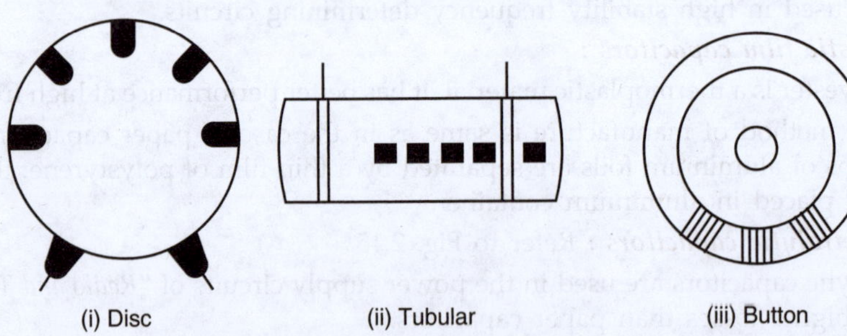

Fig. 2.16. Ceramic capacitors.

- Ceramic capacitors are used primarily as *coupling and bypass portions of radio frequency circuits* rather frequency determining elements. Specially designed ceramic capacitors are used in resonant circuits.

6. ***Air capacitors :***

These capacitors (variable) are used in *radio receivers* for tuning the receiver to a particular transmitting station.

- Such a capacitor consists of a number of semicircular plates of sheet aluminium mounted together by metal rod and capable of moving in between a number of fixed aluminium semi-circular sheets. The *capacity increases* when the rotating sheets are moving into the fixed sheets. The set of rotating sheets is called the rotor while the set of fixed sheets is called the stator (Fig. 2.17). A circular dial and a pointer is used to read the value of the capacity for any position of the rotating plates.

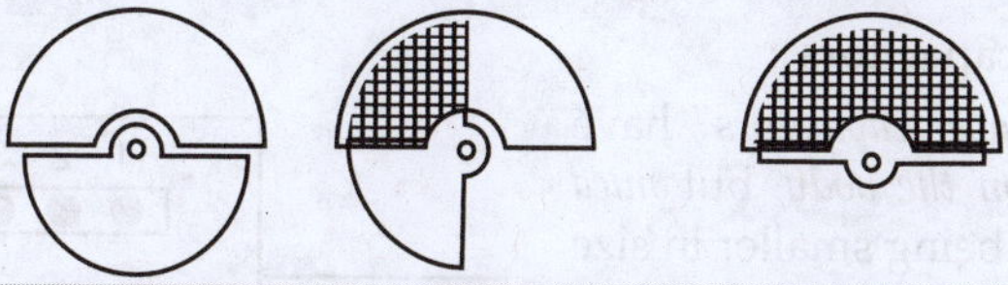

Fig. 2.17. Air capacitor—rotor and stator.

- Such condensers commonly used have a value of capacity varying almost from zero to 500 μF.

Variable capacitor :

*A capacitor whose capacity can be varied is called **'variable capacitor'**.* This is done by varying the thickness of the dielectric.

The various variable capacitors used in radio receiver are :

(*i*) ***Trimmers.*** Refer to Fig 2.18 Number of metal plates are inleaved with mica dielectric. The distance between the plates is controlled by a screw which is insulated from the plates stacked in a ceramic block.

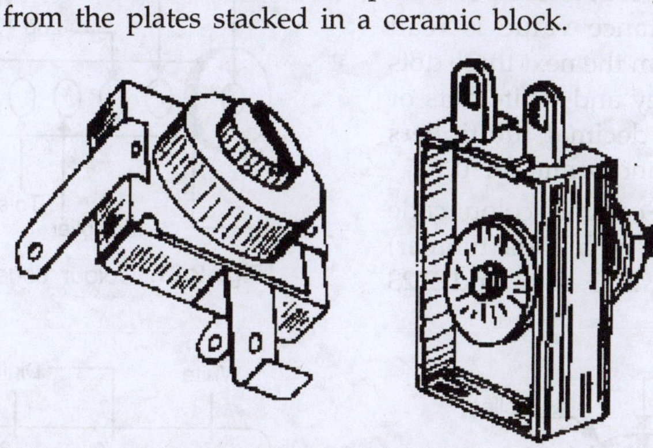

Fig. 2.18. Trimmers.

- They are available in the values of 30 pF to 70pF.

(*ii*) ***Padders.*** Refer to Fig. 2.19.

Padder is a mica capacitor (variable type). Its capacity is 600 pF.

These are continuously varying types. There are two sets of plates, fixed metal plates connected together form the stator set. Another set of movable plates form rotor. Rotor plates mesh with stator plates and can be moved with a shaft. Capacitance varies with the varying distance between the plates. Air is the dielectric. Usually two capacitors are ganged over a common shaft.

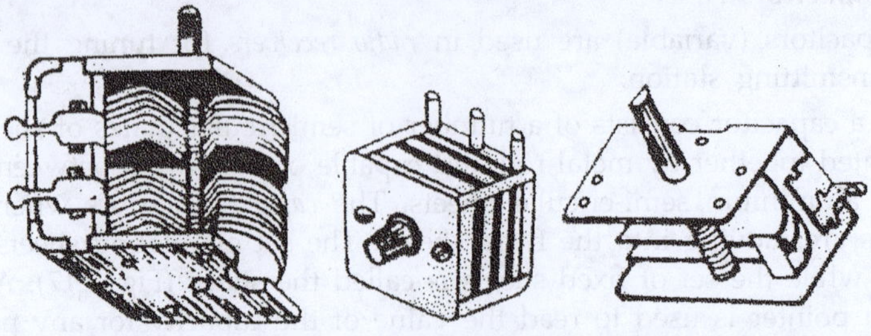

Fig. 2.19. Padders

Colour code of capacitors :

"Electrolyte and paper capacitors" have their values *printed on the body*, but *mica and tubular capacitors'* being smaller in size are colour coded. The *colour and their values are the same as in resistors.*

"Mica capacitors " (See Fig. 2.20) have six colour dots. Dots are marked from left to right in clockwise directon. Second and third dots indicate the digit and 4th dot is the multiplier. Dot 5 reads tolerance.

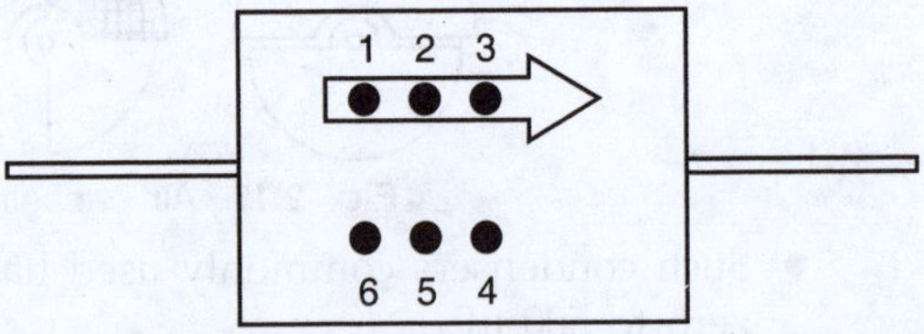

Fig. 2.20. Mica capacitor

"Ceramic capacitors" (Fig. 2.21) have colour dots or bands. The wide colour band on the left specifies temperature coefficient. Capacitance value is read from left to right from the next three dots or colour strips. Grey and white dots or strips are used as decimal multipliers with grey for 0.01 and white for 0.1.

- Colour code ceramic, colour code mica and colour code with leads (tubular) are shown in Figs. 2.21, 2.22 and 2.23 respectively.

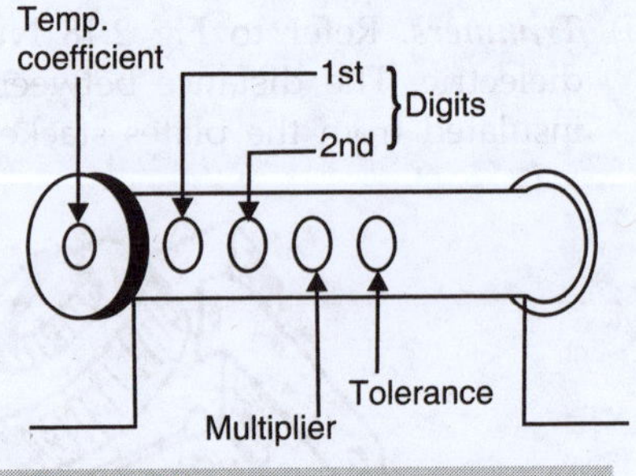

Fig. 2.21. Colour code ceramic.

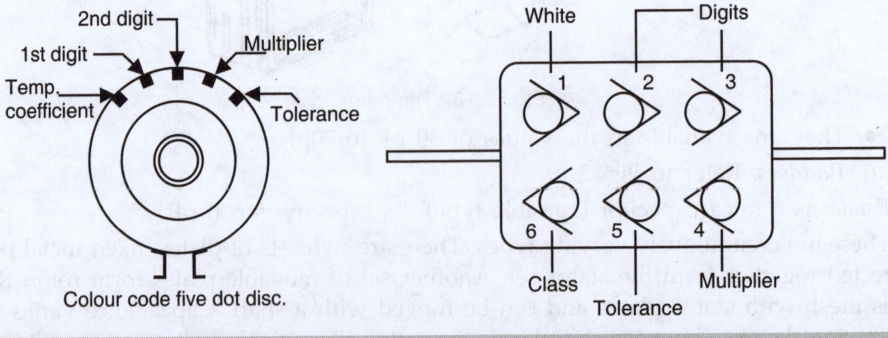

Fig. 2.22. Colour code mica.

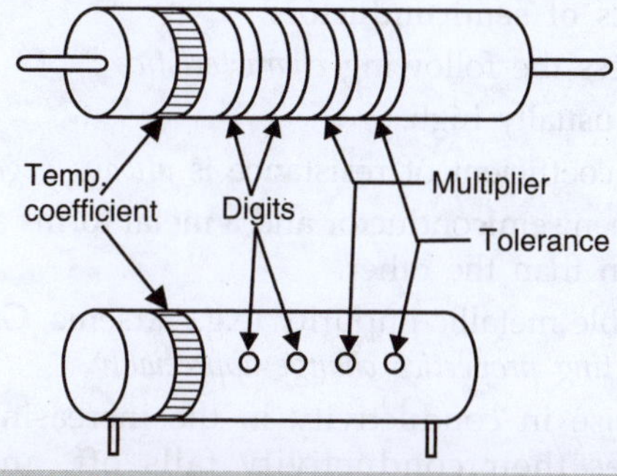

Fig. 2.23. Colour code with leads (Tubular).

Colour code chart

S.No.	Colour	First figure	Second figure	Multiplier	Tolerance
1.	Black	—	0	1	C < 10 pF
2.	Brown	1	1	10	± 1 pF
3.	Red	2	2	100	C > 10 pF
4.	Orange	3	3	1000	± 20%
5.	Yellow	4	4	10000	
6.	Green	5	5	—	
7.	Blue	6	6	—	
8.	Violet	7	7	—	
9.	Grey	8	8	0.01	
10.	White	9	9	0.1	

2.2. ELECTRONIC DEVICES

2.2.1. General Aspects

An ordinary electrical equipment enters the electronic class whenever its circuit includes electronic devices such as electron tubes or solid state devices which are formed by junctions of semiconducting materials.

Understanding electronics includes understanding of ordinary electrical devices and circuits. No matter what electronic devices are used, the equipment is still electrical. In other words, ***all that is electronic is also electrical.***

The characteristic features of electronic devices are

1. The electronic devices can rectify A.C. into D.C.
2. They can amplify input signals.
3. They can respond at speeds far beyond the speeds that one comes across in electrical and mechanical devices.
4. Some electronic devices are photosensitive. Some of these devices can produce radiations such as X-xays

2.2.2. Semiconductors

***Semiconductors** are solid materials, either non-metallic elements or compounds, which allow electrons to pass through them so that they conduct electricity in much the same way as a metal.*

2.2.2.1 Characteristics of semiconductors

Semiconductors possess the following *characteristics* :

1. The resistivity is usually high.
2. The temperature coefficient of resistance is *always negative.*
3. The contact between semiconductor and a metal forms a layer which has a higher resistance in one direction than the other.
4. When some suitable metallic impurity (e.g., Arsenic, Gallium, etc.) is added to a semiconductor, its *conducting properties change appreciably.*
5. They exhibit a rise in conductivity in the increasing temperature, with the decreasing temperatures their conductivity falls off, and at low temperatures semiconductors become dielectrics.
6. They are usually metallic in appearance but (unlike metals) are generally hard and brittle.

Both the resistivity and the contact effect are as a rule very sensitive to small changes in physical conditions, and the *great importance* of semiconductors for a wide range of uses apart from rectification depend on the *sensitiveness.*

Examples of semiconducting materials :

Of all the elements in the periodic table, *eleven* are *semiconductors* which are listed below :

S. No.	*Element*	*Symbol*	*Group in the perodic table*	*Atomic No.*
1.	Boron	B	III	15
2.	Carbon	C	IV	6
3.	Silicon	Si	IV	14
4.	Germanium	Ge	IV	32
5.	Phosphorus	P	V	15
6.	Arsenic	As	V	33
7.	Antimony	Sb	V	51
8.	Sulphur	S	VI	
9.	Sellinum	Se	VI	
10.	Tellurium	Te	VI	
11.	Iodine	I	VIII	

Examples of semiconducting *compounds* are given below :

(*i*) Alloys : Mg_3Sb_2, ZnSb, Mg_2Sn, CdSb, AlSb, InSb, GeSb.

(*ii*) Oxide : ZnO, Fe_3O_4, Fe_2O_3, Cu_2O, CuO, BaO, CaO, NiO, Al_2O_3, TiO_2, UO_2, Cr_2O_3, WO_2, MoO_3.

(*iii*) Sulphides : Cu_2S, Ag_2S, PbS, ZnS, CdS, HgS, MoS_2.

(*iv*) Halides : AgI, CuI.

(*v*) Selenides and Tellurides.

PbS is *used in photo-conductive devices, BaO in oxide coated cathodes, caesium antimonide in photomultipliers, etc.*

Atomic structure :

To understand how semiconductors work it is necessary to study briefly the structure of matter. All atoms are made of electrons, protons and neutrons. Most solid materials are classed, from the stand point of electrical conductivity, as conductors, semiconductors or insulators. *To be conductor, the substance must contain some mobile electrons—one that can move freely between the atoms.* These free electrons come only from the valence (outer) orbit of the atom. Physical force associated with the valence electrons bind adjacent atoms together. The inner electrons below the valence level, do not normally enter into the conduction process.

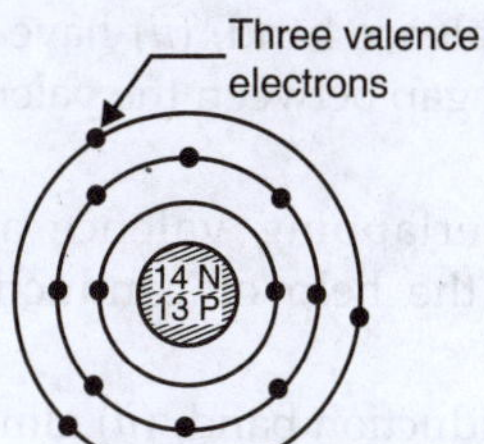

Aluminium

Fig. 2.24

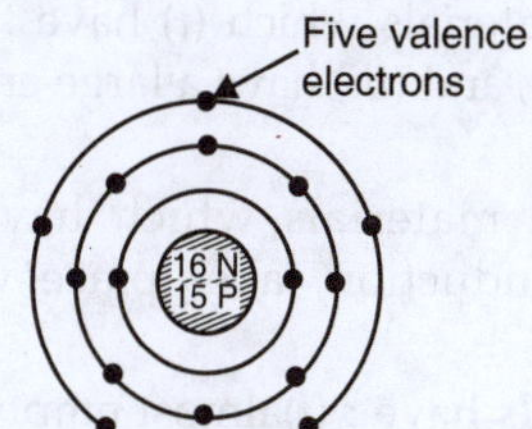

Phosphorus

Fig. 2.25

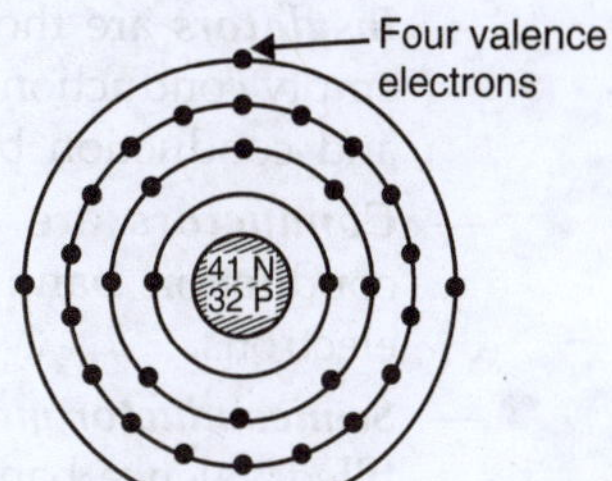

Germanium

Fig. 2.26

Conductivity depends on the number of electrons in the valence orbit. Electron diagrams for three typical elements, aluminium, phosphorus and germanium are shown in Figs. 2.24, 2.25, 2.26.

These elements can all be used in semiconductor manufacture. The *degree of conductivity* is determined as follows :

1. Atoms with *fewer than four valence electrons* are *good conductors.*
2. Atoms with *more than four valence electrons* are *poor conductors.*
3. Atoms with *four valence electrons* are *semiconductors.*

Fig. 2.24 shows aluminium which has *three valence electrons.* When there are less than four valence electrons they are loosely held so that at least one electron per atom is normally free ; hence aluminium is a good conductor. This ready availability of free electrons is also true of copper and most other metals.

Fig. 2.25 shows Phosphorus with *five valence electrons.* When there are more than four valence electrons, they are lightly held in orbit so that normally *none are free.* Hence phosphorus and similar elements are poor conductors (insulators).

Germanium (Fig. 2.26) has four *valence electrons.* This makes it neither a good conductor nor a good insulator, hence its name "semiconductor". Silicon also has four valence electrons and is a semiconductor.

Note. The energy level of an electron increases as its distance from the nucleus increases. Thus an electron in the second orbit possess more energy than electron in the first orbit ; electrons in the third orbit have higher energy than in the second orbit and so on. It follows, therefore, that electrons in the last orbit will possess very high energy. These high energy electrons are less bound to the nucleus and hence they are more mobile. *It is the mobility of last orbit electrons that they acquire the property of combining with other atoms.* Further it is due to this combining power of last orbit electrons of an atom that they are called *valence electrons.*

- Following points are worth noting:
  — ***Conduction electrons*** are those valence electrons which have gained enough energy to take part in conduction of electricity through a solid.
  — ***Valence band*** is the band of energy occupied by valence electrons. It is the highest occupied band and it may be completely or partially filled with electrons.
  — ***Conduction band*** is the higher energy band to the valence band. It is occupied by conduction electrons. It may be empty or partially filled. It is the lowest unfilled or unoccupied energy band.
  — ***Insulators*** are those materials which (*i*) have full valence band, (*ii*) have an empty conduction band, and (*iii*) have a large energy gap between the valence and conduction bands.
  — ***Conductors*** are those materials which have overlapping valence and conduction bands. Conduction takes place with the help of conduction electrons.
  — ***Semiconductor materials*** have : (*i*)almost empty conduction band, (*ii*) almost filled valence band, and (*iii*) narrow energy gap between the two.

2.2.3. Intrinsic Semiconductor

A *pure semiconductor is called "intrinsic semiconductor"*. Here no free electrons are available since all the covalent bonds are complete. A *pure semiconductor, therefore behaves as an insulator*. It exhibits a peculiar behaviour even at room temperature or with rise in temperature. The *resistance of a semiconductor decreases with increase in temperature.*

When an electric field is applied to an intrinsic semiconductor at a temperature greater than 0°K, conduction electrons move to the anode and, the holes (when an electron is liberated into the conduction band a positively charged hole is created in valence band) move to cathode. *Hence semiconductor current consists of movement of electrons in opposite direction.*

Fig. 2.27 shows the energy diagram for intrinsic (pure) semiconductor at absolute zero.

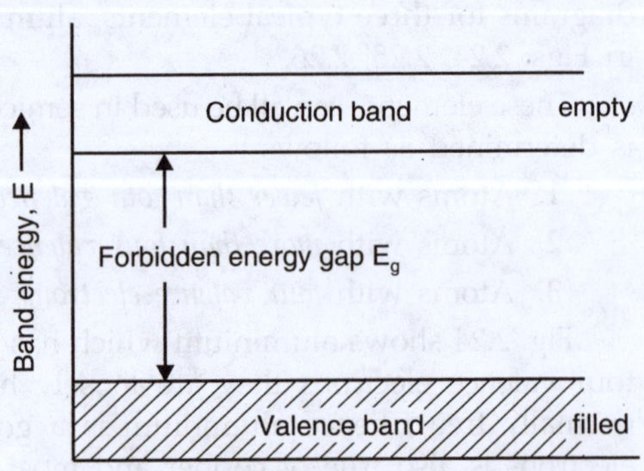

Fig. 2.27. Energy diagram for intrinsic (pure) semiconductor at absolute zero.

2.2.4. Extrinsic Semiconductor

In a pure semiconductor, which behaves like an insulator *under ordinary conditions*, if small amount of certain *metallic impurity* is *added* it attains *current conducting properties.* The impure semiconductor is then called *"impurity semiconductor" or "extrinsic semiconductor". The process of adding impurity (extremely in small amounts, about 1 part in 10^8) to a semiconductor to make it extrinsic (impurity) semiconductor is called* **Doping.**

Generally following doping agents are used :

(*i*) *Pentavalent atom* having five valence electrons (arsenic, antimony, phosphorus) ... called *donor atoms.*

(*ii*) *Trivalent atoms* having three valence electrons (gallium aluminium, boron) ... called *acceptor atoms.*

With the addition of suitable impurities to semiconductor, two type of semiconductors are :

(*i*) N-type semiconductor.
(*ii*) P-type semiconductor.

N-type semiconductor :

The presence of *even a minute quantity of impurity,* can produce N-type semiconductor. If the impurity atoms has *one valence electron more* than the semiconductor atom which it has substituted, this *extra electron* will be loosely bound to the atom. For example, an atom of *Germanium* possesses *four valence electrons* ; when it is replaced in the crystal lattice of the substance by an impurity atom of antimony (Sb) which has *five valence electrons,* the fifth valence electron (free electron) produces extrinsic N-type conductivity *even at room temperature.* Such an impurity into a semiconductor is called *donor impurity* (or donor). The conducting properties of germanium will depend upon the *amount of antimony* (*i.e.,* impurity) *added.* This means that controlled conductivity can be obtained by proper addition of impurity. Fig. 2.28 (*a*) shows the loosely bound excess electron controlled by the donor atom.

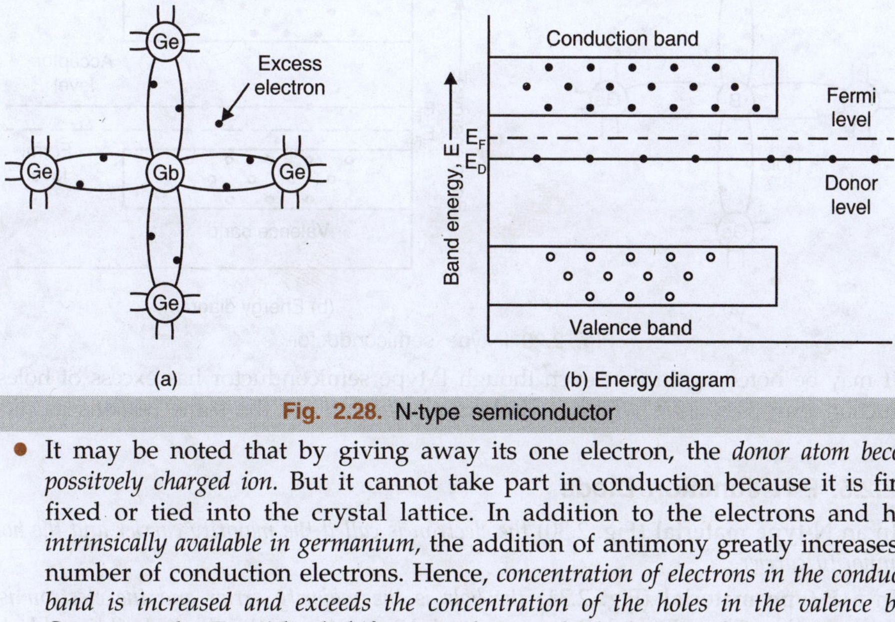

Fig. 2.28. N-type semiconductor

- It may be noted that by giving away its one electron, the *donor atom becomes possitvely charged ion.* But it cannot take part in conduction because it is firmly fixed or tied into the crystal lattice. In addition to the electrons and holes *intrinsically available in germanium,* the addition of antimony greatly increases the number of conduction electrons. Hence, *concentration of electrons in the conduction band is increased and exceeds the concentration of the holes in the valence band.* Consequently, *Fermi level shifts upwards towards the bottom of the conduction band* as shown in Fig. 2.28 (*b*). [Since the number of electrons as compared to the number of holes increases with temperature, the *position of Fermi level also changes considerably with temperature*].
- It is worth noting that even though N-type semiconductor has excess of electrons, still it is *electrically neutral.* It is so because by addition of donor impurity, number of electrons available for conduction purposes becomes more than the number of holes available intrinsically. But the total charge of the semiconductor does not change because the donor impurity brings in as much negative charge (by way of electrons) as positive charge (by way of protons).

Note. In terms of energy levels, the fifth antimony electron has as energy level (called donor level) just below the conduction band. Usually, the donor level is 0.01 eV below conduction band for germanium and 0.054 eV for silicon.

P-type semiconductor :

- P-type extrinsic semiconductor can be produced if the impurity atom has *one valence electron less* than the semiconductor atom that it has replaced in the crystal lattice. This impurity atom cannot fill all the *interatomic* bonds, and the free bond can accept an electron from the neighbouring bond ; leaving behind a vacancy of *hole*. Such an impurity is called an *acceptor impurity* (*or acceptor*). Fig. 2.29 (*a*) shows structure of P-type semiconductor (Germanium and Boron).
- In this type of semiconductor, conduction is by means of holes in the valence band. Accordingly, *holes form the majority carriers whereas electrons constitute minority carriers.* The process of conduction is called *deficit conduction.*
- Since the concentration of holes in the valence band is more than the concentration of electrons in the conduction band, Fermi level shifts nearer to the valence band [Fig. 2.29 (*b*)]. The acceptor level lies immediately above the Fermi level. *Conduction is by means of hole movement at the top of valence band, the acceptor level readily accepting electrons from the valence band.*

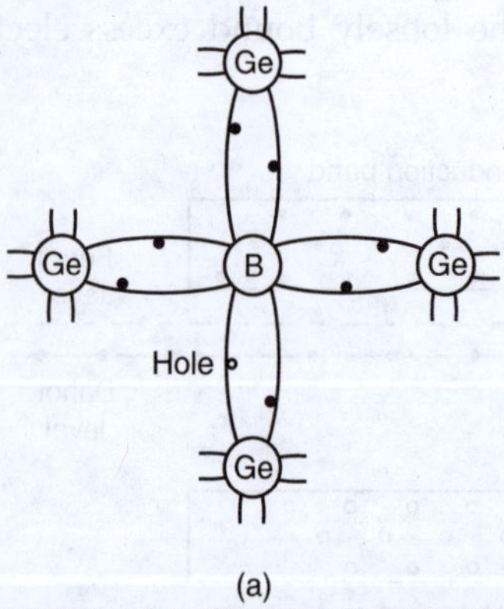

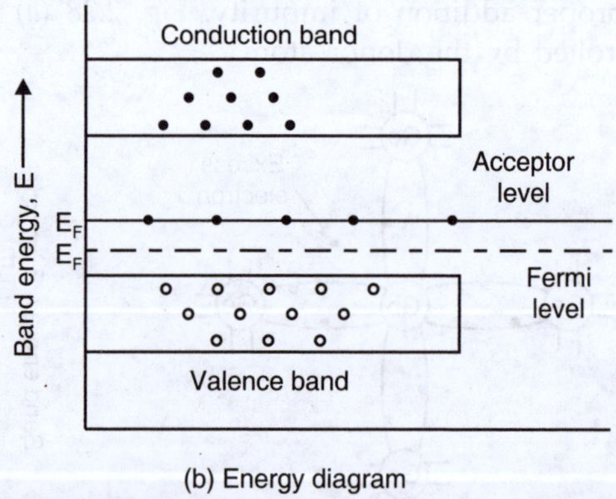

Fig. 2.29 P-type semiconductor.

It may be noted again that even though P-type semiconductor has excess of holes for conduction purposes, as a whole it is *electrically neutral* for the same reasons as dicusse earlier.

2.2.5. P-N Junction Diode

In an **N-type material** (Fig. 2.30) *the electron is called the majority carrier and the hole as the minority carrier.*

In a **P-type material** (Fig. 2.31) *the hole is the majority crrier and the electron is the minority carrier*. The N- and P-type materials represent the *basic building blocks of semiconductor devices.*

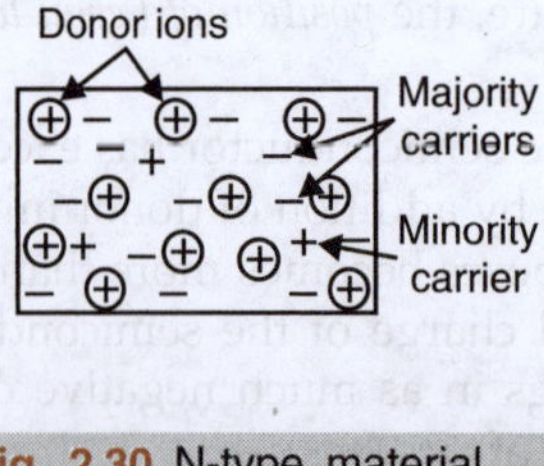

Fig. 2.30 N-type material.

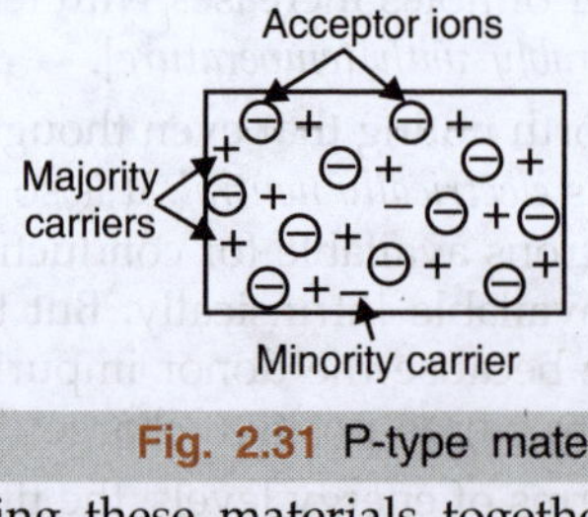

Fig. 2.31 P-type material.

The semiconductor diode is simply bringing these materials together (constructed from the same base-Ge or Si). At instant the two materials are *"joined"* the electrons and holes in the region of the junction will combine resulting in a lack of carriers in the region near the junction. *This region of unconvered positive and negative ions is called the* ***depletion region*** *due to the depletion of carriers in this region.*

Construction and types of P-N junction diodes :

The most extensively used elements in the manufacture of junction diodes are *germanium and silicon* (although some other materials are also assuming importance in recent years).

A *P-N* junction diode (known as a semiconductor or crystal diode) consists of a *P-N* junction, formed either in germanium or silicon crystal. The diode has two terminals namely *anode and cathode.* The anode refers to the *P*-type region and cathode refers the *n*-type region as shown in Fig. 2.32 (*a*)

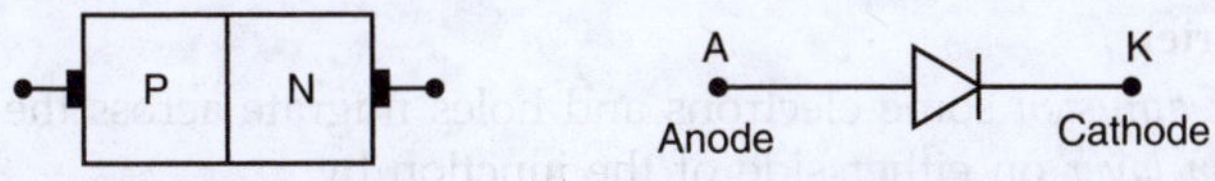

Fig. 2.32 *P-N* junction diode.

The arrow head, shown in the circuit symbol, points the direction of current flow, when it is "forward biased" (It is the same direction in which the movement of holes takes place).

The commercially available diodes, usually have some notations to identify the *P* and *N* terminals or leads. *The standard notation consists of type numbers preceded by IN,* such as *IN* 240 and *IN* 1250. Here 240 and 1250 *correspond* to colour bands. *In some diodes, the schematic symbol of a diode is painted or the colour dots are marked on the body.*

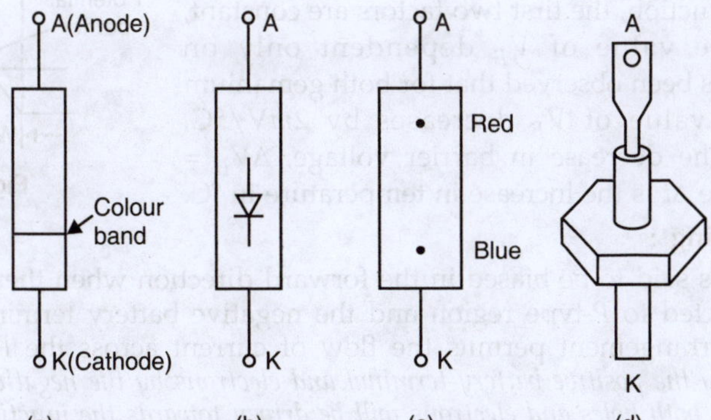

Fig. 2.33 Low, medium and high current diodes.

Fig. 2.33 shows low, medium and high current diodes.

— *Refer to Fig. 2.33* (*a*). The diode shown has a colour band located near one of the ends. The end, which is near the colour band, is identified as cathode, and other end, obviously, is the anode (*A*).

— *Refer to Fig 2.33* (*b*). The diode has a schematic symbol actually painted at its cathode (*K*) and the other end as anode.

The diodes of Fig. 2.33 (*a*) and (*b*) *can pass a forward current of 100 mA and are known as* ***low current diodes.***

— *Refer to Fig. 2.33* (*c*). The diode has colour dots marked on its body. The end lying near the blue dot is a cathode, while the other end is anode. Sometimes this diode is *shown bigger in size* than that of diodes shown in Fig. 2.33 (*a*) and (*b*). The diodes of this size can *pass a forward current of 500 mA and are known as* ***medium current diodes.***

— *Refer to Fig. 2.33 (d).* It shows a diode, which *can pass a forward current of several amperes.* Therefore it known as a ***power diode*** or a ***high current diode.***

- *The outstanding property of P-N junction / crystal diode to conduct current in one direction only permits it to be used as a* **rectifier.**

Potential barrier and biasing :

A *P-N* junction diode which consists of *P-* and *N*-type semiconductors formed together to make a *P-N* junction is shown in the Fig 2.34. The *place dividing the two zones* is known as a *"junction"*.

Potential barrier :

As a result of *diffusion* some electrons and holes migrate across the junction thereby forming a *depletion layer* on either side of the junction by neutralisation of holes in the *P*-regional and of free electrons in the *N*-region. *This diffusion of holes and electrons across the junction continues till* ***potential barrier*** *is developed in the depletion layer which then prevents further diffusion.* By the application of an external voltage this potential barrier is either *increased* or *decreased.*

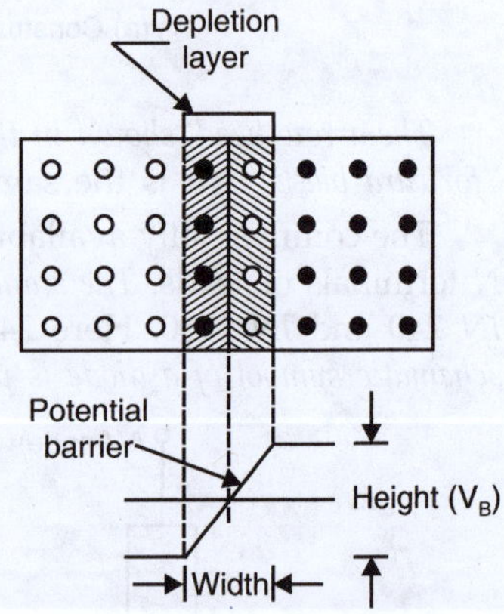

Fig. 2.34

The barrier voltage of a *P-N* junction depends upon three factors namely *density, electonic charge* and *temperature.* For a given *P-N* junction, the first two factors are constant, thus making the value of V_B dependent only on temperature. It has been observed that for both gemanium and silicon the value of V_B decreases by $2mV/°C$. Mathematically, the decrease in barrier voltage, $\Delta V_B = -0.002 \times \Delta t$, where Δt is the increase in temperature in °C.

Forward biasing :

The junction is said to be biased in the forward direction when then positive battery terminal is connected to *P*-type region and the negative battery terminal to the *N*-type (Fig. 2.35). This arrangement permits the flow of current across the *P-N* junction. The *holes are repelled by the positive battery terminal and electrons by the negative battery terminal with the result that both holes and electrons will be driven towards the junction where they will recombine.* Hence as long as the battery voltage is applied large current flows. In other words, the *forward bias lowers the potential barrier across the depletion layer* thereby *allowing more current to flow* across the junction.

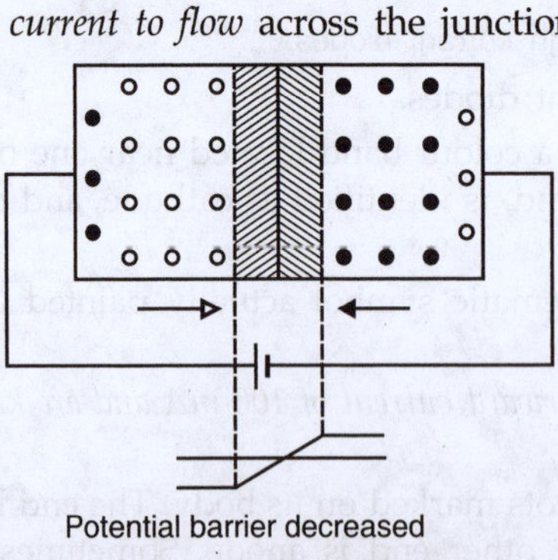

Fig. 2.35 Forward biasing.

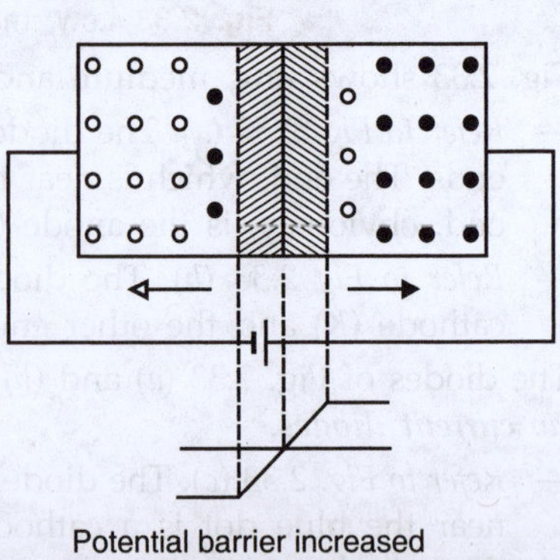

Fig. 2.36 Reverse biasing

Reverse biasing (Zener diode) :

The junction is said to be reversed biased when battery connection to the battery are reversed as shown in Fig. 2.36. In this arrangement holes are attracted by the negative battery terminal and electrons by the positive battery terminal so that both holes and electrons move *away* from the junction. Since there is no recombintion of electron-hole pairs, diode current is negligible and the junction has high resistance. Reverse biasing *increases the potential barrier* at the junction, thereby *allowing very little current to flow through* the junction.

2.2.6 Zener Diode

A properly doped P-N junction crystal diode which has a sharp breakdown voltage is known as **Zener diode.**

*The voltage-regulator diode is commonly called a **'Zener' diode.*** It is a *voltage limiting diode* that has some applications in common with the older voltage-regulator gas tubes but serves a much wider field of application, because the devices cover a wide spectrum of voltages and power levels.

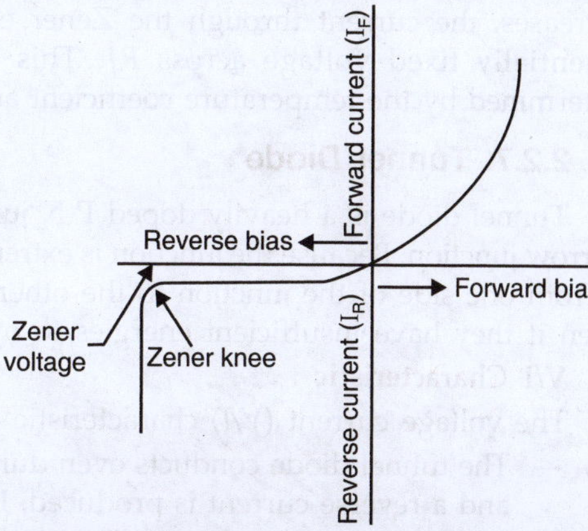

Fig. 2.37 Zener diode characteristics.

Performance/Operation :

The electrical performance of a zener diode is based on the *avalanche characteristics* of the *P-N* junction. When a source of voltage is applied to a diode in the *reversse direction* (negative to anode), a reverse current I_R is observed (see Fig 2.47). As the reverse potential is increased beyond the *"Zener knee"* avalanche breakdown becomes well developed at zener voltage V_Z. At voltage V_Z, the high counter resistance drops to a low value and the junction current increases rapidly. The current must of necessity be limited by an external resistance, since the voltage V_Z developed across the zener diode remains essentially constant. *Avalanche breakdown of the operating zener diode is not destructive as long as the rated power dissipation of the junction is not exceeded.*

Externally, the zener diode looks much like other silicon rectifying devices, and electrically it is capable of rectifying alternating current.

The following points about the *Zener diode* are worth noting :

(*i*) It looks like an ordinary diode except that it is properly doped so as to have a sharp breakdown voltage.

(*ii*) It is always reverse connected *i.e.*, it is *always reverse biased.*

(*iii*) It has sharp breakdown voltage, called Zener voltage V_Z.

(*iv*) When forward biased, its characteristics are just those of ordinary diode.

(*v*) It is not immediately burnt just because it has entered the breakdown region (The current is limited only by both external resistance and power dissipation of Zener diode).

- The location of Zener region can be controlled by varying the doping levels. An *increase in doping, producing an increase in the number of added impurities, will decrease the Zener potential.*
- Zener diodes are available having Zener potentials of 1.8 to 200 V with power ratings from $\frac{1}{4}$ to 50 W. *Because of its higher temperature and current capability,*

silicon is usually preferred in the manufacture of Zener diodes.

Applications of zener diode :

Zener diode serves in the following variety of *applications :*

1. Voltage reference or regulator element :

The primary use of a zener diode is as a *voltage reference or regulator element.* Fig. 2.48 shows the fundamental circuit for the Zener diode employed as a shunt regulator. In the circuit, diode element and load R_L draw current through the series resistance R_S. If E_{in} increases, the current through the Zener element will increase and thus maintain an essentially fixed voltage across R_L. This ability to maintain the desired voltage is determined by the temperature coefficient and the diode impedance of the zener device.

2.2.7. Tunnel Diode

Tunnel diode is a heavily doped P-N junction type germanium having an extremely narrow junction. Because the junction is extremely narrow, the electrons can tunnel through it from one side of the junction to the other. The electrons are able to tunnel through it even if they have insufficient energies to overcome the barrier.

V/I Characteristic :

The voltage current (*V/I*) characteristic of such a diode is shown in Fig. 2.38.

— The tunnel diode conducts even during the reverse bias (less than Zener voltage) and a reverse current is produced. For low forward voltages the current is high, and at a certain value of (low) voltage V_1, the current reaches its peak value. When the forward bias increases beyond V_1, the tunnel diode current begins to decrease and reaches a minimum value for a voltage V_2.

— The LM portion of the curve represents a *negative resistance characteristic of the tunnel diode.* A tunnel diode when operated in this region may be used as an *amplifier*, or *oscillator*, or as *switch for timing circuits.* When forward voltage is increased beyond the value V_2 the current starts increasing just as in a conventional diode.

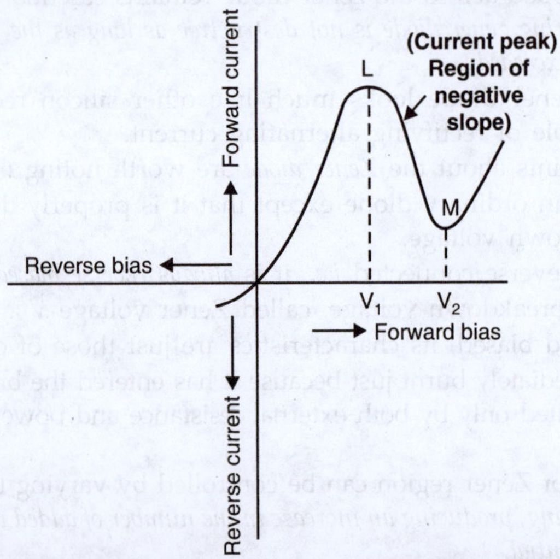

Fig. 2.38. V/I characteristic of a tunnel diode.

Advantages :

1. It is a special type of diode which can withstand very large temperature changes.

2. It can be very efficiently used in microwave region.
3. Its consumption is very low (about $\frac{1}{1000}$ th of a transistor)
4. Its cost is low.
5. It is of small size
6. It has a long life.

2.2.8. Biopolor Junction Transistor (BJT)

Introduction :

A transistor may be *defined* as follows :

— The word *transistor* was derived from the two word combination, *transfer-resistance* (Transfer + resistor ⟶ Transistor). A **transistor** *is a device to transfer a low resistance into a circuit having a high resistance.*

— A *'transistor' is a semiconductor device in which current flows in semiconductor materials.*

— When a thin layer of *P*-type or *N*-type semiconductor is between a pair of opposite types it constitutes a transistor.

- The **transistor** *is a solid state device, whose operation depends upon the flow of electric charge carriers within the solid.*

A **transistor** *is a semiconductor device having both rectifying and amplifying properties.*

"The main difference between a vacuum triode and a transistor is that while a vacuum triode is a voltage controlled device, a transistor is a current controlled device".

The transistor was invented by a team of three scientists at Bell Laboratories, USA in 1947. Although the first transistor was not a bipolar junction device, yet it was the beginning of a technological revolution that is still continuing in the twenty first century. All of the complex electronic devices and systems developed or in use today, are an outgrowth of early developments in semiconductor transistors.

The two basic types of transistors are :

1. Bipolar junction transistor (BJT)
2. Field-effect transistor (FET)

The *bipolar junction transistor is used in the following two broad area of electronics :*

(*i*) *As a* **linear amplifier** *to boost an electric signal.*

(*ii*) *As an* **electronic switch.**

P-N-P nd N-P-N transistors. To understand the basic mechanism of transistor operation the following facts need to be kept in mind.

1. Since *emitter is to provide charge carriers, it is always "forward biased".*
2. *First letter of transistor type indicates the polarity of the emitter voltage with respect to base.*
3. Collector's job is *to collect or attract those carriers through the base, hence it is always "reverse-biased".*
4. *Second letter* of transistor type *indicates the polarity of collector voltage with respect to the base.*

The above points apply both to *P-N-P* and *N-P-N* transistors.

Working of P-N-P tansistor. Fig. 2.60 shows a ***P-N-P transistor*** connected in the common-base (or grounded-base) configuration (it is so called because both the emitter and collector are returned to the base terminals). The *emitter junction is forward-biased whereas the collector junction is reverse-biased.* The holes in the emitter are repelled by the positive battery terminal towards the *P-N* or emitter junction. The potential barrier at the junction is reduced due to the forward-biased, hence holes cross the junction and enter the *N*-type base. Because the base is thin and lightly-doped, majority of the holes (about 95%) are able to drift across the base without meeting electrons to combine with. The balance of 5% of holes are lost in the base region due to recombination with electrons. The holes which after crossing the *N-P* collector junction enter the collector region are swept up by the negative collector voltage V_C.

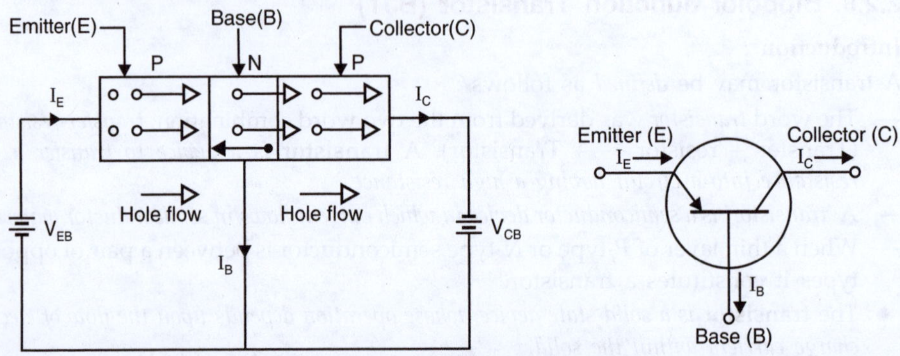

Fig. 2.39. *P-N-P* transistor.

The following points are worth noting :

1. In a *P-N-P* transistor *majority charge carriers are holes.*
2. *The collector current is always less than the emitter current because some recombination of holes and electrons take place.*

$$(I_C = I_E - I_B).$$

3. The *current amplification* (α) (or gain of P-N-P transistor) for steady conditions when connected in common base configuration is expressed as :

$$\alpha = \frac{I_C \text{ (collector current)}}{I_E \text{ (emitter current)}} < 1.$$

4. *Emitter arrow shows the direction of flow of conventional current.* Evidently, electron flow will be in the opposite direction.

Working of N-P-N transistor. Fig 2.40 shows a ***N-P-N junction transistor.*** The emitter is forward-biased and the collector reverse-biased. The electrons in the emitter region are repelled by the negative battery terminal towards the emitter or *N-P* junction. The *electrons*

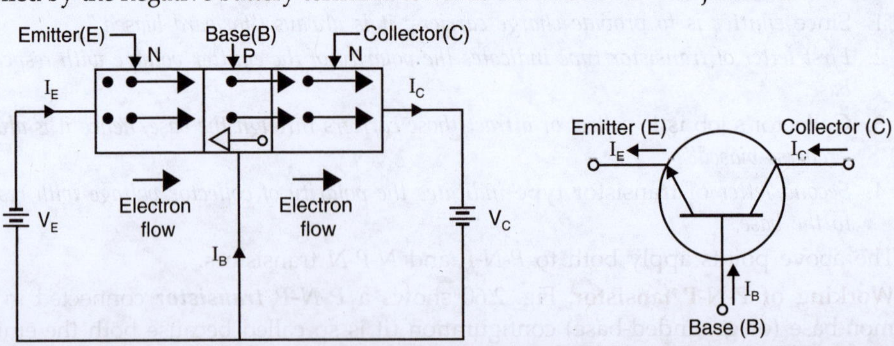

Fig. 2.40. *N-P-N* transistor.

cross over into the P-type base region because potential barrier is reduced due to forward bias. Since the base is thin and lightly doped, most of the electrons (about 95%) cross over to the collector junction and enter the collector region where they are readily swept up by the positive collector voltage V_C. Only about 5% of the emitter electrons combine with the holes in the base and are lost as charge carriers.

The following points are worth noting :

1. In a *N-P-N* transistor, *majority change carriers are electrons.*
2. I_C *(collector current) is less than* I_C (emitter current) *so that* $\alpha < 1$.
3. *Emitter arrow shows the direction of flow of conventional current.*

- The choice of *N-P-N* transistor is *made more often because majority charge carriers are electrons whose mobility is much more than that of holes.*

Note. *The junction transistors have been made in power ranges from a few milliwatts to tens of watts. The tiny junction transistor is unparalleled in that it can be made to work at power level as I microwatt.*

Transistor circuit configurations. A transistor is a three-terminal device (having three terminals namely *emitter, base* and*collector*) but we require four terminals-two for the input and two for the output for connecting it in a circuit. Hence one of the terminals of the transistor is made common to the input and output circuits. Thus there are three types of configurations for operation of a transistor. These configurations are :

(*i*) Common-base (CB) configuration.
(*ii*) Common-emitter (CE) configuration.
(*iii*) Common-collector (CC) configuration.

The term 'common' is used to denote the electrode that is common to the input and output circuits. Because the common electrodes is generally grounded, these modes of operation are frequently referred to as ground-base, ground-emitter and grounded-collector configurations as shown in Fig. 2.41 for a *N-P-N* transistor.

Each circuit configuration has specific advantages and disadvantages. It may be noted here that regrdless of circuit connection, the *emitter is always biased in the forward direction, while the collector always has a reverse biase.*

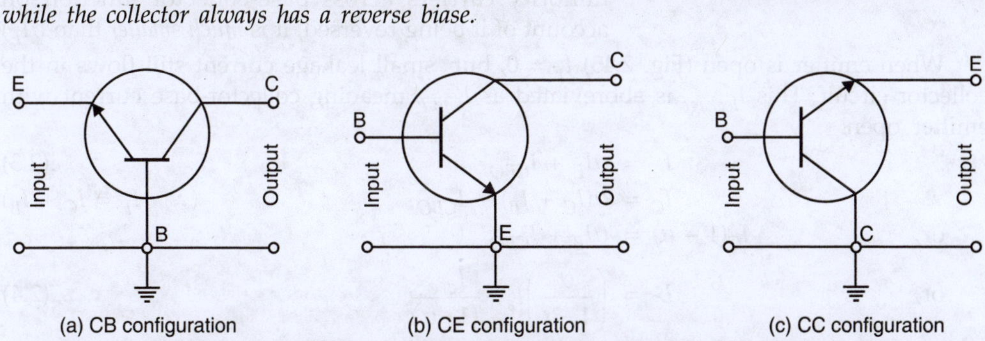

Fig. 2.41. Different circuit configurations for *N-P-N* transistor.

Common-base (CB) configuration :

In this circuit configuration, input is applied between emitter and base and output is taken from collector and base. Here, base of the transistor is common to both input and output circuits and hence the name common base configuration. A common-base configuration for *N-P-N* transistor is shown in Fig 2.42.

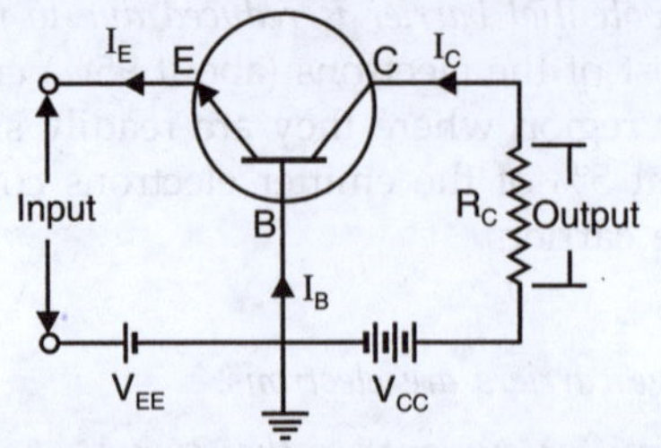

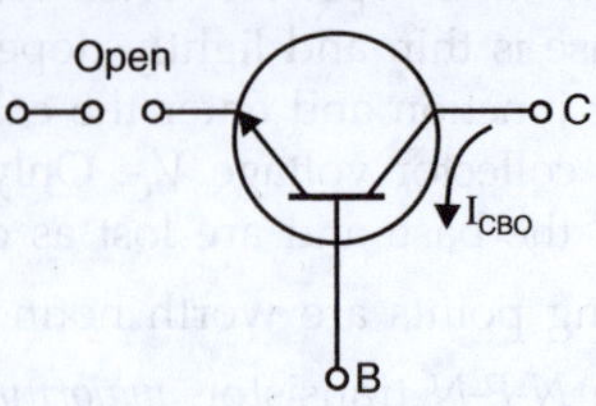

Fig. 2.42. Common-base *N-P-N* transistor. Fig. 2.43

Current amplification factor (α). *It is the ratio of output current to input current.* In CB configuration, the input current is the emitter current I_E and output current is the collector current I_C.

The ratio of change in collector current to the change in emitter current at constant collector-base voltage V_{CB} *is known as* ***current amplification factor*** *i.e.,*

$$\alpha = \frac{\Delta I_C}{\Delta I_E} \text{ at constant } V_{CB} \quad ...(2.1)$$

If *only D.C. values are considered, then* $\alpha = \dfrac{I_C}{I_E}$ ...(2.2)

α is *less than unity.* This value can be increased (not more than unity) by decreasing the base current. This is accomplished by *making the base thin and doping it lightly.*

In commercil transistors, practical value of α varies from 0.9 to 0.99.

Collector current (I_C) :

Total collector current, $I_C = \alpha I_E + I_{leakage}$

(αI_E is the part of emitter current that reaches the collector terminal)

where, I_E = Emitter current, and

$I_{leakage}$ = Leakage current (This current is due to movement of minority carriers across base-collector junction on account of it being reversed; it is *much smaller* than αI_E)

When emitter is open (Fig. 2.43) $I_E = 0$, but small leakage current still flows in the collector circuit. This $I_{leakage}$ is abbreviated as I_{CBO}, meaning collector-base current with emitter open.

$$I_C = \alpha I_E + I_{CBO} \quad ...(2.3)$$

$\therefore$ $I_C = \alpha(I_C + I_B) + I_{CBO}$ $\quad (\because \; I_E = I_C + I_B)$

or, $I_C(1 - \alpha) = \alpha I_B + I_{CBO}$

or, $$I_C = \left(\frac{\alpha}{1-\alpha}\right) I_B + \frac{I_{CBO}}{(1-\alpha)} \quad ...(2.4)$$

— In view of improved construction techniques, the magnitude of I_{CBO} for general-purpose and *low-powered transistors* (especially silicon transistors) is usually very small and *may be neglected in calculations.*

— For *high power calculations,* I_{CBO} *appears in* μA *range.*

— I_{CBO} *is temperature dependent, therefore, at high temperature it must be considered in calculations.*

2.2.9 Field-Effect Transistor (FET)

Introduction :

In an ordinary transistor both holes and electrons play part in the conduction process and so it is sometimes called a *bipolar transistor.* This ordinary transistor has the following two main *disadvantages* : (*i*) It has a low input impedance (because of forward biased emitter junction), and (*ii*) It has considerable noise level. The field-effect transistor (FET) has, by virtue of its construction and biasing, large input impedance (which may be more than 100 MΩ. The FET is generally *much less noisy* than the BJT).

Types of Field-effect Transistors :

A **field-effect transistor** (*FET*) *is a three terminal (namely* **drain, source** *and* **gate**) *semiconductor device in which current conduction is by only one type of majority carriers (electrons in case of an N-channel FET or holes in a P-channel FET).* It is also sometimes called the *unipolar transistor.*

In a broad sense, following are two main types of field-effect transistors :

1. *Junction field-effect transistor (JFET)*

(*i*) *N*-channel

(*ii*) *P*-channel.

2. *Metal oxide semiconductor field-effect transistor (MOSFET) or insulated gate field-effect transistor (IGFET).*

(*i*) Depletion type :

(*a*) *N*-channel

(*b*) *P*-channel

(*ii*) Enhancement type :

(*a*) *N*-channel

(*b*) *P*-channel

1. Junction field-effect transistors (JFET) :

The junction field-effect transistors (JFETs) can be divided depending upon their structure into the two following categories :

1. *N*-channel JFET
2. *P*-channel JFET

Construction :

- The basic construction of a *N*-channel JFET is as shown in Fig. 2.80 (*a*). It consists of an *N-type semiconductor bar with two P-type heavily doped regions diffused on opposite sides of its middle part.* The *P*-type regions form two *P-N* junctions. The *space between the junctions (i.e., N-type regions) is called a* **channel.** Both the *P-type* regions are connected internally and a single wire is taken out in the form of a terminal callled the **gate (G).** The electrical connections (called *ohmic contacts*) are made to both ends of the **N**-type semiconductor and are taken out in the form of two "terminals called **drain** (D) and **source (S)**. *The "drain (D)" is a terminal through which the electrons leave the semiconductor and "source (S)" is a terminal through which the electrons enter the semiconductor.*

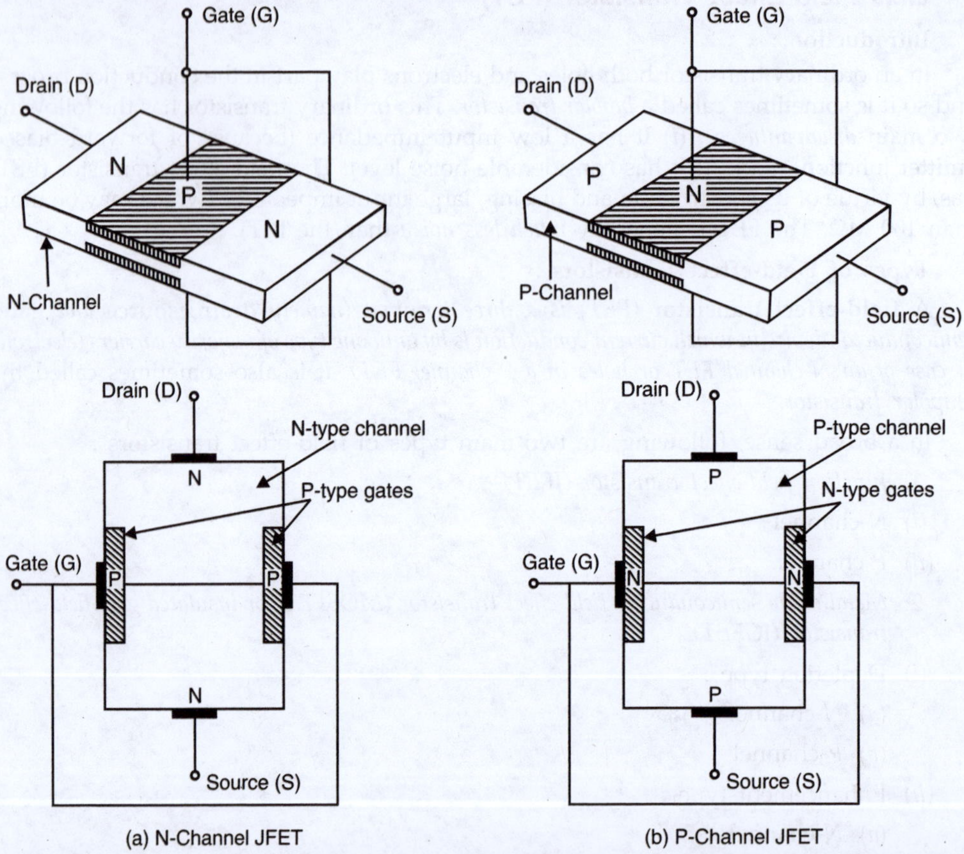

Fig. 2.44. JFETs.

Whenever a voltage is applied across the drain and source terminals, a current flows through the *N*-channel. The current consists of only one type of carriers (*i.e.,* electrons), therefore, the FET is called a *unipolar device.* (This distinguishes FET from BJT where the current consists of the flow of both the electrons and holes).

- A *P*-channel JFET is shown in Fig. 2.45 (*b*). Its construction is similar to that of *N*-channel JFET, except that it consists of a *P-channel and N-type junctions.* The current carriers in JFET are the *holes,* which flow through the *P*-type channel.

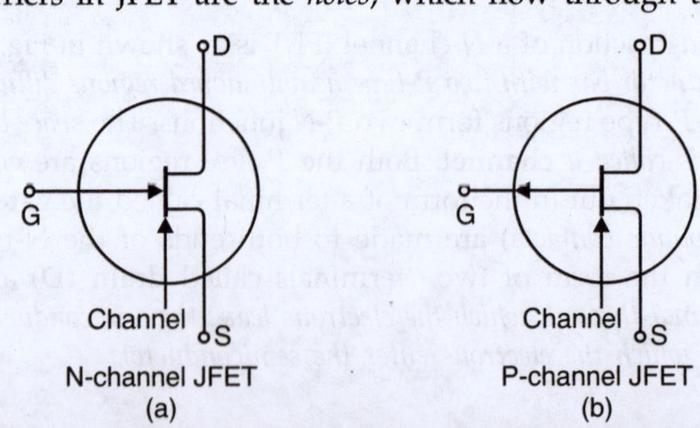

Fig. 2.45. Symbols for JFETs.

— Fig. 2.45 (*a*) shows the schematic symbol for a *N*-channel *JFET*. The arrow points *towards* the vertical line. The *vertical line* represents the *N-channel.*

— Similarly, Fig. 2.45 (*b*) shows the schematic symbol for a *P*-channel JFET.

The arrow points *away from* the vartical line. Here the vertical line represents the *P*-channel.

JFET polarities :

Fig. 2.46 (*a*) shows *N*-channel JFET polarities whereas Fig. 2.46 (*b*) shows the *P*-channel JFET polarities. It may be noted that in each case, the voltage between gate and source is such that the *gate is reverse biased.* This is the normal way of JFET connection. The drain and source terminals are interchangeable (This is generally valid for low frequencies but not for high frequencies applications).

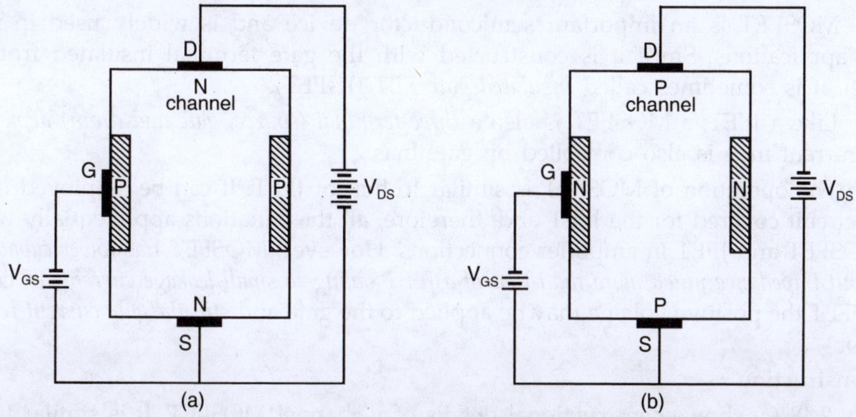

Fig. 2.46. JFET polarities.

Working :

Fig. 2.47 shows the circuit *N*-channel JFET with normal polarities. The circuit action is as follows :

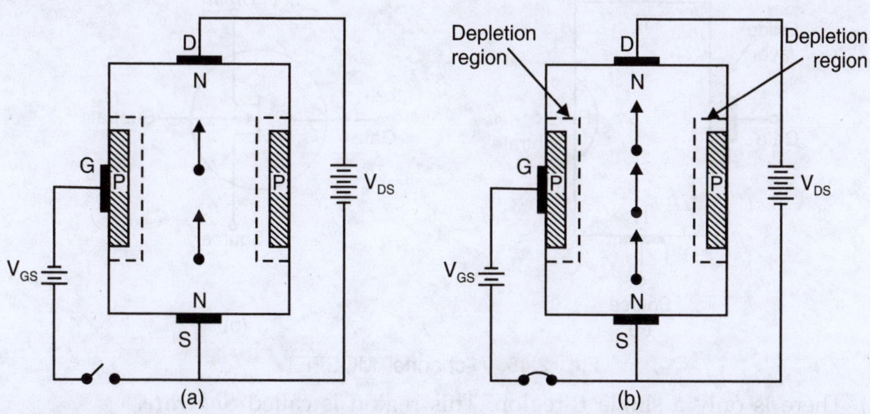

Fig. 2.47

(*a*) When a voltage V_{DS} is applied between drain and source terminals and voltage on the gate is zero [Fig. 2.47 (*a*)], the two *P-N* junctions at the sides of the bar establish deplection layers. The electrons will flow from source to drain through a channel between the depletion layers. *The size of these layers determines the width of the channel* and hence the current conduction through the bars.

(*b*) When a reverse voltge V_{GS} is applied between the gate and source [Fig. 2.47 (*b*)], the *width of the depletion layers is increased.* This *reduces* the width of conducting channel, thereby increasing the resistance of *N*-type bar. Consequently, the current from the source to drain is *decreased.* On the other hand if the reverse voltage on the gate is decreased, the width of depletion layers also decreases. This increases the width of the channel and source to drain current increases.

From the above discussion it is evident that current from source to drain can be controlled by the application of potential or electric field on the gate. It is due to this reason that this device is called field-effect transistor. Note that a *P-channel* JFET operates in the same manner as an *N*-channel JFET *except that channel current carriers will be the holes instead of electrons and the polarities of* V_{GS} *and* V_{DS} are reversed.

2. Metal oxide semiconductor FET (MOSFET) :

- MOSFET is an important semiconductor device and is widely used in many circuit applications. Since it is constructed with the gate terminal insulated from the channel, it is sometimes called *insulated gate FET* (IGFET).
- Like a JFET, a MOSFET is also a *three-terminal (source, gate and drain) device* and drain current in it is also controlled by gate bias.
- The operation of MOSFET is similar to that of JFET. It can be employed in any of the circuit covered for the JFET and, therefore, all the equations apply equally well to the MOSFET and JFET in amplifier connections. However, MOSFET has *lower capacitance and input impedance much more than that of a JFET owing to small leakage current.* In case of a MOSFET the positive voltage may be applied to the gate and *still the gate current remains the zero.*

Construction :

Fig. 2.48 (*a*) shows constructional details of *n*-channel MOSFET. It is similar to FET except with the following modifications :

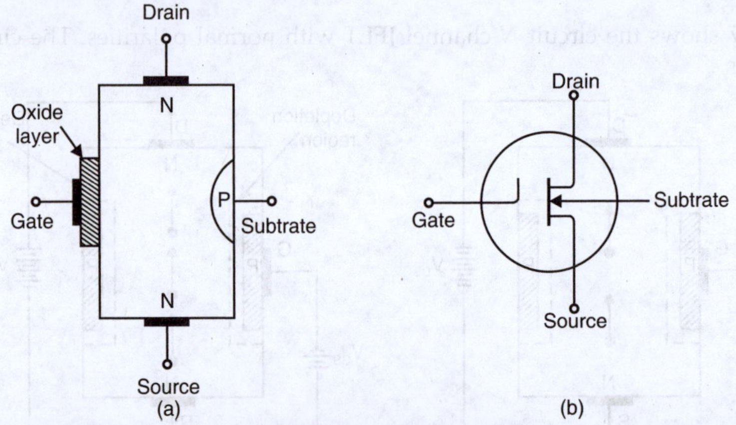

Fig. 2.48. *N*-channel MOSFET.

(*i*) There is only a single *P*-region. This region is called ***Subtrate.***

(*ii*) A thin layer of metal oxide (usually silicon oxide) is deposited over the left side of the channel. A *metalic **gate** is deposited over the oxide layers.* As silicon dioxide is an insulator, therefore, *gate is insulated from the channel.*

(*iii*) Like JFET, a MOSFET has three terminals *viz, source, gate* and *drain.*

Fig. 2.48 (*b*) shows the symbolic symbol of *N*-channel MOSFET.

Working :

Fig. 2.49 shows the MOSFET circuit. Instead of gate diode as in JFET, here gate is formed as a small capacitor. *One plate of this capacitor is the gate and the other plate is the channel with metal oxide as the dielectric.*

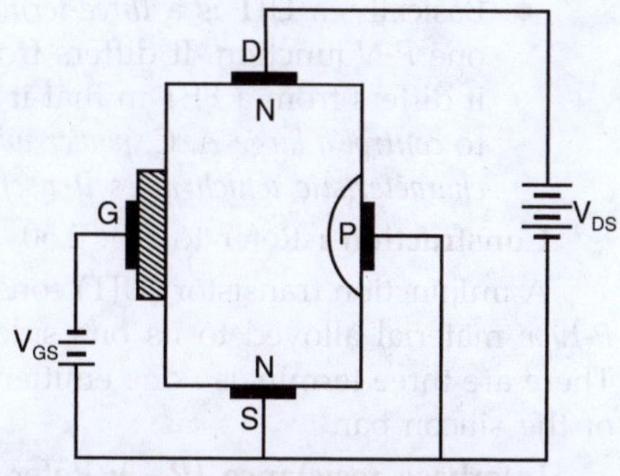

Fig. 2.49. MOSFET circuits.

— When *negative voltage* is applied to the gate, electrons accumulate on it. These electrons repel the conduction band electrons in the *N*-channel. Therefore, lesser number of conduction electrons are made available for current conduction through the channel. The greater the negative voltage on the gate, the lesser is the current conduction from source to drain.

— When the gate is given *positive voltage,* more electrons are made available in the *N*-channel. Consequently, current from source to drain *increases.*

Regarding MOSFET, the following *points* are worth noting :

- **A MOSFET, unlike the JFET, has *no gate diode.*** Therefore, it is possible *to operate the device with positive or negative gate voltage.*
- In a MOSFET, *the source to drain current is controlled by the electric field of capacitor formed at the gate.*
- As the gate forms a capacitor, therefore, negligible currents flows whether + ve or – ve voltage is applied to the gate. Consequently, the input impedance of MOSFET is *very high* (varying from 10^4 MΩ to 10^6 MΩ).

2.2.10 Unijunction Transistor (UJT)

A unijunction transistor, unlike a bipolar transistor has *only one junction.* Like other conventional transistors, it also processes the transistor action and *works like a switch.*

Its characteristics are similar to those of a silicon unilteral switch (SUS) and a complementary silicon controlled rectifier (CSCR). Its construction is, however, different and it does *not* belong to the thyristor family.

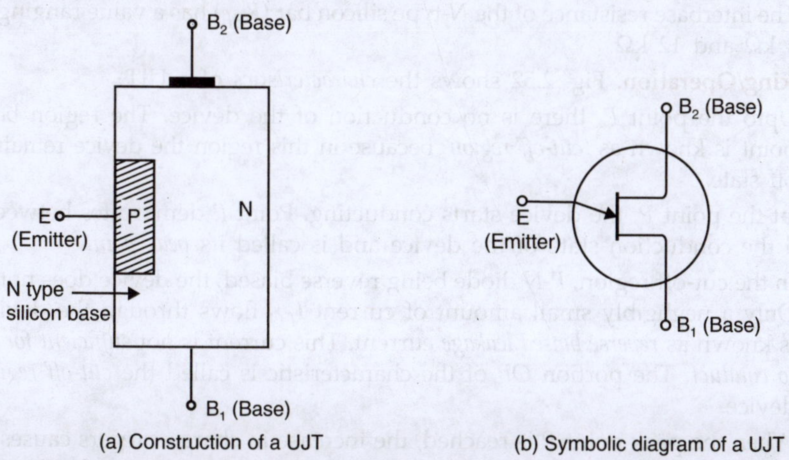

Fig. 2.50. Unijunction Transistor (UJT).

- Basically, a UJT is a *three-terminal silicon diode.* As its name indicates, it has only one *P-N* junction. It differs from an ordinary diode in that it has *three* leads and it differs from a FET in that it has *no ability to amplify*. However, it has the ability to *control a large A. C. power with a small signal.* It also exhibits a *negative resistance characteristic which mkes it useful as an oscillator.*

Construction : Refer to Fig. 2.50

A unijunction transistor (UJT) consists of a *lightly doped silicon* bar with a *heavily doped P-type* material alloyed to its one side (closer to B_2) for producing single *P-N* junction. There are three terminals : one emitter, E and two bases B_1 and B_2 at the bottom and top of the silicon bar.

Interbase resistance (R_{BB}): Refer to Fig. 2.51.

The interbase resistance (R_{BB}) is the total resistance of the silicon bar from one end to the other with emitter terminal open; from equivalent circuit (see Fig. 2.51), we have

$$R_{BB} = R_{B2} + R_{B1}$$

The point C is such that $R_{B1} > R_{B2}$ (usually R_{B1} is 60 percent of R_{BB}). R_{B1} has been shown as a variable resistor because its value varies inversely as I_E.

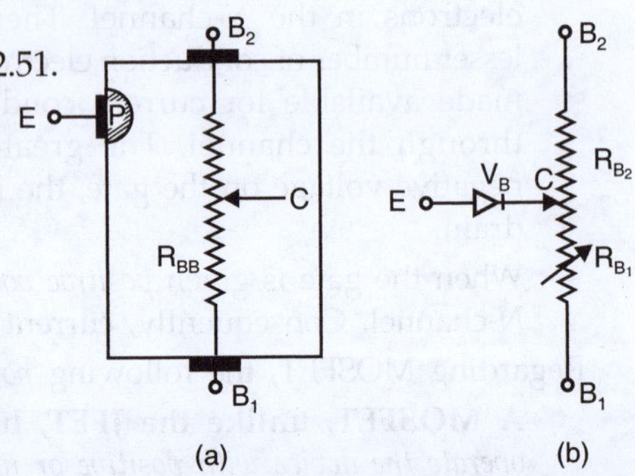

Fig. 2.51

Let the voltage drop across R_{B1} is V_C. Then,

$$V_C = V_{BB} \times \frac{R_{B_1}}{R_{B_1} + R_{B_2}} \quad \text{...using voltage binder relationship}$$

$$= \eta . V_{BB}$$

where, $$\eta = \frac{R_{B_1}}{R_{B_1} + R_{B_2}}$$

η is called the ***instrinsic stand ratio.***

— The value of η depends on two factors namely : (*i*) Construction of the UJT, and (*ii*) Spacing between the emitter junction and the two base contacts.

— The value of η is always less than unity (lies between 0.51 and 0.81)

— The interbase resistance of the *N*-type silicon bar (R_{BB}) has a value ranging between 4 kΩ and 12 kΩ.

Working/Operation. Fig. 2.52 shows the *characteristics* of a UJT.

— Upto the point *P*, there is no conduction of the device. The region before this point is known as *'cut-off region'* because in this region the device remains in cut-off state.

Just at the point *P*, the device starts conducting. Point *P* demarcates between cut-off state and the conduction state of the device and is called its *peak point.*

— In the cut-off region, *P-N* diode being reverse biased, the device does not conduct. Only a negligibly small amount of current I_{EO} flows through the device which is known as *reverse biased leakage* current. This current is not *sufficient for the device to conduct.* The portion *OP* of the characteristic is called the *cut-off region* of the device.

— When the peak point *P* is reached, the increase in charge carriers causes decrease in resistance R_{B1} and the device starts *conducting.* In the conduction state, the device depicts a negative resistance charcteristics. This means, as the emitter voltage V_E is further increased, the voltage across the device decreases, but the current increases. This region of conduction is called the *negative resistance region.* This region continues till the valley point V is reached in the characteristic. The portion PV of the characteristic is called the *negative resistance region.*

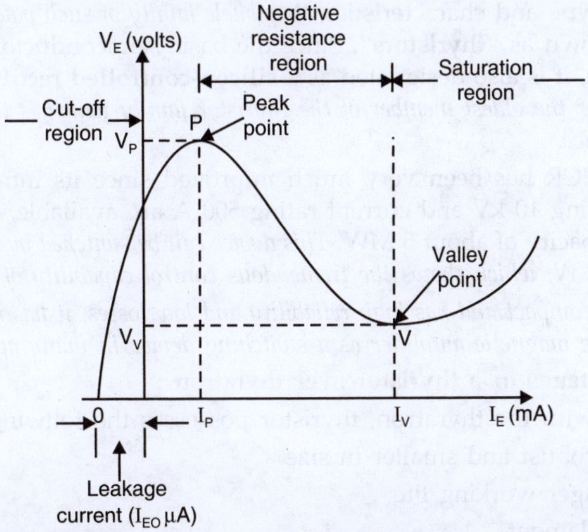

I_{EO} = Leakage current ; V_P = Peak point voltage; I_P = Peak point current; V_V = Valley point voltage; I_V = Valley point current; V_E = Emitter voltage ; I_E = Emitter current.

Fig. 2.52. Characteristics of UJT.

— After reaching the valley point, the device goes to its saturation state where further fall in the voltage across the device does not take place. The device voltage and current both reach standard values and do not change any more even if the emitter voltage is changed. This portion of the characteristic beyond the valley point is called *'saturation region'*.

A set of *V-I* characteristic for UJI can be obtained for different values of interbase voltage V_{BB}.

- It is seen that only terminals *E and* B_1 *are active terminals whereas* B_2 *is the bias terminal i.e.,* it is meant only for applying external voltage across UJT.
- Generally, *UJT is triggered into conduction by applying a suitable positive pulse at its emitter.* It can be brought *back to OFF state by applying a negative trigger pulse.*

Applications. One significant property of UJT is that it *can be triggered by* (or an output can be taken from) *any one of its three terminals.* Once triggered, the emitter current I_E of the UJT increases regeneratively till it reaches a limiting value determined by the external power supply. Owing to their particular behaviour, UJT is used in variety of circuit applications; some of these are :

1. Pulse generation;
2. Sine wave generator;
3. Saw tooth wave generator;
4. Switching;
5. Phase control;
6. Voltage or current regulated supplies;
7. Timing and trigger circuits.

2.2.11. Thyristor

2.2.11.1. Introduction :

Ample pioneering work on theory and fabrication of the *power-switching device,* which later came to be known as a *thyristor* (because its characteristics are *similar to those of the gas-tube thyratron*), was done at the Bell Laboratories in the U.S.A. The first prototype was introduced by the General Electric Company (USA) in 1957. Since then, many improvements have been made, both in the technique of its fabrication and in adapting it to numerous industrial appplications. With the development of a number of other devices of similar type and characteristics, the *whole family of such power-switching devices* has come to be known as **"thyristors".** Since the basic semiconductor material used for the device is silicon, it is also designated as a **silicon-controlled rectifier (SCR).** *The term SCR is often used for the oldest member of the thyristor family which is the most widely used power-switching device.*

The rating of SCR has been very much improved since its introduction and now SCRs of voltage rating 10 kV and current rating 500 A are available, corresponding to a power-handling capacity of about 5 MW. *This device can be switched by a low-voltage supply of about 1 A and 10 W, which shows the tremendous control capability of the device.*

Because SCR is compact and has high reliability and low losses, it has more or less replaced the thyratron and the magnetic amplifier as a switching device in many applications.

2.2.11.2. Advantages of a thyristor over thyratron :

It comparison with the thyratron, thyristor possesses the follwing *advantages* :

1. It is more robust and smaller in size.
2. It has a longer working life.
3. It has no filament.
4. The voltage drop in the forward direction is only about 1 to 2 volts, compared 10 to 15 volts for the thyratron.
5. The triggering and recovery periods are much shorter, so that it is more suitable for *high-frequency switching operations.*
6. The arc ionizing and deionizing times for a thyratron are comparatively large and so the device applications are limited to a frequency of 1 kHz. *A thyristor can operate over a much greater range of frequency.*

Comparison between transistors and thyristors :

The comparison between transistors and thyristors is given in Table 2.1.

Table 2.1. Comparison between "Transistors" and "Thyristors"

S. No.	Aspects	Transistors	Thyristors
1.	*Type of device*	3-layers, 2-junction devices	4-layer, 2-or more junction devices
2.	*Reponse*	Fast	Very fast
3.	*Efficiency*	High	Very high
4.	*Reliability*	Highly reliable	Very highly reliable
5.	*Voltage drop*	Small voltage drop	Very small voltage drop
6.	*Life*	Long life	Very long life
7.	*Power ratings*	Small to medium power ratings	Very small to very large power ratings
8.	*Conducting state*	Require a continuous flow of current to remain in conducting state	Require only small pulse for triggering and thereafter remaining in conducting state.
9.	*Power consumption*	Low power consumption	Very low power consumption
10.	*Control capability*	Low control capability	High control capability
11.	*ON, OFF timings*	Small turn-ON and turn-OFF timings	Very small turn-ON and turn-OFF timings.

2.2.11.3. Construction, operation and characteristics of a thyristor :

Construction :

- The cross-sectional view of a typical SCR is shown in Fig. 2.53. Basically, the SCR consists of a four-layer pallet of *P*-type and *N*-type semiconductor materials. Silicon is used as the intrinsic semiconductor to which the proper impurities are added. The junctions are either *diffused* or *alloyed.*
- The *planer construction* shown in Fig 2.53 (*a*) is used for *low-power* SCRs. This technique is useful for making a number of units for a single silicon wafer. Here, all the junctions are *diffused.*

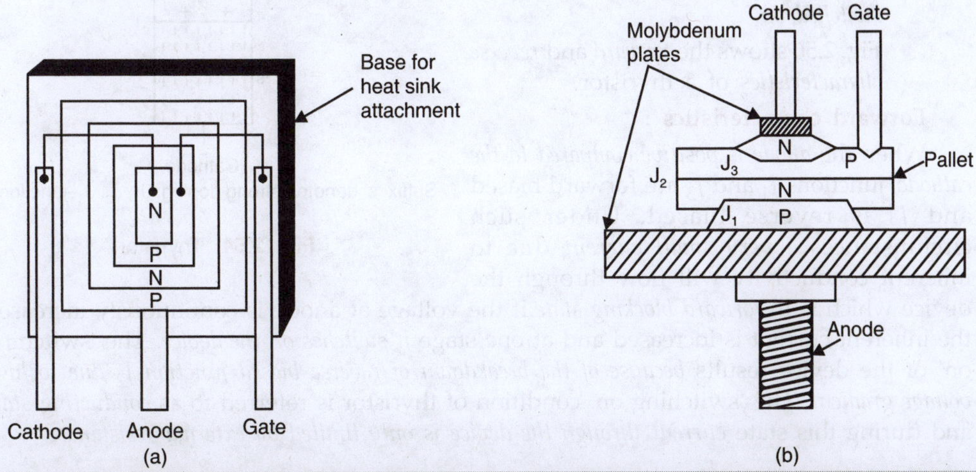

Fig. 2.53. (*a*) Planer type (all diffused), (*b*) Mesa type (alloy diffused).

- The other technique the *mesa construction* is shown in Fig. 2.53 (*b*). This is used for *high-power* SCRs. Here, the inner junction J_2 is obtained by diffusion, and then the outer two layers are alloyed to it. Because the *PNPN* pallet is required to handle large currents, it is properly braced with tungsten or molybdenum plates to provide greter mechanical strength. One of these plates is handsoldered to a copper or an aluminium stud, which is threaded for attachment to a heat sink. This provides an efficient thermal path for conducting the internal losses to the surrounding medium. The use of hand solder between the pallet and back up plates *minnimises thermal fatigue* when the SCRs are subjected to temperature-induced stresses. For medium and low-power SCRs, the pallet is mounted directly on the copper stud or casing, using soft-solder which absorbs the thermal stresses set-up by differential expansion and provides a good thermal path for heat transfer. When a larger cooling arrangement is required for highpower SCRs, the press-pack or hockey pack construction is used, which provides for double-sided air or water cooling.
- Fig. 2.54 shows the terminal configuration and symbolic diagram of a SCR.

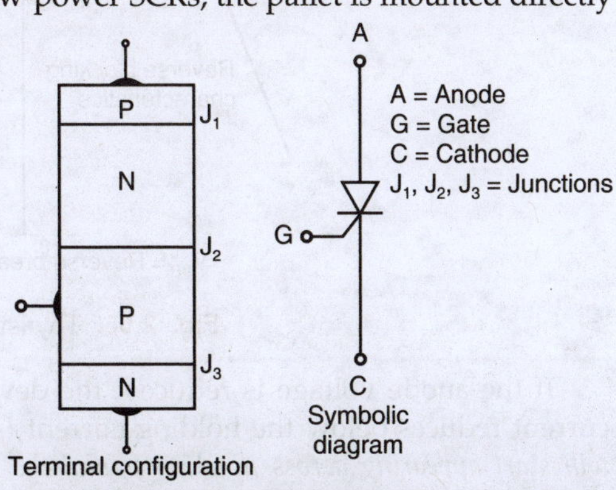

Fig. 2.54. Schematic diagram of a SCR.

- Fig. 2.55 shows a thyristor. It has four layers alternately of *P* and *N* with *outer layers heavely doped.* The control current is generally applied *via* middle *P* layer and *N* emitter. *Junctions* J_1 *and* J_3 *are forward biased while junction* J_2 *is reverse biased. If the voltage gradient between anode and cathode is too high then due to increased inherent current the thyristor may be switched on.*
- Fig. 2.56 shows the *forward* and *reverse characteristics* of a thyristor.

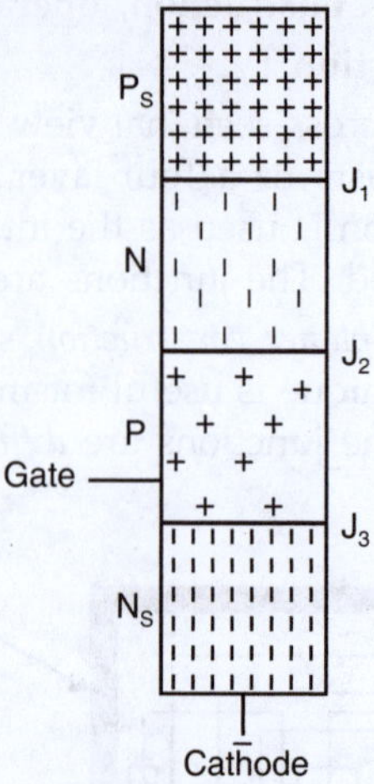

Suffix 's' denotes strong doping; J_1, J_2, J_3–junctions.

Fig. 2.54. Thyristor.

Forward characteristics :

When the *anode is positive compared to the cathode,* junctions J_1 and J_3 are forward biased and J_2 is reverse biased. Under such circumstances *a very small current* due to inherent conductivity will flow through the device which is in *forward blocking state.*If the voltage at anode is continuously increased the inherent current is increased and at one stage *it switches on the device.* The 'switching on' of the device results *because of the breakdown of reverse biased junction* J_2 *due to high voltage gradient.* The 'switching on' condition of thyristor is referred to as *conducting state* and during this state *current through the device is only limited by external resistance.*

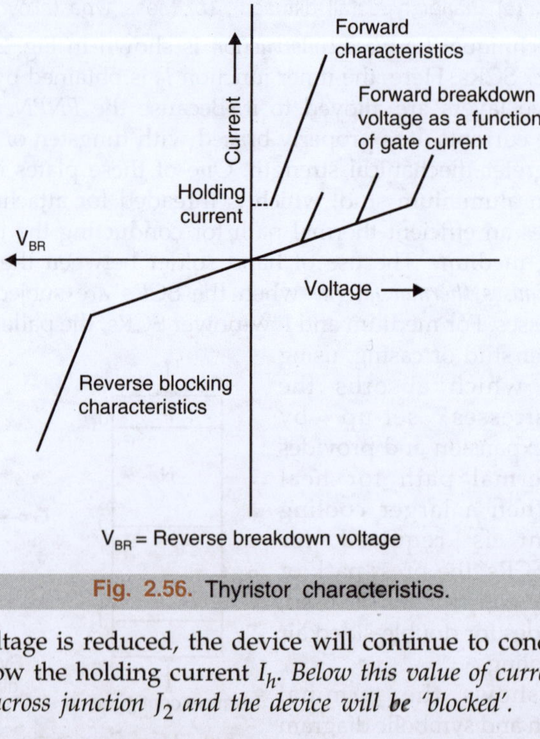

V_{BR} = Reverse breakdown voltage

Fig. 2.56. Thyristor characteristics.

If the anode voltage is reduced, the device will continue to conduct till the forward current reduces below the holding current I_h. *Below this value of current the depletion layer will start appearing across junction* J_2 *and the device will be 'blocked'.*

Reverse blocking characteristics :

If the *cathode is positive as compared to anode,* the junction J_2 *is reverse biased* and only a small current flows through the device and the characteristics are called *reverse blocking characteristics.* If the voltage is continuously increased at one stage it may result in breaking of depletion layers at junctions J_1 and J_3 and the current through the device suddenly increases to a very high value. This is called *reverse breakdown* and the voltage is called reverse breakdown voltage V_{BR}.

As the outer layers of thyristor are highly doped compared to inner layers, the thickness of depletion layers at J_2 during forward bias is much more as compared to the total thickness of depletion layer at junctions J_1 and J_3 during reverse bias. Hence, the forward breakdown voltage V_{BF} *is normally greater than reverse breakdown voltge* V_{BR}.

Thyristor Family :

There are several members in the thyristor family, some of them are mentioned below:

1. DIAC (Bidirectional Diode Thyristor)
2. TRIAC (Bidirectional Triode Thyristor)
3. SCR (Silicon Controlled Rectifier)
4. SUS (Silicon Unilateral Switch), also known as complementry SCR (CSCR)
5. SBS (Silicon Bilateral Switch)
6. SCS (Silicon Controlled Switch)
7. LASCR (Light Activated SCR)
8. LASCS (Light Activated SCS).

2.2.11.4. Typical SCR parameters :

Typical SCR parameters are given in the table 2.2

Table 2.2. Typical SCR parameters

S. No.	Parameters	Typical
1.	Forward breakover voltage, V_{FBO}	50 to 500 volts
2.	Maximum on-state voltage	About 1.5 V
3.	Peak reverse voltage, PRV	Upto 2.5 kV
4.	Holding voltage, V_H	0.5 to 20 volts
5.	Forward breakover current	Less than a few hundred μA
6.	Peak forward current	30 A to over 100 A
7.	Holding current	A few mA to few hundred mA
8.	Turn-on and turn-off times	A few tenths of μs for fast acting SCRs; A few μs for slow acting SCRs
9.	Dynamic resistance in cut-off region	A few MΩ to a few hundred MΩ
10.	Dynamic resistance in saturation region.	Lesser than 1 Ω for currents of several amperes; Lesser than 10 Ω for large currents.

2.2.11.5. Diac

Refer to Fig. 2.57.

A **Diac** *is a two terminal, three layer bi-directional device which can be switched to ON state for either polarity of the applied voltage.* It is, therefore, also known as a *'bi-directional avalanche diode'.*

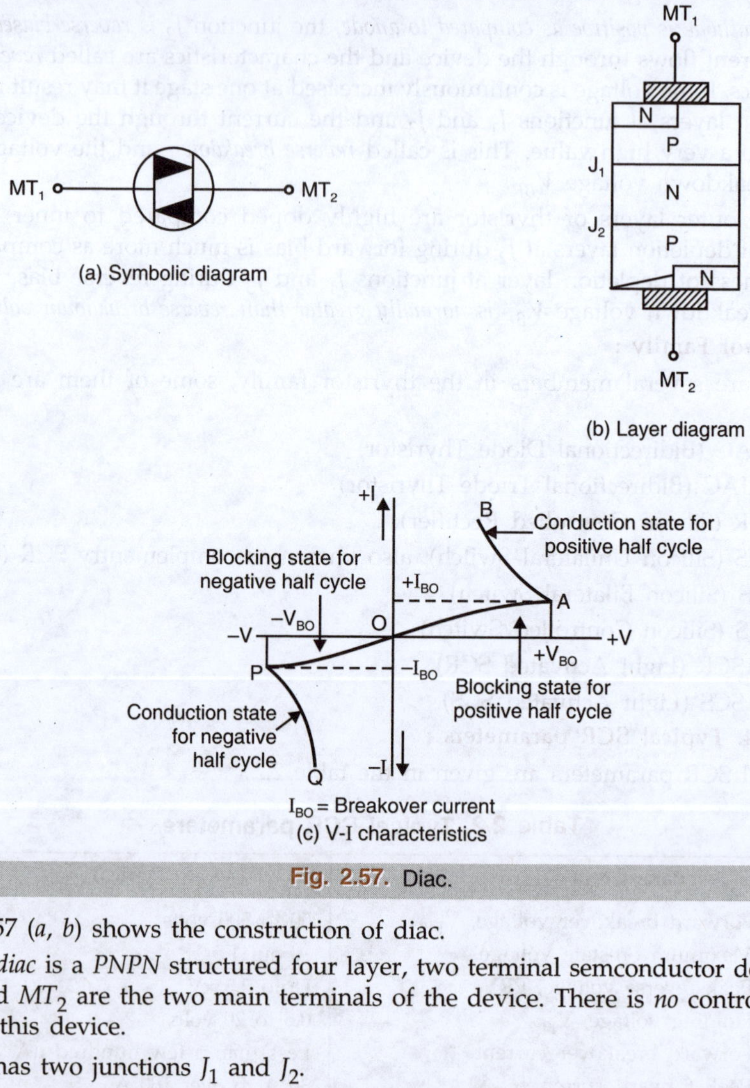

Fig. 2.57. Diac.

Fig. 2.57 (*a, b*) shows the construction of diac.

— A *diac* is a *PNPN* structured four layer, two terminal semconductor device. MT_1 and MT_2 are the two main terminals of the device. There is *no* control terminal in this device.

— It has two junctions J_1 and J_2.

— It is evident from the layer diagram (Fig. 2.57 (*b*)) that, a diac unlike a diode resembles bipolar transistor. However, the central layer of the diac is free from any connection with the terminals. The doping level at the two ends of the device is the same which leads to identical *V-I* characteristics in both Ist and IIIrd quadrants. Fig. 2.57 (*c*) shows the *V-I* characteristics of a diac.

— *When positive or negative voltage is applied across the terminals of a diac, only small leakage current will continue to flow through the device. As the applied voltage is increased, the leakage current will continue to flow until the voltage reaches the breakdown point. At this point, avalanche breakdown of the reverse biased junctions occurs and current through the device increases sharply.*

Applications. Diacs are used primarily for *triggering triacs in adjustable phase control of A.C. mains supply.*

2.2.11.6. Triac

A **triac** *is a three terminal semiconductor switching device which can control alternating current in load.*

One major difference between an SCR and triac is that whereas SCR is a unidirectional switch and can conduct in one direction only, a triac is *bi-directional switch and can conduct in either direction.*

Construction :

The **triac** is *a three terminal, four layer semiconductor device.* Its three terminals are MT_1, MT_2 and the 'Gate'. Its symbol, layer diagram and pin configuration are shown in Fig. 2.58 (*a*), (*b*) (*c*) respectively.

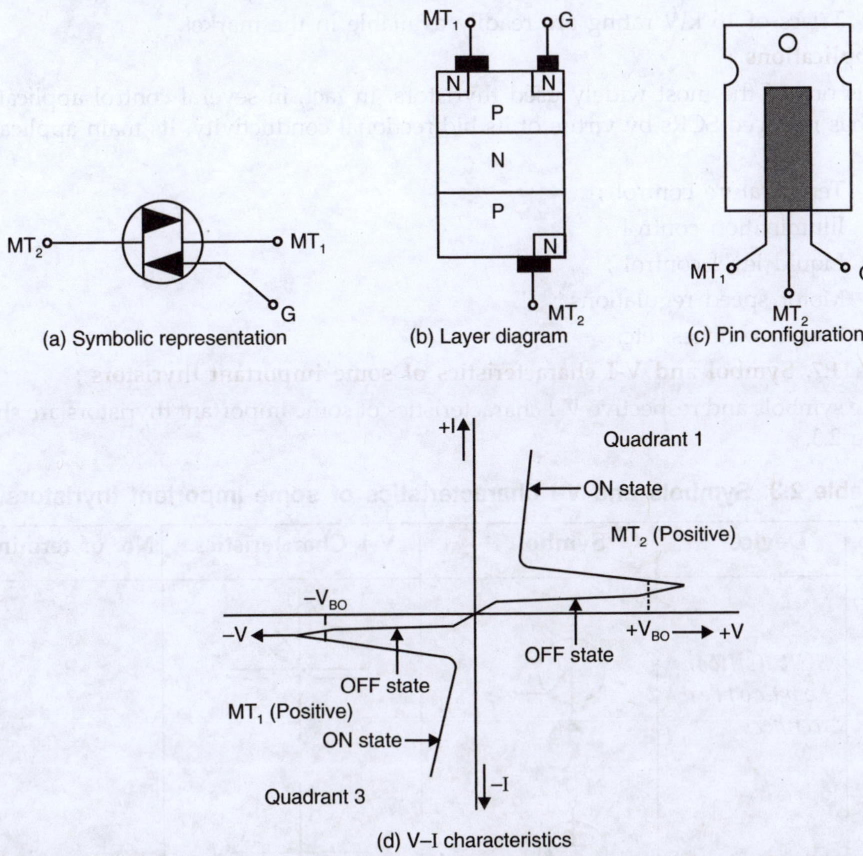

Fig. 2.58. Triac.

Working/Operation of a triac :

Fig. 2.58 shows the *V-I* characteristics of a triac.

— A triac, like an SCR, also starts conducting only when the breakover voltage is reached. Earlier to that, the leakage current which is very small in magnitude, flows through the device and therefore it remains in the OFF state.

— The 1st quadrant characteristic is just like an SCR, but 3rd quadrant characteristic of a triac is identical to its 1st quadrant, *except that, as the polarities of the main terminals change, the direction of current changes.*

— MT_2 is positive with respect to MT_1 in the 1st quadrant and it is negative in the 3rd quadrant. The device, when starts conducting, allows very heavy amount of current to flow through it. This high inrush of current must be limited by using external resistance, or it may otherwise damage the device. The **'gate'** is the *control terminal* of the device. By applying proper signal at the gate, the firing angle of the device can be changed thus, the phase control process can be changed.

- The *great advantage of triac* is that by adjusting the gate current to a proper value, any portion of both positive and negative half cycles of A.C. supply can be made to flow through the load. This *permits to adjust the transfer of A.C. power from the source to the load.*

— Its *main limitation* in comparison to SCRs is, its *low power handling capacity.*

- Triacs of 16 kW rating are readily available in the market.

Applications :

It is one of the most widely used thyristors. In fact, in several control applications, it has replaced SCRs by virtue of its bidirectional conductivity. Its main applications are:

1. Temperature control ;
2. Illumination control ;
3. Liquid level control ;
4. Motor speed regulations ;
5. Power switches, etc.

2.2.11.7. Symbol and V-I characteristics of some important thyristors :

The symbols and respective *V-I* characteristics of some important thyristors are shown in table 2.3.

Table 2.3. Symbols and V-I characteristics of some important thyristors

S.No.	Device	Symbol	V-I Charcteristics	No. of terminals
1.	*SCR (silicon controlled rectifier)*	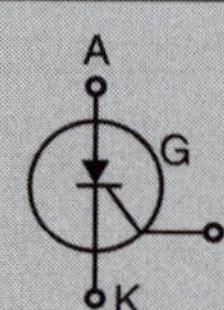	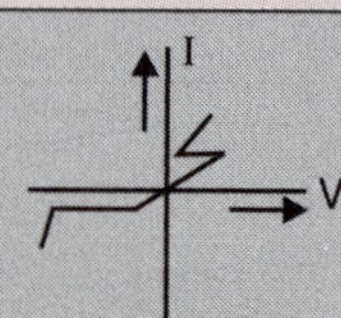	3
2.	*Diac*	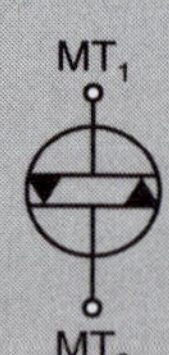	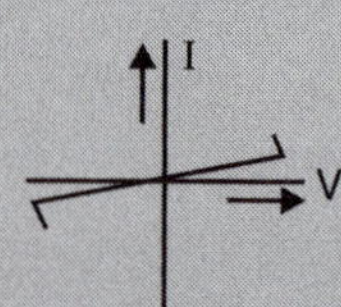	2
3.	*Triac*	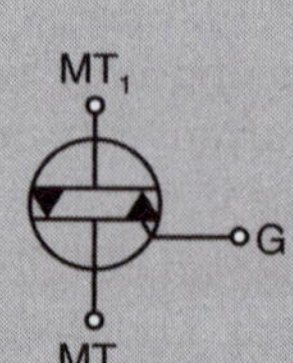	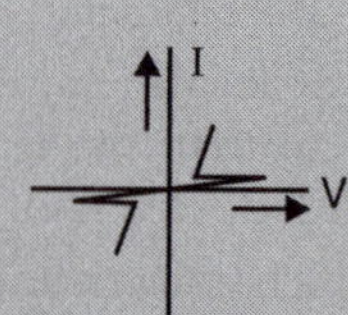	3

4.	*SCS (silicon controlled switch)*	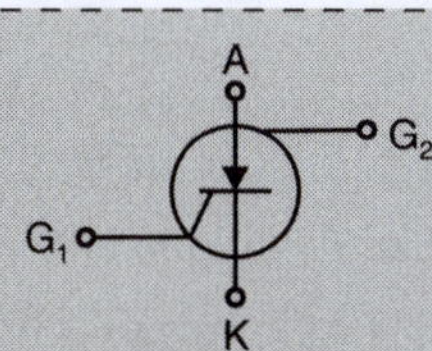	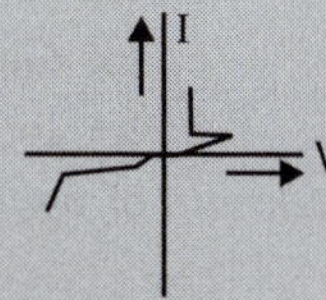	4
5.	*SUS (silicon unilateral switch)*	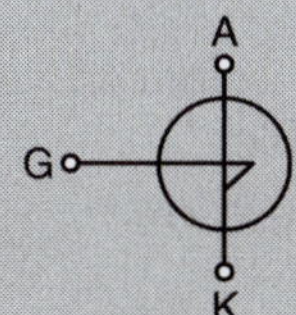	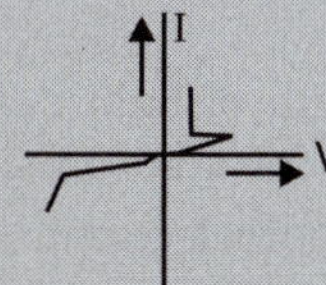	3
6.	*LASCR (light activated SCR)*	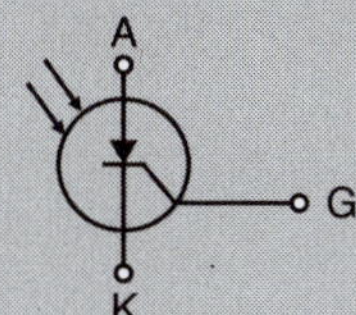	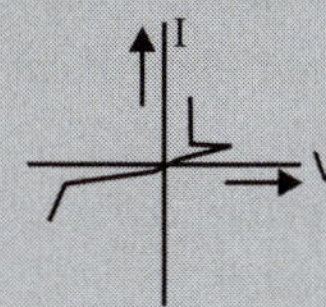	3

2.2.12 Optoelectronic Devices

Fundamentals of Light :

- As per Quantum theory, light consists of discrete packet of energy called *photons.* The energy (*E*) contained in a photon is given by;

$$E = hf$$

$$\left[\begin{array}{l}\text{where, } h = \text{Planck's constant } \left(= 6.625\times10^{-34} \text{ joule-second}\right)\text{, and} \\ \qquad\quad f = \text{frequency of light (in Hz)}\end{array}\right]$$

$$= h\times\frac{c}{\lambda}$$

or, $$\lambda = \frac{hc}{E} \text{ metres}$$

$$\left[\begin{array}{l}\text{where, } c = \text{Velocity of light } (= 3\times10^{8}\text{ m/s})\text{, and} \\ \qquad\quad \lambda = \text{Wavelength of light (metres).}\end{array}\right]$$

$$= \frac{6.625\times10^{-34}\times3\times10^{8}}{E} = \frac{19.875\times10^{-26}}{E} \quad \text{...}E \text{ in joules}$$

If E is in eV (electron – volt), then since 1 eV = 1.6×10^{-19} J

$$\therefore \quad \lambda = \frac{19.875\times10^{-26}}{E\times1.6\times10^{-19}} = \frac{12.42\times10^{-7}}{E} \text{ metre} = \frac{1.242}{E}\,\mu\text{m}$$

- When the *P-N* junction is forward biased, both the electrons as well as holes cross the junction. During this process some electrons recombine with holes, consequently some energy is lost by the electrons. The *amount of energy lost (given off in the form of light energy) is equal to the difference in energy between the conduction and valence bands, this being known as the semiconductor energy band gap* E_g.

E_g for *silicon* ...1.1 eV

E_g for Ga As ...1.43 eV

E_g for In As ...0.36 V

Example. *The wavelength of light emitted by silicon P-N junction,*

$$\lambda = \frac{1.242}{E_g} = \frac{1.242}{1.1} = 1.13\ \mu\text{m}$$

- The wavelength of light determines its colour in visible range and whether it is ultraviolet or infrared outside the visible.
- The various optoelectronic devices in use are :
  - — *Light Emitting Diode (LED)*
  - — *Liquid Crystals Displays (LCD)*
  - — *P-N junction photo diode*
  - — *Dust Sensor*
  - — *Photoconductive cell*
  - — *Phototransistor*
  - — *Photodarlington*
  - — *Photovoltaic or Solar cell*
  - — *Laser Diode*
  - — *Optical Disks*
  - — *Hologram Scanners*
  - — *Light activated SCR (LASCR)*
  - — *Optical Isolators*
  - — *Optimal Modulators etc.*

Some of these devices are discusses hence forth.

1. Light Emitting Diode (LED) :

A *P-N* junction can absorb light energy and produce electric current. The opposite process is also possible, that is a junction diode can emit light. The emission of light occurs under forward bias condition due to recombination of electrons and holes. The emitted light may be visible or invisible.

A P-N junction diode, which emits light when forward biased is known as a **light emitting diode (LED).**

The amount of light output is *directly proportional to the foward current.* Thus, higher the forward current, higher is the light output.

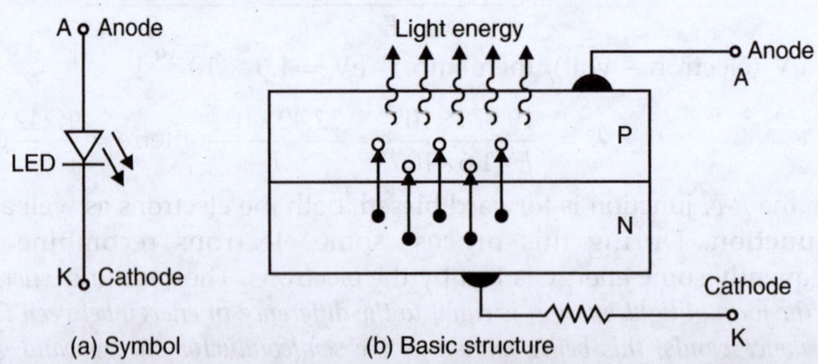

Fig. 2.59. Light emitting diode (LED).

Fig. 2.59 (*a*) shows the schematic symbol of a light emitting diode. The arrows pointing away from the diode symbol represent the *light*, which is being transmitted away from the junction.

Fig. 2.59 (*b*) shows the basic structure of a light emitting diode.

In a forward based *P-N* junction, free electrons from *N*-type material diffuse into *P*-region. Once in *P*-region these free electrons encounter holes and eventually recombine. Thus the free conduction electron fills a vacancy in valent structure and thus becomes a valence electron. In doing so the electron loses a certain amount of energy as it jumps from conduction band to the valence band. In *Si or Ge diode, the energy that recombining electrons lose is dissipated in the form of heat. But if other semiconductor material such as gallium arsenide and gallium phosphide are used to form P-N diode, the energy lost by recombining electrons is given off in the form of* **light energy.**

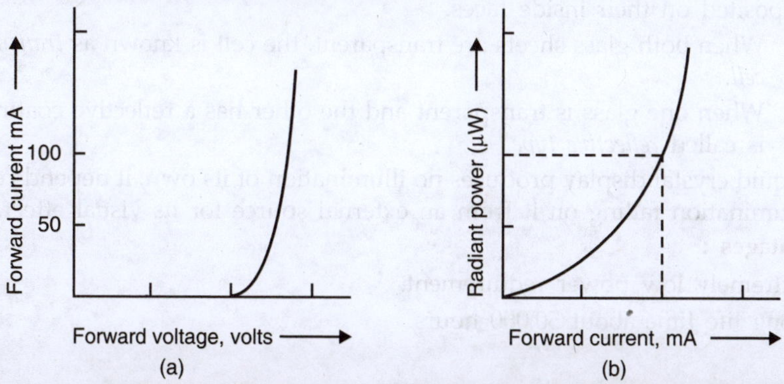

Fig. 2.60. Operating characteristics—LED.

— Diodes made of gallium arsenide (GaAs) emit infrared radiation invisible to eyes and such diodes are referred to as IRED—Infrared emitting diodes.

Fig. 2.60 shows two curves used to determine LED operating characteristics.

— Fig. 2.60 (*a*) is forward bias *V-I* curve for a typical IRED, the type used in burglar alarms. Forward bias of around 1 V is required to produce significant forward current.

— Fig. 2.60 (*b*) is a plot of radiant output power *vs* forward current. The radiant output power is rather small (μW) and *indicates a very low efficiency of electrical to radiant energy conversion.*

Applications :

Since LEDs operate at voltage levels 1.5 V to 3.3 V, they are highly compatible with solid state circuitry. Their uses include the following :

(*i*) Panel indicator

(*ii*) Digital watches

(*iii*) Calculators

(*iv*) Multimeters

(*v*) Intercoms

(*vi*) Switch boards

(*vii*) Burglar-alarm systems

(*viii*) Optical fibre communication system etc.

2. **Liquid Crystal Displays (LCD)**

- A liquid crystal is a material, usually an organic compound, which flows like a liquid at room temperature; its molecular structure has some properties normally associated with solids (*e.g.* chloesteryl nonanoate and *p*-azoxyanisole).
- When light is incident on an activated layer of a liquid crystal, it is *either absorbed or else is scattered by the disoriented molecules.*
- A liquid crystal 'cell' (Fig. 2.61) consists of a thin layer (about 10 μm) of a liquid crystal sandwiched between two glass sheets with transparent electrodes deposited on their inside faces.

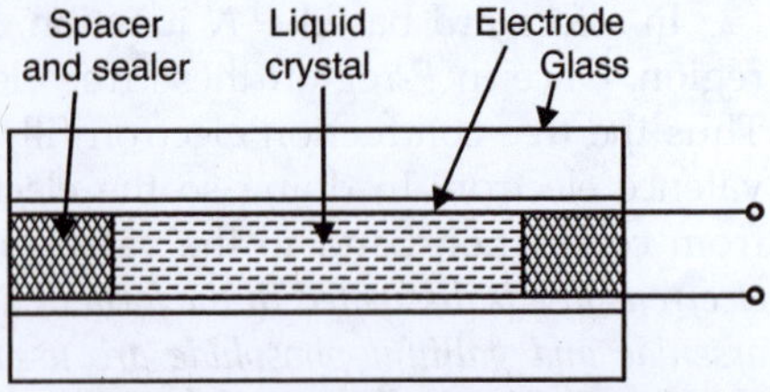

Fig. 2.61. A liquid crystal 'cell'.

  — When both glass sheets are transparent, the cell is known as *transmittive type cell.*

  — When one glass is transparent and the other has a reflective coating, the cell is called *reflective type.*

- Liquid crystal display produces no illumination of its own; it depends entirely on illumination falling on it from an external source for its visual effect.

Advantages :

1. Extremely low power requirement.
2. Long life time-about 50,000 hours.

Uses :

1. Cellular phone display.
2. Desk top LCD monitors.
3. Note book computer display.
4. Watches and portable instruments.
5. Pocket T.V. receiver.

3. **Photo-voltaic cell :**

- In this cell sensitive element is a semiconductor (not metal) which generates voltage in proportion to the light or any radiant energy incident on it. The most commonly used photo-voltaic cells are barrier layer type like *iron-selenium* cells or $Cu\text{-}CuO_2$ cells.
- Fig. 2.62 shows a typical widely used photo-voltaic cell—*Selenium cell.* It consists of a metal electrode on which a layer of selenium is deposited; on the top of this a barrier layer is formed which is coated with very thin layer of gold. The latter serves as a transluscent electrode. When light falls, a negative charge will build up on the gold electrode and a positive charge on the bottom electrode.

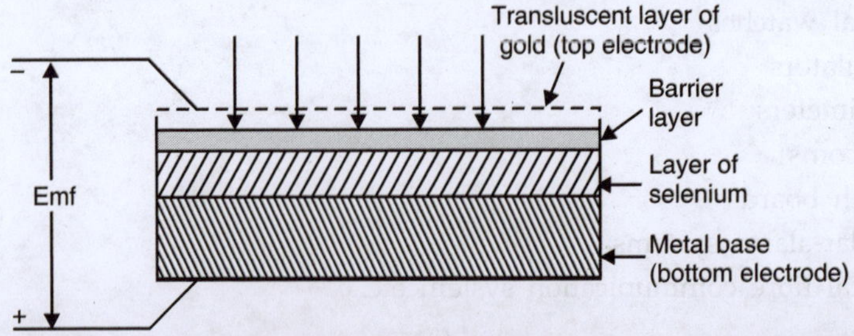

Fig. 2.62. Photo-voltaic cell.

Uses :

Photo-voltaic cells are widely used in the following fields :

(*i*) Automatic control systems.

(*ii*) Television circuits.

(*iii*) Sound motion picture and reproducing equipment.

4. P-N Junction photodiode :

- It is a two-terminal junction device which is operated first by reverse-biasing the junction and then illuminating it.
- The active diameter of these devices is about 2.5 mm but they are mounted in standard TO-5 packages with a window to allow maximum incident light.
- A photodiode can turn its current ON and OFF in nanoseconds, hence it is one of the fastest phtotodetectors.

Applications :

The following are the *fields of application* of *P-N* junction photodiode :

(*i*) Logic circuits that require stability and high speed.

(*ii*) Switching.

(*iii*) Detection (both visible and invisible).

(*iv*) Optical communication system.

(*v*) Demodulation.

(*vi*) Encoders.

(*vii*) Character recognition etc.

5. Laser diode :

The word LASER is an acronym for *Light Amplification by Stimulated Emission of Radiation.*

Laser diodes, like LED, are typical *P-N* junction devices used under a *forward bias.* Laser diodes are of the following *two types :*

1. **Surface-emitting laser diodes.** These diodes emit light in a direction *perpendicular* of the *P-N* junction plane.
2. **Edge-emitting laser diodes.** These diodes emit light in a direction *parallel* to the *P-N* junction plane.

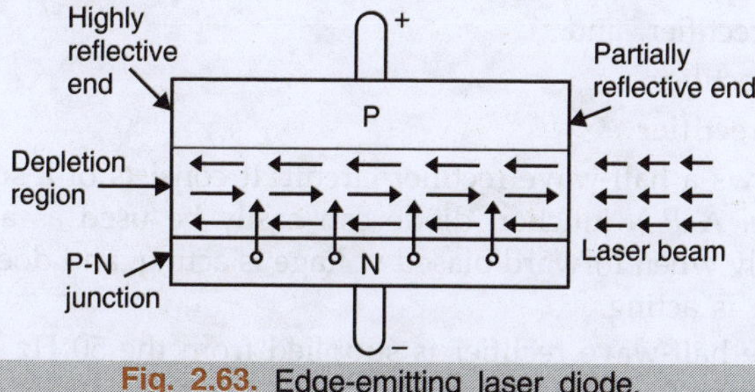

Fig. 2.63. Edge-emitting laser diode.

Fig. 2.63 shows an edge-emitting laser diode (called Fabry-Petrot type laser).

When an external voltage forward biases the *P-N* junction the electons move across the junction and usual combination takes place in the depletion region, resulting in the *production of photons.* With the increase in forward current, more photons are produced which drift at random in the depletion region. Some of these photons strike the reflective surface in the perpendicular direction. These reflected photons enter the depletion region, strike other atoms and release more photons. These photons move back and forth between reflective surfaces. The photon activity becomes so intense that at some point, a strong beam of laser light comes out of the partially reflective surface of the diode.

The beam of laser light is *coherent, monochromatic* and is *collimated.*

- The schematic symbol (Fig. 2.64) of a laser diode is similar to that of LED; a filler or lens is necessary to view the laser beam.

Fig. 2.64. Schematic symbol of a laser diode.

Applications :

1. Medical equipment used in surgery.
2. Compact disk (CD) players.
3. Laser printers.
4. Hologram scanners.
5. Parallel processing of information.
6. Parallel interconnections between computers etc.

6. Light Activated SCR (LASCR) :

- It is just an ordinary SCR except that it *can also be light-triggered.*
- Most LASCRs also have a gate terminal for being triggered by an electronic pluse just as conventional SCR. Fig. 2.65 shows the two LASCR symbols commonly used.
- These are manufactured mostly in relatively low-current ranges.

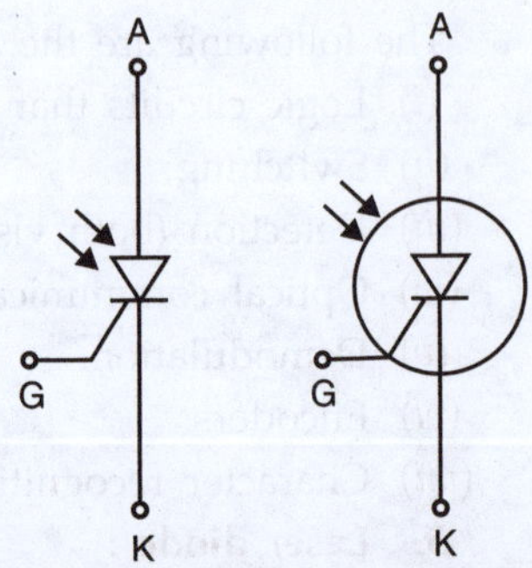

Fig. 2.65. LASCR symbols.

Applications :

1. Used for *triggering laser SCRs and triac.*
2. Used in *optical light controls, relays, motor control and a variety of computer applications.*

2.2.13. Rectifiers

A **rectifier** is *a circuit, which uses one or more diodes to convert A.C. voltage into pulsating D.C. voltage.*

A rectifier my be broadly categorized in the followign *two types :*

1. Half-wave rectifier, and
2. Full-wave rectifier.

1. Half-wave rectifier :

Fig. 2.66 (*a*) shows a half-wave rectifier circuit. It consists of a single diode in series with a load resistor. A *P-N* junction diode can easily be used as a rectifier because it conducts current only when forward biased voltage is acting, and does not conduct when reverse bias voltage is acting.

The input to the half-wave rectifier is supplied from the 50 Hz A.C. supply, whose wave form is shown in Fig. 2.66 (*b*).

Operation :

When an A.C. voltage source is connected across the junction diode as shown in Fig. 2.66 (*a*) the *positive half cycle of the input acts as forward bias voltage* and the output across the load resistance varies correspondingly. The *negative half cycle* of the input acts as a *reverse bias* and practically no current flows in the circuit. The output is, therefore, *intermittent, pulsating* and *unidirectional.*

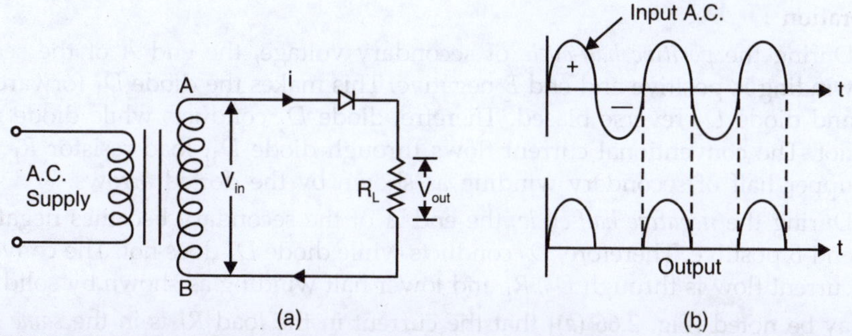

Fig. 2.66. Half-wave rectifier.

It is evident from the above discussion, that as the *circuit uses only one-half cycle of the A.C. input voltage, therefore, it is popularly known as a "half-wave rectifier".*

Disadvantages :

The main *disadvantages* of a half-wave rectifier are :

(*i*) The A.C. supply delivers power only half the time; therefore, its *output is low.*

(*ii*) The pulsating current in the load contains alternating component whose basic frequency is equal to the supply frequency. Therefore, an elaborate filtering is required to produce steady direct current.

2. Full-wave rectifier :

A ***full-wave rectifier*** *is a circuit, which allows a unidirectional current to flow through the load during the entire input cycle.* This can be achieved with *two diodes working alternately.* For the positive half-cycle of input voltage, one diode supplies current to the load and for the negative half-cycle, the other doide does so; *current being always in the same direction through the load.*

For full-wave rectification the following *two circuits* are commonly used :

1. Centre-tapped full-wave rectifier.
2. Full-wave bridge rectifier.

Centre-tapped full-wave rectifier :

Fig. 2.67 shows the circuit of a centre-tapped full-wave rectifier. The circuit uses two diodes (D_1 and D_2) which are connected to the centre-tapped secondary winding *AB* of the transformer.

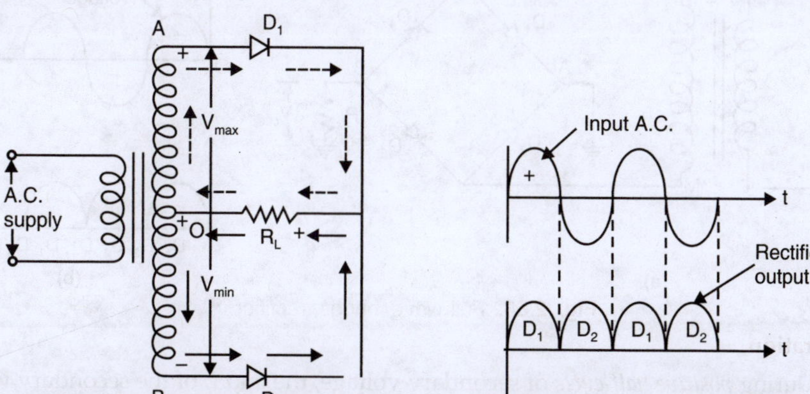

(a) Full-wave rectifier. (b) Wave forms of full-wave rectifier.

Fig. 2.67. Centre-tapped full-wave rectifier.

Operation :

- During the *positive half-cycle* of secondary voltage, the end A of the secondary winding is positive and end B negative. This makes the diode D_1 forward biased and diode D_2 reverse biased. Therefre, diode D_1 conducts while diode D_2 does not. The conventional current flows through diode D_1, load resistor R_L and the upper half of secondary winding as shown by the dotted arrows.
- During the *negative half-cycle,* the end A of the secondary becomes negative and end B positive. Therefore, D_2 conducts while diode D_1 does not. The conventional current flow is through D_2, R_L and lower half winding as shown by solid arrows.

It may be noted [Fig. 2.66 (*a*)] that the current in the load R_L is in the *same direction for both the cycles* of input A.C. voltage. Therefore, D.C. is obtained from the load R_L.

Also, Peak inverse voltage (PIV)

= Twice the maximum voltage across the half-secondary winding

i.e., $PIV = 2\,V_{max.}$

Advantages :

1. The D.C. output voltage and load current values are twice than those of half-wave rectifiers.
2. The ripple factor is much less (0.482) than that of half-wave rectifier (1.21).
3. The efficiency is twice that of half-wave rectifier.

For a full-wave rectifier, the maximum possible value of efficiency is 81.2% while that of half-wave rectifier is 40.6%.

Disadvantages :

1. The diodes used must have high peak inverse voltage.
2. It is difficult to locate the centre tap on the secondary winding.
3. The D.C. output is small as each diode utilises only one-half of transformer secondary voltage.

Full-wave bridge rectifier. It uses four diodes (D_1, D_2, D_3, D_4) across the main supply, as shown in Fig. 2.67 (*a*). The A.C. supply to be rectifier is applied to the diagonally opposite ends of the bridge through the transformer. Between other two ends of the bridge, the load resistance R_L is connected.

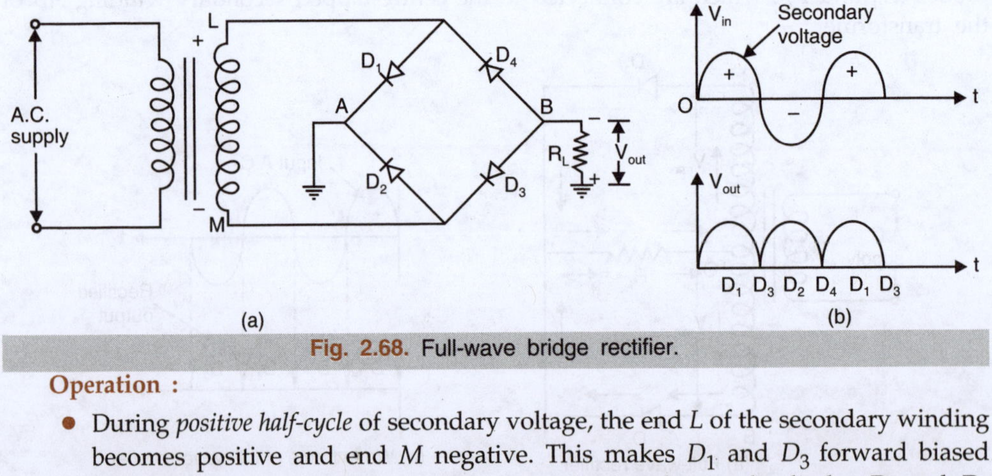

Fig. 2.68. Full-wave bridge rectifier.

Operation :

- During *positive half-cycle* of secondary voltage, the end L of the secondary winding becomes positive and end M negative. This makes D_1 and D_3 forward biased while diods D_2 and D_4 are reverse biased. Therefore, only diodes D_1 and D_3

conduct. These two diodes will be in series through the load R_L as shown in Fig. 2.69 (*a*). The current flows (dotted arrows) from *A* to *B* through R_L.

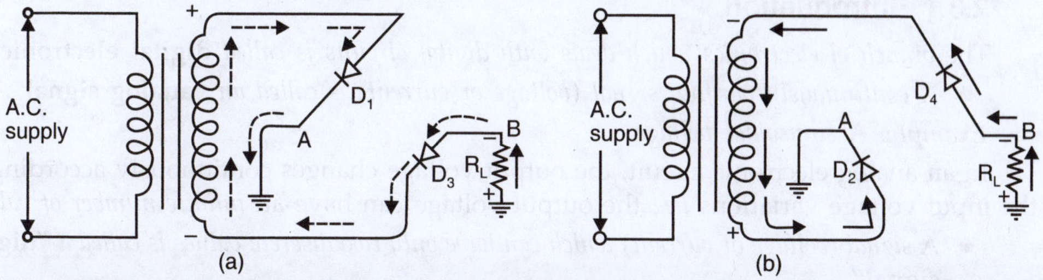

Fig. 2.69.

- During the *negative half-cycle* of the secondary voltage, end *L* becomes negative and *M* positive. This makes D_2 and D_4 forward biased whereas diodes D_1 and D_3 are reverse biased. Therefore, only diodes D_2 and D_4 conduct. These two diodes will be in series with R_L as shown in Fig. 2.69 (*b*). The current flows (*solid arrows*) from *A* to *B* through R_L *i.e.,* in the same direction as for positive half-cycle. Therefore, D.C. output is obtained across R_L.

Further it may noted that *peak inverse voltage* (*PIV*) *of each diode is equal to the maximum secondary voltage of transformer.*

Advantages :

1. It can be used with advantage in applications allowing floating input terminals *i.e.,* no output terminal is grounded.
2. The transformer is less costly as it is required to provide only half the voltage of an equivalent centre-tapped transformer used in a full-wave rectifier circuit.
3. No centre-tap is required on the transfomer.
4. The output is *twice* that of the centre-tapped circuit for the secondary voltage.

Disadvantages :

1. It uses four diodes as compared to two diodes for centre-tapped full wave rectifier.
2. Since during each half-cycle of A.C. input two diodes that conduct are in series, therefore, voltage drop in the internal resistance of the rectifying unit will be twice as great as in the centretapped circuit. This is objectionable when secondary voltage is small.
   - These days, the *bridge rectifiers are so common that manufacturers* are packing them as a single unit with bakelite or some other plastic encapsulation with external connections brought out.

Comparison of rectifiers :

The comparison of various types of rectifiers is given below :

S. No.	Aspects	Half-wave	Centre-tap	Bridge type
1.	*No. of diodes*	1	2	4
2.	*Transformer necessary*	No	Yes	No
3.	*Max. efficiency*	40.6%	81.2%	81.2%
4.	*Ripple factor*	1.21	0.48	0.48
5.	*Output frequency*	f_{in}	$2f_{in}$	$2f_{in}$
6.	*Peak inverse voltage*	V_m	$2V_m$	V_m

2.3. DIGITAL ELECTRONICS

2.3.1 Introduction

The *branch of electronics which deals with digital circuits is called* **digital electronics.**

- A *continuously varying signal (voltage or current) is called an* **"analog signal".**

Example. A *sinusoidal voltage.*

In an analog electronic circuit, the output voltage changes continuously according to the input voltage variations *i.e.,* the output voltage can have an *infinite number of values.*

- A *signal (voltage or current) which can have only two discrete values is called a* **"digital signal".**

Example. A *square wave.*

- *An electronic circuit that is designed for two-state operation is called a* **digital circuit.**

These days digital circuits are being used in many electronic products such as *video games, microwave ovens, oscilloscopes etc.*

2.3.2. Advantages and Disadvantages of Digital Electronics

The advantages and disdvantages of digital electronics are listed below :

Advantages :

1. Digital system can be normally *easily designed.*
2. Digital circuits are less *affected by noise.*
3. Storage of information is easy with digital circuits.
4. Digital circuits provide greater accuracy and precision.
5. More digital circuitry can be fabricated on integrated chips.

Disadvantages :

1. The digital circuits can handle only digital signals ; it requires encoders and decorders, due to which *cost of the equipment is increased.*
2. Under certain situations the use of only the analog techniques is simpler and economical (*e.g.* the process of signal amplification).

However, since the advantages outweigh the disadvantages, therefore, we are switching to digital techniques at a faster pace.

2.3.3. Digital Circuit

An electronic circuit that handles only a digital signal is called a **digital circuit.**

Or

An electronic circuit in which a state switches between the two states with time or with the change of the input states, and it is its state at the inputs and the outputs which has a significance is called a **digital circuit.**

"Digital" is derived from *"digitus"*. In Latin, the latter means "finger". A finger is either up or down. Similarly an electronic circuit may have one of the states as :

(*i*) 'ON' (conduction) or 'OFF' (poor conduction), or

(*ii*) *'High'* voltage or *'Low'* voltage between two terminals, or

(*iii*) *'High'* current through a circuit or *'Low'* current through a circuit, or

(*iv*) *'High'* frequency signal or *'Low'* frequency signal, or

(*v*) 'Negative' potential difference or *'Positive'* potential difference, or

(*vi*) **"1"** or **"0"** etc.

Therefore, *digital circuit is one that expresses the values in digits 1's or 0's, hence the name "digital'*. The number concept that uses only the two digits 1 and 0 is the *binary numbering system.*

- As a digital circuit is based opon the two states, it is used in dealing with binary numbers ; digital circuit is therefore used in *computers.*

Advantages of digital circuit :

1. *Noise free* as output is measured in terms of its state, not in terms of a voltage, or a current, or a frequency. A *state has a difiniteness.*
2. *Capabilities of logical decision, arithmetic and Boolean operation on the binary numbers.*

Disadvantages :

1. *Slower speed* due to greater number of components to represent a state.
2. The circuits have *complexities* also. To represent a big decimal number, a *large number of components needed.*

Advantages of "Analog circuit" :

1. More close to physical system values.
2. A voltage level may represent temperature, wind, speed etc.

Disadvantages :

Lack of definiteness, preciseness and reliability.

2.3.4. Number Systems

In the field of digital electronics and computers, the number systems are used quite frequently. However, the type of number system used in computers could be different at different stages of the usage.

In digital circuits the following *four systems of arithmetic* are often used :

1. **Decimal.** It has a *base* (or radix) *of* **10** *i.e.,* it *uses 10 different sysmbols* to represent the number.
2. **Binary.** It has a *base of* **2** *i.e.,* it uses *only two different sysmbols.*
3. **Octal.** It has a base of **8** *i.e.,* it uses *eight different symbols.*
4. **Hexadecimal.** It has *base of* **16** *i.e.,* it uses *sixteen different symbols.*

All the above mentioned systems use the same type of ***positional notation*** except that :

— Decimal system uses *powers of* **10**

— Binary system uses *powers of* **2**

— Octal system uses *powers of* **8**

— Hexadecimal system uses *powers of* **16**

- ***Decimal numbers*** *are used to represent quantities which are outside the digital system.*
- ***Binary system*** is extensively used by *digital system like digital computers which operate on binary information.*
- ***Octal system*** has certain *advantages in digital work because it requires less circuitry to get information into and out of a digital system. Moreover it is easier to read, record and print out octal numbers than binary numbers.*
- ***Hexadecimal number system*** is *particularly suited for micro-computers.*

2.3.5. Digital Coding

In digital circuits, each number of piece of information is *defined by an equivalent combination of binary digits. A complete group of these combinations which represent numbers, letters or symbols is called a* ***digital code.***

The group of 0s and 1s in the binary number can be thought of as a code representing the decimal numbers. When a decimal number is represented by its equivalent binary number, it is called a *straight binary coding.*

In modern digital equipment, *codes are used to represent and process numerical information.*

Types of codes. The various types of codes are enumerated and briefly discussed below :

1. BCD Code

It is also known as 'natural BCD' and is very convenient for representing decimal digits in digital circuits.

- It consists of four bits from 0000 to 1001 representing the decimal numbers from 0 to 9.1010 to 1111 are don't care conditions since they do not have any meaning in BCD.

2. Excess-3 Code

- The code can be derived from BCD by *adding 3 to each coded number.*

  It is useful when it is desired to obtain the 9's complement of a decimal digit represented by this code. The 9's complement is obtained simply by complementing each bit.
- This code can be *conveniently used for performing substracting operations in digital computers.*

3. Gray Code

- In this code ony one bit changes between any two successive numbers.
- It is mainly *used in the location of angular positions of a rotating shaft.*

4. Octal Code

- The octal system is a 8 base system.
- It uses 3 bits to represent one octal digit.

5. Hexadecimal Code

- The hexadecimal system is a base 16 system.
- It uses four bits to represent one hexadecimal digit.
- The hexadecimal digits are represented as 0 to 9 continued by aphabetical characters from *A* to *F*.

2.3.6. Logic Gates

General aspects :

A *digital circuit with one or more input signals but only output signal is called a* **logic gate.**

A ***logic gate*** *is an electronic circuit which makes logic decision.*

- Logic gates are the *basic building blocks* from which most of the digital systems are built up. They implement the hardware logic function based on the logical algebra developed by George Boolean which is called *Boolean algebra* in his honour.
  - — A unique characteristic of Boolean algebra is that variables used in it can assume only one of the two values *i.e., either 0 or 1.* Hence, every variable is either a 0 or a 1 (Fig. 2.69–limits on TTLIC's).
- Each gate has *distinct graphic symbol and its operation can be described by means of Boolean algebraic function.*

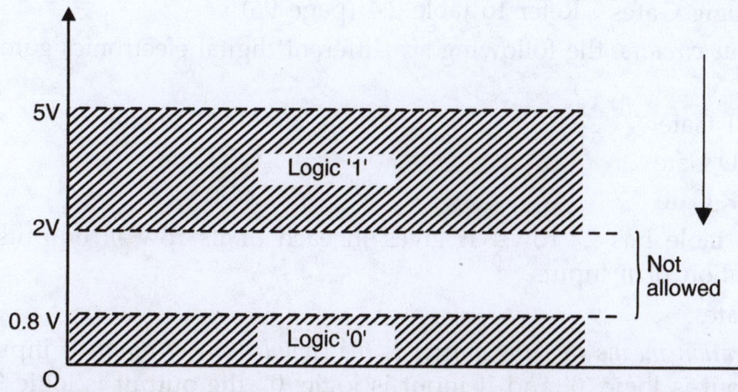

Fig. 2.70. Voltage assignment in a digital system.

- The *table which indicates output of gate for all possible combinations of input is known as a* ***truth table.***
- These gates are available today in the form of various IC families. The *most popular families are :*

(*i*) Transistor-transistor logic (TTL)
(*ii*) Emitter-coupled logic (ECL)
(*iii*) Metal-oxide-semiconductor (MOS)
(*iv*) Complementary metal-oxide-semiconductor (CMOS).

Applications of logic gates :

The following are the *fields of application of logic gates :*

1. Calculators and computers.
2. Digital measuring techniques.
3. Digital processing of communications.
4. Musical instruments.
5. Games and domestic appliances, etc.
6. The logic gates are also employed for decision making in *automatic control of machines and various industrial processes and for building more complex devices such as binary counters* etc.

Positive and negative logic :

The number symbols 0 and 1 represent, in computing systems, two possible states of a circuit or device. It does not make any difference if these two states are referred to as *'ON'* and *'OFF'*, *'Closed'* and *'Open'*, *'High'* and *'Low'*, *'Plus'* and *'Minus'* or *'True'* and *'False'* depending upon the situations. The main point is they must be symbolized by *two opposite conditions.* In *positive logic* a **'1'** *represents* : an *'ON circuit'* ; a *'Closed switch'*; a *'High voltage'*, a *Plus sign'*, *'True statement'*. *Consequently*, a **0** *represent* : an 'OFF circuit' ; an *'Open switch'*, a *'Low voltage'* ; a *'Minus sign'*, a 'False statement'.

In negative logic, the just opposite conditions prevail.

Example. A digital system has two voltage levels of 0 V and 5 V. If we say that symbol 1 stands for 5 V and symbol 0 for 0 V, then we have *positive logic system.* If on the other hand, we decide that a 1 should represent 0 V and 0 should represent 5 V, then we will get negative logic system.

Main point is that in *'positve logic'* the ***more positive*** of the two voltage levels represents the 1 while in *'negative logic'* the ***more negative*** voltage represents the 1.

Types of Logic Gates : Refer to table 2.4 (page 96)

In the *complex circuits*, the following *six* different digital electronics gates are used as *basic elements* :

1. NOT Gate
2. NAND Gate
3. AND Gate
4. OR Gate
5. NOR Gate
6. XOR Gate.

— A truth table has 2^n rows. It gives in each of its row *m* outputs for a given combination of *n* inputs.

1. **NOT Gate :**

- *Not operation means that the output is the complement of input.* If input is logic '1', the output is logic '0' and if input is logic '0', the output is logic '1'.
- Fig. 2.71 shows the *symbol of NOT Gate.* It is generally represented by a *triangle followed by a bubble* (or a bubble followed by a triangle).
- NOT gate is *used when an output is desired to be complement of the input.*
- If all inputs of NAND gates are joined it shall act as NOT gate.
- NOT gate is also *called 'inverting logic circuit.* It is also *called a 'complementing circuit'.*

2. **NAND Gate :**

- A NAND gate can said to be *basic building block* of the *all digital TTL logic gates and other digital circuits.*
- It is represented by the symbol shown in Fig. 2.72.
- Its *unique property is that output is high '1' if any of the input is at low '0' logic level.*

Let us consider two inputs with the states *A* and *B* at the NAND gate. The answer (output) $X = \overline{A.B}$. Bar denotes a NOT log operation on A.B. The meaning of A.B, called AND operation, is given in 3 below.

3. **AND Gate :**

- *A NAND gate followed by a NOT gate gives us AND gate.*
- It is represented by a symbol in Fig. 2.73. Its symbol differs from NAND only by *omission of a bubble (circle).*
- Its *unique property is that its output is '0' unless all the inputs to it are at the logic 1's.*
- A two inputs, AND gate has X = *A.B. Dot* between the two states indicates 'AND' logic operation using these.

4. **OR Gate :**

- An 'OR' operation means that the *output is '0' only if all the inputs are '0s'.*
- It is represented by a symbol shown in Fig. 2.74.
- If any of the inputs is '1' the output is '1'. A two inputs 'OR' gate has *X* = *A* + *B*. Sign + between the two states indicates an 'OR' logic operation.

5. **NOR Gate :**

- An 'OR' circuit followed by a NOT circuit gives a 'NOR' gate (Fig. 2.75).
- ts *unique property is that its output is '0' if any of input is '1'.*
- A NOR gate is a *basic building block for other types of the logic gates than TTLs.* In the TTL circuits, a NOR is fabricated in an IC by the several NANDs.

A two input NOR has $X = \overline{A + B}$.

6. **XOR Gate :**

- A XOR gate (Fig. 2.76) is called 'Exclusive OR' gate.
- Its *unique property is that the output is '1' only if odd number of the inputs at it are '1's.*
- The 'Exclusive OR' can be written as : $X = A \,.\, \overline{B} + \overline{A} \,.\, B$ or $A \oplus B$.
- Exclusive OR gate is important in the circuits *for addition of two binary numbers.*

7. **Coincidence Gate :**

- This gate (Fig. 2.77) can be written as : $X = \overline{A} \,.\, \overline{B} + A \,.\, B$.
- Output available to those states when the inputs are identical.

Basic building blocks. AND, OR and NOT gates are called *basic* building blocks or basic gates because they are *essential to realize any boolean expression.*

Universal gates. NAND and NOR gates are known as *universal gates* because any logic gate can be constructed either by using NAND gates only or by using NOR gates only.

2.3.7. Universal Gates

NAND and NOR gates are known as **universal gates.**

The AND, OR, NOT gates can be realized using only NAND or NOR gates.

— Demorgan's theorem afford a convenient method to use these two gates in *logic design.* The entire logic system can be implemented by using any of these two gates.

— These two gates are *easier to realize and consume less power than other gates.*

(*i*) **Realization of logic gates using NAND gates :**

Fig. 2.77 (*a*), (*b*), (*c*) shows realization of NOT, AND, OR gates using NAND gates respectively, which is self explanatory.

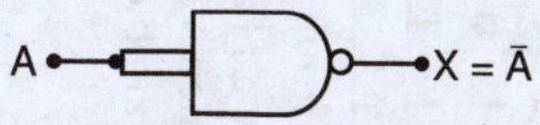

(a) Realization of NOT gate using NAND gate

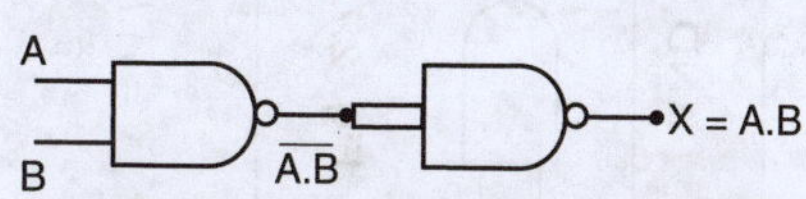

(b) Realization of AND gate using NAND gate

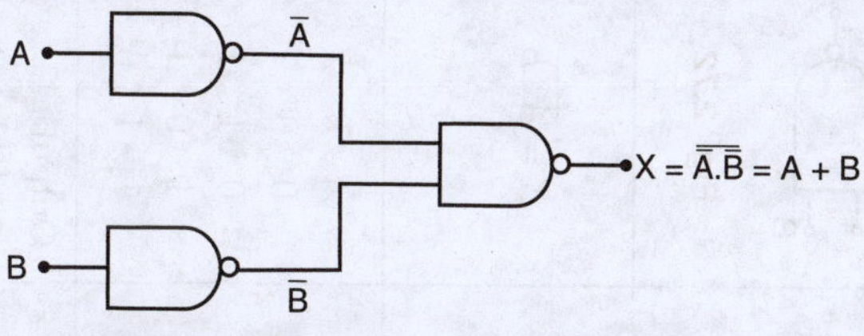

(c) Realization of OR gate using NAND gate

Fig. 2.77. Realization of NOT, AND and OR gates using NAND gates.

(*ii*) **Realization of logic gates using NOR gates :**

The realization of NOT, OR and AND gates using NOR gates is shown in Fig. 2.78 (*a*), (*b*), (*c*) respectively.

Table 2.4. **Symbols, Boolean Expressions and Truth Tables of various Logic Gates**

Logic	NOT	NAND	AND	OR	NOR	EX.OR	COINCIDENCE
Symbol	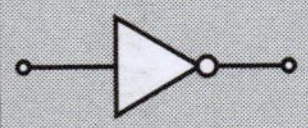 Fig. 2.71	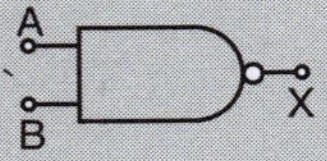 Fig. 2.72	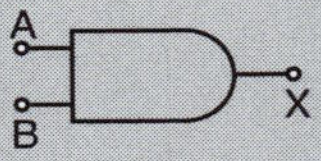 Fig. 2.73	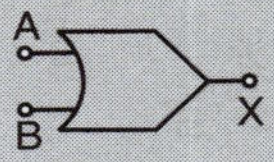 Fig. 2.74	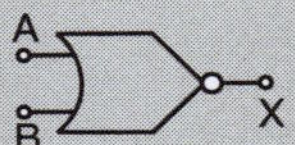 Fig. 2.75	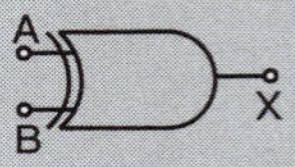 Fig. 2.76	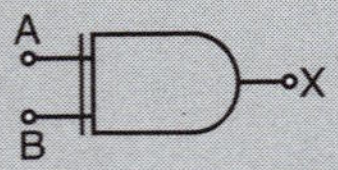 Fig. 2.77
Boolean expression	$X = \overline{A}$	$X = \overline{A \,.\, B}$	$X = A.B$	$X = A + B$	$X = \overline{A + B}$	$X = A \,.\, \overline{B} + \overline{A} \,.\, B = A \oplus B$	$X = \overline{A} \,.\, \overline{B} + A.B$
Truth table	*A X* 0 1 1 0	*A B X* 0 0 1 0 1 1 1 0 1 1 1 0	*A B X* 0 0 0 0 1 0 1 0 0 1 1 1	*A B X* 0 0 0 0 1 1 1 0 1 1 1 1	*A B X* 0 0 1 0 1 0 1 0 0 1 1 0	*A B X* 0 0 1 0 1 1 1 0 1 1 1 0	*A B X* 0 0 1 0 1 0 1 0 0 1 1 1
Definition	Output available – when there is no innput.	Output available – to all states except when all the inputs are available.	Output available– when all inputs available.	Output available– when only one or more inputs available.	Output available– when no input is available.	Output available– when the inputs are *not* identical.	Output available– to those states when the inputs are identical.

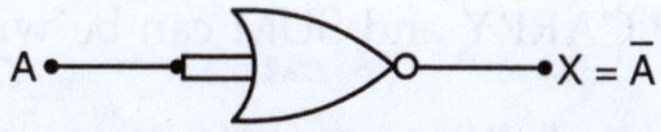

(a) Realization of NOT gate using NOR gate

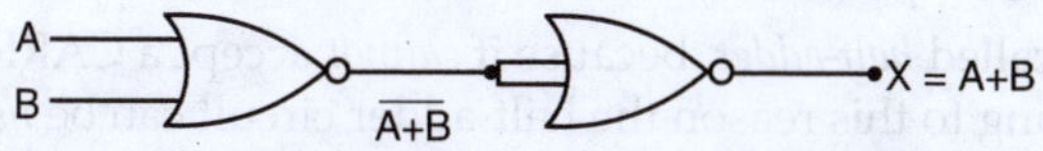

(b) Realization of OR gate using NOR gate

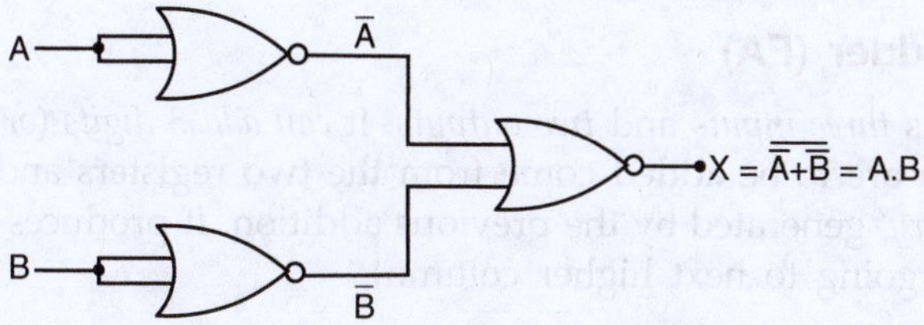

(c) Realization of AND gate using NOR gate

Fig. 2.78. Realization of NOT, OR and AND gates using NOR gates.

2.3.8. Half Adder (HA)

It is a *1-bit adder* and carries out binary addition with the help of XOR and AND gates. It has ***two** inputs and **two** outputs.*

It can add 2 binary digits at a time and produce a 2-bit data *i.e.,* 2-bit data *i.e.,* SUM and CARRY according to binary addition rules.

The circuit of a half adder is shown in Fig. 2.79. (*a*). It consists of an Ex-OR gate and AND gate. The outputs of the Ex-OR gate is called the SUM (S), while the output of the AND gate is known as CARRY (C). As the AND gate produces a *high* output only when both inputs are *high* and Ex-OR gate produces a *high* output if either input (not both) is high, the truth table of a half adder is developed by writing the truth table output of AND gate in the CARRY column and the output truth table of Ex-OR gate in SUM column. Truth table for half adder is given in table 2.5.

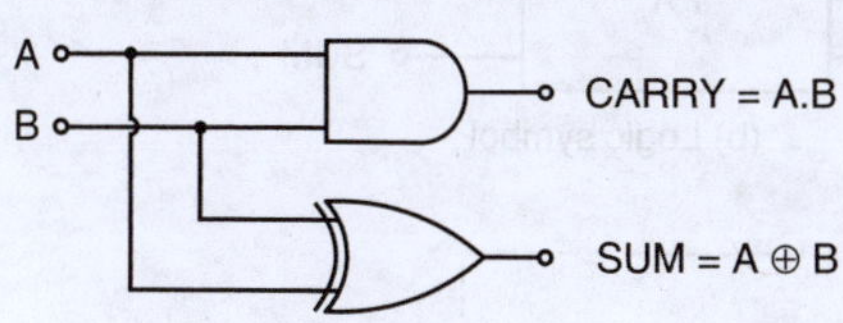

(a) Logic circuit

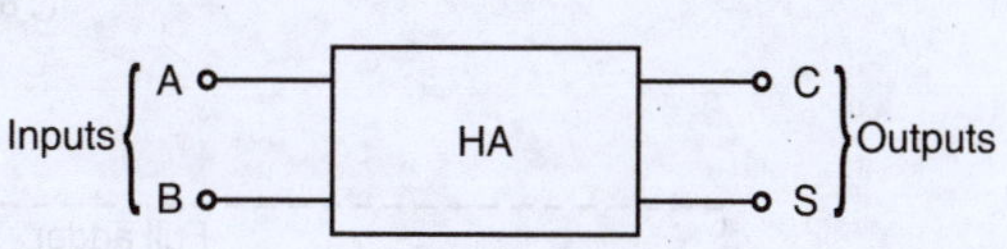

(b) Logic symbol

Fig. 2.79. Half adder.

Table 2.5. Truth table for Half Adder

Inputs		Outputs	
A	B	C	S
0	0	0	0
0	1	0	1
1	0	0	1
1	1	1	0

The logical expressions for CARRY and SUM can be written from the truth table for a half adder as follows :

$$\text{CARRY, C} = \text{A.B}$$

$$\text{SUM, S} = \text{A} \oplus \text{B}$$

- This circuit is called *half-adder*, because it *cannot* accept a CARRY-IN from previous additions. Owing to this reason the half-adder circuit can be *used for binary addition of lower most bit only.*

For higher-order columns, 3-input adder called *full adder* are used.

2.3.9. Full Adder (FA)

A full adder has *three inputs* and *two outputs*. It *can add 3 digits (or bits) at a time.* The bits A and B which are to be added come from the two registers and the third input C comes from the *'carry'* generated by the previous addition. It produces two outputs, SUM and CARRY-OUT (going to next higher column).

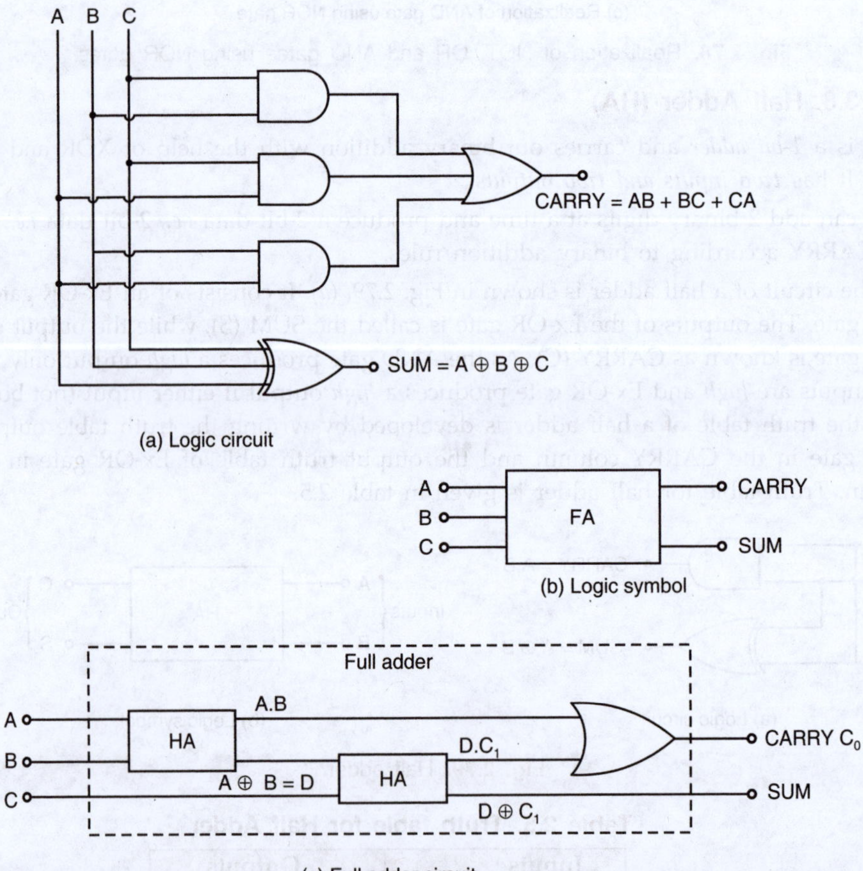

(a) Logic circuit

(b) Logic symbol

(c) Full adder circuit

Fig. 2.68. Full adder.

Table 2.6. Truth table for Full Adder

A	B	C	CARRY	SUM
0	0	0	0	0
0	0	1	0	1
0	1	0	0	1
0	1	1	1	0
1	0	0	0	1
1	0	1	1	0
1	1	0	1	0
1	1	1	1	1

A simple circuit of a full adder is shown in Fig. 2.80 (*a*), though other designs are also possible. It uses 3 AND gates, one Ex-OR gate and one OR gate. The final CARRY is given by the OR gate while the final SUM is given out by the Ex-OR gate.

Fig. 2.80(*b*) shows the logic symbol for a full adder.

Truth table for full adder for all possible inputs/outputs is given in Table 2.6. Truth table can be checked easily for its validity.

A full adder can be made by using two half adders and an OR gate. The circuit is shown in Fig. 2.80(*c*).

- The full adder can do more than a million additions per second. Besides that, it never get tired or bored or asks for a rest.

Note : *Binary additions :* Following are the four rules/cases for addition of binary numbers :

(1) 0 + 0 = 0

(2) 0 + 1 = 1

(3) 1 + 0 = 1

(4) 1 + 1 = 10 (This sum is *not ten* but *one-zero*).

2.3.10. Boolean Algebra

George Boolean in 1854 developed a mathematics now referred as *Boolean algebra.* It is the *algebra of logic presently applied to the operation of computer devices.* The rules of this algebra are based on human reasoning.

Digital circuits perform the binary arithmetic operations with binary digits 1 and 0. These operations are called logic functions or logic operations. *The algebra used to symbolically describe logic functions is called* **Boolean algebra.** Boolean algebra is *a set of rules and theorems by which logical operations can be expressed symbolically in equation form and be manipulated mathematically.*

Boolean algebra differs from ordinary algebra in that Boolean constant and variables *can have only two values :* '0' and '1' :

In Boolean algebra the following *four connecting symbols* are used :

1. Equal sign (=). In Boolean algebra the 'equal sign' refers to the standard mathematical equality. In other words, the logical value on one side of the sign is identical to the logical value on the other side of the sign.

Example. We are given two logical variables such that $A = B$. Then if $A = 1$, then $B = 1$ and if $A = 0$ then $B = 0$.

2. **Plus sign (+).** In Boolean algebra the 'plus sign' refers to logical *OR operation.*

The statement $A + B = 1$ means A ORed with B equals 1. Consequently, either $A = 1$ or $B = 1$ or both equal to 1.

3. **Multiply sign (.).** In Boolean algebra the 'multiply sign' refers to *AND operation.*

The statement $A.B = 1$ means A ANDed with B equals 1. Consequently, $A = 1$ and $B = 1$.

The function $A \,.\, B$ often written as AB, *omitting the dot* for convenience.

4. **Bar sign (–).** In Boolean algebra the 'bar sign' refers to *NOT operations.* The NOT has the effect of inverting (complementing) the logic value.

Thus if A = 1, then $\overline{A} = 0$.

2.3.11. Boolean Laws (For Outputs from Logic Inputs).

The following Laws can said to be associated with Boolean algebra :

1. **'OR' Laws**

The 'OR' Laws are described by the following equations :

$$A + 1 = 1 \qquad ...[2.5(a)]$$

$$A + 0 = A \qquad ...[2.5(b)]$$

$$A + A = A \qquad ...[2.5(c)]$$

$$A + \overline{A} = 1 \qquad ...[2.5(d)]$$

- An 'OR' operation is denoted by *plus sign.*
- 'OR' Law means :

(*i*) Any number (0 or 1) is a first input to an OR gate and another member at the second input is 1 then answer is 1,

(*ii*) If another is 0 then answer is as first input, and

(*iii*) If two inputs to an OR gate *complement* then output is '1'.

2. **'AND' Laws**

'AND' operation is denoted by the *dot* sign.

- True and true make true
- True and false make false
- False and false make false.

$$A \,.\, 1 = A \qquad ...[2.6(a)]$$

$$A \,.\, 0 = 0 \qquad ...[2.6(b)]$$

$$A \,.\, A = A \qquad ...[2.6(c)]$$

$$A \,.\, \overline{A} = 0 \qquad ...[2.6(d)]$$

3. **'NOT' Laws (Laws of Complementation)**

A NOT operation is denoted by putting a *bar* over a number.

- The NOT true means false.
- The NOT false means true.

$$\overline{1} = 0 \qquad ...[2.7(a)]$$

$$\overline{\overline{A}} = A \qquad ...[2.7(b)]$$

Eqn. [2.16(*b*)] means that if *A is inverted (complemented) and then again inverted, we get the original number.*

4. **Commutative Laws**

These Laws mean that *order of a logical operation is immaterial.*

$$A + B = B + A \quad ...[2.8(a)]$$
$$A \,.\, B = B \,.\, A \quad ...[2.8(b)]$$

5. Associative Laws

These laws *allow a grouping of the Boolean variables.*

$$A + (B + C) = (A + B) + C \quad ...[2.9(a)]$$
$$A \,.\, (B \,.\, C) = (A \,.\, B) \,.\, C \quad ...[2.9(b)]$$

6. Distributive Laws

These laws *simplify the problems in the logic disigns.*

$$A \,.\, (B + C) = (A \,.\, B) + (B \,.\, C) \quad ...[2.10(a)]$$
$$A + (B \,.\, C) = (A + B) \,.\, (A + C) \quad ...[2.10(b)]$$
$$A + (\overline{A} \,.\, B) = A + B \quad ...[2.10(c)]$$

The last two equations are typical to the Boolean algebra, and are not followed in the usual algebra.

2.3.12. De Morgan's Theorems

First theorem shows an equivalence of a NOR gate with an AND gate having bubbled inputs (Fig. 2.81), and is given by the equation :

$$\overline{A + B} = \overline{A} \,.\, \overline{B} \quad ...(2.11)$$

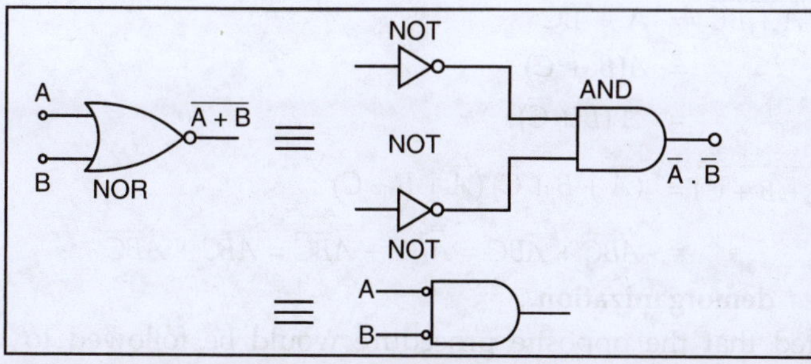

A	B	X
0	0	1
1	0	0
0	1	0
1	1	0

Fig. 2.81. De Morgan's First theorem showing an equivlence of a NOR gate (same holds for multiple inputs).

Second theorem shows an equivalence of a NAND gate with an OR having bubbled inputs as shown in Fig. 2.82 and is given by the equation :

$$\overline{A \,.\, B} = \overline{A} + \overline{B} \quad ...(2.12)$$

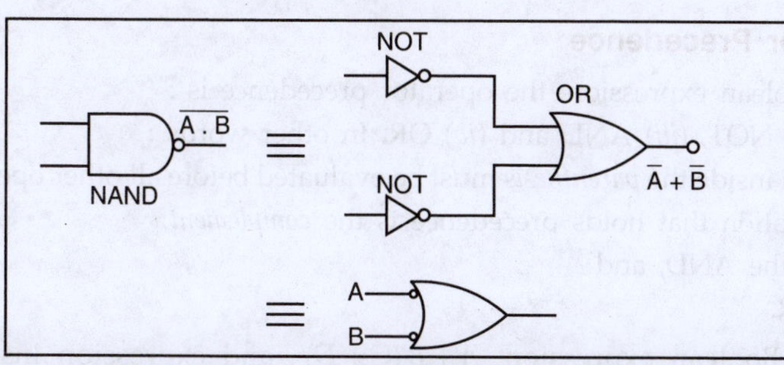

A	B	X
0	0	1
1	0	1
0	1	1
1	1	0

Fig. 2.82. De Morgan's Second theorem showing the equivalence of a NAND gate (same holds for the multiple inputs).

In fact the eqns. (2.21) and (2.12) also hold for the cases of the multiple (more than two) inputs.

$$\overline{A + B + C + ...} = \bar{A} . \bar{B} . \bar{C} \qquad ...[2.13(a)]$$

$$\overline{A . B. C .....} = \bar{A} + \bar{B} + \bar{C} \qquad ...[2.13(b)]$$

The *purpose of these theorems is to enable digital circuit designers to implement all the other logic gates with the help of either NOR gates only or NAND gates only.* For example, a NOT gate is implementable by a NAND or a NOR as shown in the left part or lower right part of Fig. 2.81 respectively. This theorem finds wide use in the digital logic circuits as these are implementable on one single basic logic gate considered as a basic building unit.

- The *'first statement'* (De Morgan's) says that *the complement of a sum equals the product of the complements.* The *'second statement'* says that the *complement of a product equals the sum of the complements.* In fact, it allows transformation from a sum-of-products form to a product-of-sum form.

— The procedure required for *taking out an expression from under a NOT sign is as follows :*

1. *Complement the given expression i.e., remove the overall NOT sign.*
2. *Change all and ANDs to ORs and all the ORs to ANDs.*
3. Complement or negate all individual variables.

Examples : (*i*)

$$\overline{A + BC} = A + BC \qquad ...Step\ 1$$

$$= A(B + C) \qquad ...Step\ 2$$

$$= \bar{A}(\bar{B} + \bar{C}) \qquad ...Step\ 3.$$

(*ii*)

$$\overline{(\bar{A} + B + \bar{C})(\bar{A} + B + C)} = (\bar{A} + B + \bar{C})(\bar{A} + B + C)$$

$$= \bar{A}B\bar{C} + \bar{A}BC = \bar{\bar{A}}\bar{B}\bar{\bar{C}} + \bar{\bar{A}}\bar{B}\bar{C} = A\bar{B}C + A\bar{B}\bar{C}$$

This *process is called* **demorganization.**

— It may be noted that the opposite procedure would be followed to *bring an expression under the NOT sign :*

Example :

$$\bar{A} + \bar{B} + \bar{C} = \bar{\bar{A}} + \bar{\bar{B}} + \bar{\bar{C}} \qquad ...Step\ 3$$

$$= A + B + C$$

$$= ABC \qquad ...Step\ 2$$

$$= \overline{ABC} \qquad ...Step\ 1$$

2.3.13. Operator Precedence

For evaluating Boolean expression, the operator precedence is :

(*i*) parenthesis, (*ii*) NOT, (*iii*) AND and (*iv*) OR. In other words :

— The expression inside the *parenthesis* must be evaluated before all other operations,

— The next operation that holds precedence is the *complement,*

— Then follows the AND, and

— Finally the OR.

Example. In the Boolean expression $A + \bar{B}(C + D)$, and expression inside the parenthesis will be evaluated first, then $\bar{B}$ will be evaluated, then the results of the two [*i.e.,* $\bar{B}$ *and* $(C + D)$] will be ANDed and finally, the result of the product ORed with *A*.

Example 2.1. *Prove the following identity : AC + ABC = AC.*

Solution. Taking the left hand expression as X, we get

$$X = AC + ABC = AC(1 + B)$$

Now, $1 + B = 1$ [Eqn. 2.5(*a*)]

$\therefore$ $X = AC.\,1 = AC$

$\therefore$ $AC + ABC = AC$ **...Proved.**

Example 2.2. *Prove the following Boolean identity : (A + B) (A + C) = A + BC.*

Solution. Putting the left hand side expression equal to X, we get

$$\begin{aligned} X &= (A + B)(A + C) \\ &= AA + AC + AB + BC && \text{...[Eqn. 2.10(a)]} \\ &= A + AC + AB + BC && [AA = A \text{ ... } (2.6(c))] \\ &= A + AB + AC + BC \\ &= A(1 + B) + AC + BC && (\because\ 1 + B = 1) \\ &= A + AC + BC \\ &= A(1 + C) + BC \\ &= A + BC && (\because\ 1 + C = 1) \end{aligned}$$

$\therefore$ $(A + B)(B + C) = A + BC.$ **...Proved.**

Example 2.3. *Prove the following identity :* $A + \overline{A}B = A + B.$

Solution. Putting the left hand expression equal to X, we get

$$\begin{aligned} X &= A + \overline{A}B = A.1 + \overline{A}B && \text{[Eqn. 2.6(a)]} \\ &= A(1 + B) + \overline{A}B && \text{[Eqn. 2.5(a)]} \\ &= A.1 + AB + AB && \text{[Eqn. 2.10(a)]} \\ &= A + BA + B\overline{A} \\ &= A + B(A + \overline{A}) && \text{[Eqn. 2.10(a)]} \\ &= A + B\,.\,1 && \text{[Eqn. 2.5(d)]} \\ &= A + B && \text{[Eqn. 2.6(a)]} \end{aligned}$$

$\therefore$ $A + \overline{A}B = A + B.$ **...Proved.**

Example 2.4. *Simplify the following Boolean expression to a minimum of literals :*

$$X = AB + \overline{A}C + BC.$$

Solution.

$$\begin{aligned} X &= AB + \overline{A}C + BC \\ &= AB + \overline{A}C + BC(A + \overline{A}) && \text{...[Eqn. 2.5 (d)]} \\ &= AB + \overline{A}C + ABC + \overline{A}BC \\ &= AB(1 + C) + \overline{A}C(1 + B) \\ &= AB + \overline{A}C && \text{...[Eqn. 2.5(a)]} \end{aligned}$$

$X = AB + \overline{A}C.$ **(Ans.)**

Example 2.5. *Simplify the following Boolean expression :*

$$AB\overline{C} + A\overline{B}\,\overline{C} + \overline{A}BC + ABC + A\overline{B}C.$$

Solution. Let, $X = AB\overline{C} + A\overline{B}\,\overline{C} + \overline{A}BC + ABC + A\overline{B}C$

Bringing together those terms which have two cammon letters, we get

$$X = ABC + AB\overline{C} + A\overline{B}\,\overline{C} + A\overline{B}C + \overline{A}BC$$

$$= AB(C + \bar{C}) + A\bar{B}(\bar{C} + C) + \bar{A}BC$$

$$= AB + A\bar{B} + \bar{A}BC \quad \text{...[Eqn. 2.5(d)]}$$

$$= A(B + \bar{B}) + \bar{A}BC$$

$$= A + \bar{A}BC = A + BC. \quad \textbf{(Ans.)} \quad \text{...[Eqn. 2.10(c)]}$$

Example 2.6. *Using Boolean algebra techniques, simplify the following expression :*

$$X = A.B.\bar{C}.\bar{D} + \bar{A}.B.\bar{C}.\bar{D} + \bar{A}.B.C.\bar{D} + A.B.C.\bar{D}.$$

Soution.

$$X = B\bar{C}\bar{D}(A + \bar{A}) + BC\bar{D}(A + \bar{A}) \text{ jlm}$$

...Taking out the common factors

$$= B\bar{C}\bar{D} + BC\bar{D} \quad \text{...[Eqn. 2.5 (d)]}$$

$$= B\bar{D} + (C + \bar{C}) \quad \text{...Again factorize}$$

$$= B\bar{D}.1 = B\bar{D} \quad \text{...(Simplified form) } \textbf{(Ans.)} \quad \text{...[Eqn. 2.5(d)]}$$

Example. 2.7. *Simplify the following expression and show the minimum gate implementation.*

$$X = A.B.\bar{C}.\bar{D} + \bar{A}.B.\bar{C}\bar{D} + B.\bar{C}.D$$

Solution.

$$X = B.\bar{C}.\bar{D}(\bar{A} + B) + B.\bar{C}.D$$

$$= B.\bar{C}.\bar{D}.1 + B.\bar{C}.D$$

...[Eqn. 2.14 (d)]

$$= B.\bar{C}.\bar{D} + B.\bar{C}.D$$

$$= B.\bar{C}.(D + \bar{D}) = B.\bar{C}.1 = B.\bar{C}$$

...[Eqn. 2.14 (d)]

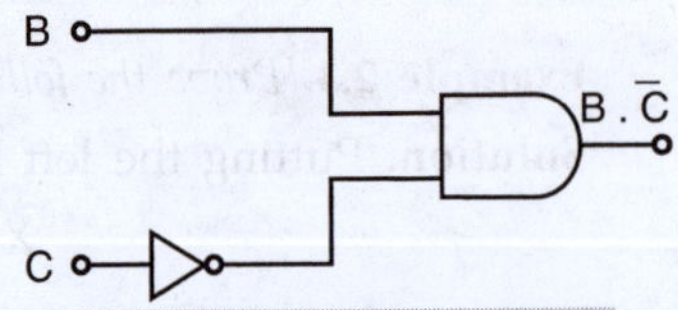

Fig. 2.83

Minimum gate implementation is shown in Fig. 2.83

Example 2.8. *Determine output expression for the circuit shown in Fig. 2.84.*

Solution. The output expression for the circuit shown in Fig. 2.84 is :

$$X = \overline{[(A + B).C.\bar{D}]}.$$

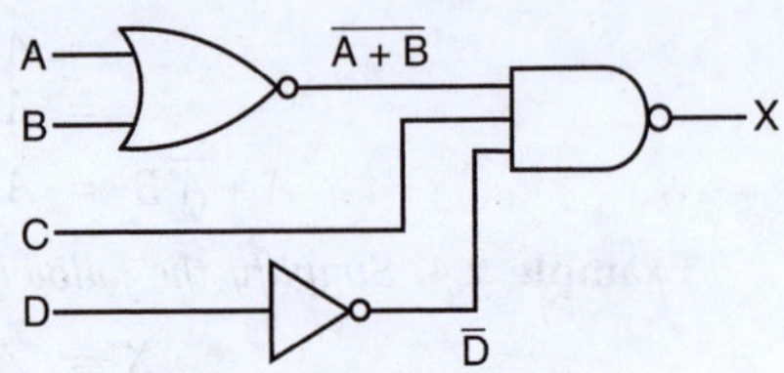

Fig. 2.84

Example 2.9. *Simplify the following Boolean expression and draw the logic circuit for simplified expression :*

$$X = \bar{B}(A + C) + C(\bar{A} + B) + AC.$$

Solution.

$$X = \bar{B}(A + C) + C(\bar{A} + B) + AC$$

$$= A\bar{B} + \bar{B}C + \bar{A}C + BC + AC$$

$$= A\bar{B} + C(\bar{B} + \bar{A} + B + A)$$

$$= A\bar{B} + C.1 = A\bar{B} + C \quad \text{...Simplified expression. } \textbf{(Ans.)}$$

Logic circuit for the simplified expression is shown in Fig. 2.85

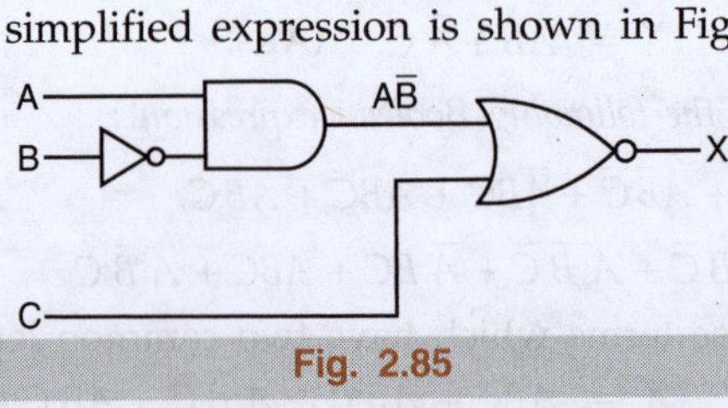

Fig. 2.85

Example 2.10. *Simplify the expression :* $(AB + C)(AB + D)$.

Solution. Let

$$
\begin{aligned}
X &= (AB + C)(AB + D) \\
&= AB\,AB + ABD + ABC + CD && \text{...[Eqn. 2.10(a)]} \\
&= AABB + ABD + ABC + CD \\
&= AB + ABD + ABC + CD && \text{...[Eqn. 2.6(c)]} \\
&= AB(1 + D) + ABC + CD \\
&= AB + ABC + CD && \text{...[Eqn. 2.5(a)]} \\
&= AB(1 + C) + CD
\end{aligned}
$$

$\therefore \quad (AB + C)(AB + D) = AB + CD.$ **(Ans.)**

Example 2.11. *Draw the logic circuit represented by the expression :*

$$X = AB + \bar{A}.\bar{B} + \bar{A}.B.C.$$

Solution. A circuit using gates can simply be designed by looking at the expression and finding out the basic gates which can be used to realize the various terms and then correct these gates appropriately.

In the given expression there are three input logical variables and X is the output.

- The *first* term $A. B$ is obtained by ANDing A with B as shown in Fig. 2.123 (*i*).
- The *second* term $\bar{A}.\bar{B}$ is obtained by using two INVERTERs and one AND gate and connecting them as shown in Fig. 2.86.

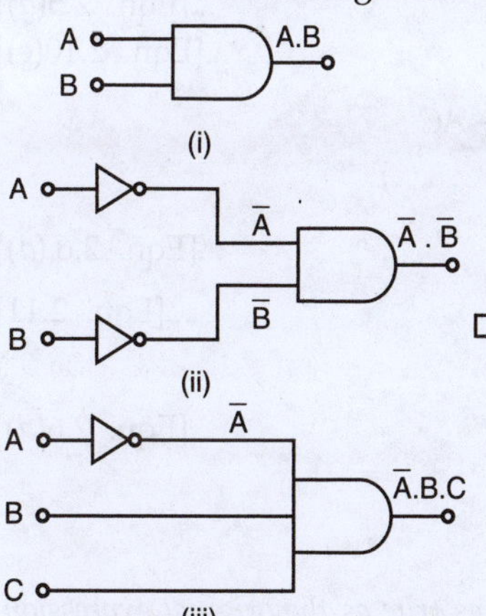

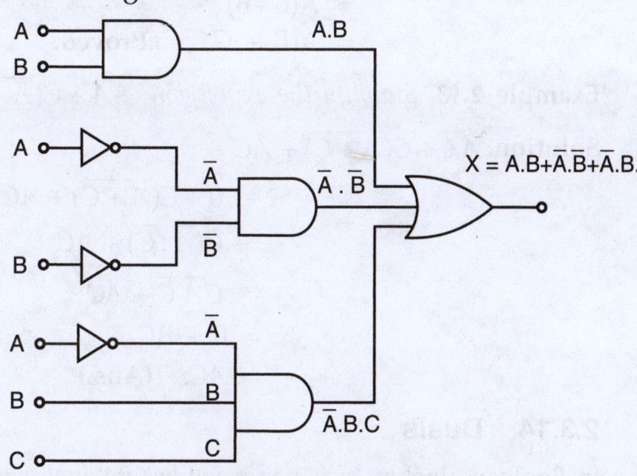

Fig. 2.86. **Fig. 2.87.** Logic gate implementation of expression $A.B + \bar{A}.\bar{B} + \bar{A}.B.C.$

- The *last* term is used by using one INVERTER, one AND gate and connecting them as shown in Fig. 2.86(*iii*).

Now, the complete logic expression is realised by ORing the three outputs of the arrangements explained above *i.e.*, by ORing. $A.B$, $\bar{A}.\bar{B}$ and $\bar{A}.B.C$. The logic gate implementation for the given expression is shown in Fig. 2.87.

Example 2.12. *Show that :*

(*i*) $(A + B)(\bar{A} + C) = AC + \bar{A}B$ (*ii*) $AB + ABC + \bar{A}\,B + A\bar{B}\,C = B + AC$

(*iii*) $ABC + A\bar{B}\,C + AB\bar{C} = A(B + C)$.

Solution. (*i*) $(A + B)(\bar{A} + C)$

$$= A\bar{A} + AC + B\bar{A} + BC \qquad (\because \quad A.\bar{A} = 0)$$

$$= AC + B\bar{A} + BC$$

Multiplying the third term by $(A+\bar{A})$, we get

$= AC + B\bar{A} + BC(A+\bar{A})$ [$A+\bar{A}$, being equal to 1 does not make any effect]

$= AC + B\bar{A} + ABC + \bar{A}\,BC$ ...[Eqn. 2.5(*d*)]

$= AC\,(1+B) + BA\,(1+C)$

$= AC + B\,\bar{A}$ ...[Eqn. 2.5(*d*)]

$= AC + \bar{A}B.$ **...Proved.**

(*ii*) $AB + ABC + \bar{A}\,B + A\bar{B}\,C$

$= AB + \bar{A}\,B + AC(B+\bar{B})$

$= B(A+\bar{A}) + AC(B+\bar{B})$

$= B + AC.$ **...Proved** ...[eqn. 2.5(*d*)]

(*iii*) $ABC + A\bar{B}C + AB\,\bar{C}$

$= AC\,(B+\bar{B}) + AB\,\bar{C}$

$= AC + AB\,\bar{C}$ ...[Eqn. 2.5(*d*)]

$= A(C + B\,\bar{C})$

$= A(C+B)$ ...[Eqn. 2.5(*d*)]

$= A(B + C).$ **...Proved.** ...[Eqn. 2.10(*c*)]

Example 2.13. *Simplify the expression* $A\bar{A} + C(\overline{A+C}) + AC.$

Solution. $A\bar{A} + C(\overline{A+C}) + AC$

$= 0 + C(\overline{A+C}) + AC$ ...[Eqn. 2.6.(*d*)]

$= C(\bar{A}\,.\,\bar{C}) + AC$ ...[Eqn. 2.11]

$= C\bar{A}\,\bar{C} + AC$

$= 0 + AC$ ...[Eqn. 2.6(*d*)]

$= AC.$ **(Ans.)**

2.3.14. Duals

In Boolean algebra *each expression has its dual which is as true as the original expression.* For getting the dual of a given Boolean expression, the procedure involves conversion of

(*i*) all 1s to 0s and all 0s to 1s.

(*ii*) all ANDs to ORs and all ORs to ANDs.

The dual so obtained is *also found to be true.*

Some of the Boolean relations and their duals are given in Table 2.7.

Table 2.7.

Relation	Dual relation
$A\,.\,0 = 0$	$A + 1 = 1$
$A\,.\,A = 1$	$A + A = A$
$A\,.\,\bar{A} = 0$	$A + \bar{A} = 1$
$A\,.\,1 = A$	$A + 0 = A$
$A\,.\,(A + B) = A$	$A + AB = 1$
$A + (\bar{A} + B) = AB$	$A + \bar{A}\,B = A + B$

Example 2.14. *Determine the Boolean expression for the logic circuit shown in Fig. 2.88. Simplify the Boolean expression using Boolean laws and De Morgan's theorem. Redraw the logic circuit using the simplified Boolean expression.*

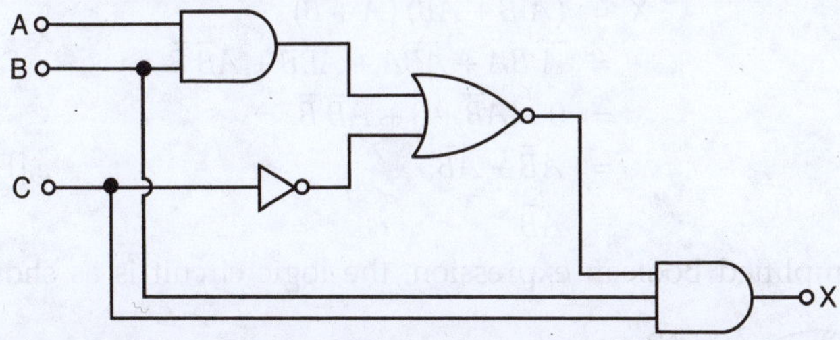

Fig. 2.88

Solution. The output of a given circuit can be obtained by *determining the output of each logic gate while working from left to right.*

With reference to Fig. 2.89, the output of the circuit is :

$$X = BC\overline{\left(AB+\overline{C}\right)}$$

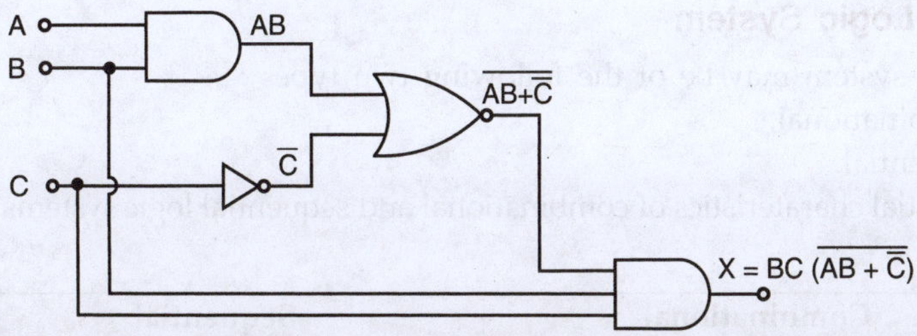

Fig. 2.89

The output X can be simplified by *De Morganizing* the term $\overline{(AB+\overline{C})}$ as follows :

$$
\begin{aligned}
BC\overline{(AB+\overline{C})} &= BC(AB+\overline{C}) && \text{...Step-1}\\
&= BC(A+B)\,.\,\overline{C} && \text{...Step-2}\\
&= BC(\overline{A}+\overline{B})\,.\,\overline{\overline{C}} && \text{...Step-3}\\
&= BC(\overline{A}+\overline{B})C && \text{...[Eqn. 2.7(b)]}\\
&= BC(\overline{A}+\overline{B}) && \text{..[Eqn. 2.6(c)]}\\
&= \overline{A}\,BC+BC\overline{B} &&\\
&= \overline{A}\,BC+0 && \text{...[Eqn. 2.15(d)]}\\
&= \overline{A}\,BC && \text{...[Eqn. 2.14(b)]}
\end{aligned}
$$

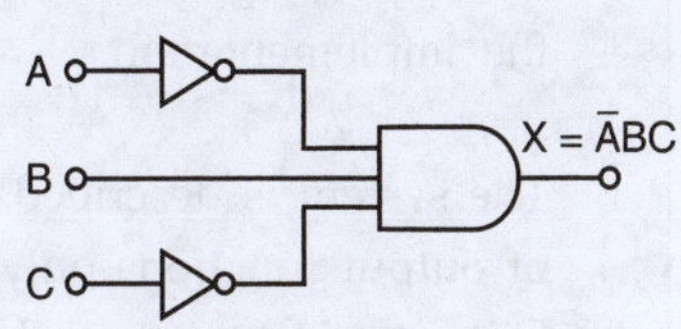

Fig. 2.90

The logic circuit with a simplified Boolean expression $X = \overline{A}\,BC$ is as shown in Fig. 2.90.

Example 2.15. *Determine the output X of a logic circuit shown in Fig. 2.91. Simplify the output expression using Boolean Laws and theorems. Redraw the logic circuit with the simplified expression.*

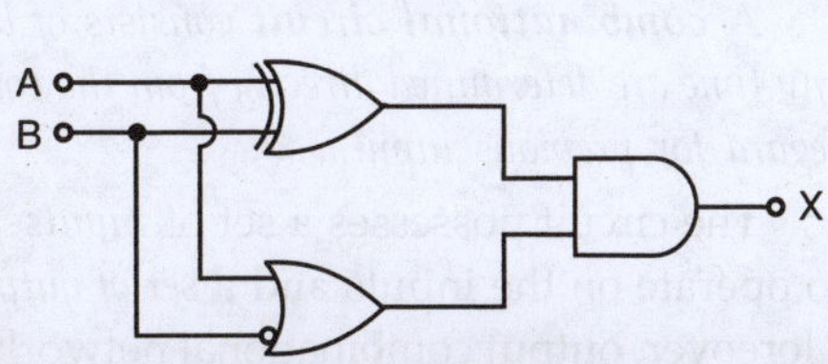

Fig. 2.91

Solution. The output of the given logic circuit can be obtained by determining the output of each logic gate while working from left to right.

As seen from Fig. 2.92, the output ;

$$X = (\bar{A} B + A\bar{B})(A + \bar{B})$$
$$= \bar{A} BA + A\bar{B}A + \bar{A}B\bar{B} + A\bar{B}\,\bar{B}$$
$$= 0 + A\bar{B} + 0 + A\bar{B}\,\bar{B} \quad \text{...[Eqn. 2.15(d)]}$$
$$= A\bar{B} + A\bar{B} \quad \text{...[Eqn. 1(b), 2.15(c)]}$$
$$= A\bar{B} \quad \text{...[Eqn. 2.14(c)]}$$

Using the simplified Boolean expression, the logic circuit is as shown in Fig. 2.93.

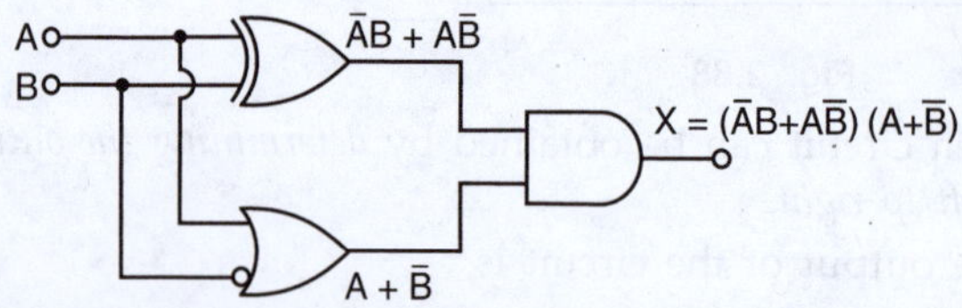

Fig. 2.92

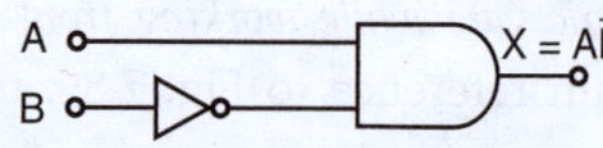

Fig. 2.93

2.3.15. Logic System

The logic system may be of the following two types :

1. Combinational.
2. Sequential.

The essential charateristics of combinational and sequential logic systems are compared as follows :

Combinational	Sequential
1. Possesses no memory or storage capacity.	Possesses memory or storage capacity.
2. Simple logic gates only carry out the implementation.	To carry out the implementation alongwith the logic gates, flip-flops, counters, registers memory cores are also used.
3. The system is described by a set of output functions only.	It is described by a set of output functions and also next state functions.
4. Output of the system depends only on the present input.	Output of the system depeds on the present input as well as on the present state of the system.

Combinational circuits :

A ***combinational circuit*** *consists of logic gates whose outputs at any time are determined directly from the combination of inputs without regard for previous input.*

The circuit possesses a set of *inputs*, a *memoryless logic network* to operate on the inputs and a *set of outputs* as shown in Fig. 2.94. Moreover, output combinational networks are used to make logical decisions and control the operation of different circuits in digital

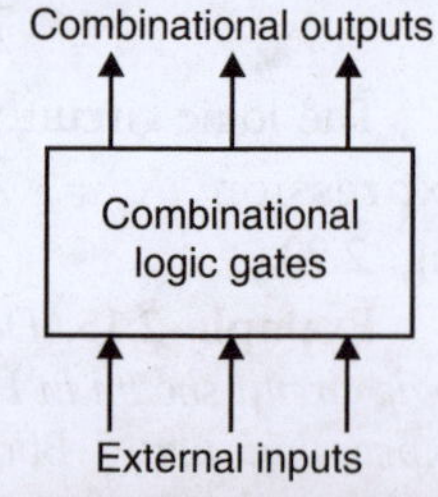

Fig. 2.94. Combinational logic circuit.

electronic systems. For a given set of input conditions, the output of such a circuit is the same. Consequently, truth table can fully describe the operation of such a circuit.

Examples. Examples of a combinational circuit are :

(*i*) Decoders (*ii*) Adders

(*iii*) Multiplexers (*iv*) Demultiplexers etc.

- **Multiplexers and demultiplexers :**
  — *Transmission of a large number of information units over a small number of lines is known as* **"Multiplexing"**.
  — **"Demultiplexing"** is a *reverse operation and denots receiving information from a small number of channels and distributing it over a large number of destinations.*

Design procedure of combinational circuit :

Following *operations* are involved in the design procedure :

1. To state the problem.
2. To determine the number of available input variables and required output variables.
3. To assign letter symbol to each input and output variable.
4. To derive the truth table that defines the required relationship between inputs and outputs.
5. To obtain the simplified Boolean function for each output.
6. To draw the logic diagram.

- A circuit that adds two bits is called a ***half adder.***
- A ***full adder*** consists of three inputs and two outputs. The outputs are designated by the symbol S for sum and C for carry.
- A ***two bit subtractor*** has two inputs X (minuend) and Y (subtrahend).
- A ***full subtractor* (FS)** is a *combinational circuit that performs a subtraction between two bits.* This circuit has *three inputs and two outputs.*

Code conversion :

- A variety of codes are used by different digital systems. It is sometimes necessary to use the output of one system as input to the other.
- A conversion circuit must be inserted between the two systems if each uses different codes for the same information.
- To convert from binary code A to binary code B, the input lines must supply the bit combinations of elements as specified by code A and the output lines must generate the corresponding bit combination of code B. A combination circuit performs this transformation by means of *logic gates.*

Comparator. A comparator is a combinational circuit that compares two number A and B and determine their relative magnitude. The outcome of the comparison is displayed in three outputs that indicate $A > B \equiv X$, $A = B = Y$, $A < B \equiv Z$.

Decoders and encoders :

— A ***decoder*** is a combination circuit that converts a binary code of n variables into m output lines, one for each discrete element of information.

— An ***encoder*** is a combination circuit that accepts m input lines, one for each element of information, and generates a binary code of n output lines.

Sequential circuits :

Such circuits have *inputs, logic network, outputs* and a *memory,* as shown in Fig. 2.95.

Their present output depends not only on their present inputs but also on the pevious logic states of the outputs.

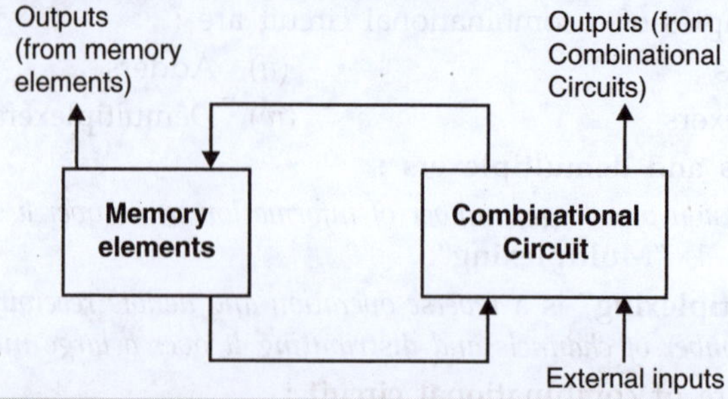

Fig. 2.95. Block diagram of a sequential circuit.

Examples. Examples of sequential circuits are :

(*i*) *Latches* (*ii*) *Flip-flops.*

The two main *types* of sequential circuits are :

1. Synchronous sequential circuits ..... referred to as *clocked-sequential circuits*
2. Asynchronous sequential circuits.

- The ***synchronous sequential circuits*** *are built to operate at a clocked rate whereas* **asynchronous ones** *are without clocking.*

2.3.16. Flip-Flop Circuits

The memory elements used in clocked sequential circuits are called **flip-flops.** These circuits are binary cells capable of storing one bit of information. It has two *outputs, one for the normal value and one for the complement value of the bit stored in it.* Binary information can enter a flip-flop in a variety of ways. Hence there different types of flip-flops.

A number of flip-flops are available in *IC* form. Some of these are *SR* (Set-Reset), *J-K* and *D* flip-flops. They are widely used as *switches, latches, counters, registers* and *memory cells* in computers.

- A salient feature of the flip-flop is that *output can exist in one of the two stable states, logic 1 and logic 0, simultaneously.* This is ensured by the appropriate crossed feedback connections associated with the most elementary form of the flip-flop known as a *latch.*

The following flip-flops will be discussed in the following articles :

1. *R-S* flip-flop.
2. Clocked *R-S* flip-flop.
3. *D* flip-flop.
4. *J-K* flip flop.
5. *T* flip-flop.

R-S flip-flop :

Fig. 2.96 shows a *R-S flip-flop using* **NOR** *gates.* There are two inputs to the flip-flops called *S* (set) and *R* (*reset*). *The* cross-coupled connection from the output of one gate and input of the other constitutes a feedback path. For that reason, the *circuit is classified as synchronous circuit.*

- A low *R* and a high *S* results in the *set* state.
- A high *R* and a low *S* give the *reset* state.
- If both *R* and *S* are high, the output becomes *indeterminate* and this is known as

'race condition'. This condition is *avoided by proper design.*

The truth table is shown in Table 2.8.

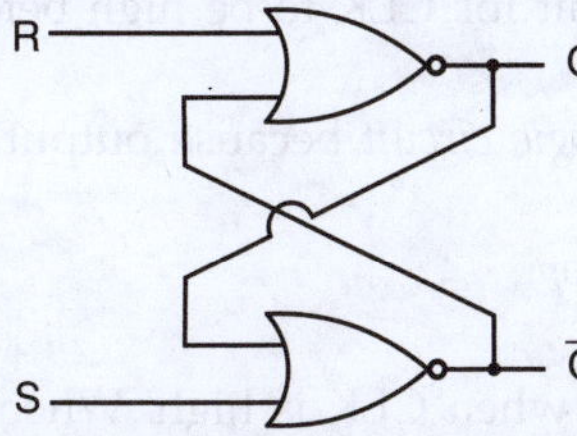

(*a*) Circuit diagram

Table 2.8. Truth table for NOR latch

R	S	Q	Comment
0	0	NC	No change
0	1	1	Set
1	0	0	Reset
1	1		Race

(*b*) Truth table

Fig. 2.96. *R-S* flip-flop using NOR gates.

Fig. 2.97 shows a *R-S* flip-flop using **NAND** *gates.* Table 2.9 shows the truth table. It is seen that the inactive and race conditions are **reversed.**

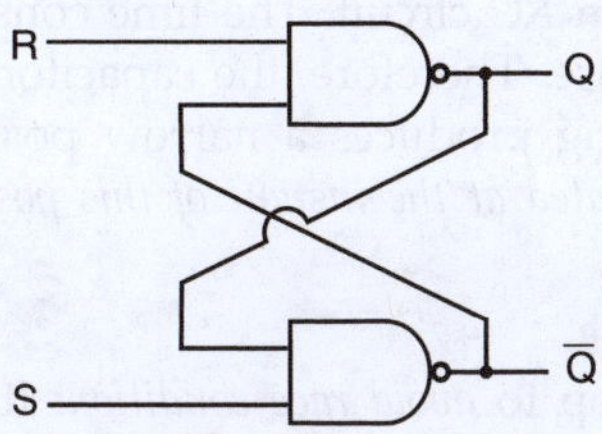

(*a*) Circuit diagram

Table 2.9. Truth table for NAND latch

R	S	Q	Comment
0	0		Race
0	1	1	Set
1	0	0	Reset
1	1	NC	No change

(*b*) Truth table

Fig. 2.97. *R-S* flip-flop using NAND gates.

- When *R* is *low,* output *Q* is *high.*
- When *R* is *high,* output *Q* is *low.*
- When both *R* and *S* are low, we get race condition which must be *avoided.*
- When both *R* and *S* are high - *no change condition.*

Clocked R-S flip-flop :

A large number of flip-flops are used in a computer. In order *to coordinate their working a square wave signal* known as *clock* is applied to the flip-flop. This *clock signal (indicated as CLK) pevents the flip-flop changing state till the right instant occurs.*

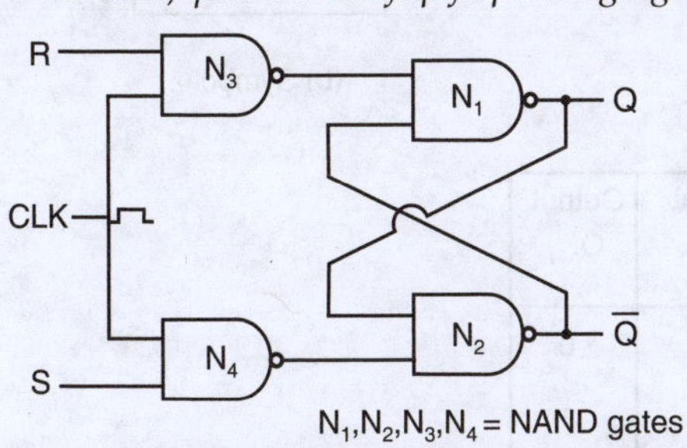

(a) Circuit diagram

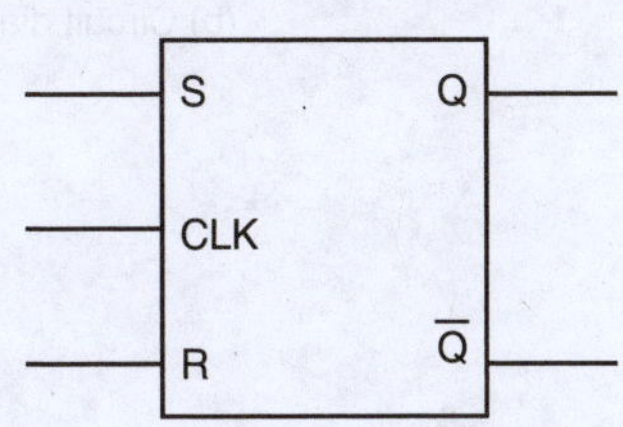

(b) Symbol

Fig. 2.98. Clocked *R-S* flip-flop

Fig. 2.98(*a*) shows a clocked R-S flip-flop using NAND gats (N_1 and N_2). This circuit uses two NAND gates N_3 and N_4 to apply CLK signal.

- When CLK is *low* the flip-flop *output Q ndicates no change.*

- If S is high and R is *low, the flip-flop must wait till CLK becomes high before Q can be set on 1.*
- If S is low and R is low, the flip-flop must wait for CLK to be high before Q is reset to low (0).

Clocked *R-S* flip-flop is a *synchroneous sequential logic circuit* because output state of the circuit changes at *discrete clocked instant of time.*

Fig. 2.98(*b*) shows a symbol for clocked *R-S flip-flop.*

Level clocking and edge triggering :

In a clocked flip-flop, the output can change state when CLK is high. When CLK is low, the output remains in the same state. Thus, the output can change state during the entire half cycle when CLK is high. This may be a *disadvantage in several situations. It is necessary that the output should change state only at one instant in the positive half cycle of the clock.* This *is known as* ***edge triggering*** *and the resulting flip-flop is known as* ***edge triggered flip-flop.***

Edge triggering can be made feasible by the use of an *RC* circuit. The time constant *RC* is made much smaller than the width of the clock pulse. Therefore, the capacitor can charge fully when CLK is high. The exponential charging produces a narrow positive voltage strike across the resistor. *The input gates are activated at the instant of this positive strike.*

D flip-flop :

A *D* flip-flop is an improvement over the *R-S* flip-flop to *avoid race condition.* It can be *level clocked or edge triggered.* The edge triggered one causes the change in output state at a unique instant.

In a *clocked R-S* flip-flop two input signals are required to drive the flip-flop which is a disadvantage with many digital circuits. In some events, both input signals become *high* which is again an undesirable condition. So these shortcomings/drawbacks of clocked *R-S* flip-flop are overcome in *D* flip-flop.

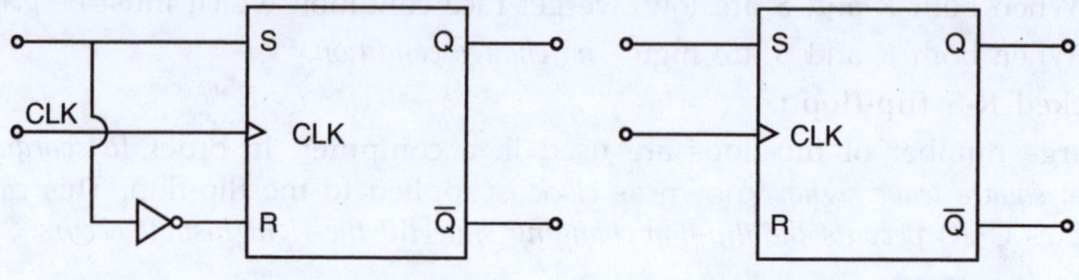

(b) Circuit diagram (b) Symbol

Input D_n	Output Q_{n+1}
0	0
1	1

(c) Truth table

Fig. 2.99. D flip-flop

Fig. 2.99(*a*), (*b*), (*c*) show the circuit diagram, symbol and truth table of *D* flip-flop respectively. It may be observed that only single data bit, *D* is required to drive the flip-flop.

— When the clock signal is at low level, data bit *D* is prevented to reach at output *Q* until clock signal becomes *high* at next pulse.

— It may be noted from the truth table that when data bit D_n is *high*, output Q_{n+1} gets at *high* level and when data bit D_n is low, Q_{n+1} gets at low level. Thus *D* flip-flop transfers the data bit *D* to *Q* as it is, and *Q* remains in the same state until the next pulse of the clock arrives.

- The flip-flop is named (*D*) flip-flop since *the transfer of data from the input to output is delayed.*
- The *D*-type flip-flop is *either used as a delay device* or *as a latch to store 1-bit of binary information.*

Edge triggered D flip-flop :

Fig. 2.100.(*a*) shows the circuit diagram and symbols of an edge triggered *D* flip-flop. The *clock provides the square wave signal.* RC circuit *converts* this signal *into strikes* so that triggering occurs at the instant of positive strike. The data bit *D drives one of the inputs.* Because of inverter, the complement $\overline{D}$ *drives the other output.* At the instant of positive strike, input *D* and its complement $\overline{D}$ cause the output *Q* to set or reset. Fig. 2.100(*b*) shows the truth table.

- When CLK is 0 or 1, the *D* input is not ***there*** and there is no change in state of *Q*.
- On the *negative* edge of the clock (marked ↓) the output remains in the same state.
- On the *positive* edge of the clock (marked ↑) *Q* changes to 0 if *D* is 0 and to 1 if *D* is 1.

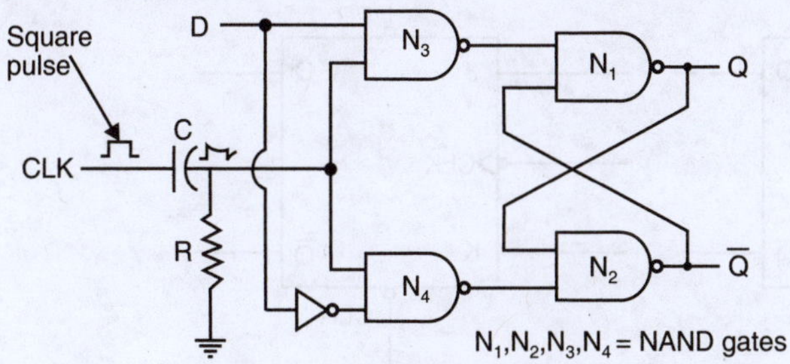

(a) Circuit diagram

CLK	D	Q
0	×	No change
1	×	No change
↓	×	No change
↑	0	0
↑	1	1

(b) Truth table

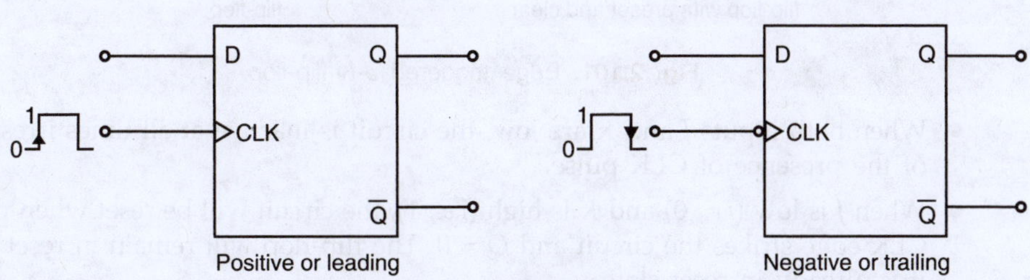

Symbols

Fig. 2.100. Edge triggered *D* flip-flop.

Edge triggered J-K flip-flop :

— *J-K* fip-flop is *very versatile* and is perhaps the most *widely used type of flip-flop.*

— The *J* and *K* designations for the inputs have no known significance except that they are adjacent letters in the alphabet.

— *J-K flip-flop functions identically to R-S flip-flop.*

The difference is that the *J-K* flip-flop has no *invalid state* as does the *R-S* flip-flop.

— It is widely used in digital devices such as *counters, registers, arithmetic logic units, and other digital systems.*

Fig. 2.101(*a*) shows the circuit diagram of a edge triggered *J-K* flip-flop used in *digital counters.* The CLK input is through an *RC* circuit with a short time constant. The *RC* circuit converts the rectangular clock pulse to narrow spikes as shown. Due to double inversion through NAND gates, the circuit is positive edge triggered.

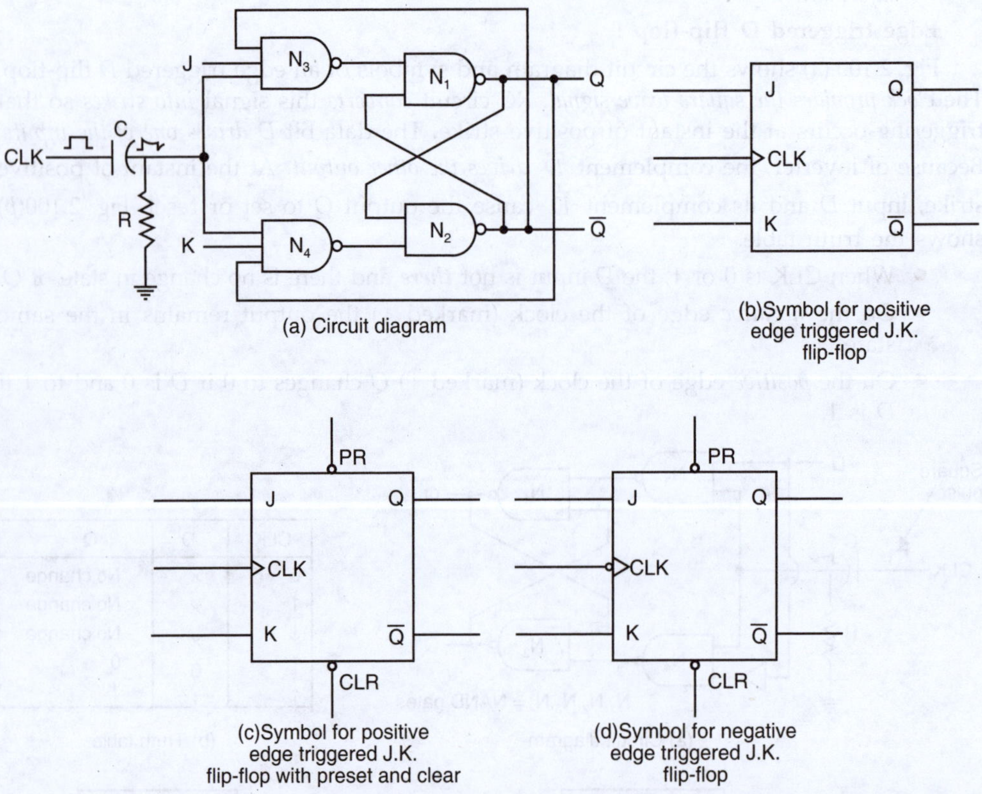

(a) Circuit diagram

(b)Symbol for positive edge triggered J.K. flip-flop

(c)Symbol for positive edge triggered J.K. flip-flop with preset and clear

(d)Symbol for negative edge triggered J.K. flip-flop

Fig. 2.101. Edge triggered *J-K* flip-flop.

- When both inputs *J* and *K* are low, the circuit is inactive at all times irrespective of the presence of CLK pulse.
- When *J* is low (*i.e.,* 0) and *K* is high (*i.e.,* 1), the circuit will be reset when positive CLK edge strikes the circuit and *Q* = 0. The flip-flop will remain in reset state if it is already in reset state.
- When *J* = 1 and *K* = 0, the circuit sets at the arrival of next positive clock edge.
- When *J* = 1 and *K* = 1, the flip-flop will toggle (means to switch to opposite state) on the next positive CLK edge. The action is illustrated in the table 2.10 :

Table 2.10. Positive edge triggered J-K flip-flop

CLK	J	K	Q
0	×	×	No change
1	×	×	No change
↓	×	×	No cahnge
X	0	0	No change
↑	0	1	0 (reset)
↑	1	0	1 (set)
↑	1	1	toggle

Using of *RC* circuit for edge triggering is not very convenient for fabrication. Actual circuits use additional NAND gates for edge triggering, such circuits are known as *direct coupled circuit*.

— Fig. 2.101(*b*) shows the symbol for positive edge triggered *J-K* flip-flop.

— Fig. 2.101(*c*) shows a positive edge *J-K* flip-flop with present (*PR*) and clear (CLR).

— Fig. 2.101(*d*) shows the symbol for negative edge triggered *J-K* flip-flop with *PR* and CLR.

The *small bubble at CLR indicates negative triggering.*

T flip-flop :

T flip-flop is basically a *J-K* flip-flop, in this circuit input terminals *J* and *K* are connected with each other and this input is named as *T*.

Fig. 2.102(*a*) and (*b*) show the circuit diagram and symbol respectively of a trailing edge triggered *T* flip-flop.

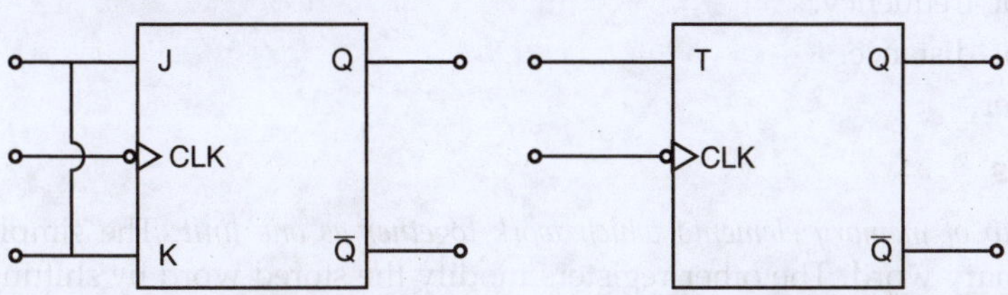

(a) Circuit diagram (b) Symbol

Input T_n	Output T_{n+1}
0	Q_n
1	$\bar{Q}_n$

(c) Truth table

Fig. 2.102. Trailing edge triggered T flip-flop.

— When ***low** level signal* is applied to the input terminal *T*, then *initial state of output of flip-flop remains the same.*

— When ***high*** level signal is applied to the input terminal *T*, output of the flip-flop toggles after arrival of every new clock pulse. So the frequency of output signal is *half* of the clock signal frequency.

This flip-flop can be treated as frequency divider or a device which takes the input frequency at the clock terminal and divide it by *two.*

2.3.17. Counters

— A ***counter** is a sequential circuit that goes through a prescribed sequence of states upon the application of input pulses.*

— The input pulses, called *count pulses,* may be clock pulse or may originate from an external source and may occur at prescribed intervals of true or random.

— The sequence of states in a counter may follow a binary count or any other sequence of states.

— They are used for counting the number of occurrences of an event and are useful for generating time sequences to control operations in a digital system.

Straight binary sequence counter. It is the simple and most straight forward. An *n*-bit binary counter has *n* flip-fops and can count in binary from 0 to 2^{n-1}.

Binary ripple couter. It consists of a series connections of *T* flip-flops without any logic gates. Each flip-flop is triggered by the output of its preceding flip-flop goes from 1 to 0. The signal propagates through the counter in a *Ripple* manner, *i.e.*, the flip-flop essentially changes once at a time in rapid succession. It is the most simplest and most straight forward. It, however, has speed limitations ; an increase in speed can be obtained by the use of a parallel or a synchronous counter.

Synchronous 3-bit binary counter. In this all flip-flops are triggered simultaneously by count pulse. The flip-flop is complemented only if its *T* input is equal to 1.

Counter-decoder circuits :

— Counters together with decoders are used to generate timing and sequencing signals that control the operation of digital systems.

— The counter-decoder can be designated to give any desired number of repeated timing sequence.

Applications of counters :

The fundamental applications of counters are given below :

1. Measurement of time interval.
2. Direct counting.
3. Measurement of speed.
4. Measurement of frequency.
5. Measurement of distance.
6. Gating a counter.

2.3.18. Registers

A **register** *is a group of memory elements which work together as one unit*. The simple registers only store a binary word. The other registers modify the stored word by shifting its bits to left or right.

The registers can be *classified* as :

(*i*) Accumulator.

(*ii*) General purpose registers.

(*iii*) Special purpose registers.

— A *counter* is a special kind of register to count the number of clock pulses arriving at the input.

2.3.19. Logic Families

The basic building block for digital systems is the *logic gate.* Logic circuits have evolved into *logic families.* Usually a system is fabricated with circuits from one logic famiy. When circuits from more than one family are to be used to implement a given function, it is necessary to ensure that output of one family is *compatible with input of the other.*

The logic famiies are *classified* as follows :

1. *Bipolar families :*

(*i*) DTL (Diode Transistor Logic)

(*ii*) TTL (Transistor Transistor Logic)

(*iii*) ECL (Emitter Coupled Logic)

2. *MOS families :*

(*i*) PMOS (P-channel MOSFET Logic)

(*ii*) NMOS (N-channel MOSFET Logic)

(*iii*) CMOS (Complementary MOSFET Logic)

Note. The PMOS and DTL are now obsolete.

2.3.20. Integrated Circuits

General aspects :

*An **integrated circuit** (IC) is a complete electronic circuit in which both the *active (e.g. transistors and FETs) and passive components (e.g. resistance, capacitors and inductors) are fabricated on a tiny single chip of silicon.*

An *IC* is different from a discrete (*i.e.,* distinct or separate) circuit, which is built by *connecting separated devices.* In this case, each device is fabricated separately and then all the devices are assembled together to make an electronic circuit. Discrete circuits have two main *disadvantages :*

(*i*) In a large circuit (*e.g.* T.V. circuit, computer circuit) there may be hundreds of components and consequently discrete assembly would occupy large space,

(*ii*) There will be hundreds of soldered points posing a considerable problem of reliability. To overcome these drawbacks of space conservation and reliability, engineers started a drive for *miniatured circuits.* This led to the development of integrated circuits.

– J.S. Kilby of Texas Instruments was the first person to develop in 1959 an integrated circuit – a single monolithic silicon chip in which active and passive components were fabricated by successive deposition, etching and diffusions. He was soon followed by Robert Noyce of fair-child who successfully fabricated a complete *IC* including the interconnections on a single silicon chip. Since then a lot of progress has been made.

Advantages and disadvantages of Integrated Circuits (ICs) :

As compared to standard printed circuits which use discrete components *IC*s have the following *advantages :*

1. *Exteremely small size* (physical)—Often the size is thousands of time smaller than a discrete circuit.

2. *Very small weight* (owing to miniaturised circuit).

3. *Reduced cost* (since many identical circuits can be built simultaneously on a single wafer).

4. *Extremely high reliability* (IC logic gate has been found to be 100 000 times more reliable than a vacuum tube and 100 times more reliable than a transistor logic gate.

5. *Increased response time and speed.*

6. *Low power consumption* (due to smaller size).

7. *Easy replacement.*

8. *Higher yield* (Because of the batch production, the yield is very high).

9. *Improved functional performance* as more complex circuits can be fabricated for achieving better characteristics.

10. *Greater ability of operating at extreme temperatures.*

Disadvantages :

1. They are *quite delicate* and *cannot withstand rough handling* or *excessive heat.*
2. They function at *fairly low voltages.*
3. They handle only *limited amount of power.*
4. If any component in an *IC* goes out of order, the whole *IC* has to be *replaced* by the new one.
5. It is not possible to produce high power *ICs* (greater than 10 W).
6. Coils or inductors *cannot* be fabricated.
7. There is a *lack of flexibility* in an *IC, i.e.,* it is not generally possible to modify the parameters within which an integrated circuit will operate.
8. In a *IC,* it is neither convenient nor economical to fabricate capacitances exceeding 30 pF. Therefore, for high values of capacitance, discrete components exterior to *IC* chip are connectted.
9. High grade *P-N-P* assembly is not possible.
10. Difficult to produce an *IC* with low noise.
11. Voltage dependence of resistors and capacitors.

Scale of Integration :

The *number of electonic circuits or components, which can be fabricated on a standard size of a silicon chip is called the* **Scale or level of integration.**

The scale of integration is generally classified on two basis : (*i*) The number of *circuits,* and (*ii*) The number of *components.*

The various types of scale of integration are :

1. *Small scale integration* **(SSI) :**

   No. of circuits per package .... Less than 12

   No. of components .... Less than 50

2. *Medium scale integration* **(MSI) :**

   No. of circuits per package .... Beteween 30 and 100

   No. of components .... Between 50 and 5000

3. *Large scale integration* **(LSI) :**

   No. of circuits per package .... Between 100 and 10 000

   No. of components .... Between 5000 and 100 000

4. *Very large scale integration* **(VLSI) :**

   No. of circuits per package .... Between 10 000 and 100 000

   No. of components .... Between 100 000 and 1 000 000

5. *Ultra large scale integration* **(ULSI) :**

   No. of circuits per package .... Between 100 000 and 1 000 000

   No. of components .... Between 1 000 000 and 10 000 000

6. *Giga scale integration* **(GSI) :**

   No. of circuits per package .... 1 000 000 or more

   No. of components .... over 100 000 000.

Classification of Integrated Circuits :

There are many ways of classifying integrated circuits but the following two classifications are important from subject point of view :

1. *Fabrication or structure :*

(*i*) Monolithic integrated circuits.

(*ii*) Thick and thin film integrated circuits.

(*iii*) Hybrid or multichip integrated circuits.

2. *Application or function :*

(*ii*) Linear (or analog) integrated circuits.

(*ii*) Non-linear (or digital) integrated circits.

Monolithic Integrated Circuits :

The word 'monolithic' means 'single stone' or more appropriately *'a single solid structure'*. In this *IC*, all circuit components (both active and passive) are fabricated variably *within a single continuous piece of silicon crystalline material called* ***water*** (or ***subtrate***). All components are atomically part of the same chip.

Monolithic *ICs* are by *far the most common type of ICs used in practice* because of :

(*i*) *Mass production ;*

(*ii*) *Lower cost ;*

(*iii*) *Higher reliability.*

Commercially available *ICs* of this type can be used is :

— *Amplifiers ;*

— *Voltage regulators ;*

— *A. M. receivers ;*

— *T. V. circuits;*

— *Computer circuits.*

Limitations of monolithic ICs :

(*i*) Low power rating.

(*ii*) Lack of flexibility in circuit design.

(*iii*) Poorer isolation between components.

(*iv*) Small range of values of passive components used in the *ICs*.

(*v*) No possibility of fabrication of inductors.

Thick and thin film Integrated Circuits :

The essential difference between thick-film and thin-film *ICs* is *not their relative thickness* but the *method of depositing the film*. Both have *similar appearance, properties and general characteristics.*

- These devices are larger than monolithic *ICs* but smaller than discretes circuits.
- These *ICs* can be used *when power requirement is comparatively higher.*

MOS Integrated Circuits :

Integrated circuits (*ICs*) based upon the active devices are of the following *two types :*

1. Bipolar *ICs* using *bipolar* active devices such as BJT.
2. Unipolar *ICs* using *unipolar* active devices like FET.

MOS *ICs* based on MOSFET structure find wide applications particularly in *digital field,* because of the following *"advantages"* over bipolar *ICs* :

(*i*) *Fabrication process is simple and cheaper* comparatively.

(*ii*) Occupy less area (the MOS IC typically occupies only 5 percent of the surface required by an expitaxial double-diffused transistor in conventional *IC* ; a MOS resistor occupies 1 percent of the area of a conventional diffused resistor).

(iii) Low power consumption.

(iv) Less costly to fabricate.

(*v*) MOS transistor has a *higher bandwidth than bipolar transistor.*

(vi) High packing density.

Disadvantage :

The major demerit of MOS *ICs* is that their *operating speed is smaller than that of* bipolar *ICs* and as such they are *not suitable for ultra high-speed applications.*

Applications :

MOS *ICs* find wide applications in LSI and VLSI *chips* such as :

— Calculator chips ;

— Memory chips ;

— Micro processors (μP) ;

— Single-chip computers.

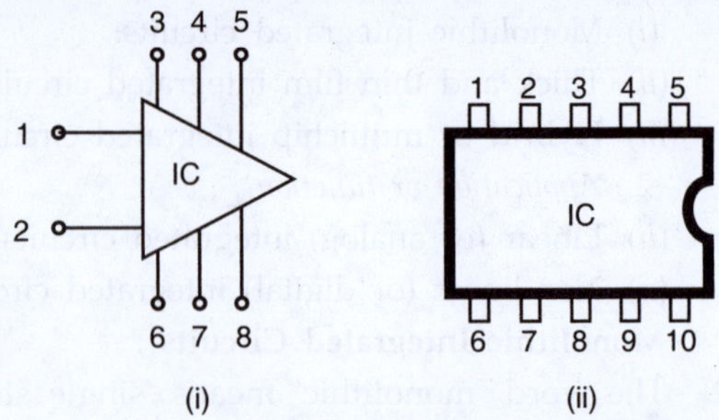

Fig. 2.103. *IC* symbols.

IC symbols :

In general, no standard symbol exist for *ICs*. Oftenly, the circuit diagram merely shows a block with numbered terminals. However, sometimes standard symbols are used for operational amplifiers or digital logic gates. Some of the symbols used with *ICs* are shown in Fig. 2.103.

- *IC* symbol does not show the internal circuit.

The Integrated Transistor Amplifier :

Fig. 2.104. (*i*) shows the schematic diagram of an integrated transistor amplifier ; the cross section view and top view of the interconnections are shown in Fig. 2.104 (*ii*) and (*iii*) respectively.

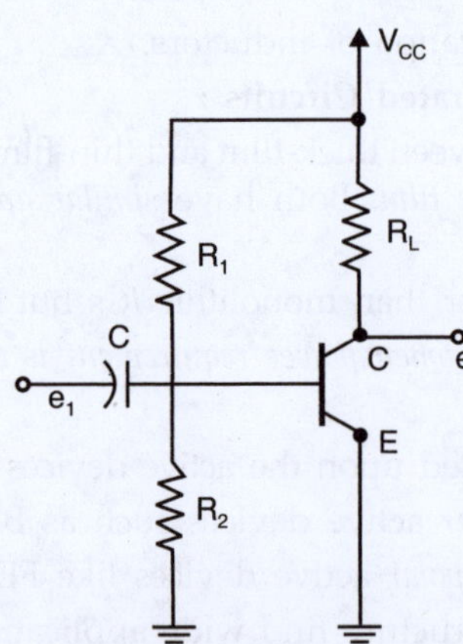

(*i*) Schematic diagram.

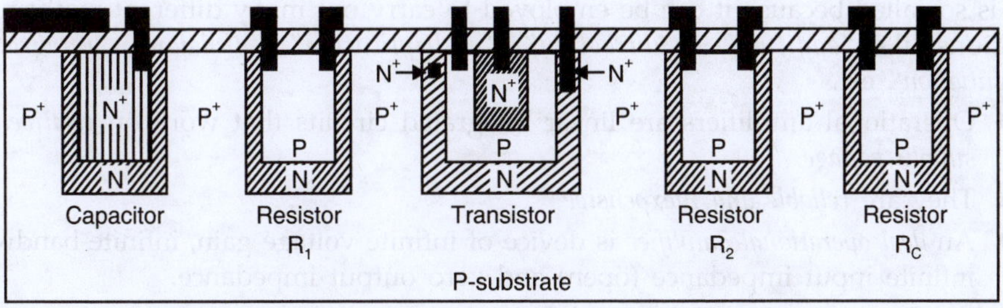

(*ii*) Cross-sectional view showing each element.

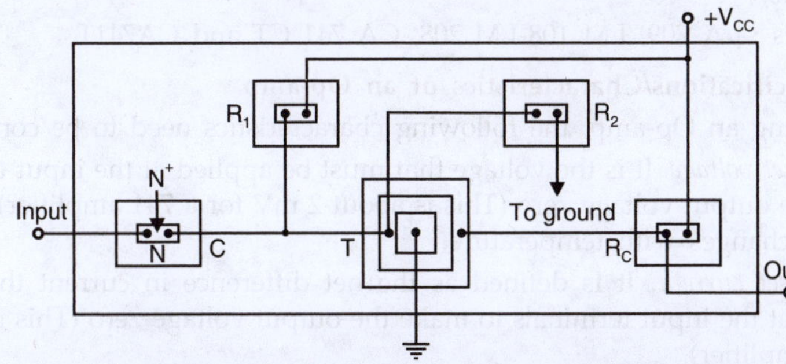

(*iii*) Top view showing the interconnections

Fig. 2.104. Integrated transistor amplifier.

The five circuit elements-one capacitor, three resistors and one transistor—and all the interconnections are created by the same masking, etching, and diffusion process. In actual *IC* the circuit elements would not appear in the tandem arrangement shown in Fig. 2.104 (*ii*) and (*iii*) ; rather the *elements would be so placed that as to make optimum use of the available space and to reduce the length of the interconnections to the minimum possible.* The tandem arrangement is used here merely for convenience and classification.

- The total area on the chip covered by this amplifier is only a very small fraction of a square mm.

Applications of ICs :

The popular applications of *ICs* are :

1. *Digital watch* ;
2. *Electronic calculator* ;
3. Pocket *PC* ;
4. Personal digital assistant (PDA) ;
5. MP3 players ;
6. Digital cameras ;
7. Mobile phones ;
8. Digital dictionaries ;
9. Digital translators ;
10. CD (compact disk) player ;
11. DVD (Digital versatile disk) players.

2.3.21. Operational Amplifiers (Op-amp)

An ***operational amplifier (Op-amp)*** *is a linear integrated circuit (IC) that has a very high voltage gain, a high input impedance and a low output impedance.*

It is so called because it can be employed to carry out many different mathematical operations like *"addition"*. *"subtraction"*, *"multiplication"*, *"division"*, *"integration"*, *"differentiation"* etc.

- Operational amplifiers are linear integrated circuits that work on *relatively low supply voltage.*
- They are *reliable and inexpensive.*
- An *deal operational amplifier* is device of infinite voltage gain, infinite bandwidth, infinite input impedance (open) and zero output impedance.
- An Op-amp may contain two dozen transistors, a dozen resistors and one or two capacitors.

  Examples : μA 709, LM 108-LM 208, CA 741 CT and CA741T.

2.3.21.1. Specifications/Characteristics of an Op-amp

While selecting an Op-amp, the following characteristics need to be considered:

1. *Input offset voltage.* It is the voltage that must be applied at the input terminals to make the output voltage zero (This is about 2 mV for a 741 amplifier). The offset voltage changes with temperature.
2. *Input offset current.* It is defined as the net difference in current that must be applied at the input terminals to make the output voltage zero (This is 20 nA for a 741 amplifier).
3. *Input check current.* It is the mean of the two input currents to make the voltage zero.
4. *Slew rate.* It is the maximum rate at which the output can change. It is expressed as volts/microseconds.
5. *Unity gain frequency.* This is the frequency at which the open loop gain of the amplifier becomes unity.
6. *Common mode rejection ratio (CMRR).* It is the ratio of desirable signals to undesirable signals.

- An Op-amp is the *basic building block* for:

— Amplifiers
— Integrators
— Summers
— Differentiators
— Comparators
— A/D and D/A converters
— Active filters
— Sample and hold amplifiers.

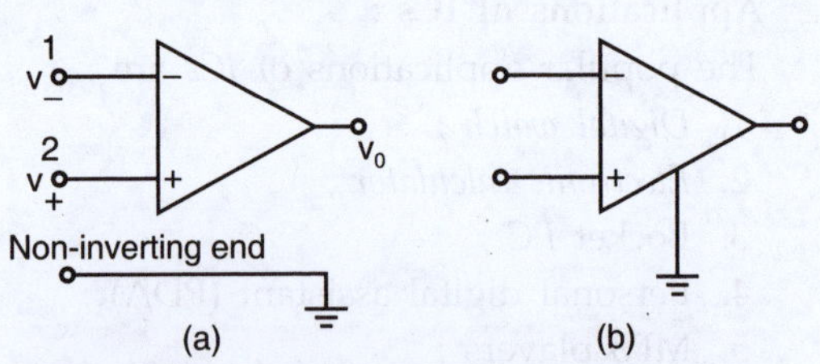

Fig. 2.105. Op-amp symbol

2.3.21.2. Op-amp description

Fig. 2.104(*a*) shows a standard symbol (a triangle having two input labelled differently and a single output) for an Op-amp, the one shown in Fig. 4.4(*b*) is also oftenly used.

One input terminal is designated by *–ve* sign, it is called *inverting end* while other input terminal is designated by a *+ve* sign, it is called *non-inverting end.*

These plus (+) and minus (–) polarities indicate *phase reversal* only. It does not mean that v_1 and v_2 [Fig. 2.105(*a*)] are negative and positive respectively. Additionally, it also *does not imply* that a positive input voltage has to be connected to the plus-marked non-inverting terminal 2 and negative input voltage to negative-marked inverting terminal 1.

In fact amplifier can be used either way up so to speak. It may also be noted that all input and output voltages are referred to a common reference usually the ground shown in Fig. 4.4(*a*).

Operational amplifier operating with –ve feedback possesses stable closed loop gain and also insensitivity to variation of supply voltage and ambient temperature.

• The voltage at the output terminal v_o, is the product of the amplifier gain G, and the voltage difference :

$$v_o = G(v_+ - v_-) \quad ...(4.9)$$

The output voltage is roughly limited to the power supply voltages V_{cc} and V_{ee}, as the voltage difference increases; if the voltage difference becomes too large, the output *saturates* near one of these values and remain constant.

• *The differential characteristic of op-amp has great importance in instrumentation because it eliminates offset voltages and noise signals common to both input terminals.* For example, nearby power lines may induce 50-cycle noise in the exterior circuitry leading to the amplifier. Such line noise is often present in identical form at both input terminals. This behaviour is known as *common-mode rejection.* If, instead, an op-amp receives the output of a voltage-sensitive Wheatstone bridge, the common offset voltages of the two voltage dividers are cancelled, and only the desired difference voltage is applied.

Limitations of Op-amp : Most of the Op-amps have a nonideal characteristic according to which they *do not completely satisfy the differential amplifying property.* With *both inputs grounded residual output voltage remains.*

— The multitude of transistors, resistors, and other elements within the Op-amp are *never perfectly matched*, so the amp output actually reaches zero at some small *non-zero* input voltage. To accommodate this input offset voltage, the common op-amp is provided with pins marked "offset null" or "balance" which provides a means for adjusting the unwanted offset voltage towards zero.

Another limitation is that the actual common-mode rejection is finite. If the two input signals each include a common-mode voltage v_{cm}, the Op-amp's actual voltage will be,

$$v_o = G(v_+ - v_-) + G_{cm}\, v_{cm} \quad ...(4.10)$$

The finite common-mode rejection is characterised by the "*common-mode rejection ratio* (CMRR)" in decibels:

$$\text{CMRR} = 20 \log_{10}\left(\frac{G}{G_{cm}}\right) \text{dB} \quad ...(4.11)$$

Since typical Op-amps have a CMRR of 60 to 120 dB, therefore, the common-mode gain is typically 10^3 to 10^6 times smaller than the differential gain, hence a *high CMRR is desirable.*

• Further, the performance of Op-amp may be limited by thermal drift. Both internal and external circuit elements may be temperature sensitive, and design of each circuit usually includes compensating features.

A wide variety of Op-amps are available, and their differences largely represent attempts to *improve*:

— Thermal stability;

— CMRR;

— Offset voltages;

— Frequency response .... These refinements, however, increase the cost.

2.3.21.3. Applications of Op-amp

Operational amplifiers may be used as the *basic components of* :

- Linear voltage amplifiers;
- Differential amplifiers;
- Integrators and differentiators;
- Voltage comparators;
- Function generators;
- Filters;
- Impedance transformers;
- Many other devices.

2.3.21.4. Op-amp circuits used in Instrumentation

Some of the commonly used Op-amp circuits are described below:

1. Inverter;
2. Adder;
3. Subtracter;
4. Multiplier and divider;
5. Integrator;
6. Differentiator;
7. Buffer amplifier;
8. Differential amplifier.

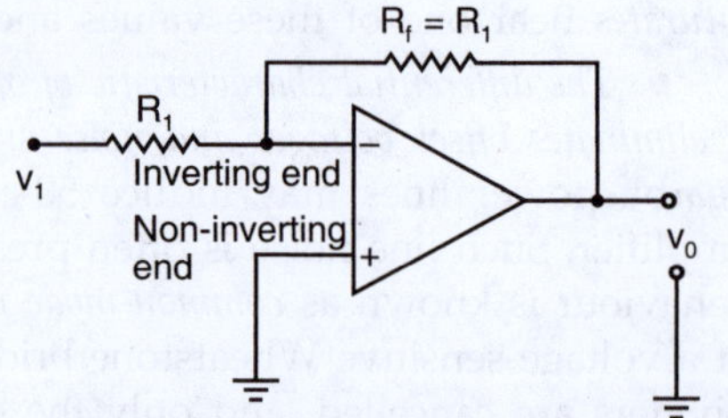

Fig. 2.106. Op-amp as an inverter.

1. Inverter. Fig. 2.106, shows the circuit of an Op-amp used as *inverter*. The feedback resistance R_f is made equal the resistance R_1, connected to the inverting end of the amplifier,

$$\text{Output voltage, } v_o = -\frac{R_f}{R_1} v_1 = -v_1 \quad (\because R_f = R_1) \qquad \text{...(4.12)}$$

Obviously, the output voltage is 180° out of phase with the input voltage.

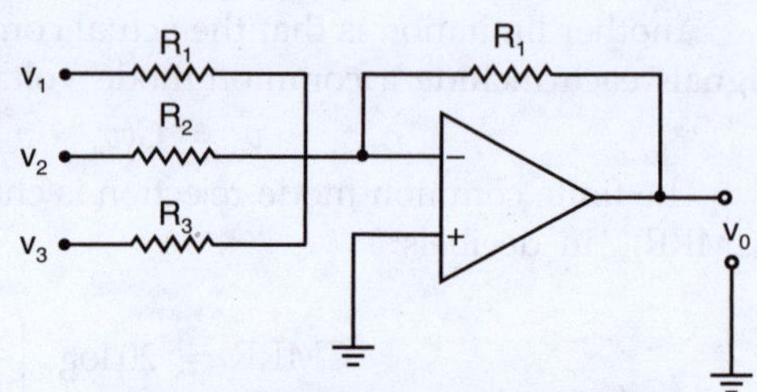

Fig. 2.107. Op-amp as an adder.

2. Adder. Fig. 2.107 shows an Op-amp circuit that performs the signals with amplification (if desired); using superposition theorem, we get

Output voltage,

$$v_o = -\left(\frac{R_f}{R_1}v_1 + \frac{R_f}{R_2}v_2 + \frac{R_f}{R_3}v_3\right) \qquad \text{...(4.13)}$$

If $R_1 = R_2 = R_3 = R_f$, then

$$v_o = -(v_1 + v_2 + v_3) \qquad \text{...(4.14)}$$

i.e., sum of the individual input voltages. The inversion that occurs cannot be avoided.

3. Subtracter. The Op-amp circuit used for subtraction of two input signals is shown in Fig. 2.108. The output of the 2nd Op-amp is given by :

$$v_o = -\left(-v_1 \frac{R_f}{R_1} \cdot \frac{R_f}{R_3} + v_2 \frac{R_f}{R_2}\right) \qquad \text{...(4.15)}$$

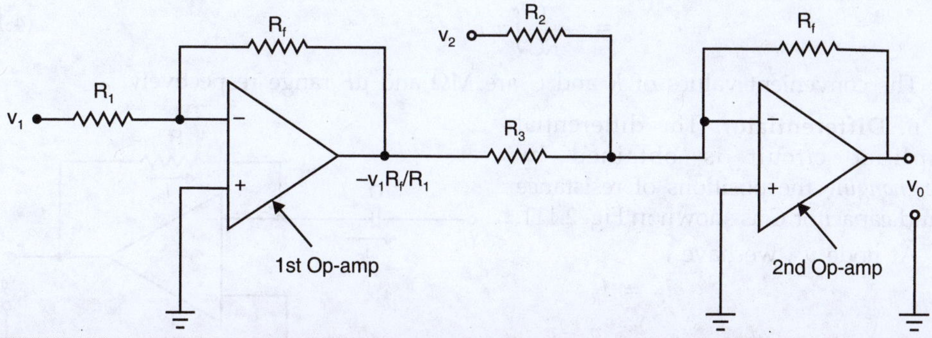

Fig. 2.108. Op-amp as a subtractor.

If $R_f = R_1 = R_2 = R_3$, the circuit acts as a *pure subtractor* and the output, in this case, is given by

$$v_o = v_1 - v_2 \quad ...(4.16)$$

4. Multiplier and divider. The output of an Op-amp in the inverting mode is given by

$$v_o = -\left(\frac{R_f}{R_1}\right) \times v_1$$

In case $R_f > R_1$, the circuit shown in Fig. 2.109 acts as a multiplier and in case $R_f < R_1$ it acts as a divider.

Thus, by choosing the values of R_f and R_1, the multiplier and the divider circuits can be designed for multiplication and division by any number.

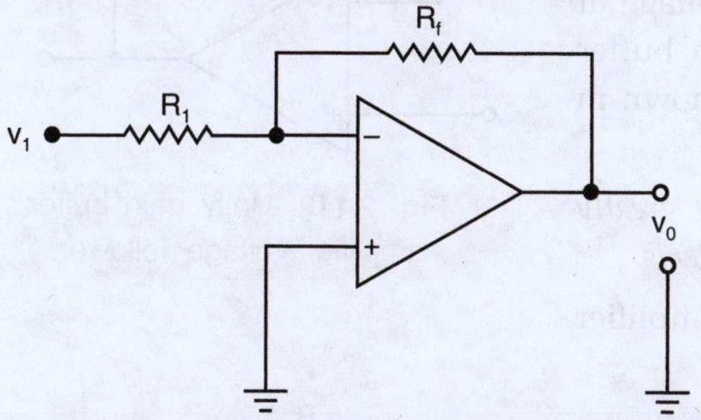

Fig. 2.109. Op-amp as multiplier and divider.

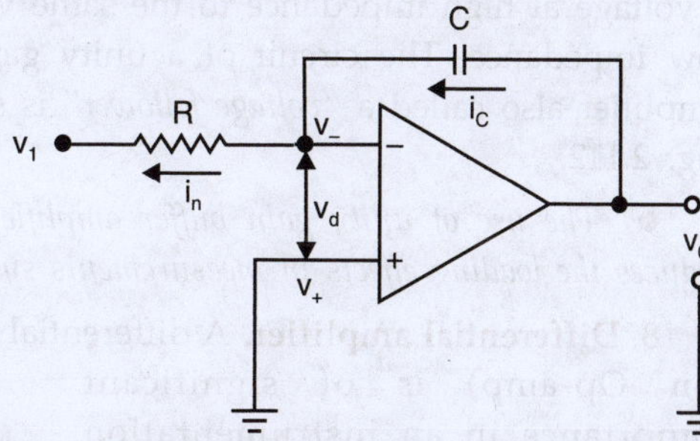

Fig. 2.110. Op-amp as an integrator.

5. Integrator. Fig. 2.110. shows a circuit in which the output voltage is proportional to the integral of the input voltage.

In order to show that the circuit shown in Fig. 4.9. acts as an integrator using KCL (Kirchoff's Current Law) at node v_-, we have,

$$i_n = i_C$$

or, $$\frac{v_- - v_1}{R} = C\frac{d}{dt}(v_o - v_-)$$

For *infinite differential gains*, $v_- = 0$

$$\therefore \quad -\frac{v_1}{R} = C\frac{d}{dt}v_o$$

By integration, we have

$$v_o = -\frac{1}{RC}\int v_1\, dt \qquad ...(4.17)$$

The convenient values of R and C are MΩ and μF range respectively.

6. Differentiator. The differential amplifier circuit is obtained by *interchanging* the positions of resistance R and capacitor C as shown in Fig. 2.111.

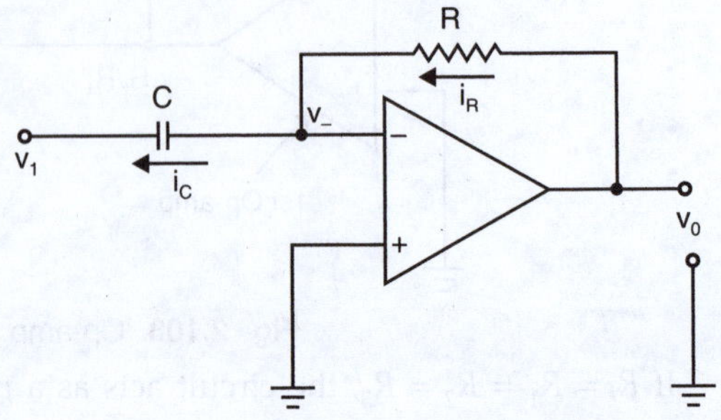

Fig. 2.111. Op-amp as differentiator.

At node v_-, we have :

$$i_C = i_R$$

$$C\frac{d}{dt}(v_- - v_1) = \frac{v_o - v_-}{R}$$

Now, $$v_- = 0$$

$$\therefore \quad -C\frac{d}{dt}(v_1) = \frac{v_o}{R}$$

or, Output voltage, $$v_o = -RC\frac{d}{dt}(v_1) \qquad ...(4.18)$$

Thus, the output voltage is equal to the differentiated input voltage.

- The *Op-amps are normally used as differentiators as they tend to decrease the signal to noise (S/N) ratio.*

7. Buffer amplifier. The buffer amplifier is essentially an impedance transformer which converts a voltage at high impedance to the same voltage at low impedance. The circuit of a unity gain buffer amplifier also called a *"voltage follower"* is shown in Fig. 2.112.

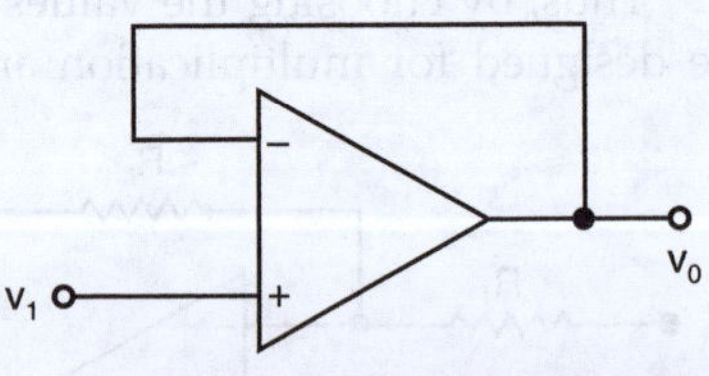

Fig. 2.112. Unity gain buffer amplifier voltage follower.

- *The use of unity gain buffer amplifiers greatly reduces the loading effects in measurements systems.*

8. Differential amplifier. A differential amplifier (an Op-amp) is of significant importance in an instrumentation system.

In its basic form it has two *inputs* and *outputs*. The signals available to the two outputs are identical except that the two are 180° out-of-phase with each other.

The output voltage of the amplifier is proportional to the difference between the two input voltages.

Fig. 2.113, shows an Op-amp used as a differential amplifier.

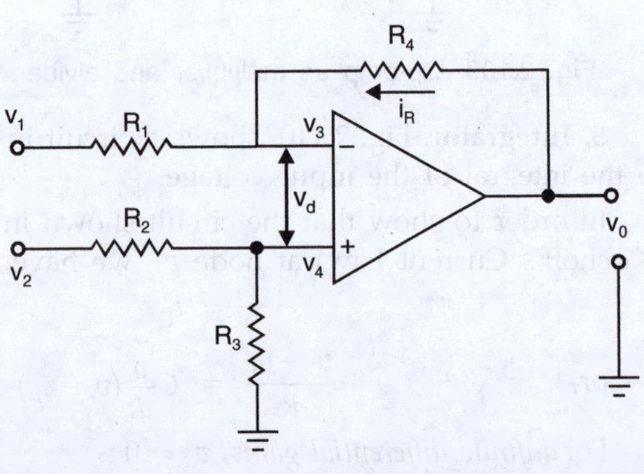

Fig. 2.113. Op-amp used as a differential amplifier.

Here, $v_o = G_d(v_+ - v_-)$

where, G_d = Differential gain.

The signal $v_d = (v_+ - v_-)$ is called *"Difference Mode Signal"* or simply *"Difference Signal"*

$$G_d = \frac{v_o}{v_+ - v_-} = \frac{v_o}{v_d} \quad ...(4.19)$$

It can be proved that,

$$v_o = G_d(v_2 - v_1) \quad ...(4.20)$$

When the *two input voltages are equal,* the output voltage is zero. Equal inputs are known as *"Common mode signals"* because the input signal is common to both inputs. However, in actual practice when equal input voltages are applied to the inputs, the *output voltage is not exactly equal to zero (difference is typically of the order of several hundred microvolts) on account of difference in response of the two inputs to common mode signals.*

Common mode gain,

$$G_{cm} = \frac{v_o}{v_{cm}} \quad ...(4.21)$$

where, G_{cm} = Common mode gain, and

v_{cm} = Common mode input signal.

The *"Common mode rejection ratio (CMRR)"* is defined as

$$CMRR = \frac{G_d}{G_{cm}} \quad ...(4.22)$$

Also, $$CMRR = 20 \log_{10}\left(\frac{G_d}{G_{cm}}\right) \text{dB} \quad ...(4.22a)$$

... when expressed in *dB.*

Advantages of differential amplifiers :

1. ***Noise immunity:***
   - These amplifiers are extensively used in equipment such as *electronic voltmeters and oscilloscopes.*
2. ***Drift immunity :***
   - The differential amplifier *has inherent capabilities of eliminating problem of drift.*
   - The differential amplifier construction is used for the *early stages of oscilloscope and electronic voltmeter amplifiers, where low drift is extremely important.*

Instrumentation amplifiers. The instrumentation amplifier is a dedicated differential amplifier with *extremely high impedance. The high common mode rejection makes this amplifier very useful in receiving small signals buried in large common-mode offsets and noise.*

These amplifiers consist of *two stages:*

— The *first stage* offers very high input impedance to both input signals and allows to set the gain with a single resistor.

— The *second stage* is a differential amplifier (unity gain) with output, negative feedback and ground connections all throughout.

HIGHLIGHTS

1. When electricity flows through open space or vacuum as in the case of lightning or vacuum tubes instead of being confined to metallic conductors, it is termed as *electronic.*

2. *Semiconductors* are solid materials, either non-metallic elements or compounds, which allow electrons to pass through them so that they conduct electricity in much the same way as a metal.
3. A pure semiconductor is called *instrinsic semiconductor.*
4. The process of adding impurity (extremely in small amounts, about 1 part in 10^8) to a semiconductor to make it extrinsic (impure) semiconductor is called *doping.*
5. The *N*- and *P*-type materials represent the basic building blocks of semiconductor devices.
6. The outstanding property of *P-N* junction diode to conduct current in one direction only permits it to be used as a *rectifier.*
7. *P-N* junction diodes usually made of germanium or silicon, are commonly used as *power rectifiers.*
8. A properly doped *P-N* junction diode which has a sharp breakdown voltage is known as *Zener diode.*
9. *Tunnel diode* is a heavily doped *P-N* junction type germanium having an extremely narrow junction.
10. A *"transistor"* is a semiconductor device having both rectifying and amplifying properties.
11. The two basic types of transistors are :
    (*i*) Bipolar junction transistor (BJT)
    (*ii*) Field-effect transistor (FET).
12. *Transistor circuit configurations :*
    (*i*) Common-base (CB) configuration
    (*ii*) Common-emitter (CE) configuration
    (*iii*) Common-collector (CC) configuration.
13. A *FET* is a three terminal (namely drain, source, gate) semiconductor device in which current conduction is by only one type of majority carriers (electrons in case of an *N*-channel FET or holes in a *P*-channel *FET*).
14. In a broad sense, following are two main types of FETs :
    (*i*) JFET (*ii*) MOSFET
15. Metal oxide semiconductor FET (MOSFET) is an important semiconductor device and is widely used in many circuit applications. It is also called insulated gate FET (IGFET).
16. The term SCR is often used for the member of the thyristor family which is the most widely used *power-switching* device.
17. A *rectifier* is a circuit which uses one or more diodes to convert A. C. voltage into pulsating D. C. voltage. A rectifier may be half-wave or full-wave.
18. The ratio of D.C. power output to the applied A.C. input power is known as *rectifier efficiency.*
19. The ratio of D.C. power output to the applied A.C. input power is known as *rectifier efficiency.*
20. The A.C. component present in the output is called a *ripple.*
21. The branch of electronics which deals with digital circuits is called *digital electronics.*
22. An electronic circuit that handles only a digital signal is called a *digital circuit.* In digital circuits the following four systems of arithmetic are often used :
    Decimal, Binary, Octal, Hexadecimal.
23. A digital circuit with one or more input signals but only one output signal is called a *logic gate.*
    In the complex circuits, the following *six* different digital electronics gates are used as basic elements :
    (*i*) NOT gate (*ii*) NAND gate
    (*ii*) AND gate (*iv*) OR gate

(v) NOR gate (vi) XOR gate.

24. The algebra used to symbolically describe logic functions is called *Boolean algebra.*

25. A *combinational circuit* consists of logic gates whose outputs at any time are determined directly from the combination of inputs without regard for previous input.

26. The *synchronous sequential circuits* are built to operate at *a clocked rate* whereas *asynchronous ones* are *without clocking*.

27. The memory elements used in clocked sequential circuits are called *flip-flops.*

28. A *counter* is a sequential circuit that goes through a prescribed sequence of states upon the application of input pulses.

29. A *integrated circuit* (IC) is a complete electronic circuit in which both the active (*e.g.* transistors and FETs) and passive components (e.g. resistors, capacitors and inductors) are fabricated on a tiny single chip of silicon.

30. The number of electronic circuits or components, which can be fabricated on a standard size of a silicon chip is called the *scale or level of integration.*
The various types of scale of integration are : SSI, MSI, LSI, VLSI, ULSI, GSI.

31. The *ICs* can be classified as follows :
(*i*) Monolithic integrated circuits.
(*ii*) Thick and thin-film integrated circuits.
(*iii*) Hybrid or multichip integrated circuits.

Or

(*i*) Linear (or analog) integrated circuits.
(*ii*) Non-linear (or digital) integrated circuits.

OBJECTIVE TYPE QUESTIONS

Choose the correct answer :

P-N junction diode

1. A *P-N* junction diode has ....
(*a*) one *P-N* junction (*b*) two *P-N* junctions
(*c*) three *P-N* junctions (*d*) none of these.

2. A crystal diode has forward resistance of the order of ......
(*a*) kΩ (*b*) Ω
(*c*) MΩ (*d*) none of these.

3. The schematic symbol for a *P–N* junction diode is
(*a*) 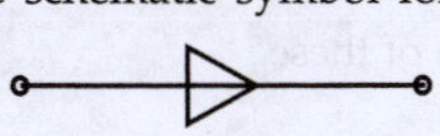
(*b*) 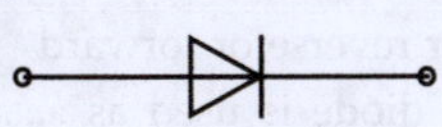
(*c*) 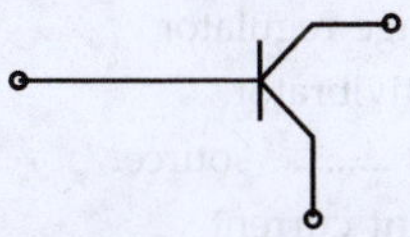
(*d*) 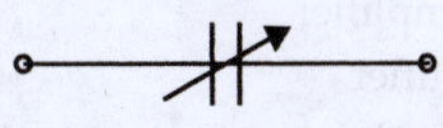

4. The resistance of a diode is equal to
(*a*) ohmic resistance of *P*- and *N*- semiconductors
(*b*) junction resistance
(*c*) reverse resistance
(*d*) algebraic sum of (*a*) and (*b*).

5. An ideal crystal diode is one which behaves as a perfect .... when forward biased.
(*a*) conductor (*b*) insulator
(*c*) resistance material (*d*) algebraic sum of (*a*) and (*b*)

6. A crystal diode is ....device.
   (*a*) non-linear (*b*) bilateral
   (*c*) linear (*d*) none of the these.
7. When a diode is reverse biased, it is equivalent to
   (*a*) an OFF switch (*b*) an ON switch
   (*c*) a high resistance (*d*) none of these.
8. The reverse current in a diode is of the order of ....
   (*a*) kA (*b*) mA
   (*c*) μA (*d*) A.
9. The conventional current in a P-N junction diode flows
   (*a*) from positive to negative (*b*) from negative to positive
   (*c*) in direction opposite to the electron flow (*d*) both (*a*) and (*c*) above.
10. The leakage current in a crystal diode is due to
    (*a*) minority carriers (*b*) majority carriers
    (*c*) junction capacitance (*d*) none of the above.
11. The cut in voltage (or knee voltage) of a silicon diode is
    (*a*) 0.2 V (*b*) 0.6 V
    (*c*) 0.8 V (*d*) 1.0 V.
12. A crystal diode is used as ....
    (*a*) an amplifier (*b*) a rectifier
    (*c*) an oscillator (*d*) A voltage regulator.
13. If the doping level of a crystal diode is increased, the breakdown voltage ....
    (*a*) remains the same (*b*) is increased
    (*c*) is decreased (*d*) none of the above.
14. The knee voltage of a crystal diode is approximately equal to ....
    (*a*) applied voltage (*b*) breakdown voltage
    (*c*) forward voltage (*d*) barrier potential
15. A Zener diode has ....
    (*a*) one *P-N* junction (*b*) two*P-N* junctions
    (*c*) three *P-N* junctions (*d*) none of these.
16. A Zener diode is always .... connected.
    (*a*) reverse (*b*) forward
    (*c*) either reverse or forward (*d*) none of these.
17. A Zener diode is used as ....
    (*a*) an amplifier (*b*) a voltage regulator
    (*c*) a rectifier (*d*) a multivibrator.
18. In the breakdown region, a Zener diode behaves like a .... source.
    (*a*) constant voltage (*b*) constant current
    (*c*) constant resistance (*d*) none of these.
19. The doping level in a Zener diode is .... that of a crystal diode.
    (*a*) the same as (*b*) less than
    (*c*) more than (*d*) none of these.
20. A Zener diode is .... device.
    (*a*) a non-linear (*b*) a linear
    (*c*) an amplifying (*d*) none of these.

Transistors (BJP, FET, etc.)

21. In P-N-P transistor, base will be of
(*a*) P material (*b*) N material
(*c*) either of the above (*d*) none of these.

22. In a transistor symbol, slant line to bar without any arrow head represents
(*a*) emitter (*b*) base
(*c*) collector (*d*) none of these.

23. In a transistor symbol, slant line to the bar with arrow head represnts
(*a*) emitter (*b*) base
(*c*) collector (*d*) none of these.

24. In a transistor highly doped part is
(*a*) emitter (*b*) base
(*c*) collector (*d*) none of these.

25. In a transistor lightly doped part is
(*a*) emitter (*b*) base
(*c*) collector (*d*) none of these.

26. In a transistor largest dimension is that of
(*a*) emitter (*b*) base
(*c*) collector (*d*) none of these.

27. A dot near the transistor pin denotes
(*a*) emitter (*b*) base
(*c*) collector (*d*) none of these.

28. In a transistor symbol, if slant line arrow head is drawn towards the bar, then the transistor is.
(*a*) P-N-P (*b*) N-P-N
(*c*) either of these (*d*) none of these.

29. A notch or a tab on the transistor cap denotes
(*a*) emitter pin (*b*) base pin
(*c*) collector pin (*d*) none of these.

30. Resistance across which of the following two pairs of transistor be nearly equal?
(*a*) Emitter base, emitter collector (*b*) Emitter base, base collector
(*c*) Base collector, collector emitter (*d*) None of these.

31. Which of the following is valid for both P-N-P as well as N-P-N transistors?
(*a*) The emitter injects holes into the base region
(*b*) The electrons are the minority carriers in the base region
(*c*) The EB junction is forward biased for active operation
(*d*) When biased in the active region, current flows into emitter terminal.

32. In a transistor with normal bias
(*a*) the emitter junction offers high resistance (*b*) the emitter junction is reverse biased
(*c*) the emitter junction has a low resistance (*d*) none of the above.

33. Which region of a transistor is lightly doped?
(*a*) Collector (*b*) Base
(*c*) Emitter (*d*) All regions are equally doped.

34. Semiconductor is a material which
(*a*) allows one type of carriers to pass through it
(*b*) has conductivity greater than insulator

(*c*) allows current to flow in one direction but not in the opposite direction
(*d*) none of these.

35. ..... is the region of a transistor which has highest conductivity.
(*a*) Base (*b*) Emitter
(*c*) Collector (*d*) Any of the above.

36. Bipolar transistor is a
(*a*) three terminal semiconductor device (*b*) three layer semiconductor device
(*c*) three junction semiconductor device (*d*) none of these.

37. Silicon controlled rectifier belongs to
(*a*) diode family (*b*) triode family
(*c*) thyristor family (*d*) non of these.

38. Current flow through a bipolar transistor is by means of
(*a*) electrons (*b*) holes
(*c*) both electrons and holes (*d*) none of these.

39. Transistor works as an open switch when emitter junction is .....biased and collector junction is ..... biased.
(*a*) forward, reverse (*b*) reverse, reverse
(*c*) reverse, forward (*d*) forward, forward.

40. Transistor works as a closed switch when emitter junction is ..... biased and collector junction is .....biased.
(*a*) forward, reverse (*b*) reverse, reverse
(*c*) reverse, forward (*d*) forward, forward.

41. Transistor works as a variable rheostat when emitter junction is ..... biased and collector junction is ..... biased.
(*a*) forward, reverse (*b*) reverse, reverse
(*c*) reverse, forward (*d*) forward, forward.

42. With two diodes connected back to back with emitter diode forward biased, and collector diode reverse biased
(*a*) emitter and collector currents are nearly equal and base current is very small
(*b*) emitter and base currents are nearlyequal and collection current is very small
(*c*) base and collector currents are nearly equal and emitter current is large
(*d*) any of the above.

43. Current base part of a transistor behaves like
(*a*) constant current source (*b*) forward biased diode
(*c*) a resistance (*d*) none of the above.

44. Majority carriers emitted by the emitter
(*a*) mostly recombine in base region
(*b*) mostly pass through the base region
(*c*) are stopped by the collector junction barrier
(*d*) recombine in the collector region.

45. The following relationship between α and β are correct *except*
(*a*) $1 - \alpha = \dfrac{1}{1+\beta}$ (*b*) $\alpha = \dfrac{\beta}{1+\beta}$
(*c*) $\beta = \dfrac{\alpha}{1-\alpha}$ (*d*) $\alpha = \dfrac{\beta}{1-\beta}$

46. A P-N-P transistor has
(*a*) only acceptor ions (*b*) only donor ions

(*c*) two-P-regions and one N-region (*d*) three P-N junctions.

47. Regarding common emitter configuration which of the following statements is *incorrect*?

(*a*) Its output resistance is very high.

(*b*) It is the only circuit which has voltage and current gains higher than unity.

(*c*) Its power gain is the best.

(*d*) It is the only configuration which provides inversion.

48. The active region of the output characteristics for a common base transistor is that in which

(*a*) emitter is forward-biased but collector is reverse-biased

(*b*) both emitter and collector are forward-biased

(*c*) collector is forward-biased and emitter is reverse-biased

(*d*) both emitter and collector are reversed-biased.

49. The set of transistor characteristics that enables α to be determined directly from the slope is ..... characteristics.

(*a*) common emitter transfer (*b*) common emitter output

(*c*) common base transfer (*d*) common base input.

50. When a common emitter transistor is cut off which of the following happens?

(*a*) Maximum voltage appears across the collector.

(*b*) Maximum collector current flows.

(*c*) Minimum voltage appears across the collector.

(*d*) Miximum voltage appears across the load resistor.

51. In amplifier circuit, biasing of transistor is necessary to

(*a*) fix the value of current amplification

(*b*) establish suitable D.C. workig conditions

(*c*) ensure that transistor is saturated

(*d*) ensure that transistor is cut off.

52. The configuration in which current gain of transistor amplifier is lowest is

(*a*) common base (*b*) common collector

(*c*) common emitter (*d*) any of these.

53. The configuration in which voltage gain of transistor amplifier is lowest is

(*a*) common base (*b*) common collector

(*c*) common emitter (*d*) any of these.

54. The configuration in which input impedance of transistor amplifier is lowest is

(*a*) common base (*b*) common collector

(*c*) common emitter (*d*) any of these.

55. The configuration in which output impedance of transistor amplifier is highest is

(*a*) common base (*b*) common collector

(*c*) common emitter (*d*) any of these.

56. The configuration which provides both high current gain and high voltage gain of transistor amplifier is

(*a*) common base (*b*) common collector

(*c*) common emitter (*d*) any of these.

57. A transistor-terminal current is positive when the

(*a*) current is due to flow of electrons (*b*) current is due to flow of holes

(*c*) electrons flow into the transistor at the terminal

(*d*) electrons flow out of the transistor at the terminal.

58. A transistor is said to be in quiescent state when

(*a*) no signal is applied to the input (*b*) no currents are flowing

(*c*) it is unbiased

(*d*) emitter junction and collector-junction biases are equal.

59. The most noticeable effect of a small increase in temperature in the CE transistor is the

(*a*) increase in the A.C. current gain (*b*) decrease in the A.C. current gain

(*c*) increases in I_{CED} (*d*) increases in the output resistance.

60. FETs have similar properties to

(*a*) thermionic valves (*b*) unijunction transistor

(*c*) P-N-P transistors (*d*) N-P-N transistors.

61. A JFET can operate in

(*a*) depletion mode only (*b*) enhancement mode only

(*c*) depletion and enhancement modes

(*d*) neither enhancement nor depletion mode.

62. In a JFET ..... is usually the point of reference.

(*a*) gate (*b*) drain

(*c*) source (*d*) either (*b*) or (*c*).

63. The primary control on drain current in a JFET, is exerted by which of the following?

(*a*) Gate reverse bias (*b*) Channel resistance

(*c*) Voltage drop across channel (*d*) Size of depletion regions.

64. For the operation of enhancement-only N-channel MOSFET, value of gate voltage has to be

(*a*) zero (*b*) low positive

(*c*) high positive (*d*) high negative.

65. Which of the following statements is *correct* regarding a JFET operating above pinch-off voltage?

(*a*) The depletion regions become smaller.

(*b*) The drain current starts decreasing.

(*c*) The drain current remains practically constant.

(*d*) The drain current increases steeply.

66. A FET can be used as a variable

(*a*) inductor (*b*) capacitor

(*c*) resistor (*d*) voltage source

(*e*) current source.

67. In FET the drain voltage above which there is no increases in the drain current is called ..... voltage

(*a*) pick off (*b*) pinch off

(*c*) breakdown (*d*) critical.

68. The operation of JFET involves a flow of

(*a*) minority carriers (*b*) majority carriers

(*c*) recombination carriers (*d*) any of the above.

69. When the positive voltage on the gate of a P-channel JFET is increased, the drain current will

(*a*) increase (*b*) decrease

(*c*) remain the same (*d*) any of the above.

70. A FET differs from a bipolar transistor as it has
(*a*) simpler fabrication (*b*) negative resistance
(*c*) high input impedance (*d*) any of the above.

71. A FET, for its operation, depends on the variation of
(*a*) forward-biased junction (*b*) reversed-biased junction
(*c*) magnetic field (*d*) The depletion-layer width with reverse voltage.

72. N-channel FETs are superior to P-channel FETs because
(*a*) they have a higher switching time
(*b*) they have a higher input impedance
(*c*) mobility of electrons is greater than that of holes
(*d*) all of the above.

73. UJT is also called a
(*a*) transistorized junction (*b*) voltage controllable device
(*c*) current controllable device (*d*) relaxation oscillator.

74. The difference between a thyristor and a silicon diode is that the thyristor
(*a*) conducts when it is triggered in addition to being forward biased
(*b*) conducts when it is forward biased
(*c*) blocks when it is reverse biased
(*d*) none of the above.

75. On stopping the gate pulse to a SCR
(*a*) it will stop conduction
(*b*) it will continue conduction in the same direction
(*c*) it will continue conduction in the opposite direction
(*d*) none of the bove will happen.

76. SCR is used for current control in
(*a*) D.C. circuit only (*b*) A.C. circuit only
(*c*) both (*a*) and (*b*) (*d*) none of these.

77. SCR is
(*a*) three layer two terminal device (*b*) three layer three terminal device
(*c*) four layer two terminal device (*d*) four layer three terminal device.

78. A triac is a ..... switch.
(*a*) undirectional (*b*) bi-directional
(*c*) either of these (*d*) none of these.

79. A diac is ..... switch.
(*a*) an A.C. (*b*) D.C.
(*c*) either of these (*d*) none of these.

80. The normal way to turn on a diac is by
(*a*) gate current (*b*) breakover voltage
(*c*) either of these (*d*) none of these.

81. A diac is equivalent to a
(*a*) triac with two gates (*b*) diode and two resistors
(*c*) pair of SCRs (*d*) pair of four-layer SCRs.

82. Regarding triac which of the following statements is *incorrect*?
(*a*) It is not particularly suited for A.C. or mains power control.
(*b*) It is a 5-layer bi-directional semiconductor device.

(*c*) Any one of its main terminals can be used either as cathode or as anode.

(*d*) It can be triggered in response to both positive and negative gate terminals.

83. ..... is the best electronic device for fast switching.

(*a*) MOSFET (*b*) JFET

(*c*) BJT (*d*) Triode.

84. Which semiconductor device acts like a diode and two resistors ?

(*a*) UJT (*b*) SCR

(*c*) diac (*d*) triac.

85. ..... is the device which acts like an N-P-N and P-N-P transistor connected base-to-base and emitter-to-collector.

(*a*) SCR (*b*) UJT

(*c*) diac (*d*) triac.

86. An SCR may be considered to be diodes back-to-back consisting of an anode, cathode and .....

(*a*) two, plate (*b*) three, plate

(*c*) three, gate (*d*) four, base.

87. A LASCR in just like a conventional SCR except that it

(*a*) has no gate terminal (*b*) can also be light-triggered

(*c*) cannot carry large current (*d*) cannot be pulse-triggered.

88. Which semiconductor device behaves like two SCRs?

(*a*) MOSFET (*b*) JFET

(*c*) UJT (*d*) Triac.

89. An SCR conducts appreciable current when

(*a*) gate is negative and anode is positive with respect to cathode

(*b*) anode is negative and gate is positive with respect to cathode

(*c*) anode and gate are both negative with respect to cathode

(*d*) anode and gate are both positive with respect to cathode.

90. An SCS has which of the following?

(*a*) One anode, one cathode and two gates (*b*) Two anodes and two gates

(*c*) Four layers and three terminals (*d*) Three layers and four terminals.

91. An SCS may be switched ON by a

(*a*) positive pulse at its anode gate G_1 (*b*) positive pulse at its cathode gate G_2

(*c*) negative pulse at its cathode (*d*) positive pulse at its anode.

92. Which of the following methods used for protecting MOSFET against damage from stray voltage developing at the gate is *incorrect*?

(*a*) Only source terminal is earthed during transit.

(*b*) Bach-to-back Zener diodes are formed into the monolithic structure of MOSFET.

(*c*) Grounding rings are used which are removed only when it is wired securely into the circuit.

(*d*) It is inserted into conducting sponge during visit.

93. Regarding MOSFET which of the following statements is *incorrect*?

(*a*) It can operate in depletion mode.

(*b*) It can operate in enhancement mode.

(*c*) It can operate in depletion and enhancement modes.

(*d*) It can operate in depletion-only-mode.

(*e*) It can operate in enhancement-only mode.

94. The main factor which differentiates a DE MOSFET from an E-only MOSFET is the absence of

(*a*) P-N junctions (*b*) insulated gate
(*c*) electrons (*d*) channel.

95. The input gate current of a FET is
(*a*) a few amperes (*b*) a few milliamperes
(*c*) a few microamperes (*d*) negligibly small.

96. Silicon devices are preferred at high temperature operations as compared to germanium because
(*a*) silicon can dissipate more power
(*b*) reverse saturation current is less in case of silicon
(*c*) silicon is more thermally stable
(*d*) all of the above.

97. Hall effect can be used to measure
(*a*) carrier concentration (*b*) electric field intensity
(*c*) magnetic field intensity (*d*) none of the above.

98. Which of the following statements is *correct* in case of a properly biased transistor?
(*a*) The emitter to base depletion region is small and collector to base depletion region is large
(*b*) The emitter to base depletion region is large and collector to base deplection region is small
(*c*) both depletion regions are small
(*d*) both depletion regions are large.

99. Ebers-Moll equations for transistors provide
(*a*) true terminal currents regardless of junction biases
(*b*) true terminal voltages dependent on junction biases
(*c*) separate input and output circuits
(*d*) all of the above.

100. In the symbols of P-N-P transistors and N-P-N transistor the arrow on the emitter shows the direction of flow of
(*a*) electrons, electrons (*b*) holes, holes
(*c*) holes, electrons (*d*) electrons, holes.

ANSWERS

1. (*a*)	2. (*b*)	3. (*b*)	4. (*d*)	5. (*a*)	6. (*a*)	7. (*a*)
8. (*c*)	9. (*d*)	10. (*a*)	11. (*b*)	12. (*b*)	13. (*c*)	14. (*d*)
15. (*a*)	16. (*a*)	17. (*b*)	18. (*a*)	19. (*c*)	20. (*a*)	21. (*b*)
22. (*c*)	23. (*a*)	24. (*a*)	25. (*b*)	26. (*c*)	27. (*c*)	28. (*a*)
29. (*a*)	30. (*b*)	31. (*c*)	32. (*c*)	33. (*b*)	34. (*b*)	35. (*b*)
36. (*b*)	37. (*c*)	38. (*c*)	39. (*b*)	40. (*d*)	41. (*a*)	42. (*b*)
43. (*a*)	44. (*b*)	45. (*d*)	46. (*c*)	47. (*a*)	48. (*a*)	49. (*c*)
50. (*a*)	51. (*a*)	52. (*a*)	53. (*b*)	54. (*a*)	55. (*a*)	56. (*c*)
57. (*d*)	58. (*a*)	59. (*c*)	60. (*a*)	61. (*a*)	62. (*c*)	63. (*a*)
64. (*c*)	65. (*c*)	66. (*c*)	67. (*b*)	68. (*b*)	69. (*b*)	70. (*c*)
71. (*d*)	72. (*c*)	73. (*b*)	74. (*a*)	75. (*b*)	76. (*c*)	77. (*d*)
78. (*b*)	79. (*a*)	80. (*b*)	81. (*d*)	82. (*a*)	83. (*c*)	84. (*a*)
85. (*a*)	86. (*c*)	87. (*a*)	88. (*d*)	89. (*d*)	90. (*a*)	91. (*b*)
92. (*a*)	93. (*d*)	94. (*d*)	95. (*d*)	96. (*b*)	97. (*c*)	98. (*a*)
99. (*a*)	100. (*b*).					

THEORETICAL QUESTIONS

1. Define a 'semiconductor'.
2. List the important characteristics of semiconductors.
3. Give examples of semiconducting materials.
4. What is the difference between a semiconductor and an insulator?
5. What is an intrinsic semiconductor?
6. What do you mean by the term *doping*?
7. How does an extrinsic semiconductor differ from an intrinsic semiconductor?
8. Explain the structure of a *P*-type semiconductor with help of neat sketches.
9. Explain briefly about 'atomic binding in semiconductors'.
10. How are holes formed in semiconductors?
11. Derive an expression for electron conductivity of a metal.
12. Derive expressions for conductivity of *N*-type and *P*-type semiconductors.
13. What do you mean by conductivity modulation ?
14. Explain briefly the following :
    (*i*) Thermistors and sensitors (*ii*) Photoconductors.
15. List the applications of semiconductor materials.
16. How is germanium prepared?
17. What is a *P-N* junction diode? How its terminals are identified?
18. Draw the *V-I* characteristics of a junction diode when it is (*a*) forward biased and (*b*) reverse biased.
19. Draw the graphical symbol of a crystal diode and explain its significance. How the polarities of function diode are identified ?
20. Draw the equivalent circuit of a crystal diode.
21. What is an ideal diode and a real diode?
22. Explain the following terms :
    (*i*) Static resistance
    (*ii*) Bulk resistance
    (*iii*) Junction resistance
    (*iv*) A.C. or dynamic resistance
    (*v*) Reverse resistance of a diode.
23. What are the important applications of a diode?
24. Write a short note on the power and current ratings of a diode.
25. What is a Zener diode ? Draw its equivalent circuit.
26. Explain briefly the applications of a Zener diode.
27. What do you understand by Zener voltage?
28. Explain why Zener diode is always operated in reverse biasing.
29. Explain how a Zener diode can stabilize the voltage across the load.
30. Explain the process of Zener breakdown.
31. Draw and explain a Zener diode voltage regulator.
32. Define the term 'Transistor'.
33. What are the various types of transistors?
34. Explain the function of emitter in the operation of a junction transistor.
35. What is the significance of arrow in the transistor symbol?
36. Why is emitter wider than collector and base?

37. Why is base made thin?
38. Draw *N-P-N* and *P-N-P* transistors.
39. Explain the working of a *P-N-P* transistor.
40. Differentiate between *P-N-P* and *N-P-N* transistors. Why are collector and emitter currents nearly equal in these transistors?
41. Define α and β of a transistor and derive the relationship between them.
42. Draw three basic configurations of *N-P-N* transistor.
43. Draw input and output characteristics of *CB* transistor configuration.
44. Draw the circuits of the various transistor configurations. List their important features. Why *CE* configuration is mainly used?
45. Explain the construction and working of a JFET.
46. What is the difference between a JFET and a BJT?
47. How will you determine the drain characteristics of JFET? What do they indicate?
48. What are the advantages and disadvantages of JFET?
49. What are the applications of FETs?
50. What is the difference between MOSFET and JFET?
51. Define the following terms for a JFET :
    (*i*) The pinch-off voltage.
    (*ii*) Channel ohmic resistance.
    (*iii*) Drain resistance.
52. Draw the *V-I* characteristics of an *N*-channel FET.
53. Discuss briefly, the construction, working, characteristics and applications of SCR.
54. Explain the forward and reverse characteristics of a thyristor.
55. What is the difference between SCR and Triac?
56. List the applications of thyristor?
57. Describe a half-wave rectifier using a crystal diode.
58. Derive and expression for the efficiency of a half-wave rectifier.
59. With neat sketch, explain the working of the following :
    (*i*) Centre-tapped full-wave rectifier.
    (*ii*) Full-wave bridge rectifier.
60. Derive an expression for the efficiency of a full-wave rectifier.
61. What is a ripple factor? What is its value for a half-wave and a full-wave rectifier?
62. What is 'digital electronics'?
63. State the advantages and disadvantages of digital electronics?
64. What is a 'digital circuit'?
65. Why binary system is preferred in 'digital system'?
66. Discuss the importance of 1's and 2's complement numbers.
67. Explain the Gray code and alphanumeric codes.
68. What is meant by a radix (or base of a number system)?
69. Draw the diagram of a clocked R-S flip-flop and give the truth table.
70. Show that a R-S flip-flop results when two NOR gates are cross-coupled.
71. What is a flip-flop? Explain the principle of operation of S-R flip-flop with truth table.
72. With the aid of a neat sketch, explain the operation of J-K flip-flop.
73. Briefly describe J-K, D- and T-type flip-flops.
84. Write short notes on "logic families".

CHAPTER

3 Concepts of Measurements and Measurement Systems

3.1. Introduction to measurement; 3.2. Significance of measurement; 3.3. Methods of measurement; 3.4. Modes of measurement; 3.5. Generalised measurement system and its functional elements; 3.6. Input-output configurations of measuring instruments and measuring systems; 3.7. Applications of measurement systems; 3.8. Static and dynamic characteristics of instruments; 3.9. Errors in measurements. *Highlights – Objective Type Questions – Theoretical Questions – Unsolved Examples.*

3.1 INTRODUCTION TO MEASUREMENT AND INSTRUMENTATION

3.1.1. Measurement

Measurement *is the act, or the result of a quantitative comparison between a predetermined standard and an unknown magnitude.* The procedure and apparatus employed for obtaining the comparison, however, must be *provable*; the procedure for this is called *calibration.*

Probably it has rightly been said: "*Whatever exists, exists in some amount*". The determination of the amount is what measurement is all about.

Measurand. *The physical quantity or the characteristic condition which is the object of measurement in an instrumentation system is variously termed as "**measurand**", "measurement variable", "instrumentation variable" or "process variable".*

The measurand may be :

— Fundamental quantity, *e.g.,* length, mass and time;

— Derived quantity, *e.g.,* speed, velocity, acceleration, pressure.

The word *measurand* is used to designate the particular physical parameter being observed and quantified; that is, the *input quantity* to the measuring process. The act of measurement produces a *result.* Fig. 3.1 shows a *fundamental measuring process.*

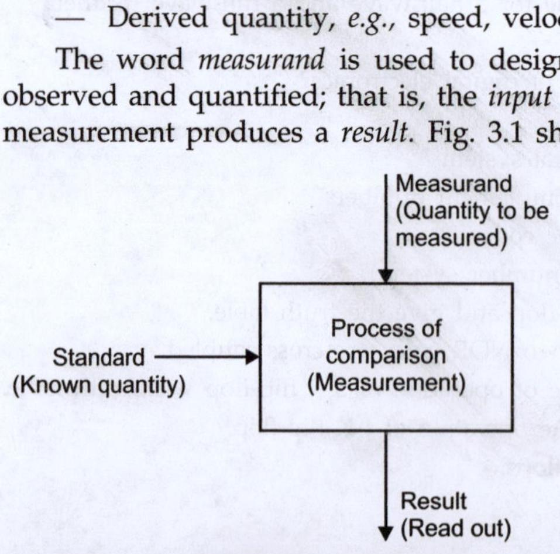

Fig. 3.1. Fundamental measuring process.

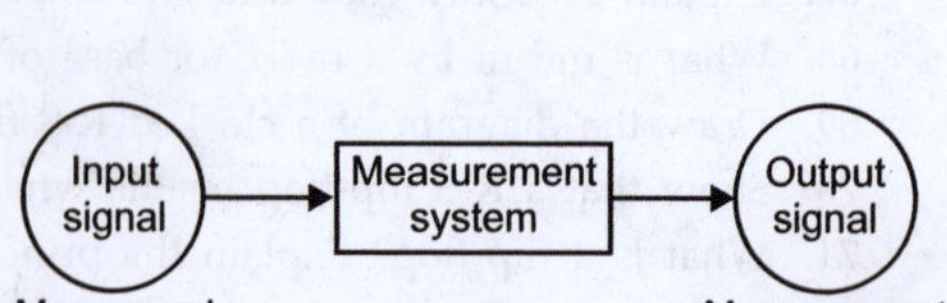

Fig. 3.2. Measurand and measurement.

— Whereas the input to a measuring system is known as *measurand*, the output is called *measurement* (See Fig. 3.2).

The *basic requirements* for getting meaningful result of measurement are:

(*i*) The standard employed for comparison purpose *must be accurately defined* and should be *commonly acceptable.*

(*ii*) The standard must be of the *same character* as the measurand and usually, but not always, is prescribed and defined by a legal or recognized agency or organisation, *e.g.*, Indian Bureau of Standards (IBS) or the International Organisation of Standards (ISO).

(*iii*) The apparatus used and method adopted for the comparison purposes must be *provable.*

- Several measurands of interest *vary with time in a continuous manner over a range of magnitude.*

For example, the speed of the automobile, as it starts from rest, has some magnitude at every instant during its motion. A sensor that responds to velocity will produce an output signal having a line variation analogous to the time change in automobile's speed. Such a signal is referred to as **analog signal** *because it is analogous to a continuous physical process.*

The analog signal, which has a value at every instant of time, becomes a **digital signal.**

— A digital signal is a *set of discrete numbers each corresponding to the value of the analog signal at a single specific instant of time.*

— The digital signal contains *no information* about the value of the analog signal at time *other than sample times.*

The mechanical quantities such as pressure, stress, strain, temperature, fluid-flow rates behave time-wise in an analog manner. However, by converting an analog signal to an equivalent digital signal for the purpose of signal conditioning and/or read out oftenly following distinct *advantages* are obtained:

(*i*) *Noise problems are reduced or sometimes eliminated completely.*

(*ii*) *Data transmission becomes simpler.*

(*iii*) *Computers are designed to process digital information, and direct numerical display or recording is more easily accomplished by manipulating digital quantities.*

3.1.2. Instrumentation

The technology of using instruments to measure and control the physical and chemical properties of materials is called ***Instrumentation.***

- When the instruments are used for the *measurement and control of industrial manufacturing, conversion or treatment process,* the term **"process instrumentation"** is used.
- *When the measuring and controlling instruments are combined so that measurements provide impulses for remote automatic action, the result is called a* ***Control system.***

3.2 SIGNIFICANCE OF MEASUREMENT

The significance of measurements is discussed below:

- Measurements play a very significant role in every branch of scientific research and engineering processes which include the following:

  — Control systems;

— Process instrumentation;

— Data reduction.

- The whole area of automation or automatic control is based on measurements. The very concept of control is based on the comparison of the actual condition and the desired performance. *The exactness of error depends on the precision and accuracy of measurement made.*
- The measurements confirm the validity of a hypothesis and also add to it its understanding. This eventually leads to new discoveries that require new and sophisticated measuring techniques.
- Through measurements a product can be designed or a process be operated with maximum efficiency, minimum cost, and with desired degree of reliability and maintainability.

3.3 METHODS OF MEASUREMENT

The broad classification of methods of meaurement is as follows:

1. Direct comparison methods.
2. Indirect comparison methods.

1. Direct comparison method:

In this method the parameter to be measured is *directly compared with either a primary or a secondary standard.*

Direct comparison is quite commonly used for measurements of length. However, for measurement of mass the problem becomes much more intricate since it is just not possible for human beings to distinguish between wide margins of mass.

- Generally, this method is not always the most accurate or the best; it is not sensitive enough also.

2. Indirect comparison method:

In this method the *comparison* is *done with a standard through the use of a calibrated system.*

These methods for measurement are used in those cases where the desired parameter to be measured is *difficult to be measured directly, but it has got some correlation with some other parameter which can be easily measured.*

Example: The elimination of bacteria in milk is directly dependent upon its temperature. Thus the bacteria elimination can be measured indirectly by *measuring the temperature of the milk.*

— In indirect measurements an *empirical relation is generally established between the measurement actually made and the results that are desired.*

- When the primary purpose of making a measurement is to determine quality of a product, then quality should be measured *directly*. However, in case direct measurement is not possible, then *indirect* measurement should be made.
- *The measurement systems used in engineering applications make use of indirect methods for measurement purposes.*

— A measurement system consists of a *transducing element* which converts the quantity to be measured with an *analogous signal*. This signal after being processed by some *intermediate means* is then fed to the end devices which provide the measurement results.

3.4 MODES OF MEASUREMENT

Following are the *three* modes of measurement :

1. Primary measurements.
2. Secondary measurements.
3. Tertiary measurements.

1. Primary measurements :

- In this case the *sought value of a parameter is determined by comparing it directly with "reference standards"*.
- There is *no conversion* of measurand in terms of length.

Examples :

(*i*) Measurement of time by counting the number of strokes of a clock.
(*ii*) Matching of two lengths when determining the length of an object with a ruler.
(*iii*) Matching of two colours when judging the temperature of a red hot steel.

2. Secondary measurements :

The indirect measurements involving 'one translation' are called ***secondary measurements.***

Examples :

(*i*) The pressure measurement by manometers.
(*ii*) The temperature measurement by mercury-in-glass thermometers.

3. Tertiary measurements:

The indirect measurements involving 'two conversions' are called ***tertiary measurements.***

Examples :

(*i*) The measurement of temperature of an object by thermocouple (Fig. 3.3). The primary signal (temperature of object) is transmitted to a thermocouple which generates a *voltage* which is a function of the temperature (first translation/ conversion). The voltage in turn is applied to a voltmeter through a pair of wires. The second conversion is then voltage into length. The tertiary signal is transmitted to brain of the observer.

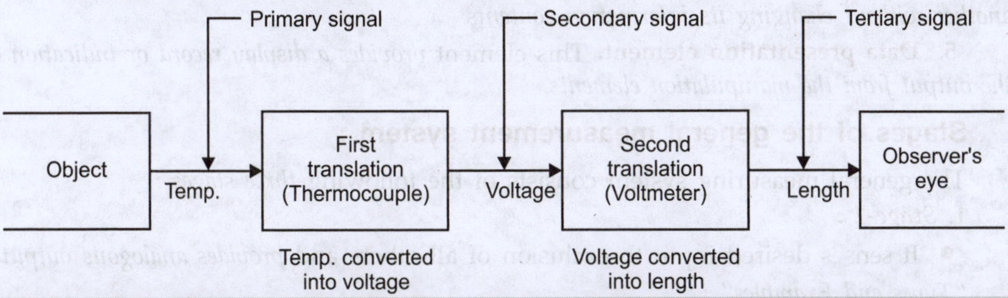

Fig. 3.3. Tertiary measurement : Measurement of temperature by a thermocouple.

(*ii*) The measurement of static pressure by a Bourdon tube pressure gauge.
(*iii*) The measurement of the speed of a rotating shaft by means of an electric tachometer.

The unit of a measuring system where translation of measurand takes place is called the *transducer or translator*.

- Majority of measurement systems are tertiary systems and they include the whole range of mechanical, electrical, pneumatic electro-mechanical and electro-pneumatic instruments.

Measurements may also be divided into following *two categories* :

1. **Contact type.** In this case the sensor of the measuring device *contacts* the controlled medium.
2. **Non-contact type.** Here the sensor *does not contact* the controlled medium.

- Non-contact measurements include *optical, radioactive and others.*

3.5 GENERALISED MEASUREMENT SYSTEM AND ITS FUNCTIONAL ELEMENTS

Fig. 3.4 shows a measurement system (generalized) with different components (called *elements*).

The various elements are :

1. **Primary sensing element.** It is an element that is sensitive to the measured variable. The sensing elements *sense the condition, state or value of the process variable by extracting a small part of energy from the measurand, and then produces an output which reflects this condition, state or value of the measurand.*

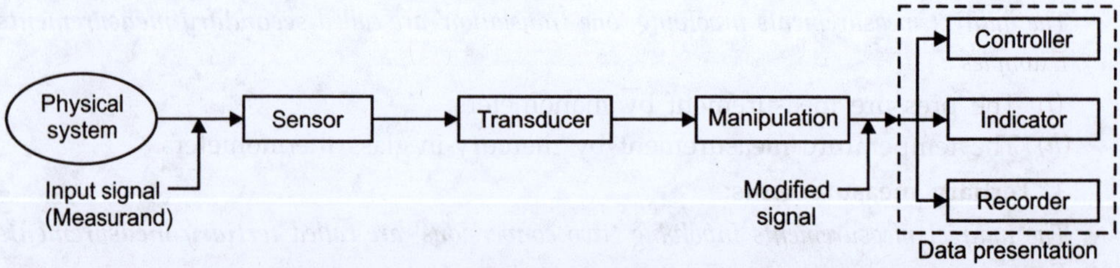

Fig. 3.4. Generalised measurement system.

2. **Variable conversion of transducer element.** This element *converts the signal from one physical form to another* without changing the information content of the signal.
3. **Manipulation element.** This element *operates* on the signal according to some *mathematical rule without changing the physical nature of the variable.*
4. **Data transmission element.** This element *transmits the signal from one location to another without changing its information contents.*
5. **Data presentation element.** This element *provides a display record or indication of the output from the manipulation elements.*

Stages of the general measurement system :

The general measuring system consists of the following *three stages* :

1. *Stage-1 :*

- It senses desired input to exclusion of all others and *provides analogous output.*

"Types and Examples" :

Mechanical : Contacting spindle, spring-mass, elastic devices (*e.g.,* Bourdon tube for pressure, proving ring for force), gyro.

Electrical : Contacts, resistance, capacitance, inductance, piezoelectric crystals and polymers, thermocouple, semiconductor junction, etc.

Optical : Photographic film, photoelectric diodes and transistors, photomultiplier tubes, holographic plates.

Hydro-pneumatic : Buoyant float, orifice, venturi, vane, propeller.

2. *Stage-2 :*

- *It modifies transduced signal into form usable by final stage.*
- It *usually increases amplitude and/or power* depending on requirement.
- It may also *selectively filter unwanted component or convert signal into pulsed form.*

"Types and Examples" :

Mechanical : Gearing, cranks, slides, connecting links, cams, etc.

Electrical : Amplifying or attenuating systems, bridges, filters, telemetering systems, various special purpose I.C. (Integrated circuit) devices.

Optical : Mirrors, lenses, optical filters, optical fibres, spatial filters (pin hole, slit).

Hydro-pneumatic: Piping, valving, dashpots, plenum chambers.

3. *Stage-3 :*

- It *provides an indication or recording* in form that can be evaluated by an unaided human sense or by a controller.
- It *records data* digitally on a computer.

"Types and Examples" :

Indicators Displace types : Moving pointer and scale, moving scale and index, light beam and scale, electron beam and scale (oscilloscope), liquid column.

Digital types : Direct alphanumeric readout.

Recorder : Digital printing, inked pen and chart, direct photography, magnetic recording (hard disk or type).

Processor and computers : Various types of computing systems, either special purpose or general, used to feed read out/recording devices and/or controlling systems.

Controllers : All types.

Illustration of measuring systems :

I. Bourdon tube pressure gauge:

Fig. 3.5 shows the simple Bourdon tube pressure gauge – an example of a simple measurement system.

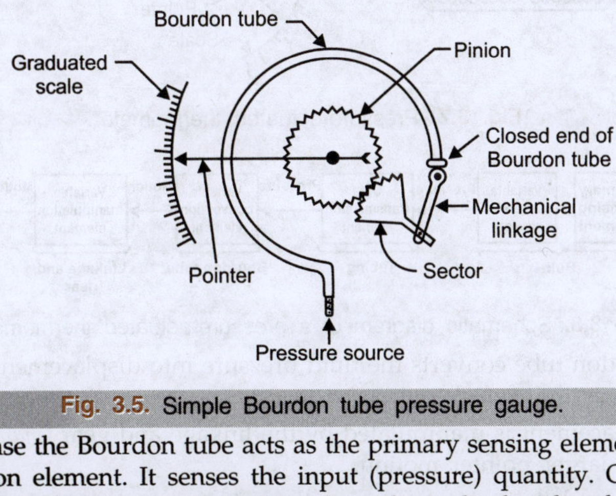

Fig. 3.5. Simple Bourdon tube pressure gauge.

— In this case the Bourdon tube acts as the primary sensing element and a variable conversion element. It senses the input (pressure) quantity. On account of the pressure, the closed end of the Bourdon tube is displaced and thus the pressure is converted into a small displacement.

— The closed end of the Bourdon tube is connected to a gearing arrangement through mechanical linkage. The gearing arrangement *amplifies* the small displacement and consequently the pointer rotates through a large angle. Thus, the mechanical linkage acts as a data transmission element while the gearing arrangement acts as a data manipulation element.

— The final data presentation stage consists of a pointer and dial arrangement which, when calibrated with known pressure inputs, gives an indication of the pressure signal applied to the Bourdon tube.

Fig. 3.6 shows the schematic diagram of this measurement system.

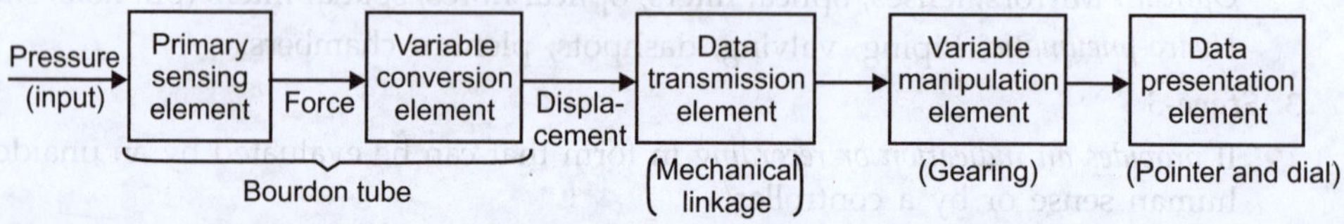

Fig. 3.6. Schematic diagram of a Bourdon tube pressure gauge.

II. Pressure-actuated thermometer :

Fig. 3.7 shows a pressure-actuated thermometer.

— The liquid-filled bulb acts as the *primary sensor* and *variable conversion element* since a temperature change results in pressure build-up within the bulb, due to the constrained thermal expansion of filling liquid.

— The pressure tubing is employed to transmit the pressure to the Bourdon tube and thus functions as the *data transmission element.*

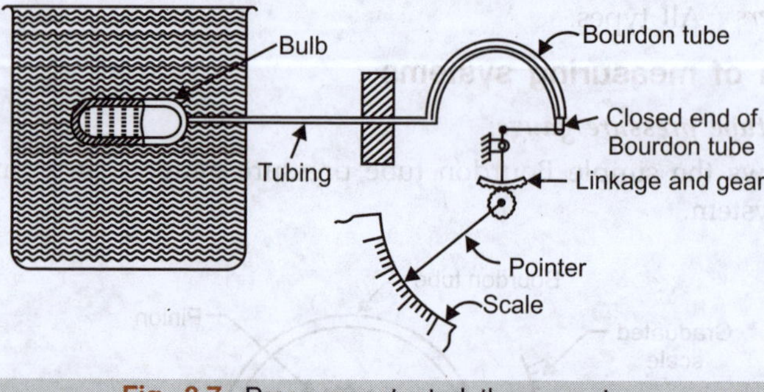

Fig. 3.7. Pressure-actuated thermometer.

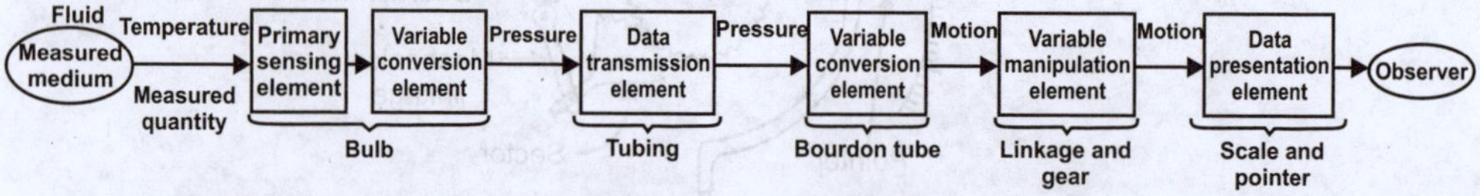

Fig. 3.8. Schematic diagram of a pressure-actuated thermometer.

— The Bourdon tube converts the fluid pressure into displacement of its tip, and as such acts as the *variable conversion element.*

— The displacement is manipulated by the linkage and gear (*manipulation elements*) to give a larger pointer motion.

— The scale and the pointer serve as the *data presentation element.*

Fig. 3.8 shows the schematic diagram of a pressure-actuated thermometer.

3.6 INPUT-OUTPUT CONFIGURATIONS OF MEASURING INSTRUMENTS AND MEASURING SYSTEMS

Fig. 3.9 shows the generalised input-output configuration of measurement systems. The various inputs to the measurement system are *classified* as follows :

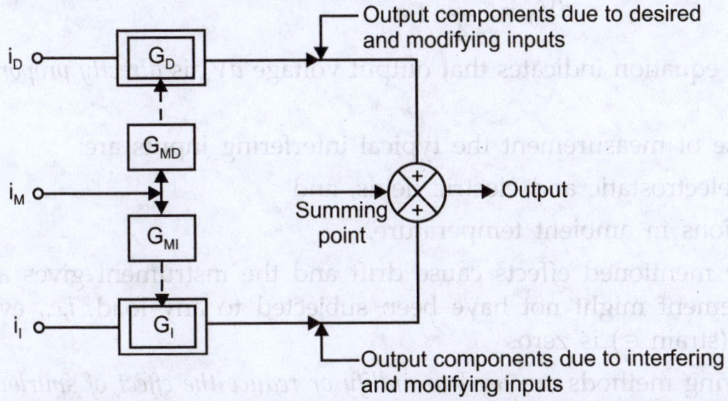

Fig. 3.9. Generalised input-output configuration of measurement systems.

1. Desired inputs.
2. Interfering inputs.
3. Modifying inputs.

1. Desired inputs :

The quantities for which the instrument or the measurement system is specifically designed to measure and respond are called ***desired inputs.***

The desired input i_D produces an output component according to an input-output relation symbolised by G_D; here G_D (*Transfer function*) represents the mathematical operation necessary to obtain the necessary output from the input.

- The transfer function G_D is a representative of a wide range of functions from a *constant in the case of linear systems to a statistical function used for statistical measurements.*

2. Interfering inputs :

The quantities to which an instrument or a measurement system becomes unintentionally sensitive are called ***interfering inputs.***

The interfering inputs i_I would produce an output component according to input-output relation symbolised by G_I.

3. Modifying inputs :

The inputs which cause a change in input-output relationships for either desired input or interfering input or for both are called ***modifying inputs.***

The modifying input i_M would cause a change in G_D and/or G_I. The specific manner in which i_M affects G_D and G_I is represented by G_{MD} and G_{MI} respectively.

For illustration, let us consider *strain measurement by a strain gauge.* When the gauge is strained due to mechanical loading, the change in output voltage is given by,

$$dV_0 = \frac{V_s}{4} \cdot \frac{dR}{R}$$

where, dV_0 = Output voltage,

V_s = Supply voltage (battery), and

dR = Change in resistance of the strain gauge which constitutes one arm of the Wheatstone bridge circuit (assuming, $R_1 = R_2 = R_3 = R_4 = R$)

In terms of strain $\in$ and strain gauge factor, F,

$$dV_0 = \frac{V_s}{4} \cdot F \in$$

The above equation indicates that output voltage dV_0 is *directly proportional to applied strain.*

In this case of measurement the typical interfering inputs are:

- Stray electrostatic and electric fields, and
- Variations in ambient temperature.

The above mentioned effects cause drift and the instrument gives an output even though the element might not have been subjected to any load, *i.e.*, even though the desired input (strain $\in$) is zero.

The following methods are *used to nullify or reduce the effect of spurious inputs on the indicated output of an instrument:*

1. Signal filtering.
2. Compensation by opposing inputs.
3. Output correction.
4. Use of high amplification and stable/accurate feedback device.
5. Use of instrument elements which are inherently sensitive only to the desired input (*Example :* The strain gauge may be made of a material with a very low temperature coefficient of resistance and temperature independent gauge factor).

3.7 APPLICATIONS OF MEASUREMENT SYSTEMS

The instruments and measurement systems are employed for the following *applications :*

1. Monitoring of processes and operations.
2. Control of processes and operations.
3. Experimental engineering analysis.

1. Monitoring of processes and operations :

In these types of applications the measuring instruments *simply indicate the value or condition of parameter under study and do not serve any control function.*

Examples :

(*i*) Water and electric energy meters installed in homes.

(*ii*) An ammeter or a voltmeter indicates the value of current or voltage being monitored at a particular instant.

2. Control of processes and operations :

There is a strong association between measurements and control. The instruments find a very useful application in automatic control systems.

- A common example is the typical refrigeration system which employs a thermostatic control. A temperature measuring device (often a bimetallic element) senses the room temperature, thus providing the information necessary for proper functioning of the control system.

3. **Experimental engineering analysis :**
   The engineering problems can be solved by theoretical as well as experimental methods; several applications require the use of both the methods.
   Following are the *uses* of experimental engineering analysis :
   (*i*) To determine system parameters, variable and performance indices.
   (*ii*) To formulate the generalised empirical relationship in cases where there is no proper theoretical backing.
   (*iii*) To test the validity of theoretical predictions.
   (*iv*) To solve mathematical relationships with the help of analogies.

3.8 STATIC AND DYNAMIC CHARACTERISTICS OF INSTRUMENTS

3.8.1. Introduction

The performance of an instrument is described by means of quantitative qualities, termed as **characteristics**; these characteristics having been broken down into the sub areas of *static characteristics* and *dynamic characteristics.*

- *The **static characteristics** pertain to a system where the quantities to be measured are constant or vary slowly with time. Performance criteria based on dynamic relations (involving rapidly varying quantities) constitute **dynamic characteristics**.*

  — The static characteristics, in a real sense, also influence the quality of measurement under dynamic conditions, but these characteristics (static) *show up as non-linear or statistical effects* in otherwise linear differential equations giving the dynamic characteristics. *These effects would make the differential equations analytically unmanageable and so the conventional approach is to treat the two aspects of the problem separately*. Thus, even though these effects influence the dynamic behaviour, the differential equations of dynamic performance generally *neglect* the effects of dry friction, backlash, hysteresis, statistical scatter etc.

- *The overall performance of an instrument is judged by a semiquantitative superimposition of the static and dynamic characteristics.*

3.8.2. Definitions Relating to Measuring Instruments

Some important terms used in connection with measuring instruments are discussed below :

1. **True or actual value.** *The actual magnitude of a signal input to a measuring system which can only be approached and never evaluated is termed as "**true or actual value**".*
2. **Indicated value.** *It is the magnitude of a variable indicated by a measuring instrument.*
3. **Correction.** *The revision applied to the critical value so that the final result obtained improves the worth of the result is called "**correction**".*
4. **Overall error.** *It is the difference of the scale reading and the true value.*
   — When the instrument is properly designed and correctly adjusted, the consistent bias in error is very rare.
5. **Range.** *The region between the limits within which an instrument is designed to operate for measuring, indicating or recording a physical quantity is called the "**range of the instrument**".*
6. **Sensitivity.** *The ratio of output response to a specified change in the output is called "**sensitivity**".*

- The minimum change in the measured variable which produces an effective response of the instrument is called *"Resolution sensitivity"*. It is also called *"discrimination"*.
- The lowest level of measured variable which produces effective response of the instrument is called *"Threshold sensitivity"*.

7. **Scale sensitivity.** *It is defined as the ratio of a change in scale reading to the corresponding change in pointer deflection.*
8. **Scale readability.** The scale readability (in analog instruments) *indicates the closeness with which the scale can be read.*
9. **Repeatability.** It is defined as the *variation of scale reading;* it is random in nature.
   - *It is a measure of closeness with which a given input can be measured over and over again.*
10. **Accuracy.** It may be defined as *conformity with or closeness to an accepted standard value (true value).*
    - Accuracy of an instrument is influenced by factors like static error, dynamic error, reproducibililty, dead zone.
11. **Uncertainty.** Uncertainty denotes the *range of error, i.e.,* the region in which one guesses the error to be.
12. **Precision.** It refers to the *degree of agreement* within a group of measurements.
    - It is usually expressed in *terms of the deviation in measurement.*
13. **Drift.** *An undesired gradual departure of the instrument output over a period of time that is unrelated to changes in input, operating conditions or lead is called "**drift**".*
14. **Linearity or non-linearity.** *Deviation of transducer output curve from a specified straight line.* The *"non-linearity"* may be: (*i*) *Terminal linearity* (deviation from a straight line through the end points; (*ii*) *Best-fit linearity* (deviation from the straight line which gives minimum errors, both plus and minus.
15. **Dead zone.** *It is the range within which variable can vary without being detected.*
16. **Dead time.** *It is the time before the instrument begins to respond after the measured quantity has been changed.*
17. **Speed of response.** *The quickness of an instrument to read the measured variable is called "**speed of response**".*
18. **Reproducibility.** *The degree of closeness with which the same value of a variable may be measured at different times is called "**reproducibility**".*
19. **Tolerance.** *It is the range of inaccuracy which can be tolerated in measurements.*
20. **Backlash.** It is defined as the *maximum distance or angle through which any part of a mechanical system may be moved in one direction without applying appreciable force or motion to the next part in a mechanical system.*
21. **Stiction (static friction).** *It is the force or torque that is necessary just to initiate motion from rest.*
22. **Noise.** It may be defined as *extraneous disturbance generated in a measuring system which conveys no meaningful information w.r.t. desired signal.*

3.8.3. Static Characteristics

Measurements of applications in which parameter of interest is more or less constant; or varies very slowly with time are called **static measurements.** *A set of criteria (e.g., "accuracy", "error", "reproducibility", "drift", "sensitivity", "dead zone") that provide meaningful description of measurements under static conditions are called* **static characteristics.**

The main *static characteristics* may be summed up as follows :

(*i*) Accuracy (*ii*) Sensitivity
(*iii*) Reproducibility (*iv*) Drift
(*v*) Static error (*vi*) Dead zone.

3.8.3.1. Accuracy, errors and correction

Accuracy :

It is the degree of correctness with which a measuring means yields the "true value" with reference to accepted engineering standards.

Accuracy is determined as the maximum amount by which the result differs from the true value (the average of an infinite number of measured values when the average deviation due to the various contributing factors tends to zero). It is *almost impossible to determine experimentally the true value. The true value is not indicated by any measurement system due to the loading effects, lags and mechanical problems (e.g., wear, hysteresis, noise pick-up etc.).*

Accuracy of the measured signal depends upon the following *factors* :

- Variation of the signal being measured;
- Intrinsic accuracy of the instrument itself;
- Accuracy of the observer;
- Whether or not the quantity is being truly impressed upon the instrument.

Errors :

The accuracy of any measurement system is measured in terms of its *errors*. Since it is impossible to measure the true value of a quantity, as such, an approximation of the *"true value"* obtained by sufficiently extended series of measurements and also taking into parameters and conditions to which corrections may be applied, we obtain what is called *"best-measured value"* of the quantity. *The difference between the best measured value and the true value of the quantity is known as* **static error.**

The absolute value of error does not indicate precisely the accuracy of measurements. It is the *relative static error* which provides the quality of measurement. The **relative static error** *is defined as the ratio of absolute static error to the true value of the quantity under measurement.*

Let, V_m = Measured value of quantity,
V_t = True value of quantity,
E_s = Static (or absolute) error of measurement, and
E_r = Relative static error.

Then,
$$E_s = V_m - V_t \quad ...(3.1)$$

and,
$$E_r = \frac{\text{Absolute error}}{\text{True value}} = \frac{E_s}{V_t} \quad ...(3.2)$$

or, %age relative error, $\% E_r = E_r \times 100$ ...(3.3)

We have,
$$V_t = V_m - E_s \quad ...\text{[From eqn. (3.1)]}$$
$$= V_m - E_r \cdot V_t \quad ...\text{[From eqn. (3.2)]}$$

or,
$$V_t + E_r \cdot V_t = V_m$$

or,
$$V_t = \frac{V_m}{1+E_r} \quad ...(3.4)$$

However, when E_r is small, $E_r << 1$,

Eqn. (3.4) may be written as :

$$V_t = V_m (1 - E_r) \quad ...(3.5)$$

- The error may be either *+ve* or *–ve*. For *+ve* static errors the instrument reads *high* whereas it reads *low* for *–ve* static errors.

Correction.

The difference between the true value and the measured value of a quantity is called **static correction** (C_s).

i.e.,
$$C_s = V_t - V_m \; (= -E_s) \quad ...(3.6)$$

Eqn. (3.6) indicates that the static correction of the instrument reading is of the same magnitude as the error, but *opposite in sign* (*i.e.*, $C_s = -E_s$).

"Accuracy" may be *expressed as* :

(*i*) Point accuracy, or

(*ii*) Percent of scale range, or

(*iii*) Percent of true value.

(*i*) *Point accuracy.* In this case, the accuracy of an instrument is stated for one or more points in a range.

Examples :

(*a*) A given thermometer may be stated to read ± 0.2°C between 50°C and 100°C;

(*b*) A scale of length may be read within ± 0.2 mm.

(*ii*) *Percent of scale range.* Here, the error is calculated on the basis of maximum value of the scale.

$$\% \text{ Error} = \left(\frac{V_m - V_t}{\text{Max. scale value}} \right) \times 100 \quad ...(3.7)$$

Eqn. (3.7) refers to the intrinsic accuracy of the instrument itself and does not include procedural or personal performance.

(*iii*) *Percent of true value.* In this case the absolute error of measurement is expressed as %age of true value of the unknown quantity.

i.e.,
$$\%\text{Error} = \left(\frac{V_m - V_t}{V_t} \right) \times 100 \quad ...(3.8)$$

— Here, with a drop in true value, the size of error decreases.

— In this type of expression of accuracy, the percentage error is the maximum for any point in the instrument range.

The percent accuracies on the basis of actual scale or full-scale reading are normally used to express *equipment accuracies* and do not include procedural or personnel performance which is actually included in the total error.

Example 3.1. *A voltmeter reads 112.68 V. If the true value of the voltage is 112.6 V, determine the following :*

(i) The static error.

(ii) The static correction for the voltmeter.

Solution. *Given :* $V_m = 112.68$ volts; $V_t = 112.6$ volts.

(*i*) **Static error, E_s :**

$$E_s = V_m - V_t = 112.68 - 112.6 = +0.08 \text{ V} \quad \textbf{(Ans.)}$$

***(ii)* Static correction, C_s :**

$$C_s = -E_s = -\mathbf{0.08\ V} \quad \textbf{(Ans.)}$$

Example 3.2. *A thermometer reads 92.35°C and the static correction given in the correction curves is – 0.07°C. Determine the true value of the temperature.*

Solution. *Given :* $V_m = 92.35°C;\ C_s = -0.07°C.$

True value, V_t :

We know that, $E_s = V_m - V_t$

or, $-C_s = V_m - V_t$ $(\because C_s = -E_s)$

or, $V_t = V_m + C_s$ ...[Eqn (3.6)]

$= 92.35 - 0.07 = \mathbf{92.28°C}$ **(Ans.)**

Example 3.3. *An analog indicating instrument with a scale range of 0–5.0V shows a voltage of 2.65 V. The true value of a voltage is 2.70 volts.*

(i) What are the values of absolute error and correction ?

(ii) Express the error as a function of the true value and full scale deflection.

Solution. *Given :* $V_m = 2.65\text{ V};\ V_t = 2.70\text{ V}$

***(i)* Absolute error, E_s :**

$$E_s = V_m - V_t = 2.65 - 2.70 = \mathbf{-0.05\ V} \quad \textbf{(Ans.)}$$

Absolute correction, C_s :

$$C_s = -E_s = -(-0.05) = \mathbf{0.05\ V} \quad \textbf{(Ans.)}$$

***(ii)* Relative error, E_r :**

As a function of the *true value* :

$$E_r = \frac{E_s}{V_t} = \frac{-0.05}{2.70} = -0.0185, \text{ or, } \mathbf{-1.85\%} \quad \textbf{(Ans.)}$$

As a function of *full scale deflection* :

$$E_s = \frac{E_s}{(5.0-0)} = \frac{-0.05}{5} = -0.01, \text{ or, } \mathbf{-1\%} \quad \textbf{(Ans.)}$$

Example 3.4. *A pressure indicator showed a reading as 42 bar on a scale range of 0–50 bar. If the true value was 41.4 bar, determine :*

(i) Static error.

(ii) Static correction.

(iii) Relative static error.

Solution. *Given :* $V_m = 42$ bar ; $V_t = 41.4$ bar ; Scale range = 0–50 bar

***(i)* Static error, E_s :**

$$E_s = V_m - V_t = 42 - 41.4 \text{ bar} = \mathbf{+\ 0.6\ bar} \quad \textbf{(Ans.)}$$

***(ii)* Static correction, C_s :**

$$C_s = -E_s = -(+\ 0.6) = \mathbf{-\ 0.6\ bar} \quad \textbf{(Ans.)}$$

***(iii)* Relative error, E_r :**

$$E_r = \frac{E_s}{V_t} = \frac{0.6}{41.4} = 0.0145 \text{ or } \mathbf{1.45\%} \quad \textbf{(Ans.)}$$

Example 3.5. *A pressure gauge of range 50 bar is stated to have an error of ± 0.15 bar when calibrated by the manufacturer. Determine :*

(i) The percentage error on the basis of maximum scale value.

(ii) The possible error as a percentage of the indicated value when a reading of 10 bar is obtained in a test.

Solution. *Given :* Pressure range = 50 bar ; Error = ± 0.15 bar

(i) Percentage error on the basis of maximum scale value

$$= \pm\frac{0.15}{50}\times 100 = \mathbf{\pm 0.3\%} \quad \textbf{(Ans.)}$$

(ii) Percentage error on the basis of indicated value of 10 bar pressure

$$= \pm\frac{0.15}{10}\times 100 = \mathbf{\pm 1.5\%} \quad \textbf{(Ans.)}$$

The above results indicate that the pressure gauge is *more unreliable at the lower end of its gauge.*

Example 3.6. *A system comprising a transmitter, a relay and a receiver element has been used to measure the pressure at a remote point. If the specified accuracy limits for transmitter, relay and receiver are within ± 0.3%, ±1.4% and ± 0.9% estimate the maximum possible error and root-square accuracy of the measurement system.*

Solution. Maximum possible error (*least accuracy*) = ± (0.3 + 1.4 + 0.9) = **± 2.6%** **(Ans.)**

Least root-square accuracy or root-sum square error

$$= \pm\sqrt{(0.3)^2 + (1.4)^2 + (0.9)^2} = \mathbf{\pm 1.69\%} \quad \textbf{(Ans.)}$$

The above results indicate that the error is possibly as large as ± 2.6% but probably not larger than ± 1.69%.

- In an instrument system, *root-square accuracy is more justifiable because it is not probable that all the elements of the system will have the greatest static error at the same point and at the same time.*

3.8.3.2. Static calibration

Static calibration *is a process by which all the static performance characteristics are obtained in one form or another.* In general, static calibration refers to a situation in which all inputs, whether desirable, interfering or modifying except one, are kept at some constant values. Then the one input under study is varied over some range of constant values, which causes the output (s) to vary over some range of constant values. Thus an *output-input relationship is developed which comprise a static calibration valid under the stated constant conditions of all the other inputs.*

- The procedure of calibration involves a comparison of the particular instrument with either :
  (i) a *primary standard.*
  (ii) a *secondary standard with a higher accuracy than the instrument to be calibrated,* or
  (iii) an *instrument of known accuracy.*

— All working instruments must be calibrated against some reference instruments *which have a higher accuracy.* The reference instruments in turn must be calibrated against instruments of still higher grade of accuracy or against primary standards or against other standards of known accuracy.

- The following **steps** are necessary in *performing a calibration :*
  1. *Examine the construction of the instrument,* and identify and list all the *possible inputs.*

2. Decide as best as possible which of the *inputs will be significant* in the application for which instrument is to be calibrated.
3. Procure apparatus that *will allow to vary all the significant inputs over the ranges considered necessary.* Procure standards to measure each input.
4. By holding some inputs constant, varying others, and recording the output(s), *develop the desired static input-output relations.*

3.8.3.3. Range and span

Range. *The difference between the largest and the smallest reading of the instrument is called the "Range of an instrument"*. The range is expressed by stating the lower and upper values.

Span *represents the algebraic difference between the upper and lower range values of the instrument.*

If the highest point of calibration is S_{max} units while the lowest is S_{min} units and that the calibration is continuous between the points, then we say that the *instrument range is between* S_{min} *and* S_{max}.

The *instrument* **span** is given by, $S_{max} - S_{min}$.

The above definitions apply both to analog as well as digital instruments.

Error calibration curve : Refer to Fig. 3.10.

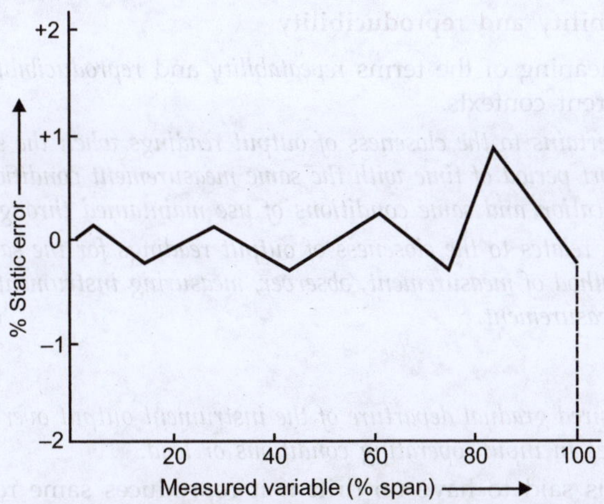

Fig. 3.10. Error calibration curve.

- *It is an error curve which can be used for correcting instrument readings.*
- The curve is plotted by calibrating the instrument against a suitable standard at a number of points on the scale.

  Examples :

  (*i*) *Range* : 2 kN/m^2 to 50 kN/m^2;
  Span : 50 – 2 = 48 kN/m^2

  (*ii*) *Range* : –5°C to 90°C ;
  Span : 90 – (–5) = 95°C.
- While determining the range of the instrument the frequency range must be taken into consideration. *"Frequency range"* is defined as *frequency over which measurements can be performed with a specified degree of accuracy.*

  Example : A moving voltmeter may have a range of 0–120 V range and 0–125 Hz frequency range with an accuracy of 0.5 or 1 per cent of full scale reading.

Example 3.7. *A pressure gauge is calibrated 5 bar to 60 bar. The accuracy is specified with ± 0.20 per cent of instrument span. What is the maximum static error ?*

Solution. Span of pressure gauge = 60 – 5 = 55 bar

$$\therefore \quad \text{Maximum static error} = \frac{\pm 0.20 \times 55}{100} = \pm \mathbf{0.11\ bar} \quad \textbf{(Ans.)}$$

3.8.3.4. Scale readability

The term scale readability is frequently used in analog type of instruments. *It indicates the closeness with which the scale of an analog type of instrument can be read.*

The readability depends upon many *factors* such as :

(*i*) Number of graduations;

(*ii*) Spacing of graduations;

(*iii*) Size of pointer;

(*iv*) Parallax effects;

(*v*) Discriminating power of the observer.

The readability, thus, is logically implied by the number of significant figures; the higher the number of significant figures, the better the readability of the instrument scale.

3.8.3.5. Repeatability and reproducibility

Although the meaning of the terms *repeatability* and *reproducibility* is same but they are applied in different contexts.

Repeatability *pertains to the closeness of output readings when the same input is applied repetitively over a short period of time with the same measurement conditions, same instrument and observer, same location and same conditions of use maintained throughout.*

Reproducibility *relates to the closeness of output readings for the same input when there are changes in the method of measurement, observer, measuring instrument, location, conditions of use and time of measurement.*

3.8.3.6. Drift

Drift *is an undesired gradual departure of the instrument output over a period of time that is unrelated to changes in input, operating conditions or load.*

An instrument is said to have no drift if it reproduces same readings at different times for same variation in measured variable.

The drift may be caused by the following *factors* :

(*i*) High mechanical stresses developed in some parts of instruments and systems.

(*ii*) Wear and tear.

(*iii*) Mechanical vibrations.

(*iv*) Temperature changes.

(*v*) Stray electric and magnetic fields.

(*vi*) Thermal e.m.fs.

Examples of drift :

- Drift occurs in *"resistance thermometers and thermocouples" due to the contamination of the metal and a change in its atomic or metallurgical structure.*
- Drift may occur in *"flow meters" due to wear and erosion of the orifice plate, nozzle or venturimeter.*

Classification of drift:

Drift may be *classified* as follows :

1. Zero drift; 2. Span drift; 3. Zonal drift.

1. **Zero drift :**

   Drift is called *zero drift if the whole of instrument calibration gradually shifts over by the same amount.*

   — It may be due to *permanent set or slippage* and can be corrected by *shifting pointer position.*

2. **Span drift :**

   If the calibration from zero upwards changes proportionally, it is called "span (or sensitivity) drift".

   — It may be due to change in spring gradient etc.

3. **Zonal drift :**

   When the drift occurs only over a portion of span of an instrument, it is called "zonal drift".

   — In industrial instruments, drift is an undesirable quantity since it is rarely apparent and cannot be easily compensated for.

- *Drift occurs very slowly and can be checked only by periodic inspection and maintenance.*

3.8.3.7. Accuracy and precision

Accuracy. *The closeness with which an instrument reading approaches the true value of the quantity being measured is called* ***accuracy***. Accuracy is determined as the maximum amount by which the result differs from the truė value. Accuracy of an instrument is influenced by factors like static error, dynamic error, reproducibility, dead zone.

Precision. The term 'precise' means clearly or sharply defined. *"Precision" is a measure of reproducibility of measurement.* It is usually expressed in terms of the deviation in measurement.

— ***Significant figure*** *is an indication of the precision of measurement.* Significant figures convey actual information regarding the magnitude of precision.

Following **examples** will illustrate the difference in meaning of the two terms *accuracy* and *precision :*

(*i*) Let us consider the case of an ammeter having high precision which can take readings to 1/150 of an ampere, but has a wrong zero adjustment. Now we can take readings to 1/150 of an ampere which are consistent and clearly defined since the ammeter is as *precise* as ever. However, the readings taken with this ammeter are *not accurate,* since they do not conform to truth because of faulty adjustment of zero.

(*ii*) Take the case of a micrometer which is normal in every respect but its anvil is displaced from its true position. When the readings are taken by this instrument, these would be clearly defined and consistent and we would say that the micrometer is as *precise* as ever. However, the readings do not conform to the truth as anvil is displaced from its true position. Thus, the readings taken by this micrometer are *not accurate.*

- It is worth noting that *precision is a necessary prerequisite to accuracy but it does not guarantee accuracy.*

3.8.3.8. Sensitivity

The ratio of the magnitude of output signal to the input signal or response of measuring system to the quantity being measured is called **sensitivity.**

It is represented by the slope of the calibration curve if the ordinates are expressed in actual units.

— When the calibration curve is linear [Fig. 3.11 (*a*)], the sensitivity is constant. However, if the calibration curve is non-linear [Fig. 3.11 (*b*)], the sensitivity is different at different points, being the slope of curve in various points.

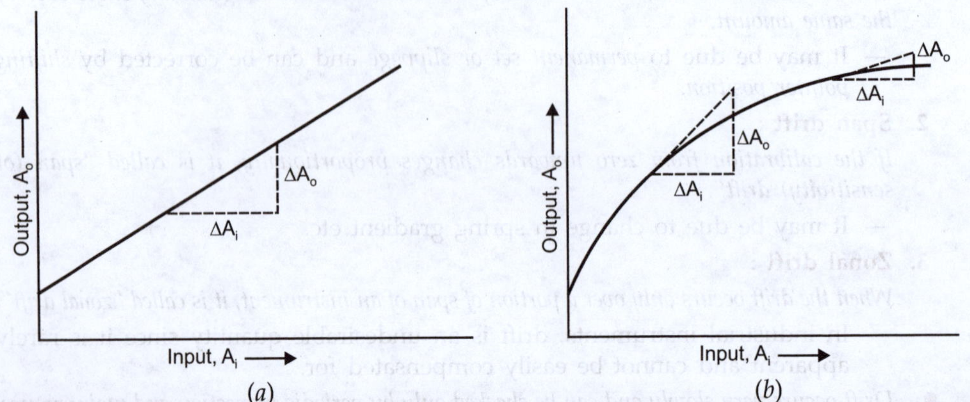

Fig. 3.11. Static sensitivity.

Static sensitivity, in general, is defined as :

$$\text{Static sensitivity } (S) = \frac{\text{Infinitesimal change in output}}{\text{Infinitesimal change in input}} = \frac{\Delta A_o}{\Delta A_i} \quad ...(3.9)$$

Similarly, *Inverse sensitivity* or *deflection factor* or *scale factor* $= \frac{\Delta A_o}{\Delta A_i}$ ...(3.10)

- *The sensitivity of an instrument should be high and thus the range should not be high in comparison to the value being measured.*
- Sensitivity has a wide range of units, and these depend upon the instrument or measurement system being investigated.

The *following points are worth noting :*

— In case the form of input to and output from the measurement system used with electrical/electronic equipment is same, the term **gain** is used instead of *sensitivity.*

— The *increase* in displacement with the optical and mechanical instruments is described by the term **amplification.**

— When the input or output signal is *changing with time*, the term **transfer function** or **transfer operator** is used (in place of the terms *sensitivity, gain or amplification*— which apparently mean the same and describe a relationship between output and input).

Example 3.8. *A pressure gauge, which has linear calibration curve, has radius of scale line as 90 mm and pressure of zero to 60 pascals is displayed over an arc of 300°. Determine the sensitivity of the gauge as a ratio of scale length to pressure.*

Solution. Full scale deflection = 300° $= 300 \times \frac{\pi}{180} = \frac{5\pi}{3}$ rad.

$$\text{Length of scale} = \frac{5\pi}{3} \times 90 = 150\,\pi \text{ mm}$$

$$\therefore \quad \textbf{Sensitivity} = \frac{\text{Scale length}}{\text{Calibration pressure}}$$

$$= \frac{150\pi}{60} = \mathbf{2.5\pi \text{ mm/Pa}} \quad \textbf{(Ans.)}$$

Example 3.9. *In a Wheatstone bridge, a change of 6Ω in the unknown arm of the bridge is required to produce a change in deflection of 2.4 mm of the galvanometer. Determine :*

(i) Sensitivity; *(ii) Deflection factor.*

Solution. *Given* : Magnitude of output response = 2.4 mm;

Magnitude of input = 6 Ω

$$\therefore \quad (i)\ \textbf{Sensitivity} = \frac{2.4}{6} = \mathbf{0.4 \text{ mm/}\Omega} \quad \textbf{(Ans.)}$$

and, $$(ii)\ \textbf{Deflection factor} = \frac{6}{2.4} = \mathbf{2.5\ \Omega\text{/mm}} \quad \textbf{(Ans.)}$$

Example 3.10. *A pressure measuring system consists of a piezoelectric transducer, a charge amplifier and an ultra violet charge recorder; their sensitivities are 6.8 pC/bar, 0.0032 V/pC and 16 mm/V respectively. Determine the deflection on the chart for a pressure change of 20 bar.*

Solution. With the elements arranged in series, the *overall sensitivity* of the measuring system (*S*)

= Sensitivity of piezoelectric transducer (S_1) × sensitivity of the charge amplifier (S_2) × sensitivity of ultra violet charge recorder (S_3).

$$= 6.8 \frac{pC}{\text{bar}} \times 0.0032 \frac{V}{pC} \times 16 \frac{\text{mm}}{V} = 0.348 \frac{\text{mm}}{\text{bar}}$$

Also, $$\textit{Sensitivity} = \frac{\text{Change of output signal}}{\text{Change of input signal}}$$

or, $$0.348 = \frac{\text{Output}}{20 \text{ bar}}$$

$$\therefore \quad \textit{Deflection on the chart} = 0.348 \times 20 = \mathbf{6.96 \text{ mm}} \quad \textbf{(Ans.)}$$

3.8.3.9. Linearity

The ability to reproduce the input characteristics symmetrically is called **linearity**. It can be expressed by the straight line equation.

The linearity is simply a measure of maximum deviation of any of the calibration points from the straight line (drawn by using the method of least square from the given calibration data). Fig. 3.12 shows the actual calibration curve and a straight line drawn from the origin using method of least squares.

Any departure from straight line relationship is non-linearity. The non-linearity may be due to the following *factors* :

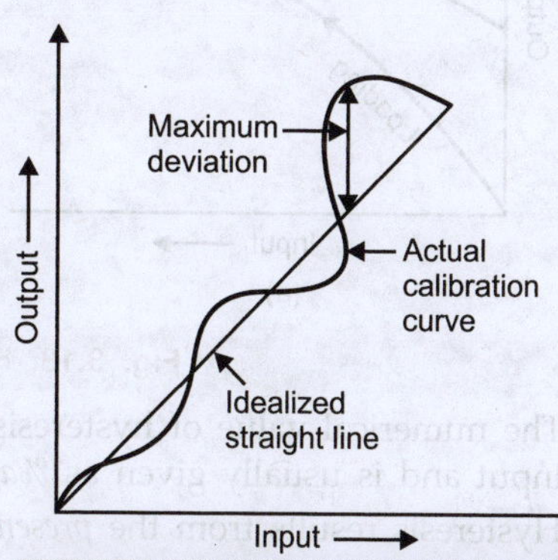

Fig. 3.12. Linearity exhibited as deviation between actual calibration curve and idealized straight line relationship.

(*i*) Viscous flow or creep;

(*ii*) Non-linear elements in the measurement device;

(*iii*) Mechanical hysteresis;

(*iv*) The elastic after-effects in the mechanical system.

The two common types of non-linearity are :

1. **Terminal linearity.** It is the deviation from a straight line through the end points.
2. **Best fit linearity.** It is the deviation from the straight line which gives minimum errors, both plus and minus.

- **Linearity** (*i.e.*, output is linearly proportional to input) is considered to be one of the best characteristics of an instrument or measurement system. This is so, because the conversion from a scale reading to the corresponding value of input quantity is very convenient. However, it may be understood that a non-linear behaviour does not essentially lead to inacuracy; an instrument having non-linear calibration curve may be highly accurate as ever.
- Since accuracy and linearity are related closely to each other, it is better to keep the non-linearity as small as possible so as to result in small errors.

3.8.3.10. Hysteresis

The maximum differences in output at any measured value within the specified range when approaching the point first with increasing and then with decreasing input may be termed as **hysteresis.**

- It is a phenomenon which shows different output effects when loading and unloading. It is *non-coincidence of loading and unloading curves.*

Figure 3.13(*a*) shows output and input curves (loading and unloading) for an instrument which has no friction due to sliding parts. The non-coincidence of loading and unloading curves is on account of *internal friction* or *hysteresis damping.*

Figure 3.13(*b*) shows the input-output relations of instruments which do not have internal friction but have external sliding friction.

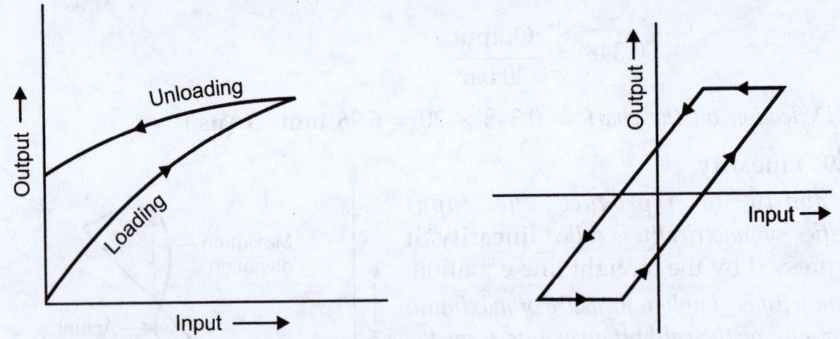

Fig. 3.13. Hysteresis effects.

- The numerical value of hysteresis can be specified in terms of either output or input and is usually given as *%age of full scale.*
- Hysteresis results from the *presence of irreversible phenomenon such as :*

— Mechanical friction;

— Slack motion in bearings;

— Magnetic and thermal effects.

3.8.3.11. Threshold and resolution

Threshold :

The minimum value below which no output change can be detected when the input of an instrument is increased gradually from zero is called the **threshold** *of the instrument.* Thus threshold defines minimum value of input which is necessary to cause a detectable change from zero output.

— In a *digital system,* threshold is the input signal necessary to cause one least significant digit of the output reading to change.

- Threshold may be caused by *backlash or internal noise.*

Resolution or discrimination :

When the input is slowly increased from some arbitrary (non-zero) input value, it is observed that the output does not change at all until a certain increment is exceeded; this increment is called ***Resolution or discrimination*** *of the instrument.* Thus resolution defines the smallest change of input for which there will be a change of output.

— In case of *analog instruments,* the resolution is determined by the observer's ability to judge the position of a pointer on a scale. Resolution is usually reckoned to be no better than ± 0.2 of the smallest division of the scale.

— In case of digital instruments, resolution is determined by the number of neon tubes taken to show the measured value.

- **Threshold** *defines the smallest measurable input* while the **resolution** *defines the smallest measurable input change.*
- *"Threshold"* and *"resolution"* may be *expressed as an actual value or as a fraction or percentage of full scale value.*

Example 3.11. *A transducer measures a range of 0–200 N force with a resolution of 0.15 per cent of full scale. What is the smallest change which can be measured by this transducer ?*

Solution. *Given :* Range of force : 0 – 200 N

Resolution : 0.15% of full scale

Smallest change which can be measured $= \frac{0.15}{100} \times 200 = \mathbf{0.3\ N}$ **(Ans.)**

Example 3.12. *Determine the resolution of a moving coil voltmeter having a uniform scale with 50 divisions; the full scale reading is 50 V and $\frac{1}{10}$ of a scale division can be estimated with a fair degree of certainty.*

Solution. One scale division $= \frac{50}{50} = 1$ V

∴ *Resolution* $= \frac{1}{10} \times 1 = \mathbf{0.1\ V}$ **(Ans.)**

Example 3.13. *Determine the resolution of a digital voltmeter which has read-out range 0 to 9999 counts and its full scale reading is 9.999 V.*

Solution. *Resolution of the digital voltmeter*

$$= \frac{1}{9999} \times 9.999$$

$= 0.001$ V or **1 mV** **(Ans.)**

3.8.3.12. Dead zone and dead time

Dead zone :

Refer to Fig. 3.14. *The largest change of input quantity for which there is no output of the instrument is termed as* ***dead zone.*** It may occur due to friction in the instrument which does not allow pointer to move till sufficient driving force is developed to overcome the friction loss.

- *"Dead zone" is caused by backlash and hysteresis in the instrument.*

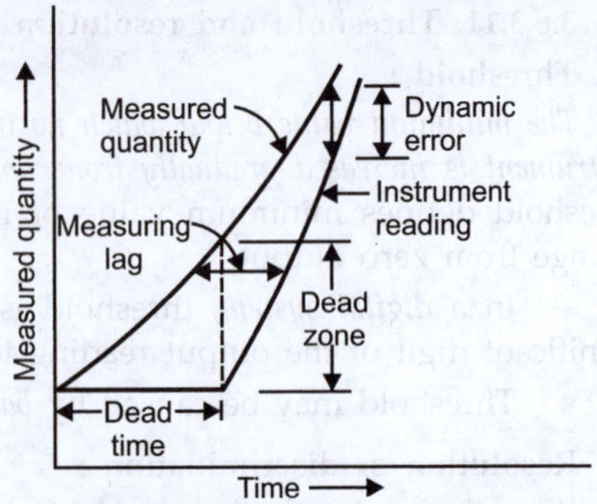

Fig. 3.14. Dead zone and dead time.

Dead time :

Refer to Fig. 3.14. *The time required by a measurement system to begin to respond to a change in the measurand is termed as* ***dead time.*** It represents the time before the instrument begins to respond after the measured quantity has been changed.

Example 3.14. *The calibration range of a certain pyrometer is 300°C to 800°C. If the dead zone in it is 0.11 per cent of span, determine the temperature change which might occur before it is detected.*

Solution. Span of the pyrometer = 800 – 300 = 500°C

$$\therefore \quad \text{Dead zone} = \frac{0.11}{100} \times 500 = 0.55°C$$

Hence, a change of 0.55°C must occur before it is detected. **(Ans.)**

3.8.3.13. Loading effects

Under ideal conditions an element used for signal sensing, conditioning, transmission and detection should not change/distort the original signal. The sensing element should not use any energy or take least energy from the process so as not to change the parameter being measured. However, under practical condition, it has been observed that the introduction of any element in a system results invariably *in extraction of the energy* from the system thereby *distorting the original signal.* This distortion may take the form of *attenuation, waveform distortion, phase shift etc.* Consequently the ideal measurements become impossible. *The incapability of the system to faithfully measure the input signal in undistorted form is called the* ***loading effect.*** This results in *loading errors.*

- The loading effects, in a measurement system, not only occur in *detector-transducer stage* but also occur in *signal conditioning and signal presentation stages* as well. The loading problem is carried right down to the basic elements themselves.

The loading effects may occur on account of both electrical and mechanical elements; these are *due to impedances of the various elements connected in a system.* The mechanical impedances may be treated similar to electrical impedances.

- *Loading effects occur due to extraction of energy from the measurand* (as earlier stated). The transfer of energy requires the specification of two variables. The physical variables which determine the flow of energy in all dynamical systems can be classified as :

(*i*) Through variables; (*ii*) Across variables.

— *"Through variables"*, also called flow variables, are those which can be specified and measured at *one point in a space.* These are extensive variables in the sense, that their magnitude depends upon the extent of the system taking part in the energy transfer.

— *"Across variables"*, also called effort variables and transvariables, are those which can be specified by *two points in a space;* usually one point is a reference point. These are intensive variables whose magnitude is independent of the medium being considered.

All objects in a measurement system or any dynamical system involve a measurand or defined relationship between an "across variable" and a "through variable".

Static stiffness and static compliance :

In some instruments when input signal is applied to them under *static conditions,* the output signal is a faithful representation of the input signal. But when these instruments are subjected to *dynamic conditions* (*i.e.,* going from one steady state to another one), they produce loading effects since energy is required to perform these operations. Under such conditions concepts of static stiffness and static compliance (instead of the concepts of impedance and admittance) need to be introduced.

It may be understood that *while impedance and admittance are concerned with* ***power drain,*** *stiffness and compliance are concerned with* ***energy drain.***

Mechanical Stiffness, in *general,* is defined as :

$$K_g = \frac{\text{Across variable}}{\int (\text{Through variable})\, dt} \qquad ...(3.11)$$

Compliance is defined as *inverse of stiffness* (as admittance is inverse of impedance) and is given by

$$C_g = \frac{1}{K_g} = \frac{\text{Through variable}}{\int (\text{Across variable})\, dt} \qquad ...(3.12)$$

Example 3.15. *(a) Discuss the concept of impedance loading and impedance matching as applied to a measurement system.*

(b) An unknown resistance is connected in series with a milliammeter of very small internal resistance. A voltmeter with internal resistance 125 kΩ is connected across the unknown resistance.

Determine the loading error when :

(i) the voltmeter and milliammeter read 180 V and 6 mA respectively.

(ii) the same voltmeter and milliammeter with another resistance read 60 V and 1.2 A respectively. Comment upon the results.

Solution. (*a*) **Impedance loading and impedance matching :**

In a generalised measurement system, the sensing element senses the condition, state or value of the process variable by *extracting a small part of energy from the measurand,* and then processes an output which reflects this condition, state or value of the measurand. Due to this energy extraction the measurand is always disturbed by the measurement act and consequently a perfect measurement becomes *theoretically impossible.*

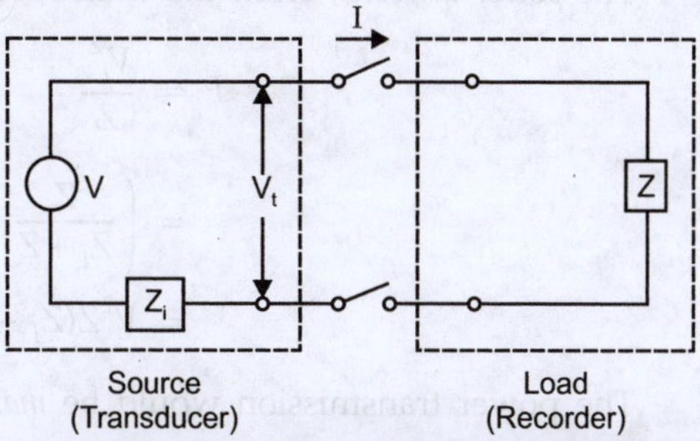

Fig. 3.15. Loading effect of recorder on the transducer element.

Examples : (*i*) The reading shown by the hand tachometer would vary with the pressure with which it is pressed against the shaft; (*ii*) A milliammeter would introduce additional resistance in the circuit and thereby change the flow of current by a significant amount.

The *interaction of system* (mentioned above) is referred to as *"loading effect"* and the *term loading when applied to a general electrical system is called* **impedance loading.** While designing an instrument efforts should be made to ensure that the *alteration due to system interaction is minimal.*

Refer to Fig. 3.15. Consider a transducer element with internal impedance, Z_i (= R_i + jX_i), R and X being the resistance and reactance respectively.

Let, V = Voltage developed in the transducer, and

V_t = Terminal voltage.

Then, when output terminals are open (*i.e.*, no-load condition),

$$V_t = V \quad \text{...(i)}$$

When the transducer is connected in series with a recorder of input impedance Z (= $R + jX$), a current flows through the circuit, the value of which is given as,

$$I = \frac{V}{Z_i + Z}$$

and,

$$V_t = V - I Z_i = V - \left(\frac{V}{Z_i + Z}\right) Z_i$$

$$= V\left[1 - \frac{Z_i}{Z_i + Z}\right] = \frac{Z}{Z_i + Z} \cdot V$$

i.e.,

$$V_t = \frac{Z}{Z_i + Z} \cdot V \quad \text{...(ii)}$$

[In ideal cases, when impedance loading is zero (*i.e.*, $Z_i = 0$), eqn. (*ii*) transforms to eqn. (*i*)]

From eqn. (*ii*), it may be observed that for the *loading effect to be minimal, the following conditions should be satisfied :*

(*i*) Z_i (transducer impedance) should be *small.*

(*ii*) Z (impedance of the measuring device) must be *high.* (It is due to this reason that vacuum tube voltmeters having an internal resistance to the tune of 10 MΩ are used in the measurement of the voltage).

The *power delivered* from the transducer to the external load,

$$P = \frac{V_t^2}{Z}$$

$$= \left(\frac{Z}{Z_i + Z} \cdot V\right)^2 \cdot \frac{1}{Z} \quad \text{...[Substituting for } V_t \text{ from eqn. (ii)]}$$

$$= V^2 Z (Z_i + Z)^{-2}$$

The power transmission would be *maximum* when, $\frac{dP}{dZ} = 0$

i.e.,

$$\frac{dP}{dZ} = V^2\left[(Z_i + Z)^{-2} \times 1 + Z \times (-2)(Z_i + Z)^{-3}\right] = 0$$

or, $$(Z_i + Z)^{-2} - 2Z(Z_i + Z)^{-3} = 0$$

or, $$1 - 2Z(Z_i + Z)^{-1} = 0$$

or, $$(Z_i + Z) - 2Z = 0$$

or, $$Z = Z_i \quad \text{...}(iii)$$

Eqn. (*iii*) indicates that maximum power will be delivered from the transducer when the impedance of the external load just *matches* its internal impedance. This aspect is referred to as ***impedance matching*** *in electrical circuits.*

If $$Z_i = R_i \quad \text{and} \quad Z = R$$

Then, $$V_t = \frac{R}{R_i + R}, \quad \text{and} \quad R = R_i.$$

(*b*) Refer to Fig. 3.16. *Given* : Internal resistance of the voltmeter, R_v = 125 kΩ;

Readings of voltmeter and ammeter in two cases :

(*i*) 180 V, 6 mA; (*ii*) 60 V, 1.2 A

(*i*) Total resistance of the circuit,

$$R_T = \frac{V}{I} = \frac{180}{6 \times 10^{-3}} = 30 \text{ k}\Omega$$

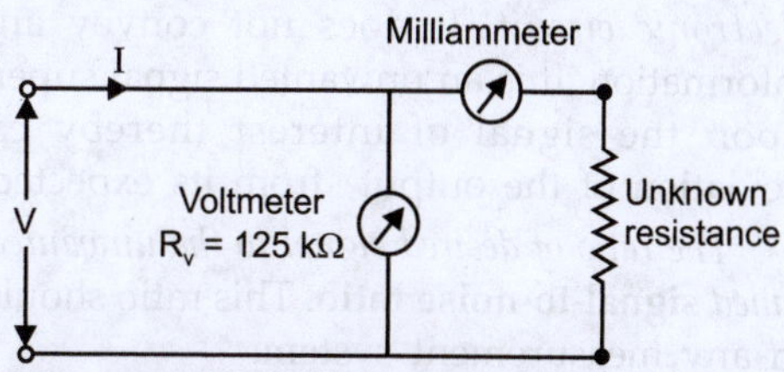

Fig. 3.16.

Since the milliammeter is stated to have negligible resistance, the total resistance as calculated above equals the apparent resistance ($R_{appt.}$) of the resistor.

i.e., $$R_{appt.} = 30 \text{ k}\Omega$$

The actual value of resistance of the resistor is obtained from the condition that voltmeter connected is parallel with the resistor, *i.e.,*

$$\frac{1}{R_T} = \frac{1}{R_{act.}} + \frac{1}{R_v}$$

or, $$\frac{1}{R_{act.}} = \frac{1}{R_T} - \frac{1}{R_v} = \frac{R_v - R_T}{R_T \times R_v}$$

or $$R_{act.} = \frac{R_T \times R_v}{R_v - R_T} = \frac{30 \times 125}{125 - 30} = 39.474 \text{ k}\Omega$$

∴ $$\textit{Percentage loading error} = \frac{R_{act.} - R_{appt.}}{R_{act.}}$$

$$= \frac{39.474 - 30}{39.474} \times 100 = \mathbf{24\%} \quad \textbf{(Ans.)}$$

(*ii*) $$\text{Total resistance, } R_T = \frac{60}{1.2} = 50 \ \Omega$$

i.e., $$R_{appt.} = 50 \ \Omega$$

$$R_{act.} = \frac{R_T \times R_v}{R_v - R_T} = \frac{50 \times 125}{125 - (50/1000)} = 50.02 \ \Omega$$

$$\therefore \quad \textit{Percentage loading error} = \frac{50.02 - 50}{50.02} \times 100 = \mathbf{0.04\%} \quad \textbf{(Ans.)}$$

From the above results it may be concluded that :

- When a voltmeter with high internal resistance is connected across two points in a *high resistance circuit*, the loading effect is appreciable and, therefore, the voltmeter gives *misleading voltage reading*.
- The same voltmeter has a *negligible loading error* when connected across a *low resistance circuit*, and it gives more *reliable voltage reading*.

3.8.3.14. Noise

Noise *is a spurious current or voltage extraneous to the current or voltage of interest in an electrical or electronic circuit*. It does not convey any useful information. It is an unwanted signal superimposed upon the signal of interest thereby causing a deviation of the output from its expected value.

The ratio of desired signal to the unwanted signal is called **signal-to-noise ratio**. This ratio should be *large* in any measurement system.

The noise may be broadly *classified* as follows :

(*i*) Generated noise;

(*ii*) Conducted noise;

(*iii*) Radiated noise.

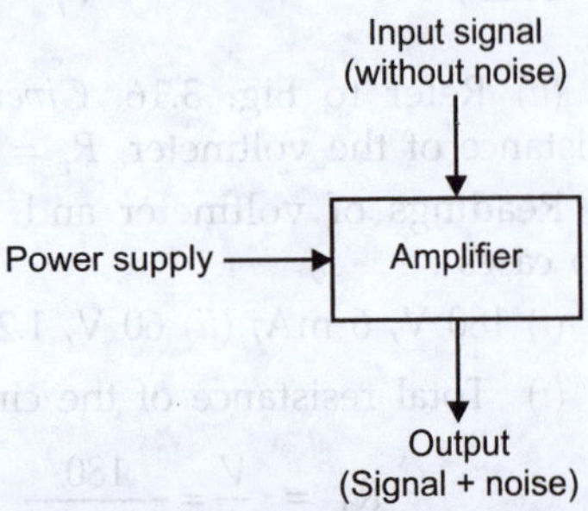

Fig. 3.17. Block diagram of an amplifier for illustration of different types of noise signals.

Refer to Fig. 3.17. It shows a block diagram of an amplifier for illustration of different types of noise signals.

(*i*) Generated noise :

- The internally generated noise is on account of components like resistors, capacitors, transistors, etc.
- The temperature dependent noise in a resistor which increases with internal heating (I^2R loss) or with an increase in the ambient temperature is called *"Johnson noise"*.
- The wide band noise, produced within a resistor, due to vibrations produced by the thermal effects, covering a wide frequency range is called *"White noise"*.

  The internally generated noise in resistor can be *reduced* by :

  (*a*) lowering the internal temperature;

  (*b*) using special film and glass substrates.

  The internally generated noise in semiconductor and vacuum tube devices on account of random movement of charges is called *"Shot noise"*. It is difficult to reduce this type of noise; however, it can be reduced to some extent by the use of selective filtering.

- The internally generated noise may also be caused by the changing electric fields in the region between plates of a capacitor and changing magnetic fields around in an inductor. In this case noise signal has a fixed frequency (since fields change in orderly manner) and as such its magnitude can be reduced by using a filter tuned to this frequency.

(*ii*) **Conducted noise :**

- This type of noise may be caused by the power supply (to the amplifier) which may have ripples or random deviations.
- 50 Hz power supply and the harmonics contained in it are one of the common sources for conducted noise.
- It can be reduced by using filters in the leads to trap out the noise.

(*iii*) **Radiated noise :** Radiated noise consists of unwanted signals (relating to electric or magnetic fields or disturbances in the environments around the amplifier) which are radiated into the interior of the amplifier.

- The electromagnetic impulses radiated from ignition wiring of spark plugs are one of the common sources of radiated noise.
- It can be *reduced by proper shielding.*

3.8.4. Dynamic Characteristics

When an instrument is required to measure a time-varying process variable, one has to be concerned with *"dynamic characteristics"*; these *characteristics quantify the dynamic relation between the input and output.*

3.8.4.1. Dynamic response

The evaluation of the ability of a system to faithfully transmit and present all the pertinent information included in the input signal and to exclude all else, is called the **response.**

The behaviour of the system when inputs vary with time (i.e., inputs are dynamic in nature) and so does the output, is called **dynamic response.** Such a response in measurement systems is found to occur in industrial, aerospace and biological applications.

The dynamic inputs are of two types :

(*i*) Steady state periodic; (*ii*) Transient.

The *"steady state periodic"* quantity is one whose magnitude has a definite repeating time, whereas the time variation of a ***"transient"*** magnitude does not repeat.

- The response of a measurement system subjected to a time varying input can be divided into the following two parts :

  (*i*) Steady state response; (*ii*) Transient response.

  "Steady state response" is simply the response when time reaches infinity.

  "Transient response" in measurement systems is defined as the part of response which goes to zero as time becomes large.

- **Periodically varying inputs may be :**

  — A *purely sinusoidal input* like the A.C. voltage with a frequency of 50 Hz.

  or

  — A complex time varying repetitive input which is a *combination of the fundamental* and *harmonics.*

When the systems are subjected to periodically varying inputs, they exhibit in their response a magnitude and phase relationship which is *different* from that of the input signal because of the *energy storage elements. The output is not a faithful representation of the input.* The harmonic contents of the input are not reproduced in the output both in magnitude and phase relationship as in the input. *The output is distorted both in terms of magnitude as well as phase relationships.*

3.8.4.2. Dynamic characteristics of a measurement system

The *dynamic characteristics* of a measurement system are :

1. Speed of response ... *Desirable*
2. Measuring lag ... *Undesirable*
3. Fidelity ... *Desirable*
4. Dynamic error ... *Undesirable*

1. Speed of response. The *"speed of response"* or *"responsiveness"* is defined as the *rapidity with which a measurement system responds to changes in the measured quantity.*

2. Measuring lag. It refers to *retardation or delay in the response of a measurement system to changes in measured quantity.*

The lag *is caused by conditions such as capacitance, inertia, or resistance.*

The measuring lags are of the following two types :

(i) ***Retardation type lag.*** In this type of measuring lag the response of the measurement system *begins immediately after a change in measured quantity has occurred.*

(ii) ***Time delay type lag.*** In this case the response of the measurement system *begins after a dead time after the application of the input.* If the measured quantity varies at a fast rate, the dead time has a severe adverse effect on the performance of the system. The measurement lags of this type are *very small* and are of the order of a fraction of a second and hence can be *ignored.*

3. Fidelity. It is defined as the *degree to which a measurement system indicates changes in the measured quantity without any dynamic error.* It refers to the ability of the system to reproduce the output in the same form as the input.

4. Dynamic error. The dynamic error, also called **"measurement error"**, is the *difference between the true value of the quantity changing with time and the value indicated by the measurement system if no static error is assumed* (see Fig. 3.14.)

- *The maximum amount by which the pointer moves beyond the steady state is called "overshoot".*

3.8.4.3. 'Dynamic' analysis of measurement systems

The dynamic behaviour of measurement systems is studied in the following two domains :

1. Time domain analysis.
2. Frequency domain analysis.

1. Time domain analysis :

In this the input signal is applied to the measurement system and the behaviour of the system is studied as a *function of time.* The dynamic response of the system to different types of inputs, which are a function of time, is analysed at different intervals of time after the application of the input signal. In most cases, the actual input signals vary in random fashion with respect to time and therefore cannot be mathematically defined. Consequently the performance of a system can be analysed (in the time domain analysis) by using the following standard *test signals/inputs* :

(i) Step input;
(ii) Ramp input;
(iii) Parabolic input;
(iv) Impulse input.

2. Frequency domain analysis :

This type of analysis of a system *pertains to the steady state response of the system to a sinusoidal input*. Here, the system is subjected to a sinusoidal input and the system response is studied with *frequency as the independent variable*.

- *Frequency response.* It is the maximum frequency of the measured variable that an instrument is capable of following without error. The usual requirement is that the frequency of measurand should not exceed 60 per cent of the natural frequency of the measuring instrument.

Standard test signals/inputs :

The most common standard inputs used for dynamic analysis are discussed below :

1. ***Step function :*** Refer to Fig. 3.18(*a*). It is a *sudden change from one steady value to another*.

It is mathematically represented by the relationship :

$$x = 0 \quad \text{at} \quad t < 0$$
$$x = x_c \quad \text{at} \quad t \geq 0$$

where x_c is a constant value of the input signal x_i.

- The *"transient response" indicates the capacity of the system to cope with change in the input signal.*

2. ***Ramp or linear function :*** Refer to Fig. 3.18(*b*). In this case, the *input varies linearly with time*.

This input is mathematically represented as :

$$x = 0 \quad \text{at} \quad t < 0$$
$$x = \psi \quad \text{at} \quad t \geq 0$$

where, ψ is the slope of the input versus time relationship.

- *The ramp-response becomes indicative of the steady state error in following the changes in the input signal.*

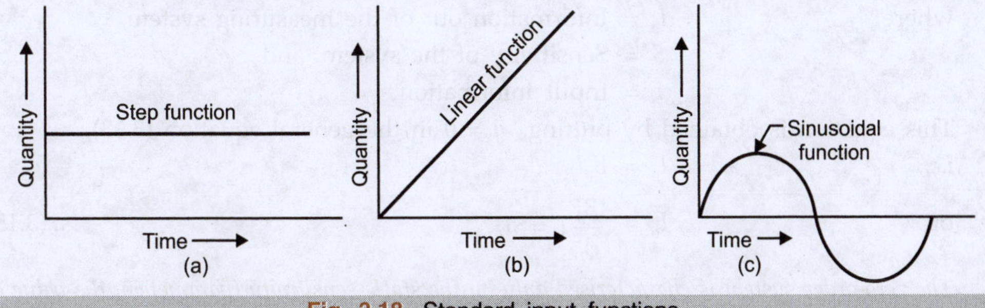

Fig. 3.18. Standard input functions.

3. ***Sinusoidal function :*** Refer to Fig. 3.18(*c*). In this case the *input varies sinusoidally with a constant maximum amplitude.*

$$I_i = A \sin \omega t$$

where, A = Amplitude, and

ω = Frequency in rad/s.

- *The frequency or harmonic response is a measure of the capability of the system to respond to inputs of cyclic nature.*

A general measurement system can be mathematically described by the following differential equation:

$$(A_nD^n + A_{n-1}D^{n-1} + ... + A_1D + A_0)I_o = (B_mD^m + B_{m-1}D^{m-1} + ... + B_1D + B_0)I_i \quad ...(3.13)$$

where, A's and B's = Constants, depending upon the physical parameters of the system,

D^k = Operative derivative of the order k,

I_o = The information out of the measurement system, and

I_i = The input information.

The order of the measurement system is generally classified by the value of the power of n.

- *Zero-order system* : $n = 0$ and $A_1, A_2, A_3 .... A_n = 0$
- *First-order system* : $n = 1$ and $A_2, A_3, A_4 .... A_n = 0$
- *Second-order system* : $n = 2$ and $A_3, A_4, A_5 .... A_n = 0$

The above method of classification is used for most of the instruments and systems.

Although general equation can be solved by various methods, we shall be using methods of D-operator for getting its solution.

3.8.4.4. Zero, first and second order systems

3.8.4.4.1. Zero-order system

Figure 3.19 shows the block diagram of a 'Zero-order system'. In this case the output of the measuring system (ideal) is directly proportional to input, no matter how the input varies. The *output is faithful reproduction of input without any distortion or time lag.*

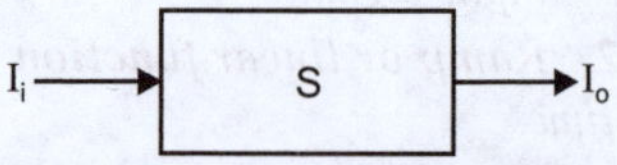

Fig. 3.19. Block diagram for Zero-order system.

The behaviour of the zero-order system is represented by the following mathematical solution :

$$I_o = S\,I_i \quad ...(3.14)$$

where, I_o = Information out of the measuring system,

S = Sensitivity of the system, and

I_i = Input information.

This equation is obtained by putting $n = 0$ in the general equation (3.13),

i.e., $A_oI_o = B_oI_i$

or $$I_o = \frac{B_o}{A_o} I_i = SI_i \quad ...(3.15)$$

The zero-order system is characterised only by the static sensitivity (parameter), the value of which is obtained through the process of static calibration.

Examples of zero-order system :

- Mechanical levers;
- Amplifiers;
- Potentiometer (It gives an output voltage which is proportional to wiper's displacement) etc.

3.8.4.4.2. First-order system

Figure 3.20 shows the block diagram of a 'First-order system'.

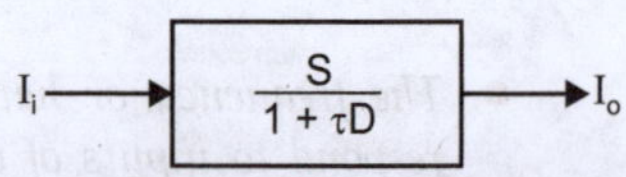

Fig. 3.20. Block diagram for First-order system.

The behaviour of a first-order system is given by following first-order differential equation :

$$A_1 \frac{dI_o}{dt} + A_o I_o = B_o I_i \qquad ...(3.16)$$

(This equation is obtained by inserting $n = 1$ in the general equation).

Eqn. (3.16) may be written in standard form as follows:

$$\frac{A_1}{A_o}\frac{dI_o}{dt} + I_o = \frac{B_o}{A_o} I_i \qquad ...(3.17)$$

or, $$\tau \frac{dI_o}{dt} + I_o = SI_i \qquad ...(3.18)$$

where, $$\tau = \frac{A_1}{A_0} = \text{Time constant, and}$$

$$S = \frac{B_o}{A_o} = \text{Sensitivity.}$$

Using D-operator, we get

$$\left[\text{where,} \quad D = \frac{d}{dt}, \quad \text{and} \quad D^2 = \frac{d^2}{dt^2}\right]$$

$$\tau DI_o + I_o = SI_i$$

or, $$I_o(\tau D + 1) = SI_i$$

or, $$\frac{I_o}{I_i} = \frac{S}{I + \tau D} \qquad ...(3.19)$$

Equation (3.19) gives the standard form of *transfer operator* for first-order system.

Examples of first-order system :

- Velocity of a true falling mass;
- Air pressure build-up in bellow;
- Measurement of temperature by mercury-in-glass thermometers;
- Thermistors and thermocouples;
- Resistance-capacitance network.

3.8.4..4.3. Second-order system

Fig. 3.21 shows the block diagram of 'Second-order system':

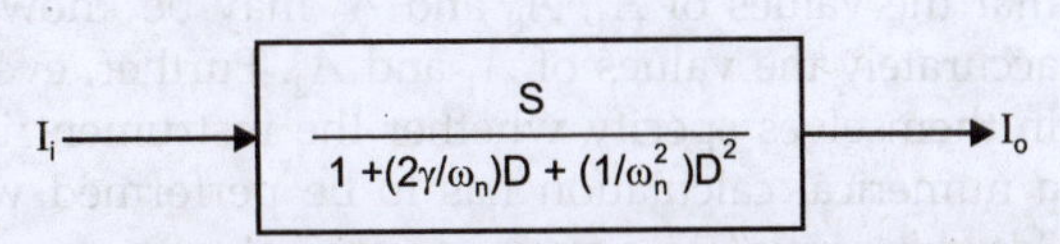

Fig. 3.21. Block diagram for Second-order system.

The behaviour of a second-order system is given by the following differential equation (obtained by putting $n = 2$ in the general equation);

$$A_2 \frac{d^2 I_o}{dt^2} + A_1 \frac{dI_o}{dt} + A_o I_o = B_o I_i \qquad ...(3.20)$$

Dividing the above equation by A_o, we have

$$\frac{A_2}{A_o}\frac{d^2 I_o}{dt^2} + \frac{A_1}{A_0}\frac{dI_o}{dt} + I_o = \frac{B_o}{A_o} I_i \qquad ...(3.21)$$

Let, $\omega_n = \sqrt{\dfrac{A_o}{A_2}}$ = Undamped natural frequency, rad/s,

$\gamma = \dfrac{A_1}{2\sqrt{A_0 A_2}}$ = Damping ratio, dimensionless, and

$S = \dfrac{B_o}{A_o}$ = Static sensitivity or steady-state gain.

Then, by substituting these values in eqn. 3.21, we get

$$\frac{I}{\omega_n^2} \cdot \frac{d^2 I_o}{dt^2} + \frac{2\gamma}{\omega_n} \cdot \frac{dI_o}{dt} + I_o = SI_i \quad ...(3.22)$$

or, in terms of D-operator, we have

$$\left(\frac{D^2}{\omega_n^2} + \frac{2\gamma}{\omega_n} D + 1 \right) I_o = SI_i$$

or,

$$\frac{I_o}{I_i} = \frac{S}{\frac{1}{\omega_n^2} D^2 + \frac{2\gamma}{\omega_n} D + 1} \quad ...(3.23)$$

Examples of second-order system :

- Piezoelectric pickup;
- Spring-mass system (used for acceleration and force measurements);
- Pen control system on X-Y plotters;
- U.V. galvanometer, etc.

Damping ratio :

In the design of instruments a term which is very frequently used is the "*damping ratio*" (γ) defined as the *ratio of the actual value of coefficient of viscous friction in movement and the value required to produce critical damping.*

i.e., $$\gamma = \frac{A_1}{2\sqrt{A_o A_2}}$$

This dimensionless term is very useful because to determine its value, it is not necessary that the values of A_1, A_0 and A_2 may be known. In practice it is not easy to determine accurately the values of A_1 and A_2. Further, even if these values are known, they do not in themselves specify whether the instrument is under, over or critically damped, since a numerical calculation has to be performed with them first. Therefore, designers find "*damping ratio*" as a very convenient measure of the amount of damping present in the movement.

The terms *damping ratio* (γ) and *underdamped natural frequency* (ω_n) immediately conjure up a physical picture of the response of an instrument and both of the quantities are very easy to measure. Thus γ and ω_n easily do away with quantities A_2, A_1 and A_0.

3.8.4.5. First-order system responses

The complete solution of an equation which describes the dynamical behaviour of a system consists of the following *two parts :*

(*i*) **Complementary function.** *It corresponds to the short time or transient response.*

(*ii*) **Particular integral.** *It refers to the long time steady state response.*

The transfer operator form of the first-order system is given by :

$$\frac{I_o}{I_i} = \frac{S}{I + \tau D}$$

When S (static sensitivity or steady state gain) equals *unity*, we get

$$(I + \tau D)I_o = I_i \quad ...(3.24)$$

Now we shall obtain the solution of this equation for different standard inputs (The solutions are not mathematically rigorous, but are practical).

Transient response (Complementary function):

The transient response from the auxiliary equation is obtained by putting input I_i equal to zero :

i.e., $$(1 + \tau D)I_{o,t} = 0 \quad ...(3.25)$$

(Subscript t refers to the transient value)

Let the solution be of the form :

$$I_{o,\,t} = A\,e^{mt}$$

where, m is an algebraic variable.

or, $$(1 + \tau D)A\,e^{mt} = 0$$

or, $$A\,e^{mt} + \tau\cdot\frac{d}{dt}(A\,e^{mt}) = 0$$

or, $$A\,e^{mt} + \tau\cdot A\,m\,e^{mt} = 0$$

$$A\,e^{mt}\,(1 + \tau\cdot m) = 0$$

$\therefore$ $$m = -\frac{1}{\tau}$$

Then, $$I_{o,\,t} = A\,e^{mt} = Ae^{-t/\tau} \quad ...(3.26)$$

The transient response of a first-order system is *same for different standard inputs.*

Steady state response (Particular integral) :

The steady state response is given by

$$(1 + \tau D)I_{o,s} = I_i \quad ...(3.27)$$

(Subscript s refers to the steady state value)

or, $$I_{o,s} = (1 + \tau D)^{-1}\,I_i$$

$$= (1 - \tau D + \text{terms in } D^2 \text{ and higher})\,I_i \quad ...(3.28)$$

1. Step input :

Since the input I_i is a step of constant magnitude; its differential equals zero, and subsequently, we get

$$I_{o,s} = (1 - \tau D)I_i = I_i \quad ...(3.29)$$

Total response = Transient response + steady state response

or $$I_o = Ae^{-t/\tau} + SI_i \quad ...(3.30)$$

The constant A is evaluated from the initial conditions as follows :

At $t = 0,\qquad I_o = 0$

$\therefore$ $0 = A + SI_i \quad$ or, $\quad A = -SI_i$

$\therefore$ $$I_o = \underbrace{-I_i\,e^{-t/\tau}}_{\text{Transient}} + \underbrace{I_i}_{\text{Steady state}}$$

or, $$I_o = I_i(1 - e^{-t/\tau}) \quad ...(3.31)$$

or, $$\frac{I_o}{I_i} = (1 - e^{-t/\tau}) \quad ...(3.32)$$

... in *non-dimensional form.*

Salient features (with step input) :

Following are the salient features of first-order system with step input :

(i) The transient response of the first-order system is *time dependent;* as the time passes, its value decreases (Refer to eqn. 3.32) and after a very long time the value becomes zero approximately. Thus magnitude of output (I_o) will be same as input (I_i) when the time is very large.

(ii) The speed of response relates to the time constant τ. A large τ indicates that response of the system is slow, whereas a small τ represents a fast system response. Thus in order to get good fidelity (*i.e.,* for accurate dynamic measurements) efforts should be made to minimise the value of τ.

(iii) Refer to Fig. 3.22, which shows the time response of a first-order system to a step-input; when $t = \tau$, $\frac{I_o}{I_i} = (1 - e^{-1}) = 0.632$. Thus, the *time constant* (τ), *for a rising exponential function, is defined as the time to reach* 63.2% *of its steady state value*. The time constant for a *decaying function* would correspond to the time taken to fall to *36.8% of its initial value.*

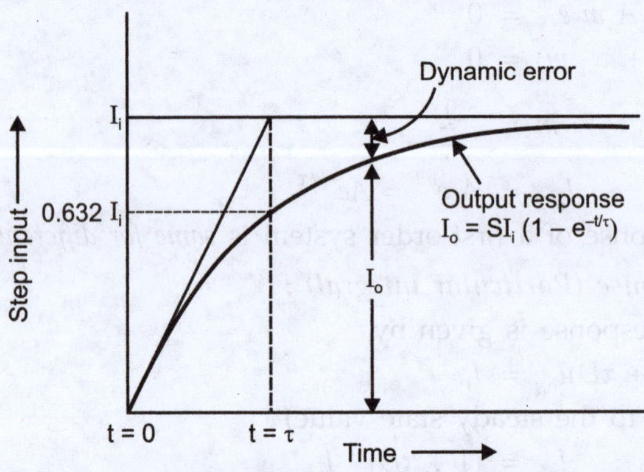

Fig. 3.22. Time response of a first-order system to step input.

(iv) Dynamic error (*i.e.,* vertical difference between the input and output response curve),

$$E_{dy.} = I_i - I_o$$
$$= I_i - I_i(1 - e^{-t/\tau})$$
$$= -I_i\, e^{-t/\tau} \quad ...(3.33)$$

or, $$\frac{E_{dy.}}{I_i} = -e^{-t/\tau} \quad ...(3.34)$$

... (*in dimensionless form*)

(v) In case the measurand has an initial value of $I_{initial}$ at $t = 0$, then the output I_o at any instant t is given by

$$I_o = I_i[1 - e^{-t/\tau}] + I_{initial}\, e^{-t/\tau}$$

or, $$I_o = I_i + (I_{\text{initial}} - I_i)\, e^{-t/\tau} \quad \text{...(3.35)}$$

(*vi*) The speed response of a system is defined in terms of *settling time* (it is time taken by the system to reach and remain within a certain percentage tolerance band of the final state value). Smaller the settling time, faster is the response. Typical values of tolerance band are 2% and 5% settling times.

For unit step input, the output reaches the values given in Table 3.1 at various intervals of time. This table also shows the per unit dynamic error at different intervals of time.

Table 3.1: Per Unit Output and Dynamic Error for a Step Input for a First-order System

t/τ	*Per unit output* (I_o/I_i)	*Per unit dynamic error* (E_{dy}/I_i)
0	0	1.000
1	0.632	0.368
2	0.865	0.135
3	0.950	0.050
4	0.982	0.018
5	0.993	0.007
∞	1.000	0

Thus, 5% settling time means that the system has reached its specified value after a time which is thrice the time constant (Table 3.1).

2. Ramp Input :

Consider that the input signal varies linearly with time so that ramp input I_i is prescribed by the relation $I_i + \psi t$, where ψ is constant. Then the governing differential equation is given by

$$(1 + \tau D)I_o = \dot{\psi} t \quad \text{...(3.36)}$$

Now, *Transient response* (complimentary function) is given as:

$$I_{o,t} = A\, e^{-t/\tau} \quad \text{... as before}$$

and, *steady state response* (particular integral) will be;

$$I_{o,s} = (1 - \tau D + \text{terms in } D^2 \text{ and higher})\, \psi t = \psi t - \tau \frac{d}{dt}(\psi \tau)$$

or, $$I_{o,s} = \psi t - \psi \tau \quad \text{...(3.37)}$$

∴ Complete response = Transient response + steady state response

$$I_o = A\, e^{-t/\tau} + (\psi t - \psi \tau) \quad \text{...(3.38)}$$

The value of constant A can be evaluated by applying the initial condition,

At, $t = 0 \quad I_0 = 0$

or, $0 = A - \psi\tau \quad \therefore \quad A = \psi\tau$

∴ $$I_o = \psi t - \psi\tau + \psi\tau \times e^{-t/\tau} = \psi(t - \tau) + \psi\tau\, e^{-t/\tau} \quad \text{...(3.39)}$$

or, $$I_o = \psi[t - \tau(1 - e^{-t/\tau})] \quad \text{...(3.39a)}$$

Fig. 3.23 shows the time response of a first-order system to a ramp input.

The dynamic error,

$$E_{dy.} = I_i - I_o$$

$$= \psi t - [\psi t - \psi\tau + \psi\tau\, e^{-t/\tau}]$$

$$= \underbrace{\psi\tau}_{\text{steady}} - \underbrace{\psi\tau\, e^{-t/\tau}}_{\text{Transient}} \qquad ...(3.40)$$

or, $$\frac{E_{dy.}}{\psi\tau} = 1 - e^{-t/\tau} \qquad ...(3.41)$$

... (in dimensionless form)

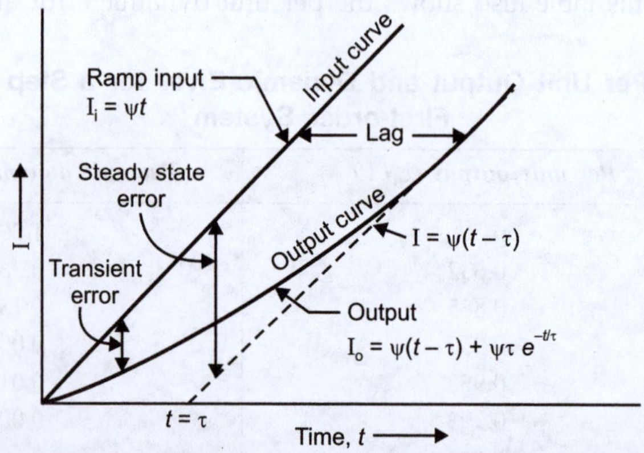

Fig. 3.23. Time response of a first-order system to a ramp input.

Salient features (with ramp input) :

(*i*) The term $\psi\tau$ being independent of time continues to exist and so it is called the steady state error. The term $\psi\tau\, e^{-t/\tau}$ gradually decreases with time and hence is called the *transient error.*

— Since the steady state error is directly proportional to τ (time constant), therefore, the larger the value of τ, the larger will be the magnitude of the error.

— When τ is made small the transient error decreases rapidly; this implies that the system attains the steady state at a faster pace.

(*ii*) The output response curve always lags behind the input curve by a constant amount known as *lag*.

3. Sinusoidal (Harmonic) input :

The frequency analysis of a system pertains to the steady state response of the system to a sinusoidal input. In this analysis, the system is subjected to a sinusoidal input and the system response is studied with frequency as the independent variable. The sinusoid is a unique input signal, and the resulting output signal for a linear system is sinusoidal in the steady state. However, the output signal differs from the input waveform in amplitude and phase.

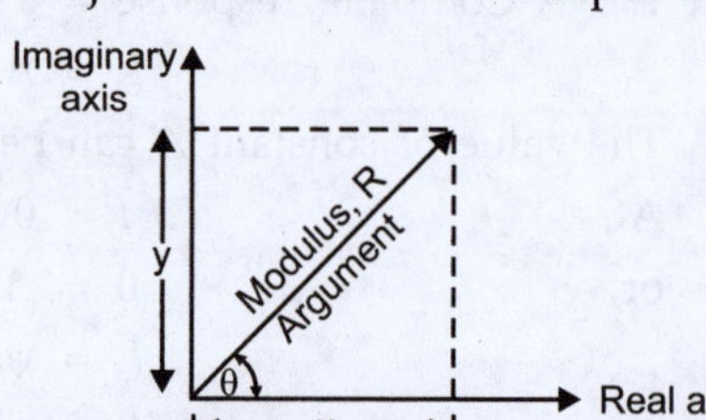

Fig. 3.24.

In order to determine the frequency response of sinusoidal input to a first-order system, let us replace the transfer operator D by a factor $j\omega$ in the input/output relationship; then we get,

$$\frac{I_o}{I_i} = \frac{1}{1 + D\tau} = \frac{1}{1 + j\omega\tau} \qquad ...(3.42)$$

where, ω = Input frequency, rad/s, and

$j = \sqrt{(-1)}$

In a frequency response the following two quantities are of interest : Refer to Fig. 3.24.

(*i*) *Amplitude ratio or modulus* $\left(\frac{I_o}{I_i}\right)$. It prescribes the size of the output amplitude relative to the input amplitude.

(*ii*) *Phase shift* of output relative to input.

For the first-order system represented by equation (3.42),

$$\text{Modulus} = \sqrt{1+(\omega\tau)^2}$$

$$\text{Argument/Phase angle} = \tan^{-1}(\omega\tau) \quad ...(3.43)$$

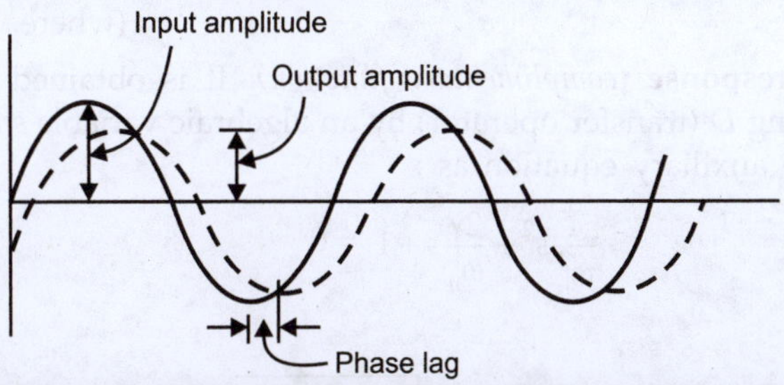

Fig. 3.25. Relationship between an input frequency and corresponding output frequency.

Salient features (with sinusoidal input): Refer to Fig. 3.25.

(*i*) When a system is subjected to a sinusoidal input with frequency ω, its output will also be sinusoidal, but the magnitude of the output amplitude necessarily may not be the same (as the input one). *The ratio of the amplitude* (often called *attenuation*) is given as :

$$\frac{I_o}{I_i} = \frac{1}{\sqrt{1+(\omega\tau)^2}} \quad ...(3.44)$$

Thus, with the increase in input frequency, the amplitude ratio decreases.

(*ii*) The output from the system may not necessarily be in phase with the input; and the phase difference is given by

$$\phi \text{ (phase angle)} = -\tan^{-1}(\omega\tau) \quad ...(3.45)$$

–ve indicates that output *lags* behind the input. When $\omega = \frac{1}{\tau}$, the phase lag is $\frac{\pi}{4}$ or 45°.

As the accuracy of an instrument measuring dynamic input depends upon the time constant, therefore, smaller the time constant, greater the accuracy; for phase shift to be small, the time period τ should be small.

(*iii*) When the input and output signals are given by the relations :

$$I_i = A \sin \omega t, \text{ and } I_o = B \sin(\omega t + \phi) = zA \sin(\omega t + \phi),$$

Then the amplitude ratio may be represented as follows :

$$K = \left|\frac{I_o}{I_i}\right| = \frac{z}{\sqrt{1+(\omega\tau)^2}} \quad ...(3.46)$$

In order to produce amplitude of sinewave without any attenuation ($K = 1$) we must use an instrument whose time constant, $\tau = \frac{\sqrt{z^2 - 1}}{\omega}$.

3.8.4.6. Second-order system responses

In case of typical second-order system having unit static sensitivity, the homogeneous equation is given by :

$$\frac{I_o}{I_i} = \frac{1}{\left(\frac{1}{\omega_n^2}\right)D^2 + \left(\frac{2\gamma}{\omega_n}\right)D + 1}$$

or,
$$\left[\left(\frac{1}{\omega_n^2}\right)D^2 + \left(\frac{2\gamma}{\omega_n}\right)D + 1\right]I_o = I_i \qquad ...(3.47)$$

(where, γ = damping ratio)

(*a*) Transient response ***(complimentary function).*** It is obtained from the auxiliary equation by replacing D (transfer operator) by an algebraic variable s and putting I_i equal to zero; we get the auxiliary equation as :

$$\frac{1}{\omega_n^2}s^2 + \frac{2\gamma}{\omega_n}s + 1 = 0$$

The roots are :

$$s_1, s_2 = \frac{\frac{-2\gamma}{\omega_n} \pm \sqrt{\left(\frac{2\gamma}{\omega_n}\right)^2 - \frac{4}{\omega_n^2}}}{\frac{2}{\omega_n^2}}$$

$$= -\gamma\omega_n \pm \sqrt{\gamma^2\omega_n^2 - \omega_n^2}$$

or,
$$s_1, s_2 = -\gamma\omega_n \pm \omega_n\sqrt{\gamma^2 - 1} \qquad ...(3.48)$$

The *transient solution* has the accepted form,

$$I_{o,t} = A\,e^{s_1 t} + B\,e^{s_2 t}$$

where, A and B = Arbitrary constants to be determined from initial conditions, and s_1, s_2 = Roots of the auxiliary equation (The roots may be real and different, real and equal or imaginary and *that determines* the *nature of transient response of the system*).

The response of the system is of the following three types depending upon the roots of the characteristic equations :

(*i*) Overdamped systems.

(*ii*) Critically damped systems.

(*iii*) Under-damped systems.

(*i*) ***Over-damped systems.*** In this case $\gamma > 1$ and the roots are *real and unequal.*

- There is *heavy damping* and the system responds to the final steady-state value without any oscillations but in a *sluggish manner.*
- *No overshoot* in step response and no *"resonance"* (resonance refers to the output signal *greater* in magnitude than the ideal output) in the *frequency response.*

- The overdamped systems, owing to their sluggish response, are usually unsuitable for several control applications.

(ii) *Critically-damped systems.* In this case $\gamma = 1$ and the roots are *real and equal.*

- The system has a *quick and smooth response to the final steady state value without any oscillations.*
- *No overshoot in the step response and no resonance in the frequency response.*

(iii) *Under-damped systems.* For an undamped system $\gamma < 1$ and roots of the characteristic equation are a complex conjugate pair; these are given as

$$s_1, s_2 = -\gamma\omega_n \pm j\omega_n\sqrt{(1-\gamma^2)}$$

$$= -\gamma\omega_n \pm j\omega_d \qquad ...(3.49)$$

where, $\omega_d = \omega_n\sqrt{(1-\gamma^2)}$; this quantity is called *"damped natural frequency"* of the system (This is the frequency at which the damped system freely oscillates when disturbed).

- Such systems *take a long time to reach steady state, but have quick initial response.*
- In these systems, there are oscillations in the step response and resonance effects in the frequency response for values of $\gamma > 0.707$.
- Majority of instruments and control are generally underdamped (*light damping*).

(b) The steady state response (*particular integral*). It is given by :

$$\left(\frac{1}{\omega_n^2}D^2 + \frac{2\gamma}{\omega_n}D + 1\right)I_{o,s} = I_i$$

or,

$$I_{o,s} = \left(1 + \frac{2\gamma}{\omega_n}D + \frac{1}{\omega_n^2}D^2\right)^{-1} I_i$$

$$= \left(1 - \frac{2\gamma}{\omega_n}D + \text{terms in } D^2 \text{and higher}\right) I_i \qquad ...(3.50)$$

1. Step input : Fig. 3.26 shows the time response of a second-order system to a step input.

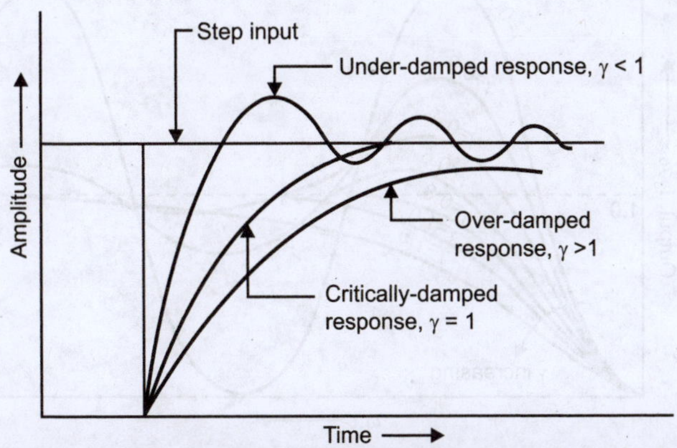

Fig. 3.26. Time response of a second-order system to a step input.

Since the input I_i is a step of constant magnitude, its differential equals zero and we have,

$$I_{o,s} = \left(1 - \frac{2\gamma}{\omega_n} D\right) I_i = I_i$$

$\therefore$ The complete response

$$I_o = I_{o,s} + I_{o,t}$$

or, $$I_o = I_i A e^{s_1 t} + B e^{s_2 t} \quad ...(3.51)$$

For the under-damped system, the complex conjugate pair of roots are given by

$$s_1, s_2 = -\gamma\omega_n \pm j\omega_d$$

$\therefore$ $$I_o = I_i + A e^{-(\gamma\omega_n + j\omega_d)t} + B e^{-(\gamma\omega_n - j\omega_d)t} \quad ...(3.52)$$

Replacing the complex exponentials by sines and cosines, we get

$$I_o = I_i + e^{-\gamma\omega_n t}(A \cos w_d t + B \sin w_d t) \quad ...(3.53)$$

By applying the initial conditions,

At $t = 0$, $I_o = 0$ and $\frac{dI_o}{dt} = 0,$

we get the values of the constants A and B as :

$$A = -I_i, \text{ and}$$

$$B = \frac{-I_i\gamma}{\sqrt{1-\gamma^2}}$$

Inserting these values in equation (3.53), we have

$$I_o = I_i\left[1 - e^{-\gamma\omega_n t}\left\{\cos\omega_d t + \frac{\gamma}{\sqrt{1-\gamma^2}}\sin\omega_d t\right\}\right] \quad ...(3.54)$$

Fig. 3.27 shows the transient response of a second-order system to a unit step function for different values of damping factor γ; *the curves indicate the overshoot and oscillations increase with a reduced damping in the system.*

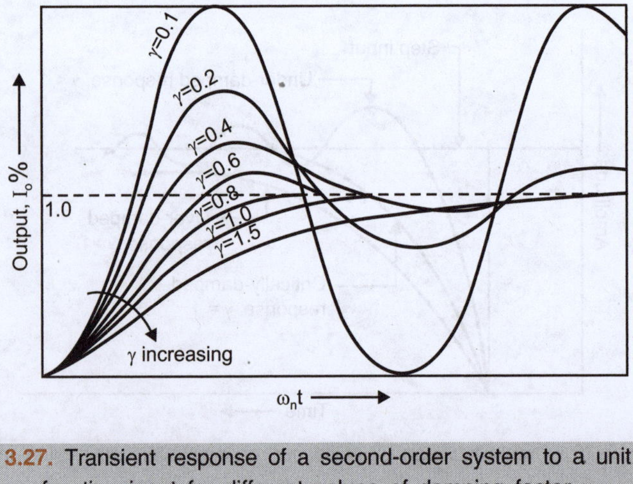

Fig. 3.27. Transient response of a second-order system to a unit step function input for different values of damping factor γ.

2. Sinusoidal (Harmonic) input :

When a sinusoidal input is given to the system, its steady state response is determined by replacing *the operator D by jω in the input/output relationship, as given below :*

$$\frac{I_o}{I_i} = \frac{1}{\frac{1}{\omega_n^2}D^2 + \frac{2\gamma}{\omega_n}D + 1} = \frac{\omega_n^2}{D^2 + 2\gamma\omega_n D + \omega_n^2}$$

$$= \frac{\omega_n^2}{(j\omega)^2 + 2\gamma\omega_n(j\omega) + \omega_n^2} = \frac{\omega_n^2}{(\omega_n^2 - \omega^2) + j(2\gamma\omega_n\omega)} \quad ...(3.55)$$

where, ω = Input frequency in rad/s, and

$j = \sqrt{-1}$

The denominator is a complex number having :

$$\text{Modulus} = \sqrt{\left[\left(\omega_n^2 - \omega^2\right)^2 + (2\gamma\omega_n\omega)^2\right]}$$

$$\text{Argument} = \tan^{-1}\left\{\frac{2\gamma\omega_n\omega}{\omega_n^2 - \omega^2}\right\} \quad ...(3.56)$$

∴ The amplitude ratio,

$$\left|\frac{I_o}{I_i}\right| = \frac{\omega_n^2}{\sqrt{\left[\left(\omega_n^2 - \omega^2\right)^2 + (2\gamma\omega_n\omega)^2\right]}} \quad ...(3.57)$$

and, the system phase lag,

$$\phi = \tan^{-1}\left\{\frac{2\gamma\omega_n\omega}{\omega_n^2 - \omega^2}\right\} \quad ...(3.58)$$

...[using eqn. (3.56)

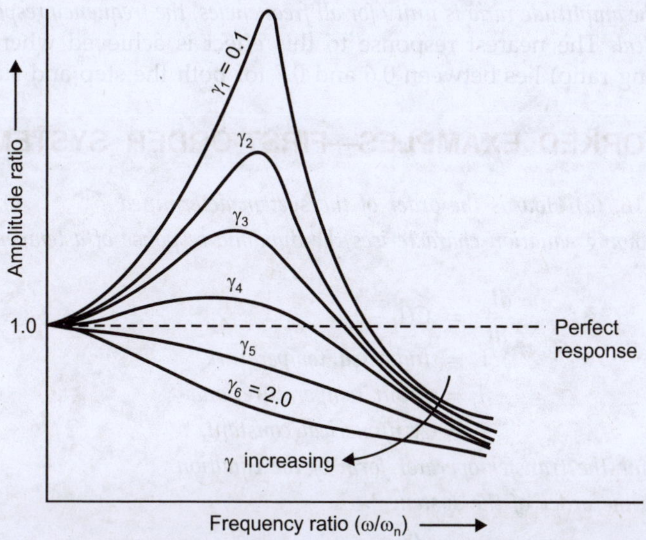

Fig. 3.28. Step response of a second-order system.

Fig. 3.28 and Fig. 3.29 show the variation of amplitude ratio and phase lag versus frequency ratio (ω/ω_n) for various values of damping ratio (γ). From these graphs we observe that the *salient features* of the *steady state response* of a second-order system are :

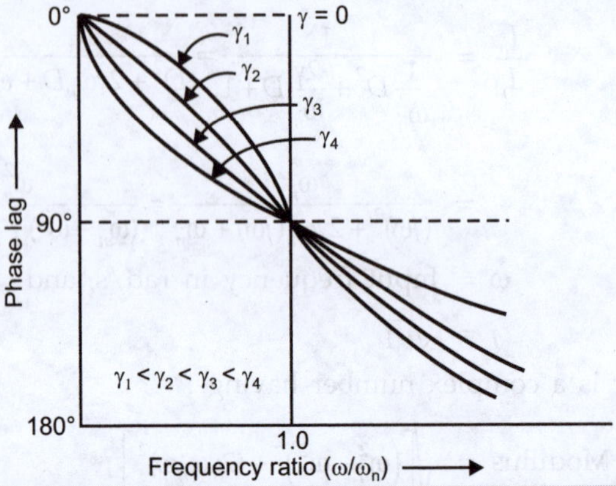

Fig. 3.29. Plot between phase lag and frequency ratio for a second-order system.

(*i*) As *frequency ratio* $\to 0$:
Amplitude ratio $\to 1$;
Phase lag $\to 0°$

(*ii*) As frequency ratio $\to \infty$:
Amplitude ratio $\to 0$;
Phase lag $\to 180°$

(*iii*) When frequency ratio = 1 :
Amplitude ratio $\to \infty$ in undamped systems ($\gamma = 0$)
Phase lag $\to -90°$ in all the systems.
This condition is known as "*resonance*" and can *result in destructive oscillations in lightly damped system.*

(*iv*) *When the amplitude ratio is unity for all frequencies, the frequency response is considered to be ideal.* The nearest response to this effect is achieved when the value of γ (damping ratio) lies between 0.6 and 0.7 for both the step and sinusoidal inputs.

WORKED EXAMPLES—FIRST-ORDER SYSTEMS

Example 3.16. *(a) How is the order of the system determined ?*

(b) The following equation characterises the dynamic response of a temperature measuring instrument :

$$\frac{dI_o}{dt} = C(I_i - I_o)$$

where, I_o = *Indicated temperature,*
I_i = *Input temperature, and*
C = *A numerical constant.*

(i) Determine the transfer operator form of the equation.

(ii) What is the order of the system ?

Solution. (*b*) *Given equation :* $\frac{dI_o}{dt} = C(I_i - I_o)$

(i) Transfer operator form :

The given equation can be rewritten as :

$$\frac{1}{C} \cdot \frac{dI_o}{dt} = I_i - I_o$$

or, $$\frac{1}{C} \cdot \frac{dI_o}{dt} + I_o = I_i$$

or, $$I \cdot \frac{dI_o}{dt} + I_o = I_i \qquad \text{(where, } \tau = \text{time constant} = \frac{1}{C}\text{)}$$

or, $$(\tau D + 1)I_o = I_i$$

$\therefore$ The transfer operator form of the equation is given by :

$$\frac{I_o}{I_i} = \frac{1}{(\tau D + 1)} \quad \textbf{(Ans.)}$$

(ii) Order of the system :

Since the highest differential in the denominator of the transfer operator is *unity*, therefore, the temperature measuring instrument has a *first-order system.*

The grouping of the measurement and control systems is done according to the *order of the highest differential in the denominator* of the system transfer operation.

Examples :

- First-order system : $\frac{I_o}{I_i} = \frac{2}{1+3D}$
- Second-order system : $\frac{I_o}{I_i} = \frac{6}{D^2+3D+4}$, or

$$\frac{I_o}{I_i} = \frac{14}{(1+0.3D)(1+0.2D)}.$$

Example 3.17. *Formulate the governing equation for a first-order system-temperature measurement by a thermal measuring element (say a thermometer or thermocouple).*

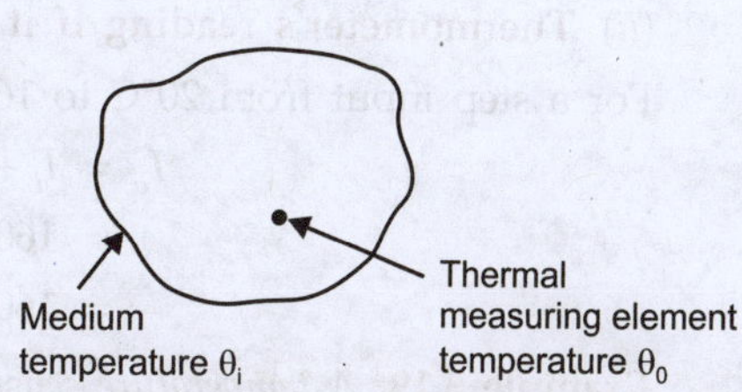

Fig. 3.30. Thermal element.

Solution. Refer to Fig. 3.30.

Let I_i = Temperature of the medium,

I_o = Temperature indicated by the thermal measuring instrument (say a thermometer or thermocouple),

A = Exposed area of the thermal measuring element,

h = Convective heat transfer coefficient,

m = Mass of thermal element, and

c = Specific heat of the element.

Then, the rate of heat flux into the element is,

$$Q = hA(\theta_i - \theta_0) \qquad ...(i)$$

The rate of enthalpy gain by the element

$$= mc\frac{d\theta_0}{dt} \qquad ...(ii)$$

Since the rate of heat flow equals the rate of enthalpy gain by the element, therefore, equating (*i*) and (*ii*) we get :

$$mc\frac{d\theta_0}{dt} = hA(\theta_i - \theta_o)$$

or,
$$\frac{mc}{hA}\cdot\frac{d\theta_o}{dt}+\theta_o = \theta_i$$

$$\tau\frac{d\theta_o}{dt}+\theta_o = \theta_i$$

where, $\tau = \frac{mc}{hA}$ is known as *time constant* of the system.

In terms of *D*-operator (where $D = \frac{d}{dt}$), we have

$$(\tau D + 1)\theta_0 = \theta_i$$

or,
$$\frac{\theta_o}{\theta_i} = \frac{1}{\tau D+1} \quad \text{...Required equation.}$$

which is an equation of *first-order*. **(Ans.)**

Example 3.18. *A thermometer, idealised as a first-order system with a time constant of 2.2 seconds, is suddenly given an input of 160°C from 0°C.*

(i) What will be the reading of the thermometer after 1.2 seconds ?

(ii) Determine its reading if it is initially held at 20°C.

Solution. *Given :* I_i = 160°C; t = 1.2*s*; t = 2.2*s*; I_{initial} = 20°C.

(*i*) **Thermometer's reading after 1.2*s* :**

We know that,
$$I_o = I_i(1 - e^{-t/\tau}) \quad \text{...[Eqn. (3.31)]}$$
$$= 160[1 - e^{-(1.2/2.2)}] = \mathbf{67.27°C} \quad \textbf{(Ans.)}$$

(*ii*) **Thermometer's reading if it was initially held at 20°C :**

For a step input from 20°C to 160°C, we have

$$I_o = I_i + (I_{\text{initial}} - I_i)\, e^{-t/\tau} \quad \text{...[Eqn. (3.35)]}$$
$$= 160 + (20 - 160)\, e^{-(1.2/2.2)}$$
$$= 160 + (20 - 160) \times 0.5796 = \mathbf{78.86°C} \quad \textbf{(Ans.)}$$

Example 3.19. *A temperature sensing device can be modelled as a first-order system with a time constant of 5 seconds. It is suddenly subjected to a step input of 30°C – 160°C. Calculate the temperature indicated by the device after 10 seconds after the start of the process.*

Solution. *Given :* τ = 5*s*; I_{initial} = 30°C, I_i = 160°C; t = 10*s*.

Temperature after 10 seconds is calculated as follows :

$$I_o = I_i + (I_{\text{initial}} - I_i)\, e^{-t/\tau} \quad \text{...[Eqn. (3.35)]}$$
$$= 160 + (30 - 160)\, e^{-10/5}$$
$$= 160 - 130 \times 0.1353 = \mathbf{142.4°C} \quad \textbf{(Ans.)}$$

Example 3.20. *A temperature sensitive transducer when subjected to sudden temperature change takes 9 seconds to reach equilibrium conditions (Three time constants). Calculate the time taken by the transducer to read half of the temperature difference.*

Solution. Time taken to reach equilibrium condition = $3\tau = 9s$ (*Given*)

$\therefore$ Time constant, $\tau = \frac{9}{3} = 3s$

Time taken by the transducer to read half of the temperature difference is calculated as follows :

$$I_o = I_i(1 - e^{t/\tau}) \quad \text{...[Eqn. (3.31)]}$$

or, $$\frac{I_o}{I_i} = 1 - e^{-t/\tau}$$

or, $$0.5 = 1 - e^{-t/3}$$

or, $$e^{-t/3} = 0.5 \quad \text{or,} \quad e^{t/3} = 2$$

$\therefore$ $$t = \mathbf{2.08s} \quad \textbf{(Ans.)}$$

Example 3.21. *A resistance temperature detector has a steady state gain of 2.96 Ω/°C and a time constant of 4.8 seconds. It is subjected to a step change of 80°C in temperature. If its stable resistance before the step change is 90Ω, write down the time-domain equation for resistance and find its value after 12 seconds of application of step input.*

Solution. Step input in terms of resistance

= Steady stage gain × change of temperature

= 0.296 × 80 = 23.68 Ω step input in terms of resistance.

In order to determine the time domain equation for resistance, the value of initial resistance must be added to it.

i.e., $$R_t = 23.68(1 - e^{-t/4.8}) + 90 \quad \textbf{(Ans.)}$$

...*Time domain equation*

$\therefore$ The value of resistance at $t = 12s$.

$$R_{12} = 23.68(1 - e^{-12/4.8}) + 90 = \mathbf{111.74\ \Omega} \quad \textbf{(Ans.)}$$

Example 3.22. *A thermometer initially at a temperature of 15°C is suddenly plunged into a liquid bath maintained at 140°C. After a time interval of 4 seconds, the thermometer indicated a reading of 75°C. Determine :*

(i) The time constant for the thermometer.

(ii) The indicated temperature after five times constant; comment upon this result.

Solution. *Given :* $I_{initial} = 15°C$; $I_i = 140°C$; $t = 4s$; $I_o = 75°C$.

(*i*) The time constant for the thermometer, τ :

When a first-order system is subjected to a step input from $I_{initial}$ to I_i, the indicated temperature is given by the relation :

$$I_o = I_i + (I_{initial} - I_i)e^{-t/\tau} \quad \text{...[Eqn. (3.35)]}$$

$$75 = 140 + (15 - 140)e^{-4/\tau}$$

or, $$e^{-4/\tau} = \frac{75-140}{15-140} = 0.52$$

or, $$e^{4/\tau} = 1.923$$

Taking logs on both sides, we get

$$\frac{4}{\tau} = 0.6539$$

$\therefore$ $$\tau = \mathbf{6.12\ s} \quad \textbf{(Ans.)}$$

(*ii*) Indicated temperature after five times constant, I_o :

Substituting the value in the above equation, we get

$$I_o = 140 + (15 - 140)\, e^{-5\tau/\tau}$$

$$= 140 + (15 - 140)e^{-5} = \mathbf{139.16°C} \quad \textbf{(Ans.)}$$

Comments : After a time interval equivalent to five times constants, the thermometer reaches the *equilibrium condition* (*i.e.*, it indicates a temperature which is approximately equal to the maximum step input).

Example 3.23. *A temperature sensitive transducer used to measure the temperature of a furnace has been idealised as a first-order system subjected to ramp input. If the maximum permissible error in temperature measurement is limited to 3.9°C, calculate the time constant of the transducer when the furnace temperature increases at the rate of 0.2°C/s.*

Solution. *Given :* Dynamic error, $E_{dy.}$ = 3.9°C; slope, ψ = 0.2°C/s.

For the ramp input, the input signal varies linearly with time ($I_i - \psi t$) and the dynamic or measurement error,

$$E_{dy.} = \underbrace{\psi\tau}_{\text{Steady}} \underbrace{-\psi\tau\, e^{t/\tau}}_{\text{Transient}} \qquad \text{...[Eqn. (3.40)]}$$

Since the transient error gradually vanishes with time, we have

$$E_{dy.} = \psi\tau$$

$$\therefore \quad \text{Time constant,} \quad \boldsymbol{\tau} = \frac{E_{dy.}}{\psi} = \frac{3.9}{0.2} = \mathbf{19.5\ s} \quad \textbf{(Ans.)}$$

Example 3.24. *A balloon (weather) carrying a temperature sensing device (of first-order, time constant 8 seconds) rises through the atmosphere at 5 m/s. It supplies information about temperature and altitude through radio signals. At 2500 m height, a temperature indication of 30°C has been received. If the temperature varies with altitude at a uniform rate of 0.011°C/m, determine the time altitude at which 30°C temperature occurs.*

Solution. Time constant of temperature sensing device, τ = 8s

Rate of rise of the balloon = 5 m/s

Temperature indicated at an altitude of 2500 m = 30°C

Rate of temperature variation with altitude = 0.011°C/m

True altitude at 30°C :

Rate of change of temperature with time $y = 0.011 \times 5 = 0.055$°C/s

For a ramp signal applied to a first-order system, the error, when transient conditions die away,

$$E_{dy.} = \psi\tau = 0.055 \times 8 = 0.44°\text{C}$$

$$\therefore \quad \text{Error in amplitude} = \frac{0.44}{0.011} = 40 \text{ m}$$

$$\therefore \quad \textit{Actual altitude} = 2500 - 40 = \mathbf{2460\ m} \quad \textbf{(Ans.)}$$

Example 3.25. *A thermometer with time constant of 50 seconds is used for measuring temperature cycling with 500 s time period. Calculate :*

(i) The ratio of output to input.

(ii) The time lag.

Solution. Time constant of Thermometer, τ = 50 s

Time period, t = 500 s

Now, frequency of temperature variation,

$$\omega = \frac{2\pi}{t} = \frac{2\pi}{500} = 0.01257 \text{ rad/s}$$

(i) The ratio of output to input, $\frac{I_o}{I_i}$:

$$\frac{I_o}{I_i} = \frac{1}{\sqrt{1+(\omega\tau)^2}}$$

$$= \frac{1}{\sqrt{1+(0.01257\times 50)^2}} = \mathbf{0.8467} \quad \textbf{(Ans.)}$$

(ii) Time lag :

$$\text{Phase shift} = \tan^{-1}(\omega\tau) = \tan^{-1}(0.01257 \times 50) = 32.15° \text{ or } 0.561 \text{ rad.}$$

$$\therefore \quad \text{Time lag} = \frac{1}{\omega}\times \text{phase shift in radian}$$

$$= \frac{1}{0.01257}\times 0.561 = \mathbf{44.63\ s} \quad \textbf{(Ans.)}$$

Example 3.26. *During a particular process, the air temperature cycle at the rate of 1 cycle every 4 minutes. The time constant of the temperature measuring device is 20 seconds. Determine :*

(i) The variation in the indicated temperature if true (input) temperature has a sinusoidal variation of ± 25°C.

(ii) The time by which the maximum reading of the thermometer lags the true maximum value.

Solution. *Given :* Air temperature variation = 1 cycle every 4 minutes

Time constant of the device, $\tau = 20$ s

Sinusoidal variation of the input, $I_i = \pm 25°C$

Now, $t = 4$ minutes

$$\therefore \quad f = \frac{1}{t} = \frac{1}{4\times 60} = \frac{1}{240} \text{ cycles/s}$$

and, $$\omega = 2\pi f = 2\pi\times\frac{1}{240} = 0.0262 \text{ rad./s}$$

(i) The variation in the indicated temperature (I_o) :

The steady state (when transient errors are omitted) solution for the sinusoidal input is given as :

$$\frac{I_o}{I_i} = \frac{1}{\sqrt{(1+\omega\tau)^2}}\sin(\omega t - \phi) \quad \text{...[Eqn. (3.44)]}$$

where, I_o = Output temperature as indicated by the thermometer,

I_i = True temperature, and

ϕ = Phase lag = $\tan^{-1}(\omega\tau)$

Substituting the value of ω and τ (calculated above) in the above equation, we get

$$\frac{I_o}{I_i} = \frac{I}{\sqrt{1+(0.0262\times 20)^2}} = 0.886$$

$$I_o = \pm 25 \times 0.886 = \mathbf{\pm 22.15°C} \quad \textbf{(Ans.)}$$

(ii) The lag (in seconds) :

$$\text{Phase lag} = \tan^{-1}(\omega\tau)$$

or, $$\phi = \tan^{-1}(0.0262 \times 20) = 27.65° \text{ or } 0.4826 \text{ rad.}$$

The lag in seconds

$$= \frac{1}{\omega} \times \text{phase lag in radian}$$

$$= \frac{1}{0.0262} \times 0.4826 = \textbf{18.42 seconds} \quad \textbf{(Ans.)}$$

Example 3.27. *A first-order instrument having an input of sine-wave pattern is required to measure signals with frequency response of 120 cycles per second. If an amplitude inaccuracy of 4 per cent is permissible, calculate the maximum time constant and time lag at 90 cycles per second.*

Solution. *Given :* Input frequency = 120 cycles per second ;

Amplitude inaccuracy permissible = 4%.

Now, $\omega = 2\pi f = 2\pi \times 120 = 754$ rad/s

Maximum time constant τ :

Corresponding to 4% amplitude inaccuracy, the amplitude ratio of output and input signals is $\left(1 - \frac{4}{100}\right) = 0.96.$

For a first-order instrument, the amplitude ratio is given as :

$$\frac{I_o}{I_i} = \frac{1}{\sqrt{1 + (\omega\tau)^2}} \quad \text{...[Eqn. (3.44)]}$$

or, $$0.96 = \frac{1}{\sqrt{(1 + 754\tau)^2}}$$

or, $$(1 + 754\tau)^2 = \left(\frac{1}{0.96}\right)^2 = 1.085$$

or, $$1 + 754\tau = 1.0416$$

∴ $$\tau = \mathbf{5.517 \times 10^{-5}\ s} \quad \textbf{(Ans.)}$$

Time lag at 90 cycles per second :

We know that for a sinusoidal input,

$$\phi = \tan^{-1}(\omega\tau) = \tan^{-1}(2\pi \times 90 \times 5.517 \times 10^{-5})$$

$$= 1.787° \quad \text{or} \quad 0.03119 \text{ rad}$$

∴ $$\text{Time lag} = \frac{1}{\omega} \times \text{phase lag in radian}$$

$$= \frac{1}{2\pi \times 90} \times 0.03119 = \mathbf{5.515 \times 10^{-5}\ s} \quad \textbf{(Ans.)}$$

Example 3.28. *The temperature of a furnace is found to vary sinusoidally between 520°C and 580°C with a periodic time of 50 seconds. A thermocouple system with a time constant of 10 seconds is employed to measure the furnace temperature. Determine :*

(i) The maximum and minimum values that will be indicated by the thermocouple.

(ii) The phase shift and the corresponding time lag between the temperature signals and the thermocouple input signals.

Solution. *Given :* Range of variation of temperature = 520° to 580°C

Periodic time, t = 50 seconds

Time constant of the thermocouple system, τ = 10 seconds.

(i) The maximum and minimum values of the temperature :

The temperature would oscillate about the mean value of $\frac{580+520}{2}$ = 550°C, with an amplitude of 30°C.

Angular frequency of oscillation,

$$\omega = 2\pi f = \frac{2\pi}{t} = \frac{2\pi}{50} = 0.1256 \text{ rad/s}$$

The amplitude ratio, after the transient effects die away, is given by :

$$\frac{I_o}{I_i} = \frac{1}{\sqrt{1+(\omega\tau)^2}} = \frac{1}{\sqrt{1+(0.1256\times10)^2}} = 0.6229$$

$$\therefore \quad I_o = 0.6229 \times 30 = 18.7°\text{C}$$

$\therefore$ *Maximum temperature* = 550 + 18.7 = **568.7°C** (Ans.)

Minimum temperature = 550 − 18.7 = **531.3°C** (Ans.)

(ii) Phase shift and time lag :

$$\text{Phase shift} = \tan^{-1}(\omega\tau)$$

$$= \tan^{-1}(0.1256 \times 10) = 51.47°, \text{ or, } 0.898 \text{ rad.}$$ **(Ans.)**

$$\text{Time lag} = \frac{1}{\omega} \times \text{phase shift in radian}$$

$$= \frac{1}{0.1256} \times 0.898 = 7.15 \text{ s}$$ **(Ans.)**

Fig. 3.31 shows the variation of actual and indicated temperatures of furnace.

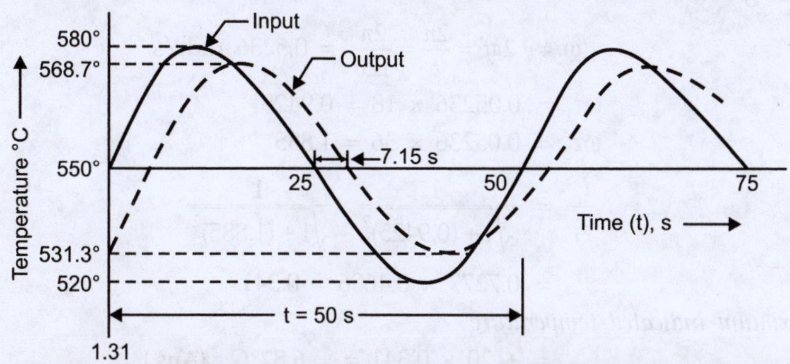

Fig. 3.31. Variation of actual and indicated temperatures of furnace.

Example 3.29. *A first-order instrument is subjected to a sinusoidal input given by relation : I_i = 0.35 sin 25t. If the instrument has a time constant of 0.3 second, develop an expression for the corresponding output.*

Solution. *Given* : Sinusoidal input relation, I_i = 0.35 sin 25 t

Time constant of the instrument, τ = 0.3 s.

Expression for the output :

The amplitude ratio, after the transient effects die away, is given as :

$$\frac{I_o}{I_i} = \frac{1}{\sqrt{1+(\omega\tau)^2}} = \frac{1}{\sqrt{1+(25\times0.3)^2}} = 0.132$$

(Here, ω = 25 ..... from the given relation)

∴ Magnitude of output,

$$I_o = 0.132 \times \text{magnitude of input}$$
$$= 0.132 \times {}^*0.35 = 0.0462 \qquad \text{(* from the given relation)}$$

Phase shift, $\phi = \tan^{-1}(\omega\tau)$

$$= \tan^{-1}(25 \times 0.3) = 82.4°$$

∴ I_o = **0.0462 sin (25*t* – 82.4°)** ...(*Required expression*) **(Ans.)**

Example 3.30. *A thermometer (its glass bulb protected by a well) is subjected to a temperature which is changing ± 20°C every 120 seconds. The measurement system may be idealised by a double capacity system with time constants of 18 seconds for the bulb and 36 seconds for the well.*

Determine the maximum value of temperature indicated by the thermometer. After how much time of the actual occurrence of temperature, it will be indicated by the instrument (i.e., time lag) ?

Solution. *Given* : Rate of change of temperature : ± 20°C every 120 seconds

Time constant for the bulb = 18 seconds

Time constant for the well = 36 seconds.

Maximum value of the temperature indicated :

The amplitude ratio (I_o/I_i) for a double capacity system is given as :

$$\frac{I_o}{I_i} = \frac{1}{\sqrt{1+(\omega\tau_1)^2}} \times \frac{1}{\sqrt{1+(\omega\tau_2)^2}}$$

and, the angle of lag, $\phi = \tan^{-1}(\omega\tau_1) + \tan^{-1}(\omega\tau_2)$

From the given data :

$$\omega = 2\pi f = \frac{2\pi}{t} = \frac{2\pi}{120} = 0.5236 \text{ rad./s}$$

$$\omega\tau_1 = 0.05236 \times 18 = 0.9425$$

$$\omega\tau_2 = 0.05236 \times 36 = 1.885$$

$$\frac{I_o}{I_i} = \frac{1}{\sqrt{1+(0.9425)^2}} \times \frac{1}{\sqrt{1+(1.885)^2}}$$

$$= 0.7277 \times 0.4686 = 0.341$$

∴ *Maximum indicated temperature*

$$= \pm 20 \times 0.341 = \pm \mathbf{6.82°C} \quad \textbf{(Ans.)}$$

Time lag :

$$\text{Angle of lag} = \tan^{-1}(0.9425) + \tan^{-1}(1.885)$$
$$= 43.30° + 62.05° = 105.35° \text{ or } 1.839 \text{ rad.}$$

$$\therefore \quad \text{Time lag (delay)} = \frac{1}{\omega} \times \text{angle of lag (in rad.)}$$

$$= \frac{1}{0.05236} \times 1.839 = \mathbf{35.12 \text{ seconds}} \quad \textbf{(Ans.)}$$

Example 3.31. *An input of 2 sin 2t + 0.5 sin10t is applied to a first-order instrument having a time constant of 0.3 second. Find the output.*

Solution. *Given* : Input equation :

$$I_i = 2 \sin 2t + 0.5 \sin 10t$$

Time constant, τ = 0.3 s

Output :

The output of the instrument to an input 2 sin 2*t* + 0.5 sin 10*t* when time constant is 0.3s, is given as:

$$I_o = \frac{1}{\sqrt{1+(2\times0.3)^2}}\sin[2t-\tan^{-1}(2\times0.3)] + \frac{1}{\sqrt{1+(10\times0.3)^2}}\sin[10t-\tan^{-1}(10\times0.3)]$$

$$= 0.857 \sin(2t - 30.96°) + 0.316 \sin(10t - 71.56°) \quad \textbf{(Ans.)}$$

Example 3.32. *A first-order system having a time constant of 0.15 second is used to measure a signal prescribed by the relation: I_i = 2 sin 2t + 0.2 cos 8t. Develop an expression for the corresponding output.*

Solution. *Given* : Relation followed by the signal : $I_i = 2 \sin 2t + 0.2 \cos 8t$

Time constant of the system, $\tau = 0.15$ s

Expression for the output :

The concept of sinusoidal transfer function can be applied for the frequency response only if input, I_i, is either a sinewave or sum of sinewaves of different frequencies and amplitudes. In order to satisfy this condition, the given input need to be rewritten in the following form :

$$I_i = 2 \sin 2t - 0.2 \sin [8t + \pi \text{ (or } 180°)]$$

Further, the first-order measuring systems are *linear one*, and therefore the output can be obtained by using the *superposition principle.*

- For the sinusoidal input : $I_i = 2 \sin 2t$

$$\frac{I_o}{I_i} = \frac{1}{\sqrt{1+(\omega\tau)^2}} = \frac{1}{\sqrt{1+(2\times0.15)^2}} = 0.9578$$

Magnitude of output = 0.9578 × 2 = 1.9156

Phase shift, $\phi = \tan^{-1}(\omega\tau) = \tan^{-1}(2 \times 0.15) = 16.7°$

∴ Output, $I_o = 1.9156 \sin(2t - 16.7°)$

- For the sinusoidal input : $I_i = -0.2 \sin(8t + \pi)$

$$\frac{I_o}{I_i} = \frac{1}{\sqrt{1+(\omega\tau)^2}} = \frac{1}{\sqrt{1+(8\times0.15)^2}} = 0.6402$$

Magnitude of output = 0.2 × 0.6402 = 0.128

Phase shift, $\phi = \tan^{-1}(\omega\tau) = \tan^{-1}(8 \times 0.15) = 50.19°$

∴ Output, $I_o = -0.128 \sin(8t + \pi - 50.19°)$

Hence, *overall output*, $\boldsymbol{I_o} = 1.9156 \sin(2t - 16.7°) - 0.128 \sin(8t + \pi - 50.19°)$ **(Ans.)**

SECOND-ORDER SYSTEMS

Example 3.33. *Formulate the governing equation for a second-order system–spring mass system with damping.*

Solution. Refer to Fig. 3.32.

Let, x_i = Input displacement,

x_o = Output displacement,

k = Stiffness of the spring,
C_d = Viscous damping coefficient, and
γ = Damping ratio.

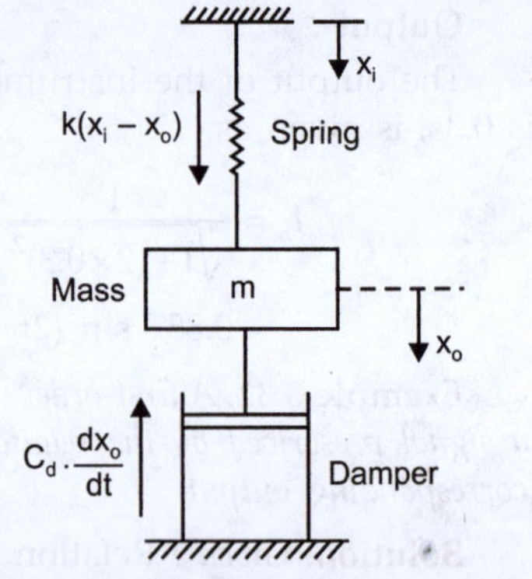

Fig. 3.32. Spring-mass system.

The *forces* acting on the mass are :

(*i*) As both ends of the springs are free to move, therefore

Spring force = Spring stiffness × displacement of one end of the spring relative to other

= $k(x_i - x_o)$, acting *downward*.

(*ii*) One end of the dashpot fixed; there is a reaction force acting in the *upward* direction.

Damping force = Damping coefficient × velocity

$$= C_d \times \frac{dx_o}{dt}$$

For translational systems, Newton's law states that,

$$\Sigma \text{ Force} = \text{Mass} \times \text{acceleration} = m\frac{d^2x_o}{dt^2}$$

or,
$$m \cdot \frac{d^2x_o}{dt^2} = k(x_i - x_o) - C_d\frac{dx_o}{dt}$$

or,
$$mD^2x_o + C_dDx_o + kx_o = kx_i \qquad \left(\text{where } D = \frac{d}{dt}, \text{ and } D^2 = \frac{d^2}{dt^2}\right)$$

or,
$$x_i = \left(\frac{m}{k}D^2 + \frac{C_d}{k}D + 1\right)x_o \qquad \text{...Required equation}$$

which is an equation of *second-order* type. **(Ans.)**

Comparing the above expression with the standard second-order form, we have :

Undamped natural frequency,

$$\omega_n = \sqrt{\frac{k}{m}}$$

and
$$\frac{2\gamma}{\omega_n} = \frac{C_d}{k}$$

∴ Damping ratio,
$$\gamma = \frac{C_d\omega_n}{2k} = \frac{C_d}{2k}\sqrt{\frac{k}{m}}$$

or,
$$\gamma = \frac{C_d}{2\sqrt{km}} \qquad \text{...Required equation.}$$

Example 3.34. *(a) Write down the expression describing the motion of linear and rotational displacement systems of the second-order.*

(b) The pen arrangement of U.V. recorder (second-order system) has a mass of 4.5g. Calculate the percentage reduction in mass if it is desired to have 15 per cent increase in natural frequency of the recorder.

Solution. *(a)* • The expression for *linear displacement* (spring-mass-damper) *system* is given by :

$$m\frac{d^2I_o}{dt^2}+C_d\frac{dI_o}{dt}+k\,I_o = k\,I_i \quad ...(i)$$

where, m = Mass (kg),
C_d = Viscous damping force (Ns/m),
k = Spring stiffness (N/m),
I_o = Output reading, and
I_i = Input reading.

• For the *rotational system*, the expression may be written as :

$$J\frac{d^2I_o}{dt^2}+C_d\frac{dI_o}{dt}+q\,I_o = q\,I_i$$

where, J = Inertia (kg m^2), and
q = Torsional stiffness.

Comparing these expressions with differential equation in the standard form,

$$\frac{1}{\omega_n^2}\cdot\frac{d^2I_o}{dt^2}+\frac{2\gamma}{\omega_n}\cdot\frac{d\,I_o}{dt}+I_o = I_i, \text{ we get}$$

Natural frequency, $\omega_n = \sqrt{\frac{k}{m}} = \sqrt{\frac{q}{J}}$, as the case may be. **(Ans.)**

(b) *Given* : m = 4.5 g; Percentage increase required = 15%.
Using subscripts 1 and 2 for initial and final values respectively, we have :

$$\omega_{n1}^2 = \frac{k}{m_1}, \text{ and, } \omega_{n2}^2 = \frac{k}{m_2}$$

$$\therefore \quad m_2 = m_1 \times \left(\frac{\omega_{n1}}{\omega_{n2}}\right)^2$$

But, $\omega_{n2} = 1.15\,\omega_{n1}$ ...(*Given*)

$$\therefore \quad m_2 = m_1 \times \left(\frac{\omega_{n1}}{1.15\,\omega_{n1}}\right)^2 = 0.756\,m_1$$

∴ *Percentage reduction in mass*

$$= \left(\frac{m_1 - m_2}{m_1}\right) = \left(\frac{1-0.756m_1}{m_1}\right)\times 100 = \mathbf{24.4\%} \quad \textbf{(Ans.)}$$

Example 3.35. *A second-order system follows the differential equation given below :*

$$\frac{d^2I_o}{dt^2}+3\frac{dI_o}{dt}+30\,I_o = 30\,I_i$$

where, I_o and I_i are output and input quantities respectively. Determine the following :

(i) Damping ratio.
(ii) Damped natural frequency.
(iii) Static sensitivity.
(iv) Time constant.

Solution. The standard form of the differential equation of a second-order system is given as :

$$\frac{1}{\omega_n^2}\cdot\frac{d^2I_o}{dt^2}+\frac{2\gamma}{\omega_n}\cdot\frac{dI_o}{dt}+I_o = k\,I_i \qquad ...(i)$$

Since the term I_o in eqn. (*i*) has a unit coefficient, therefore to recast the given equation in the standard form, let us divide it throughout by 30; we get

$$\frac{1}{30}\cdot\frac{d^2I_o}{dt^2}+\frac{1}{10}\frac{dI_o}{dt}+I_o = I_i \qquad ...(ii)$$

Comparing eqns. (*i*) and (*ii*), we get

$$\omega_n^2 = 30;\ \frac{2\gamma}{\omega_n}=\frac{1}{10}=0.1;\ k=1$$

∴ Natural frequency, $\omega_n = \sqrt{30} = 5.477$ rad./s

(*i*) Damping ratio, γ :

$$\frac{2\gamma}{\omega_n} = 0.1, \text{ or } \gamma = \frac{\omega_n}{2}\times 0.1$$

or, $$\gamma = \frac{5.477}{2}\times 0.1 = \mathbf{0.274} \quad \textbf{(Ans.)}$$

(*ii*) Damped natural frequency, ω_d :

$$\omega_d = \omega_n\sqrt{1-\gamma^2} = 5.477\sqrt{1-0.274^2} = \mathbf{5.267\ rad./s} \quad \textbf{(Ans.)}$$

(*iii*) Static sensitivity :

Static sensitivity, k = **1** **(Ans.)**

(*iv*) Time constant, τ :

$$\tau = \frac{1}{\omega_n}=\frac{1}{5.477} = \mathbf{0.1826\ s} \quad \textbf{(Ans.)}$$

Example 3.36. *The following data pertain to a torque sensing transducer connected to the shaft of an electric motor which drives a load :*

Input torque = Sinusoidal with frequency varying at 1.95 rad./s; Maximum permissible error = ± 8 per cent; Moment of inertia of load, J = 0.14 kg m^2*; Torsional constant of the shaft, q = 1.22 Nm/rad. Calculate the following :*

(*i*) *Natural frequency.*

(*ii*) *Damping ratio.*

(*iii*) *Damped natural frequency.*

(*iv*) *Time constant of the system.*

Solution. *Given :* ω = 1.95 rad/s; Max. permissible error = ± 8%; J = 0.14 kg m^2; q = 1.22 Nm/rad.

(*i*) Natural frequency, ω_n :

$$\omega_n = \sqrt{\frac{q}{J}}=\sqrt{\frac{1.22}{0.14}} = \mathbf{3.95\ rad./s} \quad \textbf{(Ans.)}$$

(*ii*) Damping ratio, γ :

Also, normalised frequency ratio, $r = \frac{\omega}{\omega_n}=\frac{1.95}{2.95}=0.661$

Corresponding to maximum permissible error of ± 8%, the amplitude ratio is 1.08. Further the amplitude ratio (*M*) is prescribed by the relation :

$$M = \frac{1}{\sqrt{(1-r^2)^2 + (2\gamma r)^2}}$$

or,
$$1.08 = \frac{1}{\sqrt{(1-0.661^2)^2 + (2\gamma \times 0.661)^2}}$$

$$= \frac{1}{\sqrt{0.317 + 1.747\gamma^2}}$$

or,
$$0.317 + 1.747\gamma^2 = \left(\frac{1}{1.08}\right)^2 = 0.857$$

∴ γ = **0.555** **(Ans.)**

(*iii*) Damped natural frequency, ω_d :

$$\omega_d = \omega_n\sqrt{1-\gamma^2}$$

$$= 2.95\sqrt{1-(0.555)^2} = \textbf{3.454 rad./s} \quad \textbf{(Ans.)}$$

(*iv*) Time constant, τ :

$$\tau = \frac{1}{\omega_n} = \frac{1}{2.95} = \textbf{0.339 s} \quad \textbf{(Ans.)}$$

Example 3.37. *(a) Discuss the time domain parameters for the response of step input to second-order system.*

(b) When a step input was given to a second-order system, the measurements revealed that the system had an overshoot of 12 per cent in a rise time of 0.22 second. Determine :

(i) Effective damping ratio.

(ii) Undamped natural frequency of the system.

Solution. (*a*) In an under-damped system, the time domain parameters of interest are (Refer to Fig. 3.33) :

(*i*) How long does the system take to practically reach its final steady value ?

(*ii*) How fast the system moves to follow the applied input ?

(*iii*) How oscillatory is the system ?

- **Settling time (t_s)** is defined as the *time required for the system to reach a steady state.*
- **Rise time (t_r)** is the *time required for the system to rise from 0 to 100% of its final value.*

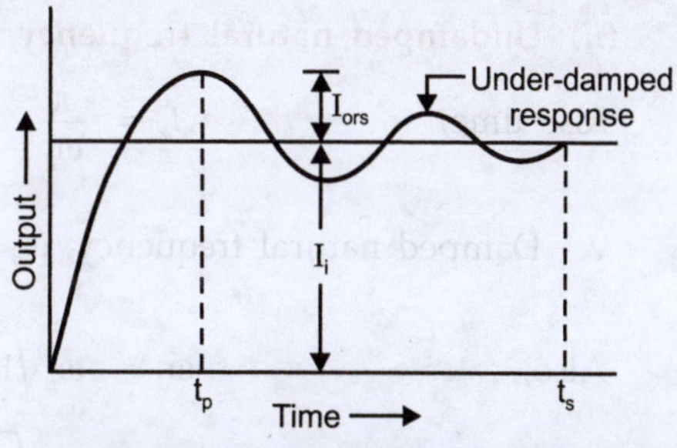

Fig. 3.33. Step response illustrating response, rise and settling times.

- **Peak time (t_p)** is the *time taken to reach the first peak (overshoot).*

The percentage overshoot $= \frac{I_{ors}}{I_i} \times 100$.

Mathematical analysis of the under-damped system would show that :

$$\textit{Rise time } (t_r) = \frac{\pi}{\omega_d}$$

$$\textit{Peak time } (t_p) = \frac{\pi}{\omega_n\sqrt{1-\gamma^2}}$$

$$\textit{Overshoot} = \exp\left[\frac{-\pi\gamma}{\sqrt{(1-\gamma^2)}}\right]$$

(*b*) *Given* : Percentage overshoot = 12%; Rise time = 0.22 s.

(*i*) Effective damping ratio, γ :

$$\text{Percentage overshoot} = \exp\left[\frac{-\pi\gamma}{\sqrt{1-\gamma^2}}\right]\times 100$$

or,
$$12 = \exp\left[\frac{-\pi\gamma}{\sqrt{1-\gamma^2}}\right]\times 100$$

Rearranging and taking logs on both sides, we get

$$\log_e\left(\frac{12}{100}\right) = \frac{-\pi\gamma}{\sqrt{1-\gamma^2}}$$

$$-2.12 = \frac{-\pi\gamma}{\sqrt{1-\gamma^2}}$$

or,
$$0.675\sqrt{1-\gamma^2} = \gamma$$

$$0.456 - 0.456\,\gamma^2 = \gamma^2 \quad \text{...(Squaring both sides)}$$

∴
$$\gamma = \sqrt{\frac{0.456}{1.456}} = \mathbf{0.56} \quad \text{(Ans.)}$$

(*ii*) Undamped natural frequency, f_n :

Rise time,
$$t_r = \frac{\pi}{\omega_d}$$

∴ Damped natural frequency,
$$\omega_d = \frac{\pi}{t_r} = \frac{\pi}{0.22} = 14.28 \text{ rad./s}$$

Also,
$$\omega_d = \omega_n\sqrt{1-\gamma^2} \quad \text{(where, } \omega_n = \text{undamped natural frequency)}$$

$$14.28 = \omega_n\sqrt{1-0.56^2}$$

or,
$$\omega_n = \frac{14.28}{\sqrt{1-0.56^2}} = 17.24 \text{ rad./s}$$

∴
$$f_n = \frac{\omega_n}{2\pi} = \frac{17.24}{2\pi} = \mathbf{2.74 \text{ Hz}} \quad \text{(Ans.)}$$

Example 3.38. *A second-order instrument has a natural frequency of 5 Hz and a damping ratio of 0.62. If the excitation frequency of the system is 7 Hz, determine the error due to the proximity of excitation frequency with the natural frequency of the system.*

Solution. *Given :* Natural frequency of the instrument, f_n = 5 Hz.

Damping ratio, γ = 0.62

Excitation frequency of the system, f = 7 Hz

Now, the ratio of excitation frequency to natural frequency

$$r = \frac{\omega}{\omega_n} = \frac{f}{f_n} = \frac{7}{5} = 1.4$$

$$\text{Amplitude ratio, } M = \frac{1}{\sqrt{(1-r^2)^2 + (2\gamma r)^2}} = \frac{1}{\sqrt{(1-1.4^2)^2 + (2\times 0.62\times 1.4)^2}} = 0.504$$

$\therefore$ The *"error"* due to proximity of excitation frequency with the natural frequency of the system

$$= (1 - 0.504) \times 100 = \mathbf{49.6\%} \quad \textbf{(Ans.)}$$

Example 3.39. *While measuring force by a diaphragm type pressure transducer (second-order system), the system is found to have a natural frequency of 800 cycles per second (cps) and a damping ratio of 62 per cent.*

Determine the frequency range over which the amplitude ratio corresponding to sinusoidal input deviates by a maximum amount of 12 per cent.

Solution. *Given :* Natural frequency of the system = 800 cps

Damping ratio, γ = 62 per cent

Maximum amount of deviation of amplitude ratio = 12 per cent.

Corresponding to a maximum deviation of 12%, the amplitude ratio (M) is either 0.88 or 1.12.

Now,
$$M = \frac{1}{\sqrt{(1-r^2)^2 + (2\gamma r)^2}}$$

where, r = Frequency ratio = $\dfrac{f}{f_n} = \dfrac{\omega}{\omega_n}$

(*i*) *When M = 1.12 :*

$$1.12 = \frac{1}{\sqrt{(1-r^2)^2 + (2\times 0.62\times r)^2}}$$

or,
$$(1 - r^2)^2 + (2 \times 0.62\ r)^2 = \left(\frac{1}{1.12}\right)^2 = 0.797$$

or,
$$1 + r^4 - 2\ r^2 + 1.538\ r^2 = 0.797$$

or,
$$r^4 - 0.462\ r^2 + 0.203 = 0$$

The solution has imaginary roots and this implies that with γ = 0.62, the *output would never be 1.12 times the output for any frequency.*

(*ii*) *When M = 0.88 :*

$$0.88 = \frac{1}{\sqrt{(1-r^2)^2 + (2\times 0.62\times r)^2}}$$

or, $(1-r^2)^2 + (2 \times 0.62\, r)^2 = \left(\frac{1}{0.88}\right)^2 = 1.29$

or, $1 + r^4 - 2r^2 + 1.538\, r^2 = 1.29$

or, $r^4 - 0.462\, r^2 - 0.29 = 0$

or, $r^2 = \frac{0.462 \pm \sqrt{(0.462)^2 + 4 \times 0.29}}{2} = 0.817$

or, $r = 0.904$

$\therefore$ $f = f_n \times r = 800 \times 0.904 = 723$ cps

Thus, the deviation remains within 12 per cent of output for the *frequency range* **0 – 723 cps** **(Ans.)**

Example 3.40. *A unit amplitude sinusoidal input signal having a frequency of 0.6 Hz is given to a second-order control system of transfer function :* $\frac{I_o}{I_i} = \frac{8}{D^2 + 4D + 20}$.

Develop an expression for the steady state response of the system and determine the corresponding output amplitude, output frequency and phase lag.

Solution. Frequency, $\omega = 2\pi f = 2\pi \times 0.6 = 3.77$ rad/s

As the sinusoidal input signal has a *unit* amplitude, therefore,

$$I_i = 1 \sin (3.77\, t)$$

Putting $D = j\omega$ in the given relation, we get

$$\frac{I_o}{I_i} = \frac{8}{(j\omega)^2 + 4j\omega + 20} = \frac{8}{(3.77j)^2 + 4 \times 3.77j + 20}$$

$$= \frac{8}{-14.21 + 15.08j + 20} = \frac{8}{5.79 + 15.08j} = \frac{1}{0.724 + 1.885j}$$

The denominator has a magnitude of

$$\sqrt{(0.724)^2 + (1.885)^2} = 2.02$$

$$\text{Phase lag, } \phi = \tan^{-1} \frac{1.885}{0.724} = 69° = 1.204 \text{ rad}$$

$$\text{Magnitude of output} = \frac{1}{2.02} \times \text{magnitude of input}$$

$$= \frac{1}{2.02} \times 1 = 0.495$$

$\therefore$ The required expression for the output in the sinusoidal form is :

$$I_o = 0.495 \sin (3.77\, t - 69°) \quad \textbf{(Ans.)}$$

Hence, *corresponding output amplitude* = **0.495** (*as compared with unit amplitude of input*) **(Ans.)**

Output frequency = **3.77** rad/s (*same as that of input signal*) **(Ans.)**

Phase lag = **69°** (*between input and output*) **(Ans.)**

3.9 ERRORS IN MEASUREMENTS

3.9.1. Introduction

Errors and uncertainties are *inherent* in the process of making any measurement and in the instrument with which the measurements are made. The study of error is important as a step in finding ways of reducing them, and also as means of estimating the reliability of final results.

The accuracy and precision of a measuring instrument depends upon its design, proper selection of material for each and every part, and skill of manufacturing.

3.9.2. Limiting (or Guarantee) Errors

In most instruments the accuracy is generated to be *within a certain percentage of full scale reading*. The manufacturer has to specify the deviations from the *nominal value* of a particular quantity. *The limits of these deviations from the specified value are defined as* **Limiting errors** *or* **Guarantee errors.**

The **relative error** is *defined as the ratio of the error to the specified (nominal) magnitude of a quantity :*

If, Q_s = The specified quantity or true quantity of a component, and

Q_a = The actual or erroneous quantity ;

Then, error $= Q_a - Q_s$ $(= \delta Q)$

and, Relative limiting error (also called accuracy),

$$E_r = \frac{Q_a - Q_s}{Q_s} = \frac{\delta Q}{Q_s}$$

Percentage limiting error,

$$\% E_r = E_r \times 100$$

If a component has a relative limiting error $\pm E_r$, then limiting values of the component, $Q_a = Q_s (1 \pm E_r)$. Thus if the specified magnitude of a resistor is 100 Ω with a limiting error of ± 4 ohms, then the magnitude of the resistor will lie between the limits 100 ± 4, *i.e.*, 96 ohms and 104 ohms. In other words, the manufacturer of the resistor guarantees its value to lie between 96 and 104 ohms.

Example 3.41. *A wattmeter having a range of 500 W has an error of ± 1.5 per cent of full scale deflection. If the true power is 50 W, what would be the range of the readings ?*

If the error is specified as percentage of true value, what would be the range of the readings ?

Solution. *Given :* Range of the wattmeter = 500 W

Error (% age of full scale deflection) = ± 1.5%

True or specified power, $Q_s = 50$ W

- When the error is specified as a *percentage of full scale deflection*, the magnitude of limiting error at full scale

$$= \pm \left(\frac{1.5}{100}\right) \times 500 = \pm 7.5 \text{ W}$$

Thus the wattmeter reading when the true reading is 50 W may be 50 ± 7.5, *i.e.*, between **42.5 and 57.5 W** **(Ans.)**

- Relative error, $E_r = \frac{7.5}{50} \times 100 = 15\%$

Now suppose the error is specified as percentage of true value, then,

$$\text{Magnitude of error} = \pm \frac{1.5}{100} \times 50 = \pm 0.75 \text{ W, and}$$

The meter may read between 50 ± 0.75 or **49.25 and 50.75 W.** **(Ans.)**

Example 3.42. *The guaranteed accuracy of a flow meter working on thermal principles is ± 3 per cent of full scale reading of 2.5 × 10^{-6} m^3/s. If the flow measured by this meter is 1.25 × 10^{-6} m^3/s, calculate the limiting error in per cent. Comment upon the results.*

Solution. *Given* : Guaranteed accuracy of flow meter,

$$E_r = \pm 3\% \text{ (of full scale reading)}$$

Full scale reading, $Q_s = 2.5 \times 10^{-6}$ m^3/s

Flow measured by the meter = 1.25 × 10^{-3} m^3/s.

Limiting error (Per cent) :

The magnitude of limiting error at full scale of the meter is,

$$\delta Q = E_r \times Q_s$$
$$= \pm 0.03 \times 2.5 \times 10^{-6}$$
$$= \pm 0.075 \times 10^{-6} \text{ m}^3/\text{s}$$

The magnitude of the flow being measured is 1.25 × 10^{-6} m^3/s;

∴ The relative error at this flow,

$$E_r' = \frac{\delta Q}{Q_s} = \frac{0.075 \times 10^{-6}}{1.25 \times 10^{-6}} = 0.06$$

∴ The value of flow being measured is,

$$Q = Q_s(1 \pm E_r') = 1.25 \times 10^{-6} (1 \pm 0.06)$$
$$= 1.25 \times 10^{-6} \pm 0.075 \times 10^{-6} \text{ m}^3/\text{s}$$

Hence, *Percentage limiting error,*

$$\% E_r = \pm \frac{0.075 \times 10^{-6}}{1.25 \times 10^{-6}} \times 100 = \pm 6\% \quad \textbf{(Ans.)}$$

Comments. From above results we observe that when the meter reading is full scale value of 2.5 × 10^{-6} m^3/s, the flow meter is guaranteed to have a limiting error of ± 3%, but when the meter is reading its half scale value of 1.25 × 10^{-6} m^3/s, the limiting error is ±6%. Thus when reading is halved, the limiting error is doubled. *This indicates that the per cent limiting error will be greater if smaller values are read by the meter.* Therefore, while selecting instruments, special care should be taken as regards the range; the values to be measured *should not lie in the lower third of the range.* Hence as far as possible, meter which will read up their scale should be used.

Resultant error of a device composed of several components :

A measuring instrument/device/system may be formed by combining several components, each having its limiting error. The limiting error of the combination can be computed in the following ways depending on how the various components are connected in the total system:

When two or more quantities, each having a limiting error, are combined, it is advantageous to be able to compute the limiting error of the combination. The limiting error can be easily found by *considering the relative increment of the function if the final result is in the form of an algebraic equation.*

(i) When the final result is either sum or difference of outputs of various components :

Let, $q_1, q_2, q_3, \ldots\ldots\ldots\ldots q_n$ = Outputs of each device, and

Q = The final result ;

Then, Q (sum or difference) $= q_1 \pm q_2 \pm q_3 \pm \ldots\ldots\ldots\ldots\ldots\ldots \pm q_n$,

and, $$\frac{\delta Q}{Q} = \pm\left[\frac{q_1}{Q}\cdot\frac{\delta q_1}{q_1} + \frac{q_2}{Q}\cdot\frac{\delta q_2}{q_2} + \frac{q_3}{Q}\cdot\frac{\delta q_3}{q_3} + \ldots\right] \quad \ldots(3.59)$$

Equation (3.59) shows that the *resultant limiting error in this case is equal to the sum of the products formed by multiplying the individual relative errors by the ratio of each term to the function.*

(ii) When the result is either product or quotient of output of various components :

In this case, $Q = q_1 \times q_2 \times q_3 \times \ldots\ q_n$

or, $Q = q_1 \div q_2 \div q_3 \div \ldots\ q_n$

or, $Q = q_1 \times q_2 \div q_3 \times q_4 \div \ldots\ q_n$

Then, $$\frac{\delta Q}{Q} = \pm\left(\frac{\delta q_1}{q_1} + \frac{\delta q_2}{q_2} + \frac{\delta q_3}{q_3} + \ldots\ldots\frac{\delta q_n}{q_n}\right) \quad \ldots(3.60)$$

In other words, the *relative limiting error of the system is the sum of relative errors of all devices/components.*

(iii) When the final output is composite factor of output of various components :

In this case, $Q = q_1^{\pm n_1} \times q_2^{\pm n_2} \times q_3^{\pm n_3} \ldots\times q_n^{\pm n_n}$

Then, $$\frac{\delta Q}{Q} = \pm\left[n_1\frac{\delta q_1}{q_1} + n_2\frac{\delta q_2}{q_2} + n_3\frac{\delta q_3}{q_3} + \ldots\right] \quad \ldots(3.61)$$

Example 3.43. *The three resistors R_1, R_2 and R_3 have the following ratings :*

$$R_1 = 25\ \Omega \pm 4\%,\ R_2 = 65\ \Omega \pm 4\%;\ R_3 = 45\ \Omega \pm 4\%$$

Determine the following :

(i) Limiting value of the resultant resistance.

(ii) Per cent limiting error of series combination of resistance.

Solution. *Given :* $R_1 = 25\ \Omega \pm 4\%$

$R_2 = 65\ \Omega \pm 4\%$

$R_3 = 45\ \Omega \pm 4\%$.

(i) Limiting value of the resultant resistance :

The values of the resistances are :

$$R_1 = 25 \pm \frac{4}{100}\times 25 = 25 \pm 1.0\ \Omega$$

$$R_2 = 65 \pm \frac{4}{100}\times 65 = 65 \pm 2.6\ \Omega$$

$$R_3 = 45 \pm \frac{4}{100}\times 45 = 45 \pm 1.8\ \Omega$$

∴ The limiting value of resultant resistance,

$$R = (25 + 65 + 45) \pm (1.0 + 2.6 + 1.8) = (135 \pm 5.4)\ \Omega$$

Magnitude of resistance = **135 Ω** and, *error* = **± 5.4 Ω** **(Ans.)**

(*ii*) Per cent limiting error of the series combination of resistance :

Per cent limiting error of the combination

$$= \pm \frac{5.4}{135} \times 100 = \pm \mathbf{4\%} \quad \textbf{(Ans.)}$$

Example 3.44. *A power meter and an ammeter are used to measure resistance of an unknown resistance. Find the limiting error in the measurement of resistance when the limiting errors in the measurement of power and current are ± 1.2% and 0.8% respectively.*

Solution. *Given :* Limiting error in the measurement of power (*P*) = $\frac{\delta P}{P} = 1.2\%$

Limiting error in the measurement of current (*I*) = $\frac{\delta I}{I} = 0.8\%$

We know that, $R = \frac{P}{I^2}$

∴ Limiting error in the measurement of resistance,

$$\frac{\delta R}{R} = \pm\left(\frac{\delta P}{P} + 2\frac{\delta I}{I}\right)$$

$$= \pm (1.2 + 2 \times 0.8) = \pm \mathbf{2.8\%} \quad \textbf{(Ans.)}$$

Example 3.45. *The unknown resistance in a Wheatstone bridge is measured by using three known resistances such that :*

$$R_4 = \frac{R_2 R_3}{R_1}$$

where, $R_1 = 50 \pm 0.5\%\ \Omega;\ R_2 = 500 \pm 0.5\%\ \Omega;\ R_3 = 440 \pm 0.5\%\ \Omega$

Determine the magnitude of the unknown resistance (R_4) and the limiting error in per cent and in ohms (for the unknown resistance).

Solution. *Given :* $R_1 = 50 \pm 0.5\%\ \Omega;\ R_2 = 500 \pm 0.5\%\ \Omega;\ R_3 = 440 \pm 0.5\%\ \Omega$

The unknown resistance is given by

$$R_4 = \frac{R_2 R_3}{R_1} = \frac{500 \times 440}{50} = \mathbf{4400\ \Omega} \quad \textbf{(Ans.)}$$

The relative limiting error of the unknown resistance is,

$$\frac{\delta R_4}{R_4} = \pm\left(\frac{\delta R_2}{R_2} + \frac{\delta R_3}{R_3} + \frac{\delta R_1}{R_1}\right)$$

$$= \pm (0.5 + 0.5 + 0.5) = \pm \mathbf{1.5\%} \quad \textbf{(Ans.)}$$

Limiting error in ohms $= \frac{1.5}{100} \times 4400 = \pm \mathbf{66\ \Omega}$ **(Ans.)**

i.e., The guaranteed values of resistance are between **4334** (*i.e.,* 4400 – 66) **Ω** and **4466** (*i.e.,* 4400 + 66) **Ω** **(Ans.)**

Example 3.46. *A cantilever beam of length l, width b and depth d is used to measure the force F. The deflection is measured using an LVDT of accuracy ± 0.1 mm. The length, width and depth are measured to an accuracy of ± 0.01 mm and their values are 45 mm, 4.5 mm and 0.9 mm respectively. The modulus of rigidity is known accurate to 0.2 %. If the deflection noted was 1.8 mm, determine the error in force measured.*

Given force, $F = \frac{3Ebd^3}{12t^3} \cdot y$

where, E = *Modulus of elasticity;* b = *width,* d = *depth, and*
l = *length of the beam ;* y = *deflection.*

Solution. The *limiting error in force F is,*

$$\frac{\delta F}{F} = \pm\left(\frac{\delta E}{E}+\frac{\delta b}{b}+3\frac{\delta d}{d}+3\frac{\delta l}{l}+\frac{\delta y}{y}\right)$$

$$= \pm\left(0.2+\frac{0.01}{4.5}+3\times\frac{0.01}{0.9}+3\times\frac{0.01}{45}+\frac{0.1}{1.8}\right)$$

$$= \pm\,(0.2 + 0.00222 + 0.03333 + 0.00067 + 0.05555)$$

$$= \pm\ \mathbf{0.2917\%}\quad \textbf{(Ans.)}$$

Example 3.47. *A dynamometer is used to measure the output power of a rotating shaft. The output power is given by the following relation :*

$$P = \frac{2\pi\times 9.81\, FLR}{t\times 10^6}\ kW$$

where, F = *Force at the end of torque arm (kg),*
L = *Length of torque arm (mm),*
R = *Number of revolutions during time t, and*
t = *Time for test run (s).*

The given test data are

F = *4.26 ± 0.02 kg;* L = *382 ± 1.2 mm;*
R = *1192 ± 1.0 revolutions;* t = *60 + 0.50 s*

The errors are limiting (absolute) errors.

(i) The magnitude of power.

(ii) The magnitude of the limiting error in the computed power.

Solution. *Given :* F = 4.26 ± 0.02 kg; L = 382 ± 1.2 mm; R = 1192 ± 1.0 revolutions; t = 60 + 0.50 s

(*i*) The magnitude of power *P* :

$$P = \frac{2\pi\times 9.81\times FLR}{t\times 10^6}\ \text{kW} \qquad \text{...(Given)}$$

Inserting the various values in the above expression, we get

$$P = \frac{2\pi\times 9.81\times 4.26\times 382\times 1192}{60\times 10^6} = \mathbf{1.993\ kW}\quad \textbf{(Ans.)}$$

(*ii*) Limiting errors in the computed power :

Total per unit error in computed power,

$$\frac{\delta P}{P} = \pm\left(\frac{\delta F}{F}+\frac{\delta L}{L}+\frac{\delta R}{R}+\frac{\delta t}{t}\right)$$

$$= \pm\left(\frac{0.02}{4.26}+\frac{1.2}{382}+\frac{1.0}{1192}+\frac{0.50}{60}\right)$$

$$= \pm\,(0.00469 + 0.00314 + 0.00084 + 0.00833) = \pm\, 0.017$$

Limiting error in computer power

$$\delta P = \pm 0.017\times P = \pm\, 0.017\times 1.993 = \mathbf{0.0339\ kW}$$

The result is written as: P = **1.993 ± 0.0339 kW** **(Ans.)**

Example 3.48. *(a) How are "known errors" treated ?*

(b) During a test the current was measured as 26.5 A flowing in a resistor of 0.12 Ω. Later on it was discovered that the ammeter reading was low by 1.1% and the marked resistance was high by 0.25%.

Determine the true power as a percentage of the power that was originally calculated.

Solution. (*a*) **Known errors.** When the error of quantity of an instrument is known, the effect of this error, when combined with errors, can be computed in a manner similar to the combinations of the limiting errors. However, the difference is that in case of known errors the *signs of relative errors are given and must be presented in the calculations.*

(*b*)

$$\text{True value of current } I = 26.5(1 + 0.011) = 26.791 \text{ A}$$

$$\text{True value of resistance } R = 0.12(1 - 0.0025) = 0.1197\ \Omega$$

$$\text{True value of power} = I^2R = (26.791)^2 \times 0.1197 = 85.916 \text{ W}$$

$$\text{Measured value of power} = (26.5)^2 \times 0.12 = 84.27 \text{ W}$$

$$\therefore \quad \frac{\text{True power}}{\text{Measured power}} \times 100 = \frac{85.916}{84.27} \times 100 = \mathbf{101.9\%} \quad \textbf{(Ans.)}$$

Alternatively :

$$\text{Power, } P = I^2R$$

$$\text{Total relative error} = \frac{\delta P}{P} = 2 \cdot \frac{\delta I}{I} + \frac{\delta R}{R}$$

$$= 2(-0.011) + 0.0025 = -0.0195$$

$$\frac{\text{True power}}{\text{Measured power}} = \frac{1}{1 + \text{relative error}}$$

$$= \frac{1}{1 - 0.0195} = \mathbf{101.9\%} \quad \textbf{(Ans.)}$$

3.9.3. Types of Errors

In measurement systems, the errors could originate from several sources. Broadly these may be *classified* as follows :

1. *Gross errors*
2. *Systematic errors :*
   (*i*) Instrumental errors;
   (*ii*) Environmental errors;
   (*iii*) Observational errors.
3. *Random errors.*

3.9.3.1. Gross errors

These errors occur due to *human mistakes* in reading instruments and recording and calculating results of measurement. Although it is probably impossible to eliminate the gross error completely, yet one should try to anticipate and correct them.

The mathematical analysis of gross errors is impossible since these may occur in different amounts. While some gross errors may be easily detected, others may go unnoticed.

These errors can be *avoided* by adopting *two means :*

(*i*) Immense care should be taken while taking the reading and recording the data.

(*ii*) Two, three, or even more readings should be taken for the quantity being measured.

• Designers of instruments to some extent can take care to avoid gross errors but much would depend upon the interest and competence of the observer.

3.9.3.2. Systematic errors

The systematic errors are repeated consistently with the repetition of the experiment and are caused by *such effects as* :

— Sensitivity shift;

— Zero off-set;

— Known non-linearity.

• These errors *cannot* be determined by direct and repetitive observations of the measurand made each time with same technique.

These errors can be located only by having repeated measurements under different conditions or with different equipment and where possible by an entirely different method.

These errors may be *instrumental, environmental or observational errors.*

1. Instrumental errors :

These errors arise *due to :*

— Inherent shortcomings in the instruments.

— Misuse of the instruments.

— Loading effects of instruments.

Instrumental errors are inherent in instruments due to construction, calibration or operation of the instruments. *These errors may cause the instrument to read too low or too high.*

Some *"examples"* of such errors are :

(*i*) Mechanical friction and wear, backlash, yielding of supports, pen or pointer drag and hysteresis of elastic members due to aging.

(*ii*) Constructional faults resulting from finite width of knife edges, excessive friction at the mating parts, lost motion due to necessary clearance in gear teeth etc.

(*iii*) Improper selection and poor maintenance of the instrument.

(*iv*) Unavoidable physical phenomenon due to friction, capillary attraction and imperfect rarefaction.

Assembly errors. These errors may be due to the following *reasons* :

(*i*) Displaced scale *i.e.,* incorrect fitting of the scale zero with respect to the actual zero position of the movement.

(*ii*) Non-uniform division of the scale.

(*iii*) Bent or distorted pointer.

The assembly errors do not change with time, so these can be easily discovered and rectified.

Misuse of instruments. A good instrument used in an unintelligent way may give erroneous results.

• *Loading effects.* In several measurement techniques, the detector/sensor utilizes the energy of the medium, thus altering slightly the parameter to be measured, resulting in errors.

2. Environmental errors :

The environmental errors are due to conditions external to the measuring device (*e.g.,* effects of pressure, temperature, humidity, dust, vibrations, external magnetic or electrostatic fields). These errors are much more troublesome than assembly errors.

These errors can be *eliminated or reduced* by taking the following measures :

(*i*) Using the instrument in controlled conditions of pressure, temperature, humidity in which it was originally assembled and calibrated.

(*ii*) Measure deviations in the local conditions from the calibrated ones and then apply a suitable correction to the instrument readings.

(*iii*) Automatic compensation for departures from the calibrated conditions by using sophisticated devices.

(*iv*) Make a complete new calibration under the local conditions.

(*v*) In case it is suspected that external magnetic or electrostatic fields can affect the readings of the instruments, magnetic or electrostatic shields may be provided.

(*vi*) Apply computed corrections. Normally, efforts are made to avoid the use of applications of computed corrections, but where these corrections are needed and are necessary, they are incorporated for the computation of the results.

3. Observational errors :

These errors occur due to *carelessness of operators.* Even when an instrument has been properly selected, carefully installed and faithfully calibrated, shortcoming in the measurement occurs due to certain failings on the part of the observer.

The observational errors may be due to the following *reasons :*

(*i*) *Parallax*—Since the parallax errors arise on account of pointer and the scale not being in the same plane, we can eliminate this error by having the pointer and the scale in the same plane.

(*ii*) Wrong scale reading, and wrong recording of data.

(*iii*) Inaccurate estimates of average reading, lack of ability to interpolate properly between graduations.

(*iv*) Incorrect conversion of units in between consecutive readings, and non-simultaneous observation of interdependent quantities.

(*v*) Tendency to read high or low, anticipate a signal and read too soon—Personal bias.

- Modern electrical instruments have *digital display of output* which completely eliminates the errors on account of human observational or sensing power as the output is in the *form of digits.*

3.9.2.3. Random errors

- The random errors are *accidental, small and independent.*
- *They vary in an unpredictable manner.*
- The magnitude and direction of these errors *cannot* be predicted from a knowledge of measurement system; however these errors are *assumed to follow the law of probabilities.*

The most common causes of these errors are :

(*i*) Friction in instrument movement.

(*ii*) Backlash in the movement.

(*iii*) Parallax errors between pointer and scale.

(*iv*) Finite dimensions of the pointer and scale divisions.

(*v*) Hysteresis in elastic members.

(*vi*) Mechanical vibrations.

3.9.4. Sources of Errors

Besides the errors so far enumerated (of the physical and mechanical nature) and the errors which necessarily result from the faulty calibration of a measuring system, the errors arise from the *following sources also :*

(*i*) **Noise.** *It is defined as any signal that does not convey useful information.*

— Noise signals/errors can be reduced to a minimum level through filtering, careful selection of components, shielding and isolation of the entire measuring system.

(*ii*) **Response time.** *It is defined as the time taken by the instrument to show 63.2% change in a reading to a step input.* This factor contributes to the uncertainty of the measurement.

(*iii*) **Design limitations.** In the design of the instrument, certain inevitable factors such as friction and resolving power lead to uncertainty of measurements.

(*iv*) **Energy exchanged by interaction.** As earlier pointed out, whenever the energy required for operating the measuring system is extracted from the measurand, the value of the latter is altered to a greater or lesser extent. This alternation is dependent upon the capacity of the system.

(*v*) **Transmission.** During the transmission of information from the primary sensing element to the indicator, the signal may be *attenuated* due to any of the following *reasons :*

— It may suffer loss through leakage ;

— It may be absorbed or otherwise consumed in the communication channel;

— It may be distorted by resonance, attenuation or delay phenomenon whose actions are selective on various signal components.

(*vi*) **Deterioration of measuring system.** The examples of some types of deterioration which occur in a measuring system and result in a source of error in measurement are :

— Change in resistance of a circuit element through strain relief ;

— Alteration of thermocouple characteristics through oxidizing or reducing atmosphere;

(*vii*) **Ambient influences on measuring systems.**

(*viii*) **Errors of observation and interpretation.**

3.9.5. Statistical Analysis of Test Data

The *systematic and random errors* are evaluated and studied by *statistical procedures* which make it possible to state from a limited group of data the most probable value of a quantity, the probable uncertainty of a single observation, and the probable limits of uncertainty of the best value that can be derived from the data.

— The object of the statistical methods based on laws of chance, which operate only on random errors and not on systematic errors, is to achieve consistency (precision) of value and not their accuracy.

The experimental data is obtained in the following *two forms of tests :*

(*i*) Single-sample test;

(*ii*) Multi-sample test.

(*i*) **Single-sample test.** *A single measurement (or succession of measurements) done under identical conditions excepting for time is known as "single-sample test".*

(*ii*) **Multi-sample test.** In a multi-sample test, *repeated measurements of a given quantity are taken under different test conditions such as using different instruments, different ways of measurement and by employing different observers.* This type of test does not materialise result by simply making measurements with the same equipment, procedure, technique and same observer.

- For getting the exact value of the quantity under measurement, tests should be conducted using as many different procedures, techniques or experiments as practicable. It should be understood that the *statistical means which help us to arrive at correct results are only valid for multi-sample tests.*

3.9.5.1. Statistical averages

When several multi-sample observations are taken experimentally there is a scatter of the data about some central value, one of the methods, presenting the result in the form of a histogram (also called a *frequency distribution curve*). Thus, histogram (Fig. 3.34) is a pictorial representation of the test data where the ordinates indicate the number of observed readings of a particular value. With more and more data taken at smaller and smaller increments, the histogram finally changes into a *smooth curve.* The smooth curve is symmetrical with respect to the central value.

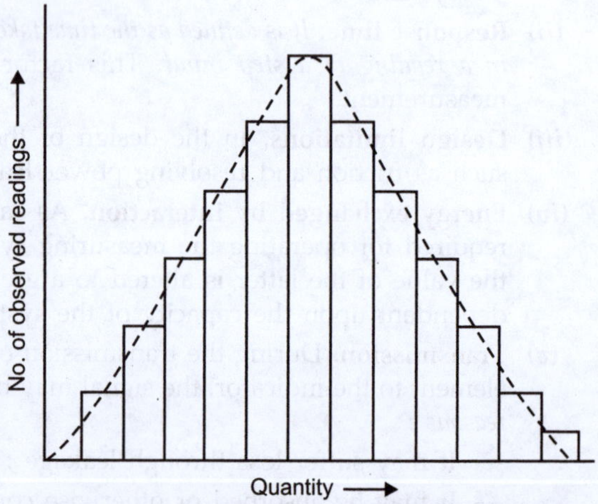

Fig. 3.34. Histogram.

(*i*) **Arithmetic mean (AM).** *The most probable value of measured variable (variate) is the arithmetic mean of the number of readings taken.* We arrive at the best approximation if the number of readings of the same quantity are very large. The mean is computed by summing all the values and dividing by the number of measurements. If $q_1, q_2, q_3 \dots q_n$ be the set of measured values of a *quantity*, then the arithmetic mean $\bar{q}$ will be given by,

$$\bar{q} = \frac{q_1 + q_2 + q_3 + \dots q_n}{n} = \frac{1}{n}\sum_{i=1}^{n} q_i \qquad \text{...(3.62)}$$

(*ii*) **Geometric mean (G.M.).** It is defined as the nth root of the product of n terms.

$$\text{Geometric mean (G.M)} = \sqrt[n]{q_1 \times q_2 \times q_3 \dots \times q_n} \qquad \text{...(3.63)}$$

(*iii*) **Median.** *The middle value of a set of an "odd" number of readings, if variables are arranged in numerical order, is called the "median".* For an *"even"* number of readings, the median is the *arithmetic average of the two central readings.*

(*iv*) **Mode.** *The value of the variable that occurs most frequently is termed as "mode".*

- For a symmetrical distribution curve, the three measures of central value, *viz., 'mean', 'median' and 'mode', are equal.*

3.9.5.2. Dispersion from mean

The property which denotes the extent to which the values are dispersed about the central value is termed as **dispersion** *or* **stored** *or* **scatter**.

A large dispersion indicates that some factors involved in the measurement process are not under close control and therefore it becomes difficult to estimate the measured quantity with confidence and definiteness.

The simplest possible measure of dispersion is the **Range** *which is the difference between greatest and least values of data.*

(*i*) **Deviation.** *The departure of the observed reading from the arithmetic mean of the group of readings is termed as* ***deviation***. Let the deviation of reading q_1 be d_1 and that of reading q_2 be d_2 etc.

Then,
$$d_1 = q_1 - \bar{q}$$
$$d_2 = q_2 - \bar{q}$$
$$\ldots \quad \ldots\ldots$$
$$d_n = q_n - \bar{q} \qquad \ldots(3.64)$$

and,
$$\bar{q} = \frac{\Sigma(q_n - d_n)}{n} \qquad \ldots(3.65)$$

$$\text{Algebraic sum of deviations} = d_1 + d_2 + d_3 + \ldots.\ d_n$$
$$= (q_1 - \bar{q}) + (q_2 - \bar{q}) + (q_3 - \bar{q}) + \ldots.(q_n - \bar{q})$$
$$= (q_1 + q_2 + q_3 + \ldots. + q_n) - n\bar{q} = 0$$
$$(\because q_1 + q_2 + q_3 + \ldots + q_n) = n\bar{q}$$

Thus, the *algebraic sum of deviations is zero.*

The above mathematical result implies that the deviations can be *positive as well as negative.*

(*ii*) **Average deviation.** *It is defined as the sum of the absolute values of deviations divided by the number of readings.*

Average deviation may be expressed as

$$\bar{d} = \frac{|d_1| + |d_2| + |d_3| + \ldots\ldots.. + |d_n|}{n}$$

$$= \frac{\Sigma|d|}{n}, \text{ or, } \sum_{i=1}^{n} \frac{|q_i - \bar{q}|}{n} \qquad \ldots(3.66)$$

The average deviation is taken as a measure of accuracy of measurement.

- The average deviation is an *indication of the precision of the instruments* used in making the measurements. *Highly precise instruments yield a low average deviation between readings.*

(*iii*) **Standard deviation.** "*Standard deviation*" (or "*mean square deviation*") of an infinite number of data is defined *as the square root of the sum of individual deviations squared divided by the number of readings.*

Thus, standard deviation

$$\sigma = \sqrt{\frac{d_1^2 + d_2^2 + d_3^2 + \ldots\ldots + d_n^2}{n}}$$

$$= \sqrt{\frac{\Sigma d^2}{n}} = \sqrt{\sum_{i=1}^{n} \frac{(q_i - \bar{q})^2}{n}} \qquad \ldots(3.67)$$

- In practice, however, the number of observations is *finite*. When the number of observations is *greater than 20*, standard deviation is denoted by the symbol σ while if it is *less than* 20, the symbol used is *s*. The standard deviation of a finite number of data is given by :

$$s = \sqrt{\frac{d_1^2 + d_2^2 + d_3^2 + ...+ d_n^2}{n-1}} = \sqrt{\frac{\Sigma d^2}{n-1}} \quad ...(3.68)$$

Variance. *The square of the standard deviations is called* ***variance***.

Variance, $V = \text{(Standard deviation)}^2$

$$= \sigma^2 = \frac{d_1^2 + d_2^2 + d_3^2 ......d_n^2}{n} = \frac{\Sigma d^2}{n} \quad ...(3.69)$$

But when the number of observation is less than 20,

Variance, $$V = s^2 = \frac{\Sigma d^2}{n-1} \quad ...(3.70)$$

3.9.5.3. Best value from a sample of readings

Let, q_b = The best value (approximately the true value) from a set of readings/data.

Then, the deviation or individual component error,

$$d_i = q_i - q_b \quad \text{where, } i = 1, 2, 3, ...n$$

This deviation may be *+ve, zero* or *–ve.*

(i) **Sum of deviations :**

$$\sum_{i=1}^{n} d_i = \sum_{i=1}^{n} (q_i - q_b) \quad ...(3.71)$$

In case of a *best value,* the sum of deviations is equal to *zero.*

i.e., $$\sum_{i=1}^{n} (q_i - q_b) = 0 \quad ...(3.72)$$

or, $$q_b = \frac{1}{n}\sum_{i=1}^{n} q_i$$, which is the arithmetic mean.

Thus, the arithmetic mean may be interpreted as the best value around which the chances of error being *+ve* or *–ve* likely to be equal.

(ii) $$d_i = q_i - q_b$$

Squaring both sides, we have

$$d_i^2 = q_i^2 - 2q_iq_b + q_b^2$$

Sum of the square of deviations,

$$\sum_{i=1}^{n} d_i^2 = \sum_{i=1}^{n} q_i^2 - 2q_b\sum_{i=1}^{n} q_i + nq_b^2 \quad ...(3.73)$$

The sum of the square of deviations, corresponding to the *best,* would be *minimum.* This condition can be obtained by setting the deviation w.r.t. q_b equal to zero;

i.e., $$\frac{d}{dq_b}\left[\sum_{i=1}^{n} d_i^2\right] = 0 \quad ...(3.74)$$

or, $$\frac{d}{dq_b}\left[\sum_{i=1}^{n} q_i^2 - 2q_b \sum_{i=1}^{n} q_i + nq_b^2\right] = 0$$

As all the q_i^2s are constant, the differentiation gives,

$$-2\sum_{i=1}^{n} q_i + 2nq_b = 0$$

or, $$q_b = \frac{1}{n}\sum_{i=1}^{n} q_i \qquad ...(3.75)$$

which is again the arithmetic mean. Obviously, *the arithmetic mean is the most probable value or the best value which one can obtain from a set of scattered measurement data.*

Example 3.49. *By using a micrometer screw the following readings were taken of a certain physical length :*

1.34, 1.38, 1.56, 1.47, 1.42, 1.44, 1.53, 1.48, 1.40, 1.59 mm

Assuming that only random errors are present, calculate the following :

(i) Arithmetic mean,

(ii) Average deviation,

(iii) Standard deviation, and

(iv) Variance.

Solution. *Given* : Readings of the micrometer screw : 1.34, 1.38, 1.56, 1.47, 1.42, 1.44, 1.53, 1.48, 1.40, 1.59 mm.

(i) **Arithmetic mean :**

We know that arithmetic mean,

$$\bar{q} = \frac{1}{n}\sum_{i=1}^{n} q_i \qquad ...(\text{Eqn. } 3.62)$$

$$= \frac{q_1 + q_2 + q_3 + q_4 + q_5 + q_6 + q_7 + q_8 + q_9 + q_{10}}{10}$$

$$= \frac{1.34 + 1.38 + 1.56 + 1.47 + 1.42 + 1.44 + 1.53 + 1.48 + 1.40 + 1.59}{10}$$

$= \mathbf{1.461\ mm}$ **(Ans.)**

(ii) **Average deviation :**

From eqn. (3.64), the *deviations* are :

$$d_1 = q_1 - \bar{q} = 1.34 - 1.461 = -0.121$$
$$d_2 = q_2 - \bar{q} = 1.38 - 1.461 = -0.081$$
$$d_3 = q_3 - \bar{q} = 1.56 - 1.461 = +0.099$$
$$d_4 = q_4 - \bar{q} = 1.47 - 1.461 = +0.009$$
$$d_5 = q_5 - \bar{q} = 1.42 - 1.461 = -0.041$$
$$d_6 = q_6 - \bar{q} = 1.44 - 1.461 = -0.021$$
$$d_7 = q_7 - \bar{q} = 1.53 - 1.461 = +0.069$$

$$d_8 = q_8 - \bar{q} = 1.48 - 1.461 = +0.019$$

$$d_9 = q_9 - \bar{q} = 1.40 - 1.461 = -0.061$$

$$d_{10} = q_{10} - \bar{q} = 1.59 - 1.461 = 0.129$$

∴ *Average deviation,*

$$\bar{d} = \frac{\Sigma|d|}{n} \left(\text{or} \sum_{i=1}^{n} \frac{|q_i - \bar{q}|}{n} \right)$$

$$= \frac{0.121 + 0.081 + 0.099 + 0.009 + 0.041 + 0.021 + 0.069 + 0.019 + 0.061 + 0.129}{10}$$

= **0.065** **(Ans.)**

It may be noted that for calculation of average deviations, the *signs of deviations are not considered.*

(iii) **Standard deviation :**

Since the number of readings is 10, which is *less than 20*, therefore equation (3.68) is used for calculation of standard deviation.

$$\text{Standard deviation, } s = \sqrt{\frac{\Sigma d^2}{n-1}} \qquad \text{...[Eqn. (3.68)]}$$

$$\text{or, } s = \sqrt{\frac{(-0.121)^2 + (-0.081)^2 + (+0.099)^2 + (0.009)^2 + (-0.041)^2 + (-0.021)^2 + (0.069)^2 + (0.019)^2 + (-0.061)^2 + (0.129)^2}{(10-1)}}$$

$$= \sqrt{\frac{0.05869}{9}} = \mathbf{0.0807\ mm} \quad \textbf{(Ans.)}$$

(iv) **Variance :**

$$\textit{Variance, } V = s^2 = (0.0807)^2 = \mathbf{0.00651\ mm^2} \quad \textbf{(Ans.)}$$

Example 3.50. *Eight different students tuned in the circuit for resonance and the values of resonant frequency in kHz were recorded as :*

412, 428, 423, 415, 426, 411, 423, 416

Calculate the following :

(i) Arithmetic mean,

(ii) Average deviation,

(iii) Standard deviation, and

(iv) Variance.

Solution. *Given* : Values of resonant frequency : 412, 428, 423, 415, 426, 411, 423, 416

(i) **Arithmetic mean :**

Arithmetic mean,

$$\bar{q} = \frac{\Sigma q}{n} = \frac{412 + 428 + 423 + 415 + 426 + 411 + 423 + 416}{8}$$

= **419.25 kHz** **(Ans.)**

(ii) **Average deviation :**

The deviations are :

$$d_1 = q_1 - \bar{q} = 412 - 419.25 = -7.25 \text{ kHz}$$

$$d_2 = q_2 - \bar{q} = 428 - 419.25 = +8.75 \text{ kHz}$$

$$d_3 = q_3 - \bar{q} = 423 - 419.25 = 3.75 \text{ kHz}$$
$$d_4 = q_4 - \bar{q} = 415 - 419.25 = -4.25 \text{ kHz}$$
$$d_5 = q_5 - \bar{q} = 426 - 419.25 = +6.75 \text{ kHz}$$
$$d_6 = q_6 - \bar{q} = 411 - 419.25 = -8.25 \text{ kHz}$$
$$d_7 = q_7 - \bar{q} = 423 - 419.25 = +3.75 \text{ kHz}$$
$$d_8 = q_8 - \bar{q} = 416 - 419.25 = -3.25 \text{ kHz}$$

∴ *Average deviation,*

$$\bar{d} = \frac{\Sigma|d|}{n}$$
$$= \frac{7.25 + 8.75 + 3.75 + 4.25 + 6.75 + 8.25 + 3.75 + 3.25}{8}$$
$$= \mathbf{5.75\ kHz} \quad \textbf{(Ans.)}$$

(For the calculation of average deviation, the *signs of deviations are not considered*)

(iii) **Standard deviation :**

Since the number of readings is 8, which is less than 20, therefore eqn. (3.68) is used for calculation of standard deviation.

$$\text{Standard deviation, } s = \sqrt{\frac{\Sigma d^2}{n-1}} \quad \text{...[Eqn (3.68)]}$$

$$\text{or, } s = \sqrt{\frac{(-7.25)^2 + (+8.75)^2 + (+3.75)^2 + (-4.25)^2 + (+6.75)^2 + (-8.25)^2 + (+3.75)^2 + (-3.25)^2}{(8-1)}}$$
$$= \mathbf{6.54\ kHz} \quad \textbf{(Ans.)}$$

(iv) **Variance :**

$$\textit{Variance, } V = s^2 = (6.54)^2 = \mathbf{42.77\ (kHz)^2} \quad \textbf{(Ans.)}$$

3.9.5.4. Normal or Gaussian curve of errors

In experimental work the random effects are normally studied with reference to the Gaussian or normal error distribution. Here, the following *assumptions* are made :

(i) All observations include a large number of small random disturbing effects, and these small random errors may be both *+ve* and *–ve*. Thus, there is a strong central tendency, *i.e., there is a greater probability of small errors and large errors of both +ve and –ve values.*

(ii) There is equal probability of *+ve* and *–ve* disturbing effects. As such the probability curve of errors plotted against the magnitude of error will be *symmetrical about the zero value.*

The law of probability states that the normal occurrence of deviations from average value of an infinite number of measurements or observations can be expressed as :

$$P = \frac{1}{\sigma\sqrt{2\pi}} e^{-x^2/2\sigma^2} \quad \text{...(3.76)}$$

where,
P = Number of readings at any deviation 'd', *i.e.,* probability of occurrence of deviation d,
σ = Standard deviation, and
x = Magnitude of deviation from mean.

Fig. 3.35 shows a Gaussian distribution curve in which P has been plotted as a function of x. The curve is symmetrical about the mean value and may be regarded as the limiting form of a histogram. The two values of standard deviation σ_1 and σ_2 ($\sigma_1 > \sigma_2$) have been illustrated in the given figure. *Larger the value of σ, the flatter the curve and hence larger the expected error of all measurements.*

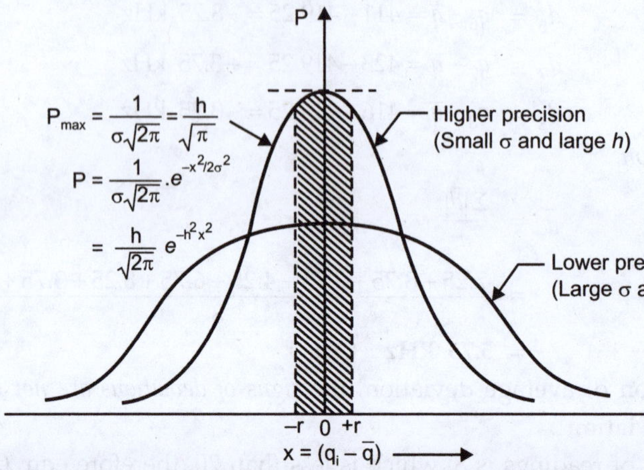

Fig. 3.35. Gaussian distribution curve.

If we integrate P between the limits $x = \pm x$, we shall obtain the area under the curve between these two values of x. Because the curve is related to the probability that the measured lies within same range, we have

$$\int_{-x}^{+x} P dx = 1.0 \qquad \text{...(3.77)}$$

i.e., area under the probability curve is *unity*.

If the total number of readings are represented by unity, then the fraction of the total number of readings lying between x_1 and x_2 will be represented by the area under the curve between these two values of deviation ;

i.e., $$n_{1-2} = \int_{-x}^{+x} P dx$$

If the area between x_1 and x_2 is 0.6, then 60 per cent of the deviations fall between x_1 and x_2.

Precision index. It may be observed from the normal distribution function that the maximum probability occurs when deviation is *zero, i.e.,* $q = \bar{q}$ and the corresponding value of probability is,

$$P_{max} = \frac{1}{\sigma\sqrt{2\pi}} = \frac{h}{\sqrt{\pi}} \qquad \text{...(3.78)}$$

where, $h = \frac{1}{\sigma\sqrt{2}}$ is called the **precision index.**

A large value of h represents high precision of the data because the probability of occurrence of variates in a given range falls off *rapidly* as the deviation increases because the variates tend to cluster (become closer) into a narrow range. On the other hand, a small value of h represents low precision because the probability of occurrences of variates in a given range falls off *gradually* as the deviation increases; this is because the variates are spread over a wide range.

Probable error : Let us consider the two points $-r$ and $+r$ marked in Fig. 3.35. These points are so located that the area bounded by the curve, the X-axis and the ordinates erected at $x = -r$ and $x = +r$ is equal to half of the total area under the curve. That is half the deviations lie between $x = \pm r$. The location of the points r can be worked out as follows :

$$\int_{-r}^{+r} P.dx = \frac{1}{2}$$

or,
$$\frac{h}{\sqrt{\pi}} \int_{-r}^{r} e^{-h^2x^2} dx = \frac{1}{2}$$

This gives *probable error,*

$$r = \frac{0.4769}{h} \qquad ...(3.79)$$

Average deviation for the normal curve :

In case of normal curve,

Average deviation,

$$\bar{d} = \int_{-\infty}^{+\infty} |x| P\,dx$$

(| x | represents the numerical value of deviation)

$$= \frac{h}{\sqrt{\pi}} \int_{-\infty}^{+\infty} e^{-h^2x^2} x.dx = \frac{2h}{\sqrt{\pi}} \int_{0}^{+\infty} e^{-h^2x^2} x.dx$$

$$= \frac{2h}{\sqrt{\pi}} \int_{0}^{+\infty} e^{-h^2z} dz \qquad \left[\begin{array}{l} \text{where, } z = x^2 \\ dz = 2x.dx \\ x.dx = \frac{dz}{2}, \text{and} \\ \int e^{ax} dx = \frac{e^{ax}}{a} \end{array}\right]$$

$$= \frac{2h}{\sqrt{\pi}} \times \frac{1}{2} \left[\frac{e^{-h^2z}}{-h^2}\right]_0^{+\infty}$$

$$= \frac{h}{\sqrt{\pi}} \left[\frac{0-1}{-h^2}\right] = \frac{h}{\sqrt{\pi}} \times \frac{1}{h^2} = \frac{1}{h\sqrt{\pi}}$$

$$= \frac{r}{0.4769\sqrt{\pi}} = \frac{r}{0.8453}$$

i.e.,
$$\bar{d} = \frac{r}{0.8453} \qquad ...(3.80)$$

Standard deviation for the normal curve :

We know that standard deviation σ is given by :

$$\sigma = \sqrt{\frac{\Sigma d^2}{n}} \qquad ...[\text{Eqn. (3.67)}]$$

Following a method similar to that followed above, we have

$$\sigma^2 = \frac{2h}{\sqrt{\pi}} \int_0^{+\infty} e^{(-h^2x^2)} x^2 . dx = \frac{1}{2h^2}$$

or,
$$\sigma = \frac{1}{\sqrt{2}h} \quad ...(3.81)$$

$$= \frac{r}{0.4769 \times \sqrt{2}}$$

$$= \frac{r}{0.6745} \quad ...(3.82)$$

From eqns. (3.80) and (3.82), we have *probable error,*

$$r = 0.8453\,\bar{d} \quad ...(3.83)$$

and,
$$r = 0.6745\ \sigma \quad ...(3.84)$$

Also, *Standard deviation of mean,*

$$\sigma_m = \frac{\sigma}{\sqrt{n}} \quad ...(3.85)$$

and, *Standard deviation of standard deviation,*

$$\sigma_\sigma = \frac{\sigma}{\sqrt{2n}} = \frac{\sigma_m}{\sqrt{2}} \quad ...(3.86)$$

3.9.5.5. Probability tables

The probability of an individual measurement to fall within the deviations x_1 and x_2 from the mean is given by,

$$P(x_1, x_2) = \frac{h}{\sqrt{\pi}} \int_{x_1}^{x_2} e^{(-h^2x^2)} dx$$

for a normal Gaussian curve.

The parameter $h = \frac{1}{\sigma\sqrt{2}}$ and the mean value $\bar{q}$ may be known from the raw data.

Since the evaluation of the integral is a cumbersome and a tedious task, hence, **probability tables** have been prepared from which the value of probability can be directly read off.

Let
$$y = \sqrt{2}hx = \frac{x}{\sigma},$$

or,
$$x = \sigma \cdot y$$

Now,
$$P(x_1, x_2) = P(y_1, y_2)$$

$$= \frac{1}{\sqrt{2\pi}} \int_{y_1}^{y_2} e^{(-y^2/2)} dy \quad ...(3.87)$$

where,
$$y_1 = \sqrt{2}hx_1$$

and,
$$y_2 = \sqrt{2}hx_2$$

Eqn. (3.87) may be written as :

$$P(y_1, y_2) = \frac{1}{\sqrt{2\pi}}\int_0^{y_2}(e)^{(-y^2/2)}dy - \frac{1}{\sqrt{2\pi}}\int_0^{y_1}e^{(-y^2/2)}.dy \qquad ...(3.88)$$

Each of the two integrals in equation (3.88) is a function of a single parameter y_1 or y_2. Therefore, probability,

$$P(0, y) = \frac{1}{\sqrt{2\pi}}\int_0^{y}(e)^{(-y^2/2)}dy \qquad ...(3.89)$$

The results of this integral are given in Table 3.2 and can be used to compute the probability between the limits x_1 and x_2. Tabulated values of the integral represent the area under the normalised Gaussian error curve for $y = 0$ and $y = y$ as shown in Fig. 3.35.

As the curve is *symmetrical* the same table may be used for *–ve* deviations replacing y by $-y$ in equation (3.89) since the value of probability is not affected by this substitution.

— Suppose we are required to find the probability that the deviation lies between $\pm\sigma$. For the *+ve* deviation σ,

$$y = \frac{q-\bar{q}}{\sigma} = \frac{\sigma}{\sigma} = 1$$

Table 3.2 indicates the probability value $P = 0.3413$ for $y = 1.0$. Due to symmetrical nature of the error curve, an equal probability would exist for the deviations to lie between $-\sigma$ and 0. Hence the probability that deviation lies between $\pm\sigma$ works out to be $2 \times 0.3413 = 0.6826$.

Table 3.2: Probability Tables

y	0	0.01	0.02	0.03	0.04	0.05	0.06	0.07	0.08	0.09
0.0	0.0000	0.0040	0.0080	0.0120	0.0160	0.0199	0.0239	0.0279	0.0319	0.0359
0.1	0.0398	0.0438	0.0478	0.0517	0.0557	0.0596	0.0636	0.0675	0.0714	0.0754
0.2	0.0793	0.0832	0.0871	0.0910	0.0948	0.0987	0.1026	0.1064	0.1103	0.1141
0.3	0.1179	0.1217	0.1255	0.1293	0.1331	0.1368	0.1406	0.1443	0.1480	0.1517
0.4	0.1554	0.1591	0.1628	0.1664	0.1700	0.1736	0.1772	0.1808	0.1844	0.1879
0.5	0.1915	0.1950	0.1985	0.2019	0.2054	0.2088	0.2123	0.2157	0.2190	0.2224
0.6	0.2258	0.2291	0.2324	0.2357	0.2389	0.2422	0.2454	0.2486	0.2518	0.2549
0.7	0.2580	0.2612	0.2642	0.2673	0.2704	0.2734	0.2764	0.2794	0.2823	0.2852
0.8	0.2881	0.2910	0.2939	0.2967	0.2996	0.3023	0.3051	0.3079	0.3106	0.3133
0.9	0.3159	0.3186	0.3212	0.3238	0.3264	0.3289	0.3315	0.3340	0.3365	0.3389
1.0	0.3413	0.3438	0.3461	0.3485	0.3508	0.3531	0.3554	0.3577	0.3599	0.3621
1.1	0.3643	0.3665	0.3686	0.3708	0.3729	0.3749	0.3770	0.3790	0.3810	0.3830
1.2	0.3849	0.3869	0.3888	0.3907	0.3925	0.3944	0.3962	0.3980	0.3997	0.4015
1.3	0.4032	0.4049	0.4066	0.4082	0.4099	0.4115	0.4131	0.4147	0.4162	0.4177
1.4	0.4192	0.4207	0.4222	0.4236	0.4251	0.4265	0.4279	0.4292	0.4306	0.4319
1.5	0.4332	0.4345	0.4357	0.4370	0.4382	0.4394	0.4406	0.4418	0.4430	0.4441
1.6	0.4452	0.4463	0.4474	0.4485	0.4495	0.4505	0.4515	0.4525	0.4535	0.4545
1.7	0.4554	0.4564	0.4573	0.4582	0.4591	0.4599	0.4608	0.4616	0.4625	0.4633
1.8	0.4641	0.4649	0.4656	0.4664	0.4671	0.4678	0.4686	0.4693	0.4700	0.4706

1.9	0.4713	0.4719	0.4726	0.4732	0.4738	0.4744	0.4750	0.4756	0.4762	0.4767
2.0	0.4773	0.4778	0.4783	0.4788	0.4793	0.4798	0.4803	0.4808	0.4812	0.4817
2.1	0.4821	0.4826	0.4830	0.4834	0.4838	0.4842	0.4846	0.4850	0.4854	0.4857
2.2	0.4861	0.4865	0.4868	0.4871	0.4875	0.4878	0.4881	0.4884	0.4887	0.4890
2.3	0.4893	0.4896	0.4898	0.4901	0.4904	0.4906	0.4909	0.4911	0.4913	0.4916
2.4	0.4918	0.4920	0.4922	0.4925	0.4927	0.4929	0.4931	0.4932	0.4934	0.4936
2.5	0.4938	0.4940	0.4941	0.4943	0.4945	0.4946	0.4948	0.4949	0.4951	0.4952
2.6	0.4953	0.4955	0.4956	0.4957	0.4959	0.4960	0.4961	0.4962	0.4963	0.4964
2.7	0.4965	0.4966	0.4967	0.4968	0.4969	0.4970	0.4971	0.4972	0.4973	0.4974
2.8	0.4974	0.4975	0.4976	0.4977	0.4977	0.4978	0.4979	0.4980	0.4980	0.4981
2.9	0.4981	0.4982	0.4983	0.4983	0.4984	0.4984	0.4985	0.4985	0.4986	0.4986
3.0	0.4987	0.4987	0.4987	0.4988	0.4988	0.4989	0.4989	0.4989	0.4990	0.4990
3.1	0.4990	0.4991	0.4991	0.4991	0.4992	0.4992	0.4992	0.4992	0.4993	0.4993
3.2	0.4993	0.4993	0.4994	0.4994	0.4994	0.4994	0.4994	0.4995	0.4995	0.4995
3.3	0.4995	0.4995	0.4996	0.4996	0.4996	0.4996	0.4996	0.4996	0.4996	0.4997
3.4	0.4997	0.4997	0.4997	0.4997	0.4997	0.4997	0.4997	0.4997	0.4998	0.4998
3.5	0.4998	0.4998	0.4998	0.4998	0.4998	0.4998	0.4998	0.4998	0.4998	0.4998
3.6	0.4998	0.4999	0.4999	0.4999	0.4999	0.4999	0.4999	0.4999	0.4999	0.4999
3.7	0.4999	0.4999	0.4999	0.4999	0.4999	0.4999	0.4999	0.4999	0.4999	0.4999
3.8	0.4999	0.4999	0.4999	0.4999	0.4999	0.4999	0.4999	0.5000	0.5000	0.5000

– Further, let it be required to find the probable error which represents the interval about the mean in which lies half of the data. This corresponds to symmetrical area about the mean which has a value of 0.5. Only half of this area lies to the right $y = 0$ and accordingly we look for the value of y in the probability table that corresponds to 0.25. The corresponding value of $y =$ **0.6745** and accordingly the probable error is given by : *Probable error* $= \pm\ 0.6745\ \sigma$.

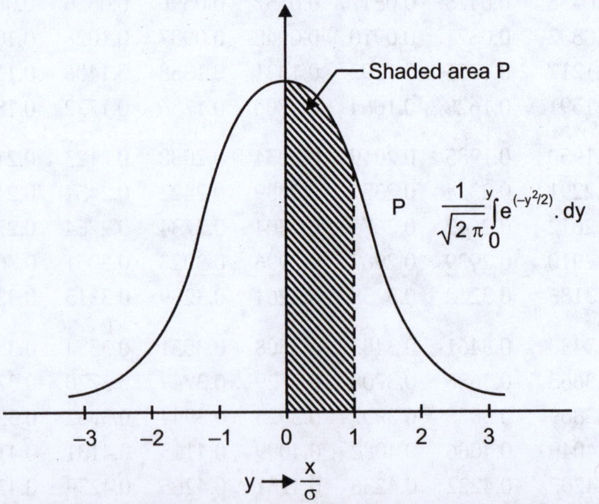

Fig. 3.36. Graphical representation of probability.

Example 3.51. *While measuring a temperature the following ten readings were recorded : 39.6, 39.9, 39.7, 39.9, 40.0, 39.8, 39.9, 39.8, 40.4 and 39.7°C.*

Calculate the following :

(i) The mean,

(ii) The standard deviation,

(iii) The probable error of one reading,

(iv) The probable error of mean, and

(v) The range.

Solution. The statistical values of the given date are computed with the help of the following table :

Observation number	*Temperature reading, °C* (*q*)	*Deviation,* $d = q - \bar{q}$	d^2
1	39.6	– 0.27	0.0729
2	39.9	+ 0.03	0.0009
3	39.7	– 0.17	0.0289
4	39.9	+ 0.03	0.0009
5	40.0	+ 0.13	0.0169
6	39.8	– 0.07	0.0049
7	39.9	+ 0.03	0.0009
8	39.8	– 0.07	0.0049
9	40.4	+ 0.53	0.2809
10	39.7	– 0.17	0.0289
$n = 10$	$\Sigma q = 398.7$ $\bar{q} = \frac{\Sigma q}{n} = \frac{398.7}{10} = 39.87$		$\Sigma d^2 = 0.441$

***(i)* The mean value of the reading :**

$$\bar{q} = \frac{\Sigma q}{n} = \frac{398.7}{10} = \mathbf{39.87°C} \quad \textbf{(Ans.)}$$

***(ii)* The standard deviation :**

$$s = \sqrt{\frac{\Sigma d^2}{(n-1)}}$$ ...since the number of observations is 10 (*i.e.,* less than 20).

$$= \sqrt{\frac{0.441}{10-1}} = \mathbf{0.22°C} \quad \textbf{(Ans.)}$$

***(iii)* Probable error of one reading :**

$$r_1 = 0.6745s = 0.6745 \times 0.22 = \mathbf{0.15°C} \quad \textbf{(Ans.)}$$

***(iv)* Probable error of mean :**

$$r_m = \frac{r_1}{\sqrt{n-1}} = \frac{0.15}{\sqrt{10-1}} = \mathbf{0.05°C} \quad \textbf{(Ans.)}$$

***(v)* Range :**

$$\text{Range} = 40.4 - 39.6 = \mathbf{0.8°C} \quad \textbf{(Ans.)}$$

Example 3.52. *In a test temperature, measurements of temperature were made 100 times with variations in apparatus and procedures. After applying the corrections the following results were obtained :*

Temperature °C	*197*	*198*	*199*	*200*	*201*	*202*	*203*	*204*	*205*
Frequency of Occurrence	*2*	*4*	*10*	*24*	*36*	*14*	*5*	*3*	*2*

Calculate the following :

(i) Arithmetic mean,

(ii) Average deviation,

(iii) Standard deviation,

(iv) Variance,

(v) Probable error of one reading,

(vi) Probable error of the mean, and

(vii) Standard deviation of the standard deviation.

Solution. The computations in tabular form are given below :

Temperature °C T	*Frequency of occurrence, f*	*T × f*	*Deviation* $d\left(=T-\frac{\Sigma T\times f}{n}\right)$	*f × d*	d^2	$f \times d^2$
197	2	394	– 3.77	– 7.54	14.213	28.426
198	4	792	– 2.77	– 11.08	7.673	30.692
199	10	1990	– 1.77	– 17.7	3.133	31.330
200	24	4800	– 0.77	– 18.48	0.593	14.232
201	36	7236	+ 0.23	+ 8.28	0.053	1.908
202	14	2828	+ 1.23	+ 17.22	1.513	21.182
203	5	1015	+ 2.23	+ 11.15	4.973	24.865
204	3	612	+ 3.23	+ 9.69	10.433	31.299
205	2	410	+ 4.23	+ 8.46	17.893	35.786
Total	n = 100	ΣT × f = 20077		Σ|f × d| = 109.6		$\Sigma f \times d^2$ = 219.72

(*i*) **Arithmetic mean :**

$$\text{Arithmetic mean} = \frac{\Sigma T\times f}{n}=\frac{20077}{100}=\mathbf{200.77°C} \quad \textbf{(Ans.)}$$

(*ii*) **Average deviation :**

$$\sigma = \frac{\Sigma|f\times d|}{n}=\frac{109.6}{100}=\mathbf{1.096°C} \quad \textbf{(Ans.)}$$

(*iii*) **Standard deviation :**

$$\textit{Standard deviation, } \sigma = \sqrt{\frac{\Sigma f\times d^2}{n}}=\sqrt{\frac{219.72}{100}}=\mathbf{1.482°C} \quad \textbf{(Ans.)}$$

(*iv*) **Variance :**

$$\text{Variance, } V = \sigma^2 = (1.482)^2 = \mathbf{2.196°C^2} \quad \textbf{(Ans.)}$$

(*v*) **Probable error of one reading :**

Probable error of one reading,

$$r_1 = 0.6745\ \sigma = 0.6745 \times 1.482 \simeq \mathbf{1°C} \quad \textbf{(Ans.)}$$

(*vi*) **Probable error of the mean :**

$$r_m = \frac{r_1}{\sqrt{n}}=\frac{1(°C)}{\sqrt{100}}=\mathbf{0.1°C} \quad \textbf{(Ans.)}$$

(*vii*) **Standard deviation of the standard deviation :**

Standard deviation of the mean,

$$\sigma_m = \frac{\sigma}{\sqrt{n}} = \frac{1.482}{\sqrt{100}} = 0.148°C$$

∴ Standard deviation of the standard deviation,

$$\sigma_\sigma = \frac{\sigma_m}{\sqrt{2}} = \frac{0.148}{\sqrt{2}} = \mathbf{0.1046°C} \quad \textbf{(Ans.)}$$

Example 3.53. *A known current of 80A is measured by an ammeter. If 40% of the readings are within 0.8A of true value, determine the following :*

(i) The standard deviation for the meter.

(ii) The probability of an error of 1.2 A.

Solution. (*i*) **Standard deviation for the meter :**

Considering a normal distribution curve, half the readings, *i.e.*, 20% of the total readings have a +*ve* error and the remaining 20% have a –*ve* error.

From probability tables, it is found that corresponding to a probability of $P(y) = 0.2$,

$$y = \frac{x}{\sigma} = 0.5248$$

∴ Standard deviation,

$$\sigma = \frac{x}{y} = \frac{0.8}{0.5248} = \mathbf{1.524} \quad \textbf{(Ans.)}$$

(*ii*) **Probability of an error of 1.2 A :**

Now, $$y = \frac{x}{\sigma} = \frac{1.2}{1.524} = 0.787$$

Corresponding to $y = 0.787$, probability is 0.2842

∴ Probability of an error of 1.2*A* = 2 × 0.2842 = 0.5684

Thus about 57% of the readings are *within ± 1.2A of the true value.* **(Ans.)**

Example 3.54. *An underdamped galvanometer was energized 100 times under the same carefully controlled experimental conditions and maximum deflection was read in each case. The readings were distributed about a mean value of 21.9 mm and had a probable error of 2.1 mm. How many of 100 readings would you estimate exceeded 25 mm ?*

Solution. Deviation, $x = 25 - 21.9 = 3.1$ mm

Probable error, $r = 2.1$ mm

From eqn. 3.82, standard deviation,

$$\sigma = \frac{r}{0.6745} = \frac{2.1}{0.6745} = 3.1$$

$$\text{Ratio, } y = \frac{x}{\sigma} = \frac{3.1}{3.1} = 1.0$$

Corresponding to ratio $y = \frac{x}{\sigma} = 1.0$, the area under the curve is 0.3413 (from table 3.2).

∴ Number of readings having a deviation within ± 3.1 mm

= 2 × 0.3413 × 100 = 68

Number of readings exceeding a deviation of ± 3.1 mm

= 100 – 68 = 32

Hence, the number of readings having a deviation of + 3.1 mm

$$= \frac{32}{2} = 16.$$

Thus the *number of readings exceeding a maximum deflection of 25 mm, i.e., (21.9 + 3.1) mm is 16.* **(Ans.)**

Example 3.55. *A machine shop was assigned the task to manufacture 10,000 aluminium rods of a nominal length of 20 mm; the rod length was stipulated neither to exceed 20.25 mm and nor to be smaller than 19.5 mm. When inspected for quality control, it was found that 1000 of the rods were too long to fit into a gauge set at 20.25 mm. Determine the number of remaining 9000 rods which will conform to the specifications. Assume that the measurement data conforms to Gaussian normal distribution curve.*

Solution. In case of normal distribution, there is equal probability of the *+ve* and *–ve* errors about the mean. Therefore 5000 rods have lengths greater than 20 mm and 5000 rods have lengths smaller than 20 mm. Since 1000 rods have length greater than 20.25 mm, therefore,

The number of rods where length lies between 20 mm and 20.25 mm

$$= 5000 - 1000 = 4000$$

∴ Probability that 4000 rods have a value greater than 20 mm and less than 20.25 mm,

$$P(y) = \frac{4000}{10,000} = 0.4$$

From probability tables, it is found that corresponding to a probability of $P(y) = 0.4$,

$$y = \frac{x}{\sigma} \simeq 1.3 \qquad \text{...(From table 3.2)}$$

∴ Standard deviation,

$$\sigma = \frac{x}{1.3} = \frac{20.25 - 20.00}{1.3} = \frac{0.25}{1.3} = 0.1923$$

Now, for rods with nominal length of 19.5 mm and 20 mm,

$$y = \frac{x}{\sigma} = \frac{(20 - 19.5)}{0.1923} = 2.60$$

Corresponding probability, $P(y) = 0.4953$ ...(From table 3.2)

∴ Number of rods that have lengths between 19.5 mm and 20 mm

$$= 10000 \times 0.4953 = 4953$$

Hence, *total number of rods whose lengths lie between specified limits of 19.5 mm to 20.25 mm,*

$$= 4000 + 4953 = \mathbf{8953} \quad \textbf{(Ans.)}$$

Example 3.56. *The speed of a hydraulic turbine model, which is running at 1500 rpm, is measured by a tachometer. The hydraulic turbine is subject to variations in speed. For a sample of 20 readings at this speed, how many readings would you expect between 1485 and 1515. Assume that the tachometer gives a normal set of deviations with precision index of 0.04 at 1500 rpm.*

Solution. *Given* : Deviation, $d = q - \bar{q} = \pm 15$ r.p.m.

and, precision index, $h = 0.04$

Standard deviation, $\sigma = \frac{1}{\sqrt{2}h}$ ...[Eqn. (3.81)]

$$= \frac{1}{\sqrt{2} \times 0.04} = 17.68$$

$$y = \frac{x}{\sigma} = \frac{q - \bar{q}}{\sigma} = \frac{15}{17.68} = 0.8484$$

Corresponding to $y = 0.8484$, as read from the probability tables

$P(y) = 0.3015$ ...(From table 3.2)

Probability of an error ± 15 r.p.m. = 2 × 0.3015 = **0.603**

Thus we expect 60% of 0.6 × 20 = **12 readings** *to lie between 1485 and 1515 r.p.m.* **Ans.**

Example 3.57. *The depth of water, which had a nominal value of 15 cm, was measured with a hook. Measurements were taken 40 times and 10 readings were found to lie outside a particular range. Assuming that the measurements conform to normal distribution with precision index 9 cm^{-1}, determine the prescribed range.*

Solution. *Given* : Nominal value of depth of water = 15 cm

Total number of times the measurements were taken = 40

No. of measurement readings found to lie outside a particular range = 10

Precision index, h = 9 cm^{-1}

Prescribed range :

Probability of falling within a particular range

$$= \frac{(40-10)}{40} = 0.75$$

Half of these measurements have a *+ve* error and half have a *–ve* error.

$$P(y) = \frac{0.75}{2} = 0.375$$

From probability tables, it is found that corresponding to a probability

$P(y) = 0.375,$

$$y = \frac{x}{\sigma} = \frac{q - \bar{q}}{\sigma} = 1.15$$ ...(From table 3.2)

Now, standard deviation, $\sigma = \frac{1}{\sqrt{2}h}$ ...[Eqn. (3.81)]

$$= \frac{1}{\sqrt{2} \times 9} = 0.0786$$

$\therefore$ $x = (q - \bar{q}) = 1.15 \times 0.0786 = 0.0904$ cm

Thus 75 per cent of depth measurements lie within the **range (15 ± 0.0904) cm. (Ans.)**

Example 3.58. *(a) A mass flow meter, using electromagnetic principle, is set to record a large number of readings. The fixed mass flow rate is 100 kg/s; half of the readings lie between 104.8 and 95.2 kg/s. The sample measurements give a straight line between the (0, 50%) points on the probability coordinates. Calculate the precision index of the instrument.*

(b) An alarm rings when the flow meter gives a reading of 88 kg/s.

(i) How many false alarms are expected in the month of November if the flow is checked automatically 4 times a day ?

(ii) What will be the value of precision index if one wants to reduce the false alarm rate by a factor of 2 ?

Solution. (*a*) The overall probability is 0.5. Considering normal distribution, 0.25 of the measurements have a *+ve* error and the rest 0.25 of the measurements have a *–ve* error.

Corresponding to a probability of $P(y) = 0.25$, from probability tables, we have

$$y = \frac{x}{\sigma} = 0.675 \quad \text{...(From table 3.2)}$$

$$\therefore \text{ Standard deviation,} \quad \sigma = \frac{x}{0.675} = \frac{4.8}{0.675} = 7.11$$

$$\textbf{Precision index,} \quad h = \frac{1}{\sqrt{2}\sigma} = \frac{1}{\sqrt{2} \times 7.11} = \mathbf{0.099} \quad \textbf{(Ans.)}$$

(*b*) (*i*) For a mass flow rate of 88 kg/s.

$$y = \frac{x}{\sigma} = \frac{100-88}{7.11} = 1.69$$

Corresponding probability $P(y) = 0.45$ ...(From table 3.2)

Thus it is expected that (0.5 – 0.45) = 0.05 or 5% of the flow measurements will fall in the range of false alarms.

Number of measurements in the month of November = 30 × 4 = 120

∴ *Expected false alarms* = 120 × 0.05 = **6** **(Ans.)**

(*ii*) Reduced number of false alarms $= \frac{1}{2} \times 6 = 3$

Probability of false alarms $= \frac{3}{120} = 0.025$ (or 2.5%)

Probability of data to lie in the tolerant band,

$$P(y) = 0.5 - 0.025 = 0.475$$

Corresponding to $P(y) = 0.475$

$$y = \left(\frac{x}{\sigma}\right) = 1.96 \quad \text{...(From table 3.2)}$$

$$\text{or, Standard deviation,} \quad \sigma = \frac{x}{1.96} = \frac{100-88}{1.96} = 6.12$$

$$\therefore \textit{ Precision index,} \quad h = \frac{1}{\sqrt{2}\sigma} = \frac{1}{\sqrt{2} \times 6.12} = \mathbf{0.1155} \quad \textbf{(Ans.)}$$

3.9.5.6. Odds

Odds *is the number of chances that a particular reading will occur when error limit is specified.* Thus the probability of occurrence can be stated in terms of *odds*.

Example. When the error limits are specified as ± 0.6745σ, the chances are that 50 per cent of observations will lie between the above limits or in other words we can say odds are 1 to 1.

The odds can be calculated as under :

$$\text{Probability of occurrence} = \frac{\text{Odds}}{\text{Odds} + 1} \quad \text{...(3.90)}$$

Certain deviations, their probability of occurrence and the corresponding odds are given in Table 3.3.

Table 3.3: Deviations, Probability and Odds

Deviation	*Probability*	*Odds*
$\pm$ 0.6745 σ	0.5000	1 to 1
$\pm$ σ	0.6826	2.15 to 1
$\pm$ 2σ	0.9546	21 to 1
$\pm$ 3σ	0.9974	369 to 1

3.9.5.7. Specification of measurement data

The results of statistical analysis of multi-sample data are normally expressed as deviations about a mean value; the deviations are expressed as follows :

(*i*) **Probable error.** The result is expressed as $\bar{q} \pm 0.6745\sigma$, which implies that 0.50 (50%) of the readings are within this limit and odds are 1 to 1. Obviously there is an even probability that a new reading will lie within these limits.

(*ii*) **Standard deviation.** The result is expressed as $\bar{q} \pm \sigma$. In this case, the error limit is the standard deviation. It implies that 0.6828 (about 68%) of the readings are within the limits σ = ± 1 and the odds are 2.15 to 1. Thus there is approximately a 2 to 1 possibility that a new reading will fall within this limit.

(*iii*) **± 2σ limit.** The result is expressed as $\bar{q} \pm 2\sigma$, which implies that 0.9546 (about 95%) of the readings are within the given limits and the odds are 21 to 1.

(*iv*) **± 3σ limit.** The result is expressed as $\bar{q} \pm 3\sigma$, which implies that 0.9974 (about 99%) of the readings are within the given limits. The odds of any observation falling within this limit are 369 to 1.

3.9.5.8. Confidence interval and confidence level

The range of deviation from the mean value within which a certain fraction of all values are expected to lie is called **confidence interval.**

The probability that the value of a randomly selected observation will lie in the range of confidence interval is termed as **confidence level.**

When the number of observations is large and their errors are random and follow Gaussian distribution, the various confidence intervals about the mean value $\bar{q}$ are given in Table 3.4.

Table 3.4: Confidence Intervals

Confidence level	*Confidence interval*	*Values lying outside confidence interval*
0.500	$\bar{q} \pm 0.674\sigma$	1 in 2
0.800	$\bar{q} \pm 1.282\sigma$	1 in 5
0.900	$\bar{q} \pm 1.645\sigma$	1 in 10
0.950	$\bar{q} \pm 1.960\sigma$	1 in 20
0.990	$\bar{q} \pm 2.576\sigma$	1 in 100
0.999	$\bar{q} \pm 3.291\sigma$	1 in 1000

If the number of observations is small (say *less* than 20) and the standard deviation is not accurately known, the confidence interval must be *broadened.* The standard deviation, in this case, is computed as :

$s = \sqrt{\frac{\Sigma d^2}{n-1}}$; this standard is multiplied by a suitable factor to establish the confidence interval.

3.9.5.9. Probability graph paper

Prior to undertaking statical analysis of the measurement data, the experimenter needs to check that the data is in conformity with the normal distribution. To accomplish this, the data is plotted on a probability graph paper wherein the abscissa is the value of a particular reading (on a linear scale) and the ordinate has a percentage of readings at or below the value of abscissa (on a logarithmic scale). A straight line is then drawn through the data points; however, while fitting the curve much weightage is not given to the points at low and high ends of the plot. *If the 50 percent reading at the ordinate exactly corresponds to averages of all the data points, the data is considered to follow the Gaussian distribution.*

3.9.5.10. Rejection of test data–Chauvenet's criterion

During experimentation, it is observed that some points are noticeably different from the majority of the data. Such points can be determined and subsequently eliminated by applying Chauvenet's criterion.

Chauvenet's criterion states as follows :

"An observed reading may be rejected if the probability of obtaining the particular deviation from the mean is less than $\frac{1}{2n}$". The quantity *n* refers to the number of observations which are large enough to presume that the data follows the Gaussian error distribution.

Let us consider that during a measurement act ten readings have been made. Then,

$$\frac{1}{2n} = \frac{1}{2 \times 10} = 0.05 \text{ or } 5\%$$

Thus readings with 95 percent of probability are to be accepted, and less than that can be rejected. Corresponding to probability

$$P(y) = \frac{0.95}{2} = 0.475; \; y = 1.96$$

Results based on such calculations for different number of readings are given in table 3.5.

Table 3.5: Chauvenet's Criterion for Rejecting Data

Number of readings	*Ratio of maximum acceptable deviation to standard deviation, d_{max}/σ*
2	1.15
3	1.38
4	1.54
5	1.65
6	1.73
7	1.80
10	1.96
15	2.13

25	2.33
50	2.57
100	2.81
300	3.14
500	3.29
1000	3.48

When applying Chauvenet's criterion, in order to eliminate any dubious data, the following *procedure* is adopted :

(*i*) Using all data points calculate the standard deviation.

(*ii*) Compare the deviation of the individual readings with standard deviation.

(*iii*) If the ratio of deviation of a reading to the standard deviation exceeds the limits given in Table 3.5, reject the readings.

(*iv*) Calculate again the mean value and the standard deviations by excluding the rejected readings from the data.

- A criterion (*based upon "confidence intervals"*) used for discarding a data point is that its deviation from the mean exceeds four times the probable error of a single reading. This results in discarding a data outside a confidence interval for a single reading at a confidence level of 0.993.

— A better criterion which does not involve the evaluation of probable error when set of data points is small and standard deviation is not accurately known, *is to discard a reading that lies outside the interval corresponding to confidence level of 0.99 for a single observation* (Refer to Tables 3.4 and 3.5). On this basis not more than one reading in 100 would lie outside their range.

— A still better method is to use confidence interval corresponding to a confidence level of 0.95 in order to scrutinize the measurement procedure adopted.

Example 3.59. *Ten copper rods, selected at random, were found to a have the following lengths in metres : 5.30, 5.73, 6.77, 5.26, 4.33, 5.45, 6.09, 5.64, 5.81, 5.75.*

Determine any reading that can be rejected by applying Chauvenet's criterion. The ratio of maximum deviation to standard deviation should not exceed 1.96.

Solution. The computations in a tabular form are given below :

S.No.	*Length q, m*	*Deviation* $d = q - \bar{q}$	d^2	$\frac{\lvert d \rvert}{s}$
1	5.30	– 0.313	0.09797	0.5
2	5.73	+ 0.117	0.01369	0.19
3	6.77	+ 1.157	1.33865	1.85
4	5.26	– 0.353	0.12461	0.56
5	4.33	– 1.283	1.64609	2.05
6	5.45	– 0.163	0.02657	0.26
7	6.09	+ 0.477	0.22753	0.76
8	5.64	+ 0.027	0.00073	0.04
9	5.81	+ 0.197	0.03881	0.31
10	5.75	+ 0.137	0.01877	0.22
$n = 10$	$\Sigma q = 56.13$ $\bar{q} = \frac{56.13}{10} = 5.613$		$\Sigma d^2 = 3.53342$	

Arithmetic mean, $\bar{q} = \frac{\Sigma q}{n} = \frac{56.13}{10} = 5.613$

Standard deviation, $s = \sqrt{\frac{\Sigma d^2}{n-1}}$ ... since the number of readings is less than 20.

$$= \sqrt{\frac{3.53342}{10}} = 0.626$$

It is given that for 10 readings the ratio of deviation to standard deviation is not to exceed 1.96 and therefore **reading number 5, *i.e.*, 4.33 m,** *should be rejected.* **(Ans.)**

3.9.5.11. Method of least squares

- Let us consider a set of n readings $q_1, q_2, q_3, ... q_n$. The sum of squares of their deviations for some mean value Q_m is,

$$S = \sum_{i=1}^{n} (q_1 - Q_m)^2$$

For minimizing S with respect to mean value Q_m, we have

$$\frac{\partial S}{\partial Q_m} = 2\left(\sum_{i=1}^{n} q_1 - nQ_m\right) = 0$$

or, $$Q_m = \frac{1}{n}\sum_{i=1}^{n} q_i$$

But, $$\frac{1}{n}\sum_{i=1}^{n} q_i = \bar{q}$$

$\therefore$ $$Q_m = \bar{q} \quad ...(3.91)$$

Thus the *mean value which minimizes the sum of the squares of the deviation is the arithmetic mean.*

- Now consider two variables u and v measured over a range of values and let us find the linear function connecting y and x.

Let the linear function be :

$$v = au + b$$

Now let us minimize the quantity,

$$S = \sum_{i=1}^{n} [v_i - (au_i + b)]^2$$

This is accomplished by taking derivatives of S w.r.t. a and b and setting them equal to zero. This gives

$$nb + a\Sigma u_i = \Sigma v_i \quad ...(i)$$

and, $$b\Sigma u_i + a\Sigma u_i^2 = \Sigma u_i\, v_i \quad ...(ii)$$

Solving, the above eqns. (*i*) and (*ii*), we get

$$a = \frac{n\Sigma u_i v_i - (\Sigma u_i)(\Sigma v_i)}{n\Sigma u_i^2 - (\Sigma u_i)^2} \quad ...(3.92)$$

$$b = \frac{(\Sigma v_i)(\Sigma u_i^2) - (\Sigma u_i v_i)(\Sigma u_i)}{n\Sigma u_i^2 - (\Sigma u_i)^2} \quad ...(3.93)$$

The standard deviations may be found as :

$$s_a = \sqrt{\frac{n}{n\Sigma u_i^2 - (\Sigma u_i)^2}}\, s_v \quad ...(3.94)$$

and

$$s_b = \sqrt{\frac{\Sigma u_i^2}{n\Sigma u_i^2 - (\Sigma u_i)^2}}\, s_v \quad ...(3.95)$$

where, $s_v = \sqrt{\frac{1}{n}\sum (au_i + b - v_i)^2}$

Standard deviation of u is,

$$s_u = \sqrt{\frac{1}{n}\sum\left(\frac{v_i - b}{a} - u_i\right)^2} = \frac{s_v}{a} \quad ...(3.96)$$

Example 3.60. *The data given below are expected to follow a linear relationship :*

$$v = au + b$$

u	1.8	4.6	6.6	9.0	11.4	13.4
v	2.2	3.2	5.2	6.4	8.0	10.0

(i) Obtain the best linear relation, and

(ii) Calculate the standard deviation.

Solution. *Given* : $v = au + b$ .... Linear relationship ;

Number of readings, $n = 6$

The computations in a tabular form are given below :

u_i	v_i	$u_i v_i$	u_i^2
1.8	2.2	3.96	3.24
4.6	3.2	14.72	21.16
6.6	5.2	34.32	43.56
9.0	6.4	57.60	81.0
11.4	8.0	91.20	129.96
13.4	10.0	134.00	179.56
$\Sigma u_i = 46.8$	$\Sigma v_i = 35$	$\Sigma u_i v_i = 335.8$	$\Sigma u_i^2 = 458.48$

(i) **Best linear relation :**

Using equations (3.92) and (3.93), we have

$$a = \frac{(6)\times(335.8) - (46.8)\times(35)}{(6)\times(458.48) - (46.8)^2} = \frac{376.8}{560.64} = 0.672$$

$$b = \frac{(35)\times(458.48) - (335.8)(46.8)}{(6)(458.48) - (46.8)^2} = \frac{331.36}{560.64} = 0.591$$

The best linear relation is : $\boldsymbol{v = 0.672\, u + 0.591}$ **(Ans.)**

(ii) Standard deviation :

The computations for finding standard deviation, in a tabular form, are given below :

u_i	v_i	$au_i + b$	$au_i + b - v_i$	$(au_i + b - v_i)^2$
1.8	2.2	1.8006	– 0.3994	0.1595
4.6	3.2	3.6822	+ 0.4822	0.2325
6.6	5.2	5.0262	– 0.1738	0.0302
9.0	6.4	6.6390	+ 0.239	-0.0571
11.4	8.0	8.2518	+ 0.2518	0.0634
13.4	10.0	9.5958	– 0.4074	0.1660
				$\Sigma(au_i + b - v_i)^2 = 0.7087$

∴ Standard deviation of v,

$$s_v = \sqrt{\frac{1}{n}\Sigma(au_i + b - v_i)^2} = \sqrt{\frac{0.7087}{6}} = \pm \mathbf{0.3436} \quad \textbf{(Ans.)}$$

From eqn. (3.96), standard deviation of u,

$$s_u = \frac{s_v}{a} = \frac{\pm 0.3436}{0.672} = \pm \mathbf{0.511} \quad \textbf{(Ans.)}$$

From eqns. (3.94 and 3.95) the standard deviations in a and b are :

$$s_a = \sqrt{\frac{n}{n\Sigma u_i^2 - (\Sigma u_i)^2}}\, s_v$$

$$= \sqrt{\frac{6}{6 \times 458.48 - (46.8)^2}} \times (\pm 0.3436) = \pm \mathbf{0.0355} \quad \textbf{(Ans.)}$$

$$s_b = \sqrt{\frac{\Sigma u_i^2}{n\Sigma(u_i^2) - (\Sigma u_i)^2}}\, s_v$$

$$= \sqrt{\frac{458.48}{6 \times 458.48 - (46.8)^2}} \times (\pm 0.3436) = \pm \mathbf{0.3107} \quad \textbf{(Ans.)}$$

Example 3.61. *(a) Enumerate the factors which should be considered while selecting an instrument for a specific measurement application.*

(b) The iron losses in a ferromagnetic material used in a transformer vary with frequency f of the supply given to the transformer. For a particular transformer these iron losses were determined at various frequencies with a constant flux density in the ferromagnetic material; the results obtained are given below :

Frequency, f (Hz) :	550	700	850	1000
Iron losses, P (mW) :	23	31	37	61

Assuming the iron losses to have general form $P = af^2 + bf$, *determine the constants a and b to achieve best fit for the four measured values, using method of least squares.*

Solution. (a) The following *factors* need to be considered for the selection of an instrument to be used for a specific measurement application :

1. Easy calibration.
2. Convenience and ease in reading the instrument.
3. The maximum and minimum values the input variables are expected to assume.

4. Nature of the input signal.
5. Non-interference with measuring system.
6. Accuracy expected from the instrument.
7. Cost criterion.
8. Specific application of the data.
9. Required form of data.
10. Negligible or well-known environmental errors.

(*b*) *Given :* General form $P = af^2 + bf$

or, $$\frac{P}{f} = af + b$$

The above relation may be expressed as

$$v = au + b$$

which is a straight line relationship between, $v = \frac{P}{f}$ and $u = f$,

i.e., $u = f$:	550	700	850	1000
$v = \frac{P}{f}$:	0.04182	0.04429	0.05529	0.061

The computations are tabulated below :

u_i	v_i	$u_i v_i$	u_i^2
550	0.04182	23	0.302×10^6
700	0.04429	31	0.49×10^6
850	0.05529	47	0.722×10^6
1000	0.0610	61	1×10^6
$\Sigma u_i = 3100$	$\Sigma v_i = 0.2024$	$\Sigma u_i v_i = 162$	$\Sigma u_i^2 = 2.514 \times 10^6$

In order to find the values of *a* and *b* for the best straight line fit using the method of least squares, let us use the equations (3.92) and (3.93) as follows :

$$a = \frac{n\Sigma u_i v_i - (\Sigma u_i)(\Sigma v_i)}{n\Sigma u_i^2 - (\Sigma u_i)^2} \quad \text{...[Eqn (3.92)]}$$

$$= \frac{4 \times 162 - (3100)(0.2024)}{4 \times 2.514 \times 10^6 - (3100)^2} = \frac{20.56}{446000} = 46.1 \times 10^{-6}$$

$$b = \frac{(\Sigma v_i)(\Sigma u_i^2) - (\Sigma u_i v_i)(\Sigma u_i)}{n\Sigma u_i^2 - (\Sigma u_i)^2} \quad \text{...[Eqn. (3.93)]}$$

$$= \frac{0.2024 \times 2.514 \times 10^6 - 162 \times 3100}{4 \times 2.514 \times 10^6 - (3100)^2}$$

$$= \frac{6633.6}{446000} = 14.87 \times 10^{-3}$$

Hence, *the best fit relationship is* :

$$P = 46.1 \times 10^{-6} f^2 + 14.87 \times 10^{-3} f \text{ mW}$$ **(Ans.)**

Example 3.62. *The final result of a measurement depends on two measurements in series, each having values of 50 and 100 units with errors of ± 2% and ± 1% such that final result is sum of both. Determine the error of final result if errors are considered as :*

(i) Limiting errors, and

(ii) Standard deviations.

Solution. Error of final result :

(i) Limiting errors :

When individual errors are limiting errors, then fractional errors of combination

$$= \pm\left(\frac{q_1}{q_1+q_2}\times\frac{\delta q_1}{q_1}+\frac{q_2}{q_1+q_2}\times\frac{\delta q_2}{q_2}\right)$$

$$= \pm\left(\frac{50}{50+100}\times 0.02+\frac{100}{50+100}\times 0.01\right)$$

= 0.0133 or **1.33%** **(Ans.)**

(ii) Standard deviations :

When individual errors are standard deviations, then errors in individual measurements are 2% of 50 and 1% of 100, *i.e.*, 1 and 1

Thus, resultant error $= \sqrt{1^2+1^2}=\sqrt{2}$

$$= \frac{\sqrt{2}}{(50+100)}\times 100\% = \mathbf{0.943\%}$$ **(Ans.)**

HIGHLIGHTS

1. *Measurement* is the act, or the result, of a quantitative comparison between a predetermined standard and an unknown magnitude. Measurements provide us with a means of describing various phenomena in quantitative terms.
2. The physical quantity or the characteristic condition which is the object of measurement in an instrumentation system is variously termed as *"measurand"*, *"measurement variable"*, *"instrumentation variable"* or *"process variable"*.
3. The technology of using instruments to measure and control the physical and chemical properties of materials is called *"instrumentation"*.
4. Methods of measurements are : (*i*) Direct method, (*ii*) Indirect method.
   — In *direct comparison method* the parameter to be measured is directly compared with either a primary or a secondary standard.
   — In *indirect comparison method* the comparison is done with a standard through the use of a calibrated system.
5. The three modes of measurement are :
   (*i*) Primary measurements, (*ii*) Secondary measurements, and (*iii*) Tertiary measurements.
6. A generalised measurement system consists of the following components/elements:
   (*i*) Primary sensing element;

(*ii*) Variable conversion or transducer element;

(*iii*) Manipulation element;

(*iv*) Data transmission element;

(*v*) Data presentation element.

7. The *static characteristics* pertain to a system where the quantities to be measured are constant or vary slowly with time. The main static characteristics are : *Accuracy, sensitivity, reproducibility, drift, static error and dead zone.*
8. The *relative static error* is defined as the ratio of absolute static error to the true value of the quantity under measurement.
9. The difference between the true value and the measured value of a quantity is called *static correction.*
10. *Static calibration* is a process by which all the static performance characteristics are obtained in one form or another.
11. The difference between the largest and the smallest reading of the instrument is called the *range* of the instrument.
12. *Span* represents the algebraic difference between the upper and lower range values of the instrument.
13. An error curve which can be used for correcting instrument readings is called *"Error calibration curve".*
14. *Scale readability* indicates the closeness with which the scale of an analog type of instrument can be read.
15. *Drift* is an undesired gradual departure of the instrument output over a period of time that is unrelated to the changes in input, operating conditions or load.
16. The closeness with which an instrument reading approaches the true value of the quantity being measured is called *accuracy.*
17. *Precision* is a measure of reproducibility of measurements.
18. The ratio of the magnitude of the output signal to the input signal or response of measuring system to the quantity being measured is called *sensitivity.*
19. The ability to reproduce the input characteristics symmetrically is called *linearity.*
20. The maximum differences in output at any measured value within the specified range when approaching the point first increasing and then with decreasing input may be termed as *hysteresis.*
21. The minimum value below which no output change can be detected when the input of an instrument is increased gradually from zero is called the *threshold* of an instrument.
22. *Threshold* defines the smallest measurable input while the *resolution* defines the smallest measurable input change.
23. The largest change of input quantity for which there is no output of the instrument is termed as *dead zone.*
24. The time required by a measurement system to begin to respond to a change in the measurement is called *dead time.*
25. The incapability of the system to faithfully measure the input signal in undistorted form is called the *loading effect.*
26. *Noise* is a spurious current or voltage extraneous to the current or voltage of interest in an electrical or electronic circuit.

27. "*Dynamic characteristics*" quantify the dynamic relation between the input and output.
28. The evaluation of the ability of a system to faithfully transmit and present all the pertinent information included in the input signal and to exclude all else, is called *response*.
29. The behaviour of the system when inputs vary with time (*i.e.*, inputs are dynamic in nature) and so does the output is called *dynamic response*.
30. Response of a measurement system may be :
    (*i*) Transient response; (*ii*) Steady state response.
31. The rapidity with which a measurement system responds to changes in the measured quantity is called the *speed of response* or "*responsiveness*".
32. *Measuring lag* refers to retardation or delay in the response of a measurement system to changes in measured quantity.
33. *Fidelity* is defined as the degree to which the measurement system indicates changes in the measured quantity without any dynamic error.
34. The *dynamic error* or *measurement error* is the difference between the true value of the quantity changing with time and the value indicated by the measurement system if no static error is assumed.
35. "Examples" of the systems :
    (*i*) *Zero-order systems* : Mechanical levers, amplifiers etc.
    (*ii*) *First-order systems* : Velocity of a free falling mass; air pressure buildup in bellows, thermistors and thermocouples etc.
    (*iii*) *Second-order systems* : Piezoelectric pick-ups, spring-mass system, pen control system on X-Y plotters etc.
36. *Damping ratio* (γ) is defined as the ratio of the actual value of coefficient of viscous friction in movement and the value required to produce critical damping.
37. The limits of the deviations from the specified value are defined as *limiting errors* or *guarantee errors*.
38. The *relative error* is defined as the ratio of the error to the specified (nominal) magnitude of a quantity.
39. The errors are classified as :
    1. Gross errors
    2. Systematic errors :
        (*i*) Instrumental errors.
        (*ii*) Environmental errors.
        (*iii*) Observational errors.
    3. Random errors.
40. *Sources of errors are* :

    Noise, Response time; Design limitations; Energy exchanged by interaction; Transmission; Determination of measuring system; Ambient influences on measuring systems, Errors of observation and interpretation.
41. *Arithmetic mean* of the number of readings taken is the most probable value of measured variable.
42. *Geometric mean* is defined as the *n*th root of the product of *n* terms.

43. *Median* is the middle value of a set of an "odd" number of readings, if variables are arranged in numerical order. For an "even" number of readings, the median is the arithmetic average of the two central readings.
44. *Mode* is the value of the variable that occurs most frequently.
45. *Dispersion* (or *spread* or *scatter*) is the property which denotes the extent to which the values are dispersed about the central value.
46. The departure of the observed readings from the arithmetic mean of the group of readings is termed as *deviation*.
47. *Standard deviation* of an infinite number of data is defined as the square root of the sum of individual deviations squared divided by the number of readings.
48. The square of the standard deviation is called *variance*.
49. The *"law of probability"* states that the normal occurrence of deviations from the average value of an infinite number of measurements or observations can be expressed as :

$$P = \frac{1}{\sigma\sqrt{2\pi}} e^{-x^2/2\sigma^2}$$

where P = Number of readings at any deviation *'d'*, *i.e.*, probability of occurrence of deviations *'d'*,
σ = Standard deviation, and
x = Magnitude of deviation from mean.

50. Chauvenet's criterion states as follows :
"An observed reading may be rejected if the probability of obtaining the particular deviation from the mean is less than $\frac{1}{2n}$". The quantity n refers to the number of observations which are large enough to presume that the data follows the Gaussian error distribution.

OBJECTIVE TYPE QUESTIONS

Choose the Correct Answer :

1. The purpose of instruments is to
   (*a*) allow measurements to be made (*b*) transmit the information
   (*c*) change signals (*d*) any of the above.
2. The measurement refers to which of the following ?
   (*a*) Primary signal (*b*) Measured variable
   (*c*) Output (*d*) All of the above.
3. The measurement of a quantity
   (*a*) is an act of comparison of an unknown quantity with another quantity.
   (*b*) is an act of comparison of an unknown quantity with a known quantity whose accuracy may be known or may not be known.
   (*c*) is an act of comparison of an unknown quantity with a predefined acceptable standard which is accurately known.
   (*d*) none of the above.
4. In present-day measurement systems
   (*a*) direct methods are commonly used.

(*b*) one of the direct methods is limited but indirect methods are commonly used.
(*c*) both direct and indirect methods are commonly used.
(*d*) all of the above.

5. The measurand is
(*a*) output (*b*) measured variable
(*c*) secondary signal (*d*) any of the above.

6. The output stage of a generalised measurement system may comprise
(*a*) manipulator (*b*) detector-transducer
(*c*) indicating or recording unit (*d*) all of the above.

7. A null type of instrument as compared to a deflection type instrument has
(*a*) a higher accuracy (*b*) a lower sensitivity
(*c*) a faster response (*d*) all of the above.

8. In an intelligent information system used for the measurement of a physical quantity
(*a*) the digital computer must perform all or nearly all the signal and information processing.
(*b*) the observer must perform all or nearly all the signal and information processing.
(*c*) the digital computer is not required.
(*d*) none of the above.

9. A high grade of slip gauges preserved in a factory and not put into general use would be a
(*a*) primary standard (*b*) secondary standard
(*c*) tertiary standard (*d*) any of the above.

10. The temperature measurement by a thermocouple is
(*a*) primary measurement (*b*) secondary measurement
(*c*) tertiary measurement (*d*) any of the above.

11. Purely mechanical instruments cannot be used for dynamic measurements because they have
(*a*) high inertia (*b*) large time constant
(*c*) higher response time (*d*) all of the above.

12. The usage of electronic instruments is becoming more extensive because they have
(*a*) a high sensitivity and reliability.
(*b*) a fast response and compatibility with digital computers.
(*c*) the capability to respond to signals from remote places.
(*d*) all of the above.

13. The intermediate stage of an instrument is used to
(*a*) sense the primary (input) signal.
(*b*) indicate or record the measurements.
(*c*) amplify and transmit the secondary signal (signal after it has been transduced).
(*d*) any of the above.

14. The function of the signal manipulating element, in a generalised measurement system, is to
(*a*) change the quantity under measurement to an analogous signal.
(*b*) change the magnitude of the input signal while retaining its identity.
(*c*) to perform non-linear operations like filtering, chopping and clipping and clamping.
(*d*) none of the above.

15. A modifying input to a measurement system can be defined as an input,
    (*a*) which changes the input-output relationship for desired inputs only.
    (*b*) which changes the input-output relationship for desired as well as interfering inputs.
    (*c*) which changes the input-output relationship for interfering inputs only.
    (*d*) none of the above.
16. The ratio of the output to input change for a given measuring system is referred to as
    (*a*) sensitivity (*b*) linearity
    (*c*) stability (*d*) none of these.
17. The largest change in the measured variable which produces no instrument response is known as
    (*a*) threshold (*b*) dynamic error
    (*c*) dead zone (*d*) none of these.
18. The conformity of the output to the true value of the measurement is called
    (*a*) precision (*b*) accuracy
    (*c*) sensitivity (*d*) none of these.
19. The resolution of a system refers to
    (*a*) smallest change in the measurand that can be measured
    (*b*) true value of the input
    (*c*) retardation of the response (*d*) none of these.
20. The least change of the measured variable which can be detected at the output of the measuring system is
    (*a*) least count (*b*) sensitivity
    (*c*) discrimination (*d*) any of these.
21. State which of the following describes the linearity of instrument :
    (*a*) Range of inaccuracy which can be tolerated.
    (*b*) Largest change in the measured value which produces no instrument response.
    (*c*) Relationship between the output and the input.
    (*d*) None of these.
22. In measurement systems, which of the following are undesirable static characteristics?
    (*a*) Sensitivity and accuracy (*b*) Drift, static error and dead zone
    (*c*) Reproducibility and non-linearity
    (*d*) Drift, static error, dead zone and non-linearity.
23. The values of static stiffness and compliance in a measurement system determine the amount of
    (*a*) power drain from a system (*b*) current drain from a system
    (*c*) energy drain from a system (*d*) none of the above.
24. In the centre zero analog ammeter having a range of –10A to +10A, there is a detectable change of the pointer from its zero position on either side of the scale only if the current reaches a value of 1 A. The ammeter has a
    (*a*) resolution of 1 A (*b*) dead zone of 1 A
    (*c*) dead zone of 2 A (*d*) none of the above.
25. The input resistance of a cathode ray oscilloscope is of the order of
    (*a*) tens of ohm (*b*) megaohms
    (*c*) kilohms (*d*) fraction of an ohm.
26. Johnson noise is
    (*a*) caused by thermal agitation of free electrons carrying currents thereby modulating the current.

(*b*) noise carried into a circuit through conductors.
(*c*) noise of an electromagnetic origin that is radiated into a circuit.
(*d*) shot noise which results from random emission of electrons across a *p-n* junction.

27. A pressure measurement instrument is calibrated between 10 bar and 250 bar. The scale span of the instrument is
(*a*) 10 bar (*b*) 250 bar
(*c*) 240 bar (*d*) 260 bar.
while the scale range is
(*a*) 10 bar (*b*) 250 bar
(*c*) 240 bar (*d*) 260 bar.

28. A pressure gauge is calibrated from 0-50 kN/m^2. It has a uniform scale with 100 scale divisions. One-fifth of a scale division can be read with certainty. The gauge has a
(*a*) resolution of 0.1 kN/m^2 (*b*) threshold of 0.1 kN/m^2
(*c*) dead zone of 0.2 kN/m^2 (*d*) a resolution of 0.5 kN/m^2.

29. A D.C. circuit can be represented by an internal voltage source of 50 V with an output resistance of 100 kΩ. In order to achieve 99 per cent accuracy for voltage measurement across its terminals, the voltage measuring device should have
(*a*) a resistance of at least of 10 MΩ (*b*) a resistance of 100 kΩ
(*c*) a resistance of at least 10 Ω (*d*) none of the above.

30. In A.C. circuits, the connection of measuring instruments causes loading errors which may affect
(*a*) only the magnitude of quantity being measured
(*b*) only the phase of the quantity being measured
(*c*) both the magnitude and phase of the quantity being measured
(*d*) magnitude, phase and also the waveform of the quantity being measured.

31. Backlash is
(*a*) gradual departure of instrument output from its calibrated value.
(*b*) the ability of the device to give identical output when repeated or measurements are made with the same input signal.
(*c*) maximum angle or distance through which any part of mechanical system may be moved in one direction without causing motion of the next part.
(*d*) unwanted signal tending to obscure the transducer signal.

32. In measurement systems, which of the following static characteristics are desirable ?
(*a*) Accuracy (*b*) Sensitivity
(*c*) Reproducibility (*d*) All of the above.

33. In force-current analogy stiffness constant K is analogous to :
(*a*) inductance (*b*) reciprocal of inductance
(*c*) capacitance (*d*) reciprocal of capacitance.

34. In force-voltage analogy viscous friction constant B is analogous to
(*a*) resistance (*b*) reciprocal of resistance
(*c*) inductance (*d*) capacitance.

35. Which of the following are the desirable dynamic characteristics of a measurement system?
(*a*) Fast response, fidelity, measuring lag and dynamic error
(*b*) Fast response and lag
(*c*) Fidelity and measuring lag
(*d*) Fast response and fidelity.

36. Which of the following errors can arise, as a result of mistakes in reading, parallax, improper instrument location and inadequate lighting ?

(*a*) Construction error (*b*) Transmission error

(*c*) Observation error (*d*) Translation error.

37. Gearing, backlash, friction between moving parts, and scale accuracies are generally known as

(*a*) instrument errors (*b*) interference errors

(*c*) calibration errors (*d*) interaction errors.

38. Errors which may be variable both in magnitude and nature (positive or negative) are classified as

(*a*) hysteresis errors (*b*) random errors

(*c*) systematic errors (*d*) interaction errors.

39. The characteristic of an observer to read either systematically higher or systematically lower than other observers is referred to as

(*a*) insensitivity (*b*) lack of precision

(*c*) lack of discrimination (*d*) operator bias.

40. The maximum probability has a large value for small values of

(*a*) standard deviation (*b*) precision index

(*c*) uncertainty (*d*) average deviation.

41. What refers to the "number of changes that a particular reading will occur when the error limit is specified" ?

(*a*) Probability (*b*) Probable error

(*c*) Odds (*d*) Confidence interval.

42. Systematic errors in a Bourdon tube pressure gauge may be caused by

(*a*) friction in the pins and gears of the amplifying mechanism

(*b*) incorrect zero setting of the pointer

(*c*) variation of atmospheric pressure

(*d*) incorrect readings of the scale due to parallax.

43. If the confidence level is 0.95, then the values laying outside the confidence interval are

(*a*) 1 in 5 (*b*) 1 in 20

(*c*) 1 in 100 (*d*) 1 in 1000.

44. Uncertainty distribution is used for

(*a*) analysis of multi-sample data

(*b*) analysis of single-sample data

(*c*) analysis of both single and multi-sample data

(*d*) none of the above.

45. The mean deviation $\bar{d}$ in terms of deviations from the mean value of n reading is

(*a*) $\frac{\Sigma|d|}{n}$ (*b*) $\frac{\Sigma d}{n}$

(*c*) $\frac{\sqrt{\Sigma d^2}}{n}$ (*d*) $\sqrt{\frac{\Sigma d^2}{n}}$.

ANSWERS

1. (a)	2. (c)	3. (c)	4. (b)	5. (b)	6. (c)	7. (a)
8. (a)	9. (b)	10. (c)	11. (d)	12. (d)	13. (c)	14. (b)
15. (b)	16. (a)	17. (c)	18. (b)	19. (a)	20. (c)	21. (c)
22. (d)	23. (c)	24. (c)	25. (b)	26. (a)	27. (c)	28. (a)
29. (a)	30. (d)	31. (c)	32. (d)	33. (b)	34. (a)	35. (d)
36. (c)	37. (a)	38. (b)	39. (d)	40. (b)	41. (c)	42. (b)
43. (b)	44. (b)	45. (a).				

THEORETICAL QUESTIONS

1. What do you understand by the term "measurement" ?
2. Describe the significance of measurements.
3. Explain briefly the term "measurand".
4. State the basic requirements for getting the meaningful result of the measurement.
5. Explain the difference between an analog signal and a digital signal.
6. Define the term "instrumentation".
7. How are standards of measurement classified ?
8. Explain clearly the difference between primary, secondary and working standards.
9. Explain briefly the following methods of complete solutions to complex problem in mechanical design :
   (i) The empirical method.
   (ii) The rational method.
   (iii) The experimental method.
10. Distinguish between direct and indirect methods of measurement.
11. Describe the mechanics and power types of measurements. Explain with suitable examples.
12. What are primary, secondary and tertiary measurements ? Explain with examples.
13. Explain the difference between the following :
    (i) Measurand and measurement.
    (ii) Measurement and instrument.
    (iii) Input and output.
14. Draw a block diagram representation of a generalised measurement system. Identify the various elements and point out the functions performed by each element/component.
15. Explain briefly various stages of the general measurement system.
16. Describe briefly the following examples of measurement systems :
    (i) Bourdon tube pressure gauge.
    (ii) Pressure-actuated thermometer.
17. What do you mean by characteristics of a measuring instrument ?
18. What are static characteristics ? How do they differ from dynamic characteristics ?
19. Define the following terms as applied to the measuring instrument :
    (i) True value
    (ii) Correction
    (iii) Range
    (iv) Sensitivity
    (v) Accuracy
    (vi) Linearity
    (vii) Dead zone
    (viii) Backlash.

Static characteristics :

20. Enumerate the main static characteristics of measuring instruments.
21. Explain briefly the terms : Accuracy, errors and correction.
22. Explain the following terms :
    (*i*) Static error (*ii*) Static correction
    (*iii*) Relative error (*iv*) Percentage relative error.
23. How can "accuracy" be expressed ? Explain briefly.
24. What do you mean by "Static calibration" ? Give the steps which are necessary in performing a calibration.
25. Differentiate between "Scale range" and "Scale span" giving suitable examples.
26. What is an "Error calibration curve" ?
27. What do you mean by "Scale readability" ?
28. Differentiate between the terms "Repeatability" and "Reproducibility".
29. Explain the term "Drift" using suitable examples.
30. How is "drift" classified ?
31. Explain briefly the following :
    (*i*) Zero drift (*ii*) Span drift (*iii*) Zonal drift.
32. Differentiate between the terms "Accuracy" and "Precision" with suitable examples.
33. What is "sensitivity" ?
34. Define the term "Deflection factor".
35. What is "linearity" ? Explain briefly.
36. What do you mean by the term "Non-linearity" ? Explain briefly "Terminal linearity" and "Best fit linearity".
37. Describe briefly the term "Hysteresis".
38. Explain clearly the difference between "Threshold" and "Resolution".
39. Explain briefly the following :
    (*i*) Dead zone; (*ii*) Dead time.
40. What do you mean by the term "loading effect" as applied to measuring instruments ? Explain briefly.
41. Explain briefly the terms "Static stiffness" and "Static compliance".
42. What is "noise" ?
43. What is signal to noise ratio ?
44. Explain briefly the following :
    (*i*) Generated noise; (*ii*) Conducted noise; (*iii*) Radiated noise.

Dynamic characteristics :

45. Define the following terms :
    (*i*) Response; (*ii*) Dynamic response of a measuring system
46. Explain briefly the following types of dynamic inputs :
    (*i*) Transient response; (*ii*) Steady state periodic.
47. Define the following terms :
    (*i*) Transient response; (*ii*) Steady state response.
48. Define the dynamic response of a system and distinguish between steady state response and transient response.
49. Define the following terms :
    (*i*) Speed of response; (*ii*) Measuring lag;
    (*iii*) Fidelity; (*iv*) Dynamic error;
    (*v*) Overshoot.

50. What are the different standard inputs for studying the dynamic response of a system ? Explain them briefly.

51. Derive the equations for time response of a first-order system when subject to :

(*i*) Unit step input; (*ii*) Unit ramp input.

52. Derive the expression for time response of a second-order underdamped system when subjected to unit ramp input. Show that the nature of the response is same as that for a unit step input.

53. Derive the expression for the magnitude and phase of a first-order system when subjected to a sinusoidal input signal. Draw the magnitude and phase versus frequency plots.

54. Derive the expressions for the frequency response of a second-order system. Draw the magnitude and phase versus frequency plots.

55. Sketch and explain the response of a first-order system when subjected to step input signal.

56. Sketch and explain the response of a first-order system subjected to ramp input signal. Indicate thereon the transient error, steady error and the time lag.

57. Describe the dynamics for sinusoidal input to a first-order system.

58. Develop an expression for the amplitude ratio and phase difference of indicated reading and the input signal which is of sinusoidal form.

59. What is the importance of time constant in appropriate measuring instruments ?

Errors in measurements :

60. What are limiting errors ? Explain briefly.

61. Explain briefly the term *"relative error"*.

62. How are errors classified ?

63. What are "Gross errors " ? How can these be avoided ?

64. What are "Systematic errors" ? Explain briefly the following systematic errors with examples :

(*i*) Instrumental errors.

(*ii*) Environmental errors.

(*iii*) Observational errors.

65. How can the systematic errors be located ?

66. What are assembly errors ? Can they be easily discovered and rectified ? If yes, why ?

67. State the factors which give rise to observational errors.

68. What are random errors ? State the common causes of their occurrence.

69. Enumerate the various sources of errors.

70. Explain briefly the following sources of errors :

(*i*) Noise (*ii*) Response time

(*iii*) Design limitations (*iv*) Energy exchanged by interaction

(*v*) Transmission (*vi*) Deterioration of measuring system.

71. What are single-sample and multi-sample tests ? Explain briefly.

72. Explain briefly the following terms :

(*i*) Arithmetic mean (*ii*) Geometric mean

(*iii*) Median (*iv*) Mode.

73. Describe briefly the following terms :

(*i*) Deviation (*ii*) Average deviation

(*iii*) Standard deviation (*iv*) Variance.

74. What is the law of probabilities ? Explain briefly.

75. Explain briefly the term "Precision index".

76. Write a short not on "Probability tables".

77. While specifying measurement data, how are the deviations expressed ?

78. Explain briefly the following terms :

(*i*) Confidence interval. (*ii*) Confidence level.

79. What is "Chauvenet's criterion" ? Expain briefly.

UNSOLVED EXAMPLES

1. A meter reads 136.6 V and the true value of the voltage is 136.52 V. Determine :

(*i*) The static error; (*ii*) The static correction for this instrument

[**Ans.** (*i*) + 0.08 V (*ii*) – 0.08 V]

2. An analog indicating instrument with a scale range of 0 – 10.0 V shows a voltage of 5.30 V. The true value of a voltage is 5.40 volts.

(*i*) What are the values of absolute error and correction ?

(*ii*) Express the error as a function of the true value and full scale deflection.

[**Ans.** (*i*) – 0.1 V, 0.1 V, (*ii*) 1.85 %, – 1 %]

3. A pressure indicator showed reading as 89 bar on a scale range of 0 – 100 bar, but the true value was 88.5 bar. Determine :

(*i*) Static error, (*ii*) Static correction, and

(*iii*) Relative static error. [**Ans.** (*i*) + 0.5 bar, (*ii*) – 0.5 bar, (*iii*) 0.565%]

4. A thermometer is calibrated 100°C to 150°C. The accuracy is specified within ± 0.25 per cent of instrument span. What is the maximum static error ? [**Ans.** ± 0.125°C]

5. In a Wheatstone bridge a change of 9 Ω in the unknown arm of the bridge is required to produce a change in the deflection of 3 mm of the galvanometer. Determine (*i*) Sensitivity, and (*ii*) Deflection factor. [**Ans.** (*i*) 0.333 mm/Ω; (*ii*) 3Ω/mm]

6. A pressure measuring system consists of a piezoelectric transducer, a charge amplifier and an ultra violet charge recorder; their sensitivities are 8.5 pC/bar, 0.004 V/pC and 20 mm/V respectively. Determine the deflection on the chart for a pressure change of 25 bar. [**Ans.** 17 mm]

7. Determine the resolution of a moving coil voltmeter having a uniform scale with 100 divisions; the full scale reading is 100 V and $\frac{1}{10}$ of a scale division can be estimated with a fair degree of certainty. [**Ans.** 0.1 V]

8. The calibration range of a certain pyrometer is 200°C to 900°C. If the dead zone in it is 0.12 per cent of the span, determine the temperature change which might occur before it is detected. [**Ans.** 0.84°C]

9. On the application of a step input of 100 bar to a pressure gauge, the pointer swings to pressure of 102.5 bar and finally comes to rest at 101.3 bar. Determine :

(*i*) The overshoot of the gauge readings and express it as a percentage of the final reading.

(*ii*) The percentage error of the gauge. [**Ans.** (*i*) 1.2 bar, 1.18%, (*ii*) 1.3%]

10. A thermometer has a time constant of 2.5 seconds. It is quickly taken from a temperature 0°C to liquid bath having a temperature 200°C. Calculate the temperature indicated by the thermometer after a time of 1.5 seconds.

Determine its reading when it was initially held at 25°C. [**Ans.** 90.24°C, 103.96°C]

11. A temperature sensitive device can be modelled as first-order system with a time constant of 6 seconds. It is suddenly subjected to a step input of 25°C – 150°C. Calculate the temperature indicated by the device after 10 seconds after the start of the process.

[**Ans.** 126.4°C]

12. A temperature sensitive transducer when subjected to sudden temperature change takes 10 seconds to reach the equilibrium conditions (five times constant). Calculate the time taken by the transducer to read half of the temperature difference. [**Ans.** 1.39s]

13. A resistance temperature detector has a steady state gain of 0.3925 Ω/°C and a time constant of 5.5 seconds. It is subjected to a step change of 75°C in temperature. If its stable resistance before the step change is 100 Ω, write down the time-domain equation for resistance and find its value after 15 seconds of applications of step input.

[**Ans.** $R_t = 29.44\ (1 - e^{-t/5.5}) + 100$, 127.5 Ω]

14. A thermometer initially at a temperature of 20°C is suddenly plunged into a liquid bath maintained at 150°C. After a time interval of 3 seconds, the thermometer indicated a temperature of 95°C. Determine.

(*i*) The time constant for the thermometer.

(*ii*) The indicated temperature after five times constant; comment upon this result.

[**Ans.** (*i*) 3.487 s, (*ii*) 149.12°C]

15. A temperature sensitive transducer used to measure the temperature of a furnace has been idealised as a first-order system subjected to ramp input. If the maximum permissible error in temperature measurement is limited to 4.2 °C, calculate the time constant of the transducer when the furnace temperature increases at the rate of 0.14/°C. [**Ans.** 30°C]

16. A weather balloon carrying a temperature sensing device (of first-order, time constant 10 seconds) rises through the atmosphere at 6m/s. It supplies information about temperature and altitude through radio signals. At 2800 m height a temperature indication of 35°C has been received. If the temperature varies with altitude at a uniform rate of 0.01°C/m, determine the true altitude at which 35°C temperature occurs. [**Ans.** 2740 m]

17. The temperature of a furnace is found to vary sinusoidally between 400°C and 450°C with a periodic time of 60 seconds. A thermocouple system with a time constant of 8 seconds is employed to measure the furnace temperature. Determine :

(*i*) The maximum and minimum values that will be indicated by the thermocouple.

(*ii*) The phase shift and corresponding time lag between the temperature signals and the thermocouple input signals. [**Ans.** (*i*) 444.2°C, 405.8 °C, (*ii*) 39.76°, 6.669s]

18. A first-order system having a time constant of 0.1 second is used to measure a signal prescribed by the relation; $I_i = 3 \sin 2t + 0.4 \cos 10t$. Develop an expression for the corresponding output. [**Ans.** $I_o = 2.94 \sin (2t - 11.31°) - 0.283 \sin (10t + \pi - 45°)$]

19. A liquid thermometer has its glass bulb protected by a well. The system can be represented by a double capacity system with time constants of 40s for the well and 20s for the bulb. The thermometer is subjected to a cyclic change of ± 10°C which occurs every 120 seconds. Find the maximum value of the indicated temperature and the delay time.

[**Ans.** ± 2.97°C; 36.8s]

20. An input of $1\sin 2t + 0.3\sin 20t$ is applied to a first-order instrument having a time constant of 0.2 second. Find the output. [**Ans.** $0.93 \sin (2t - 21.8°) + 0.073 \sin (20t - 76°)$]

21. The pen arrangement of a U.V. recorder has a mass of 4.8 g. What should be the percentage reduction in mass if it is desired to have 18 per cent increase in the natural frequency of the recorder ? Consider the pen motion to conform to a second-order system.

[**Ans.** 28.18%]

22. Determine the damping ratio, damped natural frequency, static sensitivity and time constant for a second-order system which follows the differential equation given as :

$$\frac{d^2 I_o}{dt^2} + 2\frac{dI_o}{dt} + 25 I_o = 25 I_i$$ where I_o and I_i are the output and input quantities respectively.

[**Ans.** 0.2; 4.899 rad/s; 1.0; 0.2s]

23. While measuring force by a diagram type pressure transducer (second-order system), the system is found to have a natural frequency of 900 cycles per second (cps) and damping ratio of 60 per cent.

Determine the frequency range over which the amplitude ratio corresponding to sinusoidal input deviates by a maximum amount of 10 per cent. [**Ans.** 0 – 824 cps]

24. A unit amplitude sinusoidal input signal having a frequency of 0.5 Hz is given to second-order control system of transfer function : $(I_o/I_i) = 8/(D^2 + 4D + 20)$.

Develop an expression for the steady state response of the system and determine the corresponding output, amplitude, output frequency and phase lag.

[**Ans.** I_o = 0.4587 (sin πt – 46.04°); 0.4587; π; 46.04°]

25. A torque sensing transducer is connected to the shaft of an electric motor which drives a load that has a moment of interia of 0.20 kg-m^2. If the torsional constant is 1.5 Nm/rad, what should be the damping ratio of the system when a simusoidal torque varying at frequency of 2 rad/s its to be reproduced with a maximum error of + 10 per cent ?

[**Ans.** 0.534]

26. A second-order instrument has a natural frequency of 4 Hz and a damping ratio of 0.66. If the excitation frequency of the system is 6Hz, determine the error due to the proximity of excitation frequency with the natural frequency of the system. [**Ans.** 57.3%]

Errors in measurements :

27. A wattmeter having a range of 2000 W has an error of ± 1 per cent of full scale deflection. If the true value is 200 W, what would be the range of the readings ?

If the error is specified as percentage of the true value, what would be the range of the readings ? [**Ans.** 180 to 220 W; 198 W and 202 W]

28. The guaranteed accuracy of a flow meter working on thermal principles is ± 4 per cent of full scale reading of 10.0 × 10^{-6} m^3/s. If the flow measured by this meter is 5 × 10^{-6} m^3/s, calculate the limiting error in per cent. [**Ans.** ± 8 percent]

29. A – 300 Pa manometer has a guaranteed accuracy of 2 per cent of full scale reading. The pressure measured by this instrument is 150 Pa. Calculate the limiting error in per cent.

[**Ans.** 4 per cent]

30. Three resistors have the following ratings :

R_1 = 74 Ω ± 5%, R_2 = 150 Ω ± 5%; R_3 = 100 Ω + 5%.

Determine the magnitude and limiting error in ohms and in per cent of the resistance of these resistances connected in series. [**Ans.** 324 ± 16.20; ± 5%]

31. The resistance of a circuit is found by measuring current flowing and the power fed into the circuit. Find the limiting error in the measurement of resistance when the limiting errors in the measurement of power and current are 0.9% and ± 0.6% respectively.

[**Ans.** ± 2.1%]

32. The unknown resistance in a Wheatstone bridge is measured by using three known resistances such that $R_4 = \dfrac{R_2 R_3}{R_1}$;

where, R_1 = 100 ± 0.5% Ω ; R_2 = 500 ± 5% Ω ; R_3 = 280 ± 0.5% Ω.

Determine the magnitude of the unknown resistance (R_4) and the limiting error in per cent and in ohms (for the unknown resistance). [**Ans.** 1400 Ω ; 1.5% ± 21 Ω].

33. A cantilever beam of length l, width b and depth d is used to measure the force F. The deflection is measured using an LVDT of accuracy ± 0.1 mm. The length, width and depth are measured to an accuracy of ± 0.01 mm and their values are 50 mm and 1.0 mm respectively. The modulus of rigidity is known accurate to 0.2%. If the deflection noted was 2 mm, determine the error in force measured.

Given force $F = \dfrac{3Ebd^3}{12l^3} \cdot y$

where, E = modulus of elasticity; l, b and d = length, width and depth of the beam respectively; y = deflection. [**Ans.** ± 0.2820%]

34. A dynamometer is used to measure the output power of a rotating shaft. The output power is given by the following relation : $P = \dfrac{2\pi \times 9.81\, FLR}{t \times 10^6}\, kW$

where F = force at the end of torque arm (kg); L = length of torque arm (mm); R = number of revolutions during time t; t = time for test run (s).

The given test data are : $F = 4.58 + 0.02$ kg ; $L = 397 \pm 7.3$ mm ; $R = 1202 \pm 1.0$ revolutions; $t = 60 + 0.50$ s.

The errors are limiting (absolute) errors.

Determine the magnitude of power and magnitude of the limiting error in the computed power. [**Ans.** 2.245 kW ; ± 0.038 kW]

35. During a test the current was measured as 30.4A, flowing in a resistor of 0.105 Ω. Later on its was discovered that the ammeter reading was low by 1.2% and the marked resistance was high by 0.3%. Determine the true power as a percentage of the power that was originally calculated. [**Ans.** 102.1%]

36. A circuit was tuned for resonance by eight different students, and the values of resonant frequency in KHz was recorded as 532, 548, 543, 535, 546, 531, 543 and 536. Calculate the following :

(*i*) Arithmetic mean. (*ii*) Average deviation, and

(*iii*) Variance. [**Ans.** (*i*) 539.25 kHz, (*ii*) 5.75 kHz; (*iii*) 6.54 kHz, (*iv*) 42.77 $(kHz)^2$]

37. During a test run, measurements of temperature were made 100 times with variation in apparatus and procedure. After applying corrections for known systematic errors, the following data were obtained :

Temperature °C :	397	398	399	400	401	402	403	404	405
Frequency of Occurrence :	1	3	12	23	37	16	4	2	2

Calculate the following :

(*i*) Arithmetic mean, (*ii*) Mean deviation,

(*iii*) Standard deviation, (*iv*) Probable error of one reading,

(*v*) Probable error of the mean, and

(*vi*) Standard deviation of the standard deviation.

[**Ans.** (*i*) 400.78°C, (*ii*) 1.028°C, (*iii*) 1.380°C, (*iv*) 0.93°C, (*v*) 0.093°C, (*vi*) 0.0796°C]

38. The speed of a hydraulic turbine mode, which is running at 1000 r.p.m., is measured by a tachometer. The hydraulic turbine is subjected to variations in speed. For a sample of 20 readings at this speed, how many readings would you expect between 990 and 1010 r.p.m.? Assume that the tachometer gives a normal set of deviation with precision index of 0.04 at 1000 r.p.m. [**Ans.** 9 readings]

39. The depth of water, which had a nominal value of 18 cm, was measured with a hook. Measurements were taken 24 times and 6 readings were found to lie outside a particular range. Assume that the measurements conform to normal distribution with precision index 8 cm^{-1}. Determine the prescribed range. [**Ans.** (18 ± 0.1016) cm]

40. (*a*) A mass flow meter using electromagnetic principle is set to record a large number of readings. The fixed mass flow rate is 50 kg/s; half of the readings lie between 52.4 and 47.6 kg/s. The sample measurements give a straight line between the (0, 0.50%) points on the probability coordinates. Calculate the precision index of the instrument.

(*b*) An alarm rings when the flow meter gives a reading of 44 kg/s.

(*i*) How many false alarms are expected in the month of June if the flow is checked automatically eight times a day ?

(*ii*) What will be the value of precision index if one wants to reduce the false alarm rate by a factor of two ? [**Ans.** (*a*) 0.199; (*b*) (*i*) 12, (*ii*) 0.231]

41. The data given below are expected to follow a linear relationship : $v = u + ab$

u :	0.9	2.3	3.3	4.5	5.7	6.7
v :	1.1	1.6	2.6	3.2	4.0	5.0

Using least square analysis : (*i*) obtain the best linear relation, and (*ii*) calculate the standard deviation.

[**Ans.** $v = 0.672u + 0.296$, (*ii*) $s_v = \pm 0.1716$; $s_u = \pm 0.255$; $s_a = \pm 0.0355$; $s_b = \pm 0.155$]

42. The final result of a measurement depends on two measurements in series, each having values of 25 and 50 units with errors of ±2% and ±1%, such that result is sum of both. Determine the error of final result if errors are considered as (*i*) Limiting errors, (*ii*) Standard deviations. [**Ans.** (*i*) 1.34%; (*ii*) 0.943%]

CHAPTER

4 Analog Electromechanical Instruments

4.1 Introduction to instruments; 4.2. Classification of instruments; 4.3. Selection of insruments; 4.4. Functions of instruments; 4.5. Comparison of analog and digital instruments; 4.6. Electrical instruments; 4.7. Galvanometers – D′ Arsonval galvanometer – Ballistic galvanometer – Grassot fluxmeter – Vibration galvanometer – Duddall's oscillograph; **4.8. Ammeters, Voltmeters** and **Ohmmeter** – Introduction – Types of instruments mainly used as ammeters and voltmeters – Errors in ammeters and voltmeters – *Moving iron instruments – Moving-coil instruments* – Permanent magnet moving-coil instruments – ohmmeters – *Electrodynamic or dynamometer type instruments – Hot-wire instruments – Thermocouple instruments – Rectifier instruments – Electrostatic instruments;* **4.9. Wattmeters and measurements of power** – Types of wattmeters – Dynamometer wattmeter – Induction wattmeter – Electrostatic wattmeter – Thermal wattmeter – Measurement of power in single-phase circuits – Measurement of power in 3-phase circuits – Measurement of power at high frequencies (AF/RF) – Bolometer method of power measurement – calorimetric method for measurement of large amount of RF power – Output power meter; **4.10. Energy meters and energy measurement** – General aspects – Essential characteristics of energy meters – Types of energy meters – Motor meters – Motor-driven meter – Watt-hour meter – Induction type of watt-hour meter – Induction type single-phase energy meters – Maximum demand indicators/meters – Measurement of VAh and VARh – Testing of energy meters; **4.11. Miscellaneous measuring instruments** – Power factor meters – Frequency meters – Rotational speed measurements – Synchromoscopes – Phase sequence indicators. *Highlights – Objective Type Questions – Theoretical Questions – Unsolved Examples.*

4.1 INTRODUCTION TO INSTRUMENTS

The **instrument** *serves as an extension of human faculties and enables the man to determine the value of unknown quantity or variable which his unaided human faculties cannot measure.* A measuring instrument provides information about the physical value of some variable being measured.

— The instrument would sense a physical parameter (e.g., velocity, pressure, temperature, etc.), process and translate it into a format and range which can be interpreted by the observer.

— The instrument *must also provide the controls* by which the operator can obtain, manipulate and respond to the information.

• An instrument, in *simple cases,* consists of a *single unit* which gives an output reading or signal according to the unknown variable (measurand) applied to it. Under *more complex measurement situations,* a measuring instrument may consist of *transducing elements* which convert the measurand to analogous form. The *analogous signal* is then processed by some intermediate means and then fed to the end devices for presenting the measurement results for the purpose of display, record and control.

- The instruments made by man not only retain their characteristics for extended periods of time but are also accurate and sensitive in their response.
- *Measurements* involve the use of *instruments* as a physical means of determining quantities and variables.

4.2 CLASSIFICATION OF INSTRUMENTS

The instruments may be *classified* as follows:

1. Absolute and secondary instruments.
2. Analog and digital instruments.
3. Mechanical, electrical and electronic instruments.
4. Manual and automatic instruments.
5. Self-contained and remote indicating instruments.
6. Self-operated and power-operated instruments.
7. Deflection and null output instruments.

1. Absolute and secondary instruments:

(i) Absolute instruments :

These intruments give the magnitude of the quantity under measurement in terms of physical constants of the instrument.

Examples : Tangent galvanometer and Rayleigh's current balance.

- These instruments are *seldom used except in standard institutions.*

(ii) Secondary instruments :

These instruments are so constructed that the quantity being measured can only be measured observing the output indicated by the instrument.

These instruments are *calibrated* by comparison with an absolute instrument or another secondary instrument which has already been calibrated against an absolute instrument.

Examples : Voltmeter, glass thermometer, pressure gauge etc.

- The secondary instruments find *wide use in every sphere of measurement.*

2. Analog and digital instruments:

(i) Analog instruments :

The signals of an analog unit vary in a continuous fashion and can take on infinite number of values in a given range.

Examples : Fuel gauge, ammeters and voltmeters, wrist watch, speedometer of an automobile, etc.

(ii) Digital instruments :

Signals varying in *discrete steps* and taking on a finite number of different values in a given range are *digital signals* and the corresponding instruments are of *digital type.*

Examples : Odometer of an automobile, calibrated balance of a platform scale, timer on a scoreboard, etc.

- The digital devices have the following *advantages* :
  - (*i*) High accuracy.
  - (*ii*) High speed.
  - (*iii*) Elimination of human operational errors.

- Owing to the application of digital computers for data handling, reduction and in automatic controls, the importance of *digital instrumentation* is increasing very fast. Thus, it becomes necessary to have both analog-to-digital (A/D) converters at input to the computers and digital-to-analog (D/A) converters at the output to the computers.

3. Mechanical, electrical and electronic instruments:

(*i*) ***Mechanical instruments :***

- These instruments are very reliable for static and stable conditions.
- They are *unable to respond rapidly* to the measurements of dynamic and transient conditions due to the fact that they have moving parts that are rigid, heavy and bulky and consequently have a large mass. Mass presents *inertia problems* and hence these instruments cannot faithfully follow the rapid changes which are involved in dynamic instruments.
- Most of the mechanical instruments *cause noise pollution.*

(*ii*) ***Electrical instruments :***

- The electrical methods of indicating the output of detectors *are more rapid than that of mechanical methods.*
- Unfortunately, an electrical system normally depends upon a mechanical meter as an indicating device. This mechanical movement has some inertia due to which the *frequency response* of these instruments is *poor.*

(*iii*) ***Electronic instruments :***

- Most of the scientific and industrial instruments require *very fast responses.* Such requirements cannot be met with by mechanical and electrical instruments.
- These instruments use *semiconductor devices. In electronic devices, since the only movement involved is that of electrons, the response time is extremely small owing to very small inertia of electrons.*
- With the use of electronic devices, *very weak signals can be detected by using pre-amplifiers and amplifiers.*

Comparison between mechanical and electrical/electronic instruments:

Advantages and disadvantages of mechanical instruments:

Advantages :

(*i*) Simple in design and use.
(*ii*) More durable due to rugged construction.
(*iii*) Relatively cheaper in cost.
(*iv*) For operation of the instruments, no external power supplies are required.
(*v*) Quite reliable and accurate for the measurement of parameters which are stable and non-variant with time.
(*vi*) Long history of development and successful use.

Disadvantages :

(*i*) Poor frequency response to dynamic and transient measurements.
(*ii*) Large forces are required to overcome mechanical friction.
(*iii*) Incompatibility when remote indication or control is needed.
(*iv*) Cause noise pollution.

Advantages of electrical/electronic techniques of measurements:

(*i*) Less power consumption and less load on the system being measured.
(*ii*) Non-contact measurements are possible.

(*iii*) Light, compact and more reliable instrumentation.
(*iv*) Remote indication and recording feasible.
(*v*) The amplification produced is greater than that produced by a mechanical contrivance.
(*vi*) The mathematical processing of signals like summation, differentiating and integrating is possible.
(*vii*) The frequency and transient response is good.
(*viii*) Greater flexibility.
(*ix*) Lighter in construction.

• The electronic instruments are *widely used in detection of electro-magnetically produced signals such as radio, video and microwaves.*

• The use of these instruments has made it possible to *build analog and digital computers* which are quite instrumental in the modern developments in the fields of science and technology.

4. Manual and automatic instruments :

• In case of *manual* instruments services of an operator are required.

Example : Measurement of temperature by a resistance thermometer incorporating a Wheatstone bridge in its circuit (an operator is required to indicate the temperature being measured).

• In an *automatic* type of instrument, an operator is *not* required.

Example : Measurement of temperature by mercury-in-glass thermometer.

5. Self-contained and remote indicating instruments :

• A self-contained instrument *has all its different elements in one physical assembly.*

• In a remote indicating instruments, the primary sensing element *may be located at an adequate long distance* from the secondary indicating element. Such types of instruments are *finding wide use in the modern instrumentation technology.*

6. Self-operated and power-operated instruments :

• A self-operated instrument does not require any outside power for its operation; the output energy is supplied wholly or almost wholly by the input signal.

Examples : Mercury-in-glass thermometer, Dial indicator etc.

The power operated instruments are those instruments which require some auxiliary source of power such as electricity, compressed air, hydraulic supply etc. for their operation. In such cases, the input signal supplies only an insignificant portion of the output power.

Example : Electro-mechanical measurement system shown in Fig. 4.1.

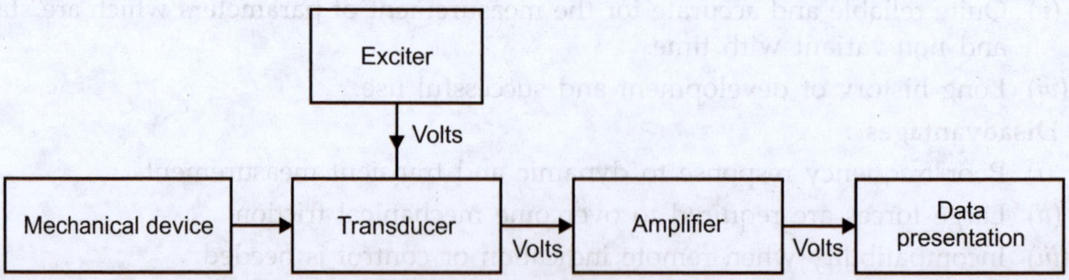

Fig. 4.1. Electro-mechanical measurement system.

7. Deflection and null output instruments :

• *In a "**deflection type**" instrument, the deflection of the instrument provides a basis for determining the quantity under measurement.* The measured quantity produces some physical effect which deflects or produces a mechanical displacement of the moving system of the instrument. An opposing effect is built in the instrument which tries to oppose the deflection or the mechanical displacement of the moving system. The balance is achieved when opposing effect *equals* the cause, producing the deflection or mechanical movement. The deflection or mechanical displacement at the point of balance then gives the value of measured quantity (which depends upon the calibration of the instrument).

Examples : Permanent magnet moving coil (PMMC) ammeter etc.

— These instruments are more suited for measurements under *dynamic conditions* than null type of instruments whose intrinsic response in slower.

• In a *"**null type**" of instrument, a zero or null indication leads to determination of the magnitude of measured quantity.* The null condition is dependent upon some other known conditions.

Example : D.C. potentiometer.

Advantages :

(*i*) The "null type" instruments are *more accurate* than deflection type instruments.

(*ii*) These instruments can be *highly sensitive* as compared with deflection type instruments.

4.3 SELECTION OF INSTRUMENTS

The following are some important factors/points which are very essential to be considered for the proper selection of a correct instrument for any application :

1. The *accuracy* expected from the instrument.

   The following most significant and fundamental rule, which is perhaps the most important rule in measurement, must always be kept in mind :

   "Never demand an accuracy of measurement higher than that which is really needed, and never forget that each degree of accuracy, if demanded, is likely to have a disproportionate effect on the complexity and cost of measuring apparatus".

2. *When are the final data required* (*i.e.* at the time of taking the measurement, or at later one)?
3. The *cost criterion* (*i.e.,* the cost of the measuring process).
4. *In what form the data should be displayed* (*i.e.,* indicating, recording, integrating etc.)?
5. Whether quantity to be measured has *constant value or is it a time variant*?

   — In the measurement of time variant quantities the manner in which the quantity to be measured changes has a significant effect on the type of measuring instruments to be used. *The physical characteristics of the measuring device must always be taken into account since these characteristics determine the amount of interaction which takes place between the measurand quantity and the instrument.*

4.4 FUNCTIONS OF INSTRUMENTS

Following are the *three* main functions of instruments :

1. Indicating function.
2. Recording function.
3. Controlling function.

1. Indicating function :

— The value of the quantity can be read by the movement of the needle on a calibrated scale provided on the instrument. Readings can be taken to any fraction within the limitations of the instrument and human eye.

Examples : (*i*) The deflection of a pointer of a *'speedometer'* indicates the speed of the automobile at that moment.

(*ii*) The use of a *'pressure gauge'* for indicating pressure.

— These days digital display (*i.e.*, the reading in numericals) is also commonly used. In fact, where highly accurate and exact readings are desired, use of digital readouts is recommended.

2. Recording function :

— In several cases the instrument continuously records, with pen and ink, the value of measured quantity against some other variable or against time. A graph paper is provided whose speed can be adjusted. A recording instrument can be used to record more than one number of quantities also, in which case, either the recording of different parameters may be done by different pens or same pen may record one parameter after other.

Example : A 'potentiometric type of recorder' used for monitoring temperature records the instantaneous values of temperatures on a strip chart recorder.

3. Controlling function :

— The controlling function is perhaps one of the most important functions particularly in the field of industrial control processes. Here, the information is used by the instrument/system to control the original measured quantity.

Examples : (*i*) *Floats* for liquid level control.

(*ii*) *Thermostats* for temperature control.

Essential requirements of a measuring instrument :

The essential requirements of a measuring instrument are :

1. Its introduction into the circuit where measurements are to be made, should not change the circuit configuration.
2. Its power consumption should be small.

4.5 COMPARISON OF ANALOG AND DIGITAL INSTRUMENTS

Qualitatively, the *"electrical indicating instruments"* are widely used for measurement of current, voltage, resistance and power. These instruments can be *classified* as follows :

1. Analog instruments.
2. Digital instruments. (For details of digital instruments refer to *chapter* **10**)

• *The deflection type instruments with a scale and movable pointer are called* ***analog instruments.***

— The measured parameter is converted into electrical signal which further actuates some electro-mechanical device to move the pointer continuously as the parameter changes.

• *The instruments which use logic circuits and techniques to obtain a measurement and then display it in numerical reading (digital) form are called* ***digital instruments.***

The digital instruments claim the following *advantages over analog instruments* :

(*i*) Greater accuracy.

(*ii*) Better resolution.

(*iii*) Automatic polarity and zeroing.

(*iv*) Easy readability.

Table 4.1 shows the comparison between analog and digital type instruments.

Table 4.1. Comparison between Analog and Digital type Instruments

S.No.	Aspects	Analog type instruments	Digital type instruments
1.	*Information form*	As the position of a pointer against a calibrated scale or dial.	As a number.
2.	*Possibility of human error*	Exists	Does not exist.
3.	*Best possible accuracy*	+ 0.25%	+ 0.005% or better.
4.	*Resolution*	One in several hundreds.	One part in several hundred thousands.
5.	*Presence of moving parts*	Moving parts involved.	These instruments can be made without moving parts.
6.	*Construction*	Simple in construction and direct reading type; can perform under unfavourable conditions.	Since these instruments involve electronics, proper environmental conditions are essential.
7.	*Rate of change of parameter*	These instruments enable the operator to judge the rate of change of parameter by seeing the needle movement.	Change of digital reading does not give any knowledge of rate of change of parameter.
8.	*Time required to observe the reading*	If exact reading is required operator takes more time as he has to guess the approximate tenths of small division.	Reading of digital meters is very fast.
9.	*Auxiliary power requirement*	These instruments require no auxiliary source of power for actuation but derive driving power for indicating system from the process.	These instruments require auxiliary source of power.
10.	*Mobility*	Can be portable also.	Usually stationary type.
	Examples :	Examples of pointer-dial output devices are : • Micrometer and platform scales; • Manometers and Bourdon-tube pressure gauge; • Mercury in glass and filled system thermometers; • Speedometer of an automobile; • Common voltmeters and ammeters etc.	Examples of Digital output devices are : • Digital ammeters and voltmeters; • Electronic and mechanical counters; • Odometers; • Yes-No light (On or Off); • Time on a scoreboard etc.

Pointer-scale analog indicators :

In analog instruments, the value of the measured parameter is indicated by positioning of the indicating pointer against a calibrated scale. This purpose can be achieved either by moving the pointer with relation to a stationary scale (fixed-scale moving-pointer indicators) or the scale may be moved with relation to a fixed reference (moving-scale fixed-pointer indicators).

Multi-pointer indicator. This type of indicator contains more than one number of pointers and above each point the identification number of the medium being measured is marked. *Usually this arrangement is used in recorders and not in indicators.*

Multi-range indicators. An instrument with multi-range indicators has *different scales for different ranges;* The choice of a particular scale is made by a selector switch.

1. *Single-point indicators:*

The *fixed-scale and movable-pointer indicators,* available in a variety of forms, are shown in Fig. 4.2(*a*).

Fig. 4.2(*b*) shows the fixed-pointer and movable-scale indicators.

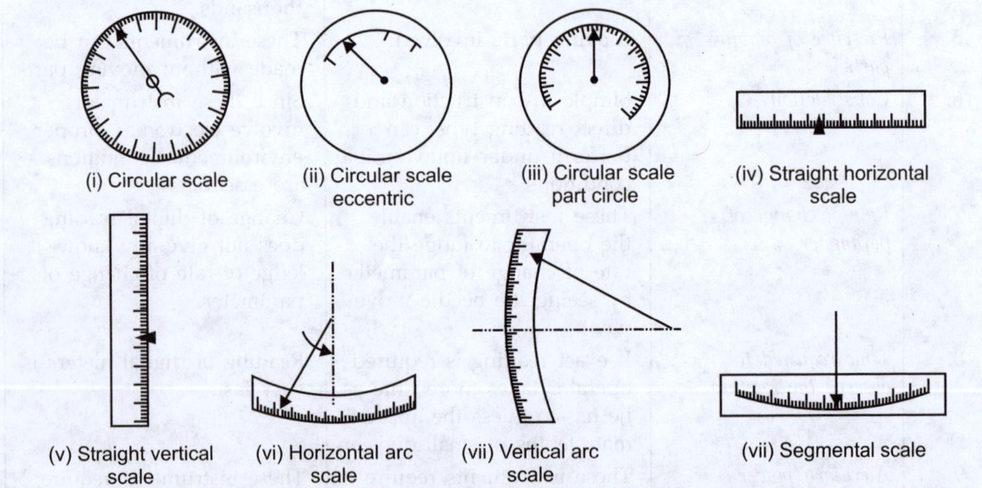

Fig. 4.2 (a). Fixed-scale and movable-pointer indicators.

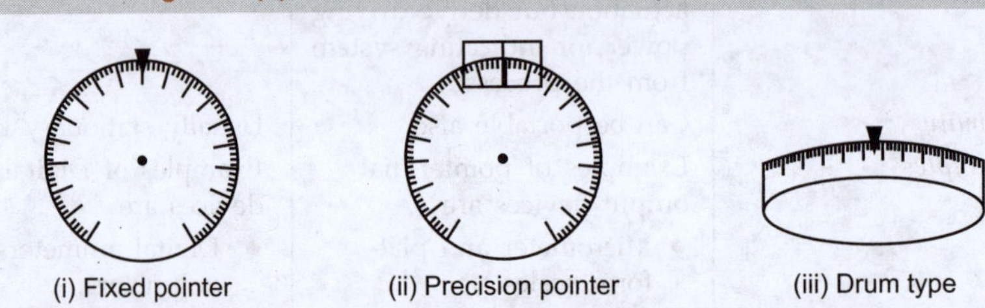

Fig. 4.2 (b). Fixed-pointer and movable-scale indicators.

The readability of graduated dials is influenced by the following factors :

- The shape and length of the pointer.
- The number, spacing, length and thickness of scale marking.
- The system of numbering of the scale marks.
- The size and design of the numerals.

2. *Multi-point, multi-pointer and multi-range indicators:*

Multi-point indicator. In this system, the indicator pointer can be connected to a number of inputs, one at a time with the help of a selector switch.

The selector switch may be operated either manually or automatically after a predetermined time. The observed reading is multiplied by a factor corresponding to the particular measurand.

Generally, such systems are confined where measurable variables are of electrical signals as the selection is accomplished by switching electrical circuits. However, gas selector switches also exist which connect one gas pipe at a time to the measuring instrument and are well-designed to avoid leakage of gas.

4.6 ELECTRICAL INSTRUMENTS

4.6.1. Introduction and Classification

The instruments used for all electrical measurements are called *measuring instruments.* They include *ammeters, voltmeters, wattmeters, energy meters* etc. The various electrical instruments may broadly be divided into two groups :

1. **Absolute instruments.** Absolute instruments are those instruments which indicate the quantity to be measured in terms of the *constants of the instrument* (dimensions, turns etc.) and in order to find out the quantity in the practical units it is necessary to *multiply such deflections with an instrument constant.*

- *No previous calibration or comparison is necessary in this case.*
- The most common absolute instrument is *tangent* galvanometer which gives the measured current in terms of tangent of the deflected angle, the radius and the number of turns of the galvanometer.
- Such insruments are rarely used (the use being merely confined within laboratories as standardising instruments).

2. **Secondary instruments.** Secondary instruments are those in which the value of electrical quantity to be measured can be determined from the *deflection of instrument only* when they have been *pre-calibrated* by comparison with an absolute instrument.

- The deflection of the instrument gives directly the quantity to be measured.
- These instruments are most generally used in everyday work.

Secondary instruments may also be *classified* as follows :

1. **Indicating instruments.** Indicating instruments are those which indicate the instantaneous value of the electrical quantity being measured at the time at which it is being measured. Their indications are given by pointers moving over calibrated scales.

Example. Ammeters, voltmeters and *wattmeters.*

2. **Recording instruments.** Recording instruments are those which give a *continuous record* of the variations of an electrical quantity over a selected period of time. The pointer in these types of instruments is an infed pen which leaves a trace on a paper put over a moving drum.

3. **Integrating instruments.** Integrating instruments are those which measure the total quantity of electricity delivered in a particular time.

Example. Ampere-hour and watt-hour meters.

Electrical measuring instruments may also be *classified* as follows :

1. **According to the quantity being measured :**

Ammeters. ...... for measuring the magnitude of current.

Voltmeters. ...... for measuring voltages.

Ohmmeters and resistance bridges. ...... for measuring resistances.

Wattmeters. ...... for power measurements.

Watt-hour meters. ...... for energy measurements.

Frequency meters. ...... for frequency measurements.

Power factor meters. ...... for power-factor measurements.

2. **According to the kind of current :**

Instruments are classified into *D.C., A.C.* and *A.C./D.C. instruments.*

3. **According to accuracy limits :**

4. **According to the principle of operation :**

Instruments are grouped into :

— Moving coil

— Moving iron

— Electrodynamic

— Induction

— Hot-wire

— Thermoelectric

— Rectifier types.

5. **According to the type of indication :**

Instruments may be :

— Indicating type

— Recording type.

6. **According to application :**

— Switch board

— Portable.

4.6.2. Electrical Principles of Operation

All electrical measuring instruments depend for their action on any of many physical effects of electric current or potential. The following are the effects generally used in their manufacture :

(*i*) **Magnetic effect.** ...... Voltmeters, ammeters, wattmetes, power factor meters etc.

(*ii*) **Thermal effect.** ...... Ammeters, voltmeters, maximum demand meters etc.

(*iii*) **Chemical effect.** ...... D.C. ampere hour meters (integrating meters).

(*iv*) **Electrostatic effect.** ...... Voltmeters which can indirectly be used as ammeters and wattmeters.

(*v*) **Electro-magnetic induction effect.** ...... Voltmeters, ammeters, wattmeters and integrating meters used in A.C. only.

- **Difference between an ammeter and a voltmeter :**

An ammeter and a voltmeter work on the same principle. The ammeter has a *low resistance* so that when it is connected in **series** with any circuit, it does not change the current. The voltmeter has a *high resistance* and it is so designed that when connected in **parallel** to the circuit for measuring voltages it does not take appreciable current.

An ammeter of low range can be used as a voltmeter by connecting an external resistance in series with it.

4.6.3. Essential Features of Indicating Instruments

Indicating instruments possess *three essential features* :

1. **Deflecting device.** ...... whereby a mechanical force is produced by the electric current, voltage or power.
2. **Controlling device.** ...... whereby the value of deflection is dependent upon the magnitude of the quantity being measured.

3. **Damping device.** ...... to prevent oscillation of the the moving system and enable the latter to reach its final position quickly.

4.6.3.1. Deflecting device

A deflecting device produces a deflecting torque which is caused by anyone of these effects: thermal effect, chemical effect, electrostatic effect etc.; with the help of this deflecting torque the needle or the pointer moves from zero position to the final position. The arrangement of the deflecting device with each type of instrument will be discussed individually.

4.6.3.2. Controlling devices

These are two types of controlling devices :

(*i*) Spring control. (*ii*) Gravity control.

(*i*) ***Spring control.*** Fig. 4.3 shows a commonly used spring control arrangement. It utilises two spiral hair springs, 1 and 2, the inner ends of which are attached to the spindle S. The outer end of spring 2 is fixed while that of 1 is attached to a lever, the adjustment of which gives zero adjustments. The two springs 1 and 2 are wound in *opposite directions* so that when the moving system is deflected, one spring winds up while the other unwinds, and the controlling torque is due to the *combined torsions* of the springs.

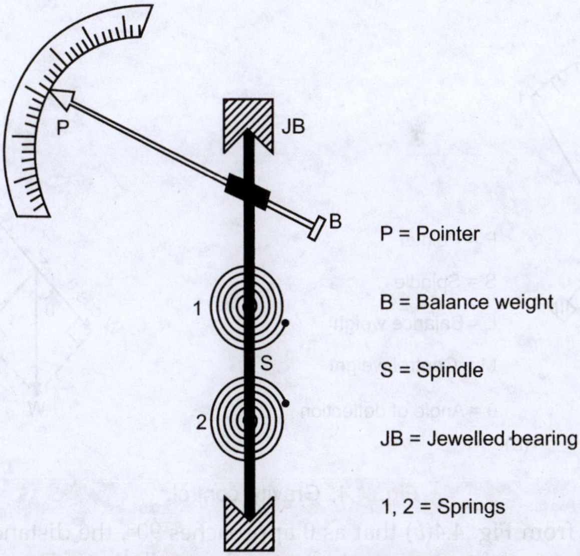

Fig. 4.3. Spring control.

Since the torsional torque of a spiral spring is proportional to the angle of twist, the controlling torque (T_c) is directly proportional to the angular deflection of the pointer (θ).

i.e., $T_c \propto \theta$

The spring material should have the following *properties* :

(*i*) It should be non-magnetic.

(*ii*) It must be of low temperature co-efficient.

(*iii*) It should have low specific resistance.

(*iv*) It should not be subjected to fatigue.

- The exact expression for controlling the torque is :

$$T_c = k\,\theta \qquad ...(4.1)$$

where k is spring constant (or control constant, or torsion constant or restoring constant). Its value is given by :

$$k = \frac{Ebt^3}{12l} \text{ N-m/rad.; the angle } \theta \text{ is in radians}$$

where, E = Young's modulus of spring material,

b = Width of spring (m),
t = Thickness of spring (m), and
l = Length of spring (m).

} *Flat spiral spring*

(*ii*) **Gravity control.** With gravity control, weights L and M are attached to the spindle S [Fig. 4.4 (*a*)], the function of L being to *balance* the weight of the pointer P. Weight M therefore provides the controlling torque. When the pointer is at zero, M hangs vertically downwards. When P is deflected through angle θ, the controlling torque is equal to (weight of $M \times$ distance d) and is therefore proportional to the *sine* of the angular deflection [Fig. 4.4 (*b*)]

i.e., $T_c \propto \sin\theta$

The degree of control is adjusted by screwing the weight up or down the carrying system.

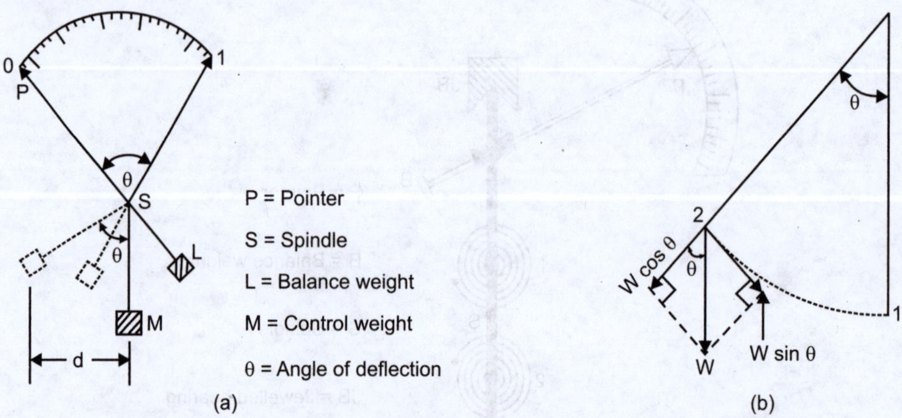

Fig. 4.4. Gravity control.

It may be seen from Fig. 4.4(*b*) that as θ approaches 90°, the distance 1-2 increases by a relatively small amount for a given change in the angle that when θ is just increasing from its zero value. Hence, *gravity-controlled instruments have scales which are not uniform but are cramped or crowded at their lower ends.*

Advantages :

1. The gravity controlled instrument is cheaper than corresponding spring-controlled instrument.
2. It is not subjected to fatigue.
3. It is unaffected by temperature.

Disdvantages:

1. Gravity control gives a cramped scale.
2. The instrument must be levelled before use.

Example 4.1. *The length, width and thickness of the control spring of an instrument are 390 mm, 0.52 mm and 0.075 mm. The Young's modulus of the spring material is 110 GN/m^2. Determine the torque exerted by the spring when it is turned through 80°.*

Solution. *Given* : l = 390 mm; b = 0.52 mm; t = 0.075 mm; T_c = 110 GN/m^2;

$\theta = 80° = 80 \times \frac{\pi}{180}$ rad. or 1.396 rad.

We know that, controlling torque,

$$T_c = \frac{Ebt^3}{12l} \cdot \theta \qquad \text{[Eqn. (4.1)]}$$

$$= \frac{(110 \times 10^9) \times (0.52 \times 10^{-3}) \times (0.075 \times 10^{-3})^3}{12 \times (390 \times 10^{-3})} \times 1.396$$

$= \mathbf{5.156 \times 10^{-6}}$ **Nm** **(Ans.)**

Example 4.2. *The toque of an ammeter varies as the square of the current through it. If a current of 5A produces a deflection of 90°, what deflection will occur for a current of 3A when the instrument is :*

(i) Spring-controlled; *(ii) Gravity controlled.* **(Jadavpur University)**

Solution. Since deflecting torque varies as (current)^2, we have $T_d \propto I^2$

For spring control : $T_c \propto \theta$ $\therefore$ $\theta \propto I^2$

For gravity control : $T_c \propto \sin\theta$ $\therefore$ $\sin\theta \propto I^2$

(i) ***For spring control :*** $90° \propto 5^2$ and $\theta° \propto 3^2$

$$\theta = 90° \times \frac{3^2}{5^2} = \mathbf{32.4°} \quad \textbf{(Ans.)}$$

(ii) ***For gravity control :*** $\sin 90° \propto 5^2$, and $\sin\theta° \propto 3^2$

$$\therefore \quad \frac{\sin\theta}{\sin 90°} = \frac{3^2}{5^2} = \frac{9}{25}$$

or, $$\sin\theta = \frac{9}{25}$$

$$\therefore \quad \theta = \sin^{-1}(9/25) = \mathbf{21.2°} \quad \textbf{(Ans.)}$$

Example 4.3. *In a spring controlled indicating instrument the control is exerted by two springs. The allowable maximum stress is 30 MN/m^2 and Young's modulus of spring material is 120 GN/m^2.*

Determine the suitable dimensions for spring if the deflecting torque corresponding to a full scale deflection of 90° is 1.2 × 10^{-3} Nm. Assume the width of spring strip as 0.6 mm.

(Roorkee University)

Solution. *Given* : Allowable maximum stress, σ_{max} = 30 MN/m^2; E = 120 GN/m^2;

T_d = 1.2 × 10^{-3} Nm; $\theta = 90°$ or $\frac{\pi}{2}$ rad.; b = 0.6 mm

Dimensions of spring, *l*, *t* :

As there are two controlling springs, the controlling torque (T_c) caused by each spring is equal to half of the deflecting torque (T_d) *i.e.*, controlling torque of each spring,

$$T_c = \frac{1.2 \times 10^{-3}}{2} = 0.6 \times 10^{-3} \text{ Nm} \qquad ...(i)$$

We know that, $T_c = \frac{Ebt^3}{12l}\theta$ [Eqn. (4.1)]

or, $$\frac{t^3}{l} = \frac{12T_c}{Eb\theta} = \frac{12\times0.6\times10^{-3}}{(120\times10^9)\times(0.6\times10^{-3})\times(\pi/2)} = 0.6366 \times 10^{-10} \quad ...(i)$$

The maximum stress (σ_{max}) developed in the spring must be well below the elastic limit of the spring material at the maximum deflection of the moving system in order to avoid failure and to preserve long period stability; it is given by the expression :

$$\frac{l}{t} = \frac{E\theta}{2\sigma_{max}}$$

or, $$\frac{l}{t} = \frac{120\times10^9\times(\pi/2)}{2\times(30\times10^6)} = 3142 \quad ...(ii)$$

Multiplying (*i*) and (*ii*), we get

$$t^2 = 0.6366 \times 10^{-10} \times 3142 = 0.2 \times 10^{-6}$$

or, *Thickness of each strip,* $t = \sqrt{0.2\times10^{-6}} \simeq 4.5 \times 10^{-4}$ m or **0.45 mm** **(Ans.)**

Substituting this value of t in eqn. (*ii*) we get

$l = 4.5 \times 10^{-4} \times 3142 =$ **1.414 m** **(Ans.)**

4.6.3.3. Damping devices

Owing to the inertia of the moving system, when subjected to the deflecting and restoring torques, number of vibrations will be produced before finally coming to rest. To-avoid this, a *damping torque* is required which opposes the motion and ceases when the pointer comes to rest. The degree of damping should be adjusted to a value which is sufficient to enable the pointer to rise quickly to its deflected position without overshooting. In that case, the instrument is said to be *dead-beat*. If the instrument is over-damped the movement is very slow (and the instrument becomes lethargic) as shown in Fig. 4.5.

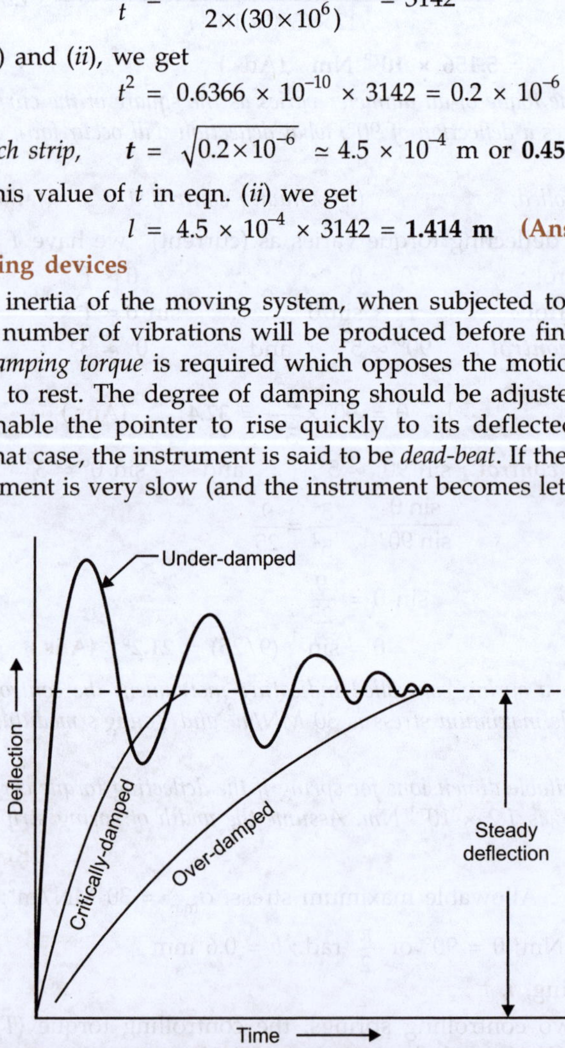

Fig. 4.5. Damping curves.

Damping can be provided by the following *methods* :

1. Air damping.

2. Eddy current damping.
3. Fluid friction damping.

1. ***Air damping.*** Fig. 4.6 shows an arrangement for obtaining air damping. It consists of a thin metal vane *MV* attached to the spindle *S*; the vane moves in a sector-shaped box *B*. *Any tendency of the moving system to oscillate is damped by the action of the air on the vane.*

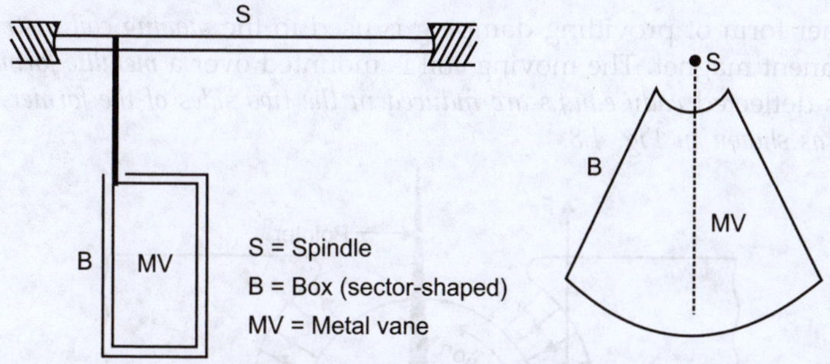

Fig. 4.6. Eddy current damping.

2. ***Eddy current damping.*** This method of damping is based on the *principle* that when a conducting, non-magnetic material is moved in a magnetic field, an e.m.f. is induced in it which causes currents called the *eddy-currents.* Due to these eddy currents, a force exists between them and the field. Due to Lenz's law this force is always in opposition to the force causing rotation of the conducting material, thus, it provides the necessary damping.

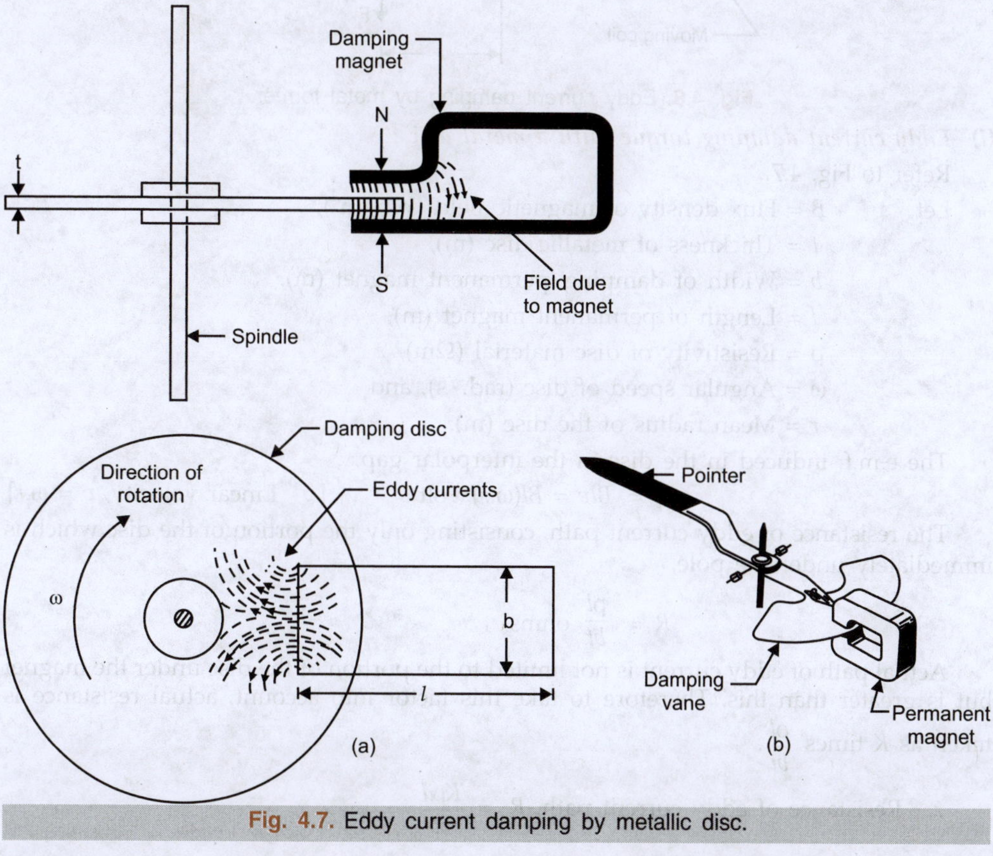

Fig. 4.7. Eddy current damping by metallic disc.

- One form of eddy-current damping is shown in Fig. 4.7. Here a copper or aluminium disc, carried by a spindle, can move between the poles of permanent magnet. If the disc moves clockwise, the e.m.f.s induced in the disc circulate eddy currents as shown dotted. It follows from Lenz's law that these currents exert a force *opposing* the motion producing them, namely the clockwise movement of the disc.
- Another form of providing damping is used in the *moving coil instruments* using permanent magnet. The moving coil is mounted over a *metallic former*. When the coil is deflected *eddy e.m.f.s are induced in the two sides of the former, causing eddy forces as shown in Fig. 4.8.*

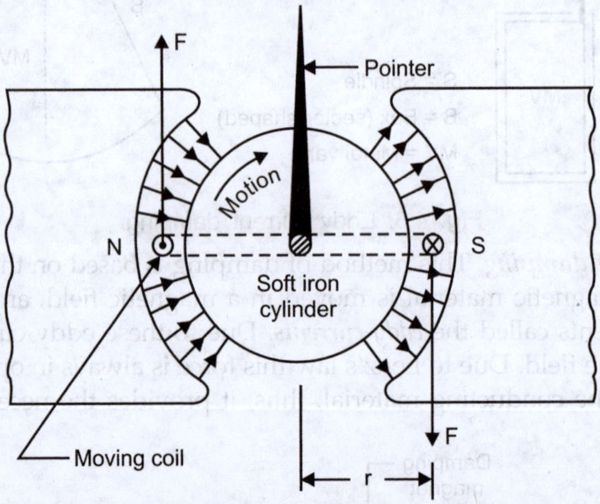

Fig. 4.8. Eddy current damping by metal former.

(i) Eddy current damping torque with a metal disc :

Refer to Fig. 4.7.

Let, B = Flux density of magnetic field (Wb/m^2),
t = Thickness of metallic disc (m),
b = Width of damping/permanent magnet (m),
l = Length of permanent magnet (m),
ρ = Resistivity of disc material (Ωm),
ω = Angular speed of disc (rad./s), and
r = Mean radius of the disc (m).

The e.m.f. induced in the disc in the interpolar gap,

$$E_e = Blv = Bl(\omega.r) \text{ volt} \qquad [\because \text{ Linear velocity, } v = \omega.r]$$

The resistance of eddy current path, consisting only the portion of the disc which is immediately under the pole,

$$R = \frac{\rho l}{bt} \text{ ohms}$$

Actual path of eddy current is not limited to the portion of the disc under the magnet but is greater than this. Therefore to take this factor into account, actual resistance is taken as K times $\frac{\rho l}{bt}$.

$\therefore$ Resistance of eddy current path, $R_e = \frac{K\rho d}{bt}$

where, K = a constant which depends upon radial position of poles, it is *always greater than unity.*

$$\text{Eddy current, } I_e = \frac{E_e}{R_e} = \frac{Bl\omega r}{K\rho l/(bt)} = \frac{Brbt\omega}{K\rho}$$

$$\textit{Damping force, } F_D = BI_e l = B \times \frac{Brbt\omega}{K\rho} \times l = \frac{B^2 rlbt\omega}{K\rho} \quad ...(4.2)$$

$$\textit{Damping torque, } T_d = F_D \times r = \frac{B^2 rlbt\omega}{K\rho} \times r = \frac{B^2 r^2 lbt\omega}{K\rho} \quad ...(4.3)$$

$$\textit{Damping constant, } k_D = \frac{T_D}{\omega} = \frac{B^2 r^2 lbt}{K\rho} \text{ Nm/rad. s}^{-1} \quad ...(4.4)$$

— The magnitude of damping torque can be varied by adjusting the radial position of the poles; damping torque (T_D) decreases with the movement of the magnet towards the edge of the disc and becomes zero when the centres of the poles are at the edge of the disc.

— The above phenomenon is employed in the induction type instruments by placing a damping magnet (permanent type) at the opposite side of the disc from the electromagnet, so that the disc *serves for damping as well as for operating purposes.*

(ii) Eddy current damping with a metal former :

Fig. 4.8, shows a metallic former moving in the field of a permanent magnet.

Let, B = Strength of magnetic field (Wb/m^2),

l = Length of the former (m),

d = Diameter of the former (m),

b = Width of the former (m),

t = Thickness of the former (m), and

ω = Angular speed of former (rad./s).

$$\text{Linear velocity of former, } v = \omega \times \frac{d}{2}$$

$$\text{E.m.f. generated in the former, } E_e = 2Blv = 2Bl\left(\omega \times \frac{d}{2}\right) = Bld\omega$$

Resistance of path of eddy currents,

$$R_e = \frac{\text{Resistivity} \times \text{Length}}{\text{Area}} = \frac{\rho \times 2(l+d)}{b \times t} = \frac{2\rho(l+d)}{bt}$$

$$\text{Eddy current, } I_e = \frac{E_e}{R_e} = \frac{Bld\omega}{2\rho(l+d)/(bt)} = \frac{Blbtd\omega}{2\rho(l+d)}$$

$$\therefore \textit{ Damping force, } \boldsymbol{F_d} = BI_e l = B \times \frac{Blbtd\omega}{2\rho(l+d)} = \frac{B^2 l^2 btd\omega}{2\rho(l+d)} \quad ...(4.5)$$

$$\textit{Damping torque, } T_D = F_D \times d = \frac{B^2 l^2 btd\omega}{2\rho(l+d)} \times d = \frac{B^2 l^2 d^2 bt\omega}{2\rho(l+d)} \quad ...(4.6)$$

$$\textit{Damping constant, } \boldsymbol{K_D} = \frac{T_D}{\omega} = \frac{B^2 l^2 d^2 bt}{2\rho(l+d)} \text{ Nm/rad. s}^{-1} \quad ...(4.7)$$

— It is obvious from eqn. (4.6) that damping torque (T_D) can be varied by varying the thickness (t) of the metal former.

3. ***Fluid friction damping.*** Fig. 4.9 shows the method of fluid friction damping. Here light vanes are attached to the spindle of the moving system. The vanes are dipped into a pot of damping oil and are completely submerged by the oil. The motion of the moving system is always opposed by the friction of the damping oil on the vanes. The damping force thus created always increases with the increase in velocity of vanes. There is no damping force when the vanes are stationary.

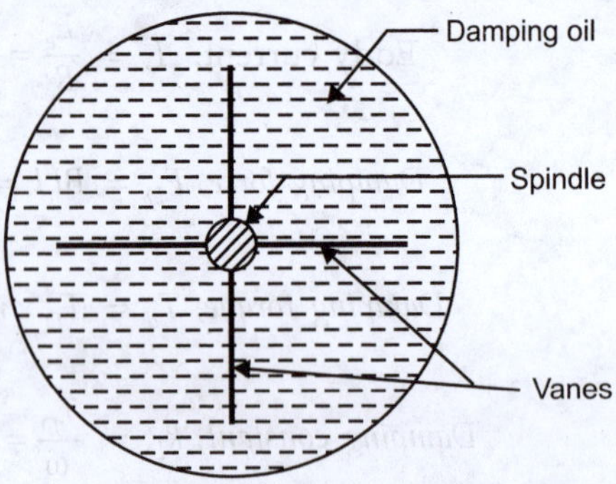

Fig. 4.9. Fluid friction damping.

The damping oil used must have the following *properties* :

(*i*) Must be a good insulator.

(*ii*) Should be non-evaporating.

(*iii*) Should not have corrosive action upon the metal of the vane.

(*iv*) The viscosity of the oil should not change with the temperature.

Though in this method of damping, no case is required as in the air friction damping but it is not much used due to the following *disadvantages* :

(*i*) Objectionable creeping of oil.

(*ii*) Using the instrument always in vertical position and its obvious unsuitability for use in portable instruments.

☞ The various type of indicating instruments used for measurement of current, voltage, power and energy along with suitability for the *type of measurement, type of control, type of damping, special features and applications are given below in tabular form :*

S.No.	*Type of instrument*	*Suitability for type of measurement*	*Type of control*	*Type of damping*	*Remarks*
1.	***Moving iron***	D.C. or A.C. measurements (current and voltage)	Spring or gravity control	Air friction	• Cheapter to manufacture. • Very accurate on A.C. or D.C., if properly designed. • *Mostly used as an indicating instrument.*
2.	***Moving coil*** (*i*) *Permanent magnet type*	D.C. or A.C. measurements (current and voltage only)	Spring control	Eddy current	• Most accurate type for D.C. measurements. • *Most widely used for measurement of D.C. voltage, current and resistance.*
	(*ii*) *Dynamometer type*	D.C. or A.C. measurements	Spring control	Air friction	• Mainly used as *wattmeter*. • Also used as standard meter for calibration and as transfer instrument.

3.	***Thermal*** *(i) Hot wire*	D.C. or A.C. measurements (current, voltage, and power)	Spring control	Eddy current	• Same calibration for both D.C. and A.C. • Free from errors due to frequency, wave form and external field when used on A.C., therefore, these are particularly used for A.C. requirements. Since its disadvantages outweigh its advantages, hence it is *now little used.*
	(ii) Thermo-couple	D.C. or A.C. measurements (current and voltage)	Spring control	Eddy current	• Free from errors due to frequency, wave form and external field when used on A.C. • *Used for measurement of current and voltage at power frequencies upto 100 MHz.*
4.	***Electrostatic***	D.C. or A.C. (voltage only)	Spring or gravity control	Air friction	• Very low power consumption and can be made to cover a large range of volts (above 500 V).
5.	***Induction***	A.C. measurements (current, voltage, power and energy) only	Spring control	Eddy current	• Expensive and degree of accuracy not high. • Mainly used for *measurement of power and energy in A.C. circuits.*
6.	***Rectifiers***	D.C. or A.C.	Spring control	Eddy current	• These are PMMC instruments used in conjunction with rectifying devices for A.C. measurements (current and voltage) from about 20 Hz to 20 kHz

4.7 GALVANOMETERS

A **galvanometer** *is an instrument used for detecting presence of small currents or voltages in a closed circuit or for measuring their magnitudes.*

However, measurement of magnitude is avoided as much as possible, for accuracy in reading a deflection is seriously limited.

Several types of galvanometers have been devised, including *moving coil* and *magnet types.* Special instruments are needed for some purposes, but the great majority of the galvanometers in use now are the *permanent-magnet moving coil type, normally referred as* "**D' Arsonval type**". This is *D.C. galvanometer.*

Other types of galvanometers are :

Ballistic galvanometer. *It is used to measure the quantity of electricity (charge) passed through it.* In *"magnetic measurements"*, this quantity of electricity is due to an instantaneous e.m.f. induced in a search coil connected across the ballistic galvanometer.

The main advantage of *moving coil instrument lies in its freedom from the effects of stray fields and the ease with which a long and uniform scale may be attained.* These are the features that have caused it to be adopted, in a modified form, for D.C. ammeters and voltmeters.

Vibration galvanometer. These galvanometers are most widely used *"tuned detectors"*. These galvanometers are of the following types :

(*i*) Moving magnet type.

(*ii*) Moving coil type.

— The moving coil type galvanometer is more generally used, the moving magnet having the disadvantage of being seriously affected by magnetic fields of the resonance frequency, unless adequately screened.

Galvanometers find their principal applications in *bridge* and *potentiometer measurements.* Therefore, a galvanometer in addition to being sensitive, should have :

— A stable zero;

— A short periodic time;

— Nearly critical damping.

4.7.1. D'Arsonval Galvanometer

4.7.1.1. Construction and Working

Fig. 4.10 shows the essential features of a D'Arsonval galvanometer :

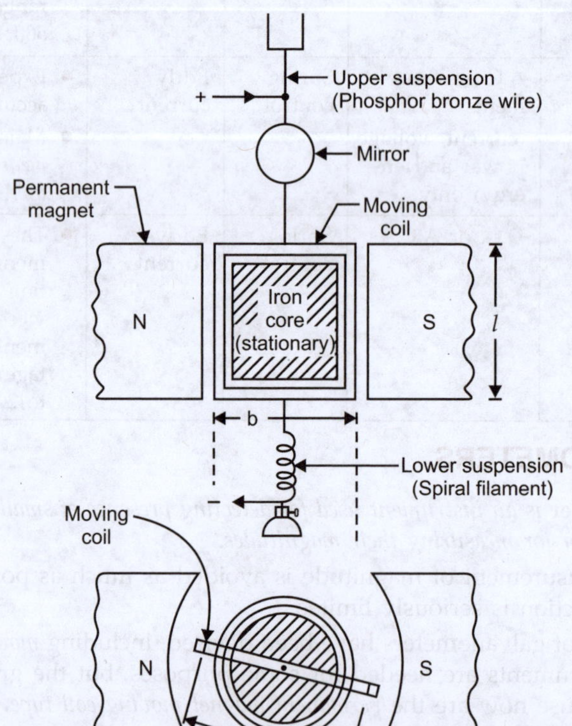

Fig. 4.10. D'Arsonval galvanometer.

Construction :

— It consists of a circular or rectangular coil of very fine insulated wire with many turns usually wound on an aluminium frame or bobbin suspended between the poles of a permanent magnet.

— The coil is suspended by a phosphor bronze filament which acts as one lead-in for the coil. The other lead-in wire is a very flexible spiral wire at the bottom of the coil. *The elasticity of the filament sets up a moderate torque in opposition to the rotation of the coil.*

— The poles of the magnet are usually cylindrical in shape and a cylindrical soft iron core is placed between the poles and coaxial with them.

The cylindrical form of poles faces and core possesses the following *advantages* :

(*i*) The length of air gap is reduced so that the amount of flux linking the coil is increased, thus increasing the sensitivity of the instrument;

(*ii*) The flux between the pole faces and core is practically radial. In radial field the deflection of coil is directly proportional to the current in the coil and, therefore, a uniform scale is obtained.

Working/Operation :

— When current flows through the coil, a deflecting force proportional to the flux density, the coil current and dimensions of the coil makes the coil to rotate on its vertical axis.

— The "*deflecting force is opposed by the restraining force*" of the suspension filament so that coil does not continue to rotate as in a motor but turns until the deflecting torque is *balanced* by the restraining force.

— Since the deflecting torque is directly proportional to the coil current, therefore, the amount of deflection of the coil indicates the magnitude of current flowing in the coil. The amount of deflection and hence the magnitude of curent flowing through the coil may be indicated by a pointer attached to the moving element and moving over a calibrated scale.

For the adjustment of coil and also for *zero adjustment*, a "*torsion head*" is provided.

— In a *sensitive galvanometer*, a small "*mirror*" is attached to the moving coil to indicate the position of the coil. This arrangement may be used in *two* ways : (*i*) To mount a telescope and scale in front of the galvanometer, so that the observer reads the scale reflected by the mirror; (*ii*) Another arrangement (more common and less tiring to the user) consists of a lamp and scale device in which a spot of light is projected on the mirror and reflected to the scale. The scale is usually mounted 1 m away from the mirror although 0.5 m is used sometimes for compactness.

- There is a "*damping torque*" present owing to production of eddy currents in the metal former on which the coil is mounted. Damping is also obtained by connecting a low resistance across the galvanometer terminals. Damping torque depends upon the resistance and we *can obtain critical damping by adjusting the value of resistance.*

4.7.1.2. Torque equation

Refer to Fig. 4.10 :

Let, N = Number of turns in coil,

B = Flux density, Wb/m^2,

I = Current through the coil,

l = Length of coil side (vertical), m

b = Width of coil, m,

θ = Deflection of pointer, rad., and

C = Spring (or controlling or restoring) constant, Nm/rad.

Force on each side of coil = BIl newtons [∵ Field is radial]

Since the coil has N turns and each turn has two sides, therefore,

$$\text{Total force} = 2NBIl \text{ newtons}$$

This force acts at a radius $\frac{b}{2}$.

∴ *Deflecting torque*, T_d = Force × radius

$$\text{or,} \quad T_d = 2NBIl \times \frac{b}{2} = NBIlb = NBIA \text{ Nm} \quad ...(4.8)$$

where, A = Area of the coil (= $l \times b$), m^2

N, B, A are constants for a galvanometer.

$$\therefore \text{ Deflecting torque, } T_d = GI \quad ...(4.9)$$

G is called the *displacement constant* (= NBA) of galvanometer. *Controlling torque* exerted by the suspension at deflection θ,

$$T_c = C\theta \quad ...(4.10)$$

For steady deflection :

$$T_d = T_c \quad ...(4.11)$$

$$GI = C\theta$$

$$\text{or,} \quad \theta = \frac{GI}{C} = \left(= \frac{NBIA}{C} \right) \quad ...[4.11(a)]$$

Several times the deflection is measured on a scale kept at a distance *r* metres from a mirror. A beam of light is reflected on the scale by a small mirror which is mounted on the moving system (Fig. 4.11). The distance *r* is usually kept as 1 metre. If the moving system rotates through angle θ, the light beam is turned through angle 2θ. Then,

Deflection *oa i.e.*, $d = (r) \times (2\theta)$ metres.

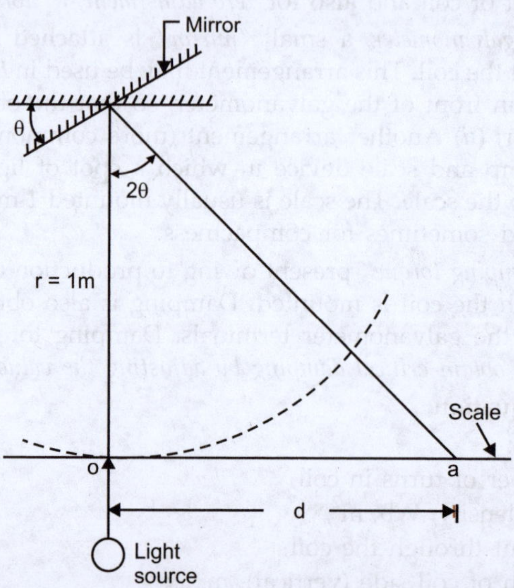

Fig. 4.11. Measurement of deflection using mirror and scale.

4.7.1.3. Galvanometer Constants

The galvanometer has *four* constants known as *"Intrinsic constants."* These are :

1. Displacement constant.
2. Inertia constant.
3. Damping constant.
4. Control or restoration constant.

1. ***Displacement constant :***

- The deflection torque is given by $T_d = GI$ (eqn. 4.9) where G is the *displacement constant* of the galvanometer and is equal to $NBlb$.
- The units of G are Nm/A.

2. ***Inertia constant :***

- Owing to inertia of moving system, a *retarding torque* is produced.
- This torque (T_j) is dependent upon the moment of inertia of moving system and the angular acceleration.

$$T_j = J \cdot \frac{d^2\theta}{dt^2} \qquad ...(4.12)$$

where, J = Moment of inertia of moving system about the axis of rotation, kg-m^2, and

θ = Deflection at any time t, rad.

3. ***Damping constant :***

- *Damping* is provided by friction due to motion of the coil air and also by induced electrical effects if a closed circuit is provided.
- *Damping torque* (T_D) is usually assumed to be proportional to the angular velocity of the system and may be expressed as :

$$T_D = D\frac{d\theta}{dt} \qquad ...(4.13)$$

where, D = Damping constant.

4. ***Control or restoration constant :***

- A *controlling torque* (T_c) is produced due to *elasticity* of the system which tries to restore the moving system back to its original position, and may be expressed as :

$$T_c = C\theta$$

where, C = Control constant, Nm/rad.

4.7.1.4. Dynamic response

The deflecting torque T_d (GI) *causes* the motion, while the inertia torque $\left(T_j = J\frac{d^2\theta}{dt^2}\right)$, damping torque $\left(T_D = D\frac{d\theta}{dt}\right)$ and controlling torque ($T_c = C\theta$) opposite the motion. The *"equation of motion"* is given as :

$$T_j + T_D + T_C = T_d$$

or,

$$\underbrace{J\frac{d^2\theta}{dt^2}}_{\text{Inertia torque}} + \underbrace{D\frac{d\theta}{dt}}_{\text{Damping torque}} + \underbrace{C\theta}_{\text{Controlling torque}} = \underbrace{GI}_{\text{Deflecting torque}} \qquad ...(4.14)$$

Eqn. (4.14) is well known *second order differential equation.* Use of Laplace transform provides an easy solution. Taking Laplace transforms of both sides and writing Laplace transform of response θ as θ(*s*), we get

$$\theta(s) = \frac{GI}{s\,(Js^2 + Ds + C)} = \frac{GI/C}{s\left[\frac{s^2}{\omega_n^2} + \frac{2\gamma s}{\omega_n} + 1\right]} \quad ...(4.15)$$

where, $\omega_n = \sqrt{\frac{C}{J}}$

and, $\gamma = \frac{D}{2J\omega_n}\left(= \frac{D}{\sqrt{2JC}}\right)$ ...[4.15(*a*)]

where, γ is known as *"damping ratio"*;

ω_n = *Natural* frequency of oscillation (rad./sec.), and

$$f_n = \frac{\omega_n}{2\pi}, \quad \text{and} \quad T = \frac{1}{f_n}$$

The characteristic equation is :

$$s^2 + 2\gamma\omega_n s + \omega_n^2 = 0 \quad ...[4.16(a)]$$

or,

$$s^2 + \frac{D}{J}s + \frac{C}{J} = 0 \quad ...[4.16(b)]$$

or,

$$s = \frac{-\frac{D}{J} \pm \sqrt{\frac{D^2}{J^2} - \frac{4C}{J}}}{2}$$

∴ The roots of characteristic equation are :

$$s_{1,2} = \frac{-D}{2J} \pm \sqrt{\frac{D^2}{4J^2} - \frac{C}{J}} \quad ...(4.17)$$

The actual response depends on the expression under the radical in eqn. (4.17)

1. If $\frac{D^2}{4J^2} < \frac{C}{J}$ or $D^2 < 4CJ$, the response is *underdamped;*
2. If $\frac{D^2}{4J^2} = \frac{C}{J}$ or $D^2 = 4CJ$, the response is *critically damped;*
3. If $\frac{D^2}{4J^2} > \frac{C}{J}$ or $D^2 > 4CJ$, the response is *overdamped.*

— When damping is *critical"*, the pointer reaches its final position $\theta_d \left(= \frac{GI}{C}\right)$ *without any oscillation.*

The ***damping ratio*** *γ is the ratio of actual value of D to the value of D for critical damping.*

i.e.,

$$\gamma = \frac{D_{actual}\ (\text{or } D)}{D_{critical}} = \frac{D}{2\sqrt{CJ}} \quad ...(4.18)$$

or, $$\frac{D}{2J} = \frac{\gamma\,(2\sqrt{CJ})}{2J} = \gamma\sqrt{\frac{C}{J}} = \gamma\omega_n \qquad ...(4.19)$$

and, $$\sqrt{\frac{C}{J} - \left(\frac{D}{2J}\right)^2} = \sqrt{\omega_n^2 - \gamma^2\omega_n^2} = \omega_n\sqrt{1-\gamma^2} \qquad ...(4.20)$$

Therefore, the roots, s_1, s_2 in eqn. (4.17) can be written as :

$$s_{1,2} = -\gamma\omega_n \pm j\omega_n\sqrt{1-\gamma^2} \qquad ...(4.21)$$

The damped frequency of oscillation ω_d is :

$$\omega_d = \omega_n\sqrt{1-\gamma^2} \qquad ...(4.22)$$

or, $$f_d = f_n\sqrt{1-\gamma^2} \qquad ...[4.22(a)]$$

Eqn. (4.21) can be written as :

$$s_{1,2} = -\gamma\omega_n \pm j\omega_d \qquad ...(4.23)$$

The *response of the three conditions of damping* is given by :

1. $$\theta = \frac{GI}{C}\left[1 - \frac{e^{-\gamma\omega_n t}}{\sqrt{1-\gamma^2}}\sin\left\{\omega_n\sqrt{(1-\gamma^2)}\,t + \sin^{-1}\sqrt{(1-\gamma^2)}\right\}\right] \qquad ...(4.24)$$

for $\gamma < 1$ *or underdamped motion.*

2. $$\theta = \frac{GI}{C}\left[1 - (1+\omega_n t)\,e^{-\omega_n t}\right] \qquad ...(4.25)$$

for $\gamma = 1$ *or critically damped motion.*

3. $$\theta = \frac{GI}{C}\left[1 - \frac{e^{-\gamma\omega_n t}}{\sqrt{\gamma^2-1}}\sinh\left\{\omega_n\sqrt{(\gamma^2-1)}\,t + \sinh^{-1}\sqrt{\gamma^2-1}\right\}\right] \qquad ...(4.26)$$

for $\gamma > 1$ *or overdamped motion.*

The system response for the underdamped, critically damped and overdamped cases is shown in Fig. 4.12.

☞ *Following observations are made :*

- When the system is *underdamped* the pointer makes a number of oscillations about the final position θ_d before setting down. Thus an underdamped response *leads to waste of time.*
- *Overdamped* response also leads to *wastage of time* because the system shows a sluggish behaviour.
- In *critical damping,* the time required for setting down to the final value θ_d is the *minimum.*

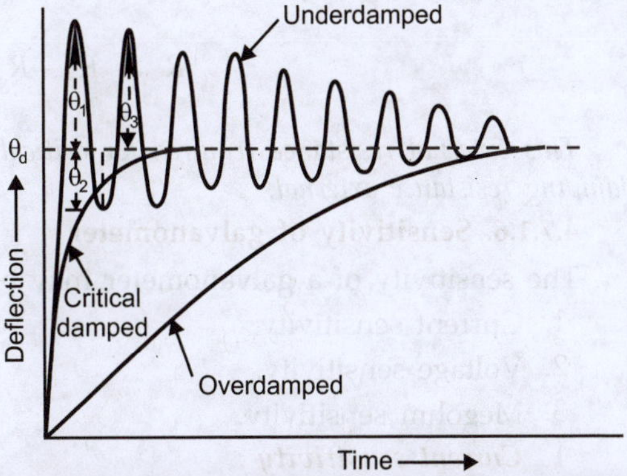

Fig. 4.12. Underdamped, critically damped and overdamped response of second order system.

4.7.1.5. Effect of galvanometer resistance on damping

Even when the galvanometer circuit is *open*, the *air friction and eddy current damping are present.*

When the galvanometer circuit is closed, a current I flows in this circuit. I is given by the equation :

$$IR = E - G\frac{d\theta}{dt} \quad ...(4.27)$$

where, R = Resistance of galvanometer circuit,

E = Applied voltage, and

$G\frac{d\theta}{dt}$ = Back e.m.f. induced in the galvanometer coil due to its motion in the magnetic field. The constant, as defined earlier, equals NBA.

Now the equation of motion is :

$$J\frac{d^2\theta}{dt^2} + D'\frac{d\theta}{dt} + C\theta = \frac{G}{R}\left(E - G\frac{d\theta}{dt}\right) \quad ...(4.28)$$

or, $$J\frac{d^2\theta}{dt^2} + \left(\frac{G^2}{R} + D'\right)\frac{d\theta}{dt} + C\theta = \frac{GE}{R} = GI \quad ...(4.29)$$

The term $\left(\frac{G^2}{R} + D'\right)$ in eqn. (4.29) replaces the term D in eqn. (4.14). If *damping constant D' is small and negligible,* the galvanometer will be *critically damped* when :

$$\frac{G^2}{R} = \sqrt{4CJ} = 2\sqrt{CJ} \quad ...(4.30)$$

∴ Resistance R for *critical damping* is given by,

$$R_c = \frac{G^2}{2\sqrt{CJ}} \quad ...(4.31)$$

If R_g is the resistance of galvanometer, the external series resistance R_e required for *critical damping* is given as :

$$R_e = R_c - R_g = \frac{G^2}{2\sqrt{CJ}} - R_g \quad ...(4.32)$$

This external resistance required for critical damping is called **CDRX** *or* **ECDR**, *critical damping resistance external.*

4.7.1.6. Sensitivity of galvanometer

The sensitivity of a galvanometer may be defined as follows :

1. Current sensitivity.
2. Voltage sensitivity.
3. Megohm sensitivity.

1. ***Current sensitivity :***

— It is defined as the *deflection produced by unit current.*

i.e., Current sensitivity, $S_I = \frac{\theta}{I}$

But, $$\theta = \frac{GI}{C} = \frac{G}{C} \text{ rad./A} \quad ...(4.33)$$

— In case of galvanometer attached with a *mirror* it may be defined as *deflection in mm* produced on a scale *at a distance of one metre from the galvanometer mirror by unit current.*

i.e. $$\textit{Current sensitivity, } S_I = \frac{d}{I \times 10^6} \text{ mm per microampere} \quad ...(4.34)$$

where, d = The deflection on scale in mm.

— Sometimes the current density is defined as the *current required to cause a deflection of one scale division.*

2. **Voltage sensitivity :**

— It is defined in terms of *deflection in scale divisions caused by unit voltage applied across the galvanometer.*

— In case of a galvanometer attached with a *mirror*, the voltage sensitivity is defined *as the deflection produced in mm on a scale at a distance of 1 m from the mirror by a unit voltage impressed on the galvanometer.*

i.e., $$\textit{Voltage sensitivity, } S_V = \frac{d}{V \times 10^6} \text{ mm}/\mu\text{V} \quad ...(4.35)$$

3. **Megohm sensitivity :**

— It is defined as the *resistance required in MΩ in the circuit to cause a deflection of one scale division on impressing 1 V.*

— In case of galvanometer attached with a *mirror* the megohm sensitivity is defined as the *resistance in MΩ placed in series with the galvanometer through which one volt of impressed voltage will cause a deflection of 1 mm on a scale at a distance of 1m from the mirror.* This is *reicprocal of current density.*

• *A sensitive galvanometer is one which produces a large deflection for a small current.* Hence, *for a sensitive galvanometer the displacement constant G should be large and control constant C should be small.*

— As displacement constant G = *NBA*, therefore, the sensitivity can be increased by increasing the number of turns, using high flux density and by increasing the dimensions of the coil. For a given coil area *A* (= *lb*) and a constant value of flux density B, the value of G can be increased by *increasing the number of turns.* If the winding area is fixed, a large number of turns can only be wound if we use *thin wire.* Hence *for a sensitive galvanometer* the value of resistance of coil R_g is large.

— The sensitivity can also be *increased* by making the control constant C *small.*

4.7.1.7. Types of D.C. galvanometer

D'Arsonval type galvanometers, which are D.C. galvanometers, are made in three general types :

1. Portable point type.
2. Laboratory reflecting type.
3. Box type.

1. ***Portable point type :***

— These are small instruments, similar in construction to other types of portable instruments. They may be pivoted or may have taut suspension.

— This type of galvanometer has sensitivity from 0.5 to 4.0 mm/μA and has a period from 2.5 to 3 seconds.

— They have CDRX from 20 to 1800 ohms and coil resistance from 12 to 250 ohms.

- *Light spot and pointer D.C. galvanometer* has sensitivities from 50.4 to 252.0 mm/μA with period of 2 seconds.

It has CDRX from 100 to 3500 ohms and coil resistance from 100 to 1000 ohms.

2. ***Laboratory reflecting type :***

— These are highly sensitive instruments.

— Reflecting lamp and scale type have sensitivities from 250 to 100,000 mm/μA and period of 5 to 40 seconds.

— They have CRDX from 900 to 1000 kΩ and a coil resistance of 80 to 800 ohms.

3. ***Box type :***

— They have taut suspension and do not require exact levelling.

— The box includes the complete optical system and, therefore, it is not necessary to make separate mounting for galvanometer, lamp and scale etc. They are *self contained.*

— Their sensitivity ranges from 200 to 2000 mm/μA having a period of 2.5 to 3 seconds.

— Their CDRX is 400 to 25000 ohms having coil resistance of 25 to 550 ohms. The characteristics of different types of galvanometers are given in Table 4.1.

Table 4.1. Characteristic of Moving Coil Galvanometers

Type of galvanometer	*Sensitivity in mm/μA*	*Period in seconds*	*Resistance in ohms*	
			CDRX	*Coil*
Pointer	0.5	2.5	20	12
	1.0	3	110	25
	4.0	3	1800	250
Light spot and pointer	50.4	2	100	10
	2520	2	3500	1000
Reflecting lamp and scale	250	5	900	80
	2000	6	10,000	650
	10,000	14	22,000	500
	100,000	40	100,000	800
Box type self contained	200	2.5	400	25
	2000	3	2500	550

4.7.1.8. Galvanometer shunt

In bridge and potentiometer measurements, during preliminary balancing work, the potential across the galvanometer circuit may be high due to unbalance. If galvanometer with sufficient sensitivity for final balance is used; the large unbalance that probably exists at the initial stage is likely to *damage the coil of the suspension.* In addition, excessive currents that may flow, send the light spot so far off the scale that the *deflection is not readable* and experimenter does not know what to do. Therefore, it is *necessary to use protective parallel resistances.* The simplex procedure is to use :

(*i*) A *high resistance in series with the galvanometer* to limit the current and therefore the deflection produced by the *available e.m.f. under conditions of unbalances;*

(ii) A *low resistance connected across the terminals of the galvanometer* thus by-passing most of the currents. This reduces the sensitivity of galvanometer circuit so that the response of the instrument to a given current is reduced.

— A *low resistance shunt will heavily overdamp the* instrument so that its motion will be *sluggish.*

— For both series and shunt resistances the *time required* to attain the final deflection is *excessive.*

— As a *general rule, a combination of series and a shunt resistance is preferable which will give the desired reduction in sensitivity and at the same time the proper damping characteristics.*

Ayrton universal shunt :

— **Fig. 4.13** shows an arrangement in which a combination of series and parallel resistances is incorporated. The lower part of the resistance (*i.e. BC* below the sliding contact) is in *parallel*, the upper part of resistance (*i.e. AB* above the sliding contact) is in *series* with the galvanometer.

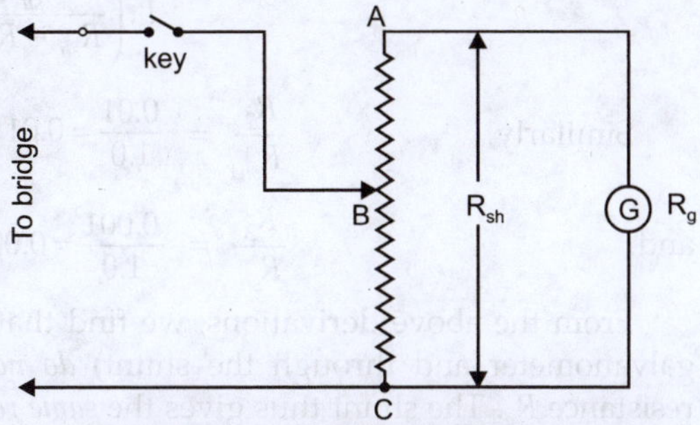

Fig. 4.13. Galvanometer shunt.

When the slider is near the lower end, a small portion of the resistance is in parallel with the galvanometer and the major part is in series with the galvanometer. The current divides itself in two parts, the major going through the shunt and only a small part through the galvanometer, thus protecting it and giving it small sensitivity.

When the sliding contact is moved up, the shunt resistance increases and the series resistance decreases. This increases the current through the galvanometer and thus the sensitivity is increased. But, there is a loss of sensitivity (as compared with galvanometer alone) with the sliding contact even at the top point, this loss, however, is not serious if R_{sh} is very high in comparison to galvanometer resistance R_g.

When key is *open* (see Fig. 4.13), there is closed circuit through the galvanometer coil and R_{sh} and this provides a damping circuit. R_{sh} should, accordingly be selected *to have a value approximately equal to that of critical damping resistance of the galvanometer* (somewhat higher is satisfactory) as the galvanometer response is excellent; if considerably too high, a fixed shunt can be added across R_{sh}. As CDRX is usually several times the coil resistance, the loss of sensitivity due to use of shunt of this value is *not serious.*

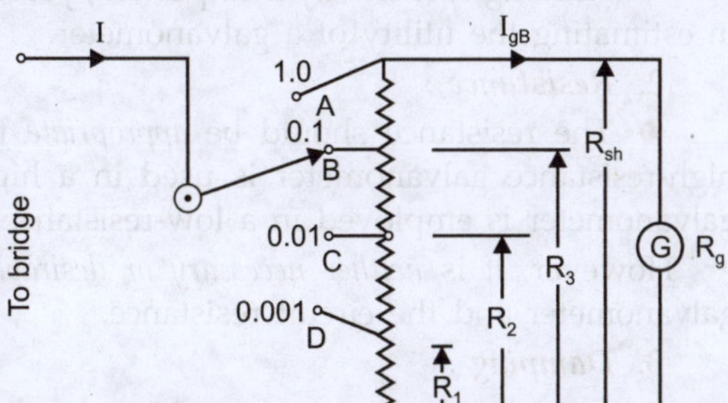

Fig. 4.14. Ayrton universal shunt.

— **Fig. 4.14** shows the construction of *"Ayrton universal shunt"*, using a dial type switch. The resistance of each shunt can be *proportioned* to give the desired result, as explained below :

When the contact is on stud *A* (or 1.0),

$$I_{gA} = I \frac{R_{sh}}{R_{sh} + R_g}, \quad \text{where } I \text{ is the total current}$$

When the contact is on stud B (or 0.1),

$$I_{gB} = I \cdot \frac{R_3}{R_3 + (R_{sh} - R_3 + R_g)} = I \cdot \frac{R_3}{R_{sh} + R_g}$$

Thus, we have,
$$\frac{I_{gB}}{I_{gA}} = \frac{I \cdot \left(\dfrac{R_3}{R_{sh} + R_g}\right)}{I \cdot \left(\dfrac{R_{sh}}{R_{sh} + R_g}\right)} = \frac{R_3}{R_{sh}} = \frac{0.1}{1.0} = 0.1$$

Similarly,
$$\frac{R_2}{R_{sh}} = \frac{0.01}{1.0} = 0.01$$

and,
$$\frac{R_3}{R_{sh}} = \frac{0.001}{1.0} = 0.001$$

From the above derivations, we find that the relative values of current (through the galvanometer and through the shunt) *do not depend* upon the value of galvanometer resistance R_g. The shunt thus gives the *same relative current values* for the various steps for all galvanometers although *not the same fraction of the total current*; therefore, it is called a *Universal shunt :*

- The *"Ayrton shunt" is very convenient in bridge measurements as the galvanometer sensitivity can be increased as bridge balance is approached.*

4.7.1.9. Selection of galvanometer

The following *features* should be considered while making choice of a galvanometer :

1. ***Sensitivity :***

- The sensitivity should be *sufficient* for the work in hand so that measurements may be made without undue fatigue and loss of time.
- Although sensitivity is important, yet it should not be the only feature considered in estimating the utility of a galvanometer.

2. ***Resistance :***

- The resistance should be *appropriate* for the measurement in hand. Usually a high-resistance galvanometer is used in a high-resistance circuit and a low-resistance galvanometer is employed in a low-resistance circuit.

However, it is *neither necessary or desirable* to have an exact matching between a galvanometer and the circuit resistance.

3. ***Damping :***

- For measurements of current and voltage, the damping should be *somewhat less than, or at most equal to critical, this will save time.*

In general, a galvanometer should be selected whose *critical damping resistance matches reasonably well with the resistance of the circuit* in which it is to be used.

4. ***Period :***

- In order that the instrument will respond quickly to the current, the time of vibration of the moving system should be *as short as possible.*

5. ***Promptness of response :***

- The promptness of response implies a *short period* and *critical damping.*

6. ***Freedom from mechanical disturbances :***

- The symmetric making and mounting the system about the axis of rotation contributes much to the system stability when it is subjected to mechanical disturbances.
- When the taut suspension galvanometers are properly balanced, they are, in general, *less expensive* than slack-suspension instruments to mechanical disturbances.

7. ***Freedom from stray-field effects :***

It is of significant importance that the indications of the instrument be not influenced by the unavoidable variations of the local field.

- The D'Arsonval galvanometer is usually relatively free from such trouble as its coil operates in a *strong field.* However, in case of instruments with weak operating field the use of an *astatic system* or of *magnetic shielding* in necessary.
- Moving magnet galvanometers are sensitive to external field and must be very carefully shielded.

8. ***Definiteness of zero reading :***

- The *zero reading* should be *definite.*
- The deflections should come promptly to their final values with no viscous actions of the suspension.

9. ***Law of deflection :***

- The deflection, throughout its useful range, should be proportional to the current.

10. ***Coil damping :***

- A suitable device should be provided for damping the coil and holding it rigidly in place during transport.
- A zero adjustment accessible from outside the case is essential.

11. *Visibility of suspended parts.*
12. *Accessibility for repairs.*
13. The *mirror* should be of the *smallest practical dimensions.*

Example 4.4. *Determine the angle through which coil turns when a deflection of 48 mm is observed on a scale of a galvanometer placed at a distance of 650 mm from the mirror.*

Solution. Refer to Fig. 4.11

Given: d = 48 mm or 0.048 m; r = 650 mm or 0.65 m.

The angle turned, θ :

The angle through which the mirror or coil turns may be found from the following expression :

$$d = r \times 2\theta$$

or,
$$\theta = \frac{d}{2r} = \frac{0.048}{2 \times 0.65} = 0.0369 \text{ rad. or } \mathbf{2.11^\circ} \quad \textbf{(Ans.)}$$

Example 4.5. *Determine the displacement constant of a suspended coil in a galvanometer of a moving coil vibrating type having 30 turns, 4.5 mm × 2.5 mm mean area and situated in a magnetic field of 0.8 tesla.*

Solution. *Given:* N = 30; A = 4.5 mm × 2.5 mm = 11.25 mm^2 or 11.25 × 10^{-6} m^2; B = 0.8 Wb/m^2

Displacement constant, G :

Displacement constant is given by :

$$G = NBA$$

$$= 30 \times 0.8 \times 11.25 \times 10^{-6} = \mathbf{270 \times 10^{-6}\ Nm/A} \quad \textbf{(Ans.)}$$

Example 4.6. *The suspended coil of a galvanometer has a 5 mm × 2.5 mm mean area and is situated in a magnetic field of 1.1 tesla. The moment of inertia of moving parts is 0.25 × 10^{-6} kg-m and the control string constant is 35 × 10^{-6} Nm/radian. If a current of 12 mA produces a deflection of 110°, calculate the number of turns of the suspended coil.*

Solution. *Given* : $A = 5 \text{ mm} \times 2.5 \text{ mm} = 5 \times 2.5 \times 10^{-6} \text{ m}^2$ or $12.5 \times 10^{-6} \text{ m}^2$;

$B = 1.1 \text{ Wb/m}^2$; $J = 0.25 \times 10^{-6}$ kg-m; $C = 35 \times 10^{-6}$ Nm/rad.; $\theta = 110°$ or 1.92 radians.

Number of turns of the suspended coil, N :

We know that for steady deflection,

$$GI = C\theta \quad \text{[Eqn. (4.11)]}$$

But,

$$G = NBA$$

$\therefore$

$$NBAI = C\theta$$

or,

$$N = \frac{C\theta}{BAI} = \frac{35 \times 10^{-6} \times 1.92}{1.1 \times 12.5 \times 10^{-6} \times 0.012} = 407 \quad \textbf{(Ans.)}$$

Example 4.7. *Fig. 4.15 shows a galvanometer system in which a wire of 0.12 mm in diameter is used as a torsion member. The wire is made of phosphor bronze for which modulus of rigidity is 40 GN/m². The polar moment of inertia for a round wire is $\frac{\pi}{32} d^4$, where d is the diameter of the wire.*

A collimated ray of light is reflected by the galvanometer mirror shown. The radius from the mirror to the scale is 180 mm.

Calculate the following :

(*i*) *Torsional constant;*

(*ii*) *Maximum allowable torque if the shear stress is limited to 100 MN/m²;*

(*iii*) *Sensitivity of the instrument in m/Nm;*

(*iv*) *Maximum possible scale deflection.*

(Roorkee University)

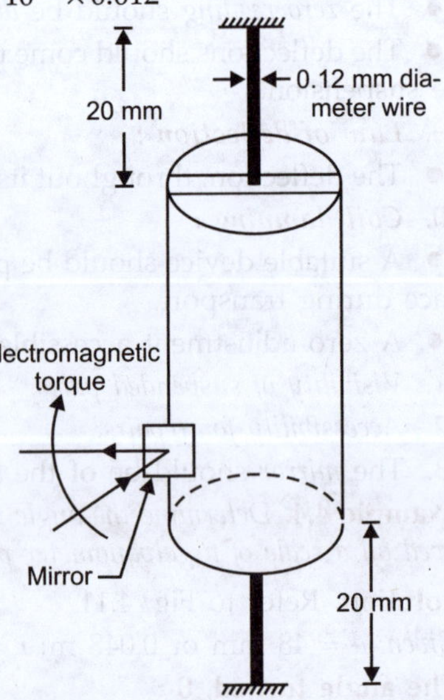

Fig. 4.15. A galvanometer moving system.

Solution. *Given* : $d = 0.12$ mm or 0.12×10^{-3} m; $l = 20$ mm $= 0.02$ m

G (modulus of rigidity) $= 40 \text{ GN/m}^2$; $r = 180$ mm or 0.18 m; τ (shear stress) $= 100 \text{ MN/m}^2$

(*i*) **Torsional constant, C :**

For a round wire :

$$\frac{T}{I_p} = \frac{\tau}{r} = \frac{G\theta}{l} \quad ...(i)$$

[where, I_p (Polar moment of inertia) $= \frac{\pi}{32} d^4$ ... *given*]

$\therefore$ Deflection per unit torque,

$$\frac{\theta}{T} = \frac{l}{GI_p}$$

Since there are two wires, one at the top and the other at the bottom, the deflection is half of that for single wire.

$$\therefore \quad \frac{\theta}{T} = \frac{l}{2GI_p} = \frac{l}{24\times(\pi/32)\,d^4} = \frac{0.02}{2\times(40\times10^9)\times(\pi/32)\times(0.12\times10^{-3})^4}$$

$$= 12280.5 \text{ rad./Nm}$$

Hence, torsional constant, $C = \frac{T}{\theta} = \frac{1}{12280.5} = \mathbf{81.43 \times 10^{-6}}$ **Nm/rad.** **(Ans.)**

(*ii*) **Maximum allowable torque, T :**

$$T = \frac{\tau I_p}{r} = \frac{\tau I_p}{(d/2)} = \frac{I\times(\pi/32)\,d^4}{(d/2)} = \frac{\tau\times\pi d^3}{16}$$

Since there are two wires, the maximum allowable torque is :

$$T = \frac{2\times\tau\times\pi d^3}{16} = \frac{\pi}{8}\tau d^3$$

$$= \frac{\pi}{8}\times(100\times10^6)\times(0.12\times10^{-3})^3$$

$$= \mathbf{67.86 \times 10^{-6}\ Nm} \quad \textbf{(Ans.)}$$

(*iii*) **Sensitivity of the instrument in m/Nm :**

The scale is kept at a radius of 0.18 m, and therefore the sensitivity of the scale is :

$$S = 2 \times (12280.5) \times 0.18 = \mathbf{4421\ m/Nm} \quad \textbf{(Ans.)}$$

A factor of 2 has been introduced because if mirror turns through an angle θ, the collimated light turns through 2θ.

(*iv*) **Maximum possible scale deflection :**

The deflection for maximum torque = $4421 \times 67.86 \times 10^{-6}$ m

= 0.3 m or **300 mm** **(Ans.)**

Example 4.8. *The coil of a moving coil galvanometer has 300 turns and is suspended in a uniform magnetic field of 0.1 Wb/m². The control constant is 0.2 × 10⁻⁶ Nm per radian. The coil is 2 cm wide and 2.5 cm high with a moment of inertia of 0.15 × 10⁻⁶ kg m². If the galvanometer resistance in 250 Ω calculate the value of resistance which when connected across the galvanometer terminals will give critical damping, assuming damping to be entirely electromagnetic.*

(Punjab University)

Solution. *Given :* $N = 300$; $B = 0.1$ Wb/m²; $C = 0.2 \times 10^{-6}$ Nm/rad; $b = 2$ cm or 0.02 m; $l = 2.5$ cm or 0.025 m; $J = 0.15 \times 10^{-6}$ kg m²; $R_g = 250\ \Omega$.

External resistance for critical damping, CDRX :

Total resistance required for critical damping,

$$R_c = \frac{G^2}{2\sqrt{CJ}} \qquad \text{[Eqn. (4.31)]}$$

or,
$$R_c = \frac{(NBA)^2}{2\sqrt{CJ}} \qquad (\because\ G = NBA)$$

$$= \frac{[300\times0.1\times(0.025\times0.02)]^2}{2\sqrt{0.2\times10^{-6}\times0.15\times10^{-6}}} = 650\ \Omega$$

∴ External resistance required for critical damping,

$$CDRX = R_c - R_g = 650 - 250 = \mathbf{400\ \Omega} \quad \textbf{(Ans.)}$$

Example 4.9. *The coil of a moving coil galvanometer is wound on a non-magnetic former whose height and width are both 25 mm. It moves in a constant field of 0.1 Wb/m^2. The moment of inertia of its moving parts is 0.3 × 10^{-6} kg-m^2 and the control spring constant is 32 × 10^{-6} Nm/rad.*

Calculate :

(i) The number of turns that must be wound on the former to produce a deflection of 140° with a current of 12 mA.

(ii) The resistance of coil to produce critical damping, all damping being assumed as electromagnetic. **(Bangalore University)**

Solution. *Given :* b = 25 mm or 0.025 m; B = 0.1 Wb/m^2; J = 0.3 × 10^{-6} kg-m^2;

$C = 32 \times 10^{-6}$ Nm/rad.; $\theta = 140° = 140 \times \frac{\pi}{180} = 2.44$ rad.; I = 12 mA or 0.012 A.

(i) **Number of turns, N :**

For steady deflection, $GI = C\theta$ [Eqn. (4.11)]

$\therefore$ Displacement constant, $G = \frac{C\theta}{I}$

$$= \frac{32 \times 10^{-6} \times 2.44}{0.012} = 6.5 \times 10^{-3} \text{ Nm/A}$$

Also, $G = NBlb$

$$\therefore \quad N = \frac{G}{Blb} = \frac{6.5 \times 10^{-3}}{0.1 \times 0.025 \times 0.025} = \textbf{104} \quad \textbf{(Ans.)}$$

(ii) **Resistance of galvanometer for critical damping, R_c :**

We know that, $$R_c = \frac{G^2}{2\sqrt{CJ}} = \frac{(6.5 \times 10^{-3})^2}{2\sqrt{32 \times 10^{-6} \times 0.3 \times 10^{-6}}} = \mathbf{6.82\ \Omega} \quad \textbf{(Ans.)}$$

Example 4.10. *In a moving coil galvanometer with free time period of 2.5 seconds, a current of 2.4 mA gives a deflection of 250 mm on a scale at a distance of 1.4 m. The moving system has a moment of inertia of 3.8 × 10^{-6} kg-m^2.*

Determine the total resistance in the galvanometer circuit for critical damping.

Solution. *Given :* T_0 (Free time period) = 2.5 s; I = 2.4 μA; d = 250 mm or 0.25 m; r = 1.4 m; J = 3.8 × 10^{-6} kg-m^2.

Total resistance in the galvanometer circuit for critical damping, R_c :

We know that, $$R_c = \frac{G^2}{2\sqrt{CJ}} \quad ...(i)$$

For steady deflection : $GI = C\theta$

or, $$G = \frac{C\theta}{I} = \frac{C \times (d/2r)}{I} \quad \left(\because \theta = \frac{d}{2r}\right)$$

or, $$G = \frac{C \times [0.25/(2 \times 1.4)]}{2.4 \times 10^{-6}} = 3.72 \times 10^4 C$$

Also, free (or natural) time period

$$T_n = 2\pi\sqrt{\frac{J}{C}}$$

or, $$C = \left(\frac{2\pi}{T_n}\right)^2 J = \left(\frac{2\pi}{2.5}\right)^2 \times 3.8\times 10^{-6} = 24 \times 10^{-6} \text{ Nm/rad.}$$

Inserting the values in eqn. (*i*), we get

$$R_c = \frac{(3.72\times 10^4 \times 24\times 10^{-6})^2}{2\sqrt{(24\times 10^{-6})\times(3.8\times 10^{-6})}} = \mathbf{41733\ \Omega} \quad \textbf{(Ans.)}$$

Example 4.11. *A moving coil galvanometer deflects 220 mm on a scale at a distance of 1.2 m from the mirror when a current of 2.2 μA passes through it. The free time period of the galvanometer is 3.8 seconds. The galvanometer is dead beat when the total resistance in the galvanometer is 19500 Ω.*

Determine the moment of inertia of the galvanometer moving system.

Solution. *Given* : d = 220 mm or 0.22 m; r = 1.2 m; I = 2.2 μA; T_0 = 3.8 s; R_c = 19500 Ω

Moment of inertia of the system, *J* :

For steady deflection : $$GI = C\theta$$

or, $$G = \frac{C\theta}{I} = \frac{C\times(d/2r)}{I}$$

or, $$G = \frac{C\times[0.22/(2\times 1.2)]}{2.2\times 10^{-6}} = 4.167 \times 10^4\ C$$

Also, free time period, $$T_n = 2\pi\sqrt{\frac{J}{C}}$$

or, $$3.8 = 2\pi\sqrt{\frac{J}{C}}$$

or, $$J = \left(\frac{3.8}{2\pi}\right)^2 \times C = 0.3657\ C$$

Critical damping resistance, $$R_c = \frac{G^2}{2\sqrt{CJ}}$$

or, $$19500 = \frac{(4.167\times 10^4 C)^2}{2\sqrt{C\times(0.3657C)}} = \frac{(4.167\times 10^4)^2\ C^2}{2\sqrt{0.3657}\times C}$$

or, $$C = \frac{19500\times 2\sqrt{0.3657}}{(4.167\times 10^4)}$$

$$= 0.1358 \times 10^{-4} \text{ Nm/rad.}$$

and, $$J = 0.3657\ C = 0.3657 \times 0.1358 \times 10^{-4}$$

$$= \mathbf{4.966 \times 10^{-6}\ kg\text{-}m^2} \quad \textbf{(Ans.)}$$

Example 4.12. *The following data relate to a moving coil galvanometer which has a former of a non-conducting material :*

The current sensitivity of the instrument = 0.012 μA/mm at 1 m

The period of undamped oscillation = 5.5 seconds

The displacement constant of the instrument = 4.8 × 10^{-3} Nm/A

Calculate :

(i) Control constant.

(ii) Inertia constant.

(iii) Total circuit resistance for instrument to be dead beat.

Solution. *Given :* Current sensitivity, S_I = 0.0012 μA/mm at 1 m; T_n = 5.5 s; $G = 4.8 \times 10^{-3}$ Nm/A.

(i) Control constant, C :

Suppose that the instrument is reflecting type. Therefore, if the final steady deflection is θ, the spot moves through an angle 2θ. The spot moves a distance 1 mm at 1 m with a current of 0.0012 μA or in other words a current of 0.0012 μA deflects the instrument through,

$$\theta = \frac{d}{2r} = \frac{(1/1000)}{2 \times 1} = 0.5 \times 10^{-3} \text{ rad.}$$

∴ Control constant,

$$C = \frac{GI}{\theta} = \frac{4.8 \times 10^{-3} \times 0.0012 \times 10^{-6}}{0.5 \times 10^{-3}}$$

$$= \mathbf{11.52 \times 10^{-9} \text{ Nm/rad.}} \quad \textbf{(Ans.)}$$

(ii) Inertia constant, *J* :

Undamped (free or natural time) period,

$$T_n = 2\pi\sqrt{\frac{J}{C}}$$

or, Inertia constant,

$$J = \frac{CT_n^2}{4\pi^2} = \frac{11.52 \times 10^{-9} \times (5.5)^2}{4\pi^2}$$

$$= \mathbf{8.827 \times 10^{-9} \text{ kg-m}^2} \quad \textbf{(Ans.)}$$

(iii) Total circuit resistance for critical damping, R_c :

Total circuit resistance for critical damping or dead beat,

$$R_c = \frac{G^2}{2\sqrt{CJ}} = \frac{(4.8 \times 10^{-3})^2}{2\sqrt{11.52 \times 10^{-9} \times 8.827 \times 10^{-9}}}$$

$$= \mathbf{1142.4 \ \Omega} \quad \textbf{(Ans.)}$$

Example 4.13. *A d'Arsonval galvanometer has a rectangular coil wound on an aluminium former of resistivity 30 × 10^{-9} Ωm. The sides of the former, each of effective length 35 mm, move in a radial field of uniform flux density 0.12 Wb/m^2 against a control torque of 20 × 10^{-6} Nm/rad. The width of the former is 25 mm, the ends being effectively outside the magnetic field, and the moment of inertia of the moving system is 90 × 10^{-9} kg-m^2. Calculate the cross-sectional area of the aluminium former if it is to provide critical damping, other sources of damping being neglected.* **(Nagpur University)**

Solution. *Given :* $r = 30 \times 10^{-9}$ Ωm; l = 35 mm or 0.035 m; b = 0.025 m; B = 0.12 Wb/m^2; $J = 90 \times 10^{-9}$ kg-m^2; control torque, $C = 20 \times 10^{-6}$ Nm/rad.

Cross-sectional area of aluminium former, *a* :

Total damping factor (D_f) in the absence of all other damping except the damping produced by *former* is given by :

$$D_f = \frac{G^2}{N^2 R_f}$$

$$= \frac{N^2 B^2 l^2 b^2}{N^2 R_f} = \frac{B^2 l^2 b^2}{R_f}$$

For critical damping, $D = D_c = 2\sqrt{CJ}$ [Eqn. (4.18)]

$$\therefore \quad \frac{B^2 l^2 b^2}{R_f} = 2\sqrt{CJ}$$

Hence, resistance of former, $R_f = \frac{B^2 l^2 b^2}{2\sqrt{CJ}}$

$$= \frac{(0.12)^2 \times (0.035)^2 \times (0.025)^2}{2\sqrt{20 \times 10^{-6} \times 90 \times 10^{-9}}} = 4108.8 \times 10^{-6}\ \Omega$$

Length of eddy currents, $L = 2(l + b)$

$$= 2(0.035 + 0.025) = 0.12 \text{ m}$$

Now, $R_f = \frac{\rho L}{a} = \frac{30 \times 10^{-9} \times 0.12}{a} = 4108.8 \times 10^{-6}\ \Omega$

$$\therefore \quad a = \frac{30 \times 10^{-9} \times 0.12}{4108.8 \times 10^{-6}} = 0.876 \times 10^{-6} \text{ m}^2 \text{ or } \mathbf{0.876 \text{ mm}^2} \quad \textbf{(Ans.)}$$

Example 4.14. *A moving coil galvanometer has the following data :*

Number of turns = *320*

Flux density = *0.12* Wb/m^2

Control constant = *0.12* × 10^{-6} *Nm/rad.*

Coil size = *21 mm* × *22 mm*

Moment of inertia = *0.2* × 10^{-6} $kg\text{-}m^2$

Galvanometer resistance (including external resistance) = *2100* Ω

Determine :

(*i*) *Total resistance for critical damping.*

(*ii*) *Damping ratio.*

(*iii*) *Frequency of undamped oscillations.*

(*iv*) *Frequency of damped oscillations.*

Solution. *Given:* $N = 320$; $B = 0.12$ Wb/m²; $C = 0.12 \times 10^{-6}$ Nm/rad.; $A = 21 \times 22 \times 10^{-6} = 4.62 \times 10^{-4}$ m²; $J = 0.2 \times 10^{-6}$ kg-m²; $R_{actual} = 2100\ \Omega$.

(*i*) Total resistance for critical damping, R_c :

$$R_c = \frac{G^2}{2\sqrt{CJ}} = \frac{(NBA)^2}{2\sqrt{CJ}}$$

$$= \frac{[320 \times 0.12 \times (4.62 \times 10^{-4})]^2}{2\sqrt{0.12 \times 10^{-6} \times 0.2 \times 10^{-6}}} = \mathbf{1015.8\ \Omega} \quad \textbf{(Ans.)}$$

(*ii*) **Damping ratio, γ :**

$$\gamma = \frac{D}{D_{critical}}$$

$$= \frac{G^2/R_{actual}}{G^2/R_{critical}} = \frac{R_{critical}}{R_{actual}} = \frac{1015.8}{2100} = \mathbf{0.4837} \quad \textbf{(Ans.)}$$

(*iii*) **Frequency of undamped oscillations, f_n :**

Frequency of undamped oscillations, f_n is given by

$$\omega_n = \sqrt{\frac{C}{J}} = \sqrt{\frac{0.12 \times 10^{-6}}{0.2 \times 10^{-6}}} = 0.7746$$

$$\therefore \quad f_n = \frac{0.7746}{2\pi} = \mathbf{0.1233\ Hz} \quad \textbf{(Ans.)}$$

(*iv*) **Frequency of damped oscillations, f_d :**

$$f_d = f_n\sqrt{1-\gamma^2} \qquad \text{[Eqn. 4.22(a)]}$$

$$= 0.1233\sqrt{1-(0.4837)^2}$$

$$= \mathbf{0.1079\ Hz} \quad \textbf{(Ans.)}$$

Example 4.15. *The following data relate to a moving coil galvanometer :*

Number of turns of the coil	*= 340*
Flux density in the magnetic field	*= 0.12 Wb/m²*
Control constant	*= 0.22 × 10^{-6} Nm/rad.*
Dimensions of the coil	*= 25 mm (wide) × 30 mm (high)*
Moment of inertia	*= 0.2 × 10^{-6} kg-m²*
Galvanometer resistance	*= 350 Ω*
Resistance connected in series with the galvanometer	*= 3150 Ω*

Determine :

(*i*) *The resistance to be connected to galvanometer for critical damping.*
(*ii*) *Relative damping.*
(*iii*) *Logarithmic decrement.*
(*iv*) *Frequency of free oscillations.*
(*v*) *First maximum deflection.*
(*vi*) *Periods of undamped and damped oscillations.* **(Roorkee University)**

Solution. *Given:* $N = 340$; $B = 0.12$ Wb/m²; $C = 0.22 \times 10^{-6}$ Nm/rad.; $A = 25$ mm × 30 mm = 750 mm² or 0.00075 m²; $J = 0.2 \times 10^{-6}$ kg-m²; $R_g = 350\ \Omega$; Resistance connected in series with galvanometer = 3150 Ω.

(*i*) **External resistance required for critical damping, CDRX :**

Total resistance required for critical damping

$$R_c = \frac{G^2}{2\sqrt{CI}} = \frac{(NBA)^2}{2\sqrt{CJ}} = \frac{(340 \times 0.12 \times 0.00075)^2}{2\sqrt{0.22 \times 10^{-6} \times 0.2 \times 10^{-6}}} = 2232\ \Omega$$

$$\therefore \quad CDRX = R_c - R_g = 2232 - 350 = \mathbf{1882\ \Omega} \quad \textbf{(Ans.)}$$

(*ii*) **Relative damping, γ :**

$$\text{Damping, } D = \frac{G^2}{R_{actual}} = \frac{G^2}{(350 + 3150)} = \frac{G^2}{3500}$$

$$\text{Critical damping, } D_c = \frac{G^2}{R_c} = \frac{G^2}{2232}$$

$$\therefore \quad \gamma = \frac{D}{D_c} = \frac{(G^2/3500)}{(G^2/2232)} = \frac{2232}{3500} = \mathbf{0.638} \quad \textbf{(Ans.)}$$

(*iii*) Logarithmic decrement, λ :

Logarithmic decrement λ is given by :

$$\lambda = \frac{\pi\gamma}{\sqrt{1-\gamma^2}} = \frac{\pi \times 0.638}{\sqrt{1-(0.638)^2}} = \mathbf{2.6} \quad \textbf{(Ans.)}$$

(*iv*) Frequency of free oscillations, f_n :

$$f_n = \frac{1}{2\pi}\sqrt{\frac{C}{J}} = \frac{1}{2\pi}\sqrt{\frac{0.22 \times 10^{-6}}{0.2 \times 10^{-6}}} = \mathbf{0.167\ Hz} \quad \textbf{(Ans.)}$$

(*v*) First maximum deflection, θ_1 :

First maximum deflection θ_1, in terms of steady deflection θ_f, is given by :

$$\theta_1 = \theta_f \left[1 + \exp.\left(-\frac{\pi\gamma}{\sqrt{1-\gamma^2}}\right)\right]$$

$$= \theta_f \left[1 + \exp.\left(-\frac{\pi \times 0.638}{\sqrt{1-0.638^2}}\right)\right] = \mathbf{1.074\ \theta_f} \quad \textbf{(Ans.)}$$

(*vi*) Periods of undamped and damped oscillations, T_n, T_d :

Frequency of damped oscillations, f_d :

$$f_d = f_n\sqrt{1-\gamma^2} \qquad \text{[Eqn. 4.22(a)]}$$

$$= 0.167\sqrt{1-(0.638)^2} = \mathbf{0.128\ Hz} \quad \textbf{(Ans.)}$$

$$\therefore \quad \text{Period of undamped oscillations, } T_n = \frac{1}{f_n} = \frac{1}{0.167} = \mathbf{5.98\ s} \quad \textbf{(Ans.)}$$

$$\text{Period of damped oscillations, } T_d = \frac{1}{f_l} = \frac{1}{0.128} = \mathbf{7.81\ s} \quad \textbf{(Ans.)}$$

Example 4.16. *The following data relate to a moving coil galvanometer :*

Number of turns of the coil	= 300
Flux density in the uniform magnetic field	= 0.06 Wb/m^2
Torsion constant	= 1.6 × 10^{-7} Nm/radian
Dimensions of the coil	= 20 mm (width) × 25 mm (height)
Moment of inertia	= 1.7 × 10^{-7} $kg\text{-}m^2$
Galvanometer resistance	= 250 Ω
Total damping	= 4.5 × 10^{-8} Nm/radian/second.

Determine :

(*i*) *The deflection of galvanometer in radian, and in mm when a current of 3 μA is flowing through the galvanometer, the scale is at a distance of 1.6 m from the mirror.*

(ii) *The current sensitivity.*
(iii) *The voltage sensitivity.*
(iv) *The megohm sensitivity.*
(v) *Frequency of oscillations.*
(vi) *Relative damping.*
(vii) *Logarithmic decrement.* **(Delhi University)**

Solution. *Given :* $N = 300$; $B = 0.06$ Wb/m^2; $C = 1.6 \times 10^{-7}$ Nm/radian; $A = 20 \times 10^{-3} \times 25 \times 10^{-3} = 0.0005$ m^2; $J = 1.7 \times 10^{-7}$ kg-m^2; $R_g = 250\ \Omega$; Total damping, $D = 4.5 \times 10^{-8}$ Nm/rad./s; $I = 3\ \mu$A; $r = 1.6$ m.

(i) **The deflection of galvanometer :**

The deflection of galvanometer in radian :

$$\theta = \frac{GI}{C} = \frac{(NBA)\,I}{C}$$

$$= \frac{300 \times 0.06 \times 0.0005 \times 3 \times 10^{-6}}{1.6 \times 10^{-7}} = \mathbf{0.1687\ rad.}\quad \textbf{(Ans.)}$$

The deflection of galvanometer in *mm*:

$$d = r \times 2\theta = 1.6 \times 2 \times 0.1687 \times 1000\ \text{mm} = \mathbf{539.8\ mm}\quad \textbf{(Ans.)}$$

(ii) **The current sensitivity, S_I :**

$$S_I = \frac{d \text{ in mm}}{I \text{ in } \mu\text{A}} = \frac{539.8}{3} = \mathbf{179.9\ mm/\mu A}\quad \textbf{(Ans.)}$$

(iii) **Voltage sensitivity, S_V :**

$$S_V = \frac{d \text{ in mm}}{V \text{ in microvolts}} = \frac{d \text{ in mm}}{R \times I \text{ in } \mu\text{A}}$$

$$= \frac{539.8}{250 \times 3} = \mathbf{0.7197\ mm/\mu V}\quad \textbf{(Ans.)}$$

(iv) **The megohm sensitivity :**

Current required for a deflection of 1 mm $= \frac{1}{179.9}\ \mu$A

$$\therefore \quad \text{Megohm sensitivity} = \frac{1}{\frac{1}{179.9} \times 10^{-6}} = \mathbf{179.9\ M\Omega/mm}\quad \textbf{(Ans.)}$$

(v) **Frequency of oscillations :**

$$\text{Frequency of oscillations} = \frac{1}{2\pi}\sqrt{\frac{C}{J} - \frac{D^2}{4J^2}}$$

$$= \frac{1}{2\pi}\sqrt{\frac{1.6 \times 10^{-7}}{1.7 \times 10^{-7}} - \frac{(4.5 \times 10^{-8})^2}{4(1.7 \times 10^{-7})^2}}$$

$$= \frac{1}{2\pi}\sqrt{0.9412 - 0.0175} = \mathbf{0.153\ Hz}\quad \textbf{(Ans.)}$$

(vi) **Relative damping, γ :**

$$\gamma = \frac{D}{\sqrt{2JC}} \qquad \text{...[Eqn. 4.15(a)]}$$

$$= \frac{4.5 \times 10^{-8}}{\sqrt{2 \times 1.7 \times 10^{-7} \times 1.6 \times 10^{-7}}} = \mathbf{0.193} \quad \textbf{(Ans.)}$$

(viii) **Logarithmic increment, λ :**

Logarithmic increment is given by :

$$\lambda = \frac{\pi\gamma}{\sqrt{1-\gamma^2}} = \frac{\pi \times 0.193}{\sqrt{1-(0.193)^2}} = \mathbf{0.618} \quad \textbf{(Ans.)}$$

Example 4.17. *The following data relate to a panel type PMMC instrument which has a full scale deflection 90° = 100 scale divisions, with a current of 1.2 mA.*

The period of oscillations = 0.6 s

In order to measure the spring constant, a small weight having a gravitational force of 98.1 × 10^6 N is placed at a distance of 110 mm from the horizontal axis of rotation with the horizontal pointer acting as lever arm; the resultant deflection = 40 divisions.

The first maximum deflection, when a current of 1.2 mA is suddenly passed through the meter = 115 divisions

The value of flux density in the air gap measured with a Hall probe = 0.26 Wb/m^2

The length and average diameter of the coil = 18 mm and 16 mm respectively.

Determine the following :

(i) Spring constant.

(ii) Moment of inertia.

(iii) Number of turns.

(iv) Damping ratio.

(v) Equivalent resistance of the short-circuited winding represented by the damping action of the former of the coil.

Solution. *Given* : Full scale deflection = 90° = 100 scale divisions, or, 1 scale division = 0.9° = 0.0157 rad.; I = 1.2 mA or 0.0012 A; T_n = 0.6 s; B = 0.26 Wb/m^2; l = 18 mm or 0.018 m; d (or b) = 16 mm or 0.016 m.

(i) **Spring constant, C :**

Restoring torque, $T_c = 98.1 \times 10^{-6} \times (110 \times 10^{-3}) = 10.79 \times 10^{-6}$ Nm

Deflection, θ = 40 div. = 40 × 0.0157 rad. = 0.628 rad.

$\therefore$ Spring constant, $C = \frac{T_c}{\theta} = \frac{10.79 \times 10^{-6}}{0.628} = \mathbf{17.18 \times 10^{-6}}$ **Nm/rad. (Ans.)**

(ii) **Moment of inertia, J :**

Period of free oscillations $T_0 = 2\pi\sqrt{\frac{J}{C}}$

$$\therefore \quad J = \frac{CT_n^2}{4\pi^2} = \frac{17.18 \times 10^{-6} \times (0.6)^2}{4\pi^2}$$

$$= \mathbf{0.1567 \times 10^{-6}\ kg\text{-}m^2} \quad \textbf{(Ans.)}$$

(iii) **Number of turns, N :**

Control torque for 40 scale divisions = 10.79 × 10^{-6} Nm (calculated above)

∴ Control torque for full scale deflection (100 div.),

$$T_c = 10.79 \times 10^{-6} \times \frac{100}{40} = 26.97 \times 10^{-6} \text{ Nm}$$

$\therefore$ Deflecting torque at full scale,

$$T_d = NBlbI = N \times 0.26 \times 0.018 \times 0.016 \times 0.0012 = 89.86 \text{ N} \times 10^{-9} \text{ Nm}$$

At equilibrium : $T_d = T_c$

or, $89.86 \text{ N} \times 10^{-9} = 26.97 \times 10^{-6}$

$\therefore$ Number of turns, $N = \dfrac{26.97 \times 10^{-6}}{89.86 \times 10^{-9}} = \mathbf{300}$ **(Ans.)**

(*iv*) **Damping ratio, γ :**

First maximum deflection, θ_1 = 115 division = 1.15 θ_f

(where θ_f = final steady deflection)

But, the value of θ_1 in terms of θ_f can be expressed as :

$$\theta_1 = \theta_f \left[1 + \exp.\left(-\frac{\pi\gamma}{\sqrt{1-\gamma^2}}\right)\right]$$

or, $$\frac{\theta_1}{\theta_f} = 1 + \exp.\left(-\frac{\pi\gamma}{\sqrt{1-\gamma^2}}\right)$$

or, $$1.15 = 1 + \exp.\left(-\frac{\pi\gamma}{\sqrt{1-\gamma^2}}\right)$$

or, $$\exp.\left(-\frac{\pi\gamma}{\sqrt{1-\gamma^2}}\right) = 0.15$$

or, $$\frac{\pi\gamma}{\sqrt{1-\gamma^2}} = 1.897$$

or, $$\frac{\gamma}{\sqrt{1-\gamma^2}} = \frac{1.897}{\pi} = 0.604$$

Squaring both sides, we get

$$\gamma^2 = 0.604^2 (1 - \gamma^2)$$

or, $$\gamma^2 (1 + 0.604^2) = 0.604^2$$

or, $$\gamma = \left(\frac{0.604^2}{1+0.604^2}\right)^{1/2} = \mathbf{0.517}$$ **(Ans.)**

(*v*) **Damping constant, *D* :**

We know that, $$\gamma = \frac{D}{2\sqrt{CJ}}$$ [Eqn. (4.18)]

or, $$D = \gamma \times 2\sqrt{CJ} = 0.517 \times 2 \times \sqrt{17.18 \times 10^{-6} \times 0.1567 \times 10^{-6}}$$

$$= \mathbf{1.696 \times 10^{-6}} \textbf{ Nm/rad. s}^{-1}$$ **(Ans.)**

(*vi*) **Resistance of former, R_f :**

Considering the former to be a single-turn coil, the damping constant is,

$$D_{\text{former}} = \frac{(Blb)^2}{R_f}$$

$$\therefore \quad R_f = \frac{(Blb)^2}{D_{former}} = \frac{(0.26 \times 0.018 \times 0.016)^2}{1.696 \times 10^{-6}} = \mathbf{3.3 \times 10^{-3}\ \Omega} \quad \textbf{(Ans.)}$$

Example 4.18. *The deflections of a galvanometer with no current and a steady state current of 6 μA are zero and 80 mm respectively. Its first maximum deflection, after a step function of voltage which produces a steady state current is applied, is 115 mm. The maximum deflection in the next cycle is 90 mm.*

Determine : (*i*) *Current sensitivity.*
(*ii*) *Logarithmic decrement.*
(*iii*) *Relative damping.*

Solution. *Given* : $I = 6\ \mu A$; $\theta_f = 80$ mm ; $\theta_1 = 115$ mm ; $\theta_3 = 90$ mm

(*i*) **Current sensitivity, S_I :**

$$S_I = \frac{\theta_f}{I} = \frac{80}{6} = \mathbf{13.33\ mm/\mu A} \quad \textbf{(Ans.)}$$

(*ii*) **Logarithmic decrement, λ :**

First maximum deflection θ_1 is given by :

$$\theta_1 = \theta_f \left[1 + \exp\left(-\frac{\pi\gamma}{\sqrt{1-\gamma^2}} \right) \right] \quad ...(i)$$

Similarly, *second maximum* deflection θ_3 is given by :

$$\theta_3 = \theta_f \left[1 + \exp\left(-\frac{3\pi\gamma}{\sqrt{1-\gamma^2}} \right) \right] \quad ...(ii)$$

Hence, from (*i*) and (*ii*), we have :

$$\frac{\theta_1 - \theta_f}{\theta_3 - \theta_f} = \exp\left(\frac{2\pi\gamma}{\sqrt{1-\gamma^2}} \right)$$

But logarithmic decrement, λ is given by :

$$\lambda = \frac{\pi\gamma}{\sqrt{1-\gamma^2}}$$

$$\therefore \quad \frac{\theta_1 - \theta_f}{\theta_3 - \theta_f} = \exp(2\lambda)$$

or, $$\frac{115-80}{90-80} = \exp(2\lambda) \quad \text{or,} \quad 3.5 = \exp(2\gamma)$$

or $$\lambda = \mathbf{0.626} \quad \textbf{(Ans.)}$$

(*iii*) **Relative damping, γ :**

$$0.626 = \frac{\pi\gamma}{\sqrt{1-\gamma^2}}, \text{ or, } \frac{\gamma}{\sqrt{1-\gamma^2}} = \frac{0.626}{\pi} = 0.1993$$

or, $\gamma^2 = (1 - \gamma^2) \times 0.1993^2$ ...squaring both sides

or, $\gamma^2(1 + 0.1993^2) = 0.1993^2$

or, $$\gamma = \left(\frac{0.1993^2}{1+0.1993^2}\right) = 0.195$$ **(Ans.)**

Example 4.19. *The Ayrton universal shunt has a total resistance of 9000 Ω and galvanometer has a resistance 3000 Ω. Determine the multiplying power of shunt for 1000 Ω, 3000 Ω, 6000 Ω tapping.*

Solution. Total shunt resistance, $R_{sh} = 9000\ \Omega$

Galvanometer resistance, $R_g = 3000\ \Omega$

$$\text{Multiplying power of shunt, } N = \frac{I}{I_g} = \frac{R_{sh} + R_g}{R_1}$$

∴ (*i*) $$N = \frac{R_{sh} + R_g}{R_1} = \frac{9000 + 3000}{1000} = 12$$ **(Ans.)**

...when $R_1 = 1000\ \Omega$

(*ii*) $$N = \frac{9000 + 3000}{3000} = 4$$ **(Ans.)**

...when $R_1 = 3000\ \Omega$

(*iii*) $$N = \frac{9000 + 3000}{6000} = 2$$ **(Ans.)**

...when $R_1 = 6000\ \Omega$

4.7.2. Ballistic Galvanometer

4.7.2.1. Construction of ballistic galvanometer

A ballistic galvanometer is also a *permanent magnet moving coil instrument and has D'Arsonval type construction.*

- *It is used to measure quantity of electricity, i.e., charge passing through it in a given time. Thus, it is an integrating instrument and is used extensively in **magnetic measurement.***

The quantity of electricity is the result of an e.m.f. induced instantaneously in a *search coil* connected to galvanometer terminals when the magnetic flux linking the coil is changed. *The 'charge' is proportional to the e.m.f. induced in the search coil.* By noting the maximum value of the deflection, the amount of charge that has passed through the galvanometer is measured. The *proportionality* between the deflection and amount of charge is *valid only* if the whole of charge has passed through the galvanometer before any appreciable deflection of the moving system occurs. Therefore, this instrument has a *large inertia.* Very often, *additional weights* are added to the moving system so that *inertia is high as compared to the restoring moment due to the suspension.* Therefore, natural period of vibration is *large,* about 10 to 15 seconds.

The damping is *small so that first deflection* (which indicates the total charge) *is large.*

- It is desirable that the instrument is *dead beat.* The galvanometer is short circuited by a key so that the moving system is brought to rest immediately.
  - — The *damping is only electromagnetic.*
  - — Air friction damping is *avoided because it is indeterminate.* Electromagnetic can be determined from the *constants of the instrument.*
- It is necessary that the moving coil is free from magnetic material.

- The terminals, coil and internal connections are of *copper* so as to *avoid thermoelectric effects at the junctions.*
- In the best ballistic galvanometer, the suspension is *non-conducting*, the *current being led into the coil by delicate spiral of very thin copper strip.*

4.7.2.2. Theory of ballistic galvanometer

In a ballistic galvanometer, as discussed earlier, the coil moves after the charge to be measured has passed through it. Obviously, during the movement of the coil, there is no current flowing through it. Hence the equation of motion is :

$$J\frac{d^2\theta}{dt^2} + D\frac{d\theta}{dt} + C\theta = 0 \qquad \text{...(4.36)}$$

where, J = Moment of inertia,

D = Damping constant, and

C = Controlling or restoring constant.

Since *damping is extremely small,* the approximate solution of the eqn. (4.36) is :

$$\theta = U_{exp.}\,[-(D/2J)t]\,\sin(\omega_0 t + \phi) \qquad \text{...(4.37)}$$

(where ω_0 (= ω_n) is the natural frequency of oscillations)

At the start of motion, when $t = 0$, $\theta = 0$ and hence $\phi = 0$.

$$\therefore \qquad \theta = U_{exp.}\,[-(D/2J)t]\,\sin\omega_0 t \qquad \text{...(4.38)}$$

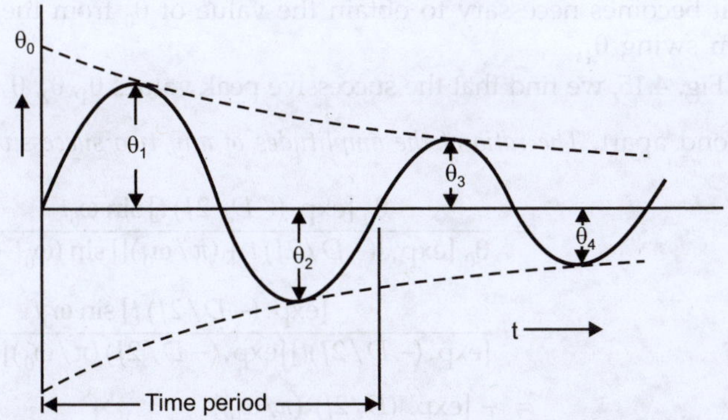

Fig. 4.16

During the passage of charge, at any instant, there will be a deflecting torque of GI acting on the coil. If t is the time taken by the whole charge to pass through, the torque impulse due to this charge is given as : $\int_0^t GI\,dt$

But, $$\int_0^t I\,dt = Q$$

$\therefore$ Torque impulse = GQ

This must be equal to the change of angular momentum produced *i.e.* $J.\ \alpha$, where α is the angular velocity of the coil at the end of impuse period.

$$\therefore \qquad GQ = J\alpha \qquad \text{...(4.39)}$$

or, $$\alpha = \frac{GQ}{J} \qquad \text{...[4.39(a)]}$$

Differentiating eqn. (4.38), we get

$$\frac{d\theta}{dt} = U\left[\exp.\{-(D/2J)\,t\}\cos\omega_0 t\,.\,\omega_0 + \sin\omega_0 t\,.\exp.\{-(D/2J)\,t\}\left(-\frac{D}{2J}\right)\right]$$

Since duration of the passage of charge is very small, at the end of the passage, $t \simeq 0$, so that from above,

$$\frac{d\theta}{dt} = U\omega_0 \qquad ...(4.40)$$

However, at this time, $\dfrac{d\theta}{dt} = \alpha = \dfrac{GQ}{J}$, $\therefore \quad \dfrac{GQ}{J} = U\omega_0$

or,

$$Q = \frac{J\omega_0}{G}U \qquad ...(4.41)$$

Now, *U being the amplitude which the oscillations would* have if the damping were *zero*, it may be called *undamped swing* θ_0.

$$\therefore \quad Q = \frac{J\omega_0}{G}\theta_0 \qquad ...(4.42)$$

$$(\text{or, } Q \propto \theta_0)$$

However, in practice, due to the presence of small amount of damping, the successive oscillations diminish exponentially (See Fig. 4.16). Even the first swing θ_1 is much less than θ_0. Hence, it becomes necessary to obtain the value of θ_0 from the observed value of first maximum swing θ_1.

Referring to Fig. 4.15, we find that the successive peak values θ_1, θ_2, θ_3 etc. are ϕ radian apart or $\dfrac{\phi}{\omega_0}$ second apart. *The ratio of the amplitudes of any two successive peaks is*

$$= \frac{\theta_0\,[\exp.(-D/2J)\,t]\sin\omega_0 t}{\theta_0\,[\exp.\{(-D/2J)\,t + (\pi/\omega_0)\}]\sin(\omega_0 t + \pi)}$$

$$= \frac{[\exp.(-D/2J)\,t]\sin\omega_0 t}{[\exp.(-D/2J)t][\exp.(-D/2J)(\pi/\omega_0)](-\sin\omega_0 t)}$$

$$= -[\exp.(D/2J)(\pi/\omega_0)]$$

$$\therefore \quad \frac{\theta_1}{\theta_2} = \frac{\theta_2}{\theta_3} = \frac{\theta_3}{\theta_4} = \text{.......}\ [\exp.(D/2J)(\pi/\omega_0)] \qquad ...(4.43)$$

Let $\exp.\,[(D/2J)(\pi/\omega_0)] = \delta^2$, where δ is called the **damping factor*

*Now, $\dfrac{\theta_1}{\theta_2} = \dfrac{\theta_2}{\theta_3} = \text{.....}\dfrac{\theta_{n-1}}{\theta_n} = \delta^2$

$$\therefore \quad \frac{\theta_1}{\theta_2}\times\frac{\theta_2}{\theta_3}\times\frac{\theta_3}{\theta_4}\times\text{......}\times\frac{\theta_{n-1}}{\theta_n} = (\delta^2)^{n-1}$$

or, $\dfrac{\theta_1}{\theta_n} = (\delta^2)^{n-1}$ or $\delta = \left(\dfrac{\theta_1}{\theta_n}\right)^{1/2(n-1)}$

Hence, δ may be obtained by observing the first and *n*th swing.

The time period of oscillation, $T_0 = \dfrac{2\pi}{\omega_0}$. If damping is very small $\theta_0 = \theta_1$, $t = \dfrac{T_0}{4} = \dfrac{\pi}{2\omega_0}$ as a very close approximation.

Putting $t = \frac{\pi}{2\omega_0}$ in eqn. (4.38), we have :

$$\theta_1 = \theta_0 \exp. [-(D/2J)(\pi/2\omega_0)] \sin(\pi/2), \text{ or, } \theta_1 = \theta_0 \cdot \delta^{-1}$$

or, $$\theta_0 = \delta\theta_1 \quad ...(4.44)$$

or, Undamped swing = Damping factor × Ist swing

Suppose, a steady current of I_s flowing through the galvanometer produces a steady deflection θ_s, then

$$C\theta_s = GI_s \quad ...(4.45)$$

or, $$G = \frac{C\theta_s}{I_s} \quad [4.45(a)]$$

Since damping is small, $\omega_0 = \sqrt{\frac{C}{J}}$

or, $$C = J\omega_0^2 = J \times \left(\frac{2\pi}{T_0}\right)^2 = 4\pi^2 \frac{J}{T_0^2}$$

Inserting this value of C in eqn. 4.45 (*a*), we get :

$$G = \frac{4\pi^2 J\theta_s}{T_0^2 I_s} \quad ...(4.46)$$

Substituting this value of G in eqn. (4.42), we have :

$$Q = \frac{J\omega_0\theta_0}{4\pi^2 J\theta_s/(T_0^2 I_s)} = \frac{J\omega_0\theta_0 \times T_0^2 I_s}{4\pi^2 J\theta_s} = \frac{J \times (2\pi/T_0)\,\theta_0 T_0^2 I_s}{4\pi^2 J\theta_s}$$

$$\left(\because \omega_0 = \frac{2\pi}{T_0}\right)$$

or $$Q = \frac{T_0}{2\pi} \cdot \frac{I_s}{\theta_s} \cdot \delta\theta_1 \quad ...(4.47)$$

Alternatively, let *quantity (D/2J) (π/ω$_0$) be called the "logarithmic decrement λ".*

Since, $$\delta^2 = e^{\lambda},$$

$$\delta = e^{\lambda/2} = 1 + (\lambda/2) + \frac{(\lambda/2)^2}{2!} + ... \simeq \left(1 + \frac{\lambda}{2}\right)$$

when λ is ***small.*

**Since $\delta^2 = e^{\lambda}$, taking logs we have : $2 \ln(\delta) = \lambda \ln(e) = \lambda$

$$\therefore \quad \frac{\lambda}{2} = \ln(\delta) = \ln\left(\frac{\theta_1}{\theta_n}\right)^{1/2(n-1)} = \frac{1}{2(n-1)} \ln\left(\frac{\theta_1}{\theta_n}\right)$$

$$\therefore \quad \lambda = \frac{1}{n-1} \ln\left(\frac{\theta_1}{\theta_n}\right) \quad ...[4.47(a)]$$

Hence, from eqn. (4.47), we have :

$$Q = \frac{T_0}{2\pi} \cdot \frac{I_s}{\theta_s}\left(1 + \frac{\lambda}{2}\right)\theta_1 \quad ...(4.48)$$

In general, eqn. (4.48) may be expressed as :

$$Q = k\theta_1 \quad ...[4.48(a)]$$

where, $k = \frac{T_0}{2\pi} \cdot \frac{I_s}{\theta_s}\left(1 + \frac{\lambda}{2}\right)$ is a constant and is found by *calibration*. Since the value of k depends upon the damping and shunting of the galvanometer it is essential that the resistance of the galvanometer circuit during calibration shall be the *same* as it is when the galvanometer is being used for testing purposes.

4.7.2.3. Calibration of a ballistic galvanometer

*The experimental determination of the constants of a ballistic galvanometer is termed as its "**calibration**".*

There are *three* methods of calibration of ballistic galvanometer :

1. Using a mutual inductance.
2. Using a capacitor.
3. Using Hibbert magnetic standard.

1. Using a mutual inductance :

This method is *most commonly used for the calibration purposes.*

Fig. 4.17(*a*) shows the schematic arrangement for the calibration of a ballistic galvanometer with mutual inductance. The mutual inductance is usually in the form of a standard solenoid with a secondary coil.

A *standard solenoid* (*primary coil*) consists of a long coil wound on a cylinder. The length of cylinder is very large as compared with its diameter. At the centre of the solenoid a *secondary coil* is placed whose length is *small*. The secondary coil is connected to the ballistic galvanometer (B.G.). When a current is passed through the solenoid, the field strength (H) at the centre is :

$$H = \frac{N_1 I}{l}\cos\theta$$

or,

$$H = \frac{N_1 I}{l} \times \frac{(l/2)}{\sqrt{(l/2)^2 + (r_1)^2}} = \frac{N_1 I}{2\sqrt{(l/2)^2 + (r_1)^2}} \quad ...(4.49)$$

where, N_1 = Number of turns of the solenoid,

I = Current through the solenoid,

l = Length of the solenoid, and

r_1 = Mean radius of solenoid.

Flux density at the centre of solenoid, $B = \mu_0 H$

Flux linking with the secondary coil,

$$\phi = B \times \text{area of secondary coil}$$

i.e.

$$\phi = B \times \pi r_2^2 = \pi B r_2^2 = \pi\mu_0 H r_2^2 \quad ...(4.50)$$

where, r_2 = Mean radius of the secondary coil.

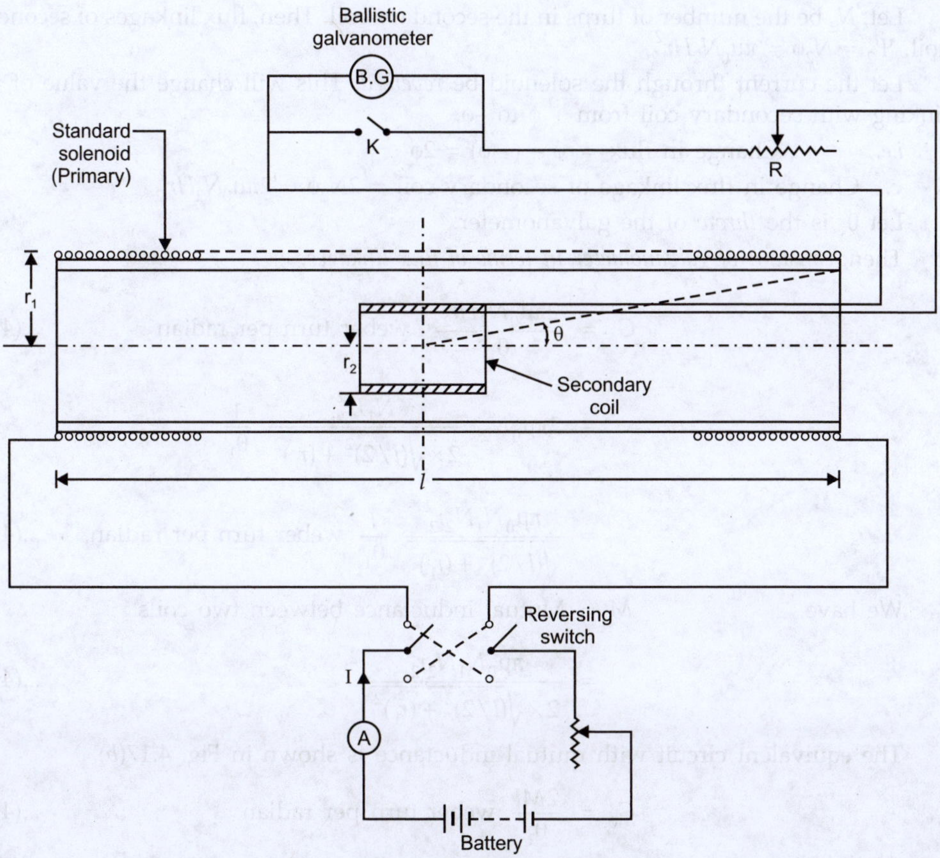

(a) Schematic arrangement

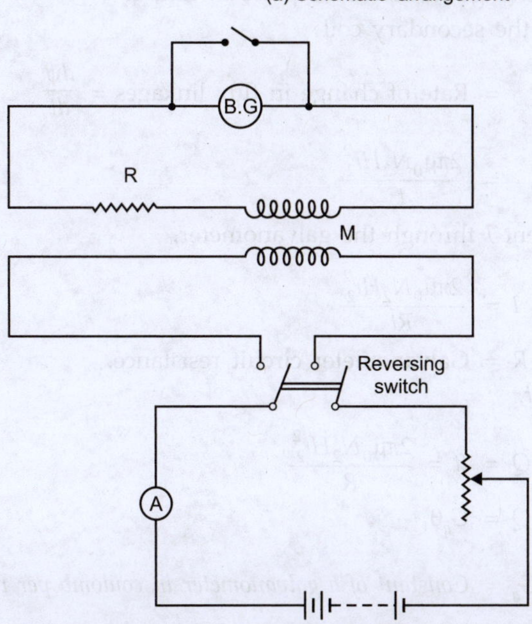

(b) Equivalent circuit

Fig. 4.17. Calibration of ballistic galvanometer with mutual inductance.

Let, N_2 be the number of turns in the secondary coil. Then, flux linkages of secondary coil, $\Psi_2 = N_2\phi = \pi\mu_0 N_2 H r_2^2$.

Let the current through the solenoid be *reversed.* This will change the value of flux linking with secondary coil from $+\phi$ to $-\phi$.

i.e., Change in flux $= \phi - (-\phi) = 2\phi$

$\therefore$ Change in flux linkage of secondary coil $= 2N_2\phi = 2\pi\mu_0 N_2 H r_2^2$

Let θ_1 is the *throw* of the galvanometer.

Then, *constant of galvanometer in terms of flux linkages,*

$$G_\phi = \frac{2\pi\mu_0 N_2 H r_2^2}{\theta_1} \text{ weber turn per radian} \quad ...(4.51)$$

$$= 2\pi\mu_0 N_2 \cdot \frac{N_1 I r_2^2}{2\times\sqrt{(l/2)^2 + (r_1)^2}} \cdot \frac{1}{\theta_1}$$

$$= \frac{\pi\mu_0 N_1 N_2 r_2^2}{\sqrt{(l/2)^2 + (r_1)^2}} \cdot \frac{I}{\theta_1} \text{ weber turn per radian.} \quad ...(4.52)$$

We have, M = Mutual inductance between two coils

$$= \frac{\pi\mu_0 N_1 N_2 r_2^2}{2\times\sqrt{(l/2)^2 + (r_1)^2}} \quad ...(4.53)$$

The equivalent circuit with mutual inductance is shown in Fig. 4.17(*b*).

$$\therefore \quad G_\phi = \frac{2MI}{\theta_1} \text{ weber turn per radian} \quad ...(4.54)$$

Let us assume that the change in flux takes place in time t.

Then, E.m.f. induced in the secondary coil

$$= \text{Rate of change in flux linkages} = \frac{d\psi}{dt}$$

$$= \frac{2\pi\mu_0 N_2 H r_2^2}{t}$$

This e.m.f. sends a current I through the galvanometer,

$$I = \frac{2\pi\mu_0 N_2 H r_2^2}{Rt}$$

where, R = Galvanometer circuit resistance.

Charge passing in time t,

$$Q = It = \frac{2\pi\mu_0 N_2 H r_2^2}{R} \quad ...(4.55)$$

But, $$Q = G_q\theta_1 \quad ...(4.56)$$

where, G_q = *Constant of a galvanometer in coulomb per radian.*

∴ Constant of galvanometer,

$$G_q = \frac{Q}{\theta_1} = \frac{2\pi\mu_0 N_2 H r_2^2}{R} \cdot \frac{1}{\theta_1} = \frac{\pi\mu_0 N_1 N_2 r_2^2 I}{R\sqrt{(l/2)^2 + (r_1)^2}} = \frac{2MI}{R\theta_1} \quad \text{...(4.57)}$$

Thus the galvanometer constant can be obtained in terms of coulomb per radian.

- *"Ballistic sensitivity"* and *"Flux linkage sensitivity"* are sometimes used in place of galvanometer constants and are defined as :

Ballistic sensitivity, $$S_q = \frac{\theta_1}{Q} \text{ rad/C} = \frac{1}{G_q} \quad \text{...(4.58)}$$

Flux linkage sensitivity, $$S_\phi = \frac{\theta_1}{\Delta\psi} \text{ rad./weber turn} = \frac{1}{G_\phi} \quad \text{...(4.59)}$$

Also, $$S_q = RS_\phi \quad \text{...(4.60)}$$

2. Using a capacitor :

In this method, a capacitor (*C*) which has been charged to a known voltage (*V*) is discharged through the galvanometer. The throw of the galvanometer is observed. Let this throw be θ.

∴ *Constant of galvanometer,* $G_q = \frac{Q}{\theta} = \frac{CV}{\theta}$ coulomb/rad.

- The *disadvantages* of this method are :

(*i*) The capacitance of condenser is difficult to be determined accurately under all the conditions.

(*ii*) The damping of the galvanometer during calibration is different from that during testing.

3. Using Hibbert magnetic standard :

This method of calibration of ballistic galvanometer is *very simple and convenient,* and so is used in the *laboratories.*

Fig. 4.18 shows Hibbert magnetic standard :

— It consists of a circular bar magnet and an iron yoke which have a narrow annular gap between them. A brass tube carrying a coil can slide through the gap.

— The brass tube is taken to its top position and is then released with the help of a trigger. It falls under gravity and it slides through the gap (See Fig. 4.18), its coil cuts through the magnetic field of a permanent magnet and therefore an e.m.f. is produced in the coil.

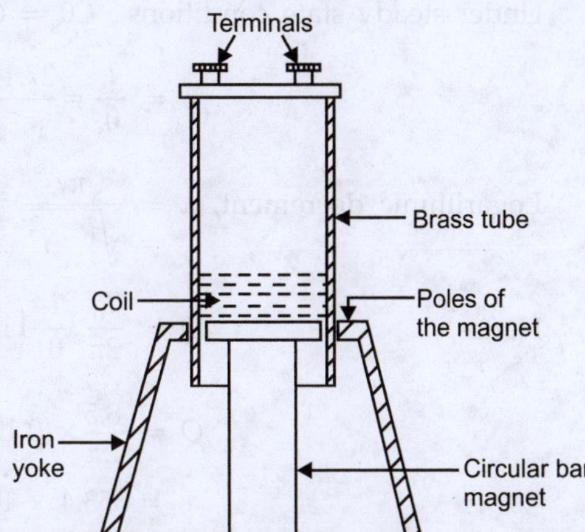

Fig. 4.18. Hibbert magnetic standard.

— The coil is connected across the terminals of the ballistic galvanometer, the *throw* of which can be observed. The rate at which coil cuts through the field is constant for a particular apparatus. The *e.m.f. per turn is constant.*

— By the use of the standard, the flux linkages which produce an observed throw of galvanometer can be known by knowing the air gap flux and number of turns in the coil. Hence the *galvanometer constant can be calculated.*

Example 4.20. *A ballistic galvanometer has a free period of 10 seconds and gives a steady deflection of 200 divisions with a steady current of 0.1 mA. A charge of 121 μC is instantaneously discharged through the galvanometer giving rise to a first maximum deflection of 100 divisions. Calculate the decrement of the resulting oscillations.* **(Bombay University)**

Solution. *Given* : $T_0 = 10$ s; $\theta_s = 200$; $I_s = 0.1$ mA; $Q = 121$ μC ; $\theta_1 = 100$

Decrement of the resulting oscillations, λ :

We know, $$Q = \frac{T_0}{2\pi} \cdot \frac{I_s}{\theta_s}\left(1 + \frac{\lambda}{2}\right)\theta_1 \qquad \text{[Eqn. (4.48)]}$$

Substituting the various values, we get

$$121 \times 10^{-6} = \frac{10}{2\pi} \times \frac{0.1 \times 10^{-3}}{200}\left(1 + \frac{\lambda}{2}\right) \times 100$$

or, $\lambda = \mathbf{1.04}$ **(Ans.)**

Example 4.21. *A ballistic galvanometer has a resistance of 140 Ω and an undamped period of 8.5 seconds. A steady e.m.f. of 3.8 mV produces a deflection of 220 mm. If the deflection produced is 720 mm and relative damping is 0.78, determine the quantity of electricity discharged from the capacitor.*

Solution. $R = 140$ Ω; $T_0 = 8.5$ s; $V = 3.8$ mV; $\theta_s = 220$ mm; $\theta_1 = 720$ mm; $\gamma = 0.78$

Quantity of electricity discharged, Q :

Steady state current, $$I_s = \frac{3.8 \times 10^{-3}}{140} = 27.14 \times 10^{-6} \text{ A}$$

Under steady state conditions : $C\theta_s = GI_s$ [Eqn. (4.45)]

$$\therefore \quad \frac{C}{G} = \frac{I_s}{\theta_s} = \frac{27.14 \times 10^{-6}}{220} = 0.123 \times 10^{-6} \text{ A/mm}$$

Logarithmic decrement, $$\lambda = \frac{\pi\gamma}{\sqrt{1-\gamma^2}} = \frac{\pi \times 0.78}{\sqrt{1-(0.78)^2}} = 3.9$$

Hence, $$Q = \frac{T_0}{2\pi} \cdot \frac{I_s}{\theta_s}\left(1 + \frac{\lambda}{2}\right)\theta_1 \qquad \text{[Eqn. (4.48)]}$$

$$Q = \frac{8.5}{2\pi} \times (0.123 \times 10^{-6})\left(1 + \frac{3.9}{2}\right) \times 720$$

$$= 353.4 \times 10^{-6} \text{ C} = \mathbf{353.4\ \mu C}$$ **(Ans.)**

Example 4.22. *A solenoid 1 m long and wound with 960 turns has a search coil of 60 turns and cross-sectional area 340 mm² at its centre. Reversing a current of 3.5 A in the solenoid causes a deflection of 4 divisions in a ballistic galvanometer connected to the search coil.*

Calculate :

(i) The galvanometer constant in flux linkages per division.

(ii) The flux linkage sensitivity.

Solution. *Given* : $l = 1$ m; $N_1 = 960$; $N_2 = 60$; $A = 340$ mm²; $\theta_1 = 4$ divisions; $I = 3.5$ A.

(*i*) **Galvanometer constant, G_ϕ :**

Flux density at the centre of solenoid $B = \dfrac{\mu_0 N_1 I}{l}$

Flux linking with search coil, $\phi = B \times A = \dfrac{\mu_0 N_1 IA}{l}$

Flux linkages of search coil, $\psi = N_2 BA = \dfrac{\mu_0 N_1 N_2 IA}{l}$

Change of flux linkages, $\Delta\psi = 2\psi = \dfrac{2\mu_0 N_1 N_2 IA}{l}$ (since current is *reversed*)

$$\text{Galvanometer constant}, G_\phi = \frac{2\psi}{\theta_1} = \frac{2\mu_0 N_1 N_2 IA}{l\,\theta_1}$$

$$= \frac{2\times 4\pi\times 10^{-7}\times 960\times 60\times 3.5\times (340\times 10^{-6})}{1\times 4}$$

$$= \mathbf{43.07 \times 10^{-6}} \textbf{ weber turn per division} \quad \textbf{(Ans.)}$$

(*ii*) **The flux linkage sensitivity, S_ϕ :**

$$S_\phi = \frac{1}{G_\phi} = \frac{1}{43.07\times 10^{-6}}$$

$$= \mathbf{23.22 \times 10^{3}} \textbf{ divisions per weber turn} \quad \textbf{(Ans.)}$$

Example 4.23. *The discharge of a capacitor through a ballistic galvanometer produces a damped frequency of 0.13 Hz and successive swings of 130, 102 and 85.4 mm.*

Calculate :

(*i*) *Logarithmic decrement.*

(*ii*) *Damping ratio.*

(*iii*) *Period of undamped and damped oscillations.*

Solution. *Given :* $f_d = 0.13$ Hz; $\theta_1 = 130$ mm; $\theta_3 = 85.4$ mm

(*i*) **Logaritmic decrement, λ :**

We know that, $$\lambda = \frac{1}{n-1}\ln\left(\frac{\theta_1}{\theta_n}\right) \qquad [4.47(a)]$$

or, $$\lambda = \frac{1}{3-1}\ln\left(\frac{130}{85.4}\right) = \mathbf{0.21} \quad \textbf{(Ans.)}$$

(*ii*) **Damping ratio, γ :**

Logarithmic decrement, $$\lambda = \frac{\pi\gamma}{\sqrt{1-\gamma^2}}$$

or, $$0.21 = \frac{\pi\gamma}{\sqrt{1-\gamma^2}}$$

Squaring both sides, we get

$$(0.21)^2\,(1-\gamma^2) = \pi^2\gamma^2$$

or, $$\gamma^2(\pi^2 + 0.21^2) = 0.21^2$$

or, $$\gamma = \left(\frac{0.21^2}{\pi^2 + 0.21^2}\right)^{1/2} = \mathbf{0.067} \quad \textbf{(Ans.)}$$

(iii) **Period of damped and undamped oscillations :**

Frequency of undamped oscillations, $$f_n = \frac{f_d}{\sqrt{1-\gamma^2}} = \frac{0.13}{\sqrt{1-0.067^2}} = 0.1303 \text{ Hz}$$

∴ Period of damped oscillations, $$T_d = \frac{1}{f_d} = \frac{1}{0.13} = \mathbf{7.69s} \quad \textbf{(Ans.)}$$

and, Period of undamped oscillations, $$T_n \text{ (or } T_0) = \frac{1}{f_n} = \frac{1}{0.1303} = \mathbf{7.67s} \quad \textbf{(Ans.)}$$

Example 4.24. *A ballistic galvanometer having 200 Ω resistance gives a throw of 80 divisions when the flux through search coil to which it is connected is reversed. The galvanometer constant is 120 μC/division. The search coil has 1500 turns, a mean area 4500 mm² and a resistance 230 Ω.*

Determine :

(i) Flux density in which the reversal of the coil takes place.

(ii) Ballistic and flux linkages sensitivities of the galvanometer.

Solution. *Given :* R_g = 200 Ω; θ_1 = 80 divisions; G_q = 120 μC/div.; N_{sc} = 1500; A_{sc} = 4500 mm² or 4500 × 10^{-6} m²; R_{sc} = 230 Ω

(i) **Flux density in which the reversal of the coil takes place, *B* :**

Resistance of galvanometer circuit, $R = R_g + R_{sc} = 200 + 30 = 230\ \Omega$

Galvanometer constant, $G_q = 120\ \mu\text{C/div.}$

Now, $$G_q = \frac{Q}{\theta_1} = \frac{It}{\theta_1} = \frac{Et}{R\theta_1}$$

But change in flux linkages $\Delta\psi = 2\psi$ since the current is reversed.

∴ $$E = \frac{2\psi}{t} = \frac{2\phi N}{t} = \frac{2BAN}{t}$$

Hence, $$G_q = \frac{2BAN}{t} \cdot \frac{t}{R\theta_1} = \frac{2BAN}{R\theta_1}$$

∴ Flux density, $$B = \frac{G_q R\theta_1}{2AN} = \frac{120 \times 10^{-6} \times 230 \times 80}{2 \times (4500 \times 10^{-6}) \times 1500} = \mathbf{0.16\ Wb/m^2} \quad \textbf{(Ans.)}$$

(ii) **The ballistic sensitivity and flux linkages sensitivity :**

Ballistic sensitivity, $$S_q = \frac{1}{G_q} = \frac{1}{120} = 8.333 \times 10^{-3} \text{ div./}\mu\text{C}$$

$$= \mathbf{8333\ div./C} \quad \textbf{(Ans.)}$$

Flux linkages sensntivity, $$S_\phi = \frac{S_q}{R}$$ ...[Eqn. (4.60)]

or, $$S_\phi = \frac{8333}{230} = \textbf{36.23 weber turn/div.} \quad \textbf{(Ans.)}$$

Example 4.25. *A galvanometer with an undamped period of 8 seconds and a current sensitivity of 1.1 mA/div. is connected to a search coil in a circuit of 3500 Ω resistance. Determine the flux linkage change in search coil to produce a first swing of 80 divisions if the logarithmic decrement is 0.18.*

Solution. T_0 (or T_n) = 8 s; Current sensitivity, $\frac{I_s}{\theta_s}$ = 1.1 μA/div.; R = 3500 Ω;

θ_1 = 80 divisions; λ = 0.18.

Change in flux linkages, Δψ :

Charge, $$Q = \frac{T_0}{2\pi}\cdot\frac{I_s}{\theta_s}\left(1+\frac{\lambda}{2}\right)\theta_1 \quad \text{[Eqn. (4.48)]}$$

$$= \frac{8}{2\pi}\times 1.1\times 10^{-6}\left(1+\frac{0.18}{2}\right)\times 80 = 122.13\times 10^{-6}\text{ C}$$

Let Δψ be the change in flux linkages in time t.

Then, $$\text{e.m.f., } E = \frac{\Delta\psi}{t}$$

Also, $$\text{charge } Q = It = \frac{E}{R}\cdot t = \frac{E}{R}\times\frac{\Delta\psi}{E} = \frac{\Delta\psi}{R}$$

$\therefore$ $\Delta\psi = RQ = 3500 \times 122.13 \times 10^{-6}$ = **427.5 × 10^{-3} Weber turn** **(Ans.)**

Example 4.26. *A ballistic galvanometer gives a first maximum deflection of 50° for a discharge of 900 μC. Find the quantity of electricity which when discharged through this galvanometer gives rise to a spot deflection of 8 divisions on a millimetre circular scale 1 metre away.*

Solution. *Given* : $\theta_1 = 50°$ or $50 \times \frac{\pi}{180}$ = 0.873 rad.; Q = 900 μC; d = 8 mm or 0.008 m;

r = 1 m

Charge Q :

Now, $$Q = G_q\,\theta_1, \quad \therefore \quad G_q = \frac{Q}{\theta_1} = \frac{900}{0.873} = 1030.9\ \mu\text{C/rad.}$$

Angle turned through by reflected ray for a swing of 8 mm,

$$\theta = \frac{d}{2r} = \frac{0.008}{2\times 1} = 0.004 \text{ rad.}$$

$\therefore$ Charge for a swing of 0.004 radian, $Q = 1030.9 \times 0.004$ = **4.12 μC** **(Ans.)**

4.7.3. Grassot Fluxmeter

This instrument is a special type of ballistic galvanometer in which the *controlling torque is very small and electromagnetic damping is heavy. It is used for measurement of flux.*

Construction :

Fig. 4.19 shows the construction of a Grassot fluxmeter :

— It consists of a moving coil of small cross-section suspended by means a single silk thread from a spring support and hanging with its parallel sides in the narrow air gap of a permanent magnet (See Fig. 4.19). The coil is formerless and the air friction damping is negligible.

— The current is led into the coil by spirals of very thin, annealed silver strips. By this construction the *controlling torque is reduced to a minimum.*

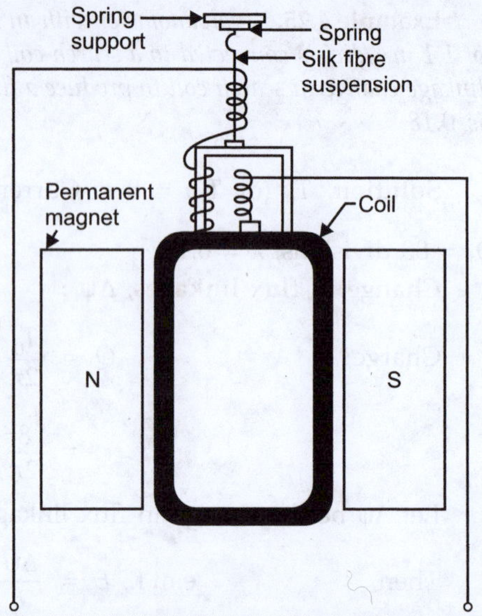

Fig. 4.19. Grassot fluxmeter.

— The instrument is usually fitted with a pointer (attached to the moving system) and a scale, although it may also be used as a reflecting instrument. The *scale is graduated in terms of flux-turns.*

Working :

— When the instrument terminals are connected to a *search coil* and the flux linking with the search coil is changed, the moving system of the flux meter is deflected and rotates through an angle which depends upon the change in flux-turns.

— The instrument coil rotates during the whole period of the flux change but stops as soon as the flux change ceases because of its high electromagnetic damping. The *high damping is obtained by reducing the resistance of the circuit and search coil to the minimum possible.* Though theoretically, the value of the resistance of the circuit comprising the instrument coil and search coil does not affect the sensitivity of the instrument but it should be kept fairly small (below 20 Ω), otherwise the damping will not be adequate.

— Theoretically, the time taken for the change in flux does not affect the deflection but in practice if the time flux change exceeds about 30 seconds serious errors are introduced due to *friction*, the *slight controlling torque due to the current leads and the effect of incorrect levelling combined with slight unbalance of the moving parts.*

Advantages :

The instrument possesses the following *advantages* :

1. It is *very portable* and has *scale calibrated directly in weber-turns.*

2. The length of time taken for the change in the flux producing the deflection need not be small. The deflection obtained, for a given change of flux interlinking with the search coil connected to the instrument, will be the *same* whether the time taken for the change be a fraction of a second or as much as one or two minutes.

Disadvantages :

The fluxmeter is *less sensitive and accurate than the ballistic galvanometer.*

Note : • In the complete absence of controlling torque, the instrument would remain in its deflected position indefinitely. Actually the pointer returns very slowly to zero, but *readings may be taken by observing the difference in deflection at the beginning and end of the change in flux to be measured without waiting for the pointer to return to zero, the scale being uniform.*

• In some modern instruments, the coil is fitted with pivots and mounted in jewelled bearings, the current being led into the coil by fine ligaments. This form of construction is rather *more robust* than the suspended instruments.

Theory of Fluxmeter :

Let, N_{sc} = Number of turns in search coil,

R_{sc}, L_{sc} = Resistance and inductance of search coil respectively,

R_{fm}, L_{fm} = Resistance and inductance of fluxmeter respectively,

ϕ = Flux linking with search coil,

i = Current in the circuit at any instant, and

θ = Deflection of the instrument at any instant.

The *equation of motion* is given as :

$$J\frac{d^2\theta}{dt^2} + D\frac{d\theta}{dt} + C\theta = Gi$$

Assuming control torque to be negligibly small and also the air friction damping is small, the equation of motion reduces to :

$$J\frac{d^2\theta}{dt^2} = Gi \qquad ...(4.61)$$

as D and C are both zero. (The effect of electromagnetic damping is considered later in expression for current i)

The e.m.f. due to any change of flux linked with search coil,

$$e_{sc} = N_{sc} = \frac{d\phi}{dt}$$

Simultaneously, due to the movement of the fluxmeter coil in the field of the magnet, a rotational e.m.f. is induced in the meter coil. This e.m.f., $e_{fm} = G \cdot \frac{d\theta}{dt}$. Besides this there are voltage drops in the resistance and inductance of the circuit. Therefore, we can write :

$$e_{sc} = e_{fm} + \left(L_{fm} + L_{sc}\right)\frac{di}{dt} + (R_{fm} + R_{sc})\, i \qquad ...(4.62)$$

or, $$N_{sc} \cdot \frac{d\phi}{dt} = G \cdot \frac{d\theta}{dt} + \left(L_{fm} + L_{sc}\right)\frac{di}{dt} + (R_{fm} + R_{sc})\, i$$

or, $$i = \frac{N_{sc} \cdot \frac{d\phi}{dt} - G \cdot \frac{d\theta}{dt} - (L_{fm} + L_{sc})\frac{di}{dt}}{(R_{fm} + R_{sc})}$$

∴ The equation of motion is :

$$J \cdot \frac{d^2\theta}{dt^2} = \frac{G}{(R_{fm} + R_{sc})}\left[N_{sc} \cdot \frac{d\phi}{dt} - G \cdot \frac{d\theta}{dt} - (L_{fm} + L_{sc})\frac{di}{dt}\right] \qquad ...(4.63)$$

or, $$N_{sc} \cdot \frac{d\phi}{dt} = \frac{J(R_{fm} + R_{sc})}{G} \cdot \frac{d^2\theta}{dt^2} + G \cdot \frac{d\theta}{dt} + (L_{fm} + L_{sc})\frac{di}{dt}$$

or, $$N_{sc} \cdot \frac{d\phi}{dt} = \frac{J(R_{fm} + R_{sc})}{G} \cdot \frac{d\omega}{dt} + G \cdot \frac{d\theta}{dt} + (L_{fm} + L_{sc})\frac{di}{dt} \qquad ...(4.64)$$

where, ω = angular velocity of moving coil at instant t.

If T is the time taken by the change in flux, then we have :

$$\int_0^T N_{sc} \cdot \frac{d\phi}{dt} \cdot dt = \int_0^T \frac{J(R_{fm} + R_{sc})}{G}\frac{d\omega}{dt} \cdot dt + \int_0^T G \cdot \frac{d\theta}{dt} \cdot dt + \int_0^T (L_{fm} + L_{sc})\frac{di}{dt} \cdot dt$$

or, $$\int_{\phi_1}^{\phi_2} N_{sc} \, . \, d\phi = \int_{\omega_1}^{\omega_2} \frac{J(R_{fm} + R_{sc})}{G} d\omega + \int_{\theta_1}^{\theta_2} G \, . \, d\theta + \int_{i_1}^{i_2} (L_{fm} + L_{sc}) \, di$$

$$\therefore \quad N_{sc} \, (\phi_2 - \phi_1) = \frac{J(R_{fm} + R_{sc})}{G} (\omega_2 - \omega_1) + G(\theta_2 - \theta_1) + (L_{fm} + L_{sc})(i_2 - i_1) \quad ...(4.65)$$

where, ϕ_1, ϕ_2 = Inter linking fluxes,

ω_1, ω_2 = Angular velocities, and

i_1, i_2 = Currents.

The suffices 1 and 2 indicate respectively values at the *beginning* and at the *end* of the change.

Now, since the angular velocities and currents are zero at both the beginning and end of the change, therefore,

$$\omega_1 = \omega_2 = 0, \quad \text{and} \quad i_1 = i_2 = 0$$

Inserting these values in eqn. (4.65), we have :

$$N_{sc} \, (\phi_2 - \phi_1) = G_2 \, (\theta_2 - \theta_1) \quad ...(4.66)$$

or, if ϕ is the change in 0 flux and θ the change in fluxmeter deflection, then,

$$\phi = \left(\frac{G}{N_{sc}} \right) \theta \quad ...(4.67)$$

When the permanent magnet field of fluxmeter is uniform for all positions of the moving coil, G is a constant. Thus the *change in the value of flux is directly proportional to the change in deflection and hence the instrument will have a uniform scale.*

Use of shunt with Fluxmeter :

When the values of flux are very high, the deflection of a fluxmeter, even when a single-turn search coil is used, may exceed the scale length. In these circumstances, the range may be increased by employing a low-resistance, non-inductive shunt.

Fig. 4.20 shows the circuit using a shunt with a fluxmeter.

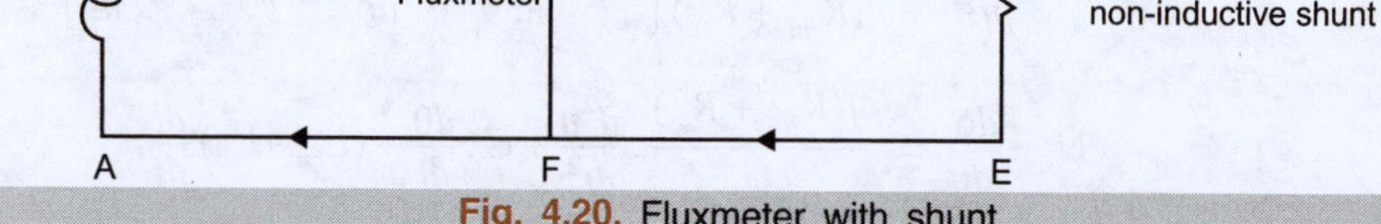

Fig. 4.20. Fluxmeter with shunt.

Let, R_{sc}, R_{fm}, R_{sh} = Resistances of search coil, fluxmeter and shunt respectively,

L_{sc}, L_{fm} = Inductances of search coil and fluxmeter respectively,

i_{sc}, i_{fm}, i_{sh} = Currents through search coil, fluxmeter and shunt respectively,

e_{sc} = e.m.f. induced in search coil because of change in flux change $= N_{sc} \dfrac{d\phi}{dt}$, and

e_{fm} = e.m.f. generated in fluxmeter coil because of rotation $= G. \dfrac{d\theta}{dt}$

- Applying Kirchhoff's law for the mesh ABCF, we get :

$$e_{sc} = i_{sc} R_{sc} + L_{sc} \frac{di_{sc}}{dt} + I_{fm} . R_{fm} + L_{fm} . \frac{di_{fm}}{dt} + e_{fm}$$

or, $$N_{sc} . \frac{d\phi}{dt} - L_{sc} \frac{di_{sc}}{dt} - L_{fm} . \frac{di_{fm}}{dt} - G . \frac{d\theta}{dt} = i_{sc}R_{sc} + i_{fm} . R_{fm}$$

Integrating as before, we have :

$$N_{sc} (\phi_2 - \phi_1) - L_{sc} (i_{sc2} - i_{sc1}) - L_{fm} (i_{fm2} - i_{fm1}) - G(\theta_2 - \theta_1)$$

$$= R_{sc} \int_0^T i_{sc} \, dt + R_{fm} \int_0^T i_{fm} \, dt \qquad ...(4.68)$$

Since the currents are zero at both the beginning and the end of the flux change, therefore,

$$i_{sc2} = i_{sc1} = 0; \; i_{fm2} = i_{fm1} = 0$$

Thus, eqn. (4.68) reduces to :

$$N_{sc} (\phi_2 - \phi_1) - G(\theta_2 - \theta_1) = R_{sc} \int_0^T i_{sc} \, dt + R_{fm} \int_0^T i_{fm} \, dt \qquad ...(4.69)$$

The equation of motion is :

$$J \frac{d^2\theta}{dt^2} = G i_{fm}$$

or, $$i_{fm} = \frac{J}{G} . \frac{d^2\theta}{dt^2} = \frac{J}{G} . \frac{d\omega}{dt} \qquad ...(4.70)$$

where ω is the angular velocity at any instant.

Hence, $$\int_0^T i_{fm} dt = \frac{J}{G} \int_0^T \frac{d\omega}{dt} . dt = \frac{J}{G} \int_{\omega_1}^{\omega_2} d\omega$$

or, $$\int_0^T i_{fm} dt = \frac{J}{G} (\omega_2 - \omega_1)$$

But, $$\omega_2 = \omega_1 = 0$$

∴ $$\int_0^T i_{fm} dt = 0$$

Hence eqn. (4.69) reduces to,

$$N_{sc} (\phi_2 - \phi_1) - G(\theta_2 - \theta_1) = R_{sc} \int_0^T i_{sc} dt \qquad ...(4.71)$$

- Applying Kirchhoff's law to mesh *CDEF*, we have :

$$e_{fm} + L_{fm} . \frac{di_{fm}}{dt} + i_{fm} R_{fm} = i_{sh} R_{sh}$$

or, $$G.\frac{d\theta}{dt}+L_{fm}.\frac{di_{fm}}{dt} = -i_{fm}R_{fm} + i_{sh}R_{sh} = -i_{fm}R_{fm} + (i_{sc} - i_{fm})R_{sh}$$

or, $$G.\frac{d\theta}{dt}+L_{fm}.\frac{di_{fm}}{dt} = -(R_{fm} + R_{sh})\, i_{fm} + i_{sc}R_{sh}$$

Integrating as before, we get :

$$G(\theta_2 - \theta_1) + L_{fm}(i_{fm2} - i_{fm1}) = -(R_{fm}+R_{sh})\int_0^T i_{fm}dt + R_{sh}\int_0^T i_{sc}dt$$

But $i_{fm2} - i_{fm1} = 0$ and $\int_0^T i_{fm}dt = 0$

Hence, $$G(\theta_2 - \theta_1) = R_{sh}\int_0^T i_{sc}dt \qquad ...(4.72)$$

or, $$\int_0^T i_{sc}dt = \frac{G(\theta_2-\theta_1)}{R_{sh}}$$

Putting this value of $\int_0^T i_{sc}dt$ in eqn. (4.71), we have :

$$N_{sc}(\phi_2 - \phi_1) - G(\theta_2 - \theta_1) = \frac{R_{sc}}{R_{sh}}G(\theta_2-\theta_1)$$

or, $$N_{sc}(\phi_2 - \phi_1) = G\frac{R_{sc}+R_{sh}}{R_{sh}}(\theta_2-\theta_1) \qquad ...(4.73)$$

If $(\phi_2 - \phi_1) = \phi$, and, $(\theta_2 - \theta_1) = \theta$

we have, $$\phi = \frac{G}{N_{sc}}.\frac{R_{sc}+R_{sh}}{R_{sh}}.\theta \qquad ...(4.74)$$

Comparison of eqn. (4.74) with corresponding eqn. (4.67) of unshunted fluxmeter shows that the *multiplying factor* for the shunted fluxmeter is : $\frac{R_{sc}+R_{sh}}{R_{sh}}$

It may be noted that this factor involves the search coil resistance (R_{sc}), but *not* the fluxmeter resistance (R_{fm}).

Difference between fluxmeter and ballistic galvanometer :

Fluxmeter differs from ballistic galvanometer in following *respects* :

1. Fluxmeter has *very small controlling torque* in comparison to ballistic galvanometer.
2. Fluxmeter has *heavy electromagnetic damping* as compared to ballistic galvanometer.
3. The fluxmeter is *less sensitive and accurate* than the ballistic galvanometer.
4. The indication of fluxmeter is *independent of time taken by the flux changes.* This is of great importance in measurement of flux linking with highly inductive circuit where the flux changes may be relatively slow rendering the use of ballistic galvanometer impractical.

Example 4.27. *The following constants relate to a certain fluxmeter :*

Air gap flux density = *0.06 Wb/m²*

Turns on moving coil = *50*

Area of moving coil = *800 mm²*

If the flux linking a 12 turn search coil of 220 mm² area connected to the fluxmeter is reversed in a uniform field of 0.6 Wb/m², calculate the deflection of the fluxmeter.

Solution. *Given* : $B_a = 0.06 \text{ Wb/m}^2$; $N_{mc} = 50$; $A_{mc} = 800 \text{ mm}^2$; $N_{sc} = 12$; $A_{sc} = 220 \text{ mm}^2$; $B_f = 0.6 \text{ Wb/m}^2$.

Deflection of the fluxmeter, θ :

Constant of fluxmeter, $G = N_{mc} \times B_a \times A_{mc} = 50 \times 0.06 \times (800 \times 10^{-6}) = 2400 \times 10^{-6}$

Flux linking with the search coil $= B_f \times A_{sc} = 0.6 \times (220 \times 10^{-6}) = 132 \times 10^{-6}$ Wb

Change in flux linking the coil, $\phi = 2 \times 132 \times 10^{-6} = 264 \times 10^{-6}$ Wb

Also, change in flux $\phi = \dfrac{G}{N_{sc}} . \theta$ [Eqn. (4.67)]

$$\therefore \quad \theta = \frac{\phi \times N_{sc}}{G} = \frac{264 \times 10^{-6} \times 12}{2400 \times 10^{-6}} = \textbf{1.32 rad. or 75.6°} \quad \textbf{(Ans.)}$$

Example 4.28. *A fluxmeter is connected to a search coil having 600 turns and a mean area of 600 mm². The search coil is placed at a centre of a solenoid 1 metre long, wound with 900 turns. When a current of 6A is reversed, there is a deflection of 30 scale divisions.*

Calculate the calibration in flux linkages per scale division.

Solution. $N_{sc} = 600$; $A_{sc} = 600 \text{ mm}^2$; $l = 1\text{m}$; $l_{sd} = 1\text{m}$; $N_{sd} = 900$; $I = 6\text{A}$; $\theta = 30$ scale divisions.

Flux linkages per scale division :

Field strength of solenoid, $H_{sd} = \dfrac{N_{sd} \times I}{l_{sd}} = \dfrac{900 \times 6}{1} = 5400$ A/m (assuming it to be very long)

Flux density, $B_{sd} = \mu_0 H_{sd} = 4\pi \times 10^{-7} \times 5400 = 6.78 \times 10^{-3} \text{ Wb/m}^2$

Flux linkages of fluxmeter search coil $= N_{sc} \times B_{sd} \times A_{sc}$

$= 600 \times 6.78 \times 10^{-3} \times (600 \times 10^{-6})$

$= 2.44 \times 10^{-3}$ Weber turn

Change in flux linkages $= 2 \times (2.44 \times 10^{-3}) = 4.88 \times 10^{-3}$ weber turn

Flux linkages per scale division $= \dfrac{4.88 \times 10^{-3}}{30} = \mathbf{0.1627 \times 10^{-3}}$ **weber turn** **(Ans.)**

Example 4.29. *A fluxmeter when used with a single turn search coil gives full scale deflection if a given uniform field linking with the search coil is reversed. It is proposed to extend the range of fluxmeter to measure flux densities six times the above value using the same search coil and method of reversals. If the resistance of search coil is 1.2 Ω, find the resistance of the shunt to be connected in parallel with the fluxmeter.*

Solution. *Given :* $R_{sc} = 1.2\ \Omega$

The range of instrument is extended $\dfrac{R_{sc} + R_{sh}}{R_{sh}}$ times by the use of a shunt

...[From eqn. 4.74]

where, R_{sc} = Resistance of search coil, and

R_{sh} = Resistance of shunt.

Now, we have : $$\frac{R_{sc}+R_{sh}}{R_{sh}} = 6$$

or, $$\frac{1.2+R_{sh}}{R_{sh}} = 6$$

∴ R_{sh} = **0.24 Ω** **(Ans.)**

Example 4.30. *A magnetic circuit excited by an m.m.f. of 8500 AT has a reluctance of 125,000 AT/Wb. A fluxmeter is used to measure the flux. If the fluxmeter has 125 divisions and flux linkages required for a deflection of one division are 0.16 × 10^{-3} weber turn, calculate the resistance of the shunt required for use with the search coil. Number of turns of search coil is 1 and its resistance is 0.03 Ω. The measurement is made by switching off the excitation.*

Solution. *Given* : m.m.f. = 8500 AT; Reluctance = 1,25,000 AT/Wb; N_{sc} = 1; R_{sc} = 0.03 Ω; Number of divisions on fluxmeter = 125; Flux linkages required for deflection of one division = 0.16 × 10^{-3} weber turn.

$$\text{Flux to be measured} = \frac{\text{m.m.f.}}{\text{reluctance}} = \frac{8500}{1,25,000} = 0.068 \text{ Wb}$$

Value of flux when the fluxmeter shows full scale deflection

$$= \frac{0.16\times10^{-3}\times125}{1} = 20 \times 10^{-3} \text{ Wb}$$

$$\text{Ratio } \frac{\text{flux to be measured}}{\text{flux for full scale deflection}} = \frac{0.068}{20\times10^{-3}} = 3.4$$

Thus, we have : $$\frac{R_{sc}+R_{sh}}{R_{sh}} = \frac{0.03+R_{sh}}{R_{sh}} = 3.4$$

∴ R_{sh} = **0.0125 Ω** **(Ans.)**

Example 4.31. *A fluxmeter is connected to a search coil of 100 turns and the mean area of the coil is 5 × 10^{-4} m^2. The search coil is placed at the centre of a standard solenoid, 1 metre long, uniformly wound with 800 turns. When a current of 5A is reversed, a deflection of 10 scale divisions is obtained with the fluxmeter. Calculate the calibration constant of the instrument in weber-turn per division.*

Solution. *Given* : N_{sc} = 100; A_{sc} = 5 × 10^{-4} m^2; l_{sd} = 1 m ; N_{sd} = 800; I = 5A;
θ = 10 scale divisions.

Instrument constant :

Since the axial length is large as compared with the mean diameter of the solenoid, therefore, the field strength in the neighbourbood of the centre of the solenoid core is *assumed to be uniform.*

$$H_{sd} = \frac{N_{sd}I}{l_{sd}} = \frac{800\times5}{1} = 4000 \text{ AT/m}$$

The flux passing through the search coil,

$$\phi = \beta_{sd} \times A_{sc} = \mu_0\mu_r H_{sd} \times A_{sc}$$

or, $$\phi = 4\pi \times 10^{-7} \times 1 \times 4000 \times 5 \times 10^{-4} = 25.13 \times 10^{-7} \text{ Wb}$$

(μ_r = 1 for air medium)

$$\text{Weber-turn} = \text{Flux} \times \text{number of turns on search coil}$$
$$= \phi \times N_{sc} = 25.13 \times 10^{-7} \times 100 = 25.13 \times 10^{-5}$$

Change in number of weber-turns on the reversal of the solenoid current

$$= 2 \times 25.13 \times 10^{-5} = 50.6 \times 10^{-5} \text{ weber-turn.}$$

Assuming uniform scale,

$$\textit{Instrument constant} = \frac{\text{Change in weber-turn on reversal of solenoid current}}{\text{Corresponding deflection}}$$

$$= \frac{50.6 \times 10^{-5}}{10}$$

$$= \mathbf{5.06 \times 10^{-5}\ Weber\text{-}turn\ per\ scale\ division} \quad \textbf{(Ans.)}$$

4.7.4. Vibration Galvanometers

Vibration galvanometers are of two types : (*i*) *Moving magnet*, (*ii*) *Moving coil*. The latter type is more generally used because moving magnet type has the *disadvantage* of being seriously affected by magnet fields of the resonant frequency (unless adequately screened). The **moving-coil** *galvanometers are not appreciably affected by such fields.* Hence only moving coil type will be discussed here.

These galvanometers are of *"D'Arsonval type"* having a moving coil suspended between the poles of a permanent magnet. When an alternating current is passed through the moving coil, an *alternating deflecting torque is produced which makes the coil vibrate with a frequency equal to the frequency of the current passing.* On account of the moving parts, the amplitude of vibrations is *small.* However, if natural frequency of the moving system is made equal to the frequency of the current, *mechanical resonance* is obtained and the moving system vibrates with a *large amplitude.*

- These galvanometers are the most widely used *"**tuned detectors**" particularly as null point detectors in A.C. bridges.* They are manufactured for various frequencies from 5 Hz upto 1000 Hz; these are most commonly used below 1200 Hz above which range they are considerably more sensitive than telephone.

Construction :

Fig. 4.21(*a*) shows the construction of a Duddell's moving coil vibration galvanometer :

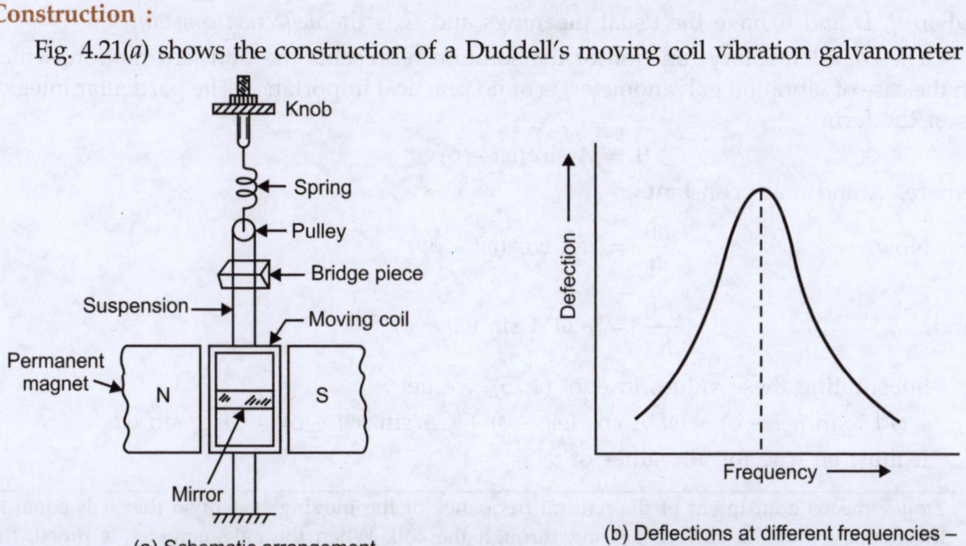

(a) Schematic arrangement

(b) Deflections at different frequencies – Resonance curve

Fig. 4.21. Vibration galvanometer.

— The moving coil (consisting of a fine bronze or platinum silver sire) is suspended between poles of a strong permanent magnet. This wire passes over a small pulley at the top and is pulled tight by a spring attached to the pulley, the tension of the spring can be adjusted by turning a milled head attached to the spring.

Generally, the suspension is rigidly held by small movable clamps on bridges, and the active length of the suspension is varied by sliding the clamps along the suspension to change the natural response frequency.

— The moving system carries a small *mirror* upon which a light beam is thrown.

Working :

— When an alternating current is passed through the coil, an alternating torque is applied to it so that the reflected spot of light on the scale is drawn out in the form of band of light. The light of this band of light is *maximum* if the natural frequency of oscillation of coil *coincides* with the supply frequency due to *resonance.*

— The *tuning of the coil is done by either changing the length of suspension or by adjusting tension in the suspension. The former is achieved by raising or lowering bridge piece against which the bifilar loop presses while the latter is achieved by turning the knob [See Fig. 4.21(*a*)].

- By making the damping very small, the *resonance curve* of the galvanometer can be made sharply-peaked [Fig. 4.21(*b*)]. In that case, the instrument discriminates sharply against frequencies other than its own natural frequency. That is, its reflection becomes very small even when the frequency of the applied current differs by a very small amount from its resonance frequency.

Theory :

Let the value of current passing through the moving coil at an instant t be :

$$i = I_{max} \sin \omega t$$

Then, the *equation of motion* of the coil is :

$$J\frac{d^2\theta}{dt^2} + D\frac{d\theta}{dt} + C\theta = Gi = GI_{max} \sin \omega t \qquad ...(4.75)$$

where J, D and C have the usual meanings and G is the *deflection constant*.

The complementary function of the solution represents the transient motion, which in the case of vibration galvanometer, is of no practical importance. The particular integral is of the form :

$$\theta = A \sin (\omega t - \phi)$$

where, A and ϕ are constant.

Now, $$\frac{d\theta}{dt} = \omega A \cos (\omega t - \phi)$$

and, $$\frac{d^2\theta}{dt^2} = -\omega^2 A \sin (\omega t - \phi)$$

Substituting these values in eqn. (4.75), we get :

$$-\omega^2 JA \sin (\omega t - \phi) + \omega DA \cos (\omega t - \phi) + CA \sin (\omega t - \phi) = GI_{max} \sin \omega t$$

It must be true for all values of i.

* *Tuning* means adjustment of the natural frequency of the moving system so that it is equal to the frequency of the current passing through the coil. When the galvanometer is tuned, the amplitude of vibration is very large and consequently a very wide band of light is observed on the scale.

When, $\omega t = \phi$ $\omega DA = GI_{max} \sin \phi$ ...(*i*)

When, $(\omega t - \phi) = \frac{\pi}{2}$ $-\omega^2 JA + CA = GI_{max} \cos \phi$ ...(*ii*)

Since the phase angle ϕ of oscillations is of *no* practical significance, it may be eliminated by squaring and adding eqns. (*i*) and (*ii*).

$$\therefore \quad \omega^2 D^2 A^2 + A^2 (C - \omega^2 J)^2 = G^2 I^2_{max}$$

or,
$$A = \frac{GI_{max}}{\sqrt{[(\omega D)^2 + (C - \omega^2 J)^2]}} \quad ...(4.76)$$

This represents the amplitude A of the resulting oscillation for a sinusoidally alternating current of peak value I_{max} flowing through the moving coil of the galvanometer.

- The *current sensitivity* of a vibration galvanometer is about 50 mm/μA with a scale 1 m away. The *sensitivity is almost inversely proportional to the frequency.*

Comparison of ballistic galvanometer with vibration galvanometer :

1. Both are of D'Arsonval type.
2. They differ in details of usage rather in fundamental ideas.
3. Ballistic galvanometers have long period of the order of 20 to 30 seconds compared to 4 to 6 seconds for vibration type.
4. In vibration galvanometer, current through it is steady as long as the switch is pressed while in ballistic case it is transitory.
5. *"Vibration galvanometers" are more suitable for use at power and low audio frequencies while ballistic type can be used for higher frequencies.*

Example 4.32. *The following data relate to a vibration galvanometer tuned to a frequency of 50 Hz :*

Inertia constant $= 0.012$ *kg-m*^{2}

Damping constant $= 0.028$ *Nm/rad.s*$^{-1}$

Calculate the ratio of sensitivity for fundamental to that for third harmonic of the galvanometer.

Solution. *Given :* $f = 50$ Hz; $J = 0.012$ kg-m^2; $D = 0.028$ N-m/rad.-s^{-1}

The amplitude of vibration of frequency ω is :

$$A = \frac{GI_{max}}{\sqrt{[\omega^2 D^2 + (C - \omega^2 J)^2]}} \quad \text{[Eqn. 4.76]}$$

∴ The ratio of amplitudes of fundamental and third harmonic is :

$$\frac{A_1}{A_3} = \frac{\sqrt{(3\omega_1 D)^2 + (C - 9\omega_1^2 J)^2}}{\sqrt{(\omega_1 D)^2 + (C - \omega_1^2 J)^2}}$$

But, for a *tuned galvanometer*, $\omega_1 = \sqrt{\frac{C}{J}}$ or $C = \omega_1^2 J$

$$\therefore \quad \frac{A_1}{A_3} = \frac{\sqrt{(3\omega_1 D)^2 + (\omega_1^2 J - 9\omega_1^2 J)^2}}{\omega_1 D}$$

$$= \frac{\sqrt{(3\omega_1 D)^2 + (8\omega_1^2 J)^2}}{\omega_1 D}$$

$$= \frac{\sqrt{(3\times\pi\times50\times0.028)^2 + \{8\times(2\pi\times50)^2\times0.012\}^2}}{2\pi\times50\times0.028}$$

$= \mathbf{1077}$ **(Ans.)**

4.7.5. Duddell's Oscillograph

This instrument is essentially a vibration galvanometer having a *low period of vibrations.* The natural frequency is about 10 kHz.

Construction :

Fig. 4.22 shows the schematic arrangement of a Duddell's oscillograph element :

— It consists of a single loop of thin phosphor-bronze strip which forms the moving system and is called a *vibrator loop*. This loop is situated in the field of a permanent magnet. Highly saturated electromagnets are used when high sensitivities are required.

— The vibrator loop is formed by passing the strip over an ivory pulley as shown in the figure.

— A spring tension adjusting device is attached to the ivory pulley

— A small mirror for reflecting light is cemented to the loop midway between the bridge pieces.

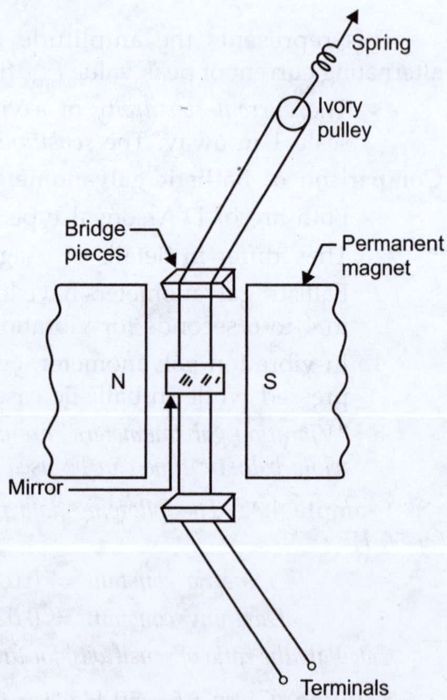

Fig. 4.22. Duddell's oscillograph element.

Such types of galvanometer can be constructed with diameter 3 mm to 6 mm and about 75 mm length and are called *pencil galvanometer*. They consist of moving element of the galvanometer and are set in a fixed frame of the permanent magnets. Two or more units per cm can be constructed and 50 or more galvanometers are included in a single oscillograph to record many waveforms simultaneously.

Working :

— When an alternating current is passed through the loop, an alternating torque acts upon the loop and it vibrates.

— When a beam of light is cast upon the mirror, it is reflected. As the loop is vibrating, the reflected beam will move backwards and forwards.

— If a wave shape is to be recorded, a photographic film is passed at a constant speed in a direction perpendicular to that of movement of the reflected beam of light. The speed of film depends upon the frequency of the wave being recorded. The wave shape is recorded on the film as a result of the two perpendicular motions *i.e.* of the film and the reflected light.

The oscillograph is calibrated by passing a known value of direct current in the loop and recording the deflection produced by the current.

— If, however, the waveform is *not to be recorded* but is *only to be observed,* the arrangement as shown in Fig 4.23 is used. The reflected beam of light from the vibrator mirror falls on the plain mirror and is further relected on to a screen. The plain mirror is rocked by a cam which is driven by a synchronous motor which in turn is supplied from the source whose waveform is to be observed. The plain mirror driven by the cam gives forward motion of the reflected beam on the screen. The forward motion is carried out for $1\frac{1}{2}$ cycles of the wave, and the plain mirror is returned to the original position during the next half cycle. During this cycle, the light is cut off by rotating shutter as shown in the figure. This is repeated continuously and the wave appears stationary on the screen on account of persistence of vision.

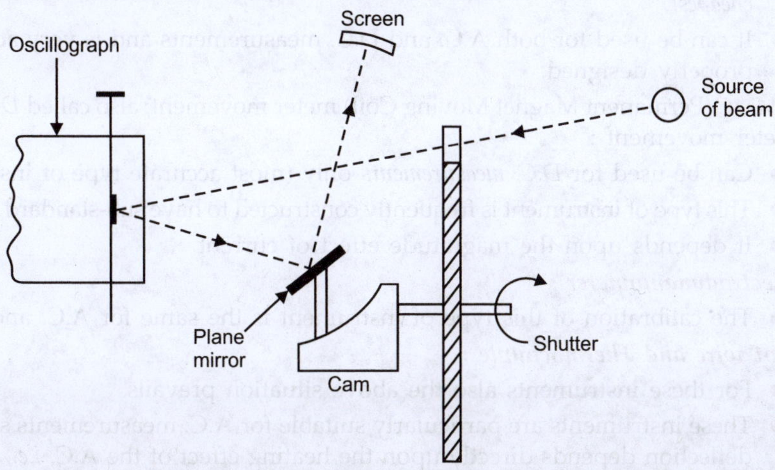

Fig. 4.23. Observation of waveform with Duddell's oscilloscope.

- Duddell's oscillographs are *very useful at low voltages and low frequencies.* They can be provided with a large member of elements and therefore a number of quantities may be *observed or recorded simultaneously.* They provide *convenient facilities for photographic purposes.*

However, their use is *limited to frequencies upto 500 Hz if an accurate record of the waveform is required.* The use of these instruments is quite *complicated.* Hence, these oscillographs are not used these days and are being replaced by cathode ray oscillographs (CROs).

4.8 ANALOG AMMETERS, VOLTMETERS AND OHMMETERS

4.8.1. Introduction

- *Analogue* ammeters and voltmeters are classified together, since there is no basic difference in their operating principles. The action of all ammeters and voltmeters, except those of the electrostatic variety, depends upon a deflecting torque principle by an electric current. In an ammeter, this torque is produced by the current to be measured, or by a definite fraction of it. In a voltmeter, it is produced by a current that is proportional to the voltage to be measured. Hence, both *voltmeters and ammeters are essential current measuring devices.*

— The ***"ammeter"*** has a *low resistance* so that when it is connected in ***series*** with any circuit, it does not change the current. Consequently, there is a small voltage drop and small power is absorbed.

The *"voltmeter"* has a *high resistance* and it is so designed that when connected in *parallel* to the circuit for measuring voltages it does *not* take appreciable current; consequently power consumed is small.

— *An ammeter of low range can be used as a voltmeter by connecting an external resistance in series with it.*

4.8.2. Types of Instruments mainly used as Ammeters and Voltmeters

The following types of instruments are mainly used as *ammeters* and *voltmeters.*

1. *Moving iron :*
   - It is the *most commonly used* form of indicating instrument, as well as the *cheapest.*
   - It can be used for both A.C. and D.C. measurements and is very accurate, if properly designed.
2. *PMMC* (Permanent Magnet Moving Coil) meter movement; also called *D' Arsonval* meter movement :
   - Can be used for *D.C. measurements* only (most accurate type of instrument).
   - This type of instrument is frequently constructed to have sub-standard accuracy.
   - It depends upon the magnitude effect of current.
3. *Electrodynamometer :*
   - The calibration of this type of instrument is the same for A.C. and D.C.
4. *Hot wire and Thermocouple :*
   - For these instruments also the above situation prevails.
   - These instruments are particularly suitable for A.C. measurements since their deflection depends directly upon the heating effect of the A.C. *i.e.,* upon the r.m.s. value of the current. Their readings are therefore *independent* of the frequency.
5. *Rectifier.*
6. *Electrostatic :*
   - These instruments used as voltmeters have the advantage that *their power consumption is exceedingly small.*
   - They can be made to cover a large range of voltages and can be constructed to have sub-standard accuracy.
7. *Induction type :*
   - The induction principle is most *generally used for watt-hour meters. This principle is not preferred for use in ammeters and voltmeters because of the comparatively high cost and inaccuracy of the instrument.*

4.8.3. Errors in Ammeters and Voltmeters

The common errors in ammeters and voltmeters are on account of the following *factors :*

(*i*) Friction.

(*ii*) Change in temperature.

(*iii*) Mechanical unbalance.

(*iv*) Variation of the electrical and mechanical properties of material employed with time.

1. Friction error :

— The *friction of pivot in the jewel* produces a frictional torque which opposes the deflection of the moving element and as a result, the pointer comes to rest a short distance from its correct position; this causes an error in readings of the instrument.

This error is more serious for sensitive instruments designed for low operating torque.

— In order to *reduce the effect of friction torque,* and consequently the error produced by it the weight of the moving system must be made as small as possible compared with the operating force, *i.e.,* the *ratio of torque to weight must be large.* In high quality instruments, the ratio of deflecting torque of full scale to weight of moving system is rarely much less than 0.01 Nm/kg.

From the point of view of a small friction torque a *vertical spindle is generally preferred to a horizontal one.*

2. Temperature error :

— The most serious error is produced by the heat generated in the instrument, or by changes in ambient (room) temperature, is *due to a change in the resistance of the working coil.* Such a change of resistance is of *little importance in ammeters* but in *voltmeters,* in which working current should be directly proportional to the applied voltage, it is *essential that the resistance of the instrument remains as nearly constant as possible.*

The power loss in the instrument should be small, and resistance coils which are likely to produce appreciable heating should be mounted, if possible, in such a position that they are *well-ventilated.*

— In order to *eliminate temperature errors, the working coil is wound with copper wire and is of comparatively low resistance.* A high *"swamping resistance"* of material whose resistance temperature coefficient is small, is connected in series with the coil, so that, although the resistance of the coil may change considerably, the *change in total resistance of circuit is small.*

Other errors resulting from heating may be caused by *expansion of control spring, or of other parts of the instrument,* although such errors are usually *small.*

3. Observational error :

— These errors are due to *mis-reading of the scale, parallex in readings* and *errors of estimation.* Precision of readings becomes difficult if the pointer has a broad tip. *Interpolation errors* may enter, particularly if the scale readings are too large or too small.

— *Parallex error* is caused by the observer *not* having his line of sight on the pointer exactly at right angles to the plane of the scale. Such an error is eliminated by providing *a mirror beneath the scale and a knife-edged pointer.*

4.8.4. Moving-Iron Instruments

Moving-iron instruments are commonly used in *laboratories and switch boards* at commercial frequencies because they are *very cheap and can be manufactured with required accuracy.*

4.8.4.1. Principle of working

The general principle of working of a moving-iron instrument (which is strictly relevant to *attraction type* of moving-iron instruments) is explained below :

— The moving element of the system consists of a plate or vane of soft iron or of high permeability steel. This iron vane is so situated that it can move in a magnetic field produced by a stationary coil.

— When the coil is excited, it becomes an electromagnet and iron vane moves in such a way so as to increase the flux of the electromagnet; this is because the vane tries to occupy a position of minimum reluctance. Thus, the force produced is always in such direction so as to increase the inductance of the coil (since inductance is *inversely proportional to reluctance of magnetic circuit of the coil*).

4.8.4.2. Theory of the moving-iron instruments

• An expression for the torque of a moving iron instrument may be derived by considering the energy relations when there is a small increment in current supplied to the instrument. When this happens there will be a *small deflection* $d\theta$ of the pointer and *work will be done.* This is given by $T_d \cdot d\theta$, where T_d is the deflecting torque of the instrument. There will also be a *change in the energy stored owing to inductance.*

Let, I = Initial current,

L = Inductance of the instrument, and

θ = Deflection.

Now, if the current increases by dI, then deflection changes by $d\theta$ and inductance by dL. In order to effect an increment dI in the current, there must be an increase in the applied voltage given by :

$$e = \frac{d}{dt}(LI) = I\frac{dL}{dt} + L \cdot \frac{dI}{dt} \qquad ...(i)$$

The *electrical energy supplied* is,

$$e\,I\,dt = I^2\,dL + IL\,dI \qquad ...(ii)$$

The stored energy changes from

$$\frac{1}{2}I^2L \text{ to } \frac{1}{2}(I+dI)^2(L+dL) \qquad ...(iii)$$

Hence the *change in stored energy* is

$$= \frac{1}{2}(I+dI)^2(L+dL) - \frac{1}{2}I^2L$$

or,

$$= \frac{1}{2}(I^2 + 2\,I\,dI + dI^2)(L+dL) - \frac{1}{2}I^2L$$

Neglecting second-and higher-order terms in small quantities, this reduces to :

$$IL\,dI + \frac{1}{2}I^2\,dL$$

Invoking the principle of conservation of energy, we have :

Electrical energy supplied = Increase in stored energy + mechanical work done

$$I^2dL + IL\,dI = IL\,dI + \frac{1}{2}I^2\,dL + T_d \cdot d\theta$$

or,

$$T_d \cdot d\theta = \frac{1}{2}I^2dL$$

or,

$$T_d = \frac{1}{2}I^2\frac{dL}{d\theta} \qquad ...(4.77)$$

where, T_d is in Nm, I in amps., L in henry, and θ in radians.

- The moving system is provided with control springs and in turn the deflecting torque T_d is balanced by the controlling torque, $T_c = C\theta$, where C is the control spring constant (Nm/rad.) and θ is the deflection (rad.).

At steady deflection state :

$$T_c = T_d$$

or, $$C\theta = \frac{1}{2} I^2 \frac{dL}{d\theta}$$

or, $$\text{Deflection, } \theta = \frac{1}{2} \frac{I^2}{C} \frac{dL}{d\theta} \qquad ...(4.78)$$

It is evident from eqn. (4.78) that the *deflection is proportional to square of the r.m.s. value of the operating current.* The deflecting torque is, therefore, *unidirectional* whatever may be the polarity of the current.

4.8.4.3. Classification of moving-iron instruments

Moving-iron instruments can be divided into two types :

1. *Attraction type* ..... in which a sheet of soft iron is *attracted towards a solenoid.*
2. *Repulsion type* ..... in which two parallel rods or strips of soft iron, magnetised inside a solenoid, are regarded as *repelling each other.*

1. Attraction type :

Fig. 4.24, shows the sectional front and an end elevation of the attracted-iron type instrument.

It consists of a solenoid (or coil) C and oval shaped soft-iron disc D in such a way that it can move in or cut of the solenoid. To this iron, a pointer P is attached so that it may deflect along with the moving iron over a graduated scale. The soft-iron disc is made of sheet metal specially shaped to give a scale as nearly uniform as possible.

— When the current to be measured (or a definite fraction of the current to be measured or proportional to the voltage to be measured) is passed through the solenoid, a magnetic field is set up inside the solenoid, which in turn magnetises the iron. Thus the soft-iron disc is attracted into the solenoid/coil, causing the spindle and the pointer to rotate. *Damping* is provided by vane V attached to the spindle and moving in an *air chamber*, and *control* is by *hair spring.*

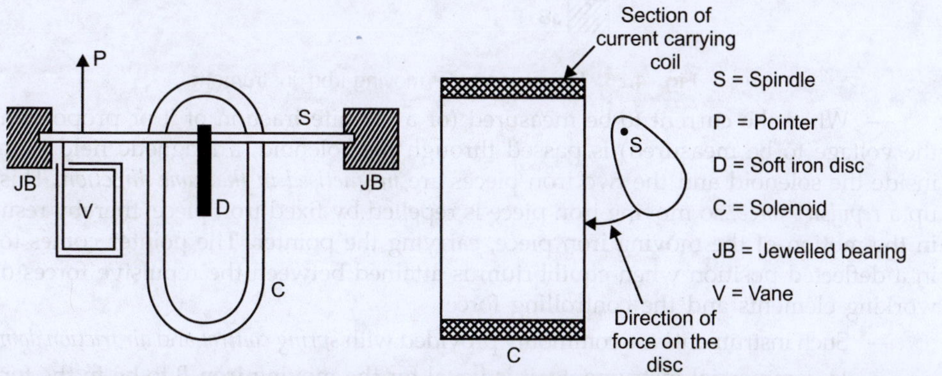

Fig. 4.24. Attraction-type moving-iron instrument

2. Repulsion type :

Repulsion-type moving-iron instrument is shown in Fig. 4.25 :

— Here, there are two irons, one fixed (*A*) and the other mounted on a short arm fixed (*B*) to the instrument spindle. The two irons lie in the magnetic field due to a solenoid/coil C. When there is no current in the coil the two iron pieces (moving one and fixed one) are almost touching each other and the pointer rests on zero position.

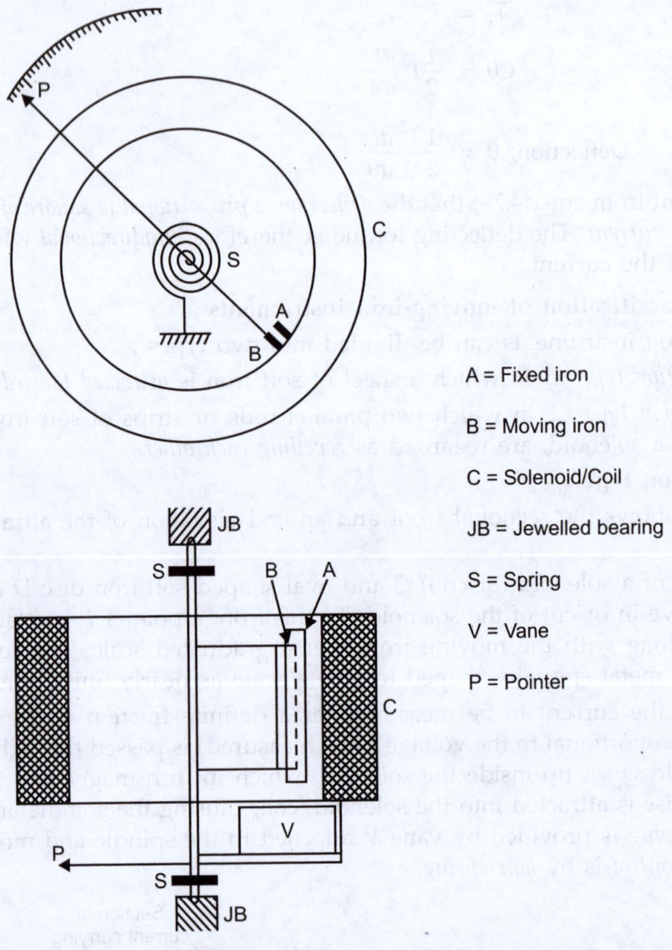

Fig. 4.25. Repulsion-type moving-iron instrument

— When the current to be measured (or a definite fraction of it or proportional to the voltage to be measured) is passed through the solenoid, a magnetic field is set up inside the solenoid and the two iron pieces are *magnetised in the same direction.* This sets up a *repulsive force,* so moving iron piece is repelled by fixed iron piece, thereby resulting in the motion of the moving iron piece, carrying the pointer. The pointer comes to rest in a deflected position when equilibrium is attained between the repulsive forces of the working elements and the controlling force.

— Such instruments are commonly provided with *spring control* and *air friction damping.*

— In commercial instruments, it is usual for the moving-iron *B* to be in the form of a *thin curved plate* and for the fixed iron *A* to be a tapered curved sheet. This construction can be arranged to give *a longer and more uniform scale* than is possible with the rods shown in Fig. 4.25.

Deflecting torque in moving-iron instruments :

In both the attraction and repulsion type moving-iron instruments it is found that for a given position of the moving system, the value of the *deflecting torque is proportional to the square of the current, so long as the iron is working below saturation*. Hence, if the current waveform is as shown in Fig. 4.26, the variation of the deflecting torque is represented by the dotted wave. If the supply frequency is 50 Hz, the torque varies between zero and a maximum 100 times a second, so that the moving system (due to its inertia) takes up a position corresponding to the mean torque, where

mean torque $\propto$ mean value of the square of the current = kI^2

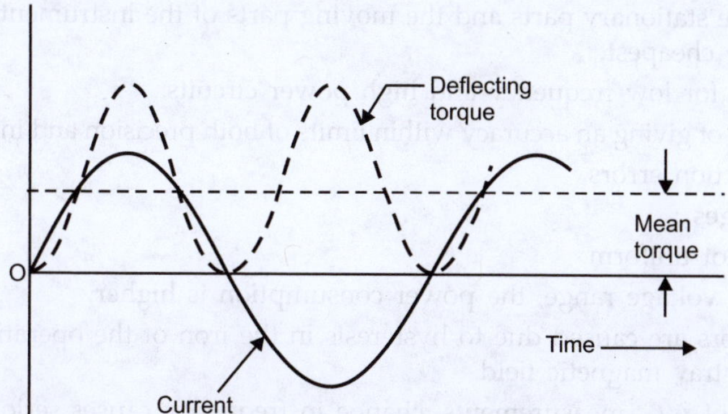

Fig. 4.26. Deflecting torque in a moving-iron instrument

where, k = a constant for a given instrument, and

I = r.m.s. value of the current.

Thus the moving-iron instrument can be used to *measure both direct current and alternating current*, and in the latter case the instrument gives the r.m.s. value of the current.

Owing to the deflecting torque being proportional to the square of the current, the scale divisions are not uniform, being cramped at the beginning and open at the upper end of the scale.

Note. For both types of instruments (attraction-type and repulsion-type), the necessary magnetic field is produced by the *ampere-turns of a current-carrying coil.*

— In case the instrument is to be used as an *ammeter*, the coil has *comparatively few turns of thick wire so* that the ammeter has *low resistance* because it is connected in *series* with the circuit.

— In case it is to be used as a *voltmeter*, the coil has *high impedance* so as to draw as small a current as possible since it is connected in *parallel with the circuit*. As current through the coil is small it has large *number of turns* in order to produce sufficient ampere-turns.

4.8.4.4. Comparison between attraction and repulsion types of instruments

Although, in general, it may said that both types of these instruments have the same advantages and are subject to same limitations, yet they differ as follows :

— An *attraction type instrument* will usually have a *lower inductance* that the corresponding repulsion type instrument, and voltmeters will therefore be *accurate over a wider range of frequency and there is a greater possibility of using shunts and ammeters.*

— Repulsion instruments, however, are *more suitable for economical production in manufacture, and a nearly uniform scale is more easily obtained.* These instruments are *much more common than the attraction type.*

4.8.4.5. Advantages, disadvantages and applications of moving-iron instruments

Advantages :

1. Can be used both in D.C. as well as in A.C. circuits.
2. Robust and simple in construction.
3. Possess high operating torque.
4. Can withstand overload momentarily.
5. Since the stationary parts and the moving parts of the instruments are simple so they are cheapest.
6. Suitable for low frequency and high power circuits.
7. Capable of giving an accuracy within limits of both precision and industrial grades.
8. Less friction errors.

Disadvantages :

1. Scales not uniform.
2. For low voltage range, the power consumption is higher.
3. The errors are caused due to hysteresis in the iron of the operating system and due to stray magnetic field.
4. In case of A.C. measurements, change in frequency causes serious error.
5. With the increase in temperature the stiffness of the spring decreases.

- There is a *difference* between D.C. and A.C. calibrations because of inductance effect of the instrument and eddy currents on A.C. The instruments should be calibrated for frequency at which these are to be tested.

Applications :

- The moving iron instruments are widely used as ammeters and voltmeters for D.C. and A.C.
- These instruments can be easily constructed for ranges of 0.1 A to 30 A without the use of a shunt and a wide range of voltages.

4.8.4.6. Errors in moving-iron instruments

The various types of errors which occur in moving-iron instruments are discussed as under :

I. Errors with both D.C. and A.C.

(i) *Errors due to hysteresis :*

— Because of hysteresis in the iron parts of the moving system, readings are higher for descending values but lower for ascending values of current (and voltage).

— The hysteresis error is almost completely *eliminated* by using Mumetal or Perm-alloy, which have negligible hysteresis loss.

(ii) *Errors due to stray fields :*

— External stray magnetic fields are liable to affect adversely the accurate functioning of the instrument.

— These errors can be *minimized* by using an iron case or a thin iron shield over the working parts.

(iii) Error due to temperature changes :

— The effect of temperature changes on moving-iron instruments arise chiefly from temperature coefficient of spring; the error may be 0.02 per cent per °C change in temperature.

— The errors in *voltmeter* are caused due to self-heating of coil and series resistance. Therefore, the series resistance should be made of material like *manganin* which has a small temperature coefficient. The value of series resistance should be very large as compared with the coil resistance in order to minimize errors due to self-heating. The series resistance is about ten times the coil resistance in case of switch board instruments.

II. Errors with A.C. only

(i) Errors due to change in frequency : The changes in frequency may cause errors due to *changes of reactance of the working coil* and also due to changes of magnitude of *eddy currents* set up in the metal parts of instrument.

(a) Reactance of instrument coil :

If the frequency of the supply changes, the reactance of the coil will change. This causes change in the current in coil of the voltmeter and hence an error in measurement. Since an ammeter is in series with the circuit, the current in its coil is the same as load current. Hence, *change in frequency causes an error in voltmeter reading only.*

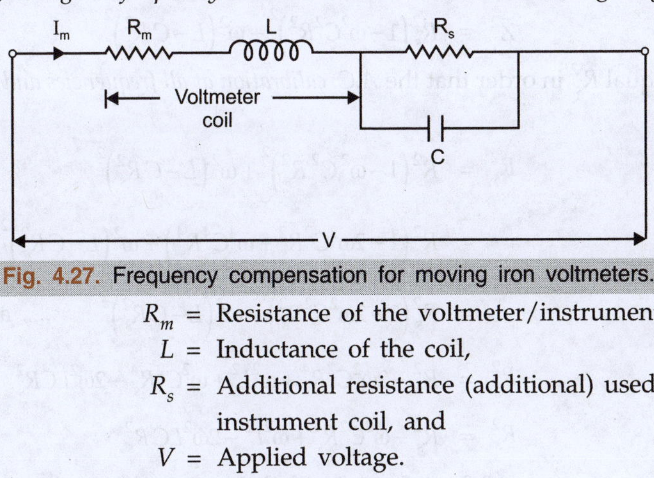

Fig. 4.27. Frequency compensation for moving iron voltmeters.

Let, R_m = Resistance of the voltmeter/instrument coil,

L = Inductance of the coil,

R_s = Additional resistance (additional) used in *series* with instrument coil, and

V = Applied voltage.

Then the current through the voltmeter coil,

$$I_m = \frac{V}{\sqrt{(R_m + R_s)^2 + \omega^2 L^2}} \quad \text{...(4.79)}$$

If the frequency is *lower* than the rated value, I_m would be higher and the error will be *positive.* If the frequency is *higher* than rated value, I_m would be lower and error will be *negative.*

One of method of compensation for frequency error is to connect a capacitor C across the series resistance R_s as shown in Fig. 4.27. The impedance of the circuit,

$$\bar{Z} = R_m + j\omega L + \frac{R_s}{1 + j\omega C R_s}$$

$$= R_m + j\omega L + \frac{R_s}{1 + j\omega C R_s} \times \frac{1 - j\omega C R_s}{1 - j\omega C R_s}$$

or, $$\bar{Z} = R_m + j\omega L + \frac{R_s - j\omega C R_s^2}{1+\omega^2 C^2 R_s^2} \quad ...(4.80)$$

The value of C can be found from eqn. (4.80) by equating Z to resistance $R_m + R_s$. Generally $R_m << R_s$. Then,

$$\bar{Z} = j\omega L + \frac{R_s - j\omega C R_s^2}{1+\omega^2 C^2 R_s^2}$$

Since $\omega C R_s << 1$, we can write

$$\bar{Z} = j\omega L + \left(R_s - j\omega C R_s^2\right)\left(1 - \omega^2 C^2 R_s^2\right)$$

$$= j\omega L + R_s - j\omega C R_s^2 - \omega^2 C^2 R_s^3 + j\omega^3 C^3 R_s^4$$

$$= R_s - \omega^2 C^2 R_s^3 + j\left[\omega L - \omega C R_s^2(1-\omega^2 C^2 R_s^2)\right]$$

$$\simeq R_s - \omega^2 C^2 R_s^3 + j\left[\omega L - \omega C R_s^2\right] \quad \text{..... as } \omega^2 C^2 R_s^2 << 1$$

$$= R_s\left(1 - \omega^2 C^2 R_s^2\right) + j\omega\left(L - C R_s^2\right)$$

or, $$\bar{Z}^2 = R_s^2\left(1-\omega^2 C^2 R_s^2\right)^2 + \omega^2\left(L - C R_s^2\right)^2$$

This must equal R_s^2 in order that the *A.C. calibration at all frequencies and D.C. calibration is the same.*

$$R_s^2 = R_s^2\left(1-\omega^2 C^2 R_s^2\right)^2 + \omega^2\left(L - C R_s^2\right)^2$$

$$= R_s^2\left(1 - 2\omega^2 C^2 R_s^2 + \omega^4 C^4 R_s^4\right) + \omega^2\left(L - C R_s^2\right)^2$$

$$= R_s^2\left(1 - 2\omega^2 C^2 R_s^2\right) + \omega^2\left(L - C R_s^2\right)^2 \quad \text{...... as } \omega^4 C^4 R_s^4 << 1$$

or, $$R_s^2 = R_s^2 - 2\omega^2 C^2 R_s^4 + \omega^2 L^2 + \omega^2 C^2 R_s^4 - 2\omega^2 L C R_s^2$$

or, $$R_s^2 = R_s^2 - \omega^2 C^2 R_s^4 + \omega^2 L^2 - 2\omega^2 L C R_s^2$$

or, $$\omega^2 L^2 - 2\omega^2 L C R_s^2 - \omega^2 C^2 R_s^4 = 0$$

or, $$L^2 - 2LC R_s^2 - C^2 R_s^4 = 0$$

or, $$L = \frac{2C R_s^2 \pm \sqrt{4C^2 R_s^4 + 4C^2 R_s^4}}{2}$$

or, $$L = \frac{2C R_s^2 \pm 2.83\, C R_s^2}{2} = 2.41 C R_s^2$$

$\therefore$ $$C = \frac{L}{2.41 R_s^2} = 0.41 \frac{L}{R_s^2} \quad ...(4.81)$$

Because of the approximations involved in deriving eqn. (4.81), the *above value of C will provide compensation only for a limited frequency range (upto about 100 Hz only).*

(*b*) *Eddy currents :* These errors are caused by eddy currents induced in the iron parts of the instruments.

Refer to Fig. 4.28.

Let, M = Mutual inductance between the instrument coil and the iron parts,

I = Current in the instrument coil,

I_e = Eddy current,

I_{e1}, I_{e2} = Component of eddy current I_e,

R_e, L_e = Resistance and inductance of eddy current path, and

$E_e = \omega MI$ = Induced voltage due to current I in the instrument coil; it lags I by 90°.

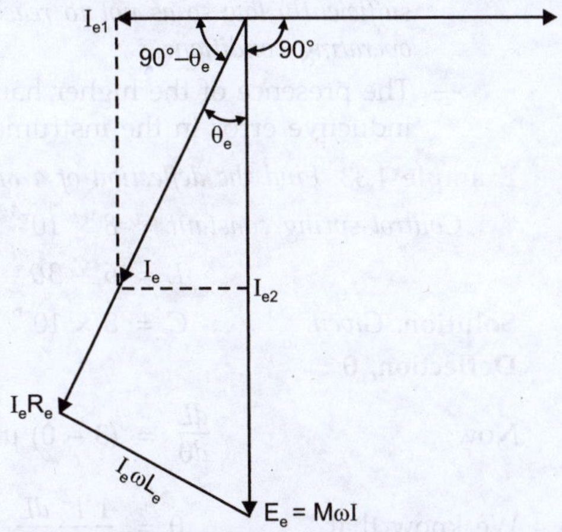

Fig. 4.28. Effect of eddy currents in moving-iron instruments.

Now,
$$I_e = \frac{E_e}{\sqrt{R_e^2 + \omega^2 L_e^2}} = \frac{\omega MI}{\sqrt{R_e^2 + \omega^2 L_e^2}} \qquad ...(4.82)$$

I_e lags E_e by an angle $\theta_e = \tan^{-1}\left(\frac{\omega L_e}{R_e}\right)$

A component of this current, $I_{e1} = I_e \cos (90° - \theta_e) = I_e \sin \theta_e$ opposes the instrument current I and sets up an opposite field thus reducing the torque on the moving system. Referring to phase diagram shown in Fig. 4.28, we have :

$$I_{e1} = I_e \sin \theta_e = \frac{\omega MI}{\sqrt{R_s^2 + \omega^2 L_e^2}} \times \frac{\omega L_e}{\sqrt{R_e^2 + \omega^2 L_e^2}}$$

$$= \frac{\omega^2 ML_e I}{R_e^2 + \omega^2 L_e^2}$$

$$\therefore \qquad I_{e1} \simeq \frac{\omega^2 ML_e I}{R_e^2} \quad \text{when } R_e >> \omega L_e \textit{ i.e., } \omega \textit{ is small}$$

$$\simeq \frac{MI}{L_e} \quad \text{when } \omega L_e >> R_e \textit{ i.e., when } \omega \textit{ is large}$$

Thus we find that *at low frequencies the eddy current error increases with square of the frequency while at high frequencies the error is practically constant.*

- *For reducing eddy current error the metal parts are placed as far out of the field as possible.*

(*ii*) *Waveform error :*

— The moving-iron instruments may be seriously affected by waveform, both on account of the change in the form of flux waveform and, in the case of voltmeters, the effect of harmonics upon the inductance.

— A moving-iron instrument with moderately saturated iron calibrated on a sine wave current and used on a peaky current wave gives readings *considerably lower than actual ones.* Contrary a flat top wave causes the instrument to read too high. This is the reason for *keeping the induction in iron sufficiently low so as not to reach the bend of the magnetisation curve under any operating condition.*

— The presence of the higher harmonics on any considerable scales causes the inductive error in the instrument and makes it to read too low.

Example 4.33. *Find the deflection of a moving-iron ammeter having the following data :*

Control spring constant = 8×10^{-6} Nm/rad.; Current = 6A,

$$L = 6 + 3\theta - 0.5\,\theta^2\ \mu H.$$

Solution. *Given:* $C = 8 \times 10^{-6}$ Nm/rad.; $I = 6A$; $L = 6 + 3\theta - 0.5\,\theta^2\ \mu H$

Deflection, θ :

Now, $$\frac{dL}{d\theta} = (3 - \theta)\ \mu H/\text{rad. or } (3 - \theta) \times 10^{-6}\ H/\text{rad.}$$

We know that, $$\theta = \frac{1}{2}\frac{I^2}{C}\frac{dL}{d\theta} \quad \text{...[Eqn. (4.78)]}$$

or, $$\theta = \frac{1}{2} \times \frac{6^2}{8 \times 10^{-6}} \times (3 - \theta) \times 10^{-6}$$

$$\theta = 2.25 \times 10^6 (3 - \theta) \times 10^{-6} = 2.25(3 - \theta)$$

or, $$\theta = \frac{2.25 \times 3}{3.25} = 2.077 \text{ rad.} = \mathbf{119°}$$ **(Ans.)**

Example 4.34. *The inductance of a moving-iron ammeter is given by the expression :*

$$L = (12 + 5\theta - 2\theta^2)\mu H,$$

where θ is the angular deflection in radians from zero position.

Determine : (i) The spring constant.

(ii) The angular deflection in radians for a current of 10 A if the deflection for a current of 5A is 30°. **(Delhi University)**

Solution. *Given :* $L = (12 + 5\theta - 2\theta^2) \times 10^{-6}$ H

(i) Spring constant, C :

Now, $$\frac{dL}{d\theta} = (5 - 4\theta) \times 10^{-6}\ H/\text{rad.} \quad \text{...(i)}$$

Also, $$\theta = \frac{1}{2}\frac{I^2}{C}\frac{dL}{d\theta} \quad \text{...[Eqn. (4.78)]}$$

or, $$\frac{dL}{d\theta} = \frac{2C\theta}{I^2} \quad \text{...(ii)}$$

From (*i*) and (*ii*), we get

$$(5 - 4\theta) \times 10^{-6} = \frac{2C\theta}{I^2}$$

Inserting $\theta = 30°$ or $\frac{\pi}{6}$ rad. and $I = 5A$ in the above expression, we get :

$$\left(5 - 4 \times \frac{\pi}{6}\right) \times 10^{-6} = \frac{2 \times C \times (\pi/6)}{5^2}$$

or, $$2.906 \times 10^{-6} = 0.0419\ C$$

∴ $$C = \frac{2.906 \times 10^{-6}}{0.0419} = \mathbf{69.36 \times 10^{-6}\ Nm/rad.}$$ **(Ans.)**

***(ii)* Angular deflection for a current of 10A, θ :**

Substituting I = 10A, in the equation we get :

$$\theta = \frac{1}{2}\times\frac{10^2}{69.36\times10^{-6}}\times(5-4\theta)\times10^{-6}$$

or, $$\theta = 0.72(5 - 4\theta) = 3.6 - 2.88\theta$$

∴ $$\theta = \frac{3.6}{3.88} \text{ rad.} = 0.928 \text{ rad. or } \mathbf{53.2°} \text{ (Ans.)}$$

Example 4.35. *The inductance of a moving iron ammeter with a full scale deflection of 90° at 1.5A is given by the expression :*

$$L = 180 + 40\theta - 4\theta^2 - \theta^3 \ \mu H$$

where, θ is the deflection in radians from the zero position, calculate :

(i) Spring constant.

(ii) The angular deflection of the pointer for a current of 1.0 A. **(Nagpur University)**

Solution: *Given :* $L = (180 + 40\theta - 4\theta^2 - \theta^3) \times 10^{-6}$ H,

$$\theta = 90° \text{ or } \frac{\pi}{2} \text{ rad. at 1.5 A current}$$

***(i)* Spring constant, *C* :**

Now, $$\frac{dL}{d\theta} = (40 - 8\theta - 3\theta^2) \times 10^{-6} \text{ H rad.} \quad ...(i)$$

Also, $$\theta = \frac{1}{2}\frac{I^2}{C}\frac{dL}{d\theta} \quad ...[\text{Eqn. (4.78)}]$$

or, $$\frac{dL}{d\theta} = \frac{2C\theta}{I^2} \quad ...(ii)$$

From *(i)* and *(ii)*, we have :

$$(40 - 8\theta - 3\theta^2) \times10^{-6} = \frac{2C\theta}{I^2}$$

Substituting θ = π rad., and I = 1.5 A in the above expression, we get,

$$\left[40-8\times\frac{\pi}{2}-3\times\left(\frac{\pi}{2}\right)^2\right]\times10^{-6} = \frac{2C\times(\pi/2)}{(1.5)^2}$$

or, $$20.03 \times 10^{-6} = 1.396 \text{ C}$$

∴ $$C = \mathbf{14.35 \times 10^{-6} \text{ Nm/rad.}} \text{ (Ans.)}$$

***(ii)* The angular deflection for a current of 1.0 A, θ :**

Substituting I = 1.0 A in the above equation, we get

$$\theta = \frac{1}{2}\times\frac{(1)^2}{14.35\times10^{-6}}\times(40-8\theta-3\theta^2)\times10^{-6}$$

$$28.7\theta = 40 - 8\theta - 3\theta^2$$

or, $$3\theta^2 + 36.7\theta - 40 = 0$$

or, $$\theta = \frac{-36.7 \pm \sqrt{(36.7)^2+4\times3\times40}}{6} = 1.007 \text{ rad.} = \mathbf{57.7°} \text{ (Ans.)}$$

Example 4.36. *The inductance of a moving-iron instrument is given by the expression :*

$$L = (0.01 + K\theta)^2\ \mu H,$$

where θ is the angular deflection in radians from zero position. The instrument angular deflections corresponding to currents of 2A and 5A are 45° and 90° respectively. Determine the value of K.

(Bangalore University)

Solution. *Given :* $L = (0.01 + K\theta)^2\ \mu H$

Value of K :

Now, $$\frac{dL}{d\theta} = 2(0.01 + K\theta)K\ \mu H \quad ...(i)$$

Also, $$\theta = \frac{1}{2}\frac{I^2}{C}\frac{dL}{d\theta} \quad ...[\text{Eqn. } (4.28)]$$

or, $$\frac{dL}{d\theta} = \frac{2C\theta}{I^2} \quad ...(ii)$$

From (*i*) and (*ii*), we have :

$$\frac{2C\theta}{I^2} = 2(0.01 + K\theta)K \times 10^{-6}$$

Substituting $\theta = 45°$ or $\frac{\pi}{4}$ radian when $I = 2A$ and $= 90°$ or $\frac{\pi}{2}$ when $I = 5A$, we get:

$$\frac{2C\times(\pi/4)}{(2)^2} = 2\left(0.01+K\times\frac{\pi}{4}\right)K\times10^{-6} \quad ...(i)$$

and, $$\frac{2C\times(\pi/2)}{(5)^2} = 2\left(0.01+K\times\frac{\pi}{2}\right)K\times10^{-6} \quad ...(ii)$$

Dividing (*i*) by (*ii*), we get :

$$\frac{50}{16} = \frac{0.01+\frac{\pi}{4}K}{0.01+\frac{\pi}{2}K}$$

or, $$50\left(0.01+\frac{\pi}{2}K\right) = 16\left(0.01+\frac{\pi}{4}K\right)$$

or, $$0.5 + 25\pi K = 0.16 + 4\pi K$$

or, $$21\pi K = -0.34$$

or, $$K = -\frac{0.34}{21\pi} = \mathbf{-5.15 \times 10^{-3}} \text{ (Ans.)}$$

Example 4.37. *The law of deflection of a moving iron ammeter is given by the expression :*

$$I = 4\theta^n$$

where θ is the deflection in radians and n is constant. The self-inductance when the meter current is zero is 10 mH. The spring constant is 0.16 Nm/rad.

(*i*) *Establish an expression for self-inductance of the meter as a function of θ and n.*

(*ii*) *Calculate the meter current and deflection that corresponds to a self inductance of 60 mH, when n = 0.75.*

(Gorakhpur University)

Solution. *Given :* $I = 4\theta^n$; $L = 10$ mH, when $I = 0$; $C = 0.16$ Nm/rad.

(*i*) **Expression for self-inductance :**

Deflection, $\theta = \frac{1}{2}\frac{I^2}{C}\frac{dL}{d\theta}$ radian ...[Eqn. (4.78)]

or, $\frac{dL}{d\theta} = \frac{2C\theta}{I^2}$

or, $dL = \frac{2C}{I^2}\theta \cdot d\theta$

Substituting $I = 4\theta^n$ in the above expression, we get :

$$dL = \frac{2C\theta}{(4\theta^n)^2}\cdot d\theta = \frac{1}{8}C\,\theta^{1-2n}\cdot d\theta$$

Integrating the above expression, we get

$$L = \frac{C\,\theta^{2-2n}}{8(2-2n)} + A = \frac{C}{16(1-n)}(\theta)^{2-2n} + A \quad ...(i)$$

where, A = constant of integration.

We have $I = 4\theta^n$; thus $\theta = 0$, when $I = 0$. It is given that when $I = 0$, $L = 10$ mH = 0.01 H

Inserting this in eqn. (*i*), we get :

$$10 \times 10^{-3} = 0 + A \quad \text{or} \quad A = 10 \times 10^{-3}$$

Hence the *expression for self-inductance as a function of θ and n* is given by :

$$L = \frac{0.16}{16(1-n)}(\theta)^{2-2n} + 0.01\,H$$

or $$L = \frac{1}{100(1-n)}(\theta)^{2-2n} + 0.01\,H \quad \textbf{(Ans.)} \quad ...(ii)$$

(*ii*) **Deflection and meter current, θ, *I* :**

When $n = 0.75$, $L = 60$ mH = 0.06 H (*Given*)

Therefore, by putting $L = 0.06$ H and $n = 0.75$ in (*ii*), we get

$$0.06 = \frac{1}{100(1-0.75)}(\theta)^{2-2\times0.75} + 0.01$$

or, $$0.05 = 0.04\,(\theta)^{0.5}$$

∴ $$\theta = \left(\frac{0.05}{0.04}\right)^2 = 1.5625 \text{ rad. or } \mathbf{89.5°} \textbf{ (Ans.)}$$

Current, $$I = 4\theta^n = 4 \times (1.5625)^{0.75} = \mathbf{5.59\ A} \textbf{ (Ans.)}$$

Example 4.38. *The full scale deflecting torque of a 10 A moving-iron ammeter is 4 × 10⁻⁵ Nm. Determine in μH/rad. the rate of change of self-inductance of the instrument at full scale.* **(Gorakhpur University)**

Solution. *Given :* $T_d = 4 \times 10^{-5}$ Nm; $I = 10$ A

We know that; $T_d = \frac{1}{2}I^2\frac{dL}{d\theta}$ ...[Eqn. (4.77)]

∴ $$\frac{dL}{d\theta} = \frac{2T_d}{I^2} = \frac{2\times4\times10^{-5}}{(10)^2} = \mathbf{0.8\ \mu H/rad.} \textbf{ (Ans.)}$$

Example 4.39. *A moving-iron voltmeter reads correctly on 250 V D.C. If 250 V, 50 Hz A.C. is applied to it, what will be the reading of the voltmeter? The instrument coil has a resistance of 500 Ω and an inductance of 1H and series non-inductive resistance is 2 kΩ.*

With what value of capacitance must the series resistance be shunted to make the meter read correctly at 50 Hz ? **(UPTU)**

Solution. *Given :* R = 500 Ω; L = 1H; R_s = 2 kΩ or 2000 Ω.

Voltmeter reading when connected to 250 V, 50 Hz supply :

Total resistance of the moving iron voltmeters $R + R_s$ = 50 + 2000 = 2500 Ω. When the instrument is used on D.C. system the inductance of the coil will not effect the instrument readings, so if instrument reads correctly, then when the instrument is used on A.C. system the reading of the voltmeter will be *affected* by *the inductance of the coil* and the readings given by the instrument will be *incorrect*.

The impedance of the instrument circuit at 50 Hz,

$$Z = \sqrt{(R+R_s)+(2\pi/L)^2}$$

$$= \sqrt{(2500)^2+(2\pi\times 50\times 1)^2} = 2520\ \Omega$$

Current drawn by the instrument, when connected to 250 V, 50 Hz A.C. supply,

$$I = \frac{V}{Z} = \frac{250}{2520}\ \text{A}$$

Since voltmeter reads correctly on D.C. supply on 250 V, so its current corresponding to 250 V

$$= \frac{V}{R+R_s} = \frac{250}{500+2000} = 0.1\ \text{A}$$

∴ *Voltmeter reading when connected to 250 V, 50 Hz supply*

$$= 250\times\frac{(250/2520)}{0.1} = \mathbf{248\ V}\quad \textbf{(Ans.)}$$

- From eqn. (4.81) value of capacitance required to make the circuit non-inductive,

$$C = 0.41\frac{L}{R_s^2} = 0.41\times\frac{1}{(2000)^2} = \mathbf{0.1\ \mu F}\quad \textbf{(Ans.)}$$

Example 4.40. *The coil of a 250 V moving-iron voltmeter has a resistance of 500 Ω and inductance of 1 H. The current taken by the instrument when placed on 250 V, D.C. supply is 0.05 A. Determine the percentage error when the instrument is placed on 250 V, A.C. supply at 100 Hz.*

Solution. Total ohmic resistance, $R = \frac{250}{0.05} = 5000\ \Omega$

(Original calibration of the instrument is with direct current)

Reactance of the coil, $X_L = 2\pi fL = 2\pi \times 100 \times 1 = 628\ \Omega$

Coil impedance, $Z = \sqrt{R^2+X_L^2} = \sqrt{(5000)^2+(628)^2} = 5039\ \Omega$

∴ Voltage reading on A.C. = $\frac{250\times 5000}{5039} = 248\ \text{V}$

∴ Error = 248 − 250 = − 2V

$$\textbf{Percentage error} = \frac{2}{250} \times 100 = \mathbf{0.8\%}\ (low) \quad \textbf{(Ans.)}$$

Example 4.41. *A 15-volt moving-iron voltmeter has a resistance of 300 Ω and an inductance of 0.12 H. Assuming that this instrument reads correctly on D.C., what will be its readings on A.C. at 15 volts when frequency is (i) 25 Hz and (ii) 100 Hz ?*

Solution. On *D.C., only ohmic resistance* is involved and the voltmeter reads correctly. But on *A.C.*, it is the *impedance* of the instrument which has to be taken into account.

(*i*) **When frequency is 25 Hz :**

Impedance at 25 Hz,

$$Z = \sqrt{R^2 + X_L^2} = \sqrt{R^2 + (2\pi f L)^2}$$

$$= \sqrt{(300)^2 + (2\pi \times 25 \times 0.12)^2} = 300.6\,\Omega$$

$$\therefore \quad \textbf{Voltmeter reading} = 15 \times \frac{300}{300.6} = \mathbf{14.97\ V} \quad \textbf{(Ans.)}$$

(*ii*) **When frequency is 100 Hz :**

Impedance at 100 Hz,

$$Z = \sqrt{(300)^2 + (2\pi \times 100 \times 0.12)^2} = 309.3\,\Omega$$

$$\therefore \quad \textbf{Voltmeter reading} = 15 \times \frac{300}{309.3} = \mathbf{14.55\ V} \quad \textbf{(Ans.)}$$

Incidently, it may be noted that as the frequency is increased, the impedance of the voltmeter is also increased. Hence, the current is decreased and, therefore, the voltmeter readings are lower.

Example 4.42. *An A.C. voltmeter with a maximum scale reading of 50 V has an inductance of 0.09 H and a total resistance of 500 Ω. The magnetising coil is wound with copper wire having a resistance of 50 Ω and the remainder of the circuit consists of a non-inductive resistance in series with it. Find the value of capacitance, that should be placed across the non-inductive series resistance, to make the instrument read correctly both on D.C. as well as on A.C. (of 50 Hz) circuits.* **(Karnataka University)**

Solution. *Given :* $L = 0.09$ H; $R + R_s = 500\ \Omega$; R (Resistance of magnetising coil) = $50\ \Omega$; $f = 50$ Hz.

The total circuit resistance, $R + R_s = 500\ \Omega$

∴ Non-inductive series resistance, $R_s = 500 - R = 500 - 50 = 450\ \Omega$

Let C be the capacitance to be placed across R_s to make the instrument read correctly on D.C. as well as on 50 Hz A.C. supply. This will be true only when,

$$C = 0.41 \frac{L}{R_s^2} \quad \text{...[Eqn.(4.81)]}$$

or, $$C = 0.41 \times \frac{0.09}{(450)^2} = 0.1822 \times 10^{-6} \text{ or } \mathbf{0.1822\ \mu F} \quad \textbf{(Ans.)}$$

Example 4.43. *The coil of a 150 V moving-iron voltmeter has a resistance of 400 Ω and an inductance of 0.75 H. The coil is made of copper which has a resistance temperature coefficient of 0.004 / °C. The current consumed by the instrument when placed on a 150 V. D.C. supply is 0.05 A. The series resistance of the voltmeter is of Manganin with a resistance temperature coefficient of 0.00015 / °C.*

Calculate :

(i) Resistance temperature coefficient of the instrument.

(ii) Alteration of the reading between D.C. and A.C. at 100 Hz.

(iii) Capacitance of the capacitor necessary to eliminate the frequency errors.

(Punjab University)

Solution. *Given :* $V = 150$ volts; $R = 400\ \Omega$; $L = 0.75$ H, $\alpha_{cu} = 0.004/°C$;
$I = 0.05$ A; $\alpha_{mang.} = 0.00015/°C$

(*i*) **Resistance temperature coefficient of instrument, $\alpha_{inst.}$:**

Total resistance of instrument $(R + R_s) = \dfrac{V}{I} = \dfrac{50}{0.05} = 3000\ \Omega$

$\therefore$ Series (swamping) resistance, $R_s = 3000 - R = 3000 - 400 = 2600\ \Omega$

Now, change in resistance of coil/°C = $0.004 \times 400 \times 1 = 1.6\ \Omega$

Change in swamping resistance/°C = $0.0015 \times 2600 \times 1 = 0.39\ \Omega$

Total change in resistance of instrument circuit/°C = $1.6 + 0.39 = 1.99\ \Omega$

$$\therefore \quad \alpha_{inst.} = \frac{\text{Change in resistance/°C}}{\text{Total resistance}} = \frac{1.99}{3000} = \mathbf{0.000633/°C}$$ **(Ans.)**

(*ii*) **Error in the reading at 100 Hz :**

Reactance of coil at 100 Hz, $X_L = 2\pi f L = 2\pi \times 100 \times 0.75 = 471.2\ \Omega$

Impedance of instrument at 100 Hz = $\sqrt{(3000)^2 + (471.2)^2} = 3037\ \Omega$

Current drawn by the instrument at 100 Hz = $\dfrac{V}{Z} = \dfrac{150}{3037} = 0.0494$ A

$\therefore$ Reading of the instrument at 100 Hz = $\dfrac{0.0494}{0.05} \times 150 = 148.2$ V

% Error in the reading = $\dfrac{148.2 - 150}{150} \times 100 = \mathbf{1.2\%}$ **(*low*) (Ans.)**

(*iii*) **Value of *C* :**

In order to eliminate frequency error, the value of capacitance to be connected across R_s,

$$C = 0.41 \frac{L}{R_s^2} \qquad \text{...[Eqn. (4.81)]}$$

or, $$C = 0.41 \times \frac{0.75}{(2600)^2} = \mathbf{0.0455\ \mu F}$$ **(Ans.)**

4.8.5. Moving-coil Instruments – Classification

The moving-coil instruments are of the following two types :

1. Permanent magnet type ...... can be used for D.C. only.
2. Dynamometer type .... can be used for both A.C. and D.C.

4.8.6. Permanent Magnet Moving-coil (PMMC) Instruments

The permanent magnet moving-coil instrument is the *most accurate type for "D.C. measurements"*.

The working principle of these instruments is the same as that of the D'Arsonval type of galvanometers, the difference being that a direct reading instrument is provided with a pointer and a scale.

4.8.6.1. Principle, construction and working

A permanent-magnet moving coil-type instrument works on the principle that *"when a current-carrying conductor is placed in a magnetic field, it is acted upon by a force which tends to move it to one side and out of the field"*.

Construction : Refer to Fig. 4.29.

- The instrument consists of a permanent magnet *M* and a rectangular coil *C* which consists of insulated copper wire wound on light aluminium frame fitted with polished steel pivots resting in jewel bearings. The magnet is made of Alnico and has soft-iron pole-pieces *PP* which are bored out cylindrically.
- The rectangular coil C is free to move in air gaps between the soft-iron pole pieces and a soft-iron cylinder *A* (central core), supported by a brass plate (not shown).

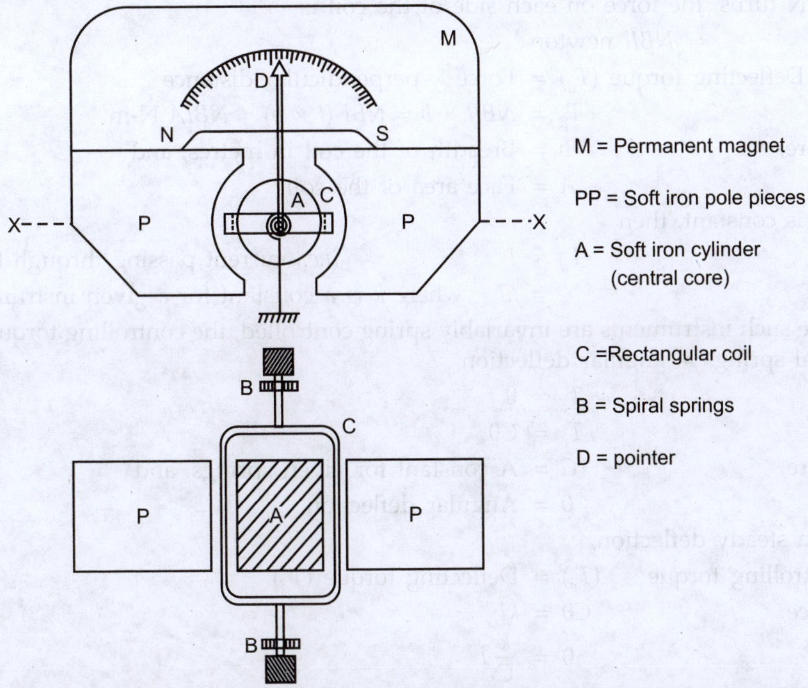

Fig. 4.29. Permanent-magnet moving-coil instrument.

The *functions of the central core A* are :

(*i*) To intensify the magnet field by reducing the length of air gap across which the magnetic flux has to pass.

(*ii*) To give a radial magnetic flux of uniform density, thereby enabling the scale to be uniformly divided.

— The movement of the coil is controlled by two phosphor bronze hair springs *BB* (one above and one below), which additionally serve the purpose of leading the current in and out of the coil. The two springs are spiralled in *opposite directions* for neutralizing the effects of changes in temperature.

— The aluminium frame not only provides support for the coil but also *provides damping by eddy currents induced in it.*

Deflecting torque. Refer to Fig. 4.30. When current is passed through the coil, forces are set up on its both sides which produce *deflection torque.* If I amperes is the current passing through the coil, the magnitude of the force (F) experienced by each of its sides is given by :

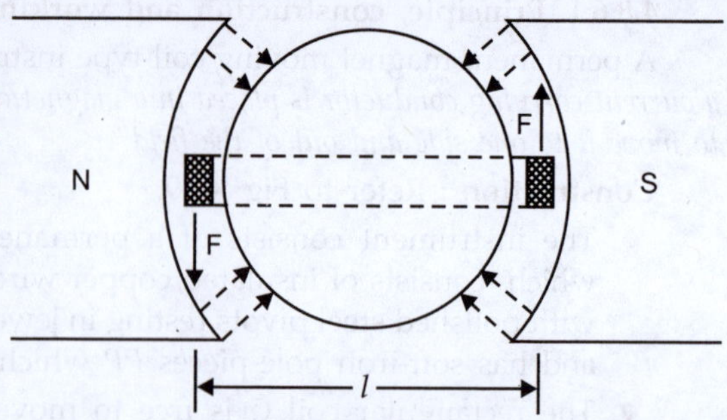

Fig. 4.30.

$$F = BIl \text{ newton}$$

where, B = flux density in Wb/m^2, and

l = length or depth of coil in metres.

For N turns, the force on each side of the coil is

$$= NBIl \text{ newton}$$

$\therefore$ Deflecting torque (T_d) = Force × perpendicular distance

or, $$T_d = NBIl \times b = NBI\,(l \times b) = NBIA \text{ N-m} \quad ...(4.83)$$

where, b = Breadth of the coil in metres, and

A = Face area of the coil.

If B is constant, then

$$T_d \propto I \quad \text{(i.e., current passing through the coil)}$$

$= kI$ where k is a constant for a given instrument.

Since such instruments are invariably spring controlled, the controlling torque (T_c) of the spiral springs ∝ angular deflection

i.e., $$T_c \propto \theta$$

or, $$T_c = C\theta$$

where, C = A constant for given springs, and

θ = Angular deflection.

For a steady deflection,

Controlling torque (T_c) = Deflecting torque (T_d)

Hence, $$C\theta = kI$$

$\therefore$ $$\theta = \frac{k}{C} I \quad ...(4.84)$$

i.e., The *deflection is proportional to current* and the *scale is therefore uniformly divided.* PMMC instrument is *unsuitable for A.C. measurements unless the current is rectified.*

Ranges :

D.C. ammeters :

(*i*) *Without shunt.........0/5 micro-amperes up to 0/30 micro-amperes.*

(*ii*) *With internal shunts........upto 0/2000 amperes.*

(*iii*) *With external shunts.......upto 0/5000 amperes.*

D.C. voltmeters :

(*i*) *Without series resistance.......0/100 milli-volts.*

(*ii*) *With series resistance.......upto 20000 or 30000 volts.*

- The PMMC system can be converted into an instrument to measure D.C. as well as A.C. quantities like current, voltage and resistance etc. It can carry a maximum current of I_m, without any modification and can withstand a maximum D.C. voltage, $V = I_m R_m$.

The various modifications of PMMC meter movement may be summed up a follows :

1. ***D.C. instruments :***
   (*i*) D.C. ammeter – By using a *shunt resistor.*
   (*ii*) D.C. voltmeter – By using *series multiplier resistor.*
   (*iii*) Ohmmeter – By using *battery and series resistor.*
2. ***A.C. instruments :***
   (*i*) Audio-frequency (AF) A.C. ammeter or voltmeter – By using a *rectifier.*
   (*ii*) Radio-frequency (RF) ammeter or voltmeter – By using a *thermocouple.*

4.8.6.2. Temperature compensation

The PMMC basic movement is not inherently insensitive to temperature, but it may be *temperature-compensated* by the appropriate use of series and shunt resistors.

- Complete temperature compensation may be accomplished by using swamping resistors as shown in Fig. 4.31. The swamping resistor is made of manganin (having temperature coefficient of practically zero) and combined with copper in the ratio of 20 : 1 to 30 : 1. The total resistance of the coil and swamping resistor increases slightly with rise in temperature but only just enough to counteract the change of springs and magnet, so that the overall temperature effect is zero.

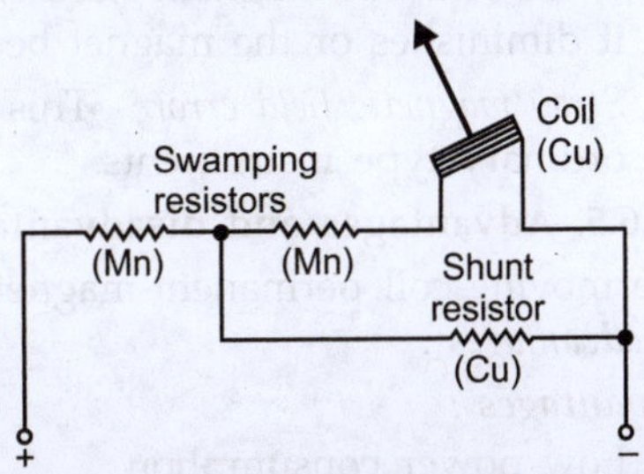

Fig. 4.31. Temperature compensatiousing series and shunt resistors.

— The main *"disadvantage"* of the use of swamping resistors is a reduction in the full-scale sensitivity of the movement; because a higher applied voltage is necessary to sustain the full-scale current.

4.8.6.3. Characteristics of PMMC meter movement :

The following *three* characteristics of PMMC meter movement will be discussed :

1. ***Full-scale deflection current (I_m) :***

- *It is the current needed to deflect the pointer all the way to the right to the last mark on the calibrated scale.*
- Typical valves of the I_m for D'Arsonval movement vary from 2 μA to 30 mA.

  — It may be remembered that for smaller currents, the number of turns in the moving coil should be *more* so that the magnetic field produced by the coil is strong enough to react with the field of permanent magnet for producing reasonable deflection of the pointer. *Fine wire* has to be used for reducing the weight of the moving coil but this increases its resistance.

  — For heavy currents, thick wire with lesser number of turns is needed so that resistance of the moving coil is comparatively less.

2. ***Internal resistance (R_m) :***

- *It is the D.C. ohmic resistance of the wire of the moving coil.*
- Typical values of R_m range from 1-2Ω for a 30 mA movement to 2 kΩ for a 50 μA movement.

3. ***Sensiivity (S) :***

- It is also known as *current-sensitivity* or *sensitivity factor.*
- It is given by the reciprocal of full scale deflection current I_m.

$$\therefore \qquad S = \frac{1}{I_m} \text{ ohm/volt} \qquad ...(4.85)$$

— The sensitivity of a meter movement depends on the *strength of the permanent magnet and number of turns in the coil.* Larger the number of turns, smaller the amount of current required to produce full-scale deflection and hence, higher the sensitivity.

4.8.6.4. Errors in PMMC instruments

The following *errors* usually occur in PMMC instruments :

1. *Frictional error* – This can be avoided by winding of coil carefully.
2. *Temperature error* – It can be avoided as discussed in Art. 4.8.3.
3. *Error due to weakening of permanent magnet* – It can be eliminated completely by weakening of mechanical control and attaching a symmetrical iron or steel needle to the coil. The magnetic needle is so arranged that magnetic control provided by it diminishes or the magnet becomes weaker.
4. *Stray magnetic field error* – This error can be avoided by using an iron case for ordinary type instruments.

4.8.6.5. Advantages and disadvantages

The moving-coil permanent-magnet type instrument have the following *advantages* and *disadvantages* :

Advantages :

(*i*) Low power consumption.
(*ii*) Their scales are uniform.
(*iii*) No hysteresis loss.
(*iv*) High torque/weight ratio.
(*v*) They have very effective and efficient eddy-current damping.
(*vi*) Range can be extended with shunts or multipliers.
(*vii*) No effect of stray magnetic field as intense polarised or unidirectional field is employed.

Disadvantages :

(*i*) Somewhat costlier as compared to moving-iron instruments.
(*ii*) Cannot be used for A.C. measurements.
(*iii*) Friction and temperature might introduce errors as in case of other instruments.
(*iv*) Some errors are set in due to the ageing of control springs and the permanent magnets.

☞ • Moving-coil permanent-magnet instrument can be *used as* :

(*i*) *"Ammeters"* — By using a *low resistance shunt.*
(*ii*) *"Voltmeters"* — By using a *high series resistance.*
(*iii*) *"Fluxmeters"* — By *eliminating the control spring.*
(*iv*) *"Ballistic galvanometers"* — By *making control springs of large moment of inertia.*

Example 4.44. *The coil of an instrument has 38 turns. The mean width and axial length of the coil are 25 mm and 20 mm respectively. If the flux density is 0.12 Wb/m², calculate the torque on the moving coil for a current of 12 mA through the coil.*

Solution. *Given :* $N = 38$; $b = 25$ mm or 0.025 m; $l = 20$ mm or 0.02 m;
$B = 0.12$ Wb/m^2; $I = 12$ mA $= 0.012$ A

Torque on the moving coil,

$$T_d = NBIlb \qquad \text{[Eqn. 4.83]}$$

$$= 38 \times 0.12 \times 0.012 \times 0.02 \times 0.025$$

$$= \mathbf{2.736 \times 10^{-5}\ Nm} \quad \textbf{(Ans.)}$$

Example 4.45. *The resistance of a moving voltmeter is 11 kΩ. The moving coil has 100 tuns and is 40 mm long and 30 mm wide. The flux density in the air gap is 0.05 Wb/m^2.*

Determine the deflection produced by 220 V if the spring control gives a deflection of 1° for a torque of 20 × 10^{-7} Nm. **(Allahabad University)**

Solution. *Given* : R = 11000 Ω; N = 100; l = 40 mm or 0.04 m; b = 30 mm or 0.03 m; B = 0.05 Wb/m^2; V = 220 V; T_d = 20 × 10^{-7} Nm/degree deflection.

Value of deflection, θ :

$$I = \frac{V}{R} = \frac{220}{11000} = 0.02 \text{ A}$$

For steady deflection state,

Deflecting torque, T_d = Controlling torque, T_c

$\therefore \quad NBIlb = 20 \times 10^{-7} \times \theta$

or, $$\theta = \frac{NBIlb}{20\times10^{-7}} = \frac{100\times0.05\times0.02\times0.04\times0.03}{20\times10^{-7}} = \mathbf{60°} \quad \textbf{(Ans.)}$$

Example 4.46. *A moving coil millivoltmeter has a resistance of 20 Ω and full scale deflection of 120° is reached when a potential difference of 100 mV is applied across its terminals. The moving coil has the effective dimensions of 3.1 cm × 2.6 cm and is wound with 120 turns. The flux density in the gap is 0.15 Wb/m^2.*

Determine the control constant of the spring and suitable diameter of copper wire for coil winding if 55 per cent of total instrument resistance is due to coil winding. ρ for copper = 1.73 × 10^{-8} Ωm. **(M.S. University of Baroda)**

Solution. *Given* : R = 20 Ω; Full scale deflection, θ = 120°; Voltage applied for full scale deflection = 100 mV = 0.1 V; l = 3.1 cm = 0.031 m; b = 2.6 cm or 0.026 m; N = 120; B = 0.15 Wb/m^2; ρ_{copper} = 1.73 × 10^{-8} Ωm.

Spring constant, *C* :

Full scale deflection current, $I = \frac{V}{R} = \frac{0.1}{20} = 0.005$ A

Full scale deflection torque, $T_d = NBIlb$

$$= 120 \times 0.15 \times 0.005 \times 0.031 \times 0.026 = 72.54 \times 10^{-6} \text{ Nm}$$

For steady deflection state, $T_d = T_c$

or, $$T_c = C\theta = 72.54 \times 10^{-6}$$

or, $$C = \frac{72.54\times10^{-6}}{120} = \mathbf{6.045 \times 10^{-7}\ Nm/degree} \quad \textbf{(Ans.)}$$

Diameter of copper wire, *d* :

Resistance of coil winding, $R = \frac{55}{100}\times 20 = 11\ \Omega$

Length of copper wire = (0.031 + 0.026) × 2 × 120 = 13.68 m

Area of cross-section of wire, $a = \frac{\rho l}{R}$

or, $$a = \frac{1.73\times10^{-8}\times13.68}{11} = 2.15 \times 10^{-8} \text{ m}^2$$

But, $a = \frac{\pi}{4} d^2$

$$\therefore \quad d = \sqrt{\frac{4a}{\pi}} = \sqrt{\frac{4 \times 2.15 \times 10^{-8}}{\pi}} = \mathbf{1.655 \times 10^{-4}\ m\ or\ 0.1655\ mm} \quad \textbf{(Ans.)}$$

Example 4.47. *The following data relate a moving coil instrument:*

Number of turns (wound on non-inductive former)	*= 200*
Width of the former	*= 20 mm*
Height of the former	*= 25 mm*
Flux density of the field	*= 0.1 Wb/m²*

Moment of the inertia of the moving parts = 3 × 10^{-7} kg-m²
Torque produced by the control spring = 90 × 10^{-7} Nm/rad.
Calculate the current in the coil to produce a deflection 120°. **(AMIE)**

Solution. *Given* : $N = 200$; $b = 20$ mm or 0.02 m; $l = 25$ mm or 0.025 m; $B = 0.1$ Wb/m²;
$T_c = 90 \times 10^{-7}$ Nm/rad.; $\theta = 120°$ or $120 \times \frac{\pi}{180} = \frac{2\pi}{3}$ rad.

Current in the coil, *I* :

For steady deflection state,

$$T_d = T_c$$

$$NBIlb = 90 \times 10^{-7} \times \frac{2\pi}{3}$$

or, $$200 \times 0.1 \times I \times 0.025 \times 0.02 = 90 \times 10^{-7} \times \frac{2\pi}{3}$$

$$I = \mathbf{1.88\ mA} \quad \textbf{(Ans.)}$$

Example 4.48. *The coil of a moving coil voltmeter is 40 mm long and 30 mm wide and has 100 turns on it. The control spring exerts a torque of 240 × 10^{-6} Nm when the deflection is 100 divisions on full scale. If the flux density of the magnetic field in the air gap is 1.0 Wb/m², calculate the resistance that must be put in series with the coil to give 1 V/division. The resistance of the voltmeter coil may be neglected.* **(Punjab University)**

Solution. *Given* : $l = 40$ mm or 0.04 m; $b = 30$ mm or 0.03 m;
$N = 100$; $T_c = 240 \times 10^{-6}$ Nm; $B = 1.0$ Wb/m²

Resistance, R :

Deflecting torque at full scale deflection,

$$T_d = NBIlb = 100 \times 1.0 \times I \times 0.04 \times 0.03 = 0.12\ I\ \text{Nm}$$

At final steady state position,

$$T_d = T_c$$

or, $$0.12\ I = 240 \times 10^{-6}$$

$\therefore$ Current at full scale deflection, $I = \frac{240 \times 10^{-6}}{0.12} = 0.002$ A

Voltage across the instrument = $IR = 0.002\ R$

$\therefore$ Volts per division $= \dfrac{0.002\,R}{100}$

This value should be equal to 1 in order to get 1 volt per division.

$$\therefore \quad \frac{0.002\,R}{100} = 1$$

or, $$R = \frac{100}{0.002} = 50000\ \Omega \text{ , or , } \mathbf{50\ k\Omega} \quad \textbf{(Ans.)}$$

4.8.6.6. Extension of range

D.C. ammeters :

Shunt resistor : Since the coil winding of a basic movement of a D.C. ammeter (PMMC galvanometer) is small and light, it *can carry only very small currents.* When it is required to measure large current, it is necessary to bypass the major part of the current through a resistance, called a *shunt,* as shown in Fig. 4.32. The resistance of the shunt can be calculated as follows.

Let, R_m = Internal resistance of the meter,

R_{sh} = Resistance of the shunt,

I_m = Current through the meter,

I_{sh} = Shunt current, and

I = Current of the circuit to be measured.

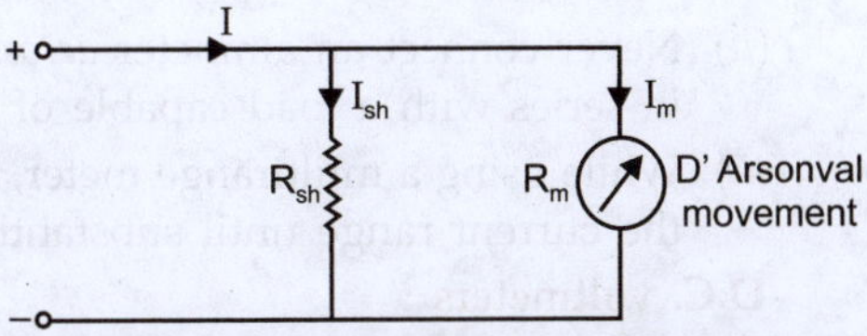

Fig. 4.32. Basic D.C. ammeter circuit.

Thus, $$I = I_m + I_{sh}$$
$$(\text{or } I_{sh} = I - I_m).$$

Also, since the voltage drop across the shunt and the meter/instrument is same,

$$I_{sh}R_{sh} = I_mR_m \qquad \text{...(4.86)}$$

or, $$R_{sh} = \frac{I_mR_m}{I_{sh}}$$

or, $$R_{sh} = \frac{I_mR_m}{(I - I_m)} \qquad \text{...(4.87)}$$

The shunt resistance with PMMC meter movement may consist of a length of constant-temperature resistance wire within the case of the instrument or it may be an external (manganin or constantan) shunt having a very low resistance.

— The *"external shunt"* consists of evenly spaced sheets of resistance material welded into a large block of heavy copper on each end of the sheets.

— The resistance material has a very low temperature coefficient, and a low thermoelectric effect exists between the resistance material and copper.

— External shunts of this type are normally used for measuring very large currents.

— **Fig. 4.33** shows a simple ***multirange ammeter*** in which the current range of the D.C. ammeter may be further

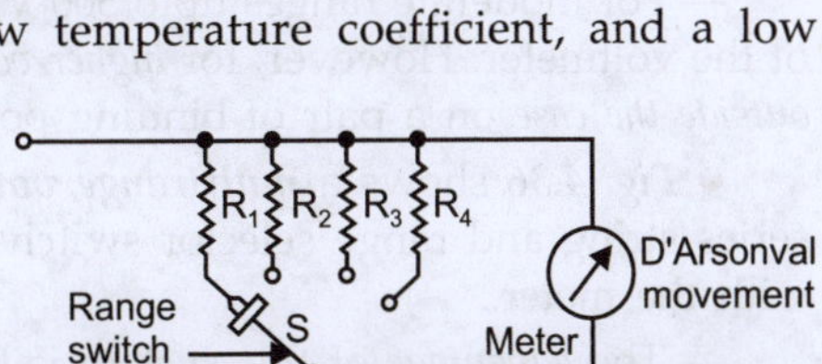

Fig. 4.33. Simple multirange ammeter.

extended by a number of shunts, selected by a *range switch.*

— *Fig. 4.34* shows a *universal or ayrton shunt,* the use of which eliminates the possibility of having the meter in the circuit without a shunt. This *advantage* is gained at the cost of a slightly higher overall meter resistance.

— The Ayrton shunt provides an excellent opportunity *to apply basic network theory to a practical circuit.*

- D.C. ammeters are commercially available in a large number of ranges, from 20 μA to 50A full-scale for a *self-contained* meter and to 500 A for a meter with *external shunt.*

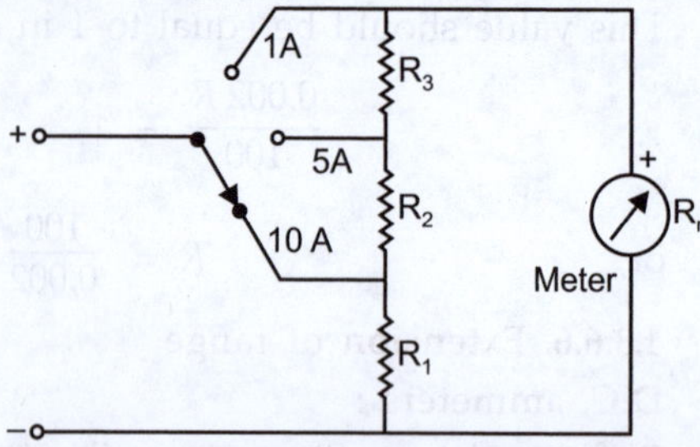

Fig. 4.34. Universal or Ayrton shunt.

While using an ammeter in measurement work, the following ***precautions*** should be observed :

(*i*) Observe the correct polarity.

(*ii*) Never connect an ammeter *across* an e.m.f. source. Always connect an ammeter in series with a load capable of limiting the current.

(*iii*) While using a multirange meter, first use the higher current range; then decrease the current range until substantial deflection is obtained.

D.C. voltmeters :

The basic D'Arsonval movement can be converted into a D.C. voltmeter by adding a series resistor or *multiplier,* as shown in Fig. 4.35. The multiplier *limits* the current through the meter so as not to *exceed* the value of the full-scale deflection current (I_{fsd}). The value of multiplier, required to extend the voltage range, is calculated from Fig. 4.35 as follows :

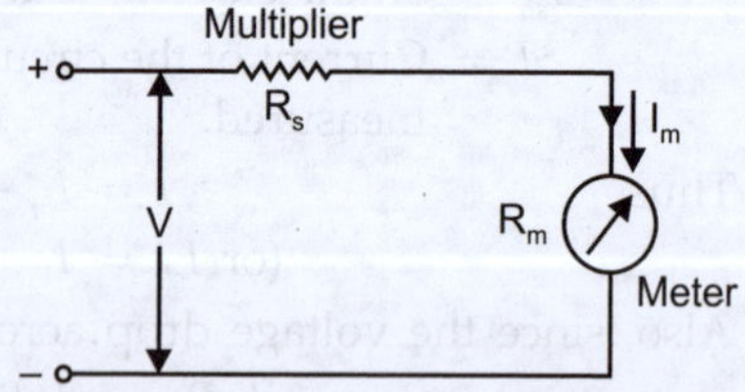

Fig. 4.35. Basic D.C. voltmeter circuit.

Let, I_m = Full scale deflection current of the meter (I_{fsd})

R_m = Internal resistance of the meter,

R_s = Multiplier resistance, and

V = Voltage of the circuit to be measured.

Now, $V = I_m (R_s + R_m)$

or, $$R_s = \frac{V - I_m R_m}{I_m} = \frac{V}{I_m} - R_m \quad ...(4.88)$$

— For moderate ranges upto 500 V, the multiplier is usually mounted *inside* the case of the voltmeter. However, for *higher voltages,* the multiplier may be mounted separately *outside the case* on a pair of binding posts *to avoid excessive heating inside the case.*

- *Fig. 4.36* shows a ***multirange voltmeter*** where the multipliers are connected in a series string and range selector switches the appropriate amount of resistance in series with the meter.

— The *advantage* of this system is that all multipliers except the first have standard resistance values and can be obtained commercially in precision tolerances. The low-range multiplier, R_4, is the only special resistor that must be manufactured to meet the specific circuit requirements.

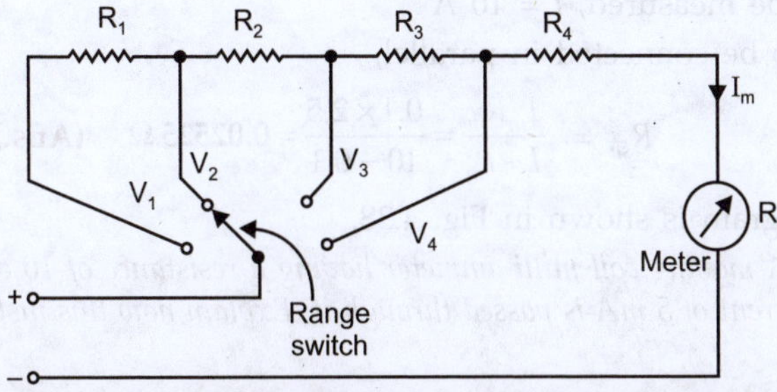

Fig. 4.36. Multirange voltmeter.

Loading effect of a voltmeter :

If the resistance of a voltmeter is *not high* as compared to the resistance of the circuit across which it is connected, the measured voltage will be *less* than the actual one. The decrease in voltage may be negligible or it may be appreciable depending on the sensitivity and input resistance of the voltmeter. *It is called **loading effect** because the voltmeter loads down the circuit across which it is connected.*

Loading effect can be minimised by using a voltmeter whose resistance is as high as possible as compared to that of the circuit across which it is connected.

— In case of an *electronic voltmeter "there is no loading problem because its input resistance is very high (10 MΩ or more).*

• It may be remembered that whereas *"accuracy"* is always required in instruments, *"sensitivity" is needed only in special applications where loading distorts which is being measured.*

Example 4.49. *A milliammeter of 2.5 ohms resistance reads upto 100 milliamperes. Calculate the resistance which is necessary to enable it to be used as :*

(i) A voltmeter reading upto 10 V.

(ii) An ammeter reading upto 10 A.

Draw the connection diagram in each case.

Solution. Resistance of the milliammeter, $R_m = 2.5\ \Omega$

Maximum current of the milliammeter, $I_m = 100$ mA = 0.1 A.

(i) Voltage to be measured, $V = 10$ volts

Resistance to be connected in series.

$$\mathbf{R_s} = \frac{V}{I_m} - R_m = \frac{10}{0.1} - 2.5 = \mathbf{97.5\ \Omega} \quad \textbf{(Ans.)}$$

Connection diagram is shown in Fig. 4.37.

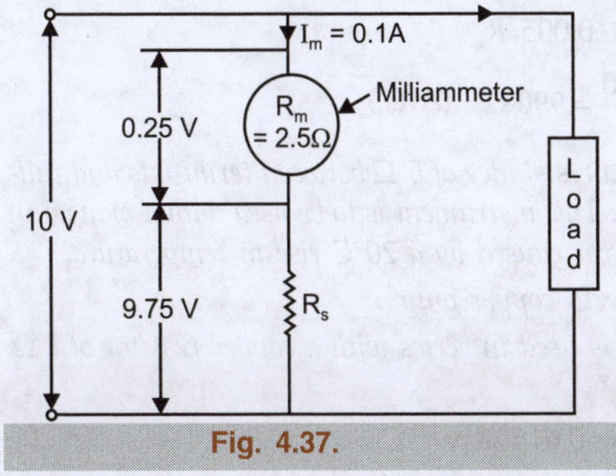

Fig. 4.37.

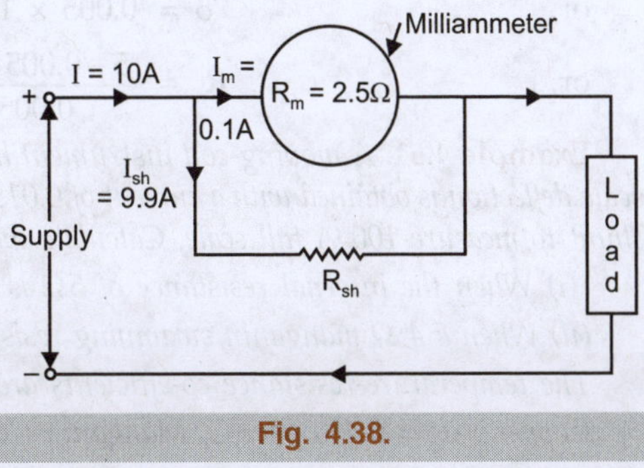

Fig. 4.38.

(*ii*) Current to be measured, I = 10 A

∴ Resistance to be connected in parallel,

$$R_{sh} = \frac{I_m R_m}{I - I_m} = \frac{0.1 \times 2.5}{10 - 0.1} = \mathbf{0.02525\ \Omega} \quad \textbf{(Ans.)}$$

Connection diagram is shown in Fig. 4.38.

Example 4.50. *A moving-coil milli-ammeter having a resistance of 10 ohms gives full scale deflection when a current of 5 mA is passed through it. Explain how this instrument can be used for measurement of :*

(i) Current upto 1 A.

(ii) Voltage upto 5 V.

Solution. Resistance of the milli-ammeter, R_m = 10 Ω

Full scale deflection current, I_m = 5 mA = 0.005 A.

(i) To measure current upto 1 A :

Resistance of the shunt, R_{sh} : Refer to Fig. 4.39.

Since voltage drop across the milliammeter and the shunt are equal

$$\therefore \quad I_m R_m = I_{sh} R_{sh} = (I - I_m)\, R_{sh}$$

$$\therefore \quad \mathbf{R_{sh}} = \frac{I_m R_m}{(I - I_m)} = \frac{0.005 \times 10}{(1 - 0.005)} = \mathbf{0.05025\ \Omega} \quad \textbf{(Ans.)}$$

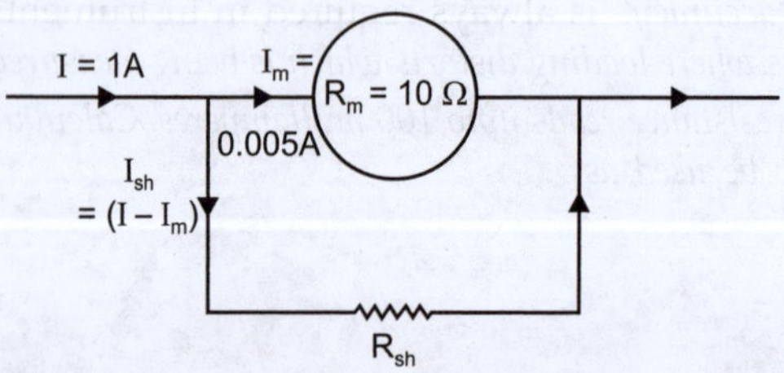

Fig. 4.39.

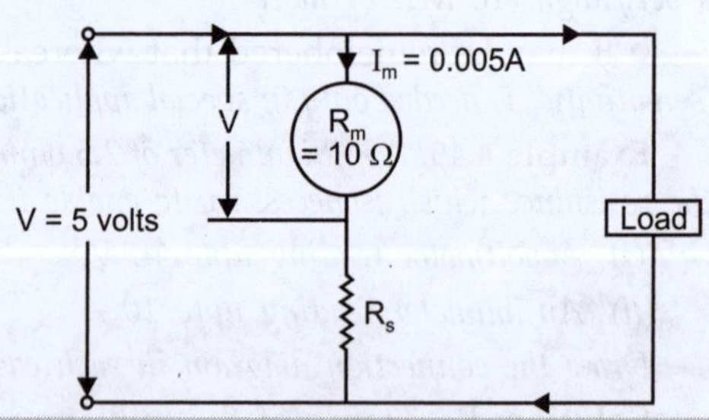

Fig. 4.40.

(ii) To measure voltage upto 5 V :

The value of external series resistance, R : Refer to Fig. 4.40.

Now, voltage across supply loads

= Voltage drop across the milliammeter + voltage drop across external series resistance R_s

$$\therefore \quad V = I_m R_m + I_m R_s$$

$$\text{or,} \quad 5 = 0.005 \times 10 + 0.005\, R_s$$

$$\text{or,} \quad R_s = \frac{5 - 0.005 \times 10}{0.005} = \mathbf{990\ \Omega} \quad \textbf{(Ans.)}$$

Example 4.51. *A moving-coil instrument has a resistance of 5 Ω between terminals and full-scale deflection is obtained with a current of 0.015 A. This instrument is to be used with a manganin shunt to measure 100 A full scale. Calculate the error caused by a 20°C rise in temperature.*

(i) When the internal resistance of 5Ω is due to copper only.

(ii) When a 4 Ω manganin swamping resistor is used in series with a copper resistor of 1 Ω.

The temperature-resistance co-efficients are :

Copper : α_c *= 0.4% per °C, Manganin :* α_m *= 0.015% per °C.*

Solution. Resistance of the instrument, $R_m = 5\Omega$

Current through the instrument, $I_m = 0.015$ A

Current to be measured, $I = 100$ A

Current through the shunt,

$$I_{sh} = I - I_m = 100 - 0.015 = 99.985 \text{ A}$$

Voltage across the shunt

$$I_m R_m = 5 \times 0.015 = 0.075 \text{ V}$$

$\therefore$ Shunt resistance, $R_{sh} = \dfrac{0.075}{99.985} = 0.00075\ \Omega \qquad [\because I_m R_m = (I - I_m) R]$

Shunt resistance after a rise of 20°C

$$= 0.00075\ (1 + 20 \times 0.00015) = 0.000752\ \Omega.$$

(*i*) The instrument resistance (which is wholly copper) after a rise of 20°C

$$= 5\ (1 + 20 \times 0.004) = 5.4\ \Omega$$

Hence, current through the instrument corresponding to 100 A in the line

$$= \frac{0.000752}{(5.4 + 0.000752)} \times 100 = 0.01392 \text{ A}$$

Reading of the instrument = $0.01392 \times 100/0.015 = 92.8$ A

$\therefore$ **Percentage error** = 100 – 92.8 = **7.2% (*low*)** **(Ans.)**

(*ii*) The instrument resistance after a rise of 20°C

$$= 1\ (1 + 20 \times 0.004) + 4\ (1 + 20 \times 0.00015)$$

$$= 1.08 + 4.012 = 5.092\ \Omega$$

Instrument current with a line current of 100 A

$$= \frac{0.000752}{(5.092 + 0.000752)} \times 100 = 0.01476 \text{ A}$$

$\therefore$ Instrument reading $= 0.01476 \times \dfrac{100}{0.015} = 98.4$ A

Percentage error = 100 – 98.4 = **1.6% (*low*)** **(Ans.)**

4.8.7. Ohmmeters

The ohmmeters may be of the following two types :

1. Series-type ohmmeter.
2. Shunt-type ohmmeter.

4.8.7.1. Series-type ohmmeter

Fig. 4.41 shows the elements of a simple single-range series ohmmeter :

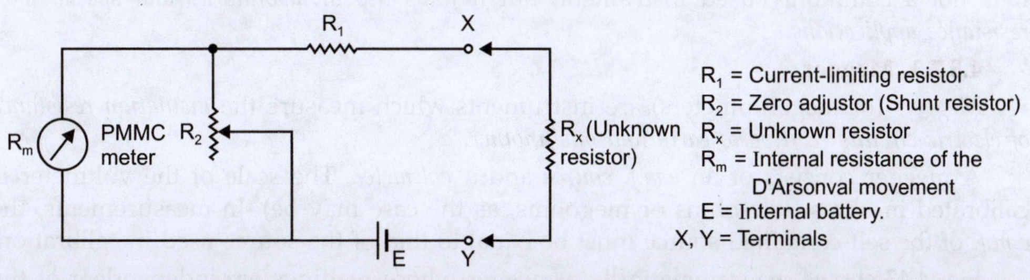

Fig. 4.41. Series-type ohmmeter.

— It consists of a D'Arsonval movement connected in series with a resistance and a battery to a pair of terminals (*X*, *Y*) to which the unknown resistance (R_x) is connected. The current through the meter then depends on the magnitude of unknown resistor, and the meter indication is proportional to the value of the unknown, provided that the calibration problems are taken into account,

— When $R_x = 0$ (*i.e.*, terminals *X* and *Y* shorted), *maximum* current flows in the circuit. Under this condition, shunt resistor R_2 is adjusted until the meter indicates full-scale current (I_{sd}); this position of the pointer is marked *"0 Ω"* on this scale.

— When $R_x = \infty$ (*i.e.*, terminals *X* and *Y* are open), the current in the circuit drops to *zero* and the meter indicates zero current, which is then marked "∞" on the scale.

Intermediate markings may be placed on the scale by connecting different known values of R_x to the instrument.

— The accuracy of these scale markings depends on the repeating accuracy of the meter and the tolerances of the calibrating resistors.

4.8.7.2. Shunt-type Ohmmeter :

Fig. 4.42. shows the circuit diagram of a shunt-type ohmmeter :

— It consists of a battery in series with an adjustable resistor R_1 and PMMC meter movement. The unknown resistor R_x is connected across terminals *X* and *Y*, in parallel with the meter. An *off-on switch* is provided to disconnect the battery from the circuit when the instrument is not used.

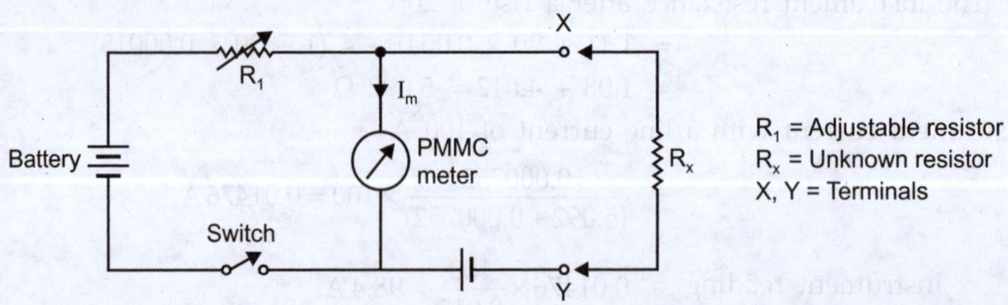

Fig. 4.42. Shunt-type ohmmeter.

— When $R_x = 0\ \Omega$ (*X* and *Y* terminals shorted), the meter current is zero.

— When $R_x = \infty$ (*X* and *Y* terminals open), the current finds a path only through the meter, and by the appropriate selection of the value R_1, the pointer can be made to read full scale.

The ohmmeter therefore has the 'zero' mark at the left hand side of the scale (no current) and the "infinite" mark on the right hand side of the scale (full-scale deflection).

• This type of ohmmeter is particularly suited to the measurement of *low-value resistors*. It is not a commonly used instrument, but it *finds use in laboratories and special low-resistance applications.*

4.8.7.3. Megger

Meggers (or megohmmeters) are instruments which measure the *insulation resistance of electric circuits relative to earth and one another.*

A megger consists of an *e.m.f. source* and a *voltmeter*. The scale of the voltmeter is calibrated in ohms (kilo-ohms or megohms, as the case may be). In measurements, the *e.m.f.* of the self-contained source must be equal to that of the source used in calibration.

Fig. 4.43 shows diagrammatically a megger whose readings are independent of the speed of the self-contained generator. The moving system incorporates two coils 1 (current coil) and 2 (pressure coil) mounted on the same shaft and *placed in the field of a permanent magnet* (now shown) 90° apart. The generator energizes the two coils over separate wires. Connected in series with one coil is a fixed resistance R_1 (or several different resistances in order to extend the range of the instrument). The unknown resistance R_x is connected in series with the other coil. *The currents in the coils interact with the magnetic field and produce opposing torques.*

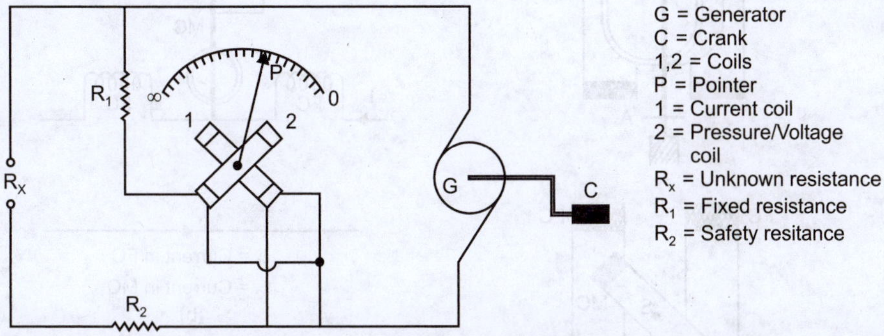

Fig. 4.43. Circuit diagram of megger.

The deflection of the moving system depends on the ratio of the currents in the coils and is independent of the applied voltage. The unknown resistance is read *directly* from the scale of the instrument. (The accuracy of measurement is unaffected by variations in the speed of the generator between 60 and 180 r.p.m.).

4.8.8. Electrodynamic or Dynamometer Type Moving Coil Instruments

4.8.8.1. Introduction

In this type of instrument, the *operating field is produced by another fixed coil* and *not by a permanent magnet.*

- These instruments are capable of service as *transfer instruments* (A *'**transfer instrument**' is one that may be calibrated with a D.C. source and then used without modification to measure A.C.*). Indeed, their principal use as ammeters and voltmetes in laboratory and measurement work is for the transfer calibration of working instruments and *as standards for calibration of other instruments a their accuracy is very high.*
- These instruments are employed as *A.C. voltmeters and ammeters both in the range of power frequencies and lower part of the audio frequency range.* They are used as *watt-meters, var-meters and with some modification as power factor meters and frequency meters.*

4.8.8.2. Construction and working

Fig. 4.44 shows a dynamometer type instrument :

The *main components* of the instrument are :

(*i*) Field system (Fixed coil).
(*ii*) Moving system (Moving coil).
(*iii*) Control system (Hair springs).
(*iv*) Damping system (Air friction damping).
(*v*) Shielding.
(*vi*) Cases and scales.

— These instruments essentially consist of *fine wire moving coil* placed in the magnetic field produced by *another fixed coil* when carrying currents.

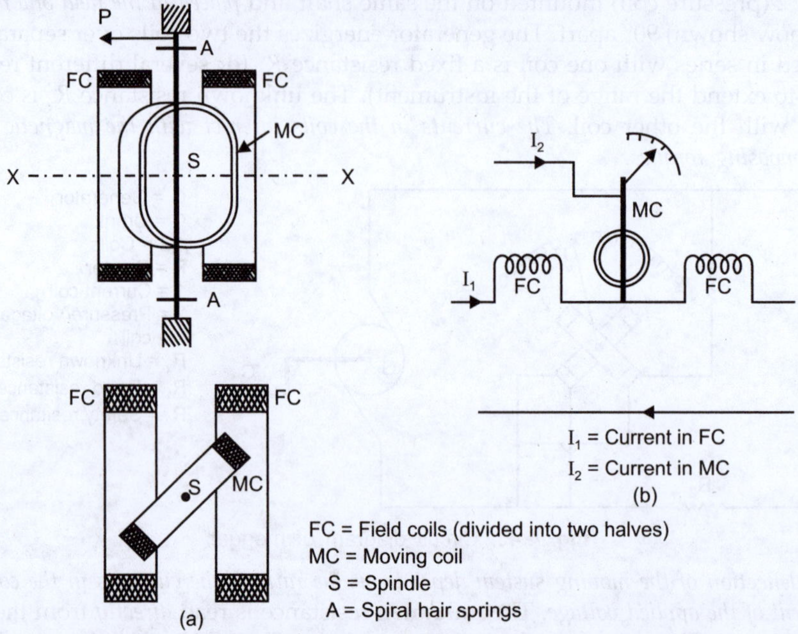

Fig. 4.44. Electrodynamic or dynamometer type instrument.

— The coils are usually *air cored* to *avoid hysteresis, eddy currents and other errors* when the instrument is used on A.C. The *fixed coil FC* is divided into *two halves* placed close together and *parallel* to each other in order to provide a *fairly uniform field*, within the range of the movement of the moving coil.

— The diagram in Fig. 4.44(*a*) shows a sectional elevation through fixed coil *FC* and the lower diagram represents a sectional plan on *XX*. The moving coil *MC* is carried by a spindle S and the *controlling torque is exerted by spiral hair spring A*, which may also serve to lead the current into and out of MC.

— In these instruments *air friction damping is used* and may be either piston type or vane type. Eddy current damping *cannot* be used as introduction of a permanent magnet for the purpose would *distort* the working magnetic field of the instrument.

— The complete assembly is surrounded by a *laminated steel shield to protect the instrument from external magnetic field* which may affect the operation of the instrument.

— Laboratory standard instruments are usually contained in *highly polished wooden* cases. The case is supported by adjustable levelling screws. A spirit level is also provided to ensure proper levelling. The *scales* are hand drawn, using machine sub-dividing equipment.

Deflecting torque. The deflecting torque is due to *interaction of the magnetic fields produced by currents in the fixed and moving coils.*

— Fig. 4.45(*a*) shows the magnetic field due to current flowing through *FC* (I_1) in the direction indicated by the dots and cross.

— Fig. 4.45(*b*) shows the magnetic field due to current (I_2) in *MC*.

— Fig. 4.45(*c*) shows the combined effect of the above magnetic fields. By combining these magnetic fields it will be seen that when currents (I_1 and I_2) flow simultaneously through *FC* and *MC*, the resultant magnetic field is *distorted* and effect is to exert a clockwise torque on *MC*.

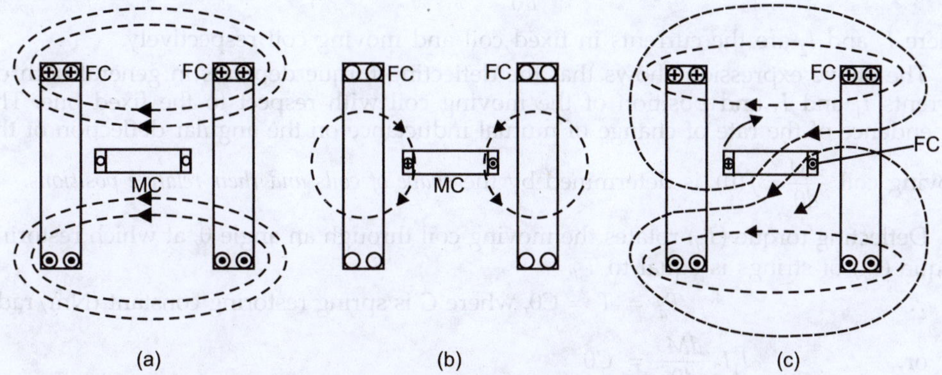

Fig. 4.45. Magnetic fields due to fixed and moving coils.

Since *MC* is carrying current (I_2) at right angles to the magnetic field produced by *FC*, *deflecting torque,*

$$T_d \propto I_1 \times I_2$$

or, $T_d = KI_1I_2$ where, K is a constant.

Since the instrument is spring-controlled, the *restoring or control torque* (T_c) is proportional to the angular deflection θ.

$$T_c \propto \theta \quad \text{or} \quad T_c = K'\theta$$

The two torques (T_d and T_c) are equal and opposite in the final deflection position.

i.e., $T_d = T_c$

or, $KI_1I_2 = K'\theta$

or, $\theta \propto I_1I_2$

or, $\theta \propto I^2$ since $I_1 \propto I$ and $I_m \propto I$

Expression for torque in terms of mutual inductance M between fixed and moving coils :

The total energy stored in the magnetic field of the fixed coil,

$$E = \frac{1}{2}L_1I_1^2 + \frac{1}{2}L_2I_2^2 + I_1I_2M \quad ...(4.89)$$

where L_1 and L_2 are the self inductances of the fixed coil and moving coil respectively.

The interaction of these currents give rise to a deflecting torque, T_d that sets the moving coil in a position for which the energy of the coil magnetic field attains maximum value.

Deflecting torque, $$T_d = \frac{dE}{d\theta} = \frac{1}{2}\cdot\frac{dL_1}{d\theta}\cdot I_1^2 + \frac{1}{2}\frac{dL_2}{d\theta}I_2^2 + I_1I_2\frac{dM}{d\theta}$$

where $d\theta$ is the increase in angular deflection at which the field energy increases by dE.

Since coil inductances L_1 and L_2 are constant, therefore, dL_1 and dL_2 each is equal to zero and T_d is given by :

$$T_d = I_1I_2\frac{dM}{d\theta} \quad ...(4.90)$$

For direct current :

$$T_d = I_1 I_2 \frac{dM}{d\theta}$$

where I_1 and I_2 are the currents in fixed coil and moving coil respectively.

The above expression shows that the deflecting torque depends in general both on currents I_1 and I_2 and position of the moving coil with respect to the fixed one. The dependence of the rate of change of mutual inductance on the angular deflection of the moving coil $\frac{dM}{d\theta} = f(\theta)$ is determined by the *shape of coils and their relative positions.*

Deflecting torque (T_d) rotates the moving coil through an angle θ, at which restoring torque (T_c) of strings is equal to T_d.

$\therefore$ $T_d = T_c = C\theta$, where C is spring restoring constant (Nm/rad.)

or, $I_1 I_2 \frac{dM}{d\theta} = C\theta$

or, Deflection, $\theta = \frac{I_1 I_2}{C} \cdot \frac{dM}{d\theta}$ ...(4.91)

For alternating current :

When the coils carry alternating currents i_1 and i_2 the instantaneous deflecting torque,

$$(T_d)_i = i_1 i_2 \frac{dM}{d\theta} \quad \text{...(4.92)}$$

The average deflecting torque over a complete cycle,

$$(T_d)_{av.} = \frac{1}{T}\int_0^T T_{di}\, dt$$

or, $$(T_d)_{av.} = \frac{dM}{d\theta}\frac{1}{T}\int_0^T i_1 i_2\, dt \quad \text{...(4.93)}$$

where, T = Time period for one complete cycle.

If currents i_1 and i_2 are *sinusoidal and are displaced by a phase angle φ, i.e.,*

$$i_1 = I_{max.1} \sin \omega t, \quad \text{and} \quad i_2 = I_{max.2} \sin(\omega t - \phi),$$

then, the *average deflecting torque,*

$$(T_d)_{av.} = \frac{dM}{d\theta}\cdot\frac{1}{T}\int_0^T i_1 i_2\, dt$$

$$= \frac{dM}{d\theta}\cdot\frac{1}{2\pi}\int_0^{2\pi} I_{max.1} \sin \omega t \,.\, I_{max.2} \sin(\omega t - \phi)\, d(\omega t)$$

or, $$(T_d)_{av.} = \frac{I_{max.1}\,.\,I_{max.2}}{2}\cos\phi\,\frac{dM}{d\theta} = I_1 I_2 \cos\phi\,\frac{dM}{d\theta} \quad \text{...(4.94)}$$

where I_1 and I_2 are r.m.s. values of current flowing in the coils.

At steady deflection state,

$$T_d = T_c$$

or, $$I_1 I_2 \cos\phi \frac{dM}{d\theta} = C\theta$$

$\therefore$ Deflection, $$\theta = \frac{I_1 I_2}{C}\cos\phi \cdot \frac{dM}{d\theta} \quad ...(4.95)$$

From the above discussion it is obvious that for a *sinusoidal alternating currents the deflecting torque and the deflection are determined by the product of r.m.s. values of coil currents and cosine of the phase angle between them.* It is clear that the *dynamometer type of instruments can be used on both D.C. and A.C.*

Use of the instrument as an ammeter. When the instrument is used as an ammeter then same current passes through both moving coil (*MC*) and fixed coils (*FC*) as shown in Fig. 4.46. In this case, $I_1 = I_2 = I$, hence $\theta \propto I^2$ or $I \propto \sqrt{\theta}$. The connections of Fig. 4.46 are used when *small currents* are to be measured.

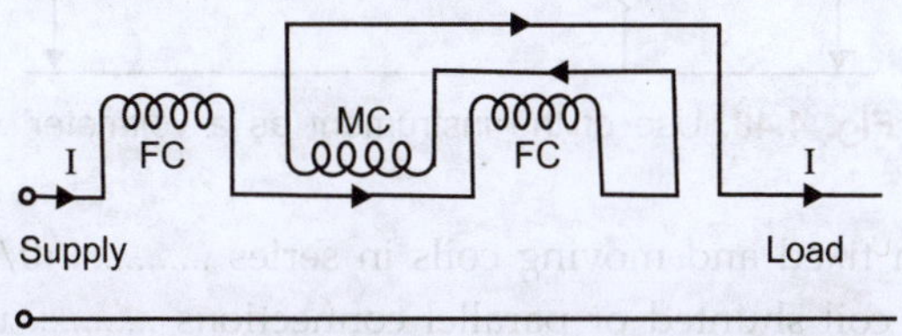

Fig. 4.46. Measurement of small currents.

In the case of heavy currents, a shunt is used to limit current through the moving coil as shown in Fig. 4.47.

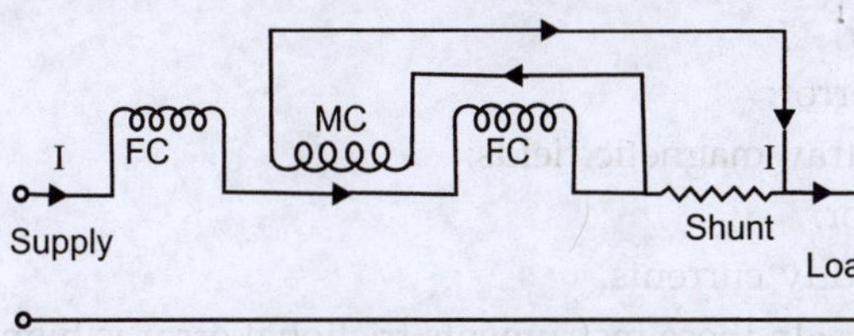

Fig. 4.47. Measurement of heavy currents.

Use of the instrument as voltmeter. When the instrument is used as a voltmeter, the fixed and moving coils are used in series along with high resistance as shown in Fig. 4.48. Here again, $I_1 = I_2 = I$,

where, $I = \frac{V}{R}$ ......in D.C. circuits

and, $I = \frac{V}{Z}$ ......in A.C. circuits

$\therefore$ $\theta \propto V \times V$ or $\theta \propto V^2$

or, $V = \sqrt{\theta}$

- Thus, whether the instrument is used as an ammeter or voltmeter, its scale is *uneven* through the whole of its range and is cramped or crowded near the zero in particular.

Note: When the dynamometer instrument is used to measure an alternating current or voltage, the moving coil due to its inertia takes up a position where the average deflecting torque over one cycle is balanced by the restoring torque of the spiral springs. For that position, the deflecting torque is proportional to the mean value of the square of current or voltage, and the instrument scale can therefore be *calibrated to read the r.m.s. value.*

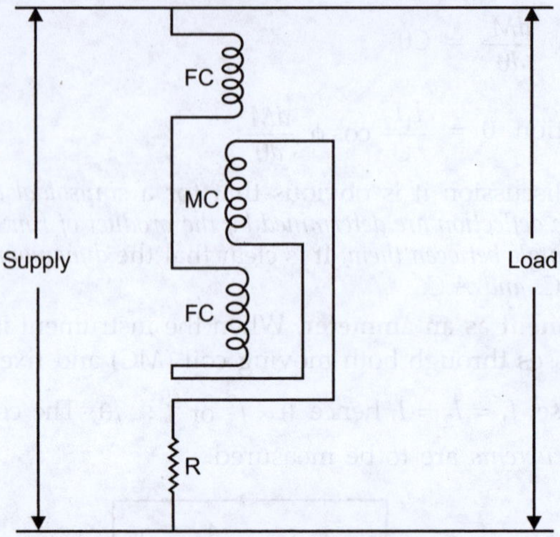

Fig. 4.48. Use of the instrument as a voltmeter.

Ranges :

Ammeter. *(i)* With fixed and moving coils in series .... 0/0.01 A – 0/0.05 A

(ii) With moving coil shunted or parallel connections .... upto 0/30 A.

Voltmeters. Upto 0–750 volts.

4.8.8.3. Errors in dynamometer type instruments

Following errors usually occur in these instruments :

(i) Frictional error;

(ii) Temperature error;

(iii) Error due to stray magnetic fields;

(iv) Frequency error;

(v) Error due to eddy currents.

1. ***Frictional error :*** In these instruments frictional error is high. So, in order to have high accuracy of measurement this error needs to be minimised to the lowest possible, this is achieved as follows :

(i) By a reasonable reduction in weight;

(ii) By proper selection of materials for bearings and spindle, and adequate polishing of these parts.

However, these measures increase the cost of the instrument and make it more sensitive to overloads and mechanical impacts.

2. ***Temperature error :***

- High currents are carried by coils which produce heat. *Self-heating of coils produces errors.*
- High accuracy instruments contain *temperature compensating resistors* which tend to *neutralise* the effects of changes in temperature.

3. ***Errors due to stray magnetic fields :*** In order to avoid the influence of stray fields a good amount of *screening is necessary.*

- In precision instruments the influence of external magnetic fields can be minimised by using *"Astatic system".*
  — *"Astatic"* dynamometer type instruments are constructed with two similar

sets of fixed and moving coils mounted on the same shaft. The two pairs of fixed coils are so connected that their *magnetic fields are in opposition. The currents in the moving coils are also opposed* and, in consequence, the deflection torques applied to the moving coils act in the same direction. Since the two fields are in opposition and, therefore, any uniform external field reduces the field of one coil and increases by an equal amount the field of the other coil, the deflecting torque produced by one coil is increased while that produced by the other coil is reduced by an equal amount and, thus *net torque on account of external magnetic field is zero.*

4. ***Frequency error :*** This error in these instruments is largely due to variation of self-reactance of coils, with frequency.

- In order to reduce frequency error in dynamometer ammeters, the ratio of currents in fixed and moving coils *i.e.* $\frac{I_1}{I_2}$ should be independent of frequency. This requires that *time constant* $\left(\frac{L}{R}\right)$ *of two circuits should be same.*
- For reducing frequency errors in voltmeters the *coil winding is made a very small part of the circuit.* This makes inductive reactance ($X_L = 2\pi fL$), a *small fraction* of total impedance, and therefore $Z = \sqrt{R^2 + (2\pi fL)^2} \approx R$. This means frequency does not affect the calibration of the instrument. In fact, dynamometer voltmeters may be used, in general, within their guaranteed accuracy from D.C. to about 125 Hz.

5. ***Error due to eddy currents :***

- The eddy currents are induced in metal parts of the instrument and develop a torque because of coupling between the moving coil and the neighbouring metal part.
- This error can be *reduced* by keeping the metal in the supports of the coil and other structural parts to the *minimum possible* and selecting the metal, to be used, of *high resistivity.*

4.8.8.4. Advantages and disadvantages

Advantages :

1. Can be used on both D.C. as well as A.C. systems (as the instrument has square law response).
2. They are free from hysteresis and eddy current errors, because of absence of iron in the operating parts of the instrument.
3. It is possible to construct ammeters upto 10 A and voltmeters upto 600 V with precision grade accuracy.
4. Very useful for accurate measurement of r.m.s. values of voltages irrespective of waveforms.
5. Owing to precision grade accuracy and same calibration for D.C. and A.C. measurements, these instruments are used as *transfer and calibration instruments.*

Disadvantages :

1. Since torque/weight ratio is small, such instruments have *low sensitivity.*
2. The *scale is not uniform* because $\theta \propto \sqrt{I}$.
3. Cost of these instruments is *higher* in comparison to those of moving-iron instruments. So, these are *only used as voltmeters and ammeters for precision measurements.*
4. *Higher frictional losses.*

5. These instruments are sensitive to overloads and mechanical impacts. Therefore, they *must be handled with care.*
6. In these instruments the operating current is large and consequently they have a *higher power consumption than PMMC instruments.*

Example 4.52. *The mutual inductance of a 25 A electrodynamometer ammeter changes uniformly at a rate of 0.0035 μH/degree. The torsion constant of the controlling spring is 10^{-6} Nm/degree. Determine the angular deflection for full scale.* **(Poona University)**

Solution. *Given* : $I_1 = I_2 = I = 25$ A; $\frac{dM}{d\theta} = 0.0035$ μH/degree; $C = 10^{-6}$ Nm/degree

Angular deflection for full scale, θ :

Now,
$$T_d = I^2 \frac{dM}{d\theta} \quad \text{[Eqn. (4.90)]}$$
... instrument being ammeter ($I_1 = I_2$)

Since spring constant is given Nm/degree, therefore, we must express $\frac{dM}{d\theta}$ in H/rad. if deflection is to be found in degrees.

$$\therefore \quad \frac{dM}{d\theta} = \frac{0.0035 \times 10^{-6}}{(\pi/180)} = 0.2 \times 10^{-6} \text{ Nm/rad.}$$

Full scale deflection torque,
$$T_d = T_c = C\theta = 10^{-6}\theta \text{ Nm}$$

Substituting this value of T_d in the above eqn., we get :
$$10^{-6}\theta = (25)^2 \times 0.2 \times 10^{-6}$$
or
$$\theta = \mathbf{125°} \quad \textbf{(Ans.)}$$

Example 4.53. *A 10 A electrodynamometer is controlled by a spring having a torsion constant of 0.1 × 10^{-6} Nm/degree. The full scale deflection is 110°. Determine the inductance of the instrument when measuring a current of 10 A. The mutual inductance at 0° deflection is 2 μH and the change in mutual inductance is linear as the deflection.* **(Nagpur University)**

Solution. *Given* : $C = 0.1 \times 10^{-6}$ Nm/degree; $\theta = 110°$, $I = 10$ A; M at 0° deflection = 10A

Inductance of the instrument :

Full scale deflecting torque, $T_d = T_c = C\theta = 0.1 \times 10^{-6} \times 110 = 11.1 \times 10^{-6}$ Nm

Also,
$$T_d = I^2 \frac{dM}{d\theta}$$

$$\therefore \quad \frac{dM}{d\theta} = \frac{T_d}{I^2} = \frac{11.1 \times 10^{-6}}{(10)^2} = 0.111 \text{ μH/rad.}$$

Change in inductance, $dM = 0.111 \times 10^{-6} \times 110 \times \frac{\pi}{180} = 0.213$ μH

∴ *Total inductance of the instrument* = Initial inductance + change in inductance

= 2 + 0.213 = **2.213 μH** **(Ans.)**

Example 4.54. *The spring constant of a 10 A dynamometer wattmeter is 10.5 × 10^{-6} Nm/radian. The variation of inductance with angular position of moving system is practically linear over the operating range, the rate of change being 0.078 mH/radian. If the full-scale deflection of the instrument is 83 degrees, calculate the current required in the voltage coil at full scale on D.C. circuit.* **(Nagpur University)**

Solution. *Given* : C = 10.5×10^{-6} Nm/rad.; $10.5 \times 10^{-6} \times \frac{\pi}{180}$ Nm/degree; $\frac{dM}{d\theta} = 0.078$ mH/rad.; θ = 83°.

Current required in voltage coil, I_2 :

$$T_d = T_c = C\theta = \left(10.5 \times 10^{-6} \times \frac{\pi}{180}\right) \times 83 = 15.21 \times 10^{-6} \text{ Nm}$$

Also, $$T_d = I_1 I_2 \frac{dM}{d\theta} \quad \text{[Eqn. (4.90)]}$$

∴ $$15.21 \times 10^{-6} = 10 \times I_2 \times 0.078$$

or, $$I_2 = 19.5 \times 10^{-6} \text{ A} \quad \text{or} \quad \mathbf{19.5\ \mu A} \quad \text{(Ans.)}$$

Example 4.55. *In a certain dynamometer ammeter the mutual inductance M varies with deflection θ (expressed in degees) as :*

$$M = -8 \cos(\theta + 30°) \text{ mH}$$

Find the deflecting torque produced by a direct current of 60 mA corresponding to a deflection of 60°.

Solution. *Given* : M = – 8 cos (θ + 30°) mH ; I = 60 mA or 0.06 A; θ = 60°

Deflecting torque, T_d :

Rate of change of mutual inductance with deflection,

$$\frac{dM}{d\theta} = \frac{d}{d\theta}[-8\cos(\theta + 30°)] = 8 \sin(\theta + 30°) \text{ mH}$$

$\frac{dM}{d\theta}$ at a deflection of 60° *i.e.*, $\left(\frac{dM}{d\theta}\right)_{\theta = 60°} = 8 \sin(60° + 30°) \text{ mH} = 8 \times 10^{-3}$ H/degree

∴ Deflecting torque, $$T_d = I^2 \frac{dM}{d\theta} = (0.06)^2 \times 8 \times 10^{-3} = \mathbf{28.8\ \mu Nm} \quad \text{(Ans.)}$$

Example 4.56. *In a torsion-head type electrodynamometer a current of 36 A requires a deflection of 90° in the head to give balance.*

Determine :

(i) The range of the instrument (maximum angle 360°)

(ii) For what current will the deflection be 180° ?

(iii) What will be the deflection for a current of 30 A ?

Solution. In torsion-head electrodynamometer,

Torque, $T \propto (\text{current})^2$

Also, torque, $T \propto \theta$ (deflection)

∴ $$\theta \propto I^2$$

or, $$I \propto \sqrt{\theta}$$

(*i*) Current corresponding to deflection of 360°,

$$I_2 = I_1 \times \sqrt{\frac{\theta_2}{\theta_1}} = 36 \times \sqrt{\frac{360}{90}} = \mathbf{72\ A} \text{ (Ans.)}$$

(*ii*) Current corresponding to deflection of 180°,

$$I_3 = I_1 \times \sqrt{\frac{\theta_2}{\theta_1}} = 36 \times \sqrt{\frac{180}{90}} = \mathbf{50.9\ A} \quad \text{(Ans.)}$$

(*iii*) Deflection corresponding to current 30 A,

$$\theta_4 = \theta_1 \times \left(\frac{I_4}{I_1}\right)^2 = 90 \times \left(\frac{30}{36}\right)^2 = \mathbf{62.5°} \quad \textbf{(Ans.)}$$

Example 4.57. *A 50 V range spring controlled, electrodynamic voltmeter having a square law scale response takes 0.05 A on D.C. for full scale deflection of 90°. The control constant is 0.5 × 10^{-6} Nm/degree and the initial mutual inductance of the instrument is 0.25 H. Find the true potential difference across the instrument when it reads 50 V at 50 Hz.*

Solution. *Given* : $V = 50$ volts; $I = 0.05$ A at $\theta = 90°$; $C = 0.5 \times 10^{-6}$ Nm/degree; Initial inductance = 0.25 H; $f = 50$ Hz.

True p.d. across the instrument :

Full scale deflection controlling torque, $T_c = 0.5 \times 10^{-6} \times 90 = 45 \times 10^{-6}$ Nm

For steady deflection state, $T_d = T_c$

$$\therefore \quad I^2 \frac{dM}{d\theta} = 45 \times 10^{-6}$$

$$\text{or,} \quad \frac{dM}{d\theta} = \frac{45 \times 10^{-6}}{I^2} = \frac{48 \times 10^{-4}}{(0.05)^2} = 18 \text{ mH/rad.}$$

Total change in mutual inductance for deflection of 90° or $\frac{\pi}{2}$ rad.,

$$dM = 18 \times 10^{-3} \times \frac{\pi}{2} = 0.02827 \text{ H}$$

∴ Total mutual inductance (M) = Initial mutual inductance + change in mutual inductance

$$= 0.25 + 0.02827 \simeq 0.2783 \text{ H}$$

Reactance of the instrument at 50 Hz = $2\pi f M = 2\pi \times 50 \times 0.2783 = 87.43\ \Omega$

Resistance of the instrument, $R = \frac{V}{I} = \frac{500}{0.05} = 1000\ \Omega$

Impedance of the instrument at 50 Hz, $Z = \sqrt{(1000)^2 + (87.43)^2} = 1003.8\ \Omega$

∴ *True p.d. across the instrument* = $IZ = 0.05 \times 1003.8 =$ **50.19 V** **(Ans.)**

Example 4.58. *A 50 V range spring controlled electrodynamic voltmeter has an initial inductance of 0.25 H, the full scale deflection torque of 0.4 × 10^{-4} Nm and full scale deflection current of 50 mA.*

Determine the difference between D.C. and 50 Hz A.C. readings at (i) 50 V and (ii) 25 V if the voltmeter inductance increases uniformly over the full scale of 90°. **(Delhi University)**

Solution. *Given* : Initial inductance = 0.25 H; $T_d = 0.4 \times 10^{-4}$ Nm; $I = 50$ mA or 0.05A

$$\text{Now,} \quad T_d = I^2 \frac{dM}{d\theta}$$

$$\text{or,} \quad 0.4 \times 10^{-4} = (0.05)^2 \times \frac{dM}{d\theta}$$

$$\text{or,} \quad \frac{dM}{d\theta} = \frac{0.4 \times 10^{-4}}{(0.05)^2} = 0.016 \text{ H/rad.}$$

Total change in inductance for full scale deflection,

$$dM = 0.016 \times \left(90 \times \frac{\pi}{180}\right) = 0.0251 \text{ H}$$

$\therefore$ Total mutual inductance, $M = 0.25 + 0.251 = 0.2751$ H

At 50 V :

The resistance of voltmeter, $R = \frac{V}{I} = \frac{50}{0.05} = 1000\ \Omega$

The impedance while measuring the voltage of 50 V at 50 Hz A.C.,

$$Z = \sqrt{(1000)^2 + (2\pi \times 50 \times 0.2751)^2} = 1003.7\ \Omega$$

$$\therefore \quad \text{Voltmeter reading} = \frac{50}{1003.7} \times 1000 = 49.8 \text{ V}$$

Hence, *difference in readings* = 50 – 49.8 = **0.2 V** **(Ans.)**

At 25 V :

Current through instrument, $I' = \frac{25}{1000} = 0.025$ A

$$\text{Deflection, } \theta' = \theta \times \frac{(I')^2}{(I)^2}, \text{ since } \frac{dM}{d\theta} \text{ is constant}$$

$$= \frac{\pi}{2} \times \left(\frac{0.025}{0.05}\right)^2 = 0.3927 \text{ rad.}$$

$$\text{Mutual inductance at 25 V} = \text{Initial inductance} + \frac{dM}{d\theta} \times \theta'$$

$$= 0.25 + 0.016 \times 0.3927 = 0.2563 \text{ H}$$

$$\text{Impedance at 25 V,} \quad Z' = \sqrt{(1000)^2 + (2\pi \times 50 \times 0.2563)^2} = 1003.24\ \Omega$$

$$\therefore \quad \text{Voltmeter reading} = \frac{25}{1003.24} \times 1000 = 24.92 \text{ V}$$

Hence, difference in reading = 25 – 24.92 = **0.08 V** **(Ans.)**

THERMAL INSTRUMENTS (HOT-WIRE AND THERMOCOUPLE)

4.8.9. Hot-wire Instruments

- *These instruments utilise the expansion of wire heated by current being measured.*
- These instruments are primarily meant for being used as ammeters but can be adopted as voltmeters by connecting a high resistance in series with them.
- The hot wire instruments are suited both for A.C. and D.C. work.

Construction. Fig. 4.49 shows the working parts of the instrument :

— It consists of platinum-iridium (It can *withstand oxidation at high temperatures*) wire *AB* stretched between a fixed end *B* and the tension-adjusting screw at *A*. The diameter of this wire is about 0.2 mm.

— A phosphor bronze wire is attached to the hot wire and in turn a fine silk thread *R* is attached to phosphor bronze wire as shown in the figure.

— The silk thread *R* passes around a pulley *E* and the other end of it is fixed to spring *S*.

— A light pointer P and thin aluminium disc L are carried by the spindle upon which the pulley is also mounted.

— M is the damping magnet; eddy current damping is used for these instruments.

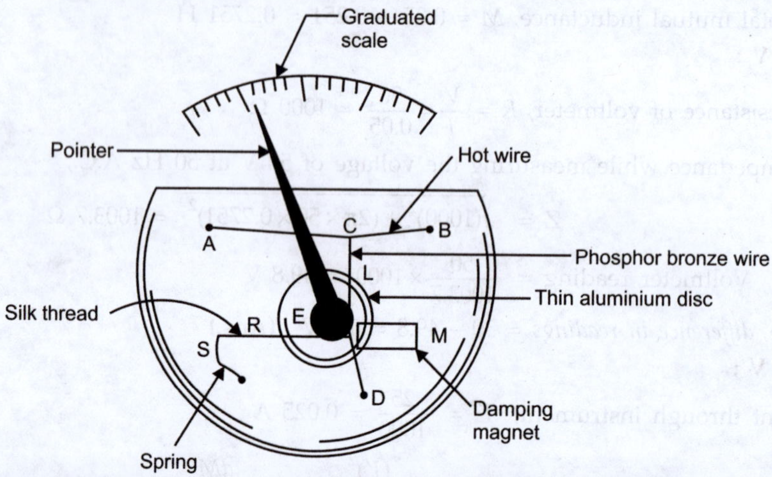

Fig. 4.49. Hot-wire instrument.

Working. When the current to be measured (or a definite fraction of it or proportional to the voltage to be measured) is passed through the wire AB, it expands due to heating effect of the current and slack in the phosphor bronze wire which is taken up by string S through the silk thread R and the moving system is rotated due to motion of the pulley E. Thus the pointer P moves over a calibrated scale.

It will be seen that the deflection of the pointer is proportional to the extension of AB which in itself proportional to I^2. Hence, *deflection is proportional to* I^2. If spring control is used then $T_c \propto \theta$.

Hence, $$\theta \propto I^2 \quad ...(4.96)$$

So, these instruments have a *square law type scale. They read the "r.m.s. value of current" and their readings are independent of its form and frequence.*

- The hot wire should be made *as thin as possible so that it may attain steady temperature as quick as possible when current flows through it.* At the same time the wire should be of *such size that it may be capable of withstanding the normal mechanical stresses* developed in the instrument due to handling, vibration, tension etc.
- In order to minimise the error due to uneven expansion between wire and its fixtures, the base of the instrument is made of material of coefficient of expansion equal to that of hot wire.

Magnification of the expansion :

- *Refer to Fig. 4.50 (a) :*

Let, l = Length of hot wire, and

δL = Increase in length of wire due to flow of current I through it.

Then, the sag in wire due to expansion,

$$S = \sqrt{\left(\frac{L+\delta L}{2}\right)^2 - \left(\frac{L}{2}\right)^2}$$

$$= \sqrt{\frac{2L\,\delta L + (\delta L)^2}{4}}$$

or, $$S = \sqrt{\frac{L\,.\,\delta L}{2}} \qquad ...(4.97)$$

neglecting (δL) being very small

The *magnification factor,*

$$M = \frac{S}{\delta L} = \frac{\sqrt{L\,.\,\delta L/2}}{\delta L} = \sqrt{\frac{L}{2\,\delta L}} \qquad ...(4.98)$$

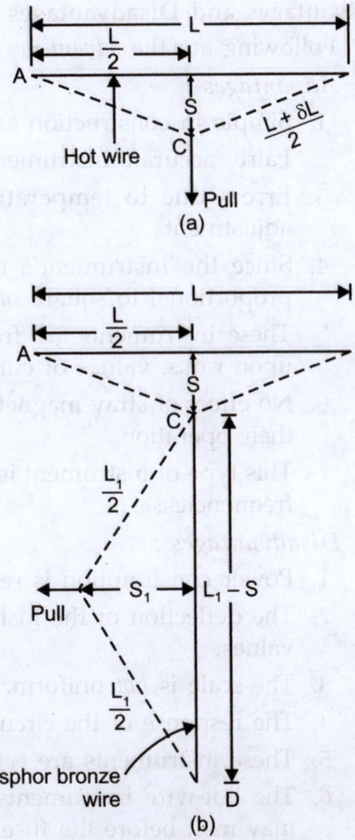

Fig. 4.50.

- *Refer to Fig. 4.50(b) :*

Further magnification is obtained by connecting a phosphor bronze wire of length L_1 to the centre of the wire *AB*.

Let the sag in this case be S_1.

Now, $$S_1 = \sqrt{\left(\frac{L_1}{2}\right)^2 - \left(\frac{L_1 - S}{2}\right)^2}$$

$$= \sqrt{\frac{L_1^2}{4} - \frac{L_1^2}{4} + \frac{2L_1 S - S^2}{4}}$$

or, $$= \sqrt{\frac{L_1 S}{2}}$$

...neglecting S^2, being very small

$$= \sqrt{\frac{L_1}{2} \times \sqrt{\frac{L.\delta L}{2}}}$$ ...substituting the value of S from eqn. (4.97)

or, $$S_1 = \sqrt{\frac{L_1}{2\sqrt{2}}}\,(L.\delta L)^{1/4} \qquad ...(4.99)$$

or $S_1 \propto (\delta L)^{1/4} \propto \sqrt{\text{current}}$ since $\delta L \propto (\text{current})^2$

The *magnification factor,*

$$M = \frac{S_1}{\delta L} = \frac{\sqrt{\frac{L_1}{2\sqrt{2}}}\,.\,(L.\delta L)^{1/4}}{\delta L} = \sqrt{\frac{L_1}{2\sqrt{2}}}\left(\frac{L}{(\delta L)^3}\right)^{1/4} \qquad ...(4.100)$$

Ranges :

Ammeter : (*i*) Without use of shunt : 0–1 A.

(*ii*) Without use of shunt by using device like dividing the wire paths into several parallel paths : 0–5 A.

Voltmeter : Upto 400 V by using high non-inductive resistance in series with the instrument.

Advantages and Disadvantages :

Following are the *advantages* and *disadvantages* of hot-wire instruments :

Advantages :

1. Simple in construction and cheap in cost.
2. Fairly accurate instruments.
3. Errors due to temperature variations can be made negligible with suitable adjustments.
4. Since the instrument's operation depends upon the heating effect which is proportional to square of current, it can be used on A.C. and D.C. both.
5. These instruments are free from waveform errors since their readings depend upon r.m.s. values of current.
6. No effect of stray magnetic fields because no magnetic effects are used to cause their operation.
7. This type of instrument is quite suitable for measurement of current at very high frequencies.

Disadvantages :

1. Power consumption is relatively high.
2. The deflection of the instrument is *not* the same for ascending and descending values.
3. The scale is *not* uniform.
4. The response of the circuit is *very slow* as the wire takes time to heat up.
5. These instruments are *very delicate.*
6. The hot-wire instruments cannot take overload because wire is very fine and may melt before the fuse can protect it.
7. Changes in room temperature causes errors which shifts the zero as a result of which it has to be reset frequently.

- In fact, although the accuracy of hot-wire instruments for measurement of current and voltage at higher frequency is higher than others, yet due its disadvantages they are *practically obsolete.*

Example 4.59. *The working wire of a single sag hot wire instrument is 15 cm long and is made-up of platinum-silver with a coefficient of linear expansion of 16 × 10⁻⁶. The temperature rise of the wire is 85°C and the sag is taken up at the centre. Find the magnification.*

(*i*) *With no initial sag;*

(*ii*) *With an initial sag of 1 mm.* **(Calcutta University)**

Solution. *Given :* $L = 15$ cm or 0.15 m; $\alpha = 16 \times 10^{-6}$; Temperature rise, $\Delta t = 85°C$

Magnification, *M* :

Increase in length of wire when heated through 85°C,

$$\delta L = \alpha L.(\Delta t) = 16 \times 10^{-6} \times 0.15 \times 85 = 2.04 \times 10^{-4} \text{ m}$$

(*i*) *With no initial sag :*

$$M = \sqrt{\frac{L}{2.\delta L}} \qquad \text{[Eqn. (4.98)]}$$

$$= \sqrt{\frac{0.15}{2 \times 2.04 \times 10^{-4}}} = \mathbf{19.17} \quad \textbf{(Ans.)}$$

(ii) *With an initial sag of 1 mm :*

$$\text{Sag, S} = \sqrt{\frac{L.\delta L}{2}} \qquad \text{[Eqn. (4.97)]}$$

$$= \sqrt{\frac{0.15 \times 2.04 \times 10^{-4}}{2}} = 3.91 \times 10^{-3} \text{ m}$$

Net increase in sag, $S' = S - \text{initial sag} = 3.91 \times 10^{-3} - 1 \times 10^{-3} = 2.91 \times 10^{-3}$ m

$$\therefore \quad M = \frac{S'}{\delta L} = \frac{2.91 \times 10^{-3}}{2.04 \times 10^{-4}} = \mathbf{14.26} \quad \textbf{(Ans.)}$$

4.8.10. Thermocouple Instruments

Working principle :

When two wires of different materials are joined together to form two junctions and if these junctions are maintained at different temperatures, a current flows around a circuit. This is known as ***Seebeck effect*** *and the system is known as a "thermocouple".* Seebeck arranged 25 metals in the order of their thermoelectric properties, these are : Bi-Ni-Co-Pd-Pt-U-Cu-Mu-Ti-Hg-Pb-Sn-Cr-Mo-Ph-Ir-Au-Ag-Zn-W-Cd-Fe-As-Sb-Te.

The current flows across the hot junction from the former to the latter metal of the above series, e.g., if copper and iron are used to form a thermocouple, the current flows from copper to iron at the hot junction and iron to copper at the cold junction (Fig. 4.51). If the copper wire is cut, an e.m.f. appears across the open circuit.

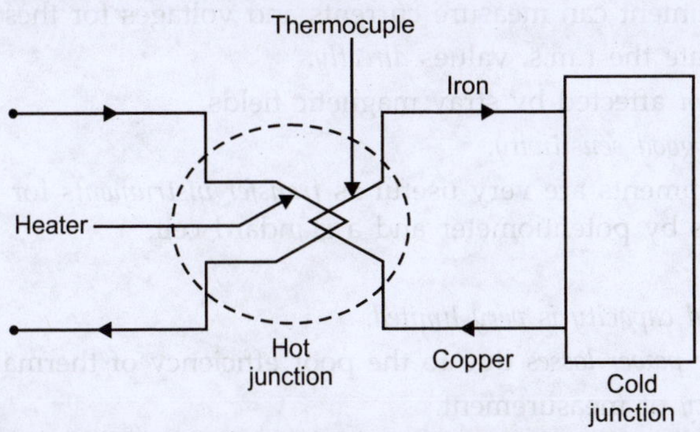

Fig. 4.51. Seebeck effect.

Construction and Working :

Fig. 4.52 shows the construction of a thermocouple instrument :

— It consists of a *heater element* (through which the current being measured is passed), *thermocouple* and *a PMMC instrument* to measure the e.m.f. produced by Seebeck effect.

— The heater wire and one junction of the thermocouple are in contact at the hot junction.

— The heater converts electrical energy to heat. A part of this heat is transferred to the thermocouple at the hot junction. Due to Seebeck effect an *e.m.f. is produced and it is measured by the PMMC instrument* (D.C. moving coil microvoltmeter). *The deflection is proportional to thermoelectric e.m.f. which in turn is proportional to the difference between the temperatures of hot and cold junctions.* This *temperature difference is proportional to* I^2.

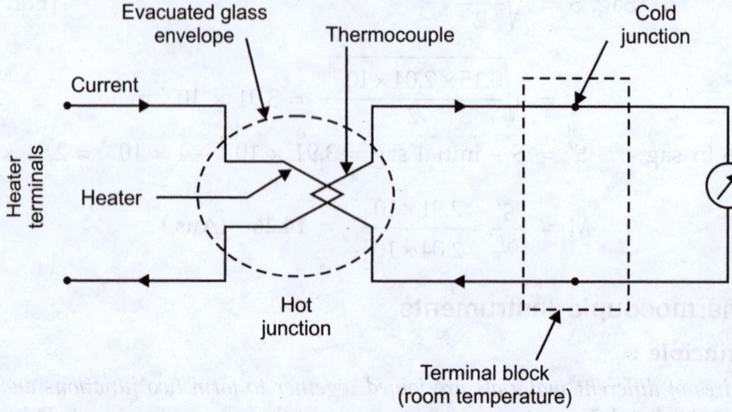

Fig. 4.52. Thermocouple instrument.

The scale of PMMC meter can be *calibrated to indicate the r.m.s. value.*

- The commonly used junctions for thermocouple ammeters are *copper-constantan* and *iron-constantan.*

Advantages and disadvantages :

Advantages :

1. These instruments are suitable for *very high frequencies* upto *about 50 MHz.* No other instrument can measure currents and voltages for these high frequencies.
2. They indicate the r.m.s. values *directly.*
3. They are *not* affected by stray magnetic fields.
4. They have *good sensitivity.*
5. These instruments are very useful as *transfer instruments* for calibration of D.C. instruments by potentiometer and a standard cell.

Disadvantages :

1. The *overload capacity is very limited.*
2. *Considerable power losses* due to the poor efficiency of thermal conversion.
3. *Low accuracy* of measurement.
4. The millivoltmeters used with thermo-elements must be necessarily more sensitive and delicate than those used with shunts, and therefore, *require careful handling.*

4.8.11. Rectifier Type Instruments

4.8.11.1 Introduction

These are not separate types of instruments, but rather a means of using a D'Arsonval movement, in conjunction with a *rectifier,* to change A.C. to D.C. Thus, a direct current movement can be adopted for use with alternating current as shown in Fig. 4.53. Rectifier type meters, using copper oxide rectifiers, are useful at low frequencies and will give good indication upto about 20 kHz.

- Rectifier instruments can operate well into the RF (radio-frequency) range with the proper use of silicon or germanium rectifiers.
- Measurement of current and voltage at several hundred MHz is possible.
- This type of meter is *more sensitive than any other type of A.C. meter.*

4.8.11.2. Rectifier type elements

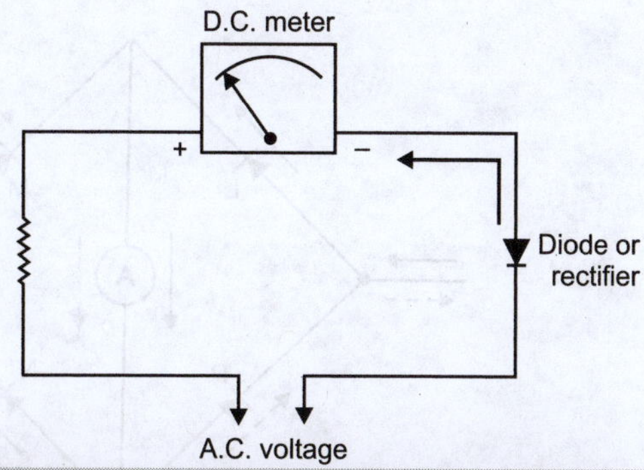

Fig. 4.53. A D.C. moving-coil meter can be used to measure A.C. voltage by putting a diode or rectifier in the meter circuit.

- *A rectifier element is used to convert A.C. to D.C. so that a unidirectional current flows through the PMMC movement.*
- The rectifier type of instruments use a copper oxide or a selenium cell, a germanium or silicon diode.
  - — *Copper oxide elements* have a peak inverse voltage (PIV) of about 2 V while *selenium elements* have a PIV of about 10 V and also they can handle only limited amount of current and hence are becoming *obsolete.*
  - — Germanium diodes have a PIV of the order of 300 V, and a current rating of 100 mA (appr.). *Low current silicon diode* rectifiers have a PIV of upto 1000 V, and a current rating of the order of 500 mA. For these reasons germanium and silicon diodes find extensive applications in rectifier instruments these days.

It may be mentioned here that though silicon diodes can carry currents as high as 85 A per rectifying elements, but these *high current silicon diodes are seldom used in indicating instruments and find wide applications in power supply circuits.*

4.8.11.3. Characteristics of rectifier instruments

The *characteristics* of rectifier instruments are given below :

1. They provide an economical and practical means of A.C. measurements in radio and communication circuits at audio frequencies (1 kHz to 10 kHz) and in other A.C. circuits where small available power makes it necessary to use sensitive low-energy instruments.
2. A rectifier instrument is, in general, approximately *fifty times* more sensitive than either an electrodynamometer or a moving-iron instrument.
3. In most instances rectifier instruments have *essentially linear scales.*
4. They commonly possess sensitivity of the order of 1000 to 2000 ohms per volt and more.
5. The power consumed by a rectifier instrument is generally several times as high as its permanent magnet moving coil (PMMC) mechanism because of the resistance of its rectifier.
6. Shunting of rectifier instruments is not practical because of the change in resistance of the rectifier with both temperature and the amount of the current.
7. Rectifier instruments are manufactured as micrometers and milliammeters in ratings from 100 μA to 15 mA for full scale deflection. The higher rating of 15 mA is more or less governed by the special size of the rectifier in comparison with the mechanism with which it is to be used.

4.8.11.4. Rectifier ammeters and voltmeters

Rectifier ammeters. The *rectifier ammeters usually* consists of four rectifier elements *L, M, N* and *P* arranged in the form of a bridge (as shown in Fig. 4.54) and A represents a moving-coil ammeter. The apex of the black triangle (shown in the bridge) indicates the direction in which the resistance is low.

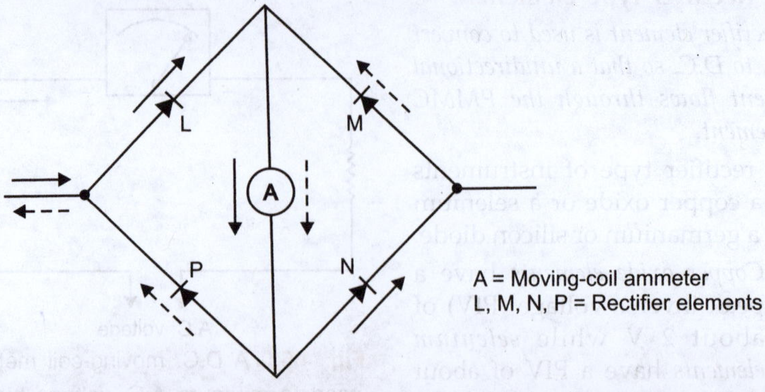

Fig. 4.54. Bridge circuit for full-wave rectification.

As shown in Fig. 4.54 during the half-cycles current flows through elements L and N from left to right as shown by the *full arrows.* During other half-cycles, the current flows through M and P, as shown by *dotted arrows.* The waveform of the current flowing through A is therefore as shown in Fig. 4.55. The deflection of moving-coil ammeter A, thus depends upon the *average value of the current;* and the scale of A can be calibrated to read the r.m.s. value of the current on the assumption that the waveform of the latter is sinusoidal with a form factor of 1.11.

Rectifier voltmeters. In a *rectifier voltmeter,* A is a milliammeter and the bridge circuit of Fig. 4.54 is connected in *series with a suitable resistor.*

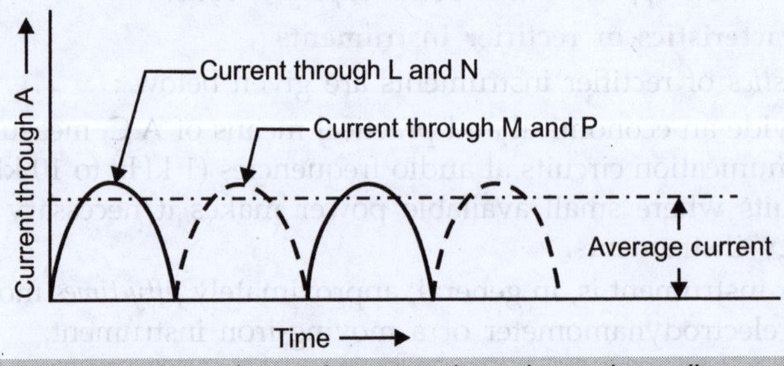

Fig. 4.55. Waveform of current through moving-coil ammeter.

Ranges :

(*i*) 0–100 μA; upto 0–100 mA

(*ii*) 0–1.0 V; upto 0–250 V without external resistance

(*iii*) Frequency range 20 Hz to 20 kHz.

☞ **Precaution :** The circuit to the instrument should *never be opened* while the rectifier is connected to the line, because this will result in *full voltage impressed across the rectifier plates, which may get damaged due to it.*

4.8.11.5. Sensitivity of rectifier type instruments

- The D.C. *sensitivity* of a rectifier type instrument is given by :

$$S_{D.C.} = \frac{1}{I_{fsd}} \ \Omega/V \qquad ...(4.101)$$

where, I_{fsd} = The current required to produce full scale deflection.

- For a half-wave rectifier with *sinusoidal input* the *A.C. sensitivity* is given as :

$$S_{A.C.} = 0.45\, S_{D.C.} \qquad ...(4.102)$$

Similarly, for a full-wave rectifier type instrument, with *sinusoidal input*, the A.C. sensitivity is given by :

$$S_{A.C.} = 0.9\, S_{D.C.} \qquad ...(4.103)$$

These relations are not valid for non-sinusoidal input waveforms such as square, triangular and sawtooth waveforms. Further these relations are *based on the assumption that the forward resistances of diodes are zero and the reverse resistance infinite.*

4.8.11.6. Extension of range of rectifier instruments as voltmeters :

Let us support we are to extend the range of a rectifier instrument which uses a *PMMC instrument having a D.C. sensitivity of* $S_{D.C.}$

Further, let v = Voltage drop across the PMMC instrument,

V = Applied voltage,

R_m = Resistance of the meter, and

R_d = Diode resistance.

Then, for D.C. operation the values of series resistance (multiplier) needed is,

$$R_s = \left(\frac{V}{I_{fsd}}\right) - R_m - R_d \qquad ...(4.104)$$

or, $$R_s = S_{D.C.} V - R_m - R_d \qquad \text{for } \textit{half-wave rectification}$$

$$= \left(\frac{V}{I_{fsd}}\right) - R_m - 2R_d$$

or, $$R_s = S_{D.C.} V - R_m - 2Rd \qquad ...(4.105)$$

...for *full-wave rectification*

For A.C. voltmeters :

$$R_s = \left(\frac{V}{I_{fsd}}\right) - R_m - R_d$$

$$= S_{A.C.} V - R_m - R_d = 0.45\, S_{D.C.} V - R_m - R_d \qquad ...(4.106)$$

...for *half-wave rectification*

or, $$R_s = \left(\frac{V}{I_{fsd}}\right) - R_m - 2R_d = 0.9\, S_{D.C.}\, V - R_m - 2R_d \qquad ...(4.107)$$

...for *full-wave rectification*

- *Following factors affect the performance of rectifier type instruments :*
  (*i*) Waveform; (*ii*) Resistance of rectifier;
  (*iii*) Temperature changes; (*iv*) Capacitance of rectifier;
  (*v*) Sensitivity.
- *"Loading effects of rectifier instruments"* :

The sensitivity of rectifier type of voltmeters, using either half or full wave rectification, is *less* than the sensitivity of D.C. voltmeters. Therefore, the *loading effects of an A.C. rectifier voltmeter is greater than that of a corresponding D.C. voltmeter.*

4.8.11.7. Advantages of rectifier instruments

1. The primary advantage of the rectifier *voltmeter* is that it is *far more sensitive* as compared to other types of voltmeter suitable for measuring A.C. voltages.

2. Metal rectifiers can be incorporated in universal instruments, such as ammeter, thereby enabling a moving-coil milliammeter to be used in combination with shunt and series resistances to measure various ranges of D.C. and D.C. *voltage,* and in combination with a *bridge rectifier and suitable resistors* to measure various ranges of A.C. and A.C. *voltage.*
3. These instruments can be employed at frequencies well above the range of other A.C. instruments of the ordinary type (from about 20 Hz to 20 kHz).
4. They incorporate the high torque, efficient damping of the moving-coil instrument with *linear scale shape, except of low ranges.*
5. They have a very low power consumption.
6. Their accuracy is about ± 5% under normal operating conditions.
7. These instruments have a much lower operating current for voltmeters (or higher value of ohm per volt) than other A.C. instruments, with the exception of electrostatic and valve voltmeters.

Example 4.60. *An alternating current is measured by a rectifier and a hot-wire ammeters and the readings are found to be 31 A and 33 A respectively.*

Determine the form factor of the current wave.

Solution. R.m.s. value of current under measurement,

$$I_{\text{r.m.s.}} = \text{Hot-wire ammeter reading} = 33 \text{ A}$$

Average value of current under measurement,

$$I_{\text{av.}} = \text{Rectifier ammeter reading} = 31 \text{ A}$$

$\therefore$ From factor, $K_f = \dfrac{I_{r.m.s.}}{I_{av.}} = \dfrac{33}{31} = \mathbf{1.064}$ **(Ans.)**

Example 4.61. *A PMMC ammeter gives reading of 35 mA when connected across two opposite corners of bridge rectifier, the other two corners of which are connected in series with a capacitor to 100 kV, 50 Hz supply.*

Determine the capacitance.

Solution. Since PMMC ammeter gives average reading, therefore, average value of current (I_{av}) = 35 mA or 0.035 A.

Assuming sinusoidal waveform *i.e.* form factor 1.11,

R.m.s. value of current ($I_{\text{r.m.s.}}$) = 1.11 × 0.035 = 0.0388 A

Neglecting resistance of the instrument and assuming the capacitance connected is of C farads,

$$I_{\text{r.m.s.}} = \frac{V}{X_C} = \frac{V}{(1/2\pi fC)} = 2\pi fCV$$

or,
$$C = \frac{I_{r.m.s.}}{2\pi fV} = \frac{0.0388}{2\pi \times 50 \times 100 \times 10^3} = 1235 \times 10^{-12} \text{ or } \mathbf{1235\ pF} \quad \textbf{(Ans.)}$$

Example 4.62. *A PMMC ammeter is connected across appropriate points of a bridge having resistance of each arm of 40 Ω and connected across a supply of voltage $v = 10 \sin \theta + 0.4 \sin 3\theta$.*

Determine the reading of the ammeter if its resistance is 20 Ω.

Solution. *Given* : Instantaneous applied voltage, $v = 10 \sin \theta + 0.4 \sin 3\theta$; Resistance of each arm = 40 Ω; Ammeter resistance = 20 Ω.

Ammeter reading, $I_{av.}$:

Total resistance to the flow of current, $R = 2 \times 40 + 20 = 100\ \Omega$

Instantaneous current through the ammeter,

$$i = \frac{V}{R} = \frac{10 \sin\theta + 0.4 \sin 3\theta}{100} = 0.1 \sin\theta + 0.004 \sin 3\theta$$

Since PMMC reads average value, therefore, reading of PMMC ammeter,

$$I_{av.} = \frac{1}{\pi}\int_0^{\pi}(0.1 \sin\theta + 0.004 \sin 3\theta)\, d\theta$$

$$= \frac{1}{\pi}\left|-0.1\cos\theta - \frac{0.004}{3}\cos 3\theta\right|_0^{\pi}$$

$$= \frac{1}{\pi}\left(0.2 + \frac{0.008}{3}\right) = 0.0645 \text{ A or } \mathbf{64.5\ mA}\ \textbf{(Ans.)}$$

Example 4.63. *A moving coil ammeter, a hot wire ammeter and a resistance of 100 Ω are connected in series with a rectifying device across a sinusoidal alternating supply of 200 V. If the rectifying device has a resistance of 100 Ω to current in one direction and of 500 Ω in the opposite direction, calculate the readings on the two ammeters.* **(UPTU)**

Solution. Reading of moving coil ammeter :

R.m.s. current in one direction,

$$I_{\text{r.m.s.1}} = \frac{200}{100 + 100} = 1\text{A}$$

Average current in one direction, $I_{av.1} = \dfrac{I_{r.m.s.1}}{K_f} = \dfrac{1}{1.11} = 0.9$ A

Similarly r.m.s. value of current in other direction,

$$I_{\text{r.m.s.2}} = \frac{200}{500 + 100} = 0.333 \text{ A}$$

Average current in other direction,

$$I_{\text{av.2}} = \frac{I_{r.m.s.2}}{K_f} = \frac{0.333}{1.11} = 0.3 \text{ A}$$

Average current over the whole cycle,

$$I_{\text{av.1}} - I_{\text{av.2}} = \frac{0.9 - 0.3}{2} = 0.3 \text{ A}$$

Hence, the *reading of moving coil ammeter* = **0.3 A** **(Ans.)**

Reading of hot-wire ammeter :

Average heating effect during *positive* half cycle

$$= (I_{\text{r.m.s.}})^2 \times R = (1)^2 \times R = R,$$

where R is the resistance of hot-wire ammeter

Similarly, average heating effect during *negative* half cycle

$$= (0.333)^2 \times R = 0.11\ R$$

Average heating effect over the complete cycle

$$= \frac{1}{2}(R + 0.11R) = 0.555\ R$$

If I is the direct current which produces the same heating effect, then

$$I^2R = 0.555\,R$$

or,
$$I = \sqrt{0.555} = 0.745 \text{ A}$$

Hence, *reading of hot-wire ammeter* = **0.745 A** **(Ans.)**

Example 4.64. *A sinusoidal alternating voltage of 250 V peak is applied to a circuit containing a rectifying device which entirely prevents current flowing in one direction, and offers a non-reactive resistance of 200 Ω to the flow of current in the other direction. Determine the readings on :*

(*i*) *Hot-wire ammeter;* (*ii*) *Moving coil in the circuit*

(Punjab University)

Solution. *Given* : V_{max} = 250 V; Resistance of rectifying device, R = 200 Ω

Maximum value of current flowing through the instruments,

$$I_{max} = \frac{V_{max}}{R} = \frac{250}{200} = 1.25 \text{ A}$$

Reading of hot-wire ammeter, $I_{r.m.s.}$:

The hot-wire ammeter *reads r.m.s. value,* thus,

$$I_{r.m.s.} = \sqrt{\frac{1}{2\pi}\int_0^{\pi} I_{max}^2 \sin^2\theta\, d\theta}$$

$$= \sqrt{\frac{I_{max}^2}{2\pi}\int_0^{\pi}\left(\frac{1-\cos 2\theta}{2}\right) d\theta}$$

$$= \sqrt{\frac{I_{max}^2}{2\pi}\times\frac{\pi}{2}} = \frac{I_{max}}{2} = \frac{1.25}{2} = \mathbf{0.625\ A} \quad \textbf{(Ans.)}$$

Reading of moving coil ammeter, $I_{av.}$:

The moving coil ammeter *reads average value,* thus,

$$I_{av.} = \frac{1}{2\pi}\int_0^{\pi} I_{max} \sin\theta\, d\theta$$

$$= \frac{I_{max}}{2\pi}\left|-\cos\theta\right|_0^{\pi} = \frac{1.25}{\pi} = \mathbf{0.398\ A} \quad \textbf{(Ans.)}$$

Example 4.65. *A permanent magnet moving coil type full-wave bridge rectifier ammeter is used to measure current in a load connected across a supply of voltage :*

$$v = 5 \sin\theta + 0.2 \sin 3\theta$$

Determine the reading of the ammeter, if its resistance is 30 Ω. Assume dynamic resistance of each diode as 35 Ω under forward biased condition. **(UPTU)**

Solution. Refer to Fig. 4.56

Given : Supply voltage, $v = 5 \sin\theta + 0.2 \sin 3\theta$; R_m = 30 Ω; Resistance of each diode (forward biased) = 35 Ω.

Reading of the ammeter :

Total resistance to the flow of current (through D_2 – PMMC – D_4 or through D_3 – PMMC – D_1),

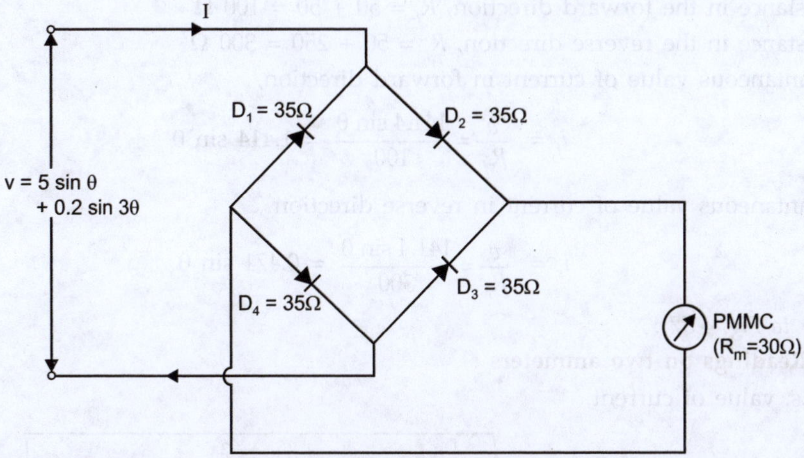

Fig. 4.56. Reading of the ammeter.

$$R = 35 + 30 + 35 = 100\ \Omega$$

Now,
$$i = \frac{v}{R} = \frac{5 \sin\theta + 0.2 \sin 3\theta}{100} = 0.05 \sin\theta + 0.002 \sin 3\theta$$

The PMMC reads *average value,* so,

$$I_{av.} = \frac{1}{\pi}\int_0^{\pi}(0.05 \sin\theta + 0.002 \sin 3\theta)\, d\theta$$

$$= \frac{1}{\pi}\left[-0.05 \cos\theta - \frac{0.002}{3}\cos 3\theta\right]_0^{\pi}$$

$$= \frac{1}{\pi}[-0.05(\cos\pi - \cos 0) - 0.000667(\cos 3\pi - \cos 0)]$$

$$= \frac{1}{\pi}[-0.05(-1-1) - 0.000667(-1-1)]$$

$$= \frac{1}{\pi}(0.1 + 1.334 \times 10^{-3}) = 0.03225 \text{ A} \quad \text{or} \quad 32.25 \text{ mA}$$

Hence, *reading of the ammeter* = **32.25 mA** **(Ans.)**

Example 4.66. *A resistance of 50 Ω is connected in series with a rectifying device, a moving coil ammeter and a dynamometer type ammeter. The circuit is connected across a sinusoidal A.C. supply of 100 V. The resistance of the rectifying device is 50 Ω in one direction and 250 Ω in the opposite direction.*

Calculate :

(*i*) *Reading on two ammeters.*

(*ii*) *Power supplied from mains.*

(*iii*) *Power dissipated in rectifying elements.* **(Anna University)**

Solution. R.m.s. value of supply voltage, $V = 100$ V

Maximum value of the voltage, $V_{max} = \sqrt{2} \times 100 = 141.4$ V

Instantaneous value of the voltage, $v = 141.4 \sin\theta$ V

Resistance in the forward direction, $R_f = 50 + 50 = 100\ \Omega$

Resistance in the reverse direction, $R_r = 50 + 250 = 300\ \Omega$

Instantaneous value of current in forward direction,

$$i_f = \frac{v}{R_f} = \frac{141.4 \sin\theta}{100} = 1.414 \sin\theta$$

Instantaneous value of current in reverse direction,

$$i_r = \frac{v}{R_r} = \frac{141.4 \sin\theta}{300} = 0.471 \sin\theta$$

Refer to Fig. 4.57 :

(i) Readings on two ammeters :

R.m.s. value of current,

$$I_{\text{r.m.s.}} = \sqrt{\frac{1}{2\pi}\left[\int_0^{\pi}(1.414\sin\theta)^2\, d\theta + \int_{\pi}^{2\pi}(0.471\sin\theta)^2\, d\theta\right]}$$

$$= \sqrt{\frac{1}{2\pi}\left[\int_0^{\pi}2\sin^2\theta\, d\theta + \int_{\pi}^{2\pi}0.222\sin^2\theta\, d\theta\right]}$$

$$= \sqrt{\frac{1}{2\pi}\left[2\int_0^{\pi}\left(\frac{1-\cos 2\theta}{2}\right)d\theta + 0.222\int_{\pi}^{2\pi}\left(\frac{1-\cos 2\theta}{2}\right)\right]}$$

$$= \sqrt{\frac{1}{2\pi}\left\{\left[\theta - \frac{\sin 2\theta}{2}\right]_0^{\pi} + 0.111\left[\theta - \frac{\sin 2\theta}{2}\right]_{\pi}^{2\pi}\right\}}$$

or, $$I_{\text{r.m.s.}} = \sqrt{\frac{1}{2\pi}\{\pi + 0.111(2\pi - \pi)\}} = 0.745\ \text{A}$$

i.e., Reading of dynamometer ammeter = **0.745 A** **(Ans.)**

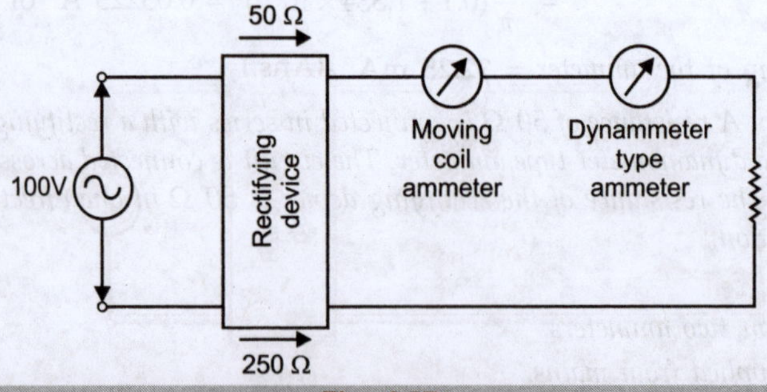

Fig. 4.57.

Average value of current,

$$I_{av.} = \frac{1}{2\pi}\left[\int_0^{\pi}1.414\sin\theta\, d\theta + \int_{\pi}^{2\pi}0.417\sin\theta\, d\theta\right]$$

$$= \frac{1}{2\pi}\left\{[-1.414\cos\theta]_0^{\pi} + [-0.417\cos\theta]_{\pi}^{2\pi}\right\}$$

$$= \frac{1}{2\pi}\left[\{-1.414(\cos\pi - \cos 0)\} + \{-0.417(\cos 2\pi - \cos\pi)\}\right]$$

$$= \frac{1}{2\pi}\left[\{-1.414(-1-1)\} + \{-0.417(1+1)\}\right]$$

$$= \frac{1}{2\pi}(2.828 - 0.834) = 0.317 \text{ A}$$

i.e., Reading of moving coil ammeter = **0.317 A** **(Ans.)**

(*ii*) **Power supplied from mains :**

Power supplied from mains = Power supplied in the forward half cycle + power supplied in the backward half cycle

$$= \frac{1}{2}\left(\frac{V^2}{R_f} + \frac{V^2}{R_r}\right) = \frac{1}{2}\left[\frac{(100)^2}{100} + \frac{(100)^2}{300}\right]$$

= **66.67 W** **(Ans.)**

(*iii*) **Power dissipated in rectifying element :**

Power consumed in 50 Ω resistor = I_2 R = $(0.745)^2 \times 50$ = 27.75 Ω

Power dissipated in rectifier = Total power supplied – power consumed in resistor

= 66.67 – 27.75 ≃ **39 W** **(Ans.)**

4.8.12. Electrostatic Instruments

4.8.12.1. Working principle

The underlying principle of operation of electrostatic instruments is the force of attraction between electric charges on neighbouring plates between which potential difference is maintained. This force gives rise to a deflecting torque. Unless the p.d. is adequately large, the force is small. Hence such instruments are *used for measurement of very high voltages.*

- The instruments are *almost always used as voltmeters and that too more as a laboratory rather than as industrial instruments.*

Such instruments *may be used with the help of external components, to measure current and power.*

4.8.12.2. Force and torque equations

Linear motion : Refer to Fig. 4.58. Let a voltage V be applied between two parallel plates, one fixed (A) and the other movable (B). The capacitance between the plates is C and the energy stored is $\frac{1}{2}CV^2$. If the voltage is increased by dV, the force of attraction (F) will increase and the movable plate will move by a distance dx.

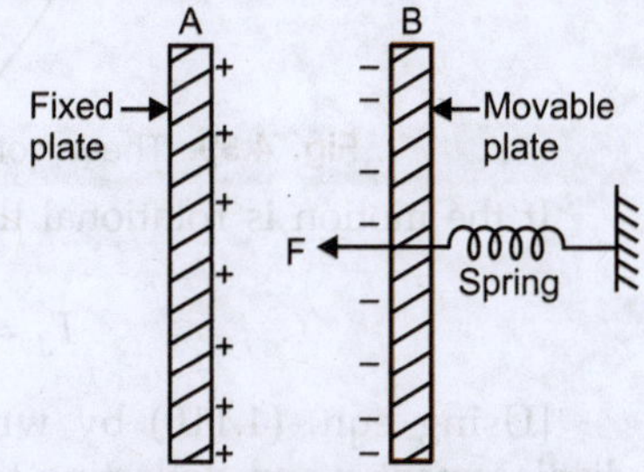

Fig. 4.58. Theory of linear motion of electrostatic instruments.

The capacitance current,

$$i = \frac{dQ}{dt} = \frac{d}{dt}(CV)$$

$$= C\frac{dV}{dt} + V\frac{dC}{dt} \qquad ...(4.108)$$

The energy input is,

$$Vi\,dt = V^2.dC + CV.dV \qquad ...(4.109)$$

The change in energy stored in electric field

$$= \frac{1}{2}(C + dC)(V + dV)^2 - \frac{1}{2}CV^2$$

$$= \frac{1}{2}(C + dC)(V^2 + 2VdV + dV^2) - \frac{1}{2}CV^2$$

$$= \frac{1}{2}\left(CV^2 + CVdV + CdV^2 + V^2dC + 2VdVdC + dC.V^2\right) - \frac{1}{2}CV^2$$

$$= \frac{1}{2}V^2dC + CV.dV$$

neglecting the higher order terms as they are small quantities.

Invoking law of conservation of energy, we have :

Energy input = Increase in energy stored + mechanical work done

or, $$V^2.dC + CV.dV = \frac{1}{2}V^2.dC + CV.dV + F.dx$$

or, $$\text{Force, } F = \frac{1}{2}V^2\frac{dC}{dx} \text{ newton} \qquad ...(4.110)$$

Rotational motion : Refer to Fig. 4.59.

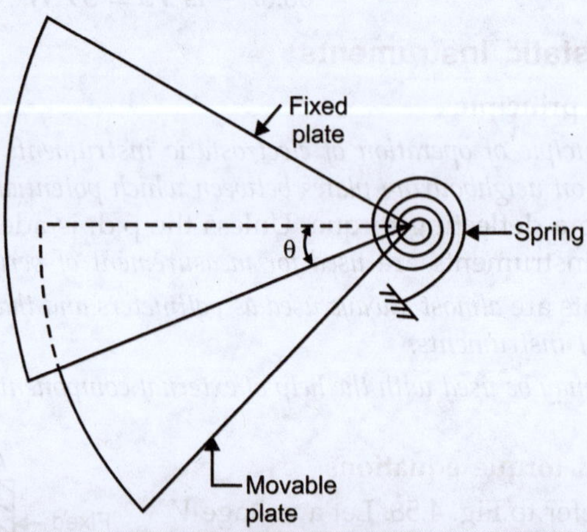

Fig. 4.59. Theory of rotational motion of electrostatic instruments.

If the motion is rotational the deflecting torque is given as :

$$T_d = \frac{1}{2}V^2\frac{dC}{d\theta} \qquad ...(4.111)$$

[Using eqn. (4.110) by writing an angular displacement θ in place of linear displacement x and deflecting torque T_d in place of force F].

When the instrument is spring controlled or has a suspension,

Control torque, $T_c = K\theta$

where, K = spring constant, and

θ = Deflection.

At steady deflection state, $T_d = T_c$

$$\therefore \quad \frac{1}{2}V^2\frac{dC}{d\theta} = K\theta$$

or, deflection, $$\theta = \frac{1}{2}\frac{V^2}{K}\frac{dC}{d\theta} \quad ...(4.112)$$

- Because the deflection is *proportional to the square of voltage* to be measured, the instrument *can be used on both A.C. and D.C.*
- Since the instrument exhibits *a square law response*, therefore, the *scale is non-uniform and is compressed at lower end.*

4.8.12.3. Classification of electrostic instruments

There are two general types of such instruments :

1. The *quadrant type* ...... Used upto 20 kV
2. The *attracted disc type* .... Used upto 500 kV.

4.8.12.4. Quadrant electrometer

Construction. Fig. 4.60(*a*) shows a *quadrant electrometer* :

— It consists of *four fixed metal double quadrants* arranged to form a *circular box*. The opposite pairs are connected together. There are *short air gaps between the quadrants*. Inside the box, a *double sector shaped needle* is suspended by a phosphor bronze or silver quartz thread. The needle is *equidistant* from top and bottom quadrants.

— The needle carries a mirror and the deflection read off by means of a *lamp and scale arrangement*.

An alternate arrangement may be mounting the needle on a spindle and the controlling troque is provided by a spring attached to the spindle. The deflecting torque moves a pointer and the deflection is read on a calibrated scale. For these instruments, *eddy current damping is used.*

— Usually, a high resistance is connected in series with an electrostatic voltmeter, particularly if the instrument is of the suspended-vane type. This resistance is used for limiting the current which can flow if there is accidental connection between a vane and a quadrant.

Principle of operation. The fixed quadrants are connected together and the voltage to be measured is applied between the fixed quadrants and the moving needle. *Due to charge accmulated, an electrostatic force is set up because of which the needle rotates. The suspension exerts a controlling torque and the needle takes up a position where the deflecting torque is equal to the controlling torque.*

Types of connections. The quadrant electrometer may be used in *two ways* :

(*i*) ***Heterostatic connection.*** This connection is shown in Fig. 4.60(*b*). A high tension (HT) battery is used for charging the needle to a potential considerably above that of the quadrants to which the negative of the voltage to be measured is connected.

(*ii*) ***Idiostatic connection.*** This connection is shown in Fig. 4.60(*c*). The needle is directly connected to one pair of quadrants.

Generally, *idiostatic connection is used.*

- It can be shown that with the four quadrants and a double sector shaped needle, as shown in Fig. 4.60, the deflecting torque (T_d) in Nm is given as :

$$T_d = \frac{\in r^2}{d}.V^2 \quad ...(4.113)$$

where, $\epsilon = \epsilon_0$ (since air is the dielectric) = Absolute permittivity = 8.854×10^{-12} F/m,

r = Radius of each quadrant m,

d = Distance of needle from the top or bottom plate, m, and

V = Voltage to be measured, volts.

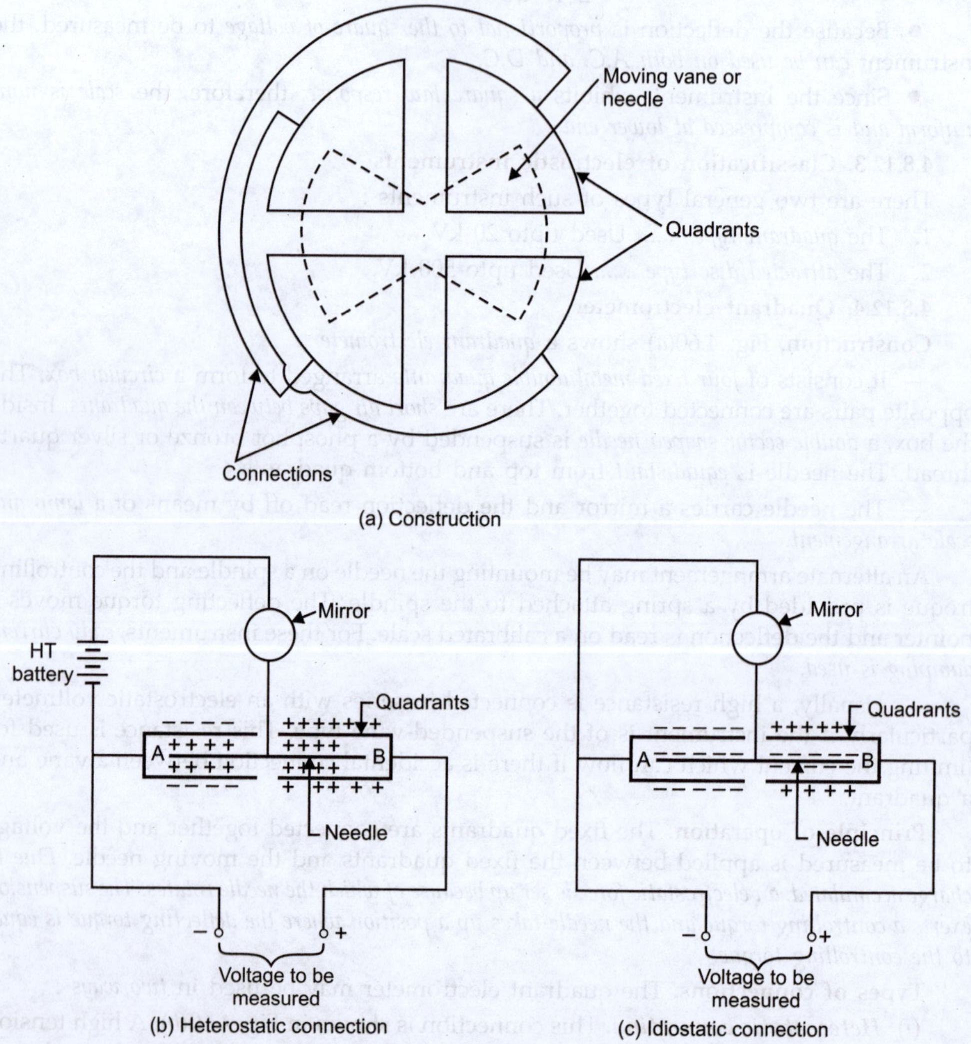

Fig. 4.60. Quadrant electrometer.

The controlling torque is provided by the spring. Thus,

$$T_c = K\theta$$

At steady deflection state,

$$T_d = T_c$$

i.e.,

$$\frac{\epsilon_0 r^2}{d} \cdot V^2 = K\theta$$

or, $$\theta = \frac{\epsilon_0 r^2}{Kd} \cdot V^2 \qquad ...(4.114)$$

Hence, with *idiostatic connection* the instrument gives a *square law response.*

Kelvin multicellular voltmeter :

The construction shown in Fig. 4.60 has a rather small deflecting torque. To increase sensitivity, a multicellular electrostatic voltmeter devised by Lord Kelvin is used.

Fig. 4.61 shows the components of a *Kelvin multicellular voltmeter* :

— It has a large number of *fixed quadrants and needles*. The large number of quadrants is needed to *give a high deflecting torque even at low voltages*.

— The moving system is suspended by a *phosphor bronze suspension to reduce bearing friction*.

— The *coach spring* provides *protection* against breaking of suspension due to vibration.

— The *torsion head* is provided for *zero adjustment*.

— Damping is done by using an *oil dashpot*.

- This voltmeter is suitable for *even low voltages in the range of 100 to 1000 V*.

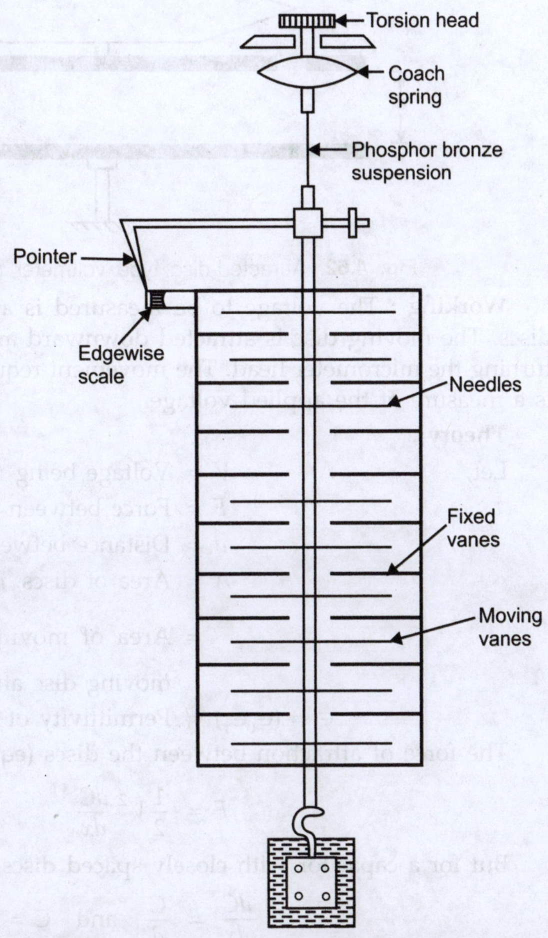

Fig. 4.61. Kelvin multicellular voltmeter.

4.8.12.5. Attracted-disc type voltmeter

Construction: This instrument is *suitable for only high voltages* (more than 10 kV).

Fig. 4.62 shows the main constructional features of an attracted-disc type voltmeter (also known as Kelvin absolute electrometer) :

— It consists of two discs, the fixed disc and the moving disc.

— The moving disc is supported by coach spring which is suspended from a micrometer head as shown.

— The moving disc is provided with a *guard ring to reduce the fringing effects.* The effective area of the moving disc is its actual area plus half the area of air gap.

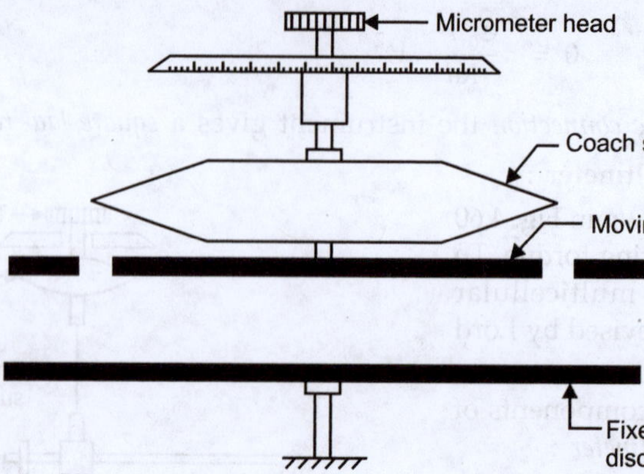

Fig. 4.62. Attracted-disc type voltmeter (Kelvin absolute electrometer)

Working : The voltage to be measured is applied between the fixed and moving discs. The moving disc is attracted downward and is brought back to zero position by turning the micrometer head. The movement required to return the disc to zero position is a measure of the applied voltage.

Theory :

Let,
V = Voltage being measured, volts,
F = Force between discs, N,
d = Distance between discs, m,
A = Area of discs, m^2,
= Area of moving disc + $\frac{1}{2}$ area of air gap between moving disc and guard ring, and
$\in = (\in_0 \in_r)$ = Permittivity of dielectric, F/m.

The force of attraction between the discs (eqn. 4.110),

$$F = \frac{1}{2} V^2 \frac{dC}{dx}$$

But for a capacitor with closely spaced discs/plates,

$$\frac{dC}{dx} = \frac{C}{d} \quad \text{and} \quad C = \frac{\in A}{d}$$

$$\therefore \quad F = \frac{1}{2} \in A . \frac{V^2}{d^2} \quad \text{...(4.115)}$$

Hence,

$$V = \sqrt{\frac{2Fd^2}{\in A}} \quad \text{...(4.116)}$$

The above theory shows that such an instrument will give an *absolute determination of potential difference*, as p.d. is given in terms of force and linear dimensions. The deflecting force is only adequate if the voltage to be measured is high..

— In order to ensure good insulation, screening, and freedom from corona effects, special construction may be necessary.

— A modern development is to enclose the working parts of the instruments within a *highly evacuated chamber*.

The superior dielectric strength of a high vacuum, in comparison with that of air at normal pressures, enables the clearance between the plates to be *reduced*, thus *giving an increased force for a given voltage. The use of a compressed-gas dielectric is an alternative means of obtaining higher dielectric strength and increased sensitivity.*

4.8.12.6. Advantages and disadvantages

Following are the *advantages and disadvantages of electrostatic instruments.*

Advantages :

1. Suitable for both A.C. and D.C.
2. Measure r.m.s. values.
3. Draw negligible power from the mains.
   — For operation on D.C., the initial current is the charging current which decreases exponentially and the final current is zero.
   — For operation on A.C., the current is inversely proportional to the impedance of the measuring circuit.
4. Free from hysteresis and eddy current errors (because there is no iron in the working system).
5. Free from frequency and waveform errors.
6. Especially suitable for high voltages.
7. Owing to negligible power loss there is no heating error.
8. Can be manufactured with first grade accuracy.
9. Cause little disturbance to circuits to which they are connected.

Disadvantages :

1. Low deflecting torque especially when the voltage required to be measured is less than 500 V.

   However, Kelvin multicellular voltmeter can be designed for use for measurements of voltages as low as 30 V.
2. Fragile and expensive and cannot be made robust.
3. Non-uniform scale.
4. Liable to friction errors.

4.8.12.7. Range extension of electrostatic voltmeters

The range of electrostatic voltmeters can be extended by the use of *multipliers* which are in form of :

1. Resistance potential divider ------ *Can be used both for direct and alternating voltages.*
2. Capacitance potential divider ------ Useful *only for alternating voltages.*

1. Resistance potential divider : This divider consists of a high non-inductive resistance across a small portion which is attached to the electrostatic voltmeter as shown in Fig. 4.63.

Let, R = The resistance of the whole of the potential divider across which the voltage V under measurement is applied, and

r = The resistance of the portion of the divider across which voltmeter is connected.

Then, the *multiplying factor* is given by :

$$\frac{V}{v} = \frac{R}{r} \qquad ...(4.117)$$

The above eqn. (4.117) is true for D.C. circuits but for A.C. circuits, the capacitance of the voltmeter (which is in parallel with *r*) has to be taken into account. Since this capacitance is variable, it is advisable to calibrate the voltmeter alongwith its multiplier.

2. **Capacitance potential divider :** In this method, the voltmeter may be connected in series with a single capacitor C and put across the voltage V which is to be measured [Fig. 4.64(*a*)] or a number of capacitor may be joined in series to form the potential divider and the voltmeter may be connected across one of the capacitors as shown in Fig. 4.64(*b*).

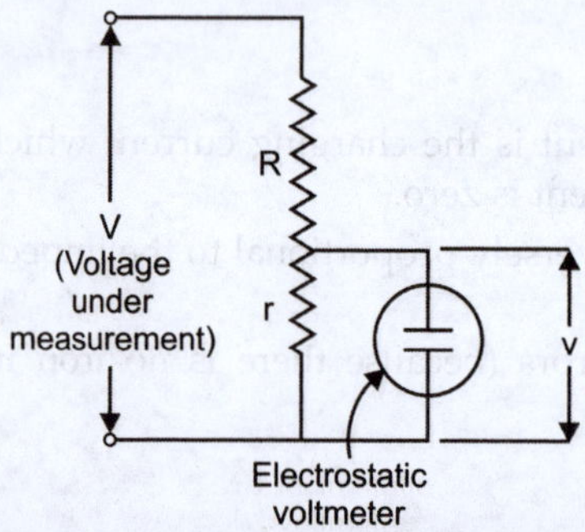

Fig. 4.63. Resistance potential divider.

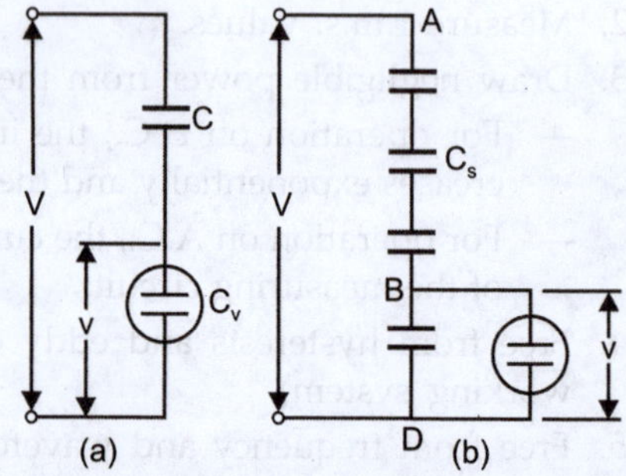

Fig. 4.64. Capacitance potential divider.

Referring to Fig. 4.64(*a*), the multiplying factor is given by :

$$\frac{V}{v} = \frac{\text{Reactance of total circuit}}{\text{Reactance of voltmeter}}$$

Now, capacitance of the total circuit $= \dfrac{CC_v}{C + C_v}$

and, its reactance $= \dfrac{1}{\omega \times \text{capacitance}} = \dfrac{C + C_v}{\omega CC_v}$

Reactance of the voltmeter $= \dfrac{1}{\omega C_v}$

$$\therefore \quad \frac{V}{v} = \frac{(C + C_v)/\omega CC_v}{1/\omega C_v} = \frac{C + C_v}{C}$$

$$\therefore \text{ Multiplying factor, } \frac{V}{v} = \frac{C + C_v}{C} = 1 + \frac{C_v}{C} \quad \text{...(4.118)}$$

Example 4.67. *An absolute electrometer uses a movable circular plate 80 mm in diameter. During a voltage measurement the distance between the plates is 3.2 mm and the force of attraction is 0.003 N. The medium is air.* $\epsilon_0 = 8.85 \times 10^{-12}$ *F/m. Determine V.*

Solution. *Given* : $A = \frac{\pi}{4} \times (0.08)^2 = 5.0265 \times 10^{-3}$ m^2; d = 3.2 mm or 0.0032 m; F = 0.03 N; $\epsilon = \epsilon_0 = 8.85 \times 10^{-12}$ F/N

We know that,
$$V = \sqrt{\frac{2Fd^2}{\epsilon A}} \quad \text{[Eqn. (4.116)]}$$

or,
$$V = \sqrt{\frac{2 \times 0.03 \times (0.0032)^2}{8.85 \times 10^{-12} \times 5.0265 \times 10^{-3}}} = \mathbf{1175\ V} \quad \textbf{(Ans.)}$$

Example 4.68. *In an electrometer, the movable plate is 110 mm in diameter. When 12 kV is applied between the movable plate and the fixed plate, the force is 0.006 N. Find the change in capacitance for 1.2 mm movement of the movable plate.*

Solution. *Given :* Diameter of movable plate = 110 mm or 0.11 m; V = 12 kV; F = 0.006 N; x = 1.2 mm.

We know that, $$F = \frac{1}{2} V^2 \frac{dC}{dx} \quad \text{[Eqn. (4.110)]}$$

$$0.006 = \frac{1}{2} \times (12 \times 10^3)^2 \times \frac{dC}{dx}$$

or, $$\frac{dC}{dx} = \frac{2 \times 0.006}{(12 \times 10^3)^2} = 83.33 \times 10^{-12} \text{ F/m}$$

∴ Change in capacitance, $$dC = 83.33 \times 10^{-12} \times (1.2 \times 10^{-3})$$
$$\simeq 0.1 \times 10^{-12} \text{ F or } \mathbf{0.1\ pF} \text{ (Ans.)}$$

Example 4.69. *An electrostatic voltmeter has full scale deflection of 90°. The control constant of the spring is 8 × 10^{-6} Nm/rad. When reading is zero, the capacitance is 12 pF. What is the capacitance when the instrument reads 1500 V and the deflection is 90°.*

Solution. *Given :* $\theta = 90° = \frac{\pi}{2}$ rad.; $K = 8 \times 10^{-6}$ Nm/rad.; C = 12 pF, when reading is zero; V = 1500 volts.

At steady deflection state,

$$T_d = T_c$$

or, $$\frac{1}{2} V^2 \frac{dC}{d\theta} = K\theta \quad \text{[Eqn. (4.111)]}$$

or, $$\frac{dC}{d\theta} = \frac{2K\theta}{V^2} = \frac{2 \times 8 \times 10^{-6} \times (\pi/2)}{(1500)^2} = 0.1117 \times 10^{-10} \text{ F/rad.}$$

Change in capacitance, $$dC = 0.1117 \times 10^{-10} \times \frac{\pi}{2} = 0.1755 \times 10^{-10} \text{ F}$$

∴ New value of capacitance $= 12 \times 10^{-12} + 0.1755 \times 10^{-10} = 29.55 \times 10^{-12}$ F
$= 29.55 \times 10^{-12}$ F or **29.55 pF** **(Ans.)**

Example 4.70. *An electrostatic voltmeter is constructed with 6 parallel, semicircular fixed plates equally-spaced at 4 mm intervals and 5 interleaved semi-circular movable plates that move in planes midway between the fixed plates, in air. The movement of the movable plates is about an axis through the centre of the circles of the plates system, perpendicular to the planes of the plates. The instrument is spring-controlled. If the radius of the movable plates is 4 cm, calculate the spring constant if 10 kV corresponds to a full-scale deflection of 100°. Neglect fringing, edge-effects and plate thickness.* **(Bombay University)**

Solution. *Given :* No. of fixed plates = 6; No. of movable plates = 5; d (Distance between fixed and movable plates) = $\frac{4}{2}$ = 2 mm or 0.002 m ; r = 4 cm or 0.04 m ; V = 10 kV ; θ = 100° or $100 \times \frac{\pi}{180}$ = 1.745 rad.

Spring constant, K :

Total number of plates (both fixed and movable) is 11, hence there are *10 parallel plate capacitors* (See Fig. 4.65)

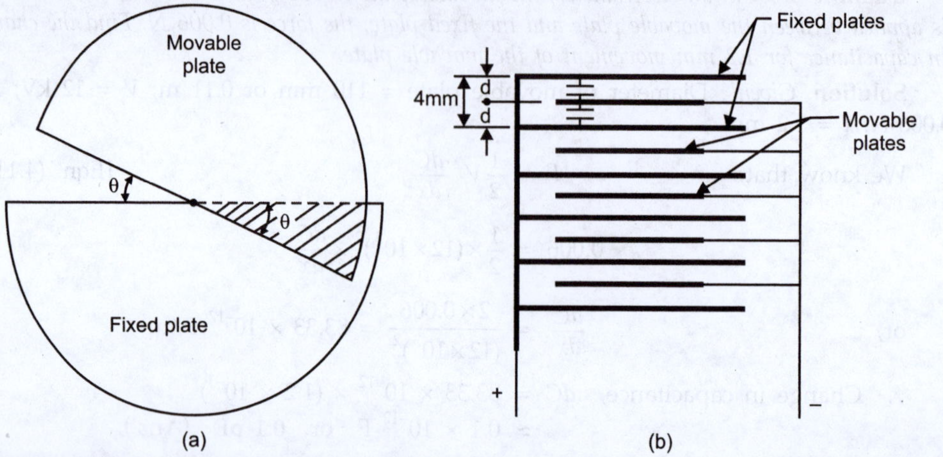

Fig. 4.65.

Suppose, the movable plates are rotated into the fixed plaes by an angle of θ radian. Then, overlap area between one fixed and one movable semi-circular plate is,

$$A = \frac{1}{2}r^2\theta = \frac{1}{2}\times 0.04^2 \times \theta = 8 \times 10^{-4}\ \theta$$

Capacitance of each of ten parallel-plate capacitors,

$$C_1 = \frac{\epsilon_0 A}{d} = \frac{8.854\times 10^{-12}\times 8\times 10^{-4}\ \theta}{2\times 10^{-3}} = 3.54 \times 10^{-12}\ \theta\ \text{F}$$

Total capacitance, $C = 10 \times C_1 = 10 \times 3.54 \times 10^{-12}\ \theta = 35.4 \times 10^{-12}\ \theta$ F

$$\therefore \quad \frac{dC}{d\theta} = 35.4 \times 10^{-12}\ \text{F/rad.}$$

Deflecting torque, $T_d = \frac{1}{2}V^2 \cdot \frac{dC}{d\theta}$ Nm [Eqn. (3.111)]

$$= \frac{1}{2}\times (10\times 10^3)^2 \times 35.4\times 10^{-12} = 17.7 \times 10^{-4}\ \text{Nm}$$

If K is the spring constant, then control torque is,

$$T_c = K\theta$$

At steady deflection state,

$$T_d = T_c$$

$$17.7 \times 10^{-4} = K \times 1.745$$

$$\therefore \quad K = \mathbf{10.14 \times 10^{-4}\ Nm/rad.}$$ **(Ans.)**

Example 4.71. *The reading '100' of a 120 V electrostatic voltmeter is to represent 10,000 volts when its range is extended by the use of a capacitor in series. If the capacitance of the voltmeter at the above reading is 70 pF, find the capacitance of the capacitor multiplier required.*

Solution. *Given* : V = 10,000 volts; v = 100 volts, C_v = 70 pF.

We know that, multiplying factor $= \frac{V}{v} = 1 + \frac{C_v}{C}$ [Eqn. 4.118)]

Inserting the various values in the above eqn., we get

$$\frac{10,000}{100} = 1 + \frac{70}{C}$$

or, $$\frac{70}{C} = \frac{10,000}{100} - 1 = 99$$

or, $$C = \frac{70}{99}\ \text{pF} = \mathbf{0.707\ pF}\ \textbf{(Ans.)}$$

Example 4.72. *A current of 0.5 + 0.3 sin ωt – 0.2 sin 2ωt is passed through the circuit shown in Fig. 4.66. Determine the reading of each instrument if* $\omega = 10^6$ *rad./s.*

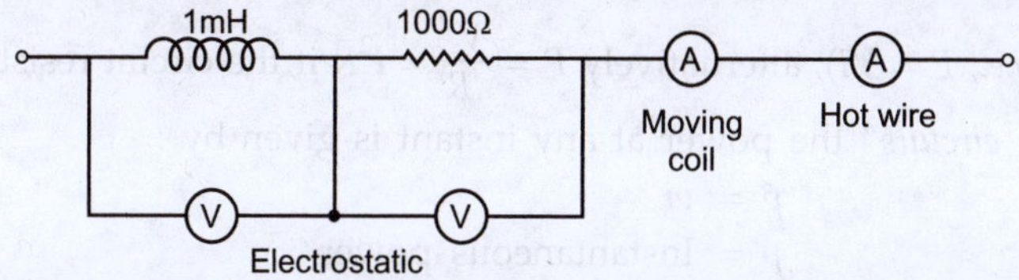

Fig. 4.66.

Solution. *Given* : $i = 0.5 + 0.3 \sin \omega t - 0.2 \sin 2\omega t$; $\omega = 10^6$ rad./s.

Reading of each instrument :

Let, the instantaneous value of current,

$$i = I_0 + I_{1max} \sin \omega t + I_{2max} \sin 2\omega t$$

or, $$i = 0.5 + 0.3 \sin \omega t - 0.2 \sin 2\omega t \quad \textit{(Given)}$$

Average value of a sinusoidal quantity over a number of cycles or over a time very much greater than the time period of a cycle is equal to zero.

∴ $$I_{av.} = i = 0.5\ \text{A}$$

Hence, the *reading of moving-coil instrument* = **0.5 A (Ans.)**

R.m.s. value of current, $$I = \sqrt{I_0^2 + \left(\frac{I_{1max}}{\sqrt{2}}\right)^2 + \left(\frac{I_{2max}}{\sqrt{2}}\right)^2}$$

$$= \sqrt{(0.5)^2 + \left(\frac{0.3}{\sqrt{2}}\right)^2 + \left(\frac{0.2}{\sqrt{2}}\right)^2} = \mathbf{0.561\ A}$$

Hence, the *reading of hot-wire instrument* = **0.561 A (Ans.)**

Reading of "electrostatic voltmeter across 1000 Ω resistance",

$$V_R = 0.561 \times 1000 = \mathbf{561\ V}\ \textbf{(Ans.)}$$

Instantaneous value of voltage across 1 mH inductor,

$$v_L = L\frac{di}{dt} = (1 \times 10^{-3}) \times \frac{d}{dt}(0.5 + 0.3 \sin \omega t - 0.2 \sin 2\omega t)$$

$$= (1 \times 10^{-3}) \times \omega(0.3 \cos \omega t - 0.4 \cos 2\omega t)$$

$$= 300 \cos \omega t - 400 \cos 2\omega t \quad ...(\because \omega = 10^6 \text{ rad./s})$$

Hence, *"reading of electrostatic voltmeter across 1 mH inductor",*

$$V_L = \sqrt{\left(\frac{300}{\sqrt{2}}\right)^2 + \left(\frac{400}{\sqrt{2}}\right)^2} \simeq 354 \text{ V} \quad \textbf{(Ans.)}$$

4.9 WATTMETERS AND MEASUREMENT OF POWER

4.9.1. Introduction

4.9.1.1. Power in D.C. and A.C. circuits

Power *may be defined as the rate at which energy is transformed or made available.*

- In *"D.C. circuits"* the power may be measured either by a wattmeter or by an ammeter and voltmeter, the product of whose readings gives the power in the circuit (*i.e.*, $P = VI$); alternatively $P = \frac{V^2}{R} = I^2R$ if the circuit resistance is known.
- In *"A.C. circuits"* the power at any instant is given by :

$$p = vi \quad ...(4.119)$$

where, p = Instantaneous power,

v = Instantaneous voltage, and

i = Instantaneous current.

Thus, if both the current and voltage waves are sinusoidal, the current lagging in phase by an angle ϕ, and

$$v = V_{max} \sin \omega t,$$

and, $$i = I_{max} \sin (\omega t - \phi),$$

Then instantaneous power p is given by,

$$p = vi$$

$$= V_{max} I_{max} \sin \omega t \cdot \sin (\omega t - \phi)$$

or, writing θ for ωt,

$$p = V_{max} I_{max} \sin \theta \sin (\theta - \phi)$$

The mean power in given by :

$$P = \frac{1}{2\pi} \int_0^{2\pi} V_{max} I_{max} \sin\theta \sin(\theta - \phi) d\theta \quad ...(4.120)$$

$$= \frac{V_{max} I_{max}}{2\pi} \int_0^{2\pi} \frac{\cos\phi - \cos(2\theta - \phi)}{2} d\theta$$

$$= \frac{V_{max} I_{max}}{4\pi} \left[\theta \cos\phi - \frac{\sin(2\theta - \phi)}{2}\right]_0^{2\pi}$$

$$= \frac{V_{max} I_{max}}{2} \cos\phi = \frac{V_{max}}{\sqrt{2}} \cdot \frac{I_{max}}{\sqrt{2}} \cdot \cos\phi$$

or, $$P = VI \cos \phi \quad ...(4.121)$$

where V and I are r.m.s. values of voltage and current.

The fact that the power factor ($\cos \phi$) is involved in the expression for the power means that a *wattmeter must be used instead of the merely an ammeter and voltmeter, since the latter method takes no account of power factor.*

4.9.1.2. Wattmeter connections/measurements in single-phase A.C. circuits

Fig. 4.67 shows a wattmeter connected in a single-phase A.C. circuit.

— The *"current coil"* of the instrument carries the load current, while the *"pressure or voltage coil"* carries a current proportional to, and in phase with, the voltage. The deflection of the wattmeter depends upon the currents on these coils and upon the power factor.

— Inductance in the voltage-coil circuit should be avoided as far as possible, since it causes the voltage-coil current to lag behind the applied voltage. *A high non-inductance resistance is connected in series with the voltage coil in order that the reactance of the coil itself shall be small in comparison with the resistance of the whole voltage-coil circuit and also, of course, to ensure that the current taken by the voltage coil shall be small.*

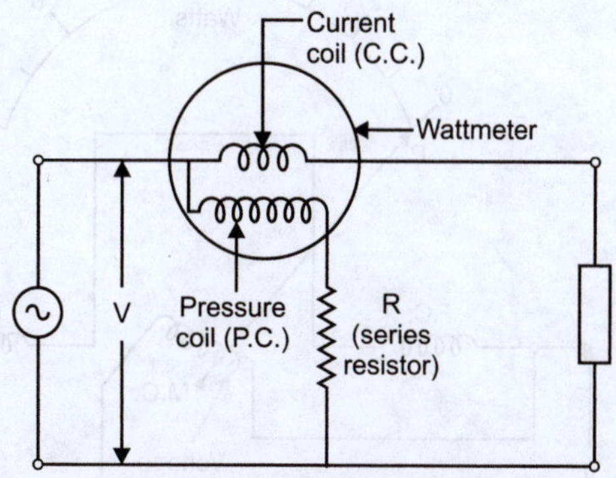

Fig. 4.67. Wattmeter connections.

WATTMETERS

4.9.2. Types of Wattmeters

A **wattmeter** is *a combination of an ammeter and a voltmeter* and, therefore, consists of two coils known as *current coil* and *pressure coil. The operating torque is produced due to interaction of fluxes on account of currents in current and pressure coils.*

There are following *three* types of wattmeters :

1. Dynamometer wattmeters. } Most commonly used
2. Induction wattmeters. }
3. Electrostatic wattmeters.

4.9.3. Dynamometer Wattmeter

4.9.3.1. Description

Fig. 4.68 shows the connection of dynamometer for measuring power. If the coils are connected so that a value of current *proportional to the load voltage* flows in one, and a value of current *proportional to the load current* flows in the other, the meter may be calibrated directly in watts. This is true because the indication depends upon the product of the two magnetic fields. The *strength of the magnetic fields depends upon the values of currents flowing* through the coils. If one current is proportional to load voltage and other current is the load current, then the meter can be calibrated in terms of watts or true power consumed by the load.

Let v = Supply voltage,

i = Load current, and

R = Resistance of the moving coil circuit.

Current through fixed coils, $i_f = i$,

Current through the moving coil, $i_m = \frac{v}{R}$

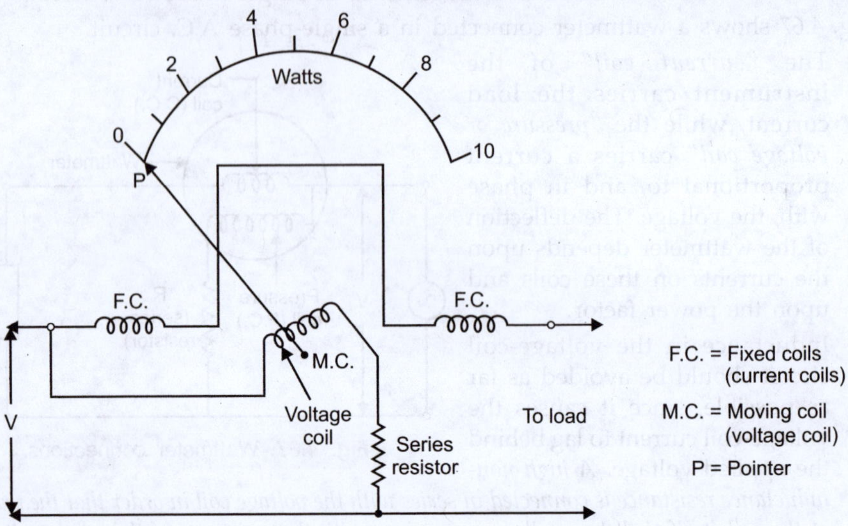

Fig. 4.68. Connection of dynamometer for measuring power

Deflecting torque, $T_d \propto i_f \times i_m \propto \dfrac{iv}{R}$ ...(4.122)

— For a *D.C. circuit the deflecting torque is thus proportional to the power.*

— For any circuit with *fluctuating torque, the instantaneous torque is proportional to instantaneous power*. In this case due to inertia of moving parts, the deflection will be proportional to the *average torque i.e.,* the deflection will be proportional to the *average power*. For sinusoidal alternating quantities the average power is $VI \cos \phi$, where

V = r.m.s. value of voltage,

I = r.m.s. value of current, and

ϕ = phase angle between V and I.

Hence an electrodynamic instrument, when connected as shown in Fig. 4.68, indicates the power, irrespective of the fact it is connected in an A.C. or D.C. circuit.

Effect of mutual inductance on the deflecting torque :

The mutual inductance between the fixed and moving coils varies as the moving coil moves and the deflecting torque is affected due to change in mutual inductance. Hence it is necessary to study the effect of change of mutual inductance on the deflecting torque.

Refer to Fig. 4.69 :

Let, i_1, i_2 = Instantaneous values of currents in two coils, and

V, I = R.m.s. values of voltage and current being measured, and

M = Mutual inductance between the current coil and potential or pressure coil.

Then, the instantaneous torque of a dynamometer type of instrument,

$$T_i = i_1 i_2 \frac{dM}{d\theta} \qquad ...(4.123)$$

Now, instantaneous value of voltage across the pressure coil circuit,

$$v = \sqrt{2}\, V \sin \omega t$$

If the pressure coil circuit has a very high resistance, it can be *treated as purely resistive*. Therefore, current i_p in the pressure coil is in phase with the voltage and its instantaneous value is,

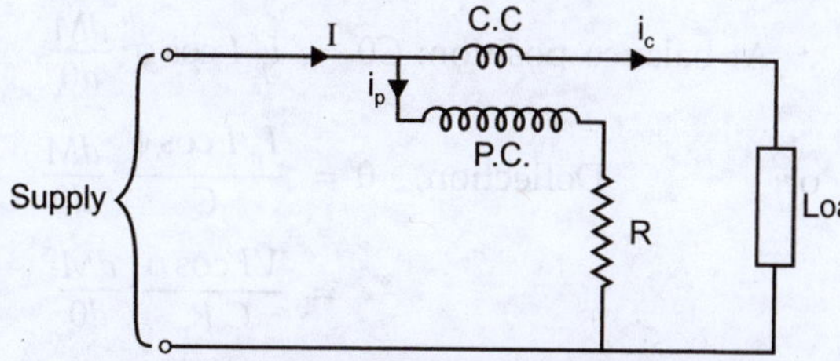

Fig. 4.69. Circuit of dynamometer wattmeter.

$$i_p = \frac{v}{R_p} = \frac{\sqrt{2} V \sin \omega t}{R_p} = \sqrt{2}\, I_p \sin \omega t$$

where, $I_p = \frac{V_p}{R_p}$ = r.m.s. value of current in pressure coil circuit, and

R_p = Resistance of pressure-coil circuit.

If the current in the current coil *lags* the voltage in phase by an angle ϕ, instantaneous value of current through current coil is,

$$i_c = \sqrt{2}\, I \sin(\omega t - \phi)$$

∴ Instantaneous torque,

$$T_i = \sqrt{2}\, I_p \sin \omega t \times \sqrt{2}\, I \sin(\omega t - \phi) \frac{dM}{d\theta}$$

$$= 2\, I_p\, I \sin \omega t \sin(\omega t - \phi) \frac{dM}{d\theta}$$

or, $$T_i = I_p\, I[\cos\phi - \cos(2\omega t - \phi)] \frac{dM}{d\theta} \quad ...(4.124)$$

From above it is obvious that there is a component of power which varies as *twice the frequency of current and voltage* (refer to term containing 2 ωt).

Average deflecting torque $(T_d)_{av.}$ is given by :

$$(T_d)_{av.} = \frac{1}{T}\int_0^T T_i \cdot d(\omega t)$$

$$= \frac{1}{T}\int I_p I[\cos\phi - \cos(2\omega t - \phi)] \frac{dM}{d\theta} \cdot d(\omega t)$$

or, $$(T_d)_{av.} = I_p\, I \cos\phi \cdot \frac{dM}{d\theta} \quad ...(4.125)$$

or, $$(T_d)_{av} = \frac{V}{R_p} \cdot I \cos\phi \frac{dM}{d\theta} \left(= \frac{P}{R_p} \cdot \frac{dM}{d\theta} \right) \quad ...(4.126)$$

Controlling torque exerted by springs, $T_c = C\theta$

where, C = Spring constant, and

θ = Final steady deflection.

Since the moving system of the instrument cannot follow the rapid variations in torque (the torque has a double frequency, component), it will take up a position at which the average deflection torque is equal to the restoring torque of the springs.

$$\therefore \text{ At balance position: } C\theta = I_p I \cos\phi \cdot \frac{dM}{d\theta} \quad ...(4.127)$$

$$\text{or, Deflection, } \theta = \frac{I_p I \cos\phi}{C} \cdot \frac{dM}{d\theta} \quad ...(4.128)$$

$$= \frac{VI \cos\phi}{C R_p} \cdot \frac{dM}{d\theta} \quad ...(4.129)$$

$$= kVI \cos\phi \cdot \frac{dM}{d\theta} \quad ...(4.130)$$

$$= \left(k \cdot \frac{dM}{d\theta}\right) \cdot P \quad ...(4.131)$$

where, P = Average power being measured = $VI \cos\phi$

and, $$k = \frac{1}{CR_p} \quad ...(4.132)$$

It is evident from eqn. (4.131) that the *deflection is directly proportional to the power being measured and the scale is essentially uniform over the range in which* $\frac{dM}{d\theta}$ *is constant.* By suitable design, the mutual inductance between fixed and moving coils can be made to vary linearly with angle over a range of 40° to 50° on either side of zero mutual inductance position. If the position of zero mutual inductance is at the midscale, the scale will be uniform over 80° to 100° which covers almost the entire range of scale.

Ranges :

(*i*) *Current circuit :* 0.25 A to 100 A with employing current transformers (CTs)

(*ii*) *Potential circuit :* 5 V to 750 V without employing potential transformers (PTs).

4.9.3.2. Types of dynamometer wattmeter

Dynamometer wattmeters may be divided into *two classes* :

1. Suspended-coil torsion instruments.
2. Pivoted-coil, direct indicating instruments.

1. ***Suspended-coil torsion wattmeters.*** These instruments are used largely as *standard wattmeters.*

— The moving, or voltage, coil is suspended from a torsion head by a metallic suspension which serves as a lead to the coil. This coil is situated entirely inside the current, or fixed coils and the winding in such that the system is astatic. Errors due to external magnetic fields are thus avoided.

— The torsions head carries a scale, and when in use, the moving coil is bought back to the zero position by turning this head, the *number of divisions turned through when multiplied by a constant for the instrument gives the power.*

— Eddy currents are eliminated as far as possible by winding the current coils of stranded wire and by using no metal parts within the region of the magnetic field of the instrument.

— The mutual inductance errors are completely eliminated by making zero position of the coil such that the angle between the planes of moving coil and fixed coil = 90°, *i.e.,* the mutual inductance between the fixed and moving coils is zero.

— The *elimination of pivot friction* makes possible the construction of *extremely sensitive and accurate* electrodynamic instruments of this pattern.

2. ***Pivoted-coil direct-indicating wattmeters.*** These instruments are commonly used as *switchboard or portable instruments.*
   — In these instruments, the *fixed coil* is wound in two halves, which are placed parallel to another at such a distance, that uniform uniform field is obtained. The *moving coil* is wound of such a size and pivoted centrally so that it does not project outside the field coils at its maximum deflection position.
   — The *springs* are provided for controlling the movement of the moving coil, which also serves as current leads to the moving coil.
   — The *damping* is provided by using a damping vane attached to the moving system and moving in a sector-shaped box.
   — The reading is indicated directly by the pointer attached to the moving system and moving over a calibrated scale.
   — The eddy current errors, within the region of the magnetic field of the instrument, are minimised by the use of non-metallic parts of high resistivity material.

4.9.3.3. Advantages and disadvantages

The advantages and disadvantages of dynamometer type wattmeters are as under :

Advantages :

1. The scale of the instrument is uniform (because deflecting torque is proportional to true power in both the cases *i.e.,* D.C. and A.C. and the instrument is spring controlled.)
2. High degree of accuracy can be obtained by careful design, hence these are used for calibration purposes.

Disadvantages :

1. The error due to the inductance of pressure coil at low power factor is very serious (unless special features are incorporated to reduce its effect).
2. Stray field may effect the reading of the instrument. To reduce it, magnetic shielding is provided by enclosing the instrument in an iron case.

Example 4.73. *A wattmeter is used to measure power in the circuit with the help of the following equation :*

$$P = E^2/R$$

where limiting values of voltage and resistance are, E = 200 V + 1% and R = 100 Ω ± 5%. Calculate :

(i) *The nominal power consumed.*

(ii) *The limiting error of power in per cent and watts.* **(UPTU)**

Solution. *Given :* Equation for power, $P = \frac{E^2}{R}$; Limiting values : $E = 200$ V + 1%;

$R = 100\ \Omega \pm 5\%$

(i) **The nominal power consumed, *P* :**

$$P = \frac{E^2}{R} = \frac{(200)^2}{1000} = \textbf{40 W (Ans.)}$$

(ii) **The limiting errors of power in per cent and watts :**

$$P = \frac{E^2}{R}$$

Taking log on both sides, we get :

$$\log_e P = 2\log_e E - \log_e R$$

Differentiating w.r.t. P, we have :

$$\frac{1}{P} = 2\cdot\frac{1}{E}\cdot\frac{dE}{dP} - \frac{1}{R}\cdot\frac{dR}{dP}$$

or, $$\frac{dP}{P} = 2\frac{dE}{E} - \frac{dR}{R}$$

Hence, the *relative limiting error of P in percent is,*

$$\frac{\delta P}{P} = \pm\left(2\frac{\delta E}{E} + \frac{\delta R}{R}\right)$$

$$= \pm(2\times 1 + 5) = \pm 7\% \text{ (Ans.)}$$

Limiting error in watts $= \pm\frac{7}{100}\times 40 = 2.8$ **W (Ans.)**

4.9.3.4. Errors in dynamometer wattmeters

The various types of *errors* in dynamometer wattmeters are :

1. Error due to pressure-coil inductance.
2. Error due to pressure-coil capacitance.
3. Error due to eddy currents.
4. Error due to power lose in pressure coil or current coil.
5. Error due to friction.
6. Error due to temperature.
7. Error due to stray fields.

1. *Error due to pressure-coil inductance :*

Let, r_p = Resistance of the pressure-coil,

L_p = Inductance of the pressure coil,

R = Resistance connected in series with the pressure coil, and

I_p = Current through the pressure coil.

Then, $$I_p = \frac{V}{\sqrt{(r_p+R)^2+\omega^2 L_p^2}},$$

lagging behind the supply voltage V by,

$$\theta = \tan^{-1}\left(\frac{\omega L_p}{r_p+R}\right)$$

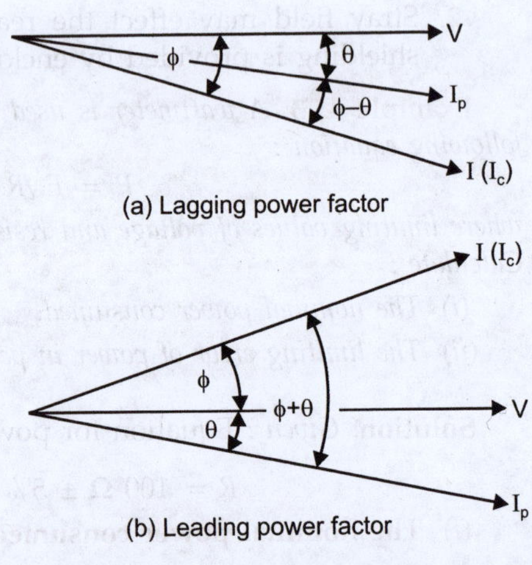

Fig. 4.70. Phasor diagram.

where, ω is the angular velocity of supply.

If ϕ is the phase angle (lagging) between load current I and supply voltage V, then phase angle between the currents of current coil (I_c) and pressure coil (I_p) will be $\phi - \theta$; then deflection of wattmeter $\alpha\ I_p\ I_c \cos(\phi - \theta)$.

or, Deflection $\alpha \ \frac{V}{Z_p} I \cos(\phi - \theta)$ ...since $I_p = \frac{V}{Z_p}$

where, Z_p is the impedance of pressure coil and $I_c = I$,

or, Deflection $\alpha \ \frac{V}{R+r_p} \ I \cos\theta \cos(\phi - \theta)$ ...since $Z_p = \frac{R+r_p}{\cos\theta}$

When $L_p = 0$, then $\theta = 0$, and the wattmeter deflection would be proportional to $\frac{V}{R+r_p} \ I \cos\phi$ at all frequencies and power factors.

$$\text{Now,} \quad \frac{\text{True power}}{\text{Power indicated by the wattmeter}} = \frac{VI\cos\phi/(R+r_p)}{VI\cos\theta\cos(\phi-\theta)/(R+r_p)}$$

$$= \frac{\cos\phi}{\cos\theta\cdot\cos(\phi-\theta)} \qquad ...(4.133)$$

$$\text{or,} \quad \text{True power} = \frac{\cos\phi}{\cos\theta\cdot\cos(\phi-\theta)} \times \text{Reading of wattmeter} \qquad ...[4.133(a)]$$

The term $\frac{\cos\phi}{\cos\theta\cdot\cos(\phi-\theta)}$ for *lagging p.f.* is known as *correction factor.*

$$\text{Correction factor for leading p.f.} = \frac{\cos\phi}{\cos\theta\cdot\cos(\phi+\theta)}$$

$$\text{Error} = \text{Actual reading} - \text{true reading}$$

$$= \frac{VI}{R+r_p}\cos\theta\cdot\cos(\phi-\theta) - \frac{VI}{R+r_p}\cos\phi$$

$$= \left[1 - \frac{\cos\phi}{\cos\theta\cos(\phi-\theta)}\right]\frac{VI}{R+r_p}\cos\theta\cos(\phi-\theta)$$

$$= \left[1 - \frac{\cos\phi}{\cos\theta\cos(\phi-\theta)}\right] \times \text{Actual reading}$$

$$= \left[1 - \frac{\cos\phi}{\cos\phi + \sin\phi\sin\theta}\right] \times \text{Actual reading}$$

... when $\cos\phi$ is *assumed equal to unity*

$$= \left(\frac{\cos\phi + \sin\phi\sin\theta - \cos\phi}{\cos\phi + \sin\phi\sin\theta}\right) \times \text{Actual reading}$$

$$= \left(\frac{\sin\phi\sin\theta}{\cos\phi + \sin\phi\sin\theta}\right) \times \text{Actual reading}$$

Dividing numerator and denominator by $\sin\phi$, we have :

$$\text{Error} = \left(\frac{\sin\theta}{\cot\phi + \sin\theta}\right) \times \text{Actual reading} \qquad ...(4.134)$$

Eqn. (4.134) indicates that *instrument gives high reading on lagging power factor and low reading on leading power factor.*

2. ***Error due to pressure coil capacitance :***

— The pressure-coil circuit may have capacitance as well as inductance, this being largely due to inter-turn capacitance in the high-value series resistor. The effect produced is similar to that of inductance in the circuit, except that the pressure-coil current tends to *lead* the applied voltage instead of to lag behind it. This causes the wattmeter to *read low, on lagging power factors* of the load, by increasing the angle between the load and voltage-coil contents.

— The effect of frequency will be, of course, to vary the phase angle between V and I_p, the *angle increasing with increase of frequency.*

— If the capacitance reactance of pressure-coil circuit is *equal* to its inductive reactance, there will be *no error* due to these effects since the two individual errors will *neutralise* one another.

3. ***Error due to eddy currents :***

— The eddy currents induced in the solid metal parts of the instrument, by the alternating magnetic field of the current coil, *alter* the magnitude and phase of this field, and so produce an error.

— The phase of the induced e.m.fs will be 90° behind the inducing flux, *i.e.*, rather more than 90° behind the main current in the current coil. The eddy current is practically in phase with its e.m.f. and this current sets up a magnetic field which, combined with that of the current coil, produces a resultant magnetic field which is less than that of the current coil alone and which also lags behind the current-coil field by a small angle.

— The eddy current error is *not easily calculable* and may be serious if care is not taken to ensure that any solid metal parts (which should be avoided as far as possible) are well removed from the current coil.

If current coil is designed for heavy currents, it should consist of *stranded conductors* in order to minimize the eddy currents flowing in the current coil itself.

4. ***Error due to power loss in pressure coil (P.C.) or current coil (C.C.).*** Fig. 4.71 shows two methods of connecting a wattmeter in circuit; *neither measures the power in the load directly, without corrections, even neglecting the errors discussed above.*

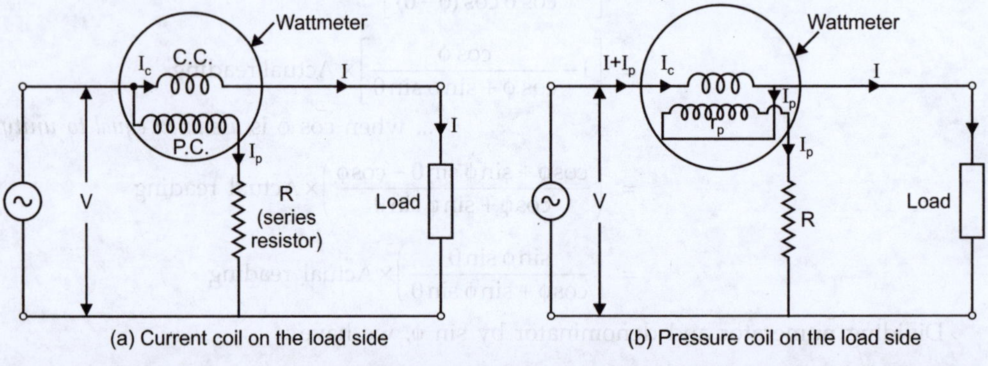

Fig. 4.71. Alternative wattmeter connections.

- In the method illustrated in Fig. 4.71(*a*), the pressure coil is connected on the "supply" side of the current coil, the voltage applied to the pressure coil is higher than that of the load on account of the voltage drop in the current coil.

Power indicated by the wattmeter

$$= \text{Power consumed by load + power loss in } \textit{current coil}$$

$$= \text{Power consumed by load} + I^2R_c \quad \text{...(4.135)}$$

where, R_c = The resistance of current coil.

- In the method illustrated in Fig. 4.71(*b*), the current coil carries the small current taken by the pressure coil, in addition to the load current.

∴ Power indicated by wattmeter

$$= \text{Power consumed by load + Power loss in } \textit{pressure coil}$$

$$= \text{Power consumed by load} + I_p^2(R + r_p) \quad \text{...(4.136)}$$

- If the load current is *small* the voltage drop in the current coil is small, so that the first method of connection introduces a *very small error*. On the other hand, *if the load current is large*, the power lost in the voltage coil will be small compared with the power in the load, and the second method of connection is *better*.

Compensation for power loss in pressure/voltage coil :

— In some wattmeters a *"compensating coil"* is used to eliminate error due to current coil carrying the pressure-coil current in addition to the load current when the connections are as in Fig. 4.71(*b*).

— This compensating coil is as *nearly as possible identical and coincident with the current coil*, so that if it were connected in series with the latter, and a current passed through the two coils-connected so that this magnitude effects were in opposition– the *resultant magnetic field would be zero*. Actually, the compensating coil is connected in series with the pressure coil, but in such a way that its magnetic effect opposes that of the current coil and *neutralises the pressure-coil component of the current in the current coil*. Thus, if no-load current flows in the instrument, the deflection should be zero, since the resultant current-coil field should be zero. The connections of this method of compensation are shown in Fig. 4.72.

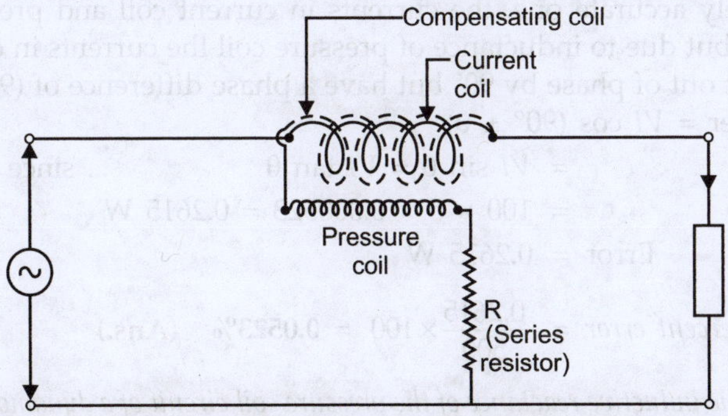

Fig. 4.72. Connections of compensating coil.

5. ***Error due to friction.*** Owing to availability of small deflecting torque, the frictional error becomes relatively important in such an instrument.

- In order *to reduce friction error* it becomes necessary that the *weight of moving system be reduced to the minimum possible on one hand and great care must be taken in the pivoting on the other hand.*

6. ***Error due to temperature.*** The indication of a wattmeter is affected by changes in room temperature. This is because any change in room temperature changes the resistance of the pressure coil and the stiffness of the springs. These effects are opposite in nature and would nearly neutralize each other, if the pressure coil circuit were composed of copper and of a resistance alloy having a negligible resistance temperature coefficient in the ratio of 1 : 10.

7. ***Errors due to stray fields.*** The dynamometer wattmeter has a relatively weak operating field and therefore, it is particularly affected by stray magnetic fields resulting in the serious errors. Hence these instruments should be *shielded* against effects of stray magnetic fields.

- *Laminated iron shields are used in portable laboratory instruments while steel cases are sometimes provided to shield the switchboard instruments.*
- Precision type of wattmeters are *not* provided with shields in order to keep down errors caused by eddy currents and to avoid the small D.C. errors caused by permanent magnet of the shields; such wattmeters use *astatic system.*

Example 4.74. *A dynamometer type wattmeter is connected across 100 V, 50 Hz supply with a load of 5 A of zero power factor. The inductance and resistance of the pressure coil are 5 mH and 3000 Ω respectively. If the voltage drop in the current coil of the wattmeter is negligible, calculate the percent error in the reading of wattmeter for full scale of 500 W.* **(AMIE)**

Solution. *Given* : Supply voltage = 100 V, 50 Hz; I = 5A, zero p.f., L_p = 5 mH or 0.005 H; $r_p + R$ = 3000 Ω; wattmeter reading for full scale = 500 W.

Percent error in wattmeter reading :

The angle θ by which the current in pressure coil lags behind the voltage

$$= \tan^{-1} \frac{\omega L_p}{(r_p + R)}$$

$$= \tan^{-1}\left(\frac{2\pi \times 50 \times 0.005}{3000}\right) = \tan^{-1} 0.000523$$

Since the power factor of load is zero, the true power is zero. If the instrument would have been completely accurate one, the currents in current coil and pressure coil were out of phase by 90° but due to inductance of pressure coil the currents in current coil and pressure coil are not out of phase by 90° but have a phase difference of (90° ± θ), and the reading of wattmeter = $VI \cos (90° \pm \theta)$

or, $= VI \sin \theta = VI \tan \theta$ ... since θ is very small

$= 100 \times 5 \times 0.000523 = 0.2615$ W

or, Error = 0.2615 W

Hence, *percent error* $= \frac{0.2615}{500} \times 100 =$ **0.0523%** **(Ans.)**

Example 4.75. *The inductive reactance of the pressure coil circuit of a dynamometer wattmeter is 0.35 percent of its resistance at normal frequency and the capacitance is negligible. Calculate the correction factor and percentage error due to reactance for load having 0.5 p.f. lagging.* **(Nagpur University)**

Solution. *Given* : $X_L = 0.0035$ R; cos φ = 0.5 lagging.

Now, $\frac{X_L}{R} = \tan \theta = 0.0035$

∴ $\theta = \tan^{-1} (0.0035) = 0.2°$

$\sin \theta = \tan \theta = \theta = 0.2°$ ...since angle θ is very small

$$\cos\theta = \cos 0.2°$$

$$\cos\phi = 0.5 \quad \text{or} \quad \phi = \cos^{-1}(0.5) = 60°$$

$$\cot\phi = 0.5773$$

$$\therefore \quad \textit{Correction factor} = \frac{\cos\phi}{\cos\theta\cdot\cos(\phi-\theta)} \qquad \text{[Eqn. (4.133)]}$$

$$= \frac{\cos 60°}{\cos(0.2°)\cos(60°-0.2°)} = \mathbf{0.994} \ \textbf{(Ans.)}$$

$$\%\text{age error} = \frac{\sin\theta}{\cot\phi+\sin\theta}\times 100 \qquad \text{[Eqn. (4.134)]}$$

$$= \frac{0.0035}{0.5773+0.0035}\times 100 = \mathbf{0.6\%} \ \textbf{(Ans.)}$$

Example 4.76. *A wattmeter has a current coil of 0.1 Ω resistance and a pressure coil of 6500 Ω resistance. Calculate the percentage errors, due to resistance only, with each of the methods of connection, when reading the input to an apparatus which takes :*

(i) 12 A at 250 V with unity power factor, and

(ii) 12 A at 250 V with 0.4 power factor. **(Punjab University)**

Solution. *Given :* Resistance of coil, R_c = 0.1 Ω; Resistance of pressure coil, R_p = 6500 Ω; I = 12 A; V = 250 V; cos ϕ = unity; 0.4.

Percentage error :

(*i*) *When p.f. is unity, i.e., cos ϕ = 1 :*

Power input to the apparatus,

$$P = VI\cos\phi$$

$$= 250 \times 12 \times 1 = 3000 \text{ W}$$

There are only *two possible ways* in which the pressure coil can be connected as shown in Figs. 4.71(*a*) and 4.71(*b*) respectively.

- *For connection 4.71(a),* the error is caused due to *power lost in the current coil.*

Power lost in current coil = I^2R_c = $(12)^2$ × 0.1 = 14.4 W

$$\therefore \quad \textit{Percentage error} = \frac{I^2R_c}{VI\cos\phi}\times 100 = \frac{14.4}{3000}\times 100 = \mathbf{0.48\%} \ \textbf{(Ans.)}$$

- *For connection 4.71(b),* the error is due to *power lost in pressure coil.*

$$\text{Power lost in pressure coil} = \frac{V^2}{R_p} = \frac{(250)^2}{6500} = 9.6 \text{ W}$$

$$\therefore \quad \textit{Percentage error} = \frac{V^2/R_p}{VI\cos\phi} = \frac{9.6}{3000}\times 100 = \mathbf{0.32\%} \ \textbf{(Ans.)}$$

(*ii*) *When p.f. is 0.4, i.e., cos ϕ = 0.4 :*

$$P = VI\cos\phi = 250 \times 12 \times 0.4 = 1200 \text{ W}$$

Power lost in current coil and pressure coil is the *same* as worked out earlier in part (*i*), since *supply voltage and current are same.*

- *Percentage error* for connection shown in *Fig. 4.71(a)*

$$= \frac{I^2R_c}{VI\cos\phi}\times 100 = \frac{14.4}{1200}\times 100 = \mathbf{1.2\%} \ \textbf{(Ans.)}$$

- *Percentage error* for connection shown in *Fig. 4.71(b)*

$$= \frac{V^2/R_p}{VI\cos\phi} = \frac{9.6}{1200} \times 100 = \textbf{0.8\%} \text{ (Ans.)}$$

Example 4.77. *A wattmeter has a current coil of 0.03 Ω resistance and a pressure coil of 6000 Ω resistance. Calculate the percentage error if the wattmeter is so connected that :*

(i) The current coil is on the load side;

(ii) The pressure coil is on the load side.

(a) If the load takes 20 A at a voltage of 220 V and 0.6 power factor in each case;

(b) What load current would give equal errors with the two connections ?

(Nagpur University)

Solution. *Given :* $R_c = 0.03\ \Omega$; $R_p = 6000\ \Omega$; $I = 20$ A; $V = 220$ V; $\cos\phi = 0.6$

(*a*) Power consumed by the load = $VI\cos\phi = 220 \times 20 \times 0.6 = 2640$ W

(*i*) *When the current coil is on the load side :*

$$\text{Power loss in current coil} = I^2R_c = (20)^2 \times 0.03 = 12 \text{ W}$$

$$\therefore \quad \textit{Percentage error} = \frac{12}{2640} \times 100 = \textbf{0.45\%} \text{ (Ans.)}$$

(*ii*) *When the pressure coil is on the load side :*

For this connection, the wattmeter measures the *loss in pressure coil.*

Now, Power loss in pressure coil

$$= \frac{V^2}{R_p} = \frac{(220)^2}{6000} = 8.07 \text{ W}$$

$$\therefore \quad \textit{Percentage error} = \frac{8.07}{2640} \times 100 = \textbf{0.31\%} \text{ (Ans.)}$$

(*b*) *For equal error for the two connections; we have :*

$$I^2R_c = \frac{V^2}{R_p}$$

or,
$$I^2 = \frac{V^2}{R_c \times R_p} = \frac{(220)^2}{0.03 \times 6000} = 268.89$$

$$\therefore \quad I = \textbf{16.4 A} \text{ (Ans.)}$$

Example 4.78. *An electrodynamic wattmeter is employed to measure power in a single-phase circuit. The load voltage is 200 V and the load current is 5 A at a lagging power factor of 0.1. The wattmeter potential coil has a resistance of 12000 Ω and inductance of 120 mH. Determine the percentage error in the wattmeter reading.* **(Kanpur University)**

Solution. *Given :* $V = 200$ volts; $I = 5$ A; $\cos\phi = 0.1$ (lagging); $R_p = 12000\ \Omega$; $L = 120$ mH = 0.12 H.

Percentage error :

Let the wattmeter be connected in the circuit as shown in Fig. 4.71(*b*).

$$\text{Load power} = VI\cos\phi = 200 \times 5 \times 0.1 = 100 \text{ W}$$

$$\phi = \cos^{-1}(0.1) = 84.26°$$

$$\text{Power lost in the coil} = \frac{V^2}{R_p} = \frac{(200)^2}{12000} = 3.33 \text{ W}$$

Neglecting the inductance of the voltage coil the reading of wattmeter would be 100 + 3.33 = 103.33 W.

Reactance of pressure coil circuit, $X_p = 2\pi \times 50 \times 0.12 = 37.7\ \Omega$

$$\theta = \tan^{-1}\left(\frac{X_p}{R_p}\right) = \tan^{-1}\left(\frac{37.7}{12000}\right) = 0.18°$$

$$\text{Wattmeter reading} = \text{True reading} \times \frac{\cos\theta \times \cos(\phi - \theta)}{\cos\phi} \quad \text{[Eqn. 4.133(a)]}$$

$$= 103.33 \times \frac{\cos(0.18°) \times \cos(84.26° - 0.18°)}{\cos(84.26°)} = 106.56\ \text{W}$$

$$\therefore \quad \textit{Percentage error} = \frac{106.56 - 100}{100} \times 100 = \mathbf{6.56\%\ (\textit{high})} \quad \textbf{(Ans.)}$$

Example 4.79. *A dynamometer wattmeter is used to measure the power factor of a 20 μF capacitor. The pressure coil of the wattmeter having a resistance 1000 Ω and an inductive reactance of 15 Ω is connected across a 50 Hz supply. The current coil of the wattmeter, a variable resistor R and the capacitor are connected in series across the same supply. The wattmeter deflection is made zero by adjusting the value of R to 1.65 Ω. If the current coil resistance is 0.1 Ω and its inductance negligible, determine the power factor of the capacitor.* **(Bombay University)**

Solution. *Given:* $C = 20\ \mu F = 20 \times 10^{-6}$ F; $R_p = 1000\ \Omega$; $X_L = 15\ \Omega$; $f = 50$ Hz; $R = 1.65\ \Omega$; $R_c = 0.1\ \Omega$.

Power factor of capacitor :

$$\text{Phase angle of pressure coil circuit, } \theta = \tan^{-1}\left(\frac{X_L}{R_p}\right) = \tan^{-1}\left(\frac{15}{1000}\right) = 0.86°$$

$$\text{Capacitance reactance, } X_C = \frac{1}{2\pi \times 50 \times (20 \times 10^{-6})} = 159.2\ \Omega.$$

Let r be the series resistance of the capacitor. Total resistance of the load circuit including resistance of current coil

$$= 0.1 + 1.65 + r = (1.75 + r)\ \Omega$$

$$\therefore \quad \phi = \text{Phase angle of load circuit} = \tan^{-1}\left(\frac{159.2}{1.75 + r}\right)$$

The load circuit has a *leading* p.f. as the reactance is *capacitance. For leading p.f.,*

Reading of wattmeter $\propto \cos\theta \cos(\phi + \theta)$

For a wattmeter reading to be zero,

$$\cos\theta \cos(\phi + \theta) = 0$$

or, $$\cos(\phi + \theta) = 0$$

or, $$\phi + \theta = 90°$$

or, $$\phi = 90 - \theta = 90° - 0.86° = 89.14°$$

or, $$\tan^{-1}\left(\frac{159.2}{1.75 + r}\right) = 89.14°$$

or, $$\frac{159.2}{1.75 + r} = \tan(89.14°) = 66.62$$

or, $$r = \frac{159.2}{66.62} - 1.75 = 0.64\ \Omega$$

Impedance of a capacitor, $Z_C = \sqrt{r^2 + (X_C)^2} = \sqrt{(0.64)^2 + (159.2)^2} = 159.2\ \Omega$

$\therefore$ *Power factor of the capacitor* $= \dfrac{r}{Z_C} = \dfrac{0.64}{159.2} = \mathbf{0.004}$ **(Ans.)**

Example 4.80. *A dynamometer wattmeter reading correctly on D.C. is used to measure power in circuit of resistance of 2 Ω and inductance of 0.25 H. The supply is 200 V at 50 Hz and the pressure coil circuit of wattmeter has a resistance of 1000 Ω and inductance of 5.6 mH. What is the actual reading of the wattmeter ?*

Neglect the impedance of the current coil circuit. Assume that pressure coil is connected on the load side of the instrument. **(Punjab University)**

Solution. *Given:* Load resistance, $R_L = 2\Omega$; Load resistance, $X_L = 2\pi fL = 2\pi \times 50 \times 0.25 = 78.5\ \Omega$; $V = 200$ volts; $R_p = 1000\ \Omega$; $L_p = 5.6$ mH $= 0.0056$ H.

Actual reading of wattmeter :

Load impedance, $\bar{Z}_L = R_L + jX_L = (2 - j78.5)\ \Omega$

Load current, $\bar{I}_L = \dfrac{V}{\bar{Z}_L} = \dfrac{200}{2 + j78.5} = (0.0648 - j2.55)\text{A}$

Pressure coil circuit impedance, $\bar{Z}_p = R_p + jX_{Lp}$

$$= 1000 + j(2\pi \times 50 \times 0.0056) = (1000 + j1.76)\ \Omega$$

Pressure coil current, $\bar{I}_p = \dfrac{V}{\bar{Z}_p} = \dfrac{200}{1000 + j1.76} = (0.2 - j0.000352)\text{A}$

$$\theta = \tan^{-1}\left(-\frac{0.000352}{0.2}\right) = -0.1°$$

Current coil current, $\bar{I}_c = \bar{I}_L + \bar{I}_p$

$$= (0.0648 - j2.55) + (0.2 - j0.000352)$$
$$= 0.2648 - j2.550352 = 2.56\ \text{A}$$

$$\phi = \tan^{-1}\left(\frac{-2.550352}{0.2648}\right) = -84.07°$$

$$\phi - \theta = -84.07 - (-0.1°) = -83.97°$$
$$\cos(\phi - \theta) = \cos(-83.97°) = 0.105$$
$$\cos\theta = \cos(-0.1°) \simeq 1$$

$\therefore$ Actual reading of wattmeter

$$= VI_c \cos\theta \cos(\phi - \theta)$$
$$= 200 \times 2.56 \times 1 \times 0.105 = \mathbf{53.76\ W}$$ **(Ans.)**

Example 4.81. *The current coil of a wattmeter is connected in series with an ammeter and an inductive load. A voltmeter and the voltage coil are connected across a 100 Hz supply. The ammeter reading is 4.5 A and the voltmeter and wattmeter readings are respectively 240 V and 23 W. The inductance of voltage circuit is 10 mH and its resistance 2000 Ω. If the voltage drop across the ammeter and current coil are negligible, what is the percentage error in the wattmeter reading ?* **(Banglore University)**

Solution. *Given:* $f = 100$ Hz; $I = 4.5$ A; $V = 240$ V; Wattmeter reading = 23 W; $L = 10$ mH = 0.01 H; $R_p = 2000\ \Omega$.

Percentage error in wattmeter reading :

Inductive reactance of pressure coil = $2\pi fL = 2\pi \times 100 \times 0.01 = 6.28\ \Omega$

Phase angle of pressure coil circuit, $\theta = \tan^{-1}\left(\frac{X_L}{R_p}\right) = \tan^{-1}\left(\frac{6.28}{2000}\right) = 0.18°$

For an inductive load, reading of wattmeter $\propto \cos\theta \cos(\phi - \theta)$

True power $\propto \cos\phi$, where $\cos\phi$ = load power factor

$$\therefore \quad \text{True power} = \frac{\cos\phi}{\cos\theta\cos(\phi-\theta)} \times \text{reading of wattmeter} \quad ...(i)$$

But, True power = $I^2R = (4.5)^2 \times R = (4.5)^2 Z \cos\phi$

where, R = Resistance of load

$$\text{Impedance of load, } Z = \frac{V}{I} = \frac{240}{4.5} = 53.33\ \Omega$$

$$\therefore \quad \text{True power} = (4.5)^2 \times 53.33 \cos\phi$$

Substituting the various values in (*i*), we get

$$(4.5)^2 \times 53.33 \cos\phi = \frac{\cos\phi}{\cos\theta\cos(\phi-\theta)} \times 23$$

$$\text{or,} \quad \cos\theta\cos(\phi-\theta) = \frac{23}{(4.5)^2 \times 53.33} = 0.0213$$

or, $\cos(0.18°)\cos(\phi - 0.18°) = 0.0213$

$$\text{or,} \quad \cos(\phi - 0.18°) = \frac{0.0213}{\cos(0.18°)} = 0.0213$$

$$\text{or,} \quad \phi = \cos^{-1}(0.0213) + 0.18° = 88.96°$$

Percentage in wattmeter reading

$$= \frac{\sin\theta}{\cot\phi + \sin\theta} \times 100 \quad \text{[Eqn. (4.134)]}$$

$$= \frac{\sin(0.18°)}{\cot(88.96°) + \sin(0.18°)} \times 100 = \mathbf{14.75\%} \ \textbf{(Ans.)}$$

Example 4.82. *The current coil of a dynamometer is connected to a 24 V D.C. source in series with a 5Ω resistor. The potential circuit is connected through an ideal rectifier in series with a 50 Hz source of 100 V. The inductance of pressure circuit and current coil resistance are negligible. Compute the reading of the wattmeter.* **(AMIE)**

Solution. Current through the current coil = $\frac{24}{5} = 4.8$ A.

The pressure coil is energised by an ideal rectifier. Therefore, the pressure coil carries current during one half cycle and in the other half cycle there is no current in it. This means there is *deflecting torque on the meter during one half cycle.*

$\therefore$ Reading of wattmeter = Average power over a cycle

$$= \frac{1}{2\pi}\left[\int_0^{\pi} vi\, d\theta\right] = \frac{1}{2\pi}\left[\int_0^{\pi}\left(\sqrt{2}\times 100\right)\sin\theta \times 4.8\, d\theta\right]$$

$$= \frac{\sqrt{2}\times 100\times 4.8}{2\pi}\left[\int_0^{\pi}\sin\theta\cdot d\theta\right]$$

$$= 108.04\left[-\cos\theta\right]_0^{\pi} = \mathbf{216\ W}\ \textbf{(Ans.)}$$

Example 4.83. *A dynamometer wattmeter is used to measure the power in an inductive load having a phase angle of 87°. The potential coil has a resistance of 1000 Ω and inductance of 6 mH. The potential coil current is negligible. The impedance of current coil is also negligible. Find the % error in wattmeter reading.*

Solution. Fig. 4.73 shows the phasor diagram. V and I are the load voltage and load current. I_p is the potential coil current. The current I is lagging the voltage by 87°. If the potential coil were purely resistive, I_p would be in phase with V and the reading of the wattmeter would be correct. Because of inductance of potential coil, I_p is lagging by $\angle\alpha$. The power factor as seen by the wattmeter is cos(87 – α)°.

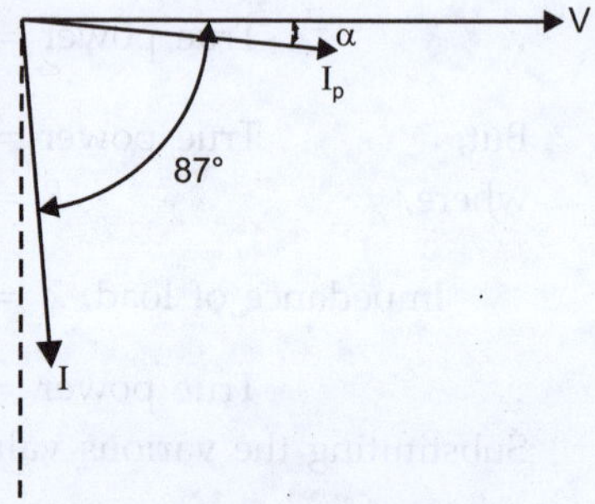

Fig. 4.73.

$$\alpha = \tan^{-1}\left(\frac{2\pi\times 50\times 6\times 10^{-3}}{1000}\right) = 0.108°$$

$$\%\ \text{error} = \frac{VI\cos(87-0.108)° - VI\cos 87°}{VI\cos 87°}\times 100 = \mathbf{3.6\%}\ \textbf{(Ans.)}$$

4.9.3.5. Low power factor electro-dynamometer type wattmeters

Ordinary electro-dynamometer wattmeter is not suitable for measuring power in circuits having low power factors due to the following *reasons:*

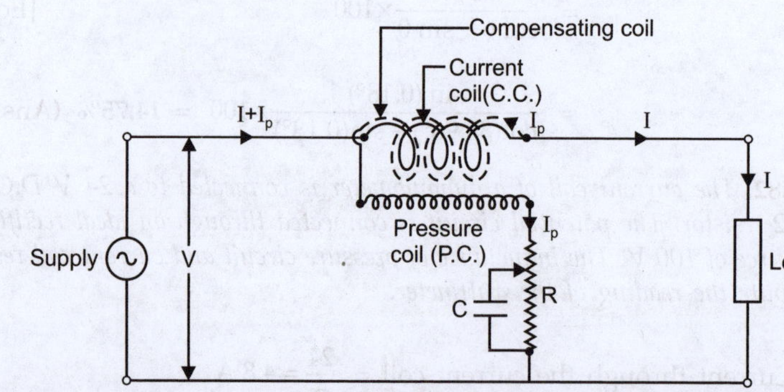

Fig. 4.74. Low power factor electro-dynamometer type wattmeter.

(*i*) Small deflecting torque on the moving system even when the current and pressure coils are fully excited.

(*ii*) Introduction of large error due to inductance of pressure coil at low power factors.

Thus, the following *special features are incorporated* in an electro-dynamometer type wattmeter to make it suitable for measurement of power in low p.f. circuits:

1. ***Pressure coil circuit :***
   — The pressure coil circuit is made of *low resistance* so that the current flowing through it is increased to give an *increased operating torque.*
   — The pressure coil current in low p.f. wattmeter may be as much as 10 times the value used for high p.f. meters.
2. ***Compensation for pressure coil current.*** This compensation is same as given in Fig. 4.72.
3. ***Compensation for inductance of pressure coil :***
   — The error caused by pressure coil inductance is $\frac{\sin\theta}{\cot\phi + \sin\theta}$ times the actual reading of wattmeter. Now, when the *p.f. is low* the value of ϕ is large and, therefore, the *error is large.*
   — The error caused by inductance of pressure coil is compensated by connecting a capacitor across a part of series resistance in the pressure coil circuit shown in Fig. 4.74.
4. ***Small control torque.*** Low p.f. wattmeters are *designed to have a small control torque so that they give full scale deflection for power factor as low as 0.1.*

4.9.3.6. Calibration of dynamometer wattmeters

The various methods of calibration of dynamometer wattmeters are enumerated and discussed as follows :

A. *D.C. calibration :*
   1. By comparison with standard wattmeter.
   2. By D.C. potentiometer.

B. *A.C. calibration :*
   1. By comparison with standard wattmeter.
   2. By A.C. potentiometer.
   3. By using resistive load.
   4. By electrostatic wattmeter.

A. D.C. calibration :

1. ***By comparison with standard wattmeter :***
   — In this method, the current coils of test wattmeter (W_T) and standard wattmeter (W_S) are connected in *series*, and the two voltage circuits are connected in *parallel.*
   — It is used to employ different sources of supply for voltage and current circuits, the *source for current being a low-voltage supply source.* This arrangement with separate supplies is known as *"fictitious load (phantom loading)"* and is almost invariably employed in wattmeter testing.

   It may be noted that an equalizing connection is required to maintain the wattmeter coils at the same voltage; this arrangement keeps the losses in the current regulating rheostat small and the total power consumed during the test is only a small fraction of the wattmeter indication.

   Fig. 4.75 shows the circuit diagram for such a test. The voltmeter and ammeter are used for checking the voltage and current, but the *calibration is a direct comparison between the readings of two wattmeters* W_T *and* W_S.

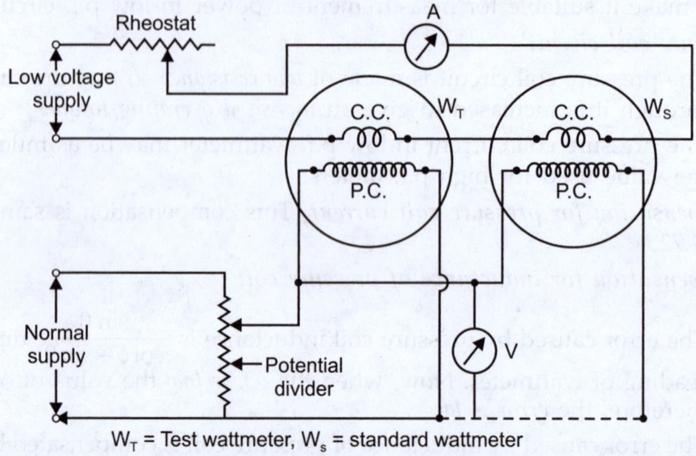

Fig. 4.75. D.C. calibration of dynamometer wattmeter with standard wattmeter.

- *Errors due to stray fields are eliminated by including reversing switch* in both supplies and taking the *mean readings* obtained with connections as shown and with both supplies *reversed.*

2. ***By D.C. potentiometer.*** Refer to Art. 9.2.

B. A.C. calibration :

1. ***By comparison with standard wattmeter.*** This method is similar to the corresponding D.C. test, the circuit being shown in Fig. 4.75.
   — The low voltage supply may be obtained conveniently from a step-down transformer.
   — The test may be carried out at various power factors by direct loading, using chokes or capacitors in addition to resistance loads.

   In fictitious loading arrangement a phase-shifting transformer may be employed for shifting the phase of the voltage applied to the circuits with respect to current in the current coils.

   - This method is *not very convenient and involves wastage of huge amount of power.*

2. ***By A.C. potentiometer.*** This method is similar to the D.C. potentiometer method (Art. 9.2) *and will normally include fictitious loading and arrangement for varying the p.f. The true power is given by VI cos* ϕ; *V, I and* ϕ *are measured by potentiometer.*

3. ***By using resistive load.*** In this method, a non-inductive resistance (standard) is used as a load and power consumed by this load is obtained from its resistance and the voltage across it. The voltage across the load may be measured with an electrostatic voltmeter.

   - This is an *accurate method of testing low range wattmeters at unity power factor.*

4. ***By electrostatic wattmeter.*** The arrangement of calibrating a wattmeter by an electrostatic wattmeter is an *expensive arrangement* (two alternators of similar electrical design coupled together and driven by a variable speed motor) and as such it is justificable where large number of tests at varying power factors are to be conducted.

4.9.4. Induction Wattmeter

4.9.4.1. Description

Induction wattmeters can be used in *A.C. circuit only* (in contrast with, dynamometer wattmeters can be used both in D.C. and A.C. circuits) and *are useful only when the frequency and supply voltage are constant.*

The *operation* of all induction instruments depends on the *production of torque due to reaction between a flux* ϕ_1 (whose magnitude depends on the current or voltage to be measured) *and eddy currents induced in a metal disc or drum by another flux* ϕ_2 (whose magnitude also depends on the current or voltage to be measured). Since the magnitude of eddy currents also depends on the flux producing them, the *instantaneous value of the deflecting torque is proportional to the square of the current or voltage under measurement and the value of mean deflecting torque is proportional to the mean square of the current or voltage.*

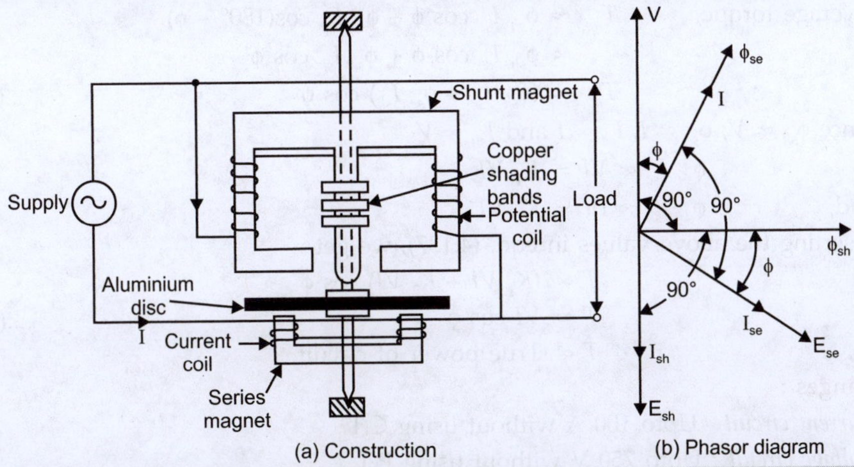

Fig. 4.76. Induction wattmeter.

Construction. Fig. 4.76(*a*) shows the construction of an induction wattmeter :

— It has two laminated electromagnets. The upper electromagnet (called magnet shunt) has two coils in series which are wound such that the direction of fluxes set up by them are same. They are connected in series with a *high resistance across the main supply.* Thus these two coils serve as *pressure coils.*

— One or more copper shading bands are fitted on the central limb of the upper electromagnet. The induced currents in these bands set up their own fluxes. The total flux of the upper electromagnet is the resultant of the main flux and the fluxes of currents in the shading bands. The positions of these bands are adjusted so that this *resultant flux lags the applied voltage by 90°.*

— The two coils of the lower electromagnet (called series magnet) are also in series and carry the circuit current. They are also wound such that their *fluxes are in the same direction.*

— A thin aluminium disc supported by jewelled bearings is mounted on the spindle. The spindle also carries a hair spring (for providing torque) and a pointer.

Theory. Fig. 4.76(*b*) shows the phasor diagram.

Let the load current *I* lag the circuit voltage *V* by an angle ϕ as shown in the figure. The flux ϕ_{sh} of shunt magnet lags behind *V* by 90° (this is done by adjusting the position of shading bands).

Assuming that hysteresis and saturation effects in the iron are negligible the series magnet flux ϕ_{se} *is taken proportional to and in phase with load current I.*

E.m.fs E_{sh} and E_{se} are induced in the disc by shunt magnet flux ϕ_{sh} and series magnet flux ϕ_{se} respectively and lag behind their respective fluxes by 90°. Eddy currents I_{sh} and I_{se} are set up by the induced e.m.fs E_{sh} and E_{se} respectively and are in phase with their respective e.m.fs.

Now, the two opposite torques given by $\phi_{sh}I_{se}$ and $\phi_{se}I_{sh}$ will act on the disc. The instantaneous values of resultant torque is the difference of the two *i.e.,*

$$T_{res.} = \phi_{sh} I_{se} - \phi_{se} I_{sh}$$

where, ϕ_{sh}, ϕ_{se}, I_{sh} and I_{se} are the *instantaneous values.*

Since the phase angle between ϕ_{sh} and I_{se} is ϕ and phase angle between ϕ_{se} and I_{sh} is $(180° - \phi)$, therefore,

Average torque, $$T_{av.} \propto \phi_{sh} I_{se} \cos \phi - \phi_{se} I_{sh} \cos(180° - \phi)$$

$$\propto \phi_{sh} I_{se} \cos \phi + \phi_{se} I_{sh} \cos \phi$$

or, $$T_{av.} \propto (\phi_{sh} I_{se} + \phi_{se} I_{sh}) \cos \phi \quad ...(4.137)$$

Since $\phi_{sh} \propto V$; $\phi_{se} \propto I$; $I_{se} \propto I$ and $I_{sh} \propto V$

$\therefore$ $$\phi_{sh} I_{se} \propto VI = K_1 VI$$

and, $$\phi_{se} I_{sh} \propto IV = K_2 VI$$

Inserting the above values in eqn. (4.137), we get

$$T \propto (K_1 VI + K_2 VI) \cos \phi$$

or, $$T \propto VI \cos \phi \quad ...(4.138)$$

or, $T \propto$ True power of circuit.

Ranges :

Current circuit : Upto 100 A without using C.T.

Voltage circuit : Upto 750 V without using P.T.

4.9.4.2. Advantages and disadvantages

Advantages :

1. Fairly long scale (extending over 300°).
2. Free from the effects of stray fields.
3. Good damping.
4. Practically free from frequency errors.

Disadvantages :

Sometimes subjected to serious temperature error because the main effect of temperature is on the resistance of the eddy current paths.

4.9.4.3. Comparison between dynamometer and induction wattmeters

The comparison between dynamometer and induction wattmeters is given in tabular form below :

S. No.	Aspects	Dynamometer wattmeters	Induction wattmeters
1.	*Working torque*	Comparatively weaker	Comparatively stronger
2.	*Power consumption*	Comparatively low	Comparatively higher
3.	*Weight of moving system*	Comparatively low	Comparatively higher

4.	*Type of scale*	Uniform	Uniform and long
5.	*Degree of accuracy*	High degree of accuracy if carefully designed.	Less accurate; it is accurate only at stated frequency and temperature.
6.	*Use*	Both D.C. and A.C. systems.	Can be used on A.C. system only.

Example 4.84. *A voltage 200 sin ωt + 50 sin (3ωt – 30°) + 40 sin(5ωt + 220°)V is applied to the pressure coil of a wattmeter. The current through the current coil is 10 sin (ωt – 10°) + 2 cos(3ωt – 150°) + 2 cos(5ωt + 150°)A. Find :*

(i) The reading of the wattmeter.

(ii) The contribution of each harmonic to the total power.

Solution. *Given:*

$$v = 200 \sin \omega t + 50 \sin (3\omega t - 30°) + 40 \sin(5\omega t + 220°)\text{V}$$

$$i = 10 \sin (\omega t - 10°) + 2 \cos(3\omega t - 150°) + 2 \cos(5\omega t + 150°)\text{A}$$

$$= 10 \sin (\omega t - 10°) + 2 \sin(3\omega t - 60°) + 2 \sin(5\omega t + 240°)\text{A}$$

(i) **The reading of the wattmeter :**

$$\text{Average power, } P = \frac{1}{2\pi}\int_0^{2\pi} ei\, d(\omega t)$$

The average value of all product terms involving different frequencies is zero. Only products of current and voltage of the same frequency need be considered to calculate power.

For nth harmonic,

$$P_n = \frac{1}{2\pi}\int_0^{2\pi} V_{n\,\max.}\, I_{n\,\max.} \sin(n\,.\,\omega t)\sin(n\,.\,\omega t - \phi_n)\, d(\omega t)$$

$$= \left(\frac{V_{n\,\max.}}{\sqrt{2}}\right)\left(\frac{I_{n\,\max.}}{\sqrt{2}}\right)\cos\phi_n$$

$$\therefore \quad P = \frac{200\times 10}{2}\cos 10° + \frac{50\times 2}{2}\times\cos 30° + \frac{40\times 2}{2}\cos 20°$$

or, $P = 984.81 + 43.30 + 37.59 = \mathbf{1065.7}$ **(Ans.)**

(ii) **The contribution of each harmonic to total power :**

$$\textit{Contribution of fundamental component} = \frac{984.81}{1065.7}\times 100 = \mathbf{92.41\%} \textbf{ (Ans.)}$$

$$\textit{Contribution of third harmonic} = \frac{43.30}{1065.7}\times 100 = \mathbf{4.063\%} \textbf{ (Ans.)}$$

$$\textit{Contribution of fifth harmonic} = \frac{37.59}{1065.7}\times 100 = \mathbf{3.527\%} \textbf{ (Ans.)}$$

Example 4.85. *A voltage 50 + 25 sin ωt is applied to a series R-L circuit having a resistance of 10 Ω and inductance 0.1 H. A wattmeter is connected in the circuit to measure power. Find the reading of the wattmeter. ω = 2π × 50 rad./sec.*

Solution. The inductance offers zero reactance to D.C.

For first harmonic,

$$\bar{Z}_1 = 10 + j(2\pi \times 50 \times 0.1) = 10 + j31.46\ \Omega = 32.97\angle 72.34°$$

$$i = \frac{50}{10} + \frac{25}{32.97} \sin(\omega t - 72.34°)$$

$$= 5 + 0.76 \sin(\omega t - 72.34°)$$

$$P = 50 \times 5 + \frac{25 \times 0.76}{2} \cos 72.34° = \mathbf{252.9\ W}\ \textbf{(Ans.)}$$

4.9.5. Electrostatic Wattmeter

These wattmeters are used for the *measurement of small amounts of power, particularly when the voltage is high and power factor low.* This type of wattmeter is also used for *measurement of dielectric loss of cables on alternating voltage and for calibration of wattmeters and energy meters.*

Electrostatic wattmeter consists of a quadrant electrometer used with a non-inductive resistor R as shown in Fig. 4.77.

Let i, e, v, v_1 and v_2 be the instantaneous load current, instantaneous potentials of the load, needle and of the two pairs of quadrants respectively, as shown in the figure.

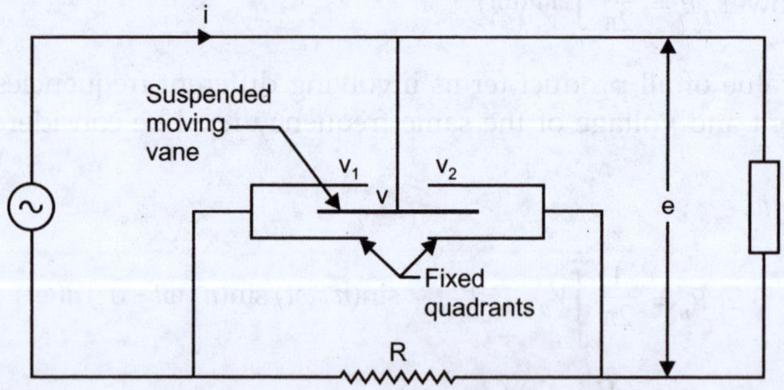

Fig. 4.77. Electrostatic wattmeter.

Instantaneous torque, $T \propto (v - v_1)^2 - (v - v_2)^2$

Since, $v_1 - v_2 = e$, the potential across load, and

$v - v_1$ = supply voltage

= voltage across load + voltage drop in non-inductive resistance

$= e + iR$

$\therefore$ Instantaneous torque, $T \propto (e + iR)^2 - e^2$

$$\propto e^2 + 2eiR + i^2R^2 - e^2$$

$$\propto 2eiR + i^2R^2$$

or, $$T \propto 2R\left(ei + \frac{i^2R}{2}\right) \quad ...(4.139)$$

i.e., Instantaneous torque is proportional to the instantaneous power in the load plus half of the power lost in non-inductive resistance.

Advantages and Disadvantages :

- Electrostatic wattmeter is a *precision instrument* and should be used as such.
- *It is free from errors on account of waveforms, frequency and eddy currents.*
- It has a *very small working torque.*

4.9.6. Thermal Wattmeter

Fig. 4.78 shows the arrangement for measuring power with thermocouples :

This method uses two similar thermocouples (**1 and 2**) whose outputs are connected in *opposition* with a galvanometer in between. R_h is the resistance of each thermocouple heating element. R is a high series resistance, and between C and D is a low resistance R_2 capable of carrying the load current i. The resistance R_2 develops a potential difference which depends upon the load current, together with the current of one heater, and the series resistance R carries the current of both heaters.

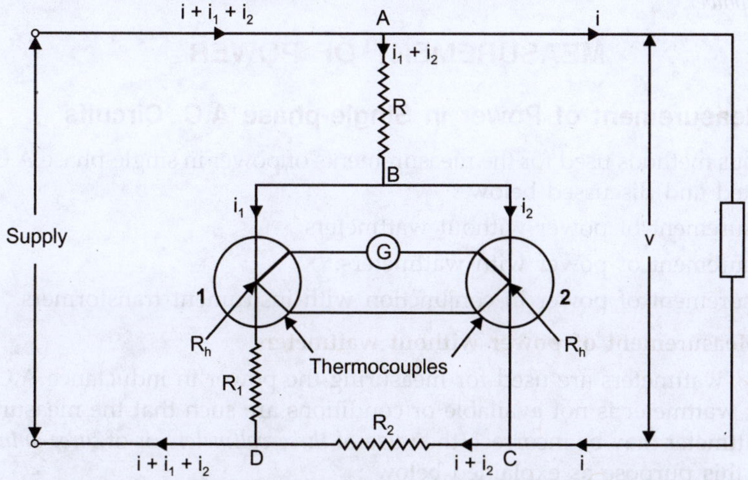

Fig. 4.78. Thermocouple/Thermal wattmeter.

If v be the instantaneous voltage at the load, then assuming identical thermocouples, we have :

$$i_1(R_h + R_1) = i_2 R_h + (i + i_2)R_2 \quad ...(4.140)$$

$$(i_1 + i_2)R + i_2 R_h = v \quad ...(4.141)$$

From (4.140) and (4.141), we get :

$$i_1 = \frac{v(R_h + R_2) + i R_2(R_h + R)}{(R + R_h)(R_h + R_1) + R(R_h + R_2)} \quad ...(4.142)$$

and,

$$i_2 = \frac{v(R_h + R_1) - i R_2 R}{(R_h + R)(R_h + R_1) + R(R_h + R_2)} \quad ...(4.143)$$

E.m.f. of *thermocouple-1*, $e_1 \propto$ heat produced by heating element, or,

$$e_1 \propto i_1^2 R_h$$

E.m.f. of *thermocouple-2*, $e_2 \propto i_2^2 R_h$

E.m.f. across galvanometer $G = e_1 - e_2$

Instantaneous deflecting torque of galvanometer,

$$T_{inst.} = K_1(e_1 - e_2) = K_2(i_1^2 - i_2^2) \quad ...(4.144)$$

where, K_1 and K_2 are constants.

Substituting the values of i_1 and i_2 in eqn. (4.144), we get

$$T_{inst.} = K(C_1\, vi + C_2 i^2)$$

where, C_1 and C_2 are constants involving the resistances.

If $R_h + R$ is not very different from R_h, then $C_2 i^2$ may be *neglected and* $T_{inst.} \propto vi$ *or instantaneous power.*

Thus, *galvanometer may be calibrated to read the power.*

— The commercial thermal wattmeters employ a number of thermocouples connected in the form of a chain in order to increase the output. C.Ts and P.Ts are also used with these instruments.

— For high frequency measurements careful shielding is required.

- The thermal wattmeters can be *used for measurement of power is several circuits and the sum of their outputs can be applied to a recording potentiometer which records the total power.*

MEASUREMENT OF POWER

4.9.7. Measurement of Power in Single-phase A.C. Circuits

The various methods used for the measurements of power in single-phase A.C. circuits are enumerated and discussed below :

1. Measurement of power without wattmeters.
2. Measurement of power with wattmeters.
3. Measurement of power in conjunction with instrument transformers.

4.9.7.1. Measurement of power without wattmeters

Generally, wattmeters are used for measuring the power in inductance A.C. circuits but in case if wattmeter is not available or conditions are such that the measurement of power by wattmeter may be incorrect, then use of *three-voltmeters* or of *three-ammeters* can be made for this purpose as explained below :

(*i*) Three-voltmeter method :

The connections are as shown in Fig. 4.79(*a*), in which V_1, V_2 and V_3 are three voltmeters and R is a non-inductive resistor which is connected in series with the load.

From the phasor diagram of Fig. 4.79(*b*) we have :

$$V_1^2 = V_2^2 + V_3^2 + 2V_2V_3 \cos\phi \qquad ...(4.145)$$

$$= V_2^2 + V_3^2 + 2(IR)V_3 \cos\phi$$

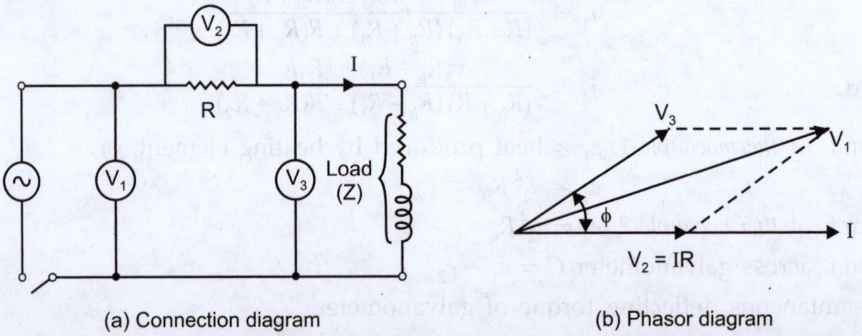

Fig. 4.79. Three-voltmeter method of measuring single-phase power.

$$= V_2^2 + V_3^2 + 2R(V_3 I \cos \phi)$$

$$= V_2^2 + V_3^2 + 2PR$$ ...since power in the inductive circuit, $P = V_3 I \cos \phi$

or, Power,
$$P = \frac{V_1^2 - V_2^2 - V_3^2}{2R} \quad ...(4.146)$$

From eqn. (4.145) *power factor* of the circuit is given by :

$$\cos \phi = \frac{V_1^2 - V_2^2 - V_3^2}{2V_2V_3} \quad ...(4.147)$$

- *The assumptions are made that the current in the resistor R is the same as the load current.*

Disadvantages :

(*i*) Supply voltage higher than normal voltage is required because an additional resistance R is connected in series with the load Z (inductive circuit).

(*ii*) Even small errors in measurement of voltages may cause serious errors in the value of power determined by this method.

(*ii*) Three-ammeter method :

The connection diagram and phasor diagram respectively are shown in Fig. 4.80(*a*) and (*b*). The current measured by ammeter A_1, is the vector sum of the load current and that taken by the non-inductive resistor R, this latter being in phase with V.

From the phasor diagram :

$$I_1^2 = I_2^2 + I_3^2 + 2I_2I_3 \cos \phi \quad ...(4.148)$$

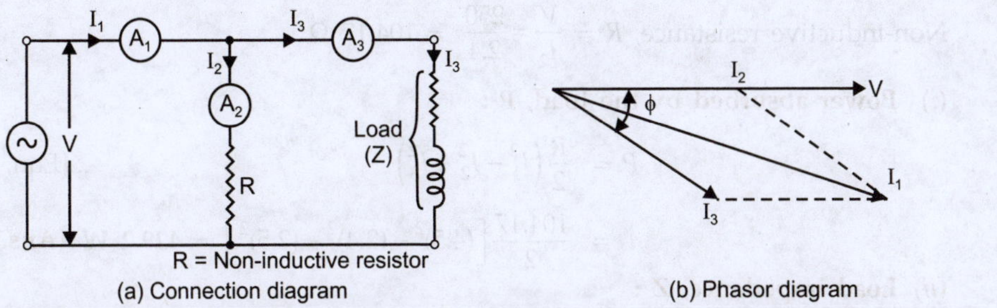

Fig. 4.80. Three-ammeter method of measuring single-phase power.

But,
$$I_2 = \frac{V}{R}$$

$\therefore$
$$I_1^2 = I_2^2 + I_3^2 + 2 \times \frac{V}{R} \times I_3 \cos \phi$$

Hence, the power $P = VI_3 \cos \phi$, is given by :

$$P = \frac{R}{2}\left(I_1^2 - I_2^2 - I_3^2\right) \quad ...(4.149)$$

Also,
$$\cos \phi = \frac{I_1^2 - I_2^2 - I_3^2}{2I_2I_3} \quad ...(4.150)$$

... From eqn. (4.148)

- The *"advantage"* of this method is that the value of power determined is *independent of supply frequency and waveforms.* The disadvantages of measurement of power by three-voltmeter method are overcome in this method.

Example 4.86. *The following readings were obtained from three voltmeters used for a single-phase power measurement :*

V_2 = 180 volts across a non-inductive resistor; V_3 = 200 volts across an inductive load; V_1 = 300 volts across the two in series.

Calculate the power factor of the inductive load. **(Gorakhpur University)**

Solution. *Given :* $V = 180$ V; $V_3 = 200$ V; $V_1 = 300$ V

Power factor, cos φ :

We know that : $$\cos \phi = \frac{V_1^2 - V_2^2 - V_3^2}{2V_2V_3} \quad \text{...[Eqn. (4.147)]}$$

or, $$\cos \phi = \frac{(300)^2 - (180)^2 - (200)^2}{2 \times 180 \times 200} = \mathbf{0.244} \text{ (Ans.)}$$

Example 4.87. *An inductive load takes a current of 2.5 A; a non-inductive resistor connected in parallel takes 2.4 A, when connected across 250 V supply. The total current taken from the supply is 4.5 A. Calculate :*

(*i*) *Power absorbed by the load.*

(*ii*) *Load impedance.*

(*iii*) *Power factor of the load.* **(Punjab University)**

Solution. *Given :* $I_3 = 2.5$ A; $I_2 = 2.4$ A; $I_1 = 4.5$ A; $V = 250$ V.

Non-inductive resistance, $$R = \frac{V}{I_2} = \frac{250}{2.4} = 104.17\ \Omega.$$

(*i*) **Power absorbed by the load, P :**

$$P = \frac{R}{2}\left(I_1^2 - I_2^2 - I_3^2\right) \quad \text{...[Eqn. (4.149)]}$$

$$= \frac{104.17}{2}\left[(4.5)^2 - (2.4)^2 - (2.5)^2\right] = \mathbf{429.2\ W} \text{ (Ans.)}$$

(*ii*) **Load impedance, Z :**

$$Z = \frac{V}{I_3} = \frac{250}{2.5} = \mathbf{100\ \Omega} \text{ (Ans.)}$$

(*iii*) **Power factor of the load, cos φ :**

$$\cos \phi = \frac{I_1^2 - I_2^2 - I_3^2}{2I_2I_3} \quad \text{...[Eqn. (4.150)]}$$

$$= \frac{(4.5)^2 - (2.4)^2 - (2.5)^2}{2 \times 2.4 \times 2.5} = \mathbf{0.687} \text{ (Ans.)}$$

4.9.7.2. Measurement of power with wattmeter

Refer to Art. 4.9.1.2.

4.9.7.3. Measurement of power in conjunction with instrument transformers

Power measurements are made in high voltage circuits connecting the wattmeter to the circuit through current and potential transformers as shown in Fig. 4.81:

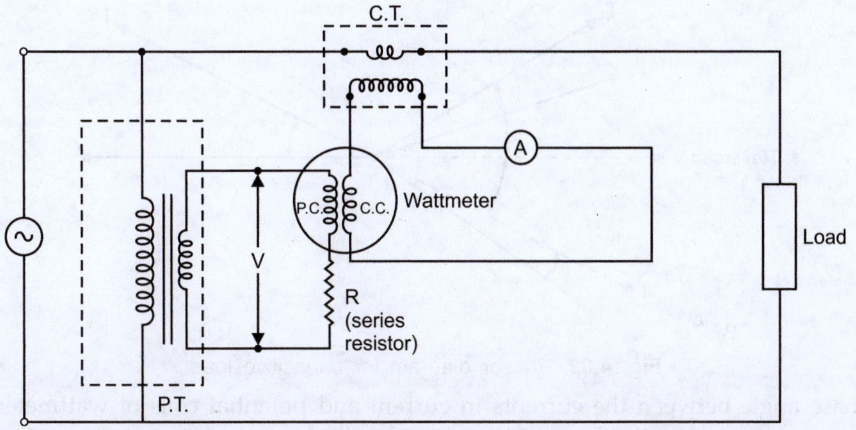

Fig. 4.81. Measurement of power with wattmeter in conjunction with instrument transformers in single-phase A.C. circuits.

— The primary winding of C.T. is connected in series with the load and the secondary winding is connected in series with an ammeter and the current coil (C.C.) of a wattmeter.

— The primary winding of the potential transformer (P.T.) is connected across the supply lines and a voltmeter and the pressure coil (P.C.) of the wattmeter are connected in parallel with the secondary winding of the transformer. One secondary terminal of each transformer and the casings are earthed.

— When the wattmeter is used in conjunction with instrument transformer, then *correction should be applied for ratio and phase angle errors of instrument transformers.*

Let, V, I and $\cos\phi$ be the load voltage, current and power factor respectively.

Further, let, current in C.C. of wattmeter = Current in secondary of C.T. = I_s

Voltage across pressure coil of wattmeter = Voltage across secondary of P.T. = V_s

Current in pressure coil of wattmeter = I_p lagging behind V_s by a small angle θ due to inductance of pressure coil.

Phase angle of P.T. = δ

Phase angle of C.T. = β

From phasor diagram shown in Figs. 4.82 and 4.83 we have :

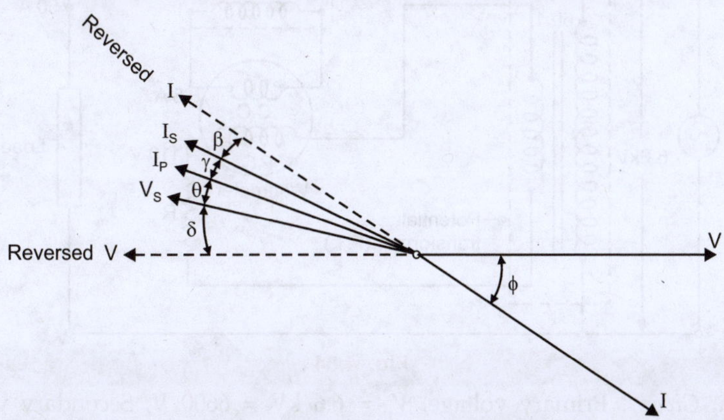

Fig. 4.82. Phase diagram for inductive load.

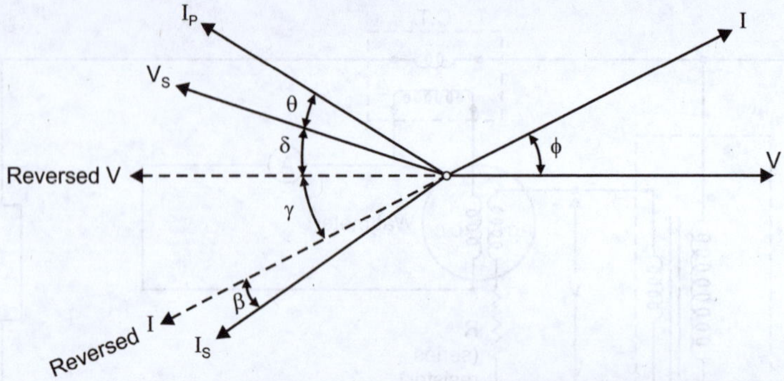

Fig. 4.83. Phasor diagram for capacitive load.

Phase angle between the currents in current and potential coils of wattmeter,

$$\gamma = \phi - \theta - \beta \pm \delta \quad \text{... for } \textit{inductive} \text{ loads} \qquad ...(4.151)$$

and, $$\gamma = \phi + \theta + \beta \pm \delta \quad \text{... for } \textit{capacitive} \text{ loads} \qquad ...(4.152)$$

Since phase angles of P.T. may be lagging or leading.

Neglecting transformation ratio errors of instrument transformers,

$$\text{Correction factor, } K = \frac{\cos\phi}{\cos\theta\cos(\phi - \theta - \beta \pm \delta)} \quad \text{...for } \textit{inductive} \text{ loads} \qquad ...(4.153)$$

and, $$K = \frac{\cos\phi}{\cos\theta\cos(\phi + \theta + \beta \pm \delta)} \quad \text{...for } \textit{capacitive} \text{ loads} \qquad ...(4.154)$$

True power = K × Actual ratio of C.T. × actual ratio of P.T. × wattmeter reading ...(4.155)

Example 4.88. *Power of a single phase 6.6 kV load drawing a current of about 50 A is required to be measured by means of a wattmeter having volt-terminals marked as 110 V and current terminals as 5 A. Draw the circuit diagram showing the connection of the wattmeter using the instrument transformers. Write suitable transformation ratio against each instrument transformer shown in the diagram.* **(UPTU)**

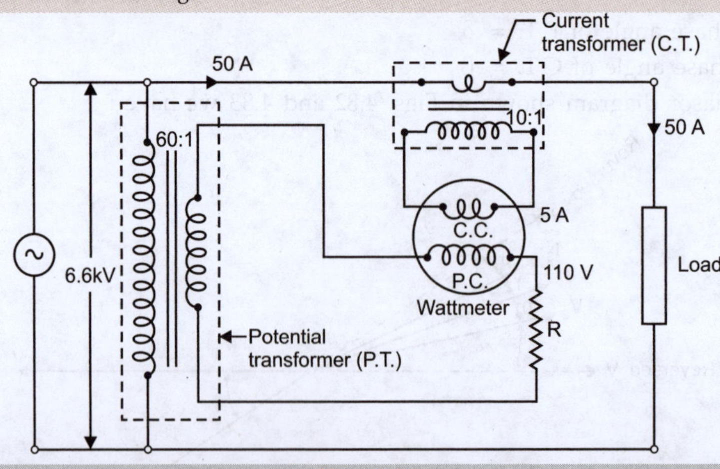

Fig. 4.84

Solution. *Given* : Primary voltage, V_p = 6.6 kV = 6600 V; Secondary voltage, V_s = 110 V; Primary current, I_p = 50 A; I_s = 5A.

$$\text{Transformation ratio of P.T.} = \frac{V_p}{V_s} = \frac{6600}{110} = 60 \textbf{ (Ans.)}$$

$$\text{Transformation ratio of C.T.} = \frac{I_p}{I_s} = \frac{50}{5} = 10 \textbf{ (Ans.)}$$

Fig. 4.84 shows the circuit diagram indicating the connection of the wattmeter using the instrument transformers. The transformation ratio against each instrument is also shown in the figure.

Example 4.89. *A 100 V, 5 A range wattmeter is used in conjunction with instrument transformers to measure power consumed by a 6 kV single-phase load taking 100 A at 0.5 p.f. (lagging). The potential and current transformers used have nominal ratios of 60 : 1 and 20 : 1, ratio errors of 1% and –0.5% and phase angle errors of –1° and +2° respectively. Assuming that there is no error in the wattmeter, find out the error in the measurement of power in the circuit due to the errors of instrument transformers.*

Solution. *Given:* $V = 6$ kV = 6000 volts; $I = 100$ A; $\cos \phi = 0.5$ (lagging) Phase angle of P.T., $\delta = -1°$; Phase angle of C.T. = 2°.

Percentage error in measurement :

$$\text{Actual power consumed} = VI \cos \phi$$

$$= 6000 \times 100 \times 0.5 = 3{,}00{,}000 \text{ W} = 300 \text{ kW}$$

Since, $\cos \phi = 0.5$ (lagging),

$\therefore$ $\phi = \cos^{-1}(0.6) = 60°$

Also, $\theta = 0°$ since wattmeter reads correctly

Now, phase angle between the currents in C.C. and P.C. of wattmeter,

$$\gamma = \phi - \theta - \beta + \delta \quad \text{...[Eqn. (4.151)]}$$

$$= 60° - 0° - 2° - 1 = 57°$$

$$\text{Correction factor, } K = \frac{\cos \phi}{\cos \theta \ \cos \gamma} \quad \text{...[Eqn. (4.153)]}$$

$$= \frac{\cos 60°}{\cos 0° \ \cos 57°} = 0.918$$

Actual transformation ratio of C.T. = 20(1 – 0.005) = 19.9

Actual transformation ratio of P.T. = 60(1 + 0.01) = 60.6

$$\text{Power indicated by wattmeter} = \frac{\text{Actual power consumed}}{K \times \text{actual ratio of P.T.} \times \text{Actual ratio of C.T.}}$$

$$= \frac{300}{0.918 \times 60.6 \times 19.9} = 0.27 \text{ kW}$$

$$\text{True reading of the wattmeter} = \frac{\text{Actual power consumed}}{\text{Nominal ratio of P.T.} \times \text{Nominal ratio of C.T.}}$$

$$= \frac{300}{60 \times 20} = 0.25 \text{ kW}$$

$$\therefore \quad \text{Percentage error} = \frac{\text{Power indicated – true reading}}{\text{True reading}} \times 100$$

$$= \frac{0.27 - 0.25}{0.25} \times 100 = 8\% \text{ (\textit{high})} \textbf{ (Ans.)}$$

4.9.8. Measurement of Power in 3-Phase Circuit

- The power in 3-phase load can be measured by using the following methods :
  1. Three-wattmeter method.
  2. Two-wattmeter method.
  3. One-wattmeter method.

4.9.8.1. Three-wattmeter method

Figs. 4.85 and 4.86 show the connection diagram for star-connected and delta-connected loads respectively. As indicated in the figures, three wattmeters are connected in each of the three phases of the load whether star or delta connected. The current coil of each wattmeter carries the current of one phase only and the pressure coil measures the phase-voltage of the phase. Hence, *each wattmeter measures the power in a single phase. The total power in the load is given by the algebraic sum of the readings of the three wattmeters.*

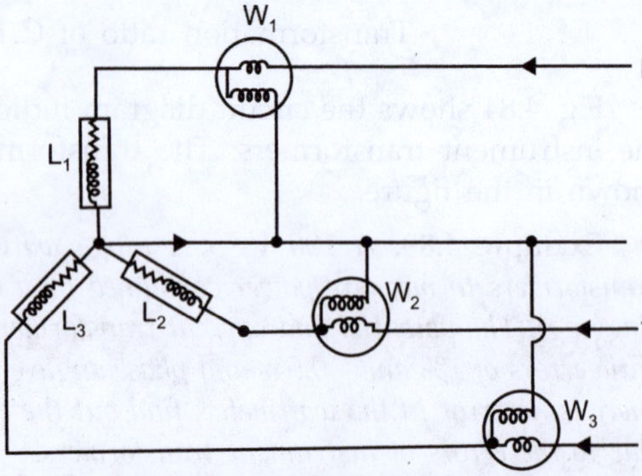

Fig. 4.85. Star-connected load.

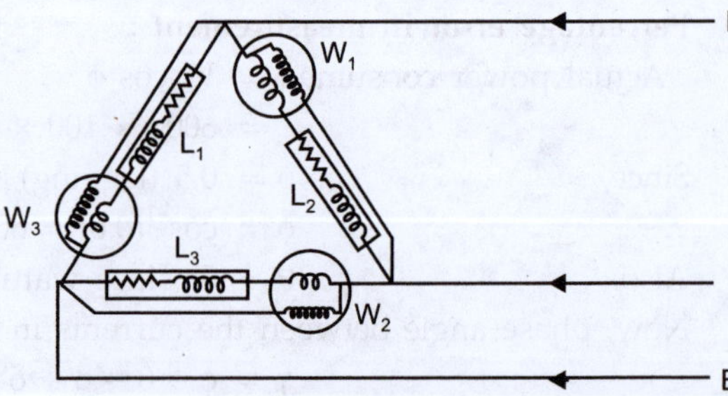

Fig. 4.86. Delta-connected load.

While using this method, following *difficulty* is met with :

- In case of star-connected load it is not always possible to get a neutral point which is required for connections (Fig. 4.85).
- In case of delta-connected load, under ordinary conditions it is not generally feasible to break into the phases of the load.

To measure power it is not necessary to use three wattmeters, two wattmeters can be used for the purpose as explained in the Article 4.9.8.2.

4.9.8.2. Two-wattmeter method – balanced load

The total power consumed by a *balanced load* can be found by using two wattmeters (Figs. 4.87 and 4.88). When load is assumed inductive in Fig. 4.87 the vector diagram for such a balanced star-connected load is shown in Fig. 4.89.

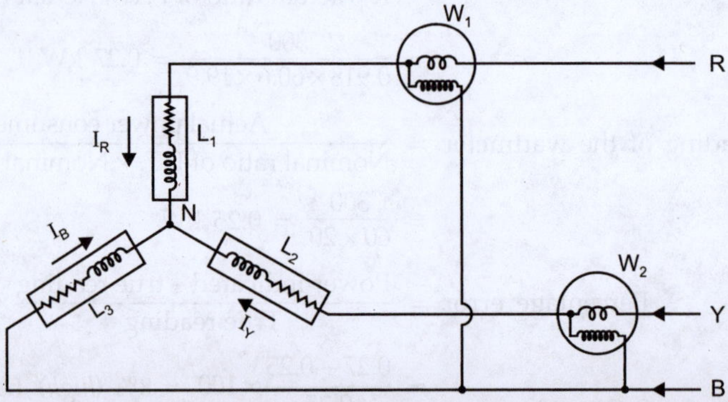

Fig. 4.87. Star-connected load

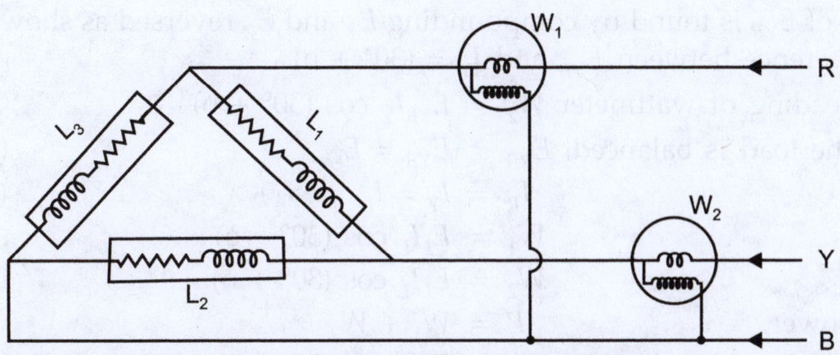

Fig. 4.88. Delta-connected load.

Let us consider the problem in terms of r.m.s. values (instead of instantaneous values).

Let, E_R, E_Y, E_B = r.m.s. values of the three phase voltages, and

I_R, I_Y, I_B = r.m.s. values of the currents.

Since these voltages and currents are assumed sinusoidal, they can be represented by vectors, the currents lagging behind their respective phase voltages by ϕ.

Refer to Fig. 4.89.

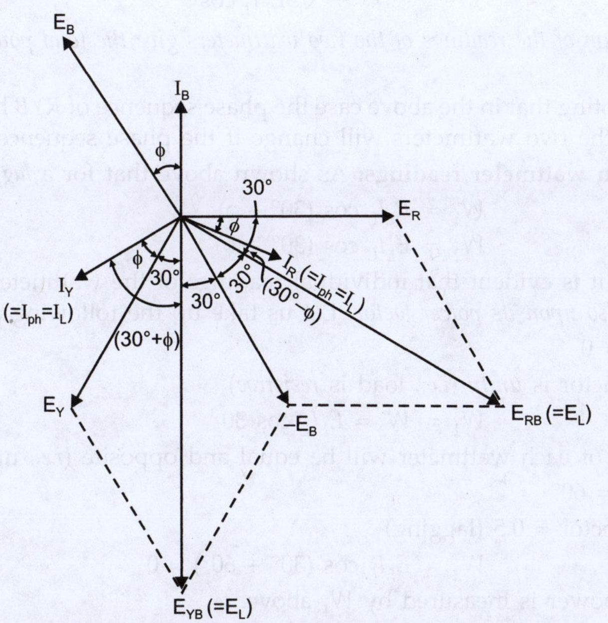

Fig. 4.89. Vector diagram – two-wattmeter method.

Current through wattmeter, $W_1 = I_R$

Potential difference across pressure coil of wattmeter W_1

$$= E_{RB} = E_R - E_B \quad \text{(Vectorially)}$$

The value of E_{RB} is found by compounding E_R and E_B reversed as shown in Fig. 4.89. It may be observed that phase difference between E_{RB} and $I_R = (30° - \phi)$.

$\therefore$ Reading of wattmeter $W_1 = E_{RB} I_R \cos (30° - \phi)$ ...(i)

Similarly, current through wattmeter $W_2 = I_Y$

Potential difference across pressure coil of wattmeter W_2

$$= E_{YB} = E_Y - E_B \quad \text{(Vectorially)}$$

The value of E_{YB} is found by compounding E_Y and E_B reversed as shown in Fig. 4.89. The phase difference between E_{YB} and $I_Y = (30° + \phi)$.

$\therefore$ Reading of wattmeter $W_2 = E_{YB} I_Y \cos(30° + \phi)$ ...(*ii*)

Since the load is balanced, $E_{RB} = E_{YB} = E_L$ (Line voltage)

and, $I_R = I_Y = I_L$ (Line current)

$\therefore$ $W_1 = E_L I_L \cos(30° - \phi)$

and, $W_2 = E_L I_L \cos(30° + \phi)$

$\therefore$ Total power,

$$P = W_1 + W_2$$
$$= E_L I_L \cos(30° - \phi) + E_L I_L \cos(30° + \phi)$$
$$= E_L I_L [\cos(30° - \phi) + \cos(30° + \phi)]$$
$$= E_L I_L [\cos 30° \cos\phi + \sin 30° \sin\phi + \cos 30° \cos\phi - \sin 30° \sin\phi]$$
$$= E_L I_L (2\cos 30° \cos\phi) = E_L I_L \times 2 \times \frac{\sqrt{3}}{2} \cos\phi$$
$$= \sqrt{3} E_L I_L \cos\phi$$

i.e.,
$$P = \sqrt{3} E_L I_L \cos\phi \qquad ...(4.156)$$

Hence the *sum of the readings of the two wattmeters give the total power consumption in the 3-phase load.*

It is worth noting that in the above case the phase sequence of *RYB* has been assumed, the readings of the two wattmeters will change if the phase sequence is reversed.

Variations in wattmeter readings. As shown above that for a *lagging power factor*

$$W_1 = E_L I_L \cos(30° - \phi)$$

and,
$$W_2 = E_L I_L \cos(30° + \phi)$$

From above it is evident that individual readings of the wattmeters *not only depend on the load but also upon its power factor.* Let us take up the following cases :

(*i*) When $\phi = 0$

i.e., power factor is *unity* (*i.e.,* load is *resistive*)

Then, $W_1 = W_2 = E_L I_L \cos 30°$

The reading of each wattmeter will be equal and opposite (*i.e.,* up-scale reading).

(*ii*) When $\phi = 60°$

i.e., power factor = 0.5 (lagging)

Then, $W_2 = E_L I_L \cos(30° + 60°) = 0$

Hence, the power is measured by W_1 above.

(*iii*) When $90° > \phi > 60°$

i.e., $0.5 > \text{p.f.} > 0$

Then W_1 is still *positive* but reading of W_2 is *reversed.* For a leading p.f., conditions are just the opposite of this. In that case, W_1 will read *negative* because the phase angle between the current and voltage is *more than* 90°. For getting the total power, the reading of W_2 is to be *subtracted* from that of W_1. Under this condition, W_2 will read *'down scale' i.e., backwards.* Hence, *to obtain a reading on W_2, it is necessary to reverse either its pressure coil or current coil, usually the former.*

All readings taken after reversal of pressure coil are to be taken as negative.

(*iv*) When $\phi = 90°$

(*i.e.,*p.f. = 0 *i.e.,* pure inductive or capacitive load)

Then, $W_1 = E_L I_L \cos(30° - 90°) = E_L I_L \sin 30°$

and, $W_2 = E_L I_L \cos(30° + 90°) = -E_L I_L \sin 30°$

These two readings are equal in magnitude but opposite in sign

$\therefore \quad W_1 + W_2 = 0$

So far we have considered *lagging angles* (taken as *positive*). Now let us discuss how the readings of wattmeters change when the power factor is *leading* one.

- For $\phi = +60°$ (lag) : $W_2 = 0$
- For $\phi = -60°$ (lead) : $W_1 = 0$

Thus we find that for angles of lead the readings of the two wattmeters are interchanged. Hence, when the power is *leading* :

$$W_1 = E_L I_L \cos(30° + \phi)$$
$$W_2 = E_L I_L \cos(30° - \phi).$$

Power factor – When the load is 'balanced'. When load is balanced with a *lagging power factor* and the voltage and currents are sinusoidal :

$$W_1 + W_2 = E_L I_L \cos(30° - \phi) + E_L I_L \cos(30° + \phi) = \sqrt{3} E_L I_L \cos\phi \quad ...(i)$$

Similarly, $W_1 - W_2 = E_L I_L \cos(30° - \phi) - E_L I_L \cos(30° + \phi)$

$$= E_L I_L \cos(2 \sin\phi \sin 30°) = E_L I_L \sin\phi \quad ...(ii)$$

Dividing (*ii*) by (*i*), we get

$$\tan\phi = \frac{\sqrt{3}(W_1 - W_2)}{(W_1 + W_2)} \quad ...(4.157)$$

For a *leading power*, this expression becomes

$$\tan\phi = -\frac{\sqrt{3}(W_1 - W_2)}{(W_1 + W_2)} \quad ...(4.158)$$

After finding tan ϕ, hence ϕ, the value of power factor cos ϕ can be found (from trigonometrical tables).

One important point which must be kept in mind is that if W_2 reading has been taken after reversing the pressure coil *i.e.*, if W_2 is *negative*, then the eqn., (4.158) becomes

$$\tan\phi = \sqrt{3}\frac{W_1 - (-W_2)}{W_1 + (-W_2)}$$

or,
$$\tan\phi = \sqrt{3}\frac{W_1 + W_2}{W_1 - W_2} \quad ...(4.159)$$

The power factor may also be expressed in terms of ratio of the readings of the two wattmeters.

Let, $\dfrac{\text{Smaller reading}}{\text{Larger reading}} = \dfrac{W_2}{W_1} = \alpha$

Then from eqn. (4.157) above, we have

$$\tan\phi = \frac{\sqrt{3}\left[1 - \left(\frac{W_2}{W_1}\right)\right]}{\left[1 + \left(\frac{W_2}{W_1}\right)\right]} = \frac{\sqrt{3}(1-\alpha)}{(1+\alpha)}$$

We know that, $\sec^2 \phi = 1 + \tan^2 \phi$

or, $\dfrac{1}{\cos^2 \phi} = 1 + \tan^2 \phi$, or, $\cos^2 \phi = \dfrac{1}{1 + \tan^2 \phi}$

$$\cos \phi = \frac{1}{\sqrt{1 + \tan^2 \phi}} = \frac{1}{\sqrt{1 + \left[\frac{\sqrt{3}(1-\alpha)}{1+\alpha}\right]^2}}$$

$$= \frac{1}{\sqrt{1 + 3\left(\frac{1-\alpha}{1+\alpha}\right)^2}} = \frac{1+\alpha}{\sqrt{(1+\alpha)^2 + 3(1-\alpha)^2}}$$

$$= \frac{1+\alpha}{\sqrt{1 + \alpha^2 + 2\alpha + 3(1 + \alpha^2 - 2\alpha)}} = \frac{1+\alpha}{\sqrt{4 + 4\alpha^2 - 4\alpha}}$$

$$= \frac{1+\alpha}{2\sqrt{1 - \alpha + \alpha^2}}$$

i.e., $$\cos \phi = \frac{1+\alpha}{2\sqrt{1 - \alpha + \alpha^2}} \quad ...(4.160)$$

If a curve is plotted between α and $\cos \phi$, then the curve obtained will be as shown in Fig. 4.90, this curve is called **watt-ratio curve.**

Reactive volt amperes *(with two wattmeters)* **:**

We know that, $\tan \phi = \dfrac{\sqrt{3}(W_1 - W_2)}{W_1 + W_2}$ [Eqn. (4.157)]

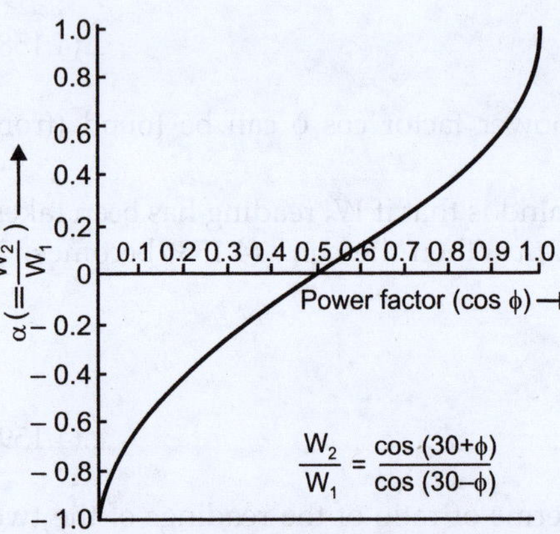

Fig. 4.90. Watt-ratio curve.

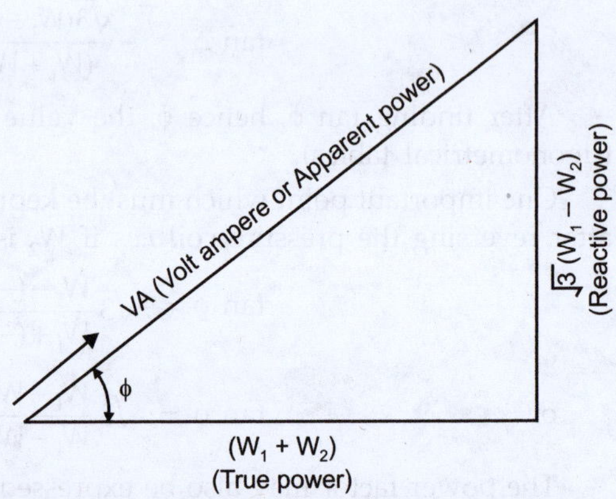

Fig. 4.91

As the tangent of the angle of lag between phase current and phase voltage of a circuit is always equal to the ratio of reactive power to the true power (Fig. 4.91) hence, in case of a balanced load, the reactive power is given by $\sqrt{3}$ *times the difference of the readings of the two wattmeters* used to measure the power of a 3-phase circuit by two-wattmeter method.

Mathematical proof is as follows :

$$\sqrt{3}\,(W_1 - W_2) = \sqrt{3}\,[(E_L I_L \cos (30° - \phi) - E_L I_L \cos (30° + \phi)]$$

$$= \sqrt{3}\, E_L I_L\, [(\cos 30° \cos \phi + \sin 30° \sin \phi) - (\cos 30° \cos \phi - \sin 30° \sin \phi)]$$

$$= \sqrt{3}\, E_L I_L\, [\cos 30° \cos \phi + \sin 30° \sin \phi - \cos 30° \cos \phi + \sin 30° \sin \phi]$$

$$= \sqrt{3}\, E_L I_L\, (2 \sin 30° \sin \phi) = \sqrt{3}\, E_L I_L \sin \phi$$

i.e., $$W_r = \sqrt{3}\,(W_1 - W_2) = \sqrt{3}\, E_L I_L \sin \phi. \quad ...(4.161)$$

4.9.8.3. Blondel's theorem

According to Blondel's theorem (for measurement of power in a *poly-phase system*) :

"If a network is supplied through N conductors, the total power is measured by summing the readings of N wattmeters so arranged that a current element of a wattmeter is in each line and the corresponding voltage element is connected between that line and a common point. If the common point is located on one of the lines, then the power may be measured by N-1 wattmeters."

- The most common applications of a general *theorem* known as ***"Blondel's theorem"*** *include the two-wattmeter method of measurement of power in 3-phase, 3-wire load circuits and three-wattmeter method of measurement of power in 3-phase, 4-wire load circuits.*

— The total power supplied in a N-wire circuit is given by the sum of the readings of (N – 1) wattmeters, so arranged that *a current coil of a wattmeter in each wire and the corresponding potential coil is connected between the wire and a common point on the system.*

— When the common point is located on one of the wires, the potential difference across the potential coil circuit of the wattmeter whose current coil is in that wire is *zero,* and the wattmeter has a zero reading. *Thus only (N – 1) wattmeters are required e.g., two wattmeters for a 3-wire circuit and three wattmeters for a 4-wire circuit.*

4.9.8.4. One-wattmeter method

In this method, the current coil is connected in any one line and the pressure coil is connected alternately between this and the other two lines as shown in Fig. 4.90. The two readings so obtained, for a balanced load, correspond to those obtained by normal two-wattmeter method.

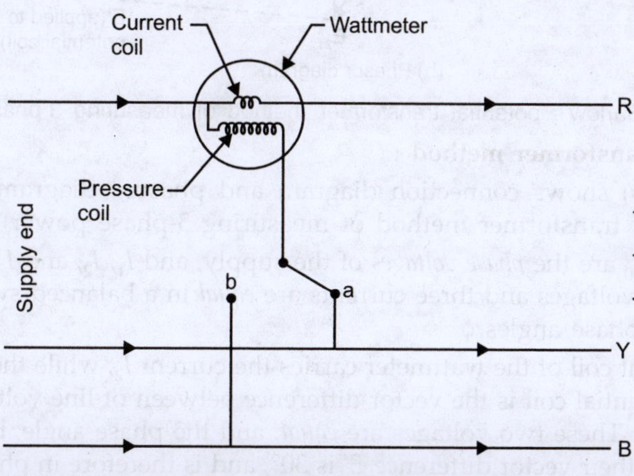

Fig. 4.92. One-wattmeter method.

This method is not of as much universal application as the two-wattmeter method because it is restricted to *fairly balanced loads only.* However, it may be conveniently applied, for instance, when it is desired to find the power input to a factory motor *in order to check the load upon the motor.*

4.9.8.5. "Barlow's method" for a balanced 3-phase circuit using instrument transformers with one wattmeter

Barlow's method is used for measuring the power in a balanced 3-phase circuit, with the *load delta-connected, which does not involve the provision of an artificial neutral point and in which the wattmeter gives the total power directly without any multiplying factor.*

There are two alternative methods of connection, one using a P.T. of 1 : 1 ratio, and the other using two currents transformers each of ratio 1 : 1.

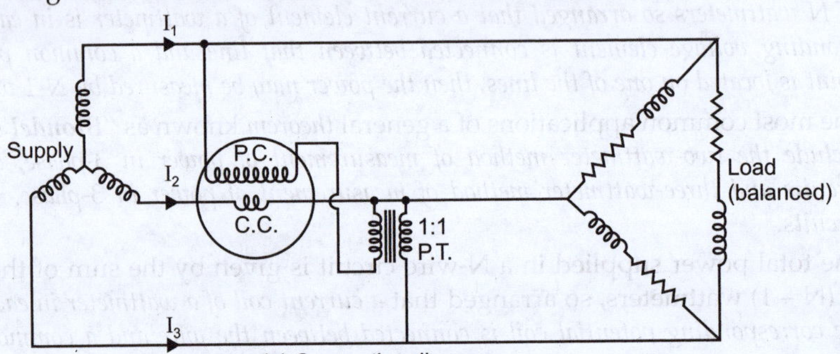

(a) Connection diagram

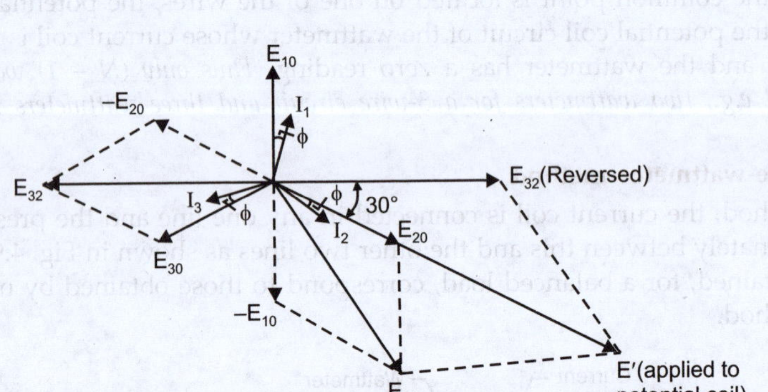

(b) Phasor diagram

Fig. 4.93. Barlow's potential transformer method of measuring 3-phase power.

1. Potential transformer method :

Fig. 4.93(*a*), (*b*) shows connection diagram and phasor diagram respectively of Barlow's potential transformer method of measuring 3-phase power :

— E_{10}, E_{20}, E_{30} are the *phase voltages* of the supply, and I_1, I_2, and I_3 the *line currents.* The three voltages and three currents are *equal* in a balanced system, as are also the three phase angles ϕ.

— The current coil of the wattmeter carries the current I_2, while the voltage applied to the potential coil is the vector difference between of line voltages E_{21}, and E_{32} (reversed). These two voltages are *equal*, and the phase angle, between either of them and their vector difference E' is 30°, and is therefore in phase with E_{20}. The phase angle between E' and I_2 is thus ϕ.

— The wattmeter measures $E'I_2 \cos \phi$.

Now, if
$$E_{10} = E_{20} = E_{30} = E, \text{ then,}$$
$$E_{21} = E_{32} = \sqrt{3}E \text{ and, thus}$$
$$E' = \sqrt{3}E_{21} = \sqrt{3} \times \sqrt{3}E = 3E$$

The power measured by the wattmeter is 3EI cos φ (since $I_1 + I_2 + I_3 = I$), which is the mean value of the total power in the balanced circuit. The ratio error and phase angle of the P.T. are, of course, *neglected* in the above reasoning.

2. Current transformer method :

The connection diagram and phasor diagram of current transformer method of measuring 3-phase power are shown respectively in Fig. 4.94(*a*), (*b*) :

— Since the load is balanced,
$$I_1 = I_2 = I_3 = I$$
and,
$$E_{10} = E_{20} = E_{30} = E$$
and, the phase angles φ are *all equal.*

— The vector E', which is the vector difference of the voltages E_{10} and E_{20}, is the voltage applied to the potential coil and equals $\sqrt{3}E$.

— The currents in the secondary circuits of current transformers 1 and 2 are I_1 reversed and I_2 reversed. The connection to current coil of the wattmeter are such that the current in it is the vector sum of I_1 and I_2 reversed and is represented by the vector I'. The value of this current is $\sqrt{3}I$, and the phase angle between E' and I' is φ. Then,

$$\text{Power measured by wattmeter} = E'I' \cos \phi$$
$$= \sqrt{3}E \times \sqrt{3}I \cos \phi = 3EI \cos \phi$$
$$= \text{Total power in load.}$$

Again, the current-transformer ratio and phase-angle errors are *neglected*.

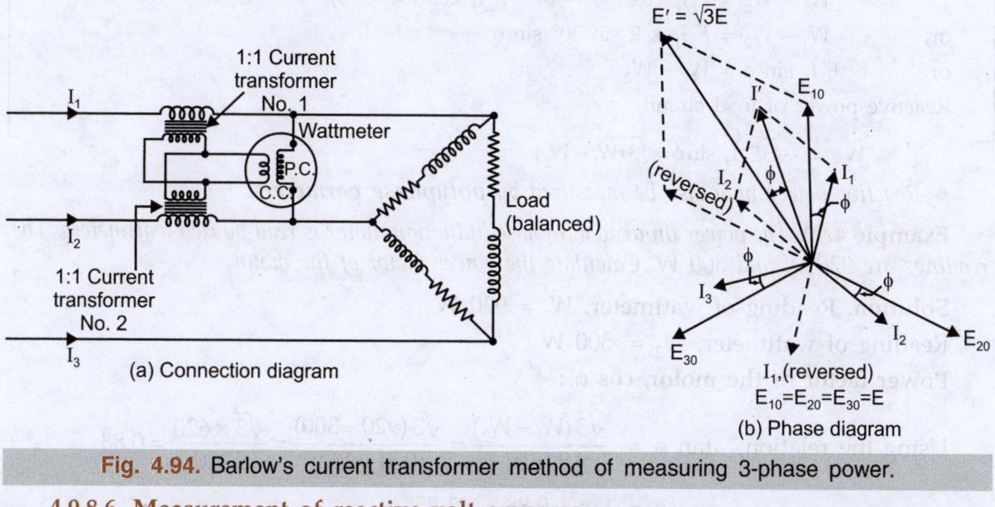

Fig. 4.94. Barlow's current transformer method of measuring 3-phase power.

4.9.8.6. Measurement of reactive volt amperes

In order to measure reactive power in a single phase circuit, a *compensated wattmeter* is used. In this wattmeter, the voltage applied to the pressure coil is 90° *out of phase with the actual voltage* and hence it will read VI cos (90° – φ) *i.e.,* VI sin φ, reactive power.

In balanced three-phase circuit the reactive power can be determined by using one wattmeter. The necessary connections are shown in Fig. 4.95. The current coil is inserted in one line and the pressure coil is connected across the other two lines.

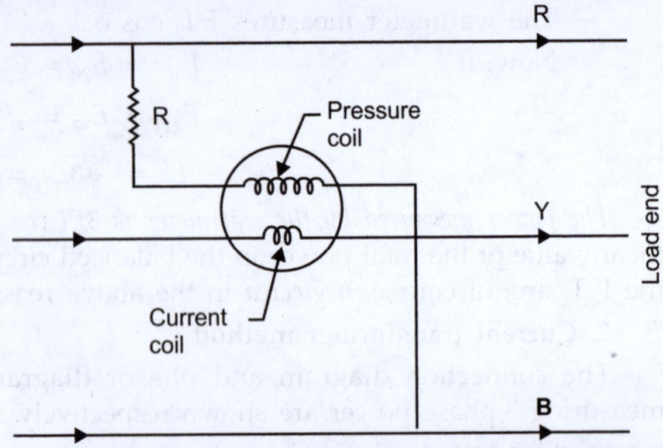

Fig. 4.95. Measurement of reactive volt-amperes.

The current following through the current coil of the wattmeter.

$= I_Y = I_{ph}$ (say)

The potential difference across the potential coil of wattmeter,

$E = E_R - E_B$

$= \sqrt{3}E_{ph}$ leading vector E_Y by 90°

$= \sqrt{3}E_{ph}$ leading the vector of current I_Y by (90° + φ)

∴ Reading of the wattmeter (Fig. 4.96)

$= \sqrt{3}E_{ph}I_{ph}\cos(90° + \phi)$

$= -\sqrt{3}E_{ph}I_{ph}\sin\phi = -W_r$

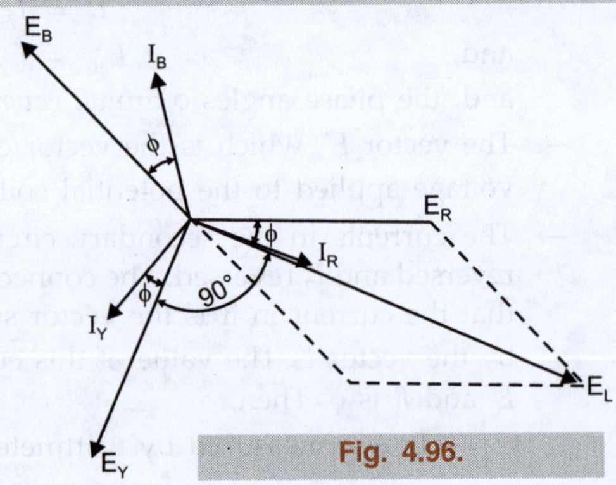

Fig. 4.96.

Total reactive power of the circuit,

$= 3E_{ph}I_{ph}\sin\phi = -\sqrt{3}W_r$

Reactive power (as earlier stated) can also be determined from two wattmeter readings connected for measurement of power.

$W_1 - W_2 = E_LI_L\cos(30° - \phi) - E_LI_L\cos(30° + \phi)$

or, $W_1 - W_2 = E_LI_L \times 2\sin 30°\sin\phi$

or, $E_LI_L\sin\phi = W_1 - W_2$

Reactive power of load circuit,

$W_r = \sqrt{3}E_LI_L\sin\phi = \sqrt{3}(W_1 - W_2).$

- *Reactive power may also be measured by* ***polyphase varmeters.***

Example 4.90. *The power input to a 3-phase induction motor is read by two wattmeters. The readings are 920 W and 300 W. Calculate the power factor of the motor.*

Solution. Reading of wattmeter, W_1 = 920 W

Reading of wattmeter, W_2 = 300 W

Power factor of the motor, cos φ :

Using the relation, $\tan\phi = \dfrac{\sqrt{3}(W_1 - W_2)}{(W_1 + W_2)} = \dfrac{\sqrt{3}(920 - 300)}{(920 + 300)} = \dfrac{\sqrt{3} \times 620}{1220} = 0.88$

∴ $\phi = \tan^{-1} 0.88 = 41.35°$

Power factor of the motor,

$\cos\phi = \cos 41.35° = $ **0.75 (lag) (Ans.)**

Example 4.91. *While performing a load test on a 3-phase wound-rotor induction motor by two-wattmeter method, the readings obtained on two wattmeters were + 14.2 kW and – 6.1 kW and the line voltage was 440 V. Calculate :*

(i) True power drawn by the motor, (ii) Power factor, and (iii) Line current.

Solution. Reading of wattmeter, $W_1 = 14.2$ kW

Reading of wattmeter, $W_2 = -6.1$ kW

Line voltage, $E_L = 440$ V

(*i*) **True power drawn by the motor**

$$= 14.2 - 6.1 = \mathbf{8.1\ kW} \quad \textbf{(Ans.)}$$

(*ii*)
$$\tan\phi = \frac{\sqrt{3}(W_1 - W_2)}{(W_1 + W_2)} = \frac{\sqrt{3}[14.2 - (-6.1)]}{[14.2 + (-6.1)]} = \frac{\sqrt{3}\times 20.3}{8.1} = 4.34$$

$$\therefore \quad \phi = \tan^{-1} 4.34 = 77°$$

$\therefore$ **Power factor,** $\cos\phi = \cos 77° = \mathbf{0.2249\ (lag).}$ **(Ans.)**

(*iii*) **Line current, I_L :**

Using the relation, $P = \sqrt{3}E_L I_L \cos\phi$

$$8.1 \times 1000 = \sqrt{3}\times 440 \times I_L \times 0.2249$$

$$\therefore \quad I_L = \frac{8.1\times 1000}{\sqrt{3}\times 440\times 0.2249} = 47.26\text{ A}$$

Hence, *line current* = **47.26 A** **(Ans.)**

Example 4.92. *A 3-phase, 440 V motor load has a power factor of 0.6. Two wattmeters connected to measure the power show the input to be 25 kW. Find the reading on each instrument.*

Solution. Input power, = 25 kW

Line voltage, $E_L = 440$ V

Power factor of the motor load, $\cos\phi = 0.6$

W_1 ; W_2 :

Using the relation : $\tan\phi = \dfrac{\sqrt{3}(W_1 - W_2)}{W_1 + W_2}$ ...(*i*)

Now, $\cos\phi = 0.6, \quad \therefore \phi = 53.13°$

$\therefore \quad \tan\phi = \tan 53.13° = 1.333$

Also $W_1 + W_2 = 25$ kW (*Given*) ...(*ii*)

Substituting these values in (*i*), we get

$$1.333 = \frac{\sqrt{3}(W_1 - W_2)}{25}$$

$$\therefore \quad W_1 - W_2 = \frac{1.333\times 25}{\sqrt{3}} = 19.24\text{ kW} \quad ...(iii)$$

From (*ii*) and (*iii*), we get

$$\mathbf{W_1 = 22.12\ kW;}$$

and $\mathbf{W_2 = 2.88\ kW}$ **(Ans.)**

Example 4.93. *In a 3-phase circuit, two wattmeters used to measure power indicate 1200 W and 600 W respectively. Find the power factor of the circuit :*

(i) When both wattmeter readings are positive.

(ii) When the latter is obtained by reversing the current coil connections.

Solution. (*i*) *When both wattmeter readings are positive :*

Reading of wattmeter, $W_1 = 1200$ W

Reading of wattmeter, $W_2 = 600$ W

We know that, $$\tan\phi = \frac{\sqrt{3}\,(W_1 - W_2)}{W_1 + W_2} = \frac{\sqrt{3}(1200 - 600)}{(1200 + 600)} = 0.577$$

$$\phi = \tan^{-1} 0.577 = 30°$$

Power factor $= \cos\phi = \cos 30° =$ **0.866 (lag). (Ans.)**

(*ii*) *When the reading of wattmeter W_2 is obtained by reversing the coil connection :*

Reading of wattmeter, $W_1 = 1200$ W

Reading of wattmeter, $W_2 = -600$ W

We know that, $$\tan\phi = \frac{\sqrt{3}\,(W_1 - W_2)}{W_1 + W_2} = \frac{\sqrt{3}(1200 - (-600))}{(1200 + (-600))} = \frac{\sqrt{3} \times 1800}{600}$$

$$= 5.196$$

or $$\phi = \tan^{-1} 5.196 = 79.1°$$

Hence, **power factor** $= \cos\phi = \cos 79.1° =$ **0.1889. (Ans.)**

Example 4.94. *In order to measure the power input and the power factor of an over-excited synchronous motor two wattmeters are used. If the meters indicate (– 3.5 kW) and (+ 8.0 kW) respectively, calculate :*

(*i*) *Power factor of the motor;* (*ii*) *Power input to the motor.*

Solution. (*i*) *Since an over-excited synchronous motor runs with a leading power factor,* we should use the relation,

$$\tan\phi = -\frac{\sqrt{3}(W_1 - W_2)}{(W_1 + W_2)}$$

Moreover it is W_1 that gives negative reading and not W_2. Hence $W_1 = -3.5$ kW.

$\therefore$ $$\tan\phi = -\frac{\sqrt{3}(-3.5 - 8)}{(-3.5 + 8)} = \frac{\sqrt{3} \times 11.5}{5.5} = 3.62$$

$\therefore$ $$\phi = \tan^{-1} 3.62 = 74.6° \text{ (lead)}$$

$\therefore$ **Power factor** $= \cos\phi = \cos 74.6° =$ **0.2655 (lead). (Ans.)**

(*ii*) **Power input** $= W_1 + W_2 = -3.5 + 8 =$ **4.5 W (Ans.)**

Example 4.95. *Two wattmeters are used to measure power input to a 1.5 kV, 50 Hz, 3-phase motor running on full-load at an efficiency of 85 per cent. Their readings are 250 kW and 80 kW respectively. Calculate :*

(*i*) *Input,* (*ii*) *Power factor,*

(*iii*) *Line current, and* (*iv*) *Output.*

Solution. *Since the motor is running at full-load, its power factor must be greater than 0.5. Hence, W_2 reading is positive.*

$\therefore$ $W_1 = +250$ kW and $W_2 = +80$ kW

(*i*) **Input** $= W_1 + W_2 = 250 + 80 =$ **330 kW. (Ans.)**

(*ii*) $$\tan\phi = \frac{\sqrt{3}\,(W_1 - W_2)}{W_1 + W_2} = \frac{\sqrt{3}(250 - 80)}{(250 + 80)} = 0.892$$

$\therefore$ $$\phi = \tan^{-1} 0.892 = 41.74°$$

and, **power factor** $= \cos\phi = \cos 41.74° =$ **0.746 (lag). (Ans.)**

(*iii*) Power, $P = \sqrt{3}\, E_L I_L \cos\phi$

$\therefore \quad I_L = \dfrac{P}{\sqrt{3}\, E_L \cos\phi} = \dfrac{330 \times 1000}{\sqrt{3} \times 1.5 \times 1000 \times 0.746} = 170.27$

Hence, **line current = 170.27 A. (Ans.)**

(*iv*) **Output** = Input × efficiency = 330 × 0.85 = **280.5 kW. (Ans.)**

Example 4.96. *Two wattmeters are used to measure power input to a synchronous motor. Each of them indicates 60 kW. If the power factor be changed to 0.866 leading, determine the reading of the two wattmeters, the total input power remaining the same.*

Solution. Reading of wattmeter, $W_1 = 60$ kW

Reading of wattmeter, $W_2 = 60$ kW.

First case. In the first case, *both wattmeters read equal and positive. Hence, motor must be running at unity power factor.*

Second case. *Power factor is 0.866 leading.*

In this case : $W_1 = E_L I_L \cos(30° + \phi)$

$W_2 = E_L I_L \cos(30° - \phi)$

$\therefore \quad W_1 + W_2 = \sqrt{3}\, E_L I_L \cos\phi$

$W_1 - W_2 = -E_L I_L \sin\phi$

$\therefore \quad \tan\phi = -\dfrac{\sqrt{3}(W_1 - W_2)}{(W_1 + W_2)}$

Since, $\cos\phi = 0.866$ (*Given*)

$\therefore \quad \phi = \cos^{-1} 0.866 = 30°$

and, $\tan\phi = \tan 30° = \dfrac{1}{\sqrt{3}}$

$\therefore \quad \dfrac{1}{\sqrt{3}} = -\dfrac{\sqrt{3}(W_1 - W_2)}{120}$ [$\because W_1 + W_2 = 120$ kW (*Given*)]

$\therefore \quad W_1 - W_2 = -\dfrac{120}{3} = -40$ ...(*i*)

But $W_1 + W_2 = 120$ ...(*ii*)

From (*i*) and (*ii*), we get **W_1 = 40 kW, W_2 = 80 kW. (Ans.)**

Example 4.97. *Two wattmeters connected to read the total power in a 3-phase system supplying a balanced load read 10.5 kW and –2.5 kW respectively. Calculate the total power and power factor.*

Also, explain the significance of (*i*) *equal wattmeter readings and* (*ii*) *a zero reading on one wattmeter.*

Solution. *Given :* $W_1 = 10.5$ kW; $W_2 = -2.5$ kW.

For two wattmeter method,

total power = $W_1 + W_2 = 10.5 + (-2.5) =$ **8.0 kW. (Ans.)**

We know that, $\tan\phi = \dfrac{\sqrt{3}(W_1 - W_2)}{W_1 + W_2} = \dfrac{\sqrt{3}\,[10.5 - (-2.5)]}{[10.5 + (-2.5)]} = 2.8145$

or, $\phi = 70.44°$

$\therefore$ **Power factor,** $\cos\phi = \cos 70.44° =$ **0.335. (Ans.)**

(*i*) For readings of the two wattmeters to be equal,

$$W_1 = W_2$$

or, $\sqrt{3}E_L I_L \cos(30° - \phi) = \sqrt{3}E_L I_L \cos(30° + \phi)$

or, $\cos(30° - \phi) = \cos(30° + \phi)$

$\therefore$ $\phi = 0°$ or $\cos\phi = 1$

i.e., for *unity power factor, the reading of two wattmeters are equal.* **(Ans.)**

(*ii*) As reading of wattmeters are

$$W_1 = \sqrt{3}E_L I_L \cos(30° - \phi), \text{ and}$$

$$W_2 = \sqrt{3}E_L I_L \cos(30° + \phi),$$

for *reading of one of the wattmeters to be zero* ϕ *must be* 60°, *which makes* $W_2 = \sqrt{3}E_L I_L \cos 90° =$ **0. (Ans.)**

Example 4.98. *A star connected 3-phase load consists of three similar impedances. When load is connected to a 3-phase 400 V, 50 Hz supply it takes 30A line current at a power factor of 0.8 lagging.*

(*i*) *Calculate the total power taken by the load.*

(*ii*) *Calculate the resistance and reactance of each phase of the load.*

(*iii*) *If power is measured by two-wattmeter method, then calculate the reading of each wattmeter.*

Solution. *Given :* $E_L = 400$ V; $f = 50$ Hz; $I_L = (I_{ph}) = 30$ A, $\cos\phi = 0.8$ lagging.

(*i*) **Total power taken by the load, *P* :**

$$P = \sqrt{3}E_L I_L \cos\phi$$

$$= \sqrt{3} \times 400 \times 30 \times 0.8 = 16627 \text{ W or } \mathbf{16.627\ kW.}$$ **(Ans.)**

(*ii*) **Resistance and reactance of each phase of the load, *R, X* :**

$$E_{ph} = \frac{E_L}{\sqrt{3}} = \frac{400}{\sqrt{3}} = 230.95 \text{ V}$$

$\therefore$ Load impedance, $Z = \frac{E_{ph}}{I_{ph}} = \frac{230.95}{30} = 7.7\ \Omega$

$\therefore$ Load resistance, $R = Z\cos\phi = 7.7 \times 0.8 =$ **6.16 Ω. (Ans.)**

and, load resistance, $X = Z\sin\phi = 7.7 \times 0.6 =$ **4.62 Ω. (Ans.)**

(*iii*) **Reading of each wattmeter, W_1, W_2 :**

Reading of wattmeter 1, $W_1 = E_L I_L \cos(30° + \phi)$

Reading of wattmeter 2, $W_2 = E_L I_L \cos(30° - \phi)$

where, $\phi = \cos^{-1} 0.8 = 36.9°$

$\therefore$ $W_1 = 400 \times 30 \cos(30° + 36.9°) = 4708$ W or **4.708 kW. (Ans.)**

and, $W_2 = 400 \times 30 \cos(30° - 36.9°) = 11913$ W or **11.913 kW. (Ans.)**

Example 4.99. *A 3-phase, 3-wire, 415 V system supplies a balanced load of 20 A at a power factor 0.8 lag. The current coil of wattmeter 1 is in phase R and of wattmeter 2 in phase B. Calculate :*

(*i*) *The reading on 1 when its voltage coil is across R and Y ;*
(*ii*) *The reading on 2 when its voltage coil is across B and Y ;*
(*iii*) *The reading on 1 when its voltage coil is across Y and B.*

Solution. *Given* : $E_L = 415$ V, $I_L = 20$ A, $\cos\phi = 0.8$ (lag) or $\phi = \cos^{-1} 0.8 = 36.87°$ (lag)

(*i*) *The reading on 1 when its voltage coil is across R and Y and current coil is in R phase*

$$= E_L I_L \cos(30° - \phi) = 415 \times 20 \times \cos(30° - 36.87°)$$
$$= 8240 \text{ W} \quad \text{or} \quad \mathbf{8.24\ kW} \quad \textbf{(Ans.)}$$

(*ii*) *Reading on wattmeter 2 when its voltage coil is across B and Y and current coil is in phase B*

$$= E_L I_L \cos(30° + \phi) = 415 \times 20 \cos(30° + 36.87°)$$
$$= 3260 \text{ W} \quad \text{or} \quad \mathbf{3.26\ kW} \quad \textbf{(Ans.)}$$

(*iii*) *When the current coil is in one phase (phasor R) and the voltage coil is across other two phases (Y and R),* the wattmeter reads $\frac{\text{total } kVAR}{\sqrt{3}}$ and,

$$\text{Total } kVAR = \sqrt{3}(W_1 - W_2)$$

$$\therefore \quad \text{Wattmeter reading} = \frac{\text{Total } kVAR}{\sqrt{3}} = \frac{\sqrt{3}(W_1 - W_2)}{\sqrt{3}}$$
$$= W_1 - W_2 = 8240 - 3260 = \mathbf{4980\ VAR} \quad \textbf{(Ans.)}$$

Example 4.100. *The ratio of the readings of two wattmeters connected to measure power in a balanced 3–φ, 3 wire load is 5 : 3. The load is known to be inductive with a lagging power factor. Calculate the power factor of the load.* **(Punjab University)**

Solution. Let the ratio of wattmeter reading be *r i.e.,* $\frac{W_2}{W_1} = r$

Now,
$$\tan\phi = \sqrt{3}\left(\frac{W_1 - W_2}{W_1 + W_2}\right) = \sqrt{3}\left[\frac{1 - \frac{W_2}{W_1}}{1 + \frac{W_2}{W_1}}\right] = \sqrt{3}\left(\frac{1-r}{1+r}\right)$$

and, power factor,
$$\cos\phi = \frac{1}{\sec\phi} = \frac{1}{\sqrt{1 + \tan^2\phi}} = \frac{1}{\sqrt{1 + 3\left(\frac{1-r}{1+r}\right)^2}}$$

Substituting $r = \frac{W_2}{W_1} = \frac{3}{5} = 0.6$ in the above expression, we get

$$\cos\phi = \frac{1}{\sqrt{1 + 3\left(\frac{1-0.6}{1+0.6}\right)^2}} = \mathbf{0.918\ (lag)} \quad \textbf{(Ans.)}$$

Example 4.101. *A symmetrical star-connected three-phase load is made of three identical coils having an internal resistance of 12 Ω and inductance of 142 mH. The power taken by this load when connected to a balanced 3-phase, three wire supply of 400 V, 50 Hz is measured by* *using the two-wattmeter method. The wattmeters used are similar and are of type of 400 V, 5 A, UPF, class 1 accuracy.*

(i) Determine expected readings of W_1 and W_2.

(ii) Compute the total power.

Solution. Line voltage, $E_L = 400$ V

Phase voltage, $E_{ph} = \frac{400}{\sqrt{3}} = 231$ V

Coil resistance, $R_{ph} = 12\ \Omega$

Coil inductive reactance, $X_{ph} = 2\pi f L_{ph} = 2\pi \times 50 \times 142 \times 10^{-3} = 44.61\ \Omega$

Coil impedance, $Z_{ph} = \sqrt{R_{ph}^2 + X_{ph}^2} = \sqrt{12^2 + 44.61^2} = 46.2\ \Omega$

Coil phase angle, $\phi = \cos^{-1}\frac{R_{ph}}{Z_{ph}} = \cos^{-1}\left(\frac{12}{46.2}\right) = 79.94°$ (lagging)

Phase current, $I_{ph} = \frac{E_{ph}}{Z_{ph}} = \frac{231}{46.2} = 5$ A

(i) **Readings of wattmeters W_1, W_2 :**

Readings of wattmeter, $W_1 = E_L I_L \cos(30° - \phi)$

$= 400 \times 5.0 \cos(30° - 74.94°) =$ **1415.7 W** **(Ans.)**

$(\because\ I_L = I_{ph} = 5.0 \text{ A})$

Reading of wattmeter, $W_2 = E_L I_L \cos(30° + \phi)$

$= 400 \times 5.0 \times \cos(30° + 74.24°) =$ **– 515.6 W** **(Ans.)**

(ii) **Total power :**

Total power $= W_1 + W_2$

$= 1415.7 + (-515.6) =$ **900.1 W** **(Ans.)**

Example 4.102. *Two wattmeters are being used to measure power of a balanced load of 30 A at power factor 0.8 lagging being supplied by a 3-phase, 3-wire, 440 V supply. Calculate :*

(i) Power consumed *(ii) Reading of wattmeter No. 1*

(iii) Reading of wattmeter No. 2.

Solution. Line voltage, $E_L = 440$ V

Line current $I_L = 30$ A

Power factor, $\cos\phi = 0.8$

Phase angle, $\phi = \cos^{-1} 0.8 = 36.87°$

(i) **Power consumed,** $P = \sqrt{3}\, E_L I_L \cos\phi = \sqrt{3} \times 440 \times 30 \times 0.8$

$= 18290$ W or **18.29 kW** **(Ans.)**

(ii) **Reading of wattmeter No. 1**

$= E_L I_L \cos(30° - \phi) = 440 \times 30 \cos(30° - 36.87°)$

$= 13105$ W or **13.105 kW** **(Ans.)**

(iii) **Reading of wattmeter No. 2**

$= E_L I_L \cos(30° + \phi) = 440 \times 30 \times \cos(30° + 36.87°)$

$= 5185$ W or **5.185 kW (app.)** **(Ans.)**

4.9.8.7. Polyphase meters

*A **polyphase meter** is used to measure power in a polyphase (3-phase) system.*

It works on the *principle that two single-phase wattmeters are enough to measure power in 3-phase system (balanced or unbalanced).*

It is a double element wattmeter consisting of two separate elements mounted one above the other in one and same case. The moving coils of each unit are rigidly coupled together axially by a rod of insulating material.

The instruments give the total power directly because the total deflecting torque action on the moving system is equal to the algebraic sum of torques produced by the two separate working systems.

— For avoiding mutual interference between the units, these are magnetically shielded from each other by placing a nickel iron plate between them.

— The electrical connections of this type of meter for measurement of power in a 3-phase system are identical with those of two-wattmeter method of measuring power in 3-phase system.

- The *"advantage"* of such an instrument is that the total power in circuit is given by *direct indication* instead of indirectly by addition of readings of the two wattmeters.

Its *"disadvantage"* is that it is *not* as accurate as a single-phase wattmeter.

4.9.9. Measurement of Power at High Frequencies (AF/RF)

Quite often we are required to measure the output power of electronic devices such as amplifiers and oscillators. In such devices the frequencies involved are generally *audio frequencies* (up to 15 kHz appr.) or *radio frequencies* (above 15 kHz). The dynamometer type of wattmeters are not suitable for measurement of power at high frequencies since they are designed to measure power at power frequencies.

Out of a few methods of power measurements at high frequencies, the simplest one is to connect a *"dummy load"* to the output terminals of the device under test. Then either the voltage across the dummy load or the current through the dummy load is measured using instruments like *rectifier or thermocouple types since these instruments can work at high frequencies*. If R is the resistance of dummy load, then power output can be calculated as follows :

Power, $$P = \frac{V^2}{R} \quad ...(4.162)$$

where, V = Voltage across the dummy load, and

I = Current through the dummy load.

Fig. 4.97. shows the basic circuit for power measurement at *"audio frequencies"*.

— The resistance R (non-inductive type) which forms the *dummy load* is connected across the output terminals of the device under test. The value of R should be nearly equal to that of output impedance of the device.

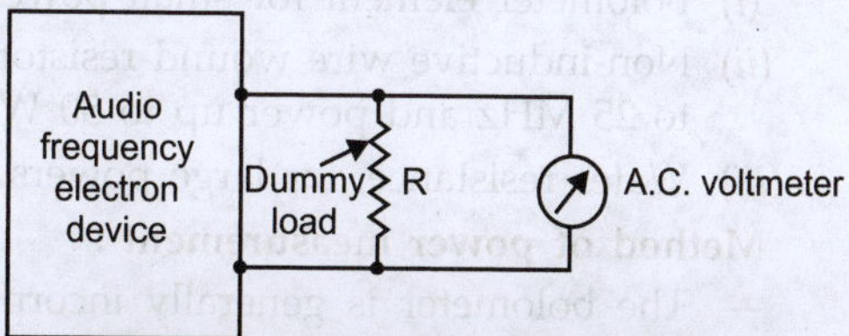

Fig. 4.97. Power measurement at Afrequencies.

— The A.C. voltmeter (a high resistance rectifier type of instrument) when used with a fixed value of R, its scale can be directly calibrated in terms of power output at audio frequencies. Such A.C.

voltmeters are provided with *decibel scales.*

Fig. 4.98 shows the basic circuit for power measurement at *radio frequencies* (RF)

— The output tank of amplifier dummy or oscillator (consisting of inductance L_1 and capacitance C) is transmitted to the dummy load consisting of inductance L_4, capacitance C_1 and resistance *R* in series with a thermocouple ammeter *A*, by means of an inductive link (L_2, L_3) that couples L_1 and L_2 and L_3 and L_4.

— Using capacitor C_1, the dummy load is tuned to resonance at the resonant frequency of the tank circuit. Thus at resonant frequency the dummy load is only *R*.

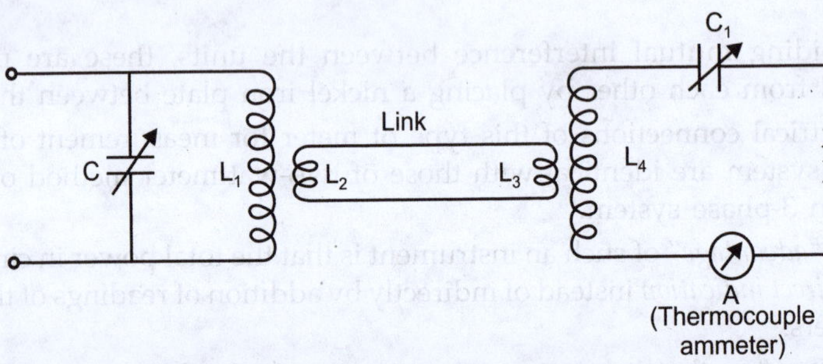

Fig. 4.98. Measurement of power at radio frequencies.

— A is directly calibrated to read the value of power.

— For the measurement of power at high frequencies, *effective shielding must be used.*

4.9.10. Bolometer method of power measurement

At *AF* and *RF*, it is easier to measure voltage, current and impedance, than to measure power. In this range direct measurement of power is not done. The power is calculated from the relation : $P = \frac{V^2}{R} = I^2R$ (as stated earlier) .

At *'microwave frequencies', voltage, current and impedance are difficult to determine.* Their values *may vary* at different points in the circuit, and they are affected appreciably by small changes in geometry. Therefore, in this range, *direct measurement of power is more accurate* and the actual load is replaced by a *dummy load.*

The *requirements* of a dummy load are :

(*i*) Low skin effect;

(*ii*) Low reactances;

(*iii*) Ability to dissipate the required amount of power; and

(*iv*) A suitable value equal to that of the actual load.

The dummy load may take any of the following *forms* :

(*i*) Bolometer element for small powers;

(*ii*) Non-inductive wire wound resistors, cooled by air, or water, for frequencies up to 25 MHz and power up to 50 W; and

(*iii*) Water resistance, for large powers up to 50 W etc.

Method of power measurement :

— The bolometer is generally incorporated into a bridge network, so that small changes in resistance can be readily detected and the power measurement performed by substitution of low frequency power.

— In one method of power measurement, the bridge is initially balanced with low frequency (or bias) power. When *RF* power is applied to bolometer, *low frequency power is withdrawn until the bridge is balanced again.*

In any case, whatever bridge technique is used, the *underlying assumption of measurement is that equal amount of low frequency and RF power produce the same heating effect and resistance changes in the bolometer.*

Bolometer :

— *Bolometric measurements* are *based on the dissipation of the RF power in a small temperature sensitive resistive element, called a "**Bolometer**".*

— The bolometer may be a *short ultra thin wire* having a positive temperature coefficient (PTC) of resistance, called *"Baretter"*, or a bead of semiconductor having a negative temperature coefficient (NTC) called *"Thermistor"*. Both Baretters and Thermistors can measure small powers, of the order of a *fraction of microwatts*. They can also indicate or monitor large amounts of power by inserting a directed compiler.

— The RF power to be measured heats the bolometer and causes a change in its electrical resistance, which serves as an indication of the magnitude of power.

Bolometer bridge :

One of the most commonly used *absorption type meters* is called *"bolometer bridge"*.

It is basically a bridge circuit where one of the arms contains a temperature-sensitive resistor (R_T), as shown in Fig. 4.99.

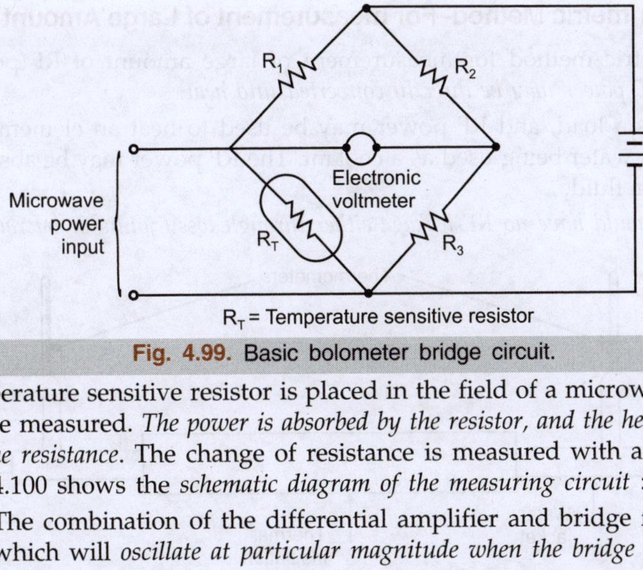

Fig. 4.99. Basic bolometer bridge circuit.

The temperature sensitive resistor is placed in the field of a microwave signal whose power is to be measured. *The power is absorbed by the resistor, and the heat generated causes a change in the resistance.* The change of resistance is measured with a bridge circuit.

Fig. 4.100 shows the *schematic diagram of the measuring circuit* :

— The combination of the differential amplifier and bridge form an oscillator which will *oscillate at particular magnitude when the bridge is unbalanced.*

— The sensitive resistive element (R_T) absorbs the power and heats itself until the bridge is very nearly balanced. The D.C. bias is then adjusted until *exact balance is achieved.*

— Now the element is placed in the microwave field. The element absorbs the power and is heated, which causes a change in its resistance, thereby unbalancing the bridge.

— *The imbalance is in a direction opposite to that of the initial cold resistance. This action causes the output from the oscillator to decrease so as to effect a balance.* The electronic voltmeter circuit measures the amount of this power decrease from the oscillator and displays the measurement as a *power increase which is that supplied by the microwave field.*

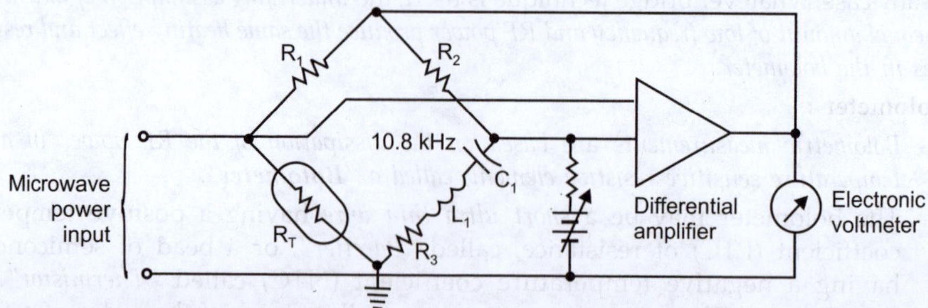

Fig. 4.100. Simplified diagram of the measurement circuit of a modern bolometer type power meter.

Uses of bolomoter :

(*i*) The bolometer is customarily used for power measurements in the range 0.01 – 10 mW.

(*ii*) A comparison-type bolometer is quite suitable in the range of 10 mW to 10 W.

(*iii*) Power measurements above 10 W can be made by *using attenuators and a low-power bolometer.*

4.9.11. Calorimetric Method–For measurement of Large Amount of RF Power

The calorimetric method for measurement of large amount of RF power uses the *principle that a RF power may be directly converted into heat.*

Water acts as a load, and RF power may be used to heat an element such as long transmission line, water being used as a coolant. The RF power may be absorbed directly in the calorimeter fluid.

The *system should have no RF leakage either through lossy joints or by radiation.*

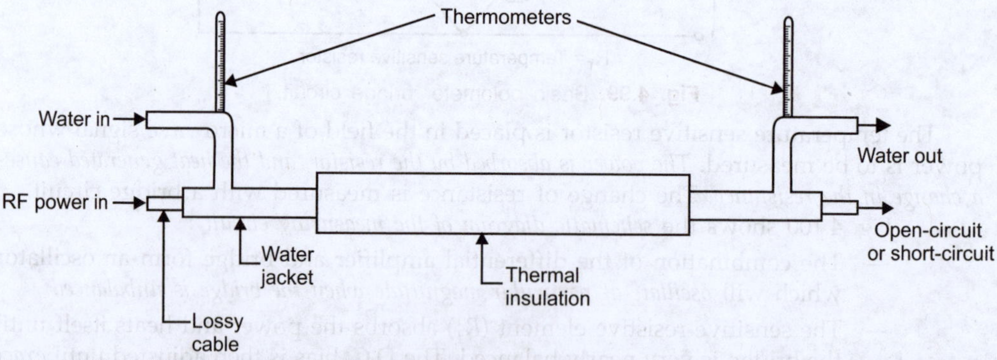

Fig. 4.101. Measurement of power in a lossy cable using the calorimetric method.

The power radiated (*P*) is given by the relation :

$$P = m \times c_p \times \Delta t \quad \text{...(4.163)}$$

where,

m = Mass flow rate of the fluid (water),

c_p = Specific heat of fluid, and

Δt = Temperature rise.

Alternatively, the relationship between temperature rise and power dissipated may be established by dissipating a known amount of A.C. (50Hz) or D.C. power in the calorimeter system.

- Fig. 4.101 shows an arrangement for the power measurement in a lossy cable using the calorimetric method. Here the difference in temperature is measured by thermometers.

4.9.12. Ouput power meter

An *"output power wattmeter"* is designed to directly measure the output in an arbitrary load. In this instrument *a set of resistive loads* are provided to be selected for *power measurements.* Besides this, the instrument *can be used to measure impedance and frequency response characteristics.*

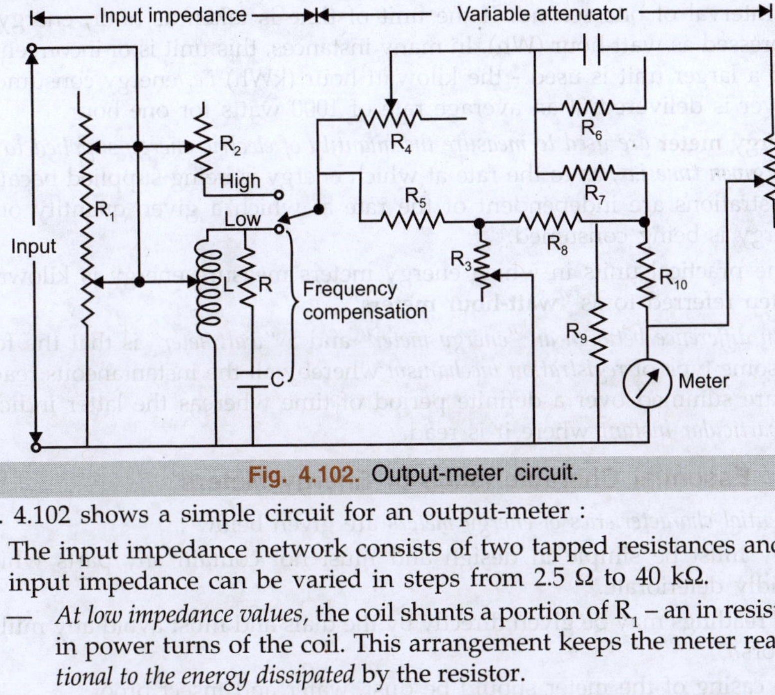

Fig. 4.102. Output-meter circuit.

Fig. 4.102 shows a simple circuit for an output-meter :

- The input impedance network consists of two tapped resistances and a coil. The input impedance can be varied in steps from 2.5 Ω to 40 kΩ.
  — *At low impedance values,* the coil shunts a portion of R_1 – an in resistance results in power turns of the coil. This arrangement keeps the meter reading *proportional to the energy dissipated* by the resistor.
  — At *high impedance values,* the coil is replaced by another tapped resistance R_2.
- The R-C frequency compensator is in parallel with the coil. At low frequency, capacitive reactance X_C is high; it decreases as frequency increases. This *compensates* for the losses of the coil at low frequencies.
  — The frequency range is generally between 20 Hz and 20 kHz.
- Resistance R_3 is the control in a T-network. This amounts to a variable meter shunt, which is used to *extend the range of the meter.*
  — The lowest meter range is normally 5 mW, but it can be extended in decade steps of 50 mW, 500 mW etc.
- The rest of the circuit is a *combination of a calibration and frequency compensation network.*
- This type of meter may have a midscale accuracy of ± 2% at 1 kHz. The accuracy, over the frequency range of 20Hz – 15kHz, may be within ± 2% of the 1kHz value.
- The input meter is subjected to a waveform error when the input is other than sinusoidal.

- *Practical instruments for measuring the power output of oscillators, amplifiers, transformers, transducers and low frequency lines use an input impedance of a tapped transformer with 48 impedance setting.*

This meter can also be used :

(*i*) To measure output impedance by adjusting the maximum power.

(*ii*) To check the frequency response characteristics of audio frequency devices.

4.10 ENERGY METERS AND MEASUREMENT OF ENERGY

4.10.1. General Aspects

- **Energy** *is the total power delivered or consumed after a time internal.* Thus energy is the *product of power and time.* Its unit is *joule* or *watt-second* which is 1 watt over an interval of one second. If the unit of time is taken as hour, energy is then expressed as watt-hour (Wh). In many instances, this unit is of inconvenient size and a larger unit is used - the kilowatt hour (kWh) *i.e.* energy consumed when power is delivered at an average rate of 1000 watts for one hour.
- **Energy meter** *are used to measure the quantity of electric energy supplied to a circuit in a given time i.e.,* as to the rate at which energy is being supplied because their registrations are independent of the rate at which a given quantity of electric energy is being consumed.

Since the practical units in which energy meters measure energy is kilowatt-hour, they are often referred to as "**watt-hour meters**".

The main difference between an "energy meter" and a *"wattmeter"* is that the former is fitted with some type of *registration mechanism* whereby all the instantaneous readings of the power are summed over a definite period of time whereas the latter indicates the value at a *particular instant* where it is read.

4.10.2. Essential Characteristics of Energy Meters

The *essential characteristics of energy meters* are given below :

1. They must be simple in design and must not contain any parts which may rapidly deteriorate.
2. The readings may be given directly by the dials and must avoid any multiplying factors.
3. The casing of the meter should be dust, water and insect proof.
4. Permanency of calibration is a prime requisite and to attain it, the friction at the pivots etc., and retarding torque of the magnetic brakes must remain constant. The magnets should be so placed that they are not affected in their strength by the magnetic field of the current coil.
5. The friction losses should be minimum (being unavoidable) and must remain practically constant over long periods of time. This requires that the moving parts should be light, and the jewels and pivots should be of best quality and kept in good order.
6. There should be minimum possible friction in the counter device. The torque of the meter should be high so that the unavoidable irregularities in friction may not cause inaccuracies.
7. The energy meter should maintain its accuracy under reasonably varying conditions of voltage and load.
8. The energy loss in the meter itself must be small.

4.10.3. Types of Energy Meters

Energy meters are generally of the following three types :

1. *Electrolytic meters* ..... Their operation depends on electrolytic action.
2. *Motor meters* ..... They are really small electric motors.
3. *Clock meters* ...... They function as clock mechanisms.

Here only motor meters will be discussed.

4.10.4. Motor Meters

The motor meters can be used in *D.C. as well as in A.C. circuits.* In principle the motor meter is a small motor of D.C. or A.C. type whose *instantaneous speed of rotation is proportional to the circuit current in case of an ampere-hour meter and to the power of the circuit in case of a watt-hour meter.*

The following are the essential parts of the motor meters :

1. **An operating torque system.** It produces a torque and causes the moving system to rotate continuously.

2. **A braking device.** It is usually a permanent magnet, known as *brake magnet.* This brake magnet induces current in some part of the moving system which in turn produces the braking torque. Thus the braking torque is *proportional to the induced currents* whereas the induced currents are *proportional to the speed of the moving system* (and hence the *braking torque is proportional to the speed of the moving system* (*disc*). When the braking torque is equal to the driving torque the moving system attains a steady speed.

3. **Revolution registering device.** This device is obtained by having a worm cut on the spindle of the instrument. The worm engages with a pinion and thus drives the train of wheels and registers ampere-hours and watt-hours directly.

Types of motor meters. The various types of motor meters are :

(*i*) *Mercury motor meters* ------ Normally used on D.C. circuits.

(*ii*) *Commutator motor meters* ------ Can be used both for D.C. as well A.C. work

(*iii*) *Induction motor/energy meters* ------ Used only on A.C. circuit

4.10.5. Motor-driven Meter—Watt-hour Meter

The motor-driven meter shown in Fig. 4.103 can be used on direct or alternating current. It contains a small motor and an aluminium retarding disc. *The field winding is connected in series with the load, and the field strength is proportional to the load current.*

The armature is connected *across* the source, and the *current in the armature is proportional to the source or line voltage.* The torque produced in the armature is *proportional to the power* consumed by the load.

The armature shaft drives a series of counters that are calibrated in watt-hours. The power that is used can be read directly from the dials.

The aluminium disc attached to the armature is used to control the armature speed. The disc turns is used can be read direcly from the dials.

The aluminium disc attached to the armature is used to control the armature speed. The disc turns in a magnetic field produced by the permanent magnets, and the retarding force increases as the rotation increases and stops when the disc stops. The retarding force is produced by the aluminium conductor cutting through the lines of force of the permanent magnets. This is a form of magnetic damping.

The meter must overcome the friction of the bearing and indicators at *very light loads.* A portion of the field is produced by the armature current (coil in series with the armature

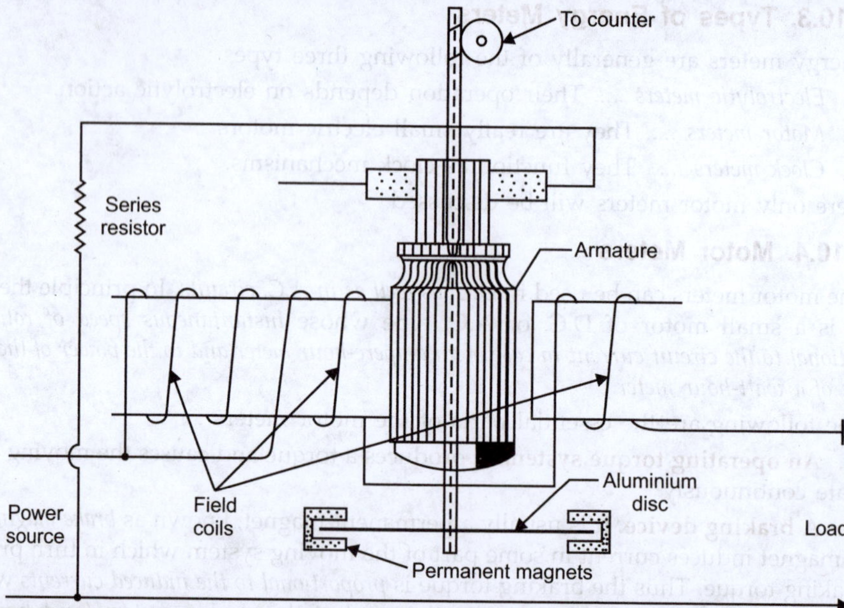

Fig. 4.103. Motor-driven meter (watt-hour meter) designed to operate on direct or alternating current.

winding). This coil is (called as compensating coil) wound to aid the field and is adjusted to the point where it just overcomes the meter friction.

4.10.6. Induction Type Watt-hour Meter

This is the *most commonly used meter on A.C. circuits for measurement of energy.*

Advantages :

(*i*) Simple in operation.

(*ii*) High torque/weight ratio.

(*iii*) Cheap in cost.

(*iv*) Correct registration even at very low power factor.

(*v*) Unaffected by temperature variations.

(*vi*) More accurate than commutator type energy meter on light loads (owing to absence of a commutator with its accompanying friction).

4.10.7. Induction type Single-phase Energy Meters

These are, by far, the most common form of A.C. meters met with in every-day *domestic* and *industrial installations.* These meters measure electric energy in kWh.

4.10.7.1. Construction and theory of operation

The principle of these meters is practically the same as that of induction watt meters. Instead of the control spring and the pointer of the wattmeter, the watt-hour meter, (energy meter) employs a *brake magnet and a counter attached to the spindle.* Just like other watt-hour meters, the eddy currents induced in the aluminium disc by the brake magnet due to the revolution of the disc, are utilised to control the continuously rotating disc.

Construction. The construction of a typical meter of this type is shown in Fig. 4.104(*a*); the brake-magnet and recording wheel-train being omitted for clearances. It consists of the following :

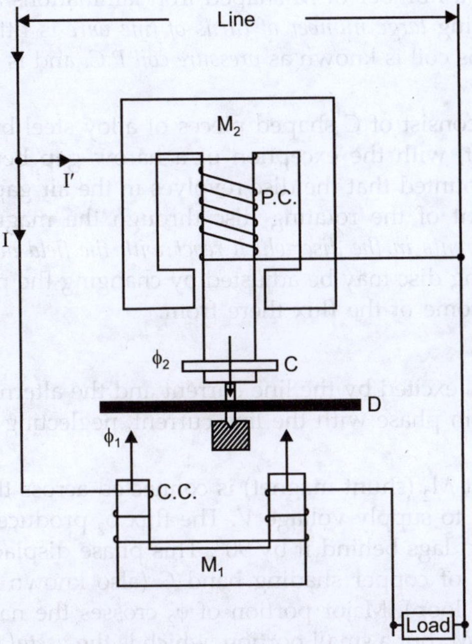

(a) Construction and principle of operation

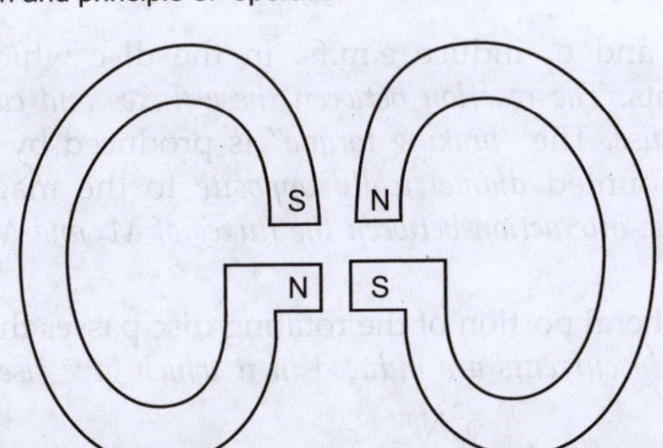

Fig. 4.104. Induction type single-phase energy meter.

(i) Series magnet M_1;

(ii) Shunt magnet M_2;

(iii) Brake magnet;

(iv) A rotating disc.

The *series electromagnet* M_1 consists of a number of U-shaped iron laminations assembled together to form a core, wound with a *few turns of a heavy gauge wire*. This wound coil is known as *current coil* and is connected in one of the lines and in series with the load to be metered. The series electromagnet is energized and sets up a magnetic field cutting through the rotating disc, when load current flows through the current coil C.C. The rotating disc is an aluminium disc mounted on a vertical spindle and supported on a sapphire cup contained in a bottom screw. The bottom pivot, which is usually removable, is of hardened steel, and the end, which is hemispherical in shape, rests in the sapphire cup. The top pivot (not shown) merely serves to maintain the spindle in a vertical position under working condition and does not support any weight or exert any appreciable thrust in any direction.

The *shunt magnet* M_2 consists of a number of M shaped iron laminations assembled together to form a core. A core having *large number of turns of fine wire* is fitted on the middle limb of the shunt magnet, this coil is known as *pressure coil P.C.* and is connected across the supply mains.

The *brake magnet* [Fig. 4.104(*b*)] consist of C shaped pieces of alloy steel bent round to form a complete magnetic circuit, with the exception of a narow gap between the poles. The pair of magnets are so mounted that the disc revolves in the air gap between the polar extremities. The movement of the rotating disc through the magnetic field crossing the air gap *sets up eddy currents in the disc which react with the field and exerts a braking effect*. The speed of the rotating disc may be adjusted by changing the position of the brake magnets or by diverting some of the flux there from.

Theory of operation :

— Magnet M_1 (series magnet) is excited by the line current and the alternating flux ϕ_1, produced is proportional to and in phase with the line current, neglecting of course the hysteresis and iron saturation.

The winding of the other magnet M_2 (shunt magnet) is connected across the supply line and carries current proportional to supply voltage V. The flux ϕ_2 produced by it is proportional to supply voltage V and lags behind it by 90°. This phase displacement of exact 90° is achieved by adjustment of copper shading band C (also known as power factor compensator or compensating loop). Major portion of ϕ_2 crosses the narrow gap between the centre and side limbs of M_2 but a small portion, which is the *useful flux passes through disc D*.

— Two fluxes ϕ_1 and ϕ_2 induce e.m.fs. in the disc which further produce the circulatory eddy currents. *The reaction between these fluxes and eddy currents produces the "driving torque" on the disc.* The *"braking torque"* is produced by a pair of magnets [Fig. 4.104(*b*)], which are mounted *diametrically opposite* to the magnets M_1 and M_2. *This arrangement minimises the interaction between the fluxes of M_1 and M_2 and that of the braking magnet.*

— When the peripheral portion of the rotating disc passes through the air gap of the braking magnet, the *eddy currents are induced in it which give rise to the necessary torque.*

The braking torque, $$T_b \propto \frac{\phi_b^2 N}{R} \quad ...(4.164)$$

where,

ϕ_b = The flux of braking magnet,

N = The speed of the rotating disc, and

R = The resistance of the eddy current path.

If ϕ_b and R are constant, then,

$$T_b \propto N \quad ...(4.165)$$

— The register mechanism is either of pointer type or of cyclometer type. In the pointer type mechanism, the pinion on the rotor shaft drives, with the help of a suitable train of reduction gears, a series of five or six pointers rotating on dials marked with ten equal divisions. The gearing between different pointers is such that each pointer advances by $\frac{1}{10}$th of a revolution for a complete revolution of the adjacent pointer on the main rotor disc in the train of gearing as shown in Fig. 4.105.

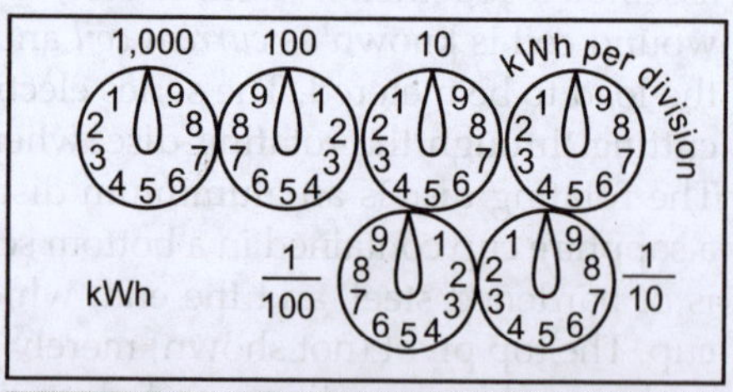

Fig. 4.105. Pointer type register.

- It can be derived that the driving torque, is given by :

$$T_d \propto \omega\, \phi_{1max}\, \phi_{2max} \sin \alpha \qquad ...(4.166)$$

where, ϕ_{1max} and ϕ_{2max} = The maximum fluxes produced by the magnets M_1 and M_2, and

α = The angle between the *fluxes* = $(90 - \phi)°$ ...(See Fig. 4.106)

Assuming that fluxes are proportional to the current, we have :

Current through the winding of M_1 = I ....the line current

Current through the winding of M_2 = $\dfrac{V}{\omega L}$

where, L = Inductance of the pressure/voltage coil (The pressure coil is made highly inductive with negligible resistance; hence $\phi_2 \propto \dfrac{V}{\omega L}$)

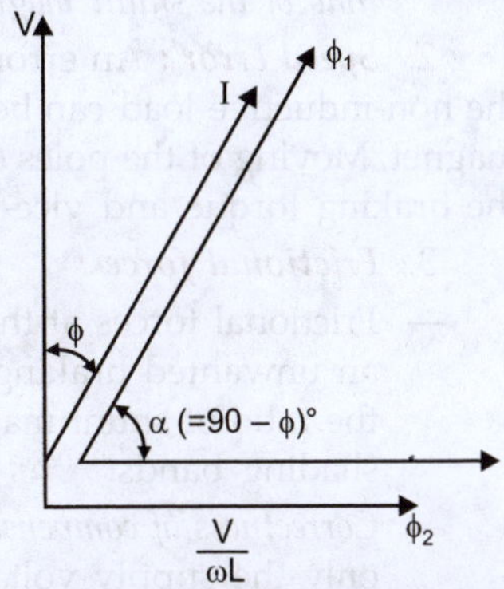

Fig. 4.106.

By inserting various values in eqn. (4.166), we get :

$$T_d \propto \omega . I . \frac{V}{\omega L} \sin (90 - \phi)°$$

or, $T_d \propto VI \cos \phi \propto$ Power ...(4.167)

Also, $T_b \propto N$

The disc achieves a steady speed N when the two torques are equal *i.e.*, when,

$$T_d = T_b$$

$\therefore$ $N \propto$ power ...(4.168)

Hence, in a given period of time, the number of revolutions $\int_0^t N dt$ *is proportional to* $\int_0^t W . dt$ *i.e. the electrical energy consumed.*

4.10.7.2. Errors, compensation and adjustments

I. Sources of errors :

The various sources of errors in an induction-type energy meter are given below :

(*i*) *Incorrect magnitude of the fluxes.* These may arise from abnormal voltages and load currents.

(*ii*) *Incorrect phase relation of fluxes.* These may arise from defective lagging, abnormal frequencies, changes in the iron losses etc.

(*iii*) *Unsymmetrical magnetic structure.* The disc may go on rotating while no current is being drawn but pressure coils alone are excited.

(*iv*) *Changes in the resistance of the disc.* It may occur due to changes in temperature.

(*v*) *Changes in the strength of the drag magnets.* It may be due to temperature or ageing.

(*vi*) Phase-angle errors due to lowering of power factor.

(*vii*) Abnormal friction of moving parts.

(*viii*) Badly distorted waveform.

(*ix*) Changes in the retarding torque due to the disc moving through the field of the current coils.

Errors and **compensations** are discussed below :

1. ***Phase error :***

— Normally the flux due to shunt magnet does *not* lag behind the supply voltage by exactly 90° due to the fact that the coil has some resistance. Consequently the *torque is not zero at zero power factor*. This is known as *"phase error"*.

— This error is *compensated by means of adjustable copper bands placed over the central limb of the shunt magnet.*

2. ***Speed error :*** An error in the speed of the motor (*i.e.* speed error) when tested on the non-inductive load can be eliminated by correctly adjusting the position of the brake magnet. Moving of the poles of the braking magnet towards the centre of the disc reduces the braking torque and vice-versa.

3. ***Frictional force :***

— Frictional forces at the rotor bearings and in the register mechanism give rise to an unwanted braking torque on the disc rotor. This can be reduced by making the ratio of shunt magnet flux and series magnet flux large with the help of two shading bands.

— *Correctness of compensation* is achieved when the rotor does not run on load with only the supply voltage connected.

4. ***Creeping :***

— The slow but continuous rotation of motor when only the voltage *i.e.*, the pressure coil excited but no current flowing in the circuit is called *"creeping"*. It may be *caused due to friction compensation, vibration, stray magnetic field or the voltage supply in excess of the normal.*

— To overcome this creeping on no load, *two holes are drilled in the disc on the diameter, i.e., on the opposite side of the spindle.* This causes sufficient distortion of the field to prevent rotation when one of the holes comes under one of the poles of shunt magnet.

5. ***Temperature error :*** The errors due to temperature variations of the various instruments are usually *small* because the various effects produced tend to neutralise one another.

6. ***Overload compensation :***

— Under load conditions, the disc revolves continuously in the field of the series magnet and, therefore, there is a dynamically induced e.m.f. in the disc because of this rotation. This e.m.f. causes eddy currents which interact with the field of the series magnet to produce a braking torque. This braking torque is obviously *proportional to the square of the load current* and it assumes increasing importance as the load rises. Thus at *high values of load current the registration tends to be lower than the actual.*

— To *minimise the self braking action, the full-load speed of the disc is kept as low as possible* and the current coil flux is made *small* in comparison to the potential coil flux so that the dynamically induced e.m.f. responsible for self braking action is *very small* as compared with statically induced e.m.f. produced by flux alteration at the rate of 50 Hz and which is responsible for the production of driving torque.

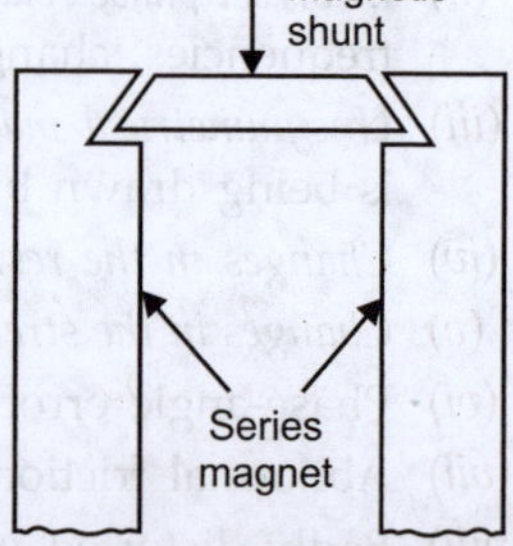

Fig. 4.107. Overload compensation device.

— It is also customary to add an *overload compensator*, which usually takes the form of a saturable magnetic shunt for the series magnet core as shown in Fig. 4.107. The magnetic shunt, at high values of load current, *diverts some of the series magnetic flux and so compensation for self-braking action.*

7. **Voltage compensation :**

— Voltage variations may introduce errors due to non-linear magnetic characteristics of the shunt magnet core and also due to self braking torque which is approximately proportional to the square of the supply voltage.

Since a certain amount of variation is allowed in the declared voltage of supply, therefore, energy meters must be compensated for this variation.

— *Compensation for variations* in supply voltage is provided by making use of a *saturable magnetic shunt which diverts a greater proportion of the flux into the active path when the voltage rises.* The compensation can be conveniently provided by *increasing the reluctance of the side limbs of the shunt magnet; this is done by providing holes in the side limbs.*

II. Energy meter adjustments :

In energy meter, some adjustments are carried out so that they give correct reading and their errors are within allowable limits. Following is the sequence of these adjustments :

(*i*) *Preliminary light load adjustment :*

— The disc is so positioned that the holes are *not* underneath the electromagnets.

— Rated voltage is applied to the voltage coil with no current through the current coil.

— The light load device is adjusted until the disc just fails to start.

(*ii*) *Full load unity factor adjustment :*

— The pressure/voltage coil is connected across the rated supply voltage and rated full current at unity p.f. is passed through the current coils.

— The position of the brake magnet is adjusted to vary the braking torque so that the meter revolves at the correct speed with required limits of errors.

(*iii*) *Lag adjustment :*

— The pressure coil is connected across rated supply voltage and rated full load current is passed through the current coil at 0.5 p.f. lagging.

— The lag device is adjusted till the meter runs at correct speed.

(*iv*) With rated supply voltage, rated full current and unity p.f., the speed of the motor is checked and full load unity p.f. and low p.f. adjustments are repeated until the desired accuracy limits are reached for both the conditions.

(*v*) *Light load adjustment :*

— Rated supply voltage is applied across the pressure coil and a very low current (about 5% of full load) is passed through the meter at unity p.f.

— This adjustment is done so that the meter runs at correct speed.

(*vi*) Full load-unity p.f. and light load adjustments are again carried out until the speed is correct for both loads *i.e.*, full load as well as light load.

(*vii*) The performance is rechecked at 0.5 p.f. lagging.

(*viii*) *Creep adjustment :*

— As a final check on light load adjustment, the pressure coil is excited by 110% of rated voltage with zero load current.

— If the light load adjustment is correct, the meter should not creep under the above conditions.

4.10.7.3. Polyphase energy meters

In polyphase circuits, the energy can be measured by a group of single-phase energy meters connected as required by Blondel's theorem; the *total energy consumed being the sum of the readings of all energy meters (–ve sign being used for the reading of any meter which runs backward).*

However, this is never done in practice as it would be both more expensive and more troublesome than the use of polyphase meters.

In *commercial measurement, polyphase meters are used.*

- As in the case of wattmeters for measuring power in polyphase circuits, polyphase energy meters are also constructed by combining two or more single-phase energy meters in one unit.

Polyphase meters may be *'multi-disc type'* or *'single disc type'*. In the multi-disc type, *each element drives a separate disc.* In the single disc type meter, *all the elements drive the same disc.* This disc is slotted or laminated in sector to prevent interaction between eddy currents produced by one element with the flux produced by another element.

Two element energy meter :

Fig. 4.108, shows a two element energy meter used for measurement of energy in 3-phase, 3-wire circuit :

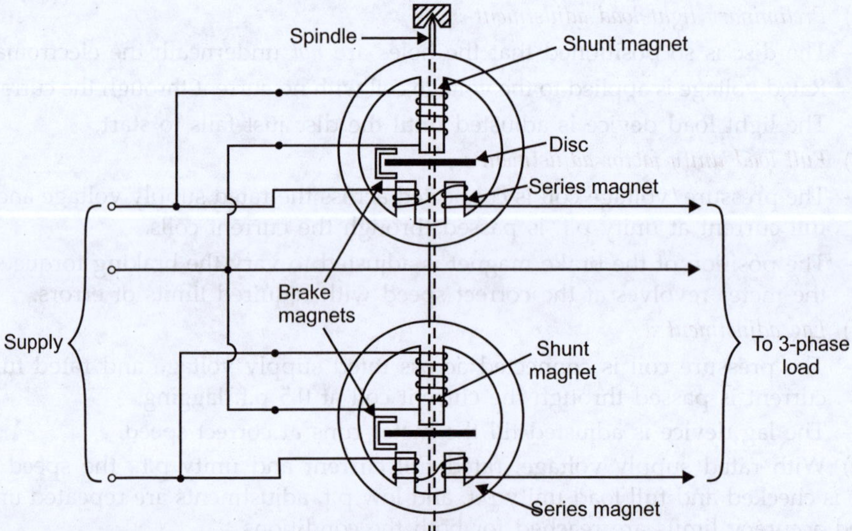

Fig. 4.108. Two element energy meter.

— It consists of two discs mouned on common spindle.

— The moving system drives a single gear train. Each unit is provided with its own shading ring, shading bands and brake magnet for adjustments.

— It is essential *that the driving torque of two elements be exactly the same for the same power passing through them.* This condition is checked by *energising the coils from a single circuit,* the two pressure coils being connected in parallel and the circuit coils in series, with one reversed to oppose the other. This arrangement causes the two torques to be in opposition and if the two torques are *equal,* the disc will not rotate. If there is slight motion, the magnetic shunt is adjusted till the speed of the discs become zero.

— The lag, light load and unity p.f. adjustments are made independently for each element.

— For measurement of energy in 3-phase, 4-wire unbalanced system three element energy meters are used.

- Polyphase energy meters are subjected to *errors due to interaction between the elements,* this may be reduced by suitable positioning and screening of the elements. They also have *relatively large friction error owing to the greater weight of the rotating parts.*

These meters are little *cheaper* than the corresponding number of single-phase meters and have the clear advantage of *giving the total power on a single register.*

- When the currents and voltages of the circuits to be measured are high then use of instrument transformers (C.T.s and P.T.s) is made with energy meters just as with wattmeters. If the line voltage exceeds 440 V CTs and PTs are almost invariably used irrespective of the magnitude of current in order to *insulate meter from the line.*

Example 4.103. *A 5 A, 230 V meter on full load unity power factor test makes 60 revolutions in 360 seconds. If the normal disc speed is 520 revolutions per kWh, what is the percentage error ?*

(Punjab University)

Solution. Energy consumed in 360 seconds

$$= \frac{VI \cos\phi \times t}{3600 \times 1000} \text{ kWh}$$

$$= \frac{230 \times 5 \times 1 \times 360}{3600 \times 1000} = 0.115 \text{ kWh}$$

where, t is in seconds.

Energy recorded by the meter

$$= \frac{60}{520} = 0.11538 \text{ kWh}$$

$$\therefore \text{ \%age error } = \frac{\text{Energy registered/recorded} - \text{energy actually consumed}}{\text{Energy actually consumed}} \times 100$$

$$= \frac{0.11538 - 0.115}{0.115} \times 100 = \mathbf{0.33\%}\ (\textit{fast}). \quad \textbf{(Ans.)}$$

Example 4.104. *The constant of a 230 V, 50 Hz, single-phase energy meter is 185 revolutions per kWh. The meter takes 190 seconds for 10 revolutions while supplying a non-inductive load of 4.5 A at normal voltage. What is the percentage error of the instrument ?*

Solution. Energy consumed in 190 seconds

$$= \frac{VI \cos\phi}{1000} \times t = \frac{230 \times 4.5 \times 1}{1000} \times \frac{190}{3600} = 0.0546 \text{ kWh}$$

[cos ϕ = 1, since load supplied is non-inductive]

$$\text{Energy registered by the meter} = \frac{10}{185} = 0.054 \text{ kWh}$$

$$\therefore \quad \text{\%age error} = \frac{0.054 - 0.0546}{(0.054)} = \mathbf{0.06\%}\ (\textit{slow}). \quad \textbf{(Ans.)}$$

Example 4.105. *The name plate of a meter reads "1 kWh = 15000 revolutions". In a check up, the meter completed 150 revolutions during 45 seconds. Calculate the power in the circuit.*

Solution. Power metered in 150 revolutions

$$= 1 \times 150/1500 = 0.01 \text{ kWh}$$

If P kilowatt is the power in the circuit, then energy consumed in 45 seconds

$$= \frac{P \times 45}{3600} \text{ kWh} = 0.0125\, P \text{ kWh}$$

Equating the two amounts of energy, we have

$$0.0125\, P = 0.01$$

$$\therefore \quad P = 0.8 \text{ kW} = \mathbf{800\ W} \quad \textbf{(Ans.)}$$

Example 4.106. *A 230 V ampere-hour type meter is connected to a 230 V D.C. supply. If the meter completes 225 revolutions in 10 minutes when carrying 14 A, calculate :*

(i) The kWh registered by the meter.

(ii) The percentage error of the meter above or below the original calibration.

The timing constant of the meter is 40 A-s/revolution.

Solution. During 225 revolutions the meter would register 40 × 225 A-s or coloumbs. Since time taken is 10 minutes or 600 seconds it corresponds to a current of $\frac{40 \times 225}{600} = 15$ A.

(*i*) Energy recorded by the meter $= \frac{VIt}{1000}$ kWh where, t is in hour

$$= \frac{230 \times 15}{1000} \times \frac{10}{60} = 0.575 \text{ kWh}$$

(*ii*) Actual energy consumed $= \frac{230 \times 14}{1000} \times \frac{10}{60} = 0.5367$ kWh

$$\textbf{Percentage error} = \frac{0.575 - 0.5367}{0.5367} \times 100 = \mathbf{7.14\%}\ (\textit{fast}) \quad \textbf{(Ans.)}$$

Example 4.107. *A single phase kWh meter makes 500 revolutions per kWh. It is found on testing as making 40 revolutions in 58.1 seconds at 5 kW full load. Find out the percentage error.* **(Nagpur University)**

Solution. Number of revolutions made by the meter in one hour testing

$$= \frac{40}{58.1} \times 3600 = 2478.5 \text{ rev.}$$

These revolutions correspond to an energy of 5 × 1 kWh

$$\therefore \quad \text{Number of revolution per kWh} = \frac{2478.5}{5} = 495.7 \text{ rev.}$$

Hence, $\quad$ Percentage error $= \frac{500 - 495.7}{495.7} \times 100 = \mathbf{0.867\%}$ **(Ans.)**

Example 4.108. *An energy meter is designed to make 100 revolutions of the disc for one unit of energy. Calculate the number of revolutions made by it when connected to a load carrying 40 A at 230 V and 0.4 power factor for an hour. If it actually makes 360 revolutions, find the percentage error.* **(Allahabad University)**

Solution. Energy consumed in one hour $= 230 \times 40 \times 0.4 \times \dfrac{1}{1000} = 3.68$ kWh

Number of revolutions the meter should make when it is correct

$$= 3.68 \times 100 = 368 \text{ rev.}$$

Number of revolutions actually made = 360 rev.

$\therefore$ Percentage error $= \dfrac{360-368}{368} \times 100 = -2.17\%$ *i.e.* **2.17%** (*slow*) **(Ans.)**

Example 4.109. *A 50 A, 230 V energy meter in a full load test makes 61 revolutions in 37 seconds. If the normal speed of the disc is 520 revolutions/kWh, compute the percentage error.* **(Nagpur University)**

Solution. Assuming unity power factor :

Energy consumed in kWh in 37 seconds $= \dfrac{230 \times 50 \times 37}{3600 \times 1000}$

$$= 0.1182 \text{ kWh}$$

Number of revolutions corresponding to this energy

$$= 520 \times 0.1182 = 61.464 \text{ rev.}$$

The meter makes 61 revolutions,

$\therefore$ % *Error* $= \dfrac{61 - 61.464}{61.464} =$ **0.755%** (*slow*) **(Ans.)**

Example 4.110. *The constant of a 25 A, 220 V meter is 500 rev./kWh. During a test at full load of 4.4 kW, the disc makes 50 revolutions in 83 seconds. Calculate the meter error.* **(UPTU)**

Solution. Number of revolutions which the meter should make in one hour at full load

$$= \frac{(4.4 \times 1000) \times 1 \times 500}{1000} = 2200 \text{ rev.}$$

This corresponds to speed of $\dfrac{2200}{60} = 36.7$ r.p.m.

Correct time for 50 revolutions $= \dfrac{50}{(36.7/60)} = 81.7$ s

Hence, meter is *slow* by 83 – 81.7 = 1.3 s

$\therefore$ % *Error* $= \dfrac{1.3}{81.7} \times 100 =$ **1.59%** (*Slow*) **(Ans.)**

Example 4.111. *A 16 A ampere-hour meter with a dial marked in kWh, has an error of +2.5% when used on 250 V circuit. Find the percentage error in the registration of the meter if it is connected for an hour in series with a load taking 3.2 kW at 200 V.* **(Nagpur University)**

Solution. Number of revolutions which the meter should make in one hour at full load

$$= \frac{16 \times 250}{1000} = 4 \text{ kWh}$$

Correct reading $= 4 + \dfrac{2.5}{100} \times 4 = 4.1$ kWh

Meter constant on a 3.2 kW load at 200 V $= \frac{3.2 \times 1000}{200} = 16$ A

Since on the given load meter constant is same as the normal current of meter, hence in one hour it would give a correct reading of 4.1 kWh.

But actual load = 3.2 kWh

$\therefore$ Error = 4.1 – 3.2 = 0.9 kWh

Hence, % *Error* $= \frac{0.9}{3.2} \times 100$ = **28.12%** **(Ans.)**

Example 4.112. *The disc of an energy meter makes 600 revolutions per unit of energy. When a 1 kW load is connected the disc rotates at 10.2 r.p.m. If the load is on for 12 hours how many units are recorded as error ?* **(Allahabad University)**

Solution. Energy actually consumed = 1 (kW) × 12 (h) = 12 kWh

Total number of revolutions made by the disc during the period of 12 hours

= (10.2 × 60) × 12 = 7344 rev.

Since 600 revolutions record 1 kWh, therefore, energy recorded by the meter

$= \frac{7344}{600} = 12.24$ kWh

Hence, *extra* unit recorded as *error* = 12.24 – 12 = **0.24 kWh** **(Ans.)**

Example 4.113. A *D.C. ampere-hour meter is rated at 5A, 250 V. The declared constant is 5A- s/rev. Express the constant in revolutions per kWh. Also calculate the full load speed of the meter.*

Solution. Meter constant = 5A-s/rev.

= 5 × 250 W-s/rev. with rated voltage of 250 V

$= \frac{5 \times 250}{1000 \times 3600}$ kWh/rev. $= \frac{1}{2880}$ kWh/rev.

$\therefore$ *Meter constant in terms of revolution per kWh* = **2880** **(Ans.)**

Energy consumption in 1 min. at full load $= \frac{5 \times 250 \times (1/60)}{1000} = \frac{1}{48}$ kWh

$\therefore$ *Full load speed* = Meter constant (rev./kWh) × Energy consumption (kWh/min.)

$= 2880 \times \frac{1}{48}$ = **60 r.p.m.** **(Ans.)**

Example 4.114. *A meter whose constant is 750 revolutions per kWh makes 15 revolutions in 30 seconds. Determine the load in kW.* **(Allahabad University)**

Solution. Energy consumed in 30 seconds

$$= \frac{\text{No. of revolutions made}}{\text{Meter constant in rev./kWh}} = \frac{15}{750} = 0.02 \text{ kWh}$$

$\therefore$ *Load in kW* $= \frac{\text{Energy consumed in kWh}}{\text{Time in hours}} = \frac{0.02}{(30/3600)}$ = **2.4 kW** **(Ans.)**

Example 4.115. *A correctly adjusted single phase 240 V watt-hour meter has a meter constant of 600 revolutions per kWh. Determine the speed of the disc for a current of 8A at a lagging power factor of 0.6.* **(Punjab University)**

Solution. *Given* : V = 240 volts ; I = 8A ; cos ϕ = 0.6 ; meter constant = 600 rev./kWh.

Speed of the disc, N :

Energy supplied per minute $= \frac{VI \cos\phi}{1000} \times t$ kWh

$$= \frac{240 \times 8 \times 0.6}{1000} \times \frac{1}{60} = 0.0192 \text{ kWh/min.}$$

∴ *Speed of the disc, N* = Energy supplied/min. × meter constant (rev./kWh)

= 0.0192 × 600 = **11.52 rev./min.** **(Ans.)**

Example 4.116. *A 230 V, 50 Hz single-phase energy meter has a constant of 200 revolutions per kWh. While supplying a non-inductive load of 4.4 A at normal voltage, the meter takes 3 minutes for 10 revolutions. Calculate the percentage error of the instrument.* **(Bombay University)**

Solution. Energy consumed in 3 minutes

$$= \frac{VI\cos\phi \times t}{1000} \text{ kWh}$$

$$= \frac{230 \times 4.4 \times 1 \times (3/60)}{1000} = 0.0506 \text{ kWh}$$

[∵ Power factor (cos ϕ) for non-inductive load is unity]

$$\text{Energy consumption registered} = \frac{\text{No. of revolutions made}}{\text{Meter constant in rev./kWh}}$$

$$= \frac{10}{200} = 0.05 \text{ kWh}$$

$$\therefore \quad \%\ Error = \frac{\text{Energy registered} - \text{Energy consumed}}{\text{Energy consumed}} \times 100$$

$$= \frac{0.05 - 0.0506}{0.0506} \times 100 = \mathbf{-\ 1.186\%} \quad \textbf{(Ans.)}$$

Example 4.117. *A 230 V, single-phase domestic energy meter has a constant load of 4A passing through it for 6 hours at unity power factor. If the meter disc makes 2,208 revolutions during this period, what is the meter constant in revolutions/kWh ?*

Calculate the power factor of the load if the number of revolutions made by the meter are 1,472 when operating at 230 V and 5A for 4 hours. **(AMIE)**

Solution. Energy consumed in 6 hours

$$= \frac{VI\cos\phi \times t}{1000} = \frac{230 \times 4 \times 1 \times 6}{1000} = 5.52 \text{ kWh}$$

$$\therefore \quad \textit{Meter constant} = \frac{2208}{5.52} = \textbf{400 rev./kWh} \quad \textbf{(Ans.)}$$

In the *second case* :

$$\text{Energy consumed} = \frac{\text{Revolutions made}}{\text{Meter constant}}$$

$$= \frac{1472}{400} = 3.68 \text{ kWh}$$

$$\text{Also,} \quad \text{Energy consumed} = \frac{VI\cos\phi \times t}{1000}$$

$$\therefore \quad 3.68 = \frac{230 \times 5 \times \cos\phi \times 4}{1000}$$

$$\text{or,} \quad \cos\phi = \frac{3.68 \times 1000}{230 \times 5 \times 4} = \mathbf{0.8} \quad \textbf{(Ans.)}$$

Example 4.118. *In a test run of 30 minutes duration with a constant current of 5A, the ampere-hour meter registers 525 Wh. The supply voltage is 220 V. Calculate the error and state whether the meter is fast or slow.* **(Punjab University)**

Solution. *Given* : t = 30 minutes or 0.5 h ; I = 5 A; Consumption registered by the meter = 525 Wh or 0.525 kWh; Supply voltage, V = 220 volts.

% Error :

$$\text{Energy supplied} = \frac{VIt}{1000} = \frac{220 \times 5 \times 0.5}{1000} = 0.55 \text{ kWh}$$

$$\therefore \quad \%\ Error = \frac{\text{Energy registered} - \text{Energy consumed}}{\text{Energy consumed}} \times 100$$

$$= \frac{0.525 - 0.55}{0.55} \times 100 = -4.54\% \text{ i.e. } \mathbf{4.54\%} \text{ (slow)} \quad \text{(Ans.)}$$

Example 4.119. *The meter constant of a 230 V, 10 A wattmeter is 1800 revolutions per kWh. The meter is tested at half load and rated voltage and unity power factor. The meter is found to make 80 revolutions in 138 seconds. Determine the meter error at half load.*

(Bangalore Univeristy)

Solution. Actual energy consumed at half load during 69 seconds

$$\frac{VI \cos \phi \,.\, t}{1000} = \frac{230 \times (10/2) \times 1 \times (138/3600)}{1000} = 44.08 \times 10^{-3} \text{ kWh}$$

$$\text{Energy recorded} = \frac{\text{No. of revolutions made}}{\text{Rev./kWh}} = \frac{80}{1800} = 44.44 \times 10^{-3} \text{ kWh}$$

$$\therefore \quad \%\ Error = \frac{44.44 \times 10^{-3} - 44.08 \times 10^{-3}}{44.08 \times 10^{-3}} \times 100 = \mathbf{0.817\%} \text{ (fast)} \quad \text{(Ans.)}$$

Example 4.120. *An A.C. energy meter is tested for half hour run at a supply voltage of 230 V and a load current of 10 A at 0.8 lag power factor. The dial reading at the start was 57.35 and at the end of the test was 58.25. The meter constant is 1200 revolutions per kWh. The meter made 890 revolutions in 30 minutes run.*

Calculate : (i) Error in registration.

(ii) Error in r.p.m. of the meter. **(AMIE)**

Solution. *Given :* Supply voltage, V = 230 volts ; I = 10 A ; $\cos \phi$ = 0.8 lag p.f.

Energy consumption recorded by the meter = 58.25 – 57.35 = 0.9 kWh

(*i*) Error in registration :

Actual energy consumed in half an hour

$$= \frac{VI \cos \phi \,.\, t}{1000} = \frac{230 \times 10 \times 0.8 \times (30/60)}{1000} = 0.92 \text{ kWh}$$

$$\therefore \quad Error\ in\ registration = \text{Energy recorded} - \text{Energy actually consumed}$$

$$= 0.90 - 0.92 = -0.02 \text{ kWh} \quad \text{i.e., } \mathbf{0.02\ kWh} \text{ (less)} \quad \text{(Ans.)}$$

(*ii*) Error in r.p.m. of the metre :

Actual revolutions required to be made by the meter for an energy consumption of 0.90 kWh = 1200 × 0.90 = 1080 rev.

Revolutions made by the meter = 890

$$\therefore \quad Error\ in\ r.p.m. = \frac{890 - 1080}{30} = \mathbf{-6.33\ r.p.m.} \quad \text{(Ans.)}$$

Example 4.121. *A single-phase energy meter of the induction type is rated 230 V, 10A, 50 Hz and has a meter constant of 600 rev./kWh when correctly adjusted. If 'quadrature' adjustment is slightly disturbed so that the lag is 85°, calculate the percentage error at full load, 0.8 power factor lag.* **(Bombay University)**

Solution. *Given* : Supply voltage, $V = 230$ volts, $I = 10$ A ; $f = 50$ Hz ; Meter constant = 600 rev./kWh ; $\cos \phi = 0.8$ (lag)

Percentage error at full load :

Deflecting torque, $T_d \propto \sin \alpha$, where α is the angle between the alternating fluxes due to series and shunt magnets.

When the voltage flux lagging adjustment is disturbed so that the phase angle between the voltage flux and the voltage is *less than 90°; the error is introduced.*

$$\phi = \cos^{-1}(0.8) = 36.87°$$

$$\alpha' = 85° - 36.87 = 48.13°, \text{ instead of } \alpha = 90° - 36.87° = 53.13°$$

$$\therefore \quad \% \; Error = \frac{\sin \alpha' - \sin \alpha}{\sin \alpha} \times 100$$

$$= \frac{\sin 48.13° - \sin 53.13°}{\sin 53.13°} \times 100 \simeq \mathbf{-6.92\%} \quad \textbf{(Ans.)}$$

Example 4.122. *A 3-phase, 2 element energy meter has a constant of 0.2 revolutions of disc per kWh. The meter is being used with a P.T. of ratio 22 kV/220 V and C.T. of ratio 100/5A.*

If the line voltage is 220 V, current is 10 A, time to complete 10 revolutions is 30 seconds on unity power factor, determine the error expressed as a percentage of the correct reading. **(Gorakhpur University)**

Solution. Actual energy consumed during the test period

$$= \sqrt{3} \times \text{ratio of P.T.} \times \text{ratio of C.T.} \times V_s I_s \cos \phi \times t \times 10^{-3} \text{ kWh}$$

$$= \sqrt{3} \times \frac{22000}{220} \times \frac{100}{5} \times 220 \times 10 \times 1 \times \frac{30}{3600} \times 10^{-3} = 63.5 \text{ kWh}$$

Energy recorded by the meter during the test period

$$= \frac{\text{No. of revolutions made}}{\text{Meter constant in rev./kWh}} = \frac{10}{0.2} = 50 \text{ kWh}$$

$$\% \; Error = \frac{50 - 63.5}{62.5} \times 100 = -21.3\% \quad i.e. \; \mathbf{21.3\% \; (\mathit{low})} \quad \textbf{(Ans.)}$$

4.10.8. Maximum Demand Indicators/Meters

4.10.8.1. Introduction

Firstly, we shall discuss how to determine the cost of generating electric energy and hence the rates or tarrifs of charging for this energy. The cost consists of :

(*i*) Standing charges which are independent of the output.

(*ii*) Running or operating charges which are proportional to output.

— *The size or capacity of the generating plant and hence necessary capital investment is determined by the* ***maximum demand*** *of the system supplied by the generating plant.*

— The demand of a system is its load requirement, usually in kW or kVA, averaged over a suitable and specified interval of time of short duration. Thus the *"maximum demand" of an installation is defined as the greatest of all the demands which have occured during a given period, such as a day, a month or a year.*

— The instrument which measures the maximum demand is called maximum demand indicator or meter.

— The primary requirement of maximum demand meters is that they shall record maximum power taken by a consumer during a particular period. These meters should be so designed that any sudden momentary increase in loads such as due to short circuits on the system or due to high starting current of motors are not taken into account. Therefore, *these meters must be designed to record the average power over successive predetermined periods; these periods may be of 15 to 30 minutes duration.*

4.10.8.2. Types of maximum demand indicators

Maximum demand indicators are of the following *four* types :

1. Recording demand indicators.
2. Average demand indicators (Merz price maximum demand indicator)
3. Thermal type maximum demand indicator.
4. Digital maximum demand indicators.

Out of all these, only average demand indicators will be described here.

4.10.8.3. Average demand indicator (Merz Price maximum demand indicator)

This indicator is not in the *form of a separate instrument*, but is a fitting which can be used with any type of motor on clock meter to indicate the maximum consumption during a half-hour or other period throughout a quarter of year (or other period between consecutive readings of the instrument).

- *This indicator can be used to record either maximum current or maximum power.*

Construction and operation : Refer to Fig. 4.109.

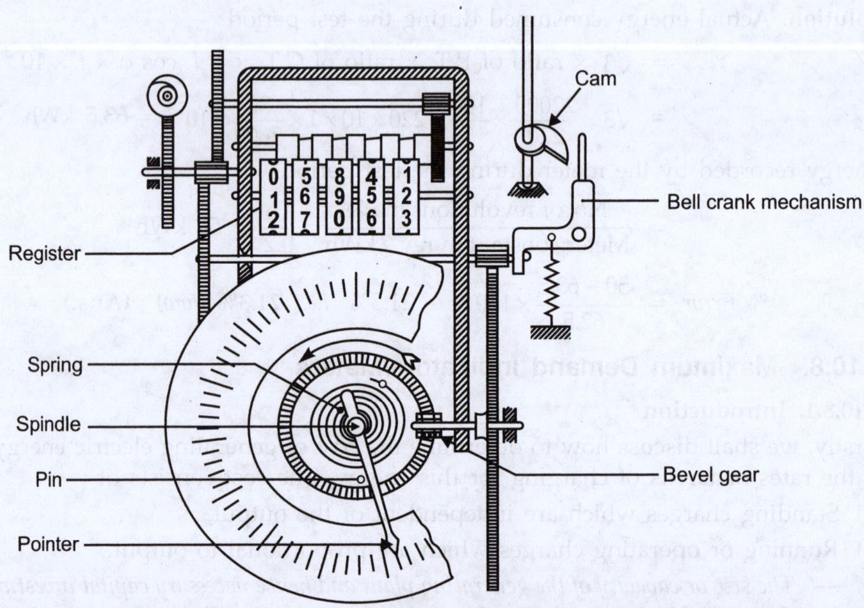

Fig. 4.109. Merz Price maximum demand indicator.

— It consists of a separate dial fitted inside the instrument, the pointer of which is driven by the spindle of the moving system of the meter through a train of gears and pin.

— Normally, the pointer is driven forward for half an hour period or other period known as integrating time period and the energy consumed during this period is indicated on the dial by the pointer. At the end of the time period, by means of a time switch and an operating mechanism the pointer is disconnected from the gearing, returned to zero and reconnected to the gearing.

— A maximum demand pointer is advanced by the driving pointer, but is *held by friction when not being driven.* This pointer will be advanced to a new position when one half-hour's energy consumption *exceeds* than that in any preceding half hour, but will not be affected during the half periods when the energy consumption is *lower* than that corresponding to its existing position.

— There are a number of variations in details of construction. For instance, the cam may be replaced by an electromagnetic relay and a clutch may be substituted for the bell crank releasing device.

The *average demand* can be calculated as follows :

$$\text{Average demand in kW} = \frac{\text{Maximum energy recorded over a time interval in kWh}}{\text{Time interval in hours}} \quad ...(4.169)$$

- *These meters can also be used to measure the maximum demand in terms of kVAh or kVARh by attaching to a suitable meter element which will measure such quantities.*

Advantages :

(*i*) Uniform scale.

(*ii*) More accurate than the thermal type.

Disadvantages :

(*i*) More expensive than thermal type, since it involves mechanisms of greater complexity.

(*ii*) The serious disadvantage is that if the maximum demand occurs sometime after the start of one time interval and continues over only a part of the next interval, the real maximum is *not* indicated since it is split up into two different time intervals (thermal lagged meters do not suffer from this disadvantage since their indication does not depend upon preselected times as their integration is continuous).

4.10.9. Measurement of VAh and VARh

If a circuit has a supply voltage V, a load current I, and cos ϕ as the p.f. then we can write :

$$(VI)^2 = (VI)^2 (\cos^2\phi + \sin^2\phi)$$

or, $$(VI)^2 = (VI \cos \phi)^2 + (VI \sin \phi)^2$$

Multiplying both sides by h^2 where h is the time in hours,

$$(VIh)^2 = (VIh \cos \phi)^2 + (VIh \sin \phi)^2 \quad ...(4.170)$$

$$(VAh)^2 = (Wh)^2 + (VARh)^2 \quad ...(4.171)$$

where, VAh = Volt-ampere hours,

Wh = Watt-hours, and

$VARh$ = volt-ampere hours *reactive*

Expressing the eqn. (4.171) in kilo-units, use have :

$$(kVAh)^2 = (kWh)^2 + (kVARh)^2 \quad ...(4.172)$$

Eqns. (4.170), (4.171), and (3.172) hold *good only for the condition that cos ϕ is maintained constant;* in practice, this condition is not usually met. Thus measurement of *VAh* or *kVAh* is not a simple matter. *Meters required for these measurements are a combination of watt-hour meters and reactive volt-ampere hour meters.*

4.10.9.1. Measurement of volt-ampere hours reactive (VARh)

Sometimes the metering of volt-ampere hours *reactive* (*VARh*) of a circuit is necessary in connection with electricity tariffs. For this, a meter is required whose motoring action is proportional to *VI* sin ϕ or *VI* cos (90° – ϕ), where ϕ is the power factor angle. The measurement of *VARh* may, therefore, be affected by employing a watt-hour meter, in which either the voltage flux or current flux is given a phase displacement of 90°. Thus an induction watt-hour meter with voltage flux in phase with the voltage and current flux in phase with the current will register *VARh*.

- In order to measure VARh in single-phase circuits, use of specially compensated watt-hour meter is made. *Phase compensation is accomplished by suitable combination of resistance, capacitance and inductance.*

Normal induction type energy meter can be used to register *VARh* with the help of following schemes, these schemes apply to 3-phase, 3-wire system :

1. Single element method.
2. Crossed phase method.
3. Auto-transformer method.

1. Single element method :

Fig. 4.110 illustrate the single element method of measuring *VARh* in a 3-phase, 3-wire system.

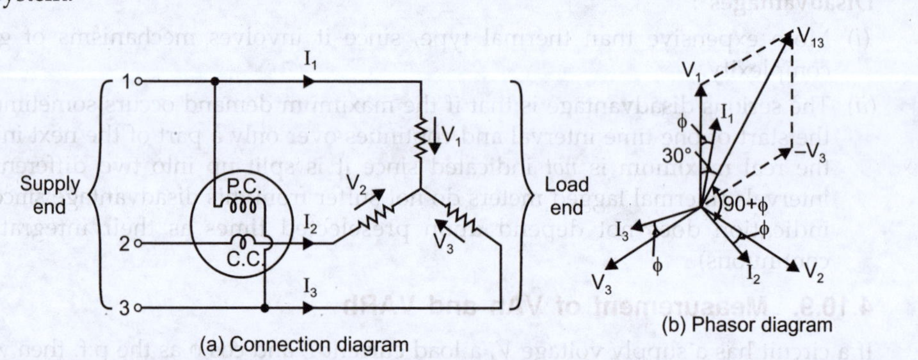

Fig. 4.110. Single element mehod of measuring VARh in a 3-phase, 3-wire system.

A normal single-phase induction type energy meter is used. The current coil is connected in series with one of the phase while the pressure coil is connected across the other two phases as shown in Fig. 4.110(*a*). This connection is same as used for measurement of reactive power.

$$\text{Total number of revolutions} = K \int V_{13} I_2 \cos(90 + \phi)^\circ \, dt$$

$$= -\sqrt{3}K \int VI \sin\phi \, dt$$

$$= -\sqrt{3}K \times (\text{VARh of each phase}) \quad ...(4.173)$$

Thus, if a system is balanced the meter registers the total *VARh*.

- The *disadvantage* of this method is that it *can only be used for balanced loads.*

2. **Crossed phase method :**

This method can be employed for *unbalanced currents in a 3-phase, 3-wire system.*

Fig. 4.111, illustrates the cross phase method of measuring *VARh* in a 3-phase, 3-wire system :

— In this method a 2-element polyphase energy meter is used. A neutral point is obtained by a *choke* as shown in Fig. 4.111(*a*).

— Each current is associated with a voltage differing in phase from the voltage normally used for power or energy measurements [See Fig. 4.111(*b*)].

— In this method, we must ensure that torque acting on the two elements (Element-1 and Element-2) are *opposite* in direction.

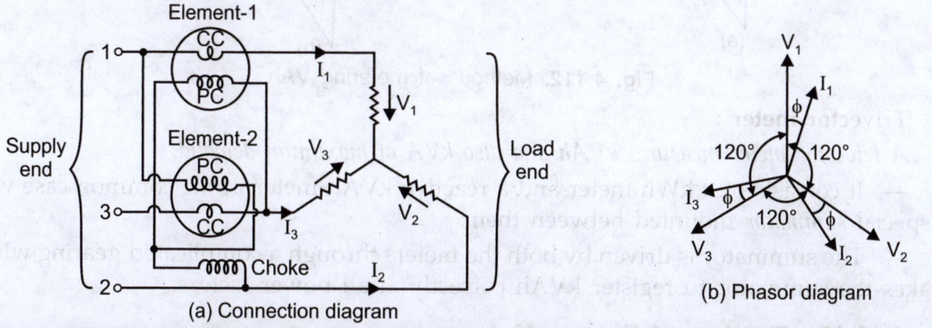

Fig. 4.111. Crossed phase method of measuring VARh in 3-phase, 3-wire system.

— Since the voltage across the pressure coil is $\frac{1}{\sqrt{3}}$ of that used for power or energy measurements, evidently the registration requires a multiplier of value $\sqrt{3}$, this is evident from the following :

Now, Total number of revolutions $= K\int[V_3 I_1 \cos(120° + \phi) - V_1 I_3 \cos(120° - \phi)]\, dt$

In use of balanced currents, total number of revolutions

$$= K\int VI\,[\cos(120° + \phi) - \cos(120° - \phi)]\, dt$$

$$= K\int(\sqrt{3}\, VI \sin\phi)\, dt \qquad ...(4.174)$$

3. **Auto-transformer method :**

This method uses a two element watt-hour meter with a phase shifting arramgement employing two open delta-connected auto-transformers.

4.10.9.2. Measurement of volt-ampere hours (VAh)

Fig. 4.112 illustrate the two alternative methods of metering *VAh*.

- In **Fig. 4.112(*a*)** meter *L* is the watt-hour meter while meter *M* registers volt-ampere hours. The meter *M* may be maximum demand indicator which gives *VA* demand.

  The reading of meter *B* is in *VAh*

- In **Fig. 4.112(*b*)** meter *L* records watt-hours (*Wh*) while meter *N* registers volt-ampere hours reactive (*VARh*).

Then, $VAh = \sqrt{(Wh)^2 + (VARh)^2}$

$$= \sqrt{(\text{Reading of meter } L)^2 + (\text{Reading of meter } N)^2} \quad \text{...(4.175)}$$

This principle of vector addition is utilised for the constructin of a meter which measures *VAh* directly.

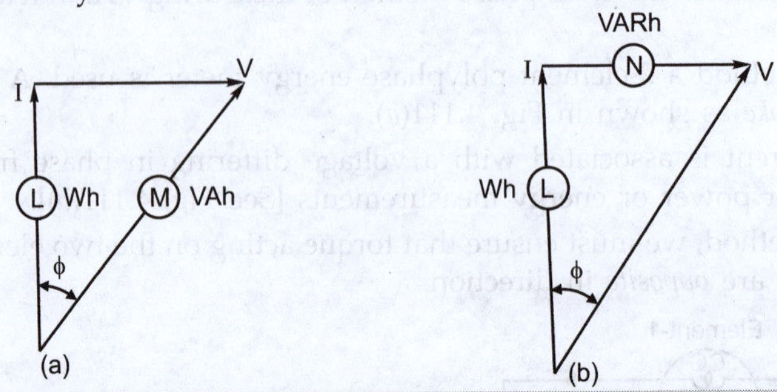

Fig. 4.112. Methods of metering VAh.

Trivector meter :

A trivector meter measures kVAh and also kVA of maximum demand.

— It consists of a kWh meter and a reactive kVAh meter in the common case with a special *summator* mounted between them.

— The summator is driven by both the meters through a complicated gearing which makes the summator to register kVAh correctly at all power factors.

4.10.10. Testing of Energy Meters

4.10.10.1. General aspects

Prior to the installation of energy meters in domestic or commercial premises, it is essential by law, to test the accuracy of the meter. The errors, under specified conditions, must lie within permissible limits.

A.C. energy meter must be tested for the following *conditions :*

(*i*) At 5% of marked current with unity p.f.

(*ii*) At 100% or 125% of marked current with unity p.f.

(*iii*) At one intermediate load with unity p.f.

(*iv*) At marked current and 0.5 lagging p.f.

Besides this, the meters should also be subjected to following *tests* :

(*i*) *Starting test.* The meter should start and run at 0.5% of marked value of current and with normal voltage (Accuracy not to be checked at this current).

(*ii*) *Creep test.* The meter should not *creep* (should not revolve through more than one revolution) with an applied voltage of 110% of its marked value and with the current circuit open.

4.10.10.2. Apparatus required for meter testing

The approved apparatus for testing of energy meters is listed below :

A. Standard apparatus :

(*i*) D.C. potentiometer, calibrated by National Physical Laboratory.

B. Substandard apparatus :

(*i*) Indicating wattmeter

(*ii*) Rotating meter

(*iii*) Ammeter

(*iv*) Voltmeter

(*v*) Electrolytic ampere-hour meter.

- *Time standard* – Ship's chronometer or pendulum-type clock.
- *Time substandard* – Stop-watch or other suitable tuning device.

4.10.10.3. Methods of testing motor meters

Three main methods of testing motor meters are described below :

Method A : "Long-period dial tests," using substandard rotating meters.

— In this method, a rotating substandard meter is used to measure the amount of energy passing through the meter being tested in a given time.

— The current circuits of two meters are connected in *series* and thus carry the same current while the voltage circuits are connected in *parallel* across the same supply.

— The meters are started and stopped *simultaneously.*

— The registrations of two meters are compared at the end of the test period (which is an hour or even more) and the error is calculated as under :

Let, K_x = Registration of test meter, in kWh

K_s = Registration of substandard meter, in kWh,

Then, $$\%\ \text{Error} = \frac{K_x - K_s}{K_s} \times 100 \qquad ...(4.176)$$

This is the *most rigorous* of the three methods of testing.

Advantages :

(*i*) Reliability only on dial indications of the test meter.

(*ii*) The complete elimination of errors due to starting or stopping of substandard stop-watches.

(*iii*) Avoidance of any necessity for holding the load absolutely constant.

Disadvantage. The main diadvantage is the length of time required for a single accurate test and thus for the complete set of tests.

- This method is *unsuitable if adjustments to the meters have to be made during testing.*

Method B : "Short period test" using substandard or rotating meters.

This test extends over a *short period* and therefore it is known as a *short period test.*

— A rotating substandard meter is used in this method also.

— The current coils of the substandard meter and the meter under test are connected in series, and their pressure coils are connected in *parallel.*

— Necessary provision is made to start and stop the two meters simultaneously.

- In this method either *"revolutions"* of the disc of meter under test are compared with those made by the disc of substandard meter during the same time or *"speeds"* of the test and the substandard meters are directly compared.

Let, R_x = Number of revolutions per kWh for the meter under test,

R_s = Number of revolutions per kWh for the substandard meter,

R_{xt} = Total number of revolutions made by the meter under test,

R_{st} = Total number of revolutions made by the substandard meter during the test interval,

N_x = Speed of disc of the meter under test,

N_s = Speed of disc of the substandard meter, and

t = Time interval in hours.

(i) *When revolutions are compared :*

Energy recorded by the meter under test during test interval $= \dfrac{R_{xt}}{R_x}$

Energy recorded by the substandard meter during test interval $= \dfrac{R_{st}}{R_s}$

$$\therefore \quad \% \text{ Error} = \frac{(R_{xt}/R_x) - (R_{st}/R_s)}{(R_{st}/R_s)} \times 100 \quad \text{...(4.177)}$$

(ii) *When speeds are compared :*

Energy recorded by the meter under test $= \dfrac{N_x . t}{R_x}$

Energy recorded by the substandard meter $= \dfrac{N_s . t}{R_s}$

$$\therefore \quad \% \text{ Error} = \frac{(N_x.t/R_x) - (N_s.t/R_s)}{(N_s t/R_s)} = \left(\frac{N_x R_s}{N_s R_x} - 1\right) \times 100 \quad \text{...(4.178)}$$

- An additional requirement of this method is one *"long period dial test" extending over an hour at full load. This is done to check the registering mechanism and also the effect of time.* The meter is then again subjected to *short time test at full load to check for any effect due to heating.*

Advantages :

(i) Less time required comparatively.

(ii) The complete elimination of error owing to starting or stopping of substandard or stop watches.

(iii) Avoidance of any necessity for holding the load absolutely constant.

Method C : *Precision test*

— In this method precision grade indicating instruments are used as reference standards.

— There indicating instruments are connected in the circuit of meter under test.

— The current and voltage are held constant during the test.

— The number of revolutions made by the meter disc during the test are recorded. The time taken is also measured.

Now, Energy recorded by the meter under test $= \dfrac{R_{xt}}{R_x}$ kWh

Energy computed from the readings of the indicating instrument = kW × t

where, R_{xt} = Total number of revolutions made by the disc of meter under test,

R_x = Number of revolutions per kWh for the meter under test, and

t = Time in hour

$$\therefore \quad \% \text{ Error} = \frac{(R_{xt}/R_x)}{kW \times t} \times 100 \qquad ...(4.179)$$

- In addition to this test, a 'long period dial test' is required to check the registration mechanism.

Caution : Before starting any of the test mentioned above on a watt-hour meter, its potential circuit must be connected to the supply for one hour in order to enable the self-heating of the potential coil to stabilise.

4.10.10.4. Phantom or Fictitious loading

When the meter under test has high current rating, a test with *actual loading arrangements would involve a considerable waste of power. This can be avoided by using phantom or fictitious loading.*

In phantom loading arrangement, the pressure circuit is supplied from a circuit of required normal voltage and the current circuit from a separate low voltage supply (The rated current can be circulated through the current circuit since the impedance of this circuit is very low). The total power supplied for the test, in this arrangement, is the sum of power supplied due to the small pressure coil current at normal voltage and that due to the circuit current supplied at low voltage. Thus, the total power required for testing the meter with phantom loading is *comparatively very small.*

- For testing of meter of high capacity while in service, the pressure circuit of the meter is supplied from the line in the normal way while the current circuit of the meter is removed from the consumer load circuit and supplied from a source of low voltage.

4.10.10.5. Testing of A.C. meters

— When A.C. meters are to be tested, it is not only necessary to vary the voltage and load current, but, in addition, the *relative phases of the two quantities must be altered;* in other words the *p.f. of the load must be variable also.*

— Variation of the p.f. may be brought about by making up the load circuit with variable resistors, inductors and capacitors, and adjusting these as required. This method is, however, inconvenient in most cases, since it involves the waste of the power in the load and necessitates several fairly expensive pieces of apparatus, these objections being particularly apparent if the capacity of the meters is high.

- Due to this reason the load employed for testing purposes is usually a *fictitious one,* as in the case of D.C. meter testing. Provision must, of course, be made for *varying the phase of the current in the current circuit relative to that of voltage circuit in addition to the variation of the magnitude of the current.*

— This phase of current is usually altered in one of the following *two* ways :

(*i*) Using two similar alternators coupled together and driven by a motor, with the *stator of one of them capable of rotation through any desired angle relative to the position of the stator of the other.*

(*ii*) Using a *phase-shifting transformer.*

- **Fig. 4.113** shows the connections for the testing of the a single-phase meter X, using two coupled alternators A_1 and A_2 for the supply to the voltage-and current-coil circuits respectively.

— The stator of A_1 is capable of rotation for phase variation.

— A_2 is connected first of all to a step-down transformer T whose secondary then supplies the current coils of the two meters *X* and *S*, the latter being either a substandard indicating wattmeter or a standard watt-hour meter, according to the method of test employed.

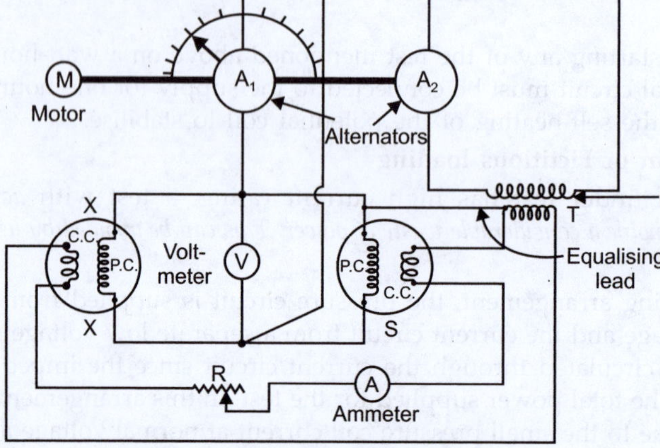

X = Meter under test; S = Standard meter;
A_1, A_2 = Coupled alternators.

Fig. 4.113. Connections for the testing of single-phase meters.

— The variable resistor *R* is for adjustment of the current in the current-coil circuit.

— *V* and *A* are a voltmeter and ammeter for the measurement of the load voltage and current.

A phase-shifting transformer may, of course, be used instead of the two alternators.

— The *equilising lead* shown is necessary *when a fictitious load is used,* in order to ensure that p.d. between the current and voltage coils of the meters is zero, as it is when they are in service.

After making adjustment of power factor the load as desired for the test, the errors of the meters under test can be determined either by timing a certain number of revolutions and measuring the load by an indicating wattmeter or by comparing their rates of revolutions with that of a standard watt-hour meter connected in the circuit so that it may measure the same load.

It may noted that the *link between the current and voltage coils in each meter must be disconnected for this test.*

- Fig. 4.114 shows the connections for the testing of 3-phase, 3-wire meters :
  - — In this method, 3-phase supply is used for the following two purposes : (*i*) To supply the stator of a phase-shifting transformer; (*ii*) To supply three step-down transformers, two of whose secondaries supply the current coils of the meter.
  - — A regulating choke coil is connected in the primary circuits of these transformers.
  - — The motor of the phase-shifting transformer supplies two auto-transformers having variable secondary tappings. These supply the voltage coils of the meters, the latter consisting of a double standard wattmeter and one or more meters to be tested. These are connected as shown.
  - — A_1, A_2 and A_3 are ammeters and V_1 and V_2 voltmeters, for adjustment of the load.

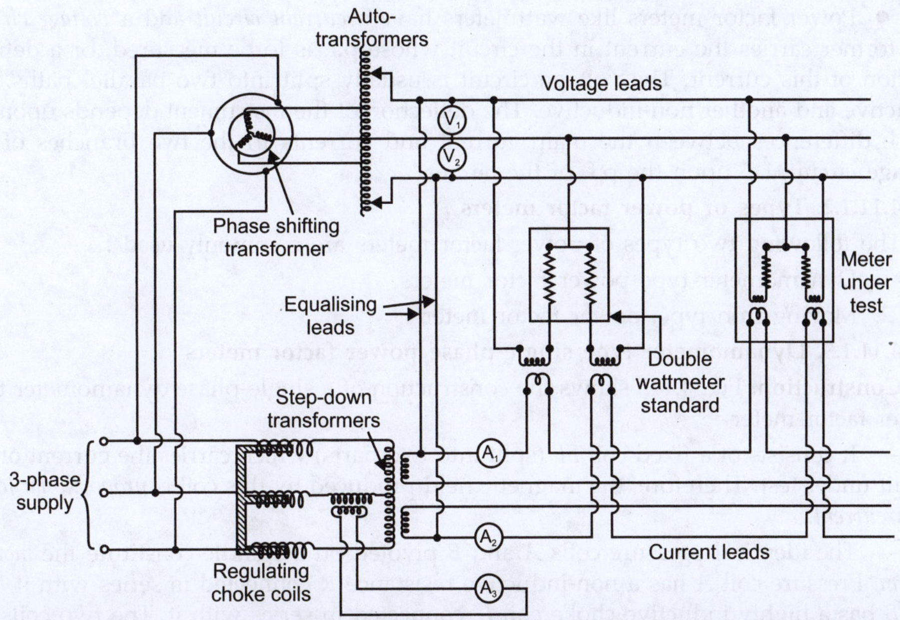

Fig. 4.114. Connections for testing of 3-phase, 3-wire energy meters.

4.10.10.6. Measurement of kVA

For charging the industrial consumer on power factor tariff, in which consumer is penalized for low p.f. of his load, *maximum demand kVA indicator is required for measurement of maximum demand in kVA.*

One type of this instrument is essentially an induction type A.C. ampere-hour meter consisting of a shaded pole, a continuously rotating disc and a braking magnet of permanent magnet type. It is used with a timing device of the Merz type and records the maximum ampere hours during successive time intervals of 15 minutes. The dial of the instrument is *graduated in terms of kVA* corresponding to the circuit rated voltage. *The readings of the instrument are independent of p.f. of the load* and frequency of supply within such variations of the latter might be expected in the normal supply system.

4.11. MISCELLANEOUS MEASURING INSTRUMENTS

Under this heading we shall discuss the following instruments :

1. Power factor meters.
2. Frequency meters.
3. Rotational speed measurements.
4. Synchroscope.
5. Phase sequence indicators.

4.11.1. Power Factor Meters

4.11.1.1. Introduction

- *Power-factor meters or phase meters* indicate the power factor of a circuit *directly*, as distinct from instruments from whose readings the power factor may be obtained by *dividing the watts supplied by the volt-amperes in the circuit.*

The accuracy obtained with the use of power factor meters is sufficient for most purposes other than high precision testing.

- Power factor meters like wattmeters have a *current circuit* and a *voltage circuit.* The former carries the current in the circuit whose p.f. is to be measured, or a definite fraction of this current. The voltage circuit is usually split into two parallel paths, one inductive and another non-inductive. The deflection of the instrument depends upon the phase differences between the main current and currents in the two branches of the voltage circuit, i.e. upon the p.f. of the circuit.

4.11.1.2. Types of power factor meters

The following two types of power-factor meters are commonly used :

1. Dynamometer type power factor meters.
2. Moving iron type, power factor meter.

4.11.1.3. Dynamometer type single-phase power factor meters

Construction. Fig. 4.115 shows the construction of a single-phase dynamometer type power factor meter :

— It consists of a fixed coil *FF* (split into two parts) which carries the current of the circuit under test. Therefore, the magnetic field produced by this coil is *proportional to the main current.*

— The identical pressure coils *A* and *B* pivoted on a spindle constitute the *moving system.* Pressure coil *A* has a non-inductive resistance *R* connected in series with it, and coil *B* has a highly inductive choke coil *L,* connected in series with it. The two coils are *connected across the voltage of the circuit. The values of R and L are so adjusted that the two coils carry the same value of current at normal frequency, i.e.,* $R = \omega L$.

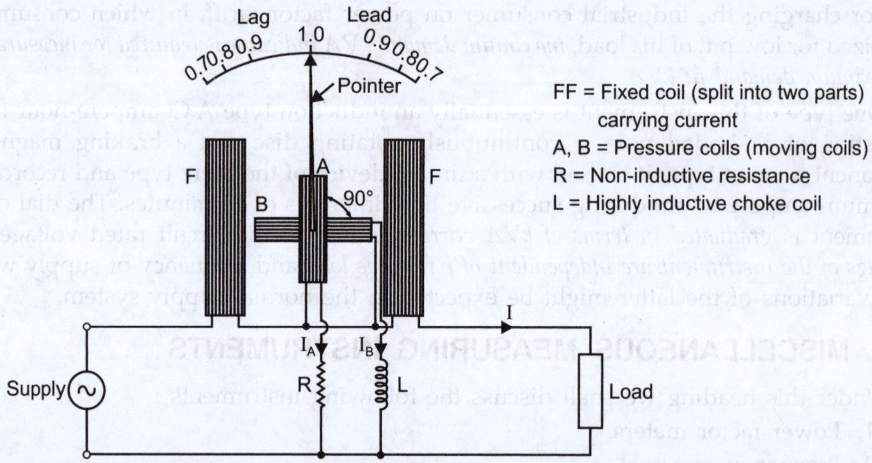

Fig. 4.115. Single-phase dynamometer type power factor meter.

— The current through the coil *A* is in phase with the circuit voltage while that through the coil *B* lags the voltage by an angle δ which is nearly equal to 90°. The angle between the planes of coils is also made equal to δ. Connections to moving coils are made through thin silver or gold ligaments which are *extremely flexible* and this gives a *minimum control effect on the moving system.*

Theory. Let us assume (to simplify the problem) that the current through the coil *B* lags the voltage by exactly 90°. Also that the angle between planes of coils is exactly 90° (*i.e.,* δ = 90°).

Fig. 4.116 illustrates the *action of the instrument.*

Let us assume that the field of the two fixed coils is uniform and in the direction arrowed. The torque on each coil for a given coil current will be maximum when the coil is parallel to the field, *i.e.*, along XX. Suppose that, when the system power factor angle is ϕ, the coils take up a position of equilibrium displaced θ from the vertical. Then the torques due to the two coils must be *equal and opposite.* Now, since the current in coil A is in phase with the system voltage, and the field in which it moves is proportional to the system current, then coil A is essentially a wattmeter movement displaced $90° - \theta$ from the maximum torque position. Then the torque of A,

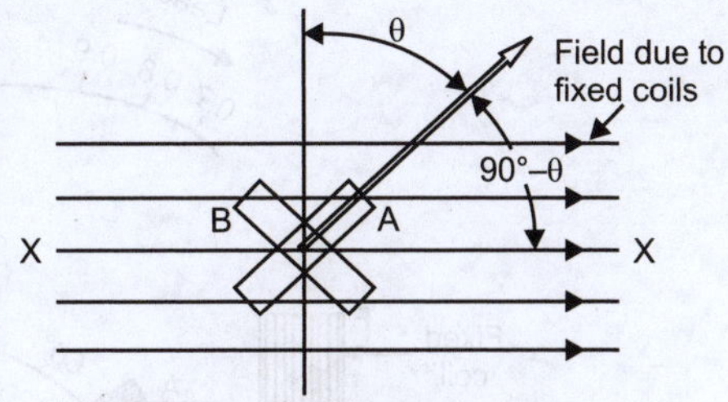

Fig. 4.116.

$$T_A = kVI \cos \phi \cos (90° - \theta) \quad ...(4.180)$$

where, k is a constant.

Similarly, since the current in coil B lags 90° on the system voltage, coil B is in a *sine-meter movement* displaced θ from the maximum torque position. Then torque on B is,

$$T_B = kVI \sin \phi \cos \theta \quad ...(4.181)$$

Hence, in the equilibrium position,

$$T_A = T_B$$

i.e., $kVI \cos \phi \cos [90° - \theta] = kVI \sin \phi \cos \theta$

or, $\cos \phi \sin \theta = \sin \phi \cos \theta$

or, $\tan \theta = \tan \phi$

or, $\theta = \phi$

Therefore, the *deflection of the instrument is a measure of phase angle of the circuit.*

- It may be noted that the instrument must be designed for, and calibrated at, the frequency of the supply on which it is to be used. In case the meter is *used for any other frequency or if supply contains harmonics it will give rise to serious errors in the indication due to change in the value of reactance of choke coil.*

4.11.1.4. Dynamometer-type power factor meter for balanced 3-phase load

Fig. 4.117, shows the connections of a 3-phase power-factor meter, the *readings of which are correct only when the load is balanced :*

— In this instrument the two moving coils are fixed with their planes 120° apart, and are connected across two different phases of the supply circuit, the fixed coils being connected in the third phase and carrying the current in the line. There is now no necessity for phase-splitting by artificial means, since the required phase displacement between the currents in the two moving coils can be obtained from the supply itself, as shown.

— *Provided the two moving coils are 120° apart, the angle through which the pointer is deflected from the unity power factor position is equal to the phase angle of the circuit.*

- The 3-phase instrument gives indications which are *independent of frequency and waveform,* since the currents in the two moving coils are both affected in the *same way* by any change of frequency.

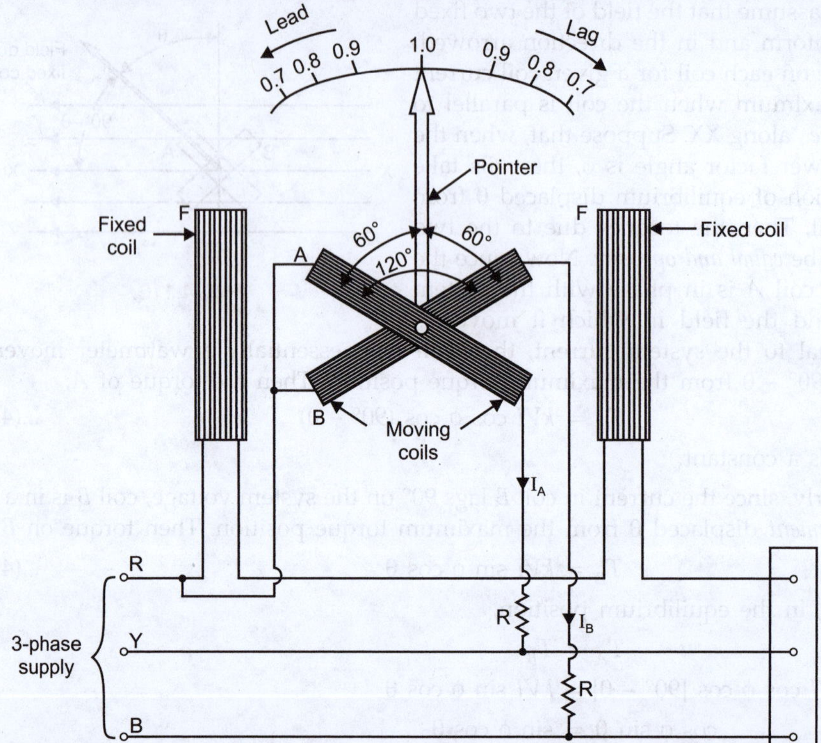

Fig. 4.117. Dynamometer-type power factor meter for balanced 3-phase load.

Advantage : The dynamometer type instrument is *more accurate* than moving iron type.

Disadvantages :

(*i*) Limited scale arc.

(*ii*) The connecting ligaments or threads to the moving coils are required, which subject the moving system of the instrument to a minimum of restraint, this is difficult to accomplish to a satisfactory degree due to low working torque obtainable with this instrument.

4.11.1.5. Moving-iron power factor meters

There are two types of moving iron power factor meters :

1. Rotating field type.
2. Alternating field type.

1. Rotating field type for 3-phase loads :

Fig. 4.118 illustrates the principle of the Westing house instrument of this type :

— A rotating field is set up by the three coils A_1, A_2 and A_3, which are supplied from the 3-phase mains through current transformers.

— The coil *B* is placed at the centre of this system of coils, and is connected in series with a resistance, across two lines of the supply. Inside coil *B* is a short pivoted iron rod, with sector-shaped iron pieces S_1 and S_2 at its ends. Damping vanes and a pointer are also carried on the same spindle. The moving system is shown in Fig. 4.118(*b*).

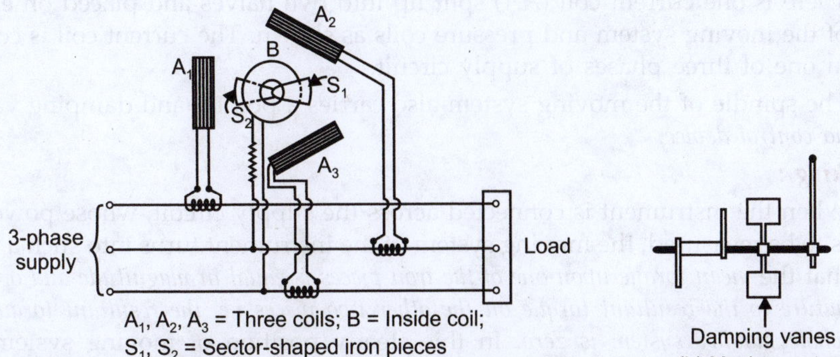

(a) Connection diagram (b) Moving system

Fig. 4.118. Rotating field type moving iron power factor meter for a balanced 3-phase load.

There are no controlling forces.

— The moving-iron system carries an alternating flux. It does not tend to rotate continuously, but sets in a definite position which depends upon the relative phases of the current in coil B, which is practically in phase with the voltage, and of the currents in coil A.

— The deflection of the moving system is approximately equal to the angle of phase displacement between current and voltage in the 3-phase circuit.

— Owing to the presence of iron in the instrument and the dependence of the reactance of coil B upon frequency it becomes necessary to calibrate the instrument at the normal frequency.

2. Alternating field type :

Fig. 4.119 shows an alternating field type (Lipman) moving iron power factor meter :

Construction :

— It consists of three sets of moving irons (B_1, B_2, B_3) mounted on a single spindle one above the other with non-magnetic separating pieces (*S*) in between them and spaced at 120° in different parallel planes.

— Each iron piece is magnetised by a separate coil known as *pressure coil*. The three pressure coils (C_1, C_2, C_3) are star-connected to the 3-phases of supply.

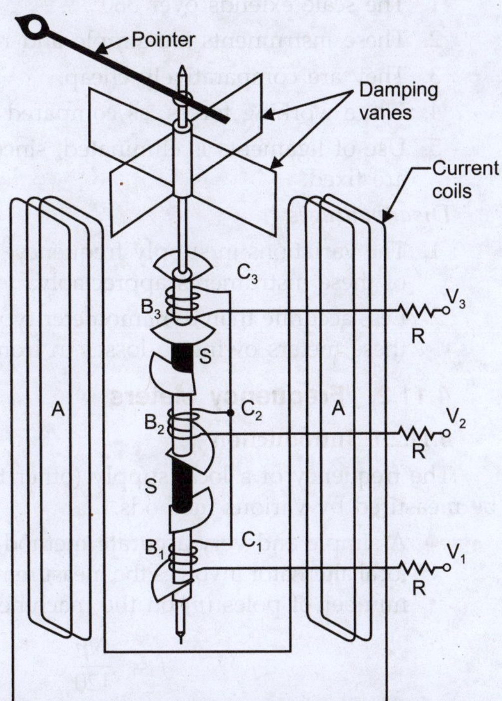

Fig. 4.119. Alternating field type (Lipman) moving iron power factor meter.

— There is one current coil (*AA*) split up into two halves and placed on each side of the moving system and pressure coils as shown. The current coil is connected in one of three phases of supply circuit.

— The spindle of the moving system also carries a pointer and damping vanes but *no control device.*

Working :

— When the instrument is connected across the supply circuit, whose power factor is to be measured, the moving system of the instrument turns into such a position that the *mean torque upon one of the iron pieces is equal in magnitude and opposite in nature to the resultant torque on the other two pieces i.e. the resultant torque on the whole moving system is zero.* In this steady position of moving system of the instrument, the deflection of that iron piece, which is magnetised by the same phase as the current coil, is *equal* to the phase angle between the voltages and currents of 3-phase supply circuit provided that the effects of iron losses and of inductance of pressure coils are negligible.

— As the three pressure coils are at different levels, no resultant rotating magnetic field is set up by them hence any tendency for the moving system to rotate continuously following the *"drag"* of such a field is *eliminated.*

- The arranement shown in Fig. 4.119 is *suitable for 3-phase balance system but it can be suitably modified for use on unbalanced 3-phase circuits and for two phases and single-phase circuits.*

Advantages :

1. The scale extends over 360°.
2. These instruments are simple and robust in construction.
3. They are comparatively cheap.
4. Large working forces, as compared with those in dynamometer type.
5. Use of ligaments is eliminated, since all the coils in a moving iron instruments are fixed.

Disadvantages :

1. The variations in supply frequency, voltage and waveform affect the calibration of these instruments appreciably.
2. Less accurate than dynamometer type instruments, since errors are introduced in these meters owing to losses in iron parts.

4.11.2. Frequency Meters

4.11.2.1. Introduction

The frequency of a local supply (other than that of the controlled grid system) can be measured by various methods.

- A simple and very accurate method of determining frequency of supply from a local alternator involves the measurement of speed (*N*) of the machine and knowing number of poles (*p*) on the machine, the frequency (*f*) supply is given by :

$$f = \frac{Np}{120} \qquad ...(4.182)$$

- Certain A.C. circuits of bridge type are also used for the measurement of frequency over a wide range but most of them are used when the *source of alternating supply is not accessible or where the frequency is in the audible range.*

- Commercial frequency (*i.e.* frequency of the order of 50–100 Hz) is measured by portable commercial instruments, known as *frequency meters* or *indicators*. Although the accuracy of such instruments is less than that of above methods but sufficient for normal uses.

4.11.2.2. Types of frequency meters

The different types of frequency meters are :

1. Mechanical resonance type.
2. Electrical resonance type
3. Weston type.
4. Ratiometer type.
5. Saturable core type.
6. Digital frequency meters.

- Measurement of audio and radio frequencies will also be discussed.

4.11.2.3. Mechanical resonance type (Vibrating reed type) frequency meter

This meter works on the *principle of the mechanical resonance* of thin flat steel reeds arranged alongside and close to an electromagnet. This meter is also known as *vibrating reed frequency meter.*

Construction :

— It consists of a number of thin steel strips called *reeds*. These reeds are placed in a row alongside and close to an electromagnet as shown in Fig. 4.120(*a*).

— The electromagnet has a laminated iron core and its coil is connected in series with a resistance, across the supply whose frequency is to be measured.

— Each reed is approximately about 4 mm wide and 0.5 mm thick. All the reeds are not exactly similar to each other; they *have either slightly different dimensions* or *carry different weights or flags at their tops.*

The natural frequency of vibration of the reeds depends upon their weights and dimensions.

— The reeds are so designed and arranged that the natural frequency of one reed differs by one or half a cycle from that of reed next to it. So if a frequency meter has a range of 47–53 Hz then natural frequency of first reed will be 47 Hz, 47.5 Hz for second reed, 48 Hz for the third reed and so on.

The reeds are fixed at the bottom end and are free at the top end. Since the reeds on a frequency meter are arranged to be viewed end on, they have a portion bent over at free end to *serve as a flag* as shown in Fig. 4.120(*b*). The flags are painted *white* to provide *maximum contrast* against their *black* background.

Working :

— When the instrument is connected across the supply whose frequency is to be measured, an alternating flux is set up. Due to this alternating field an attractive force is experienced upon the reeds after every half cycle. Consequently the reeds tend to vibrate but only the reed whose actual frequency is double of supply frequency will be in resonance and vibrate most (with maximum amplitude). Normally the vibration of other reeds is so slight as to be unobservable.

— If two adjacent reeds vibrate with equal amplitude then the supply frequency will be half way between the frequencies of the two adjacent reeds. The usual range of frequency meters of this type is about 6 Hz (say from 47 Hz to 53 Hz).

Fig. 4.120(*c*) illustrates the *vibration of reeds at different frequencies.*

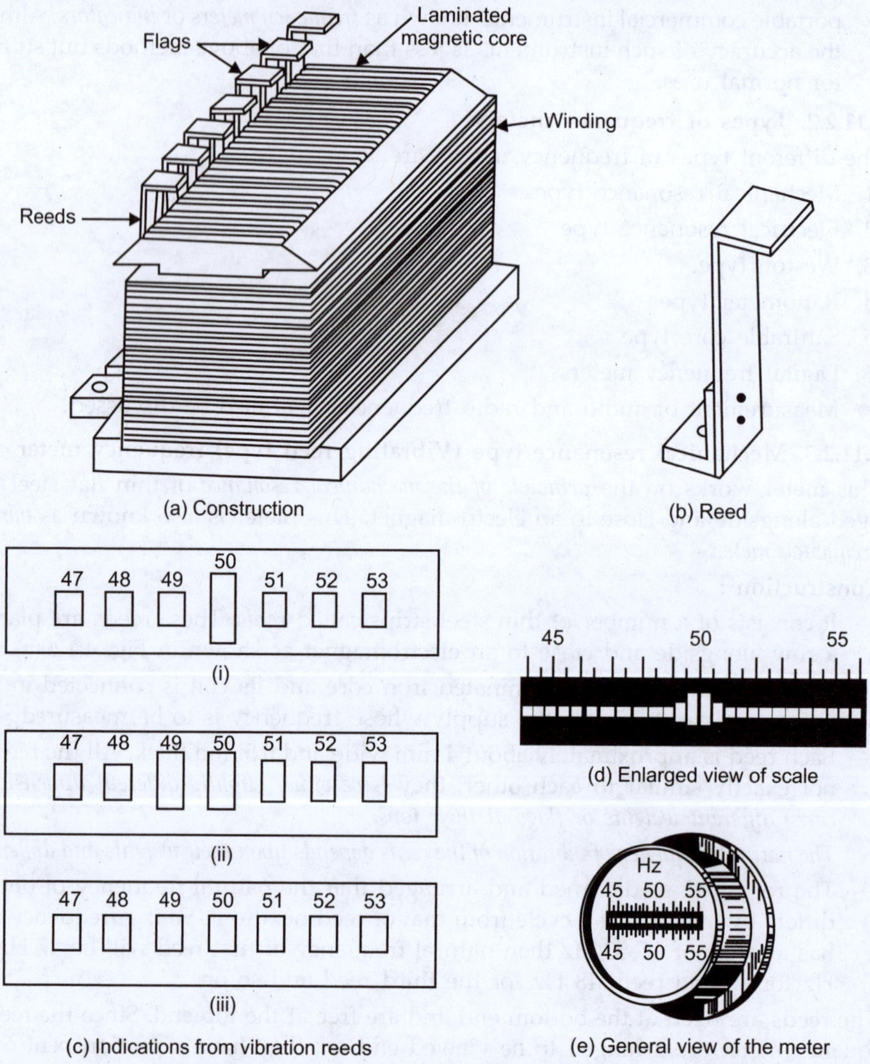

Fig. 4.120. Vibrating reed type frequency meter.

Fig. 4.120(c)–(i) illustrates that when supply frequency is 50 Hz, the reed having natural frequency of 50 Hz vibrates with maximum amplitude.

Fig. 4.120(c)–(ii) illustrates that when frequency is 49.5 Hz then reeds having natural frequencies of 49 Hz and 50 Hz will vibrate with equal and maximum amplitude.

Fig. 4.120(c)–(iii) illustrates that neither of the reeds has considerable amplitude, which indicates that either magnetic coil is *unexcited* or frequency under measurement is *out of range.*

- *The frequency range of a given set of reeds may be doubled by polarising the magnets.* The polarisation may be accomplished by using a D.C. winding in addition to the A.C. winding or by using a permanent magnet.

Advantage. *The indications of the meter are independent of the waveform of the applied voltage and of the magnitude of the voltage,* provided that the voltage should be high enough to provide sufficient amplitude for the reed vibration so as to make its reading reliable.

Disadvantage. These instruments cannot read much closer than half the frequency difference between adjacent reeds. Thus they *cannot be used for precision measurements.*

4.11.2.4. Electrical resonance type frequency meter

Fig. 4.121 shows a very simple form of indicating frequency meter, whose action is dependent upon *electrical resonance.*

Construction :

— It consists of a *laminated iron core of varying cross-section, a fixed magnetising coil* mounted at one end of the iron core and a *moving coil* with a pointer attached to it and pivoted so that it can move freely over the iron core as shown.

— The moving coil is connected across a capacitor C and the fixed coil is connected across the supply circuit, whose frequency is to be measured.

Working :

— When the magnetising coil is connected across the supply circuit, current *I* flows through it and a flux ϕ in phase with current is set up in the iron core and an e.m.f. *e* lagging behind the flux ϕ by 90° is induced in the moving coil.

The current in the moving coil be lagging or leading the induced e.m.f. *e*, depending upon whether the circuit of the moving coil is largely inductive or capacitive, as shown in Fig. 4.121(*b–i*) and (*b–ii*) respectively. The current in the moving coil will be *in phase* with *e* [Fig. 4.121 (*b–iii*)] when inductance reactance (X_L) is *equal* to capacitive reactance (X_C).

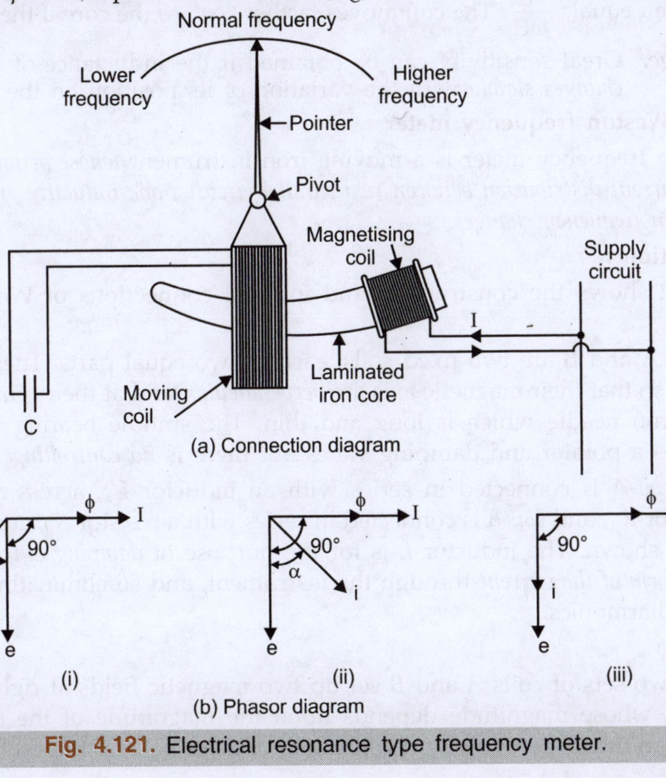

Fig. 4.121. Electrical resonance type frequency meter.

— The torque on the moving system (which has *no controlling device*) is given by :

$$T \propto Ii \cos (90° \pm \alpha) \quad ...(4.183)$$

where, I = The magnetising coil current,
i = The moving coil current, and
α = The phase angle between current and e.m.f. induced in the moving coil.

From the eqn. (4.183) it is evident that the torque on the moving system will be *zero* when X_L (or ωL)= $X_C \left(\text{or } \frac{1}{\omega C}\right)$

- Referring now to the actual operation of the instrument, the capacitive reactance $\frac{1}{\omega C}\left(\text{or } \frac{1}{2\pi fC}\right)$ is *constant for any given frequency.* The inductive reactance ωL ($2\pi fL$) depends upon the *position of the pivoted coil on the core. The nearer this coil approaches the magnetising coil the greater is its inductance.* The pivoted coil is pulled towards the magnetising coil, therefore until $\omega L = \frac{1}{\omega C}$ when, as already seen, the torque is *zero.* Obviously the *circuit of the moving coil is in resonance when the torque is zero.*

The value of the capacitor C is so chosen that the pivoted coil takes up a convenient mean position when the frequency is at its normal value. If the frequency increases $\frac{1}{\omega C}$ decreases, and the coil moves *"off"* the core, so that its inductance is reduced to the point where ωL again equals $\frac{1}{\omega C}$. The coil moves farther *"on"* to the core if the frequency falls.

Advantage. Great sensitivity can be obtained if the inductance of the moving coil *changes slowly* with the variation of its position on the core.

4.11.2.5. Weston frequency meter

A Weston frequency meter is a moving iron instrument *whose action depends on the variation in current distribution between two parallel circuits, one inductive and the other non-inductive, when frequency changes.*

Construction :

Fig. 4.122 shows the construction and internal connections of Weston frequency meter :

— Coils A and B are two fixed coils, each in two equal parts. These two coils are fixed so that their magnetic axis are *perpendicular*, and at their centres is a pivoted soft-iron needle which is long and thin. The spindle bearing the needle also carries a pointer and damping vanes but there is *no controlling device.*

— The *coil-A* is connected in series with an inductor L_A, across a non-inductive resistor R_A, and *coil B* is connected in series with a resistor R_B, across an inductor L_B as shown. The inductor L is for the purpose of *damping out harmonics in the waveform of the current* through the instrument, and so eliminating errors due to such harmonics.

Working :

— The two sets of coils A and B set up two magnetic fields at right angle to each other, whose magnitude depends upon the magnitude of the current flowing through the respective sets of coils. Both of these fields act upon the pivoted iron needle, so the position taken by the soft-iron needle depends upon the magnitude of two fields or of currents in coils *A* and *B*. The values of various resistances and inductances are so chosen that *for normal frequency of supply,* the voltage drop across non-inductive resistor R_A and inductance L_B sends equal currents through the two pairs of coils *A* and *B* so that the needle takes such a position that the *pointer is at the centre* as shown in the figure.

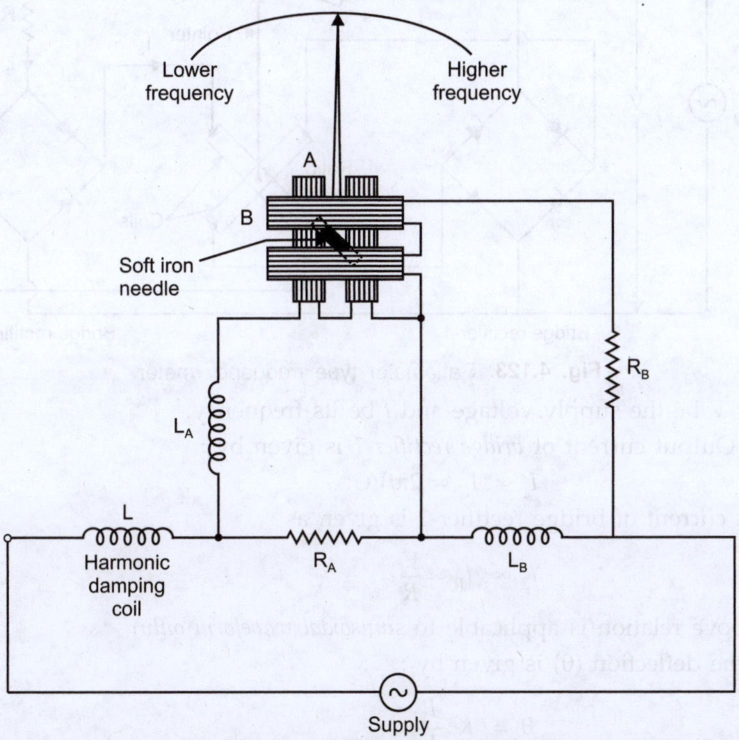

Fig. 4.122. Weston frequency meter.

— When the supply frequency *increases* above the normal one, the voltage across R_A remains more or less same and reactance of coil L_A increases due to increase in frequency, so the current through coil-A *decreases.* Simultaneously the current through the coil-B *increases* on account of increase of reactance of inductive coil L_B resulting in increase of voltage drop across the coil-B. Thus the magnitude of field of coil-A decreases and that of coil-B increases and the needle sets itself more in line with axis of coil-B; the pointer, therefore, swings over to the right indicating increased frequency.

Decrease of frequency obviously has the opposite effect, and these variations of frequency are followed by the pointer as shown.

4.11.2.6. Ratiometer type frequency meter

Fig. 4.123 shows the construction and connection diagram for a ratiometer type frequency meter :

— This type of frequency meter consists of a *ratiometer which gives a clear relationship between the current ratio and the deflection.*

— The two coils of this ratiometer are fed with rectified output currents of two separate bridge rectifiers as shown.

— The input sides of the two bridge rectifiers are connected to A.C. supply whose frequency is to be measured. Input sides of one of the bridges rectifiers has a series capacitance C and the other has series resistance *R*.

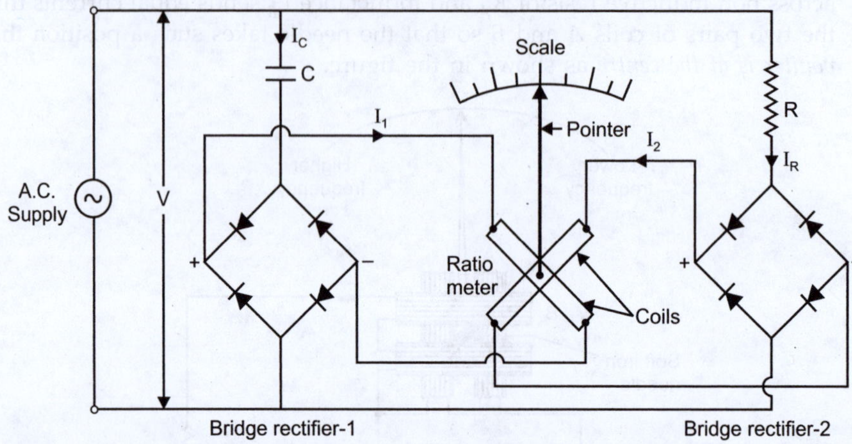

Fig. 4.123. Ratiometer type frequency meter.

- Let *V* be the supply voltage and *f* be its frequency.

Then, Output current of *bridge rectifier-1* is given by :

$$I_1 \propto I_C \propto 2\pi fVC \quad ...(4.184)$$

Output current of bridge rectifier-2 is given as :

$$I_2 \propto I_R \propto \frac{V}{R} \quad ...(4.185)$$

(The above relation is applicable to *sinusoidal waveform only*)

Then, the deflection (θ) is given by :

$$\theta = K.\frac{I_1}{I_2}$$

$$= K_1 \times \frac{2\pi fVC}{(V/R)} = 2\pi K_1 CRf$$

Since K_1, C and *R* are constants.

$$\therefore \quad \theta = K_2 f \quad ...(4.186)$$

where K_2 = a constant $2\pi K_1 CR$

— Thus, the instrument has a *linear scale of frequency* provided the ratiometer is so designed that the *deflection is directly proportional to the ratio of two currents.*

— Since the supply voltage *V* does not appear in the expression for deflection, the instrument *may be used for a fairly wide range of voltage below the maximum specified*. However, the voltage *should not be too low* otherwise distortions are introduced which make the meter read wrongly.

- This meter may be employed *upto a frequency of 5 kHz.*

4.11.2.7. Saturable core frequency meter

Fig. 4.124 shows the construction and connection diagram of a saturable core frequency meter.

Construction :

— This instrument has a saturable core transformer T as its primary detector.

— The core assembly consists of the following parts :

(*i*) An outer core made of magnetic material;

(*ii*) Leg A made of non-magnetic material;

(*iii*) Leg B made of magnetic material, it *saturates at low values of m.m.f.*

The cross-section of leg A being same as that of B.

— The primary winding P is wound around legs A and B. The secondary winding S consists of two coils, one around leg A and the other around B as shown in the figure. The two coils are connected in *series. The e.m.fs. induced in these coils oppose each other.*

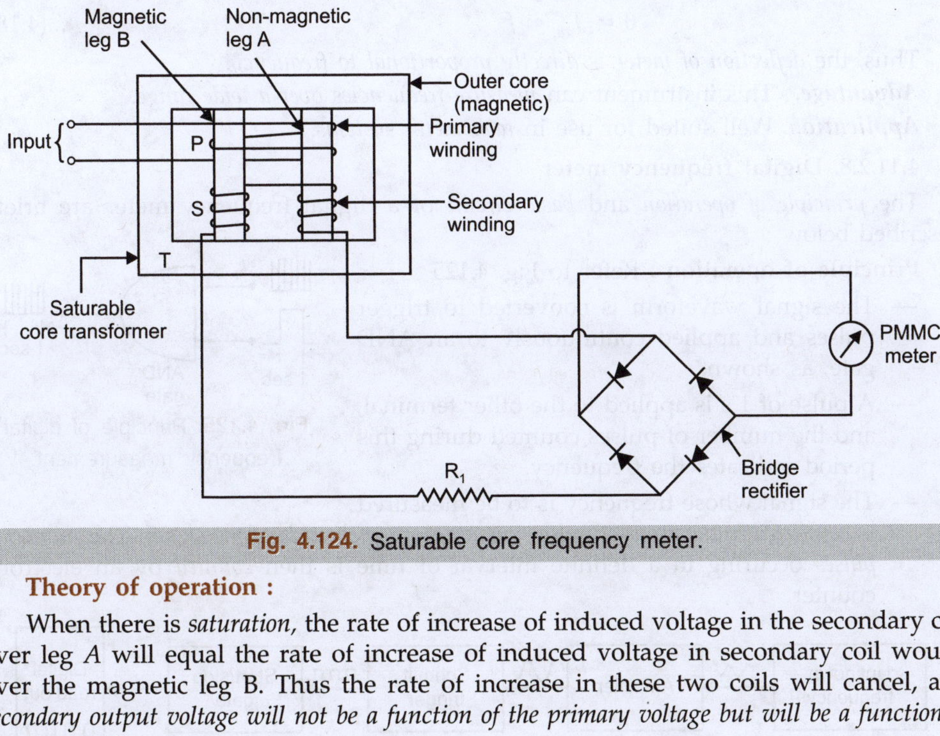

Fig. 4.124. Saturable core frequency meter.

Theory of operation :

When there is *saturation*, the rate of increase of induced voltage in the secondary coil over leg A will equal the rate of increase of induced voltage in secondary coil wound over the magnetic leg B. Thus the rate of increase in these two coils will cancel, and *secondary output voltage will not be a function of the primary voltage but will be a function of frequency only.*

If ϕ is the difference between the flux of leg A and leg B, then the induced e.m.f. in the secondary winding is given by :

$$e = N \frac{d\phi}{dt}$$

Now, the average secondary voltage per half cycle,

$$E_{av.} = \frac{\omega}{\pi} \int_0^{\pi/\omega} e.dt$$

$$= \frac{\omega}{\pi} N \int_{-\phi_m}^{+\phi_m} d\phi$$

or, $$E_{av.} = 4Nf\phi_{max}$$

$\therefore$ $$f = \frac{E_{av.}}{4N\phi_{max}} = kE_{av.} \quad ...(4.187)$$

where k is a constant after the magnetic leg B becomes saturated.

The output of this circuit is rectified and the current in the D.C. meter is :

$$I_{av.} = \frac{E_{av.}}{R}$$

where R includes the resistance of the bridge rectifier, the meter and the resistance R_1.

$\therefore$ $$f = kRI_{av.} \quad ...(4.188)$$

The deflection θ is directly proportional to current $I_{av.}$ as current is detected by PMMC meter, or

$$\theta \propto I_{av.} \propto f \quad ...(4.189)$$

Thus, the *deflection of meter is directly proportional to frequency.*

Advantage. This instrument can *measure frequencies over a wide range.*

Application. Well suited for use in *tachometer systems.*

4.11.2.8. Digital frequency meter

The *principle of operation* and *basic circuit* of a digital frequency meter are briefly described below :

Principle of operation : Refer to Fig. 4.125

— The signal waveform is converted to trigger pulses and applied continuously to an AND gate, as shown.

— A pulse of 1 s is applied to the other terminal, and the number of pulses counted during this period indicates the frequency.

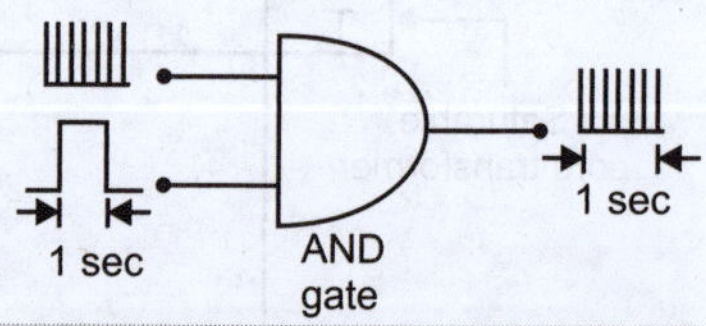

Fig. 4.125. Principle of digital frequency measurement.

— The signal whose frequency is to be measured is *converted into a train of pulses one pulse for each cycle of the signal.* The *number of pulses* occuring in a definite interval of time is then *counted* by an electronic counter.

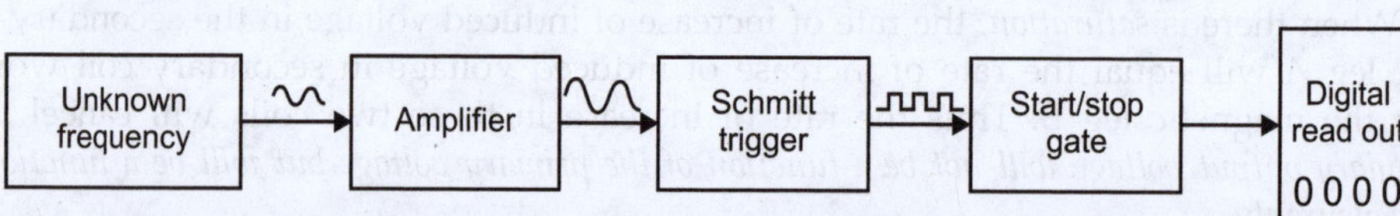

Fig. 4.126. Basic circuit of a digital frequency meter.

Since each pulse represents the cycle of the unknown signal the number of counts is a direct indication of the frequency of the unknown signal.

Since electronic counters have a high speed of operation, high frequency signals can be measured.

Basic circuit :

Fig. 4.126 shows the block diagram of a basic circuit of a digital frequency meter :

— The signal may be amplified before being applied to Schmitt trigger.

— The Schmitt trigger converts the input signal into square wave with fast rise and fall times, which is then differentiated and clipped. As a result, the output from a Schmitt trigger is a train of pulses, one pulse for each cycle of the signal.

— The output pulses from the Schmitt trigger are fed to 'Start'/'Stop' gate. When this gate is *enabled,* the input pulses pass through this gate and are fed directly to the electronic counter, which counts the number of pulses. When this gate is *disabled,* the counter stops counting the incoming pulses. The counter displays the number of pulses that have passed through it in the time interval between start and stop. If this interval is known, the unknown frequency can be measured.

4.11.2.9. Measurement of Audio and Radio Frequencies

The audio and radio frequencies (AF and RF) can be measured by using the following techniques/methods :

1. Bridge technique.
2. Heterodyne technique.
3. Wavemeters.
4. Interpolation technique.

Some of these methods/techniques are described below :

1. Bridge technique :

An unknown frequency can be determined by applying it to bridge in which the balance dependent upon frequency and then calculating the frequency from the circuit constants is required to give balance. Such arrangements find considerable use for measuring AF.

Any type of bridge or null network in which balance depends upon frequency can be used. Typical examples are : *Wien, Hay* and *resonance bridges* and the *null network.*

- Fig. 4.127 shows *"null network"* for measuring AF; the frequency is calculated using the relation :

$$\omega = \frac{1}{R_2}\sqrt{\frac{2}{C_1C_2}} \qquad ...(4.190)$$

- The *"Wien bridge"* is particularly satisfactory for measuring AF because it can be brought into balance merely by varying resistive elements containing no inductance, and can be conveniently proportioned to cover a wide frequency range (for details refer to Art. 6.3.11). An accuracy in the order of 0.5 to 1.0 % can be readily obtained in frequency bridges. In the frequency range 300 to 500 Hz, balance can be mostly easily made by the use of telephone receivers. At frequencies outside this range some indicating arrangements, such as an amplifier and VTVM combination, is normally required.

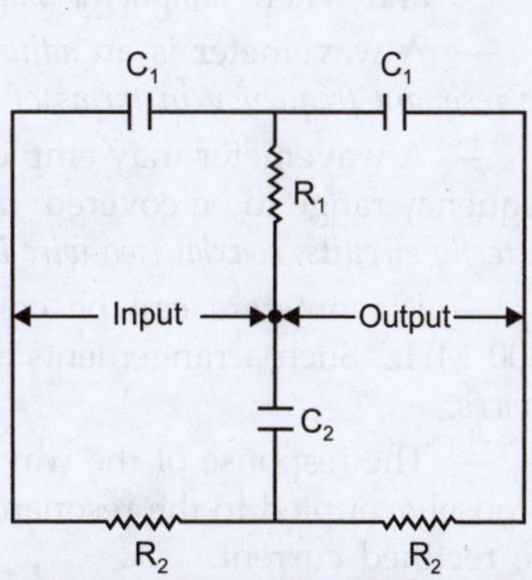

Fig. 4.127. Null network for measuring AF.

2. Heterodyne technique :

- The basis of heterodyne frequency meter is a *stable and tunable oscillator.*
- Heterodyne frequency meters are commonly provided with a *detector, audio amplifier,* and *headphones* as shown in Fig. 4.128 :

— Energy from the *oscillator* is supplied to the *detector* by internal coupling, and in addition provision is made whereby energy from an external source may be superimposed upon the detector input. The frequency of the heterodyne meter can then be made the same as the unknown frequency by adjusting so that the difference (or beat) frequency of the two oscillators as observed on the headphones is reduced to zero.

— These frequency meters are sometimes provided with a *crystal oscillator calibrator*. This is simple but well-designed crystal oscillator utilizng a *low temperature coefficient crystal to ensure good frequency stability*. The crystal oscillator produces a known frequency which can be used to check one or more points on the frequency calibration of the heterodyne meter with the aid of *detector audio amplifier*.

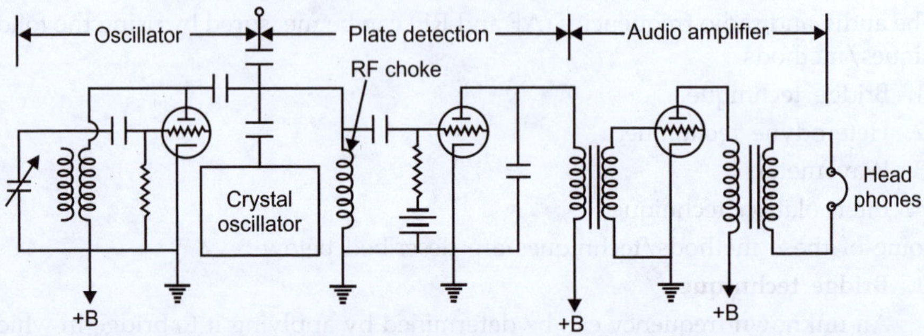

Fig. 4.128. Heterodyne frequency meter.

Uses : Heterodyne frequency meters are employed for the following *purposes* :

1. To produce known frequencies.
2. To measure an unknown frequency by comparison with the controllable and known frequency of heterodyne frequency meter.
3. To interpolate between fixed frequencies.

3. Wavemeters :

Wavemeters are used to measure frequency when high *accuracy is not required* and where *simplicity and portability are important*.

— A wavemeter is an *adjustable resonant circuit provided with a calibration that gives the resonant frequency in terms of setting of the tuning adjustments.*

— A wavemeter may employ any type of resonant circuit that is convenient for the frequency range to be covered; thus resonant circuits based on *lumped constants including butterfly circuits, coaxial two-wire lines, and cavities,* all find use in wavemeter applications.

— Wavemeters can be constructed with lumped circuit elements up to about 1200 MHz. Such arrangements are normally employed either as *absorption or as reaction devices*.

— The response of the wavemeter is commonly indicated by a crystal detector that is loosely coupled to the resonant circuit and associated with a micrometer that indicates the rectified current.

— Arrangements employing a fixed inductance and variable capacitance are suitable for use from the lowest radio frequencies up to about 100 MHz; this entire range can be covered by single variable condenser provided with a suitable set of plug-in coils as shown in Fig. 4.129. In this the coupling between the crystal and resonant circuit is made small by making capacitance C_1 small and C_2 large.

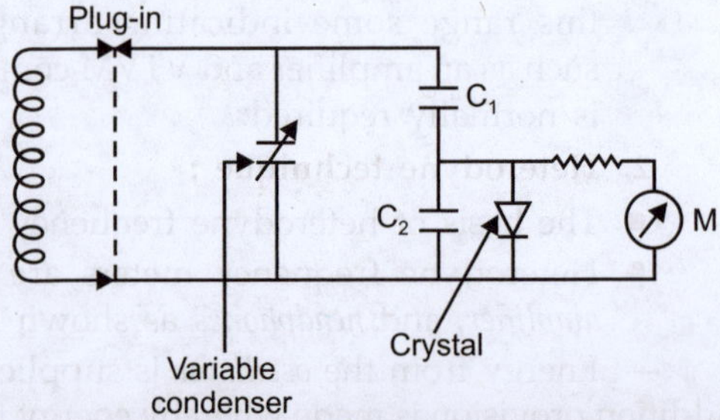

Fig. 4.129. Tuned-circuit type wavemeter.

— A variation of the inductance-capacitance wavemeter that is useful at *very high frequencies consists in varying both the coil and the condenser simultaneously*. This increases the frequency range that can be covered with a single coil and condenser combination, and is particularly useful at frequencies of the order of 50 to 400 MHz.

- The *accuracy* obtainable with such type of wavemeters depends upon the circuit Q and upon details of construction. In general, accuracies of the order of 1% or, better are readily obtained but it is quite difficult to make the accuracy as great as 0.1 %.

Cavity wavemeters :

- The *resonant cavities are extensively used as wavemeter at* ***microwave frequencies****. They have high accuracy, mechanical simplicity, and a large physical size in proportion to the wavelength being measured.*
- Practical cavity wavemeters are always in the form of cylinders which are tunable by means of a piston that varies the cavity length, as shown in Fig. 4.130.

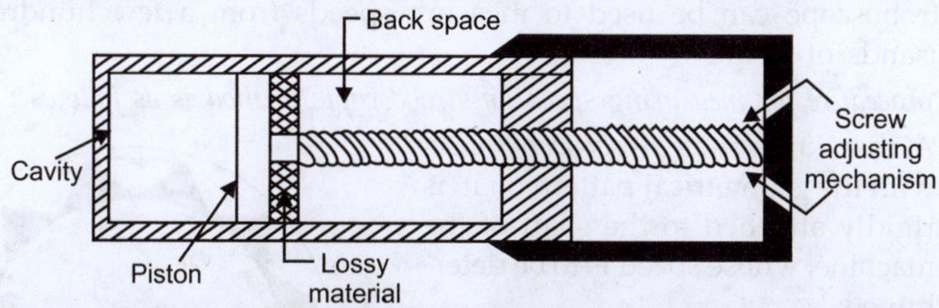

Fig. 4.130. Simple cylindrical cavity.

- Coupling to a wavemeter is accomplished by a loop or an orifice. A loop is used when co-axial systems are used to excite a cavity or carry power from it, while orifice is normally used in a waveguides system.
- The accuracy with which frequency can be determined using a well made cavity wavemeter is *quite high* (Precision of the order of 1 part in 1000 is easy to obtain and 1 part in 100,000 can be obtained by careful mechanical design combined with temperature compensation and correction).

Frequency standards :

The fundamental standard of frequencies is the period of rotation of earth. It can be measured with great accuracy by astronomical methods and is taken as a standard frequency of 1 cycle/day. The error in the measurement is approximately 164 μs. All standards of frequency must ultimately be referred to this fundamental source for calibration purpose.

Practical standards can be *classified* as follows :

(*i*) *Primary standards.* A primary standard of frequency is an oscillator which generates a frequency that is constant over a long period and is checked directly with the earth's rotation.

(*ii*) *Secondary standards.* These standards of frequency are *very stable oscillators* which have their frequencies checked periodically against the primary standards.

- Recent developments in microwave spectroscopy have introduced the *possibility of using spectrum lines as frequency standard.*

4.11.3. Rotational Speed Measurements

4.11.3.1. Introduction

The rotational speed of a machine can be measured by using electrical methods or methods involving electrical-operated measuring equipment. If this machine is an alternator the speed is a measure of the frequency of the A.C. voltage generated, the number of the poles of the alternator being known.

One simple and direct method is to use a small generator with a permanent-magnet field, and to *measure, with a voltmeter, the generated voltage. This is proportional to the speed,* and *the moving-coil voltmeter, with a uniform scale, can be calibrated directly in rotational speed (r.p.m. or r.p.s.).*

4.11.3.2. Stroboscopic method for measurement of speed

***Stroboscope** is a device which emits periodic flashes of light of very short duration.*

— The stroboscopes are generally *calibrated to read the speed directly in r.p.m.* The frequencies of the flashes is adjusted until the body appears stationary and the speed is read directly from the dial which is employed for adjusting the frequency. When the adjustment is calibrated to read the speed directly, the stroboscope is called *"Tachometer"*.

— A stroboscope can be used to measure speeds from a few hundred to many thousands of r.p.m.

- *The procedure for measuring speed by stroboscopic method is as follows :*

  — A disc known as *stroboscopic disc*, with the geometrical pattern on it, is rigidly attached to the shaft of the machine, whose speed is to be determined.

  — The disc is illuminated by a series by rapid flashes, each being of very short direction in comparison to time interval between successive flashes. The series of flashes for illuminating the stroboscopic disc are obtained with the help of a stroboscope.

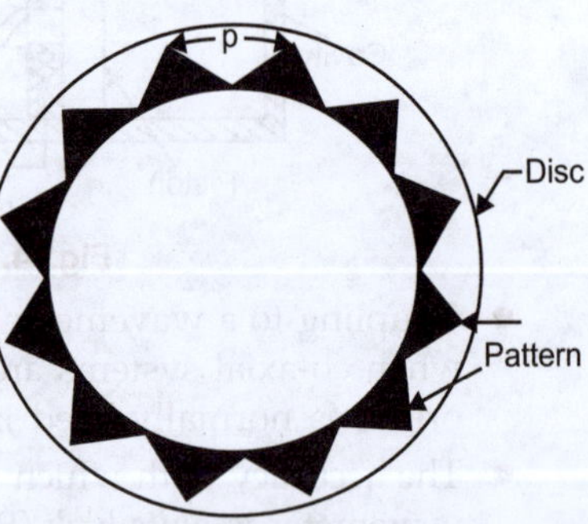

Fig. 4.131. Disc with a geometrical pattern.

  — The frequency of the flashes is adjusted such that disc seems to be stationary. This will happens when the point adjacent to that viewed in the preceeding flash, has taken up the same position and since all points are exactly identical, the eye cannot mark the change. In other words, the disc will *seem to be stationary when the disc moves one point pitch during the time interval between the two successive flashes*. Same effect will be caused if disc moves through $2p, 3p, ... np$ between the successive flashes.

Let, n = Number of points of the disc,

F = Number of flashes per minute,

N = The speed (primary) of the machine in r.p.m., and

d = Diameter of the circular pattern.

The pitch of the points, $p = \dfrac{\pi d}{n}$ ...(*i*)

The distance moved by the disc during the successive flashes $= \dfrac{\pi d N}{F} = p.$ ...(*ii*)

From (*i*) and (*ii*), we get

$$\frac{\pi d}{n} = \frac{\pi d N}{F}$$

or, $$N = \frac{F}{n}$$ ...(4.191)

Thus, if frequency of flashes is known, the rotating speed (N) of the shaft can be determined.

- The fact that the pattern will be stationary even if the actual speed is a whole multiple of N, causes no confusion since the approximate speed can be measured by another but less accurate method. So, the difference between N and its multiple can be easily known *e.g.* if the speed determined by stroboscopic method is 300 and with a less accurate method 890, the accurate speed of the machine is 900 r.p.m., multiplying factor being 3.

When the actual speed is a whole number of N then N is known as the primary speed of the machine.

- A neon lamp with a tripping circuit is generally used with stroboscope to provide periodic flashes of light of very short duration.

Disadvantages :

(*i*) This method is not as accurate as the methods using digital meters.

(*ii*) The stroboscope cannot be used for measuring speed in surroundings having ambient light above a certain level.

Applications :

(*i*) This method is especially suitable for measuring the speed of *small-power rating machines.*

(*ii*) Also suitable for determining the speed of machine parts which are *not readily accessible or where it is not practible to use mechanical methods or where the speed is variable.*

4.11.4. Synchroscopes

4.11.4.1. Introduction

It may be logically assumed that one alternator is placed in parallel with one or more other alternators only when an additional load requires it. Those alternators already carrying load are known as the *running machines,* while that which is to be placed in the system is known as the *incoming machine.*

At the time of synchronising the following *conditions* must be met :

1. The effective voltage of the incoming alternator *must be exactly equal to that of the others, or of the busbars connecting them.*
2. The *phase rotation,* or sequence of the running and incoming alternators, *must be the same.*
3. The individual phase voltages which are to be connected to each other *must be in exact phase opposition.* This is the same as saying that D.C. generators must be connected + to + and – to –.
4. The *frequencies should be the same,* although it is more desirable that the frequencies at the instant of paralleling be almost, but not quite, identical.

- *A* **synchroscope** *is used to determine the correct instant for closing the switch which connects an alternator with another alternator* (*or to the power station busbars*).
  — The voltages can be checked with the help of a voltmeter.
  — For a 3-phase machine the phase sequence of the two should be the same. This condition is verified by a *phase sequence indicator*.

— *The function of a synchroscope is to indicate the difference in phase and frequency of voltages of the busbar and the incoming machine.*

Types of synchroscopes :

There are two types synchroscopes :

1. *Moving iron type.*
2. *Electrodynamic (Weston) type.*

Both are special forms of respective power factor meters.

4.11.4.2. Moving iron type synchroscope

Fig. 4.132 shows the construction and connection diagram of a moving iron type synchroscope, which is due to Lipman :

Construction :

— It consists of a *fixed coil A* divided into two parts. This coil is designed for a small value of current and is connected in series with a resistance across two phases of the busbar.

— There are two iron cylinders C_1 and C_2 mounted on the spindles. Each iron cylinder is provided with two iron vanes whose axes are 180° out with each other.

— The iron cylinders are excited by two pressure coils P_1 and P_2 which are connected to two phases of the incoming machine one of the coils has series resistance R and the other has a series inductance L_1. This is done *to create an artificial phase difference of 90° between the currents of two pressure coils.*

— There are *no control springs.*

— The pointer (provided in the instrument) moves over a dial marked *"Fast"* and *"Slow"*

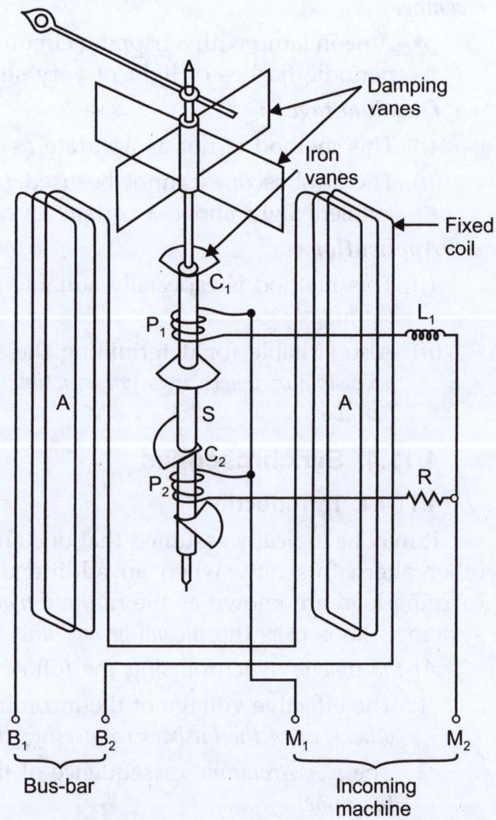

A = Fixed coil; C_1, C_2 = Iron cylinders; P_1, P_2 = Pressure coils

Fig. 4.132. Moving-iron alternating-field synchroscope.

Working :

— When the frequency of the incoming machine is equal to that of busbar, the synchroscope works as a single-phase power factor meter, and so indicates the phase difference between the two e.m.fs.

— When the frequency of the incoming machine is different from the frequency of busbar, the spindle rotates continuously at a speed in r.p.s. equal to difference between the frequencies of the two systems. The direction of rotation depends upon whether the incoming machine is running too fast or too slow.

— The instrument has 360° scale. The displacement from the zero position indicates the phase difference between the e.m.fs. When pointer of the instrument is stationary at the *centre or zero position, the parallel switch should be closed.*

- The moving iron type synchroscopes are commonly used because of the following *reasons* :
  (*i*) They are cheap and robust.
  (*ii*) They are inherently suitable for continuous operation.
  (*iii*) No slip rings are necessary since there are no electrical connections to the moving system.

4.11.4.3. Electro-dynamometer (Weston) type synchroscope

Weston synchroscope is an instrument of dynamometer type and is *spring-controlled.* Fig. 4.133 shows the construction and connection diagram of this instrument.

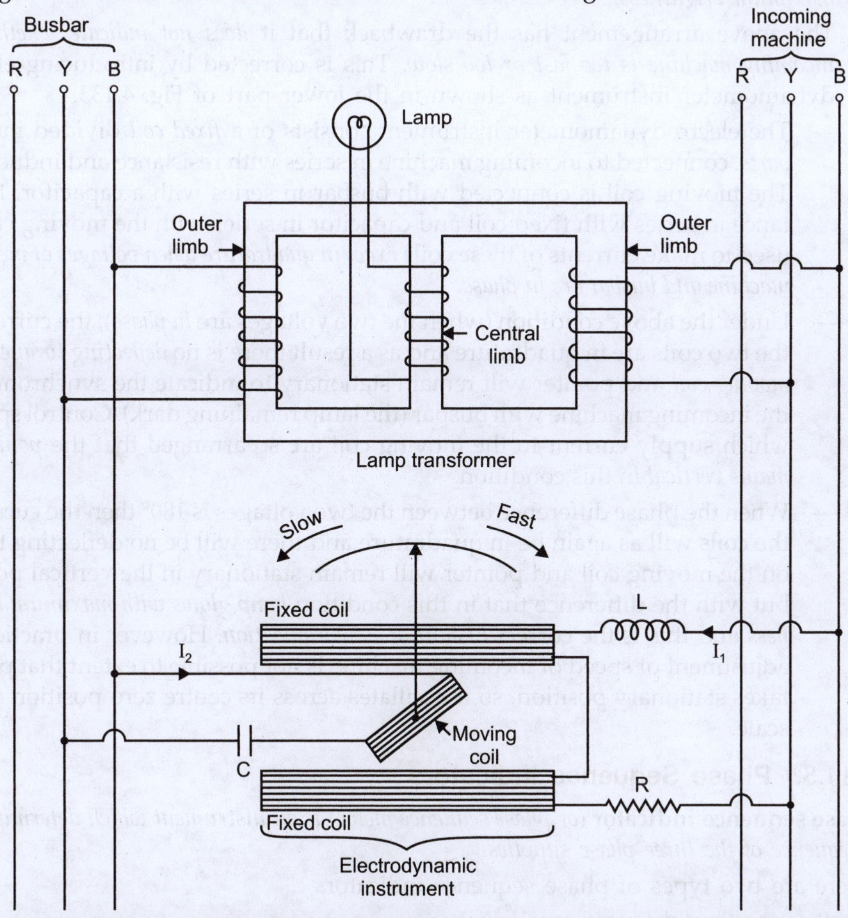

Fig. 4.133. Electro-dynamometer (Weston) type synchroscope.

- It consists of a three limbed transformer (also called lamp transformer). The winding on one of the outer limbs is excited from bus-bars and that on the other limb by incoming machine. The winding on the central limb is connected to a lamp.
  — The two fluxes produced by the outer limbs are forced through the central limb. *The resultant flux through the central limb is equal to the phasor sum of these fluxes.* This resultant flux induces an e.m.f. in the windings of the central limb.

— The windings on the two outer limbs are so arranged that when the bus-bar and the incoming machine voltages are in phase, the two fluxes through the central limb are *additive* and thus e.m.f. induced in the central limb winding is *maximum*. Hence under these conditions the *lamp glows with maximum brightness*.

— When the two voltages are *180° out of phase* with each other the resultant flux is *zero* and hence *no e.m.f. is induced* in the central limb *winding*, consequently the lamp *does not glow at all and is dark*.

— When the frequency of incoming machine is different from that of busbar, the lamp will flicker (*i.e.* will be bright and dark alternately). *The frequency of flickering is equal to the difference in frequencies of the busbar and the incoming machine.*

The correct instant of synchronising is when the lamp is flickering at a very slow rate and is at its maximum brightness.

- The above arrangement has the drawback that it *does not indicate whether the incoming machine is too fast or too slow.* This is corrected by introducing electro-dynamometer instrument as shown in the lower part of Fig. 4.133.

  — The electrodynamometer instrument consists of a *fixed coil* divided into *two parts*, connected to incoming machine in series with resistance and inductance. The moving coil is connected with busbar in series with a capacitor. Inductance in series with fixed coil and capacitor in series with the moving coil are used to make currents of these coils *exact in quadrature when voltages of incoming machine and busbar are in phase.*

  — Under the above condition (when the two voltages are *in phase*), the currents in the two coils are in quadrature and as a result there is no *deflecting torque on the moving coil* and pointer will remain stationary to indicate the synchronism of the incoming machine with busbar (the lamp remaining dark). Control springs, which supply current to the moving coil are so arranged that the *pointer remains vertical* in this condition.

  — When the phase difference between the two voltages is 180° then the current in the coils will as again be in quadrature and there will be no deflecting torque on the moving coil and pointer will remain stationary in the vertical position but with the difference that in this condition *lamp glows with maximum brightness* and this is the correct *instant of synchronisation.* However in practice, the adjustment of speed of incoming machine is not possible to extent that pointer takes stationary position, so it oscillates across its centre zero position on the scale.

4.11.5. Phase Sequence Indicators

Phase sequence indicator (*or phase sequence meter*) *is an instrument which determines the phase sequence of the three phase supplies.*

There are two types of phase sequence indicators :

1. *Rotating type phase sequence indicator.*
2. *Static type sequence indicator.*

4.11.5.1. Rotating type phase sequence indicator

This indicator works on the *principle similar to that of three-phase induction motors.*

— It consists of three coils mounted 120° apart (*i.e.* in star connection) in space. The three ends of the coils are brought out and connected to three terminals marked R Y B as shown in Fig. 4.134.

— These star-connected coils are excited by the supply whose frequency is to be determined. On the top of the coils is mounted an aluminium disc.

— The coils produce a rotating magnetic field and eddy current e.m.fs. are induced in the disc. These e.m.fs. cause eddy currents to flow in the aluminium disc. *A torque is produced due to the interaction of the eddy currents with the field.* Due to this torque the disc revolves and its *direction of rotation depends upon the phase sequence of the supply.*

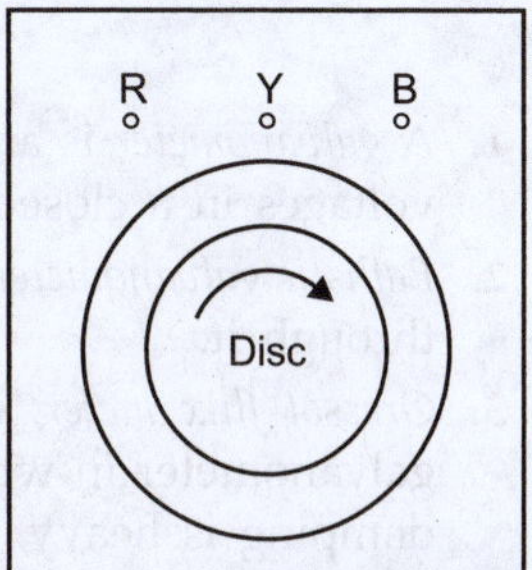

Fig. 4.134. Phase sequence indicator.

— The direction of the rotation of the disc is indicated by an arrow. *When the direction of the rotation is the same as that indicated by the arrow head, the phase sequence of the supply is the same as marked on the terminals of the instrument.* However, if the disc revolves *opposite* to the direction indicated by the arrow head, the sequence of suply is *opposite* to that marked on the terminals.

4.11.5.2. Static type sequence indicator

A static type sequence indicator may have the following *two* types of arrangement :

- One arrangement consists of two lamps and an inductor as shown in Fig. 4.135.
  - — When the phase sequence is RYB, lamp-1 will be dim and lamp-2 will glow brightly.
  - — In case, the phase sequence is RBY, lamp-1 will glow brightly and lamp-2 will be dim.
- In another arrangement, similar phase sequence indicator consists of two neon lamps and a capacitor basically as shown in Fig. 4.135(*a*). In practice, a resistor must be added in series with each lamp *to limit the current* as shown in Fig. 4.135(*b*).

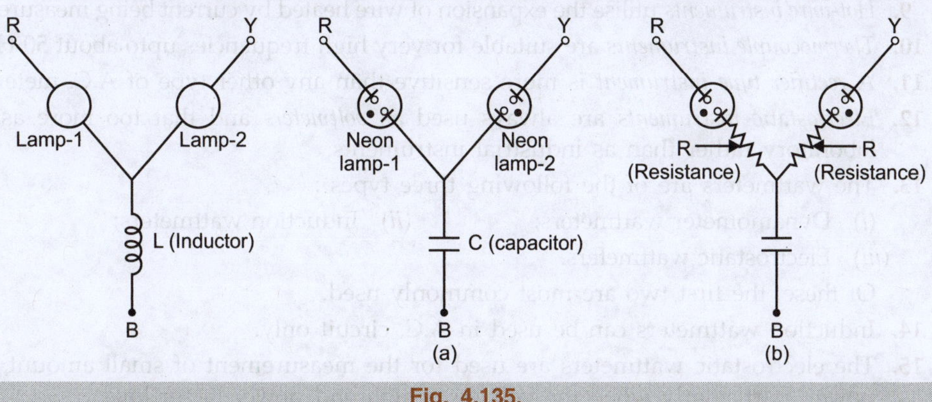

Fig. 4.135.

— When the phase sequence is RYB, *lamp-1 will glow* and *lamp-2 will be dark*, since when neon lamps are used as indicators, the lamp which has the lower voltage will not glow at all, as the voltage across it is lower than the breakdown voltage of the lamp.

HIGHLIGHTS

1. A *galvanometer* is an instrument used for detecting presence of small currents or voltages in a closed circuit or for measuring their magnitude.
2. *Ballistic galvanometer* is used to measure the quantity of electricity (charge) passed through it.
3. *Grassot flux meter,* used for measurement of flux, is a special type of ballistic galvanometer in which the controlling torque is very small and electromagnetic damping is heavy.
4. *Vibration galvanometers* are of D'Arsonval type having a moving coil suspended between the pieces of a permanent magnet. When an alternating current is passed through the moving coil, an alternating deflecting torque is produced which makes the coil vibrate with a frequency equal to the frequency of the current passing.

   These galvanometers are the most widely used "tuned detectors" particularly as null point detectors in the A.C. bridges.
5. *Duddell's oscillograph* is essentially a vibration galvanometer having a low period of vibrations.
6. *Moving iron instruments* are commonly used in laboratories and switch boards at commercial frequencies because they are very cheap and can be manufactured with required accuracy.
7. The permanent moving-coil instrument is the most accurate type for D.C. measurements.
8. *Meggers* (megohm meters) are instruments which measure the insulation resistance of electric circuits relative to earth and one another.
9. *Hot-wire instruments* utilise the expansion of wire heated by current being measured.
10. *Thermocouple instruments* are suitable for very high frequencies upto about 50 Hz.
11. A *rectifier type instrument* is more sensitive than any other type of A.C. meter.
12. *Electrostatic instruments* are always used as *voltmeters* and that too more as a laboratory rather than as industrial instruments.
13. The wattmeters are of the following three types :

    (*i*) Dynamometer wattmeters; (*ii*) Induction wattmeters;

    (*iii*) Electrostatic wattmeters.

    Of these, the first two are most commonly used.
14. Induction wattmeters can be used in A.C. circuit only.
15. The electrostatic wattmeters are used for the measurement of small amount of power, particularly when the voltage is high and power factor low.
16. According to *Blondel's theorem* :

    "If a network is supplied through N conductors, the total power is measured by summing the readings of N wattmeters so arranged that a current element of a wattmeter is in each line and the corresponding voltage element is connected between that line and a common point. If the common point is located on one of the lines, then the power may be measured by N-1 wattmeters".
17. A *polyphase meter* is used to measure power in a polyphase (3-phase) system.
18. *Energy meters* are used to measure the quantity of electric energy supplied to a circuit in a given time.

19. The various types of energy meters are :

(*i*) *Electrolytic meters* ...... Their operation depends on electrolytic action.

(*ii*) *Motor meters* ...... There are really small electric motors

— Mercury motor meters ...... Normally used on D.C. circuits.

— Commutator motor motors ...... Can be used both for D.C. as well as A.C. work

— Induction motor/energy meters ...... used only on A.C. circuits.

(*iii*) *Clock meters* ...... They function as clock mechanisms.

OBJECTIVE TYPE QUESTIONS

Choose the Correct Answer :

Galvanometers

1. A d'Arsonval galvanometer uses a light and scale arrangement. The light source is placed 1 m away from the moving system of galvanometer. The arrangement uses a circular scale calibrated in mm. The deflection indicated by the scale is

(*a*) 1000 mm (*b*) 2000 mm

(*c*) 500 mm (*d*) none of the above.

if a current of 1 mA is passed through the coil. The spring stiffness is 2×10^{-6} Nm/rad and the displacement constant is 2 Nm/A.

2. The time period of free oscillations in a galvanometer having a relative damping of 0.6 is 2 s. The frequency of damped oscillations is :

(*a*) 0.5 rad/s (*b*) 0.3 rad/s

(*c*) 0.4 rad/s (*d*) none of the above.

3. The resistance required for critical damping in a circuit is 1000 Ω. The galvanometer circuit has a resistance of 800 Ω. Is the galvanometer circuit

(*a*) underdamped ? (*b*) undamped ?

(*c*) overdamped ? (*d*) none of the above.

4. A galvanometer has a ratio of 0.9 for damped frequency oscillations to undamped frequency oscillations. Suppose moment of inertia, stiffness constant and damping constant are made twice their original value, what would be the new ratio of damped frequency oscillations to undamped frequency oscillations,

(*a*) 0.9 (*b*) 1.11

(*c*) 4 (*d*) 2.

5. In d'Arsonval galvanometer, an iron core is usually used between the permanent magnet pole faces. This is used so that

(*a*) flux density in the air gap becomes high thereby a large deflecting torque is produced.

(*b*) the effect of stray magnetic fields is reduced.

(*c*) moment of inertia of moving parts becomes smaller.

(*d*) none of the above.

6. Sometimes, the d'Arsonval galvanometers, do not use ferromagnetic cores between poles of the permanent magnet. In this case

(*a*) the flux density becomes smaller resulting in low deflecting torque.

(*b*) the dimensions of the moving coil can be made smaller thereby reducing the moment of inertia.

(*c*) the magnetic field may not be radial resulting in a non-uniform scale even if spring control is used.

(*d*) all of the above.

7. A current of 2 μA is passed though the moving coil of an undamped d'Arsonval galvanometer which has a displacement constant of 2 Nm/A and a control constant of 10×10^{-4} Nm/rad. The moving oscillates with an amplitude of :
(*a*) 0.2 rad. (*b*) 0.4 rad.
(*c*) 0.8 rad. (*d*) none of the above.

8. If the damping in a d'Arsonval galvanometer is only due to electromagnetic effects, the resistance required for critical damping is :
(*a*) $G^2/\sqrt{CJ}$ (*b*) $G/\sqrt{CJ}$
(*c*) $G/2\sqrt{CJ}$ (*d*) $G^2/2\sqrt{CJ}$

9. The relative damping in a galvanometer is 0.8. Its logarithmic decrement is approximately :
(*a*) 0.48 (*b*) 1.25
(*c*) 4.19 (*d*) −4.19.

10. In a critically damped galvanometer the deflection at a time 0.9 times the time of free oscillations after a current is passed through the moving coil is approximately
(*a*) 0.986 times the final deflection.
(*b*) 0.901 times the final deflection.
(*c*) 0.866 times the final deflection.
(*d*) none of the above.

11. A d'Arsonval galvanometer uses a lamp and scale arrangement. Its current sensitivity is 250 mm/μA. If the resistance of the coil is 100 Ω and the external resistance required for critical damping is 900 Ω, (which is connected in the circuit), the voltage sensitivity is
(*a*) 0.25 mm/μV (*b*) 2.5 mm/V
(*c*) 25 mm/μV (*d*) 250 mm/μV.

12. Ayrton shunt is used in d'Arsonval galvanometers so as to limit the current in the galvanometer coil to its maximum permissible value. The relative value of current through the galvanometer coil and the shunt
(*a*) depends upon the value of resistance of galvanometer coil only.
(*b*) depends upon the values of resistance of galvanometer coil and the shunt.
(*c*) does not depend upon the value of resistance of galvanometer coil.
(*d*) none of the above.

13. A Ballistic galvanometer should be designed with
(*a*) a large period of natural oscillations and a negligble damping contant.
(*b*) a small period of natural oscillations and a high damping constant.
(*c*) a large period of natural oscillations and a high damping factor.
(*d*) a small period of natural oscillations and a low damping factor.

14. A Duddell's oscillograph will give no amplitude disortion and phase displacement if its
(*a*) moment of inertia and stiffnes constant are zero.
(*b*) stiffness constant and damping factor are zero.
(*c*) moment of inertia and damping factor are zero.
(*d*) all of the above.

15. In a Duddell's oscilloscope the phase displacements of fundamental and 13th harmonic are calculated to be 3° and 36° respectively. The oscillograph will show the 13th harmonic to be
(*a*) 39° ahead of its true position.
(*b*) 38° ahead of its true position.
(*c*) 3° ahead of its true position.
(*d*) 3° behind its true position.

16. A vibration galvanometer is to be tuned to a frequency of 50 Hz. The ratio of its control constant to inertia constant should be

(*a*) 98696 (*b*) 2500

(*c*) 10.132×10^{-6} (*d*) none of the above.

17. A vibration galvanometer is tuned

(*a*) by changing the length and tension of vibrating coil.

(*b*) by attaching weight to the vibrating coil.

(*c*) by changing its damping constant.

(*d*) all of the above.

18. A Duddell's oscillograph can be used for frequencies

(*a*) upto 50 Hz (*b*) upto 500 Hz

(*c*) above 500 Hz (*d*) upto 10 kHz.

19. In a fluxmeter

(*a*) the controlling torque is produced by weights attached to moving coil.

(*b*) the controlling torque is produced by springs.

(*c*) there is no controlling torque.

(*d*) none of the above.

20. In an unshunted flux meter, the sensitivity is dependent upon

(*a*) resistance of moving coil.

(*b*) resistance of search coil.

(*c*) resistance of both moving coil and search coil.

(*d*) none of the above.

Ammeters, Voltmeters, Wattmeters and Energy meters

21. The heater wire of thermocouple instruments is made very thin in order to have

(*a*) a high value of resistance.

(*b*) to reduce skin effects at high frequencies.

(*c*) to reduce the weight of the instrument.

(*d*) to decrease the over-ranging capacity of the instrument.

22. Electrostatic type instruments are primarily used as

(*a*) ammeters (*b*) wattmeters

(*c*) voltmeters (*d*) ohmmeters.

23. Creeping in a single-phase induction type energy meter may be due to :

(*a*) overcompensation for friction

(*b*) overvoltage

(*c*) vibrations

(*d*) all of the above.

24. A Merz Price maximum demand Indicator Indicates :

(*a*) maximum demand

(*b*) average maximum demand over a specified period of time

(*c*) maximum energy consumption

(*d*) all of the above.

25. The range of an electrostic voltmeter can be extended by using

(*a*) a capacitor in series with the voltmeter whose capacitance is greater than the capacitance of the voltmeter.

(*b*) a capacitor in series with the voltmeter whose capacitance is smaller than the capacitance of the voltmeter.

(*c*) a resistor in series with the voltmeter.

(*d*) an inductor in series with the voltmeter.

26. The sensitivity of a PMMC instrument is 10 kΩ/V. If this instrument is used in a rectifier type voltmeter with half-wave rectification, what would be the sensitivity ?

(*a*) 10 kΩ/V (*b*) 9 kΩ/V
(*c*) 4.5 kΩ/V (*d*) 22.2 kΩ/V.

27. The high torque to weight ratio in an analog indicating instrument indicates

(*a*) high friction loss (*b*) low friction loss
(*c*) nothing as regards friction loss (*d*) none of the above.

28. A PMMC instrument produces a deflection of 90° when a current 2 mA passes through its moving coil. The instrument is spring controlled. Supposing the instrument is redesigned with its spring constant as before, and its permanent magnet is replaced with a new permanent magnet which produces twice the flux density as before, what would be the deflection if a current of 1 mA is passed through the coil ? The other features of the PMMC instrument remain the same.

(*a*) 180° (*b*) 90°
(*c*) 45° (*d*) 135°.

29. A PMMC meter rated at 50 μA is used in a rectifier type of instrument which uses full wave rectification. What is the sensitivity on sinusoidal A.C. ?

(*a*) 20 kΩ/V (*b*) 9 kΩ/V
(*c*) 22.2 kΩ/V (*d*) 18 kΩ/V.

30. A reverse biased diode in addition to a forward biased diode is used in rectifier type voltmeters using half-wave rectification. The use of the reverse biased diode is :

(*a*) it does not allow any current to flow through the meter during negative half cycle
(*b*) it short circuits the meter during the negative half cycle
(*c*) it does not allow the reverse leakage current to flow through the meter during the negative half cycle
(*d*) all of the above.

31. VAh metering can be done by using

(*a*) a ball and disc friction gearing.
(*b*) Trivector meter.
(*c*) Bridge connected rectifiers.
(*d*) all of the above.

32. Light load adjustments for induction type energy meters are usually done at

(*a*) 10% of full load current.
(*b*) 5% of full load current.
(*c*) 50% of full load current.
(*d*) 1% of full load current.

33. In order to reduce errors on account of temperature, instruments are provided with shunts and series connected swamping resistance. In order that errors on account of temperature changes be low, the swamping resistance

(*a*) should be made of a material having a high resistance temperature co-efficient.
(*b*) should be made of a material having a low resistance temperature co-efficient with the value of swamping resistance equal to meter resistance.
(*c*) should be made of a material having low resistance temperature co-efficient and should have a value of about 20 to 30 times that of meter resistance.
(*d*) should have an infinite value.

34. A 4-range milliammeter having ranges of 0—10 mA, 0—50 mA, 0—100 mA and 0—500 mA is used for measurement of current whose magnitude is not known. In order to prevent damage to the instrument, the selector switch of the ammeter should be first placed at

(*a*) 0—50 mA range (*b*) 0—100 mA range
(*c*) 0—10 mA range (*d*) 0—500 mA range.

35. The braking torque provided by a permanent magnet in a single-phase energy meter is proportional to the
(*a*) square of the flux of the permanent magnet.
(*b*) speed of the meter.
(*c*) distance of the permanent magnet from the centre of the revolving disc.
(*d*) all of the above.

36. The braking torque provided by a permanent magnet in a single-phase energy meter can be changed by
(*a*) providing a magnetic shunt and changing its position.
(*b*) changing the distance of the permanent magnet from the centre of the revolving disc.
(*c*) both (*a*) and (*b*).
(*d*) none of the above.

37. Which meter has the highest accuracy in the prescribed limit of frequency range :
(*a*) PMMC (*b*) Moving iron
(*c*) Electrodynamometer (*d*) Rectifier.

38. Which instrument is the cheapest disregarding the accuracy ?
(*a*) PMMC (*b*) Moving Iron
(*c*) Electrodynamometer (*d*) Rectifier.

39. Which instrument has the highest frequency range with accuracy within reasonable limits ?
(*a*) Moving Iron (*b*) Electrodynamometer
(*c*) Thermocouple (*d*) Rectifier.

40. In the single-phase induction meter, in order to obtain true value of energy, the shunt magnet flux should lag behind the applied voltage by
(*a*) 90° (*b*) 0°
(*c*) 45° (*d*) none of the above.

41. In an induction type of meter, maximum torque is produced when the phase angle, between the two fluxes is
(*a*) 0° (*b*) 45°
(*c*) 60° (*d*) 90°.

42. In an induction type meter, maximum torque is obtained when the parameters of rotating disc are :
(*a*) $R = 0$ (*b*) $X = 0$
(*c*) $R = X$ (*d*) none of the above.
where R and X are respectively the resistance and reactance of eddy current paths in the disc.

43. A 1 mA ammeter has a resistance of 100 Ω. It is to be converted to a 1 A ammeter. The value of shunt resistance is :
(*a*) 0.001 Ω (*b*) 0.1001 Ω
(*c*) 100000 Ω (*d*) 100 Ω.

44. Swamping resistance is connected
(*a*) in series with the shunt to reduce temperature error in shunted ammeters.
(*b*) in series with the ammeters to reduce errors on account of friction.
(*c*) in series with meter and have a high resistance temperature co-efficient in order to reduce temperature errors in ammeters.
(*d*) in series with the meter and have a negligible resistance co-efficient in order to reduce temprature errors in shunted ammeters.

45. A quadrant type electrostatic instrument uses two types of connections (*i*) Heterostatic and (*ii*) Idiostatic. An external battery is used
(*a*) for Idiostatic connection
(*b*) for Heterostatic connection
(*c*) for both Idiostatic and Heterostatic connections
(*d*) none of the above.

46. The size of electrostatic voltmeters is large as compared with that of corresponding electromagnetic voltmeters because
(*a*) they are used for measurement of high voltages.
(*b*) energy density in an electostatic field is consideraly smaller than that in an electromagnetic field.
(*c*) their frequency range is smaller.
(*d*) they have a high dielectric loss.

47. In a single-phase induction type energy meter, the lag adjustment is done
(*a*) to make the current coil flux to lag 90° behind the applied voltage.
(*b*) to make the pressure coil flux to lag 90° behind the applied voltage.
(*c*) to bring the pressure coil flux in phase with the applied voltage.
(*d*) none of the above.

48. In a circuit of a single-phase induction energy meter, the pressure coil lags the voltage by 88°, the errors while measuring power in two circuits having power factors of unity and 0.5 lagging are respectively are :
(*a*) – 0.061%, + 6.1% (*b*) + 0.061%, – 6.1%
(*c*) – 0.061%, – 6.1% (*d*) –6.1%, – 6.1%.

49. Horizontally mounted moving iron instruments use
(*a*) eddy current damping (*b*) electromagnetic damping
(*c*) fluid friction damping (*d*) air friction damping.

50. The reason why eddy current damping cannot be used in moving iron instruments is :
(*a*) they have a strong operating magnetic field.
(*b*) they are not normally used in vertical position.
(*c*) they need a large damping force which can only be provided by air friction.
(*d*) they have a very weak operating magnetic field and introduction of a permanent magnet required for eddy current damping would distort the operating magnetic field.

51. Phantom loading for testing of energy meters is used
(*a*) to isolate the current and potential circuits.
(*b*) to reduce power loss during testing.
(*c*) for meters having low current ratings.
(*d*) to test meters having a large current rating for which loads may not be available in the laboratory. This also reduces power losses during testing.

52. In an induction type energy meter, compensation for static friction is provided by
(*a*) shading bands which are actuated to provide a constant torque irrespective of load.
(*b*) lag circuits.
(*c*) drilling holes in the disc.
(*d*) none of the above.

53. An hypothetical electrostatic voltmeter with spring control is designed with one fixed and another movable plate. Both plates are in the form of a circle with one plate over the other so as to produce a rotatry motion. If a voltage is applied between them, no rotatory motion is produced because
(*a*) there is no change in the voltage.

(*b*) there is no change in the capacitance irrespective of the relative position of the two plates.
(*c*) the spring force is very large.
(*d*) the applied voltage is small to produce any noticeable deflection.

54. An electrostatic voltmeter draws a small value of current on d.c.
(*a*) under steady state condition respective of the applied voltage.
(*b*) when switched on irrespective of the applied voltage.
(*c*) when measuring low voltages.
(*d*) when measuring high voltages.

55. In a household single-phase single phase induction type wattmeter, the meter can be reversed by
(*a*) reversing the supply terminals.
(*b*) reversing the load terminals.
(*c*) opening the meter connections and reversing either the potential coil terminals or current coil terminals.
(*d*) opening the meter and reversing connections of both current and potential coil circuits.

56. If an induction type energy meter runs fast, it can be slowed down by
(*a*) lag adjustment.
(*b*) light load adjustment.
(*c*) by adjusting the position of braking magnet and making it come closer to the centre of the disc.
(*d*) by adjusting the position of braking magnet and making it move away from the centre of the disc.

57. The PMMC meters used in rectifier type voltmeters are provided with shunts even at the expense of lowering sensitivity. Why ?
(*a*) The shunting of the meter causes the circuit current to be high thereby making the rectifier work in the linear portion of its $v - i$ characteristics.
(*b*) The shunt prevents the meter against sudden overloads.
(*c*) The provision of shunt increases the sensitivity of the circuit.
(*d*) All of the above.

58. Why are multimeters provided with separate scale for low a.c. voltages ?
(*a*) To improve the readability of the scale.
(*b*) To have high accuracy.
(*c*) To take into account the high value of resistance of rectifier at low voltages (and currents) and also the fact that at low voltages (and currents) the value of rectifier resistance is not constant but varies considerably even for small changes in voltages (or current).
(*d*) None of the above.

59. A voltmeter has a resistance of 2000 Ω. When it is connected across a d.c. circuit its power consumption is 2 mW. Suppose this voltmeter is replaced by a voltmeter of 4000 Ω resistance, the power consumption will be :
(*a*) 4 mW (*b*) 1 mW
(*c*) 2 mW (*d*) none of the above.
The circuit conditions remain undisturbed.

60. A milliammeter of resistance 100 Ω is connected in series with a circuit. Its power consumption is 0.1 mW. Supposing it is replaced with a milliammeter of 200 Ω resistance the power consumption will be
(*a*) 0.2 mW (*b*) 0.05 mW
(*c*) 0.1 mW (*d*) none of the above.

61. A 1 mA d'Arsonval movement has a resistance of 100 Ω. It is to be converted to a 10 V voltmeter. The value of multiplier resistance is

(*a*) 999 Ω (*b*) 9999 Ω
(*c*) 9900 Ω (*d*) 990 Ω.

62. A d'Arsonval movement is rated at 50 μA. Its sensitivity is
(*a*) 20000 Ω/V (*b*) 200000 Ω/V
(*c*) 200 Ω/V (*d*) cannot be determined.

63. An Astatic movement in an electrodynamometer type of instrument is used for
(*a*) eliminating errors on account of external magnetic fields.
(*b*) increasing the operating torque of the instruments.
(*c*) providing eddy current damping.
(*d*) none of the above.

64. In order that an electrodynamometer type of instrument exhibits a pure square law response, the meter range should be limited to
(*a*) –45° to + 45° about the position for zero mutual inductance between fixed and moving coils.
(*b*) –45° to + 45° about the position for maximum mutual inductance between fixed and moving coils.
(*c*) –22.5° to 22.5° about the position for zero inductance between fixed and moving coils.
(*d*) –90° to 90° about the position for maximum mutual inductance between fixed and moving coils.

65. A d'Arsonval movement has a sensitivity of 40,000 Ω/V and its lateral resistance is 4000 Ω. The resistance of multiplier to convert it to 1 V voltmeter is :
(*a*) 44000 Ω (*b*) 36000 Ω
(*c*) 3600 Ω (*d*) none of the above.

66. A PMMC instrument is spring controlled. The control spring stiffness decreases by about 0.04 percent per °C rise in temperature and the strength of magnets goes down by 0.02 percent per °C rise in temperature. The rise in temperature is 10°C. The deflection
(*a*) decreases by about 0.2 percent (*b*) decreases by about 0.4 percent
(*c*) increases by about 0.4 percent (*d*) increases by about 0.2 percent.

67. Moving iron type of instrument can be used as
(*a*) standard instruments for calibration of other instruments.
(*b*) transfer type instruments.
(*c*) indicator type instruments as on panels.
(*d*) all of the above.

68. Moving iron instruments when measuring voltages or currents
(*a*) indicate the same values of the measurement for both ascending and descending values.
(*b*) indicate higher value of measurand for ascending values.
(*c*) indicate higher value of measurand for descending values.
(*d*) none of the above.

69. An electrodynamometer type of instruments finds its major use as
(*a*) standard instrument only.
(*b*) transfer instrument only.
(*c*) both as standard and transfer instrument.
(*d*) an indicator type of instrument.

70. The operating magnetic field in an electrodynamometer type of instrument has a flux density typically about
(*a*) 1 Wb/m^2 (*b*) 0.5 Wb/m^2
(*c*) 0.05 Wb/m^2 (*d*) 0.005 Wb/m^2.

71. The power consumption of PMMC instruments is typically about
(*a*) 0.25 W to 2 W (*b*) 0.25 mW to 2 mW
(*c*) 25 μW to 200 μW (*d*) none of the above.

72. In a series type ohmmeter
(*a*) 0 marking is on the left hand side of scale while ∞ marking is on the right hand side.
(*b*) 0 marking is on the right hand side of scale and ∞ infinity marking on the left hand side.
(*c*) any of the above two markings can be on left or right side of the scale.
(*d*) 0 marking is in the middle of scale.

73. The frequency range of moving iron instruments is
(*a*) audio frequency band 20 Hz to 20 kHz.
(*b*) very low frequency band 10 Hz to 30 kHz.
(*c*) low frequency band 30 Hz to 300 kHz.
(*d*) power frequencies 0 to 125 Hz.

74. Spring controlled moving iron instruments exhibit a square law response resulting in a non-linear scale. The shape of the scale can be made almost linear by
(*a*) keeping rate of change of inductance, L, with deflection, θ, as constant.
(*b*) keeping $\frac{1}{\theta} \cdot \frac{dL}{d\theta}$ as constant.
(*c*) keeping $\frac{1}{K\theta}$ as constant where K is the spring constant.
(*d*) keeping $\theta \cdot \frac{dL}{d\theta}$ as constant.

75. A megger is used for measurement of
(*a*) low valued resistances.
(*b*) medium valued resistances.
(*c*) high valued resistances, particularly insulation resistance.
(*d*) all of the above.

76. The controlling torque in a megger is provided by
(*a*) springs
(*b*) weights attached to the moving system
(*c*) it does not need any controlling torque
(*d*) none of the above.

77. A voltage of 200 V at 5 Hz is applied to an electrodynamometer type of instrument which is spring controlled. The indication on the instruments is
(*a*) 200 V (*b*) 0 V
(*c*) the instrument follows the variations in voltage and does not give a steady response
(*d*) upto 1 MHz.

78. Thermocouple instruments can be used for a frequency range
(*a*) upto 100 Hz (*b*) upto 5000 Hz
(*c*) upto 1 MHz (*d*) 50 MHz and above.

79. A series type ohmmeter is designed with a d'Arsonval movement having full scale current of 0.5 mA and an internal resistance. It uses a battery of 3V and has a half scale resistance of 3000 Ω. The battery voltage goes down to 2.7 V. What is the value of shunt resistance required so that full scale current is obtained if the meter terminals are short circuited.
(*a*) 62.5 Ω (*b*) 2700 Ω
(*c*) 55 Ω (*d*) 130 Ω.

80. A make break switch is provided to disconnect the battery when the meter is not in use in

(*a*) both series and shunt type ohmmeters. (*b*) only in series type ohmmeters.

(*c*) only in shunt type ohmmeters. (*d*) none of the above.

81. A moving iron instrument can be used for current and voltage measurements

(*a*) in a.c. circuits only (*b*) in d.c. circuits only

(*c*) in both a.c. and d.c. circuits for any value of frequency (in case of a.c. circuits)

(*d*) in both a.c. and d.c. circuits for frequencies upto about 125 Hz (in case of a.c. circuits).

82. In spring controlled moving iron instruments, the scale is

(*a*) uniform

(*b*) cramped at the lower end and expanded at the upper end

(*c*) expanded at the lower end and cramped at the upper end

(*d*) cramped both at the lower and the upper ends.

83. The moving iron voltmeters indicate

(*a*) the same value for d.c. and a.c. voltages.

(*b*) lower values for a.c. voltages than for corresponding d.c. voltages.

(*c*) higher values for a.c. voltages than for corresponding d.c. voltages.

(*d*) none of the above.

84. A moving iron voltmeter reads low for a.c. voltages than for corresponding values of d.c. voltages. The meter can be made to read equally for both a.c. and d.c. voltages

(*a*) if the resistance of the multiplier is made very high.

(*b*) if the inductance of the coil is made small.

(*c*) if the resistance of the coil is made very large.

(*d*) if the multiplier resistance is shunted by a capacitor of appropriate value to make the circuit non-inductive.

85. The power in a 3-phase 4-wire circuit can be measured by using

(*a*) 2 wattmeters (*b*) 4 wattmeters

(*c*) 3 wattmeters (*d*) 1 wattmeters.

86. The power in a 3-phase circuit is measured with the help of 2 wattmeters. The readings of one of wattmeters is positive and that of the other is negative. The magnitude of readings is different. It can be concluded that the power factor of the circuit is

(*a*) unity (*b*) zero (lagging)

(*c*) 0.5 (lagging) (*d*) less than 0.5 (lagging)

87. A capacitor is connected across a portion of resistance of the multiplier in order to make the pressure coil circuit non-inductive. The value of this resistance is r while the total resistance and inductance of pressure circuit are respectively R_p and L. The value of capacitance C is :

(*a*) $0.41\ L/r^2$ (*b*) $0.41\ L/R_p^{\ 2}$

(*c*) $L/R_p^{\ 2}$ (*d*) L/r^2.

88. When measuring power with an electrodynamometer wattmeter in a circuit where the load currents is small

(*a*) the current coil should be connected on the load side

(*b*) the pressure coil should be connected on the load side

(*c*) it is immaterial whether current coil or pressure coil is connected on the load side.

89. The power in a d.c. circuit is measured with the help of an ammeter and a voltmeter. The voltmeter is connected on the load side. The power indicated by the product of readings of two instruments (VI) is :

(*a*) the power consumed in the load.

(*b*) the sum of power consumed by load and the voltmeter.

(*c*) the sum of power consumed by load and the ammeter.

(*d*) none of the above.

90. When measuring power with an electrodynamometer wattmeter in a circuit where the load current is large

(*a*) the current coil should be connected on the load side.

(*b*) the pressure coil should be connected on the load side.

(*c*) the pressure coil should be connected on the supply side.

(*d*) it is immaterial whether the pressure coil or the current coil is on the load side.

91. When measuring power with an electrodynamometer wattmeter in a circuit having a low power factor

(*a*) the current coil should be connected on the load side.

(*b*) the current coil should be connected on the supply side.

(*c*) the pressure coil should be connected on the load side.

(*d*) a compensated wattmeter with pressure coil connected on the load side should be used.

92. In electrodynamometer type wattmeters, current coils designated for carring heavy currents use standard wire or laminated conductors

(*a*) to reduce iron loses. (*b*) to reduce hysteresis losses.

(*c*) to reduce eddy current losses in conductors.

(*d*) all of the above.

93. In electrodynamometer type of wattmeters, the inductance of presure coil circuit produces error

(*a*) which is constant irrespective of the power factor of the load.

(*b*) which is higher at low power factors.

(*c*) which is lower at low power factors.

(*d*) none of the above.

94. In an electrodynamometer type of wattmeters

(*a*) the current coil is made fixed. (*b*) the pressre coil is made fixed.

(*c*) any of the two coils i.e. current coil or pressure coil can be made fixed.

(*d*) both the coils should be movable.

ANSWERS

1. (*b*)	**2.** (*c*)	**3.** (*c*)	**4.** (*a*)	**5.** (*a*)	**6.** (*d*)	**7.** (*c*)
8. (*d*)	**9.** (*c*)	**10.** (*a*)	**11.** (*a*)	**12.** (*c*)	**13.** (*a*)	**14.** (*c*)
15. (*c*)	**16.** (*a*)	**17.** (*a*)	**18.** (*b*)	**19.** (*c*)	**20.** (*d*)	**21.** (*b*)
22. (*c*)	**23.** (*d*)	**24.** (*b*)	**25.** (*b*)	**27.** (*c*)	**27.** (*b*)	**28.** (*b*)
29. (*d*)	**30.** (*d*)	**31.** (*d*)	**32.** (*b*)	**33.** (*c*)	**34.** (*d*)	**35.** (*d*)
37. (*c*)	**37.** (*c*)	**38.** (*b*)	**39.** (*c*)	**40.** (*a*)	**41.** (*d*)	**42.** (*b*)
43. (*b*)	**44.** (*d*)	**45.** (*b*)	**46.** (*b*)	**47.** (*b*)	**48.** (*c*)	**49.** (*d*)
50. (*d*)	**51.** (*d*)	**52.** (*a*)	**53.** (*b*)	**54.** (*a*)	**55.** (*c*)	**56.** (*d*)
57. (*a*)	**58.** (*c*)	**59.** (*b*)	**50.** (*a*)	**61.** (*c*)	**62.** (*a*)	**63.** (*a*)
64. (*a*)	**65.** (*b*)	**66.** (*d*)	**67.** (*c*)	**68.** (*c*)	**69.** (*c*)	**70.** (*d*)
71. (*c*)	**72.** (*b*)	**73.** (*d*)	**74.** (*d*)	**75.** (*c*)	**76.** (*c*)	**77.** (*c*)
78. (*d*)	**79.** (*a*)	**80.** (*c*)	**81.** (*d*)	**82.** (*b*)	**83.** (*b*)	**84.** (*d*)
85. (*c*)	**86.** (*d*)	**87.** (*d*)	**88.** (*a*)	**89.** (*b*)	**90.** (*b*)	**91.** (*d*)
92. (*c*)	**93.** (*b*)	**94.** (*a*).				

THEORETICAL QUESTIONS

Galvanometers

1. What is the purpose of an instrument ?
2. How are instruments classified ?
3. Give the comparison between mechanical and electrical/electronic instruments.
4. What factors should be taken into considerations while selecting the instruments ?
5. Explain briefly the functions of instruments.
6. Give the comparison between analog and digital instruments.
7. How are electrical instruments classified ?
8. Discuss briefly essential features of indicating instruments.
9. Explain briefly the following damping methods :
   (*i*) Air damping; (*ii*) Eddy current damping;
   (*iii*) Fluid friction damping;
10. What is a galvanometer ?
11. List the various types of galvanometers.
12. Describe the constuctional details and principle of operation of a D'Arsonval galvanometer. Derive the expression for steady state deflection.
13. How are the constants of a D'Arsonval galvanometer determined ? Why the knowledge of these constants is necessary ?
14. Explain with the help of a neat sketch the construction, and give the theory of operation of a D'Arsonval galvanometer.
15. Name the four intrinsic constants of a galvanometer.
16. Explain briefly the effect of galvanometer resistance on damping.
17. Discuss briefly the following sensitivities of galvanometer :
   (*i*) Current sensitivity;
   (*ii*) Voltage sensitivity;
   (*iii*) Megohm sensitivity.
18. Explain briefly the following three general types of galvanometers :
   (*i*) Portable point type.
   (*ii*) Laboratory reflecting type.
   (*iii*) Box type.
19. Explain the function of Ayrton Universal shunt used for D'Arsonval galvanometers. Prove that the current through the galvanometer does not depend upon the resistance of the galvanometer in case an Ayrton shunt is used.
20. Define the terms "current sensitivity", "voltage sensitivity" and "megohm sensitivity" as applied to D'Arsonval galvanometers. Explain how current sensitivity of a galvanometer can be increased.
21. Dscribe the construction and working of a ballistic galvanometer. Explain the different in constructional details of a ballistic galvanometer and a D'Asonval galvanometer.
22. Prove that is a ballistic galvanometer, the charge is proportional to first swing of the moving coil.
23. Describe the different methods used for calibration of a ballistic galvanometer.
24. Describe the construction and working of an unshunted fluxmeter. Prove that the flux is proportional to the deflection of the fluxmeter.
25. Describe the function of a shunted fluxmeter. Prove that the deflection of a shunted fluxmeter is independent of the resistance of its coil.

26. Explain the construction and working of a vibration galvanometer. Explain the term 'tuning' and describe low a vibration galvanometer is tuned.

27. Derive the expression for steady state deflection of a vibration galvanometer when a sinusoidal voltage is applied to its coil.

28. Give the comparison of ballistic galvanometer with vibration galvanometer.

29. Explain the construction and working of Duddell's oscillograph. List is applications.

Ammeters, Voltmeters and Ohmmeters

30. Explain the working principle of a moving iron indicating instrument. Show that this type of instrument can be used both of d.c. and a.c. measurements. Indicate also the errors involed.

31. Describe the constructional details and working of a moving iron attraction type meter. Derive its torque equation.

32. Derive the expression for torque equation for a moving iron instrument and comment upon the nature of scale.

33. With the help of neat diagrams explain the working principle of a repulsion type moving iron instrument.

34. Explain the construction and operating principle of moving iron instruments. List the errors and methods of compensation.

35. What is essential difference between a moving coil and a moving iron instrument ?

36. Explain the construction of a PMMC meter with the help of a neat sketch. How are different forces produced ? Derive the torque equation. Enlist the advantages and disadvantages of PMMC metres.

37. Describe the construction and working of a PMMC type of instrument. Derive the expression for deflection for a PMMC ammeter if it is (*i*) spring controlled (*ii*) gavity controlled. Comment upon the shape of scale in both cases.

38. Describe the principle of operation, construction, main features and applications of a moving coil instrument.

39. Give a sketch showing the construction of a moving coil voltmeter. Derive an expression for the torque of moving coil instruments.

40. State the causes of change of accuracy in moving coil instruments with change of temperature. Explain how compensation is made in ammeters for change of electrical resistance of the moving coil with change of temperature.

41. Explain with the help of a neat sketch the construction of an electrodynamic type moving coil ammeter. Give the equation for torque of the ammeter and the principal advantages and disadvantages of these instruments. State the causes of change of accuracy with change of temperature and frequency.

42. Describe in details the construction and principle of operation of a transfer instrument generally used in the laboratory for measurement of current.

43. How are thermal instruments classified ? Explain their principles of operation. Explain the operation of a hot wire ammeter with neat diagrams.

44. Draw an illustrative diagram and explain the working of a hot wire instrument. Why is it not normally used as a standard meter ?

45. Describe briefly the constructional features of a thermocouple instrument and the principle of operation. Point out the advantages and disadvantages of this type over the other types.

46. Draw the circuit diagram and explain the principle of operation of a rectifier type voltmeter employing a bridge rectifier.

47. Explain with the help of a neat diagram the principle of operation of a rectifier ammeter. Also briefly discuss the factors which influence the performance of a rectifier ammeter. Specify the suitable current and frequency range of a rectifier ammeter.

48. Describe the construction and working principle of the electrostatic voltmeters.

49. Give the theory of quadrant electrometer. Explain how the instrument can be used for measurement of voltage.

50. Describe the construction of a quadrant type of electrostatic voltmeter. Derive the expression for deflection when idiostatic type of connections are used. The meter is spring controlled.

51. Derive an expression for the force of attraction between the plates in a parallel plate electrostatic voltmeter.

Wattmeters

52. Draw the possible methods of connecting the pressure coil of a wattmeter and compare the errors. Explain the meaning of "compensation winding" in a wattmeter and show how it helps to reduce the error.

53. Discuss the erors introduced in the dynamometer type wattmeter due to the resitance of current and potential coils. How is the error due to resistance of presure coil compensated for by using a compensating coil ?

54. Discuss the various types of errors and their methods of compensation in the dynamometer type wattmeters.

55. Describe the three-ammeter method for measurement of power and power factor in a single phase circuit. Derive the expressions for power and power factor.

56. With neat connection and vector diagrams explain how power can be measured with three ammeters in a single-phase a.c. circuit.

57. Give connection and vector diagrams for the 3-voltmeter method of power measurement.

58. Show with the help of neat diagram how would you extend range of a wattmeter using C.T. and P.T.

59. Draw the circuit diagram of a measurement set-up using a curent transformer, a potential transformer and a wattmeter.

60. Explain how you measure the total power in a 3-phase circuit with the help of two-wattmetes only.

61. Draw the necessary circuit diagram for measurement of three phase power by two wattmeter method. Make necessary derivations. In case of balanced load, discuss the effects of the following load power factors on the two wattmeter readings : Zero, 0.5 and unity.

62. Deduce an expression for the power factor of a balanced three phase load with the help of two wattmeter readings. What will be the power factors :

(*i*) When the readings of the two ammeters are equal.

(*ii*) When the reading of one of the wattmeter is zero.

(*iii*) When the readings of two wattmeters are equal and opposite.

63. Show that in two-wattmeter method of 3-phase power measurement the sum of the readings of the two wattmeters gives the total power consumed in a 3-phase circuit. Hence prove,

$$\tan \Phi = \sqrt{3}\,\frac{W_1 - W_2}{W_1 + W_2}$$

where Φ is the phase angle of the load and W_1 and W_2 are the readings of the wattmeters.

64. What is Blondel's theorem ? Explain with vector diagram how the total power and power factor of a balanced 3-phase 3-wire circuit, can be measured by means of two wattmetes, proving the formula employed.

65. With the help of circuit and phasor diagrams and using usual notations show that the total power in a 3-pase, 3-wire, star-connected balanced load can be measured with the help of only one wattmeter.

66. Show that the power in a 3-phase system is measured by the use of (*i*) only one wattmeter and (*ii*) two wattmeters. Indicate how the power factor is determined. Comment on the accuracy of the measuremens when the load is unbalanced and the supply is a four-wire system.
67. Explain the construction and principle of operation of a dynamometer type wattmeter. How it can be made to read d.c. as well as a.c. ?
68. Explain the construction and working of an electro-dynamometer type of wattmeter.
69. Describe the constructional details and working of a single phase electro-dynamometer type of wattmeter. Derive the expression for deflection for a.c. operation if the instrument is spring controlled.
70. What is special feature of a wattmeter suitable for working on low power factor circuits? Explain.
71. Describe the working principle and construction of an induction type wattmeter. How does this instrument compare with the dynamometer type ?
72. What are the errors in induction type wattmeter, and how are they compensated ?
73. Describe the construction of an induction wattmeter with the help of a neat diagram. What is the function of copper shading band ?

    Explain how torque is produced in an induction type wattmeter. Prove that the torque is proportional to the power in the circuit.
74. Explain the theory of electrostatic wattmeter and discuss its merits over other wattmeters.
75. Draw a circuit diagram of thermal wattmeter and explain its principle of operation.
76. Explain with neat diagram as to how wattmeters are tested.
77. Describe, with connection diagram, the method of calibrating a dynamometer type wattmeter by means of an electrostatic meter.

Energy meters and Measurement of energy

78. Give the essential components of motor meters. Describe the errors common to all meters.
79. Describe, with a neat sketch, the construction and working of a mercury motor meter. How it can be used as a watt-hour meter ?
80. Describe with a neat sketch the construction and working of Elihu Thomson commutator meter. How is braking torque provided in this instrument ? Give its merits also.
81. Describe with the help of diagrams the construction and working of a domestic single phase energy meter.
82. Describe with neat sketch the working of single-phase energy meter. Discuss the friction compensation and creep prevention in an energy meter.
83. Explain working principle of a single-phase induction type energy meter with the help of connection diagam. Give the various sources of errors and their compensation.
84. Describe the construction and working of a single-phase induction type energy meter. Show that the total number of revolutions made by its disc during a particular time is proportional to the energy consumed.
85. Draw a sketch showing the essential features of construction of a single-phase energy meter. Give the theory of operation of the instrument explaining clearly the function of the quadrature loops of the shunt magnet. Describe the mechanism of friction compensation.
86. Describe the expresion for torque for a single-phase induction type of energy meter. Explain that this meter can measure the true energy consumed only if the phase angle between the applied voltage and the pressure coil flux is 90°. Describe a method for making this adjustment.
87. Draw the connection diagram of a 3-phase energy meter and explain its working. How do you correct it, if it is found to be moving fast?
88. Draw the circuit diagram for the measurement of the energy in a 3-phase, 3-wire system using a 2-element energy meter. Is the above arrangement suitable for the measurement of energy in a 3-phase, 4-wire system. Explain.

89. What are the errors in energy meter and how they are compensaed in multi-element induction type energy meter ?

    How these energy meter errors will compare with that of watt-meter ?

90. Classify different types of integrating electricity meters. What type of meter is used for measuring reactive kVAh ? How does it differ from a meter used for measuring kWh on the same circuit ?

91. Explain the advantages of using "phantom load" in meter testing. Describe with the aid of diagram how a single-phase energy meter is tested and calibrated with the help of rotating sub-standard watt-hour meter.

92. Explain briefly how would you carry out single phase meter testing in the laboratory.

93. How will you test the single-phase energy meter ? Explain the procedure of adjustments of different compensation systems.

94. Explain, with the help of a neat diagram the principle and working of a thermal type maximum demand indicator.

95. Explain the working of Merz Price demand indicator.

96. Explain how the following adjustments are made in a single-phase induction type energy meter :
    - (*i*) Lag adjustment
    - (*ii*) Adjustment for friction compensation
    - (*iii*) Creep
    - (*iv*) Overload compensation
    - (*v*) Temperature compensation.

97. Explain the sources of errors in single-phase induction type energy meters.

98. Describe the construction and working of a two element induction energy meter.

99. Explain the different types of tariffs used. Explain the underlying principle for using different types of tariffs for different types of consumers.

100. Describe the construction and working of a Merz Price maximum demand indicator.

101. Explain the different types of methods used for VAh metering.

102. What is a Trivector meter ?

103. What is Phantom loading ? Explain with an example how is it more advantageous than testing with direct loading.

104. Describe a circuit for testing of a single-phase induction type energy meter at different loads and power factors.

105. Enumerate the different types of tests that are carried out on single-phase Induction type watt-hour meters. Describe the Long period dial test.

106. Describe the crossed phase method for measurement of VARh in 3-phase 3-wire circuits.

Miscellaneous Instruments

107. Explain the principle of operation of the moving iron power factor meter.

108. How the power factor of a single phase circuit is measured ? Describe one type of meter for this purpose.

109. Explain with a neat sketch the construction and principle of a dynamometer type single phase power factor meter. How would the accuracy of such an instrument be affected by frequency and waveform variation ?

110. Discuss the construction, principle and theory of one type of frequency meter.

111. What methods can be used for the measurement of frequency ? Explain any one frequency meter with its advantages and disadvantages.

112. Explain the working of a frequency meter, which depends for its action on the phenomenon of electrical resonance.

113. Describe the working of a frequency meter, which depends on mechanical resonance for its action.
114. Describe the construction and give the theory of operation of one type of frequency meter. Explain clearly whether the instrument performance will be affected or not by normal changes in the supply voltage.
115. Draw a neat single sketch of a Weston synchroscope and explain its working.
116. Explain clearly and briefly the construction and working of a synchroscope.
117. Draw synchroscope internal circuit diagram and explain the working and uses. In a synchroscope it is observed that the pointer is revolving once in every second. What is frequency of the incoming machine ?
118. What are the uses of synchroscope ?
119. Describe the construction and working of rotating and static type phase sequence indicators.

UNSOLVED EXAMPLES

Galvanometers

1. The scale of a galvanometer is placed at a distance of 600 mm. A deflection of 42 mm is observed. What is the angle through which the coil has turned ? [**Ans.** 2°]
2. Determine the displacement constant of a suspended coil in a galvanometer of moving coil vibrating type having 28 turns, 4 mm × 2 mm mean area and situated in a magnetic field of 0.8 tesla. [**Ans.** 179.2×10^{-6} Nm/A]
3. Fig. 4.10 shows a galvanometer system in which a wire of 0.1 mm in diameter is used as a torsion member. The wire is made of phosphor bronze for which modulus of rigidity is 40 GN/m^2. The polar moment of inertia for a round wire is $\frac{\pi}{32} d^4$, where d is the diameter of the wire.

   A collimated ray of light is reflected by the galvanometer mirror shown. The radius from the mirror to the scale is 200 mm.

   Calculate : (*i*) Torsional constant; (*ii*) Maximum allowable torque if the shear stress is limited to 100 MN/m^2; (*iii*) Sesntivity of the instrument in m/Nm; (*iv*) Maximum possible scale deflection.

   [**Ans.** (*i*) 52.6×10^{-6} Nm/rad; (*ii*) 39.3×10^{-6} Nm; (*iii*) 7600 m/Nm; (*iv*) 298.7 mm]
4. The coil of a moving coil galvanometer is wound on a non-magnetic former whose height and width are both 20 mm. It moves in a constan field of 0.12 Wb/m^2. The moment of inertia of its moving parts is 0.25×10^{-6} kg-m^2, and the control spring constant is 30×10^{-6} Nm/rad. Calculate : (*i*) The number of turns that must be wound on the coil to produce a deflection of 150° with a current of 10 mA, and (*ii*) The resistance of the coil to produce critical damping, all damping being assumed as electromagnetic.

   [**Ans.** (*i*) 164; (*ii*) 11.28 Ω]
5. In a moving coil galvanometer with free time period of 3 seconds, a current of 2.5 μA gives a deflection of 300 mm on a scale at a distance of 1.5 m. The moving system has a moment of inertia of 4×10^{-6} kg-m^2. Determine the total resistance in the galvanometer circuit for critical damping. [**Ans.** 29,400 Ω]
6. A moving coil galvanometer deflects 200 mm on a scale at a distance of 1 m from the mirror when a current of 2 μA is passed through it. The free time period of the galvanometer is 4 seconds. The galvanoeter is dead beat when the total resistance in galvanometer circuit is 20,000 Ω.

   Determine the moment of inertia of the galvanometer moving system.

   [**Ans.** 4.13×10^{-6} kg-m^2]

7. The following data relate to a moving coil galvanometer which has a former of a non-conducting material :
The current sensitivity of the instrument = 0.001 μA/mm at 1 m; The period of undamped oscillation = 6 seconds; The displacement constant of the instrument = 5×10^{-3} Nm/A.
Calculate : (*i*) Control constant;
(*ii*) Inertia constant;
(*iii*) Total circuit resistance for instrument to be dead beat.
[**Ans.** (*i*) 10×10^{-9} Nm/rad;
(*ii*) 9.12×10^{-9} kg-m^2; (*iii*) 1309 Ω]

8. A d'Arsonval galvanometer has a rectangular coil wound on an aluminium former of resistivity 27×10^{-9} Ωm. The sides of the former, each of effective length 30 mm, move in a radial field of uniform flux density 0.15 Wb/m^2 against a control torque of 18×10^{-6} Nm/rad. The width of the former is 20 mm, the ends being effectively outside the magnetic field, and the moment of inertia of the moving system is 80×10^{-9} kg-m^2. Determine the cross-sectional area of the aluminium former if it is to provide critical damping, other sources of damping being neglected. [**Ans.** 0.8 mm^2]

9. The following data relate to a moving coil galvanometer :
Number of turns of the coil = 250; Flux density in the uniform magnetic field = 0.05 Wb/m^2; Torsion constant = 1.5×10^{-7} Nm per radian; Dimensions of the coil = 15 mm (width) × 20 mm (height); Moment of inertia = 1.5×10^{-7} kg-m^2; Galvanometer resistance = 200 Ω; Total damping = 4×10^{-8} Nm/radian/second.
Determine :
(*i*) The deflection of galvanometer in radian and in mm when a current of 2.5 μA is flowing through the galvanometer, the scale is at a distance of 1.5 m from the mirror.
(*ii*) The curent sensitivity.
(*iii*) The voltage sensitivity.
(*iv*) The megohm sensitivity.
(*v*) Frequency of oscillations.
(*vi*) Relative damping.
(*vii*) Logarithmic decrement.
[**Ans.** (*i*) 0.0625 radian; 187.5 mm; (*ii*) 75 mm/μA;
(*iii*) 0.375 mm/μV; (*iv*) 75 MΩ/mm;
(*v*) 0.158 Hz; (*vi*) 0.189; (*vii*) 0.6]

10. The following data relate to a moving coil galvanometer :
Number of turns of the coil = 300; Flux density in the magneic field = 0.1 Wb/m^2; Control constant = 0.2×0^{-6} Nm/rad; Dimensions of the coil = 20 mm (wide) × 25 mm (high); Moment of inertia = 0.15×10^{-6} kg-m^2; Galvanometer resistance = 300 Ω; Resistance connected in series with the galvanometer = 1000 Ω.
Determine :
(*i*) The resistance to be connected to galvanometer for critical damping.
(*ii*) Relative damping.
(*iii*) Logarithmic decrement.
(*iv*) Frequency of free oscillations.
(*v*) First maximum deflection.
(*vi*) Period of undamped and damped oscillations.
[**Ans.** (*i*) 350 Ω; (*ii*) 0.5; (*iii*) 1.814; (*iv*) 0.184 Hz;
(*v*) 1.163 θ_f ; (*vi*) 5.44s ; 6.3 s]

11. The Ayston universal shunt has a total resistance of 6,000 Ω and galvanometer has a resistance of 2000 Ω. Determine the multiplying power of shunt for 1000 Ω, 2000 Ω and 3000 Ω tapping. [**Ans.** 8, 4, 2.667]

12. A ballistic galvanometer has a resistance of 150 Ω and an undamped period of 7.5 seconds. A steady e.m.f. of 3.5 mV produces a deflection of 210 mm. Determine the quantitiy of electricity discharged from a capacitor if the deflection produced is 750 mm. The relative damping is 0.8. [Ans. 307.5 μC]

13. A solenoid 1m long and wound with 800 turns has a search coil of 50 turns and cross-sectional area 300 mm^2 at its centre. Reversing a current of 4A in the solenoid causes a deflection of 5 divisions in a ballistic galvanometer connected to the search coil. Calculate : (*i*) The galvanometer constant in flux linkages per division; (*ii*) The flux linkage sensitivity.
[Ans. (*i*) 24.1×10^{-6} weber turns per divisions ;
(*ii*) 41.5×10^{-3} divisions per weber turn]

14. The discharge of a capacitor through a ballistic galvanometer produces a damped frequency of 0.125 Hz and successive swings of 120, 96 and 76.8 mm.
Calculate : (*i*) The damping ratio; (*ii*) The logarithmic decrement; (*iii*) The period of undamped and damped oscillations. [Ans. (*i*) 0.0708; (*ii*) 0.223; (*iii*) 7.98s; 8s]

15. A moving coil ballistic galvanometer of 150 Ω resistance gives a throw of 75 dicisions when the flux through a search coil to which it is connected is reversed. Find the flux density in which the reversal of the coil takes place, given that the galvanometer constant = 110 μC/division, and the search coil has 1400 turns, a mean area of 5000 mm^2 and a resistance of 20 Ω.
Calculate also the ballistic and flux linkages sensitivities of the galvanometer.
[Ans. 0.1 Wb/m^2; 9100 div./C; 60.67 weber turns/div.]

16. A galvanometer with an undamped period of 9 seconds and a current sensitivity of 1 μA/div. is connected to a search coil in a circuit of 4000 Ω resistance. Determine the flux linkage change in the search coil to produce a first swing of 100 divisions if the logarithmic decrement is 0.2. [Ans. 632×10^{-3} weber turn]

17. A ballistic galvanometer gives a first maximum deflection of 60° for a discharge of 1000 μC. Find the quantity of electricity which when discharged through this galvanometer gives rise to a spot deflection of 10 divisions on a millimeter circular scale 1 metre away.
[Ans. 4.77 μC]

18. A certain fluxmeter has the following constants :
Air gap flux density = 0.05 Wb/m^2; Turns on moving coil = 40, Area of moving coil = 750 mm^2.
If the flux linking a 10 turn search coil of 200 mm^2 area connected to the fluxmeter is reversed in a uniform field of 0.5 Wb/m^2, calculate the deflection of the fluxmeter.
[Ans. 76.5°]

19. A fluxmeter when used with single turn search coil gives full scale deflection if a given uniform field linking with the search coil is reversed. It is proposed to extend the range of fluxmeter to measure flux densities four times the above value using the same search coil and method of reversals. If the resistance of search coil is 1.1 Ω, find the resistance of shunt to be connected in parallel with the fluxmeter. [Ans. 0.367 Ω.]

20. A magnetic circuit excited by an m.m.f. of 800 AT has a reluctance of 1,20,000 AT/Wb. A fluxmeter is used to measure the flux. If the fluxmeter has 120 divisions and flux linkages required for deflection of one division are 0.15×10^{-3} weber turn, calculate the resistance of the shunt required for use with the search coil. Number of tuns of search coil is one and its resistance is 0.025 Ω. The measurement is made by switching off the excitation. [Ans. 0.00926 Ω]

21. The following data relate to a vibration galvanometer tuned to a frequency of 50 Hz : Inertia constant = 0.01 kg-m^2; Damping constant = 0.025 Nm/rad.-s^{-1} Calculate the ratio of sensitivity for fundamental to that for third harmonic for the galvanometer. **[Ans.** 1000]

Ammeters, Voltmeter and Ohmmeters

22. The coil of instrument has 42½ turns. The mean width of the coil is 25 mm and the axial length of the coil is 20 mm. If the flux density is 0.1 Wb/m^2, calculate the torque on the moving coil for a current of 15 mA through the coil. **[Ans.** 3.187 × 10^{-5} Nm]

23. A moving coil voltmeter with a resistance of 20 Ω gives a full scale deflection of 120° when a potential difference of 100 mV is applied across it. The moving coil has dimensions of 30 mm × 25 mm and is wound with 100 turns. The control spring constant is 0.378 × 10^{-6} Nm/degree. Find the flux density in the gap. Find also the diameter of copper wire of coil winding if 30 percent of instrument resistance is due to coil winding. The specific resistance for copper = 1.7 × 10^{-8} Ωm. **[Ans.** 0.12 Wb/m^2; 0.2 mm]

24. The inductance of a moving iron ammeter in μH is given by the expresion :

$$L = 20 + 10\,\theta - 3\theta^2$$

where, θ is the deflection in radians from zero position.

Determine the deflection of the ammeter for a current of 10 A if the spring constant is 8 × 10^{-6} Nm/rad. **[Ans.** 93°]

25. For a certain dynamometer ammeter the mutual inductance M varies with deflection θ (expressed in degrees) as M = – 6 cos (θ + 30°) mH. Find the deflecting torque produced by a direct current of 50 mA corresponding to a deflection of 60°. **[Ans.** 15 μNm]

26. A PMMC meter gives reading of 40 mA when connected across two opposite corners of a bridge rectifier, the other two corners of which are connected in series with a capacitor to 100 kV, 50 Hz supply. Determine the capacitance. **[Ans.** 1413 pF]

27. A PMMC meter is connected across appropriate points of bridge having resistance of each arm of 35 Ω and connected across a supply of voltage, $v = 5 \sin\theta + 0.2 \sin 3\theta$. Determine the reading of the ammeter if its resistance is 30 Ω. **[Ans.** 32.25 mA]

28. An absolute electrometer uses a movable circular plate 70 mm in diameter. During a voltage measurement, the distance between the plates is 3.2 mm and the force of attraction is 0.003 N. The medium is air. $\epsilon_0 = 8.85 \times 10^{-12}$ F/m. Find V.

[Ans. 1343.1 V]

29. A moving-coil instrument gives full-scale deflection with 15 mA and has a resistance of 5 Ω. Calculate the resistance to be connected :

(*i*) in parallel to enable the instrument to read upto 1 A.

(*ii*) in series to enable it to read upto 100 V. **[Ans.** (*i*) 0.076 Ω, (*ii*) 6661.7 Ω]

30. A moving-coil instrument has a resistance of 10 ohms and gives a full-scale deflection when carrying 50 mA. Show how it can be adopted to measure voltage upto 750 volts and current upto 100 amperes. **[Ans.** R = 14990 Ω, R_s = 0.005 Ω]

31. A 20 V moving-iron voltmeter reads correctly when put on D.C. and the instrument has a resistance of 800 ohms and inductance of 0.15 H. Find out the reading on 20 V A.C. mains (*i*) at 250 Hz (*ii*) 50 Hz. **[Ans.** (*i*) 18.6 V, (*ii*) 19.87 V]

32. A 15 V moving-iron voltmeter has a resistance of 500 Ω and inductance of 0.12 H. Assuming that this instrument reads correctly on D.C. what will be its reading on A.C. at 15 V when the frequency is (*i*) 25 Hz and (*ii*) 100 Hz ?

[Ans. (*i*) 14.99 V, (*ii*) 14.83 V]

33. The total resistance of a moving-iron voltmeter is 1000 Ω and coil has an inductance of 0.765 H. The instrument is calibrated with a full-scale deflection of 50 V D.C. Calculate the percentage error when the instrument is used on (*i*) 25 Hz supply, (*ii*) 50 Hz supply, the applied voltage being 50 V in each case. [**Ans.** (*i*) 0.72%, (*ii*) 36%]

34. In a moving-coil instrument, the moving coil has 40 turns and is of square shape with a mean length of 40 mm along each side. The coil hangs in a uniform radial field of 0.08 Wb/m^2. Find the turning moment of the coil when it is carrying a current of 10 mA. [**Ans.** 512×10^{-7} Nm]

35. A moving-coil instrument has a resistance of 5 Ω between terminals and full scale deflection is obtained with a current of 0.015 A. The instrument is to be used with a manganin shunt to measure 100 A full scale. Calculate the error caused by a 10°C rise in temperature. (*i*) when the internal resistance of 5 Ω is due to copper only, (*ii*) when a 4 Ω manganin swamping resistor is used in series with a copper resistor of 1 Ω.
Take : α_{copper} = 0.004°C and $\alpha_{manganin}$ = 0.00015/°C. [**Ans.** (*i*) 3.7%, (*ii*) 0.8%]

Wattmeters

36. The volt circuit of an electrodynamic wattmeter has an inductance of 8 mH and a resistance of 2000 Ω. What is the percentage error of the instrument when measuring an inductive load having phase angle of 89° and supply frequency is 50 Hz. Neglect the impedance of the current coil and assume that the volt circuit current is negligible as compared with the load current. [**Ans.** 7.2%]

37. A dynamometer type wattmeter with its voltage coil connected across the load side of the instrument reads 250 watts. If the load voltage be 200 volts, what power is being taken by load ? The voltage coil branch has a resistance of 2000 Ω. [**Ans.** 230 W]

38. An electro-dynamic wattmeter has a voltage circuit of resistance of 8,000 Ω and inductance of 63.6 mH which is connected directly across a load carrying 8A at a 50 Hz voltage of 240 V and pf of 0.1 lagging. Estimate the percentage error in the wattmeter reading caused by the loading and inductance of the voltage circuit. [**Ans.** 6.46% (high)]

39. The reading on the ammeters connected for the 3-ammeter method of power measurement are 2.5 A, 4A and 5.6 A in the non-inductive resistor, the load and the main respectively. The terminal voltage is 300 volts. Calculate the value of (*i*) non-inductive resistor (*ii*) the load impedance (*iii*) the power absorbed by the load and (iv) power factor of the load. [**Ans.** (*i*) 120 Ω (*ii*) 75 Ω (*iii*) 546.6 W (*iv*) 0.455]

40. You are supplied with a wattmeter, the current coil of which can take a maximum current of 5A and whose voltage coil can take a maximum voltage of 150 V. You are required to measure the total power of 10 kW in a 250 V, single phase supply at 0.8 pf. What additional equipment you need ? Connect them in proper circuit and find out the multiplying factor which has to be used.
[**Ans.** CT of ratio 50/5A, PT of ratio 250/150V; 16.667]

41. The power in a single phase high voltage circuit is measured by using instrument transformers with voltmeter, ammeter and wattmeter. Observed readings of the instruments (assuming no error) are 115V, 4.5 A, 200W. Characteristics of the transformers are PT : nominal ratio 11500/115V, ratio correction factor 0.995, phase angle–25 minutes; CT : nominal ratio 25/5A, ratio correction factor 0.997, phase angle +15 minutes. Neglecting the voltage coil phase angle in the wattmeter, calculate the true power.[**Ans.** 99.86 kW]

42. Two wattmeters connected to read total power in a three phase system supplying a balanced load gives readings of 8.5 kW and –3.5 kW respectively. Calculate total power and power factor. [**Ans.** 5 kW; 0.2339 (agging)]

43. Two wattmetes used to measure the power input in a 3-Φ circuit indicate 1000W and 5000W respectively. Find the power factor of the circuit (*i*) when both wattmeter readings are positive, (*ii*) when the latter is obtained by reversing the current coil connections. [**Ans.** 0.866 (lag); 0.189 (lag)]

44. Two wattmeters connected to measure the input in a balanced 3-phase circuit indicate 2000 watts and 500 watts respectively. Find power factor
    (*i*) when both readings are obtained in same direction of connecting switch.
    (*ii*) when second reading is obtained after reversing the connection of current coil of the instrument. [**Ans.** (*i*) 0.694 (lag); (*ii*) 0.325 (lag)]

45. The power input to a 2,000V, 50 Hz, 3-phase motor running on full load at an efficiency of 90% is measured by two wattmeters which indicate 300 kW and 100 kW respectively. Calculate (*i*) input (*ii*) power factor (*iii*) line current and (*iv*) hp output. [**Ans.** (*i*) 400 kW (*ii*) 0.756(lag) (*iii*) 152A (*iv*) 490(metric)]

46. A 30phase motor draws a line current of 50A from 220V source while starting. The pf is 0.4. Find the readings of the two wattmeters connected to measure power. [**Ans.** 8.85 kW; (–)1.23 kW]

47. A balanced load is supplied from a 3-Φ, 400V, 3-wire system whose power is measured by two wattmeters. If the total power supplied is 26 kW at 0.75 power factor lagging, find the readings of the each of the two wattmeters. [**Ans.** 19.62 kW; 6.38 kW]

48. The power flowing in a 3-Φ, 3-wire balanced load system is measured by the two-wattmeter method. The reading of wattmeter A is 5000 watts and on wattmeter B is –1,000 watts. (*a*) what is the power factor of the system ? (*b*) If the voltage of the circuit is 440 volts, what is the value of capacitance which must be introduced into each phase to cause the whole of the power measured to appear on wattmeter A? [**Ans.** (*a*) 0.359 (*b*) 5.4 Ω]

49. In a particular measurement, the wattmeter readings were 5 kW and 1 kW. Calculate the power and power factor, if (*a*) both meters read direct, (*b*) one of the meters has been reversed. [**Ans.** (*a*) 6 kW; 0.656 (*b*) 4 kW, 0.359]

50. In balanced three phase system power is measured by two-wattmeter method and the ratio of two wattmeter readings is 2 : 1. Determine the power factor of the system and deduce the relation used. [**Ans.** 0.866]

51. In a dynamometer wattmeter the moving coil has 500 turns of means diameter 3 cm. Estimate the torque if the axis of the field and moving coils are at (*a*) 60° (*b*) 90° when the density in the field coils is 15 mWb/m^2, the current in the moving coil is 0.5 A and power being measured has a power factor of 0.866. [**Ans.** (*a*) 199×10^{-6} Nm (*b*) 230×10^{-6} Nm]

Energy meters

52. In a test of 30 minute duration with a constant current of 5 A such a meter registered 0.50 kWh. If the meter is subsequently used on a 200 V supply, determine its error, and state whether the meter is running fast or slow. [**Ans.** No error]

53. A d.c. ampere-hour meter is rated at 10 A, 250 V. The declared constant is 5 A-s/revolution. Express this constant in revolutions per kWh. Also determine the full load speed of the meter. (**Ans.** 2,880 revolution per kWh; 2 revolutions per second)

54. A meter, whose constant is 600 revolutions per kWh, makes 5 revolutions in 20 seconds. Calculate the load in kW. [**Ans.** 1.5 kW]

55. An energy meter is designed to make 100 revolutions of the disc for one unit of energy. Calculate the number of revolutions made by it when connected to a load carrying 20 A at 230 volts at 0.8 pf for an hour. If it actually makes 360 revolutions; find the percentage error. [**Ans.** 368 revolutions, 2.175% slow]

56. The constant of a 3-phase 2-element energy meter is 0.24 revolution of disc per hour. If the meter is normally used with a PT of ratio 22 kV/220 V and a CT of ratio 500/5 A, determine the error expressed as a percentage of the correct reading, from the following test results :

Line voltage 220 V, current 5.25 A, power factor 1.0, time to complete 80 revolutions 61 seconds. [**Ans.** 1.65% slow]

57. The declared constant of a 5 A, 220 V d.c. watthour meter is 3275 revolutions per kWh. Calculate the speed of the disc at full load.

In a test run at half load, the meter takes 59.5 s to complete 30 revolutions. Calculate the error of the meter. [**Ans.** 1 r.p.s., 0.84% fast]

58. A 50 A, 230 V meter on full load test makes 61 revolutions in 37 s. If the normal disc speed is 520 revolutions per kWh, find the percentage error. [**Ans.** 0.75% low]

59. A wattour meter is calibrated to measure energy on a 250 V supply. On test a steady current of 15 A is passed through it for 5 hours at unity power factor. If the meter readings before and after the test are 8234.21 kWh and 8253.13 kWh respectively, calculate the percentage error.

If the spindle turns through 290 revolutions during 5 minutes when a current of 20 A is passing through the meter at 250 V and 0.87 power factor, calculate the meter constant.

[**Ans.** 0.9% high ; 800 revolutions/kWh]

CHAPTER

5 Instrument Transformers

5.1. Introduction; 5.2. Terms relating to instrument transformers; 5.3. Current transformer (CT) – General aspects – Relationships in a current transformer – Errors in current transformer – Characteristics of current transformers – Clip-on type current transformer; 5.4. Potential transformer (PT) – General aspects – Comparison of current and potential transformers – Relationships in a potential transformer – Characteristics of potential transformers – Reduction of errors in instrument transformers 5.5. Capacitance potential transformers; 5.6. Testing of instrument transformers – Testing of current transformers – Mutual inductance method (absolute) – Arnold method (comparison) – Testing of PTs – Absolute null method – Comparison method– *Highlights – Objective Type Questions – Theoretical Questions – Unsolved Examples.*

5.1 INTRODUCTION

For the measurement of *large currents in D.C. circuits,* it is usual to use low-range ammeters with suitable *shunts*. For measuring *high voltages,* low range voltmeters are used with *high resistance connected in series with them*. However, this method is neither convenient nor practical for alternating current and voltage instruments. For this purpose, *specially constructed accurate-ratio **"instrument transformers"** are employed in conjunction with standard low-range A.C. instruments. Their purpose is to reduce the line current or supply voltage to a value small enough to be easily measured with meters of moderate size and capacity.* In other words, they are used for *extending the range of A.C. ammeters and voltmeters.*

The following are the *two basic advantages* inherent in the method using instrument transformers for measurement purposes :

(*i*) Standard rated instruments may be used.

(*ii*) Operating personnel coming in contact with the instruments are not subjected to high voltage and current of the lines, and so there is less danger to them. Even, with a low-voltage system, instrument transformers are used for measuring large currents, so that heavy loads to the instrument panel and to the ammeter and other current terminals are avoided.

The principle of the instrument transformer is fundamentally the same as that of the power transformer.

The instrument transformers are *classified* as follows :

1. *Current transformers (CT)* : These are used for the measurement of *"current"*.
2. *Potential transformers (PT)* : These are used for the measurement of *"voltage"*.

These transformers are extensively used for *precise measurements.*

• The instrument transformers are widely *used* in protection circuits of power systems *for the operation of over-current, under-voltage, earth-fault* and *various other relays.*

Fig. 5.1 shows the wiring diagrams for current and potential instrument transformers.

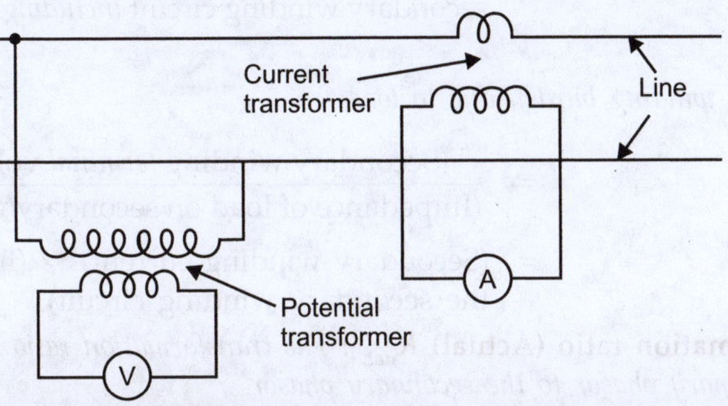

Fig. 5.1. Wiring diagram for current and potential instrument transformers.

Advantages and disadvantages of instrument transformers :

The advantages and disadvantages of instrument transformers as *compared with ammeter shunts and voltmeter multipliers* are as follows :

Advantages :

1. The instrument transformer enables the single range instrument to cover a *large* current or voltage range (normal ranges : 0-5A; 0-110V).
2. It is possible to locate the measuring instruments far away from the high voltage circuit by using long leads connecting the instrument transformer to the instrument.
3. When instruments are employed in conjunction with instrument transformers, their readings *do not depend* upon their constants *viz.*, R, L and C as in the case of shunts and multipliers.
4. Many instruments can be operated from a single instrument transformer.
5. Small power loss.
4. By employing current transformer with a suitable split and hinged core, upon which the secondary winding is wound, the current in a heavy current busbar can be measured without breaking the current circuit.

Disadvantages : The instrument transformers *can not* be used for D.C. measurements.

- In using instrument transformers *for current (or voltage) measurements*, we must know the *ratio of primary current (or voltage) to the secondary current (or voltage)*. These ratios give the *multiplying factor* for finding the primary values from the instrument readings on the secondary side.

However, for *energy or power measurements*, it is essential to know not only the transformation ratio *but also the phase angle between primary and secondary currents (or voltages) because it necessitates further correction to the meter reading.*

5.2 TERMS RELATING TO INSTRUMENT TRANSFORMERS

The various terms relating to instrument transformers are discussed below :

1. **Burden of an instrument transformer :** The load across the secondary winding can be conveniently expressed as the output in volt-ampere at the rated secondary voltage. *The rated burden is the volt-ampere loading which is permissible without errors exceeding the limits for the particular class of accuracy.*

$$\textit{Total secondary burden} = \frac{(\text{Secondary winding } \textit{induced} \text{ voltage})^2}{(\text{Impedance of the secondary winding circuit } \textit{including} \text{ impedance of secondary winding})}$$

= (Secondary winding current)^2 × (Impedance of secondary winding circuit *including* secondary winding) ...(5.1)

Secondary winding burden due to load

$$= \frac{(\text{Secondary winding } \textit{terminal} \text{ voltage})^2}{(\text{Impedance of load on secondary winding})}$$

= (Secondary winding current)^2 × (Impedance of *load* in the secondary winding circuit) ...(5.2)

2. **Transformation ratio (Actual) $K_{act.}$** : *The transformation ratio is defined as the ratio of the primary phasor to the secondary phasor.*

   Transformer ratio (Actual) $K_{act.}$

$$= \frac{|\text{Primary phasor}|}{|\text{Secondary phasor}|} \quad ...(5.3)$$

$$= \frac{\text{Primary winding current (Actual)}}{\text{Secondary winding current (Actual)}} \text{ for a } CT \quad ...(5.4)$$

$$= \frac{\text{Primary winding voltage (Actual)}}{\text{Secondary winding voltage (Actual)}} \text{ for a } PT \quad ...(5.5)$$

3. **Nominal transformation ratio ($K_{nom.}$)** : *It is the ratio of rated primary winding current (or voltage) to the rated secondary winding current (or voltage).*

$$i.e., \quad K_{nom.} = \frac{\text{Rated primary winding current}}{\text{Rated secondary winding current}} \text{ ... for a } CT \quad ...(5.6)$$

$$= \frac{\text{Rated primary winding voltage}}{\text{Rated secondary winding voltage}} \text{ ... for a } PT \quad ...(5.7)$$

4. **Turns ratio (K_t)** : This is defined as under :

$$\text{Turns ratio, } K_t = \frac{\text{Number of turns of secondary winding}}{\text{Number of turns of primary winding}} \text{ ...for a } CT \quad ...(5.8)$$

$$= \frac{\text{Number of turns of primary winding}}{\text{Number of turns of secondary winding}} \text{ ...for a } PT \quad ...(5.9)$$

- *Normally,* $K_{nom.} = K_t$

5. **Ratio correction factor (*RCF*)** : The ratio correction factor is *the transformation (actual) ratio divided by nominal ratio.*

   In other words,

$$RCF = \frac{K_{act.}}{K_{nom.}}$$

$$\text{or,} \quad K_{act.} = RCF \times K_{nom.} \quad ...(5.10)$$

- The ratio marked on transformers is their *nominal ratio.*

5.3 CURRENT TRANSFORMER (CT)

5.3.1. General Aspects

- Just as a shunt extends the range of a D.C. ammeter, so does the current transformer perform the same function in A.C. circuits. Thus a *high magnitude alternating current can be easily measured by a combination of a current transformer and a low range ammeter.*

- The primary of a current transformer (CT) consists of a few turns of thick cross-section connected in *series* with the high current line. Very often the primary is just one turn formed by taking the line conductor through the secondary winding (Fig. 5.2). The secondary winding consists of a large number of turns of fine wire designed for either 5A or 1A rating. Thus a current transformer is *step-up* transformer. The current transformer has the secondary effectively short-circuited through the low impedance of the ammeter. Fig. 5.3 shows the *current transformer connections.*

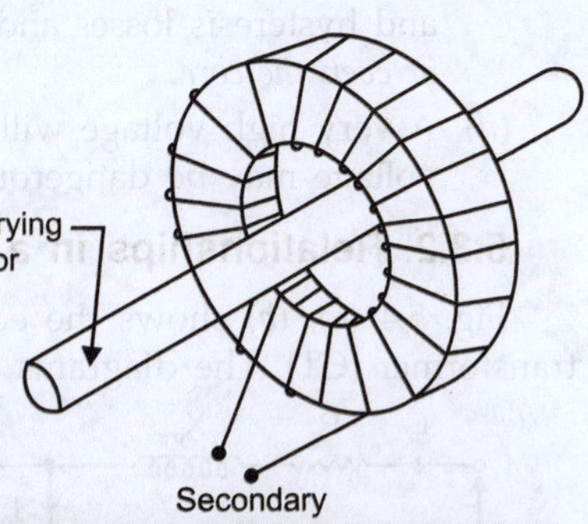

Fig. 5.2. Line conductor acting as primary.

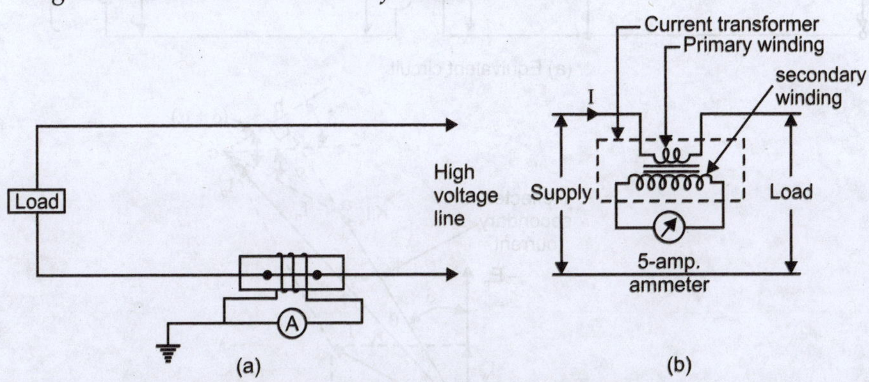

Fig. 5.3. Current transformer connections.

— *Core material* for CT is *usually low-loss silicon steel, mumetal.*

— Core should have *high permeability* and *low loss.*

— To keep losses low, flux density is kept low (≈ 0.1 tesla).

— Core lamination can be *rectangular or ring shaped.* Ring type core is jointless but difficult to wind.

— Windings must be rigid to withstand extremely large mechanical forces produced by short circuit currents.

— For low and medium voltages, tape and varnish insulations are used while for high voltage working CT may be compound filled or oil immersed.

- The *current transformer ratio is not equal to the ratio of secondary to primary turns, mainly because of the effect of the magnetising current.* The primary current can be thought of as the sum of two currents, the first to balance secondary current so that primary and secondary m.m.fs may balance and the second being the no-load current I_0. The component I_0 besides being responsible for a slight error in the current ratio, is also responsible for a phase angle error. The transformer *must be carefully designed to minimise the ratio and phase angle errors.*

☞ - It may be noted that *current transformer* ***must never be operated on open-circuit*** *for the* following two *reasons* :

(*i*) There will be no secondary m.m.f. and since the primary current (and m.m.f.) is fixed, the core flux will increase enormously. This will cause large eddy current and hysteresis losses and the *resulting high temperature may damage the insulation or even the core.*

(*ii*) A very high voltage will be induced in the multi-turn secondary and this high voltage may be dangerous both to life and to the insulation.

5.3.2. Relationships in a Current Transformer

Fig. 5.4 (*a*), (*b*) shows the equivalent circuit and the phasor diagram of a current transformer (CT). The diagrams are same as for any other transformer. Here,

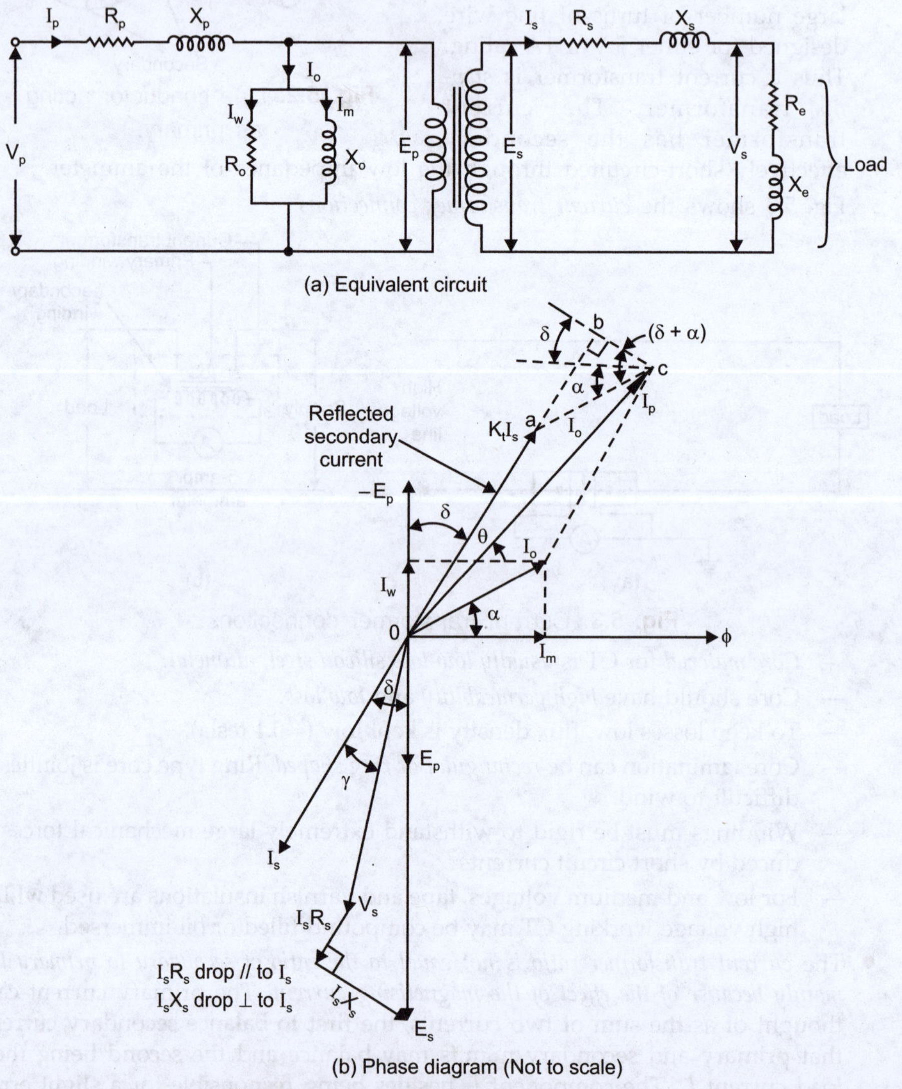

Fig. 5.4. Current transformer (CT).

ϕ = Working flux,

$$K_t = \text{Turns ratio} = \frac{\text{Number of secondary winding turns}}{\text{Number of primary winding turns}},$$

R_p = Resistance of primary winding,
X_p = Reactance of primary winding,
R_s = Resistance of secondary winding,
X_s = Reactance of secondary winding,
R_e = Resistance of external burden *i.e.,* load on secondary,
X_e = Reactance of external burden *i.e.,* load on secondary,
E_p = Primary induced voltage,
E_s = Secondary induced voltage,
V_p = Primary supply voltage,
V_s = Secondary terminal voltage,
I_0 = Exciting current,
I_p = Primary winding current,
I_s = Secondary winding current,
I_m = Magnetising component of I_0
I_w = Loss component of I_0.
α = Angle between exciting current I_0 and working flux ϕ,
θ = Phase angle of the transformer,
δ = Angle between secondary winding induced voltage (E_s) and secondary winding current (I_s)
= Phase angle of total impedance of secondary burden.

$$= \tan^{-1}\left(\frac{X_s + X_e}{R_s + R_e}\right), \text{ and}$$

γ = Phase angle of load or burden *i.e.,* $R_e + jX_e$

$$= \tan^{-1}\frac{X_e}{R_e}$$

Actual transformer ratio (current) ($K_{act.}$) :

Consider the phasor diagram of CT with a lagging power factor, as shown in Fig. 5.4 (*b*).

Consider the right-angled triangle *obc* :

$$(oc)^2 = (ob)^2 + (bc)^2$$
$$= (oa + ab)^2 + (bc)^2$$

or,
$$I_p^2 = \left[K_t I_s + I_o \sin(\delta+\alpha)\right]^2 + \left[I_o \cos(\delta+\alpha)\right]^2$$
$$= K_t^2 I_s^2 + 2K_t I_s I_o \sin(\delta+\alpha) + I_o^2 \sin^2(\delta+\alpha) + I_o^2 \cos^2(\delta+\alpha)$$
$$= K_t^2 I_s^2 + 2K_t I_s I_o \sin(\delta+\alpha) + I_o^2 [\sin^2(\delta+\alpha) + \cos^2(\delta+\alpha)]$$
$$= K_t^2 I_s^2 + 2K_t I_s I_o \sin(\delta+\alpha) + I_o^2$$

or,
$$I_p = \sqrt{K_t^2 I_s^2 + 2K_t I_s I_o \sin(\delta+\alpha) + I_o^2} \quad \text{...(5.11)}$$

∴ Actual transformation (current) ratio,

$$K_{act.} = \frac{I_p}{I_s} = \frac{\sqrt{K_t^2 I_s^2 + 2K_t I_s I_o \sin(\delta+\alpha) + I_o^2}}{I_s} \quad \text{...(5.12)}$$

In a well designed transformer $I_o << K_t I_s$. Usually I_o is less than 1% of I_p and I_p is, therefore, very nearly equal to $K_t I_s$.

∴ Eqn. (5.12) can be written as :

$$K_{act.} = \frac{\sqrt{K_t^2 I_s^2 + 2K_t I_s I_o \sin(\delta+\alpha) + I_o^2 \sin^2(\delta+\alpha)}}{I_s}$$

... Adjusting I_o^2 as $I_o^2 \sin^2(\delta+\alpha)$

or, $$K_{act.} = \frac{K_t I_s + I_o \sin(\delta+\alpha)}{I_s} = K_t + \frac{I_o}{I_s}\sin(\delta+\alpha) \quad ...(5.13)$$

Although (5.13) is approximate, yet it is *sufficiently accurate for all practical purposes.* The above *theory is applicable when the burden is "inductive" which is normally the case.* Eqn. (5.13) can be further expanded as,

$$K_{act.} = K_t + \frac{I_o}{I_s}(\sin\delta\cos\alpha + \cos\delta\sin\alpha)$$

$$= K_t + \frac{I_o \cos\alpha \,.\, \sin\delta + I_o \sin\alpha \,.\cos\delta}{I_s}$$

or, $$K_{act.} = K_t + \frac{I_m \sin\delta + I_w \cos\delta}{I_s} \quad ...(5.14)$$

$(\because I_o \cos\alpha = I_m; I_o \sin\alpha = I_w)$

Note that δ is *positive* for *lagging* p.f. load while *negative* for *leading* p.f. load.

Phase angle (θ) :

The angle by which the secondary current phasor, when reversed, differs in phase from the primary current, is known as the ***phase angle*** *of the transformer.*

This angle is taken to be +ve if the secondary current reversed leads the primary current. The angle is taken as –ve if the secondary current reversed lags behind the primary current.

The angle between I_s reversed and I_p is θ. Therefore, the phase angle is θ.

From the phasor diagram [Fig. 5.4 (*b*)], we have :

$$\tan\theta = \frac{bc}{ob} = \frac{bc}{oa+ab} = \frac{I_o\cos(\delta+\alpha)}{K_t I_s + I_o \sin(\delta+\alpha)}$$

As θ is very small, we can write,

$$\theta = \frac{I_o\cos(\delta+\alpha)}{K_t I_s + I_o \sin(\delta+\alpha)} \text{ rad.} \quad ...(5.15)$$

Since $I_o << K_t I_s$, therefore, we can neglect the term $I_o \sin(\delta+\alpha)$

$$\theta \simeq \frac{I_o\cos(\delta+\alpha)}{K_t I_s} \text{ rad.} \quad ...(5.16)$$

$$\simeq \frac{I_o\cos\delta\cos\alpha - I_o\sin\delta\sin\alpha}{K_t I_s}$$

$$\simeq \frac{I_m\cos\delta - I_w\sin\delta}{K_t I_s} \text{ rad.} \quad ...(5.17)$$

$$\simeq \frac{180}{\pi}\left(\frac{I_m\cos\delta - I_w\sin\delta}{K_t I_s}\right) \text{ degrees} \quad ...(5.18)$$

5.3.3. Errors in Current Transformer

It is evident from the eqn. 5.14 that the value of transformation ratio (actual ratio) is *not* equal to the turns ratio (K_t). Also the value is not constant and depends upon magnetising and loss components of the exciting current, the secondary winding load current and its power factor. This means the secondary winding current is not a constant fraction of the primary winding current but depends upon the above mentioned factors. This introduces *considerable errors* into current measurements.

In power measurements, due to the use of CT transformers, the following two types of errors are introduced :

1. *Ratio error* : Due to actual transformation ratio being *different* from the turns ratio (K_t).

2. *Phase angle error* : Due to secondary winding current *not* being exactly 180° out of phase with the primary winding current (difference being angle θ).

1. **Ratio error.** It is defined as :

$$\text{Percentage ratio error} = \frac{\text{Nominal ratio} - \text{Actual ratio}}{\text{Actual ratio}} \times 100$$

$$= \frac{K_{nom.} - K_{act.}}{K_{act.}} \times 100 \qquad \text{...(5.19)}$$

2. **Phase angle error.** This error is denoted by angle θ, by which the phase difference between primary and secondary is different from 180°. It is given by :

$$\theta = \frac{180}{\pi}\left[\frac{I_m \cos\delta - I_w \sin\delta}{K_t I_s}\right] \text{degrees}$$

- ***Approximate formulae for errors :***

Usually, the *instrument burden* is largely resistive with some inductance and, therefore, *δ is +ve and is generally small.*

Hence, sin δ ≃ 0 and cos δ ≃ 1

Thus, from eqns. (5.14) and (5.18), we have :

$$K_{act.} = K_t + \frac{I_w}{I_s} \qquad \text{...(5.20)}$$

and,

$$\theta = \left(\frac{180}{\pi}\right)\left(\frac{I_m}{K_t I_s}\right) \qquad \text{...(5.21)}$$

But $I_p \simeq K_t I_s$, and, therefore, eqns. (5.20) and (5.21) can be rewritten as :

$$K_{act.} \simeq K_t + \frac{K_t I_w}{I_p} \simeq K_t\left(1 + \frac{I_w}{I_p}\right) \qquad \text{...(5.22)}$$

and,

$$\theta = \left(\frac{180}{\pi}\right)\left(\frac{I_m}{I_p}\right) \qquad \text{...(5.23)}$$

5.3.4. Characteristics of Current Transformers

The performance of current transformers under different operating conditions may be estimated by considering the phasor diagram. Brief statements of the performance variation with load conditions are as follows :

1. **Effect of change in power factor of secondary burden :**
The power factor of the secondary circuit depends on the power factor of the burden on secondary, this directly affects the two errors of the transformer as under :
   - *Ratio error* : As δ is positive for all *inductive* loads, therefore, sin (δ + ∝) is positive and hence actual ratio $K_{act.}$ is always *greater* than the turns ratio K_t [refer to eqn. (5.13)]. δ is negative for *capacitive* burdens and, consequently $K_{act.}$ is *less* than K_t.
   - *Phase angle error* : When load is *inductive* and δ is small positive then θ is positive. But as δ approaches 90° as load becomes highly inductive then θ becomes negative. In case of *capacitive* load, δ is negative and θ is always positive.
2. **Effect of change in burden on secondary :**
   — An increase in secondary winding circuit burden impedance means an increase in volt ampere rating. This necessitates an increase in the secondary winding induced voltage which can be generated by an increased flux and flux density. Therefore, both magnetising component I_m and loss component I_w are increased; subsequently *errors also increase*.
   — A great burden impedance, in general, not only increases the transformation ratio, but also *shifts* the phase angle between primary winding current and secondary winding current reversed to *more positive values*.
3. **Effect of change in primary winding current (I_p) :**
   — When the primary current I_p changes, the secondary current I_s changes proportionately.
   — At *low values* of I_p (or I_s), the exciting current component I_m and I_w are a greater proportion of I_p and, therefore, the errors are *greater*.
   — When I_p increases, I_s also increases and there is a *decrease in ratio error and phase angle* [refer to eqns. (5.22) and (5.23)].
4. **Effect of change in frequency :** The effect of increase in frequency is similar to that produced by a decrease in secondary burden, as increase in frequency will produce a corresponding *decrease in flux for the given e.m.f.* Consequently, there is reduction in I_m and I_w and hence *errors also get reduced*.

5.3.5. Clip-on Type Current Transformer

The main feature of this type of transformer is that *it can be used instantaneously for measurement of current in a current carrying conductor* (cable, insulated bus-bar or fuse link) *without interrupting the circuit for connection of ammeter in series with the circuit*.

Fig. 5.5, shows the schematic diagram of a clip-on type current transformer.

— It consists of a laminated core which is so arranged that it can be opened out at a hinged section by merely pressing a trigger-like projection. When the core is thus opened, it permits the admission of very heavy current-carrying busbars or feeders where upon the trigger is released and the core is tightly closed by a spring.

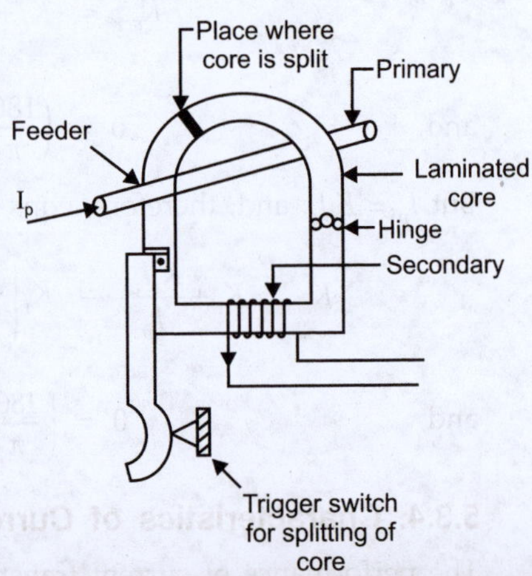

Fig. 5.5. Schematic diagram of a clip-on type current transformer.

— The current-carrying conductor of feeder acts as a single-turn primary whereas the secondary is connected across the standard ammeter conveniently mounted in the handle itself; the ammeter gives the value of current flowing through the conductor.

- The instrument can be used in A.C. circuits only.
- If can be operated by one hand.
- The accuracy claimed by the manufacturers is within 3% of full-scale with the conductor in a central position in the core.
- There transformers are designed for a voltage of 5 kV between the primary and secondary windings.

Example 5.1. *The following data relate to a current transformer :*

Number of primary turns = single turn
Number of secondary turns = 240
Secondary winding current = 5 A
External burden = 1.2 Ω (non-inductive)
Magnetomotive force (mmf) = 96 AT
for setting up flux in the core
Cross-section of the core = 1200 mm²
Supply frequency = 50 Hz

Neglecting the effects of magnetic leakage, iron losses and I^2R losses, calculate :

(i) Actual transformation ratio.
(ii) Phase angle.
(iii) Maximum flux density in the core.

Solution. *Given* : $N_p = 1$; $N_s = 240$; $I_s = 5$ A; $R_e = 1.2\ \Omega$; mmf = 96 AT; $A_c = 1200$ mm²; $f = 50$ Hz.

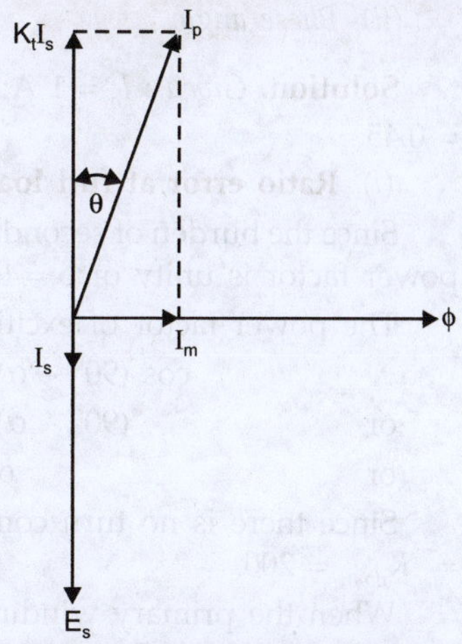

Fig. 5.6. Phasor diagram.

(*i*) **Actual transformation ratio ($K_{act.}$) :**

$$\text{Turns ratio, } K_t = \frac{N_s}{N_p} = \frac{240}{1} = 240$$

Voltage induced in the secondary winding,

E_s = Current in the secondary winding (I_s) × impedance of secondary winding (R_e)
= 5 × 1.2 = 6.0 V

As the secondary burden is purely resistive, the I_s is in phase with E_s, and the secondary power factor is unity or $\delta = 0$ (Refer to Figs. 5.6 and 5.4).

The loss component (I_w) of no-load current (I_o) is to be neglected and, therefore, $I_w = 0$

$$\therefore \quad I_o = \sqrt{I_m^2 + I_w^2} = I_m = \text{magnetising component}$$

and, $\alpha = 0$

$$\text{Also,} \quad I_m = \frac{mmf}{N_p} = \frac{96}{1} = 96\,A$$

Now, reflected secondary winding current = K_tI_s = 240 × 5 = 1200 A

Referring to phasor diagram (Fig. 5.6), we have :

$$\text{Primary current, } I_p = \sqrt{(K_t I_s)^2 + (I_m)^2}$$

$$= \sqrt{(1200)^2 + (96)^2} = 1203.8\,A$$

$$\therefore \text{ Actual transformation ratio, } K_{act.} = \frac{I_p}{I_s} = \frac{1203.8}{5} = \mathbf{240.76} \quad \textbf{(Ans.)}$$

(ii) **Phase angle, θ :**

From Fig. 5.6, we get :

$$\tan\theta = \frac{I_m}{K_t I_s} = \frac{96}{1200} = 0.08$$

$$\therefore \quad \theta = \tan^{-1}(0.08) = \mathbf{4° \; 34'} \quad \textbf{(Ans.)}$$

(iii) **Maximum flux density in the core, B_{max} :**

We know that : $E_s = 4.44\, f\, \phi_{max}\, N_s$

$$\therefore \quad \phi_{max} = \frac{E_s}{4.44\, f N_s} = \frac{6.0}{4.44 \times 50 \times 240} = 0.1126 \times 10^{-3}\, Wb$$

$$\text{Hence,} \quad B_{max} = \frac{\phi_{max}}{A_c} = \frac{0.1126 \times 10^{-3}}{1200 \times 10^{-6}} = \mathbf{0.0938\ Wb/m^2} \quad \textbf{(Ans.)}$$

Example 5.2. *A ring core current transformer of ratio 1000/5 A is operating at full primary current with a secondary burden of non-inductive resistance of 1.1 Ω. Its exciting current is 1 A at a power factor of 0.45. Calculate :*

(i) The ratio error at full load, assuming that there has been no compensation.

(ii) Phase angle.

Solution. *Given* : $I_o = 1$ A; Nominal ratio, $K_{nom.} = \frac{1000}{5} = 200; R_e = 1.1\Omega$, Power factor = 0.45.

(i) **Ratio error at full load :**

Since the burden of secondary winding is purely resistive therefore, secondary winding power factor is unity or δ = 0.

The power factor of exciting current is 0.45,

$$\therefore \quad \cos(90° - \alpha) = 0.45$$

$$\text{or} \quad (90° - \alpha) = \cos^{-1}(0.45) = 63.26°$$

$$\text{or} \quad \alpha = 90° - 63.26° = 24.74°$$

Since there is no turn compensation, the turns ratio is equal to nominal ratio or $K_t = K_{nom.} = 200$.

When the primary winding carries rated current of 1000 A, the secondary winding carries a current of 5 A.

Rated secondary winding current, $I_s = 5$ A

$$\therefore \quad K_t I_s = 200 \times 5 = 1000 \text{ A.}$$

Actual transformation ratio,

$$K_{act.} = K_t + \frac{I_o}{I_s}\sin(\delta + \alpha) \qquad \text{...[Eqn. (5.13)]}$$

$$= 200 + \frac{1}{5}\sin(0 + 26.74°) = 200.09$$

$$\therefore \quad \text{Ratio error} = \frac{\text{Nominal ratio} - \text{Actual ratio}}{\text{Actual ratio}} \times 100 = \frac{K_n - K_T}{R} \times 100$$

$$= \frac{200 - 200.09}{200.09} \times 100 = \mathbf{-0.045\%} \quad \textbf{(Ans.)}$$

(*ii*) **Phase angle, θ :** Refer to Fig. 5.4.

$$\theta = \frac{180}{\pi}\left[\frac{I_o \cos(\delta + \alpha)}{K_t I_s}\right] \text{degrees} \qquad \text{...[Eqn. (5.16)]}$$

$$= \frac{180}{\pi}\left[\frac{1 \times \cos(0 + 26.74°)}{200 \times 5}\right] = \mathbf{3'4''} \quad \textbf{(Ans.)}$$

Example 5.3. *The primary winding of a 1000/5A, 50 Hz current transformer has a single turn. Its secondary burden consists of a non-inductive impedance of 1.4 Ω. If the iron loss in the core is 1.4 W at full load and magnetising mmf is 80 AT, calculate :*

(i) Flux in the core.

(ii) Ratio error at full load.

Neglect leakage reactance.

Solution. *Given* : Nominal ratio, $K_{nom.} = \frac{1000}{5} = 200$; $N_p = 1$; Secondary impedance = 1.4Ω; mmf = 80 AT; Iron loss in the core = 1.4 W.

(*i*) **Flux in the core, ϕ_{max} :**

Let us *assume* turns ratio to be equal to nominal, in the absence of any other data.

Then, Turns ratio, $K_t = K_{nom.} = 200$.

$$\therefore \quad N_s = K_t N_p = 200 \times 1 = 200 \qquad (\because N_p = 1)$$

Now, secondary induced voltage, $E_s = 5 \times 1.4 = 7$ V

Also, $E_s = 4.44 f \phi_{max} N_s$

$$\therefore \quad \phi_{max} = \frac{E_s}{4.44 \times f \times N_s} = \frac{7}{4.44 \times 50 \times 200} = \mathbf{0.1576 \times 10^{-3}\ Wb} \ \textbf{(Ans.)}$$

(*ii*) **Ratio error at full load :**

$$\text{Primary induced voltage, } E_p = \frac{E_s}{K_t} = \frac{7}{200}\ \text{V}$$

$$\text{Loss component of exciting current, } I_w = \frac{\text{Iron loss}}{E_p} = \frac{1.4}{(7/200)} = 40\ \text{A}$$

$$\text{Magnetising component, } I_m = \frac{mmf}{N_p} = \frac{80}{1} = 80\ \text{A}$$

As the burden is purely resistive, therefore, the secondary load angle is zero or δ = 0.

We know that,

$$\text{Actual ratio, } K_{act.} = K_t + \frac{I_m \sin\delta + I_w \cos\delta}{I_s} \qquad \text{...[Eqn. (5.14)]}$$

$$= 200 + \frac{40}{5} = 208\ \text{V}$$

$$\therefore \quad \text{Ratio error} = \frac{K_{nom.} - K_{act.}}{K_{act.}} \times 100$$

$$= \frac{200 - 208}{208} \times 100 = -\mathbf{3.846\%} \quad \textbf{(Ans.)}$$

Example 5.4. *A current transformer with a bar primary has 250 turns in its secondary winding. The resistance and reactance of the secondary circuit are 1.4 Ω and 1.1 Ω respectively including the transformer winding. When 5A currents flows in the secondary winding, the magnetising mmf is 80 AT and the iron loss is 1.1 W. Determine the following :*

(i) Ratio error.

(ii) Phase angle error.

Solution. *Given* : $N_p = 1$; $N_s = 250$; $R_p = 1.4\ \Omega$; $X_s = 1.1\ \Omega$; $I_s = 5\text{A}$; Magnetising mmf = 80 AT; Iron loss = 1.1 W.

(i) **Ratio error :**

$$\text{Turns ratio, } K_t = \frac{N_s}{N_p} = \frac{250}{1} = 250$$

$$\text{Secondary circuit impedance} = \sqrt{(1.4)^2 + (1.1)^2} = 1.78\ \Omega$$

For secondary winding circuit :

$$\cos\delta = \frac{1.4}{1.78} = 0.7865; \quad \sin\delta = \frac{1.1}{1.78} = 0.618$$

$$\text{Secondary induced voltage, } E_s = 5 \times 1.78 = 8.9\ \text{V}$$

$$\text{Primary induced voltage, } E_p = \frac{E_s}{250} = \frac{8.9}{250} = 0.0356\ \text{V}$$

Loss of component referred to primary winding,

$$I_w = \frac{\text{Iron loss}}{E_p} = \frac{1.1}{0.0356} = 30.9\ \text{A}$$

$$\text{Magnetising current, } I_m = \frac{mmf}{N_p} = \frac{80}{1} = 80\text{A}$$

$$\text{Actual transformation ratio, } K_{act.} = K_t + \frac{I_m \sin\delta + I_w \cos\delta}{I_s} \quad \text{...[Eqn. (5.14)]}$$

$$= 250 + \frac{80 \times 0.618 + 30.9 \times 0.7865}{5} = 264.75$$

Assuming, $K_{nom.} = K_t = 250$, since no other contrary information is given, we have :

$$\text{Ratio error} = \frac{K_{nom.} - K_{act.}}{K_{act.}} \times 100$$

$$= \frac{250 - 264.75}{264.75} \times 100 = -\mathbf{5.57\%} \quad \textbf{(Ans.)}$$

(ii) **Phase angle error, θ :**

$$\text{We know that,} \quad \theta = \frac{180}{\pi}\left[\frac{I_m \cos\delta - I_w \sin\delta}{K_t I_s}\right] \text{degrees} \quad \text{...[Eqn. (5.18)]}$$

$$= \frac{180}{\pi}\left[\frac{80 \times 0.7865 - 30.9 \times 0.618}{250 \times 5}\right] = \mathbf{2.01°} \quad \textbf{(Ans.)}$$

Example 5.5. *The secondary winding burden of current transformer, having a bar primary and 300 secondary turns, is an ammeter of resistance 1.0 Ω and reactance 0.55 Ω. The resistance and reactance of secondary winding are 0.3 Ω and 0.25 Ω respectively. The core requires the equivalent of an mmf of 90 AT for magnetisation and 45 AT for core losses. Determine the following :*

(i) The primary winding current and actual transformation ratio when the ammeter in the secondary winding circuit indicates 5 A.

(ii) The number of turns which could be reduced in the secondary winding so that the ratio error be zero for this condition.

Solution. *Given* : $N_p = 1$; $N_s = 300$; $R_e = 1.0$ Ω; $X_e = 0.55$ Ω; $R_s = 0.3$ Ω; $X_s = 0.25$ Ω; mmf for magnetisation = 90 AT; mmf for core loss = 45 AT; $I_s = 5$ A.

(i) I_p ; $K_{act.}$:

Total secondary circuit resistance = $R_s + R_e = 0.3 + 1.0 = 1.3$ Ω

Total secondary circuit reactance = $X_s + X_e = 0.25 + 0.55 = 0.8$ Ω

$$\text{Secondary circuit phase angle, } \delta = \tan^{-1}\left(\frac{0.8}{1.3}\right) = 31.6°$$

$$\cos\delta = \cos 31.6° = 0.8517, \text{ and}$$

or,
$$\sin\delta = \sin 31.6° = 0.524$$

$$\text{Turn ratio, } K_t = \frac{N_s}{N_p} = \frac{300}{1} = 300$$

$$\text{Magnetising current, } I_m = \frac{\text{Magnetising } mmf}{N_p} = \frac{90}{1} = 90\text{ A}$$

$$\text{Loss component, } I_w = \frac{mmf \text{ equivalent to iron loss}}{N_p} = \frac{45}{1} = 45\text{A}$$

∴
$$\textit{Actual ratio, } K_{act} = K_t + \frac{I_m \sin\delta + I_w \cos\delta}{I_s} \quad \text{...[Eqn. (5.14)]}$$

$$= 300 + \frac{90 \times 0.524 + 45 \times 0.8517}{5} = \mathbf{317.1} \quad \textbf{(Ans.)}$$

$$\text{Primary current, } I_p = K_{act.} \times I_s = 317.1 \times 5 = \mathbf{1585.5\ A} \quad \textbf{(Ans.)}$$

(ii) **Number of turns to be reduced :**

For eliminating the ratio error, we must reduce the secondary winding turns or in other words we must *reduce the turn ratio.*

For zero ratio errors :

$$K_{nom.} = K_{act.} \quad (\text{where, } K_{nom.} = K_t = 300)$$

or,
$$300 = K_t' + \frac{I_m \sin\delta + I_w \cos\delta}{I_s}$$

or,
$$300 = K_t' + \frac{90 \times 0.524 + 45 \times 0.8517}{5} = K_t' + 17.1$$

or
$$K_t' = 282.9$$

Hence secondary winding turns, $N_s = K_t' \times N_p = 282.9 \times 1 = 282.9.$

$\therefore$ *Reduction in secondary winding turns* = 300 – 282.9 ≃ **17** **(Ans.)**

Example 5.6. *A current transformer has a bar primary, 100 secondary winding turns, and 100/1 nominal ratio. It is operating on an external burden of 1.45 Ω non-inductive, the secondary winding resistance being 0.25 Ω. The primary winding exciting current is 1.8 A, lagging 38.4° to the secondary voltage reversed. With 1A current flowing in the secondary winding, calculate :*

(i) The actual ratio of primary winding current to secondary winding current;

(ii) The phase angle between these currents.

Solution. *Given* : $N_p = 1$; $N_s = 100$; $K_{nom.} = 100/1$; $R_e = 1.45\ \Omega$; $R_s = 0.25\ \Omega$; $I_o = 1.8\text{A}$; Lagging angle, with secondary voltage reversed = 38.4°; $I_s = 1\text{A}$.

(*i*) **The actual ratio, $K_{act.}$:**

$$\text{Turn ratio, } K_t = \frac{N_s}{N_p} = \frac{100}{1} = 100$$

$$\text{Total secondary circuit resistance} = 0.25 + 1.45 = 1.7\ \Omega$$

$$\text{Secondary circuit reactance} = 0$$

$\therefore$ The secondary p.f. is unity and its phase angle, $\delta = 0$

Now, α = Angle between I_o and ϕ

= 90° – (angle between I_o and E_s reversed)

= 90° – 38.4° = 51.6°

Also,
$$K_{act.} = K_t + \frac{I_o}{I_s}\sin(\delta + \alpha) \quad \text{...[Eqn. (5.13)]}$$

$$= 100 + \frac{1.8}{1.0}\sin(0 + 51.6°) = \mathbf{101.41} \quad \textbf{(Ans.)}$$

(*ii*) **Phase angle, θ :**

$$\theta = \frac{180}{\pi} \times \frac{I_o \cos(\delta + \alpha)}{K_t I_s} \text{ degrees} \quad \text{...[Eqn. (5.16)]}$$

$$= \frac{180}{\pi} \times \frac{1.8\cos(0 + 51.6°)}{100 \times 1} = \mathbf{0.641°} \quad \textbf{(Ans.)}$$

Example 5.7. *A current transformer rated as 1000/5A, 25 VA has a turn ratio of 1 : 200. The core loss and magnetising components of the primary current are 5A and 8A under rated conditions. Determine the ratio and phase angle errors for the rated burden and rated secondary current at 0.8 p.f. lagging and 0.8 p.f. leading.*

Neglect the resistance and leakage reactance of secondary winding.

Solution. *Given* : $N_p = 1$, $N_s = 200$ *i.e.*, $K_t = \frac{200}{1} = 200$; Magnetising current, $I_m = 8\text{A}$;

Loss component, $I_w = 5\text{A}$; $\cos\phi = 0.8$ lagging and leading.

Case-1 : *0.8 p.f. lagging* :

For lagging p.f. the secondary phase angle δ is + ve.

$$\cos\delta = \text{secondary p.f.} = 0.8,\ \ \sin\delta = \sqrt{1^2 - (0.8)^2} = 0.6$$

$$\text{Actual ratio, } K_{act.} = K_t + \frac{I_m \sin\delta + I_w \cos\delta}{I_s} \quad \text{...[Eqn. (5.14)]}$$

$$= 200 + \frac{8 \times 0.6 + 5 \times 0.8}{5} = 201.76$$

$$\therefore \quad \textit{Ratio error} = \frac{K_{nom.} - K_{act.}}{K_{act.}} \times 100$$

$$= \frac{200 - 201.76}{201.76} = \mathbf{-0.872\%} \quad \textbf{(Ans.)} \quad (\because K_{nom.} = K_t = 200)$$

$$\textit{Phase angle, } \theta = \frac{180}{\pi}\left[\frac{I_m \cos\delta - I_w \sin\delta}{K_t I_s}\right] \text{degrees} \quad \text{...[Eqn. (5.18)]}$$

$$= \frac{180}{\pi}\left[\frac{8 \times 0.8 - 5 \times 0.6}{200 \times 5}\right] = \mathbf{0.1948°} \quad \textbf{(Ans.)}$$

Case-2 : *0.8 p.f. leading* :

For leading p.f. δ is –ve. cos δ = 0.8 and sin δ = – 0.4.

$$\text{Now,} \quad \text{Actual ratio, } K_{act.} = K_t + \frac{I_m \sin\delta + I_w \cos\delta}{I_s}$$

$$= 200 + \frac{8 \times (-0.6) + 5 \times 0.8}{5} = 199.84$$

$$\therefore \quad \textit{Ratio error} = \frac{K_{nom.} - K_{act.}}{K_{act.}} \times 100$$

$$= \frac{200 - 199.84}{199.84} \times 100 = \mathbf{+0.08\%} \quad \textbf{(Ans.)}$$

$$\textit{Phase angle, } \theta = \frac{180}{\pi}\left[\frac{I_m \cos\delta - I_w \sin\delta}{K_t I_s}\right] \text{degrees}$$

$$= \frac{180}{\pi}\left[\frac{8 \times 0.8 - 5(-0.6)}{200 \times 5}\right] = \mathbf{0.538°} \quad \textbf{(Ans.)}$$

Example 5.8. *A 500/5A, 50 Hz current transformer has a one turn primary and an output of 20 VA. The number of secondary turns is 99, and the resistance and leakage reactance of the secondary winding are 0.4 Ω and 0.35 Ω respectively. At rated load with non-inductance burden, the inphase and quadrature components (referred to the flux) of the exciting mmf are 6AT and 8AT respectively.*

Calculate the current and phase angle errors.

Solution. *Given* : $N_p = 1$; $N_s = 99$; ∴ $K_t = 99$; $K_{nom.} = \frac{500}{5} = 100$; $R_s = 0.4\ \Omega$; $X_s = 0.35\ \Omega$; Magnetising mmf = 6 AT; Loss mmf = 8 AT.

(*i*) **Ratio (current) error :**

$$\text{Magnetising current, } I_m = \frac{\text{Magnetising mmf}}{\text{Primary winding turns}} = \frac{6}{1} = 6\ \text{A}$$

$$\text{Loss component, } I_w = \frac{\text{Loss mmf}}{\text{Primary winding turns}} = \frac{8}{1} = 8\ \text{A}$$

Resistance of external burden (purely resistive – given), R_e

$$= \frac{VA}{I_s^2} = \frac{20}{5^2} = 0.8\ \Omega$$

Reactance of external burden, $X_e = 0\ \Omega$

Resistance of total secondary circuit burden

$$= R_s + R_e = 0.4 + 0.8 = 1.2\ \Omega$$

Reactance of total secondary circuit burden

$$= X_s + X_e = 0.35 + 0 = 0.35\ \Omega$$

$$\text{Secondary phase angle, } \delta = \tan^{-1}\frac{0.35}{1.2} = 16.26°$$

$\therefore \cos\delta = \cos 16.26° = 0.96$ and $\sin\delta = \sin 16.26° = 0.28$

$$\text{Actual ratio, } K_{act.} = K_t + \frac{I_m \sin\delta + I_w \cos\delta}{I_s}$$

$$= 99 + \frac{6 \times 0.28 + 8 \times 0.96}{5} = 100.87$$

$$\textit{Ratio (current) error} = \frac{K_{nom.} - K_{act.}}{K_{act.}} \times 100$$

$$= \frac{100 - 100.87}{100.87} \times 100 = \mathbf{-0.86\%} \quad \textbf{(Ans.)}$$

(*ii*) **Phase angle error, θ :**

$$\theta = \frac{180}{\pi}\left[\frac{I_m \cos\delta - I_w \sin\delta}{K_t I_s}\right] \text{degrees}$$

$$= \frac{180}{\pi}\left[\frac{6 \times 0.96 - 8 \times 0.28}{99 \times 5}\right] = \mathbf{0.407°} \quad \textbf{(Ans.)}$$

Example 5.9. *A 100/5A current transformer, at its rated load of 20VA, has an iron loss of 0.18 W and a magnetising current of 1.4 A. It is supplying rated output to a meter having a ratio of resistance to reactance of 4. Calculate :*

(i) Ratio error.

(ii) Phase angle.

Solution. *Given* : $K_{nom.} = \frac{100}{5} = 20$; Burden of secondary circuit = 20 VA; Iron loss = 0.18 W; I_m = 1.4 A; $\frac{X}{R} = 4$.

(*i*) **Ratio error :**

Assuming $K_{nom.} = K_t = 20$ ... Any other data being absent.

Neglecting the secondary winding burden,

the total burden of the secondary circuit = Burden of the meter.

$\therefore$ Burden of secondary circuit = 20 VA

$$\text{Voltage across primary winding, } E_p = \frac{VA}{I_p} = \frac{20}{100} = 0.2V$$

$$\text{Loss component of exciting current, } I_w = \frac{\text{Iron loss}}{E_p} = \frac{0.18}{0.2} = 0.9\text{ A}$$

Secondary circuit phase angle, $\delta = \tan^{-1}\left(\frac{X}{R}\right)$

or, $\delta = \tan^{-1}\left(\frac{1}{4}\right) = 14.04°$

$\therefore$ $\cos\delta = 0.97$ and $\sin\delta = 0.2426$

$$\text{Actual ratio, } K_{act.} = K_t + \frac{I_m \sin\delta + I_w \cos\delta}{I_s}$$

$$= 20 + \frac{1.4 \times 0.2426 + 0.9 \times 0.97}{5} = 20.24$$

$$\therefore \quad \textit{Ratio error} = \frac{K_{nom.} - K_{act.}}{K_{act.}} \times 100$$

$$= \frac{20 - 20.24}{20.24} \times 100 = -\mathbf{1.186\%} \quad \textbf{(Ans.)}$$

(*ii*) **Phase angle, θ :**

$$\theta = \frac{180}{\pi}\left[\frac{I_m \cos\delta - I_w \sin\delta}{K_t I_s}\right] \text{ degree}$$

or, $$\theta = \frac{180}{\pi}\left[\frac{1.4 \times 0.97 - 0.9 \times 0.2426}{20 \times 5}\right] = \mathbf{0.653°} \quad \textbf{(Ans.)}$$

5.4 POTENTIAL TRANSFORMER (PT)

5.4.1. General Aspects

A **potential transformer** *is a step down transformer used along with a low range voltmeter for measuring a high voltage.*

- The primary is connected across the high voltage supply and the secondary to the voltmeter or potential coil of the wattmeter.
- Since the voltmeter (or potential coil) impedance is very high, the secondary current is very small and the potential transformer behaves as an ordinary two winding transformer operating on no-load.
- Fig. 5.7. shows a potential transformer used to measure the voltage of a circuit. It may be noted that the secondary is grounded. *This is done so that if the insulation breaks down, the high voltage does not endanger personnel who may be reading the meters.*
- These transformers are *made with high quality iron core operating at very low flux densities so that the magnetising current may be very small.*
- Careful design ensures minimum variation of voltage ratio with load and minimum phase shift between input and output voltages.

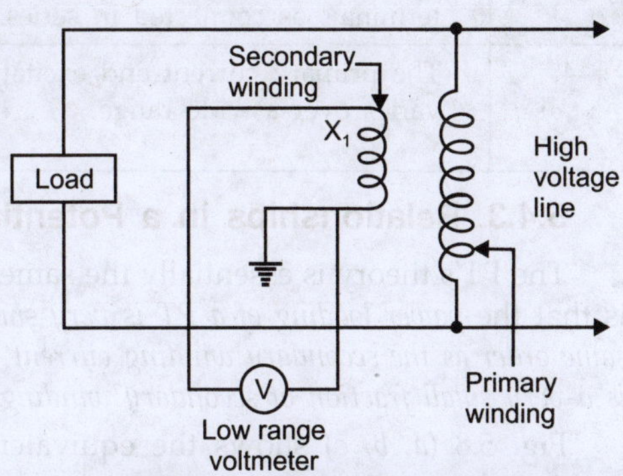

Fig. 5.7. Potential transformer connections.

- Potential transformer secondary is commonly designed for an output of **110 V.**
- PTs are usually of the following types :
  - — Upto voltage of 5000 V ..... Usually of *dry type.*
  - — Between 5000 V and 1,38000 V .... *Either dry type or oil immersed.*
  - — Voltages exceeding 1,38000 V .... *Always immersed type.*
- As a rule, the PTs are of *shell type* because such a construction develops a *high degree of accuracy.*
- Potential transformer (PT) is different from power transformer. In power transformer current is 5% of full-load current, while in *case of PT, secondary current is of the same order as magnetising current.* Thus *PT has a small volt-ampere ratings (say about 15 VA).*
- The primary and secondary winding of the PT are *coaxial* in order to reduce the *leakage reactance to a minimum.* The low voltage winding (secondary winding) is placed next to the core.
- Oil filled bushings are used for oil filled PTs to minimise the overall size of the transformer. When neither side of the line is at ground potential, two bushings are used.

5.4.2. Comparison of Current and Potential Transformers

The comparison between current and potential transformers is given in the table below :

S.No.	Current transformer (CT)	Potential transformer (PT)
1.	Secondary must *always be shorted.*	Secondary is *nearly under open circuit conditions.*
2.	The winding carries *full-line current.*	The winding is impressed with *full-line voltage.*
3.	The primary current is *independent* of the secondary circuit conditions.	The primary current depends on the secondary circuit conditions.
4.	It can be treated as series transformer under short circuit conditions.	It can be treated as parallel transformer under open circuit secondary.
5.	A small voltage exists across its terminals as connected in series.	Full line voltage appears across its terminals.
4.	The primary current and excitation varies over a wide range.	The line voltage is almost constant hence exciting current and flux density varies over a limited range.

5.4.3. Relationships in a Potential Transformer (PT)

The PT's theory is essentially the same as that of a power transformer. The difference is that the *power loading of a PT is very small and consequently the exciting current is of the same order as the secondary winding current, while in a power transformer the exciting current is a very small fraction of secondary winding load current.*

Fig. 5.8 (*a, b, c*) shows the equivalent circuit, phasor diagram and enlarged and concise phasor diagram (referred to primary) of a potential transformer. Here,

$$\phi = \text{Working flux},$$

$$I_m = \text{Magnetising current},$$

I_w = Iron loss component of no load (exciting) current,
I_o = No load (exciting) current,
E_s = Secondary winding induced voltage,

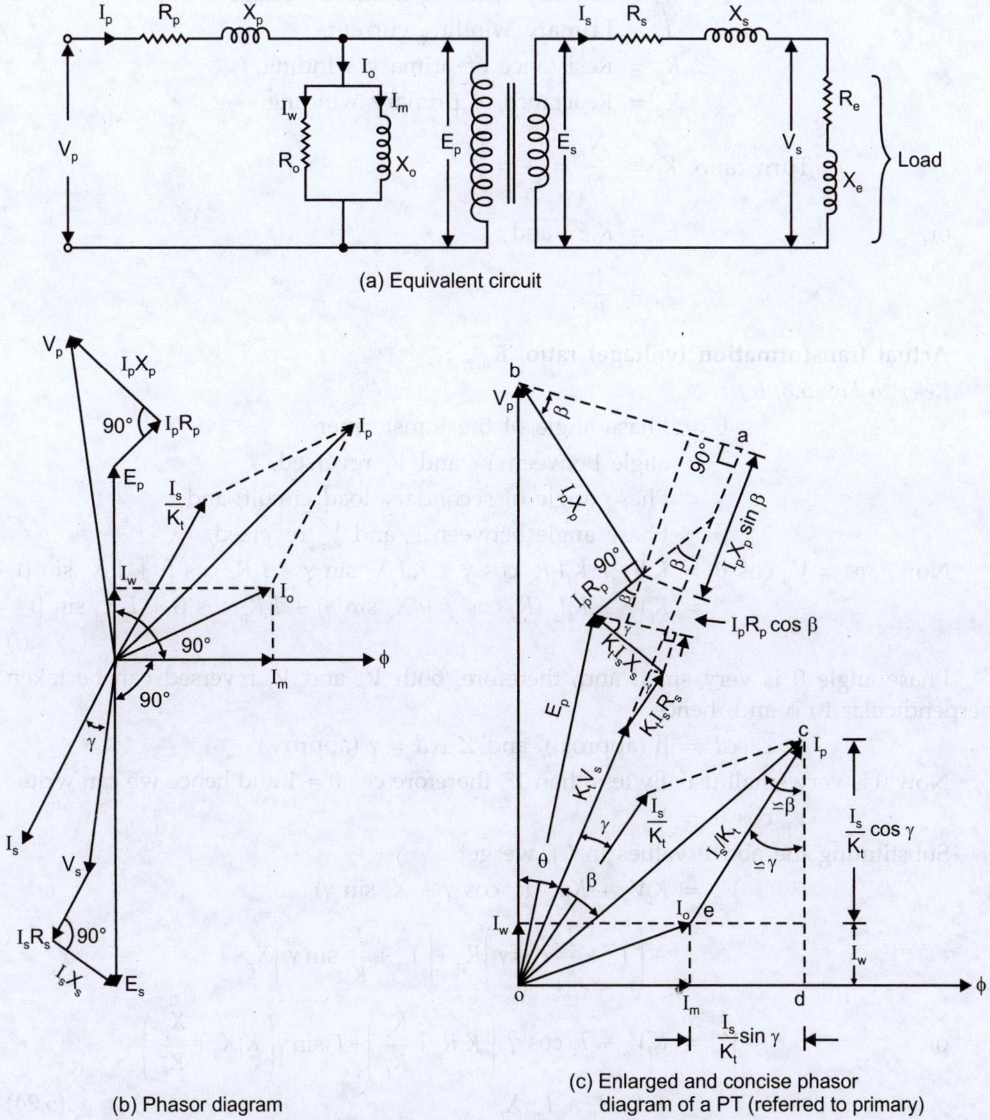

Fig. 5.8. Potential transformer (PT).

V_s = Secondary winding terminal voltage,
I_s = Secondary winding current,
R_s = Resistance of secondary winding,
X_s = Reactance of secondary winding,
R_e = Resistance of secondary load circuit,
X_e = Reactance of secondary load circuit,
γ = Phasor angle of secondary load circuit,

$$= \tan^{-1}\left(\frac{X_e}{R_e}\right)$$

E_p = Primary winding induced voltage,

I_p = Primary winding current,

R_p = Resistance of primary winding,

X_p = Reactance of primary winding.

$$\text{Turn ratio, } K_t = \frac{N_p}{N_s} = \frac{E_p}{E_s} = \frac{I_s}{I_p}$$

or, $E_p = K_t E_s$, and

$$I_p = \frac{I_s}{K_t}$$

Actual transformation (voltage) ratio, $K_{act.}$:

Refer to Fig. 5.8. (c) :

θ = Phase angle of the transformer

= angle between V_p and V_s reversed,

γ = Phase angle of secondary load circuit, and

β = Phase angle between I_p and V_s reversed,

Now, $oa = V_p \cos\theta = K_t V_s + K_t I_s R_s \cos\gamma + K_t I_s X_s \sin\gamma + I_p R_p \cos\beta + I_p X_p \sin\beta$

$$= K_t V_s + K_t I_s (R_s \cos\gamma + X_s \sin\gamma) + I_p R_p \cos\beta + I_p X_p \sin\beta \quad ...(i)$$

Phase angle θ is very small and, therefore, both V_p and V_s reversed can be taken perpendicular to ϕ and, hence

$$\angle ocd = \beta \text{ (approx.), and } \angle ecd = \gamma \text{ (approx.)}$$

Now θ is very small usually less than 1°, therefore $\cos\theta = 1$ and hence we can write :

$$V_p \cos\theta = V_p$$

Substituting the above values in (*i*), we get :

$$V_p = K_t V_s + K_t I_s (R_s \cos\gamma + X_s \sin\gamma) + \left(I_w + \frac{I_s}{K_t}\cos\gamma\right)R_p + \left(I_m + \frac{I_s}{K_t}\sin\gamma\right)X_p$$

or, $$V_p = K_t V_s + I_s \cos\gamma\left(K_t R_s + \frac{R_p}{K_t}\right) + I_s \sin\gamma\left(K_t X_s + \frac{X_p}{K_t}\right) + I_w R_p + I_m X_p \quad ...(5.24)$$

$$= K_t V_s + \frac{I_s}{K_t}\cos\gamma\left(K_t^2 R_s + R_p\right) + \frac{I_s}{K_t}\sin\gamma\left(K_t^2 X_s + X_p\right) + I_w R_p + I_m X_p$$

$$= K_t V_s + \frac{I_s}{K_t}\cos\gamma\,(R_p)_{eq} + \frac{I_s}{K_t}\sin\gamma\,(X_p)_{eq} + I_w R_p + I_m X_p$$

[where, $K_t^2 R_s + R_p = (R_p)_{eq}$ = equivalent resistance of transformer referred to 'primary'
and, $K_t^2 X_s + X_p = (X_p)_{eq}$ = equivalent reactance of transformer referred to 'primary']

$$\therefore \quad V_p = K_t V_s + \frac{I_s}{K_t}\left[(R_p)_{eq}\cos\gamma + (X_p)_{eq}\sin\gamma\right] + I_w R_p + I_m X_p \quad ...(5.25)$$

$\therefore$ *Actual transformation (voltage) ratio,*

$$K_{act.} = \frac{V_p}{V_s} = K_t + \frac{\frac{I_s}{K_t}\left[(R_p)_{eq}\cos\gamma + (X_p)_{eq}\sin\gamma\right] + I_w R_p + I_m X_p}{V_s} \quad ...(5.26)$$

Eqn. (5.24) can be rewritten as :

$$V_p = K_t V_s + K_t I_s \cos\gamma\left(R_s + \frac{R_p}{K_t^2}\right) + K_t I_s \sin\gamma\left(X_s + \frac{X_p}{K_t^2}\right) + I_w R_p + I_m X_p$$

$$= K_t V_s + K_t I_s \cos\gamma\,(R_s)_{eq} + K_t I_s \sin\gamma\,(X_s)_{eq} + I_w R_p + I_m X_p$$

[where, $R_s + \frac{R_p}{K_t^2} = (R_s)_{eq}$ = equivalent resistance of transformer *referred to 'secondary'*

and, $X_s + \frac{X_p}{K_t^2} = (X_s)_{eq}$ = equivalent reactance of transformer *referred to 'secondary'*]

$$\therefore \quad V_p = K_t V_s + K_t I_s\,[(R_s)_{eq}\cos\gamma + (X_s)_{eq}\sin\gamma) + I_w R_p + I_m X_p \quad ...(5.27)$$

$\therefore$ *Actual transformation (voltage) ratio,*

$$K_{act.} = \frac{V_p}{V_s} = K_t + \frac{K_t I_s\left[(R_s)_{eq}\cos\gamma + (X_s)_{eq}\sin\gamma\right] + I_w R_p + I_m X_p}{V_s} \quad ...(5.28)$$

Using eqns. 5.26 and 5.28, the difference between $K_{act.}$ and K_t is :

$$K_{act.} - K_t = \frac{\frac{I_s}{K_t}\left[(R_p)_{eq}\cos\gamma + (X_p)_{eq}\sin\gamma\right] + I_w R_p + I_m X_p}{V_s} \quad ...(5.29)$$

$$= \frac{K_t I_s\left[(R_s)_{eq}\cos\gamma + (X_s)_{eq}\sin\gamma\right] + I_w R_p + I_m X_p}{V_s} \quad ...(5.30)$$

Phase angle, θ :

From the phasor diagram [Fig. 5.8 (*c*)], we have :

$$\tan\theta = \frac{ab}{oa} = \frac{I_p X_p \cos\beta - I_p R_p \sin\beta + K_t I_s X_s \cos\gamma - K_t I_s R_s \sin\gamma}{K_t V_s + K_t I_s R_s \cos\gamma + K_t I_s X_s \sin\gamma + I_p R_p \cos\beta + I_p X_p \sin\beta}$$

The terms in denominator involving I_p and I_s are small and, therefore, they can be neglected as compared with $K_t V_s$

$$\therefore \quad \tan\theta = \frac{I_p X_p \cos\beta - I_p R_p \sin\beta + K_t I_s X_s \cos\gamma - K_t I_s R_s \sin\gamma}{K_t V_s}$$

$$= \frac{X_p\left(I_w + \frac{I_s}{K_t}\cos\gamma\right) - R_p\left(I_m + \frac{I_s}{K_t}\sin\gamma\right) + K_t I_s X_s \cos\gamma - K_t I_s R_s \sin\gamma}{K_t V_s}$$

$$= \frac{I_s \cos\gamma\left(\frac{X_p}{K_t} + K_t X_s\right) - I_s \sin\gamma\left(\frac{R_p}{K_t} + K_t R_s\right) + I_w X_p - I_m R_p}{K_t V_s}$$

$$= \frac{\frac{I_s \cos\gamma}{K_t}(X_p + K_t^2 X_s) - \frac{I_s \sin\gamma}{K_t}(R_p + K_t^2 R_s) + I_w X_p - I_m R_p}{K_t V_s}$$

$$= \frac{\frac{I_s \cos\gamma}{K_t}(X_p)_{eq} - \frac{I_s \sin\gamma}{K_t}(R_p)_{eq} + I_w X_p - I_m R_p}{K_t V_s}$$

$$= \frac{\frac{I_s}{K_t}\left[(X_p)_{eq}\cos\gamma - (R_p)_{eq}\sin\gamma\right] + I_w X_p - I_m R_p}{K_t V_s}$$

Since θ is small, therefore, θ = tan θ

$$\therefore \quad \theta = \frac{\frac{I_s}{K_t}\left[(X_p)_{eq}\cos\gamma - (R_p)_{eq}\sin\gamma\right] + I_w X_p - I_m R_p}{K_t V_s} \text{ rad.} \qquad ...(5.31)$$

In terms of quantities *referred to secondary,*

$$(R_s)_{eq} = \frac{R_p}{K_t^2} \text{ and } (X_s)_{eq} = \frac{X_p}{K_t^2}$$

$$\therefore \quad \theta = \frac{K_t I_s\left[(X_s)_{eq}\cos\gamma - (R_s)_{eq}\sin\gamma\right] + I_w X_p - I_m R_p}{K_t V_s}$$

or,

$$\theta = \frac{I_s}{V_s}\left[(X_s)_{eq}\cos\gamma - (R_s)_{eq}\sin\gamma\right] + \frac{I_w X_p - I_m R_p}{K_t V_s} \text{ rad.} \qquad ...(5.32)$$

Ratio (voltage) error. The actual ratio of transformation varies with operating conditions and the error in secondary voltage may be defined as :

$$\textit{Percentage ratio error} = \frac{K_{nom.} - K_{act.}}{K_{act.}} \times 100 \qquad ...(5.33)$$

Phase angle error. In an ideal transformer there should not be any phase difference between primary winding voltage and the secondary winding voltage *reversed*. However, in an actual transformer there exists a phase difference between V_p and V_s reversed.

Phase angle error = θ, is given by : same as given in eqns. 5.31 and 5.32. The phase angle is taken as *+ve* when secondary winding voltage reversed *leads* the primary winding voltage. The angle is *–ve* when secondary winding voltage reversed *lags* the primary winding voltage.

- It may be remembered that *while measuring 'voltage', ratio error is only important while for 'power measurements' both ratio and phase angle errors are involved.*

5.4.4. Characteristics of Potential Transformers

1. **Effect of change in Secondary Burden :**
   - With the increase in secondary burden, the secondary current is increased and consequently the primary current is increased. Both the primary and secondary voltage drops increase and thus for a given value of V_p, there is a decrease in the value of V_s and hence the *actual ratio increases as the burden increases.* This variation of error is *almost linear* with change in burden.
   - Further, owing to increased voltage drop, V_p advances while V_s retards in phase. As a result, the *phase angle* between V_p and V_s reversed *increases* and becomes *more negative as the current increases.*
2. **Effect of power factor of secondary burden :**
   - When the power factor of a secondary circuit burden is reduced, angle γ increases. As a consequence, current I_p shifts towards I_o [see Fig. 5.8 (*b*)]. The voltages V_p and V_s come more nearly into phase with E_p and E_s respectively since the voltage drops are almost constant. The results in an increase in V_p relative to E_p. But, since V_p is constant, therefore, E_p reduces relative to V_p; V_s also reduces relative to E_s. Thus the *transformation ratio increases as the p.f. of secondary burden reduces.*
   - While V_s advances in phase and V_p retards hence *negative phase angle reduces as lagging p.f. of the secondary decreases.*
3. **Effect of change in frequency :**
   - The flux is inversely proportional to frequency, for a constant voltage. Thus, with increase in frequency, flux is reduced and, therefore I_m and I_w are decreased leading to *decrease in voltage ratio*. The decrease is not so much, as with increase in frequency the leakage reactance increases and, therefore, leakage reactance drops are increased giving an *increase in the ratio*. Thus changes in voltage ratio (due to change in frequency) are dependent upon relative values of I_o and leakage reactances (as they produce opposite effects).
   - With the *increase in frequency*, the *phase angle is increased* since both the effects advance V_s and the increase in secondary reactance retards V_s.
4. **Effect of primary voltage.** The variations of errors due to change in primary voltage are of no importance, as supply-system voltages do not vary widely.

5.4.5. Reduction of Errors in Instrument Transformers

The errors (ratio and phase angle) in instrument transformer may be reduced by using the following methods :

1. ***Reduction of magnetising and loss components (of exciting current I_o).*** From the equations (derived earlier) of ratio and phase angle errors it is evident that these errors depend on I_m and I_w (components of I_o). These components can be *reduced* by :
   (*i*) Using materials of high permeability; (*ii*) Keeping the flux density in the core to low value; (*iii*) Choosing low reactance core; (*iv*) Using a core of large cross-section; (*v*) Providing smaller magnetic paths to the flux; (*vi*) Designing the assembly and interleaving the core with proper precautions.
2. ***Reduction of resistance and leakage reactance.*** The errors depend on the voltage drops which in turn depend on the values of resistance and leakage reactance.
   - The *resistance* can be reduced by using thick conductors (*i.e.*, of large cross-section) and using the smallest length of mean turn.

- The *leakage reactance* of windings depends upon the leakage fluxes of primary and secondary windings and, therefore, windings *must be kept as close as possible.*

— Further, a high flux density means a high flux in the core and so the windings have lesser number of turns, which obviously results in smaller leakage reactance of the windings.

3. ***Turns compensation*** **:**

- At no load (when $I_s = 0$) the actual ratio ($K_{act.}$), in case of PT, *exceeds* the turns ratio (K_t) by an amount $\frac{(I_w R_p + I_m X_p)}{V_s}$ (Refer to eqn. 5.26). The ratio further increases with an inductive or resistive load because of voltage drops in resistance and leakage reactance of these windings. If $K_t = K_{nom.}$, the $K_{act.}$ differs from the nominal ratio and *errors creep* in.

  The remedy lies in making the K_t less $K_{nom.}$ This can be accomplished by either *reducing* the number of turns of primary winding or *increasing* the number of secondary winding turns. By doing so, it becomes possible to make the actual ratio $K_{act.}$ equal to $K_{nom.}$ for one particular value of burden. At the same time, there is a general reduction in the ratio error over the entire working range of the burden. This is called *"turns compensation"*.

- *The phase angle θ, practically, remains unaffected due to a small change in* K_t.

Example 5.10. *The following data relate to a 1000/100 volts potential transformer :*

Primary resistance = 84.4 Ω

Primary reactance = 62.5 Ω

Secondary resistance = 0.78 Ω

Total equivalent reactance = 102 Ω

No-load current = 0.03 A at 0.42 power factor

Calculate :

(i) Phase angle error at no load.

(ii) Burden in VA at unity power factor at which the phase angle will be zero.

Solution. *Given :* $K_t = \frac{1000}{100} = 10$; $R_p = 84.4\ \Omega$; $X_p = 62.5\ \Omega$; $R_s = 0.78\ \Omega$; $X_e = 102\ \Omega$; $I_o = 0.03$ A at 0.42 p.f.

(i) Phase angle error at no load, θ :

No load p.f. $\cos\phi_o = 0.42$;

$$\therefore \quad \sin\phi_o = \sqrt{1-(0.42)^2} = 0.9075$$

Now,
$$I_w = I_o \cos\phi_o = 0.03 \times 0.42 = 0.0126 \text{ A}$$
$$I_m = I_o \sin\phi_o = 0.03 \times 0.9075 = 0.0272 \text{ A}$$

$$\text{Phase angle, } \theta = \frac{\frac{I_s}{K_t}\left[(X_p)_{eq}\cos\gamma - (R_p)_{eq}\sin\gamma\right] + I_w X_p - I_m R_p}{K_t V_s} \text{ rad} \quad \text{...[Eqn. (5.31)]}$$

At no load, $I_s = 0$

or, $$\theta = \frac{I_w X_p - I_m R_p}{K_t V_s} \text{rad}$$

or, $$\theta = \frac{0.0126 \times 62.5 - 0.0272 \times 86.4}{10 \times 100} = \mathbf{-0.00156\ rad} \quad \textbf{(Ans.)}$$

(ii) **Burden in VA :**

$$\theta = 0 \qquad ...(Given)$$

At unity p.f., $\cos \gamma = 1$, $\sin \gamma = 0$

$$\theta = \frac{\frac{I_s}{K_t}\left[(X_p)_{eq} \cos \gamma - (R_p)_{eq} \sin \gamma\right] + I_w X_p - I_m R_p}{K_t V_s} \text{rad.}$$

Substituting the values, we have :

or, $$\theta = \frac{\frac{I_s}{10}(102 \times 1 - 0) + 0.0126 \times 62.5 - 0.0272 \times 86.4}{10 \times 100}$$

or, $$\theta = 10.2\, I_s - 1.563$$

$\therefore$ $$I_s = \frac{1.563}{10.2} = 0.1532\, A$$

$\therefore$ Burden load in $VA = V_s I_s = 100 \times 0.1532 =$ **15.32 VA** **(Ans.)**

Example 5.11. *A single-phase potential transformer has a turns ratio of 3810/63. The nominal secondary voltage is 63 V and the total equivalent resistance and leakage reactance referred to secondary are 2 Ω and 1 Ω respectively. Calculate the ratio and phase angle errors when the transformer is supplying a burden of 100 + j 200 Ω.* **(PTU – Dec. 2005)**

Solution. *Given* : $K_t (= K_{nom.}) = \frac{3810}{63} = 60.476$; $V_s = 63$ V; $(R_s)_{eq} = 2\Omega$; $(X_s)_{eq} = 1\ \Omega$;

Burden = 100 + j 200 Ω.

Ratio error :

Neglecting no load components of exciting current,

Transformation (actual) ratio,

$$K_{act.} = K_t + \frac{K_t I_s \left[(R_s)_{eq} \cos \gamma + (X_s)_{eq} \sin \gamma\right]}{V_s}$$

or, $$K_{act.} = K_t + \frac{K_t \left[(R_s)_{eq} \cos \gamma + (X_s)_{eq} \sin \gamma\right]}{(V_s / I_s)} \qquad ...[Eqn. (5.28)]$$

$$\text{Burden} = R_e + jX_e = 100 + j\, 200\ \Omega \qquad ...(i)$$

$\therefore$ $$\gamma = \tan^{-1}\left(\frac{X_e}{R_e}\right) = \tan^{-1}\left(\frac{200}{100}\right) = 63.43°$$

Also, $$\frac{V_s}{I_s} = Z_s = \sqrt{100^2 + 200^2} = 223.6\,\Omega$$

Substituting the various values in eqn. (*i*), we get :

$$K_{act.} = 60.476 + \frac{60.476[2 \times \cos 63.43° + \sin 63.43°]}{223.6} = 60.9598$$

$$\therefore \qquad \% \text{ Ratio error} = \frac{K_{nom.} - K_{act.}}{K_{act.}} \times 100$$

$$= \frac{60.476 - 60.9598}{60.9598} \times 100 = -\mathbf{0.7936\%} \quad \textbf{(Ans.)}$$

Phase angle error, θ :

$$\theta = \frac{I_s}{V_s}\left[(X_s)_{eq} \cos\gamma - (R_s)_{eq} \sin\gamma\right] \text{ rad.} \qquad \text{...[Eqn. (5.32)]}$$

$$= \frac{1}{223.6}[1 \times \cos 63.43° - 2 \sin 63.43°] \text{ rad.}$$

$$= -5.999 \times 10^{-3} \text{rad.} = -5.999 \times 10^{-3} \times \frac{180}{\pi} \text{degrees}$$

$$= -\mathbf{0.3437°} \quad \textbf{(Ans.)}$$

5.5 CAPACITIVE POTENTIAL TRANSFORMERS

The conventional electro-magnetic type of potential transformer becomes *expensive* at voltages above about 100 kV (phase), due to insulation requirements. A capacitive voltage transformer is a less expensive alternative, this transformer is illustrated in Fig. 5.9.

— It consists of a capacitance potential divider used in conjunction with a conventional auxiliary transformer.

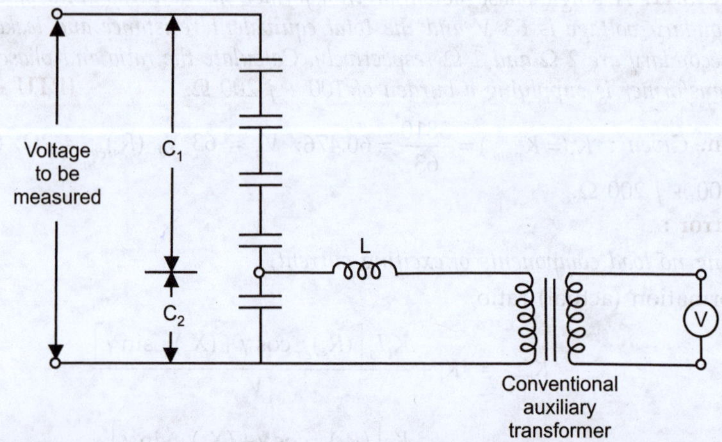

Fig. 5.9. Capacitive potential transformer.

— The capacitance potential divider *steps down the voltage to be measured*, say to about 10 kV. The output voltage of the capacitance divider is *further stepped down by the auxiliary transformer* to the desired secondary voltage, say 110 V.

— The auxiliary transformer consists of an inductance L, which may consist wholly or partly of leakage inductance of the windings of auxiliary transformer. The value of this L *may be adjusted to* $1/[\omega^2(C_1 + C_2)]$ *so that the voltage drop due to the current drained from the divider is largely compensated. Thus the overall voltage transformation ratio is independent of burden.*

The overall ratio is the product of the divider and transformer ratio.

— *The auxiliary transformer must be of very small ratio error and phase angle to give satisfactory performance of the complete unit.*

5.6 TESTING OF INSTRUMENT TRANSFORMERS

The methods of testing of instrument transformers *i.e.*, finding their ratio and phase angle errors may be broadly classified into the following two groups :

1. **Absolute methods :**
   - In these methods, the errors of transformer under test are determined in terms of constants – resistance, inductance and capacitance of the testing circuit.
2. **Comparison method :**
   - In this case, the transformer (under test) errors are compared with those of a standard current transformer whose errors are known.

The above two methods, *according to their measurement techniques*, can be *classified* as follows :

1. ***Deflection methods :***
   - These methods *employ the deflections of suitable instruments* like electrodynometer wattmeters for measuring quantities related to the phasors under consideration or to their deflection. The required ratio and phase angles are then found out from the magnitudes of deflections.
   - The deflection methods, in some cases, may be made direct reading.
2. ***Null methods :***
   - These methods employ a network in which *appropriate phasor quantities are balanced against one another. The ratio and phase angle errors are then found from the magnitudes of deflection.*
   - The null methods may be made direct reading in terms of calibrated scales on the adjustable elements in the network.

5.6.1. Testing of Current Transformers (CTs)

The following two methods of testing of CTs will be discussed :

1. Mutual inductance method ... *Absolute method*
2. Arnold method ... *Comparison method*

5.6.1.1. Mutual inductance method (*Absolute*)

This absolute method of testing CTs uses *null technique.*

Fig. 5.10 shows the circuit and phasor diagrams for the mutual inductance method :

— R_p and R_s are low resistances, non-inductive shunts. R_p is fixed while R_s is variable and has a slide wire for fine adjustment of the resistance.

— The voltage drop across R_p is matched against voltage drop across R_s. The balance conditions are indicated by a vibration galvanometer (V.G.) placed in the circuit.

— If there is no phase difference between I_p and I_s, the virbration galvanometer will indicate zero deflection when;

$$I_pR_p = I_sR_s, \quad \text{or,} \quad \frac{I_p}{I_s} = \frac{R_s}{R_p} \qquad ...(5.34)$$

Thus, R_p and R_s should be so selected that the ratio $\frac{R_s}{R_p}$ is nearly equal to nominal ratio ($K_{nom.}$) of the current transformer.

— The adjustment of resistance R_s is so made that two voltage drops are *equal*. In order *to obtain zero deflection, the magnitude as well as phase of the voltage drops should be same*. Consequently, a mutual inductance M is put *to compensate for phase* *difference between I_p and I_s as without any phase compensating device it is impossible to obtain the balance with resistance alone.*

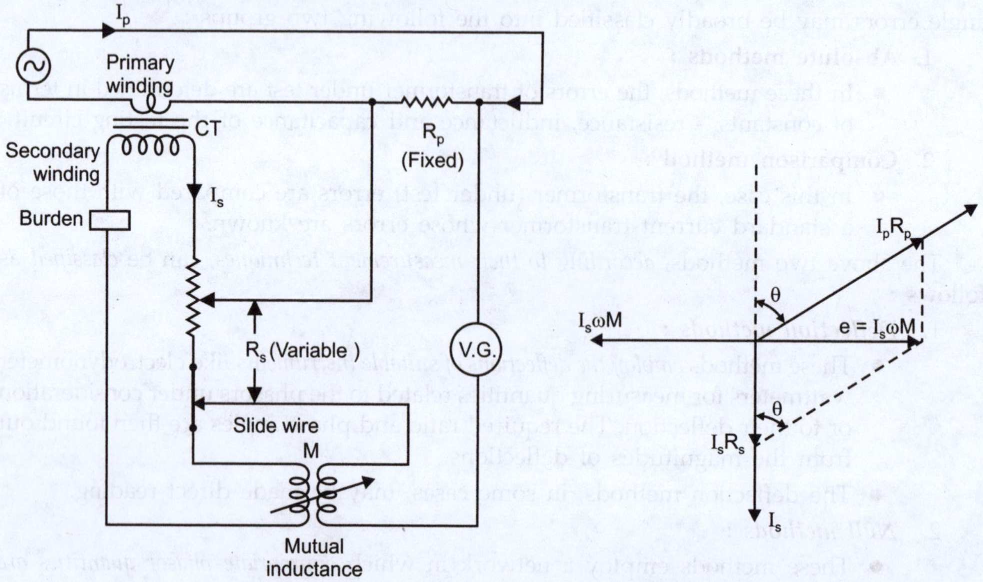

(a) Circuit diagram (b) Phasor diagram (under balance conditions)

Fig. 5.10. Mutual inductance method.

Referring to Fig. 5.10(*b*) representing phasor diagram under balance conditions, we get :

$$\tan\theta = \frac{I_s\omega M}{I_sR_s} = \frac{\omega M}{R_s} \quad ...(5.35)$$

and,
$$\cos\theta = \frac{I_sR_s}{I_pR_p} \quad ...(5.36)$$

From the above relations, we have :

$$\textit{Phase angle, } \theta \simeq \frac{\omega M}{R_s} \text{ rad.} \quad ...(5.37)$$

$$\textit{Actual ratio, } K_{act.} = \frac{I_p}{I_s} = \frac{R_s}{R_p\cos\theta} \simeq \frac{R_s}{R_p} \quad ...(5.38)$$

...θ being very small

- While stating the burden at which errors have been measured, the secondary load circuit, which includes R_s, impedance of primary winding of mutual inductance and impedance marked burden, must be taken into account.

5.6.1.2. Arnold method (*Comparison*)

This comparison method of CT testing involves *null techniques*.

- This method provides *very accurate results.*

Fig. 5.11 shows the circuit and phasor diagrams for Arnold method of CT testing:

— In this method, CT under test (*X*) is *compared with a known standard CT* (*S*) *of the same nominal ratio* ($K_{nom.}$)

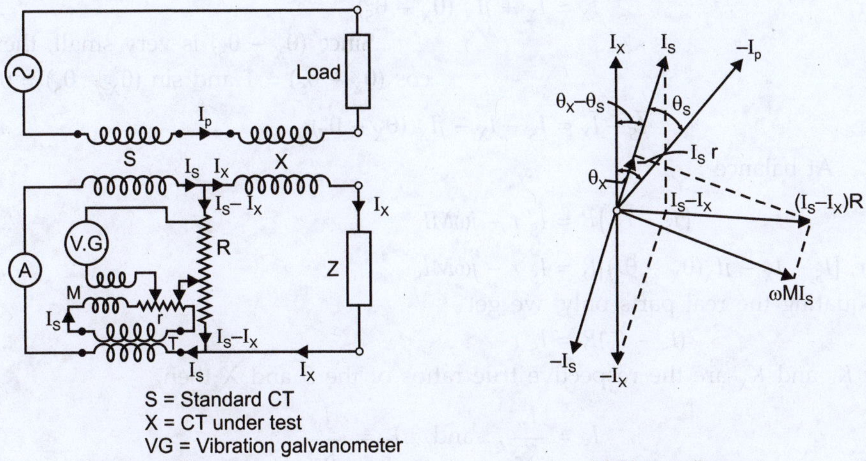

(a) Circuit diagram (b) Phasor diagram

Fig. 5.11. Arnold method of CT testing.

— *S* and *X* are connected with primaries in series and, secondaries also in series. The secondary circuit is completed by an ammeter the primary of a 5/5 isolating CT, *T* and a suitable burden *Z* for the CT under test.

— *R* is a non-inductive resistance; its value is decided according to the required sensitivity and range.

— *M* is a variable mutual inductor with a range of ± 2.4 μH and *r* is a variable resistance ranging ± 500 μΩ.

— Since the inductive inteference in measuring circuit causes error during measurement, therefore, it is essential to eliminate it. The *elimination procedure* involves the following *steps*:

(*i*) The leads of resistance *R* are disconnected and short-circuited, *r* and *M* are set at zero and full rated primary current is allowed to flow.

(*ii*) The connecting leads and apparatus of the measuring circuit are so arranged that vibration galvanometer shows no deflection.

But doing so, the circuit becomes free from inductive interference.

— The vector difference of two secondary currents ($I_S - I_X$) flows through the non-inductive resistance *R*. The potential difference across *R* is balanced against e.m.f. in the secondary of mutual inductance *M* and the voltage drop across the portion *r* of the low resistance centre tapped slide wire.

— The current in the primary mutual inductance *M* and slide wire is due to current transformer *T*. As *T* is of unity ratio so the current flowing in the primary of mutual inductance *M* and slide wire is equal to secondary current of *S* (standard CT). Balance is indicated by the vibration galvanometer and is obtained by adjustment of *M* and *r*.

• If θ_x and θ_s are the respective phase angles of the test and standard current transformers, then the angles between the *reversed* vector of I_p and the vectors I_S and I_X will be as shown in Fig. 5.11(*b*).

Taking I_S as the referene vector, we have :

$$\bar{I}_S = I_S + j0 \quad ...(5.39)$$

$$\bar{I}_X = I_X \cos(\theta_X - \theta_S) + jI_X \sin(\theta_X - \theta_S) \quad ...(5.40)$$

or, $$\bar{I}_X \simeq I_X + jI_X\,(\theta_X - \theta_S) \quad ...(5.41)$$

...Since $(\theta_X - \theta_S)$ is very small, therefore,
$\cos(\theta_X - \theta_S) \simeq 1$ and $\sin(\theta_X - \theta_S) \simeq \theta_X - \theta_S$

or, $$\bar{I}_S - \bar{I}_X = I_S - I_X - jI_X\,(\theta_X - \theta_S) \quad ...(5.42)$$

∴ At balance :

$$(\bar{I}_S - \bar{I}_X)R = I_S\, r - j\omega M I_S \quad ...(5.43)$$

or, $$[I_S - I_X - jI_X(\theta_X - \theta_S)]R = I_S\, r - j\omega M I_S \quad ...(5.44)$$

Equating the real parts only, we get

$$(I_S - I_X)R = I_S\, r \quad ...(5.45)$$

If K_S and K_X are the respective true ratios of the S and X then,

$$I_S = \frac{I_p}{K_S}, \text{ and, } I_X = \frac{I_p}{K_X}$$

Inserting these values in eqn. (5.45) we get :

$$\left(\frac{I_p}{K_S} - \frac{I_p}{K_X}\right)R = \frac{I_p}{K_S}r$$

or, $$1 - \frac{K_S}{K_X} = \frac{r}{R}$$

or, $$\frac{K_S}{K_X} = 1 - \frac{r}{R}$$

or, $$\frac{K_S}{K_X} = \frac{1}{1-(r/R)} = 1 + \frac{r}{R} \quad ...(5.46)$$

...since $\frac{r}{R}$ is very small.

Hence, K_X can be determined from the known values of K_S, r and R. Now, equating imaginary parts of eqn (5.44) we get :

$$-jI_X\, R(\theta_X - \theta_S) = -j\omega M I_S \quad ...(5.47)$$

or, $$(\theta_X - \theta_S) = \frac{\omega M I_S}{R I_X} \simeq \frac{\omega M}{R} \quad ...(5.48)$$

...since $\frac{I_S}{I_X}$ is very *close to unity*.

5.6.2. Testing of Potential Transformers (PTs)

The following tests are conducted on the potential transformers, for the determination of the ratio and phase angle errors :

5.6.2.1. Absolute null methods

The principle of most absolute methods of testing potential transformers is to balance the secondary voltage against a fraction of the primary voltage obtained from a potential divider and the errors are measured in terms of circuit constants.

Fig. 5.12 shows the circuit and phasor diagrams for such a method :

— The burden (Z) with which the transformer is to be tested is connected across the secondary winding, and the normal primary voltage at normal frequency, is applied to the primary winding. One end of the secondary winding is connected to one end of the primary winding.

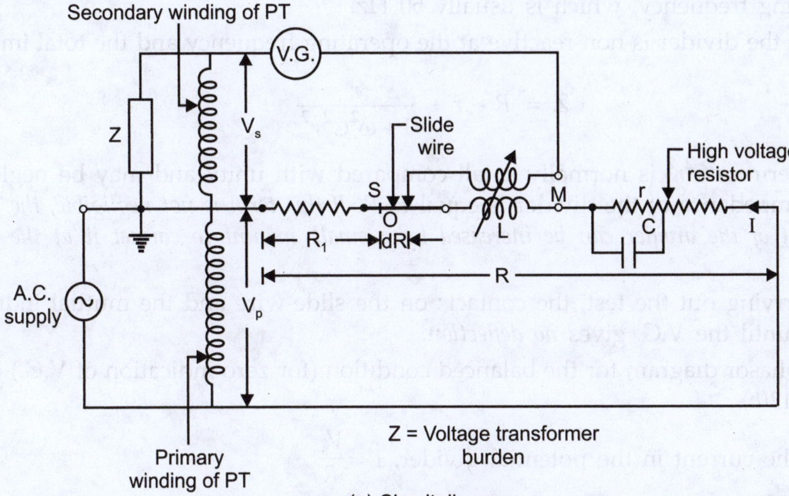

(a) Circuit diagram

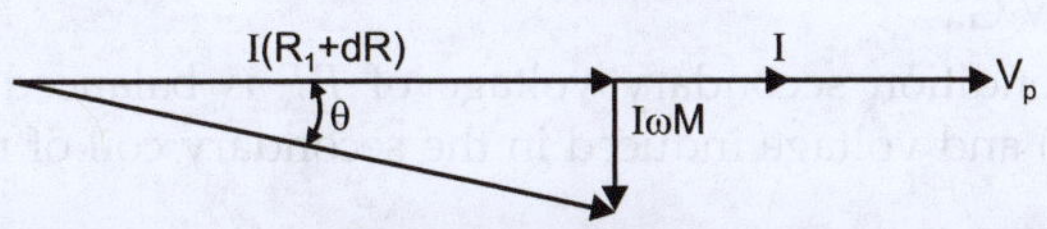

(b) Phasor diagram

Fig. 5.12. Absolute method of testing potential transformers.

— A non-inductive and non-capacitive potential divider is connected across the primary winding. This divider consists of fixed resistance R_1 in series with slide wire *S*, which is connected to *h.v.* resistance *r* in series with a primary coil of a variable mutual inductor *M*. A variable capacitor C shunts a small portion of the resistance of the potential divider.

— V.G. is a vibration galvanometer. The resistances are so chosen that nominal ratio ($K_{nom.}$) of the test transformer is equal to $\frac{R}{R_1}$.

L is the inductance of the primary winding of the mutual inductor and will cause a phase error in the divider ; the capacitor *C* compensates for this error in the following way :

The impedance of the whole divider,

$$\bar{Z} = R - r + j\omega L + \frac{r}{1 + j\omega Cr} \qquad ...(5.49)$$

or,

$$\bar{Z} = R - r + j\omega L + \frac{r}{1 + j\omega Cr} \times \frac{1 - j\omega Cr}{1 - j\omega Cr}$$

$$= R - r + j\omega L + \frac{r(1 - j\omega Cr)}{1 + \omega^2 C^2 r^2}$$

$$= R - r + \frac{r}{1 + \omega^2 C^2 r^2} + j\omega \left(L - \frac{Cr^2}{1 + \omega^2 C^2 r^2} \right) \qquad ...(5.50)$$

Now, in this divider, the values of C anf r are so chosen that the term $\frac{Cr^2}{1+\omega^2C^2r^2} = L$ at operating frequency, which is usually 50 Hz.

Now, the divider is non-reactive at the operating frequency and the total impedance,

$$Z = R - r + \frac{r}{1+\omega^2C^2r^2} \quad ...(5.51)$$

The term $\omega^2C^2r^2$ is normally small compared with unity and may be neglected, so that the impedance of the divider is equal to R. *If this term is not negligible, the resistance of the part of the divider can be increased by a small amount to correct it at the operating frequency.*

In carrying out the test, the contacts on the slide-wire and the mutual inductor are adjusted until the V.G. gives *no deflection.*

The phasor diagram for the balanced condition (for zero indication of V.G.) is shown in Fig. 5.12(*b*) :

The current in the potential divider, $I = \frac{V_s}{R}$

Let dF be the resistance of slide wire inserted and M be the mutual inductor to have zero deflection of V.G.

In balanced, condition secondary voltage of PT is balanced by the voltage across resistance ($R_1 + dR$) and voltage induced in the secondary coil of mutual inductor M (see Fig. 5.12b).

$$\therefore \quad \bar{V}_s = I(R_1 + dR) - jI\omega M \quad ...(5.52)$$

The phase angle θ of the PT under test is usually very small, therefore, it can be neglected and,

$$V_s = I(R_1 + dR)$$

$$\text{or,} \quad V_s = \frac{V_p}{R}(R_1 + dR) \quad ...(5.53)$$

Thus, the transformer ratio is, $\frac{V_p}{V_s} = \frac{R}{R_1 + dR}$

- The *ratio error* (of PT) $= \frac{K_{nom.} - K_{act.}}{K_{act.}}$

$$= \frac{\frac{R}{R_1} - \frac{R}{R_1 + dR}}{\frac{R}{R_1 + dR}} = \frac{R_1 + dR}{R_1} - 1 = \frac{dR}{R_1} \quad ...(5.54)$$

R_1 is a fixed quantity in the circuit so that ratio error of PT can be obtained directly from the slide wire calibrated in values of ratio error.

- *Phase angle,* θ. From the phasor diagram (Fig. 5.12b),

$$\tan\theta = \frac{I\omega M}{I(R_1 + dR)} = \frac{\omega M}{R_1 + dR} \quad ...(5.55)$$

Since dR is very small in comparison to R_1, therefore, it can be neglected and θ being very small, $\tan\theta$ can be taken equal to θ.

$\therefore$ *Phase angle error* (of PT), $\theta = \dfrac{\omega M}{R_1}$ ...(5.56)

Thus *mutual inductor M can be calibrated directly in terms of the phase angle error.*

5.6.2.2. Comparison method

In this method, the PT under test (X) with burden across the secondary is *compared* with standard PT (S) whose errors are small and known.

Fig. 5.13 shows the circuit diagram and phasor diagram respectively for comparison method of testing potential transformers :

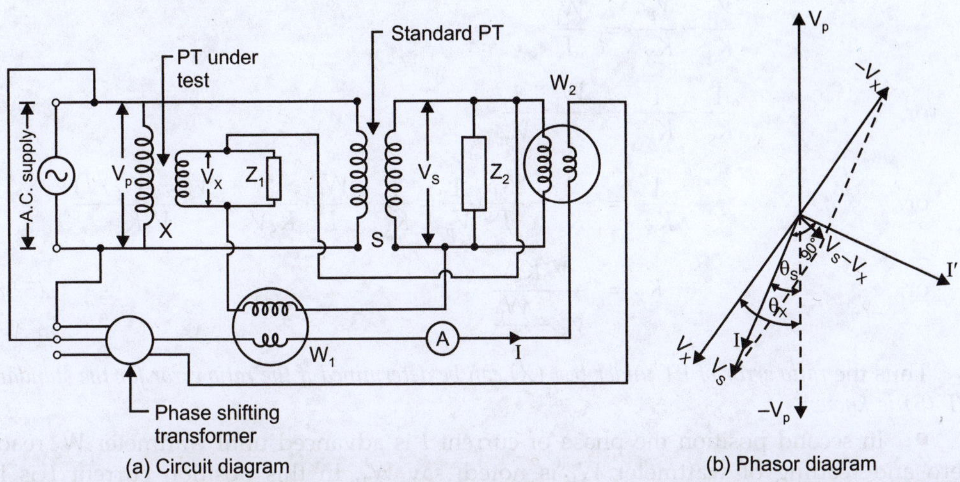

Fig. 5.13. Comparison method of testing potential transformers.

— The primary windings of both the PTs (*i.e.*, S and X) are connected in parallel; the secondary windings are so connected that the voltage acting across the potential coil of wattmeter W_1 is the vector difference of the two secondary voltages V_S and V_X. The potential coil of wattmeter W_2 is directly connected across the secondary of the standard PT (S).

— The current coils of both the wattmeters are connected in series so as to carry the same current and are *supplied from a phase shifting transformer.*

— Burden impedances Z_1, and Z_2 are adjusted to the required values.

— The wattmeter W_2 acts as a phase indicating device and also indicates the voltages at which transformers are tested.

The phasor diagram is shown in Fig. 5.13(b).

- During experimentation, since the phase shifting transformer is first adjusted to give the maximum value, therefore, the current I will be in phase with secondary voltage of standard transformer (S), V_S as shown in phasor diagram (Fig. 5.13b) and reading of the wattmeter $W_2 = V_S I$. The *reading of wattmeter W_1 for this setting* is also noted which will be equal to the *product of current I and the component of the voltage which is in phase with this current*. Since in this position V_S and V_X are nearly in phase opposition across the potential coil of wattmeter W_1, so the voltage component of voltage circuit in phase with currents I is $(V_S - V_X)\cos(\theta_X - \theta_S)$, and

$$W_1 = I\,[(V_S - V_X)\cos(\theta_X - \theta_S)] \quad ...(5.57)$$

$$\simeq I\,(V_S - V_X)$$

...since $\sin(\theta_X - \theta_S)$ is very small, therefore, $\cos(\theta_X - \theta_S) \simeq 1$.

or, $$V_S - V_X = \frac{W_1}{I}$$

If R_S and R_X are true voltage transformation ratios of S and X potential transformers, then

$$\frac{V_p}{V_S} = K_S, \text{ and, } \frac{V_p}{V_X} = K_X$$

Inserting these values in eqn. (5.57), we get

$$\frac{V_p}{K_S} - \frac{V_p}{K_X} = \frac{W_1}{I}$$

or, $$\frac{1}{K_S} - \frac{1}{K_X} = \frac{W_1}{I} \cdot \frac{I}{V_p}$$

or, $$\frac{1}{K_X} = \frac{1}{K_S} - \frac{W_1}{I} \cdot \frac{1}{V_p} = \frac{1}{K_S} - \frac{W_1}{I} \cdot \frac{1}{K_S V_S} = \frac{V_S - (W_1/I)}{V_S K_S}$$

or, $$K_X = \frac{V_S K_S}{V_S - \dfrac{W_1}{I}} \qquad \text{..(5.58)}$$

Thus the *ratio error of PT under test (X) can be determined if the ratio error for the standard PT (S) is known.*

- In second position the phase of current I is advanced until wattmeter W_2 reads zero and reading of wattmeter W_1 is noted, say W'_1. In this position current I is in quadrature with V_S and new reading of wattmeter W_1 is given as :

$$W'_1 = IV_X \cos [90° - (\theta_X - \theta_S)] \qquad \text{...(5.59)}$$

$$= IV_x \sin (\theta_X - \theta_S)$$

or, $$\sin (\theta_X - \theta_S) = \frac{W'_1}{IV_x}$$

Since $V_X = V_S$ and $(\theta_X - \theta_S)$ is very small, therefore, $\sin (\theta_X - \theta_S) = \theta_X - \theta_S$

$\therefore$ $$\theta_X - \theta_S = \frac{W'_1}{IV_s} \qquad \text{...(5.60)}$$

Thus, *the phase angle of PT under test (X) can be determined if the phase angle for standard PT (S) is known.*

HIGHLIGHTS

1. The principle of the instrument transformer is fundamentally the same as that of the power transformer.
2. Current transformers (CTs) are used for the measurement of *'current'*.
3. Potential transformers (PTs) are used for the measurement of *'voltage'*.
4. CT and PT are extensively used for precise measurements.
5. Ratio Correction Factor (RCF) is the ratio of actual transformation ratio to nominal ratio (*i.e.*, RCF = $\frac{K_{act.}}{K_{nom.}}$).

4. The *CT must never be operated on open circuit* because : (*i*) There will be no secondary mmf and since the primary current (and mmf) is fixed, the core flux will increase enormously. This will cause large eddy current and hysteresis losses and the resulting high temperature may damage the insulation or even the core; (*ii*) A very high voltage will be induced in the multi-turn secondary and this high voltage may be dangerous both to life and to the insulation.
7. A potential transformer (PT) is a *step down transformer* used along with a low range voltmeter for measuring a high voltage.
8. PT has a small volt-ampere ratings (say 15 VA).

OBJECTIVE TYPE QUESTIONS

Choose the Correct Answer :

1. The ratio of transformation in the case of potential transformers
   (*a*) increases with increase in power factor of secondary burden.
   (*b*) remains constant irrespective of the power factor of secondary burden.
   (*c*) decreases with increase in power factor of secondary burden.
   (*d*) none of the above.
2. In case of potential transformers
   (*a*) the phase angle error is always positive.
   (*b*) the phase angle error is always negative.
   (*c*) the phase angle error is usually zero.
   (*d*) the phase angle error is positive when the secondary winding voltage reversed leads the primary winding voltage and is negative when the secondary winding voltage reversed lags behind the primary winding voltage.
3. The disadvantages of using multipliers with voltmeters at high voltages are :
   (*a*) The power consumption of multipliers becomes large at large voltages.
   (*b*) The multipliers at high voltages have to be shielded in order to prevent capacitive currents.
   (*c*) The metering circuit is not electrically isolated from the power circuit.
   (*d*) all of the above.
4. The nominal ratio of a current transformer is
   (*a*) primary winding current/secondary winding current
   (*b*) rated primary winding current/rated secondary winding current
   (*c*) number of secondary winding turns/number of primary winding turns
   (*d*) all of the above.
5. The errors in current transformers can be reduced by designing them with
   (*a*) high permeability and low loss core materials, avoiding any joints in the core and also keeping the flux density to a low value.
   (*b*) using primary and secondary windings as close to each other as possible.
   (*c*) using large cross-sections for both primary and secondary winding conductors.
   (*d*) all of the above.
6. Capacitive potential transformers are used
   (*a*) for primary winding phase voltages above 100 kV.
   (*b*) for keeping the value of transformation ratio constant irrespective of the burden by making certain adjustment.

(c) because they are cheaper than the electromagnetic transformers above a certain voltage range.
(d) all of the above.

7. The size of potential transformers
(a) is the same as that of power transformers of the same VA rating.
(b) is much greater than that of power transformers of the same VA rating because they are designed for low ratio and phase angle errors which require large sized cores and winding conductors.
(c) is smaller than that of power transformers of the same VA rating.
(d) none of the above.

8. The transformation ratio in the case of a potential transformer is defined as ratio of
(a) primary winding voltage/secondary winding voltage
(b) rated primary winding voltage/rated secondary winding voltage
(c) number of turns of primary winding/number of turns of secondary winding
(d) all of the above.

9. The burden of current transformers is expressed in terms of
(a) secondary winding current
(b) VA rating of transformer
(c) voltage, current and power factor of secondary winding circuit
(d) none of the above.

10. The current in the primary winding of a current transformer depends upon
(a) burden of the secondary winding of transformer.
(b) load connected to the system in which the C.T. is installed.
(c) both burdens of the transformer secondary winding and load connected to the system.
(d) none of the above.

11. Turns compensation is used in current transformers primarily for reduction of
(a) phase angle error.
(b) both ratio and phase angle errors.
(c) ratio error, reduction in phase angle error is incidental.
(d) none of the above.

12. The advantages of instrument transformers are :
(a) The readings of instruments used in conjunction with them do not depend upon their resistance, inductance etc.
(b) The ratings of instrument transformers have been standardized and the ratings of instruments used in conjunction with them also get standardized. Therefore, there is reduction of cost and ease in replacement.
(c) The metering circuit is electrically isolated from the power circuit thereby providing safety to operating personnel.
(d) all of the above.

13. The disadvantages of shunts for use at high currents are :
(a) It is difficult to achieve good accuracy with shunts.
(b) Power consumption of the shunts is large.
(c) The metering circuit is not electrically isolated from the power circuit.
(d) All of the above.

14. When the secondary winding of a current transformer is open-circuited with the primary winding energized
(a) the whole of the primary current produces large value of flux in the core (limited to

only saturation) thereby producing a large voltage in the secondary winding.

(*b*) the large voltage may act as safety hazard for the operators and may even rupture the insulation.

(*c*) when the large magnetizing force is taken off it leaves a large value of residual magnetism.

(*d*) all of the above.

15. Clamp on meters are used because

(*a*) with their use it is possible to measure current flowing in a line without breaking the circuit.

(*b*) their accuracy is high.

(*c*) it is not possible to measure voltage without them.

(*d*) all of the above.

16. A short-circuiting link is provided on the secondary side of a current transformer because :

(*a*) when the secondary winding of the CT is short-circuited by the link with the primary winding energized a very high current flows on the primary side.

(*b*) when the secondary winding of the CT is short circuited by the link with the primary winding energized, it is possible to make any adjustments in the secondary winding circuit like replacing a faulty ammeter.

(*c*) when the short-circuiting link is opened with the secondary winding open-circuited, the current on the primary side falls to almost zero.

(*d*) all of the above.

17. When the secondary winding of a potential transformer is suddenly open circuited with primary winding excited

(*a*) large voltages are produced in the secondary winding which may be a safety hazard to operating personnel.

(*b*) the large voltages so produced may rupture the insulation.

(*c*) the primary winding draws only the no load current.

(*d*) none of the above.

18. The ratio and phase angle errors in potential transformers may be reduced by

(*a*) increasing the exciting current.

(*b*) increasing the resistance and leakage reactance in the transformer.

(*c*) not employing turns compensation.

(*d*) none of the above.

ANSWERS

1.	(*c*)	**2.**	(*d*)	**3.**	(*d*)	**4.**	(*b*)	**5.**	(*a*)	**4.**	(*d*)	**7.**	(*b*)
8.	(*a*)	**9.**	(*b*)	**10.**	(*b*)	**11.**	(*c*)	**12.**	(*d*)	**13.**	(*d*)	**14.**	(*d*)
15.	(*a*)	**14.**	(*b*)	**17.**	(*c*)	**18.**	(*d*).						

THEORETICAL QUESTIONS

1. Explain the use of instrument transformers.

2. What are the basic advantages inherent in the method of using instrument transformers for measurement purposes ?

3. List the advantages and disadvantages of instrument transformers.

4. Draw the wiring diagrams for current and potential transformers.

5. What are the advantages of instrument transformers over ammeter shunts and voltmeter multipliers ?

4. Explain the meaning of the term "burden" in the instrument transformer.
5. What is meant by "ratio error" and "phase angle error" in CTs ?
6. What is meant by turns compensation and why is it done ?
7. Derive expressions for the ratio and phase angle errors of a current transformer.
8. Draw the phasor diagram of a CT and discuss the effect of variation of power factor of the secondary burden upon the performance.
9. What is phase angle error and on which factors does it depend ?
10. Define the following terms as applied to instrument transformers :
    (*i*) Burden of an instrument transformer. (*ii*) Actual transformation ratio.
    (*iii*) Nominal transformation ratio. (*iv*) Turn ratio.
11. What is "Ratio correction factor" ? Explain briefly.
12. State the reasons why current transformer must never be operated on open circuit.
13. Write a short note on 'clip-on type current transformer'.
14. How does a PT differ from a power transformer ?
15. Explain the use of PT with the help of a connection diagram.
16. Give the comparison between CT and PT.
17. Derive the expressions for 'actual transformation (voltage), ratio and phase angle' in case of a potential transformer.
18. Explain the following in relation to a PT :
    (*i*) Effect of change in secondary burden.
    (*ii*) Effect of power factor of secondary burden.
    (*iii*) Effect of change in frequency.
19. How errors can be reduced in instrument transformers ?
20. Describe the working of a capacitive type potential transformer. Describe how the transformation ratio can be made independent of the burden of the transformers.
21. Describe the following methods for testing of a current transformer:
    (*i*) Mutual inductance method; (*ii*) Arnold comparison method.
22. Explain the 'Absolute null method' for testing of a potential transformer.

UNSOLVED EXAMPLES

Current transformer (CT)

1. The turns of primary and secondary windings of a current transformer are 1 and 200 respectively. The secondary winding supplies a current of 5A to a non-inductive burden of 1 Ω resistance. The requisite flux is set up in the core by an mmf of 80 AT. The frequency is 50 Hz and net cross-section of the core is 1000 mm^2. Neglecting the effects of magnetic leakage, iron losses and I^2R losses, calculate :
    (*i*) The actual transformation ratio. (*ii*) Phase angle.
    (*iii*) Maximum flux density in the core. **[Ans.** (*i*) 200.64; (*ii*) 4°34′; (*iii*) 0.1126 Wb/m^2]
2. A ring core current transformer of ratio 1000/5A is operating at full primary current, with a secondary burden of non-inductive resistance of 1 Ω. Its exciting current is 1 A at a p.f. of 0.4.

    Calculate : (*i*) The ratio error at full load, assuming there has been no compensation; (*ii*) Phase angle. **[Ans.** (*i*) – 0.04%; (*ii*) 3′9″.]
3. The primary winding of a 1000/5A, 50 Hz current transformer has a single turn. Its secondary burden consists of a non-inductive impedance of 1.6 Ω. If the iron loss in the core is 1.5 W at full load and magnetising mmf is 100 AT, calculate : (*i*) Flux in the core; (*ii*) Ratio error at full load. Neglect leakage reactance.
    [Ans. (*i*) 0.18 × 10^{-3} Wb; (*ii*) – 3.61%]

4. A current transformer with a bar primary has 300 turns in its secondary winding. The resistance and reactance of the secondary circuit are 1.5 Ω and 1.0 Ω respectively including the transformer winding. With 5A flowing in the secondary winding, the magnetising mmf is 100 AT and the iron loss is 1.2 W. Calculate :

   (*i*) Ratio error; (*ii*) Phase angle error. **[Ans.** (*i*) – 5.54%; (*ii*) 2.34°]

5. The secondary winding burden of a current transformer, having a bar primary and 200 secondary winding turns, is an ammeter of resistance 1.2 Ω and reactance 0.5 Ω. The resistance and reactance of secondary winding are 0.2 Ω and 0.3 Ω respectively. The core requires the equivalent of an mmf of 100 AT for magnetisation and 50 AT for core losses. Determine the following :

   (*i*) The primary winding current and actual transformation ratio when the ammeter in the secondary winding indicates 5 A.

   (*ii*) The number of turns which could be reduced in the secondary winding so that the ratio error be zero for this condition.

6. A current transformer has a bar primary, 100 secondary winding turns and 100/1 nominal ratio. It is operating on an external burden of 1.6 Ω non-inductive, the secondary winding resistance being 0.2 Ω. The primary winding current is 1.9 A, lagging 40.6° to the secondary voltage reversed. With 1 A current flowing in the secondary winding, calculate :

   (*i*) The actual ratio of primary winding current to secondary winding current;

   (*ii*) The phase angle between these currents. **[Ans.** (*i*) 101.44; (*ii*) 0.708°]

7. The current transformer rated as 1000/5A, 25 VA has a turn ratio of 1 : 200. The core loss and magnetising components of the primary current are 4A and 7A under rated conditions. Determine the ratio and phase angle errors for the rated burden and the rated secondary current at 0.8 p.f. lagging and 0.8 p.f. leading.

   Neglect the resistance and leakage reactance of secondary winding.

   [Ans. – 0.24%, + 0.185°; + 0.603%, 0.46°]

8. A 500/5A, 50 Hz current transformer has a one turn primary and an output of 15 VA. The number of secondary turns is 98 and the resistance and leakage reactance of the secondary winding are 0.35 Ω and 0.3 Ω respectively. At rated load with non-inductive burden, the in-phase and quadrature components (referred to the flux) of the exciting mmf are 13 AT and 10 AT respectively. Calculate the current and phase angle errors.

   [Ans. – 0.38%; 0.537°]

9. A 100/5A current transformer, at its rated load of 25 VA, has an iron loss of 0.2 W and a magnetising current of 1.5 A. It is supplying rated output to a meter having a ratio of resistance to reactance of 5. Calculate it ratio error and phase angle.

   [Ans. – 1.075%; 0.75°]

Potential transformer (PT)

10. The following data relate to a 1000/100 volts potential transformer :
    Primary resistance = 94.5 Ω; Primary reactance = 66.2 Ω; Secondary resistance = 0.86 Ω; Total equivalent reactance = 110 Ω. Calculate :

    (*i*) Phase angle error at no load;

    (*ii*) Burden in VA at unity p.f. at which the phase angle will be zero. **[Ans.** –4.1′; 10.9 VA]

CHAPTER

6 Electronic Instruments

6.1 INTRODUCTION

"Instrument" *may be defined as a device for measuring the value or magnitude of a quantity or variable.* **"Measurement"** *involves using an instrument as a physical means of determining a quantity or variable.*

For any measuring instrument, the *primary requirements* are :

1. Power consumption should be as small as possible.
2. No alteration in the circuit conditions with the introduction of instrument in the circuit.
   - Most of the scientific and industrial instruments require *very fast responses.* Such requirements cannot be met with by mechanical and electrical instruments. The *"electronic instruments"* use *semiconductor devices.* In the electronic devices, since the only movement involved is that of electrons, *the response time is extremely small owing to very small inertia of electrons.* With the use of electronic devices *very weak signals can be detected by using pre-amplifiers and amplifiers.*

- The electronic instruments are based on electrical or electronic principles for their measurement.
- These instruments are either indicating or recording type instruments just like the electrical measuring instruments. In addition to indicating and recording a quantity, some instruments also control a quantity e.g. tachogenerator may be used for indicating or recording the speed as well for controlling the speed in a speed controlled motor.

6.2 ESSENTIALS OF AN ELECTRONIC INSTRUMENT

An electronic instrument is made up of the following three elements (See Fig. 6.1).
1. Transducer; 2. Signal modifier; 3. Indicative device.

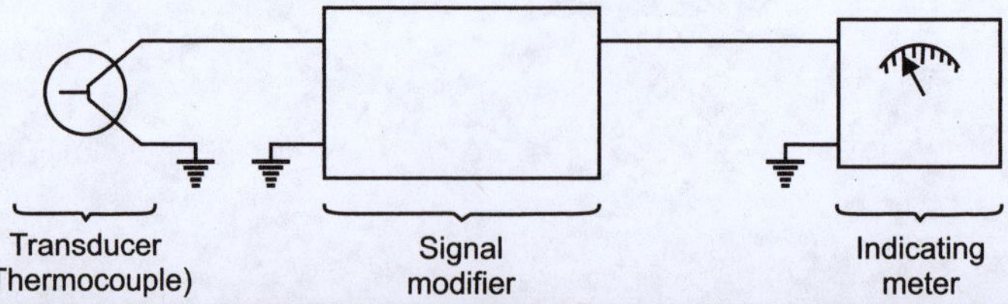

Fig. 6.1. An electronic instrument.

1. **Transducer :**
   - This sensing element is required only for measuring non-electrical quantities e.g. pressure, temperature. Its function is to convert the non-electrical quantity into an electrical signal.
   - It is not required when the quantity being measured is already in the electrical form.
2. **Signal modifier :**
   - The function of this element is to make the incoming signal suitable for application to the indicating device.

   Example. The signal may need *amplification* before it can be properly displayed.

   — Other types of signal modifiers are : Voltage dividers for reducing the amount of signal applied to the indicating device or wave shaping circuits such as *filters, rectifiers* or *choppers* etc.
3. **Indicating device :**
   - The indicating device is usually a deflection type meter (See F.g. 6.1) for general purpose instruments like voltmeters, ammeters, or ohm meters.
   - For digital readout instruments, the indicating device is of *digital design.*

6.3 ADVANTAGES OF ELECTRONIC INSTRUMENTS

Following are the *advantages of electrical/electronic techniques of measurements :*

1. Less power consumption and less load on the system being measured.
2. Non-contact measurements are possible.
3. Light, compact and more reliable instrumentation.
4. Remote indication and recording feasible.
5. The amplification produced is greater than that produced by a mechanical contrivance.
6. The mathematical processing of signals like summation, differentiating and integrating is possible.
7. The frequency and transient response is good.
8. Greater flexibility.
9. Lighter in construction.

- The electronic instruments are *widely used in detection of electro-magnetically produced signals such as radio, video and microwaves.*
- The use of these instruments has made it possible to *build analog and digital computers* which are quite instrumental in the modern developments in the fields of science and technology.

6.4 ELECTRONIC VOLTMETERS

6.4.1 Introduction

"Electronic voltmeters" use *rectifiers, amplifiers and other circuits to generate a current proportional to the voltage being measured.* The current so produced is measured by a permanent moving coil instrument (normally abbreviated as PMMC or D'Arsonval movement).

— Any measurement using PMMC movement must draw a current of 50 μA from the measured quantity for its operation for full scale deflection if conventional voltmeters are used. This would produce *great loading effects* specially in electronic and communication circuits. *"Electronic voltmeters"* avoid the loading errors by supplying power required for measurement by using external circuits like *'amplifiers'*. The amplifiers not only supply power for the operation but also make it possible for low level signals, which produce a current less than 50 μA for full scale deflection, to be detected which otherwise cannot be detected in the absence of amplifiers.

6.4.2. Advantages of Electronic Voltmeters

The electronic voltmeters claim the following *advantages over conventional ones* :

1. Low power consumption.
2. High accuracy.
3. High frequency range.
4. Compactness and portability.
5. Low level signal detection.
6. Less loading effect.
7. High input impedance.
8. High sensitivity.

The *comparison* between electronic voltmeters and conventional analog meters is given in table 6.1. below :

Table 6.1. Comparison between Electronic Voltmeters and Conventional Analog Voltmeters

S.No.	*Aspects*	*Electronic voltmeters*	*Conventional analog voltmeters*
1.	*Power consumption*	Low	High
2.	*Type of construction*	Not rugged	Rugged
3.	*Use of amplifiers*	Yes	No
4.	*Voltage range*	Higher	Limited
5.	*Loading effect*	Not present	Present
6.	*Low level signals*	Can be detected	Cannot be detected
7.	*Sensitivity*	High	Low
8.	*Frequency range*	High	Limited
9.	*Input impedance*	Very high	Low
10.	*Instrument range*	More	Less

6.4.3. Types of Electronic Voltmeters

The multimeter, also called volt-ohm-milliammeter (VOM), can be used to measure voltages but it lacks both sensitivity and high input resistances; moreover, its input resistance is different for each range. The *electronic voltmeter* (EVM), on the other hand, has *input resistance ranging from 100 MΩ to 1000 MΩ, thus producing less loading of the circuit under test than the VOM*. Another advantage of EVM is that its *input remains constant over all ranges.*

Following two types of voltmeters are used these days :

1. Analog voltmeters.
2. Digital voltmeters.

- A "digital instrument" is one which uses internal circuitry of digital design. A *"digital readout instrument" is* one whose measuring circuitry is of analog design but the indicating device is of digital design.

The electronic voltmeters are also *classified according to technology used*, as follows:

1. Vacuum-tube voltmeter (VTVM) :
   - It uses vacuum tubes with deflection meter movement.

2. Digital voltmeters :
   - Transistor voltmeter (TVM).
   - FET voltmeter (FETVM)

6.4.4. Vacuum Tube Voltmeters (VTVM)

Introduction :

R.A. Heising, in 1915, used a **valve** to measure the voltage. Most of the pioneer work in the united Kingdom was done by E.B. Moullin Oxford University. In fact, at one time, valve voltmeters were usually referred to as Moullin voltmeters. In U.S.A. these meters are known as **'Vacuum Tube Voltmeters'** usually abbreviated as 'V.T.V.M.'.

— *'Valve–voltmeters'* essential consists of a *thermionic valve which has a millimeter connected in its anode circuit*. The voltage which is to be measured is normally applied to its control grid circuit, which imposes very little load on the circuit, even at a high frequency. Although the basic arrangement of a voltmeter has limited range, this can be extended by the use of a potential divider.

— The valve voltmeter *takes practically no power at all from the source under test*; this factor is an important one in the measurement of voltages in a *radio circuit*. An ordinary moving coil meter, no matter how sensitive it may be, always draws some power from the circuit under test. In circuits where a plenty of power is available this is not serious, but when dealing with a circuit in which *even a load of a few microamperes could seriously affect the accuracy of the reading, a valve voltmeter should be used.*

Description :

- A **vacuum tube voltmeter** is a useful measuring device to measure *A.C. Voltages over a wide frequency range.*

  — It can work on high frequencies, without losing any efficiency, even if it is calibrated at such a low frequency as 50 Hz.

  — It consumes practically no power.

- It works on the principle that a *detector can be used as a voltmeter.*

The voltage to be measured is first rectified and the rectifier plate current is used to measure the applied voltage, by having a D.C. indicating instrument in the plate circuit of the tube. Usually an *amplifier stage is also added, which in addition to increasing sensitivity of the instrument, also increases the input impedance.* Because of this increase in input impedance, the power drawn by the instrument *reduces* nearly to *nil*. This makes the measurements *more accurate* and the instrument can be used over a wide voltage range with the same high sensitivity.

Types of vacuum type voltmeter :

The different types of V.T.V.M. are given below :

1. D.C. vacuum tube voltmeter
2. Diode rectifier-amplifier meter
3. Voltmeter
4. Plate circuit rectification meter
5. Gridh rectification meter
6. Amplifier type meter.

True R.M.S. voltmeter :

Fig. 6.2. shows one type of commercial true r.m.s. instrument :

- In this instrument the input signal is amplified and fed into the directly-heated filament of high vacuum diode valve, thus furnishing part of the filament heating power. The filament must have a small thermal time constant if the instrument is to respond to rapid changes in signal level.

- After amplification of A.C. input signal a cathode follower is used to obtain an impedance match between the amplifier and the low-impedance filament. The signal current will tend to raise the filament temperature, which will cause the anode current to rise and the anode-to-cathode voltage to fall. This drop in voltage is amplified by the D.C. amplifier. The output of this amplifier supplies the component of current I_{fb} (feedback current) to the filament.

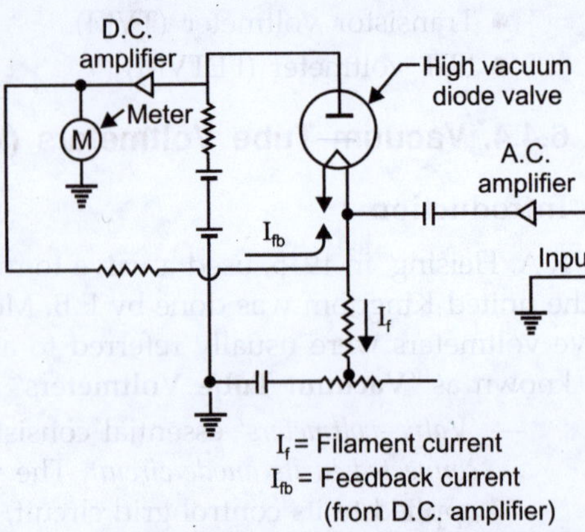

Fig. 6.2. True r.m.s. voltmeter.

- The feedback circuit is adjusted so that the decrease in $I_{fb}^2 R_f$ equals the signal power, $I_s^2 R_f$, in the filament, where I_s is the *r.m.s. signal current.*
- The meter reads the *output voltage of the amplifier, which is proportional to the feedback current* I_{fb}. It can be shown that the feedback current I_{fb} *is proportional to the square of the r.m.s. signal current*. The meter reading is, *therefore, proportional to the r.m.s. signal current.*
- The meter will have a *linear power scale.*

Peak-reading Electronic Voltmeter :

In Fig. 6.3. is shown the basic circuit of a peak-reading electronic voltmeter. The diode is placed in a probe for high-frequency response and low-level signals. The diode permits the A.C. signal to charge C to the peak value of the signal voltage. The network R_1C_1 is a simple filter which removes all ripple voltage from the D.C. that is delivered to the D.C. amplifier in the instrument.

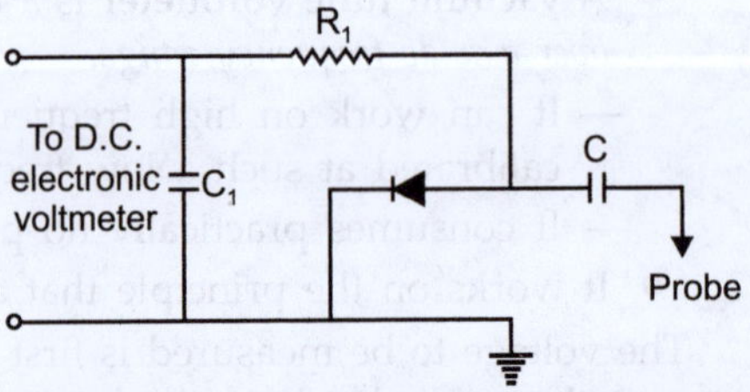

Fig. 6.3. Peak-reading electronic voltmeter.

Advantages-Applications of valve voltmeters :

1. An important advantage of the valve voltmeter is that it may be designed to cover *practically any frequency;* commercially manufactured instruments are suitable for frequencies upto 50 MHz and above.
2. The valve voltmeter will often be used on circuits where the voltage is not very high so that input impedance can only be compared with moving-coil meter which has been adjusted to appropriate range. This means that it is usually impossible to determine accurately a low voltage on the high voltage range of a moving-coil meter. On the other hand, the valve voltmeter *maintains high impedance over all its ranges and can be used for low-voltage R.F. measurement.*
3. A thermionic valve is also used in the diode peak voltmeter. When an instrument of this type is connected to an A.C. source, rectification takes place in the diode each half-wave surge charging a capacitor in the output circuit to the peak value

of the wave. Provided the voltmeter in the circuit is of sufficiently high resistance and is large enough to prevent a loss of charge through the meter when the rectifier is not passing current the reading on the meter will indicate the *peak voltage, irrespective of the waveform*.

- *"Electronic voltmeters"* are made in the following ***two basic forms*** :

1. ***A.C. amplifier types*** ... in the A.C. amplifier configuration, the voltage to be measured is applied directly to the input terminals of a wide-band gain-stabilized amplifer, whose output is rectified and applied to a moving-coil meter.
2. ***D.C. amplifier types*** ... in the D.C. amplifer construction, the input voltage is applied directly to a diode detector, and the resulting D.C. is amplified and monitored.

— It is generally accepted that the A.C. *amplifier type is usually the more sensitive of the two, but the usable range is restricted by the amplifier band-width.*

— The D.C. amplifer form permits voltage measurement upto frequencies approaching the resonance of the diode unit. Sensitivity however, is restricted by the linearity of the diode and stability of the D.C. amplifer. The first of these limitations is reduced considerably by the use of semiconductor diodes, which maintain their low forward resistance down to very small voltage levels. High gain D.C. amplification is obtained by the use of an A.C. coupled amplifer in conjunction with a chopper and synchronous detector.

Note : VTVMs are useful for the measurement of D.C. and A.C. voltages. They are till today used in laboratories.

6.4.5. Transistor Voltmeters (TVM)

Transistor voltmeter claims the following *advantages* over the vacuum tube voltmeter :

1. It does not require any warm up time.
2. It is comparatively compact and is light in weight and as such suited for field work.
3. Being battery operated, it can be used where a source of external power is not available.

- A transistor offers a low input resistance. That is why in early days hybrid (valve + transistor) configuration was used to provide high input resistance. However, now-a-days field effect transistors (FET) which offers as high input resistance as valve are extensively used.

D.C. VOLTMETERS

6.4.5.1 Transistorised voltmeter using cascaded transistors

Fig. 6.4 shows a TVM using cascaded transistors (TVM using a single transistor has *low sensitivity*) :

— It consists of two transistors Q_1 and Q_2 directly coupled in a complementary configuration (i.e. one is NPN type while other is PNP type); they form a *direct coupled amplifier*.

— The input voltage is applied through a network of resistances. The range switch selects the resistor that is in series with the transistor for the desired range of voltage. It can be seen from Fig. 6.4 that for the 0 to 1V range the value of resistance is 0.2 MΩ *i.e.* 200 K Ω; this means sensitivity is 200 kΩ/V neglecting small resistance offered by the transistor Q_1. Thus a current of $\frac{1}{200 \times 1000}$ a.m.p. or 5 μA is drawn from the circuit. The values of the other resistors are so chosen that the same current (*i.e.* 5μA) is drawn from other range.

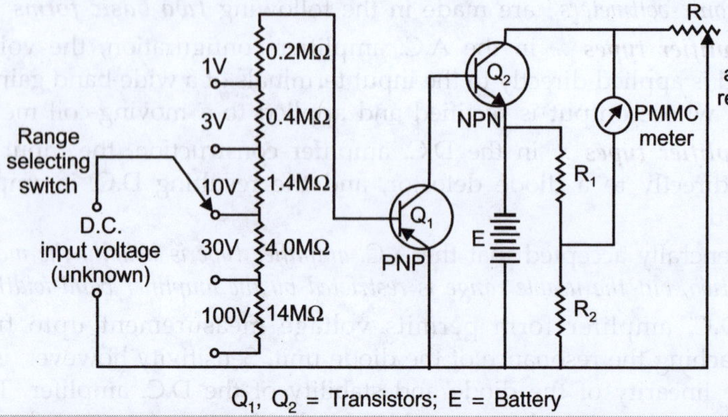

Fig. 6.4. TVM using cascaded transistors.

— A variable resistance R is put in the circuit for zero adjustment of the PMMC. It controls the bucking current from the supply E to buck out the quiescent current.

- The *drawback* of such a voltmeter is that it has to work under specified ambient temperature to get the required accuracy otherwise excessive drift problem occurs during operation.

6.4.5.2. Transistorised voltmeter using FET as input stage

Fig. 6.5. shows a TVM D.C. voltmeter with a *FET* (field effect transistor) forming the *input stage* of a direct coupled amplifier :

— In this voltmeter, voltage to be measured is firstly attenuated with range selection switch to keep the input voltage of amplifier within its level.

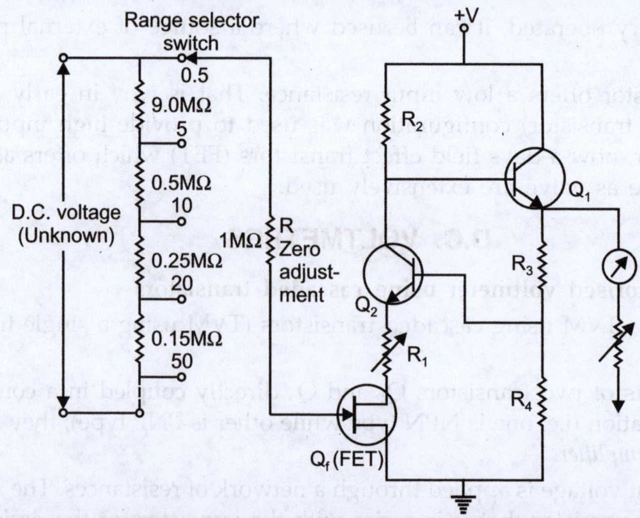

Fig. 6.5. VTM using FET as input stage.

— *FET* is used in the input stage of amplifier because of its *high impedance so that it does not load the circuit of which voltage is to be measured and it also keeps the sensitivity of voltmeter very high*. As FET is a voltage controlled device so resistance network of attenuator is put in *shunt* in the circuit.

— The two transistors Q_1 and Q_2 form a *direct coupled D.C. amplifier* using the PMMC meter movement. Provided that the amplifier operates within the limits of its dynamic range, the deflection of meter is *directly proportional* to the magnitude of the input applied voltage. The gain of this amplifier permits the instrument to be used in the *mV range*.

Further, accidental overhead voltages *do not burn out the meter* since the amplifier saturates and thus limits the maximum current through the meter to safe value.

— The input impedance of this voltmeter is high enough to make it necessary to correct for the loading effect of the circuit under measurement.

- Voltmeter used for mV ranges require a high gain D.C. amplifier in order to supply sufficient current for driving the meter movement. For overcoming the drift problems usually encountered with direct coupled D.C. amplifiers, the D.C. input is converted into A.C. voltage, before it is applied to an A.C. amplifier. The amplified A.C. voltage is then converted back into a D.C. voltage.

— In order to avoid the drift problems of D.C. amplifiers, *chopper type D.C. amplifiers* are commonly used in high sensitivity voltmeters.

Advantages :

1. Very high sensitivity (as much as 100 times the normal PMMC meter).
2. No damage of the meter due to the overload.
3. No loading effect, since the amount of power drawn is very low.
4. High input impedance (to isolate meter from the measurement circuit).
5. The voltages in the mV range can be measured.

6.4.5.3. Balanced bridge TVM

This is also called *"difference-amplifier type TVM"*.

A basic configuration of a balanced bridge TVM is shown in Fig. 6.6 :

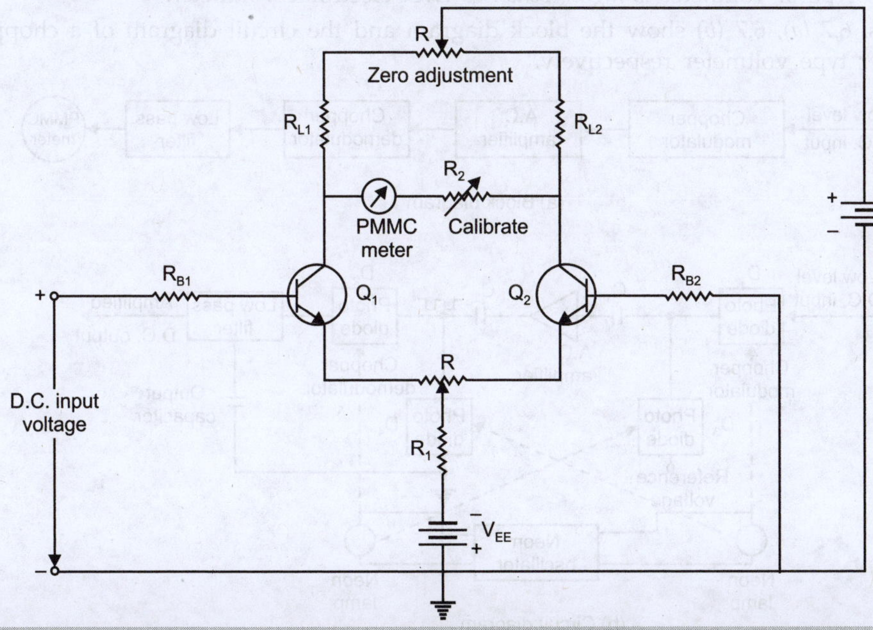

Fig. 6.6. Balanced bridge TVM.

— The bridge arms consist of resistances R_{L1}, R_{L2} and internal collector resistances of two identical transistors Q_1 and Q_2.

— When a positive voltage is applied to the base of Q_1, an increased emitter current flows. In turn, Q_2 draws decreased emitter current due to increased voltage drop across R_1. Thus the bridge is unbalanced and a current proportional to input voltage flows through the meter which can be calibrated to read the voltage supply directly.

— Resistor R is used for *zero adjustment*. The purpose of this adjustment is to equalise both halves of the difference amplifier so that there is no current through the meter under the absence of D.C. input voltage.

— Resistance R_2 is used for calibration purpose – It does accurate selling of the full-scale reading of the ammeter reading. In order to get several different full-scale ranges a range switch may be used.

Advantages :

1. When transistors Q_1 and Q_2 are similar then with the change in ambient temperature, β valves of both the transistors will change equally so there will be *no effect on collector current* in the changed ambient temperature condition.

   Also, with change in ambient temperature, saturation current of both the transistors will change equally, and, therefore, the *meter reading will remain unaffected.*

2. Any fluctuation in power supply V_{CC} and V_{EE} does not affect its performance as it *changes the biasing of both transistors equally.*

Disadvantage :

The main disadvantage of this type of voltmeter is its comparatively *low input resistance.* However a high resistance comparable to that of VTVM can be achieved by designing the voltmeter with FET as bridge circuit component.

6.4.5.4. Chopper amplifier type voltmeter

This type of voltmeter is *highly sensitive* D.C. electronic voltmeter.

Figs. 6.7 (*a*), 6.7 (*b*) show the block diagram and the circuit diagram of a chopper amplifier type voltmeter respectively.

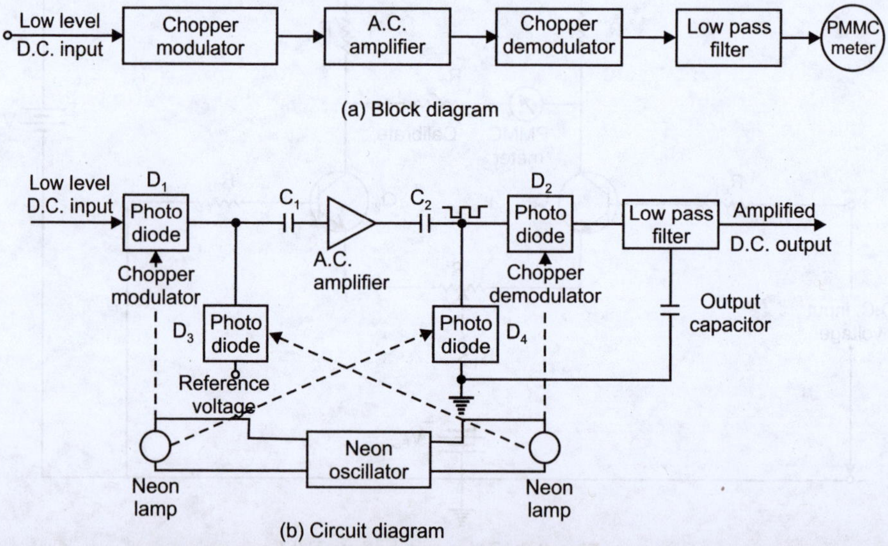

Fig. 6.7. Chopper amplifier type voltmeter.

Fig. 6.7 (*a*) : Firstly D.C. input voltage is converted into A.C. voltage by chopper modulator and then it is supplied to an A.C. amplifier. Output of amplifier is then demodulated to a D.C. voltage proportional to the original input voltage. Modulator chopper and demodulator chopper act in synchronism.

Fig. 6.7. (*b*) :

— The D.C. input voltage is chopped to a low frequency, 200 to 300 Hz A.C.

— It is passed through a blocking capacitor, amplified and then passed through another blocking capacitor so as to remove D.C. drift or offset from the amplified signal. By suitable choice of C_1 and C_2 it is possible to design a sensitive D.C. voltmeter which is drift free. This has high D.C. input impedance and high degree of accuracy.

— Such sensitive voltmeter employs photodiodes D_1, D_2, D_3 and D_4 for modulation (that is conversion from D.C. to A.C.) and demodulation (that is conversion from A.C. to D.C.). A photoconductor has a low resistance (say of the order of few hundred ohms) when it is illuminated by an incandescent or a neon lamp. Its resistance sharply increases to several megaohms when it is not illuminated. An oscillator drives two neon bulbs into illumination on alternate half cycles of oscillation. Each neon lamp illuminates one photodiode in the input circuit of the amplifier and in the output circuit as shown in Fig. 6.7 (b). Together they act like switch across the input to the amplifier alternately opening and closing at a rate determined by the frequency of the neon oscillator.

— The input signal to the amplifier is a square wave voltage whose amplitude is *proportional to the input voltage level and its frequency is equal to the frequency of oscillator*. The A.C. amplifier gives an amplified square wave in magnitude.

— The photodiodes in the amplifier *output circuit*, operating in antisynchronism with the input chopper, produces a *demodulating* action yielding D.C. output voltage.

— This D.C. signal is passed through a *low pass filter to remove any residual A.C. components*. This amplified drift free D.C. output is then fed to PMMC movement for measurement.

- A commercial available instrumentary using a photo chopper amplifier has an input impedance of 100 M Ω, a resolution of 0.1 μV and input ranges from 3 μ V full scale to 1 k V full scale with an accuracy of ±2% of full-scale deflection.

Note. The input impedance of chopper amplifier D.C. voltmeter is high except on very low input ranges. To overcome this high impedance problem an arrangement for *nulling* is included in meter circuit. A control in the front panel enables to *null the input voltage with a bucking voltage.* When the bucking voltage is equal to the input voltage a null is indicated and the meter shows an infinite impedance and therefore the loading effects are totally eliminated. The input voltage is then removed, and bucking voltage which is equal to the input voltage is shown by the meter.

Advantages :

1. The input impedance of chopper amplifier is usually of the order of 10 M Ω or higher, except on very low input ranges.
2. The drift in an ordinary D.C. amplifier is of the order of m V. The full scale range of an ordinary D.C. amplifier is limited to measuring input signal of 1 – 100 mV. In a chopper modulator system with the use of A.C. amplifier, drift can be cut down by a factor of 100, thus allowing an input signal range of about 0.01 mV = 10 μV full scale to be handled.

6.4.6. Differential voltmeter (D.C.)

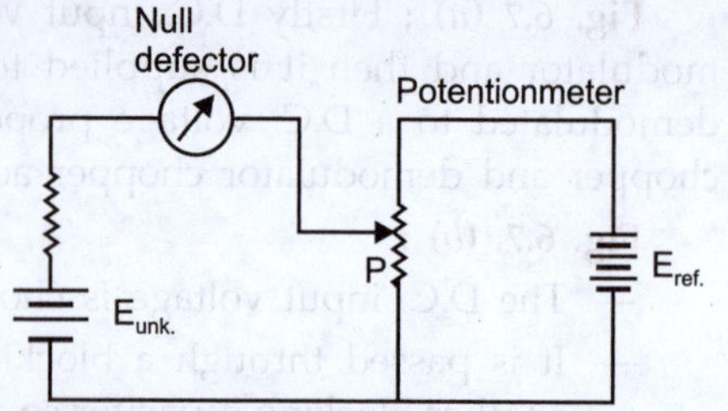

Fig. 6.8. Basic circuit of a differenital voltmeter.

- *This is one of the most common and accurate methods of measuring unknown voltages.*
- In this voltmeter the unknown voltage is *compared* with a known voltage. The principle of operation of this voltmeter is similar to a potentiometer that is why it is often called a *potentiometric voltmeter*.

Fig. 6.8 shows the *basic circuit of a differential voltmeter* :

— A precision resistance potential divider P is used to divide down an accurately known reference voltage E_{ref}.

— The adjustments of P are done till the output voltage *equals* the input/unknown voltage E_{unk}.

— Under balance conditions, neither the input source nor the reference delivers any current to the null detector, and thus voltmeter offers an infinite impedance to the source under test.

— The reference source usually consists of a low voltage D.C. standard or a low voltage zener-controlled D.C. source. In order to measure high voltages, a high voltage reference source may be used.

- The ***drawback*** of this system is that due to use of potential divider, the input resistance of the voltmeter is no longer infinite and this meter has a *loading effect which is undesirable.*

Direct current FETVM using difference amplifer :

Fig. 6.9. shows the schematic diagram of FETVM using difference amplifier.

— The two FETs (Q_1 and Q_2) are identical so that increase in the current of one FET is offset by corresponding decrease in source current of the other. The two FETs form the lower arms of the balanced bridge circuit whereas the two drain resistors R_D form the upper arms. The meter movement is connected across the drain terminals of the FETs.

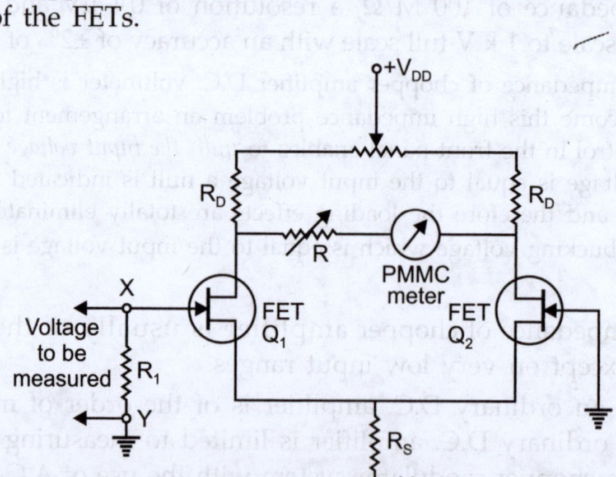

Fig. 6.9. FETVM using difference amplifier.

— The circuit is balanced under zero-input-voltage condition provided the two FFTs are identical. In that case there would no current through the PMMC meter. Zero-adjustment potentiometer is used to get null deflection in case there is a small current through the meter under zero-signal condition.

— The variable resistor R is used to adjust the full-scale calibration.

— When positive voltage is applied to the gate of Q_1, some current flows through the meter. The magnitude of this current is found to be *proportional to the voltage being measured*. Hence the meter is calibrated is volts to indicate input voltage.

Examples 6.1. *Given a difference-amplifier type FETVM shown in Fig. 6.10, find the ammeter current. If is given that g_m = 0.005 siemens, FETs have resistance of 100 kΩ and the meter's resistance is 50Ω.* **(M.S. University, Baroda)**

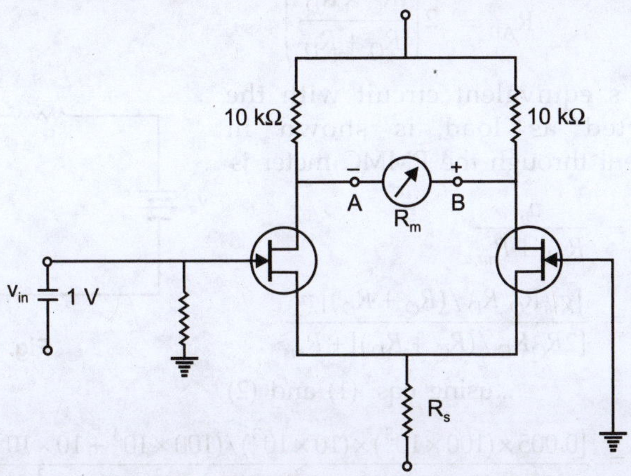

Fig. 6.10.

Solution. *Given* : g_m (Transconductance) = 0.005 siemens : $R_{Q_1} = R_{Q_2} = R_Q$ = 100 kΩ; R_m = 50 Ω; R_D (Drain resistance) = 10 kΩ

Let us use Thevenin's theorem to solve the problem.

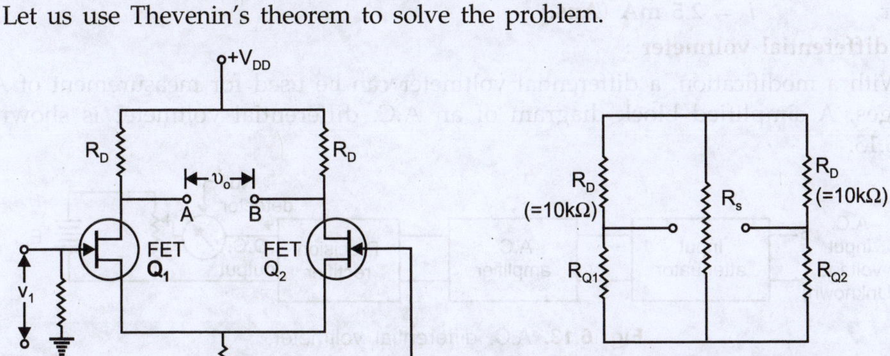

Fig. 6.11. Thevenin's equivalent circuit.

When we remove the ammeter (considering it as the load) the circuit will become as shown in Fig. 6.11 (a). The output voltage is the voltage gain of single FET times the difference between the input of Q_1 and input of Q_2. Since the input voltage of Q_2 is zero as shown in Fig. 6.11 (a) the output voltage under open circuit conditions is,

$$v_o = g_m\left(\frac{R_Q R_D}{R_Q + R_D}\right)v_1 \quad ...(1)$$

Where, R_D = Drain resistor,

R_Q = Resistance of FET.

In order to find the resistance of Thevenin source generator looking into terminals A and B, v_1 and V_{DD} are made zero. Under this condition both FETs have a resistance of R_Q as shown in Fig. 6.11 (b). From this balance circuit and assuming R_S to be large, the resistance looking into terminals A and B is

$$R_{AB} = 2\left(\frac{R_Q \times R_D}{R_Q + R_D}\right) \quad ...(2)$$

The Thevenin's equivalent circuit with the ammeter connected as load is shown in Fig. 6.12. The current through the PMMC meter is

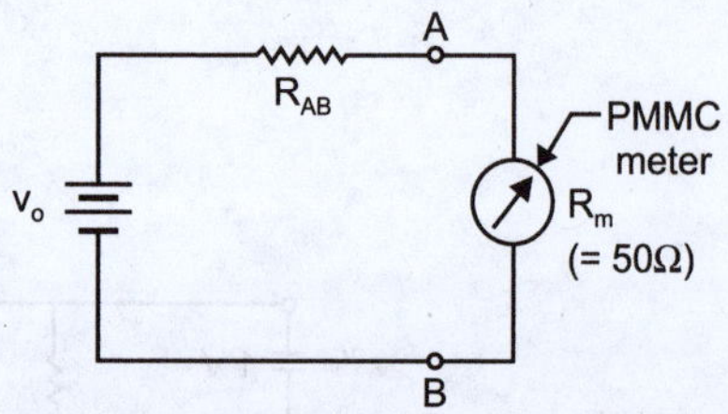

Fig. 6.12.

$$i = \frac{v_o}{R_{AB} + R_m}$$

or,
$$i = \frac{[g_m R_Q R_D/(R_Q + R_D)]v_1}{[2R_Q R_D/(R_Q + R_D)] + R_m}$$

...using eqs. (1) and (2)

$$= \frac{[0.005 \times (100 \times 10^3) \times (10 \times 10^3)/(100 \times 10^3 + 10 \times 10^3)] \times 1}{[2 \times (100 \times 10^3)(10 \times 10^3)/(100 \times 10^3 + 10 \times 10^3)] + 50}$$

$$= \frac{[0.005 \times 10^9/110{,}000] \times 1}{[2 \times 10^9/110{,}000] + 50} = \frac{45.45}{18231.8} = 2.5 \times 10^{-3}\ \text{A}$$

or, i = **2.5 mA (Ans.)**

A.C. differential voltmeter :

With a modification, a differential voltmeter can be used for measurement of A.C. voltages. A simplified block diagram of an A.C. differential voltmeter is shown in Fig. 6.13:

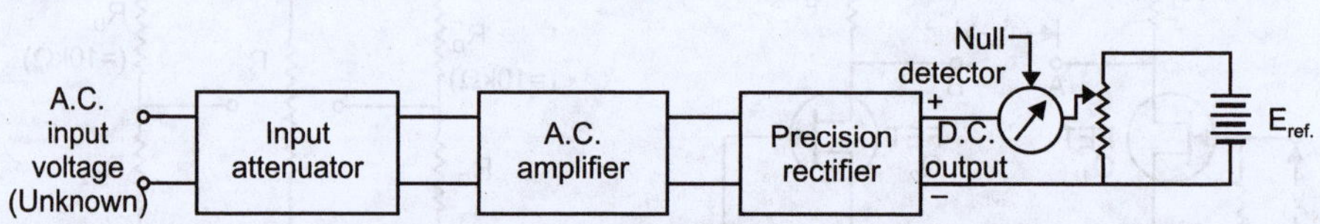

Fig. 6.13. A.C. differential voltmeter.

— An unknown A.C. voltage to be measured is applied at the input.

— This voltage is applied to an *attenuator*, consisting of number of resistors used in voltage divider mode.

— The output of attenuator is given to an A.C. amplifier which amplifies the small signals.

— The amplified A.C. voltage is then converted to D.C. voltage, by using a precision *rectifier*. Once D.C. output is proportional to the A.C. input voltage obtained, the remaining function of the voltmeter is exactly same as D.C. differential voltmeter as discussed above.

6.4.7. Rectifier Type A.C. Voltmeter

The PMMC (or D' Arsonval) meter movement can be used for measuring alternating voltage (and current) by inserting a rectifier in the measuring circuit. Such meters are more widely used than either (costly but more accurate) dynamometer type or more delicate thermal and hotwire type.

1. A.C. voltmeter using half-wave rectifier :

Fig. 6.14 shows an A.C. voltmeter using half-wave rectifier. Here, a half-wave diode rectifier has been combined in series with a D.C. meter movement.

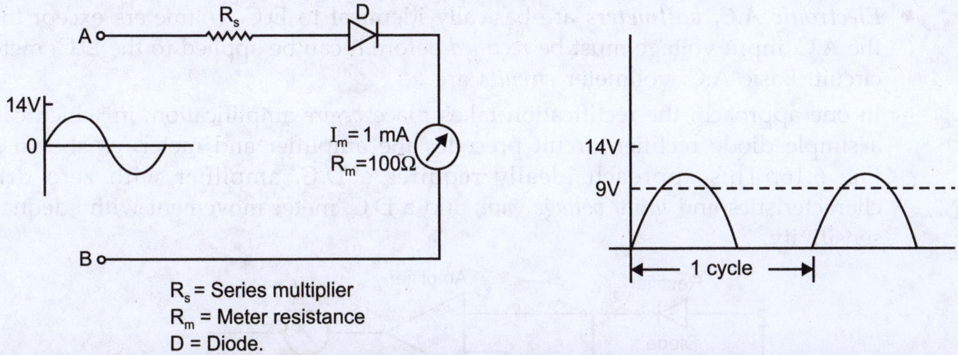

Fig. 6.14. A.C. Voltmeter using half-wave rectifier.

- When used as a D.C. voltmeter (*i.e.*, with rectifier) it would have say, for example, a range of 10 V. However, if an A.C. voltage of r.m.s. value 10 V is applied across input terminals *AB*, it would read 4.5 V; it is explained below :

— An A.C. voltage of r.m.s. value 10 V has a peak valve of $10 \times \sqrt{2} = 14$ V and an average value of 0.636 × 14 = 9 V. Since in the half-wave rectified output, one half-cycle is absent, the average for full cycle = $\frac{9}{2} = 4.5$ V. The meter movement will, therefore, read 4.5 V *i.e.* 45% of the D.C. value.

— It may also be noted that *A.C. sensitivity of a half-wave A.C. meter is only 45 percent* of *the D.C. sensitivity* (*i.e.* A. C. voltage is *less sensitive* than D.C. voltmeter).

- The value of series multiplier (R_S) for A.C. voltmeters way be obtained as follows:

$$R_S = \frac{E_{D.C.}}{I_{D.C.}} - R_m$$

or,

$$R_S = \frac{0.45\, E_{r.m.s}}{I_{D.C.}} - R_m$$

where, I_{DC} = Full scale deflection current.

2. A.C. voltmeter using full-wave rectifier :

Fig. 6.15 shows an A.C. voltmeter using full-wave rectifier.

- In this case, obviously, the meter reading would be 90% of r.m.s. input voltage *i.e.*, 90% of the D.C. value.

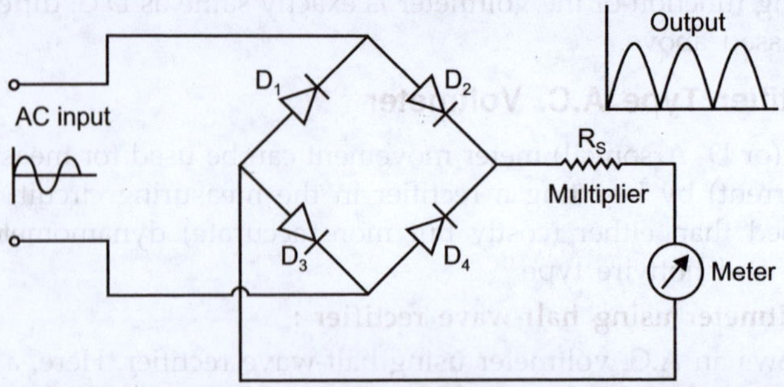

Fig. 6.15. A.C. voltmeter using full-wave rectifier.

- ***Electronic A.C. voltmeters*** are basically identical to D.C. voltmeters except that the A.C. input voltage must be *rectified* before it can be applied to the D.C. meter circuit. Basic A.C. voltmeter *circuits* are :

— In one approach, the rectification takes place *before* amplification, in which case a simple diode rectifier circuit precedes the amplifier and meter, as shown in Fig. 6.16. This approach ideally requires a D.C. amplifier with zero drift characteristics and *unity voltage gain*, and a D.C. meter movement with adequate sensitivity.

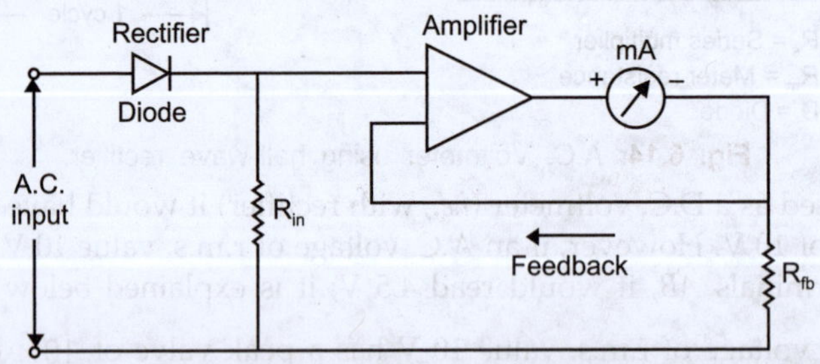

Fig. 6.16. D.C. mode of operation.

— In another approach the A.C. signal is rectified after amplification (See fig. 6.17) where full-wave rectification takes place in the meter circuit connected to the output terminals of the A.C. amplifier. This approach generally requires an A.C. amplifier with *high open-loop gain* and *large amounts of negative feedback* to overcome the nonlinearity of the rectifier diodes.

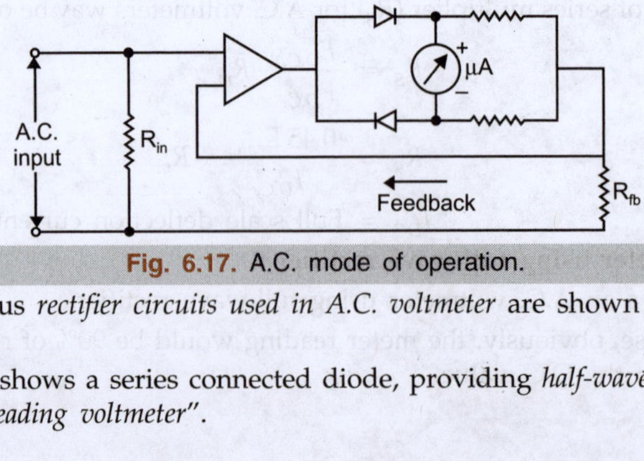

Fig. 6.17. A.C. mode of operation.

- The various *rectifier circuits used in A.C. voltmeter* are shown in Figs. 6.18, 6.19, and 6.20.

— **Fig. 6.18.** shows a series connected diode, providing *half-wave rectification* for an *"average-reading voltmeter"*.

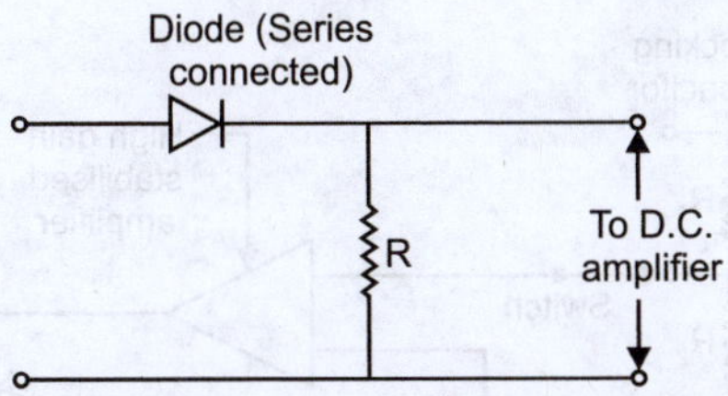

Fig. 6.18

— **Fig. 6.19.** shows four diodes in a bridge circuit for full-wave rectification and application to an *"average-reading meter."*

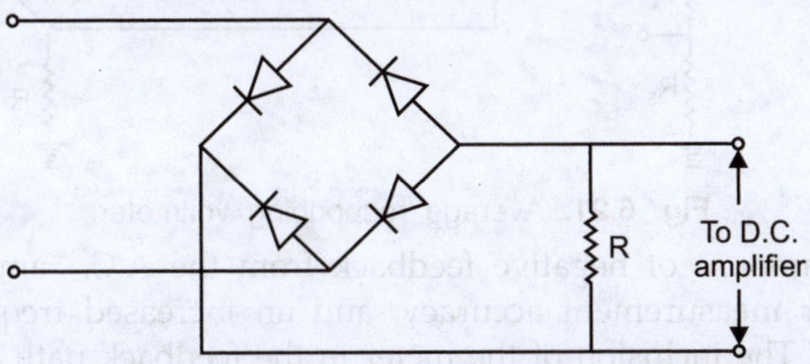

Fig. 6.19

— **Fig. 6.20** shows shunt-connected diode used in a *"peak-reading voltmeter"*. The rectifier diode charges the small capacitor to the peak of the applied input voltage.

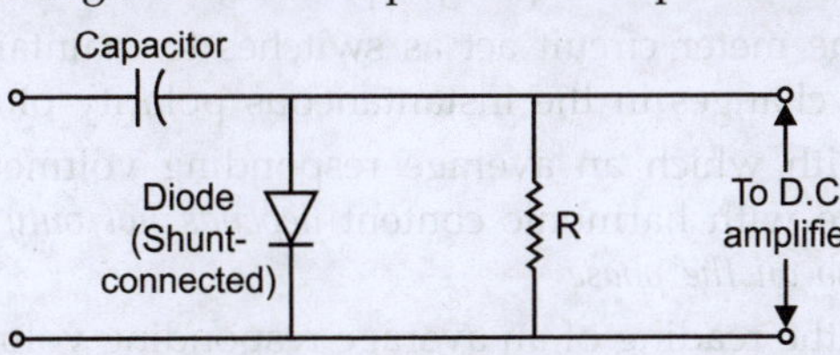

Fig. 6.20

- Usually A.C. voltmeters are of **average-responding** *type, with the meter scale calibrated in terms of the r.m.s. value of a sine wave.* In electronics since so many wave forms are sinusoidal, this is an entirely satisfactory solution and certainly *much less expensive* than a true-responding voltmeter. However, this meter will read *high or low* depending on the form factor of the wave, in case of nonsinusoidal wave-forms (Refer to examples 6.2 and 6.3).

6.4.8. Average Responding Voltmeter

Fig. 6.21 shows the block diagram of a typical average responding voltmeter :

— The input/applied wave form is *amplified* in a high gain stabilised amplifier and then *rectified* and fed to a D.C. mA calibrated in terms of r.m.s. input voltage. In this meter instrument, the rectified current is *averaged by a filter* to produce a steady deflection of the pointer of the meter.

— The input *blocking capacitor* (preceding the high gain amplifier) *excludes* the D.C. component (in the applied voltage) from the measurement.

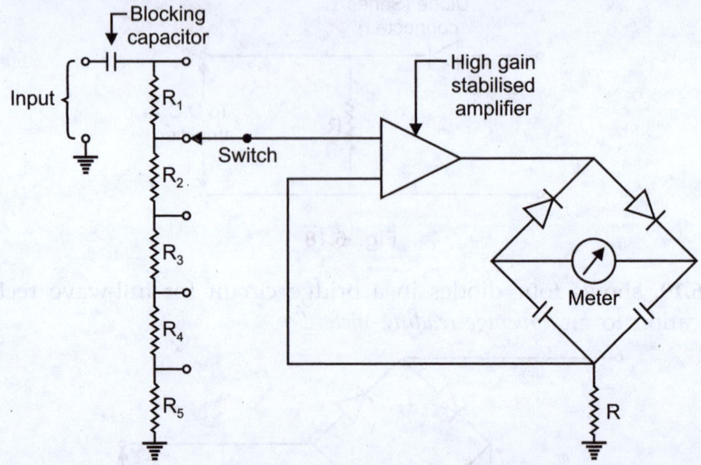

Fig. 6.21. Average responding voltmeter.

— The large amount of negative feedback from the A.C. amplifier ensures gain stability for measurement accuracy, and an increased frequency range of the instrument. The inclusion of the meter in the feedback path minimises the effect of diode non-linearity and meter impedance variations on performance of the circuit.

— The *capacitors* in the meter circuit tend to act as storage or filter capacitors for the rectifier diodes as well as coupling capacitors for the feedback signal.

— The *diodes* in the meter circuit act as switches to maintain unidirectional meter current despite changes in the instantaneous polarity of the input voltage.

- The *accuracy* with which an average responding voltmeter indicates the r.m.s. value of a wave with harmonic content *depends not only on the amplitude of the harmonic but also on the phase*

  — The *errors* in the reading of an average responding voltmeter may be *due to the application of complex waveforms.*

6.4.9. Peak Responding Voltmeter

The main difference between peak responding voltmeters and average responding voltmeters is the *use of storage capacitors with the rectifying diode in the former case.* The capacitor charges through the diode to the peak value of the applied voltage and the meter circuit then responds to voltage of the capacitor.

Fig. 6.22 (*a*) and (*b*) shows the *two* most common types of peak responding voltmeters:

— Fig. 6.22 (a) shows a *D.C. coupled peak voltmeter;* here the capacitor charges to the total peak voltage above ground reference.

In this type of meter, the reading of the meter is affected by the presence of D.C. with A.C. voltage.

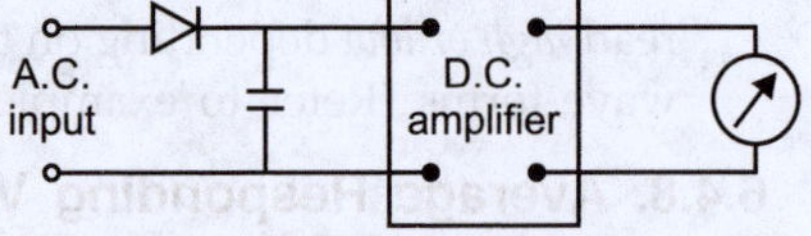

(a) D.C. coupled peak voltmeter

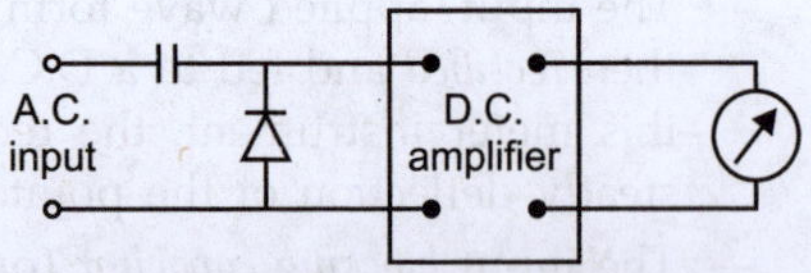

(b) A.C. coupled peak voltmeter

Fig. 6.22. Peak responding voltmeter.

— Fig. 6.22 (b) shows an *A.C. coupled peak voltmeter.*

In both the circuits, the capacitor discharges very slowly through the high impedance input of D.C. amplifier, so that a negligible small amount of current supplied by the circuit under test keeps the capacitor charged to the peak A.C. voltage. *The D.C. amplifier is used in the peak responding meter to develop the necessary meter current.*

Advantages. The main advantage of a peak responding voltmeter is that the rectifying diode and the storage capacitor may be taken out of the instrument and placed in the probe when no A.C. pre-amplification is required. The measured A.C. signal then travels no farther than diode. The meter is then able to measure frequencies upto 100s of MHz with a minimum of circuit loading.

Disadvantages :

1. Error caused due to harnomic distortion in the input waveforms.
2. Limited sensitivity of the instrument due to imperfect characteristics of diode.

Example 6.2. *A sawtooth voltage has a peak value of 80 V and a time period of 2.0 s as shown in Fig. 6.23. Calculate the error when measuring this voltage with an average reading voltmeter calibrated in terms of r.m.s. value of a sinusoidal wave.*

Solution. The given sawtooth wavefrom can be mathematically expressed as :

$$e = mt, \text{ where } m \text{ is the slope of ramp}$$

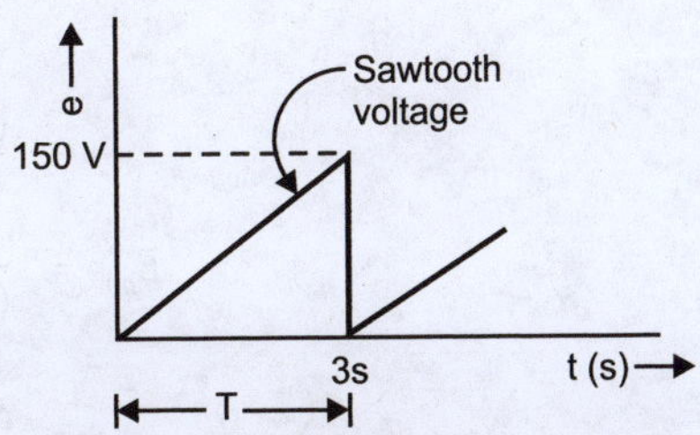

Fig. 6.23.

$$\therefore \quad m = \frac{e}{t} = \frac{150}{3} = 50$$

$$\text{or} \quad e = 50t \qquad 0 \le t \le T$$

$$\therefore \quad E_{r.m.s} = \sqrt{\frac{1}{T}\int_0^T e^2 . dt}$$

$$= \sqrt{\frac{1}{3}\int_0^3 (50t)^2 dt} = \sqrt{\frac{2500}{3}\left|\frac{t^3}{3}\right|_0^3}$$

$$= \sqrt{\frac{2500}{9}(3^3 - 0)} = 50\sqrt{3} \text{ V}$$

$$E_{av} = \frac{1}{T}\int_0^T e . dt$$

$$= \frac{1}{3}\int_0^3 50\, t\, dt = \frac{50}{3}\left|\frac{t^2}{2}\right|_0^3 = \frac{50}{6}(3^2 - 0) = 75 \text{ V}$$

$$\therefore \quad \text{Form factor, } k_f = \frac{E_{r.m.s}}{E_{av}} = \frac{50\sqrt{3}}{75} = 1.155$$

The ratio of the two form factors is,

$$\frac{(k_f)_{\text{sine wave}}}{(k_f)_{\text{sawtooth}}} = \frac{1.11}{1.155} = 0.961$$

Thus, the meter indication is **low** by 0.961.

$$\therefore \quad \text{Percentage error} = \frac{0.961-1}{1} \times 100 = -\mathbf{3.9\%} \ \textbf{(Ans.)}$$

Example 6.3. *Calculate the error when a symmetrical square-wave voltage as shown in Fig. 6.24, is applied to an average reading A.C. voltmeter with scale calibrated in terms of r.m.s. value of a sinusoidal wave.*

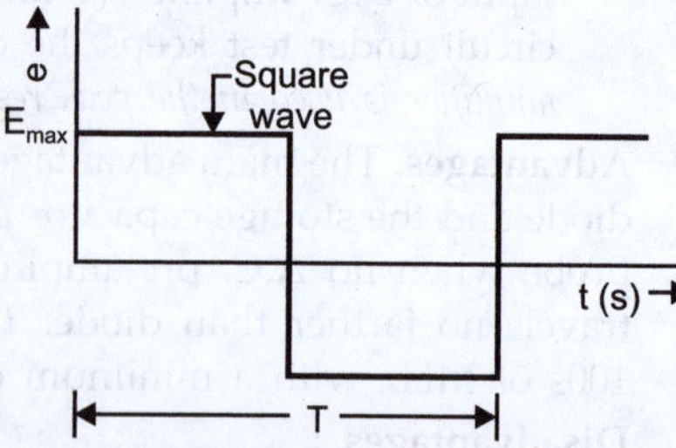

Fig. 6.24

Solution. The r.m.s. value of the square voltage is given as :

$$E_{r.m.s.} = \sqrt{\frac{1}{T}\int_0^T e^2\, dt}$$

$$= \sqrt{\frac{1}{T}\int_0^T E_{\max}^2\, dt} = \sqrt{\frac{1}{T} E_{\max}^2 \, |t|_0^T}$$

$$= \sqrt{\frac{1}{T} \times E_{\max}^2\,(T-0)} = E_{\max}$$

$$E_{av.} = \frac{1}{(T/2)}\int_0^{T/2} e.\, dt$$

$$= \frac{2}{T} E_{\max} . |t|^{T/2} = \frac{2}{T} E_{\max}\left(\frac{T}{2} - 0\right) = E_{\max}$$

$$\therefore \quad \text{Form factor, } k_f = \frac{E_{r.m.s}}{E_{av}} = \frac{E_{max}}{E_{max}} = 1$$

The ratio of the two form factors is,

$$\frac{(k_f)_{sine\ wave}}{(k_f)_{square\ wave}} = \frac{1.11}{1} = 1.11$$

Thus, the meter indication is **high** by a factor 1.11.

$$\therefore \quad \text{Percentage error} = \frac{1.11-1}{1} \times 100 = \mathbf{11\%} \quad \textbf{(Ans.)}$$

6.4.10. True R.M.S.-reading Voltmeter

The average reading voltmeter can be converted into an r.m.s. reading voltmeter by calibrating the meter scale in terms of the r.m.s. value of a *sine wave.* However, if the waveform is not sinusoidal the reading causes considerable error. *Complex waveforms can be accurately measured by r.m.s.-reading voltmeter.*

This type of voltmeter provides an output by sensing the heating power of the waveform. *Since the heating power of the waveform is proportional to the square of the r.m.s. value of the voltage, the scale of the instrument can be calibrated to read the r.m.s. directly.* In practice, the heating power is measured by measuring an amplified version of the input waveform to the heater element of thermocouple whose output voltage is proportional to the square of the r.m.s. value of input voltage.

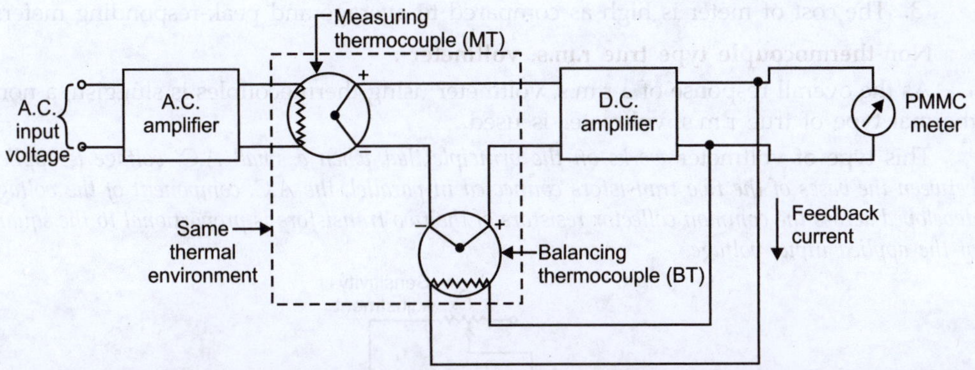

Fig. 6.25. Block diagram of a true r.m.s.-reading voltmeter.

Fig. 6.25. shows the block diagram of a true r.m.s.-reading voltmeter :

— The r.m.s.-reading voltmeter consists of two thermocouples MT and BT as shown in Fig. 6.25; MT is called measuring thermocouple and BT the balancing thermocouple. BT is used in the feedback loop to cancel out the non-linear effect of the MT. These two thermocouples MT and BT form part of the bridge circuits connected to the input of a D.C. amplifier.

— The unknown A.C. voltage is amplified and applied to the heating element of MT. As already stated, heating produces an output voltage which *upsets the balance of bridge.*

— This *unbalanced voltage* is *amplified by D.C. amplifier* and is fed back to the heater element of the BT.

— Balance is re-established when the voltage output of the thermocouple currents in the heater element of BT is equivalent to the A.C. current in the heating element of MT. D.C. current is directly proportional to the r.m.s. value of input voltage.

— The PMMC meter is calibrated to read the r.m.s. voltage directly.

- A typical laboratory-type r.m.s.-reading voltmeter provides accurate r.m.s.reading of complex waveforms having a *crest factor* (ratio of peak value to r.m.s. value) of 10/1.
  - — At 10 per cent of full-scale meter deflection, where there is less chance of amplifier saturation, waveforms with crest factors as high as 100/1 could be accommodated.
  - — With the use of good instruments, voltages throughout a range of 100μV to 300V within a frequency range of 10 Hz to 10 MHz may be measured.

Advantages :

1. Sensitivities in the mV region is possible.
2. The non-linear behaviour is avoided by using two thermocouples (MT and BT) in the same thermal environment.
3. The true r.m.s. value measured is independent of the A.C. input waveform, if the peak amplitude of A.C. input is within the dynamic range of the A.C. amplifier.

Limitations :

1. The overall meter response is sluggish (since the response of thermocouples is slow).
2. In case of highly non-linear waveforms, the crest factor puts the limitation on the meter reading.

3. The cost of meter is high as compared to average and peak-responding meters.

Non-thermocouple type true r.m.s. voltmeter :

As the overall response of a r.m.s. voltmeter using thermocouples is sluggish, a non-thermal type of true r.m.s. voltmeter is used.

This type of voltmeter *works on the principle that when a small A.C. voltage is applied between the bases of the two transistors connected in parallel, the A.C. component of the voltage developed across the common collector resistors of the two transistors is proportional to the square of the applied input voltage.*

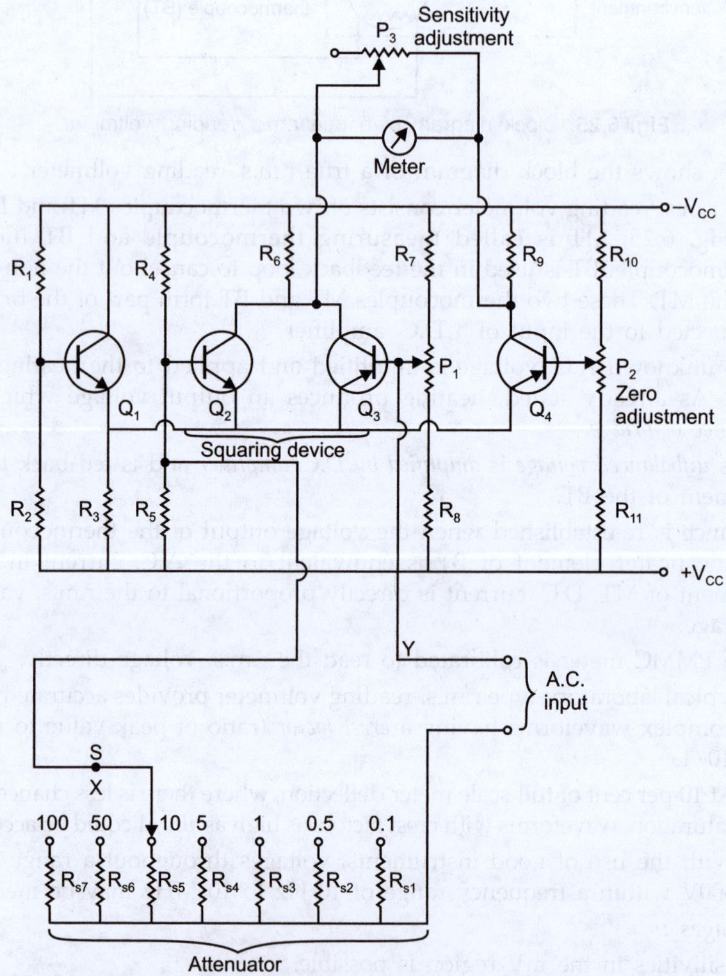

Q_1, Q_2, Q_3, Q_4 = Transistors; P_1, P_2, P_3 = Potentiometers; S = Switch

Fig. 6.26. Non-thermocouple type true r.m.s. voltmeter.

Fig. 6.26 shows a non-thermocouple type transistorised true R.M.S. voltmeter, using a squaring device and other components :

— The transistors Q_2 and Q_3 constitute the *squaring device*. The resistance R_6 is common collector resistance. The potentiometer P_1 is used for *bias setting*.

— The transistor Q_4 constitutes the other side of the bridge. The potentiometer P_2 is used for bias setting which is nothing but *zero adjustment*. It adjusts zero for zero A.C. input *i.e.* shorting the terminals X and Y. Under this condition the bridge is balanced and drops across R_6 and R_9 are equal.

— The function of transistor Q_1 is to improve temperature stability through emitter resistance R_3. The resistances R_1 and R_2 are used to adjust the biasing to Q_1 such that drop across R_3 is 0.7V for silicon transistors.

— A.C. input is applied through an attenuator consisting of resistances R_{S1} and R_{S7}. The proper range is selected through the switch S (a range selector).

On the application of input, the bridge is unbalanced and *drop between resistances R_6 and R_9 is proportional to r.m.s. value of input;* this is given to the meter. The potentiometer P_3 is used to adjust the sensitivity of the meter.

6.4.11. Solid State Voltmeter

Fig. 6.27 shows the circuit of an electronic voltmeter using an IC op amp 741C :

— This is *directly coupled very high gain amplifier.*

— The gain of the op-amp can be adjusted to any suitable lower value by providing appropriate resistance between its output terminal, Pin No. 6, and inverting input, Pin No.2, to provide a negative feedback.

— The ratio $\frac{R_2}{R_1}$ determines the gain *i.e.* 101 in this case, provided by the op-amp.

— The 0.1 μF capacitor across the 100 kΩ resistance R_2 is for *stability under stray pick-ups.*

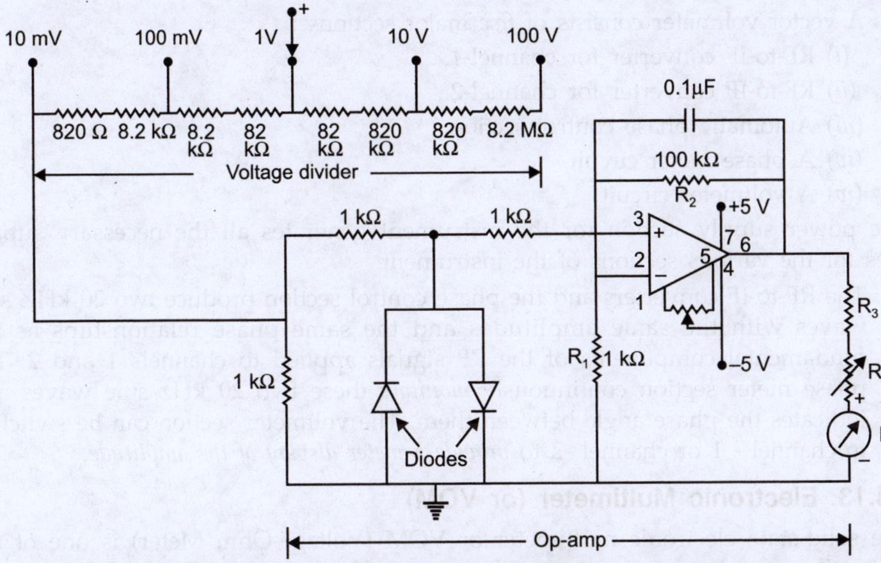

Fig. 6.27. Solid state mV voltmeter using opamp.

— Terminals 1 and 5 are called *offset null terminals*. A 10 kΩ potentiometer is connected between these two offset null terminals with its centre tap connected to a –5V supply. This potentiometer is called *"zero set "* and is used for adjusting zero output for zero input conditions.

— The two diodes used are for *IC protection*. Under normal conditions, they are non-conducting, as the maximum voltage across them is 10 mV. If an excessive voltage, say more than 100 mV appears across them, then depending upon the polarity of the voltage, one of the diode conducts and protects the IC. A μA scale of 50-1000μA full scale deflection can be used as an indicator. R_4 is adjusted to get maximum full scale deflection.

6.4.12. Vector Voltmeter

A ***vector voltmeter*** *measures the amplitude of a signal at two points in a circuit and simultaneously measures the phase difference between the voltage waveforms at these two points.*

This voltmeter has wide applications especially in situations where other methods are very difficult or time consuming. The vector voltmeter is useful in VHF applications and can be used successfully in measurements such as :

(*i*) Complex insertion loss.

(*ii*) Amplifier gain and phase shift.

(*iii*) Two-port network parameters.

(*iv*) Filter transfer functions.

(*v*) Radio frequency distortion.

- This voltmeter basically converts two RF signals of the same fundamental frequency (from 1MHz to 16Hz) to two IF signals with 20 kHz fundamental frequencies. These I.F. signals have the same amplitudes, waveforms and phase relationships as the original RF signals. Consequently, the fundamental components of the IF signals have the same amplitude and phase relationships as the fundamental components of the RF signals. *These fundamental components are filtered from the IF signals and are measured by a voltmeter and a phase meter.*

  A vector voltmeter consists of *five* major sections :

  (*i*) RF-to-IF converter for channel-1.

  (*ii*) RF-to-IF converter for channel-2.

  (*iii*) Automatic phase control section.

  (*iv*) A phase meter circuit.

  (*iv*) A voltmeter circuit.

The power supply section for the instrument generates all the necessary supply voltages for the various sections of the instrument.

- The RF-to-IF converters and the phase control section produce two 20 kHZ sine waves with the same amplitudes and the same phase relationships as the fundamental components of the RF signals applied to channels 1 and 2. The phase meter section continuously *monitors* these two 20 kHz sine waves and indicates the phase angle between them. The voltmeter section can be switched to channel - 1 or channel -2 to *provide a meter display of the amplitude.*

6.4.13. Electronic Multimeter (or VOM)

The solid-state electronic multimeter or VOM (Voltage-Ohm Meter) is one of the most versatile general-purpose shop instruments *capable of measuring D.C. and A.C. voltages as well as current and resistance.*

An electronic multimeter generally consists of the following *elements* :

1. Balanced-bridge D.C. amplifier and indicating meter.
2. Input attenuator or Range switch...*To limit the magnitude of the input voltage to the desired value.*
3. Rectifier section... *To convert an A.C. input voltage to a proportional D.C. value.*

4. Internal battery and additional circuitry... *To provide the capability of resistance measurement.*
5. "Function switch" ... *To select the various measurement functions of the instrument such as voltage, current or resistance.*

The instrument is also usually provided with a built-in power supply for operation on A.C. mains and, in most cases, one or more batteries for operation as *portable test instrument.*

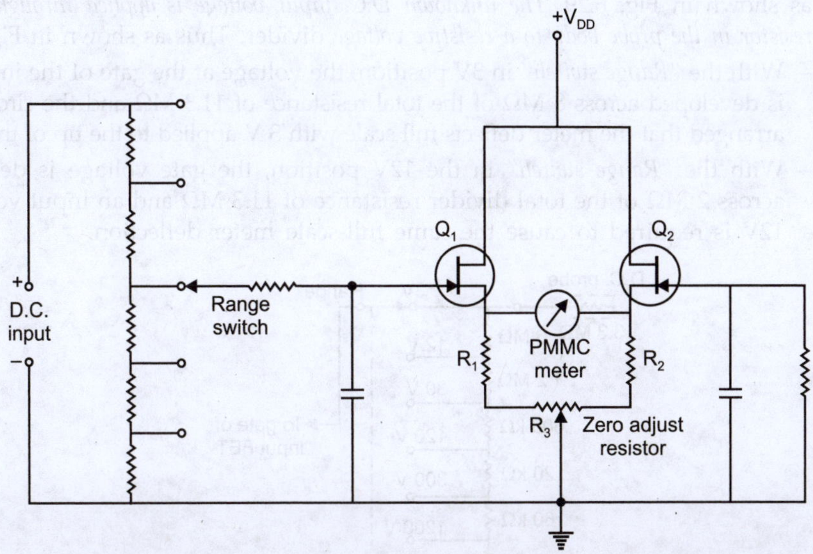

Q_1, Q_2 = FETs ; R_1, R_2 = Source resistors, R_3 = Zero adjust resistor.

Fig. 6.28. Balanced-bridge D.C. amplifier with input attenuator and indicating meter.

Fig. 6.28 shows the schematic diagram of a balanced-bridge D.C. amplifier using field effect-transistors (FETs). (This circuit also applies to a bridge amplifier with ordinary bipolar transistors (BJTs) :

— Q_1 and Q_2 are the two *FETs;* these should be reasonably well matched for current gain to ensure thermal stability of the circuit.

— The two *FETs,* Q_1 and Q_2 constitute the *upper* arms of a bridge circuit, whereas, source resistors R_1 and R_2, together with *"Zero"* adjust resistor R_3 form the *lower* bridge arms.

— The PMMC meter is connected between the source terminals of the FETs, representing two opposite corners of the bridge.

Working :

Voltage measurement :

- Where there is no input signal, the gate terminals of the FETs are at ground potential and the transistors operate under identical quiescent conditions. Under such a condition the bridge is balanced and meter reads zero. However, in practice, small differences in the operating characteristics of the transistors, and slight tolerance differences in the various resistors, cause a certain amount of unbalance in the drain currents, and the meter shows a small deflection from zero. This current is reduced to zero by adjust control resistor R_3 and now the bridge is balanced.

- On the application of the voltage to the gate of input FET Q_1, its drain current increases which causes the voltage at the source terminal to rise. The resulting unbalance between Q_1 and Q_2 source voltages is indicated by the PMMC meter, whose scale is calibrated to agree with the magnitude of the applied input voltage.
- The maximum voltage which can be applied to the gate of Q_1 is determined by the operating range of FET and is usually on the order of a few volts. The range of input voltages can easily be extended by an input attenuator a *"Range switch"*, as shown in Fig. 6.29. *The unknown D.C. input voltage is applied through a large resistor in the probe body to a resistive voltage* divider. Thus as shown in Fig. 6.29 :
  — With the *"Range switch"* in 3V position, the voltage at the gate of the input FET is developed across 8 MΩ of the total resistance of 11.3 MΩ and the circuit is so arranged that the meter deflects full scale with 3 V applied to the tip of the probe.
  — With the *"Range switch"* in the 12V position, the gate voltage is developed across 2 MΩ of the total divider resistance of 11.3 MΩ and an input voltage of 12V is required to cause the same full-scale meter deflection.

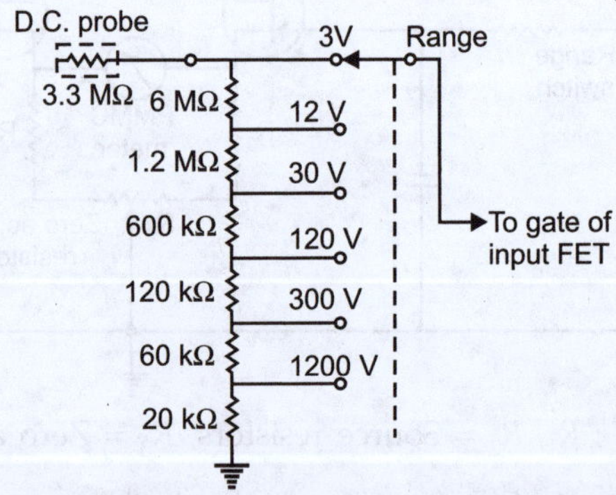

Fig. 6.29. Typical input voltage attenuator for a multimeter.

Resistance measurement :

In an electronic multimeter, when the *function switch* is placed in the 'OHMS' position the unknown resistor is connected in series with an internal battery, and the PMMC meter simply measures the voltage drop across the unknown.

Fig. 6.30 shows a typical circuit, where a separate divider network, used only for resistance measurements, provides for a number of differents resistance ranges.

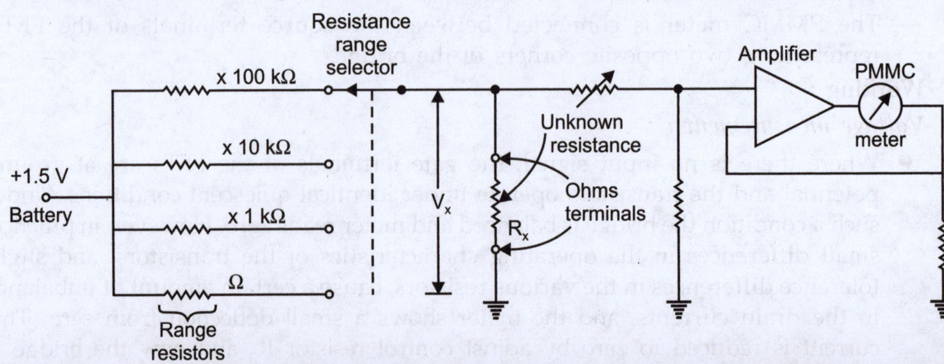

Fig. 6.30. Resistance range selector circuit of an electronic multimeter.

— When an unknown resistor R_x is connected to the OHMS terminals of the multimeter, the 1.5 V battery supplies current through one of the range resistors and unknown resistor to ground.

— Voltage drop V_x (across R_x) is applied to the input of the bridge amplifier which causes a deflection on the meter. Since the voltage drop across R_x is *directly proportional to its resistance, the meter scale can be calibrated in terms of resistance.*

If may be noted that the resistance scale of a multimeter reads increasing resistance from left to right, opposite the way resistance scale reads on conventional meters. It is expected because electronic multimeter reads a *larger resistance as a higher voltage, whereas the ordinary multimeter indicates a higher resistance as a smaller current.*

Advantages and disadvantages :

Following are the *advantages* and *disadvantages* of an electronic multimeter :

Advantages :

1. Simple circuit.
2. Less costly.
3. High input impedance.
4. High frequency range.
5. Less electric noise.
6. Rugged construction.

Disadvantages :

1. Poor reliability and repeatability.
2. Poor resolution.
3. Size is not compact.
4. Accuracy not high.
5. Interfacing of the output with external devices is difficult.

6.4.14. Selection of an Analog Voltmeter

The selection of the most appropriate instrument for a particular voltage measurement depends on the performance required in a given situation. Some important considerations needed in selecting an analog voltmeter (*an analog voltmeter always measures average value; it is calibrated to read r.m.s. or peak values*) are as follows :

1. Input impedance :

- The input resistance/impedance should be at least of an order of magnitude higher than the impedance of the circuit under measurement to avoid loading effects.

2. Voltage ranges :

- The voltage ranges on the meter scale may be in the 1-3-10 sequence with 10 db of separation, or in the 1.5-5-15 sequence, or in a single scale calibrated in decibels.

In any case, the *scale divisions should be compatible with the accuracy of the instrument :*

— An instrument with an accuracy of 1% or less should also have a mirror-backed scale to reduce parallax and improve accuracy.

3. Decibels :

- In case of measurements that cover a *wide range* of voltages, use of the decibel scale can be very effective.

— This kind of measurement is found, for example, in the frequency response curve of an amplifier or filter, where the output voltage is measured as a *function of the frequency of the applied input voltage.*

— All voltmeters with *db* scales are calibrated in *dbm*, referenced to some particular impedance.

4. Sensitivity versus bandwidth :

- Noise is a function of bandwidth.

— A voltmeter with a broad bandwidth will pick up and generate more noise than one operating over a narrow range of frequencies.

5. Battery operation :

- A voltmeter powered by an internal battery is essential for field work.

6. A.C. current measurements :

- A sensitive A.C. voltmeter and a series resistance may be used to measure currents.
- In the usual case, however, an A.C. *current probe* is used which enables the operator to measure an A.C. current without disturbing the circuit under test.

In view of the above considerations, the following *general guidelines* can be stated :

(*i*) *For D.C. measurements* :	.... Select the meter with the broadest capability meeting the requirements of the circuit.
(*ii*) *For A.C. measurements* : [involving sine waves with only modest amounts of distortion (less than 10%)]	.... The average-responding voltmeter provides the best accuracy and most sensitivity per unit investment.
(*iii*) *For high frequency measurements :* (greater than 10 MHz)	.... The peak-responding voltmeter with a diode-probe input is the most economical choice.
(*iv*) *For measurements involving the effective power of waveforms which depart from the true sinusoidal form* :	.... The r.m.s. responding voltmeter is the appropriate choice.

HIGHLIGHTS

1. Essentials of an electronic instrument are :
   (*i*) Transducer; (*ii*) Signal modifier; (*iii*) Indicating device.
2. Valve voltmeter essentially consists of a thermionic valve which has a millimeter connected in its anode circuit.
3. The various electronic voltmeters widely used are :
   (*i*) Transistorised voltmeter using cascaded transistors;
   (*ii*) Transistorised voltmeter using FET as input stage;
   (*iii*) Balanced bridge T.V.M;
   (*iv*) Chopper amplifier type voltmeter;
   (*v*) Differential voltmeter (D.C);
   (*vi*) Rectifier type A.C. voltmeter;

(*vii*) Averge responding voltmeter;
(*viii*) Peak reading voltmeter;
(*ix*) True R.M.S. reading voltmeter,
(*x*) Solid state voltmeter; and
(*xi*) Vector voltmeter.

OBJECTIVE TYPE QUESTIONS

Choose the Correct Answers :

1. A true rms reading voltmeter uses two thermocouples in order
(*a*) to increase sensitivity
(*b*) that the second thermocouple cancels out the no-linear effects of the first thermocouple
(*c*) to prevent drift in the d.c. amplifier
(*d*) all of the above.

2. In an electronic ohmmeter, an OP-AMP is used as a
(*a*) Summer
(*b*) Multiplier
(*c*) Buffer amplifier
(*d*) Integrator.

3. An average reading VTVM uses one diode connected in series with an external resistance. A high value of series resistance is used so that the instrument should have
(*a*) a high input impedance
(*b*) a linear *v-i* characteristics
(*c*) low power consumption
(*d*) all of the above.

4. In a compensated shunt type peak reading VTVM a bucking battery with a series adjustable resistance is used for
(*a*) providing power for the circuit
(*b*) zero setting
(*c*) the purpose of amplifiying low level signals
(*d*) all of the above.

5. The power consumption of a d.c. voltmeter using a direct coupled amplifier when measuring a voltage 0.5 V is of the order of
(*a*) a few watt
(*b*) a few milliwatt
(*c*) a few microwatt
(*d*) a few nanowatt.

6. A chopper stabilized A.C. amplifier may use
(*a*) an electromechanical chopper
(*b*) bipolar transistors as choppers
(*c*) MOS field effect transistors as choppers
(*d*) light activated devices as choppers
(*e*) all of the above.

7. A voltmeter with a broad bandwidth has
   (*a*) low noise level and high sensitivity
   (*b*) high noise level and high sensitivity
   (*c*) high noise level and low sensitivity
   (*d*) low noise level and low sensitivity.
8. In circuits of RF voltmeters
   (*a*) vacuum tube diodes are used
   (*b*) conventional PN junction diodes are used
   (*c*) schottky barrier or point contact type diodes are used
   (*d*) all of the above.
9. In a differential amplifier using two FETs, a resistance has to be used to reduce the meter current to zero even when no voltage is applied to the circuit. This is necessitated on account of
   (*a*) mismatch between characteristics of the FETs
   (*b*) difference between tolerance values of resistors used in the circuit even though they are marked nominally equal.
   (*c*) variations in the operating voltage of the circuit.
   (*d*) both (*a*) and (*b*).
10. A PMMC meter has an internal resistance 200 Ω and the current required for its full scale deflection is 50 μA. The meter is capable of measuring, on its own, a maximum voltage of
   (*a*) 5 mV
   (*b*) 10 mV
   (*c*) 5 μV
   (*d*) 10 μV.
11. A PMMC meter has an internal resistance of 500 Ω and the current required for its full scale deflection is 100 μA.The power consumed by the meter is
   (*a*) 5 μ W
   (*b*) 5 mW
   (*c*) 5 W
   (*d*) 0.5 μW.
12. Electronic voltmeters which use rectifiers employ negative feedback. This is done
   (*a*) to increase the overall gain
   (*b*) to improve stability
   (*c*) to overcome nonlinearity of diodes
   (*d*) none of the above.
13. A voltage having symmetrical square waveshape is applied to an average reading voltmeter with scale calibrated in terms of rms value of a sinusoidal wave which has an error of
   (*a*) –3.9%
   (*b*) + 3.9%
   (*c*) – 11%
   (*d*) +11%.
14. The resistances, R_D, connected in the bridge circuit of a difference amplifier type of Electronic Voltmeter are each 10 kΩ. The value of R_D is much smaller than the value of A.C. drain resistance, r_d, of FETs. The transconductance of FETs is 0.005 mho. The value of

current through the meter when an input voltage of 1V is applied to the gate of left FET with the gate of right hand FET grounded, and the meter resistance almost negligible is:

(*a*) 5 mA

(*b*) 2.5 mA

(*c*) 10 mA

(*d*) 1 mA.

15. In a source follower type electronic voltmeter, the transconductance of FETs is 0.005 mho and the resistance of the meter is 100Ω. The approximate value of current through the meter, when the input voltage is 1V, is :

(*a*) 2.5 mA

(*b*) 5 mA

(*c*) 2 mA

(*d*) 1 mA

ANSWERS

1. (*b*)	**2.** (*c*)	**3.** (*d*)	**4.** (*b*)	**5.** (*d*)	**6.** (*e*)
7. (*c*)	**8.** (*c*)	**9.** (*d*)	**10.** (*b*)	**11.** (*a*)	**12.** (*c*)
13. (*d*)	**14.** (*b*)	**15.** (*c*)			

THEORETICAL QUESTIONS

1. What is an *"instrument"* ?

2. What do you mean by the term *"Measurement"*?

3. What are the primary requirements for any measuring instrument ?

4. What are the essentials of an electronic instrument ? Explain each of them briefly.

5. Explain briefly the following elements of electronic instrument :

(*a*) Transducer ;

(*b*) Signal modifier;

(*c*) Indicating device.

6. State the advantages of electronic instruments.

7. What are electronic voltmeters ? Explain briefly.

8. What are the advantages of electronic voltmeters over conventional ones ?

9. Give the comparison between *'Electronic voltmeters'* and *'Conventional analog voltmeters'*.

10. Enumerate the various types of electronic voltmeters.

11. How are electronic voltmeters classified ?

12. Explain briefly *'Vacuum Tube Voltmeters'* (VTVM).

13. List the various types of vacuum type voltmeters.

14. What are the advantages-applications of valve voltmeters ?

15. State the advantages of transistor voltmeter over the vacuum tube voltmeter.

16. Explain briefly any two of the following electronic voltmeters :

(*a*) Transistorised voltmeter using cascaded transistors.

(*b*) Transistorised voltmeter using FET as input stage.

(*c*) Balanced bridge TVM.

17. Describe briefly the following :
    (*a*) Chopper amplifier type voltmeter.
    (*b*) Differential voltmeter (D.C.)
18. Explain briefly 'Rectifier type A.C. voltmeter'.
19. Describe briefly with neat sketches any two of the following :
    (*a*) Average responding voltmeter
    (*b*) Peak responding voltmeter
    (*c*) True r.m.s.-reading voltmeter.
20. What are the advantages and limitations of true r.m.s. reading voltmeter ?
21. What is a *'Solid state voltmeter'* ? Explain briefly.
22. Discuss briefly *'Vector voltmeter'*.
23. Explain with a neat circuit diagram *"Electronic multimeter"* (or VOM).
24. List some important considerations needed in selecting an analog voltmeter.

CHAPTER

7 Cathode Ray Oscilloscope (CRO)

7.1. INTRODUCTION

Cathode ray oscilloscope *(CRO) is a device that allows the amplitude of electrical signals, whether they be voltage, current, power, etc., to be displayed primarily as a function of time.*

It is a very useful and versatile laboratory instrument used for display, measurement and analysis of waveforms and other phenomena in electrical and electronic circuits.

— It is basically an *electron beam voltmeter*.

- The oscilloscope depends on the movement of an *electron beam* to impinge on a phosphor surface, which produces a *visible spot*. If the electron beam is deflected in either of two orthogonal axes, such as the familiar *X* and *Y* axes used in conventional graph construction, the luminous spot can be used to create *two-dimensional displays*. Typically, the *X*-axis of the oscilloscope is deflected at a constant rate, relative to time, and the vertical or *Y*-axis is deflected in response to an input stimulus such as a voltage. This produces the time-dependence variation of the input voltage, which is very important to the design and development of electronic circuits.
- The oscilloscope is *capable of displaying events that take place over periods of microseconds and nanoseconds.*
- CROs *operate on voltages*. However it is possible to convert current, strain, acceleration, pressure and other physical quantities into voltages with the help of transducers and thus to present visual representations of a wide variety of dynamic phenomena on CROs.

These oscilloscopes have been evolved continuously, and they are now available which can measure frequencies upto 1GHz and observe events as small as 20 Hz in duration.

- With some oscilloscopes several additional features, such as digital multimeters and counters, are available. Several oscilloscopes are *microprocessor controlled*. They have the ability to calculate several features, such as rise time or pulse width of the measured waveform, and to display these values along with the display of waveforms. They are *easier to use and internal routines often act as a guide for the user, and display a warning if there is any error in setting.*

Although, most oscilloscopes are monochromatic, *colour oscilloscopes* are finding increasing applications in computers and in television.

- Most present day oscilloscopes are *capable of accepting two or more inputs displaying them simultaneously*. This may be achieved through using a *split beam* or *by using a multiple beam tube.*

7.2. APPLICATIONS OF CRO

There are variety of applications in which CRO is used. Some of these applications are:

1. Tracing of an actual waveform of current or voltage.
2. Determination of amplitude of a variable quantity.
3. Comparison of phase and frequency.
4. Measurement of capacitance and inductance.
5. In televisions.
6. In radar.
7. For finding B-H curves for hysteresis loop.
8. For engine pressure analysis.
9. For studying the heart beats, nervous reactions etc.
10. For tracing transistor curves.
11. For determining the modulation characteristics and to detect the standing waves in transmission lines.
12. It can be used to check the diodes and the faulty components in the various circuits.

7.3. BLOCK DIAGRAM OF AN OSCILLOSCOPE

Fig. 7.1. shows the block diagram of general-purpose oscilloscope. Its consists of the following major systems :

1. ***Cathode ray tube :***
   — It generates the electron beam, accelerates the beam to a high velocity, deflects the beam to create the image, and contains a phosphor screen where the electron beam eventually becomes visible.

   — It displays the quantity being measured.
2. ***Vertical amplifier.*** It amplifies the signal waveform to be viewed.
3. ***Horizontal amplifier.*** It is fed with a sawtooth voltage which is then applied to horizontal deflection plates.
4. ***Sweep or time base generator.*** It produces sawtooth voltage waveform used for horizontal deflection of the electron beam.
5. ***Trigger circuit.*** It produces trigger pulses to start horizontal sweep.
6. ***High and low-voltage supply :***

— The power supply block provides the voltages required by the cathode ray tube to generate and accelerate the beam, as well as to supply the required operating voltages for the other circuits of the oscilloscope.

— Relatively high voltages are required by cathode ray tubes, on the order of a few thousand volts, for acceleration, as well as a low voltage for the heater of the electron gun, which emits the electrons. Supply voltages for other circuits are various values, usually not more than a few hundred volts.

7. ***Delay line.*** High frequency oscilloscopes always include delay line in the vertical amplifiers. *The purpose of such lines is to delay the vertical signal enough to keep it from reaching the CRT deflection plates before the horizontal sweep circuits are running.*

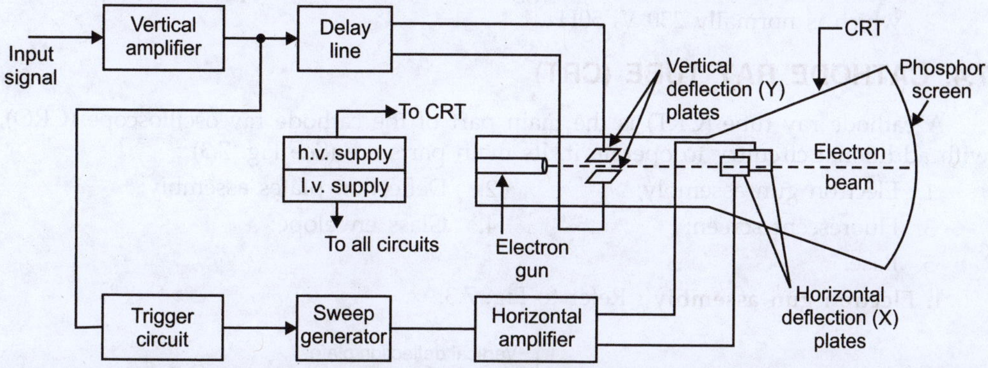

Fig. 7.1. Block diagram of a general purpose CRO.

Working of a CRO :

- When signal is to be displayed or viewed on the screen it is applied across the Y-plates of a cathode ray tube (CRT). But to see its waveform or pattern, it is essential to spread it to horizontally from left to right. This is achieved by applying a sawtooth voltage wave to X-plates.

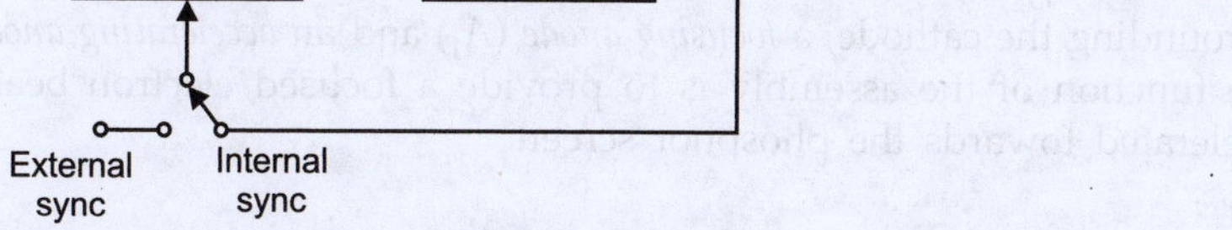

Fig. 7.2

Under these conditions the electron beam would move uniformly from left to right thereby graphing vertical vibrations of the input signal versus time. Due to repetitive tracing of the viewed waveform, we get a continuous display because of persistence of vision.

- However, to get a stable stationary display on the screen it is essential to *synchronize the horizontal sweeping of the beam* (sync) with the input signal across the Y-plates. The signal will be properly synced only when *its frequency equals the sweep generator frequency.*

The usual method of synchronizing the input signal is to *use a portion of the input signal to trigger the sweep generator so that the frequency of the sweep signal is locked or synchronized to the input signal.* It is called internal sync because the synchronization is obtained by internal wiring connection as shown in Fig. 7.2.

— In the *external synchronization,* an external trigger is used to trigger the signal being measured.

— In the *line synchronization,* the trigger is obtained from the power supply to CRO which is normally 230 V, 50Hz.

7.4. CATHODE RAY TUBE (CRT)

A cathode ray tube (CRT) is the main part of the cathode ray oscilloscope (CRO), with additional circuitry to operate it. Its main parts are (See Fig. 7.3) :

1. Electron gun assembly;
2. Deflection plates assembly;
3. Fluorescent screen;
4. Glass envelope;
5. Base.

1. Electron gun assembly : Refer to Fig. 7.3.

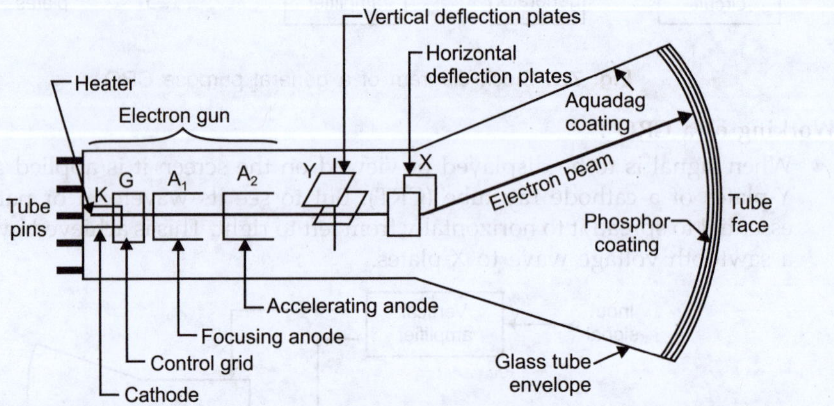

Fig. 7.3. Cathode ray tube.

- The electron gun assembly consists of an indirectly heated *cathode* a *control grid* surrounding the cathode, a *focusing anode* (A_1) and an *accelerating anode* (A_2). The sole function of the assembly is to provide a focused electron beam which is accelerated towards the phosphor screen.

Cathode :

- The *'cathode'* is a nickel cylinder coated with an oxide coating of barium and strontium and emits plenty of electrons, when heated. The emitting surface of the cathode should be as small as possible, theoretically a *point.*
- Rate of emission of electrons (or say the intensity of electron beam) depends on the magnitude of cathode current, which can be controlled by the control grid in a manner similar to a conventional vacuum tube.

- The typical values of current and voltage required by an indirectly heated cathode are 600 mA at 6.3 V (A.C. or D.C.). The special low power designs use 140 mA at 1.5 V.

Control grid :

- The *'control grid'* is usually a metal cylinder covered at one end but with a small hole in the cover. This is usually a metal cap of low permeability steel, about 15 mm in diameter and 15 mm long. An aperture of approximate 0.25 mm is drilled in the grid cap for the passage of electrons through it.
- The grid is kept at *negative potential* (variable) w.r.t. cathode and its function is to vary the electron emission and so the brilliancy of the spot on the phosphor screen. The hole in the grid is provided to allow passage for electrons through it and concentrate the beam of electrons along the axis of tube.

Focusing and accelerating anodes :

- Electron beam comes out from the control grid through a small hole in it and enters a *pre-accelerating anode,* which is a hollow cylinder in shape and is at a potential of few hundred volts more positive than the cathode so as to accelerate the electron beam in the electric field.
- The electron beam is then focused on the screen by an electrostatic lens consisting of two more cylindrical anodes called the *'focusing anode'* and *'accelerating anode'* apart from the pre-accelerating anode. The function of these anodes is to concentrate and focus the beam on the screen and also to accelerate the speed of electrons.
- The pre-accelerating and accelerating anodes are connected to a common positive high voltage of about 1500V; the focusing anode is connected to a lower adjustment voltage of 500 V.

2. Deflection plates assembly :

- The electron beam, after leaving the electron gun, passes through two pairs of deflection plates. One pair of plates is mounted horizontally and produces an electric field in the vertical plane. This pair produces a vertical deflection and is thus called *Vertical deflection plates* or *Y-plates*. The other pair of plates is mounted vertically and produces a horizontal deflection. This pair of plates is called *Horizontal deflection plates or X-plates.*

  The plates are flared so as to allow the beam to pass through them without striking the plates.
- When a sinusoidal voltage is applied to Y-plates, the beam will be moved up and down according to the variation of plate potential. If one frequency of variation is more than 16 Hz the deflection will be a vertical line in the centre of the screen. In case the sinusoidal voltage is applied to X-plates and frequency of variation is more than 16 Hz the deflection will be a horizontal line. If potentials are applied to both sets of plates simultaneously, the deflection will be an oblique line. The amount of deflection is in proportion to the voltage applied to the pair of plates.

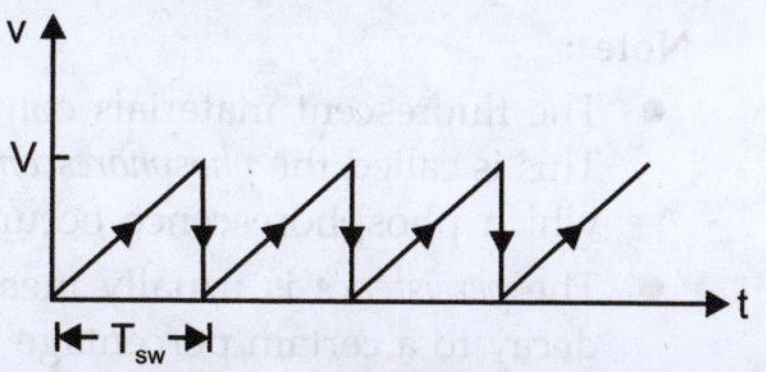

Fig. 7.4

It must be remembered that :

— The signal to be displayed on the screen is always applied across the Y-plates.

— The voltage applied across X-plates is a ramp voltage *i.e.* a voltage which increases linearly with time. It has a *'sawtooth waveform'* as shown in Fig. 7.4. It is also called *horizontal time-base or sweep voltage*. It has a sweep time of T_{sw}.

3. Fluorescent screen :

- The end wall of the CRT, called *'screen'*, is coated with *phosphor*. When struck by high-energy electrons, it glows. In other words, it absorbs the kinetic energy of the electrons and converts it into light–the process being known as *fluorescence*. That is why the screen is called *fluorescent screen*. The colour of the emitted light depends on the type of phosphor used.

The intensity of light emitted from the CRT screen, called *luminance*, depends upon the following *factors* :

(*i*) The number of bombarding electrons striking the screen per second.

(*ii*) The energy with which the bombarding electrons strike the screen which is determined by the accelerating potential (more the accelerating potential more the luminance).

(*iii*) The time the beam strikes a given area of the phosphor *i.e.*, sweep speed.

(*iv*) The physical characteristics of the phosphor itself.

- Among the fluorescent materials used are :
  (*i*) *Zinc or orthosilicate* ... for giving a green trace very suitable for visual observations;
  (*ii*) *Calcium tungstate* ... For giving blue and ultra-violet radiations very suitable for photography;
  (*iii*) *Zinc sulphide* (with other materials) ... For giving a white light suitable for T.V.;
  (*iv*) *Zinc phosphate* ... It gives a pronounced after glow and is useful when studying transient phenomena because the trace persists for short while after the transient has disappeared.
- The phosphor of the oscilloscope is designated as follows :

P_1 = Green medium; P_2 = Blue green medium;
P_5 = Blue very short; P_{11} = Blue short ect.

— The material P_{31} is used commonly for general purpose oscilloscopes due to the following *characteristics* :

(*i*) High illumination level. (*ii*) High writing speed.

(*iii*) High *burn resistance (to avoid accidental damage)

(*iv*) It gives colour to which human eye response is maximum.

(*v*) It gives short persistence required to avoid multiple image display.

* When an electron beam strikes the screen its K.E. is absorbed by the phosphor material. This creates heating. The heat produced gives rise to *"phosphor burn"* which is damaging and may cause phosphor destruction. So the phosphor material should have a *high burn resistance*.

Note :

- The fluorescent materials continue to emit light even after radiation exposure is cut off. This is called the *phosphorescence* characteristics of the materials. The length of time during which phosphorescence occurs is called the *persistence* of the phosphor.
- The *persistence* is usually measured in terms of the time required for the CRT image to decay to a certain percentage (usually 10%) of the original light output. It is expressed as short, medium and long.

(*i*) *"Short persistence"*. It lasts for *microseconds*. It is required for extremely high speed phenomena, to avoid smearing and interference caused when one image persists and overlaps with next one.

(*ii*) *"Medium persistence"*. It lasts for *milliseconds*. These traces are mostly used for general purpose applications.

(*iii*) *"Long persistence"*. It lasts for seconds to several minutes. These traces are used for transients as they keep the fast transient on the screen for observation after the transient has disappeared.

4. Glass envelope :

- The glass envelope is funnel-shaped having a phosphor-coated screen at its flared end. It is *highly-evacuated in order to permit the electron beam to traverse the tube easily*.
- The inside part of the flared part of the tube is coated with a conducting graphite layer called *"aquadag"* which is maintained at the same potential as the second anode. This layer performs the two functions : (*i*) It accelerates the electron beam after it passes between the deflecting plates, and (*ii*) It collects the electrons produced by secondary emission when electron beam strikes the screen. Hence it prevents the formation of negative charge on the screen.

— In some tubes, particularly CRTs with magnetic focusing (such as TV picture tubes), the accelerating anode is dispersed with entirely and the conductive coating is used as the final accelerating anode.

- *Horizontal and vertical marks* an marked on the screen of the CRT to provide user a correct measurement. These marks, *usually in rectangular form*, are called *graticule*.

— The graticule may be *external*, *internal* and *projected*.

— The graticules are generally laid out in an 8 × 10 pattern as shown in Fig. 7.5. with 8 major divisions along the vertical axis and 10 major divisions along the horizontal axis. The tick marks on the centre graticule represent minor divisions or subdivisions.

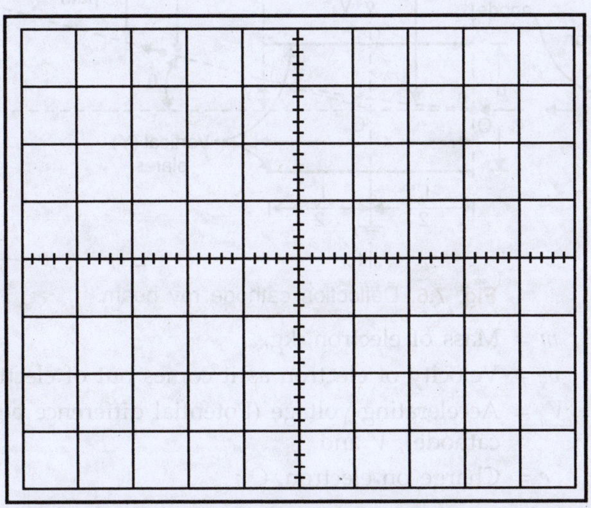

Fig. 7.5. The CRT graticule.

Size : Size refers to *screen diameter*. CRTs for Oscilloscopes are available in sizes 1, 2, 3, 5 and 7 inches. 3 inch tube is most common for portable instruments.

Example : *A CRT having a number* $5GP_1$ *is 5 inch tube with a medium persistence green trace. The first number indicates that it is a 5 inch Tube and* P_1 *indicates a medium persistence green trace.*

— Both round and rectangular CRTs are found is oscilloscopes. Vertical viewing size is 8 cms and horizontal is 10 cms (See Fig. 7.5).

- **Operating voltages :**

  The CRT requires a heating voltage of 6.3 V A.C. or D.C. at 600 mA. Several D.C. voltages are listed below; the voltages vary with the types of tube used :

  (*i*) Negative grid (control) voltage : –14 V to –200 V

  (*ii*) Positive anode A_1 (Focusing anode) : –100 V to –1100 V

  (*iii*) Positive anode A_2 (Accelerating anode) : 600 V to 6000 V

  A_3 : 200 V to 20000 V in some cases

5. Base. The base is provided to the CRT through which the connections are made to the various parts.

7.5. ELECTROSTATIC DEFLECTION

Fig. 7.6. shows the upward deflection of an electron beam when it passes between the vertical or *Y*-plates of a CRT. The beam deflects upwards because the upper *Y*-plate has been made positive w.r.t. the lower plate. *Reversing* the polarity of the *"applied voltage"* would, obviously, cause the beam to deflect downwards.

The velocity of an electron coming out of the electron gun can be found by equating the kinetic and potential energies.

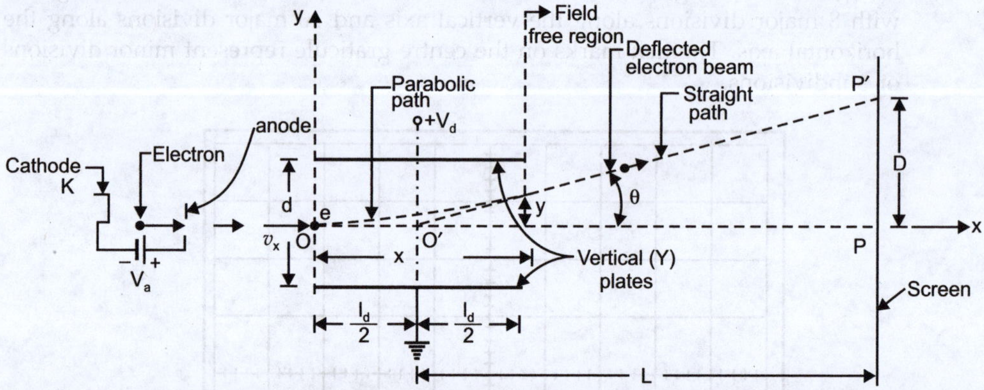

Fig. 7.6. Deflection cathode ray beam.

If,

m = Mass of electron, kg,

v_x = Velocity of electron as it comes out of electron gun, *m*/*s*,

V_a = Accelerating voltage (Potential difference between anode and cathode), *V* and

e = Charge on electron, C ;

Then,

$$\frac{1}{2} m v_x^2 = eV_a$$

or,

$$v_x = \sqrt{\frac{2eV_a}{m}} \qquad ...(7.1)$$

This velocity v_x *remains constant* because the fields due to X-and Y-plates act at right angles to each other.

Fig. 7.6. shows the *electrostatic deflection* due to the Y-plates.

Let,
d = Distance between Y-plates,
l_d = Length of each Y-plate,
L = Distance from O′ to screen,
V_d = Deflecting voltage (*i.e.*, potential difference between Y-plates), and
D = Deflection on the screen.

(The straight line of travel of the electron is tangent to the parabola at $x = l_d$)

Now, electric field between Y-plates

$$= \frac{V_d}{d} \text{ volts/m} \quad ...(7.2)$$

$$\therefore \quad \text{Force on electron} = \frac{eV_d}{d} \text{ newtons} \quad ...(7.3)$$

Also, force on electron = $m \times$ acceleration in y-direction (*i.e.* a_y)

$$\therefore \quad a_y = \frac{eV_d}{m\,d} m/s^2 \quad ...(7.4)$$

The time taken by the electron to traverse the length l is $\frac{l_d}{v_x}$. Since the initial velocity in the y-direction is zero, v_y, *i.e.* velocity in y-direction when electron leaves the field between plates, is

$$v_y = 0 + a_y.t$$

$$= \frac{eV_d}{md}.\frac{l_d}{v_x}$$

$$i.e. \quad v_y = \frac{eV_d}{md}\frac{l_d}{v_x} \quad ...(7.1)$$

After the electron leaves the field of the Y-plates, it moves in a *straight line* because it is in field-free space/region. *During the motion between the deflecting plates, a uniform acceleration is acting on the electron and therefore, its path is parabolic.* If the straight line is extended backward, it can be shown that it passes through O′, midway between the plates.

The deflection of the electron when it leaves the field is,

$$y = (\text{Average velocity in } y\text{-direction}) \times \frac{l_d}{v_x}$$

$$= \frac{1}{2}.\frac{eV_d}{md}\frac{l_d}{v_x} \; x \frac{l_d}{v_x}$$

$$\text{or,} \quad y = \frac{1}{2}\left(\frac{eV_d}{md}\right)\left(\frac{l_d}{V_x}\right)^2 \quad ...(7.6)$$

$$\text{From Fig. 7.36;} \quad \frac{y}{D} = \frac{(l_d/2)}{L} \text{ or } \frac{l_d}{2L}$$

or, $$D = \frac{2L}{l_d}.y = \frac{2L}{l_d} \times \frac{1}{2} \times \frac{eV_d}{md} \times \frac{l_d^2}{v_x^2}$$

i.e. $$D = \frac{L\,e\,V_d\,l_d}{m\,d\,v_x^2} \qquad \text{...(7.7)}$$

Substituting the values of v_x from eqn. (7.1), we get

$$D = \frac{LeV_d l_d}{m\,d} \times \frac{m}{2eV_a} = \frac{LV_d l_d}{2dV_a}$$

i.e. $$D = \frac{LV_d\,l_d}{2\,dV_a} \qquad \text{...(7.8)}$$

It is seen from eqn. (7.8) that D (vertical deflection on screen) is proportional to V_d (deflection voltage). This direct proportionality indicates that CRT may be used as a *linear voltage-indicating device*. The deflection voltage, although *assumed fixed* in the above discussion, usually varies and the image on the screen follows the variations of the deflection voltage in a linear manner.

- **Electrostatic deflection sensitivity (S)** *is defined as the deflection per volt of the deflection voltage.*

  i.e., $$S = \frac{D}{V_d} = \frac{L\,l_d}{2\,d\,V_a} \qquad \text{...(7.9)}$$

  Its units are m/V.

- **Deflection factor (G)** of a cathode ray tube is given by the *reciprocal of the sensitivity*.

  i.e. $$G = \frac{1}{S} = \frac{2dV_a}{L\,l_d} \qquad \text{...(7.10)}$$

  Its units are V/m.

— It is seen that S is independent of the deflection voltage V_d and the ratio $\frac{e}{m}$.

Moreover, S varies inversely with V_a (acceleration voltage); thus by decreasing V_a, sensitivity can be increased.

On the other hand high V_a value, produces a bright spot. But for high V_a, high V_d is required and such a beam which is highly accelerated is *difficult to deflect*. Such a beam is called *"hard beam"*.

— The typical values of deflection sensitivities are 1.0 mm/V to 0.1 mm/V corresponding to the deflection factors of 10 V/cm to 100 V/cm respectively.

7.6. MAGNETIC DEFLECTION

When a conductor of length l carrying a current I is placed in a magnetic field of flux density B, a force acts on the conductor. This force F is given by (assuming the directions of I and B to be perpendicular),

$$F = BIl \text{ newtons} \qquad \text{...(7.11)}$$

The direction F is perpendicular to the directions of I and B. If I and B are not perpendicular, only the component of I perpendicular to B contributes to the force.

Let the length l contains n electrons; if these electrons take t seconds to travel length

l then the charge per second *i.e.* current is given as :

$$I = \frac{n\,e}{t} \qquad ...(7.12)$$

where e is the charge on an electron, C.

$\therefore$ Force on the n electrons is,

$$F = \frac{B\,n\,e\,l}{t} \qquad ...(7.13)$$

$\frac{l}{t}$ is the drift speed (*i.e.* v) of the electrons.

$\therefore$ Force *per electron* $= B\ e\ v$ ...(7.14)

Equ. (7.14) indicates that if v is zero, F is zero. This means that if electron is at rest, it does not experience any force and it remains at rest. If the velocity is along the direction of magnetic flux, there is no force on the electron and it continues to move at the same velocity.

Refer to **Fig. 7.7.** When an electron enters a magnetic field, having flux density directed into the page, with a speed v the force F is perpendicular to v and therefore the speed v remains constant. This forces causes the electron to move in a circular motion with constant speed v. If r is the radius of the circular path, the acceleration towards the centre is $\frac{v^2}{r}$.

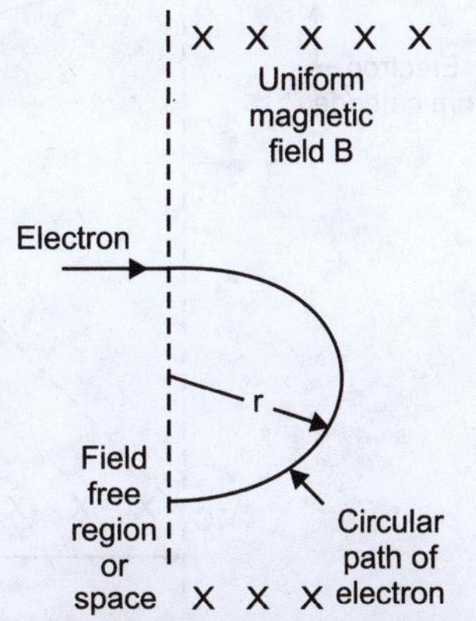

Fig. 7.7. Electron moving in a magnetic field.

Then, $$F = \frac{mv^2}{r} = B\,e\,v$$

or, $$r = \frac{mv}{eB} \qquad ...(7.15)$$

Also, angular velocity,

$$\omega = \frac{v}{r} = \frac{v}{(mv/eB)} = \frac{eB}{m} \text{ rad./sec.} \qquad ...(7.16)$$

and, Time period T (*i.e.*, to complete one revolution) is given as :

$$T = \frac{2\pi}{\omega} = \frac{2\pi}{(eB/m)} = \frac{2\pi m}{eB} \text{ rad./sec.} \qquad ...(7.17)$$

From above discussion, we find that the radius of the circular path is directly proportional to v. ω and T do not depend on r or v. Thus, the fast moving particles travel in a bigger circle while the slow moving particles travel in smaller circle.

The *time period for both the slow and the fast particles is the same*. This result forms the basis of operation of *cyclotron, magnetic focusing apparatus and many other apparatus*.

- A magnetic field can be used for deflection of electron beam in a cathode ray tube.

Fig. 7.8 shows a transverse magnetic field of flux density B for detecting an electron beam. The field is pointing into page. The electron moves in a straight line from cathode to boundary A of the magnetic field. In the magnetic field region, the electron experiences a force F (eqn. 7.14) and moves in a circular path AB. The radius of the circular path is r (eqn. 7.15) and the centre is at point C. The angle of deflection θ is very small and is given by,

$$\theta \simeq \frac{l}{r} \qquad \text{...(7.18)}$$

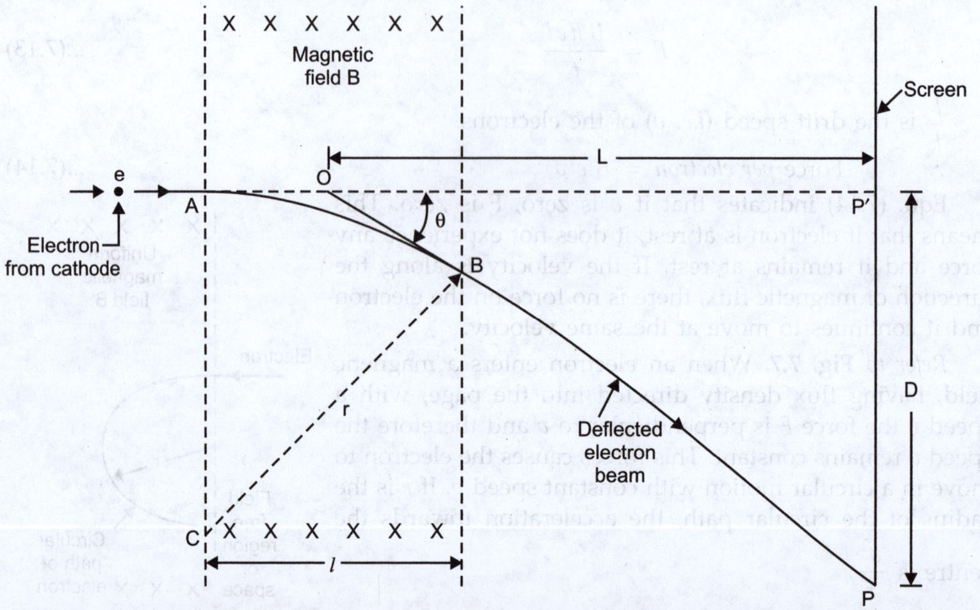

Fig. 7.8. Deflection of electron beam by a magnetic field.

As distance L is very large as compared to l, therefore, the straight line PB when projected backward passes through point O at the centre of the magnetic field. The deflection D is given as :

$$D = L\tan\theta \simeq L\theta = \frac{Ll}{r} = \frac{LleB}{mv} \qquad \text{...(7.19)}$$

The velocity v is v_x as given by equ. (7.1). Substituting this value in eqn. (7.19), we get :

$$D = \frac{LleB}{\sqrt{\frac{2eV_a}{m}}} = BLl\left(\sqrt{\frac{e}{2mV_a}}\right) \qquad \text{...(7.20)}$$

Magnetic deflection sensitivity (S) *is defined as the deflection on the screen in meters per tesla of magnetic flux density.*

i.e.
$$S = \frac{D}{B} = Ll\sqrt{\frac{e}{2mV_a}} \qquad \text{...(7.21)}$$

- It is seen that magnetic deflection and magnetic deflection sensitivity are inversely proportional to $\sqrt{V_a}$. Therefore, magnetic deflection is preferable in tubes with high accelerating voltage like *TV tube* and *radar* (while *electrostatic deflection is prefered for general purpose oscilloscope*).

Comparison between deflection methods :

The comparison between 'electrostatic deflection' and 'magnetic deflection' method is given below :

S.N.	Aspects	Electrostatic deflection	Magnetic deflection
1.	*Method used to obtain deflection.*	By the application of voltage to the plates.	By controlling the magnetic field by varying the current through the coils.
2.	*Relationship between the deflection and V_a.*	Inversely proportional to V_a.	Invesely proportional to $\sqrt{V_a}$.
3.	*Size of the beam for given V_a.*	Small	Large (wide).
4.	*Sensitivity (S) formula.*	$S = \frac{Ll_d}{2dV_a}$	$S = Li\sqrt{\frac{e}{2mV_a}}$
5.	*Dependance of deflection on the ratio e/m.*	No	Yes
6.	*Size of tubes necessary for given display area.*	Longer tubes.	Shorter tubes (can be built for the given display area).
7.	*Effect on the spot brightness with reduction in V_a (for greater sensitivity)*	Reduced brightness.	Increased brightness.
8.	*Applications.*	General propose oscilloscopes.	T.V. and radar.

- ***Differences between CRT and television picture tube.***

1. The CRT used in CRO uses electrostatic method of focussing as compared to a TV picture tube which employs electromagnetic envelope.
2. In CRT used in CRO, connection to accelerating anode is brought out through pin in the base along with other electrode connections as shown in Fig. 7.33. However in TV picture tube the lead in from accelerating anode is brought out through the sides of glass envelope.
3. In CRT used in CRO, EHT, (*i.e.*, extra high tension transformer) to accelerating anode is given from mains supply whereas in TV picture tube EHT supplied to accelerating anode is generated in the line output stage during flyback. During flyback the anode current of line output stage is suddenly cut off and anode voltage generates a spike of 2 to 3 kV amplitude. By increasing the number of turns on the primary winding to form an auto-transformer of say 1 : 5 turns pulses of 15 kV or more is obtained. After rectification and smoothing, EHT is obtained and applied to accelerating anode.

This method of applying EHT has an advantage because if input to line and valve fails, no EHT is produced and thus the tube is saved.

Example 7.1. *A CRT has an anode voltage of 2 kV and parallel deflecting plates 2 cm long and 5 mm apart. The screen is 30 cm from the centre of the plates. Find the input voltage required to deflect the beam through 3 cm. The input voltage is applied to the deflecting plates through amplifiers having an overall gain of 100.* **(Nagpur University S - 2005)**

Solution. *Given :* $V_a = 2$ kV or 2000 V; $l_d = 2$ cm or 0.02 m; $d = 5$ mm or 0.005 m; L = 30 cm or 0.3 m; D = 3 cm or 0.03 m; Overall gain = 100.

We know that, $$D = \frac{L l_d V_d}{2 d V_a} \quad \text{...[Eqn. (7.1)]}$$

or, $$V_d = \frac{2 d V_a D}{L l_d}$$

$$= \frac{2 \times 0.005 \times 2000 \times 0.03}{0.3 \times 0.02} = 100\text{V}$$

∴ Input voltage required for deflection of 3 cm

$$= \frac{V_d}{gain} = \frac{100}{100} = \mathbf{1\ V} \quad \textbf{(Ans.)}$$

Example 7.2. *An electrostatically deflected cathode ray tube has plane parallel deflecting plates which are 2.5 cm long and 0.5 cm apart, and the distance from their centre to the screen is 20 cm. The electron beam is accelerated by a potential difference of 2500 V and is projected centrally between the plates.*

Calculate the deflection voltage required to cause the beam to deflect through 3 cm and find the corresponding deflection sensitivity of the screen. **(PTU Dec. - 2005)**

Solution. *Given :* $l_d = 2.5$ cm = 0.025 m; $d = 0.5$ cm = 0.005 m; $L = 20$ cm = 0.2 m; $V_a = 2500$ V ; $D = 3$ cm = 0.03 m

$$V_d \ ; \ S :$$

We know that, $$D = \frac{L l_d V_d}{2 d V_a} \quad \text{...[Eqn. (7.8)]}$$

or, *Deflection voltage,* $$V_d = \frac{D \times 2 d V_a}{L l_d} = \frac{0.03 \times 2 \times 0.005 \times 2500}{0.2 \times 0.025} = \mathbf{150\ V} \quad \textbf{(Ans.)}$$

∴ *Deflection sensitivity,* $$S = \frac{D}{V_d} = \frac{0.03}{150} \text{ m/V or } \mathbf{0.2\ mm/V} \quad \textbf{(Ans.)}$$

Example 7.3. *The horizontal deflection plates in a CRT are 20 mm long and 5 mm apart. The centre of the plates is 20 cm from the screen. Accelerating voltage is 2500 V. Determine the deflection sensitivity.* **(UPTU - 2004)**

Solution. *Given:* $l_d = 20$ mm = 0.02 m; $d = 5$ mm = 0.005 m; $L = 20$ cm = 0.2 m; $V_a = 2500$ V.

We know that, deflection sensitivity,

$$S = \frac{L l_d}{2 d V_a} \quad \text{...[Eqn. (7.9)]}$$

or, $$S = \frac{0.2 \times 0.02}{2 \times 0.005 \times 2500} \text{ m/V or } \mathbf{0.16\ mm/V} \quad \textbf{(Ans.)}$$

Example 7.4. *In a CRT, the anode to cathode voltage is 2500 V. The parallel deflection plates are 1.5 cm long and spaced 5 mm. The screen is 50 cm from the centre of the deflecting plates. If mass of electron is 9.109×10^{-31} kg and charge on electron 1.602×10^{-19}C, find :*

(*i*) *The beam speed.*

(*ii*) *The deflection sensitivity of the tube.*

Solution. *Given* : V_a = 2500 V; l_d = 1.5 cm = 0.015 m; d = 5 mm = 0.005 m;
L = 50 cm = 0.5 m; $m = 9.109 \times 10^{-31}$ kg; $e = 1.602 \times 10^{-9}$ C.

(*i*) **The beam speed, v :**

We know that, $$v = \sqrt{\frac{2eV_a}{m}}$$ ...(Eqn. (7.1)]

$$= \sqrt{\frac{2 \times 1.602 \times 10^{-19} \times 2500}{9.109 \times 10^{-31}}}$$

$$= \mathbf{29.65 \times 10^6\ m/s}$$ **(Ans.)**

(*ii*) **Deflection sensitivity, S :**

$$S = \frac{Ll_d}{2dV_a}$$ ...[Eqn. (7.2)]

$$= \frac{0.5 \times 0.015}{2 \times 0.005 \times 2500}\ m/V$$

or, $$= \mathbf{0.3\ mm/V}$$ **(Ans.)**

Example 7.5. *A cathode-ray tube, with magnetic deflection, has its screen 22 cm from the centre of the magnetic field. The width of the uniform magnetic field is 3.3 cm and the final anode potential is 6 kV. Calculate the density of the magnetic field to produce a deflection of 4.4 cm on the screen.*

Take : Mass of electron = 9.107×10^{-31} kg; charge on electron = 1.6×10^{-19} C.

Solution. *Given* : L = 22 cm = 0.22 m; l = 3.3 cm = 0.033 m; V_a = 6 kV;
D = 4.4 cm = 0.044 m; $m = 9.107 \times 10^{-31}$ kg; $e = 1.6 \times 10^{-19}$ C.

Density of magnetic field, B :

We know that, $$D = BLl\left(\sqrt{\frac{e}{2mV_a}}\right)$$ ...[(Eqn. (7.20)]

Substituting the values, we get:

$$0.044 = B \times 0.22 \times 0.033 \left(\sqrt{\frac{1.6 \times 10^{-19}}{2 \times 9.107 \times 10^{-31} \times (6 \times 10^3)}}\right)$$

or, $$0.044 = B \times 27.779$$

or, $$B = 1.584 \times 10^{-3}\ \text{Wb/m}^2$$ or **1.584 m Wb/m²** **(Ans.)**

Example 7.6. *In a cathode-ray tube, the electron beam is displaced vertically by a magnetic field of flux density 1.8×10^{-4} Wb/m². The length of the magnetic field along the tube axis is same as that of the electrostatic deflection plates. The final anode voltage is 800 V.*

Calculate the voltage which should be applied to the Y-deflection plates 10 mm apart, to return the spot back to the centre of the screen.

Take : Mass of electron = 9.107×10^{-31} kg; charge on electron = 1.6×10^{-19} C.

Solution. *Given* : $B = 1.8 \times 10^{-4}$ Wb/m²; V_a= 800 V; d = 10 mm = 0.01 m.

Deflection voltage, V_d :

Case I. *Magnetic deflection* :

In this case, deflection, $D = BLl\left(\sqrt{\dfrac{e}{2mV_a}}\right)$ ...[Eqn. (7.20)]

$$= 1.8\times10^{-4}\,Ll\left(\sqrt{\frac{1.6\times10^{-19}}{2\times9.107\times10^{-31}\times800}}\right)$$

or, $D = 1.886\ lL$ metres ...(1)

Case II. *Electrostatic deflection* :

In this case, $D = \dfrac{LV_d\,l}{2dV_a}$ ...[Eqn. (7.8)]

or, $D = \dfrac{lLV_d}{2\times0.01\times800} = 0.0625\ l\ M\ Vd$ metres ...(2)

Equating (1) and (2), we get

$$1.886\ lL = 0.0625\ lLV_d$$

or, $V_d = \dfrac{1.886}{0.0625}$

$= \mathbf{30.176\ V}$ **(Ans.)**

7.7. VERTICAL AMPLIFIER

- The input signal is fed to the *Y*-deflecting plates through a vertical amplifier.
- The function of vertical amplifier is to *amplify the input signal.*
- It may consist of several stages in cascade and generally, operates with fixed gain. The output stage of this amplifier is invariably a push-pull amplifier. It is necessary to adjust the magnitude of the input signal so that a large signal may not cause the display to go out of the screen. Therefore, the input signal is fed to the vertical amplifier through an *attenuator*.
- The *"attenuator"* is a network of resistances and capacitances and it *reduces the input signal* so that the display of large input signals *may not go out of the screen.*

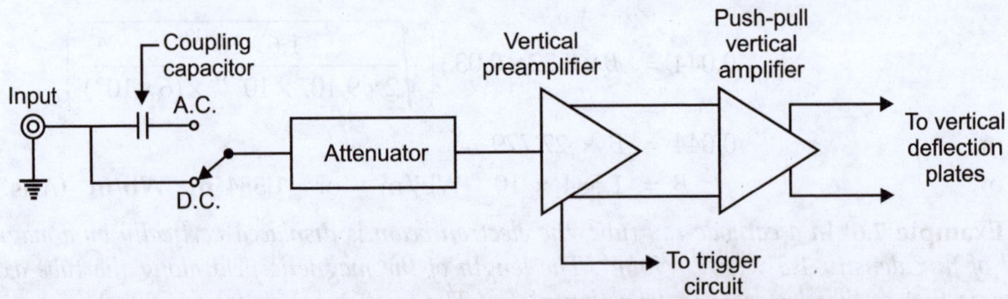

Fig. 7.9. Block diagram of the vertical section of an oscilloscope.

Fig. 7.9 shows the block diagram of a complete *vertical deflection system.*

— The input convector feeds an input attenuator, after which follows the vertical amplifier. The input impedance of an oscilloscope is rather high, being on the order of 1 MΩ, which is desirable for measuring voltages in high impedance circuits.

— The attenuator sets the sensitivity of the oscilloscope in the common 1-2-5 sequence. As an example, the input attenuator could provide for 10, 20, 50, 100, 200 mV, etc. per centimetre.

— Practically all oscilloscopes provide a switchable input coupling capacitor. This is provided so that measurements of A.C. signals may be viewed in the presence of high D.C. voltages by including the coupling capacitor. When D.C. measurements are to be made, the capacitor may be removed. The value of the capacitor is chosen so that the frequency response of the oscilloscope is preserved down to a few Hz.

- *The vertical amplifier is the principal factor in determining the sensitivity and bandwidth of an oscilloscope.*
  — The gain of the vertical amplifier determines the smallest signal that the oscilloscope can satisfactorily reproduce on the CRT screen. The sensitivity of an oscilloscope is directly proportional to gain of the vertical amplifier that is, as gain increases sensitivity increases, which allows us to observe smaller-amplitude signal.
  — The vertical sensitivity of an oscilloscope is the smallest deflection factor that can be selected with the rotary switch.
  — The *bandwidth* of an oscilloscope determines the range of frequencies that can be observed with the instrument. The bandwidth of an oscilloscope is the range of frequencies over which the gain of the vertical amplifier is within 3 dB of the midband frequency gain.

7.8. TIME BASE (SWEEP OR RAMP) GENERATOR

If the horizontal deflecting plates X-plates are not energised, the electron beam would draw a vertical line on the screen.

- The *function of the time base generator* is to drive the beam at a steady speed across the screen and, when it reaches the right-hand side of the screen, the beam is made to fly back to the starting position on the left-hand side of the screen. The time base generator produces a *saw-tooth wave* (Fig. 7.10) of *the same frequency as the* input signal to the *Y*-plates.

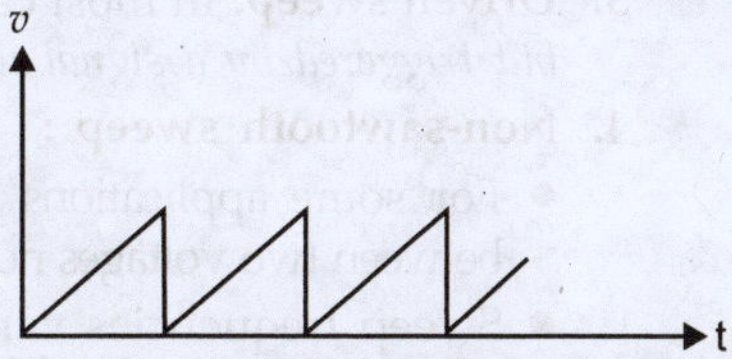

Fig. 7.10. Saw-tooth voltage wave.

- The horizontal deflection of the beam is proportional to the instantaneous voltage of the saw-tooth wave, therefore, the beam sweeps at a uniform rate across the screen in the horizontal direction, running to the starting position almost instantaneously and sweeping again.
- The time base generator gives an output sweep signal of adjustable frequency so that the *Y* input signals of a wide range of frequencies may be displayed on the screen.

There are many applications of a CRO in which a time base generator is *not* needed and it is necessary to apply an external input signal to the X-plates (through a switch).

7.9. TYPES OF SWEEPS

There are *four* types of sweeps :

1. **Free running sweep :**
   - In very basic oscilloscopes the sweep generator is continuously charging and discharging a capacitor. One ramp voltage is followed *immediately* by another, hence the sawtooth pattern appears. A sweep generator operating in this manner is said to be *"free running"*.

- In order to present a stationary display on the screen, the sweep generator signal must be forced to run in *synchronization* with a vertical input signal. In basic oscilloscopes this is accomplished by carefully adjusting the sweep frequency to a value very close to the exact frequency of the vertical input signal, or submultiple of this frequency. With signals at the same frequency, an *internal synchronization pulse will lock the sweep generator into the vertical input signal*. This method of synchronization has some serious limitations when an attempt is made to observe *low-amplitude* signals. However, the most serious limitation is probably the inability of the instrument to maintain synchronization when the amplitude or frequency of the vertical signal is not constant, such as voice or music signals.

2. **Triggered sweep :**
   - A waveform to be observed on the CRO may not be periodic but may perhaps occur at irregular intervals. In this case it is desirable that the sweep circuit remain inoperative and the sweep be initiated by the waveform under examination. In some cases the waveform may be periodic, but it may be that the interesting part of the waveform is of very short duration compared to the period of the waveform. Under such cases a *triggered switch is used.*
   - In triggered sweep or single sweep, the spot is swept once across the screen in response to a trigger signal.
   - The *triggered sweep is used for examination of transient or one time signals and waveform is photographed for record.* The trigger can be obtained from the signal under investigation or by an external source.
3. **Driven sweep.** In most cases, a *'driven sweep'* is used where the sweep is *recurrent but triggered by a signal under test.*
4. **Non-sawtooth sweep :**
   - For some applications like comparison of two frequencies or finding phase shift between two voltages non-sawtooth sweep voltages are utilised for the sweep circuit.
   - Sweep frequencies vary with the type of the oscilloscope.

A laboratory oscilloscope may have sweep frequencies upto several MHz; a simple oscilloscope for audio work has an upper limit 100 kHz. Most T.V. services require a sweep voltage frequency upto 1 MHz.

7.10. HORIZONTAL AMPLIFIER

- The function of the horizontal amplifier is to *amplify the signal applied to the X-plates.*
- This amplifier consists of several stages in the cascade.
- For measurement of *voltages and currents*, the input to the horizontal amplifier is a *sawtooth wave*. For *frequency and phase* measurements, the input is an *external signal*. The horizontal amplifier should be capable of amplifying both these inputs without distortion.

The input impedance of most general purpose oscilloscopes is about 1 MΩ shunted by a capacitance of about 20 pF.

7.11. DELAY LINE

The *purpose of a "delay line" is to delay the vertical signal enough to keep it from reaching the CRT deflection plates before the horizontal sweep circuits are running.*

- In an oscilloscope all electronic circuitry (attenuators, amplifiers, pulse shapers, etc.) causes a certain amount of line delay in the transmission of signal voltages to the deflection plates. Almost all of this delay is created in circuits that *switch*, *shape* or *generate*.

Referring to Fig. 7.11, we observe that the horizontal signal (time base, or sweep voltage) is initiated or *triggered*, by a portion of the output signal applied to the vertical plates of the CRT. In horizontal channel, the signal processing consists of generating and shaping a trigger pulse (trigger pick off) that starts the sweep generator, whose output is fed to the horizontal amplifier and then to the horizontal deflection plates. This whole process consumes time on of the order of 80 ns or so. Thus, to allow the operator to observe the leading edge of the signal waveform, the signal drive for the vertical CRT plates must therefore be *delayed* by at least the same amount of time. This is the function of the vertical delay line. In the Fig. 7.11 is shown a 200 ns delay line which has been added to the vertical channel. It may be remembered that a trigger pick off must *precede* the delay line, although the delay line can appear almost anywhere along the vertical signal path.

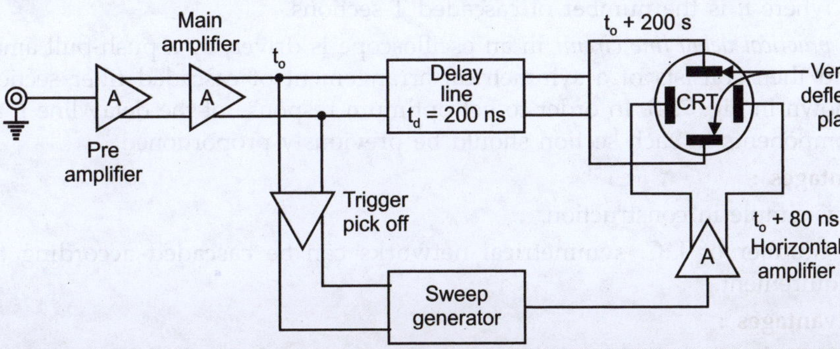

Fig. 7.11. Delay line circuit.

In CRO the following two types of delay lines are used :

1. Lumped-parameter delay line. 2. Distributed-parameter delay line.

1. Lumped-parameter delay line :

- This type of delay line consists of a cascaded symmetrical LC networks, such as the so called T-section of Fig. 7.12.

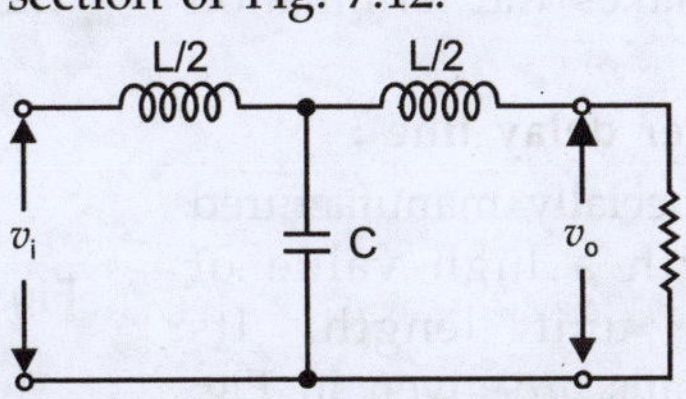

Fig. 7.12. T-filter section.

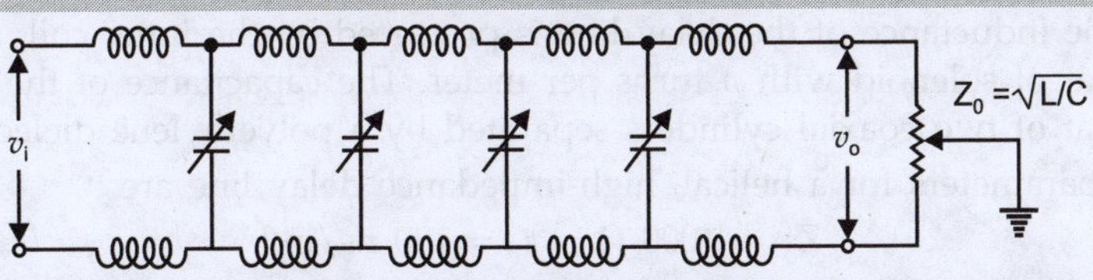

Fig. 7.13. Practical delay line.

- Each section is capable of delaying the signal by 3 to 6 ns.
- The T-section is terminated into its characteristic impedance Z_0. So, it acts as

a low-pass filter, its cut-off frequency is given as :

$$f_c = \frac{1}{\pi\sqrt{LC}} \quad ...(7.22)$$

If the spectrum of input signal v_i consists of frequencies much less than the cut-off frequency, output signal v_o will be a faithful reproduction of v_i but delayed by a time,

$$t_s \simeq \frac{1}{\pi f_c} = \sqrt{LC} \quad ...(7.23)$$

where, t_s is the delay time for a single T-section. However, when a number of T-sections are cascaded into a so-called *lumped-parameter* delay line, the total delay time,

$$t_d = n\, t_s \quad ...(7.24)$$

where n is the number of cascaded T-sections.

- A *practical delay line circuit* in an oscilloscope is driven by a push-pull amplifier and then consists of a symmetrical arrangement of cascaded filter sections as shown in Fig. 7.13. In order to get optimum response of the delay line L and C components in each section should be previously proportioned.

Advantages :

(*i*) It is simple in construction.

(*ii*) A number of L.C. symmetrical networks can be cascaded according to the requirement.

Disadvantages :

(*i*) Since it has a sharp cut-off frequency therefore, as the input frequency increases the amplitude and phase *distortion* takes place.

(*ii*) In order to get better transient response the delay line should be matched to the characteristic impedance at the input and output side. This makes the termination circuiry *complex*.

2. Distributed-parameter delay line :

- It consists of a specially manufactured coaxial cable with a high value of inductance per unit length. Its constructional details are given in Fig. 7.14.

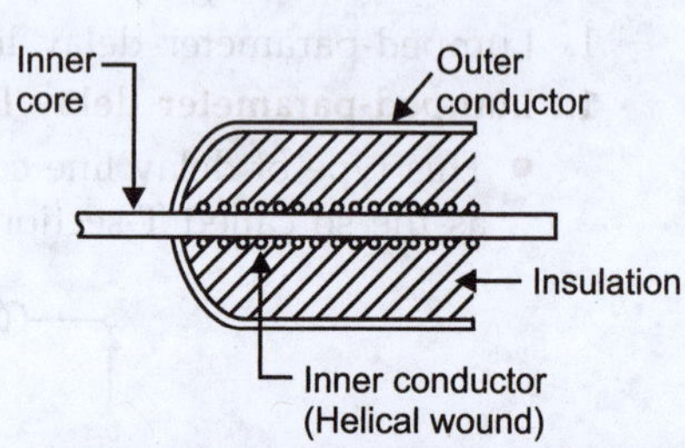

Fig. 7.14. Distributed parameter delay line.

- The inductance of the delay line is produced by the inner coil, and it equals that of solenoid with n turns per meter. The capacitance of the delay line is that of two coaxial cylinders separated by a polyethylene dielectric.

Typical parameters for a helical, high-impedance delay line are :

$$Z_o = 1000\ \Omega\ ;\ t_d = 180\ \text{ns/m.}$$

Advantages :

(*i*) Occupies less space.

(*ii*) It does not require careful adjustment of a lumped-parameter delay line.

7.12. BASIC CONTROLS OF CRO

Fig. 7.15. shows the detailed block diagram of a general purpose oscilloscope, exhibiting a number of controls required for the proper functioning of CRO.

1. Vertical (Y) deflection system :

- The signals to be examined are usually applied to the vertical or Y-deflection plates through an attenuator and a number of amplifier stages. Vertical amplifier is required because the signals are not strong enough to produce measurable deflection on CRT screen.

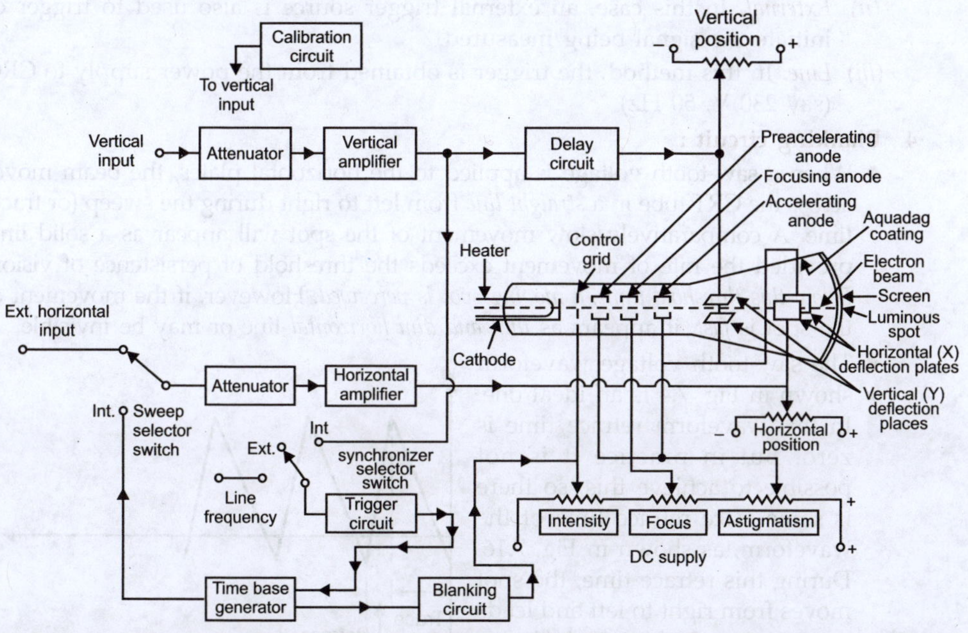

Fig. 7.15. Detailed block diagram of a general purpose oscilloscope.

— *The amplifier response must be wide enough to pass faithfully the entire band of frequencies to be measured.*

- The high voltage signals (to be examined) must be attenuated to bring them as within the range of vertical amplifies. The output of the vertical amplifier is also applied to the synchronizing amplifier through the *synchronizer selector switch in internal position (Int.)*. This permits the horizontal sweep circuit to be triggered by the signal being investigated.

2. Horizontal (X) deflection system :

- The horizontal or X-deflection plates are fed by a sweep voltage that provides a *time base*. The horizontal plates are supplied through an amplifier, but they can be fed directly when voltages are of sufficient magnitudes.
- When external signals are to be applied to the horizontal deflection system, they can also be fed through the horizontal amplifier, *via* the sweep selector switch which is in the external position (Ext.). When the sweep selector switch is in the internal position, the horizontal amplifier receives an input from the saw tooth sweep generator which is triggered by the synchronizing amplifier.

3. Synchronization :

- In order to obtain a *stationary pattern* synchronization [irrespective of the type of sweep (e.g. free running, triggered sweep, driven sweep, non saw-tooth sweep] has to be done. This requires that the time base be operated at a submultiple frequency of the signal under measurement (applied to *Y*-plates).

The following three usual sources for synchronization can be selected by synchronizing selector :

(*i*) *Internal.* Here the trigger is obtained from the signal being measured through the vertical amplifier.

(*ii*) *External.* In this case, an external trigger source is also used to trigger or initiate the signal being measured.

(*iii*) *Line.* In this method, the trigger is obtained from the power supply to CRO (say 230 V, 50 Hz)

4. Blanking circuit :

- When a saw-tooth voltage is applied to the horizontal plates, the beam moves across the CRT tube in a *straight line* from left to right during the sweep (or trace) time. A comparatively slow movement of the spot will appear as a solid line, provided the rate of movement exceeds the threshold of persistence of vision. *Below this threshold limit, a moving spot is perceived.* However, if the movement of the spot is *fast*, it appears as *thin and dim horizontal* line or may be invisible.
- The saw-tooth voltage waveform shown in Fig. 7.4 is an ideal one. In this waveform retrace time is zero. But in practice it is not possible to achieve this, so there is some trace retrace time of the waveform, as shown in Fig. 7.16. During this retrace time, the spot moves from right to left and leads to confusion. *So this trace is blanked out by supplying a high -ve voltage to the grid during the retrace time.* The blanking voltage is usually developed/triggered by time-base generator.

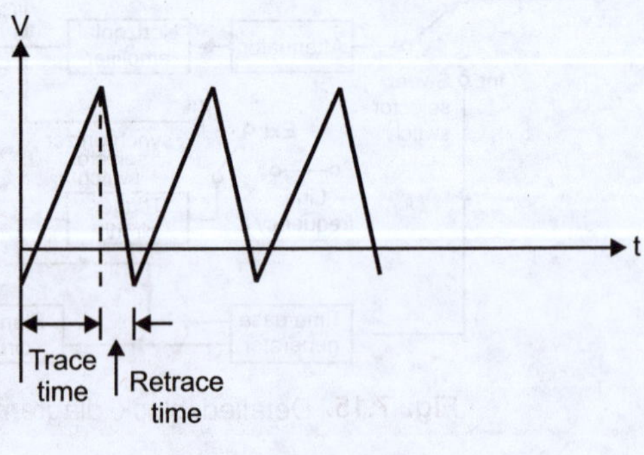

Fig 7.16. Sawtooth voltage waveform.

5. Positioning controls. These controls involve two knobs–one for controlling the horizontal position and another for controlling the vertical position.

— The spot can be moved to *left or right i.e.*, horizontally with the help of a knob, which regulates the D.C. potential applied to the *X*-deflection plates in addition to the usual saw-tooth wave. Similarly the spot can be moved *up and down i.e.* vertically with the help of another knob, which regulates the D.C. potential applied to the *Y*-deflection plates in addition to the signal.

6. Intensity (Z-axis) modulation :

- The intensity modulation is done by inserting a signal between the ground and the cathode/control grid.
- It is applied during normally visible portion of the trace.
- It can be used for *brightening the display*.
  — The periodically brightened spots may be used as markers for time calibration of the main waveform.

7. Intensity control :

- The electron-beam intensity is varied by the *intensity control potentiometer* which changes the grid potential w.r.t. cathode. The grid potential determines the amount of electrons leaving the cathode and thus controls the intensity of the beam.
- It must be ensured that the electron beam does not form burning spots on the screen. A stationary spot should be kept on very low intensity. If the intensity is kept high, the spot must be kept moving. If a *"halo"* appears around the spot, the intensity is too high; the intensity should be turned before turning the oscilloscope on.

8. Focus control :

- The middle anode, in the electron gun of a CRT, is kept at a lower potential w.r.t. other two anodes; it acts like an electrostatic lens and focal length of this lens can be varied by varying the potential of the middle anode w.r.t. other two anodes. So the focusing of an electron beam is done by varying the potential of the middle anode with the help of a potentiometer, as shown in Fig. 7.15.
- *By increasing the +ve potential applied to the focusing anode the electron beam can be narrowed and the spot on the screen can be made a pin point.*

9. Astigmatism:

- *Astigmatism* is an additional focusing control (in most modern oscilloscopes) and is analogous to as astigmatism in optical lenses.
- In order to focus the spot correctly, it is necessary to stop it near the centre of the screen by switching off the time base and adjusting the X and Y positioning controls. The spot is then made as *sharp as possible by successive adjustment of focus and astigmatism controls.*

10. Calibration circuit :

- Laboratory oscilloscopes normally have an internally generated and stablised voltage of known amplitude which is used for calibration purposes.
- The calibrating voltage usually has a square waveform.

7.13. DUAL TRACE OSCILLOSCOPE

The comparison of two or more voltages is invariably necessary in the analysis and study of many electronic circuits and systems. This is possible by using more than one oscilloscope but in such a case it is difficult to trigger the sweep of each oscilloscope precisely at the same time. A common and less costly method to solve this problem is to use dual trace or multitrace oscilloscopes. In this method, the same electron beam is *used to generate "two traces" which can be deflected from two independent* vertical sources.

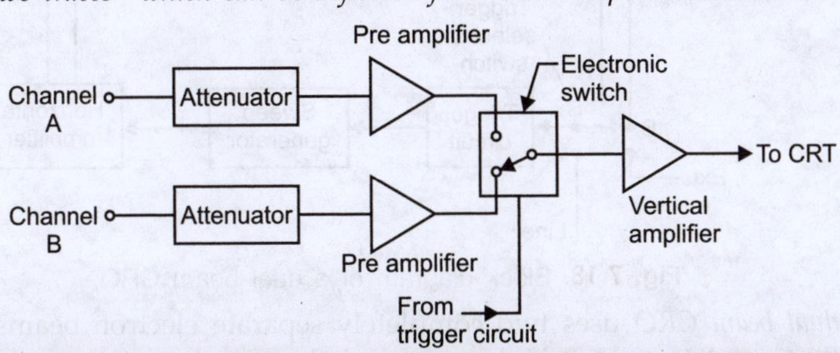

Fig. 7.17. Block diagram of a dual trace oscilloscopes.

Dual trace oscilloscopes are used extensively by *industrial firms and research laboratories.*

These oscilloscopes produce a dual-trace display by means of electronic switching of two separate input signals.

Fig. 7.17 shows the block diagram of a dual trace oscilloscope :

- There are two vertical input circuits marked channels *A* and *B* with identical preamplifiers.
- The outputs of these preamplifiers are fed to an electronic switch which alternately connects them to the main vertical amplifier of the oscilloscope.
- In this type of oscilloscope there is *only one electron beam.*
- There are two common operating modes for the electronic switch, called *alternate* and *chop*; these are selected from the instrument's front panel.
  - In ***alternate mode*** the electronic switch alternates between channels A and B, letting each through for one cycle of the horizontal sweep. The display is blanked during the flyback and hold-off periods, as in conventional oscilloscopes provided the sweep speed is much greater than the decay time of the CRT phosphor, the screen will show a steady display of both the waveforms at channels *A* and *B*. *This mode can not be used for displaying very low frequency signals.*
  - In the ***chop mode***, the electronic switch free runs at a high frequency of the order of 100 kHz to 300 kHz. The result is that small segments from channels *A* and *B* are connected alternately to the vertical amplifier, and displayed on the screen. Provided the chopping rate is much faster than the horizontal sweep rate, the display will show a continuous line for each channel. If the sweep rate approaches the chopping rate then the individual segments will be visible and the alternate mode should now be used.

7.14. DUAL BEAM OSCILLOSCOPE

Fig. 7.18 illustrates a dual beam oscilloscope.

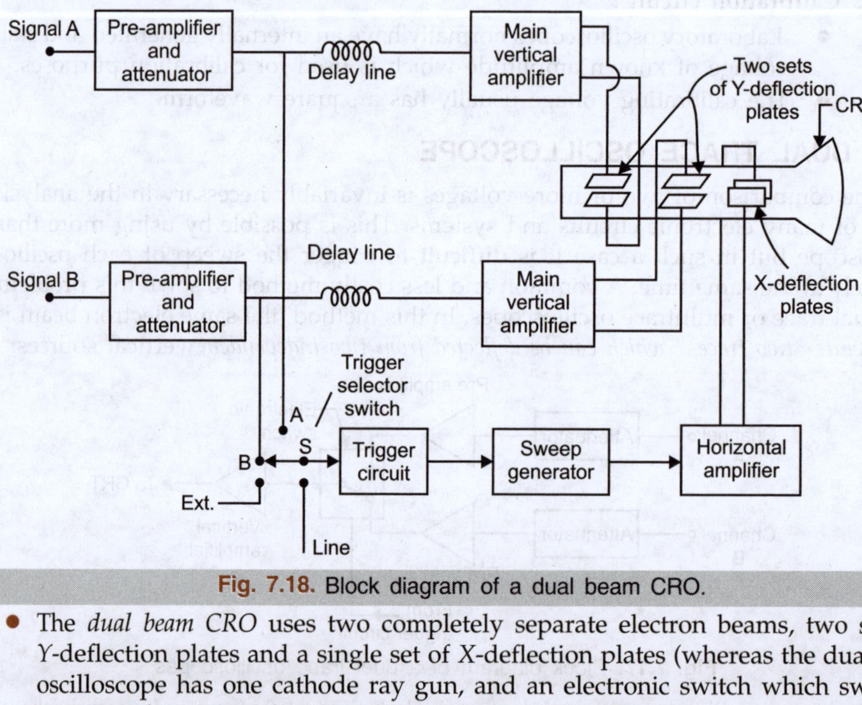

Fig. 7.18. Block diagram of a dual beam CRO.

- The *dual beam CRO* uses two completely separate electron beams, two sets of *Y*-deflection plates and a single set of *X*-deflection plates (whereas the dual trace oscilloscope has one cathode ray gun, and an electronic switch which switches two signals to a single vertical amplifier).

- Only one beam can be synchronized at one time, since the sweep is the same for both signals, *i.e.*, a common time base is used for both beams. Therefore, the signals must have the same frequency or must be related harmonically, in order to obtain both beams locked on the CRT screen, *e.g.*, the input signal of an amplifier can be used as signal A and its output signal as signal B.
- ***Multiple beam oscilloscope*** has a *single tube but several beam producing systems inside. Each system has separate vertical deflecting pair of plates and generally a common time base system*. The triggering can be done internally using either of the multiple inputs or externally by an external signal or line voltages.

Comparison of dual trace and dual beam oscilloscopes :

S. No.	*Aspects*	*Dual trace CRO*	*Dual beam CRO*
1.	*No. of electron beams used.*	One (to generate two traces)	Two
2.	*No. of vertical amplifiers used.*	One	Two
3.	*Display of two signals.*	Not simultaneous (the two signals are not displayed simultaneously in real time but appears to be displayed simultaneously.	Simultaneous
4.	*Switching between traces.*	Difficult (since beam is shared between the two signals)	Easy
5.	*Frequencies of signals.*	Different	Same frequency or their frequencies must be integer multiples of each other
6.	*Size and weight.*	Less	More
7.	*Modes of operation.*	Alternate and chop	The two different types are using double gun tube or split beam using single electron gun

7.15. STORAGE OSCILLOSCOPE

In the conventional CRT the persistence of the phosphor ranges from a *few milliseconds to several seconds*, so that an event that occurs only once will disappear from the screen after a relatively short period of time.

- A "*storage CRT can retain the display much longer, upto several hours after the image was first written on the phosphor*. This *retention* feature can also be useful when displaying the waveform of a very low-frequency signal. In the conventional (non-storage) oscilloscope, the start of such a display would fade before the end is written.
- A storage oscilloscope uses the phenomenon of "*secondary electron emission*" *to build up and store electrostatic charges on the surface of an insulated target*.
  — Storage CRT tubes can be classified as : *Bistable tubes* and *halftone tubes*. The "*bistable tube*" will either store or not store an event and produces only one

level of image brightness. The *"halftone tube"* can retain an image for varying lengths of time (variable persistence) and at different levels of image brightness.

- The storage oscilloscopes are specially useful in the following cases :
  (*i*) For real-time observation of events that occur only once.
  (*ii*) For displaying the waveform of a very low-frequency signal.

7.16. SAMPLING OSCILLOSCOPE

Above the range of 50-300 MHz, *sampling techniques* have to be employed to obtain suitable display and CRO employing such sampling methods are called *"sampling oscilloscopes"*. The display is made up from as many 1000 dots of luminescence. The vertical deflection for each dot is obtained from progressively later points in each successive cycles of input waveform as shown in Fig. 7.19. The horizontal deflection of the electron beam is obtained by applications of a staircase waveform to X-deflection plates. Fig. 7.20 shows the display of dots.

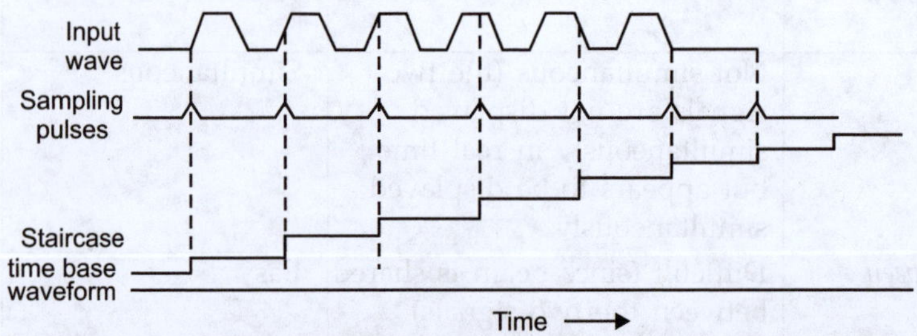

Fig. 7.19. Sampling principle.

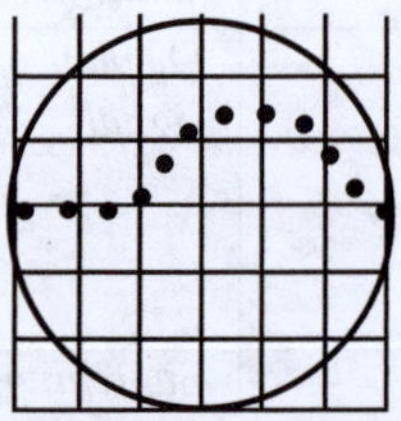

Fig. 7.20. Display of dots.

Fig. 7.21 shows the block diagram of a sampling oscilloscope :

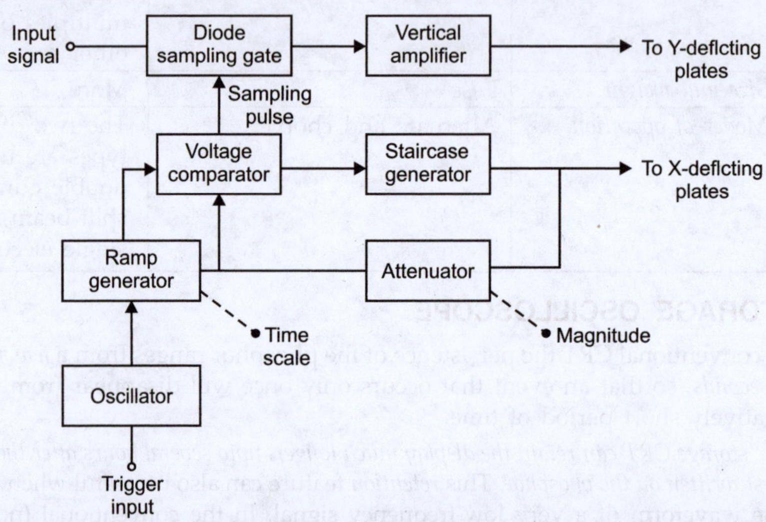

Fig. 7.21. Sampling oscilloscope.

— The input is applied to the *diode sampling gate*.

— At the start of each sampling cycle a trigger input pulse is generated which activates the blocking *oscillator*. The oscillator output is given to the *ramp generator* which generates the linear ramp signal.

— Since the sampling must be synchronized with the input signal frequency, the signal is delayed in the *vertical amplifier*.

— The *staircase generator* produces a staircase waveform which is applied to an *attenuator*. The attenuator controls the magnitude of the staircase signal and then it is applied to a *voltage comparator*. Another input to the voltage comparator is the output of the ramp generator. The voltage comparator compares the two signals and produces the output pulse when the two voltages are equal. This is nothing but a *sampling pulse* which is applied to sampling gate through the gate control circuitry. This pulse opens the diode gate and the sample is taken in. This sampled signal is then applied to the vertical amplifier and the *Y*-deflecting plates.

— The output of the staircase generator is also applied to the *X*-deflecting plates. During each step of staircase the spot moves on the screen. *The comparator output advances the staircase output through one step*. After the certain number of pulses about 100 or so, the staircase generator resets. The smaller the size of the steps of the staircase generator, larger is the number of samples and higher is the resolution of image.

— The sampling oscilloscope responds and stores rapid bits of information and displays them continuously. It is this quantity of sampling oscilloscopes which has limited sensitivity, bandwidth and small display size. The *sampling method transforms the input signal into lower frequency range, where low frequency circuitry is then capable of producing an effective display.*

Advantages :

1. The display produced is clear.
2. High speed electrical signals can be produced.
3. Very high frequency performance can be achieved.
4. By controlling the size of the steps of the staircase generator, the number of samples and hence the resolution can be controlled.
5. The sampling technique allows the design of the oscilloscope with wide bandwidth, high sensitivity even for low duty cycle pulses.

Limitation. This oscilloscope *cannot be used to display* the *transient waveforms*.

7.17. DIGITAL READ OUT OSCILLOSCOPE

This oscilloscope provides readout of the signal information such as voltage, time etc., in addition to the conventional CRT display.

It consists of a *high-speed laboratory CRO and an electronic counter, both in one cabinet*.

Fig. 7.22. shows the block diagram of a digital CRO for voltage measurement :

— The input signal is first sampled using sampling circuit.

— The sampled signal is given as one output to each start and step comparators. When input is sampled, the sampling circuit advances the sampling position by a fixed amount. This process is known as *strobing*. The selected sweep time per cm control and number of samples taken per cm decide the equivalent time between two samples.

— The CRT trace is used to identify 0% and 100% zone positions. The positions can be shifted anywhere on the display. The potential divider is used which taps the voltage between 0% and 100% levels. The 0% level is used to produce a pulse for opening of gate while the 100% is used to produce a pulse of closing of gate.

— The coincidence of any of the input waveforms with the selected percentage point is sensed by the voltage comparator. When 0% level gives start pulse, clock gate opens and counter starts counting the pulses. When 100% level gives stop pulse, clock gate closes and counter stops counting.

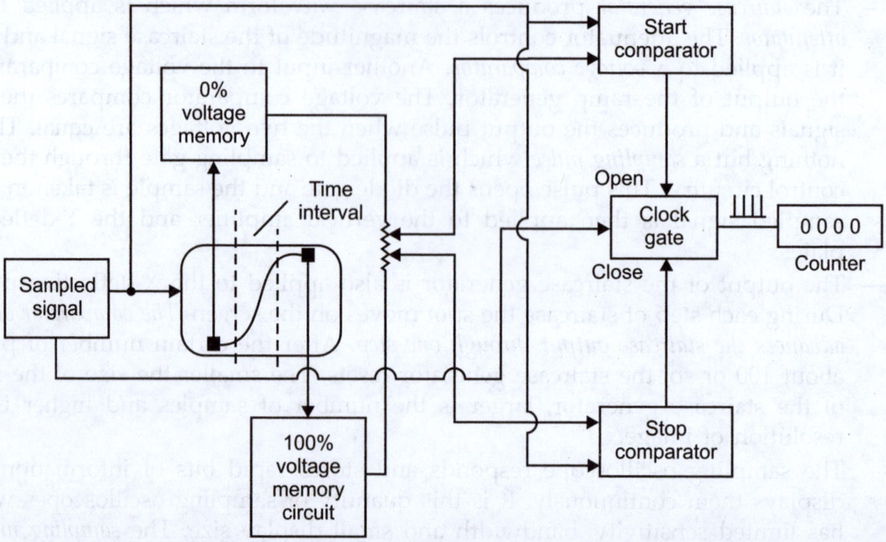

Fig. 7.22. Block diagram of a digital CRO for voltage measurement.

— *The number of pulses counted by a counter are proportional to actual sample taken.* The digital readout is obtained by using Nixie display tube.

7.18. DIGITAL STORAGE OSCILLOSCOPE

Following are the distinct *disadvantages of the storage cathode ray tube* :

(*i*) The storage tube can preserve a stored waveform for a finite amount of time; eventually, waveform is lost. The power to the storage tube must be present as long as the image is to be stored.

(*ii*) The trace of a storage tube is, generally, not as fine as a normal cathode ray tube.

(*iii*) The writing rate of the storage tube is less than a conventional cathode ray tube, which limits the speed of the storage oscilloscope.

(*iv*) The storage cathode ray tube is considerably more expensive than a conventional tube and requires additional power supplies.

(*v*) Only one image can be stored. If two traces are to be compared, they must be superimposed on the same screen and displayed together.

The *"digital storage oscilloscope"* is a superior method of trace storage. In this technique, the waveform to be stored is *digitized, stored in a digital memory, and retrieved for display on the storage oscilloscope.* The stored display can be displayed indefinitely as long as power is applied to the memory, which can be supplied with a small battery.

Fig. 7.23 shows the block diagram of a digial storage oscilloscope :

— The input is amplified and attenuated with input amplifiers (as in any oscilloscope).

— The digital storage oscilloscope uses the same types of input circuitry and oscilloscope probes as a conventional oscilloscope, and many digital storage oscilloscopes can operate in a conventional mode, bypassing the digitizing and storing features.

— The output of the input signal amplifiers feeds an A/D (analog-to-digital) converter. The main requirement of the A/D converter is its speed, while accuracy and resolution are of secondary impedance.

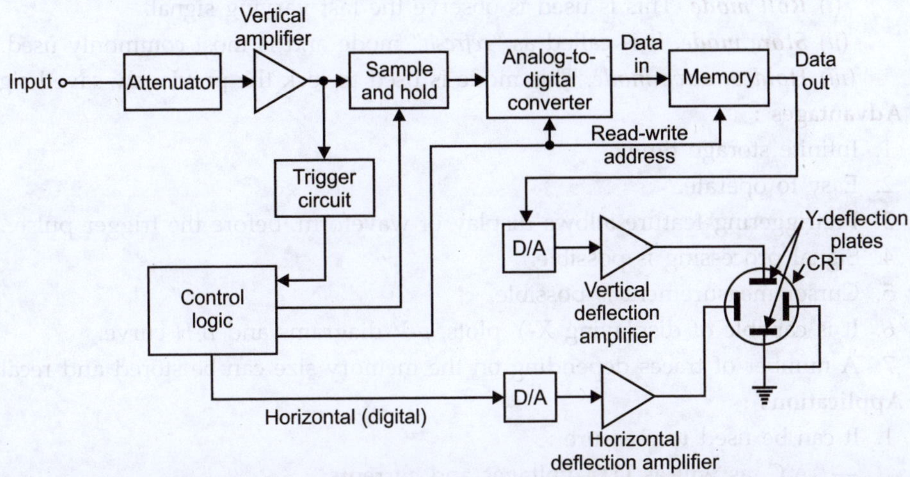

Fig. 7.23. Digital storage oscilloscope.

— The digitized output is in the binary form and not BCD form. Generally, successive approximation type of ADC is used in digital storage oscilloscope. If a 12 bit converter is used then accuracy of about 0.1% and resolution of about 0.025% is obtained, while if a 10 bit ADC is used then the resolution of about 0.1% is obtained.

— Digitizing the analog signal means to take samples of the input signal at periodic intervals of time. The sampling rate should be greater than twice the highest frequency present in the input signal, so that the sampling theory is satisfied. This ensures that aliasing effect will not occur and there is no loss of information. The sampling rate may be as high as 100,000 samples per second. For such high sampling rates, a fast conversion A/D convertor is required. *'Flash' A/D converters* are used at the cost of decreased resolution with increase in sampling rate.

— The size of the memory is related to the amount of horizontal segments of the trace that can be divided into one sweep of the time base, *e.g.*, if memory size is 1K then the trace will be divided into 1000 horizontal segments. If a 10 bit ADC is used then the frequency response of 25 kHz is obtained; the total memoery storage capacity is 4096 for a single channel, 2048 for two channels, 1024 for four channels each.

— The selection of the sampling rate and memory size depends on the type of waveform being recorded. As the analog signal is sampled at suitable rates, the data points are readable on to the memory. Once the signal is stored, the required manipulations can be done.

— A *continuous storage oscilloscope* consists of a feature called ***pretrigger view***. This indicates that the events that occurred before the trigger input signal was applied, are displayed. This selection is a percentage selection.

This mode is useful when failure occurs.

— *Single shot events* one sometimes transient in nature and hence they are quickly lost. Unless the wavefrom is stored the observer cannot see such events. *Such events are stored in the memory of the digital storage oscilloscope and memory can be rapidly read and on the screen a continuous waveform can be displayed.*

- The digital storage oscilloscope has the following threee modes of operation:
  (*i*) ***Roll mode.*** This is used is observe the fast varying signal.
  (*ii*) ***Store mode.*** It is called as *"refresh"* mode and is most commonly used.
  (*iii*) ***Hold or save mode.*** This mode is used to lock the previously saved signal.

Advantages :

1. Infinite storage time.
2. Easy to operate.
3. Pretriggering feature allows display of waveform, before the trigger pulse.
4. Signal processing is possible.
5. Cursor measurement is possible.
6. It is capable of displaying *X-Y* plots, *p-V* diagrams and *B-H* curve.
7. A number of traces depending on the memory size can be stored and recalled.

Applications :

1. It can be used to measure :
   — A.C. as will as D.C. voltages and currents;
   — Frequency, time period, time interval between two signals etc.,
   — Inductance, capacitance.
2. It is used to give the visual representation for a target of radar such as aeroplane, ship etc.
3. In medical fields, it is used to display cardiograms that are useful for diagnosis of heart of the patient.
4. It is used to observe the radiation pattern generated by the transmitting antenna.
5. It can be used to check the faulty components in various circuits. It can also be used to analyse TV waveforms.
6. It can be used to determine the modulation characteristics and detect the standing waves in transmission lines.

7.19. HIGH FREQUENCY OSCILLOSCOPE

A general purpose oscilloscope has an upper frequency limits of about *300 kHz*. However, there are many applications which involve measurement or examination of signals whose frequencies are much higher than this upper limit. For such high-frequency (HF) applications conventional HF oscilloscopes are built. These HF oscilloscopes can display signals with frequencies upto *500 MHz*. They differ in their internal construction from low-frequency (LF) oscilloscopes by their special HF CRTs and high-frequency amplifiers.

Fig. 7.24 shows a high frequency CRT.

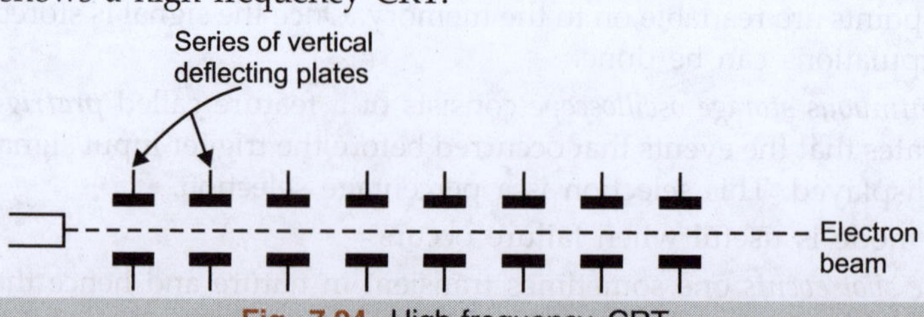

Fig. 7.24. High-frequency CRT.

— In an ordinary cathode ray oscilloscope, there is only one pair of vertical deflection plates. When the HF signal is to be displayed, the electron beam does not get sufficient time to pick up the instantaneous level of the signal. Also, at high frequencies the number of electrons striking the screen in a given time and the intensity of the beam is reduced. Hence, instead of one set of vertical deflection plates, a *series of vertical deflection plates* are employed.

— The vertical deflection plates are so *shaped* and *spaced that an electron moving along CRT receives from each set of plates an additional deflection force in proper time sequence.* This synchronization is obtained by making the signal travel from one plate to the next plate at the *same speed as the transit time of electrons.*

— The signal is applied to each pair of plates, and as the electron beam moves the signal also moves through the delay times. The time delays are so arranged that the same electrons are deflected by the input signal. Thus the electron beam picks up the level of the input signal. *The time delays between the plates correspond exactly to the electron transit times.* In addition, new fluorescent materials have now been developed to increase the brightness at high frequencies.

7.20. ACCESSORIES OF CATHODE RAY OSCILLOSCOPES

The CRO is one of the most useful instruments in the electronic industry. Its usefulness is further extended by the provision of accessories or auxiliary equipment. Source of the accessories are enumerated and described below :

1. Probes
2. Electronic switch
3. Calibrators
4. Cameras.

7.20.1. Probes

The *function of a probe is to connect the test circuit to the oscilloscope without altering, loading, or otherwise disturbing the test circuit.*

Probes are of the following types :

1. Direct probe;
2. Isolation probe;
3. High impedance or 10:1 probe;
4. Active probes;
5. Current probe.

1. ***Direct probe :***
   - It is the simplest of all the probes and uses a shielded co-axial cable.
   - It avoids stray pick-ups which may create problems when low level signals are being measured.
   - It is usually used for low frequency or low impedance circuits.
   - While using the shielded probe, the shunt capacitance of the probe and cable is added to the input impedance and capacity of the oscilloscope and acts to lower response of the oscilloscope to high impedance and high frequency circuits.

2. ***Isolation probe :***
   - An isolation probe is employed *to avoid the undesirable circuit loading effect of the shielded probe.*

- It is made by placing a carbon resistor in series with the test lead, as shown in Fig. 7.25.

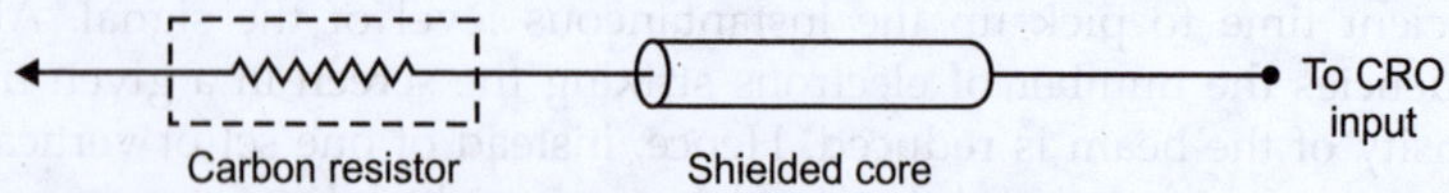

Fig. 7.25. Isolation probe.

- This probe causes a slight change in the amplitude of the waveform and a slight change in the wave shape. To avoid this possibility, a high impedance compensated probe, called a *low capacitance* or a 10 : 1 probe is employed.

3. *High Impedance or 10:1 Probe :*

These probes are used *to increase the input resistance and reduce the effective input capacitance of an oscilloscope.*

Fig. 7.26 (*a*, *b*) shows the layout of a high impedance or 10:1 probe and its equivalent circuit.

The capacitor is adjusted so that the elements of the bridge are balanced. Under conditions of balance we have :

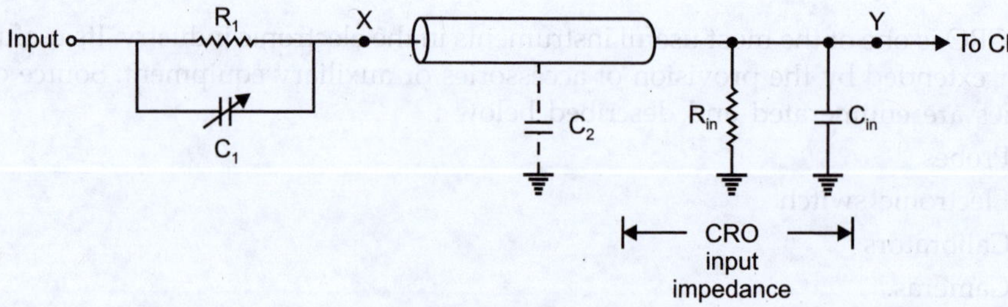

(a) Lay out of a 10:1 probe.

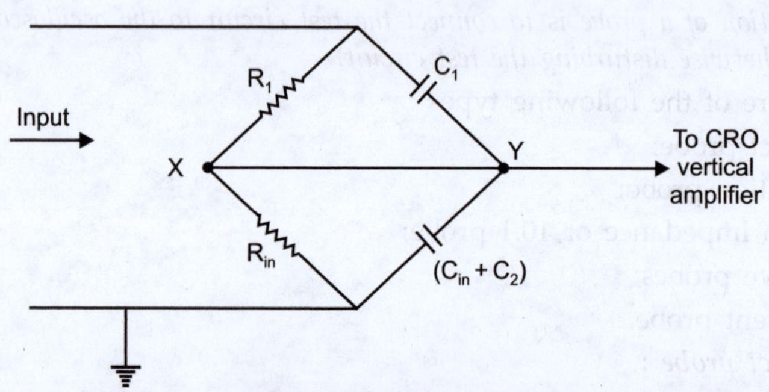

(b) Equivalent circuit of 10:1 probe.

Fig. 7.26. High impedance or 10:1 probe.

$$R_1 X_{(C_{in}+C_2)} = R_{in} X_{(C_1)}$$

or, $$\frac{R_1}{\omega (C_{in} + C_2)} = \frac{R_{in}}{\omega C_1}$$

or, $$R_1 C_1 = R_{in} (C_{in} + C_2) \qquad ...(7.25)$$

Therefore, X and Y are equipotential and the effect of the probe is equivalent to placing a potential divider consisting of R_1 and R_{in} across, the input circuit.

- The attenuation of the signal is 10:1, *i.e.*, $\frac{R_1 + R_{in}}{R_{in}} = 10:1$ over a wide frequency range. Hence, it is called a compensated 10:1 probe.
- Capacitor C_1 is adjusted by connecting the probe tip to a square-wave of 1 kHz and observing the CRT display. When the CRT display has *optimum response* the value of C_1 is deemed to be appropriate.

Then,
$$V_{out} = 0.1\, V_{in} = \frac{V_{in} \times R_{in}}{R_1 + R_{in}} \qquad ...(7.26)$$

4. *Active probes :*

- These probes are designed *to provide an efficient method of coupling high frequency, fast rise time signals to the CRO input.*
- Usually active probes have *very high input impedance with less attenuation than passive probes.*
- Active devices may be diodes, FETs, BJTs, etc.
- These probes are *more expensive and bulky than passive probes, but they are useful for small signal measurements, because their attenuation is less.*

Fig. 7.27 shows a basic circuit of an active probe using a FET and BJT.

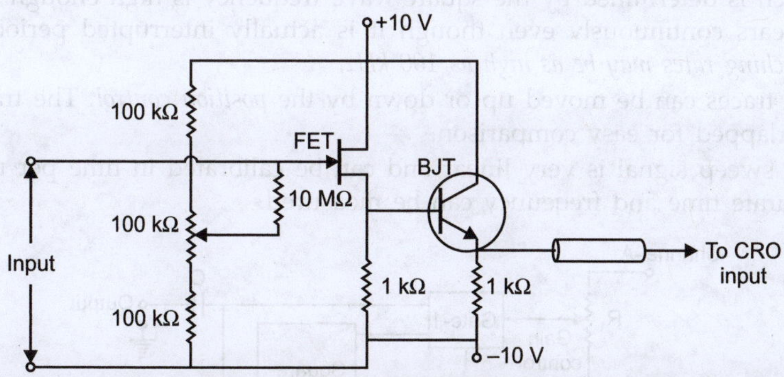

Fig. 7.27. FET probe.

- The FET is used as the active element to amplify the input signal. Although voltage gain of FET follower circuit shown is unity, the follower circuit provides a power gain so that input impedance can be increased.
- To be effective the FET must be mounted directly in the voltage probe tip, so that the capacitance of the interconnecting cable can be eliminated. This requires that the power for the FET be supplied from the oscilloscope to the FET in the probe tip. The FET voltage follower drives a coaxial cable but instead of the cable connecting directly to the high input impedance of the oscilloscope it is terminated in its characteristic impedance.
- Because there is no signal attenuation between FET amplifier and probe tip, the range of the signals that can be handled by the FET probe is limited to the dynamic range of the FET amplifier and is typically *less than a few volts*. This is a *significant drawback of the FET probe*. Therefore, for *handling a large dynamic range, external attenuators are added at the probe tip.*
- *Active probes have limited use because the FET probe effectively becomes an attenuator probe. So the oscilloscopes are typically used with 10:1 attenuator probe.*

5. *Current probe :*

- It is a very valuable probe and can be clamped around a wire carrying current *without any physical contact to the probe,* allowing the oscilloscope to be employed to measure the magnitude of the current with a frequency from 0(D.C.) to 50 Hz.
- The current sensor consits of two parts : A *conventional transformer* for transforming alternating current to voltage, and a *Hall effect device* for converting direct current to a voltage.

7.20.2. Electronic switch

The electronic switch is an extremely useful accessory for displaying the two signals simultaneously on single trace oscilloscopes. Fig. 7.28 shows the block diagram of an electronic switch :

— Basically, the electronic switch is a *square-wave generator.*

— Two signals which are to be displayed are each applied to the separate gain control and gate stage. The gate stages are alternately biased to cut-off by square-wave signals from the square-wave generator. Thus only one gate stage is in a condition to pass its signal at any given time.

— The output of both stages are applied directly to the oscilloscope input. Amplitudes of the two signals are adjusted by their gain controls. When the switching rate, which is determined by the square-wave frequency is high enough each signal appears continuously even though it is actually interrupted periodically. *The switching rates may be as high as 100 kHz.*

— The traces can be moved up or down by the *position control*. The traces can be overlapped for easy comparison.

— The sweep signal is very linear and can be calibrated in time per mm so that accurate time and freqeuncy can be measured.

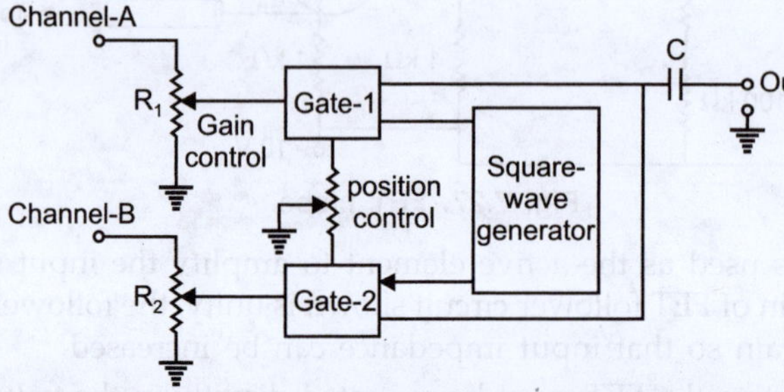

Fig. 7.28. Block diagram of electronic switch.

7.20.3. Calibrators

Several oscilloscopes have a *built-in reference source of voltage* which has usually a square waveform with a frequency of 1 kHz. The voltage may have a single magnitude or may be of several selectable magnitudes. *Provision of this facility enables the oscilloscope amplifiers and time base to be calibrated and checked for accuracy.*

7.20.4. Cameras

The use of photography is the best method of permanent records of oscilloscope traces for analysis.

- The special cameras used for keeping the records are of the following two types :
  (*i*) 35 mm film; (*ii*) Polaroid film.

- *35 mm unit* is usually prefocused and bolted over the tube face. Some cameras have a fixed lens aperture, the exposure is simply controlled by hand operation of flapper shutter.
- *Polaroid film* photographic technique approaches *instant photography*. A permanent record is obtained with 10–20s. However, this process is comparatively *expensive*.

7.21. CRO MEASUREMENTS

CRO can be used to measure various characteristics of an input signal and the properties of the output signal. The various parameters which can be measured by using CRO are

(i) Voltage;	*(ii)* Current;
(iii) Period;	*(iv)* Frequency;
(v) Phase;	*(vi)* Amplitude;
(vii) Peak-to peak value	*(viii)* Duty cycle etc.

7.21.1. Voltage measurement

The procedure for voltage measurement is as follows :

— The voltage to be measured is applied to Y-deflection plates through the vertical amplifier. Generally, a coaxial cable is used for connection.

— The X-deflection plates are excited by the time-base generator.

— To aid the measurement, a set of squares is marked on the transparent screen cover. Using these markings and the gain of the vertical amplifier, the peak-to-peak (p-p) value (and hence r-m.s. value) of the voltage wave can be measured.

— After noting down the selection in *volts/division* from the front panel (selected for measurement) the peak-to-peak value, amplitude and r.m.s value of sinusoidal signal can be obtained as given below :

(i) Peak-to-peak value, $V_{p-p} = \frac{\text{Volts}}{\text{div.}} \times \text{(number of divisions)}$

(The volts/div. is nothing but *deflection sensitivity* of CRO)

(ii) Amplitude, $V_{max} = \frac{V_{p-p}}{2}$

(iii) R.M.S. value, $V_{r.m.s.} = \frac{V_{p-p}}{2\sqrt{2}}$ (for sinusoidal signals only)

- It must be remembered that an oscilloscope is very complex device and that the accuracy of measurements is not better than ± 3%. It is basically a *display device so that we can observe the waveform in detail and find out the extent of harmonics in the wave.*

Example 7.7. *For the waveform shown in Fig. 7.29, if vertical attenuation is 3 mV/div., determine :*

(i) Peak-to-peak value,
(ii) Amplitude, and
(iii) R.M.S. value of the signal.

Solution. Vertical attenuation = 3 mV/div.

From the figure, it is evident that the screen is divided such that one part is subdivided into 5 units.

$\therefore$ 1 subdivision $= \frac{1}{5} = 0.2$ units (see Fig. 7.29)

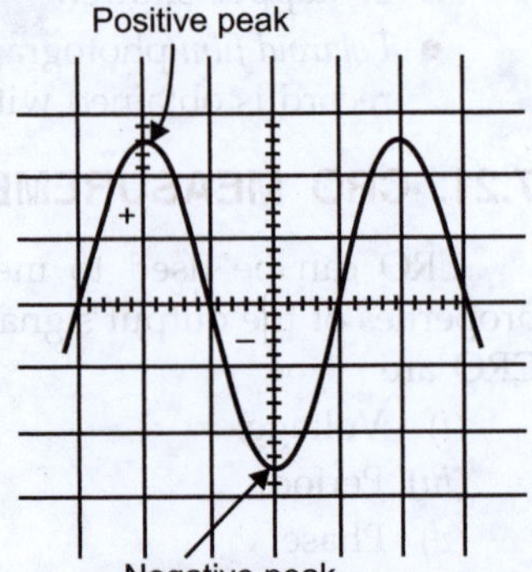

Fig. 7.29.

(*i*) **Peak-to-peak value-V_{p-p} :**

It can be observed that positive peak of the signal corresponds to two full divisions and three subdivisions. Hence positive peak is $2 + 3 \times 0.2 = 2.6$ units. While the negative peak also corresponds to 2.6 units.

$$\therefore \quad V_{p-p} = 2.6 + 2.6 = 5.2 \text{ divisions}$$

Also,
$$V_{p-p} = \frac{\text{Volts}}{\text{division}} \times \text{No. of divisions} = 3 \times 5.2 = \mathbf{15.6\ mV}\ \textbf{(Ans.)}$$

(*ii*) **Amplitude, V_{max} :** $V_{max} = \frac{V_{p-p}}{2} = \frac{15.6}{2} = \mathbf{7.8\ mV}$ **(Ans.)**

(*iii*) **R.m.s. value, $V_{r.m.s.}$:** $V_{r.m.s.} = \frac{V_{max}}{\sqrt{2}} = \frac{7.8}{\sqrt{2}} = \mathbf{5.515\ mV}$ **(Ans.)**

7.21.2 Current measurement

A CRO has a very high input impedance and cannot be used for direct measurement of current. However, the current can be measured in terms of voltage drop across a standard resistance as shown in Fig. 7.30; *i.e.*, $I = \frac{V \text{ (measured on CRO)}}{R}$.

- For measurement of *high frequency currents, the shunt should be non-inductive.*

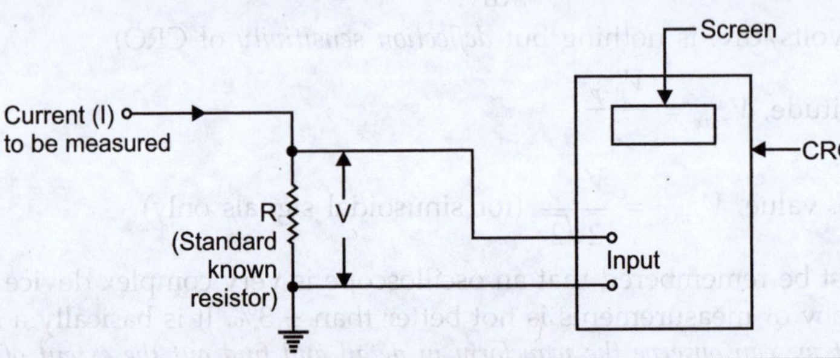

Fig. 7.30. Current measurement.

7.21.3. Time period measurement

For the measurement of time period (*T*) of the waveform, it is displayed on the screen such that *one complete cycle is visible on the screen*. The accuracy increases if a single cycle occupies as much as the horizontal distance on the screen.

After noting the time/division selected on the front panel, the time period of the waveform can be obtained as :

Time period, $$T = \frac{\text{Time}}{\text{Division}} \times \text{number of division occupied by } \textit{one cycle}.$$

The frequency (f) is then given as :

$$f = \frac{1}{T}$$ (This is the method of frequency measurement without Lissajous pattern)

7.21.4. Lissajous figures/patterns

Lissajous patterns (named in the honour of French scientist who first obtained them geometrically and optically) *are formed when two sine waves are applied simultaneously to the vertical and horizontal deflecting plates of CRO.* The two sine waves may be obtained from two audio oscillators. Obviously, in this case, a *sine wave sweeps a sine-wave input signal.*

- *The shape of the Lissajous pattern depends on the frequency and phase relationship of the two sine waves.*
- Two sine waves of the *same* frequency and amplitude may produce a line, an ellipse or a circle depending on their *phase difference* as shown in Fig. 7.31.

  In general, the shape of Lissajous figures depend on :

(*i*) Amplitude ;
(*ii*) Phase difference ;
(*iii*) Ratio of frequency of the two waves.

Uses. The Lissajous figures are used for the following purposes :

1. To determine an unknown frequency by comparing it with a known frequency.
2. To check audio oscillator with a known-frequency signal.
3. To check audio amplifiers and feedback networks for phase shift.

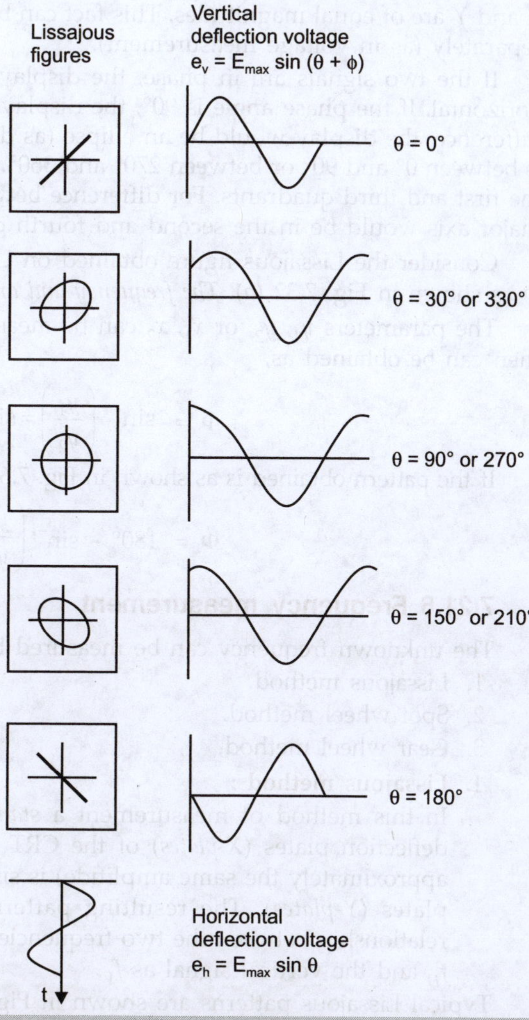

Fig. 7.31. Lissajous patterns for same frequency, different phase shift.

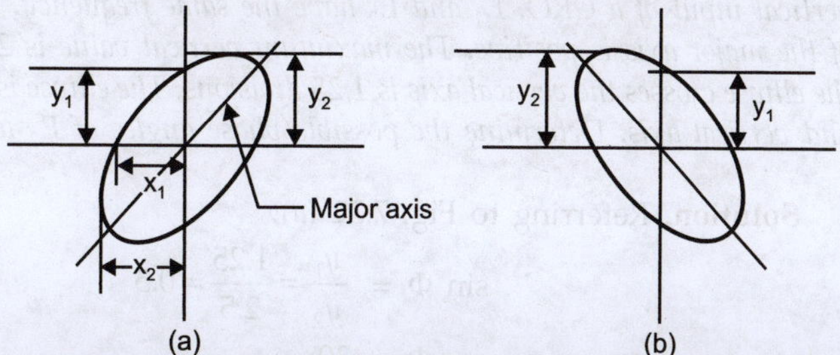

Fig. 7.32. Measurement of phase angles by using CRO.

7.21.5. Phase-angle measurement

An oscilloscope can be used to find the phase angle between the two sinusoidal quantities of the same frequency. One of the signals is fed to the *Y*-plates. The time-base generator is switched out and the second signal is fed to the *X*-plates. It is necessary that *X* and *Y* are of equal magnitudes. This fact can be checked by measuring the amplitudes separately (as in voltage measurement).

If the two signals are in phase, the display would be a straight line at 45° to the horizontal. If the phase angle is 90°, the display would be a circle. For any other phase difference, the display would be an ellipse (as discussed earlier). If the phase difference is between 0° and 90° or between 270° and 360°, the ellipse would have its major axis in the first and third quadrants. For difference between 90° and 180° or 180° and 270°, the major axis would be in the second and fourth quadrants.

Consider the Lissajous figure obtained on CRO with an unknown phase difference Φ as shown in Fig. 7.32 (*a*). *The frequency and amplitude of the two waves are same.*

The parameters y_1, y_2, or x_1, x_2 can be measured (See Fig. 7.32 (*a*); the phase angle then can be obtained as,

$$\Phi = \sin^{-1}\left(\frac{y_1}{y_2}\right) = \sin^{-1}\left(\frac{x_1}{x_2}\right)$$

If the pattern obtained is as shown in Fig. 7.62 (*b*), then the phase angle Φ is given by:

$$\Phi = 180° - \sin^{-1}\left(\frac{y_1}{y_2}\right)$$

7.21.6 Frequency measurement

The unknown frequency can be measured by the following methods :

1. Lissajous method.
2. Spot wheel method.
3. Gear wheel method.

1. Lissajous method :

In this method of measurement a *standard frequency* is applied to one set of deflection plates (*X-plates*) of the CRT tube while the unknown frequency (of approximately the same amplifide) is simultaneously applied to the other set of plates (*Y-plates*). The resulting patterns depend on the integral and phase relationship between the two frequencies. The horizontal signal is designated as f_H and the vertical signal as f_V.

Typical Lissajous patterns are shown in Fig. 7.33 (*a* & *b*) for sinusoidal frequencies which are equal, integral and in ratio.

Example 7.8. *Voltage E_1 is applied to the horizontal input and voltage E_2 is applied to the vertical input of a CRO. E_1 and E_2 have the same frequency. The trace is an ellipse. The slope of the major axis is positive. The maximum vertical value is 2.5 divisions and the point where the ellipse crosses the vertical axis is 1.25 divisions. The ellipse is symmetrical about the horizontal and vertical axis. Determine the possible phase angles of E_2 with respect to E_1.*

(Nagpur University, W-2005)

Solution. Referring to Fig. 7.32 (*a*),

$$\sin \Phi = \frac{y_1}{y_2} = \frac{1.25}{2.5} = 0.5$$

$$\therefore \quad \Phi = 30°$$

Thus possible phases are **30° or 330° (Ans.)**

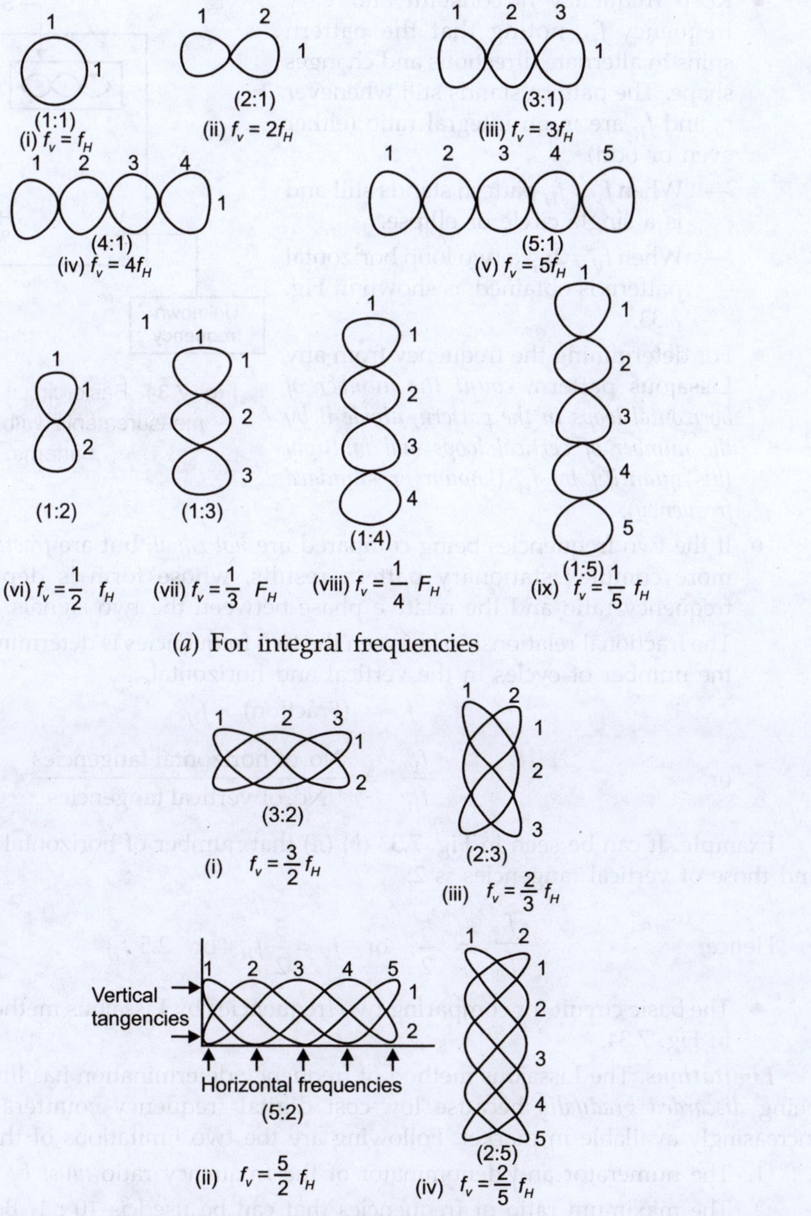

(*a*) For integral frequencies

(*b*) For non-integral frequencies

Fig. 7.33. Lissajous patterns.

Measurement procedure :

- Set up the oscilloscope and switch off the internal sweep (change to Ext.)
- Switch off sync. control.
- Connect the signal source as shown in Fig. 7.34.
- Set the horizontal and vertical gain control for the desired width and height of the pattern.

- Keep frequency f_V constant and vary frequency f_H, noting that the pattern spins in alternate directions and changes shape. The pattern stands still whenever f_V and f_H are in an integral ratio (either even or odd)
  — When $f_V = f_H$ pattern stands still and is a single circle or ellipse.
  — When $f_V = 2f_H$, a two loop horizontal pattern is obtained as shown in Fig. 7.33.
- For determining the frequency from any Lissajous pattern, *count the number of horizontal loops in the pattern, divide it by the number of vertical loops and multiply this quantity by* f_H *(known or standard frequency).*

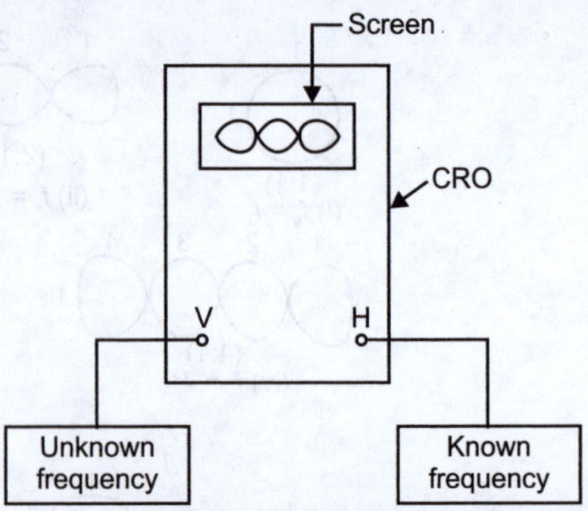

Fig. 7.34. Basic circuit for frequency measurements with Lissajous patterns.

- If the two frequencies being compared are *not equal*, but are *fractionally related*, a more complex stationary pattern results, whose form is dependent on the frequency ratio and the relative phase between the two signals.

  The fractional relationship between the two frequencies is determined by counting the number of cycles in the vertical and horizontal.

$$f_V = (\text{Fraction}) \times f_H$$

or,
$$\frac{f_V}{f_H} = \frac{\text{No. of horizontal tangencies}}{\text{No. of vertical tangencies}}$$

Example. It can be seen in Fig. 7.33 (*b*) (*ii*) that number of horizontal tangencies is 5 and those of vertical tangencies is 2;

Hence,
$$\frac{f_V}{f_H} = \frac{5}{2}, \text{ or } f_V = \frac{5}{2} f_H \text{ or } 2.5\, f_H.$$

- The basic circuit for comparing two frequencies by Lissajous method is illustrated in Fig. 7.34.

Limitations. The Lissajous method of frequency determination has limitations and is being *discarded gradually* because low-cost digital frequency counters are becoming increasingly available in market. Following are the two limitations of this method :

1. The numerator and denominator of the frequency ratio *must be whole number*.
2. The maximum ratio of frequencies that can be used is 10 : 1. Beyond that, the Lissajous patterns become *too complex* to *analyse*.

2. Spot wheel method :

With the increase of ratio of the two frequencies being compared, the Lissajous pattern becomes more complicated hence more difficult to interpret. In such cases, it is advantageous to use spot wheel method of display.

Fig. 7.35 illustrates the *spot wheel method of frequency determination* :

— *Fig. 7.35 (a)* shows the *basic circuit* for frequency measurement, whereby an intensity modulated circular or elliptical pattern is produced on the face of the CRT.

— The *lesser* of the two frequencies being compared is applied to the deflecting plates of the CRT through resistance-capacitance phase shifter as illuspraited, thereby producing the circular or elliptical sweep.

— The higher frequency is applied to the control grid of the CRT (biased slightly below cut off), as shown in *Fig. 7.35 (b)*, thus modulating the intensity of beam. As a consequence, the pattern appears as a series of alternate bright and dark spots, as shown in *Fig. 7.35 (c)*, where the ratio of high to low frequencies is given by the *number of bright spots*.

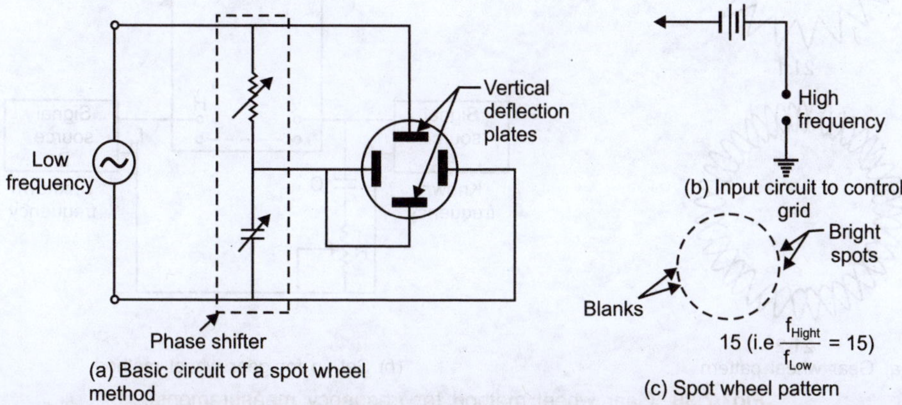

Fig. 7.35. Spot wheel method for frequency measurements.

— When the voltages across *R* and C are equal, the pattern is circle; if they are unequal, the pattern is elliptical.

— The variation of the grid-cathode potential changes the density of the electron stream within the CRT and *determines the intensity*. When the grid bias of the CRT reaches cut-off potential, there is no electron stream and fluorescence at the screen cannot occur. *This property is used for the comparison of two frequencies, provided their ratio is an integer. The high frequency signal is applied to the grid.* Sufficient grid must be present so that the negative peaks of the sine wave cut off the electron beam.

— The high frequency signal need not be a pure sine wave, it can be a square wave.

Fig. 7.35(*c*) shows that the frequency ratio is 15 : 1 $\left(i.e. \frac{f_{High}}{f_{Low}} = 15\right)$ and there are 15 blanks in the pattern :

3. Gear wheel method :

When the number of loops in a Lissajous figure is large, the accurate counting becomes difficult.

- A test method that uses a *modulated ring pattern* in place of looped figure and permits a higher count is shown in Fig. 7.36 (a). This pattern is also called a 'gear wheel' or 'toothed wheel', because of its shape. The unknown frequency is determined by multiplying the known frequency by the number of teeth in the pattern.
- Fig. 7.36 (*b*) shows the set-up for gear wheel method for frequency measurements :

— In this method, a phase shift network (RC) introduces a 90° phase shift between the horizontal and vertical channels of the oscilloscope, which is needed to produce a ring or circle pattern with a known frequency f_V. This ring is *modulated* by the voltage from the unknown frequency.

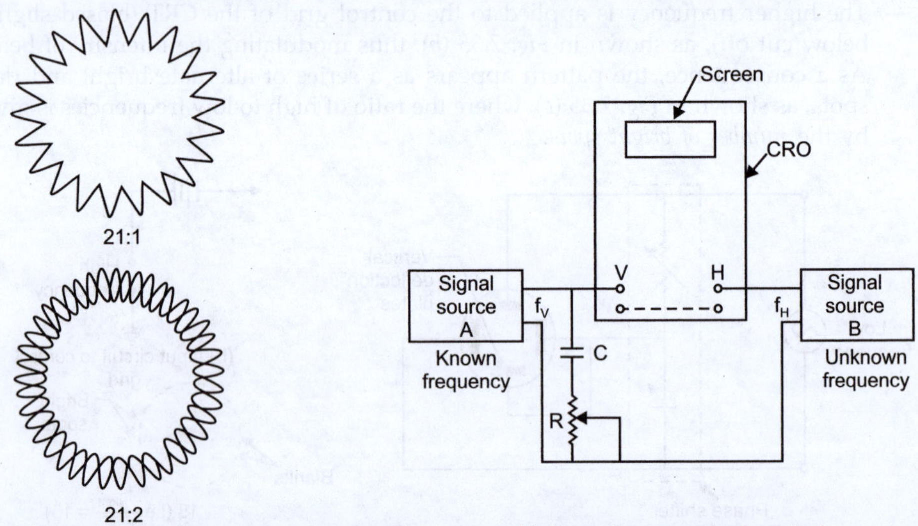

(*a*) Gear wheel pattern (*b*) Set-up for gear wheel method

Fig. 7.36. Gear wheel method for frequency measurements.

— When the *voltages across* and *R* and *C* are unequal, the pattern is elliptical. If they are equal, the pattern is circle.

— *The unknown frequency must be higher than the known frequency and the amplitude of the unknown must be below that of the known to prevent distortion.*

Measurement procedure :

- After setting up the oscilloscope, switch off the and Sync–control and connect the circuit as shown in Fig. 7.36 (b). Switch off the signal *B* temporarily.
- *Switch on the signal source A* and adjust *R* to obtain a ring pattern on the screen. Adjust the horizontal and vertical gain controls to spread the ring over the screen as much as possible.
- *Switch on the signal source B*, noting that the ring becomes *toothed by the signal*. Adjust the known frequency f_V to stop the ring from rotating. Adjust the amplitude of f_H to obtain a distinct teeth pattern. The unknown frequency is then given by :

$$f_H = nf_V$$

where f_H is an integral multiple (odd or even) of f_V.

— *The wheel rotates or spins if f_H is not exact multiple of f_V, therefore adjust the known frequency to stop the wheel from rotating.*

7.21.7. Dielectric loss measurement

Fig. 7.37 (*a*) shows the circuit diagram or measurement of dielectric loss by cathode ray oscilloscope :

- C is capacitor formed by dielectric material under test and C_s is the *standard capacitor of negligible loss;* these capacitors C and C_s are connected in series across A.C. supply source.
- The voltage across the capacitor C or a known fraction of it applied to the X-deflection plates and the voltage across the standard capacitor C_s is applied to

the Y-deflection plates of CRO. There will be phase difference between the voltages across capacitors C and C_s, because voltage across standard capacitor C_s will lag 90° behind the screen flowing through it but the voltage across capacitor C will not *lag exactly 90°* behind the current flowing through it *owing to its losses*.

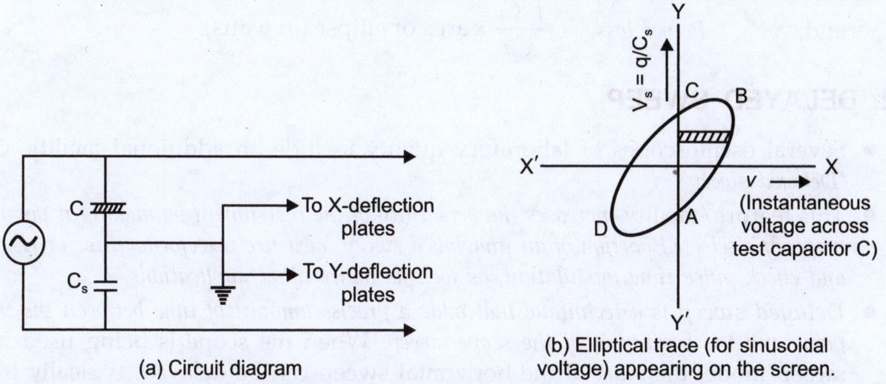

Fig. 7.37. Dielectric loss measurement by CRO.

- When the voltage is sinusoidal then the trace appearing on the screen of CRT will be an ellipse, as shown in Fig. 7.37 (*b*), and the *area of this trace will represent the power loss per cycle of the capacitor C under test*. More the power loss in the test capacitor C, more will be the phase difference of the two voltages and so more will be the area of the trace.

The work done on the test capacitor C over any period of time is given as :

$$dw = \int vidt = v\int \frac{dq}{dt}.dt = \int v.\,dq$$

where, v = Instantaneous voltage across the test capacitor C and

q = Instantaneous charge on the standard capacitor C_s.

If the deflection sensitivity CRO is x and y mm/V in X-and Y-axes respectively, then the area of the hatched section shown in the figure will be :

$$x.\ v.\ y.\frac{dq}{C_s}\ mm^2$$

— When the *spot moves from point A to point B*, the area between curve AB and the Y-axis is

$$\int x.v.y.\frac{dq}{C_s} = \frac{xy}{C_s}\int v.dq = \frac{xy}{C_s}\times \text{work done on capacitor C}$$

— When the *spot moves from point B to point C*, both the voltages across the capacitor C and charge on the capacitor C_s are being reduced so that *energy is being restored* to the circuit and this amount of energy is represented by the curve BC and the Y-axis. But the area enclosed by curve ABC and the Y-axis represents the work done on the dielectric and *not returned to the circuit*. Similarly the area enclosed by CDA represents work done during the remaining half cycle, so that the *area of the ellipse represents the work done on the dielectric per cycle*.

Since area of ellipse = $\frac{xy}{C_x}\times$ work done on capacitor C, therefore:

Work done or dilectric loss per cycle

$$= \frac{C_s}{x.y} \times \text{area of ellipse}$$

and, *Power loss* $= \frac{f.C_s}{x.y} \times$ area of ellipse (in watts).

7.22. DELAYED SWEEP

- Several oscilloscopes of laboratory quality include an additional facility, called *'Delayed sweep'*.
- This feature/facility *increases the versatility of the instrument by making it possible to magnify a selected portion of an undelayed sweep, measure waveform jitter or rise time, and check pulse time modulation, as well as many other applications.*
- ***Delayed sweep*** *is a technique that adds a precise amount of time between the trigger point and the beginning of the scope sweep.* When the scope is being used in the sweep mode, the start of the horizontal sweep can be delayed, typically from a few μs to perhaps 10 seconds or more. Delayed sweep operation allows the users to view a small segment of the waveform, e.g. an oscillation or ringing that occurs during a small portion of a low frequency waveform.

The following are the most common approaches used by the oscilloscope manufacturers for delayed sweep operations :

(*i*) Intensified sweep, where the delayed sweep acts as a positional magnifier.

(*ii*) Normal triggering sweep after the desired time delay, which is set from the panel controls.

(*iii*) A Delay Plus Trigger mode, where a visual indication, such as light, indicates that the delay time has elapsed and sweep is ready to be triggered.

HIGHLIGHTS

1. The *cathode ray oscilloscope* is a device that allows the amplitudes of electrical signals, whether they be voltage, current, power, etc., to be displayed primarily as a function of time.
2. Cathode ray tube (CRT) consists of the following parts :
   (*i*) Electron gun assembly;
   (*ii*) Deflection plates assembly;
   (*iii*) Fluorescent screen;
   (*iv*) Glass envelope
   (*v*) Base
3. *Electrostatic deflection :*

$$v_x = \sqrt{\frac{2eV_a}{m}} \quad \text{...(i)}$$

$$D = \frac{LV_d l_d}{2dV_a} \quad \text{...(ii)}$$

where, v_x = Velocity of electron as it comes out of electron gun, m/s,

e = Charge on electron, C,
m = Mass of electron, kg,
V_a = Accelerating voltage,
D = Deflection on the screen,
V_d = Deflection voltage,
l_d = Length of each Y-plate,
d = Distance between Y-plates, and
L = Distance of middle of Y-plate to the screen.

Electroatic deflection sensitivity, $S = \frac{D}{V_d}$; its units are m/V.

4. Magnetic deflection :

$$D = BLl\left(\sqrt{\frac{e}{2mV_a}}\right)$$

where, B = Flux density.

5. The function of *vertical amplifier* is to amplify the input signal.

6. The function of *time base generator* is to drive the beam at a steady speed across the screen and, when it reaches the right-hand side of the screen the beam is made to fly back to the starting position on the left-hand side of the screen. This generator produces a *saw-tooth wave of the same frequency as the input signal* to the Y-plates.

7. The purpose of *"delay line"* is to delay the vertical signal enough to keep it from reaching the CRT deflection plates before the horizontal sweep circuits are running.

8. Basic controls of CRO are :

(*i*) Vertical (Y) deflection system; (*ii*) Horizontal (X) deflection system; (*iii*) Synchronization; (*iv*) Blanking circuit; (*v*) Positioning controls; (*vi*) Intensity (Z-axis) modulation; (*vii*) Intensity control; (*viii*) Focus control; (*ix*) Astigmatism; (*x*) Calibration circuit.

9. A *"storage oscilloscope"* uses the phenomenon of *"secondary electron emission"* to build up and store electrostatic charges on the surface of an insulated target.

10. *"Digital readout oscilloscope"* consists of a *high-speed laboratory CRO and electronic counter, both in one cabinet*.

11. The function of a *probe* is to connect the test circuit to the oscilloscope without altering, loading, or otherwise disturbing the test circuit.

12. The *"electronic switch"* is an extremely useful accessory for *displaying the two signals simultaneously on single trace oscilloscopes*.

13. The various parameters which can be measured by using CRO are :

(*i*) Voltage; (*ii*) Current; (*iii*) Period; (*iv*) Frequency; (*v*) Phase : (*vi*) Amplitude; (*vii*) Peak-to-peak value ; (*viii*) Duty cycle etc.

14. *Lissajous figures/patterns* are formed when two sine waves are applied simultaneously to the vertical and horizontal deflecting plates of CRO.

15. The unknown frequency can be measured by the following methods :

(*i*) Lissajous method; (*ii*) Spotwheel method : (*iii*) Gearwheel method.

16. *"Delayed sweep"* is a technique that adds a precise amount of time between the trigger point and the beginning of the scope sweep.

17. D.C. bridges are used to measure the *resistance*.

18. A.C. bridges are used to measure the *impedances* consisting of *inductances* and *capacitances*.

19. The simplest form of bridge is for the purpose of measuring resistance and is called the *Wheatstone bridge*.

20. $\bar{Z}_1 \bar{Z}_4 = \bar{Z}_2 \bar{Z}_3$ is the general equation for balance of the A.C. bridge.

21. Maxwell bridge is limited to the measurement of low Q values (1-10).

22. Hay bridge is more convenient for measuring high-Q coils.

23. Schering bridge is widely used for the measurement of unknown capacitors, dielectric loss and power factor.

24. Q-meter is an instrument designed to measure some electrical properties of coils and capacitors.

25. Bolometric measurements are based on the dissipation of the *RF* power in a small temperature sensitive resistive element, called a *bolometer*.

26. An *output power wattmeter* is designed to directly measure the output in an arbitrary load.

OBJECTIVE TYPE QUESTIONS

Choose the Correct Answers :

1. A thin aluminium film is usually deposited on the non-viewing side of the phosphor because

(*a*) it acts as a heat-sink and prevents phosphor burn

(*b*) the light scatter from the phosphor is reduced

(*c*) it does not allow the screen to be negatively charged

(*d*) all of the above.

2. During the retrace time, the electrons forming the horizontal beam

(*a*) move from left to right on the screen

(*b*) move from right to the left on the screen

(*c*) move from bottom to top of screen

(*d*) move from top to bottom of screen.

3. The deflection of an electron beam on a CRT screen is 10 mm. Suppose the pre-accelerating anode voltage is halved and the potential between deflecting plates is doubled. The deflection of the electron beam will be :

(*a*) 80 mm (*b*) 40 mm

(*c*) 20 mm (*d*) 10 mm.

4. If the distance of screen from a CRT to centre of deflection plates is 15 cm, the length of deflection plates is 2 cm, the distance between plates is 1 cm and the accelerating voltage is 500 V, the deflection sensitivity is :

(*a*) 33.2 V/cm (*b*) 0.03 cm/V

(*c*) 66.4 V/cm (*d*) 0.015 cm/V.

5. Post-acceleration is needed in a CRO if the frequency of the signal is

(*a*) less than 1 MHz (*b*) more than 1 MHz

(*c*) more than 10 MHz (*d*) more than 10 Hz.

6. In a digital storage oscilloscope, the input signals are

(*a*) directly applied to the oscilloscope

(*b*) multiplexed, converted to digitial form and stored and applied to oscilloscope

(c) multiplexed, converted to digital form and stored, converted to analog form and applied to oscilloscope

(d) applied to amplifier, stored as analog signals, multiplexed, converted to digital form, stored in digital form, converted to analog form,and applied to CRO through an amplifier.

7. The source of emission of electrons in a CRT is

(a) PN junction diode

(b) a barium and strontium oxide coated cathode

(c) accelerating anodes

(b) post-accelerating anodes.

8. In a CRT the focusing anode is located

(a) between pre-accelerating and accelerating anodes

(b) after accelerating anode

(c) before pre-accelerating anode

(d) none of the above.

9. The horizontal amplifier should be designed for

(a) high frequency signals with a fast rise time

(b) high amplitude signals with a slow rise time

(c) high amplitude signals with a fast rise time

(d) low amplitude signals with a fast rise time.

10. A vertical amplifier for a CRO can be designed for

(a) only a high gain

(b) Only a broad bandwidth

(c) a constant gain times bandwidth product

(d) all of the above.

11. In an oscilloscope, when the unknown signal applied to the vertical plates is being synchronized with the sweep signal applied to horizontal plates, the pattern seen on the CRO screen moves towards the right. It means that

(a) the frequency of the signal is lower than that of the sweep signal

(b) the frequency of the signal is greater than that of the sweep signal

(c) the frequency of the signal is equal to the frequency of that sweep signal

(d) none of the above.

12. P_1 phosphor material is used for display in CRTs for

(a) photographic applications (b) general purpose applications

(c) television applications (d) all of the above.

13. An aquadag is used in a CRO to collect

(a) primary electrons

(b) secondary emission electrons

(c) both primary and secondary emission electrons

(d) none of the above.

ANSWERS

1. (d)	**2.** (b)	**3.** (b)	**4.** (b)	**5.** (c)	**6.** (d)
7. (b)	**8.** (a)	**9.** (b)	**10.** (c)	**11.** (a)	**12.** (b)
13. (b)					

THEORETICAL QUESTIONS

1. What is a 'CRO'? Explain briefly.
2. What are the applications of 'CRO'?
3. Draw the block diagram of an oscilloscope and explain briefly its major systems.
4. Explain briefly the working of a CRO.
5. What is a 'CRT' ? Explain briefly.
6. With the help of a neat diagram, explain briefly the main parts of a cathode ray tube (CRT).
7. Derive an expression for vertical deflection of an electron beam in a 'CRT'.
8. Describe the different types of phosphor materials used in a CRO and list their applications.
9. Explain the functioning of a time-base generator in a CRO.
10. What are the different types of amplifiers used for CROs? Describe the basis on which they are classified.
11. Describe in details the vertical amplifier used in a CRO.
12. Describe the different types of sweeps used in a CRO. Explain their fields of application.
13. Describe the function of attenuators in CROs.
14. Describe the following :
    (*a*) Blanking circuit (*b*) Astigmatism control
    (*c*) Sources of synchronization (*d*) Z-axis modulation.
15. What do you understand by "*Delay line*"? Explain briefly.
16. Enumerate the basic controls of CRO and explain them briefly.
17. Describe briefly the following :
    (*a*) Dual trace oscilloscope (*b*) Dual beam oscilloscope.
18. Give the comparison between '*dual trace*' and '*dual beam*' oscilloscopes.
19. Discuss briefly any two of the oscilloscopes :
    (*a*) Sampling oscilloscope (*b*) Digital readout oscilloscope
    (*c*) High frequency oscilloscope
20. What is a '*probe*' ?
21. Explain briefly high impedance (or 10 : 1) probe.
22. What is an "*Electronic switch*"? Explain its working with the help of a block diagram.
23. Discuss the advantages and disadvantages of analog and digital type of oscilloscopes.
24. Discuss in detail the '*delayed sweep*'.
25. Describe briefly how the following measurements can be made with the use of CRO.
    (*a*) Frequency; (*b*) Phase angle;
    (*c*) Voltage.
26. Explain briefly any two of the following methods of frequency measurement :
    (*a*) Lissajous figures (*b*) Spot wheel method
    (*c*) Gear wheel method.

UNSOLVED EXAMPLES

1. In a CRT, the anode to cathode voltage is 2000V. The parallel deflector plates are 1.5 cm long and spaced 5 mm. The screen is 50 cm from the centre of the deflection plates. Find the following :

   (*a*) The beam speed.

   (*b*) The deflection sensitivity of the tube.

   Take : Mass of electron = 9.109×10^{-31} kg : charge on electron = 1.602×10^{-19} C.

   [**Ans.** (*i*) 26.5×10^6 m/s; (*ii*) 0.325 mm/V]

2. A cathode ray tube, with magnetic deflection, has its screen 20 cm from the centre of the magnetic field. The width of the uniform magnetic field is 3 cm and the final anode potential is 6 kV. Calculate the density of magnetic field to produce a deflection of 4 cm on the screen.

   Take : Mass of electron = 9.107×10^{-31} kg : charge on electron = 1.6×10^{-19}C.

   [**Ans.** 1.7423 Wb/m^2]

3. In a cathode-ray tube the electron beam is displaced vertically by a magnetic field of flux density 2×10^{-4} wb/m^2. The length of the magnetic field along the tube axis is the same as that of the electrostatic deflection plates. The final anode voltage is 800 V. Calculate the voltage which should be applied to the *Y*-deflection plates 1 cm apart, to return the spot back to the centre of the screen. [**Ans.** 33.53 V]

CHAPTER

8 Digital Instruments

8.1. General aspects; 8.2. Analog and digital systems; 8.3. Basic concepts of digital instruments – Introduction – Binary counters – Decimal counting units (DCUs) – Display devices – Frequency counters – Period counters – Characteristics of digital meters – Analog-to-digital (A/D) and digital-to-analog conversion **8.4. Digital voltmeters (DVMs)** – Introduction – Advantages of DVMs – Characteristic features of DVMs – Applications of DVMs – Classification of digital voltmeters – Ramp-type DVM – Dual-slope integrating type DVM – Integrating type DVM (voltage-to-frequency conversion) – Successive-approximation DVM – General specifications of a DVM; **8.5. Digital LCR meter; 8.6. Digital multimeter (DMM); 8.7. Microprocessor-based instruments-intelligent instruments; 8.8. Microprocessor-based RAM type DVM; 8.9. IEEE-488 general purpose interface bus (GPIB) instruments** – *Highlights – Objective Type questions – Theoretical Questions.*

8.1 GENERAL ASPECTS

- *A **digital instrument** can be considered as a counter which counts the pulses in a predetermined time.* Digital transducers whose output is in the form of pulses are used to monitor the desired parameter.

"Accuracy" of a digital instrument is dependent on the number of pulses generated by the transducer because the fraction of pulse cannot be generated and in counting there can be ambiguity of only one pulse at start/stop. Hence *more are pulses* corresponding to a measure less the possibility of error corresponding to one pulse and *more the accuracy.*

— The digital instruments use *logic circuits and techniques* for carrying out measurements of quantities.

— The information in the electronic digital read-out (display) devices is *presented as a series of digits on the tubes, screen, or printed on a piece of paper.* The relevant characters (letters of alphabet from A to Z, numerals from 0 to 9, punctuation mark and other symbols in common use) can be generated by :

(*i*) Semiconductor light emitting diodes (LED).

(*ii*) Liquid crystal displays (LCD).

(*iii*) Numerical indicators tubes (NIT).

(*iv*) Hot filament or bar tubes.

— Digital instruments provide better resolution than analog instruments. A typical value of *resolution* of a digital instrument may be one part in 10^6 *i.e.,* 1 μV can be read on the 1-V input range.

— *The **digital meters** work on the principle of "quantization".*

The analog quantity to be measured is first subdivided or quantized into a number of small intervals upto many decimal places. The objective of the digital instrument is then to determine in which portion of the subdivision the measurand can thus be identified as an integral multiple of the smallest unit called the *quantum*, chosen for subdivision.

The measuring procedure thus *reduces to one of counting the number of quanta present in the measurand.*

— Fig. 8.1. shows the building block of a digital instrument. Any digital instrument invariably consists of a an analog-to-digital converter (ADC) in its input stage. The display block may be analog or digital in nature. If an analog read out is desired, it becomes necessary to include a stage involving digital-to-analog (D/A) conversion.

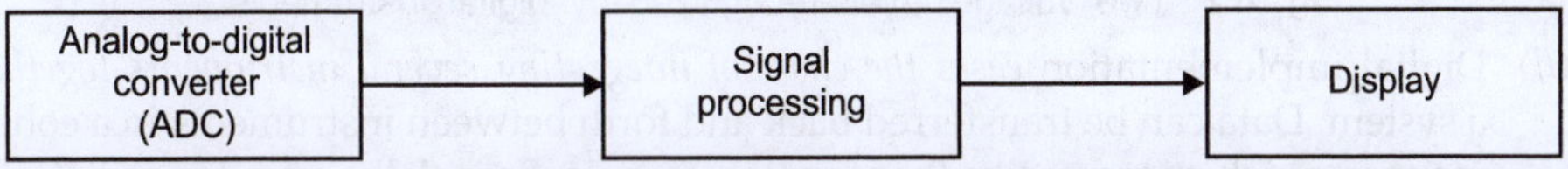

Fig. 8.1. Building block of a digital instrument.

8.2 ANALOG AND DIGITAL SYSTEMS

An ***analog system*** comprises devices that manipulate the physical quantities represented in *analog form.* Examples of analog systems are : Automobile speedometer, Magnetic tape recording and playback equipment etc.

A ***digital system*** is a combination of devices designed for manipulating physical quantities or information represented in *digital form.* Examples of digital systems are : Digital computers, typewriters, digital watches etc.

Advantages of digital systems/techniques :

1. Easier to design.
2. Easy storage of information.
3. Greater accuracy and precision.
4. Operation is programmable.
5. Digital circuits are less affected by the noise.

- Digital system may consist of the following *components :*

(*i*) Resistors; (*ii*) Capacitors;
(*iii*) Transistors; (*iv*) Linear ICs;
(*v*) Digital ICs; (*vi*) Digital devices;
(*vii*) Analog-to-digital converters (ADC) (*viii*) Digital-to-analog converters (DAC).

— *Data is digital form facilitates various operations that are normally required in signal processing.*

— An increase in the availability and type of computer facilities and a decrease in the cost of various methods required for digital systems is accelerating the development of digital instrumentation for measurement and signal processing.

8.3 BASIC CONCEPTS OF DIGITAL INSTRUMENTS

8.3.1. Introduction

A truly digital instrument uses *binary* (*i.e.,* base 2) *logic elements* such as *gates, flip-flops,* and so on, to perform the measurement function. These circuits operate with only two voltage levels (Fig. 8.2), and these represent the binary digits 0 and 1.

Digital implementation entails the following *"advantages" :*

(*i*) The instrument is more *easily modified*, especially where the complexity is in embedded software.

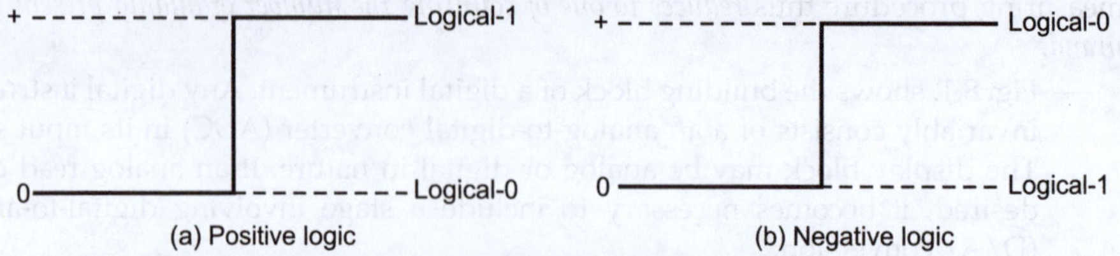

Fig. 8.2. Two voltage levels-recognized by digital instruments.

(*ii*) Digital implementation *eases the chore of integrating several instruments together* in a system. Data can be transferred back and forth between instruments in a coherent manner much more easily than analog signals, especially over long distances.

— For larger instrumentation systems, the use of general purpose interface bus (IEEE-488 GPIB) makes the chore of integrating instruments easy.

(*iii*) Even though software digital solutions are also very complex as in the case of analog solutions, yet the *hardware implications are minimal*. One can add RAM or ROM or use a hard drive to store a larger, more complex program that would be impossibly difficult to implement in analog hardware.

8.3.2. Binary Counters

The binary counter forms the heart of many "discrete logic" digital implementations.

- *A **binary counter** consists of a chain of locked flip-flops, each state is designed to count by 2.*
- *A **decoder** is used to indicate the state of the count* by examining the unique *Q* and not-*Q* states at the output of each stage.
- A 4-bit binary counter will count from 0000_2 to 1111_2 (*i.e.,* 0_{10} to 15_{10}) while a *"binary coded decimal"* **(BCD)** counter counts only from 0000_2 to 1001_2 (0_{10} to 9_{10}). Table 8.1 shows both binary and BCD coding, the latter of which is a subset of 4-bit binary coding.

Table 8.1. Binary and BCD Codes

Decimal	Binary	BCD		
0	0000		0000	
1	0001		0001	
2	0010		0010	
3	0011		0011	
4	0100		0100	
5	0101		0101	
6	0110		0110	
7	0111		0111	
8	1000		1000	
9	1001		1001	
10	1010	0001	0000	
11	1011	0001	0001	
12	1100	0001	0010	
13	1101	0001	0011	Two 4-bit
14	1110	0001	0100	counters
15	1111	0001	0101	required

8.3.3. Decimal Counting Unit (DCUs)

Decimal counting unit *is the circuit of any electronic counter.* It consists of the following components :

(*i*) Decade counter;

(*ii*) Decoder/display-driver;

(*iii*) Data latch.

Fig. 8.3. shows the block diagram of an example of a decimal counting unit (DCU) :

— This circuit is constructed of TTL integrated circuit logic elements. The decade counter uses a chain of flip-flops, but it also contains some internal gates that sense the tenth state (*i.e.*, "9") and reset the counter to 0000 if another input pulse is received. In fact, now, there are several MSI and LSI integrated circuits available that contain all circuitry needed to form a complete electronic counter of several decades.

— The four output lines from the *"decade counter"* are coded in 8421 BCD format and they change state with every pulse.

— *"Input"* and *"reset"* are other two lines, in which the input line accepts the pulses being counted, while the reset line will reset the counter to 0000 when brought to ***High***. For normal counting the line is kept ***Low***, and to clear the counter it is brought momentarily ***High***.

— The *"decoder"* is used to convert the 4-bit BCD code to the code needed to correctly drive the read-out device.

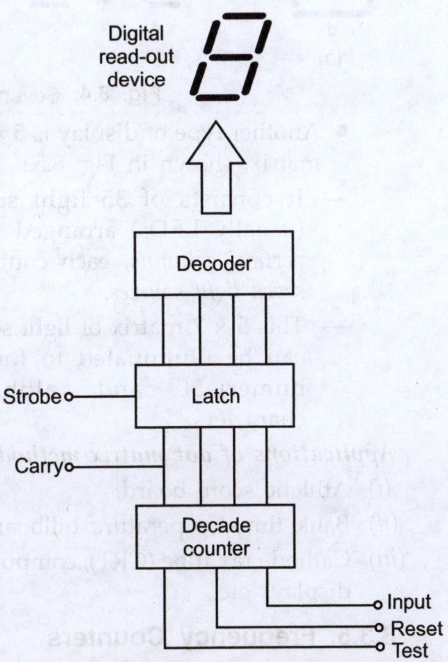

Fig. 8.3. Decimal decode counter.

— A quad *"latch"* circuit is used to eliminate the rolling effect (this effect is created because the display changes with the count).

— The *"strobe"* line, in practice, is held low while the count is being accumulated, and at the end of the count period the strobe line is brought high momentarily, just long enough to transfer the count data from the input lines to the output lines.

- The individual DCU can count only from 0 to 9, but if two DCUs are used in cascade, the assembly can count from 0 to 99, and three DCUs can count from 0 to 999 etc.; the assembly is known as ***decimal counting assembly (DCA).***

8.3.4. Display Devices

- In modern equipment the most common display is the *seven-segment* readout shown in Fig. 8.4(*a*). This type of display consists of seven illuminated bars arranged in a *figure-eight pattern.* It is possible to *form all ten decimal digits by lighting appropriate bars* [Fig. 8.4(*b*)]. A BCD-to-seven-segment decoder will ground those terminals *a* through *g* required to form the decimal digit represented in BCD form at the decoder input. Probably the most popular types as of this writing are the light-emitting diode (LED) and liquid crystal diode (LCD) displays.

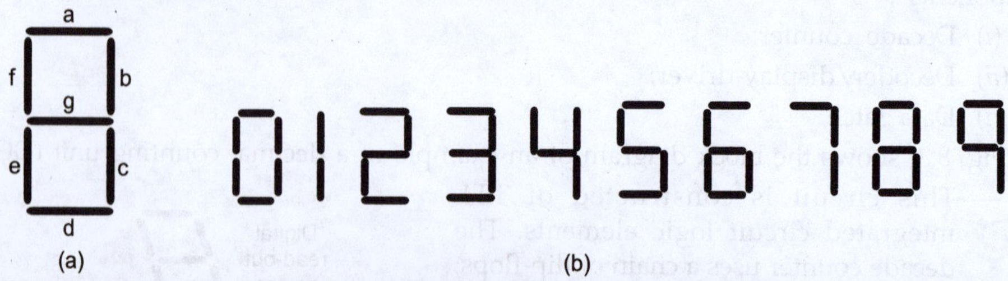

Fig. 8.4. Seven-segment decimal readout.

- Another type of display is 5 × 7 dot matrix, shown in Fig. 8.5.:
  — It consists of 35 light sources (usually LEDs) arranged in *five vertical columns,* each containing *seven light sources.*
  — This 5 × 7 matrix of light sources can be illuminated to form all numerical and alphabetic characters.

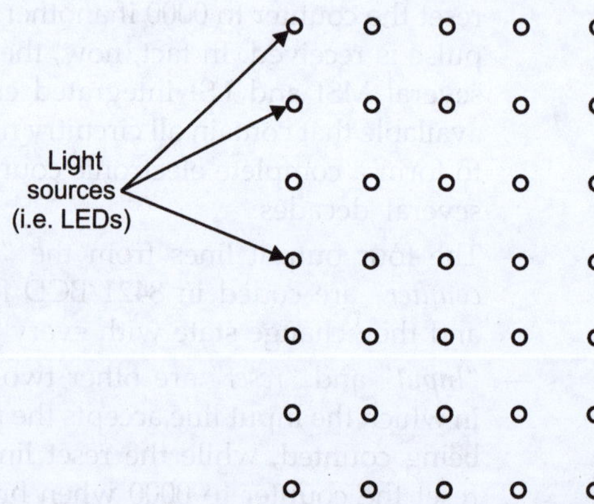

Fig. 8.5. 5 × 7 dot matrix.

Applications of dot matrix method :

(*i*) Athletic score board;
(*ii*) Bank time-temperature billboards;
(*iii*) Cathode ray tube (CRT), computer TV displays etc.

8.3.5. Frequency Counters

Refer to Fig. 8.6 :

- **Fig. 8.6(*a*)** shows the basic block diagram of a **"frequency counter"**. It consists of the following *sections :*

(*i*) Decade counter assembly (DCA); (*ii*) Main gate;
(*iii*) Trigger; (*iv*) Input amplifier;
(*v*) Main gate flip-flop; (*vi*) Time base;
(*vii*) Display clock.

— The frequency counter measures *"events per unit of time (EPUT)"* (*i.e.,* cycles per second), so the DCA must be tuned on only for a given period of time (*e.g.,* 0.1, 1, or 10 s).

— The main gate, main gate flip-flop, and time base sections are used *to allow input pulses into the DCA* for the designed period of time.

— The time base section consists of a crystal *oscillator* that produces pulses at precise rate such as 100 kHz, 1 MHz, 4 MHz, 10 MHz, and so on. A chain of *decade dividers* is used to reduce the crystal oscillator frequency *to a lower frequency.* The time base output frequency will be 10 Hz for 0.1 s, 1 Hz for 1 s and 0.1 Hz for 10-s measuring periods.

- **Fig. 8.6(*b*)** shows the timing diagram for one complete interval of EPUT counter.

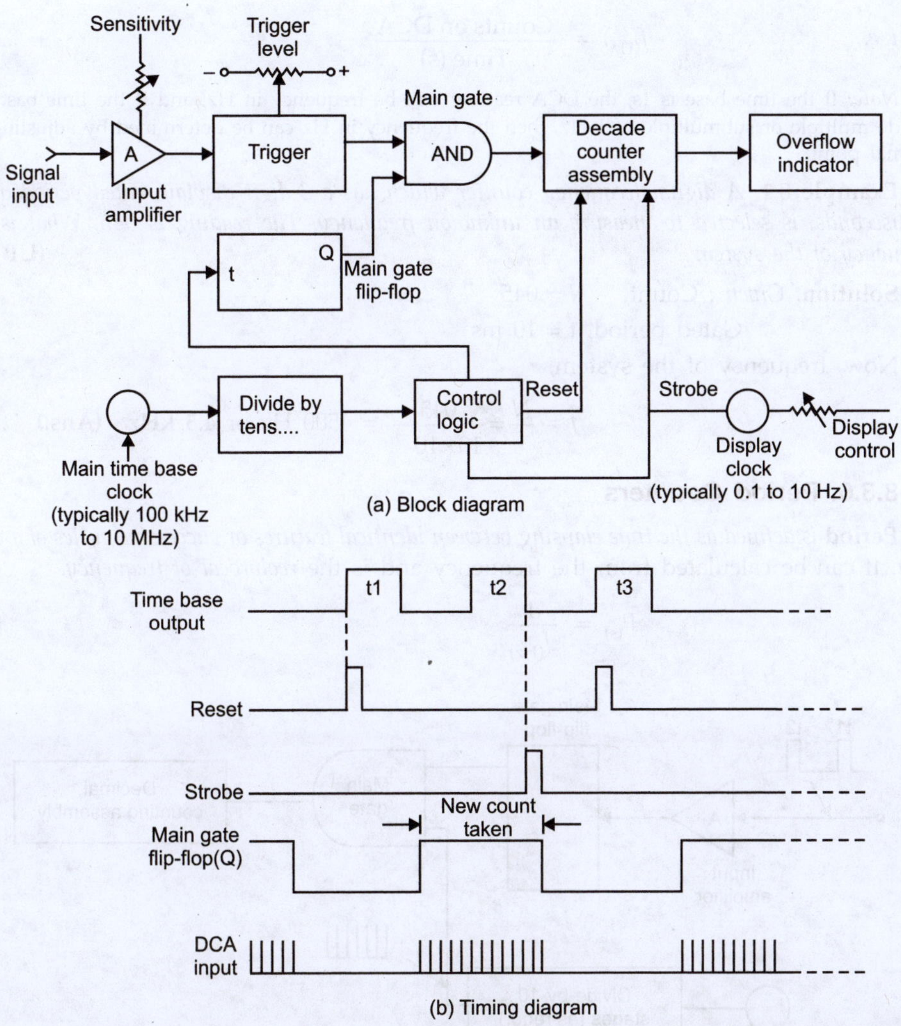

Fig. 8.6. Frequency counter.

— Pulses *t1*, *t2* and *t3* are output from the time base section. When pulse *t1* goes high, the control logic section generates a short pulse to reset the DCA to zero. When *t1* goes low again, the *Q* output of the J-K main gate flip-flop will go high. The main (AND) gate has one input tied to the *Q* output of the flip-flop and the other input is tied to the signal being counted. As a result, the main gate passes input pulses to the DCA *only* when the *Q* terminal of the flip-flop is high.

— Until the negative transition of *t2* occurs, the flip-flop remains set. At that time the *Q* output of the flip-flop drops low, turning off the flow of pulses into the DCA and causing the control logic section to generate a strobe pulse. This pulse tells the DCU latches to transfer data from the counter to the decoders. The display, then, shows *only completed count cycles and will hold the previous count until the end of the next interval.*

The frequency of the input signal is the number of counts accommodated on the DCA divided by the time base period in seconds.

i. e.,

$$f_{(Hz)} = \frac{\text{Counts on DCA}}{\text{Time (s)}} \quad ...(8.1)$$

Note: If the time base is 1s, the DCA reading will be frequency in Hz, and if the time base is a decade multiple or submultiple of 1 Hz, then the frequency in Hz can be determined by adjusting the decimal point.

Example 8.1. *A digital frequency counter which has a 3-digit display, gated period of 10 milliseconds, is selected to measure an unknown frequency. The reading is 045. What is the frequency of the system ?* **(UPTU)**

Solution. *Given* : Count, $N = 045$

Gated period, t = 10 ms

Now, frequency of the system,

$$f = \frac{N}{t} = \frac{045}{10 \times 10^{-3}} = 4500 \text{ Hz or } \mathbf{4.5\ kHz} \quad \textbf{(Ans.)}$$

8.3.6. Period Counters

Period *is defined as the time elapsing between identical features or successive cycles of a wave form.* It can be calculated from the frequency and is the *reciprocal of frequency.*

$$P_{(s)} = \frac{1}{f_{(Hz)}} \quad ...(8.2)$$

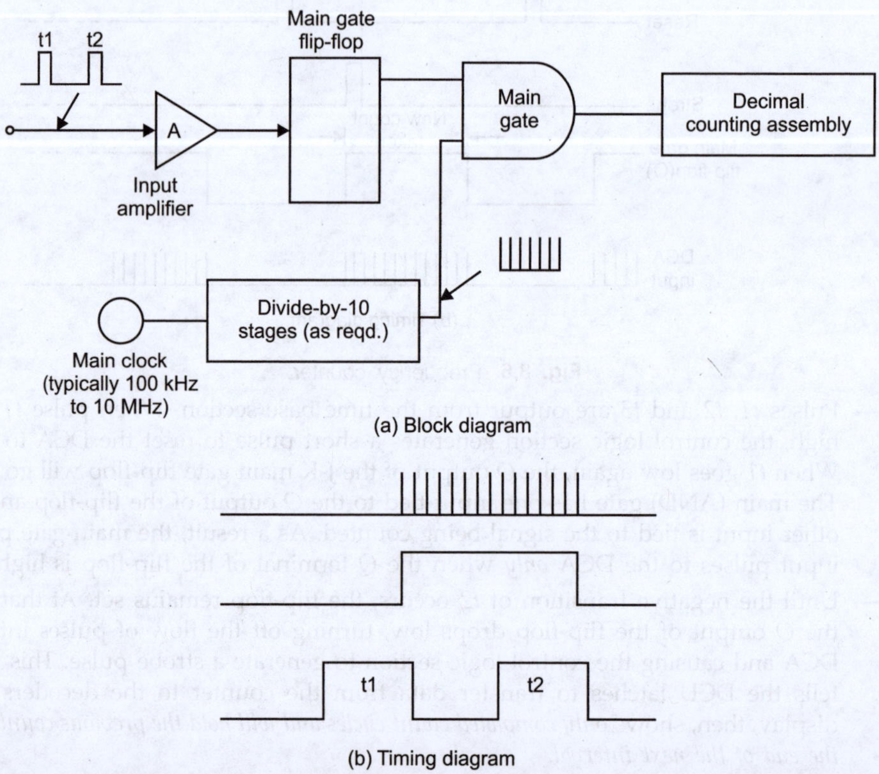

Fig. 8.7. Period counter.

Refer to Fig. 8.7.

- A period counter can be made by *reversing* the roles of the input stage and time base; the input amplifier is connected to the main gate flip-flop, and the time base is connected to the main gate as shown in **Fig. 8.7(*a*).**
- **Fig. 8.7(*b*)** shows the timing diagram for a period counter. The flip-flop is set (*i.e.*, Q goes high) on the negative-going transitions of $t1$, and this allows the gate to pass pulses from the time base to DCA.

The time base frequency determines the time interval represented by each pulse. The period is given by the count accumulated by the DCA and the time base frequency :

$$P_{(s)} = \frac{\text{DCA count}}{f_{(\text{Hz})}} \qquad ...(8.3)$$

Note: A time base frequency of 1000 Hz yields a period counter resolution of $\frac{1}{1000}$ s or 1 ms; similarly, one 10 kHz time base yields 0.1 ms resolution.

8.3.7. Characteristics of Digital Meters

Following are the few specifications which characterise digital meters :

1. Resolution; 2. Sensitivity; 3. Accuracy.

1. **Resolution.** *It is defined as the number of digit positions or simply the number of digits used in a meter.*

If number of full digits is n, then resolution,

$$R = \frac{1}{10^n} \qquad ...(8.4)$$

For $n = 4$, $$R = \frac{1}{10^4} = 0.0001 \text{ or } 0.01\%.$$

- A three-digit display on the digital meter for 0-1 V range will be able to indicate from 000 to 999 mV, with *smallest increment (resolution)* of 1 mV.

2. **Sensitivity.** *It is the smallest change in input which a digital meter is able to detect.* Thus, it is the full-scale value of the lowest voltage range multiplied by the resolution of the meter. In other words,

$$\text{Sensitivity, } S = (fs)_{min} \times R \qquad ...(8.5)$$

where, $(fs)_{min}$ = Lowest full-scale value of digital meter, and

R = Resolution is decimal.

3. **Accuracy.** In the accuracy specifications of digital meters, the following two quantities are included :

(*i*) A percentage of range.

(*ii*) A percentage of reading.

Following solved examples, illustrate these characteristics :

Example 8.2. *What is the resolution of $3\frac{1}{2}$ digit display on 1 V and 10 V ranges? A $3\frac{1}{2}$ digital voltmeter has an accuracy specification of ± 0.5% of reading ± 1 digit. What is possible error in volt when the instrument is reading 2.00 V on 5-V scale.*

Solution. • Number of full digits on $3\frac{1}{2}$ digit display, $n = 3$

$\therefore$ Resolution, $R = \dfrac{1}{10^n} = \dfrac{1}{10^3} = 0.001$

It means, the meter cannot distinguish between values that differ from each other by less than 0.001 of full scale (*fs*).

For full-scale range of 1 V:

The *resolution* = (*fs*) × *R* = 1 × 0.001 = **0.001 V** **(Ans.)**

For full-scale range of 10 V :

The *resolution* = 10 × 0.001 = **0.01 V** **(Ans.)**

- The display for 2 V reading on 10 V scale of $3\frac{1}{2}$ digital meter would be 0.0200.

The digit in the least significant digit (LSD) has a value of :

$$(fs) \times R = 5 \times \frac{1}{10^3} = 0.005 \text{ V}$$

$$\textit{Total possible error} = \frac{0.5}{100} \times \text{reading} + \text{value of digit in LSD}$$

$$= \frac{0.5}{100} \times 2 + 0.005 = 0.01 + 0.005 = \mathbf{0.015\ V} \quad \textbf{(Ans.)}$$

Example 8.3. *What is the resolution of a $4\frac{1}{2}$ digit display? How would 16.95 V be displayed on a 10 V range and 0.6564 V on a 1 V and 10 V ranges ?*

Solution. • Number of full digits on $4\frac{1}{2}$ digit display, $n = 4$

$\therefore$ *Resolution*, $R = \dfrac{1}{10^n} = \dfrac{1}{10^4} = \mathbf{0.0001}$ **(Ans.)**

- There are 5 digit faces in $4\frac{1}{2}$ digit display, so *16.95 would be displayed as* **16.950** **(Ans.)**

Resolution on 1 V range is 0.000 × 1 = 0.0001 V.

Any reading upto the *4th decimal can be displayed.*

Hence, *0.6564 will be displayed as* **0.6564** **(Ans.)**

Resolution on 10 V range is 0.0001 × 10 = 0.001 V.

Hence, decimals upto *third decimal place can be displayed.*

Hence, *0.6564 will be displayed as* **0.656** **(Ans.)**

Example 8.4. *Explain briefly the $3\frac{1}{2}$-digit display in a digital meter.*

Solution. The number of digit positions in a digital meter determines the resolution. Hence a 3-digit display on digital voltmeter (DVM) for 0–1 V range will indicate values from 0–999 mV with a smallest increment of 1 mV.

Normally, a *fourth* digit capable of indicating 0 or 1 (hence called a *half digit* is placed to the left. This permits the digital meter to read values above

Fig. 8.8. 3½-digit display.

999 upto 1999, to give overlap between ranges for convenience, a process called *"over-ranging"*. This type of display is called a $3\frac{1}{2}$ *digit display*, shown in Fig. 8.8.

8.3.8 Analog-to-Digital (A/D) and Digital-to-Analog Conversion

Refer to Article 9.4

8.4 DIGITAL VOLTMETERS (DVM)

8.4.1. Introduction

A **digital voltmeter** *(DVM) displays measurements of D.C. or A.C. voltage as discrete numerals* instead of pointer deflections on a continuous scale as in analog instruments.

- It is a *versatile and accurate instrument* that is employed in many laboratory measurement applications.
- Owing to the development and perfection of IC modules, the size, power requirements and cost of the digital voltmeter have been drastically reduced and, therefore, DVMs can actively compete with conventional analog instruments, both in price and portability.
- The various varieties of DVM differ in the following ways :
  - (*i*) Number of digits;
  - (*ii*) Number of measurements;
  - (*iii*) Accuracy;
  - (*iv*) Speed of reading;
  - (*v*) Digital output of several types.

8.4.2. Advantages of DVMs

Following are the *advantages of DVM over other voltmeters* :

1. Higher accuracy and resolution.
2. Greater speed.
3. No parallax.
4. Reduced human error.
5. Compatibility with other digital equipment for further processing and recording.

8.4.3. Characteristic Features of DVMs

The characteristic features of digital voltmeters (DVMs) are :

1. *Input range :* From ± 1.000 V to ± 1000 V with automatic range selection and overload indication.
2. *Absolute accuracy :* As high as ± 0.005% of the reading.
3. *Resolution :* 1 part in million (1 μV reading can be read or measured on 1 V range)
4. *Stability :* Short-term, 0.002% of the reading for a 24-hr period; long term, 0.008% of the reading for a 6-month period.
5. *Input resistance :* Typically 10 MΩ;
   Input capacitance : 40 pF
6. *Calibration :* Internally from stabilised reference sources, independent of measuring circuit.
7. *Output signals :* In BCD (Binary-coded-decimal) form, for print output and further digital processing.
   - *Optional features* may include additional circuitry to measure current, ohms and voltage ratio.

8.4.4. Applications of DVMs

DVMs are often used in *"data processing systems"* or *"data logging systems"*.

— In such systems, a number of analog input signals are scanned sequentially by an electronic system and then each signal is *converted to an equivalent digital value by the A/D (analog-to-digital) converter in the DVM.*

— The digital value is then transmitted to a printer along with the information about the input line from which the signal has been derived.

— The whole data is then printed out.

In this way, *a large number of input signals can be automatically scanned or processed and their values either printed or logged.*

8.4.5. Classification of Digital Voltmeters (DVMs)

The digital voltmeters can broadly be *classified* as follows :

1. Ramp-type DVM.
2. Dual-slope integrating type DVM.
3. Integrating type DVM (Voltage-to-frequency conversion).
4. Successive–approximation DVM.

8.4.6. Ramp-type DVM

The *principle of operation* of the ramp-type DVM is based *on the measurement of the time it takes for a linear ramp voltage to rise from 0 V to the level of input voltage, or decrease from the level of the input voltage to zero.* This interval of time is measured with an *electronic time-interval counter*, and the count is displayed as a *number of digits* on electronic indicating tubes.

- *Fig. 8.9* shows a ramp-type DVM.

Fig. 8.9(*a*) shows the *"voltage-to-time conversion"* using gated clock pulses.

— At the start of the measuring cycle, a ramp voltage is initiated; this voltage can be positive-going or negative going. The negative-going ramp, shown in Fig. 8.9(*a*) is continuously compared with the unknown input-voltage.

— At the instant that the ramp voltage equals the unknown voltage, a coincidence circuit, or comparator, generates a pulse which opens a gate **[See Fig. 8.9(*b*)]**. The ramp voltage continues to decrease with time until it finally reaches 0 V (or ground potential) and a second comparator generates an output pulse which closes the gate.

— An oscillator generates clock pulses which are allowed to pass through the gate to a number of decade counting units (DCUs) which totalise the number of pulses passed through the gate.

— The *decimal number*, displayed by the indicator tubes associated with the DCUs, *is a measure of the magnitude of the input voltage.*

— The sample-rate multivibrator (MV) determines the rate at which the measurement cycles are initiated. The sample-rate circuit provides an initiating pulse for the ramp generator to start its next ramp voltage. At the same time, a reset pulse is generated which returns all the DCUs to their *zero state*, removing the display momentarily from the indicator tubes.

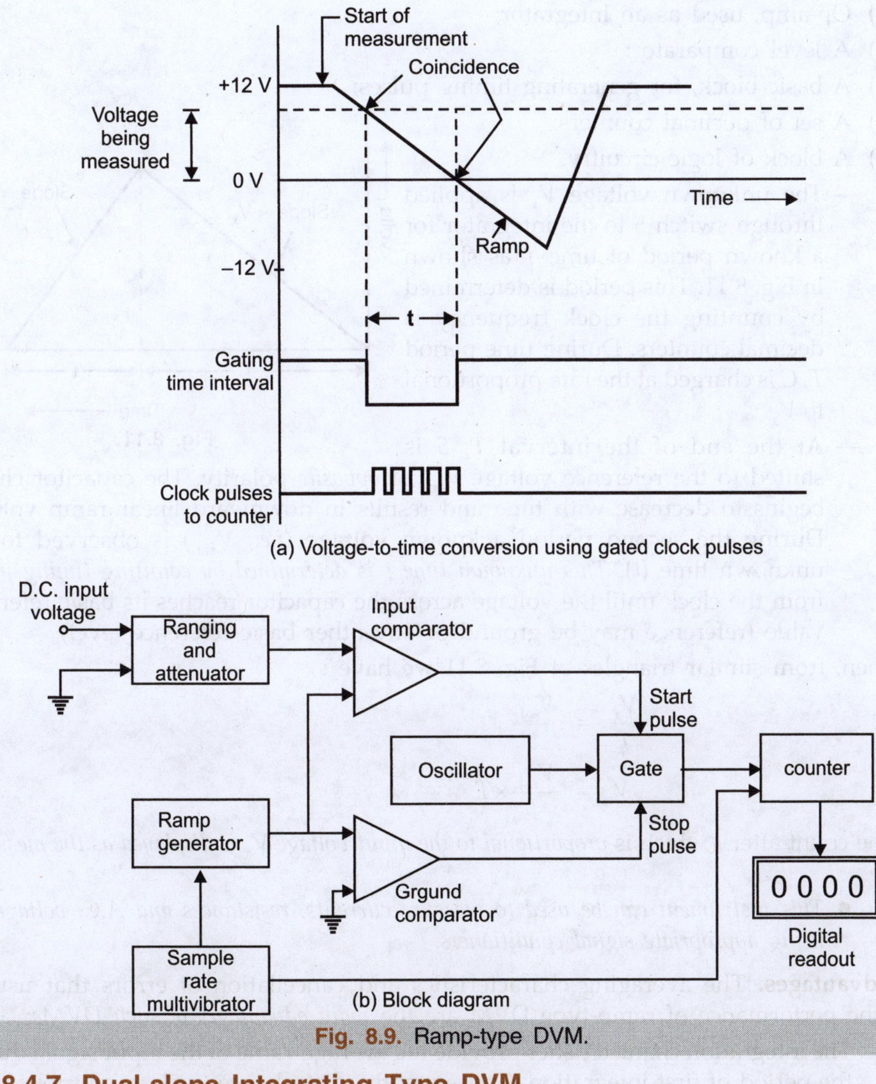

Fig. 8.9. Ramp-type DVM.

8.4.7. Dual-slope Integrating Type DVM

Fig. 8.10 shows the block diagram of a dual-slope integrating type DVM. It consists of the following *five* basic blocks:

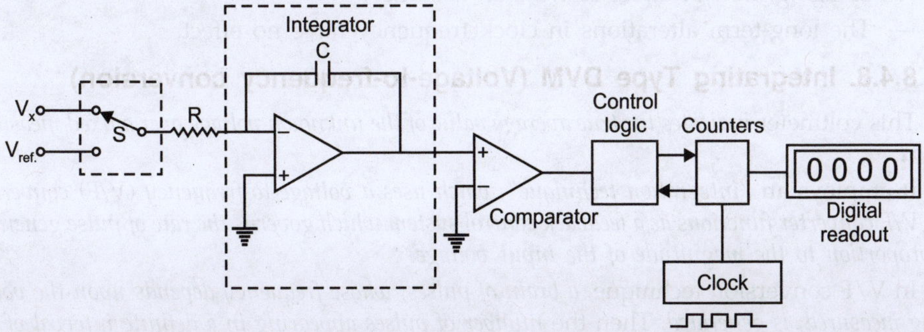

Fig. 8.10. Block diagram of a dual-slope integrating type DVM.

(*i*) Op-amp, used as an integrator;
(*ii*) A level comparator;
(*iii*) A basic block, for generating timing pulses;
(*iv*) A set of decimal counters;
(*v*) A block of logic circuitry.

— The unknown voltage V_x is applied through switch S to the integrator for a known period of time T as shown in Fig. 8.11. This period is determined by counting the clock frequency is decimal counters. During time period T, C is charged at the rate proportional to V_x.

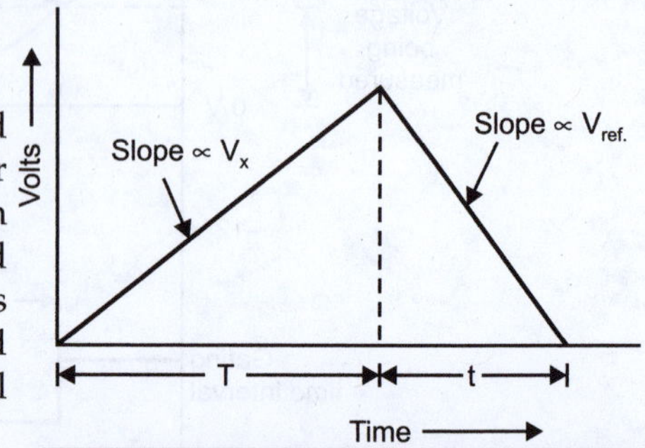

Fig. 8.11.

— At the end of the interval T, S is shifted to the reference voltage $V_{ref.}$ of *opposite* polarity. The capacitor charge begins to decrease with time and results in downward linear ramp voltage. During the second period a known voltage (*i.e.*, $V_{ref.}$) is observed for an unknown time (t). *This unknown time t is determined by counting timing pulses* from the clock until the voltage across the capacitor reaches its basic reference value (reference may be ground or any other basic reference level).

Then, from similar triangles of Fig. 8.11 we have :

$$\frac{V_x}{T} = \frac{V_{ref.}}{t}$$

or,
$$V_x = \frac{V_{ref.}}{t} \times T \qquad ...(8.5)$$

The count after t which is *proportional to the input voltage V_x is displayed as the measured voltage.*

- *This instrument can be used to measure currents, resistances and A.C. voltages by using appropriate signal conditioners.*

Advantages. The averaging characteristics and cancellation of errors that usually limit the performance of ramp-type DVM are the *main advantages of such DVMs:*

— The integration characteristics provide the average value of the input signal during the period of first integration. Consequently, disturbance, such as spurious noise pulses, are *minimised.*

— Long-term drifts in the time constant, as may result from temperature variations or aging, do not affect conversion accuracy.

— The long-term alterations in clock frequency have no effect.

8.4.8. Integrating Type DVM (Voltage-to-frequency conversion)

This voltmeter *indicates the true average value of the unknown voltage over a fixed measuring period.*

It employs an *"integration technique" which uses a voltage-to-frequency (V/F) conversion. The V/F converter functions as a feedback control system which governs the rate of pulse generation in proportion to the magnitude of the input voltage.*

In V/F conversion technique, *a brain of pulses, whose frequency depends upon the voltage being measured, is generated.* Then the *number of pulses appearing in a definite interval of time is measured.*

The basic requirement of this technique is *"operational integrator"*, the circuit diagram and symbolic representation of which is shown in [Fig. 8.12(*a*, *b*)], it is a modification of operational amplifier. The operational integrator consists of an amplifier (having high gain) with a capacitor C connected between output and input. If V_i is the input voltage, then V_o, of the operational integration is given by :

$$V_o = \frac{1}{RC}\int V_i \cdot dt \qquad ...(8.6)$$

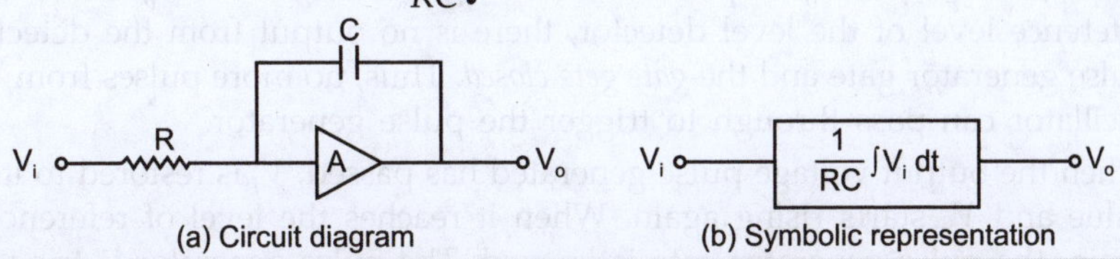

Fig. 8.12. Operational integrator.

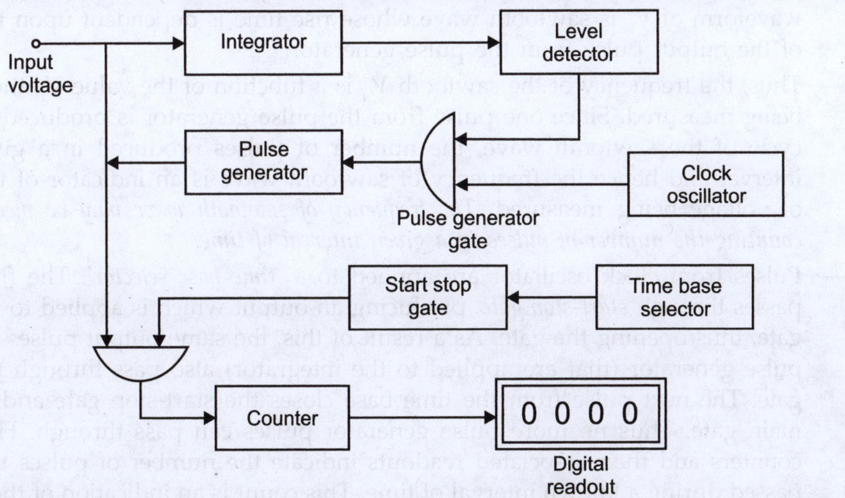

Fig. 8.13. Block diagram of an integrating type DVM (Voltage-to-frequency conversion).

If the input voltage is constant, we have :

$$V_o = -\frac{V_i}{RC} \cdot t \qquad ...(8.7)$$

Thus if a constant input voltage V_i is applied, an output voltage V_o is produced which rises at a uniform rate and has a polarity *opposite* to that of input voltage. That is, for a constant input voltage the integrator produces a ramp output voltage of opposite polarity such that the greater the input voltage, the sharper is the rate of rise, or step of the output voltage.

Fig. 8.13. shows the *block diagram* of this voltmeter :

— The unknown voltage, V_i is applied to the input of the integrator and the output voltage V_o starts to rise. *The slope of V_o is determined by the value of input voltage V_i.* This voltage is fed to a level detector, and, when V_o reaches a certain reference level, the *detector* sends a pulse to *pulse generator gate.* The level detector is a device similar to a voltage comparator.

— The output voltage from integrator, V_o is compared with the fixed voltage of an internal reference source and when V_o reaches that level, the level detector produces an *output pulse.* This output pulse opens the pulse level gate, permitting pulse from a fixed frequency clock oscillator to pass through the *pulse generator*. This generator is similar to schmitt trigger that *produces an output pulse of fixed amplitude and width for every pulse it receives.*

— The output pulse whose polarity is *opposite* to that of V_i and has greater amplitude is fed back to the input of the integrator. Hence, the net output to the integrator is now of reversed polarity (opposite to that of V_i). As a result of this reversed output, the output V_o drops back to its original level. Since V_o is now below the reference level of the level detector, there is no output from the detector to the pulse generator gate and the *gate gets closed.* Thus, no more pulses from the clock oscillator can pass through to trigger the pulse generator.

— When the output voltage pulse generated has passed, V_i is restored to its original value and V_o starts rising again. When it reaches the level of reference voltage again, the pulse generator gate is opened. The pulse generator is triggered by a pulse from the clock generator and the entire cycle is repeated again. Thus, the waveform of V_o is sawtooth wave whose rise time is dependent upon the value of the output pulse from the pulse generator.

— Thus, the frequency of the sawtooth V_n is a function of the value V_i, the voltage being measured. Since one pulse from the pulse generator is produced for each cycle of the sawtooth wave, the number of pulses produced in a given time interval and hence the frequency of sawtooth wave is an indicator of the value of voltage being measured. *The frequency of sawtooth wave may be measured by counting the number of pulses in a given interval of time.*

— Pulses from clock oscillator are applied to a *'time base selector'*. The first pulse passes through *start-stop gate,* producing an output which is applied to the main gate, thus opening the gate. As a result of this, the same output pulses from the pulse generator (that are applied to the integrator) also pass through the main gate. The next pulse from the time base closes the start-stop gate and also the main gate. Thus no more pulse generator pulses can pass through. Hence the counters and their associated readouts indicate the number of pulses that have passed during a known interval of time. This count is an indication of the voltage being measured. *In order to make the counter read directly in terms of voltage the amplitude and width of pulse generator can be suitably adjusted.*

- The drawback of this DVM is that it *needs excellent characteristics in the linearity of the ramp.* The A.C. noise and supply noise are averaged out.

8.4.9. Successive-approximation DVM

Fig. 8.14, shows the block diagram of a successive-approximation DVM, in which the electronic implementation of the successive-approximation technique is illustrated :

— A *"digital-to-analog (D/A) converter"* is used to provide the estimates. The "equal to or greater than" or "less than" decision is made by the *"comparator"*. The D/A converter provides the estimate and is compared to the input signal.

— A special shift register called a *"successive-approximation register (SAR)"* is used to control the D/A converter and consequently the estimates. At the *beginning* of the conversion all the outputs from the SAR are at *logic zero.* If the estimate is *greater than the input,* the comparator output is *high* and the first SAR output reverses state and the second output changes to a logic "one". If the comparator output is *low,* indicating that the estimate is *lower than the input signal* the first-output remains in the logic one state and the second output assumes the logic one state. *This continues to all the states until the conversion is complete.*

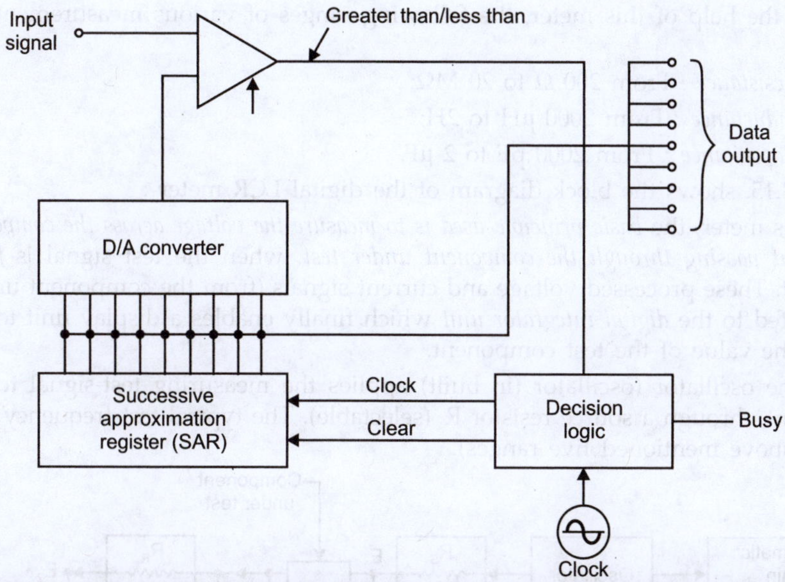

Fig. 8.14. Block diagram of a successive-approximation DVM.

— The above sequence of events are performed electronically. *For an N-bit conversion after N clocks, the actual value of the input is known*. The least significant bit is the state of the comparator. In some systems an additional clock is employed to store the last bit in the SAR and thus N + 1 clocks are required for a conversion.

8.4.10. General Specifications of a DVM

Following are the general specifications of a DVM :

(*i*) *Display* : 3 ½ digits, LCD; (*ii*) *Unit Annuciation*: mV, V, mA, Ω, kΩ, MΩ, buzzer, B (low battery); MANU (Manual), ac and ⊣▷⊢ (diode test); (*iii*) *Max. Indication*: 1999 or –1999; (*iv*) *Over-range indication*: only (1) or (–1) displayed at the MSB position; (*v*) *Polarity*: AUTO negative polarity indication; (*vi*) *Zero adjustment* : Automatic; (*vii*) *Functions*: DC volts, AC volts, DC amps, AC amps, Ohms, continuity test, diode test; (*viii*) *Ranging*: Sepectable automatic or mannual; (*ix*) *Automatic*: Instrument automatically selects maximum range for measurement and display. Auto ranging operates on all functions except for dc or ac current; (*x*) Manual switch selection as desired; (*xi*) *Sampling rate* : 2 samples/s, nominal; (*xii*); Low battery: β mark on LCD readout; (*xiii*) *Temperature*: Operating 0°C – 40°C, <80% RH(Relative humidity); Storage – 20°C – 60°C, < 70% RH; (*xiv*) *Power* : Two AR size 1.5 V batteries, Life 2000 hours typically with zinc-carbon; (*xv*) *Standard accessories;* Probe red-black; safety fuse 250 – 0.2A; (*xvi*) Size: 160(L) × 80(B) × 30(H); (*xvii*) *Weight*: 250 g without batteries; (*xviii*) *Input impedance*: 11 MΩ – 1000 MΩ; (*xix*) Accuracy : ± 0.5% – 0.7% or ± 5 digits for dc; 1.0% reading or ± 5 digits for ac at 40 – 500 kHz.

8.5 DIGITAL LCR METER

This type of meter is used *to measure the resistance, inductance, capacitance and dissipation factor*. The desired function can be selected by using a *rotary switch*. The various ranges available are :

(*i*) 200 μH/pF/Ω; (*ii*) 2000 μH/pF/Ω; (*iii*) 20 mH/nF/kΩ; (*iv*) 200 mH/nF/kΩ; (*v*) 2H/μF/MΩ.

With the help of this meter, the following ranges of various measurements can be made:

- *Resistance* : From 200 Ω to 20 MΩ;
- *Inductance* : From 2000 μH to 2H;
- *Capacitance* : From 2000 pF to 2 μF.

Fig. 8.15. shows the block diagram of the digital LCR meter :

In this meter, the *basic principle used is to measure the voltage across the component and the current passing through the component under test*, when the test signal is *fed to the component.* These processed voltage and current signals (from the component under test) are then fed to the *digital integrator unit* which finally enables a display unit to provide directly the value of the test component.

— The oscillator (oscillator (in built) applies the measuring test-signal to the test component, through a source resistor R_s (selectable). The typical test frequency is 1 kHz (for the above mentioned five ranges).

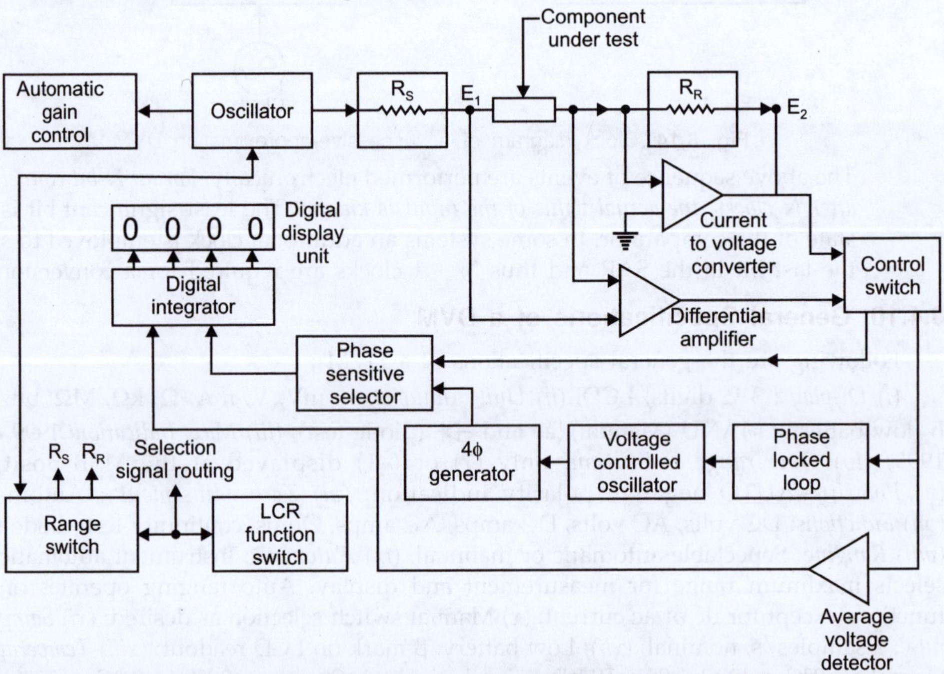

Fig. 8.15. Block diagram of a digitial LCR meter.

— The signal current then flows to current-to-voltage converter, which is essentially an op-amp with range resistor R_R, connected in the feedback path.

— The op-amp drives the junction of component and R_R to a virtual ground and hence R_R does not change the current through the component. Thus *signal current develops a voltage E_2 proportional to the current through the component.*

The voltage E_1 and E_2 are the vector quantities and as such they define the characteristics of the component at a particular test frequency and signal level.

Mathematically : $E_1 \propto V$ and $E_2 \propto I$

Thus, *Capacitance,* $C \propto \frac{I}{V} \propto \frac{E_2}{E_1}$; *Inductance,* $L \propto \frac{V}{I} \propto \frac{E_1}{E_2}$.

The above ratios are adopted in measurement modes and are obtained by dual-slope integration method. For *inductance* measurements, a *series* equivalent circuit of an inductor is assumed while for capacitance measurements, a parallel equivalent circuit of capacitor is assumed for five ranges.

— The values of R_S and R_R are selected, based on the impedance of the unknown component. For the measurement of *inductance,* the component impedance is usually low and hence R_S is chosen much higher. This achieves a *constant current drive* to the component. The R_S decides the value of current. For *capacitance* measurement, the component impedance is high and R_S is chosen to be much low. Thus *constant voltage drive* is provided to the test component.

— The voltage E_1 is applied as one input to *'differential amplifier'*; E_1 is then fed to a *control switch,* along with E_2. The greater of the two is fed to *'average voltage detector (AVD)'* and lesser to the *'phase sensitive detector (PSD)'*.

— The signal given to AVD is also given to *'phase locked loop (PLL)'* and voltage controlled oscillator (VCD). These two produce the clock signals, which are locked in phase with the reference signal. The clock signal is then divided into phase shifted 90° and 270° signals. These are given to PSD, which detector the phase angle.

— The D.C. voltage outputs, from AVD and PSD are then given to *'digital integrator unit'* and final to *'digital display unit'* where the component value is displayed.

- For wide range of values, the basic accuracy of this meter is ± 0.15 percent of the reading.

— The range selection is fully automatic.

— The display used is normally 4 digit LEDs with automatic decimal point.

— The inductors and capacitors are automatically differentiated.

Commercial digital LCR meter :

Important features available in commercial digital LCR meter are :

(*i*) Auto-ranging and auto-computing facility.
(*ii*) High basic accuracy, upto 0.2 percent.
(*iii*) High frequency range, from 0.1 to 1 kHz.
(*iv*) Hold function.
(*v*) Multipoint measurement techniques.
(*vi*) Series or parallel modes of operation.

- Most of these meters are *"microprocessor controlled"* and multipoint measurement technique ensures better accuracies in measurement. Apart from the basic measurement of LCR and Q they can also be used for series and parallel measurement *i.e.,* they can measure series and parallel components of L and C.

— The microprocessor configuration provides operation flexibility *i.e.,* comparison with previous recorded values, storage of these values for future references (memory function).

— The user software provides user prompt facility by means of flashing LED for respective functions and to select appropriate, test frequencies to achieve optimum results.

— The *'Hold function'* is useful for selecting the components within a specified range.

8.6 DIGITAL MULTIMETER (DMM)

Digital multimeter offers *increased versatility due to its additional capability to measure A.C. voltage and current, D.C. current, and resistance.*

Fig. 8.16, shows the block diagram of a digital multimeter (DMM) :

- In the *"A.C. voltage mode"*, the applied input is fed through a calibrated, compensated attenuator, to a precision full-wave rectifier circuit followed by a ripple reduction filter. The resulting D.C. is fed to ADC and the subsequent display system.
- For *current measurements,* the drop across an internal calibrated shunt is measured, directly by the ADC in the *"D.C. current mode"*, and after A.C. to D.C. conversion in the *"A.C. current mode"*. This drop is often in the range of 200 mV (corresponding to full scale).
  - — Due to the lack of precision in the A.C.–D.C. conversions, the accuracy in the A.C. range is in general of the order of 0.2 to 0.5%. In addition, the measurement range is often limited to about 50 Hz at the lower frequency end due to the ripple in the rectified signal becoming a non-negligible percentage of the display and hence results in fluctuation of the displayed number. At the higher frequency end, deterioration of the performance of the AC/DC converter limits the accuracy.
  - — The A.C. measurement, is *often average reading, r.m.s. calibrated.*

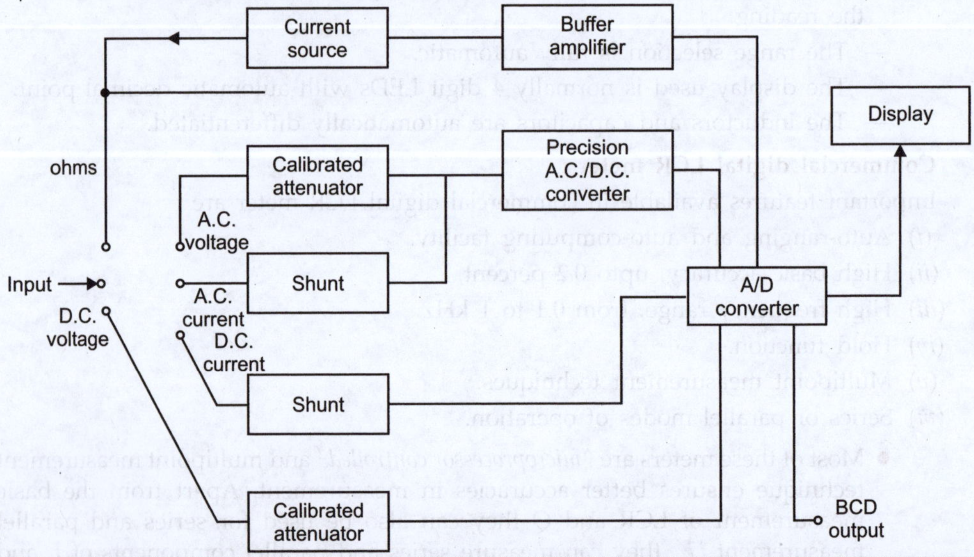

Fig. 8.16. Block diagram of a digital multimeter.

- In the *"resistance range"* the digital multimeter operates by measuring the voltage across the externally connected resistance, resulting from a current forced through it from a calibrated internal current source.
  - — The accuracy of resistance measurement is of the order of 0.1 to 0.5% depending on the accuracy and stability of the internal current sources. The accuracy may be *proper* in the *highest range* which is often about 10 to 20 MΩ. In the *lowest range,* the full scale may be ≃ 200 Ω with a resolution of about 0.01 Ω for a $4\frac{1}{2}$ digit digital multimeter. In this range of resistance

measurement, the effect of the lead resistance will have to be carefully considered.

Comparison between "Analog" and "Digital" Multimeters

S. No.	Aspects	Analog multimeters	Digital multimeters
1.	*Power supply*	Not required	Required
2.	*Visual indication of changes in the reading*	Better	Not that much better
3.	*Effect of electric noise*	Less	More
4.	*Isolation problems*	Less	More
5.	*Accuracy*	Less	High
6.	*Interface of the output with external equipment*	Not possible	Possible
7.	*Construction*	Simple	Complicated
8.	*Size*	Bigger	Smaller
9.	*Cost*	Less	More
10.	*Nature of output*	Ambiguous–many times	Unambiguous reading

8.7 MICROPROCESSOR-BASED INSTRUMENTS–INTELLIGENT INSTRUMENTS

"Intelligent instruments", a new class of instruments, involve the *use of microprocessors (µP) as an integral part of the measuring instruments.*

Fig. 8.17. shows a block diagram of *microprocessor based impedance measuring instrument :*

— The operation makes interface with the instrument via the IEEE 488 bus to allow control by, or to make the measurement available to a large external computer system.

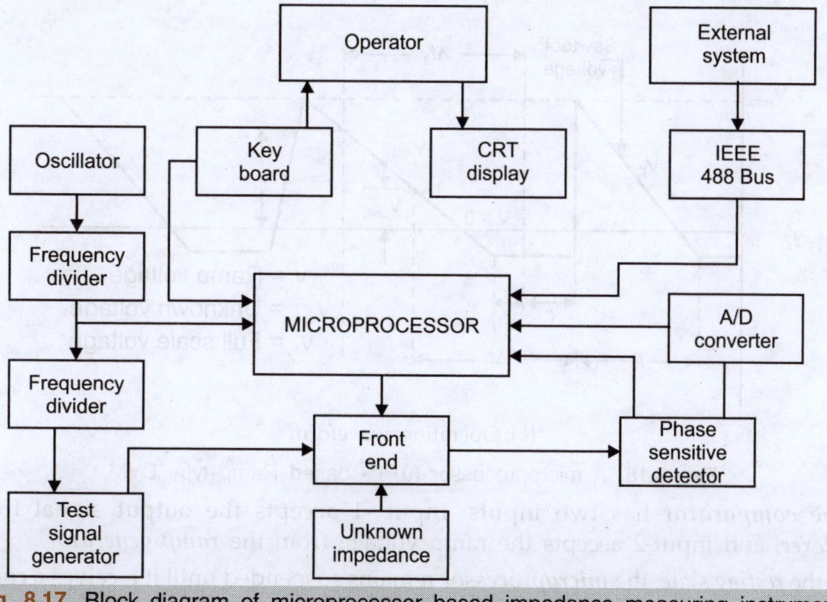

Fig. 8.17. Block diagram of microprocessor based impedance measuring instrument.

— The *frequency division of the oscillator signal* provides the timing clock signal and the A.C. test signal.

— The *front end circuit* applies the test signal to the unknown impedance and a standard impedance provides an output signal, proportional to the voltage across each, to the *phase sensitive detector* (PSD).

— PSD, which is also controlled by the microprocessor, converts the A.C. inputs of the impedance in vector form to a D.C. output.

— The A/D converter provides the digital data, which is used by the microprocessor to compute the value of the unknown impedance. This is then displayed on the CRT or sent as output to the IEEE 488 bus.

8.8 MICROPROCESSOR-BASED RAM TYPE DVM

Fig. 8.18 (a) and (b) shows the basic *block diagram* and *operating waveform* of a micro processor-based Ramp type DVM.

Description: The multiplexer has the following inputs:

(*i*) Input-1. Connected to the ground potential;

(*ii*) Input-2. The unknown input;

(*iii*) Input-3. Reference voltage input

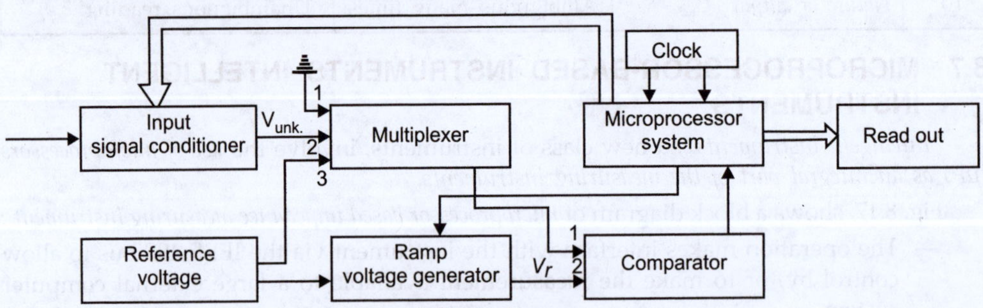

(*a*) Block diagram

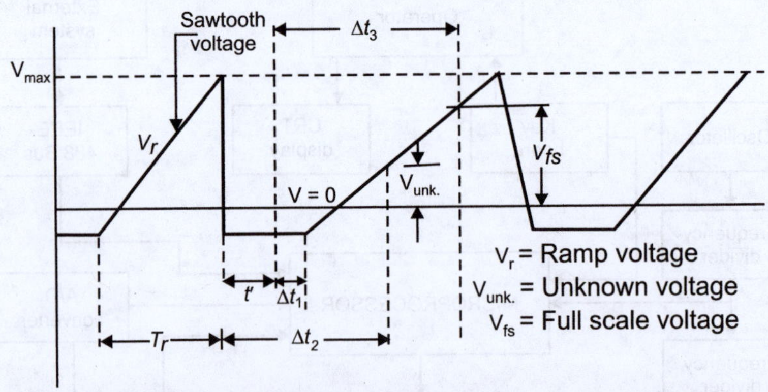

(*b*) Operating waveform

Fig. 8.18. A microprocessor (μP) - based Ramp type DVM

The ***comparator*** has two inputs: Input -1 accepts the output signal from the ***multiplexer***, and input-2 accepts the ramp voltage from the ***ramp generator***.

In the *resting state*, the ***microprocessor*** remains suspended until it receives a command to start conversion. During the resting period, the μP regularly sends reset signals to the *ramp generator*. Whenever the ramp generator is reset, its capacitor discharges. It produces a ramp (i.e. a saw tooth voltage) whose duration T_r, and amplitude V_{max} remain constant.

- As and when a conversion command arrives at the *microprocessor* at a time t_1, the multiplexer first connects Input-1 of the comparator to its Input-1 (*i.e.* ground potential) and brings the former to the ground potential.

– After a pause when Input-2 voltage arriving from the ramp generator becomes equal to Input-1 of the comparator, the comparator sends a signal to the microprocessor, that ramp voltage is zero. The microprocessor measures this time interval Δt_1 [Fig. 518(*b*)] by counting the number of clock pulses supplied by the clock generator during this interval. Let the count be N_1, which is there stored by the microprocessor.

– Then microprocessor sends a command due to which the comparator Input-1 is connected to Input-2 of the multiplexer; this connects the unknown voltage, $V_{unk.}$ to the Input-1 of the comparator. When the ramp voltage equals the unknown voltage, the comparator sends a signal to the microprocessor to measure the time internal Δt_2 . The count N_2, during this interval is also stored.

– Similarly, the next command from the microprocessor causes the comparator Input-1 to be connected to the Input-3 of the multiplexer, which is full scale reference voltage (V_{fs}). At the instant, when the ramp voltage equals V_{fs}, a pulse is sent to the microprocessor from the comparator output to measure this time internal, Δt_3 [Fig. 8.18 (*b*)]. The count N_3, during this interval is also stored.

The unknown voltge, $V_{unk.}$ is determined from the following equation:

$$V_{unk.} = C.\frac{(N_2 - N_1)}{(N_3 - N_1)}$$

where, C = Coefficient, it depends on the characteristics of the instrument and the units selected, and

N_1, N_2, N_3 = The count representing the zero drift, the unknown voltage and full scale voltage respectively.

In this method, the zero drift has practically no effect on the result, on account of the slope variation of the ramp.

Advantages :

1. The instrument's *accuracy is not affected* by the time and temperature instabilities of the element values of the circuit.

2. Good repeatability in switching instants in the presence of noise and interference.

3. The size of scale remains constant due to zero drift connection and maximum count.

Disadvantage : The suppression of noise and interference cannot be achieved.

8.9 IEEE-488 GENERAL PURPOSE INTERFACE BUS (GPIB) INSTRUMENTS

Now-a-days automatic test equipment is one of the leading methods for testing electronic equipment in factory production and troubleshooting situations. *The basic method is to use a programmable digital computer to control a bank of test instruments.* The bank of instruments can be configured for a special purpose or for general use.

The IEEE-488 bus or General purpose interface bus (GDIB) is a tool that is based on the IEEE (Institution of Electrical and Electronics Engineers) specifications.

The digital signals on IEEE-488 bus are generally similar to TTL (transistor-transistor logic), *i.e.,* a logic LOW is less than 0.8 V and a logic HIGH is greater than 2.0 V.

The digital signals can be connected to the instruments through a multiconductor cable upto 20 metres is length provided that an instrument load is placed every 2 metres.

Most IEEE-488 bus/GPIB systems operate unrestricted to 250 bilobytes per second or faster with some restrictions.

There are *two basic configurations* for the IEEE-488/GPIB system :

1. *Linear.* In this type of configuration a tap-off to the next instrument is taken from the previous one in series.
2. *Star.* In this case, the instruments are connected from a central point.

Fig. 8.19. shows the basic structure of IEEE-488/GPIB system. The figure indicates the following four different devices :

(*i*) Computer;
(*ii*) Frequency counter;
(*iii*) Signal generator;
(*iv*) Digital multimeter.

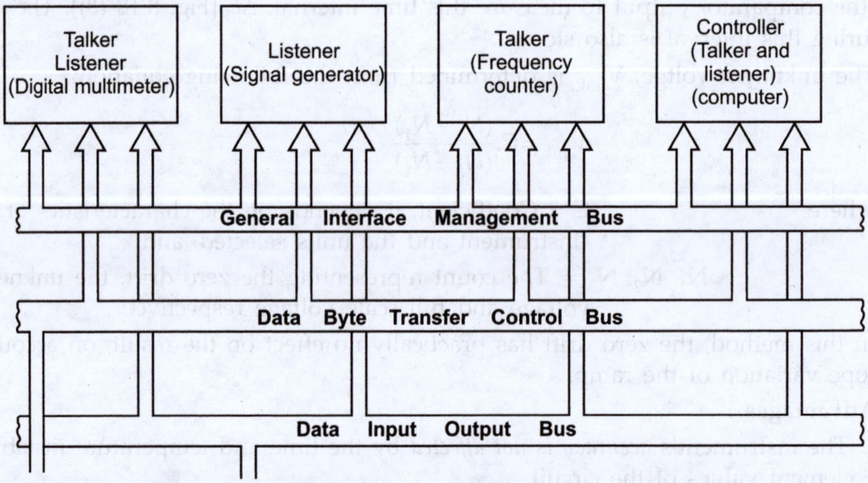

Fig. 8.19. Basic structure of IEEE-488/GPIB system.

The IEEE-488/GPIB system itself consists of *three major buses* :

(*i*) *General Interface Management (GIM) bus.* This bus co-ordinates the whole system and ensures an orderly flow of data over the data input/output (DIO) bus.

(*ii*) *Data I/O (DIO) bus.* This bus is a bi-directional 8-bit data bus that carries data, interface messages and device-dependent messages between the *controllers, talkers* and *listeners.* This bus sends asynchronously in byte-serial format.

(*iii*) *Data Byte Transfer (DBT) bus.* This bus controls the sending of data along the DIO bus.

- The signals defined for the above three buses are implemented as conductors in a system interface cable.

- Each IEEE-488/GPIB system compatible instrument will have a female 36-pin Amphenol-style connector on the rear panel.
- The devices connected to IEEE-488 GPIB system (*i.e.*, computer, frequency counter, signal generator and digital multimeter) are categorised as : *Controller; Listener and/or Talker.*

***(i)* Controller :**

— Its function is to communicate device addresses and other inface buses to instrument in the system.

***(ii)* Listener :**

— Its function is to receive commands from other instruments (usually the controller) when the correct address is placed on the base.

— The listener acts on the message received but does not send back any data to the controller.

***(iii)* Talker :**

— Its function is to respond to the message sent to it by the controller.

- ***V* × *I* bus** is another widely growing platform for instrumentation systems. It uses a mainframe chasis with a maximum of 13 slots to hold *modular instruments on plug-in boards.*

— The *V* × *I* backplane includes the 32-bit VME data bus, high performance instrumentation, buses for precision timing and synchronisation between instrument components, standardised initialisation and resource management to ease configuration.

HIGHLIGHTS

1. A *digital instrument* can be considered as a counter which counts the pulses in a predetermined time.
2. The digital meters work on the principle of *"quantization"*,
3. Digital voltmeters (DVM) are broadly classified as :
   (i) Ramp-type DVM;
   (ii) Dual-slope integrating type DVM;
   (iii) Integrating type DVM (Voltage-to-frequency);
   (iv) Successive-approximation DVM.
4. Intelligent instruments involve the use of microprocessor as an integral part of the measuring instruments.

OBJECTIVE TYPE QUESTIONS

Choose the Correct Answer :

1. The D' Arsonval meter movement can be converted into an audio-frequency A.C. ammeter by adding a .... to it.
   (a) thermocouple *(b)* rectifier
   (c) chopper *(d)* transducer.
2. In a linear meter, half-scale deflection occurs when there is .... per cent of the rated current through its coil
   (a) 100 *(b)* 25 *(c)* 50 *(d)* 75.

3. It is desired to convert a 0–1000 μA meter movement, with an internal resistance of 100 Ω into a 0–100 mA meter. The required value of shunt resistant is about

(*a*) 1 Ω (*b*) 10 Ω
(*c*) 99 Ω (*d*) 100 Ω.

4. Loading effect is principally caused by .... instruments.

(*a*) high resistance (*b*) low-sensitivity
(*c*) high-sensitivity (*d*) high-range.

5. A multimeter is used to measure

(*a*) resistance (*b*) current
(*c*) voltage (*d*) all of the above.

6. Digital instruments are those which

(*a*) have numerical readout (*b*) use LED or LCD displays
(*c*) have a circuitry of digital design (*d*) use deflection type meter movement.

7. The main difference between the electronic and electrical instruments is that an electronic instrument contains.

(*a*) an electronic device (*b*) a transducer
(*c*) a digital readout (*d*) electrons.

8. A sinusoidal voltage of rms value 10 V is applied to a D'Arsonval movement connected in series with a half-wave rectifier. It will show a reading of .... volt.

(*a*) 9 (*b*) 4.5
(*c*) 10 (*d*) 7.7

9. A VTVM produces negligible loading effect on a circuit under test primarily because

(*a*) it virtually draws no current from the circuit
(*b*) of its very high internal resistance
(*c*) it uses high vacuum tubes
(*d*) it is a null deflection instrument.

10. In a 3½ digit voltmeter, the largest number that can be read is

(*a*) 0999 (*b*) 1999
(*c*) 4999 (*d*) 9999.

11. A 3½ digit voltmeter having a resolution of 100 mV can be used to measure maximum voltage of

(*a*) 100 V (*b*) 200 V
(*c*) 1000 V (*d*) 5000 V.

12. The essential elements of an electronic instrument are

(*a*) transducer (*b*) signal conditioner
(*c*) indicating device (*d*) all of the above.

13. The current sensitivity of a meter is expressed in

(*a*) ampere (*b*) ohm/ampere
(*c*) ohm/volt (*d*) ampere/division.

14. The basic meter movement can be converted into an ohmmeter by connecting a .... with it.

(*a*) high resistance in series (*b*) low resistance in parallel
(*c*) battery in series (*d*) battery and a variable resistance in series.

15. A 0–1 mA meter has a sensitivity of

(*a*) 1 kΩ/V (*b*) 1 mA
(*c*) 1 kΩ (*d*) 1000 A.

16. A moving coil instrument has a resistance of 10 Ω and takes 40 mA to produce full-scale deflection. The shunt resistance required to convert this instrument for use as an ammeter of range 0 to 2A is

(*a*) 0.1021 Ω (*b*) 0.2041 Ω

(*c*) 0.2561 Ω (*d*) 0.4210 Ω.

17. A moving coil ammeter has a fixed shunt of 0.02 ohm resistance. If the coil resistance of the meter is 1000 Ω, a potential difference of 500 mV is required across it for full-scale deflection. Under this condition, the current in the shunt would be

(*a*) 2.5 A (*b*) 2.5 A

(*c*) 0.25 A (*d*) 0.25 A.

ANSWERS

1. (*b*)	**2.** (*c*)	**3.** (*a*)	**4.** (*b*)	**5.** (*d*)	**6.** (*c*)	**7.** (*a*)
8. (*b*)	**9.** (*b*)	**10.** (*b*)	**11.** (*c*)	**12.** (*d*)	**13.** (*c*)	**14.** (*d*)
15. (*a*)	**16.** (*b*)	**17.** (*b*)				

THEORETICAL QUESTIONS

1. What is a digital instrument? Explain.
2. Explain briefly the principle on which the digital meters work.
3. What are the building blocks of a digital instrument? Explain briefly.
4. How does a digital system differ from an analog system? Explain.
5. What are the advantages of digital systems/techniques?
6. Enumerate the components of a digital system.
7. What are the advantages of digital implementation?
8. Explain briefly the following :
   (*i*) Binary counters;
   (*ii*) Decimal counting units (DCUs);
   (*iii*) Display devices.
9. Explain with a block diagram the working of a 'frequency counter'.
10. What do you mean by the term 'Period'. How is it measured by a 'Period counter'?
11. Explain briefly the following characteristics of digital meters :
    (*i*) Resolution;
    (*ii*) Sensitivity;
    (*iii*) Accuracy.
12. What is the function of a digital voltmeter?
13. State the advantages of digital voltmeters.
14. State the characteristic features of DVMs.
15. List the applications of DVMs.
16. Give the classification of digital voltmeters.
17. Explain briefly with neat diagrams the working of any two of the following instruments :
    (*i*) Ramp-type DVM;
    (*ii*) Dual-slope integrating type DVM;
    (*iii*) Integrating type DVM (Voltage-to-frequency conversion)
18. With the help of a block diagram, describe briefly the working of a successive-approximation digital voltmeter.
19. What is the function of a digital LCR meter? Explain its working with the help of a neat block diagram.
20. State the important features available in a commercial digital LCR meter.
21. Describe briefly a "digital multimeter" with the help of a block diagram.
22. Give the comparison between analog and digital multimeters.
23. Give the description of a microprocessor based impedance measuring instruments.
24. Discuss briefly the working of IEEE-488 bus system.

CHAPTER

9 Potentiometers

9.1. Introduction; 9.2. D.C. potentiometers – Simple D.C. potentiometer – Direct reading potentiometer – Crompton potentiometer – Practical applications of D.C. potentiometers – Measurement of resistance – Measurement of current – Measurement of high e.m.fs (say 250 V) – Measurement of power – Calibration of voltmeter – Calibration of ammeter – Calibration of wattmeter – Protection of standard cell and galvanometer – Self balancing potentiometers. **9.3. A.C. potentiometers** – Standardisation of A.C. potentiometers – Classification of A.C. potentiometers – Drysdale polar potentiometer – Gall co-ordinate potentiometer – Advantages and disadvantages of A.C. potentiometers – Application of A.C. potentiometers – Measurement of self inductance – Calibration of voltmeter – Calibration of ammeter – Calibration of wattmeter – *Highlights – Objective Type Questions – Theoretical Questions – Unsolved Examples.*

9.1 INTRODUCTION

A **potentiometer** *is an instrument used for measuring and comparing the e.m.fs of different cells and for calibrating and standardising voltmeters, ammeters, etc.*

Since a potentiometer measures voltage, it *can also be used to determine current* simply by measuring the voltage drop produced by the unknown current passing through a known standard resistance.

- Some *significant features/advantages* of potentiometers are :

(*i*) Measurements using comparison methods are capable of *high degree of accuracy* because the result obtained does not depend on the actual deflection of a pointer, as is the case in deflection methods, but only upon the accuracy with which the voltage of the reference source is known.

(*ii*) A potentiometer makes use of a *balance or null condition, no current flows and hence no power is consumed* in the circuit containing the unknown e.m.f. when the instrument is balanced. Thus the determination of voltage by a potentiometer is quite *independent of the source resistance.*

Uses of potentiometer. The following are the *applications/uses of potentiometers* :

1. Measurement of small e.m.fs. (upto 2 volts).
2. Comparison of e.m.fs. of two cells.
3. Measurement of high e.m.fs. (say 250 volts).
4. Measurement of resistance.
5. Measurement of current.
6. Calibration of ammeter.
7. Calibration of voltmeter.

9.2 D.C. POTENTIOMETERS

9.2.1. Simple D.C. Potentiometer

- For the *accurate measurement of potential difference, current and resistance the potentiometer is one of the most useful instruments.*
- *Its principle of action is that an unknown e.m.f. or p.d. is measured by balancing it, wholly or in part, against a known difference of potential.*
- Fig. 9.1, shows a *simple D.C. potentiometer.*

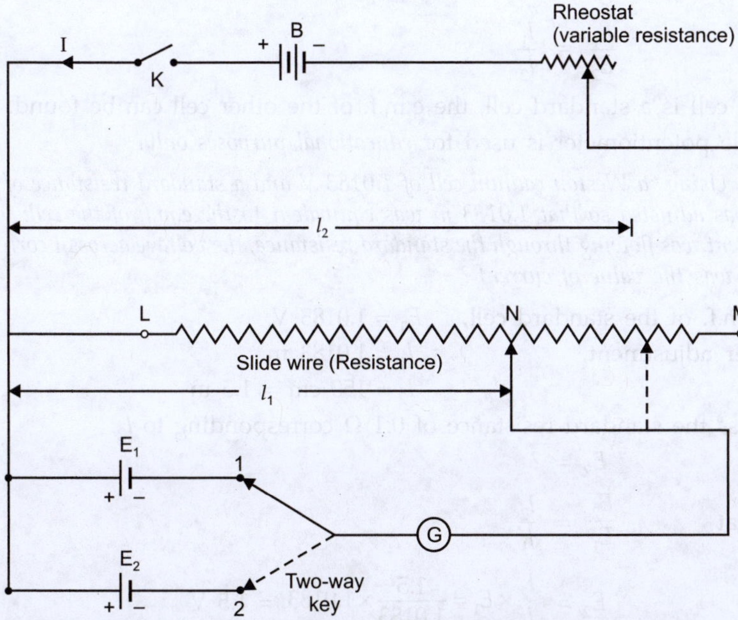

Fig. 9.1. Simple D.C. potentiometer.

Construction :

— In its simplest form, it consists of a German silver or manganin wire usually one metre long and stretched between two terminals as shown.

— This wire is connected in series with a suitable rheostat and battery B which sends a steady current through the resistance wire LM. As the wire is of uniform cross-section throughout, the fall in potential across it is uniform and the drop between any two points is *proportional to the distance between them.*

— As seen, the battery voltage is spread over the rheostat and the resistance wire LM. As we go along LM, there is a progressive fall of potential. If r is the resistance per cm of this wire, l is the length, then for a current I amperes, the fall of potential over the whole length of wire is Irl volts.

Working :

— The two cells whose e.m.fs. are to be compared are joined as shown in Fig. 9.1, always remembering the *positive terminals of the cells and battery must be joined together.*

— The cells can be joined with the galvanometer in turn through a two-way key. The other end of the galvanometer is connected to a movable contact on LM. By this movable contact, a point like *M* is found when there is no current in and hence no deflection of galvanometer G. Then it means that the e.m.f. of the cell just balances the potential fall LN due to the battery current passing through it.

Suppose that the balance or null point for first cell of e.m.f. E_1 occurs at a length l_1 as measured from point L.

Then, $$E_1 = Ir\,l_1 \quad ...(i)$$

Similarly, if the balance point is at l_2 for the other cell,

then, $$E_2 = Ir\,l_2 \quad ...(ii)$$

Dividing (*i*) by (*ii*), we get :

$$\frac{E_1}{E_2} = \frac{l_1}{l_2} \quad ...(9.1)$$

If one of the cell is a standard cell, the e.m.f. of the other cell can be found.

- This simple potentiometer is used for *educational purposes only.*

Example 9.1. *Using a Weston cadium cell of 1.0183 V and a standard resistance of 0.1 Ω a potentiometer was adjusted so that 1.0183 m was equivalent to the e.m.f. of the cell; when a certain direct current was flowing through the standard resistance, the voltage across I correspond to 150 cm. What was the value of current ?*

Solution: E.m.f. of the standard cell, $E_1 = 1.0183$ V

Potentiometer adjustment, $l_1 = 1.0183$ m

and, $l_2 = 150$ cm $= 1.5$ m

Voltage across the standard resistance of 0.1 Ω corresponding to l_2,

$$E_2 = ?$$

We know that $$\frac{E_2}{E_1} = \frac{l_2}{l_1}$$

$$\therefore \quad E_2 = \frac{l_2}{l_1} \times E_1 = \frac{1.5}{1.0183} \times 1.0183 = 1.5 \text{ V}$$

Current flowing through the standard resistance,

$$I_2 = \frac{E_2}{R} = \frac{1.5}{0.1} = \mathbf{15\ A.}\ \textbf{(Ans.)}$$

9.2.2. Direct-reading Potentiometer

The simple potentiometer (described above) in its commercial form is so calibrated that the readings of the potentiometer give the voltage ***directly***, thereby eliminating tedious arithmetical calculations and so saving appreciable time.

Fig. 9.2, shows a *direct-reading potentiometer* :

— The resistance *R* consists of 14 equal resistances joined in series, the resistance of each unit being equal to that of the whole slide wire *S*, which is divided into 100 equal parts.

— The current of the battery *B* is controlled by the slide wire resistance W.

Standardisation of the potentiometer : Refer to Fig. 9.2. The procedure for standardising the potentiometer is as follows :

— A standard cell *i.e.*, Weston cadium cell of e.m.f. 1.0183 V is connected to sliding contacts *L* and *M* through a sensitive galvanometer *G*.

— First, *L* is put on stud 10 and *M* on 18.3 divisions on *S* and then *W* is adjusted for zero deflection in *G*. In that case, potential difference between *L* and *M* is

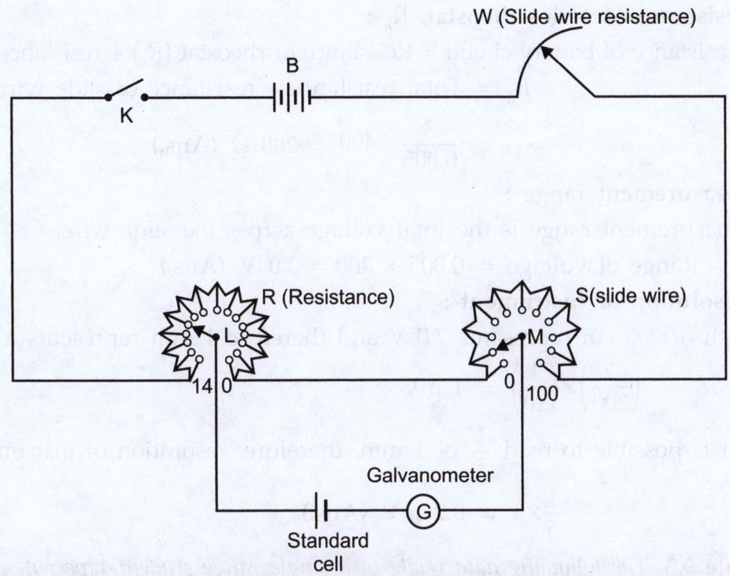

Fig. 9.2. Direct-reading potentiometer.

equal to cell voltage, *i.e.,* 1.0183 V so that potential drop on each resistance of R is $\frac{1}{10} = 0.1$ V and every division of S represents $\frac{0.1}{100} = 0.001$ V. After standardising this way, the *position of W is not to be changed in any case otherwise the whole adjustment would go wrong*. After this the instrument becomes *direct reading*.

Suppose in a subsequent experiment, for balance, L is moved to stud no. 6 and M to 75 divisions, then voltage would be $= (6 \times 0.1) + (75 \times 0.001) = 0.675$ V.

- It may be noted that since most potentiometers have 14 steps on R, it is usually not possible to measure p.ds. exceeding 1.5 V. For measuring *higher voltages* it is necessary to use a ***volt-ratio box.***

Example 9.2. *A basic slide wire potentiometer has a working battery voltage of 3.0 V with negligible resistance. The resistance of the slide wire is 400 Ω and its length is 200 cm. A 200 cm scale is placed along the slide wire. The slide wire has 1 mm scale divisions and it is possible to read upto 1/5 of a division. The instrument is standardised with 1.018 V standard cell with sliding contact at the 101.8 cm mark on scale. Calculate :*

(*i*) *Working current.*

(*ii*) *Resistance of series rheostat.*

(*iii*) *Measurement range.*

(*iv*) *Resolution of instrument.* **(Nagpur University)**

Solution: (*i*) Working current, I_m :

Because the instrument is standardised with an e.m.f. of 1.018 V with sliding contact at 101.8 cm, it is obvious that a length 101.8 cm represents a voltage of 1.018 V.

$$\text{Resistance of 101.8 cm length of wire} = \frac{101.8}{200} \times 400 = 203.6\ \Omega.$$

$$\therefore \quad \text{Working current, } I_m = \frac{101.8}{203.6} = 0.005 \text{ A} \quad \text{or} \quad \mathbf{5\ mA}\ \textbf{(Ans.)}$$

(*ii*) **Resistance of series rheostat, R_h :**

Total resistance of battery circuit = Resistance of rheostat (R_h) + resistance of slide wire

$$\therefore \quad R_h = \text{Total resistance} - \text{resistance of slide wire}$$

$$= \frac{3}{0.005} - 400 = \mathbf{200\ \Omega} \ \textbf{(Ans.)}$$

(*iii*) **Measurement range :**

The measurement range is the total voltage across the slide wire.

$$\therefore \quad \text{Range of voltage} = 0.005 \times 400 = \mathbf{2.0\ V} \ \textbf{(Ans.)}$$

(*iv*) **Resolution of instrument :**

A length of 200 cm represents 2.0 V and therefore 1 mm represents a voltage of

$$\left(\frac{2}{200}\right) \times \left(\frac{1}{10}\right) = 1 \text{ mV}$$

Since it is possible to read $\frac{1}{5}$ of 1 mm, therefore, resolution of instrument is

$$\frac{1}{5} \times 1 = \mathbf{0.2\ mV} \ \textbf{(Ans.)}$$

Example 9.3. *The following data relate to a single range student type potentiometer :*
18 step dial switch–each step represents 0.1 V;
The dial resistors are 10 Ω;
The slide wire of the potentiometer is circular; and has 11 turns and a resistance of 11 Ω each;
The slide wire has 100 divisions and interpolation can be done to one fourth of a division.
The working battery has a voltage of 6.0 V and negligible internal resistance.
Calculate :

(*i*) *Measuring range of instrument.*
(*ii*) *Resolution of instrument.*
(*iii*) *Working current.*
(*iv*) *Setting of rheostat.* **(Roorkee University)**

Solution: (*i*) **Measuring range of instrument :**

Total resistance of measuring circuit, R_m

= Resistance of dial + resistance of slide wire

or, $R_m = 18 \times 10 + 11 = 191\ \Omega$

Since each step of 10 Ω represents a voltage drop of 0.1 V, range of instrument

$$= 191\left(\frac{0.1}{10}\right) = 1.91 \text{ V.}$$

(*ii*) **Resolution of instrument :**

The slide wire has a resistance of 11 Ω and therefore voltage drop across slide wire

$$= 11 \times \left(\frac{0.1}{10}\right) = 0.11 \text{ V}$$

The slide wire has 11 turns and therefore voltage drop across each turn

$$= \frac{0.11}{11} = 0.01 \text{ V}$$

Each turn is divided into 100 divisions and therefore each divisions represents a voltage drop of $\frac{0.01}{100} = 0.0001$ V.

Since each turn can be interpolated to $\frac{1}{4}$ of a division,

$\therefore$ *Resolution* of instrument $= \frac{1}{4} \times 0.0001\text{ V} = 0.000025 = \mathbf{25\ \mu V}$ **(Ans.)**

(iii) **Working current, I_m :**

The voltage drop across working circuit of potentiometer is 1.91 V and the resistance is 191 Ω.

$\therefore$ Working current, $I_m = \frac{1.91}{191} = 0.01\text{ A} = \mathbf{10\ mA}$ **(Ans.)**

(iv) **Setting of rheostat :**

Total resistance across battery circuit $= \frac{6}{0.01} = 600\,\Omega$.

The resistance of potentiometer circuit is 191 Ω.

$\therefore$ Resistance of series rheostat, $R_h = 600 - 191 = \mathbf{409\ \Omega}$ **(Ans.)**

9.2.3. Crompton Potentiometer

Fig. 9.3 shows the general arrangement of a crompton potentiometer :

— It consists of a graduated slide-wire AC connected in series with fourteen (or more) coils each of which has a resistance equal to that of the slide-wire (of the order of 10 ohms).

— There are two moving contacts P_1 and P_2 sliding over the slide-wire; and the studs of the resistance coils respectively.

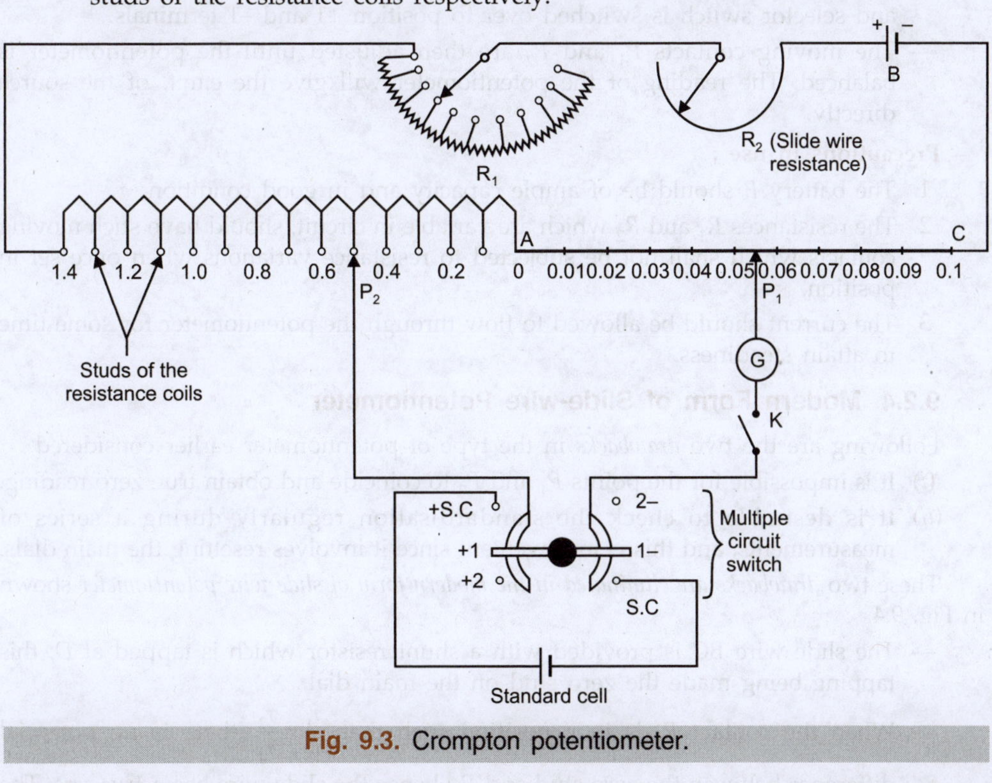

Fig. 9.3. Crompton potentiometer.

— B is the supply battery (2 V), and R_1 and R_2 are two variable resistors, the former consisting of a number of coils for *coarse adjustment* of the potentiometer current, and the latter taking the form of a slide-wire for fine adjustment.

— The galvanometer G is connected in series with a key K, and multiple circuit switch, by means of which either the standard cell or e.m.f. to be measured can be connected in the galvanometer circuit.

— The terminals to which the apparatus under test is connected are marked *+ve* and *–ve* to avoid the possibility of damage to the potentiometer due to the wrong polarity being used. The standard cell and supply battery terminals are marked similarly.

Standardisation. Before putting the potentiometer in use, it is standardised, *i.e.*, made direct reading by adjustment of the current from the supply battery as follows :

— A Weston type standard cell S.C., whose e.m.f. is 1.0183 V is connected to the terminals of selector switch marked as S.C.

— The moving contact P_2 is set at 1.00 stud of the coils and moving contact P_1 is set at 0.0183 on the slide wire.

— Then the resistances R_1 and R_2 are so adjusted that there is no deflection in the galvanometer when they key K is pressed. The potentiometer is now said to be *standardised*. After standardisation the positions of R_1 and R_2 *should not be disturbed in any case* otherwise the whole adjustment would go wrong.

Measurement of unknown e.m.fs. The unknown e.m.fs. can be measured as follows:

— The source whose e.m.f. is to be measured is connected to terminals +1 and –1 and selector switch is switched over to position +1 and –1 terminals.

— The moving contacts P_1 and P_2 are then adjusted until the potentiometer is balanced. The reading of the potentiometer will give the e.m.f. of the source directly.

Precautions in use :

1. The battery B should be of ample capacity and in good condition.
2. The resistances R_1 and R_2 which are variable in circuit, should have such moving contacts which shall not be subjected to resistance variations when once set in position.
3. The current should be allowed to flow through the potentiometer for some time to attain steadiness.

9.2.4. Modern Form of Slide-wire Potentiometer

Following are the two *drawbacks* in the type of potentiometer earlier considered :

(*i*) It is impossible for the points P_1 and P_2 to coincide and obtain true zero reading.

(*ii*) It is desirable to check the standardisation regularly during a series of measurements, and this is inconvenient since it involves resetting the main dials.

These two *drawbacks are eliminated in the modern form of slide wire potentiometer* shown in Fig. 9.4.

— The slide wire BC is provided with a shunt resistor which is tapped at D, this tapping being made the zero stud on the main dial.

— When the contact P_1 is in a position such that $\frac{r_1}{r_3} = \frac{r_2}{r_4}$ there is no potential difference between the zero stud and P_1; hence the slide-wire has a *true zero*. The

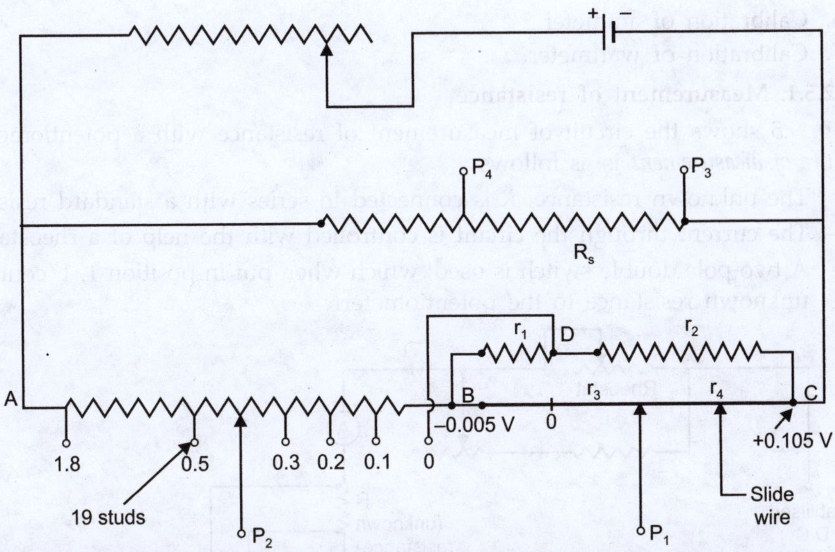

Fig. 9.4. Circuit diagram for modern form of slide-wire potentiometer.

slider can travel a short distance to the left of the zero position, giving a small negative reading, and movement to the right gives positive readings. The range of the slide wire is usually from –0.005 V to 0.105 V, and the resistance of the slide-wire and its shunt is slightly greater than that of one step on the main dial.

Standardisation :

— Standardisation in this potentiometer is carried out with the aid of the shunt resistor R_s between A and C. This resistor is tapped at points P_3 and P_4, which are so chosen that a potential difference equal to the e.m.f. of a standard cell exists between them when the potentiometer is correctly standardised. Thus the potentiometer may be standardised independently of the main dial by connecting a galvanometer and standard cell between P_3 and P_4 and adjusting the potentiometer current until balance is achieved. This circuit is known as an *independent standardising circuit.*

— Sometimes for quick standardisation, instead of shunt resistor R_s, a *standard coil* in series with the slide wire is provided. The resistance of this coil is so chosen that the current flowing in the slide wire in standardised position gives the voltage drop across the standard coil *equal to the e.m.f. of the standard cell.* Now the potentiometer can be standardised independently of the main dial by connecting a galvanometer and standard cell across this coil terminals and then adjusting the potentiometer current with rheostat until balance is obtained.

9.2.5. Practical Applications of D.C. Potentiometers

Some of the important applications of D.C. potentiometers are enumerated and described below :

1. Measurement of resistance.
2. Measurement of current.
3. Measurement of high e.m.fs. (say 250 V).
4. Measurement of power.
5. Calibration of voltmeter.

6. Calibration of ammeter.
7. Calibration of wattmeter.

9.2.5.1. Measurement of resistance

Fig. 9.5 shows the circuit of measurement of resistance with a potentiometer. The *procedure of measurement* is as follows :

— The unknown resistance, R is connected in series with a standard resistor S.

— The current through the circuit is controlled with the help of a rheostat.

— A two-pole double switch is used; which when put in position 1, 1′ connects the unknown resistance to the potentiometer.

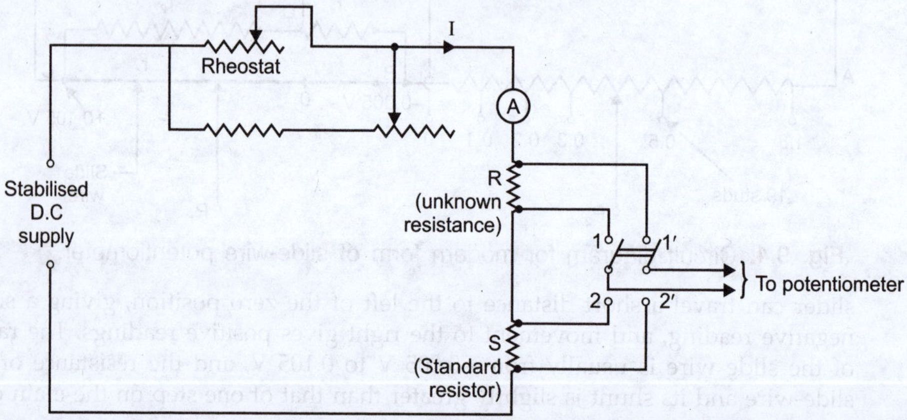

Fig. 9.5. Measurement of resistance with potentiometer.

If the reading of the potentiometer is V_R, then,

$$V_R = IR \quad \text{...(i)}$$

Now the switch is thrown to position 2, 2′, this connects the standard resistor S to the potentiometer. If the reading of potentiometer is V_S, then,

$$V_S = IS \quad \text{...(ii)}$$

Dividing (*i*) by (*ii*), we get

$$\frac{V_R}{V_S} = \frac{IR}{IS}$$

or,

$$R = \frac{V_R}{V_S} \cdot S \quad \text{...(9.2)}$$

The value of R can be accurately known since the value of standard resistance S is accurately known.

Following points are *worth noting* :

(*i*) It is desirable that the current flowing through the circuit be so adjusted that the value of voltage drop across each resistor is of the order of 1 V.

— The ammeter inserted in the circuit is merely for indicating that whether the current flowing through the circuit is within the capacity of the resistors or not otherwise the exact value of current flowing need not not be known.

(*ii*) The resistance of the *standard resistor* S, which must be accurately known, should be of the *same order as the resistance*, R under the measurement.

- The potentiometer method of measurement of resistance is *suitable for measurement of "low resistances"*.

9.2.5.2. Measurement of current

Fig. 9.6 shows the circuit of measurement of current with a potentiometer. The *procedure* is as follows :

— The unknown current *I*, whose value is to be determined, is passed through a standard resistor *S* as shown. The standard resistance should be of such a value that *voltage drop* across it caused by flow of current to be measured, may *not exceed* the range of the potentiometer.

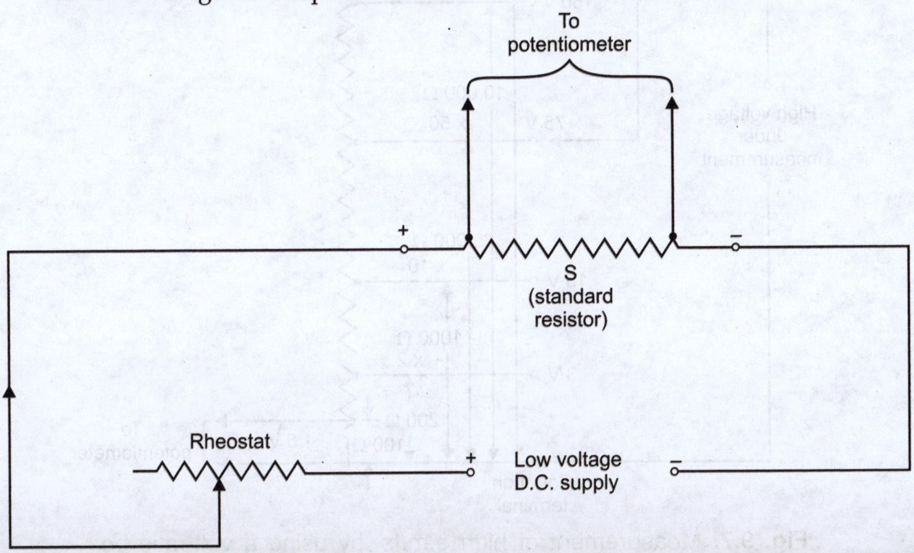

Fig. 9.6. Measurement of current with potentiometer.

— Voltage drop across *S* measured in volts divided by the value of *S* in ohms gives the value of unknown current in amperes.

i.e., Unknown current (in amps.), $I = \dfrac{\text{Voltage drop across } S \text{ (measured in volts)}}{\text{Value of } S\text{, in ohms}}$ ...(9.3)

9.2.5.3. Measurement of high e.m.fs. (say 250 V)

A potentiometer *cannot* be used to measure high e.m.fs. of the order of 250 V *directly*, as it is quite beyond its range.

The voltage above the direct range of potentiometer (normally 1.5 V) can be measured by using a *volt-ratio box* in conjunction with the potentiometer. The volt-ratio box consists of a simple resistance potential divider with various tappings on the input side, as shown in Fig. 9.7. Each input terminal is marked with the maximum voltage which can be applied and with the corresponding multiplying factor for the potentiometer scale.

The *procedure* of measurement is as follows :

— High e.m.f. to be measured is applied across the suitable input terminals of volt-ratio box and leads to the potentiometer are taken from *two tapping points* meant for this purpose.

— The p.d. across these two points is measured on potentiometer as earlier described.

Let, v = Potential difference (p.d.) measured by potentiometer, and

k = Multiplying factor,

Then, Voltage to be measured, $V = kv$ volts ...(9.4)

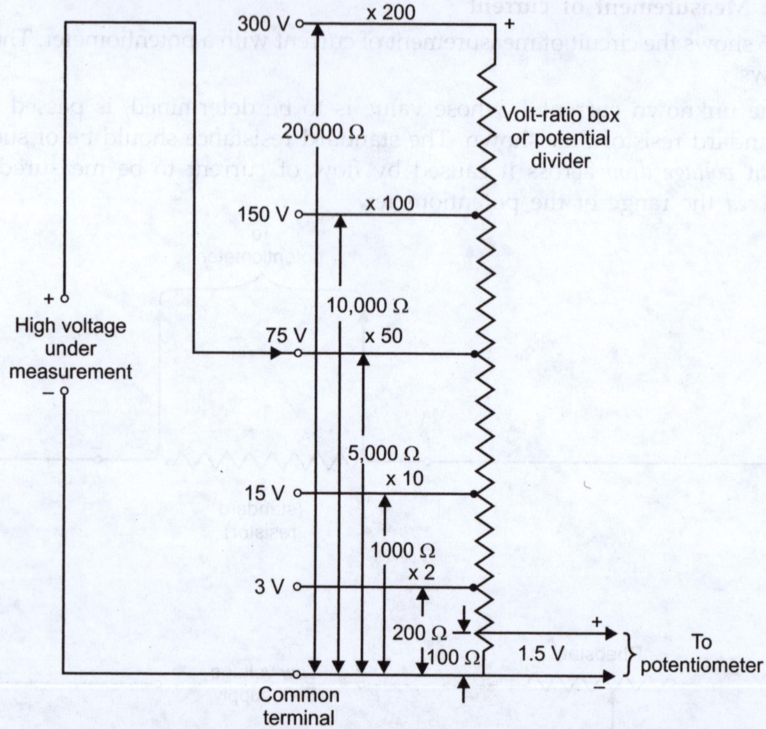

Fig. 9.7. Measurement of high e.m.fs. by using a volt-ratio box in conjunction with the potentiometer.

9.2.5.4. Measurement of power

Fig. 9.8, shows the circuit for measurement of power with a potentiometer. The *procedure* of measurement is as follows :

— The measurements are made, one across the standard resistance *S*, connected in series with the load and other across at output terminals of volt-ratio box.

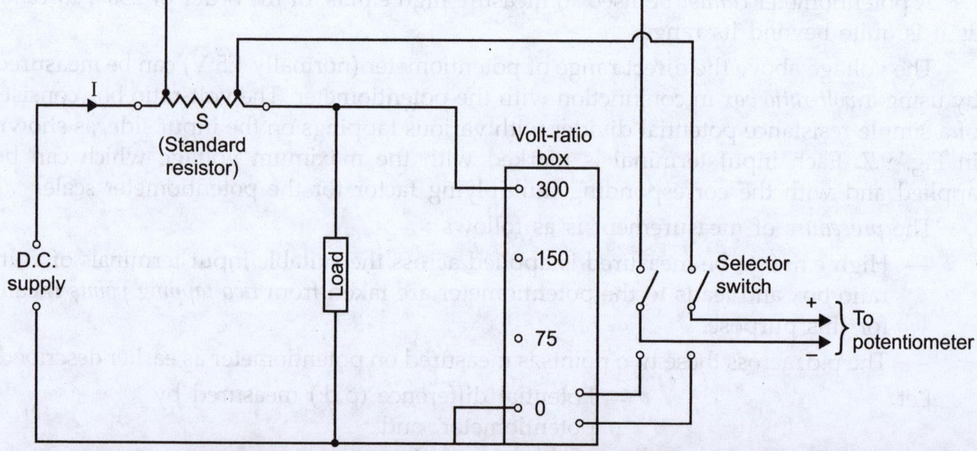

Fig. 9.8. Measurement of power with potentiometer.

— The current is computed from the voltage drop across the standard resistance and the voltage across the load is computed from the potentiometer reading across the output terminals of volt-ratio box.

If V_S is the reading of potentiometer when connected across S and V_R is the reading of potentiometer when connected across volt-ratio box, then :

$$\text{Current through load, } I = \frac{V_S}{S}, \text{ and}$$

$$\text{Voltage across load, } V = k\,V_R, \quad \text{where } k \text{ is the multiplying factor of volt-ratio box.}$$

$$\therefore \quad \text{Power consumed, } P = VI = kV_R \times \frac{V_S}{S} \qquad \text{...(9.5)}$$

9.2.5.5. Calibration of voltmeter

Fig. 9.9 shows the circuit for calibration of a voltmeter. The *procedure* of calibration is as following :

— The primary requirement in this calibration process is that a *suitable stable D.C. voltage supply* is available otherwise any changes in the supply voltage will cause a corresponding change in the calibration of voltmeter.

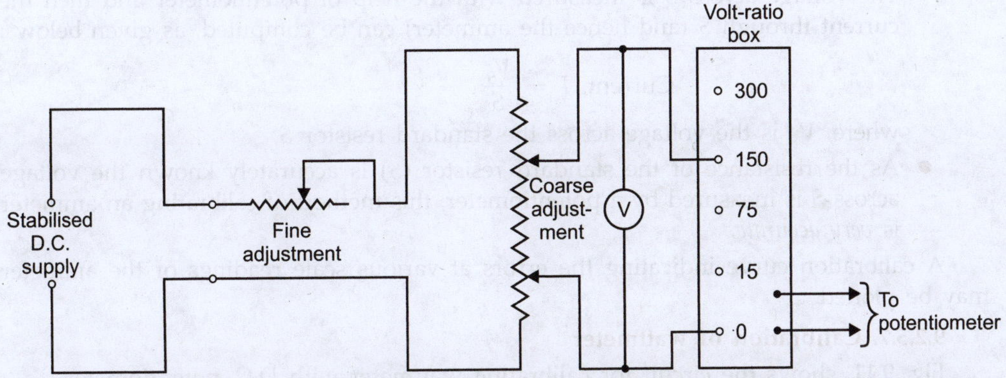

Fig. 9.9. Calibration of voltmeter with potentiometer.

— The potential divider network consists of two rheostats, one for *coarse* and the other for *fine* control of calibrating voltage. These controls are connected to the supply source and with the help of these controls it is possible to *adjust the voltage* so that the pointer coincides exactly with a major division of the voltmeter.

— The voltage across the voltmeter is stepped down to a value suitable for application to a potentiometer, with the help of a volt-ratio box.

In order to get accurate measurements, it is necessary to *measure voltages near the maximum range of the potentiometer*, as far as possible.

— The potentiometer measures the *true value of voltage*. If the reading of the potentiometer does not agree with the voltmeter reading, a *–ve* or *+ve* error is indicated. A calibration curve may be drawn with the help of the readings of voltmeter and potentiometer.

9.2.5.6. Calibration of ammeter

Fig. 9.10. shows the circuit for calibrating an ammeter. The *procedure* of calibration is as follows :

— A standard resistor, S, of suitable value and sufficient current carrying capacity is placed in *series* with the ammeter under calibration.

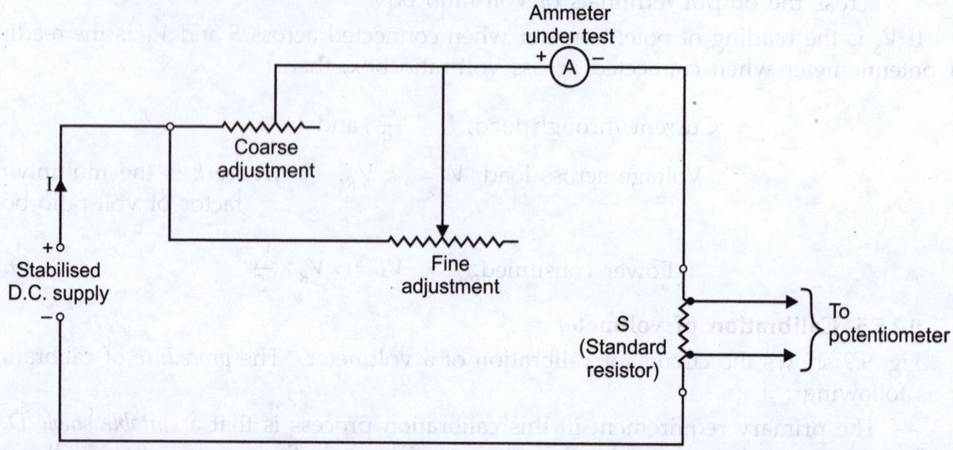

Fig. 9.10. Calibration of an ammeter with potentiometer.

— The voltage across S is measured with the help of potentiometer and then the current through S (and hence the ammeter) can be computed, as given below :

$$\text{Current, } I = \frac{V_S}{S}$$

where, V_S is the voltage across the standard resistor S.

- As the resistance of the standard resistor (S) is accurately known the voltage across S is measured by a potentiometer, this method of calibrating an ammeter is *very accurate*.

A calibration curve indicating the errors at various scale readings of the ammeter may be plotted.

9.2.5.7. Calibration of wattmeter

Fig. 9.11. shows the circuit for calibrating wattmeter with D.C. potentiometer :

— The current coil (C.C.) of wattmeter under calibration is supplied from low voltage supply and potential coil (P.C.) from the normal supply through potential divider.

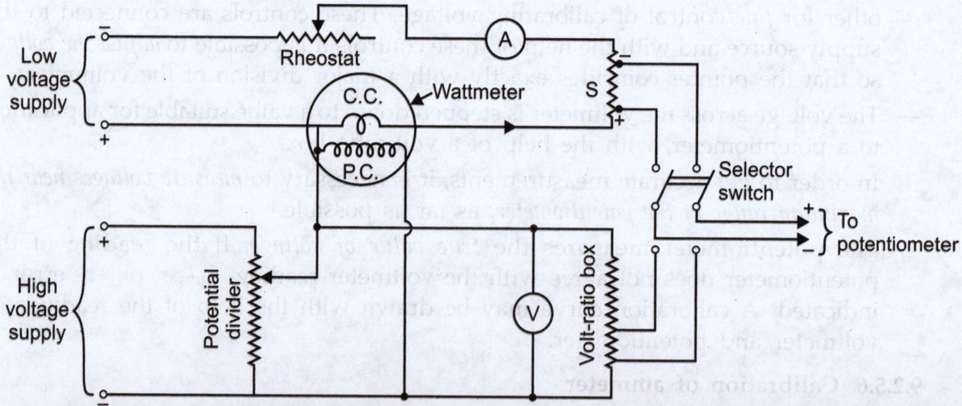

Fig. 9.11. Calibration of wattmeter with D.C. potentiometer.

— The voltage V across the potential coil and the current I through current coil are measured in turn by the potentiometer employing a volt-ratio box and standard resistor, S, as shown.

The true power is then VI and the wattmeter reading may be compared with this value.

This method of calibration of wattmeter is quite *precise*.

Example 9.4. *A simple slide wire is used for the measurement of current in a circuit. The voltage across a standard resistance of 1.0 Ω is balanced at 75 cm. Find the magnitude of the current if the standard cell having an e.m.f. of 1.45 volts is balanced at 50 cm.* **(AMIE)**

Solution: *Given* : $S = 1.0\ \Omega$; $l_1 = 75$ cm; $E = 1.45$ volts; $l_2 = 50$ cm.

Magnitude of current, *I* :

Voltage per cm length of potentiometer wire,

$$v = \frac{E}{l_2} = \frac{1.45}{50} = 0.029 \text{ V}$$

The p.d. which balances at 75 cm,

$$V = v \times l_1 = 0.029 \times 75 = 2.175 \text{ V}$$

$$\therefore \quad \text{Magnitude of current, } I = \frac{2.175}{1.0} = \mathbf{2.175\ A} \text{ (Ans.)}$$

Example 9.5. *The e.m.f. of standard cell used for standardisation is 1.0183 volts. If the balance is obtained at a length of 55 cm, determine :*

(i) The e.m.f. of the cell which balances at 70 cm.

(ii) The current flowing through a standard resistance of 2 Ω if the p.d. across it balances at 65 cm.

(iii) The voltage of a supply main which is reduced by a volt-ratio box to one hundredth and balance is obtained at 85 cm.

(iv) The percentage error in ammeter reading 0.35 ampere when balance is obtained at 45 cm with the p.d. across a 2.5 Ω resistor in the ammeter circuit.

(v) The percentage error in a voltmeter reading 1.42 volts when balance is obtained at 75 cm.

Solution: *Given* : E.m.f. of the standard cell = 1.0183 V

The voltage drop per cm length of potentiometer wire,

$$v = \frac{1.0183}{55} = 0.0185145$$

(i) The e.m.f. of a cell which balances at 70 cm

$$= v \cdot l = 0.0185145 \times 70 = \mathbf{1.29601\ volts} \text{ (Ans.)}$$

(ii) The potential difference which balance at 65 cm

$$= v \cdot l = 0.0185145 \times 65 = 1.20344 \text{ V}$$

Magnitude of the standard resistor, $S = 2\ \Omega$

$$\therefore \quad \textit{Current flowing through 2 Ω resistance} = \frac{V}{S} = \frac{1.20344}{2} \text{ volts} = \mathbf{0.60172\ A} \text{ (Ans.)}$$

(iii) The p.d. which balances at 85 cm,

$$V = v \cdot l = 0.0185145 \times 85 = 1.57373 \text{ volts (Ans.)}$$

$$\therefore \quad \textit{Voltage of supply main} = V \times \text{Ratio of volt-ratio box}$$

$$= 1.57373 \times 100 = \mathbf{157.373\ volts} \text{ (Ans.)}$$

(*iv*) The p.d. which balances at 45 cm,

$$V = v \,.\, l = 0.0185145 \times 45 = 0.83315 \text{ volts}$$

Current through 2.5 Ω resistance,

$$I = \frac{V}{S} = \frac{0.83315}{2.5} = \textbf{0.33326 A} \text{ (Ans.)}$$

∴ *Percentage error in ammeter reading*

$$= \frac{0.35 - 0.33326}{0.33326} \times 100 = \textbf{5\% high} \text{ (Ans.)}$$

(*v*) The p.d. which balances at 75 cm

$$= 0.0185145 \times 75 = 1.38859 \text{ volts}$$

$$\text{Voltmeter reading} = 1.42 \text{ volts}$$

∴ Percentage error in voltmeter reading

$$= \frac{1.42 - 1.38859}{1.38859} \times 100 = \textbf{2.26\% high} \text{ (Ans.)}$$

Example 9.6. *A Crompton's potentiometer consists of a resistance dial having 15 steps of 10 Ω each and a series connected slide wire of 10 Ω which is divided into 100 divisions. If the working current of the potentiometer is 10 mA and each division of slide wire can be read accurately upto $\frac{1}{5}$ of its span, calculate the resolution of the potentiometer in volt.*

(Punjab University)

Solution. Resolution of the potentiometer :

Total resistance of the potentiometer,

$$R = \text{Resistance of the dial + resistance of wire}$$

$$= 15 \times 10 + 10 = 160 \ \Omega$$

$$\text{Working current, } I = 10 \text{ mA} = 0.01 \text{ A}$$

Voltage range of the potentiometer = Working current × total resistance of potentiometer

$$= 0.01 \times 160 = 1.6 \text{ V}$$

Voltage drop across slide-wire = Working current × slide-wire resistance

$$= 0.01 \times 10 = 0.1 \text{ V}$$

Since slide wire has 100 divisions, therefore, each division represents $\frac{0.1}{100}$ or 0.001 V

As each division of slide wire can be read accurately upto $\frac{1}{5}$ of its span, therefore,

$$\textit{Resolution of potentiometer} = \frac{0.01}{5} = \textbf{0.0002 V} \text{ (Ans.)}$$

Example 9.7. *The following readings were obtained during the measurement of a low resistance using a potentiometer :*

Voltage drop across a 0.1 Ω standard resistance = 1.0439 V

Voltage drop across the low resistance under test = 0.4305 V

Calculate the value of unknown resistance, current and power lost in it.

Solution. *Given* : $S = 0.1\ \Omega$; $V_S = 1.0439$ V; $V_R = 0.4305$ V

Resistance of unknown resistor, $R = \frac{V_R}{V_S} \times S$ [Eqn. (9.2)]

$$= \frac{0.4305}{1.0439} \times 0.1 = \mathbf{0.041239\ \Omega} \quad \textbf{(Ans.)}$$

Current through the resistor, $I = \dfrac{V_S}{S} = \dfrac{1.0439}{0.1} = 10.439 \text{ A}$

Power loss in unknown resistance, $P = I^2R = (10.439)^2 \times 0.041239 = \mathbf{4.494\ W}$ **(Ans.)**

Example 9.8. *The following readings were obtained during the measurement of low resistance by means of a potentiometer :*

Voltage drop across a 0.1 Ω standard resistance connected in series with the unknown resistance = 1.03606 V

Voltage drop across low resistance under test = 0.85621 V

The resistance of standard resistor at the temperature of test is 0.10016 Ω. Upon setting the potentiometer dial to zero and breaking the current passing through the unknown resistance, the thermal e.m.f. of latter produced a galvanometer deflection equivalent to 24 μV, the direction of deflection being the same as that produced by an increase of the potentiometer reading during the voltage measurement.

Calculate the unknown resistance. **(Bengalore University)**

Solution. Unknown resistance, *R* :

Indicated voltage drop across the unknown resistance = 0.85621 V

Thermal e.m.f. with unknown resistance = 24 μV = 0.000024 V

∴ Actual voltage drop across unknown resistance,

$$V_R = 0.85621 - 0.000024 = 0.856186 \text{ V}$$

∴ Unknown resistance, $R = \dfrac{V_R}{V_S} \times S = \dfrac{0.856186}{1.03606} \times 0.10016 = \mathbf{0.08277\ \Omega}$ **(Ans.)**

Example 9.9. *The current is measured through a standard resistance of 0.1 ± 0.1 Ω% (standard deviation) by a potentiometer that is accurate to ± 0.0001 V (standard deviation). The voltage across the resistance is measured to be 0.25643 V.*

What is the current and to what accuracy it has been determined ?

Solution. *Current,* $I = \dfrac{V_S}{S} = \dfrac{0.25643}{0.1} = \mathbf{2.5643\ A}$ **(Ans.)**

The fractional standard deviation of current,

$$\frac{\delta I}{I} = \sqrt{\left(\frac{\delta V}{V}\right)^2 + \left(\frac{\delta S}{S}\right)^2}$$

$$= \sqrt{\left(\frac{0.0001}{0.25643}\right)^2 + (0.001)^2} = 0.00107$$

or, *percentage accuracy of measurement of current* = ± 0.00107 × 100 = **± 0.107%** **(Ans.)**

9.2.6. Protection of Standard Cell and Galvanometer

— If the potentiometer is unbalanced, current flows through the galvanometer circuit, which may be excessively high if the working point is far away from balance conditions.

In a standard cell, the current must be limited to 0.1 mA otherwise it will be damaged.

— Normally, a *galvanometer is protected against overcurrents by using a shunt of low resistance.* However, if this practice is followed in a potentiometer, it will *not only fail to protect the standard cell but also increase the current in the standard cell during the standardising process* (since the circuit resistance decreases considerably on account of low resistance put in parallel with the galvanometer). Thus a *series resistance is used to limit the current in both the standard cell and galvanometer. This resistance should be sufficiently large to protect the circuit if the polarity of the standard cell should accidently be reversed.*

Usually, a *series resistance of at least 20,000 Ω be inserted in the standard cell to protect it against overcurrents.*

9.2.7. Self-balancing Potentiometers

The self-balancing potentiometers find wide applications in *"industry"* due to their following *special features* :

1. Their action is automatic and thus the constant attention of an operator is eliminated.
2. They draw a curve of quantity being measured with the help of recording mechanism.
3. They can be mounted on a switchboard or panel and thus acts as monitoring devices for the quantity under measurements.
   - These potentiometers are very useful for *instrumentation purposes.*
   - They are also used in *multiplying and dividing networks in analogue computers and in special function generators.*

9.3 A.C. POTENTIOMETERS

The potentiometer method is an exceedingly useful one for the accurate measurement of alternating currents and voltages, since such measurements are not easily carried out by other methods.

• The principle of A.C. potentiometer is the *same* as that of D.C. potentiometer, the *most important difference in operation being that* :

"Whereas in the D.C. potentiometer *only the magnitudes* of the "unknown" e.m.f. and slide-wire voltage drop must be made equal to obtain balance, in the A.C. potentiometer the *phases of the two voltages, as well as their magnitudes, must be equal for balance to be obtained.*"

This condition obviously necessitates modification of the potentiometer as constructed for D.C. work, and means that the operation is somewhat more complicated.

The following *factors* must be considered for the *operation of an A.C. potentiometer* :

(*i*) In all A.C. potentiometers, the potentiometer circuit must be supplied from the *same source as the voltage or current being measured.*

(*ii*) The sources of A.C. supply should be *free from harmonics* because in the presence of harmonic it may not be possible to achieve a balance. The supply source should be as *sinusoidal as possible.*

(*iii*) There being no reference source (the reference source in D.C. being a standard cell or a Zener source) the absolute accuracy with which an A.C. voltage can be measured in an A.C. potentiometer cannot be comparable with corresponding type of D.C. measurement.

(*iv*) Extraneous or stray e.m.fs. picked up from stray fields or couplings between portions of the potentiometer circuit, which seriously affect the result, must be eliminated, compensated for or measured (since they may add vectorially to the e.m.f. being measured).

- In order to avoid error in readings, the slide wire and resistance coils should be *non-reactive.*

9.3.1. Standardisation of A.C. Potentiometers

— The A.C. potentiometers are made direct reading type, *i.e.*, the readings are read off directly from the dial settings. In order to do this, the A.C. potentiometer must be standardised as is done in case of D.C. instruments.

— The standardisation is done with the help of a standard D.C. source, *i.e.*, a standard cell or a Zener source and a *"transfer instrument"*. This transfer instrument is usually an electrodynamometer milliammeter, so constructed that its response to alternating currents is the same as its D.C. response. Such an instrument *can be calibrated on D.C. and then brought to the same setting on A.C.* Alternatively a *thermocouple type of instrument may be used as a transfer instrument.*

9.3.2. Classification of A.C. Potentiometers

A.C. potentiometers are usually *classified* according to the manner in which the instrument dials (or scales) present the value of unknown voltage. There are two general types of A.C. potentiometers :

1. Polar potentiometers :

— In these potentiometers, the unknown e.m.f. is measured in *"polar form" i.e.*, in *terms of its magnitude and relative phase.* The magnitude is given by one scale and its phase angle with respect to some reference vector is given by the second scale.

— Provision is made for reading phase angles upto 360°.

— The voltage is read in the form $V\angle\theta$ (See Fig. 9.12).

- *Example :* Drysdale instruments.

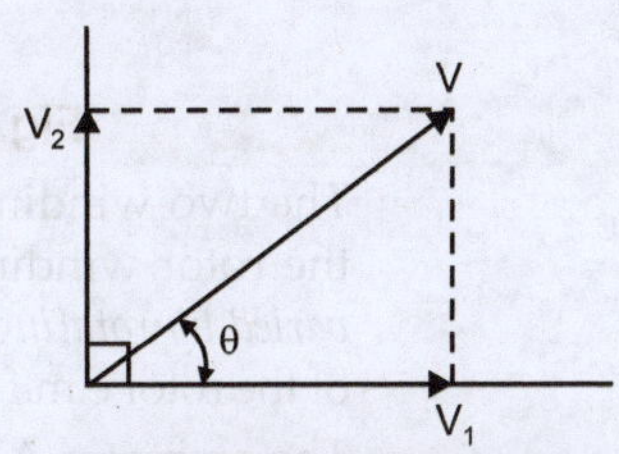

Fig. 9.12.

2. Co-ordinate potentiometers :

— In these instruments, the unknown e.m.f. is measured in *terms of its components along and perpendicular to a standard axis.* The two components of the unknown e.m.f. are given directly by two different scales known as *"in-phase"* (V_1) and *"quadrature"* (V_2) scales.

— Provision is made in these instruments to read both +ve and –ve values of voltages so that all angles upto 360° are covered.

— Voltage, $V = \sqrt{V_1^2 + V_2^2}$; $\theta = \tan^{-1}\left(\dfrac{V_2}{V_1}\right)$

- *Examples :* Gall-Tinsley and Campbell-Larsen type instruments.

9.3.3. Drysdale Polar Potentiometer

Fig. 9.13 shows a Drysdale polar potentiometer :

— For A.C. measurements, the slide wire MN is supplied from a phase shifting circuit so arranged that the *magnitude of the voltage supplied by it remains constant while its phase can be varied through 360°*. Consequently, slide-wire current *I* can be maintained constant in magnitude but varied in phase.

— The *"phase shifting circuit"* consists of two stator coils supplied from the same source in parallel, their currents are made to differ by 90° by using well-known *phase-splitting technique*.

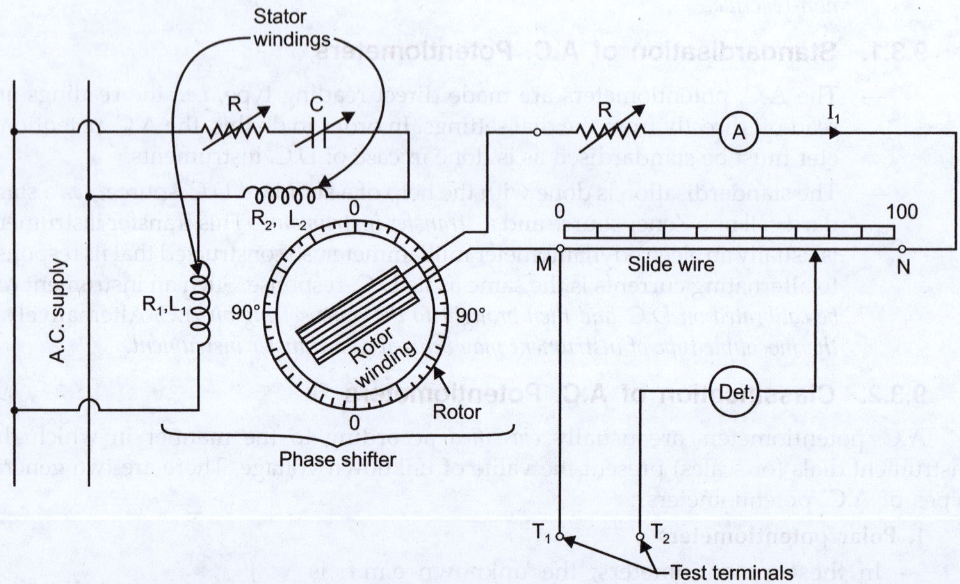

Fig. 9.13. Drysdale polar potentiometer.

The two windings produce rotating flux which induces a secondary e.m.f. in the rotor winding which is of *constant magnitude but the phase of which can be varied by rotating the rotor in any position either manually or otherwise*. The phase of the rotor e.m.f. is read from the circular graduated dial provided for the purpose.

— The ammeter A in the slide-wire circuit is of electrodynamic or thermal type.

— *Before using it for A.C. measurements, the potentiometer is first calibrated by using D.C. supply for slide wire and standard cell for test terminals T_1 and T_2.*

— The unknown alternating voltage to be measured is applied across test terminals T_1 and T_2 and the balance is effected by the *alternate adjustment of the slide-wire contact and the position of phase-shifting rotor. The slide-wire reading represents the magnitude of the test voltage and the voltage phase-shifter reading gives the phase with reference to an arbitrary reference vector.*

9.3.4. Gall Co-ordinate Potentiometer

Fig. 9.14 shows a Gall co-ordinate potentiometer :

— This potentiometer uses two slide-wires *XY* and *MN* with their currents I_1 and I_2 having a *mutual phase difference of 90°*.

— The two currents are obtained from single-phase supply through isolating transformers, the *circuit for "quadrature" slide wire MN incorporating a phase shift arrangement*.

— Before using the instrument, the current I_1 is first standardised as described for Drysdale potentiometer (Art. 9.3.3). Next current I_2 is standardised with the help

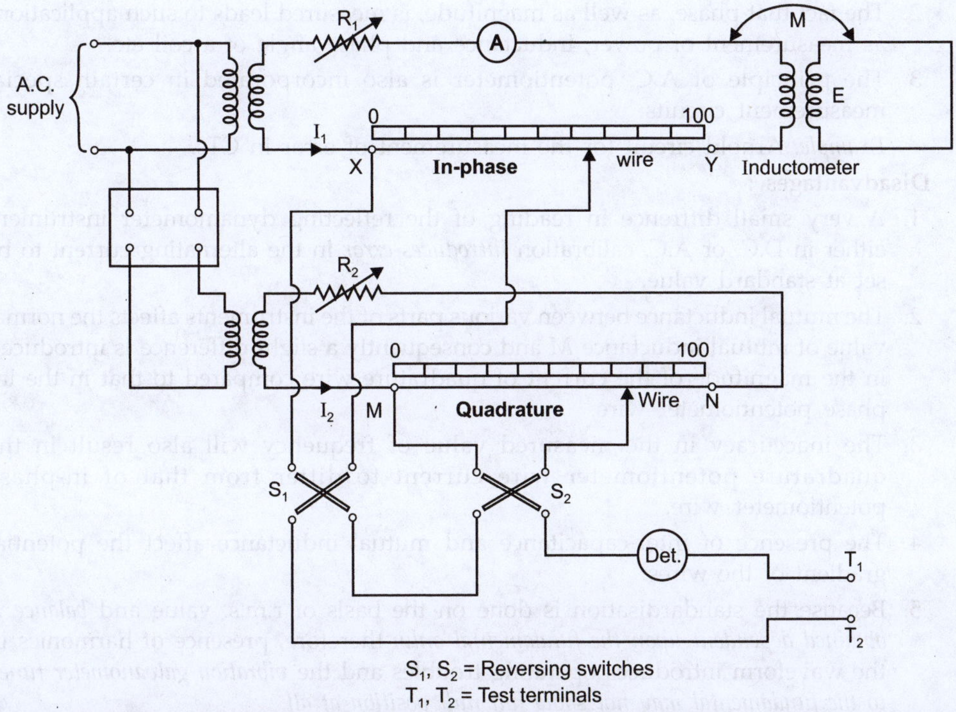

Fig. 9.14. Gall co-ordinator potentiometer.

of the mutually induced e.m.f. in inductometer secondary. This e.m.f. $E = \omega MI_1$, and is in quadrature phase with I_1.

— Now, E is balanced against the voltage drop on the slide-wire MN. This balance will be obtained only when I_2 is of current magnitude and is in exact quadrature with I_1. *Balance is obtained with the help of phase-shifter and rheostat* R_2.

— The *unknown voltage* is applied across the test terminals T_1 and T_2.

— 'Slide-wire MN measures that component of unknown voltage which *is in phase with* I_2. Similarly, slide-wire XY measures that component of the unknown voltage which is *in phase with* I_1. Since I_1 and I_2 are in quadrature, the two measured values are quadrature components of the unknown voltage. If V_1 and V_2 are these values, then :

Unknown voltage, $V = \sqrt{V_1^2 + V_2^2}$, and, $\theta = \tan^{-1}\left(\frac{V_2}{V_1}\right)$ w.r.t. I_1

— Reversing switches S_1 and S_2 are used for measuring both +ve and –ve in-phase and quadrature components of the unknown voltage.

9.3.5. Advantages and Disadvantages of A.C. Potentiometers

A.C. potentiometers entail the following *advantages* and *disadvantages* :

Advantages :

1. *Very versatile instrument.* By employing shunts and volt-ratio boxes its use may be extended to cover current, voltage and resistance measurements over a wide range of values.

2. The fact that phase, as well as magnitude, is measured leads to such applications as measurement of power, inductance and phase angle of a coil etc.
3. The principle of A.C. potentiometer is also incorporated in certain special measurement circuits.

   Example. Arnold circuit for the measurement of error in CTs.

Disadvantages :

1. A very small diffrence in reading of the reflecting dynamometer instrument either in D.C. or A.C. calibration *introduces error* in the alternating current to be set at standard value.
2. The mutual inductance between various parts of the instruments affects the normal value of mutual inductance M and consequently a slight difference is introduced in the magnitude of the current of quadrature wire compared to that in the in-phase potentiometer wire.
3. The inaccuracy in the measured value of frequency will also result in the quadrature potentiometer wire current to differ from that of in-phase potentiometer wire.
4. The presence of inter-capacitance and mutual inductance affect the potential gradient of the wires.
5. Because the standardisation is done on the basis of r.m.s. value and *balance is obtained dependent upon the fundamental only,* therefore, presence of harmonics in the waveform introduces operating troubles and the *vibration galvanometer tuned to the fundamental may not show full null position at all.*

   — The indicating instruments of high quality may give very nearly as high accuracy and be much simpler to use for measurement of current, voltage and power.

   — For the measurement of electrical properties of components, A.C. bridges can be arranged to give higher accuracy and freedom from frequency errors.

9.3.6. Applications of A.C. Potentiometers

Out of the several applications, only the following important applications of A.C. potentiometers will be described here :

1. Measurement of self inductance.
2. Calibration of voltmeter.
3. Calibration of ammeter
4. Calibration of wattmeter.

9.3.6.1. Measurement of self inductance

Fig. 9.15 (a, b) shows the circuit diagram and vector diagram for the measurement of inductance and resistance of a coil under test by using an A.C. potentiometer.

— A standard non-inductive resistor is connected in series with the coil under test, and two potential differences V_1 and V_2 are measured in magnitude and phase by the potentiometer.

Refer to Fig. 9.15(b) :

Voltage drop across standard resistor R_s, $V_2 = IR_s$

where, I = Current flowing through the coil and R_s, and

R_s = Resistance of the standard non-inductive resistor.

or, $$I = \frac{V_2}{R_s} \quad ...(9.6)$$

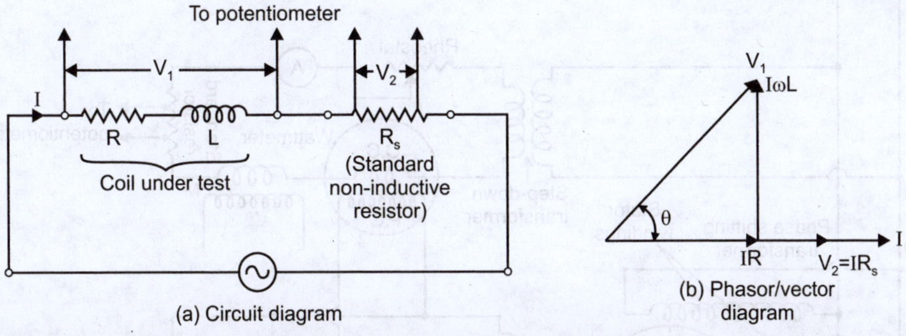

Fig. 9.15. Measurement of self inductance.

Voltage drop across inductive coil = V_1 (By measurement)

Phase angle between current and voltage drop across the coil = θ.

Voltage drop due to resistance of coil, $IR = V_1 \cos\theta$

$$\therefore \quad \text{Resistance of the coil, } R = \frac{V_1 \cos\theta}{I} = \frac{V_1 \cos\theta}{(V_2 / R_s)} = \frac{V_1 R_s \cos\theta}{V_2} \quad \ldots(9.7)$$

Voltage drop due to inductance of coil, $I\omega L = V_1 \sin\theta$

$$\therefore \quad \textit{Inductance of coil, } L = \frac{V_1 \sin\theta}{I\omega} = \frac{V_1 \sin\theta}{\omega(V_2 / R_s)} = \frac{V_1 R_s \sin\theta}{V_2 \omega} \quad \ldots(9.8)$$

- The *accuracy obtained* by this method largely depends on the *accuracy of the potentiometer*.

9.3.6.2. Calibration of voltmeter

The method of calibration of an A.C. voltmeter by using A.C. potentiometer is similar to that adopted for calibration of D.C. voltmeter by using D.C. potentiometer (refer to Art. 9.2.5.5).

9.3.6.3. Calibration of ammeter

The method of calibration of an A.C. ammeter is similar to D.C. potentiometer method for D.C. ammeter *i.e.* A.C. ammeter under calibration is connected in series with a control resistance (non-inductive variable resistor), for varying the current, and a non-inductive standard resistor and voltage drop across standard resistor is measured on A.C. potentiometer.

However, the standardising A.C. potentiometer *involves the use of a suitable transfer instrument.*

9.3.6.4. Calibration of wattmeter

The method of calibration of wattmeter by A.C. potentiometer is similar to D.C. potentiometer method (refer to Art. 9.3.5.7).

Fig. 9.16 shows the circuit for calibration of wattmeter by A.C. potentiometer :

— The current coil (C.C.) of the wattmeter is supplied through a stepdown transformer and the potential coil (P.C.) from the secondary of a variable transformer whose primary is supplied from the rotor of a phase shifting transformer.

— The voltage V across the P.C. of wattmeter and current I through the C.C. of wattmeter are measured by the potentiometer, employing a volt-ratio box and a standard resistor as shown.

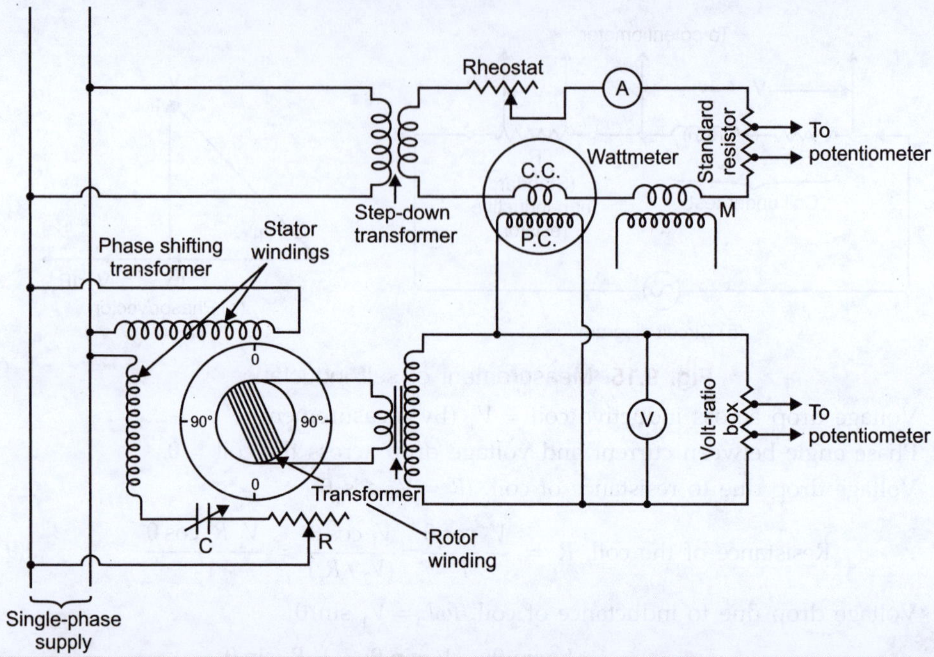

Fig. 9.16. Calibration of wattmeter by A.C. potentiometer.

— The *power factor cos φ is varied by rotating the phase-shift rotor,* the phase angle between voltage and current, φ being given by the reading on the dial of the phase-shifter. The *power is then VI cos φ* and the wattmeter reading may be compared with this value.

— A *small mutual inductance M is included to ensure accuracy of measurement of zero p.f.*

Example 9.10. *The following readings were obtained during the measuement of power by a polar potentiometer :*

Voltage across 0.2 Ω standard resistance in series with the load = 1.52 ∠35°

Voltage across 200 : 1 potential divider across the line = 1.43 ∠53°

Determine the current, voltage, power and power factor of the load.

Solution. (*i*) Current through the load, $\bar{I} = \dfrac{1.52\angle 35^\circ}{0.2} = 7.6\angle 35^\circ$ A

Magnitude of the current, **I = 7.6 A (Ans.)**

(*ii*) Voltage across the load, $\bar{V} = 200\,(1.43\angle 53^\circ) = 286\angle 53^\circ$ V

Magnitude of voltage, **V = 286 V (Ans.)**

(*iii*) Phase angle of load = 53° – 35° = 18°

∴ *Power factor of the load,* **cos φ** = cos 18° = **0.951 lagging (Ans.)**

(*iv*) *Power consumed by the load,*

$$P = VI\cos\phi = 286 \times 7.6 \times 0.951 = \mathbf{2067\ W}$$ **(Ans.)**

Example 9.11. *The following results were obtained for determination of impedance of a coil by using a co-ordinate type potentiometer :*

Voltage across 1.0 Ω resistor in series with coil

= + 0.2404 V in-phase dial and 0.0935 V on quadrature dial

Voltage across 10 : 1 potential divider used with the coil

= + 0.3409 V on in-phase dial and + 0.2343 V on quadrature dial

Calculate the resistance and reactance of the coil.

Solution. Current through the coil, $\bar{I} = \dfrac{(+0.2404 - j\,0.0935)}{1.0} = (0.2404 - j\,0.0935)$ A

Voltage across the coil, $\bar{V} = 10\,(0.3409 + j\,0.2343) = (3.409 + j\,2.343)$ V

$\therefore$ Impedance of the coil, $\bar{Z} = \dfrac{\bar{V}}{\bar{I}} = \dfrac{3.409 + j\,2.343}{0.2404 - j\,0.0935} = \dfrac{4.136\,\angle 34.5°}{0.258\,\angle -21.2°}$

$= 16.03\,\angle 55.7° = 9.03 + j13.24)\ \Omega$

$\therefore$ Resistance of coil, R = **9.03 Ω** **(Ans.)**

Reactance of coil, X = **13.24 Ω** **(Ans.)**

Example 9.12. *The following readings were obtained during measurement of inductance of a coil on an A.C. potentiometer :*

Voltage drop across 0.1 Ω standard resistor conneced in series with the coil

= 0.613 ∠12° 6′ V

Voltage across the test coil through a 100 : 1 volt-ratio box

= 0.781 ∠50° 48′

Frequency = 50 Hz

Determine the value of inductance of the coil. **(U.P. Tech. University)**

Solution. Current through the coil, $\bar{I} = \dfrac{0.613\,\angle 12°6'}{0.1} = 6.13\,\angle 12°\,6'$ A

Voltage across the coil, $\bar{V} = 100 \times 0.781\,\angle 50°\,48' = 78.1\,\angle 50°\,48'$ V

$\therefore$ Impedance of the coil, $\bar{Z} = \dfrac{\bar{V}}{\bar{I}} = \dfrac{78.1\angle 50°48'}{6.13\angle 12°6'} = 12.74\angle 38°42'\ \Omega$

Resistance of the coil, $R = 12.74 \cos 38°42' = 9.94\ \Omega$

Reactance of the coil, $X = 12.74 \sin 38°42' = 7.96\ \Omega$

$\therefore$ Inductance of coil, $L = \dfrac{X}{2\pi f} = \dfrac{7.96}{2 \times \pi \times 50} = 0.0253$ H or **25.3 mH** **(Ans.)**

Example 9.13. *The current taken by a small iron core choke coil is measured by a rectangular co-ordinate A.C. potentiometer. A 1.0 Ω non-inductive resistance is connected in series with the choke coil. The voltages measured across the resistance and the coil are (0.8 – j 0.75) V and (1.2 + j 0.3) V respectively. Determine the iron loss in the coil.*

Assume the voltage and current to be sinusoidal. **(AMIE)**

Solution. Standard resistance, $R_s = 1.0\ \Omega$

Voltage drop across the resistance, $V_s = (0.8 - j\,0.75)$ V

Current through the coil, $\bar{I} = \dfrac{0.8 - j\,0.75}{1.0} = (0.8 - j\,0.75)$ A

$= 1.0966\,\angle -43.15°$

Voltage across the coil, $\bar{V}_{coil} = (1.2 + j\,0.3)$ V $= 1.237\,\angle 14°$ V

Phase angle between voltage across the coil and the current in the coil

$= 14° - (-43.15°) = 57.15°$

Power loss in the coil $= VI \cos \phi$

$= 1.237 \times 1.0966 \times \cos 57.15° = 0.7358$ W

Assuming copper loss in the coil to be negligible and hence taking the total losses occuring in the coil as iron loss, we have :

Iron loss in the coil = **0.7358 W (Ans.)**

Example 9.14. *In calibrating a wattmeter with the help of an A.C. rectangular co-ordinate potentiometer (as shown in Fig. 9.17) using a 300 : 1 volt-ratio box and a 0.1 Ω standard resistor the voltage across the resistor was found to be $V_1 = (0.35 - j\,1.0)$ V and low voltage tap of volt-ratio box gave a reading of $V_2 = (0.80 + j\,0.15)$ V.*

Determine :

(i) True reading of the wattmeter.

(ii) Load power factor. **(AMIE)**

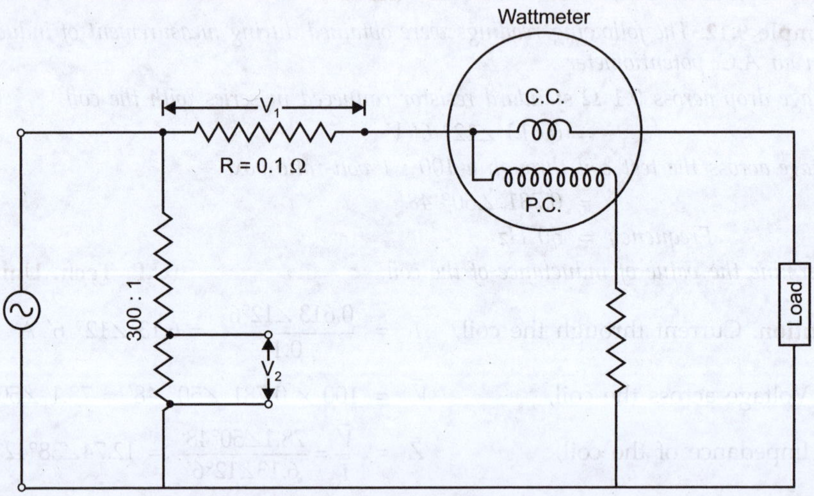

Fig. 9.17.

Solution. ***(i)* True reading of the wattmeter :**

$$\bar{V}_1 = (0.35 - j\,0.1)\text{ V}$$

$$\bar{I} = \frac{\bar{V}_1}{R} = \frac{0.35 - j\,0.1}{0.1} = 3.5 - j\,1.0$$

$$\bar{V}_2 = (0.80 + j\,0.15)\text{ V}$$

Supply voltage, $\bar{V} = 300\ \bar{V}_2 = 300(0.80 + j\,0.15) = (240 + j\,45)$

Now, product of $\bar{V}$ and $\bar{I}$ (Apparent power) $= (\bar{V}) \times (\bar{I}) = (240 + j\,45) \times (3.5 - j\,1.0)$

$= (244.2\ \angle 10.62°)\ (3.64 \angle{-15.94})$

$$888.89\ \angle 26.56° = \underbrace{795.1}_{\text{True or real power}} + \underbrace{j\,397.4}_{\text{Reactive power}}$$

$\therefore$ True reading of the wattmeter (true or real power) = **795.1 W (Ans.)**

***(ii)* Power factor of the load circuit,**

$\cos \phi = \cos 26.56° =$ **0.894 (Ans.)**

HIGHLIGHTS

1. A *potentiometer* is an instrument used for measuring and comparing the e.m.fs. of different cells and for calibrating and standardising voltmeters, ammeters, etc.
2. Simple potentiometer is used for educational purposes only.
3. Some of the important *applications* of D.C. potentiometers are :
   (*i*) Measurement of resistance;
   (*ii*) Measurement of current;
   (*iii*) Measure of high e.m.fs. (say 250 V);
   (*iv*) Measurement of power;
   (*v*) Calibration of voltmeter;
   (*vi*) Calibration of ammeter;
   (*vii*) Calibration of wattmeter.
4. A.C. potentiometers are *classified* as follows :
   (*i*) *Polar potentiometers.* In these potentiometers, the unknown e.m.f. is measured in *polar form i.e.* in terms of its magnitude and relative phase. The magnitude is given by one scale and its phase angle with respect to some reference vector is given by the second scale. Provision is made for relative phase angle upto 360°.
   Example. Drysdale instrument.
   (*ii*) *Co-ordinate potentiometers.* In these potentiometers, the unknown e.m.f. is measured in terms of its components along and perpendicular to a standard axis. The two components of the known e.m.f. are given directly by two different scales known as 'in-phase' and 'quadrature' scales.
   Examples. Gall-Tinsley and Campbell-Larsen type instruments.

OBJECTIVE TYPE QUESTIONS

Choose the Correct Answer :

1. Standardization of potentiometers is done in order that, they become
   (*a*) accurate (*b*) precise
   (*b*) accurate and direct reading (*d*) accurate and precise.
2. A Crompton's potentiometer is provided with a dial having 15 steps of 10 Ω each and a slide wire of 10 Ω resistance. The slide wire is divided into 100 divisions and one fifth of a division can be read with certainty. The working current of potentiometer is 10 mA. The range and resolution of the potentiometer are respectively :
   (*a*) 1.6 V, 0.2 mV (*b*) 1.6 V, 0.5 mV
   (*c*) 1.5 V, 0.1 mV (*d*) 1.6 V, 0.1 mV.
3. The readings of a polar type A.C. potentiometer when measuring the reactance of a coil are $I = 12\angle 13.8°$, $V = 27.8\ \angle 29.7°$. The reactance of the coil is
   (*a*) 2.317 Ω (*b*) 0.634 Ω
   (*c*) 2.22 Ω (*d*) – 2.22 Ω.
4. A co-ordinate type potentiometer is used for determination of impedance of a coil and the results obtained are : voltage across 1 Ω resistor in series with coil, 0.24 V on in-phase dial and – 0.09 V on quadrature dial,
   Voltage across 10 : 1 potential divider used with coil + 0.37 V on in-phase dial and + 0.32 on quadrature dial. The resistance of coil is :

(a) 9.14 Ω (b) 2.34 Ω
(c) 10.76 Ω (d) none of the above.

5. A potentiometer is basically a
(a) deflectional type instrument (b) null type instrument
(c) deflection as well as null type instrument (d) a digital instrument.

6. When a potentiometer is used for measurement of voltage of an unknown source, the power consumed in the circuit of the unknown source under null conditions
(a) is very high (b) is high
(c) is small (d) is ideally zero.

7. A volt-ratio box is designed with a resistance 100 Ω/V. It has input terminals of 300 V, 150 V and 75 V, and an output terminal of 1.5 V which is connected to the potentiometer. Supposing a voltage of 270 V is connected to the 300 V terminal, what is voltage indicated by the potentiometer and what is the power consumption under null conditions ?
(a) 1.5 V, 3 W (b) 1.35 V, 3 W
(c) 1.5 V, 2.43 W (d) 1.35 V, 2.43 W.

8. The standardisation of A.C. potentiometers is done by
(a) directly using A.C. standard voltage source.
(b) using D.C. standard sources and transfer instruments.
(c) using D.C. standard sources and d'Arsonval galvanometer.
(d) using A.C. standard sources and transfer instruments.

ANSWERS

1. (c) 2. (a) 3. (b) 4. (a) 5. (b) 6. (d) 7. (d)
8. (b).

THEORETICAL QUESTIONS

1. What is a potentiometer ?
2. Describe the construction and working of a simple D.C. potentiometer.
3. Describe with the help of a diagram the principle and working of one form of D.C. potentiometer. Explain how this potentiometer is standardised.
4. Give some significant features/advantages of potentiometers.
5. List the applications/uses of potentiometers.
6. Draw a connection diagram of Crompton potentiometer and bring out its salient features. How is it standardised ?
7. Draw a neat sketch of a modern D.C. potentiometer and discuss how the potentiometer is standardised.
8. Draw the circuit of a D.C. potentiometer. Explain how you can calibrate the same against a standard cell. Discuss the effect of room temperature on this calibration.
9. Explain how the potentiometer can be used for measurement of resistance.
10. Name the different types of potentiometer and describe one out of them.
11. Briefly explain how you will calibrate a given D.C. ammeter in the laboratory using a direct reading D.C. potentiometer.
12. Explain how the potentiometer may be used for precise measurement of voltage (240 V D.C.).
13. What is a volt-ratio box ? Explain briefly.

14. Describe with the help of suitable diagrams, how a D.C. potentiometer can be used for :
    (*i*) Calibration of a voltmeter;
    (*ii*) Calibration of an ammeter;
    (*iii*) Calibration of wattmeter;
    (*iv*) Determination of an unknown reistance.
15. What are the applications of self-balancing potentiometers ?
16. How does an A.C. potentiometer differ from a D.C. potentiometer ?
17. What factors must be considered for the operation of an A.C. potentiometer ?
18. What are the problems associated with an A.C. potentiometer ? Describe the working of any one A.C. potentiometer.
19. Name the different forms of A.C. potentiometers and bring out the difference between them.
20. Describe the construction and working of a polar type potentiometer. Explain the method for standardising it.
21. Describe with the help of diagram, the principle and construction of polar potentiometer.
22. Describe the construction and working of a co-ordinate type of potentiometer. Explain the procedure for standardisation of the potentiometer. Describe how this potentiometer can be used for :
    (*i*) Measurement of self inductance of a coil.
    (*ii*) Calibration of a voltmeter and A.C. energy meters.
23. Explain the principle of a rectangular type co-ordinate A.C. potentiometer. Draw a diagram of scheme of connections and describe how the potentiometer is standardised.
24. What are the advantages and disadvantages/limitations of A.C. potentiometers ?
25. Describe briefly the applications of A.C. potentiometer.

UNSOLVED EXAMPLES

1. The e.m.f. of a standard cell used for standardisation is 1.0186 volts. If the balance is obtained at 60 cm determine :
    (*i*) The e.m.f. of the cell which balances at 75 cm.
    (*ii*) The current flowing through a standard resistance of 2 Ω if the p.d. across it balances at 66 cm.
    (*iii*) The voltage of the supply main which is reduced by a volt-ratio box to one hundredth and balance is obtained at 84 cm.
    (*iv*) The percentage error in ammeter reading 0.28 ampere when balance is obtained at 40 cm with p.d. across a 2.5 Ω resistance in the ammeter circuit.
    (*v*) The percentage error in a voltmeter reading 1.25 volts when balance is obtained at 72 cm.

    [**Ans.** (*i*) 1.27325 V; (*ii*) 0.56023 A; (*iii*) 142.604 V; (*iv*) 3% high; (*v*) 2.25% high]
2. The following readings were obtained during the measurement of a low resistance using a potentiometer :

    Voltage drop across a 0.1 Ω standard resistance = 1.0235 V

    Voltage drop across the low resistance under test = 0.4221 V

    Calculate the value of unknown resistance, current and power lost in it.

    [**Ans.** 0.041208 Ω; 10.235 A; 4.316 W]
3. The following readings were obtained during measurement of low resistance by means of a potentiometer :

    Voltage drop across a 0.1 Ω standard resistance connected in series with the unknown resistance = 1.01575 V.

Voltage drop across low resistance under test = 0.83942 V

The resistance of standard resistor at the temperature of test is 0.10014 Ω. Upon setting the potentiometer dials to zero and breaking the current passing through the unknown resistance, the thermal e.m.f. of latter produced a galvanometer deflection equivalent to 23 μV, the direction of the deflection being the same as that produced by an increase of the potentiometer reading during the voltage measurement.

Calculate the unknown resistance. **[Ans.** 0.08275 Ω]

4. The current is measured through a standard resistance of 0.1 ± 0.1 Ω% (standard deviation) by a potentiometer that is accurate to ± 0.0001 V (standard deviation). The voltage across the resistance is measured to be 0.2514 V. To what accuracy current has been determined ?

[Ans. ± 0.108%]

5. The following readings were obtained during the measurement of power by a polar potentiometer :

Voltage across a 0.2 Ω standard resistance in series with the load = $1.46\angle 32°$

Voltage across a 200 : 1 potential divider across the line = $1.37\angle 56°$

Determine the current, voltage, power and power factor of the load.

[Ans. 7.3 A; 274 V; 1827 W; 0.9135]

6. A co-ordinate type potentiometer is used for determination of impedance of a coil and the results obtained are :

Voltage across 1.0 Ω resistor in series with coil = 0.238 V on in-phase dial and

− 0.085 V on quadrature dial

Voltage across a 10 : 1 potential divider used with coil

= + 0.3375 V on in-phase dial and

+ 0.232 V on quadrature dial

Calculate the resistance and reactance of the coil. **[Ans.** R = 9.49 Ω; X = 13.13 Ω]

CHAPTER

10 Measurement of Resistance, Inductance and Capacitance

10.1. Introduction; 10.2. Measurement of resistance – Classification of resistances – **Measurement of *low resistance*** – Ammeter – voltmeter method – potentiometer method – Kelvin double bridge method – Measurement of ***medium resistance*** – Ammeter-voltmeter method – Substitution method – Wheatstone bridge – Carey-foster slide-wire bridge method – Measurement of ***high resistances*** – Direct deflection method – Loss of charge method – Megohm bridge – Megger – Measurement of *"insulation resistance"* when power is *on* – two-wire system – Three-wire system (Russell's method) – Measurement of earth resistance – *"Location of cable faults"* – Murray loop test – Varley loop test – Location of faults in high-volage cables; **10.3. Measurement of Inductance – A.C. bridges** – A.C. bridge-general form – Maxwell bridge – Hay bridge – Anderson bridge – Owen bridge – De sauty bridge – Schering bridge – Heaviside mutual inductance bridge – Carey Foster (Heydweiller) bridge – Campbell bridge – Wien bridge – Resonance bridge – Wagner earthing device – Types of detectors – Sources of errors in bridge circuits and the elimination of errors – *Highlights – Objective Type Questions – Theoretical Questions – Unsolved Examples.*

10.1 INTRODUCTION

For several years the precision measurements of component values have been made by using various forms of *bridges*.

- **Bridge circuits** *are extensively used for measuring component values such as* **R** (*Resistance*), **L** (*Inductance*) and **C** (*Capacitance*).
- A "bridge circuit" in its simplest form consists of a network of four resistance arms forming a closed circuit, with a D.C. source of current applied to two opposite junctions and a current detector connected to the other two junctions.
- Since the bridge circuit merely compares the value of an unknown component with that of an accurately known component (a standard), its measurement accuracy can be very high. This is because the readout of this comparison is based on the null indication at bridge balance, and is essentially independent of the characteristics of the null detector. The *measurement accuracy is, therefore, directly related to the accuracy of the bridge component and not to that of the null detector used.*

Advantages of the bridge circuits are :

(*i*) High measurement accuracy (as the measurement is done by comparing the unknown value with the standard value).

(*ii*) The accuracy is independent of null detector's characteristics (it depends on the values of the components).

(*iii*) The balance equation is independent of the magnitude of the input voltage or its source impedance, the sensitivity and impedance of the null detector or any impedance shunting the detector.

(*iv*) The interchange of the source and detector does not affect the balance condition.

(*v*) The bridge circuit can be used in control circuits.

Types of bridges :

Bridges are of the following two types :

1. ***D.C. bridges :***

- These bridges are used to measure the *"resistances."*
- They use the *D.C. voltage* as the excitation voltage.

Examples : (*i*) Wheatstone bridges.
(*ii*) Kelvin bridges.

2. ***A.C. bridges :***

- These bridges are used to measure the *"impedances"* consisting of inductances and capacitances.
- They use the *alternating voltage* as the excitation voltage.

Examples : (*i*) Maxwell bridge;
(*ii*) Hay bridge;
(*iii*) Schering bridge;
(*iv*) Anderson bridge etc.

10.2 MEASUREMENT OF RESISTANCE

10.2.1. Classification of Resistance

From the point of view of measurement, resistances can be *classified* as follows :

1. ***Low resistances.*** All resistances of the *order of 1 Ω and under* may be classified as low resistances.
   — Such resistances, in practice, may be met with in the *armatures and series winding of large machines, in ammeter shunts, cable lengths, contacts etc.*
2. ***Medium resistances.*** This class includes resistances from about *1 Ω upwards to about 100 kΩ.*
   — In practice, the majority of the pieces of electrical apparatus used have resistances which lie between these limits.
3. ***High resistances.*** Resistances of the order of *100 kΩ and upwards* are classified as high resistances.

Measurements of high resistances are required for determination of :

(*i*) Insulation resistance of components and built-up electrical equipment of all types.

(*ii*) Resistance of high resistance circuit elements.

(*iii*) Volume resistivity and surface resistivity of a material.

10.2.2. Measurement of Low Resistances

Following methods are used for the measurement of *low resistances* :

1. Ammeter-voltmeter method.
2. Potentiometer method.
3. Kelvin double bridge method.

10.2.2.1. Ammeter-voltmeter method

- This method, which is the *simplest* of all, is in *very common use* for the measurement of *low resistances* when an *accuracy of the order of 1% is sufficient.*
  — However, it must be realised that it is essentially a comparatively *rough* method, the accuracy being limited by those of the ammeter and voltmeter used, even if corrections are made for the "shunting" effect of the voltmeter.

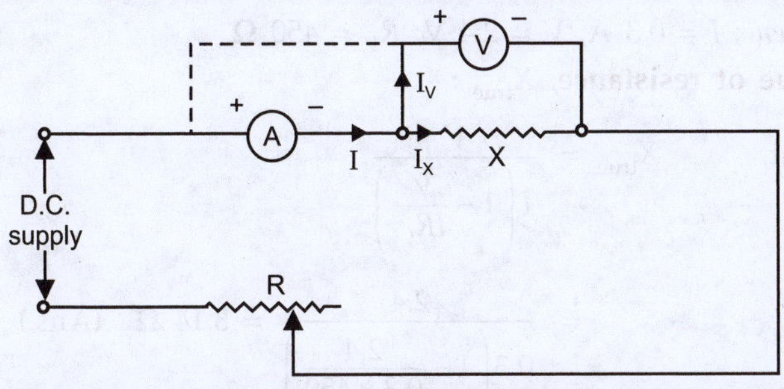

Fig 10.1. Ammeter-voltmeter method of measuring low resistances.

- In this method, current through the resistor (X) under test and the potential drop across it are simultaneously measured. The readings are obtained by ammeter and voltmeter respectively. The attainable accuracy depends primarily on the accuracy and ranges of the instruments employed for measurement of current and voltage.

From Fig. 10.1, it may be observed that there are *two ways* in which the ammeter and voltmeter may be connected for measurement, and that, in either case the resistance of one of the instruments will affect the measurement to some extent.

Case 1 : *When the voltmeter is connected directly across the resistor,* the ammeter measures current flowing through the unknown resistance X and the voltmeter. In this case :

Current through ammeter = Current through unknown resistance (X) + Current through voltmeter

or, $$I = I_X + I_V \quad ...(10.1)$$

or, $$I_X = I - I_V$$

∴ *True value* of unknown resistance,

$$X_{\text{true}} = \frac{V}{I_X} = \frac{V}{I - I_V} = \frac{V}{I - \frac{V}{R_V}} = \frac{V}{I\left(1 - \frac{V}{IR_V}\right)} \quad ...(10.2)$$

where, V = The voltmeter reading,

R_V = The resistance of the voltmeter, and

I = The current indicated by the ammeter.

Case 2 : *When the ammeter is connected so that it indicates only the current flowing through the unknown resistance,* the voltmeter measures voltage drop across the ammeter and unknown resistance X. In this case :

$$V = IR_A + IX = I(R_A + X)$$

or, $$X_{\text{true}} = \frac{V}{I} - R_A \quad ...(10.3)$$

where, R_A = The resistance of the ammeter.

Example 10.1. *The ammeter-voltmeter method is used to measure a resistance. With the voltmeter connected across the resistance the readings on the ammeter and voltmeter are 0.3 A and 2.4 V respectively. The resistance of the voltmeter is 450 Ω.*

Calculate :

(i) True value of resistance.

(ii) Percentage error in the value of resistance, if the voltmeter current is ignored.

Solution. *Given* : $I = 0.3$ A; $V = 2.4$ V; $R_V = 450\ \Omega$

(i) True value of resistance, X_{true} :

We know that,
$$X_{true} = \frac{V}{I\left(1 - \frac{V}{IR_V}\right)} \quad \text{[Eqn. (10.2)]}$$

$$= \frac{2.4}{0.3\left(1 - \frac{2.4}{0.3 \times 450}\right)} = \mathbf{8.14\ \Omega} \quad \textbf{(Ans.)}$$

(ii) Percentage error in the value of resistance :

Measured value of unknown resistance,

$$X_m = \frac{V}{I} \quad \text{....ignoring } I_V$$

$$= \frac{2.4}{0.3} = 8\ \Omega$$

$$\therefore \quad \%\ \text{age error} = \frac{X_m - X_{true}}{X_{true}} \times 100 = \frac{8 - 8.14}{8.14} \times 100$$

$$= -1.72\%\ \textit{i.e.}\ \mathbf{1.72\%\ (low)} \quad \textbf{(Ans.)}$$

10.2.2.2. Potentiometer method

In this method the unknown resistance is *compared with a standard resistance of the same order of magnitude.*

Fig. 10.2 shows the circuit diagram for this method :

— The unknown resistance X, an ammeter A, a rheostat R to limit the current and a standard resistance S are connected, all in series with a *low voltage, high current supply source.* The value of S should be accurately known and of the same order of resistance and must be of the same or higher current rating than the one under test (X).

— The current flowing through the circuit is adjusted so that potential difference, if possible, across each of the resistor is about 1 V.

— The voltage drop across both the unknown resistor X and standard resistor S are measured by a potentiometer. The ratio of the two potentiometer readings gives the ratio of X to S.

Mathematically,
$$\frac{X}{S} = \frac{\text{Potentiometer reading across } X}{\text{Potentiometer reading across } S} = \frac{V_X}{V_S} \quad ...(10.4)$$

As the accuracy of this method depends on there being no change in current between the two readings, therefore, it is necessary that source of supply of current through the circuit be *extremely stable.* When the necessary precautions are taken, and a good potentiometer and sensitive galvanometer are used, very high accuracy can be obtained.

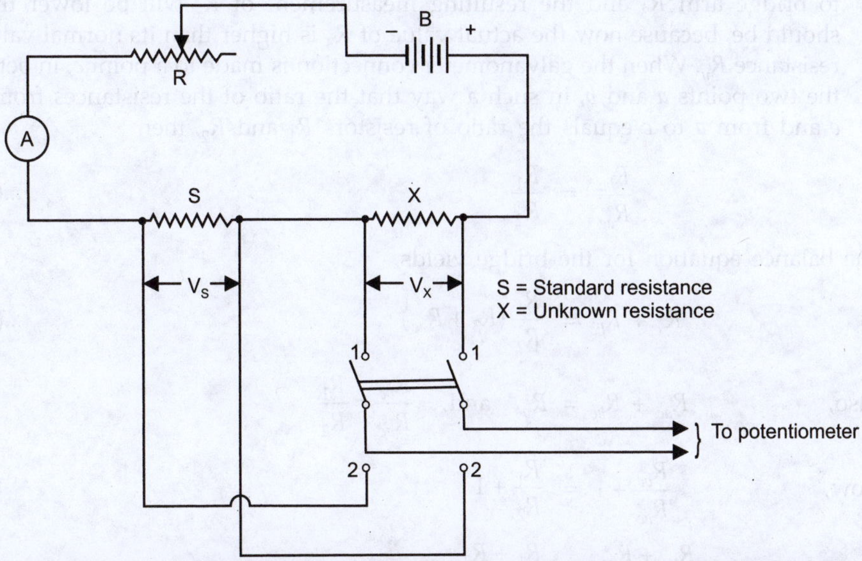

Fig. 10.2. Potentiometer method of measuring low resistances.

Example 10.2. *In a measurement of resistance by potentiometer, the voltage drops across a resistor under test and across 0.024 Ω standard resistor were found to be 0.885 V and 1.1 V respectively. Determine the value of resistor under test.*

Solution. *Given* : $S = 0.024\ \Omega$; $V_X = 0.885$ V; $V_S = 1.1$ V

Resistance of resistor under test, X :

We know that, $$\frac{X}{S} = \frac{V_X}{V_S}$$ [Eqn. (10.4)]

or, $$X = \frac{SV_X}{V_S} = \frac{0.024 \times 0.885}{1.1} = \mathbf{0.0193\ \Omega}$$ **(Ans.)**

10.2.2.3. Kelvin double bridge method

Firstly we shall discuss Kelvin bridge and thereafter Kelvin double bridge method will be described in detail.

Kelvin bridge :

This bridge is a modified version of Wheatstone bridge and provides greatly increased accuracy in the measurement of *low-value resistances*, generally below 1 Ω.

Fig. 10.3 shows the schematic of Kelvin bridge :

— R_y represents the *resistance of connecting lead* from R_3 to R_x

— Two galvanometer connections are available, to point *a* or to point *b*. When the galvanometer is connected to point *a*, the resistance R_y (of connecting lead) is added to the unknown resistor R_x, resulting in too high an indication for R_x. When connection is made to *b*, R_y is added

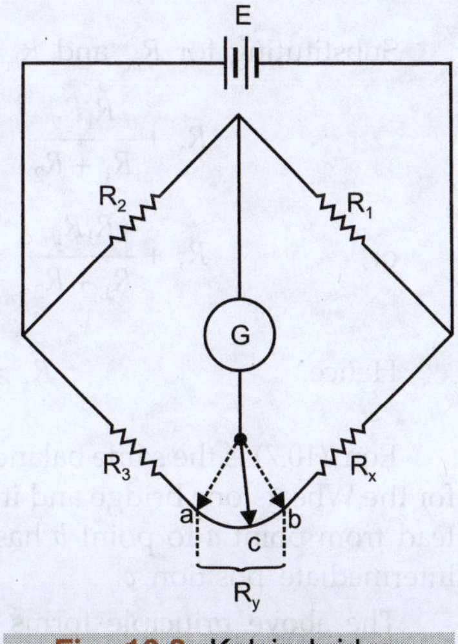

Fig. 10.3. Kelvin bridge.

to bridge arm R_3 and the resulting measurement of R_x will be lower than it should be, because now the actual value of R_3 is higher than its normal value by resistance R_y. When the galvanometer connection is made to a point *c*, in between the two points *a* and *b*, in such a way that the ratio of the resistances from *b* to *c* and from *a* to *c* equals the ratio of resistors R_1 and R_2, then

$$\frac{R_{bc}}{R_{ac}} = \frac{R_1}{R_2} \quad \text{...(10.5)}$$

The balance equation for the bridge yields,

$$R_x + R_{bc} = \frac{R_1}{R_2}(R_3 + R_{ac}) \quad \text{...(10.6)}$$

Also, $R_{ac} + R_{bc} = R_y$, and, $\frac{R_{bc}}{R_{ac}} = \frac{R_1}{R_2}$

Now,
$$\frac{R_{bc}}{R_{ac}} + 1 = \frac{R_1}{R_2} + 1$$

or,
$$\frac{R_{bc} + R_{ac}}{R_{ac}} = \frac{R_1 + R_2}{R_2}$$

or,
$$\frac{R_y}{R_{ac}} = \frac{R_1 + R_2}{R_2}$$

or,
$$R_{ac} = \frac{R_2 R_y}{R_1 + R_2} \quad \text{...}(i)$$

and,
$$R_{bc} = R_y - R_{ac}$$
$$= R_y - \frac{R_2 R_y}{R_1 + R_2}$$

$\therefore$
$$R_{bc} = \frac{R_1 R_y + R_2 R_y - R_2 R_y}{(R_1 + R_2)} = \frac{R_1 R_y}{R_1 + R_2} \quad \text{....}(ii)$$

Substituting for R_{ac} and R_{bc} in eqn. (10.6), we get

$$R_x + \frac{R_1 R_y}{R_1 + R_2} = \frac{R_1}{R_2}\left(R_3 + \frac{R_2 R_y}{R_1 + R_2}\right)$$

or,
$$R_x + \frac{R_1 R_y}{R_1 + R_2} = \frac{R_1 R_3}{R_2} + \frac{R_1 R_y}{(R_1 + R_2)}$$

Hence,
$$R_x = \frac{R_1 R_3}{R_2} \quad \text{...(10.7)}$$

Eqn. (10.7) is the same balance equation [Eqn. (10.24), suffices to be noted] as developed for the Wheatstone bridge and it indicates that the effect of the resistance of the connecting lead from point *a* to point *b* has been eliminated by connecting the galvanometer to the intermediate position *c*.

The above principle forms the basis for construction of the Kelvin double bridge, commonly known as *Kelvin bridge*.

Kelvin double bridge :

This bridge is so called (*i.e.* double bridge) since the circuit contains a *second set of ratio arms* (l and m) as shown in Fig. 10.4. This second set of ratio arms connects the galvanometer to a point f at the appropriate potential between c and d and it eliminates the effect of the yoke resistance R_y. The ratio of the resistances of arms l and m is the same as the ratio of R_1 and R_2.

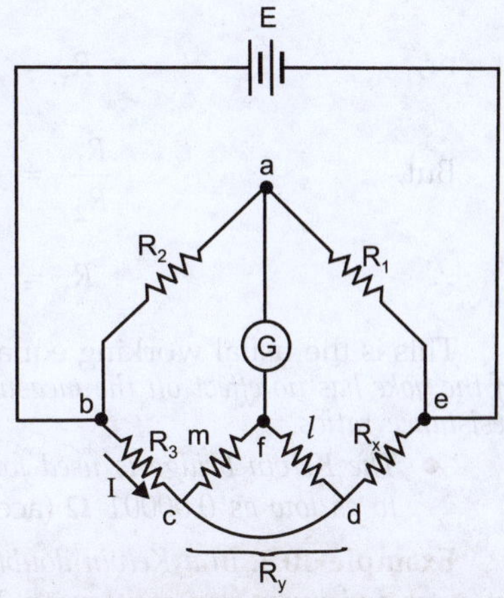

Fig. 10.4. Kelvin double bridge.

The galvanometer will read zero when the potential at a equals the potential at f.

i.e., $$E_{ab} = E_{bcf}$$

But, $$E_{ab} = \frac{R_2}{R_1 + R_2} \times E \quad ...(10.8)$$

and, $$E = I\left[R_3 + R_x + \frac{(l+m)\,R_y}{(l+m)+R_y}\right] \quad ...(10.9)$$

Substituting for E in eqn. (10.8), we get

$$E_{ab} = \frac{R_2}{R_1 + R_2} \times I\left[R_3 + R_x + \frac{(l+m)\,R_y}{(l+m)+R_y}\right] \quad ...(10.10)$$

Similarly, $$E_{bcf} = I\left[R_3 + \frac{m}{l+m}\left\{\frac{(l+m)\,R_y}{(l+m)+R_y}\right\}\right] \quad ...(10.11)$$

But, $$E_{ab} = E_{bcf}$$

$$\therefore \quad \frac{IR_2}{R_1+R_2}\left[R_3 + R_x + \frac{(l+m)\,R_y}{(l+m)+R_y}\right] = I\left[R_3 + \frac{m}{l+m}\left\{\frac{(l+m)\,R_y}{(l+m)+R_y}\right\}\right]$$

or, $$R_3 + R_x + \frac{(l+m)\,R_y}{(l+m)+R_y} = \frac{R_1+R_2}{R_2}\left[R_3 + \frac{m\,R_y}{(l+m)+R_y}\right]$$

or, $$R_3 + R_x + \frac{(l+m)\,R_y}{(l+m)+R_y} = \left(\frac{R_1}{R_2}+1\right)\left(R_3 + \frac{mR_y}{l+m+R_y}\right)$$

or, $$R_x + \frac{(l+m)\,R_y}{l+m+R_y} + R_3 = \frac{R_1R_3}{R_2} + R_3 + \frac{m\,R_1R_y}{R_2\,(l+m+R_y)} + \frac{m\,R_y}{l+m+R_y}$$

or, $$R_x = \frac{R_1R_3}{R_2} + \frac{m\,R_1R_y}{R_2\,(l+m+R_y)} + \frac{m\,R_y}{l+m+R_y} - \frac{(l+m)\,R_y}{l+m+R_y}$$

or, $$R_x = \frac{R_1R_3}{R_2} + \frac{m\,R_1R_y}{R_2\,(l+m+R_y)} + \frac{m\,R_y - l\,R_y - m\,R_y}{l+m+R_y}$$

or, $$R_x = \frac{R_1R_3}{R_2} + \frac{m\,R_1R_y}{R_2\,(a+b+R_y)} - \frac{l\,R_y}{l+m+R_y}$$

or, $$R_x = \frac{R_1 R_3}{R_2} + \frac{m R_y}{l + m + R_y}\left(\frac{R_1}{R_2} - \frac{l}{m}\right) \quad ...(10.12)$$

But, $$\frac{R_1}{R_2} = \frac{l}{m}$$

$\therefore$ $$R_x = \frac{R_1 R_3}{R_2}$$

This is the usual working equation for the Kelvin bridge. *It indicates that the resistance of the yoke has no effect on the measurement, provided that the two sets of ratio arms have equal resistance ratios.*

- *The Kelvin bridge is used for measuring very low resistances, from approximately 1 Ω to as low as 0.00001 Ω* (accuracy varying from ± 0.05 % to ± 0.2%).

Example 10.3. *In a Kelvin double bridge, there is error due to mismatch between the ratios of outer and inner arm resistances. The following data relate to this bridge :*

Standard resistance = 100.03μ Ω; Inner ratio arms = 100.31Ω and 200Ω; Outer ratio arms = 100.24Ω and 200Ω; The resistance of connecting leads from standard to unknown resistor is 680μ Ω.

Calculate the unknown resistance.

Solution. Refer to Fig. 10.4. *Given* : R_3 = 100.03 μ Ω; l = 100.31 Ω; m = 200 Ω; R_1 = 100.24 Ω; R_2 = 200 Ω; R_y = 680μ Ω

Unknown resistance, R_x :

We know that, $$R_x = \frac{R_1 R_3}{R_2} + \frac{m R_y}{l + m + R_y}\left(\frac{R_1}{R_2} - \frac{l}{m}\right) \quad \text{[Eqn. (10.12)]}$$

Inserting the various values, we get

$$R_x = \frac{100.24 \times 100.03 \times 10^{-6}}{200} + \frac{200 \times 680 \times 10^{-6}}{100.31 + 200 + 680 \times 10^{-6}}\left(\frac{100.24}{200} - \frac{100.31}{200}\right)$$

$$= 50.135 \times 10^{-6} + 4.528 \times 10^{-4} (-3.5 \times 10^{-4})$$

$$= 49.97 \times 10^{-6}\ \Omega = \mathbf{49.97\ \mu\Omega} \quad \textbf{(Ans.)}$$

Kelvin bridge ohmmeter :

- This is a modified form of Kelvin bridge and is *intended for the rapid measurement of the winding resistances of machines and transformers, and for the measurement of contact and earth conductor resistances.*
- The accuracy is of the order of ± 0.2 per cent.
- This instrument is *direct reading* and balance is obtained by rotating a single dial.
- Six standard resistors are used and these range from 0.01 Ω to 1000 Ω. The ratio dial is calibrated from 0.5 Ω to 10.5 Ω using 100 Ω standard resistor. The overall range of the instrument is from 0.00005 Ω to 105 Ω.

10.2.3. Measurement of Medium Resistances

The methods used for measurement of *medium resistances* are :

1. Ammeter-voltmeter method.
2. Substitution method.
3. Wheatstone bridge.
4. Carey-Foster slide-wire bridge method.

10.2.3.1. Ammeter-voltmeter method

This method has already been explained in Art. 10.2.2.1.

10.2.3.2. Substitution method

The connections for measurement of resistance by substitution are shown in Fig. 10.5. Let R be a variable resistance which can be changed in small steps, say of 0.1 Ω.

(i) Refer to Fig. 10.5 *(a)*. First resistance X is put in the circuit and the value of current noted.

Then resistance X is *removed* and it is *substituted* by a known variable resistance R which is varied so that the value of the current is same in both the cases. *This value of R is equal to the unknown resistance.*

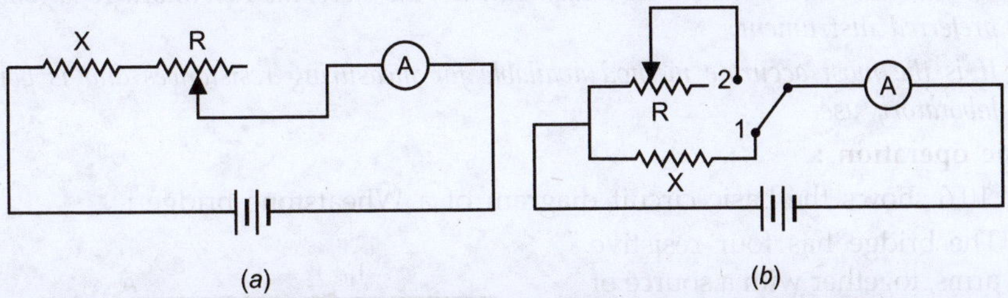

Fig. 10.5. Measurement of resistance by substitution.

(ii) If R is of fixed value, note the *readings* of the ammeter for the following cases :

(a) For resistances X and R in series.

(b) When resistance X is removed.

Let the readings of the ammeter for these cases be I_1 and I_2

∴ $$I_1 = \frac{V}{R+X} \quad ...(10.13)$$

and, $$I_2 = \frac{V}{R} \quad ...(10.14)$$

$$\frac{I_2}{I_1} = \frac{R+X}{R} = 1 + \frac{X}{R} \quad ...(10.15)$$

or, $$\frac{X}{R} = \frac{I_2}{I_1} - 1$$

or, $$X = R\left(\frac{I_2 - I_1}{I_1}\right) \quad ...(10.16)$$

(iii) Refer to Fig. 10.5*(b)*. The two-way switch first makes contact with 1 and then with 2 and let these readings be I_1 and I_2.

$$I_1 = \frac{V}{X} \text{ (where } X \text{ is in the circuit)} \quad ...(10.17)$$

$$I_2 = \frac{V}{R} \text{ (when } R \text{ is in the circuit)} \quad ...(10.18)$$

∴ $$\frac{I_2}{I_1} = \frac{X}{R}$$

or, $$X = R \times \frac{I_2}{I_1} \quad ...(10.19)$$

10.2.3.3. Wheatstone bridge

The *simplest form of bridge is for the purpose of measuring resistance and is called the* **Wheatstone bridge.**

- The Wheatstone bridge is widely used for precision measurement of resistance from *approximately 1Ω to the low megaohm range.*

— The *"upper limit"* is set by the reduction in sensitivity to unbalance, caused by high resistance values, thus reducing the galvanometer current.

— The *"lower limit"* is set by the resistance of the connecting leads and the contact resistance at the binding posts. The resistance of the leads could be calculated or measured, and the final result modified, but *contact resistance is very hard to compute or measure. For low resistance measurements, therefore, the Kelvin bridge is generally the preferred instrument.*

- *It is the most accurate method available for measuring resistances and is popular for laboratory use.*

Basic operation :

Fig. 10.6 shows the basic circuit diagram of a Wheatstone bridge :

— The bridge has four resistive arms, together with a source of e.m.f. (a battery) and null detector, usually a galvanometer (G) or other sensitive current meter.

— The current through the galvanometer depends on the potential difference between the points C and D. The bridge is said to be *balanced* when the potential difference across the galvanometer is *zero volt* so that there is no current through the galvanometer. Hence the bridge is balanced when potential difference between points C and D is equal,

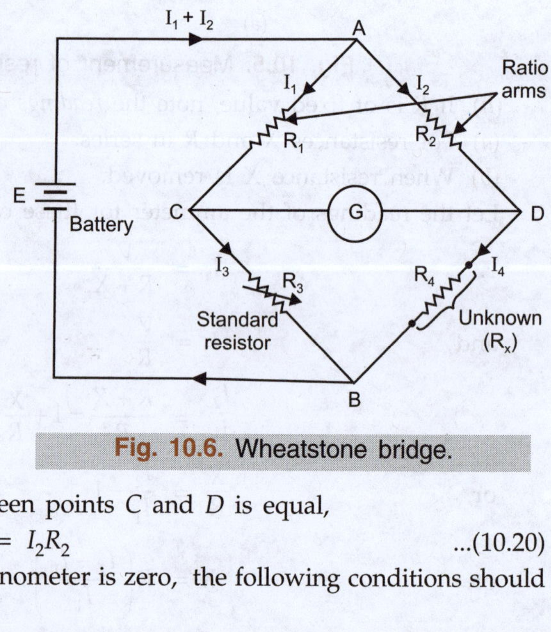

Fig. 10.6. Wheatstone bridge.

i.e. $$I_1R_1 = I_2R_2 \quad ...(10.20)$$

When the current through the galvanometer is zero, the following conditions should be satisfied :

$$I_1 = I_3 = \frac{E}{R_1 + R_3} \quad ...(10.21)$$

and, $$I_2 = I_4 = \frac{E}{R_2 + R_4} \quad ...(10.22)$$

Substituting the values of I_1 and I_2 in Eqn. (10.20), we get

$$\frac{E}{R_1 + R_3} \times R_1 = \frac{E}{R_2 + R_4} \times R_2$$

or, $$\frac{R_1}{R_1 + R_3} = \frac{R_2}{R_2 + R_4}$$

$$\text{or,} \quad R_1(R_2 + R_4) = R_2 (R_1 + R_3)$$

$$\text{or,} \quad R_1R_2 + R_1R_4 = R_1R_2 + R_2R_3$$

$$\text{or} \quad R_1R_4 = R_2R_3 \qquad ...(10.23)$$

Eqn. (10.23) is the well known expression for balance of the Wheatstone bridge. When the three of the resistors are known, the fourth may be determined. Thus if R_4 is the unknown resistor, its resistance R_x can be found as follows :

$$R_x = \frac{R_2}{R_1} . R_3 \qquad ...(10.24)$$

The resistors R_2 and R_1 are called *ratio arms* while the resistor R_3 is called the *standard arm* of the bridge.

When the null detector has sufficient sensitivity to indicate the balance position of the bridge with the required degree of position, the measurement of the unknown resistance R_x is *independent* of the characteristics or calibration of the null-detecting galvanometer.

Measurement errors :

The various sources of measurement errors are :

(*i*) The limiting errors of three known resistances–main measurement error.

(*ii*) Insufficient sensitivity of the null detector.

(*iii*) Changes in resistance of the bridge arms due to heating effect of the current through the resistors.

— In order to check this error the power dissipation in the bridge arms must be computed in advance, particularly when low-resistance values are to be measured, and the *current must be limited to a safe value.*

(*iv*) Thermal e.m.fs. in the bridge circuit or the galvanometer circuit.

— To prevent thermal effects, the more sensitive galvanometers sometimes have copper coils and copper suspension systems to avoid having dissimilar metals in contact with one another and generating thermal e.m.fs.

(*v*) Errors due to the resistance of leads and contacts exterior to the actual bridge circuit. These errors can be *reduced* by using a Kelvin bridge.

Applications of Wheatstone bridge :

Following are the *applications* of Wheatstone bridge :

1. To measure the D.C. resistance of various types of wire either for the purpose of quality control of the wire itself, or of some assembly in which it is used.

Example : Measurement of the resistance of motor windings, transformers, solenoids, and relay coils.

2. This bridge is also used extensively by telephone companies and others to *locate cable faults.*

Limitations :

The Wheatstone bridge entails the following *limitations* :

1. In case of measurement of low resistance, the resistance of the leads and contacts becomes significant, resulting in introduction of an error. This error can be *eliminated* by using Kelvin's double bridge.
2. While measuring high resistances, the resistance of the bridge becomes so large that the galvanometer becomes *insensitive to imbalance.* Consequently, a power supply has to replace the battery and a D.C. VTVM replaces the galvanometer. The Wheatstone bridge *cannot* be used for measuring high resistances in megaohms.

3. The *change in resistance* of the bridge arms due to the heating effect of current through the resistance–the temperature rise causes a change in the value of the resistance, and *excessive current may cause a permanent change in value.*

Example 10.4. *Fig. 10.6 consists of the following parameters :*

$R_1 = 20\ k\ \Omega$; $R_2 = 30\ k\ \Omega$; $R_3 = 80\ k\ \Omega$.

Determine the unknown resistance R_x.

Solution. Using the bridge balance equation, we have :

$$R_1 R_x = R_2 R_3$$

or,
$$R_x = \frac{R_2 R_3}{R_1} = \frac{30 \times 80}{20}$$

$$= \mathbf{120\ k\ \Omega}\ \textbf{(Ans.)}$$

Example 10.5. *Each of the ratio arms of a laboratory type Wheatstone bridge has guaranteed accuracy of ± 0.05%, while the standard arm has a guaranteed accuracy of ± 0.1%. The ratio arms are both set at 1000 Ω and bridge is balanced with standard arm adjusted to 3154 Ω.*

Determine the upper and the lower limits of the unknown resistance, based upon the guaranteed accuracies of the known bridge arms. **(Punjab University)**

Solution. Refer to Fig. 10.6. *Given* : $R_1 = R_2 = 1000\ \Omega$, $R_3 = 3154\ \Omega$

We know that :
$$\frac{R_1}{R_2} = \frac{R_3}{R_x}$$

or, Value of unknown resistance,
$$R_x = \frac{R_2 \times R_3}{R_1}$$

$$= \frac{1000 \times 3154}{1000} = 3154\ \Omega \qquad \text{[Eqn. (10.24)]}$$

∴ Percentage error in determination of R_x,

$$\frac{\delta R_x}{R_x} = \pm \frac{\delta R_1}{R_1} \pm \frac{\delta R_2}{R_2} \pm \frac{\delta R_3}{R_3}$$

$$= \pm 0.05 \pm 0.05 \pm 0.1 = \pm 0.2\%$$

∴ *Limiting value of* R_x = 3154 ± 0.2% = **3147** *to* **3160 Ω** **(Ans.)**

Wheatstone bridge with digital readout :

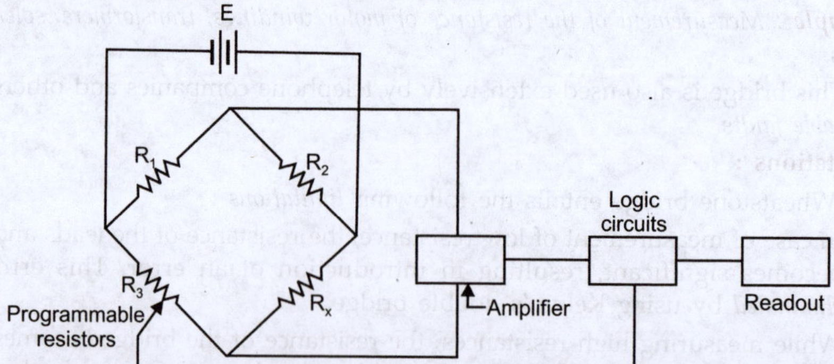

Fig. 10.7. Block diagram of Wheatstone bridge with digital readout.

Fig. 10.7 shows the block diagram of Wheatstone bridge with digital readout. Here a logic circuit is used to provide a signal to R_3, reuse the null, and provide a digital readout representing the value of R_x.

10.2.3.4. Carey-Foster slide-wire bridge method

This bridge circuit is an elaboration of the Wheatstone bridge and is *specially suited for comparing two nearly equal resistances.*

Fig. 10.8, shows the connection for Carey-Foster bridge :

— R_1 and R_2 are nominal ratio arms, R_x the resistance under test, and R_3 the standard resistance.

— A slide-wire of length L and uniform cross-section is included between R_x and R_3 as shown.

— Resistances R_1 and R_2 are first adjusted so that the ratio $\frac{R_1}{R_2}$ is approximately equal to the ratio $\frac{R_x}{R_3}$. Exact balance is obtained by adjustment of the sliding contact on the slide-wire. Let l_1 be the distance of the sliding contact from the left-hand end of the slide-wire. The resistances R_x *and* R_3 *are then interchanged* and balance again obtained. Let the distance now be l_2.

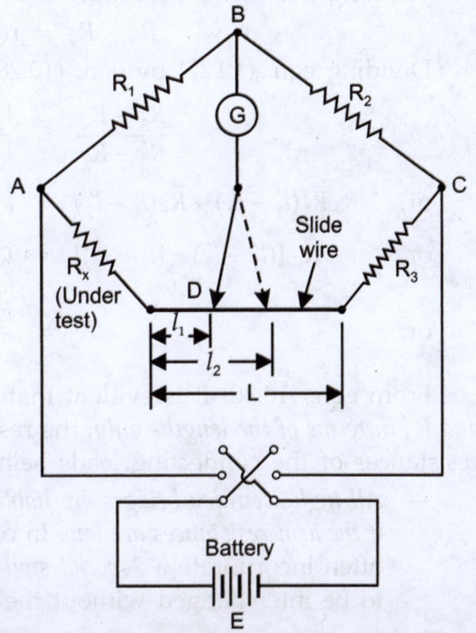

Fig. 10.8. Carey-Foster slide-wire bridge.

Then, for the *first balance,* $$\frac{R_1}{R_2} = \frac{R_x + l_1 r}{R_3 + (L - l_1)r} \quad ...(10.25)$$

where, r is the resistance per unit length of the slide-wire

For the *second balance,* $$\frac{R_1}{R_2} = \frac{R_3 + l_2 r}{R_x + (L - l_2)r} \quad ...(10.26)$$

Comparing eqns. (10.25) and (10.26), we get :

$$\frac{R_x + l_1 r}{R_3 + (L - l_1)r} = \frac{R_3 + l_2 r}{R_x + (L - l_2)r}$$

or, $$\frac{R_x + l_1 r}{R_3 + (L - l_1)r} + 1 = \frac{R_3 + l_2 r}{R_x + (L - l_2)r} + 1$$

or, $$\frac{R_x + l_1 r + R_3 + Lr - l_1 r}{R_3 + (L - l_1)r} = \frac{R_3 + l_2 r + R_x + Lr - l_2 r}{R_x + (L - l_2)r}$$

or, $$\frac{R_x + R_3 + Lr}{R_3 + (L - l_1)r} = \frac{R_3 + R_x + Lr}{R_x + (L - l_2)r}$$

or, $$R_3 + (L - l_1)r = R_x + (L - l_2)r$$

or, $$R_3 + Lr - l_1 r = R_x + Lr - l_2 r$$

or, $$R_3 - R_x = r(l_1 - l_2) \qquad ...(10.27)$$

Thus the difference between R_3 and R_x is obtained from the resistance of the slide-wire between the two balance points.

The slide wire is calibrated *i.e.* r is obtained by shunting R_3 by a known high resistance, thus reducing the effective value to R_3'. Now balance points l_1' and l_2' are then obtained by repeating the whole procedure previously described, so that :

$$R_3' - R_x = r(l_1' - l_2') \qquad ...(10.28)$$

Dividing eqn. (10.27) by eqn. (10.28), we get

$$\frac{R_3 - R_x}{R_3' - R_x} = \frac{l_1 - l_2}{l_1' - l_2'} \qquad ...(10.29)$$

or, $$R_3(l_1' - l_2') - R_x(l_1' - l_2') = R_3'(l_1 - l_2) - R_x(l_1 - l_2)$$

or, $$R_x\,[(l_1' - l_2') - (l_1 - l_2)] = R_3\,(l_1' - l_2') - R_3'(l_1 - l_2)$$

or, $$R_x = \frac{R_3\,(l_1' - l_2') - R_3'(l_1 - l_2)}{(l_1' - l_2' - l_1 + l_2)} \qquad ...(10.30)$$

From eqn. (10.30) it is evident that this method gives a *direct comparison between R_x and R_3 in terms of the lengths only,* the resistances of R_1 and R_2, contact resistances and the resistances of the connecting leads being eliminated.

— *All high-accuracy bridges are liable to error owing to thermoelectric e.m.fs. particularly if the arm resistances are low.* In order to *minimise this trouble,* Carey-Foster bridges often incorporate a *"special switch,"* which enables the connections to R_x and R_3 to be interchanged without the necessity for handling the coils.

☞ • *The methods used for measurement of medium resistances are unsuitable for measurement of low resistances i.e. resistances having a value under 1 Ω. The reason being that the resistance of leads and contacts, though small, are appreciable in comparison in the case of low resistances; hence special techniques have to be used for the measurement of low resistances.*

10.2.4. Measurement of High Resistances

The various methods employed for the measurement of *high resistances* are :

1. Direct deflection method.
2. Loss of charge method.
3. Megohm bridge.
4. Megger (Refer to Art. 4.8.7.3)

10.2.4.1. Direct deflection method

- In this method a very sensitive and high resistance (1 kΩ or more) moving coil galvanometer is connected in series with the resistance to be measured, and to a battery supply. The deflection of the galvanometer gives a measure of the insulation resistance. This is *very simple but rough method of measuring insulation resistance.*
- Fig. 10.9 illustrates the direct deflection method for measuring insulation resistance of a cable :
  — The galvanometer G measures the current I_R between conductor core and metal sheath; leakage currents I_L over the surface of the insulating material are carried by the guard wire wound on the insulation and does not flow through the galvanometer [Fig. 10.9 (*a*)].

— The cables without metal sheath can be tested in a similar way if the cable, except the end or ends on which connections are made, is immersed in water in a tank [Fig. 10.9 (*b*)] The water and tank then form the return path for the current. The cable is immersed in saline water for about 24 hours and temperature in kept constant (at about 20°C) and then the measurement is taken as in Fig. 10.9(*b*). The resistance of the cable, $R = \frac{V}{I_R}$.

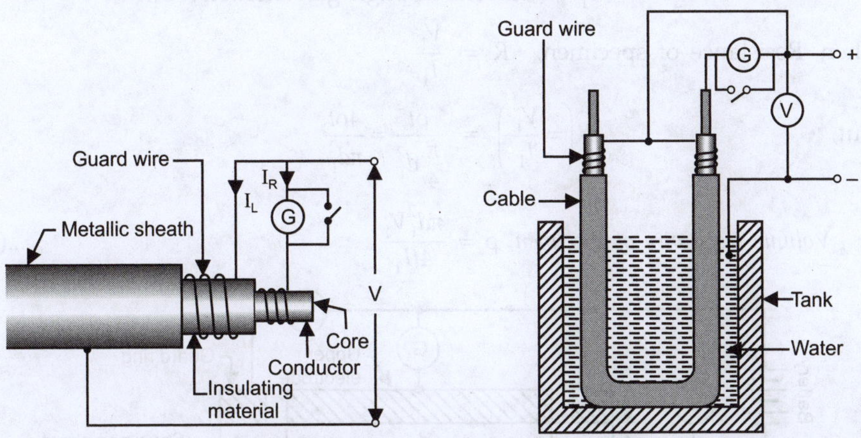

(a) Cable having sheath. (b) Cable having no conducting sheath.

Fig. 10.9. Measurement of insulation resistance of cable.

The deflection of the galvanometer, in some cases, is observed and its scale is afterwards calibrated by replacing the insulation by a standard high resistance (usually 1 MΩ), the galvanometer shunt being varied, as required to give a deflection of the same order as before.

— While conducting tests on cables, the galvanometer should be short-circuited before applying the voltage. *The short-circuiting connection is removed only after sufficient time has elapsed so that charging and absorption currents cease to flow.*

The galvanometer (of D'Arsonval type) employed should be *very sensitive* (at least 1000 mm/microampere at a scale distance of 1 metre), should be *of high resistance,* and also have *uniform scale.* The galvanometer and its circuit, together with the keys, *must be well insulated to prevent leakage currents.*

— The battery should be of about 500 V and its e.m.f. should remain constant throughout the test.

Measurement of volume and surface resistivities :

The insulation resistance of insulating material samples available in *sheet form* is often measured by the *direct deflection method.* In such cases, *volume resistivity* and *surface resistivity* of the material are measured.

Fig. 10.10 shows the schematic diagram for measurement of volume and surface resistivities of a specimen of insulating material. The specimen is provided with tin foil or colloidal graphite electrodes; the upper electrode having a guard ring.

Volume resistivity :

In order to measure *"volume resistivity"*, which is *in fact the specific resistance,* readings of the voltage applied and the current through the galvanometer are taken. Leakage currents over the specimen's edge will flow between the guard ring and the lower electrode and hence will *not* introduce error into the measurement.

The *volume resistivity* is calculated as under :

Let, d_1 = Diameter of upper electrode,

t = Thickness of the specimen sheet,

V_1 = Voltmeter reading, and

I_1 = Current through galvanometer G_1.

Then, Resistance of specimen, $R = \frac{V_1}{I_1}$

But, $$R\left(=\frac{V_1}{I_1}\right) = \frac{\rho t}{\frac{\pi}{4}d_1^2} = \frac{4\rho t}{\pi d_1^2}$$

$\therefore$ *Volume resistivity of specimen,* $$\rho = \frac{\pi d_1^2 V_1}{4tI_1} \quad ...(10.31)$$

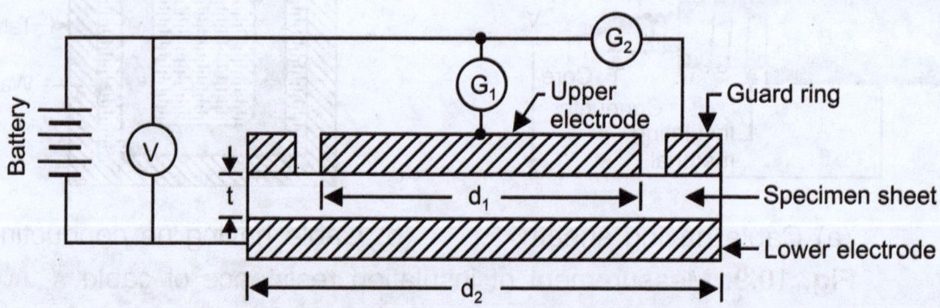

Fig. 10.10. Measurement of volume and surface resistivities of insulating material specimen.

— The resistivity of a thin layer of dielectric materials is different from volume resistivity, not only because of an *adherent humidity layer but also of contamination, chemical alterations, absorption of gases, or structural modification.*

The resistance R_t between two electrodes embedded in or attached to a dielectric medium is given by :

$$\frac{1}{R_t} = \frac{1}{R_v} + \frac{1}{R_s} \quad ...(10.32)$$

where, R_v = Volume resistance, and

R_s = Surface resistance.

— R_v can be measured separately from R_s by means of guard rings.

Surface resistivity :

The "*surface resistivity*" *is defined as the resistance between opposite edges of a square area of the surface of the material.* This quantity depends upon the general condition of the surface and upon humidity and is not a constant.

— In order to measure surface resistivity, the galvanometer is placed in position G_2 (See Fig. 10.10). In this position, the *galvanometer measures the leakage current and current flowing between upper and lower electrodes will be eliminated from measurement.*

Let, d_2 = Diameter of lower electrode disc,

V_2 = Voltmeter reading, and

I_2 = Current through the galvanometer G_2.

$\therefore$ Surface resistance, $R_s = \dfrac{V_2}{I_2}$

The leakage current flows along a path of length t and width πd_2 and, therefore, surface resistivity,

$$\rho_s = \frac{R_s \times \pi d_2}{t} = \frac{\pi d_2}{t} \cdot \frac{V_2}{I_2} \qquad ...(10.33)$$

- Other forms of *specimen and electrodes* (*e.g.* the electrodes and guard ring may be mercury, either placed in specially machined recesses, in moulded insulating materials, or retained by metal rings on the surface of sheet materials) are also used.

10.2.4.2. Loss of charge method

Fig. 10.11 shows the connection diagram for loss of charge method for measurement of high resistance :

C is a capacitor of known capacitance, V is electrostatic voltmeter, R_{leak} is the total leakage resistance of the capacitor and voltmeter and R_x is the resistance to be measured.

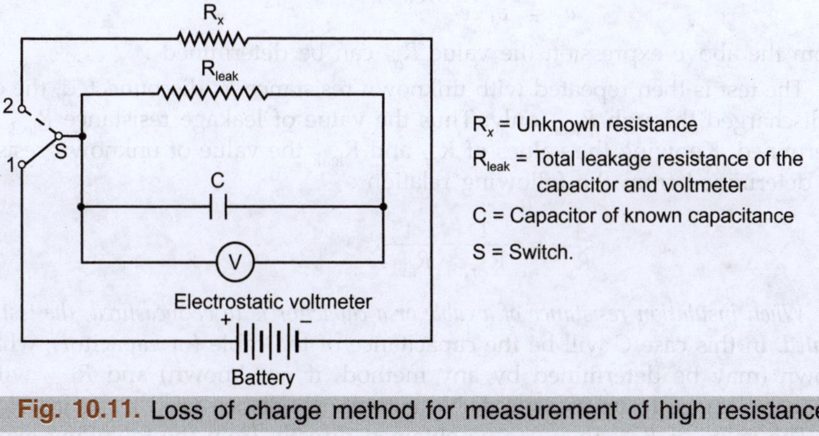

Fig. 10.11. Loss of charge method for measurement of high resistance.

Working :

— The capacitor C is first charged by means of a battery to some suitable voltage by putting switch S on stud-1 and then allowed to discharge through the resistance R_x and R_{leak} by throwing switch S to stud-2. The time taken 't' for the potential difference to fall from V_1 to V_2 during discharge is observed by a stop watch.

Let, $R_{eff.}$ = Effective/equivalent resistance of R_x and R_{leak} connected in parallel.

— If at any instant, say time 't', the voltage across the discharging capacitor is v volts, the charge on the discharging capacitor is q couloumbs and the capacity of the capacitor is C farads then current (i) at this instant,

$$i = -\frac{dq}{dt} = -C\frac{dv}{dt} \qquad ...(10.34)$$

Also,
$$i = \frac{\text{Potential drop across } R_{eff.}}{R_{eff.}} = \frac{v}{R_{eff.}} \qquad ...(10.35)$$

Comparing eqns. (10.34) and (10.35), we get :

$$\frac{v}{R_{eff.}} = -C \cdot \frac{dv}{dt}$$

or, $$\frac{dv}{v} = -\frac{dt}{CR_{eff.}} \quad ...(10.36)$$

Integrating both sides, we have :

$$\int_{v_1}^{v_2} \frac{dv}{v} = \int_0^t -\frac{dt}{CR_{eff.}}$$

or, $$[\ln v]_{v_1}^{v_2} = -\frac{t}{CR_{eff.}}$$

or, $$l_n\left(\frac{v_2}{v_1}\right) = -\frac{t}{CR_{eff.}}$$

or, $$\frac{v_2}{v_1} = e^{(-t/CR_{eff.})}$$

or, $$v_2 = v_1 \cdot e^{(-t/CR_{eff.})} \quad ...(10.37)$$

From the above expression the value $R_{eff.}$ can be determined :

— The test is then repeated with unknown resistance R_x disconnected, the capacitor being discharged through R_{leak} only. Thus the value of leakage resistance R_{leak} can also be determined. Knowing the values of $R_{eff.}$ and R_{leak} the value of unknown resistance R_x can be determined from the following relation :

$$\frac{1}{R_x} = \frac{1}{R_{eff.}} - \frac{1}{R_{leak}} \quad ...(10.38)$$

— *When insulation resistance of a cable or a capacitor is to be measured, the test need not be repeated.* In this case C will be the capacitance of the cable (or capacitor), which must be known (may be determined by any method, if not known) and R_{leak} will be the insulation resistance to be measured. No external resistance need to be inserted in the circuit. The value of R_{leak} then can be obtained directly from the following expression :

$$v_2 = v_1\, e^{(-t/CR_{leak})} \quad ...(10.39)$$

where, v_1 = Voltage at the start instant of charging, and

v_2 = Voltage across the object under test after time t seconds.

- This measurement method is associated with several *difficulties and errors.* Some important among them are :

(i) Effect of temperature upon insulation resistance :

— Usually, the resistance of insulating materials falls rapidly with increasing temperature, so temperature change and operating temperature play very important roles in the determination of the resistance.

— The value of insulation resistance should always be stated together with the temperature at which the test was carried out.

(ii) Leakage and absorption effects :

— Since the leakage currents are of comparable magnitude to the current under measurement, therefore, to eliminate leakage, the capacitor and voltmeter having exceedingly high resistance must be employed.

— The current which flows through the resistance is due to resistance as well as absorption. In order to eliminate absorption effect the observations should be made over a long period (several hours).

(iii) Effect of time of electrification :

— All insulating materials have absorption effects. Although the absorption current falls fairly steeply in the beginning, yet its fall becomes gradual thereafter.

— Thus the determination of insulation resistance by the ratio of applied voltage to the current flowing through insulation depends upon the time of application of the voltage.

Example 10.6. *The value of a high resistance is measured by loss of charge method. A capacitor having a capacitance of 2.5 μF is charged to a potential of 500 V D.C. and is discharged through the high resistance. An electrostatic voltmeter, kept across the high resistance, reads the voltage as 300 V at the end of 60 seconds.*

Calculate the value of high resistance. **(AMIE)**

Solution. *Given :* C = 2.5 μF = 2.5 × 10^{-6} F; v_1 = 500 V; v_2 = 300 V; t = 60 s

Value of unknown resistance, R_x :

We know that,
$$v_2 = v_1 e^{(-t/CR_x)} \quad \text{[Eqn. (10.37)]}$$

or,
$$\frac{v_2}{v_1} = e^{(-t/CR_x)} = \frac{1}{e^{(t/CR_x)}}$$

or,
$$e^{(t/CR_x)} = \frac{v_1}{v_2}$$

Taking log on both sides, we get :

$$t/CR_x = \ln\left(\frac{v_1}{v_2}\right)$$

or,
$$R_x = \frac{t}{C\ln\left(\frac{v_1}{v_2}\right)}$$

$$= \frac{60}{2.5\times10^{-6}\times\ln\left(\frac{500}{300}\right)} = \mathbf{46.98\ M\Omega} \quad \textbf{(Ans.)}$$

Example 10.7. *A 2.5 μF capacitor is charged to a potential difference of 450 volts. The capacitor is disconnected from the supply and the reading on an electrostatic voltmeter in parallel with the capacitor is observed to fall to 280 volts in 15.2 minutes. The test is repeated with a resistance R in parallel with the capacitor and the voltmeter. The voltmeter's reading is now found to fall from 450 V to 280 V in 10.8 minutes.*

Calculate the value of R.

Solution. *Given :* C = 2.5 μF = 2.5 × 10^{-6} F; v_1 = 450 V; v_2 = 280 V; t_1 = 15.2 min. = 15.2 × 60 = 912 s; t_2 = 10.8 × 60 = 646 s

Value of *R* :

Case-1 : Total leakage resistance of capacitor and voltmeter,

$$R_{\text{leak}} = \frac{t}{C\ln\left(\frac{v_1}{v_2}\right)} = \frac{912}{2.5\times10^{-6}\ \ln(450/280)} \simeq 768.9\ M\Omega$$

Case-2 : Equivalent resistance of R_{leak} and insulation resistance R to be measured,

$$R_{eff.} = \frac{t}{C \ln\left(\frac{v_1}{v_2}\right)} = \frac{648}{2.5 \times 10^{-6} \ln\left(\frac{450}{280}\right)} = 546.3 \text{ M}\Omega$$

Unknown resistance R can be determined from the following relation :

$$\frac{1}{R} = \frac{1}{R_{eff.}} - \frac{1}{R_{leak}} \quad \text{[Eqn. (10.38)]}$$

or,
$$R = \frac{R_{eff.} \times R_{leak}}{R_{leak} - R_{eff.}} \cdot \frac{546.3 \times 768.9}{(768.9 - 546.3)} = \mathbf{1887\ M\Omega} \quad \textbf{(Ans.)}$$

Example 10.8. *The following results were obtained by loss of charge method of testing a cable :*
Discharged immediately after charging, the deflection = 200 divisions
Discharged 30 seconds after charging, the deflection = 125 divisions
Discharged 30 seconds after charging,
when in parallel with a resistance of 10 MΩ, the deflection = 100 divisions
Calculate insulation resistance of the cable. **(Roorkee University)**

Solution.
$$v_2 = v_1 \cdot e^{(-t/CR)} \quad \text{[Eqn. (10.37)]}$$

or,
$$CR = \frac{t}{\ln\left(\frac{v_1}{v_2}\right)} \quad \text{...(i)}$$

where, v_1 = Voltage at $t = 0$,
v_2 = Voltage at time t seconds,
C = Capacitance of the cable, and
R_x = Resistance of the cable to be calculated.

Assuming 1 scale division deflection of galvanometer = 1V, then,

Case-1 : $v_1 = 200$ V, when $t = 0$
$v_2 = 125$ V, when $t = 30$ s

Substituting these values in eqn. (*i*) we have :

or
$$CR_x = \frac{t}{\ln\left(\frac{v_1}{v_2}\right)} = \frac{30}{\ln\left(\frac{200}{125}\right)} = 63.83 \quad \text{...(i)}$$

Case-2 : Effective resistance in MΩ,

$$R_{eff.} = \frac{1}{\frac{1}{R_x} + \frac{1}{10}} \quad \text{[From Eqn. (10.38)]}$$

$$= \frac{10R_x}{10 + R_x}$$

and,
$$CR_{eff.} = \frac{30}{\ln\left(\frac{v_1}{v_2}\right)} = \frac{30}{\ln\left(\frac{200}{100}\right)} = 43.28 \quad \text{...(ii)}$$

Dividing (*ii*) by (*i*), we get :

$$\frac{R_{eff.}}{R_x} = \frac{43.28}{63.83} = 0.678$$

or, $$\frac{10R_x}{(10+R_x)\,R_x} = 0.678$$

or, $$10 = (10 + R_x) \times 0.678$$

or, $$R_x = \frac{10}{0.678} - 10 = \mathbf{4.75\ M\Omega} \quad \textbf{(Ans.)}$$

10.2.4.3. Megohm bridge

- Fig. 10.12 shows the circuit of a completely self-contained Megohm bridge; it includes the following :
  - — Power supplies;
  - — Bridge members;
  - — Amplifiers;
  - — Indicating instrument.
- Sensitivity for balancing against high resistance is obtained by use of adjustable high voltage supplies of 500 V or 1000 V and the use of a sensitive null indicating arrangement such as a high gain amplifier with an electronic voltmeter or a CRO.

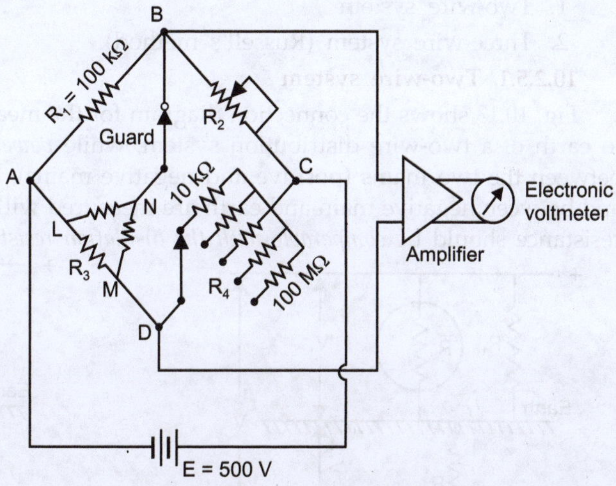

Fig. 10.12. Megohm bridge.

  - — The dial on R_2 is calibrated 1--10–100–1000 MΩ with main decade 1–10 occupying greater part of the dial space.
  - — Since unknown resistance $R_3 = \frac{R_1 R_4}{R_2}$, the arm R_2 is made tapered, so that the dial calibration is approximately logarithmic in main decade 1–10.
  - — Arm R_4 gives five multipliers 0.1, 1, 10, 100 and 1000.
  - — The junction of ratio arms R_1 and R_2 is brought on the main panel and is designated as 'Guard' terminal.

Example 10.9. *A test voltage is applied for several minutes between the conductor of a 100 metres long cable and earth. A reflecting galvanometer connected in series reads 280 scale divisions, the universal shunt being 2.5. With a standard resistance of 1 MΩ in the circuit the scale reading is 350, the value of the universal shunt being 1000.*

Calculate the insulation resistance of the cable. **(Bangalore University)**

Solution. Let, I_R = Current through the circuit with unknown resistance R connected, and

I_S = Curent through the circuit with standard resistance S connected.

Then, $I_R \propto$ galvanometer deflection × universal shunt multiplier

or, $I_R \propto 280 \times 2.5 \propto 700$

and, $I_S \propto 350 \times 1000 \propto 3{,}50{,}000$

Since current is inversely proportional to the resistance of the circuit, the circuit voltage remaining the same,

$$R = \frac{I_S}{I_R} \times S = \frac{3{,}50{,}000}{700} \times 1\ \text{M}\Omega = \mathbf{500\ M\Omega} \quad \textbf{(Ans.)}$$

10.2.5. Measurement of Insulation Resistance When Power is On

Following two cases will be discussed :

1. Two-wire system.
2. Three-wire system (Russell's method).

10.2.5.1. Two-wire system

Fig. 10.13 shows the connection diagram for the measurement of insulation resistance to earth of a two-wire distribution system, while *power is on*. In this case, the voltages between the two mains (positive and negative mains), between positive main and earth and between negative main and earth are measured with a *high resistance voltmeter* whose resistance should be *comparable with the insulation resistance under test*.

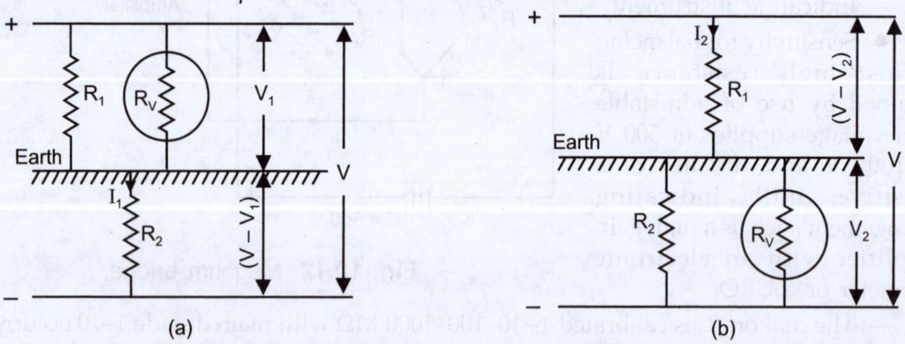

Fig. 10.13. Measurement of insulation resistance of 2-wire live mains.

Let, V = Voltage between positive and negative mains,
V_1 = Voltage between positive main and earth,
V_2 = Voltage between negative main and earth, and } From measurements
R_V = Resistance of voltmeter (known),
R_1 = Insulation resistance between positive outer and earth, and
R_2 = Insulation resistance between negative outer and earth.

Case-1 : Let I_1 be the current flowing from positive main to negative main in the first case. Then,

$$V_1 = I_1 \times \left(\frac{1}{\frac{1}{R_1} + \frac{1}{R_V}}\right)$$ ... Since R_1 and R_V are in parallel across positive main and earth

or, $$V_1 = I_1 \times \left(\frac{R_1 R_V}{R_1 + R_V}\right) \quad ...(i)$$

and, $$V - V_1 = I_1 R_2 \quad ...(ii)$$

Dividing (*ii*) by (*i*), we get :

$$\frac{V - V_1}{V_1} = \frac{R_2 (R_1 + R_V)}{R_1 R_V}$$

or, $$\frac{V}{V_1} - 1 = \frac{R_2 (R_1 + R_V)}{R_1 R_V}$$

or, $$\frac{V}{V_1} = \frac{R_2 (R_1 + R_V)}{R_1 R_V} + 1$$

or, $$\frac{V}{V_1} = \frac{R_1 R_V + R_2 (R_1 + R_V)}{R_1 R_V} = \frac{R_1 R_V + R_1 R_2 + R_2 R_V}{R_1 R_V}$$

or, $$\frac{V}{V_1} = \frac{R_1 R_2 + R_V (R_1 + R_2)}{R_1 R_V} \quad ...(iii)$$

Case-2 : Let I_2 be the current flowing from positive main to negative main. Then,

$$V_2 = I_2 \times \left(\frac{R_2 R_V}{R_2 + R_V} \right) \quad ...(iv)$$

and $$V - V_2 = I_2 R_1 \quad ...(v)$$

Dividing (*v*) by (*iv*), we have :

$$\frac{V - V_2}{V_2} = \frac{R_1 (R_2 + R_V)}{R_2 R_V}$$

or, $$\frac{V}{V_2} - 1 = \frac{R_1 (R_2 + R_V)}{R_2 R_V}$$

or, $$\frac{V}{V_2} = \frac{R_2 R_V + R_1 (R_2 + R_V)}{R_2 R_V}$$

or, $$\frac{V}{V_2} = \frac{R_1 R_2 + R_V (R_1 + R_2)}{R_2 R_V} \quad ...(vi)$$

Dividing (*iii*) by (*vi*), we get :

$$\frac{V_2}{V_1} = \frac{R_2}{R_1}$$

or, $$R_2 = R_1 \frac{V_2}{V_1}$$

Substituting this value of R_2 in (*iii*), we get :

$$\frac{V}{V_1} = \frac{R_1 \times R_1 \cdot \frac{V_2}{V_1} + R_V \left(R_1 + R_1 \cdot \frac{V_2}{V_1} \right)}{R_1 R_V}$$

or, $$\frac{V}{V_1} = \frac{R_1 \cdot \frac{V_2}{V_1} (R_1 + R_V) + R_1 R_V}{R_1 R_V}$$

$$\text{or,} \qquad \frac{V}{V_1} = \left(\frac{R_1 + R_V}{R_V}\right)\frac{V_2}{V_1} + 1$$

$$\text{or,} \qquad R_V V = R_1 V_2 + R_V V_2 + R_V V_1$$

$$\text{or,} \qquad R_1 V_2 = R_V (V - V_1 - V_2)$$

$$\text{or} \qquad \boldsymbol{R_1} = R_V \left(\frac{V - V_1 - V_2}{V_2}\right) \qquad ...(10.39)$$

$$\text{Similarly,} \qquad \boldsymbol{R_2} = R_V \left(\frac{V - V_1 - V_2}{V_1}\right) \qquad ...(10.40)$$

- The *combined insulation resistance* of the system can be determined by assuming that the two mains are bunched together so that the resultant vlaue is the combined resistance of R_1 and R_2 in *parallel*. Then,

$$\textit{Insulation resistance of the system} = \frac{R_1 R_2}{R_1 + R_2}$$

10.2.5.2. Three-wire system (Russell's method)

Fig. 10.14 shows the connection diagrams for measurement of insulation resistance of three-wire live mains.

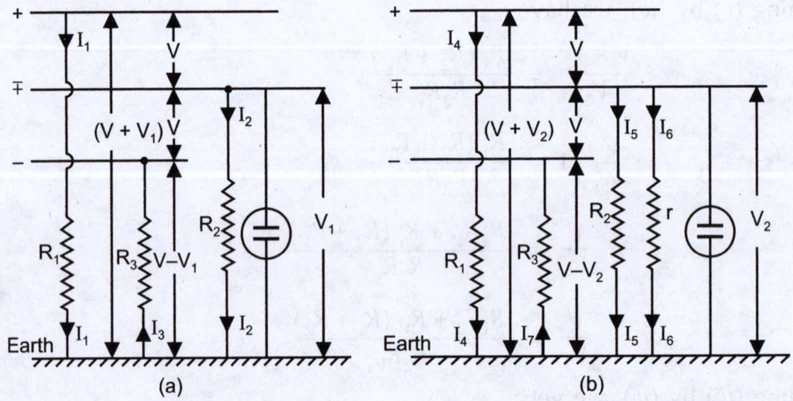

Fig. 10.14. Measurement of insulation resistance of 3-wire line mains.

— In this case an *electrostatic voltmeter* is used and connection of middle wire to earth is temporarily removed.

— Firstly, the voltmeter is connected between the middle wire and earth [See Fig. 10.14(*a*)] and reading of voltmeter is noted; let it be V_1. Then, a known resistance r of a fairly high value (about 2 kΩ) is connected between the middle wire and earth and the new reading of the voltmeter is noted; let it be V_2. Since the resistance of an electrostatic voltmeter can be taken as infinity, therefore, no current will flow through it.

Using Kirchhoff's laws, we get :

In case of *first test* :

$$I_1 + I_2 = I_3$$

$$\text{or,} \qquad \frac{V + V_1}{R_1} + \frac{V_1}{R_2} = \frac{V - V_1}{R_3} \qquad ...(i)$$

In case of *second test* : Refer to Fig. 10.14(*b*).

$$I_4 + I_5 + I_6 = I_7$$

or,
$$\frac{V+V_2}{R_1} + \frac{V_2}{R_2} + \frac{V_2}{r} = \frac{V-V_2}{R_3} \quad ...(ii)$$

Subtracting (*ii*) from (*i*), we get :

$$\left(\frac{V+V_1}{R_1} - \frac{V+V_2}{R_1}\right) + \left(\frac{V_1}{R_2} - \frac{V_2}{R_2}\right) - \frac{V_2}{r} = \left(\frac{V-V_1}{R_3} - \frac{V-V_2}{R_3}\right)$$

$$\frac{V_1 - V_2}{R_1} + \frac{V_1 - V_2}{R_2} - \frac{V_2}{r} = \frac{V_2 - V_1}{R_3}$$

or,
$$(V_1 - V_2)\left(\frac{1}{R_1} + \frac{1}{R_2} + \frac{1}{R_3}\right) = \frac{V_2}{r} \quad ...(iii)$$

The combined insulation resistance of the system is given by the resultant of the resistances R_1, R_2 and R_3 connected in parallel.

i.e.
$$R = \frac{1}{\frac{1}{R_1} + \frac{1}{R_2} + \frac{1}{R_3}}$$

Hence, the expression (*ii*) reduces to,

$$\frac{V_1 - V_2}{R} = \frac{V_2}{r}$$

or,
$$(V_1 - V_2)r = V_2 R$$

or,
$$\frac{V_1 - V_2}{V_2} = \frac{R}{r}$$

or,
$$R = r\left(\frac{V_1}{V_2} - 1\right) \quad ...(10.41)$$

10.2.6. Measurement of Earth Resistance

In any electrical system, the provision of an earth electrode is made due to the *reasons* mentioned below :

1. It is of significant importance that all the parts of electrical equipment (such as switches, circuit breakers, machines, tanks of transformers etc.) must be at earth potential and therefore they *must be connected to an earthed electrode.* When all these parts are connected to an earth electrode, a continuous low resistance path is available for leakage currents to flow to earth; these currents operate the protective devices and thus the faulty circuit is isolated in case a fault occurs.

Thus, *the purpose of earthing is to protect the various parts of the installation, as well as persons working against damage in case the insulation of a system fails at any point.*

2. The purpose of an earth electrode is to ensure that in the event of overvoltage on the system due to lightning discharges or other system faults, those parts of equipment which are normally *dead* as far as voltages are concerned do not attain dangerously high potentials.

3. In case of 3-phase circuit the neutral of the system is earthed so that the potential of the circuit with respect to earth is stabilised.

- The resistance of any earthing system depends on the following *main factors* :
  (*i*) Shape and material of earth electrode or electrodes used.
  (*ii*) Depth in the soil at which the electrodes are burried.
  (*iii*) Specific resistance of soil surrounding and in the neighbourhood of electrodes.

Megger earth tester :

The earth resistance can be measured with the help of *Earth tester* or a *Megger earth tester.*

The *Megger earth tester* (Fig. 10.15) is essentially a *direct reading ohmmeter and a hand driven generator which supplies the testing current.*

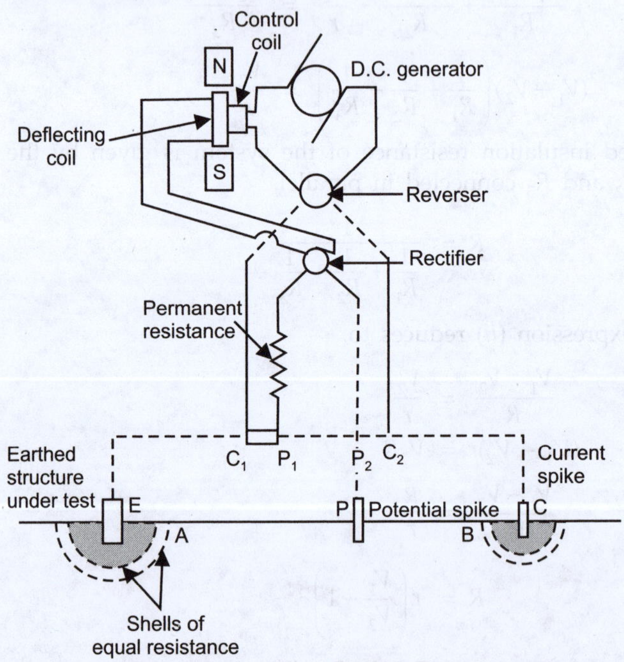

Fig. 10.15. Megger tester circuit containing direct current shown by full lines; circuit containing A.C. current shown by dotted lines.

Construction :

— The ohmmeter consists of two coils (current coil and potential coil) mounted at a fixed angle to each other on a common axle. The current coil carries current proportional to the current flowing in the test circuit, while the potential coil carries current proportional to the potential across the resistance under test. Thus the potential coil acts as a voltmeter while current coil acts as an ammeter. Since the *deflection* of the needle is proportional to the ratio of the current in the two coils, it gives *resistance directly.*

— The hand-operated generator produces the direct current, but to eliminate the effect of electrolytic e.m.f., it is necessary to pass alternating current through the soil, so to change the D.C. into an alternating supply a rotary current reverser is mounted on the same shaft of the generator. The alternating current in the soil will produce an alternating drop in the soil but the potential to be applied across the moving coil must be direct because the ohmmeter in a moving coil instrument works on D.C. alone, so for changing the alternating drop into the direct drop a synchronous rotary rectifier is also attached as shown in Fig. 10.15. Although the instrument is immune to stray currents, yet while taking readings sometimes it will be observed that the needle vibrates; this happens only at the instant when the stray alternating current has the same frequency as the frequency generated. For such occasions either *increase the speed of the generator a little or lower its speed.*

Working :

- For *measurement of earth resistance* two spikes acting as current and potential electrodes are driven into the ground at a suitable distance (C_2 at a distance of about 25 m from earth electrode under test). P_1 and C_1 terminals of the meter are shorted and are connected to the earth electrode under test.
- First spike P_2 is fixed between the spike C_2 and the earth electrode and the resistance is determined; then this spike taken 3 to 4 m near to the earth electrode and then for the same distance near to electrode C_2 and the two readings are again noted. The average of these three readings is the required earth resistance.

Specifications required for earthing as per ISI :

Following are *recommended specifications as per I.S.I.* for providing good earthing :

1. The earthing electrode should be situated at a place at least $1\frac{1}{2}$ metres away from the building (outside) whose installation system is being earthed.
2. The earth wire should be of same material as that of earth electrode used.
3. The minimum sectional area of earth lead wire should not be less than 0.02 sq. inch (No. 8 S.W.G.) and not more than 0.1 sq. inch.
4. The size of earth conductor as a general rule should not be less than half of the section of the line conductor.
5. The earth wire should be taken through G.I. pipe of 12 mm dia. for at least 32 cm length above and below ground surface to the earth electrode to safeguard against mechanical wear and tear.
6. Loose earth and coal salt mixture should be filled around the earth electrode for effective earthing.
7. The earth wire connected to the earth electrode *shall not be* necessarily run along the whole wiring system. All the earth wire run along the sub-circuits should be terminated and hooked firmly at the main board and from where the main earth wires should be run to the earth electrode. The loop earth wires should be of 14 S.W.G. copper wire.
8. All the joints in the earth wire should be firmly done with nuts and bolts of the same material as of earth wire.

10.2.7. Location of Cable Faults

Out of several methods available for locating the faults in underground cables, the *two commonly used methods are :*

1. Murray loop test.
2. Varley loop test.

- Both these tests use the principle of Wheatstone bridge for locating the faults.
- These tests can be employed to locate the *earth fault* or *short circuit fault in* underground cables *provided that a sound cable runs along the faulty cable.*

10.2.7.1. Murray loop test

This is the *most common and accurate method* of locating earth or short-circuit faults in the underground cables.

(i) **Earth fault.** Fig. 10.16 shows the circuit diagram for locating the earth fault by Murray loop test. *AB* is the sound cable and *CD* is the faulty cable. *F* is the point at which earth fault occurs. The far end *D* of the faulty cable is joined to the far end *B* of the sound cable through a low resistance link. Two *variable* resistances *P* and *Q* are joined to ends *A* and *C* respectively and serve as the ratio arms of the Wheatstone bridge.

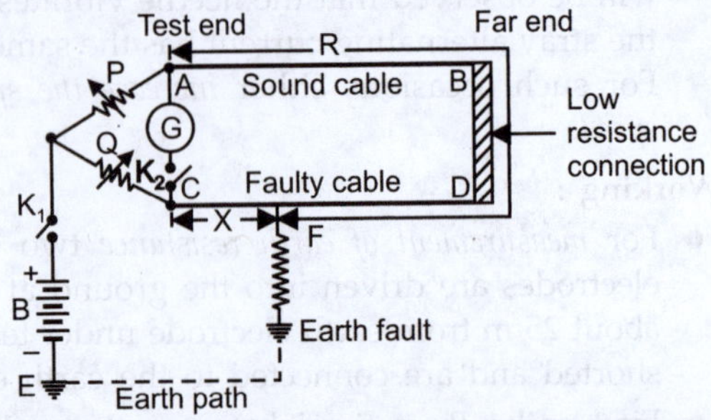

Fig. 10.16. Murray loop test for locating earth fault.

Let, R = Resistance of the conductor loop up to the fault from the test end, and

X = Resistance of the other length of loop.

P, Q, R and *X* constitute the four arms of the Wheatstone bridge. The resistances *P* and *Q* are varied till the galvanometer (*G*) *indicates zero deflection.*

When the bridge is balanced, we have

$$\frac{P}{Q} = \frac{R}{X}, \quad \text{or,} \quad \frac{P}{Q} + 1 = \frac{R}{X} + 1,$$

or,
$$\frac{P+Q}{Q} = \frac{R+X}{X}$$

If r is the resistance of each cable, then

$$R + X = 2r$$

$$\frac{P+Q}{Q} = \frac{2r}{X}, \quad \text{or,} \quad X = \frac{Q}{P+Q} \times 2r \qquad \text{...(10.42)}$$

If l is the length of each cable in metres, then resistance per metre length of the cable $= \frac{r}{l}$

∴ Distance of fault point (*F*) from the test end,

$$d = \frac{X}{r/l} = \frac{Q}{P+Q} \times 2r \times \frac{l}{r} = \frac{Q}{P+Q} \times 2l$$

or,
$$d = \frac{Q}{P+Q} \times \text{(loop length) metres} \qquad \text{...(10.43)}$$

It may be noted that resistance of the fault is in the battery circuit and not in bridge circuit. Therefore, *fault resistance does not affect the balancing of the bridge. However, in case the fault resistance is high, the sensitivity of the bridge is reduced.*

(ii) **Short-circuit fault.** Fig. 10.17 shows the circuit diagram for locating the short-circuit fault by Murray loop test. Again, *P, Q, R* and *X* constitute the arms of the bridge. The fault resistance is in battery circuit and not in the bridge circuit. The bridge is balanced by varying the resistances *P* and *Q*. When the bridge is balanced, we have :

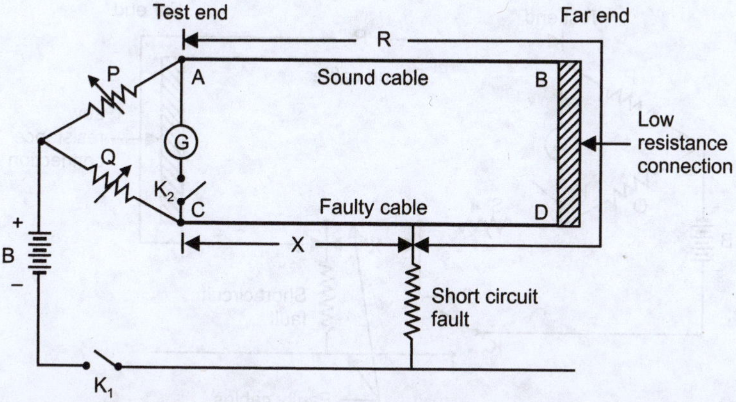

Fig. 10.17. Murray loop test for locating short-circuit fault.

$$\frac{P}{Q} = \frac{R}{X}, \quad \text{or,} \quad \frac{P+Q}{Q} = \frac{R+X}{X} = \frac{2r}{X}$$

$$X = \frac{Q}{P+Q} \times 2r$$

or, $$X = \frac{Q}{P+Q} \times (\text{loop length}) \text{ metres} \quad ...(10.44)$$

Thus the position of the fault is located.

10.2.7.2. Varley loop test

This test which is also used to locate earth fault or short-circuit fault in the underground cables uses Wheatstone bridge principle.

It *differs* from Murray loop test in that here the *ratio arms P and Q have fixed resistances.* Balance is obtained by adjusting the *variable resistance* S connected to the test end of the faulty cable. The circuit diagrams for locating earth fault and short-circuit fault by Varley loop test are shown in Figs. 10.18 and 10.19 respectively.

- The key K_2 is first thrown to position 1. The variable resistance S is varied till the bridge is balanced for resistance value of S_1. Then,

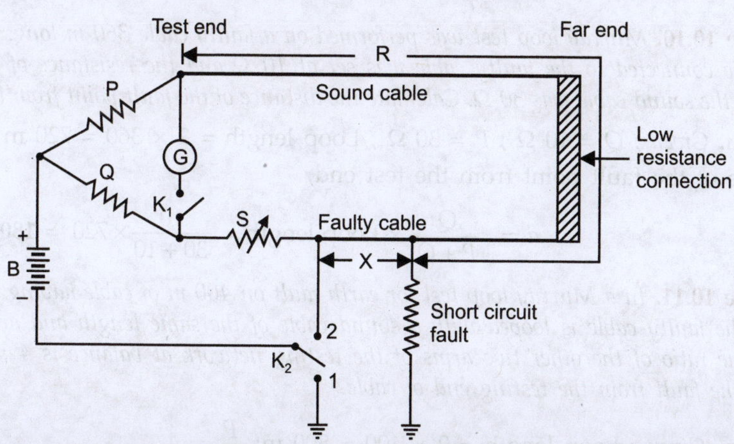

Fig. 10.18. Varley loop test for locating earth fault.

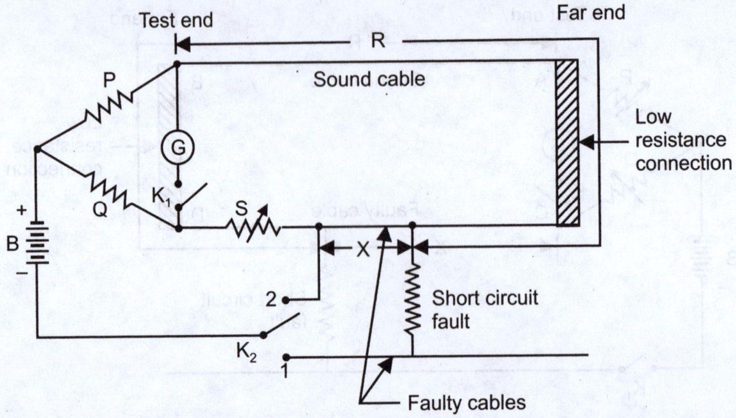

Fig. 10.19. Varley loop test for locating short-circuit fault.

$$\frac{P}{Q} = \frac{R}{X + S_1}, \quad \text{or,} \quad \frac{P+Q}{Q} = \frac{R + X + S_1}{X + S_1}$$

or,
$$X = \frac{Q(R + X) - PS_1}{P + Q} \qquad \text{...(10.45)}$$

- The key is now thrown to position 2 and bridge is balanced with the new value of resistance S_2. Then,

$$\frac{P}{Q} = \frac{R + X}{S_2} \quad \text{or,} \quad Q\,(R + X) = PS_2$$

Since the values of P, Q, S_1 and S_2 are known, the value of X can be determined.

$$\text{Loop resistance} = R + X = \frac{P}{Q} S_2$$

If r is the resistance of the cable per metre length, then,

Distance of fault from the test end,

$$d = \frac{X}{r} \text{ metres.} \qquad \text{...(10.46)}$$

Example 10.10. *Murray loop test was performed on a faulty cable 360 m long. At balance, the resistance connected to the faulty cable was set at 10 Ω and the resistance of the resistor connected to the sound cable was 30 Ω. Calculate the distance of the fault point from the test end.*

Solution. *Given* : $Q = 10\ \Omega$; $P = 30\ \Omega$; Loop length = 2 × 360 = 720 m

Distance of the fault point from the test end,

$$d = \frac{Q}{P+Q} \times \text{Loop length} = \frac{10}{30 + 10} \times 720 = \mathbf{180\ m} \quad \textbf{(Ans.)}$$

Example 10.11. *In a Murray loop test for earth fault on 400 m of cable having a resistance 1.5 Ω/km, the faulty cable is looped with a sound cable of the same length and area of cross-section. If the ratio of the other two arms of the testing network at balance is 4 : 1, find the distance of the fault from the testing end of cable.*

Solution. *Given* : Loop length = 2 × 400 = 800 m; $\frac{P}{Q} = 4$

$$\frac{P}{Q} = \frac{4}{1} \quad \text{or} \quad \frac{P+Q}{Q} = \frac{4+1}{1} = 5$$

Distance of fault from test end,

$$d = \frac{Q}{P+Q} \times \text{loop length} \qquad \text{[Eqn. (10.43)]}$$

$$= \frac{1}{5} \times 800 = \mathbf{160\ m.} \quad \textbf{(Ans.)}$$

Example 10.12. *In Murray loop test for a fault to earth on a 400 m length of cable having a resistance of 1.5 Ω/km, the faulty cable is looped with a sound cable of the same length but having a resistance of 3.5 Ω/km. The resistances of the other two arms of the testing network at balance are in the ratio 3 : 1. Calculate the distance of the fault from the testing end of the cable.*

Solution.

$$\frac{P}{Q} = \frac{3}{1}, \quad \text{or,} \quad \frac{P+Q}{Q} = \frac{3+1}{1} = 4$$

$$\text{Loop resistance} = \frac{1.5}{1000} \times 400 + \frac{3.5}{1000} \times 400 = 2\ \Omega$$

Resistance of the faulty cable from the test end up to fault point,

$$X = \frac{Q}{P+Q} \times \text{loop resistance} = \frac{1}{4} \times 2 = 0.5\ \Omega$$

Distance of fault point from the testing end,

$$d = \frac{X}{15/1000} = \frac{0.5}{15/1000} = \mathbf{333.33\ m} \quad \textbf{(Ans.)}$$

Example 10.13. *Varley loop test is conducted to locate an earth fault on a 25 km long cable. The resistance per km of single conductor is 15 Ω. The loop is completed with a similar sound conductor. At balance, the variable resistance connected to the faulty conductor is 150 Ω. The fixed resistors have equal values. Calculate the distance of the fault from the test end.*

Solution. Resistance of faulty cable from test end to fault point,

$$X = \frac{Q(R+X) - PS}{P+Q} \qquad \text{[Eqn. (10.45))]}$$

Here, $P = Q\ ;\ S = 150\ \Omega\ ;\ R + X = 25\ (15 + 15) = 750\ \Omega$

$$\therefore \quad X = \frac{Q \times 750 - Q \times 150}{Q+Q} = 300\ \Omega$$

The resistance per km = 15 Ω

∴ Distance of fault from the test end,

$$d = \frac{X}{15} = \frac{300}{15} = \mathbf{20\ km.} \quad \textbf{(Ans.)}$$

10.2.7.3. Location of faults in high-voltage cables

As the resistance presented by high-voltage cable under fault condition is very high, therefore, such faults cannot be located with accuracy by methods already described earlier.

In high-voltage cables breakdown occurs at the time of application of the high voltage especially when the oil is in the cable and cable joints breaks down and therefore presents a low resistance. However, when the applied voltage is withdrawn the oil *recovers its insulating property and the cable again presents high insulation resistance. The fault resistance may be* ***reduced*** *by a continued application of the high voltage so as to carbonise or burn out the insulation at the point of the fault. After the insulation resistance is so reduced, ordinary loop-tests may then be applied to locate the fault position.*

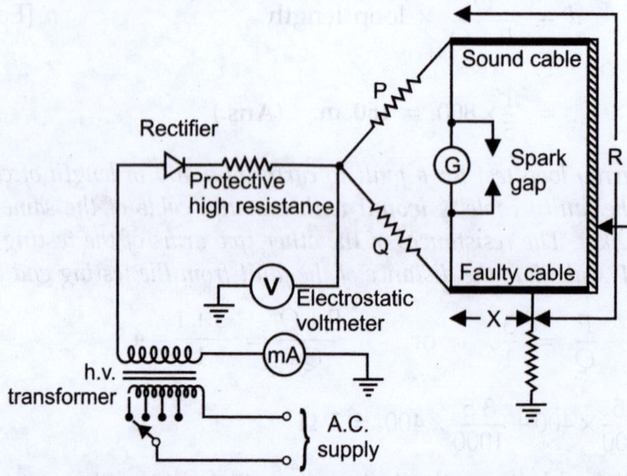

Fig. 10.20. Location of faults in high voltage cables.

- Fig. 10.20 shows a more satisfactory method for location of faults in high voltages. This method involves performance of Murray loop test at high voltage obtained from either half-wave or full-wave rectifier :
  — The rectified supply at about 60 kV is obtained from a high-vacuum diode; a higher voltage, if desired, may be obtained with two valves.
  — The rectifier is fed from a step-up h.v. transformer.
  — A current limiting high resistance and a milliammeter are included in the supply circuit for the limitation and measurement of the fault current respectively.
  — An electrostatic voltmeter is employed to measure the voltage at the junction point of the ratio arms. A loop is made connecting together the faulty and sound cables with the help of a short-circuiting link. In order to vary the ratio arms (*P*, *Q*) a highly insulated slide-wire, having a sliding contact which can be operated by a long insulating handle is used.
  — The spark gap is used to protect the galvanometer.

Test procedure : The test is conducted as follows :

— The circuit is energised and the tap at the primary of the h.v. transformer is adjusted to give a low fault current of the order of 5 mA at the time of cable breakdown.

— The bridge is balanced by varying P/Q ratio.

— The tap of primary of h.v. transformer is then changed to increase the fault current to about 50 mA.

— The bridge is *rebalanced*.

It may be mentioned here that at the time of breakdown, a very large voltage occurs across the protective high resistance and therefore the voltage across the cable is low. However, if the fault clears, the voltage across the cable would rise suddenly and it is on this account that highly insulated slide-wire is used for operator's safety.

— The position of the fault can be obtained using equations given for Murray loop test.

10.3. MEASUREMENT OF INDUCTANCE AND CAPACITANCE

A.C. BRIDGES

10.3.1. A.C. Bridge—General form

An A.C. bridge in its basic form consists of the following elements/components :

1. *Four bridge arms.*

2. *A source of excitation.* The power source supplies an A.C. voltage to the bridge at the desired frequency.

- For power measurements at low frequencies, the power line may serve as the source of excitation.
- At higher frequencies, the electronic oscillators are universally used as the source of supply due to the following *reasons* :

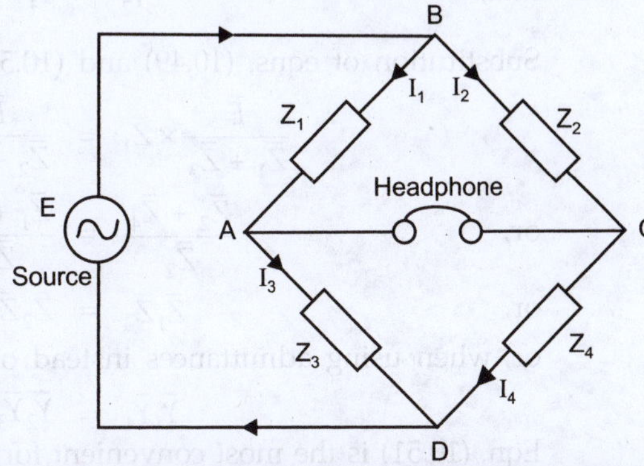

Fig. 10.21. General form of the A.C. bridge

(*i*) The output waveform is very close to sine wave.

(*ii*) The output frequency is very stable.

(*iii*) The output frequency can be determined with accuracy and is also easily adjustable.

(*iv*) The output power is adequate to drive the logic circuits.

— A typical oscillator has a frequency range of 40 Hz to 125 kHz with power output of 7W.

3. *Null detector.* It must respond to A.C. unbalance currents. The null/balance detectors commonly used for A.C. bridges are :

(*i*) *Headphones*. These are used as detectors at the frequencies of 250 Hz to 3 to 4 kHz.

(*ii*) *Tunable amplifier circuits* :

— A tuned detector is the most sensitive detector while working with single frequency.

— Tunable amplifier detectors are used for frequency range of 10 Hz to 100 Hz.

(*iii*) *Vibration galvanometers.* These are useful for low audio frequency range from 5 Hz to 1000 kHz but are commonly used below 200 Hz.

General equation for bridge balance :

Fig. 10.21 shows the general form of an A.C. bridge.

- The four bridge arms $\bar{Z}_1, \bar{Z}_2, \bar{Z}_3$ and $\bar{Z}_4$ (Bar *i.e.* – indicates complex notation) are indicated as unspecified impedances and detector is represented by headphones (*cheapest and effective form*).
- The balance condition is reached when the detector response is zero, or indicates a null. Balance adjustment to obtain a null response is made by varying one or more of the bridge arms.
- The *condition for bridge balance* requires that potential difference from *A* to *C* in Fig. 10.21 be zero. This will be the case when the voltage drop from *B* to *A* equals the voltage drop from *B* to *C* in both *magnitude* and *phase*. In complex notation we can write :

$$\bar{E}_{BA} = \bar{E}_{BC} \quad ...(10.47)$$

or,
$$\bar{I}_1\bar{Z}_1 = \bar{I}_2\bar{Z}_2 \quad ...(10.48)$$

Also at balance,
$$\bar{I}_1 = \bar{I}_3 = \frac{\bar{E}}{\bar{Z}_1 + \bar{Z}_3} \quad ...(10.49)$$

and,
$$\bar{I}_2 = \bar{I}_4 = \frac{\bar{E}}{\bar{Z}_2 + \bar{Z}_4} \quad ...(10.50)$$

Substitution of eqns. (10.49) and (10.50) into eqn. (10.48) yields :

$$\frac{\bar{E}}{\bar{Z}_1 + \bar{Z}_3} \times \bar{Z}_1 = \frac{\bar{E}}{\bar{Z}_2 + \bar{Z}_4} \times \bar{Z}_2$$

or,
$$\frac{\bar{Z}_2 + \bar{Z}_4}{\bar{Z}_2} = \frac{\bar{Z}_1 + \bar{Z}_3}{\bar{Z}_1} \quad \text{or} \quad 1 + \frac{\bar{Z}_4}{\bar{Z}_2} = 1 + \frac{\bar{Z}_3}{\bar{Z}_1}$$

or,
$$\bar{Z}_1\bar{Z}_4 = \bar{Z}_2\bar{Z}_3 \quad ...(10.51)$$

or, when using admittances instead of impedance, we have

$$\bar{Y}_1\bar{Y}_4 = \bar{Y}_2\bar{Y}_3 \quad ...(10.52)$$

Eqn. (10.51) is the most convenient form in most cases and is the *general equation for balance* of the A.C. bridge. Eqn. (10.52) is useful when *dealing wih parallel components in bridge arms.*

— Eqn. (10.51) states the product of the impedances of one pair of opposite arms must be equal to the product of impedances of the other pair of opposite arms, with the impedances expressed in complex notation. If the impedance is written in the form $\bar{Z} = Z \angle\theta$, where Z represents the magnitude and θ the phase angle of the complex impedance, Eqn. (10.52) can be written in the form

$$(Z_1 \angle\theta_1)(Z_4 \angle\theta_4) = (Z_2 \angle\theta_2)(Z_3 \angle\theta_3) \quad ...(10.53)$$

or,
$$Z_1Z_4 \angle(\theta_1 + \theta_4) = Z_2Z_3 \angle(\theta_2 + \theta_3) \quad ...(10.54)$$

— Eqn. (10.54) shows that following *two conditions* must be met *simultaneously* when balancing an A.C. bridge :

(*i*) The magnitudes of the impedances should satisfy the relationship :

$$Z_1Z_4 = Z_2Z_3 \quad ...(10.55)$$

(*ii*) The phase angles of the impedances should satisfy the relationship :

$$\angle\theta_1 + \angle\theta_4 = \angle\theta_2 + \angle\theta_3 \quad ...(10.56)$$

Example 10.14. *The following data relate to the basic A.C. bridge of Fig. 10.21 :*

$\bar{Z}_1 = 50\angle 80°$ *(inductive impedance);*

$\bar{Z}_2 = 125\Omega$ *(pure resistance) ;*

$\bar{Z}_3 = 200\angle 30°$ *(inductive impedance);*

$\bar{Z}_4 =$ *Unknown*

Determine the constants of the unknown arm :

Solution. *Given* : $\bar{Z}_1 = 50 \angle 80°$; $\bar{Z}_2 = 125\ \Omega$, $\bar{Z}_3 = 200 \angle 30°$

$\bar{Z}_4$:

We know that :
$$Z_1Z_4 = Z_2Z_3 \quad ...(10.55)$$

or,
$$Z_4 = \frac{Z_2Z_3}{Z_1} = \frac{125 \times 200}{50} = 500\Omega$$

Also, $\angle\theta_1 + \angle\theta_4 = \angle\theta_2 + \angle\theta_3$ ...(10.56)

or, $80° + \theta_4 = 0° + 30°$

or, $\theta_4 = 0° + 30° - 80° = -50°$

Hence, $\bar{Z}_4 = \mathbf{500\angle{-50°}\ \Omega}$,

indicating that we are dealing with a *capacitive* element, possibly consisting of a series combination of a resistor and a capacitor. **(Ans.)**

Example 10.15. *The following data relate to the balanced A.C. bridge shown in Fig. 10.22.*

Arm AB, R_1 = 225 Ω; Arm BC, R_2 = 150 Ω in series with C_2 = 0.53 μ F; arm CD, unknown; arm DA, R_3 = 100 Ω in series with L = 7.95 mH. The oscillator frequency = 1 kHz.

Calculate the constants of arm CD.

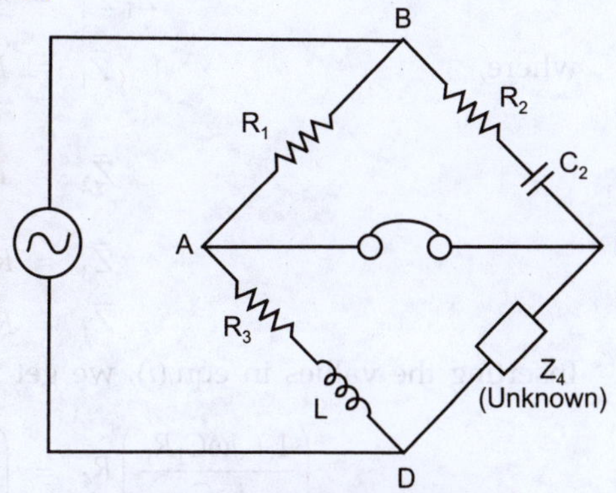

Fig. 10.22.

Solution. *Given* : f = 1 kHz, R_1 = 225 Ω; R_2 = 150 Ω, C = 0.53 μ *F* or $\omega\, C_2 = 2\,\pi\, f \times 0.53 \times 10^{-6} = 2\pi \times 1000 \times 0.53 \times 10^{-6}\ \Omega = 3.33 \times 10^{-3}\ \Omega$; R_3 = 100 Ω, *L* = 7.95 mH or $\omega L = 2\,\pi\, f\, L = 2\pi \times 1000 \times 7.95 \times 10^{-3} \simeq 50\Omega$.

Arm *CD*, $\bar{Z}_4$:

Now,
$$\bar{Z}_1 = R_1 = 225\ \Omega$$

$$\bar{Z}_2 = R_2 - \frac{j}{\omega C_2} = 150 - \frac{j}{3.33\times10^{-3}} = (150 - j\,300)\ \Omega$$

$$\bar{Z}_3 = R_3 + j\omega L = (100 + j\,50)\ \Omega$$

$$\bar{Z}_4 = \text{Unknown}$$

The general equation for bridge balance states that :

$$\bar{Z}_1\,\bar{Z}_4 = \bar{Z}_2\,\bar{Z}_3$$

or,
$$\bar{Z}_4 = \frac{\bar{Z}_2\,\bar{Z}_3}{\bar{Z}_1} = \frac{(150 - j\,300)\,(100 + j\,50)}{225}$$

$$= \frac{(15000 + j\,7500 - j\,30000 + 15000)}{225}$$

$$= \frac{30000 - j\,22500}{225} = \mathbf{133.33 - j\,100}$$

i.e.
$$R_4 = 133.33\ \Omega \text{ and } X_{C_4} = \frac{1}{\omega\, C_4} = \frac{1}{2\,\pi\, f\, C_4} = 100$$

or,
$$C_4 = \frac{1}{2\,\pi\times1000\times100} = 1.59\times10^{-6}\ F \text{ or } 1.59\ \mu F$$

Hence, $\mathbf{R_4 = 133.33\ \Omega}$, $\mathbf{C_4 = 1.59\ \mu F}$ **(Ans.)**

Example 10.16. *In a four-arm bridge network (Fig. 10.23), the arm AB consists of an imperfect condenser, BC and CD are non-reactive resistances of 1000 Ω each and DA is a standard capacitor of 0.0115 μF capacitance in series with a resistance of 140 Ω. If the bridge* *is balanced for frequency ω = 7500 radians/sec., find the shunt-less resistance and capacitance of the imperfect condenser.* **(UPTU, 2004-05)**

Solution. Refer to Fig. 10.23 : *Given* : $\omega = 7500$ rad/s; $R_2 = 140\ \Omega$; $R_3 = R_4 = 1000\ \Omega$; $C_2 = 0.0115\ \mu$ F.

We know that for balance,

$$\bar{Z}_1 \bar{Z}_4 = \bar{Z}_2 \bar{Z}_3 \quad ...(i)$$

where,

$$\bar{Z}_1 = R_1 + \frac{1}{j\omega C_1} = \frac{1 + j\omega C_1 R_1}{j\omega C_1}$$

$$\bar{Z}_2 = R_2 + \frac{1}{j\omega C_2} = \frac{1 + j\omega C_2 R_2}{j\omega C_2}$$

$$\bar{Z}_3 = R_3$$

$$\bar{Z}_4 = R_4$$

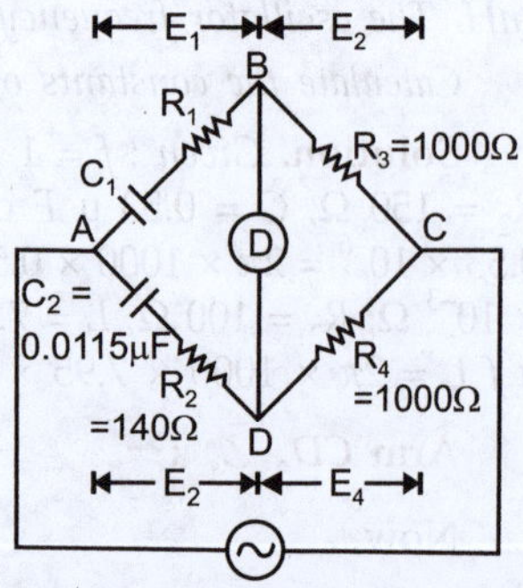

Fig. 10.23.

Inserting the values in eqn.(*i*), we get

$$\left(\frac{1 + j\omega C_1 R_1}{j\omega C_1}\right) R_4 = \left(\frac{1 + j\omega C_2 R_2}{j\omega C_2}\right) R_3$$

or,

$$j\omega C_2 R_4 = \omega^2 R_1 C_1 C_2 R_4$$

$$= j\omega C_1 R_3 - \omega^2 C_1 C_2 R_2 R_3$$

Comparing the *real part*, we have :

$$\omega^2 R_1 C_1 C_2 R_4 = \omega^2 C_1 C_2 R_2 R_3$$

$$R_1 = \frac{R_2 R_3}{R_4} \quad ...(ii)$$

or,

$$\boldsymbol{R_1} = \frac{140 \times 1000}{1000} = \mathbf{140\ \Omega} \textbf{ (Ans.)}$$

Comparing the *imaginary part*, we get :

$$\omega C_2 R_4 = \omega C_1 R_3$$

$$C_2 R_4 = C_1 R_3 \quad ...(iii)$$

or,

$$\boldsymbol{C_1} = \frac{C_2 R_4}{R_3} = \frac{(0.0115 \times 10^{-6}) \times 1000}{1000}$$

$$= 0.0115 \times 10^{-6} \text{ F or } \mathbf{0.0115\ \mu\ F} \textbf{ (Ans.)}$$

MEASUREMENT OF INDUCTANCE

10.3.2. Maxwell Bridge (*for measurement of* "inductance")

Maxwell bridge, shown in Fig. 10.24 *measures an unknown inductance in terms of a known capacitor* :

— The use of standard arm offers the advantage of compactness and easy shielding.

— The capacitor is almost a loss-less component.

— One arm has a resistance R_1 in parallel with C_1, and hence it is easier to write the balance equation using the admittance of arm 1 instead of the impedance.

The general equation for bridge balance is,

$$\bar{Z}_1 \bar{Z}_x = \bar{Z}_2 \bar{Z}_3$$

or, $$\bar{Z}_x = \frac{\bar{Z}_2 \bar{Z}_3}{\bar{Z}_1} = \bar{Z}_2 \bar{Z}_3 \bar{Y}_1 \quad ...(10.57)$$

where, $\bar{Z}_1 = R_1$

in parallel with C_1 *i.e.* $\bar{Y}_1 = \frac{1}{\bar{Z}_1}$

$$\bar{Y}_1 = \frac{1}{R_1} + j\omega C_1$$

$$\bar{Z}_2 = R_2$$

$$\bar{Z}_3 = R_3$$

$\bar{Z}_x = R_x$ in series with L_x

$(= R_x + j\omega L_x)$

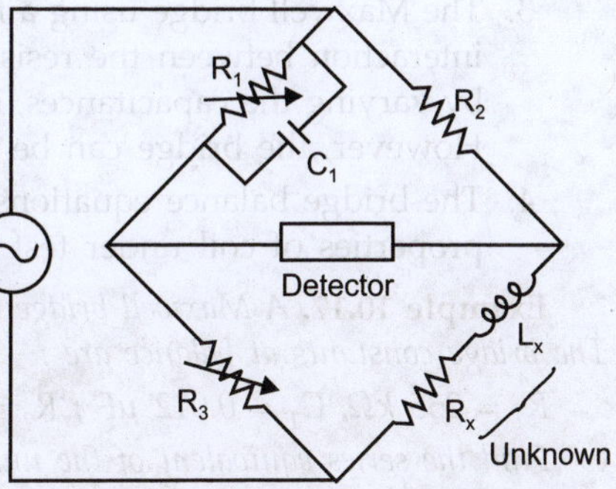

Fig. 10.24. Maxwell bridge for inductance measurements.

Substituting these values in eqn. (10.57), we get

$$R_x + j\omega L_x = R_2 R_3 \left(\frac{1}{R_1} + j\omega C_1\right)$$

or, $$R_x + j\omega L_x = \frac{R_2 R_3}{R_1} + j\omega C_1 R_2 R_3$$

Equating *real terms* and *imaginary terms*, we have

$$R_x = \frac{R_2 R_3}{R_1}; \quad ...(10.58)$$

$$L_x = C_1 R_2 R_3$$

To obtain the bridge balance, first R_3 is adjusted for inductive balance and R_1 is adjusted for resistive balance.

The *"quality factor"* of the coil is given by,

$$Q = \frac{\omega L_x}{R_x} = \frac{\omega.\, C_1 R_2 R_3}{(R_2 R_3 / R_1)} = \frac{\omega.C.\, R_2 R_3 R_1}{R_2 R_3}$$

i.e. $$Q = \omega\, C_1 R_1 \quad ...(10.59)$$

- Maxwell bridge is limited to the measurement of *low Q values (1-10).*
- The bridge is *particularly suited for inductance measurements,* since comparison with a capacitor is more ideal than with another inductance. Commercial bridges measure from *1-1000 H, with ± 2 % error* (If Q is very large, R_1 becomes excessively large and it is impractical to obtain a satisfactory variable standard resistance in the range of values required).

Advantages :

1. The balance equation is independent of losses associated with inductance.
2. The measurement is independent of the excitation *frequency.*
3. The scale of the resistance can be calibrated to read inductance directly.
4. A wide range of inductance at power and audio frequencies can be measured.

Disadvantages :

1. High Q values cannot be measured.
2. It cannot be used for measurement of very low Q values, because of balance coverage problem.
3. The Maxwell bridge using a fixed capacitor has the disadvantage that there is an interaction between the resistance and reactance balances. This can be avoided by varying the capacitances, instead of R_2 and R_3, to obtain a reactance balance. However, the bridge can be made to read directly in Q.
4. The bridge balance equations are independent of frequency. But practically, the properties of coil under test vary with frequency which can cause error.

Example 10.17. *A Maxwell bridge (Fig. 10.24) is used to measure an inductive impedance. The bridge constants at balance are :*

$R_1 = 235\ k\Omega,\ C_1 = 0.012\ \mu F\ ;\ R_2 = 2.5\ k\Omega;\ R_3 = 50\ k\Omega$

Find the series equivalent of the unknown impedance.

Solution. Refer to Fig. 10.26.

$R_x : ; L_x :$

We know that : $R_x = \dfrac{R_2 R_3}{R_1} = \dfrac{2.5 \times 50}{235}$

$= \mathbf{0.53\ k\ \Omega}$ **(Ans.)**

$L_x = C_1 R_2 R_3 = 0.012 \times 10^{-6} \times (2.5 \times 10^3) \times (50 \times 10^3)$

$= \mathbf{1.5\ H}$ **(Ans.)**

The *equivalent series circuit* is shown in Fig. 10.25.

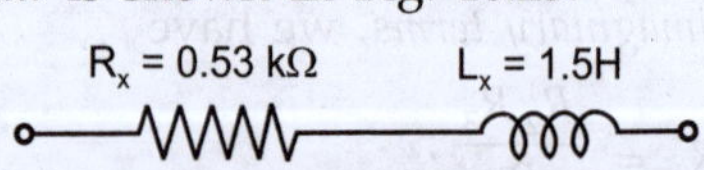

Fig. 10.25.

10.3.3. Hay Bridge (*For measurement of* "inductance")

The Hay bridge shown in Fig. 10.26 differs from the Maxwell bridge by having resistor R_1 in series with standard capacitor C_1 instead of in parallel. It is apparent that for large phase angles, R_1 should have a very low value. This bridge is, thus, *more convenient for measuring high – Q coils.*

The general equation for bridge balance is,

$$\bar{Z}_1 \bar{Z}_x = \bar{Z}_2 \bar{Z}_3 \qquad ...(i)$$

where,

$$\bar{Z}_1 = R_1 - \frac{j}{\omega C_1};$$

$$\bar{Z}_2 = R_2\ ;$$

$$\bar{Z}_3 = R_3\ ;$$

$$\bar{Z}_x = R_x + j\omega L_x$$

Substituting these values in eqn. (*i*), we get :

$$\left(R_1 - \frac{j}{\omega C_1}\right)(R_x + j\omega L_x) = R_2 R_3$$

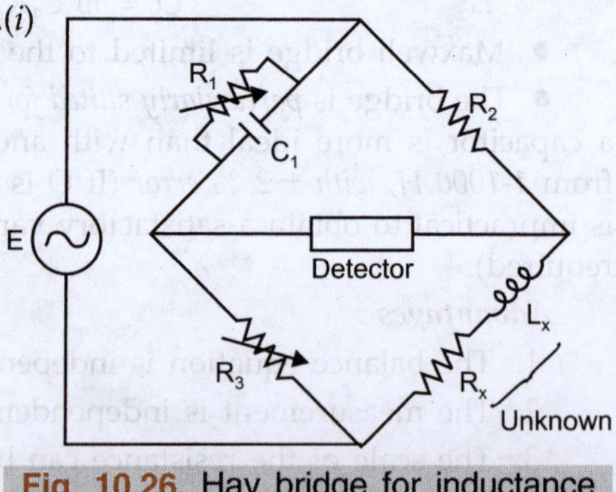

Fig. 10.26. Hay bridge for inductance measurements.

$$R_1R_x + j\omega L_x R_1 - \frac{jR_x}{\omega C_1} + \frac{L_x}{C_1} = R_2 R_3$$

Equating the *real* and *imaginary* terms, we get :

$$R_1 R_x + \frac{L_x}{C_1} = R_2 R_3 \quad ...(10.60)$$

and,
$$\frac{R_x}{\omega C_1} = \omega L_x R_1 \quad ...(10.61)$$

or,
$$R_x = \omega^2 L_x C_1 R_1 \quad ...[10.61\ (a)]$$

Substituting for R_x in eqn. (10.60), we have

$$R_1(\omega^2 L_x C_1 R_1) + \frac{L_x}{C_1} = R_2 R_3$$

or,
$$\omega^2 R_1^2 C_1 L_x + \frac{L_x}{C_1} = R_2 R_3$$

Multiplying both sides by C_1, we get

$$\omega^2 R_1^2\, C_1^2\, L_x + L_x = R_2 R_3 C_1$$

or,
$$L_x\,(1 + \omega^2 R_1^2\, C_1^2\, L_x) = R_2 R_3 C_1$$

or,
$$L_x = \frac{R_2 R_3 C_1}{1 + \omega^2 R_1^2\, C_1^2} \quad ...(10.62)$$

Substituting for L_x in eqn. (10.61*a*), we have

$$R_x = \frac{\omega^2 C_1^2\, R_1 R_2 R_3}{1 + \omega^2 R_1^2\, C_1^2} \quad ...(10.63)$$

The term ω appears in the expressions for both L_x and R_x. This indicates that the bridge is *frequency sensitive.*

— This bridge is also used in the *measurement of incremental inductance*. The inductance balance equation depends on the losses of the inductor (or Q) and also on the operating frequency.

- A *commercial bridge measures from 1 μH – 100H with ± 2 % error.*
- An inconvenient feature of this bridge is that equation giving the balance condition for inductance contains the multiplier $\frac{1}{(1+1/Q^2)}$ (where $Q = \frac{1}{\omega R_1 C_1}$). The inductance balance thus depends on its Q and frequency.

$$\therefore \quad L_x = \frac{R_2 R_3 C_1}{1 + (1/Q^2)} \quad ...(10.64)$$

For a value of Q greater than 10, the term $\frac{1}{Q^2}$ will be smaller than $\frac{1}{100}$ and can therefore be *neglected*. Then,

$L_x = R_2 R_3 C_1$, which is the same as Maxwell's equation. But for inductors with a Q less than 10, the $\frac{1}{Q^2}$ term cannot be neglected. Hence *this bridge is not suited for measurements of coils having Q < 10.*

Example 10.18. *Find the series equivalent inductance and resistance of the network that causes an opposite angle (Hay bridge) to null the following bridge arms (Fig. 10.27)*

$\omega = 3000$ *rad/s ;* $R_2 = 9$ *kΩ ;* $R_1 = 1.8$ *kΩ ;* $C_1 = 0.9$ *μF,* $R_3 = 0.9$ *kΩ.*

Solution. We know that,

$$R_x = \frac{\omega^2 C_1^2 R_1 R_2 R_3}{1 + \omega^2 R_1^2 C_1^2} \quad ...(10.63)$$

$$= \frac{(3000)^2 (0.9 \times 10^{-6})^2 \times (1.8 \times 1000)(9 \times 1000)(0.9 \times 1000)}{1 + (3000)^2 (1.8 \times 1000)^2 (0.9 \times 10^{-6})^2}$$

$$= \frac{106.27 \times 10^3}{1 + 14.58}$$

$$= 6822 \; \Omega \quad \text{or} \quad 10.822 \; k\Omega$$

Also,

$$L_x = \frac{R_2 R_3 C_1}{1 + \omega^2 R_1^2 C_1^2} \quad ...(10.62)$$

$$= \frac{(9 \times 1000)(0.9 \times 1000) \times (0.9 \times 10^{-6})}{1 + (3000)^2 (1.8 \times 1000)^2 (0.9 \times 10^{-6})^2}$$

$$= \frac{7.29}{1 + 14.58} = 0.468 \text{H}$$

i.e. R_x = **6.822 kΩ;** L_x = **0.468 H (Ans.)**

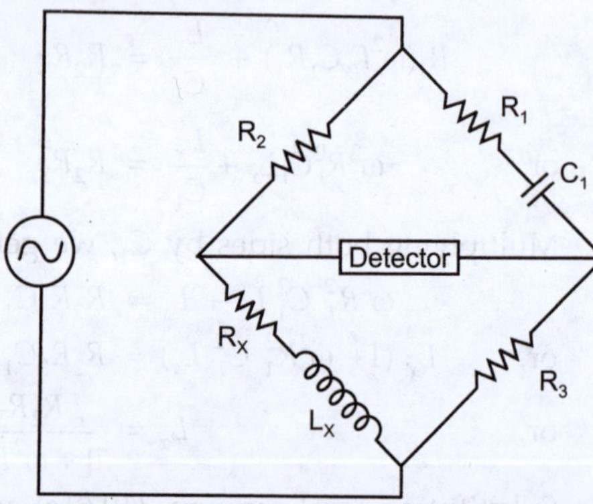

Fig. 10.27.

10.3.4. Anderson Bridge (For *measurement of* "Inductance")

This method is one of the commonest and best bridge methods for precise measurement of inductance over a wide range.

— Here, the self-inductance is *measured in terms of a standard capacitor.*

Fig. 10.28 shows the circuit and the phasor diagrams of the bridge for *balanced conditions.*

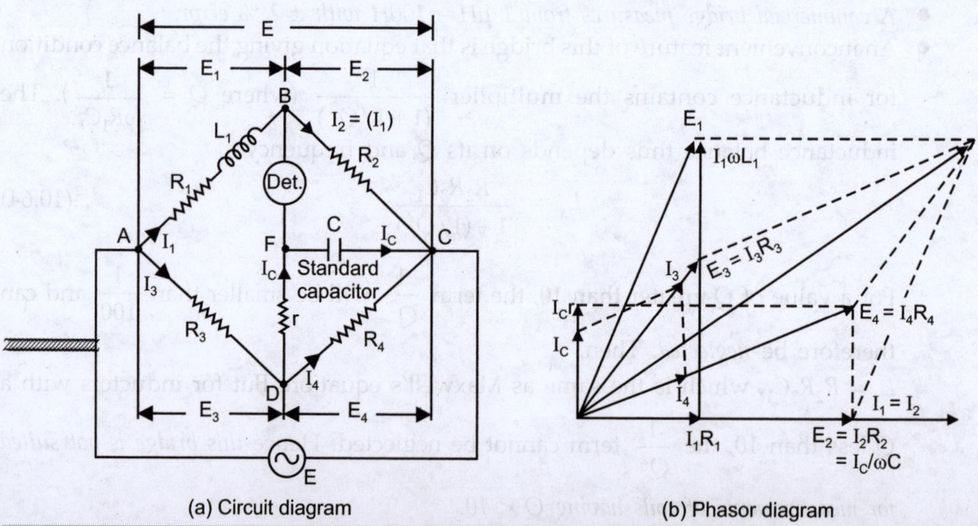

Fig. 10.28. Anderson bridge.

Let, L_1 = Self-inductance to be measured,

R_1 = Resistance of the coil under test,

R_2, R_3, R_4, r = Known non-inductive resistances, and

C = Standard known capacitor.

At balance :

$$\bar{I}_1 = \bar{I}_2 \quad \text{and,} \quad \bar{I}_3 = \bar{I}_C + \bar{I}_4$$

Now,
$$\bar{I}_1 R_2 = I_C \times \frac{1}{j\omega C}$$

or,
$$I_C = \bar{I}_1 j\omega C R_2 \qquad \text{...(i)}$$

Other balance equations are :

$$\bar{I}_1(R_1 + j\omega L_1) = \bar{I}_2 R_3 + I_C \cdot r \qquad \text{...(ii)}$$

and,
$$\bar{I}_C \cdot r + \bar{I}_C \cdot \frac{1}{j\omega C} = (\bar{I}_3 - \bar{I}_C)\, R_4$$

or,
$$I_C\left(r + \frac{1}{j\omega C}\right) = (\bar{I}_3 - \bar{I}_C)\, R_4 \qquad \text{...(iii)}$$

Substituting the value of $\bar{I}_C$ from (*i*) in eqns. (*ii*) and (*iii*), we get :

$$\bar{I}_1\,(R_1 + j\omega L_1) = \bar{I}_3 R_3 + \bar{I}_1 j\omega C R_2 \cdot r$$

or,
$$\bar{I}_1\,(R_1 + j\omega L_1 - j\omega C R_2 \cdot r) = \bar{I}_3 R_3 \qquad \text{...(iv)}$$

and,
$$\bar{I}_1 j\omega C R_2\left(1 + \frac{1}{j\omega C}\right) = (\bar{I}_3 - \bar{I}_1 \cdot j\omega C R_2)\, R_4$$

or,
$$\bar{I}_1 j\omega C R_2 r + \bar{I}_1 R_2 = \bar{I}_3 R_4 - \bar{I}_1 j\omega C R_2 R_4$$

or,
$$\bar{I}_1\,(j\omega C R_2 r + j\omega C R_2 R_4 + R_2) = \bar{I}_3 R_4 \qquad \text{...(v)}$$

Substituting the value of R_4 from (*v*) in (*iv*), we get

$$\bar{I}_1\,(R_1 + j\omega L_1 - j\omega C R_2 \cdot r) = \bar{I}_1\left(\frac{j\omega C R_2 R_3 r}{R_4} + j\omega C R_2 R_3 + \frac{R_2 R_3}{R_4}\right)$$

Equating the *real* and *imaginary terms*, we get :

$$R_1 = \frac{R_2 R_3}{R_4} \qquad \text{...(10.65)}$$

$$L_1 - C R_2 r = \frac{C R_2 R_3 r}{R_4} + C R_2 R_3$$

or,
$$L_1 = C R_2 r + \frac{C R_2 R_3 r}{R_4} + C R_2 R_3$$

or,
$$L_1 = C R_2\left[r\left(1 + \frac{R_3}{R_4}\right) + R_3\right]$$

or, $$L_1 = \frac{CR_2}{R_4}\left[r(R_3 + R_4) + R_3R_4\right] \quad ...(10.66)$$

Advantages :

1. A fixed capacitor can be used instead of a variable capacitor as in the case of a Maxwell bridge.
2. It is much easier to obtain balance in case of Anderson bridge than a Maxwell bridge for low Q-coils.
3. This bridge may be used for accurate determination of capacitance in terms of inductance.

Disadvantages/Limitations :

1. It is more complex than its prototype Maxwell bridge.
2. The shielding of the bridge becomes more difficult due to an additional junction point.
   - In view of the above complications of the Anderson bridge, in all the cases where a variable capacitor is permissible the more simple Maxwell bridge is used (instead of Anderson bridge).

Example 10.19. *Fig. 10.29 shows the connection of an Anderson bridge for measuring the inductance L and resistance R of the coil. Find R and L, if balance is obtained when Q = S = 1,000 Ω, P = 500 Ω, k = 100 Ω and C = 0.5 μF.* **(AMIE)**

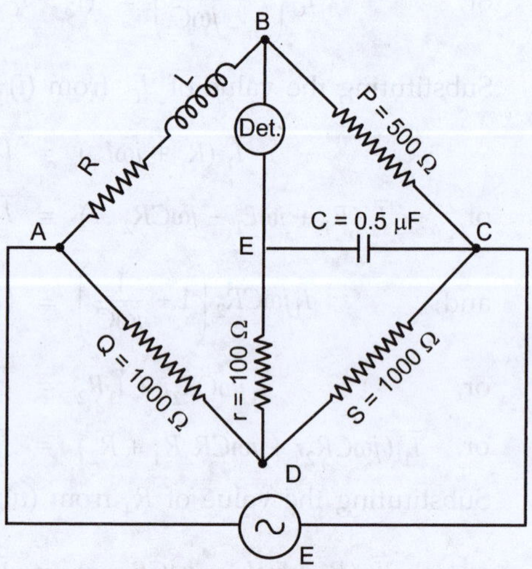

Fig. 10.29.

Solution. *Given* : Q = S = 1,000 Ω, P = 500 Ω, r = 100 Ω, C = 0.5 μF.

R, L :

In the balance condition of bridge :

$$R = \frac{P \times Q}{S} \quad \text{[Eqn. (10.65)]}$$

$$= \frac{500 \times 1000}{1000} = 500\ \Omega \quad \textbf{(Ans.)}$$

$$L = \frac{CP}{S}[r(Q+S) + QS] \quad \text{[Eqn. (10.67)]}$$

$$= \frac{0.5 \times 10^{-5} \times 500}{1000}[100(1000 + 1000) + 1000 \times 1000]$$

$$= \mathbf{0.3\ H} \quad \textbf{(Ans.)}$$

Example 10.20. *The following data relate to Anderson bridge shown in Fig. 10.30.*

The arms BC, CD and DA consist of resistances having values 1000 Ω, 1000 Ω and 500 Ω respectively.

A resistance of 100 Ω and a capacitance of 3 μF are connected respectively to the arms DF and CF.

An A.C. supply of 100 Hz is applied between the terminals A and C and a detector is connected between the terminals B and F.

The detector indicates null under the above conditions. Determine the values of R and L connected to the arm AB. **(UPTU)**

Solution. *Given* : $R_2 = 1000\ \Omega$; $R_3 = 500\ \Omega$; $R_4 = 1000\ \Omega$; $r = 100\ \Omega$; $C = 3\ \mu F$.

R, L :

At balance condition of the bridge :

We know that :

$$R = \frac{R_2 R_3}{R_4} \qquad \text{[Eqn. (10.65)]}$$

$$= \frac{1000 \times 500}{1000} = \mathbf{500\ \Omega} \quad \textbf{(Ans.)}$$

$$L = \frac{CR_2}{R_4}[r(R_3 + R_4) + R_3 R_4] \qquad \text{[Eqn. (10.66)]}$$

$$= \frac{3 \times 10^{-6} \times 1000}{1000}[100(500 + 1000) + 500 \times 1000]$$

$$= \mathbf{1.95\ H} \quad \textbf{(Ans.)}$$

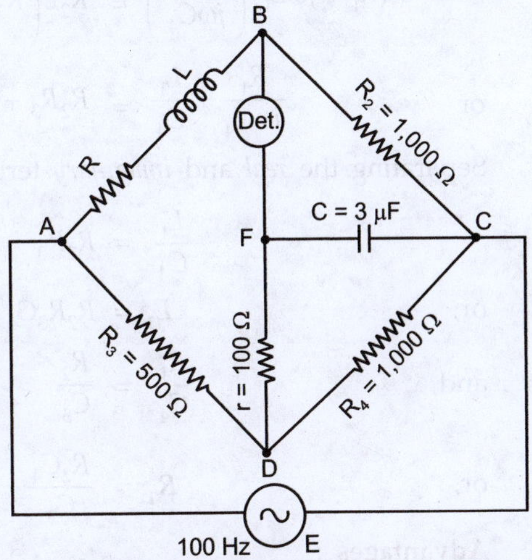

Fig. 10.30.

10.3.5. Owen Bridge (For *measurement of* "Inductance")

This bridge may be used *for measurement of an inductance in terms of capacitance.*

Fig. 10.31 shows the circuit and phasor diagrams, for this bridge, under balance condition.

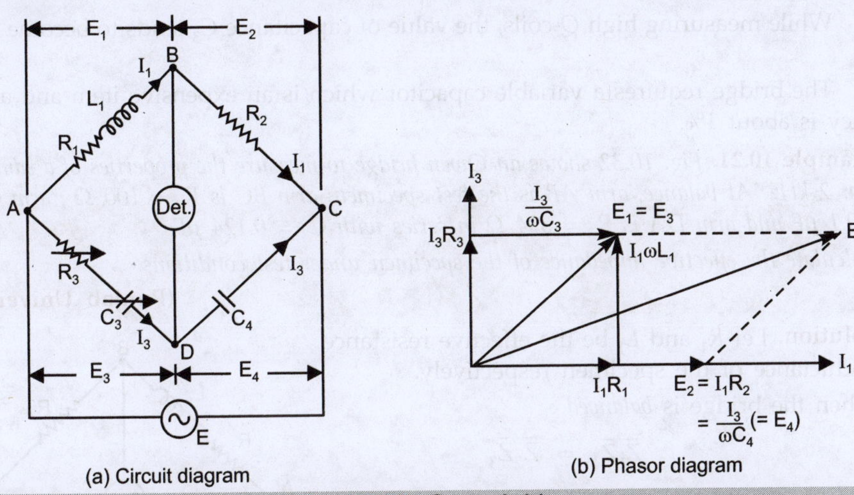

Fig. 10.31. Owen bridge.

Let, L_1 = Unknown self-inductance of resistance R_1,

R_2 = Fixed non-inductive resistance,

R_3 = Variable non-inductive resistance,

C_3 = Variable standard capacitor, and

C_4 = Fixed standard capacitor.

When the bridge is balanced :

$$\bar{Z}_1 . \bar{Z}_4 = \bar{Z}_2 . \bar{Z}_3$$

$$(R_1 + j\omega L_1)\left(\frac{1}{j\omega C_4}\right) = R_2 \cdot \left(R_3 + \frac{1}{j\omega C_3}\right)$$

or $$\frac{R_1}{j\omega C_4} + \frac{L_1}{C_4} = R_2R_3 + \frac{R_2}{j\omega C_3}$$

Separating the *real* and *imaginary* terms, we have :

$$\frac{L_1}{C_4} = R_2R_3$$

or, $$L_1 = R_2R_3C_4 \qquad ...(10.67)$$

and, $$\frac{R_1}{C_4} = \frac{R_2}{C_3}$$

or, $$R_1 = \frac{R_2C_4}{C_3} \qquad ...(10.68)$$

Advantages :

1. The balance equations are quite simple and do not contain any frequency component.
2. Since R_3 and C_3, the variable elements, are in the same arm, convergence to balance condition is much easier.
3. The bridge can be used over a wide range of measurement of inductances.

Disadvantages :

1. While measuring high *Q*-coils, the value of capacitance C_3 tends to become rather large.
2. The bridge requires a variable capacitor which is an expensive item and also its accuracy is about 1%.

Example 10.21. *Fig. 10.32 shows an Owen bridge to measure the properties of a sample of sheet at 2 kHz. At balance, arm AB is the test specimen; arm BC is $R_2 = 100\ \Omega$; arm CD is $C_4 = 0.1\ \mu F$ and arm DA is $R_3 = 834\ \Omega$ in series with $C_3 = 0.124\ \mu F$.*

Calculate the effective impedance of the specimen under test conditions.

(Punjab University)

Solution. Let R_1 and L_1 be the effective resistance and inductance of the specimen respectively.

When the bridge is *balanced* :

$$\bar{Z}_1\bar{Z}_4 = \bar{Z}_2\bar{Z}_3$$

or, $$(R_1 + j\omega L_1)\left(\frac{1}{j\omega C_4}\right) = R_2\left(R_3 + \frac{1}{j\omega C_3}\right)$$

or, $$R_1 \cdot \frac{1}{j\omega C_4} + \frac{L_1}{C_4} = R_2R_3 + R_2 \cdot \frac{1}{j\omega C_3}$$

Equating *real* and *imaginary* terms, we get :

$$\frac{L_1}{C_4} = R_2R_3$$

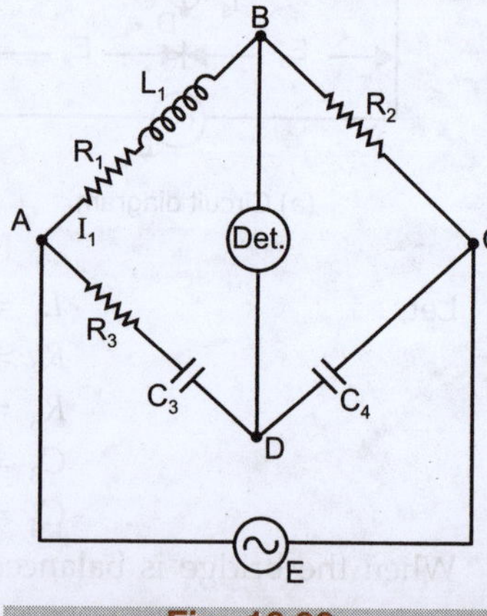

Fig. 10.32.

or, $L_1 = R_2 R_3 C_4 = 100 \times 834 \times 0.1 \times 10^{-6}$

$= 8.34 \times 10^{-3}$ H or **8.34 mH** **(Ans.)**

and, $\frac{R_1}{C_4} = \frac{R_2}{C_3}$

or, $R_1 = R_2 \times \frac{C_4}{C_3} = 100 \times \frac{0.1}{0.124}$ = **80.6 Ω** **(Ans.)**

Reactance of specimen at 2 kHz,

$$X_1 = 2\pi f L_1 = 2\pi \times (2 \times 1000) \times 8.34 \times 10^{-3} = 104.8\ \Omega$$

∴ Impedance of the specimen, $Z_1 = \sqrt{R_1^2 + X_1^2}$

$= \sqrt{(80.6)^2 + (104.8)^2}$ = **132.2 Ω** **(Ans.)**

MEASUREMENT OF CAPACITANCE

10.3.6. De Sauty Bridge (For *measurement of* "Capacitance")

This bridge is the simplest method of *comparing two capacitances.*

Fig. 10.33(*a* & *b*) shows the circuit and phasor diagrams of the bridge.

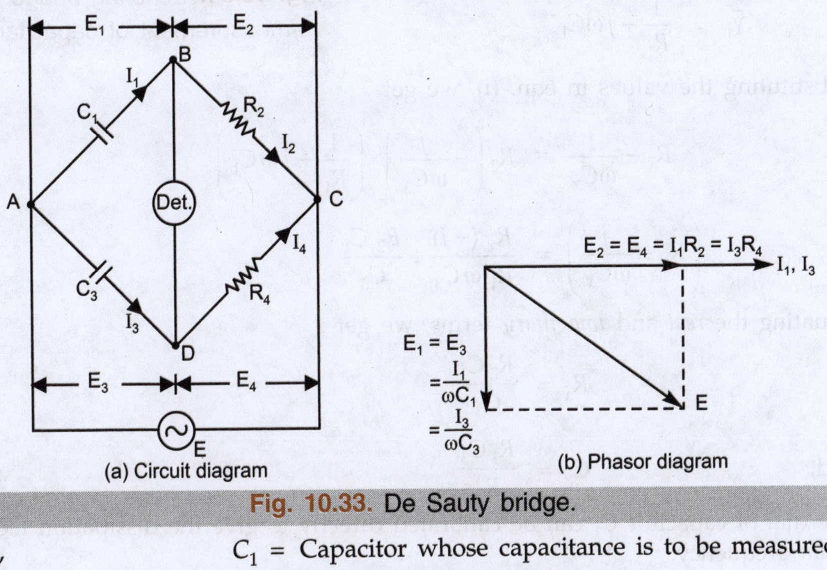

Fig. 10.33. De Sauty bridge.

Let, C_1 = Capacitor whose capacitance is to be measured,

C_3 = Standard capacitor, and

R_3, R_4 = Non-inductive resistors.

At balance : $I_1 = I_2$; $I_3 = I_4$

and, $\left(\frac{1}{j\omega C_1}\right) R_4 = \left(\frac{1}{j\omega C_3}\right) R_2$

or, $\frac{R_4}{C_1} = \frac{R_2}{C_3}$

or, $$C_1 = \frac{C_3 R_4}{R_2} \quad ...(10.69)$$

10.3.7. Schering Bridge (*for Measurement of* "Capacitance")

It is one of the most important A.C. bridges, widely used for the *measurement of unknown capacitors, dielectric loss and power factor.*

Fig. 10.34 shows the basic circuit arrangement of Schering bridge. The capacitor C_3 is a high quality mica capacitor (low-loss) for general measurements, or an air capacitor (having a very stable value and a very small electric field) for insulation measurement.

The general equation for balance is,

$$\bar{Z}_1 \bar{Z}_x = \bar{Z}_2 \bar{Z}_3$$

or, $$\bar{Z}_x = \frac{\bar{Z}_2 \bar{Z}_3}{\bar{Z}_1} = \bar{Z}_2 \bar{Z}_3 \bar{Y}_1 \quad ...(i)$$

where, $$\bar{Z}_x = R_x - \frac{j}{\omega C_x}$$

$$\bar{Z}_2 = R_2$$

$$\bar{Z}_3 = -\frac{j}{\omega C_3}$$

$$\bar{Y}_1 = \frac{1}{R_1} + j\omega C_1$$

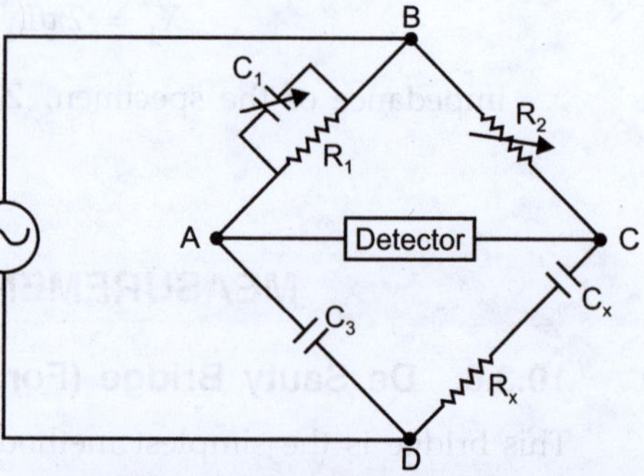

Fig. 10.34. Schering bridge for the measurement of capacitance.

Substituting the values in eqn. (*i*), we get

$$R_x - \frac{j}{\omega C_x} = R_2\left(-\frac{j}{\omega C_3}\right) \times \left(\frac{1}{R_1} + j\,\omega\,C_1\right)$$

or $$\left(R_x - \frac{j}{\omega C_x}\right) = \frac{R_x\,(-j)}{R_1\,\omega\,C_3} + \frac{R_2\,C_1}{C_3}$$

Equating the *real* and *imaginary* terms, we get :

$$R_x = \frac{R_2\,C_1}{C_3} \quad ...(10.70)$$

and, $$C_x = \frac{R_1\,C_3}{R_2} \quad ...(10.71)$$

The dial of capacitor C_1 can be calibrated directly to give the dissipation factor at a particular frequency.

- The lower junction of the bridge is grounded. At the frequency normally used on this bridge, the reactances of capacitors C_3 and C_x are much higher than the resistances of R_1 and R_2. Hence if the junction of R_1 and R_2 is grounded, the detector is effectively at ground potential. This reduces any stray-capacitance effect, and makes the bridge more stable.

Power factor. The *power factor* (p.f.) of a series R.C. combination is defined as *the cosine of the phase angle of the circuit.*

$\therefore$ The power factor of the unknown, p.f. $= \dfrac{R_x}{Z_x}$

For phase angles very close to 90°, the reactance is almost equal to the impedance and we can approximate the power factor to,

$$\text{p.f.} \simeq \frac{R_x}{X_x} = \frac{R_x}{\frac{1}{\omega C_x}} = \omega R_x C_x \quad ...(10.72)$$

Dissipation factor (D). The *dissipation factor* of a series *R-C* circuit is defined *as the cotangent of the phase angle.*

i.e.,
$$D = \frac{R_x}{X_x} = \omega C_x R_x \quad ...(10.73)$$

Also, since the quality of a coil is defined by $Q = \frac{X_L}{R_L}$, we find that the dissipation factor, *D*, is the *reciprocal of the quality factor, Q,*

i.e.
$$D = \frac{1}{Q} \quad ...(10.74)$$

The dissipation factor tells us something about the quality of the capacitor; *i.e.,* how close the phase angle of the capacitor is to the ideal value of 90°.

Substituting the value of C_x in eqn. (10.71) and of R_x in eqn. (10.70) in the eqn. (10.74), we get

$$D = \omega \times \frac{R_1 C_3}{R_2} \times \frac{R_2 C_1}{C_3} = \omega R_1 C_1 \quad ...(10.75)$$

— When resistor R_1 in the Schering bridge of Fig. 10.34 has a fixed value, the dial of capacitor C_1 may be calibrated directly in dissipation factor *D*. This is the usual practice in a Schering bridge. The appearance of the term ω in the expression for the dissipation factor [eqn. (10.75)] means, of course, that the calibration of the C_1 dial holds for only one particular frequency at which dial is calibrated. A different frequency can be used, provided that a correction is made by multiplying the C_1 dial reading by the ratio of the two frequencies.

- Commercial units *measure from 100 pF–1μF, with ± 2% accuracy.*
- The Schering bridge is *widely used for testing small capacitors at low voltages with very high precision.*

Example 10.22. *The Schering bridge shown in Fig. 10.34 has the following constants :*
Arm AB ... Capacitor of 0.4 μF in parallel with 1.5 kΩ resistance;
Arm BC ... Resistance of 3 kΩ;
Arm CD ... Unknown capacitor C_x and R_x in series
Arm DA ... Capacitor of 0.4 μF;
Frequency ... 1 kHz.
Determine the following :
(i) R_x and C_x.
(ii) Dissipation factor.

Solution. *Given:* R_1 = 1.5 kΩ; C_1 = 0.4 μF ; R_2 = 3 kΩ; C_3 = 0.4 μF.

(i) ***R_x; C_x :***

We know that,
$$R_x = \frac{R_2 C_1}{C_3} = \frac{3k\Omega \times 0.4\,\mu F}{0.4\mu F} \quad \text{[Eqn. (10.70)]}$$
$$= \mathbf{3\ k\Omega} \quad \textbf{(Ans.)}$$

and, $$C_x = \frac{R_1 C_3}{R_2} = \frac{1.5\,k\Omega \times 0.4\,\mu F}{3\,k\Omega}$$
$$= \mathbf{0.2\ \mu F}\ \textbf{(Ans.)}$$

(*ii*) **Dissipation factor, *D* :**

$$D = \omega\, C_x\, R_x \qquad \text{[Eqn. (10.73)]}$$
$$= 2\,\pi\, f.\, C_x.\, R_x$$
$$= 2\,\pi \times 1000 \times 0.2 \times 10^{-6} \times (3 \times 1000) = 3.77$$

i.e. $$D = \mathbf{3.77}\ \textbf{(Ans.)}$$

Example 10.23. *A bakelite sheet of 5 mm thickness is tested at 50 Hz between the electrodes 12 cm in diameter. The Schering bridge (Fig. 10.35) used has an air capacitor C_2 of 106 pF, a non-reactive resistance of R_4 of $\left(\frac{1000}{\pi}\right)$ Ω in parallel with a variable capacitor C_4 and a non-reactive variable resistance R_3. Balance is obtained with C_4 = 0.55μF and R_3 = 270Ω.*

Determine the following :

(*i*) *Capacitance ;*
(*ii*) *Power factor, and*
(*iii*) *Relative permittivity of the sheet.*

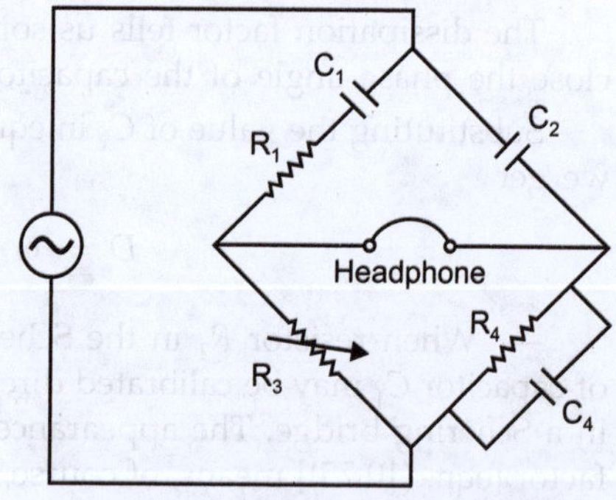

Fig. 10.35.

Solution. Refer to Fig. 10.35.

Given :
$$C_2 = 106\text{pF}$$
$$= 106 \times 10^{-12}\ \text{F};$$
$$R_4 = \frac{1000}{\pi}\,\Omega\,;$$
$$C_4 = 0.55\mu\ \text{F}$$
$$= 0.55 \times 10^{-6}\ \text{F};$$
$$R_3 = 270\ \Omega$$

(*i*) **Capacitance *C* :**

When the bridge is balanced, we have

$$R_1 = \frac{R_3\, C_4}{C_2} \qquad \text{...[Eqn. (10.70)]}$$
$$= \frac{270 \times 0.55 \times 10^{-6}}{106 \times 10^{-12}} = 1.4\ \text{M}\,\Omega$$

and, $$C_1 = \frac{R_4\, C_2}{R_3} \qquad \text{...[Eqn. (10.71)]}$$
$$= \frac{(1000/\pi) \times 106 \times 10^{-12}}{270}$$
$$= \mathbf{124.9\ pF}\ \textbf{(Ans.)}$$

(*ii*) **Power factor :**

$$\text{Power factor} = \omega R_1 C_1 \qquad \text{...[(Eqn. (10.72)]}$$
$$= 2\pi \times 50 \times (1.4 \times 10^{6}) \times (124.9 \times 10^{-12})$$
$$= \mathbf{0.0549}\ \textbf{(Ans.)}$$

(iii) **Relative permittivity ϵ_r :**

We know that $$C = \frac{\epsilon_0 \epsilon_r A}{t} \quad ...(i)$$

where, $C (= C_1)$ = Capacitance,

$\epsilon_o = 8.854 \times 10^{-12}$

A = Area of the electrodes

$$= \frac{\pi}{4} \times \left(\frac{12}{100}\right)^2 = 0.0113\, m^2, \text{ and}$$

t = Thickness of bakelite sheet

= 5 mm or 0.005 m.

Substituting the values in eqn. (*i*), we get

$$124.9 \times 10^{-12} = \frac{8.854 \times 10^{-12} \times \epsilon_2 \times 0.0113}{0.005}$$

or $$\epsilon_r = \frac{124.9 \times 10^{-12} \times 0.005}{8.854 \times 10^{-12} \times 0.0113}$$

= **6.24 (Ans.)**

Example 10.24. *The four arms of an A.C. bridge network are as follows :*

Arm AB : an unknown impedance; arm BC: a standard capacitor C_3 of 1000 pF; arm CD : a non-inductive resistor R_4 of 100 Ω in parallel with a capacitor C_4 of 0.01 μF; arm DA : a non-inductive resistor R_2 of 1000 Ω.

The A.C. supply is connected across terminals B, D and the supply frequency is 50 Hz. If the bridge is balanced with the above values, determine the components of the unknown impedance, while deriving the balance conditions. **(AMIE)**

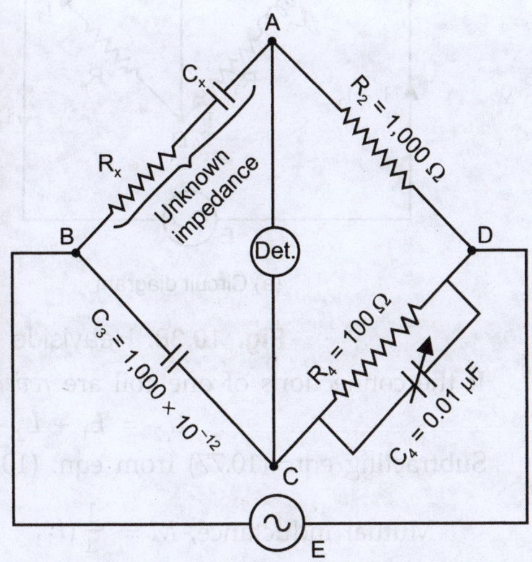

Fig. 10.36.

Solution. The configuration of the bridge is that of *Schering bridge* (Fig. 10.36).

Given : $R_2 = 1000\ \Omega$; $C_3 = 1000 \times 10^{-12}$ pF;

$R_4 = 100\ \Omega$; $C_4 = 0.01\ \mu F$.

We know that, $$R_x = \frac{R_2 C_4}{C_3}$$

$$= \frac{1000 \times 0.01 \times 10^{-6}}{1000 \times 10^{-12}}$$

= **0.01 MΩ (Ans.)**

and, $$C_x = \frac{R_4 C_3}{R_2} = \frac{100 \times 1000 \times 10^{-12}}{1000}$$

= **100 pF (Ans.)**

Measurement of Mutual Inductance

10.3.8. Heaviside Mutual Inductance Bridge

Measurement of mutual inductance. If the terminals of two coils whose mutual inductance is to be measured are available for series connections, measurement may be made by one of the self-inductance bridges discussed earlier. If the connections are made so that the magnetic fields of two coils are additive as shown in Fig. 10.37(a), the effective inductance of the two coils in series is,

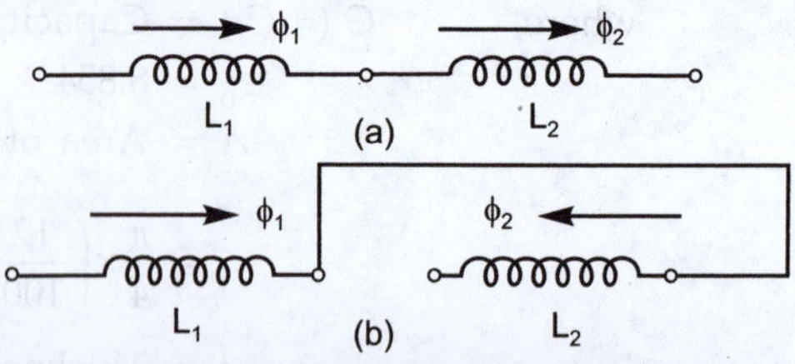

Fig. 10.37. Positive and negative couplings for mutual inductors.

$$L_{e1} = L_1 + L_2 + 2M \quad ...(10.76)$$

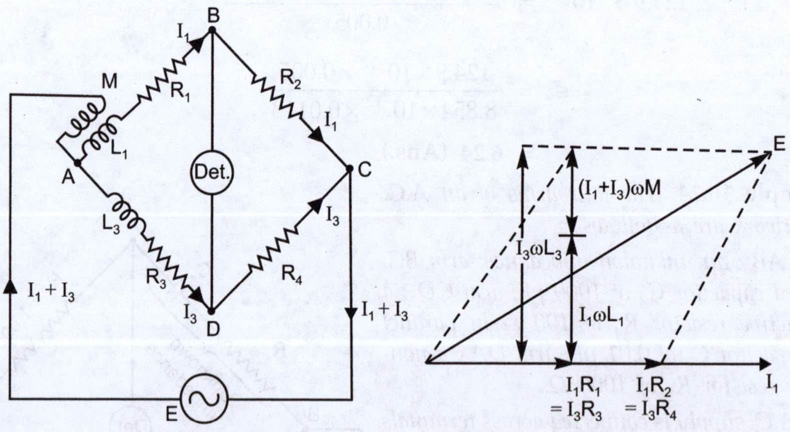

(a) Circuit diagram (b) Phasor diagram

Fig. 10.38. Heaviside mutual inductance bridge.

If the connections of one coil are *reversed* as shown in Fig. 10.37(b), then

$$L_{e2} = L_1 + L_2 - 2M \quad ...(10.77)$$

Subtracting eqn. (10.77) from eqn. (10.76), we get

$$\text{Mutual inductance, } M = \frac{1}{4}(L_{e1} - L_{e2}) \quad ...(10.78)$$

Hence mutual inductance is obtained as one-fourth of difference of self-inductance measured with *series additive* and *series subtractive* connections.

This method is of advantage only when fairly high coupling between the two coils is obtained otherwise it results in poor accuracy due to nearly equal terms of eqn. (10.78).

- Fig. 10.38 shows the circuit and phasor diagrams of Heaviside mutual inductance bridge, which *measures mutual inductance in terms of a known self-inductance.*

Let,

M = Unknown mutual inductance,

L_1 = Self-inductance of secondary of mutual inductance,

L_3 = Known self-inductance, and

R_2, R_3, R_3, R_4 = Non-inductive resistors.

At balance voltage drop between B and C must equal the voltage drop between D and C. Also voltage drop across A–B–C equals the voltage drop across A–D–C. Thus we have the following equations at balance :

$$\bar{I}_1 R_2 = \bar{I}_2 R_4 \quad ...(i)$$

and, $$(\bar{I}_1 + \bar{I}_2)(j\omega M) + \bar{I}_1 (R_1 + R_2 + j\omega L_1) = \bar{I}_2 (R_3 + R_4 + j\omega L_3) \quad ...(ii)$$

Inserting the value of $\bar{I}_1 = \dfrac{\bar{I}_1 R_4}{R_2}$ from (*i*) in (*ii*), we get

$$\left(\frac{\bar{I}_2 R_4}{R_2} + \bar{I}_2\right)(j\omega M) + \frac{\bar{I}_2 R_4}{R_2}(R_1 + R_2 + j\omega L_1) = \bar{I}_2 (R_3 + R_4 + j\omega L_3)$$

or, $$I_2\left(\frac{R_4}{R_2} + 1\right)(j\omega M) + \frac{\bar{I}_2 R_4}{R_2}(R_1 + R_2 + j\omega L_1) = \bar{I}_2 (R_3 + R_4 + j\omega L_3)$$

or, $$j\omega M\left(\frac{R_4}{R_2} + 1\right) + \frac{R_4 R_1}{R_2} + R_4 + j\omega L_1 \frac{R_4}{R_2} = R_3 + R_4 + j\omega L_3$$

Separating the *real* and *imaginary* terms, we have :

$$\frac{R_4 R_1}{R_2} + R_4 = R_3 + R_4$$

or, $$\frac{R_4 R_1}{R_2} = R_3$$

or, $$R_1 = \frac{R_2 R_3}{R_4} \quad ...(10.79)$$

and, $$\left(\frac{R_4}{R_2} + 1\right) M + L_1 \frac{R_4}{R_2} = L_3$$

$$\left(\frac{R_4}{R_2} + 1\right) M = L_3 - \frac{L_1 R_4}{R_2}$$

or, $$\left(\frac{R_4 + R_2}{R_2}\right) M = \frac{R_2 L_3 - R_4 L_1}{R_2}$$

or, $$M = \frac{R_2 L_3 - R_4 L_1}{R_2 + R_4} \quad ...(10.80)$$

It is clear from eqn. (10.80), that L_1, the self-inductance of the secondary of the mutual inductor must be known in order that M be measured by this method.

In case, $R_2 = R_4$, we get, $$M = \frac{L_2 - L_1}{2} \quad ...(10.81)$$

and $$R_1 = R_3 \quad ...(10.82)$$

- This method *can be used for measurement of self-inductance.* Let L_3 be the self-inductance to *be determined.*

Then, from eqns. (10.79) and (10.80), we get :

$$L_3 = \frac{M(R_2 + R_4) + R_4 L_1}{R_2}$$

$$= M\left(1 + \frac{R_4}{R_2}\right) + \frac{R_4}{R_2} L_1 \qquad ...(10.83)$$

and,
$$R_3 = \frac{R_1 R_4}{R_2} \qquad ...(10.84)$$

In case $R_4 = R_2$, we have

$$L_3 = 2M + L_1$$

and
$$R_3 = R_1$$

Campbell's Modification of Heaviside bridge :

This bridge is used to meaure self-inductance in terms of a mutual inductance.

Fig. 10.39 shows a modified Heaviside bridge; this modification is due to Campbell :

— In this case an additional balancing coil L, R is included in arm AD in series with the inductor under test.

— An additional resistance r is put in arm AB. Balance is obtained by varying M and r.

— A short-circuiting switch S is placed across the coil R_3, L_3 *under measurement.*

— Two sets of readings are taken, one with switch being open and the other with switch being closed. Let values of M and r be M_1 and r_1 with switch open and M_2, r_2 with switch closed.

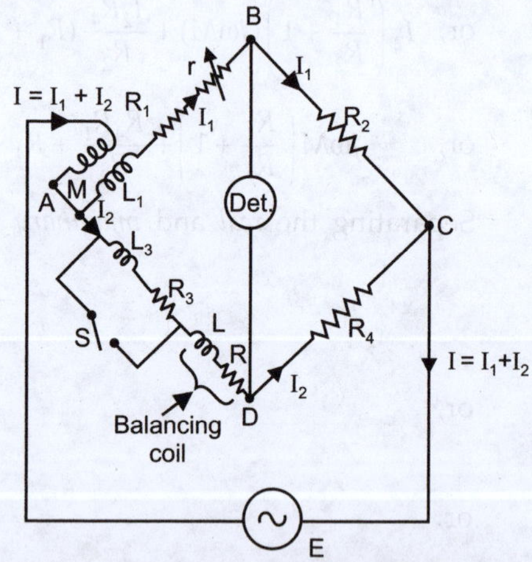

Fig. 10.39. Campbell's modification of Heaviside bridge.

∴ With switch *open*, from eqn. (10.83), we have :

$$L_3 + L = \frac{M_1 (R_2 + R_4) + R_4 L_1}{R_2} \qquad ...(i)$$

and, with switch *closed*,

$$L = \frac{M_2 (R_2 + R_4) + R_4 L_1}{R_2} \qquad ...(ii)$$

Subtracting (*ii*) from (*i*), we get :

$$L_3 = (M_1 - M_2)\left(1 + \frac{R_4}{R_2}\right) \qquad ...(10.85)$$

With switch open, from eqn. (10.84), we have :

$$R_3 + R = (R_1 + r_1) \times \frac{R_4}{R_2} \qquad ...(iii)$$

and, with switch closed, $R = (R_1 + r_2)\dfrac{R_4}{R_2}$ ...(*iv*)

Subtracting (*iv*) from (*iii*), we get :

$$R_3 = (r_1 - r_2) \times \frac{R_4}{R_2} \qquad ...(10.86)$$

- When we have "*equal ratio arms*", $R_2 = R_4$

and, therefore from eqns. (10.85) and (10.86), we get

$$\boldsymbol{L_3} = 2(M_1 - M_2) \qquad ...(10.87)$$

and $\quad \boldsymbol{R_2} = r_1 - r_2 \qquad ...(10.88)$

10.3.9. Carey Foster (Heydweiller) Bridge

Basically, Carey Foster used this bridge, but it was subsequently modified by Heydweiller for use on A.C.

This bridges is *used* for the following purposes :

(*i*) *To measure the capacitance in terms of a standard mutual inductance.* The bridge in this case is known as *Carey Foster bridge.*

(*ii*) *To measure the mutual inductance in terms of a standard capacitance.* The bridge in this case is known as *Heydweiller bridge.*

Fig. 10.40 shows the circuit and phasor diagrams for the bridge under balance conditions.

The unusual feature of this bridge is that one of its arms, *AD*, is *short-circuited* and therefore the potential drop across this arm is *zero.* For obtaining the balance, the potential drop across arm *AB should also be equal to zero* and for this reason *negative coupling is needed for the mutual inductance.*

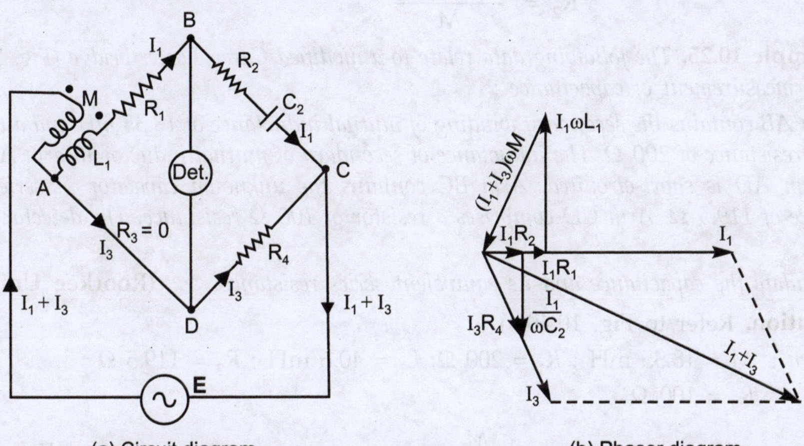

Fig. 10.40. Carey Foster (Heydweiller) bridge.

When the bridge is balanced :

$$\bar{I}_1 (R_1 + j\omega L_1) - (\bar{I}_1 + \bar{I}_3)\, j\omega M = 0$$

or, $$\bar{I}_1 (R_1 + j\omega L_1 - j\omega M) = \bar{I}_3 j\omega M \qquad ...(i)$$

and, $$\bar{I}_1 \left(R_2 + \frac{1}{j\omega C_2} \right) = \bar{I}_3 R_4 \qquad ...(ii)$$

Substituting the value of $\bar{I}_3$ from (*ii*) in (*i*), we get :

$$\bar{I}_1 (R_1 + j\omega L_1 - j\omega M) = \frac{\bar{I}_1}{R_4}\left(R_2 + \frac{1}{j\omega C_2}\right) j\omega M$$

or,
$$R_1 + j\omega L_1 - j\omega M = \frac{R_2}{R_4} j\omega M + \frac{M}{R_4 C_2}$$

Equating the *real* and *imaginary* terms, we get :

$$R_1 = \frac{M}{R_4 C_2}$$

or,
$$\boldsymbol{M} = R_1 R_4 C_2 \quad ...(10.89)$$

and,
$$L_1 = M\left(1 + \frac{R_2}{R_4}\right)$$

or,
$$L_1 = \frac{M(R_2 + R_4)}{R_4} = R_1 C_2 (R_2 + R_4) \quad ...(10.90)$$

- If the bridge is used for *measurement of capacitance*, eqns. (10.89) and (10.90) may be written as :

$$\boldsymbol{C_2} = \frac{M}{R_1 R_4} \quad ...(10.91)$$

and,
$$\boldsymbol{R_2} = \frac{R_4(L - M)}{M} \quad ...(10.92)$$

Example 10.25. *The following data relate to a modified Carey Foster bridge [Fig. 10.40(a)], used for measurement of capacitance :*

Arm AB contains the secondary winding of mutual inductance of 18.35 mH and a total non-reactive resistance of 200 Ω. The inductance of secondary of mutual inductor in arm AB is 40.6 mH. Arm AD is short-circuited. Arm BC contains the unknown capacitor in series with a resistance of 119.5 Ω. Arm CD comprises a resistor of 100 Ω resistance. The detector is across BD.

Calculate the capacitance and its equivalent series resistance. **(Roorkee University)**

Solution. Refer to Fig. 10.40(*a*).

Given : $M = 18.35$ mH ; $R_1 = 200\ \Omega$; $L_1 = 40.6$ mH ; $R_2 = 119.5\ \Omega$; $R_4 = 100\ \Omega$.

We know that :
$$\boldsymbol{C_2} = \frac{M}{R_1 R_4} \quad \text{[Eqn. (10.91)]}$$

$$= \frac{18.35 \times 10^{-3}}{200 \times 100} F = \mathbf{0.9175\ \mu F} \text{ (Ans.)}$$

and,
$$\boldsymbol{R_2} = \frac{R_4 (L - M)}{M}$$

$$= \frac{100(40.6 - 18.35)}{18.35} = 121.25\ \Omega \quad \text{[Eqn. (10.92)]}$$

∴ *Series resistance of capacitor* = 121.25 − 119.5 = **1.75 Ω** **(Ans.)**

10.3.10. Campbell Bridge (*For measurement of "Mutual Inductance"*)

This bridge *measures an unknown mutual inductance in terms of a standard mutual inductance.*

Fig. 10.41 shows the circuit diagram of Campbell bridge.

Let, M_1 = *Unknown* mutual inductance,

L_1 = Self-inductance of secondary of mutual inductance M_1,

M_2 = Variable standard mutual inductance,

L_2 = Self-inductance of secondary of mutual inductance M_2, and

R_1, R_2, R_3, R_4 = Non-inductive resistance.

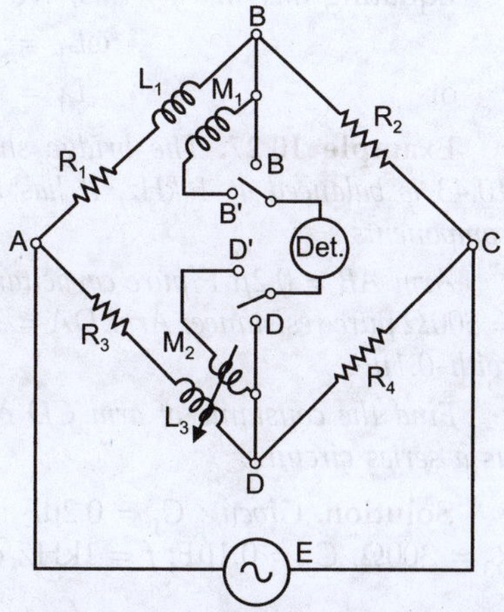

Fig. 10.41. Campbell's bridge.

For the balancing process, the following *two steps* are required :

Step 1. Detector is connected between B and D. The circuit now becomes a simple self-inductance comparison bridge. Then, condition for balance is :

$$\frac{L_1}{L_2} = \frac{R_1}{R_3} = \frac{R_2}{R_4} \quad ...(10.93)$$

The bridge may be balanced by adjustment of R_2 (or R_4) and R_1 (or R_3).

Step 2. Detector is connected between B' and D'. Keeping the adjustments as in step 1 above, the variable inductance M_2 is varied to get balance. Then,

$$\frac{M_1}{M_2} = \frac{R_2}{R_4}$$

or,
$$M_1 = M_2 \cdot \frac{R_2}{R_4} \quad ...(10.94)$$

Additional Worked Examples on Bridges

Example 10.26. *In a bridge shown in Fig. 10.42 the balance was obtained with L_1 = 52.6 mH and R_2 = 1.68Ω. Find the resistance and inductance of the coil.*

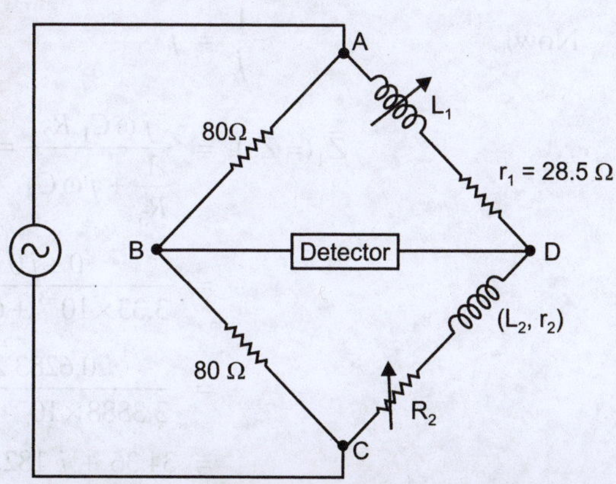

Fig. 10.42.

Solution : *Given* : L_1 = 52.6 mH; R_2 = 1.68Ω ; r_1 = 28.5Ω.

Refer to Fig. 10.42.

r_2; L_2;

Balance of the bridge was obtained when L_1 = 52.6 mH and R_2 = 1.68Ω

At balance :

$80\,(r_1 + j\,\omega L_1) = 80\,[(R_2 + r_2) + j\omega L_2]$

or, $r_1 + j\omega L_1 = (R_2 + r_2) + j\omega L_2$

Equating the *real terms*, we get,

$$r_1 = R_2 + r_2$$

$$\therefore \quad r_2 = r_1 - R_2 = 28.5 - 1.68 = \mathbf{26.82\ \Omega}\ \textbf{(Ans.)}$$

Equating *imaginary terms*, we have :

$$\omega L_1 = \omega L_2$$

or $$L_1 = L_2 = \mathbf{52.6\ \mathit{m}\,H}\ \textbf{(Ans.)}$$

Example 10.27. *The bridge shown in Fig. 10.43 is balanced at 1 kHz. It has the following components :*

Arm AB = 0.2μ F pure capacitance; Arm BC = 500Ω pure resistance; Arm DA = 300Ω parallel with 0.1μF.

Find the constants of arm CD considering it as a series circuit.

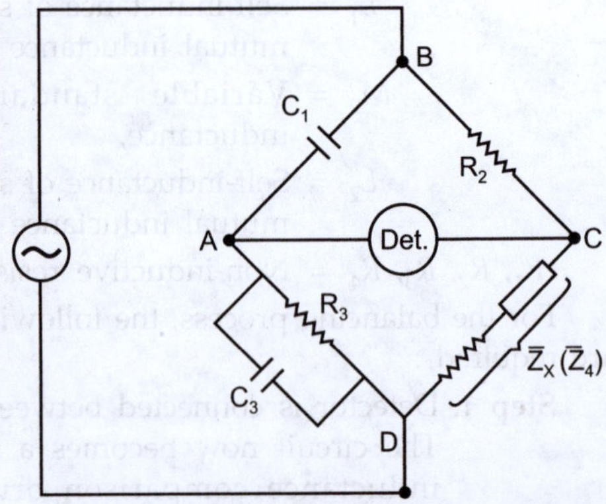

Fig. 10.43.

Solution. *Given* : $C_1 = 0.2\mu F$: $R_2 = 500\Omega$; $R_3 = 300\Omega$, $C_3 = 0.1\mu F$; f = 1kHz or 1000 Hz.

Now, $$\bar{Z}_1 = 0 - j\left(\frac{1}{\omega C_1}\right)$$

$$\bar{Z}_2 = R_2$$

$$\bar{Z}_3 = R_3 \parallel X_{C3}$$

or, $$\bar{Y}_3 = \frac{1}{R_3} + j\left(\frac{1}{X_{C3}}\right) = \frac{1}{R_3} + j(\omega C_3)$$

$$\bar{Z}_4 = \bar{Z}_x$$

The basic balance equation is given by :

$$\bar{Z}_1\,\bar{Z}_4 = \bar{Z}_2\,\bar{Z}_3$$

$$\therefore \quad \bar{Z}_4 = \frac{\bar{Z}_2\,\bar{Z}_3}{\bar{Z}_1} = \frac{\bar{Z}_2}{\bar{Z}_1\,\bar{Y}_3} = \frac{R_2}{-\dfrac{j}{\omega C_1}\left[\dfrac{1}{R_3} + j(\omega C_3)\right]}$$

Now, $$-\frac{1}{j} = j$$

$$\therefore \quad \bar{Z}_4 (= \bar{Z}_x) = \frac{j\,\omega C_1 R_2}{\dfrac{1}{R_3} + j\,\omega C_3} = \frac{0 + j(2\pi \times 1000) \times 0.2 \times 10^{-6} \times 500}{\dfrac{1}{300} + j(2\pi \times 1000) \times 0.1 \times 10^{-6}}$$

$$= \frac{0 + j\,0.6283}{3.33 \times 10^{-3} + 6.283 \times 10^{-4}}$$

$$= \frac{0.6283 \angle 90^\circ}{3.3888 \times 10^{-3} \angle +10.68^\circ} = 185.4 \angle 79.32^\circ$$

$$= 34.36 + j\,182.19\ \Omega$$

$$\therefore \quad \bar{Z}_x = (34.36 + j\,182.19)\ \Omega$$

or, $R_x = \mathbf{34.36\ \Omega}$ **(Ans.)**

and, $X_{L_x} = 2\pi f L_x = 182.19$

$$\therefore \quad L_x = \frac{182.19}{2\pi \times 1000} \simeq 0.029\text{ H or } \mathbf{29\ mH}\ \textbf{(Ans.)}$$

Example 10.28. *The following data relate to the bridge shown in Fig. 10.44.*

$R_1 = 200\ \Omega$; $R_2 = 200\ \Omega$; $C_2 = 5\mu F$; $R_3 = 500\Omega$; $C_3 = 0.2\mu F$.

Find the constants of Z_x considering it as a series circuit.

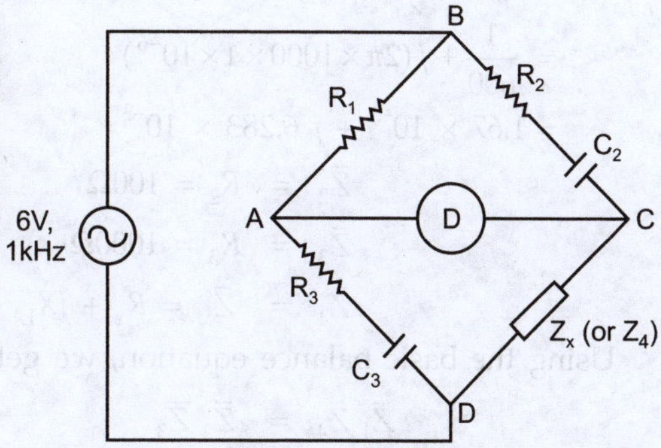

Fig. 10.44.

Solution. Refer to Fig. 10.44.

$$\bar{Z}_1 = R_1 = 200\Omega$$

$$\bar{Z}_2 = R_2 - j\left(\frac{1}{\omega C_2}\right)$$

$$= 200 - j\left[\frac{1}{2\pi \times 1 \times 10^3 \times 5 \times 10^{-6}}\right] = 200 - j\ 31.83\ \Omega$$

$$\bar{Z}_3 = R_3 - j\left(\frac{1}{\omega C_3}\right)$$

$$= 500 - j\left[\frac{1}{2\pi \times 1 \times 10^3 \times 0.2 \times 10^{-6}}\right] = 500 - j\ 795.77\ \Omega$$

When the bridge is balanced,

$$\bar{Z}_1 \bar{Z}_4 = \bar{Z}_2 \bar{Z}_3$$

$$\therefore \quad \bar{Z}_4 \text{ (or } \bar{Z}_x) = \frac{\bar{Z}_2 \bar{Z}_3}{\bar{Z}_1} = \frac{(200 - j\,31.83)(500 - j\,795.77)}{200}$$

$$= \frac{(202.517 \angle -9.04^\circ)(939.81 \angle -57.86^\circ)}{200}$$

or, $\bar{Z}_x = 951.64 \angle -66.9^\circ = (373.36 - j\ 875.3)\ \Omega$

$\therefore$ $R_x = \mathbf{373.36\ \Omega}$ **(Ans.)**

and, $X_C = \dfrac{1}{2\pi f C_x} = 875.3$

or, $C_x = \dfrac{1}{2\pi \times 1 \times 10^3 \times 875.3} = \mathbf{0.1818\mu F}$ **(Ans.)**

Example 10.29. *The following data relate to the bridge shown in Fig. 10.45.*

$R_1 = 600\Omega$; $C_1 = 1\mu F$; $R_2 = 100\Omega$; $R_3 = 1000\Omega$

Find the values of R_x and L_x at balance.

Solution. Refer to Fig. 10.45.

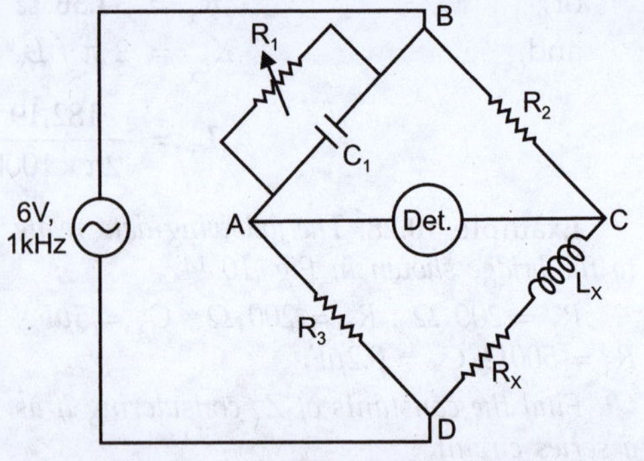

Fig. 10.45.

$$\bar{Z}_1 = R_1 \parallel X_{C_1}$$

or, $$\bar{Y}_1 = \frac{1}{R_1} + j\omega C_1$$

$$= \frac{1}{600} + j(2\pi \times 1000 \times 1 \times 10^{-6})$$

$$= 1.67 \times 10^{-3} + j\,6.283 \times 10^{-3}$$

$$\bar{Z}_2 = R_2 = 100\Omega$$

$$\bar{Z}_3 = R_3 = 1000\Omega$$

$$\bar{Z}_4 = \bar{Z}_x = R_x + jX_L$$

Using the basic balance equation, we get :

$$\bar{Z}_1 \bar{Z}_4 = \bar{Z}_2 \bar{Z}_3$$

or, $$\bar{Z}_4 (= \bar{Z}_x) = \frac{\bar{Z}_2 \bar{Z}_3}{\bar{Z}_1} = \bar{Z}_2 \bar{Z}_3 \bar{Y}_1$$

$$= 100 \times 1000 \times [1.67 \times 10^{-3} + j\,6.283 \times 10^{-3}]$$

$$= (167 + j\,628.3)\ \Omega = (R_x + j\,X_L)\ \Omega$$

∴ $\mathbf{R_x = 167\ \Omega}$ **(Ans.)**

$$X_{L_x} = 628.3\ \Omega = 2\pi f L_x$$

∴ $L_x = \frac{628.3}{2\pi \times 1000} \simeq$ **0.1 H (Ans.)**

10.3.11. Wien Bridge (*For measurement of* "Frequency")

The Wien bridge is *primarily known as a "frequency" determining bridge*. It can also be used for the *measurement of an unknown capacitor with great accuracy.*

Fig. 10.46 shows a Wien bridge. It has a series *R-C* combination in one arm and a parallel combination in the adjoining arm.

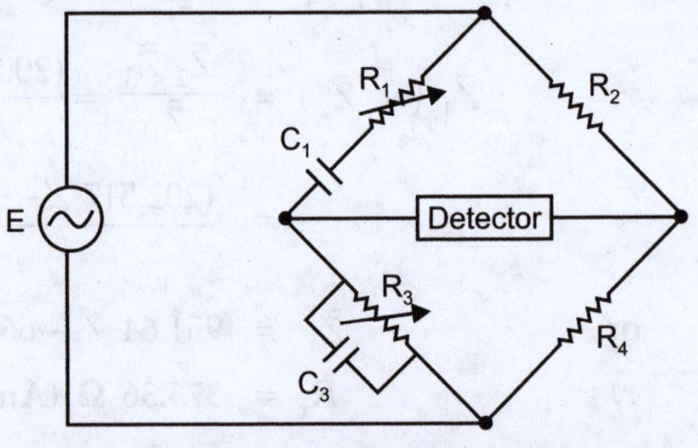

Fig. 10.46. Wien bridge for measuring frequency.

The impedance of one arm is,

$$\bar{Z}_1 = R_1 - \frac{j}{\omega C_1}$$

The admittance of parallel arm is

$$\bar{Y}_3 = \frac{1}{R_3} + j\omega C_3$$

Using the bridge balance equation, we have

$$\bar{Z}_1 \bar{Z}_4 = \bar{Z}_2 \bar{Z}_3$$

or, $$\bar{Z}_1 \bar{Z}_4 = \bar{Z}_2 \times \frac{1}{\bar{Y}_3}$$

or, $$\bar{Z}_2 = \bar{Z}_1 \bar{Z}_4 \bar{Y}_3$$

or, $$R_2 = \left(R_1 - \frac{j}{\omega C_1}\right) R_4 \left(\frac{1}{R_3} + j\omega C_3\right) \quad ...(10.95)$$

Expanding this expression, we get

$$R_2 = \frac{R_1 R_4}{R_3} + j\omega R_1 R_4 C_3 - \frac{jR_4}{\omega C_1 R_3} + \frac{R_4 C_3}{C_1}$$

or, $$R_2 = \left(\frac{R_1 R_4}{R_3} + \frac{R_4 C_3}{C_1}\right) - j\left(\frac{R_4}{\omega C_1 R_3} - \omega C_3 R_1 R_4\right) \quad ...(10.96)$$

Equating the *real terms*, we obtain

$$R_2 = \frac{R_1 R_4}{R_3} + \frac{R_4 C_3}{C_1}$$

or, $$\frac{R_2}{R_4} = \frac{R_1}{R_3} + \frac{C_3}{C_1} \quad ...(10.97)$$

Equating the *imaginary terms*, we have

$$\omega C_3 R_1 R_4 = \frac{R_4}{\omega C_1 R_3} \quad ...(10.98)$$

or, $$\omega^2 = \frac{1}{C_1 C_3 R_1 R_3} \quad ...[10.98\ (a)]$$

or, $$\omega = \frac{1}{\sqrt{C_1 C_3 R_1 R_3}}$$

Since, $\omega = 2\pi f$, therefore,

$$2\pi f = \frac{1}{\sqrt{C_1 C_3 R_1 R_3}}$$

or, $$f = \frac{1}{2\pi\sqrt{C_1 C_3 R_1 R_3}} \quad ...(10.99)$$

We observe that the two conditions for bridge balance, (eqns. 10.97, 10.99), result in an expression determining the required resistance ratio $\frac{R_2}{R_4}$, and another expression determining the frequency of the applied voltage. In other words, if we satisfy eqn. (10.97), and also excite the bridge with a frequency described by eqn. (10.99), the bridge will be in balance.

In most Wien bridge circuits, the components are chosen such that $R_1 = R_3$ and $C_1 = C_3$; this reduces eqn. (10.97) to $\frac{R_2}{R_4} = 2$ and eqn. (10.99) to

$$f = \frac{1}{2\pi RC} \quad ...(10.100)$$

which is *general expression* for the frequency of the Wien bridge.

— In a practical bridge, capacitors C_1 and C_3 are *fixed* capacitors, and resistors R_1 and R_3 are variable resistors *controlled by a common shaft*. Provided now that $R_2 = 2R_4$, the bridge may be used as a frequency-determining device balanced by a single *control. This control may be calibrated directly in terms of frequency.*

- By the use of this bridge an *accuracy of 0.5% - 1% can be easily obtained.*

- Because it is frequency sensitive, it is difficult to balance unless the waveform of the applied voltage is purely sinusoidal.

Applications :

1. The bridge is employed for *measuring frequency in the audio range.*

— The audio range is normally divided into 20Hz - 200Hz - 2kHz - 20kHz ranges. In this case, the resistances can be used for range changing and capacitors C_1 and C_3 for fine frequency control within the range.

2. The bridge can also be used for *measuring capacitances;* in this case operational frequency must be known.
3. The bridge is also used in a harmonic distortion analyser, as a notch filler, and in audio frequency and radio frequency oscillators as a frequency determining element.

Example 10.30. *Determine the equivalent parallel resistance and capacitance that causes a Wien bridge (Fig. 10.46) to null with the following component values :*

$R_1 = 2.8k\Omega$; $C_1 = 4.8\mu F$; $R_2 = 20k\Omega$; $R_4 = 80k\Omega$; $f = 2$ kHz;

Solution. *Given* : f = 2kHz or $\omega = 2\pi f = 2\pi \times (2 \times 1000) = 12.57 \times 10^3$ rad/s.

$$\mathbf{R_3 = ?,\ C_3 = ?}$$

We know that :
$$\frac{R_2}{R_4} = \frac{R_1}{R_3} + \frac{C_3}{C_1} \quad \text{...(Eqn. 10.97)}$$

and
$$\omega^2 = \frac{1}{C_1 C_3 R_1 R_3} \quad \text{...(Eqn. 10.98 (a))}$$

or,
$$C_3 = \frac{1}{w^2 C_1 R_1 R_3}$$

Substituting for C_3 in eqn. (10.97), we get

$$\frac{R_2}{R_4} = \frac{R_1}{R_3} + \frac{1}{\omega^2 C_1^2 R_1 R_3}$$

$$\frac{R_2}{R_4} = \frac{1}{R_3}\left(R_1 + \frac{1}{\omega^2 C_1^2 R_1}\right)$$

or,
$$R_3 = \frac{R_4}{R_2}\left(R_1 + \frac{1}{\omega^2 C_1^2 R_1}\right) \quad \text{...}(i)$$

$$= \frac{80\times10^3}{20\times10^3}\left[2.8\times10^3 + \frac{1}{(12.57\times10^3)^2 (4.8\times10^{-6})^2 (2.8\times10^3)}\right]$$

$$= 4\ (2.8 \times 10^3 + 0.098) = \mathbf{11.2\ k\Omega}\ \textbf{(Ans.)}$$

Now,
$$C_3 = \frac{1}{\omega^2 C_1 R_1 R_3} \quad \text{Rearranging (Eqn. 10.98}a)$$

$$= \frac{1}{(12.57\times10^3)^2 \times(4.8\times10^{-6})\times(2.8\times10^3)\times11.2\times10^3}$$

$$= \mathbf{42.04\ pF}\ \textbf{(Ans.)}$$

10.3.12. Resonance Bridge

Fig. 10.47 shows a resonance bridge, consisting of a series resonance circuit :

— The series resonance circuit is formed by R_4, C_4 and L_4 in series.

— All the other arms consist of resistors only.

Using the bridge balance equation, we have

$$Z_1 Z_4 = Z_2 Z_3$$

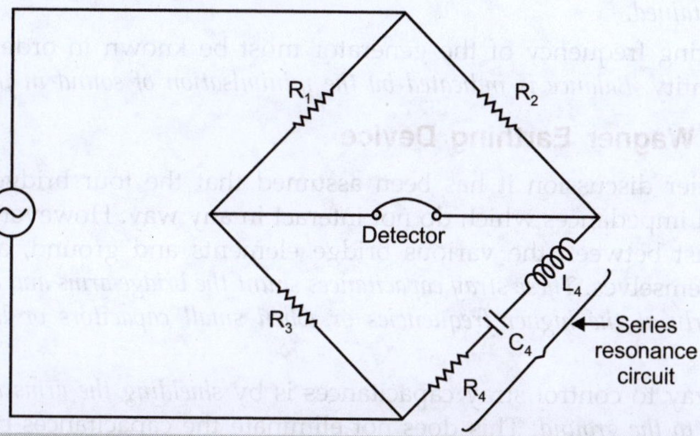

Fig. 10.47. Resonance bridge

Here, $\bar{Z}_1 = R_1$; $\bar{Z}_2 = R_2$; $\bar{Z}_3 = R_3$; $\bar{Z}_4 = R_4 + j\omega L_4 - \frac{j}{wC_4}$

Substituting the values in the balance equation, we get

$$R_1\left(R_4 + j\omega L_4 - \frac{j}{\omega C_4}\right) = R_2 R_3$$

or, $$R_1 R_4 + j\,\omega\, L_4 R_1 - \frac{jR_1}{\omega C_4} = R_2 R_3$$

Equating the *real* and *imaginary terms*, we have

$$R_1 R_4 = R_2 R_3$$

$\therefore$ $$R_4 = \frac{R_2 R_3}{R_1}$$

and, $$\omega L_4 - \frac{1}{\omega C_4} = 0$$

or, $$\omega^2 = \frac{1}{L_4 C_4}$$

or, $$(2\pi f)^2 = \frac{1}{L_4 C_4} \quad \text{or} \quad 2\pi f = \frac{1}{\sqrt{L_4 C_4}}$$

or *Resonance frequency*, $$f = \frac{1}{2\pi\sqrt{L_4 C_4}} \qquad ...(10.101)$$

- This bridge can be *used to measure unknown inductances or capacitances*. The resistor R_4 can be measured by keeping a fixed ratio $\frac{R_3}{R_1}$ and using a standard variable resistance to obtain balance.
  — If an *"inductance"* is *being measured, a standard capacitor is varied unitl balance is obtained.*
  — If a *"capacitance"* is *being measured, a standard inductor is varied until balance is obtained.*

The operating frequency of the generator must be known in order to calculate the unknown quantity. *Balance is indicated by the minimisation of sound in the headphones.*

10.3.13. Wagner Earthing Device

In the earlier discussion it has been assumed that the four bridge arms consist of simple lumped impedances which do not interact in any way. However, in practice, *stray capacitances* exist between the various bridge elements and ground, and also between bridge arms themselves. *These stray capacitances shunt the bridge arms and cause measurement error, particularly at the higher frequencies or when small capacitors or large inductors are measured.*

- One way to control stray capacitances is by *shielding the arms and connecting the shields to the ground*. This does not eliminate the capacitances but at least makes them constant in value, and they can therefore be *compensated*.
- One of the most widely used methods for eliminating some of the effects of stray capacitance in a bridge circuit is the *Wagner earth (ground) connection*. This circuit *eliminates the troublesome capacitance which exists between the detector terminals and the ground.*

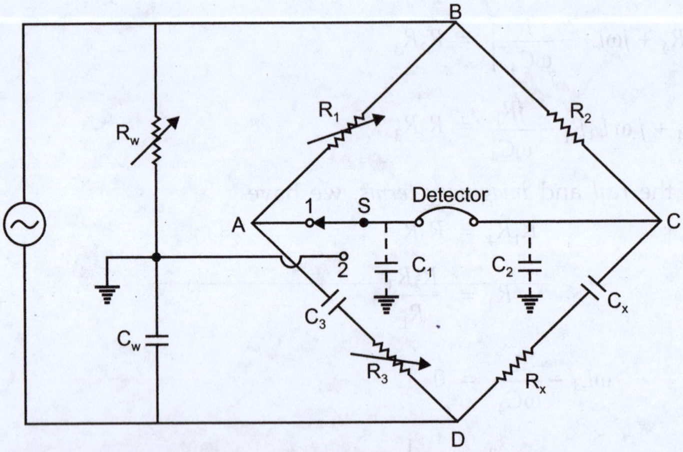

Fig. 10.48. Wagner earthing device.

Fig. 10.48 shows the circuit of a capacitance bridge. C_1 and C_2 represent the stray capacitances. The oscillator is removed from its usual ground connection and bridged by a series combination of resistor R_w and capacitor C_w. The junction of R_w and C_w is grounded and is called *Wagner ground connection*.

The initial adjustment of the bridge involves the following *procedure* :

— The detector is connected to point 1, and R_1 is adjusted for null or minimum sound in the headphones.

— The switch is then thrown to position 2, which connects the detector to Wagner ground point. Resistor R_w is now adjusted for minimum sound.

— When the switch is thrown to position 1 again, some unbalance will probably exist. Resistors R_1 and R_3 are then adjusted for minimum detector response, and the switch is again thrown to position 2.

This procedure is repeated till a null is reached on both switch positions; under this condition points 1 and 2 are at the same potential and this is *ground potential*. Stray capacitances C_1 and C_2 are then effectively shorted out and have no effect on normal bridge balance.

The capacitances from point *B* to *D* to ground are also eliminated by the addition of Wagner ground connection, since current through these capacitances will enter the Wagner ground connection.

The capacitances across the bridge arms are *not* eliminated by this earthing device and they will still *affect* the accuracy of the measurement.

- Because the addition of Wagner earthing device does not affect the balance conditions, the *measurement procedure* remains *unchanged*.

10.3.14. Types of Detectors

Following types of detectors are used in various applications :

1. *For low frequency* ...	Vibration galvanometer – A moving magnet vibration galvanometer has a range of 1.5 k Hz
2. *For ordinary laboratory work at frequencies up to a few 100 Hz*...	Moving coil type of instrument; it has a high sensitivity.
3. *For higher frequencies (> 800 Hz)* ...	Telephone (headphone)–*best detector* (Vibration galvanometers and headphones have no phase selectivity, *i.e.*, they do not indicate whether it is resistance or reactance adjustments that are required).
4. *For frequencies above about 3 kHZ, and particularly at high AF or RF*...	Heterodyne or beat tone detector.

☞ Following points are worth noting :

- The A.C. galvanometer and separately excited dynamometer are phase selective, and are best suited at low frequencies.
- An electrodynamometer can also be used as an A.C. detector.
- Electrometers are used as detectors because small capacitances possess a very large impedance when used with A.C. circuits at low frequency, and when measured in the bridge they form a high impedance branch. Hence, this detector is used to *increase sensitivity*.
- In several cases, especially in bridges for routine use, a pointer instrument is used. It is advantageous if it can be made phase selective. These pointer instruments are generally moving coil milliammeters operated with some arrangement of copper oxide rectifiers (frequency range 40 Hz - 1 kHz)
- *Modern bridge techniques employ the amplifier as a regular feature.*
- With all detectors, the impedance should be selected to suit that of the bridge. A higher sensitivity can be obtained by using an interbridge transformer.

— When *headphones* are used as detectors, precautions should be taken *to eliminate capacitance effects* between the observer and the headphones.

10.3.15. Sources of Errors in Bridge Circuits and the Elimination of Errors

Factors causing errors :

The following *factors cause errors in A.C. bridge circuit* :

1. Stray-conductance effects, due to imperfect insulation.
2. Mutual-inductance effects, due to magnetic coupling between various components of the bridge.
3. Stray-capacitance effects, due to electrostatic fields between conductors at different potentials.
4. 'Residues' in components – *e.g.*, the existence of small amount of series inductance or shunt capacitance in nominally non-reactive resistors.

Methods of reducing bridge errors :

The important *methods of reducing bridge errors are* :

1. *Use of high-quality components.*
2. *Proper bridge lay-out.*

The bridge lay-out should take care of the following *requirements* :

— All the leads to a single corner should be brought as nearly as possible to the same point.

— A pair of leads to a component should not form a large loop; if possible the leads should be co-axial.

— If there is more than one inductor in the bridge, leads may have to be quite long.

By doing so, the stray mutual inductance coupling and lead self-inductances are reduced to minimum and the stray capacitances are definitely located across individual arms of bridge; they can be measured and compensated for relatively easily.

3. *Operation of bridge at sufficient sensitivity.* In order to obtain a high degree of accuracy *e.g.* 1 part in 10^4 or better, it is necessary that a bridge must be operated at sufficient sensitivity which implies that the change in voltage at the detector terminals from the balance conditions should be large when the detector is considered open.

4. *Use of insulating stands.* When the insulation between various components/elements of a bridge is not good, trouble may arise because of leakage currents from one arm to another. This is specially true in case of high impedances bridges. It can be avoided by mounting the various bridge components and other pieces of apparatus on insulating stands.

5. In order to *avoid 'eddy current errors'* the presence of large conducting masses near the bridge work should be avoided.

6. *Residual errors.* Residues mean small inherent inductance or capacitance of the resistors. In precise work it becomes necessary to evaluate them in order to eliminate them or compensate for errors due to these. The self-inductance is important only when the coils used are multi-turn coils and supply used is of high frequency.

7. *Frequency and waveform errors.* These errors are overcome by employing *wave filters* which eliminate the unwanted harmonics from the source or by employing *tuned detectors* in place of headphones such as vibration galvanometers which do not respond to harmonics and respond readily only to the fundamental for which they are tuned.

10.3.16. Precautions to be taken when using a Bridge

When using a bridge, the following *precautions* should be taken to obtain accurate readings :

1. The layout of the leads should be such that no loops or long lengths enclosing magnetic flux are produced, with consequent stray inductance errors.
2. The self-capacitance of the leads, with a large L, is more important than their inductance, so they should be spaced relatively far apart.
3. When measuring a capacitor, it is important to keep capacitance as low as possible. Owing to this reason the leads should *not be too close together* and should be made of fine wire.
4. In case of very precise inductive and capacitive measurements, leads are encased in metal tubes to shield them from mutual electromagnetic action, and are so designed as to completely shield the bridge.

10.4. Q-METER, IMPEDANCE BRIDGE, LCR AND RX METERS

10.4.1. Q-Meter

The overall efficiency of the coils and capacitors intended for radio frequency (RF) applications is best evaluated using the Q value.

Q-meter *is an instrument designed to measure some electrical properties of coils and capacitors.* The *principle* of the Q-meter is based on *'series resonance'*, the voltage drop across the coil or capacitor is Q times the applied voltage (where *Q is the ratio of reactance to resistance,* $\frac{X_L}{R}$). If a fixed voltage is applied to the circuit, a *voltmeter across the capacitor can be calibrated to read Q directly.*

At resonance : $X_L = X_C$; $V_L = IX_L$; $V_C = IX_C$, $V = IR$

where,
X_L = Inductive reactance,
X_C = Capacitive reactance,
R = Coil resistance,
I = Circuit current,
V_L = Inductive voltage,
V_C = Capacitor voltage, and
V = Applied voltage.

Therefore,
$$Q = \frac{X_L}{R} = \frac{X_C}{R}\left(=\frac{1}{\omega CR}\right) = \frac{V_C}{V} \quad ...(10.102)$$

In the above eqn. (10.102), if *V is kept constant, the voltage across the capacitor can be measured by a voltmeter calibrated to read directly in terms of Q.*

Fig. 10.49 shows the basic Q-meter circuit :

- A wide-range *oscillator* with frequency range from 50 kHz to 50 MHz is used as a power supply to the circuit. The output of the oscillator is shorted by a low-value resistance, Rsh usually of the order of 0.02Ω. So it introduces almost no resistance into the oscillatory circuit and represents a voltage source with a very small or almost negligible internal resistance. The voltage across the low-value shunt resistance R_{sh}, V is measured by a *thermocouple meter* and the voltage across the capacitor V_C is measured by an electronic voltmeter.

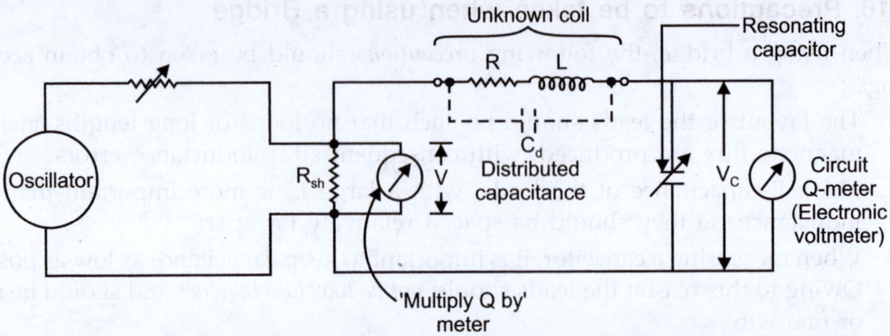

Fig. 10.49. Circuit diagram of a Q-meter.

- The inductance (L) of the coil can be determined by connecting it to the test terminals of the instrument. The circuit is tuned to resonance by varying either the capacitance or the oscillatory frequency. If the capacitance is varied, the oscillator frequency is adjusted to a given frequency and resonance is obtained. If the capacitance is pre-set to a desired value, the oscillator frequency is varied until resonance occurs. The Q *reading on the output meter must be multiplied by index setting or the "Multiply Q by" switch to obtain the actual value.*

  The inductance of the coil can be calculated from known values of the coil frequency and resonating capacitor (C).

$$X_L = X_C$$

or, $$2\pi f L = \frac{1}{2\pi f C} \quad \text{or} \quad 4\pi^2 f^2 LC = 1$$

or $$L = \frac{1}{(2\pi f)^2 C} \qquad \text{...(10.103)}$$

- The Q indicated is *not the actual* Q, because the losses of the resonating capacitor, voltmeter and inserted resistance are all included in the measuring circuit. The actual Q of the measured coil is *somewhat greater than the indicated* Q. *This difference is negligible except where the resistance of the coil is relatively small compared to the inserted resistance* R_{sh}.

Measurement of low impedance components :

Low-impedance components, such as low-value resistors, small coils, and large capacitors, are measured in series with the measuring circuit, as shown in Fig. 10.50. The component to be measured, Z_x, is connected in series with a stable *work coil* usually provided with the instrument across the test terminals.

Two measurements are made :

(*i*) In the first measurement the unknown is short-circuited by a small *shorting-strap* and the circuit is resonated, establishing a reference condition. The values of the tuning capacitor (C_1) and the indicated Q (Q_1) are noted.

(*ii*) In the *second* measurement, the shorting strap is removed and the circuit is retuned, giving a new value for the tuning capacitor (C_2) and a change in the Q value from Q_1 to Q_2.

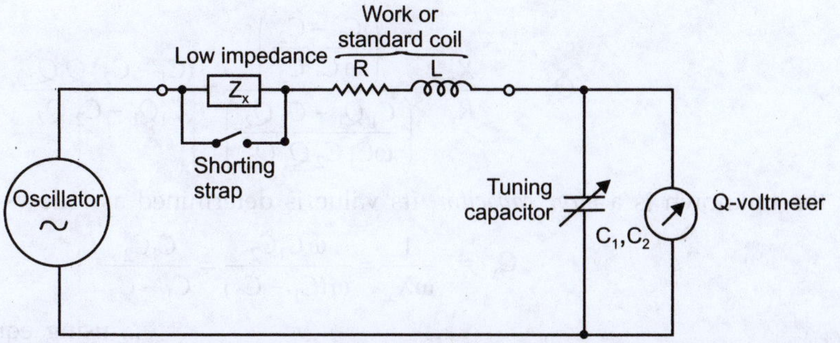

Fig. 10.50. Q-meter measurement of a low-impedance component in the *series* connection.

For the second measurement the reactance of the unknown (X_x) can be expressed in terms of the *new* value of the tuning capacitor (C_2) and the *in-circuit* value of the inductor (L), this yields

$$X_x = X_{C2} - X_L$$

or,

$$X_x = X_{C2} - X_{C1} \quad \text{...(10.104)}$$

$$(\because X_L = X_{C1})$$

$$= \frac{1}{\omega C_2} - \frac{1}{\omega C_1}$$

or,

$$X_x = \frac{C_1 - C_2}{\omega C_1 C_2} \quad \text{...(10.105)}$$

X_x is inductive if $C_1 > C_2$ and capacitive if $C_1 < C_2$.

The resistive component of the unknown impedance R_x can be found in terms of reactance X_x and the indicated values of circuit Q, since,

$$R_1 = \frac{X_1}{Q_1}, \text{ and } R_2 = \frac{X_2}{Q_2}$$

Also,

$$R_x = R_2 - R_1 = \frac{1}{\omega C_2 Q_2} - \frac{1}{\omega C_1 Q_1}$$

or,

$$R_x = \frac{C_1 R_1 - C_2 Q_2}{\omega C_1 C_2 Q_1 Q_2} \quad \text{...(10.106)}$$

— If the unknown is *purely resistive*, the setting of the tuning capacitor would *not* have changed in the measuring process, and $C_1 = C_2$. The equation for resistance, then, reduces to,

$$R_x = \frac{Q_1 - Q_2}{\omega C_1 Q_1 Q_2} = \frac{\Delta Q}{\omega C_1 Q_1 Q_2} \quad \text{...(10.107)}$$

— If the unknown is a *small inductor*, the value of the inductance is found from eqn. (10.107) as follows :

$$L_x = \frac{C_1 - C_2}{\omega^2 C_1 C_2} \quad \text{...(10.108)}$$

$$(\because X_x = \omega L_x)$$

- The Q of the coil can be determined by the definition,

i.e.,
$$Q_x = \frac{X_x}{R_x} = \frac{\left(\dfrac{C_1 - C_2}{\omega C_1 C_2}\right)}{\left(\dfrac{C_1 Q_1 - C_2 Q_2}{\omega C_1 C_2 Q_1 Q_2}\right)} = \frac{(C_1 - C_2) Q_1 Q_2}{C_1 Q_1 - C_2 Q_2} \quad \text{...(10.109)}$$

— If the unknown is a *large capacitor*, its value is determined as follows :

$$C_x = \frac{1}{\omega X_x} = \frac{\omega C_1 C_2}{\omega (C_1 - C_2)} = \frac{C_1 C_2}{C_1 - C_2} \quad \text{...(10.110)}$$

... using eqn. (10.108)

• The Q of the capacitor can be found by using eqn. (10.109).

Measurement of high impedance components :

High impedance components, such as high-value resistors, certain inductors, and small capacitors, are measured by connecting them in *parallel* with the measuring circuits, as shown in Fig. 10.51.

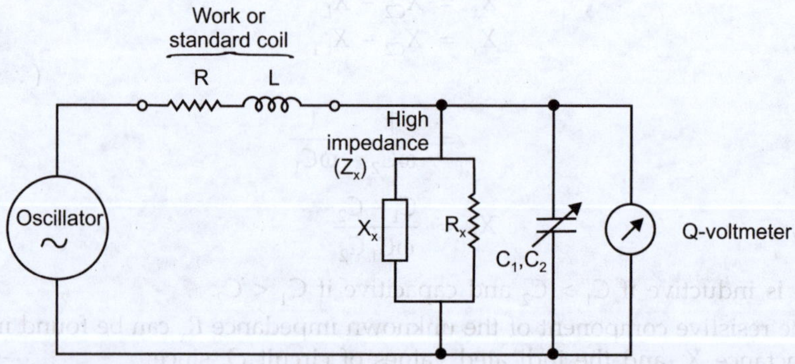

Fig. 10.51. Q-measurement of a high-impedance component in the parallel connection.

Prior to connecting the unknown, the circuit is resonated, by using a suitable work coil, to establish reference values for Q and C (Q_1 and C_1). Then, when the unknown component is connected to the circuit, the capacitor is readjusted for resonance, and a *new value* for the tuning capacitance (C_2) is obtained and a change in the value of circuit Q (ΔQ) from Q_1 to Q_2.

The unknown impedance is measured in terms of its parallel components X_x and R_x. At the initial resonance condition, when the unknown is *not yet* connected into the circuit, the working coil (L) is tuned by the capacitor (C_1), and we have

$$\omega L = \frac{1}{\omega C_1} \quad \text{...(10.111)}$$

and,
$$Q_1 = \frac{\omega L}{R} = \frac{1}{\omega C_1 R} \quad \text{...(10.112)}$$

When the unknown impedance (Z_x) is connected into the circuit and the capacitor is tuned for resonance, the reactance of the working coil (X_L) equals the parallel reactances of the tuning capacitor (X_{C_2}) and the unknown (X_x), and we have,

$$X_L = \frac{X_{C_2}\, X_x}{X_{C_2} + X_x}$$

or, $$X_L\,(X_{C_2} + X_x) = X_{C_2}\, X_x$$

or, $$X_L\, X_{C_2} + X_L\, X_x = X_{C_2}\, X_x$$

or, $$X_x\,(X_{C_2} - X_L) = X_L\, X_{C_2}$$

or, $$X_x = \frac{X_L\, X_{C_2}}{(X_{C_2} - X_L)} = \frac{\frac{1}{\omega C_1}\frac{1}{\omega C_2}}{\frac{1}{\omega C_2} - \frac{1}{\omega C_1}} = \frac{1}{\omega\,(C_1 - C_2)} \quad \text{...(10.113)}$$

— If the unknown is *inductive*, $X_x = \omega L_x$, eqn. (10.103) yields the value of the unknown inductance :

$$L_x = \frac{1}{\omega^2\,(C_1 - C_2)} \quad \text{...(10.114)}$$

—If the unknown is *capacitive*, $X_x = \dfrac{1}{\omega C_x}$ the eqn. (10.113) yields the value of the unknown capacitor.

$$C_x = C_1 - C_2$$

In a *parallel resonant circuit* the total resistance (R_t) at resonance is equal to the product of the circuit Q and the reactance of the coil.

i.e. $$R_t = Q_2 X_L = Q_2 X_{C_1} = \frac{Q_2}{\omega C_1} \quad \text{...(10.115)}$$

[$\because X_L = X_{C_1}$ ... eqn. (10.111)]

The *resistance* (R_x) of the unknown impedance can be determined by comaring the conductances in the circuit (see Fig. 10.51).

Let, G_t = Total conductance of the resonant circuit,

G_x = Conductance of the unknown impedance, and

G_L = Conductance of the working coil.

Then, $$G_t = G_X + G_L$$

or $$G_x = G_t - G_L \quad \text{...(10.116)}$$

Also, $$G_t = \frac{1}{R_t} = \frac{\omega C_1}{Q_2} \quad \text{... using eqn. (10.115)}$$

$$\therefore \quad \frac{1}{R_x} = \frac{\omega C_1}{Q_2} - \frac{R}{R^2 + \omega^2 L^2}$$

[$\because$ Conductance of the working coil, $G_L = R/(R^2 + \omega^2 L^2)$]

$$= \frac{\omega C_1}{Q_2} - \frac{1}{R}\left[\frac{1}{1 + \frac{\omega^2 L^2}{R^2}}\right] = \frac{\omega C_1}{Q_2} - \frac{1}{R}\left(\frac{1}{1 + Q_1^2}\right)$$

Normally value of Q_1 is very high so $(1 + Q_1^2)$ can be taken equal to Q_1^2.

$$\therefore \quad \frac{1}{R_x} = \frac{\omega C_1}{Q_2} - \frac{1}{RQ_1^2}$$

$$= \frac{\omega C_1}{Q_2} - \frac{1}{RQ_1 \times \dfrac{1}{\omega C_1 R}}$$

$$= \frac{\omega C_1}{Q_2} - \frac{\omega C_1}{Q_1} \quad \text{...[using eqn. (10.112)]}$$

$$= \frac{\omega C_1 Q_1 - \omega_2 C_1 Q_1}{Q_1 Q_2}$$

or
$$R_x = \frac{Q_1 Q_2}{\omega C_1 (Q_1 - Q_2)} \quad \text{...(10.117)}$$

The Q of the unknown impedance,

$$Q_x = \frac{R_x}{X_x} = \frac{\dfrac{Q_1 Q_2}{\omega C_1 (Q_1 - Q_2)}}{\dfrac{1}{\omega (C_1 - C_2)}} = \frac{(C_1 - C_2) Q_1 Q_2}{C_1 (Q_1 - Q_2)} \quad \text{...(10.118)}$$

...(using eqn. 10.117 and 10.113, respectively)

Sources of error :

The various errors occurring in measurements are discussed below :

1. Error due to distributed capacitance :

The *distributed capacitance or self-capacitance* of the measuring circuit is the most important factor which affects accuracy of the measurement. This source of error affects the actual value of Q and the inductance of the coil.

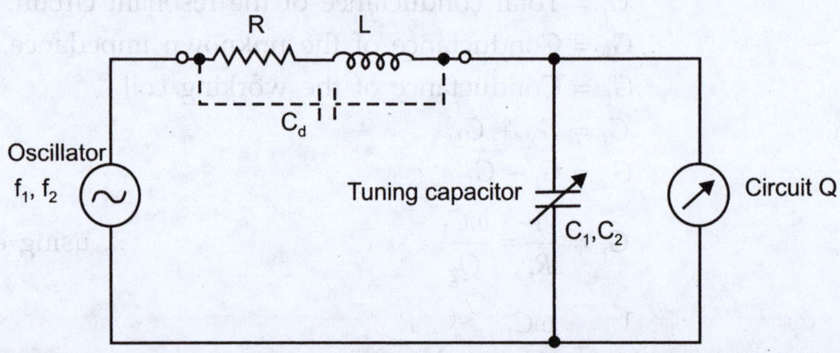

Fig. 10.52. Determination of the distributed capacitance of an inductor.

Distributed or self-capacitance C_d of a coil can be measured by making two measurements at different frequencies. The circuit diagram is shown in Fig. 10.52. The coil under test in connected directly to the test terminals of the measuring circuit (See Fig. 10.52). The tuning capacitor is set to a high value, preferably to its maximum position, and the circuit is resonated by adjusting the oscillatory frequency. Resonance is indicated by maximum deflection on the *"Circuit Q"* meter. The values of the tuning capacitor (C_1) and the oscillatory frequency (f_1) are noted. The frequency is then increased to twice its original value ($f_2 = 2f_1$) and the circuit is retuned by adjusting the resonating capacitor (C_2). For these two resonant conditions, we have :

$$f_1 = \frac{1}{2\pi\sqrt{L(C_1 + C_d)}} \qquad ...(10.119)$$

and,
$$f_2 = \frac{1}{2\pi\sqrt{L(C_2 + C_d)}} \qquad ...(10.120)$$

Since,
$$f_2 = 2f_1$$

$$\therefore \quad \frac{1}{2\pi\sqrt{L(C_2 + C_d)}} = 2 \times \frac{1}{2\pi\sqrt{L(C_1 + C_d)}}$$

Squaring both sides and simplifying, we get

$$\frac{1}{C_2 + C_d} = \frac{4}{C_1 + C_d}$$

or,
$$(C_1 + C_d) = 4(C_2 + C_d) \quad \text{or} \quad (4C_2 + 4C_d)$$

or,
$$C_d = \frac{C_1 - 4C_2}{3} \qquad ...(10.121)$$

The *distributed capacitance reduces the effective value of Q* of a coil and the difference between the measured value and actual value depends on the value of the distributed capacitance and the resonating capacitor. It can be shown that,

$$\text{True } Q = Q_e\left(\frac{C + C_d}{C}\right) \qquad ...(10.122)$$

where, Q_e = Effective Q of the coil,
C = Resonating capacitance, and
C_d = Distributed capacitance.

2. ***Error due to residual inductance of the instrument :***
   - This type of error is very small, of the order of 0.015μH.
   - It affects the measurement of only very small inductors, say inductors of value lesser than 0.5μ H.
3. ***Error due to shunt resistor :***
   - The error due to shunt resistor R_{sh} is *very small* and can be neglected as in case of measuring Q of a coil of the resistance 20 Ω with a shunt resistor of 0.02 Ω.
   - However, the shunt resistor R_{sh} of the Q-meter circuit introduces error when resistance of the test coil is *comparable to the shunt resistor*, say, for example, a coil of resistance 0.2 Ω is measured with shunt resistance of 0.2 Ω.
4. ***Error due to conductance of Q voltmeter :***
   - The *conductance* of the Q voltmeter has a slight shunting effect on the tuning capacitor at high frequencies, but this effect can usually be *neglected*.

Applications : Some of the specialized uses of this instrument are *to measure* :

(*i*) Q of a coil;
(*ii*) Inductance and capacitance;
(*iii*) Distributed capacitance of a coil;
(*iv*) Q and power factor of a dielectric material;
(*v*) Mutual inductance of coupled circuits;
(*vi*) Coefficient of coupling;
(*vii*) Critical coupling;

(*viii*) Reactance and effective resistance of an inductor at operating frequency;

(*ix*) Bandwidth of a tuned circuit etc.

- It has been very rightly said that *if ever an RF problem arises, a Q meter can always provide the answer.*

Example 10.31. *The following data were obtained during the measurement of the distributed or self-capacitance of a coil :*

First measurement : $f_1 = 2$ *MHz and* $C_1 = 420$ *pF;*

Second measurement : $f_2 = 4$ *MHz and* $C_2 = 90$ *pF.*

Find the distributed capacitance, C_d.

Solution. *Given* : $f_1 = 2$ MHz ; $C_1 = 420$ pF; f_2 $(= 2f_1) = 4$MHz, $C_2 = 90$ pF.

We know that, $$C_d = \frac{C_1 - 4C_2}{3} \quad \text{...(Eqn. 10.121)}$$

or, $$C_d = \frac{420 - 4 \times 90}{3} = \mathbf{20pF} \textbf{ (Ans.)}$$

Example 10.32. *Calculate the value of distributed capacitance of a coil when the following measurements are made :*

At frequency $f_1 = 2$ *MHz, the tuning capacitor is set at 410pF.*

At frequency, $f_2 = 5$ *MHz, the tuning capacitor is tuned at 50pF.*

Solution. *Given* : $f_1 = 2$MHz, $C_1 = 410$pF; $f_2 = 5$MHz, $C_2 = 50$pF.

Distributed capacitance C_d :

We know that : $$f_1 = \frac{1}{2\pi\sqrt{L(C_1 + C_d)}} \quad \text{...[Eqn. (10.119)]}$$

and, $$f_2 = \frac{1}{2\pi\sqrt{L(C_2 + C_d)}} \quad \text{...[Eqn.(10.120)]}$$

Now, $$\frac{f_2}{f_1} = \frac{5}{2} = 2.5 \text{ or } f_2 = 2.5 f_1$$

$$\therefore \quad \frac{1}{2\pi\sqrt{L(C_2 + C_d)}} = 2.5 \times \frac{1}{2\pi\sqrt{L(C_1 + C_d)}}$$

or, $$\frac{1}{C_2 + C_d} = \frac{6.25}{C_1 + C_d} \quad \text{...squaring both sides and simplifying}$$

or, $$(C_1 + C_d) = 6.25(C_2 + C_d)$$

or, $$C_d = \frac{C_1 - 6.25\, C_2}{5.25} = \frac{410 - 6.25 \times 50}{5.25}$$

$$= \mathbf{18.57pF} \textbf{ (Ans.)}$$

Example 10.33. *In a Q-meter, after connecting a standard coil, the resonance is obtained with a frequency* f_1 *with resonating capacitor set at* C_1*; the indicated Q-factor is* Q_1*. An unknown impedance is connected in series with the standard coil and the resonance is re-established by reseting the resonating capacitor at* C_2*, the corresponding Q-factor being* Q_2.

Determine the resistive and reactive components of the unknown impedance when $C_1 = 190$*pF,* $Q_1 = 75$; $C_2 = 170$*pF and* $Q_2 = 45$*. The frequency is 200 kHz.*

Solution. *Given* : C_1 = 190pF; Q_1 = 75; C_2 = 170pF; Q_2 = 45; f = 200kHz.

Resistive (R_x) and reactive components (X_x) :

Resistive component (R_x) of unknown impedance (Z_x) is given by :

$$R_x = \frac{C_1Q_1 - C_2Q_2}{\omega C_1C_2Q_1Q_2} \quad \text{...[Eqn. (10.106)]}$$

or,
$$R_x = \frac{190\times10^{-12}\times75 - 170\times10^{-12}\times45}{(2\pi\times200\times1000)\times190\times10^{-12}\times170\times10^{-12}\times75\times45}$$

$= \mathbf{48.18\Omega}$ **(Ans.)**

Reactive component (X_x) is given by :

$$X_x = \frac{C_1 - C_2}{\omega\, C_1\, C_2} \quad \text{...[Eqn. (10.105)]}$$

or,
$$X_x = \frac{190\times10^{-12} - 170\times10^{-12}}{(2\pi\times200\times1000)\times190\times10^{-12}\times170\times10^{-12}}$$

$= \mathbf{492.74\ \Omega}$ **(Ans.)**

Example 10.34. *A coil of resistance 4Ω is connected in the circuit shown in Fig. 10.88 and resonance occurs when oscillator frequency is 500 kHz and the resonating capacitor is set at 120 pF. Determine the percentage error introduced in the calculated value of Q if a resistance of 0.02 Ω is used across the oscillatory circuit.*

Solution. *Given* : R = 4Ω; f = 500kHz; C = 110pF; Resistance used across the oscillatory circuit = 0.02 Ω.

Percentage error :

The true or effective Q of the coil,

$$Q_{true} = \frac{1}{\omega CR} = \frac{1}{(2\pi\times500\times10^3)\times120\times10^{-12}\times4}$$

$= 663.14$

The indicated or calculated Q of the coil,

$$Q_{indicated} = \frac{1}{\omega C\,(R+0.02)} = \frac{1}{(2\pi\times500\times10^3)\times120\times10^{-12}\times(4+0.02)}$$

$= 659.85$

$$\therefore \quad \text{Percentage error} = \frac{Q_{true} - Q_{indicated}}{Q_{true}}\times100$$

$$= \frac{663.14 - 659.85}{663.14}\times100 \simeq \mathbf{0.5\ \%}\ \textbf{(Ans.)}$$

Example 10.35. *The following results were obtained during tests conducted, using a Q-meter, on a radio-tuning coil to establish the distributed/self-capacitance :*

(*i*) *With the radio coil connected normally, resonance was obtained at 600 kHz with the tuning capacitor of the Q-meter set at 100pF.*

(*ii*) *With a standard inductor connected in place of radio coil, resonance was obtained at 2 MHz and this condition was not changed when the radio coil was connected in parallel with the standard inductor.*

Determine the self-capacitance of the radio coil.

Solution. *Given* : $C_1 = 100\text{pF}$; $f_1 = 600\text{kHz}$; $f_2 = 2\text{MHz}$.

Self-capacitance of the coil, C_d :

If L and C_d are the inductance and self-capacitance of the coil respectively, then from the *first test* we have :

$$L = \frac{1}{\omega_1^2 (C_1 + C_d)} \qquad \text{...}(i) \qquad \text{...(From eqn. 10.121)}$$

where, ω_1 = Angular frequency, and

C_1 = Tuning capacitor setting to produce resonance in first setting.

In the *second test* the reason for resonance condition of the standard inductor being unchanged is that a very high impedance has been placed in parallel, this being the resonant circuit of the radio coil and its own self-capacitance, *i.e.*, the radio coil in parallel with its own self-capacitance is resonant at 2 MHz. Thus, we can write,

$$L = \frac{1}{\omega_2^2 C_d} \qquad \text{...}(ii)$$

Equating expressions (*i*) and (*ii*), we have :

$$\frac{1}{\omega_1^2 (C_1 + C_d)} = \frac{1}{\omega_2^2 C_d}$$

or,
$$\omega_2^2 C_d = \omega_1^2 C_1 + \omega_1^2 C_d$$

or,
$$C_d = \frac{\omega_1^2 C_1}{(\omega_2^2 - w_1^2)} = \frac{f_1^2 C_1}{f_2^2 - f_1^2} \qquad (\because \omega = 2\pi f)$$

Substituting the value, we get

$$C_d = \frac{(600 \times 1000)^2 \times 100 \times 10^{-12}}{(2 \times 10^6)^2 - (600 \times 1000)^2}$$

$$= \mathbf{9.89pF} \textbf{ (Ans.)}$$

Example 10.36. *A circuit consisting of an unknown coil, a resistance and a variable capacitor connected in series is tuned to resonance using a Q-meter. If the frequency is 400 kHz, the resonating capacitor is set at 220 pF, the resistance is 0.8 Ω and the Q-meter indicates 110, determine the effective inductance and resistance of the unknown coil.*

Solution. *Given* : $f = 400\text{kHz}$; $C = 200\text{pF}$; $R_{sh} = 0.8\Omega$; $Q = 110$.

Inductance (*L*) ; resistance (*R*) :

Under *resonant conditions,* we have :

$$L_{coil} = \frac{1}{\omega^2 C} = \frac{1}{(2\pi \times 400 \times 1000)^2 \times 220 \times 10^{-12}}$$

$$= \mathbf{719.6\ \mu H} \textbf{ (Ans.)}$$

$$R_{coil} = \frac{\omega L}{Q} - R_{sh} = \frac{2\pi \times 400 \times 1000 \times 719.6 \times 10^{-6}}{110} - 0.8$$

$$= \mathbf{16.44\ \Omega} \textbf{ (Ans.)}$$

10.4.2. Impedance (Or R-L-C) Bridge

The impedance bridge is employed for the measurement of :

(*i*) Resistance; (*ii*) Inductance;

(*iii*) Capcitance; (*iv*) Dissipation factor;

(*v*) Quality factor.

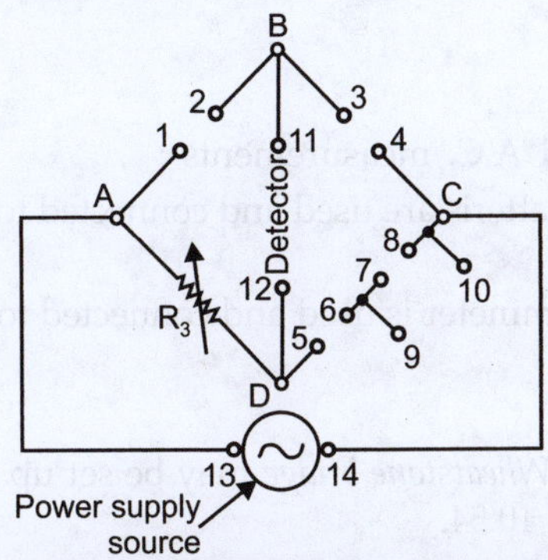

Fig. 10.53. Skelton type R-L-C ridge.

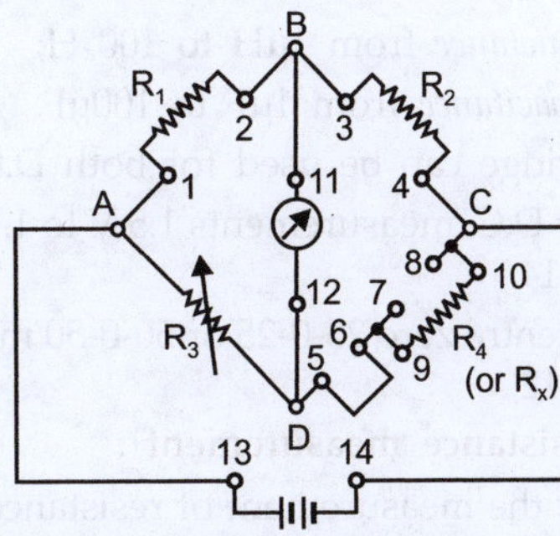

Fig. 10.54. Wheatstone bridge (Resistance measurement)

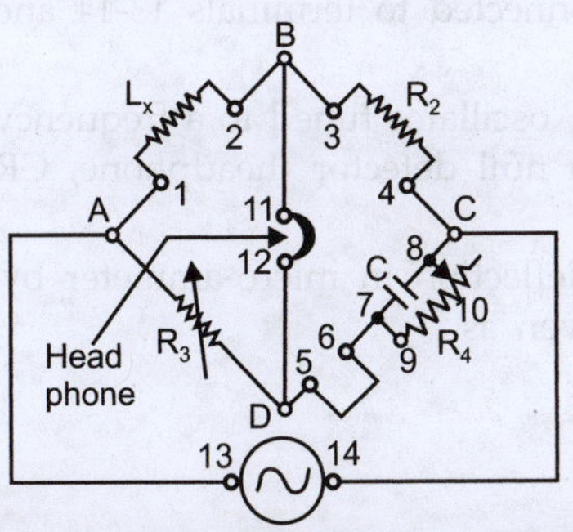

Fig. 10.55. Inductance measurement.

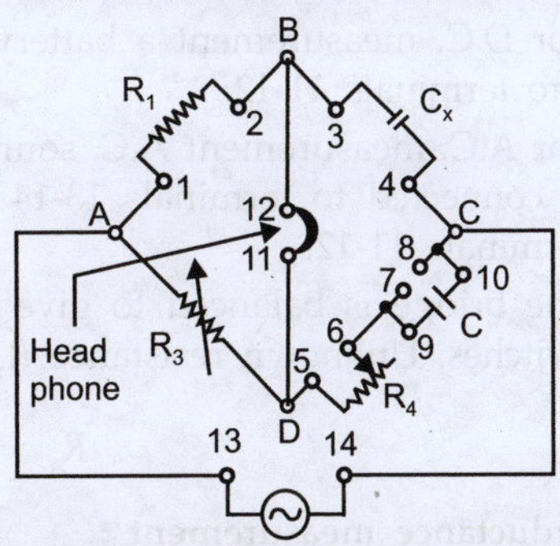

Fig. 10.56. Capacitance measurement.

This bridge consists of the following components/elements :

(*i*) ***Signal sources;***

(*ii*) ***Bridges*** :

(*a*) *Wheatstone bridge* ... For measurement of resistance;

(*b*) *Maxwell and Hay bridges* ... For measurement of inductance and quality factors;

(*c*) *De Sauty bridge* ... For measurement of capacitance and dissipation factors.

(*iii*) ***Imbalance amplifier;***

(*iv*) ***A metered detector** in addition to power supply.*

— A sensitive centre zero-micro-ammeter (25μA-0.25μA) is used to indicate the balance condition.

— Provision is also made for application of external signal sources and connecting external detectors.

- Fig. 10.53 shows the schematic circuit diagram of a *skelton type bridge* employed for measurement of *R*, *L* and *C* :

— R_3 is a decade resistance box.

— The other arms are completed by connecting the unknown and standard components to 1-2, 3-4, 5-6, 7-8 and 9-10, a null detector to 11-12 and a source of supply to 13-14.

This bridge can be used to measure :

(a) *Resistance* from 100mΩ to 10MΩ;

(b) *Inductance* from 1μH to 100 H;

(c) *Capacitance* from 1μF to 100μF.

This bridge can be used for both D.C. and A.C. measurements :

— For D.C. measurements 1.5 V to 1.75 V batteris are used and connected to terminals 13-14.

— A centre zero 25-0-25 or 50-0-50 micro-ammeter is used and connected to terminals 11-12.

1. **Resistance measurement :**

- For the measurement of resistance, the *Wheatstone bridge* may be set up by proper arrangement of bridge arms, as shown in Fig. 10.54.
- The standard resistors R_1 and R_2 are connected to the terminals 1-2 and 3-4 respectively, unknown resistance R_x is connected to terminals 9-10 and a heavy wire jumper is connected to terminals 5-6.
- For D.C. measurement a battery is connected to terminals 13-14 and a micro-ammeter to terminals 11-12.

For A.C. measurement A.C. source (AF oscillator tuned to a frequency of 1kHZ) is connected to terminals 13-14 and a null detector (headphone, CRO etc.) to terminals 11-12.

- The bridge is balanced to give null deflection in micro-ammeter by adjusting decade switches. Unknown resistance R_x is given as :

$$R_x = \frac{R_2R_3}{R_1} \quad ...(10.123)$$

2. **Inductance measurement :**

- In order to measure unknown inductance L_x, standard resistor R_2, standard capacitor C, 100Ω rheostat R_4, heavy wire jumper, A.C. supply source and detector are connected to terminals 1-2, 3-4, 7-8, 9-10, 5-6, 13-14 and 11-12 respectively as shown in Fig. 10.55.
- The bridge is balanced to give null deflection by adjusting the decade switches and rheostat R_4. Then :

Unknown inductance, $L_x = R_2R_3C$ henrys

where R_2 and R_3 are in ohms and C is in farads.

3. **Capacitance measurement :**

- For the measurement of capacitor, standard resistor R_1, unknown capacitance C_x, a 100Ω rheostat R_4, standard capacitor C, source of A.C. supply and detector are connected to terminals 1-2, 3-4, 5-6, 9-10, 13-14 and 11-12 respectively as shown in Fig. 10.56.
- The bridge is balanced by adjusting decade switches and rheostat R_4. Then :

Unknown capacitance, $C_x = \frac{CR_3}{R_1}$ μF

where R_1 and R_3 are in ohms and C is in μF.

10.4.3. LCR Meter

These meters are used to measure values of inductance (*L*), capacitance (*C*) and resistance (*R*) *directly*.

LCR meters are provided with switching arrangement for the measurement of *L*, *C* and *R*.

Fig. 10.57. illustrates a typical LCR meter with a block diagram and important controls on the face panel of the meter.

— It consists of a permanent magnet moving coil (PMMC) instrument.

— The scale is calibrated in terms of *R*, *L* and *C*.

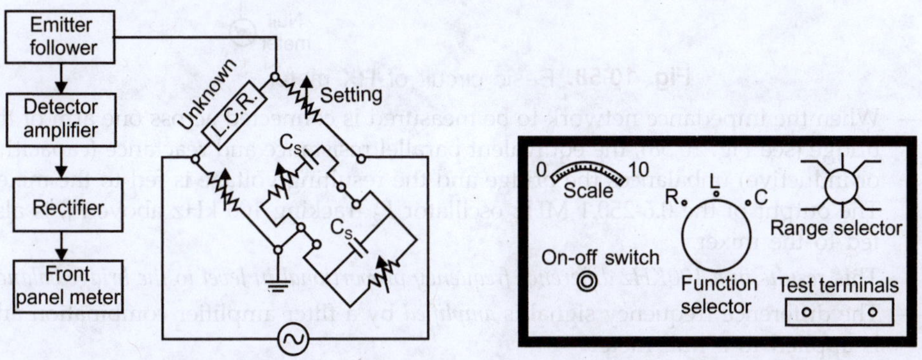

(a) Block diagram (b) Important controls on the face panel of LCR meter

Fig. 10.57. Typical LCR meter.

— When the component under test (*R*, *L* or *C*) is connected across the *"test terminals"* of the meter, the pointer of the meter indicates the value of the component *directly*.

Basically the meter is an electronic instrument containing various electronic components fabricated on a PCB (printed circuit board) which needs D.C. supply. For this purpose, the meter also has a stepdown transformeter and a rectifier. However, a sinusoidal single-phase supply is given to the meter from outside.

- The *use of this meter* is as follows :

— The meter is connected to the supply and ON/OFF switch is made 'ON'.

— The glow of the indicator confirms that the meter is properly connected with the supply.

— Now the position of function selector is adjusted at L, C or R depending upon the component to be tested. Also suitable range is selected by range selector.

— The unknown component is connected across the test terminals of the meter and the value is read on the scale.

10.4.4. RX Meter

This meter is used *to measure resistive and reactive components of a parallel Z network. It is especially useful in the design of RF tank circuits but can be used with any parallel 'Z'.*

Fig. 10.58 shows the basic circuit of a typical *RX* meter :

— It consists of two variable frequency oscillators that track each other at frequency *100 kHz apart*. The output of a 0.5-250MHz oscillator F_1 is fed into a bridge.

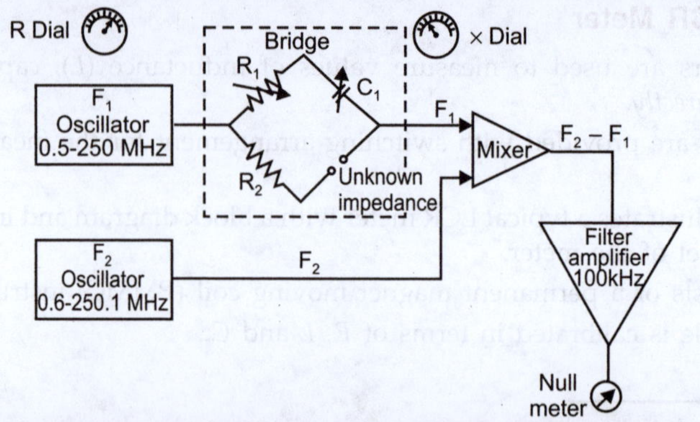

Fig. 10.58. Basic circuit of RX meter.

— When the impedance network to be measured is connected across one arm of the bridge (see Fig. 10.58), the equivalent parallel resistance and reactance (capacitive or inductive) unbalances the bridge and the resulting voltage is fed to the mixer. The output of the 0.6-250.1 MHz oscillator F_2 tracking 100 kHz above F_1, is also fed to the mixer.

This *results in a 100KHz difference frequency proportional in level to the bridge balance.*

— The difference frequency signal is *amplified* by a filter amplifier combination and is applied to a null meter.

— When the bridge resistive and reactive controls are nulled, their respective dials accurately indicate the parallel impedance components of the network under test.

10.5. MEASUREMENT OF RESISTANCE (EFFECTIVE), INDUCTANCE AND CAPACITANCE AND IMPEDANCE AT HIGH FREQUENCIES

- Although some of the methods described earlier *e.g.*, bridge methods can be used to measure *R*, *L* and *C* at high frequencies, say *0.1 to 100 MHz* but these methods do not give correct results owing to the fact that they do not take into consideration the special effects which are inherent in any measurement at high frequencies. Some of the special effects are :

  (*i*) High reactance due to stray series inductance and its variation with frequency; (*ii*) Low reactance due to stray shunt capacitance and its variation with frequency; (*iii*) Increase of effective resistance on account of proximity effects ; (*iv*) Increase of effective resistance with frequency on account of skin effects; (*v*) Increase of effective resistance due to production of eddy currents; and (*vi*) Variation of dielectric loss of capacitors with frequency.
- As errors to the tune of 1 to 5% are acceptable in measurements in this frequency range, the methods used involve the use of very simple apparatus which may include :

  (*i*) A set of inductors from 1.0 μ H to 10mH; (*ii*) A variable air capacitor of maximum capacitance 1000 pF; (*iii*) A decade resistance box of resistance up to 100Ω; and (*iv*) A thermoelectric ammeter or an electronic voltmeter in order to indicate resistance conditions.

10.5.1. Measurement of Resistance (Effective)

Following are the *two methods* of measurement of effective resistance (at high frequency) of a series R-L-C resonant circuit :

1. Resistance variation method.
2. Reactance variation method.

1. Resistance variation method :

The effective series resistance of a resonant circuit is measured by *adding extra resistance from outside.*

Fig. 10.59. illustrates the resistance variation method :

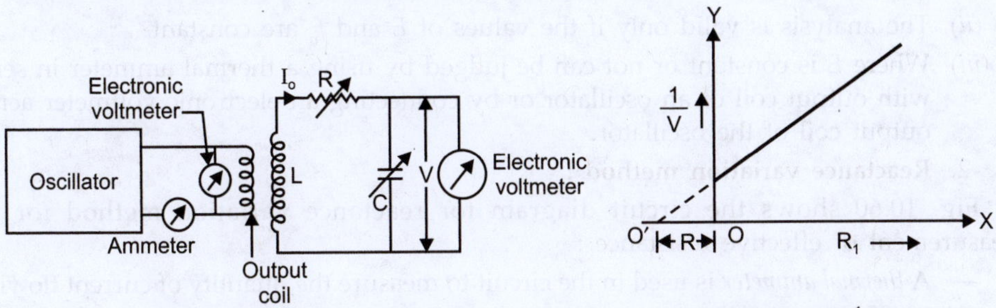

(a) Circuit diagram (b) Plot between $\frac{1}{V}$ and R_1

Fig. 10.59. Resistance variation method.

Let, E = Induced e.m.f. in the resonant circuit;

I_o = Current in the resonant circuit,

R_1 = Additional resistance inserted in the resonant circuit

R = Effective resistance of resonant circuit excluding resistance R_1

L = Inductance of coil,

C = Capacitance of coil,

V = Reading of voltmeter connected across C, and

f_o = Resonant frequency.

At resonance (achieved by varying capacitor) :

$$X_L = X_C$$

$\therefore$ Total resistance $= R + R_1$

and,
$$I_0 = \frac{E}{R + R_1} \quad \text{...(10.124)}$$

$$V = L_o \times X_C = \frac{1}{2\pi f_o C} \quad \text{...(10.125)}$$

or,
$$V = \frac{E}{2\pi f_o C (R + R_1)} \quad \text{...[10.125(a)]}$$

or,
$$R + R_1 = \frac{E}{2\pi f_o CV} \quad \text{...(10.126)}$$

If the values of E, f_o and C are kept constant, we get,

$$R + R_1 = \frac{K}{V}$$

where, k = a constant $= \frac{E}{2\pi f_o C}$ ...(10.127)

Thus, under the above conditions, if a curve is plotted between $\frac{1}{V}$ and R_1, we get a straight line as shown in Fig. 10.59 (*b*). The effective resistance is given by the intercept 00′ provided that the loss resistance of capacitor and the resistance of the leads is *negligible*.

Conclusion from the above analysis :

(*i*) An increase in the value of R_1 decreases V and increases the value of $\frac{1}{V}$.

(*ii*) The analysis is valid only if the values of E and f_o are constant.

(*iii*) Where E is constant or not can be judged by using a thermal ammeter in series with output coil of an oscillator or by connecting an electronic voltmeter across output coil of the oscillator.

2. Reactance variation method :

Fig. 10.60 shows the circuit diagram for reactance variation method for the measurement of effective resistance :

— A *thermal ammeter* is used in the circuit to measure the quantity of current flowing.

— By varying the circuit capacitance (*C*) or frequency (*f*), the circuit reactance is varied till the ammeter reads maximum value. This occurs only when the circuit is under resonant conditions.

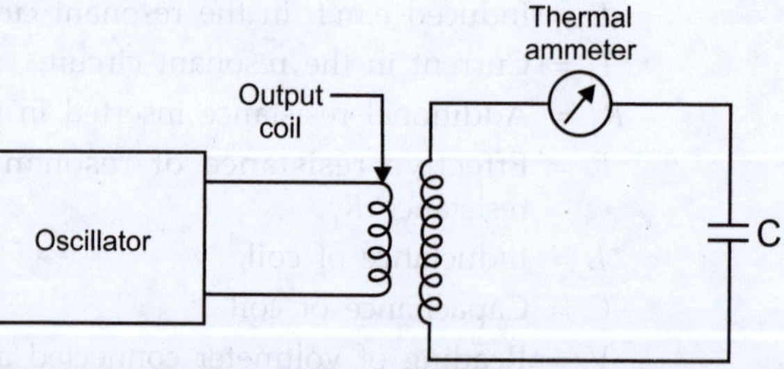

Fig. 10.60. Reactance variation method.

The maximum value of current,

$$I_o = \frac{E}{R} \qquad ...(10.128)$$

where, E = E.m.f. induced, and

R = Effective resistance.

The reactance is then adjusted so that current $I = \frac{I_o}{\sqrt{2}}$

$$\therefore \quad I = \frac{E}{\sqrt{R^2 + X^2}} = \frac{I_o}{\sqrt{2}} = \frac{E}{\sqrt{2}\,R}$$

or,

$$\sqrt{R^2 + X^2} = \sqrt{2}\,R$$

Squaring boths sides, we get

$$R^2 + X^2 = 2R^2$$

or,

$$X = \pm R \qquad ...(10.129)$$

Thus, there are *two values* of reactance X which give a current, $I = \frac{I_o}{\sqrt{2}}$. The difference of these two values or reactance gives twice the value of effective resistance, *i.e.*, 2R.

Following are the two ways of varying the reactance :

(i) Variation of capacitance (C) and

(ii) Variation of frequency (*f*).

(i) ***Variation of capacitance (C) :***

In this method the value of frequecny (*f*) is kept constant and capacitance (C) is varied.

Let the value of capacitance which gives I_0 be C_0 and the two values of capacitance which give a current of $\frac{I_0}{\sqrt{2}}$ be C_1 and C_2.

Then, $$\frac{1}{\omega C_2} - \frac{1}{\omega C_1} = 2R$$

or, $$\frac{1}{\omega}\left(\frac{C_1 - C_2}{C_1 C_2}\right) = 2R$$

or, $$R = \frac{C_1 - C_2}{2\omega\, C_1 C_2} = \frac{C_1 - C_2}{4\pi f C_1 C_2} \quad ...(10.130)$$

When the circuit loss is small and the resonance is peaked and symmetrical,

$$C_1 C_2 \simeq C_0^2 \quad ...(10.131)$$

$\therefore$ $$R = \frac{C_1 - C_2}{2\omega C_0^2} = \frac{C_1 - C_2}{4\pi f C_0^2} \quad ...(10.132)$$

Also, for a series resonant circuit,

$$Q = \frac{\omega L}{R} = \frac{1}{\omega C_0 R} \quad ...(10.133)$$

or, $$Q = \frac{1}{\omega C_0 \left\{\frac{C_1 - C_2}{\omega C_0^2}\right\}}$$

... Substituting for *R* from eqn. (10.132)

or, $$Q = \frac{2C_0}{C_1 - C_2} \quad ...(10.134)$$

— The above eqn. (10.134) indicates that the circuit *Q* can be determined in terms of two values of capacitance.

— The accuracy of this method depends upon the accuracy with which the difference $C_1 – C_2$ can be known. In order to determine this difference accurately, a micrometer capacitor is put in parallel with the main capacitor.

(ii) ***Variation of frequency :***

In this method the capacitance (C) is kept constant and the frequency (*f*) is *varied*. Let ω_0 be the angular frequency to give maximum current I_0 and ω_1 and ω_2 be the two angular frequencies which give a value of current $I = \frac{I_0}{\sqrt{2}}$.

Then, $$\left(\omega_1 L - \frac{1}{\omega_1 C}\right) - \left(\omega_2 L - \frac{1}{\omega_2 C}\right) = 2R$$

or, $$(\omega_1 - \omega_2) L + \frac{1}{C}\left(\frac{1}{\omega_2} - \frac{1}{\omega_1}\right) = 2R$$

or, $$(\omega_1 - \omega_2) L + \frac{1}{C}\left(\frac{\omega_1 - \omega_2}{\omega_1 \omega_2}\right) = 2R$$

or, $$(\omega_1 - \omega_2) L + \omega_0^2 L\left[\frac{\omega_1 - \omega_2}{\omega_1 \omega_2}\right] = 2R \qquad \left[\because \frac{1}{C} = \omega_0^2 L, \text{ as } X_L = X_C \text{ at resonance frequency } \omega_0\right]$$

Also, for a peaked and symmetrical resonance curve $\omega_0^2 = \omega_1 \omega_2$

$\therefore$ $$(\omega_1 - \omega_2) L + \omega_0^2 L\left(\frac{\omega_1 - \omega_2}{\omega_0^2}\right) = 2R$$

or, $$(\omega_1 - \omega_2) L + (\omega_1 - \omega_2) = 2R$$

or, $$R = (\omega_1 - \omega_2) L \quad \text{or} \quad \frac{(\omega_1 - \omega_2)}{(1/L)} \qquad ...(10.135)$$

$$\left(\because \frac{1}{L} = \omega_0^2 C\right)$$

or, $$R = \frac{\omega_1 - \omega_2}{\omega_0^2 C} \qquad ...(10.136)$$

and, $$Q = \frac{\omega_0 L}{R} = \frac{\omega_0}{\omega_1 - \omega_2} \qquad ...(10.137)$$

...(using eqn. (10.135)

- The *limitation* of this method is that it is quite difficult to *keep the value of E constant with change of frequency*. However, error introduced on this account is *quite small.*

Example 10.37. *A coil of inductance L is tuned by a capacitance C to resonate 2 MHz. A standard capacitor of 210pF is tuned to resonate with the same coil at 4 MHz. It the resonance is indicated by a VTVM of capacitance 6pF in parallel with the circuit, calculate the values of L and C.*

Solution. *Given* : Capacitance of standard resonant capacitor, C_s = 210 pF; Capacitance of voltmeter, C_v = 6 pF; Resonant frequencies = 2 MHz and 4 MHz respectively.

Values of L and C :

We know that at resonance :

$$X_L = X_C$$

or, $$2\pi f L = \frac{1}{2\pi f C} \quad \text{or} \quad C = \frac{1}{4\pi^2 f^2 L}$$

Now, we have $$C + C_v = \frac{1}{4\pi^2 L (2 \times 10^6)^2}$$

or, $$(C + 6) \times 10^{-12} = \frac{1}{16\pi^2 L \times 10^{12}} \quad \text{...(i)}$$

and, $$(C_s + C_v) = \frac{1}{4\pi^2 L\,(4\times10^6)^2}$$

or, $$(210 + 6) \times 10^{-12} = \frac{1}{64\pi^2\, L\times10^{12}} \quad \text{...(ii)}$$

or, $$L = \frac{1}{(210+6)\times10^{-12}\times64\pi^2\times10^{12}}$$

$$= \mathbf{7.33\mu H} \text{ (Ans.)}$$

Substituting for L in eqn. (*i*), we get

$$(C + 6) \times 10^{-12} = \frac{1}{16\pi^2\times7.33\times10^{-6}\times10^{12}}$$

or, $$C = \mathbf{857.9pF} \text{ (Ans.)}$$

Example 10.38. *A coil of inductance L and resistance R, a calibrated variable capacitor C and a thermojunction, all in series, are put across a constant r.m.s. voltage of frequency 4MHz. The uncalibrated ammeter is found to show the same scale reading I for the two values of C namely 40 and 48pF, one being below and the other above the capacitance C_0 needed to produce maximum current. The capacitor is now adjusted to a value C_0, and an added series resistance of 60Ω is found to reduce the current previous value of I.*

Determine the values of L and R.

Solution. *Given* : C_1 = 40 pF ; C_2 = 48 pF (C_1 and C_2 are the two values of capacitance (C) which produce the same current);

f = 4MHz; Additional series resistance, R' = 60Ω

Values of *L* and *R* :

Let the resonance curve be peaked and symmetrical and therefore capacitance required for resonance is :

$$C_o = \frac{C_1 + C_2}{2} = \frac{40+48}{2} = 44\text{pF}$$

At resonance, $$L = \frac{1}{4\pi^2 f^2 C_o} = \frac{1}{4\pi^2\times(4\times10^6)^2\times44\times10^{-12}}\ \text{H}$$

$$= \mathbf{35.98MH} \text{ (Ans.)}$$

The value of current when the capacitance in the circuit is C_1,

$$I = \frac{E}{\sqrt{R^2 + \left(\omega L - \frac{1}{\omega C_1}\right)^2}}$$

or, $$I = \frac{E}{\sqrt{R^2 + \left(2\pi\times4\times10^6\times35.9\times10^{-6} - \frac{1}{2\pi\times4\times10^6\times48\times10^{-12}}\right)^2}}$$

or, $$I = \frac{E}{\sqrt{R^2 + [(902.26 - 828.93)]^2}} = \frac{E}{\sqrt{R^2 + 5377}} \quad ...(i)$$

Also, $$I = \frac{E}{R + R'} = \frac{E}{R + 60} \quad ...(ii)$$

Equating eqns. (*i*) and (*ii*) we get :

$$\frac{E}{\sqrt{R^2 + 5377}} = \frac{E}{R + 60}$$

or, $$R + 60 = \sqrt{R^2 + 5377}$$

Squaring both sides, we get

$$R^2 + 120R + 3600 = R^2 + 5377$$

$$\therefore \quad R = \mathbf{14.81\Omega} \text{ (Ans.)}$$

Example 10.39. *A circuit having an effective capacitance of 160 pF is tuned to a frequency of 1.2 MHz. In this circuit the current falls to 70.7 percent of its resonant value when the frequency of an e.m.f. of constant magnitude injected in series with the circuit deviates from the resonant frequency by 6 kHz.*

Calculate the Q factor and effective resistance of the circuit.

Solution. *Given* : C = 160pF ; f_o = 1.2MHz

Values of *Q* and *R* (effective) :

We know that, $$Q = \frac{\omega_o}{\omega_1 - \omega_2} = \frac{f_o}{f_1 - f_2} \quad ...[\text{Eqn. } (10.137)]$$

The two frequencies at which current is 70.2% or $\frac{1}{\sqrt{2}}$ time the current at resonant frequency are :

$$f_1 = 1.2 \times 10^6 + 6 \times 10^3$$

and, $$f_2 = 1.2 \times 10^6 - 6 \times 10^3$$

$$\therefore \quad f_1 - f_2 = 12 \times 10^3 \text{ Hz}$$

Hence, $$Q = \frac{1.2 \times 10^6}{12 \times 10^3} = \mathbf{100} \text{ (Ans.)}$$

Now, $$R = \frac{\omega_1 - \omega_2}{\omega_0^2 C} \quad ...[\text{Eqn. } (10.136)]$$

or, $$R = \frac{f_1 - f_2}{2\pi f_0^2 \times C} = \frac{12 \times 10^3}{2\pi \times (1.2 \times 10^6)^2 \times 160 \times 10^{-12}}$$

$$= \mathbf{8.29\ \Omega} \text{ (Ans.)}$$

10.5.2. Measurement of Inductance

Fig. 10.61 (*a*) shows the circuit diagram for the measurement of inductance of a coil :

— The coil under test is loosely coupled to the output coil of a variable frequency oscillator.

— The oscillator's dial is calibrated in terms of frequency.

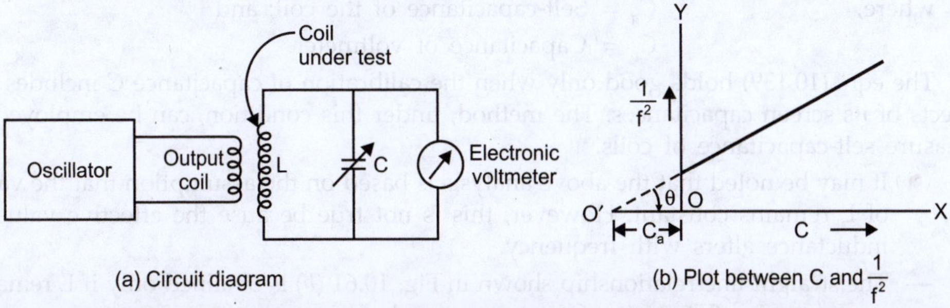

Fig. 10.61. Measurement of inductance.

— A variable capacitor, *C*, is used for obtaining resonance conditions which are indicated by an electronic voltmeter connected across it.

At resonance : $X_L = X_C$

or, $$\omega L = \frac{1}{\omega C}$$

or, $$LC = \frac{1}{\omega^2} = \frac{1}{4\pi^2 f^2} \quad ...(10.138)$$

Thus knowing the values of *f* and *C*, the value of *L* can be calculated.

The value of the inductance (L) obtained by this method is *rarely accurate* due to the following reasons :

— The *additional capacitances* which are introduced into the circuit are *not taken into account*.

— Errors are caused due to the fact that dial setting of capacitor may not be accurately known.

In order to get better accuracy, several readings are take by varying the frequency, *f*.

If C_a is the value of additional capacitance, then for resonance at any frequency, we have :

$$L\,(C + C_a) = \frac{1}{4\pi^2 f^2} \quad ...(10.139)$$

or, $$\frac{1}{f^2} = 4\pi^2 L\,(C + C_a)$$

$$= 4\pi^2 LC + 4\pi^2 LC_a \quad ...[10.139\,(a)]$$

From the eqn. (10.139) it is obvious that if a curve is plotted between $\frac{1}{f^2}$ and C, we get a straight line as shown in Fig. 10.61 (*b*). The line produced back cuts the *x*-axis at *D*. The slope of the line, $\tan\theta = 4\pi^2 L$

$$\therefore \quad L = \frac{\tan\theta}{4\pi^2} \quad ...(10.140)$$

Thus by knowing value of slope of the line, the value of *L* can be obtained.

Also, the value of additional capacitance C_a is equal to the intercept 00′.

Here $C_a = C_d + C_v$ ...(10.141)

where, C_d = Self-capacitance of the coil; and

C_v = Capacitance of voltmeter.

The eqn. (10.139) holds good only when the calibration of capacitance C includes the effects of its screen capacitances. The method, under this condition, can be employed to measure self-capacitance of coils.

- It may be noted that the above analysis is based on the assumption that the value of L remains constant. However, this is not true because the effective value of inductance alters with frequency.
  - — The straight line relationship shown in Fig. 10.61 (*b*) is obtained only if L remains constant over the range of frequency used.
  - — When the value of L changes with frequency, a curve, instead of a straight line, is obtained. Therefore, in order to determine the value of L at a particular frequency a tangent is drawn to the curve at a value of the frequency where the value of L is required. *The value of L is obtained from the slope by using the eqn. (10.140).*

Example 10.40. *A capacitance of 200 pF produces resonance with a coil at a frequency of* $\frac{2}{\pi} \times 10^6$ *Hz, while at the second harmonic of this frequency resonance is produced by a capacitance of 40pF. Neglecting the effect of voltmeter capacitance and other stray capacitance, calculate :*

(*i*) *The self-capacitance of the coil;*

(*ii*) *The inductance of the coil,*

Solution. *Given :* $C_1 = 200\text{pF}$; $f = \frac{2}{\pi} \times 10^6 \text{Hz}$; $C_2 = 40\text{pF}$; f_2 (second harmonic) $= 2f_1 = \frac{4}{\pi} \times 10^6 \text{Hz}$

(*i*) Self-capacitance of the coil, C_d :

At resonance : $$L = \frac{1}{4\pi^2 f^2 (C + C_d)}$$ ...[Eqn. (10.139)]

Therefore, with the given data, we have :

$$L = \frac{1}{4\pi^2 \left(\frac{2}{\pi}\right)^2 \times 10^{12} (200 + C_d) \times 10^{-12}}$$

$$= \frac{1}{16(200 + C_d)} \quad ...(i)$$

and, $$L = \frac{1}{4\pi^2 \left(\frac{4}{\pi}\right)^2 \times 10^{12} (40 + C_d) \times 10^{-12}}$$

$$= \frac{1}{64(40 + C_d)} \quad ...(ii)$$

From (*i*) and (*ii*), we get :

$$\frac{1}{16(200+C_d)} = \frac{1}{64(40+C_d)}$$

or, $$64\,(40 + C_d) = 16\,(200 + C_d)$$

$$160 + 4C_d = 200 + C_d$$

or, $$C_d = \mathbf{13.33\ pF}\ \textbf{(Ans.)}$$

Substituting the value of C_d in eqn. (*i*), we get

$$L = \frac{1}{16(200+13.33)}\ \text{H}$$

$$= \mathbf{292.97\mu\ H}\ \textbf{(Ans.)}$$

10.5.3. Measurement of Capacitance

The capacitance at high frequencies is normally measured by using *"substitution method"*.

Fig. 10.62 shows a simple circuit for measurement of capacitance :

— The unknown capacitance C is connected in parallel with a variable standard capacitor C_s, which is of the *same order* as the unknown capacitance.

— Keeping the switch open, resonance is obtained by varying the standard capacitance C_s. The resonance condition is indicated by an electronic voltmeter connected across C_s.

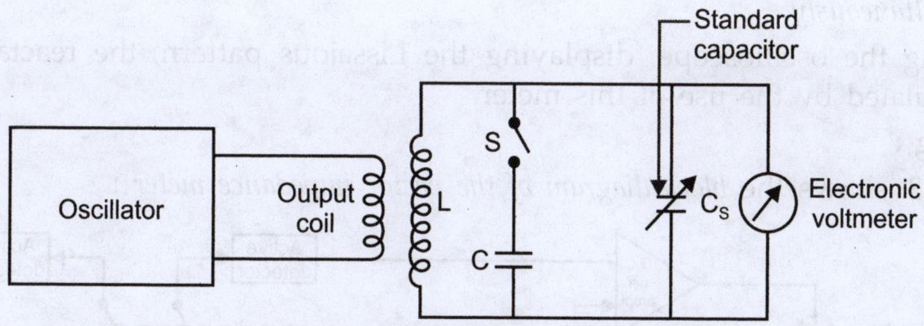

Fig. 10.62. Measurement of capacitance.

If f = frequency at resonance (in Hz), L = inductance of coil coupled with the output coil of the oscillator (in H), and C_s = setting of standard capacitor with switch S open (in F), then we have :

$$C_{s1} = \frac{1}{4\pi^2 f^2 L} \qquad \text{...(10.142)}$$

— Now, close the switch S to put unknown capacitance C in parallel with C_S. Keeping the frequency f the same as earlier, obtain the resonant conditions by varying C_s. If C_{se} = setting of standard capacitor with switch closed, then we have :

$$C + C_{S2} = \frac{1}{4\pi^2 f^2 L} \qquad \text{...(10.143)}$$

From eqn. (10.142) and (10.143), we get :

$$C + C_{S2} = C_{S1}$$

or, Unknown capacitance, $$C = C_{S1} - C_{S2} \qquad \text{...(10.144)}$$

The eqn. (10.144) indicates that the value of unknown capacitance can be directly known from the *difference of the two settings* of the standard capacitor C_S.

10.5.4. Measurement of Impedance

The measuring of impedance means to measure its magnitude and the phase angle. At the frequencies below 100MHz, the impedance magnitude can be determined by voltage and current measurements. The phase difference between the voltage and current waveforms indicates whether the impedance is inductive or capacitive in nature. To specify the component values precisely, many measurements at several different frequencies are necessary. At high frequencies *such measurements become tedious, time consuming and lengthy. For such measurements complex impedance measurement meters are used.*

Two such meters are discussed below :

1. Vector impedance meter :

The meter used for complex impedance measurements at high frequencies is called **vector impedance meter.**

Applications :

(*i*) It is used to measure the impedances over wide frequency range, from 400 kHz to 110 MHz.

(*ii*) It can also be used to obtain *sweep frequency plots, i.e.,* magnitude and phase angle against frequency; these plots provide complete information within the frequency band of interest.

(*iii*) This meter can measure the magnitude and the phase angle of the impedance *simultaneously.*

(*iv*) Using the oscilloscope, displaying the Lissajous pattern, the reactance can be calculated by the use of this meter.

Working :

Fig. 10.63. shows the *block diagram of the vector impedance meter* :

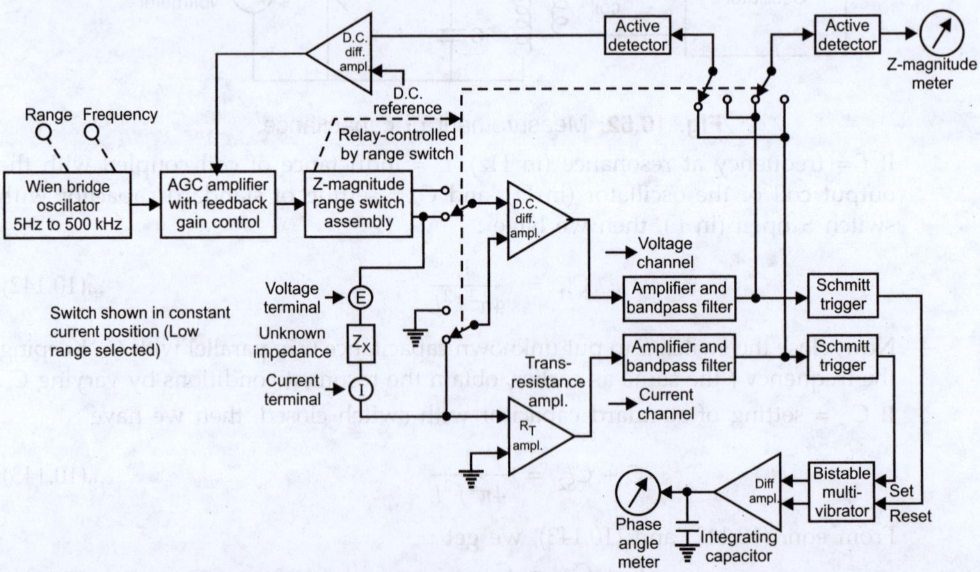

Fig. 10.63. Block diagram of the vector impedance meter.

- Two measurements take place

(*i*) The magnitude of the impedance is determined by measuring the current through the unknown component when a known voltage is applied across it, or by measuring the voltage across the component when a known current is passed through it.

(*ii*) The *phase angle* is found by determining the phase difference between the voltage across the component and the current through the component.

- As shown in Fig. 10.63, the instrument contains a *signal source* (Wien bridge oscillator) with two front panel controls to select the frequency range and to continuously adjust the selected frequency. The output of the oscillator is fed to an *AGC amplifier* which permits accurate gain adjustment by means of its feedback voltage. This gain adjustment is an internal control actuated by the setting of the *impedance range switch*, to which AGC amplifier output is connected.

The impedance range switch allows operation of the instrument in the following two modes :

(*i*) Constant-current mode.

(*ii*) Constnat-voltage mode.

Constant-current mode :

— In this mode the three lower ranges (×1, ×10, and 100) operate.

— Here, the unknown component is connected across the input of the A.C. differential amplifier; the current supplied (to the unknown) depends on the setting of the impedance range switch. This current is held constant by the action transresistance or R_T amplifier, which converts the current through the unknown to a voltage output equal to the current times its feedback. The output R_T amplifier is fed to a detector circuit and compared to a D.C. reference voltage. The resulting control voltage regulates the gain of the AGC amplifier and hence the voltage applied to the impedance range switch.

— The output of the A.C. differential amplifier is applied to an amplifier and filter section consisting of high-and low-band filters that are changed with the frequency range to restrict the amplifier bandwidth.

— The filter output is applied to the detector that drives the Z magnitude meter. Since the current is maintained constant, the Z magnitude meter which measures the voltage across the unknown component deflects proportional to the magnitude of the impedance.

Constant-voltage mode :

— In this mode, the two inputs to the differential amplifier are switched. The terminal which was connected to the input of the transresistance amplifier in constant-current mode is now grounded. The other input of the differential amplifier, which was connected to the voltage terminal of the unknown component, is now connected to a point on the *Z-magnitude range switch which is held at constant potential*. The voltage terminal of the unknown is connected to this same point of constant potential, or depending on the setting of the Z-magnitude range switch, to a decimal fraction of this voltage. *In any case, the voltage across the unknown is held at a constant level.*

— The current through the unknown is applied to the transresistance amplifier which again produces an output voltage proportional to its input current.

— The roles of the A.C. differential amplifier and transresistance amplifier are now *reversed*. The voltage output of the R_T amplifier is applied to the detector and then to the Z-magnitude meter. The output voltage of the differential amplifier controls the gain of the AGC amplifier in the same manner that the R_T amplifier did in the constant current meter.

Phase angle measurements :

— The phase angle measurements are carried out simultaneously with the magnitude measurements.

— The outputs of both the voltage channel and the current channel are amplified and each ouput is connected to a schmitt trigger circuit.

— The Schmsitt trigger circuits produce a positive-going spike every time the input sine wave goes through zero crossing. These positive spikes are applied to a *binary phase detector circuit*.

The phase detector consists of a bistable multivibrator, a differential amplifier, and an integrating capacitor.

— The positive-going pulse from the constant-current channel *sets* the multivibrator and the pulse from the constant-voltage channel *resets* the multivibrator. The "*set*" time of the multivibtrator is, therefore, determined by the *zero crossings of the voltage and current waveforms*.

— The "*set*" and "*reset*" outputs of the multivibrator are applied to the differential amplifier, which applies the difference voltage to an integrating capacitor.

— *The capacitor voltage is directly proportional to the zero-crossing time interval and is applied to the phase-angle meter which then indicates the phase difference in degrees between the voltage and current waveforms.*

- The vector impedance meter is usually *calibrated* by connecting standard components (e.g., standard resistors or capacitors) to the input terminals.

2. **RF vector impedance meters :**

This meter uses the *phase locked sampling system* to obtain and operate in a frequency range of 0.5 MHz to 108 MHz.

Fig. 10.64 shows the block digram of the the RF vector impedance meter :

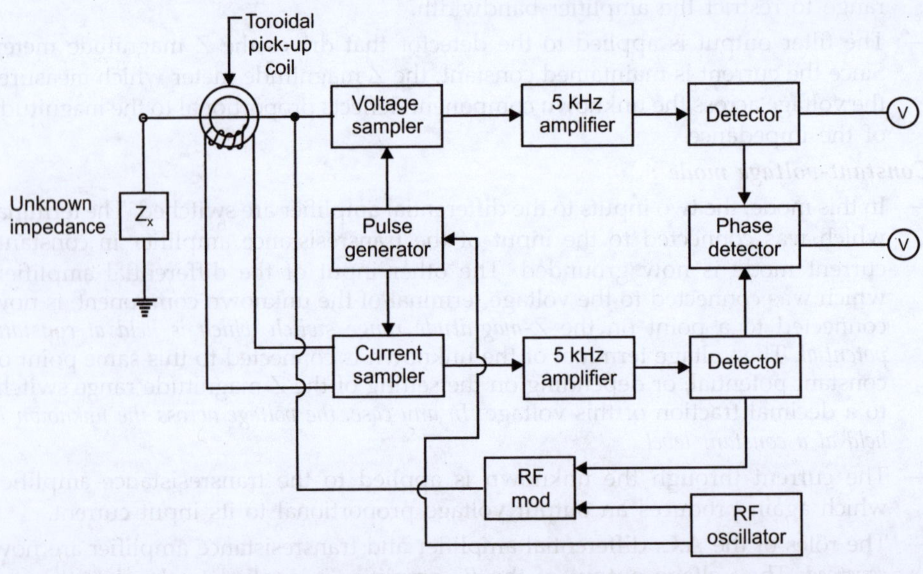

Fig. 10.64. RF vector impedance meter.

— The RF current is supplied to the unknown impedance.

— The current is measured by toroidal pick-up coil. It is then sampled, amplified, detected and then applied to the modulator. This forms an *automatic current level control*.

— The voltage is similarly sampled and because the current is held constant, it is a *measure of the magnitude of impedance*.

— The *phase detector* measures the phase angle θ.

- This instrument (commercially available) popularly known as HP 4815 is manufactured by Hewlett-Packard company.

10.5.5. Automatic Bridges

The "automatic bridges" provide an automatic readout without adjustment of balance controls (as in conventional bridges). In some cases they also provide a *Binary Coded Digital (BCD) readout* to external equipment.

These bridges are similar in operation to digital meters.

Fig. 10.65 shows a typical automatic bridge circuit :

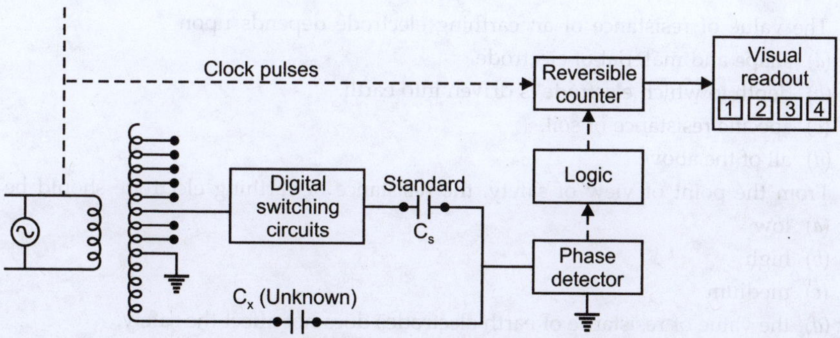

Fig. 10.65. Typical automatic bridge circuit.

— The circuit is transformer-ratio-arm bridge for the automatic measurement of capacitance.

— The circuit is in balance when the currents through the standard capacitor (C_s) and the unknown capacitor (C_x) are equal, so that the current in phase detector is zero.

— The range is chosen automatically by relays that select decade taps on the ratio transformer.

— The phase detector determines whether the current passing the unknown arm of the bridge is higher or lower than that through the standard arm and produces an error signal that indicates whether more or less voltage is required on the standard capacitor to reach a balance. This information is used by a decade counter that controls the voltage on the standard capacitor through electronic switching circuits.

— *At balance*, the value of the unknown is displayed on an *in-line digital readout* that *indicates capacitance. This information is also presented in BCD form for use with printers and other data handling equipments.*

HIGHLIGHTS

1. Resistances are classified as follows :
   (*i*) Low resistances — of the order of 1 Ω and under
   (*ii*) Medium resistances — From about 1 Ω upwards to about 100 kΩ
   (*iii*) High resistances — Of the order of 100 kΩ and upwards
2. When balancing an A.C. bridge, the following two conditions must be met simultaneously
   (*i*) The magnitudes of the impedances should satisfy the relationship :
   $$Z_1Z_4 = Z_2Z_3$$
   (*ii*) The phase angle of impedances should satisfy the relationship :
   $$\angle\theta_1 + \angle\theta_4 = \angle\theta_2 + \angle\theta_3$$

OBJECTIVE TYPE QUESTIONS

Choose the Correct Answer :

1. The value of resistance of an earthing electrode depends upon
   (*a*) shape and material of electrode.
   (*b*) depth to which electrode is driven into earth.
   (*c*) specific resistance of soil.
   (*d*) all of the above.
2. From the point of view of safety, the resistance of earthing electrode should be :
   (*a*) low
   (*b*) high
   (*c*) medium
   (*d*) the value of resistance of earth electrodes does not affect the safety.
3. A Wheatstone bridge cannot be used for precision measurements because errors are introduced into it on account of
   (*a*) resistance of connecting leads (*b*) thermo-electric emfs
   (*c*) contact resistances (*d*) all of the above.
4. Low resistances are provided with four terminals
   (*a*) to facilitate the connection of current and potential circuits.
   (*b*) in order that the resistance value becomes definite irrespective of the nature of contacts at the current terminals.
   (*c*) to eliminate the effect of thermo-electric emfs.
   (*d*) to eliminate the effect of leads.
5. In a Kelvin's double bridge two sets of readings are taken when measuring a low resistance, one with the current in one direction and the other with direction of current reversed. This is done to
   (*a*) eliminate the effect of contact resistance.
   (*b*) eliminate the effect of resistance of leads.
   (*c*) correct for changes in battery voltage.
   (*d*) eliminate the effect of thermo-electric emfs.
6. A resistance of value 10 Ω approximately is to be measured by ammeter-voltmeter method with resistance of ammeter as 0.02 Ω and that of voltmeter as 5000 Ω. The resistance should be measured

(a) by connecting the ammeter on the side of unknown resistance as this connection gives better accuracy.
(b) by connecting the voltmeter on the side of unknown resistance as this connection gives better accuracy.
(c) by any of the two connections, as both of them give equal accuracy.
(d) none of the above.

7. Equal resistances of 100 Ω each are connected in each arm of a Wheatstone bridge which is supplied by a 2 V battery source. The galvanometer of negligible resistance connected to the bridge can sense as low current as 1 μA. The smallest value of resistance that can be measured is
(a) 20 mΩ (b) 2 μΩ
(c) 20 μΩ (d) none of the above.

8. A Wheatstone bridge has ratio arms of 1000 Ω and 100 Ω resistances, the standard resistance arms consist 4 decade resistance boxes of 1000, 100, 10, 1 Ω steps. The maximum and minimum values of unknown resistance which can be determined with this setup are
(a) 111100 Ω, 1 Ω (b) 11110 Ω, 10 Ω
(c) 111100 Ω, 10 Ω (d) none of the above.

9. High resistances are provided with a guard terminal. This guard terminal is used to
(a) bypass the leakage current.
(b) guard the resistance against stray electrostatic fields.
(c) guard the resistance against overloads.
(d) none of the above.

10. When measuring insulation resistance of cables using d.c. source, the galvanometer used should be initially short-circuited because
(a) cables have a low value of initial resistance.
(b) cables have a high value of capacitance which draws a high value of charging current.
(c) cables have a low value of capacitance which draws a high value of charging current.
(d) none of the above.

11. The advantage of Varley loop tests over Murray loop tests is
(a) they can be used for localizing of short circuit faults.
(b) they can be used for localizing of earth fault.
(c) the loop resistance can be experimentally determined.
(d) their accuracy is higher.

12. When localizing ground fault with the help of loop tests, the resistance of the fault
(a) affects the balance conditions.
(b) affects the value of cable resistance.
(c) affects the sensitivity of the bridge.
(d) all of the above.

13. A bridge circuit works at a frequency of 2 kHz. The following can be used as detectors for detection of null conditions in the bridge :
(a) Vibration galvanometers and headphones.
(b) Headphones and tunable amplifiers.
(c) Vibration galvanometers and tunable amplifiers.
(d) Vibration galvanometers, headphones and tunable amplifiers.

14. Wagner's Earth Devices are used in A.C. bridge circuits for
(a) eliminating the effect of earth capacitances.
(b) eliminating the effect of inter-component capacitances.

(c) eliminating the effect of stray electrostatic fields.
(d) shielding the bridge elements.

15. Maxwell's inductance-capacitance bridge is used for measurement of inductance of
(a) low Q coils
(b) medium Q coils
(c) high Q coils
(d) low and medium Q coils.

16. The advantage of Hay's bridge over Maxwell's inductance-capacitance bridge is because
(a) its equations for balance do not contain any frequency term.
(b) it can be used for measurement of inductance of high Q coils.
(c) it can be used for measurement of inductance of low Q coils.
(d) none of the above.

17. The equations under balance conditions for a bridge are
$R_1 = R_2R_3/R_4$ and $L_1 = R_2R_3C_4$
where R_1 and L_1 are respectively unknown resistance and inductance. In order to achieve converging balance
(a) R_2 and R_3 should be chosen as variables
(b) R_2 and C_4 should be chosen as variables
(c) R_4 and C_4 should be chosen as variables
(d) R_3 and C_4 should be chosen as variables.

18. Frequency can be measured by using
(a) Maxwell bridge
(b) Schering bridge
(c) Heaviside Campbell bridge
(d) Wien bridge.

19. The resonant frequency of an RLC series circuit is 1.5 MHz with the tuning capacitor set at 150 pF. The bandwidth is 10 kHz. The effective resistance of the circuit is :
(a) 29.5 Ω
(b) 14.75 Ω
(c) 9.4 Ω
(d) 4.7 Ω

ANSWERS

1. (d)	2. (a)	3. (d)	4. (b)	5. (d)	6. (c)	7. (a)
8. (c)	9. (a)	10. (b)	11. (c)	12. (c)	13. (b)	14. (a)
15. (b)	16. (b)	17. (c)	18. (d)	19. (d).		

THEORETICAL QUESTIONS

Measurement of Resistances

1. How are resistances classified ?
2. Explain what do you mean by low, medium and high resistances. Suggest various suitable methods for measuring them giving justification.
3. State the various methods of measurement of low resistance. Why is the voltmeter-ammeter method unsuitable for the precise measurement of low resistance ?
4. What do you understand by low, medium and high resistance ? Describe Kelvin's double bridge method for measurement of small resistances.
5. What are the problems associated with measurement of low resistances ? How are they overcome through use of Kelvin's double bridge ? Derive the expression for the unknown resistance in the case of Kelvin's double bridge.
6. Explain the procedure of measuring a low resistance with the help of Kelvin's double bridge. Derive the relation for finding unknown resistance.
7. Explain the theory and working principle of Kelvin's double bridge method for measurement of low resistance.
8. Explain with the help of a connection diagram how you would determine the value of low resistance shunt by Kelvin's double arm bridge method. What are the important precautions that are to be taken in such an experiment ?

9. Describe Kelvin's double bridge for the comparison of two low resistances. Give the theory of the bridge and arrangement necessary in order that the greatest precision possible may be obtained.
10. Describe Carey Foster's slide wire bridge for the measurement of medium resistances.
11. Describe a method for measurement of insulation resistance of a cable.
12. What are the special problems associated with the measurement of high resistances of the order of mega ohms ?
13. Describe with the help of diagram the loss of charge method for determining the insulation resistance of a length of cable.
14. Describe price-guard wire method of measuring high resistance with the help of diagram.
15. Describe the loss of charge method for the determination of high resistance.
16. Describe the 'loss of charge' method of measuring high resistance and prove the formula used. Mention the possible errors and suggest methods to minimise these.
17. Explain the loss of charge method for the measurement of insulation resistance of a cable.
18. Describe and explain with the help of neat sketches the construction and working of megger.
19. Explain how will you measure insulation resistance (i) 2-wire D.C. live main and (ii) 3-wire D.C. live main.
20. What is a *"bridge circuit"* ?
21. What are the applications of bridge circuits?
22. What are the advantages of bridge circuits ?
23. What are the types of bridges ?
24. Explain briefly how a 'Wheatstone bridge' is used for measurement of resistance.
25. What are the sources of measurement errors in a Wheatstone bridge?
26. State the limitations of "Wheatstone bridge".
27. Explain briefly the following bridges :
    (*a*) Kelvin bridge; (*b*) Kevin double bridge.
28. Draw the block diagram of Wheatstone bridge with digital readout and explain it briefly.
29. Draw the general form of an A.C. bridge and derive the general equation for its balance.
30. Explain briefly how *"inductance"* can be measured by using a 'Maxwell bridge'.
31. What are the advantages and disadvantages of a Maxwell bridge ?
32. Discuss briefly how "Hay bridge" can be used for the measurement of inductance.
33. With the help of circuit diagram explain how capacitance can be measured by the use of a 'Schering bridge'.
34. Describe briefly how Wien bridge can be used for the measurement of frequency.
35. State the applications of a Wien bridge.
36. Explain briefly *"Resonance bridge"* with the help of a neat circuit diagram. Name the parameters which can be measured by this bridge.
37. Explain with a neat circuit diagram the working of a "Wagner earthing device".
38. How are detectors classified ?
39. What precautions should be taken while using a bridge ?
40. Draw a neat diagram of the Anderson bridge. Deduce the equations when the bridge is under balance condition. Explain clearly how you may measure self-inductance by using this bridge. Draw the phasor diagram of the voltages and currents of the bridge arms at balance.
41. Prove that in Wien bridge the value of the resistance of the capacitor at balance is given by :

$$R_1 = \frac{R_3\,(1+\omega^2 R_2^2 C_2^2)}{\omega^2 R_2 R_4 C_2^2}$$

where ω, C_2, R_2, R_3, R_4 are known values with usual notation.
What are the uses of the above bridge in electrical engineering ?

42. Describe with connection diagram the Heaviside Campbell bridge for precise measurement of self-inductance of a coil. Deduce the formula used.
43. Discuss the common sources of error in A.C. bridges. How are they eliminated ?
44. Write short notes on the following :
    (*i*) De Sauty bridge
    (*ii*) Carey Foster bridge
    (*iii*) Wagner earthing.

UNSOLVED EXAMPLES

1. A resistance is measured by the ammeter-voltmeter method. With the voltmeter connected across the resistance the readings on the ammeter and voltmeter are 0.4 A and 3.2 V respectively. The resistance of the voltmeter is 500 Ω.
   Calculate :.
   (*i*) True value of resistance.
   (*ii*) Percentage error in the value of the resistance if the voltmeter current is ignored.
   [Ans. (*i*) 8.13 Ω ; (*ii*) –1.6%]
2. In a measurement of resistance by potentiometer, the voltage drops across a resistor under test and across a 0.02 Ω standard resistor were found to be 0.735 V and 0.98 V respectively. Determine the value of resistor under test. **[Ans.** 0.015 Ω]
3. Determine the insulation resistance of a short length of cable in which voltage falls from 125 to 100 volts in 25 seconds. The capacity of the condenser is 600×10^{-12} F.
   [Ans. 1,86,726 MΩ]
4. The insulation resistance of 2 metre cable was measured by the loss of charge method. The voltage across the standard capacitor of 0.003 μF falls from 222 V to 155 V in one minute. Calculate the insulation resistance of the cable. Derive the formula used. What will be the insulation resistance if the length of the cable is doubled ?
   [Ans. 55,670 MΩ]
5. Calculate the insulation resistance of a short length of cable which was tested for insulation resistance by 'loss of charge' method. An electrostatic voltmeter of infinite resistance is connected between the cable and earth forming a joint capacitance of 0.0006 μF. It is observed that after charging voltage falls from 300 volts to 100 volts in one minute.
   [Ans. 91,024 MΩ]
6. Calculate the insulation resistance of a length of cable in which voltage falls from 100 to 80 volts in 20 seconds, the capacity being 0.0003 μF. **[Ans.** 299,000 MΩ]
7. A cable is tested by loss of charge method using a ballistic galvanometer with the following results : discharged immediately after electrification, deflection 250 divisions; discharged 25 seconds after electrification (*i*) deflection 150 divisions (*ii*) When in parallel with a resistance of 7 MΩ deflection 100 divisions. Calculate the insulation resistance of the cable. **[Ans.** 5.56 MΩ]
8. Fig. 10.6 consists of the following parameters :
   R_1 = 10 kΩ; R_2 = 15 kΩ ; R_3 = 40 kΩ
   Determine the unknown resistance R_x. **[Ans.** 60 kΩ]
9. In a Kelvin double bridge, there is error due to mismatch between the ratios of outer and inner arm resistances. The following data relate to this bridge :
   Standard resistance = 100.03 μΩ; Inner ratio arms = 100.31 Ω and 200 Ω; Outer ratio arms = 100.24 Ω and 200 Ω; The resistance of the connecting leads from standard to unknown resistance is 700 μΩ.
   Calculate the unknown resistance. **[Ans.** 49.97 μΩ]
10. The impedances of the basic A.C. bridge of Fig. 10.21 are given as follows :
    $\bar{Z}_1$ = 100 Ω∠80° (inductive impedance); $\bar{Z}_2$ = 250 Ω (pure resistance);
    $\bar{Z}_3$ = 400 Ω ∠30° (inductive impedance); $\bar{Z}_4$ = Unknown
    Determine the constants of the unknown arm. **[Ans.** 1000 ∠–50°]

11. A Maxwell bridge (Fig. 10.24) is used to measure an inductive impedance. The bridge constants at balance are :

R_1 = 470 kΩ; C_1 = 0.01 μF; R_2 = 5.1 kΩ ; R_3 = 100 kΩ

Find the series equivalent of the unknown impedance. [**Ans.** R_x = 1.09 kΩ; L_x = 5.1 H]

12. The Schering bridge shown in Fig. 10.34 has the following constants :

Arm AB ... Capacitor of 0.5 μF in parallel with 1 kΩ resistance;

Arm BC ... Resistance of 3 kΩ;

Arm CD ... Unknown capacitor C_x and R_x in series

Arm DA ... Capacitor of 0.5 μF

Frequency ... 1 kHz

Determine the following :

(*a*) Unknown resistance and capacitance.

(*b*) Dissipation factor.

[**Ans.** (*a*) 2Ω, 0.25 μF; (*b*) 3.1416]

13. An A.C. bridge is connected as shown in Fig. 10.49. Determine the resistance and inductance of AB. [**Ans.** 400 Ω ; 1.12 H]

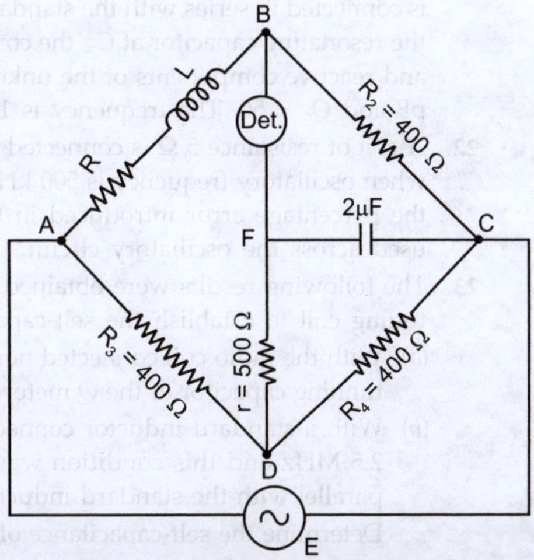

Fig. 10.49.

14. A dielectric sheet of thickness 1 mm is tested at 50 Hz between two electrodes of 10 cm diameter. The Schering bridge employed has a standard compressed air capacitor C_3 of 100 pF; a non-inductive resistor R_4 of 350 Ω in parallel with a variable capacitor C_4 and a non-inductive variable resistor R_2. At balance C_4 = 0.4 μF, R_2 = 250 Ω.

Calculate power factor and permittivity of the sheet. [**Ans.** 0.04398 ; 2]

15. A capacitor is tested by a Schering bridge. It forms one arm *AB* of the bridge. The other arms are : *AD* – a non-reactive resistance of 100 Ω ; *DC* – a non-reactive resistance of 300 Ω shunted by a capacitor of 0.5 μF ; *BC* – a standard loss-free capacitor of 100 pF, the supply frequency is 50 Hz. The bridge is at balance with the above components. Deduce the balance contions of the bridge and hence find out the capacitance and power factor of the capacitor under test. [**Ans.** 298.7×10^{-12} F ; 0.0471]

16. A sample of bakelite was tested by the Schering bridge method at 25 kV, 50 Hz. Balance was obtained with a standard condenser of 109 pF capacitance, a condenser of capacitance 0.5 μF in parallel with a non-reactive resistor of 309 Ω and a non-reactive resistor of 100 Ω. Determine the equivalent series resistance and the power factor of the specimen. [**Ans.** 0.458 MΩ ; 0.0485]

17. A balanced 1 kHz bridge has the following configuration :

Arm *AB* : R_1 = 1000 Ω in parallel with C_1 = 0.053 μF ; *BC* : R_2 = 1500 Ω in series with C_2 = 0.53 μF ; *CD* : the unknown, *DA* : pure capacitance C_4 = 0.265 μF.

Determine *R* and *L* or *C* constants of the unknown. Draw the phasor diagram of the bridge at above frequencey. [**Ans.** *R* = 119.46 Ω ; 0.1656 μF]

18. Determine the equivalent parallel resistance and capacitance that causes a Wien bridge (Fig. 10.46) to null with the following component values :

R_1 = 3.1 kΩ; C_1= 5.2 μF ; R_2 = 25 kΩ; R_4 = 100 kΩ; *f* = 2.5 kHz

[**Ans.** 12.4 kΩ ; 20.3 pF]

19. The following data were obtained during the measurement of the distributed or self-capacitance of a coil :

First measurement : f_1 = 2 MHz and C_1 = 460 pF;
Second measurement : f_2 = 4 MHz and C_2 = 100 pF.
Find the distributed capacitance, C_d. **[Ans.** 20 pF]

20. Calculate the value of distributed capacitance of a coil when the following measurements are made :
At frequency f_1 = 2MHz, the tuning capacitor is set at 450 pF;
At frequency f_2 = 5MHz, the tuning capacitor is tuned at 60 pF. **[Ans.** 14.28 pF]

21. In a *Q* meter, after connecting a standard coil, the resoanance is obtained with a frequency f_1 with resonating capacitor set at C_1; the indicated Q-factor is Q_1. An unknown impedance is connected in series with the standard coil and the resonance is re-established by resetting the resonating capacitor at C_2, the corresponding Q-factor being Q_2. Determine the resistive and reactive components of the unknown impedance when C_1 = 208 pF, Q_1 = 80, C_2 = 184 pF and Q_2 = 50. The frequency is 165 kHz. **[Ans.** 46.88 Ω, 605 Ω]

22. A coil of resistance 5 Ω is connected in the circuit shown in Fig. 2.88 and resonance occurs when oscillatory frequency is 500 kHz and resonating capacitor is set at 120 pF. Determine the percentage error introduced in the calculated value of *Q* if a resistance of 0.02 Ω is used across the oscillatory circuit. **[Ans.** 0.4 %]

23. The following results were obtained during tests conducted, using a *Q* meter, on a radio-tuning coil to establish the self-capacitance :
(*a*) With the radio coil connected normally resonance was obtained at 800 kHz with the tunning capacitor of the *Q* meter set at 95 pF.
(*b*) With a standard inductor connected in place of radio coil, resonance was obtained at 2.5 MHz and this condition was not changed when the radio coil was connected in parallel with the standard inductor.
Determine the self-capacitance of the radio coil. **[Ans.** 10.84 pF]

24. A circuit consisting of an unknown coil, a resistance and a variable capacitor connected in series is tuned to resonance using a *Q* meter. If the frequency is 450 kHz and the resonating capacitor is set at 250 pF, determine the effective inductance and resistance of unknown coil. The *Q* meter indicates 105 and the resistance is 0.75 Ω.
[Ans. 500 μH; 12.76 Ω]

25. A coil of inductance L is tuned by a capacitance C to resontde at 1 MHz. A standard capacitor of 230 pF is tuned to resonate in the same coil at 2 MHz. If the resonance is indicated by a VTVM of capcitance 8 pF in parallel with the circuit, calculate the values of *L* and C. **[Ans.** 26.6 μH; 944 pF]

26. A coil of inductance *L* and resistance *R*, a calibrated variable capacitor C and a thermojunction, all in series, are put across a constant r.m.s. voltage of frequency 3 MHz. The uncalibrated ammeter is found to show the same scale reading I for the two values of C–namely 37 pF and 45 pF, one being below and the other above capacitance C_O needed to produce maximum current. The capacitor is now adjusted to a value C_O, and an added series resistance of of 75 Ω is found to reduce the current to the previous value of *I*. Determine the values of *L* and *R*. **[Ans.** 69.2 μH ; 70.8 Ω]

27. A circuit having an effective capacitance of 150 pF is tuned to a frequency of 1.5 MHz. In this circuit the current falls to 70.70 per cent of its resonant value when the frequency of an e.m.f. of constant magnitude injected in series with the circuit deviates from the resonant frequency by 5 kHz. Calculate the *Q* factor and effective resistance of the circuit.
[Ans. 150; 4.7 Ω]

28. A capacitance of 250 pF produces resonance with a coil at a frequency $\left(\frac{2}{\pi}\right)\times 10^6\ Hz$, while at the second harmonic of this frequency resonance is produced by a capacitance of 50 pF. Calculate : (*i*) The self-capacitance, and (*ii*) The inductance of the coil.
Neglect the effect of voltmeter capacitance and also other stray capacitances.
[Ans. 16.67 pF; 234 μH]

CHAPTER

11 Magnetic Measurements

11.1. Introduction to magnetic materials – Terms connected with magnetic materials – Magnetic parameters – Classification of magnetic materials – Principal ferromagnetic elements – Ferromagnetism – Magnetic domains. **11.2. Classification of magnetic measurements; 11.3. Ballistic (D.C.) tests** – Measurement of flux/flux density – Measurement of value of magnetising force (H) – Determination of B-H curve – Determination of hysteresis loop – Determination of leakage factor in dynamo-electric machinery – Magnetic potentiometer – Testing bar specimens of magnetic materials – Bar specimens versus ring specimens. **11.4. Permeameters** – Hopkinson permeameter (Bar and Yoke method) – Ewing double bar permeameter – Illiovici permeameter – Burrows permeameter – Fahy's simplex permeameter – The National Physical Laboratory (England) form of permeameter. **11.5. A.C. magnetic testing; 11.6. Methods of measurement of Iron losses** – Wattmeter method – A.C. bridge methods – A.C. potentiometer method – Oscillographic method. **11.7. Methods of measurement of air gap flux and field strength** – Measurement of air gap flux – Measurement of field strength by withdrawal of search coil – By nuclear magnetic resonance – Use of bismuth spiral – Use of mumetal wires – Use of Hall effect devices. **11.8. Testing of permanent magnets; 11.9. Testing of Feebly magnetic materials; 11.10. Precautions in Magnetic Testing** – *Highlights – Objective Type Questions – Theoretical Questions – Unsolved Examples.*

11.1 INTRODUCTION TO MAGNETIC MATERIALS

- ***Magnetic materials** are those materials in which a state of magnetisation can be induced.* Such materials when magnetised create a magnetic field in the surrounding space.
- In a material all the molecules contain electrons which orbit around the nucleus; these orbits are therefore, equivalent to circulating currents and so develop an m.m.f. Each m.m.f. due to an individual orbit, in most molecules, is neutralised by an opposite one. But in magnetic materials such as iron and steel *there are a number of unneutralised orbits,* such that a resultant axis of m.m.f. exists which *produces a magnetic dipole.*
- *In unmagnetised specimens,* because of mutual attraction and repulsion among the dipoles, the *molecular m.m.f. axes lie along continuous closed paths, and no external magnetic effect can be detected.*

In *magnetised specimens,* the dipoles line up parallel with the exciting m.m.f. and when the exciting m.m.f. is removed, a *number of dipoles may remain aligned in the direction of the external field* and thus exhibit *permanent magnetism.*

- The *readiness of a material to accept magnetism is expressed by its permeability.* For all materials except a few magnetic ones, the permeability is that of free space and is constant. It is denoted by μ_0 ($= 4\pi \times 10^{-7}$).

For magnetic materials the permeability equals μ_0 times the relative permeability which is denoted by μ_r(*i.e.*, $\mu = \mu_0\mu_r$).

The *relative permeability varies with the degree of magnetisation of the material and may have a value* as high as 2500.

11.1.1. Terms Connected with Magnetic Materials

1. **Magnetic force.** It is the force exerted by one magnet on another either to attract it or repel it.
2. **Unit pole strength.** It is defined as the strength of that pole which when placed in vacuum at a distance of one metre from a similar and equal pole, repels it with a force of one newton.
3. **Magnetic flux density.** It is usually represented by letter B and is defined as the flux or lines of force passing per unit area through any substance through a plane at right angles to the direction of magnetic flux; it is measured in Wb/m^2 (or T, *i.e.,* Tesla).
4. **Magnetic field strength.** It may be defined in the following *two ways* :
   (*i*) Field strength at any point within a magnetic field is the number of lines of force passing through a unit area round the point considered and held perpendicular to the lines.
   (*ii*) Field strength at any point within a magnetic field is the force exerted by a unit north pole at that point.
5. **Relative permeability.** It is the ratio of the flux density (B) produced in that material to the flux density produced in vacuum by the same magnetising force (H). It is denoted by μ_r.
6. **Absolute permeability.** Absolute permeability of the material is the ratio of flux density in that material to the magnetising force producing that flux density and is denoted by μ : $\mu = \mu_0\mu_r$ where μ_0 is the permeability of free space having a value of $4\pi \times 10^{-7}$ H/m.
7. **Magnetic potential.** The magnetic potential at any point within a magnetic field is measured by the work done in carrying a unit north pole from infinity to that point against the force of magnetic field.
8. **Intensity of magnetisation.** It is defined as the pole strength per unit area of the bar or magnetic moment per unit volume of the bar. It is denoted by letter I.
9. **Susceptibility.** It is denoted by letter χ and is defined as the ratio of intensity of magnetisation (I) to magnetising force (H). In order words $\chi = \dfrac{I}{H}$.
10. **Magnetomotive force.** It is that force which drives or tends to drive the flux through a magnetic circuit. In short it is written as m.m.f. It is the product of number of conductors (N) and current (I) passing through them, *i.e.,* m.m.f. = NI.
11. **Magnetic reluctance.** It is that property of the material which opposes the production of magnetic flux in it.
12. **Coercive force.** It may be defined as the demagnetising force which is necessary to neutralise completely the magnetism in an electromagnetic after the value of magnetising force becomes zero.
13. **Remanence.** It is defined as the magnetic flux density which still persists in magnetic material even when the magnetising force is completely removed. It is expressed in Wb/m^2.

14. **Retentivity.** It is that property of magnetic material which is measured by its maximum value of the residual induction.

15. **Core loss, incremental.** The core loss in a magnetic material when subjected simultaneously to a D.C. biasing magnetising force and an alternating magnetising force.

16. **Core loss, specific.** The power expended per unit mass of magnetic material in which there is symmetrical, sinusoidally varying induction of a specified maximum value, *B*, at a specified frequency, *f*.

17. **Core loss, total.** The power expended in a magnetic specimen in which there is cyclically alternating induction.

Note. *Measurements of core loss are normally made with sinusoidally alternating induction, or the results are corrected for deviations from the sinusoidal condition.*

18. **Curie temperature.** The temperature above which a ferromagnetic material becomes paramagnetic.

19. **Cyclically magnetised condition.** A magnetic material is in a cyclically magnetised condition when, after having been subjected to a sufficient number of identical cycles of demagnetising force. It follows identical hysteresis loops on successive cycles which are not necessarily symmetrical with respect to the origin of the axes.

20. **Hysteresis loop, normal.** A closed curve obtained with a ferromagnetic material by plotting (usually to rectangular co-ordinates) corresponding D.C. value of magnetic induction *B* for ordinates and magnetising force *H* for abscissa when the material is passing through a complete cycle between equal definite limits of either magnetising force $\pm H_m$ or magnetic induction $\pm B_m$. In general the normal hysteresis loop has inversion symmetry with respect to the origin of the *B* and *H* axes, but this may not be true for special materials.

21. **Hysteresis loss, incremental.** The hysteresis loss occurring in a ferromagnetic material when subjected simultaneously to constant biasing magnetic force and cyclic excitation.

22. **Hysteresis loss, normal.** (*i*) The power expended in a ferromagnetic material as a result of hysteresis, when the material is subjected to an SCM (symmetrically cyclically magnetised) excitation. (*ii*) The energy loss per cycle in a magnetic material as a result of magnetic hysteresis when the induction is cyclic (but not necessarily periodic).

23. **Magnetic hysteresis.** The property of a ferromagnetic magnetic material revealed by the lack of correspondence between the changes in induction resulting from increasing magnetising force and from decreasing magnetising force.

24. **Induction, intrinsic.** The vector difference between the magnetic induction in a magnetic material and the magnetic induction that would exist in a vacuum under the influence of the same magnetising force.

25. **Induction, normal.** The maximum induction in a magnetic material that is in symmetrically cyclically magnetised condition.

Note. *Normal induction is a magnetostatic parameter usually measured by ballistic methods.*

26. **Induction curve, normal.** A curve of a previously demagnetised specimen depicting the relation between normal induction and corresponding ascending values of magnetising force. This curve starts at the origin of *B* and *H* axes.

27. **Magnetic field of induction.** A state of a region such that a conductor carrying a current in the region would be subjected to mechanical force and an e.m.f. would be induced in an elementary loop rotated with respect to the field in such a manner as to change the flux linkage.

28. **Ageing coefficient.** The percentage change in specific magnetic property resulting from a specified ageing treatment.

29. **Magnetic ageing.** The change in magnetic properties of a material resulting from metallurgical change. This term applies whether the change results from a continued normal or specified accelerated ageing condition.

11.1.2. Magnetic Parameters

The various *magnetic parameters* are described below :

1. Magnetic moment :

It has been observed that a coil carrying a current *I*, when located in a region of uniform magnetic flux density, will experience, a *torque* whose magnitude depends on the *area of the coil, the current* and the *component of flux density in the plane of the coil.* The magnetic dipole moment is defined by the following relation :

$$p_m = I \times A \quad \text{...(11.1)}$$

where, p_m = magnetic dipole moment–Ampere metre² (A-m²) and is a vector quantity,

I = Current, and

A = Area of the coil.

A *permanent bar magnet* also experiences a torque when placed in a uniform field, which tends to align it with the field. If the magnet is supposed to have free poles of opposite kind, we can define the dipoles moment, *as the product of pole strength and distance between them.*

2. Magnetisation :

It has been found that all materials are affected by the presence of a magnetic field, in that they are found experimentally to *acquire magnetic moments.* The *magnitude of this moment per unit volume, is called the magnetisation of the medium* and is described by the vector *M*.

The magnetisation *M* of a material may be expressed in terms of its elementary magnetic dipole moments p_m by

$$M = Np_m \quad \text{...[11.1(a)]}$$

where, *N* is the number of magnetic dipoles per unit volume.

To prove for a magnetic material : $B = \mu_0(M + H)$

Consider a solenoid.

Let, l = Length of the solenoid,

A = Area of cross-section of the solenoid,

N = Number of turns, and

I = Current through the solenoid.

If the solenoid is placed in vacuum the flux density will be :

$$B_0 = \mu_0 H \quad \text{...(11.2)}$$

$$\left(\text{where } H = \frac{NI}{l},\ H \text{ being magnetising force}\right)$$

If the vacuum is replaced by a homogeneous magnetic medium the flux density will increase to

$$B = \mu_0 \mu_r H \quad ...(11.3)$$

Thus, there will be an increase in flux density of

$$B - B_0 = \mu_0(\mu_r - 1) H = \mu_0 H' \quad ...(11.4)$$

Now this increase in flux density can be thought of due to an increase in H by H' rather than due to the change in the medium. If I' is the corresponding increase in current then we have

$$H' = \frac{NI'}{l} = \frac{NI' A}{lA} \quad ...(11.5)$$

The right hand side of eqn. (11.5) represents the *total magnetic dipole moment per unit volume and is the magnetisation of the medium* denoted by the letter M. Hence

$$H' = (\mu_r - 1) H = M$$

or,

$$\frac{M}{H} = \chi = \mu_r - 1 \quad ...(11.6)$$

where χ is a *dimensionless quantity* defined as the *magnetic susceptibility* of the medium. Under this condition

$$B = B_0 + \mu_0 H = \mu_0(1 + \chi)H \quad ...(11.7)$$

$$= \mu_0\left(1 + \frac{M}{H}\right)H$$

i.e.,

$$B = \mu_0(M + H) \quad ...(11.8)$$

3. Magnetic susceptibility :

The eqn. (11.7) can be rewritten as

$$B = \mu_0(1 + \chi)H = \mu_0 \mu_r H \quad ...(11.9)$$

where $\mu_r = (1 + \chi)$ is defined as the *relative permeability* of the medium.

- The *magnetic susceptibility* depends on the *nature of the magnetic material and on its state, i.e.*, temperature etc. The susceptibility of a sample may change on *cold working*. If copper is cold worked, the susceptibility may change from a –ve to a +ve value and on annealing after cold work, it may again become –ve.
- The susceptibility may be determined by *measuring the force exerted on a magnetic material* when it is placed in a magnetic field. The susceptibility of a ferromagnetic substance is very strongly dependent on the field strength. Therefore, the magnetic properties of ferromagnetic materials are best described by giving the magnetisation or the flux density as a function of the field strength.
- The *M–H* and *B–H* curves of the ferromagnetic materials are very similar in nature. The slope of the former when divided by the magnetic space constant, μ_0 gives the *magnetic susceptibility* and the slope of the latter gives the *magnetic permeability*, $\mu(= \mu_0 \mu_r)$.
- For ferromagnetic materials the *susceptibility* (or, the permeability) is *not constant.* It has a low value at weak fields. As the field intensity is increased, the susceptibility increases, reaches a maximum value and then begins to drop, *ultimately reaching a constant value on saturation.*

11.1.3. Classification of Magnetic Materials

All materials possess magnetic properties to a greater or lesser degree and these are determined by the facts that (*i*) a magnetic field exerts forces and torques on the bodies, (*ii*) a body placed in a magnetic field distorts the field. The magnetic properties of the materials are characterised by their relative permeabilities. In accordance with the value of relative permeability the materials may be classified in the following *three* ways :

1. Ferromagnetic materials :

The relative permeabilities of these materials are *much greater than unity* and are dependent on the field strengths. They attract the lines of force strongly [Fig. 11.1(*c*)]. The principal ferromagnetic elements are *iron, cobalt, nickel.* Gadolinium however, also comes under this classification. These have high susceptibility.

2. Paramagnetic materials :

These have relative permeability *slightly greater than unity* and are *magnetised slightly.* They attract the lines of forces weakly [Fig. 4.1(*b*)]. Aluminium, platinum and oxygen belong to this category.

3. Diamagnetic materials :

The relative permeability of these materials is *slightly less than unity.* They *repel the lines of force slightly* [Fig. 4.1(*a*)]. The examples are bismuth, silver, copper and hydrogen.

Fig. 4.2. shows the arrangement of atomic magnetic moment.

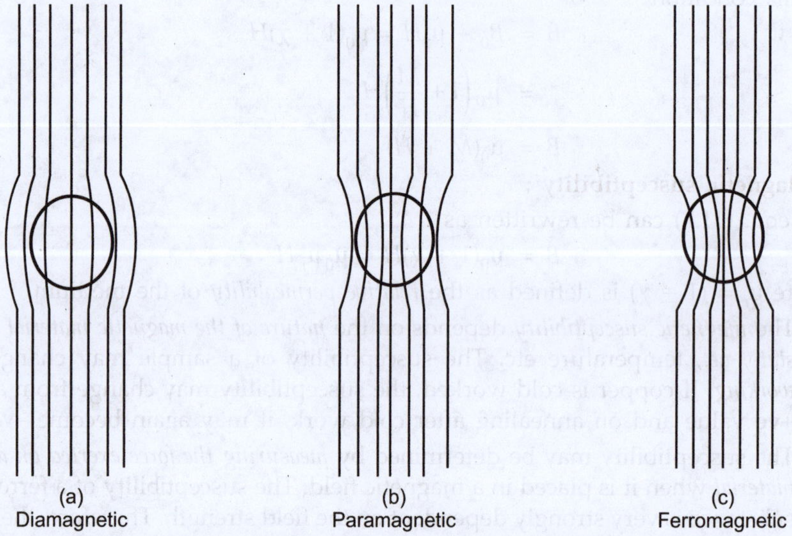

Fig. 11.1

11.1.4. Principal Ferromagnetic Elements

The principal ferromagnetic elements are discussed below :

1. Iron :

- It has the highest susceptibility comparatively. This property added with low cost makes iron most suitable for commercial purposes.
- Its magnetic properties are considerably affected by the presence of traces of carbon, oxygen and nitrogen, therefore, these undesirable elements should be reduced to the possible attainable limits. High purity means high permeability and reduced hysteresis loss. Its *permeability is to the tune of 2000.*

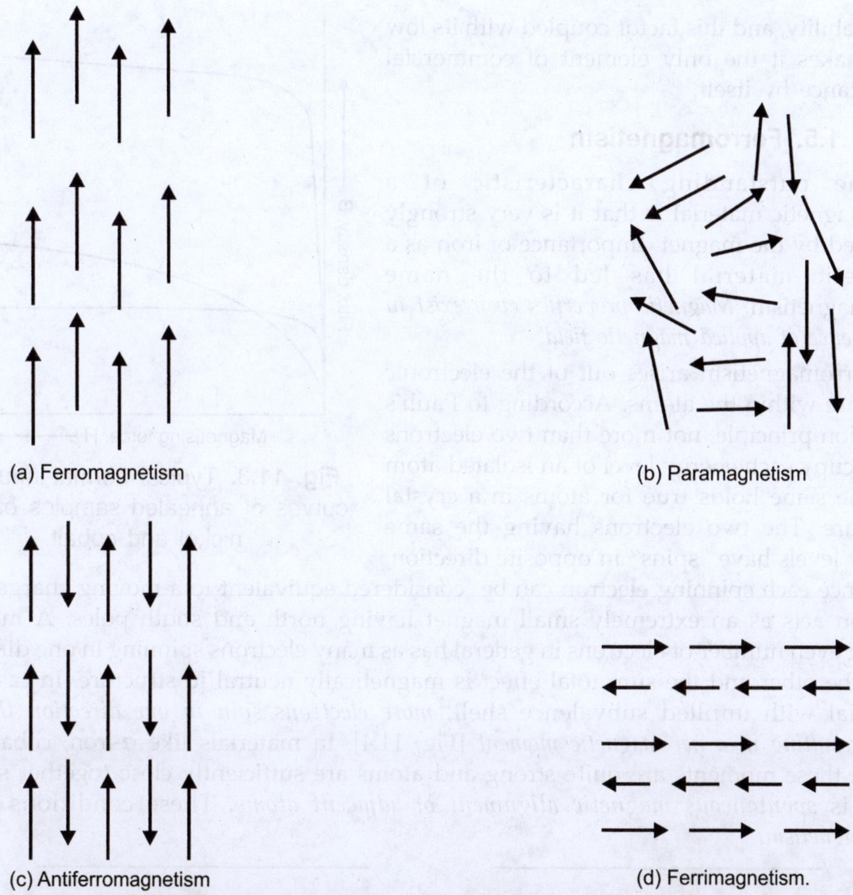

Fig. 11.2. Arrangement of atomic magnetic moment.

2. Nickel :

- It is also one of the important ferromagnetic elements. When it is heated, it remains ferromagnetic upto 395°C and beyond that it gets converted to paramagnetic.
- Its properties as ferromagnetic material are *considerably improved when alloyed with iron and cobalt. Its permeability is 300.*

3. Cobalt :

- Its permeability is 250 and increases with the temperature upto 300°C. It loses its magnetic properties at about 1130°C.
- When in pure state it does not prove to be an important ferromagnetic material.
- Its cost is adequately high.

The only elements which are strongly ferromagnetic at normal temperature are iron, cobalt and nickel. In Fig. 11.3 are shown typical *normal-induction curves* of annealed samples of iron, nickel and cobalt of comparatively high purity. These curves are given only for the purpose of general comparison and should not be considered at representing critical values. Small variations in the degree of purity or in the annealing procedure lead to substantial differences in normal induction. It may be noted that iron has greatest permeability, and this factor coupled with its low cost makes it the only element of commercial importance by itself.

11.1.5. Ferromagnetism

The outstanding characteristic of a ferromagnetic material is that it is very strongly attracted by the magnet. Importance of iron as a magnetic material has led to the name ferromagnetism. *Magnetic properties even exist in the absence of applied magnetic field.*

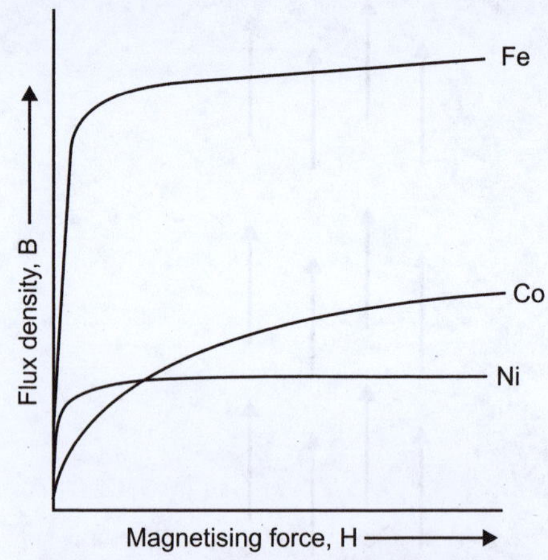

Fig. 11.3. Typical normal-induction curves of annealed samples of iron, nickel and cobalt.

Ferromagnetism arises out of the electronic structure within the atoms. According to Pauli's exclusion principle, not more than two electrons can occupy each energy level of an isolated atom and the same holds true for atoms in a crystal structure. The two electrons having the same energy levels have "spins" in opposite direction, and since each spinning electron can be considered equivalent to a moving charge, each electron acts as an extremely small magnet having north and south poles. A material having even number of electrons in general has as many electrons spinning in one direction as in the other and the sum total effect is magnetically neutral in structure. In case of a material with unfilled subvalence shell, *more electrons spin in one direction than in other resulting in a net magnetic moment* [Fig. 11.4]. In materials like α-iron, cobalt and nickel, these moments are quite strong and atoms are sufficiently close together so that there is *spontaneous magnetic alignment of adjacent atoms.* These conditions cause *ferromagnetism.*

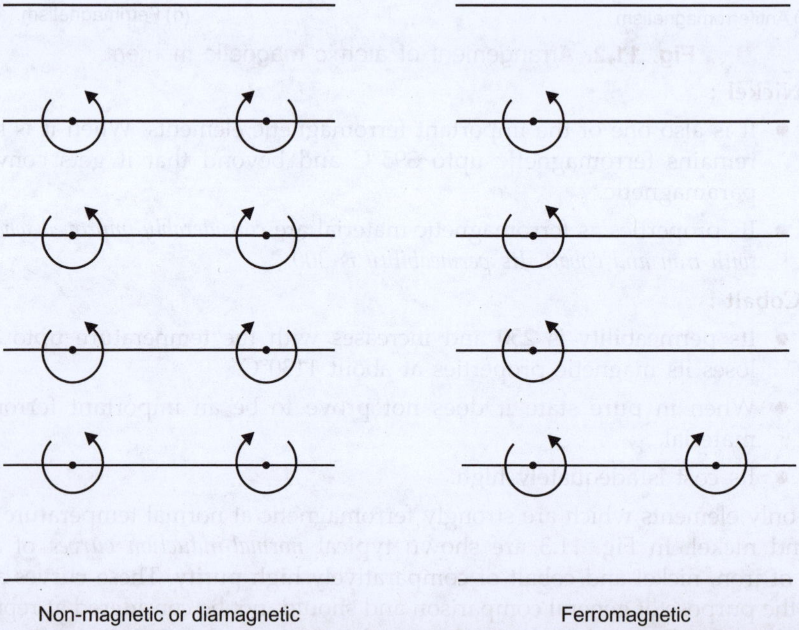

Fig. 11.4

11.1.6. Magnetic Domains

A specimen of ferromagnetic material, according to Weiss, consists of a large number or *regions or domains* which are *permanently magnetised.* The atomic moments in the individual domains are all aligned parallel to one another *at temperatures far below the Curie point. Each domain is magnetically saturated and has a net magnetic moment.* However, the direction of the permanent magnetisation varies from domain to domain, and consequently the resultant magnetisation may be zero, or nearly zero. *Above the Curie temperature, the domains may disrupt and the material may lose its ferromagnetic properties.*

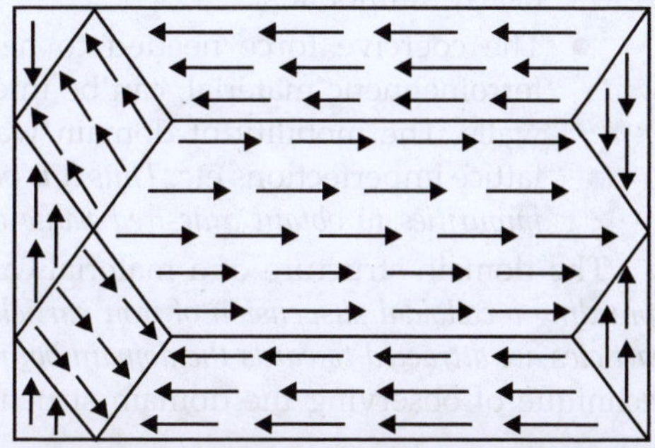

Fig. 11.5. Domains in an unmagnetised ferromagnetic specimen.

- Fig. 11.5. shows domains in an unmagnetised ferromagnetic substance.
- The domains exist in single crystal as well as in polycrystalline samples. The domains are separated by domain walls in which the spin direction gradually changes from the preferred direction of one domain to the preferred direction of the neighbouring domain. A schematic arrangement of the domains with zero resultant magnetic moment is shown in Fig. 11.6.

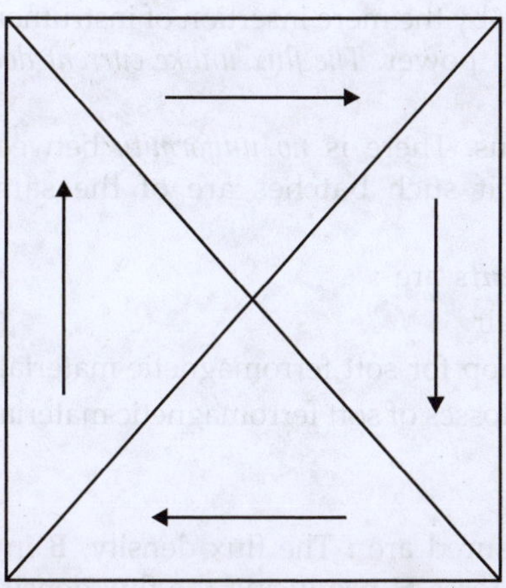

Fig. 11.6. Domain configuration according to Weiss hypothesis giving zero resultant magnetisation (unmagnetised crystal).

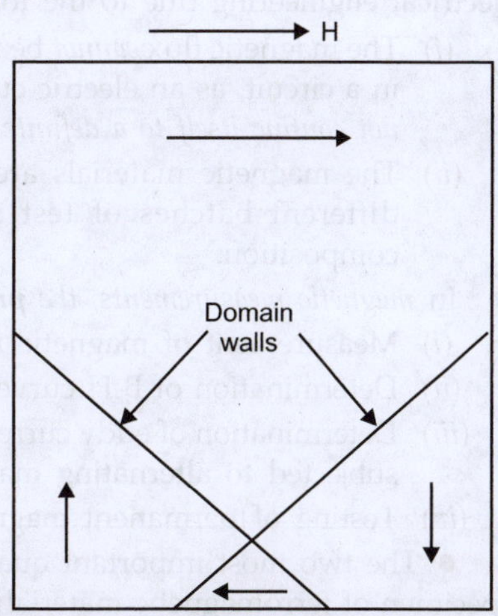

Fig. 11.7. Magnetisation by domain growth.

When an external field H is applied, say in horizontal direction, the domains having parallel orientation of spontaneous magnetisation grow in size *at the expense of other domains* that are magnetised in other directions by virtue of a motion of domain walls. This condition is shown in Fig. 11.7. The motion of domain walls can be considered as stealing of the neighbouring dipoles from other domains and aligning them in the direction of the external field so that the preferred domain increases in size. As the external field is increased progressively, a stage is reached when the *whole material becomes one single domain giving rise to saturation of the magnetisation.* Thus the *hysteresis results due to motion of domain walls and also due to domain rotation to some extent.*

- The *spontaneous magnetisation takes place in a material in certain preferred directions.* When an external field is applied in an arbitrary direction, some domains which are easy to rotate, orient themselves in the direction of the field and become nearly immobile.
- The coercive force needed to neutralise the spontaneous magnetisation in a ferromagnetic material, can be interpreted in terms of the *mobility of* the domain walls. The mobility of domain walls depends on the impurities in the material lattice imperfections etc. *Thus it is possible to design a material by adding or subtracting impurities to obtain a desired value of coercive force.*

The domain structure of a material can be *observed* under a powerful microscope *by spreading a colloidal suspension of iron particles on nicely prepared surface of the material. These particles are attracted towards the domain boundaries where the magnetisation is quite large.* This technique of observing the domain structure is known as the *magnetic powder technique.*

11.2 CLASSIFICATION OF MAGNETIC MEASUREMENTS

Measurements of various magnetic phenomena are called **magnetic measurements.**

The magnetic measurements and a thorough knowledge of characteristics of magnetic materials are of significant importance in designing and manufacturing electrical equipment.

The magnetic measurements are *more inaccurate* than other types of measurements in electrical engineering due to the following ***reasons :***

(*i*) The magnetic flux *cannot* be measured directly, by the mere insertion of instrument in a circuit, as an electric current, voltage and power. *The flux unlike current does not confine itself to a definite path.*

(*ii*) The magnetic materials are *not* homogeneous. There is *no uniformity* between different batches of test specimens even if such batches are of the same composition.

In *magnetic measurements, the* ***principal requirements*** are :

(*i*) Measurement of magnetic field strength in air.

(*ii*) Determination of B-H curve and hysteresis loop for soft ferromagnetic materials.

(*iii*) Determination of eddy current and hysteresis losses of soft ferromagnetic materials subjected to alternating magnetic fields.

(*iv*) Testing of permanent magnets.

- The two most important quantities to be measured are : The flux density, **B** in a specimen of ferromagnetic material and magnetising force **H** producing the flux density.

— The flux density B is measured by a *ballistic galvanometer or a fluxmeter* which is a special type of ballistic galvanometer. Ballistic galvanometer or fluxmeter does not measure B in a magnetic material directly, instead measures the changes in the flux.

— Magnetising force H is measured by *permeameter.*

Types of tests :

The study of magnetic measurements has been divided into three categories :

1. ***D.C. tests.*** These tests are often termed as *"ballistic tests".*

    — These are used for *determination of B.H. curves and hysteresis loops* of ferromagnetic materials.

    — These tests provide an adjustable m.m.f. on the magnetic circuit and a ballistic galvanometer or flux meter for measurement of flux density.

2. ***A.C. tests.*** These tests may be carried at power, audio or radio frequencies.
   — A.C. tests are used to determine iron losses or core losses in strip (sheet) material when it is subjected to alternating field.
   — These losses can be separated into hysteresis losses and eddy current losses by a separate method involving variation of frequency and form factor.
   — The strip material is assembled as a close magnetic circuit in the form of square. There are two types of such squares : (*i*) Epstein square; and (*ii*) Lloyd Fisher square.
3. ***Steady state tests.***
   — These are used to measure the *steady state value of flux in the air gap of magnetic circuit.*

- The above tests are normally carried out on *ring* specimens of the ferromagnetic materials although *bar* specimens are much easier to construct. However, some difficulties and accuracies arise in performing the test on bar specimens; *"permeaters" overcome these difficulties and inaccuracies in testing bar specimens of magnetic materials.*

11.3 BALLISTIC (OR D.C.) TESTS

These tests are used for :

1. Determination of flux/flux density in a specimen.
2. Determination of B.H. curves.
3. Plotting of hysteresis loop.

11.3.1. Measurement of Flux/Flux Density

Fig. 11.8 shows the connection diagram for the measurement of flux/flux density in a *ring specimen.*

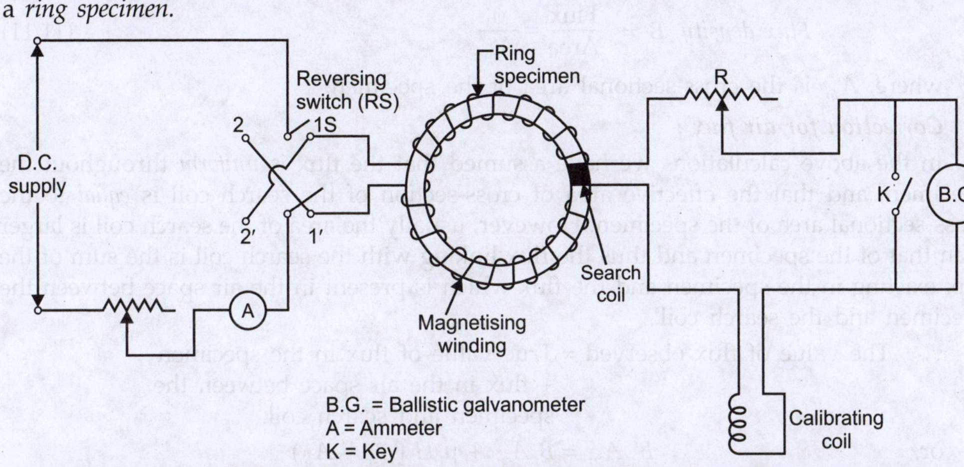

Fig. 11.8. Measurement of flux/flux density in ring specimens.

— The ring specimen is wound with a magnetising winding which carries a current *I*.
— A search coil (known as **"B coil"**) of convenient number of turns is wound on the specimen and connected through a resistance and calibrating coil, to a ballistic galvanometer as shown.

— The current through the magnetising coil is reversed and therefore the flux linkages of the search coil change inducing an e.m.f. e in it. This e.m.f. sends a current through the ballistic galvanometer causing it to deflect.

Let, ϕ = Flux linking with search coil,
R = Resistance of the ballistic galvanometer circuit,
N = Number of turns in the search coil, and
t = Time taken to reverse the flux.

∴ Average e.m.f. induced in the search coil,

$$e = N \cdot \frac{d\phi}{dt} = \frac{2N\phi}{t}$$

Average current through the ballistic galvanometer is,

$$i = \frac{2N\phi}{Rt}$$

Charge passing is : $Q = it = \frac{2N\phi}{Rt} \times t = \frac{2N\phi}{R}$

Let θ_1 be the throw of the galvanometer and G_q be the constant of galvanometer expressed in coulomb per unit deflection.

Charge indicated by the galvanometer = $G_q\,\theta_1$.

$$\therefore \quad \frac{2N\phi}{R} = G_q \cdot \theta_1$$

or, $$\text{Flux, } \boldsymbol{\phi} = \frac{R \cdot G_q \cdot \theta_1}{2N} \qquad ...(11.10)$$

In a uniform field with search coil turns at right angles to the flux density vector, we have,

$$\textit{Flux density, } \mathbf{B} = \frac{\text{Flux}}{\text{Area}} = \frac{\phi}{A_{sp.}} \qquad ...(11.11)$$

where, $A_{sp.}$ is the cross-sectional area of the specimen.

Correction for air flux :

In the above calculations we have assumed that the flux is *uniform* throughout the specimen, and that the effective area of cross-section of the search coil is *equal* to the cross-sectional area of the specimen. However, usually the area of the search coil is larger than that of the specimen and thus the flux linking with the search coil is the sum of the flux existing in the specimen and the flux which is present in the air space between the specimen and the search coil.

∴ The value of flux observed = True value of flux in the specimen + flux in the air space between the specimen and search coil.

or, $$B' A_{sp.} = B\, A_{sp.} + \mu_0 H\,(A_{sc} - A_{sp})$$

∴ *True value of flux density,* $$B = B' - \mu_0 H\left(\frac{A_{sc}}{A_{sp}} - 1\right) \qquad ...(11.12)$$

where, B' = Observed or apparent value of flux density, Wb/m^2,
B = True value of flux density in specimen, Wb/m^2,
A_{sp} = Cross-sectional area of speciment, m^2, and
A_{sc} = Cross-sectional area of search coil, m^2.

11.3.2. Measurement of Value of Magnetising Force (H)

The value of H *inside a specimen* can be calculated from the formula, given by :

$$H = \frac{NI}{l} \text{ AT/m} \qquad ...(11.13)$$

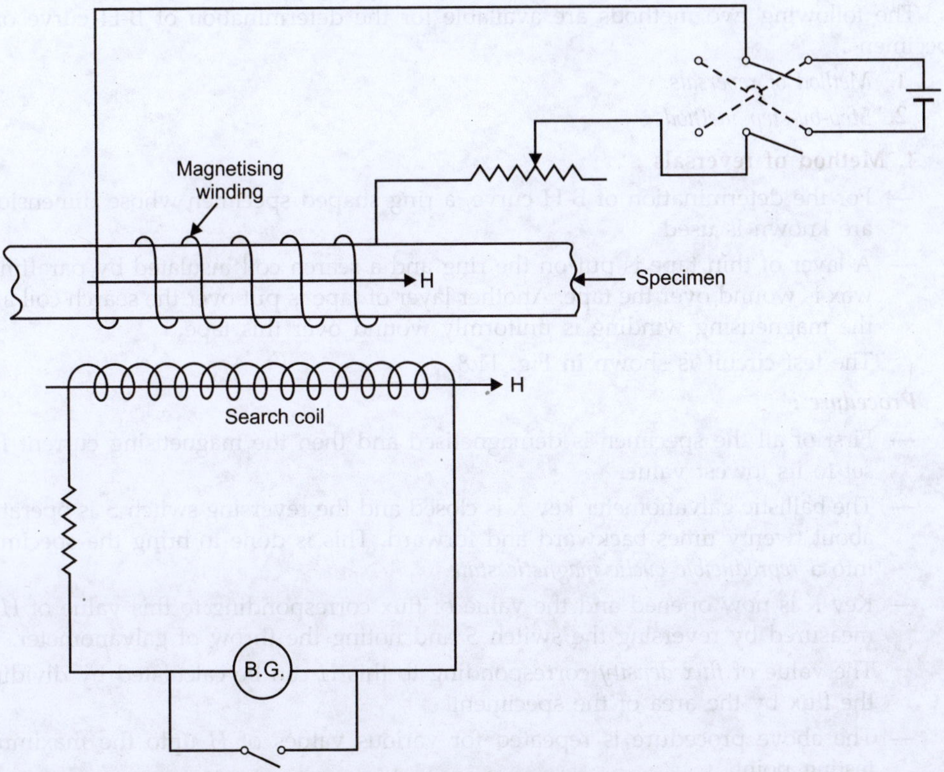

Fig. 11.9. Determination of magnetising force H.

where, N = Number of turns in the specimen,

I = Current flowing through the specimen, and

l = Mean circumference of the ring, *m*.

H can also be determined from measurements made *outside the specimen* from the equation,

$$H = \frac{B_0}{\mu_0} \qquad ...(11.14)$$

where, $\mu_0 = 4\pi \times 10^{-7}$ henry/metre and the value of B_0 can be determined by the method already described.

That is *H cannot be measured directly.*

Thus the magnetising force of a constant magnetic field can be measured *indirectly by a ballistic galvanometer and a search coil* as shown in Fig. 11.9.

— If magnetising force is to be determined in the air gap, the search coil (called **H-coil**) is placed in the air gap itself ; while magnetising force H within the speciment can be determined by measuring the H on its surface, since the tangential components of the field are of equal magnitude for both sides on the interface.

— The search coil as positioned in Fig. 11.9 measures the value of flux density B_0 in air in a similar manner described earlier. After finding out the value of B_0, H can be readily determined from eqn. 11.14.

11.3.3. Determination of B-H Curve

The following two methods are available for the determination of B-H curve of a specimen :

1. *Method of reversals.*
2. *Step-by-step method.*

1. Method of reversals :

— For the determination of B-H curve, a ring shaped specimen whose dimensions are known is used.

— A layer of thin tape is put on the ring and a search coil insulated by paraffined wax is wound over the tape. Another layer of tape is put over the search coil and the magnetising winding is uniformly wound over this tape.

The test circuit is shown in Fig. 11.8.

Procedure :

— First of all the specimen is demagnetised and then the magnetising current I is set to its lowest value.

— The ballistic galvanometer key K is closed and the reversing switch S is operated about twenty times backward and forward. This is done to bring the specimen into a *reproducible cyclic magnetic state.*

— Key K is now opened and the value of flux corresponding to this value of H is measured by reversing the switch S and noting the throw of galvanometer.

The value of *flux density* corresponding to this H can be calculated by dividing the flux by the area of the speciment.

— The above procedure is repeated for various values of H upto the maximum testing point.

The B-H curve may be plotted from the measured values of B corresponding to the various values of H.

2. Step-by-step method :

The special feature of this method of determination of B-H curve is that there is *no reversal of magnetising current.*

The connection diagram of this method remains the same as for the previous method (see Fig. 11.8) except that the direct current is now supplied to the magnetising coil through a *potential divider having a number of tapping* as shown in Fig. 11.9. The tappings are so arranged that the value of magnetising current I and so magnetising force H is increased upto the desired maximum value, in a number of suitable steps.

Procedure :

— The tapping switch S_2 is set on tapping 1 and switch S_1 is closed. The throw of galvanometer corresponding to this increase in flux density in the specimen, from zero to some value B_1 is observed. The value of B_1 can be calculated from the throw of the galvanometer. The value of the corresponding magnetising force H_1 may be calculated from the value of current flowing in the magnetising winding at tapping 1.

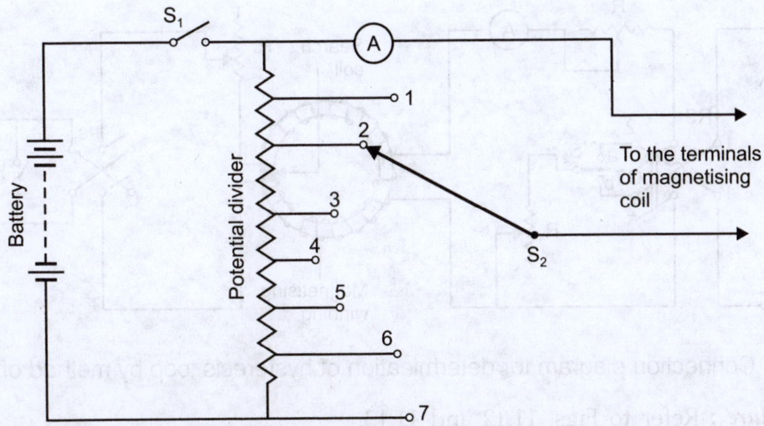

Fig. 11.10

— The magnetising force is then increased suddenly to H_2 by suddenly changing the position of switch S_2 from tapping 1 to tapping 2 and the corresponding increase in flux density is determined from the galvanometer throw observed. The flux density B_2 corresponding to magnetising force H_2 will be equal to B_1 *plus* increase in flux density (δB) determined from the galvanometer throw.

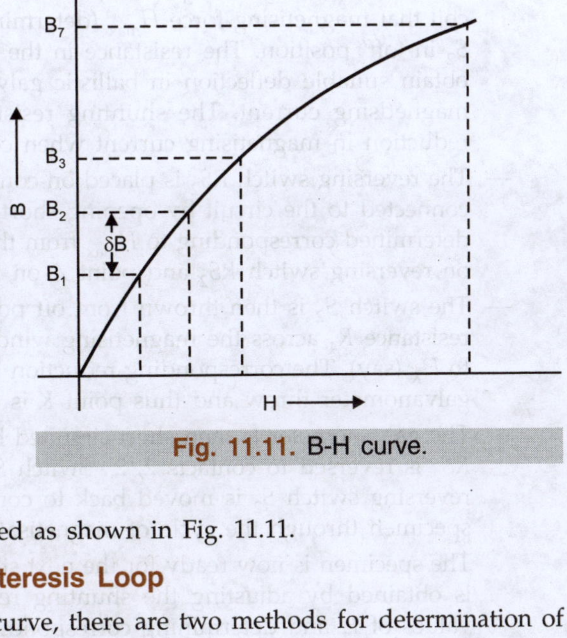

Fig. 11.11. B-H curve.

— The process is repeated for other values of H to the maximum point and complete B-H curve is obtained as shown in Fig. 11.11.

11.3.4. Determination of Hysteresis Loop

Similar to determination of B-H curve, there are two methods for determination of hysteresis loop of a magnetic material specimen :

1. *Method of reversals.*
2. *Step-by-step method.*

1. Method of reversals. Fig. 11.12 shows the connection diagram for determination of hysteresis loop by method of reversals.

R_1, R_2 and R_3 are the *variable resistances* for adjusting the resistances in the ballistic galvanometer and magnetising coil circuits, R_4 is a variable *shunting resistance,* which can be connected across the magnetising coil by means of switch S_2 thus reducing the magnetising current from its maximum value down to any desired value depending upon the value of R_4.

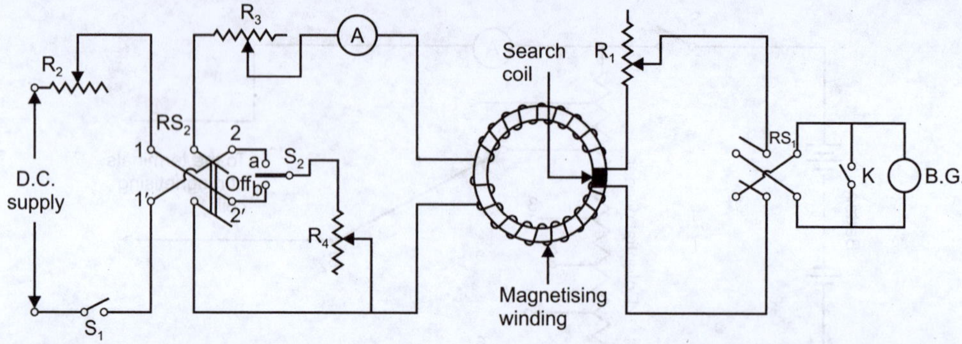

Fig. 11.12. Connection diagram for determination of hysteresis loop by method of reversals.

Procedure : Refer to Figs. 11.12 and 11.13.

— The value of magnetising force H_{max} required to produce flux density B_{max} to be used during the test is obtained from the B-H curve of the specimen.

— The resistances R_2 and R_3 are adjusted to give such a current in the magnetising coil that magnetising force H_{max} (determined from B-H curve) is produced with S_2 in 'off' position. The resistance in the galvanometer circuit R_1 is adjusted to obtain suitable deflection in ballistic galvanometer on reversing the maximum magnetising current. The shunting resistance R_4 is adjusted to give required reduction in magnetising current when connected across magnetising winding.

— The reversing switch RS_2 is placed on contacts 1, 1′ and ballistic galvanometer is connected to the circuit by opening short-circuiting key *K*. The value of B_{max} is determined corresponding to H_{max} from the deflection of galvanometer observed on reversing switch RS_2 and point *A* on the hysteresis loop is obtained.

— The switch S_2 is then thrown from off position to contact *b* in order to connect resistance R_4 across the magnetising winding and reduce the magnetising force to H_K (say). The corresponding reduction in flux density δB is obtained from the galvanometer throw and thus point *K* is obtained on the loop.

— The galvanometer is then short-circuited by closing key *K* and reversing switch RS_2 is reversed to contacts 2, 2′. Switch S_2 is moved to the "off" position and reversing switch S_2 is moved back to contacts 1, 1′. This procedure passes the specimen through the cycle of magnetisation and back to the point A.

The specimen is now ready for the next step in the test. The part AC of the loop is obtained by adjusting the shunting resistance R_4 to give different reduced values of *H* and determining corresponding reduction in *B*.

— In order to obtain part *CDE* of the loop, the galvanometer is short-circuited, switch S_2 is put in the "off" position, and reversing switch RS_2 moved back again to contacts 1, 1′. Now place switch S_2 on contact *a*, open the key *K* and rapidly reverse RS_2 to contacts 2, 2′. This causes the magnetising force to change from $+H_{max}$ to $-H_L$ (say). From the throw of the galvanometer change in flux density δB′ can be obtained. Thus point *L* on the hysteresis loop can be located. The magnetisation of the specimen is brought back to point *A* by reversing switch RS_2 on to contacts, 1, 1′ with key *K* closed. By continuing this procedure, other parts on part *CDE* of the hysteresis loop are obtained. Thus part *ACDE* of the loop can be traced.

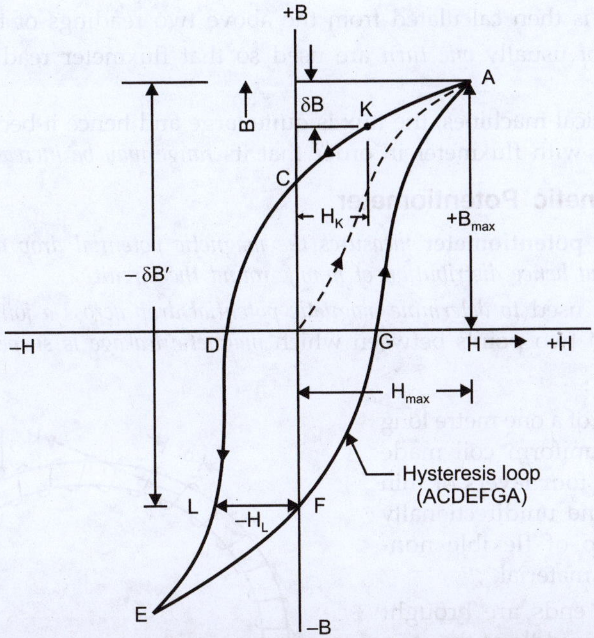

Fig. 11.13. Hysteresis loop of a magnetic material.

— The part *EFGA* of the loop may be obtained by drawing in the reverse of the part *ACDE* as the two halves are identical.

11.3.5. Determination of Leakage Factor in Dynamo-Electric Machinery

In the dynamo-electric machine, the *leakage factor* is defined as :

$$\text{Leakage factor} = \frac{\text{Total flux}}{\text{Useful flux}}$$

where *"useful flux" is the flux in the armature (i.e., flux crossing the air gap)* and *"total flux"* in the pole bodies which in turn is equal to the useful flux *plus* leakage flux existing on the pole body at its root.

Thus in order to measure the leakage factor, we have to measure flux in the pole bodies and flux in the armature. *For measurement of leakage factor the 'flux meter' is used, as ballistic galvanometer is unsuitable due to high inductance of the field system of the machine.*

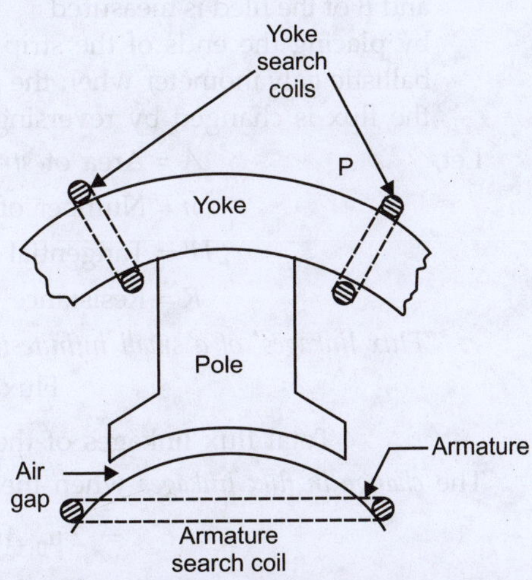

Fig. 11.14. Measurement of leakage factor.

— The yoke carries *half* of the total flux and therefore it is possible to measure the value of *"total flux per pole" by using two search coils on the yoke and connecting them in series across the fluxmeter.*

The armature is kept stationary and another search coil is put on it. The coil is so positioned that it embraces the useful flux per pole. *This search coil is then connected to the fluxmeter and this way the "useful flux" is measured.*

Leakage factor is then calculated from the above two readings of the fluxmeter.

Search coils of usually *one turn* are used so that fluxmeter reading gives the flux directly.

In large electrical machines, the flux is quite large and hence it becomes necessary to use a *shunt* along with fluxmeter in order that its *range may be increased.*

11.3.6. Magnetic Potentiometer

The **magnetic potentiometer** *measures the magnetic potential drop along any part of a magnetic circuit and hence distribution of m.m.f. round the circuit.*

It can also be used *to determine magnetic potential drop across a joint* in the magnetic circuit or between two points between which *magnetic leakage is suspected.*

Construction :

— It consists of a one metre long flat and uniform coil made of two or four layers of thin wire wound unidirectionally on a strip of flexible non-magnetic material.

— The coils ends are brought out at the middle of the strip as shown in Fig. 11.15 and connected to a ballistic galvanometer.

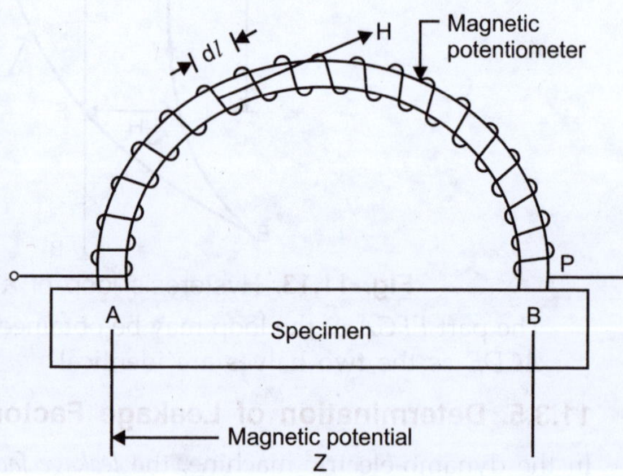

Fig. 11.15. Magnetic potentiometer.

Working :

— The magnetic potential difference between points A and B of the filed is measured by placing the ends of the strip at these points and observing the throw of the ballistic galvanometer when the flux through the specimen is changed, assuming the flux is changed by reversing the field.

Let, A = Area of *strip*, m^2,

n = Number of turns per unit length of the strip,

H' = Tangential component of magnetising force, AT/m, and

R = Resistance of the ballistic galvanometer.

∴ *'Flux linkages' of a small infinitesimal portion of strip* of length dl are given as :

$$\text{Flux turns} = (\mu_0 H'A)n \cdot dl = \mu_0 H'An \cdot dl$$

$$\text{Total flux linkages of the strip} = \int \mu_0 H'Andl = \mu_0 An \int H'dl$$

The *change in flux linkages* when the current in the magnetising winding is *reversed*

$$= 2\,\mu_0\, An \int H'dl$$

But, $\int H' \cdot dl = Z$ = Magnetic potential difference between A and B.

$$\therefore \text{Change in flux linkages} = 2\,\mu_0\, An\, Z$$

$$\text{Now, Change } Q = it = \frac{e}{R}t = \frac{2\mu_0 AnZ}{R \times t} \cdot t = \frac{2\mu_0 An}{R} Z \text{ coulomb} \quad ...(i)$$

Also, Change indicated by galvanometer,

$$Q = G_q \, \theta_1 \text{ coulomb} \qquad ...(ii)$$

Equating eqns. (*i*), and (*ii*) we have :

$$\frac{2\mu_0 \, AnZ}{R} = G_q \cdot \theta_1$$

or, Magnetic potential, $$Z = \frac{RG_q \, \theta_1}{2\mu_0 \, An} \qquad ...(11.15)$$

The value of G_q (galvanometer constant) can be found with the help of a calibrating circuit.

- The above analysis is correct only if the portion *AB* of the specimen does not carry any part of the magnetising winding. Since it is practically impossible to achieve uniformity of *H* when this condition is satisfied, direct measurement of *H* by this method is *rarely used.*

11.3.7. Testing Bar Specimens of Magnetic Materials

The preparation of ring specimens is difficult while *bar specimens are easier to construct.* However, if the methods of measurements described above are applied to bar specimens some difficulties and inaccuracies are encountered. This is because when we use bar specimens, the return circuit for the flux is through air whose *resistance is very high.* In addition when the bar is magnetised poles are produced at its ends, and these poles produce, inside the rod a magnetising force which acts from North-pole to South which is in opposition to the applied magnetising force. This is called *self-demagnetisation or end effects.*

Thus the value of *H* of the magnetising force which *is effective in producing the flux in specimen* is *less* than the applied magnetising force as calculated from the m.m.f. of the magnetising winding. Thus the *true magnetising force H* is given by :

$$H = H_a - H_d$$

where, H_a = Applied magnetising force

$$= \frac{NI}{l} \qquad \text{...for a } \textit{long solenoid}$$

$$= \frac{NI}{\sqrt{l^2 + d^2}} \qquad \text{...for a } \textit{sort solenoid} \text{ of length } l \text{ and diameter } d.$$

H_d = The *magnetising force due to self-demagnetisation or end effects.*

- For obtaining accuracy in measurements with bar specimens, we must apply *corrections for the demagnetising force or reduce the effect to negligible proportions.* The self-demagnetising force H_d, however, is somewhat an uncertain quantity.

— Since the effect of H_d is small when the ratio $\frac{l}{d}$ of the bar is large, the dimensions of the specimen should be so chosen as to make $\frac{l}{d}$ *as large as possible.* It has been observed that the demagnetising has negligible effect on the value of magnetising force when the ratio $\frac{l}{d}$ of bar specimens is 25 or more.

11.3.8. Bar Specimens Versus Ring Specimens

- The ballistic methods which use specimen in the form of a ring give the *highest accuracy* under proper conditions.

- The *main advantage of "ring specimens" is that the end effects are absent in them and they are free from errors resulting from magnetic leakage.*
- The magnetising winding in ring specimen, however, has to be wound *uniformly* on the ring otherwise the m.m.f. distribution will not be uniform which may lead to magnetic leakage resulting in errors in the measurements.

— The magnetising force is calculated in terms of mean diameter of the ring and the m.m.f. in the magnetising winding. But in the inner circumference of the ring is smaller than the outer, hence the magnetising force and the flux density are not uniform over the cross-section of the ring. *If the ratio of mean diameter to the radial width is more than eight, both magnetising force and flux density are reasonably uniform over the cross-section of the specimen and consequently the error involved is not appreciable.*

Types of ring specimens. There are three types of ring specimen used, depending upon the material to be tested :

1. *Laminated type :*

— This is the normal form of "isotropic sheet materials."

— The ring is assembled with the directions of rolling of individual laminations distributed radially so that the test should yield result corresponding to *mean properties of the material.*

2. *Clock-spring type :*

— A long strip of material is wound in the form of a spiral to form a ring.

— This type is used for testing *"anisotropic materials" i.e.,* grain oriented cold rolled steel.

3. *Solid type :*

— Solid rings are prepared either by casting or forging.

Uses of ring specimens. The use of ring specimen in magnetic testing is limited owing to following *reasons* :

(*i*) The labour involved in preparing them, because the magnetising winding has to be put with hand for each individual specimen.

(*ii*) The value of the magnetising force is limited by heating.

The ring specimens are therefore used for the following purposes :

1. Testing of small quantities of materials of very high permeability.
2. Testing powder or dust core materials.

11.4 PERMEAMETERS

— Normally it is not a practical arrangement to a have $\frac{l}{d}$ as large as possible, for a bar specimen to be tested. ***Permeameters*** have been developed *to overcome the difficulties and inaccuracies encountered in performing measurements on bar specimens of magnetic materials.* Thus pemeameters test bar specimens under controlled field conditions.

— The permeameter generally consists of a fixed steel frame to which bar specimens can be connected.

— There are various types of permeameters differing widely in the arrangements of magnetising and search coils and in the means of guarding against leakage fluxes and m.m.f. drop at joints.

— Most of the permeameters incorporate modifications of the *bar and yoke* arrangements invented by Hopkinson.

The following permeameters will be discussed :

1. Hopkinson permeameter (Bar and yoke method).
2. Ewing double bar permeameter.
3. Illiovici permeameter.
4. Burrow's permeameter.
5. Fahy's simplex permeameter.
6. The National Physical Laboratory (England) form of permeameter.

11.4.1. Hopkinson Permeameter (Bar and Yoke Method)

The bar and yoke method is generally employed for the testing of bar specimens in which the demagnetising effect of free poles is largely eliminated by the use of heavy section yokes. Bars give the advantage of simple arrangement for the application of the magnetising force.

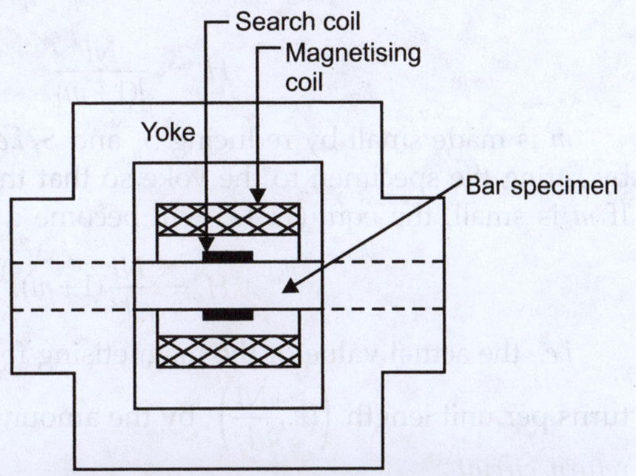

Fig. 11.16. Bar and yoke method (Hopkinson permeameter).

Fig. 11.16 shows the schematic arrangement of bar and yoke method :

— A test coil is wound upon the central part of a bar specimen.

— The bar is clamped between two halves of a massive iron yoke whose reluctance is *low* compared to that of the bar specimen. This yoke provides a return path for the flux.

— The specimen is wound with a magnetising winding as shown.

Let, N = Number of turns on the magnetising winding,

I = Current in the magnetising winding,

l = Length of the bar specimen between two halves of the yoke,

A_{sp} = Cross-sectional area of specimen,

μ_{sp} = Permeability of the specimen when the magnetising current is I,

S_y = Reluctance of the yoke,

S_j = Reluctance of the joints between the bar specimen and the yoke, and

ϕ = Flux in the magnetic circuit.

Then, Reluctance of specimen,

$$S_{sp} = \frac{l}{\mu_{sp} A_{sp}}$$

$$\therefore \quad \text{Flux, } \phi = \frac{\text{m.m.f.}}{\text{Reluctance of magnetic circuit}}$$

$$= \frac{NI}{S_y + S_j + (l/\mu_{sp} A_{sp})}$$

Flux density in the specimen,

$$B = \frac{\phi}{A_{sp}} = \frac{NI}{A_{sp}\left[S_y + S_j + (l/\mu_{sp} A_{sp})\right]}$$

$$\text{Magnetising force, } H = \frac{B}{\mu_{sp}} = \frac{NI}{\mu_{sp} A_{sp}\left[S_y + S_j + (l/\mu_{sp} A_{sp})\right]}$$

Let,

$$m = \frac{\text{Reluctance of yoke and joints}}{\text{Reluctance of specimen}}$$

$$= \frac{S_y + S_j}{(l/\mu_{sp} A_{sp})} = \frac{\mu_{sp} A_{sp}}{l}(S_y + S_j) \quad ...(11.16)$$

$$\therefore \quad H = \frac{NI}{l(1+m)} \quad ...(11.17)$$

m is made small by reducing S_y and S_j *i.e.*, by making the yokes of large section and by fitting the specimen to the yoke so that the *air gaps at the joints are as small as possible.* If m is small, the eqn. (11.17) will become :

$$H = \frac{NI}{l}(1+m)^{-1} = \frac{NI}{l}(1-m+....) = \frac{NI}{l}(1-m) \quad ...(11.18)$$

i.e., the actual value of the magnetising force is *less* than that computed from ampere-turns per unit length $\left(i.e., \frac{NI}{l}\right)$, by the amount $\frac{m\,NI}{l}$ which *depends on the reluctance of the return circuit*.

The flux density B may be measured by ballistic galvanometer as explained earlier.

11.4.2. Ewing Double Bar Permeameter

Fig. 11.17 shows an Ewing double bar permeameter :

— It consists of two exactly similar bar specimens of the material under test and two pairs of magnetising coils, one pair being exactly of half the length of the other pair. The number of turns per metre axial length for both pairs of coil is the *same.*

— The bar specimens fit tightly into the holes of two yokes. The yokes are made of annealed soft iron. The position these yokes on the bars can be adjusted. The

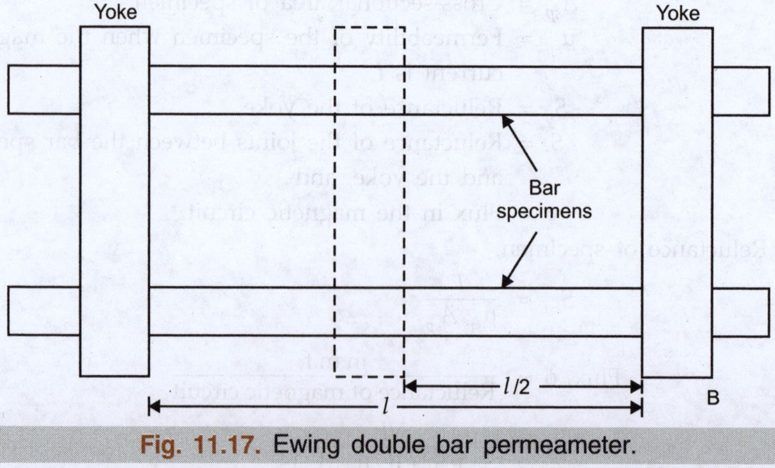

Fig. 11.17. Ewing double bar permeameter.

purpose of the arrangement of bar specimens and yokes (shown in Fig. 11.17) is the elimination of the reluctance of yoke and joints.

— The tests are made, one with length of specimen l and the other with length $\frac{l}{2}$, assuming that for both positions, the reluctance of yokes and the joints is the *same* for a given flux density.

Let,
n = Number of turns per unit length of magnetising coils,
I_1 = Current in the coils when the specimen length is l,
I_2 = Current in the coils when the specimen length is $\frac{l}{2}$,
H_1 = Apparent magnetising force when length is l,
H_2 = Apparent magnetising force when length is $\frac{l}{2}$,
a = m.m.f. required for the yokes and joints in each case, and
B = Flux density in the specimen (same in both the cases).

Then,
$$H_1 = \frac{nlI_1}{l} = nI_1,$$

and,
$$H_2 = \frac{n(l/2)\cdot I_2}{(l/2)} = nI_2$$

If H is the true magnetising force in the iron for a flux density B, then, we have :

$$H_1 \cdot l = H \cdot l + a \quad \text{...(i)}$$

or,
$$H_2 \cdot l/2 = H \cdot l/2 + a \quad \text{...(ii)}$$

($\because$ Total applied m.m.f. = m.m.f. required for iron + m.m.f. required for yoke and joints).

Subtracting expressions (*i*) from (*ii*), we get

$$a = (H_2 - H_1)l$$

Inserting this value of a in (*i*), we have :

$$H_1 l = Hl + (H_2 - H_1)l$$

or,
$$H = \frac{H_1 l - (H_2 - H_1)l}{l} = \frac{H_1 l - H_2 l + H_1 l}{l} = 2H_1 - H_2$$

i.e., True magnetising force,

$$H = 2H_1 - H_2 \quad \text{...(11.19)}$$

The flux density B corresponding to this value of H is measured with the help of a ballistic galvanometer in the usual way.

The complete test is made by first obtaining and plotting a B-H curve for the specimens with a length l, the apparent values of magnetising force being plotted. The specimens are then demagnetised, and a second B-H curve is plotted with a specimen length $\frac{l}{2}$, the two curves being plotted on the same axis. The B-H curve is then obtained from these two curves, *determining the true value of magnetising force corresponding to any value of flux density* from eqn. (11.19).

Disadvantages. The disadvantages of this method are :

1. It requires two carefully prepared specimens and hence *lengthy in operation.*
2. The reluctance of yokes and joints for two different positions of the yokes are *not* quite the same unless the specimens are very precisely machined.

11.4.3. Illiovici Permeameter

Fig. 11.18 shows the construction of Illiovici permeameter :

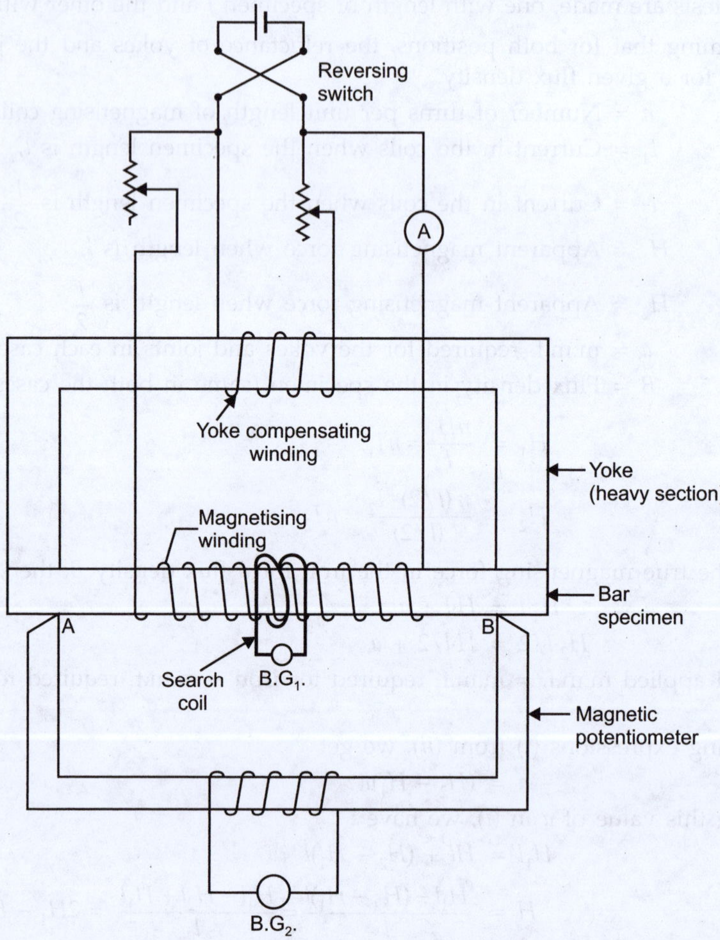

Fig. 11.18. Illiovici permeameter.

- It consists of a bar specimen clamped against a yoke of heavy section.
- A magnetising winding and a search coil (connected across a ballistic galvanometer $B.G_1$.) are wound on the specimen.
- A magnetic potentiometer is connected to a section AB of the specimen.
- A compensating winding is wound on the yoke; this winding is connected in parallel with the magnetising winding across a reversing switch.
- When the magnetic potentiometer indicates *no* magnetic potential drop, the m.m.f. for part AB is provided by the *magnetising winding*, while for the remaining part of the specimen, the yoke and the joints, is provided by the *compensating winding*. This condition is obtained by adjusting the current in the main winding to the required test value and then adjusting the current in the compensating winding so that ballistic galvanometer $B.G_2$. shows no throw when currents are reversed.

Under this condition there is *no flux* through the *magnetic potentiometer and therefore *no magnetic potential difference* exists between A and B.

Thus the magnetising force (H) for length AB is given as :

$$H = \frac{NI}{l}$$

where, N = Number of turns in the magnetising winding,

I = Current in the magnetising winding, A, and

l = Length AB, m.

The *flux density* in the specimen is found by noting the throw of ballistic galvanometer $B.G_1$. when the currents in the magnetising and the corresponding windings are *reversed-simultaneously.*

- In this apparatus, specimen can be tested upto field H = *35000 AT/m.*

Advantages :

1. Simplicity in construction.
2. Independent of the reluctance of yoke.

Disadvantages :

1. Owing to the lack of symmetry of arrangement the results obtained are *not* satisfactory.
2. It does not incorporate a direct test of uniformity of magnetisation along the specimen.
3. *Operation is complicated* on account of readjustment of compensating winding for each reading.
4. Errors may also be caused on account of some leakage flux between yoke and specimen and the magnetic *potentiometer may be affected by leakage flux.*

11.4.4. Burrow's Permeameter

In this permeameter, the effect of magnetic leakage at the joints between yoke and specimen is eliminated by the use of a number of compensating coils which apply compensating m.m.fs. at different parts of the magnetic circuit, these m.m.fs. being just sufficient to drive the flux through the reluctance of the part upon which the coils are placed.

Fig. 11.19 shows the arrangement of the magnetic circuit and coils.

— S_1 is the bar specimen to be tested, S_2 being a bar of similar dimensions to S_1.

— These bars are surrounded by magnetising windings M_1 and M_2, which are uniformly wound along the lengths of the bars.

— A_1, A_2, B_1 and B_2 are *compensating coils* for the elimination of leakage effects at the joints between the two bars and the massive yoke, in which the bars fit.

— c and c' are two exactly *similar search coils* wound at the *centres* of the two bars, while d_1 and d_2 are two *similar search coils* wound in the *position* shown, on the test bar, and each having exactly *half* the number of turns of search coil c. Coils d_1 and d_2 are connected *permanently in series*, as are also the four coils A_1, A_2, B_1, B_2.

---

* In general, the cores of magnetic potentiometer should be of non-magnetic materials but in Illiovici permeameter *ferromagnetic cores are allowable since it is used as a null detector and not for actual measurement.*

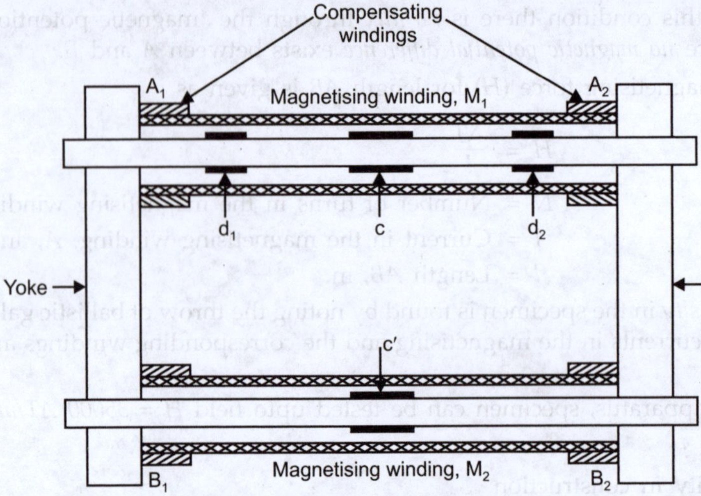

Fig. 11.19. Burrow's permeameter.

The dimensions of the coils M_1, M_2 are such that H in the specimen is approximately 10^4 times the current in the windings, the maximum value of H for which the apparatus is used being about *40,000 AT/m.*

— The dimensions of the bar specimens are approximately 30 cm long and 1 cm in diameter.

— The coils A_1, A_2, B_1, B_2 are supplied from a separate battery supply, coil M_1 from another separate supply, and M_2 from another.

— In carrying out the test it is necessary, first of all, to ensure that, for a given value of the current in M_1–*i.e.*, of H in the test bar—the flux threading all four search coils c, c', d_1 and d_2, is the *same, the current in the compensating coils, and in M_2, being adjusted until this condition is fulfilled.* If the flux threading coils d_1 and d_2 is the *same* as that threading c and c', there can be *no* appreciable leakage of flux through the air in the neighbourhood of the joints. This *means that the m.m.f. for the joints is being supplied by the compensating coils, and that the m.m.f. in coil M_1 is used up merely in driving the flux through the bar specimen S_1 inside it.*

Thus, H in the specimen is given by :

$$H = \frac{NI}{l}$$

where, N = Number of turns on coil M_1,

I = Current in M_1, amperes, and

l = Length of specimen in metres.

In some cases, it may be necessary to make corrections for the fact that the magnetised solenoids are not infinitely long, but such corrections are usually negligible.

- Following is the *"procedure" for obtaining equal flux threading all four search coils :*

— After demagnetising the specimen, the current in the magnetising coil M_1 is set at the required testing value. Search coils c and c' are then connected in series, but in opposition, to a ballistic galvanometer.

— The currents in coils M_1 and M_2 are then simultaneously *reversed.* A throw will be observed on the galvanometer. The current in M_2 is *adjusted* until *no throw* is obtained when the two currents are *reversed.* Since the search coils *c* and *c′* have equal number of turns, this means that *equal fluxes* are now threading through them.

— Next search coil *c* is connected in series with but in opposition to coils d_1 and d_2, and then to the ballistic galvanometer. The current in the compensating coils A_1, A_2, B_1, B_2 is then adjusted until *no* galvanometer throw is obtained upon simultaneous reversal of the currents in these coils and in M_1 and M_2. Then, since d_1 and d_2 together have the same number of turns as coil *c*, the flux threading all three coils is the same.

— The *"flux density"* corresponding to the value of *H* in coil M_1 (obtained as described above) can be measured by connecting coil *c alone* to the ballistic galvanometer, and noting the throw when the currents in the two magnetising coils and the compensating coils are *simultaneously reversed.*

Advantages :

1. This method results in *uniform conditions along the specimen* provided that the material itself is uniform.
2. This method can be used *for magnetic testing of many kinds and also for calibration of specimens that serve as reference standards in some of the simpler methods.*

11.4.5. Fahy's Simplex Permeameter

Fig. 11.20 shows the outline of construction of Fahy's simplex permeameter :

— Two iron clamping posts clamp the specimen against a laminated steel yoke. The yoke carries the magnetising winding, and the specimen is surrounded by a search coil for the measurement of *flux density* in it.

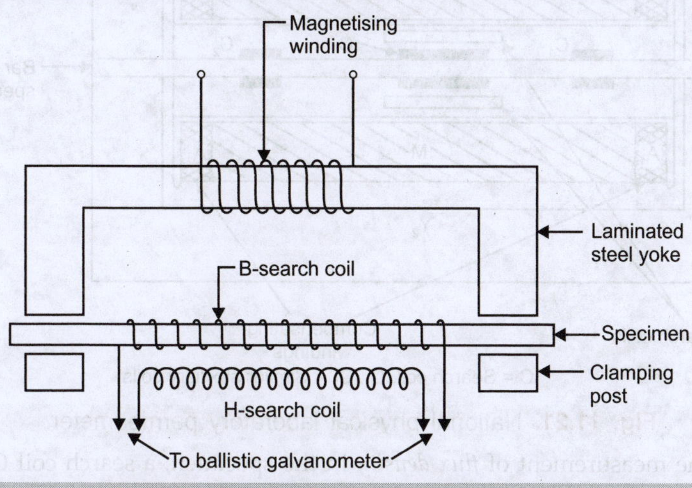

Fig. 11.20. Fahy's simplex permeameter.

— The magnetising force *H* acting on the specimen is measured, like the flux density, by a ballistic galvanometer, utilising an air-used search coil of several thousand turns located between the two clamping posts as shown. The values of *H* so measured are corrected by the calibration of *H*-search coil, utilising a specimen of known magnetic characteristics in place of the test specimen.

- An hysteresis loop may be determined equally well by application of the method for the ring specimen described earlier. Some Fahy meters have been equipped with a flip coil for *H* for use in hysteresis loop determination.

Advantages :

1. *Simpler* in both construction and operation as compared to Burrows permeameter (while its accuracy is of the same order as Burrows apparatus).
2. *Less sensitive* to the effects of magnetic inhomogeneities in the specimen (since the *B* coil is wound uniformly over the entire length of the specimen and, therefore, it tends to measure the average flux density in the specimen; eventually the effects of local magnetic irregularities are diluted).
3. *Only test specimen is required.*

11.4.6. The National Physical Laboratory (England) Form of Permeameter

This is one of the most accurate forms of permeameter and is comparatively easy to use in practice.

The arrangement of the permeameter is shown schematically in Fig. 11.21 :

— The specimen is clamped between two similar rectangular yokes Y_1 and Y_2. To keep the flux path of low reluctance, cross-sectional area of these yokes is taken *large.*

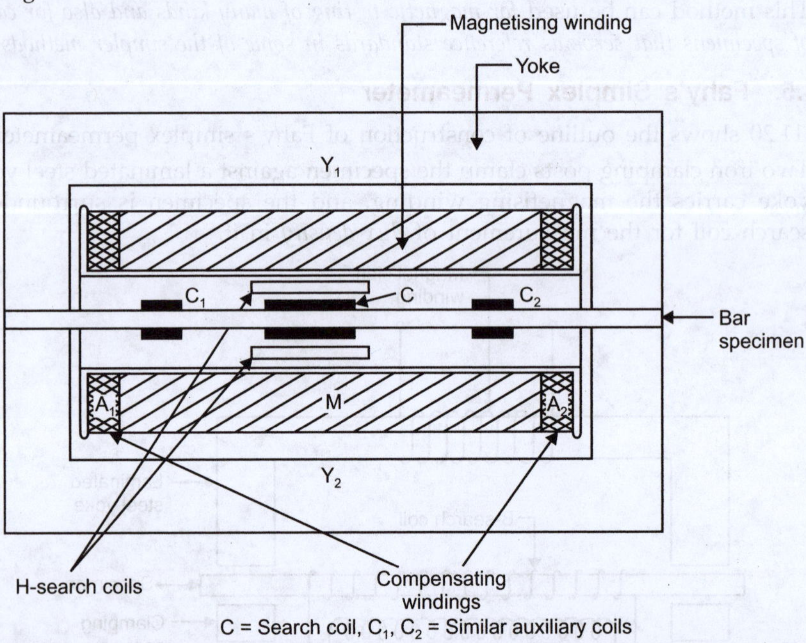

Fig. 11.21. National physical laboratory permeameter.

— For the measurement of *flux density* in the specimen, a search coil C is bound in middle part of the specimen.

— Two similar auxiliary coils C_1 and C_2, each having *half* the number of turns of that in search coil C, are bound at either side of the specimen. Coils C_1 and C_2 are used to check the uniformity in flux in the specimen (in a similar way as described earlier in Burrows permeameter).

— The main magnetising winding M is bound such that it surrounds the specimen. Both ends of M are provided with compensating windings A_1 and A_2, which supply necessary m.m.fs. to reluctance of the joints.

— Now series combination of windings C_1 and C_2 is connected in series opposition with search coil C and then connected with ballistic galvanometer. Current in the main magnetising winding is set for the required value of H. Current in the compensating windings is so adjusted that there is *no throw* in the ballistic galvanometer, during simultaneous reversal of current in these windings A_1 and A_2 and main winding M. This condition shows the *uniform distribution of flux density in the specimen.* Now this flux density is measured with the help of ballistic galvanometer by connecting it to search coil C.

— **H-coils** are provided for the measurement of value of H in the specimen. Measurement of H is done by connecting H *coils* to the ballistic galvanometer.

Significant features. This permeameter may be taken as typical of modern precision permeameters because of its following *features* :

1. It gives accurate results over a wide range of magnetising force. This range extends from *10 AT/m to 300,000 AT/m.*
2. It uses an *H-coil* for determination of value of magnetising force H instead of using a calculated value of H and these values are normally erroneous.
3. It is capable of even testing circular cross-section specimens by addition of shaped poles pieces.
4. The field strength can be doubled by mounting a short specimen between massive soft iron pole pieces.

Example 11.1. *A solenoid 50 cm long and 2.2 cm in diameter is uniformly wound with 500 turns of insulated wire.*

(*i*) *Find the magnetic field strength at the centre of the solenoid when carrying a current of 1.8 amperes.*

(*ii*) *If a secondary coil is wound round the central part of the solenoid, calculate the flux passing through it.*

Solution. *Given :* l = 50 cm or 0.5 m; d = 2.2 cm or 0.022 m,

N = 500; I = 1.8 A.

(*i*) **Magnetic field strength at the centre, H :**

$$H = \frac{NI}{l} = \frac{500 \times 1.8}{0.5} = \mathbf{1800\ AT/m}\ \textbf{(Ans.)}$$

(*ii*) **Flux passing through the secondary coil, ϕ :**

$$\phi = \mu_0\, \mu_r\, HA$$

$$= 4\pi \times 10^{-7} \times 1 \times 1800 \times \left(\frac{\pi}{4} \times 0.022^2\right)$$

$$= \mathbf{85.98 \times 10^{-8}\ Wb}\ \textbf{(Ans.)}$$

(assuming air-cored solenoid *i.e.*, μ_r = 1)

Example 11.2. *A ballistic galvanometer having a circuit resistance of 5000 Ω and a constant of 0.1 μC per scale division is connected in turn (i) to a coil of 4 turns wound round the field coil of a D.C. machine and (ii) to a 6 turns coil placed on the armature surface embracing the total flux per pole entering armature. When normal field current is broken the galvanometer readings are 226 and 272 divisions respectively.*

Calculate :

(i) Flux per pole, and

(ii) Leakage coefficient. **(Punjab University)**

Solution. *Given :* $R = 5000\ \Omega$; $G_q = 0.1\ \mu C$ per scale division.

(*i*) **Total flux per pole, ϕ :**

Electricity passed through the galvanometer,

$$Q = G_q \theta \quad ...(i)$$

Electricity passed through the galvanometer, when current is broken also,

$$Q = \frac{\phi N}{R} \quad ...(ii)$$

where, ϕ = Flux per pole,

N = Number of turns on search coil, and

R = Circuit resistance of the galvanometer.

Comparing expression (*i*) and (*ii*), we get

$$G_q \theta = \frac{\phi N}{R}$$

or,
$$\phi = \frac{R G_q \theta}{N}$$

$$= \frac{5000 \times 0.1 \times 10^{-6} \times 226}{4} = \mathbf{0.02825\ Wb}\ \textbf{(Ans.)}$$

(*ii*) **Leakage coefficient :**

Useful flux = Flux measured by galvanometer when connected to a search coil wound on the armature surface

$$= \frac{5000 \times 0.1 \times 10^{-6} \times 272}{6} = 0.02266\ \text{Wb}$$

$\therefore$ *Leakage coefficient* $= \frac{\text{Total flux}}{\text{Useful flux}} = \frac{0.02825}{0.02266} = \mathbf{1.247}$ **(Ans.)**

Example 11.3. *A moving coil ballistic galvanometer of 150 ohms resistance gives a throw of 75 divisions when the flux through a search coil to which it is connected is reversed. Find the flux density given that the galvanometer constant is 110 μC per scale division and the search coil has 1400 turns, a mean area of 50 cm² and a resistance of 20 Ω.* **(Madras University)**

Solution. *Given :* $R = 150 + 20 = 170\ \Omega$; $\theta = 75$ divisions; $G_q = 110\ \mu C$ per scale division; $N_{sc} = 1400$; $A = 50\ cm^2 = 50 \times 10^{-4}\ m^2$.

Flux density, B :

We know that,
$$B = \frac{R \cdot G_q \cdot \theta_1}{2\,NA} \quad ...[\text{Eqns, (11.10, 11.11)}]$$

or,
$$B = \frac{170 \times 110 \times 10^{-6} \times 75}{2 \times 1400 \times 50 \times 10^{-4}} = \mathbf{0.1\ Wb/m^2}\ \textbf{(Ans.)}$$

Example 11.4. *An iron ring of 400 mm² cross-sectional area with a mean length of 1.2 m is wound with a magnetising winding of 120 turns. A secondary coil with 240 turns of wire is connected to a ballistic galvanometer having a constant of 1.1 μC per scale division, the total* *resistance of the secondary circuit being 2200 Ω. On reversing a current of 12 A in the magnetising winding, the galvanometer shows a deflection of 100 scale divisions. Calculate :*

(i) The flux density in the specimen;

(ii) The value of permeability, at this flux density.

Solution. *Given* : Area, $A = 400\ mm^2 = 400 \times 10^{-6}\ m^2$; $l = 1.2$ m; $N_1 = 120$; $N_2 = 240$; $G_q = 1.1\ \mu C$ per scale division; $R = 2200\ \Omega$; $I = 12$ A; $\theta_1 = 100$ scale divisions.

(i) The flux density in the specimen, B :

$$\text{Total m.m.f. of coil} = N_1 I_1 = 120 \times 12 \text{ AT}$$

$$\text{Magnetising force, } H = \frac{N_1 I_1}{l} = \frac{120 \times 12}{1.2} = 1200 \text{ AT/m}$$

Charge through the galvanometer, $Q = G_q\ \theta_1 = 1.1 \times 100 = 110\ \mu C$ ...(*i*)

Let ϕ is the flux passing through the ring,

Then, Flux linkages of search coil,

$$\psi = N_2 \phi = 240\ \phi$$

Change in flux linkages of search coil due to reversal of current,

$$d\psi = 2\psi = 480\ \phi$$

Average e.m.f. induced in the search coil,

$$e = \frac{d\psi}{dt} = \frac{480\ \phi}{dt}, \text{ where, } dt = \text{time of reversal.}$$

∴ Current through the search coil,

$$Q = i.dt = \frac{e}{R} \cdot dt$$

$$= \frac{480\ \phi}{dt\ R} \cdot dt = \frac{480\ \phi}{R} = \frac{480\ \phi}{2200} \quad ...(ii)$$

From (*i*) and (*ii*), we have :

$$\frac{480\ \phi}{2200} = 110 \times 10^{-6}$$

$$\text{or,} \quad \phi = \frac{110 \times 10^{-6} \times 2200}{480} = 504 \times 10^{-6} \text{ Wb}$$

∴ Flux density in the specimen,

$$B = \frac{\phi}{A} = \frac{504 \times 10^{-6}}{400 \times 10^{-6}} = \mathbf{1.26\ Wb/m^2} \text{ (Ans.)}$$

(ii) The value of permeability, μ_r :

We know that, $B = \mu_0\ \mu_r\ H$

$$\therefore \quad \mu_r = \frac{B}{\mu_0 H} = \frac{1.26}{4\pi \times 10^{-7} \times 1200} = \mathbf{836} \text{ (Ans.)}$$

Example 11.5. *An iron ring with a mean diameter of 0.12 m and a cross-section of 35.5 mm² is wound with a magnetising winding of 350 turns and a secondary winding of 250 turns. On reversing a current of 12 A in the magnetising winding, a ballistic galvanometer gives a throw of 280 scale divisions, while a Hibbert magnetic standard with 12 turns and a flux of 0.28 × 10⁻³ Wb gives a reading of 105 scale divisions, other conditions remaining the same.*

Find the relative permeability of the specimen. **(Rorkee University)**

Solution. *Given* : Mean diameters of iron ring, $D = 0.12$ m; $A = 35.5$ mm^2; $N_1 = 350$; $N_2 = 250$, $I = 12$ A; $\theta = 280$ scale divisions; Turns on Hibbert magnetic standard, $N_{st} = 12$; $\phi_{st} = 0.28 \times 10^{-3}$ Wb; Throw of galvanometer with Hibbert's standard, $\theta_{st} = 105$ scale divisions.

Relative permeability of the specimen, μ_r :

In a Hibbert's magnetic standard, the coil slides through the gap of a permanent magnet.

∴ Change in flux linkages of the coil when it goes down the Hibbert's standard

$$= \text{Flux } (\phi_{st}) \times \text{turns of the Hibbert's standard } (N_{st})$$

$$= (0.28 \times 10^{-3}) \times 12 = 3.36 \times 10^{-3} \text{ weber turns}$$

Constant of galvanometer,

$$G_\phi = \frac{3.36 \times 10^{-3}}{105} = 32 \times 10^{-6} \text{ Weber turns/division.}$$

The galvanometer shows a deflection of 280 scale divisions when connected to search coil.

∴ Change in flux linkages of search coil,

$$d\psi = 280 \times 32 \times 10^{-6} = 8960 \times 10^{-6} \text{ Weber turn}$$

This change is caused by the reversal of flux.

$$\therefore \text{ Flux linkages of coil, } \psi = \frac{d\psi}{2} = \frac{8960 \times 10^{-6}}{2} = 4480 \times 10^{-6} \text{ weber turn}$$

$$\text{Flux linking with search coil, } \phi = \frac{\text{Flux linkage of coil}}{\text{Turns of coil}}$$

$$= \frac{4480 \times 10^{-6}}{250} = 17.92 \times 10^{-6} \text{ Wb}$$

$$\text{Flux density in the ring specimen, } B = \frac{17.92 \times 10^{-6}}{35.5 \times 10^{-6}} = 0.505 \text{ Wb/m}^2$$

$$\text{Length of flux path in ring, } l = \pi D = \pi \times 0.12 = 0.377 \text{ m}$$

$$\text{Total m.m.f. of magnetism winding} = N_1 I = 350 \times 12 = 4200$$

$$\text{Magnetising force, } H = \frac{N_1 I}{l} = \frac{4200}{0.377} = 11141 \text{ AT/m}$$

$$\therefore \text{ Relative permeability of specimen, } \mu_r = \frac{B}{\mu_0 H} = \frac{0.505}{4\pi \times 10^{-7} \times 11141} = \mathbf{36.07} \textbf{ (Ans.)}$$

Example 11.6. *A permeameter, used to test a ferromagnetic material, is provided with a coil wound over the specimen, this coil being known as "B-coil" and another coil wound near the surface of the coil called "H-coil". These two coils are connected to two separate ballistic galvanometers having a constant of 28×10^{-6} weber turn per scale division. The testing is done by reversing the current of the magnetising winding wound over the specimen. The reading of the ballistic galvanometer connected to 'B'-coil is 100 divisions and that connected to 'H'- coil is 10 divisions.*

The data of the two coils is :

B-coil :	*Turns = 100;*	*Area = 60 mm^2*
H-coil :	*Turns = 10,000;*	*Area = 6 mm^2*

Calculate :

(i) Flux density in specimen.

(ii) Magnetising force.

(iii) Relative permeability of the specimen.

Solution. *(i)* **Flux density in speciment, B_{sp} :**

Let ϕ and N represent the flux and the number of turns respectively.

Then, Flux linkage, $\psi = N\phi$

Change in flux linkages due to reversal of current,

$$d\psi = 2\psi = 2\,N\phi$$

Flux linkages indicated by galvanometer $= G_\phi\,\theta_1$

$$\therefore \quad 2\,N\phi = G_\phi\,\theta_1$$

or, $$\text{Flux } \phi = \frac{G_\phi\,\theta_1}{2N}, \text{ and}$$

Flux density $$B = \frac{\phi}{A} = \frac{G_\phi\,\theta_1}{2NA}$$

Since *'B-coil'* is wound over the specimen, therefore it gives flux density in the *specimen*,

$\therefore$ Flux density in specimen,

$$B_{sp} = \frac{G_\phi\,\theta_1}{2NA} = \frac{28\times10^{-6}\times100}{2\times100\times60\times10^{-6}} = \mathbf{0.233\ Wb/m^2}\ \textbf{(Ans.)}$$

(ii) **Magnetising force, H :**

Since *'H'*-coil is put at the surface of the coil, therefore, it indicates the flux density in the *air*.

$$\text{Flux density in air, } B_{air} = \frac{28\times10^{-6}\times10}{2\times10000\times6\times10^{-6}} = 0.00233\ \text{Wb/m}^2$$

$$\text{Magnetising force, } H = \frac{B_{air}}{\mu_0} = \frac{0.00233}{4\pi\times10^{-7}} = \mathbf{1854\ AT/m}\ \textbf{(Ans.)}$$

(iii) **Relative permeability of the specimen, μ_r :**

$$\mu_r = \frac{B_{sp}}{\mu_0\,H} = \frac{0.233}{4\pi\times10^{-7}\times1854} = \mathbf{100}\ \textbf{(Ans.)}$$

Example 11.7. *The constant of a given potentiometer is obtained by aid of a coil of 320 turns in which a current of 0.7A is reversed. The resulting throw of galvanometer is 160 scale divisions. It is then used to measure the magnetic potential difference between two points and throw is 310 divisions.*

Calculate the magnetic potential difference.

Solution. *Given :* $N = 320$; $I = 0.7$ A; $\theta_1 = 160$ scale divisions; $\theta_2 = 310$ scale divisions.

Magnetic potential difference :

$$\text{M.m.f. of coil} = NI = 320 \times 0.7 = 224 \text{ AT}$$

Change in m.m.f. due to reversal of current in the coil = 2 × 224 = 448 AT

A deflection of 160 scale divisions is caused by an m.m.f. of 448 AT,

$$\therefore \quad \text{m.m.f. causing a deflection of 310 divisions} = \frac{310}{160}\times448 = 868 \text{ AT}$$

Hence, *magnetic potential difference* = **868 AT (Ans.)**

11.5 A.C. MAGNETIC TESTING

The A.C. magnetic testing is carried out for the following *purposes* :

(*i*) To determine the iron losses in magnetic materials at different values of flux density and frequency;

(*ii*) To separate two components of iron losses, *i.e.*, eddy current losses and hysteresis losses.

Iron losses :

When ferromagnetic materials are subjected to an alternating field, power loss due to *hysteresis effects* and *eddy currents* occurs.

(*i*) ***Hysteresis loss*** (P_h). This loss depends upon the *frequency*, and *maximum flux density* of the magnetic field to which the specimen is subjected. It is expressed as :

$$P_h = K_h f B_{max}^k \text{ watts per m}^3 \text{ or per kg} \quad ...(11.20)$$

where, K_h = Hysteresis coefficient for the given material,

f = Frequency of alternation, Hz,

B_{max} = Maximum flux density of the alternating magnetic field, Wb/m^2,

k = Steinmetz coefficient; its value varies from 1.6 to 2.

(*ii*) ***Eddy current loss*** (P_e). This loss depends upon the *wave shape of the variation of flux with time, frequency* and *maximum flux density* of the magnetic field to which the specimen is subjected.

This loss is given by the expression :

$$P_e = K_e K_f^2 f^2 B_{max}^2 \text{ watts per m}^3 \text{ or per kg} \quad ...(11.21)$$

where, K_e = Eddy current loss constant for the given material, and

K_f = Form factor of alternating value of flux.

— The *"hysteresis loss"* for a given specimen can be easily determined by plotting hysteresis loop with direct current supply. But when the ferromagnic materials are subjected to alternating currents, the actual hysteresis loss differs from that determined by plotting hysteresis loop.

— The *"eddy current loss"* occurs only when the magnetic materials are subjected to alternating currents. Hence the ferromagnetic materials used in manufacture of all A.C. electrical machinery and other equipment are tested with alternating currents only.

• *"Total iron loss"* is the sum of hysteresis and eddy current losses and is given by the relation :

$$P_i = P_h + P_e$$

or,

$$P_i = K_h f B_{max}^k + K_e K_f^2 f^2 B_{max}^2 \text{ watts per m}^3 \text{ or per kg} \quad ...(11.22)$$

The total iron loss can be calculated with the aid of above expression but it *gives good results only in the case of static machinery and not in case of dynamo electric machinery. The reason being that this expression is applicable when the field is simply alternating, but when the reversals of flux are due to rotational magnetic field, the losses do not follow the same laws.* Therefore, in practice, ***iron loss curves*** *are used which represent the loss in watt per kg as a function of flux density.*

Fig. 11.22 shows loss curves for thickness t = 0.5 mm and t = 0.3 mm for a particular frequency say 50 Hz.

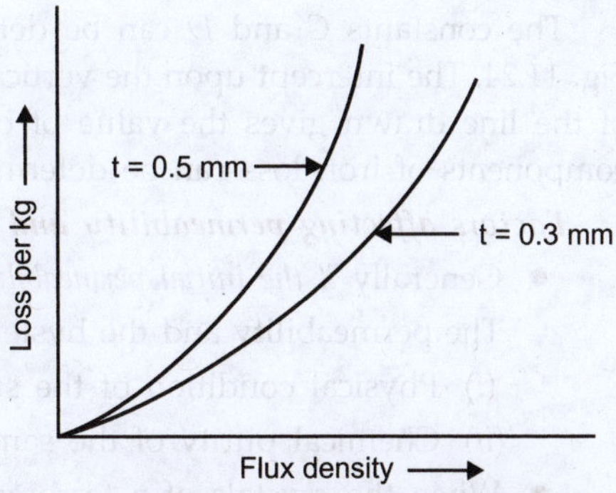

Fig. 11.22. Loss curves.

Separation of iron losses :

As the hysteresis and eddy current losses have different laws of variation with both frequency and form factor, K_f, it is *possible to separate the losses by variation of frequency or form factor if B_{max} can be maintained constant.*

1. *Variation of frequency.* In this test the form factor K_f and maximum flux density B_{max} remain constant. When the *frequency is varied* the total iron loss may be expressed as :

$$P_i = Af + Bf^2 \quad ...(11.23)$$

where, $$A = K_h \; B_{max}^k \quad ...(11.24)$$

and, $$B = Ke \; K_f^2 \, B_{max}^2 \quad ...(11.25)$$

Both A and B being constants for the test.

Now, we have : $$\frac{P_i}{f} = A + Bf \quad ...(11.26)$$

These constants may be determined by plotting $\frac{P_i}{f}$ against f, as shown in Fig. 11.23. The intercept on the vertical axis gives value of constant A and slope of the line gives value of constant B. Thus knowing constants A and B the hysteresis and eddy current loss components for any given frequency can be determined.

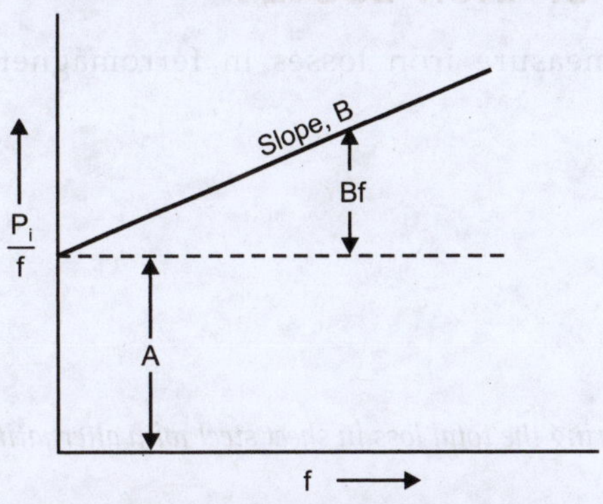

Fig. 11.23. Variation of iron loss with frequency.

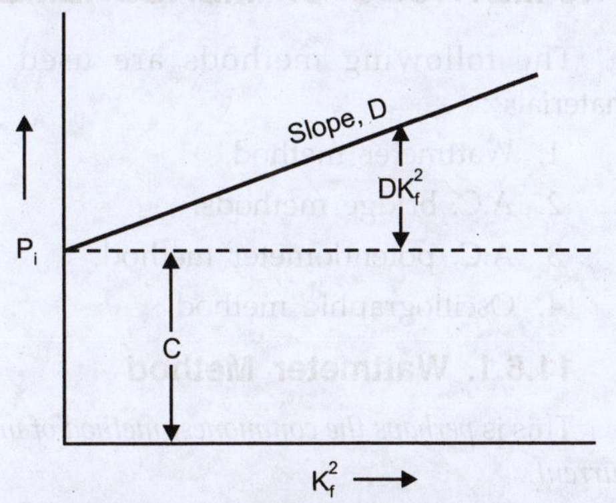

Fig. 11.24. Variation of iron loss with K_f^2.

2. *Variation of form factor.* The value of maximum flux density B_{max} and frequency f are kept constant in this test. When form factor K_f is varied, the total iron loss may be written as :

$$P_i = C + D \, K_f^2 \quad ...(11.27)$$

where, $$C = K_h \, f \, B_{max}^k \quad ...(11.28)$$

and, $$D = K_e \, f^2 \, B_{max}^2 \quad ...(11.29)$$

Both C and D being constant for the test.

The constants C and D can be determined by plotting P_i against K_f^2, as shown in Fig. 11.24. The intercept upon the vertical axis gives the value of constant C and the slope of the line drawn gives the value of constant D. Knowing the constants C and D the components of iron loss can be determined.

Factors affecting permeability and hysteresis loss :

- Generally *if the initial permeability is high, the hysteresis loss is low and vice versa.* The permeability and the hysteresis loss depend upon the following conditions :
  (*i*) Physical condition of the sample.
  (*ii*) Chemical purity of the sample.
- When the crystals of a ferromagnetic material are *cold worked,* they experience *deformation* as a result of which the material has *very poor magnetic properties.* Due to internal strains on the domains, greater magnetic field is required to give a definite magnetisation, as a consequence the *permeability decreases and the hysteresis loss is increased.* A material which has *suffered magnetic damage due to cold work may be treated to a sufficiently high temperature when the magnetic properties will be restored.*
- The impurity content of the material exercises a limit on the highest magnetic permeability and the lowest hysteresis loss that can be obtained. *Impurities bring harm to the magnetic properties by affecting the regular geometric pattern of the crystals.*
- The main impurities in the magnetic materials used for transformer cores and electrical machinery are *carbon, sulphur, oxygen* and *nitrogen. Carbon is the most detrimental* and the amount of carbon is commercial materials is kept *as low as 0.01 per cent.*

11.6 METHODS OF MEASUREMENT OF IRON LOSSES

The following methods are used to measure iron losses in ferromagnetic materials :

1. Wattmeter method.
2. A.C. bridge methods.
3. A.C. potentiometer method.
4. Oscillographic method.

11.6.1. Wattmeter Method

This is *perhaps the commonest method of measuring the total loss in sheet steel with alternating current.*

— The sheet material to be tested is arranged in the form of a ***magnetic square*** [of which there are several forms, Epstein being the originator of the arrangement] with fixed magnetising and search coils into which the strips of sheet material to be tested are slipped and clamped.

— The clamping arrangement should be such that *air gaps* in the circuits are reduced to a *minimum possible.*

— Each winding is divided into *four parts,* each part being on one side of the square.

— The windings extend as far as possible along the whole length of the side and a reasonable approximation to *uniform flux density is obtained.*

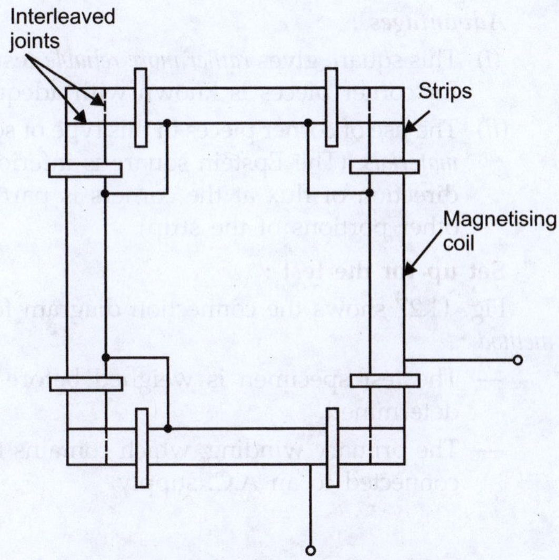

Fig. 11.25. Epstein square.

1. Epstein square. Fig. 11.25 shows the arrangement of an Epstein square :

— It consists of four stacks of strips. These stacks are bound and then taped.

— The individual strips are insulated from each other and each strip is in the plane of the square.

— The stacks are slipped into four magnetising coils with the strips projecting beyond the coils.

— The ends of the four strips are interleaved (as in the construction of transformer core) and clamped at corners.

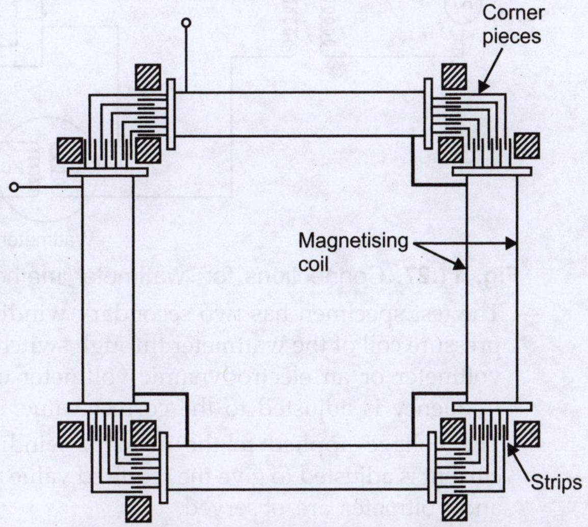

Fig. 11.26. Lloyd-Fisher square.

2. Lloyd-Fisher square. Fig. 11.26 shows the construction of Lloyd-Fisher square :

— Each strip (usually 0.25 m long and 50 to 60 mm wide) is perpendicular to the plane of the square. The strips are built up into four stacks which are placed inside four similar magnetising coils of large cross-sectional area. These four coils are connected in series to form the *primary winding*.

— Each magnetising coil has two similar single layer coils underneath it; these are called *secondary coils*. These secondary coils are connected in series in group of four, one from each core, to form *two separate secondary windings*.

— The magnetic circuit is completed by bringing the four stacks together in the form of a square and joining them at the corners. The corner joints are made by a set of standard right angled corner pieces.

— There is an overlapping of corner pieces and strips at the corners due to which cross-section of iron is doubled at the corners; therefore, a correction must be applied for this. Also the measured losses has to be *corrected* for loss in the corner pieces.

Advantages :

(*i*) This square gives *rather more reliable* results than Epstein square, in case allowance for corner pieces is known with adequate accuracy.

(*ii*) The use of corner pieces in this type of square makes it *superior for testing anisotropic materials.* (The Epstein square is inferior for tests on anisotropic materials as the direction of flux at the corners is partially perpendicular to the path of flux in other portions of the strip).

Set up for the test :

Fig. 11.27 shows the connection diagram for finding the total iron loss by *wattmeter method* :

— The test specimen is weighed before assembly and its cross-sectional area is determined.

— The primary winding, which contains the current coil (C.C.) of the wattmeter, is connected to an A.C. supply.

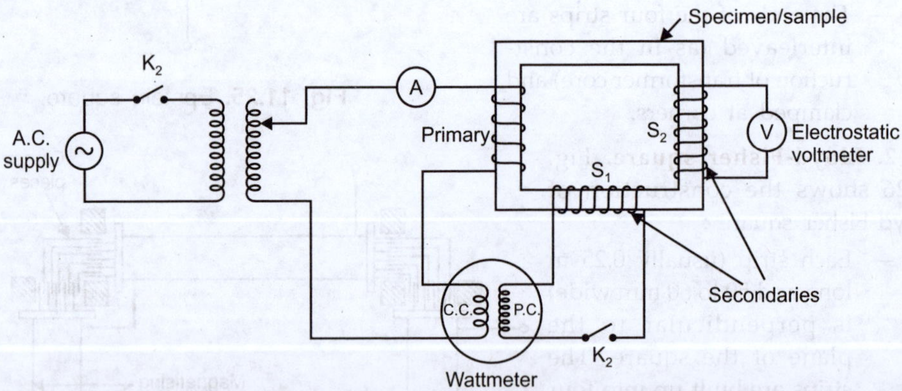

Fig. 11.27. Connections for "wattmeter method" of iron loss measurement.

— The test specimen has two secondary windings S_1 and S_2, S_1 is connected to the pressure coil of the wattmeter through switch K_2. S_2 is connected to an electrostatic voltmeter or an electrodynamic voltmeter of very high impedance. The supply frequency is adjusted to the correct value.

— The voltage applied to the primary winding is adjusted, till the magnetising current is adjusted to give the required value of B_{max}. The readings of the wattmeter and voltmeter are observed.

Theory of the method :

The voltage induced in secondary winding S_2, whose r.m.s. value is measured by the voltmeter, is given by the expression :

$$E = 4\, K_f\, \phi_{max} \cdot f \cdot N_2 \text{ volts}$$

or,
$$E = 4\, K_f B_{max} \cdot A \cdot f N_2 \text{ volts} \quad ...(11.30)$$

where, K_f = Form factor (1.11 for sinusoidal waves),

ϕ_{max} = Maximum value of flux, Wb,

B_{max} = Maximum flux density Wb/m^2,

A = Cross-section of the specimen, m^2,

f = Frequency, Hz, and

N_2 = Number of turns on the secondary winding.

$$\therefore \qquad B_{max} = \frac{E}{4\,K_f \cdot A \cdot f \cdot N_2} \qquad ...[11.30(a)]$$

The expression may need to be corrected, especially at high values of B_{max}, for the fact that the coil S_2 encloses some air flux as well as the flux in the specimen/sample, since the cross-sectional area of the coil must be greater than that of specimen itself.

Let, A_c = Cross-sectional area of the coil,

A_{sp} = Cross-sectional area of the specimen,

H_{max} = The maximum magnetising force (equal to that in the air space within the coil)

$B_{max\,(act.)}$ = Actual maximum flux density in the specimen.

Then, the *total flux within the coil is :*

$$B_{max(act.)} \cdot A_{sp} + \mu_0\, H_{max}\,(A_c - A_{sp}) = B_{max(app.)} \cdot A_{sp}$$

where, $B_{max\,(app.)}$ is the *apparent maximum flux* density in the specimen.

Thus, $E = 4\,K_f \cdot f \cdot N_2\,[B_{max(act.)} \cdot A_{sp} + \mu_0\, H_{max}\,(A_c - A_{sp})]$

or, $E = 4\,K_f \cdot f \cdot N_2\, A_{sp}\, B_{max(app.)}$ volts ...(11.31)

where, $$B_{max(app.)} = B_{max(act.)} + \mu_0\, H_{max}\left(\frac{A_c - A_{sp}}{A_{sp}}\right) \qquad ...(11.32)$$

H_{max} may be *determined from the permeability curve of the specimen.*

As regards ***power loss in iron :***

Let, P_i = Total iron loss, W,

P = Wattmeter reading, W,

V = Voltage applied to wattmeter voltage (or pressure) coil,

E = Voltmeter reading, V

= Voltage induced in coil S_1, since S_1 and S_2 have equal number of turns.

r_p = Resistance of wattmeter voltage (or pressure) coil, Ω,

r_c = Resistance of coil S_1, Ω, and

I_p = Current in the voltage (or pressure) coil circuit, A.

Then, $E = I_p(r_p + r_c)$ ...(*i*)

$V = I_p r_p$ ...(*ii*)

∴ *Total iron loss in specimen + total copper loss in the secondary winding circuit*

$$= P \cdot \frac{E}{V} = P\left[\frac{I_p(r_p + r_c)}{I_p\, r_p}\right] = P\left(\frac{r_p + r_c}{r_p}\right)$$

$$= P\left(1 + \frac{r_c}{r_p}\right) \qquad ...(iii)$$

Again, the *copper losses* in r_p and r_c are :

$$I_p^2(r_p + r_c) = \left(\frac{E}{r_p + r_c}\right)^2 (r_p + r_c) = \frac{E^2}{r_p + r_c} \qquad ...(iv)$$

$\therefore$ *Total iron loss,* $P_i = P\left(1+\frac{r_c}{r_p}\right) - \frac{E^2}{r_p + r_c}$ watts ...(11.33)

...[Subtracting (*iv*) from (*iii*)]

- *Specific iron loss i.e.,* iron loss per kg can be calculated by dividing the total iron loss by the weight of the specimen.
- *The hysteresis and eddy current components of the loss can be graphically determined from the results of power measurements* such as the above, *at different frequencies.*

Measurement of iron loss at different from factors :

Fig. 11.28, shows the connections for the measurement of power loss at different form factors.

— The form factor is obtained from the readings of two voltmeters V_1 and V_2. The voltmeter V_1 (may be an electrostatic or electrodynamic type) is connected across S_1 and it reads the *r.m.s. value* of the voltage. Voltmeter V_2 (a moving coil instrument) is connected across S_2 through a bridge rectifier and, therefore, it reads the *average value of the voltage.*

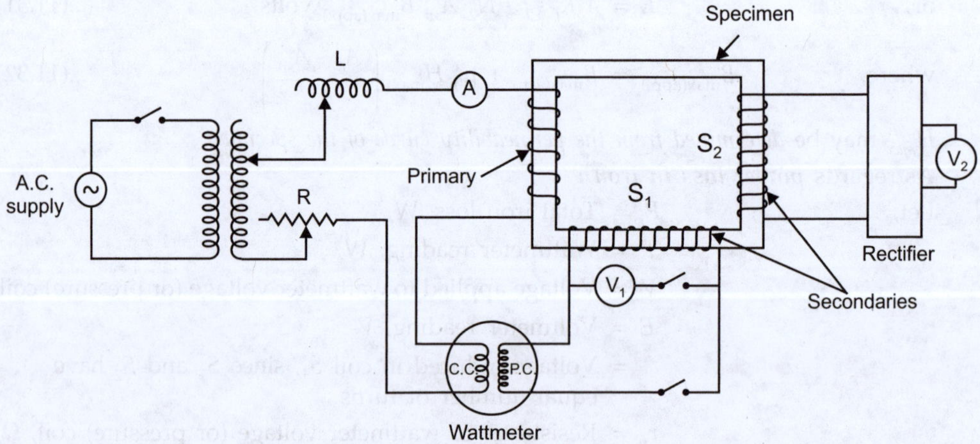

Fig. 11.28. Measurement of power loss at different form factors—Wattmeter method.

$$\text{Form factor, } K_f = \frac{\text{R.m.s. value}}{\text{Average value}} = \frac{\text{Reading of voltmeter } V_1}{\text{Reading of voltmeter } V_2}$$

(since S_1 and S_2 have *equal* number of turns)

The variation of form factor is obtained by adjusting the values of series resistance (R) and inductance (L).

Wattmeter method for testing ring specimens :

Since sheet steel in the form of rings is used for making current transformers and also spiral cores are used for construction of current transformers and variacs, therefore it becomes necessary to measure iron loss in ring specimens. The methods employed and experimental set up is the same as that for strips.

Fig. 11.29 shows the connection diagram for measurement :

— S_1, S_2 are the secondary windings;

— Voltmeter V_1 reads the *r.m.s. value;*

— Voltmeter V_2 reads the *average value* of voltage.

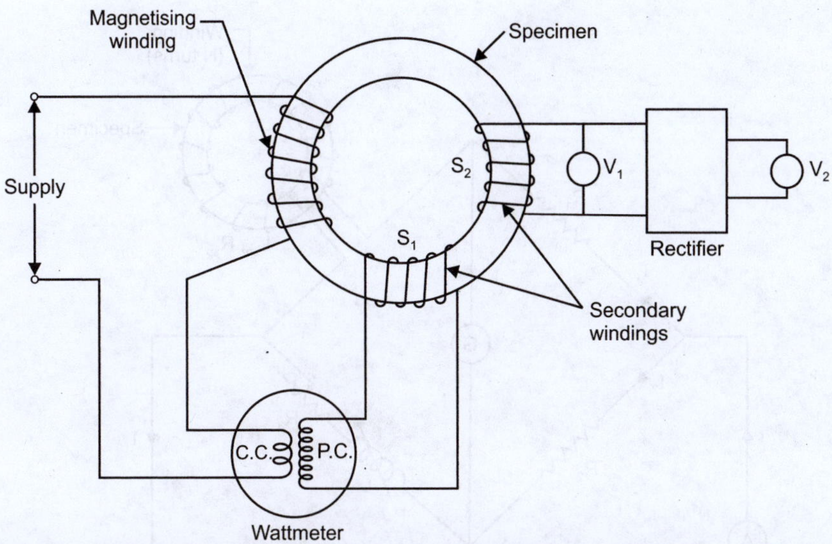

Fig. 11.29. Wattmeter method testing of ring specimen.

- The wattmeter method described above is *not sufficiently sensitive for testing sheet materials working at low flux densities frequently employed in communication equipment.*

11.6.2. A.C. Bridge Methods

— The A.C. bridge network can be used for the measurement of iron loss and effective permeability of magnetic samples.

— Such methods are *very useful when the available samples are small, and when the test is to be carried out at commercial or audio frequencies, and with fairly low values of flux density.*

1. Maxwell bridge :

Fig. 11.30 shows the connections of the Maxwell bridge as applied to *iron loss and permeability measurements :*

— R_1, R_2 and R_3 (variable) are resistors. In series with R_3, in the same arm, is a variable inductor L of resistance r. It may be necessary to connect R_3 in series with the winding on the specimen if the resistance of the latter is small.

— The specimen, in ring form, is wound with a winding whose inductance is L_{sp} and effective resistance R_{sp}, this effective resistance containing an iron-loss component. R_w is the actual resistance of the winding on the ring.

— G is the vibration galvanometer or telephone, and A an ammeter.

— The supply is from an alternator having a pure sine waveform.

— The bridge is balanced by adjustment of L and R_3.

Theory :

When the bridge is balanced :

$$I_1R_1 = I_2R_2$$

and, $$I_1(R_{sp} + j\omega L_{sp}) = I_2[(r + R_3) + j\omega L]$$

where, $$\omega = 2\pi f$$

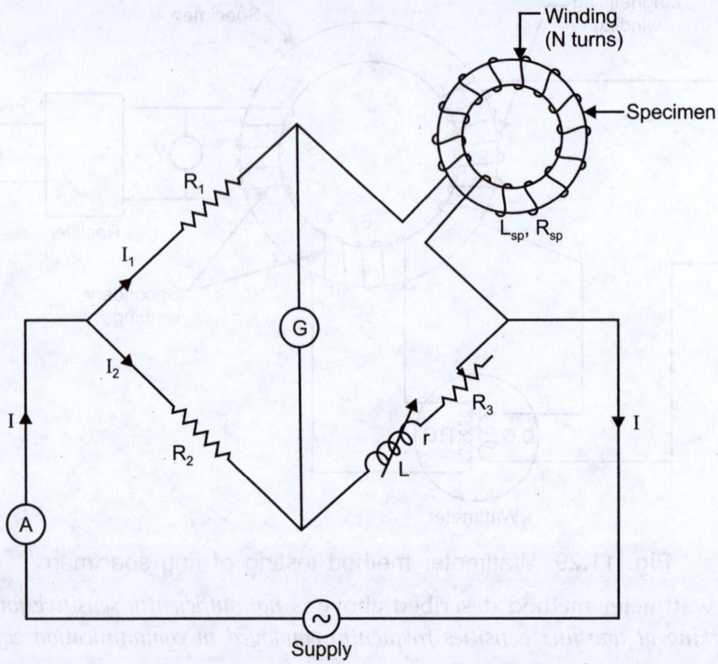

Fig. 11.30. Maxwell bridge for iron loss measurements.

Then, $$\frac{I_1}{I_2} = \frac{R_2}{R_1} \qquad ...(i)$$

$$I_1 R_{sp} = I_2(r + R_3) \qquad ...(ii)$$

and, $$I_1 L_{sp} = I_2 L \qquad ...(iii)$$

From (*i*) and (*ii*), $$\frac{R_2}{R_1} = \frac{r + R_3}{R_{sp}}$$

or, $$R_{sp} = (r + R_3)\frac{R_1}{R_2} \qquad ...(iv)$$

From (*i*) and (*iii*), $$L_{sp} = L \cdot \frac{R_1}{R_2} \qquad ...(v)$$

- The *iron loss* (P_i) *in the sample* is given by :

$$\text{Iron loss, } P_i = I_1^2 R_{sp} - I_1^2 R_w \text{ or } I_1^2(R_{sp} - R_w) \qquad ...(vi)$$

Now, current $I = I_1 + I_2$ (I_1 and I_2 being in phase)

Thus, $$I = I_1 + \frac{R_1}{R_2} I_1 = I_1\left(\frac{R_1 + R_2}{R_2}\right)$$

or, $$I_1 = I \cdot \left(\frac{R_2}{R_1 + R_2}\right) \qquad ...(vii)$$

Hence, the iron loss is given by :

$$P_i = I^2 \left(\frac{R_2}{R_1 + R_2}\right)^2 (R_{sp} - R_w) \qquad ...(11.34)$$

[Using eqn. (*vi*) and (*vii*)]

Let, N = Number of turns on the specimen,

l = Length of mean circumference of specimen, m,

a = Cross-section of specimen, m^2, and

μ_r = Relative permeability of specimen.

Then, the inductance is given as :

$$L_{sp} = \frac{NI_1}{l/(a\mu_0\mu_r)} \times \frac{N}{I_1} = \frac{N^2 a}{l}\mu_0\mu_r \text{ henry} \qquad ...(11.35)$$

From which μ_r *may be calculated when* L_{sp} *has been measured.*

2. Campbell bridge method :

This is one of the best A.C. bridges used for the measurement of iron loss.

Fig. 11.31(*a*) shows the connections for campbell bridge method :

— The ring specimen carries two windings, a primary and a secondary, having N_1 and N_2 turns respectively, and having the same cross-sectional area.

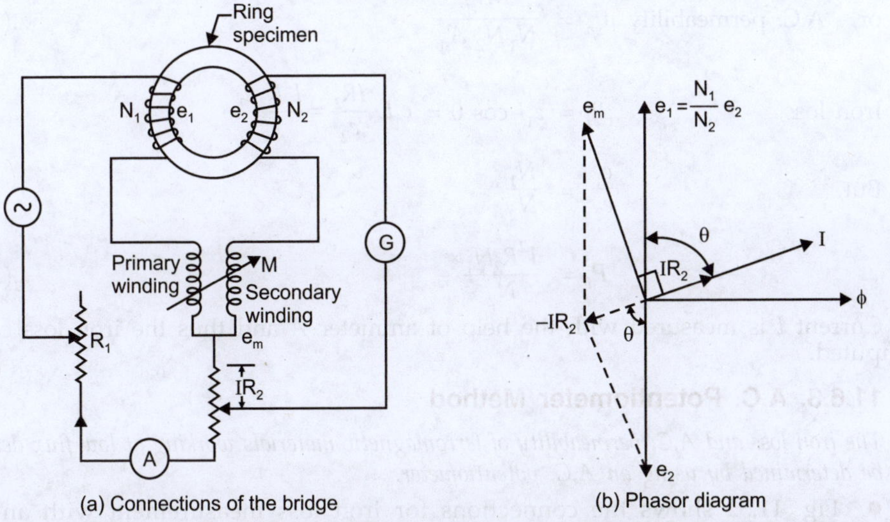

Fig. 11.31. Campbell bridge method.

— M is a variable mutual inductor connected as shown. R_1 and R_2 are variable resistors, and A is an ammeter. G is a vibration galvanometer or telephone according to the supply frequency. The waveform of the supply voltage should be sinusoidal, and R_1 should be made *sufficiently large* to ensure that current waveform is also *sinusoidal.*

— M and R_2 are adjusted until the galvanometer shows no deflection or until the telephone gives minimum sound.

- *Under balance conditions* the phasor diagram is as shown in Fig. 11.31(*b*) :

— ϕ is the flux in the specimen, and I the current in the primary winding, leading the flux by a small angle on account of iron loss.

— e_1, e_2, and e_m are the *induced voltages* in the primary winding, the secondary winding, and the secondary of the mutual inductor M respectively. Balance is obtained when the vector sum e_m and e_2 is equal, and opposite, to the vector IR_2 representing the voltage drop in resistor R_2.

Let M' = Mutual inductance between the primary and secondary windings of the specimen,

$$e_2 = \omega M'I, \text{ and}$$

$$e_m = \omega MI.$$

Since IR_2 is very small, we can write,

$$e_2 \simeq e_m$$

or, $$\omega M'I = \omega MI \quad \text{or} \quad M' \simeq M$$

Flux in the specimen, $$\phi = \frac{N_1 I}{l_{sp}/(\mu_{sp} A_{sp})} = \frac{N_1 I \mu_{sp} A_{sp}}{l_{sp}}$$

Mutual inductance between primary and secondary windings of specimen is :

$$M = \frac{N_2 \phi}{I} = \frac{N_1 N_2 \mu_{sp} A_{sp}}{l_{sp}}$$

or, A.C. permeability $$\mu_{sp} = \frac{M l_{sp}}{N_1 N_2 A_{sp}} \quad ...(11.36)$$

Iron loss, $$P_i = e_1 I \cos\theta = e_1 I \cdot \frac{IR_2}{e_2} = \frac{I^2 R_2 e_1}{e_2}$$

But, $$\frac{e_1}{e_2} = \frac{N_1}{N_2}$$

$\therefore$ $$P_i = \frac{I^2 R_2 N_1}{N_2} \quad ...(11.37)$$

Current I is measured with the help of ammeter A and thus the iron loss can be computed.

11.6.3. A.C. Potentiometer Method

The iron loss and A.C. permeability of ferromagnetic materials working at low flux densities can be determined by using an A.C. potentiometer.

- Fig. 11.32 shows the connections for iron loss measurement with an A.C. potentiometer.
  - — The ring specimen carries two windings. The primary windings has N_1 turns and the secondary N_2 turns, the supply to the former being through a regulating transformer, from an alternator (having sinusoidal voltage waveform) which also supplies current to the two potentiometer slide-wire circuits.
  - — A variable resistor R and a standard resistor are connected in series with the primary winding, the latter being for the purpose of measuring the current in the primary winding by measuring the magnitude and phase of the voltage drop across it.
  - — G is a vibration galvanometer.
- The flux density B_{max} in the specimen is obtained from the expression :

$$B_{max} = \frac{E_2}{4K_f \cdot A \cdot f \cdot N_2}$$

where, E_2 = E.m.f. (r.m.s. value) induced in the secondary winding,

K_f = Form factor = 1.11 for sinusoidal supply,

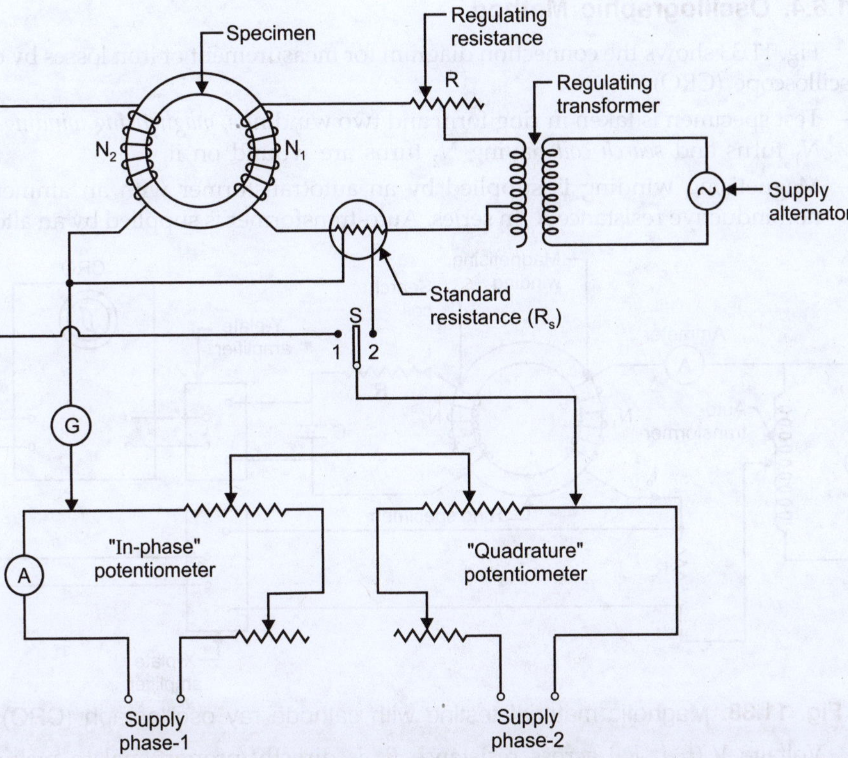

Fig. 11.32. Iron-loss measurement by A.C. potentiometer

A = Cross-sectional area of the specimen,

f = Supply frequency, and

N_2 = Number of turns in the secondary winding.

— The potentiometer is standardised first, and then the supply is given to the primary winding of the specimen.

— The voltage E_2 is measured by moving the switch S on to contact 1, setting the *"quadrature"* potentiometer at *zero*, and adjusting the *"in-phase"* potentiometer until the vibration galvanometer shows no deflection. The setting of the "in-phase" potentiometer for balance then gives the value of voltage E_2 directly.

— The switch S is then thrown over to contact 2, and both the "in-phase' and "quadrature" potentiometers are adjusted to give zero galvanometer deflection. The reading of the *"in-phase"* potentiometer then gives the value $I_e R_s$, where I_e is the loss component of the current in the primary winding and R_s is the value of the standard resistance (I_e = reading of in-phase potentiometer/R_s). The reading of the *"quadrature"* potentiometer gives the value $I_m R_s$ where I_m is the magnetising component of the primary current (I_m = reading of quadrature potentiometer/R_s).

$$\text{Iron loss} = I_e \cdot E_1 = I_e \cdot E_2 \cdot \frac{N_1}{N_2}$$

- This method is *very suitable for A.C. magnetic testing at low flux densities since magnetising and iron loss components of exciting current are measured separately.*

11.6.4. Oscillographic Method

- Fig. 11.33 shows the connection diagram for measurement of iron losses by cathode ray oscilloscope (CRO) :
  — Test specimen is taken in ring form and two windings, *magnetising winding* having N_1 turns and *search coil* having N_2 turns are wound on it.
  — Magnetising winding is supplied by an autotransformer with an ammeter and non-inductive resistance R_p in series. Auto-transformer is supplied by an alternator.

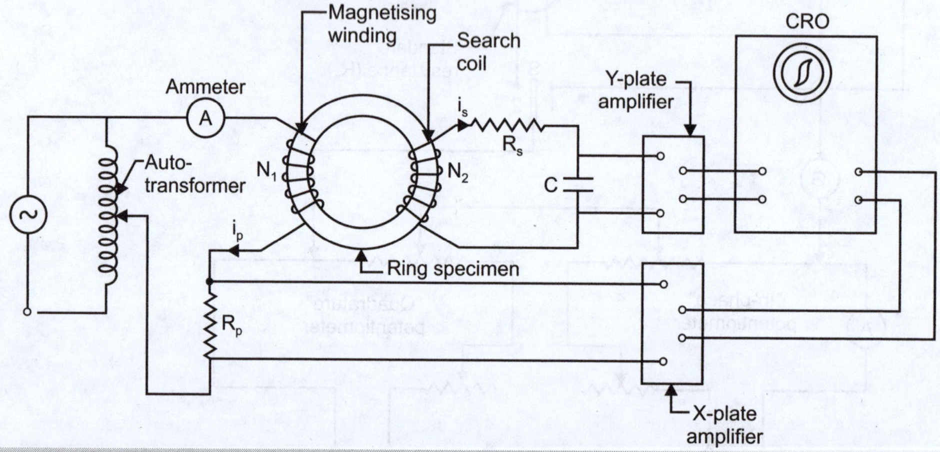

Fig. 11.33. Magnetic material testing with cathode ray oscillograph (CRO).

- Voltage $V_p(= i_p\ R_p)$ across resistance R_p is directly proportional to magnetising current i_p and, consequently, to the strength of magnetic field in the test specimen. This voltage V_p is applied to the *X-plates* of the CRO through an amplifier.

Output of search coil is supplied to a capacitor C with a high resistance R_s in series. The e.m.f. induced in the search coil,

$$e_{sc} = N_2 \cdot \frac{d\phi}{dt} = N_2 \cdot A_{sc} \frac{dB}{dt}$$

when resistance R_s is higher than that of the remaining section of the circuit, the current in the measuring circuit is :

$$i_s = \frac{e_{sc}}{R_s} = N_2 \cdot \left(\frac{A_{sc}}{R_s}\right)\left(\frac{dB}{dt}\right)$$

The voltage across capacitor C is :

$$V_c = \frac{1}{C}\int i_s\, dt = \frac{1}{C}\frac{N_2 A_{sp}}{R_s}\int dB = \frac{1}{C}\frac{N_2 A_{sp}}{R_s} B \quad ...(11.38)$$

Therefore voltage V_c across the capacitor C is proportional to flux density in the test specimen. This voltage V_c is applied to *Y-plates* of the CRO through an amplifier.

The *X-deflection* is proportional to the voltage drop across R_p carrying the magnetising current, the deflection is therefore proportional to the *magnetising force H.*

The *Y-deflection* is proportional to the voltage V_c and therefore to the *flux density B* in the specimen.

(*i*) ***Plotting of hysteresis loop.*** According to the arrangement shown in the figure, the X-deflection is proportional to H and the Y-deflection is proportional to B. Thus the figure traced on the screen will be the *hysteresis loop* of the material.

(*ii*) ***Plotting B-H curve.*** When the peak flux density in the specimen is slowly and continuously varied (*e.g.*, by varying the excition of the alternator supplying the magnetising current) the locus of the tips of the hysteresis loops gives the *B-H curve* for the material.

(*iii*) ***Determination of hysteresis loop.*** The hysteresis loss for the material may be estimated :

(*a*) From the area of the loop either by comparison with a loop for a standard specimen, or,

(*b*) By calculation from the *X* and *Y* sensitivities of the CRO and the circuit constants. This method, however is *very tedious*.

11.7 METHODS OF MEASUREMENT OF AIR GAP FLUX AND FIELD STRENGTH

11.7.1. Measurement of Air Gap Flux

Some of the suitable methods for the measurement of *air gap flux* are :

1. Deflection of a pivoted magnetic needle—The *classical magnetometer method.*
2. The rapid rotation of search coil through 180°—*e.g., the classical earth conductor.*
3. Continuous rotation of search coil fitted with a commutator to give a D.C. output.
4. Continuous rotation of search coil fitted with slip rings to give an A.C. output.
5. The force on a currentcarrying conductor in the field.
6. The effect of the field on the magnetisation of a small piece of soft-iron excited by A.C. in a magnetising winding.

11.7.2. Measurement of Field Strength

The following methods of measurement of field strength will be described :

1. By withdrawal of search coil.
2. By nuclear magnetic resonance.
3. Use of bismuth spiral.
4. Use of mumetal wires.
5. Use of Hall effect devices.

11.7.2.1. By withdrawal of search coil

— In this method a search coil is placed in a magnetic field. The terminals of the search coil are connected to a ballistic galvanometer.

— This search coil is suddenly withdrawn from the magnetic field, as a consequence the flux linking with the search coil changes. This induces an e.m.f. in the search coil which causes the galvanometer to deflect. The flux density in the field may be known from the throw of the galvanometer.

11.7.2.2. By nuclear magnetic resonance

— When a high frequency alternating magnetic field is superimposed on atomic nuclei located in a constant magnetic field, resonance absorption of energy from the high-frequency field is liable to occur at a specific ratio of the constant field strength *H* to frequency *f* of the alternating field.

— According to laws of quantum mechanics the magnetic moment of, say, the hydrogen nucleus (proton) may be oriented either in the same direction as that of external magnetic field or in *opposition to it*.

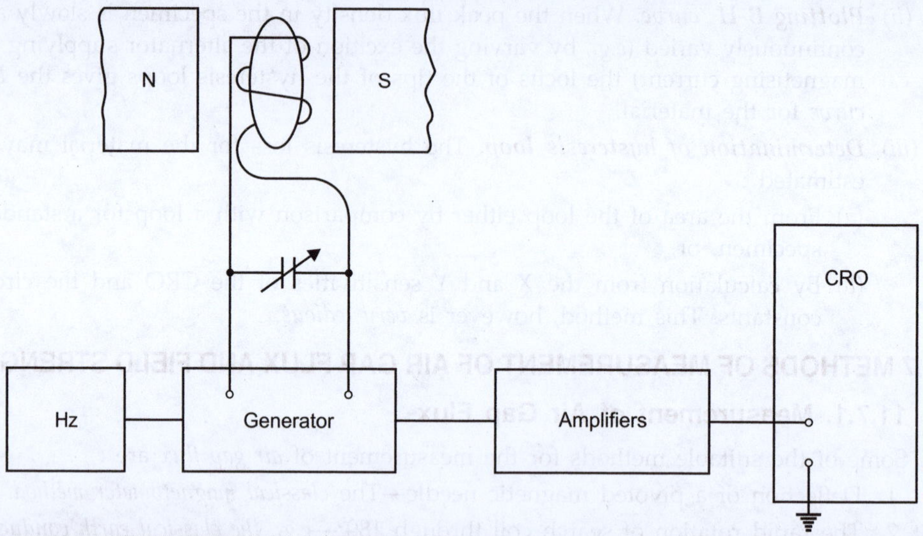

Fig. 11.34. Measurement of field strength by nuclear magnetic resonance.

— The orientation of the proton magnetic moment can be changed by application of an energy quantum hf, where h is Planck's constant and f is the frequency. The difference between these two energy states of the proton amounts to $2m_pH$ where m_p is the proton magnetic moment.

Thus, $$hf = 2m_pH$$

or, $$f = \frac{2m_pH}{h}$$

Now, $$\omega = 2\pi f = 2\pi \times \left(\frac{2m_p H}{h}\right) = \frac{4\pi m_p H}{h} = \gamma_p H$$

or, $$H = \frac{2\pi f}{\gamma_p} \quad ...(11.39)$$

where, γ_p = *"Gyromagnetic ratio" of the nucleus i.e., the ratio of the magnetic moment to the mechanical moment.*

The value of γ_p of a number of substances is known to the nearest thousandth of a percent.

According to this method the magnetic strength of a constant field is determined by measuring the frequency at which nuclear magnetic resonance occurs.

- **Fig. 11.34** shows the block diagram of the test installation for the measurement of field strength by nuclear magnetic resonance :

— A proton sample, say, an aqueous solution of $Fe(NO_3)_3$ contained in a glass ampule is surrounded by a small coil and is inserted in the magnetic field to be measured, the axis of the coil being at right angles to the field.

— The coil and capacitor constitute an oscillatory circuit, which is connected to a high frequency oscillator.

— *Resonance is detected by a sharp change in the amplitude of the resonance curve traced on the screen of CRO at a change in the oscillator frequency.*

The resonance frequency is measured by means of a *frequency meter or wavemeter.*

- This method is *most suitable for measurement of fields having a magnetic strength of 800 to 1.6 × 10^6 AT/m*. The *accuracy* of measurement lies within *±0.01%*.

11.7.2.3. Use of bismuth spiral

— It has been observed that *resistance of bismuth wire increases when it is placed in strong magnetic field*. This property of bismuth wire is used in instruments employed for measuring flux density of magnetic field.

— Bismuth wire is taken in spiral form and its resistance, under the influence of magnetic field is measured with a Wheatstone bridge, as shown in Fig. 11.35.

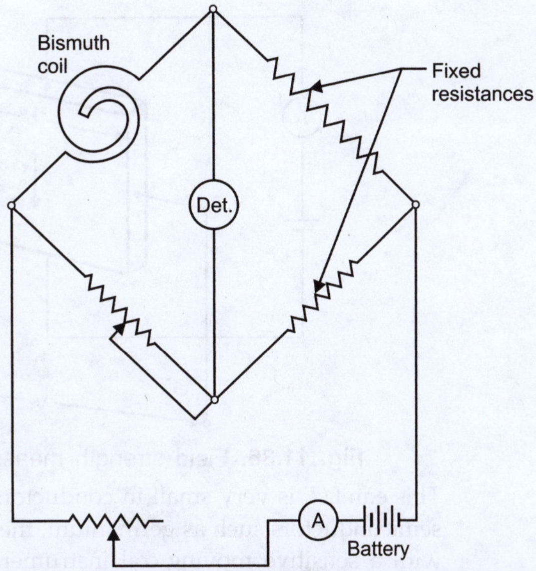

Fig. 11.35. Measurement of flux density with bismuth spiral.

— The change in the value of resistance is converted into the flux density *by calibration*.

Advantages :

(*i*) It is simple to use.

(*ii*) It takes a small area, being in spiral form, for flux measurement.

(*iii*) With this method, even flux density in air gap between poles and armature of D.C. machine can be measured.

Disadvantage. The change in resistance under influence of magnetic field is very small and a large change takes place with even small change in temperature.

11.7.2.4. Use of mumetal wires

— This method is based on the principle that *impedance of mumetal wire changes when it is kept in magnetic field.*

— The change in impedance is measured with A.C. bridges and *with proper calibration, change in impedance is converted into the value of magnetic field strength.*

— This method is free from disadvantages associated with bismuth spiral method.

- *Because of high sensitivity it is used in modern instrumentation.*

11.7.2.5. Use of Hall effect devices

- When a current carrying conductor is placed in a magnetic field a *transverse effect* (discovered by Hall in 1879) is noted. This called the *Hall effect.*

— Hall found that : *"When a magnetic field is applied at right angles to the direction of the electric current, an electric field is set up which is perpendicular to both the direction of electric current and the applied magnetic field."*

In other words : *"When any specimen carrying a current I is placed in the transverse magnetic field B, then an electric field E is induced in the specimen in the direction perpendicular to both I and B (the phenomenon is known as the Hall effect)."*

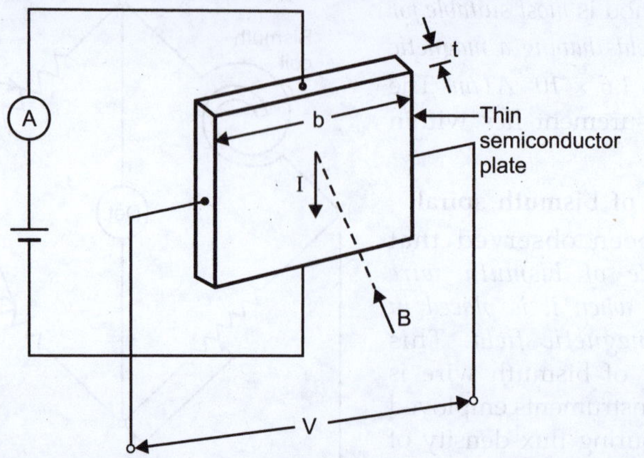

Fig. 11.36. Field strength measurement using Hall effect.

This e.m.f. E is very small in conductors and is difficult to measure, but in some semiconductors such as germanium, the e.m.f. is sufficiently large to be measured with a sensitive moving-coil instrument.

— This phenomenon now forms the *basis* of some simple and elegant field measuring instruments.

- Consider the thin plate of semiconductor illustrated in Fig. 11.36.

The current is made up from electrons and positive holes travelling in opposite directions, and since current is rate of flow of charge, it can be expressed as :

$$I = N.q.v.b.t \quad ...(11.40)$$

where, N = Number of charge carriers per unit volume,

q = Charge on each charge carrier,

v = Average velocity of the charge carriers, and

b, t = Breadth and thickness of the semiconductor plate respectively.

Now, if there is a magnetic field, of flux density B, crossing the thin plate, there will be a force on each charge, given by the expression :

$$F_B = q.v.B \quad ..(11.41)$$

where, B is flux density of the magnetic field.

This force F_B will deflect the carriers to one side in accordance with Fleming's left hand rule. Thus, there is tendency for charge to *accumulate* on one side of the semiconductor, but *existence of this charge establishes on electric field which, in turn, exercises an opposing force on the carriers*. Finally a state of equilibrium is reached when the force due to the magnetic field is equal to that due to electric field and there is then no accumulation. The polarity of the electric field depends upon whether conduction is due to negative electrons or positive holes of the charge.

If the electric-field strength is E, then the force on the charge carriers is given by :

$$F_E = q \cdot E \quad ...(11.42)$$

Then, *at equilibrium*,

$$F_E = F_B$$

or, $$q{\cdot}E = q{\cdot}v{\cdot}B$$

or, $$E = v{\cdot}B$$

or, $$v = \frac{E}{B}$$

Substituting this value of v in eqn. (11.32), we get

$$I = Nq \times \frac{E}{B} \times b \times t = Nq\, bt \cdot \frac{E}{B}$$

or, $$E = \frac{IB}{Nqbt} \quad ...(11.43)$$

Now, since the electric field is uniform throughout the breadth of semiconductor, so potential difference, V at the opposite edge of semiconductor,

$$V = Eb$$

or, $$V = \frac{IB}{Nqbt} \times b = \frac{IB}{Nqt} \quad ...(11.44)$$

The term $\frac{1}{Nq}$ is called the *Hall coefficient* of material. It can been seen from eqn. (11.44) that for a *given current* the *voltage* appearing at the opposite edges of semiconductor is *directly proportional to the flux density* of given magnetic field.

11.8 TESTING OF PERMANENT MAGNETS

The method of testing permanent magnet depends upon the shape of the magnet which in turn varies greatly according to the purpose for which the magnet is to be used.

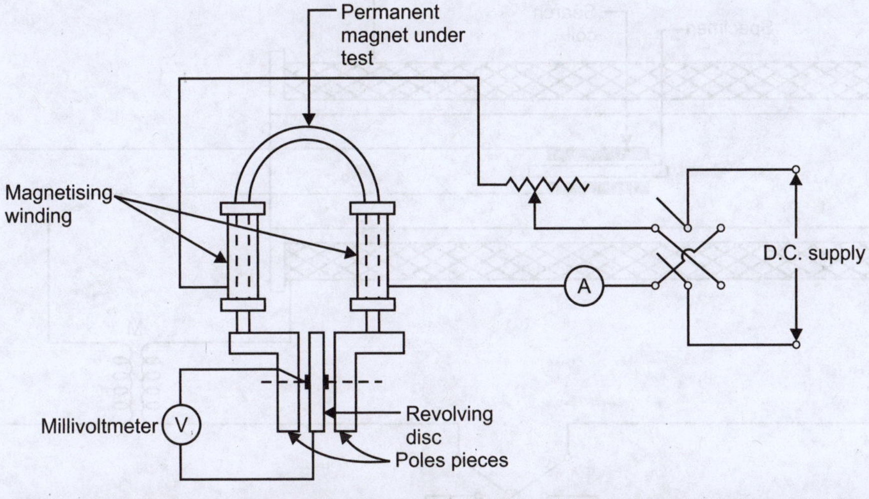

Fig. 11.37. Testing permanent magnet by batteridge apparatus.

Because of the self-demagnetising force in permanent magnets, it is necessary, in testing such magnets, *to measure H by means of search coils, of small cross-section, laid flat against the surface of the magnet.* These search coils give the value of the air flux density at the magnet surface at different points, and thus measure H for these points. The flux density B is measured by search coils in the ordinary way.

Fig. 11.37 shows ***Batteridge apparatus*** for magnet testing :

— In this method of magnet testing, the magnet to be tested is placed with its straight ends inside two magnetising coils and so that its ends press against two poles pieces, in the air gap between which a thin iron disc, plated with copper is mounted.

— The disc is mounted in ball bearings and clearances between it and the pole pieces are small. It is driven round at constant speed by a small motor, and has two small brushes making contact with its spindle and its rim.

Working. When a current flows in the magnetising windings, flux crosses the gap between the pole pieces, and the flux density in the gap will be proportional to that in the magnet. An e.m.f. will be induced in the revolving disc, this e.m.f. being proportional to the gap flux density and being measured by the millivoltmeter *V*. Thus the voltmeter reading gives a measure of flux density in the specimen, while *H* is obtained by the usual formula from the constants of magnetising coils and from the current in these coils.

Some leakage exists between the two arms of the magnet, but with proper design of apparatus its effect upon the results is small.

11.9 TESTING OF FEEBLY MAGNETIC MATERIALS

Fig. 11.38 shows the apparatus employed for measurements upon material which is distinctly magnetic, but whose permeability is low (of the order of *unity*) :

— The magnetising winding has a search coil at its centre, this search coil being fixed so that the specimen can be slipped inside it or withdrawn without disturbing the search coil itself.

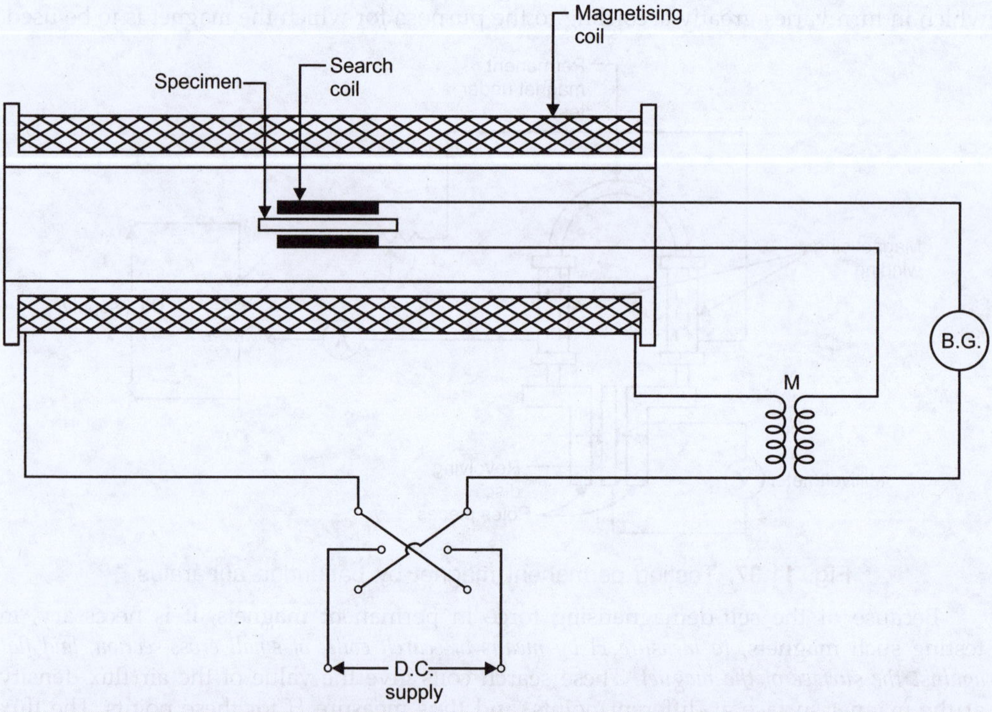

Fig. 11.38. Apparatus for testing feebly magnetic materials.

— The search coil is connected through a ballistic galvanometer (B.G.), to the secondary of a mutual inductor, the primary of which is connected in the magnetising circuit.

— As the material under test is only feebly magnetic, it is assumed that there is no demagnetisation due to end effect.

Testing procedure :

— Current is passed through the magnetising coil to give the required field strength H, which is given by $H = \frac{NI}{l}$ (NI being the number of ampere-turns on the magnetising coil and l its axial length), since the demagnetising effect of the ends of the specimen is negligible.

— Before placing the specimen inside the search coil the mutual inductance M is adjusted to give no galvanometer throw when the magnetising current is reversed. The specimen is then inserted, and the galvanometer throw upon reversal of the current observed. The flux density (B_f) obtained from this throw will be that due to the ferric induction in the specimen.

When the field strength is H, the total flux density,

$$B = B_f + \mu_0 \cdot H,$$

and, for the relative permeability μ_r, we have

$$\mu_0 \mu_r H = B_f + \mu_0 H \quad \text{for the relative permeability } \mu_r$$

or, *Relative permeability,* $$\mu_r = \frac{B_f + \mu_0 H}{\mu_0 H} = \left(\frac{B_f}{\mu_0 H} + 1\right) \quad ...(11.45)$$

11.10 PRECAUTIONS IN MAGNETIC TESTING

1. Where the inductance is measured by means of a stationary test coil surrounding the specimen, the space between the coil and the test specimen should be as small as possible. However, this space cannot be reduced to zero, and for precision work the necessary correction must be made. The test-coil area A can be determined by noting the deflection of a ballistic galvanometer when the test coil is placed in a standard solenoid and the current in the solenoid is reversed.
2. Before induction measurements are made, the *specimen should be carefully demagnetised.* This is best done by first magnetising to a value well above the maximum at which measurements will be made; the current is then gradually reduced to zero, being rapidly reversed meanwhile. Alternating current of 25 Hz frequency is convenient where much work is to be done.
3. In loss measurements, the temperature of the specimen should be carefully noted, because this affects the eddy current loss. The exciting winding should therefore, the *sufficiently large to avoid heating of the specimen.* In precision work, the apparatus is placed in an oil bath.
4. In tests of *sheet materials, the strips should not be too narrow,* because of the hardening at the edges due to cutting. This effect is negligible with a width of 5 cm. Care should be taken that burrs are removed from the edges and that the only insulation between sheets is the natural scale or oxide. The test specimen should be composed of strips cut from the sheets in both directions.
5. The readings of a ballistic galvanometer, should be kept at about the same magnitude by varying the resistance in series with it. The observational error is thus kept constant.

Example 11.8. *(a) What do you mean by eddy current losses ? Explain briefly.*

(b) The iron loss in the sample is 300 watts at 50 Hz with the eddy current component 5 times as big as the hysteresis loss component. At what frequency will the iron loss be doubled if the flux density is kept same ? **(Punjab University)**

Solution. (*a*) **Eddy current losses :**

- The term "eddy currents" is applied to those electric currents which circulate within a mass of conducting material when the latter is situated in a varying magnetic field. The conducting material may be considered as consisting of large number of closed conducting paths, each of which behaves like a short circuited winding. The varying magnetic field induces eddy e.m.fs. in these closed elemental paths giving rise to eddy currents. These eddy currents produce loss in power resulting in heating of materials. This loss is of considerable importance as it affects the efficiency and heating of electrical machines.
- *The eddy currents produce a magnetic field of their own which opposes the main magnetic field. As the effect of eddy currents is not uniform over the cross-section of the material, this results in a flux distribution which is not uniform, the flux density in the outer portions being greater than that at the centre. Thus there is a reduction in effective cross-section. The effect of eddy currents upon flux distribution is chiefly of importance in transformers where the material otherwise would be worked at a uniform flux density.*
- The magnetic materials used for varying magnetic fields are laminated (made up of thin sheets insulated from each other) so as to reduce eddy currents and associated losses, as by laminating, the *area of path of eddy currents is reduced giving rise to a large value of resistance.*

(*b*) *Given :* $P_i = 300$ W; $f = 50$ Hz; $P_e = 5\,P_h$;

Frequency, f':

Total iron loss, $P_i = P_h + P_e = 300$

or, $P_h + 5P_h = 300$

or, $P_h = 50$ W

and $P_e = 5\,P_h = 5 \times 50 = 250$ W

Let iron loss be *doubled* at supply frequency f'.

Since hysteresis loss is *proportional to the frequency*,

$\therefore$ New hysteresis loss, $P_h' = \dfrac{f'}{f} \times P_h = \dfrac{f'}{50} \times 50 = f'$ watts

Also, as eddy current loss various as *square of supply frequency*,

$\therefore$ New eddy current loss,

$$P_e' = \left(\frac{f'}{f}\right)^2 \times P_e$$

$$= \left(\frac{f'}{50}\right)^2 \times 250 = 0.1\,f'^2$$

Total iron loss, at new frequency f',

$$P_i' = P_h' + P_e' = 2P \qquad \text{...(Given)}$$

or, $f' + 0.1f'^2 = 2 \times 300 = 600$

or, $f'^2 + 10f' - 6000 = 0$

or, $$f' = \frac{-10 \pm \sqrt{10^2 + 4 \times 6000}}{2} = 72.6 \text{ Hz} \quad \textbf{(Ans.)}$$

Example 11.9. *(a) Derive an expression for eddy current loss in thin sheets.*

(b) Calculate the loss per kg in a specimen of alloy steel for a maximum flux density of 1.1 Wb/m² and a frequency of 50 Hz (sinusoidal), using 0.5 mm thick sheets. The resistivity of alloy steel is 30×10^{-8} Ωm. The density is 7800 kg/m³. Hysteresis loss in each cycle is 380 W-s/m³.

Solution. (*a*) **Eddy current loss in thin sheets :**

Fig. 11.39 shows a thin plate of thickness *t* and width *b*, thickness being *considerably smaller than the width.* Let us suppose that this sheet carries a flux, $B = B_{max} \sin \omega t$, the field running parallel to the axis of the sheet. Eddy currents would flow in the sheet in the elemental paths as shown in Fig. 11.39.

Let us consider 1 metre axial length and also 1 metre width of sheet. Take an elemental path of thickness *dx* at a distance *x* from the axis. Flux enclosed by 1 metre length of this path.

$$\phi = B(\text{flux density}) \times 2x \times 1 \text{ (area)} = 2xB$$
$$= 2xB \sin \omega t$$

Instantaneous values of eddy e.m.f. induced in the elemental path,

$$e_x = \frac{d\phi}{dt} = \frac{d}{dt}(2x B_{max} \sin \omega t)$$
$$= 2xB_{max} \cos \omega t = 2x\omega B_{max} \cos \omega t.$$

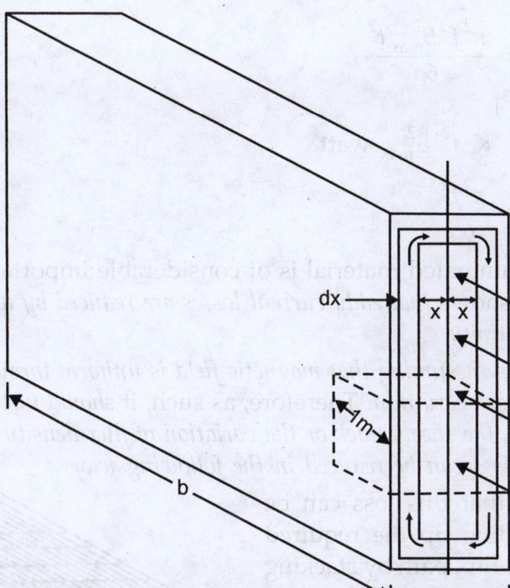

Fig. 11.39. Eddy currents in lamination.

R.M.S. value of eddy e.m.f. in the elemental path,

$$E_{r.m.s.} = \frac{1}{\sqrt{2}} \times E_{max}$$
$$= \frac{1}{\sqrt{2}} \times 2x\omega B_{max} \quad [\because E_{max} = 2x\omega B_{max} \text{ i.e., when } \cos \omega t = 1]$$
$$= \sqrt{2}\, x\omega B_{max} = \sqrt{2}x \times 2\pi f B_{max} = 2\sqrt{2}x\pi f B_{max}$$

If ρ is the resistivity of material, the resistance of each elemental path,

$$R_{xe} = \frac{2\times\rho}{1\times dx} = \frac{2\rho}{dx}$$

(Neglecting portions of path parallel to edges of plates).

∴ Eddy current in element path,

$$I_{xe} = \frac{E_{xe}}{R_{xe}} = \frac{2\sqrt{2}x\pi f B_{max}}{2\rho/dx} = \frac{\sqrt{2}\pi f B_{max} x dx}{\rho}$$

Eddy current loss in this path with 1 metre length and 1 metre width is

$$P_{xe} = I_{xe}^2 R_{xe} = \frac{2\pi^2 f^2 B_{max}^2 (xdx)^2}{\rho^2} \times \frac{2\rho}{dx}$$

$$= \frac{4\pi^2 f^2 B_{max}^2}{\rho} \cdot x^2\, dx$$

∴ Total eddy current loss in a plate 1 metre wide, 1 metre long and t metre thick is

$$= \int_{x=0}^{x=t/2} \frac{4\pi^2 f^2 B_{max}^2}{\rho} \cdot x^2\, dx = \frac{\pi^2 f^2 B_{max}^2}{6\rho} t^3 \text{ watt} \quad ...(i)$$

Eddy current loss *per unit volume*

$$= \frac{\pi^2 f^2 B_{max}^2\, t^2}{6\rho} \quad ...(ii)$$

(the volume being t m^3)

$$= K_e f^2 B_{max}^2 \text{ watt} \quad ...(iii)$$

where, $$K_e = \frac{\pi^2 t^2}{6\rho} \quad ...(iv)$$

- The thickness t of laminated material is of considerable importance as regards to eddy current loss. Eqn. (*ii*) shows *that eddy current losses are reduced by using thinner plates and a material of higher resistivity.*
- Eqn. (*ii*) is derived by *assuming that magnetic field is uniform throughout and parallel to the axis and also the sheets are very thin.* Therefore, as such, it *should **not** be applied to cases where frequency is **abnormal**, the sheets thick or the variation of flux density is **not sinusoidal** with time. The eddy current loss can be reduced in the following ways :*

1. It has been found that this loss can be minimised by building up the required cross-section for the flux path by stacking thin pieces known as *"laminations"* as shown in Fig. 11.40. Since the e.m.fs. set up in the material by the varying flux are usually of small magnitude, the natural oxide on the surface of the sheet iron or steel from which the laminations are punched will effectively insulate the laminations from one another, and thus limit each eddy current path to a single lamination.

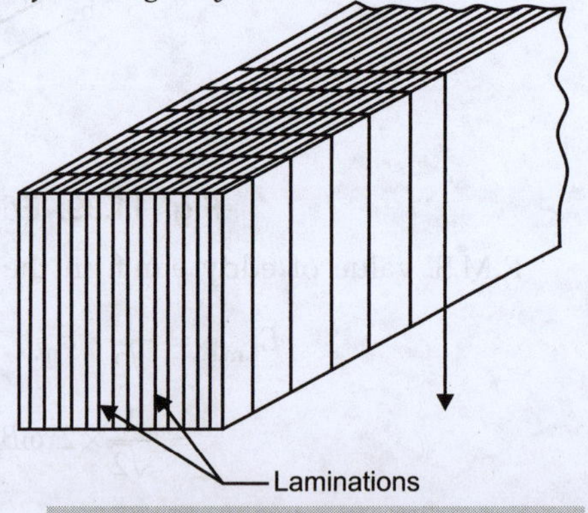

Fig. 11.40

2. This loss may also be reduced *by grinding the ferromagnetic material to a powder and mixing it with a binder that effectively insulates the particles from one another.* This mixture is then formed under pressure into the desired shape and heat treated. Magnetic cores for use of *communication equipment* are frequently made by this process.

- It may be noted that *iron losses* (hysteresis and eddy current losses) *if allowed to take place unchecked, not only reduce the efficiency of electrical equipment but also raise the temperature of the core.* Hence these losses should be kept *as small as is economically possible.*

(*b*) *Given :* B_{max} = 1.1 Wb/m^2; f = 50 Hz; t = 0.5 mm; ρ = 30 × 10^{-8} Ωm

Density = 7800 kg/m^3; P_h/cycle = 380 W-s/m^3.

Total iron loss :

Eddy current loss is given by

$$P_e = \frac{\pi^2 f^2 B_{max}^2 t^2}{6\rho} \text{ watt per m}^3$$

$$= \frac{\pi^2 f^2 B_{max}^2 t^2}{6\rho \times 7800} \text{ watt/kg}$$

$$= \frac{\pi^2 \times 50^2 \times 1.1^2 \times (0.5/1000)^2}{6 \times 30 \times 10^{-8} \times 7800} = 0.5316 \text{ W/kg}$$

$$\textit{Hysteresis loss, } P_h = \frac{380 \times 50}{7800} = 2.4359 \text{ W/kg at 50 Hz}$$

$$\textit{Total iron loss} = P_e + P_h = 0.5316 + 2.4539$$

$$= \mathbf{2.9855\ W/kg}$$ **(Ans.)**

Example 11.10. *(a) Discuss briefly the effect of the following factors on core loss : (i) Rivets and bolts; (ii) Burrs; (iii) Pressure.*

(b) Specimen of cold rolled grain oriented 0.3 mm thick stampings has a resistivity of 4.8 × 10^{-7} Ωm. The published hysteresis loop is essentially rectangular in form, with a coercive force of 12 AT/m for all values of peak flux densities upto 1.6 Wb/m^2. The manufacturer quotes the iron loss in material as 1.2 W/kg, with a sinusoidal flux density of 1.0 Wb/m^2 (peak) at 100 Hz. Calculate the loss in the material from its properties and compare it with the quoted value. The density is 7650 kg/m^3.

Solution. (*a*) The effects of rivets and bolts, burrs and pressure on core loss are described below :

(*i*) ***Rivets and bolts :*** Laminations must be held together. A common method is to use rivets or bolts, but they *tend to short circuit lamination insulation.* Rivets and bolts very often form part of a closed magnetic loop which, under A.C. operation, may *cause a serious loss of electrical energy.*

(*ii*) ***Burrs :*** As dies become worn, the burrs that they form on punched edges of lamination stock become larger and larger. Thus burrs are not always removed by deburring operation, and they *cut through sheet insulation forming contacts between laminations and allow excessive eddy currents to flow.* These are commonly referred to as *intersheet eddy currents.* Machining operations that are performed on stacked laminations, such as the grinding of induction motor stators or broaching of solenoid faces, often smear the laminations together, forming good connections between sheets. Many schemes, such as treating the faces with acid or alkali, have been tried in an attempt to eat away the metal particles after laminations have been assembled into apparatus, but this attempted cure often causes other troubles, such as rusting. Burning of the burrs with a flame has also been attempted, with varying success.

(iii) ***Pressure :*** Mechanical strains set up in magnetic materials either from *cold working during rolling or punching or distortion during final assembly operations,* affect magnetic properties in several ways. The hysteresis and eddy current losses as well as the magnetising current are increased. *The increased hysteresis loss due to punching depends upon the material, its thickness, and the configuration of the grains.* The loss may be as high as 300 percent of the loss in the unpunched sheet. The strains produced by punching operations increase the loss primarily in the narrow section next to the punched edge. The percent increase in hysteresis loss is much lower for high flux densities than for low ones.

The *effect of tension on silicon steel is to increase the permeability until high flux densities are reached. The effect of compression, however, is much more marked than that of tension and reduced the permeability considerably.*

(b) Given : $t = 0.3$ mm; $\rho = 4.8 \times 10^{-7}$ Ωm; Density = 7650 kg/m^3

Total iron loss :

The material is subjected to a peak flux density of 1 Wb/m^2 and, therefore, the value of coercive force is 12 AT/m for all densities. The hysteresis loop is thus a rectangle with sides 2 × 12 = 24 AT/m and 2 × 1 = 2 Wb/m^2.

$$\text{Loss per cycle} = \text{Area of hysteresis loop} = 24 \times 2 = 48 \text{ W-s/m}^3$$

$$\text{Hysteresis loss/kg at 100 Hz} = \frac{48 \times 100}{7650} = 0.627 \text{ W/kg}$$

Eddy current loss,

$$P_e = \frac{\pi^2 f^2 B_{max}^2 t^2}{6\rho} \text{ W/m}^3 = \frac{\pi^2 f^2 B_{max}^2 t^2}{6\rho \times 7650} \text{ W/kg}$$

$$= \frac{\pi^2 \times (100)^2 \times (1.0)^2 \times (0.3/1000)^2}{6 \times 4.8 \times 10^{-7} \times 7650} = 0.403 \text{ W/kg}$$

Total iron loss,

$$P_i = P_h + P_e = 0.627 + 0.403$$

$$= \mathbf{1.03\ W/kg}$$ **(Ans.)**

The calculated loss is *smaller* than the quoted. **(Ans.)**

Example 11.11. *The measured values of iron loss of a magnetic specimen of weight 13 kg are 17.2 W and 28.9 W at 40 Hz and 60 Hz respectively at a constant peak flux density. Determine the values of hysteresis and eddy current losses in W/kg at 50 Hz for the same value of flux density.* **(AMIE)**

Solution. *Given :* $W = 13$ kg; $P_i = 17.2$ W at 40 Hz; $P_i = 28.9$ W at 60 Hz.

P_h, P_e at f = 50 Hz

At a given flux density, $P_h \propto f$ and $P_e \propto f^2$

Let, $P_h = Af$, and $P_e = Bf^2$

where, A and B are constants.

Total iron loss, $P_i = P_h + P_e = Af + Bf^2$ per kg

Substituting P_i = 17.2 W when f = 40 Hz, and

P_i = 28.9 W when f = 60 Hz, we get:

$$\frac{17.2}{13} = 40A + (40)^2B \quad ...(i)$$

$$\frac{28.9}{13} = 60A + (60)^2B \quad ...(ii)$$

Solving expressions (*i*) and (*ii*), we get

$$A = 0.0254;\ B = 0.000194$$

∴ When f = 50 Hz,

$$P_h = Af = 0.0254 \times 50 = \mathbf{1.27\ W\ per\ kg}$$ **(Ans.)**

and, $$P_e = Bf^2 = 0.000194 \times (50)^2 = \mathbf{0.485\ W\ per\ kg}$$ **(Ans.)**

Example 11.12. *A laminated iron cylinder is rotated in a magnetic field. The iron loss is 500 W at 480 r.p.m. and 700 W at 600 r.p.m. Find the loss if the laminations were twice as thick, the induction density increased by 25% and the speed were 700 r.p.m. Take steinmetz coefficient as 1.6. Assume the waveform to be sinusoidal.*

Solution. *Given* : Iron loss at 480 r.p.m. = 500 W; Iron loss at 600 r.p.m. = 700 W; Steinmetz coefficient = 1.6

The frequency of pulsations is proportional to rotational speed.

Let N be the speed in r.p.m. and V_i be the volume of iron.

$$\text{Eddy current loss, } P_e = \frac{\pi^2 f^2 B_{max}^2\, t^2}{6\rho} V_i$$

$$= \frac{\pi^2 Z^2 N^2 B_{max}^2\, t^2}{6\rho} V_i = K_e' N^2 \quad ...(i)$$

(where $f = ZN$ and Z is constant)

$$\text{Hysteresis loss, } P_h = K_h f\, B_{max}^{1.6}\, V_i = K_h\, ZN\, B_{max}^{1.6}\, V_i = K_h' N \quad ...(ii)$$

where K_e' and K_h' are constants if B_{max} and t remain constant.

$$\therefore \quad \text{Total iron loss, } P_i = K_e' N^2 + K_h' N \quad ...(iii)$$

Substituting the value of P_i and N in the expression (*iii*), we get

$$500 = K_e' (480)^2 + K_h' (480)$$

or, $$1.04 = 480 K_e' + K_h' \quad ...(1)$$

$$700 = K_e'(600)^2 + K_h' (600)$$

or, $$1.166 = 600 K_e' + K_h' \quad ...(2)$$

Subtracting (1) from (2), we get

$$0.126 = 120 K_e'$$

$$\therefore \quad K_e' = 0.00105$$

and, $$K_h' = 0.536$$

Total loss with B_{max} increased by 25%, thickness increased to twice the original value and N = 700 r.p.m.,

$$P_i = 0.00105 \times (700)^2 \times (1.25)^2 \times 2^2 + 0.536 \times 700 \times (1.25)^{1.6}$$

$$= 3215.6 + 536.2 = \mathbf{3751.8\ W}$$ **(Ans.)**

Example 11.13. *In a power loss test on a 15 kg specimen of sheet steel laminations, the maximum flux density and wave form factor are maintained constant and the following results are obtained :*

Frequency (Hz) :	*25*	*40*	*50*	*60*	*80*
Total loss (watts) :	*18.5*	*36*	*50*	*66*	*104*

Calculate the eddy current and hysteresis loss per kg at 50 Hz. **(AMIE)**

Solution. At a given flux density and waveform factor, total iron losses (P_i) are given as :

$$P_i = Af + Bf^2$$

or, $$\frac{P_i}{f} = A + Bf$$

The value of $\frac{P_i}{f}$ for various value of f are :

f :	25	40	50	60	80
$\frac{P_i}{f}$:	0.74	0.9	1.0	1.1	1.3

A graph is plotted between $\frac{P_i}{f}$ and f is shown in Fig. 11.41.

From the graph $A = 0.5$, and, $B = 0.01$.

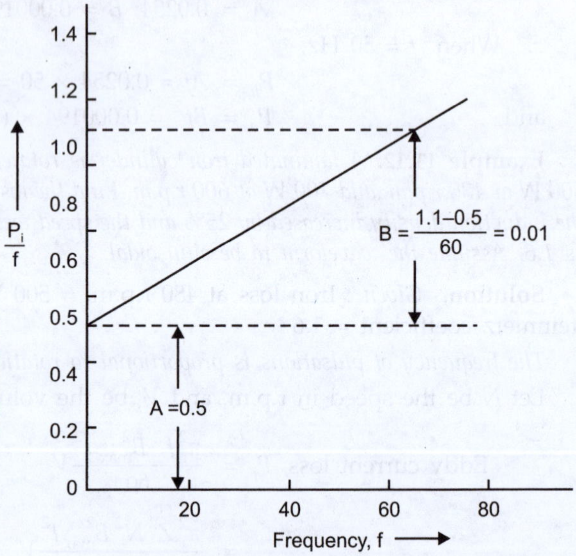

Fig. 11.41

$$\text{Eddy current loss at 50 Hz} = Bf^2 = 0.01 \times (50)^2 = 25 \text{ W}$$

$\therefore$ $$\textit{Eddy current loss per kg} = \frac{25}{15} = \mathbf{1.67\ W/kg} \text{ (Ans.)}$$

$$\text{Hysteresis loss at 50 Hz} = Af = 0.5 \times 50 = 25 \text{ W}$$

$\therefore$ $$\text{Hysteresis loss per kg} = \frac{25}{15} = \mathbf{1.67\ W/kg} \text{ (Ans.)}$$

Example 11.14. *The following data relate to a test conducted on a specimen of iron stampings weighing 12 kg by an Epstein square shown in Fig. 11.25 :*

Cross-sectional area of the specimen = $1700\ m^2$

Number of turns on each of the two windings S_1 *and* S_2 = 520

Resistance of each of the windings S_1 *and* S_2 = 45 Ω

Resistance of pressure coil of the wattmeter = 75,000 Ω

Frequency of sinusoidal voltage supply = 50 Hz

Current indicated by an ammeter in the primary circuit = 0.4 A

Voltage indicated by the voltmeter across a secondary winding = 250 V

Calculate the value of flux density in the specimen and iron loss/kg at this flux density if the wattmeter connected on the primary side indicates 85 W.

Solution. *Given* : Weight of specimen = 12 kg; A_{sp} = 1700 mm^2; N_2 =520; r_c = 45 Ω; r_p = 75,000 Ω; E = 250 V; P = 85 W.

Flux density in the specimen, B_{max} :

$$E = 4.44\, f\, B_{max}\, A_{sp} \times N_2$$

or,
$$B_{max} = \frac{E}{4.44\, f\, A_{sp} N_2} = \frac{250}{4.44 \times 50 \times 1700 \times 10^{-6} \times 520}$$

$$= \mathbf{1.27\ Wb/m^2}\ \textbf{(Ans.)}$$

Iron loss/kg, P_i :

Iron loss in stampings, $$P_i = P\left(1 + \frac{r_c}{r_p}\right) - \frac{E^2}{r_p + r_c}$$ ...[Eqn. (11.33)]

$$= 85\left(1 + \frac{45}{75,000}\right) - \frac{(250)^2}{75,000 + 45} = 84.22 \text{ W}$$

∴ *Iron loss/kg* $= \dfrac{84.22}{12}$ = **7.02 W/kg (Ans.)**

Example 11.15. *Calculate the percentage change in the hysteresis and eddy current losses of a choke coil if other factors remain constant :*

(*i*) *The form factor of the applied voltage decreases by 15%;*

(*ii*) *The frequency of the applied voltage increases by 15%.*

Neglect resistance of the choke coil.

Solution. Percentage change in P_h and P_e :

(*i*) *When the form factor of the applied voltage decreases by 75% :*

Since P_h is independent of form factor of supply voltage, so remains the same *i.e.*, percentage change in P_h *is zero* **(Ans.)**

P_e varies as square of the form factor of supply voltage,

$$\therefore \quad \frac{P_{e2}}{P_{e1}} = \left(\frac{K_{f2}}{K_{f1}}\right)^2 = (0.85)^2 = 0.7225$$

or, $P_{e2} = 0.7225\ P_{e1}$ (∵ Form factor decreases by 15%)

Hence, *percentage change in* P_e

$$= \frac{P_{e2} - P_{e1}}{P_{e1}} \times 100 = \frac{0.7225 P_{e2} - P_{e1}}{P_{e1}} \times 100$$

$$= \textbf{27.75\% } (\textit{decrease}) \textbf{ (Ans.)}$$

(*ii*) *When the frequency of the applied voltage increases by 15% :*

Now supply frequency, $f_2 = 1.15\, f_1$

Since, $P_h \propto f$

$$\therefore \quad \frac{P_{h2}}{P_{h1}} = \frac{f_2}{f_1} = 1.15$$

or, $P_{h2} = 1.15\ P_{h1}$

$\therefore$ *Percentage change in* $P_h = \dfrac{P_{h2} - P_{h1}}{P_{h1}} \times 100$

$= \dfrac{1.15\,P_{h1} - P_{h1}}{P_{h1}} \times 100 =$ **15%** *(increase)* **(Ans.)**

Since, $P_e \propto f^2$

$\therefore \quad \dfrac{P_{e2}}{P_{e1}} = \left(\dfrac{f_2}{f_1}\right)^2 = (1.15)^2 = 1.3225$

or, $P_{e2} = 1.3225\, P_{e1}$

$\therefore$ *Percentage change in* $P_e = \dfrac{P_{e2} - P_{e1}}{P_{e1}} \times 100$

$= \dfrac{1.3225\,P_{e1} - P_{e1}}{P_{e1}} \times 100 =$ **32.25%** *(increase)* **(Ans.)**

Example 11.16. *While testing a sample of sheet steel stampings at a frequency of 50 Hz the following results obtained :*

Volts	*4.9*	*69.3*	*91.8*	*100.5*	*110.5*	*118.0*
Amperes	*0.2*	*0.3*	*0.46*	*0.52*	*0.64*	*0.77*
Watts	*9.5*	*16.8*	*27.5*	*32.5*	*39.0*	*44.8*

Mean width of plates = 3 cm; Mean thickness = 0.0489 cm; Number of plates = 51; Total weight = 10 kg; Number of magnetising turns on coil = 600.

Allowing 4 watts for copper loss in the magnetising winding calculate the iron loss in watts per kg at a maximum flux density of 1 Wb/m² at a frequency of 50 Hz.

(Punjab University)

Solution. *Given :* b = 3 cm or 0.03 m; t = 0.0489 cm or 0.000489 m; N_p = 51; Total weight = 10 kg; N = 600; Copper loss = 4 watts; B_{max} = 1 Wb/m²; f = 50 Hz.

Iron loss/kg :

Mean area of plates,

$A_p = N_p \times b \times t = 51 \times 0.03 \times 0.000489 = 7.48 \times 10^{-4}\ \text{m}^2$

Induced voltage, $E = 4.44\, f\, \phi_{max}\, N$

$= 4.44 \times 50 \times (1 \times 7.48 \times 10^{-4}) \times 600 \simeq 100$ V

From the graph plotted (Fig. 11.42), corresponding to a voltage of 100 V, power supplied is 30.5 W.

Now, Total losses = Iron loss + copper loss

$\therefore$ Iron loss = Total loss – copper loss = 30.5 – 4 = 26.5 W

Hence, *Iron loss/kg* $= \dfrac{26.5}{10} =$ **2.65 W (Ans.)**

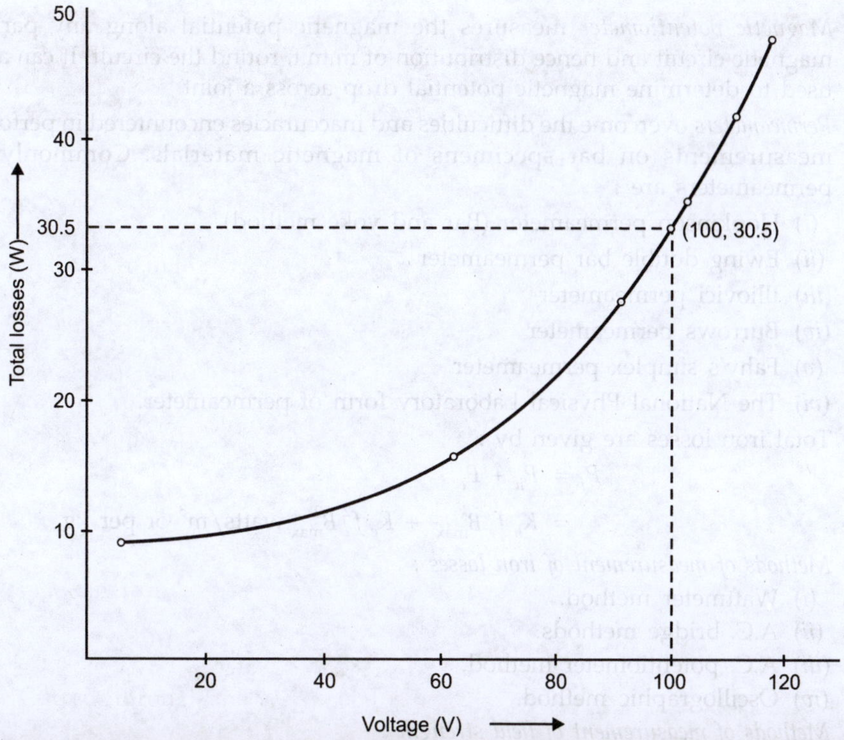

Fig. 11.42

Example 11.17. *An iron cored coil and a standard 2.5 Ω resistor are connected in series and the potential difference is measured across each on an A.C. potentiometer of co-ordinate type. The readings are 3.5 ±j 1.1 V for the coil and 4.6–j1.8 V for the resistor.*

Calculate the magnetising component and loss component of current through the coil.

Solution. I_e, I_m :

$$\text{Current through the coil, } I = \frac{4.6 - j1.8}{2.5} = 1.84 - j0.72 = 1.976 \angle -21.4°$$

$$\text{Voltage across the coil, } V = 3.5 + j1.1 = 3.67\angle 17.45°$$

$$\text{Angle between } V \text{ and } I, \ \phi = 17.45° - (-21.4°) = 38.85°$$

$$\text{Loss component of current, } I_e = I \cos \phi = 1.976 \cos 38.85°$$

$$= \mathbf{1.54 \ A} \ \textbf{(Ans.)}$$

Magnetising component of current,

$$I_m = I \sin \phi = 1.976 \sin 38.85°$$

$$= \mathbf{1.24 \ A} \ \textbf{(Ans.)}$$

HIGHLIGHTS

1. Measurements of various magnetic phenomena are called *magnetic measurements.*
2. *Types of tests* : (*i*) D.C. tests; (*ii*) A.C. tests; (*iii*) Steady state tests.
3. Ballistic (or D.C.) tests are used for :

   (*i*) Determination of flux/flux density; (*ii*) Determination of B.H curves; (*iii*) Plotting of hysteresis loop.

4. *Magnetic potentiometer* measures the magnetic potential along any part of a magnetic circuit and hence distribution of m.m.f. round the circuit. It can also be used to determine magnetic potential drop across a joint.
5. *Permeameters* overcome the difficulties and inaccuracies encountered in performing measurements on bar specimens of magnetic materials. Commonly used permeameters are :
   (*i*) Hopkinson permeameter (Bar and yoke method)
   (*ii*) Ewing double bar permeameter
   (*iii*) Illiovici permeameter
   (*iv*) Burrows permeameter
   (*v*) Fahy's simplex permeameter
   (*vi*) The National Physical Laboratory form of permeameter.
6. Total iron losses are given by :

$$P_i = P_h + P_e$$
$$= K_h f B_{max}^k + K_e f^2 B_{max}^2 \text{ watts/m}^3 \text{ or per kg.}$$

7. *Methods of measurement of iron losses :*
   (*i*) Wattmeter method.
   (*ii*) A.C. bridge methods.
   (*iii*) A.C. potentiometer method.
   (*iv*) Oscillographic method.
8. *Methods of measurement of field strength :*
   (*i*) By withdrawal of search coil.
   (*ii*) By nuclear magnetic resonance.
   (*iii*) Use of bismuth spiral.
   (*iv*) Use of mumetal wires.
   (*v*) Use of Hall effect devices.

OBJECTIVE TYPE QUESTIONS

Choose the Correct Answer :

1. Iron losses in a specimen of iron are determined by using a co-ordinate type of potentiometer. The loss component of no load current is read by :
   (*a*) In-phase potentiometer
   (*b*) Quadrature potentiometer
   (*c*) Either by in-phase or quadrature potentiometer
   (*d*) None of the above.
2. Magnetizing force is measured by Nuclear Magnetic Resonance method. The expression for magnetising force is
   (*a*) $H = f/\nu_p$ (*b*) $H = 2\pi f/h$
   (*c*) $H = fh/4\pi$ (*d*) $H = 2\pi f/\nu_p$
   f = frequency in Hz, h = Planck's constant and ν_p = gyrometric ratio.
3. The reason for using ring type specimens for ballistic tests is :
   (*a*) they are free from and effects.
   (*b*) they require less labour to prepare as compared with bar type specimens.
   (*c*) they can be used for large quantities of materials of very small permeability.
   (*d*) all of the above.

4. Ballistic tests are used in magnetic measurements for
(*a*) determination of *B-H* curve of the specimen only.
(*b*) determination of hysteresis loop of the specimen only.
(*c*) determination of flux density, magnetizing force and *B-H* curve and hysteresis loop of the specimen.
(*d*) finding out iron losses in the specimen.

5. In a Lloyd-Fisher square used for determination of iron loss in a specimen of iron laminations, the current coil of the wattmeter is connected in the primary circuit, while the pressure coil is connected in the secondary winding circuit. This is done so that
(*a*) the I^2R losses in the current coil are not included in wattmeter reading.
(*b*) the I^2R losses in the pressure coil circuit are not included in the wattmeter reading.
(*c*) the I^2R losses both in current coil and pressure coil circuit are not included in the wattmeter reading.
(*d*) the wattmeter reads only the iron losses in the specimen of laminations.

6. The true value of flux density in a ring specimen is :

(*a*) $B = B' + \mu_0 H\left(\frac{A_c}{A_s} - 1\right)$ (*b*) $B = B' - \mu_0 H\left(\frac{A_c}{A_s} - 1\right)$

(*c*) $B = B' + \mu_0 H\left(\frac{A_s}{A_c} - 1\right)$ (*d*) $B = B' - \mu_0 H\left(\frac{A_s}{A_c} - 1\right)$

where B' = measured value of flux density, H = magnetizing force, A_s = area of cross-section of specimen, A_c = area of cross-section of coil producing magnetizing force.

7. Burrow's permeameter is preferred to other permeameters as it
(*a*) incorporates arrangement for uniform distribution of mmf over the entire length of the specimen.
(*b*) incorporates arrangements for elimination of leakage flux at the joints.
(*c*) is very accurate and can serve as reference standard.
(*d*) all of the above.

8. In an Epstein square, used for finding the iron loss in a stack of laminations, the laminations are so arranged that
(*a*) plane of each lamination is perpendicular to the plane of the square.
(*b*) plane of each lamination is parallel to the plane of the square.
(*c*) the plane of laminations may be either parallel or perpendicular to the plane of the square.
(*d*) the plane of laminations may be at any angle to the plane of square.

9. Permanent magnets are tested by
(*a*) ballistic methods.
(*b*) using an electric circuit having a mutual inductance.
(*c*) potentiometric methods.
(*d*) Betteridge apparatus.

10. The iron losses when measured in a specimen of laminations at a constant maximum flux density of 1.3 Wb/m^2 are : 20 W at 40 Hz and 40 W at 60 Hz. What are the hysteresis and eddy current losses at 50 Hz and a flux density of 1.3 Wb/m^2.
(*a*) 10 W, 20 W (*b*) 7.38 W, 212.8 W
(*c*) 8.38 W, 20.8 W (*d*) 5 W, 25 W

ANSWERS

1. (*a*) **2.** (*d*) **3.** (*d*) **4.** (*c*) **5.** (*a*) **6.** (*b*) **7.** (*d*)
8. (*b*) **9.** (*c*) **10.** (*d*).

THEORETICAL QUESTIONS

1. Define the following terms relating to magnetic materials :
   (*i*) Magnetic field strength; (*ii*) Magnetic potential; (*iii*) Coercive force; (*iv*) Remanence; (*v*) Curie temperature.
2. How are magnetic materials classified ?
3. Give the classification of magnetic measurements.
4. Discuss the procedure for the determination of flux density in a ring specimen.
5. Explain the method of determination of magnetising force *H*.
6. Describe with the help of a neat diagram, a method for the determination of B-H curve of a magnetic sample. Point out various sources of errors and methods of minimising them.
7. Describe a method for the measurement of B-H curve of a magnetic substance of a bar form.
8. Give the theory underlying the use of magnetic potentiometer.
9. How is the leakage factor in a dynamo-electric machinery determined ?
10. Write a short note on "Testing bar specimens of magnetic materials".
11. Give the comparison between bar specimens and ring specimens in relation to magnetic measurements.
12. Give the advantages and disadvantages of ring and bar specimens used in magnetic testing of materials.
13. What are permeameters ?
14. Describe the principle and construction of a parameter for the determination of B-H curve of a sample.
15. Describe Fahy's simplex permeameter.
16. Describe the 'Lloyd Fisher square' method of measuring iron losses in ferromagnetic material.
17. Describe the method for the measurement of total loss in sheet steel with alternating current. Give the theory of the method.
18. Explain with a neat circuit diagram the Epstein or Lloyd Fisher square method of testing core loss of laminated sheet steel. Make necessary derivations and explain how eddy current and hysteresis losses may be evaluated separately. Can you explain why a separate sine wave generator may be required for this test.
19. Explain how transformer steel available in strips of 25 cm length and 5 cm width is tested for iron losses by the wattmeter method. Indicate the method of stacking the strips.
20. Explain the "method of reversals" for experimental determination of a magnetic specimen.
21. Describe the method for determination of B-H curve of a magnetic materials using :
   (*i*) Method of reversals, and (*ii*) Step-by-step method.
22. Describe the construction and working of the following types of permeameters :
   (*i*) Illiovici permeameter; (*ii*) Burrow's permeameter.
23. Explain the methods of separation of iron losses into their two components : eddy current and hysteresis losses if the maximum value of flux density is maintained constant and (*i*) frequency is varied keeping the form factor constant; (*ii*) form factor is varied keeping the frequency constant.
24. Explain how the value of A.C. permeability of magnetic materials is determined through the use of :
   (*i*) Maxwell's bridge, and (*ii*) Campbell's bridge.
25. Describe how magnetising and loss components of no load current of a transformer be determined by using an A.C. potentiometer.

26. What is 'Nuclear magnetic resonance"? Explain how the strength of a magnetic field in air can be determined using this principle.
27. Explain the oscillographic method of determination of hysteresis loss of a magnetic material.
28. Explain a method for testing of permanent magnets.
29. Describe a method for testing of feebly magnetic materials.
30. What precautions should be taken in magnetic testing ?

UNSOLVED EXAMPLES

1. A solenoid 60 cm long and 2.5 cm is diameter is uniformly wound with 600 turns of insulated wire.
   (*i*) Find the magnetic field strength at the centre of the solenoid when carrying a current of 1 ampere;
   (*ii*) If a secondary coil is wound round the central part of the solenoid, calculate the flux passing through it. [**Ans.** 123.2×10^{-8} Wb]
2. An iron ring of 350 mm^2 cross-sectional area with a mean length of 1 m is wound with a magnetising winding of 100 turns. A secondary coil with 200 turns of wire is connected to a ballistic galvanometer having a constant of 1 μC per scale division, the total resistance of the secondary circuit being 2000 Ω. On reversing a current of 10 A in the magnetising winding, the galvanometers shows a deflection of 100 scale divisions.
   Calculate : (*i*) The flux density in the specimen; (*ii*) The value of permeability at this flux density. [**Ans.** (*i*) 1.429 Wb/m^2; (*ii*) 1137]
3. An iron ring has a mean diameter of 0.1 m and a cross-section of 33.5 mm^2. It is wound with a magnetising winding of 320 turns and secondary winding of 220 turns. On reversing a current of 10 A in the magnetising winding, a ballistic galvanometer gives a throw of 272 scale divisions, while a Hibbert magnetic standard with 10 turns and a flux of 0.25 $\times 10^{-3}$ Wb gives a reading of 102 scale divisions, other conditions remaining the same. Find the relative permeability of the specimen. [**Ans.** 35.3]
4. A ferromagnetic material is tested with the help of a permeameter provided with a coil wound over the specimen, this coil being known as "B-coil" and another coil wound near the surface of the coil called "H-coil". These two coils are connected to two separate ballistic galvanometers having a constant of 25×10^{-6} Weber turn per scale division. The testing is done by reversing the current of the magnetising winding wound over the specimen. The reading of the ballistic galvanometer connected to 'B'-coil is 100 divisions and that connected to 'H'-coil is 10 divisions. The data of the two coils is :

   'B'-coil : Turns = 100; Area = 50 mm^2

   'H'-coil : Turns = 10000; Area = 5 mm^2

   Find : (*i*) Flux density in the specimen; (*ii*) Magnetising force; (*iii*) Relative permeability of the specimen. [**Ans.** (*i*) 0.25 Wb/m^2; (*ii*) 2000 AT/m; (*iii*) 100]
5. The constant of a given magnetic potentiometer is obtained by aid of a coil of 300 turns in which a current of 0.6 A is reversed. The resulting throw of the galvanometer is 157 scale divisions. It is then used to measure the magnetic potential difference between two points and throw is 304 divisions. Calculate the magnetic potential difference. [**Ans.** 697 AT]
6. Calculate the loss of energy caused by hysteresis in one hour in 50 kg of iron when subjected to cyclic magnetic changes. The frequency is 25 Hz, the area of the hysteresis loop represents 240 joules/m^3 and the density of iron is 7800 kg/m^3. [**Ans.** 138240J]
7. Calculate the loss per kg in a specimen of alloy steel for a maximum flux density of 1.2 Wb/m^2 and a frequency of 50 Hz, using 0.5 mm thick sheets. The resistivity of alloy steel is 30×10^{-8} ohm metre. The density is 7800 kg/m^3. Hysteresis loss in each cycle is 400 W-s/m^3. [**Ans.** 3.2 W/kg]

8. A specimen of cold rolled grain oriented 0.3 mm thick stamping has a resistivity of 5 × 10^{-7} Ωm. The published hysteresis loop is essentially rectangular in form, with a coercive force of 12 AT/m for all values of peak flux densities upto 1.6 Wb/m^2. The manufacturer quotes the iron loss in the material as 1.2 W/kg, with a sinusoidal flux density of 1.0 Wb/m^2 (peak) at 100 Hz. Calculate the loss in the material from its properties and compare it with the quoted value. The mass density is 7650 kg/m^3. **[Ans.** 1.015 W/kg]

9. A laminated iron cylinder is rotated in a magnetic field. The iron loss is 500 W at 480 r.p.m. and 700 W at 600 r.p.m. Find the loss if the laminations were twice as thick, the induction density increased by 20% and speed were 720 r.p.m. Take Steinmetz coefficient as 1.6. **[Ans.** 3622 W]

10. A specimen of iron stampings weighing 10 kg and having a cross-sectional area of 1680 mm^2 is tested by an Epstein square shown in Fig. 11.25. Each of the two windings S_1 and S_2 have 515 turns. A sinusoidal voltage supply at 50 Hz is given to the primary winding of Epstein square and an ammeter connected in the primary circuit indicates a current of 0.35 A and the electrostatic voltmeter connected across a secondary winding indicates a voltage of 250 V. The resistance of each of the windings S_1 and S_2 is 40 Ω. The resistance of the pressure coil of the wattmeter is 80,000 Ω.

Calculate the value of flux density in the specimen and the iron loss/kg at this flux density if the wattmeter connected on the primary side indicates 80 W.

[Ans. 1.3 Wb/m^2; 7.926 W/kg]

11. Calculate the percentage change in the hysteresis and eddy current losses of choke coil if other factors remain constant :

(*i*) The form factor of the applied voltage decreases by 20%.

(*ii*) The frequency of the applied voltage increases by 20%.

[Ans. (*i*) Zero, 36% decrease; (*ii*) 20% increase; 44% increase)

12. An iron-cored coil and a standard 2 Ω resistor are connected in series and the potential difference is measured across each of an A.C. potentiometer of co-ordinate type. The readings are 3.6 + *j*1.2 V for the coil and 4.8 – *j*2 V for the resistor. Calculate the magnetising component and loss component of current through the coil.

[Ans. I_e = 1.96 A; I_m = 1.72 A]

CHAPTER

12 High Voltage Measurements

12.1 Introduction; 12.2. High voltage tests – Sustained low-frequency tests – Constant direct-current tests – High-frequency tests – Surge or impulse tests; **12.3. Testing apparatus** – High voltage transformers – Apparatus for voltage regulation – Control gear and protective devices – Apparatus for voltage measurement – Low-frequency high-voltage tests – High voltage D.C. testing of cables – High-frequency high-voltage tests – Surge or impulse testing; **12.4. High voltage testing of insulating materials, porcelain insulators and insulating oil.** *Highlights – Objective Type Questions – Theoretical Questions.*

12.1 INTRODUCTION

- The rapid advance in the use of high voltages (to the tune of 220 kV and 440 kV) for transmission purposes has necessarily been accompanied by extensive research upon high-voltage phenomena. The subject of high voltage testing is, therefore, a very important one at the present time, and includes a large variety of testing methods.
- A high voltage encountered in practice involve a range of frequencies from zero, in the case of D.C. to the very high frequencies of lightning impulses.
- In measurement of high voltages, some methods measure its *peak value* while other methods measure *r.m.s. value.* For convenience, a *mean value is sometimes measured instead of an r.m.s. value.*
  - — ***Peak values** govern the maximum voltage stress and have an important bearing on ultimate breakdown or the corona formation.*
  - — ***R.m.s. values** have a more direct bearing on dielectric loss;* but through internal heating effect that also has a close connection with dielectric strength.

High voltage sources. These are of two types : *A.C.* and *D.C.* high voltage sources.

- ***A.C. sources*** can be categorised into *high* frequencies and *low* frequencies sources.
  - — *"High frequency high voltage A.C. sources"* are *lightning* and *high voltage transmission line sparks.*
  - — *"Low frequency high voltage A.C. sources"* are set up by *transformers.*
- ***High voltage D.C. sources*** can be further divided into *low current* and *high current sources.*
  - — *"High voltage high current D.C. sources"* are *A.C. step-up transformer plus rectifier.*
  - — High voltage low current D.C. sources are *high frequency oscillator plus rectifier.* This type of voltage is required for X-ray and CROs.

Crest factor. The crest factor of a waveform is defined as the ratio of *peak voltage to r.m.s. voltage.*

i.e. $$\text{Crest factor} = \frac{\text{Peak voltage}}{\text{R.m.s. voltage}} \qquad ...(12.1)$$

The value of crest factor for sinusoidal waveform is 1.414.

When the crest factor of a waveform is known, its r.m.s. voltage can be determined if its peak voltage is given with the help of eqn. (12.1). Similarly peak voltage of a waveform can be evaluated if its r.m.s. voltage is known with the knowledge of crest factor of a waveform. Thus, *the knowledge of crest factor is quie helpful in high voltage measurements.*

12.2 HIGH VOLTAGE TESTS

Although there are a large number of methods of high voltage testing, but due to limitation of space we shall be considering a few important methods only.

The high voltage tests may be *classified* as follows :

1. Sustained low-frequency tests.
2. Constant direct-current tests.
3. High-frequency tests.
4. Surge or impulse tests.

12.2.1. Sustained Low-frequency Tests

- These tests are *commonest* of all high-voltage tests. They are conducted at the power frequency *i.e.,* 50 Hz.
- In these tests, test voltage should be of *sine waveform.*
- Voltages of 2 kV or 3 kV are used for routine voltage tests upon motors, switchgear, and other apparatus after manufacture, and in some cases after installation, in accordance with specifications.

  These tests are also carried out on :

  — Specimens of insulation for the determination of dielectric constant and dielectric loss.

  — Supply mains for routine testing.

  — High voltage transformers, porcelain insulators and high voltage cables, etc. for works testing. The test voltage in such cases (and research work) may be of the order of 2000 kV.

12.2.2. Constant Direct-current Tests

- In order to transmit large amounts of power efficiently, high transmission voltages are essential.
- The cables and other equipment employed for transmission of electric power at extra high voltage, according to standard specifications, are required to withstand a test-voltage of at least 10 kV above the maximum voltage to which it is intended to be subjected in use continuously, for at least 30 minutes for insulation stability. Such tests may be performed with D.C. voltage or A.C. voltage. However *D.C. testing is preferred to A.C. testing* because of the following *reasons* :

  — With an A.C. voltage supply, the high voltage transformer required for the purpose would have to be of a *very large capacity* owing to the heavy charging (capacitive) currents drawn by the lines or the cables. Consequently, a *large expenditure* has to be incurred for procuring the high voltage testing transformer.

  — Furthermore, *great difficulties* will be experienced in *transporting* the testing transformer to the testing site due to its large size.

12.2.3. High-frequency Tests

— The high-frequency tests, at frequencies (varying from several kHz to MHz or more) are important in work upon insulators for radio purposes where such frequencies are usual.

— It has been observed that in the case of porcelain insulators in service on transmission lines, breakdown, or flashover occurs in most cases as a result of *high-frequency disturbances in the line,* these being either due to *switching operations* or *external causes.* Thus, *insulators which may be satisfactory under high voltage of low frequency are not necessarily satisfactory under the different conditions existing when the frequency is of much* higher order.

Therefore, to ensure reliable operation of insulators under disturbances caused by switching and arcing earth must be subjected to high frequency tests.

12.2.4. Surge or Impulse Tests

— The surge or impulse tests are carried out in order to investigate the influence of *"surges"* in a transmission line upon breakdown of the line insulators and of the end turns of transformers connected to the line.

— Such surges are usually produced by lightning in the neighbourhood of the transmission line, and are the result of the sudden discharge of the electricity on the suspended particles of a thunder cloud, which takes the form of a lightning stroke.

An indirect stroke (direct stroke being rare) occurs when a charged cloud above and near to the line, electrostatically induces charges of opposite sign to that of its own charge, in the line. These charges are bound as long as the cloud remains near without discharging its electricity by lightning stroke. But if the cloud is suddenly discharged, as it is when the lightning stroke occurs, the induced charges in the line are no longer bound, but travel with the velocity of light along the line to equalize the potential at all points on the line. This means that a steep-fronted voltage wave travels along the line, its forms being as shown in Fig. 12.1. The wave is characterised by a steep rise of voltage upto its peak value followed by a slower drop to zero voltage (The probable maximum voltage appearing on a line due to direct stroke is 10 to 15 MV and that by indirect stroke is 500 kV). The increase in voltage occurs in a very short time of the order of *one millionth of a second.* Thus, very severe stresses are imposed upon the line insulators and upon the transformer windings, in case the wave is allowed to reach them. Such surges invariably cause rupture of insulators and other equipment.

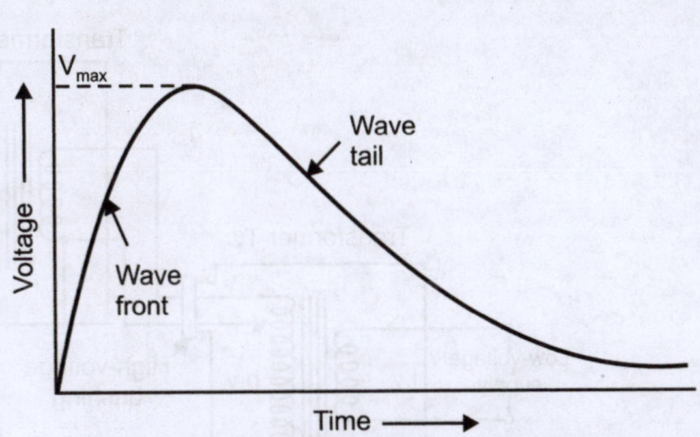

Fig. 12.1. Surge waveform.

— Impulse test is necessary to test the insulation of the power system so that voltage surges, due to switching or lightning strokes etc. are not permitted to change the insulation of power system equipment.

- *The testing plant for this purpose of testing must be capable of producing a high unidirectional voltage, suddenly applied to the apparatus under test.*

12.3. TESTING APPARATUS

For all types of high-voltage tests, certain apparatus is common. Special apparatus is required, in addition, for direct current, high frequency, and impulse tests.

The *following apparatus is usually essential for all tests* :

1. High-voltage transformers.
2. Apparatus for voltage regulation.
3. Control gear and high-voltage connections, including safety and protective devices.
4. Apparatus for voltage measurement.

12.3.1. High Voltage Transformers

— These transformers need to be carefully designed since they are subjected to transient voltages and surges during their normal operation when the insulation under test breaks down. *To withstand the internal streses set up by such disturbances the transformer insulation must be carefully proportioned.*

— The transformers are usually *single-phase core type,* and are *oil immersed,* Bakelite cylinders being used for insulation between high- and low-voltage windings.

— Transformers for cable testing, and certain other purposes, may also be required to give a considerable current, and the question of regulation and cooling must therefore be carefully considered.

— *In insulation testing the current taken from the transformer, when the test specimen breaks down, is limited by the insertion of water resistances in the test circuit, so that for such purposes the transformer need not have a large kVA capacity.*

— Equipments giving a maximum voltage upto *500 kV have usually a single high-voltage transformer. For the higher voltages, 1000 kV and upwards, two or more transformers are generally used,* connected *in "cascade" This arrangement is found to be more convenient than the single transformer for very high voltages, owing to the very large size and high cost of a single transformer for the purpose.*

The general arrangement of connecting transformers in cascade is shown in Fig. 12.2. :

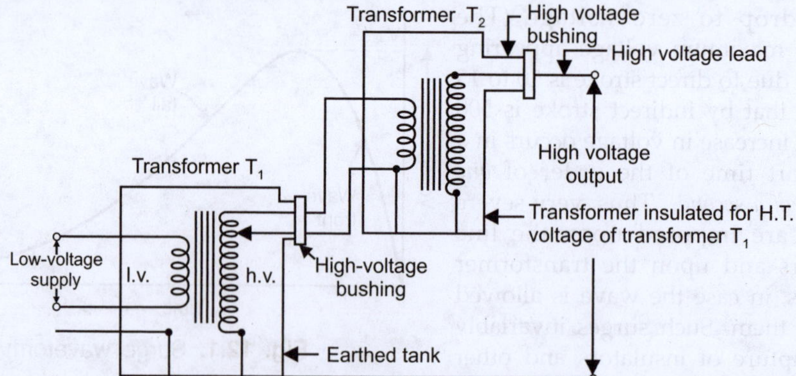

Fig. 12.2. Cascade connected transformers.

— The low-voltage supply is connected to the primary winding of transformer T_1, the tank of which is earthed. One end of the *h.v.* secondary winding of this transformer is conneted to the earthed tank. From the other end of the secondary winding, a lead passes through a high voltage bushing, which provides insulation for the full secondary voltage between this lead and the tank. Through this bushing also, runs a second lead from a tapping point on the secondary winding, the voltage between this point and high voltage end of the secondary winding being that required for the primary winding of transformer T_2.

— One end of the secondary winding of transformer T_2 is connected to its tank, which is insulated from earth for the full secondary voltage of transformer T_1. The other end of the secondary winding provides the high-voltage terminal of the equipment, *the total voltage-to-earth obtainable being approximately the sum of the secondary voltages of the two transformers.*

12.3.2. Apparatus for Voltage Regulation

The *requirements of voltage-regulating apparatus are* :

(*i*) The regulation of voltage should be *smooth* – to avoid surges on *h.v.* side of the transformer, and also for accuracy of voltage measurement.

(*ii*) The voltage-regulating apparatus *shall not cause distortion of the waveform.*

The secondary voltage of the transformer is regulated by variation of the voltage applied to its primary winding. This may be done as follows :

1. By variation of the alternator field current.
2. By resistance, or inductance control.
3. By induction regulators.
4. By a tapped transformer.

12.3.2.1. Variation of the alternator field current

This method is applicable when a separate alternator is used to supply the testing transformer. Except in the case of comparatively small plants for routine testing, a separate alternator is generally used.

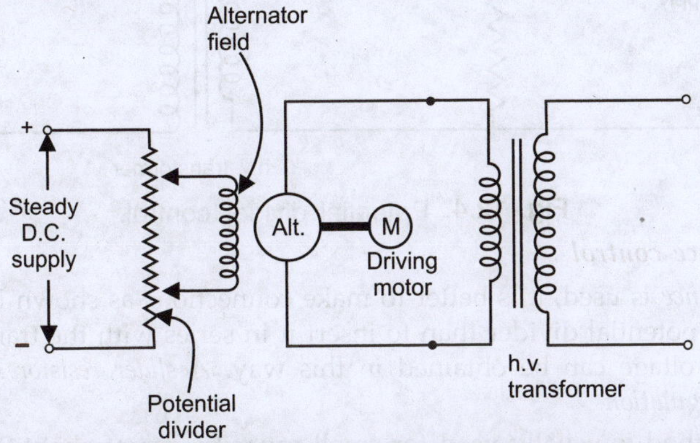

Fig. 12.3. Alternator field control method of voltage control.

The alternator should have a voltage waveform which is as nearly as possible *sinusoidal* on load, and the distortion under load conditions should be small. This is achieved by *making the air gap of alternator large, by special design of the armature windings; and by suitable regulation of the number, and shape, of the slots.*

In the case of *fairly small machines,* the field current of the alternator is varied *directly,* but when a large alternator is used the field current of the exciter may be varied, thus varying the alternator field current indirectly. Fig. 12.3. shows the connections for *direct variation of the alternator field current* :

— A potential divider, connected across a steady D.C. supply is used, and the connections are arranged so that a small field current in reverse direction to the normal one may be obtained. By this arrangement, zero voltage may be obtained by neutralisation of the residual magnetism in the field.

— Smooth and gradual voltage regulation is obtained by a special design of the potential divider, which should have a *large number of turns and some provision for very gradual and steady movement of the sliding contact.*

This method of voltage control has following *"advantages"* :

1. Convenient and simple.
2. Smooth regulation of voltage from zero to full voltage.
3. Output waveform is not distorted due to absence of impedance used in series of circuit for regulation.
4. Freedom from any disturbance caused by frequent short circuiting of secondary winding of h.v. transformer when insulation of power system under test breaks down.

12.3.2.2. Resistance or inductance control

In the case of small equipments, when a separate alternator is not used, resistance or inductance must be inserted in the A.C. supply circuit for voltage regulation.

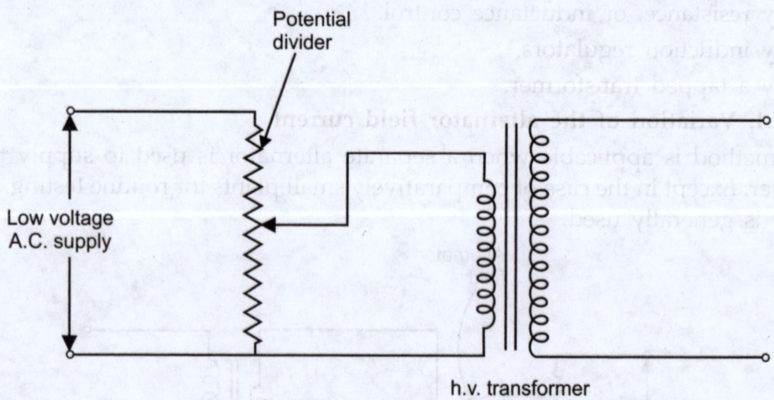

Fig. 12.4. Potential divider control.

1. ***Resistance control :***

When *resistance* is used, it is better to make connections as shown in Fig. 12.4, using the resistor as a potential divider than to insert it in series with the transformer primary winding. Zero voltage can be obtained in this way. *A slider resistor should be used for smooth voltage regulation.*

- This method is usually used for small capacity sets of about 2 to 3 kV rating.

Advantages :

(*i*) Cheap and convenient.
(*ii*) Smooth voltage variation is obtained.
(*iii*) Small distortion of voltage waveform.

Disadvantages :

(*i*) Large power loss in potential divider.
(*ii*) Because of very large size and heavy cost of the resistance potential divider, the arrangement cannot be employed for large capacity plants.

2. ***Inductance control :***

Voltage regulation may be obtained by connecting a choke coil in series with the primary winding of the h.v. testing transformer. The arrangement is shown in Fig. 12.5 wherein a *choke of variable inductance is used. Voltage regulation is obtained by the withdrawal or insertion of the iron core of the coil.*

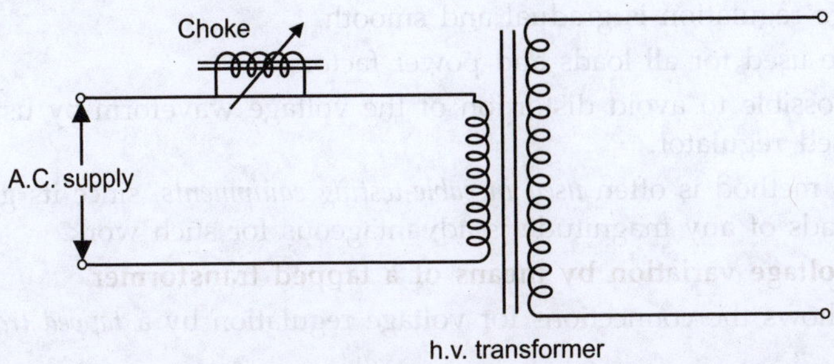

Fig. 12.5. Voltage variation by using series reactance.

Advantages :

This method is *more efficient* than resistance method, on account of its low power factor (and hence considerably smaller power loss).

Disadvantages :

(*i*) Large coil size is required when large power is to be handled.

(*ii*) Owing to non-linear characteristics of the iron core (of the choke coil), the input voltage waveform is distorted.

(*iii*) As test specimens are insulating materials, the power factor of load is leading and, therefore, the use of choke coil results in an *increase in the input voltage to the transformer with increase in load.*

12.3.2.3. Induction regulator method

Fig. 12.6 shows the circuit diagram for induction-regulator control.

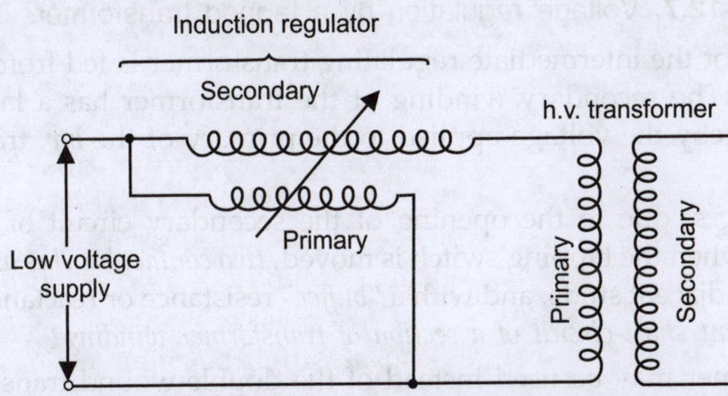

Fig. 12.6. Induction-regulator control.

— An *induction regulator is, essentially, a transformer, or mutual inductor, the secondary voltage of which can be varied by the rotation of the primary winding through any required angle upto 180°.

* Single phase induction regulator is similar in construction to single-phase induction motor except that the rotor is *not allowed to rotate continuously* but can be turned through one pole pitch by hand wheel or small rotor.

— If the primary and secondary windings have equal number of turns, by rotation of primary winding, the voltage induced in the secondary winding may be varied from $-V$ to $+V$ (where V is the low voltage supply). Thus the voltage applied to *h.v.* transformer primary way be *varied from zero to 2V*.

Advantages :

(*i*) Voltage regulation is gradual and smooth.

(*ii*) Can be used for all loads and power factors.

(*iii*) It is possible to avoid distortion of the voltage waveform by using a carefully designed regulator.

Uses. This method is often *used in cable-testing equipments*, since its gradual voltage variation at loads of any magnitude is advantageous for such work.

12.3.2.4. Voltage variation by means of a tapped transformer

Fig. 12.7 shows the connections for voltage regulation by a *tapped transformer* :

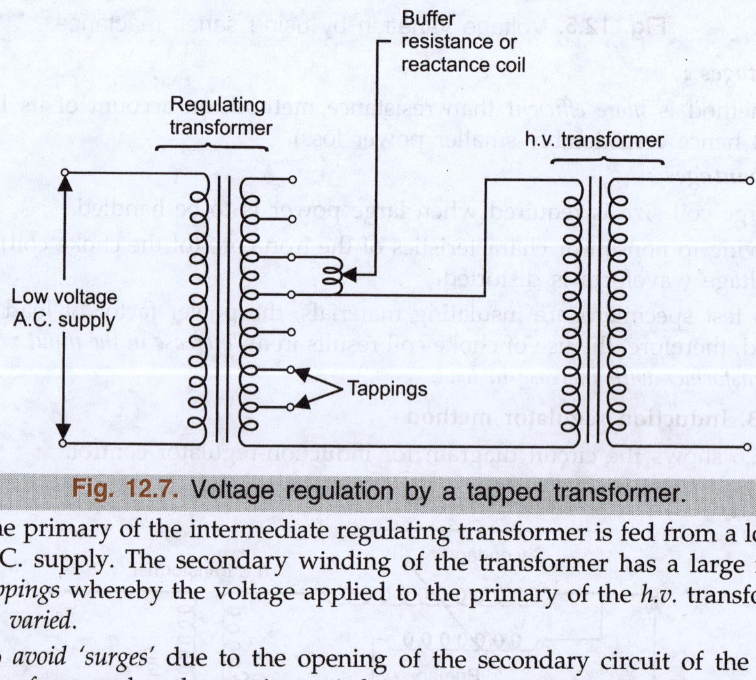

Fig. 12.7. Voltage regulation by a tapped transformer.

— The primary of the intermediate regulating transformer is fed from a low voltage A.C. supply. The secondary winding of the transformer has a large number of *tappings* whereby the voltage applied to the primary of the *h.v.* transformer may be *varied*.

— To *avoid 'surges'* due to the opening of the secondary circuit of the regulating transformer when the tapping switch is moved, *two contact brushes are used*, making contact with adjacent studs, and with a *"buffer"* resistance or reactance coil between them to *prevent short-circuit of a section of transformer winding*.

An auto-transformer may be used instead of the double-wound transformer shown.

— *In order to obtain smooth voltage regulation a large number of tappings should be used.*

Advantages :

(*i*) High efficiency.

(*ii*) Very small distortion of voltage waveform.

Disadvantages :

(*i*) Unless a large number of tappings is used, the voltage regulation is not smooth.

(*ii*) When the power to be handled is large, the equipment becomes expensive.

12.3.3. Control Gear and Protective Devices

On the ***low-voltage side*** (*i.e.* in the primary circuit) of the high voltage transformer the following protective devices should be provided :

(*i*) *Main switch :* To isolate the testing apparatus from the supply.

(*ii*) *Fuses.*

(*iii*) *A circuit breaker :* To protect the testing apparatus from damage in case of supply failure.

(*iv*) *Over-voltage relay :* To protect the apparatus from the effects of over voltages.

(*v*) *Interlocks :* To guard against the supply to the transformer being switched on except at a low value.

(*vi*) *Earth connections :* To provide safety and to protect the apparatus.

On the ***high-voltage side*** of the transformer, the following protective devices should be provided :

(*i*) *Protective resistances :* To protect the high voltage transformer from the effects of the surges [a high resistance (about 0.5 Ω/V) or a choke is connected in series with the secondary winding of h.v. transformer].

(*ii*) *Sphere gap :* To provide protection in the event of excessive voltage being applied to the circuit (In case of over voltage, the sphere gap breaks down and thus bypasses the surge).

Corona. During high voltage testing at power frequencies, some difficulties arise which are primarily due to *corona effect*.

— *"Corona"* is a phenomenon which occurs *when the voltage gradient at the surface of a high voltage conductor exceeds the breakdown strength of air.* This results in ionisation of surrounding air accompanied by a faint glow and hissing sound.

— Corona results in *power loss* and *distortion of waveform.*

Thus, it influences the breakdown and flashover voltages on account of surges in the circuit and because of presence of ionized air in the test room.

— The voltage at which corona occurs depends upon the *diameter of the high-voltage conductor,* the voltage increasing with increasing diameter. Thus to avoid such effects, the conductors in the high-voltage circuit should be made *much larger* than is necessitated by current-carrying considerations.

Fig. 12.8 shows a simplified diagram of connections for a *high-voltage testing equipment having a separate supply alternator.*

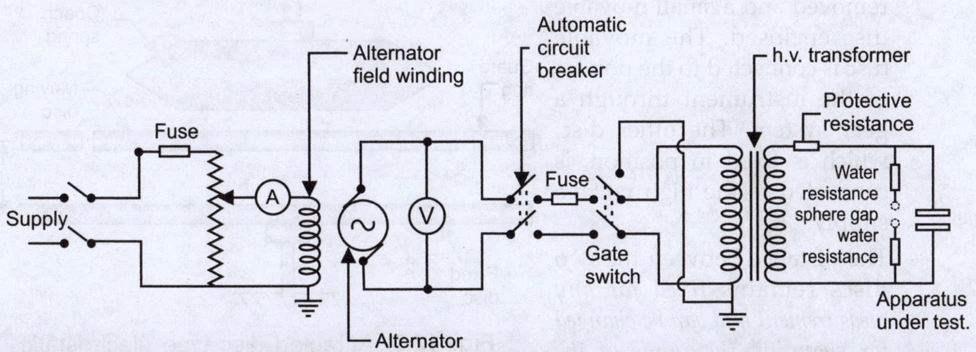

Fig. 12.8. Circuit for h.v. testing at power frequencies.

12.3.4. Apparatus for Voltage Measurement

Because of the difficulty of designing electrostatic voltmeters, for the measurement of extra high voltages, which will be free from errors due to corona effects within the instrument and to external electrostatic fields, a number of special methods of measurement have been devised for the purpose. Some methods must be available, also, for calibrating such voltmeters when constructed.

The methods of measurement of voltage are divided into *two categories* :

1. Methods used for measurement of *r.m.s. values of voltages.*
2. Methods of measurement of *peak values of the voltages.*

However, if voltage under test is of sinusoidal shape then the peak value can be determined by multiplying r.m.s. value of voltage with peak factor or *vice versa.*

12.3.4.1. Measurement of r.m.s. values of voltages

The r.m.s. value of voltage can be measured by the following *methods/instruments* :

1. Electrostatic voltmeters.
2. Transformation ratio method.
3. Potential divider methods.
4. Standard impedance methods.
5. Potential transformer method.
6. Sphere gaps.

1. Electrostatic voltmeter :

The electrostatic voltmeters have been discussed in chapter 3. However, the design of this type of meter has to be modified for high voltage use.

The following instruments may be used for measuring r.m.s. values of high voltages :

(*i*) Attracted disc voltmeters.
(*ii*) Attracted sphere voltmeter.
(*iii*) Oscillating ellipsoid voltmeter.

(i) Attracted disc voltmeters :

Fig. 12.9 shows one form of these instruments :

— This voltmeter consists of two hollow metal mushroom shaped discs. One of the discs has a portion of its centre removed and a small movable disc enclosed. The movable disc is connected to the pointer of the instrument through a gear system. The other disc, which is fixed in position, is connected to the high voltage supply.

— The distance between the two discs remains *fixed for any measurement and can be changed for changing the range of the instrument.*

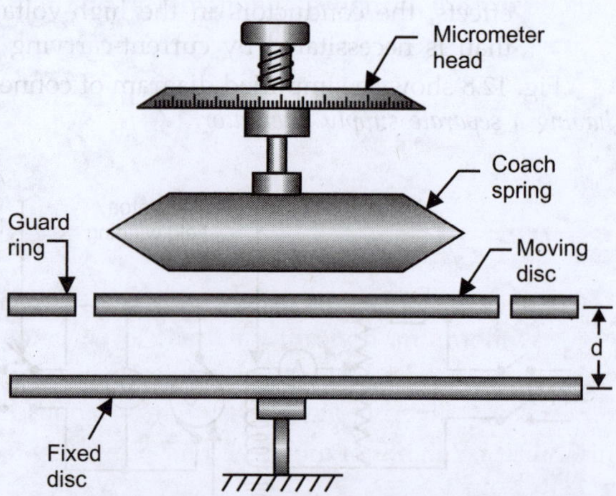

Fig. 12.9. Attracted disc type electrostatic voltmeter.

— Normally, a scale is provided on the movable disc stand which gives the proper range of the multiplying factor with which the deflection is to be multiplied to obtain the value of high voltage. *The disc's large size provides adequate protection for the working parts of the instrument against the electrostatic disturbances.*

- These meters are designed to measure *upto 200 kV directly* by connecting them across the source and *can be used for measuring higher voltage in conjunction with potential dividers.*

(ii) ***Attracted sphere voltmeter :***

— In this voltmeter, the voltage is measured based on the *measurement of electrostatic force between two spheres.*

— Forces of 4.2 to 7.5 N can be produced with direct voltage which will flashover gap of 100 mm and 500 mm between spheres of 1 m diameter. The attractive forces between the spheres can be measured with a stiff elastic support. *The errors involved are less than 1 percent for voltage as high as 1 MV.*

(iii) ***Oscillating ellipsoid voltmeter :***

— In this type of voltmeter, the voltage to be measured is applied to a pair of discharge free parallel plate pair of electrodes lying in vertical planes and a small conducting electrode of accurate prolate spheroidal shape is suspended at its mid-point by a long unspun silk thread so that it hangs in the uniform field between the central parts of the plates with its long horizontal axis.

— On the application of the voltage, an electrostatic field is produced due to which charges of opposite polarity and of calculable magnitude accumulate on the two ends of the ellipsoid. The ellipsoid, therefore, tends to align its long axis along the direction of the field. The *ellipsoid oscillates about its final position on account of the moment of inertia.*

— The controlling torque produced by the silk thread is *very small.* Therefore, *frequency of oscillations is approximately proportional to the applied voltage.* The constant of proportionality can be calculated from the mass and dimensions of the ellipsoid and the geometry of the plate electrodes.

- This voltmeter has the *drawback* that the average *electrostatic stress is much smaller* than that which can be obtained by the two voltmeters described above. This is *due to concentration of the stress at the tips of the ellipsoid.*

2. **Transformation ratio method :**

This method of measurement of high voltage is *very simple but approximate.*

- In this method, the primary voltage of the high voltage testing transformer is measured with a low voltage meter (See Fig. 12.10) and the secondary voltage is obtained as follows :

$$V_s = KV_p \qquad ...(12.2)$$

where, V_s = Terminal voltage of secondary winding,

K = Turns ratio, and

V_p = Terminal voltage of primary winding.

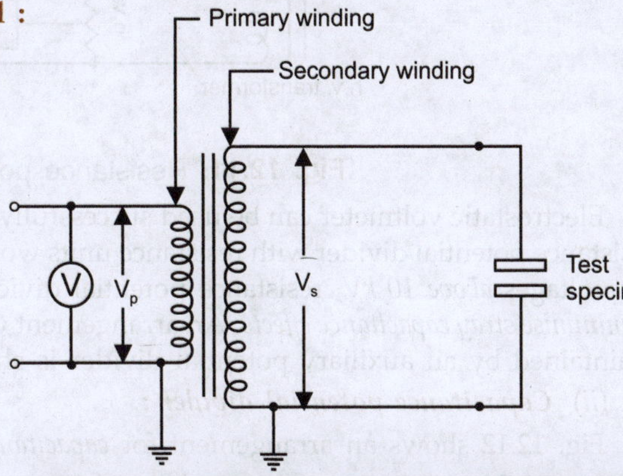

Fig. 12.10. Measurement of r.m.s. voltage by transformation method.

It is assumed, in this method, that K *remains constant* (hence the method though simple, gives wrong results).

- In some high voltage transformers, a separate voltage coil is provided whose voltage is measured with the help of a low voltage voltmeter. The secondary voltage (V_s) in this case is given by :

$$V_s = \left(\frac{N_s}{N_c}\right).V_c \quad ...(12.3)$$

where, N_s = Number of turns of secondary winding,
N_c = Number of turns of the coil, and
V_c = Voltage across coil.

— This method cannot be used for cascade arrangement of transformers. It gives an accuracy of 1 to 2 per cent.

— The advantage of this system lies in its convenience as a means of taking continuous reading after being calibrated by comparison with one of direct systems of measurements.

3. **Potential divider methods :**

In such methods, either a *high value resistor* or an *air capacitor* must be used, *connected across the high-voltage winding, a definite fraction of the total secondary voltage being measured by means of a low-voltage electrostatic voltmeter.*

(i) ***Resistance potential divider :***

Fig. 12.11 shows an arrangement for measurement of high voltage by resistance potential divider which uses a high voltage transformer.

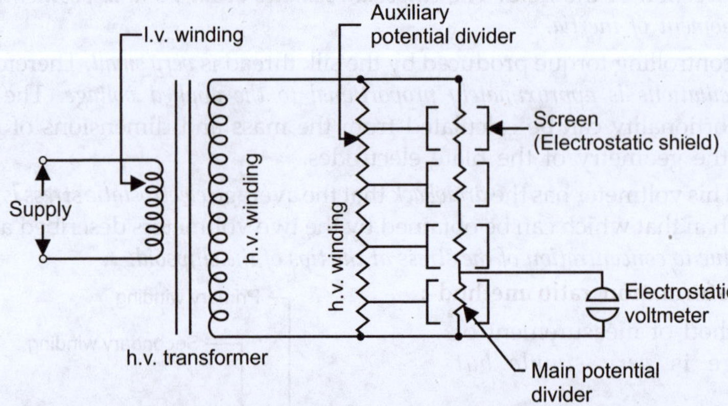

Fig. 12.11. Resistance potential divider.

Electrostatic voltmeter can be used successfully upto about 10 kV, with an unshielded resistance potential divider with resistance units wound on flat insulating sheets. However, for voltages *above 10 kV*, resistance potential divider *requires careful electrostatic shielding to minimise stray capacitance effects.* An arrangement where the shield potentials are correctly maintained by an auxiliary potential divider is shown in Fig. 12.10.

(ii) ***Capacitance potential divider :***

Fig. 12.12 shows an arrangement for *capacitance potential divider* :

— In this method, an electrostatic voltmeter, whose capacitance is large compared with that of the high-voltage standard capacitor, is connected in series with the latter, and is shunted by a capacitance of much greater value than its own.

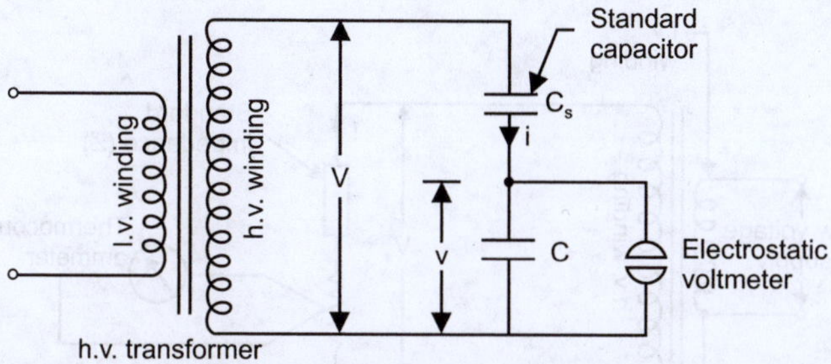

Fig. 12.12. Capacitance potential divider.

— The *capacitance of the standard capacitor must be accurately known,* and it is important that it shall be *free from dielectric loss.* For the latter reason, *air capacitors* (usually from 50 to 100 μμF) are always used for the purpose.

Let, C_s = Capacitance of the standard capacitor,

C = Capacitance of the capacitor which shunts the electrostatic voltmeter,

v = Voltmeter reading,

V = Voltage of the high voltage circuit, and

i = Current passing through the capacitors.

Then,
$$v = \frac{i}{\omega C} \quad \text{...(12.4)}$$

(where $\omega = 2\pi f$, f being the supply frequency)

Neglecting the capacitance of the voltmeter, the effective capacitance of C_s and C, in series is $\frac{C_s C}{C_s + C}$, and the voltage V of the high voltage circuit is given as :

$$V = \frac{i}{\omega\left(\frac{C_s C}{C_s + C}\right)}$$

or,
$$V = \frac{C_s + C}{C_s} \cdot v \quad \text{...(12.5)}$$

[Using eqn. (12.4)]

- This method *measures r.m.s. voltage.*

Advantages : The use of capacitance potential divider as compared to resistance potential divider for use at high voltage has the following *advantages* :

(i) Simple in construction.

(ii) Free from heating effects.

(iii) Comparatively free from effects of stray magnetic fields and can be easily shunted.

4. Standard impedance methods :

Fig. 12.13 shows the connections for the measurement of voltage by a standard impedance method :

— In this method, a very high standard impedance is connected across the h.v. terminals of the transformer. The impedance may be of a known value of resistance of high value or an air capacitor.

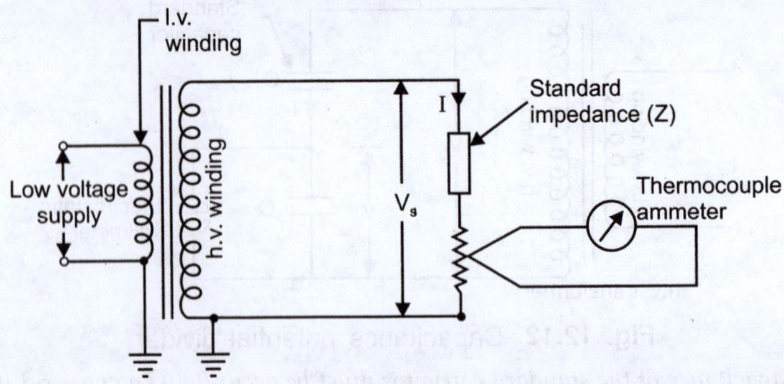

Fig. 12.13. Measurement of voltage by standard impedance method.

— The current flowing through the standard impedance may be measured by means of a thermocouple ammeter by measuring voltage across a known resistance; using an electrostatic voltmeter or by using a valve voltmeter.

Secondary voltage, $V_s = I \times Z$

where, Z = Impedance connected across the *h.v.* winding terminals

— A small heater of fine wire is connected in series with standard impedance, through which current flows. This current heats up the wire along with the thermo-junction, which is made up of two dissimilar metals. This induces e.m.f. in the thermo-junction circuit and the magnitude of the induced e.m.f. depends of the temperature of the heater. So r.m.s. value of A.C. current flowing through the standard impedance can be conveniently measured by measuring the induced e.m.f. in the thermocouple. After measuring the current through the standard impedance, the r.m.s. value of high voltage is obtained by multiplying it by the value of standard impedance.

5. **Potential transformer method :**

Modern potential transformers claim the following *advantages* over the potential dividers :

(*i*) The indicating instrument is isolated from the high voltage circuit.

(*ii*) The instrument impedance has negligible effect on the transformation ratio, within limits.

— Two state reduction by means of a combination of a potential divider and potential transformer proves very useful for the measurement of very high voltages.

— Potential transformers may be used with electrodynamometer, moving iron, or electrostatic instruments for measurement of r.m.s. values of A.C. voltages upto 200 kV.

The range may be extended by capacitor potential transformers.

6. **Sphere gaps :**

The *breakdown voltage of a spark gap between two metallic spheres may be used for the measurement of voltages upto the highest encountered in high voltage testing.*

— In this method of measurement of extra high voltage, two equal-sized metal spheres made of brass, bronze, steel, aluminium, aluminium alloy or copper (usually of copper), separated by an air gap, are mounted vertically one above the other, with the lower sphere earthed (See Fig. 12.14).

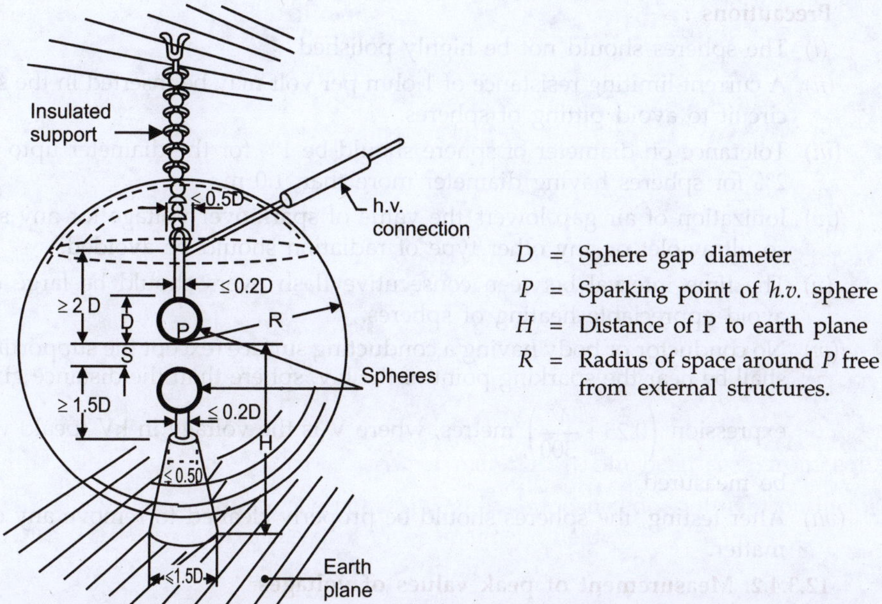

Fig. 12.14. Measurement of voltage with sphere gaps.

— The voltage between the spheres is revised till a spark passes between the two spheres. The value of voltage required to *spark over* (breakdown) *depends upon the dielectric strength of air, the size of spheres the distance between the spheres, humidity and temperature.*

— Calibration tables have been compiled on the basis of experimental work with different sizes of spheres, different gap lengths at specified conditions of pressure and temperature. The breakdown voltages corresponding to actual working conditions may be directly read off from the calibration tables. However, *corrections must be applied for temperature and pressure conditions.*

- *The spark-over voltages given in the calibration tables are in terms of r.m.s. voltage for sine wave. To obtain the peak value of the voltage under measurement, the r.m.s. values given are multiplied by the peak factor,* $\sqrt{2}$.

Advantages :

(*i*) The method is cheap, simple and reliable.

(*ii*) Peak voltages of values ranging from 2 kV to 2500 kV can be measured by this method and calibration of high voltage voltmeters and voltage measuring devices can be easily done.

(*iii*) Sphere gaps may also be used for measurement of voltage in surge (impulse) tests.

Disadvantages/Limitations :

(*i*) This method does not provide a continuous record of voltage.

(*ii*) The voltage measuremens carried out by sphere gaps can be relied upon to be accurate within 2 to 3 per cent, provided suitable precautions are taken. This is not a high accuracy in comparison with other electrical measurements. However, even this accuracy is acceptable owing to difficult conditions of working in *h.v.* testing.

Precautions :

(*i*) The spheres should not be highly polished.

(*ii*) A current limiting resistance of 1 ohm per volt may be inserted in the sphere gap circuit to avoid pitting of spheres.

(*iii*) Tolerance on diameter of sphere should be 1% for the diameter upto 1.0 m and 2% for spheres having diameter more than 1.0 m.

(*iv*) Ionization of air gap lowers the value of spark-over voltage for any sphere gap so ultraviolet or any other type of radiation should be avoided.

(*v*) The time interval between consecutive flash overs should be large enough to avoid appreciable heating of spheres.

(*vi*) No conductor or body having a conducting surface (except the supporting shanks) shall be near the sparking point of the h.v. sphere than the distance given by the expression $\left(0.25+\frac{V}{300}\right)$ metres, where V is the voltage in kV (peak) which is to be measured.

(*vii*) After testing, the spheres should be properly cleaned to remove any extraneous matter.

12.3.4.2. Measurement of peak values of voltages

— Due to the fact that the *maximum peak* value of applied voltage is that which produces the actual breakdown stress in the material under test, it is important that this value should be known in most cases. If methods of voltage measurement are known which give r.m.s. voltage values to be used, the *peak voltage may be obtained from the peak factor of the voltage wave.*

— It is often more satisfactory, however, to use some method of voltage measurement *which gives the peak value of the voltage directly.* From the fact that the breakdown of air is involved in the *"sphere-gap method"*, this is obviously one method which *depends upon the peak values,* and the calibration curves for sphere-gaps, already referred to, give such peak values.

Other methods of measurement of peak voltage are :

1. Rectified capacitor-charging current method.
2. Ryall crest voltmeter.
3. Ionic wind voltmeter.

1. Rectified capacitor-charging current method :

This method is based upon the fact that the *average charging current of a standard air capacitor is proportional to peak voltage.* The charging current is therefore rectified and its average value is then measured by a D.C. moving-coil permanent-magnet milliammeter or by a D'Arsonval galvanometer.

Fig. 12.15(*a*) shows the connections of an arrangement for peak voltage measurement using a full-wave rectification :

— C_s is a high voltage compressed gas capacitor across which practically whole of the voltage is dropped.

— The rectifier bridge (specially designed) is used for the rectification of alternating currents.

— The average charging current is measured by the milliammeter.

Assuming the impedance of the rectifier to be negligible, as compared to that of capacitor, the impedance offered to the path of current is thus on account of capacitor only and thus the charging current leads the applied voltage by 90°.

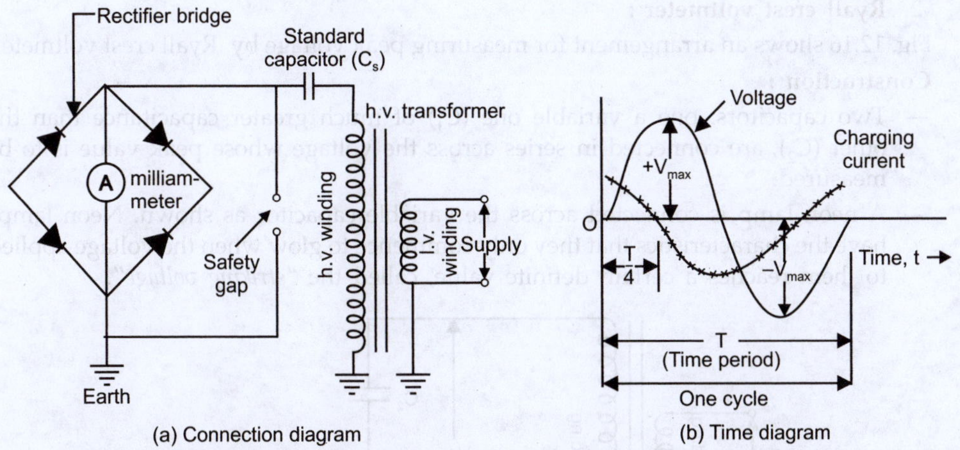

Fig. 12.15. Rectified capacitor-charging current method.

If v is the voltage at any instant, then,

$$v = V_{max} \sin \omega t,$$

where, V_{max} is the peak voltage.

Charging current at any instant,

$$i = \frac{V_{max}}{(1/\omega C)} \sin\left(\omega t + \frac{\pi}{2}\right) = V_{max}\ \omega C \cos \omega t \quad ...(12.6)$$

The average charging current [See Fig. 12.15(*b*)],

$$I_{av} = \frac{1}{(T/4)} \int_0^{(T/4)} V_{max}\ \omega C \cos \omega t\, dt = 4 V_{max} f C$$

or,

$$V_{max} = \frac{I_{av.}}{4 C f} \quad ...(12.7)$$

$$\left(\because T = \frac{1}{f}\right)$$

Thus, if the values of capacitance of the capacitor and frequency f (which must be constant throughout the test) are known, the peak value of an alternating voltage can be determined.

The *accuracy of voltage measurement* by this method depends upon the following *factors/requirements :*

(*i*) Impedance of the rectifier should be very small compared to the impedance of the capacitor.

(*ii*) Voltage to be measured should be sinusoidal.

(*iii*) Milliammeter should indicate the correct reading of the average current.

(*iv*) Rectifier should suppress fully the one-half wave of the alternating wave.

- This method gives satisfactory result for all types of wave shapes only when positive and negative halves of the wave shapes are *equal* and wave shape has *only one peak in each half cycle.*

2. Ryall crest voltmeter :

Fig. 12.16 shows an arrangement for measuring peak voltage by Ryall crest voltmeter.

Construction :

— Two capacitors, one a variable one (C_1) of much greater capacitance than the other (C_2), are connected in series across the voltage whose peak value is to be measured.

— A neon lamp is connected across the variable capacitor as shown. Neon lamps have the characteristics that they only commence to glow when the voltage applied to them reaches a certain definite value, called the *"striking voltage"*.

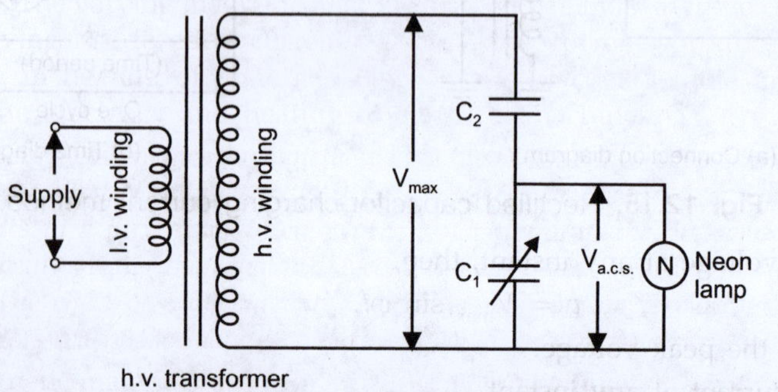

Fig. 12.16. Ryall crest voltmeter.

Working :

— The capacitance of capacitor C_1 is gradually reduced so that there is an increased voltage across it. This voltage is increased until the lamp strikes. Then peak voltage is given by :

$$V_{max} = V_{a.c.s.}\left(1+\frac{C_1}{C_2}\right) \quad ...(12.8)$$

where, $V_{a.c.s.}$ = A.C. striking voltage of the lamp (which is a constant quantity if the type of lamp is suitably chosen)

— The accuracy of the method depends upon the accuracy with which the values of C_1 and C_2 and $V_{a.c.s.}$ for neon lamp are determined.

- *This method can be used for measurement of peak value of voltage upto 150 kV (maximum).*

3. Ionic wind voltmeter :

This voltmeter which is used for direct measurement of high voltage, can measure both r.m.s. and peak values of A.C. voltages as well as D.C. voltages.

Its *principle of working* is as follows :

"When a highly-charged point is situated in air or other gas, a movement of the air surrounding the point is observed. This is referred to as the *"electric wind"*, and is brought about by the repulsion of ions from the surface of the point by the intense electrostatic field. These ions, colliding with uncharged molecules of air in the neighbourhood of the point, carry the latter with them and setup the *"electric wind"*. If an earthed electrode is placed *near the highly-charged one, an intense electric field exists of course, between the two, and a similar wind is observed also at the earth electrode."*

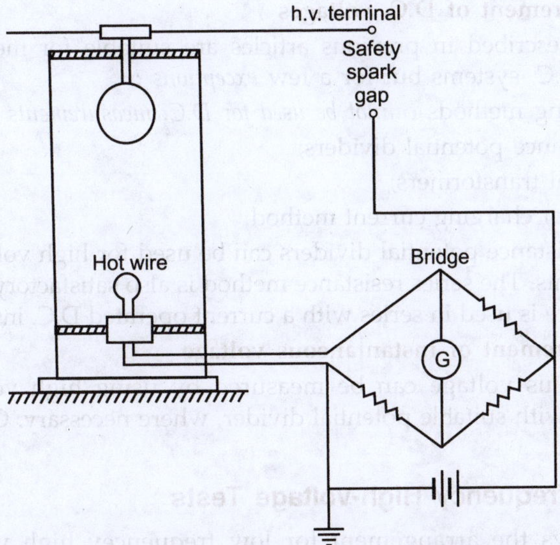

Fig. 12.17. Ionic wind voltmeter.

Fig. 12.17 shows the set up of the *ionic wind voltmeter* :

— A hot wire, of platinum-gold alloy, included in one arm of a wheatstone bridge network, is used as the earthed electrode of a high-voltage gap.

— Before the high voltage is applied to the gap the bridge network is balanced *i.e.*, galvanometer G is set to zero.

— When the voltage across the gap exceeds a certain value, called the "*threshold voltage", the *electric wind cools the hot wire and hence reduces its resistance.* A reduction of 25% in resistance is obtainable under suitable conditions, and this is *sufficient to cause an appreciable out-of-balance voltage* in the bridge network, as indicated by the galvanometer deflection.

— The *cooling and hence the galvanometer deflection, depends upon the potential gradient (and hence upon the applied voltage), the temperature of the electrode, the nature and pressure of gas in which the electrostatic field exists, and upon the frequency.*

— When the voltage is alternating, only that portion of the voltage wave which lies above the threshold value is effective from the viewpoint of cooling of the hot wire. The area of the effective portion of the voltage wave and hence cooling, varies for different wave shapes, even though the waves have the same r.m.s. value. Therefore, voltage waveform influences the instrument readings.

Calibration of the instrument is on a sine waveform.

Advantages :

(*i*) Robust in construction.

(*ii*) The high voltage may be measured by an observer at some distance from the charged conductors.

(*iii*) Freedom from disturbance by weather and temperature conditions.

- These voltmeters are used upto *300 kV*.

* The "*Threshold voltage*" is the voltage which is required before the potential gradient is sufficient for ionization to commence.

12.3.4.3. Measurement of D.C. voltages

The methods described in previous articles are suitable for measurement of high voltage D.C. and A.C. systems but for a few *exceptions :*

— The following methods *cannot be used for D.C. measurements :*

(*i*) Capacitance potential dividers;

(*ii*) Potential transformers;

(*iii*) Capacitor charging current method.

- Resistance potential dividers can be used for high voltage D.C. measurements. The series resistance method is also satisfactory when a high resistance is used in series with a current operated D.C. instrument.

12.3.4.4. Measurement of instantaneous voltage

The instantaneous voltage can be measured by using high voltage cathode ray oscilloscope (CRO) with suitable potential divider, where necessary. CRO is discussed in detail in chapter 11.

12.3.5. Low-frequency High-voltage Tests

Fig. 12.18 shows the arrangement for low frequencey high voltage testing (on insulating materials).

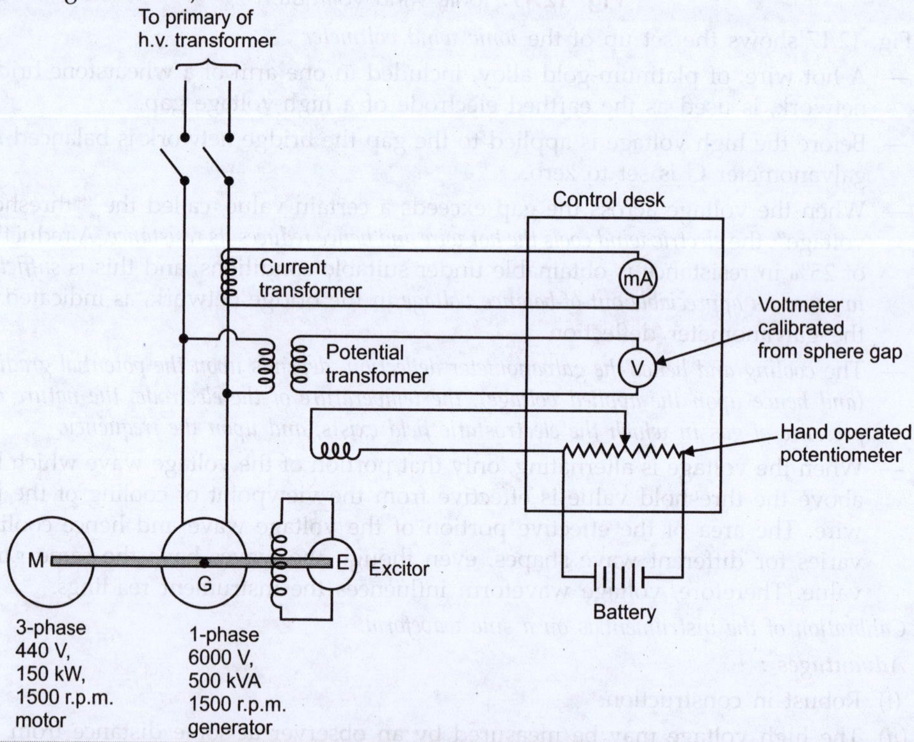

Fig. 12.18. Low-frequency high-voltage testing.

12.3.6. High Voltage D.C. Testing of Cables

Fig. 12.19 shows the connections for D.C. high-voltage cable tests.

In this test, current passes through the rectifier only for half wave of the voltage cycle. During this period, capacitor formed by the sheath and core is *charged to a voltage equal to that of maximum value,* V_{max} of high voltage transformer secondary voltage.

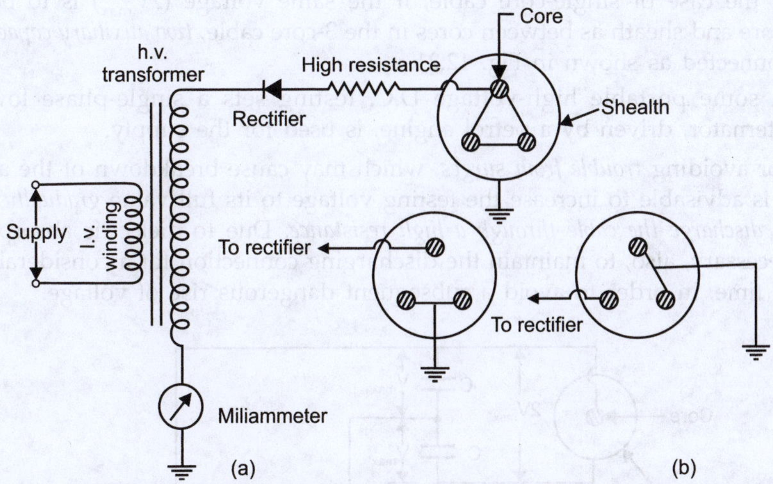

Fig. 12.19. Connections for high-voltage cables (3-core) tests.

During next half cycle, rectifier *stops conduction* and voltage of the cable capacitor remains *same.* The milliammeter, mA shows the reading of charging and leakage currents of the cable capacitor.

A high resistance of 0.25 MΩ is connected in series with the circuit to protect it against surges.

Tests are usually made upon three-core cables with connections as shown in Fig. 12.19(*a*) as well as with the connections of Fig. 12.19(*b*).

Two rectifiers are used for test voltage upto *200 kV*.

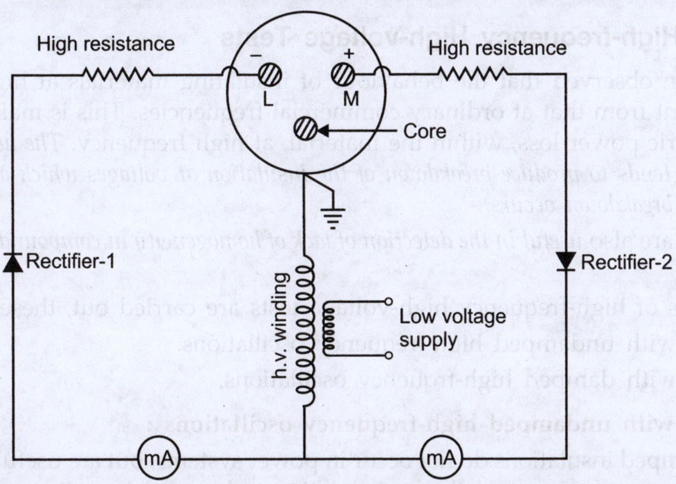

Fig. 12.20. Connections for high-voltage D.C. test upto 200 kV.

— Fig. 12.20 shows the connections for testing a 3-core cable with two rectifiers. In this case, test is being performed between cores *L* and *M*. During first half of the voltage cycle, rectifier-1 conducts and core *L* of the cable is charged to a potential V_{max} below earth and rectifier-2 does not permit any current flow through it. During next half wave of the voltage cycle rectifier-2 conducts and core *M* of the cable is charged to potential V_{max} above earth. Thus D.C. voltage between the cores is $2V_{max}$ and from each core to earth is V_{max}.

— In the case of single-core cable, if the same voltage ($2V_{max}$) is to be applied between core and sheath as between cores in the 3-core cable, *two auxiliary capacitors must be used,* connected as shown in Fig. 12.21.

- In some portable high-voltage D.C. testing sets a single-phase low-voltage alternator, driven by a petrol engine, is used for the supply.
- For avoiding *trouble from surges,* which may cause breakdown of the apparatus, it is advisable to increase the testing voltage to its full value *gradually,* and also to *discharge the cable through a high resistance.* Due to dielectric absorption, it is necessary, also, to maintain the discharging connection for a considerable length of time, in order to avoid a subsequent dangerous rise of voltage.

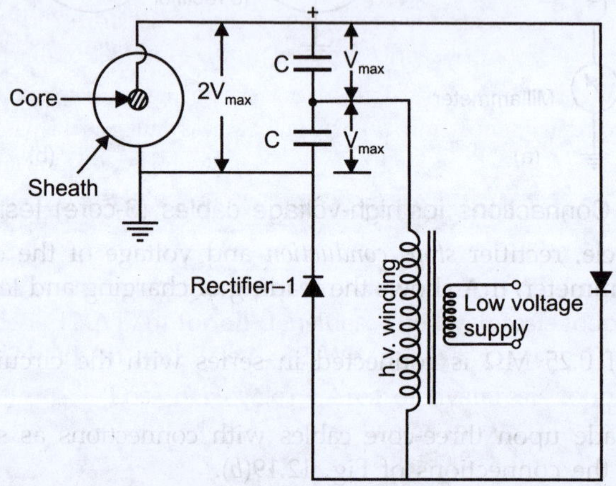

Fig. 12.21. Connections for high voltage D.C. testing for 1-core cables.

12.3.7. High-frequency High-voltage Tests

It has been observed that the behaviour of insulating materials at high frequencies is very different from that at ordinary commercial frequencies. This is mainly due to the greater dielectric power loss, within the material, at high frequency. *The heat produced by this power loss tends to produce breakdown of the insulation at voltages which are smaller than those at which breakdown occurs.*

Such tests are also *useful in the detection of lack of homogeneity in compound-filled porcelain insulators.*

Two kinds of high-frequency high-voltage tests are carried out, these are :

1. Tests with undamped high-frequency oscillations.
2. Tests with damped high-frquency oscillations.

1. Tests with undamped high-frequency oscillations :

— Undamped insulations do not occur in power systems, but are useful for insulation testing purposes, especially for insulation to be used in radio work.

— High frequency alternators have been used for frequency upto 30 kHz, and high-frequency arc generators have also been used, but these have the *disadvantage* that it is *difficult to obtain smooth voltage variation.*

— For the production of high-frequency high-voltages, usually electronic circuits are used; the voltage obtainable is 150 kV, and the frequency 100 kHz.

2. **Tests with damped high-frequency oscillations :**

Fig. 12.22. shows a circuit for high-voltages, usually electronic circuits are used; the voltage obtainable is 150 kV, and the frequency 100 kHz.

Circuit details :

— Damped high-frequency oscillations are obtained by the use of a Tesla coil, together with a circuit containing a quenched spark-gap as shown.

— The Tesla coil constitutes the h.v. tansformer. It consists of two air-cored coils which are placed concentrically. The high voltage secondary coil has a large number of turns, and is wound on a frame of insulating material, the insulation between turns being air, or in some cases, oil.

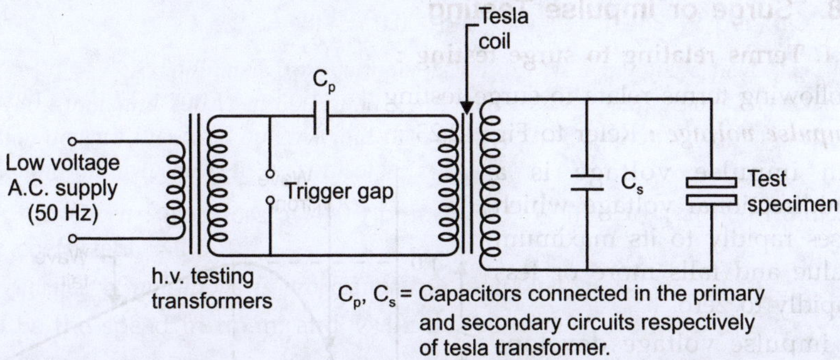

Fig. 12.22. Circuit for high-voltage tests with damped high-frequency oscillations.

If the Testla coil is oil-immersed, the spacing between turns can be made smaller than when air is the insulation. The primary winding has only a few turns, wound on an insulating frame.

— The supply is usually 50 Hz A.C. to the primary a *h.v.* testing transformer.

— Two capacitors C_p (air-cored) and C_s (made up of sphere gap) are connected in the primary and secondary circuits respectively, of the Tesla transformer. A trigger gap for sparking is also induced in the primary circuit of Tesla coil as shown.

Working :

— When the primary of Tesla coil is connected to A.C. supply, the capacitor C_p is charged to some maximum voltage depending upon the voltage on the secondary side of the supply transformer and upon the setting of the "trigger gap". The trigger gap across the secondary breaks at the pre-set value of voltage, the capacitor C_p discharges and thus high frequency damped oscillations are set up in the circuit containing C_p, the spark gap and primary of the Tesla coil. The spark gap will be conducting due to formation of arc across it during the time interval of dying out of oscillations.

— The time taken for the train of oscillations to die away is very small fraction of a second. The charge and discharge of primary capacitor C_p occurs twice per cycle of voltage (*i.e.* if supply frequency is 50 Hz then 100 trains of damped oscillations per second will be set up).

The frequency of the oscillations will be very high, usually of the order of *100 kHz and is given by :

* The highest frequency from which any trouble can be expected in a transmission system is about 80 kHz and hence 100 kHz is the maximum frequency required for testing.

$$f = \frac{1}{2\pi\sqrt{L_p C_p}} \text{ Hz} \quad ...(12.8)$$

where L_p and C_p are the inductance and capacitance respectively of the oscillatory circuit.

As the primary winding of the Tesla coil carries an oscillatory current, oscillations are induced in the secondary winding. If the primary and secondary windings are *tuned i.e.* $L_pC_p = L_sC_s$ the frequency of oscillations in the secondary winding will be the *same* as in the primary. These damped high-frequency oscillations are impressed upon the test specimen as shown in Fig. 12.22.

The voltage is measured by a sphere gap, while a CRO may be used to record the waveform of secondary voltages of the Tesla coil.

12.3.8. Surge or Impulse Testing

12.3.8.1. Terms relating to surge testing :

The following terms relate to surge testing :

1. ***Impulse voltage :*** Refer to Fig. 12.23 :

- An impulse voltage is a unidirectional voltage which rises rapidly to its maximum value and falls more or less rapidly to zero.

If an impulse voltage develops without causing a flashover or puncture, it called a *"full impulse voltage"*, however, if a flashover or puncture occurs, it is termed as a *"chopped impulse voltage"* (See Fig. 12.24).

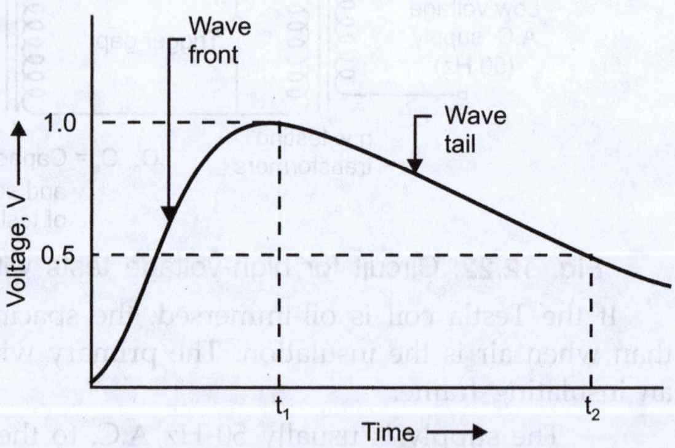

Fig. 12.23. Full surge wave.

- The impulse voltage is conveniently expresed as t_1/t_2 wave :

Time interval t_1 is the time taken by the impulse voltage to rise to its peak value and t_2 is time taken to fall to 50% of its peak value (See Fig. 12.23).

- Normally *1/50 μs impulse voltages are used for testing*. This voltage indicates that it *rises to its peak value in 1 μs and falls to 50% of its peak value in 50 μs.*

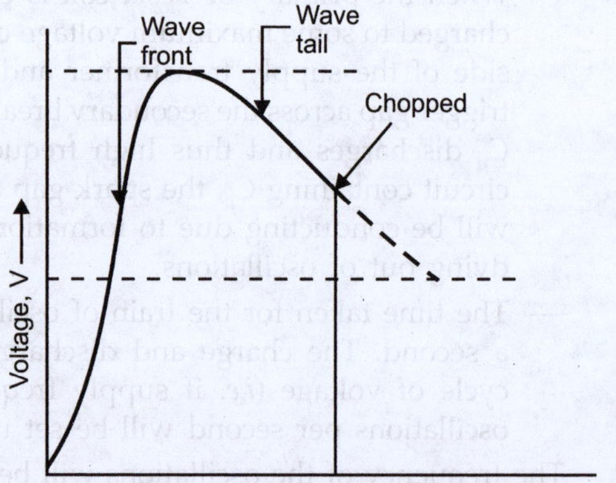

Fig. 12.24. Chopped surge curve.

2. ***Impulse flashover voltage :***

- 50% impulse flashover voltage – it is the peak value of that impulse voltage which causes flashover of the object under test for about half the number of applied impulses.

3. ***Impulse puncture voltage :***

- It is the peak value of that impulse voltage which causes the puncture of the specimen under test *when puncture occurs on wave tail and is the value of the voltage at the instant of puncture when puncture occurs on the wavefront.*

4. ***Impulse ratio for flashover :*** It is the *ratio of impulse flashover voltage to the peak value of power frequency flashover voltage.*

12.3.8.2. Basic impulse test circuit

Fig. 12.25 shows a basic impulse test circuit from which test impulse surge voltage can be obtained :

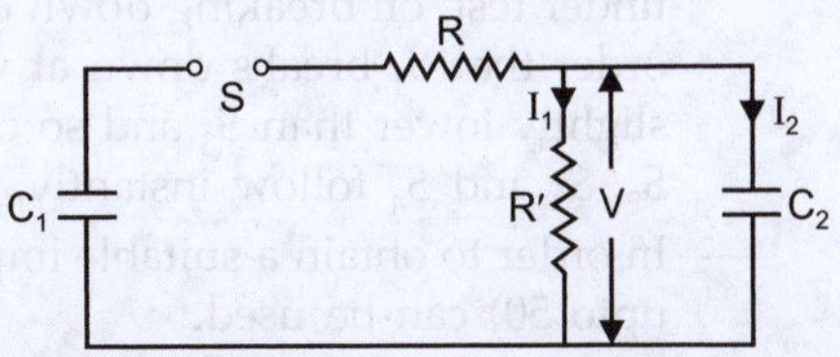

Fig. 12.25. Basic impulse test circuit.

— Capacitor C_1 is first charged from a D.C. source till the breakdown of spark-gap occurs. Now the voltage to which C_1 is charged is available.

— The discharge takes place in such a *short time* that *energy loss in the resistors is negligible* and whole of the energy stored in capacitor C_1 is suddenly discharged through the test specimen on breaking down of trigger gap. Thus the specimen is subjected to the required impulse voltage.

In the circuit diagram, *R* is the wavefront control resistor of a few hundred ohms and *R′* the wave tail control resistor of the order of 100 kΩ.

12.3.8.3. Impulse generators

Impulse generators can be single-stage or of multi-stage construction. The choice between a single- or multi-stage generator depends upon the output voltage desired and the type of equipment providing D.C. input to the generator.

Single-stage impulse generators have been built for voltages upto 200 kV but for voltage above 200 kV, use of multi-stage generators is usually made.

Where expense is not the main consideration, ***single-stage generators*** may be used upto a voltage output of 200 kV (or even higher).

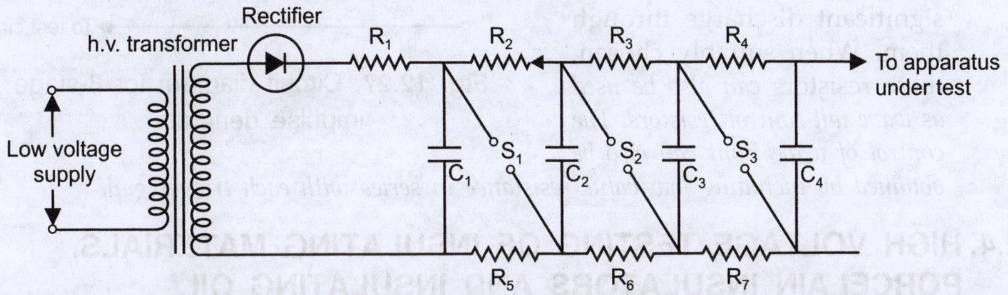

Fig. 12.26. Connections of multi-stage surge generator – Marx circuit.

In ***multi-stage impulse generators*** the capacitor C_1 of Fig. 12.25 consists of *n*-sections *connected in parallel during "charging" and in series during "discharging"*. Thus the output maximum voltage is about *n*-times of the charging voltage. The change over from parallel to series connection is *affected automatically by breakdown of trigger gaps.*

Fig. 12.26 shows the circuit employing the principle of charging capacitors in parallel and discharging them in series, this circuit is known as *"Mark circuit"* :

— In this method of obtaining impulse/surge voltage number of capacitors (C_1, C_2, C_3, C_4 etc.) connected in *parallel* are charged from a high voltage transformer through a rectifying device, as shown in the Fig. 12.26.

— At a certain voltage, which depends upon the setting of the trigger gaps, the trigger gaps breakdown and connect the capcitors in *series* simultaneously. Thus the voltage equal to the sum of the voltages, to which each of the capacitors is charged, is applied to the apparatus under test.

— The discharge takes place so rapidly that energy loss in resistors is negligible and whole of the energy stored in the capacitors is discharged through the apparatus under test, on breaking down of trigger gaps. The trigger gaps are set in such an order that S_1 breaks down at voltage slightly lower than S_2 and S_2 at a voltage slightly lower than S_3 and so on. Thus the trigger gap S_1 breaks down first, and S_2, S_3 and S_4 follow instantly.

— In order to obtain a suitable impulse voltage, any number of trigger gaps (usually upto 50) can be used.

- Fig. 12.27 shows a simplified circuit for 3-stage impulse generator, used now-a-days.

— The capacitors C_1, C_2 and C_3 are charged in parallel through high resistance resistors R_1, R_2, R_3, R_4 and R_5.

— On breakdown of spark gaps S_1, S_2 and S_3, the points L and P, M and Q, N and T etc. are connected together and therefore all the capacitors are connected in series and discharge through the circuit or apparatus under test.

— The resistors R_1, R_2, R_3, R_4 and R_5 are taken of such a high value that there is no significant discharge through them. When suitably chosen, these resistors *can also be used as wave tail control resistors. The control of wave front can also be obtained by including low value resistance in series with each trigger gap.*

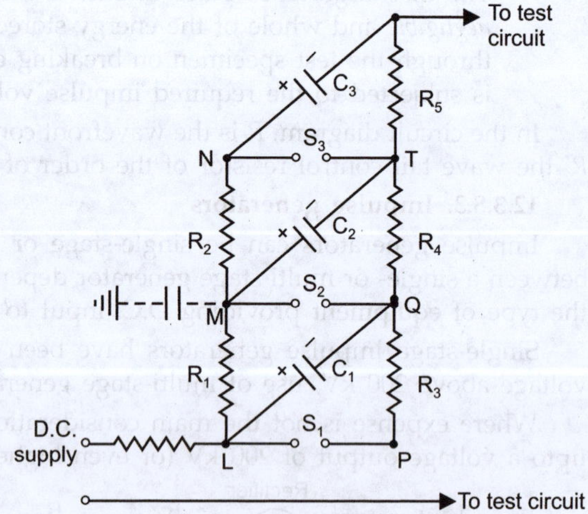

Fig. 12.27. Circuit diagram for 3-stage impulse generator.

12.4. HIGH VOLTAGE TESTING OF INSULATING MATERIALS, PORCELAIN INSULATORS AND INSULATING OIL

12.4.1. Testing of Insulating Materials

- The *"dielectric strength" of an insulating material is the minimum voltage gradient at which electrical failure or breakdown of the material occurs under prescribed conditions.* Although breakdown of insulation may be caused by leakage current, it is normally caused by voltage stress exceeding the dielectric strength of the material. The dielectric strength is expressed in terms of *voltage per unit thickness.*
- Dielectric strength varies with structure homogeneity and quality of sample, thickness of sample, shape of electrodes, waveform of test voltage and duration and rate of application of voltage, moisture content of specimen and heat capacity

of electrodes. Dielectric strength is *not* proportional to thickness but is relatively greater for thin sheets. It is also a *function of time of application*, being greater, lesser is the time.

- Measurement of capacitance, permittivity, dielectric loss and loss angle of insulating materials is very important. These measurements are *usually made at high voltages*. All these quantities vary with voltage used for measurements and for some applications, values measured at low voltages are of no use at all. *Normally,* **"Schering bridge"** *is used for determination of these quantities.*

12.4.2. High Voltage Tests on Porcelain Insulators

The following tests are applied to *porcelain insulators* are :

1. Power frequency tests.
2. Flashover tests.
3. Sample tests.
4. Routine tests.

1. ***Power frequency tests :***

- In this test, frequency used is between 25 to 100 Hz of approximately sine waveform with peak value of voltage not to exceed 1.45 times the r.m.s. value.

2. ***Flashover test :***

- In this test, porcelain insulators are designed in such a way that their *puncture voltage is higher than the flashover voltage.* Thus in case of sudden over-voltage, the air at insulator surface breaks down earlier and thus the insulator is not punctured and is saved from total destruction. Therefore, flashover voltage tests are very significant in the case of porcelain insulators.
- The flashover voltage for any insulator depends upon :
  — Barometric pressure ;
  — Temperature ;
  — Shape of the electrostatic field ;
  — Humidity ;
  — Nature of contact between electrodes and insulators.

The various *"flash-over tests"* applied are :

(i) *50 percent dry impulse flashover tests with 1/50 μs impulse voltage.*

(ii) *Dry flashover and dry one minute test.* In this test the voltage is gradually raised to a specified value in approximately 10 seconds and is maintained for one minute. The voltage is then gradually raised until flashover occurs.

(iii) *Wet flashover and one minute rain test.* During this test, the insulator is sprayed constantly with water and the same procedure as for dry flashover and dry one minute test is followed.

3. ***Sample tests :***

The sample tests include the following :

(i) *Mechanical strength test :*

— This test is performed on pin-type insulators to determine their ultimate mechanical strength.

— *During performance of this test the pin-type insulator is mounted on a steel pin and load* is applied to the tie-wire groove by means of a steel bridle. This applied load is then gradually increased till the failure takes place.

— Suspension type insulators are tested in pure tension.

(ii) Temperature cycle test :

— To perform this test, the insulator is subjected to several temperature cycles by immersing it alternately in hot and cold water.

— The insulator or its glaze must not be affected and after drying it out it must withstand the usual flashover test.

(iii) Puncture test :

— This test is performed to determine puncture voltage.

— In this test the insulator is totally immersed in oil and is subjected to the voltage being gradually increased to a certain specified figure or till puncture takes place.

4. ***Routine tests :***

— These tests are applied to all the insulators.

— The insulators are initially subjected to a low voltage which is rapidly increased till a flash-over occurs every few seconds. The voltage is maintained at this value for a minimum of 5 minutes or if failure occurs, for five minutes after the last punctured piece has been removed. The voltage is gradually reduced to $\frac{1}{3}$rd of its value before supply is switched off.

— *The porcelain insulators used for communication purposes are tested with high frequency undamped oscillations.*

12.4.3. Testing of Insulating Oil (Liquid insulant)

The dielectric strength of liquid insulant is tested in a special equipment comprising a cell made up of glass or plastic. The *cell material is transparent and non-absorbent.* Two polished electrodes made up of copper, brass, bronze, or stainless steel are used. These electrodes are either spherical (12.5 to 13 mm dia.) or of the shape shown in the Fig. 12.28 IS 6792—1972. The electrodes are arranged horizontally and the gap between them is accurately adjusted to 2.5 ± 0.1 mm. A variable high voltage supply through a step-up transformer is connected to the electrodes. A voltmeter will indicate the breakdown voltage. The meter will give r.m.s. voltage in kV.

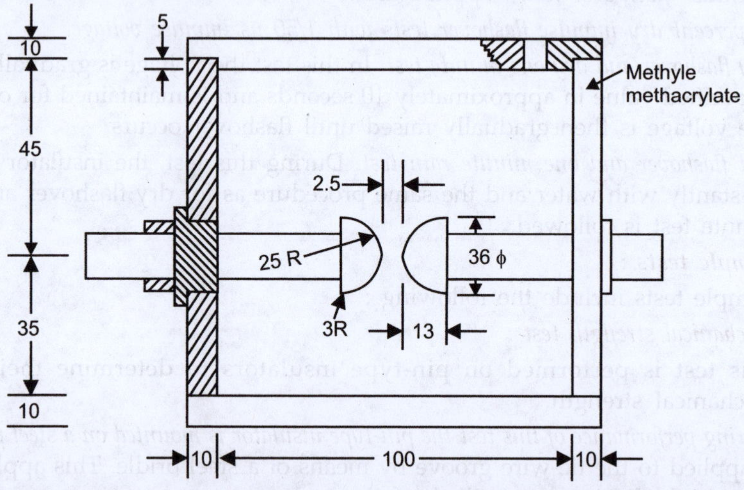

Fig. 12.28. Testing Electric strength of liquid insulant.

The test consists of increasing the test voltage at a uniform rate of 2 kV/sec. from zero upto the value producing breakdown. *The breakdown voltage is the voltage reached during the test at the time the first spark occurs between the electrodes.* The test is carried out six times on the same cell filling (oil) and the *arithmetic mean of six breakdown voltages is taken to be the breakdown voltage of the oil.*

In order to obtain accurate and reliable results attention should be paid to the following facts :

- The distance between the end faces of the two electrodes should be accurately set.
- The electrodes should be wiped, clean and rinsed out with clean dry oil or to be dried with thermostat.
- Right before the test the jar containing the oil should slowly turned upside down several times to mix the oil (it should not be shaken to avoid possible air bubbles).
- After each test the oil should be gently stirred with a clean dry glass stick.

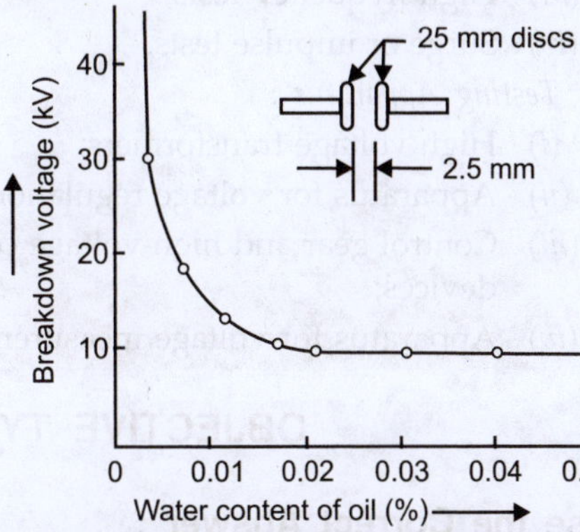

Fig. 12.29. Breakdown voltage of transformer oil as a function of the water content.

- Fig. 12.29 shows the breakdown voltage of transformer oil as a function of the water content.
- Fig. 12.30 shows the temperatures dependence of breakdown voltage of transformer oil, dry oil (1), oil containing traces of water (2).

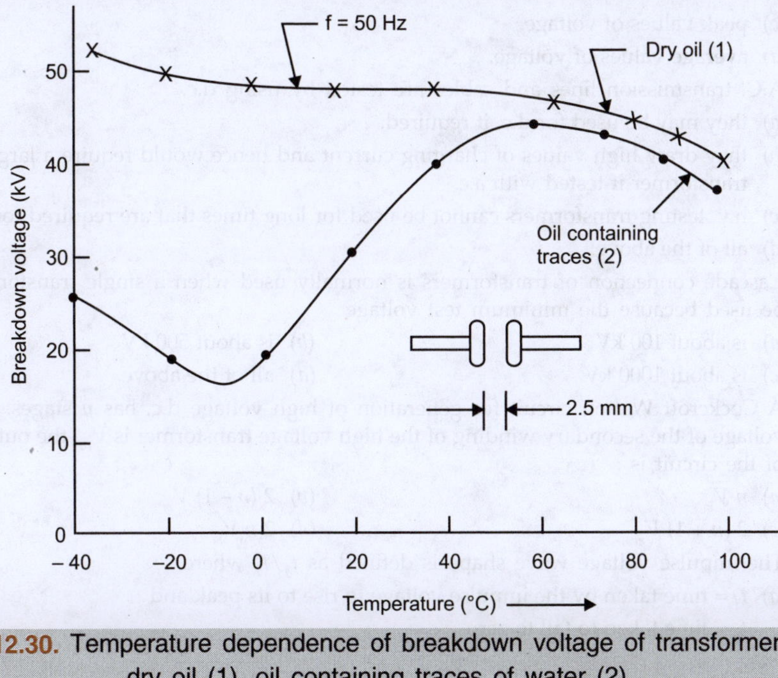

Fig. 12.30. Temperature dependence of breakdown voltage of transformer oil, dry oil (1), oil containing traces of water (2).

HIGHLIGHTS

1. High voltage tests may be classified as follows :
   (*i*) Sustained low-frequency tests;
   (*ii*) Constant direct-current tests;
   (*iii*) High-frequency tests;
   (*iv*) Surge or impulse tests.
2. *Testing Apparatus :*
   (*i*) High voltage transformers;
   (*ii*) Apparatus for voltage regulation;
   (*iii*) Control gear and high-voltage connections, including safety and protective devices;
   (*iv*) Apparatus for voltage measurement.

OBJECTIVE TYPE QUESTIONS

Choose the Correct Answer :

1. The voltage control circuits do not use resistance potential dividers because
   (*a*) they involve a large power loss.
   (*b*) they cause distortion of waveform.
   (*c*) they do not give a smooth variation of voltage.
   (*d*) they have non-linear characteristics.
2. Sphere gaps are used for measurement of
   (*a*) instantaneous values of voltage.
   (*b*) r.m.s. values of voltage.
   (*c*) peak values of voltage.
   (*d*) average values of voltage.
3. A.C. transmission lines and cables are tested by using d.c.
   (*a*) they may be used for d.c. if required.
   (*b*) they draw high values of charging current and hence would require a large sized h.v. transformer if tested with a.c.
   (*c*) h.v. testing transformers cannot be used for long times that are required for testing.
   (*d*) all of the above.
4. Cascade connection of transformers is normally used when a single transformer cannot be used because the minimum test voltage
   (*a*) is about 100 kV (*b*) is about 500 kV
   (*c*) is about 1000 kV (*d*) all of the above.
5. A Cockcroft Walton circuit for generation of high voltage d.c. has *n* stages. If the peak voltage of the secondary winding of the high voltage transformer is V_m, the output voltage of the circuit is :
   (*a*) $n V_m$ (*b*) $2(n-1) V_m$
   (*c*) $2(n+1) V_m$ (*d*) $2 n V_m$
6. The impulse voltage wave shape is defined as t_1/t_2 where
   (*a*) t_1 = time taken by the impulse voltage to rise to its peak and
   t_2 = time taken to fall to zero

(*b*) t_1 = time taken by the impulse voltage to rise to its peak and
t_2 = time taken to fall to 50% of its peak

(*c*) t_1 = time taken by the impulse voltage to rise to 50% of its peak and
t_2 = time taken to fall to zero

(*d*) none of the above.

7. The electric strength of new insulating oil should have a minimum strength (r.m.s.) of

(*a*) 30 kV/mm (*b*) 3 kV/mm

(*c*) 50 kV/mm (*d*) 5 kV/mm

8. High-frequency high-voltage tests are carried out on porcelain insulators because

(*a*) they can withstand an a.c. voltage at low frequency but fail to withstand that voltage when applied at high frequency.

(*b*) the insulators, though used on power frequency circuits, may be subjected to high frequency voltages on account of disturbances caused by switching which produce high dielectric loss and heating that may cause failure of insulator at a lower voltage.

(*c*) the insulators may be used for communication purposes wherein the frequencies may be of the order of MHz.

(*d*) all of the above.

9. If the sizes of power transformers and high voltage testing transformers of equal kVA rating are compared :

(*a*) the size of a h.v. testing transformer is greater than that of a power transformer

(*b*) the size of a h.v. testing transformer is smaller than that of a power transformer

(*c*) the size of a h.v. transformer is the same as that of a power tansformer.

10. In Marx's circuit,

(*a*) the capacitors are charged in series and discharged in series.

(*b*) the capacitors are charged in parallel and discharged in parallel.

(*c*) are charged in series and discharged in parallel.

(*d*) are charged in parallel and dicharged in series.

ANSWERS

1. (*a*) **2.** (*c*) **3.** (*b*) **4.** (*b*) **5.** (*d*) **6.** (*b*) **7.** (*b*)
8. (*d*) **9.** (*a*) **10.** (*d*).

THEORETICAL QUESTIONS

1. Write a brief account of different types of high voltage tests usually performed in a h.t. laboratory.

2. How are high voltages classified ?

3. Explain briefly any two of the following high voltage tests :

(*i*) Sustained low-frequency tests.

(*ii*) Constant direct-current tests.

(*iii*) High-frequency tests.

(*iv*) Surge or impulse tests.

4. What are the apparatus essential for all high voltage tests ? What are the methods of voltage regulation in such tests ? Describe these methods in detail.

5. With a neat diagram of connections explain the working of the cascade connected transformers for h.t. testing. Why is it considered advantageous to use this method rather than a single transformer for obtaining high voltages ?

6. Why is high voltage testing important for electrical power apparatus and insulating materials ? Describe briefly the high voltage tests done on electrical apparatus.
7. Explain the reasons for connecting h.v. transformers in cascade. Explain with the help of a diagram, the cascade connections of two transformers.
8. Describe the different methods of voltage control used for h.v. test circuits. Explain their applications, advantages and disadvantages.
9. Explain the following potential divider methods for measurement of r.m.s. value of high voltages.
   (*i*) Resistance potential divider.
   (*ii*) Capacitive potential divider.
   (*iii*) Compensated capacitive potential divider.
10. How will you measure the peak value of an A.C. high voltage ?
11. Explain briefly the use of sphere gaps for high voltage measurement bringing out their advantages and disadvantages in such measurement.
12. Explain, how can you measure high voltage by sphere gap method. Discuss the effect of humidity, if any, on your measurement. Do you measure the r.m.s. value of the voltage or the peak value ? Can you suggest any limitations of this method ?
13. Give the construction, principle and theory of electrostatic voltmeters for the measurement of high voltage.
14. Show that by measuring the rectified charging current of a standard condenser, the peak value of power frequency voltage can be measured.
15. Describe one method of measurement of high voltage of the order of 100 kV. Give the theory of the method.
16. Describe the working, advantages and disadvantages of the following types of voltmeters used for measurement of peak values of high voltages.
   (*i*) Ryall crest voltmeter;
   (*ii*) Ionic wind voltmeter.
17. Explain why transmission lines and cables, which are used for power transmission and distribution of A.C. voltages are tested with high D.C. voltage. Describe the circuits used for high voltage D.C. testing of cables. Describe why is it necessary to establish an equivalence between D.C. and A.C. test voltages ?
18. Describe the circuit of a single-stage impulse generator.
19. Describe a high voltage impulse generator.
20. Explain, with the help of neat diagrams, the working of a surge generator.
21. How is a standard impulse voltage wave shape specified ? Explain clearly with suitable connection diagram the functioning of a multi-stage surge generator. How is the shape of the same controlled ?
22. Explain briefly how the testing of electrical strength of insulating oil is carried out.
23. Explain the terms 'standard impulse voltage' and 'impulse ratio'. Describe in detail the method of generating impulse voltage using Marx circuit for surge or impulse testing.
24. Explain the nature of surge waveform. Describe a high voltage multistage impulse generator.
25. Give a brief account of tests to which porcelain insulators are subjected before commissioning.
26. What is meant by routine testing of porcelain insulators ?
27. Write short notes on :
   (*i*) High voltage testing.
   (*ii*) Crest voltmeter.
   (*iii*) Sphere gap voltmeter.

CHAPTER

13 Illumination

13.1. Introduction; 13.2. Definitions; 13.3. Laws of illumination or luminance; 13.4. Polar curves; 13.5. Photometry – Photometer heads – Photocells (for photometry measurements) – Distribution photometry – Measurement of M.S.C.P. by integrating sphere – Measurement of brightness or luminance – Measurement of illumination – Illumination photometers. *Highlights – Objective Type Questions – Theoretical Questions – Unsolved Examples.*

13.1 INTRODUCTION

Light is *a form of electromagnetic energy radiated from a body which is capable of being perceived by the human eye.*

- The sensation of light results from a flow of energy into the eye and the light will appear to vary if the rate of this flow of energy varies. Light radiations form only a very small part of the complete range of electromagnetic radiations. Light can be of different colours, which depend on the wavelength of the radiation causing it.

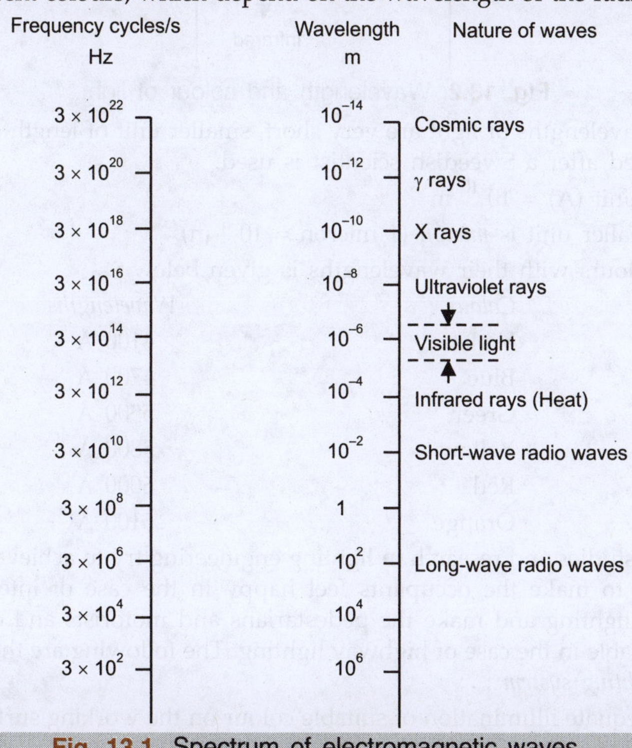

Fig. 13.1. Spectrum of electromagnetic waves.

- Light can be described as a *vibratory motion,* which is transmitted in the form of waves through space. Visible light travels in the form of transverse waves of electromagnetic oscillations. The speed of all electromagnetic waves is 3×10^8 m/s in free space. The wavelength and frequency are different for different waves. The velocity with which these waves travel is related to the wavelength and frequency, by the relation : $v = \lambda f$.

The complete range of waves along with their frequency and wavelength is illustrated in Fig. 13.1.

Fig. 13.2 shows the light emitted together with its colour for the wavelengths within the visible spectrum.

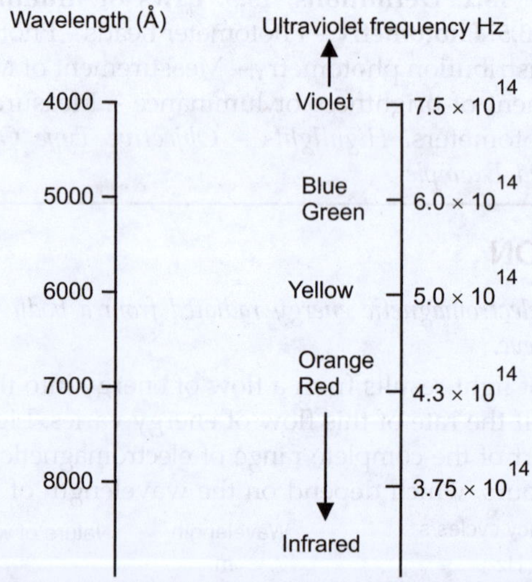

Fig. 13.2. Wavelength and colour of light.

Since the wavelengths of light are very short, smaller unit of length called Angstrom Unit (AU) named after a Sweedish scientist is used.

Angstrom Unit (Å) = 10^{-10} m

Another smaller unit is *micron* (1 micron = 10^{-6} m).

A list of colours with their wavelengths is given below :

Colour	*Wavelengths*
Violet	4100 Å
Blue	4700 Å
Green	5500 Å
Yellow	5800 Å
Red	6000 Å
Orange	6100 Å

- All the studies and research in lighting engineering try to achieve a good lighting scheme to make the occupants feel happy in the case of interior design and factory lighting and make the pedestarians and motorists and other road users comfortable in the case of highway lighting. The following are the *essentials of any good lighting system* :

(*i*) Adequate illumination of suitable colour on the working surfaces.

(*ii*) Good maintenance.

(*iii*) Avoidance of hard shadows.

(*iv*) Avoidance of glare.

- The *aim of artificial lighting* is to supplement the daylight or to replace it in modern offices, homes, industries, workplaces etc. ***Good illumination** ensures increased production, effectivity of work and reduced accidents.*

13.2. DEFINITIONS

1. Solid angle :

- *A* plane angle is subtended at a point and is enclosed by two straight lines lying in the same plane. [Fig. 13.3(*a*)]. Its magnitude is given by,

$$\theta = \frac{\text{Arc}}{\text{Radius}} \text{ 'radians'}$$

The largest angle subtended at a point is 2π radians.

A *radian* is the angle subtended by an arc of a circle whose length equals the radius of the circle.

- ***Solid angle** is the angle generated by the surface passing through the point in space and the periphery of the area [Fig. 13.3(b)].* It is denoted by ω, expressed in *'steradians'* and is given by the *ratio of the area of the surface to the square of the distance between the area and the point,*

i.e., $$\omega = \frac{\text{Area}}{(\text{Radius})^2} = \frac{A}{r^2}$$

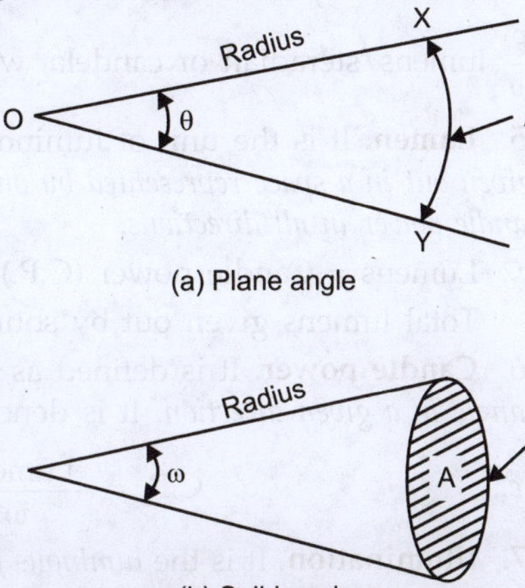

Fig. 13.3. Plane angle and solid angle.

The largest solid angle subtended at a point is due to a sphere at its centre, and is equal to $\frac{4\pi r^2 \text{ (Area of sphere)}}{r^2} = 4\pi$ steradians.

Relationship between ω and θ is obtained as follows :

Consider a curved surface of a spherical segment *ABC* of height *h* and radius *r* (Fig. 13.4)

Surface area of segment $ABC = 2\pi rh$

Here, $h\ (BD) = OB - OD = r - r\cos\frac{\theta}{2} = r\left(1 - \cos\frac{\theta}{2}\right)$

$\therefore$ Surface area of segment $ABC = 2\pi r^2\left(1 - \cos\frac{\theta}{2}\right)$

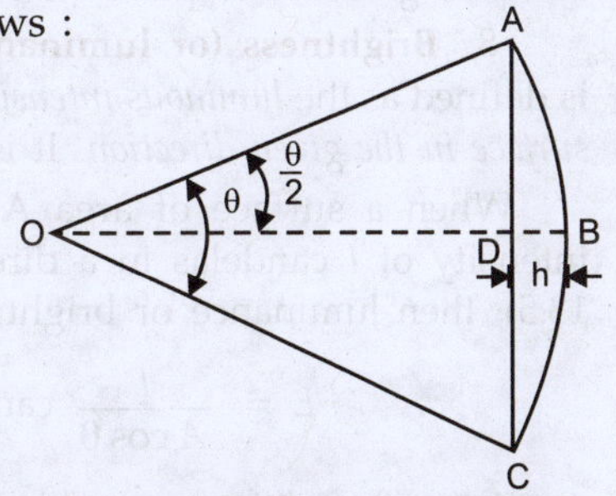

Fig. 13.4. Relation between solid and plane angles.

$$\text{Solid angle, } \omega = \frac{\text{Surface area}}{(\text{Radius})^2} = \frac{2\pi r^2\left(1 - \cos\frac{\theta}{2}\right)}{r^2}$$

$$= 2\pi\left(1 - \cos\frac{\theta}{2}\right) \qquad ...(13.1)$$

2. **Light.** *The radiant energy from a hot body which produces the visual sensation upon the human eye is called* ***light.***

It is denoted by the symbol Q, expressed in lumen-hours (analogous to watt-hours)

3. **Luminous flux.** *The total quantity of light energy emitted per second from a luminous body is called* ***luminous flux.***

It is represented by the symbol F and measured in *lumens.* The concept of luminous flux assists us to specify the output and efficiency of a given light source.

4. **Luminous intensity.** Luminous intensity in a given direction is the *luminous flux emitted by the source per unit solid angle.*

It is denoted by the symbol I and is measured in *'candela'* (cd) or *lumens/stearadian i.e.,*

$I = \frac{F}{\omega}$ lumens/steradian or candela, where ω in the solid angle.

5. **Lumen.** It is the unit of luminous flux and is defined as the *amount of luminous flux given out in a space represented by one unit solid angle by a source having an intensity of one candle power in all directions.*

i.e., Lumens = Candle power (C.P.) × Solid angle (ω)

Total lumens given out by source of one candela is 4π lumens.

6. **Candle power.** It is defined as the *number of lumens emitted by a source in a unit solid angle in a given direction.* It is denoted by symbol C.P.

i.e.,
$$\text{C.P.} = \frac{\text{Lumens}}{\omega}$$

7. **Illumination.** It is the *luminous flux received by a surface per unit area.* It is denoted by symbol E and is measured in *'lumens per square metre' or lux or 'metre-candle' (i.e.,* $E = \frac{F}{A}$ lumens/m^2 or lux, where A is the area of the surface).

Illumination differs from light very much, though generally these terms are used more or less synonymously. Strictly speaking *light is the cause and illumination is the result of the light* on the surfaces it falls. Thus illumination makes surfaces more or less bright with a certain colour and it is this brightness and colour which the eye sees and interprets as something useful or pleasant or otherwise.

8. **Brightness (or luminance).** Brightness of a surface is defined as the *luminous intensity per unit projected area of the surface in the given direction.* It is denoted by the symbol L.

When a surface of area A has an effective luminous intensity of I candelas in a direction θ to the normal (Fig. 13.5), then luminance or brightness of that surface,

$$L = \frac{I}{A\cos\theta} \text{ candela/square metre (Cd/m}^2\text{)}$$

A *"uniform diffuse source"* is one in which the *intensity per unit projected area is the same from all directions of view.*

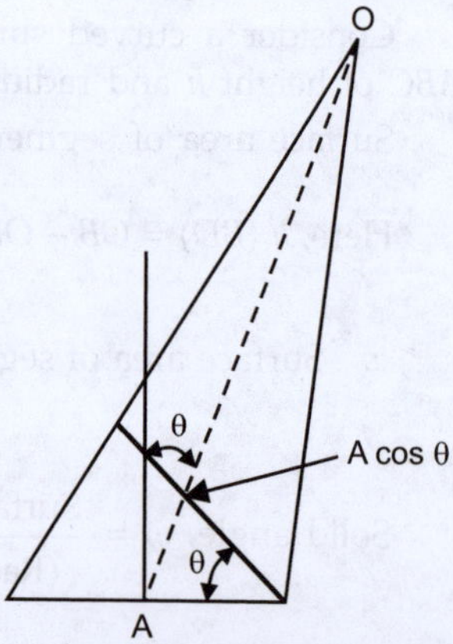

Fig. 13.5.

- ***Relation beween I, L and E :***

Consider a uniform diffuse spherical source with radius r metres and luminous intensity I candela.

Then,
$$L = \frac{I}{\pi r^2}$$

and, $$E = \frac{I}{4\pi r^2} \times 4\pi = \frac{I}{r^2}$$

$$\therefore \quad E = \pi L \qquad ...(13.2)$$

9. **Mean horizontal candle power (M.H.C.P.).** It is defined as the *mean of candle power in all directions in the horizontal plane containing the source of light.*

10. **Mean spherical candle power (M.S.C.P.).** It is defined as the *mean of candle powers in all directions and in all planes from the source of light.*

11. **Mean hemi-spherical candle power.** It is defined as the *mean of all candle powers in all directions above or below the horizontal plane passing through the source of light.*

12. **Reduction factor.** Reduction factor of a source of light is the *ratio of its mean spherical candle power to its mean horizontal candle power, i.e.,*

$$\text{Reduction factor} = \frac{\text{M.S.C.P.}}{\text{M.S.H.P.}}.$$

13. **Lamp efficiency.** It is defined as the *ratio of the luminous flux to the power input.* It is expressed in *lumens per watt.*

14. **Specific consumption.** It is defined as the *ratio of power input to the average candle power.* It is expressed in *watts per candle.*

15. **Space-height ratio.** It is defined as the *ratio of horizontal distance between adjacent lamps and height of their mountings.*

16. **Utilisation factor (UF).** The *ratio of total lumens reaching the working plane to total lumens given out by the lamp* is called *utilisation factor (or co-efficient of utilisation).*

17. **Maintenance factor (MF).** It is the *ratio of illumination under normal working conditions to the illumination when the things are perfectly clean.*

18. **Depreciation factor.** This is merely the reverse of the maintenance factor and is defined as the *ratio of initial metre-candles to the ultimate maintained metre-candles on the working plane.* It value is *more than unity.*

19. **Waste light factor.** Whenever a surface is illuminated by a number of sources of light, there is always a certain amount of wastage of light on account of overlapping and falling of light outside at the edges of the surface. The effect is taken into account by multiplying the theoretical value of lumens required by 1.2 for rectangular areas and 1.5 for irregular areas and objects such as statues, monuments etc.

20. **Absorption factor.** In the places where atmosphere is full of smoke fumes, such as in foundries, there is a possibility of absorption of light. *The ratio of total lumens available after absorption to the total lumens emitted by the source of light is called the* ***absorption factor.*** Its values varies from unity for clean atmosphere to 0.5 for foundries.

21. **Beam factors.** The ratio of lumens in the beam of a projector to the lumens given out by lamps is called the *beam factor.* This factor takes into account the absorption of light by reflector and front glass of the projector lamp. Its value varies from 0.3 to 0.6.

22. **Reflection factor.** When a ray of light impinges on a surface, it is reflected from the surface at an angle of incidence, as shown in Fig. 13.6. A certain portion of incident light is absorbed by the

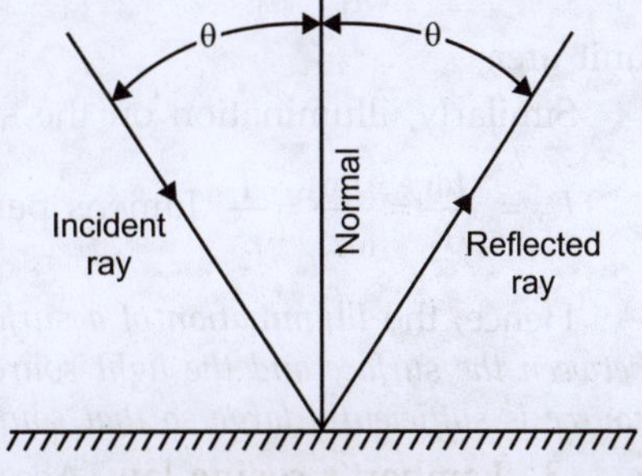

Fig. 13.6.

surface. The ratio of reflected light to the incident light is called the *'reflection factor'*. It is *always less than unity.*

23. **Glare.** The opening of the pupil in the human eye is controlled by the iris. If a bright object comes into the view of the eye, large amount of light produces an intense image on the retina and the iris automatically protects the eye by contracting the pupil, thus reducing the intensity of the image. When the eye is towards another object which is less bright as compared to the bright object already in the field of view, the iris will contract reducing the amount of light received on the retina from every object in the field of view and making it difficult to see the object desired. At the same time, the portion of the retina which received image of the bright object remains fatigued. This phenomenon is called ***"Glare"***, and is familiar in connection with motor-car head lights.

In other words,"**glare**" *may be defined as the brightness within the field of vision of such a character that causes annoyance, discomfort, interference with vision or eye-fatigue.*

13.3. LAWS OF ILLUMINATION OR LUMINANCE

The illumination (E) of a surface depends upon the following *factors* (The source is assumed to be a point source or otherwise sufficiently away from the surface to be regarded as such).

1. E is *directly proportional to the luminous intensity* (I) of the source. In other words, $E \propto I$.

2. **Inverse square law.** The *illumination of a surface is inversely proportional to the square of the distance of the surface from the source.*

In other words, $E \propto \dfrac{1}{r^2}$.

Proof : Consider surface areas A_1 and A_2 at distances r_1 and r_2 respectively from the point source S of luminous intensity I and *normal* to the rays as shown in Fig. 13.7. Let the solid angle subtended be ω.

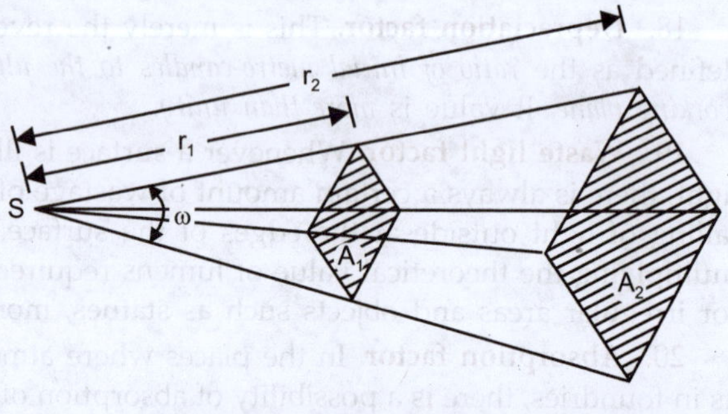

Fig. 13.7. Inverse square law.

Total luminous flux radiated

$= I\omega$ lumens

Illumination of the surface of area A_1,

$$E_1 = \frac{I\omega}{A_1} = \frac{I\omega}{\omega r_1^2} = \frac{I}{r_1^2} \text{ lumens per unit area}$$

Similarly, illumination on the surface of area A_2,

$$E_2 = \frac{I\omega}{A_2} = \frac{I\omega}{\omega r_2^2} = \frac{I}{r_2^2} \text{ lumens per unit area.}$$

Hence, the *illumination of a surface is inversely proportional to the square of the distance between the surface and the light source provided that the distance between the surface and the source is sufficiently large so that source can be regarded as a point source.*

3. **Lambert's cosine law.** According to this law, *E is directly proportional to the cosine of the angle made by the normal to the illuminated surface with the direction of the incident flux.*

As shown in Fig. 13.8, let F be the flux incident on the surface of area A when in position 1. When this surface is turned back through an angle θ, then flux incident on it is $F \cos \theta$. Hence, illumination of the surface when in position 1 is $E_1 = \frac{F}{A}$. But when in position 2, $E_2 = \frac{F \cos \theta}{A}$.

$$\therefore \quad E_2 = E_1 \cos \theta$$

Combining all these factors together, we get

$$E = \frac{I \cos \theta}{r^2}.$$

The unit is lumens per unit area.

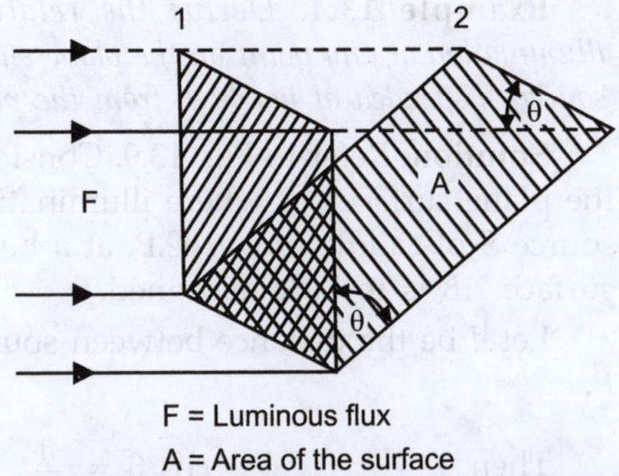

Fig. 13.8. Lambert's cosine law.

Example 13.1. *A 250 V lamp has a total flux of 1500 lumens and takes a current of 0.4 A. Calculate :*

(i) Lumens per watt. *(ii) M.S.C.P. per watt.*

Solution. *Given* : V = 250 volts ; F = 1500 lumens ; I = 0.4 A

Mean spherical candle power of lamp,

$$\text{M.S.C.P.} = \frac{F}{4\pi} = \frac{1500}{4\pi} = 119.4$$

(i) **Lumens per watt** $= \frac{\text{Output of lamp in lumens}}{\text{Wattage of lamp in watts}} = \frac{1500}{250 \times 0.4}$ **= 15. (Ans.)**

(ii) **M.S.C.P. per watt** $= \frac{\text{M.S.C.P. of lamp}}{\text{Wattage of lamp}} = \frac{119.4}{250 \times 0.4}$ **= 1.194. (Ans.)**

Example 13.2. *A 0.3 metre diameter diffusing sphere of opal glass having 15% absorption encloses an incandescent lamp with a luminous flux of 4500 lumens. Calculate average luminance of the sphere.*

Solution. *Given* : d = 0.3 m ; Lamp = 4500 lumens

Flux 1 emitted by the sphere, F_{sphere} = (1 – 0.15) × 4500 = 3825 lumens

Surface area of the sphere $= 40\pi r^2 = 4\pi \times \left(\frac{0.3}{2}\right)^2 = 0.2827 \text{ m}^2$

$\therefore$ Average luminance of sphere $= \frac{F_{sphere}}{\text{Surface area of sphere}}$

$= \frac{3825}{0.2827}$ **= 13530 lumens/m². (Ans.)**

Example 13.3. *The candle power of a lamp is 120. A plane surface is placed at a distance of 2.5 metres from this lamp. Calculate the illumination on the surface when it is (i) normal, (ii) inclined to 45°, and (iii) Parallel to rays.*

Solution. *(i)* $E = \frac{C.P.}{d^2} = \frac{120}{(2.5)^2}$ **= 19.2 lux. (Ans.)**

(ii) $E = \frac{C.P.}{d^2} \times \cos 45° = \frac{120}{(2.5)^2} \times \cos 45°$ **= 13.58 lux. (Ans.)**

(iii) E = 0, since the rays of light are parallel to the surface, they cannot illuminate it. **(Ans.)**

Example 13.4. *Derive the relation to find the illumination at any point on the plane surface due to light source suspended at height h from the plane surface.*

Solution. Refer to Fig. 13.9. Consider a point *P* on the plane surface *AB* where illumination due to light source *S* of candle power C.P. at a height h from the surface *AB* is to be determined.

Let *d* be the distance between source *S* and point *P*.

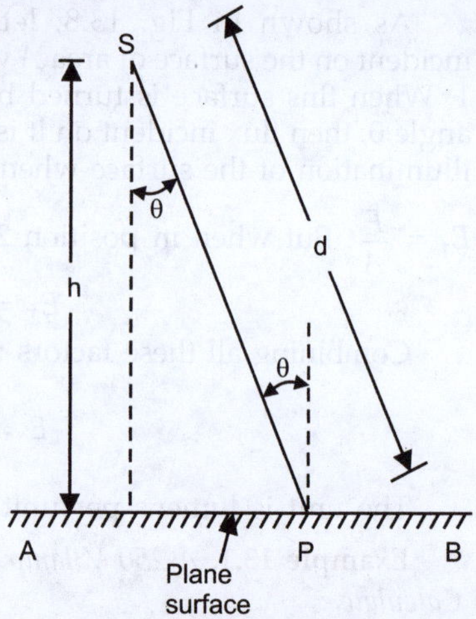

Fig. 13.9.

Then, $\cos\theta = \frac{h}{d}$, or, $d = \frac{h}{\cos\theta}$

Illumination at point *P*, by laws of illumination

$$= \frac{\text{C.P.}}{d^2}\cos\theta = \frac{\text{C.P.}}{(h/\cos\theta)^2}\cos\theta$$

$$= \frac{\text{C.P.}}{h^2}\cos^3\theta,$$

and illumination at any point *O*, *vertically below the source of light* $= \frac{\text{C.P.}}{h^2}$.

Hence, *illumination at any point on a plane is* $\cos^3\theta$*-times of illumination at a point just vertically below the light source, where θ is the angle between the normal to the surface at the point and rays of light.*

Example 13.5. *A 500 W lamp having M.S.C.P. of 800 is suspended 3 m above the working plane :*

(*i*) *Illumination directly below the lamp at the working plane.*

(*ii*) *Lamp efficiency.*

(*iii*) *Illumination at a point 2.4 m away on the horizontal plane from vertically below the lamp.*

Solution. Wattage of the lamp = 500 W

M.S.C.P. of the lamp, $I = 800$

Height of the lamp, $h = 3$ m

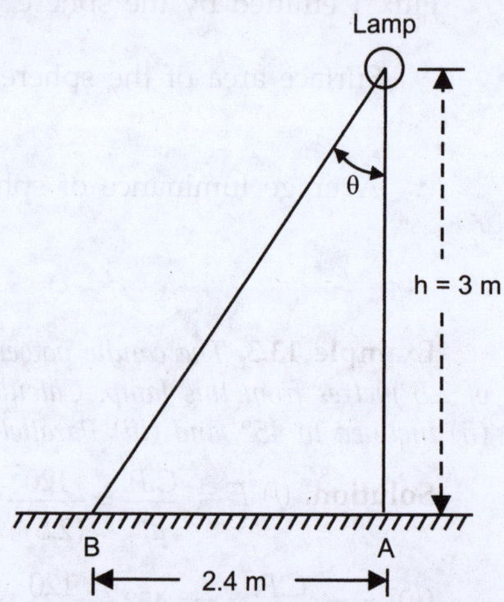

Fig. 13.10.

(*i*) Illumination directly below the lamp at the working plane :

Illumination directly below the lamp,

$$E_A = \frac{I}{h^2} = \frac{800}{3^2} = \mathbf{88.9\ lux.} \quad \textbf{(Ans.)}$$

(*ii*) Lamp efficiency :

$$\text{Lamp efficiency} = \frac{\text{Luminous flux}}{\text{Power input}}$$

$$= \frac{4\pi \times \text{M.S.C.P.}}{500}$$

$$= \frac{4\pi \times 800}{500}$$

$$= \mathbf{20.1\ lumens/watt.} \quad \textbf{(Ans.)}$$

(iii) **Illumination at a point 2.4 m away :**

Illumination at a point 2.4 m away on the horizontal plane from vertically below the lamp,

$$E_B = \frac{I}{h^2}\cos^3\theta$$

Here,
$$\cos\theta = \frac{3}{\sqrt{3^2 + 2.4^2}} = 0.7808$$

$$\therefore \quad E_B = \frac{800}{3^2}\times(0.7808)^3 = \mathbf{42.3\ lux.}\quad \textbf{(Ans.)}$$

Example 13.6. *A lamp with reflector is mounted 10 m above the centre of a circular area of 20 m diameter. If the combination of the lamp and reflector gives a uniform C.P. of 800 over the circular area, determine the maximum and minimum illumination produced on the area.*

(Punjab University)

Solution. Candle power of the lamp, C.P. = 800

Height of the lamp, h = 10 m

Diameter of the circular area = 20 m

The *maximum illumination* will occur *directly below the lamp i.e.,* at point *A* and

$$= \frac{\text{C.P.}}{h^2} = \frac{800}{10^2} = \mathbf{8\ lux.}\quad \textbf{(Ans.)}$$

The minimum illumination will occur at the periphery of the circular area *i.e.,* at point *B* and

$$= \frac{\text{C.P.}}{h^2}\cos^3\theta$$

$$= \frac{800}{10^2}\times\left(\frac{10}{\sqrt{10^2+10^2}}\right)^3 = \mathbf{2.83\ lux.}\quad \textbf{(Ans.)}$$

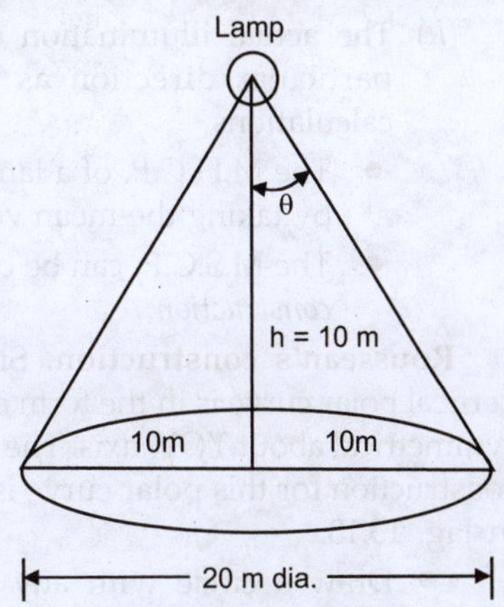

Fig. 13.11.

13.4. POLAR CURVES

The luminous intensity for most lamps or sources of light is not the same in all directions, because of their un-symmetrical shape. Often it is necessary to know the distribution of light in various directions to ascertain how the candle power of a light source varies in different directions. The luminous intensity in all the directions can be represented by ***polar curves.***

— If the luminous intensity *i.e.,* candle power is measured in a *horizontal plane* about a vertical axis and a curve is plotted between the candle power and the angular position, a *"horizontal polar curve"* or diagram is obtained.

— If the candle power is measured at angular position in a vertical plane, a polar curve in the vertical plane, called *"vertical polar curve"*, is obtained.

Fig. 13.12 shows typical polar curves for an ordinary lamp.

The *polar curves are used to determine the following :*

(i) The mean horizontal candle power (M.H.C.P.) and mean spherical candle power (M.S.C.P.).

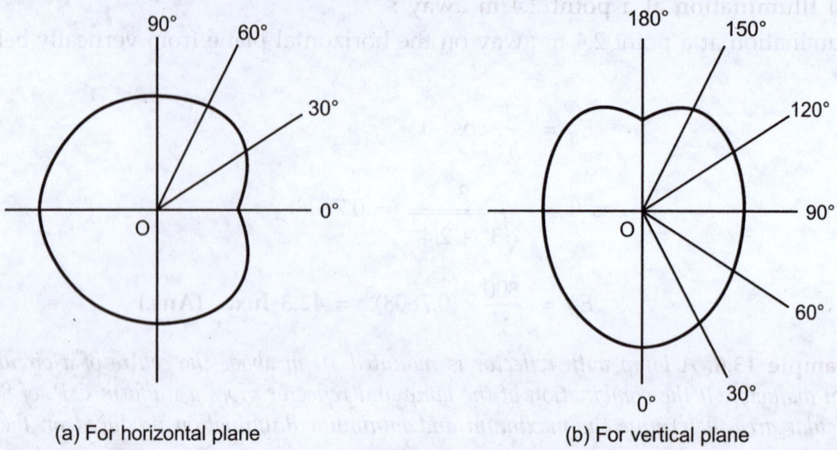

Fig. 13.12. Polar curves.

(*ii*) The actual illumination of a surface by employing the candle power in that particular direction as read from the vertical polar curve in illumination calculations.

- The M.H.C.P. of a lamp can be determined from the horizontal polar curve by taking the mean value of the candle power in a horizontal direction.
- The M.S.C.P. can be determined from the vertical polar curve by *Roussean's construction.*

Roussean's construction. Suppose the vertical polar curve is in the form of two lobes symmetrical about *YOY′* axis. The Roussean's construction for this polar curve is illustrated in Fig. 13.13.

- Draw a circle with any convenient radius with centre *O* as the centre.
- Draw *PQ* parallel to *YOY′* and equal to the vertical diameter of the circle.
- Draw any line *OEA* meeting the polar curve in *E* and the circle in *A*. Let the projection be *S*.
- At S erect an ordinate ST = OE.
- By similar construction draw other ordinates. The curve PSQUTVP obtained by joining these ordinates is known as *Roussean's curve.* The *mean ordinate of this curve gives the mean spherical candle power* (M.S.C.P.).

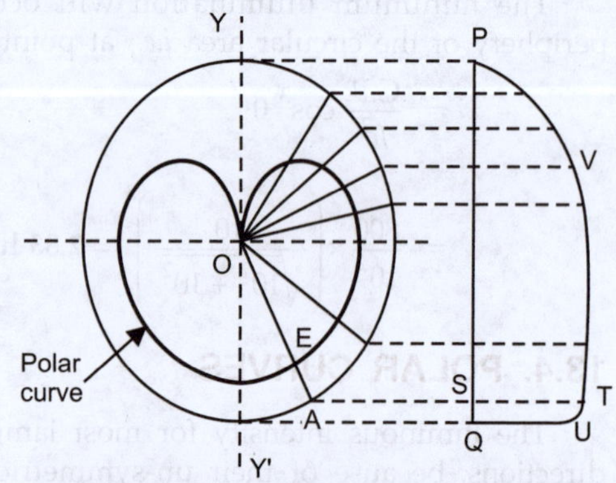

Fig. 13.13. Roussean's construction.

$$\text{The mean ordinate of the curve} = \frac{\text{Area PSQUTVP}}{\text{Length of } PQ}$$

The area under the curve can either be determined on a graph paper or by using Simpson's rule.

13.5. PHOTOMETRY

We shall now discuss the *comparison and measurement of candle powers.* The candle power of a source in any direction is measured by comparison with a standard or substandard source employing photometer bench and some form of photometer.

The *"photometer bench"* (See Fig. 13.14) consists of two steel rods (2 to 3 metres long) which carry the stands or saddles for holding the two sources, the carriage for the photometer head and for any other apparatus employed in making measurement. One of the steel rods carries a brass strip with graduate scale in mm.

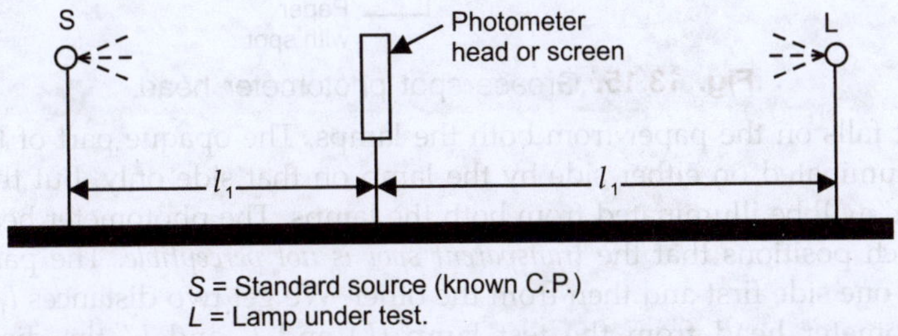

Fig. 13.14. Photometer bench for measurement of candle power.

The bench should be rigid so that source being compared may be free from vibrations and the carriage holding the photometer head should be capable of moving smoothly and with very little effort. The photometer head acts as a *screen* for comparison of illumination of the standard source and the source under test.

The photometer head or screen is moved in between the two fixed sources until the illumination on both the sides of the screen is same. If the distances of the standard source S and lamp (source) under test L from photometer head are l_1 and l_2 respectively, then applying inverse square law, we get

$$\frac{\text{Candle power of } L}{\text{Candle power of } S} = \frac{l_2^2}{l_1^2}$$

$$\therefore \quad \text{Candle power of } L = \text{Candle power of } S \times \frac{l_2^2}{l_1^2} \qquad \text{...(13.3)}$$

In order to eliminate errors due to reflected light the experiment is performed in dark room with dead black walls and ceiling.

- In this measurement since the square of distances are involved, therefore, distances should be accurately measured. For obtaining distance exactly, two points are determined at which there is a perceptible difference in illumination from the two sides and the point half way between them is taken as the position of equal illumination.

13.5.1. Photometer Heads

Although, quite a few types of photometer heads are in use, yet most common in use are *Bunsen* and *Lummer-Brodhun type.* These heads give quite accurate results when the lamps of the same or very similar colour are to be compared. However, when two lamps with different colours are to be compared, a *"Flicker photometer"* gives better results.

Bunsen head. It is also called the *"Grease spot photometer"*. It consists of a piece of thin opaque paper at the centre of which is a transluscent "spot", made by treating the paper with oil or wax.

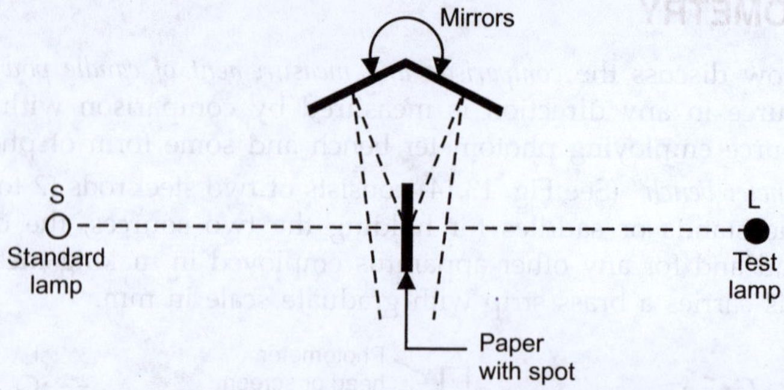

Fig. 13.15. Grease spot photometer head.

- Light falls on the paper from both the lamps. The opaque part of the paper will be illuminated on either side by the lamp on that side only. But the transparent "spot" will be illuminated from both the lamps. The photometer head is adjusted at such positions that the *transparent spot is not perceptible.* The paper is viewed from one side first and then from the other. We get two distances l_1 and l_1' of the photometer head from the test lamp (L) and l_2 and l_2' the distances of the photometer head from the standard lamp (S). The C.P. (candle power) of the test lamp is calculated from the following relation :

$$\frac{\text{C.P. of test lamp } (L)}{\text{C.P. of the standard lamp } (S)} = \frac{l_1 l_1'}{l_2 l_2'}$$

- In another method, one side of the paper is illuminated from a fixed lamp. On the other side of the photometer first the test lamp (L) is used and then the standard lamp (S) is used to make the spot to vanish. If l_1 and l_2 are these distances, then we have,

$$\frac{\text{C.P. of test lamp } (L)}{\text{C.P. of standard lamp } (S)} = \frac{l_1^2}{l_2^2}$$

The *use of two mirrors with the photometer head is perhaps the most accurate method* (Fig. 13.15). The two sides of the spot can be viewed simultaneously and the position of the head for equal "contrast" in illumination between the opaque part and the transparent spot of the paper on the two sides is located. The C.P. of the test lamp (L) can then be calculated by the expression given earlier.

Lummer-Brodhun photometer head. There are two types of Lummer-Brodhun photometer heads :

(*i*) Equality-of-brightness type.

(*ii*) Contrast type.

Contrast type is more accurate and, therefore *more used in photometric measurements.*

Flicker photometers :

- These photometers are used *when the two sources giving light of different colours are to be compared.*
- They operate on the fact that if two illuminated surfaces are presented to the eye with rapid alternations, the *flicker disappears when the surfaces are of equal brightness.* The speed of alternation should be kept as low as possible at which the disappearance of the flicker can be obtained for the small variation in brightness.

- These photometers are not affected by the difference in colour of two lights to be compared as much as photometers of the steady comparison types are affected. This is because the colour difference between two alternating fields of light disappears at a lower speed of alternation than the speed at which difference in brightness appears.

13.5.2. Photocells (for photometric measurements)

Except for the measurement of luminance of sources, measurement by visual comparisons have been replaced by *'photocell'* because of the following facts :

(*i*) The more complex procedures and the apparatus involved in visual measurements are avoided.

(*ii*) These cells give *more accurate and faster measurements.*

(*iii*) The measurements are consistent.

The two types of photocells used in photometric measurements are :

1. Photovoltaic cells
2. Photoemissive cells.

1. **Photovoltaic cell.** Fig. 13.16 shows the construction of a photovoltaic cell (also known as the barrier layer or rectified cell). It consists of a base plate made of either steel or aluminium and carries a layer of metallic selenium which is light sensitive. An electrically conducting layer of cadmium oxide is applied by sputtering over the selenium layer. The layer is *sufficiently thin* to allow light to reach the selenium and is electrically continuous as it acts as a negative pole. The negative contact is formed of a strip of wood's metal sprayed on to the edge of the top surface. The base plate forms the positive contact. The front surface of the cell is protected by a transparent varnish.

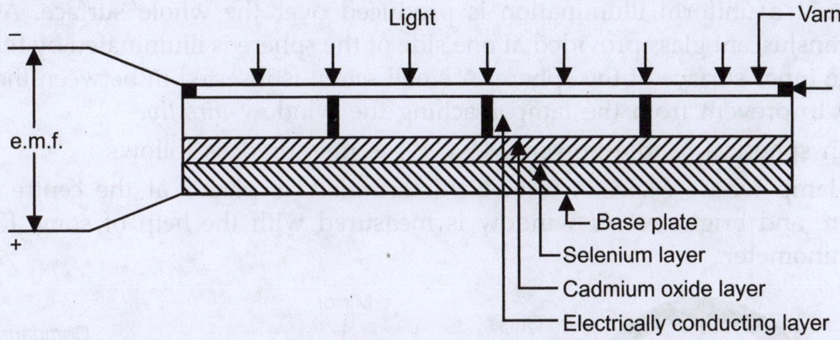

Fig. 13.16. Photovoltaic cell.

As the light falls on the upper surface of selenium, electrons are released from the surface which maintain a flow of current through the external circuit connected between the +ve and –ve contacts.

- *In photometry, the current output of a photocell should be proportional to the illumination which is achieved by keeping the external resistance at a low value.* For achieving greater accuracy, the illumination should not be allowed to exceed 275 lumens/m^2.
- The smaller the size of the cell (of course compatible with obtaining sufficient current to be properly measured) a better linear relation is obtained between current and illumination. This is because, for such a cell, the resistance of the electrically conducting film is at a minimum. Also, the current being small, the voltage drop due to the circuit resistance will be kept low.

2. **Photoemissive cell.** In order to get greater precision in terms of linearity and stability it is better to use photoemissive cell rather than photovoltaic cell. However, the circuit involved is more complex and requires the use of some sort of valve amplifiers.

13.5.3. Distribution Photometry

- ***Distribution photometry** assists in finding out the variations of luminous intensity around a lighting fitting so that polar curves and other diagrams for the fitting can be obtained and also the performance of the fitting is studied.*

Following points are important for proper distribution photometry :

(*i*) The *temperature of the test area should be kept stable* (since the output flux of some sources like fluorescent tubes depends upon the temperature of the surroundings).

(*ii*) *Only that light which is to be measured should be allowed to reach the photocell.*

(*iii*) *Optical path should be made sufficiently long* so that the inverse square law may be applied with a reasonable degree of accuracy.

- In a very simple ***distribution photometer,*** the photocell fitting moves around the fitting on a semicircular track and its position is controlled by means of a flexible cable. A complete series of curves at different angles of azimuth can be obtained by rotating the fitting about a vertical axis. The mounting of the fitting on the apparatus is facilitated by mounting a gear system on the gentry which turns the fittings so that it can be brought down to easy access.

13.5.4. Measurement of M.S.C.P. by Integrating Sphere

The ***integrating sphere*** consists of a hollow sphere whose diameter (one or more than one metre) is large compared to the lamp to be tested, having a smooth inner surface with a uniform coating of white paint. If the lamp is hung inside the sphere, the light is so diffused that a uniform illumination is produced over the whole surface. A small window of transluscent glass provided at one side of the sphere is illuminated by reflected light from the inner surface of the sphere. A small *screen* is inserted in between the lamp and window to prevent from the lamp reaching the window *directly.*

The mean spherical candle power (M.S.C.P.) is measured as follows :

- The lamp whose M.S.C.P. is to be determined, is placed at the centre of the sphere and brightness of window is measured with the help of some form of illuminometer.

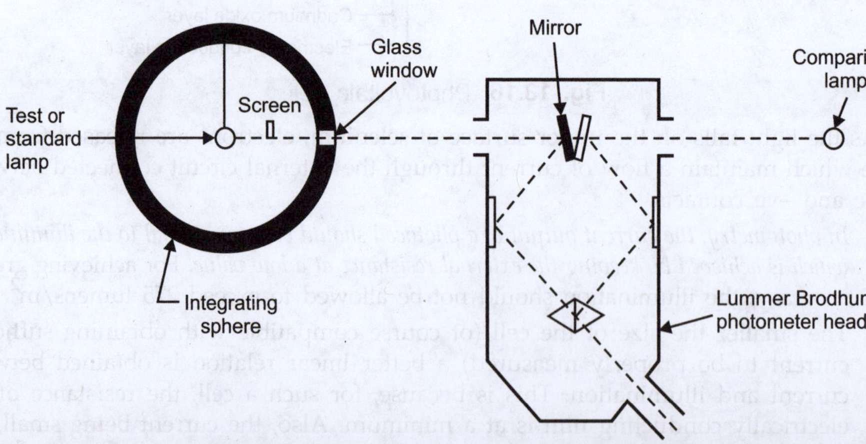

Fig. 13.17. Measurement of M.S.C.P. using the integrating sphere.

- The test lamp is replaced by a standard lamp whose M.S.C.P. is known and brightness of the window is again measured.

Since the M.S.C.P. of the two lamps are proportional to the respective brightness of the window, therefore, the M.S.C.P. of the lamp under test can be determined by knowing the M.S.C.P. of standard lamp and brightness of window in the two cases.

Fig. 13.17 shows an arrangement in which a Lummer-Brodhun photometer head is used in which a mirror is placed as indicated :

— The light from the glass window is reflected by the mirror as shown.

— Another lamp is used for comparison and for obtaining balance by shifting the position of the comparison lamp from the photometer head.

— We get two distances of comparison lamp from the photometer head : *First distance* when the glass window is illuminated by the test lamp and the *second distance* when the glass window is illuminated by the standard or sub-standard lamp.

— The squares of the corresponding distances of the comparison lamp from the photometer head are *inversely proportional to the M.S.C.Ps. of the lamps.*

13.5.5. Measurement of Illumination – Illumination Photometers

The simplest and best method of measuring illumination is by a *portable type illumination photometer calibrated to read directly in "lux (lumens/m^2)".*

— No shadow should be allowed to fall on the meter during measurements and meter should be held perfectly horizontal and stationary for accurate results.

— The meter should be *colour-corrected type* so that it can be used for all kinds of light sources.

- A number of such instruments is use are :

  Trotter Illumination Photometer, Macbeth Illuminometer, Holophane lumeter etc.

— These are portable photometers which are built *compact and light in weight* and are used for measurements of illumination in *homes, offices, commercial houses* and *industries.*

— These photometers *measure the intensity of illumination or illuminance of a source and not the candle power of the luminous sources.*

13.5.5.1. Trotter illumination photometer

This photometer works on *Lambert's cosine law.*

Fig. 13.18 shows the Trotter illumination photometer :

Construction :

— It consists of a small lamp *L* supplied with 4*V* portable battery *B*, which throws light on an inclined mirror *M* from which it is reflected to a white matt surface S_1 of high diffusion quality which can be *tilted.*

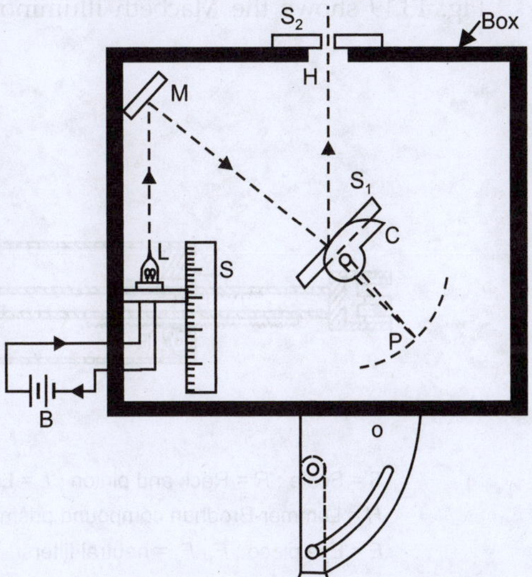

S_2 = Test plate; H = slot; S_1 = Tilting surface; B = Battery; L = Lamp; S = Pre-calibrated scale; M = Mirror; C = Cam; P = Pointer

Fig. 13.18. Trotter illumination photometer.

— The top of the box containing the lamp and the tilting surface S_1 carries a test plate S_2 which receives the illumination to be measured. The test plate S_2 is exactly the same with diffusing material as S_1. Like S_1, it is also of high diffusing quality so that *intensity variations in different directions obey "cosine law"*. It is necessary that *test plate should be made out of a material which does not introduce errors due to selective absorption.*

— The test plate has a slot H cut in it, through which the surface S_1 can be viewed.

Working :

— For measurement, S_2 is placed where illumination is required.

— Then the surface S_1 is tilted by means of a cam C, through a roller attached to the tilting surface, until its illumination as seen through slot H is the same as that of the test plate S_2.

— A pointer P moving over a pre-calibrated scale S then gives the illumination of S_2 in lumens/m^2.

— The scale S is then calibrated by producing known intensities of illumination on S_2 from a standard lamp and marking the position of pointer for photometric balance between S_1 and S_2. Thus, it is not necessary to know the candle power of the lamp although for constancy of candle power it must be aged before use in the photometer.

— The purpose of cam C is to provide a *fairly graduated scale.*

- The range of this photometer is normally from *0.1 to 42 lumens/m²*.

13.5.5.2. Macbeth illuminometer

It is a sensitive photometric device and is one of the most accurate instruments of this type. It is very simple and easy to operate.

Its operation is based upon the *inverse square law.*

Fig. 13.19 shows the Macbeth illuminometer.

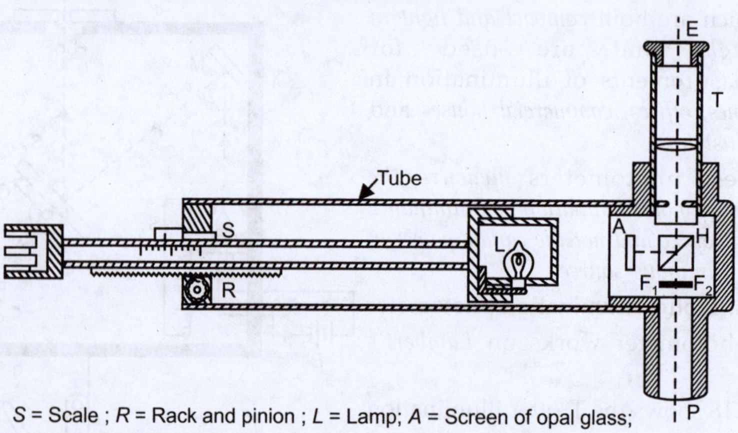

S = Scale ; *R* = Rack and pinion ; *L* = Lamp; *A* = Screen of opal glass;
H = Lummer-Brodhun compound prism ; *P* = Tube ; *T* = Telescope ;
E = Eye piece ; F_1, F_2 = neutral filters

Fig. 13.19. Macbeth illuminometer.

Construction :

— A Lummer-Brodhun compound prism H is mounted in a box and viewed through a telescope *T*.

— On the opposite side of the prism from the telescope is a tube *P*, which is pointed towards a test plate of opal glass, situated at the point at which the illumination is to be measured.

— *A* is a screen of opal glass, which is illuminated by a lamp *L* in an enclosure having an aperture in the side nearest the screen *A*.

Working :

— For measurement, the lamp *L* is moved along a tube which is at right angles to the telescope, by a rack and pinion *R*, until balance is obtained at the photometer head. A scale *S* then gives the illumination of the test plate *directly*.

— The lamp *L* is supplied from a dry battery or accumulator, and a milliammeter and rheostat are provided with instrument for adjustment of the lamp current to its correct value (*i.e.* that which makes the scale *S* direct reading).

- The normal range of this photometer is from *5 to 250 lumens/m²*. Its range can be extended either way by inserting neutral filters F_1 and F_2.

13.5.5.3. Holophane lumeter

The average brightness or luminance of a source in a given direction is given by the *ratio of its intensity to the projected area of the source in that direction.* For achieving greater sensitivity it is necessary to use some kind of optical system to collect more light flux.

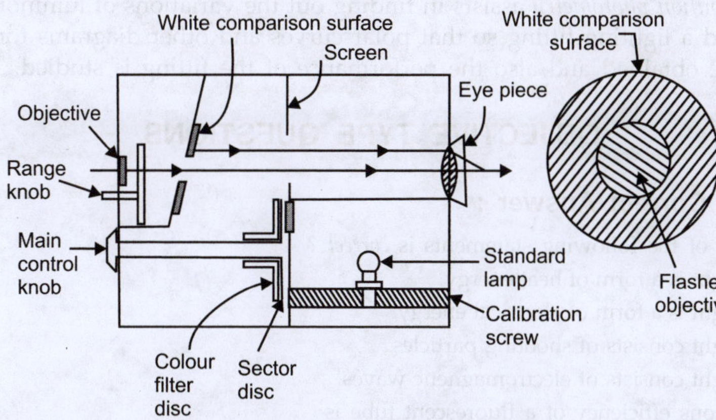

Fig. 13.20. Holophane lumeter.

Fig. 13.20 shows the schematic arrangement of the *"holophane lumeter"*:

— In this lumeter, the light from the source falls on the objective lens which is viewed through the eye piece. The source is, therefore, seen as a uniformly bright *disc* having the luminance of the source (multiplied by the percentage loss in the optical system).

— This *disc* is surrounded by a ring of light reflected by the white comparison surface. The luminance of this ring is matched against the object by adjusting the range and main control knobs which are graduated.

— A standard lamp placed in the whitened compartment, illuminates the white screen. This lamp is moved with the help of a calibration screw only when the instrument requires recalibration.

— Light reaches the white comparison surface after passing through the opal window and amount of this light is adjusted by adjusting the sector disc by means of main control knob.

HIGHLIGHTS

1. *Light* is defined as the radiant energy from a hot body which produces the visual sensation upon the human eye.
2. *Law of inverse square.* The illumination of a surface is inversely proportional to the square of the distance between the surface and the light source provided that the distance between the surface and the source is sufficiently large so that source can be regarded as a point source.
3. *Lambert's cosine law.* According to this law, the illumination of any point on a surface is proportional to the cosine of the angle between the normal at that point and the direction of luminous flux.
4. The luminous intensity in all the directions can be represented by *polar curves.*
5. Most commonly used photometer heads are :
   Bunsen head ; Lummer-Brodhun photometer head.
6. Flicker photometers are used when the two sources giving light of different colours are to be compared.
7. The two types of photocells used in photometric measurements are :
   (*i*) Photovoltaic cells; (*ii*) Photoemissive cell
8. *Distribution photometry* assists in finding out the variations of luminous intensity around a lighting fitting so that polar curves and other diagrams for the fitting can be obtained and also the performance of the fitting is studied.

OBJECTIVE TYPE QUESTIONS

Choose the Correct Answer :

1. Which of the following statements is *correct* ?
   (*a*) Light is a form of heat energy.
   (*b*) Light is a form of electrical energy.
   (*c*) Light consists of shooting particles.
   (*d*) Light consists of electromagnetic waves.
2. Luminous efficiency of a fluorescent tube is
   (*a*) 10 lumens/watt (*b*) 20 lumens/watt
   (*c*) 40 lumens/watt (*d*) 60 lumens/watt.
3. Candela is the unit of which of the following ?
   (*a*) Wavelength (*b*) Luminous intensity
   (*c*) Luminous flux (*d*) Frequency.
4. Colour of light depends upon
   (*a*) frequency (*b*) wave length
   (*c*) both (*a*) and (*b*) (*d*) speed of light.
5. Illumination of one lumen per sq. metre is called ......
   (*a*) lumen metre (*b*) lux
   (*c*) foot candle (*d*) candela.
6. A solid angle is expressed in terms of ......
   (*a*) radians/metre (*b*) radians
   (*c*) steradians (*d*) degrees.

7. The unit of luminous flux is ......

(*a*) watt/m^2 (*b*) lumen

(*c*) lumen/m^2 (*d*) watt.

8. In houses, the illumination is in the range of

(*a*) 2–5 lumens/watt (*b*) 10–20 lumens/watt

(*c*) 35–45 lumens/watt (*d*) 60–65 lumens/watt.

9. "The illumination is directly proportional to the cosine of the angle made by the normal to the illuminated surface with the direction of the incident flux."

Above statement is associated with

(*a*) Lambert's cosine law (*b*) Planck's law

(*c*) Bunsen's law of the illumination (*d*) Macbeth's law of illumination.

10. Desired illumination of level of the working plane depends upon

(*a*) age group of observers.

(*b*) whether the object is stationary or moving.

(*c*) size of the object to be seen and its distance from the observer.

(*d*) whether the object is to be seen for longer duration or shorter duration of time.

(*e*) all above factors.

11. Which of the following instruments is used for the comparison of candle powers of different sources ?

(*a*) Radiometer (*b*) Bunsen meter

(*c*) Photometer (*d*) Candle meter.

12. ...... photometer is used for comparing the lights of different colours ?

(*a*) Grease spot (*b*) Bunsen

(*c*) Lummer brodhun (*d*) Guilds flicker.

13. Luminous flux is

(*a*) the light energy radiated by sun.

(*b*) the part of light energy, radiated by sun, which is received on the earth.

(*c*) the rate of energy radiation in the form of light waves.

(*d*) none of the above.

14. Stroboscopic effect due to use of discharge lamps in workshops results in moving machinery appearing

(*a*) stationary

(*b*) stationary running slow

(*c*) stationary running in reverse direction

(*d*) all of the above.

15. When a electric bulb is broken it produces bang, this is due to

(*a*) vacuum inside the bulb

(*b*) pressure of air in the bulb

(*c*) pressure inside is equal to that outside

(*d*) none of the above.

16. Due to which of the following reasons the light of a tube appears cooler than that of a bulb ?

(*a*) Tungsten is not used in the tube.

(*b*) Tube is painted with milky colour.

(*c*) Tube consumes less power.

(*d*) Surface area of the tube is more than that of bulb.

(*e*) none of the above.

17. Soft shadows are produced by
(*a*) using surface source of light instead of point source of light.
(*b*) increasing the number of lamps.
(*c*) both (*a*) and (*b*).
(*d*) none of the above.

18. The level of illumination on a surface *least* depends on
(*a*) ambient temperature (*b*) candle power of the source
(*c*) distance of the source (*d*) type of reflector used.

19. The errors involved in photometric measurements are of the order of
(*a*) 1% (*b*) 1 – 5%
(*c*) 10 – 20% (*d*) 40 – 50%.

20. The desirable properties of a laser beam are :
(*a*) it has a high radiant energy
(*b*) it can be polarised
(*c*) it is monochromatic and coherent
(*d*) all of the above.

21. A photoconductive cell may be called as :
(*a*) photoresistive cell
(*b*) photo resistor
(*c*) light dependent resistor
(*d*) all of the above.

22. The resistance of a photoresistive cell
(*a*) increase with increase in light intensity
(*b*) remains constant irrespective of the light intensity incident upon it
(*c*) decreases with increase in light intensity in a linear manner
(*d*) decreases with increase in light intensity in an exponential manner.

23. The response of human eye under normal light conditions is called
(*a*) stopopic vision and it peaks at a wavelength of 555 nm
(*b*) stopopic vision and it peaks at a wavelength of 507 nm
(*c*) photopic vision and it peaks at a wavelength of 507 nm
(*d*) photopic vision and it peaks at a wavelength of 555 nm.

24. A light source may be considered as a point source if its diameter is
(*a*) less than 10% of the distance between it and the detector
(*b*) less than 1% of the distance between it and the detector
(*c*) more than 10% of the distance between it and the detector
(*d*) more than 1% of the distance between it and the detector.

25. In case of measurement of intensity of a test source, a standard source is used. The photometric head used when comparing two lights of different colours, is
(*a*) Bunsen photometer head.
(*b*) Constast type Lummer Brodhun photometer head.
(*c*) Comparison type Lummer Brodhun photometer head.
(*d*) Flicker photometer head.

26. The current through a photoemissive tube depends upon :
(*a*) intensity of light incident upon it.
(*b*) colour and wavelength of incident light.
(*c*) voltage applied between cathode and anode of tube.
(*d*) all of the above.

27. The luminous sensitivity of gas filled photoelectric cells varies between
(*a*) 10 to 100 μA/lm (*b*) 40 to 150 μA/lm
(*c*) 0.002 to 0.1 μA/μW (*d*) 0.01 to 0.15 μA/μW.

28. Photodiodes
(*a*) do not have any response in darkness
(*b*) have a response even in darkness
(*c*) have a leakage current even in darkness which is called dark current that doubles for every 10°C increase in the ambient temperature which decreases the signal to noise ratio.
(*d*) have a leakage current even in darkness which is called dark current that doubles for every increase of 5°C in temperature and which increases the signal to noise ratio.

29. The sensitivity of a photovoltaic cell
(*a*) is almost constant when it is short circuited.
(*b*) decreases with increase of load resistance.
(*c*) is almost negligible when the load resistance is about 10000 Ω.
(*d*) all of the above
(*e*) none of the above.

ANSWERS

1. (*d*)	**2.** (*d*)	**3.** (*b*)	**4.** (*c*)	**5.** (*b*)	**6.** (*c*)	**7.** (*b*)
8. (*d*)	**9.** (*a*)	**10.** (*e*)	**11.** (*c*)	**12.** (*d*)	**13.** (*c*)	**14.** (*d*)
15. (*a*)	**16.** (*a*)	**17.** (*c*)	**18.** (*a*)	**19.** (*c*)	**20.** (*d*)	**21.** (*d*)
22. (*d*)	**23.** (*d*)	**24.** (*a*)	**25.** (*d*)	**26.** (*d*)	**27.** (*b*)	**28.** (*c*)
29. (*d*).						

THEORETICAL QUESTIONS

1. What is the basic nature of light ? Explain.
2. Define the following terms :
Luminous flux, Lumen, Illumination, Lamp efficiency.
3. Explain briefly the following :
(*i*) Space-height ratio (*ii*) Utilisation factor
(*iii*) Depreciation factor.
4. What is a solid angle ?
5. State the laws of illumination.
6. What is a polar curve ? How is it useful to an illumination engineer ?
7. What do you understand by polar curves ? Explain Rousseau's construction for calculating M.S.C.P. of a lamp.
8. Explain briefly the laws of illumination or luminance.
9. Discuss the following laws of illumination :
(*i*) Inverse square law.
(*ii*) Lambert's cosine law.
10. What do you mean by the term 'photometry' ?
11. Explain briefly with a neat diagram "Photometer bench" for measurement of candle power.
12. Describe the working of the "Grease spot photometer".
13. Write a short note on "Flicker photometers".
14. Explain with a neat sketch the working of a "photovoltaic cell" used in photometry measurements.

15. Write short notes on the following :
   (i) Photoemissive cell.
   (ii) Distribution photometry.
16. Describe with a neat diagram the measurement of M.S.C.P. (mean spherical candle power) by integrating sphere.
17. Explain briefly the working of a *"holophane lumeter"* with a neat sketch.
18. Write a short note on *"Measurement of illumination"*.
19. Explain briefly any two of the following illumination photometers :
   (i) Trotter illumination photometer.
   (ii) Macbeth illuminometer.
   (iii) Holophane lumeter.

UNSOLVED EXAMPLES

1. A 250 V lamp has a total flux of 3000 lumens and takes a current of 0.8 A. Calculate (i) lumens/watt, and (ii) M.S.C.P./watt. [**Ans.** (i) 15 ; (ii) 1.2]
2. A 0.4 metre diameter diffusing sphere of opal glass, having 20% absorption, encloses an incandescent lamp with a luminous flux of 4850 lumens. Calculate the average luminance of the sphere. [**Ans.** 7720 lumens/m^2]
3. A filament lamp of 500 W is suspended at a height of 5 metres above working plane and gives uniform illumination over an area of 8 m diameter. Assuming efficiency of reflector as 60% and efficiency of lamp as 0.9 watt per candle power, determine the illumination on the working plane. [**Ans.** 83.33 lumens/m^2]
4. The candle power of a lamp is 100. A plane surface is placed at a distance of 2 metres from this lamp. Calculate the illumination on the surface when it is (i) normal, (ii) inclined to 45°, and (iii) parallel to the rays. [**Ans.** (i) 25 lux, (ii) 17.67 lux (iii) zero]
5. A 500 W lamp having M.S.C.P. of 100 is suspended 2.7 metres above the working plane. Calculate :
   (i) Illumination directly below the lamp at the working plane.
   (ii) Lamp efficiency.
   (iii) Illumination at a point 2.5 metres away on the horizontal plane from vertically below the lamp.

   [**Ans.** (i) 137.17 lux ; (ii) 25.13 homens/watt ; (iii) 54.2 lux]
6. A lamp with reflector is mounted 12 m above the centre of a circular area of 24 m diameter. If the combination of the lamp and reflector gives a uniform C.P. of 1000 over the circular area, determine the maximum and minimum illumination produced on the area. [**Ans.** 6.94 lux ; 2.46 lux]
7. A lamp having a uniform C.P. of 200 in all directions is provided with a reflector which directs 60% of the total light uniformly on to a circular area of 10 m diameter. The lamp is hung 6 m above the area. Calculate :
   (i) The illumination at the centre and edge of the surface with and without the reflector.
   (ii) The average illumination over the area without the reflector.

   [**Ans.** (i) 5.56 lux, 2.52 lux (without reflector), 19.2 lux (without reflector), (ii) 3.7 lux]

CHAPTER

14 Signal Generators

14.1.General aspects; 14.2. Conventional standard signal generator; 14.3. Modern signal generator; 14.4. Audio Frequency (AF) and Square wave generator; 14.5. Function generator; 14.6. Pulse and square generators; 14.7. Sweep generator; 14.8. Frequency synthesizer; 14.9. Other signal generators; *Highlights – Objective Type questions – Theoretical Questions.*

14.1 GENERAL ASPECTS

A **signal generator** *is an instrument that provides a controlled output waveform or signal for use in testing or aligning, or in measurements on other circuits or equipments.*

- The signal generators provide a variety of waveforms for testing electronic circuits, *usually at low powers.*
- The term ***"Oscillator"*** is used to describe an instrument that *provides only a sinusoidal output signal,* and the term ***"Function generator"*** to describe an instrument that provides several output waveforms including *sine wave, square wave, triangular wave and pulse trains, as well as an amplitude modulated waveform.*
  — It may be emphasized that, although we speak of oscillators as generators of signals, no energy is created; it is simply converted from a D.C. source into A.C. energy as some specific frequency.
- In various instrumentation systems, the signal at audio frequency as well as at radio frequency are required. In most of the cases, the signal at particular frequency is generated *using an oscillator which provides fixed frequency signal.* Some other variable frequency oscillators are also available.
- The audio frequency (AF) and radio frequency (RF) signal generators are the equipments which provide signals for general test purposes. These two are *specially designed to cover a wide frequency band from few Hz to several GHz.* The frequency band limits are listed in table 14.1.

Table 14.1. Frequency Band Limits

S. No.	Band	Approximate range
1.	Audio frequency (AF)	20 Hz–20 kHz
2.	Radio frequency (RF)	Above 30 kHz
3.	Very low frequency (VLF)	15–100 kHz
4.	Low frequency (LF)	100–500 kHz
5.	Broadcast	0.5–1.5 MHz
6.	Video	0–5 MHz
7.	High frequency (HF)	1.5–30 MHz
8.	Very high frequency (VHF)	30–300 MHz
9.	Ultra high frequency (UHF)	300–3000 MHz
10.	Microwave	Beyond 3000 MHz (or 3 GHz)

Characteristics of signal generators :

Although there are various types of signal generators, yet the following *characteristics* are common to all the types :

1. The frequency of the signal should be well known and stable.
2. The amplitude of the output signal should be controllable from very low values to relatively large values.
3. The signal should be free from distortion.

Applications :

Signal generators are used for the following *purposes* :

1. To provide appropriate signal for calibration, testing and trouble shooting of the amplifier circuits used in communications, electronics such as radio and television amplifiers.
2. To be employed as power source for the measurement of gain, bandwidth, signal-to-noise (S/N) ratio and standing wave ratio (SWR) and other properties of circuits.
3. To measure the characteristics of antennas and transmission lines.

Classification of signal generators :

The signal generators can be *classified* into the following categories :

1. Audio generators.
2. Function generators.
3. Pulse generators.
4. RF generators.
5. Frequency synthesizers.
6. Other signal generators.

The various types of generators are described in the following articles.

14.2 CONVENTIONAL STANDARD SIGNAL GENERATOR

A *sine-wave generator* represents the largest single category of signal generators. This instrument covers the frequency range from a few Hz to many GHz. Fig. 14.1 shows the simplest form of the sine-wave generator :

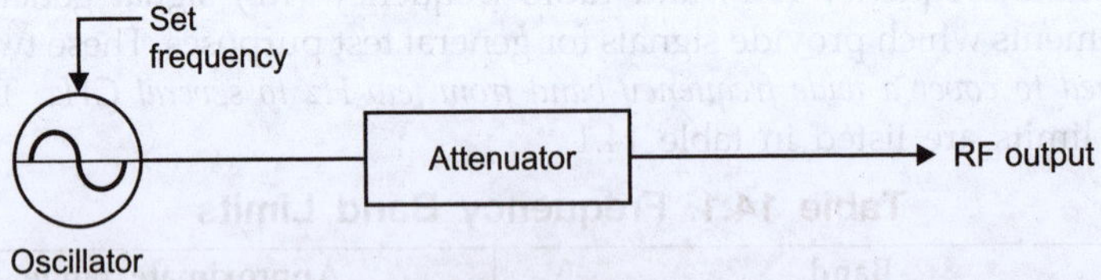

Fig. 14.1. Block diagram of a simple sine-wave generator.

— It consists of two basic blocks, an *"oscillator"* and an *"attenuator"*.

— *The frequency accuracy and stability and freedom from distortion depends on the design of the oscillator, while the amplitude accuracy depends on the attenuator's design.*

Conventional standard signal generator *is basically a radio frequency (RF) signal generator and produces known and controllable voltages*. It is extensively used in the *testing of radio receivers and transmitters*.

- The generator's output is *frequency modulated or amplitude modulated.*

— The *frequency modulation* is possible using a carrier signal from RF oscillator.

— The *amplitude modulation* can be done by using internal sine-wave oscillator.

The modulation may be done by a sine wave, square wave, triangular wave or a pulse also.

Fig. 14.2, shows the block diagram of a *conventional standard signal generator* :

— The carrier frequency (sine wave) is produced by a *stable RF oscillator* using LC tank circuit. The frequency of oscillation is shown on frequency control and the vernier dial setting.

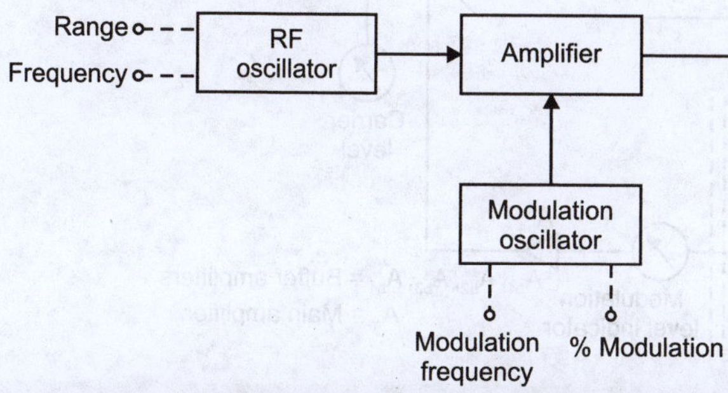

Fig. 14.2. Block diagram of a conventional standard signal generator.

— Amplitude modulation is obtained by an *internal or an external fixed frequency oscillators.*

— Modulation is done in output amplifier circuit which delivers its output (*i.e.,* the modulated carrier) to an *attenuator.*

— The output voltage is observed is an *output meter* and the *attenuation output setting.*

- Regulated power supplies are used for frequency stability at all ranges and to maintain it, temperature compensation is used. Buffer amplifiers are employed to isolate the oscillator circuit from the output in high frequency oscillator.

Advantages :

1. The output voltage of the conventional standard signal generator can be controlled as per the requirement.
2. The output obtained is stable.

Disadvantages :

1. The stability of the frequency is limited, due to LC tank circuit.
2. This type of generator takes some time to stabilise at new frequency when the range is changed.
3. The isolation of the master oscillation from output circuit is necessary in high frequency oscillators.

14.3 MODERN SIGNAL GENERATOR

The order to improve the frequency stability, a *single master oscillator is optimally designed for the highest frequency range* and frequency dividers are switched in to produce lower ranges. The *master oscillators is made insensitive to temperature variations and to the influence of the succeeding stages by careful circuit design.* Temperature compensation devices are used for any temperature changes.

Fig. 14.3. shows the block diagram of a modern signal generator (Laboratory type) :

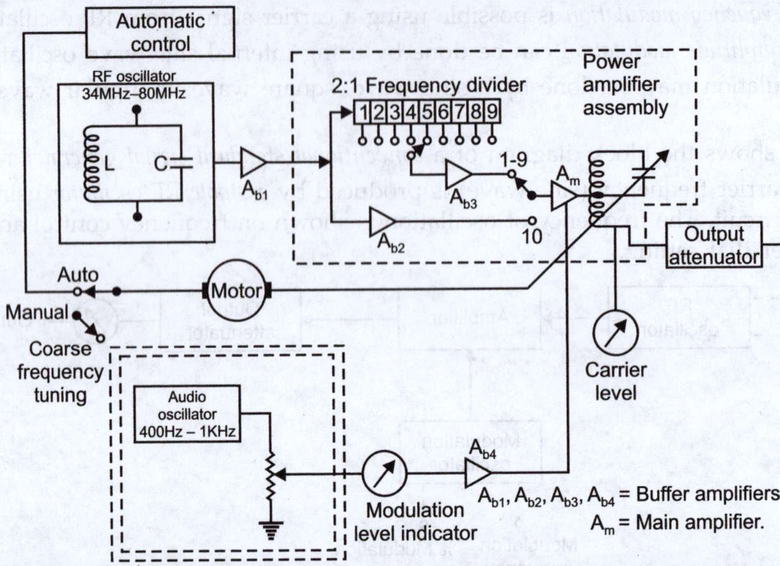

Fig. 14.3. Block diagram of a modern signal generator.

— RF oscillator (consisting of LC tank circuit) is used as a *master oscillator;* it gives a frequency range from 34 MHz-80 MHz. This signal is connected to untuned buffer amplifier A_{b1}. The output of this amplifier is given to frequency divider circuit (Fig. 14.3); the circuit consists of 9 frequency dividers. The frequency division can be done by using flip-flops to get ratio of 2 : 1.

— The LC tank circuit and frequency divider provides a carrier signal.

— The additional buffer amplifiers Ab_2 and Ab_3 provide a very high degree of isolation between the master oscillator and the power amplifier, and almost eliminates all the frequency effects (distortion) between the input and output circuits, caused by loading.

— The *modulating signal is provided by an audio oscillator;* the frequency of this oscillator is in the range of 400 Hz – 1 kHz. The modulation is done at the power amplifier stage. The level of modulation can be adjusted upto 95% by using a control device.

— The lowest frequency range produced by the cascaded frequency divider is the highest frequency divided by 2^9 or 512. Thus, the frequency stability of the highest range is imparted to the lowest frequency ranges.

— The master oscillator is turned automatically or manually. In an automatic controller for tuning master oscillator, a motor driven variable capacitor is used. This system is extensively used in programmable automatic frequency control devices. The oscillator can be fine tuned by means of a large rotary switch with each division corresponding to 0.01% of main dial setting.

— 1 MHz crystal oscillator provides the internal calibration.

— The small power consumption of the instruments makes output with very low ripple.

— *The supply voltage of the master oscillator is regulated by temperature compensated reference circuit.*

— The output of the main amplifier (A_m) is given to an output attenuator. The *"attenuator" controls the amplitude level and provides the required stable RF output.*

Advantages :

1. Very low power consumption.
2. Owing to less power consumption, the instrument shows good regulation and crystal stability.
3. The instrument provides excellent Q stability (Q-factor/quality factor) with very low ripple.
4. Due to the provision of temperature compensation, the temperature variation does not change output.
5. As the same oscillator is used on all the bands, range switching effects are eliminated.
6. The use of buffer amplifiers provides good isolation between the master oscillator and main power amplifier, eliminating loading effect completely.
7. High degree of isolation eliminates distortion between the input and output circuits.
8. Since the same master oscillator is used to get various frequency ranges, the stability is improved even at the highest frequency range.

Disadvantage : *Increased cost* (as compared to conventional signal generator) due to complex circuitry.

Comparison between Standard and Modern signal generators :

The comparison between standard and modern signal generators is given in the table 14.2 below :

S. No.	Aspects	Standard signal generator	Modern signal generator
1.	*Frequency stability*	Limited	Maintained over entire frequency range
2.	*Frequency range*	Small	Wide range possible
3.	*Regulation*	Poor	Excellent
4.	*Construction*	Simple	Complicated
5.	*Cost*	Low	High
6.	*Time taken to stabilise at new frequency*	Takes time	Very quick
7.	*Loading and distortion effects*	More	Less
8.	*Temperature compensation*	Not provided	Provided
9.	*Automatic tuning with motor*	Not available	Possible
10.	*Q-stability*	Less with high ripple	Good with low ripple

14.4 AUDIO FREQUENCY (AF) AND SQUARE WAVE GENERATOR

Fig. 14.4. shows the block diagram of AF sine and square wave generator :

— In this generator, a *Wien bridge oscillator* is used; this oscillator is *best* for the audio frequency range. The frequency of oscillations can be changed by varying the capacitance in the oscillator. The frequency can also be changed in steps by switching in resistors of different values.

— The output of the oscillator goes to the function switch. Depending upon the position of the switch, we get output as sine wave output or square wave output.

— In the square wave generation section, the output of the Wien bridge oscillator is fed to square wave shaper circuit which uses schmitt trigger circuit.

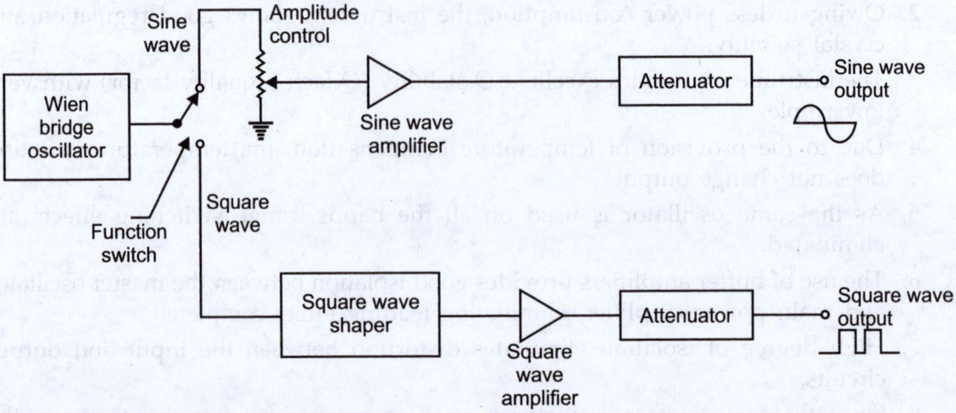

Fig. 14.4. Block diagram of AF sine and square wave generator.

— The *attenuators* in both the sections are used to control output signal level. The signal level, before attenuation, is made very high using sine wave amplifier and square wave amplifier.

Specifications :

Following are the specifications of a typical AF sine and square wave generator :

(*i*) *Frequency range* : 10 Hz to 1 MHz.

— Frequency is variable almost 5 decades continuously.

(*ii*) *Amplitude (sine wave)* : 5 mV to 5 V (r.m.s.)

— The output is taken through a push-pull amplifier.

— For low output, the impedance is 600 Ω.

(*iii*) *Amplitude (square wave)* : 0 to 20 V (Peak)

— Possible to adjust the symmetry of the square wave from 30 to 70%.

(*iv*) *Power required* : 7 W (at 220 V, 50 Hz).

14.5 FUNCTION GENERATOR

A **function generator** *is an instrument that has the capability of producing different types of waveform as its output signal.* The most common output waveforms are :

— Sine waves;

— Triangular waves;

— Square waves;

— Sawtooth waves.

The frequencies of above waves may be adjusted from a fraction of a hertz to several hundred kHz (A very common frequency range for function generators is 0.01 Hz to 3 MHz).

- The *triangular-wave and sawtooth-wave outputs* of function generators are commonly used for those applications which *need a signal that increases or reduces at a specific linear rate.* They are also used in *driving sweep oscillators in oscilloscopes and the X-axis of X-Y recorders.*

- One of the useful features of the function generator is that *it can be phase locked to an external signal source.* One function generator can be phase locked with second function generator; the *two output signals can be displaced in phase by an adjustable amount.* Furthermore, *the fundamental frequency of one generator* can be *phase locked to a harmonic of another generator. By adjusting the amplitude and phase of the harmonic, almost any waveform can be generated.*
- When a function generator is phase locked to a standard frequency of the source, then all the output waveforms of the generator will have *same accuracy and stability as that of a standard source.*

Fig. 14.5. shows the block diagram of a typical function generator :

— In this generator the frequency is controlled by varying the magnitude of current that drives the *integrator.*

— The frequency controlled voltage regulates *two* current supply sources. The upper current supply source supplies constant current to the integrator whose voltage rises linearly with time according to output signal equation $v_{out} = -\frac{1}{C}\int_0^t i\,dt$. An increase or decrease in the current increases or decreases the slope of the output voltage and thus controls the frequency.

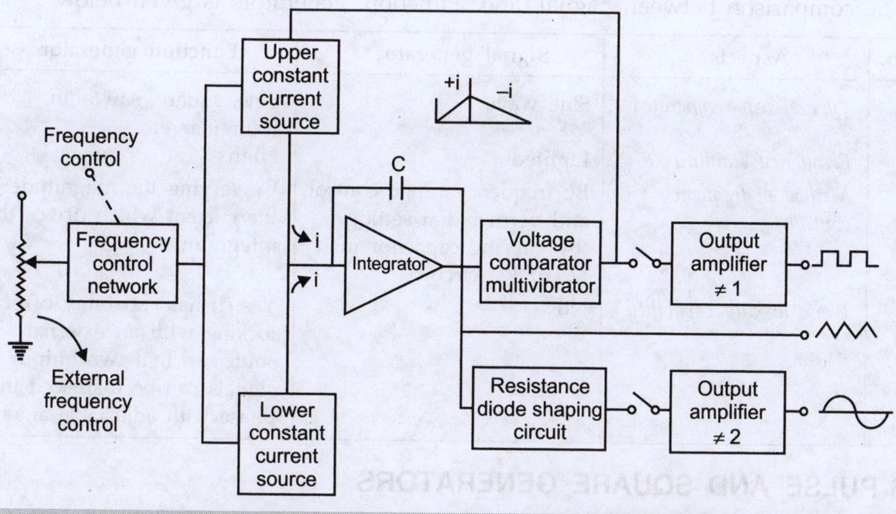

Fig. 14.5. Block diagram of a typical function generator.

— The *voltage comparator multivibrator* changes state at a predetermined maximum level of the integrator output voltage. This change cuts-off the current supply from the upper supply source and switches to the lower supply source. The *lower supply source supplies a reverse current to the integrator* so that its *output drops linearly with time.* When the output attains a predetermined level, the voltage comparator again changes state and switch 'ON' to upper supply source.

— *The output of the "integrator" is a "triangular-wave"* whose frequency depends on the upper constant current source. The *"voltage comparator" delivers square wave output voltage of the same frequency as that of input triangular waveform. The triangular wave is synthesised into sine wave using diode resistance network.* In this shaper circuit, the slope of triangular wave is changed as its amplitude changes; this results in a *sine wave with less than 1 per cent distortion.*

Specifications :

Following are the typical important specifications of a function generator :

(i) *Output waveforms* : Sine, square, triangular, ramp, pulse, AM and FM modulated, arbitrary.

(ii) *Frequency range* : 0.001 Hz to 20 MHz.

(iii) *Frequency stability* : 0.05%.

(iv) *Distortion* : –55 dB below 50 kHz, –40 dB above 50 kHz.

(v) *Output amplitude (open circuit)* : 10 $V_{p\text{-}p}$.

(vi) *Impedance* : 50 Ω.

Applications :

Function generator is used for the following purposes :

1. To test the bandwidth of the audio frequency amplifier—This method is commonly known as ***square wave testing.***
2. To be used as an important instrument in the troubleshooting of different analog and digital circuits and instruments.
3. To act as a source for the alignment of receivers.

Comparison between "Signal generator" and "Function generator :

The comparison between 'Signal' and 'Function' generators is given below :

S. No.	Aspects	Signal generator	Function generator
1.	*Type of wave generated*	Sine wave	Sine, square, sawtooth, triangular etc.
2.	*Frequency stability*	Limited	High
3.	*Method of frequency control*	By frequency range control and vernier dial setting (by varying capacitor in LC or RC circuit).	By varying the magnitude of the current which drives the integrator.
4.	*Phase locking capability*	No	Yes (It has capability of phase locking with an external source so that two output signals can be displaced in phase with adjustable amount).

14.6 PULSE AND SQUARE GENERATORS

14.6.1. Generator Aspects

These generators are *often used with an oscilloscope as the measuring device.*

The primary difference between a pulse generator and a square-wave generator relates to the *"duty cycle"*.

Duty cycle *is defined as the ratio of the average value of the pulse over one cycle to the peak value of the pulse.*

Or

It is also defined *as the ratio of the pulse width and the period or pulse repetition time.*

i.e.,
$$\text{Duty cycle} = \frac{\text{Pulse width}}{\text{Period}}$$

- The duty cycle of *'square-wave generators'* equals 0.5 or 50% since they produce an output voltage with equal *"on"* and *"off"* times.

- The duty cycle of a *'pulse generator'* may vary.
  — Very short duration pulses give a low duty cycle, and the pulse generator generally can supply *more power* during its *"on"* period than a square-wave generator can.
  — *Short duration pulses reduce the power dissipation in the component under test.*
- Square-wave generators are *used* :
  (*i*) Whenever the low-frequency characteristics of a system are being investigated, *e.g., testing audio systems.*
  (*ii*) When the transient response of a system requires some time to settle down.
- A ***"pulse"*** *differs from a* ***"square wave"*** *in that it needs neither baseline nor left-right symmetry.*

Fig. 14.6 shows one popular approach to constructing a *pulse generator*. A monostable multivibrator (*i.e.*, one-shot) follows a square wave oscillator. The pulse repetition rate is set by the square wave frequency. The one-shot triggers on the leading edge of the square wave and produces one output pulse for each input cycle. The duration of each output pulse is set by the *on-time* of the one-shot and it may be very short or may approach the period of the square wave.

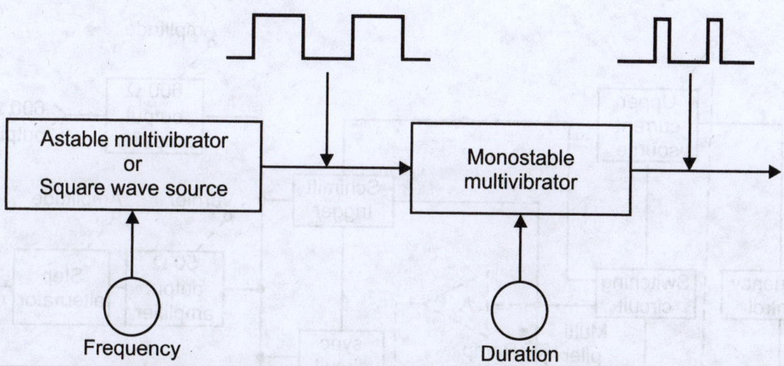

Fig. 14.6. Block diagram of a pulse generator.

— A typical pulse generator will allow the user to select the repetition rate, duration, and amplitude, and also the number of pulses to be output in a given burst.

Circuits used in pulse generation :

The circuits used in pulse generation fall into the following two categories :

1. *Passive, or pulse-shaping circuits.*
   - In these circuits, a sine-wave oscillator is used as the basic generator and its output is passed through a pulse-shaping circuit to obtain the desired waveform.

   Example : An approximate square waveform may be obtained by first amplifying and then clipping a sine wave.
2. *Active, or pulse-generating circuit.*
   - Active generators are usually of the *relaxation type.* The relaxation oscillator uses the charge-and-discharge action of capacitor to control the conduction of a vacuum tube or a transistor.
   - Some common forms of relaxation oscillators are :
     (*i*) *Multivibrators.*
     (*ii*) *Blocking oscillators.*

14.6.2. Laboratory Square-wave and Pulse Generator

Fig. 14.7 shows the block diagram of a typical general-purpose generator providing negative pulses of variable frequency, duty cycle, and amplitude.

Features :

— The frequency range of the instrument is covered in seven decade steps from 1 Hz to 10 MHz, with a linearly calibrated dial for continuous adjustment on all ranges.

— The duty cycle can be varied from 25% to 75%.

— Two independent outputs are available :

(*i*) A 50 Ω source that supplies pulses with rise-and-fall times of 5 ns at 5 V peak amplitude;

(*ii*) A 600 Ω source that supplies pulses with rise-and-fall times of 70 ns at 30 V peak amplitude.

— The instrument can be operated as a free-running generator or it can be synchronised with external signals.

— Trigger out pulses for synchronisation are also available.

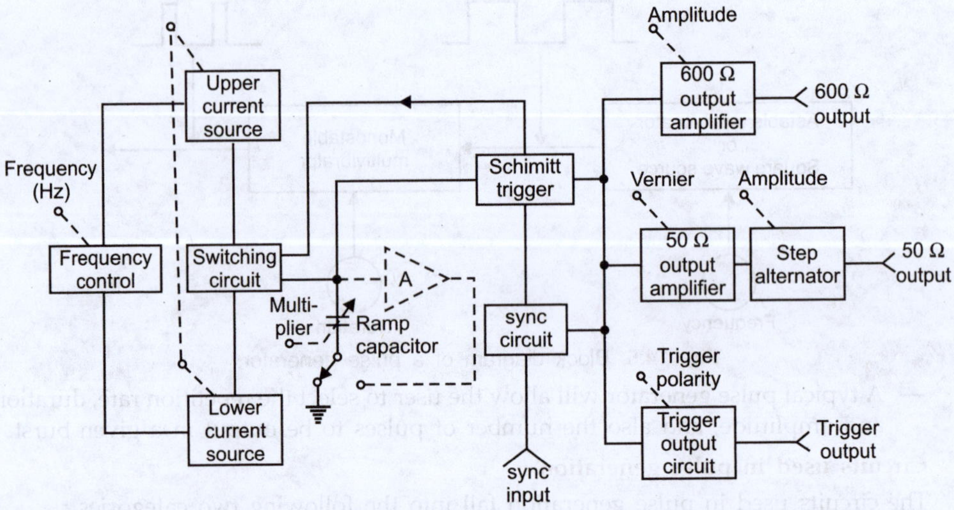

Fig. 14.7. Block diagram of a general purpose generator.

Working :

- The *upper* current source supplies a constant current to the ramp capacitor. This charges the capacitor at a constant rate, and the ramp voltage increases linearly. When the positive slope of the ramp voltage reaches the upper limit set by internal circuit components, the Schmitt trigger (a bistable vibrator) *changes state.* The trigger circuit output becomes *negative,* reversing the condition of the current control switch, and the capacitor starts discharging. The discharge rate is linear, controlled by *lower* current source. When negative ramp reaches a predetermined lower level, the Schmitt trigger switches back to its original state. This now provides a positive trigger circuit output that reverses the condition of the current switch again, cutting off the lower current source and switching on the upper current source. This gives one cycle of operation. The entire process is repetitive and the Schmitt trigger circuit provides negative pulses at a continuous rate.

- *The Schmitt circuit's output is passed to the trigger output circuit and to the 50 Ω and 600 Ω amplifiers.*
  - — The trigger output circuit differentiates the square-wave output from the Schmitt trigger, inverts the resulting pulse, and *provides a positive triggering pulse.*
  - — The 50 Ω amplifier is provided with an output attenuator to *allow a vernier control of the signal output voltage.*
- Besides its free running mode of operation, the generator can be synchronized or locked into an external signal (by triggering the Schmitt circuit by an external synchronization pulse).

The regulated voltages for all stages of the instrument are provided by an internal supply (by which the unit is powered).

14.7 SWEEP GENERATOR

A **sweep generator** *provides a sinusoidal output voltage whose frequency varies smoothly and continuously over an entire frequency band, usually at audio rate.*

The frequency modulation process may be accomplished *electronically* or *mechanically.*

— In the *electronic* process of frequency modulation the modulating voltage is used to vary the reactance of the oscillator tank circuit component.

— The *mechanical* process involves the use of a motor driven capacitor, as provided for in a modern laboratory type signal generator.

Fig. 14.8 shows a basic block diagram of a sweep generator :

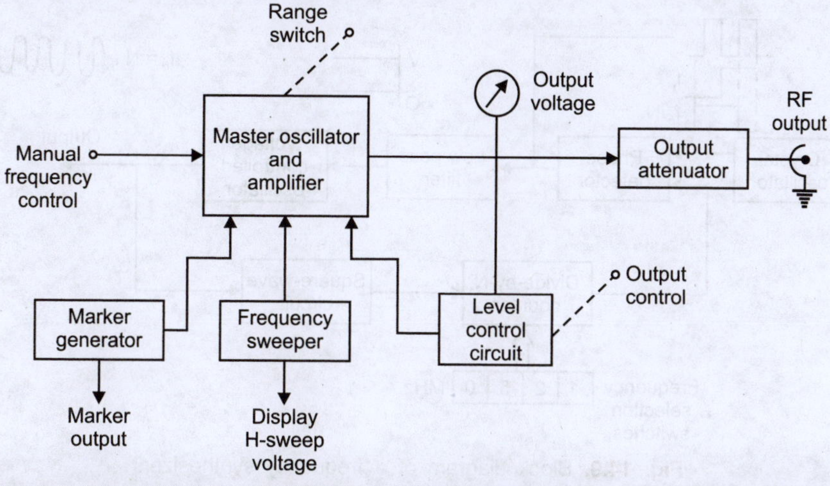

Fig. 14.8. Sweep generator.

— The *frequency sweeper* provides a variable modulating voltage which *causes the capacitance of the master oscillator to vary* (may be of the order 20 sweeps/sec). A manual control permits independent adjustment of the oscillator resonant frequency. The frequency sweeper provides a *varying sweep voltage for synchronisation* to drive the horizontal deflection plates of the CRO. Thus the amplitude of the response of a test device will be locked and displayed on the screen.

— For the identification of a frequency interval, a *marker generator* provides half sinusoidal waveforms at any frequency within the sweep range. The marker voltage can be added to the sweep voltage of the CRO during alternate cycles of the sweep voltage, and appears superimposed on the response curve.

— The *automatic level control circuit* is a closed loop feedback system which *monitors the RF level at some point in the measurement system. This circuit holds the power delivered to the load or test circuit constant and independent of frequency and impedance changes.*

A constant power level prevents any source mismatch and also provides a constant readout calibration with frequency.

14.8 FREQUENCY SYNTHESIZER

The **frequency synthesizer** *is another type of RF generator that uses phase locked loop (PLL) to generate output frequencies over a wide range.* The most common range is from 1 MHz to 160 MHz.

Fig. 14.9 shows a simple block diagram of a frequency synthesizer; its major components are :

(*i*) Voltage controlled oscillator (VCO);
(*ii*) Divide-by-N counter;
(*iii*) Phase detector;
(*iv*) Crystal oscillator;
(*v*) Low-pass filter;
(*vi*) Square-wave circuit.

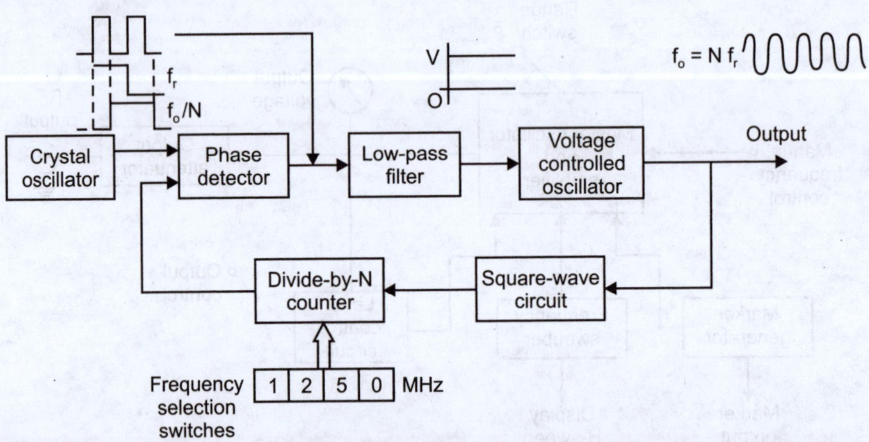

Fig. 14.9. Block diagram of a frequency synthesizer.

Working :

— The output of the *'crystal oscillator'* (a reference frequency, f_r), is fed into one input of a phase detector. The other input of a phase detector has another squarewave applied to it as shown in the figure. *The frequencies of these two square waves are identical but there is a phase difference (ϕ) between them.*

— The output of the *'phase detector'* is a pulse wave waveform with pulse width controlled by the phase difference.

— The output of the phase detector is applied to the low-pass filter which converts it into a D.C. voltage, V.

— The D.C. voltage, V is used as a control voltage for the voltage-controlled oscillator (VCO) and it determines the output frequency of VCO.

— The output of VCO is fed to a circuit that converts it into a square wave for triggering a digital divide-by-N counter. The divide-by-N counter divides the VCO frequency by a number set by a bank of switches. These switches may be push buttons with digital readout or they may be thumb-wheel type which indicate their position numerically. The switches are connected in such a way that the displayed number is the *factor N by which the output frequency is divided before being applied to the phase detector.*

The switches permit the user to obtain frequency which is *any integer multiple of the crystal oscillator frequency.*

- *The frequency synthesizer is used in almost the same areas as the RF signal generators.*

14.9 OTHER SIGNAL GENERATORS

Some of the signal generators which do not fit well in various pre-established categories are described below :

1. **RM Makers :**
   - These devices are usually crystal controlled and have a fixed output frequency for use as a reference.
   - These are used for calibrating TV signals.
2. **Digitally programmable test-oscillators :**
   - These instruments can have extremely wide frequency range although some versions have much narrow range also.
   - The set frequency can be programmed through the front panel keyboard or via a computer interface input such as IEEE-488 general purpose interface bus (commonly known as GPIB).
3. **Arbitrary waveform generators :**
   - These instruments permit the user *to design and generate virtually any desired form.*
   - This type of generator is quite useful to perform a variety of tests on communication equipment.

Applications :

These generators are extensively used in the following areas :

(*i*) High-speed low filter data and clock pulse generation.

(*ii*) Mixed signal design and test.

(*iii*) Communications design and test for producing *arbitrary IF based signals* and *standard waveforms for communications.*

(*iv*) Read word simulations.

(*v*) Disk drive read/write design and test.

HIGHLIGHTS

1. Audio signal generators usually produce sine and square waves over the range of at least 20 Hz and 20 kHz. Some will operate down to 1 Hz and upto 100 kHz.

2. The function generator generally produces sine, square, and triangle waveforms at frequencies from fractional Hz to the MHz range.
3. Audio and function generators use 600 Ω output impedance, while RF generators have a 50 Ω output impedance.
   Proper operation of signal generators depends upon using a load matched to the output impedance.
4. RF signal generators operating over 1 GHz are usually called *"microwave generators"*.

OBJECTIVE TYPE QUESTIONS

Choose the Correct Answer :

1. A triangular waveshape is obtained :
   (*a*) by integrating a square wave (*b*) by differentiating a sine wave
   (*c*) by differentiating a square wave (*d*) by integrating a sine wave.
2. An stable multivibrator
   (*a*) gives one output pulse for every two input pulses
   (*b*) gives a timed output pulse for a trigger input
   (*c*) gives a train of output pulses for a trigger input
   (*d*) gives four output pulses for a single input pulse.
3. In signal generators
   (*a*) energy is created
   (*b*) energy is generated
   (*c*) energy is converted from a simple D.C. source into A.C. energy at some specific frequency
   (*d*) all of the above.
4. A video frequency oscillator has a frequency range of :
   (*a*) 20 Hz to 20 kHz (*b*) 20 kHz to 30 MHz
   (*c*) D.C. to 5 MHz (*d*) 2 MHz to 30 MHz.
5. The desirable characteristics of a signal source are :
   (*a*) The frequency of the signal should be well known and stable.
   (*b*) The amplitude of a signal source should be controllable from very small to relatively large values.
   (*c*) The output signal should be free from distortion.
   (*d*) All of the above.
6. A voltage controlled oscillator (VCO) is an instrument :
   (*a*) whose frequency is dependent upon the amplitude of input signal.
   (*b*) whose frequency is independent of the amplitude of the input signal.
   (*c*) whose frequency is dependent upon the frequency of the input signal.
   (*d*) none of the above.
7. A random noise generator produces a signal
   (*a*) whose amplitude varies randomly (*b*) which has no periodic frequency
   (*c*) has an unpredictable power spectrum (*d*) all of the above.

ANSWERS

1. (*a*)	**2.** (*c*)	**3.** (*c*)	**4.** (*c*)	**5.** (*d*)	**6.** (*a*)	**7.** (*a*)

THEORETICAL QUESTIONS

1. What is a signal generator ? Explain briefly.
2. State the common characteristics of signal generators.
3. How are signal generators classified ?
4. Explain briefly, with neat diagrams, the following signal generators :
   (*i*) Conventional standard signal generator.
   (*ii*) Modern signal generator.
5. List the advantages and disadvantages of a modern signal generator.
6. Draw the block diagram of an audio frequency and square wave generator and explain its working.
7. Give the comparison between standard and modern signal generators.
8. What is a function generator ? Explain briefly.
9. Describe the working of 'Function generator' with a neat block diagram.
10. List the typical important specifications of a function generator.
11. Give the comparison between "Signal generator" and "Function generator".
12. What do you mean by "Duty cycle" ? Explain briefly.
13. How does a "pulse" differ from a "square wave" ?
14. Draw a simple diagram of a pulse generator and explain it briefly.
15. Explain briefly circuits used in pulse generation.
16. Explain briefly, with the help of a block diagram, the working of the laboratory square-wave and pulse generator.
17. Explain briefly the following :
    (*i*) Sweep generator.
    (*ii*) Frequency synthesizer.
18. Write short notes on :
    (*i*) RM makers;
    (*ii*) Digitally programmable test oscillators;
    (*iii*) Arbitrary waveform generators.

CHAPTER

15 Signal Analysis

15.1. Introduction; 15.2. Wave analysers – Introduction – Resonant or basic wave analyser – Frequency selective wave analyser – Heterodyne wave analyser – Applications of wave analyser; **15.3. Harmonic distortion analysers** – Harmonic distortion – Fundamental-suppression distortion analyser – Heterodyne harmonic analyser or wavemeter; **15.4. Spectrum analysers** – Introduction – Swept superheterodyne spectrum analyser – Real time spectrum analyser – Fourier analyser – Applications of spectrum analyser – *Highlights – Objective Type Questions – Theoretical Questions*

15.1 INTRODUCTION

It can be shown mathematically that any complex waveform is made up of a *fundamental* and *its harmonics*.

In several applications, the analysis of electrical signals is required. The different instruments which are used for signal analysis are :

(*i*) Wave analysers; (*ii*) Distortion analysers;
(*iii*) Spectrum analysers; (*iv*) Audio analysers;
(*v*) Modulation analysers.

All signal analysis instruments *measure the basic frequency properties of a signal*, but they use different techniques to do so.

(*i*) **A wave analyser.** *It is a voltmeter which can be accurately tuned to measure the amplitude of a single frequency, within a band of about 10 Hz to 40 MHz.*

(*ii*) **Distortion analysers.** These instruments operate over a range of 5 Hz to 1 MHz and *give a measure of the energy present in a signal outside a specified frequency band.* They, therefore, *tune out the fundamental signal and give an indication of the harmonics.*

(*iii*) **Spectrum analyser.** It sweeps the signal frequency band and *displays a plot of amplitude versus frequency*. It has an operating range of about 0.02 Hz to 250 GHz.

(*iv*) **Audio analyser.** It is similar to a distortion analyser but *can measure additional function, such as noise.*

(*v*) **Modulation analysers.** These instruments tune to the required signal and *recover the whole amplitude modulated (AM), frequency modulated (FM) or phase modulation (PM) envelope for display or analysis.*

15.2 WAVE ANALYSERS

15.2.1. Introduction

A **wave analyser** *is an instrument to measure relative amplitudes of single frequency components in a complex waveform.*

Basically, a wave analyser is a *frequency selective voltmeter* which is tuned to the frequency of one signal while rejecting all other signal components. The desired frequency is selected by a frequency calibrated dial to the input of maximum amplitude. The amplitude is displayed by CRO or a suitable voltmeter.

There are *two* types of wave analysers :

1. Resonant (or Basic) wave analyser.
2. Based upon the frequency ranges.
   (*i*) Frequency selective wave analyser.
   (*ii*) Heterodyne wave analyser.

15.2.2. Resonant or Basic Wave Analyser

Fig. 15.1. shows a resonant or basic wave analyser. It consists of the following components :

(*i*) Primary detector—a simple L-C circuit;

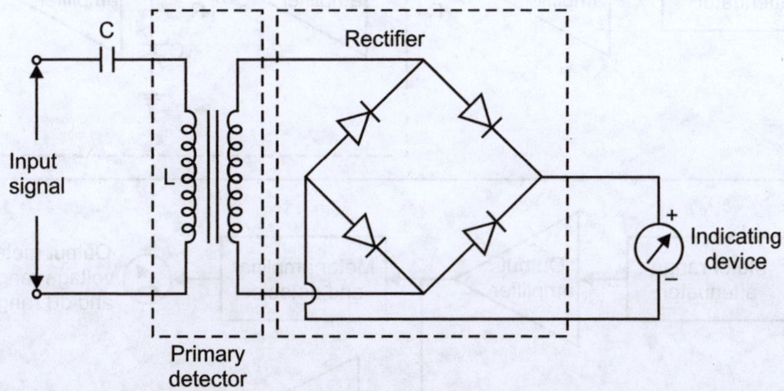

Fig. 15.1. Resonant or basic wave analyser.

(*ii*) A full-wave rectifier—intermediate stage;

(*iii*) Indicating device—A simple D.C. voltmeter calibrated to indicate the peak value of the sinusoidal input voltage.

— The *'L-C circuit'* is adjusted for resonance at the frequency of the particular harmonic component to be measured. Since L-C circuit is tuned to a single frequency, *it passes only the frequency to which it is tuned and rejects all other frequencies.*

— The *'full-wave rectifier'* output provides the *average value* of the input signal.

— A number of tuned filters, connected to the *'indicating device'* through a selector switch, would be required for a useful wave analyser.

15.2.3. Frequency Selective Wave Analyser

This wave analyser is used for measurements in the *audio frequency range* (*i.e.*, from 20 Hz to 20 kHz). It consists of a *very narrow pass-band filter section that can be tuned to a particular frequency.*

Fig. 15.2. shows the block diagram of frequency selective analyser :

— The waveform to be analysed is passed through an adjustable *'attenuator'*. This acts as a range multiplier and permits a large range of signal amplitudes to be analysed without loading the amplifier.

— A *'driver amplifier'* feeds the attenuated waveform to a *'high-Q active filter'*. The filter consists of a cascaded arrangement of RC resonant sections and filter amplifiers. The passband of the total filter section is covered in decade steps over the entire audio range by switching capacitors in the RC sections. For selecting the frequency ranges, generally close-tolerance polystyrene capacitors are used. Precision potentiometers are used to tune the filter to any desired frequency within the selective band.

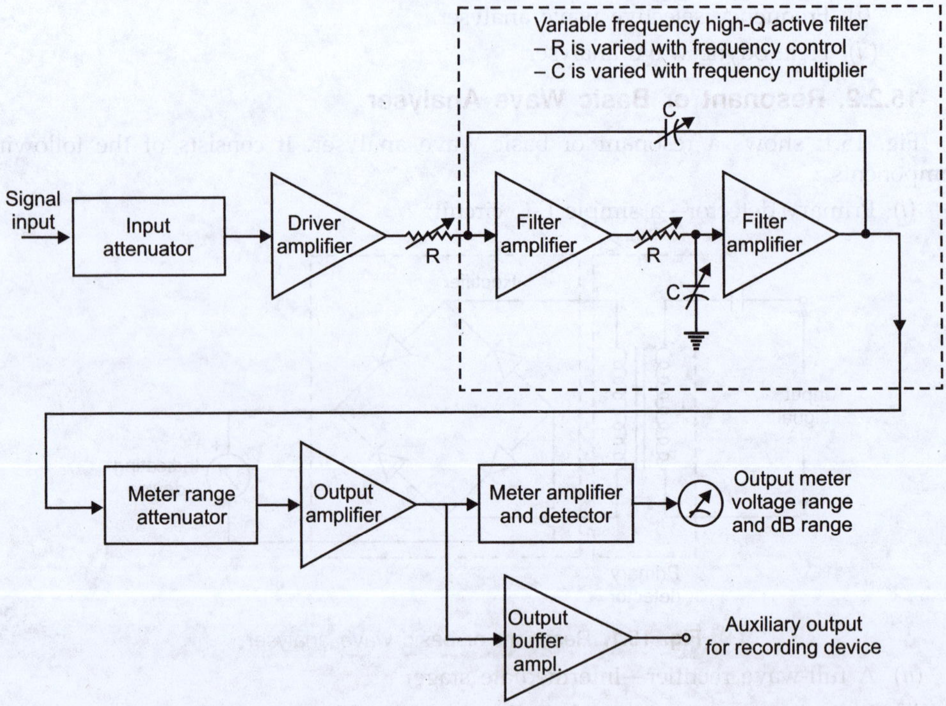

Fig. 15.2. Block diagram of frequency selective wave analyser (audio-range).

— A *'final amplifier stage'* supplies the *selected signal* to the meter circuit and to an untuned buffer amplifier.

— The buffer amplifier can be used to drive a recorder or an electronic counter.

— The meter is driven by an average-type detector and usually has several voltage ranges as well as a decibel scale.

- The *bandwidth of the instrument is very narrow, typically about 1% of the selected frequency.*

Fig. 15.3 shows the attenuation characteristics of a frequency selective wave analyser.

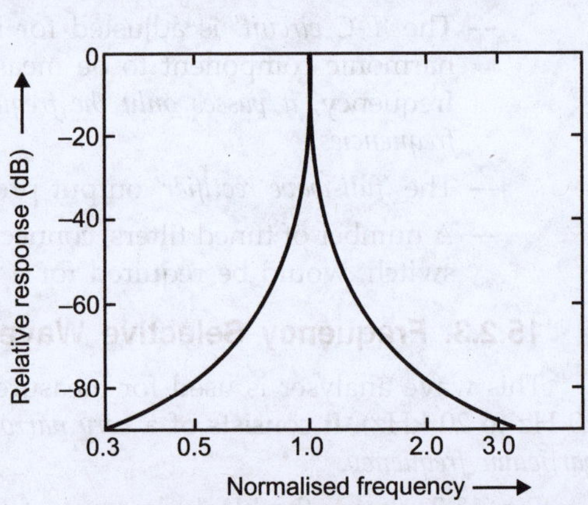

Fig. 15.3. Attenuation characteristics of frequency-selective wave analyser.

15.2.4. Heterodyne Wave Analyser

An ordinary wave analyser cannot be used for measurements in the RF range and above MHz range. Hence special type of wave analysers working on the principle of, *heterodyning* (mixing) are used. These wave analysers are known as *'Heterodyne wave analysers'*.

Fig. 15.4. shows the functional block diagram of the heterodyne wave analyser. Its operating frequency range is from 10 kHz to 18 MHz in 18 overlapping bands selected by the frequency range control of the local oscillator. The bandwidth is controlled by an active filter and can be selected at 200, 1,000 and 3,000 Hz.

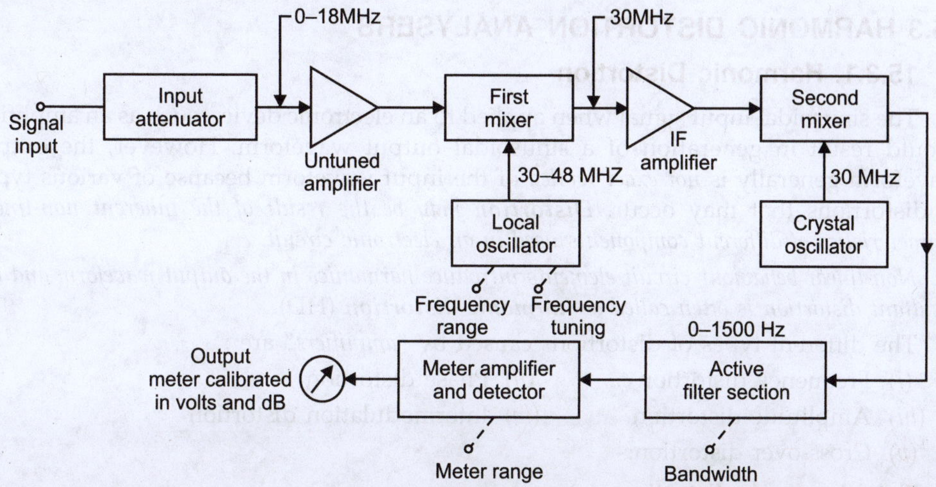

Fig. 15.4. Functional block diagram of the heterodyne wave analyser.

Working :

— The input signal enters the instrument through a probe connector that contains a *unity-gain isolation amplifier*.

— After appropriate attenuation, the input signal is heterodyned in the mixer stage with the signal from a local oscillator.

— The output of the mixer forms an intermediated frequency that is uniformly amplified by the 30 MHz *IF amplifier*. This amplified signal is then mixed again with a 30 MHz crystal oscillator signal, which results in information *centered on a zero frequency*.

— An *active filter* with controlled bandwidth and symmetrical slopes of 72 dB per octave then passes the selected component to the meter amplifier and detector circuit. The output from the meter detector can be read off on a decibel-calibrated scale or may be applied to a recording device.

- In a wave analyser, good frequency stability is achieved by using *frequency synthesisers*, which have high accuracy and resolution, or by an *automatic frequency control (AFC)*. In an AFC system the local oscillator locks to the signal, and so eliminates the drift between them.

15.2.5. Applications of Wave Analyser

The various applications of wave analyser are :

1. To carry out complete harmonic analysis.

2. To measure relative amplitudes of single frequency components in a complex waveform.
3. To measure the amplitude in the presence of noise and other interfering signals.
4. To be used to separate and display a large number of harmonics (say 50).
5. To measure the signal energy with well defined bandwidth.
6. To measure the harmonic distortion of an amplifier; the contribution of each harmonic to the total distortion also can be determined.
7. The wave analysers are applied *industrially in the field of reduction of sound and vibrations generated by rotating electrical machines and apparatus.*

15.3 HARMONIC DISTORTION ANALYSERS

15.3.1. Harmonic Distortion

The sinusoidal input signal when applied to an electronic device, such as an amplifier, should result in generation of a sinusoidal output waveform. However, the output waveform generally is *not exact replica* of the input waveform because of various types of distortions that may occur. ***Distortion** may be the result of the inherent non-linear characteristics of different components used in an electronic circuit.*

Non-linear behaviour circuit elements introduce harmonics in the output waveform and the resultant distortion is often called as **harmonic distortion** (HD).

The different types of distortions caused by *"amplifiers"* are :

(*i*) Frequency distortion (*ii*) Phase distortion

(*iii*) Amplitude distortion (*iv*) Intermodulation distortion

(*v*) Cross-over distortion.

Total harmonic distortion :

A *non-linear system* produces harmonics of an input sinewave, the *harmonics consisting of sine waves with frequencies which are multiples of the fundamental of the input signal. Total harmonic distortion (THD)* is given by,

$$\text{THD} = \frac{\left[\Sigma\,(\text{Harmonics})^2\right]^{1/2}}{\text{Fundamental}} \qquad \text{...(15.1)}$$

A *measure of distortion* represented by a particular harmonic is simply the *ratio of amplitude of harmonic to that of the fundamental harmonic distortion (HD)* and is represented by,

$$D_2 = \frac{E_2}{E_1},\quad D_3 = \frac{E_3}{E_1},\quad D_4 = \frac{E_4}{E_1}$$

where D_n (n = 2, 3, 4, ...) is distortion of nth harmonic and E_n represents the amplitude of nth harmonic; E_1 is the amplitude of *fundamental.*

THD is defined as :

$$\text{THD} = \sqrt{D_2^2 + D_3^2 + D_4^2 + \ldots\ldots D_n^2} \qquad \text{...(15.2)}$$

$$= \sqrt{\left(\frac{E_2}{E_1}\right)^2 + \left(\frac{E_3}{E_1}\right)^2 + \left(\frac{E_4}{E_1}\right)^2 + \ldots\ldots \left(\frac{E_n}{E_1}\right)^2}$$

or, $$\text{THD} = \frac{\sqrt{E_2^2 + E_3^2 + E_4^2 + ... + E_n^2}}{E_1} \quad ...(15.3)$$

$\therefore$ $$\text{Percentage harmonic distortion} = \sqrt{D_2^2 + D_3^2 + D_4^2 + ... D_n^2} \times 100 \quad ...(15.4)$$

$$= \frac{\sqrt{E_2^2 + E_3^2 + E_4^2 + ... + E_n^2}}{E_1} \times 100 \quad ...(15.5)$$

Some of the methods used to measure the harmonic distortion factor are described below.

15.3.2. Fundamental-Suppression Distortion Analyser

This method of distortion measurement is used when it is important *to measure total harmonic distortion* rather than the distortion caused by each component. Hence, the input waveform is applied to a network that *suppresses or rejects the fundamental frequency but passes all the harmonic-frequency components for subsequent measurement.*

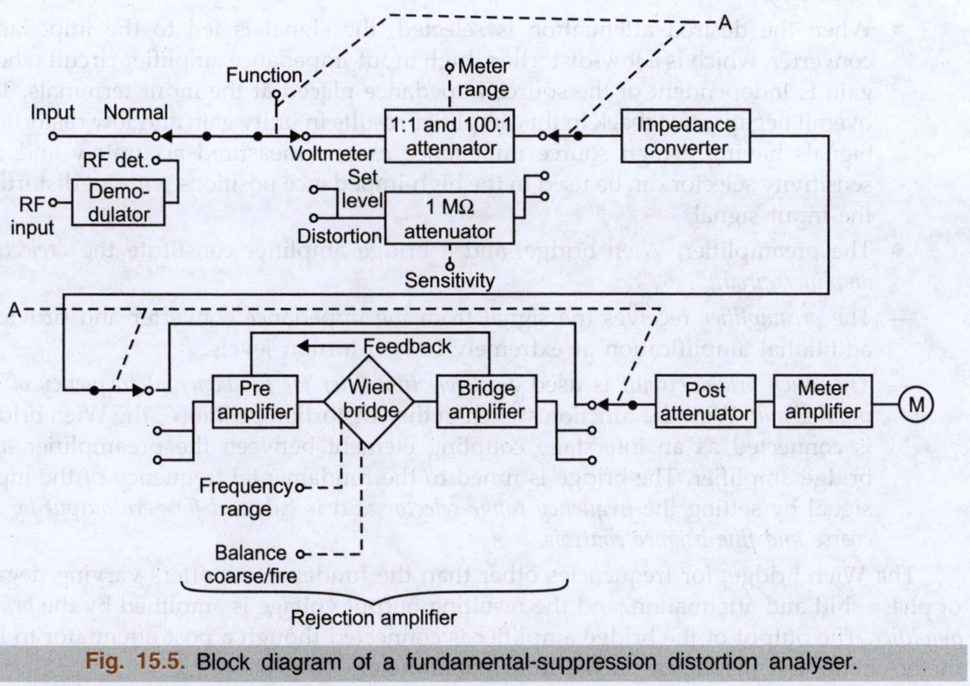

Fig. 15.5. Block diagram of a fundamental-suppression distortion analyser.

Fig. 15.5 shows the block diagram of a fundamental-suppression distortion analyser. The instrument consists of the following components:

(*i*) *Impedance converter* : It provides a low-noise, high-impedance input circuit, independent of the signal source at the input terminals to the instrument.

(*ii*) *Rejection amplifier* : It rejects the fundamental frequency of the input signal and passes the remaining frequency components on to the metering circuit where the harmonic distortion is measured.

(*iii*) *Metering circuit* : It provides a visual indication of a total harmonic distortion in terms of a percentage of total input voltage.

(*iv*) *The power supply.*

Operation/Working. The following two modes of operation are possible :

1. ***Voltmeter mode :***

- The instrument operates as a conventional A.C. voltmeter (a very convenient feature) when the function switch is in the *"voltmeter" position.*
- In this mode the input signal is applied to the impedance converter circuit through the 1 : 1 and 100 : 1 attenuator, which selects the appropriate meter range.
- The output of the impedance converter then bypasses the rejection amplifier (as shown dotted) and the signal is applied directly to the metering circuit. The meter measures the r.m.s. value of the A.C. input voltage.

2. ***Distortion mode :***

- When the instrument is in distortion mode, the function switch is in the *"distortion"* position and the rejection amplifier is the part of the circuit, and distortion measurements are made.
- In this case the input signal is applied to a 1 MΩ input attenuator that provides 50 dB attenuation in 10 dB steps, controlled by a front panel switch marked *sensitivity.*
- When the desired attenuation is selected, the signal is fed to the impedance converter, which is a low-distortion, high input-impedance amplifier circuit whose gain is independent of the source impedance placed at the input terminals. The overall negative feedback in this amplifier results in unity gain and low distortion. Signals having a high source impedance can be measured accurately and the sensitivity selector can be used in the high-impedance positions without distorting the input signal.
- The preamplifier, Wien bridge, and a bridge amplifier constitute the *"rejection amplifier circuit".*

— The *preamplifier* receives the signal from the impedance converter and provides additional amplification at extremely low distortion levels.

— The *Wien bridge circuit* is used as a *rejection filter for fundamental frequency of the input signal.* With the function switch in the "distortion position", the Wien bridge is connected as an interstage coupling element between the preamplifier and bridge amplifier. The bridge is tuned to the fundamental frequency of the input signal by setting the *frequency range selector* and is *balanced for zero output by the coarse and fine balance controls.*

The Wien bridge, for frequencies other than the fundamental, offers varying degree of phase shift and attenuation, and the resulting output voltage is amplified by the *bridge amplifier*. The output of the bridge amplifier is connected though a post attenuator to the meter circuit and displayed on the front panel meter.

Advantages :

This instrument has the following two main *advantages :*

1. The harmonic distortion produced by the instrument itself is negligibly small.
2. Selectivity requirements are not severe because only the fundamental frequency component is to be suppressed.

15.3.3. Heterodyne Harmonic Analyser or Wavemeter

This type of analyser makes use of a *highly selective, fixed frequency filter.*

Fig. 15.6. shows the block diagram of a heterodyne harmonic analyser :

— The input signal is mixed with output of variable frequency oscillator in a mixer circuit.

— After the successful mixing, output of mixer supplies the sum and the difference of each harmonic beating against the oscillator frequency. Now for each harmonic component, either the sum or difference of frequencies is made equal to the frequency of the filter by varying the frequency of oscillator.

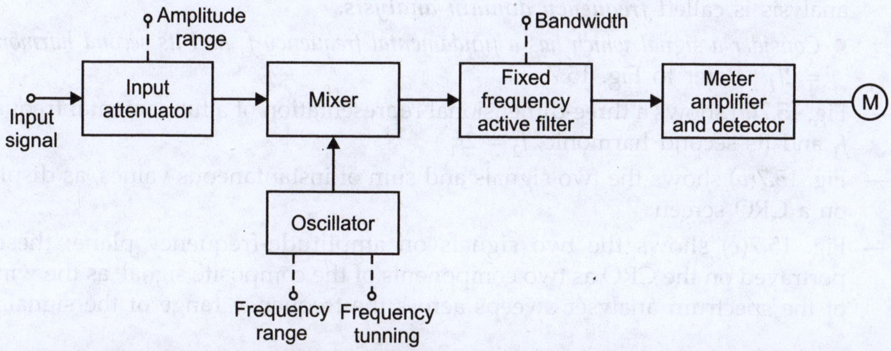

Fig. 15.6. Block diagram of heterodyne harmonic analyser or wavemeter.

— Only the mixer output signal that can pass through the filter is that signal whose frequency is equal to the frequency of the filter. *The output of the mixer is the resultant of the variable frequency oscillator output and harmonic component and the frequency of the resultant output from mixer is equal to the filter frequency.*

— The output of mixer has r.m.s. value which is proportional to the r.m.s. value of that harmonic component. So when the output of mixer is supplied to a true r.m.s. reading voltmeter through amplifier, the voltmeter reads the r.m.s. value of that harmonic component. In this way, the instrument *provides r.m.s. value of each harmonic component.*

15.4 SPECTRUM ANALYSERS

15.4.1. Introduction

Spectrum analyser *is an instrument which graphically presents an energy distribution of the signal as a function of frequency on its CRT.*

Spectrum analysis *is defined as the study of energy distribution across the frequency spectrum of a given electrical signal.*

The spectrum analysis of a signal provides the information about the following particulars/parameters :

(*i*) Bandwidth.
(*ii*) Effects of different types of modulation and spurious signal generation.
(*iii*) Frequency stability.
(*iv*) The component levels.
(*v*) Spectral purity.
(*vi*) Attenuation.

The spectrum analysis, an account of instrumentation limitations and capabilities, is divided into the following *two* major categories :

(*i*) Audio frequency (AF) analysis.
(*ii*) Radio frequency (RF) analysis.

The *"RF spectrum analysis"* covers a frequency range of 10 MHz to 40 GHz, and hence is more important, because *it includes the vast majority of communication, navigation, radar, and industrial instrumentation frequency bands.*

— In case of *normal oscilloscope* the *signal is plotted as a function of time i.e.,* amplitudes versus time plot is obtained. Such an analysis is called ***time domain analysis***.

— In case of *"spectrum analyser"*, the *signal is plotted as a function of frequency.* This analysis is called ***frequency domain analysis.***

- *Consider a signal which has a fundamental frequency f_1 and its second harmonic, $f_2 = 2f_1$* : Refer to Fig. 15.7.

— Fig. 15.7(*a*) shows a three-dimensional representation of a fundamental frequency, f_1 and its second harmonic, $f_2 = 2f_1$.

— Fig. 15.7(*b*) shows the two signals and sum of instantaneous values, as displayed on a CRO screen.

— Fig. 15.7(*c*) shows the two signals on amplitude-frequency plane; these are portrayed on the CRO as two components of the composite signal, as the window of the spectrum analyser sweeps across the frequency range of the signal.

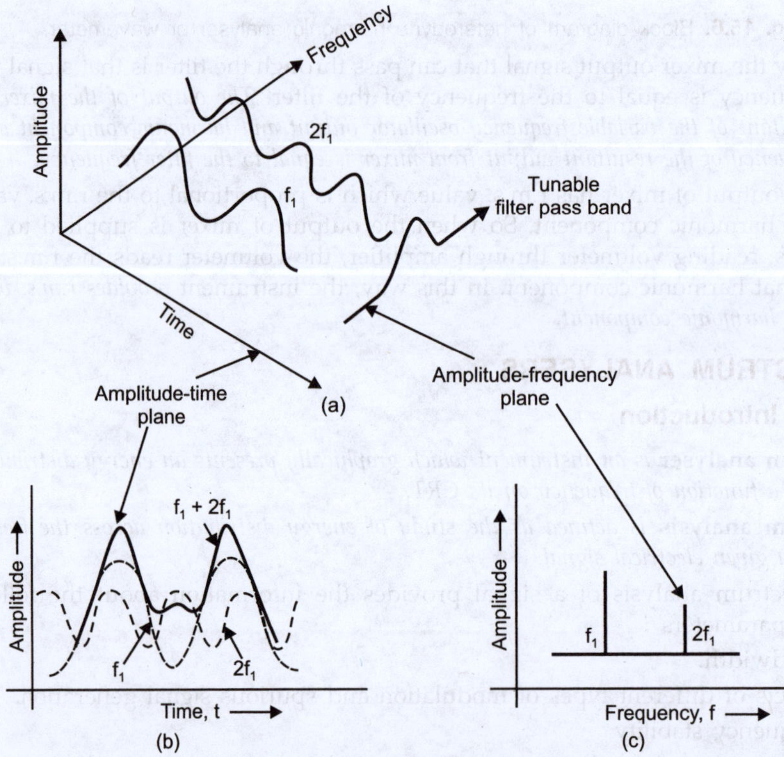

Fig. 15.7. A three-dimensional presentation of amplitude, frequency, and time.

- Spectrum analysers use either a parallel filter bank or swept frequency technique.

— In a *parallel filter bank analyser*, the frequency range is covered by a series of filters whose central frequencies and bandwidth (BW) are so selected that they overlap each other, as shown in Fig. 15.8.

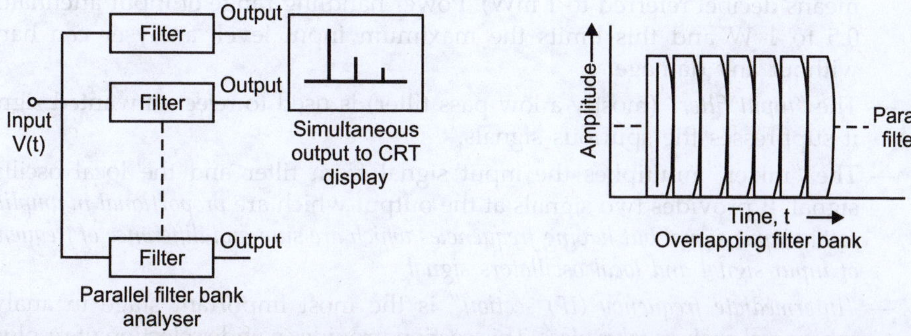

Fig. 15.8. Spectrum analyser (Parallel filter bank analyser).

— Typically, an audio analyser will have 32 of these filters, each covering one third of an octave.

For wide band narrow resolution analysis, particularly at RF or microwave signals, the *swept frequency technique* is preferred.

Classification of spectrum analysers :

According to the technique used, the spectrum analysers can be *classified* as follows :

1. ***Scanning type.*** These analysers use *swept technique.*
2. ***Non-scanning type.*** These are called *real time analysers.*

15.4.2. Swept Superheterodyne Spectrum Analyser

This analyser is so called because it uses a swept receiver of superheterodyne type.

Fig. 15.9 shows the block diagram of swept heterodyne spectrum analyser. The basic blocks of this analyser are :

(*i*) Wideband input mixer;

(*ii*) Swept local oscillator (for driving wideband mixer);

(*iii*) Resolution bandwidth filter (for deciding intermediate frequency);

(*iv*) Detector and video filter;

(*v*) Display.

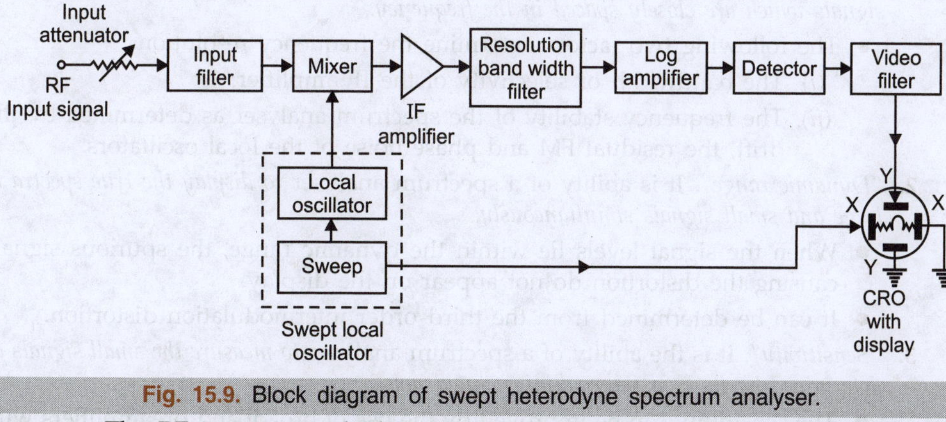

Fig. 15.9. Block diagram of swept heterodyne spectrum analyser.

— The RF input is applied to an input attenuator, which decides the level of input signal so as to keep it within the operating range of other blocks of the instrument. Normally spectrum analysers can handle 0 to 10 dBm (One dBm means decibel referred to 1 mW). Power handling range of input attenuator is 0.5 to 1 W and this limits the maximum input level, analyser can handle without any damage.

— The *"input filter"* (mostly a low pass filter) is used to reject unwanted signals; it suppresses the spurious signals.

— The "mixer" multiplies the input signal from filter and the local oscillator signal. It provides two signals at the output which are *proportional in amplitude to the input signal but having frequencies which are sum and difference of frequencies of input signal and local oscillators signal.*

— *"Intermediate frequency (IF) section"* is the most important stage in analyser where *real analysis takes place*. This section provides a wide selection of resolution width filters; these filters are described by their 3-dB bandwidth. These filters decide the resolving power of the analyser; the narrow the filter, the close two signals may be and still can be seen as separate responses. The filters are normally tuned to a narrow band of 2 GHz and thus the input signals which are separated from the local oscillator frequency by 2 GHz only are converted to the intermediate frequency band.

— The filter output is given to the *"log amplifier"*, which processes the incoming signal in a logarithmic fashion. The *logarithmic processing permits a large range of incoming signals to be measured and compared.*

— The signal from log amplifier (which is compressed one) is received by the *"detector"* and its output is given to the *"video filter"*. These filters are used *for post filtering or averaging the detector output*. The bandwidth setting of video filter is same or larger than resolution bandwidth filter. Averaging removes the pure signal remains and the random noise. The output of the video filter is given to the CRO for display purpose.

The *"swept local oscillator"* puts a limit on the stability and spectral purity in several performance areas.

Characteristics of spectrum analyser :

The characteristics of spectrum analyser are :

1. *"Frequency resolution"*. It is defined as the *ability of spectrum analyser to separate the signals which are closely spaced in the frequency.*
   - The following two factors determine the frequency resolution :
     (*i*) The bandwidth or selectivity of the IF amplifier.
     (*ii*) The frequency stability of the spectrum analyser as determined by the drift, the residual FM and phase noise of the local oscillators.
2. *"Dynamic range"*. It is ability of a spectrum analyser *to display the true spectra of large and small signals simultaneously.*
   - When the signal levels lie within the dynamic range, the spurious signals causing the distortion do not appear on the display.
   - It can be determined from the third order intermodulation distortion.
3. *"Sensitivity"*. It is the ability of a spectrum analyser *to measure the small signals as determined by its own internally generated noise.*
   - The sensitivity can be improved by the use of broadband preamplifiers with low noise figures, lower than that of spectrum analyser.
4. *"Sweep desensitization"*. It is an effect caused by scanning a spectrum analyser too

fast; it results in loss of amplitude, calibration, sensitivity and resolution.

- To avoid the sweep desensitization, the scan velocity in Hz/second must not exceed the square of 3-dB bandwidth of the IF filter in Hz.

15.4.3. Real Time Spectrum Analyser

A real time spectrum analyser is the one *which presents the effect of changes in all the input frequencies, on its spectrum display.* It is *used at audio frequencies.*

Fig. 15.10 shows the block diagram of a real time spectrum analyser :

— It consists of 24 channels and covers a range from 50 Hz to 10 kHz.

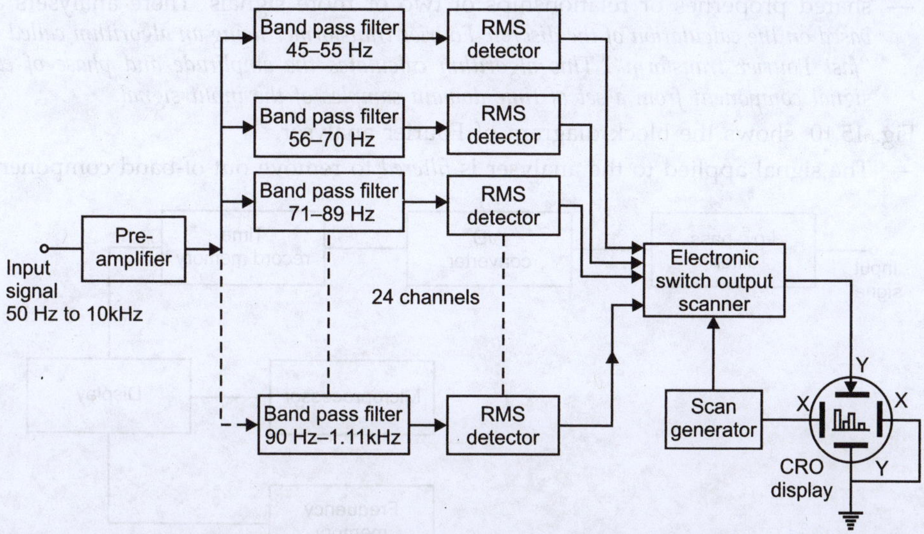

Fig. 15.10. Block diagram of real time spectrum analyser.

— Each channel consists of band pass filter and RMS detector. Since a number of band pass filters are used, it is called as a *multifilter real time spectrum analyser.*

— The electronic switch is used to scan the output of each filter sequentially.

— The output from electronic switch output scanner is given to the vertical deflecting plates of the CRO.

— The scan generator provides the horizontal deflection on CRO. The output of a scan generator is a sawtooth output which is synchronised with the electronic switch.

- In this analyser the composite amplitude of signal within the bandwidth of each filter is displayed as *a function of the frequency range of the filter.*
- These analysers are *only used in audio frequency applications* and can use as many as 32 filters.

Advantages :

1. Good resolution can be obtained by reducing the bandwidth of each filter.
2. The display rate can be increased at least upto 3 Hz.

Disadvantages :

1. Poor resolution.
2. Resolution cannot be adjusted. Hence, it is not possible to obtain the magnified portion of the display.

3. The total system is complex.
4. The frequency axis of the display is inflexible.

15.4.4. Fourier Analyser

Fourier analyser *uses digital signal processing techniques to provide measurement that is beyond the capabilities of spectrum analyser.* Some of the other applications are measurement of :

— very low-frequency or closely-spaced signals;

— random signals observed by noise;

— shared properties or relationships of two or more signals. There analysers are *based on the calculation of the discrete Fourier transformer using an algorithm called the "fast Fourier transform". This algorithm calculates the amplitude and phase of each signal component from a set of time domain samples of the input signal.*

Fig. 15.10. shows the block diagram of Fourier analyser.

— The signal applied to the analyser is *filtered* to remove out-of-band components.

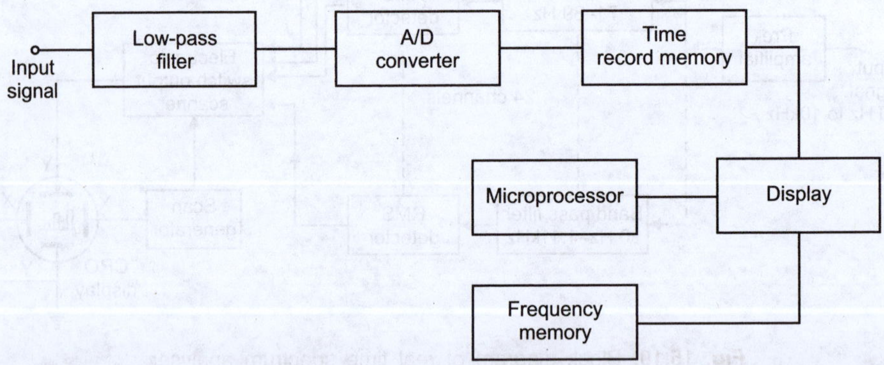

Fig. 15.11. Block diagram of Fourier analyser.

— The signal out of the *'low-pass filter'* is applied to the *'A/D converter'*, which samples and digitizes it at a regular time intervals until a full set of samples called a *time record* has been collected.

— The *'microprocessor'* then *performs the desired series of computations on the time data to obtain the frequency domain results.* These results, which are stored in the memory, *can be displayed on a CRT or recorded permanently with a recorder or plotter.*

- Since Fourier analysers are primarily digital instruments, interfacing to a computer or other digital system *is relatively simple.* Remote programming and transforming data via a bus or other interfacing systems can considerably *expand the range* of possible applications for the analyser. *By virtue of the digital nature of the analyser, it will provide a high degree of accuracy, stability, and repeatability in spectrum analysis.*

15.4.5. Applications of Spectrum Analyser

The applications of spectrum analyser are :

1. Amplitude modulation.
2. Frequency modulation.
3. Pulse modulation.
4. Measurement of harmonic distortion.

5. Noise measurement.
6. Determination of the frequency drift of a signal.
7. Analysis of air and water pollution.
8. Measurement of antenna pattern.
9. Tuning a parametric amplifier.
10. In the field of electronic testing related to troubleshooting and quality control.
11. In the fields of *biomedical, radars,* and *oceanography.*

HIGHLIGHTS

1. Following instruments are used for signal analysis :

   Wave analysers; Distortion analysers; Spectrum analysers; Audio analysers; Modulation analysers.
2. A *wave analyser* is an instrument to measure relative amplitudes of single frequency components in a complex waveform.
3. *Spectrum analyser* is an instrument which graphically presents an energy distribution of the signal as a function of frequency on its CRT.

OBJECTIVE TYPE QUESTIONS

Fill in the blanks or say "Yes" or "No" :

1. It can be shown mathematically that any complex waveform is made up of a fundamental and its ....
2. Wave analysers are used in low RF range, below 50 MHz and down through the AF range.
3. Harmonic .... analysers measure the total harmonic current in the waveforms.
4. The wave analyser must have extremely high input distortion.
5. .... analyser is an instrument to measure relative amplitudes of single frequency components in a complex waveform.
6. Frequency selective wave analyser is used for measurements in the .... frequency range.
7. Heterodyne harmonic analyser makes use of a highly selective, fixed frequency filter.
8. .... analysis is defined as the study of energy distribution across the frequency spectrum of a given electrical signal.
9. .... analyser is an instrument which graphically presents an energy distribution of the signal as a function of frequency on its CRT.
10. Spectrum analysers use either a parallel filter bank or .... frequency technique.

ANSWERS

1. harmonics	**2.** Yes	**3.** distortion	**4.** No
5. Wave	**6.** audio	**7.** Yes	**8.** Spectrum
9. Spectrum	**10.** swept.		

THEORETICAL QUESTIONS

1. Name the instruments which are used for signal analysis.
2. What is a "wave analyser"? Explain.
3. Discuss briefly a resonant or basic wave analyser.
4. Explain briefly, with a neat block diagram, the working of a frequency selective wave analyser.
5. Describe with the help of a block diagram the working of a heterodyne wave analyser.
6. What are the applications of wave analyser?
7. What do you mean by the term 'harmonic distortion'?
8. What is total harmonic distortion?
9. Explain briefly "Fundamental-suppression distortion analyser"? State its advantages also.
10. With the help of a block diagram, give the description of "Heterodyne harmonic analyser or wavemeter".
11. What is a "Spectrum analyser"?
12. What do you mean by "Spectrum analysis"?
13. Which information is provided by the spectrum analysis of a signal?
14. How are spectrum analysers classified?
15. Explain the working of a swept superheterodyne spectrum analyser, with a neat block diagram.
16. What are the characteristics of spectrum analyser?
17. Describe briefly "Real time spectrum analyser". State its advantages and disadvantages.
18. List the applications of spectrum analyser.
19. Write short notes on the following :
    (*i*) Wave analyser;
    (*ii*) Spectrum analyser;
    (*iii*) Harmonic distortion analyser.

PART – II
ELECTRICAL AND ELECTRONIC INSTRUMENTATION

CHAPTER

16 Sensors and Transducers

16.1. Introduction, 16.2. Definition of Transducer, 16.3. Classification of Transducers – Basic requirements of a transducer, – Transducer sensitivity, – Specifications for transducers, **16.4.Advantages and Disadvantages of Electrical Transducers, 16.5. Transducers Actuating Mechanisms, 16.6. Resistance Transducers** – Linear and angular motion potentiometers – Thermistors and resistance thermometers – Wire resistance strain gauges, **16.7. Variable Inductance Transducers** – Self-generating type – Passive type, **16.8. Capacitive Transducers** – Capacitive transducers – using change in area of plates – Capacitive transducer – using change in distance between the plates – Capacitive tachometer, **16.9. Piezoelectric Transducers,** – Piezoelectric materials – Desirable properties of piezoelectric materials – Working of a piezoelectric device – Advantages and disadvantages of piezoelectric transducers – Piezoelectric accelerometer, **16.10. Hall Effect Transducers** – Hall effect – Hall effect transducers, **16.11.Thermoelectric Transducers, 16.12. Photoelectric Transducers** – Principle of operation – Applications – Classification – Photoemissive cells – Photovoltaic cell – Photoconductive cell –Photoelectric tachometer, **16.13. Strain Gauges** – Introduction – Types of strain gauges– Theory of strain gauges – Strain gauge circuits, **16.14. Load Cells** – Hydraulic load cell – Pneumatic load cell – Strain-gauge load cells, **16.15. Proximity Sensors** – Eddy current proximity sensor – Capacitive proximity sensor – Inductive proximity switch, **16.16. Pneumatic Sensors, 16.17. Light Sensors, 16.18. Tactile Sensor, 16.19. Smart Sensor, 16.20. Fibre Optic Transducers** – Introduction – Types of optical fiber sensor configurations, **16.21. Digit Transducers** – Introduction – Frequency domain transducers – Digital encoders – Digital Pressure Transducer, **16.22. Recent Trends–Smart Pressure Transmitters 16.23. Selection of Sensors, 16.24. Rotary Variable Differential Transformer (RVDT), 16.25. Synchros and Resolvers, 16.26. Induction Potentiometers, 16.27. Micro Electromechanical Systems (MEMS)** – Introduction – Advantages of MEMS – Applications of MEMS – Description of MEMS – Manufacturing of MEMS– MEMS Accelerometer–Vibration microsensor – MEMS Humidity microsensor– *Highlights – Objective Type questions – Theoretical Questions – Unsolved Examples.*

16.1 INTRODUCTION

The *primary sensing element* (*sensor*) is the first and foremost requirement for measurement and automatic controls. The ***sensors*** *sense the condition, state or value of the process variable and produce an output which reflects this condition, state or value. The* **transducers** *transform the energy of the process variable to an output of some other type of energy which is able to operate some control device.* Sometimes a *secondary transducer* may be employed to transform the output of the primary sensor to still another type of energy.

Examples :

- In the ordinary *dial indicator* the indicating spindle acts as a sensor/detector for displacement. It simply performs the function of sensor/detector and nothing else.
- The function of a *Bourdon tube* of a pressure gauge is *twofold*: Firstly to sense the pressure and secondly to give the resulting effect or output in the form of displacement. Here the tube acts a sensor/detector transducer.
- In a *compressive load cell*, the platform detects the force and gives an output in the form of deflection. This deflection may be further converted into an electrical output by strain gauges (called *secondary transducer*).

For the measurement of particular quantity, different types of sensors and transducers are available and the choice of a suitable unit depends upon the static and dynamic performance characteristics.

16.2 DEFINITION OF TRANSDUCER

A broad definition of a transducer is as follows:

"A **transducer** *is a device which converts the energy from one form to another".*

Most of the transducers either convert electrical energy into mechanical displacement and/or convert some non-electrical physical quantity (*e.g.*, force, sound, temperature etc.) to an electrical signal.

A transducer performs the following *functions* in an electronic instrumentation system :

1. *Detects or senses the presence, magnitude and changes in physical quantity being measured.*
2. *Provides a proportional electrical output signal* (see Fig. 3.1.)

- A **transducer** can be broadly defined as a *device which converts a non-electrical quantity into an electrical quantity.*
- An **electrical transducer** *is a sensing device by which a physical, mechanical or optical quantity to be measured is transformed directly, with a suitable mechanism, into an electrical voltage or current proportional to the input measurand.*

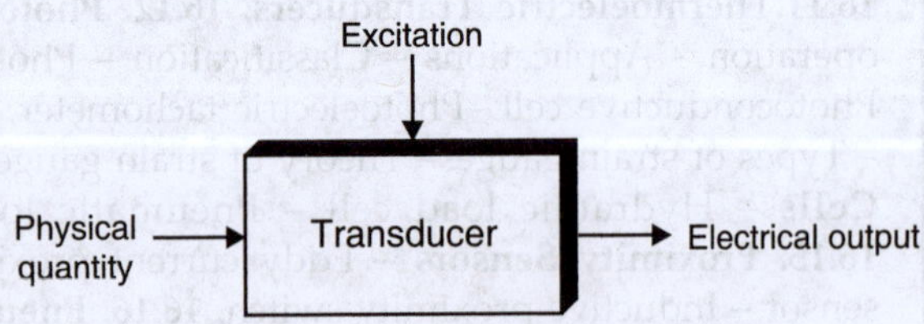

Fig. 16.1. Transducer.

- An **inverse transducer** *is defined as a device which converts an electrical quantity into a non-electrical quantity.* It is a precision actuator which has an electrical input and a low power non-electrical output. A *piezoelectric* crystal acts as an inverse transducer because when a voltage is applied across its surfaces, it changes its dimensions causing a mechanical displacement.
- The significant parameters which *dictate the transducer capability* are :

  (*i*) Linearity; (*ii*) Repeatability;
  (*iii*) Resolution; (*iv*) Reliability.

16.3 CLASSIFICATION OF TRANSDUCERS

A. Transducers are broadly *classified* into two groups as follows:

1. ***Active transducers.*** They are also known as *self-generating type transducers.* These transducers *develop their own voltage or current. The energy required for production of an output signal is obtained from the physical phenomenon being measured.*

Examples: *Thermocouples and thermopiles, piezoelectric pick-up, photovoltaic cell.*

2. ***Passive transduces.*** *They are known as externally-powered transducer.* These transducers *derive the power required for the energy conversion from an external power source.* However, they *may absorb some energy from the physical phenomenon under study.*

   Examples: *Resistance thermometers and thermistors, potentiometric devices, differential transformer, photoemission cell etc.*

B. Classification based on the type of output :

1. ***Analogue transducers.*** These transducers convert the input physical phenomenon into an analogous output which is a continuous function of time.

   Examples: *Strain gauge, a thermocouple, a thermistor or an LVDT* (linear voltage differential transformer).

2. ***Digital transducers.*** These transducers convert the input physical phenomenon into an electrical output which may be in form of pulse.

C. Classification based on electrical principle involved :

1. ***Variable-resistance type :***
   (*i*) Strain and pressure gauges. (*ii*) Thermistors, resistance thermometers.
   (*iii*) Photoconductive cell etc.
2. ***Variable-inductance type :***
   (*i*) Linear voltage differential transformer (LVDT).
   (*ii*) Reluctance pick-up. (*iii*) Eddy current gauge.
3. ***Variable-capacitance type :***
   (*i*) Capacitor microphone. (*ii*) Pressure gauge.
   (*iii*) Dielectric gauge.
4. ***Voltage-generating type :***
   (*i*) Thermocouple. (*ii*) Photovoltaic cell.
   (*iii*) Rotational motion tachometer. (*iv*) Piezoelectric pick-up.
5. ***Voltage-divider type :***
   (*i*) Potentiometer position censor. (*ii*) Pressure-actuated voltage divider.

Table 16.1. shows the *measurements versus transduction methods.*

- While *describing a particular transducer* the information must be available about the following aspects :
  (*i*) The measurand.
  (*ii*) The sensing element which responds directly to the measurand.
  (*iii*) The principle of operation of the transducer and where the output of the transducer originates.
  (*iv*) The useful range.
- While *selecting a dector-transducer element*, the following major consideration need to be looked into:
  (*i*) Mechanical suitability in terms of
  — Physical size, weight and shape;
  — Mounting arrangement;
  — Ruggedness.
  (*ii*) Electrical suitability in terms of:
  — Sensitivity;
  — Frequency response;
  — Ease of signal transmission.

(*iii*) Environmental suitability in terms of
— Sensitivity to temperature and self-heating effects;
— Magnetic fields;
— Vibration; dust and humidity;
— Supply frequency etc.

(*iv*) Transducer performance in terms of calibration accuracy.

(*v*) Desired measurement accuracy and range, power requirements, overload protection and vulnerability to sudden failure.

(*vi*) Purchase aspects.

Table 16.1. Measurements versus Transduction Methods

S. No.	Quantity to be measured	Type of transducer	S. No.	Quantity to be measured	Type of transducer
1.	***Displacement***	— Resistive	3.	***Velocity***	— Resistive
		— Inductive			— Inductive
		— Capacitive			— Capacitive
		— Piezoelectric			— Piezoelectric
		— Magnetoelectric			— Photoelectric
		— Radioactive			— Magnetoelectric
		— Electron tube.			— Radioactive
2.	***Thickness***	— Inductive			— Electron tube.
		— Capacitive	4.	***Acceleration***	— Resistive
		— Piezoelectric			— Inductive
		— Photoelectric			— Capacitive
		— Radioactive.			— Piezoelectric
		Magnetoelectric			— Magnetoelectric
		— Electron tube.			— Radioactive.
5.	***Mass***	— Inductive	9.	***Level***	— Resistive
		— Piezoelectric			— Capacitive
		— Magnetoelectric			— Piezoelectric
		— Radioactive.			— Photoelectric
6.	***Force***	— Resistive			— Radioactive.
		— Inductive	10.	***Temperature***	— Resistive
		— Piezoelectric			— Photoelectric
		— Radioactive.			— Thermoelectric
7.	***Pressure***	— Resistive			— Radioactive.
		— Inductive	11.	***Humidity***	— Resistive
		— Capacitive			— Capacitive.
		— Piezoelectric	12.	***Viscosity***	— Resistive
		— Thermoelectric			— Capacitive
		— Magnetoelectric			— Piezoelectric
		— Radioactive			— Magnetostrictive.
		— Electron tube.			
8.	***Flow***	— Resistive			
		— Inductive			
		— Capacitive			
		— Piezoelectric			

16.3.1. Basic Requirements of a Transducer

Transducers should meet the following *basic requirements* :

1. Ruggedness;
2. Linearity;
3. Repeatability;
4. High output signal quality;
5. High reliability and stability;
6. Good dynamic response;
7. No residual deformation;
8. No hysteresis.

16.3.2. Transducer Sensitivity

The relationship between the measurand and the transducer output signal is referred to as **transducer sensitivity**.

i.e., Transducer sensitivity $= \dfrac{\text{Output signal increment}}{\text{Measuand increment}}$

Sensitivity of a transducer should be *usually as high as possible* since then it becomes easier to take the measurements.

16.3.3. Specifications for Transducers

While selecting the proper transducer for any applications, or ordering the transducers, the following *specifications* should be thoroughly considered:

(*i*) Ranges available.
(*ii*) Squaring system.
(*iii*) Sensitivity.
(*iv*) Maximum working temperature.
(*v*) Method of cooling employed.
(*vi*) Mounting details.
(*vii*) Maximum depth.
(*viii*) Linearity and hysteresis.
(*ix*) Output for zero input.
(*x*) Temperature coefficient of zero drift.
(*xi*) Natural frequency.

16.4 ADVANTAGES AND DISADVANTAGES OF ELECTRICAL TRANSDUCERS

The advantages and disadvantages of electrical transducers are as follows :

Advantages :

1. Very small power is required for controlling the electrical or electronic system.
2. The electrical output can be amplified to any desired level.
3. The effect of friction is reduced to the minimum possible.
4. Mass-inertia effects are reduced to minimum possible (In fact when dealing with electrical or electronic signals the inertia effects are because of electrons, which have negligible mass).
5. The size and shape of the transducers can be suitably designed to achieve the optimum weight and volume.
6. The output can be indicated and recorded remotely at a distance from the sensing medium. Also, more than one indicator can be actuated simultaneously.
7. The output can be modified to meet the requirements of the indicating or controlling equipment.
8. The signals can be conditioned or mixed to obtain any combination with outputs of similar transducers or control signals as in air data computer or adaptive control systems.

Example. The measurement of 'Mach number' with two measurands.

9. The contour design and dimensions can be so chosen as not to disturb the measurand phenomena as in the case of turbulence measurements.

Modern digital computers have made the use of electrical transducers absolutely essential.

Disadvantages :

1. Low reliability in comparison to mechanical types, due to ageing and drift of the active components.
2. Sensing elements and associated signal conditioners are comparatively expensive.
3. In some cases, the accuracy and resolution attainable are not as high as in mechanical devices. The best accuracy available is of the order of 0.01 percent.

16.5 TRANSDUCERS ACTUATING MECHANISMS

Transducers are also known as *gauges, pick-ups and signal generators.* Most of the pick-ups have following *two basic elements :*

(*i*) Activating device.

(*ii*) Transducing element.

Fig. 16.2. shows some typical actuating mechanisms.

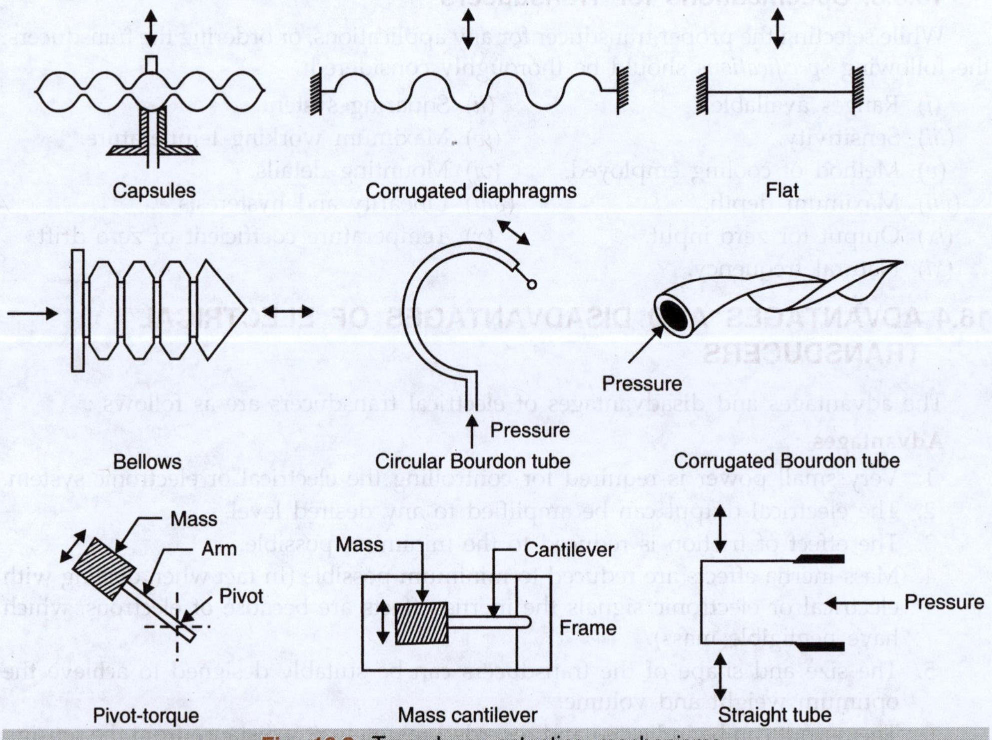

Fig. 16.2. Transducer actuating mechanisms.

16.6 RESISTANCE TRANSDUCERS

The resistance of a metal conductor is expressed by a simple equation that involves a few physical quantities. The relationship is given by $R = \frac{\rho L}{A}$;

where,
R = Resistance, Ω,
ρ = Resistivity of conductor materials, Ω-*m*,
L = Length of conductor, *m*, and
A = Cross-sectional area of the conductor, m^2.

Any method of varying one of the quantities involved in the above relationship can be the designed basis of an *electrical resistance transducer.* There are a number of ways in which resistance can be changed by a physical phenomenon.

The *translational and rotational "potentiometers"* which work on the basis of change in the value of resistance with change in length of the conductor can be used for measurement of translational or rotary displacements.

*"**Strain gauges**" work on the principle that the resistance of a conductor or a semiconductor changes when strained.* This property can be used for measurement of displacement, force and pressure.

The resistivity of materials changes with the change of temperature thus causing a change of resistance. This property may be used for measurement of *"temperature".*

In a *resistance transducer* an indication of measured physical quantity is given-by *a change in the resistance.* It may be *classified* (as discussed above) as follows :

1. Mechanically varied resistance — *Potentiometer*
2. Thermal resistance change — *Resistance thermometers*
3. Resistivity change — *Resistance strain gauge.*

16.6.1. Linear and Angular Motion Potentiometers

Such potentiometers *convert the linear motion or the angular motion of a rotating shaft into changes in resistance.* The device is a variable resistor whose resistance is varied by the movement of a slider over a resistance element.

— Translatory devices have strokes from 2.5 mm to 5 mm.

— Rotational devices have full scale from 10° to 60° full turn.

The potentiometer shown in Figs. 16.3 and 16.4 form a part of the bridge circuit whose output voltage is changed by the slider position.

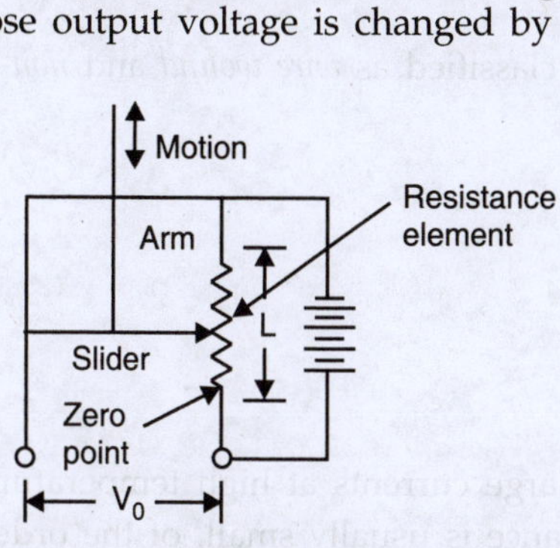

Fig. 16.3. Linear motion potentiometer.

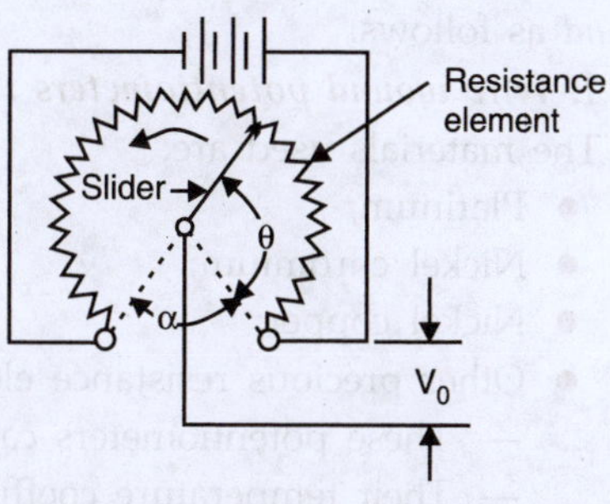

Fig. 16.4. Rotary motion potentiometer.

— The slider is powered by the mechanical part on which the linear displacement or angular measurement are to be made.

— Due to arm movement, the slider moves over the resistance element and thus shorts out a portion of the resistance. *The change in resistance in the potentiometer is then an indication of the amount of motion and the direction of movement is indicated by whether the resistance is increasing or decreasing.* The unbalanced voltage is measured directly or fed into an amplifier and recorded.

The potentiometers are used in many transducers designed to measure :

(*i*) Pressure (*ii*) Force

(*iii*) Acceleration (*iv*) Liquid level.

The potentiometers have the folllowing *advantages* and *disadvantages* :

Advantages :

(i) High output.
(ii) Less expensive.
(iii) Available in different sizes, shapes and ranges.
(iv) Simple to operate.
(v) Their electrical efficiency is very high.
(vi) Rugged construction.
(vii) Insensitivity towards vibration and temperature.

Disadvantages :

(i) Limited life due to early wear of the sliding ram.
(ii) The output tends to be noisy and erratic in high speed operation or when in high vibration environment.
(iii) In wire wound potentiometers the resolution is limited while in cermet and metal film potentiometers, the resolution is infinite.

Power rating of potentiometers :

The potentiometers are designed with a definite power rating which is related directly to their heat dissipating capacity. The manufacturer normally designs a series of potentiometers of single turn with a diameter of 50 mm with a wide range of ohmic values ranging from *100 Ω to 10 kΩ* in *steps of 100 Ω.* These potentiometers are essentially of the *same size* and of the *same mechanical configuration.* They have the *same heat transfer capabilities.* Their rating is typically *5 W at an ambient temperature of 21°C.* This limits their input excitation voltage.

Materials used for potentiometers :

The materials used for potentiometers may be classified as *wire wound* and *non-wire wound* as follows:

1. Wire wound potentiometers :

The materials used are:

- Platinum;
- Nickel chromium;
- Nickel copper;
- Other precious resistance elements.
  - — These potentiometers carry relatively large currents at high temperatures.
  - — Their temperature coefficients of resistance is usually small, of the order of 20×10^{-6}/°C or less.
  - — Their resolution is about 0.025 – 0.05 mm and is limited by the number of turns that can be accommodated on the body.
  - — The response is limited to about 5 Hz.
  - — The maximum speed with which a wire wound potentiometer may be turned is about 300 r.p.m.

2. Non-wire wound potentiometers :

These are also called *continuous potentiometers.*

The *materials* used are :

- Cermet;
- Hot moulded carbon;

- Carbon film;
- Thin metal film.
  — These potentiometers provide *improved resolution and life* (since resolution is no longer limited by the number of turns that can be wrapped onto a body)
  — A continuous potentiometer may be turned at a speed of 2000 r.p.m.
  — These are *more sensitive to temperature changes* and have a *higher wiper contact resistance*, which is variable and can *carry only moderate currents.*

Example 16.1. *Explain briefly the types of errors encountered in a transducer.*

Solution. The types of errors encountered in a transducer are briefly discussed below:

1. Scale errors.
2. Dynamic errors.
3. Noise and drift errors.

1. ***Scale errors.*** *Calibration* in general may be defined as the *process for determination, by measurement or comparison with a standard, of the correct value of each scale reading on the measuring instrument.* It is the determination of the settings of a control device that correspond to particular values of voltage, current, frequency, pressure or some other output.

2. ***Dynamic errors.*** *Fidelity* of an instrument system is defined as the *degree of closeness with which the system records the signal which is impressed upon it.* It refers to the ability of the system to reproduce the output in the same form as the input. "Dynamic error" is the *difference between the true value of a quantity changing with time and the value indicated by the instrument if no static error is assumed.*

3. ***Noise and drift errors.*** *Drift* is undesired change or a gradual variation in output over a period of time that is unrelated to changes in output, operating conditions or load. Drift for a measuring device can either be systematic, random, or some combination of the two. For most devices, drift is measured and specified as a percentage of output span.

Example 16.2. *A linear resistance potentiometer is 50 mm long and is uniformly wound with wire having a resistance of 10000 Ω. Under normal conditions, the slider is at the centre of the potentiometer. Find the linear displacement when the resistances of the potentiometer as measured by a Wheatstone bridge for two cases are (i) 3850 Ω, (ii) 7560 Ω.*

Are the two displacements in the same direction? If it is possible to measure a minimum value of 10 Ω resistance with the above arrangements, find the resolution of the potentiometer in mm. **(Anna University)**

Solution. Under normal condition, it is given that the slider is at the centre of the potentiometer, hence under normal position the resistance of the potentiometer

$$= \frac{10000}{2} = 5000\ \Omega$$

The resistance of the potentiometer wire per unit length

$$= \frac{10000}{50} = 200\ \Omega/\text{mm}$$

(i) Change in resistance of potentiometer from its normal position

$$= 5000 - 3850 = 1150\ \Omega$$

$$\therefore \quad \textbf{Displacement} = \frac{1150}{200} = \textbf{5.75 mm (Ans.)}$$

(*ii*) Change in resistance of potentiometer from its normal position

$= 7560 - 5000 = 2650\ \Omega$ (in the *opposite direction*)

$\therefore$ **Displacement** $= \frac{2560}{200} =$ **12.8 mm. (Ans.)**

The two displacements are in the *opposite direction.*

Resolution of the potentiometer

$$= \frac{\text{Min. measurable resistance}}{\Omega/\text{mm}} = \frac{10}{200} = \mathbf{0.05\ mm.}\ \textbf{(Ans.)}$$

16.6.2. Thermistors and Resistance Thermometers

These transducers are thermally sensitive variable resistors made of certain conducting and ceramic-like semiconducting materials. They are used *as temperature detecting elements used to sense temperature for the purpose of measurements and control.*

Thermistors are essentially semiconductors which behave as resistors with a *high negative temperature coefficient of resistance.* The high sensitivity to temperature changes makes the thermistors extremely useful for precision temperature (–60°C to + 15°C) measurements, control and compensation. Their resistance ranges from 0.5 Ω to 0.75 MΩ.

Thermistors are composed of sintered mixture of metallic oxides such as manganese, nickel, cobalt, copper, iron and uranium.

Fig. 16.5. shows the commercial forms of thermistors.

Applications of thermistors :

1. *Measurements of temperature* .... (major application).
2. Temperature compensation in complex electronic equipment, magnetic amplifiers and instrumentation equipment.
3. Measurement of power at high frequencies.
4. Vacuum measurements.
5. Measurements of level, flow and pressure of liquid.
6. Measurement of thermal conductivity.

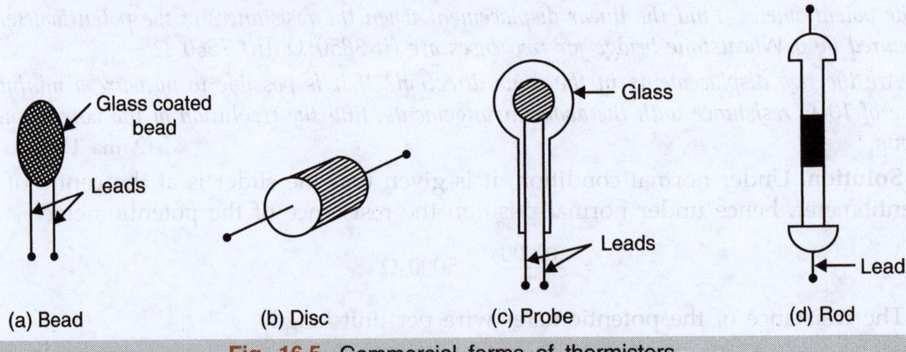

Fig. 16.5. Commercial forms of thermistors.

16.6.3. Wire Resistance Strain Gauges

Refer to article 16.14.

Salient features of thermistors :

1. The thermistors are compact, rugged and inexpensive.
2. They have good stability, when properly aged.

3. Their response time can vary from a fraction of a second to minutes depending on the size of the detecting mass and thermal capacity of the thermistors. It varies *inversely* with dissipation factor.
4. The upper limit of temperature for thermistors is dependent on physical changes in the material or solder used in attaching the electrical connections and is usually 400°C or less.
5. These can be installed at a distance from their associated measuring circuits if elements of high resistances are used such that the resistance of leads is negligible.
6. The measuring current should be maintained to as low a value as possible so that self-heating of thermistors is avoided otherwise errors are introduced on account of change of resistance caused by self-heating.

- *For more details refer to* **Art. 12.2.**

Example 16.3. *(a) A thermistor has a resistance temperature coefficient of –5% over a temperature range of 25°C to 50°C. If the resistance of the thermistor is 100 W at 25° C, what is the resistance at 35°C?*

(b) Suggest a complete instrumentation scheme in block diagram form to measure the temperature in a closed oven with the help of thermistor.

Solution. (*a*) $R_{35} = R_{25}[1 + \alpha(35 - 25) = 100[1 - 0.05(35 - 25)] = 50\ \Omega$ **(Ans.)**

(*b*) Fig. 16.6. shows the complete instrumentation scheme for the measurement of temperature with the help of a thermistor. Thermistor is mounted in the oven at a place where temperature is to be sensed. With the increase in temperature, resistance of the thermistor decreases causing imbalance in Wheatstone bridge circuit whose output balance voltage is amplified by signal conditioning device, the amplified output when connected to a suitable output device gives the value of the temperature of the oven.

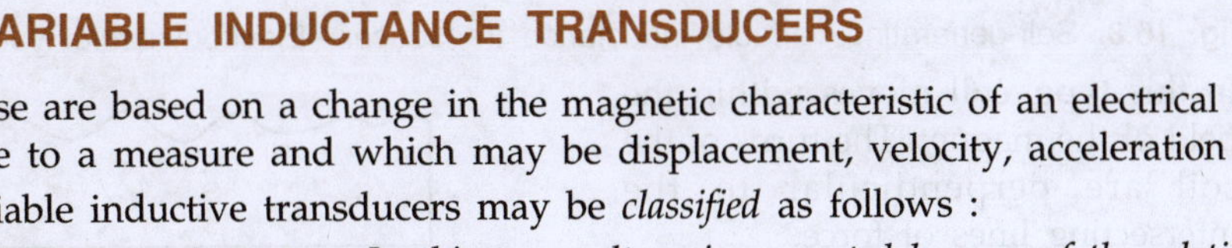

Fig. 16.6.

16.7 VARIABLE INDUCTANCE TRANSDUCERS

These are based on a change in the magnetic characteristic of an electrical circuit in response to a measure and which may be displacement, velocity, acceleration etc.

Variable inductive transducers may be *classified* as follows :

1. Self-generating type. *In this type voltage is generated because of the relative motion between a conductor and a magnetic field.*

These may be further *classified* as follows:

(*i*) Electromagnetic type.

(*ii*) Electrodynamic type.

(*iii*) Eddy current type.

2. Passive type. In this type the *motion of an object results in changes in the inductance of the coils of the transducer.*

These may be further *classified* as follows :

(*i*) Variable reluctance.

(ii) Mutual inductance.
(iii) Differential transfer type.

16.7.1. Self-generating Type

16.7.1.1. Electromagnetic type

Fig. 16.7 shows an electromagnetic type of self-generating variable inductance transducer :

— It consists of a permanent magnet core on which a coil is directly wound.
— When a plate of iron or other ferromagnetic material is moved with respect to the magnet, the flux field expands or collapses and a voltage is induced in the coil.
— This device is used for *indication of angular speed*. The measurements of speed can be made with great accuracy when the pick-up is placed near the teeth of a rotating gear.

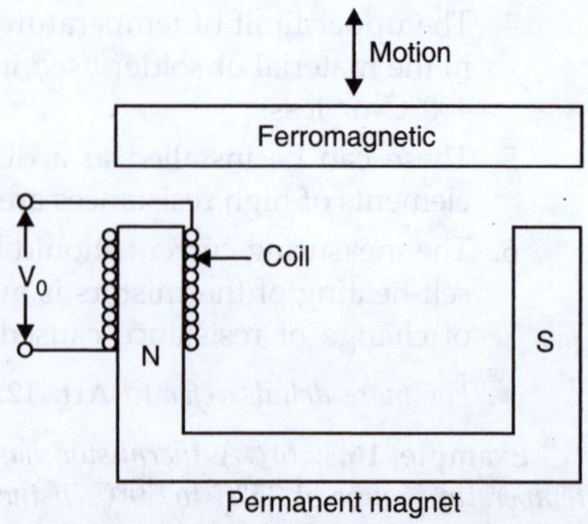

Fig. 16.7. Self-generating variable inductance transducer—Electromagnetic type.

16.7.1.2. Electrodynamic type

This type of transducer, linear and rotational is shown in Fig. 16.8.

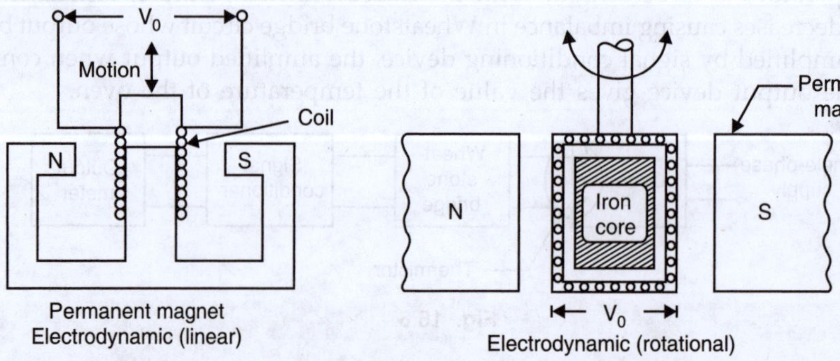

Fig. 16.8. Self-generating variable inductance transducer—Electrodynamic type.

— In this type, coil moves within the field of the magnet. The turns of the coil are perpendicular to the intersecting lines of force.
— When the coil moves it induces a voltage which at any moment is *proportional to the velocity of the coil.*

- The principle of these transducers is used in the *magnetic flow meters.*

16.7.1.3. Eddy current type

Fig. 16.9 shows an eddy current type of self-generating variable inductance transducer.

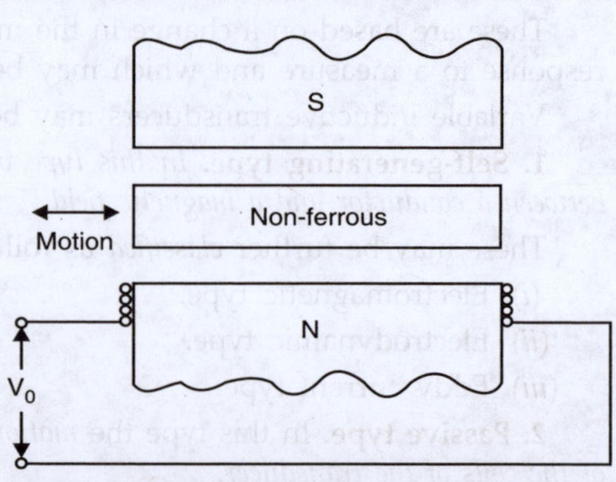

Fig. 16.9. Self-generating variable inductance transducer—Eddy current type.

Eddy current or drag cup tachometer :

In this type of tachometer (Fig. 16.10) the test-shaft rotates a permanent magnet and this induces eddy currents in a drag cup of disc held close to the magnet. The eddy currents produce a torque which rotates the cup against the torque of a spring. The disc turns in the direction of rotating magnetic field until the torque developed equals that of the spring. A pointer attached to the cup indicates the rotational speed on a calibrated scale.

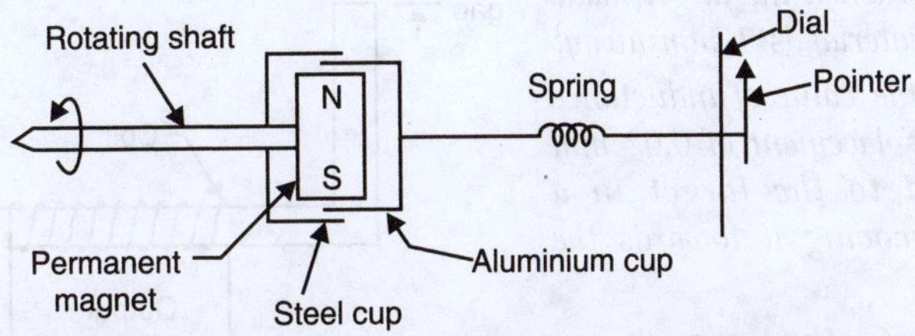

Fig. 16.10. Eddy current or drag type tachometer.

- The *automobile speedometers* operate on this principle.
- These tachometers are used for measuring rotational speeds upto 12000 r.p.m. with an accuracy of ± 3 per cent.

16.7.2. Passive Type

16.7.2.1. Variable reluctance transducer

In these transducers (comprising of a magnetic field and core with a gap between the core and the fixed coils) a change in the reluctance of the magnetic circuit by a mechanical input results in a similar change both in the inductance and inductive reactance of the coils. The change in inductance is then measured by suitable circuitry and related to the value of mechanical input.

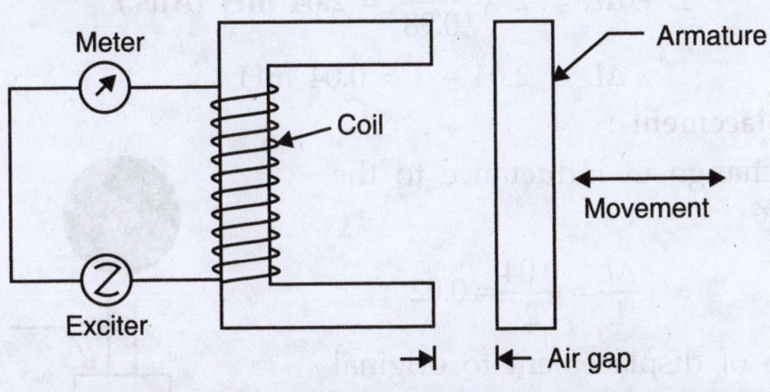

Fig. 16.11. Variable reluctance transducer.

The magnetic circuit reactance may be changed by affecting a change :

(*i*) in the *air gap* or

(*ii*) in the *amount/type of core material.*

— Transducers which make use of *air gap change* are referred as *reluctance type.*

— Transducers which utilize a *variable core* are referred as *permeance type.*

A variable *reluctance transducer* is shown in Fig. 16.11. Here the inductance of a single coil is changed through the variable air gap. The change in inductance may be calibrated in terms of movement of the armature.

This principle of variable reluctance is used for the measurement of dynamic quantities such as :

(*i*) Pressure (*ii*) Force
(*iii*) Displacement (*iv*) Acceleration
(*v*) Angular position etc.

Example 16.4. *Fig. 16.12 shows a variable reluctance type proximity inductive transducer in which the coil has inductance of 2 mH when the target made of ferromagnetic material is 1 mm away.*

(*i*) *Calculate the value of inductance when a displacement of 0.02 mm is applied to the target in a direction moving it towards the core.*

(*ii*) *Show that the change in inductance is linearly proportional to the displacement. Neglect the reluctance of the iron parts.*

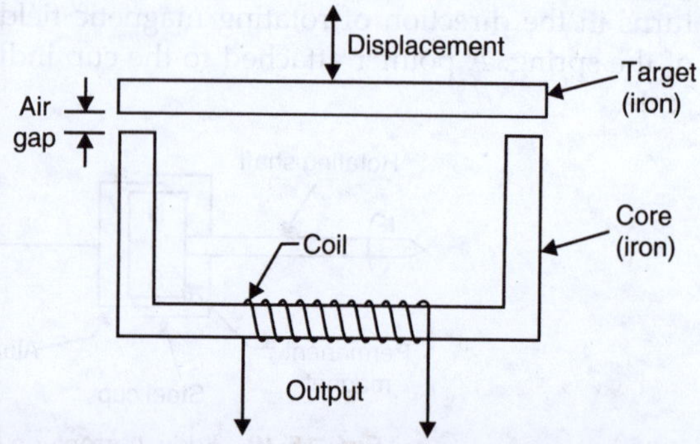

Fig. 16.12. Variable reluctance inductive transducer.

Solution. Inductance with air gap length of 1.00 mm, L = 2 mH

(*i*) **Value of inductance when a displacement of 0.02 mm is applied :**

Length of air gap when a displacement of 0.02 mm is applied towards the core

$$= 1.00 - 0.02 = 0.98 \text{ mm}$$

Now, the inductance is inversely proportional to the length of air gap as the reluctance of flux paths through iron are neglected. *Since the gap length decreases the inductance increases by ΔL.*

$$\therefore \quad L + \Delta L = 2 \times \frac{1}{0.98} = \textbf{2.04 mH (Ans.)}$$

or, $$\Delta L = 2.04 - 2 = 0.04 \text{ mH}$$

(*ii*) **ΔL α displacement :**

The ratio of change in inductance to the original inductance

$$= \frac{\Delta L}{L} = \frac{0.04}{2} = 0.02$$

Also, the ratio of displacement to original gap length

$$= \frac{0.02}{1} = 0.02$$

Hence the ΔL α **displacement** .... *Proved.*

- This relationship, however, is *true of only very small values of displacement.*

Variable permeance transducer :

Fig. 16.13 shows a *variable permeance transducer* in which the inductance of coil is changed by *varying the core material.*

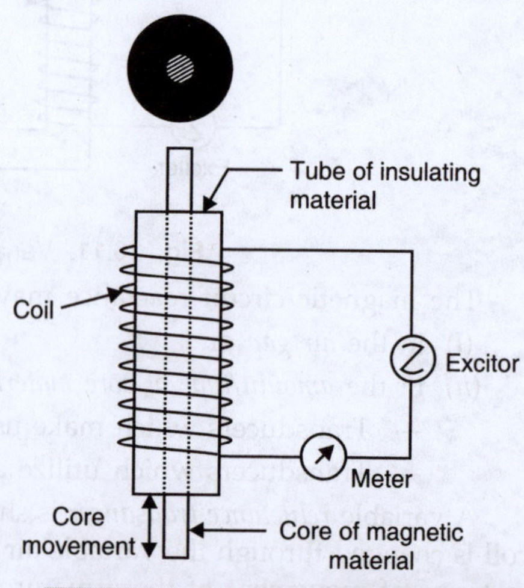

Fig. 16.13. Variable permeance transducer (self inductance arrangements).

— The transducer consists of a coil of many turns of wire wound on a tube or insulating material with a moveable core of magnetic material.

— When the coil is energized and the core enters the solenoid cell, the *inductance of the coil increases in proportion to the amount of metal within the coil.*

It is primarily used for the measurement of :

(*i*) Displacement; (*ii*) Strain; (*iii*) Force.

16.7.2.2. Mutual inductance transducer

A two-coil mutual inductance transducer is illustrated in Fig. 16.14. It consists of an energising coil *X* and a pick-up coil *Y*. A change in the position of the armature by a mechanical input changes the air gap. This causes a change in the output from coil *Y*, which may be used as measure of the displacement of the armature *i.e.*, the mechanical input.

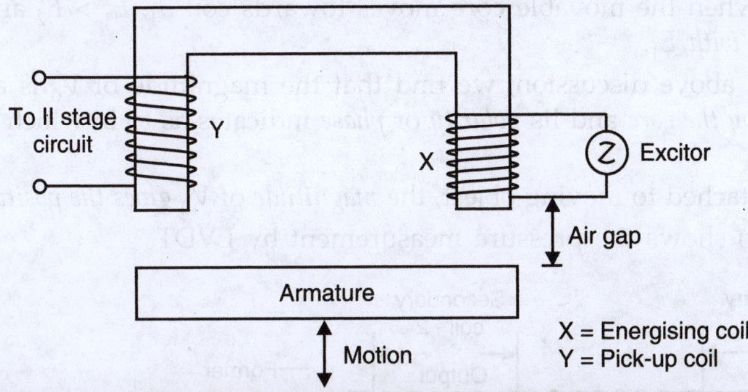

Fig. 16.14. Mutual inductance transducer.

16.7.2.3. Linear-variable-differential transformer (LVDT)

LVDT *is a passive inductive transducer and is commonly employed to measure force (or weight, pressure and acceleration etc. which depend on force) in terms of the amount and direction of displacement of an object.*

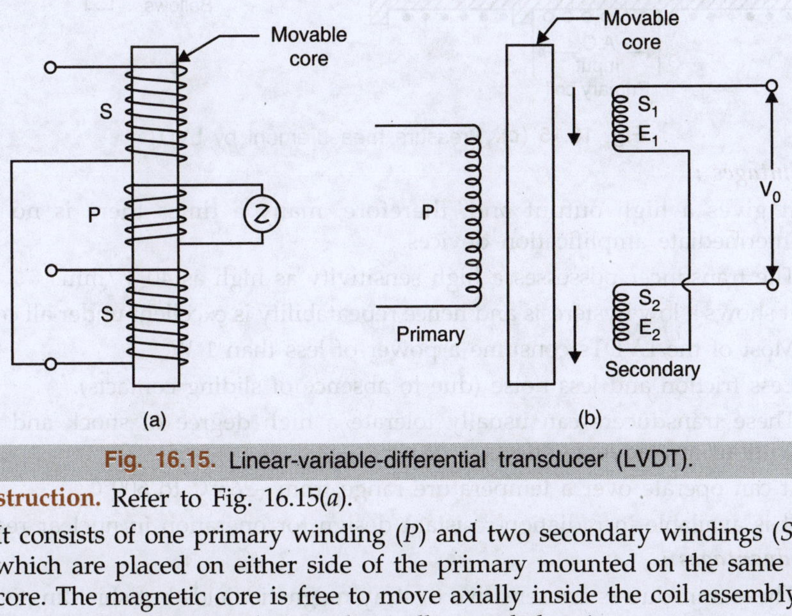

Fig. 16.15. Linear-variable-differential transducer (LVDT).

Construction. Refer to Fig. 16.15(*a*).

— It consists of one primary winding (*P*) and two secondary windings (S_1 and S_2) which are placed on either side of the primary mounted on the same magnetic core. The magnetic core is free to move axially inside the coil assembly and the motion being measured is mechanically coupled to it.

— The two secondaries S_1 and S_2 have equal number of turns but are connected in series opposition so that e.m.fs, (E_1 and E_2) induced in them are 180° out of phase with each other and hence, cancel each other out. [See Fig. 16.15 (*b*)]

— The primary is energised from a suitable A.C. source.

Working :

— When the core is in the centre (*called reference position*) the induced voltages E_1 and E_2 are *equal and opposite*. Hence they cancel out and the output voltages V_0 is zero.

— When the external applied force moves the core towards coil S_2, E_2 is increased but E_1 is *decreased* in magnitude though they are still antiphase with each other. The net voltage available is $(E_2 - E_1)$ and is *in phase with* E_2.

Similarly, when the movable core moves towards coil S_1, $E_1 > E_2$ and $V_0 = E_1 - E_2$ and is in *phase with* E_1.

Thus, from above discussion, we find that the magnitude of V_0 is a *function of the distance moved by the core* and its *polarity* or *phase* indicates as to in which direction it has moved.

If core is attached to moving object, the *magnitude* of V_0 *gives the position of that object*. Fig. 16.15(*c*) shows the pressure measurement by LVDT.

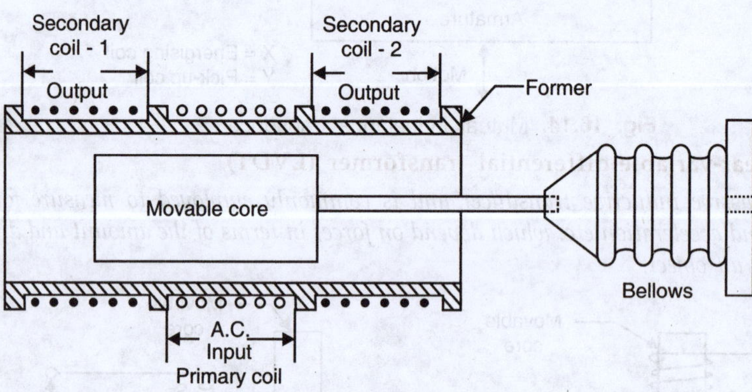

Fig. 16.15 (*c*). Pressure measurement by LVDT.

Advantages :

1. It gives a high output and, therefore, many a times there is no need for intermediate amplification devices.
2. The transducer possesses a high sensitivity as high as 40 V/mm.
3. It shows a low hysteresis and hence repeatability is excellent under all conditions.
4. Most of the LVDTs consume a power of less than 1 W.
5. Less friction and less noise (due to absence of sliding contacts).
6. These transducers can usually tolerate a high degree of shock and vibration without any adverse effects.
7. It can operate over a temperature range from –265°C to 600°C.
8. It is available in radiation-resistant design for operation in nuclear reactors.

Disadvantages :

1. These transducers are sensitive to stray magnetic fields but shielding is possible. This is done by providing magnetic shields with longitudinal slots.

2. Relatively large displacements are required for appreciable differential output.
3. The receiving instrument must be selected to operate on A.C. signals or demodulator network must be used if a D.C. output is required.
4. Several times, the transducer performance is affected by vibrations.
5. The dynamic response is limited mechanically by the mass of core and electrically by the frequency of applied voltage. The frequency of the carrier should be at least ten times the highest frequency component to be measured.

Applications:

1. Measurement of material thickness in hot strip or slab steel mills.
2. In accelerometers.
3. Jet engine controls in close proximity to exhaust gases.

Note. LVDT *is not suited for fast dynamic measurements on account of mass of the core.*

Example 16.5. *In a linear voltage differential transformer (LVDT) the output voltage is 1.8 V at maximum displacement. At a certain load the deviation from linearity is maximum and it is ± 0.0045 V from a straight line through the origin. Find the linearity at the given load.*

Solution. *Given :* The output voltage of LVDT at maximum displacement = 1.8 V

The deviation from a straight line through the origin = ± 0.0045 V

$$\therefore \quad \text{\%age linearity} = \pm\frac{0.0045}{1.8}\times 100 = \mathbf{\pm 0.25\%} \text{ (Ans.)}$$

Example 16.6. *The output of a LVDT is connected to a 4 V voltmeter through an amplifier whose amplification factor is 500. An output of 1.8 mV appears across the terminals of LVDT when the core moves through a distance of 0.6 mm. If the millivoltmeter scale has 100 divisions and the scale can be read to $\frac{1}{4}$ of a division, calculate:*

(i) The sensitivity of LVDT.

(ii) The resolution of the instrument in mm.

Solution. (*i*) The sensitivity of LVDT :

$$\text{The sensitivity of LVDT} = \frac{\text{Output voltage}}{\text{Displacement}} = \frac{1.8}{0.6} = \mathbf{3\ mV/mm} \text{ (Ans.)}$$

(*ii*) The resolution of the instrument :

Sensitivity of measurement = Amplification factor × sensitivity of LVDT

$$= 500 \times 3 = 1500 \text{ mV/mm}$$

$$\text{1 scale division} = \frac{4}{100}\text{ V} = 40 \text{ mV}$$

Minimum voltage that can be read on the voltmeter

$$= \frac{1}{4}\times 40 = 10 \text{ mV}$$

∴ Resolution of the instrument

$$= 10\times\left(\frac{1}{1500}\right) = \mathbf{0.0067\ mm} \text{ (Ans.)}$$

16.8 CAPACITIVE TRANSDUCERS

The principle of operation of capacitive transducers is based upon the familiar equation for capacitance of a parallel plate capacitor :

$$\text{Capacitance, } C = \frac{\epsilon A}{d} = \frac{\epsilon_r \epsilon_o A}{d} \quad ...(16.1)$$

where, $\epsilon = \epsilon_r \epsilon_o$ = Permittivity of medium, F/m,

ϵ_r = Relative permittivity, (for air $\epsilon_r = 1$),

ϵ_o = Permittivity of free space = 8.85×10^{-12} F/m,

A = Overlapping area of plates, and

d = Distance between the two plates.

Any physical quantity which can cause a change in ϵ, A or d can be measured by the *capacitance gauge.*

The *displacement is measured by measuring the change in capacitance brought about by :*

(*i*) Change in area, or

(*ii*) Change in distance between the plates.

The change in capacitance on account of change in dielectric is used to measure change in liquid and gas levels.

16.8.1. Capacitive Transducers—Using Change in Area of Plates

Figure 16.16(*a*), (*b*) shows the elementary diagrams of the arrangements of a capacitive transducer where capacitance change occurs because of change in the area of plates.

Since capacitance is directly proportional to the effective area of the plates, response of such a system is linear.

Fig. 16.16(*c*) shows variation of the capacitance.

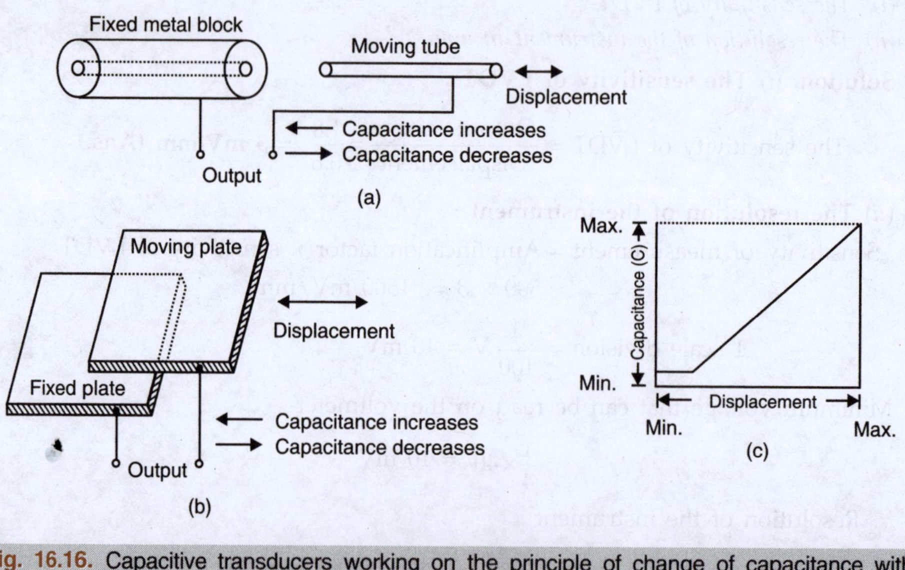

Fig. 16.16. Capacitive transducers working on the principle of change of capacitance with change of area.

16.8.2. Capacitive Transducer–Using Change in Distance Between the Plates

Fig. 16.17 shows the basic form of a capacitive transducer utilizing the effect of *change of capacitance with change in distance between the plates.*

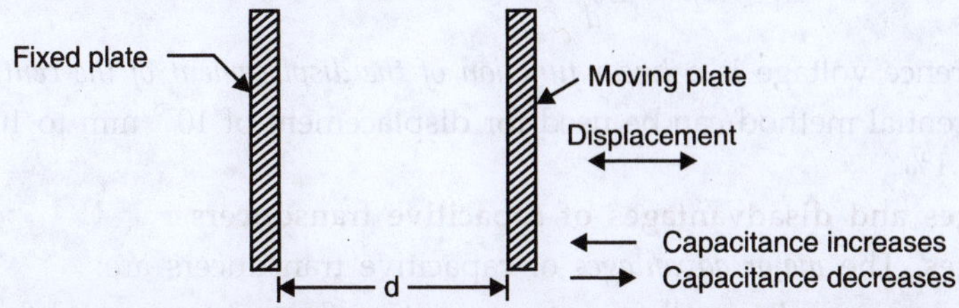

Fig. 16.17. Capacitive transducer.

One is a *fixed plate* and the displacement to be measured is applied to the other plate which is *movable.* Since, the capacitance, C varies inversely as the distance between the plates the response of this transducer is *not linear.*

Differential capacitor system:

In a differential capacitor system, let the normal position of the central plate be represented by a solid line as shown in Fig. 16.18. The capacitances C_1 and C_2 are then identical.

i.e., $$C_1 = C_2 = C = \frac{\in A}{d} \quad ...(16.2)$$

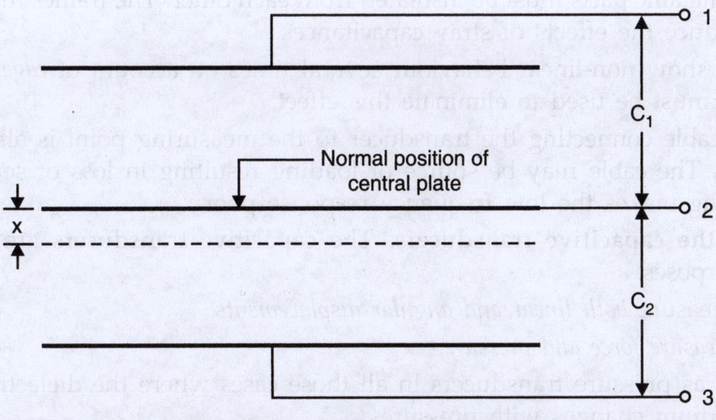

Fig. 16.18. Differential capacitor system.

When the central plate is displaced parallel to itself through a distance x, the capacitances are

$$C_1 = \frac{\in A}{d+x}, C_2 = \frac{\in A}{d-x} \quad ...(16.3)$$

For an alternating voltage E applied between the terminals 1 and 2, the voltages across C_1 and C_2 are given by

$$E_1 = \frac{EC_2}{C_1+C_2} = E\frac{d+x}{2d}$$

and, $$E_2 = \frac{EC_1}{C_1+C_2} = E\frac{d-x}{2d} \quad ...(16.4)$$

When the differential measurement circuit is fed with output from the terminals pairs 1 and 3, and 2 and 3, the difference voltage would be recorded.

$$E_1 - E_2 = E\frac{x}{d} \quad \text{...(16.5)}$$

The difference voltage is a *linear function of the displacement of the central plate.*

The differential method can be used for displacement of 10^{-8} mm to 10 mm with an accuracy of 0.1%.

Advantages and disadvantages of capacitive transducers :

Advantages. The *major advantages* of capacitive transducers are:

1. Require extremely small force for operation (hence very useful for use in small systems).
2. Extremely sensitive.
3. Require small power for operation.
4. High input impedance; therefore, loading effects are minimum.
5. Frequency response is good.
6. A resolution of the order of 2.5×10^{-3} mm can be obtained.
7. Can be used for applications where stray magnetic fields render the inductive transducers useless.

Disadvantages. The *principal disadvantages* of capacitive transducers are:

1. The metallic parts must be insulated from each other. The frames must be earthed to reduce the effects of stray capacitances.
2. They show non-linear behaviour several times on account of *edge effects ; guard rings* must be used to eliminate this effect.
3. The cable connecting the transducer to the measuring point is also a source of error. The cable may be source of loading resulting in loss of sensitivity. Also loading makes the low frequency response poor.

Uses of the capacitive transducers. The capacitive transducers are used for the following purposes:

1. To measure *both linear and angular displacements.*
2. To *measure force and pressure.*
3. Used as pressure transducers in all those cases where the dielectric constant of a medium changes with pressures.
4. To measure humidity in gases.
5. Used in conjunction with mechanical modifiers for measurement of *volume, density, weight, input level* etc.

Example 16.7. *A parallel plate capacitive transducer uses plates of area 300 mm² which are separated by a distance 0.2 mm.*

(i) Determine the value of capacitance when the dielectric is air having a permittivity of 8.85×10^{-12} F/m.

(ii) Determine the change in capacitance if a linear displacement reduces the distance between the plates to 0.18 mm. Also determine the ratio of per unit change of capacitance to per unit change of displacement.

(iii) If a mica sheet 0.01 mm thick is inserted in the gap, calculate the value of original capacitance and change in capacitance for the same displacement. Also calculate the ratio

of per unit change in capacitance to per unit change in displacement. The dielectric constant of mica is 8.

Solution. *Given* : $A = 300\ mm^2 = 300 \times 10^{-6}\ m^2$; $d = 0.2$ mm; $\epsilon_0 = 8.85 \times 10^{-12}$ F/m; ϵ_r (mica) = 8

***(i)* Value of capacitance:**

Value of capacitance, $C = \dfrac{\epsilon_0 \epsilon_r A}{d} = \dfrac{\epsilon_0 A}{d}$ $\quad (\because \epsilon_r = 1)$

$$= \frac{8.85 \times 10^{-12} \times 300 \times 10^{-6}}{0.2 \times 10^{-3}} \text{F} = \mathbf{13.275\ pF} \text{ (Ans.)}$$

***(ii)* Change in capacitance, ΔC:**

Change is displacement $\Delta d = 0.2 - 0.18 = 0.02$ mm.

Capacitance after application of displacement,

$$C + \Delta C = \frac{8.85 \times 10^{-12} \times 300 \times 10^{-6}}{0.18 \times 10^{-3}} \text{F} = 14.75 \text{ pF}$$

Change in capacitance, $\Delta C = 14.75 - 13.275 =$ **1.475 pF (Ans.)**

Ratio of per unit change of capacitance to per unit change of displacement,

$$\frac{\Delta C / C}{\Delta d / d} = \frac{(1.475/13.275)}{(0.02/0.2)} = \mathbf{1.111} \text{ (Ans.)}$$

***(iii)* C, ΔC, Ratio (ΔC/C)/(Δd/d) when mica sheet is inserted:**

Initially, the displacement between the plates is 0.2 mm. Since the thickness of mica is 0.01 mm, the length of air between the plates = 0.2 – 0.01 = 0.19 mm.

Initial capacitance of transducer,

$$C = \frac{\epsilon_0 A}{\dfrac{d_1}{\epsilon_{r_1}} + \dfrac{d_2}{\epsilon_{r_2}}}$$

or,

$$C = \frac{8.85 \times 10^{-12} \times 300 \times 10^{-6}}{\left(\dfrac{0.19}{1} + \dfrac{0.01}{8}\right) \times 10^{-3}} \text{F} = \mathbf{13.88\ pF} \text{ (Ans.)}$$

When a displacement of 0.02 mm is applied, the length of air gap is reduced to 0.19 – 0.02 = 0.17 mm.

∴ Capacitance with displacement applied

$$= \frac{8.85 \times 10^{-12} \times 300 \times 10^{-6}}{\left(\dfrac{0.17}{1} + \dfrac{0.01}{8}\right) \times 10^{-3}} \text{F} = 15.5 \text{ pF}$$

Change in capacitance, $\Delta C = 15.5 - 13.88 =$ **1.62 pF (Ans.)**

$$\text{Ratio } \frac{\Delta C / C}{\Delta d / d} = \frac{(1.62/13.88)}{(0.02/0.2)} = \mathbf{1.167} \text{ (Ans.)}$$

16.8.3. Capacitive Tachometer

Refer to Fig. 16.19. A capacitive pick-up tachometer consists of a vane attached to one end of the rotating machine shaft. When the shaft rotates between the field capacitive plates, there occurs a change in the capacitance. The capacitor forms a part of an oscillator tank so that *number of frequency changes per unit of time is a measure of the shaft speed.* The pulses thus produced are *amplified* and *squared,* and may then be *fed to frequency measuring unit* or to a digital counter so as to provide a digital analog of the shaft rotation.

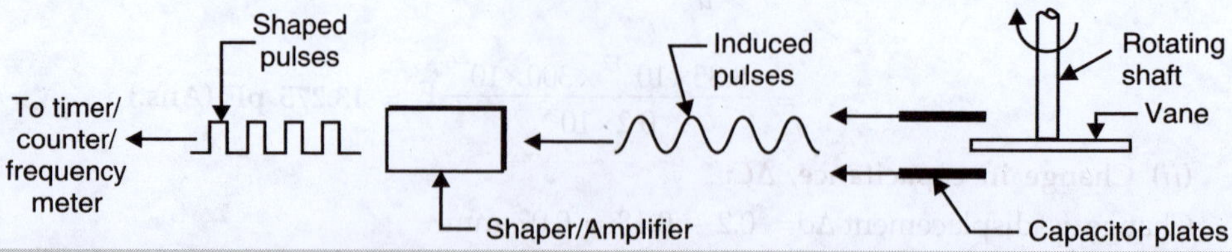

Fig. 16.19. Capacitive pick-up tachometer.

16.9 PIEZOELECTRIC TRANSDUCERS

16.9.1. Piezoelectric Materials

*A **"piezoelectric material"** is one in which an electric potential appears across certain surfaces of a crystal if the dimensions of the crystals are changed by the application of a mechanical force.* This potential is produced by the displacement of external charges. The effect is reversible, *i.e.,* conversely, if a varying potential is applied to the proper axis of the crystal, it will change the dimensions of the crystal thereby deforming it. This effect is known as **piezoelectric effect.**

Elements exhibiting piezoelectric qualities are sometimes known as *electro-resistive elements.* Common piezoelectric materials are : *Ammonium dihydrogen phosphate, Rochelle salt, lithium sulphate, dipotassium tartrate, potassium dihydrogen phosphate, quartz and ceramics A and B.*

There are two main groups of piezoelectric crystals:

1. ***Natural crystals*** ..... such as *quartz and tourmaline.*
2. ***Synthetic crystals*** ..... such as *Rochelle salt, lithium sulphate, dipotassium tartrate etc.*

16.9.2. Desirable Properties of Piezoelectric Materials

The *desirable properties* of piezoelectric materials are :

(*i*) Stability.

(*ii*) High output.

(*iii*) Insensitivity to temperature and humidity.

(*iv*) The ability to be formed into most desirable shape.

Natural crystals entail the following *advantages* :

(*i*) Higher mechanical and thermal stability.

(*ii*) Ability to withstand higher stresses.

(*iii*) Low leakage.

(*iv*) Good frequency response.

Synthetic materials, in general, have a *higher voltage sensitivity.*

- *"Quartz"* is the most stable piezoelectric material. However, its output is quite small.

- *"Rochelle"* salt provides the highest output but it can be worked over a limited humidity range and has to be protected against moisture. The highest temperature is limited to 45°C.
- *"Barium titanate"* has the advantage that it can be formed into a variety of shapes and sizes since it is polycrystalline. It has also a higher dielectric constant.

16.9.3. Working of a Piezoelectric Device

A typical mode of operation of a piezoelectric device employed for measuring varying force applied to a simple plate is shown in Fig. 16.20. *The magnitude and polarity of the induced charge on the crystal surface is proportional to the magnitude and direction of the applied force.* The charge at the electrode gives rise to voltage (E), given by,

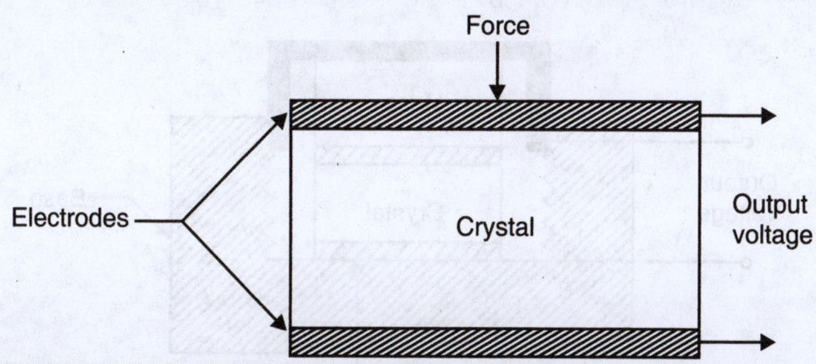

Fig. 16.20. Piezoelectric transducer.

$$E = \frac{gtF}{A} = gtp \quad ...(16.6)$$

where, g = Voltage sensitivity in Vm/N, $\left[g = \frac{K}{t}, \; k = \text{Piezoelectric constant}, \; t = \text{Thickness of the crystal.} \right]$

F = Force in N (newton),

A = Area of the crystal in m^2, and

p = Pressure $\left(= \frac{F}{A}\right)$ in N/m^2.

16.9.4. Advantages and Disadvantages of Piezoelectric Transducers

Advantages :

1. High frequency response.
2. Small size.
3. High output.
4. Rugged construction.
5. Neglible phase shift.

Disadvantages :

1. Output affected by changes in temperature.
2. Cannot measure static conditions.

Applications :

These transducers find the following *fields of application:*

1. Accelerometers.

2. Pressure cells.
3. Force cells.
4. Ceramic microphones.
5. Phonograph pick-up.
6. Cartridges.
7. Industrial cleansing apparatus.
8. Under-water detection system.

16.9.5. Piezoelectric Accelerometer

Refer to Fig. 16.21. A piezoelectric accelerator is probably the *simplest* and *most commonly used transducer for measuring acceleration.*

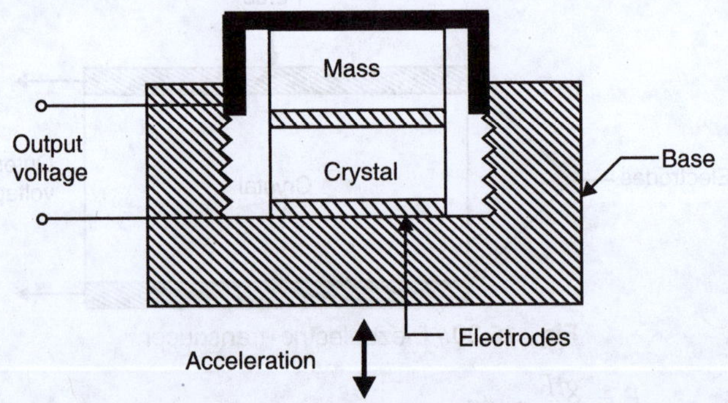

Fig. 16.21. Piezoelectric accelerometer.

Construction. It consists of a *piezoelectric crystal* sandwitched between two *electrodes* and has *mass* placed on it. The unit is fastened to the base whose acceleration characteristics are to be obtained. The *can* threaded to the base acts as a spring and squeezes the mass against the crystal. *Mass exerts a force on the crystal and a certain voltage output is generated.*

Working. When the base is accelerated *downward* inertial reaction force on the base acts upward against the top of the can. This relieves stress on the crystal. According to Newton's second law of motion, *force* = *mass* × *acceleration,* since the mass is a fixed quantity, the *decrease in force is proportional to the acceleration.* Similarly, an acceleration in the upward direction would increase the force on the crystal in proportion to the acceleration. The resulting change in the output voltage is recorded and correlated to the acceleration imposed on the base.

Advantages :

1. Small size and a small weight.
2. High output impedance.
3. Can measure acceleration from a fraction of g to thousands of g.
4. High sensitivity.
5. High frequency response (10 Hz to 50 kHz).

Disadvantages :

1. Unsuitable for applications where the input frequency is lower than 10 Hz.
2. Subject to hysteresis errors.
3. Sensitive to temperature changes.

Example 16.8. *A 2.5 mm thick quartz piezoelectric crystal having a voltage intensity of 0.055 Vm/N is subjected to a pressure of 1.4 MN/m². If the permittivity of quartz is 40.6 × 10⁻¹² F/m, calculate:*

(i) Voltage output.

(ii) Charge sensitivity of the crystal.

Solution. *Given:* $t = 2.5$ mm or 2.5×10^{-3} m; $g = 0.055$ Vm/N; $p = 1.4$ MN/m²; $\in (= \in_o \in_r) = 40.6 \times 10^{-12}$ F

(i) Voltage output, *E*:

$$E = gtp \quad \text{...[Eqn. (16.6)]}$$

$$= 0.055 \times 2.5 \times 10^{-3} \times 1.4 \times 10^6 = \mathbf{192.5\ V\ (Ans.)}$$

(ii) Charge sensitivity of the crystal:

$$\text{Charge sensitivity} = \in_o \in_r g = \in . g$$

$$= 40.6 \times 10^{-12} \times 0.055 \text{ C/N}$$

$$= \mathbf{2.233\ pC/N\ (Ans.)}$$

Example 16.9. *A piezoelectric crystal measuring 6 mm × 6 mm × 1.8 mm is used to measure a force. Its voltage sensitivity is 0.055 Vm/N. Calculate the force if voltage developed is 120 V.*

Solution. *Given:* $A = 6 \text{ mm} \times 6 \text{ mm} = 36 \times 10^{-6}$ m²; $t = 1.8$ mm or 1.8×10^{-3} m;

$g = 0.055$ Vm/N; $E = 120$ V

Force *F*:

We know that,

$$E = gtp \quad \text{...[Eqn. (16.6)]}$$

$$120 = 0.055 \times (1.8 \times 10^{-3}) \times p$$

or

$$p = \frac{120}{0.055 \times 1.8 \times 10^{-3}} \text{ N/m}^2 = 1.212 \text{ MN/m}^2$$

∴ Force

$$F = p \times A = 1.212 \times 10^6 \times 36 \times 10^{-6} = \mathbf{43.63\ N\ (Ans.)}$$

Example 16.10. *The following data relate to a barium titanate pick-up:*

Dimensions	*6 mm × 6 mm × 1.5 mm*
Force acting on the pick-up	*6N*
The charge sensitivity of the crystal	*150 pC/N*
Permittivity	*12.5 × 10⁻⁹ F/m*
Modulus of elasticity	*12 × 10⁶ N/m²*

Calculate the following:

(i) The strain.

(ii) The charge and capacitance.

Solution. *Given:* $A = 6 \times 6 = 36$ mm² or 36×10^{-6} m²; $t = 1.5$ mm or 1.5×10^{-3} m; $\in = 12.5 \times 10^{-9}$ F/m; $F = 6$ N; d(charge sensitivity) = 150 pC/N; Modulus of elasticity, $E = 12 \times 10^6$ N/m².

(i) The strain, *e*:

$$\text{Pressure, } p = \frac{F}{A} = \frac{6}{36 \times 10^{-6}} \text{ N/m}^2 = 0.167 \text{ MN/m}^2$$

$$\textbf{Strain, } e = \frac{\text{Stress}}{\text{Young's modulus}} = \frac{0.167 \times 10^6}{12 \times 10^6} = \mathbf{0.0139\ (Ans.)}$$

(*ii*) Charge and capacitance; Q, C:

$$\text{Voltage sensitivity, } g = \frac{d}{\epsilon_0 \epsilon_r} = \frac{d}{\epsilon} = \frac{150\times10^{-12}}{12.5\times10^{-9}} = 12 \times 10^{-3} \text{ Vm/N}$$

$$\therefore \quad \text{Voltage generated, } E = gtp$$

$$= 12 \times 10^{-3} \times 1.5 \times 10^{-3} \times 0.167 \times 10^{6} = 3\text{V}$$

$$\text{Hence, charge, } Q = d \times F = 150 \times 10^{-12} \times 6 \text{ C} = \mathbf{900\ pC} \text{ (Ans.)}$$

$$\text{and, } \quad \text{Capacitance, } C = \frac{900\times10^{-12}}{3}\text{F} = \mathbf{300\ pF} \text{ (Ans.)}$$

16.10 HALL EFFECT TRANSDUCERS

16.10.1. Hall Effect

When a current carrying conductor is placed in a magnetic field, a *transverse effect* is noted. This effect is called Hall effect (discovered by Hall in 1879). Hall found that: *"When a magnetic field is applied at right angles to the direction of electric current an electric field is set up which is perpendicular to both the direction of electric current and the applied magnetic field".*

In other words:

"When any specimen carrying a current I is placed in the transverse magnetic field B, then an electric field E is induced in the specimen in the direction perpendicular to both I and B. The phenomenon is known as **Hall effect".**

In the Fig. 16.22 is shown a specimen bar carrying a current *I* in the positive-X direction. Let a magnetic field *B*, be applied in the positive-Z direction. Then according to Hall effect, a force gets exerted on the charge carriers (whether electrons or holes) in the negative-*Y* direction. This current *I* may be due to holes moving in the positive-X or due to *free electrons* moving in the negative-X direction through the *semiconductor specimen.* Hence irrespective of the nature of the charge carriers, whether holes or electrons, these charge carriers get passed *downwards towards face* 1 of the specimen shown in Fig. 16.22.

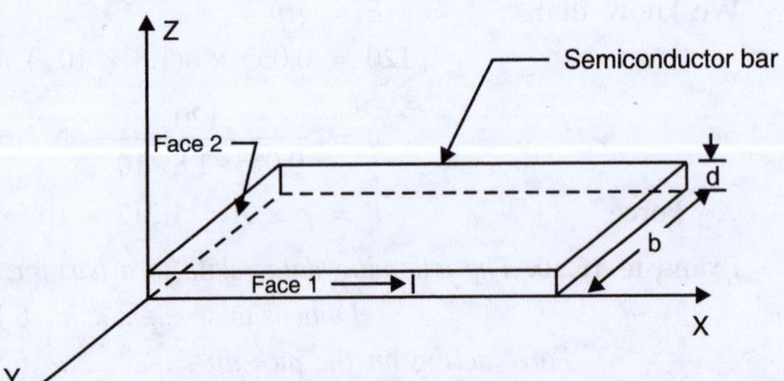

Fig. 16.22. Current carrying semiconductor bar subject to transverse magnetic field.

The current, in an N-type specimen, is carried almost fully by electrons. These electrons, as a result of Hall effect, accumulate on side 1 which surface then gets negatively charged relative to side 2. Consequently, a potential difference develops between surfaces 1 and 2 and is called the *'**Hall voltage**'* (V_H). This Hall voltage in an N-type semiconductor is positive at surface 2. On the other hand, in a P-type specimen, the Hall voltage is positive at surface 1. These two results have been verified experimentally.

The Polarity of Hall voltage enables us to determine experimentally whether the semiconductor specimen is of N-type or P-type.

The magnitude of Hall voltage (V_H) is given by the expression :

$$V_H = \frac{R_H BI}{b} \quad ...(16.7)$$

where, R_H = Hall coefficient,
B = Magnetic field strength,
I = Current carried by the specimen, and
b = Width of the specimen along the magnetic field.

The *Hall effect* may be used for:

1. Determining whether a semiconductor is N-type or P-type.
2. Determining the carrier concentration.
3. Calculating the mobility having measured the conductivity.
4. *Magnetic field meter.* The Hall voltage V_H for a given current is proportional to B. Hence measurement of V_H measures the magnetic field B.
5. *Hall effect multiplier.* The instrument gives an output proportional to the product of two signals. Thus if current I is made proportional to one input and if B is proportional to the second input, then *Hall voltage V_H is proportional to the product of the two inputs.*

16.10.2. Hall Effect Transducers

Hall effect transducers are the transducers in which Hall effect is utilised to measure various electrical or non-electrical quantities.

Commercial Hall effect transducers are made from germanium or other semiconductor materials.

The following are the *applications of Hall effect transducers* :

1. Displacement measurement:

Hall effect transducer may be used to measure a linear displacement or to locate a structural element is cases *where it is possible to change the magnetic field strength by variation in the geometry of a magnetic structure.*

Fig. 16.23 shows the arrangement of Hall-effect displacement (linear) transducer . The Hall effect element is located in the gap, adjacent to the permanent magnet. The field strength produced in the gap due to the permanent member is changed by changing the position of a ferromagnetic plate. The voltage output of the Hall effect element is proportional to the field strength in the gap which is function of the position (*i.e.,* displacement) of ferromagnetic plate with respect to the structure.

- With this method the *displacements as small as 0.025 mm can be measured.*

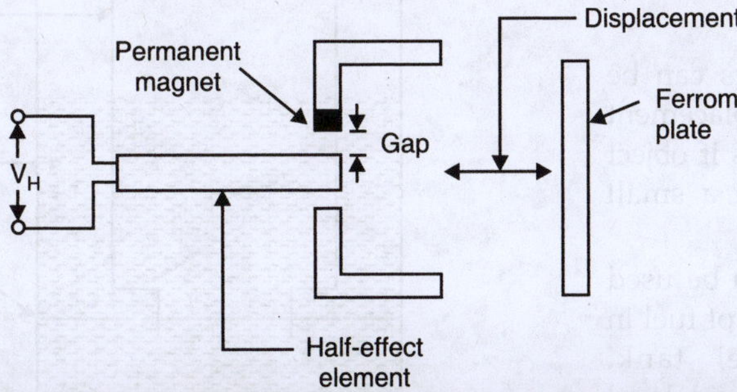

Fig. 16.23. Hall-effect displacement transducer.

2. Current measurement:

Hall effect transducer can be used to measure current in a conductor without interrupting the circuit and without making electrical connection between the conductor circuit and the meter.

When a D.C. or A.C. current flows through the conductor, it sets up a magnetic field around. This *magnetic field is proportional to the current.* A Hall effect transducer is inserted in a slotted ferromagnetic tube which acts as a magnetic concentrator. *The voltage produced at the output terminals is proportional to the magnetic field strength and hence is proportional to the current, flowing through the conductor.*

— The magnetic concentrator can be omitted at high current levels since the magnetic field is fairly strong in the vicinity of the Hall element and thus can cause output voltages which can be detected easily.

- This method can be used *to measure current from less than a mA to thousands of amperes.*

3. Magnetic flux measurement:

The magnetic flux can be measured by using Hall effect transducer. Here, a semiconductor plate is inserted into the magnetic field which is to be measured. The magnetic lines of force are perpendicular to the semiconductor. The transducer gives an output voltage which is *proportional to the magnetic field intensity (B).*

Following are the *advantages* and *disadvantages* of the system:

Advantages:

(*i*) The system requires a very small space in the direction of the magnetic field and thus the Hall effect element can be inserted in narrow gaps for magnetic measurements in air spaces.

(*ii*) The Hall effect element gives out a continuous electric signal in direct response to the magnetic field strength.

Disadvantages:

High sensitivity to temperature variations, and Hall coefficient may vary from plate to plate which may need individual calibration in each case.

4. Fluid level measurement:

Hall effect sensors can be used as position, displacement and proximity sensors if object is being sensed with a small permanent magnet.

Such a sensor can be used to determine the level of fuel in an automobile fuel tank. Fig. 16.24 shows a fluid level detector.

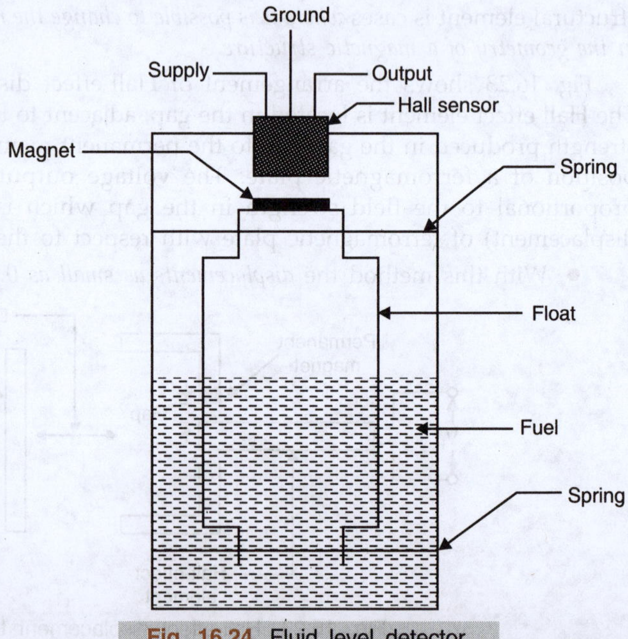

Fig. 16.24. Fluid level detector.

A magnet is attached to float and as the level of fuel changes and so the float distance from the Hall sensor changes. The result is a Hall voltage output which is a measure of the distance of the float from the sensor and hence the level of the fuel in the tank.

Example 16.11. *The resistivity of semiconductor material was known to be 0.00912 Ω m at room temperature. The flux density in the Hall model was 0.45 Wb/m².*

Calculate the Hall angle for a Hall co-efficient of 3.55×10^{-4} m³/coloumb.

Solution. Refer to Fig. 16.25.

Resistivity of the semiconductor material, ρ = 0.00912 Ωm

Flux density in the Hall model,

$$B = 0.48 \text{ Wb/m}^2$$

Hall co-efficient,

$$R_H = 3.55 \times 10^{-4} \text{ m}^3/\text{C}$$

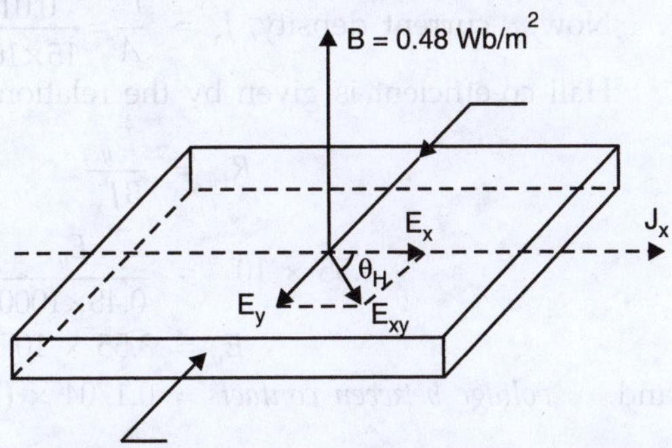

Fig. 16.25

Hall angle, θ_H:

Resistivity, $$\rho = \frac{E_x}{J_x}$$

$$0.00912 = \frac{E_x}{J_x}$$

$$E_x = 0.00912\, J_x$$

Also, $$R_H = \frac{e_y}{B\, J_x}$$

$$3.55 \times 10^{-4} = \frac{e_y}{0.48 J_x}$$

$$E_y = 3.55 \times 10^{-4} \times 0.48\, J_x = 1.704 \times 10^{-4}\, J_x$$

Now, $$\tan \theta_H = \frac{E_y}{E_x}$$

$$\frac{1.704 \times 10^{-4} J_x}{0.00912\, J_x} = 0.01868$$

$$\theta_H = 1° \, 4' \text{ (Ans.)}$$

Example 16.12. *Fig. 16.26 shows a specimen of silicon doped semiconductor having the Hall co-efficient of 3.55×10^{-4} m³/coloumb. Calculate the voltage between contacts when a current of 15 mA is flowing.*

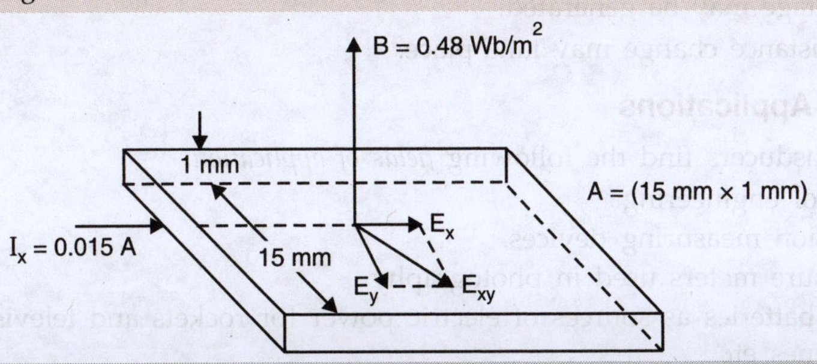

Fig. 16.26

Solution.

Hall co-efficient, $R_H = 3.55 \times 10^{-4}\ m^3/C$

Current, $I = 15\ mA = 0.015\ A$

Area, $A = 15\ mm \times 1\ mm = 15 \times 10^{-6}\ m^2$

Flux density, $B = 0.48\ Wb/m^2$.

Voltage between contacts:

Now, current density, $J_x = \frac{I_x}{A} = \frac{0.015}{15 \times 10^{-6}} = 1000\ A/m^2$

Hall co-efficient is given by the relation:

$$R_H = \frac{E_y}{BJ_x}$$

$$3.55 \times 10^{-4} = \frac{E_y}{0.48 \times 1000}$$

$$E_y = 3.55 \times 10^{-4} \times 0.48 \times 1000 = 0.1704\ V/m$$

and, *voltage between contacts* $= 0.1704 \times (15 \times 10^{-3}) =$ **0.002556 V (Ans.)**

16.11 THERMOELECTRIC TRANSDUCERS

Two dissimilar metal conductors when joined at the ends and the two junctions kept at different temperatures, then a small e.m.f. is produced in the circuit. *The magnitude of this voltage depends upon the materials of conductors and the temperature difference between the two junctions.* This thermoelectric effect is used in **thermocouples** for the *measurement of temperature.*

Any number of combination of metals may be used. Two commonly employed combinations are:

1. *Iron and constantan* (an alloy of copper and nickel).
2. *Chromel* (an alloy of chromium and nickel) and *alumel* (an alloy of aluminium and nickel).

- For more details refer to **Article 12.25.**

16.12 PHOTOELECTRIC TRANSDUCERS

16.12.1. Principle of Operation

The photoelectric transducers *operate on the principle that when light strikes special combination of materials then following may result:*

(*i*) Electrons may flow.

(*ii*) A voltage may be generated.

(*iii*) A resistance change may take place.

16.12.2. Applications

These transducers find the following *fields of application:*

1. Control engineering.
2. Precision measuring devices.
3. Exposure meters used in photography.
4. Solar batteries as sources of electric power for rockets and television, counting machines etc.
5. Satellites used in space research.

16.12.3. Classification

Photoelectric transducers may be *grouped as follows:*

1. Photoemissive cell.
2. Photovoltaic cell.
3. Photoconductive cell.

16.12.4. Photoemissive Cell

This cell is also known as *photo tube.* It is based on the emission of electrons from a metal cathode (or photo-sensitive surface) when it is exposed to light or other radiation.

Refer to Fig. 16.27. It consists of two metallic electrodes (*i.e.,* a cathode and an anode) supported in an evacuated glass bulb fitted with a base like a thermionic valve. The cathode is either semi-cylindrical or V-shaped and is made of a metal coated with an emissive material. The anode is in the form of a thin wire facing the cathode.

When the light falls on the cathode photo-electrons are emitted which are attracted by the positive anode. Subsequently current is produced whose magnitude (for a given cathode) depends on (*i*) intensity of incident radiation and (*ii*) anode cathode voltage.

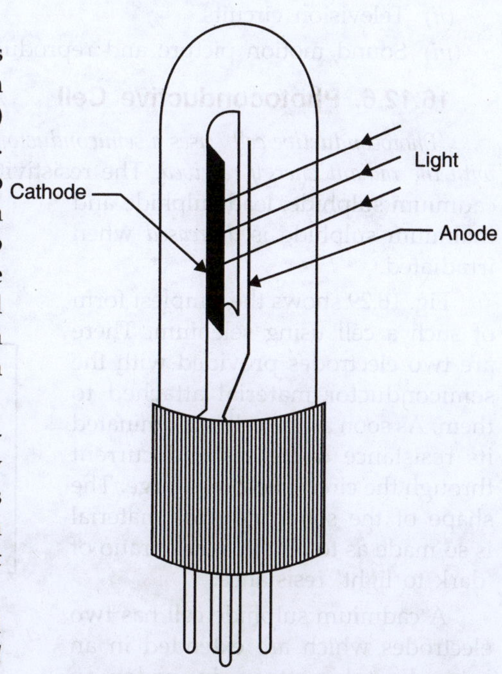

Fig. 16.27. Photoemissive cell.

Photo-emissive cell finds use in: (*i*) field of *photometry and calorimetry,* (*ii*) *sound reproduction* from a *motor-picture film,* (*iii*) *'on and off' circuits* and *other circuits* concerning the counting or sorting of objects on a conveyor belt, automatic opening of a door etc.

16.12.5. Photovoltaic Cell

In this cell sensitive element is a *semiconductor* (not metal) which generates voltage in proportion to the light or any radiant energy incident on it. The most commonly used photo-voltaic cells are barrier layer type like iron-selenium cells or Cu-CuO_2 cells.

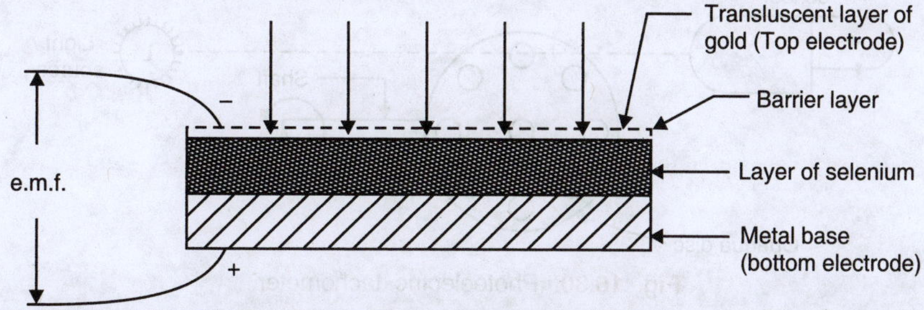

Fig. 16.28. Photovoltaic cell.

Fig. 16.28 shows a typical widely used photo-voltaic cell—"Selenium cell". It consists of a metal electrode on which a layer of selenium is deposited; on the top of this a barrier layer is formed which is coated with a very thin layer of gold. The latter serves as a transluscent electrode through which light can impinge on the layer below. Under the influence of this light, a negative charge will build up on the gold electrode and a positive charge on the bottom electrode.

Photo-voltaic cells are widely used in the following *fields:*

(*i*) Automatic control systems.

(*ii*) Television circuits.

(*iii*) Sound motion picture and reproducing equipment.

16.12.6. Photoconductive Cell

"Photoconductive cell" uses a semiconductor material whose resistance changes in accordance with the radiant energy received. The resistivity of semiconductor materials like selenium, cadmium sulphide, lead sulphide and thalmium sulphide is *decreased* when irradiated.

Fig. 16.29 shows the simplest form of such a cell using selenium. There are two electrodes provided with the semiconductor material attached to them. As soon as the cell is illuminated its resistance decreases and current through the circuit becomes large. The shape of the semiconductor material is so made as to obtain a large ratio of 'dark to light' resistance.

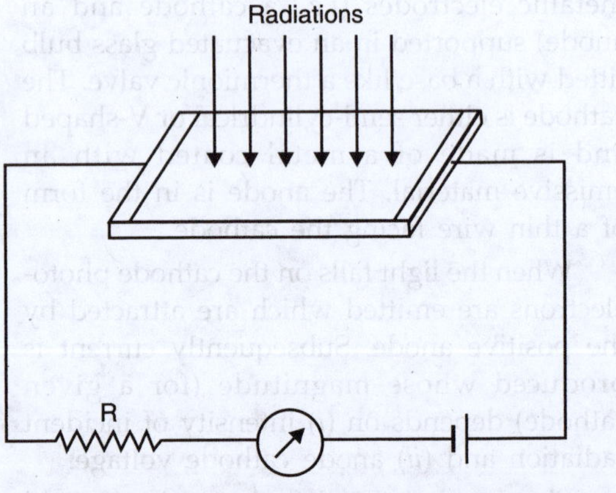

Fig. 16.29. Photoconductive cell.

A cadmium sulphide cell has two electrodes which are extended in an inter-digital pattern in order to increase the contact area with the sensitive material. It has high 'dark to light' ratio.

Photoconductive cells are generally used for *detecting ships and aircrafts by the radiations given out by their exhausts or (funnels) and for telephony by modulated infrared lights.*

16.12.7. Photoelectric Tachometer

Fig. 16.30 shows a photoelectric tachometer.

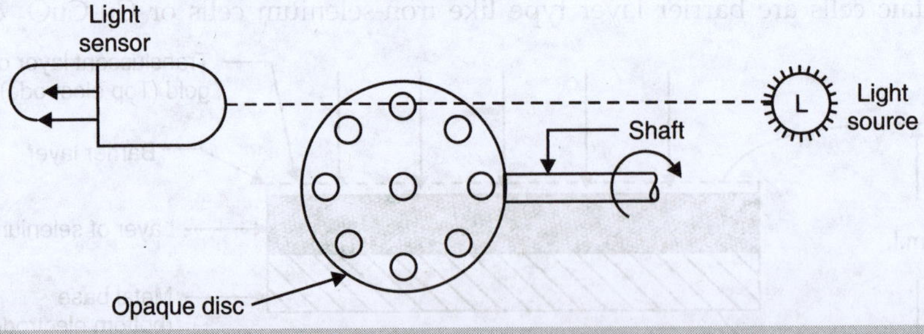

Fig. 16.30. Photoelectric tachometer.

- It consists of an *opaque disc* mounted on the shaft whose speed is to be measured. The disc has a number of equivalent holes around the periphery. On one side of the disc there is a source of light (*L*) while on the other side there is a light sensor (may be a photosensitive device or phototube) in line with it (light-source).

- On the rotation of the disc, holes and opaque portions of the disc come alternatory in between the light source and the light sensor. When a hole comes in between the *two*, light passes through the holes and falls on the light sensor, with the result that an *output pulse is generated.* But when the opaque portion of the disc comes in between, the light from the source is blocked and hence there is no pulse output. Thus *whenever a hole comes in line with the light source and sensor, a pulse is generated.* These pulses are counted/measured through an *electric counter.*

The number of pulses generated depends upon the following *factors* :

(*i*) The number of holes in the disc;

(*ii*) The shaft speed.

Since the number of holes are fixed, therefore, the number of pulses generated depends on the speed of the shaft only. The electronic counter may therefore be calibrated in terms of speed (r.p.m.)

Advantages. It is a *digital instrument.*

Disadvantages. It is required to replace the light source periodically and if the grating period is small then errors might creep in the output.

16.13 STRAIN GAUGES

16.13.1. Introduction

When a metal conductor is stretched or compressed, its resistance changes on account of the fact that both length and diameter of conductor change. The value of resistivity of the conductor also changes. When it is strained its property is called ***piezo-resistance****.* Therefore, resistance strain gauges are also known as ***piezo-resistive gauges.***

The **strain** *gauge is a measurement transducer for measuring strain and associated stress in experimental stress analysis.* Secondly many other detectors, and transducers, notably the load cells, torque meters, diaphragm type pressure gauges, temperature sensors, accelerometers and flow meters, employ *a strain gauge as a secondary transducer.*

16.13.2. Type of Strain Gauges

Four types of strain gauges are:

1. Wire-wound strain gauges.
2. Foil-type strain gauges.
3. Semiconductor strain gauges.
4. Capacitive strain gauges.

16.13.2.1. Wire-wound strain gauges

There are two main classes of wire-wound strain gauges:

1. Bonded strain gauge.
2. Unbonded strain gauge.

Bonded strain gauge:

It is composed of fine wire, wound and cemented on a resilient insulating support, usually a wafer unit. Such units may be mounted upon or incorporated in mechanical elements or structures whose deformations under stress are to be determined. While there are no limits to the basic values which may be selected for strain-gauge resistances, a typical example may be taken as of the order or 100 to 500 Ω.

Fig. 16.31 shows the commonly used form of resistance wire strain gauges.

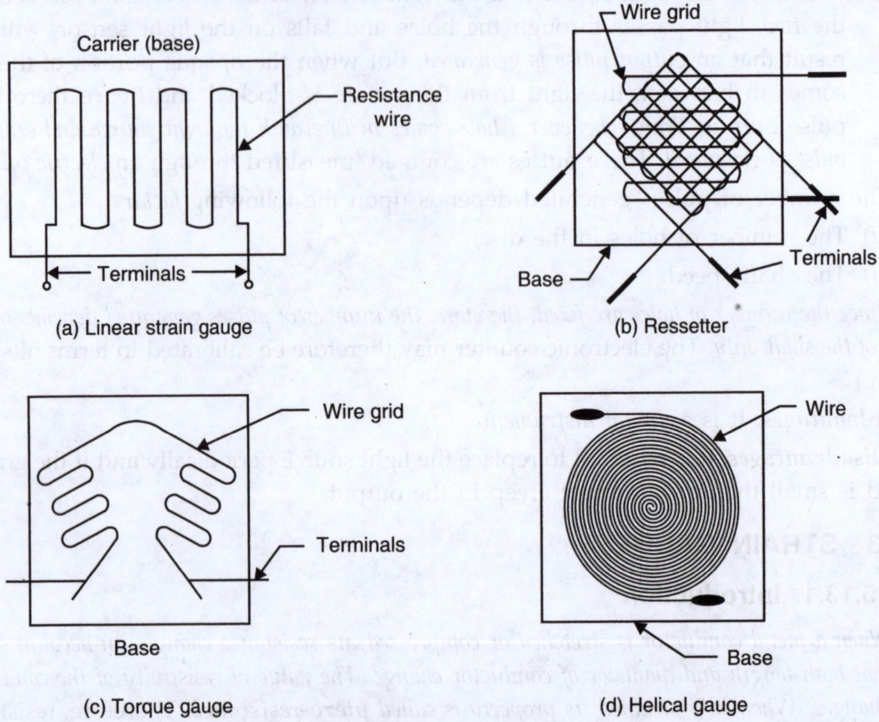

Fig. 16.31. Resistance wire strain gauges.

Unbonded strain gauge:

Fig. 16.32 shows an *unbonded strain gauge*. *M* and *N* are attached by rods *m* and *n*, respectively, to points between which *displacement is to be measured*. Pick-up and measuring networks are energized from similar but isolated source. Unbalance originating in pick-up is detected and balanced by servo-actuated measuring network, *providing a reading of strain on graduated scale.*

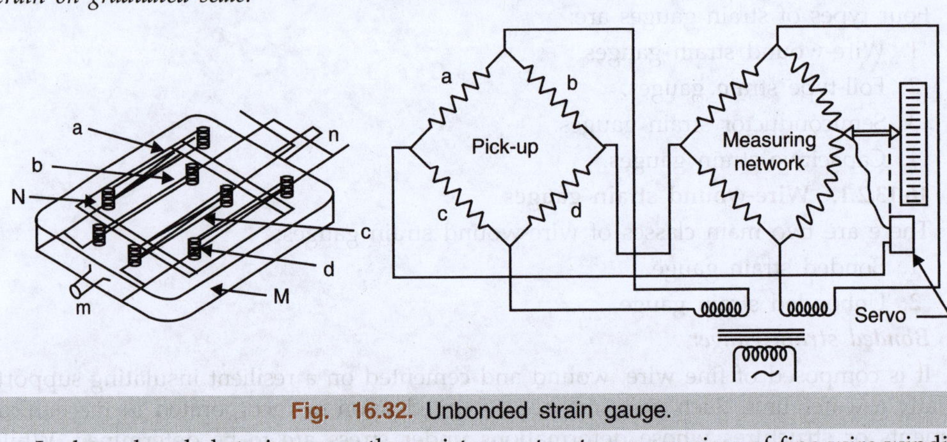

Fig. 16.32. Unbonded strain gauge.

In the unbonded strain gauge the resistance structure comprises of fine wire winding stretched between insulating supports mounted alternately on the two members between which displacement is to be measured (see Fig. 16.32). These wires comprise the four arms of a Wheatstone-bridge network of *which two opposite arms are tightened and the other two slackened by the displacement.*

— *While a bonded gauge tends to respond to the average strain in the surface to which it is cemented, the unbonded form measures displacement between the two points to which the respective supports are attached.*

— *Unbonded wire strain gauges* are usually operated on input potentials ranging upto 35 V direct or alternating current. Under conditions of extreme balance, corresponding to full operating range, the open-circuit e.m.f. may be of the order of 8 to 10 mV and closed circuit current upto 100 μA.

Strain gauges for use on A.C. circuits are supplied in both capacitive and inductive forms, wherein the corresponding characteristics of A.C. circuit components are varied by the displacement to be measured.

Requirements/Characteristics of resistance wire strain gauges:

The resistance wire strain gauges should have the following characteristics to have excellent and reproducible results.

1. The strain gauge should have a *high value of gauge factor.* A high value of gauge factor indicates a large change in resistance for a particular strain resulting in high sensitivity.
2. The *resistance of strain gauge should be as high as possible* since this minimizes the effects of undesirable variations of resistance in the measurement circuit.
3. The strain gauges should have a *low resistance temperature co-efficient.* This is essential to minimize errors on account of temperature variations which affect the accuracy of measurements.
4. The strain gauge *should not have any hysteresis effect in its response.*
5. In order to maintain constancy of calibration over the entire range of strain gauge, it should have *linear characteristics i.e.,* the variations in resistance should be a linear function of the strain.
6. The strain gauges are frequently used for dynamic measurements and hence their frequency response should be good. The linearity should be maintained within accuracy limits over the entire frequency range.

16.13.2.2. Foil strain gauges

In these gauges *the strain is sensed with the help of metal foil.* Foil gauges have a *much greater dissipation capacity as compared with wire wound gauges on account of their greater surface area for the same volume.* Due to this reason they can be employed for *higher operating temperature range.*

In these gauges, the bonding is better due to large surface area of the foil. The *bonded foil* gauges *find* a wider field of action.

Fig. 16.33 shows a typical foil gauge.

- The characteristics of foil type strain gauges are similar to those of wire wound strain gauges and their gauge factors are typically the same as that of wire wound strain gauges.
- The resistance value of foil gauges which are commercially available is between 50 and 1000 Ω.
- The maximum gauge current is about 30 mA, maximum strain for ± 1 percent, linearity is about ± 0.3 percent and strain variation upto at least 50 kHz can be measured.

Advantages :

1. Excellent strain reproducibility.
2. Low hysteresis and creep effects.

3. Superior mechanical stability under prolonged strain and higher temperature conditions (owing to larger ratio of surface area to cross-sectional area), so these gauges can be used for higher currents.
4. Longer life (since terminals have no stress concentration due to absence of joints).
5. Can be fabricated economically on a mass scale.

Uses. The foil strain gauges are employed for both *stress analysis as well as for construction of transducers.*

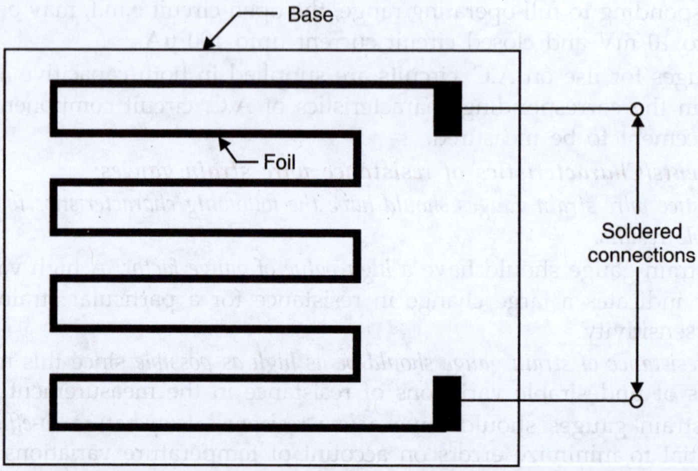

Fig. 16.33. Foil gauge.

16.13.2.3. Semiconductor strain gauges

- *Semiconductor strain gauges depend for their action upon piezo-resistive effect i.e., the change in value of the resistance due to change in resistivity.*
- These gauges are used where a *very high gauge factor and small envelope are required.*

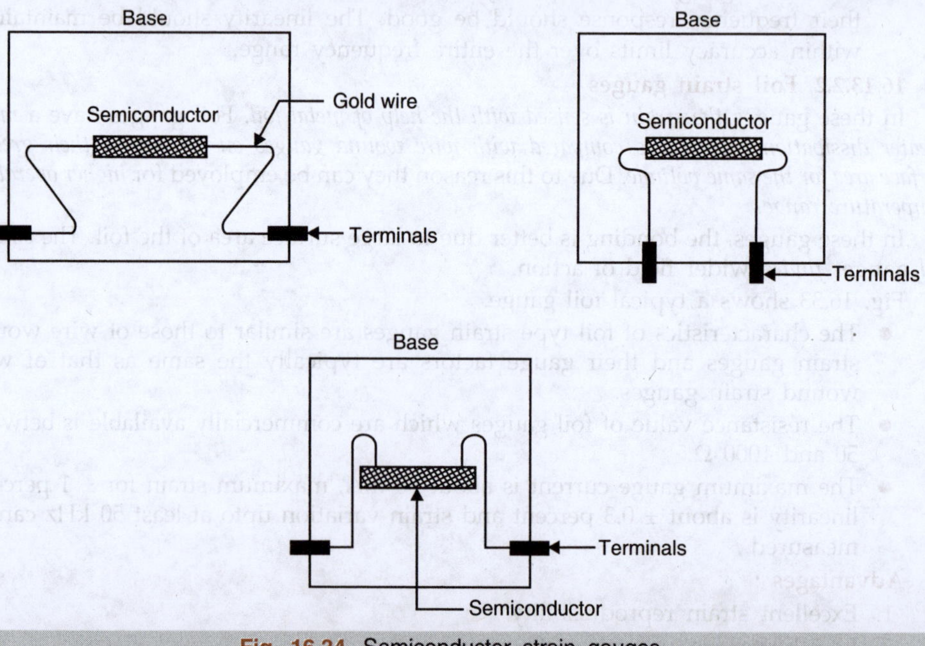

Fig. 16.34. Semiconductor strain gauges.

- For semiconductor strain gauges semiconducting materials such as *silicon* and *germanium* are used.
- A typical strain gauge consists of a *strain sensitive crystal material and leads that are sandwiched in a protective matrix*. The production of these gauges employs conventional semiconductor technology using semiconducting wafers or filaments which have a thickness of 0.05 mm and bonding them on suitable insulating substances, such as teflon. Gold, lead are generally applied for making the contacts.

Fig. 16.34 shows some typical semiconductor strain gauges.

Advantages:

1. These gauges have high gauge factor.
2. They are chemically inert and have low cross-sensitivity.
3. They are almost free from hysteresis and creep effects because of integrated format of the actual gauge itself.
4. Fatigue life is in excess of 10×10^5 operations and the frequency response is upto 10^{12} Hz.
5. These gauges can be very small ranging in length from 0.7 to 7 mm. *They are very useful for measurement of local strains.*

Disadvantages:

1. The *major and serious disadvantage is that these gauges are very sensitive to change in temperature.*
2. Linearity of these gauges is *poor.*

16.13.2.4. Capacitive strain gauges

Fig. 16.35 shows a capacitive strain gauge. It uses the *principle of variation of capacitance with variation of distance between electrodes.* The electrodes are flexible metal strips of about 0.1 mm thickness. The strain to be measured is applied to the top plate. This changes the distance between the curved electrodes resulting in change of capacitance.

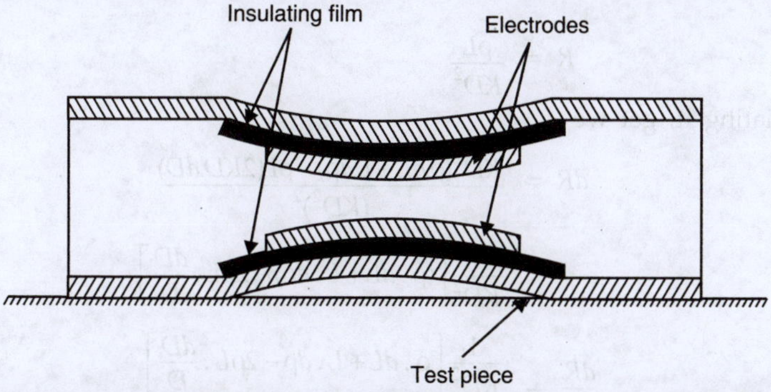

Fig. 16.35. Capacitive strain gauge.

The strain-capacitance relationship, in general, is *not linear* but variations in dimensions and shape allow gauge characteristics to be chosen so as to match the range of capacitance to be measured with a good degree of accuracy.

- A capacitance strain gauge has a capacitance of about 0.5 pF.
- Its overall size is 5 mm × 17 mm × 1 mm.
- It uses a polyamide film of insulating material.
- It can be used upto a temperature of 300°C.

16.13.3. Theory of Strain Gauges

When a strain gauge is subjected to tension (*i.e.,* positive strain) its length *increases* while its cross-sectional area *decreases.* Since the resistance of a conductor is proportional to its length and inversely proportional to its area of cross-section, the resistance of the gauge increases with positive strain. The change in the value of resistance of strained conductor is more than what can be accounted for an increase in resistance due to dimensional changes. The extra change in the value of resistance is attributed to a *change in the value of resistivity of a conductor when strained; this property is known as* ***piezo-resistive effect.***

Strain gauges are most commonly used in Wheatstone bridge circuits to measure the change of resistance of grid of wire for calibration proposes; the "**gauge factor**" *is defined as the ratio of per unit change in resistance to per unit change in length.*

i.e.,
$$\text{Gauge factor } (G_f) = \frac{\Delta R/R}{\Delta L/L} \quad \text{...(16.8)}$$

where, ΔR = Corresponding change in resistance R, and

ΔL = Change in length per unit length L.

The resistance of the wire of strain gauge R is given by :

$$R = \frac{\rho L}{A}$$

where, ρ = Resistivity of the material of wire (of strain gauge),

L = Length of the wire, and

A = Cross-sectional area of the wire,

= KD^2, K and D being a constant and diameter of the wire respectively.

As earlier stated, when the wire is strained its length increases and lateral dimension is reduced as a function of Poisson's ratio (m); consequently there is an increase in resistance.

Now,
$$R = \frac{\rho L}{KD^2}$$

Differentiating it, get we

$$dR = \frac{KD^2(\rho.dL + L.d\rho) - \rho L(2KD.dD)}{(KD^2)^2}$$

$$= \frac{1}{KD^2}\left[(\rho\,.\,dL + L\,.\,d\rho) - 2\rho L\,.\,\frac{dD}{D}\right]$$

or,
$$\frac{dR}{R} = \frac{\frac{1}{KD^2}\left[\rho\,.\,dL + L\,.\,d\rho - 2\rho L\,.\,\frac{dD}{D}\right]}{\frac{\rho L}{KD^2}}$$

$$= \frac{dL}{L} + \frac{d\rho}{\rho} - 2\frac{dD}{D}$$

Now, Poisson's ratio,
$$\mu = \frac{\text{Lateral strain}}{\text{Longitudinal strain}} = \frac{-dD/D}{dL/L}$$

or,
$$\frac{dD}{D} = -\mu \times \frac{dL}{L}$$

For small variations, the above relationship can be written as:

$$\frac{\Delta R}{R} = \frac{\Delta L}{L} + 2\mu\frac{\Delta L}{L} + \frac{\Delta \rho}{\rho} \qquad ...(16.9)$$

Gauge factor, $$G_f = \frac{\Delta R/R}{\Delta L/L}$$

or, $$\frac{\Delta R}{R} = G_f \cdot \frac{\Delta L}{L} = G_f \times e \qquad ...(16.10)$$

where, e = strain = $\frac{\Delta L}{L}$

The *gauge factor* can be written as:

$$G_f = 1 + 2\mu + \frac{\Delta \rho/\rho}{e} \qquad ...(16.11)$$

$$= 1 \quad + \quad 2\mu \quad + \quad \frac{\Delta \rho/\rho}{e}$$

= Resistance change due to change of length + Resistance change due to change in area + Resistance change due to piezo-resistive effect

or, $$G_f = 1 + 2\mu + \frac{\Delta \rho/\rho}{\Delta L/L} \qquad ...(16.12)$$

The strain is usually expressed in terms of *microstrain;* 1 micro strain = 1 μm/m

If the change in the value of resistisity of a material when strained is *neglected,* the gauge factor can be rewritten as:

$$G_f = 1 + 2\mu \qquad ...(16.13)$$

Eqn. (16.13) is valid only when *piezo-resistive effect* (*i.e.,* change in resistisity due to strain) is *almost negligible.*

- The Poission's ratio for all metals lies between 0 and 0.5. This give G_f as 2 approximately. In case of wire wound strain gauges where the common value for Poisson's ratio is 0.3, the value of G_f amounts to 1.6.
- The value of the gauge factor varies from material to material but it is generally assumed that it remains constant in the working range of strain required. *Its value is determined experimentally.*
- *Knowing the gauge factor (G_f), the strain in the member can be directly found out by the change of resistance.*

Properties of gauge materials:

The grid material for its proper functioning must possess the following desirable *properties:*

1. High resistivity.
2. High gauge factor.
3. High mechanical strength.
4. High electrical stability.
5. Low temperature sensitivity.
6. Low hysteresis.
7. Low thermal e.m.f. when joined with other materials.

8. Good corrosion resistance.
9. Good weldability.

Adhesing techniques:

For proper mounting of the strain gauges, the following *steps* should be strictly followed:

1. Before mounting the strain gauge on the surface, the surface must be preferably cleaned by emery cloth and base material exposed.
2. Remove the various traces of grease or oil etc. by using a solvent like acetone.
3. Swab the back of the strain gauge by cotton dipped in acetone once, to ensure that the back is free from grease etc.
4. Apply a generous quantity of cement to the cleaned resistance and then simply place the cleaned gauge on it and excess cement worked out. Make sure that there is no bubble between the surface and the gauge, if any one is there, that should be removed. Avoid using heavy pressure, otherwise cement may puncture the paper and short the grid.
5. Allow the gauge to sit for at least eight or ten hours before using it. If possible a slight weight might be placed by keeping a strong rubber on the gauge.
6. After cement has been fully cured, check the continuity of wire by an ohmmeter and weld the electric leads.

Example 16.13. *The gauge factor of a resistance wire strain gauge using a soft iron wire of small diameter is 4.2. Neglecting the piezo-resistive effect, calculate the Poisson's ratio.*

Solution. *Given:* $G_f = 4.2$

When the *piezo-resistive effect* is neglected, the gauge factor is given by:

$$G_f = 1 + 2\mu \qquad \text{...[Eqn. (16.13)]}$$

or,

$$4.2 = 1 + 2\mu$$

$$\therefore \quad \mu = \frac{4.2-1}{2} = \mathbf{1.6} \text{ (Ans.)}$$

Example 16.14. *A simple electrical strain gauge of resistance 120 Ω and having a gauge factor of 2 is bonded to steel having an elastic limit stress of 400 MN/m² and modulus of elasticity is 200 GN/m². Calculate the change in resistance,*

(i) due to a change in stress equal to $\frac{1}{10}$ *of the elastic range;*

(ii) due to change of temperature of 20°C if the material is advance alloy. The resistance temperature coefficient of advance alloy is $20 \times 10^{-6}/°C$.

Solution. *Given:* $R = 120\ \Omega$; $G_f = 2$; Elastic limit stress = 400 MN/m²; Modulus of elasticity = 200 GN/m²; Resistance temperature coefficient, $\alpha_0 = 20 \times 10^{-6}/°C$.

Change in resistance:

(i)

$$\text{Change in stress} = \frac{1}{10} \times 400 \text{ MN/m}^2 = 40 \times 10^6 \text{ N/m}^2$$

$$\text{Modulus of elasticity} = 200 \text{ GN/m}^2 = 200 \times 10^9 \text{ N/m}^2$$

$$\text{Strain, } e = \frac{\text{Stress}}{\text{Modulus of elasticity}} = \frac{40\times10^6}{200\times10^9} = \frac{1}{5}\times10^{-3}$$

$$\text{Gauge factor } G_f = \frac{\text{Per unit change in resistance}}{\text{Per unit change in length}}$$

or, $$G_f = \frac{\Delta R/R}{e} \quad \text{or} \quad \Delta R = R\, G_f\, e$$

or, $$\Delta R = 120 \times 2 \times \frac{1}{5} \times 10^{-3} = 48 \times 10^{-3}\ \Omega = \mathbf{45\ m\Omega}\ \textbf{(Ans.)}$$

(ii) $$R_{t2} = R_{t1}\,[1 + \alpha_0(t_2 - t_1)]$$

∴ Change in resistance $R_{t2} - R_{t1} = R_{t1}\ \alpha_0(t_2 - t_1)$

or, $$\Delta R = R_{t2} - R_{t1} = 120 \times 20 \times 10^{-6} \times (20)$$
$$= 48 \times 10^{-3}\ \Omega = \mathbf{48\ m\Omega}\ \textbf{(Ans.)}$$

Example 16.15. *A strain gauge is bonded to a beam which is 12 cm long and has a cross-sectional area of 3.8 cm^2. The unstrained resistance and gauge factor of the strain gauge are 220 Ω and 2.2 respectively. On the application of load the resistance of the gauge changes by 0.015 Ω. If the modulus of elasticity for steel is 207 GN/m^2, calculate:*

(i) The change in length of the steel beam.

(ii) The amount of force applied to the beam.

Solution. *Given:* L = 12 cm = 0.12 m; A = 3.8 cm^2 = 3.8 × 10^{-4} m^2; R = 220 Ω; G_f = 2.2; ΔR = 0.015 Ω; E = 207 GN/m^2.

***(i)* The Change in length of steel beam. ΔL:**

$$\text{Gauge factor, } G_f = \frac{\Delta R/R}{\Delta L/L}$$

∴ $$\Delta L = \frac{(\Delta R/R).L}{G_f} = \frac{(0.015/220)\times 0.12}{2.2} = \mathbf{3.72 \times 10^{-6}\ m}\ \textbf{(Ans.)}$$

***(ii)* The amount of force applied to the beam, *F*:**

$$E = \frac{\text{Stress}}{\text{Strain}} = \frac{\sigma}{e}$$

∴ $$\sigma = E \times e = E \times \frac{\Delta L}{L}$$
$$= (207 \times 10^9) \times \frac{3.72 \times 10^{-6}}{0.12} = 6.417 \times 10^6\ \text{N/m}^2$$

∴ $$\text{Force, } F = \sigma \cdot A = 6.417 \times 10^6 \times 3.8 \times 10^{-4}\ \text{N} = \mathbf{2.438\ kN}\ \textbf{(Ans.)}$$

16.13.4. Strain-gauge Circuits

The following strain gauge circuits will be discussed :

1. Ballast circuit.
2. Wheatstone bridge circuit.
   (i) Balanced (null) condition
   (ii) Unbalanced (deflection) condition
   — Quarter bridge
   — Half bridge
   — Full bridge.

16.13.4.1. Ballast-circuit (Voltage-sensitive potentiometric circuit)

Fig. 16.36 shows a ballast circuit-voltage-sensitive potentiometric circuit. Here,

v_i = Input supply voltage,

v_0 = Output voltage,

R_b = Ballast resistance, and

R_g = Resistance of the unstrained resistance gauge.

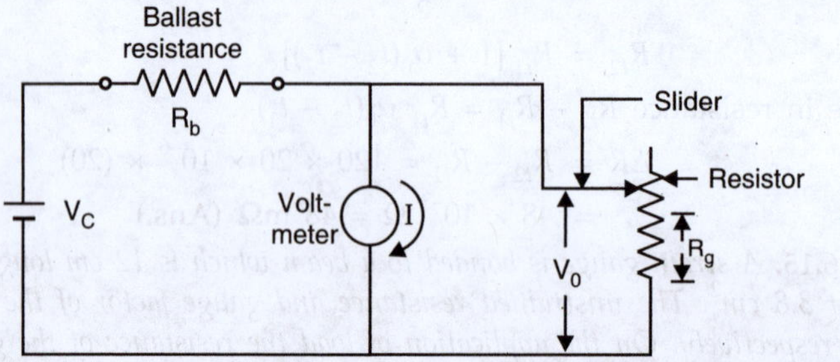

Fig. 16.36. Ballast circuit-voltage-sensitive potentiometric circuit.

The output voltage, when *no stress* is applied to the strain gauge, is given by

$$v_0 = \left(\frac{R_g}{R_g + R_b}\right) v_i \qquad ...(16.14)$$

When the gauge is strained, the gauge resistance changes to $(R_g + dR_g)$ and the output voltage becomes,

$$v_0 + dv_0 = \left[\frac{(R_g + dR_g)}{(R_g + dR_g) + R_b}\right] v_i \qquad ...(16.15)$$

∴ The *change in the output voltage,*

$$dv_0 = \left[\frac{(R_g + dR_g)}{(R_g + dR_g) + R_b} - \frac{R_g}{R_g + R_b}\right] v_i$$

$$= \left[\frac{(R_g + dR_g)(R_g + R_b) - R_g\{(R_g + dR_g) + R_b\}}{\{(R_g + dR_g) + R_b\}(R_g + R_b)}\right] v_i$$

$$= \left[\frac{\left(R_g^2 + R_gR_b + dR_gR_g + dR_gR_b - R_g^2 - dR_gR_g - R_gR_b\right)}{(R_g + R_b)^2}\right] v_i$$

$$...(\because dR_g > R_g)$$

$$= \left[\frac{dR_g \cdot R_b}{(R_g + R_b)^2}\right] v_i = \frac{R_g \cdot R_b}{(R_g + R_b)^2} \cdot \frac{dR_g}{R_g} v_i \qquad ...(16.16)$$

... Multiplying numerator and denominator by R_g.

Also, *condition of maximum sensitivity* is given by : $R_g = R_b$

Hence, $$dv_0 = \frac{v_i}{4} \cdot \frac{dR_g}{R_g}$$

Also, $$\frac{dR_g}{R_g} = G_f \cdot e$$

$$\therefore \quad dv_0 = \left(\frac{G_f}{4}\right) e\, v_i \text{ ...where, } G_f \text{ denotes the gauge factor ...(16.16a)}$$

From eqn. (16.16*a*) it is evident that *change in output-voltage when gauge is strained is directly proportional to strain.*

- The ballast circuit is used for *dynamic strain measurements where static strain components are ignored.*

Limitations of potentiometric circuits:

(*i*) No possibility of compensation for temperature variations.

(*ii*) High sensitivity to A.C. interference giving hum due to ground loops, induction from high current lines and poor connections.

16.13.4.2. Wheatstone bridge circuit

The wheatstone bridge technique can be used in the following *two ways*: (*i*) Null mode; (*ii*) Deflection mode.

1. Null mode:

Refer to Fig. 16.37. The resistances, with *no straining*, are so arranged that $v_B = v_D$ and the galvanometer gives zero deflection.

Then, $$\frac{R_1}{R_3} = \frac{R_2}{R_4} \quad \text{...(16.17)}$$

where, $R_1 = R_g$ = Unstrained resistance of the gauge.

In measurement of strains, generally R_1 is the strain gauge, R_2 and R_4 are the fixed resistances and R_3 is a variable resistor.

When the gauge is strained, its resistance R_1 changes by an amount dR_1. This change unbalances the bridge resulting into the deflection of the galvanometer. The balance is then regained by adjusting R_3 by an amount dR_3. The rebalanced conditions gives:

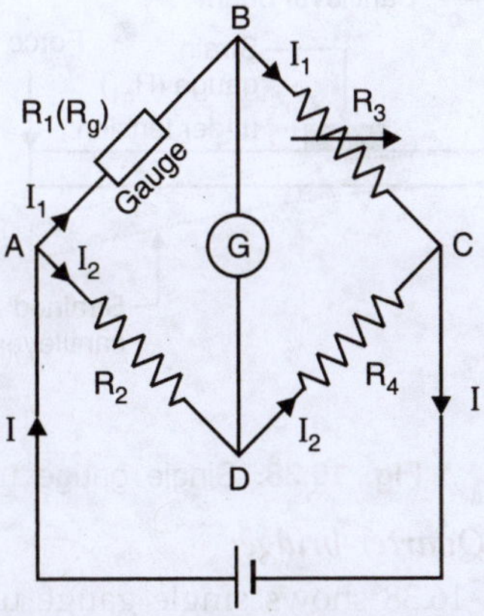

Fig. 16.37. Wheatstone bridge circuit.

$$\frac{R_1 + dR_1}{R_3 + dR_3} = \frac{R_2}{R_4}$$

or, $$R_1 + dR_1 = (R_3 + dR_3)\,\frac{R_2}{R_4}$$

or, $$R_1 + dR_1 = R_3 \times \frac{R_2}{R_4} + dR_3 \times \frac{R_2}{R_4}$$

or, $$R_1 + dR_1 = R_1 + dR_3 \times \frac{R_2}{R_4} \qquad \left[\because\ R_1 = \frac{R_3}{R_4} \times R_2 \text{ from eqn. (7.17)}\right]$$

or, $$dR_1 = dR_3 \times \left(\frac{R_2}{R_4}\right) \quad \text{...(16.18)}$$

If the resistances, of all the limbs of the wheatstone bridge are equal, then

$$R_1 = R_2 = R_3 = R_4 = R_g$$

and, $$dR_1 = dR_3 \qquad ...(16.19)$$

The change in resistance dR_1, in terms of strain, is given by

$$dR_1 = G_f\, e\, R_g,$$ where G_f is the gauge factor and e is the strain

∴ $$dR_3 = G_f\, e\, R_g \qquad ...(16.20)$$

Eqn. (16.20) indicates that the *change in the value of resistance R_3 is direct measurement of strain.*

2. Deflection mode:

Initially the bridge resistances are so adjusted that the bridge is in balanced. The equilibrium gets disturbed when the gauges are strained. Then, the voltage v_0 is measured under this unbalanced condition.

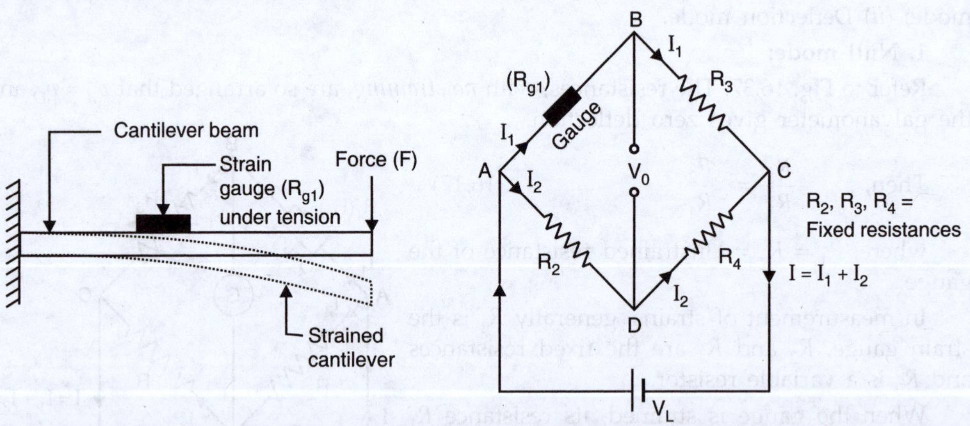

Fig. 16.38. Single gauge used for strain measurement (Quarter-bridge).

(i) Quarter-bridge:

Fig. 16.38 shows single gauge used for strain measurement (quarter-bridge). In this arrangement only one strain gauge is used and the other three elements of the bridge are fixed resistors.

Let us assume that the galvanometer (measuring instrument) has infinite impedance and therefore no current flows through it. Then,

Current flowing through the limbs AB and BC, $$I_1 = \frac{v_i}{R_{g1} + R_3} \qquad ...(16.21)$$

Voltage drop in limb AB (or voltage at terminal B),

$$v_{AB} = I_1\, R_{g1} = \frac{R_{g1}}{R_{g1} + R_3} \cdot v_i \qquad ...(16.22)$$

Similarly, $$I_2 = \frac{v_i}{R_2 + R_4} \qquad ...(16.23)$$

and, $$v_{AD} = \frac{R_2}{R_2 + R_4} \cdot v_i \qquad ...(16.24)$$

Initially, $R_{g1} = R_2 = R_3 = R_4 = R$

$\therefore$ $$v_{AB} = v_{AD} = \frac{v_i}{2} \quad ...(16.25)$$

and, v_0 = Voltage across the terminals B and D

$$= v_{AB} - v_{AD} = 0$$

Obviously, the bridge is balanced under *unstrained conditions.*

When the gauge is strained (see Fig. 16.38), the resistance R_{g1} changes by an amount dR_{g1}. Then,

$$v_{AB} = \left[\frac{R_{g1}+dR_{g1}}{(R_{g1}+dR_{g1})+R_3}\right]v_i = \left(\frac{R+dR}{2R+dR}\right)v_i$$

$[\because Rg_1 = R_3 = R \text{ and } dR_{g1} = dR]$

$$v_{AD} = \left(\frac{R_2}{R_2+R_4}\right)v_i = \frac{v_i}{2} \qquad (\because R_2 = R_4 = R)$$

The changed output voltage,

$$v_0 + dv_0 = \left(\frac{R+dR}{2R+dR} - \frac{1}{2}\right)v_i$$

$$= \left(\frac{2R+2dR-2R-dR}{2(2R+dR)}\right)v_i = \left(\frac{dR}{4R+2dR}\right)v_i$$

Since $dR << R$ and $v_i = 0$ (under unstrained conditions), therefore

$$dv_0 = \frac{v_i}{4}\cdot\frac{dR}{R} \quad ...(16.26)$$

or, $$dv_0 = \left(\frac{G_f}{4}\right)e \cdot v_i \quad ...(16.27)$$

...in terms of gauge factor G_f and applied strain e

From eqn. (16.27) it is obvious that the *output voltage is directly proportional to the applied strain.*

(ii) Half-bridge:

Fig. 16.39, shows two gauges used for strain measurement (Half-bridge). In this arrangement two of the bridge elements are strain gauges and the other two are fixed resistors. The strain gauge-1 is bonded to the upper surface of the cantilever beam and a second strain gauge-3 is bonded to the lower surface and located precisely underneath the gauge-1. These gauges are connected electrically to form adjacent limbs of the Wheatstone bridge circuit.

The temperature effects are cancelled out by having $R_2 = R_4$ and using *two identical gauges* in the opposite arms of the bridge.

Suppose, $R_{g1} = R_{g3} = R_2 = R_4 = R$

Under *no strain conditions:*

$$v_{AB} = v_{AD} = \frac{v_i}{2};\ v_B = v_D \text{ and } v_0 = 0.$$

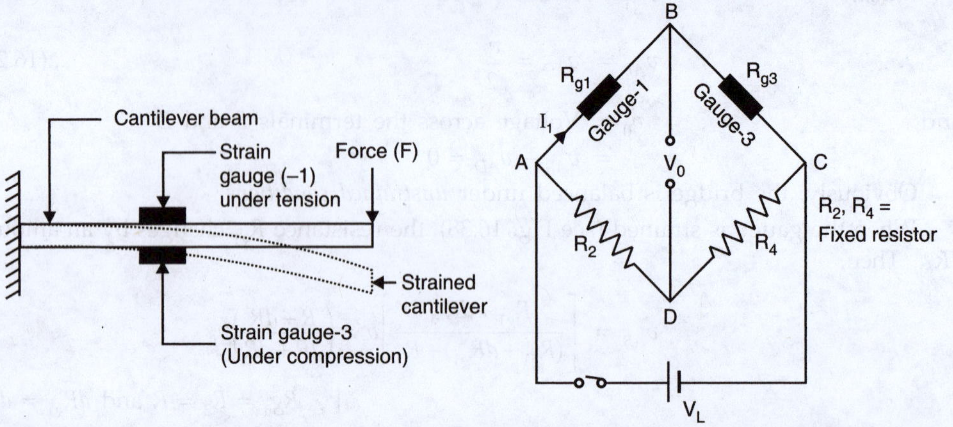

Fig. 16.39. Two gauges used for strain measurement (Half-bridge).

On the application of load to the cantilever beam, the resistance of the gauge R_{g1} *increases* due to *tensile load* whilst R_{g3} *decreases* due to equal *compressive strain* so that,

$$\text{Resistance of gauge–1} = R_{g1} + dR_{g1}$$

and,
$$\text{Resistance of gauge–3} = R_{g3} - dR_{g3}$$

Now,
$$v_{AB} = \frac{R_{g1}}{R_{g1} + R_{g3}} \cdot v_i$$

$$= \frac{R + dR}{(R + dR) + (R - dR)} v_i = \frac{R + dR}{2R} \cdot v_i \qquad ...(16.28)$$

and,
$$v_{AD} = \frac{R_2}{R_2 + R_4} v_i = \frac{v_i}{2} \qquad ...(16.29) \quad (\because R_2 = R_4)$$

The changed output voltage,

$$v_0 + dv_0 = \frac{R + dR}{2R} . v_i - \frac{v_i}{2}$$

$$= v_i\left(\frac{R + dR}{2R} - \frac{1}{2}\right) = v_i\left(\frac{2R + 2dR - 2R}{4R}\right)$$

or,
$$v_0 + dv_0 = \frac{v_i}{2} . \frac{dR}{R} \qquad ...(16.30)$$

Since under unstrained conditions $v_0 = 0$, therefore, change in output voltage due to applied strain becomes,

$$dv_0 = \frac{v_i}{2} \cdot \frac{dR}{R}$$

or,
$$dv_0 = \left(\frac{G_f}{2}\right) e \cdot v_i \qquad ...(16.31)$$

which is *twice* the output of Wheatstone bridge using one gauge only.

The eqn. (16.31) can be rewritten as:

$$dv_0 = \frac{v_i}{4}\left[\frac{dR}{R} - \left(\frac{-dR}{R}\right)\right]$$

$$\text{or,} \quad dv_0 = \frac{v_i}{4}\left[\begin{Bmatrix}\text{Fractional change in} \\ \text{resistance of gauge in limb } AB\end{Bmatrix} - \begin{Bmatrix}\text{Fractional change in} \\ \text{resistance of gauge in limb } BC\end{Bmatrix}\right] \quad ...(16.32)$$

The –ve sign with fractional change in resistance of the gauge in limb *BC* is due to the fact that compressive and tensile strains are of *opposite signs.*

In general, for the *two gauges connected in the adjacent limbs of a bridge circuit,* we have:

$$dv_0 = \frac{v_i}{4}\left(\frac{dR_{g1}}{R} - \frac{dR_{g3}}{R}\right) \quad ...(16.33)$$

Thus, *when both the gauges are mounted on the top of the cantilever beam, the two effects cancel each other and the output voltage is zero.*

(iii) Full-bridge:

Fig. 16.40, shows four gauges used for stain measurement (Full-bridge). In this arrangement all the four elements of the bridge are strain gauges.

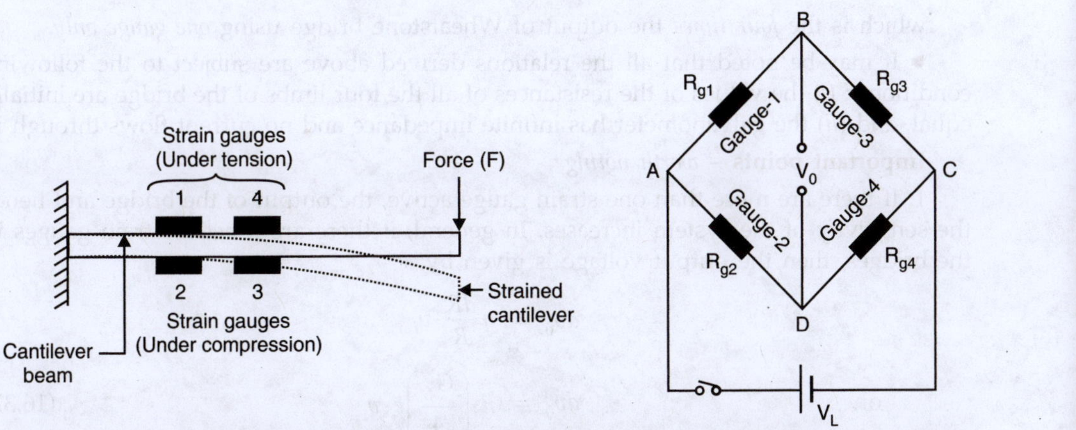

Fig. 16.40. Four gauges used for strain measurement (Full-bridge).

All the four gauges are similar and have equal resistances when unstained, *i.e.,*

$$R_{g1} = R_{g2} = R_{g3} = R_{g4} = R.$$

Under no-strain condition : $v_{AB} = v_{AD} = \frac{v_i}{2}$; $v_B = v_D$ and $v_0 = 0$.

When the load is applied to the cantilever beam, the resistance R_{g1} and R_{g4} *increase* due to *tensile load* whilstresistances R_{g2} and R_{g3} *decrease* due to equal *compressive strain.* When strained the resistances of the various gauges are

$$R_{g1} = R_{g4} = R + dR \text{ (tension)}$$

$$R_{g2} = R_{g3} = R - dR \text{ (compression)}$$

$$\text{and,} \quad v_{AB} = \frac{R_{g1}}{R_{g1} + R_{g3}} \cdot v_i$$

$$\text{or,} \quad v_{AB} = \frac{R + dR}{(R + dR) + (R - dR)} v_i = \frac{R + dR}{2R} \cdot v_i \quad ...(16.34)$$

and, $$v_{AD} = \frac{R_{g2}}{R_{g2}+R_{g4}} v_i$$

or, $$v_{AD} = \frac{R-dR}{(R-dR)+(R+dR)} \cdot v_i = \frac{R-dR}{2R} \cdot v_i \quad \text{...(16.35)}$$

The changed output voltage,

$$v_0 + dv_0 = \frac{R+dR}{2R} \cdot v_i - \frac{R-dR}{2R} \cdot v_i$$

$$= v_i \left(\frac{R+dR}{2R} - \frac{R-dR}{2R} \right) = v_i \frac{dR}{R}$$

Since the output voltage under unstrained conditions, $v_0 = 0$, therefore, change in output voltage due to applied strain becomes,

$$dv_0 = v_i \frac{dR}{R}$$

i.e., $$dv_0 = G_f \, e . v_i \quad \text{...(16.36)}$$

which is the *four times* the output of Wheatstone bridge using *one gauge only*.

- It may be noted that all the relations derived above are subject to the following conditions : *(i)* the values of the resistances of all the four limbs of the bridge are initially equal, and *(ii)* the galvanometer has infinite impedance and no current flows through it.

Important points – *worth noting* :

1. If there are more than one strain gauge active, the output of the bridge and hence the sensitivity of the system increases. In general, if there are *n* active strain gauges in the bridges, then the output voltage is given by

$$dv_0 = n \frac{dR}{R} \cdot v_i$$

or, $$dv_0 = n \cdot \left(\frac{G_f}{4} \right) e \cdot v_i \quad \text{...(16.37)}$$

(G_f and *e* are gauge factor and strain respectively).

The increased bridge output is expressed in terms of *"bridge constant"* (it represents the ratio of the actual bridge output to that if only one gauge were effective). The bridge constants for the three arrangements discussed above are 1, 2 and 4 respectively.

2. *High gauge sensitivity* can be obtained with :

(i) High gauge factor: It depends upon :

— The gauge material;

— The configuration of the gauge wire;

— The mechanical loading.

- In general the *foil* and *wire gauges* have gauge factor of about 2 and semiconductor gauges have typical values of about – 100 to + 200 (approx).

(ii) Large excitation voltage. It depends upon current or power rating of the gauge; typical values being 15 mA and 15 mW respectively.

16.13.4.3. Temperature compensation

In order to measure static strains accurately, it is absolutely necessary that adequate temperature compensation may be provided in the Wheatstone bridge. The errors due to change of temperature arise due to the following two *reasons* :

(*i*) Resistance change of the wire in the strain gauge with change in temperature.

(*ii*) Differential expansion existing between the gauges and the metal to which they are bonded, if the temperature coefficients of the two are not same.

Temperature effects may be dealt with as follows :

(*i*) *By compensation or cancellation.* This method is extensively used for both metallic as well as semiconductor gauge.

(*ii*) *By evaluation as a part of the reduction problem.* This method is employed only for semiconductor gauges.

Temperature compensation or cancellation techniques :

Temperature compensation may be provided by :

- Adjacent arm balancing or compensating gauge.
- Self-temperature compensation.
- Use of special external control circuitry.

1. Adjacent arm compensating gauge :

(*i*) ***Use of dummy gauge.*** The arrangement is shown in Fig. 16.41. The active strain gauge is installed on the test specimen (cantilever beam in this case) while the dummy or compensating gauge is installed on a like piece of material and is *not subjected to strain.*

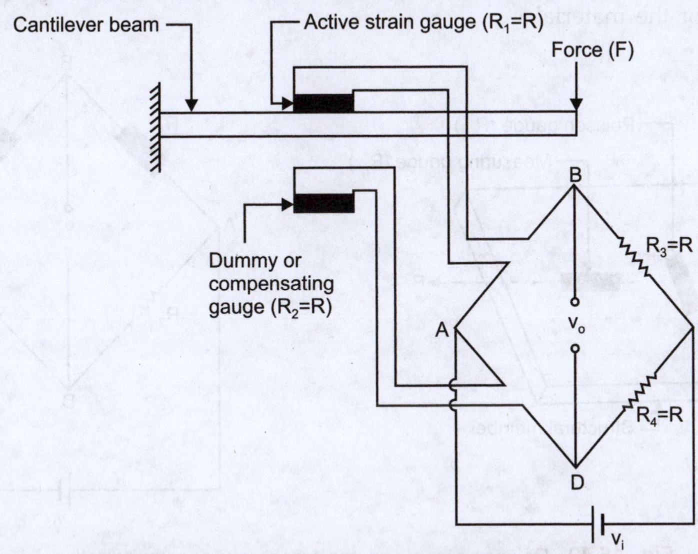

Fig. 16.41. Use of dummy gauge for temperature compensation.

The bridge is initially balanced and therefore $\frac{R_1}{R_2} = \frac{R_3}{R_4}$. Suppose a change in temperature occurs, and the resistances of active and dummy gauges change by dR_1 and dR_2 respectively. In order that there should be no output the bridge should be balanced and this requires,

$$\frac{R_1 + dR_1}{R_3} = \frac{R_2 + dR_2}{R_4}$$

or,
$$\frac{dR_1}{R_3} = \frac{dR_2}{R_4}$$

Let $R_3 = R_4 = R$ and therefore for bridge balance dR_1 should be equal dR_2. Now if active and dummy gauges are identical $R_1 = R_2 = R$ and if dummy gauge is placed in the same environments as the active strain gauge $dR_1 = dR_2$, therefore there will be no output due to changes in ambient temperature.

(ii) ***Use of two active gauges in adjacent arms.*** In some applications, where equal and opposite strains are known to exist, it is possible to attach similar gauges in such a way that one gauge experiences a positive strain and the other a negative strain (see Fig. 16.41).

- The arrangement shown in Fig. 16.41, *not only gives increased sensitivity but also eliminates the effects caused by temperature variations.*

(iii) ***Use of four active gauges.*** This arrangement is illustrated in Fig. 16.42.

- Four active-arm bridges are extensively used *when strain gauges are used as secondary transducers to give maximum sensitivity combined with full temperature compensation.* The effect of increasing the number of active gauges is the same as if a low impedance detector is used.

(iv) ***Poisson's method.*** In this method of temperature compensation two active gauges R_{g1} and R_{g3} bonded at right angles are used as shown in Fig. 16.42. Here gauge Rg_1 is in tension but the gauge Rg_3 is reduced in length depending upon Poisson's ratio of the material.

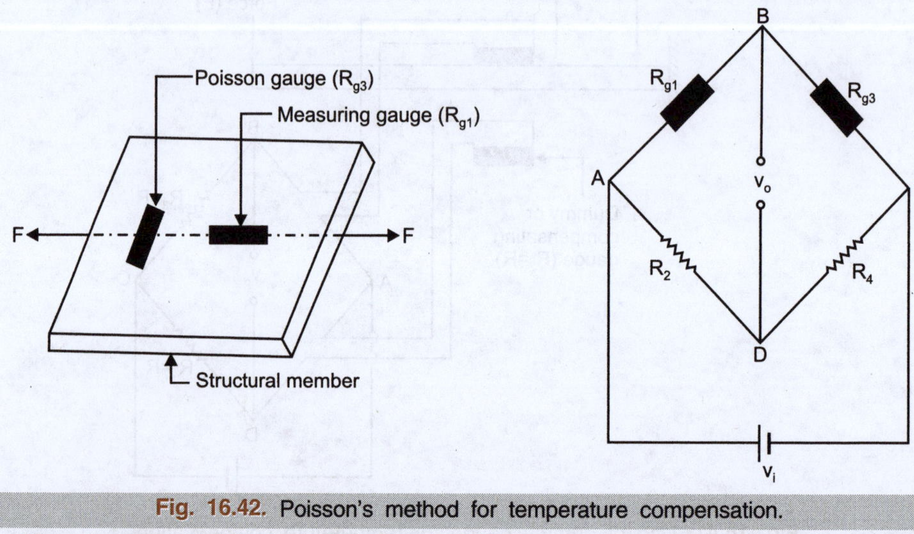

Fig. 16.42. Poisson's method for temperature compensation.

The *advantage* of this method is that *in addition to temperature compensation, the output of the bridge is increased by a factor (1 + μ) as compared to a bridge using a single active gauge.* The bridge sensitivity is increased by a factor (1 + μ) accordingly.

- This arrangement finds application *where it is not convenient or possible to mount the gauge in corresponding position on the underside of the test-member.*

2. **Self temperature compensation :**

In several applications it may be difficult or impossible to obtain temperature compensation by using adjacent arm compensating gauge acting either as a dummy gauge or a compensating gauge. Under these circumstances, it is desirable to use *self temperature compensation.*

Following two types of self-compensated gauges are available :

(i) Selected melt gauge :

- This method of self-compensation is based upon the fact that through proper manipulation of alloy and processing, usually through cold working, some control over the temperature sensitivity of the grid material may be obtained.
- Both wire and foil grid materials may be prepared which show very low apparent strain variation with temperature over a certain temperature range.

(ii) Dual element-gauge :

- In this case, two wire elements are connected in series to make one strain gauge assembly.
- *The two elements have different temperature characteristics and are so chosen that the net strain produced on account of temperature changes is minimised when the gauge is mounted on the test surface.*

16.13.4.4. Strain-gauge calibration

In most applications, a theoretical objection to the use of strain gauges is that it is impossible to check the accuracy of the readings obtained from them. Once the strain gauge is bonded in its measuring position, it cannot be removed, transferred or subjected to a known strain or calibration.

The calibration, in most of the cases, is done by introducing an accurately known sample of the variable and then observing response of the system. But this practice cannot be extended to bonded resistance strain-gauge because of the nature of transducer. In the case of strain gauges, the manufacturer has to maintain strict quality control and ensure that all the strain gauges produced by him are identical. Generally the gauge factors are usually quoted with a tolerance limit of about ± 0.2 percent or less. Thus if gauge factor and gauge resistance are known, any resistance strain-gauge system can be easily calibrated.

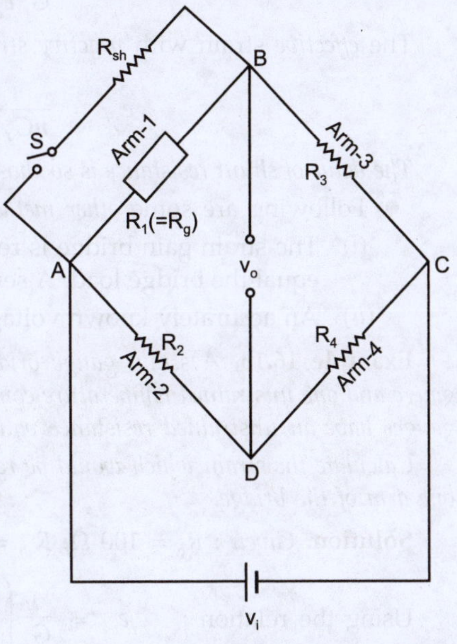

Fig. 16.43. Calibration of strain gauges.

The method of strain gauge calibration consists of determining the response of the system to the introduction of a known small resistance change at the gauge and of calculating an equivalent strain therefrom. The small resistance change is introduced by shunting a relatively high value precision resistance R_{sh} across the gauge (gauge resistance R_g) and to introduce this resistance R_{sh} in parallel to R_g, a switch S is used, as shown in Fig. 16.43.

Resistance of arm-1 of the bridge *before* switch S is closed = R_g

Resistance of arm-1 of the bridge *after* switch is closed

= Resistance of parallel combination of R_g and R_{sh}

$$= \frac{R_g \cdot R_{sh}}{R_g + R_{sh}}$$

∴ The change in resistance of arm-1 resulting from introduction of R_{sh},

$$dR = R_g - \frac{R_g \cdot R_{sh}}{R_g + R_{sh}} = \frac{R_g^2}{R_g + R_{sh}} \quad ...(16.38)$$

Also, strain $e = \frac{1}{G_f} \cdot \frac{dR}{R}$, where G_f is the gauge factor.

Substituting dR_g for dR in eqn. (16.38), we get the *equivalent strain* as,

$$e_{eq} = \frac{1}{G_f} \cdot \frac{dR_g}{R_g} = \frac{1}{G_f} \cdot \frac{1}{R_g} \cdot \frac{R_g^2}{(R_g + R_{sh})} = \frac{1}{G_f}\left(\frac{R_g}{R_g + R_{sh}}\right) \quad ...(16.39)$$

∴ Shunt resistance, $$R_{sh} = \frac{R_g(1 - G_f \cdot e_{eq})}{G_f \cdot e_{eq}} \quad ...(16.40)$$

In practice $G_f \cdot e_{eq} << 1$ and therefore the term $G_f \cdot e_{eq}$ in the numerator of eqn. 16.40 can be *neglected without-introducing any significant error.*

Then, $$R_{sh} = \frac{R_g}{G_f\, e_{eq}} \quad ...(16.41)$$

The *effective* strain with *n-active strain gauges* in the bridge is given as :

$$e_{eff.} = \frac{R_g}{nG_f\,(R_g + R_{sh})} \quad ...(16.42)$$

The value of shunt resistance is so chosen that the read-out system gives a full scale deflection.

- Following are some *other methods of electrical calibration :*
  (*i*) The strain gain bridge is replaced with a substitute load, initially adjusted to equal the bridge load. A series resistance is then used for calibration.
  (*ii*) An accurately known voltage is injected into the bridge circuit.

Example 16.16. *A strain gauge bridge consists of two fixed 100 Ω resistors; one active gauge and one unstrained temperature compensating gauge connected in adjacent arms. The two gauges have an unstrained resistance value of 100 Ω each. The gauge factor is 2.1.*

Calculate the strain which would be represented by 80000 Ω calibration resistance shunting one arm of the bridge.

Solution. *Given* : $R_g = 100\ \Omega$; $R_{sh} = 80000\ \Omega$; $G_f = 2.1$.

Using the relation : $$e_{eq} = \frac{1}{G_f}\left(\frac{R_g}{R_g + R_{sh}}\right) \quad ...(16.39)$$

or, $$e_{eq} = \frac{1}{2.1}\left(\frac{100}{100 + 80000}\right)$$

$= 5.945 \times 10^{-4}$ or **594.5 microstrain (Ans.)**

Example 16.17. *In a four arm bridge circuit four strain gauges are mounted on a simple flat tensile specimen arranged for complete temperature compensation and maximum sensitivity. A calibration resistor of 100 × 10³ Ω is connected across one of the strain gauges. Gauge resistances are each 200 Ω and each gauge has a gauge factor of 2.0.*

Calculate the axial strain in the specimen, if the strain gauge read-out has 140 divisions when calibration switch is closed and 220 divisions when the load is applied.

Solution. *Given* : $n = 4$; $R_g = 200\ \Omega$; $R_{sh} = 100 \times 10^3\ \Omega$;

$$\text{Effective strain, } e_{eff.} = \frac{R_g}{n\,G_f\,(R_g + R_{sh})} \qquad ...(16.42)$$

$$= \frac{200}{4 \times 2.0(200 + 100 \times 10^3)} = 249.5 \times 10^{-6}$$

When the calibration switch is closed, the read out gives a reading of 140 divisions.

$$\therefore \qquad 1 \text{ divisions} = \frac{249.5 \times 10^{-6}}{140} = 1.782 \times 10^{-6} = 1.782 \text{ microstrain.}$$

When the *strain gauge is loaded, the strain* = 1.782 × 220 = **392.04 microstrain (Ans.)**

16.13.4.5. Stress-strain relationships and strain gauge rosettes

Here, we shall discuss cases of uniaxial and biaxial strain measurements :

1. Simple uni-axial stresses. In this case, the axial stress (σ) can be calculated from the observed strain (*e*) using the modulus of elasticity, *E*, of the material used (within the elastic limits) as follows :

$$\sigma = e \, . \, E \qquad ...(16.43)$$

2. Biaxial stresses. Fig. 16.44 shows a cylindrical pressure vessel (closed at both ends) subjected to biaxial stresses. Consider an element subjected to stresses σ_x and σ_y applied simultaneously.

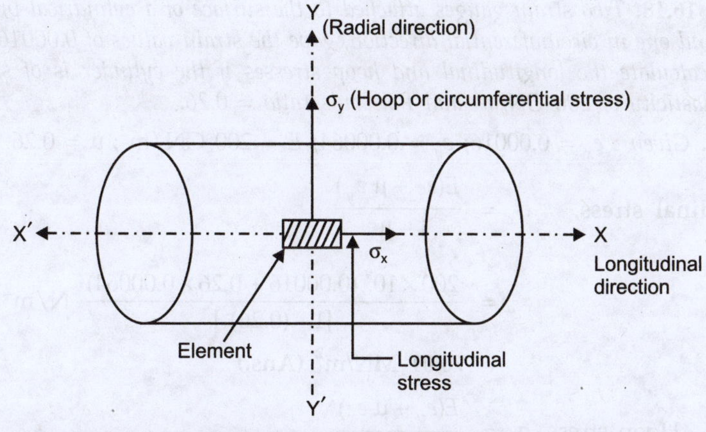

Fig. 16.44. A free surface subjected to biaxial stresses.

Strain produced by longitudinal stress σ_x, $e_x = \frac{\sigma_x}{E}$ ...in X or longitudinal direction.

At the same time, because of Poisson's ratio effect, the strain produced in *Y* or radial direction $= -\frac{\mu\,\sigma_x}{E}$, where μ is the Poisson's ratio.

Now, the stress σ_y applied in Y or radial direction will produce a strain equal to $\frac{\sigma_y}{E}$ in Y-direction and $-\frac{\mu\,\sigma_y}{E}$ in X or longitudinal direction.

The net strains in the two directions are :

$$e_x = \frac{\sigma_x}{E} - \frac{\mu\,\sigma_y}{E} = \frac{1}{E}(\sigma_x - \mu\,\sigma_y) \qquad ...(16.44)$$

and,
$$e_y = \frac{\sigma_y}{E} - \frac{\mu\,\sigma_x}{E} = \frac{1}{E}(\sigma_y - \mu\,\sigma_x) \qquad ...(16.45)$$

∴ We get expressions for the stresses as :

$$\sigma_x = \frac{E(e_x + \mu\,e_y)}{1-\mu^2} \qquad ...(16.46)$$

and.
$$\sigma_y = \frac{E(e_y + \mu\,e_x)}{1-\mu^2} \qquad ...(16.47)$$

If a third stress σ_3 acts in the third orthogonal direction, then the stress-strain relationships are given as :

$$e_x = \frac{1}{E}[\sigma_x - \mu(\sigma_y + \sigma_z)] \qquad ...(16.48)$$

$$e_y = \frac{1}{E}[\sigma_y - \mu(\sigma_z + \sigma_x)] \qquad ...(16.49)$$

$$e_z = \frac{1}{E}[\sigma_z - \mu(\sigma_x + \sigma_y)] \qquad ...(16.50)$$

Example 16.18. *Two strain gauges attached to the surface of a cylindrical pressure vessel, one in axial and one in circumferential direction, gave the strain values of 0.00016 and 0.00064 respectively, calculate the longitudinal and hoop stresses if the cylinder is of steel having a modulus of elasticity of 200 GN/m² and Poisson's ratio = 0.26.*

Solution. *Given* : e_x = 0.00016; e_y = 0.00064; E = 200 GN/m²; μ = 0.26.

Longitudinal stress,
$$\sigma_x = \frac{E(e_x + \mu\,e_y)}{1-\mu^2} \qquad ...(16.46)$$

$$= \frac{200 \times 10^9\,(0.00016 + 0.26 \times 0.00064)}{[1-(0.26)^2]}\ \text{N/m}^2$$

$$= \mathbf{70.01\ MN/m^2}\ \textbf{(Ans.)}$$

Hoop stress,
$$\sigma_y = \frac{E(e_y + \mu\,e_x)}{1-\mu^2}$$

$$= \frac{200 \times 10^9\,(0.00064 + 0.26 \times 0.00016)}{[1-(0.26)^2]}\ \text{N/m}^2$$

$$= \mathbf{146.2\ MN/m^2}\ \textbf{(Ans.)}$$

Example 16.19. *A simple tension member having an area of 110 mm*^{2} *is subjected to 25 kN. Strains of 1540 and – 420 micro-strain are measured in axial and transverse directions respectively. Calculate :*

(*i*) *Modulus of elasticity.*

(*ii*) *Poisson's ratio.*

Solution. Given : Area, $A = 110 \text{ mm}^2 = 110 \times 10^{-6} \text{ m}^2$; Load, $P = 25$ kN;

$$e_x = 1540;\ e_y = -420 \text{ micro-strain.}$$

(*i*) Modulus of elasticity, *E* :

$$\text{Axial stress, } \sigma_x = \frac{P}{A} = \frac{25 \times 10^3}{110 \times 10^{-6}} = 227.27 \times 10^6 \text{ N/m}^2$$

Transverse stress, $\sigma_y = 0$

$$\text{Axial stress, } e_x = \frac{1}{E}(\sigma_x - \mu\,\sigma_y) = \frac{\sigma_x}{E}$$

$$\therefore \quad E = \frac{\sigma_x}{e_x} = \frac{227.27 \times 10^6}{1540 \times 10^{-6}} \text{ N/m}^2 = 147.6 \times 10^9 \text{ N/m}^2$$

$$= \mathbf{147.6\ GN/m^2}\ \textbf{(Ans.)}$$

(*ii*) Poisson's ratio, μ :

Strain in the transverse direction,

$$e_y = \frac{1}{E}(\sigma_y - \mu\,\sigma_x) \qquad \text{...(16.45)}$$

$$-420 \times 10^{-6} = \frac{1}{147.6 \times 10^9}(0 - \mu \times 227.27 \times 10^6)$$

$$\text{or,} \quad \mu = \frac{420 \times 10^{-6} \times 147.6 \times 10^9}{227.27 \times 10^6} = \mathbf{0.2728}\ \textbf{(Ans.)}$$

Strain gauge rosettes :

"Rosette gauges" are used for measuring strain in complex parts. These gauges have *three or four separate grids with various angular orientations* and they can be cemented to the part with no particular attention being paid to the overall gauge orientation. The resultant strain on each of the grids can then be recorded and the true magnitudes and directions of the significant surface strains in the part can be calculated from these data.

Depending on the arrangement of the grids, we have the following types of rosettes :

- Rectangular rosettes.
- Delta rosettes.
- T-delta rosettes.

The formulae for calculation of principal strains and stresses for various arrangements are given below :

1. Two element rosette. Refer to Fig. 16.45.

- *Principal strains* : e_1 ; e_2
- *Principal stresses* :

$$\sigma_1 = \frac{E}{1 - \mu^2}(e_1 + \mu\, e_2) \qquad \text{...(16.51)}$$

$$\sigma_2 = \frac{E}{1-\mu^2}(e_2 + \mu\, e_1) \qquad ...(16.52)$$

- *Maximum shear stress :*

$$\tau_{max} = \frac{E}{2(1-\mu)}(e_1 - e_2) \qquad ...(16.53)$$

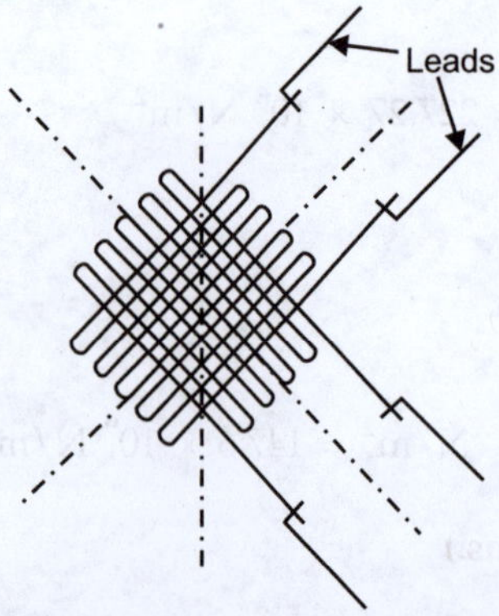

Fig. 16.45. Two element rosette. Two grid elements are mounted in between two thin papers.

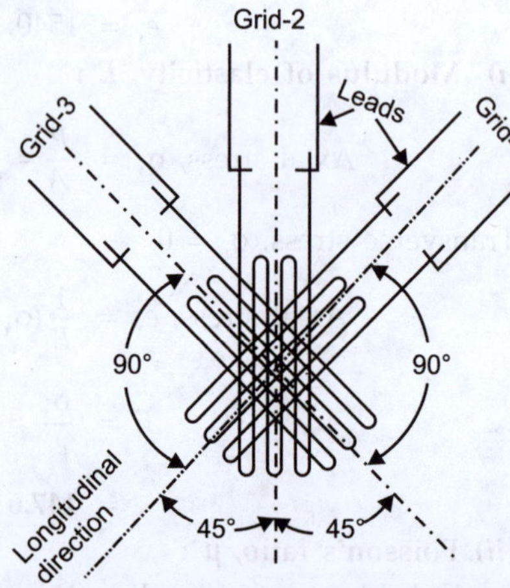

Fig. 16.46. Rectangular rosette. Grids do not make electrical contact. Grids are kept in-between two thin papers.

2. **Three-element rectangular rosette.** Refer to Fig. 16.46.
   - *Principal strains :*

$$e_{max}, e_{min} = \frac{e_1 + e_3}{2} \pm \frac{1}{\sqrt{2}}\left[(e_1 - e_2)^2 + (e_2 - e_3)^2\right]^{1/2} \qquad ...(16.54)$$

   - *Principal stresses :*

$$\sigma_{max}, \sigma_{min} = \frac{E(e_1 + e_3)}{2(1-\mu)} \pm \frac{E}{\sqrt{2}(1+\mu)}\left[(e_1 - e_2)^2 + (e_2 - e_3)^2\right]^{1/2} \qquad ...(16.55)$$

   - *Maximum shear stress :*

$$\tau_{max} = \frac{E}{\sqrt{2}(1+\mu)}\left[(e_1 - e_2)^2 + (e_2 - e_3)^2\right]^{1/2} \qquad ...(16.56)$$

The orientation of principal stress is given by

$$\tan 2\theta = \frac{2\,e_2 - e_1 - e_3}{e_1 - e_3} \qquad ...(16.57)$$

This is the axis at which τ_{max} occurs.

3. **Three-element delta rosette.** Refer to Fig. 16.47.
   - *Principal strains :*

$$e_{max}, e_{min} = \frac{e_1 + e_2 + e_3}{3} \pm \frac{\sqrt{2}}{3}\left[(e_1 - e_2)^2 + (e_2 - e_3)^2 + (e_3 - e_1)^2\right]^{1/2} \qquad ...(16.58)$$

- *Principal stresses :*

$$\sigma_{max}, \sigma_{min} = \frac{E(e_1 + e_2 + e_3)}{3(1-\mu)} \pm \frac{\sqrt{2}E}{3(1+\mu)}\left[(e_1 - e_2)^2 + (e_2 - e_3)^2 + (e_3 - e_1)^2\right]^{1/2} \quad ...(16.59)$$

- *Maximum shear stress :*

$$\tau_{max} = \frac{\sqrt{2}E}{3(1+\mu)}\left[(e_1 - e_2)^2 + (e_2 - e_3)^2 + (e_3 - e_1)^2\right]^{1/2} \quad ...(16.60)$$

The orientation of principal stress is given by

$$\tan 2\theta = \frac{\sqrt{3}(e_3 - e_2)}{2\,e_1 - e_2 - e_3} \quad ...(16.61)$$

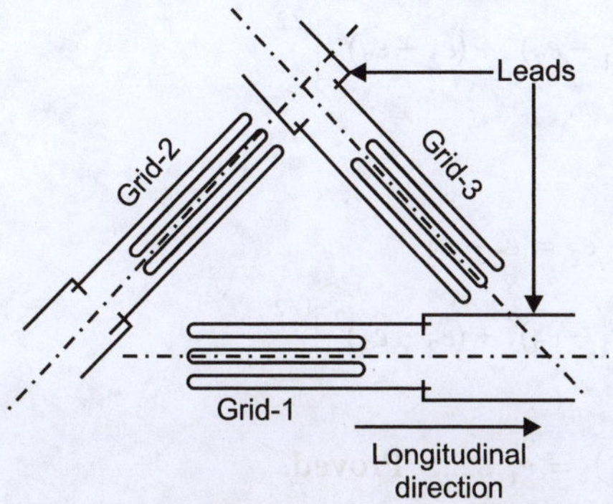

Fig. 16.47. Delta rosette.

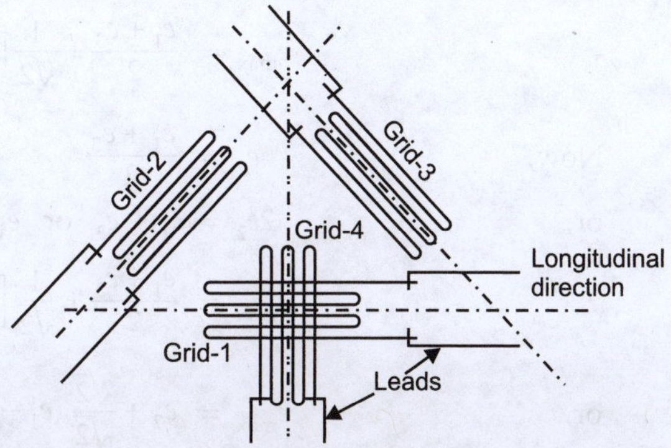

Fig. 16.48. T-delta rosette. Girds do not make electrical contacts.

4. T-delta rosette. Refer to Fig. 16.48.

- *Principal strains :*

$$e_{max}, e_{min} = \frac{1}{2}\left[e_1 + e_4 \pm (e_1 - e_4)^2 + \frac{4}{3}(e_2 - e_3)^2\right]^{1/2} \quad ...(16.62)$$

- *Principal stresses :*

$$\sigma_{max}, \sigma_{min} = \frac{E}{2}\left[\frac{e_1 + e_4}{1-\mu} \pm \frac{1}{1+\mu}\sqrt{(e_1 - e_4)^2 + \frac{4}{3}(e_2 - e_3)^2}\right] \quad ...(16.63)$$

- *Maximum shear stress :*

$$\tau_{max} = \frac{E}{2(1+\mu)}\sqrt{(e_1 - e_4)^2 + \frac{4}{3}(e_2 - e_3)^2} \quad ...(16.64)$$

The principal axis is located at θ according to,

$$\tan 2\theta = \frac{2}{\sqrt{3}}\left(\frac{e_3 - e_2}{e_1 - e_4}\right) \quad ...(16.65)$$

The T-delta rosette has the following *advantages* over others due to the fourth gauge element (three strain measurements are essential to completely define any stress situation) :

(*i*) The fourth gauge gives a *check on the results* obtained from other three elements.

(*ii*) Due to four readings, averaging effect results in *better accuracy.*

Example 16.20. *In a rectangular rosette, the maximum and minimum strains are given by*

$$e_{max}, e_{min} = \frac{e_1 + e_3}{2} \pm \frac{1}{\sqrt{2}}\left[(e_1 - e_2)^2 + (e_2 - e_3)^2\right]^{1/2}$$

and then are located at an angle, $\tan 2\theta = \dfrac{2e_2 - e_1 - e_3}{e_1 - e_3}$

Show that : $e_{max} = e_1$ *and* $e_{min} = e_3$ *when* $\theta = 0$.

Solution. When $\theta = 0$, $\tan 2\theta = 0$ and hence,

$$\frac{2e_2 - e_1 - e_3}{e_1 - e_3} = 0 \quad \text{or} \quad e_2 = \frac{e_1 + e_3}{2}$$

$$\therefore \qquad e_{\max} = \frac{e_1 + e_3}{2} + \frac{1}{\sqrt{2}}\left[(e_1 - e_2)^2 + (e_2 - e_3)^2\right]^{1/2}$$

Now, $$e_2 = \frac{e_1 + e_3}{2}$$

or, $$2e_2 = e_1 + e_3 \quad \text{or} \quad e_1 - e_2 = e_2 - e_3$$

$$\therefore \qquad e_{\max} = \frac{e_1 + e_3}{2} + \frac{1}{\sqrt{2}}\left[(e_1 - e_2)^2 + (e_2 - e_3)^2\right]^{1/2}$$

or, $$e_{\max} = e_2 + \frac{\sqrt{2}}{\sqrt{2}}(e_1 - e_2) = e_1 \text{ ....}$$ **Proved.**

Similarly, we can get, $e_{\min} = e_3$.

Example 16.21. *A rectangular rosette is mounted on a steel plate having a modulus of elasticity 200 GN/m*^{2} *and Poisson's ratio 0.3. The three strains measured are :*

$e_1 = 60$ *microstrain;* $\quad e_2 = 48$ *microstrain;* $\quad e_3 = -12$ *microstrain;*

Calculate the following :

(*i*) *The principal strains.*

(*ii*) *The principal stresses.*

(*iii*) *The maximum shear stress*

(*iv*) *Location of the principal planes.*

Solution. *Given :* $e_1 = 60$ microstrain; $e_2 = 48$ microstrain;
$e_3 = -12$ microstrain; $E = 200$ GN/m^2; $\mu = 0.3$.

(*i*) The Principal strains, e_{max} ; e_{min} :

$$e_{\max}, e_{\min} = \frac{e_1 + e_3}{2} \pm \frac{1}{\sqrt{2}}\left[(e_1 - e_2)^2 + (e_2 - e_3)^2\right]^{1/2} \qquad \text{...(16.54)}$$

$$= \frac{60 - 12}{2} \pm \frac{1}{\sqrt{2}}\left[(60 - 48)^2 + (48 + 12)^2\right]^{1/2}$$

or, $$e_{\max}, e_{\min} = 24 \pm \frac{61.19}{\sqrt{2}} = \mathbf{67.27 \times 10^{-6}, -19.27 \times 10^{-6}}$$ **(Ans.)**

(ii) The principal stresses, σ_{max} ; σ_{min} :

$$\sigma_{max}, \sigma_{min} = \frac{E(e_1 + e_3)}{2(1-\mu)} \pm \frac{E}{\sqrt{2}\,(1+\mu)}\left[(e_1 - e_2)^2 + (e_2 - e_3)^2\right]^{1/2} \quad ...(16.55)$$

$$= \frac{200 \times 10^9\,(60 \times 10^{-6} - 12 \times 10^{-6})}{2(1-0.3)} \pm \frac{200 \times 10^9}{\sqrt{2}(1+0.3)}$$

$$[(60 \times 10^{-6} - 48 \times 10^{-6})^2 + (48 \times 10^{-6} + 12 \times 10^{-6})^2]^{1/2}$$

or, $$\sigma_{max}, \sigma_{min} = (6.857 \pm 6.656) \times 10^6 \text{ N/m}^2$$

$$= \mathbf{13.513\ MN/m^2,\ 0.201\ MN/m^2}$$ **(Ans.)**

(iii) The maximum shear stress, τ_{max} :

$$\tau_{max} = \frac{E}{\sqrt{2}\,(1+\mu)}\left[(e_1 - e_2)^2 + (e_2 - e_3)^2\right]^{1/2} \quad ...(16.56)$$

$$= \frac{200 \times 10^9}{\sqrt{2}(1+0.3)}[(60 \times 10^{-6} - 48 \times 10^{-6})^2 + (48 \times 10^{-6} + 12 \times 10^{-6})^{1/2}$$

or, $$\tau_{max} = 6.656 \times 10^6 \text{ N/m}^2 \text{ or } \mathbf{6.656\ MN/m^2}$$ **(Ans.)**

(iv) Location of the principal planes, θ :

$$\tan 2\theta = \frac{2\,e_2 - e_1 - e_3}{e_1 - e_3} \quad ...(16.57)$$

$$= \frac{2 \times 48 - 60 + 12}{60 + 12} = 0.667$$

∴ $$2\theta = \tan^{-1}(0.667) = 33.7° \text{ or } 213.7°$$

or, $$\theta = \mathbf{16.85° \text{ or } 106.85°}$$ **(Ans.)**

16.14 LOAD CELLS

Load cells *are elastic devices that can be used for measurement of force through indirect methods i.e., through use of secondary transducers.*

Load cells utilize an elastic member as the primary transducer and strain gauges as secondary transducer. When the combination of the *strain gauge–elastic member* is used for weighing, it is called a *"load cell"*.

While designing load cells using strain gauges the following *factors* should be considered :

(*i*) Stiffness of the elastic element.

(*ii*) Optimum positioning of gauges on the element.

(*iii*) Provision for compensation of the temperature.

When large loads are to be measured, the direct tensile-compressive member may be used, whereas, in case of small loads, strain amplification provided by bending may be used with advantage.

16.14.1. Hydraulic Load Cell

Fig. 16.49 shows a *hydraulic load cell.*

Here the force variable is impressed upon a diaphragm which deflects and thereby transmits the force to a liquid. The liquid medium contained in a confined space has a preload pressure of 2 bar. On the application of the force the liquid pressure increases and *equals the force magnitude divided by the effective area of the diaphragm*. The pressure is transmitted to and read on an accurate pressure gauge calibrated directly on force units.

- These cells have been used to measure loads upto about 25 MN (with an *accuracy of 0.1% of full scale*); *resolution is about 0.02 per cent*.

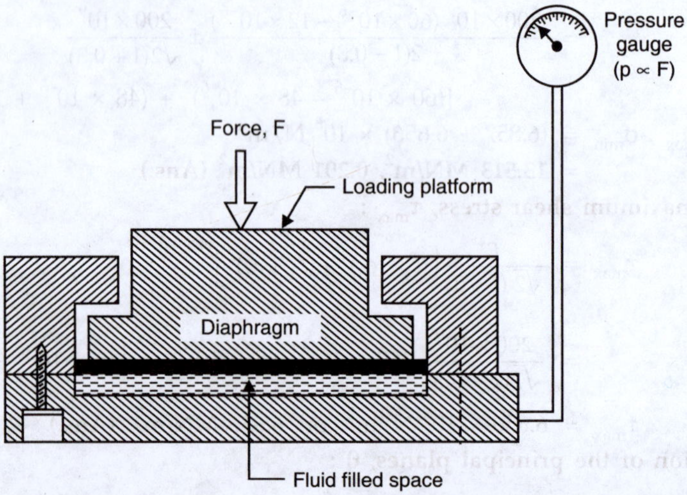

Fig. 16.49. Hydraulic load cell.

16.14.2. Pneumatic Load Cell

This cell operates on the *force-balance principle*. It employs a nozzle-flapper transducer similar to the conventional relay system. For any constant applied force, the system attains equilibrium at a specific nozzle opening and *corresponding pressure is indicated by the height of mercury column in a manometer*.

- The commercially available load cells (operating on this principle) can measure loads upto 25 kN with an accuracy of 0.5% of full scale.

16.14.3. Strain-Gauge Load Cells

These cells convert weight or force into electrical outputs which are provided by the strain gauges; these outputs can be connected to various measuring instruments for indicating, recording and controlling the weight or force.

Usually the *strain gauges are directly applied to the force-developing device and the device is calibrated against strain-gauge output.*

- These are excellent force-measuring devices, particularly for *transient and non-steady forces*.
- These are used in conjunction with CRO (for display purposes) for measurement of *rapidly changing loads*.

Construction and working of the load cell :

Fig. 16.50 shows a simple strain gauge load cell. It consists of a steel cylinder, on which are mounted four identical strain gauges. The gauges R_{g1} and R_{g4} are along the direction of applied load and the gauges R_{g2} and R_{g3} are attached circumferentially to gauges R_{g1} and R_{g4}. All the four gauges are connected electrically to the four limbs of a Wheatstone bridge circuit.

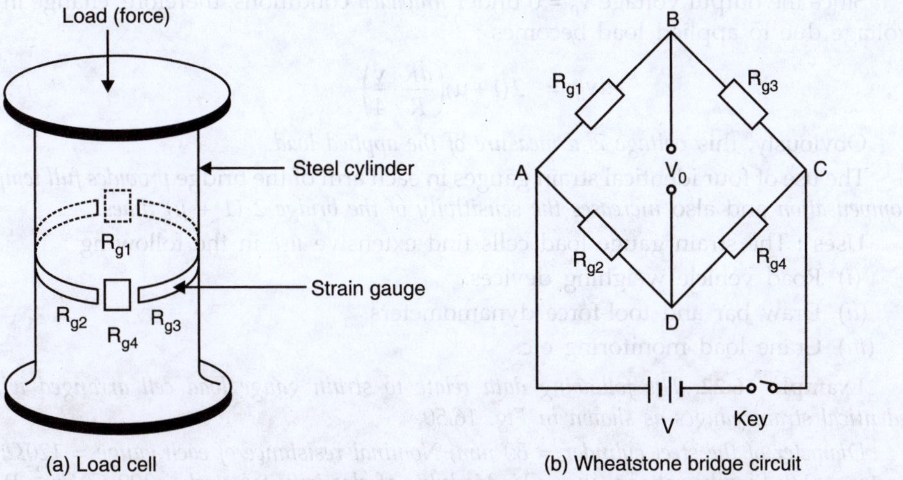

Fig. 16.50. Strain gauge load cell.

When there is *no load on the cell*, all the four gauges have the same resistance (*i.e.*, $R_{g1} = R_{g2} = R_{g3} = R_{g4}$). Obviously the terminals *B* and *D* are at the same potential, the bridge is balanced and the output voltage is zero.

i.e., $$\nu_{AB} = \nu_{AD} = \frac{\nu}{2} \quad \text{...(16.66)}$$

and, $$\nu_{AB} - \nu_{AD} = \nu_0 = 0 \quad \text{...(16.67)}$$

On the application of a compressive load to the unit, the *vertical gauges* (R_{g1} and R_{g4}) undergo compression (*i.e.* negative strain) and, therefore, there is *decrease in resistance*. The *circumferential gauges* R_{g2} and R_{g3}, simultaneously, undergo tension (*i.e.*, positive strain) leading to *increase in resistance*. The two strains are not equal; these are related to each other by a factor, μ, the Poisson's ratio. Thus when strained, the resistances of various gauges are :

$$R_{g1} = R_{g4} = R - dR \quad \text{...(compression)}$$
$$R_{g2} = R_{g3} = R + dR \quad \text{...(tension)}$$

Potential at terminal *B*, $$\nu_{AB} = \frac{R_{g1}}{R_{g1} + R_{g3}}\nu = \frac{R - dR}{(R - dR) + (R + \mu.dR)} \times \nu$$
$$= \frac{R - dR}{2R - dR(1 - \mu)} \times \nu \quad \text{...}(i)$$

Potential at terminal *D*, $$\nu_{AD} = \frac{R_{g2}}{R_{g2} + R_{g4}}\nu = \frac{R + \mu.dR}{(R + \mu.dR) + (R - dR)} \times \nu$$
$$= \frac{R + \mu.dR}{2R - dR(1 - \mu)} \times \nu \quad \text{...}(ii)$$

The changed output voltage,

$$\nu_0 + d\nu_0 = \frac{R - dR}{2R - dR(1 - \mu)} \times \nu - \frac{R + \mu.dR}{2R - dR(1 - \mu)} \times \nu$$

...[Using (*i*) and (*ii*)]

$$= \frac{dR(1 + \mu)}{2R} = 2(1 + \mu)\left(\frac{dR}{R} \cdot \frac{\nu}{4}\right) \quad \text{...(16.68)}$$

...*in magnitude*

Since the output voltage $v_0 = 0$ under *unloaded* conditions, therefore, change in output voltage due to applied load becomes :

$$dv_0 = 2(1+\mu)\left(\frac{dR}{R}\cdot\frac{v}{4}\right) \quad \text{...(16.69)}$$

Obviously, this *voltage is a measure of the applied load.*

The use of four identical strain gauges in each arm of the bridge *provides full temperature compensation* and also *increases the sensitivity of the bridge 2 (1 + μ) times.*

Uses : The strain gauge load cells find extensive *use* in the following :

(*i*) Road vehicle weighing devices.

(*ii*) Draw bar and tool-force dynamometers.

(*iii*) Crane load monitoring etc.

Example 16.22. *The following data relate to strain gauge load cell arranged with four identical strain gauges as shown in Fig. 16.50.*

Diameter of the steel cylinder = 60 mm; Nominal resistance of each gauge = 120Ω; Gauge factor = 2.0; Supply voltage (v) = 6V; Modulus of elasticity for steel = 200 GN/m²; Poisson's ratio = 0.3.

Calculate the sensitivity of the load cell.

Solution. *Given* : $d = 60$ mm $= 0.06$ m; R_g (each gauge) $= 120\ \Omega$; $G_f = 2.0$, $v = 6$V; $E = 200$ GN/m²; $\mu = 0.3$.

Sensitivity of the load cell :

Consider a load of 1 kN applied to the load cell.

$$\text{Stress } (\sigma) = \frac{\text{Load}}{\text{Cross-sectional area}} = \frac{1\times10^3}{\frac{\pi}{4}\times(0.06)^2} = 0.3537\times10^6\,\text{N/m}^2$$

$$\text{Strain, } e = \frac{\text{Stress }(\sigma)}{\text{Modulus of elasticity (E)}} = \frac{0.3537\times10^6}{200\times10^9} = 1.7685\times10^{-6}$$

Fraction change in resistance,

$$\frac{dR}{R} = G_f \times e = 2.0 \times 1.7685 \times 10^{-6} = 3.537 \times 10^{-6}$$

$$\text{Output voltage, } dv_0 = 2(1+\mu)\left(\frac{dR}{R}\cdot\frac{v}{4}\right) \quad \text{...[Eqn. (16.68)]}$$

$$= 2(1+0.3)\left(3.537\times10^{-6}\times\frac{6}{4}\right) = 13.79\times10^{-6}\,\text{V} = 13.79\,\mu\text{V}$$

Hence, the *sensitivity of the load cell* = **13.79 μV/kN** **(Ans.)**

16.15 PROXIMITY SENSORS

A ***proximity sensor*** consists of an *element that changes either its state or an analog signal when it is close to, but often not actually touching, an object.*

Magnetic, electrical capacitance, inductance, and eddy current methods are particularly suited to the design of a proximity sensor.

- A ***photoemitter-detector pairs*** represents another approach, where *interruption or reflection of a beam of light is used to detect an object in a non-contact manner.* The emitter and detecter are usually a phototransistor and a photodiode.

Common ***applications*** for proximity sensors and limit switches include :

(*i*) Counting moving objects;

(*ii*) Limiting the traverse of a mechanism.

16.15.1. Eddy Current Proximity Sensors

Working principle :

When a coil is supplied with an alternating current an alternating magnetic field is produced. If there is a metal object in close proximity to this attending magnetic field, then eddy currents are induced in it. The eddy currents themselves produce a magnetic field which *distorts* the magnetic field responsible for their production. Consequently, the *impedance of the coil changes and so the amplitude of the alternating current. This change, at some preset level, can be used to trigger a switch.*

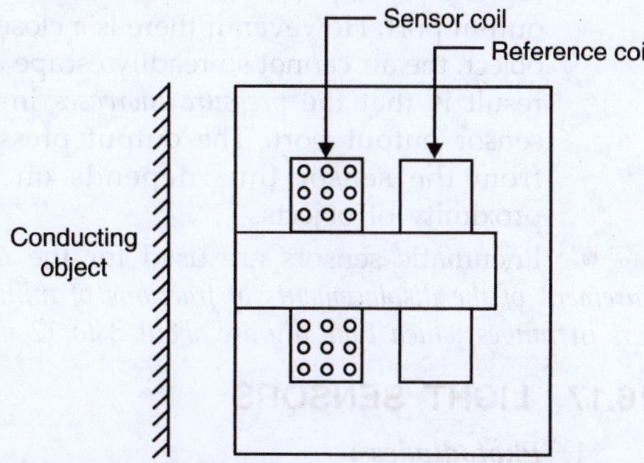

Fig. 16.51. Eddy current proximity sensor.

Fig. 16.51. shows the basic form of an eddy current proximity sensor. It is *used for the detection of non-magnetic but conductive materials.*

Advantages :

(*i*) Small in size.

(*ii*) Relatively inexpensive.

(*iii*) High flexibility.

(*iv*) High sensitivity to small displacements.

16.15.2. Capacitance Proximity Sensor

Fig. 16.52. shows a schematic diagram of a capacitance proximity sensor.

— It consists of a simple plate (one of the forms), with the object (*earthed*) acting as the other plate.

— As the object approaches the sensor, separation between the plate of the capacitor and object changes which becomes significant as the object is close to the sensor.

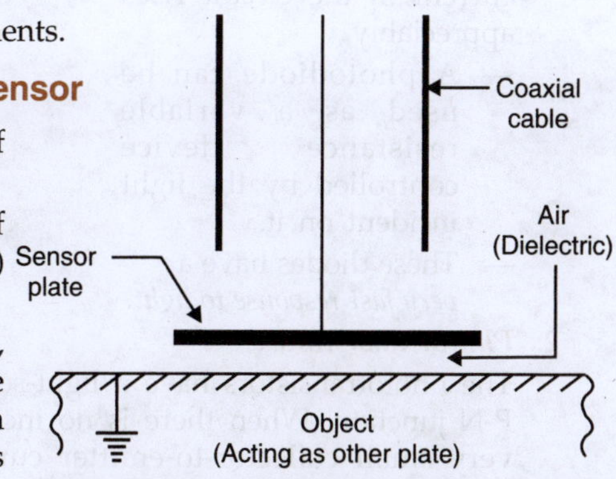

Fig. 16.52. Capacitance proximity sensor.

16.15.3. Inductive Proximity Switch

— An inductive proximity switch consists of a coil wound round a core.

— When the end of the coil is close to a metal object its *inductance changes*. This change can be monitored by its effect on a resonant circuit and the change used to trigger a switch.

- It can only be used *for the detection of metal objects* and is best with ferrous metals.

16.16 PNEUMATIC SENSORS

These sensors involve the use of compressed air, displacement or proximity of an object being transformed into a *change in air pressure.*

Fig. 16.53. shown the basic form of a pxeumatic sensor.

— Low pressure air is allowed to escape through a port in front of the sensor.

— This escaping air, in the absence of any close by object, escapes and in doing so also *reduces the pressure* in the nearby sensor output port. However, if there is a close by object, the air cannot so readily escape and the result is that the *pressure increases* in the sensor output port. The output pressure from the sensor thus depends on the proximity of objects.

- Pneumatic sensors are used for the *measurement of the displacements of fractions of millimeters in ranges which typically are about 3 to 12 mm.*

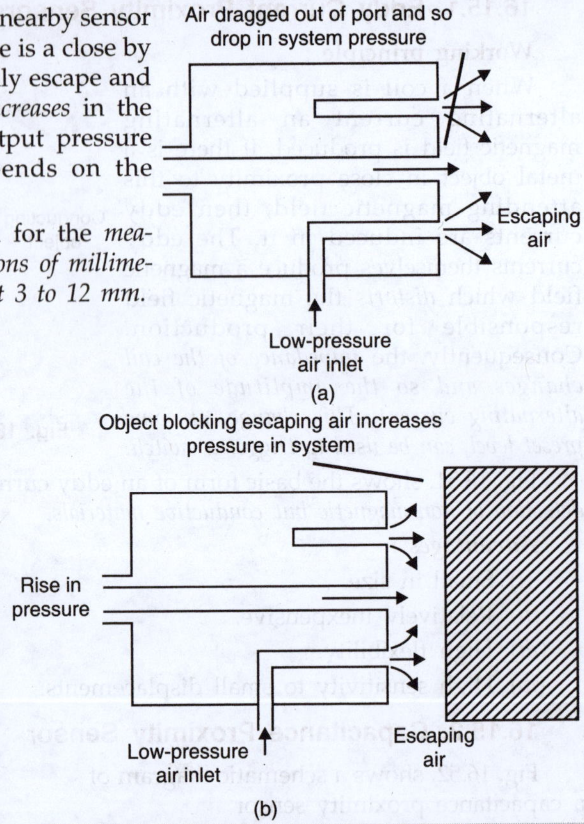

Fig. 16.53. Pneumatic proximity sensor.

16.17 LIGHT SENSORS

1. ***Photodiodes :***

"*Photodiodes*" are semiconductor junction diodes which are connected into a circuit in reverse bias, so giving a very high resistance, so that when light, falls on the junction the diode resistance drops and the current in the circuit rises appreciably

— A photodiode can be used as a variable resistance device controlled by the light incident on it.

— These diodes have a *very fast response to light*.

2. ***Phototransistors* :**

The phototransistors have a light-sensitive collector-base P-N junction. When there is no incident light there is a very small collector-to-emitter current. When light is incident, a base current is produced that is *directly proportional to the light intensity*. This leads to the production of a collector current which is then a *measure of the light intensity*.

— Phototransistors are often available as integrated packages with the phototransistor connected in a Darlington arrangement with a conventional transistor (Fig. 16.54). Since this arrangement gives a higher current gain, the device gives a *much greater collector current for a given light intensity*.

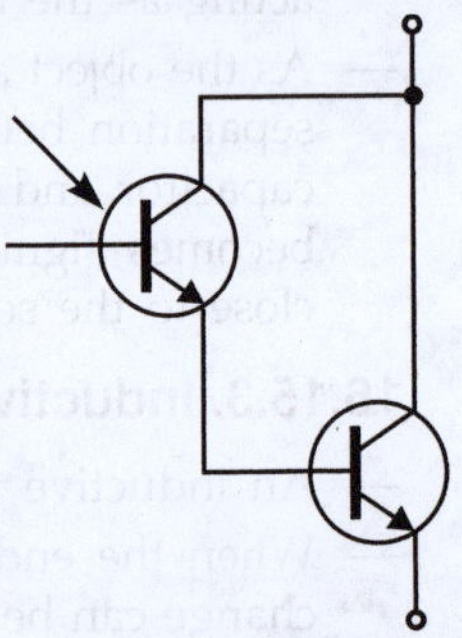

Fig. 16.54. Photo Darlington.

3. ***Photoresistor* :**

It has a *resistance* which depends on the intensity of the light falling on it, *decreasing linearly as the intensity increases.*

— The cadmium sulphide photoresistor is most responsive to light having wavelengths shorter than about 515 nm and the cadmium selinide photoresistor for wavelengths less than about 700 nm.

- An array of light sensors is often required in a small space in order to determine the variations of light intensity across that space.

16.18 TACTILE SENSOR

Human sense of touch is a marvelous phenomenon and scientists have a real challenge in developing robot fingers that can actually "feel" the difference between various textures and surface shapes.

One of the forms of *tactile sensor uses polyvinylidene fluoride* (PVF_2) *sheet which has piezo-electric qualities*. A soft film is sand wiched between two layers of PVF_2. The soft film transmits vibrations. The principle used in this case is both piezo-electric and its reverse effect.

A.C. voltage is supplied to the lower layer (i.e., which is a piezo-electric material) and as a result generates vibrations (Reverse of piezo-electric effect). These vibrations are picked up by the soft film and transmitted to the upper layer which converts these vibrations to alternate voltage (piezo-electric effect).

16.19 SMART SENSORS

The sensors having decision-making and communication logic added to the basic sensors, are called **smart or intelligent sensors.**

In such sensors, *other features* included are :

(*i*) Compensation for interferring inputs ;

(*ii*) Linearisation;

(*iii*) Self test;

(*iv*) Calibration facility.

— Whenever possible, there are usually microcomputers and other elements on the same chip.

- Fig. 16.55 shows the functional diagram for *data acquisition system using the smart pressure sensor*:

— When a piezoresistive sensor is used, as primary sensor for pressure, *the main interferring input is temperature and so a secondary temperature sensor is used to compensate for the effect of temperature on resistance change.*

— The memory of microcontroller has calibration data stored in it. The microcontroller is essentially a microprocessor with input/output (I/O) facility.

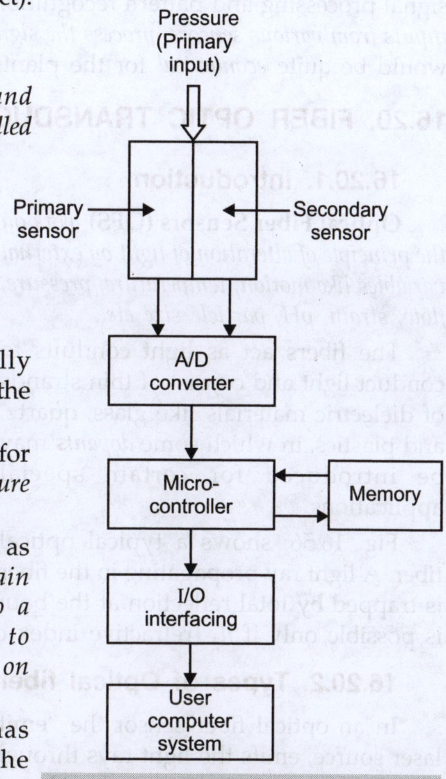

Fig. 16.55. Data acquisition system using smart pressure sensor.

— The IEEE 1451.2 standard for smart sensors includes the following *modules* :

(*i*) STIM (Smart Transducer Interface Module)

(*ii*) NCAP (Network Application Processor)

(*iii*) TEDS (Transducer Electronic Data Sheet)

(*iv*) TII (Transducer Independent Interface)

— The TEDS contain information of the transducer's identification (viz. make, model, specification, calibration, location etc.)

— Any transducer can be interfaced through STIM, performing the functions of signal conditioning, linearisation, signal processing, patterns recognition of monitored signals etc.

— The TII interfaces STIM with NCAP. It is a 10-wire serial I/O bus that defines data transport from STIM to NCAP, address and data transport from NCAP to STIM, latching data, performing triggering functions, data transport acknowledge, interrupt used by STIM to request services of NCAP etc. NCAP is a device used for supporting a network interface, and may provide local intelligence including the possible use of a microcontroller. It is the smart sensor's window to the outside world. *Any control network may be connected to a group of sensors, with an appropriate NCAP.*

A useful application of the above unit is for a condition monitoring system for a plant, to monitor the health of machines and process, to detect possible damage. Instead of using a dedicated computer with spectrum analyses or data acquisition system or signal processing and pattern recognition, *a smart unit on the lines of IEEE 1451.2 can accept inputs from various sensors, process the signals and diagonise fault if any and communicate* ; this would be quite *economical* for the plants.

16.20. FIBER OPTIC TRANSDUCERS

16.20.1. Introduction

Optical Fiber Sensors (OFS) *work on the principle of alteration of light by external variables like motion, temperature, pressure, flow, strain, pH, particle size etc.*

The fibers act as light conduits to conduct light and consist of thin strands of dielectric materials like glass, quartz, and plastics, in which some *dopants* may be introduced for certain special applications.

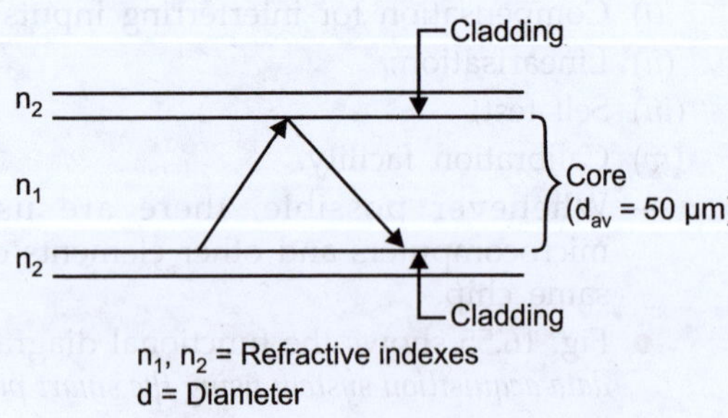

Fig. 16.56. A typical optical fiber.

Fig. 16.56, shows a typical optical fiber. A light ray propagating in the fiber is trapped by total reflection at the boundary of the core and the cladding material. This is possible only if n_2 (refractive index of the material) < n_1 (refractive index of core)

16.20.2. Types of Optical fiber sensor configurations

In an optical fiber sensor the "emitters", which may be a light emitting diode or a laser source, emits the light rays through the fiber, which get modulated due to the outer signal which is to be measured. The output fiber is connected to the "detector", which converts the optical energy into electrical energy. These detectors work on the principle of creation of an electron-hole pair in semiconductors or the release of the electrons from the cathode of the photomultiplier tube.

Basically, following are the *two types of optical fiber sensor configurations* :

1. Extrinsic (or incoherent) sensors :

- In this type of sensor (Fig.16.57) the *intensity indulation of light takes place outside the fiber.*

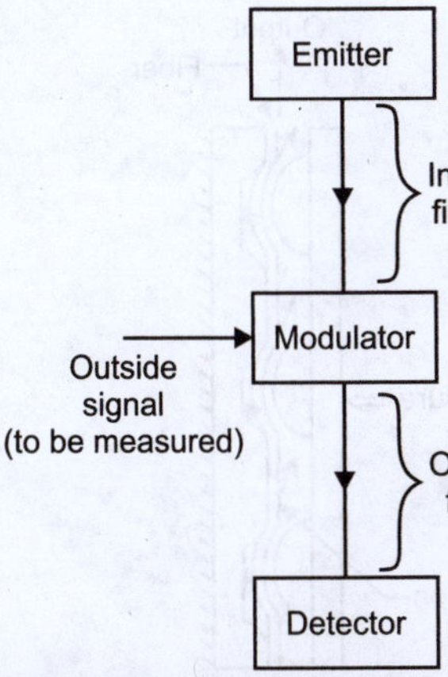

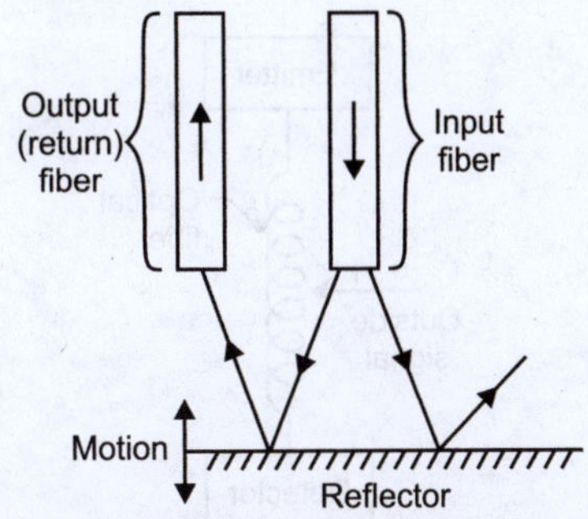

Fig. 16.57. Extrinsic sensor configuration. **Fig. 16.58.** Example of an extrinsic type sensor.

- Fig. 16.58, shows an extrinsic (or external) intensity modulator type of sensor in which the position of the reflector due to motion to be measured may change, thus *changing the light intensity in the output fiber*, which is detected by a detector.
- Fig. 16.59, shows a *typical fiber optic flow sensor* – an extrinsic type sensor.

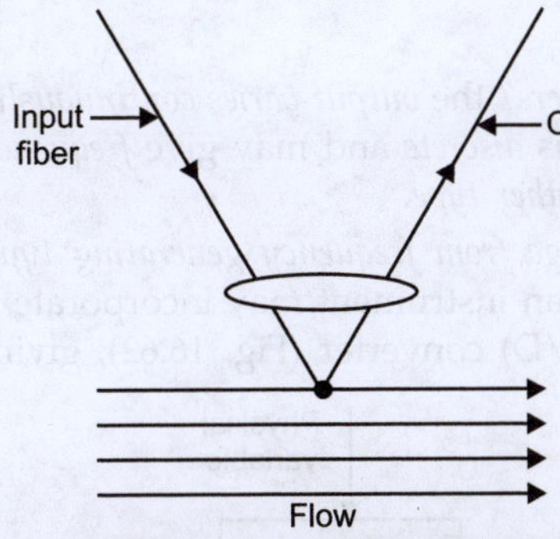

Fg. 16.59. A typical fiber optic flow sensor (extrinsic type).

— In this type of flow sensor, the frequency of light waves scattered from the particles in the moving fluid is Doppler shifted, the *frequency shift being proportional to velocity*.

— The input optical fiber carries light, which is focused on the fluid flow.

— The output fiber carries the light scattered by the particles in the flowing liquid and particle velocity is found from the modulated frequency spectrum.

2. Intrinsic (or coherent) sensor : Refer to Fig. 16.60

- In this case, the *intensity modulation of light takes place within the fiber*.
- Fig. 16.61 shows an example of an intrinsic type sensor :

— Here, the fiber bending due to pressure, which is the input variable (to be measured), *induces radiation losses, changing the intensity at the output*. This happens within the fiber itself and hence the configuration is called as intrinsic type. This causes radiation of light even at small deformation and is also called *micro-bend sensor*.

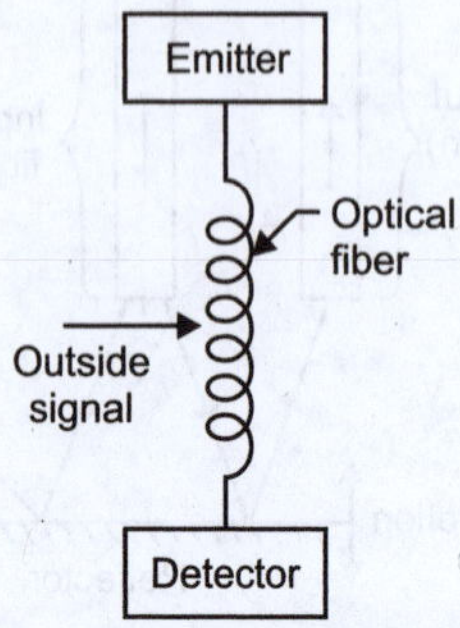

Fig. 16.60. Intrinsic sensor configuration.

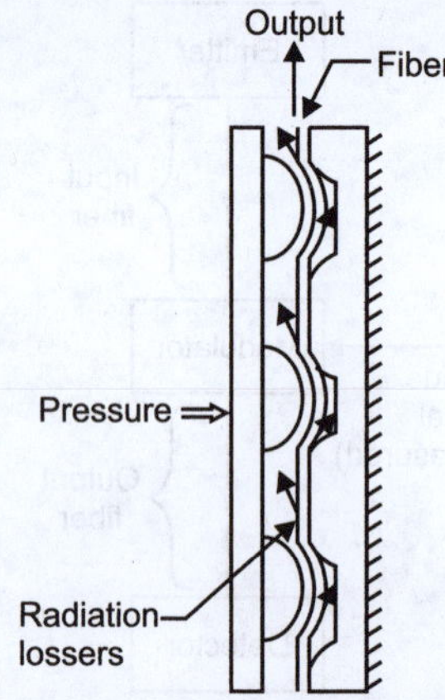

Fig. 16.61. Example of an intrinsic type sensor.

— In this type of sensor configuration, apart from the intensity modulation, *there may be phase or frequency modulation*, which after detection may be employed for many applications.

16.21 DIGITAL TRANSDUCERS

16.21.1. Introduction

Whereas in "*analog transducers*" the *output varies continuously according to the input*, the output in "**digital** *transducers*" is *discrete* and may give *frequency type output or a digitally coded output, of binary or some other type.*

- Digital transducers *range from frequency generating types of transducers to digital encoders.* Alternatively, an instrument may incorporate an analog transducer and an analog-to-digital (A/D) converter (Fig. 16.62), giving a digital output.

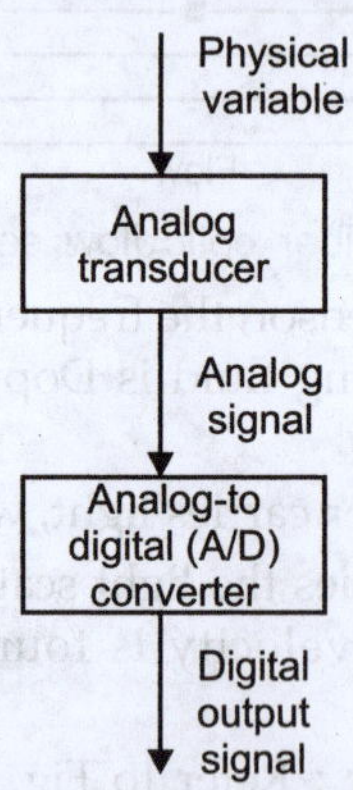

Fig. 16.62. Digital instrument incorporating A/D converter.

Advantages :

The main advantages of *digital transducers* are :

1. It is possible to obtain increased accuracy in pulse count.
2. It becomes easy to use digital computers along with the transducers, for data manipulation.
3. There are ergonomic advantages in presenting digital data.

4. It is easy to transmit digital signals without distortion and external noise (since digital signals are not dependent on signal amplitudes).

16.21.2. Frequency Domain Transducers

The *output* in these transducers is in the form of *pulses or sinusoidal wave forms, the frequency of which is a measure of the magnitude of the physical variable.*

The following three types of frequency domain transducer are described :

1. Electromagnetic frequency domain transducer.
2. Opto-electrical frequency domain transducer.
3. Vibrating string transducer.

1. **Electromagnetic frequency domain transducer :**

Fig. 16.63 shows an electromagnetic frequency domain transducer which can employed for the *measurement of speed.*

— It consists of a permanent magnent on a solenoid.

— A gear of ferromagnetic material is mounted on the rotating shaft whose speed is to be measured.

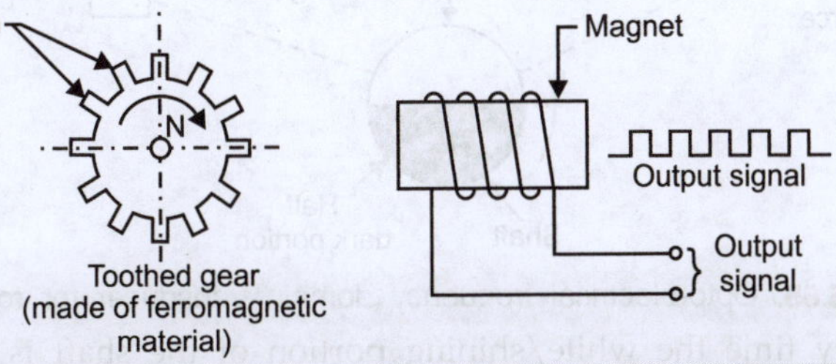

Fig. 16.63. Electromagnetic frequency domain transducer.

— As each gear tooth passes in front of the magnet, the gap length changes. Consequently there is a change in the flux density and a voltage pulse is induced in the coil. *Pulse frequency equal speed N times the number of teeth T.* The form of the output signal is shown in Fig. 16.63.

Thus, pulse frequency is a measure of rotational speed.

2. **Opto-electrical frequency domain transducer :**

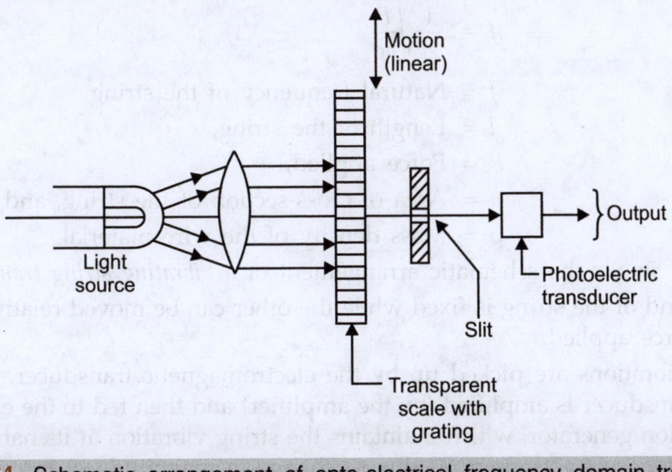

Fig. 16.64. Schematic arrangement of opto-electrical frequency domain transducer.

Fig. 16.64, shows schematic arrangement of this type of transducer for *linear motion* :

— The arrangement includes a transparent scale with a grating.

— The moving object is attached to the transparent scale.

— Light from a source passes through the scale and the slit and then falls on a photoelectric transducers. The slit width is such that a motion equal to the pitch of the grating produces one complete cycle of light and darkness at the photoelectric cell. Thus, a pulse output is obtained. From the number of pulses, the change in motion of the scale and the object attached to it can be determined.

- Fig. 16.65, shows opto-electrical frequency domain transducer for measurement of *rotational speed of a shaft* :

  — The shaft has half dark and half white or shining portion.

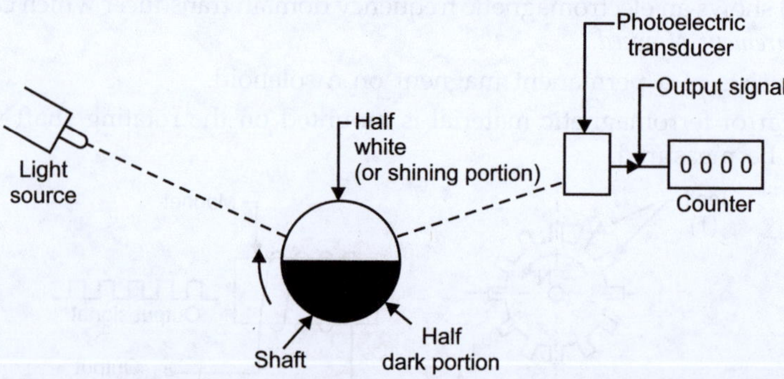

Fig. 16.65. Opto-electrical frequency domain transducer for rotary motion.

  — Every time the white/shining portion of the shaft is in front of the light source, the reflected light, falling on the photoelectric transducer, gives an electrical pulse output. The frequency of the pulses is thus a measure of speed of rotation of the shaft.

3. Vibrating string transducer :

This type of transducer is essentially employed to measure the *force* applied to a metal string, which is kept-vibrating, the *frequency* of which is dependent on the force applied. The natural frequency of the string is given by :

$$f = \frac{1}{2l}\sqrt{\frac{P}{a\rho}}$$

where, f = Natural frequency of the string,

l = Length of the string,

P = Force applied,

a = Area of cross-section of the string, and

ρ = Mass density of the wire material.

Fig. 16.66, shows the schematic arrangement of a *vibrating string transducer* :

— One end of the string is fixed while the other can be moved relative to it, due to the force applied.

— The vibrations are picked up by the electromagnetic transducer. The output of the transducer is amplified (by the amplifier) and then fed to the electromagnetic vibration generator, which maintains the string vibration at its natural frequency f. Frequency f gets changed due to change in magnitude of force P.

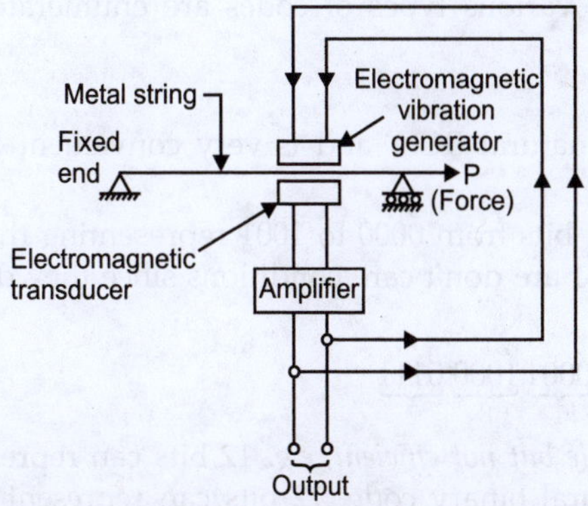

Fig. 16.66. Vibrating string transducer.

— The frequency is measured by a *frequency counter and is a measure of the force applied to the string*.

- This transducer can be used for *measurement of force and displacement*.
- Vibrating diaphragm and cylinder types of transducers have been developed to measure *pressure.*

16.21.3. Digital Encoders

*The **digital encoders** convert analog motion (rotary or linear) to a digital output form.*

16.21.3.1. Introduction to digital coding

In digital circuits, each number of piece of information is defined by an equivalent combination of binary digits. *A complete group of these combinations which represent numbers, letters or symbols is called a **digital code**.*

The group of 0s and 1s in the binary number can be thought of as a code representing the decimal numbers. When a decimal number is represented by its equivalent binary number, it is called *a straight binary coding*.

In modern digital equipment, *codes are used to represent and process numerical information.*

The digital output is usually required to be in *binary form*, especially for use with a *digital computer*. Unlike the decimal system where there are ten possible states for a digit (0-9), the binary system has two states, viz. ON or OFF or 1 and 0, respectively. To form a number, these are arranged with least significant bit (LSB) on the right and most significant bit (MSB) on the left. In natural or straight binary code, the arrangement is as shown in Fig. 16.67.

Thus, 11101 in binary system would correspond to the number 29 in the decimal system, as shown in Fig. 16.67. This arrangement uses 5 bits and this can specify any number upto *31*, in steps of 1 or the resolution is $\frac{1}{31}$. More bits would be needed for smaller resolution. *Each bit may be considered to be replaced by a switch, whose state ON would correspond to the state 1 and OFF to state 0.* Thus, a number of bits or switches can specify any desired number.

2^4 (MSB) $\quad 2^3 \quad 2^2 \quad 2^1 \quad 2^0$ (LSB)

1 1 1 0 1 $= 1(2^4) + 1(2^3) + 1(2^2) + 0(2^1) + 1(2^0) = 29$

Fig. 16.67. Binary code arrangement.

Types of codes. The various types of codes are enumerated and briefly discussed below :

1. BCB Code

It is also known as 'natural BCD' and is very convenient for representing decimal digits in digital circuits.

- It consists of four bits from 0000 to 1001 representing the decimal numbers from 0 to 9.1010 to 1111 are don't care conditions since they do not have any meaning in BCD.

  Example : $187 = \underbrace{0001}_{1}\,\underbrace{1000}_{8}\,\underbrace{0111}_{7}$

- This code is *simple but not efficient, e.g.* 12 bits can represent a number upto 999, while in the natural binary code, 12 bits can represent number upto 4095.

2. Excess-3 Code

- The code can be derived from BCD by *adding 3 to each coded number.*

  It is useful when it is desired to obtain the 9's complement of a decimal digit represented by this code. The 9's complement is obtained simply by complementing each bit.

- This code can be *conveniently used for performing subtracting operations in digital computers.*

3. Gray Code

- In this code any *one bit changes between any two successive numbers.*
- This code is used to overcome the inherent disadvantage of the straight binary code. In a straight binary code, a change of number 1 might result in a change of number of bits. Thus a change of number 7 to 8 would need a change in all the four bits as 7 is represented by 0111 and 8 by 1000. Thus an error of one bit in a large digital number, in practice, would result in a large error in decimal reconversion. In the case of a *digital transducers*, an increment in the measured variable would thus have to change all the bits, and any error in a bit, would change the number considerably. This is *overcome in the Gray code*, in which with a change from one value to the next, only one bit is changed at a time. The *conversion from natural binary to Gray code is done as follows, keeping the same number of bits* :

  (*i*) MSB is unchanged during conversion ; (*ii*) Gray code bit for other bits in the natural binary is *same* if the digit to the *left* in the natural binary is zero ; (*iii*) The bit is changed, if the bit to the left is 1. *Example* : Fig. 16.68 shows the application of these rules for changing the number 10011 in natural binary code by 11010 in Gray Code.

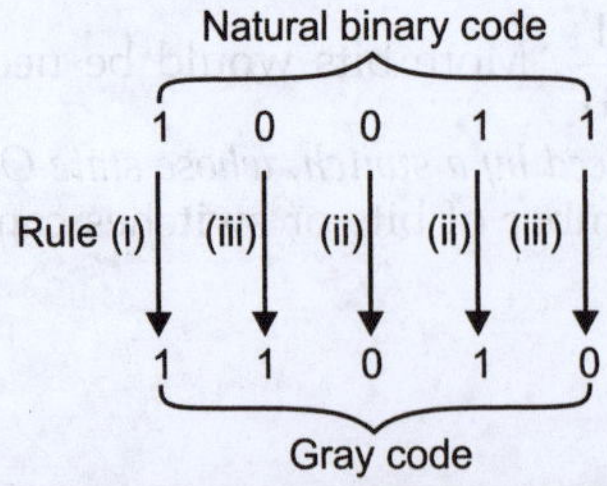

Fig. 16.68. Natural binary to gray code conversion.

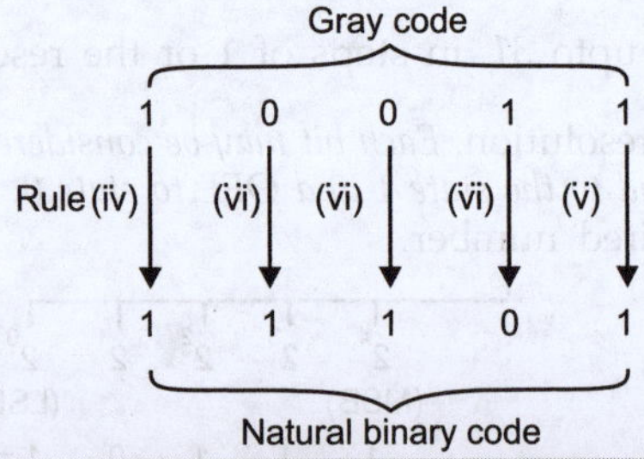

Fig. 16.69. Gray Code to natural binary conversion.

While changing from Gray to natural code, the following rules apply :
(iv) MSB is unchanged during conversion ; *(v)* for every other bit in Gray code, natural binary is *same*, if the number of 1s to the left is even ; *(vi)* The bit is *changed* if the number of 1s to the left is *odd*. *Example* : 10011 in Gray code would be changed to 11101 in natural binary code, as shown in Fig. 16.69.

- Gray code is mainly used in the *location of angular positions of a rotating shaft*.

4. Octal Code

- The octal system is a 8 base system.

It uses 3 bits to represent one octal digit.

5. Hexadecimal Code

- The hexadecimal system is a base 16 system.
- It uses four bits to represent one hexadecimal digit.
- The hexadecimal digits are represented as 0 to 9 continued by alphabetical characters from *A* to *F*.

16.21.3.2. Description of ditigal encoders

In the *contact type* arrangement there is direct contact between the brushes and encoder disc. The disadvantage of such a system is the wear of brushes and disc (due to friction between the brushes and disc)

1. Optical type encoder :

Fig. 16.70. illustrates the arrangement of an *optical type encoder* :

— It is a non-contact encoder.

— The disc has transparent and opaque areas, corresponding to the conducting and non-conducting one respectively, in the contacting type encoder disc.

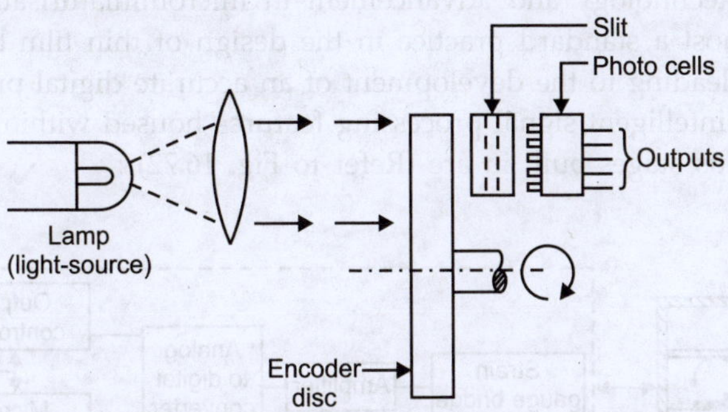

Fig. 16.70. Optical type encoder.

— A light source is used along with slits and photo cells. The photo cell, corresponding to a particular track, would produce an electrical output if the transparent portion is in front of the slit and light source, giving state ON (or 1) while no electrical output from a cell would correspond to OFF (or 0) state.

2. Magnetic type encoder :

Fig. 16.71 shows a magnetic encoder circuit :

— It is also a non-contact type encoder.

— This encoder uses a number of small toroidal magnets with coils around them.

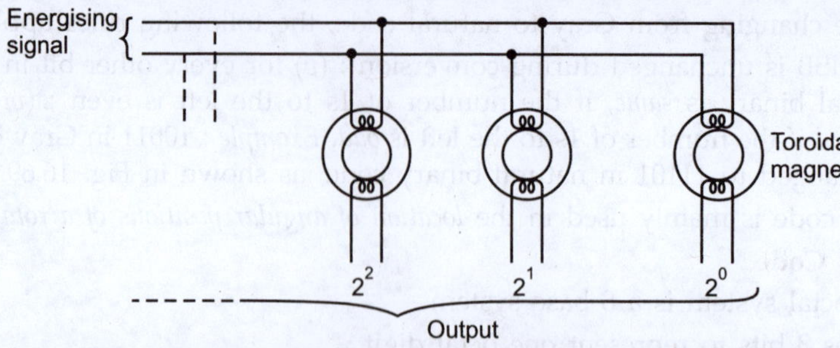

Fig. 16.71. Magnetic encoder circuit.

— The area corresponding to the conducting and non-conducting ones in the contact type encoder disc form the non-magnetic and magnetic areas on the disc. The presence or absence of such areas is detected by coils which are in close proximity to each track on the disc.

— One of the coils in each toroidal magnet is energised with high frequency A.C. carrier signal. If the toroid is over a non-magnetic area of the encoder disc, a voltage due to transformer action is induced in the output coils (state 1) while a magnetic area would saturate the magnetic circuit and output is very small (state 0). The signals on the output windings are *demodulated* to remove the high frequency carrier signal.

16.21.4. Digital Pressure Transducer

These days it has become feasible to incorporate the signal conditioner and associated electronic circuitry within the body of the transducer itself, owing to the advent of integrated circuit technology and advancement in microminiaturisation innovation. It has now been almost a standard practice in the design of thin film based and surface mounted devices, leading to the development of an accurate digital pressure transducer with a number of intelligent signal processing features housed within a relatively small volume. The various stages built in are (Refer to Fig. 16.72) :

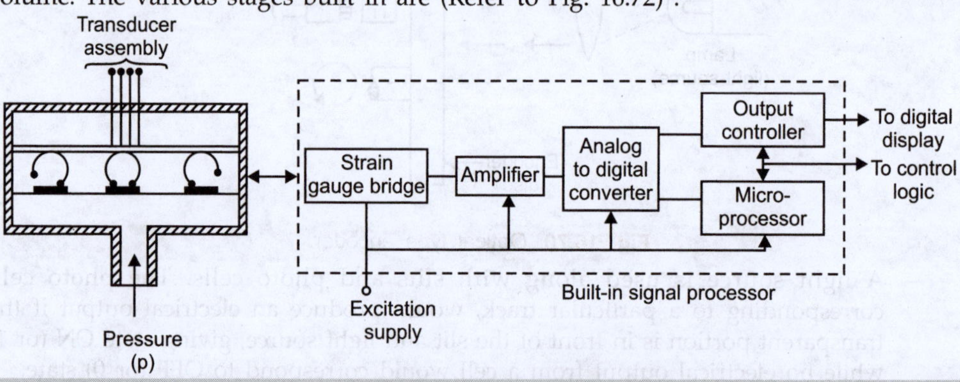

Fig. 16.72. Digital pressure transducer.

- Preamplifier ;
- Analog to digital (A/D) converter ;
- Output processor, controlled with a microprocessor (μP)

  The final electrical output can be directly connected to a digital display or modified to give a direct output to 0 to +/– 5V or 4 to 20 mA.

16.22 RECENT TRENDS–SMART PRESSURE TRANSMITTERS

The microprocessors are now being used in *transmitters* also; as a consequence of the availability of computing power the transmitters have become more intelligent.

The output in case of smart transmitters is 4–20 mA on 2-wire but with the added capability of digital communication from a hand-held interface connected anywhere on 4–20 mA signal, the remote adjustment of the transmitter data base and acquisition diagnostic information to minimise loop downtime is possible. It has *high rangeability and much better performance*.

The transmitter senses all the *three parameters, 'differential pressure', 'static pressure'* and *'temperature'*. The meter body is pre-programmed in manufacturing to characterise the unit for linearity, static pressure and temperature effects, and it *computes a highly repeatable and accurate pressure measurement*. These characteristics are held in PROM memory and being specific to one meter are kept with the meter body. The combination of characterised meter body and digital electronics has enabled a quantum leap forward in performance.

- The *rangeability* to smart transmitters is very high (400 : 1). Thus only 3 sensors would be required to cover the entire range of 2.5 millibar to 700 bar.
- The *reliability* is very high due to use of minimum number of components and protection against all foreseeable damping like radio frequency, reverse polarity, overpressure, surge voltage and lightning.

Advantages of digital transmitters :

The major advantages of digital (so called "*smart*") transmitters over their conventional analog counterparts are :

(*i*) Increased rangeability (400 : 1 against 6 : 1 of analog transmitters).
(*ii*) Higher accuracy.
(*iii*) Self-diagnostic facilities.
(*iv*) Almost no drift with time.
(*v*) Reduced cabling cost due to the use of a field bus cuts.
(*vi*) Better noise immunity.
(*vii*) Economical, because of improved overall performance.
(*viii*) Ambient temperature compensation.
(*ix*) Remote adjustability of range, damping, polarity etc. (This makes the commissioning of the entire system simpler).

16.23 SELECTION OF SENSORS

A number of factors need to be considered for selecting a sensor for a particular application are :

1. The nature of the measurement required *e.g.,*
   — The variable to be measured, its nomial value, the range of values;
   — The accuracy required;
   — The required speed of measurement;
   — The reliability required;
   — The environmental conditions under which the measurement is to be made.
2. The nature of the output required from the sensor, thus determining the signal conditioning requirements in order to give suitable output signals from the measurement.

Then possible sensors can be identified taking into account such factors as their range, accuracy, linearity, speed of response, reliability, maintainability, life, power supply requirements, ruggedness, cost, availability.

16.24 ROTARY VARIABLE DIFFERENTIAL TRANSFORMER (RVDT)

RVDT is used for the measurements of *angular displacements.* It *converts the angular displacements into electrical signals.* Its operating principle is similar to that of LVDT.

Fig. 16.73, shows the constructional details of a RVDT :

— The construction of RVDT is similar to that of LVDT (*i.e.* one primary winding P and two secondary windings S_1 and S_2), the only dfference is that the core is cam-shaped and rotated inside the windings through a shaft.

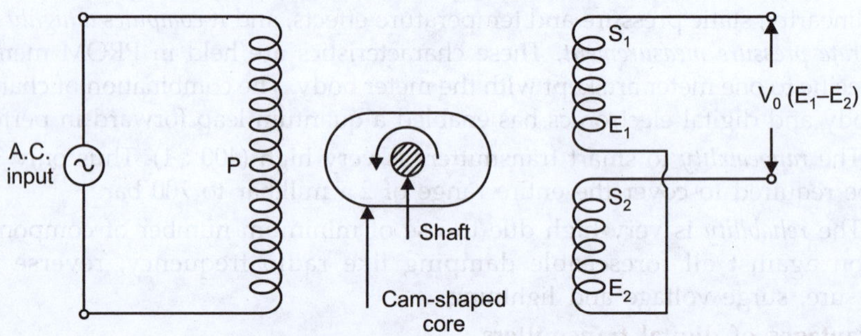

Fig. 16.73. Rotary variable differential transformer–Constructional details

— The core is rotated in clockwise and anticlockwise directions with respect to *"Null"* position. When the core is at *"Null"* position the voltage is *zero* because the voltages induced in both the secondary windings S_1 and S_2 are *equal and in opposite directions.*

— When any angular displacement to be measured (w.r.t. Null position) is applied to the shaft, differential voltage gets developed across the secondary windings at the output. When the core is rotated in clockwise direction, the voltage gets increased. If the core is rotated in anticlockwise direction, the voltage increases. The output voltage is increasing in both the directions, but there is opposition with respect to each other according to the position of shaft (clockwise and anticlockwise).

16.25 SYNCHROS AND RESOLVERS

Basically, *"synchros and resolvers" are a class of A.C. electromechanical variable-coupling transformer devices primarily employed for angular data transmission and computing systems.*

Synchro :

— The basic electromagnetic structure of a synchro consists of a *wound rotor* and a *wound stator,* concentrically arranged to give adjustable mutual coupling between the windings of the two members.

— The rotor is wound with a *single-phase* concentrated winding and the stator is wound with *three-phase distributed windings connected in a Y-configuration.*

— The rotor is excited with an A.C. signal, and by the transformer action between the rotor and stator windings, voltages are induced in the stator coils. *The magnitudes of these voltages depend upon the angular position of the rotor with respect to the stator.*

Fig. 16.74, shows a typical practical set-up in which a pair of synchros are used – one acts as a transmitter and the other as a receiver.

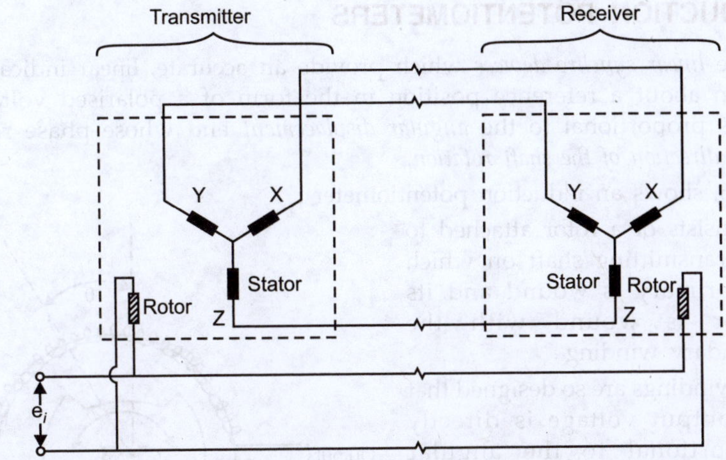

Fig. 16.74. Practical set-up of a pair of synchros.

Resolver :

— An inductive resolver is essentially a variable transformer with a *rotary electromagnetic coupling between the primary and secondary windings.*

— In a typical angle-positioning resolver, two windings are provided at *right angles to one another* for both stator and rotor. The windings are located in such a manner that output voltage has an amplitude *proportional to the sine or cosine of the input shaft position of the rotor with respect to the stator*. Depending upon the designer's particular application, either the rotor or stator may be used as the primary.

Fig. 16.75, shows the electrical connections of a resolver.

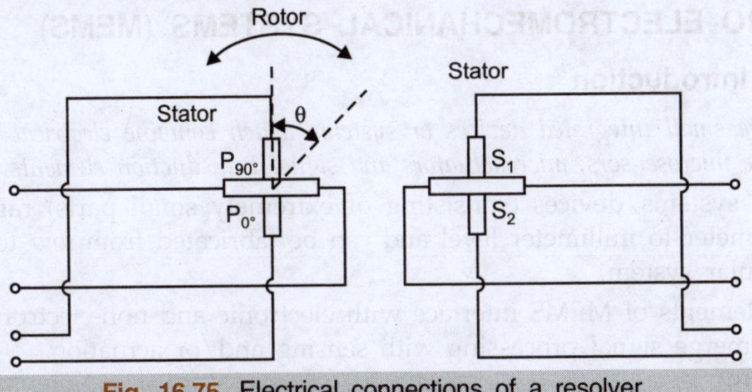

Fig. 16.75. Electrical connections of a resolver.

Advantages of synchros and resolvers :

The main advantages of synchros and resolvers are :

1. High reliability.
2. High accuracy (0.01 percent feasible).
3. Infinite resolution.
4. Useful operating angle of 360° and capable of continuous rotations.
5. No rotational wear, except at slip rings.
6. Relatively insensitive characteristics to stray cable capacitances.
7. The operational speeds of the system are quite high.

16.26 INDUCTION POTENTIOMETERS

These are *linear synchro devices* which provide an accurate, linear indication of the shaft rotation about a reference position in the form of a polarised voltage whose magnitude is proportional to the *angular displacement* and whose phase relationship indicates the *direction of the shaft rotation.*

Fig. 16.76, shows an induction potentiometer :

— It consists of a rotor attached to the transmitting shaft on which the primary is wound and its stator is wound with the secondary winding.

— The windings are so designed that the output voltage is directly proportional to the angular position of the rotor.

— Good linearity is feasible for angular displacements in the range of ± 45° only, even though the shaft rotation is continuous.

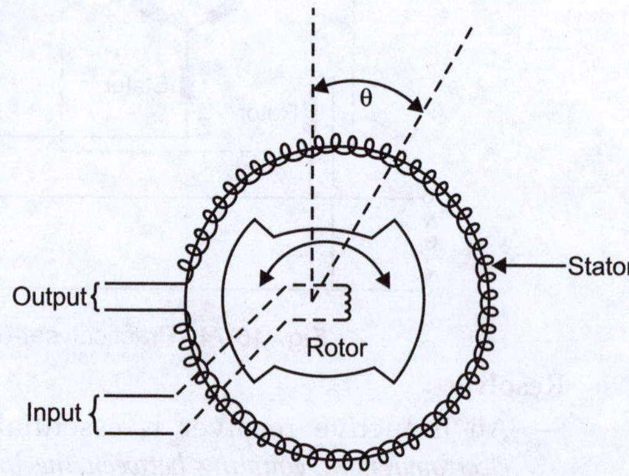

Fig. 16.76. Induction potentiometer.

Some of the *errors* that can arise are due to : (*i*) Mechanical asymmetry, (*ii*) Variable output impedance, and (*iii*) Thermal effects.

These devices are analogous to resistance potentiometers but since they are induction-type components, they have less restraining forces acting upon the rotors and hence are *capable of providing higher resolution.*

16.27 MICRO ELECTROMECHANICAL SYSTEMS (MEMS)

16.27.1. Introduction

MEMS *are small integrated devices or systems which combine electrical and mechanical components like microsensors, microactuators and signal transduction elements.*

— These systems/devices (consisting of extremely small parts) range from sub-micrometer to millimeter level and can be fabricated from few to millions in a particular system.

— The elements of MEMS interface with electronic and non-electronic signals and often merge signal processing with sensing and/or actuation.

- *"MEMS" are used to sense, control and activate mechanical processes on the microscale and function individually or in arrays.*

16.27.2. Advantages of MEMS

MEMS claim the following advantages over conventional electromechanical systems :

1. Better stability and higher accuracy in the performance.
2. Can be produced on mass scale (once the fabrication design and process is finalised).
3. As these devices are available in very small sizes, they can be easily inserted with other applications (*Miniaturization*).
4. Compact integration of comprehensive functionality.

5. Depending upon the requirements, the MEMS can be designed for various signals (*e.g.*, mechanical, optical, electrical, biological etc.)

16.27.3. Applications of MEMS

The various fields of application of MEMS are :

1. Consumer products.
2. Process and manufacturing industries.
3. Aerospace industry.
4. Automobile engineering.
5. Telecommunication engineering.
6. Medical field.

16.27.4. Description of MEMS

Fig. 16.77 shows the general block diagram of MEMS :

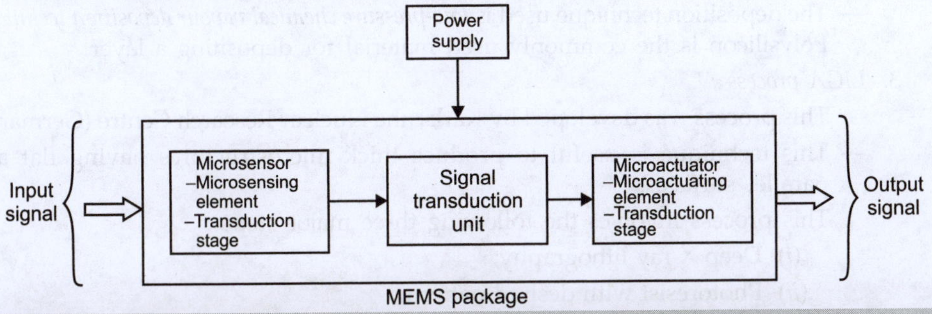

Fig. 16.77. General block diagram of MEMS.

1. ***Input signals :***

- These signals are received from various sources/systems such as mechanical, electrical, optical, chemical etc.
- They are admitted/supplied to the MEMS package.

2. ***MEMS package :***

- It is semiconductor chip (IC), which can be fabricated by suitable micromachining technique.
- It consists of the following components :

(*i*) *Microsensors :*

— Microsensing element;

— Signal processing hardware.

These sensors are used to *sense* different physical parameters such as pressure, temperature, force, humidity, sound etc.

(*ii*) *Microactuators :*

— Microactuating unit;

— Transduction stage.

The microactuator is an element that *converts electrical signals into physical signals. Example:* Microswitches, microvalves, micropumps, microgrippers, etc.

3. ***Output signals :***

- These are the signals interfaced with the system to be controlled or monitored.

Owing to monolithic integration of sensors, actuators and signal transduction unit, closed loop control system can be developed.

16.27.5. Manufacturing of MEMS

The technique employed for manufacturing MEMS is called micromachining or micromanufacturing.

The various techniques of micromachining are :

1. *Bulk micromachining :*
   - — It is used for production of microsensors and accelerometers.
   - — In this method, isotropic and anisotropic etching techniques are used.
2. *Surface micromachining :*
   - — In this method, MEMS are produced by depositing, patterning and etching a sequence of thin films, typical 1 to 100 μm thick.
   - — The deposition technique used is *low-pressure chemical vapour deposition technique.* Polysilicon is the commonly used material for depositing a layer.
3. *LIGA process :*
   - — This process was developed by Karlsruhe Nuclear Research Centre (Germany).
   - — This technique is useful to produce thick microstructures having flat and parallel surfaces.

     This process involves the following three major steps :

     (*i*) Deep X-ray lithography;

     (*ii*) Photoresist with desired pattern;

     (*iii*) Electroplating.
   - — It is widely used to produce *microgears* and *microturbines.*

16.27.6. MEMS Accelerometer—Vibration Microsensor

The most common method of observing the conditions of a rotating machine is the vibration analysis. *"Vibration" is measured by its frequency and amplitude.* The most common type of vibration sensor is **accelerometer**. The accelerometers are designed for the purpose of measurement of shock and vibrations.

- *"Piezoelectric-type accelerometer"* has the highest frequency response and high output voltage, making it convenient for measurement of shock and vibration.
  - — The most suitable piezoelectric material for fabrication of MEMS accelerometer is PZT (lead zirconate titanate oxide).
  - — The piezoelectric sensor is a self-generating device which is ideal for sensing dynamic events such as shocks and vibrations.

    Following are the advantages of PZT sensor :

    (*i*) Fast response;

    (*ii*) Better stability;

    (*iii*) High frequency bandwidth;

    (*iv*) High charge sensitivity (upto 110 pF/N)

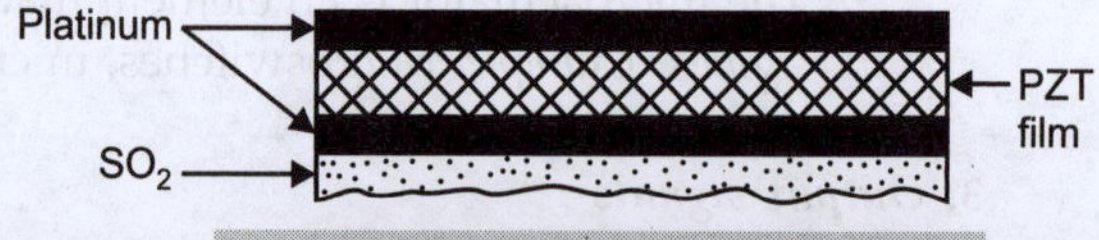

Fig. 16.78. Construction of MEMS accelerometer.

  - — In its *construction* (Fig. 16.78), a membrane on silicon wafer with SO_2 layer is fabricated by bulk micromachining. The film of PZT is sandwiched between platinum electrodes by deposition technique.

- When the accelerometer is connected to the vibrating surface, according to the working principle, the *PZT film experiences a force and an electric charge is developed across it. This charge variation can be converted into voltage variation at the output by charge amplifier.*

Applications. The MEMS accelerators is suitable for Z measuring and analysing the vibration levels at rotating machines, pumps, compressors and turbines; these measurements may be useful for controlling the vibrations to check noise and damage of rotating machines.

16.27.7. MEMS Humidity Microsensor

Fig. 16.79 shows the construction of a MEMS humidity microsensor. For such a sensor barium titanate ($BaTiO_3$) is used because it undergoes strong changes in its dielectric constant when exposed to humid atmosphere.

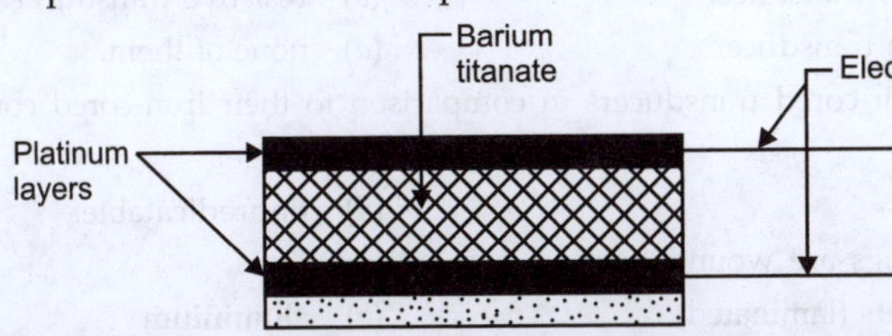

Fig. 16.79. Construction of a MEMS humidity microsensor.

$BaTiO_3$ (a synthetically produced piezoelectric material-ferroelectric ceramic) has high permittivity and with induced polarization it is conveniently used in sensors. It advantages are : Good mechanical and thermal stability, low leakage and good frequency response, large voltage sensitivity and large dielectric constant (2000–5000 appr.). A nanometric $BaTiO_3$ film can be fabricated on a silicon (SiO_2/Si) substrate.

The capacitance of the capacitor formed by placing $BaTiO_3$ between silver-palladium and nickel electrodes layer is given by :

$$C = \frac{K \epsilon_0 A}{d} \quad ...(16.70)$$

where, K = Dielectric constant,

ϵ_0 = Permittivity of free space = 8.85×10^{-12} F/m,

A = Surface area, and

d = Distance between the electrode layers.

When assembly is exposed to *humidity, the capacitance C changes due to change in the value of K. This change in C is calibrated into voltage using charge amplifiers.*

HIGHLIGHTS

1. The technology of using instruments to measure and control the physical and chemical properties of materials is called *instrumentation*.
2. Modes of measurements are :

   (*i*) Primary measurements (*ii*) Secondary measurements

   (*iii*) Tertiary measurements.

3. A transducer is a device which converts the energy from one from to another.
4. Transducers may be classified as follows :
   A. (*i*) Active transducers (*ii*) Passive transducers
   B. (*i*) Variable-resistance type (*ii*) Digital transducers.
   C. (*i*) Variable-resistance type (*ii*) Variable-inductance type
   (*iii*) Variable-capacitance type (*iv*) Voltage-generating type
   (*v*) Voltage-divider type.

OBJECTIVE TYPE QUESTIONS

Choose the Correct Answer :

1. LVDT is a
   (*a*) capacitive transducer (*b*) resistive transducer
   (*c*) inductive transducer (*d*) none of them.
2. The size of air-cored transducers in comparison to their iron-cored counter parts is
   (*a*) smaller (*b*) bigger
   (*c*) same (*d*) unpredicatable.
3. LVDT windings are wound on
   (*a*) steel sheets (laminated) (*b*) aluminium
   (*c*) ferrite (*d*) copper.
4. Piezoelectric crystals are used for measurement of .... changes.
   (*a*) static (*b*) dynamic
   (*c*) static and dynamic (*d*) any of these.
5. Piezoelectric crystals produce an e.m.f.
   (*a*) when external mechanical force is applied
   (*b*) when external magnetic field is applied
   (*c*) when radiant energy stimulates the crystal
   (*d*) when the junction of two such crystals is heated.
6. Piezoelectric crystals are used for the measurement of
   (*a*) temperature (*b*) velocity
   (*c*) sound (*d*) none of the above.
7. In the given circuit, how much the voltmeter will read ?
   (*a*) 0 V (*b*) 10 V
   (*c*) 4 V (*d*) 5 V
   (*e*) 3.33 V.

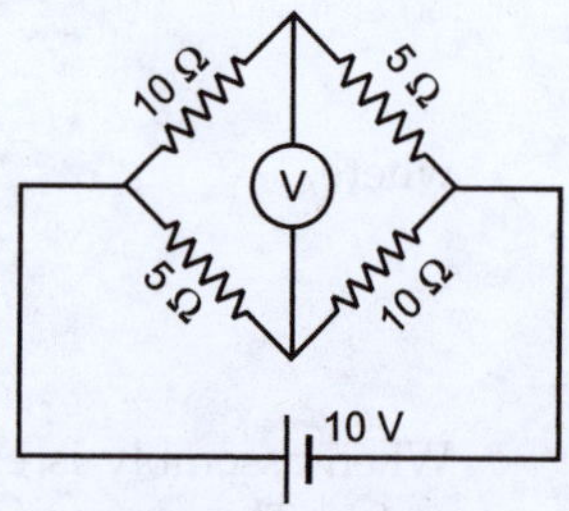

Fig. 16.80

8. Capacitive transducers operate upon the principle (*s*) of
   (*a*) variation of over-lapping area of plates
   (*b*) variation of separation of plates
   (*c*) variation of relative permittivity of dielectric material between two plates
   (*d*) all of the above.
9. Capacitive transducers are normally employed for .... measurements.
   (*a*) static (*b*) dynamic
   (*c*) both static and dynamic (*d*) transient.

10. The thermo-electric effect was first observed by

(a) Seebeck (b) Thomas Young
(c) Pirani (d) Thermus.

11. Thermocouples are .... transducers.

(a) active (b) passive
(c) adhesive (d) both (a) and (c).

12. Nitro-cellulose cement is used in strain gauges as

(a) carrier (b) base
(c) adhesive (d) lead.

13. A resistance thermometer is basically a/an

(a) active transducer (b) passive transducer
(c) potentiometer (d) none of these.

14. Platinum resistance thermometer can be used upto

(a) 200°C (b) 850°C
(c) 1200°C (d) 1500°C.

15. Which of the following should be incorporated in the RTD to make a temperature sensitive bridge most sensitive to temperature ?

(a) Platinum (b) Nickel
(c) Thermistor (d) Copper.

16. Bourdon tubes have the advantages of

(a) high accuracy and good dynamic response
(b) high sensitivity and good repeatability
(c) not being prone to shock vibrations
(d) not being susceptible to hysteresis.

17. A transducer is basically a device which converts

(a) mechanical energy into electrical
(b) energy or information from one form to another
(c) mechanical displacement into electrical
(d) none of these.

18. The gauge factor of a strain gauge is given as

(a) $G = \dfrac{\Delta R/R}{\Delta l/l}$ (b) $G = \dfrac{\Delta l/l}{\Delta R/R}$

(c) $G = \dfrac{\Delta R/R}{\Delta D/D}$ (d) none of these.

19. The gauge factor G and the Poisson's ratio μ are related as

(a) $g = 1 + \mu$ (b) $G = \mu$

(c) $G = 2 + \mu$ (d) $G = \dfrac{1+\mu}{2}$.

20. Thermocouples are generally used for accurate temperature measurement upto

(a) 350°C (b) 550°C
(c) 1400°C (d) 3500°C.

21. For surface temperature measurement one can use

(a) strain gauges (b) diaphragm
(c) RTD (d) thermocouple.

22. LVDT can be used for

(a) vibration measurement (b) angular velocity measurement
(c) force measurement in beam (d) load measurement on a column.

23. The principles of operation of LVDT is based on variation of
(*a*) self inductance (*b*) mutual inductance
(*c*) reluctance (*d*) permeance.

24. An LVDT has an output in the form of
(*a*) linear displacement of core (*b*) pulse
(*c*) rotary movement of core (*d*) none of the above.

25. Hall effect transducers have the drawbacks of
(*a*) high-sensitivity to temperature variation
(*b*) variation of Hall's coefficient from plate to plate
(*c*) poor resolution
(*d*) both (*a*) and (*b*).

26. A Hall's effect pick-up can be used for measuring
(*a*) pressure (*b*) magnetic flux
(*c*) relative humidity (*d*) current.

27. Self generating transducers are .... transducers.
(*a*) active (*b*) passive
(*c*) secondary (*d*) inverse.

28. The transducer that converts the input signal into the output signal, which is a continuous function of time, is known as .... transducer.
(*a*) active (*b*) passive
(*c*) analog (*d*) digital.

29. A transducer that converts measurand into the form of pulse is called the .... transducer.
(*a*) active (*b*) analog
(*c*) digital (*d*) pulse.

30. Certain types of materials generate an electrostatic charge or voltage when mechanical force is applied across them. Such materials are called the
(*a*) Piezo-electric (*b*) photo-electric
(*c*) thermo-electric (*d*) none of these.

31. Piezoelectric transdcuers are .... transducers.
(*a*) active (*b*) passive
(*c*) inverse (*d*) both (*a*) and (*c*).

32. Piezoelectric transducers work when we apply to it
(*a*) heat (*b*) mechanical force
(*c*) vibrations (*d*) illumination.

33. In semiconductor strain gauges, the change in resistance on application of strain is mainly on account of change in
(*a*) length of wire (*b*) diameter of wire
(*c*) resistivity (*d*) both (*a*) and (*b*).

34. The drawbacks of semiconductor strain gauges are
(*a*) low fatigue life (*b*) poor linearity
(*c*) that they are expensive and brittle (*d*) none of these.

35. In a resistance potentiometer, the non-linearity .... the ratio of potentiometer to load resistance
(*a*) decreases with the increase in (*b*) increases with the increase in
(*c*) is independent of (*d*) none of the above.

36. High value pot resistance leads to
(*a*) low sensitivity (*b*) high sensitivity
(*c*) low non-linearity (*d*) less error.

37. A strain gauge is a passive transducer and is employed for converting
(*a*) mechanical displacement into a change of resistance
(*b*) pressure into a change of resistance
(*c*) force into displacement
(*d*) none of the above.

38. Semiconductor strain gauges have
(*a*) high gauge factor (–100 to 150) (*b*) very small size (0.7 to 7 mm)
(*c*) higher fatigue value (*d*) all of these.

39. The strain gauges should have low
(*a*) gauge factor (*b*) resistance
(*c*) resistance temperature coefficient (*d*) all of the above.

40. The carrier material employed with strain gauges at room temperature is
(*a*) impregnated paper (*b*) bakelite
(*c*) aluminium foil (*d*) epoxy.

41. Seismic transducer is used for measurement of
(*a*) linear velocity (*b*) angular velocity
(*c*) acceleration (*d*) pressure.

42. The accelerometer using LVDT has the advantage of
(*a*) high natural frequency (*b*) being contactless device
(*c*) better resolution (*d*) all of these.

43. Bonded strain gauges are
(*a*) exclusively used for construction of transducers
(*b*) exclusively used for stress analysis
(*c*) used for both stress analysis and for construction of transducers
(*d*) none of the above.

44. A load cell is essentially a
(*a*) strain gauge (*b*) thermistor
(*c*) resistive potentiometer (*d*) inductive transducer.

45. A load cell is an electro-mechanical device and is widely used for measurement of
(*a*) static forces (*b*) dynamic forces
(*c*) temperature (*d*) both (*a*) and (*b*).

46. Which of the following can be used for pressure measurement ?
(*a*) Thermometer (*b*) Pyrometer
(*c*) Bolometer (*d*) Piezoelectric crystal.

47. Radiation pyrometers are used for measuring temperature in the range of
(*a*) 500 – 1000°C (*b*) 1000–2000°C
(*c*) 1200 – 3500°C (*d*) above 4000°C.

48. The best method of measuring the temperature of hot bodies radiating energy in the visible spectrum is
(*a*) thermocouple (*b*) optical pyrometer
(*c*) RTD (*d*) thermistor.

49. Pirani gauge is used for measuring .... pressure.
(*a*) very high (*b*) high
(*c*) very low (*d*) atmospheric.

50. Pirani gauge are used for measurement of pressure ranging from
(a) 10^{-4} to 1 torr (b) 1 to 10 torrs
(c) 10 to 100 torrs (d) above 100 torrs.

51. The ionization vacuum gauge, in construction, is similar to a
(a) vacuum diode (b) vacuum triode
(c) thyratron (d) none of these.

52. The device used for measuring temperatures exceeding 1500°C is
(a) radiation pyrometer (b) RTD
(c) thermocouple (d) bimetallic thermometer.

53. The most suitable device for measuring temperature of a furnace is
(a) RTD (b) thermistor
(c) optical pyrometer (d) bimetallic thermometer.

54. Which of the following devices cannot be used for measurement of temperature ?
(a) RTD (b) Thermocouple
(c) LVDT (d) Pyrometer.

55. Which of the following is not the drawback of radiation pyrometers ?
(a) Their initial as well as installation costs are high
(b) Poor precision and slow response
(c) They need maintenance
(d) Each pyrometer needs individual calibration.

56. Pyrometer is used to measure
(a) strain (b) pressure
(c) displacement (d) temperature.

57. The device used for measuring low pressure, of the order of 10^{-2} torr, is
(a) strain gauge (b) Pirani gauge
(c) ionization gauge (d) any of these.

58. Moving-coil pick-up is used for measuring
(a) linear velocity (b) vibrations
(c) displacement (d) pressure.

59. Electronic counters are used for measuring
(a) linear velocity (b) angular velocity
(c) acceleration (d) pressure.

60. Angular velocity is measured by
(a) strain gauge (b) solar cell
(c) A.C. tacho-generator (d) none of the above.

61. A Wheatstone bridge circuit using strain gauges can be used for measuring
(a) static strains (b) dynamic strains
(c) both (a) and (b) (d) none of these.

62. Dummy strain gauges are used for
(a) calibration of strain gauges (b) compensation of temperature variations
(c) increasing bridge sensitivity (d) all of the above.

63. In measurements using two strain gauges, the second strain gauge is provided for
(a) temperature compensation (b) increasing sensitivity
(c) stability (d) linearity
(e) both (a) and (b).

64. Which of the following additional devices is required for measuring pressure with an LVDT ?

(*a*) Bellows (*b*) Bourdon tubes

(*c*) Bolometer (*d*) Rotameter

(*e*) either (*a*) or (*b*).

65. Which of the following devices cannot be used for measurement of pressure ?

(*a*) LVDT (*b*) RTD

(*c*) Piezoelectric transducers (*d*) Piezoresistive transducers.

66. The transducers used for measurement of linear displacement are

(*a*) strain gauges and resistive potentiometers

(*b*) LVDTs, capacitive transducers and Hall effect transducers

(*c*) thermocouples, thermistors and RTDs

(*d*) both (*a*) and (*b*).

67. Rotational displacement can be measured by

(*a*) strain gauges (*b*) resistive potentiometers

(*c*) reluctance transducers (*d*) both (*b*) and (*c*).

68. Temperature compensation, in bridge circuit arrangement, is affected by

(*a*) using dummy strain gauges

(*b*) using strain gauges of smaller gauge factor

(*c*) reversing strain gauges

(*d*) any of these.

69. Which one of the following devices cannot be used to measure pressure ?

(*a*) Strain gauge (*b*) LVDT

(*c*) Piezoelectric crystal (*d*) Pyrometer.

70. Which of the following additional is required for measuring pressure with piezoelectric crystal ?

(*a*) Bellows (*b*) Strain gauges

(*c*) Rotameter (*d*) RTD.

ANSWERS

1. (*c*)	**2.** (*b*)	**3.** (*c*)	**4.** (*b*)	**5.** (*a*)	**6.** (*a*)	**7.** (*e*)
8. (*d*)	**9.** (*c*)	**10.** (*a*)	**11.** (*a*)	**12.** (*e*)	**13.** (*b*)	**14.** (*b*)
15. (*c*)	**16.** (*b*)	**17.** (*b*)	**18.** (*a*)	**19.** (*c*)	**20.** (*c*)	**21.** (*d*)
22. (*c*)	**23.** (*b*)	**24.** (*a*)	**25.** (*d*)	**26.** (*d*)	**27.** (*a*)	**28.** (*c*)
29. (*c*)	**30.** (*a*)	**31.** (*d*)	**32.** (*b*)	**33.** (*c*)	**34.** (*c*)	**35.** (*b*)
36. (*b*)	**37.** (*a*)	**38.** (*d*)	**39.** (*c*)	**40.** (*a*)	**41.** (*c*)	**42.** (*d*)
43. (*c*)	**44.** (*a*)	**45.** (*d*)	**46.** (*d*)	**47.** (*c*)	**48.** (*b*)	**49.** (*c*)
50. (*a*)	**51.** (*b*)	**52.** (*a*)	**53.** (*c*)	**54.** (*c*)	**55.** (*b*)	**56.** (*d*)
57. (*b*)	**58.** (*a*)	**59.** (*b*)	**60.** (*c*)	**61.** (*c*)	**62.** (*b*)	**63.** (*e*)
64. (*e*)	**65.** (*b*)	**66.** (*d*)	**67.** (*d*)	**68.** (*a*)	**69.** (*d*)	**70.** (*a*).

THEORETICAL QUESTIONS

1. Define the term "instrumentation".
2. List the various modes of measurement.
3. Enumerate the elements of a measurement system.
4. What is transducer ?
5. What are the functions of a transducer in an electronic instrumentation system ?
6. How are transducers classified ?
7. What are the advantages of electromechanical transducers ?
8. Explain briefly with diagrams important transducer actuating mechanisms.
9. Describe briefly the following :
   (*i*) Thermistors and resistance thermometers.
   (*ii*) Wire resistance strain gauges.
10. Give the classification of variable inductance transducers.
11. Explain briefly any two of the following transducers :
    (*i*) Self-generating variable inductance transducer – Electromagnetic type
    (*ii*) Variable reluctance transducer
    (*iii*) Mutual inductance transducer
    (*iv*) Linear-variable-differential transformer (LVDT).
12. What is the principle on which a capacitive transducer works ?
13. What are the advantages and disadvantages of capacitive transducers ?
14. Give the applications of capacitive transducers.
15. What is a piezoelectric transducer ? List the advantages and disadvantages of piezoelectric transducers.
16. How are photoelectric transducers classified ?
17. Explain briefly the following :
    (*i*) Photoemissive cell (*ii*) Photoconductive cell
    (*iii*) Photovoltaic cell.
18. What is a strain gauge ?
19. Explain briefly with neat diagrams, any two of the following :
    (*i*) Wire-wound strain gauges (*ii*) Foil-type strain gauges
    (*iii*) Semiconductor strain gauges (*iv*) Capacitive strain gauges.
20. What is a load cell ?
21. Explain briefly any two of the following :
    (*i*) Hydraulic load cell.
    (*ii*) Pneumatic load cell.
    (*iii*) Strain-gauge load cell.
22. Describe briefly the following sensors :
    (*i*) Eddy current proximity sensor.
    (*ii*) Capacitance proximity sensor.
23. Write a short note on "Pneumatic sensor".

UNSOLVED EXAMPLES

1. A linear resistance potentiometer is 50 mm long and is uniformly wound with a wire having a resistance of 10000 Ω. Under normal conditions, the slider is at the centre of the potentiometer.

(*i*) Find the linear displacements when the resistances of the potentiometer are measured by a Wheatstone bridge for two cases are : (*a*) 3800 ohms and (*b*) 7500 ohms.

(*ii*) If it is possible to measure a minimum value of 12 ohms resistance with the above arrangement, find the resolution of the potentiometer in mm.

[**Ans.** (*i*) 6 mm, 12.5 mm; (*ii*) 0.06 mm]

2. In a linear voltage differential transformer the output voltage is 2.0 V at maximum displacement. At a certain load, the deviation from linearity is maximum and it is ± 0.004 V from a straight line through origin, find the linearity at the given load.

[**Ans.** ±2%]

3. The output of a LVDT is connected to a 5V voltmeter through an amplifier whose amplification factor is 250. An output of 2 mV appears across the terminals of LVDT when the core moves through a distance of 0.5 mm. If the multimeter has 100 divisions and the scale can be read to a $\frac{1}{5}$ of a division. Calculate :

(*i*) The sensitivity of LVDT, and

(*ii*) The resolution of the instrument in mm, [**Ans.** (*i*) 4 m V/mm, (*ii*) 0.01 mm]

4. A parallel plate capacitive transducer uses plates of area 250 mm^2 which are separated by a distance 0.2 mm.

(*i*) Calculate the value of capacitance when the dielectric is air having a permittivity of 8.85×10^{-12}F/m.

(*ii*) Calculate the change in capacitance if a linear displacement reduces the distance between the plates to 0.18 mm. Also calculate the ratio of per unit change of capacitance to per unit change of displacement.

(*iii*) If a mica sheet 0.01 mm thick is inserted in the gap, calculate the value of original capacitance and change in capacitance for the same displacement. Also calculate the ratio of per unit change of capacitance to per unit change in displacement. The dielectric constant of mica is 8.

[**Ans.** (*i*) 11.06 pF; (*ii*) 1.23 pF, 1.11; (*iii*) 11.57 pF, 1.35 pF, 1.167]

5. A capacitive transducer uses two quartz diaphragms of area 675 mm^2 separated by a distance of 3.8 mm. A pressure of 850 kN/m^2 when applied to the top diaphragm produces a deflection of 0.55 mm. The capacitance is 330 pF when no pressure is applied to the diaphragms. Determine the value of capacitance after the application of a pressure of 850 kN/m^2.

[**Ans.** 385.8 kN/m^2]

6. A capacitive transducer, used in pressure measuring instrument has a spacing of 4.2 mm between its diaphragms. A pressure of 600 kN/m^2 produces an average deflection of 0.28 mm of the diaphragm of the transducer. A transducer which has a capacitance of 250 pF before the application of pressure is connected in an oscillation circuit having a frequency of 120 kHz. Determine the change in frequency of oscillator after the application of pressure to the transducer.

[**Ans.** 4.1 kHz approx]

7. A 2 mm thick quartz piezoelectric crystal having a voltage intensity of 0.055 Vm/N is subjected to a pressure of 1.8 MN/m^2. Calculate the voltage output and charge density of the crystal. Take the permittivity of quartz as 40.6×10^{-12} F/m.

[**Ans.** 198 V, 2.23 pC/N]

8. A piezoelectric material measuring 5 mm × 5 mm × 1.5 mm is used to measure a force. Its voltage sensitivity is 0.055 Vm/N. Calculate the force if voltage developed is 110V.

[**Ans.** 33N]

9. The following data relate to a barium titrate pick-up :

Dimensions : 5 mm × 5 mm × 1.25 mm; Force acting on the pick-up = 5N. The charge sensitivity of the crystal = 150 pC/N; Permittivity = 12.5×10^{-9} F/m; Modulus of elasticity = 12×10^6 N/m^2. Calculate strain, charge and capacitance.

[**Ans.** 0.0167 : 750 pC; 250 pF]

10. A strain of 5 micro-strain is caused in a structural member when subjected to a compressive force. Two separate strain gauges are attached to the structural member, one is nickel wire strain gauge (gauge factor = –12.1) and other is nichrome wire strain gauge (gauge factor = 2). If the resistance of strain gauges before being strained is 130 Ω, calculate the change in the value of resistance of the gauges after they are strained.

   [**Ans.** 7.865 mΩ (increase); 1.3 mΩ (decrease)]

11. A strain gauge is bonded to a beam which is 10 cm long and has a cross-sectional area of 4 cm^2. The unstrained resistance and gauge factor of the strain gauge are 220Ω and 2.2 respectively. On the application of load the resistance of the gauge changes by 0.013Ω. If the modulus of elasticity for steel is 207 GN/m^2, calculate : (*i*) the change in length of the steel beam, and (*ii*) the amount of force applied to the beam.

   [**Ans.** (*i*) 2.68×10^{-6} m; (*ii*) 2.219 kN]

12. A single strain gauge having resistance 144 Ω is mounted on a steel cantilever beam at a distance of 0.15 m from the free end. The beam dimensions are 25 cm (length) × 2.0 cm (width) × 0.3 cm (depth). An unknown force applied at the free end produces a deflection of 12.7 mm of the end. If the change in gauge resistance is found to be 0.1824Ω, calculate the gauge factor. Take Young's modulus for steel as 200 GN/m^2. [**Ans.** 2.3]

CHAPTER

17 Signal Conditioning

17.1 Introduction-General measurement system components-signal conditioning and its necessity-Processes adopted in signal conditioning-mechanical amplification and electrical signal conditioning **17.2 Functions of signal conditioning equipment; 17.3 Amplification 17.4 Types of amplifiers; 17.5 Mechanical amplifiers; 17.6 Fluid amplifiers; 17.7 Optical amplifiers; 17.8 Electrical and electronic amplifiers**-Desirable characteristics of electronic amplifiers-electronic amplification of gain-A.C. and D.C. amplifiers-modulated and unmodulated signals-integrated circuit (ICs)-Operational amplifiers (Op-amp)-attenuators-filters-input circuitry; *Highlights–Objective Type Questions – Theoretical Questions*

17.1 INTRODUCTION

17.1.1. General Measurement System Components

Fig. 17.1 shows the general measurement system components :

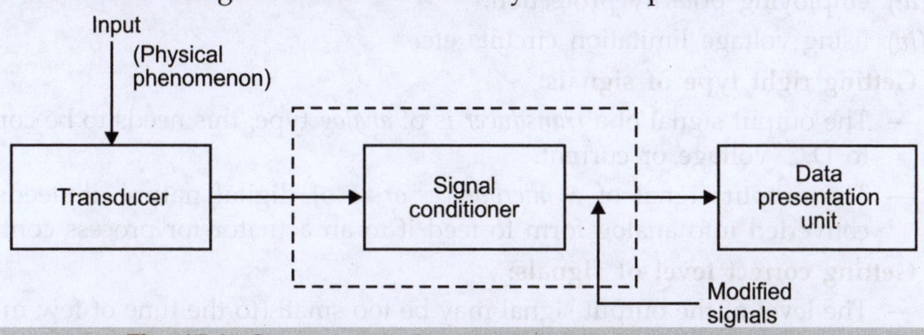

Fig. 17.1. Components of a general measurement system.

— The *"first stage" of the instrumentation or measurement system which detects the measurand (which is basically a physical quantity) is termed as* **detector-transducer stage.** In this stage, in most of the cases, the quantity is detected and is transduced into an electrical form.

— The output from the first stage *needs certain modifications* before it becomes compatible with the data presentation stage. The necessary modification is carried out in the *"intermediate stage"*, more commonly referred to as the **signal conditioning stage**.

— The *"last stage"* of the measurement system may consist of indicating, recording, displaying, data processing elements or may consist of control elements.

• *Measurement of dynamic mechanical quantities* places special requirements on the elements in the signal conditioning stage.

Large amplifications, as well good transient response, are often desired, both of which are difficult to obtain by mechanical hydraulic, or pneumatic methods. Consequently, *electrical or electronic elements are usually required.*

17.1.2. Signal Conditioning and its Necessity

Signal conditioning *may be defined as the process of modifying the output signals from the transducer into usable and satisfactory signal using amplification, attenuation, non-linearisation, linearisation or multiplication by another function.*

The necessity of signal conditioning may be due to following *reasons* :

1. Signals may be too noisy due to electromagnetic interference.
2. Signals may be too small, usually is mV range.
3. Signals may be non-linear and require to be converted into digital form.
4. Signals may be analog one and require to be converted into digital form.
5. Signals may be digital one and need to be converted into analog signals.
6. It may be required to improve the quality of digital signals.

17.1.3. Processes Adopted in Signal Conditioning

Following processes are usually adopted in signal conditioning :

1. **Protection.** The range of the output signals from the transducer may be so high that it may damage the next unit or element which needs to be protected.

   Example: If a high voltage/current signals are fed to the microprocessor, it will get damaged. The microprocessor can be protected by :

   (*i*) employing a series of current limiting resistors, fuses to break if current is too high;

   (*ii*) using a step down transformer if the voltage is too high;

   (*iii*) employing polarity protection;

   (*iv*) using voltage limitation circuits etc.

2. **Getting right type of signals:**

   — The output signal of a *transducer* is of *analog* type, this needs to be converted to D.C. voltage or current.

   — The output signal of a *microprocessor* is of digital nature, it needs to be converted into analog form to feed it to an actuator for process controlling.

3. **Getting correct level of signals:**

   — The level of the output signal may be too small (to the tune of few mV), this needs to be amplified for feeding it into an analog-to-digital converter. It may also be difficult to measure such low level signals.

   — For amplification operational amplifiers (Op-amp) are widely used.

4. **Elimination of interferences :**

   — Some undesired signals or disturbances (say noise disturbance due to electromagnetic interference) may be associated with the output signals, these need to be eliminated.

   — Such interferences are eliminated by the use of ***filters***.

5. **Manipulation of signals:** The output signals may be non-linear in nature, these need to be linearised and vice versa.

17.1.4. Mechanical Amplification and Electrical Signal Conditioning

Limitations/disadvantages of mechanical amplification:

In the field of dynamic measurements, strictly, mechanical systems are much more uncommon than they were in the years past, largely because of several inherent *disadvantages.* Mechanical amplification by the elements such as linkages, gearing, or cams (these elements present design problems of immense magnitudes particularly if dynamic inputs are to be handled) is quite limited because of the following *reasons :*

(*i*) When amplification is required *frictional forces are also amplified, resulting in considerable undesirable signal loading.* These effects, coupled with backlash and elastic deformations, result in *poor response.*

(*ii*) Initial loading results in reduced frequency response and in certain cases, depending on the particular configuration of the system, *phase response is also a problem.*

Advantages of electrical signal conditioning:

In several detector-transducer combinations which provide an output in electrical form, it is convenient to perform further signal conditioning electrically.

- Such conditioning may typically *include:*
  — Converting resistance changes to voltage changes;
  — Subtracting offset voltages;
  — Increasing signal voltages;
  — Removing unwanted frequency components.
- Electrical methods are also preferred for their ease of *power amplification.*
  — *Additional power may be fed into the system to provide a greater output power than input by the users of "power amplifiers"*, which have no important mechanical counterpart in most instrumentation. (It is true that hydraulic and pneumatic systems may be set up to increase signal power; however, their use is limited to relatively slow-acting control applications, primarily in the fields of chemical processing and electric power generation). This technology is of a particular value when recording procedures employ stylus-type recorders, mirror galvanometers, or magnetic-disc methods.

17.2 FUNCTIONS OF SIGNAL CONDITIONING EQUIPMENT

The signal conditioning equipment may be required to perform the following *functions on the transduced signal:*

1. Amplification
2. Modification or modulation
3. Impedance matching
4. Data processing
5. Data transmission.

1. **Amplification.** It means *enhancement of the signal level* which is often in the low level range. The amplification system must bring the level of transducer signal to a value adequate enough to make it useful for *conversion, processing, indicating* and *recording.*
2. **Modification or modulation.** It means to *change the form of signal.* The signal may be *smoothened, linearised, filtered or converted into digital form.*
3. **Impedance matching.** The signal conditioning equipment arranges the input, and output impedances of the matching device so as to prevent loading of the transducer and to maintain a high signal level at the recorder.
4. **Data processing.** To carry out mathematical operations (*e.g.*, addition, subtraction, differentiation, integration etc.) before indication or recording of data.
5. **Data transmission.** To transmit signal from one location to another without changing the contents of the information.

The whole task of signal conditioning requires the following:

(*i*) Ingenuity;

(*ii*) Proper selection of components;

(*iii*) Faithful reproduction of signal.

- The elements of signal conditioning are designed in such a fashion as to be *insensitive to all extraneous inputs. The accuracy, range and dynamic response are all designed to be compatible with the detector transducer.*
- In several situations the "*signal conditioning*" or "*data acquisition equipment*" is an *excitation* and *amplification system for passive transducers.* It may be an *amplification system for active transducers.* In both the applications, the transducer output is brought upto adequate level to make it useful for *conversion, processing, indicating* and *recording.*
- In case of "*passive transducers*" (*e.g.,* strain gauges, potentiometer resistance thermometers, inductive and capacitive transducers) excitation is needed because these transducers do not generate their own voltage or current; the excitation is provided from external sources.
- The "*active transducers*" (*e.g.,* technogenerators, thermocouples, inductive pick-ups and piezoelectric crystals) do not require excitation from an external source since they produce their own electrical output. However, these signals have a low voltage level and as such they need to be amplified.

The excitation sources may be:

- D.C. voltage source.
- A.C. voltage source.

Figures 17.2 and 17.3 show D.C. and A.C. signal conditioning systems respectively.

D.C. signal conditioning system:

Refer to Fig. 17.2. The resistance transducers like strain gauges constitute one or more than one arm of a Wheatstone bridge which is excited by an isolated D.C. source. The bridge can be balanced by a potentiometer and can also be calibrated for unbalanced conditions.

Characteristics of a D.C. amplifier:

(*i*) It should have extremely good thermal and long term stability.

(*ii*) It may require balanced differential inputs giving high mode rejection ratio (CMRR); CMRR is a *measure of ratio of desired signal to undesired signal, this value should be as high as possible.*

Advantages:

(*i*) D.C. amplifier is easy to calibrate at low frequencies.

(*ii*) It is able to recover from an overload condition unlike its A.C. counterpart.

Disadvantages:

- The major disadvantage of a D.C. amplifier is that it suffers from the *problem of drift.* As a result, the *low frequency spurious signals come out as data information.* This problem is overcome by the *use of the drift amplifiers.*
- The D.C. amplifier is followed by a lowpass filter which eliminates high frequency components or noise from the data signal.

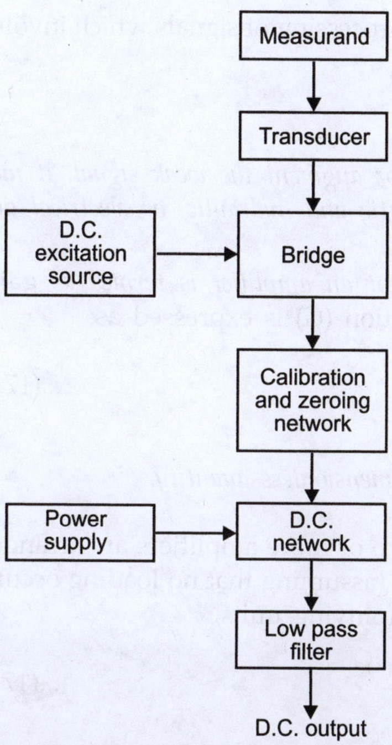

Fig. 17.2. D.C. signal conditioning system.

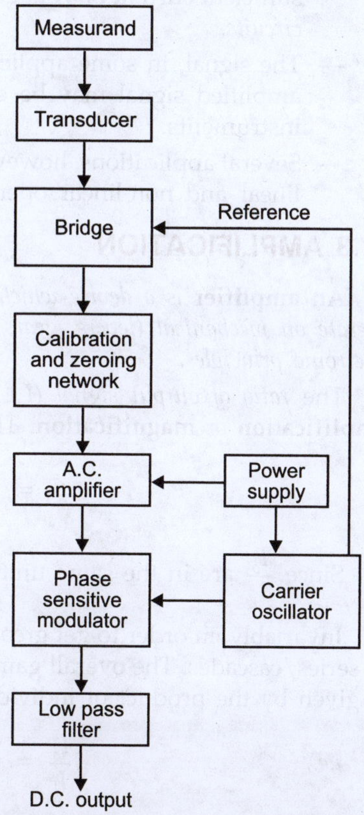

Fig. 17.3. A.C. signal conditioning system.

Uses. D.C. systems are generally used for common resistance transducers such as *potentiometers and resistance strain gauges.*

A.C. signal conditioning system:

Refer to Fig. 17.3. The problems which are encountered in D.C. systems are overcome through *carrier type A.C. signal conditioning system.*

The transducer parameter variations amplitude modulate the carrier frequencies at the bridge output and the waveform is *amplified and demodulated.* The demodulation is phase sensitive so that polarity of D.C. output indicates the direction of the parameter change in the bridge output.

In carrier systems, it is *very easy to obtain very high rejection of mains frequency pick-up.*

— Active filters can be used to reject this frequency and prevent overloading of A.C. amplifier.

— The carrier frequency components of the data signal are filtered out by the phase-sensitive demodulators.

Uses. A.C. systems are used for variable reactance transducers and for systems where signals have to be transmitted long via cables to connect the transducers to the signal conditioning equipment.

— The physical quantities like pressure, temperature, acceleration, strain etc. after having being *transduced* into their analogous electrical form and *amplified* to sufficient current or voltage levels (say 1 to 10 V) are further *processed by electronic circuits.*

— The signal, in some applications, does not need any further processing and the amplified signal may be directly applied to indicating or recording or control instruments.

— Several applications, however, involve further processing of signals which involve linear and non-linear operations.

17.3 AMPLIFICATION

An **amplifier** *is a device which is used to increase or augment the weak signal. It may operate on mechanical (levers, gears etc.) optical, pneumatic and hydraulic, or electrical and electronic principles.*

The *ratio of output signal* (I_o) *to input signal* (I_i) *for an amplifier is termed as* **gain, amplification** *or* **magnification**. The gain of amplification (*G*) is expressed as:

$$G = \frac{I_o}{I_i} \qquad ...(17.1)$$

Since $\frac{I_o}{I_i}$ are in the same units, the gain *G* is a *dimensionless quantity.*

Invariably, in order to get greater magnification, two or more amplifiers are arranged in series/cascades. The overall gain of the arrangement (assuming that no loading occurs) is given by the product of individual gains of the amplifying units,

i.e.,
$$\frac{I_o}{I_i} = G_1 \cdot G_2 \cdot G_3 \cdots\cdots \qquad ...(17.2)$$

17.4 TYPES OF AMPLIFIERS

The amplifiers, on the basis of principle of working, may be *categorised* as follows:

1. Mechanical amplifiers.
2. Fluid amplifiers.
3. Optical amplifiers.
4. Electrical and electronic amplifiers.

17.5 MECHANICAL AMPLIFIERS

The mechanical amplifiers may be further *classified* as follows:

(i) *Simple and compound levers :* The compound lever has two or more levers linked together so that output from one lever provides the input to the other.

Example. The *Huggenberger extensometer* is one of the most popular and *accurate mechanical amplifier*. It uses a system of compound levers to give *very high magnification* to the order of *2000 or even more.*

(ii) *Simple and compound gears :* The simple and compound gear trains are used quite frequently *to provide mechanical amplification of either angular displacement or rotary speed.*

A *"compound gear train"* gives *greater modification with the additional advantage of no change in the direction of input signal.*

Examples. The gear trains are used for the magnification of displacement in the *Bourdon tube pressure gauge* and in the *dial-test indicator* where linear movement is translated into rotation by means of rack and pinion.

Limitations of mechanical amplification:

The mechanical amplification, as earlier stated, usually suffers from the errors caused by the following *factors.*

(*i*) Internal loading;

(*ii*) Friction at the mating parts;

(*iii*) Elastic deformation;

(*iv*) Backlash.

17.6 FLUID AMPLIFIERS

Fluid amplifiers may be *classified* as follows:

(*i*) ***Hydraulic amplifier :*** When a small displacement is applied to a piston operating inside a cylinder containing some liquid, there occurs a large displacement of the liquid in the output tube which has a small diameter.

Example. This principle is employed in the *mercury-in-glass thermometer* and the *single-column manometers.*

(*ii*) ***Pneumatic amplifier :*** Pneumatic methods are extensively used and can be applied to any type of measurement.

17.7 OPTICAL AMPLIFIERS

In optical amplification, a ray of light strikes a mirror with an angle of incidence i and gets reflected with angle of reflection equal to the angle of incidence. When the mirror rotates through an angle θ, the angle of incidence change to $(i + \theta)$. Before rotation of the mirror, the angle between the incident ray and reflected ray is $2i$ and after rotation it is $2(i + \theta)$. Obviously there is angular magnification of 2θ between the incident and reflected rays. In order to get a greater magnification, more number of mirrors surfaces may be used.

Examples. This principle to amplify the input signals is used in the following cases.

— Optical levers;

— U.V. galvanometers;

— Mechanical-pointer galvanometers.

17.8 ELECTRICAL AND ELECTRONIC AMPLIFIERS

The electrical amplifiers are used to *increase the magnitude of weak voltage or current signals resulting from electromechanical transducers.*

17.8.1. Desirable Characteristics of Electronic Amplifiers

The following are the desirable characteristics of electronic amplifiers:

(*i*) *High input impedance* so that its loading effect on the transducer in minimum.

(*ii*) *Low output impedance* so that the amplifier is *not unduly loaded by the display or recording device.*

(*iii*) *Frequency response* should be as *good as that of the transducer.*

17.8.2. Electronic Amplification of Gain

- The following are the several *generalities* that can be listed for the ideal (but non-existent) electronic amplifier:

(*i*) Infinite gain (lower gain can be obtained by adding attenuation circuits).

(*ii*) Infinite input impedance; no input current, hence no load on the previous stage or device.

(*iii*) Zero output impedance (low noise).

(*iv*) Instant response (wide frequency bandwidth).

(*v*) Zero output for zero input.

(*vi*) Ability to ignore or reject, extraneous inputs.

Of course, none of these aims can be completely realized, it is often possible to approach them, and their assumption simplifies circuit analysis.

- In an electronic amplifier, separate power is provided so that the output power may exceed the input if that is required.

Here, if v_i = Input voltage,

i_i = Input current,

v_o = Output voltage, and

i_o = Output current,

Then :
$$\text{Gain} = \frac{\text{Power output}}{\text{Power input}} = \frac{v_o i_o}{v_i i_i} \qquad ...(17.3)$$

$$\text{Voltage amplification} = \frac{\text{Voltage output}}{\text{Voltage input}} = \frac{v_o}{v_i} \qquad ...(17.4)$$

$$\text{Current amplification} = \frac{\text{Current output}}{\text{Current input}} = \frac{i_o}{i_i} \qquad ...(17.5)$$

- Another way of expressing *power gain* is through the use of *decibel*.

The common logarithm (log to the base 10) of power gain is known as *bel power gain.*

$$\text{Power gain} = \log_{10}\left(\frac{P_o}{P_i}\right) \text{bel}$$

$$1 \text{ bel} = 10 \, dB$$

$$\therefore \quad \text{Power gain} = 10 \log_{10}\left(\frac{P_o}{P_i}\right) \text{dB} \qquad ...(17.6)$$

If the two powers are developed in the same resistance or equal resistance, then

$$P_i = \frac{V_i^2}{R} = I_i^2 R$$

$$P_o = \frac{V_0^2}{R} = I_o^2 R$$

$$\therefore \quad \textit{Voltage gain} = 10 \log_{10} \frac{V_0^2/R}{V_i^2/R} = 20 \log_{10} \frac{V_o}{V_i} \text{ dB} \qquad ...(17.7)$$

$$\text{Current gain} = 10\log_{10}\frac{I_o^2 R}{I_i^2 R} = 20\log_{10}\frac{I_o}{I_i}\,dB \qquad ...(17.8)$$

Example 17.1. *A three-stage amplifier has a first voltage gain of 100, second stage voltage gain of 200 and third stage voltage gain of 400. Find the total voltage gain in dB.*

Solution. First-stage voltage gain in *dB*

$$= 20\log_{10} 100 = 20 \times 2 = 40$$

Second-stage voltage gain in *dB*

$$= 20\log_{10} 200 = 20 \times 2.3 = 46$$

Third-stage voltage gain in *dB*

$$= 20\log_{10} 400 = 20 \times 2.6 = 52$$

Total voltage gain = 40 + 46 + 52 = **138 *dB* (Ans.)**

Example 17.2. *(i) A multistage amplifier employs five stages each of which has a power gain of 30. What is the total gain of the amplifier in dB?*

(ii) If a negative feedback of 10 dB is employed, find the resultant gain.

Solution. Absolute gain of each stage = 30

Number of stages = 5

(*i*) *Power gain of one stage* = $10\log_{10} 30$ dB = 14.77

∴ Total *power gain* = 5 × 14.77 = **73.85 dB (Ans.)**

(*ii*) *Resultant power gain with negative feedback*

= 73.85 – 10 = **63.85 dB (Ans.)**

17.8.3. A.C. and D.C. Amplifiers

The instrumentation systems usually employ the following two types of electronic amplifiers.

(*i*) A.C. coupled amplifiers.

(*ii*) D.C. coupled amplifiers.

- For an *"A.C. amplifiers"* bandwidth is the range of frequencies between which gain or amplitude ratio is constant to within – 3dB (3dB down points). This corresponds to the frequencies at which the voltage output amplitude falls by 29.3% to 70.7% of the maximum value.
  - — The *"A.C. amplifiers"* are *only capable of dealing with rapid, repetitive signals but are usually simpler and cheaper* when compared with their D.C. counterparts.
  - — In an "A.C. amplifier system" the amplifier drift and spurious noise are *not significant;* the mains frequency pick-up *rejection is very high.*
- The *"D.C. amplifiers"* are capable of *amplifying static, slowly changing or rapid-repetitive input signals.*
  - — The "D.C. amplifier systems" are *easy to calibrate at low frequencies, and have the ability to recover rapidly from overload conditions.*

17.8.4. Modulated and Unmodulated Signals

The measurands may be *"pure"* in the sense that analog electrical signal contains *nothing more than the real time variation of the measurand information itself.*

On the other hand, the signal may be *"mixed"* with a **carrier** which consists of a *voltage oscillation at some frequency higher than that of the signal.* A common rule of thumb is that the *frequency ratio should be at least 10:1.*

The measurand affects the *carrier* by varying either its *amplitude* or its *frequency:*

— In the *former* case the *carrier frequency is held constant and its amplitude is varied by the measurand. This process is known as* **Amplitude modulation** (or **AM**).

— In the *latter* case the *carrier amplitude is held constant and its frequency is varied by the measurand.* This is known as **Frequency modulation** (or **FM**).

The most familiar use of AM and FM transfer of signals is in AM and FM *radio broadcasting.*

When *"modulation"* is used in instrumentation *"amplitude modulation"* (AM) is the *more common form.*

— Nearly any mechanical signal from a passive pick-up can be transduced into an analogous AM form. Sensors based on either *inductance* (*e.g.* differential transformer) or *capacitance* (*e.g.* capacitance pick-up for liquid level) require an A.C. excitation.

In addition, however, resistance-type sensors may also use an A.C. excitation, as with some strain gauge circuits.

It is required to extract signal information from the modulated carrier.

- This operation, *when AM is used,* may take several forms:
  - — The simplest is merely to display the entire signal using an *oscilloscope or oscillograph,* and then to *"read"* the result from the envelope of the carrier.
  - — More commonly, the *mixed signal and carrier are* ***"demodulated"*** by *"rectification* and *filtering".*
- FM demodulation is more complex operation and may be accomplished through the use of
  - — Frequency discrimination,
  - — Ratio detection, or
  - — IC phase-locked loops.

17.8.5. Integrated Circuits (ICs)

The integrated circuits (ICs), as the name implies, are groups of circuit elements combined to perform specific purposes. For the most part the elements consist of *transistors, diodes, resistors,* and, to lesser extent, *capacitors,* all connected and packaged in convenient plug-in units.

ICs from the building blocks are used to construct more complex circuits such as :

— Differential amplifiers;
— Mixers (for combining signals);
— Timers;
— Filters;
— Audio preamps;
— Auto-power amplifiers;
— Voltage references;
— Regulators and comparators;
— Several digital devices.

17.8.6. Operational Amplifiers (Op-amp)

An ***operational amplifier (Op-amp)*** *is a linear integrated circuit (IC) that has a very high voltage gain, a high input impedance and a low output impedance.*

It is so called because it can be employed to carry out many different mathematical operations like *"addition"*. *"subtraction"*, *"multiplication"*, *"division"*, *"integration"*, *"differentiation"* etc.

- Operational amplifiers are linear integrated circuits that work on *relatively low supply voltage.*
- They are *reliable and inexpensive.*
- An *deal operational amplifier* is device of infinite voltage gain, infinite bandwidth, infinite input impedance (open) and zero output impedance.
- An Op-amp may contain two dozen transistors, a dozen resistors and one or two capacitors.

  Examples : μA 709, LM 108-LM 208, CA 741 CT and CA741T.

17.8.6.1. Specifications/Characteristics of an Op-amp

While selecting an Op-amp, the following characteristics need to be considered:

1. *Input offset voltage.* It is the voltage that must be applied at the input terminals to make the output voltage zero (This is about 2 mV for a 741 amplifier). The offset voltage changes with temperature.
2. *Input offset current.* It is defined as the net difference in current that must be applied at the input terminals to make the output voltage zero (This is 20 nA for a 741 amplifier).
3. *Input check current.* It is the mean of the two input currents to make the voltage zero.
4. *Slew rate.* It is the maximum rate at which the output can change. It is expressed as volts/microseconds.
5. *Unity gain frequency.* This is the frequency at which the open loop gain of the amplifier becomes unity.
6. *Common mode rejection ratio (CMRR).* It is the ratio of desirable signals to undesirable signals.

- An Op-amp is the *basic building block* for:

— Amplifiers
— Integrators
— Summers
— Differentiators
— Comparators
— A/D and D/A converters
— Active filters
— Sample and hold amplifiers.

17.8.6.2. Op-amp description

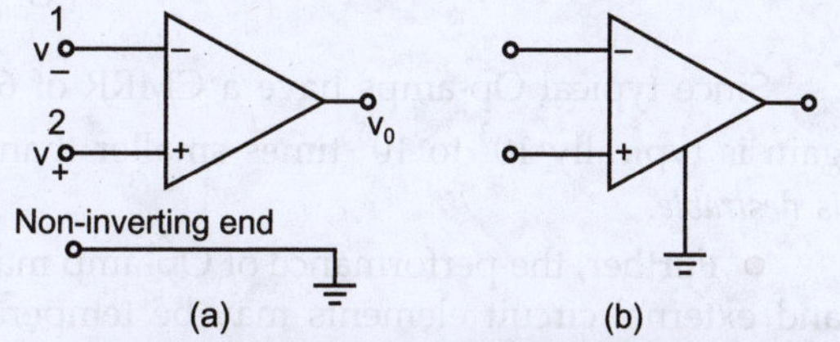

Fig. 17.4. Op-amp symbol.

Fig. 17.4(*a*) shows a standard symbol (a triangle having two input labelled differently and a single output) for an Op-amp, the one shown in Fig. 17.4(*b*) is also oftenly used.

One input terminal is designated by *–ve* sign, it is called *inverting end* while other input terminal is designated by a *+ve* sign, it is called *non-inverting end.*

These plus (+) and minus (–) polarities indicate *phase reversal* only. It does not mean that v_1 and v_2 [Fig. 17.4(*a*)] are negative and positive respectively. Additionally, it also *does not imply* that a positive input voltage has to be connected to the plus-marked non-inverting terminal 2 and negative input voltage to negative-marked inverting terminal 1. *In fact amplifier can be used either way up so to speak.* It may also be noted that all input and output voltages are referred to a common reference usually the ground shown in Fig. 17.4(*a*).

Operational amplifier operating with –ve feedback possesses stable closed loop gain and also insensitivity to variation of supply voltage and ambient temperature.

- The voltage at the output terminal v_o, is the product of the amplifier gain G, and the voltage difference :

$$v_o = G(v_+ - v_-) \qquad ...(17.9)$$

The output voltage is roughly limited to the power supply voltages V_{cc} and V_{ee}, as the voltage difference increases; if the voltage difference becomes too large, the output *saturates* near one of these values and remains constant.

- *The differential characteristic of op-amp has great importance in instrumentation because it eliminates offset voltages and noise signals common to both input terminals.* For example, nearby power lines may induce 50-cycle noise in the exterior circuitry leading to the amplifier. Such line noise is often present in identical form at both input terminals. This behaviour is known as *common-mode rejection.* If, instead, an op-amp receives the output of a voltage-sensitive Wheatstone bridge, the common offset voltages of the two voltage dividers are cancelled, and only the desired difference voltage is applied.

Limitations of Op-amp : Most of the Op-amps have a nonideal characteristic according to which they *do not completely satisfy the differential amplifying property.* With *both inputs grounded residual output voltage remains.*

— The multitude of transistors, resistors, and other elements within the Op-amp are *never perfectly matched,* so the amp output actually reaches zero at some small *non-zero* input voltage. To accommodate this input offset voltage, the common Op-amp is provided with pins marked "offset null" or "balance" which provides a means for adjusting the unwanted offset voltage towards zero.

Another limitation is that the actual common-mode rejection is finite. If the two input signals each include a common-mode voltage v_{cm}, the Op-amp's actual voltage will be,

$$v_o = G(v_+ - v_-) + G_{cm}\, v_{cm} \qquad ...(17.10)$$

The finite common-mode rejection is characterised by the "*common-mode rejection ratio* (CMRR)" in decibels:

$$\text{CMRR} = 20 \log_{10}\left(\frac{G}{G_{cm}}\right) \text{dB} \qquad ...(17.11)$$

Since typical Op-amps have a CMRR of 60 to 120 dB, therefore, the common-mode gain is typically 10^3 to 10^6 times smaller than the differential gain, hence a *high CMRR is desirable.*

- Further, the performance of Op-amp may be limited by thermal drift. Both internal and external circuit elements may be temperature sensitive, and design of each circuit usually includes compensating features.

A wide variety of Op-amps are available, and their differences largely represent attempts to *improve:*

— Thermal stability;
— CMRR;
— Offset voltages;
— Frequency response .... These refinements, however, increase the cost.

17.8.6.3. Applications of Op-amp

Operational amplifiers may be used as the *basic components of* :

- Linear voltage amplifiers;
- Integrators and differentiators;
- Function generators;
- Impedance transformers;
- Differential amplifiers;
- Voltage comparators;
- Filters;
- Many other devices.

17.8.6.4. Op-amp circuits used in Instrumentation

Some of the commonly used Op-amp circuits are described below:

1. Inverter;
2. Adder;
3. Subtracter;
4. Multiplier and divider;
5. Integrator;
6. Differentiator;
7. Buffer amplifier;
8. Differential amplifier.

1. Inverter. Fig. 17.5, shows the circuit of an Op-amp used as *inverter*. The feedback resistance R_f is made equal to the resistance R_1, connected to the inverting end of the amplifier,

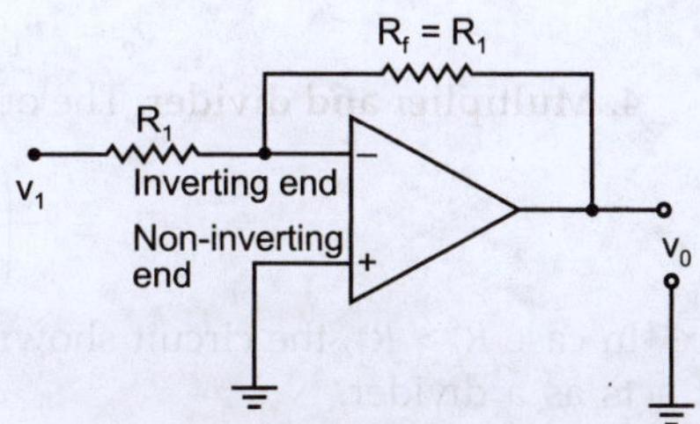

Fig. 17.5. Op-amp as an inverter.

$$\text{Output voltage, } v_o = -\frac{R_f}{R_1}v_1 = -v_1 \qquad \text{...(17.12)}$$

$$(\because R_f = R_1)$$

Obviously, the output voltage is 180° out of phase with the input voltage.

2. Adder. Fig. 17.6 shows an Op-amp circuit that performs the signals with amplification (if desired); using superposition theorem, we get

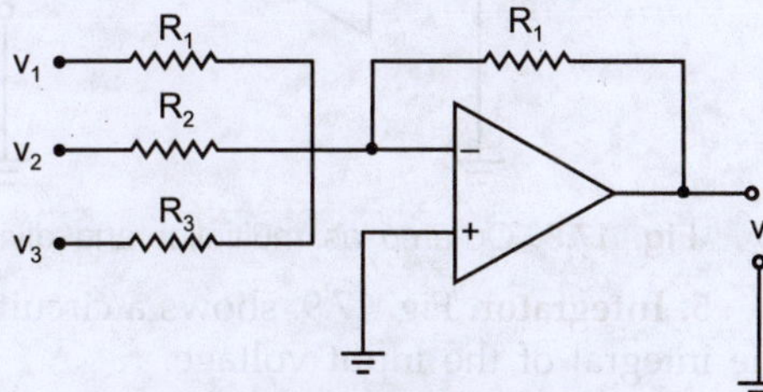

Fig. 17.6. Op-amp as an adder.

Output voltage,

$$v_o = -\left(\frac{R_f}{R_1}v_1 + \frac{R_f}{R_2}v_2 + \frac{R_f}{R_3}v_3\right) \qquad \text{...(17.13)}$$

If $R_1 = R_2 = R_3 = R_f$, then

$$v_o = -(v_1 + v_2 + v_3) \qquad \text{...(17.14)}$$

i.e., sum of the individual input voltages. The inversion that occurs cannot be avoided.

3. Subtracter. The Op-amp circuit used for subtraction of two input signals is shown in Fig. 17.7. The output of the 2nd Op-amp is given by :

$$v_o = -\left(-v_1\frac{R_f}{R_1}\cdot\frac{R_f}{R_3}+v_2\frac{R_f}{R_2}\right) \quad ...(17.15)$$

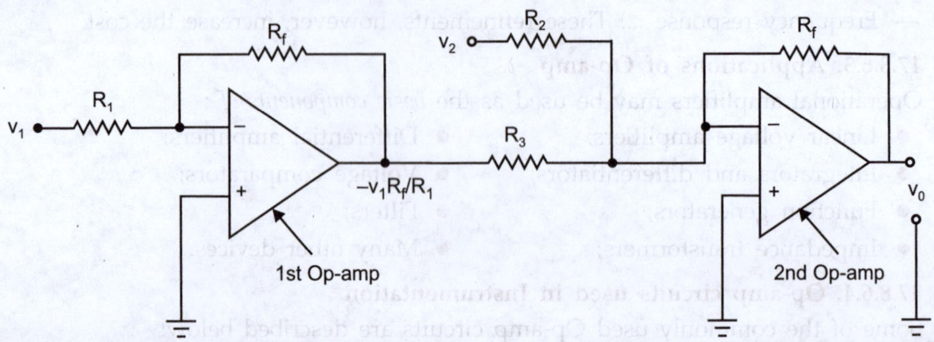

Fig. 17.7. Op-amp as a subtractor.

If $R_f = R_1 = R_2 = R_3$, the circuit acts as a *pure subtractor* and the output, in this case, is given by

$$v_o = v_1 - v_2 \quad ...(17.16)$$

4. Multiplier and divider. The output of an Op-amp in the inverting mode is given by

$$v_o = -\left(\frac{R_f}{R_1}\right)\times v_1$$

In case $R_f > R_1$, the circuit shown in Fig. 17.8 acts as a multiplier and in case $R_f < R_1$ it acts as a divider.

Thus, by choosing the values of R_f and R_1, the multiplier and the divider circuits can be designed for multiplication and division by any number.

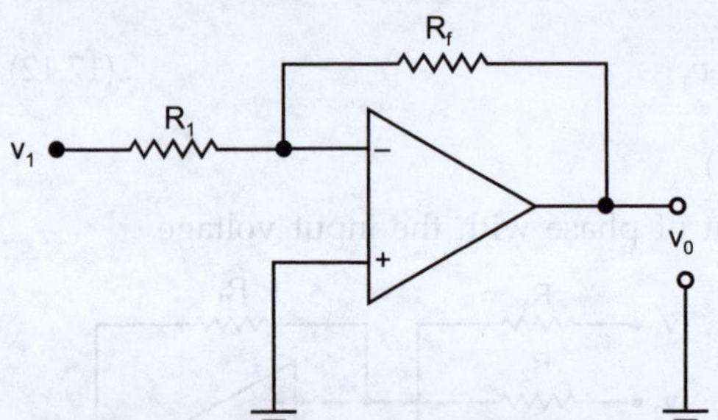

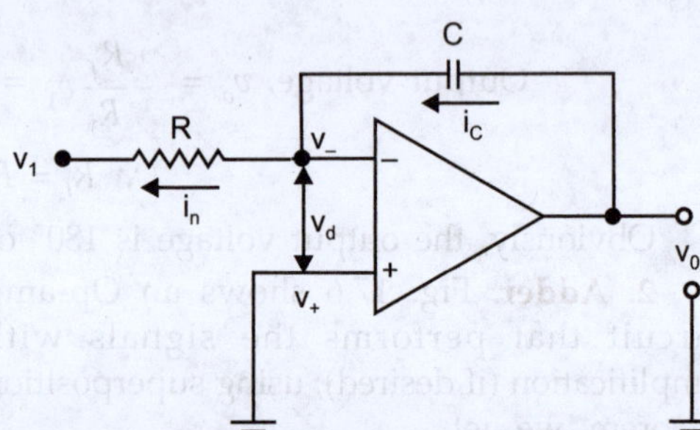

Fig. 17.8. Op-amp as multiplier and divider. **Fig. 17.9.** Op-amp as an integrator.

5. Integrator. Fig. 17.9. shows a circuit in which the output voltage is proportional to the integral of the input voltage.

In order to show that the circuit shown in Fig. 17.9. acts as an integrator using KCL (Kirchoff's Current Law) at node v_-, we have,

$$i_n = i_C$$

or,
$$\frac{v_- - v_1}{R} = C\frac{d}{dt}(v_o - v_-)$$

For *infinite differential gains*, $v_- = 0$

$$\therefore \quad -\frac{v_1}{R} = C\frac{d}{dt}v_o$$

By integration, we have

$$v_o = -\frac{1}{RC}\int v_1\,dt \qquad ...(17.17)$$

The convenient values of R and C are MΩ and μF range respectively.

6. Differentiator. The differential amplifier circuit is obtained by *interchanging* the positions of resistance R and capacitor C as shown in Fig. 17.10.

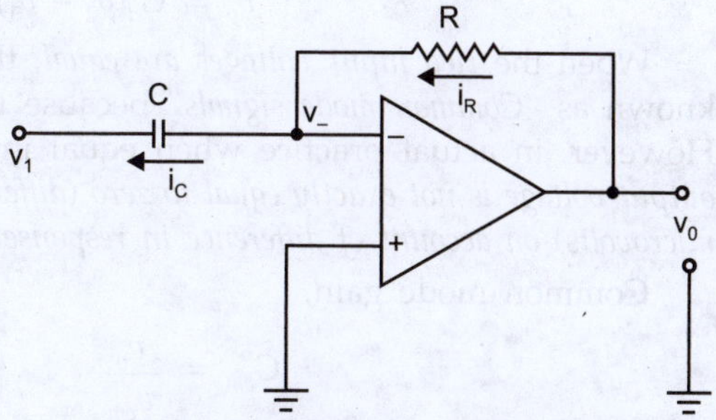

Fig. 17.10. Op-amp as differentiator.

At node v_-, we have :

$$i_C = i_R$$

$$C\frac{d}{dt}(v_- - v_1) = \frac{v_o - v_-}{R}$$

Now, $\quad v_- = 0$

$$\therefore \quad -C\frac{d}{dt}(v_1) = \frac{v_o}{R}$$

or, $\quad$ Output voltage, $v_o = -RC\frac{d}{dt}(v_1) \qquad ...(17.18)$

Thus, the output voltage is equal to the differentiated input voltage.

- The *Op-amps are normally used as differentiators as they tend to decrease the signal to noise (S/N) ratio.*

7. Buffer amplifier. The buffer amplifier is essentially an impedance transformer which converts a voltage at high impedance to the same voltage at low impedance. The circuit of a unity gain buffer amplifier also called a *"voltage follower"* is shown in Fig. 17.11.

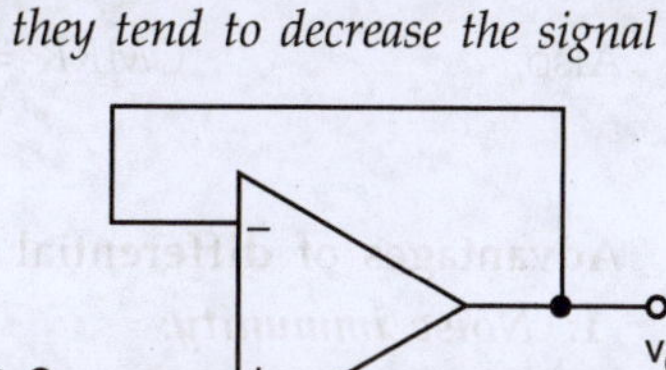

Fig. 17.11. Unity gain buffer amplifier voltage follower.

- *The use of unity gain buffer amplifiers greatly reduces the loading effects in measurements systems.*

8. Differential amplifier. A differential amplifier (an Op-amp) is of significant importance in an instrumentation system.

In its basic form it has two *inputs* and *outputs*. The signals available to the two outputs are identical except that the two are 180° out-of-phase with each other.

The output voltage of the amplifier is proportional to the difference between the two input voltages.

Fig. 17.12, shows an Op-amp used as a differential amplifier.

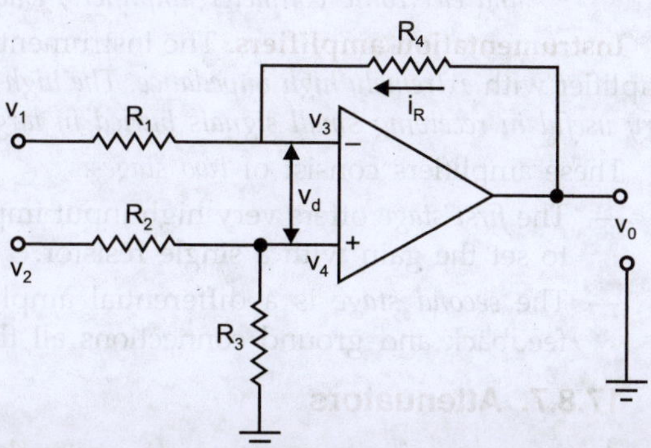

Fig. 17.12. Op-amp used as a differential amplifier.

Here, $v_o = G_d(v_+ - v_-)$

where, G_d = Differential gain.

The signal $v_d = (v_+ - v_-)$ is called *"Difference Mode Signal"* or simply *"Difference Signal"*

$$G_d = \frac{v_o}{v_+ - v_-} = \frac{v_o}{v_d} \quad ...(17.19)$$

It can be proved that,

$$v_o = G_d(v_2 - v_1) \quad ...(17.20)$$

When the *two input voltages are equal*, the output voltage is zero. Equal inputs are known as *"Common mode signals"* because the input signal is common to both inputs. However, in actual practice when equal input voltages are applied to the inputs, the *output voltage is not exactly equal to zero (difference is typically of the order of several hundred microvolts) on account of difference in response of the two inputs to common mode signals.*

Common mode gain,

$$G_{cm} = \frac{v_o}{v_{cm}} \quad ...(17.21)$$

where, G_{cm} = Common mode gain, and

v_{cm} = Common mode input signal.

The *"Common mode rejection ratio (CMRR)"* is defined as

$$CMRR = \frac{G_d}{G_{cm}} \quad ...(17.22)$$

Also,

$$CMRR = 20 \log_{10}\left(\frac{G_d}{G_{cm}}\right) \text{ dB} \quad ...(17.22a)$$

... when expressed in *dB*.

Advantages of differential amplifiers :

1. ***Noise immunity:***
   - These amplifiers are extensively used in equipment such as *electronic voltmeters and oscilloscopes.*
2. ***Drift immunity :***
   - The differential amplifier *has inherent capabilities of eliminating problem of drift.*
   - The differential amplifier construction is used for the *early stages of oscilloscope and electronic voltmeter amplifiers, where low drift is extremely important.*

Instrumentation amplifiers. The instrumentation amplifier is a dedicated differential amplifier with *extremely high impedance. The high common mode rejection makes this amplifier very useful in receiving small signals buried in large common-mode offsets and noise.*

These amplifiers consist of *two stages:*

— The *first stage* offers very high input impedance to both input signals and allows to set the gain with a single resistor.

— The *second stage* is a differential amplifier (unity gain) with output, negative feedback and ground connections all throughout.

17.8.7. Attenuators

An **attenuator** *is a two-port resistive network and is used to reduce the signal level by a given amount.*

In a number of applications, it is necessary to introduce a specified loss between the source and a matched load without altering the impedance relationship. Attenuators may be used for this purpose.

Attenuator may be *symmetrical* or *asymmetrical*, and can be either *fixed or variable. A fixed attenuator with constant attenuation* is called a **pad**.

- Variable attenuators are used as control volumes in radio broadcasting sections.
- Attenuators are also used in *laboratory to obtain small value of voltage or current for testing circuits.*

The attenuation is expressed in **decibels (dB)** or, in *naper*. The attenuation offered by a network in decibels is given by

$$\text{Attenuation in } dB = 10\log_{10}\left(\frac{P_i}{P_o}\right) \quad ...(17.23)$$

where, P_i is the input power and P_o is the output power.

The attenuators may be of the following types:

1. Resistance attenuator.
2. Symmetrical T-attenuator.
3. L-type attenuator.
4. π-type attenuator.

17.8.8. Filters

Filtering *is the process of attenuating unwanted components of a measurement while permitting the desired component to pass.*

The **filter** *is an electronic circuit which can pass or stop a particular band of frequencies through it.* The filters was first designed by G.A. Campbell and D.Z. Zobel at Bell laboratories.

The band of frequencies which will pass through filter is called the ***pass band*** and the band of all remaining frequencies is called ***attenuation band***. In case of **ideal filter**, all frequencies of pass band will pass without suffering from any attenuation while the band of all remaining frequencies of attenuation band will be suppressed completely.

Classification of filters:

The filters may be *classified* as follows:

A. On the basis of passing and attenuating of frequencies:

1. **Low pass filters :**
   - These are those filters which pass only low frequencies through them and which reject all high frequencies above the cut-off frequencies.
   - A low pass filter is also called *"lag network"* because it causes a phase lag in the output signal.
   - This type of filter is also called *"integrating network"*.
2. **High pass filters :**
   - These are those filters which pass only high frequencies through them and which reject all low frequencies below the cut-off frequency.
   - The high pass filter is a differentiating network and is also called as *"lead network"* because it will cause a phase lead in the output signal.
3. **Band pass filters :**
   - These are those filters which pass a band of frequencies through them and which reject all other frequencies to pass through them.

4. **Band stop filters :**
   - These filters, which are also known as *"band elimination filters"*, are those which reject a band of frequencies to pass through them and which allow the other frequencies to pass though them.

B. On the basis of relation between series and shunt impedances :

1. **Constant filters** (*or prototype filters*). In this filter the series impedance z_1 and stunt impedance z_2 are interrelated by the relation:

$$z_1 z_2 = k^2, \qquad ...(17.24)$$

where, k is a constant independent of frequency.

2. ***m*-derived filters.** These filters do not have the product of series and shunt impedances equal to k^2, but have the same characteristic impedance as the corresponding k section, with sharper attenuation characteristic.
   - Fig. 17.13(*a*) shows some terminology as applied to a low pass filter (Similar terms are applicable to the high pass and notch or band reject filters, respectively) while in the Fig. 17.13(*b*) are shown the band-pass filter characteristics.

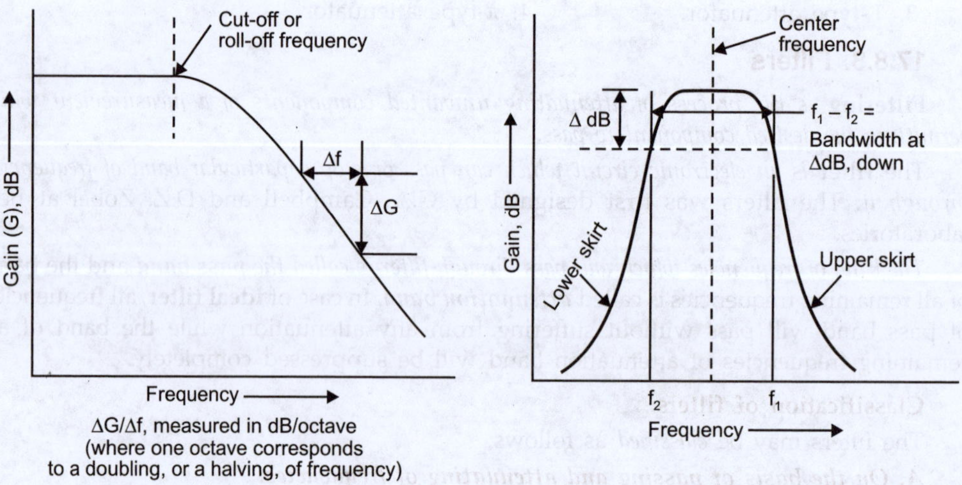

Fig. 17.13

- Fig. 17.14 shows the ideal characteristics of filters.

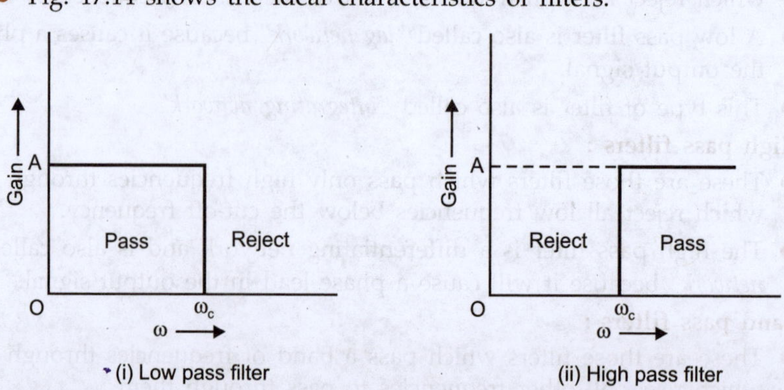

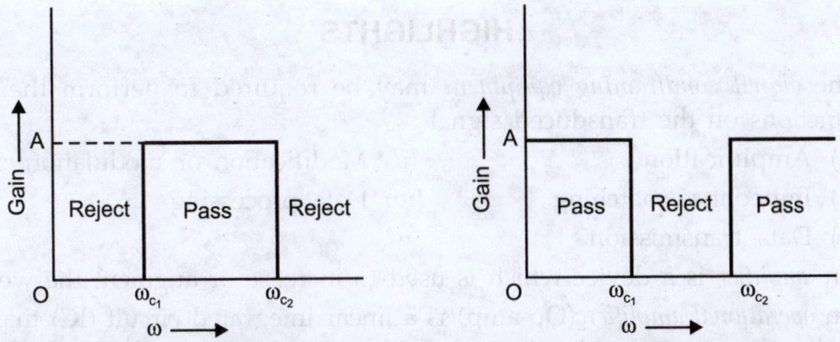

(iii) Band pass filter (iv) Band stop filter

Fig. 17.14. Ideal characteristics of filters.

17.8.9. Input Circuitry

Although not all-inclusive, the following types of input circuits are used for signal conditioning of electrical transducers :

1. Simple current-sensitive circuits.
2. Ballast circuits.
3. Voltage-dividing circuits.
4. Bridge circuits.
5. Resonant circuits.
6. Amplifier input circuits.

1. 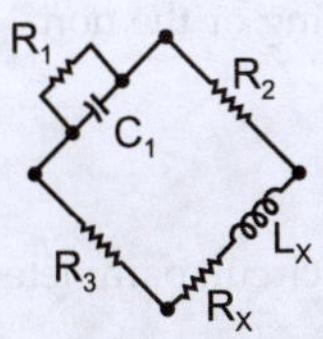 Maxwell circuit *Measures L (best for $\omega L_x/R_x < 10$)* *Balance equations :* $L_x = R_2 R_3 C_1$ $R_x = \frac{R_2 R_3}{R_1}$	2. 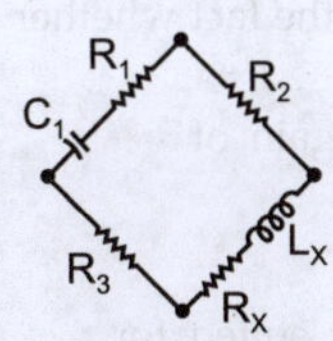 Hay circuit *Measures L (best of $\omega L_x/R_x > 10$)* *Balance equations :* $L_x = \frac{R_2 R_3 C_1}{1 + \omega^2 C_1^2 R_1^2}$ $R_x = \frac{\omega^2 C_1^2 R_1 R_2 R_3}{1 + \omega^2 C_1^2 R_1^2}$
3. 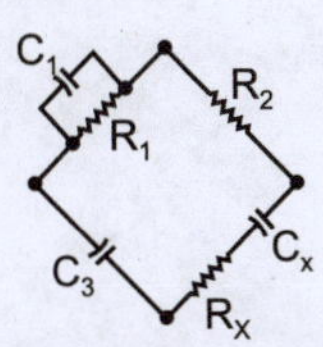 Schering circuit *Measures C* *Balance equations :* $C_x = C_3 \frac{R_1}{R_2}$ $R_x = R_2 \frac{C_1}{C_3}$	4. 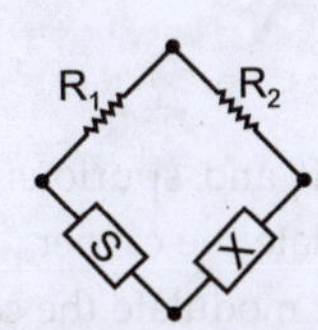 Comparison with series constants *Meausres L or C* *Balance equations :* $R_x = R_s \frac{R_1}{R_2}$ If inductive, $L_x = L_s \frac{R_2}{R_1}$ If capacitive, $C_x = C_s \frac{R_1}{R_2}$
5. 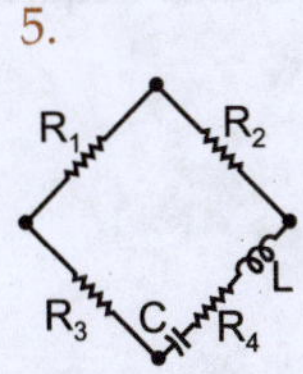 Resonant circuit *Measures L or C (f known), f (L and C known)* *Balance equations :* $X_L = X_C$ or $L.\ C = \frac{1}{\omega^2}$ $f = \frac{1}{2\pi\sqrt{LC}}$	6. 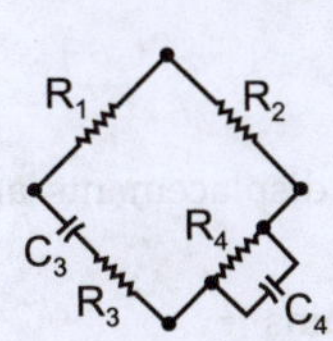 Wien or RC frequency bridge *Measures f* *Balance equations :* $f = \frac{1}{2\pi\sqrt{R_3 R_4 C_3 C_4}}$ $\frac{R_1}{R_2} = \frac{R_3}{R_4} + \frac{C_4}{C_3}$

Fig. 17.15. Impedance bridge arrangements.

HIGHLIGHTS

1. The *signal conditioning equipment* may be required to perform the following functions on the transduced signal :
   (*i*) Amplification (*ii*) Modification or modulation
   (*iii*) Impedance matching (*iv*) Data processing
   (*v*) Data transmission.
2. An *amplifier* is a device which is used to increase or augment the weak signal.
3. An *operational amplifier* (Op-amp) is a linear integrated circuit (IC) that has very high voltage gain, a high input impedance and a low output impedance.
4. *Filtering* is process of attenuating unwanted components of a measurement while permitting the desired component to pass.

OBJECTIVE TYPE QUESTIONS

Choose the Correct Answer :

1. The closed loop gain of an Op-amp is dependent upon whether the Op-amp is used
   (*a*) in inverting mode (*b*) in non-inverting mode
   (*c*) is independent of the fact whether the input is connected to inverting or non-inverting terminal.
   (*d*) is dependent upon the fact whether the input is connected to inverting or the non-inverting terminal.
2. A buffer amplifier has gain of
   (*a*) infinity (*b*) zero
   (*c*) unity (*d*) dependent upon the circuit parameters.
3. A.C. amplifiers are best suited for
   (*a*) steady-state signals (*b*) low frequency signals
   (*c*) rapidly varying signals (*d*) none of these.
4. The amplifier drift and spurious noise signals are not significant in
   (*a*) a.c. amplifiers (*b*) d.c. amplifiers
   (*c*) charge amplifiers (*d*) none of these.
5. In a carrier system, drift and spurious signals are important
   (*a*) because they modulate the carrier
   (*b*) because they do not modulate the carrier
   (*c*) because it is easier to achieve a stable carrier than a stabilized d.c. source.
   (*d*) none of the above.
6. When using d.c. signal conditioning system, with a carrier of 3 kHz, the data frequency should be limited to :
   (*a*) 1 kHz (*b*) 5 Hz
   (*c*) 600 Hz (*d*) 2 MHz.
7. The input and output displacements are of opposite phase in
   (*a*) simple lever (*b*) compound lever
   (*c*) compound gear trains (*d*) none of these.
8. What is the desirable feature in an electronic amplifier?
   (*a*) High output impedance (*b*) Low input impedance
   (*c*) good frequency response (*d*) All of these.

9. Charge amplifiers are used in order to amplify the output signals of
   (a) inductive (b) capacitive
   (c) resistive
   (d) piezoelectric and capacitive transducers.
10. Filters that transmit all frequencies below a defined cut-off frequency are known as
    (a) low-pass filters (b) high-pass filters
    (c) band-pass filters (d) any of these.
11. Excitation and amplification systems are needed
    (a) for active transducers only (b) for passive transducers only
    (c) for both active and passive transducers
    (d) for both passive and output transducers.
12. A d.c. amplifier
    (a) needs to have a balanced differential inputs with a high common mode rejection ratio (CMRR) to give very good thermal and long term stability.
    (b) easy to calibrate at low frequencies and has ability to recover from overload conditions.
    (c) is immune to drift and low frequency spurious signals come out as data information.
    (d) is followed by a low pass filter to eliminate high frequency components including noise from the data signal.
    (e) all of the above.
13. The output from frequency-modulation system is
    (a) a.c. voltage (b) d.c. voltage
    (c) a.c. and d.c. voltage (d) any of these.
14. The data transmission with synchro systems employs .... telemetering to convey the requisite information.
    (a) frequency (b) position
    (c) impulses (d) voltage.
15. When using a.c. signal conditioning system for capacitive transducers, the carrier frequencies
    (a) range between 50 Hz and 20 kHz (b) should be of the order of 0.5 MHz
    (c) should be of the order to 20 MHz (d) none of the above.
16. An a.c. signal conditioning system is normally used for
    (a) resistive transducers like strain gauges (b) inductive and capacitive transducers
    (c) piezoelectric transducers (d) all of the above.
17. The overall gain or amplification of a system of two amplifiers arranged in series is
    (a) $G_1 + G_2$ (b) $G_1 - G_2$
    (c) $G_1 \times G_2$
    (d) $\frac{G_1}{G_2}$ where G_1 and G_2 are the two gains expressed as pure numbers.
18. The properties of an ideal Op-amp are :
    (a) It should have zero input impedance (b) It should have high input impedance
    (c) It should have a zero open loop gain (d) None of the above.
19. The moving iron voltmeters indicate :
    (a) the same value of d.c. and a.c. voltages.
    (b) lower values for a.c. voltages than the corresponding values of d.c. voltages.
    (c) higher value for a.c. voltages than the corresponding values of d.c. voltages.
    (d) none of the above.

ANSWERS

Choose the correct answer

1. (*a*)	2. (*c*)	3. (*c*)	4. (*a*)	5. (*a*)	6. (*c*)	7. (*a*)
8. (*c*)	9. (*d*)	10. (*a*)	11. (*d*)	12. (*e*)	13. (*b*)	14. (*b*)
15. (*b*)	16. (*b*)	17. (*c*)	18. (*b*)	19. (*b*).		

THEORETICAL QUESTIONS

1. What do you mean by the following terms as applied to instrumentation or measurement system?
   (*i*) Detector-transducer stage.
   (*ii*) Signal conditioning stage.
2. State the limitations of mechanical amplification.
3. What are the advantages of electrical signal conditioning?
4. Explain briefly the following functions of signal conditioning equipment:
   (*i*) Amplification (*ii*) Modification or modulation
   (*iii*) Impedance matching (*iv*) Data processing
   (*v*) Data transmission.
5. Explain briefly the following:
   (*i*) D.C. signal conditioning system.
   (*ii*) A.C. signal conditioning system.
6. Describe briefly the term "Amplification".
7. Explain briefly any two of the following amplifiers:
   (*i*) Mechanical amplifiers (*ii*) Fluid amplifiers
   (*iii*) Electrical and electronic amplifiers.
8. State the generalities that can be listed for an ideal electronic amplifier.
9. What are A.C. and D.C. amplifiers? Explain briefly.
10. What do you mean by "Modulated and unmodulated signals"?
11. What is an Op-amp? State its limitations as well.
12. Explain briefly the term "Common-mode rejection ratio (CMRR).
13. State the applications of Op-amp.
14. Enumerate some of the commonly used Op-amp circuits.
15. Explain briefly the following:
    (*i*) Buffer amplifier. (*ii*) Differential amplifier.
16. State the advantages of differential amplifiers.
17. What is an attenuator? How are the attenuators classified?
18. What do you mean by the terms "Filtering" and "Filter"?
19. How are filters classified?

CHAPTER

18 Data Acquisition Systems and Conversion

18.1 Introduction; 18.2 Objectives and configuration of data acquisition system **18.3 Data acquisition systems**-Analog and automated data acquisition systems-single channel data acquisition system-Multi-channel data acquisition system-applications of data acquisition systems; **18.4 Data conversion**-Analog-to-digital conversion (ADC)-Digital-to-analog (D/A) conversion. *Highlights – Objective Type Questions – Theoretical Questions.*

18.1 INTRODUCTION

Data acquisition system (DAS) *may be defined as a system used for data processing, data conversion, data transmission and data storage.*

A typical data acquisition consists of individual sensors with necessary signal conditioning, multiplexing, data conversion, data processing, data handling and associated transmission, storage and display systems.

— *The analog data is usually acquired and converted into digital form for the purpose of processing, transmission, display, and storage.* Data may be transmitted over long distance (from one location to another one) or short distances (from a test centre to a nearby computer).

— The data my be displayed on a digital panel meter or as a part of CRT presentation. The same may be stored temporarily (in either raw or processed form) for immediate use or permanently for ready reference later.

Fig. 18.1 shows a generalised data acquisition system.

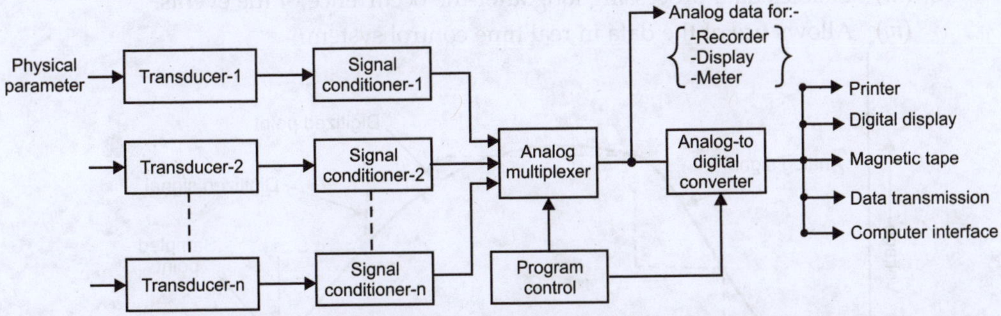

Fig. 18.1. Generalised data acquisition system (DAS).

- An **analog data acquisition system** typically consists of same or all of the following elements.
  - (*i*) Transducers
  - (*ii*) Signal conditioners

(*iii*) Calibrating equipment

(*iv*) Integrating equipment

(*v*) Visual devices

(*vi*) Graphic recording instruments

(*vii*) Magnetic tape instrumentation

(*viii*) Analog computers

(*ix*) High speed cameras and TV equipment.

- A ***digital data acquisition system*** may include some or all of the following elements :

(*i*) Transducer.

(*ii*) Signal conditioner.

(*iii*) Scanner or multiplexer.

(*iv*) Signal converter.

(*v*) Analog-to-digital converter (ADC).

(*vi*) Auxiliary equipment and system programmer.

(*vii*) Digital recorder.

(*viii*) Digital printer.

Now-a-days, in mechatronic and measurement systems, *microprocessors, microcontrollers, single-board computers, and personal computers* are widely used. As such, it is increasingly important for engineers to understand how to directly access information and analog data from the surrounding environment with these devices.

Consider a signal from a sensor as illustrated by the analog signal in Fig. 18.2. In this case there are *two options* :

- Firstly, one could record the signal with an analog device such as chart recorder (which physically plots the signal on the paper) or display *it with an oscilloscope.*
- Secondly, *the data may be stored by using a microprocessor or computer. This process is called computer* ***"data acquisition"*** and entails the following *merits*:

(*i*) Can result in greater data accuracy.

(*ii*) Provides more compact storage of the data.

(*iii*) Enables data processing long after the occurrence of the events.

(*iv*) Allows use of the data in real time control system.

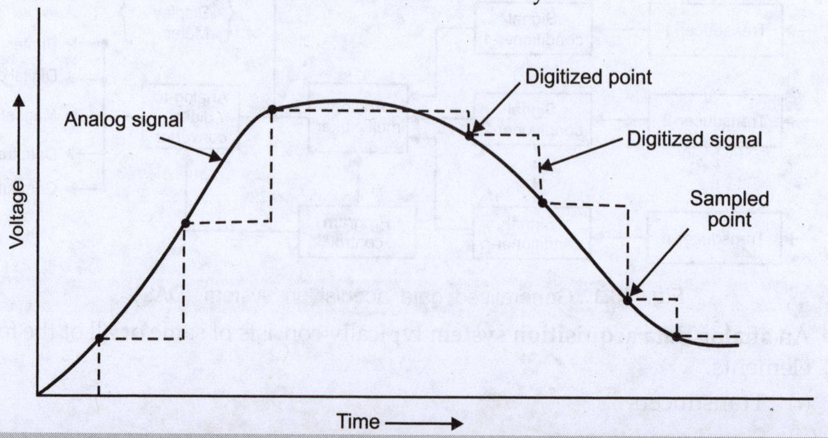

Fig. 18.2. Analog signal and sampled equivalent.

In order to input analog data to a digital circuit or microprocessor, the analog data *must be transformed into digital values.* The first step is to *numerically evaluate the signal at discrete instants in time. This process is called **"sampling"***, and the result is *"digitized signal"* composed of discrete values corresponding to each sample (See Fig. 18.2).

18.2 OBJECTIVES AND CONFIGURATION OF DATA ACQUISITION SYSTEM (DAS)

Objectives of DAS :

The objectives of a data acquisition system may be summed up as follows :

1. To be reliable, flexible and capable of being expanded for future requirements.
2. To acquire the necesary data, at correct speed and at correct time.
3. Down time not be more than 0.1 percent.
4. To be able to compute unit performance indices using on-line, real-time data.
5. To maintain on-line optimum and safe operations, it must monitor the complete plant operation.
6. To make use of all data efficiently to inform the operator about the state of the plant.
7. To be able to collect, summarise and store data for diagnosis of operation and record purpose.
8. To provide an effective human communication system and be able to identify problem areas, thereby minimising unit availability and maximising unit through point at minimum cost.

Configuration of DAS :

The important factors that decide the configuration and the sub-systems of a data acquisition system are as follows :

(*i*) The number of channels to the monitored.
(*ii*) Sampling rate per channel.
(*iii*) Signal conditioning requirement of each channel.
(*iv*) Resolution and accuracy.
(*v*) Cost.

The various general configurations include :

1. *Signal channel possibilities :*
   (*i*) Direct conversion.
   (*ii*) Pre-amplification and direct conversion.
   (*iii*) Sample and hold, and conversion.
   (*iv*) Pre-amplification, sample and hold, and conversion.
   (*v*) Pre-amplification, signal conditioning, and any of the above.
2. *Multi-channel possibilities :*
   (*i*) Multiplexing the outputs of single channel converters.
   (*ii*) Multiplexing the outputs of sample-hold circuits.
   (*iii*) Multiplexing the inputs of the sample-hold circuits.
   (*iv*) Multiplexing low level data.
3. *Noise-reduction options :*
   (*i*) Filtering.
   (*ii*) Integrating converters and digital processing.

18.3 DATA ACQUISITION SYSTEMS

Data acquisition *is the process of using output signals and inputting that into a computer.* The output signal may be one that originates from direct measurement of electrical quantities such as voltage, frequency, resistance etc. or that originates from sensors.

The description of various types of data acquisition systems is given in the following articles.

18.3.1. Analog and Automated Data Acquisition Systems

Fig. 18.3, shows the block diagram of elements of ***analog data acquisition system***:

— This system consists of a sensor-transducer the output of which is connected to DAC board (this is a PCB) through a signal conditioning unit. The DAC board is plugged to a computer. The DAC board consists of a *multiplexer, amplifier, ADC, register* and *control circuitry*, the output of control circuitry connected to a computer system.

— A software is employed to control the acquisition of data through DAC. When the program requires input from a particular sensor, it activates the DAC board by sending control word to the control and status register. The control word indicates what type of operation the board has to carry out.

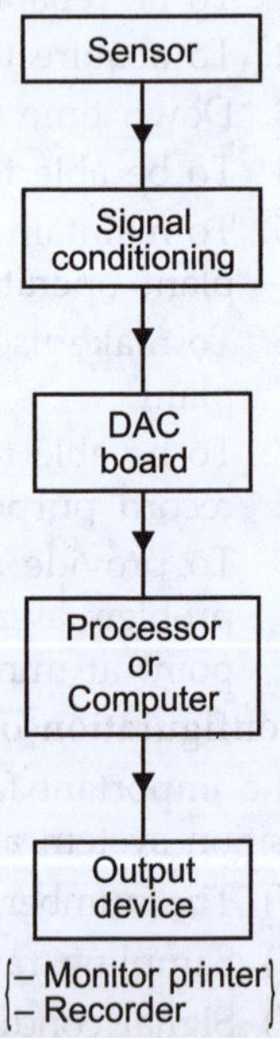

Fig. 18.3. Block diagram of analog data acquisition system.

Automated data acquisition systems may take the following forms:

1. Data loggers;
2. Computer with plug-in boards.

1. Data loggers :

— A *data logger* can monitor the inputs from a large number of sensors.

— Inputs from individual sensors, after suitable signal conditioning, are fed into the *multiplexer*. The multiplexer is used *to select one signal* which is then fed, after amplification, to the analog-to-digital converter.

— The digital signal is then processed by a microprocessor. The microprocessor is able to carry out simple arithmetic operations, perhaps taking the average of a number of measurements.

— The output of the system might be displayed on a digital meter that indicates the output and channel number, used to give a permanent record with a printer, stored on a floppy disc or transferred to perhaps a computer for analysis.

- As data loggers are often used with *thermocouples*, there are often special inputs for thermocouples, these providing cold junction compensation and linearisation. The multiplexer can be switched to each sensor in turn and so the output consists of a sequence of samples. Scanning of the inputs can be selected by programming the microprocessor to switch the multiplexer to just sample a single channel, carry out a single scan of all channels, a continuous scan of all channels, or perhaps carry out a periodic span of all channels.

2. Computer with plug-in boards :

Fig. 18.4, shows the basic elements of a data acquisition system using plug-in boards with a computer.

The signal conditioning prior to the inputs to the board depends on the sensors concerned.

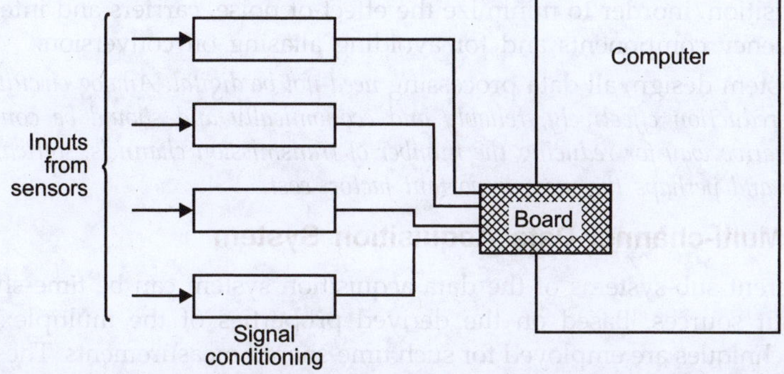

Fig. 18.4. Data acquisition system (Using plug-in-boards with a computer).

Examples:

(*i*) *Thermocouples* — Amplification, cold junction compensation and linearisation;

(*ii*) *Strain gauges* — Wheatstone bridge, voltage supply for bridge and linearisation;

(*iii*) *RTDs* — Current supply, circuitry and linearisation.

18.3.2. Single Channel Data Acquisition System

Fig. 18.5 shows the block diagram of a single channel data acquisition system.

— The system consists of a *"signal conditioner"* followed by an *"analog-to-digital (A/D) converter,"* performing repetitive conversions at a free running, internally determined rate.

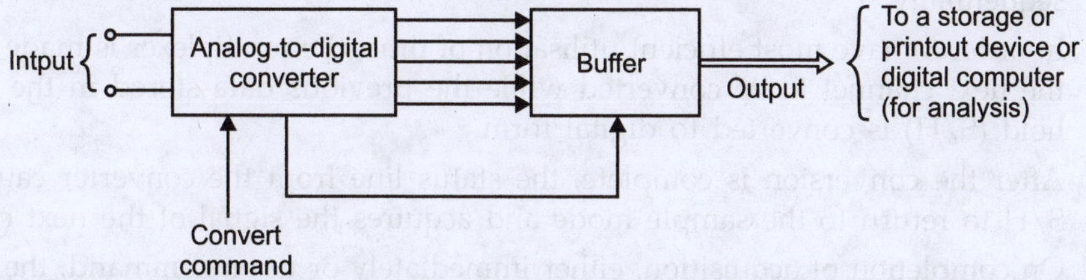

Fig. 18.5. Single channel data acquisition system.

— The digital outputs from the "buffer" are further fed to a storage or printout device or to a digital computer for analysis.

Example : Digital panel meter (DPM)

— Analog-to-digital converters used in this system applications are usually designed to receive external commands to *convert and hold.*

Although these converters based on dual-slope techniques are useful for conversion of low frequency data, such as from thermocouples, especially in the presence of noise, the most popular type of converter for data system applications is the *"successive approximation"* type since it is *capable of high resolution and high speed at moderate cost.*

- Invariably it is observed that the signal level is very low compared to input requirement. In such cases, the amplification of the input signal is done to bring its level to match the input requirements. This is called *pre-amplification.* Preamplifiers can be coupled with active filters before the process of data acquisition, inorder to minimize the effect of noise, carriers and interfering high-frequency components and for avoiding aliasing on conversion.
- In system design, all data processing *need not be digital. Analog circuits can perform data reduction effectively, reliably and economically and should be considered as an alternative way for reducing the number of transmission channels, software complexity, noise and perhaps the most important factor, cost.*

18.3.3. Multi-channel Data Acquisition System

The different sub-systems of the data acquisition system can be time-shared by two or more input sources. Based on the derived properties of the multiplexed system a number of techniques are employed for such time-shared measurements. These techniques are enumerated and discussed below :

1. Multi-channel analog multiplexed system.
2. Multiplexing the outputs of sample-holds.
3. Multiplexing after A/D conversion.
4. Multiplexing low-level data.

1. Multi-channel analog multiplexed system :

Fig. 18.6 shows a conventional multi-channel data system having a single analog-to-digital (A/D) converter preceded by a multiplexer :

— In this system the individual signals are applied directly, or after preamplification and/or signal conditioning wherever necessary, to the multiplexer.

— These analog signals are then converted to digital signals by the A/D converter, sequentially.

— In order to have most efficient utilisation of time, the multiplexer is made to seek the next channel to be converted while the previous data stored in the sample hold (S/H) is converted to digital form.

— After the conversion is complete, the status line from the converter causes the S/H to return to the sample mode and acquires the signal of the next channel.

— On completion of acquisition, either immediately or upon command, the S/H is switched to the hold mode; a conversion begins again and the multiplexer switches on to the subsequent channel.

In cases where the signal variations are extremely slow, sufficient accuracy in measurement can be achieved even without the S/H.

- The *advantage* of this system is that it has a lower cost due to the sharing of a majority of sub-system.
- This method is relatively *slower than systems where the S/H output or even A/D converter outputs are multiplexed.*

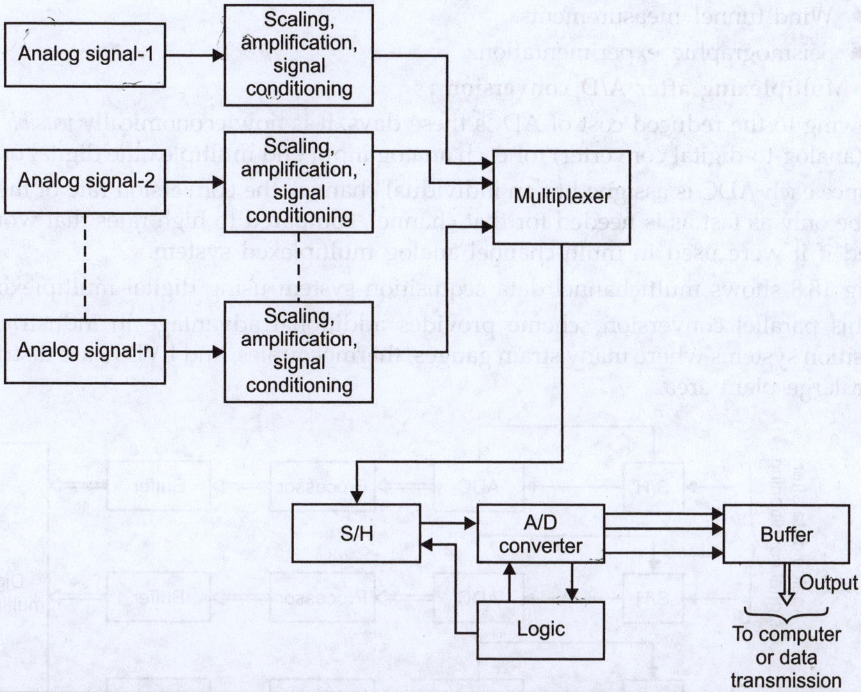

Fig. 18.6. Multichannel analog multiplexed system using a single A/D converter.

2. Multiplexing the outputs of sample-holds :

Fig. 18.7 shows a simultaneous sampled system multiplexer :

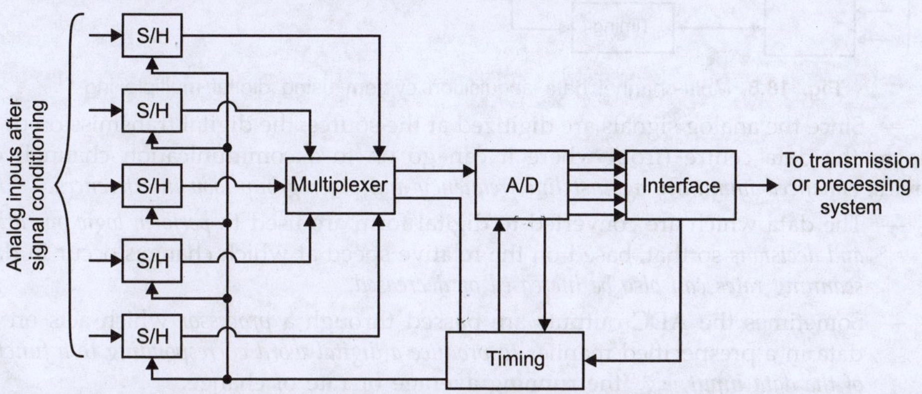

Fig. 18.7. Simultaneous sampled system multiplexer.

— The technique of multiplexing the outputs of the sample-hold (S/H) is particularly useful when a large number of channels are to be monitored at the same time (*i.e.*, synchronously) but at moderate speeds.

— An individual S/H is assigned to each channel, and they are updated synchronously by a timing circuit.

— The outputs of S/H outputs are connected to an A/D converter through a multiplexer resulting in sequential read-out of the outputs.

This approach may be used in the following applications :

- Radar and fire control systems.

- Wind-tunnel measurements.
- Seismographic experimentation.

3. Multiplexing after A/D conversion :

Owing to the reduced cost of ADCs these days, it is now economically *feasible to use ADC* (analog-to-digital converter) for each analog input and multiplex the digital outputs.

Since each ADC is assigned to an individual channel, the conversion rate of the ADC need be only as fast as is needed for that channel, compared to high rates that would be needed if it were used in multi-channel analog multiplexed system.

Fig 18.8 shows multichannel data acquisition system using digital multiplexing :

This parallel conversion scheme provides additional advantage in industrial data acquisition systems where many strain gauges, thermocouples, and LVDTs are distributed over a large plant area.

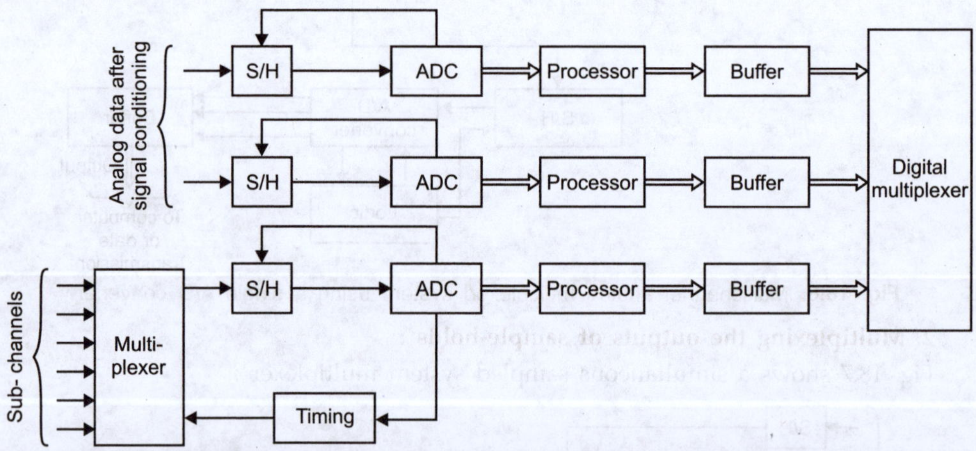

Fig. 18.8. Multi-channel data acquisition system using digital multiplexing.

— Since the analog signals are digitized at the source, the digital transmission of the data to the data centre (from where it can go on to a communication channel) can provide *enhanced immunity against line frequency and other group-loop interferences*.

— The data which are converted to digital form are used to *perform logic operations and decisions* so that, based on the relative speed at which changes occur in data, *scanning rates can also be increased or decreased.*

— Sometimes the ADC outputs are passed through a *processor* which acts on the data in a prespecified manner *to produce a digital word corresponding to a function of the data input, e.g.,* the running average or rate of change.

4. Multiplexing low-level data :

- A "low-level multiplexing system" enables the use of a *single high-quality data amplifier for handling multi-channel low-level inputs.*
- This technique can be attractive when a large number of channels (250 or more) all having low-level outputs need to be monitored at moderate speeds. *A typical application is a 200 channel stress measurement system in a transmission tower set-up.*

18.3.4. Applications of Data Acquisition Systems

Data acquisition systems find wide use in a variety of "*industrial and scientific areas*", including the *aerospace*, *biomedical* and *telemetry industries*. The type of data acquisition system to be used depends upon the application and the intended use of recorded input data.

- The *"analog data acquisition systems"* are generally used when wide bandwidth is required or when lower accuracy can be tolerated.
- *"Digital data acquisition systems"* are used when the physical quantity being monitored has a *narrow bandwidth* (i.e., when the quantity varies slowly). These systems are also used when *high accuracy and low per channel cost is required.* These systems are in general, *more complex than analog systems, both in terms of instrumentation involved and the volume and complexity of input data they can handle.*

18.4 DATA CONVERSION

18.4.1. Analog-to-Digital (A/D) Conversion

18.4.1.1. Digital signals

The majority of sensors supply the output which tends to be in analog form. Where a *microprocessor* is used as part of the measurement or control system, the analog output from the sensor *has to be converted into a digital form* before it can be used as an input to the microprocessor. Similarly, most actuators operate with analog inputs and so the digital output from a microprocessor has to be converted into an analog form before it can be used as input by the actuator.

18.4.1.2. ADC process

The *"analog-to-digital conversion" process* changes a sampled analog voltage into digital form. This process, conceptually involves the following *two steps:*

(i) **Quantizing.** It is defined as the *transformation of a continuous analog input into a set of discrete output states.*

(ii) **Coding.** It is *assignment of a digital code word or number to each output state.*

Procedure of conversion:

Analog-to-digital conversion involves converting analog signals into binary words. Fig. 18.9 shows the basic elements of analog-to-digital conversion. The procedure of A/D conversion is that a *clock supplies regular time signal pulses to ADC (analog-to-digital converter) and every time it receives a pulse it samples the analog signal.* The types of signals involved at various stages of analog-to-digital conversion are shown in Fig. 18.10:

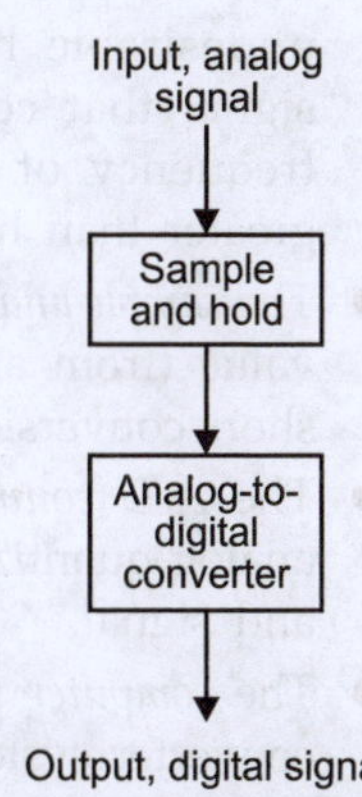

Fig. 18.9. Basic elements of A/D conversion.

- Fig. 18.10(*a*), shows the analog signal;
- Fig. 18.10(*b*), shows the clock signals which supply the time signals at which the sampling occurs;
- Fig. 18.10(*c*), shows a series of various pulses which is the result of sampling (sampled signal);
- Fig. 18.10(*d*), shows the sampled and held signal which is obtained by using a *sample and hold unit.* (This unit is necessary because A/D converter requires a finite amount of time, termed the *'conversion time'*, to convert analog signal into a digital one) which holds each sampled value until the next pulse occurs.

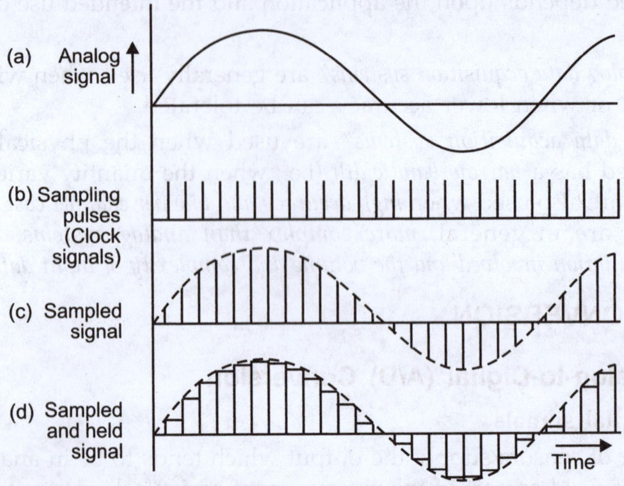

Fig. 18.10. Signals : (*a*) Analog; (*b*) Clock; (*c*) Sampled; (*d*) Sampled and held.

18.4.1.3. Components used in A/D Conversion

In order to acquire an analog voltage for digital processing, it is imperative to properly select the following components and apply them this sequence :

(*i*) Buffer amplifier; (*ii*) Low-pass filter;
(*iii*) Sample and hold amplifier; (*iv*) Analog-to-digital (A/D) converter;
(*v*) Computer.

Figure 18.11, shows the components used in A/D conversion:

- The *buffer amplifier* provides a signal in a range close to but not exceeding the full input voltage range of the A/D converter.
- The *low-pass filter* is necessary to remove any undesirable high-frequency components in the signal that could produce aliasing. The cut-off frequency of the low-pass filter should not be greater than half the sampling rate.
- The *sample and hold amplifier* maintains a fixed input value (from an instantaneous sample) during the short conversion time of the A/D converter.
- The *A/D converter* should have a resolution and analog quantization size appropriate to the system and signal.
- The *computer* must be properly interfaced to A/D converter system to store and process the data.
- The analog-to-digital conversion process requires a small but finite interval of time that must be taken into consideration when assessing the accuracy of the results.
- The *conversion time depends on the design of the converter, the method used for conversion, and the speed of the components used in the electronic design.*

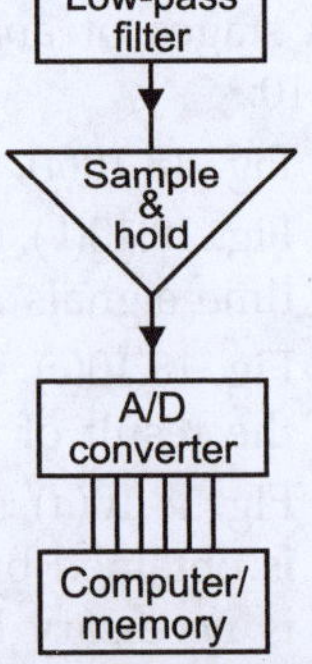

Fig. 18.11. Components used in A/D conversion.

— Owing to the continuous change in the analog signals, the uncertainty about when in the sample time window the conversion occurs causes corresponding uncertainty in the digital value. This is of significant importance if there is no sample and hold amplifier on the A/D input. The term *"aperture time"* refers to the duration of the time window and is associated with any error in the digital output due to changes in the input during this time.

— In a signal, sampling at or about Nyquist frequency will yield the correct frequency components. In order to obtain accurate amplitude resolution, we must have an A/D converter with an adequately small aperture time.

18.4.1.4. Analog-to-Digital (A/D) Converter

An ***analog-to-digital (A/D)*** *converter is an electronic device that converts an analog voltage to a digital code.* The output of the A/D converter can be directly interfaced to digital devices such as microcontroller and computers.

The ***"resolution"*** *of an A/D converter is the number of bits used to digitally approximate the analog value of the input.* The number of possible states N is equal to the number of bit combinations that can be output from the converter : $N = 2^n$.

where, n = The number of bits.

The number of analog *"decision points"* that occur in the process of quantizing is $(N - 1)$. The *"analog quantization size"* Q is defined as the *full scale range of the A/D converter by the number of output states.*

Design Principles:

Analog-to-Digital (A/D) converters *are designed based on a number of different principles; these are:*

(*i*) Successive approximations.

(*ii*) Flash or parallel encoding.

(*iii*) Single-slope and dual-slope integration.

(*iv*) Switched capacitor.

(*v*) Delta sigma.

Some of these are discussed below:

(i) Successive approximation A/D converter :

The successive approximation A/D converter is very widely used because :

- It is fast in operation;
- It has high resolution;
- It is less expensive.

The various subsystems involved in this type of converter are shown in Fig. 18.12.

— The *"clock"* generates a voltage, emitting a regular sequence of pulses which are counted, in a *binary manner*, and the resulting binary word is converted into an analog voltage by a *"DAC"* (*digital-to-analog converter*). This voltage rises in steps and is *compared with the 'analog input voltage'* from the sensor.

— When the clock-generated voltage passes the input analog voltage the pulses from the clock are stopped from being counted by a *"gate"* being closed. The output from the counter at that time is then a digital representation of the analog voltage.

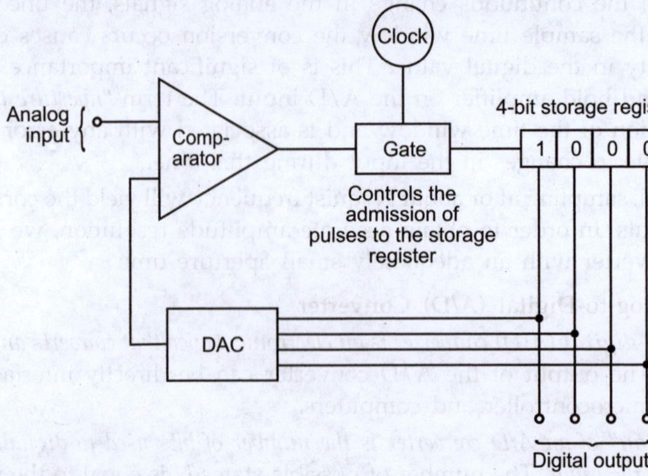

Fig. 18.12. Successive approximation analog-to-digital converter (ADC).

Note: When frequency of the clock is f, the time taken between the pulses is $\frac{1}{f}$; hence the time taken to generate the word, *i.e.*, the conversion time is $\frac{n}{f}$.

(ii) Flash A/D converter :

*The fastest type of A/D converter is known as a **flash converter**.*

Fig. 18.13, shows a flash ADC:

— For an n-bit converter, 2^n-1 separate voltage comparators are used in parallel, with each having the analog input voltage as one input.

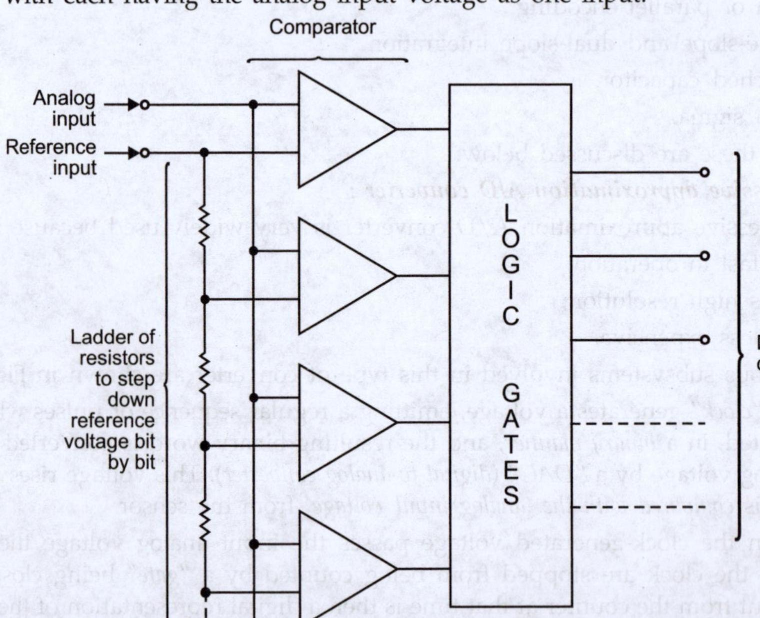

Fig. 18.13. Flash analog-to-digital converter (ADC).

— A *reference voltage* is applied to a ladder of resistors so that the voltage applied as the other input to each comparator is one bit *larger* in size than the voltage applied to the previous comparator in the ladder. Thus when the analog voltage is applied to the ADC, all these comparators for which the analog voltage is greater than the reference voltage of a comparator will give a *high* output and those for which it is less will be *low.*

— The resulting outputs are fed in parallel to a *logic system* which translates into a digital word.

(iii) Single-slope and dual-slope integration :

(a) Single-slope or ramp or voltage-to-time A/D converter :

Fig. 18.14 shows the schematic of a ramp ADC :

— A ramp analog-to-digital converter (ADC) involves an analog voltage which is increased at constant rate (and hence called *ramp* voltage) and is applied to a comparator where it is compared with the analog voltage from the sensor. The time consumed by the ramp voltage to increase to the value of the sensor voltage will depend on the size of the sampled analog voltage.

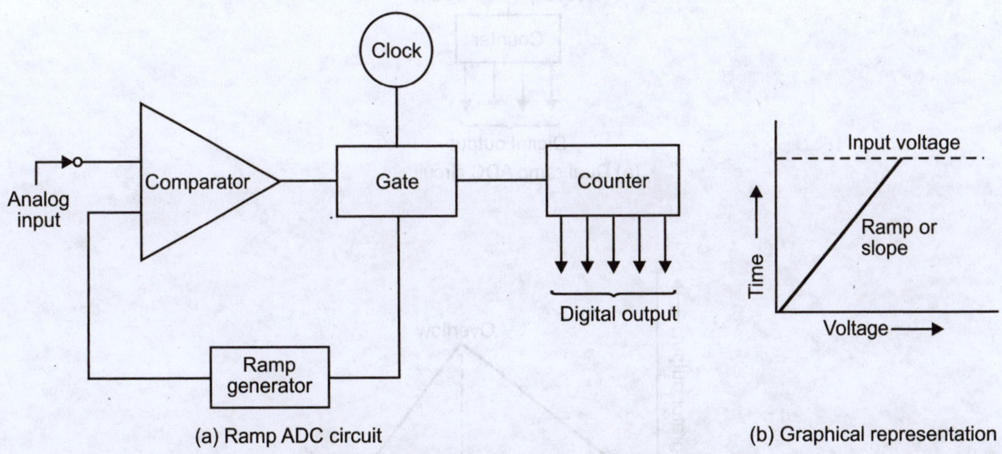

Fig. 18.14. Single slope or ramp ADC.

— When the ramp voltage starts, a gate is opened which starts a binary counter counting the regular pulses from a clock. When the two voltages are equal, the gate closes and the word indicated by the counter is the digital representation of the sampled analog voltage.

(b) Dual-slope integration A/D converter or dual ramp converter :

This type of converter, as compared to single ramp converter, is more commonly used.

Fig. 18.15, shows the dual slope/dual ramp *ADC*. The analog input voltage and the reference input voltage are successively connected to the integrator with the help of a switch. These two voltages (analog input and reference input) *must be of opposite polarities.*

— The fixed voltage is integrated for a fixed sample time.

— The integrated value is then discharged at a fixed rate and the time to do this is measured by a counter. The count is then a measure of the analog input voltage.

- The *"advantage"* of these converters is that they have *excellent noise rejection* because the integral action averages out random negative and positive contributions over the sampling period.
- Their *"limitation"* is that they are *very slow* in operation.

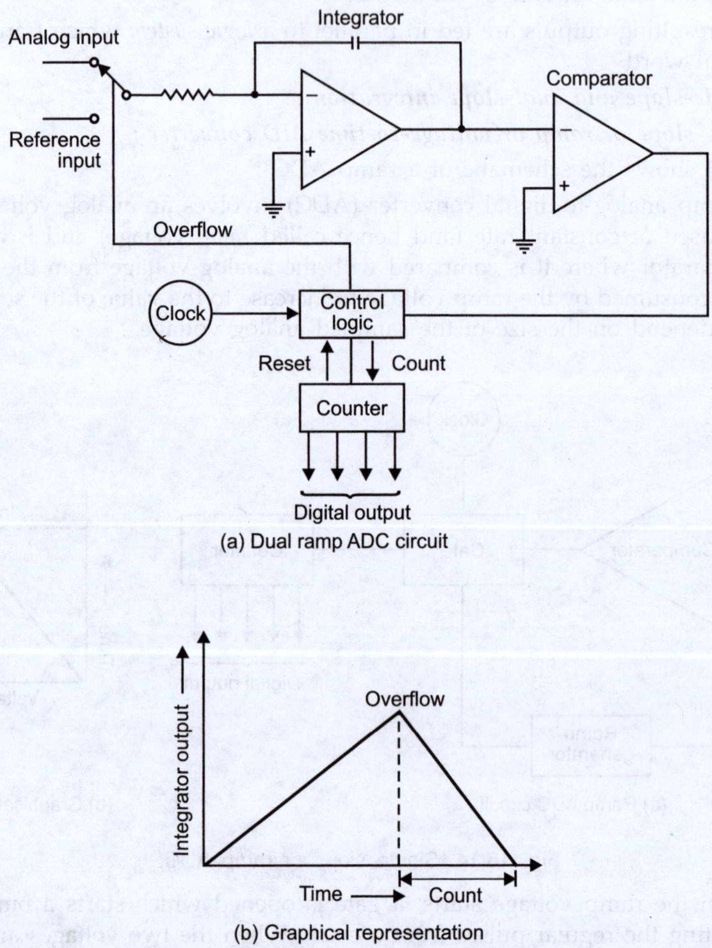

Fig. 18.15. Dual slope/dual ramp A/D converter.

Multiplexers :

The ***"multiplexer"*** *is essentially an electronic switching device which enables each of the inputs to be sampled in turn.*

A *"multiplexer"* is a circuit that is able to have inputs of data from a number of sources and then, by selecting an input channel, give an output from just one of them.

- In *applications* where there is a need for measurement to be made at a number of different locations, rather than use a separate ADC and microprocessor of each measurement, a *"multiplexer"* can be used *to select each input in turn and switch it through a single ADC and microprocessor* (Fig. 18.16).

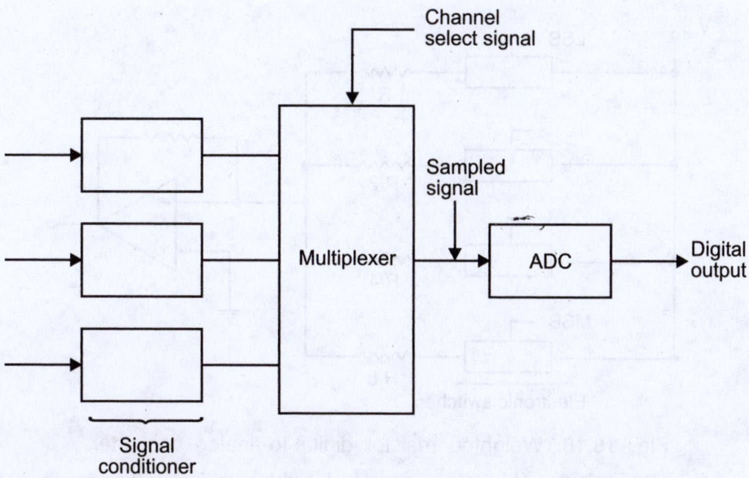

Fig. 18.16. Multiplexer.

Digital multiplexer :

Fig. 18.17 shows a two channel multiplexer. The logic level applied to the select input determines which AND gate is enabled so that its data input passes through the *OR* gate to the output.

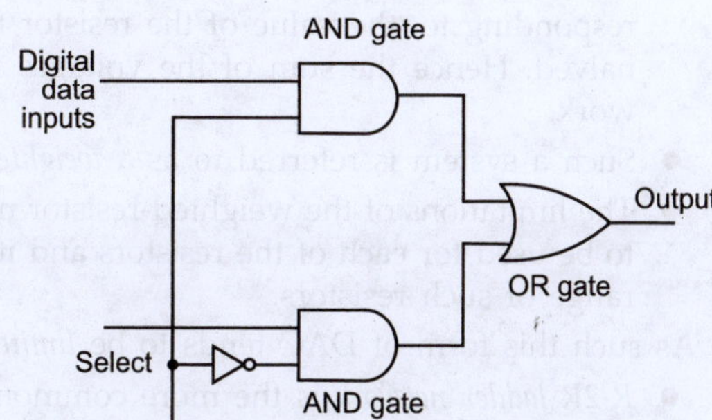

Fig. 18.17. Two channel multiplexer.

A number of forms of multiplexers are available in *IC* packages.

- ***"Demultiplexer"*** is similar to multiplexer but with *reversed action*. It accepts a digital signal through its one input and then channelises it to a particular output selected by binary value at the control port.

18.4.2. Digital-to-Analog (D/A) Conversion

Invariably we have to reverse the process of analog-to-digital (A/D) conversion by changing a digital value to an analog value. This is called ***digital-to-analog (D/A)*** *conversion.*

A D/A converter (DAC) *allows a computer or other digital device to interface with external analog circuits and devices.*

- The input to a digital-to-analog converter (DAC) is a *binary word;* the output is an analog signal that represents the weighted sum of the non-zero bits represented by the word.

  Example: An output of 0010 must give an analog output which is twice that given by an input of 0001.

Digital-to-Analog Converters :

— Fig. 18.18, shows a simple form of DAC using a summing amplifier to form the weighted sum of all the non-zero bits in the input word.

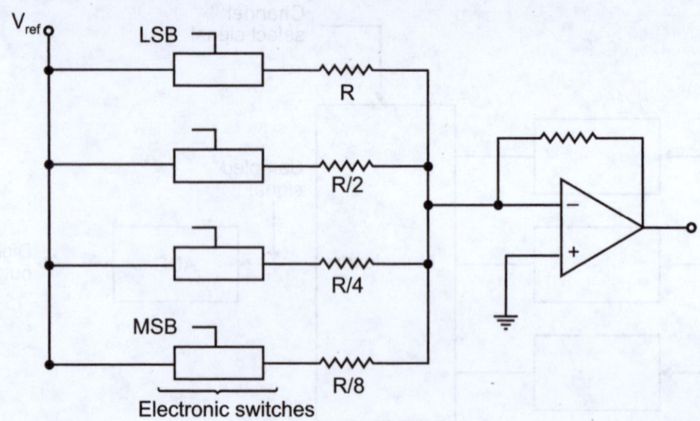

Fig. 18.18. Weighted resistor digital-to-analog converter.

— The reference voltage (V_{ref}) is connected to the resistors by means of electronic switches which respond to binary 1.

— The values of input resistances depend on which bit in the word a switch is responding to, the value of the resistor for successive bits from the LSB being halved. Hence the sum of the voltages is a weighted sum of the digits in the work.

- Such a system is referred to as a *weighted-resistor network*.
- The limitations of the weighted-resistor network is that accurate resistances have to be used for each of the resistors and it is difficult to obtain the required wide range of such resistors.

As such this form of DAC tends to be *limited to 4-bit-conversions*.

- *R-2R ladder network* is the more commonly used version (Fig. 18.19).
  — This version overcomes the problem of obtaining accurate resistances over a wide range of values, *only two values being required*.
  — The output voltage is generated by switching sections of the ladder to either the reference voltage or 0 V according to whether there is a 1 or 0 in the digital input.

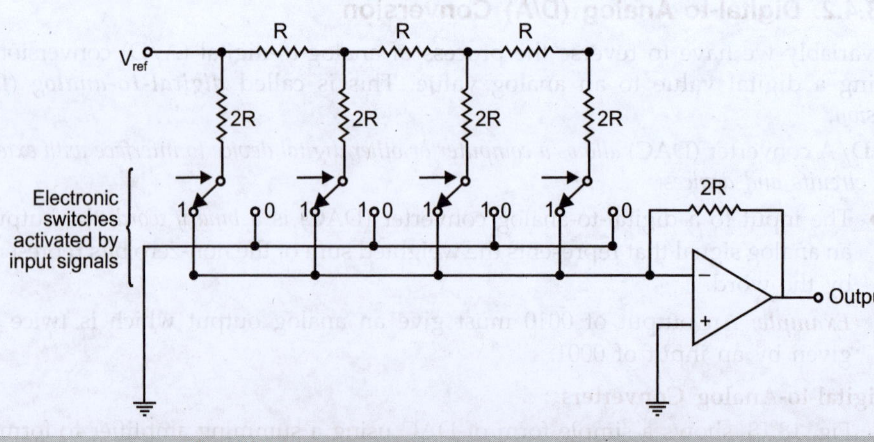

Fig. 18.19. R-2R ladder digital-to-analog converter.

- Fig. 18.20 shows computer control hardware, illustrating the *roles that A/D and D/A converters play in a* ***mechatronic control system***.
  — An analog voltage signal from a *sensor* (*e.g.*, a thermocouple) is converted to a digital value.
  — The *computer* uses this value in a control algorithm, and the computer outputs an analog signal to an *actuator* (*e.g.*, an electric motor) to cause some change in the system being controlled.

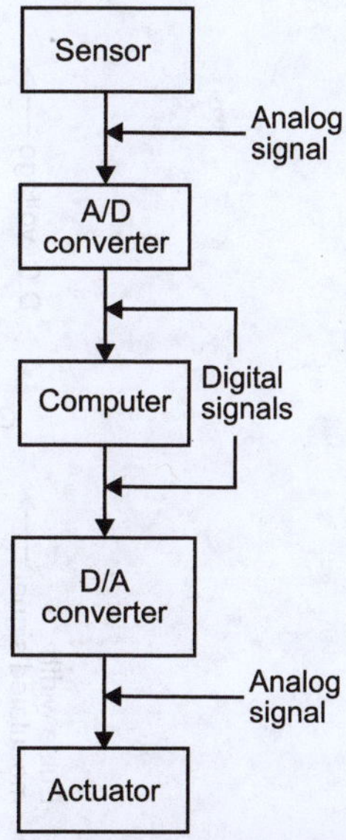

Fig. 18.20. Computer control hardware.

Pulse-Modulation:

While dealing with the transmission of low-level D.C. signals from sensors, a problem that is encountered is that the gain of Op-amp (operational amplifier) used to amplify them may drift causing a drift in the output. This problem can be solved if the signal is a *sequence of pulses rather than a continuous-time signal.* This can be achieved in the following *two* ways:

1. Pulse amplitude modulation (PAM).
2. Pulse width modulation (PWM).

1. Pulse amplitude modulation:

— In this method of conversion, D.C. signal (Fig. 18.21(*a*)] is chopped in the way as shown in Fig. 18.21(*b*). The output from the chopper is a chain of pulses, the heights of which depends on the D.C. level of the input signal. This process is called *"pulse amplitude modulation"*.

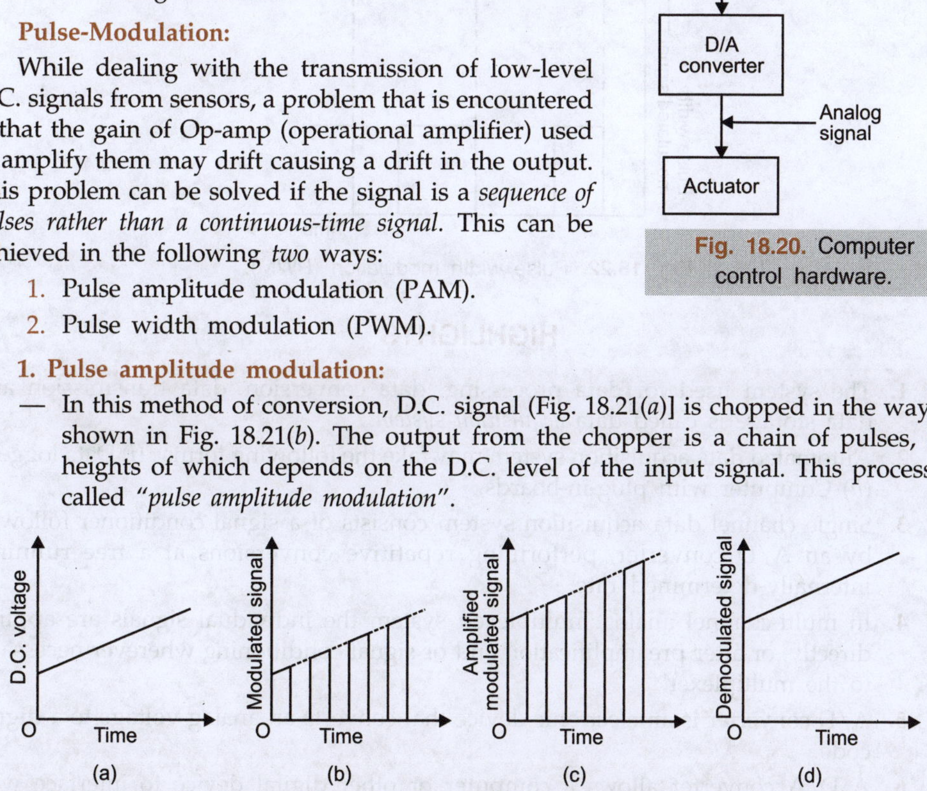

Fig. 18.21. Pulse amplitude modulation.

— After amplification and any other signal conditioning, the modulated signal [Fig. 18.21(*c*)] can be demodulated [Fig. 18.21(*d*)] to give a D.C. output.

2. Pulse width modulation (PWM) :

This type of modulation is used where the width, *i.e., duration of a pulse rather than its amplitude* depends on the size of the voltage, as shown in Fig. 18.22.

— PWM is widely used with control systems as a *means of controlling the average value of a D.C. voltage.*

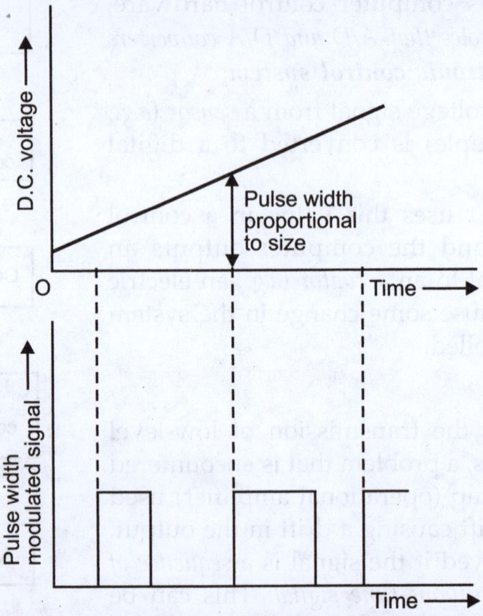

Fig. 18.22. Pulse width modulation (PWM).

HIGHLIGHTS

1. The system used for data processing, data conversion, data transmission and data storage is called data *acquisition system*.
2. Automated data acquisition system may take the following forms : (*i*) Data loggers; (*ii*) Computer with plug-in-boards.
3. Single channel data acquisition system consists of a signal conditioner followed by an A/D converter, performing repetitive conversions at a free running, internally-determined rate.
4. In multi-channel analog multiplexed system the individual signals are applied directly, or after preamplification and or signal conditioning wherever necessary, to the multiplexer.
5. A/D converter is an electronic device that converts an analog voltage to a digital code.
6. A D/A converter allows a computer or other digital device to interface with external analog circuits and devices.

OBJECTIVE TYPE QUESTIONS

Fill in the Blanks or Say "Yes" or "No"

1. .... is the process of using output signal and inputing that into a computer.
2. A data logger cannot monitor the inputs from a large number of sensors.
3. Digital panel meter (DPM) is an example of .... channel data acquisition system.
4. A low level multiplexing system enables the use of a single high-quality data amplifier for handling multi-channel low level inputs.
5. The analog data acquisition systems are generally used when .... bandwidth is required or when .... accuracy can be tolerated.

6. Digital data acquisition systems are less complex than analog systems.
7. ADC process changes a sampled analog voltage into digital form.
8. .... is defined as the transformation of a continuous analog input into a set of discrete output states.
9. .... is the assignment of a digital code word or number to each output state.
10. The faster type of A/D converter is known as .... converter.
11. A .... is essentially an electronic switching device which enables each of the inputs to be sampled in turn.

ANSWERS

1. Data acquisition	**2.** No	**3.** Single	**4.** Yes
5. Wide, lower	**6.** No	**7.** Yes	**8.** Quantizing
9. Coding	**10.** flash	**11.** multiplexer	

THEORETICAL QUESTIONS

1. What is meant by data acquisition ?
2. What is a data acquisition system ?
3. Draw the block diagram of a generalised data acquisition system and explain it briefly.
4. Explain the different components of a data acquisition system.
5. What are the objectives of a data acquisition system ?
6. What are the factors that decide the configuration of a data acquisition system ?
7. With the help of suitable block diagram, explain analog data acquisition system.
8. Write a note on "Automated data acquisition systems."
9. Explain briefly a digital acquisition system.
10. Explain with a neat diagram the working of a single-channel data acquisition system.
11. Explain the working of a multi-channel analog multiplexed data acquisition system.
12. How is low-level multiplexing achieved ?
13. Give the applications of data acquisition systems.
14. Explain briefly the procedure of analog-to-digital (A/D) conversion.
15. Define the following terms :
    (*i*) Quantizing. (*ii*) Coding.
16. Describe briefly the components use in A/D conversion.
17. Explain briefly the following
    (*i*) Successive approximation A/D converter
    (*ii*) Flash A/D converter.
18. What is a muliplexer ? Explain briefly.
19. Explain briefly the following :
    (*i*) Weighted resistor D/A converter.
    (*ii*) R-2R ladder D/A converter.
20. Discuss briefly the following :
    (*i*) Pulse amplitude modulation (PAM)
    (*ii*) Pulse width modulation (PWM).

CHAPTER

19 Data Transmission and Telemetry

19.1 Data/Signal transmission; Mechanical transmission-Hydraulic transmission-Pneumatic transmission-Magnetic transmission-Electric type of transmitters-Converters; **19.2 Telemetry**-Definition-General telemetering system; Types of telemetry systems-Voltage telemetering system-Current telemetering system-Position (or ratio) telemetering system-RF telemetry system – Introduction to types of signals and transmission paths-Modulation and demodulation-FM telemetry system-PAM telemetry system; *Highlights; Objective Type Questions; Theoretical Questions.*

19.1 DATA/SIGNAL TRANSMISSION

The terms **"Data transmission"** *and* **"Telemetry"** *refer to the process by which the measurand is transferred to a remote location for the purpose of being processed, recorded and displayed.*

The terms *"measuring devices"* and *"transmitters"* generally go by side by side and it is very difficult to make any distinction between them.

A measuring device converts a primary indication into some form of energy that can easily be displayed on a scale; some transmitters also do the same things. In the *stricter sense* **"transmitters"** *could be considered as devices which transmit the value of the primary variable at a considerable distance from the primary element.* If transmission is to be carried over *very long distances*, then devices are known as **"telemeters"**.

For transmission purposes, the measured variable is converted into a transmittable signal (either pneumatic or electrical), so that it can be received by a remote indicating, recording, or controlling device. The *selection of transmission device* depends upon the *nature of the variable and the distance the signal is required to be sent.*

For data transmission various methods have been developed; the choice of a particular method depends upon :

(*i*) The physical variable; (*ii*) The distance involved.

- The **hydraulic and pneumatic methods** are employed for transmission over a *short distance.*
  - — The pneumatic type transmission devices are generally suitable for transmission upto maximum distance of 200 m.
- The **electrical/electronic methods** are suitable *equally for short as well as long distance transmission.*
  - — Generally short transmission is carried out on own communication connections between sending and receiving devices.
  - — The *telemeters* which are designed for *long-distance transmission* may be designed to transmission over their own wire or over phone wires or by microwave.

19.1.1. Mechanical Transmission

The *"rack and pinion arrangement"* and the *"gear trains"* as used in *Bourdon-tube pressure* gauge and *dial indicator gauge* constitute mechanical transmission. They *amplify the displacement* and also *transmit the signal to a pointer which moves across a calibrated dial.*

19.1.2. Hydraulic Transmission

Fig. 19.1. shows the hydraulic method of transmission, which is commonly used. Here four bellows are employed, two at the transmission end and two at the receiving end. The four bellows are connected by an impulse pipeline and the whole system is filled with liquid. When the actuating link, on the transmission end, is operated by the measurand, then one bellow is expanded and other is contracted. This expansion and contraction is communicated to receiving end, which moves the receiving pointer an equal amount. The purpose of using two bellows on either side is to compensate for changes in ambient temperature.

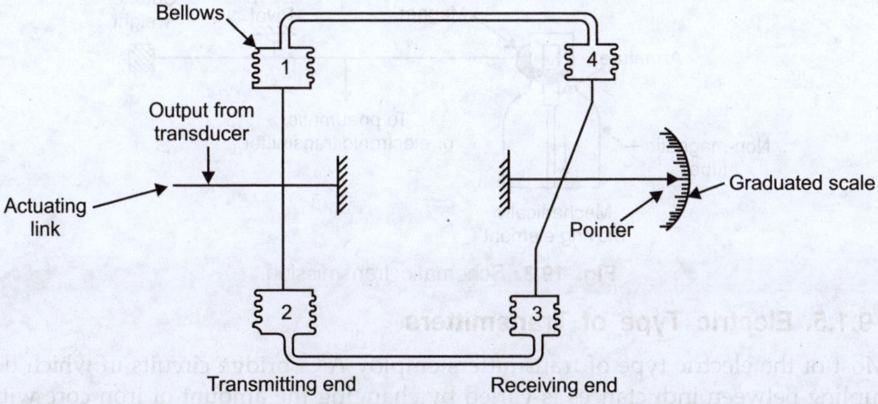

Fig. 19.1. Hydraulic method of transmission.

19.1.3. Pneumatic Transmission

Fig. 19.2, shows the one of the pneumatic methods of transmission **(Flapper nozzle mechanism).**

It consists of an open nozzle which is supplied with air through a restriction/orifice (its diameter being *smaller* than nozzle diameter for proper functioning). In front of the nozzle there is a flapper which is positioned by the measuring element. The force on the flapper is produced by a transducer which converts the measurand into linear displacement. The flapper is pivoted about a point and the other end, it contains some balancing counter weight.

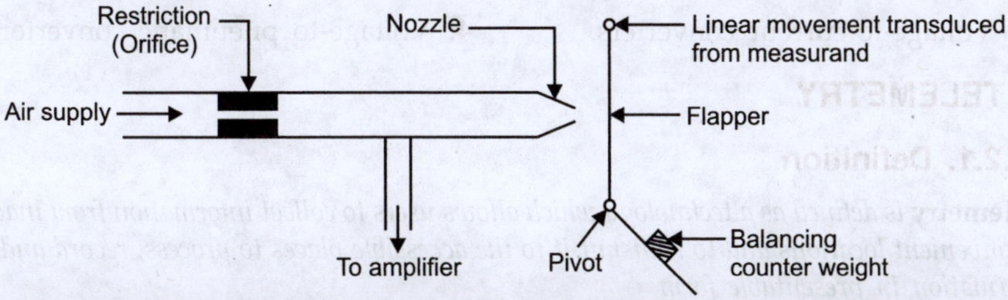

Fig. 19.2. Schematics of pneumatic transmission–Flapper nozzle mechanism.

When the flapper is moved against, the nozzle, the air cannot escape and maximum air passes to the *amplifier*, and when flapper is moved away from the nozzle, minimum air passes to the amplifier as most of the air escapes to atmosphere. Thus, the movement of flapper from one extreme position to another serves to control the amplifier, which produces an air pressure proportional to the measurand of adequate strength for transmission over the required distance.

19.1.4. Magnetic Transmission

Fig. 19.3, shows the schematics of magnetic transmission. In this arrangement/device, an armature is attached at the end of the mechanical moving part whose movement is to be transmitted outside the armature moving inside a non-magnetic tube. A magnet is placed around the armature outside the tube. The magnet follows the movement of the armature and repositions a pneumatic transmitter. The magnet movement could also be utilised to operate an electronic transmitter.

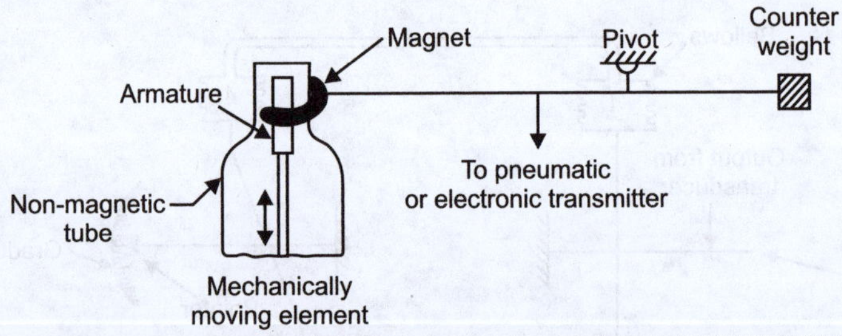

Fig. 19.3. Schematic transmission.

19.1.5. Electric Type of Transmitters

Most of the electric type of transmitters employ A.C. bridge circuits in which degree of coupling between inductances is varied by changing the amount of iron core within a coil. The common *examples* are:

1. Wheatstone bridge transmitter.
2. Inductance bridge.
3. Impedance bridge.
4. Differential transformer.
5. Self synchronous motor (Selsyn)
6. Resistance manometers.

19.1.6. Converters

The converters are series of transducers which play an important role in the modern instrumentation, linking electrical (voltage and current based) and pneumatic systems together.

Following are the most commonly used converters :

1. Current-to-pneumatic converters.
2. Pneumatic-to-current converters.
3. Voltage-to-current converters.
4. Voltage-to pneumatic converters.

19.2 TELEMETRY

19.2.1. Definition

Telemetry *is defined as a technology which allows users to collect information from inaccessible and inconvenient locations and to transmit it to the accessible places to process, record and display the information in presentable form.*

Or

Telemetry *(as the name implies) means measuring at a distance.*

19.2.2. General Telemetring System

Fig. 19.4, shows the block diagram of a general telemetring.

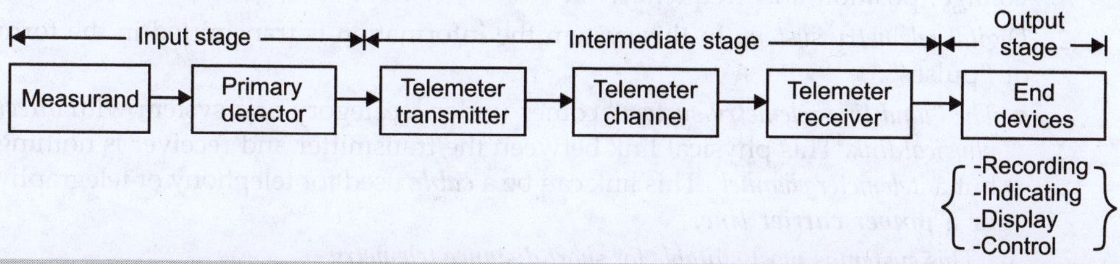

Fig. 19.4. Block diagram of general telemetring system.

- In the 'primary stage', the "measurand" is detected by the *"primary detector"* and sent to the channel transmitter.
- The 'intermediate stage' consists of the following three elements :
  - (*i*) Telemeter transmitter;
  - (*ii*) Telemeter channel;
  - (*iii*) Telemeter receiver.

The *"telemeter transmitter"* converts the output of the primary detector into an electrical signal and transmits it to the *"telemetring channel"*. This signal (in electrical format) is received by a *"telemeter receiver"* placed at a remote location. This receiver converts the signal into a usable form and is indicated or recorded or displayed by the *"end device"* (output stage) which is calibrated in terms of the measurand. The control device is provided to take corrective action on measurand through feed back loop to control the output.

For telemetry, some basic instruments are required for :

— translating the measurand into a suitable electrical signal;

— processing and transmission of signal over the channel;

— receiving the signal and converting it into a readily comprehensible reading.

19.2.3. Types of Telemetry Systems

The telemetry systems can be *classified* as follows :

I. According to the link between the telemeter transmitter and the telemeter receiver :

1. *Land line (D.C.) telemetry system :*
   - (*i*) Voltage telemetering system.
   - (*ii*) Current telemetering system.
   - (*iii*) Position (or ratio) telemetering system.
2. *Ratio frequency (RF) telemetry system :*
   - — Frequency telemetering system
   - — Pulse telemetering system.
   - **A.C.** telemetering system is employed *both for land line and radio-frequency airborne telemetry techniques.* A.C. telemetry is employed for sensors, which provide an A.C. output or voltage to frequency converters. The data is available in the form of current or voltage, which is usually very weak for such transmission. Hence *it is modulated by an A.C. carrier waves generated by an oscillator.*

II. According to the form in which the information is transmitted :

1. *Analog telemetry system.* Here information is transmitted in the form of current, voltage, position and frequency.
2. *Digital telemetry system.* In this system the information is transmitted in the form of "pulses".

- *The "land line telemetry system"* comes under a category of a system with *direct physical link.* This physical link between the transmitter and receiver is nothing but a *'telemeter channel'*. This link can be a ***cable*** used for telephony or telegraphy or a ***power carrier line.***
  – *This system is most suitable for short distance telemetry.*
- In *"radio frequency (RF) telemetry system"* there is *no physical link* between the telemeter transmitter and the telemeter receiver. The link between the transmitting end and the receiving end is established through the ***radio links.***

Advantages and disadvantages of land line telemetry system :

Advantages :

(*i*) Very effective system for short-distance transmission (as the telemeter or communication channel can be established with the help of a simple cable or transmission line).

(*ii*) Very simple circuitry is required for the measurement set-up.

(*iii*) The information can be transmitted (in the form of voltage, current and position) easily by the use of modern electronic circuitry available now-a-days.

(*iv*) A wide variety of detectors (or the primary sensing elements) are available which produce electrical signal (voltage or current) in relation with the variable being measured at the system's input stage.

Disadvantages :

(*i*) Limited frequency response.

(*ii*) Signal multiplexing is very difficult.

(*iii*) The distortions in the transmission links are directly incorporated in the main system.

(*iv*) The information transmitted through the link is affected by the EMI effects of main frequency from nearby cable.

(*v*) The effects of thermoelectric emfs are significant in case of telemetry systems transmitting D.C. signals.

19.2.4. Voltage Telemetering System

- In these systems the *measurand is converted to A.C. or D.C. voltage.*
- For such systems, the *self-balancing potentiometers are the usual receivers.*

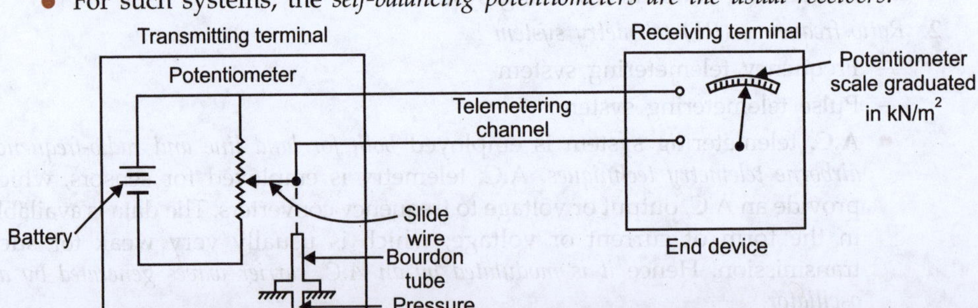

Fig. 19.5. Voltage telemetering system.

Fig. 19.5, shows a voltage telemetering system :

Construction :

— It consists of a slide-wire potentiometer connected in series with a battery at the transmitting terminal. The sliding contact is connected to the bourdon tube used for pressure measurement.

— At the receiving terminal a null balance D.C. potentiometer or a recorder is used.

— In between the transmitting terminal and the receiving terminal, a pair of wires form a telemeter channel.

Working/Operation :

— With the change of pressure in the system, the bourdon tube actuates the slider of the potentiometer. Consequently, the voltage changes at the transmitting end.

— This voltage is carried by the telemetering channel to the receiving terminal, where it is measured with the help of null balanced D.C. potentiometer indicator calibrated in terms of pressure scale (kN/m^2) or recorded if required. Some systems make use of the deflection type indicators to measure voltage; these indicators are calibrated for the line resistance. Basically, a null balance D.C. potentiometer reduces the current carried by the telemetering channel with negligible resistance.

Such systems use primary sensing elements (detectors) such as *microphones, tachometers, bourdon tubes or differential transformers as all produce voltage signal.*

Disadvantage : These systems are affected by line resistance, leakage, interfering sources nearly, noise and require higher-quality circuits than current systems, especially for low voltages.

- The voltage telemetering system is *limited for transmission upto 300 metres distances.*
- A voltage telemetering system is *very much suited for adding several input voltages in series* provided the measurement is *linear*. However, this system needs a relatively *more expensive receiving terminal.* This system is normally *not suitable* to the use of many receivers at the same time.

19.2.5. Current Telemetering System

Fig. 19.6 shows a basic current telemetering system :

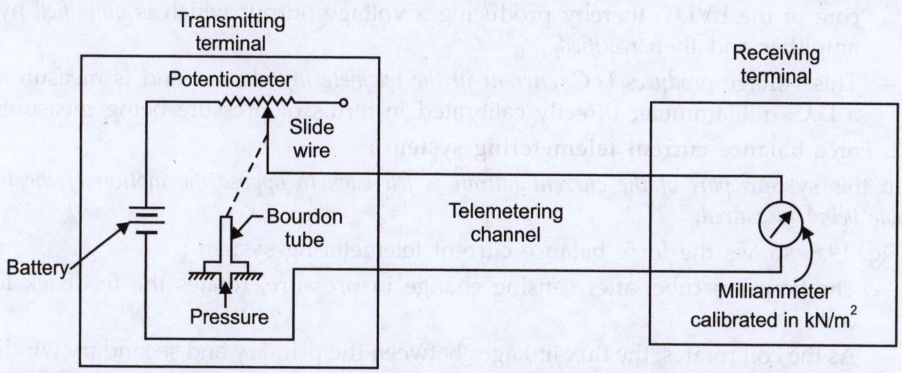

Fig. 19.6. Basic current telemetering system.

Construction :

— It consists of a slide-wire potentiometer in series with a battery.

— The slider is connected to the bourdon tube which measures pressure.

— At the receiving terminal, a multimeter is connected in series, which is calibrated in terms of pressure scale (kN/m^2).

Working/Operation :

— When the pressure in the system changes, the bourdon tube moves the sliding contact thereby changing the current at the transmitting terminal.

— This current passes to the receiving terminal through the pairs of wires (telemetering channel) and at the receiving terminal it is measured by the milliammeter.

These days, the following *improved versions* of this system are used :

1. Motion balance system; 2. Force balance system

1. Motion balance current telemetering system :

In this system slide-wire potentiometer is replaced by a position detector like inductor transducer or capacitor transducer.

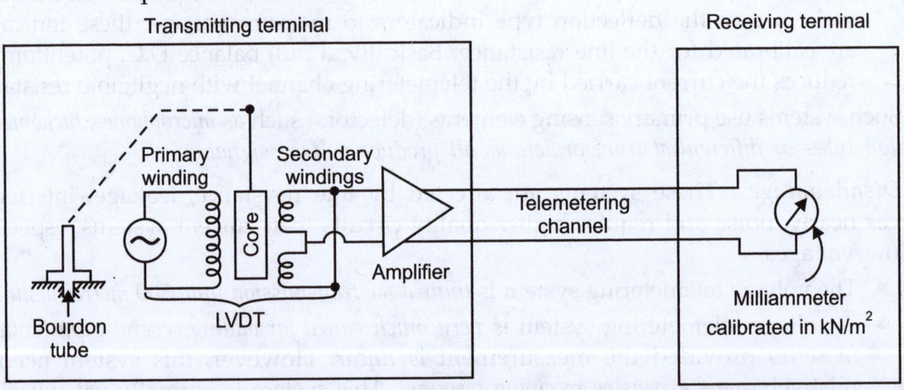

Fig. 19.7. Motion balance current telemetering system.

Fig. 19.7. shows a motion balance current telemetering system where a linear variable differential transducer (LVDT) has been used as inductive transducer :

— The pressure acting on the bourdon tube causes a displacement which moves the core of the LVDT, thereby producing a voltage output which is *amplified* by an amplifier and then *rectified.*

— This voltage *produces D.C. current in the telemetering channel* and is measured by a D.C. milliammeter directly calibrated in terms of pressure being measured.

2. Force balance current telemetering system :

In this system *part of the current output is fed back to oppose the motion of the input variable being measured.*

Fig. 19.8. shows the force balance current telemetering system :

— The bourdon tube, after sensing change in pressure, rotates the feedback force coil.

— As the coil rotates, the flux linkages between the primary and secondary windings change. This change in flux linkages varies the amplitude of the amplifier. The output is connected to the feedback force coil which produces a force opposing the bourdon tube input.

This system of current telemetering *increases the accuracy as small variations are required which result in better linearity.*

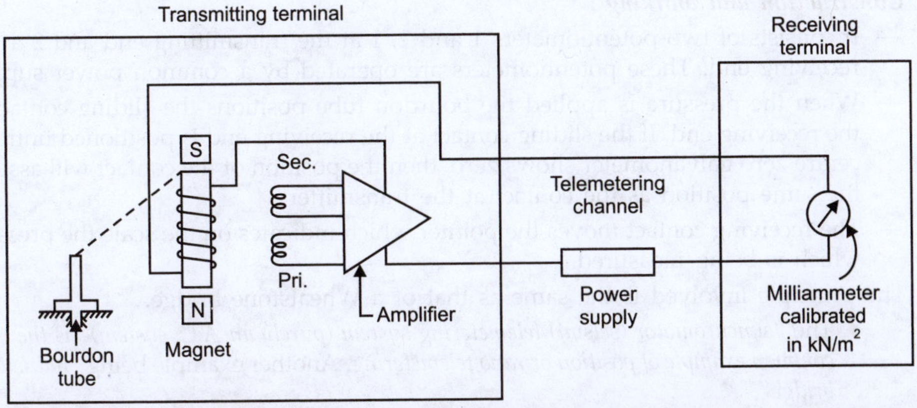

Fig. 19.8. Force balance current telemetering system.

Advantages of current telemetering system :

(*i*) The current systems can develop higher voltages than most voltage systems and, consequently, it can be made more immune to the effect of thermal and inductance voltages in the interconnecting leads as well as line resistance.

(*ii*) Simple D.C. milliammeters can be used with special calibration for line resistance.

(*iii*) Several receivers can be operated simultaneously.

(*iv*) The received signals can be added or subtracted directly.

(*v*) Changes in line resistance are compensated by basic feedback method.

(*vi*) The response of the system to an input change is almost instantaneous.

(*vii*) The energy level is adequately high to minimise the effects of extraneous voltages.

- This system is *also not suitable for long distance* since the current output is varied by means of an adjustable resistance in the line.

19.2.6. Position or (Ratio) Telemetering System

This system transmits and reproduces the measured variable by positioning variable resistors or other electrical components in a *bridge circuit* form so as to produce proportional changes at both the transmitter and the receiver ends.

Fig. 19.9 shows a position telemetering *bridge type* system :

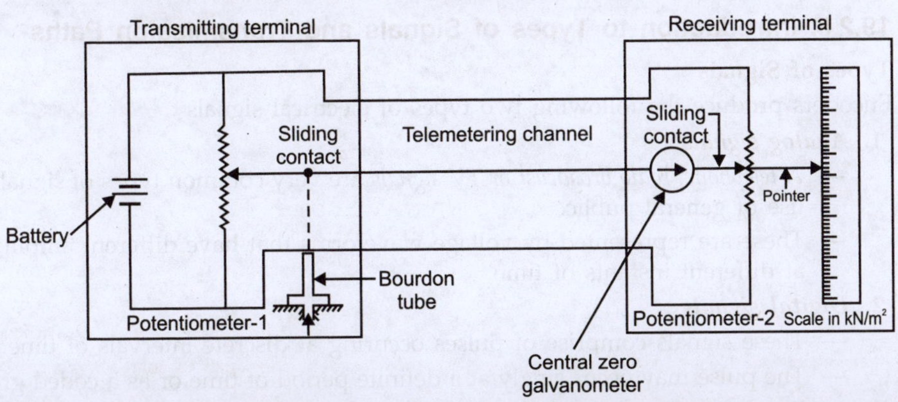

Fig. 19.9. Position telemetering bridge type system.

Construction and working :

— It consists of two potentiometers 1 and 2, 1 at the transmitting end and 2 at the receiving end. These potentiometers are operated by a common power supply.

— When the pressure is applied the bourdon tube positions the sliding contact at the receiving end. If the sliding contact at the receiving end is positioned until the centre zero galvanometer shows zero, then the position of the contact will assume the same position as the contact at the transmitter.

— The receiving contact moves the pointer which indicates on the scale the pressure which is being measured.

The principle involved is the same as that of a Wheatstone bridge.

- The *"synchromotor (selsyn) telemetering system (purely an A.C. system)" is the most common example of position or ratio telemetering*. Another example being *"inductance bridge"*.

— In this system *angular input displacement is converted into relative magnitude of three phase A.C. voltages.*

Advantages :

(*i*) Requires no intermediate amplifiers or conversions.

(*ii*) Relatively inexpensive.

(*iii*) Minimum moving parts, so the maintenance is low.

(*iv*) Instantaneous response.

(*v*) Power taken for their operation directly from the line.

Limitation : These systems are *affected by excessive line resistance.*

19.2.7. Radio Frequency (RF) Telemetry System

In this telemetry system the *link between the transmitting end and the receiving end is established through the "radio links"* (there being no physical link).

Example : The controlling of aircraft on test flights, rockets and spacecrafts.

- The RF telemetry systems are *more suitable for transmission of data over distances more than 1 km.*
- In this system, along with certain band of the radio frequency spectrum, a microwave link above 4 MHz is also allocated. This is because the radio waves at these frequencies travel in straight lines with some repeaters, located every 30 km to 60 km, on high buildings or towers.

19.2.8. Introduction to Types of Signals and Transmission Paths

Types of Signals :

Encoders produce the following two types of electrical signals :

1. *Analog signals :*

— A *telephone, radio broadcast or TV signals* are very common types of signals for use of general public.

— These are represented by voltage waveforms that have different amplitudes at different instants of time.

2. *Digital signals :*

— These signals comprise of pulses occuring at discrete intervals of time.

— The pulse may occur singly at a definite period of time or as a coded group.

— These signals play a very important role in the transmission and reception of coded messages.

— Simplest types of digital signals are the Telegraph and Teleprinter signals but sometimes analog messages are also converted into digital form before being transmitted for certain reasons.

Transmission Paths :

The transmission of message can be carried out by means of the following :

1. **Transmission lines** ... termed as *Line Communications*.
2. **Radio waves** ... termed as *Radio Communications*.

1. Line communications :

- The most common form of line communication is *telephone* and *telex networks*.
- These are carried out using overhead lines on poles and also by use of buried cables.
- In UHF range and above line communications are carried out by the use of **wave guides.** The following points are worth noting :
  (*i*) In *wave guides* the signal propagation takes place by means of *radio waves* provided the cross-sectional dimensions of the proper wave guide rectangular hollow pipe maintain a proper relation with a wavelength of the wave.
  (*ii*) *Wave guides* are employed as transmission lines for microwave communication in the frequency range 2 GHz to 11 GHz.
  (*iii*) *Wave guides* do not transmit signals *below cut-off frequency* and can have band-width in excess of 20% of their operating frequencies.
  (*iv*) *Optical wave guides* have been developed in the form of glass fibres to carry light modulated signals at frequencies of the order of 100 THz (1 Tera hertz = 10^{12} hertz) from *lasers*. This will have a very vast capacity to carry signals.

2. Radio communications :

- In this type of communication propagation of signals through atmosphere is used.
- Non-directional antennas are employed for broadcast transmission and reception while directional antennas are employed in radio telephony and other point-to-point links.
- Radio waves are commonly termed as unbounded waves but generally they are bounded by the surface of earth and various ionospheric conducting layers that lie between a height of 50 km and 400 km above ground.

Major Radio Frequency Bands

	Name	*Frequencies*	
1.	Very low frequency (VLF)	Below 30 kHz	... Radio location equipment
2.	Low frequency (LF)	30 kHz to 300 kHz	... Wartime radio navigation
3.	Medium frequency (MF)	300 kHz to 3 MHz	... Includes AM radio broadcast band
4.	High frequency (HF)	3 MHz to 30 MHz	... Radio
5.	Very high frequency (VHF)	30 MHz to 300 MHz	... Includes FM broadcast band and television VHF channels

6. Ultra high frequency (UHF)	300 MHz to 3 GHz	...	Includes television UHF channels
7. Super high frequency (SHF)	3 GHz to 30 GHz	...	Satellite communications
8. Extremely high frequency (EHF)	30 GHz to 300 GHz	...	Satellite communications

- In a *radio receiver* by providing an *RF amplifier* the following *advantages* are obtained :

(*i*) Signal-to-noise ratio is improved.

(*ii*) Availability of greater gain and better sensitivity.

(*iii*) Image frequency rejection is improved.

(*iv*) Spurious frequencies are prevented.

(*v*) Better selectivity.

(*vi*) Improved coupling of the receiver to the antenna.

19.2.9. Modulation and Demodulation

19.2.9.1. Introduction

Several signals out of different types of signals that are generally encountered in communication systems have frequency spectra that is not suitable for direct transmission especially when atmosphere is used as the transmission channel. In such a case, the frequency spectra of the signal may be translated by *modulating high frequency carrier wave with signal.*

In order to transmit and receive the intelligence (code, voice, music etc.) successfully, the following two processes are essential :

(*i*) Modulation; (*ii*) Demodulation.

19.2.9.2. Modulation :

Definition :

Modulation may be defined as follows :

- ***Modulation*** *is the process of combining the low-frequency signal with a very high-frequency radio wave called 'carrier wave (CW)'.* The resultant wave is called *modulated carrier wave.* This job is done at the transmitting station.

Or

Modulation *is a process in electronic circuits by which the characteristics of one wave form (carrier) is modified by the variations in another wave (audio signal).*

Or

Modulation *is the process of combining an audio-frequency (AF) signal with a radio frequency (RF) carrier wave (Fig. 19.10).* AF signal is called a *modulating wave* and the resulting wave produced is called *modulated wave.* During modulation, some characteristic of the carrier wave is varied in time with the modulating signal and is accomplished by combining the two.

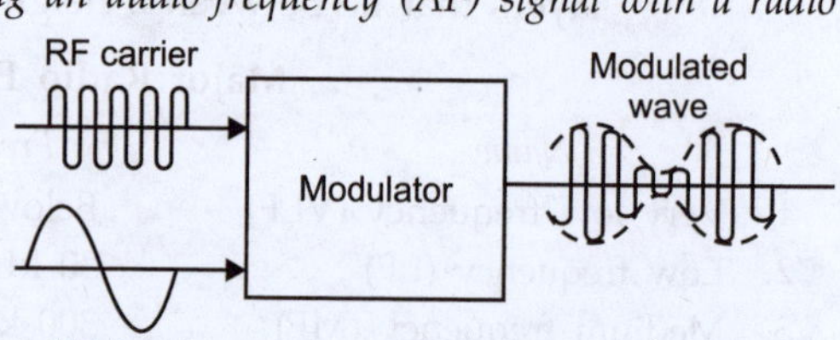

Fig. 19.10

19.2.9.3. Need of modulation

In electronic communication, the modulation is necessary for the following *reasons :*

1. Modulation increases operating range.

2. It reduces the size of transmitting and receiving antennas.
3. It permits transmission without wire.
4. It is extremely difficult to radiate low frequency signals through earth's atmosphere in the form of electromagnetic energy.

19.2.9.4. Methods of modulation

For a sinusoidal carrier wave, the mathematical expression is given as

$$e = E_c \sin (\omega_c t + \phi) = E_c \sin (2\pi f_c t + \phi) \quad ...(1)$$

Thus, the waveform can be varied by any of its following three factors or parameters :

(*i*) E_c – The *amplitude* ;

(*ii*) f_c – The *frequency* ;

(*iii*) ϕ – The *phase*.

Accordingly, there are three types of sine-wave modulations known as :

1. Amplitude Modulation (AM)
2. Frequency Modulation (FM)
3. Phase Modulation (FM).

1. Amplitude modulation (AM) :

The process by which the amplitude of a carrier wave is varied in accordance with the modulating signal is called **amplitude modulation.**

The process of amplitude modulation is shown graphically in Fig. 19.11.

— For the sake of simplicity, the AF/message signal has been assumed sinusoidal [Fig. 19.11(*a*)].

— The carrier wave by which it is desired to transmit the AF signal is shown in [Fig. 19.11(*b*)].

— The resultant wave called *modulated wave* is shown in [Fig. 19.11(*c*)].

The function of the modulation is to mix these two waves in such a way that (*a*) is transmitted along with (*b*).

— The fluctuations in the amplitude of the carrier wave depend on the *signal amplitude* and the rate at which these fluctuations take place depends on the *frequency* of the audio signal.

— *All stations are broadcasting on the standard broadcast band (550 – 1550 kHz) use AM modulation.*

— **Percent modulation** *indicates the degree to which the AF signal modulates the carrier wave.*

— *Methods of amplitude modulation are :*

(*i*) Amplifier modulation.

(*ii*) Oscillator modulation.

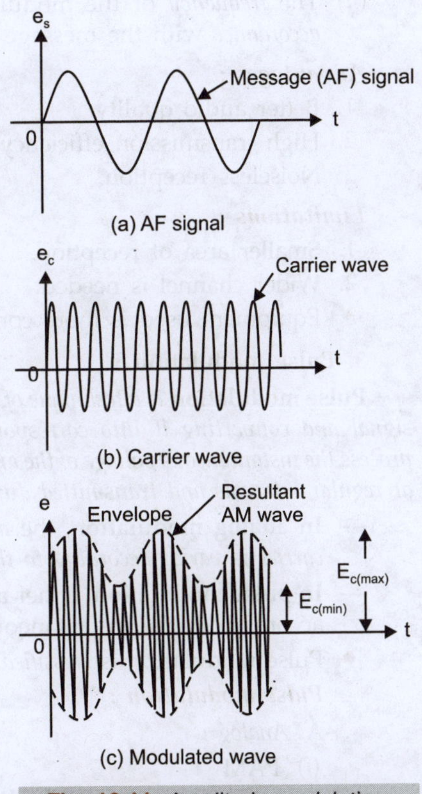

Fig. 19.11. Amplitude modulation.

Limitations of amplitude modulation :

Following are the *limitations* of amplitude modulation :

1. Smaller operating range.
2. Poor efficiency.
3. Poor audio quality.
4. Noisy reception.

2. Frequency modulation (FM) :

The process by which the frequency of a carrier wave is varied in accordance with the modulating signal is called **frequency modulation.**

The process of frequency modulation is shown graphically in Fig. 19.12.

The three waves shown in the figure are : (*a*) Signal; (*b*) Carrier; (*c*) The resultant frequency modulated wave.

When a signal of frequency f_s is modulated with carrier wave of frequency f_c, a resultant modulated wave is produced.

The following points are worth noting :

(*i*) The *amplitude* of the modulated wave is the *same* as that of the carrier wave.

(*ii*) The *frequency* of the modulated wave *varies in accordance* with the message signal.

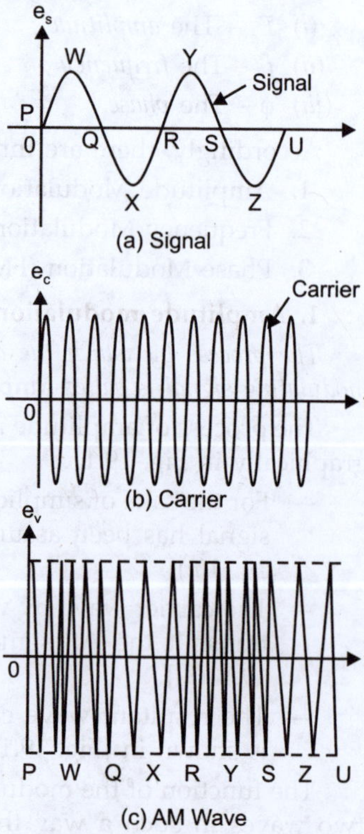

Fig. 19.12. Frequency modulation.

Advantages :

1. Better audio quality.
2. High transmission efficiency.
3. Noiseless reception.

Limitations :

1. Smaller area of reception.
2. Wider channel is needed.
3. Equipment used is more complex and costly.

3. Pulse modulation :

Pulse modulation is *a technique of modulating the analog signal and converting it into corresponding values. In this process the instantaneous voltage of the analog signal is sampled at regular intervals and transmitted during these sampling periods only.*

— **In analog modulation**, the *amplitude or frequency or phase (any one parameter) of carrier is varied according to the instantaneous voltage of signal.*

— **In pulse modulation**, either amplitude or width or position of the pulse is varied according to the instantaneous voltage of signal (except for PCM).

- Pulse modulation is *classified* as follows :

Pulse modulation :

A. *Analog :*

(*i*) PAM

(*ii*) PTM → PDM, PWM
           → PPM

B. *Digital* :

(*i*) PCM, (*ii*) Delta-modulation.

- **Sampling theorem :** This theorem was developed by *Nyquist.*

  It states : *"If sampling rate is any pulse modulating system exceeds twice the maximum signal frequency, the original signal can be constructed at the receiver with minimum distortion."*

  — If f_m is the maximum signal frequency then sampling rate is greater than $2f_m$.

  — The minimum samplings rate is called the *Nyquist rate.*

  — *The reciprocal of the sampling rate* (or sampling frequency) is called ***sampling line*** or ***Nyquist interval.***

- **Pulse Amplitude Modulation (PAM) :**

  — PAM can be *generated* by using an AND gate.

  — PAM can be *demodulated* by passing through low pass filter with cut-off frequency as the highest signal frequency.

  — Generally FM is used to modulate the pulses so that it can be transmitted; such a system is called PAM-FM.

- **Pulse Width Modulation (PWM) :**

  — It is also called Pulse Duration Modulation (PDM) or Pulse Length Modulation (PLM).

  — It can be generated using a monostable multivibrator.

  — It can be demodulated by feeding the PWM signal to an integrating circuit.

  — In PWM the width of the pulse is changed in accordance with the instantaneous value of the modulating signal with amplitude remaining constant.

  — The most oftenly used integrating circuit is the loudspeaker itself.

- **Pulse Position Modulation (PPM) :**

  — It can be generated in the similar way as PWM but the pulse width is kept *constant* from the starting point of occurrence of pulse.

  — It can be *demodulated* by converting into PWM using a flip-flop.

  — In PPM the position of the pulse or the time of occurrence of the pulses is changed in accordance with the instantaneous magnitude of the modulating signal.

- **Pulse Code Modulation (PCM) :**

  — In this type of modulation, digital equivalents of the instantaneous voltage levels of the signals are transmitted in the form of pulses.

  — The total amplitude range which the signal may occupy is divided into a number of standard levels; this process is called *Quantizing.* The level actually set is the nearest standard level. The *quantization levels depend upon number of bits per sample.*

  — PCM encoder functions similar to A/D converter; PCM decoder functions similar to D/A converter.

19.2.9.5. Digital modulation techniques

In digital communications, the modulating signal consists of binary data or an *M*-ary encoded version of it. This data is used to modulate a carrier wave (usually sinusoidal) with fixed frequency. In fact the input data may represent the digital computer outputs or PCM waves, generated by digitizing voice or video signals. The channel may be a telephone channel, microwave radio link, satellite channel or an optical fiber.

In ***digital communication***, *the modulation process involves switching or keying the amplitude, frequency or phase of the carrier in accordance with the input data.* Thus, there are three basic modulation techniques for the transmission of digital data; they are known as :

(*i*) Amplitude-shift keying (ASK).

(*ii*) Frequency-shift keying (FSK).

(*iii*) Phase-shift keying (PSK).

The above techniques can be viewed as special cases of amplitude modulation, frequency modulation and phase modulation respectively.

Types of digital modulation techniques :

Various types of digital modulation techniques are :

I. *Coherent digital modulation techniques :*

1. Coherent binary modulation techniques.
2. Coherent binary amplitude shift keying or on-off keying.
3. Coherent demodulation of binary ASK.
4. Binary phase shift keying (BPSK).
5. Coherent binary frequency shift keying (BFSK).

II. *Non-coherent binary modulation techniques :*

1. Non-coherent binary ASK.
2. Non-coherent detection of FSK.
3. Differential phase shift keying (DPSK).
4. Quadrature phase shift keying (QPSK).

19.2.9.6. Demodulation or detection

*The process of extracting the low frequency modulating signal from the modulated wave is known as **demodulation** or **detection**.*

- The demodulation of an *'AM wave'* involves *two operations :*
  (*i*) Rectification of the modulated wave ;
  (*ii*) Elimination of the RF component of the modulated wave.
- The demodulation of an *'FM wave'* involves *three operations :*
  (*i*) Conversion of frequency changes produced by modulating signal into corresponding amplitude changes ;
  (*ii*) Rectification of the modulating signal ;
  (*iii*) Elimination of RF component of the modulated wave.

19.2.9.7. Noise

- **Noise** may be defined, in an electrical sense, *as any extraneous form of energy tending to interfere with the proper and easy reception and reproduction of those signals which it is desirable to receive.*
- **Noise Power :**

— Noise power,

(*i*) is *directly proportional to temperature ;*

(*ii*) is *directly proportional to the bandwidth of system ;*

(*iii*) is *independent of resistance value.*

— Unit of *noise power is watts.*

— Units of *noise power density are watts/hertz.*

— The *average thermal noise power is constant as long as the bandwidth is fixed.*

— ***Thermal noise voltage*** is *proportional to temperature, bandwidth and resistance.*

- **Noise figure :** *It is the ratio of S/N ratio at input to S/N ratio at output.* Ideal noise figures is '1' or 0 dB. Noise figure is *always greater than 1. Higher the noise figure of the amplifier, noiser the amplifier.*
- **Classification of noise :**

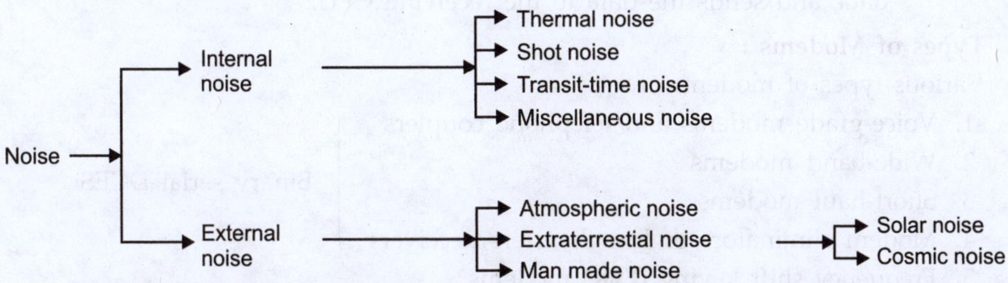

All receivers have their basic components consisting of tubes or transistors. The different types of noise present in these devices are thermal noise, shot noise, partition noise etc.

— In superheterodyne receivers, mixers are much noiser than amplifiers. It is so, because, the plate current of the partition noise is greatly increased in mixers.

— At short wave frequencies, the image frequency rejection is very poor and this introduces noise.

— A low value of transductance also helps in increasing the noise.

Thus, the effect of noise is maximum in mixers of all the receivers.

19.2.9.8. Modems

- Modem is an *acronym for MODulator DEModulator.*
- *A* **modem** *is a device that converts data from digital computer signals to analog signals that can be sent over a phone line.* This is called *"modulation". The analog signals are then converted back into digital data by the receiving modem.* This is called *"demodulation".*
- Fig. 19.13 shows a modem interfacing block diagram :

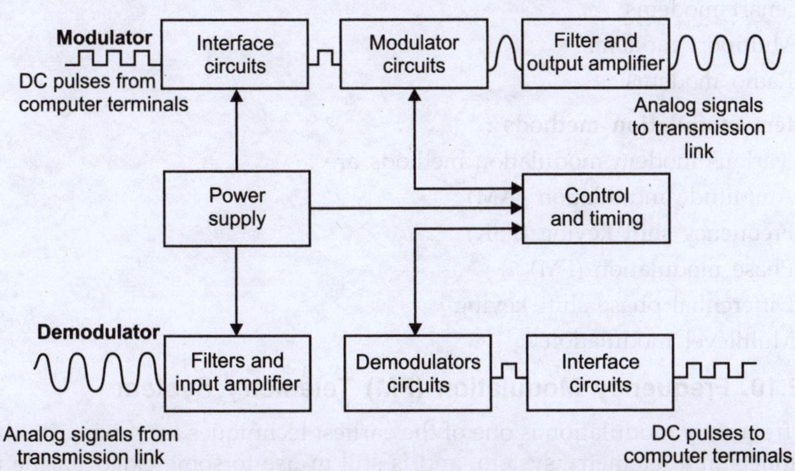

Fig. 19.13. Modem interfacing block diagram.

— A modem is fed with digital information in the form of ones and zeros, from the CPU (central processing unit). The modem then analyzes this information and converts it into analog signals, that can be sent over a telephone line.

— Another modem then receives these signals, converts them back into digital data, and sends the data to the receiving CPU.

Types of Modems :

Various types of modems are :

1. Voice-grade modems and telephone couplers
2. Wide-band modems
3. Short-haul modems
4. Modem eliminators or line diverters/receivers

(Items 1–4: Binary serial DCES)

5. Frequency shift keying (FSK) modems
   — Bell 103 modem
   — V 21 modem
   — Bell 202 modem
6. Phase shift keying (PSK) modems
   — 201 modem
   — V.26 modem
   — Bell 212 modem
   — V.22 modem
   — V.22 bis modem
   — V.27 modem
   — V.29 modem
   — V.32 modem
   — V.32 bis modem
   — V.FC modem
   — V.34 modem
7. Data modems
8. Smart modems
9. Multiport modems
10. Radio modems.

Modem modulation methods :

The various modem modulation methods are :

1. Amplitude modulation (AM)
2. Frequency shift keying (FSK)
3. Phase modulation (FM)
4. Differential phase shift keying
5. Multilevel modulation.

19.2.10. Frequency Modulation (FM) Telemetry System

The frequency modulation is one of the earliest techniques for mixing (multiplexing) data channels in a telemetry system, and is still in use to some extent in the telemetry field.

Fig. 19.14 shows a basic frequency modulation (FM) telemetry system :

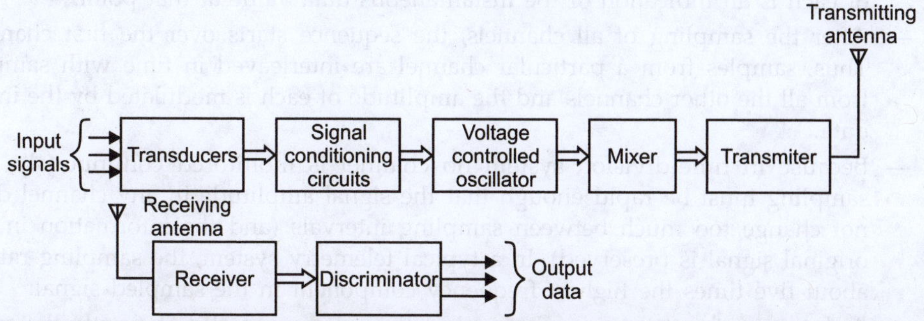

Fig. 19.14. Basic FM telemetry system.

— The output of each transducer/signal conditioner modulates the frequency of a voltage-controlled subcarrier oscillator. Many oscillators, each operating in a dedicated part of the frequency spectrum, are mixed for radio transmission. As each voltage controlled oscillator is assigned with a separate frequency of the frequency spectrum, so each signal can be modulated without interfering with other signals.

— At the receiving station, a FM demodulator (usually called *discriminator*) is tuned to the frequency of each subcarrier. Thus, when the measurement value changes at the source, the output voltages of the discriminator change accordingly.

19.2.11. Pulse Amplitude Modulation (PAM) Telemetry System

The primary limitation of the FM telemetry system is that the *capacity of channels offered is less.* Thus, to have still higher capacity of channel within a linked bandwidth, the PAM telemetry system is widely used which makes use of the *"time division multiplexing"* technique.

Fig. 19.15 shows a basic time division multiplexing system :

— All channels use same frequency band of the allotted frequency spectrum but *not at the same time.*

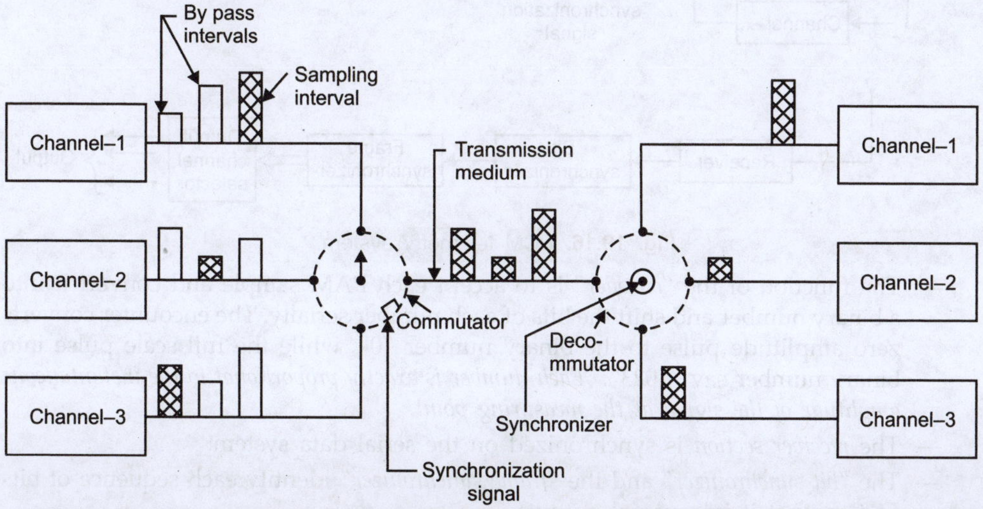

Fig. 19.15. Basic time division multiplexing system.

— The *commutator* samples the signal in each channel in sequence, and the amplitude of each is an indication of the instantaneous data value at that point.

— After the sampling of all channels, the sequence starts over the first channel. Thus, samples from a particular channel are interleaved in time with samples from all the other channels and the amplitude of each is modulated by the input data.

— Because in time-division system no channel is monitored continuously, the sampling must be rapid enough that the signal amplitude in any channel does not change too much between sampling intervals (and the information in the original signal is preserved). In a typical telemetry system, the sampling rate is about five times the highest frequency component in the sampled signal.

— A *decommutator* at the receiving end of the system operates at exactly the same frequency as the commutator and distributes the parts of the multiplexed signal to the proper output channels.

Because a time-division system is *based on precise timing*, it is of significant importance that the *demodulator be synchronized exactly with the commutator.*

19.2.12. Pulse Code Modulation (PCM) Telemetry System

This system is so named, since the measured data is represented in the form of *binary weighted codes.*

This system also makes use of time division multiplexing system (as used in PAM); however, in this system an *encoder is used after the commutator stage.*

Fig. 19.16 shows the block diagram of the PCM telemetry system :

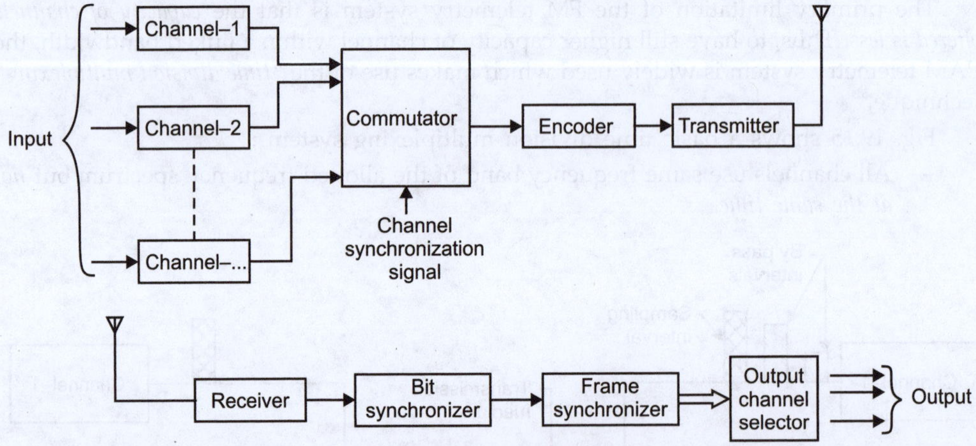

Fig. 19.16. PCM telemetry system.

— The function of the *"encoder"* is to accept each PAM sample and convert it into a binary number and shift the bits of each number serially. The encounter converts zero amplitude pulse to the binary number "0", while the full scale pulse into binary number say "1023". *Each number is exactly proportional to the instantaneous amplitude of the signal at the measuring point.*

— The *receiver section* is synchronized on the serial data system.

— The *"bit synchronizer"* and the *"frame synchronizer"* identify each sequence of bits and convert it into useful outputs.

Comparison of FM, PAM and PCM

S. No.	Aspects	FM	PAM	PCM
1.	*Accuracy*	Poor	Poor	Excellent
2.	*Efficiency in use*	Medium	Best	Worst
3.	*Percentage of use in practical systems*	20%	15%	65%

HIGHLIGHTS

1. *Telemetry* is defined as a technology which allows users to collect information from inaccessible and inconvenient locations and to transmit it to the accessible places to process, record and display the information in presentable form.
2. The RF telemetry systems are more suitable for transmission of data over distances *more* than 1 km.
3. In pulse code modulation (PCM), digital equivalents of the instantaneous voltage levels of the signals are transmitted in the form of pulses.
4. A *'modem'* is a device that converts data from digital computer signals to analog signals that can be sent over a phone line. This is called modulation. The signals are then converted back into digital one by receiving modem. This is called 'demodulation'.

OBJECTIVE TYPE QUESTIONS

Choose the Correct Answer :

1. The FM telemetry as compared with AM telemetry requires a channel that is
   (*a*) smaller than what is required for AM telemetry
   (*b*) equal to that of AM telemetry
   (*c*) 10 times of that required for AM telemetry
   (*d*) 100 times of that required for AM telemetry.
2. FM systems operate at
   (*a*) VHF and UHF (*b*) only VHF
   (*c*) only MF (*d*) MF and HF.
3. FM systems as compared to AM systems are
   (*a*) less affected by noise
   (*b*) equally affected by noise
   (*c*) more affected by noise
   (*d*) highly affected by noise since they operate at VHF and UHF.
4. Time division multiplexing requires
   (*a*) constant data transmission
   (*b*) transmission of data samples
   (*c*) transmission of data at random
   (*d*) transmission of data of only one measurand.
5. Modem is an acronym of
   (*a*) modulation (*b*) demodulation
   (*c*) modulation and demodulation (*d*) all of the above

6. Time division multiplexing is used when the data to be transmitted
   (*a*) is slow changing
   (*b*) has a small bandwidth
   (*c*) is slow changing and has a low bandwidth
   (*d*) is fast changing and has a high bandwidth.

ANSWERS

1. (*c*)	2. (*a*)	3. (*a*)	4. (*b*)	5. (*c*)	6. (*c*)

THEORETICAL QUESTIONS

1. Define the terms "Data transmission" and "Telemetry'.
2. What is the difference between a 'transmitter' and a 'telemeter' ?
3. Explain briefly any two of the following :
   (*i*) Mechanical transmission.
   (*ii*) Hydraulic transmission.
   (*iii*) Pneumatic transmission.
   (*iv*) Magnetic transmission.
4. Discuss briefly the working of a general telemetering system, with the help of a block diagram.
5. How are telemetry systems classified ?
6. List the advantages and disadvantages of land line telemetry system.
7. Describe briefly the following :
   (*i*) Voltage telemetering system.
   (*ii*) Current telemetering system.
8. Explain briefly 'Force balance current telemetering system'.
9. State the advantages of current telemetering system.
10. Discuss the working of a 'position telemetering system' with the help of a neat diagram.
11. Write a short note on 'RF telemetry system'.
12. What do you mean by the terms 'Modulation' and 'Demodulation' ?
13. Explain briefly the following :
    (*i*) Amplitude modulation.
    (*ii*) Frequency modulation.
14. What do you mean by the following :
    PAM; PWM; PPM; PCM.
15. What is a modem? Explain.
16. Explain briefly the following telemetry systems :
    (*i*) Frequency modulation (FM) telemetry system.
    (*ii*) Pulse amplitude modulation (PAM) telemetry system.

CHAPTER

20 Display Devices and Recorders

20.1 Introduction; 20.2 Electrical indicating instruments-Analog instruments-Digital instruments **20.3 Digital display units :** Cathode ray tube (CRT)-Light emitting diode (LED)-Liquid crystal display (LCD)-Cold cathode displays or Nixies-Electro-luminescent (EL) displays-Incandescent display-Electrophoretic image display (EID)-Liquid vapour display (LVD)-Comparison of various display devices; **20.4 Recorders :** Introduction-types of recorders-strip chart recorders-circular chart recorders-X-Y recorders-oscillographic recorders-magnetic tape recorders. Printers; *Highlights; Objective Type Questions; Theoretical Questions.*

20.1 INTRODUCTION

The main purpose of any measurement system is to provide information concerning the state and condition of the physical phenomenon being investigated. The measuring systems may be activated either directly from the measuring means (*e.g.*, bellows, pressure spiral etc., to which indicating pointer is attached directly through lever and leverage system) or by means of servo-operated system (null-balance system which incorporates a feedback circuit in a closed loop). The last stage of a measurement system is the *data presentation stage;* if the results of the system are meaningful they must be displayed for *instant observation by a display device* or *for storage for observation at a later stage by a recorder* (The data presentation devices may be called as *"output devices"*).

The following factors decide about the *choice between the display devices and recorders* :

(*i*) The information content of the output.

(*ii*) The expected use of the output.

The output devices may be *categorized* as follows :

1. Single number output devices.
2. Time domain output devices.

1. Single number output devices :

Such devices indicate the value of some particular quantity under condition such that the value to be measured can be regarded as time variant over the timer interval during which measurements are made; thus a *single number will represent measurement.*

"Indicating instruments" and *"Digital display units"* belong to this class.

2. Time domain output devices :

The indicating instruments or the digital display (suitable only when the *output varies at a very slow rate*) do not serve the purpose *when the values of the quantity are to be taken as a function of time.*

For fast changing outputs (where signal waveform or shape is the desired information) ... *"Cathode ray oscilloscope (CRO)"* is used.

For keeping a *permanent record of the variation of the output with time ..... "Cathode ray tube photographs", direct writing recorders" strips chart recorders," magnetic tape-recorders* etc. are used.

The machine interpretable outputs can be had from :

— Magnetic tapes;

— Punched paper tapes;

— Punched cards;

— Pulsed signals.

- The information available from an instrument may take the following forms :

(*i*) *Quantitative information* (*e.g.*, angular spread in r.p.m; force in newtons).

(*ii*) *Qualitative information* (*e.g.*, the approximate value or direction of change of some variable, a check reading).

(*iii*) Status information (*e.g.*, On/Off, in/out).

(*iv*) Alphanumeric and symbolic information (*e.g.*, the labels and instructions; letters A to Z, the numerals 0 to 9 punctuation marks and various other simple symbols can be generated and intepreted)

- *A good display, functionally, is one which permits the best combination of speed accuracy and sensitivity when transferring the necessary information from the instrument to the operator.*

20.2 ELECTRICAL INDICATING INSTRUMENTS

Qualitatively the electrical indicating instruments are widely used for measurements of current, voltage, resistance and power. These instruments can be *classified* as follows :

1. Analog instruments
2. Digital instruments.

Analog type of meters use scale and needle (pointer) type arrangement to display the value of the measured parameter. *The measured parameter is converted into electrical signal which further actuates some electromechanical device to move the pointer continuously as the parameter changes.*

Digital type meters *indicate the reading in exact numerals.*

20.2.1. Analog Instruments

Analog indicators *are basically current-operated devices, working on the "d'Arsonval type galvanometer principle"*. The device consists essentially of a circular or rectangular coil of a large number of turns of fine wire, suspended between the poles of a permanent magnet. The coil is mounted in position with a jewel bearing and is coupled to a pointer-indicator moving over a graduated scale. The restoring force for equilibrium is provided through a control spring. The sensitivity and the torque produced depend upon the coil dimensions, number of turns, flux density, and the control spring constant. The resulting deflection (δ) is given as :

$$\delta = KIB \cos \theta \quad ...(20.1)$$

where, K = Constant of proportionality,

I = Current through the coil,

B = Flux density, and

θ = Angle between the plane of the coil and magnetic field when no current is flowing.

The basic movement can be adapted to measure D.C. voltages with a series resistor and to measure A.C. voltages with a rectifying diode and resistor.

Pointer-scale analog indicators :

In analog instruments the value of the measured parameter is indicated by positioning of the indicating pointer against a calibrated scale. The purpose can be achieved either by moving the pointer with relation to a stationary scale (fixed-scale moving-pointer indicators) or the scale may be moved with relation to a fixed reference (moving-scale fixed-pointer indicators).

1. *Single-point indicators :*

The *fixed-scale and movable-pointer indicators,* available in a variety of forms, are shown in 20.1. Fig. 20.2 shows the fixed-pointer and movable-scale indicators.

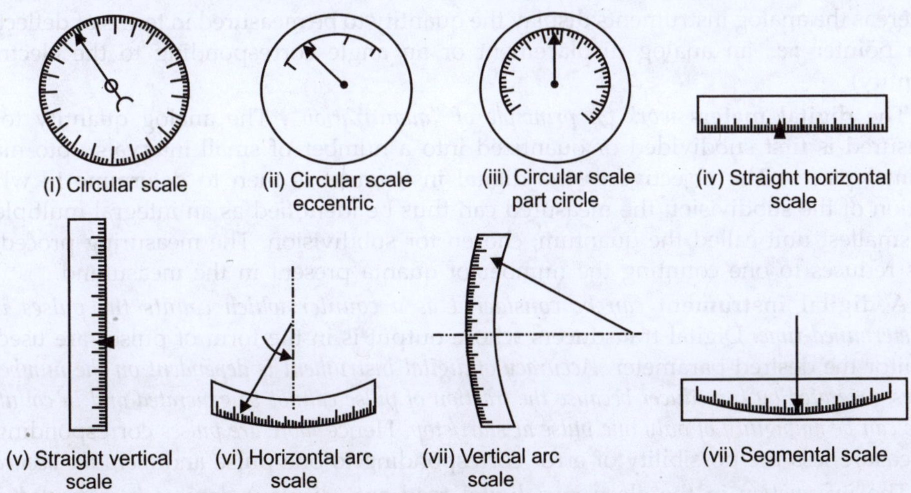

Fig. 20.1. Fixed-scale and movable-pointer indicators.

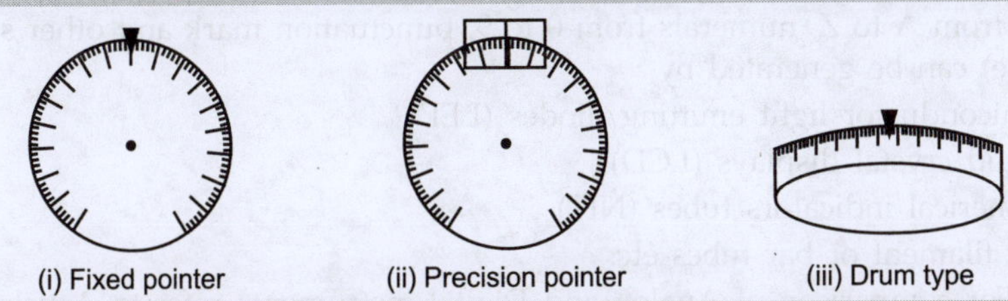

Fig. 20.2. Fixed-pointer and movable-scale indicators.

The readability of graduated dials is influenced by the following factors :

- The shape and length of the pointer.
- The number, spacing, length and thickness of scale marking.
- The system of numbering of the scale marks.
- The size and design of the numerals.

2. *Multi-point, multi-pointer and multi–range indicators :*

Mult-point indicator. In this system the indicator pointer can be connected to a number of inputs, one at a time with the help of a selector switch.

The selector switch may be operated either manually or automatically after a pre-determined time. The observed reading is multiplied by a factor corresponding to the particular measurand.

Generally such systems are confined where measurable variables are of electrical signals as the selection is accomplished by switching electrical circuits. However, gas selector switches also exist which connect one gas pipe at a time to the measuring instrument and are well designed to avoid leakage of gas.

Multi-pointer indicator. This type of indicator contains more than one number of pointers and above each point the identification number of the medium being measured is marked. *Usually this arrangement is used in recorders and not in indicators.*

Multi-range indicators. An instrument with multi-range indicators has *different scales for different ranges;* the choice of a particular scale is made by a selector switch.

20.2.2. Digital Instruments

The **digital instruments** *indicate the value of the measurand in the form of decimal number"* (whereas the analog instruments display the quantity to be measured in terms of deflection of a pointer *i.e.,* an analog displacement or an angle corresponding to the electrical quantity).

The **digital meters** *work on principle of "quantization".* The analog quantity to be measured is first subdivided or quantized into a number of small intervals upto many decimal places. The objective of the digital instrument is then to determine in which portion of the subdivision the measured can thus be identified as an integral multiple of the smallest unit called the quantum, chosen for subdivision. The measuring procedure thus reduces to one counting the number of quanta present in the measurand.

A **digital instrument** *can be considered as a counter which counts the pulses in a predetermined time.* Digital transducers whose output is in the form of pulses are used to monitor the desired parameter. *Accuracy of digital instrument is dependent on the number of pulses generated by transducer because the fraction of pulse cannot be generated and in counting there can be ambiguity of only one pulse at start/stop.* Hence *more are pulses* corresponding to a measure less the possibility of error corresponding to one pulse and *more the accuracy.*

The *information* in the electronic digital read-out (display) devices is *presented as a series of digits on tubes, screen or printed on a piece of paper.* The relevant characters (letters of alphabet from A to Z, numerals from 0 to 9, punctuation mark and other symbols in common use) can be generated by :

(*i*) Semiconductor light emitting diodes (LED)
(*ii*) Liquid crystal displays (LCD)
(*iii*) Numerical indicators tubes (NIT)
(*iv*) Hot filament or bar tubes etc.

Note : For comparison of Analog and Digital instruments refer to Article 1.13.

20.3 DIGITAL DISPLAY UNITS

- The digital display units can be classified as follows :

1. Based on the format used :

(*i*) *Planar* (*i.e.,* the entire read-out characters and in the *same plane*)
— Illuminated "segmental type"
— Illuminated "dot matrix type"
— Displays using *rear projection* and *gaseous discharge type* segmental displays.

(*ii*) *Non-planar :* Here the characters are displayed in *different planes.*
— *Gaseous discharge tubes like **nixies**.*
— *Displays using "illuminated lucite sheets."*

In general displays may be *classified* as follows :

2. Based on methods of conversion of electrical data into visible light :

(*i*) *Active displays :*

— CRTs

— Gas discharge plasma

— LEDs etc.

(*ii*) *Passive displays :*

— Light controllers

— LCDs

— EPIDs etc.

3. Based on the display size and physical dimensions :

(*i*) *Symbolic displays :*

— Alpha numeric

— Nixie tubes

— LEDs etc.

(*ii*) *Console displays :*

— CRTs

— LEDs etc.

(*iii*) *Large screen display :*

— Enlarged projection system

4. Based on resolution and legibility of characters :

(*i*) *Simple single element indicator.*

(*ii*) *Multi-element displays.*

5. Based on the applications :

(*i*) *Analog displays :*

— Bar graph displays (CRT)

(*ii*) *Digital displays :*

— Nixies

— Alphanumeric

— LEDs etc.

In digital electronic field, the *commonly used displays are* :

1. Cathode ray tube (CRT)
2. Light emitting diode (LED)
3. Liquid crystal display (LCD)
4. Gas discharge plasma displays (cold cathode display or *Nixies*)
5. Electro-luminescent (EL) displays
6. Incandescent display
7. Electrophoretic image displays (EPID)
8. Liquid vapour display (LVD).

20.3.1. Cathode Ray Tube (CRT)

Cathode ray tube (CRT) is the *usually commonly used 'read out system' for displaying large quantities of alphanumeric data.* It forms the basis of CROs and TV systems conventionally.

- The generation system of characters on CRTs require relatively simple electronic circuitry. A typical CRT display has *easy facilities for the control of digit size* by controlling the deflection sensitivity of the system. The number of characters displayed can be changed with the help of time shared deflection and modulator circuits.
- The intensity and brightness can be realized with different gray scales and the display can have different colours depending on the phosphor used in the screen; generally the phosphor is chosen to be green or white.
- CRTs of storage type facilitate showing a stationary pattern on the screen without flickering display. It is also possible to retain the pattern for a long time independent of the phosphor persistence.

20.3.2. Light Emitting Diode (LED)

A P-N junction can absorb light energy and produce electric current. The opposite process is also possible, that is a junction diode can emit light. The emission of light occurs under forward bias condition due to recombination of electrons and holes. The emitted light may be visible or invisible.

A P-N *junction diode, which emits light when forward biased is known as a* **light emitting diode (LED).**

The amount of light output is *directly proportional to the forward current.* Thus, higher the forward current, higher is the light output.

Fig. 20.3 (*a*) shows the schematic symbol of a light emitting diode. The arrows pointing away from the diode symbol represent the *light*, which is being transmitted away from the junction.

Fig. 20.3 (*b*) shows the basic structure of a light emitting diode.

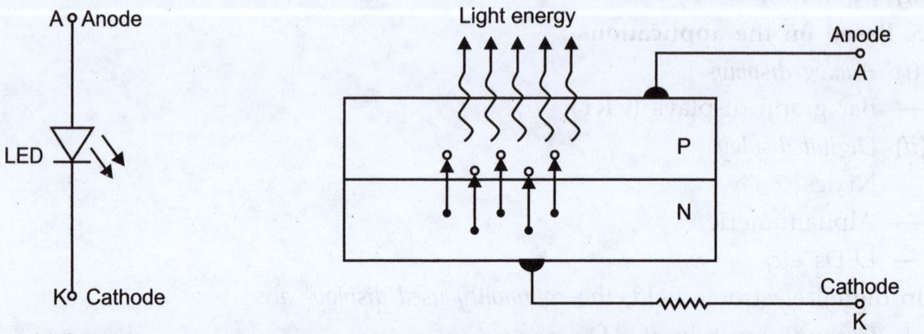

Fig. 20.3. Light emitting diode (LED).

In a forward based *P-N* junction, free electrons from *N*-type material diffuse into P-region. Once in *P*-region these free electrons encounter holes and eventually recombine. Thus the free conduction electron fills a vacancy in covalent structure and thus becomes a valence electron. In doing so the electron loses a certain amount of energy as it jumps form conduction band to the valence band. In Si or Ge *diode, the energy that recombining electrons lose is dissipated in the form of heat. But if other semiconductor material such as gallium arsenide and gallium phosphide are used to form P-N diode, the energy lost by recombining electrons is given off in the form of* **light energy**.

- The buffer requirements for CRT are complex and have to fulfil many functions such as :

(*i*) The buffer must translate data from digital to analog form, which determines the beam position and intensity;

(*ii*) The buffer must refresh the image on the tube face in excess of thirty times a second to avoid flicker;

(*iii*) The buffer must maintain sufficient accuracy to hit any one place out of about five million possible places on the tube face;

(*iv*) To reduce the amount of data needed from the computers to drive the roster and vector CRTs, the buffer must be able to fill in details of the intensity and movement needed for the continual control of the beam and to combine this with data needed to refresh the image.

— Diodes made from gallium arsenide (GaAs) emit infrared radiation invisible to eyes and such diodes are referred to as IRED—Infrared emitting diodes.

Fig. 20.4 shows two curves used to determine LED **operating characteristics**.

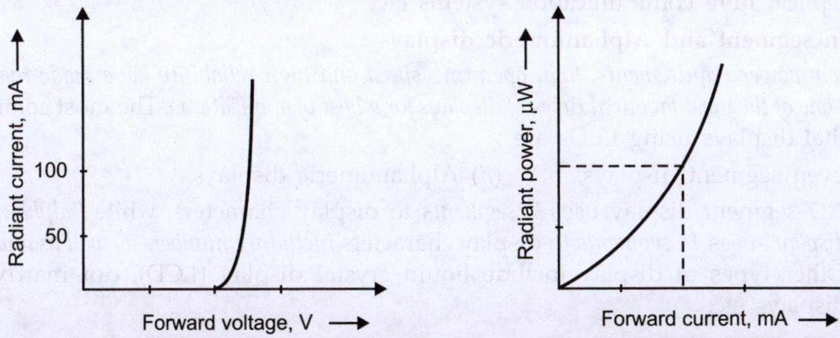

Fig. 20.4. Operating characteristics-LED.

— Fig. 20.4 (*a*) is forward bias V-I curve for a typical IRED, the type used in burglar alarms. Forward bias of around 1 V is required to produce significant forward current.

— Fig. 20.4 (*b*) is a plot of radiant output power *vs* forward current. The radiant output power is rather small (μW) and *indicates a very low efficiency of electrical to radiant energy conversion.*

- They have *fast switching characteristics.*

Advantages :

1. LEDs are *miniature in size* and they can be stacked together to form numeric and alphanumeric display in high density matrix.
2. The light output from an LED is a function of current flowing thought it. Therefore, intensity of light emitted from LEDs can be *smoothly controlled.*
3. LEDs have a *high efficiency* as emitted of electromagnetic radiation. They require moderate power for their operation. They are useful where miniaturization in the low D.C. power are important.
4. They are very useful where *dynamic operation* of large number of arrays is involved.
5. *Economical* and *high degree of reliability.*
6. Being rugged, they *can withstand shocks and vibrations.*
7. *Long life.*
8. They *occupy small area.*
9. *Available in different colours* like red, yellow, green and amber.

Disadvantage :

LEDs are not suited for large areas display, primarily because of their high cost.

Applications :

Since LEDs operate at voltage levels 1.5 V to 3.3 V, they are highly compatible with solid state circuitry. Their *uses* include the following :

(*i*) Panel indicator.
(*ii*) Digital watches.
(*iii*) Multimeters.
(*iv*) Calculators.
(*v*) Intercoms.
(*vi*) Switch boards.
(*vii*) Burglar-alarm systems.
(*viii*) Optical fibre communication systems etc.

Seven segment and Alphanumeric displays :

The low power requirements, high operating speed and high reliability have made the LEDs in general one of the most favoured types of displays for a host of applications. The most commonly used digital displays using LEDs are :

(*i*) Seven segment displays; (*ii*) Alphanumeric displays.

- A 7-segment display uses *7 segments* to display characters, while *"alphanumeric display"* uses *14 segments* to display characters *including numbers as well as alphabets*. Other types of display include liquid crystal display (LCD), dot matrix type displays etc.

1. Seven segment display :

- In segmental displays, it is usual to employ a single LED for each segment [Fig. 20.5 (*a*)].
- For conventional seven segment LED displays [including the decimal point (DP) *i.e.*, the 8th segment], the wiring pattern is simplified by making one terminal common to all LEDs and other terminals corresponding to different segments. The terminals can be either of common anode (CA) form or common cathode (CC) form, shown in Fig. 20.5 (*b*) and (*c*).

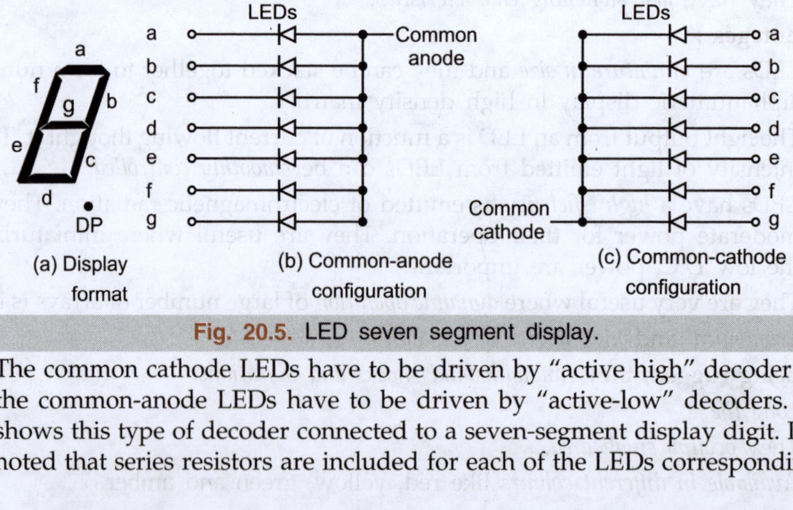

Fig. 20.5. LED seven segment display.

- The common cathode LEDs have to be driven by "active high" decoder ICs, and the common-anode LEDs have to be driven by "active-low" decoders. Fig. 20.6 shows this type of decoder connected to a seven-segment display digit. It may be noted that series resistors are included for each of the LEDs corresponding to the seven segments. Further, an *additional LED corresponding to the decimal point (DP) is also provided in the display,* which again has a current limiting series resistance. The typical value of the series resistance is 150 Ω for a segment current of about 20 mA.

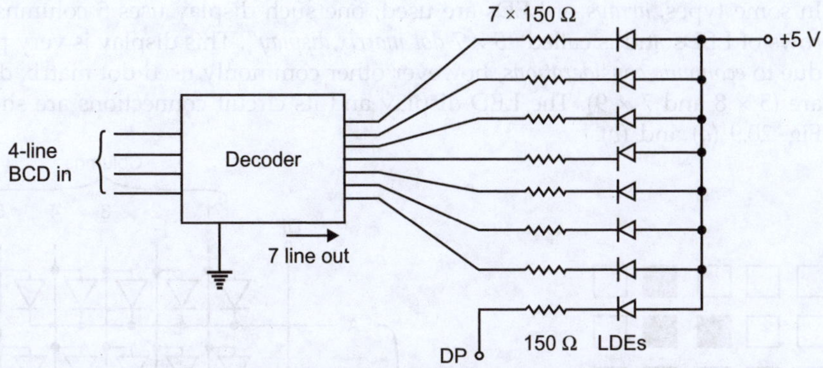

Fig. 20.6. LED seven segment display driver.

- By forward biasing different LEDs, we can display the digits 0 through 9. For *example* to display a zero, the LEDs a, b, c, d, e and f are forward biased. To light up 5, we need to forward bias segments a, f, g, c, d. Thus in a seven segment display, depending upon the digit to be displayed, the *particular set of LEDs is forward biased*. The various digits from 0 to 9 which can be displayed using seven segment display are shown in Fig. 20.7.

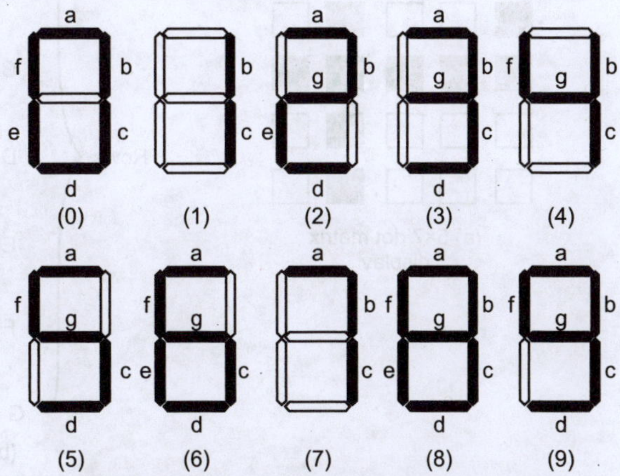

Fig. 20.7. Various digits displayed with seven-segment display.

2. Alphanumeric displays :

- The alphanumeric displays are used to display alphabets as well as numeric characters in response to the electrical inputs. Arrays of LEDs as well as CRTs are extremely useful in this application. "Read only memories" that are used to facilitate generation of alphanumeric characters, and long serial shift registers for memories are now available in MOS/LSI form.
- In such displays, the LEDs are arranged in various ways. One such alphanumeric display using 18 LEDs, called **eighteen-segment display** is shown in Fig. 20.8.

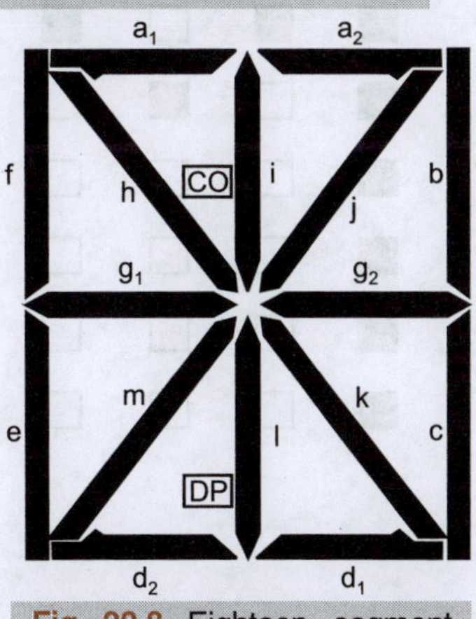

Fig. 20.8. Eighteen- segment display.

— Due to the diagonally arranged LEDs such as j, h, m, k the characters like Y, K, M etc. can be displayed with such a display. Separate LEDs are used for Decimal point (DP) and Colon Operator (CO).

— The various alphanumeric characters are displayed by lighting up the LEDs located in specific positions.

— In some types, *arrays* of LEDs are used; one such display uses 5 columns and 7 rows of LEDs and is called "*5 × 7 dot matrix display*". This display is very popular due to *economic considerations*, however other commonly used dot matrix displays are (5 × 8 and 7 × 9). The LED display and its circuit connections are shown in Fig. 20.9 (*a*) and (*b*).

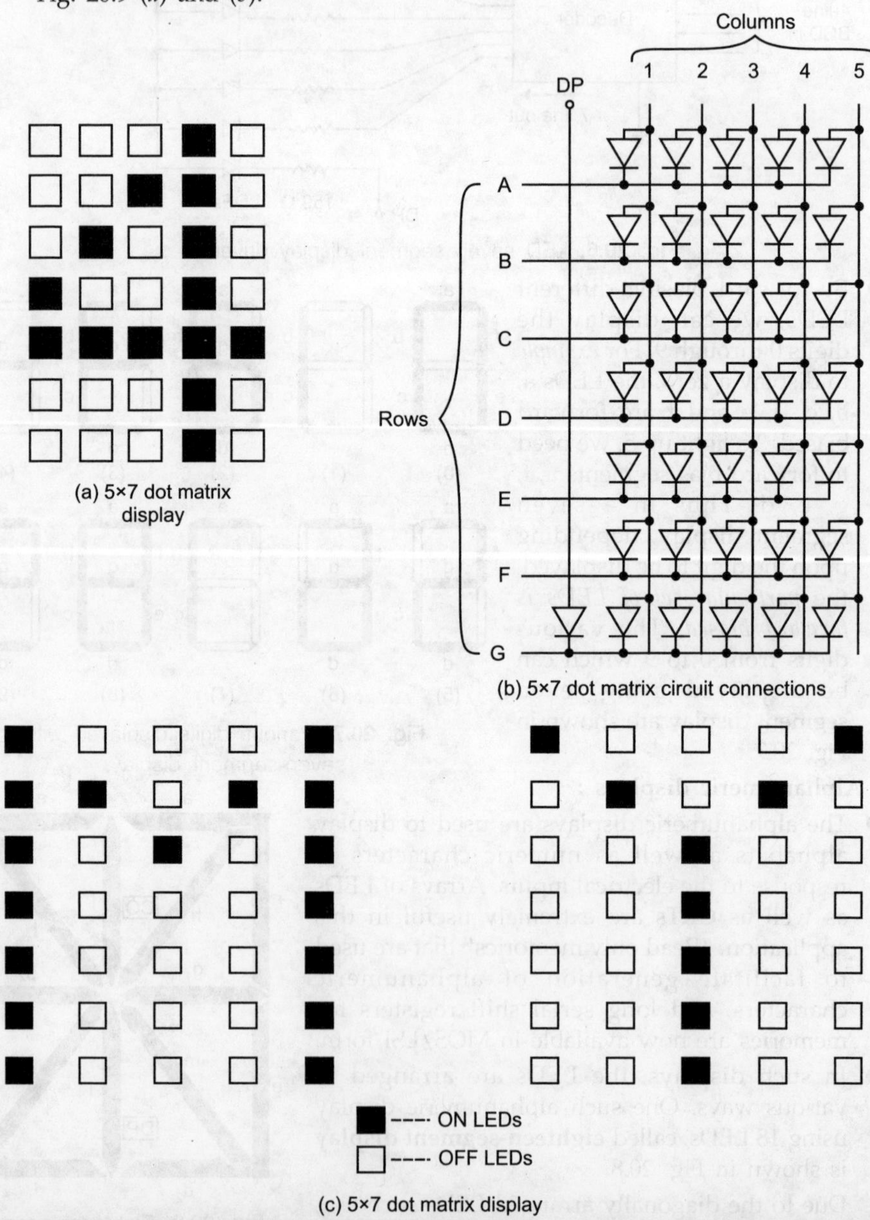

Fig. 20.9. 5 × 7 dot matrix display.

- Depending upon the required character, the corresponding LEDs are switched on, in the display. Fig. 20.9(*c*) shows the display of characters M and Y with the help of 5 × 7 dot matrix display.
- In order to improve the display's brightness, a dynamic system is used. In this arrangement the LEDs are not lit continuously but are sequentially lit by scanning in a "vertical strobe" or horizontal strobe" mode (This is similar to running lights used in modern advertisements).

— In the *"vertical strobe mode"*, a *single row is selected at a time*, the appropriate LEDs are energised in that row, and then the signal is applied to next row.

— In the *"horizontal strobe mode"*, a *single column is selected at a time.*

- Alphanumeric displays using LEDs employ a number of square and oblong emitting areas, arranged either as dot matrix or segmented bar matrix. The alphanumeric LEDs are normally lard out on a single slice of semiconductor material, all the chips being enclosed in a package, similar to an IC, except that the packaging compound is transparent and not opaque.

Bar-graph display :

The **bar graph display** is basically an array of LEDs, LCDs, or gas-discharge panels, arranged in an $n \times m$ *matrix* form, by which the signal magnitude is indicated in a linear or angular format.

The light-emitting diode bar-graph consists of a number of LED arrays (row-wise) arranged in a single or multiple column. The packed density may be as much as 50 to 200 devices per cm^2, and the display signs can be as large as 5 cm × 10 cm.

The *main advantages* of this type of display are :

(*i*) Solid-state nature (without any moving parts)

(*ii*) Good speed of response.

- Such displays find wide applications in *process instrumentation systems.*

20.3.3. Liquid Crystal Displays (LCDs)

LCDs are passive displays characterised by *very low power consumption and good contrast ratio.*

- A liquid crystal is a material, usually an organic compound, which flows like a liquid at room temperature; its molecular structure has some properties normally associated with solids (*e.g.*, cholesterol nonanoate and p-azoxyanisole).
- When light is incident on an activated layer of a liquid crystal, it is *either absorbed or else is scattered by the disoriented molecules.*
- A liquid crystal 'cell' (Fig. 20.10) consists of a thin layer (about 10 μm) of a liquid crystal sandwiched between two glass sheets with transparent electrodes deposited on their inside faces.

— When both glass sheets are transparent, the cell is known as *transmittive type cell.*

— When one glass is transparent and the other has a reflective coating, the cell is called *reflective type.*

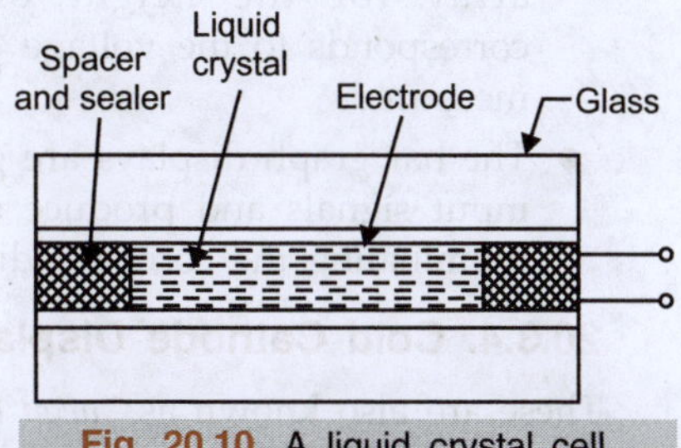

Fig. 20.10. A liquid crystal cell.

Features of LCDs :

(*i*) They are light scattering.

(*ii*) They can operate in a reflective or transmissive configuration.

(*iii*) They do not actively generate light and depend for their operation on *ambient or back lighting*.

(*iv*) The power required by an LCD to scatter or absorb light is extremely small of the order of a few μW/cm. They operate at low voltage ranging from 1 to 15 V.

(*v*) The response time of LCDs typically range between 10 and 100 ms.

Advantages :

1. Extremely low power requirement.
2. Low cost
3. Long life time–about 50,000 hours.

Disadvantages :

1. LCDs are *very slow devices*. They have a turn-on time of a few milli-seconds, and a turn-off time of tens of milliseconds.
2. They occupy a large area.
3. When used on D.C. their life span is quite small. Therefore, they are used with A.C. supplies having a frequency less than 500 Hz.
4. Poor visibility in low ambient temperature.
5. Limited temperature range of operation (0–60°C).
6. Low reliability.

LCDs are usually of the *'seven segment type' for numeric use* and have one common back electrode and seven transparent front electrodes characters.

Uses :

1. Cellular phone display.
2. Desk top LCD monitors.
3. Note book computer display.
4. Watches and portable instruments.
5. Pocket T.V. receiver.

Bar graph displays :

- These displays are *analog displays* which are an alternative to conventional D'Arsonval moving coil meters.
- They use a closely packed linear array or column of display elements, *i.e.*, "DOT-LEDs", which are independently driven so that the array (or the height of the column) corresponds to the voltage or current being measured.

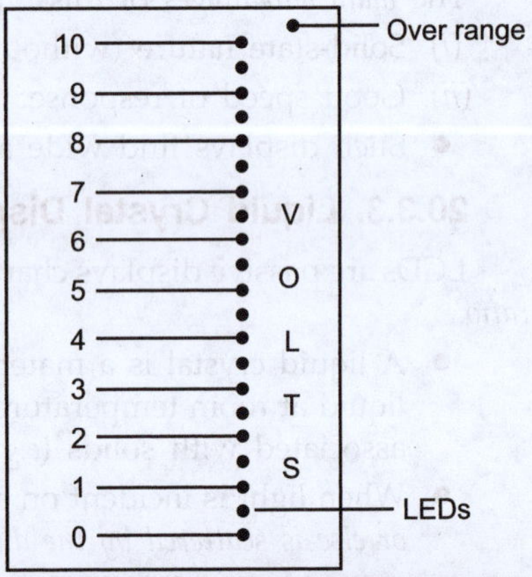

Fig. 20.11. Analog meter using bar graph of LEDs.

- The bar graph displays are generally used in the panel meters to accept analog input signals and produce an equivalent display of the input signal level by illuminating the corresponding LEDs as depicted in Fig. 20.11.

20.3.4. Cold Cathode Displays or Nixies

These are also known as *"neon or gas-discharge displays"*.

These are the most well known type of alphanumeric displays.

Their operation is based on the emission of light in a cold cathode gas filled tube under breakdown condition. The cold cathode indicators are called **Nixie tubes**.

Fig. 20.12 shows the construction of the nixie tube.

— It consists of 10 cathodes (each in the shape of a numeral from 0 to 9) and one anode, all are made of *thin wires*. The anode is in the form of *thin frame*.

— The electrodes are enclosed in a glass filled envelope with connecting pins at the bottom. Neon gas is usually emplayed and it gives an orange-red glow when activated. However, other colours are available when different gases are used.

— During normal operation, the anode is returned to *positive* supply through a suitable current limiting resistor, the value of supply being *greater than the worst-case breakdown voltage of the gas within the tube. The gas in the vicinity of the appropriate cathode glows when the cathode is switched to ground potential.*

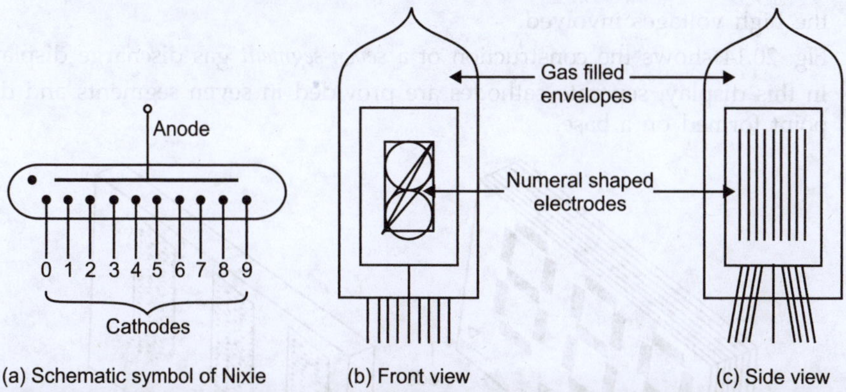

Fig. 20.12. Nixie tube.

— Because 10 cathodes have to be associated with a single anode inside the glass bulb, they have necessarily to be stacked in *different planes. This requires different voltages for different cathodes to enable the glow discharge.*

- Several nixie tubes possess *dot-cathodes* either on the left or right of the character to serve as decimal points.
- The standard nixie is not the only format used with cold cathode technology–both bar and dot matrix versions are available.

— The bar types have a cathode which forms the segment and operates in a fashion similar to the standard neon tube. *Identical supply voltage and drivers are required.* In the dot type display, each dot is in matrix fashion and operates as an individual glow discharge light source. The required dots are selected by an X-Y addressing array of thin film metal lines.

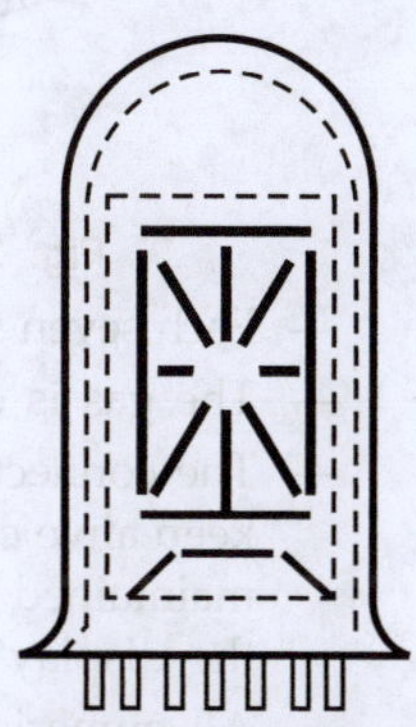

Fig. 20.13. Fifteen segment nixie.

Fig. 20.13 shows a special nixie having 15 cathodes constituting 15 segments that be used to *produce numeric as well as alphanumeric characters.*

Characteristics of nixie tubes :

1. The numerals are usually large, typically 15-30 mm high and appear in the same base line for in-line read-out.
2. These are single digit devices with or without a decimal point.
3. These can also display alphabetical symbols.

4. They are either side viewing or top viewing.
5. Most of the nixie tubes require D.C. supply of 150-220V, and the selected cathode carries current in the range of 1-5 mA.
6. A nixie tube can be pulse operated and hence can be used in multiplexed displays.

Segmental gas discharge displays :

- These displays work on the principle of gas discharge glow, similar to the case of nixie tubes.
- They are available in *seven-segment* or *fourteen-segment* format for display of numeric and alphanumeric characters.
- Integrated circuit (ICs) have been developed to drive these devices and to develop the high voltages involved.

  Fig. 20.14 shows the construction of a *seven-segment* gas discharge display.

— In this display, separate cathodes are provided in seven segments and decimal point formed on a base.

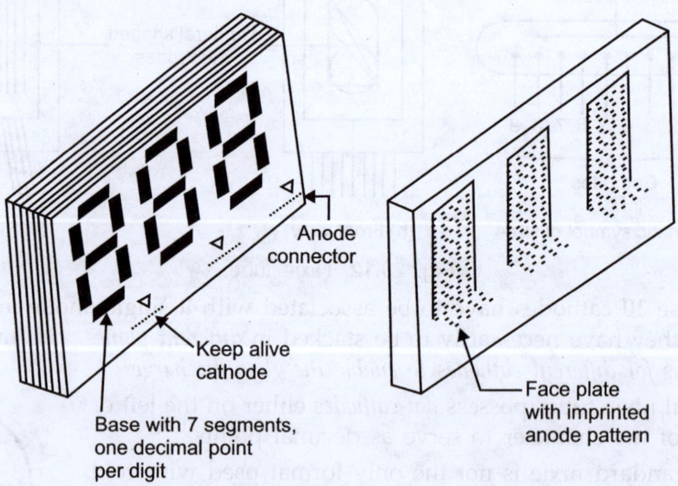

Fig. 20.14. Seven-segmented display using gaseous discharge.

— Each seven segment group has a single anode deposited on the covering face plate.
— The gas is contained in the space between anodes and cathodes.
— The connections are made with the help of pins for all electrodes at the rear. A keep alive cathode is also enclosed with each group of segments. A 50 μA current maintained through the keep alive cathode provides a source of ions; this improves the display's switching speed.
— All numeric characters can be displayed by activating the appropriate segment.
— Display panels of rows or columns of such characters can be easily constructed by extending a single character.
– *The power requirements of such devices are more or loss in the same range as these for nixie tubes.*

- A *major advantage* is that the power consumed is extremely small as we can have bright displays with the tubes currents as low as 200 μA.
- The *main disadvantage* of such displays is the *high voltage required to operate them;* this voltage is between 150-200 V. This high voltage transistors must be employed as switches for the cathodes.

20.3.5. Electro-luminescent (EL) Displays

These displays are important means of light generation.

They can be fabricated using poly crystalline semiconductors, which are essentially *phosphor powder or film type structures.*

- The **powder type** consists of powder phosphor with some binding material e.g., organic liquids deposited on a glass sheet.

The glass has transparent conductive segments (*e.g.*, seven-segment displays) or dots (dot matrix display) along with the required conductive leads on the side on which the phosphor is coated for electrical connections.

A metallic electrode, usually aluminium, is placed over the phosphor in a pressure cell by vacuum evaporation, so as to from an electrical connection on the other side of the phosphor. The resulting device is *capacitive,* because of poor conduction paths in phosphor. An A.C. field applied across the chosen segment (or dot) and aluminium electrode *excites the phosphor resulting in the emission of light*.

- In **film type** *structures* the electro-luminescent powder structure is replaced by a polycrystalline phosphor film which is deposited on a glass substrate using a vacuum or pressure cell.

These devices can be operated by A.C. as well as by D.C.

20.3.6. Incandescent Display

Incandescence has been a basic process of producing light for several decades.

- Such displays using *sixteen* segments as well as *5 × 7 dot matrix formats* fabricated using thin film micro electronics are now available for alphanumeric characters.
- These displays are characterised by *simple technology, bright output and compatibility with ICs but at very low operating speeds.*

20.3.7. Electrophoretic Image Display (EPID)

Electrophoresis is *the movement of charged pigment particles suspended in a liquid under the influence of an electric field.* This phenomenon has been utilised in EPIDs.

- These displays are characterised by *large character size, low power dissipation and internal memory.*

The relatively slow speed of these displays is a *major limitation,* particularly for use as a *dynamic display.*

- The life span of an EPID is a few thousand hours.
- The panel of EPID makes use of the electrophoretic migration of charged particles in suspension. The extremely thin suspension which largely contains the pigment particles and suspending liquid, is sandwiched between a pair of electrodes one of which is transparent. The application of a D.C. electric field, across the electrodes, moves the particles electrophoretically towards either electrode, the movement depending mainly on the polarity of the charge on the particles. The reflective colour of the suspension layer changes on account of this migration.

It is usual to have the transparent electrode as a common electrode. The back electrodes are generally segmented. During the normal operation of display the transparent electrode is maintained at ground potential and segmented electrodes at the back are given different potentials.

- EPID panels generally follow a segmented character format–*typically seven segments for numeric characters.*

EPID panel has a *memory* because the pigment particles deposited on an electrode surface remain there even after the applied voltage is removed.

- In order to achieve a *desired colour display*, the colour combinations of both the pigment particles and suspending liquid can be used.

20.3.8. Liquid Vapour Display (LVD)

These displays (latest in economical display technology) *employ a new reflective passive delay principle and depend on the presence of ambient lights for their operation.*

Fig. 20.15 shows the structure of a typical LVD cell.

— It consists of a transparent volatile liquid encased between two glass plates and side spacers.

— The rear glass plate has a black background and the front glass surface in contact with the liquid is roughened, so that the liquid wets it.

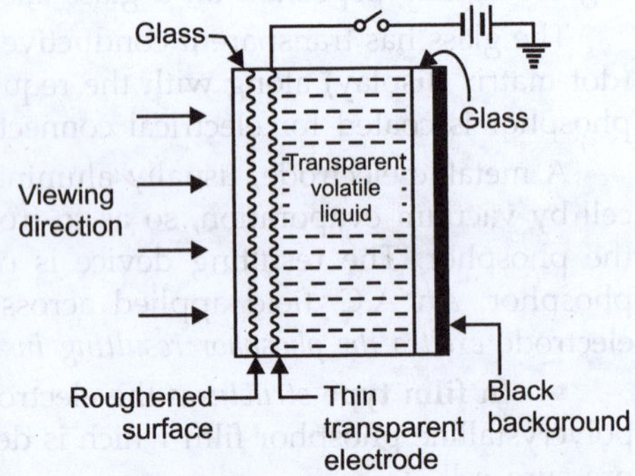

Fig. 20.15. Structure of LVD cell.

— *The transparent electrode is heated by using a voltage drive, which is the basis for the display function.*

— When the display is in 'OFF' condition with no voltage applied across the transparent electrode the viewer sees the background through the front transparent glass electrode and the liquid.

— In order to obtain an 'ON' condition of the display a voltage is applied to the transparent electrode. This causes sufficient heat in the electrode which evaporates the liquid in contact with it, and a combination of vapour film and vapour bubbles is formed around the roughened glass. As the evaporative index of vapour is approximately one, there is discontinuity established at the interface between the front glass plate and the liquid which *gives rise to light scattering.* This makes it a simple display device.

The organic liquid used in this type of display should have : (*i*) Refractive index close to that of the glass plate; (*ii*) Minimum energy for vaporising the liquid in contact with the roughened surface.

- LVD has a *better contrast ratio than LCD.*
- The *speed of operation of LVDs is low.*

20.3.9. Comparison of Various Display Devices

The comparison of various display devices is given in table 20.1 below :

Table 20.1. Comparison of display devices

S. No.	Display devices	Advantages	Disadvantages	Applications
1.	*CRTs*	— Bright — Efficient	— Bulky — High voltage	— Large display — Small and large group viewing

		— Uniform — Planar display -all colours — High reliability.	— Non-digital address — High initial cost.	— Console display.
2.	*LEDs*	— Bright — Efficient — Red, yellow, amber, green colours — Compatible with ICs — Small size.	— High cost per element — Limited reliability — Low switching speed.	—Indicators and small displays — Individual viewing — Flat panel.
3.	*LCDs*	— Good contrast in bright ambient light — Low power — Compatible with ICs — Low cost element.	— Limited temperature range (0–60°C) — Limited reliability — A.C. operation necessary — Low switching speed.	— do —
4.	*NIXIEs*	— Bright — Range of colours — Low cost element — Compatible with ICs.	— High drive power.	— Indicators — Small, medium and large displays — Small group viewing.
5.	*ELs*	— Low cost element — Many columns.	— Not compatible with ICs.	— Indicator and small display — Flat panel.

20.4 RECORDERS

20.4.1. Introduction

A **recorder** *records electrical and non-electrical quantities as a function of time.* The record may show how one variable varies with respect to another, or how the input signal varies with time.

The record serves the following *objectives* :

(*i*) It preserves the details of measurement at a particular time.
(*ii*) It provides at a glance the overall picture of the performance of unit.
(*iii*) It provides immediate reflection on the action taken by the operator.

20.4.2. Types of Recorders

In an instrumentation system, one of the important considerations is the method by which the data required is recorded. The recording method should be consistent with the type of system. If we are dealing with a wholly analog system, then *analog recording techniques* should be used. While, on the other hand, if the system has a digital output, *digital recording devices* are employed.

Two types of recorders are:

1. *Analog recorders :*

(*i*) *Graphic recorders*

(*a*) Strip chart recorders
- Galvanometer type
- Null type
  - — Potentiometric recorders
  - — Bridge recorders
  - — LVDT recorders

(*b*) Circular chart recorders

(*c*) X-Y recorders

A *"graphic recorder"* may be defined as an instrument, which draws a graph that relates two or more variables to time or to each other. A *"graph"* is an analog representation of discrete information.

(*ii*) *Oscillographic recorders.*

(*iii*) *Magnetic tape recorders.*

2. *Digital recorders.*

The above recorders are discussed briefly below :

20.4.3. Strip Chart Recorders

Fig. 20.16 shows the basic constructional features of a strip chart recorder.

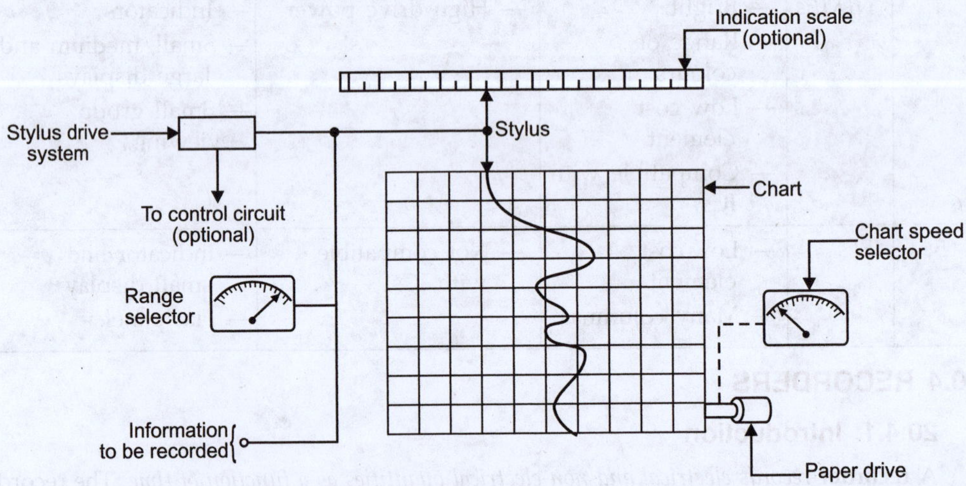

Fig. 20.16. Strip chart recorder.

A strip chart consists of the following:

- A long wall of graph paper moving *vertically*.
- A system for driving the paper at some selected speed.
- A stylus for marking paper on the moving graph paper (Most recorders use a pointer attached to the stylus, which (pointer) moves over a calibrated scale thus showing instantaneous value of the quantity being measured).
- A stylus driving system which moves the stylus in nearby exact replica or analog

of the quantity being measured (A spring wound mechanism may be used but in most of the recorders a *synchronous motor* is used for driving the paper).

Marking mechanisms. The most commonly used mechanisms employed for marking marks on the paper are :

(*i*) Marking with ink filled stylus.
(*ii*) Marking with heated stylus.
(*iii*) Chopper bar.
(*iv*) Electric stylus marking.
(*v*) Electrostatic stylus.
(*vi*) Optical marking method.

Tracing systems. For producing graphic representations, the following two types of tracing systems are used:

(*i*) Curvilinear system.
(*ii*) Rectilinear system.

20.4.3.1. Galvanometer type strip chart recorders

— This type of recorder operates on the *"deflection principle"*.

— The deflection is produced by a galvanometer, (d'Arsonval) which produces a torque on account of a current passing through its coil. This current is proportional to the quantity being measured. (See Fig. 20.17)

— These recorders can work on ranges for a few mA to several mA or from a few mV to several mV.

— The moving galvanometer type recorder is *comparatively inexpensive instrument*, having a narrow bandwidth of 0 to 10 Hz. It has a sensitivity of 0.4 mV/mm or from a chart of 100 mm width a full scale deflection of 40 mV is obtained. Linear amplifiers are used for measurement of smaller voltages.

— This type of recorder is not useful for recording fast variations in either current or voltage or power.

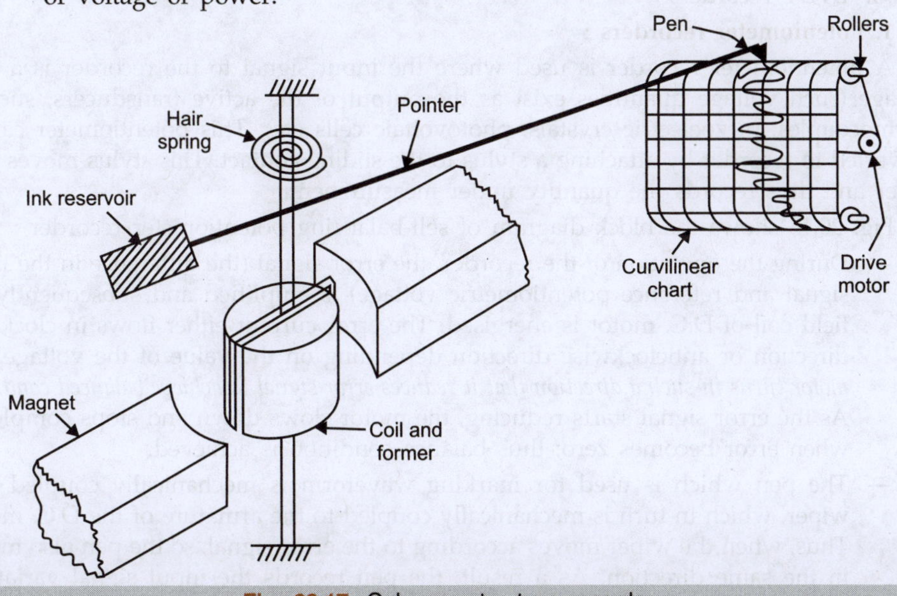

Fig. 20.17. Galvanometer type recorder.

- Some recorders contain a timing mechanism that prints a series of small dots along the edge of the paper chart, as the paper moves through the recorder. This time marker produces *one mark per second*. These types of recorders are mostly used as ***optical recorders***, and contain a light source provided by either an ultra violet or tungsten lamp. *A small mirror is connected to the galvanometer movement and this light beam is focussed on the mirror. The beam reflected from the mirror is focussed into a spot on a light sensitive paper. As the current passes through the coil the mirror deflects. The movement of the light beam is affected by the deflection of the small mirror, and the spot on the paper also varies for the same reason, thus tracing the waveform on the paper.*

20.4.3.2. Null type strip chart recorders

The main disadvantage of a galvanometer type recorder is that it has a *low input impedance and a limited sensitivity.* This disadvantage can be overcome by using an *amplifier* between the input terminals and the display or indicating instruments. This amplifier provides a high input impedance and improved sensitivity *at the cost of low accuracy.* For improving the accuracy of the instrument the input signal is *compared with a reference voltage using a potentiometer circuit.*

- These recorders *operate on the principle of "self balancing" or "null conditions".*

When the input signal given by the sensor or transducer is applied to the measuring unit of the recorder the balanced condition of the instrument gets disturbed. This unbalanced signal produces *error signal*. This operates on some device which converts this error signal to some action so that the balanced condition of the instrument is achieved. Now, this action is nothing but the magnitude of the error signal which is *directly proportional to the change in input signal.*

The null type strip chart recorders are of the following *types* :

1. Potentiometer recorders.
2. Bridge recorders.
3. LVDT recorders.

1. Poteniometer recorders :

A potentiometer recorder is used where the input signal to the recorder is a D.C. voltage (such voltage quantities exist as the output of the active transducers, such as thermocouples, piezoelectric crystals, photovoltaic cells etc). This potentiometer can be converted to recorder by attaching a stylus to the sliding contact. This stylus moves over paper and thus records the quantity under measurement.

Fig. 20.18 shows the block diagram of self-balancing potentiometer recorder :

— During the operation of the recorder, the error signal (the difference in the input signal and reference potentiometric voltage) is amplified and subsequently the field coil of D.C. motor is energised. The error current either flows in clockwise direction or anticlockwise direction depending on the value of the voltage. *The motor turns in such a direction that it reduces error signal to achieve balanced condition.* As the error signal starts reducing, the motor slows down and stops completely when error becomes zero; thus balance condition is achieved.

— The pen which is used for marking waveform is mechanically coupled to a wiper, which in turn is mechanically coupled to the armature of the D.C. motor. Thus, when the wiper moves according to the error signal, so the pen also moves in the same direction. As a result, the pen records the input signal variations moving across the paper.

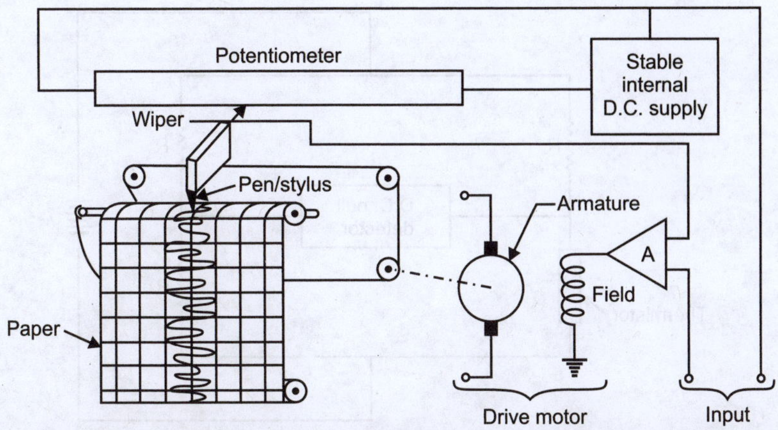

Fig. 20.18. Block diagram of self-balancing potentiometer recorders.

- Such recorders have a very high input impedance, infinity at balance conditions, and a high sensitivity of the order of 4V/mm with an error of less than ± 0.25% with a bandwidth of 0.8 Hz.
- The most common application of potentiometric recorder is for *recording and control of process temperatures.* Self balancing potentiometers are unduly used in industry because of the following *reasons:*

(*i*) Their action is automatic and thus eliminate the constant operation of an operator.

(*ii*) They draw a curve of the quantity of being measured with the help of recording mechanism.

(*iii*) They can be mounted on switchboard or panel and thus act as mounting devices for the quantity under measurement.

Single-point and multi-point recorders:

- *Instruments that record changes of only one measured variable are called* **single-point recorders**.
- A **multi-point recorder** may have as many as 24 inputs, with traces displaced in 6 colours.

2. Bridge recorders :

These recorders are used when the signals to be monitored are due to variation in some electrical parameter of a passive transducer such as *variation in resistance which occurs in strain gauges, thermistors and photoconductive cells.*

Temperature is one of the non-electrical quantities which can be measured by this type of recorder.

Fig. 20.19 shows a bridge recorder using a *thermistor*.

— The resistance changes in the thermistor cause corresponding changes is in the bridge output. These changes are applied to the detector.

— The bridge balance (null) can be restored by varying the resistance of another arm of the bridge, while recording in terms of current, voltage or temperature. Based on the kind of voltage supplied to the bridge, the output can be chosen to be D.C. or A.C.

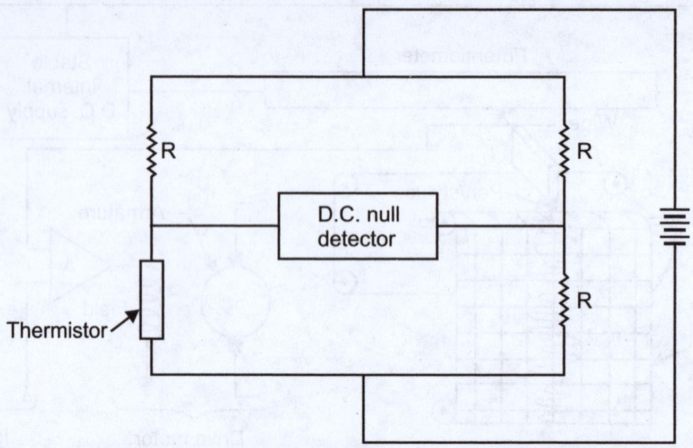

Fig. 20.19. Bridge recorder (using thermistor).

3. LVDT recorder

This recorder is used to record *force, pressure* etc. which can be translated into displacement with the help of a force assuming device.

Fig. 20.20 shows a LVDT recorder.

— It consists of linear voltage differential transformers (LVDT) T_1 and T_2.

— The displacement to be recorded is applied to the soft iron of differential transformer T_1. T_2 is associated with the recording circuit.

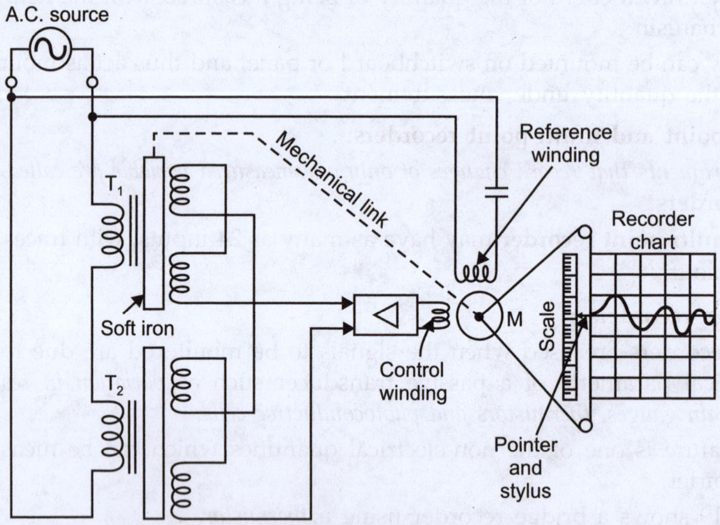

Fig. 20.20. LVDT recorder.

— The primary windings of T_1 and T_2 are connected in series to the same A.C. source. The secondary windings are connected in *series opposition*. Thus at the balance condition (*i.e.,* null condition), the output voltages of the transformers are equal and opposite and the net output (A.C. signal source) is *zero.*

— Now if a displacement is applied to the core of the transformer T_1 the output voltages of the two transformer will be different and a different output voltage is obtained. The differential output voltage upsets the balance and an A.C. error signal (with reference to the phase of the voltage of A.C. supply) depends upon the direction of the applied displacement.

— The amplified A.C. signal is then sent to the control winding of the servometer M causing it to rotate in a direction that moves the iron core of transformer T_2 to *rebalance the circuit*. As the motor rotates, it also moves the stylus across the chart of the recorder and the pointer across the scale. When the balance is restored the error signal vanishes and the motor stops. The position of the stylus reflects the new position of the movable core (*i.e.*, the displacement) and the trace produced by the stylus on the paper shows the change from the previous position. This pointer indicates the new position on the scale which is calibrated directly in terms of displacement.

20.4.3.3. Advantages and disadvantages of strip chart recorders

Advantages :

(*i*) The rate of movement of the chart can easily be changed to spread out the trace of the variable being observed.

(*ii*) Data conversion is easier when rectangular coordinates are used.

(*iii*) These recorders require the use of servo-mechanisms to position the pointer or pen. Therefore more than adequate power is available, there being no real limitations on the weight of the pen, pressure between pen and paper or length of the pointer.

(*iv*) Relatively large amount of paper can be inserted at one time in the form of a nell.

(*v*) Many more separate variables can be recorded on a strip chart than on circular chart.

Disadvantages :

(*i*) The mechanism is considerably more complicated than is required to drive a circular chart.

(*ii*) Observing behaviour several hours or days back is not as easy as picking out one circular chart which covers the desired period of time.

20.4.3.4. Applications of strip chart recorder

Following are some of the important applications of strip chart recorders :

1. Temperature recorder. The two primary methods used for recording temperature are :

(*i*) Thermocouple methods :

— This method utilises commercially available thermocouples that cover a wide range of temperatures.

— The thermocouples serve very well as temperature sensing elements and are readily compatible with strip-chart recorders.

(*ii*) The resistance method.

2. Sound level recording. Invariably it is required to obtain a record of sound level over a period of time near hospitals, schools or residences, airports etc. It can be done by with an ordinary microphone and a strip-chart recorder, provided the output signal from the microphone is of adequate amplitude to drive the recorder.

Fig. 20.21 shows a set up for obtaining sound level data using a sound level meter and a strip chart recorder.

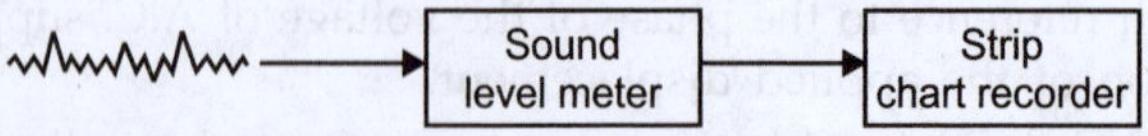

Fig. 20.21. Set-up for measuring sound level as a function of time.

3. Recording amplifier drift. Transistor amplifiers are sensitive to temperature changes. The changes in temperature *cause the bias voltage of the transistor to change*, thereby changing the operating or quiescent point, generally by a small amount. This change of the point is called *drift*. Because this is generally a slow process a strip chart recorder may be used to monitor and record the drift by connecting the recorder to the output of the amplifier.

20.4.3.5. Comparison between potentiometric recorder and galvanometeric recorder

The comparison between potentiometric and galvanometric recorders is given in table 20.2 below.

Table 20.2. Comparison between Potentiometric recorder and Galvanometric recorder

S. No.	Aspects	Potentiometric recorder	Galvanometric recorder
1.	*Working principle*	— *Self balancing or null conditions*. It compares the input signal with the reference and *produces error signal.*	— *D'arsonval movement.* — It produced deflection when current passes through coil. This produces a *torque* which is used for driving a pointer and recording input signal.
2.	*Recording mechanism*	— The *error signal* produced according to the difference between the input signal and reference is *amplified.* — The current corresponding to the error signal drives the motor which moves the pen coupled with the wiper.	— The deflection of the galvanometer is proportional to the current through the coil (which is proportional to the input signal measured). — Due to the deflection, torque is produced, which moves the pointer to which pen-ink system is fitted.
3.	*Response*	Faster	Slower
4.	*Bandwidth*	Higher–of the order of few hundreds of Hz.	Smaller of the order of 0-10 Hz.
5.	*Input impedance*	Very high, hence loading problems are eliminated.	Very low, hence loading problems are significant.
6.	*Sensitivity*	Higher–equal to 4V/mm for error less than ± 0.25%	Lower–equal to 0.4 V/mm.
7.	*Applications*	Used for recording the output from the transducer in the form of voltage, current, change in resistance or inductance or capacitance, pressure temperature etc.	Commonly used as optical recorder which uses ultraviolet source.

20.4.4. Circular Chart Recorders

As the name implies, the data in this type of recorder is *recorded on a flat circular chart*. These recorders were developed mainly to take advantage of the availability and convenience of spring-wound clock and synchronous motor movements to drive the chart in a circular direction.

— The circular chart used here has concentric circles ruled on it to form its scales. In addition, there are printed arcs extending from the centre of the chart to the paper's edge. As the pen of the recorder is moved it swings along these arcs; these arcs are called the *"time arcs"*.

— The speed of the rotation of chart is usually one revolution per 24 hours or per seven days or any other speed, which can be conveniently obtained by using synchronous motor with suitable gear assembly.

The radial position of the pen at any time indicates the instantaneous value of the quantity under measurement.

— Chart diameter is limited to a maximum of 0.3 m, speed of the chart is also limited, resolution along the scale length is usually non-uniform and the charts do not run for a long period.

— Magnitude of several variables can be recorded on a single chart which makes it easy and convenient to analyse the interrelationship of various measurements and also it saves the panel mounting space.

The various drives for circular charts are *classifieds* as follows :

(*i*) Mechanical (spring clock drive).

(*ii*) Pneumatic (air lock drive).

(*iii*) Electric (synchronous regulated D.C. motor or motor wound spring).

(*iv*) Dual powered drive (duplex) *i.e.*, a synchronous motor and spring clock mechanical drive.

(*v*) Externally controlled drives.

- *Circular chart recorders are particularly suitable for direct actuation by a number of mechanical sensors such as bellows, bourdon tubes etc.*

The comparison between circular chart and strip recorders is given in table 20.3.

Table 20.3. Comparison between Circular chart and Strip chart recorders

S. No.	Aspects	Circular chart recorders	Strip chart recorders
1.	*Handling and storing*	Easy	Very easy
2.	*Shape and size of chart*	Circular, varying in size from 100 mm to 250 mm diameter.	Curvilinear type, available in the form of long strips usually rolled on to a drum.
3.	*Usable recording area*	40 to 50% area of chart is calibrated and rest is the space covered by mechanism involved.	90% or more of the chart width is usable recording area, very small portion being taken up by punched holes for guide purposes.
4.	*Amount of information that can be carried.*	Strictly limited amount.	It can be packed with information.
5.	*Exhibition of information*	It shows all the information recorded at a glance.	It needs to be unrolled to see past records.

6.	*Facility to record*	It is possible to simultaneously record on the full-chart range upto four separate variables.	It is possible to record upto 4 to 6 points simultaneously and thus afford saving a lot of panel space.
7.	*Cost*	Low initial cost.	Cost though high is justified, considering its versatility, predictive diagnostic capability, invaluable tool for analysing the overall dynamic response.
8.	*Range of chart speeds*	Usually the circular chart moves at one constant speed and high speed phenomenon cannot be recorded.	It is possible to have wide range of chart speeds and records fast changing phenomenon.
9.	*Chart speed*	The chart speed is limited and as such recording cycle takes longer time for multiple points.	The availability of wide range of chart speeds enables the recording of greater number of points and at a much higher speed than is practical with circular chart instruments.
10.	*Type of operators required*	Less skilled operators can do the job since it is easy to adjust and repair instruments.	Skilled operators are required.

20.4.5. X-Y Recorders

A **X-Y recorder** is *an instrument which gives a graphic record of the relationship between two variables.*

This system has a pen which can be positioned along the two axes with the writing paper remaining stationary. There are two amplifier units, one amplifier actuates the pen in the Y-direction as the input signal is applied, while the second amplifier actuates the pen in X-direction. The movements of the pen in X-and Y-directions are automatically controlled by means of a motor, pulleys and a linear potentiometer. Obviously, trace of the marking pen will be due to the combined effects of two signals applied simultaneously. In these recorders, *an e.m.f is plotted as a function of another e.m.f.* There are many variations of X-Y recorders. With the help of these *recorders and appropriate tranducers a physical quantity may be plotted against another physical quantity.*

Fig. 20.22 shows a block diagram of a typical analog X-Y recorder :

— A signal enters in each of the two channels.

— The signals are *attenuated* to the inherent full scale range of the recorder (often 0.5 mV). The signal then passes to a *balance circuit* where it is compared with an internal reference voltage.

– The *error signal* (*i.e.* the difference between the input signal voltage and the reference voltage) is fed to a *"**chopper**"* which converts D.C. signal to an A.C. signal.

— The signal is then *amplified* in order to actuate a *servomotor* which is used to balance the system and hold it in balance as the value of the quantity being recorded changes.

The action described above takes place in both the axes simultaneously. Thus we get a record of one variable with respect of another.

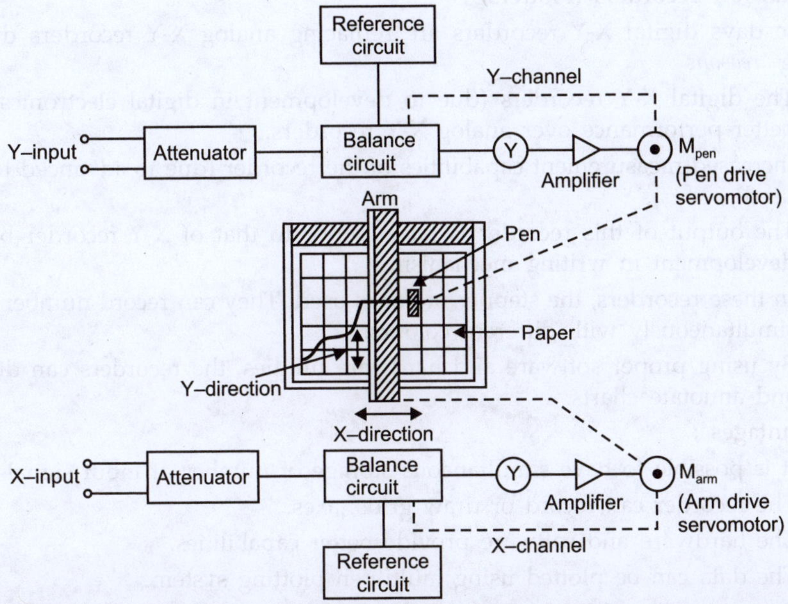

Fig. 20.22. X-Y recorder.

Advantages :

1. The instantaneous relationship between two physical quantities can be recorded.
2. The relationship between either electrical or nonelectrical quantities can be recorded.
3. In modern types of recorders, zero offset adjustments are available.

Applications. A few examples in which use of X-Y recorders is made are as under :

1. Plotting of stress-strain curves, hysteris curves and vibrations amplitude against swept frequency.
2. Pressure-volume diagrams for I.C. engines.
3. Pressure-flow studies for lungs.
4. Lift drag wind tunnel tests.
5. Electrical characteristics of materials such as resistance versus temperature and plotting the output from electronic calculators and computers.
6. Speed-torque characteristics of motor.
7. Regulation curves of power supplies.
8. Plotting of characteristics of vacuum tubes, zener diodes, rectifiers and transistors etc.

Specifications : Some important specifications of X-Y recorder are :

(*i*) Sensitivity .... 10 μV/mm.
(*ii*) Slewing speed .... 1.5 m/s
(*iii*) Frequency response .... About 6 Hz both the axes
(*iv*) Chart size .... 250 mm × 180 mm
(*v*) Accuracy .... ± 0.3 per cent.

Digital X-Y recorders (Plotters) :

These days digital X-Y recorders are replacing analog X-Y recorders due to the following *reasons* :

- The digital X-Y recorders (due to development in digital electronics) provide better performance over analog X-Y recorders.
- Increased measurement capabilities of the recorder (due to advanced techniques used).
- The output of this recorder is far superior to that of X-Y recorder because of development in writing mechanism.
- In these recorders, the stepper motor is used. They can record number of inputs simultaneously with different colours.
- By using proper software and hardware utilities, the recorders can draw grids and annotate charts.

Advantages :

(*i*) It is possible to have simultaneous storage of number of input signals.
(*ii*) The recorder can record or draw grids, axes.
(*iii*) The hardware and software provide better capabilities.
(*iv*) The data can be plotted using multi-pen plotting system.

- The comparison between strip chart recorder and analog X-Y recorders is given in table 20.4 below :

Table 20.4. Comparison between "Strip chart recorder"and "Analog X-Y recorder".

S.No.	Strip chart recorder	X-Y recorder (analog)
1.	Also called as *X-t plotter*.	Also called X-Y plotter.
2.	Input variable is plotted as a function of time.	One input variable is plotted as a function of the other.
3.	The paper is kept *rotating*.	The paper is held *stationary*.
4.	Generally a self-balancing potentiometer is used to plot the emf as a function of time.	One self-balancing potentiometer controls the position of roll while the other controls the position of the recording pen.
5.	The zero offset adjustments are *not available*.	The zero adjustments *are available*.

20.4.6. Oscillographic Recorders

The **oscillographic recorders** *are instruments which record information on paper or film with ink, powder, electrostatic charge, liquid, electric arc or light.*

Although galvanometer recorders, pen recorders, X-Y recorders, strip chart recorders and CRT recorders all may be considered oscillographic recorders, yet *here word 'oscillograph'* implies either a *galvanometer or a CRT recorder–the recorders having bandwidths more than 20 kHz.*

20.4.6.1. Ultraviolet (U.V.) recorders:

These recorders are basically *electro-mechanical oscillographic recorders* and modified version of Duddel's oscillographs.

An ultraviolet recorder consists of a number of galvanometer (moving coil) elements mounted in a single magnet block as shown in Fig. 20.23. A paper sensitive to ultraviolet light is used for producing a trace for the purpose of recording. The u.v. light is projected on the paper with the help of mirrors attached to the moving coils.

Working. When a current is passed through the moving (galvanometer) coil, it deflects under the influence of the magnetic field of the permanent magnet. The ultraviolet light falling on the mirrors is deflected and projected on to the u.v. light sensitive paper through a lens and mirror system. The paper is driven past the moving high spot and thus a trace of variation of current with respect to time is produced.

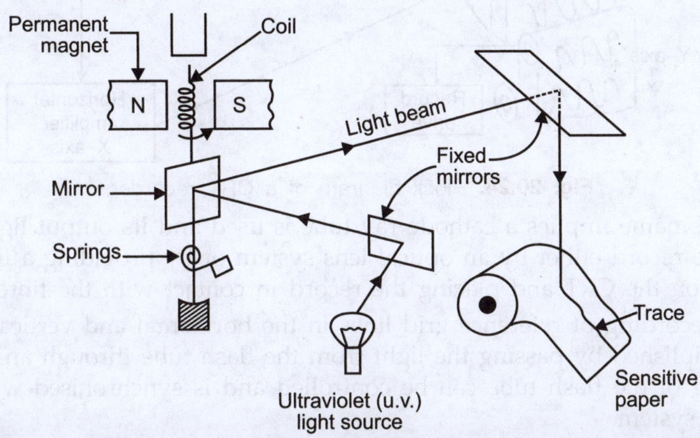

Fig. 20.23. Ultraviolet (u.v.) recorder.

The recorder, in addition to the input currents, may have the following additional traces:

(*i*) Grid lines.

(*ii*) Timing lines.

(*iii*) Trace identification.

- The ultraviolet (u.v.) recorders, compared to the mechanical and pen recorders, have *better frequency and response characteristics*; the typical values are :

  Frequency response = 0 to 300 Hz; 0 to –12 kHz (maximum)

  Response time = 16 μs

  — The maximum frequency that may be recorded depends upon the frequency response of the galvanometer used. When high frequency signals are to be recorded, the marking paper is moved with sufficient speed so as to spread out the trace along the time-axis. *The u.v. recorders have an additional advantage of multi-trace recording.*

- *"Typical applications"* of U.V. recorders are in recording :

  (*i*) Regulation transients of generators.

  (*ii*) Output of transducers.

  (*iii*) Control system performance.

- These recorders are also used for *recording the magnitude of low frequency signals which cannot be measured with analog (pointers) type instruments.*

20.4.6.2. Cathode ray tube (CRT) recorders

These recorders are *four-dimensional display and recording devices.*

Fig. 20.24 shows the block diagram of a CRT recorder.

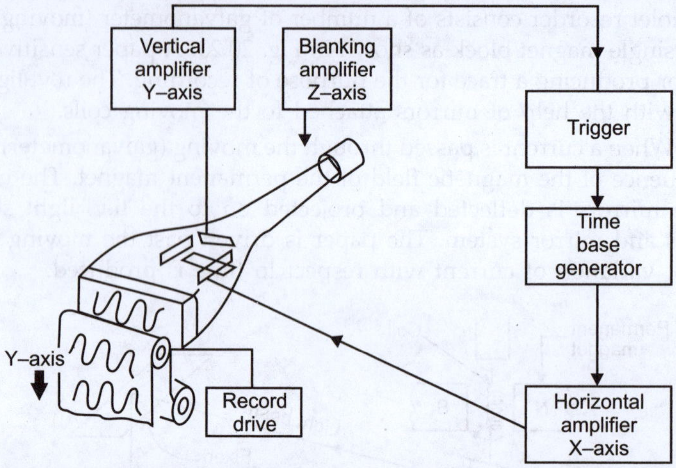

Fig. 20.24. Block diagram of a CRT recorder.

— As the name implies a cathode ray tube is used and its output light is focussed on the record either by an optical lens system or by providing a fibre optic face plate on the CRT and placing the record in contact with the fibre optics.

— The recording of reference grid lines in the horizontal and vertical directions is accomplished by passing the light from the flash tube through an aperture. The timing of the flash tube can be controlled and is synchronised with the record drive system.

- The CRT recorders can record data upto 1000 times faster than galvanometer recorders. Further because of four axes they have capability of recording continuously.

20.4.7. Magnetic Tape Recorders

These recorders have response characteristics which enable them to be used at *higher frequencies*; hence they find an *extensive use in Instrumentation systems.*

20.4.7.1. Construction and operating principle

Construction :

A magnetic tape recorder consists of the following *basic components*:

1. Recording head.
2. Magnetic tape.
3. Reproducing head.
4. Tape transport mechanism.
5. Conditioning devices.

1. Recording head :

Fig. 20.25 shows the construction of a magnetic tape recording head :

— Its construction is similar to that of transformer having a toroidal core with a coil.

— The fine air gap of lengths 5-15 μm is shunted by the passing magnetic tape.

— When the current used for recording is passed through the coil wound around magnetic core, it produces magnetic flux.

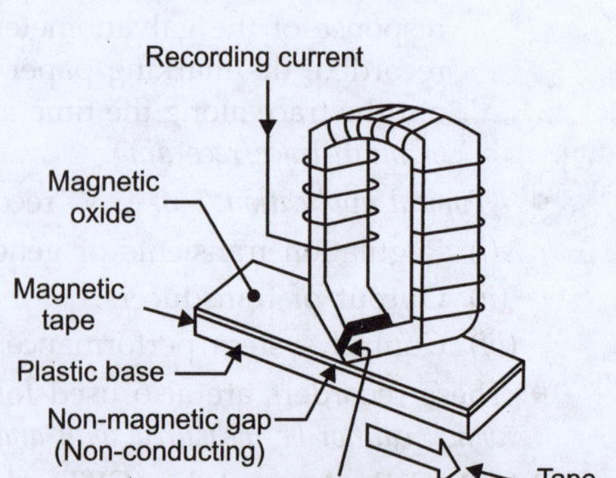

Fig. 20.25. Magnetic tape recording head.

— As the magnetic tape having iron oxide particles passes the head, the magnetic flux produced gets linked with the iron oxide particles and these particles get magnetised. The state of magnetization of the oxide as it leaves the gap is retained, thus the actual recording takes place at the trailing edge of the gap.

— Any signal recorded on the tape appears as a *magnetic pattern dispersed in space along the tape, similar to the original coil current variation with time.*

2. *Magnetic tape :*

— It is made of a thin sheet of tough, dimensionally stable plastic, one side of which is coated with a magnetic material. Typically, the plastic base is polyvinyl chloride (PVC) or polyethylene terephthalate, for example *mylar*.

— The magnetic coating consists of a dispersion of very small particles of iron oxide (Fe_2O_3) in a plastic binder.

— A typical tape is 12.7 mm wide and 25.4 μm thick.

— The magnetic particles conform to the magnetic pattern and retain it.

3. *Reproducing head :*

— It is similar to that of a recording head in appearance.

— Its *function is to detect the stored magnetic pattern and to convert it back to original electrical signal.*

4. *Tape transport mechanism :*

Fig. 20.26 shows a simple tape transport mechanism. This mechanism moves the magnetic tape along the recording head or reproducing head with a constant speed.

— The magnetic tape is wound on two reels; one is called as supply reel and other is called as tape-up reel. Both the reels rotate in same direction. The fast winding of the tape or the reversing of the tape is done by using special arrangements.

— The rollers are used to drive and guide the tape.

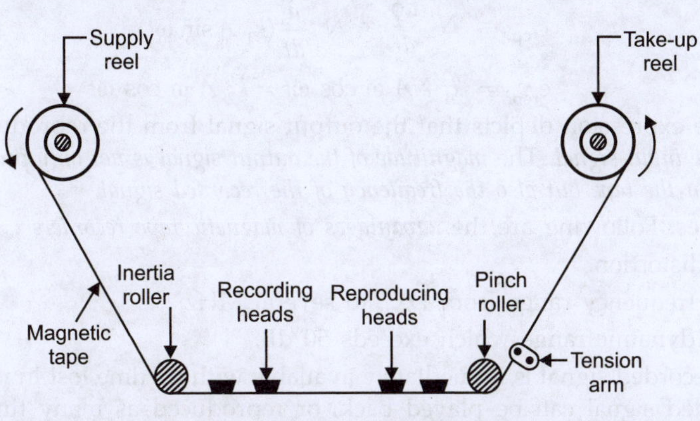

Fig. 20.26. Tape transport mechanism.

The tape transport mechanism must perform the following *tasks* :

(*i*) To handle the tape without straining and wearing it.

(*ii*) To guide the tape across magnetic heads with great precision.

(*iii*) To maintain uniform and sufficient gap between the tape and heads.

(*iv*) To maintain proper tension of magnetic tape.

5. *Conditioning devices :*

- These devices consists of *'amplifiers'* and *'filters' to modify signal to a format that can be properly recorded on a tape.*
  — *Amplifiers* amplify the signal to be recorded while *filters* remove unwanted ripple quantities.

Operating principle of tape recorders :

The *operating principle* of tape recorders is as follows :

When a magnetic tape is passed through a recording head, the signal to be recorded appears as some magnetic pattern on the tape. *This magnetic pattern is in accordance with the variations of original recording current. The recorded signal can be reproduced back by passing the same tape through a reproducing head where the voltage in induced corresponding to the magnetic patten on the tape.* The induced voltage depends on the direction of magnetization and its magnitude on the tape. The e.m.f thus produced is proportional to the rate of change of magnitude of magnetisation

$$\therefore \qquad i.e., \quad e \propto N \cdot \frac{d\phi}{dt},$$

where, N is the number of turns of the winding put on reproducing head.

Since the voltage in the reproducing head is proportional to $\frac{d\phi}{dt}$, the reproducing head acts as a differentiator.

Let us assume that the orignal signal is A sinωt. The current in the recording head winding and the flux produced will be proportional to this voltage.

$$\therefore \qquad \phi = k_1 \, A \sin \omega t, \qquad \text{where } k_1 = \text{constant}$$

Assuming that the tape retains this flux pattern and regenerates it in the reproducing head, the voltage induced in the reproducing head winding,

$$e_{rep.} = N \cdot \frac{d\phi}{dt} = N \cdot \frac{d}{dt}(k_1 \, A \sin \omega t)$$

$$\text{or,} \qquad e_{rep.} = k_1 \, NA \, \omega \cos \omega t = k_2 \, A \, \omega \cos \omega t \qquad \text{...(20.1)}$$

The above expression depicts that the output signal from the reproducing head is a *derivative of the input signal.* The *magnitude of the output signal is not only proportional to the flux recorded on the tape but also the frequency of the recorded signal.*

Advantages. Following are the *advantages of magnetic tape recorders* :

1. Low distortion.
2. Wide frequency range from D.C. to several MHz.
3. Wide dynamic range which exceeds 50 dB.
4. The recorded signal is immediately available with no time lost in processing. The recorded signal can be played back, or reproduced as many times as desired without loss of signal.
5. Multi-channel recording possible.
6. Exceedingly high density of data points giving simplified storage and handling.

Applications. These recorders are available in various sizes and in various forms (*e.g.*, portable, semi-portable and rack mounted); due to their unique features they have wide applications, some of these are :

1. Communication surveillance and spying.
2. Data recording and analysis on missiles, aircrafts and satellites.

3. Medical research and patient monitoring.
4. Industrial research and production monitoring and control.

20.4.7.2. Recording methods

Following are the three methods of magnetic tape recording which are used for instrumentation purposes :

1. Direct recording.
2. Frequency modulation (FM) recording.
3. Pulse Duration Modulation (PDM) recording.

1. Direct reading :

It is the simplest method of recording and *usually requires one tape track for each channel. The signal to be recorded is amplified, mixed with a high frequency bias and fed directly to the recording head as a varying electric current.*

Advantages:

(*i*) This requires simple electronic circuits.
(*ii*) This recording process requires only simple, moderately priced electronic circuitry.
(*iii*) It has a wide frequency response ranging from 50 Hz to about 2 MHz for a tape speed of 3.05m/s. It provides the *greatest bandwidth obtainable from a given recorder.*
(*iv*) It has a good dynamic response and takes overload without increase in distortion.
(*v*) It can be used for recording voice and multiplexing a number of channels of information into one channel of tape recording.
(*vi*) It is used to record signal where information is contained in the relation between frequency and amplitude, such as spectrum analysis of noise.

Limitations :

(*i*) This recorder is only used when maximum bandwidth is required.
(*ii*) This recording process is characterised by some amplitude instability caused by random surface inhomogeneities in the tape coating. Some portion may not be perfectly recorded owing to dirt or poor manufacture.

2. Frequency modulation (FM) recording :

This system is generally used when a more accurate response to D.C. voltage is required.

Fig. 20.27 shows a FM recording system :

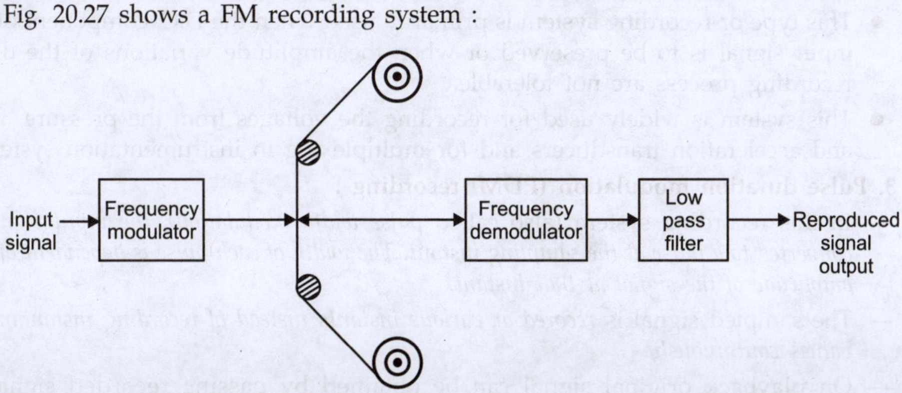

Fig. 20.27. Frequency modulation (FM) system.

— In this system a carrier oscillator frequency f_c called the centre frequency, is modulated by the level of the input signal. The centre frequency is selected with respect to the tape speed and frequency deviation selected for the tape recorders is ± 40% about the carrier frequency.

— When there is no input signal, *i.e.*, zero input, the modulation is at centre frequency f_c. If a positive input signal is applied, the frequency deviates from the centre frequency by some amount in a certain direction, the application of a *negative* input voltage deviates the centre frequency in the *opposite direction.*

— The output of the modulation, which is fed to the tape, is a signal of constant frequency for D.C. inputs, and varying frequency for A.C. inputs. The variation of frequency is directly proportional to the amplitude of the input signal.

— On playback, the output of the reproducing head is demodulated and fed through a *"low pass filter" which removes the carrier and other unwanted frequencies* reproduced due to the modulation process.

— The *"frequency demodulator"* converts the difference between the centre frequency and the frequency on the tape, to a voltage proportional to the difference in the frequencies. This system can thus record frequencies from D.C. to several thousand hertz. Residual carrier signals and out of band noise are removed by a low pass filter.

Advantages :

1. Primarily useful when the D.C. component of the input signal is to be preserved.
2. No drop-out effect due to inhomogeneities of the tape material.
3. Wide frequency range, can record from D.C. voltages to several KHz.
4. Independent of amplitude variations and accurately reproduces the waveform of the input signal.

Disadvantages :

1. Limited frequency range.
2. High tape speed is required.
3. Extremely sensitive to tape speed fluctuations.
4. Recording circuitry is more complicated than that of direct recording system.
5. Requires a high quality tape transport and speed control mechanisms.
6. Comparatively expensive.

Applications :

- This type or recording system is primarily used when the D.C. component of the input signal is to be preserved or when the amplitude variations of the direct recording process are not tolerable.
- This system is widely used for recording the voltages from the pressure, force and acceleration transducers and for multiplexing in instrumentation systems.

3. Pulse duration modulation (PDM) recording :

— In this recording system (also called *pulse width modulation), the input signal is converted to a pulse at the sampling instant. The width of each pulse is dependent on the amplitude of the signal at that instant.*

— The sampled signal is *recored at various instants instead of recording instantaneous values continuously.*

— On playback original signal can be obtained by passing recorded signal to appropriate filter.

Advantages :

1. High accuracy.
2. High signal-to-noise ratio.
3. The system is capable to record information from a large number of channels simultaneously.

Disadvantages :

1. Limited frequency response.
2. Low reliability.
3. Complex circuitry.

Applications : It is used in instrument systems for special applications such as for *simultaneous recording of large number of slowly changing variables.*

20.4.7.3. Magnetic recording to store data on the floppy discs and hard discs of computers

The use of *'magnetic recording' is restored to store data on the floppy disc and hard discs of computers.*

The basic principles are that *a recording head, which responds to the input signal, produces corresponding magnetic patterns on a thin layer of magnetic material and a read head gives an output by converting the magnetic patterns on the magnetic material to electrical signals.* Besides these heads the systems require a transport system which moves the magnetic material in a controlled way under the heads.

1. Magnetic recording codes :

- In digital recording, the signals are recorded as a *coded combination of bits.*
- A *bit cell* is the element of the magnetic coating where the magnetism is either completely saturated in one direction or completely saturated in the reverse direction. Saturation is when the magnetising field has been increased to such an extent that the magnetic material has reached its maximum amount of magnetic flux and further increases in magnetising current produce no further change.
- For getting proper flux reversals, some of the commonly used methods (involving encoding) are :

  (*i*) Phase encoding (PE);

  (*ii*) Non-return-to-zero (NRZ);

  (*iii*) Frequency modulation (FM);

  (*iv*) Modified frequency modulation (MFM);

  (*v*) Run length limited (RLL).

Optimum code is the one that allows the bits to be packed as close as possible and which can be read without error. The read head can locate reversals quite easily but they must not be too close together.

— The RLL code has the advantage of being *more compact* than the other codes, PE and FM taking up the most space.

— MFM and NRZ take up the same amount of space. *NRZ* has the disadvantage of, unlike the other codes, *not being self locking.*

2. Magnetic discs :

"Digital recording" is commonly done *on a floppy or hard disc.*

The digital data is stored on the disc surface along concentric circles called *tracks*, a single disc having many such tracks. A single read-write head is used for each disc surface and heads are moved, by means of a mechanical actuator, backwards and forwards to access different tracks. The disc is spun by the drive and the read/write heads *read or write data into a track.*

The 3.5 *"floppy disc"* used in the personal computer can store 1.4 *Mbytes* of data.

"Hard discs". These are sealed units with data stored on the disc surface along concentric circles.

— A hard disc assembly has more than one such disc and the data is stored on magnetic coatings on both sides of the discs.

— The discs are rotated at high speeds and tracks assessed by moving the read-write heads.

— Large amounts of data can be stored on such assemblies of discs; storage of the order of many *G bytes* are now common.

20.4.8. Printers

The *printers provide a record of data on paper.* Such printers are available in the following *versions* :

1. The dot matrix printer.
2. The ink jet printer.
3. The laser printer.

1. The dot matrix printer :

Fig. 20.28 shows the head mechanism of a dot matrix printer.

— It consists of either 9 or 24 pins in a vertical line.

— Each pin is controlled by an electromagnet which when turned on propels the pin onto the inking ribbon. This transfers a small bob of ink onto the paper behind the ribbon. A character is formed by moving the print head in horizontal lines back-and-forth across the *paper* and *firing* the appropriate pins.

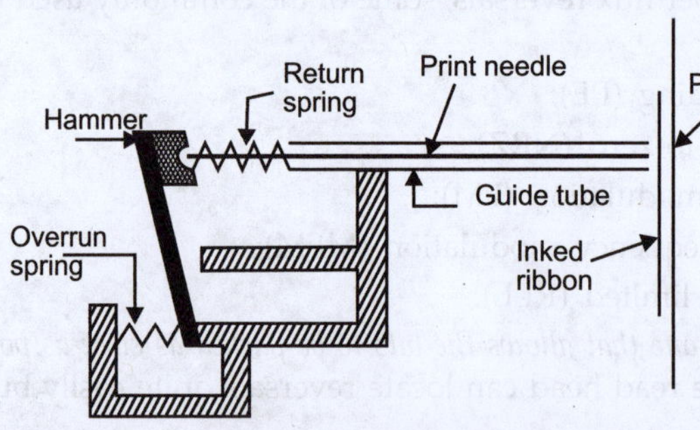

Fig. 20.28. Head mechanism of a dot matrix printer.

2. The ink jet printer :

This type of the printer uses a conductive ink which is forced through a small nozzle to produce a jet of very small drops of ink of constant diameter at a constant frequency.

— In one form a constant stream of ink passes along a tube and is pulsed to form fine drops by a piezoelectric crystal which vibrates at a frequency of about 100 kHz. (Fig. 20.29).

— In another form is used a small heater in the print head with vaporized ink in a capillary tube, so producing gas bubbles which push out 'ink drops'.

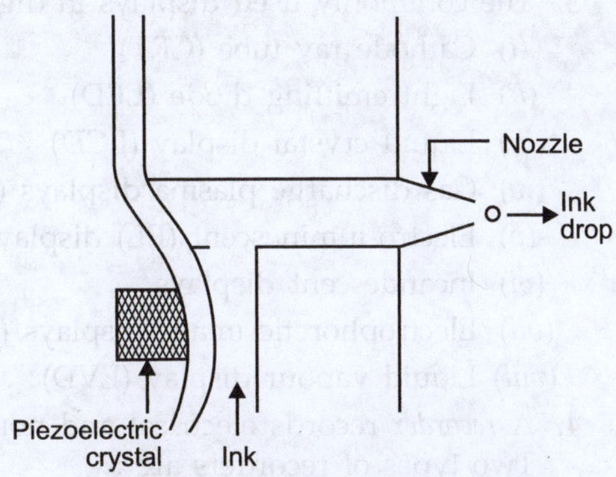

Fig. 20.29. Production of stream of drops.

3. The laser printer :

Fig. 20.30 shows the basic elements of a laser printer.

— It has a photosensitive drum which is coated with a selenium-based light sensitive material. The selenium, in the dark, has a high resistance and consequently becomes charged as it passes close to the charging wire; this is a wire at a high voltage and off which charge leaks.

— A light beam is made to scan along the length of the drum by a small rotating eight-sided mirror. When light strikes the selenium its resistance drops and it can no longer remain charged. By controlling the brightness of the beam of light the points on the drum can be discharged or left charged.

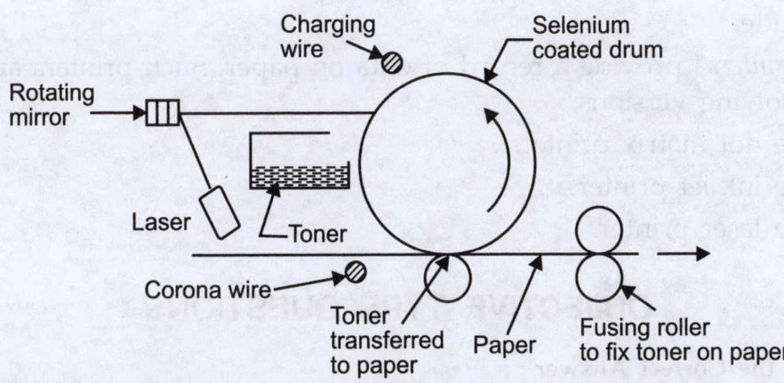

Fig. 20.30. Laser printer's basic elements.

— As the drum passes the toner reservoir, the charged areas attract particles of toner which thus stick to the areas that have not been exposed to light and do not stick on the areas that have been exposed to light.

— The paper is given a charge as it passes another charging wire, the so called corona wire, so that as it passes close to the drum it attracts the toner off the drum. A hot fusing roller is then used to melt the toner particles so that, after passing between rollers, they finely adhere to paper.

HIGHLIGHTS

1. *"Analog indicators"* are basically current-operated devices working on the d'Arsonval type galvanometer principle.
2. *"Digital type meters"* indicate the reading in exact numerals.

3. The commonly used displays in digital electronic field are :
   (*i*) Cathode ray tube (CRT)
   (*ii*) Light emitting diode (LED)
   (*iii*) Liquid crystal display (LCD)
   (*iv*) Gas discharge plasma displays (cold cathode displays or Nixies)
   (*v*) Electro-luminescent (EL) displays.
   (*vi*) Incandescent displays
   (*vii*) Electrophoretic image displays (EPID)
   (*viii*) Liquid vapour display (LVD).
4. A *recorder* records electrical and non-electrical quantities as a function of time. Two types of recorders are :
   (*i*) Analog recorders.
   (*ii*) Digital recorders.
5. Instruments that record change of only one measured variable are called *single point recorders.* A *multi-point recorder* may have as many as 24 inputs, with traces displaced in 6 colours.
6. A *X-Y recorder* is an instrument which gives a graphic record of the relationship between two variables.
7. *Cathode ray oscilloscope (CRO)* is an instrument which presents signal waveforms visually. It is also useful for comparing two signals in phase, frequency or amplitude.
8. The *"printers"* provide a record of data on paper. Such printers are available in the following versions :
   (*i*) The dot matrix printers.
   (*ii*) The ink jet printer.
   (*iii*) The laser printer.

OBJECTIVE TYPE QUESTIONS

Choose the Correct Answer :

1. The switching time of LEDs is of the order of
   (*a*) 1 s (*b*) 1 ms
   (*c*) 1 μs (*d*) 1 ns.
2. LEDs emit light
   (*a*) only in red colour (*b*) only in yellow colour
   (*c*) only in green colour (*d*) in red, green, yellow and amber colour
3. An LCD requires a power of
   (*a*) 20 W (*b*) 20 mW
   (*c*) 20 μW (*d*) 20 nW.
   to light up one segment of a seven-segmented display system.
4. X-Y recorders
   (*a*) record one quantity with respect to another quantity
   (*b*) record one quantity on X axis with respect to time on Y axis
   (*c*) record one qunatity on Y axis with respect to time on X axis.
   (*d*) any of the above.

5. In a direct magnetic tape recording system, the reproducing head and amplifier characteristics are

(*a*) – 6 db/octave. +6 db/octave (*b*) – 6 db/octave, + 20 db/decade

(*b*) + 6 db/octave, + – 6 db/octave (*d*) – 20 db/decade, + 20 db/decade.

6. A Nixie tube requires

(*a*) 10 cathodes (*b*) 14 cathodes

(*c*) 15 cathodes (*d*) 20 cathodes.

7. Nixie tubes require

(*a*) 150 V – 220 V (*b*) 100 V – 150 V

(*c*) 150 mV – 220 mV (*d*) 100 mV – 150 mV

to produce glow discharge.

8. If an information is required to be stored over a short interval of time

(*a*) a single number devices should be used

(*b*) a storage type oscilloscope should be used

(*c*) a CRO with photographic equipment should be used

(*d*) a direct writing recorder or a magnetic tape recorder should be used.

9. The machine interpretable outputs from an analog transducer can be had from

(*a*) magnetic tapes (*b*) punched cards and tapes

(*c*) teletypewriter (*d*) all of the above.

10. Digital instruments have input inpedance of the order of

(*a*) Ω (*b*) kΩ

(*c*) MΩ (*d*) mΩ.

11. The advantages of F.M. magnetic tape recording are :

(*a*) It can record from d.c. to several kHz.

(*b*) It is free from dropout effects.

(*c*) It is independent of amplitude variations and accurately reproduces the waveform of input signal.

(*d*) All of the above.

12. The turn on and turn off times of a LCD are of the order of

(*a*) 1 s (*b*) 1 ms

(*c*) 10 μs (*d*) 10 ns.

13. The power requirement of an LED is

(*a*) 40 mW per numeral (*b*) 40 μW per numeral

(*b*) 10 W per numeral (*d*) 10 μW per numeral.

14. If four flip-flops are connected in cascade, they serve as

(*a*) 2 : 1 scalar (*b*) 4 : 1 scalar

(*c*) 8 : 1 scalar (*d*) 16 : 1 scalar.

15. The LEDs, for their display require

(*a*) a voltage of 1.2 V and a current of 20 mA

(*b*) a voltage of 25 V and a current of 20 mA

(*c*) a voltage of 1.2 V and a current of 100 mA

(*d*) a voltage of 25 V and a current of 100 mA.

16. Which of the following is the visual display unit?

(*a*) Cathode any oscilloscope (*b*) U.V. recorder

(*c*) Storage oscilloscope (*d*) Moving coil oscillograph.

17. Which of the following units has a high frequency response but presents difficulty in getting a permanent record?

(a) Servo recorder (b) Moving coil oscillograph
(c) X-Y recorder (d) Cathode ray oscilloscope.

18. The advantages of F.M. magnetic tape recording are :
(a) It can record from d.c. to several kHz. (b) It is free from dropout effects.
(c) It is independent of amplitude and accurately reproduces the waveform of input signal.
(d) All of the above.

19. The source of emission of electrons in a CRT is :
(a) PN junction diode
(b) A barium and strontium oxide coated cathode
(c) Accelerating anodes (d) Post accelerating anodes.

20. The pointer-scale instruments have a
(a) very high (b) very low
(c) linear (d) stable frequencies response.

21. The operation of a moving-coil current recording instruments is based on
(a) photo-electric principle (b) D' Arsonval principle
(c) piezo-electric principle (d) thermo-electric principle.

22. The time bases of an oscilloscope are generated by
(a) horizontal amplifier (b) vertical amplifier
(c) sweep generators (d) storage oscilloscope.

23. Which of the following devices requires a matching network to avoid overloading of the signal source and prevent damage from excessive current?
(a) U.V. recorder (b) X-Y recorder
(c) Storage oscilloscope (d) Servo recorder.

24. X-Y recorders record a quantity
(a) with respect to another quantity
(b) on X-axis with respect to time on Y-axis.
(c) on Y-axis with respect to time on X-axis
(d) any of these.

25. In an electrodynamometer type of wattmeter
(a) the current coil is made fixed (b) the pressure coil is made fixed
(c) any of the two coils can be made fixed (d) both the coils should be movable.

26. An average reading VTVM uses one diode with an external series resistance. A high value of series resistance is used so that the instrument should have
(a) a high input impedance (b) a linear *v-i* characteristics
(c) low power consumption (d) all of the above.

27. In a CRT the focusing anode is located
(a) between pre-accelerating and accelerating anode
(b) after accelerating anode
(c) before pre-accelerating anode (d) none of the above.

28. Post acceleration is needed in a CRO if the frequency of the signals is
(a) less than 1 MHz (b) more than 1 MHz
(c) more than 10 MHz (d) more than 10 Hz.

29. The slewing speed in an *X-Y* recorder refers to
(a) time base
(b) maximum constant velocity that the marking pen achieves
(c) frequency response

(*d*) relationship between inputs to X and Y channels.

30. Which of the following recorder/display units has the highest frequency response and least response time?

(*a*) X-Y plotters (*b*) U.V. recorders

(*c*) Pen recorders (*d*) CRO.

31. An LCD requires a power of approximately

(*a*) 20 W (*b*) 20 mW

(*c*) 20 μW (*d*) 20 nW.

ANSWERS

1.	(*d*)	**2.**	(*d*)	**3.**	(*c*)	**4.**	(*a*)	**5.**	(*c*)	**6.**	(*c*)	**7.**	(*a*)
8.	(*b*)	**9.**	(*d*)	**10.**	(*c*)	**11.**	(*d*)	**12.**	(*c*)	**13.**	(*a*)	**14.**	(*d*)
15.	(*a*)	**16.**	(*a*)	**17.**	(*d*)	**18.**	(*d*)	**19.**	(*b*)	**20.**	(*b*)	**21.**	(*b*)
22.	(*c*)	**23.**	(*a*)	**24.**	(*a*)	**25.**	(*a*)	**26.**	(*d*)	**27.**	(*a*)	**28.**	(*c*)
29.	(*b*)	**30.**	(*d*)	**31.**	(*c*).								

THEORETICAL QUESTIONS

1. What are the factors that decide about the choice between the display devices and recorders ?

2. How are output devices categorized ?

3. What are the differences between the analog type and digital type instruments.

4. How are digital display units classified ?

5. Name the displays which are commonly used in digital electronic field.

6. Explain briefly the following dispalys :

(*i*) Cathode ray tube (CRT).

(*ii*) Light emitting diode (LED).

7. State the advantages, disadvantges and applications of LEDs.

8. Discuss briefly the various types of displays.

9. Light emitting diodes, liquid crystal diodes play very important role in digital display devices. Describe them in detail. What are their advantages and disadvantages? Compare.

10. Briefly discuss the use of LED and LCD as display devices in instrumentation. Comment on their relative merits and demerits.

11. State the difference between an analog and a digital indicator.

12. Explain a seven-segment LED display.

13. Discuss with a neat diagram a method of realising a seven segment numeric display using LEDs.

14. What are the operating principles of LCD display ?

15. What are the advantages of LCD display over Nixie tube and LED display ?

16. Compare the relative performance of the following display devices in numeric display applications :

(*i*) Flat panel alphanumeric CRT.

(*ii*) Liquid vapour display.

(*iii*) Electrophoretic image display.

(*iv*) Nixie tubes.

17. What is the function of the display or recording element of a generalised measurement system?

18. How does a display unit differ from a recorder?
19. How are the output devices categorized? Explain briefly.
20. How can we get machine interpretable outputs?
21. List the different forms in which the display is available from an instrument.
22. How are electrical indicating instruments classified?
23. How analog display meters differ from digital type meters?
24. Give a comparison between analog type and digital type instruments.
25. Give four examples each of the analog type and digital type instrumentation.
26. Explain briefly the following:
    (*i*) Single-point indicators.
    (*ii*) Multi-point multi-pointer and multi-range indicators.
27. Explain briefly the essential features of indicating instruments.
28. Describe briefly any two of the following :
    (*i*) Moving-iron instruments; (*ii*) Moving-coil instruments;
    (*iii*) Rectifier instruments.
29. What are advantages of electronic voltmeters?
30. What are digital instruments? State their principle of operation.
31. Explain briefly any two of the following:
    (*i*) Semiconductor light emitting diodes (LED);
    (*ii*) Liquid crystal displays;
    (*iii*) Hot filament or bar tubes;
    (*iv*) Numerical indicator tubes (NIT).
32. What is a recorder?
33. Elaborate the difference between a display unit and a recorder.
34. What is meant by a direct reading instrument?
35. Explain the functioning of a basic type of strip chart recorder. Enumerate the different types of marking mechanisms used in it.
36. Distinguish between single point and multi-point recorders.
37. What is a X-Y recorder? State its applications.
38. Explain the working of an ultraviolet (U.V) recorder. State its applications.
39. What are the basic components of a magnetic tape recorder for instrumentation applications? List its advantages and disadvantages.
40. Explain with neat diagram the construction and working of a cathode ray oscilloscope (CRO).
41. Describe the graphic recorders.
42. What is a strip chart recorder? Why is it called X-t recorder too? Describe the working of a basic type of strip chart recorder.
43. Explain with the help of a diagram the CRT recorder.
44. Explain with the help of a diagram the working of a galvanometer oscillographic recorder.
45. Describe the FM magnetic tape recording. Give its merits, demerits and applications.
46. What is pulse duration modulation (PDM)? Give its merits and demerits.
47. Describe the digital tape recorders.
48. Write short notes on the following :
    (*i*) Potentiometer type recorder.
    (*ii*) Circular chart recorders.
    (*iii*) X-Y recorders.

CHAPTER

21 Measurement of Non-Electrical Quantities

21.1 PRESSURE MEASUREMENT

21.1.1. Definition of Pressure

Pressure *is defined as a force per unit area.*

Pressures are exerted by gases, vapours and liquids. The instruments that we generally use, however, record pressure as the difference between the two pressures. Thus, it is the *difference between the pressure exerted by a fluid of interest and the ambient atmospheric pressure.* Such devices indicate the pressure either above or below that of the atmosphere.

Atmospheric pressure :

The atmospheric air exerts a normal pressure upon all surfaces with which it is in contact, and it is known as **atmospheric pressure.** The atmospheric pressure is also known as *Barometric pressure.*

The atmospheric pressure at sea level (above absolute zero) is called *standard atmospheric pressure.*

Note: The local atmospheric pressure may be a little lower than these values, if the place under question is higher than sea level, and higher values if the place is lower than sea level, due to the corresponding decrease of the column of air standing, respectively.

Gauge pressure :

It is the pressure, measured with the help of pressure measuring instrument, in which atmospheric pressure is taken as datum.

The *atmospheric pressure* on the scale is marked as *zero.*

Gauges record the pressure above or below the local atmospheric pressure, since they measure the difference in pressure of the liquid to which they are connected and that of surrounding air.

If the pressure of the liquid is below the local atmospheric pressure, then the gauge is designated as *'vacuum gauge'* and the recorded value indicates the amount by which the pressure of the liquid is *below* local atmospheric pressure, *i.e., negative pressure.*

(*Vacuum pressure* is defined as the pressure *below the atmospheric pressure*).

Absolute pressure :

It is necessary to establish an absolute pressure scale which is independent of the changes in atmospheric pressure. *A pressure of absolute zero can exist only in complete vacuum.*

Any pressure measured above the absolute zero of pressure is termed as an **'absolute pressure'.**

A schematic diagram showing the gauge pressure, vacuum pressure and the absolute pressure is given in Fig. 21.1.

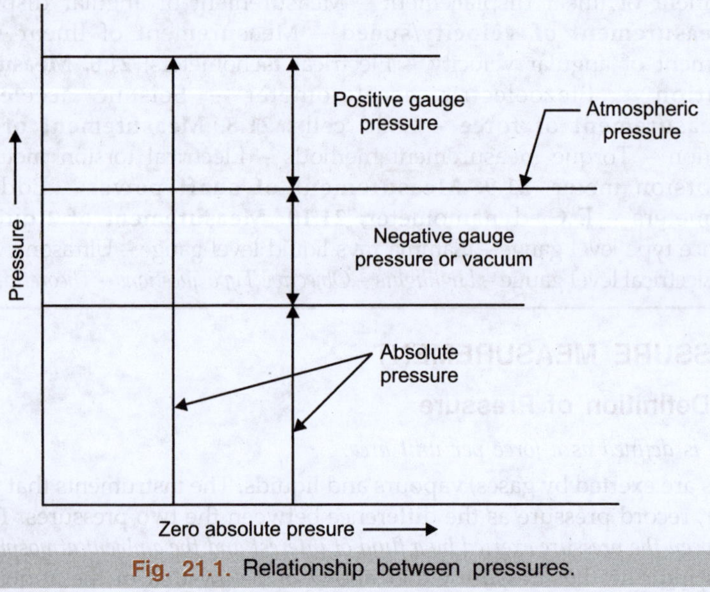

Fig. 21.1. Relationship between pressures.

Mathematically :

1. Absolute pressure = Atmospheric pressure + gauge pressure

i.e., $$p_{abs.} = p_{atm.} + p_{gauge.}$$

2. Vacuum pressure = Atmospheric pressure – absolute pressure.

Static pressure (p_s) and total pressure (p_t) :

***Static pressure** is defined as the force per unit area acting on the wall by a fluid at rest or flowing parallel to the wall in a pipeline.*

***Total or stagnation pressure** is defined as the pressure that would be obtained if the fluid stream were brought to rest isentropically.* For an incompressible fluid or gas flowing at low velocities, the *dynamic pressure* equals $\frac{V^2}{2g}$ where, V is the velocity of the fluid flow.

Thus, Total pressure = Static pressure + dynamic (or velocity) pressure

$$\therefore \quad p_t = p_s + \frac{V^2}{2g} \qquad ...(21.1)$$

21.1.2. Units for Pressure and Pressure Measuring Instruments

The fundamental S.I. units of pressure is pressure is newton per square meter (N/m^2). This is also known as *Pascal* (P_a).

"Low pressures" are often expressed in terms of *mm of water* or *mm of mercury*. This is an abbreviated way of saying that the pressure is such that will support a liquid column of stated height.

- *Standard atmospheric pressure has the following equivalent values :*
  101.3 kN/m^2 or 101.3 kPa; 10.3 m of water; 760 mm of mercury; 1013 mb (millibar); $\simeq$ 1 bar = 100 kPa = 10^5 N/m^2.

(When the local atmospheric pressure is not given in a problem, it is taken as 100 kN/m^2 or 10 m of water for simplicity of calculations.)

Pressure measuring instruments :

The pressure measurement instruments can be *categorized* as follows :

1. *Instruments for measuring low pressures (below 1 mm of Hg) :*
   - Manometers.
   - Low pressure gauges.
2. *Instrument for medium and high pressures (between 1 mm of Hg to 1000 atmospheres) :*
   - Bourdon tube.
   - Diaphragm gauges.
   - Bellow pressure gauge.
   - Dead-weight pressure gauge.
3. *Instrument for measuring low vacuum and ultra high vacuum (760 torr to 10^{-9} torr and beyond; 1 torr = 1 mm of Hg) :*
   - Mcleod.
   - Thermal conductivity.
   - Ionization gauges.
4. *Instruments for measuring very high pressures (1000 atmospheres and above) :*
   - Bourdon tube.
   - Diaphragm gauges.
   - Electrical resistance pressure gauges.
5. *Instruments for measuring varying pressure :*
   - Engine indicator.
   - Cathode ray oscilloscope (CRO).

Fig. 21.2. shows the summary of applicable range of pressure gauges.

Some of these instruments are discussed in the following Articles.

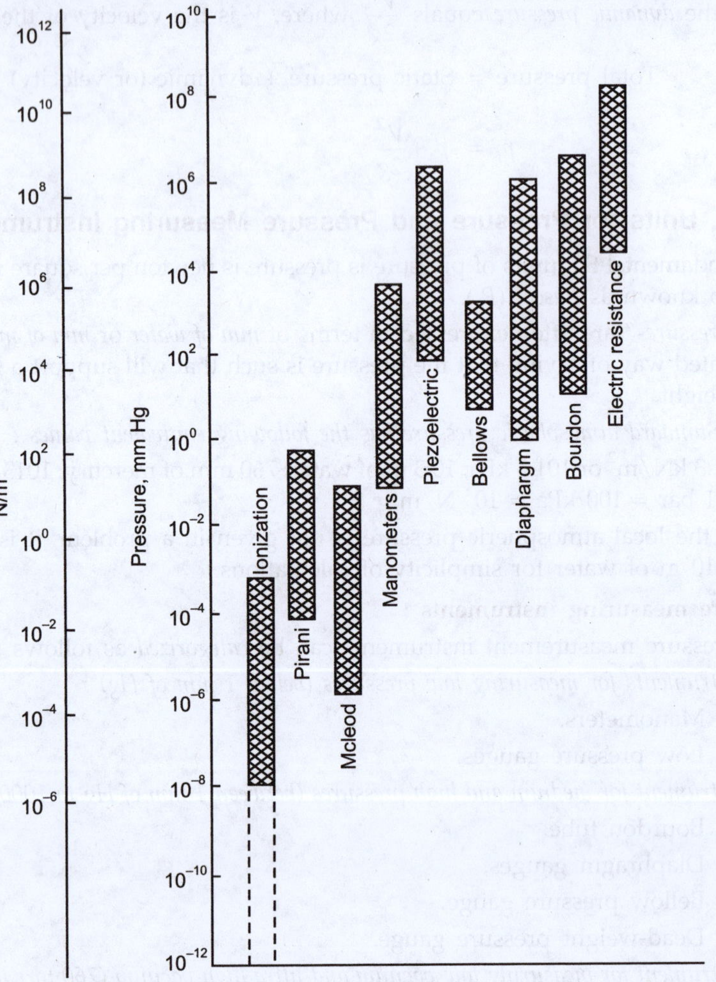

Fig. 21.2. Summary of applicable range of pressure gauges.

21.1.3. Electrical Pressure Transducers

A **transducer** *is a device for converting a signal of one sort into the other type.* In other words, a **pressure gauge** *is a transducer for converting the distortion of a Bourdon spring into an indication through use of a calibrated dial* and so on. In case of electrical transducers for measuring pressure, various transformations which are needed *to convert pressure to some sort of usable electrical signal* can be :

1. The change of resistance of an elastic material such as a wire.
2. The change in capacitance when the distance between capacitor plates is varied or the dielectric is changed.
3. The conversion of a changing pressure into electrical voltage signals.
4. Creation of a voltage by the movement of a coil within a magnetic field.
5. Generation of a voltage or current through the use of photoelectric effects.
6. Torque balance system (electromagnetic).

The following **pressure transducers** will be discussed :

1. Resistance type pressure transducers.
2. Pressure voltage type.
3. Inductive type.
4. Capacitive type.
5. Carbon pile type.
6. Piezoelectric type.
7. Photoelectric type.
8. Electromagnetic type.

21.1.3.1. Resistance type pressure transducers

Resistance type devices used in pressure transducers include;

1. Strain gauges.
2. Moving contacts.

1. *Strain gauge :*

A strain gauge is simply a fine wire in the form a grid (Fig. 21.3). When the grid is distorted, the resistance (R) of the wire changes according the formula :

$$R = k\frac{L}{A} \qquad ...(21.2)$$

where k = A constant for the particular kind of wire

L = Length of the wire, and

A = Cross-section area.

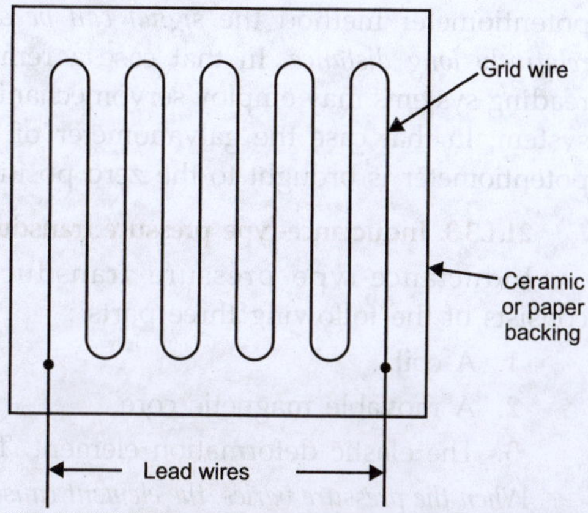

Fig. 21.3. Strain gauge.

As the strain gauge is distorted by the elastic deformation element, its *length is increased and its cross-section area is reduced.* Both of these changes *increase the resistance.*

- Since little distortion is required to change the resistance of a strain gauge through its total range, this type of transducer can be used to detect *very small movement* and, therefore, *very small pressure changes.*
- Since current flows in the strain gauge, some heat will be generated which must be compensated for. One method of compensation employs a second strain gauge placed close to the measuring unit. These two are used in a bridge circuit so that the *heating effect of the two resistive elements is cancelled out*, and the only resistance change in the bridge is due to the deformation.

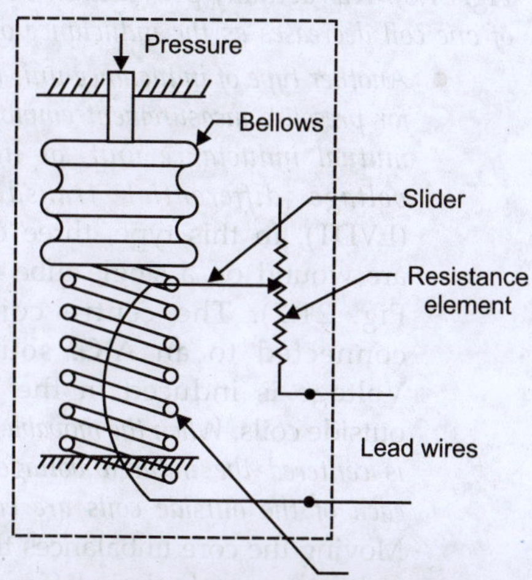

Fig. 21.4. Moving contact type resistance transducer.

Most often, strain gauge bridges are A.C. bridges.

2. *Moving contact type resistance transducer :*

This type of transducer is most often used with

bellows because of the force required. Fig. 21.4 shows a typical circuit for such a transducer.

21.1.3.2. Pressure-voltage transducer (potentiometer devices)

One of the simplest of pressure-voltage transducer uses a potentiometer. *The potentiometer is used to convert a pressure signal into an electrical voltage as shown* in Fig. 21.5.

The potentiometer should have *little friction.* Levers may be used to match the required movement of the potentiometer with that of the bellows, the signal applied to potentiometer may be relatively larger of the order of several volts. By the use of potentiometer method the *signal can be sent relatively long distance.* In that case a remote reading systems may employ servomechanism system. In that case the galvanometer of the potentiometer is brought to the zero position.

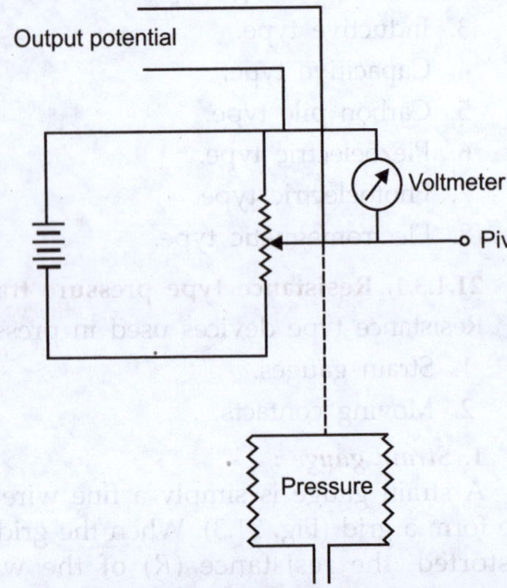

Fig. 21.5. Pressure-voltage transducer.

21.1.3.3. Inductance-type pressure transducer

Inductance-type pressure transducers consists of the following three parts :

1. A coil.
2. A movable magnetic core.
3. The elastic deformation element. The element is attached to the core.

When the pressure varies, the element causes the core to move within the coil. An alternating current is passed through the coil, and as the core moves the inductance of the coil changes. *The current passing through the coil increases as the inductance decreases.* This type of transducer is used in a *current-sensitive coil.*

To get increased sensitivity the coil may be divided into two, using a centre tap (See Fig. 21.6). This actually provides two coils. As the core moves inside the coils, the *inductance of one coil decreases as the inductance of the other increases.*

- *Another type of inductance unit used for pressure measurement employs a mutual inductance unit, or linear voltage differential transducer* (LVDT). In this type, three coils are wound on a single tube (See Fig. 21.7). The centre coil is connected to an A.C. source. Voltage is induced in the two outside coils. *When the movable core is centered the induced voltages in each of the outside coils are equal.* Moving the core unbalances these voltages, and the difference between them can be measured.

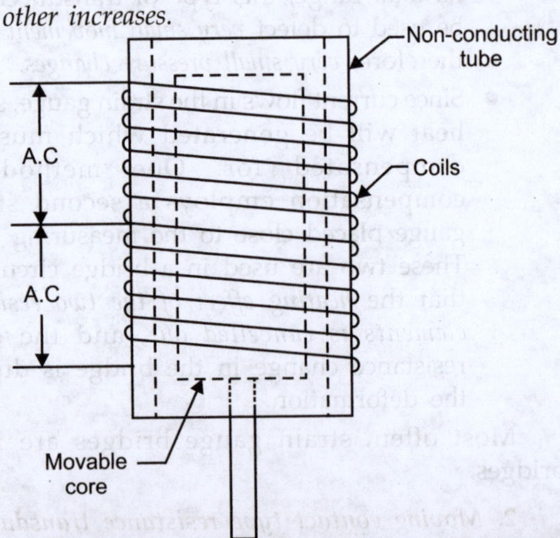

Fig. 21.6.

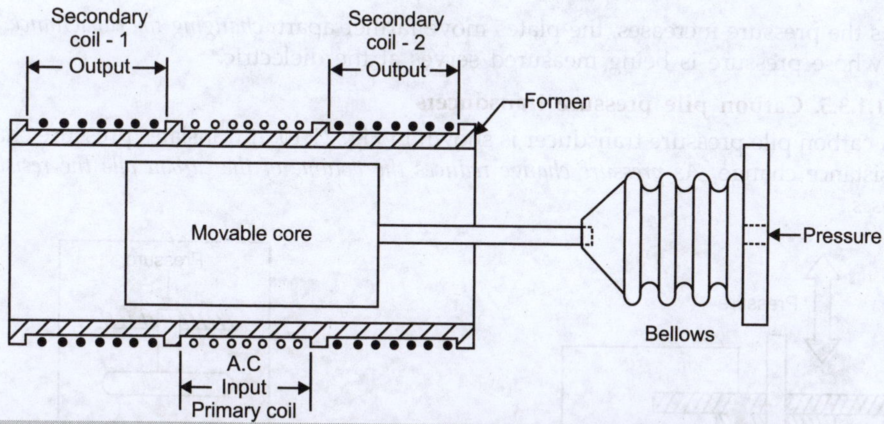

Fig. 21.7. Linear voltage differential transducer (LVDT).

Reluctance-type pressure transducer :

Similar to the inductance type pressure transducer is the reluctance-type. Fig. 21.8 shows a reluctance-type pressure trans-ducer. Here the electrical energy is provided by the field of a permanent magnet rather than A.C. source.

For pressure measurement, the elastic deformation element moves an armature closer to or farther from the permanent magnet. *The reluctance of a coil wrapped around the permanent magnet varies with the movement.*

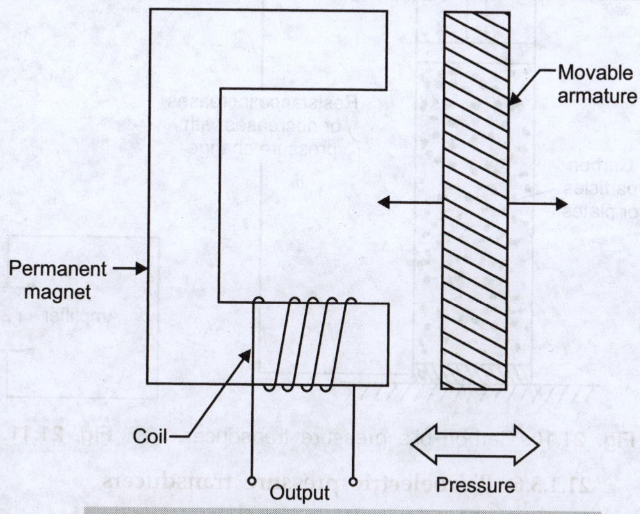

Fig. 21.8. Reluctance-type pressure transducer.

21.1.3.4. Capacitive-type pressure transducer

Capacitive pressure transducers consists of two conductive plates and a dielectric (See Fig. 21.9).

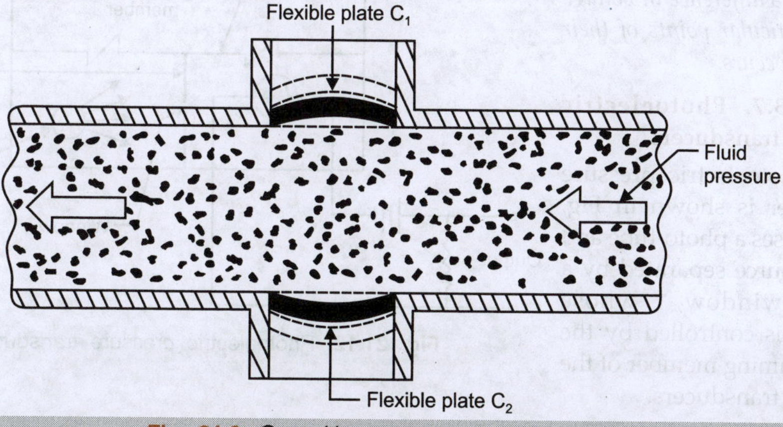

Fig. 21.9. Capacitive-type pressure transducer.

As the pressure increases, the plates move farther apart, *changing the capacitance.* The fluid whose pressure is being measured serves at the dielectric.

21.1.3.5. Carbon pile pressure transducers

A carbon pile pressure transducer is shown in Fig. 21.10. It translates pressure change to resistance change. *As pressure change reduces the volume of the carbon pile the resistance decreases.*

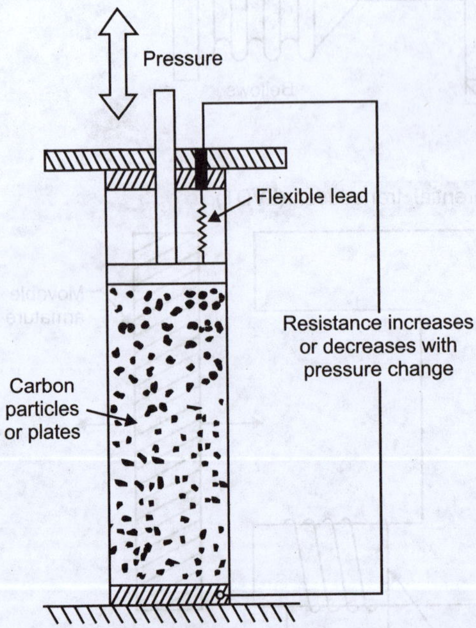

Fig. 21.10. Carbon pile pressure transducer.

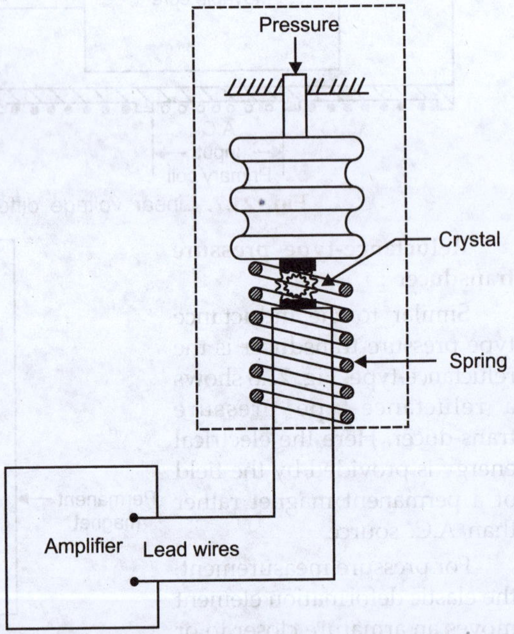

Fig. 21.11. Piezoelectric pressure transducer.

21.1.3.6. Piezoelectric pressure transducers

Fig. 21.11 shows a piezoelectric pressure transducer. As the pressure is applied to crystals to certain materials a *difference in voltage across particular points of their structure occurs.*

21.1.3.7. Photoelectric pressure transducers

A photoelectric pressure transducer is shown in Fig. 21.12. It uses a photo-tube and a light source separated by a small window, whose aperture is controlled by the force summing member of the pressure transducers.

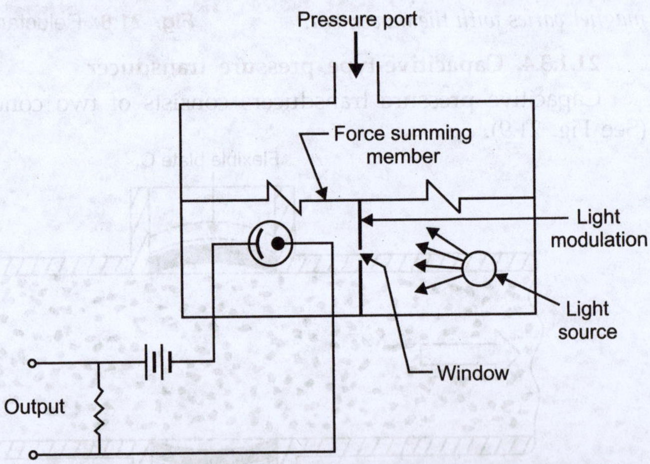

Fig. 21.12. Photoelectric pressure transducer.

21.1.3.8. Electromagnetic pressure transducers

For many years a technique has been used which consists of coil in plate or grid circuit of an oscillating tube (one that is generating alternating current) to deliver current to some other circuit. The energy may be transferred into some other winding having a transformer action. The coil has to deliver more power to the secondary winding, the additional power must come from the winding in the plate. The increase in plate current causes relays to operate at some predetermined values of current. It can also be used to develop voltages across known resistors. It can cause electromagnets to establish forces used for measuring the force applied which is obtained by pressure bellows.

21.1.4. Measurement of High Pressures

Conventional pressure-measuring devices, such as strain-gauge pressure cells and Bourdon-tube gauges may be used at pressure as high as 70000 MPa. *Bourdon tubes* for such pressures are nearly round in section and have a high ratio of wall thickness to diameter. They are, therefore, quite *stiff*, and *deflection per turn is small*. For this reason, high-pressure Bourdon tubes are often made of a number of turns.

For the measurement of very high pressure (above 70000 MPa), the *"electrical gauges" based on the principle of change of resistance with change of pressure are used.*

It is known that resistance of fine wires changes with pressure according to the following linear relationship.

$$R = R_1(1 + b\,\Delta p) \qquad ...(21.3)$$

where, R = Resistance at a pressure of 1 atm., ohms,

b = The pressure coefficient of resistance, and

Δp = The gauge pressure.

A pressure transducer based upon this principle is called a **Bridgman gauge.**

Fig. 21.13 shows a Bridgman gauge. The sensing element does not actually contact the process medium but is separated therefrom by a kerosene-filled bellows. One end of the sensing coil is connected to a central terminal, as shown, while the other is grounded, thereby completing the necessary electrical circuit.

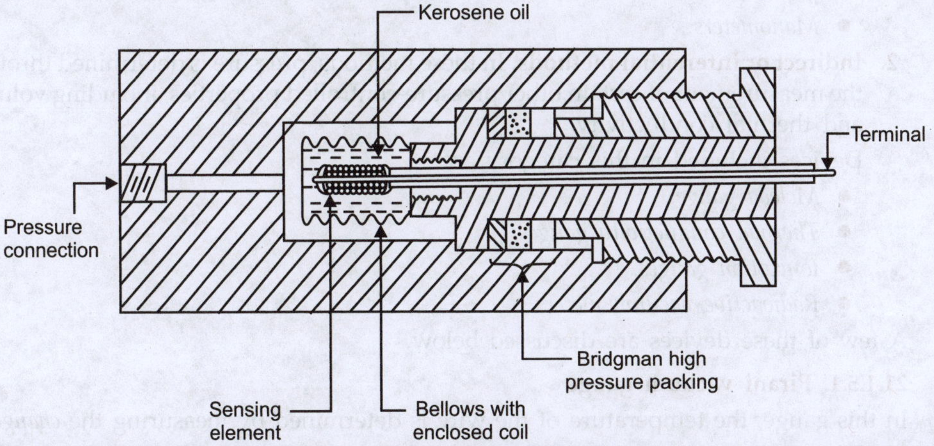

Fig. 21.13. Bridgman gauge.

The fine wire used in a typical gauge is made of manganin or gold chrome. Manganin has a pressure coefficient of resistance 25×10^{-12} Pa^{-1} and the total resistance of the wire is 100 ohms.

Although pressure coefficient resistance of gold is 10×10^{-12} Pa^{-1} (it gives a change in resistance which is $\frac{1}{4}$th of that of manganin for the same change in pressure), it is preferred to manganin in many applications as it has *resistance temperature coefficient* which is one half of that of manganin and, therefore, its use results in *lower errors on account of temperature changes.*

This gauge claims the following **advantages :**

1. Commercially available with scale upto pressures of 15 MPa with accuracies of 0.1 to 0.5% when properly calibrated.
2. Good dynamic response (the resistance wire itself can respond to variations in MHz range but the overall frequency response of the pressure measurement system is limited to *much lower values because of the acoustic response of the transmitting fluid).*

21.1.5. Measurement of Low Pressure (Vacuum)

In general, pressures below atmospheric may be called **low pressures** or **vacuums.**

It is impossible to reach an absolute pressure of zero, since a *positive* magnitude of absolute pressure exists at all times, even in a vacuum.

The *micrometer, which is one-millionth of a meter (0.001 mm) of mercury column, is a common unit of "low pressure".*

Very low pressure ...... any pressure *below* 1 mm of mercury (or torr).

Ultra low pressure ...... *less than* a nanometer (10^{-3} μm).

The two basic methods of measurement of low pressure are :

1. **Direct methods.** Here, the displacement deflection caused by the pressure is measured and is correlated to the pressure. The following devices are included in this *category* :
   - *Spiral Bourdon tubes.*
   - *Flat and corrugated diaphragms.*
   - *Capsules.*
   - *Manometers.*
2. **Indirect or inferential methods.** In these methods, pressure is determined through the measurement of certain other pressure-controlled properties, including volume and thermal conductivity.

   Devices included in this category are :
   - *Mcleod gauge.*
   - *Thermal conductivity gauges.*
   - *Ionization gauges.*
   - *Radioactive vacuum meters.*

A few of these devices are discussed below.

21.1.5.1. Pirani vacuum gauge

In this gauge, the temperature of the wire is determined by measuring the *change of resistance.*

The pirani gauge employs a single filament (in the form of four coiled wires of tungsten or platinum connected in parallel) enclosed in a glass tube/chamber, whose pressure is to be measured (See Fig. 21.14).

As the surrounding pressure changes, the filament temperature and hence its resistance also changes. A *compensating cell* is also employed to minimise variation caused by ambient temperature changes (its filament material is same as that of the measuring cell). The resistance change of the filament in the measuring cell is measured by the use of a resistance bridge which is *calibrated in terms of pressure.*

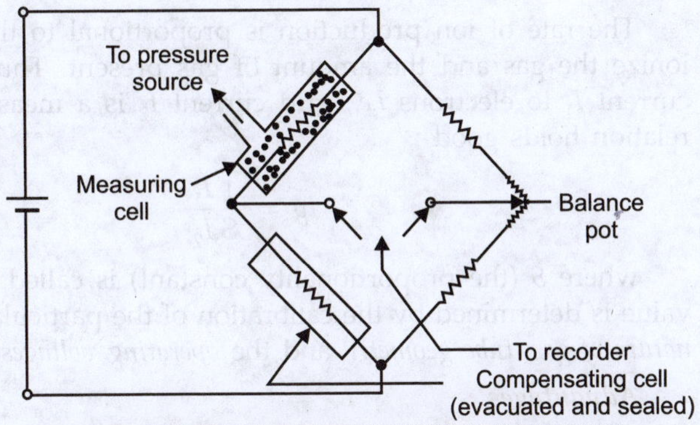

Fig. 21.14. Pirani vacuum gauge.

These gauges cover range from about 10^{-5} to 1 torr.

Advantages :

(*i*) These gauges are rugged, inexpensive and usually more accurate than thermocouple gauges.

(*ii*) The pressure reading range is wider.

(*iii*) Fast response to changes in pressure.

(*iv*) Possibility of process control and remote reading.

Disadvantages :

(*i*) Calibration is non-linear and varies from one gas to another.

(*ii*) Require calibration against some pressure standards.

(*iii*) Poor transient response.

(*iv*) Operation requires electrical power.

21.1.5.2. Ionization gauges

These gauges are used for measurement of pressures as low as 0.000001 micron (one billionth of a mm of Hg); the maximum pressure which an ionization gauge can measure is about 1 micron.

The ionization gauge (Fig. 21.15) is very similar to the ordinary electronic tube. This gauge therefore, mainly consists of an envelope which is evacuated by the pressure to be measured and contains a heated filament, a positively based grid and a negative based plate. The grid draws electrons from the heated filament, and collision between them and gas molecules causes ionization (Ionization is the process of knocking off an electron from an atom and thus producing a free electron and a positively charged ion) of the molecules. As the plate is negatively charged, the positively charged ions are attracted to the plate of the tube, causing current (I_1) flow in the external circuit; the electrons are collected by the grid and a current I_2 is produced in the grid circuit.

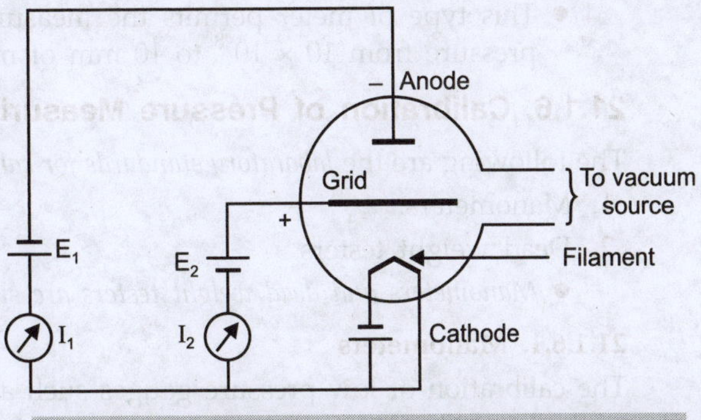

Fig. 21.15. Hot-filament ionization gauge.

The rate of ion production is proportional to the number of electrons available to ionize the gas and the amount of gas present. Thus, the ratio of *+ve* ions, *i.e.,* anode current I_1 to electrons *i.e.,* grid current I_2 is a measure of gas pressure. The following relation holds good :

$$p = \frac{1}{S}\frac{I_1}{I_2} \quad \text{...(21.4)}$$

where S (the proportionality constant) is called the *sensitivity* of the gauge and its value is determined by the calibration of the particular gauge. Sensitivity is a function of *nature of gas, tube geometry* and the *operating voltages.*

Advantages :

(*i*) Fast response to changes in pressure.

(*ii*) Possibility of process control and remote indication.

(*iii*) Constant sensitivity for a given gas over a wide range of pressures.

(*iv*) Wide pressure range.

Disadvantages :

(*i*) Some gases get decomposed by hot filament.

(*ii*) Contamination of the measured gas by gases forced out of the hot filament.

(*iii*) The filament can burn out quickly if exposed to air when hot or before the pressure is at low value (This disadvantage is eliminated by using a radioactive source of alpha-particles as the ionizing agent).

(*iv*) High cost.

(*v*) Complex electric circuitry.

(*vi*) These gauges must be provided with automatic cut out in order to protect them in case of system leak or break.

(*vii*) These gauges must be carefully installed—in particular absorbed gases must be eliminated.

(*viii*) Calibration varies with gases.

21.1.5.3. Radioactive vacuum meters

A radioactive vacuum meter works on the principle that *the current flowing through an ionization chamber depends upon the pressure, or more precisely upon the density of the gas in the chamber.*

- This type of meter permits the measurement and recording of the absolute pressure from 10×10^{-8} to 10 mm of mercury.

21.1.6. Calibration of Pressure Measuring Equipment

The following are the *laboratory standards for calibrating pressure measuring instruments :*

1. Manometers.
2. Dead weight testers.

- *Manometers and dead weight testers are suitable for static calibration only.*

21.1.6.1. Manometers

The calibration of low pressure gauges such as *manometers* and *straight bellows types* is usually done by comparison method using ***precision manometer***. For calibration, simple U-tube manometers with antiparallax devices and moving reservoir type manometers with vernier scales are employed.

- Manometers are used upto 200 kN/m^2; *above this pressure dead weight testers are used.*

21.1.6.2. Dead weight testers

This tester is a *primary standard for pressure measurement.* It provides good calibration facility over a wide pressure range (0.00007 to 70 N/mm^2 gauge in steps as small as 0.01% of range with a calibration uncertainty of 0.01 to 0.05% of the reading).

A dead weight tester is shown in Fig. 21.16. It consists of an accurately machined and finished piston which is inserted into a close fitting cylinder; both of known cross-sectional areas. At the top of the piston is provided a platform to hold standard weights of known accuracy.

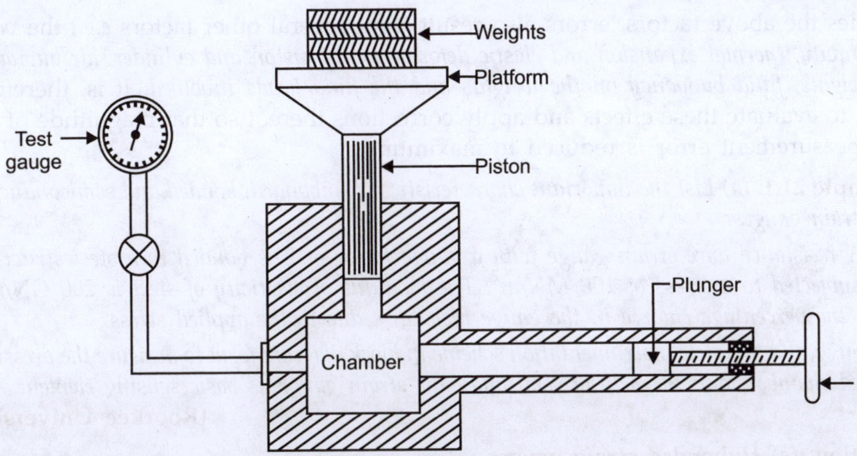

Fig. 21.16. Dead weight tester.

Working :

- The chamber and the cylinder of the tester are filled with clean oil.
- The gauge to be tested is fixed at its appropriate place and piston is inserted in the cylinder.
- A weight is placed on the piston. Consequently the pressure is exerted on the oil and this pressure is transmitted to the gauge deflecting its pointer. In order to get the desired setting of the gauge pointer, the weights are adjusted.
- Then pressure is varied by forcing the plunger in, by moving the handle. The movement of the handle is continued till enough pressure is built up to lift the piston-weight combination. The piston-weight combination *floats freely under these conditions.* Thus the equilibrium is established with *fluid force balanced against the gravitational force of the weights plus the friction drag.*

i.e., $$p \cdot A = Mg + F_d$$

Hence, pressure, $$p = \frac{Mg + F_d}{A} \quad \text{...(21.5)}$$

where, M = Mass, kg

g = Acceleration due to gravity, m/s^2

F_d = Friction drag, N, and

A = Effect area of piston-cylinder combination, m^2.

The effect of equivalent area (A) depends on such factors as *piston cylinder clearance, pressure level, temperature,* and is normally taken as the *mean of the cylinder and piston areas :*

Many kinds of pressure-measuring devices may be calibrated such as :

— *Engine indicators;*

— *Industrial pressure gauges;*

— *Piezoelectrical pressure transducers etc.*

The following *factors affect the accuracy of the dead weight tester :*

- Friction between the piston and the cylinder.
- Uncertainty of value of effective area A.
- Uncertainty of gravitational constant g.

Besides the above factors, errors also result from several other factors *e.g.,* the *value of local gravity, thermal expansion and elastic deformation of piston and cylinder, air buouancy on the weights, fluid buoyancy on the weights and the fluid heads involved.* It is, therefore, desirable to evaluate these effects and apply corrections thereof so that magnitude of the overall measurement error is reduced to maximum.

Example 21.1. *(a) List the important characteristics of unbonded, bonded and semiconductor type of strain gauges.*

(b) A resistance wire strain gauge with a gauge factor of 2 is bonded to a steel structure member subjected to a stress of 100 MN/m^2. The modulus of elasticity of steel is 200 GN/m^2. Calculate the percentage change in the gauge resistance due to the applied stress.

(c) Suggest a complete instrumentation scheme in block diagram form to measure the pressure in a closed chamber with the help of semiconductor strain gauge as basic sensing element.

(Roorkee University)

Solution. (*a*) ***Unbonded strain gauges :***

- The range of unbonded strain gauges is ± 0.15% strain with accuracies of better than 0.1% and linearity within 1%.
- The gauges are not cemented but simply screwed at the desired location hence they can be detached and used again.
- Mainly used for measurement of force, pressure, acceleration rather than for the measurement of displacement because of its massive structure.

Bonded strain gauges :

- These gauges are bonded directly into the surface of the structure member which is being examined.
- Any changes in strain in the member are directly transmitted to the gauge material.

Semiconductor strain gauges :

(*i*) These gauges can measure very small strain, of the order of 0.01 micron, owing to high gauge factor of about ± 130.

(*ii*) Very suitable for measuring *local strains* as they are available in the size of 1.0 mm.

(*iii*) Almost free from hysteresis and creep effects.

(*iv*) Frequency response upto 10^{12} Hz.

(*v*) Chemically inert and has good fatigue.

- As for *demerits*, these gauges have :

  — Poor linearity,

— Higher cost, and

— High sensitivity to temperature changes.

(*b*) *Given :* $G_f = 2;\ \sigma = 100\ MN/m^2;\ E = 200\ GN/m^2$

$$\text{Strain} = \frac{\text{Stress}}{\text{Modulus of elasticity}} = \frac{100 \times 10^6}{200 \times 10^9} = 500 \times 10^{-6}$$

$$\frac{\Delta R}{R} = \text{Gauge factor} \times \text{strain} = 2 \times 500 \times 10^{-6} = 0.001$$

Percentage change in resistance = 0.001 × 100 = **0.1%** **(Ans.)**

(*c*) The four semiconductor strain gauges will be mounted on the four walls of the closed chamber so as to sense the pressure exerted on the walls of the chamber. They will be converted in a bridge arrangement as shown in the Fig. 21.17. Output of the bridge will be amplified and given to the permanent magnet moving coil type instrument calibrated directly in terms of pressure.

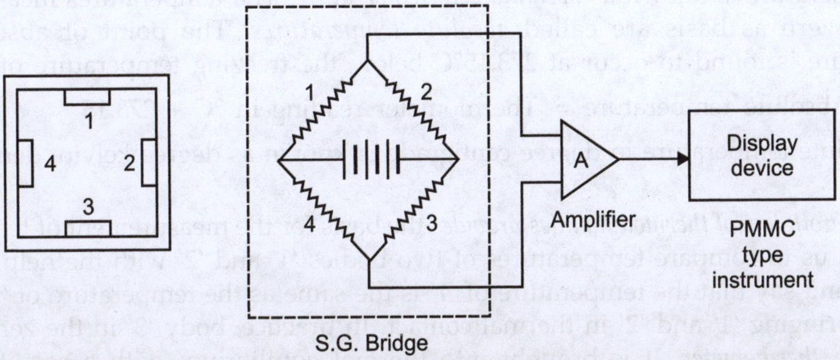

Fig. 21.17

Example 21.2. *Describe with a suitable diagram an instrumentation scheme to measure gas pressure inside an enclosed vessel using an LVDT and bellows.*

Solution. Fig. 21.18(*a*) and (*b*) shows using bellows and LVDT for measurement of pressure or force. The bellows and Bourdon tube act as Primary transducer, *i.e.*, Force summing device. The force of pressure applied to the bellows moves a core inside LVDT. The output voltage of LVDT is a measure of linear motion. The linear motion is measure of pressure. Thus force or pressure is a measure of output voltage of LVDT. In Bourdon tube circular motion is converted into linear motion which moves the core inside LVDT.

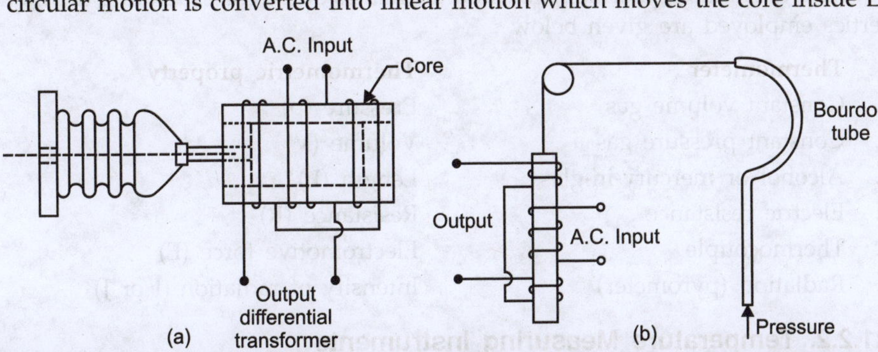

Fig. 21.18.

21.2 TEMPERATURE MEASUREMENT

21.2.1. Introduction

The **temperature** *is a thermal state of a body which distinguishes a hot body from a cold body.* The temperature of a body is proportional to the stored molecular energy *i.e.,* the average molecular kinetic energy of the molecules in a system (A particular molecule does not have a temperature, it has energy; the gas as a system has temperature).

The temperature may also be *defined* as :

- *The measure of the mean K.E. of the molecules of a substance.*
- *The degree of hotness and coldness of a body or an environment measured on a definite scale.*
- *The driving force or potential causing flow of energy as heat.*

Instruments for measuring ordinary temperatures are known as ***thermometers*** *and those for measuring high temperatures are known as* ***pyrometers.***

It has been found that a gas will not occupy any volume at a particular temperature. This temperature is known as *absolute zero temperature.* The temperatures measured with absolute zero as basis are called *absolute temperatures.* The point of absolute zero temperature is found to occur at 273.15°C below the freezing temperature of water.

$$\text{Absolute temperature} = \text{Thermometer reading in °C} + 273.15$$

Absolute temperature in degree centigrade is known as degree kelvin, denoted by K (SI unit).

The *zeroth law of thermodynamics provides* the basis for the measurement of temperature. It enables us to compare temperatures of two bodies '1' and '2' with the help of a third body '3' and say that the temperature of '1' is the same as the temperature of '2' without actually bringing '1' and '2' in thermal contact. In practice, body '3' in the zeroth law is called the *thermometer.* It is brought into thermal equilibrium with a set of standard temperature of body '2' and is thus calibrated. Later, when any other body '1' is brought in thermal communication with the thermometer, we say that the body '1' has attained equality of temperature with the thermometer and hence with body '2'. This way, the body '1' has the temperature of body '2' given for example by, say the height of mercury column in thermometer '3'. The height of mercury column in a thermometer, therefore, becomes a *thermometric property.*

These are *six other methods of temperature measurement* which utilize various other properties of materials, these are functions of temperature, as thermometric properties.

Six different kinds of thermometers and the names of the corresponding thermometric properties employed are given below :

	Thermometer	Thermometric property
1.	Constant volume gas	Pressure (p)
2.	Constant pressure gas	Volume (V)
3.	Alcohol or mercury-in-glass	Length (L)
4.	Electric resistance	Resistance (R)
5.	Thermocouple	Electromotive force (E)
6.	Radiation (pyrometer)	Intensity of radiation (I or J)

21.2.2. Temperature Measuring Instruments

The temperature measuring instruments are based on changes in a broad range of physical properties, among which are the following :

1. Changes in physical dimensions :

(*i*) *Liquid-in-glass thermometers.*

(*ii*) *Bimetallic elements.*

- *"Liquid-in-glass thermometer"* continues to appear in both laboratory and household situations, primarily because of its *ease of use and low cost.*
- *"Bimetallic elements"* are used in various low-frequency, low cost applications.

2. Changes in glass pressure or vapour pressure :

(*i*) *Constant-volume gas thermometers.*

(*ii*) *Pressure thermometers (gas, vapour and liquid filled).*

3. Change in electrical properties :

(*i*) *Resistance thermometers (RTD, PRT).*

(*ii*) *Thermistors.*

(*iii*) *Thermocouples.*

(*iv*) *Semiconductor-junction sensors*

(*a*) Diodes.

(*b*) Integrated circuits.

- *"Electrical sensors"* are widely used particularly when automatic or remote recording is desired.

4. Changes in emitted thermal radiation :

(*i*) *Thermal and photon sensors.*

(*ii*) *Total-radiation pyrometers.*

(*iii*) *Optical and two-colour pyrometers.*

(*iv*) *Infrared pyrometers.*

- "Radiant sensors" are used for non-contact temperature sensing, either in high-temperature applications like combustors or for infra red sensing at low temperatures.

5. Changes in chemical phase :

(*i*) *Fusible indicators.*

(*ii*) *Liquid crystals.*

(*iii*) *Temperature-reference (fixed-point) cells.*

The temperature measuring instruments may also be *classified* into *two broad categories* as follows :

1. Non-electrical methods :

(*i*) By using change in volume of a liquid when its temperature is changed.

(*ii*) By using change in pressure of a gas when its temperature is changed.

(*iii*) By using changes in the vapour pressure when the temperature is changed.

2. Electrical methods :

(*i*) By thermocouples.

(*ii*) By change in resistance of material with change in temperature.

(*iii*) By comparing the colours of filament and the object whose temperature is to be found out.

(*iv*) By ascertaining the energy received by radiation.

The **thermometers** may also be *classified* as follows :

1. *Expansion thermometers :*
   (*i*) Liquid-in-glass thermometers (*ii*) Bimetallic thermometers.
2. *Pressure thermometers :*
   (*i*) Vapour pressure thermometers (*ii*) Liquid-filled thermometers
   (*iii*) Gas-filled thermometers.
3. *Resistance thermometers*
4. *Thermocouple thermometers*
5. *Radiation pyrometers*
6. *Optical pyrometers.*

Table 21.1 gives the summary of temperature range of different instruments.

Some of these thermometers are discussed in the following Articles.

21.2.3. Electrical Resistance Thermometer or Resistance Temperature Detector (RTD)

Electrical methods of temperature measurement are very convenient due to the following *reasons* :

(*i*) They furnish signal that is easily detected, amplified or used for control purposes.

(*ii*) They are usually quite accurate when properly calibrated and compensated.

Electrical resistance :

The change of electrical resistance of a wire with temperature changes is another effect of the increased molecular capacity as caused by heat. Resistance thermometers utilise this characteristic and the temperature measurements are conveniently made by the change of resistance of suitable metals of known characteristics. The thermometers of this type are used over a range of temperature from –200 to 650°C. With the increase of temperature the electrical resistance of most of the metals increases in almost direct proportion to the temperature rise. Therefore, *if the electrical resistance of a wire of known and calibrated material is measured, the temperature of the wire can be determined. The resistance element should be of a suitable metal having high temperature coefficient, reproducibility, stability and ability to resist corrosion and oxidation.*

In case of variation of resistance R($R = \frac{\rho l}{a}$, where ρ is the resistivity, l length and a is the cross-sectional area of the electrical conductor) with temperature for most of the metals, the following *quadratic* relationship gives good accuracy :

$$R = R_0(1 + \alpha t + \beta t^2) \quad \text{...(21.6)}$$

where, R_0 = Resistance at temperature t = 0, and

α, β = Constants.

For a narrow range, the following *line* relationship may be used with sufficiently good accuracy.

$$R = R_0(1 + \alpha t) \quad \text{...(21.7)}$$

where, α = Resistance temperature coefficient, /°C.

Resistance thermometer construction :

The thermometer *consists of the following parts :*

(*i*) A resistance element or bulb.

Table 21.1. Summary of Temperature Range of Different Instruments

Type of Instrument	Liquid in glass thermo-ammeter	Bimetallic thermometer	Pressure thermometers			Resistance thermometers			Thermocouples				Radiation pyrometers	Optical pyrometers
	Mercury		*Vapour pressure*	*Liquid filled*	*Gas filled*	*Copper*	*Nickel*	*Platinum*	*Iron Constantan*	*Copper Constantan*	*Chromel Alumel*	*Platinum Platinum Rhodium*	–	–
Low temprature limit	–39°C	–75°C	–250°C	–87°C	–269°C	–140°C	–184°C	–184°C	–196°C	–250°C	–184°C	0°C	about room temp.	760°C
High temperature limit	600°C	540°C	316°C	650°C	540°C	120°C	316°C	760°C	760°C	350°C	1260°C	1540°C	as high as desired.	as high as desired.
Remarks	Low price, fairly accurate, easily broken.	Easier to read than liquid-in-glass thermometers.	Normally scale non-linear	To be compensated for variation in ambient temperatures	Requires large bulb, linear scale	Very sensitive. Suitable for narrow range Instruments.			Widely used in Industry	Constantan is 60% Cu, 40% Ni	Chromel is 20% Cr, 80% Ni Alumel is 2% Al, 98% Ni	Expensive	—	—

(*ii*) Suitable electrical leads.

(*iii*) An indicating-recording or resistance measuring instrument.

The ***resistance element*** is usually in the form of a coil of very fine platinum, nickel or copper wound non-conductively onto an insulating ceramic former, which is protected externally by a metal sheath.

— A *laboratory type* of resistance thermometer is often wound on a crossed *mica former* (ceramic former in case of industrial type) and enclosed in pyrex tube. The tube may be evacuated or filled with an inert gas to protect the metal wire.

It should be ensured that the resistance wire is *free from mechanical stresses.* A metal which has been strained will suffer a change in the resistance characteristics; the metal is therefore usually *annealed* at a temperature higher than that at which it is to operate.

Leads are taken out of the thermometer for the measurement of the changes in resistance in order to determine the value of temperature.

The change in resistance is usually, measured by a "*Wheatstone bridge*" which may be used either in null (balanced) condition or in the deflection out of balance condition. *The null condition suffices for steady state measurements whereas, transient conditions usually require the use of deflection mode.*

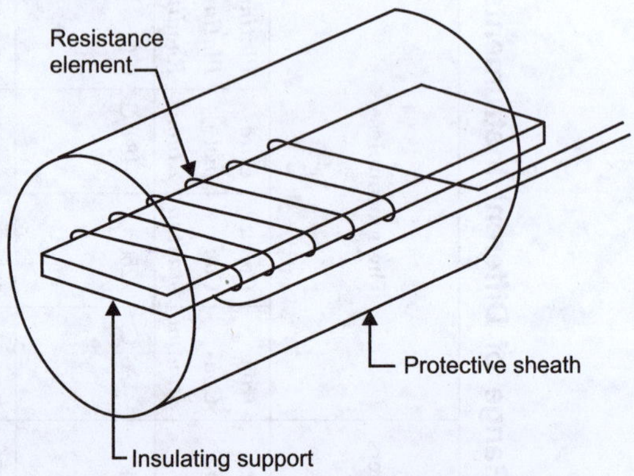

Fig. 21.19. Section illustrating the construction of a simple RTD.

Fig. 21.19, shows the section illustrating the construction of RTD.

Desirable properties of the metal used for fabricating sensing elements :

The metal to be used for fabricating sensing elements should satisfy the following *properties :*

(*i*) Linearity of resistance.

(*ii*) Resistance to corrosion and oxidation, under the temperature range of interest.

(*iii*) Should be available in conditions providing reproducible and consistent results.

(*iv*) Relatively change in resistance with temperature in order to produce a resistance thermometer with good sensitivity.

(*v*) High sensitivity so that the unit can be fabricated in a compact and convenient size.

(*vi*) No change of phase or state within a reasonable temperature change.

- *Industrial resistance thermometers often referred to as resistance temperature detectors (RTD) are usually made* with elements of *platinum, nickel* and *copper.*

— For precise measurement, platinum is preferred because it is physically stable and has high electrical resistance characteristics.

- The platinum resistance thermometer, owing to its *accuracy, stability* and *sensitivity,* has been used to define International Temperature Scale from the boiling point of oxygen (–182.9°C) to the freezing point of antimony (630.5°C).

Advantages and disadvantages of resistance thermometers :

Advantages :

(*i*) The resistance thermometer resists corrosion and is physically stable.

(*ii*) The measurement is very accurate.

(*iii*) The distance between the resistance element and the recording element can be made much larger than is possible with pressure thermometers.

(*iv*) It has lot of flexibility with regard to choice of measuring equipment and interchangeability of elements and assembly of components.

(*v*) It requires no reference junction like thermocouples and as such is more effective at room temperature.

(*vi*) The temperature sensitive resistance element can be easily installed and replaced.

(*vii*) Resistance elements can be used to measure differential temperature.

(*viii*) High working signal level, simplicity of lead wire and termination schemes compared with a thermocouple.

(*ix*) The limits of error of a resistance element are claimed to be ± 0.25 per cent of the scale range.

(*x*) The accuracy of the measuring circuit can be easily checked by substituting a standard resistor for the resistance element.

(*xi*) Self-heating effect in the conductor can be reduced to negligible with proper design.

(*xii*) The resistance element response time is of the order of 2 to 10 seconds, depending on the construction.

Disadvantages :

(*i*) The resistance thermometers cost more.

(*ii*) They suffer from time lag.

(*iii*) Possibility of current leakage between the resistance element and the ground.

(*iv*) Resistance change due to temperature changes of measuring resistors.

21.2.4. Thermistors

A **thermistor** *(contraction of the term "Thermal Resistor") is a thermally sensitive variable resistor made of a ceramic like semiconducting material.*

Thermistors, unlike metals, respond *negatively* to temperature; as the *temperature rises, the thermal resistance decreases.*

Thermistors are often composed of oxides of manganese, nickel, and cobalt in formulations having resistivities of 100 to 450,000 Ω cm; in *cryogenic applications,* doped germanium and carbon-impregnated glass are used. The metallic oxides are milled, mixed in appropriate proportions, are pressed into the desired shape with appropriate binders and finally sintered. The electrical terminals are either embedded before sintering or baked afterwards.

The electrical characteristics of thermistors are controlled by varying the type of oxide used and physical size and configuration of the thermistor.

Thermistors may be shaped in the form of beads, disc, washer etc. (Refer to Art. 7.6.2).

Thermistors may be used bare but are usually glass coated or positioned under a thin metal cap. The change in resistance is measured by using circuitry similar to that of metal conductors.

Advantages :

(*i*) Fairly less cost.
(*ii*) Availability in very small sizes.
(*iii*) Fast thermal response.
(*iv*) High sensitivity.
(*v*) Easy adaptability to electrical readout devices.

Comparison between Metal Resistors and Thermistors

S. No.	Aspects	Metal resistors	Thermistors
1.	*Resistance change*	Positive (*i.e.*, increase in resistance with temperature increase).	Negative (*i.e.*, reduced resistance with temperature rise).
2.	*Temp.-resistance relationship*	Approximately linear.	$R = R_0 \exp\left[\beta\left(\frac{1}{T} - \frac{1}{T_0}\right)\right]$, where R is the resistance at temperature $T(K)$, R_0 is the resistance at reference temperature $T_0(K)$ and β is a constant which depends on the thermistor formulation or grade.
3.	*Practical operating range*	–160°C to 600°C (for resistance thermometer)	–100°C to 300°C (for thermistors)
4.	*Stability*	More stable, provide better reproducibility with low hysteresis.	Less stable comparatively.

- Owing the inherently *high sensitivity* possessed by thermistors :
  - — Very simple electrical circuitry can be used to measure the temperature.
  - — Ordinary ohmmeters maybe used within the limits of accuracy of the meter itself.
  - — More often one of the various forms of resistance bridge is used, either in the null-balanced form or as a deflection bridge.
  - — Simple ballast circuits are also usable.
  - — In some cases, special linearizing circuits are used to obtain an output voltage that varies linearly with temperature.

Factors to be considered while selecting resistance thermometers :

The following *factors* need to be considered while making selection of resistance thermometers :

1. The *platinum resistance thermometer* should be selected for measuring temperatures upto *300° to 400°C;* for a temperature range of *0°C to 180°C, copper resistance thermometer* is employed.
2. *Adeque length* of the working part of resistance thermometer and also the protruded part of thermometer well should be chosen, since maximum accuracy of measuring is achieved when sensitive element is in the *middle of the stream.*
3. The metal for *preserving casing* should be chosen on the basis of pressure and physical phenomena of the medium. Preserving casings of standard resistance thermometers are made of the following metals :

   (*i*) Brass; (*ii*) Ordinary carbon steel;
   (*iii*) Stainless steel; (*iv*) Alloyed steel.

4. Most of the resistance thermometers cannot operate at places with high vibration or shock-type load. While choosing ratiometer, the following points should be given due considerations :
   - Correspondence of range between ratiometer and resistance thermometer.
   - Value of permissible resistance of leads; ratiometers generally operate with lead resistances of 5 Ω to 15 Ω.
   - Correspondence of scale of an instrument to measured temperatures; an *instrument with a standard scale nearest to the upper limit of measuring should be chosen.*
   - Necessity of a low voltage (4 V) D.C. power supply.

21.2.5. Thermocouple Thermometers

For higher range of temperatures *i.e.,* above 650°C, filled thermometers are unsuitable. For higher range of temperatures, *thermocouples* and *pyrometers* are used.

21.2.5.1. Thermoelectric effects

When direct current passes through a pair of thermocouples with junction maintained at different temperatures, the following *five* effects are observed :

1. Seebeck effect.
2. Peltier effect.
3. Thomson effect.
4. Joulean effect.
5. Conduction effect.

21.2.5.2. Thermocouple and thermoelectric pyrometer

A thermocouple, in its simplest form, consists of *two dissimilar metals or alloys which develop e.m.f. when the reference and measuring junctions are at different temperatures.* The reference junction or cold junction is usually maintained at some constant temperature, such as 0°C.

Fig. 21.20, shows a simple circuit of a *thermocouple and temperature measuring device.* In many industrial installations the instruments are equipped with *automatic compensating devices for temperature changes of the reference junction, thus eliminating the necessity of maintaining this junction at constant temperature.*

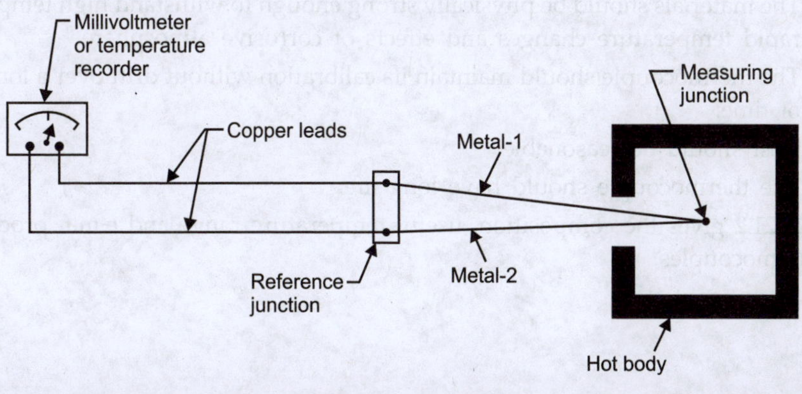

Fig. 21.20. Thermocouple.

Fig. 21.21, shows the schematic diagram of *thermoelectric pyrometer.*

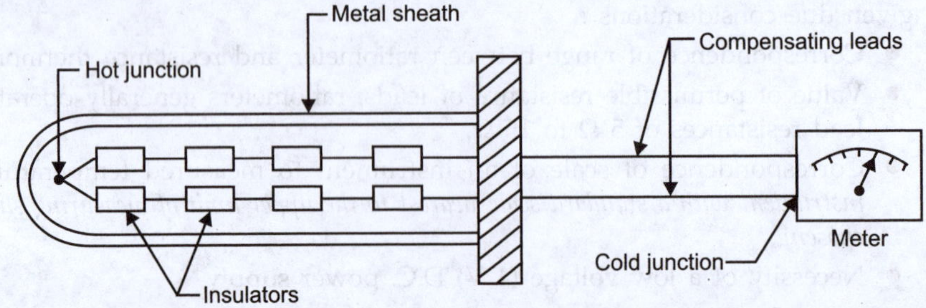

Fig. 21.21. Schematic diagram of a thermoelectric pyrometer.

Compensating leads :

These are a pair of leads wires made of such material which when connected to a thermocouple, behave such that the effective reference junction or junctions are removed to the other end of the leads. Its material should be such that it has the *same e.m.f. temperature characteristics* as the thermocouple over the range of the ambient temperature and is at the same time *cheap* also. The *resistance* of the compensating leads *should be as low as possible.* Usually they are covered with some protective material suitable for the installation.

Connectors or extension leads :

These are generally copper conductors having a *negligible resistance* and are used to connect the compensating leads further to the indicators.

- *In order to prevent the unwanted e.m.f. or unwanted ground currents,* the *thermocouple leads and extension leads must be properly insulated.*

21.2.5.3. Thermocouple materials

The thermocouple materials should possess the following *properties :*

1. The temperature–e.m.f. relationship should be reasonably linear and reproducible.
2. Adequate thermo e.m.f. per degree of temperature change to facilitate detection and measurement.
3. The materials should be physically strong enough to withstand high temperatures, rapid temperature changes and effects of corrosive atmosphere.
4. The thermocouple should maintain its calibration without drift over a long period of time.
5. Cost should be reasonable.
6. The thermocouple should have long life.

Table 21.2 gives the composition, useful temperature range and e.m.f. produced for some thermocouples.

Table 21.2. Composition, useful temperature range and e.m.f. produced for some thermocouples

S. No.	Thermocouple	Composition	Temperature (°C)		Thermoelectric power		Remarks
			Useful range	Max.	°C	Millivolt	
1.	*Platinum vs platinum-rhodium*	Pure platinum vs Pt + 10% or 13% RH	400 to 1450	1700	0 500 1000 1500	0.0 4.219 9.569 15.498	Used for high temperature measurements.
2.	*Chromel vs alumel*	90% Ni + 10% Cr vs 95% Ni + 5% (Al + Sn) Ma	–200 to 1200	1450	–200 0 300 600 900 1200	–5.75 0.0 12.21 24.90 37.36 48.85	High resistance to oxidation.
3.	*Iron vs constantan*	Pure iron vs 45–60% Cu + 55–40% Ni	–200 to 750	1000	–200 0 300 600 900	–8.27 0.0 16.59 33.27 52.29	—
4.	*Copper vs constantan*	Pure copper vs Cu-Ni constantan	–200 to 350	600	–200 0 200 400	–5.539 0.0 9.285 20.865	Not suitable in air due to excesssive oxidation.

Series combination (Thermopile):

In a series arrangement, the total output from *n* thermocouples *equals the sum of the individual e.m.fs.*

- The series arrangement provides *considerably more sensitivity* and is frequently used where it is desired to obtain a *substantially large e.m.f. for measurement of small temperature differences between the two junctions.*

Parallel combination :

- This arrangement may be used to *obtain an average temperature of a number of points, as in determining the average temperature of gas flowing through a large area passage.*
- The total potential indicated by the potentiometer will be an *average of the junction potentials.*

21.2.5.4. Advantages and disadvantages of thermocouples

The following are the *advantages* and *disadvantages* of the thermocouples :

Advantages :

1. Thermocouples have a *better response.*
2. They have a *higher range of temperature measurements.*
3. The sensing element of a thermocouple can be *easily installed.*
4. *Cheaper* than resistance thermometers.

5. *Very convenient* for measuring the temperature at *one particular point* in a piece of apparatus.

Disadvantages :

1. *Lower accuracy*, and as such they cannot be used for precision work.
2. In order to ensure long life, they need to be *amply protected.* For preventing contamination of thermocouple, when precious metals like platinum or its alloys are being used, the protecting tube has to be made chemically inert and vacuum tight.
3. The *circuitry* for thermocouple is *very complex.*

Example 21.3. *Suggest a complete instrumentation scheme to measure high temperature of a furnace using thermocouple. What are the precautions to be taken up while carrying out this measurement ?* **(Punjab University)**

Solution. The system comprises of Platinum-Rhodium-Platinum type thermocouples (suitable upto 2200°C) in the form of parallel thermopile for averaging. Reference junction compensation is done by means of an isothermal block.

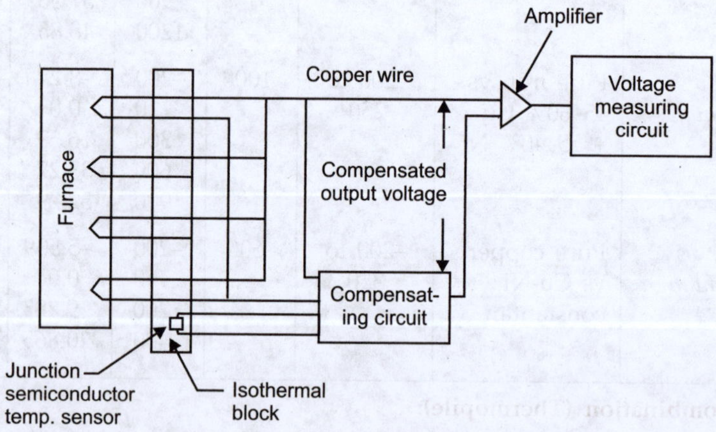

Fig. 21.22

For greater accuracy a junction semiconductor temperature sensor may be used as shown in Fig. 21.22, for taking care of the drift caused by ambient temperature variation. The compensated output voltage is amplified and measured with the help of a suitable voltage measuring circuit, directly calibrated in terms of temperatures.

Precautions :

1. Reference junction compensation is essential.
2. The thermocouple leads should be as short as possible.
3. The compensating circuit should also provide compensation for any changes in the ambient temperature.

21.2.6. Radiation Pyrometers

The principle, that the *radiant energy emitted from a body increases with temperature,* is used in measuring temperature particularly in the *higher ranges.*

Radiation pyrometers are used in the following *situations :*

(*i*) High temperatures.

(*ii*) Where it is not possible to contact the material whose temperature is to be measured.

(iii) Where furnace atmosphere is detrimental to thermocouples and causes erratic measurements and short life.

Following two *"principles"* are used for construction of radiation temperature measuring devices :

(i) ***Total radiation pyrometer.*** In this case the *total radiant energy* from a heated body is measured.

(ii) ***Selective (or partial) radiation pyrometer.*** In this case, we measure the spectral radiant intensity of the radiant energy from the heated body at a *given wavelength.*

Since it is difficult to build a practical pyrometer that is responsive to radiations of all wavelengths or to spectral radiation of one wavelength, therefore, *practical radiation pyrometers response to a wide band of radiation of approximately 0.1 to 8.0 microns width in the visible and infrared band of radiation spectrum;* the actual width of this band, however, depends entirely on the *physical construction of the radiation receiver.*

Total radiation pyrometer :

A device which measures the total intensity of radiation emitted from a body is called total ***radiation pyrometer.***

The elements of a *total radiation pyrometer* are illustrated in Fig. 21.23. It collects the radiation from an object (hot body) whose temperature is required. A mirror is used to focus this radiation on a thermocouple. This energy which is concentrated on the thermocouple raises its temperature and in turn generates an e.m.f. This e.m.f. is then measured either by the galvanometer or potentiometer method. *Thus rise of temperature is a function of the amount of radiation emitted from the object.*

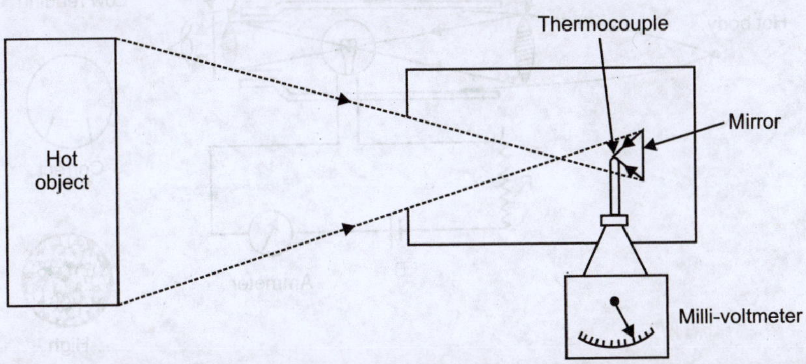

Fig. 21.23. A schematic diagram of radiation pyrometer.

Advantages and disadvantages of radiation pyrometers :

Following are the *advantages and disadvantages of radiation pyrometers :*

Advantages :

(i) Can be used to measure the temperature of an object which may be either stationary or moving, and so adaptable to continuous industrial processing.

(ii) High speed of response.

(iii) Accuracy ± 2% of scale.

(iv) Primarily used to measure temperatures in the range 700–2000°C, where resistance thermometers and thermocouples cannot be used.

(v) Direct contact is not necessary with the object whose temperature is to be measured.

(*vi*) Suitable for measuring temperatures where the atmospheric or other environmental conditions prevent satisfactory operation of other temperature sensing devices.

(*vii*) Relatively independent of the distance between the measuring element and the heated body.

Disadvantages :

(*i*) In the situations where the temperature may be high due to operating conditions, *cooling is required* to protect the instrument from overheating.

(*ii*) The presence of dust and dirt on the mirrors or lens causes the instrument to read too low.

21.2.7. Optical Pyrometers

An **optical pyrometer** *works on the principle that matters glow above 480°C and the colour of visible radiation is proportional to the temperature of the glowing matter.* The amount of light radiated from the glowing matter (solid or liquid) is measured and employed to determine the temperature.

Fig. 21.24, shows a *disappearing filament pyrometer.*

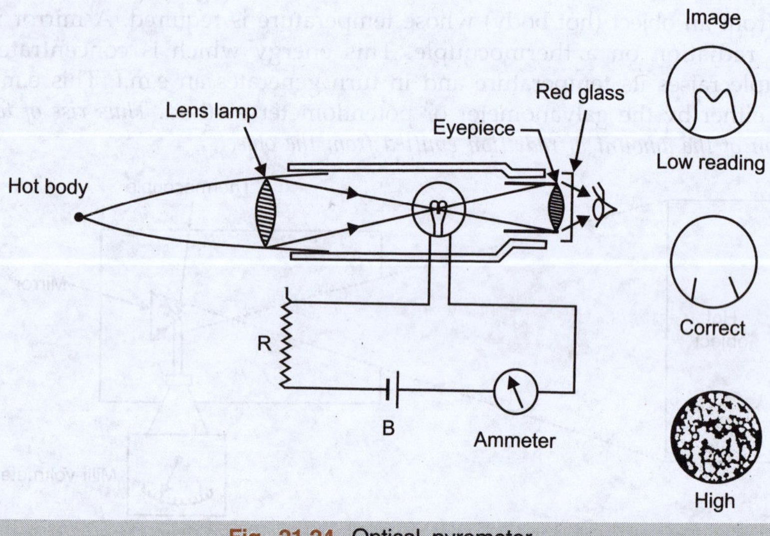

Fig. 21.24. Optical pyrometer.

Operation :

— The optical pyrometer is sighted at the hot body and focused.

— In the beginning filament will appear dark as compared to the background which is bright (being hot).

— By varying the resistance (*R*) in the filament circuit more and more current is fed into it, till filament becomes equally bright as the background and hence disappears.

— The current flowing in the filament at this stage is measured with the help of an ammeter which is calibrated directly in terms of temperature.

— If the filament current is further increased, the filament appears brighter as compared to the background which then looks dark.

— An optical pyrometer can measure temperature ranging from 700 to 4000°C.

Uses :

The optical pyrometer is widely used for *accurate measurement of temperature of :*

— furnaces;

— molten metals.

— other heated materials.

- *This pyrometer has been accepted as the standard means for determining temperatures on the International Temperature Scale from the gold point and upwards.*

Advantages :

(*i*) Excellent accuracy within ±5°C for the operating range 700–3000°C.

(*ii*) No direct contact is necessary with the object whose temperature is to be measured. Thus this type of pyrometer can be used in situations where the measuring target is remote and inaccessible such as furnace interiors, molten metals etc.

(*iii*) Measurement is independent of the distance between the target and measuring instrument.

(*iv*) The skill in operating the thermometer can be acquired readily.

Disadvantages :

(*i*) The lower measuring temperature is limited to 700°C (The eye is insensitive to wavelength characteristics below this temperature).

(*ii*) Owing to the manual null-balance operation of this pyrometer it is not suitable for continuous reading or automatic control applications.

21.3 FLOW MEASUREMENT

21.3.1. Introduction

In almost all the flow situations of engineering importance measurements of fluid velocity, flow rate and flow quantity with varying degree of accuracy are a fundamental necessity. Flow of material in a process or system can be measured by a variety of methods depending upon the *material and its condition, the type of flow, the volume, the pressure and temperature range, the accuracy needed and the control required.*

The accurate measurement of flow presents many and varied problems. The flowing medium may be *liquid, a gas, a granular solid.* The flow may be a *laminar or turbulent, steady-state or transient.* In addition, there are several different approaches to the problem of flow measurement.

Flow characteristics. When fluids move at very low velocities through uniform conduits, the motions of individual particles are generally along lines paralleling the conduits' walls. The velocity of particles is greatest at the centre and theoretically zero at the wall of conduit with the velocity distribution as shown in Fig. 21.25(*a*), such a flow is called *"laminar"*.

On increasing the flow, a point is reached where the particles motion becomes more

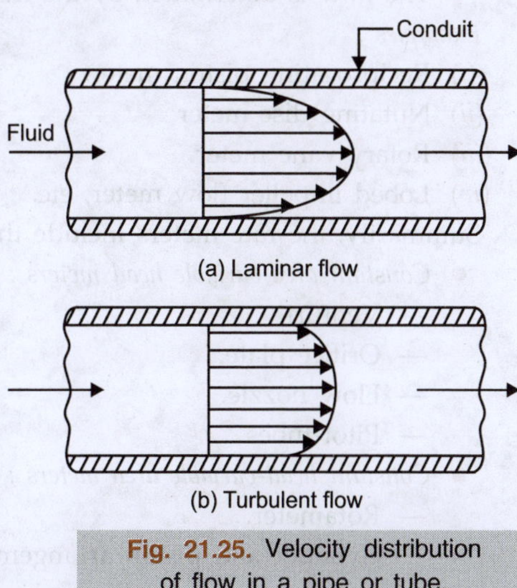

Fig. 21.25. Velocity distribution of flow in a pipe or tube.

random and complex and flow at *higher rates* is referred to as *"turbulent"* (The approximate velocity at which the change occurs is called the *critical velocity*). The corresponding time-average velocity distribution across a circular tube is shown in Fig. 21.25(*b*).

It has been observed that the critical velocity is a function of several factors that may be put in a dimensionless form called the *Reynolds number, Re,* its value for a circular pipe is given as

$$Re = \frac{\rho VD}{\mu} \quad \text{...(21.8)}$$

where,
ρ = Density of fluid, kg/m^3,
V = Velocity flow, m/s,
D = Diameter of pipe, m, and
μ = Dynamic viscosity of fluid, Ns/m^2.

When, $Re < 2000$ .... The flow is *laminar* (or viscous)
$Re > 4000$ .... The flow is turbulent

Re between 2000 and 4000 (transition zone) .... The flow is unpredictable

21.3.2. Flow Measurement Methods/Devices

The flow measurement devices are *classified* as follows :

1. Rate meters
2. Quantity meters.

1. Rate meters. They measure either the volumetric flow rate directly or use meters that measure velocity and the volume flow rate can then be calculated with the help of cross-section.

(*i*) ***Inferential type.*** Inferential methods *imply that the flow is not directly measured but is inferred from the measurement of other quantities (e.g., pressure, temperature, etc.).*

(*ii*) ***Absolute or positive displacement type.*** These meters basically *capture and release a fixed volume of fluid by some type of pumping action.* They normally count the number of cycles that occur and indicate or register an integrated flow volume. The flow is determined by the frequency of the cycle.

Examples :

(*i*) Reciprocating piston meter.
(*ii*) Nutating disc meter.
(*iii*) Rotary vane meter.
(*iv*) Lobed impeller flow meter, etc.

Summarily, the rate meters include the following :

- *Constant area-variable head meters :*
  — Venturi.
  — Orifice plate.
  — Flow nozzle.
  — Pitot tubes.
- *Constant head-variable area meters ;*
  — Rotameter.
  — Cylinder and piston arrangement.

- *Variable head and variable area meters :*
  — Weirs.
  — Flumes.
- *Variable velocity meters :*
  — Current meters.
  — Anemometers.
- *Variable force meters.*
  — Vane.

The term *obstruction meter* applies to all the above mentioned devices.

Besides the above mentioned rate meters following are some *special meters :*

— Electromagnetic flow meter ;

— Ultrasonic flow meter, etc.

2. Quantity meters. These meters may be designed for the measurement of either weight or volume. They may be absolute or displacement type. These are generally cited as *positive meters.*

Examples :

(*i*) *Weight meters :*
— Weighters and traps.

(*ii*) *Volumetric meters :*
— Tank-rotating impeller piston pumps.
— Nutating piston pump for liquids.
— Bellows and liquid sealed drums for gases.

21.3.3. Variable Head Flow for Incompressible Fluids

When the fluid flows through the obstruction meters, such as *venturimeter, orifice meter and flow nozzle,* the fluid accelerates and reduction in pressure takes place. The pressure difference before and after the obstruction is measured by means of a differential pressure sensor and is related to the rate of flow.

21.3.3.1. Venturimeter

A ***venturimeter*** *is an instrument used to measure the rate of discharge in a pipeline and is often fixed permanently at different sections of the pipeline to know the discharge there.*

It is one of the most important practical applications of Bernoulli's theorem. It has been named after the 15th century Italian *engineer Venturi.*

The Venturimeter may be used in any position, horizontal vertical or inclined. In order to measure pressure differentials U-tube manometers are used.

The size of a venturimeter is expressed in terms of inlet and throat diameters. For example, a 300 × 150 mm venturimeter fits in a 300 mm diameter pipe and has a throat diameter of 150 mm.

A venturimeter consists of following *three* parts :

(*i*) A short converging part,

(*ii*) Throat, and

(*iii*) Diverging part,

Expression for rate of flow :

Fig. 21.26 shows a venturimeter fitted in *horizontal pipe* through which a fluid is flowing.

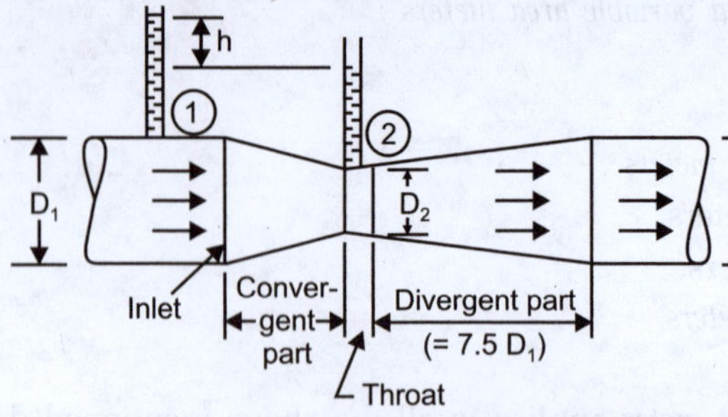

$\left(\text{Throat ratio } \frac{D_2}{D_1} \text{ varies from } \frac{1}{4} \text{ to } \frac{3}{4}\right)$

Fig. 21.26. Venturimeter.

Let, D_1 = Diameter at inlet or at section 1,

A_1 = Area at inlet $\left(= \frac{\pi}{4} D_1^2\right)$,

p_1 = Pressure at section 1, Nm^2,

w_1 = Specific weight of fluid at inlet, N/m^3

V_1 = Velocity of fluid at section 1 and

D_2, A_2, p_2, w_2 and V_2 are the corresponding values at section 2.

Applying Bernoulli's equation at sections 1 and 2, we get

$$\frac{p_1}{w_1} + \frac{V_1^2}{2g} + z_1 = \frac{p_2}{w_2} + \frac{V_2^2}{2g} + z_2$$

Here, $z_1 = z_2$ ... since the pipe is *horizontal*.

$$\frac{p_1}{w_1} + \frac{V_1^2}{2g} = \frac{p_2}{w_2} + \frac{V_2^2}{2g}$$

or,
$$\frac{p_1 - p_2}{w} = \frac{V_2^2}{2g} - \frac{V_1^2}{2g} \quad ...(21.9)$$

... For incompressible fluid, $\rho_1 = \rho_2 = \rho$ and $w_1 = w_2 = w$

$\frac{p_1 - p_2}{w}$ = Difference of pressure heads at sections 1 and 2 is equal to h; called *pressure head or piezometer* head.

It is measured by a differential manometer, pressure gauge or any other pressure measuring device.

The continuity equation gives,

$$A_1 V_1 = A_2 V_2$$

or,
$$V_1 = \frac{A_2 V_2}{A_1}$$

Substituting for V_1 in equation (21.9), we have

$$\frac{p_1 - p_2}{w} = \frac{V_2^2}{2g} - \frac{A_2^2 V_2^2}{A_1^2 \times 2g}$$

$$= \frac{V_2^2}{2g}\left(1-\frac{A_2^2}{A_1^2}\right) = \frac{V_2^2}{2g}\left[1-(A_2/A_1)^2\right]$$

or $$V_2 = \frac{1}{\sqrt{1-(A_2/A_1)^2}}\sqrt{\frac{2g}{w}(p_1-p_2)} \quad ...(21.10)$$

For a given venturimeter, A_1 and A_2 are constants and therefore,

$$V_2 = M\sqrt{\frac{2g}{w}(p_1-p_2)}$$

where, $M = \frac{1}{\sqrt{1-(A_2/A_1)^2}}$ is a constant called *"velocity approach factor"*

The value of M *is always greater than unity.*

Equation 21.10 has been derived on the basis that there are no losses. However, in actual practice, this is not true and actual velocity is obtained from the theoretical by multiplying the latter by a factor C_v, called *"coefficient of velocity"*.

$$(V_2)_{\text{actual}} = C_v M\sqrt{\frac{2g}{w}(p_1-p_2)} \quad ...(21.11)$$

Now, Discharge = area × velocity

or $$Q = A_2 C_v M\sqrt{\frac{2g}{w}(p_1-p_2)} \quad ...(21.12)$$

Sometimes, a factor, C_v called *"coefficient of contraction"* is taken into consideration for finding out the actual discharge.

∴ $$Q_{\text{actual}} = C_c C_v M A_2\sqrt{\frac{2g}{w}(p_1-p_2)} \quad ...(21.13)$$

or $$Q_{\text{actual}} = C_d M A_2\sqrt{\frac{2g}{w}(p_1-p_2)} \quad ...(21.14)$$

where, $C_d = C_c C_v$ is called *"coefficient of discharge"*. For venturimeters, the factors C_d and M are applied separately but for nozzles and Orifice plates these factors are combined into a single factor known as *"flow coefficient"* and designated as K.

∴ $$Q_{\text{actual}} = KA_2\sqrt{\frac{2g}{w}(p_1-p_2)}$$

... For nozzles and orifice plates ...(21.15)

In order to take into account the *changes in flow rate on account of changes in temperature*, a factor E is used. Then the actual discharge is essentially obtained from the following relation :

$$Q_{\text{actual}} = C_d MEA_2\sqrt{\frac{2g}{w}(p_1-p_2)} \quad ...(21.16)$$

where, E is the *"thermal expansion correction factor"*

— The value of C_d is the *less than unity* (varies between 0.96 to 0.98). Due to variation of C_d venturimeters are not suitable for *very low velocities*.

Value of 'h' by differential U-tube manometer :

Case I. *Differential manometer containing a liquid heavier than the liquid flowing through the pipe.*

Let, S_{hl} = Sp. gravity of heavier liquid,

S_p = Sp. gravity of liquid flowing through pipe, and

y = Difference of the heavier liquid column in U-tube

Then, $$h = y\left[\frac{S_{hl}}{S_p} - 1\right] \quad ...(21.17)$$

Case II. *Differential manometer containing a liquid lighter than the liquid through the pipe.*

Let, S_{ll} = Sp. gravity of lighter liquid,

S_p = gravity of liquid flowing through pipe, and

y = Difference of lighter liquid column in U-tube.

Then, $$h = y\left[1 - \frac{S_{hl}}{S_p}\right] \quad ...(21.18)$$

Advantages :

(*i*) Low head loss (about 10% of differential pressure head)

(*ii*) Due to their extensive use over a long period in the past, their characteristics are well established through practice and consequently their behaviour can be predicted perfectly.

(*iii*) On account of low loss, they have a high coefficient of discharge (0.96 to 0.98) and as such they can be used to measure high flow rates in pipes having diameters of the order of a few meters.

Disadvantages :

(*i*) High cost.

(*ii*) Owing to their large size they are unsuitable for applications where there is a limited space.

Uses : Venturimeters are used for measurement of flow of :

— Water

— Fluids used in industrial processes;

— Industrial wastes;

— Gases;

— Slurries and dirty liquids;

— Suspended particles.

21.3.3.2. Orifice plate meter

An orifice plate is *defined as plate in which an aperture has been cut*. This device is cheaper than venturimeter and is used for measuring the discharge of fluid through a pipe. It also works on the same principle of a venturimeter. It is more frequently used than all other primary elements combined together.

The orifice plates are of the following *four* types :

1. Concentric;
2. Eccentric;

3. Segmental;
4. Quadrant types.

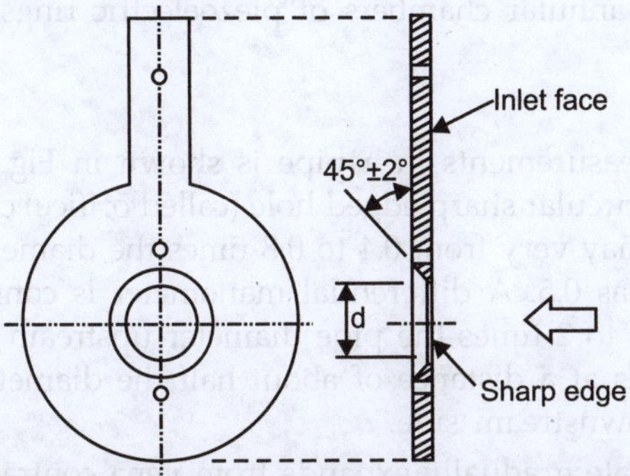

Fig. 21.27. Concentration orifice plate with 45° bevelled edges.

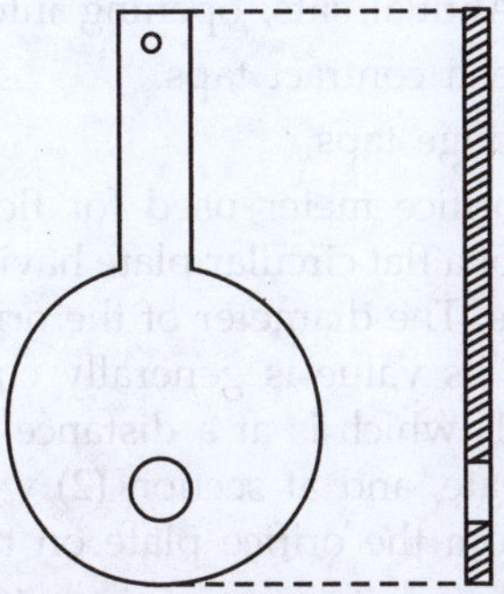

Fig. 21.28. Eccentric orifice plate.

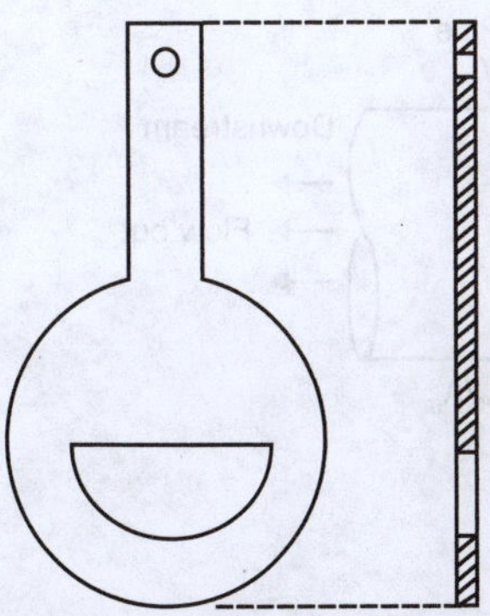

Fig. 21.29. Segmental orifice plate.

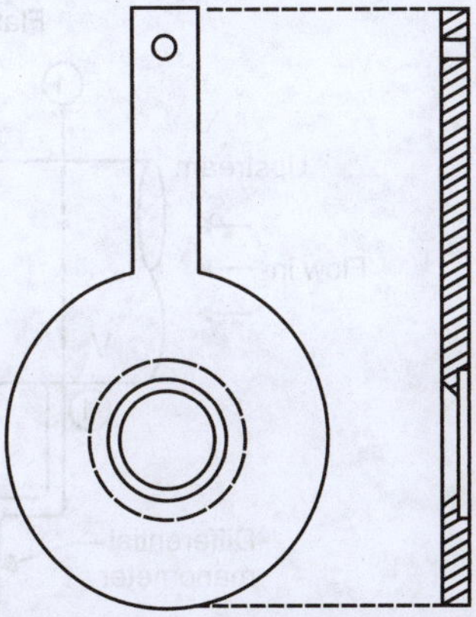

Fig. 21.30. Quadrant edge orifice plate.

1. The "*concentric plate*" has a circular hole located *in the centre* (Fig. 21.27).
2. The "*eccentric plate*" has a circular hole located below its centre (Fig. 21.28).
3. The "*segmental plate*" has a hole that is only *partly circular* located *below its centre* (Fig 21.29).
4. The "*quadrant edge orifice plate*" is so constructed that the edge is rounded so as to form a quarter circle (Fig. 21.30).

Uses : The type of the orifice plate used depends upon the *characteristics of the fluid that* is to be metered; *the eccentric and segmental types are used for fluids containing solids.*

Orifice plates are made from *steel, stainless steel, monel, phosphor bronze and other such materials, that can withstand the corrosive effect of the fluid.*

— Its thickness is only sufficient to withstand the buckling forces caused by the pressure differential.

— The circular holes are carefully made with 90% square sharp edge.

The *orifice plates can also be* **classified** according to the disposition of pressure taps (upstream and downstream) to measure the differential pressure across it. Following types of *pressure taps* are generally employed.

1. *Corner taps.*
   (*i*) Single tappings (Individual tapes)
   (*ii*) Annual slits, opening into the annular chambers of piezoelectric rings.
2. Vena contract taps.
3. Flange taps

An orifice meter used for flow measurements in a pipe is shown in Fig. 21.31. It consists of a flat circular plate having a circular sharp edged hole (called orifice) concentric with pipe. The diameter of the orifice may very from 0.4 to 0.8 times the diameter of the pipe but its value is generally chosen as 0.5. A differential manometer is connected at section (1) which is at a distance to 1.5 to 2 times the pipe diameter upstream from the orifice plate, and at section (2) which is at a distance of about half the diameter of the orifice from the orifice plate on the downstream side.

The jet coming out of the orifice plate gradually expands from vena contracta to fill the pipe. Fig. 21.31 shows the location of vena contracta point. A part of the K.E. of the jet is converted into eddy currents causing dissipation of energy and loss of head.

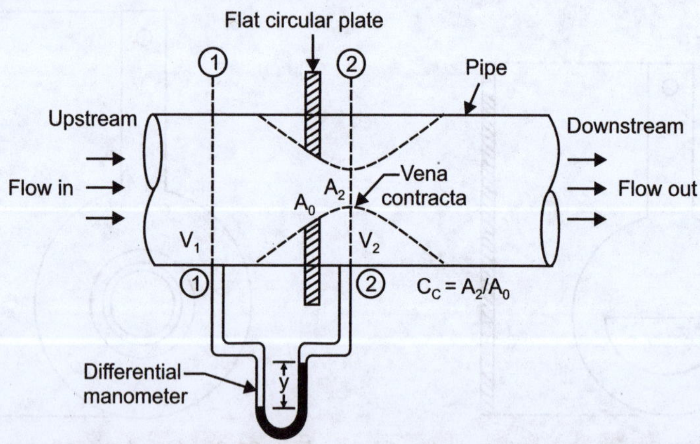

Fig. 21.31. Orifice meter.

Let, A_1 = Area of pipe at section (1),

V_1 = Velocity at section (1)

p_1 = Pressure at section (1), and

A_2, V_2 and p_2 are corresponding values at section (2).

Applying Bernoulli's equation at section (1) and (2) we get,

$$\frac{p_1}{w} + \frac{V_1^2}{2g} + z_1 = \frac{p_2}{w} + \frac{V_2^2}{2g} + z_2$$

Here $z_1 = z_2$, since the pipe is horizontal.

$$\therefore \quad \frac{p_1}{w} + \frac{V_1^2}{2g} = \frac{p_2}{w} + \frac{V_2^2}{2g}$$

or

$$V_{2(\text{actual})} = C_v M \sqrt{\frac{2g}{w}(p_1 - p_2)} \quad \text{...(21.19)}$$

$$= \frac{C_v}{\sqrt{1-(A_2/A_1)^2}} \sqrt{\frac{2g}{w}(p_1 - p_2)}$$

As the jet comes out of the orifice, it contracts to a minimum area A_0 at the vena contracta. Area at the vena contracta, $A_2 = C_v A_o$.

$$\therefore \quad V_{2(\text{actual})} = \frac{C_v}{\sqrt{1-(C_c A_0 / A_1)^2}} \sqrt{\frac{2g}{w}(p_1 - p_2)} \qquad \text{...(21.20)}$$

Actual flow rate, $Q_{\text{actual}} = A_2 V_{2(\text{actual})} = C_c A_0 V_{2(\text{actual})}$

$$= C_v C_c \frac{A_0}{\sqrt{1-(C_c A_0 / A_1)^2}} \sqrt{\frac{2g}{w}(p_1 - p_2)} \qquad \text{...(21.21)}$$

$$= C_d M E A_0 \sqrt{\frac{2g}{w}(p_1 - p_2)} \qquad \text{...(21.22)}$$

$$= K E A_0 \sqrt{\frac{2g}{w}(p_1 - p_2)} \qquad \text{...(21.23)}$$

where, $$K = C_d M = \frac{C_d}{\sqrt{1-(C_c A_0 / A_1)^2}}$$

where K is called the *"orifice flow constant"*.

Advantages :

(*i*) Low initial cost.

(*ii*) Easy to install.

(*iii*) Simple and less expensive maintenance as compared with a venturimeter.

(*iv*) Requires less space as compared to a venturimeter.

(*v*) The behaviour is predictable and the results are reproducible with good accuracy.

(*v*) Can be used in wide range of pipe sizes (1.25 cm to 150 cm)

Disadvantages :

(*i*) Low coefficient of discharge (about 0.6 as compared to almost unity for venturimeters), consequently their sensitivity is much lower than that of venturimeter.

(*ii*) Poor pressure recovery. The overall pressure loss varies form 40 to 90% of the differential pressure.

(*iii*) Susceptible to inaccuracies resulting from erosion, corrosion, and sealing.

(*iv*) Since the flat orifices tend to clog, they are not suitable for measurement of slurries or entrained particles.

(*v*) Owing to the lower physical strength of the orifices, they are likely to be damaged by pressure transients.

Difference between a venturimeter of an orificemeter :

• A **venturimeter** is a device which is inserted into pipeline to measure incompressible fluid flow rates. It consists of a convergent section which reduces the diameter to between one half to one-fourth of the pipe diameters. This is followed by a divergent section. The pressure difference between the position just before the venturi and at the throat of the venturi is measured by a differental manometer. *The working of the venturi is based on the Bernoulli's principle, that is when the velocity head increases in an accelerated flow, there is a corresponding reduction in the piezometric head.*

• The **orificemeter** is an opening, usually round, located in the side wall of the tank or reservoir for measuring the flow of a liquid. The main feature of the orifice meter is that *most of the potential energy of the liquid is converted into kinetic energy of the free jet issuing through the orifice.*

The main points of difference between a venturimeter and orificemeter are :

1. The venturimeter can be used for *measuring the flow rates of all incompressibly flows* (gases with low pressure variations, as well as liquids), whereas orifice meters are generally used for measuring the *flow rates of liquids.*
2. Venturimeter is installed in *pipeline only*, and the accelerated flow through the apparatus of subsequently decelerated to the original velocity at the outlet of the venturimeter. The flow continues through the pipeline. In the orifice meter the entire potential energy of the fluid converted to kinetic energy, and the jet *discharges freely into the open atmosphere.*
3. In venturimeter, the flow *velocity is measured by noting the pressure difference between the inlet and the throat of the venturimeter,* whereas in the orificemeter the discharge velocity measured by using *pitot tube or by trajectory method.*

21.3.3.3. Flow nozzles

In engineering practice, the nozzles are used for the creation of jets and streams for all purposes as well as for fluid metering. When placed in or at the end of a pipeline as metering devices, they are called as *"Flow nozzles"*.

In fact, a flow nozzle is nothing else but a *venturimeter without it diverging cones.* It differs from an orifice meter in the sense that it is made curved so that jet contraction is not so large as in an orifice-meter.

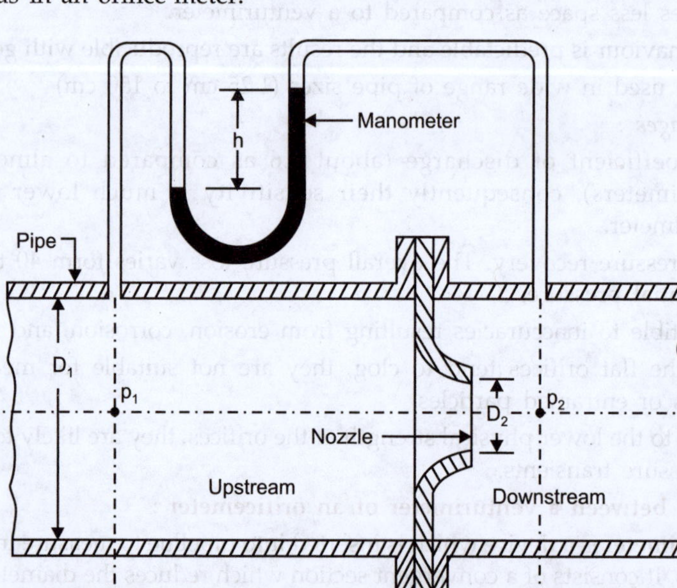

Fig. 21.32. Flow measurement through the use of a nozzle.

Fig. 21.32 shows flow measurement by the use of a nozzle. The flow nozzle comprises a smooth, gradual contraction to throat followed by a free, uncontrolled expansion back to the original pipe flow area. In a nozzle the flow is guided smoothly to the section of minimum area. The equation for discharge through a nozzle is given by

$$Q_{actual} = KEA_2 \sqrt{\frac{2g}{w}(p_1 - p_2)} \quad ...(21.25)$$

where,
$$K = \frac{C_d}{\sqrt{1-(A_2/A_1)^2}} = \frac{C_d}{\sqrt{1-(D_2/D_1)^4}} \quad ...(21.26)$$

Since the device contains no provision for an orderly transformation of velocity pressure into static pressure, the nozzle has a pressure loss of from 80 to 90 per cent of the differential pressure obtained.

The *discharge coefficient of a flow nozzle is dependent upon the following factors :*

(*i*) The smoothness of approach to tangency.

(*ii*) The length of cylindrical portion of the nozzle.

(*iii*) The location of the pressure taps. The pipe wall taps located at one pipe diameter upstream and half pipe downstream from the inlet face of the nozzle give best result.

The flow nozzles are usually made of *gunmetal, stainless steel or monel metal.*

Advantages :

(*i*) They have higher coefficient of discharge than that of orifice meters.

(*ii*) Cheaper and easier to install.

(*iii*) Physical length is less compared to venturimeter.

(*iv*) Quite useful for measurement of flow of fluids containing solids that settle.

(*v*) Widely used for high pressure/temperature steam flow.

(*vi*) Being more rugged and more resistant to erosion than sharp edged orifice, they can be used for flow measurements at high velocities.

Disadvantages :

(*i*) Owing to poor pressure recovery, they are not useful for applications where pressure heads are small or where pressure recovery is a must.

(*ii*) As compared to orificemeter, it is expensive and difficult to install.

(*iii*) Limited moderate pipe sizes, not available above 120 cm.

(*iv*) Higher maintenance.

21.3.4. Pitot Tubes

In kinematics and dynamics of fluid flow, flow velocity constitutes an important parameter and as such its measurement is quite essential. *Velocity measurements are made for the following purposes :*

(*i*) To determine the rate of flow (volumetric)

(*ii*) To locate separation points from a knowledge of velocity distribution.

(*iii*) To forecast the weather from wind velocity measurements.

- Instruments like the *"pitot tube"* and *"hot wire anemometers"* measure the *"local velocity"* at point in the channel or ducts through which the fluid is flowing.
- Instruments like the *cup and vane anemometers, the current and turbine meters* measure the *"average velocity" of flow.*

Description of pitot tubes:

The velocity distribution across a section can be obtained with the help of a *pitot-tube,* which is one of the most accurate devices for velocity measurement.

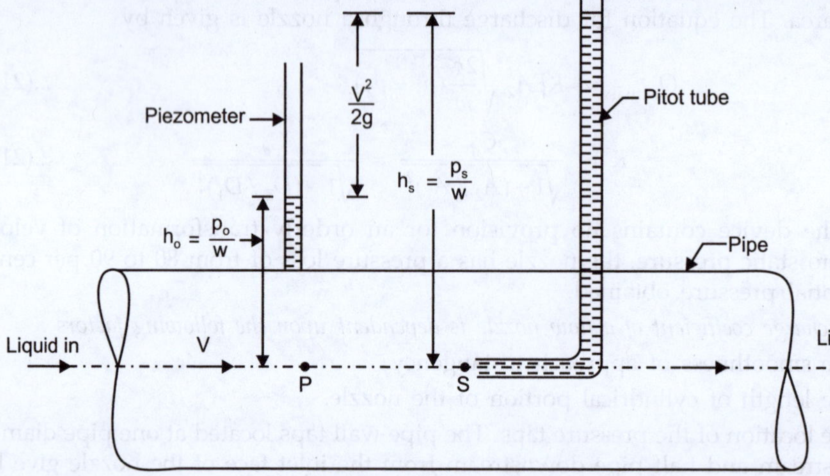

Fig. 21.33. Pitot tube.

It consists of a glass tube in the form of a 90° bend of short length open at both its ends. It is placed in the flow with its bent leg directed upstream so that a stagnation point is created immediately in front of the opening (Fig. 21.33). The kinetic energy at this point gets converted into pressure energy causing the liquid to rise in the vertical limb, to a height equal to the stagnation pressure.

Applying Bernaulli's between stagnation point (S) and a point (P) in the undistributed flow at the same horizontal plane we get,

$$\frac{p_o}{w} + \frac{V^2}{2g} = \frac{p_s}{w}$$

or, $$h_o + \frac{V^2}{2g} = h_s$$

or, $$V = \sqrt{2g(h_s - h_o)} \text{ or } \sqrt{2g\Delta h} \quad \text{...(21.27)}$$

where, p_o = Pressure at point 'P' *i.e.*, static pressure,

V = Velocity at point 'P' *i.e.*, free flow velocity.

p_s = Stagnation pressure at point 'S', and

Δh = Dynamic pressure

= Difference between stagnation pressure head (h_s) and static pressure head (h_o)

The height of liquid rise in the pitot tube indicates the stagnation pressure head. The static pressure head may be measured separately with a piezometer (Fig. 21.34).

Both the static pressure as well as stagnation pressure can be measured in a device known as *Pitot-static tube* (Fig. 21.34).

It consists of two concentric pitot-tubes with an annular space in between as shown in the figure. The outer tube has additional two or more holes drilled perpendicular to the direction of flow and thus the liquid level in it gives the static head, while the inner tube works as normal pitot-tube. If a differential manometer is connected to the tubes of a pitot static tube it will measure the dynamic pressure head. If y is manometer difference, then

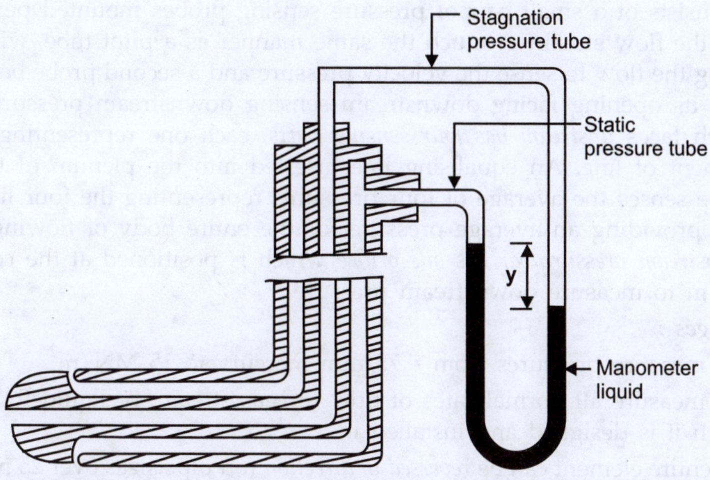

Fig. 21.34. Pitot-static tube.

$$\Delta h = y\left(\frac{S_m}{S} - 1\right) \quad ...(21.28)$$

where, S_m = Specific gravity of manometric liquid, and

S = Specific gravity of the liquid flowing through the pipe

When a pitot-tube is placed in the fluid stream the flow along its outer surface gets accelerated and causes the static pressure to decrease. Also, the stem, which is perpendicular to the flow direction, tends to produce an excess pressure head. In order to take these effects into account equation (21.27) is modified to give the actual velocity as :

$$V = C\sqrt{2g\Delta h} \quad ...(21.29)$$

where, C = A corrective coefficient which takes into account the effect of stem and bent leg.

The most commonly used form of Pitot-static tube known as Prandle-Pitot tube is so designed that the effect of stem and bent leg cancel each other *i.e.,* C = 1.

Advantages :

(*i*) Negligible loss of head when pitot tube is inserted in the pipe.

(*ii*) Less costly as compared to venturimeter or orificemeter.

(*iii*) Easy to install and remove from the pipeline.

Disadvantages :

(*i*) Poor accuracy.

(*ii*) Unsuitable for dirty or sticky liquids.

(*iii*) Requires high flow velocities of about 15 m/s to produce measurable heads.

(*iv*) Sensitive to upstream disturbance, to Reynolds number changes and to dimensions errors.

(*v*) Cannot be used for industrial applications which require an instant readout.

Annubar tube :

- An ***Annubar tube*** is a *commercial version of pitot tube and is applicable to liquid, steam and gas flow measurements.* Its simple and unique design is based upon the fundamental Bernoulli's equation for flow balance.

- It consists of a small pair of pressure sensing probes mounted perpendicularly into the flow stream in much the same manner as a pitot tube, with one probe facing the flow to sense the velocity pressure and a second probe behind the first with its opening facing downstream sensing downstream pressure. The probe which faces *upstream has four sensing parts,* each one representing an annubar segment of line. An equalising line inserted into the plenum of the upstream probe senses the average of four pressures representing the four line segments, thus providing an average pressure for the entire body of flowing liquid. The *downstream pressure bar has one orifice* which is positioned at the centre of flow stream to measure downstream pressure.

Advantages :

(*i*) Can measure pressures from – 75 mm mercury to 15 MN/m^2.

(*ii*) Can measure all normal rates of flow for gas stream, or liquid for the pipe for which it is designed and installed in.

(*iii*) The entire element can be *reversed in direction* (on pipe sizes over 25 mm diameter) to detect reverse flows or act as an off position without a system shut down.

(*iv*) Can measure a wide range of flow rates without deviating from a square root output signal which is compatible with all standard makes of flow instrumentation and control.

(*v*) Relatively less expensive as a flow detector.

(*vi*) Relatively simple to install and light in weight.

(*vii*) Reliable for flow measurements through pipes or circular ducts from 21.5 mm to 225 mm in diameter.

(*viii*) The flow resisitive effect of this meter is almost negligible.

(*ix*) These units are available as complete systems with elements, including, recording and control instrumenation.

21.3.5. Weirs and Flumes

These devices are commonly used for *open channel measurements.*

Weir : A weir is essentially a dam with a notched opening at the top through which the liquids flows.

The weir make use of *rectangular, V and tapezoidal notches.* The rate of flow is simply measured by measuring the head of water above the lowest point of the weir opening through which the liquid flows. This height is measured by means of a float installed in a box called *stilling well,* which is a part of the total structure. The float is so placed that is not disturbed by the velocity of flow or by turbulence of stream.

- The discharge through a *rectangular notch,*

$$Q = \frac{2}{3} C_d L \sqrt{2g} (H)^{\frac{3}{2}} \quad ...(21.30)$$

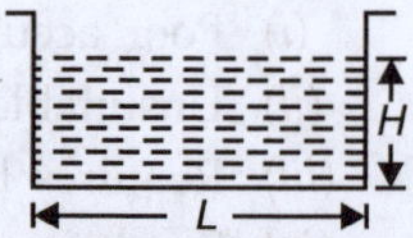

Fig. 21.35. Rectangular notch/wear.

where, C_d = Coefficient of discharge (it varies with the dimensions of the weir and is usually less than 0.7),

L = Length of weir, and

H = Height of water above sill of the notch/wear.

— The rectangular notch is the *most commonly used notch on account of its simplicity, ease of construction and accuracy.*

- The discharge through a V-notch,

$$Q = \frac{8}{15} C_d \sqrt{2g}(H)^{\frac{5}{2}} \tan\frac{\theta}{2} \quad ...(21.31)$$

where, θ = angle of notch.

C_d varies with θ and H, kinematic viscosity and surface tension of liquids; typical values are 0.58 to 0.7.

— The V-notch offers the widest range for a single size, the small opening at the lower end can accommodate small flows while the top portion for large flows. However, *because of its shape, it gives the highest loss of head.*

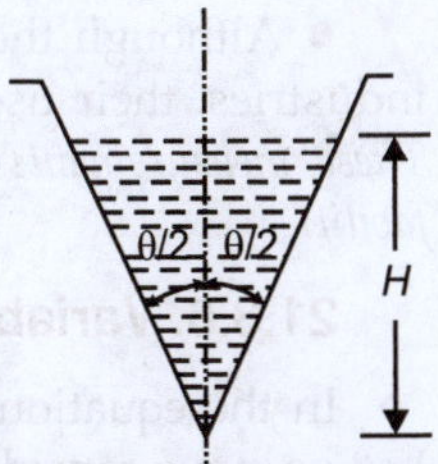

Fig. 21.36. Triangular notch/wear.

- The discharge through a *trapezoidal notch* (combination of the rectangular and triangular notches),

$$Q = \frac{2}{3} C_d L \sqrt{2g}(H)^{\frac{3}{2}} + \frac{8}{15} C_d \sqrt{2g}\,(H)^{\frac{5}{2}} \tan\frac{\theta}{2} \quad ...(21.32)$$

Flume : A *"flume" is an open channel whose sides are arranged sharply to form a throat, followed by a fairly long expansion section.* Stilling wells are provided for measuring the head at both entrance and throat as shown in Fig. 21.38. In this case,

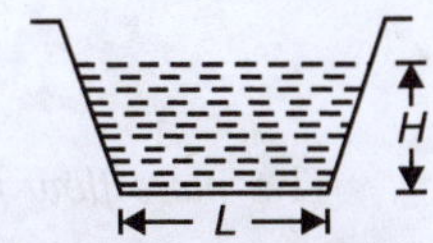

Fig. 21.37. Tapezoidal notch/wear.

$$Q_{actual} = \frac{CA_2}{\sqrt{1+(A_2/A_1)^2}}\sqrt{2gh} \quad ...(21.33)$$

C = Venturi flume coefficient (0.95 to 1), and

$h = h_1 - h_2$

When, $h_2 = \frac{2}{3}h$, *maximum value of flow occurs.*

Then, $$Q_{max} = 1.7 b_2 (H)^{\frac{3}{2}} \quad ...(21.34)$$

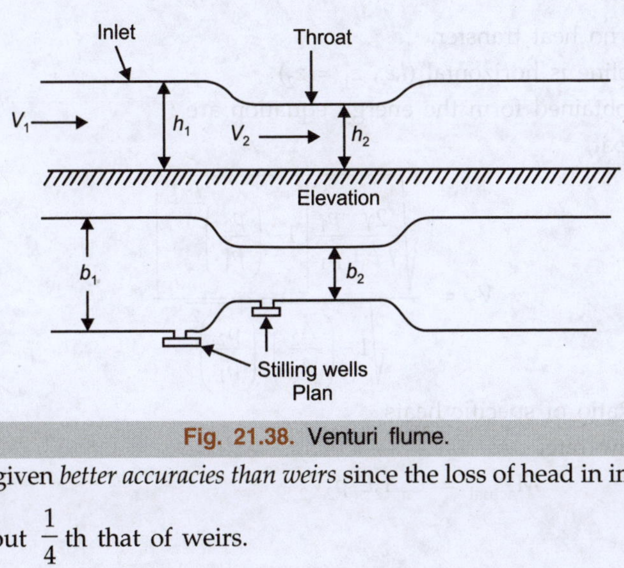

Fig. 21.38. Venturi flume.

The flumes given *better accuracies than weirs* since the loss of head in improved versions of flumes is about $\frac{1}{4}$ th that of weirs.

• Although the weirs and flumes have not previously found application in process industries, their use is increasing with the increasing demand for *water pollution control. These measurements are a must when large volumes of water are handled in waste treatment facilities.*

21.3.6. Variable Head Meters for Compressible Fluids

In the equations derived so far we have assumed that the density remains constant. Let us now consider the flow of *compressible fluids* in which *density is not constant but changes with pressure.* In order to account for changes in density with pressure, a factor Y, called "**expansion factor**" is introduced. This factor is so chosen that the mass flow rate calculated for incompressible fluids when multiplied by this factor gives the flow rate for compressible fluids.

The volumetric flow rate for a *compressible fluid* for venturies (eqn. 21.16) is given by

$$Q_{actual} = C_d M E A_2 \sqrt{\frac{2g}{w}(p_1 - p_2)}$$

$$= C_d M E A_2 \sqrt{\frac{2}{\rho}(p_1 - p_2)} \quad ...(21.35)$$

The *mass flow rate* considering the density at the inlet is,

$$m = C_d M E A_2 Y \sqrt{\frac{2}{\rho_1}(p_1 - p_2)} \quad ...(21.36)$$

The expansion factor Y can be taken as unity for liquids and for atmospheric air in the range $\left(\frac{p_1 - p_2}{\rho}\right)$ less than 0.1 cm of water.

The value of Y can be calculated theoretically for an *adiabatic flow process.* The following *assumptions* are made :

(*i*) The flow is steady and hence the mass entering is equal to mass leaving or $m_1 = m_2$.

(*ii*) There is no heat transfer.

(*iii*) The pipeline is horizontal (*i.e.*, $z_1 = z_2$)

The results obtained form the energy equation are :

Velocity at exit,

$$V_2 = \frac{\sqrt{\frac{2\gamma}{\gamma - 1}\frac{p_1}{\rho_1}\left[1 - \left(\frac{p_2}{p_1}\right)^{\frac{\gamma-1}{\gamma}}\right]}}{\sqrt{1 - \left(\frac{A_2}{A_1}\right)^2\left(\frac{\rho_2}{\rho_1}\right)^2}}$$

where, γ = Ratio of specific heats.

The mass flow rate,

$$M_{actual} = C_d A_2 E \, \rho_2 V_2$$

$$= C_d A_2 E \rho_2 \frac{\sqrt{\frac{2\gamma}{\gamma-1}\frac{p_1}{\rho_1}\left[1-\left(\frac{p_2}{p_1}\right)^{\frac{\gamma-1}{\gamma}}\right]}}{\sqrt{1-\left(\frac{A_2}{A_1}\right)^2\left(\frac{\rho_2}{\rho_1}\right)^2}} \quad ...(21.37)$$

Example 21.4. *The flow of cooling water in a manufacturing process is measured by a horizontal venturimeter with 200 mm inlet and 100 mm throat. The U-tube mercury manometer connected between inlet and throat of venturi shows a differential pressure of 220 mm of mercury. Calculate the water flow rate if coefficient of discharge is 0.98, specific gravity of mercury 13.6, and density of water 1000 kg/m*^{3}.

Solution. Given : D_1 = 200 mm = 0.2 m; D_2 = 100 mm = 0.1 m;
h = 220 mm; C_d = 0.98; Sp. gr. of mercury = 13.6;
ρ_{water} = 1000 kg/m^3

Water flow rate, Q_{actual} :

Differential pressure head,

$$(p_1 - p_2) = gh\,(\rho_{Hg} - \rho_{water})$$
$$= 9.81 \times 220 \times 10^{-3}\,(13.6 - 1) \times 1000$$
$$= 27.193 \times 10^3 \text{ N/m}^2$$

Velocity approach factor,

$$M = \frac{1}{\sqrt{1-\left(\frac{A_2}{A_1}\right)^2}}$$

(where, $A_1 = \frac{\pi}{4}D_1^2$ and $A_2 = \frac{\pi}{4}D_2^2$)

$$= \frac{1}{\sqrt{1-\left(\frac{D_2}{D_1}\right)^4}} = \frac{1}{\sqrt{1-\left(\frac{100}{200}\right)^4}} = 1.0328$$

The actual discharge through a venturimeter is given by,

$$Q_{actual} = C_d M A_2 = \sqrt{\frac{2g}{w}(p_1 - p_2)} \quad ...(21.14)$$

...(assuming thermal expansion factor E = 1)

$$= 0.98 \times 1.0328 \times \frac{\pi}{4} \times (0.1)^2 \sqrt{\frac{2 \times 9.81}{9.81 \times 1000} \times 27.193 \times 10^3}$$

$$= \mathbf{0.0586\ m^3/s.}$$ **(Ans.)**

Example 21.5. *The following data relate to an inclined venturimeter :*

Diameter of the pipeline = 400 mm

Inclination of the pipeline with horizontal = 30°

Throat diameter = 200 mm

The distance between the mouth and throat of the meter = 600 mm

Sp. gravity of oil flowing through the pipeline = 0.7

Sp. gravity of heavy liquid (U-tube) = 13.6

Reading of the differential manometer = 50 mm

The coefficient of the meter = 0.98

Determine the rate of flow in the pipeline. **(Punjab University)**

Solution. Diameter at inlet, D_1 = 400 mm = 0.4 m

$\therefore$ Area of inlet, $A_1 = \frac{\pi}{4} \times 0.4^2 = 0.1257 \text{ m}^2$

Throat diameter, D_2 = 200 mm = 0.2 m

$\therefore$ Area of throat, $A_2 = \frac{\pi}{4} \times (0.2)^2 = 0.0314 \text{ m}^2$

Reading of the differential manometer (U-tube),

$$y = 50 \text{ mm} = 0.05 \text{ m}$$

Difference of pressure head h is given by

$$h = y\left[\frac{S_{hl}}{S_p} - 1\right]$$

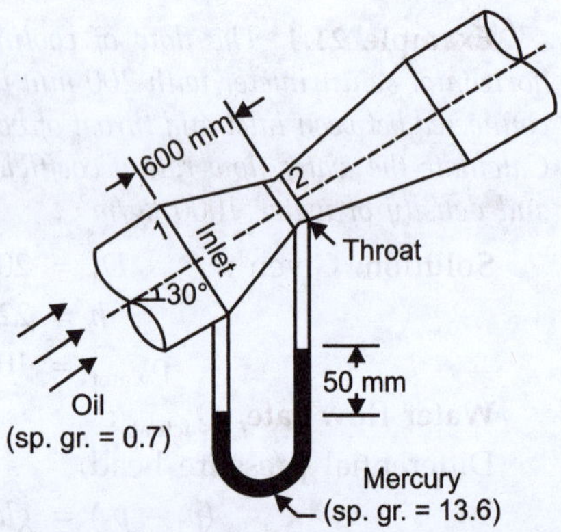

Fig. 21.38.

where, S_{hl} = Sp. of gravity of heavy liquid (*i.e.*, mercury) in U-tube = 13.6, and

S_p = Sp. gravity of liquid (*i.e.*, oil) flowing through the pipe = 0.7.

$$\therefore \qquad h = 0.05\left(\frac{13.6}{0.7} - 1\right) = 0.92 \text{ m of oil}$$

Now, applying Bernoulli's equation at sections '1' and '2', we get

$$\frac{p_1}{w} + z_1 + \frac{V_1^2}{2g} = \frac{p_2}{w} + z_2 + \frac{V_2^2}{2g} \qquad ...(i)$$

or, $\left(\frac{p_1}{w} + z_1\right) - \left(\frac{p_2}{w} + z_2\right) + \frac{V_1^2}{2g} - \frac{V_2^2}{2g} = 0$

But, $\left(\frac{p_1}{w} + z_1\right) - \left(\frac{p_2}{w} + z_2\right) = h$

or, $\left(\frac{p_1}{w} - \frac{p_2}{w}\right) + (z_1 - z_2) = h$

It may be noted that *differential gauge reading will include in itself the difference of pressure h and the difference of datum head.*

Thus eqn. (*i*) reduces to :

$$h + \frac{V_1^2}{2g} - \frac{V_2^2}{2g} = 0 \qquad ...(ii)$$

Applying continuity at sections '1' and '2' we get

$$A_1 V_1 = A_2 V_2$$

or, $$V_1 = \frac{A_2 V_2}{A_1} = \frac{\frac{\pi}{4} \times 0.2^2}{\frac{\pi}{4} \times 0.4^2} \times V_2 = \frac{V_2}{4}$$

Substituting the value of V_1 and h in equation (*ii*) we get

$$0.92 + \frac{V_2^2}{16 \times 2g} - \frac{V_2^2}{2g} = 0$$

or, $$\frac{V_2^2}{2g}\left(1 - \frac{1}{16}\right) = 0.92 \quad \text{or} \quad V_2^2 \times \frac{15}{16} = 0.92$$

or, $$V_2^2 = \frac{0.92 \times 2 \times 9.81 \times 16}{15} = 19.25 \text{ or } V_2 = 4.38 \text{ m/s}$$

Rate of flow of oil,

$$Q = A_2V_2 = 0.0314 \times 4.38 = \mathbf{0.1375\ m^3/s}\ \textbf{(Ans.)}$$

Example 21.6. *Water flows at the rate of 0.015 m³/s through a 100 mm diameter orifice used in a 200 mm pipe. What is the difference in pressure head between upstream section, and vena contracta section. Take coefficient of contraction and discharge as 0.60 each.*

Solution. *Given* : $Q = 0.015\ \text{m}^3/\text{s};\ D_0 = 100 \text{ mm} = 0.1 \text{ m}$

$D_1 = 200 \text{ mm} = 0.2 \text{ m};\ C_c = 0.6;\ C_d = 0.6$

$$A_0 = \frac{\pi}{4}D_0^2 = \frac{\pi}{4} \times 0.1^2 = 0.007854 \text{ m}^2$$

$$A_1 = \frac{\pi}{4}D_1^2 = \frac{\pi}{4} \times 0.2^2 = 0.03141 \text{ m}^2$$

Difference in pressure head :

The flow rate through an orificemeter is given by

$$Q = KEA_0\sqrt{\frac{2g}{w}(p_1 - p_2)} \quad \text{...[Eqn. (21.23)]}$$

Here, $$K = C_dM = \frac{C_d}{\sqrt{1 - \left(\frac{C_0A_0}{A_1}\right)^2}} = \frac{0.6}{\sqrt{1 - \left(0.6 \times \frac{0.007854}{0.03141}\right)^2}} = 0.6069$$

Assuming E (thermal expansion factor) as 1 and substituting the various values, we get,

$$0.015 = 0.6069 \times 0.007854\sqrt{\frac{2 \times 9.81}{9810}(p_1 - p_2)}$$

$$= 0.00021317\sqrt{(p_1 - p_2)}$$

$\therefore$ $$(p_1 - p_2) = \mathbf{4951.5\ N/m^2 \text{ or } 0.5047\ m\ of\ water}\ \textbf{(Ans.)}$$

Example 21.7. *A venturi tube of throat diameter 60 mm has a discharge coefficient of 0.97 and with a flow rate of 1.2m³/s, the pressure differential is 15.5 N/m². Determine the flow rate when an orifice of 60 mm is used in the same pipe. The discharge coefficient of orifice is 0.6 and pressure differential in same.*

Solution. Let the measurements through venturimeter and orifice are denoted by subscripts v and o respectvely.

Then, $$Q_v = C_vMA_2\sqrt{\frac{2g}{w}(p_1 - p_2)}$$

and, $$Q_o = C_o M A_2 \sqrt{\frac{2g}{w}(p_1 - p_2)}$$

∴ Discharge through the orifice,

$$Q_o = \frac{C_o}{C_v} \times Q_v = \frac{0.6}{0.97} \times 1.2 = \mathbf{0.742\ m^3/s}\ \textbf{(Ans.)}$$

Example 21.8. *A sub-marine fitted with a pitot tube moves horizontally in sea. Its axis is 12 m below the surface of water. The pitot tube fixed in front of the sub-marine and along its axis is connected to the two limbs of a U-tube containing mercury, the reading of which is found to be 200 mm. Find the speed of the sub-marine.*

Take the specific gravity of sea water = 1.025 times fresh water.

Solution. Reading of the manometer, y = 200 mm = 0.2 m of mercury

Sp. gravity of mercury, S_{hl} = 13.6

Sp. gravity of sea water, S_l = 1.025

To find the head, (h) using the relation : $h = y\left(\frac{S_{hl}}{S_l} - 1\right)$, we have

$$h = 0.2\left(\frac{13.6}{1.025} - 1\right) = 2.45$$

∴ Velocity of the sub-marine,

$$V = \sqrt{2gh} = \sqrt{2 \times 9.81 \times 2.45}$$

$$= 6.93\ \text{m/s or}\ \mathbf{24.9\ km/h}\ \textbf{(Ans.)}$$

21.3.7. Rotameter and Elbow Meter

21.3.7.1. Rotameter

*A **rotameter** is a constant-pressure drop, variable area flow meter.*

Construction. Refer to Fig. 21.39. It consists of a tapered metering glass tube, inside of which is located a rotor or active element (float) of the meter. The tube is provided with inlet and outlet connections. The specific gravity of the float or bob material is *higher* than that of the fluid to be metered. On a part of the float spherical slots are cut which cause it (float) to rotate slowly about the axis of the tube and keep it centered. Owing to this spinning accumulation of any sediment on the top or sides of float is checked. However, the stability of the bob may also be ensured by using a guide along which the float-would slide.

Working. When the rate of flow increases the float rises in the tube and consequently there is an increase in the annular area between the float and the tube. Thus, the float rides higher or lower depending on the rate of flow.

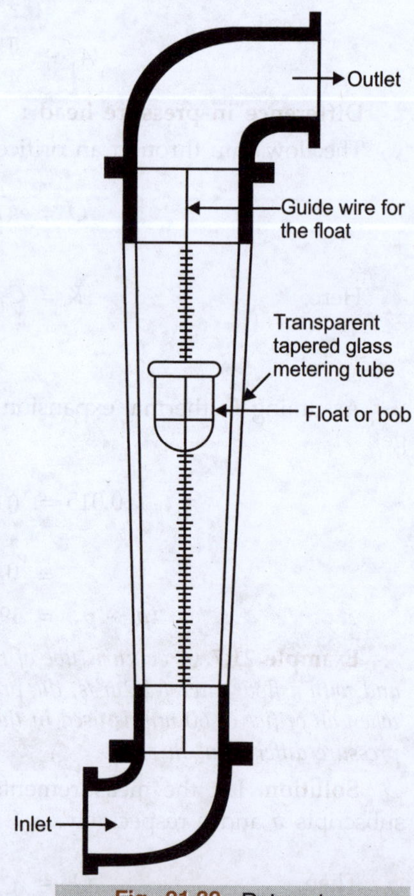

Fig. 21.39. Rotameter.

The discharge through a rotameter is given by

$$Q = C_d A_{ann.} \left[2gV_{fl} \frac{(\rho_{fl} - \rho_f)}{A_f \rho_f} \right]^{\frac{1}{2}} \quad ...(21.38)$$

where,
Q = Volume flow rate,
C_d = Co-efficient of discharge,
$A_{ann.}$ = Annular area between float and tube,
V_{fl} = Volume of float,
ρ_{fl} = Density of float material,
ρ_f = Density of fluid, and
A_f = Maximum cross-sectional area of fluid.

As the flow area $A_{ann.}$ is a function of height of float in the tube, the flow rate scale can be engraved on the tube corresponding to a particular float.

Advantages :

1. Simpler in operation.
2. Handling and installation easy.
3. Wide variety of corrosive fluids can be handled.
4. Low cost relatively.
5. Possibility of convenient and visible flow comparisons by mounting several rotameters side by side.
6. Easily equipped with data transmission and recording devices.
7. Rotameters are immune from viscosity effects upto a certain range.
8. They are particularly useful for measurement of low flow rates.
9. They can be used for measuring liquids, gases and vapours.

Limitations :

1. Mounted vertically, limited to small pipe sizes and capacities.
2. Less accurate, compared to venturimeter and orificemeter.
3. Glass tube subject to breakage.

21.3.7.2. Elbow meter

When liquid flows around a pipe bend there is an increase in pressure with radius, *i.e., the pressure at the outer wall of the bend is more than that at the inner wall. This difference of pressure which exists between the outside and inside of the bend is used for measurement of discharge in a pipeline.*

As shown in Fig. 21.40, the pipe bend is provided with two pressure tappings, one each at the inner and outer walls of the bend. These tappings are connected to the links of U-tube manometer.

As per literature, the following relation between velocity and pressure difference is available :

$$K\frac{V^2}{2g} = \left(\frac{p_o}{w} + z_o\right) - \left(\frac{p_i}{w} + z_i\right)$$

where, K = constant (depends upon the shape and size of the band), *ranges from* 1.3 to 3.2, and

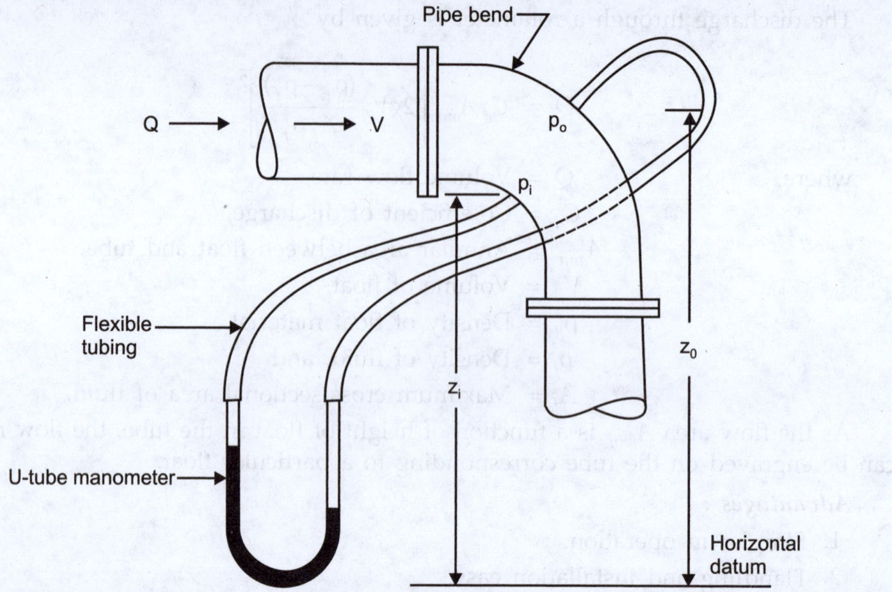

Fig. 21.40. Elbow meter.

V = Velocity of flow,

Suffices o and i represent the conditions at the outer and inner walls of the pipe bend.

or,
$$V = \frac{1}{\sqrt{K}}\sqrt{2g}\sqrt{\left(\frac{p_o}{w}+z_o\right)-\left(\frac{p_i}{w}+z_i\right)}$$

$\therefore$
$$\text{Discharge, } Q = AV = C_d A\sqrt{2g}\sqrt{\left(\frac{p_o}{w}+z_o\right)-\left(\frac{p_i}{w}+z_i\right)} \quad \text{...(21.39)}$$

where, $C_d = \frac{1}{\sqrt{K}}$ = Coefficient of discharge, and

A = Cross-sectional area of the pipe.

C_d varies between 0.56 and 0.88.

The following empirical relation has been suggested :

$$C_d = \sqrt{\frac{R}{D}}$$

where, R = Radius of the pipe bend, and

and D = Diameter of the pipe.

- An elbow meter can be conveniently used for the measurement of discharge in pipes which are fitted with elbows and bends.
- Its accuracy, with proper calibration, approaches that of a venturimeter or nozzle.

21.3.8. Electromagnetic Flow Meters

These meters are particularly suitable for the flow measurements of *slurries, sludge and any electrically conducting fluid.*

- Flow of an *electrically conducting fluid* may be determined by measuring the e.m.f. developed between a pair of electrodes set in opposite sides of an insulating conduit due to the movement of the liquid through a magnetic field established transversely of the conduit and perpendicular to both the flow and the line joining the electrodes (See Fig. 21.41). *By using an alternating field the effects of electrode polarization may be eliminated. Null measurement of the generated voltage renders the apparatus independent of the resistance of the liquid.*

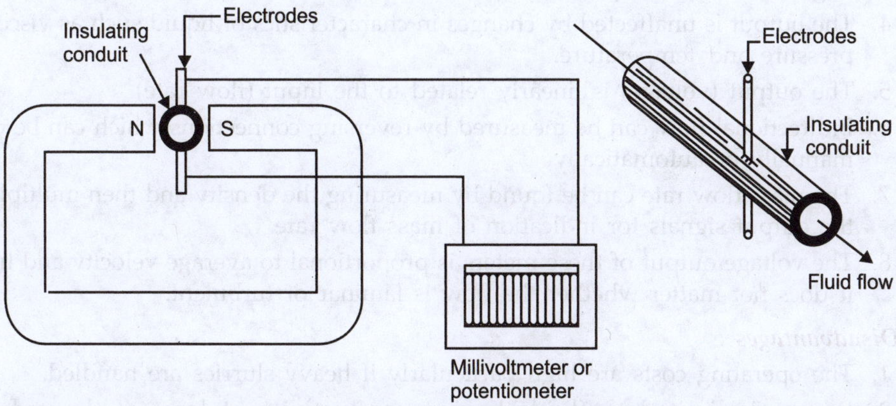

Fig. 21.41. Electromagnetic flow meter.

Theory : Refer to Fig. 21.41.

Consider the flow of a conductive liquid through a non-metallic insulating pipe located in the air gap of a permenant magnet (or an electromagnet which may be either A.C. or D.C.). Fluid motion relative to the field generates an induced *e.m.f. i.e.,* the value of which as per Faraday's law is given as :

$$e = Blv \times 10^{-8} \text{ volts} \quad ...(21.40)$$

where, B = Magnetic flux density (in V-s/cm^2)

l = Length of the conductor (in cm) and

v = Velocity of the conductor (in cm/s).

The effective length of the conductor corresponds to the inner diameter of the pipe and the velocity of the conductor is proportional to the mean flow velocity. The volume flow rate is give by,

$$Q = \frac{\pi}{4}d^2 v; \text{ the eqn. (21.40) then gets modified to :}$$

$$e = B \times d \times \frac{4Q}{\pi d^2} \times 10^{-8} \left(\because\ l = d \text{ and } v = \frac{4Q}{\pi d^2} \text{ from flow equation}\right)$$

or,
$$e = \frac{4QB}{\pi d} \times 10^{-8} \text{ volts} \quad ...(21.41)$$

This shows volume flow rate Q is directly proportional to the induced *e.m.f.* 'e' so long as the flux density remains unaltered.

The *e.m.f.* in the conductor is picked up by means of two electrodes set in the opposite sides of an insulating and non-magnetic pipe. This is to prevent short-circuiting of the induced voltage, and to allow magnetic field to penetrate the liquid. The electrodes are mounted flush with inside faces of the pipe and thus making a direct contact with the flowing liquid.

The *e.m.f.* generated by the instrument, however, is very small (typically of the order of 1 mV) and the resistance of the fluid is often very high. Therefore, the output of the instrument is generally amplified suitably.

Advantages :

1. No obstruction to flow.
2. Can handle slurries, greasy materials and liquids containing suspended matter.
3. High accuracy and reliability.
4. The output is unaffected by changes in characteristics of liquid such as viscosity, pressure and temperature.
5. The output (voltage) is linearly related to the input (flow rate).
6. Bidirectional flow can be measured by reversing connections which can be done manually or automatically.
7. The mass flow rate can be found by measuring the density and then multiplying the output signals for indication of mass flow rate.
8. The voltage output of these meters is proportional to average velocity and hence it does not matter whether the flow is laminar or turbulent.

Disadvantages :

1. The operating costs are high particularly if heavy slurries are handled.
2. Its use is limited to fluids having conductivity at least of the order of 0.05 μmho/cm.
3. They must be full at all the times because it sees velocity as analogues to volume flow rate.

Application. This type of meter in its various forms has been widely used for metering *corrosive acids, cement slurries, sewage, paper pulp, detergents etc. which present difficult handling problems.*

21.3.9. Ultrasonic Flow Meters

*The **working principle** of these meters is based on an apparent change in velocity of propagation of sound waves in a fluid with a change in the velocity of the fluid flow. The method uses transmitters and receivers of acoustic energy. "Piezoelectric crystals" are used as both transmitters and receivers.*

Fig. 21.42 illustrates an ultrasonic flow meter for measurement of flow.

Refer to Fig. 21.42(i) : *Travel time difference method–Single transmitter-receiver system.*

–When flow velocity (v) is zero, the transit time of the pulse from the transmitter to the receiver is given by :

$$t_0 = \frac{d}{c} \qquad \text{...(21.42)}$$

where, d = Distance between the transmitter and receiver, and

c = Velocity of sound in fluid.

Now, when the fluid moves with velocity v, the transit time becomes :

$$t = \frac{d}{c+v} \qquad \text{...(21.43)}$$

$$= \frac{d(c-v)}{(c+v)(c-v)} = \frac{d(c-v)}{c^2 - v^2}$$

...Multiplying numerator and denominator by (c–v)

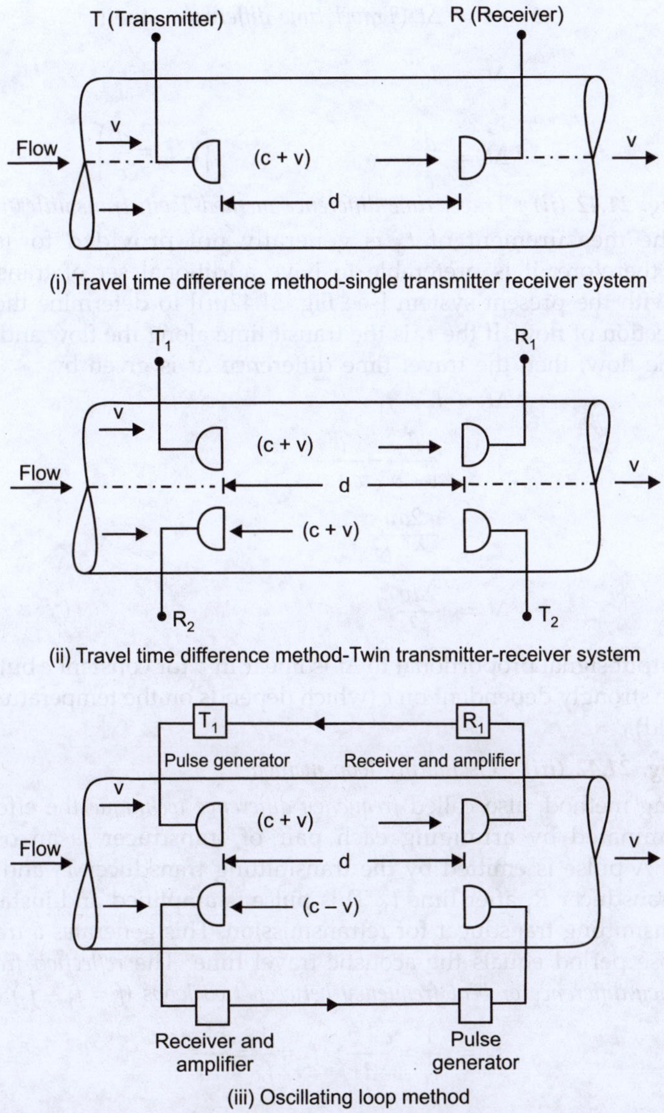

Fig. 21.42. Ultrasonic flow meter.

For any reasonable flow rate in liquids, the relationship $v << c$ holds and there is negligible error in writing, then

$$t = \frac{d(c-v)}{c^2}$$

$$= \frac{d}{c}\left(1-\frac{v}{c}\right)$$

or, $$t = t_0\left(1-\frac{v}{c}\right) \qquad ...(21.44)$$

or, $$t_0 - t = t_0\frac{v}{c}$$

Let $t_0 - t = \Delta t$ (*Travel time difference*), then

$$\Delta t = t_0 \frac{v}{c}$$

or,
$$\Delta t = \frac{dv}{c^2} \qquad \left(\because t_0 = \frac{d}{c}\right) \qquad ...(21.45)$$

Refer to Fig. 21.42 (ii) : Travel time difference method-Twin transmitter-receiver method :

Because the measurement of t_0 is generally not provided for in the present measurement, therefore, it is preferable to have additional set of transmitter-receiver system along with the present system [see Fig. 21.42(*ii*)] to determine the transmit time against the direction of flow. If the t_1 is the transit time along the flow and t_2 the transmit time against the flow, then the travel time difference Δt is given by :

$$\Delta t = t_2 - t_1$$

$$= \frac{d}{c-v} - \frac{d}{c+v}$$

$$= \frac{2dv}{c^2 - v^2}$$

or,
$$\Delta t \simeq \frac{2dv}{c^2} \qquad (\because v << c) \qquad ...(21.46)$$

Thus an output signal proportional to Δt is linear in v for constant c but the calibration constant will be strongly dependent on c (which depends on the temperature and pressure of the flow field).

Refer to Fig. 21.42 (iii) : Oscillating loop method :

In oscillating method, also called *frequency difference technique* the effect of the sonic velocity is eliminated by arranging each pair of transducer is an oscillating loop [Fig. 21.42(*iii*)]. A pulse is emitted by the transmitting transducer T_1 and is received by the receiving transducer R_1 after time t_1. This pulse is amplified and instantaneously fed back to the transmitting transducer for retransmission. This generates a train of pulses in each path whose period equals the acoustic travel time. The *reflection frequencies* (f_1, f_2) and the *frequency difference or beat frequency between two loops* $(f = f_1 - f_2)$ are given as :

$$f_1 = \frac{1}{t_1} = \frac{c+v}{d}; \; f_2 = \frac{1}{t_2} = \frac{c-v}{d} \qquad ...(21.47)$$

$$f = f_1 - f_2 = \frac{c+v}{d} - \frac{c-v}{d} = \frac{2v}{d} \qquad ...(21.48)$$

The frequency difference is thus proportional to v and independent of sonic velocity c.

— Since frequency difference is an easily measured quantity, this method has considerable appeal.

— The *drawback* of this method is that the frequency difference is small, and the long counting interval needed for a given resolution results in *slow response time.*

Advantages : The following are the *advantages* and *disadvantages* of ultrasonic flow meter :

1. Absence of moving parts.
2. No abstraction to flow.
3. Excellent dynamic response.
4. High accuracy (of the order of ±2% of the full scale values).

5. Linear relationship between output and input.
6. Output is insensitive to variations in viscosity, density and temperature.
7. Suitable for metering both liquids and gases.
8. The output being electrical in nature, the read out can be analog or digital.

Disadvantages :

1. Complex construction.
2. Relatively high costs.
3. Use of commercial meters is generally limited to liquids.

Applications : This device finds special applications, namely *measurement* of :

— Ocean occurents;
— Water flows in large conduits;
— Flows of various bio-medical and industrial fluids etc.

21.3.10. Thermal Flow Meters

This type of flow meter is also useful for the measurement of corrosive liquids. Fig. 21.43 shows a thermal flow meter.

— It consists of a heater placed in the fluid stream, with two temperature sensors such as thermocouples or resistance thermometers.

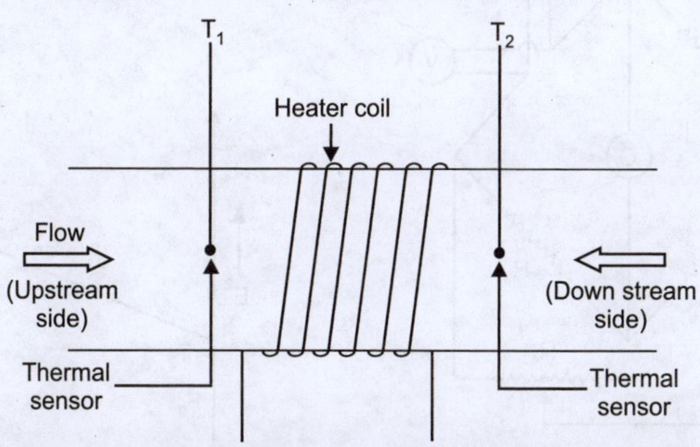

Fig. 21.43. Thermal flow meter.

— The temperature sensors are located symmetrically with respect to the heater, one at upstream side and the other at the downstream side, at short distance from the heater.

Under *no* flow conditions, the temperatures T_1 and T_2 of the sensors will the *same*, and the differential signal measured on the detector will be *zero*. When the *flow increases*, T_1 will *fall* and T_2 will *rise* until a steady state is reached. The heat transferred to the fluid by the heater is given by :

$$Q = m\, c_p\, (T_2 - T_1) \qquad ...(21.49)$$

where, Q = Heat transferred,

m = Mass flow rate of the fluid,

c_p = Specific heat of the fluid,

T_1 = Temperature of the fluid before heat is transferred to it, and

T_2 = Temperature of the fluid after heat is transferred to it by the heater coil

Using the above equation, the value of m (*i.e.*, mass flow rate) can be obtained by measuring the temperature difference $(T_2 - T_1)$ and the heat transferred to the fluid Q in terms of constant *power input to the heater*, and knowing the specific heat of c_p of the fluid.

- The dynamic response of this type of flow meter is dependent on the flow conditions, and is usually poor.

21.3.11. Wire Anemometers

These devices are hot wire resistance transducer wire filaments which are used for measurement of flow rates of fluids.

By suitably locating this transducer (in the form of wire filaments) the flow rates of non-conducting liquids in open channels and closed pipes and of gases in closed pipes can be conveniently measured. The transducer which is in the form of probe is located at the centre of pipe with direction of wire perpendicular to the direction of fluid flow.

The flow sensing element is a short length of 5 μm diameter platinum-tungsten wire welded between the prongs of the probe and heated electrically as a part of Wheatstone bridge. [Fig. 21.44].

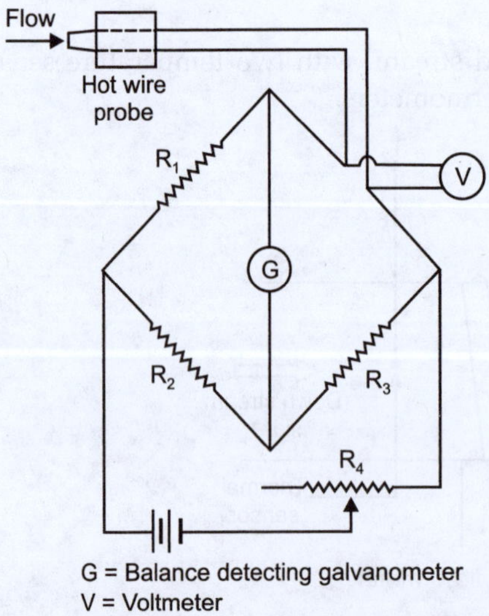

Fig. 21.44. Hot wire bridge circuit.

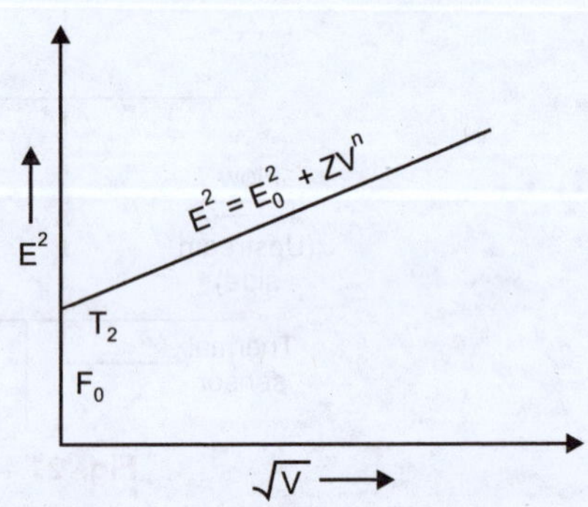

Fig. 21.45. Response of hot wire.

When the probe is introduced into the flowing fluid, it tends to be cooled by the instantaneous velocity and consequently there is a tendency for the electrical resistance to decrease. The rate of cooling of wire depends on the (*i*) dimensions and physical properties of the wire, (*ii*) difference of temperature between the wire and the fluid, (*iii*) physical properties of the fluid, and (*iv*) stream velocity under measurement. For a simple hot wire anemometer the *first three conditions are effectively constant and the instrument response is then a direct measure of the velocity.*

Hot wire anemometers are commonly used in two different modes namely :

1. Constant current mode; 2. Constant temperature mode.

1. Constant current mode :

— In this mode the heating current *i.e.*, voltage across the bridge is kept constant.

— Initially the circuit is adjusted such that the galvanometer reads zero when the heated wire lies in stationary air.

— *When the air flows, the hot wire cools, the resistance changes and the galvanometer deflects. The galvanometer deflections are amplified, measured and correlated with air velocity by previous calibrations.*

2. Constant temperature mode :

— In this mode operating resistance of the wire and hence its temperature is maintained constant.

— In the event of the tendency of the hot-wire to cool by the flowing fluid, the external bridge voltage is applied to the wire to maintain a sensibly constant temperature. *The bride voltage is varied so as to bring the galvanometer needle to zero; the reading on the voltmeter is recorded and correlated with the air velocity.*

— The response of a hot wire placed normal to the flow velocity V (Fig. 21.45) is described by the relationship.

$$E_2 = E_0^2 + ZV^n \quad \text{...(21.50)}$$

where, E = Instantaneous value of the bridge voltage,

E_0, Z = Constants depending on the odd resistance of the wire and on the properties of the fluid medium.

The values of these constants for a given wire are found by calibrating the wire in a uniform flow field against a standard pitot-static probe. Generally with exponent n = 0.5, the response of the hot wire in the form of a plot between E^2 and V^n is a straight line.

- *The constant temperature design has the inherent advantage that the wire is protected against burnout.* However, the associated electronic circuitry, specially the D.C. feedback amplifier has an upper limit of the frequency response beyond which the system becomes *unstable* and the *wire temperature oscillates.*
- A modification of the hot wire anemometer is the ***hot film transducer*** which generally enhances its utility. These transducers have great mechanical strength and maybe used for very high temperatures by constructing them with internal cooling water passages.

— They are extremely sensitive to fluctuations in the liquid flow velocity and have been employed for measurements involving frequencies as high as 50 kHz. For such measurements sophisticated electronic instrumentation is required.

Advantages and limitations. The following are the *advantages* and *disadvantages* of hot wire anemometer.

Advantages :

1. Suitable for measurement in both gases and liquids (In liquids, the probe has to be modified as the hot film probe).
2. Since the instrument's output is electrical, therefore, the read-out can be analog or digital.
3. Owing to very small probe size there is neither much disturbance nor pressure loss to the flow.
4. The instrument exhibits excellent dynamic characteristics, since its time constant is of the order of 10^{-4} to 10^{-6} s.
5. The instrument has extended measuring range from very low velocities to supersonic velocities.
6. The instrument shows excellent accuracy of the order of ±0.1% in the measurement of mean velocities and ±2% in the measurement of turbulence levels.

Disadvantages :

1. Expensive circuitry.
2. Needs skilled operations.
3. Highly non-linear operating characteristics.
4. The instrument needs frequent calibrations (since the accumulations of dirt/dust on the hot wire changes its sensitivity).
5. The hot wire does not sense whether the flow is in the forward or reverse direction. Its output depends on the cooling of the wire which is same in both the cases.
6. Since the wire is fragile and dust particle striking the hot wire can break the same, therefore, hot wire use and operation has to be done with exceptionally clean fluids.

21.3.12. Laser Doppler Anemometer

This type of anemometer measures the instantaneous velocities of gases or liquids flowing in a transparent (usually glass) channel.

The working principle of this device is based on the Doppler shift in frequency of the light-scattered by an object moving relative to the radiating source.

Basically the technique consists of focusing laser beams at the point in the fluid where the velocity is to be measured. At this focal point, a photo-detector senses the laser light scattered from the fluid or fluid particles entrained in the fluid. Signal processing of photo-detector output yields the magnitude of Doppler frequency shift which is *directly proportional to the instantaneous velocity of the flow.*

A LASER (Light Amplification by the Stimulated Emission of Radiation) beam is employed, instead of a monochromatic light beam, due to the following *reasons :*

(*i*) Laser beam is coherent.

(*ii*) The laser provides much higher quality of monochromatic (single frequency) light source.

(*iii*) Its wavelength (unlike a monochromatic beam) is affected minimally by changes in the ambient pressure, temperature or humidity.

(*iv*) Its frequency is very stable and as such it can detect the relative Doppler frequency shift ($\Delta f/f$) of the order of 10^{-7} for flow velocities of the order 45 km/h.

The following are some of the materials known as active media suitable for production of laser beams :

Ruby; Nd-glass; Nd-Yale; CO_2 gas; Neon gas; Ionised argon gas; Helium neon gas; Semiconductor crystal gallium arsenide.

Advantages and Disadvantages : The following are the advantages of laser Doppler anemometer over the usual pitot tube and the hot wire velocity metering techniques :

Advantages :

1. Very high accuracies (of the order of ± 0.2 percent).
2. Suitable for measurements in both gas and liquid flows.
3. No transfer functions involved, *i.e.*, the output voltage of the instrument is directly proportional to the instantaneous velocities rather than its inference from pressure difference (pitot tubes) or heat transfer phenomenon (hot wire).
4. It is possible to have measurements in miniature size of volume of the order 0.2 mm side cube.

5. Measurements are of non-contact type (*i.e.*, no physical object is inserted in the flow field and thus flow is undisturbed by the measurement).
6. In this instrument very high frequency response of the order of MHz is possible; consequently, its accuracy in the determination of fluctuating velocities is far superior to that of hot-wire anemometer because of practically no time lag in the measurements.

Disadvantages :

1. High cost.
2. Requires a high degree of experience and skill in operation.
3. Unsuitable for clean flows; for such flows, the tracer particles have to be seeded in the flow for scattering the incident light.
4. Need for a transparent channel.

21.4 MEASUREMENT OF DISPLACEMENT

21.4.1. Measurement of Linear Displacement

Linear displacement may be measured by the following *transducers* :

1. Resistive potentiometers.
2. Strain gauges.
3. Variable inductance transducers.
4. Linear variable differential transducers (LVDT).
5. Capacitive transducers.
6. Piezoelectric transducers.
7. Hall effect transducers.
8. Electronic transducers.
9. Ionization transducers.
10. Digital transducers.
11. Acoustic transducers.

• *Mechanical displacement* may be converted into an electrical variable by the simple expedient of adjusting resistance in an electrical circuit. A slide-wire resistor, having a movable contact attached to the part whose displacement is to be measured, may be connected through a 2-conductor circuit to a steady-voltage source in series with an ammeter (or milliammeter) calibrated in terms of the displacement. If the resistor is connected as a voltage divider, the need for a regulated supply is eliminated, and with a 3-conductor circuit the display instrument may be a *ratio-meter or potentiometer*. Such combinations are common and are available for both D.C. and A.C. operations.

• Where deflections are small, of the order of 0.25 mm, measurement may be made by use of a *differential transformer*.

Measurement of small displacements :

The linear or differential transformer is the *popular means for measuring small displacements of the order of 0.25 mm.* This device (See Fig. 21.46) is generally produced with a single primary winding and two secondaries, all disposed along a common axis and having in the common magnetic circuit a movable iron core longitudinally displacable with the motion to be measured. The secondaries may be connected additively or differentially and may be included in the circuit of a null type instrument based either by shifting the core of a similar transformer excited from the same source or by the use of a slide wire potentiometer.

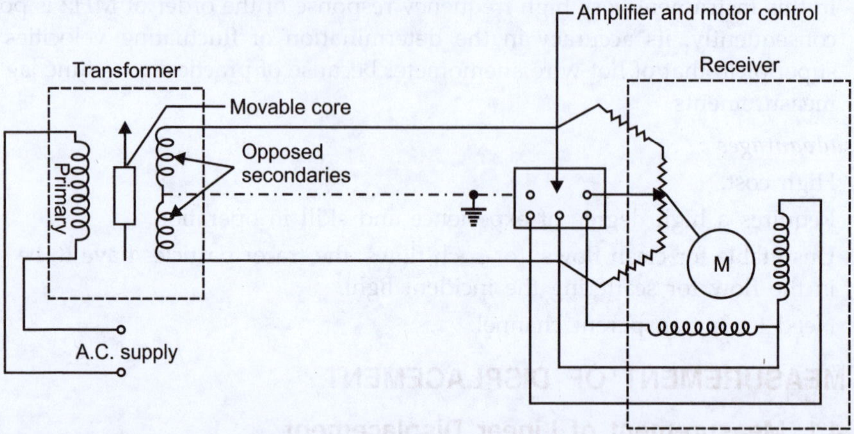

Fig. 21.46. Schematic diagram of differential transformer transducer with servo-activated receiver.

Linear transformer are regularly supplied for operation at all frequencies upto 30 kHz. The sensitivity, of course, increases with the frequency.

Linear transformers may be interconnected in a great variety of arrangements to perform computations or to express desired mathematical functions of measured variables.

21.4.2. Measurement of Angular Displacement

The following transducers are used for angular displacement :

1. Resistive potentiometer.
2. Variable inductive transducers.
3. Synchros.
4. Capacitive transducers.
5. Rotary variable differential transformer (RVDT).
6. Variable reluctance transducer.

Variable reluctance transducer :

Fig. 21.47, shows a variable reluctance transducer for *measurement of angular displacement.*

L_1 and L_2 are the two identical coils connected in series opposition. These coils are so placed/positioned with respect to a cam shaped iron core that, when the core is in the *null position, the impedances of both the coils are same.* Hence, voltage across the two coils are the same, though they are 180° out of phase. Any angular displacement from the null position results in a *differential voltage output proportional to the amount of displacement.* The clockwise

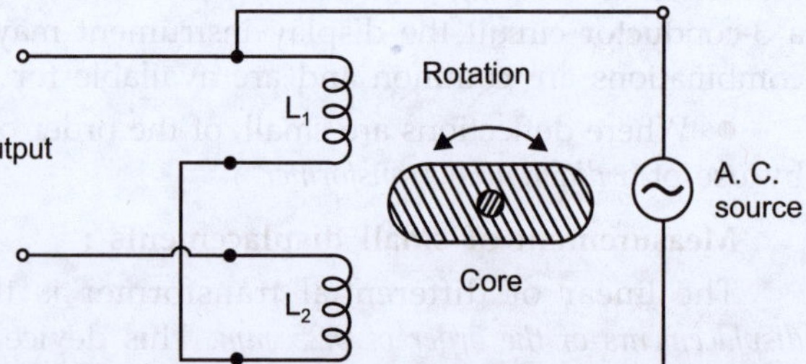

Fig. 21.47. Variable reluctance transducer for measurement of angular displacement.

and anticlockwise angular displacements produce different outputs which are *opposite in phase.* Thus the angular displacement may be measured in magnitude and direction.

In these transducers, angular changes from null position to ± 45° may be sensed.

21.5 MEASUREMENT OF VELOCITY/SPEED

21.5.1. Measurement of Linear Velocity

The following types of transducers are used (in the analogue methods) for measurement of *linear velocity* (first derivative of displacement).

1. Electro-magnetic transducers :
   (*i*) Moving magnet type.
   (*ii*) Moving coil type.
2. Seismic type transducers.

Electro-magnetic transducers :

An electro-magnetic transducer utilizes the voltage produced in a coil on account of change in flux linkages resulting from change in reluctance. This is the most commonly used transducer for measurement of linear velocities.

(*i*) Moving magnet type. A moving type electro-magnetic transducer uses *permanent magnet which provides a constant polarizing field.* Fig. 21.48, shows such a transducer. It consists of rod rigidly coupled to the device whose velocity is being measured. The rod is a permanent magnet which is surrounded by a coil.

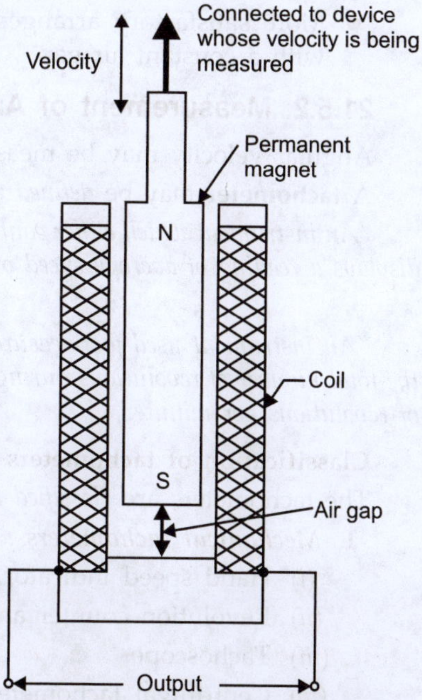

Fig. 21.48. Moving magnet type transducer.

Working. When the magnet moves a voltage is induced in the coil which is *directly proportional to the velocity.* The voltage induced in the coil (placed in a magnet field) is given by

$$e = BANV \quad ...(21.51)$$

i.e.,

$$e \propto V \quad ...(21.52)$$

where, B = Flux density, Wb/m^2,
A = Area of coil, m^2,
N = Number of turns of coil, and
V = Relative velocity of magnet with respect of coil.

The *direction of motion is determined by the polarity of the output voltage.*

Advantages :

1. The output voltage is linearly proportional to velocity.
2. Negligible maintenance (due to absence of mechanical surfaces/contacts).
3. Can be employed as event markers which are robust and inexpensive to manufacture.

Disadvantages :

1. Limited frequency response.
2. Unsuitable for measurement of vibrations.
3. Stray magnetic fields adversely affect their performance.

***(ii)* Moving coil type.** This type of transducer operates essentially through the *action of a coil moving in a magnetic field; the voltage generated in the coil being proportional to the velocity of the coil.*

The *advantages* of this type of transducer are :

- Reduced stray magnetic fields effects due to the whole device being contained in an antimagnetic case.
- More satisfactory arrangement, as the system forms a closed magnetic circuit with a constant air gap.

21.5.2. Measurement of Angular Velocity

Angular velocity may be measured with the help of a *"tachometer"*.

A **tachometer** may be *defined* as follows :

"An instrument which either continuously indicates the value of rotary speed or continuously displays a reading of average speed over rapidly operated short-intervals of time."

Or

"An instrument used for measurement of angular velocity, as of shaft, either by registering the total number of revolutions during the period of contact, or by indicating directly the number of revolutions per minute."

Classification of tachometers :

The tachometers are *classified* as follows :

I. ***Mechanical tachometers :***

(*i*) Hand speed indicator.

(*ii*) Revolution counter and timer.

(*iii*) Tachoscope.

(*iv*) Centrifugal tachometers.

(*v*) Vibrating reed tachometers.

II. ***Electrical tachometers :***

(*i*) D.C. tachometer generators.

(*ii*) A.C. tachometer generators.

(*iii*) Photoelectric tachometers (or Speed-meters)

(*iv*) Eddy current or drag-cup tachometer.

(*v*) Capacitive tachometer.

(*vi*) Stroboscopic tachometers.

Some of these meters are discussed below :

21.5.3. Electrical Tachometers

Electrical tachometers are preferred, as compared to mechanical tachometers, in view of advantages they offer as *electrical transducers.*

1. D.C. tachometer generator :

In a D.C. generator the e.m.f. generated depends upon the following two factors :

(*i*) Field excitation (*ii*) Speed.

If for the field system permanent magnet pole pieces are used, then the generated voltage depends only on the speed. Hence the *speed can be computed by measuring the generated e.m.f.*

A D.C. tachometer generator is shown in Fig. 21.49.

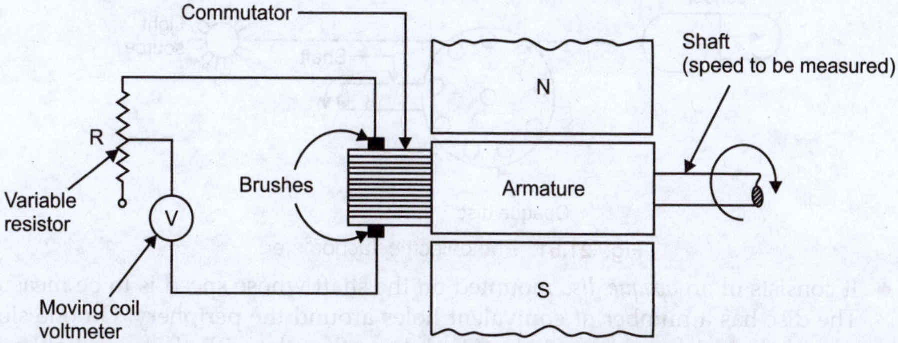

Fig. 21.49. D.C. tachometer generator.

- The shaft whose speed is to be measured is coupled to the armature.
- A moving coil voltmeter is connected across the brushes to measure the generated voltage. The variable resistance *R* is incorporated to limit the current through the voltmeter.

Since voltage is proportional to speed, the voltmeter may be calibrated in terms of speed. (r.p.m.).

2. *A.C. tachometer generator :*

The inherent demerits associated with D.C. tachometer generator, due to the provision of commutor and brushes, are eliminated in A.C. tachometer generator.

An A.C. tachometer generator is shown in Fig. 21.50.

- It consists of, like an alternator, a stationary armature (stator) and a rotating field system (rotor). Owing to the generation of e.m.f. in a stationary coil on a stator, commutation problems no longer exist.
- The alternating e.m.f. induced in the stationary coil is *rectified* and the output D.C. voltage is measured with the help of a moving coil voltmeter (*V*).
- The *ripple content* of the rectified voltage is *smoothened* by the capacitor filter (C).
- As the speed depends on both the amplitude of the voltage and frequency, any one of them can be used as a measure of the speed. In an A.C. tachometer, it is the *induced voltage* that is considered as the required parameter.

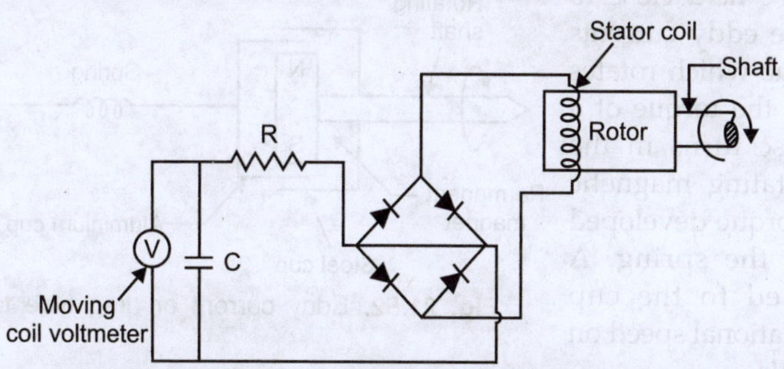

Fig. 21.50. A.C. tachometer generator.

3. *Photoelectric tachometer :*

Fig. 21.51, shows a photoelectric tachometer.

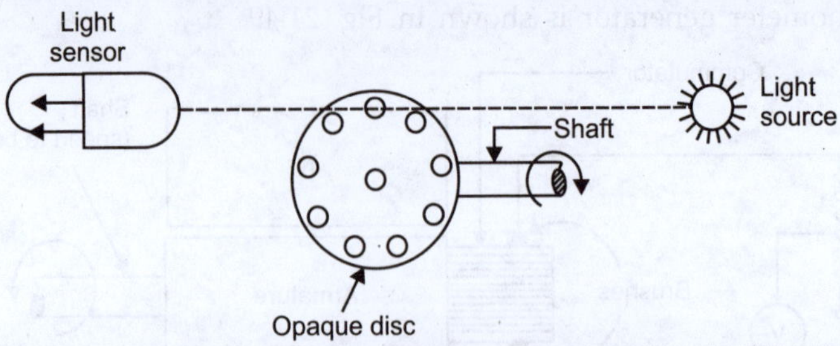

Fig. 21.51. Photoelectric tachometer.

- It consists of an *opaque disc* mounted on the shaft whose speed is to be measured. The disc has a number of equivalent holes around the periphery. On one side of the disc there is a source of light (L) while on the other side there is a light sensor (maybe a photosensitive device or photo-tube) in line with it (light-source).
- On the rotation of the disc, holes and opaque portions of the disc come alternately in between the light source and the light sensor. When a hole comes in between the *two,* light passes through the holes and falls on the light sensor, with the result that an *output pulse is generated.* But when the opaque portion of the disc comes in between, the light from the source is blocked and hence there is no pulse output. Thus *whenever a hole comes in line with the light source and sensor, a pulse is generated.* These pulses are counted/measured through an *electric counter.*

The number of pulses generated depends upon the following *factors :*

(*i*) The number of holes in the disc ;

(*ii*) The shaft speed.

Since the number of holes are fixed, therefore, the number of pulses generated depends on the speed of the shaft only. The electronic counter may therefore the calibrated in terms of speed (r.p.m.).

Advantage. It is a *digital instrument.*

Disadvantage. It is required to replace the light source periodically and if the grating period is small then errors might creep in the output.

4. *Eddy current or drag cup tachometer :*

In this type of tachometer (Fig. 21.52) the test-shaft rotates a permanent magnet and this induces eddy currents in a drag cup or disc held close to the magnet. The eddy currents produce a torque which rotates the cup against the torque of a spring. The disc turns in the direction of rotating magnetic field until the torque developed equals that of the spring. A pointer attached to the cup indicates the rotational speed on a calibrated scale.

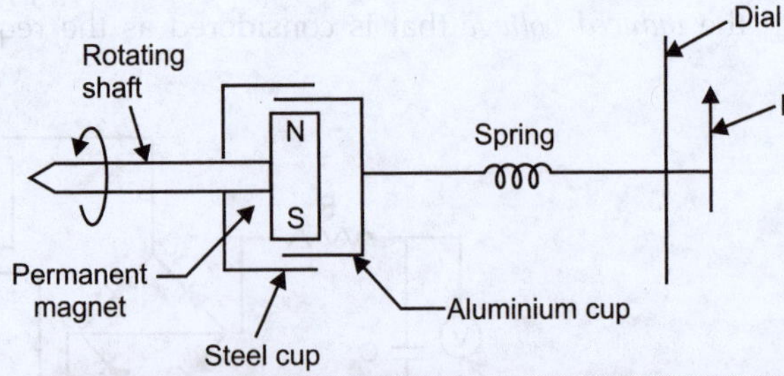

Fig. 21.52. Eddy current or drag type tachometer.

- The *automobile speedometers* operate on this principle.
- These tachometers are used for measuring rotational speeds upto 12000 r.p.m. with an accuracy of ± 3 per cent.

5. *Capacitive tachometer :*

Refer to Article. 7.8.3.

6. *Stroboscopic tachometer :*

Its working principle is to synchronize a flashing light with rotation of shaft, making it appear standstill. Usually, a reference mark is put on the rotating body and the speed of oscillation of flash produced from stroboscope itself is a measure of r.p.m. of rotating body, *without requiring any contact.*

- These tachometers are most oftenly used in *experimental work.*
- Its *disadvantage* is that it cannot be used where ambient lighting is above a certain value; for it to be most effective, the surrounding light must be subdued.

Example 21.9. *Draw an instrumentation scheme to measure the speed of a motor cycle using a magnetic transducer and explain the principle behind its operation.*

Solution. The block diagramatic representation of the instrumentation scheme for the measurement of the speed of the motorcycle using a magnetic transducer is given as follows (Fig. 21.53):

The basic principle involved for the above measurement is that the speed of the motor cycle is sensed with the help of an *A.C. tachogenerator* which is coupled with the motor cycle wheel. *The output of the generator is voltage which is direct indication of speed.* The other blocks show the necessary circuit for displaying the output in the desired form.

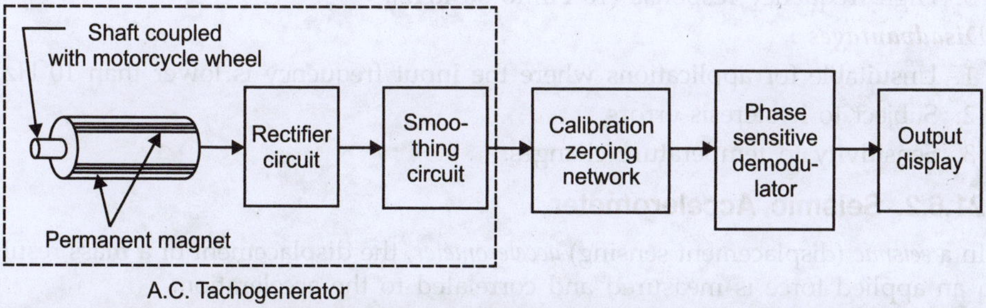

Fig. 21.53. Block diagramatic scheme for measurement of motor cycle speed using magnetic transducer.

21.6 MEASUREMENT OF ACCELERATION

The acceleration of a moving body is generally measured by means of sensors called **accelerometers.**

For measurement of acceleration, generally following types of accelerometers are used :

1. Piezoelectric type
2. Seismic type.

21.6.1. Piezoelectric Accelerometer

Refer to Fig. 21.54. A piezoelectric accelerator is probably the *simplest* and *most commonly used transducer for measuring acceleration.*

Construction. It consists of a *piezoelectric crystal* sandwiched between two *electrodes* and has *mass* placed on it. The unit is fastened to the base whose acceleration characteristics are to be obtained. The *can* threaded to the base acts as a spring and squeezes the mass against the crystal. *Mass exerts a force on the crystal and a certain voltage output is generated.*

Working. When the base is accelerated *downward* inertial reaction force on the base acts upward against the top of the can. This relieves stress on the crystal. According to Newton's second law of motion, *force = mass × acceleration*, since the mass is fixed quantity, the *decrease in force is proportional to the acceleration*. Similarly, an acceleration in the upward direction would increase the force on the crystal in proportion to the acceleration. The resulting change in the output voltage is recorded and correlated to the acceleration imposed on the base.

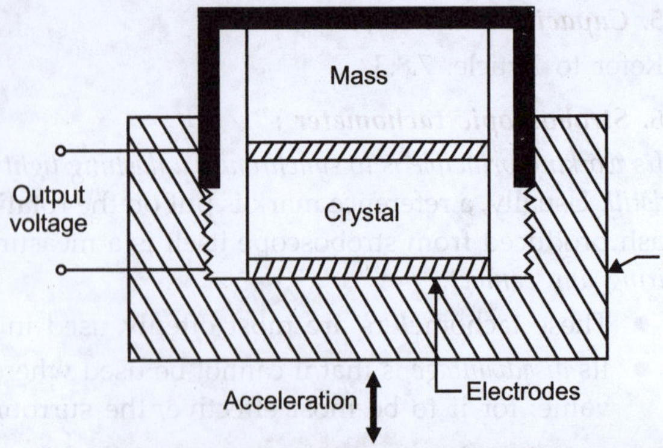

Fig. 21.54. Piezoelectric accelerometer.

Advantages :

1. Small size and a small weight.
2. High output impedance.
3. Can measure acceleration from a fraction of *g* to thousands of *g*.
4. High sensitivity.
5. High frequency response (10 Hz to 50 kHz).

Disadvantages :

1. Unsuitable for applications where the input frequency is lower than 10 Hz.
2. Subject to hysteresis errors.
3. Sensitivity to temperature changes.

21.6.2. Seismic Accelerometer

In a *seismic* (displacement sensing) *accelerometer*, the displacement of a mass resulting from an applied force is measured and correlated to the acceleration.

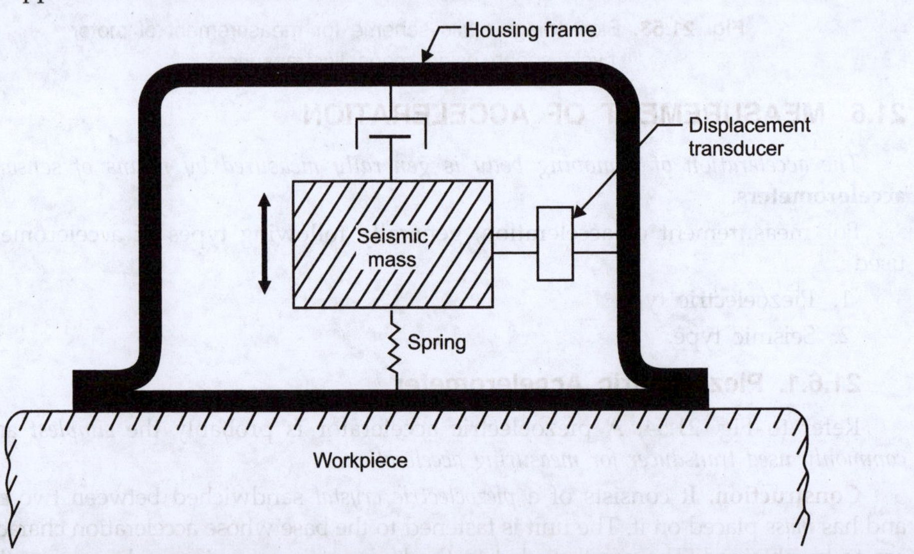

Fig. 21.54. Schematic diagram of a seismic transducer.

A schematic diagram of this accelerator is shown in Fig. 21.55. The mass is connected through the parallel spring and damper arrangement to the housing frame. The housing frame is connected to the source of vibrations whose characteristics are to be measured. *The mass has tendency to remain fixed in its spatial position so that the vibrational motion is registered as a relative displacement between mass and housing frame. This displacement is sensed and indicated by an appropriate transducer.*

- This accelerometer may also be used as a *vibration sensor* (vibration refers to the repeated cyclic oscillations of a system).

21.6.3. Strain Gauge Acceleration (Vibration) Sensor

Fig. 21.56 shows a strain gauge acceleration/vibration sensor :

— It consists of a Wheatstone bridge in which the semiconductor strain gauges (R_1 and R_2) one connected in two arms, as shown in the Fig. 21.56. These gauges measure a change in electrical resistance, which is proportional to force due to *acceleration*. Since the acceleration is related to vibrations, the *meter is calibrated to read the voltage (V_o) proportional to acceleration.* The function of *capacitor* C is to remove the D.C. signal from the output signal.

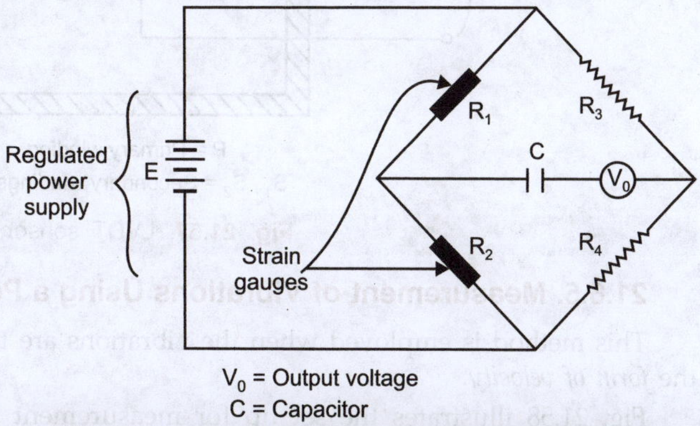

Fig. 21.56. Strain gauge acceleration/vibration sensor.

— The value of V_0 (voltage drop across the bridge) is recorded or indicated by the indicator.

- The range of this method is : Vibration amplitude : 5 to 1000 g; Vibration frequency : 0 to 1000 Hz.

21.6.4. LVDT Acceleration/Vibration Sensor

This sensor measures vibrations in terms of acceleration.

Fig. 21.57 shows LVDT sensor. It produces an A.C. voltage proportional to displacement of movable ferromagnetic core.

— LVDT's core acts as the mass. When flexible reeds are connected to the core, springs action is produced.

— When vibrations to be measured/monitored are applied to the core, it moves in the upward and downward directions according to the vibration. As a result, A.C. voltage is induced in the secondary windings of LVDT, the magnitude of which is proportional to the amplitude of vibrations. This voltage if required, can be rectified for induction or measurement of vibration amplitude.

- This sensor operates on *higher frequencies* because the mass of LVDT is *small with natural higher frequencies.*
- This system provides *better resolution* and the *operation is frictionless.*

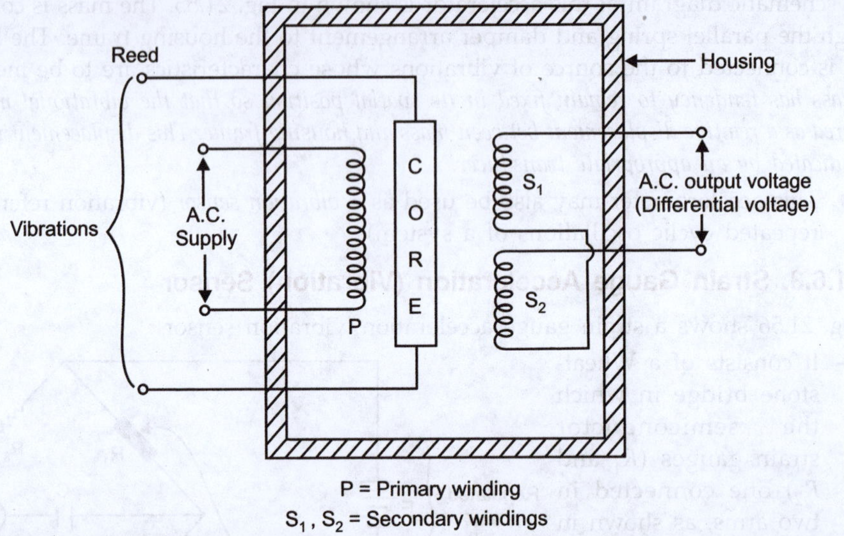

Fig. 21.57. LVDT sensor.

21.6.5. Measurement of Vibrations Using a Permanent Magnet and a Coil

This method is employed when the vibrations are to be measured or determined in the *form of velocity*.

Fig. 21.58 illustrates the set up for measurement of vibration using a permanent magnet and a coil :

— The system consists of an electrical *coil* suspended by a *spring* in the magnetic field of a *permanent magnet*. The whole assembly is mounted in a *housing*, which is *placed in contact with the vibrating surface*. The coil has low natural frequency by which it moves with housing.

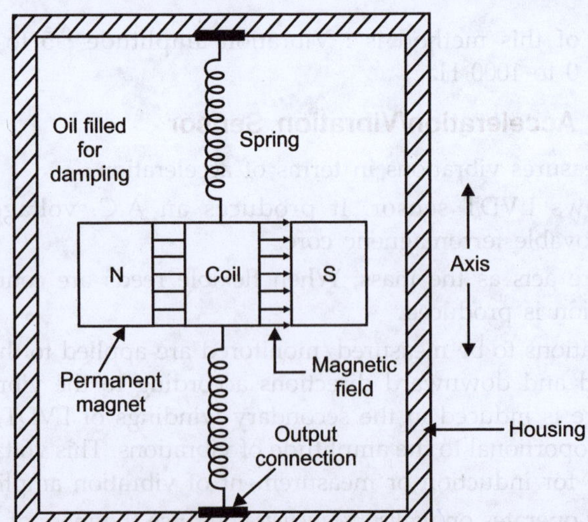

Fig. 21.58. Measurement of vibrations using a permanent magnet and a coil.

— When vibration velocity is applied to the coil, a voltage is induced in it. The amplitude of voltage is proportional to vibration velocity, with frequency equal to the vibration frequency.

— The induced voltage is monitored or displayed on a read-out.

21.7 MEASUREMENT OF FORCE

Force *is defined as any cause that produces, stops or changes the motion of a body or tends to produce these effects.*

An unknown force may be measured by the methods incorporating the following principles.

1. *Balancing the force against a known gravitational force on a standard mass.*
   ...Scales and balances
2. *Translating the force to a fluid pressure and then measuring the resulting pressure.*
   ...Hydraulic and pneumatic load cells
3. *Applying the force to some elastic member and then measuring the resulting deflection.*
   ...Proving ring
4. *Applying the force to a known mass and then measuring the resulting acceleration.*
5. *Balancing the force against a magnetic force developed by interaction of a magnet and a current carrying coil.*

21.7.1. Load Cells

Refer to Art 16.14.

Piezoelectric load cells :

The principle of working of a piezoelectric load cell is that when *pressure is applied* across certain crystals (quartz etc.) then an *electrical charge proportional to the pressure is developed across certain well-defined faces of the crystal.*

- These are used for measuring forces, particularly of a *transient or rapidly changing variety.*
- Suitable for measuring forces from 5 N to 20 kN.

21.8 MEASUREMENT OF TORQUE

21.8.1. Introduction

To transmit energy by rotation it is necessary to apply a turning force. In case of a shaft if the force is applied tangentially and in the plane of transverse cross-section the *torque or twisting moment* may be calculated by multiplying the force with the radius of the shaft. If the shaft is subjected to two opposite turning moments it is said to be in *pure torsion* and it will exhibit the tendency of shearing off at every cross-section which is perpendicular to the longitudinal axis.

The measurement of torque is carried out because of the *reasons* : *(i)* It is of considerable interest for its own sake; *(ii)* It is required to obtain load information necessary for stress and deflection analysis; *(iii)* It is often associated with determination of mechanical power.

21.8.2. Torque Measurement Methods

Torque may be measured by the following *methods* :

1. Gravity balance method.
2. Mechanical torsion meter.

3. Optical torsion meter.
4. Electrical torsion meter.
5. Strain-gauge torsion meter.

Some of these meters are discussed below.

21.8.2.1. Electrical torsion meter

Fig. 21.59, shows such a system using two photoelectric or magnetic transducers, involving two sets of measurements :

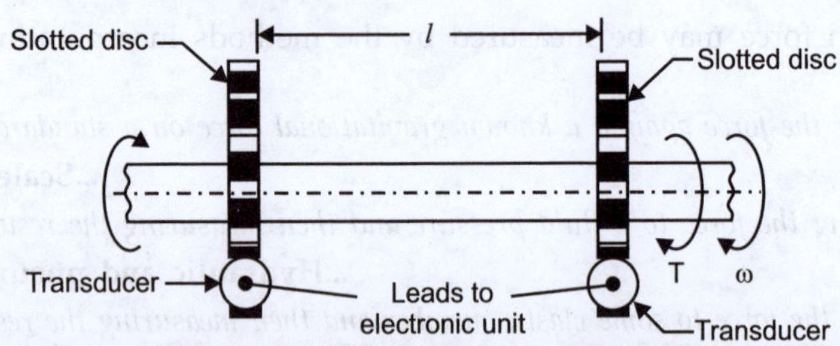

Fig. 21.59. Electrical torsion meter.

(*i*) A count of the pulse from either slotted wheel; this count gives the *frequency or shaft speed.*

(*ii*) A measure of the time between pulses from the two wheels, this signal is proportional to the twist θ and hence *T*, in the shaft.

By combining the above two signals *T* and ω the power being transmitted by the shaft can be calculated.

21.8.2.2. Strain-gauge torsion meter

Fig. 21.60, shows a general configuration of a strain gauge bridge circuit widely employed for *torque measurement from a rotating shaft.*

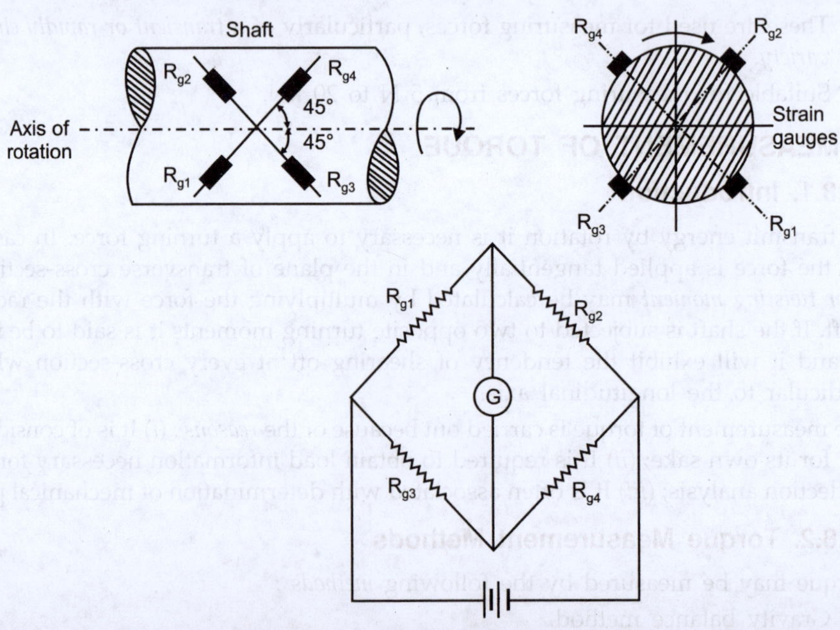

Fig. 21.60. Strain gauge torsion-meter.

Construction. Four bonded-wire strain gauges are mounted on a 45° helix with the axis of rotation; and are placed in *pairs diametrically opposite.* When the gauges are accurately placed and have matched characteristics, the system is temperature compensated and insensitive to bending and thrust or pull effects. Any change in the gauge circuit then results only from torsional deflection.

Working. When the shaft is under torsion, gauges 1 and 4 will *elongate* as a result of the tensile component of a pure shear stress on one diagonal axis, while 2 and 3 will *contract* due to compressive component on the other diagonal axis. These tensile and compressive principal strains can be measured and hence shaft torque calculated.

Advantages. This arrangement has the following *advantages* :

1. Fully temperature compensated.
2. Gives the maximum sensitivity for a given torque.
3. Provides automatic compensation for bending and axial loads.

Limitation :

The main limitation/difficulty associated with the use of this arrangement is the connection of the bridge to its power source and display arrangement. *Slip rings* are used for this purpose. These are conducting rings attached to the shaft, but insulated from it, with one of the slip rings connected to each of the bridge terminals. Slip rings are mercury filled and transmit the signal to a stationary member where it is amplified and displayed or recorded.

- *Good results from the strain gauge method are available only when the shaft experiences sufficient strain to produce a measurable output signal.* In case the torque is small the gauge bridge will not give adequate output signal for measurement. In such cases a *flexible coupling* is introduced in the shaft. The coupling incorporates one or more elastic members (elastic member may be a different shaft or a commercial torque meter) to which the gauge may be attached. The elastic members are so designed that they *produce sufficiently large deflection even under light load conditions.* This large deflection produces large strains resulting in large output of the strain gauge bridge.

21.9 MEASUREMENT OF SHAFT POWER

A **dynamometer** *is a device used to measure the torque being exerted along a rotating shaft so as to determine the shaft power input or output of power-generating, transmitting and absorbing machinery.*

The dynameters are *classified* as follows :

1. Absorption dynamometers. They work on the principle that *the power measured is converted into heat by friction or by other means.* The power absorbed is lost as heat and dissipated to the surroundings, where it serves no useful purpose.

Examples :

(*i*) Mechanical brakes :

- Prony brakes.
- Rope brakes.

(*ii*) Hydraulic or fluid friction brakes.

(*iii*) Eddy current dynamometer.

2. Transmission dynamometers. They work on the principle that the *power transmitted either to or from the dynamometer is not absorbed or dissipated.* After measurement, the energy is conveyed to the surroundings in a useful mechanical or electrical form.

Examples :

(*i*) Torsion and belt dynamometers.

(*ii*) Epicyclic train dynamometers.

(*iii*) Strain gauge dynamometers.

3. Driving dynamometers. These instruments *measure power and also supply energy to operate the tested devices.*

Example : Electric cradled dynamometer (a typical example).

Some of these dynamometers are discussed below:

21.9.1. Eddy Current Dynamometer

An **eddy current dynamometer** *(absorption) operates on the principle that when an isolated conductor cuts across magnetic flux, voltage is induced and local currents (called "eddy currents") flow in a short circular path within the conductor. These eddy currents get dissipated in the form of heat.*

Fig. 21.61, shows an eddy current dynamometer. It consists of a *toothed steel rotor* mounted on the shaft of the test-engine, it rotates inside a *smooth bared cast iron stator;* the clearance between the stator and rotor being very small. The stator carries an *exciting coil,* which is energised by D.C. supply from an external source. The stator is cradled on antifriction bearings and is provided with a brake arm to which is attached a spring balance or a scale pan.

When the dynamo is operated, the rotor turns and causes a constant change in flux density on all stator points. Consequently eddy currents are induced which oppose the rotation of the rotor. The moment of resistance is measured by the brake arm and then torque and shaft power may be calculated.

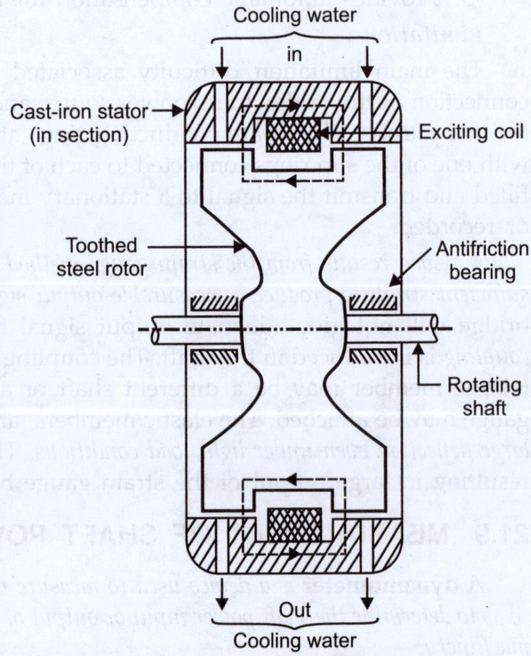

Fig. 21.61. Eddy current dynamometer.

The mechanical power supplied to the dynamometer shaft is converted to heat which is carried away partly by air circulation (induced by the rotor teeth) and partly by water circulation (See Fig. 21.61).

- Usual power limit of the dynamometer = 300 H.P.; Speed limit = 6000 r.p.m. (approx.)

Advantages :

(*i*) Good control at low rotating speeds.

(*ii*) Comparatively small size for a given capacity.

(*iii*) Suitable for a higher speed range.

21.9.2. D.C. Dynamometers

D.C. electric type dynamometer is the most *versatile and accurate dynamometer.*

This is a *cradled dynamometer* and is widely used for power and torque measurements of :

- Internal combustion engines.
- Small steam turbines.
- Pumps and other mechanical equipment.

The basic arrangement of a D.C. dynamometer is shown in Fig. 21.62. The D.C. machine acts as both motor and generator. It works as a *generator* when it is coupled to the machine under test which is a power generating machine. The *power generated by the generator is dissipated in resistance grids* or may be recovered as a useful power.

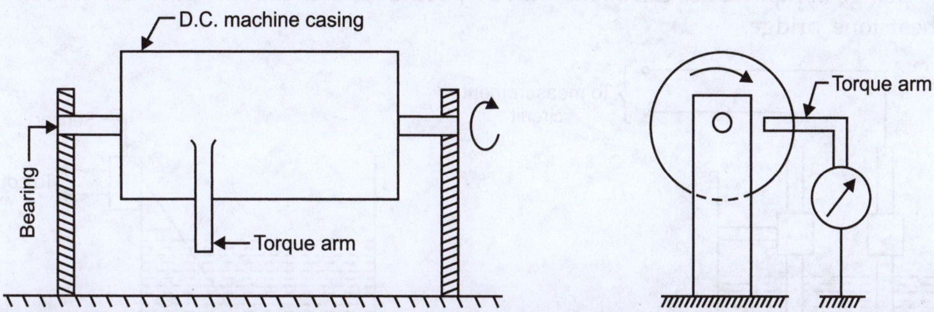

Fig. 21.62. D.C. dynamometer.

Thus, the D.C. machine can act both as *absorption dynamometer* and a *transmission dynamometer.* Force and torque is measured through a force/torque arm attached to the casing of the D.C. machine. By controlling the field excitation of the generator the output voltage and hence the power can be controlled.

When the D.C. machine is made to work as a *D.C. motor* it may act as *driving dynamometer.* The D.C. motor may drive a test machine which is a power absorbing device like a pump. The output power of a pump may be smoothly controlled by controlling the speed of the D.C. motor by varying either the armature voltage or field current.

21.10 MEASUREMENT OF LIQUID LEVEL

Liquid level *refers to the position or height of a liquid surface above datum line.* In order to ascertain the quantity of the liquid held in a container, *level measurements are made.*

Liquid level is an important process variable in industrial plants that use large quantities of liquids such as water, chemicals and solvents. For example, there is need to monitor liquid level in boilers, distillation columns etc.

The liquid level measurements can be grouped into the following *two categories* :

1. *Direct methods :*
   (*i*) Sight glass level gauge.
   (*ii*) Float gauges.
   — Float and tape liquid level gauge.
   — Float and shaft liquid level gauge.
2. *Indirect methods :*
   (*i*) Hydrostatic pressure devices.
   (*ii*) Capacitance type level gauge.

(*iii*) Gamma-rays liquid level gauge.

(*iv*) Ultrasonic liquid level gauge.

Some of these devices are discussed below.

21.10.1. Capacitance Type Level Gauge

Fig. 21.63, shows a capacitance type liquid level gauge. It consists of two parallel metal electrodes held firmly at a known fixed distance apart. *The dielectric constant between the electrodes varies with the liquid level and so does the capacitance of the system.* Greater the level the larger will be the capacitance and less the level, the smaller will be the capacitance. The change in capacitance can be measured by some suitable circuitry such as a capacitance Wheatstone bridge.

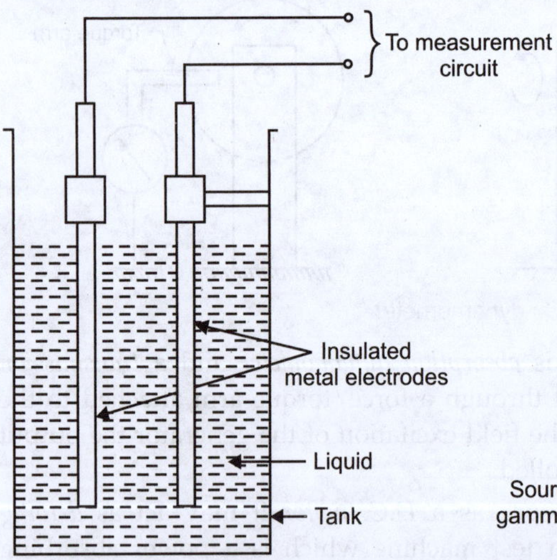

Fig. 21.63. Capacitance type liquid level gauge.

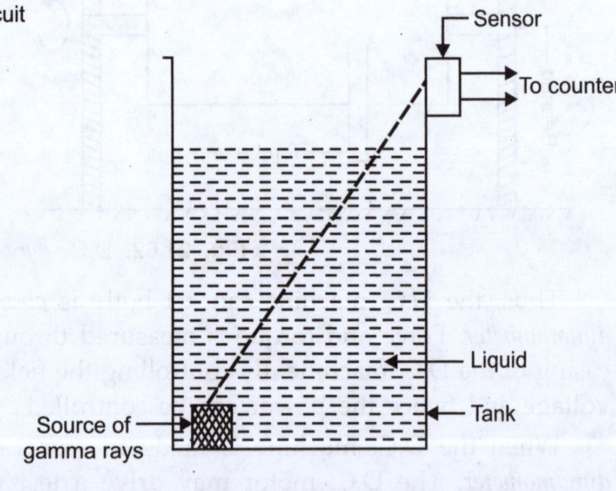

Fig. 21.64. Gamma-rays liquid level gauge.

- *This method can also be used for measuring the level of powdered solids, liquefied gases and liquid metals.*

21.10.2. Gamma-rays Liquid Level Gauge

This gauge works on the principle that the intensity of gamma rays reaching a sensor of gamma rays (*e.g., Geiger Miller tube) varies as the height of liquid column placed between the gamma-rays source and the detector.*

Fig. 21.64, shows the schematic arrangement of the gamma-rays liquid level gauge. As the gamma rays can penetrate the tank walls, the tube senses the rays and greater the intensity of the rays, the greater will be the output of the tube. *The intensity of rays will depend upon the liquid level.* When the tank is empty maximum radiation will reach the Geiger Miller tube. Higher the liquid level, the greater the absorption of gamma rays and consequently lower the detector output. The output is in the form of *pulses* and can be counted by a *counter*, the counter, therefore, may be calibrated in terms of the liquid level.

21.10.3. Ultrasonic Liquid Level Gauge

An ultrasonic liquid level gauge works on the *principle of reflection of an acoustic wave from a liquid surface.*

Fig. 21.65, shows the schematic arrangement of an ultrasonic liquid level gauge. The ultrasonic wave sent by a transmitter *'T'*, is reflected at the surface of liquid and is received by a receiver *'R'*. With the change of liquid level the transient time of the wave also changes. The liquid level can be measured either by the *time taken* by the reflected wave to reach the receiver or on the basis of *change of phase* of the wave. Piezoelectric crystals (such as quartz or barium titanate) may be used as transmitter and receiver.

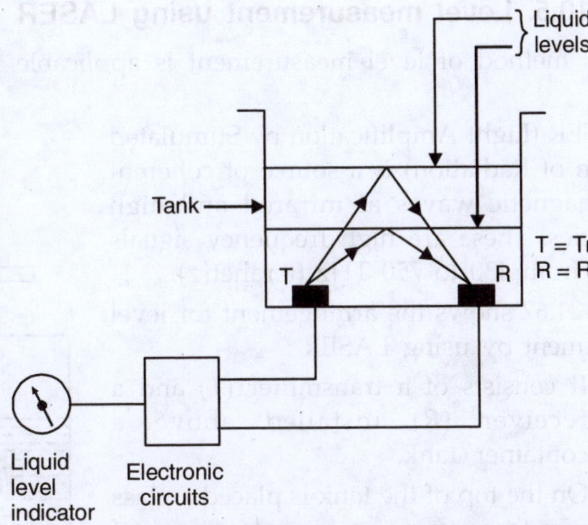

Fig. 21.65. Ultrasonic liquid level gauge.

- Although the gauge works on a simple principle, yet actual instrument involves complex circuitry.

21.10.4. Electrical Level Gauges

A simpler way of measuring the level using electrical circuit is the use of *triode tube.* Fig. 21.66 shows the principle of electrical transducers. If the level in tank is below the required level *G*, there is no complete circuit to the grid of the tube, which then accumulates a charge of electrons. This makes it very negative in effect and reduces the tube plate current to a very small amount. Under these circumstances, very little current in the relay coil completes the alarm circuit. The alarm thus indicates the dangerously low level in the boiler tank. A red light signal also serves the same purpose. On the other hand if the level of water rises, it completes the circuit between G and other terminal *G'*. The electrons can then be removed from the grid circuit. As the grid is made less negative, the plate current will continue to rise until sufficient current flows in the relay coil to operate it. The triode tube can well be *replaced by transistors.*

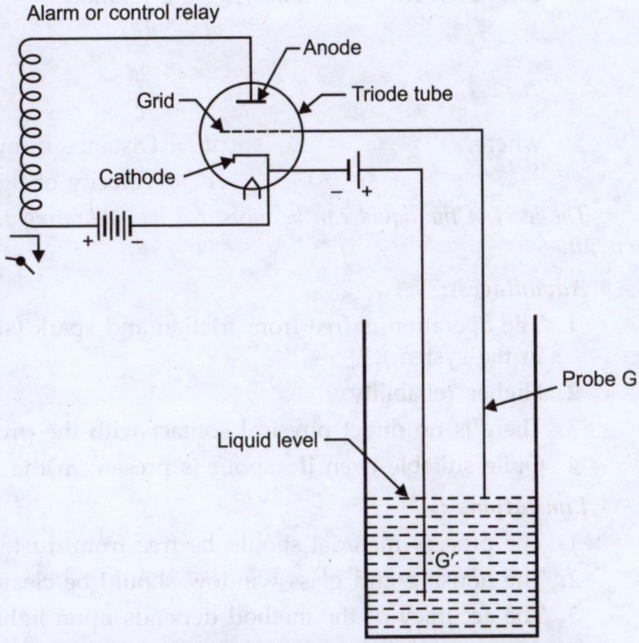

Fig. 21.66. Electrical level gauge principle.

21.10.5. Level measurement using LASER

This method of level measurement is applicable for *solids, liquids with transparent vapours.*

LASER (Light Amplification by Stimulated Emission of Radiation) is a source of coherent electromagnetic waves at infrared and high frequencies. These are high frequency signals ranging from 430 to 750 THz (terahetrz).

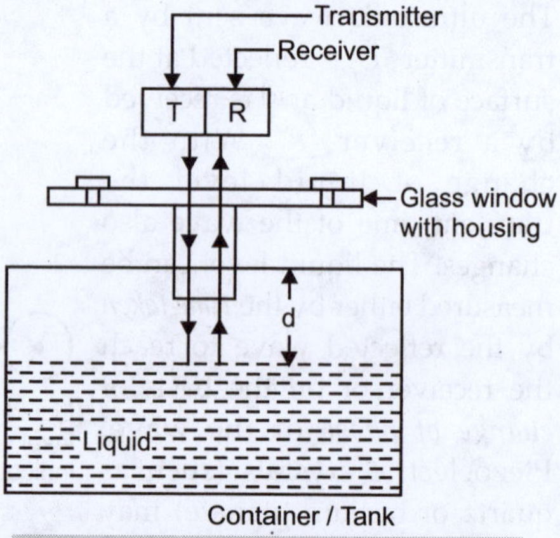

Fig. 21.67. Level measurement by using LASER.

Fig. 21.67 shows the arrangement for level measurement by using LASER :

— It consists of a transmitter (*T*) and a receiver (*R*) installed above a container/tank.

— On the top of the tank is placed a glass window, using proper housing of plastic or rubber.

— A sharp beam of laser is focused towards the liquid level by modulation from the *transmitter* end. When this signal reaches at the surface of the liquid, it gets reflected and is received by the *"receiver"*. In order to measure the liquid level, the time period (*t*) from reflection of the signal to reaching the receiver is calculated from the following the relation :

$$t = f(d) \quad \text{...(21.53)}$$

$$\therefore \quad t = \frac{2d}{c} \quad \text{...(21.54)}$$

where, d = Distance from the liquid level, and

c = Velocity of light ($= 3 \times 10^8$ m/s).

The level of the liquid can be measured by calibrating time period in pulses using counter circuit.

Advantages :

1. The operation is free from friction and spark (since no moving part is involved in the system).
2. Higher reliability.
3. There is no direct physical contact with the process material.
4. Quite suitable even if vapour is present in the process material.

Limitations :

1. The process material should be free from dust.
2. The housing and glass window should be clean and free from dust.
3. The accuracy of the method depends upon light reflection characteristics of the process material.

Applications. This method is used for measuring the *levels of* :

(*i*) Hazardous and toxic liquids and acids;

(*ii*) Liquid aluminium, liquid polyethylene, brown tobacco, etc.;

(*iii*) Wound pulp in paper industries.

21.10.6. Microwave Level Detector

In this type of detector electromagnetic radio waves in microwave X-band and K-band of frequency ranges of 8–12.5 GHz and 18–26.5 GHz respectively are used. In order to get better performance solid state devices are employed to generate and detect microwave signals. Microwave signals do not pass through metals but can pass through the vessels having walls of fiber glass, plastic and ceramic.

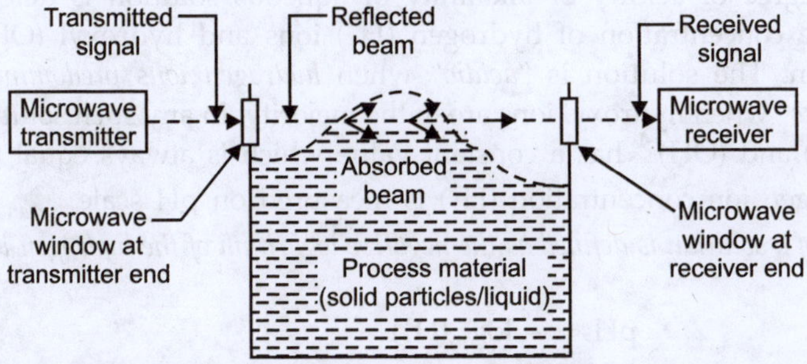

Fig. 21.68. Microwave method.

Fig. 21.68 illustrates the arrangement for detecting or measuring the level by microwave method :

— It consists of a microwave transmitter, which transmits the signals either in X-band or K-band through window inside the vessel. The windows are installed at transmitter and receiver ends; these may be made of plastic, teflon, quartz, mica etc.

— When air or vapour is present inside the vessel, strong microwave signals are received at the receiver and detected by the microwave detector. However, with the increase in the level of process material, the transmission of microwaves is retarded and these signals get reflected leading to reduced signal strength detected by the detector. Thus the level of process material can be detected.

- This system is applicable for solids *i.e.,* small size particles, liquids having dielectric upto 10.

Advantages :

1. Not affected by dust, foam and mist.
2. Applicable for conductive as well as non-conductive process materials.
3. The operation is free of friction and spark (since no moving part is involved).

Limitations :

1. Construction is complex.
2. Suitable for limited dielectric constant of process material.
3. Method is expensive.

Applications. It is useful to detect/measure *levels* of :

(*i*) Light powdered materials such as dry saw dust;
(*ii*) Toxic, hazardous chemicals or process materials;
(*iii*) Coal, minerals, and vegetable pulp.

21.11 MEASUREMENT OF pH VALUE

— Since *acids* and *bases* are of significant importance in various fields like biology, medicine, food technology, water sewage treatment and the chemical industry, therefore measurement of their concentration and strength are most common.

The scale on which the acidity or alkalinity is measured is the **pH scale**.

— The degree of acidity or alkalinity of aqueous solution is determined by the relative concentration of hydrogen (H^+) ions and hydroxyl $(OH)^-$ ions in the solution. The solution is *"acidic"* when *hydrogen ions predominate* which it is *"alkaline"* when hydroxyl ions are in the majority. In any such solution the product of (H^+) and $(OH)^{-1}$ has a constant value which is always equal to 10^{-14}.

— Hydrogen ion concentration (H^+) is measured on pH scale.

pH value *of a solution is defined as the negative logarithm of the hydrogen ion concentration or,*

$$pH = -\log_{10}(H^+) \quad ...(21.55)$$

This scale *ranges* from *0 to 14*.

In neutral solution, the concentration of both (H^+) and $(OH)^{-1}$ ions are equal *i.e.*, both are 10^{-7}.

$\therefore$ For a neutral solution : $pH = -\log(10^{-7}) = 7$

Thus a neutral solution like pure water has a pH value of 7.

For *"acidic solutions"* the pH value lies *between 0 to 7*. In case of *"alkaline solutions"* the pH value lies in *between 7 to 14*.

- The theory of pH measurement involves the breakdown of certain chemical compounds when combined with water. When these compounds breakdown, tiny electrically charged particles are released. Some of these are positively charged particles–hydrogen ions (H^+) and some are negatively charged particles–hydroxyl ions $(OH)^-$. The pH of a solution can be accurately measured by means of an *electronic pH meter*. pH meters usually cover a range of 0 to 14 pH units; however it is possible to make measurements beyond these limits in very concentrated solutions.

pH cell :

Fig 21.69 shows a pH cell :

— The pH cell consists of two electrodes (which are immersed into test solution). One of the electrodes called the *"reference electrode"*, is at a constant potential regardless of the pH value of the solution under test. The potential of the other electrode, called the *"measuring electrode"* is determined by the pH value of solution. *Thus the potential difference between the two electrodes depends upon the pH value of the test solution.*

— The *'reference electrode'* is made of glass and consists of an inner assembly containing a solution of calomel (mercury chloride) and mercury. This assembly is surrounded by a larger glass tube, and the space between the two contains an accurate solution of KCl (potassium chloride). A tiny opening in the bottom of the electrode permits KCl to diffuse very slowly into the test solution. In this way the electrical contact is made between this solution and calomel solution of the electrode. Thus the test solution sets the potential level of the reference electrode.

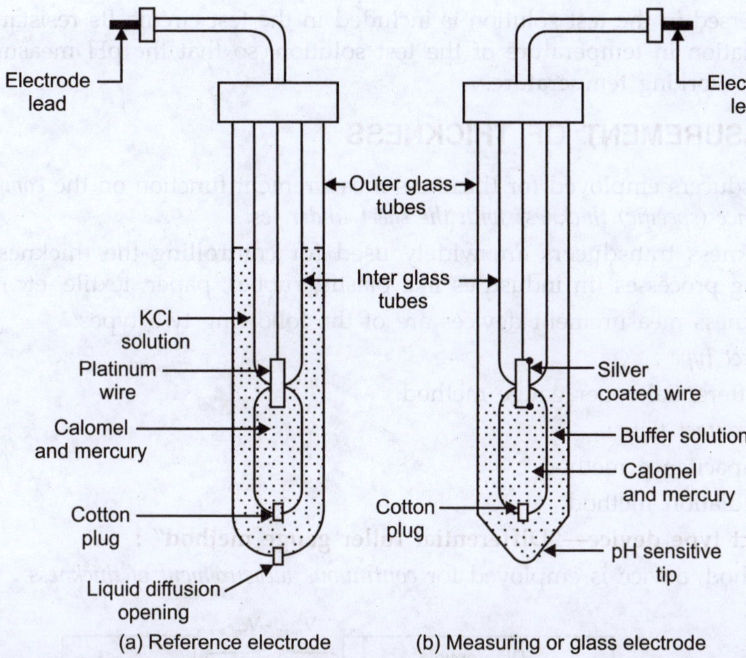

Fig. 21.69. pH cell.

In the *measuring electrode* the mercury calomel element is surround by a buffer solution of known and constant pH. There is no opening at the bottom of the outer tube, instead it tapers down to a tip made of thin glass of special composition. *At this tip a potential difference is developed between the buffer solution and the test solution because of difference in the pH value of the two solutions. Since the pH value of the buffer solution is constant the net potential of this electrode is a function of the pH value of the test solution.*

Fig. 21.70 shows the simplified circuit diagram for electronic pH measurement.

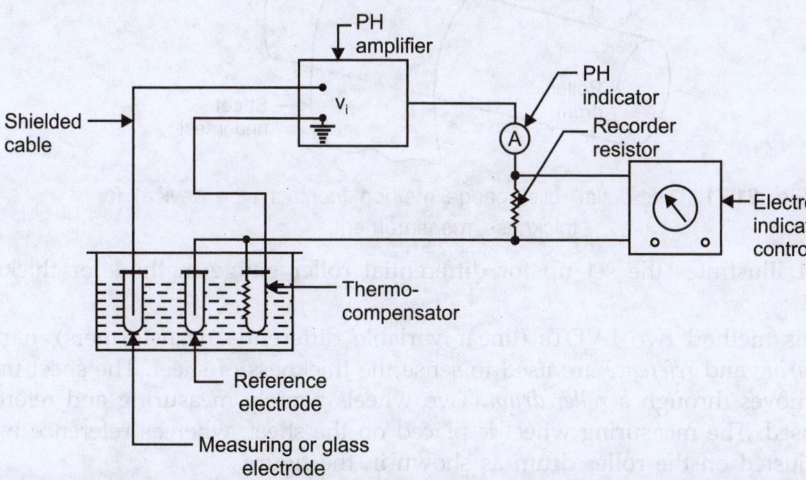

Fig. 21.70. Simplified circuit diagram for electronic pH measurement.

The voltage developed between the measuring electrode and the reference electrode is applied as the input voltage, v_i to a null-balance millimeter. A temperature compensating resistor immersed in the test solution is included in the test circuit. Its resistance varies with the variation in temperature of the test solution, so that the pH measurement is correct at the working temperature.

21.12 MEASUREMENT OF THICKNESS

The transducers employed for thickness measurement function on the *comparision of sheet of reference (specific) thickness with the sheet under test.*

The thickness transducers are widely used for controlling the thickness during manufacturing processes (in industries like plastic, rubber, paper, textile, etc.)

The thickness measurement devices are of the following two types.

1. *Contact type :*
   - Differential roller gauge method.
2. *Non-contact type :*
   - Capacitance method.
   - Radiation method.

1. Contact type device—"Differential roller gauge method" :

This method/device is employed for *continuous measurement of thickness.*

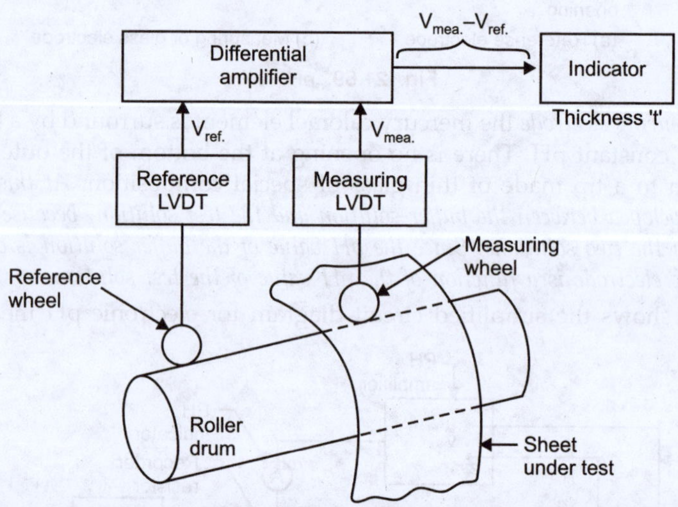

Fig. 21.71. Differential roller gauge method (contact type device) for thickness measurement.

Fig. 21.71 illustrates the set up for differential roller gauge method for thickness measurement :

— In this method two LVDTs (linear variable differential transformers), namely *measuring* and *reference* are used to sense the thickness of sheet. The sheet under test moves through a *roller drum*. Two wheels namely measuring and reference are used. The measuring wheel is placed on the sheet; whereas reference wheel is adjusted on the roller drum as shown in the figure.

— The thickness of sheet is determined by the difference of secondary voltages of both the LVDTs due to linear displacements of core or armature positions. The thickness t' of the sheet is determined from the relationship.

$$t = C(V_{mea.} - V_{ref.}) \quad ...(21.56)$$

where, C = The calibration constant,

$V_{mea.}$ = The output voltage of the measuring LVDT, and

$V_{ref.}$ = The output voltage of the reference LVDT.

— The *indicator* indicates the thickness.

- The measurement range of the operating thickness is 0.0025 mm to several cm.

2. Non-contact type devices :

(*i*) ***Capacitance method :*** This method is normally employed for measuring thickness of *insulating materials.*

Fig. 21.72 shows the set-up for capacitance method (non-contact type) for thickness measurement.

— In this method two metal electrodes are placed on the two sides of the insulating material sheet under test. This arrangement functions as a *parallel plate capacitor.*

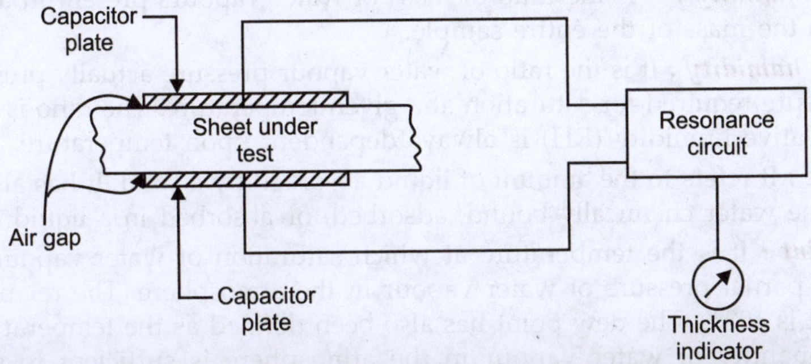

Fig. 21.72. Capacitance method (non-contact type) for thickness measurement.

— The capacitance of the capacitor varies according to or in proportion with the thickness of the sheet under test. This *change is capacitance is calibrated in terms of thickness by means of resonance circuit arrangements.* Owing to changes in the value of the capacitance, the resonance ($X_L = X_C$) gets altered which is indicated by the thickness indicator.

(*ii*) ***Radiation method :*** In this method thickness is measured by using nuclear radiations (α, β, γ, X-rays) radiated or generated from radiation source.

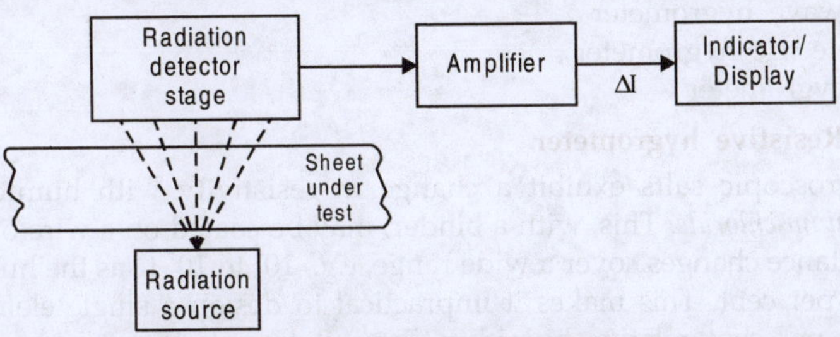

Fig. 21.73. Radiation method.

Fig. 21.73 shows nuclear radiation method for thickness measurement :

— The sheet under test is passed through the radiation source and radiation detector stage. The sheet's thickness is determined *in terms of radioactive decay or attenuation of radiations*, the relationship is given by :

$$\Delta I = I_0[1 - e^{(-\mu t)}] \quad \text{...(21.57)}$$

where, ΔI = The change in ionization current when absorber is inserted,

I_0 = The ionization current without the absorber,

m = The absorber coefficient, $cm^2/\mu g$,

t = The thickness in $\mu g/cm^2$.

— The indicator/display is calibrated to read the thickness value.

21.13 MEASUREMENT OF HUMIDITY

21.13.1. Definitions

Humidity : It is the measure of water vapour present in a gas. It is usually measured as absolute humidity, relative humidity or dew point temperature.

Absolute humidity : It is the mass of water vapour present per unit volume.

Specific humidity : It is the ratio of mass of water vapours present in a sample of a moist gas to the mass of the entire sample.

Relative humidity : It is the ratio of water vapour pressure actually present to water vapour pressure required for saturation at a given temperature. The ratio is expressed in per cent. Relative humidity (RH) is always dependent upon temperature.

Moisture : It refers to the amount of liquid absorbed by a solid. It has also been used to refer to the water chemically bound, adsorbed, or absorbed in a liquid.

Dew point : It is the temperature at which saturation of water vapour pressure is equal to the partial pressure of water vapour in the atmosphere. The relative humidity at dew point is 100%. The dew point has also been defined as the temperature at which the actual quantity of water vapour in the atmosphere is sufficient to saturate this atmosphere with water vapour.

21.13.2. Hygrometers

The *"hygrometer" is an instrument which measures humidity directly.* It can be calibrated in terms of absolute humidity. More frequently its output is used to indicate relative humidity.

The various types of hygrometers used are enumerated and described as follows :

1. Resistive hygrometer
2. Capacitance hygrometer
3. Microwave hygrometer
4. Piezoelectric hygrometer
5. Hair hygrometer

21.13.2.1. Resistive hygrometer

Some hygroscopic salts exhibit a change in resistivity with humidity. The most common is *lithium chloride.* This, with a binder, may be coated on a wire or on electrodes. Resulting resistance changes cover a wide range, *e.g.,* 10^4 to 10^9 Ω as the humidity changes from 100 to 0 per cent. This makes it impractical to design a single element to operate from 1 to 100 per cent relative humidity. Instead *several elements* are used, each in a narrow range, with provision for switching elements. Resistance is measured either with a Wheatstone bridge or by a combination of current and voltage measurements.

Most of these must not be exposed to conditions of 100 per cent humidity as the resulting condensation may damage the device. Either they must be operated in a constant temperature environment or temperature corrections must be made. These are accurate to within ±2.5 per cent or ±1.5 percent in some cases. Response times are typically of the order of a few seconds. These are currently the most common electronic hygrometers.

A typical resistance hygrometer is shown in Fig. 21.74. Two metal grids are bonded to a sheet of plastic.

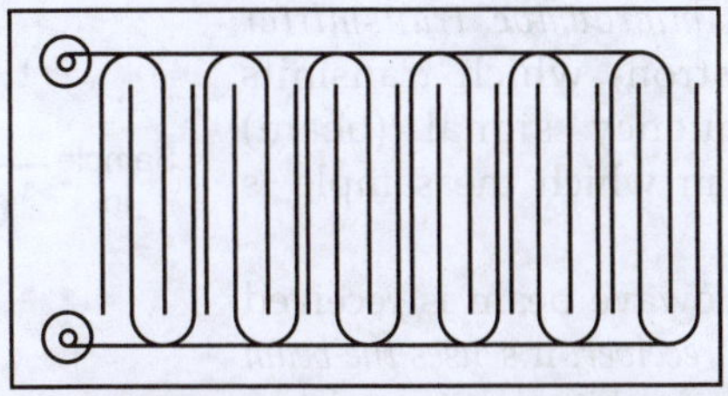

Fig. 21.74. Resistive hygrometer.

The whole is coated with a layer of moisture sensitive chemical such as lithium chloride. The resistance changes when this chemical is exposed to variations in humidity. The higher the relative humidity, the more moisture the lithium chloride will absorb, and the lower will be its resistance.

The resistance of the sensing unit is a measure of the relative humidity. Resistance should be measured by applying A.C. to the Wheatstone bridge. D.C. voltage is not applied because it tends to breakdown the lithium chloride to its lithium and chlorine atoms. The current flow is a measure of the resistance and hence of the relative humidity.

21.13.2.2. Capacitance hygrometer

Fig. 21.75 shows a capacitance hygrometer :

— It consists of a conductive aluminium substance over which a layer of porous aluminium oxide is deposited.

— The aluminium oxide is coated with a thin film of gold.

— The aluminium substrate and the gold film act as *electrodes.* These electrodes together with the aluminium oxide as the *delectric,* form a capacitor.

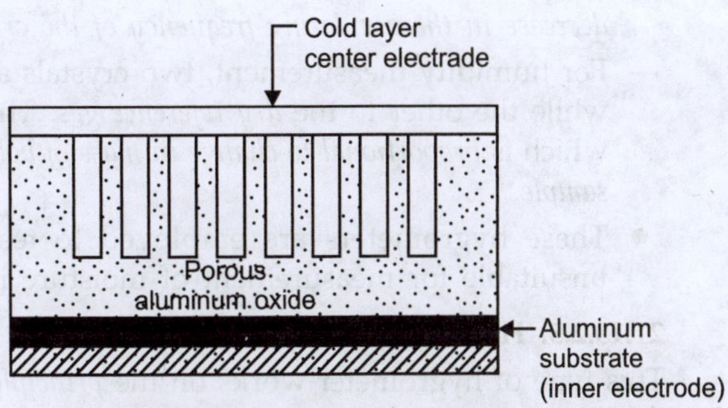

Fig. 21.75. Capacitance hygrometer.

— When this sensor is exposed to air, *the aluminium oxide will absorb the water vapours through the gold layer that changes the dielectric constant of the aluminium oxide. It results in a change in the capacitance which is measured by introducing frequency determining oscillator and other circuits.* The capacitance of the sensor changes with the variation of rate of water content.

— These days, instead of aluminium oxide dielectric, *hygroscopic polymer dielectric is used.*

- It is *small in size* and wide humidity measurement range detects the low dew points. Such a hygrometer is *unsuitable for polar material* (*e.g.,* alcohol) since the capacitor gets short circuited at 15 kHz frequency.

21.13.2.3. Microwave absorption hydrometer

In this type of hygrometer microwave frequencies to the tune of 20 to 22 GHz are used; in this frequency range water molecule produces *molecular resonance.*

Fig. 21.76 shows a microwave absorption hygrometer :

— It consists of a *microwave transmitter* (magnetron, klystron) which transmits microwave frequency signal (beam) through the *pipe* in which the sample is flowing.

— The reflected microwave beam is received by the microwave *receiver; it senses the beam attenuation and phase shift due to moisture content in the sample.*

— The *display/readout exhibits the mass of moisture per unit volume.*

- This hygrometer is suitable for moisture content measurements in *slurries and pastes.* It is unsuitable for measurement of moisture in gases.

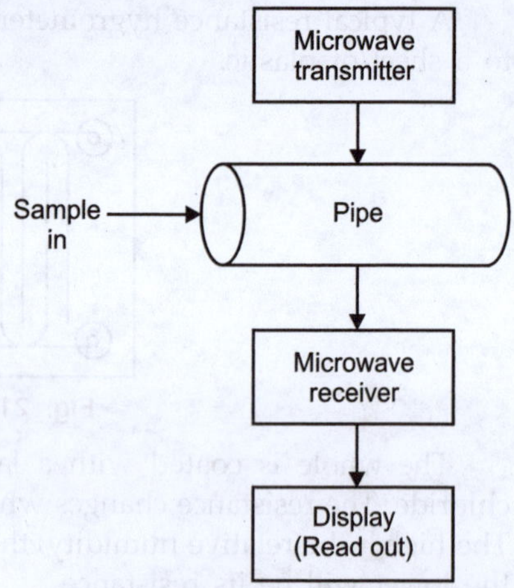

Fig. 21.76. Microwave absorption hygrometer.

21.13.2.4. Piezoelectric hygrometer

— This type of hygrometer makes use of a piezoelectric crystals like *quartz* crystals, which are coated with hygroscopic material. When this crystal is exposed to the sample, its coating absorbs water content. It results in *increase in the mass and decrease in the oscillating frequency of the crystal.*

— For humidity measurement, two crystals are used. One is exposed to *wet sample* while the other to the *dry reference gas.* Then frequency difference is determined which is *proportional to change in mass of both the crystals and moisture content in the sample.*

- These hygrometers are employed to test *liquid samples.* They are, however, unsuitable for measurement of moisture in liquid phase samples.

21.13.2.5. Hair hygrometer

This type of hygrometer works on the *principle that when hair absorbs the moisture, its length changes.*

In the hair hygrometer a strand of hair is attached to an arm through mechanical linkage. The arm is connected to the pointer, which slides over the calibrated scale. A spring is used to maintain proper position of the arm and stand. During moisture measurements when hair absorbs moisture, its length changes. This change in length is applied to the arm which results in sliding of the pointer over the calibrated scale.

- This hygrometer is *small in size, less costly* and can be *installed easily.* However, it is less accurate and requires periodic maintenance.
- It is used in industries where high precision of measurement is not necessary.

21.13.3. Psychrometers

The *"psychrometer" is a humidity measuring instrument which utilizes one wet bulb thermometer and one dry bulb thermometer.* The dry bulb thermometer measures the ambient temperature and the wet bulb thermometer measures temperature reduction due to evaporative cooling. A wick, a porous ceramic sleeve, or a similar device saturated with water is in close physical contact with the "bulb" (sensing portion) of the wet bulb thermometer to keep it moist. Relative humidity is determined from the two temperature readings and a reading of the barometric pressure usually with the help of a **Psychrometric Chart**. At any given ambient temperature, the relative humidity decreases as the difference between dry bulb and wet bulb readings increases.

Some of the psychrometers are described below :

1. *Sling psychrometer* : This device consists of two similar thermometers mounted side by side as shown in Fig. 21.77. The bulb of one of the thermometers is covered by a wet cloth.

When a thin film of water is brought into close contact with air, a portion of it evaporates, and the temperature of the remaining water is lowered due to heat required by the evaporation process. The amount of evaporation and consequent lowering of the temperature depends upon the capability of the air to absorb the moisture. Thus if the relative humidity is lower, air has small moisture and can therefore absorb more. Hence the evaporation is faster and the temperature of the remaining water will be lower.

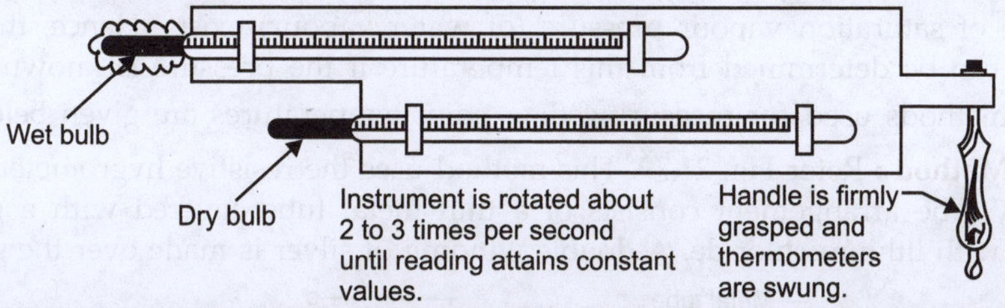

Fig. 21.77. Sling psychrometer.

In the sling psychrometer, a handle is provided whereby the thermometer may be whirled, so that the bulbs continuously come in contact with fresh air. If this is not done the stagnant moisture-laden air will engulf the thermometer thereby causing errors.

The dry bulb thermometer is unaffected by the moving air and registers the ambient temperature. The wet bulb thermometer shows a lower temperature due to evaporation. Since the cooling effect is a function of the relative humidity of air, the relative humidity may be determined by comparing the readings of the two thermometers. Relative humidity can directly be read off from psychrometric charts.

Psychrometric charts *graphically describe the relationship between the properties of moist air i.e., the dry bulb, the wet bulb and dew point temperatures of the mixture and humidity.*

2. *Commercial type psychrometer* : A commercial type psychrometer is shown in Fig. 21.78. The filled system thermometer bulbs are placed adjacent to each other and air

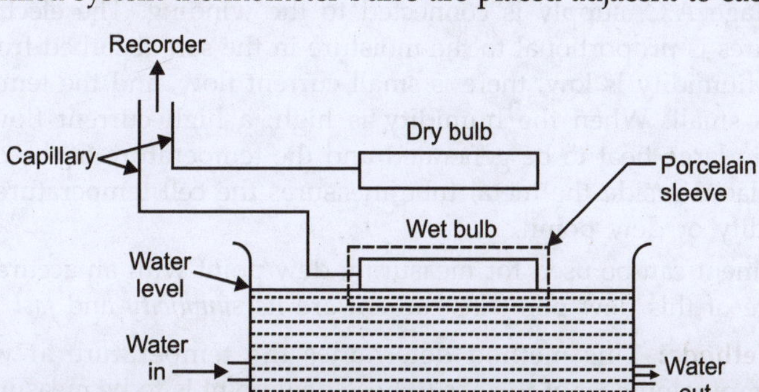

Fig. 21.78. Commercial type psychrometer.

is blown across the assembly or they may be placed where there is strong current of air (nearly 5 m/s). One of the bulbs is coated with a knitted or woven cotton which is suspended into a water reservoir whose level is controlled so that the required conditions are satisfied. The temperatures are recorded by suitable recorders.

The two thermometers are connected to a recorder through capillary tubes. In order to avoid errors arising on account of contamination of the wick the bulb may be enclosed in a porcelain sleeve and water is pumped into the annular space at a constant rate. *These instruments are not suited for instantaneous monitoring and control of humidity.*

A variation in the indication and recording arrangement is also possible if the filled in thermometers are repleced by resistance thermometers or thermocouples.

21.13.4. Dew Point Measurements

The *'dew point' is a discrete temperature. Specially, it is that temperature at which liquid water and water vapour, or ice and water vapour are in equilibrium.* At this temperature only one value of saturation vapour pressure (of water vapour) exists. Hence, the absolute humidity can be determined from this temperature if the pressure is known.

Two methods used for measuring dew point temperatures are given below :

First Method : Refer Fig. 21.79. This method uses the resistive hygrometer described previously. The arrangement consists of a thin metal tube covered with a glass cloth saturated with lithium chloride. A double winding of silver is made over the glass cloth.

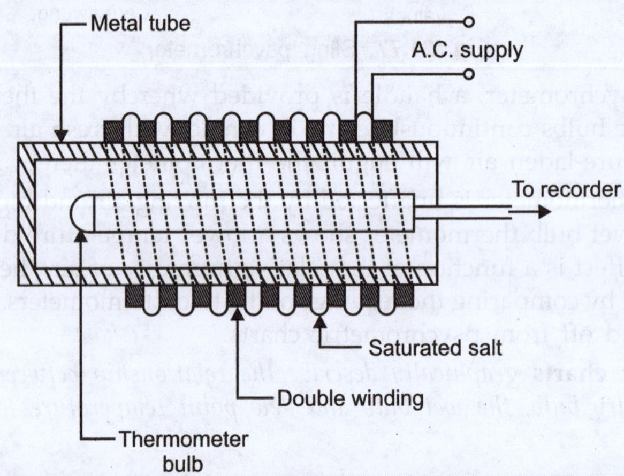

Fig. 21.79. Dew point recorder.

A low voltage A.C. supply is connected to the winding. The electric conductivity between the wires is proportional to the moisture in the salt absorbed from surrounding gas. When the humidity is low, there is small current flow, and the temperature rise of the winding is small. When the humidity is high, a high current flows through the winding causing large heat to be generated and the temperature of the cell is high. The thermometer placed inside the metal tube measures the cell temperature, and therefore absolute humidity or dew point.

This instrument can be used for measuring dew point with an accuracy of ±2%. The main advantage of this dew point instrument are its *simplicity* and *fast response.*

Second Method : This method determines the temperature at which moisture condenses by impinging a jet of the gas (whose dew point is to be measured) on a mirror and finding the exact temperature at which the surface becomes cloudy. This method is shown in Fig. 21.80. Automatic control can be used by having feedback.

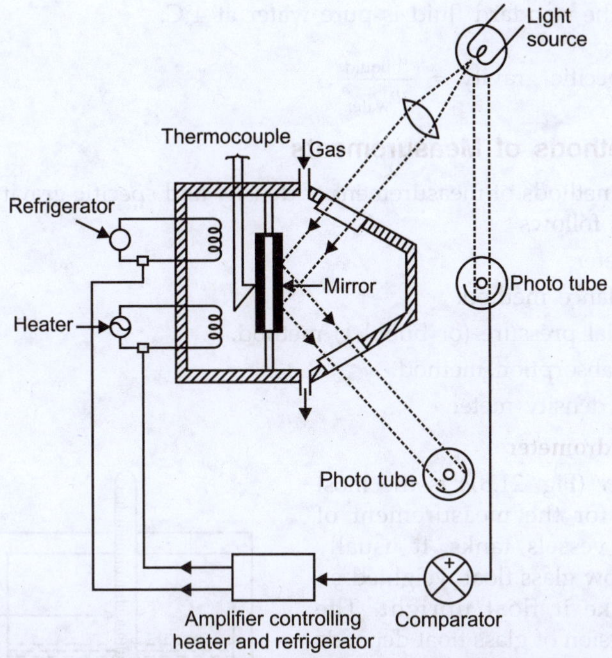

Fig. 21.80. Schematic diagram of dew point recorder.

The gas is supplied continuously. This dew point recorder is provided with arrangements for both cooling and heating of the mirror. Light falling on the mirror from a standard light source is reflected on a photo-tube. Another photo-tube receives light straight from the source. The outputs of tubes are then compared by a comparator. If the mirror is foggy due to condensation an error signal is sent and the heater starts functioning and the refrigerator is turned off. If the condensation has not started, the heater is shut off and the refrigerator turned is on. The cycle continues. Thus, the temperature of the mirror tells about the temperature necessary to condense the gas. The temperature is sensed by a thermocouple. The temperature is measured and recorded by a suitable device.

21.14 MEASUREMENT OF DENSITY AND SPECIFIC GRAVITY

21.14.1. Definitions

Mass density. *It is defined as the mass per unit volume* $\left(\frac{m}{V}\right)$ *at a standard temperature and pressure. It is usually denoted by* ρ *(rho). Its units are* kg/m^3 *i.e.,*

$$\rho = \frac{m}{V} \qquad ...(21.58)$$

Weight density (or specific weight). *It is defined as the weight per unit volume at the standard temperature and pressure.* It is usually denoted by w.

$$w = \rho g \qquad ...(21.59)$$

Specific gravity. It is defined as the *ratio of specific weight of the liquid to the specific weight of a standard liquid*. It is dimensionless and has no units. It is denoted by S.

For liquids, the standard fluid is pure water at 4°C.

$$\therefore \qquad \text{Specific gravity} = \frac{w_{\text{liquid}}}{w_{\text{water}}}.$$

21.14.2. Methods of Measurements

The various methods of measurement of density and specific gravity are enumerated and discussed as follows :

1. Hydrometer.
2. Force balance method.
3. Differential pressure (or bubbler) method.
4. Nuclear absorption method.
5. Acoustic density meter.

21.14.2.1. Hydrometer

An *hydrometer* (Fig. 21.81) is the most popular device for the measurement of liquids in open vessels/tanks. It usually consists of a hollow glass float weighted on one side to make it float upright. The amount of immersion of glass float depends on the density of the liquid. The glass float is usually calibrated in terms of specific gravity (ratio between the density of liquid under measurement to that of the water). Thus the specific gravity of the liquid can be easily determined by noting the amount of immersion from the scale on the glass float.

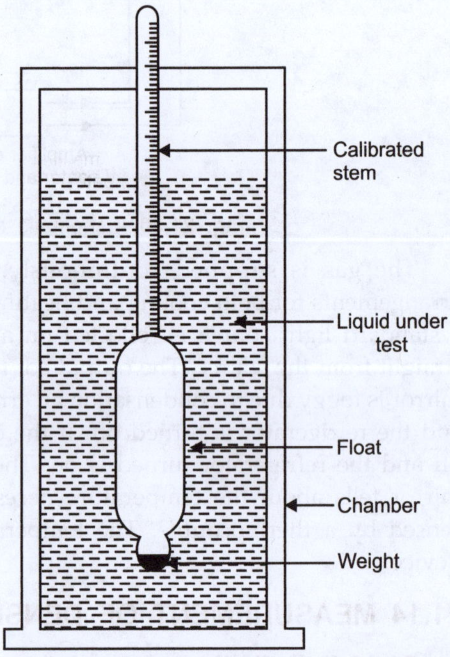

Fig. 21.81. Hydrometer.

Hydrometers are widely used where automatic operation is not required.

- ***Photoelectric hydrometer.*** In order to have *continuous monitoring of the density* of a liquid, we use a *"photoelectric hydrometer"*. As opaque stem of the hydrometer rises or falls, it cuts off light rays passing through a slit into a phototube. The output of the phototube is determined by the position of the stem and hence the density of the liquid.

- ***Density measurement using LVDT.*** In the arrangement (Fig. 21.82) an iron cone is provided at the centre of the hydrometer and the body of the hydrometer chamber is surrounded by the windings of the linear voltage differential transformer (LVDT). The sample flows continuously through the hydrometer chamber and the hydrometer will rise or fall in accordance with the changes in specific gravity of the liquid. The output of the differential transformer changes with the position of the iron core. A potentiometer is used to detect any unbalance in the circuit caused by the output from the differential transformer and to convert this signal into a specific gravity/density reading.

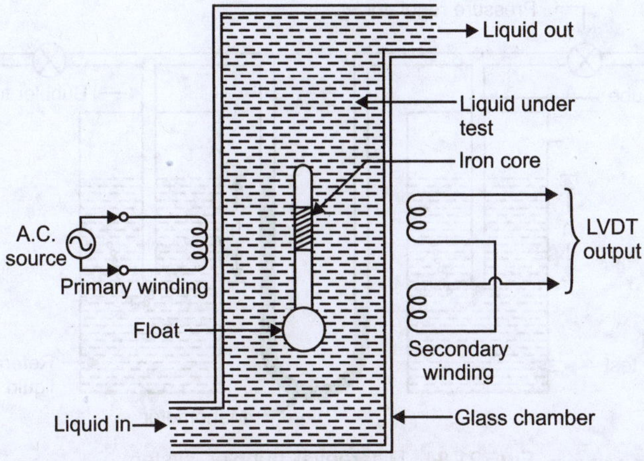

Fig. 21.82. Hydrometer-LVDT type.

This method is specially useful for *dynamic measurement of density*. Since the output signal is a voltage, it can be used for display and control of density from a remote location.

21.14.2.2. Force balance method

The schematic arrangement of the force balance method is shown in Fig. 21.83. The liquid under test flows through a tube loop of the meter. The tube loop is suspended from a pivoted beam which carries a weight. The system is balanced when the tube is full of sample liquid.

When a liquid of density different from the sample liquid is passed, the beam gets deflected. The beam is connected to a force balance system which keeps the tube loop in balance with the weight. *The force (found in terms of the current-fed to arm electromagnet) is measured on an indicator which is calibrated directly in terms of density.*

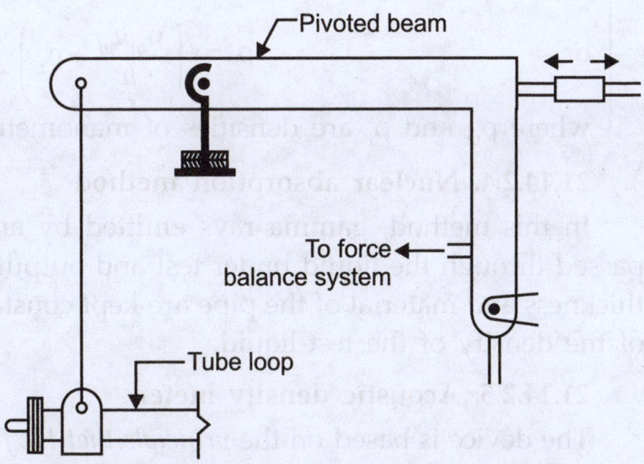

Fig. 21.83. Force balance system.

21.14.2.3. Differential pressure (or bubbler) method

This method utilizes the hydrostatic principle :

$$p = \rho g h \qquad ...(21.60)$$

This means that for a constant depth of liquid from a free surface, the *pressure exerted is directly proportional to density of the liquid* (ρ).

Fig. 21.84 shows a differential bubbler system. It consists of two containers, one containing reference liquid of known density and other connected to the liquid under test in which level/height is maintained constant by an overflow tube. Two bubbler tubes are immersed to equal depths in two containers and compressed air at regulated pressure is supplied. The air pressures to bubbler tubes are so adjusted *that bubbles are released slowly*.

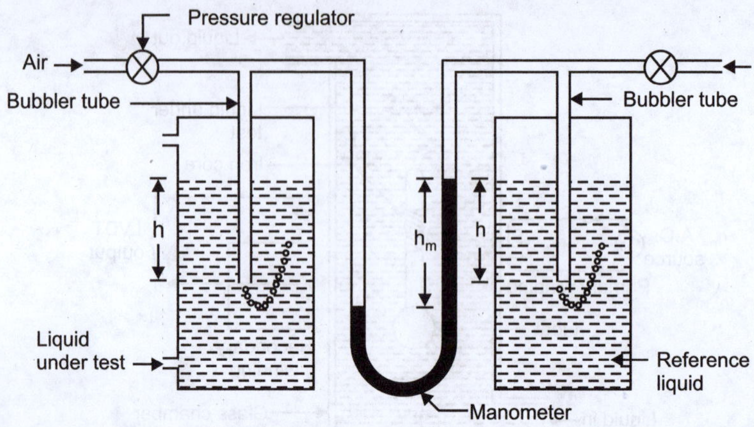

Fig. 21.84. Differential bubbler system.

- The two pressures are then proportional to the density of the liquids.
- The reading 'h_m' of differential manometer is indicative of the specific gravity of the liquid under test.

From the hydrostatic balance for the system, we have

$$\rho g h = \rho_m g h_m + \rho_r g h$$

or,

$$\rho = \left(\frac{\rho_m h_m}{h} + \rho_r\right)$$

where ρ_m and ρ_r are densities of manometric and reference liquids respectively.

21.14.2.4. Nuclear absorption method

In this method, gamma rays emitted by an isotope of cesium-137 or cobalt-60 are passed through the liquid under test and output signal detected. If other parameters like thickness and material of the pipe are kept constant, the output signal becomes a measure of the density of the test-liquid.

21.14.2.5. Acoustic density meter

The device is based on the *principle that the frequency of oscillation of a tube is dependent on its mass.* When the tube contains a liquid, then its mass is dependent on the density of the liquid. Hence, a change in density of the liquid produces a change in its acoustic resonant frequency.

21.15 MEASUREMENT OF VISCOSITY

21.15.1. Introduction to Viscosity

Viscosity *may be defined as the property of a fluid which determines its resistance to shearing stresses.* It is a measure of the internal fluid friction which causes resistance to flow. It is primarily *due to cohesion and molecular momentum exchange between fluid layers* and as flow occurs, these effects appear as shearing stresses between the moving layers of fluid.

An *ideal fluid has no viscosity.* There is no fluid which can be classified as a perfectly ideal fluid. However, the fluids with very little viscosity are sometimes considered as ideal fluids.

Viscosity of fluids is due to *cohesion and interaction between particles.*

Refer to Fig. 21.85. When two layers of fluid, at a distance '*dy*' apart, move one over the other at different velocities, say u and $u + du$, the viscosity together with relative velocity causes a shear stress acting between the fluid layers. The top layer causes a shear stress on the adjacent lower layer while the lower layer causes a shear stress on the adjacent top layer. This shear stress is proportional to the rate of change of velocity with respect to y. It is denoted by τ (called Tau).

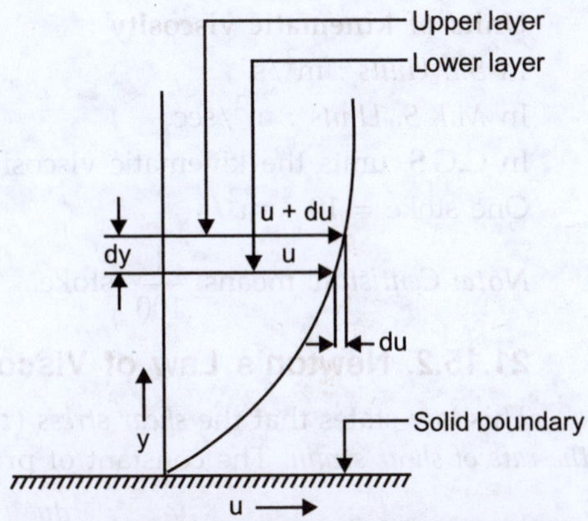

Fig. 21.85. Velocity variation near a solid boundary.

Mathematically $\tau \propto \dfrac{du}{dy}$

or, $$t = \mu \cdot \frac{du}{dy} \qquad ...(21.61)$$

where μ = Constant of proportionality and is known as *coefficient of dynamic viscosity* or *only viscosity, and*

$\dfrac{du}{dy}$ = Rate of shear stress or rate of shear deformation or velocity gradient.

From Fig. 21.85, we have

$$\mu = \frac{\pi}{\left[\dfrac{du}{dy}\right]} \qquad ...(21.62)$$

Thus viscosity may also be defined as the *shear stress required to produce unit rate of shear strain.*

Units of viscosity :

In *S.I. Units* : N.s/m^2

In *M.K.S. Units* : Kg$_f$. sec/m^2

$$\left[\therefore \mu = \frac{\text{force/area}}{(\text{length/time}) \times \dfrac{1}{\text{length}}} = \frac{\text{force/length}^2}{\dfrac{1}{\text{time}}} = \frac{\text{force} \times \text{time}}{(\text{length})^2}\right]$$

The unit of viscosity in C.G.S. is also called *poise* = $\dfrac{\text{dyne} - \text{sec.}}{\text{cm}^2}$.

One poise = $\dfrac{1}{10}$ N.s/m^2.

Note: The viscosity of water at 20°C is $\dfrac{1}{100}$ poise or one centipoise.

Kinematic viscosity : *Kinematic viscosity* is defined as the *ratio between the dynamic viscosity and density of fluid.* It is denoted by ν (called nu).

$$\text{Mathematically, } \nu = \frac{\text{Viscosity}}{\text{Density}} = \frac{\mu}{\rho} \qquad ...(21.63)$$

Units of kinematic viscosity :

In *S.I. Units :* m^2/s

In *M.K.S. Units :* $m^2/sec.$

In C.G.S. units the kinematic viscosity is also known as stoke (= $cm^2/sec.$)

One stoke = 10^{-4} m^2/s

Note: Centistoke means $\frac{1}{100}$ stoke.

21.15.2. Newton's Law of Viscosity

This law states that the *shear stress* (τ) *on a fluid element layer is directly proportional to the rate of shear strain.* The constant of proportionality is called the *co-efficient of viscosity.*

$$\text{Mathematically, } \tau = \mu\frac{du}{dy} \qquad \text{...(21.64)}$$

The fluids which follow this law are known as *Newtonian fluids.*

21.15.3. Viscosity Measurement Methods

The following methods are used to measure viscosity

1. Rotating cylinder method
2. Falling sphere method
3. Capillary tube method
4. Efflux viscometers.

} Mechanical methods

The devices used for measurement of viscosity are known as ***'Viscometers'***.

5. Electrical method.

21.15.3.1. Rotating cylinder method

In this method Newton's law of viscosity is used to measure the viscosity of a fluid. Fig. 21.86, shows a rotating cylinder viscometer. It consists of two concentric cylinders, the annular space between them is fitted with the liquid whose viscosity is to be determined. The outer cylinder is rotated at a constant angular velocity ω with respect to the inner stationary cylinder. The torque transmitted by the enclosed liquid to the stationary cylinder is measured by the torsional strain of the restraining spring attached to the top of the inner cylinder.

According to Newton's law of viscosity

$$\text{Shear stress } (\tau) = \mu \times \text{velocity gradient} \left(\frac{du}{dy}\right)$$

Since the annular space $t = (R_2 - R_1)$ is quite small, (where R_2 and R_1 are the radii of the outer and inner cylinders respectively), the velocity gradient,

$$\frac{du}{dy} = \frac{V}{t} = \frac{2\pi R_2 N}{60t}$$

(where, N = the rotational speed of the outer cylinder in r.p.m.)

$$\therefore \quad \text{Shear stress } (\tau) = \mu \times \frac{2\pi R_2 N}{60t}$$

$$\text{Viscous drag} = \text{Shear stress} \times \text{area}$$

$$= \mu \times \frac{2\pi R_2 N}{60t} \times 2\pi R_1 h, \text{ where } h = \text{height of liquid.}$$

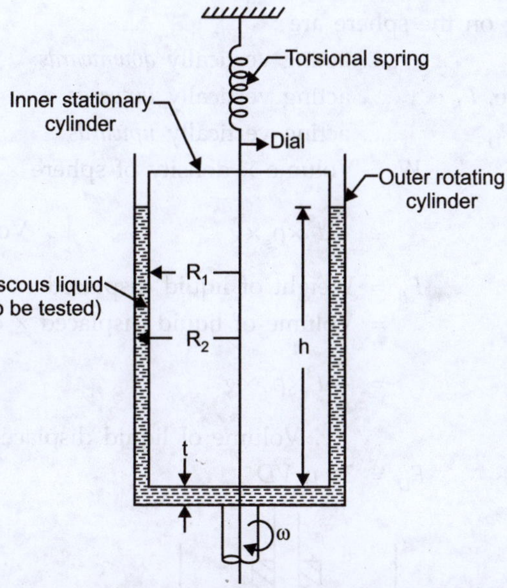

Fig. 21.86. Rotating cylinder viscometer.

Viscous torque = Viscous drag × radius

$$= \left(\mu \times \frac{2\pi R_2 N}{60t} \times 2\pi R_1 h\right) \times R_1 = \frac{\mu\pi^2 R_1^2 R_2 hN}{15t}$$

Viscous torque must equal to the torque T exerted by the torquemeter.

$$\therefore \quad T = \frac{\mu\pi^2 R_1^2 R_2 \cdot hN}{15t}$$

or, $$\mu = \frac{15\,Tt}{\pi^2 R_1^2 R_2 hN} \quad ...(21.65)$$

Thus a rotational type viscometer can be calibrated to directly give μ for given speed of rotation N.

21.15.3.2. Falling sphere method

Falling sphere method of measuring viscosity of a liquid is based on *Stokes' Law.*

Fig. 21.87, shows a falling sphere viscometer. A small spherical ball is released into the liquid to be tested and it accelerates under the gravitational force until it reaches a maximum/terminal velocity V when the *buoyant and viscous drag forces balance the gravity force.*

Let,
d = Diameter of the spherical ball,
l = Distance travelled by the sphere in viscous liquid,
t = Time taken by the sphere to cover distance l,
ρ_s = Density of sphere,
ρ_f = Density of fluid/liquid,
W = Weight of sphere,
F_B = Buoyant force, acting on the sphere
F_D = Drag force, and
μ = Dynamic viscosity of the liquid under test.

The forces acting on the sphere are :

1. Weight W ...... acting vertically *downwards.*
2. Buoyant force, F_B ...... acting vertically *upwards.*
3. Drag force, F_D ...... acting vertically *upwards*

$$W = \text{Volume} \times \text{density of sphere} \times g$$

$$= \frac{\pi}{6} d^3 \times \rho_s \times g \qquad \left(\because \text{Volume of sphere} = \frac{\pi}{6} d^3\right)$$

$$F_B = \text{Weight of liquid displaced}$$

$$= \text{Volume of liquid displaced} \times \text{density of liquid} \times g$$

$$= \frac{\pi}{6} d^3 \times \rho_f \times g$$

(∴ Volume of liquid displaced = Volume of sphere)

$$F_D = 3\pi\ \mu\ VD \qquad \text{...(Stokes, law)}$$

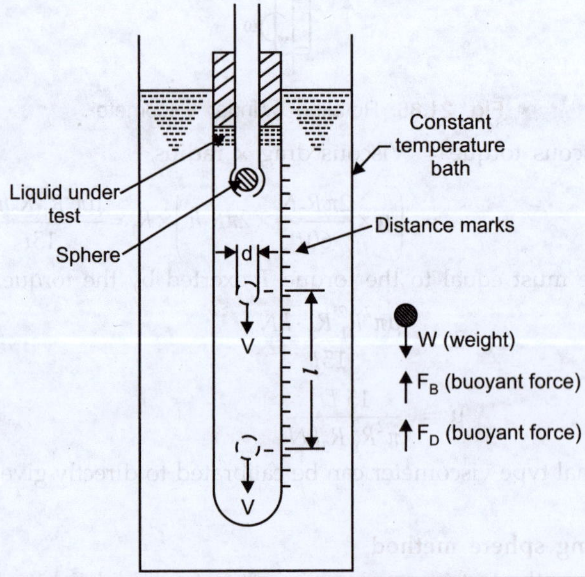

Fig. 21.87. Falling sphere viscometer.

For equilibrium

$$F_D + F_B = W$$

or,
$$F_D = W - F_B$$

$$3\pi\mu\ V_d = \frac{\pi}{6} d^3 \times \rho_s \times g - \frac{\pi}{6} d^3 \times \rho_f \times g \qquad \text{...(substituting the values)}$$

$$= \frac{\pi}{6} d^3 \times g(\rho_s - \rho_f)$$

or,
$$\mu = \frac{\pi}{6} \cdot \frac{d^3 \times g(\rho_s - \rho_f)}{3\pi V d}$$

or,
$$\mu = \frac{gd^2}{18V}(\rho_s - \rho_f) \qquad \text{...(21.66)}$$

From the above equation the value of dynamic viscosity μ can be determined.

Note: Stokes' law is essentially valid for Reynolds number below 0.1 where the wall has no effect on the terminal fall viscosity V.

21.15.3.3. Capillary tube method

This method makes use of Hagen-Poiseuille equation for laminar flow through circular tubes.

Fig. 21.88 shows a capillary tube viscometer. It consists of a tank in which the liquid whose viscosity is to be determined is filled. A capillary tube of diameter D and length L is attached horizontally very close to the bottom of tank. The tube is allowed to discharge freely into the atmosphere. The liquid is collected in a measuring tank for a given time. Then the rate of liquid collected in the tank per second is determined. The pressure head is measured at a point far away from the tank.

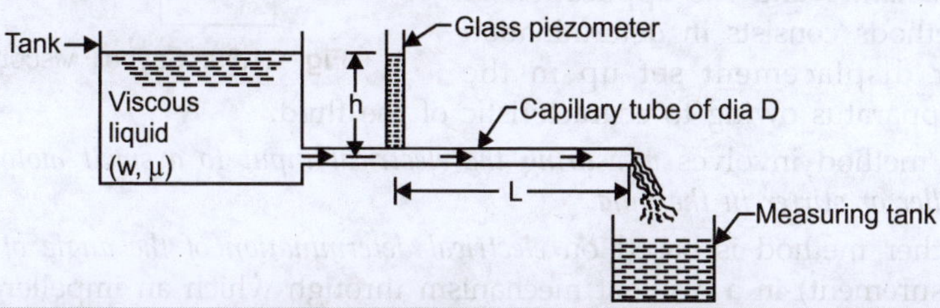

Fig. 21.88. Capillary tube viscometer.

Let, h = Difference of pressure head for length L

μ = Co-efficient of viscosity, and

w = Weight density of the liquid.

According to Hagen-Poiseuilli's equation, we have

$$h = \frac{32\,\mu\,\bar{u}L}{wD^2}$$

But, $$\bar{u} = \frac{Q}{\frac{\pi}{4}D^2}$$ (where, Q = Discharge of liquid through the tube)

$$\therefore \quad h = \frac{32\mu \times \frac{Q}{(\pi/4)\times D^2}\times L}{wD^2} = \frac{128\,\mu\,QL}{\pi w D^4}$$

or, $$\mu = \frac{\pi w h D^4}{128\,QL} \quad ...(21.67)$$

21.15.3.4. Efflux viscometers

In Efflux viscometers the viscosity is determined by noting the time of efflux, under specific conditions, for a fixed volume of fluid through a specific capillary or aperture (standardized nozzle or orifice).

The operation of *Saybolt, Redwood* and *Engler* is based on this principle.

Fig. 21.89 shows a **Saybolt viscometer**. It consists of a tank in the bottom of which is a short capillary tube. The tank is filled with the liquid whose viscosity is to be determined; a constant temperature bath surrounds this tank to ensure that the liquid remains at uniform constant temperature during the test run. The viscosity is determined by measuring the time required for 60 cm^3 of liquid at a known temperature to flow out of the reservoir through the tube. The initial level of the liquid in the reservoir is previously adjusted to a standard height. From the time measurement, the kinematic viscosity of the liquid can be determined by the use of empirical formula or the calibration chart.

- These viscometers (laboratory instruments) are widely employed in *petroleum and allied industries.*

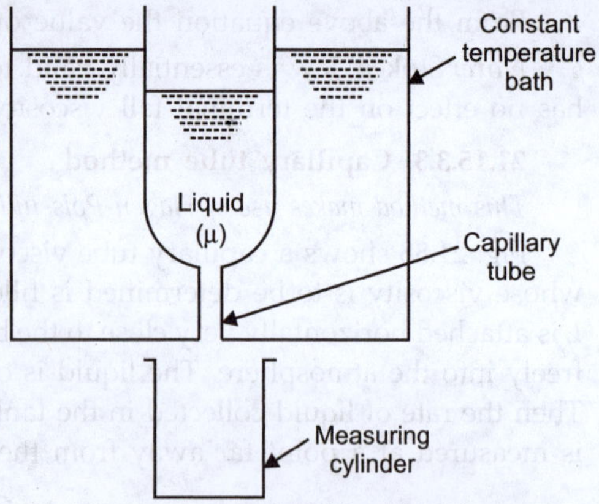

Fig. 21.89. Saybolt viscometer.

21.15.3.5. Electrical method

Viscosity measurement is essentially *mechanical in nature* and the application of electrical methods consists in determination of stress or displacement set up in the measuring apparatus owing to characteristic of the fluid.

- One method involves *measuring the electrical input to a small motor driving an impeller or stirrer in the fluid.*
- Another method is based on *electrical determination of the angle of lag* (torque measurement) in a resilient mechanism through which an impeller is driven.
- A further method utilizes *mangetostriction to produce longitudinal oscillations* in a steel rod carrying a diaphragm immersed in the liquid. Determination of the electrical loading on the exciting circuit provides a measure of viscosity.

21.16 GAS ANALYSERS–MEASUREMENT OF CO_2, O_2 AND CO

The gas analysis is carried out to determine the concentration of one or more components of a gas mixture. The common types of analysers used for the measurement of CO_2 and O_2 are described as follows :

21.16.1. Measurement of Carbon-dioxide (CO_2)

Since gases differ in their ability to conduct the heat, therefore, the *proportion of various constituents of flue gases can be determined by measuring their thermal conductivities.* If the thermal conductivity of air is considered unit of 0°C, the thermal conductivity of CO_2 = 0.585; H_2 = 7.35; N_2 = 1.015; O_2 = 1.007 respectively.

Fig. 21.90 shows the commonly used method for measuring CO_2 content in flue gases.

— It consists of hot wire thermal conductivity gas analysis cell which comprises of *two chambers*, each containing a wire filament as shown in Fig. 21.90. The gas is passed through one chamber whose conductivity is to be measured and other chamber is filled with a reference gas like air and sealed.

— The platinum wire elements are passed through the cells (as shown in Fig. 21.90) and form a Wheatstone bridge circuit. Under normal working conditions, these elements are heated by the bridge current.

— The temperature of the filament in the measuring chamber rises as the conductivity of the gases passing around it decreases, as the heat of the filament cannot be carried by the gases. Thus the temperature and hence the resistance of the two platinum wire elements changes and unbalances the bridge and provides an *unbalanced voltage which is proportional to the percentage of CO_2.* This unbalanced voltage can be used for measuring the percentage of CO_2 in flue gases.

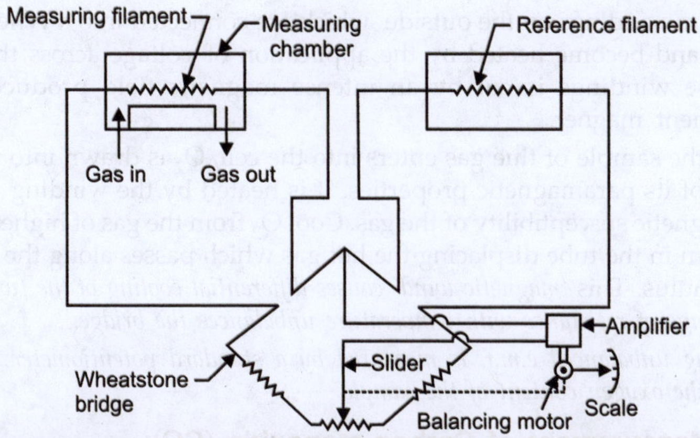

Fig. 21.90. Electrical circuit for the measurement of CO_2 content in the gases.

— To compensate the error which would arise from variation in temperature of the gas, another resistance is inserted in the circuit (not shown).

This type of instrument gives accurate values *if the proportion of H_2 present is small and constant*.

21.16.2. Measurement of Oxygen (O_2)

Although there are other methods to measure O_2, we shall discuss here only *"paramagnetic oxygen analyser"*. This analyser works on the principle of *paramagnetic properties of O_2*. Oxygen is the only paramagnetic component in the flue gases. It has very high magnetic permeance compared with other gases. The magnetic permeance of paramagnetic cases *decreases with an increase in temperature*. The gas molecules near the heated body in the magnetic field lose part of their magnetic properties and are pushed out of the magnetic field by cooler molecules. The temperature of the cooler molecules entered in the magnetic field where heated body is situated, is increased and in turn they are also pushed out of magnetic field by other molecules. In this way, a convection flow known as *'Magnetic Wind'* is created which cools the heated body. The magnetic wind increases with an increase in O_2 content in the flue gases and cooling of the heated body is intensified. *The temperature change of heated body alters its electric resistance which indicates the concentration of O_2 in the gas mixture*. Therefore, the O_2 content can be measured by measuring an electric resistance of the heated body which can be one of the arms of Wheatstone bridge.

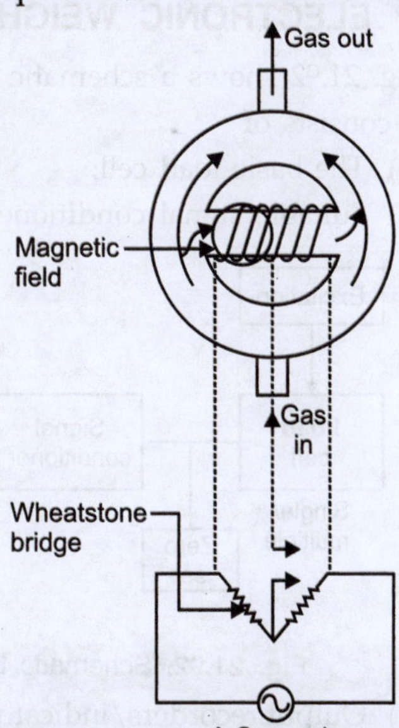

Fig. 21.91. Paramagnetic oxygen analyser.

Fig. 21.91 shows the electric circuit used for O_2 measurement using the magnetic wind type principle.

— The gas is aspirated into the measuring cell through an annulus with an horizontal by-pass the tube (as shown in Fig. 21.91). This by-pass tube has two identical adjacent platinum windings on the outside which are connected into a Wheatstone bridge circuit and become heated by the application of voltage across the bridge. One of these windings is cut by an intense magnetic field produced by a large permanent magnet.

— When the sample of flue gas enters into the cell, O_2 is drawn into the by-pass by virtue of its paramagnetic properties. It is heated by the winding which reduces the magnetic susceptibility of the gas. Cool O_2 from the gas of higher susceptibility is drawn in the tube displacing the hot gas which passes along the tube back into the annulus. This *'magnetic wind' causes differential cooling of the two windings and the change of resistance with temperature unbalances the bridge.*

The resulting unbalanced e.m.f. is measured by a standard potentiometer recorder and is proportional to the oxygen content of the sample.

21.16.3. Measurement of Carbon-monoxide (CO)

A carbon-monoxide analyser works on the principle of non-dispersive infra-red absorption and is suitable for all biatomic molecules. The principle of the apparatus is based on molecular specific absorption of bands of infra-red radiation. *The different radiation intensities, depending on the gas concentration are first converted in the receiver chamber into flow pulses and then into electrical signals.* The system first shoots beams of infra-red and visible light through the gases in the smoke stack. A molecule of a particular gas hungrily absorbs light energy at particular wavelength. By comparing the original intensities of the beam (I_0) with the amount of light that reaches the detector (I), the analyser determines the relative amounts of CO in the gases. The resulting signal is amplified and transmitted to the control room. The concentration of the gas is measured in terms of the ratio as log (I/I_0). Therefore, any light level change, darking of the window, scattering of particulates or water droplets in the gas stream, effects both I and I_0 equally and maintains the ratio unchanged.

21.17 ELECTRONIC WEIGHING SYSTEM

Fig. 21.92 shows a schematic block diagram of electronic weighing system :

It consists of :

(*i*) The basic load cell;

(*ii*) Suitable signal conditioners;

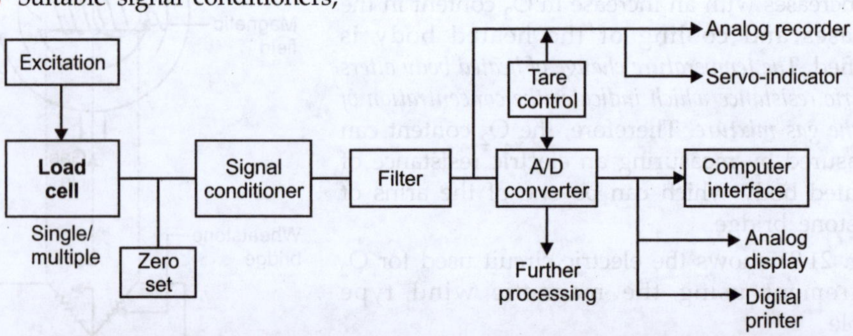

Fig. 21.92. Schematic block diagram of electronic weighing machine.

(*iii*) Output recorders/indicators giving both the analog and digital outputs for further processing.

— The signals from the load cells (which may be upto 4 or 8 cells) are added and amplified to give an output of 0 to 5 V or 4 to 20 mA for process instrumentation.

This is further converted into digital format with an A/D converter so that the output can be indicated or printed out in a digital format used for processing.

— Taring facility can be provided whenever necessary.

- The electronic system can be designed computer based with suitable software to include all the functions such as totalising, averaging, counting, processing or any other such control operation.

Applications. The electronic weighing systems find uses in :

— Platform weighers;
— Truck weighing;
— Crane weighing;
— Tank and hopper weighing;
— Weighing of vehicles in motion;
— Weigh feeters;
— General purpose commodity or product weighing;
— Bulk load sensing (as encountered in chemical plants, steel plants etc.)

Advantages : The main advantages of an electronic weighing system when compared with mechanical systems are :

1. High dependability and ruggedness.
2. The accuracy is quite good even under most adverse and hostile environmental conditions.
3. Small size and compactness.
4. Rapid weighing (useful for in-motion weighing), since response speed is high.
5. Both analog and digital output with print-out facility, remote indication, and parallel display.
6. Computer capability for processing.
7. Availability of tare facility.
8. Excellent flexibility for monitoring multiple loads.

21.18 MICROPROCESSOR-BASED INSTRUMENTATION

In microprocessor-based instruments, microprocessor (μp) forms one of the auxiliary functional elements of the instruments. In the word microprocessor *"micro"* refers to its microminiature size/dimensions, while *"processor"* signifies its vast potential to perform complex computations at fantastically high speeds, together with pre-programmed logic/software which enhances significantly the capabilities and effectiveness of the instruments.

These days, highly versatile microprocessor chips designed and developed on Very Large Scale Integration (VLSI *i.e.,* 10^7 circuits/cm^2) have provided very convenient means of adding Artificial Intelligence (AI) in many of our every day facilities, *examples* being 'automatic washing machines', 'burglar alarms' automatic teller machines (ATMs) etc.

Microprocessor-based instruments are commonly termed as *smart or intelligent instruments*. The micro-miniaturisation has resulted in offering these instruments of the size of a pocket calculator.

Microprocessor, by itself, is an operational computer. In order to shape it in the form of a *digital computer*, it is incorporated with additional circuits for memory and input/output devices.

A computer-based measurement system can process all the inputs and present the data in real time. Fig. 21.93 shows a *typical digital computer-based measurement system.*

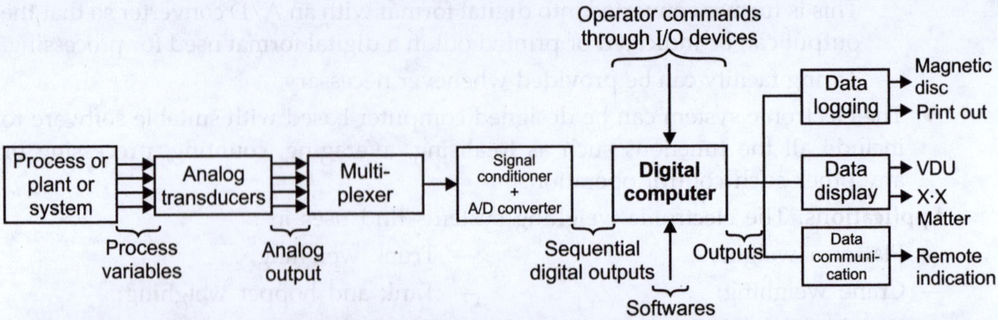

Fig. 21.93. A typical digital computer-based measurement system.

— A 'sequential list of instructions' (termed as a computer program) is fed to a digital computer for suitable processing/manipulation of the data.

— Invariably, Artificial Intelligent (AI) in the form of Artificial Neutral Networks (ANNs) or Fuzzy Logic (FL) may be incorporated. By doing so, the decision-making tasks in various processes are usually accomplished by the computer itself and not by any human operator.

Advantages and disadvantages of computer-based instrumentation systems :

The computer-based instrumentation systems entail the following *advantages* and *disadvantages*.

Advantages :

1. Lower costs, higher accuracy and more flexibility.
2. Can be battery operated, since they have low consumption of the order of 40 mW.
3. Can be adjusted/programmed with a remote control.
4. Do not require skilled operation due to the incorporation of user-friendly subroutines.
5. Suitably programmed to automatically carry out the mundane tasks of drifts correction, noise reduction/elimination, non-linearity correction, range and span adjustments, etc.
6. These instruments have compact, rugged and reliable signal conditional conditioning and display. They can perform in tough, industrial, consumer, military, automobile and environmental conditions.
7. Owing to their compact size, they are mostly of portable type.
8. The measurement, processing and display of the data of the process variables are done in real time *i.e.*, it is one-line type process information in the desired form and units.

Disadvantages :

1. These instruments are prone to virus problems and consequently they suddenly become sick and inoperative.
2. They invariably need the processing data in the digital form.
3. They cannot modify the programmes themselves.
4. The commercial computer softwares become absolute very fast and periodically updating the same involves extra expenditure.

HIGHLIGHTS

Pressure Measurement

1. *Pressure* is defined as a force per unit area.
2. *Gauges* record the pressure above or below the local atmospheric pressure, since they measure the difference in pressure of the liquid to which they are connected and that of surrounding air.
3. *Manometers* are defined as the devices used for measuring the pressure at a point in a fluid by balancing the column of fluid by the same or another column of liquid. They are treated as *standards* for pressure and differential pressure measurements because they are virtually free from any friction or hysteresis effects.
4. Some important *mechanical gauges* are :
   (*i*) Bourdon tube pressure gauge.
   (*ii*) Diaphragm gauge
   (*iii*) Vacuum gauge.
5. For the measurement of *very high pressures* (above 70000 MPa), the "*electrical gauges*" based on the principle of change of resistance with change of pressure are used.
6. The two basic methods of measurement of low pressure are :
   (*i*) *Direct methods.* The following devices belong to this category :
   — Spiral bourdon gauges;
   — Flat and corrugated diaphragms;
   — Capsules;
   — Manometers.
   (*ii*) *Indirect methods.* Devices included in this category are :
   — Mcleod gauge;
   — Thermal conductivity gauges;
   — Ionization gauges;
   — Radioactive vacuum meters.
7. The following are the laboratory standards for calibrating pressure measuring instruments :
   (*i*) Manometers.
   (*ii*) Dead weight testers.

Temperature measurement

8. The *temperature* is a thermal state of a body which distinguishes a hot body from a cold body.
9. *Electrical sensors* are widely used particularly when automatic or remote recording is desired.
10. In a *bimetallic thermometer* differential expansion of bimetallic strips is used to indicate the temperature.
11. Gas filled thermometers are generally used for pressures below 35 bar.

12. The sources of static errors in filled systems are :
    (*i*) The ambient temperature effect.
    (*ii*) The heat effect.
    (*iii*) The barometric effect.
    (*iv*) The immersion effect.
13. A *thermistor* is a thermally sensitive variable resistor made of a ceramic-like semiconducting material.
14. A *thermocouple* in its simplest form, consists of two dissimilar metals or alloys which develop e.m.f. when the reference and measuring junctions are at different temperatures.
15. A device which measures the total intensity of radiation emitted from a body is called total *radiation pyrometer*.
16. An *optical pyrometer* works on the principle that matters glow above 480°C and the colour of visible radiation is proportional to the temperature of the glowing matter.

Flow measurement

17. The flow measurement devices are classified as follows :
    (*i*) Rate meters
    (*ii*) Quantity meters.
18. A *venturimeter* is an instrument used to measure the rate of discharge in a pipeline and is often fixed permanently at different sections of the pipeline to know the discharge there.
19. A *Pitot tube* consists of a glass tube in the form of a 90° bend of short length open at both its ends. In a pitot-static tube both the static pressure as well as stagnation pressure can be measured.
20. A *rotameter* is a constant-pressure drop, variable area flow meter.
21. *Electromagnetic flow meters* are particularly suitable for the flow measurements of slurries, sludge and any electrically conducting fluid.
22. *Hot wire anemometers* are hot wire resistance transducers which are used for measurement of flow rates of fluids.
23. The working principles of *swirl meter* depends upon votex precession.
24. The working principle of *ultrasonic flow meters* is based on an apparent change in the velocity of propagation of sound waves in a fluid with a change in the velocity of the fluid flow.
25. *Lobed impeller flow meter* is primarily used for metering gases.

Measurement of displacement, velocity, force, torque etc.

26. A *tachometer* is an instrument which either continuously indicates the value of rotary speed or continuously displays a reading of average speed over rapidly operated short intervals of time.
27. The acceleration of a moving body is generally measured by means of sensors called *accelerometers*.
28. An unknown force may measured by : (*i*) Scales and balances; (*ii*) Springs; (*iii*) Proving rings; (*iv*) Load cells.

29. *Load cells* are elastic devices that can be used for measurement of force through indirect methods *i.e.,* through the use of secondary transducers.

30. Torque may be measured by the following methods :

(*i*) Gravity balance method (*ii*) Mechanical torsion meter

(*iii*) Optical torsion meter (*iv*) Electrical torsion meter.

(*v*) Strain-gauge torsion meter.

31. A *dynamometer* is a device used to measure the torque being exerted along a rotating shaft so as to determine the shaft power input or output of power-generating, transmitting and absorbing machinery. The dynamometers are classified as :

(*i*) Absorption dynamometers. (*ii*) Transmission dynamometers.

(*iii*) Driving dynamometers.

OBJECTIVE TYPE QUESTIONS

Choose the Correct Answer :

Pressure measurement

1. Bellows type of gauges have the advantage that

(*a*) they can be used for low, medium and high pressures

(*b*) they can be used for measurement of dynamic pressures

(*c*) they do not need any temperature compensating devices

(*d*) drift and hysteresis allow their use in applications having ±2% of full span accuracy.

2. The disadvantages of strain gauge pressure transducers are:

(*a*) They have a poor frequency response.

(*b*) They do not have a continuous resolution.

(*c*) They cannot be powered by a.c. sources.

(*d*) They have a low output.

3. Which of the following types of Bourdon tube shape has a small tip travel and necessitates amplification?

(*a*) C-type (*b*) Spiral

(*c*) Helical shaped (*d*) All of these.

4. Manometers measure unknown pressure by

(*a*) measuring liquid levels

(*b*) measuring height of liquid columns

(*c*) balancing the unknown force produced by pressure against a known force

(*d*) noting the deflection of a pointer.

5. In a manometer, pressure sensitivity is

(*a*) $1/g\,\rho_m$ (*b*) $g\,\rho_m$

(*c*) $g\,\rho_m\,\rho_f$ (*d*) $1/g\,\rho_m\,\rho_f$

where g = gravitational force and ρ_m and ρ_f are respectively specific gravities of manometric fluid and the transmitting fluid.

6. In thermal conductivity gauges, the major source of error is heat lost on account of

(*a*) radiation (*b*) conduction

(*c*) convection (*d*) both conduction and radiation.

7. The desirable property of a manometric fluid is
   (a) high viscosity (b) high co-efficient of thermal expansion
   (c) low vapour pressure (d) corrosiveness and sickness.
8. The advantages of manometers are their
   (a) higher accuracy, repeatability, wide range of filling liquids and their ability to work as primary standards
   (b) fragility and portability
   (c) ability to work in hazardous environments
   (d) ability to work under all kinds of atmospheric conditions and at all altitudes.
9. Piezometer is used for measuring :
   (a) gauge pressure (b) vacuum
   (c) absolute pressure
   (d) pressure difference between two fluids
10. A U-tube differential manometer is used inverted when
    (a) pressure difference is small (b) pressure difference is large
    (c) cannot be used inverted (d) any of these.
11. Capacitive type pressure transducers have
    (a) good frequency response, suitability for both dynamic and static measurements and continuous resolution
    (b) large temperature range
    (c) good thermal stability
    (d) high input impedance.
12. A typical piezoelectric pressure transducer has a range of
    (a) 0–10 kPa (b) 0–100 MPa
    (c) 0–700 kPa (d) 0–20 MPa.
13. In thermocouple gauges, the temperature of heated strip lies between
    (a) 0°C–100°C (b) 50°C–400°C
    (c) 200°C–1000°C (d) 10°C–50°C.
14. In order to avoid errors on account of changes in ambient temperature Pirani gauges
    (a) are enclosed in temperature controlled chambers
    (b) use one active and one dummy element
    (c) use bridge measuring circuits
    (d) none of the above.
15. Vacuum pressure is
    (a) equal to gauge pressure (b) equal to atmospheric pressure
    (c) lower than atmospheric pressure (d) equal to absolute pressure.
16. Velocity pressure is difference of
    (a) dynamic pressure and static pressure
    (b) stagnation pressure and static pressure
    (c) impact pressure and dynamic pressure
    (d) impact pressure and stagnation pressure.
17. Bridman gauges are used for measurement of
    (a) vacuum (b) medium pressure
    (c) high pressures (d) very high pressures.
18. Bridman gauges use
    (a) Monel and stainless steel (b) Manganin and gold chrome
    (c) Phosphor bronze (d) Beryllium copper.

19. A well-type manometer is used in preference to simple U-tube manometer to obtain

(*a*) better accuracy (*b*) better precision

(*c*) a constant zero (*d*) higher sensitivity.

20. Which of the following cannot be used for negative pressures?

(*a*) Piezometer (*b*) Pirani gauge

(*c*) U-tube manometer (*d*) Bourdon gauge.

21. Well type of manometers have

(*a*) poor accuracy (*b*) good accuracy

(*c*) low sensitivity

(*d*) high sensitivity as compared to U-tube manometers having two legs of the same area of cross-section.

22. In barometers the error caused by mercury vapour pressures is of the order of

(*a*) 1 kN/m^2 (*b*) 1 N/m^2

(*c*) 0.1 N/m^2 (*d*) 10 N/m^2.

23. The sensitivity of an ionization gauge is defined as

(*a*) I_i/pI_e (*b*) p/I_iI_e

(*c*) I_e/pI_i (*d*) none of the above.

where I_i = ion current, I_e = electron current and p = gas pressure.

24. The accuracy of dead weight testers is affected by

(*a*) friction force between the piston and the cylinder

(*b*) uncertainty of the value of effective area of piston

(*c*) uncertainty of the value of gravitational constant

(*d*) all of the above.

25. Signal conditioning is carried out by the capillary tubes which convert gas pressure into a mercury height. This statement pertains to

(*a*) bourdon tube pressure gauge (*b*) pirani gauge

(*c*) Mcleod gauge (*d*) diaphragm pressure transducer.

26. The angular error in a Bourdon tube refers to

(*a*) pointer or hand becomes loose on the spindle

(*b*) gauge tends to give a progressively higher or lower reading

(*c*) gauge reads correctly both at maximum and minimum readings but gives an inaccurate reading at the mid point

(*d*) any of these.

27. A capillary tube when used for damping of pressure signals is called a

(*a*) low pass filter (*b*) high pass filter

(*c*) acoustical filter (*d*) digital filter.

28. McLeod gauge

(*a*) can be used for pressure below 0.1 × 10^{-3} torr.

(*b*) gives continuous output.

(*c*) is sensitive to condensed vapours that may be present in the sample of the gas whose pressure is being measured.

(*d*) cannot be used as standard for vacuum measurements.

29. Inductive pressure transducers

(*a*) have poor dynamic response (*b*) have high sensitivity

(*c*) are sensitive to vibrations (*d*) none of the above.

30. Pressure transducers using LVDTs have

(*a*) poor dynamic response (*b*) linear response over a large range
(*c*) low sensitivity (*d*) all of the above.

Flow measurement

31. Which of the following meter is used for measuring flow of clean fluids only?

(*a*) Ultrasonic flow meter (*b*) Turbine flow meter
(*c*) Laser doppler anemometer (*d*) Hot wire anemometer

32. Which of the following flow meters is capable of measuring the rate of flow as well as totalized flow?

(*a*) Nutating disc flow meter (*b*) Electromagnetic flow meter
(*c*) Orifice meter (*d*) Lobed impeller flow meter

33. Annubar tubes can be used for measurement of flow rates of

(*a*) liquids only (*b*) gases only
(*c*) liquids and gases (*d*) liquids, gases and steam.

34. The meter which is suitable for flow totalization is

(*a*) turbine meter (*b*) venturimeter
(*c*) ultrasonic flow meters (*d*) orifice plate.

35. The fluid flow between the electrodes of an electromagnetic flow meter generates an e.m.f. which is a function of

(*a*) dynamic pressure (*b*) discharge
(*c*) flow velocity (*d*) all of these.

36. A quantitative estimate of density changes can be made by

(*a*) interferometer (*b*) shadograph
(*c*) schlieren system (*d*) any of these.

37. The size of a venturimeter is expressed as 300 × 150 mm. It means that

(*a*) the diameter of the upstream pipe is 300 mm and that of downstream pipe 150 mm
(*b*) the diameter of the pipe is 300 mm and that of throat is 150 mm
(*c*) the diameter of pipe is 150 mm and that of throat is 300 mm
(*d*) none of the above.

38. Venturimeters have

(*a*) high head loss, low co-efficient of discharge and small size
(*b*) low head loss, low co-efficient of discharge and small size
(*c*) low head loss, low co-efficient of discharge and large size
(*d*) low head loss, high co-efficient of discharge and large size.

39. In an ultrasonic flow meter, the error caused by velocity of sound propagation in medium can be eliminated by

(*a*) measuring transit time
(*b*) measuring phase shift
(*c*) measuring difference of transit times in the direction of flow and against the direction of flow
(*d*) measuring difference of frequency in the direction of flow and against the direction of flow.

40. Electromagnetic flow meters

(*a*) produce no head loss, can be used for measurement of flow of corrosive liquids and cannot be used for measurement of bidirectional flow.
(*b*) produce large heads loss, cannot be used for measurement of flow of corrosive liquids and cannot be used for measurement of bidirectional flow

(*c*) produce no head loss, cannot be used for measurement of flow of corrosive liquids and can be used for measurement of bidirectional flow.

(*d*) produce no head loss, can be used for measurement of flow of corrosive liquids and can be used for measurement of bidirectional flow.

41. A.C. excitation is used in electromagnetic flow meters because

(*a*) there is polarization, there are thermocouple effects and there is distortion of fluid velocity profile

(*b*) there is no polarization, there are no thermocouple effects and there is no distortion of fluid velocity profile

(*c*) there is polarization, there are no thermocouple effects and there is no distortion of fluid velocity profile

(*d*) there is no polarization, there are thermocouple effects and there is no distortion of fluid velocity profile.

42. A pitot-static tube measures

(*a*) static pressure (*b*) dynamic pressure

(*c*) total pressure

(*d*) difference between total and static pressure.

43. Which of the following instruments is a rate meter?

(*a*) Venturimeter (*b*) Hot wire anemometer

(*c*) Nutating disk meter (*d*) Current meter.

44. A flow meter that measures flow rates which are independent of density is

(*a*) rotameter (*b*) electromagnetic flow meter

(*c*) venturimeter (*d*) orifice meter.

45. Rotameter is a

(*a*) drag force flow meter (*b*) variable area flow meter

(*c*) variable head flow meter (*d*) rotating propeller type flow meter.

Temperature measurement

46. Which of the following units are employed where it is not possible to have a physical contact with the system whose temperature is to be measured?

(*a*) Thermocouples (*b*) Resistance thermometers

(*c*) Radiation pyrometers (*d*) Any of these.

47. A parallel combination of thermocouples is used for the measurement of

(*a*) small temperature differences between the two junctions

(*b*) large temperature differences between the two junctions

(*c*) average temperature of a number of points

(*d*) all of these.

48. Thermopile is

(*a*) a combination of a number of thermocouples connected in series

(*b*) a combination of a number of thermocouples connected in parallel

(*c*) a combination of a number of thermocouples some of which are connected in series and some in parallel

(*d*) none of the above.

49. Reference junction compensation in thermocouples can be provided through use of

(*a*) hardware only (*b*) software only

(*c*) both hardware and software (*d*) none of these.

50. In pressure thermometers when mercury is used, the bulb and capillary should be made of

(*a*) copper (*b*) alloys of copper
(*c*) stainless steel (*d*) none of these.

51. Partial immersion liquid-in-glass thermometers give errors because

(*a*) thermometer is not fully immersed
(*b*) the temperature of the emergent portion is higher than that of the immersed portion
(*c*) the temperature of the emergent portion is lower than that of the immersed portion
(*d*) the temperature of the emergent portion is different than that of the immersed portion.

52. Recording is not possible with

(*a*) liquid-in-glass thermometers (*b*) thermocouples
(*c*) filled in system thermometers (*d*) all of these.

53. The emf of a thermocouple with junctions at t_1 and t_3 is the algebraic sum of the emf's of two couples of the same material with junctions at t_1 and t_2, and t_2 and t_3 respectively. This statement is known as

(*a*) law of homogeneous circuit
(*b*) law of successive or intermediate temperatures
(*c*) law of intermediate metals
(*d*) none of these.

54. The advantages of thermistor are :

(*a*) Resistance of thermistor is high and therefore error due to resistance of leads is negligible.
(*b*) Errors due to self-heating are small.
(*c*) Its sensitivity is about 10^6 as high as that of platinum resistance thermometers.
(*d*) all of the above.

55. Law of intermediate metals in thermocouples allows them to

(*a*) use reference junction compensation
(*b*) use meters for measurement of emf without disturbing the circuit conditions
(*c*) use extension wires of materials other than the one used for making thermocouples
(*d*) both (*b*) and (*c*).

56. If the temperature of a radiating body is 1500°C, the wavelength at which maximum radiant energy occurs is

(*a*) 2.8 μm (*b*) 1.63 μm
(*c*) 1 μm (*d*) 20.8 μm.

57. Radiation pyrometers are used in the temperature range of :

(*a*) 0–500°C (*b*) 500–1000°C
(*c*) –250 to 500°C (*d*) 1200–2500°C.

58. Which metal/non-metal has the highest temperature range?

(*a*) Semiconductors (*b*) Nickel
(*c*) Copper (*d*) Platinum.

59. Which arrangement has the manual null-balance range?

(*a*) Optical pyrometer (*b*) Resistance thermometer
(*c*) Liquid-in-glass thermometer (*d*) None of these.

60. For the same temperature range, the volume of bulb of a gas-filled thermometer is

(*a*) greater than that of a liquid-filled or a vapour-filled thermometer
(*b*) smaller than that of a liquid-filled or a vapour filled thermometer

(*c*) greater than that of a liquid-filled thermometer but lesser than that of vapour-filled thermometer

(*d*) lesser than that of a liquid-filled thermometer but greater than that of vapour-filled thermometer.

61. In filled systems used for temperature measurement, full compensation is used for compensating the effect of

(*a*) changes in ambient temperature, head and barometric pressure

(*b*) changes in ambient temperature where the receiving element is located

(*c*) changes in ambient temperature where both the receiving element and the capillary tube are located

(*d*) all of the above.

62. The electrical output from a thermocouple circuit is detected by using

(*a*) Wheatstone bridge (*b*) current sensitive device

(*c*) voltage-balancing circuit (*d*) none of these.

63. Which arrangement has the capability to measure the temperature of an object which may be either stationary or moving?

(*a*) Thermocouples. (*b*) Thermistors.

(*c*) Total radiation pyrometer. (*d*) Filled-in-system thermometers.

64. Self-compensating capillary tube is used in the filled-in-system thermometer to eliminate or minimise

(*a*) elevation effect (*b*) immersion effect

(*c*) barometric effect (*d*) temperature effect.

65. Mercury in used in liquid-filled systems as it gives

(*a*) wide temperature range

(*b*) high sensitivity

(*c*) wide temperature range and approximately linear scale

(*d*) wide temperature range and high sensitivity.

Measurement of displacement, velocity, force etc.

66. Elastic elements used for measurement of force give

(*a*) high sensitivity and slow response if they are stiff

(*b*) low sensitivity and fast response if they are stiff

(*c*) low sensitivity and slow response if they are stiff

(*d*) none of the above.

67. Piezoelectric type of load cells can be used for measurement of

(*a*) dynamic forces only.

(*b*) dynamic forces and static forces provided the load cells have a small time constant.

(*c*) dynamic forces and static forces provided that the load cells have a large time constant.

(*d*) none of the above.

68. The average speed measurements are given by

(*a*) centrifugal tachometer (*b*) drag cup tachometer

(*c*) revolution counter and a timer (*d*) stroboscope.

69. Torque of rotating shafts is measured by mounting a strain gauge bridge on the shaft. The supply to the bridge circuit is given to it and the output from it taken from by using

(*a*) only slip rings (*b*) only rotary transformers

(*c*) slip rings, rotary transformers, telemetry equipment

(*d*) none of the above.

70. A sink is

(*a*) a transmission dynamometer (*b*) a driving dynamometer
(*c*) an absorption dynamometer (*d*) any dynamometer.

71. The stroboscopic method of speed measurement has the advantage that

(*a*) the method is simple and straight forward
(*b*) the multiples of angular speed can be measured
(*c*) a stationary image can be observed
(*d*) there is no need of any physical contact between the instrument and the rotating shaft.

72. The seismic mass transducer for the measurement of acceleration and vibration is a

(*a*) zero order instrument (*b*) first order instrument
(*c*) second order instrument (*d*) none of these.

73. Hydraulic load cells are available which have a maximum capacity of

(*a*) 10 kN (*b*) 50 kN
(*c*) 10 MN (*d*) 50 MN.

74. Pendulum type of scales used for measurement of force work on the principle of

(*a*) comparing the unknown force with a known force
(*b*) converting the unknown force to power and comparing it with known power
(*c*) converting the unknown force to torque and comparing it with a known torque.
(*d*) all of the above.

75. Tachometers are used to measure

(*a*) displacement (*b*) angular velocity
(*c*) vibration (*d*) time.

76. The speed of a sealed compressor unit can be measured by a

(*a*) stroboscope (*b*) vibrating reed tachometer
(*c*) capacitive pick-up tachometer (*d*) tachoscope.

ANSWERS

1. (*a*)	**2.** (*d*)	**3.** (*a*)	**4.** (*c*)	**5.** (*a*)	**6.** (*a*)	**7.** (*c*)
8. (*a*)	**9.** (*a*)	**10.** (*a*)	**11.** (*a*)	**12.** (*d*)	**13.** (*b*)	**14.** (*b*)
15. (*c*)	**16.** (*b*)	**17.** (*d*)	**18.** (*b*)	**19.** (*d*)	**20.** (*a*)	**21.** (*d*)
22. (*b*)	**23.** (*c*)	**24.** (*d*)	**25.** (*c*)	**26.** (*c*)	**27.** (*c*)	**28.** (*c*)
29. (*b*)	**30.** (*a*)	**31.** (*d*)	**32.** (*d*)	**33.** (*d*)	**34.** (*a*)	**35.** (*c*)
36. (*a*)	**37.** (*b*)	**38.** (*d*)	**39.** (*d*)	**40.** (*d*)	**41.** (*b*)	**42.** (*d*)
43. (*a*)	**44.** (*b*)	**45.** (*b*)	**46.** (*c*)	**47.** (*c*)	**48.** (*a*)	**49.** (*c*)
50. (*c*)	**51.** (*d*)	**52.** (*a*)	**53.** (*b*)	**54.** (*d*)	**55.** (*d*)	**56.** (*b*)
57. (*d*)	**58.** (*d*)	**59.** (*a*)	**60.** (*a*)	**61.** (*c*)	**62.** (*c*)	**63.** (*c*)
64. (*d*)	**65.** (*c*)	**66.** (*b*)	**67.** (*c*)	**68.** (*c*)	**69.** (*c*)	**70.** (*c*)
71. (*d*)	**72.** (*c*)	**73.** (*d*)	**74.** (*c*)	**75.** (*b*)	**76.** (*b*)	

THEORETICAL QUESTIONS

Pressure measurement

1. Define the following terms :

(*i*) Pressure (*ii*) Atmospheric pressure

(*iii*) Gauge pressure (*iv*) Absolute pressure
(*v*) Static pressure (*vi*) Total or stagnation pressure.

2. What are the units of pressure?
3. How are pressure measurement instruments classified?
4. What are "manometers"? How are these classified?
5. Why are manometers treated as standards for pressure and differential pressure measurements?
6. How does a differential manometer differ from a simple manometer?
7. List the advantages and limitations of manometers.
8. What are the desirable characteristics of a manometer liquid?
9. Explain briefly the following manometric liquids :
   (*i*) Mercury; (*ii*) Water; (*iii*) Alcohol and kerosene.
10. On what principle do the "Bell gauges" operate? Name the different types of Bell gauges used in practice.
11. Explain briefly any two of the following mechanical gauges :
    (*i*) Bourdon tube pressure gauge (*ii*) Diaphragm gauge
    (*iii*) Vacuum gauge.
12. Explain briefly the following types of pressure elements :
    (*i*) Bourdon tube (*ii*) Diaphragm
    (*iii*) Bellows.
13. State the advantages and disadvantages of "Bourdon tube element."
14. What are the advantages and disadvantages of "Diaphragm elements"?
15. List the advantages and disadvantages of bellow gauges/elements.
16. Explain briefly any three of the following pressure transducers :
    (*i*) Resistance type pressure transducers; (*ii*) Pressure voltage type;
    (*iii*) Capacitance type; (*iv*) Piezoelectric type;
    (*v*) Electromagnetic type.
17. How are very high pressures measured? Explain briefly with a neat and clean sketch the construction and working of a Baridgman gauge used for measuring high pressures.
18. What do you mean by low pressures or vacuums?
19. Draw diagrams to show how LVDTs can be used with bellows, elements and Bourdon tubes for measurement of pressure. Give their advantages and disadvantages.
20. Explain how capacitive type of transducers can be used for pressure measurement.
21. Describe the applications of piezoelectric transducers for measurement of pressure. List their advantages and disadvantages.
22. Describe the construction, working and theory of McLeod gauge for measurement of vacuum. List its advantages and disadvantages.
23. Explain briefly the construction and working of a thermocouple vacuum gauge. State its advantages and disadvantages.
24. Describe briefly the construction and working of a Pirani gauge. Give its advantages and disadvantages.
25. Give the construction and working of a hot-filament ionization gauge. List its advantages and disadvantages.
26. (*a*) Explain how calibration of pressure measuring equipment is carried out.
    (*b*) Describe the construction and working of a "Dead Weight Tester". State the factors which affect the accuracy of dead weight testers.

Temperature measurement

27. Define the term "temperature".
28. How are temperature measuring instruments classified?
29. Explain briefly the following expansion thermometers :
    (*i*) Liquid-in-glass thermometers.
    (*ii*) Bimetallic thermometers.
30. State the advantages, disadvantages and applications of bimetallic thermometers.
31. Discuss briefly any two of the following pressure thermometers :
    (*i*) Vapour pressure thermometer.
    (*ii*) Liquid-filled thermometer.
    (*iii*) Gas-filled thermometer.
32. State the advantages and limitations of filled-systems.
33. Explain briefly the following sources of static errors in filled systems :
    (*i*) The ambient temperature effect.
    (*ii*) The head effect.
    (*iii*) The barometric effect.
    (*iv*) The immersion effect.
34. Write a short note on "Response speed of filled-systems".
35. Explain briefly the construction and working of a resistance thermometer.
36. What are the advantages and disadvantages of resistance thermometers?
37. What are thermistors? What are their advantages?
38. "Give a comparison between "thermistors" and "metal resistors".
39. List the factors which should be considered while selecting resistance thermometers.
40. What is a "Seebeck effect"?
41. Explain with a neat diagram the construction and working of a thermoelectric pyrometer.
42. What are the desirable properties of thermoelectric materials?
43. State the advantages and disadvantages of thermocouples.
44. Explain briefly the working of a radiation pyrometer.
45. List the advantages and disadvantages of radiation pyrometers.
46. Describe with a neat sketch the working of an optical pyrometer.
47. What are the advantages and disadvantages of an optical pyrometer?
48. State the precautions which should be taken while making temperature measurements.
49. Explain the factors which influence the response of a temperature sensing device.

Flow measurement

50. List the applications of flow measurements.
51. What are the different factors which influence the choice of method used for the measurement of flow?
52. How are the flow measurements classified?
53. What is the difference between the "Rate meters" and "Quantity meters"?
54. What is the significance of term "inferential" as applied to flow meters?
55. Certain meters are known as variable head meters. Explain clearly what is meant by the designation variable head.
56. Describe briefly the following :
    (*i*) Rotameter; (*ii*) Elbowmeter.
57. Explain with a neat sketch the working of any electromagnetic flow meter.
58. What are hot wire anemometers?

Measurement of displacement, velocity, force etc.

59. How can linear displacement be measured? Explain briefly.
60. Explain briefly how small displacement can be measured by a linear or differential transformers.
61. Describe briefly with a neat sketch how angular displacement is measured by a variable reluctance transducer.
62. Name the transducers which are used for the measurement of linear velocity.
63. Explain with a neat diagram the method of measuring linear velocity with the use of moving magnet type transducer.
64. How is angular velocity measured?
65. What is a tachometer?
66. How are tachometers classified?
67. Why are electrical tachometers preferred to mechanical tachometers?
68. Explain briefly any two of the following electrical tachometers :
    (*i*) D.C. tachometer generator (*ii*) A.C. tachometer generator
    (*iii*) Photoelectric tachometer.
69. Give the description of an eddy current tachometer, with a neat sketch.
70. Describe briefly a capacitive pick-up tachometer.
71. What are accelerometers?
72. Explain briefly the following :
    (*i*) Piezoelectric accelerometer (*ii*) Seismic accelerometer.
73. How is an unknown force measured?
74. What is a load cell?
75. Describe briefly the construction and working of strain gauge load cell. State its field of application as well.
76. What is the principle of working of a piezoelectric load cell?
77. What is 'torque'? How can it be measured?
78. Explain briefly any two of the following torque measurement methods :
    (*i*) Gravity balance method. (*ii*) Mechanical torsion meter.
    (*iii*) Electrical torsion meter.
79. Describe briefly how a strain gauge torsion meter is used to measure the torque.
80. What is a dynamometer?
81. How are dynamometers classified?
82. Explain briefly the following :
    (*i*) Prony brake dynamometer. (*ii*) Rope brake dynamometer.
83. Describe briefly any two of the following dynamometers :
    (*i*) Hydraulic dynamometer. (*ii*) Eddy current dynamometer.
    (*iii*) Belt transmission dynamometer.
84. Explain with a neat diagram the construction and working of a D.C. dynamometer.

CHAPTER

22 Microprocessors and Microcontrollers

22.1 Computers-Brief description: History and development of computers – Definition of a computer – Characteristics of a computer – Classification of computers – Analog computers – Digital computers – Differences between analog and digital computers – Block diagram of a digital computer – Rating of chips – Computer peripherals – Storage devices – Hardware, software and liveware – Translators – Computer languages – Computer programming process for writing programs – Computing elements of analog computers; **22.2 Microprocessors:** Evolution of microprocessor (μP) – Microprocessor systems – Intel 8085 microprocessor – Microprocessor programming – Digital logic – Types of microprocessors – Processing speed of a processor; **22.3 Organisation of Microcomputers; 22.4 Microcontrollers:** Introduction – Characteristics of microcontrollers – Applications – Classification of microcontrollers – Description of a typical full-featured microcontroller – 8051 microcontroller – Microchip controllers – Comparision of microprocessors and microcontrollers **22.5 Programmable Logic Controllers:** Introduction – Advantages of a PLC – Selection of a PLC – Uses of PLCs – Special features – Architecture/ basic internal structure – Input/output processing and programming. **22.6 Microprocessor-based instrument system – Intelligent instrument**. *Highlights – Objective Type Questions – Theoretical Questions.*

22.1 COMPUTER—BRIEF DESCRIPTION

22.1.1. History and Development of Computers

- Charles Babbage (an English Mathematician) was responsible for conceiving the concept of the Modern computer, and is called "Father of Computers".
- He designed the early computer called *"Difference Engine"* in the year 1822, with which reliable tables could be produced. In 1833 he improved upon the machine and put forth new idea of *"Analytical Engine"* which could perform the basic arithmetic functions automatically. In this machine punched cards were used as input/output devices for basic input and output.

The concept of use of punched cards was developed further by Horman Hollerith in the year 1889.

- Leonards Torres demonstrated a *digital calculating machine* in Paris in 1920.
- In 1944 Prof. Howard Aiken (Howard University) developed Electromechanical calculators known as Mark-I. This machine could handle about a sequence of 5 arithmetic operations by using memory for previous results.
- On the basis of research done for U.S. army during the World War-II in 1946, the first electronic computer, ENiAC (Electric Numerical Integer and Computer) was designed in 1946. This computer was about 15 metres long and 2 metre high and weighed about 50 tons. It consumed about 200 kW power. This machine did not have any facility for storing program.

- In 1949, the *concept of stored program was adopted.*
- In 1951, was introduced the commercial version of stored program computer UNIVAC-(Universal Automatic Computer)—the first digital computer.

Generations of computers:

First generation .... Developed during the years 1951–1959.

- These computers are *"based" on* ***"Vacuum tubes"***.
- Very slow in operation (10^3 operations/sec.)
- Big in size and unreliable.
- Short span of life.
- Frequent breakdowns.
- High power consumption and great amount of heat generation.
- Small primitive memories and no auxiliary storage.
- Limited programming capabilities.

Examples: UNIVAC-1 and IBM 650.

Second generation. Developed during the years 1960–1965.

- These computers are *based on* **"Transistors"**.
- Faster in operation, comparatively (10^6 operations/sec.)
- Smaller in size.
- More reliable.
- Consume less power.
- Generate less heat than vacuum tubes.
- Auxiliary memory in the form of magnetic tape was introduced.

Examples: UNIVAC 1107, IBM 7090, CDC 1604, Honeywell 800 etc.

Third generation. .... Introduced during 1965–1970, also being used presently.

- These computers are based on ***"Integrated circuits"***, based on silicon technology.
- Much more smaller in size.
- More reliable.
- Faster in operation (10^9 operation/sec).
- Less expensive.
- Employ higher capacity internal storage.
- Wide range of peripheral used.
- Make use of new concepts like *multi-programming, multi-processing, high level languages.*

Examples: IBM-360/370, Honeywell 6000.

Fourth generation .... Introduced in 70s.

- These computers are based on VLSI (Very large scale integration) chips and microprocessors chips.
- Possess high processing power.
- Low maintenance.
- Faster in operation.
- High reliability.
- Very low power consumption.
- Less expensive.
- Small size.

This generation also includes the following :

— Microcomputers;

— Office automation systems;

— Distributed processing systems.

Fifth generation .... Introduced during late 1990s.

- These computers use optic fibre technology to handle *Artificial Intelligence, Expert Systems, Robotics* etc.
- Possess very high processing speeds.
- More reliable.

22.1.2. Definition of a Computer

A **computer** is *a machine that processes data according to set of instructions stored within the machine.*

- It *receives data as input, processes the data, i.e.,* performs arithmetic and logical operations on the same and *produces output in the desired form on output device as per the instructions coded in the program.*
- The processing function of the computer is *directed by the stored program, a set of codes instructions stored in the memory unit, which guides the sequence of steps to be followed during processing.*

22.1.3. Characteristics of a Computer

The following are the *characteristics* which make a computer an indispensable unit :

1. Speed
2. Consistency
3. Accuracy
4. Flexibility
5. Reliability
6. Large storage capacity
7. Automatic operation
8. Diligent
9. No emotional ego and psychological problems.

Limitations of a computer :

A computer entails the following *limitations* :

1. It does not work on itself, a set of instructions is required for its operation.
2. It cannot take decision on its own, it has to be programmed as per requirements.
3. It is not intelligent, it has to be instructed in detail for the performance of each and every task.
4. It cannot learn by experience, as human beings do.

22.1.4. Classification of Computers

The computers may be *classified* as follows:

1. **On the basis of the type of data :**

   (*i*) *Analog computers* (These computers process the data in analog form).

   (*ii*) *Digital computers* (These computers process the data in digital form).

2. On the basis of the size and capacity :

(*i*) *Microcomputers*

(*ii*) *Mini computers*

(*iii*) *Main frame*

(*iv*) *Super computers.*

3. On the basis of the type of application :

(*i*) *Special purpose computers*

(*ii*) *General purpose computers.*

4. On the basis of the number of users :

(*i*) *Single user computers*

(*ii*) *Multi-user computers.*

5. On the basis of the number of processors :

(*i*) *Single processor computers.*

(*ii*) *Multiprocessor computers.*

6. On the basis of the type of instructions set :

(*i*) *Complex Instruction Set Computers* (CISC).

(*ii*) *Reduced Instruction Set Computers* (RISC).

22.1.5. Analog Computers

- The *principle of operation* of analog computers is *to create a physical analog of mathematical problems.*
- Measure physical variables *continuously.*
- Use signals as input (which may be supplied by devices like barometers, speedometers, thermometers etc.).
- The *result* given by an analog computer is *not very precise, accurate and consistent.*
- These computers find limited applications.

Example: Speedometer of a vehicle (here speed varies continuously).

22.1.6. Digital Computers

- The digital computers *accept digits and alphabets as input.*
- Receive data in the form of discrete signals representing ON (high) or OFF (low) voltage.
- The data input can be represented as sets of 0's and 1's representing low and high respectively.
- The digital computers convert data into discrete form before operating on it.
- The *most important characteristic* of a digital computer is that it is general purpose device *capable of being used in a number of different applications.* By changing the stored program, the same machine can be used to implement totally different tasks.

Example: Digital watches.

Digital computers may be further classified based upon : (*i*) Purpose of use (*e.g.,* General purpose, special purpose); (*ii*) Size and capabilities.

On the basis of *size and capabilities,* the digital computers are classified as :

1. Super computers.
2. Mainframe computers.

3. Medium sized computers.
4. Mini computers.
5. Micro computers.

1. Super computers :

- These computers are the fastest (speed of calculations upto 1.2 billion instructions per second) and have very high processing speeds.
- They are very large in size and most powerful and costliest.
- Their fields of applications include processing weather data, geological data, genetic engineering etc.
- Word length : 64 bits and more.
- These computers can receive input from more than 1000 individual work stations.

Example: PARAM (a super computer developed in India).

2. Main frame computers :

- These are large scale general purpose computer systems.
- Possess large storage capacities in several million words.
- Secondary storage directly accessible—of the order of several billion words.
- Can support a large number of terminals (upto 100 or more).
- Faster in operation (100 million instructions/sec. approx).
- Accept all types of high level languages.
- Word length—16 or 32 or 64 bits.

3. Medium sized computers :

- Mini versions of mainframe computers.
- They have smaller power than mainframes.
- Processing speeds relatively high with support for about 200 remote systems.

4. Mini computers :

- These are general purpose computer systems.
- Reduced storage capacity and performance (as compared to main frame).
- CPU speed—few million instructions/sec.
- Word length—16 or 32 bits.
- Can accept all types of high level languages.
- Can support upto about 20 terminals.

Note: In view of fast development in electronics it is difficult to draw a line of demarcation between small main frame computers and large mini-computers.

5. Micro-computers :

- These are small sizer computers *utilising microprocessors.* These are popularly known as personal computer (PC).
- CPU is usually contained on one chip.
- Possess low storage capacity (maximum being 256 K words).
- Slow in operation (10^5 instructions/sec.).

- Usually provided with *video display unit, floppy drive and printer*. Some micro-computers can support hard disc also.
- Maximum word length is 16 bits; however most of these use 8-bit words.
- Commonly used language—BASIC. However these computers can also accept other high level languages viz., PASCAL, FORTRAN etc.

Note: **A single chip microcomputer* consists of a single chip on which the central processing unit, input/output and memory units are integrated. This is used for *industrial applications* and also in *product calculators.*

*Its *advantage is the reduction in cost and size, increase in performance and reliability.*

22.1.7. Differences between Analog and Digital Computers

The differences between analog and digital computers are given in Table 22.1.

Table 22.1. Differences between Analog and Digital Computers

S. No.	Digital computer	Analog computer
1.	It performs calculations by counting and thus counts directly. It is the most *versatile machine.*	It processes work electronically by *analogy*. It does not produce number but produces its results *in the form of graph.* It is *more efficient in continuous calculations.*
2.	It operates on inputs which are on-off type (being digits 0 or 1) and its output is in the form of signals.	It accepts variable electrical signals (analog values) as inputs, and its output is also in the form of analog electrical signals.
3.	It is based on counting operation.	It operates by measuring analog signals.

These days *digital computers* are being *widely used.*

A *hybrid computer* is combination of both analog and digital computers. It is used for *simulation applications.*

22.1.8. Block Diagram of a Digital Computer

Figure 22.1 shows a block diagram of a typical digital computer.

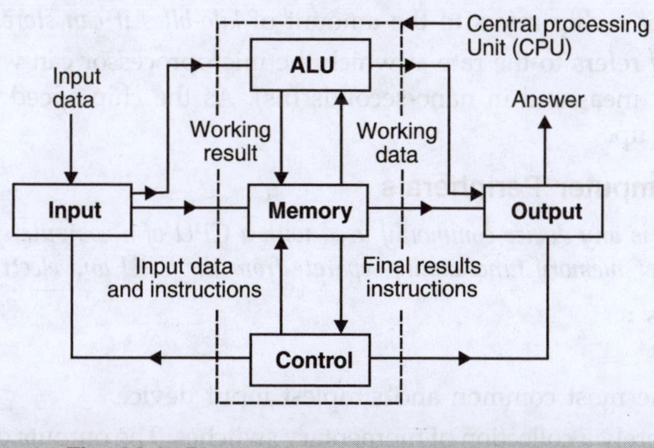

Fig. 22.1. Block diagram of a digital computer.

The following are the *five basics elements* of a computer system :

1. **Input :**
   - The data and instructions are first recorded on a machine readable medium, like punched card, and then fed into the computer via a device that codes them in a manner which is suited to conversion into electrical pulses before entering memory.
   - The input supplies data to the computer in digital (binary) form.
2. **Memory :**
   - The memory section within the computer is where data are stored or memorized.
   - Problem to be solved, inputs for the problem, a program of instruction, working data, intermediate results and final results are *types of memory data.*
   - The memory section holds data between high speed computer operation and slower input and output devices.
3. **Arithmetic Logic Unit (ALU) :**
   - ALU performs necessary arithmetical operations on the data and ensures that instructions are obeyed.
   - It also performs *logical operations.*
   - The ALU combined with control unit is called *Central Processing Unit (CPU).*
4. **Control Unit :**
   - It fetches instructions from main memory, interprets them and issues the necessary signals to the components making up the system.
   - It issues commands for all hardware operations necessary in obeying instructions.
5. **Output :**
   - The output is the path for data out of the computer and may include devices for reading out answers.

22.1.9. Rating of Chips

Chips are rated in terms of their *capacity* and *speed.*

- *Capacity* of a chip refers to the amount of *kilo-bites it can store.*
- *Chip speed* refers to the rate at which the microprocessor can write to the chip. It is usually measured in nano-seconds (ns). As the chip speed increases, its cost also goes up.

22.1.10. Computer Peripherals

A **peripheral** *is any device commonly used with a CPU of a computer for input or output of information or for memory functionally separate from the CPU and electronically detachable.*

Input devices :

1. ***Keyboard :***
   - It is the most common and simplest input device.
   - It is merely a collection of momentary switches. The outputs of the key switches are fed to electronic circuitry known as *keyboard encodes* which converts them into binary coded values. The values are then fed into the computer which

interprets the key which was pressed. Thus the function of the key changes with the type of work we are doing.

2. *Mouse :*

- It is a pointing device and its size is about the size of palm.
- It is a hand-held device that *controls a pointer on the screen.*
- It rolls on a small ball. A mouse has one or more buttons on the top. When the user moves the mouse over a flat surface, the screen cursor moves in the direction of the mouse movement.

3. *Digitizer (or Graphic tablet) :*

- It is similar to light pen.
- It consists of a glass plate on which digitizing tablet is moved.
- It is used for fine drawing works and for image manipulation applications such as Auto-cad.

4. *Optical Mark Reader (OMR) :*

- OMR is being used for reading the answer sheet by means of light. It can read upto 150 documents per minute; when on-line with respect to the computer system, can read upto 2000 documents per minute.
- OMR can also be used for such applications as *order writing payroll, inventory control, insurance, questionnaires* etc.

5. *Magnetic Ink Character Reader (MICR) :*

- MICR uses a special ink to print character. These characters can be decoded by special magnetic devices.
- This system is employed *by banks for processing cheques.*

6. *Scanner :*

- It is used for getting existing graphical image (like photographs, mats, etc.) into computer.
- Once the graphical image is scanned and brought into the computer user can include them into documents or can edit them.

7. *Light pen :*

- It consists of a pen like device and photoelectric cell.
- It is used to draw pictures on the screen.
- When light pen is in contact with screen, the electron beam activates the photoelectic cell which in turn sends signals into the computer and ultimately a mark is made on the screen where light pen contacted the screen.

8. *Joy-stick :*

- It is screen-pointing device.
- A stick is present with a button at the top. It can be held in the hand and bent in any one of the four directions. As the stick is moved, the action on the screen changes in the appropriate direction.
- A joy-stick is *widely used for playing computer games.*

9. *Touch screen :*

- The touch screen technique involves beam and ultrasonic waves.

- By using touch screen we can issue command to the computer by touching the screen.
- Limited amount of data can be entered via a terminal or a microcomputer that has a touch screen.

10. ***Compact Disk Read Only Memory (CDROM) :***
- It is a 120 mm diameter disc with a polycarbonate subtrate, a reflective metalised layer on one side, with a protective lacquer finish.
- Here a laser beam is used to burn a small hole or pit which represent binary '1'. The absence of pit represents '0'. In this way digital information is stored on the disc in large quantities (in Giga Bytes).

11. ***Voice Recognition System or Voice Synthesizer :***
- Voice recognition techniques, along with several other techniques, are used to convert the voice signals to appropriate words and device the correct meanings of words. There has been a limited success in this area and these days devices are available commercially to recognize and interpret human voices.

Output devices :

1. ***Printer :***
- A printer is *device that produces copies of text and graphics on paper.*
- The printers are classified/categorised as follows:

A. *Impact printers :*

(*i*) Solid font

(*ii*) Dot matrix.

B. *Non-impact printers :*

(*i*) Thermal printer

(*ii*) Inkjet printer

(*iii*) Laser printer

(*iv*) Electrographic printer

(*v*) Electrostatic printer.

2. ***Plotters :***
- Plotters are those devices which *reproduce drawings using pens that are attached to movable arms.*
- Platting in different colours is possible.

3. ***Monitors or Visual Display Unit (VDU) :***
- A monitor is a television like device, which is used to display information, output and input data.
- It consists of a cathade ray tube (CRT), on which the information is displayed. When the user processes any key on the keyboard, the keyboard encoder generates code of that key which is depressed. This code is then fed to the computer; from there VDU system takes that code and displays it on the screen.

22.1.11. Storage Devices

The memory devices in a memory unit (which stores the data, instructions and intermediate results) may be of the following types :

1. **Internal storage device** ..... also known as *main* or *primary* storage device.

The primary storage devices currently in use in computers are :

(*i*) Magnetic core memory device

(*ii*) Thin film memory device

(*iii*) Thin rod memory device

(*iv*) Plated wire memory device.

2. Auxiliary storage device

The popular *secondary memory devices* are :

(*i*) Magnetic tape drive

(*ii*) Magnetic disk drive

(*iii*) Magnetic drum

(*iv*) Floppy disk

(*v*) Winchester disk.

Methods of Input to Backing Stores :

The following methods are generally used :

(*i*) Key-to-tape

(*ii*) Key-to-cassette/cartridge

(*iii*) Key-to-disk/diskette

Memory. The memory is used to store information/data so that it can be retrieved whenever required. There are mainly two types of memories :

1. Primary memory
2. Secondary memory.

1. Primary memory :

- It is also known as core memory, main memory, RAM (Random Access Memory).
- It is constructed using purely semiconductor devices, data is stored in the form voltages.
- It is a volatile memory whereas ROM (Read Only Memories) are non-volatile memories.

2. Secondary memory :

- Secondary memory, also known as *auxiliary memory*, is used to store large volumes of data.
- Data is stored in the form of magnetic energy and can be stored (in the secondary memory) for large periods.

Difference between Read Only Memory (ROM) and Random Access Memory (RAM).

ROM :

- As the name implies ROM is a memory unit that performs the read operation only; it *does not have a write capability.* This implies that the binary *information stored in a ROM is made permanent during the hardware production of the unit and cannot be altered by writing different words into it.* Whereas a *RAM is a general-purpose device whose contents can be altered during the computational process.*
- ROM is a type of memory chip that we can read only and we cannot write on it.
- ROM provides permanent storage for program instructions.

- The most important ROM chip in any computer is ROM BIOS (Basic Input/ Output System).
- ROM is most oftenly used in microprocessors that always execute the same program such as boot strap loader.

Disadvantage of ROM :

(*i*) A ROM is prepared by the manufacturer and cannot be altered once the chip has been made.

(*ii*) It is slow.

The *ROM memory* may be **classified** as follows:

(*i*) *Programmable Read Only Memory (PROM).* Here, the information can be altered, but *not as easily as in the ordinary memory.* Once the operations to be performed have been written into a PROM chip, they are permanent and cannot be changed.

(*ii*) *Erasable Programmable Read Only Memory (EPROM).* This type of ROM can be erased and programmed with the help of special equipment. It has a window at its top, which if exposed to ultraviolet light, allows data to be erased.

(*iii*) *Electrically Erasable Programmable ROM (EEPROM).* In order to erase and reprogramme this type of ROM, it is required to be removed from the socket.

(*iv*) *Flash EPROM.* It is the latest type of ROM. A manufacturer can make changes to the flash EPROM while it remains in the PC, by running a special program.

RAM :

- This memory is so named since memory registers can be accessed for information transfer as required.
- RAM chip is made with Metal Oxide Semiconductor (MOS).
- RAM chips may be classified as :
  - (*i*) *Dynamic Ram:* It provides volatile storage (*i.e.*, the data stored is lost in the event of a power failure).
  - (*ii*) *Static RAM:* These chips are more complicated and take up more space for a given storage capacity than dynamic RMA chips. These chips are also volatile in nature but as long as they are supplied with power, they need not require special regenerator circuits to retain the stored data.
- *Static RAM chips are thus used in specialised applications while Dynamic RAM chips are used in the primary locations.*
- Owing to the volatile nature of these storage elements, a back up *Uninterrupted Power System (UPS) is often installed along with larger computer systems.*

22.1.12. Hardware, Software and Liveware

Hardware :

The set of *physical components, modules and peripherals comprising a computer system* is called ***Hardware.***

Apart from wires and nut bolts, the major hardware components of computer are :

(*i*) Input-output devices

(*ii*) Control unit

(*iii*) Memory

(*iv*) ALU.

Software :

The software is a *set of programs required for data processing activities of the computer.* In other words, the program written in any one of the computer languages, is called *software.*

System software includes the following :

(*i*) Operating systems

(*ii*) Language processors (assembles, compilers, interpreters)

(*iii*) Utility program

(*iv*) Subroutine program.

Liveware :

All persons concerned with computers, *i.e.*, complier, programmer, etc. are called *liveware.*

22.1.13. Translators

A *translator* is a software program which converts statements written in one language into another *e.g.*, converting assembly language to machine code etc. The assembly language program is called *'source program'* and the machine code program is called *'object program'*.

There are three types of *translators* :

1. Assembler 2. Compiler 3. Interpreter.

22.1.14. Computer Languages

1. ***Machine language.*** It is a programming language in which the instructions are in a form which allows the computer to perform them immediately, without any further translation. Instructions in machine language are *in the form of a binary code*, also called machine code and are known as machine *instructions.*
2. ***Low level language.*** Low level languages are machine-oriented languages in which each instruction corresponds or resembles a machine instruction. The low level language must be translated into machine language before use.
3. ***High level language.*** The development of high level language was intended to overcome main limitations of level language. The high level languages have an extensive vocabulary of word, symbols and sentences.

Different types of high level languages are :

(*i*) *Commercial languages.* ... The most well commercial language is COBOL (Commercial Business Oriented Language).

(*ii*) *Scientific language.* ... The most well-known languages among this group are :

(*a*) ALGOL (Arithmetic Oriented Language)

(*b*) FORTRAN (Formula Translation)

(*c*) BASIC (Beginner All Purpose Symbolic Instruction Code).

(*iii*) *Special purpose language.*

(*iv*) *Command language.*

(*v*) *Multipurpose language.*

22.1.15. Computer Programming Process for Writing Programs

The complete computer programming process followed by programmer for writing comprises the following **steps** :

1. Analysis 2. Flow charting 3. Coding
4. Debugging 5. Documentation 6. Production.

22.1.16. Computing Elements of Analog Computers

1. *Attenuators* ...... are used to multiply a variable quantity by a constant.
2. *Summing amplifiers* ...... are used to add or subtract variables as required.
3. *Servo multipliers* ...... are used to multiply two variables.
4. *Function generators* ...... are used to simulate the arbitrary behaviour of variables.
5. *Integrating amplifiers* ...... are used to integrate a variable with respect to time.

22.2 MICROPROCESSORS

22.2.1. Microprocessor—General Aspects

22.2.1.1. Definition and brief description

A ***microprocessor****, usually abbreviated as* μP, *is a large scale integration (LSI) chip that is capable of performing arithmetic and logic functions as defined by a given programme.* This system by itself does not form an operational computer, and additional circuit for memory and input/output must be supplied and interfaced with the system. The software (it is the programme for controlling the operation of the microprocessor itself) is also to be provided.

Or

A state machine on a single IC chip with very large scale integration, capable at a desired instant of working as per programme or an instruction of a programme, and which is driven by a clock of frequency of 1 MHz or more, is called a ***microprocessor****.* Such machine is also called a central processing unit (CPU). A CPU forms main part of a computer.

The ***microprocessor*** consists of the following *three segments* (See Fig. 22.2).

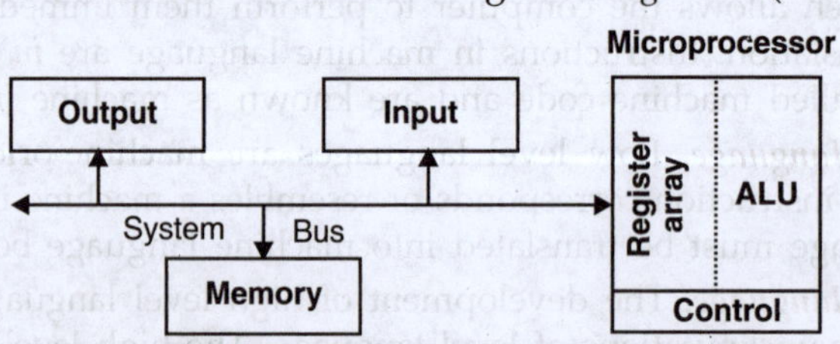

Fig. 22.2. Block diagram of a microcomputer.

1. **Arithmetic/Logic Unit (ALU).** In this area of the microprocessor, computing functions are performed on data. The ALU performs arithmetic operations such as addition and subtraction, and logic operations such as AND, OR and exclusive OR. Results are stored *either* in registers or in memory or sent to output devices.
2. **Register unit.** This area of the microprocessor consists of various registers. The registers are used primarily to store data temporarily during the execution of a program. Some of the registers are accessible to the user through instructions.
3. **Control unit.** The control unit provides the necessary timing and control signals to all the operations in the microcomputer. It controls the flow of data between the microprocessor and peripherals including memory.

In short a *microprocessor performs the following functions* :

- Communicates with all peripherals (memory and I/O) using system.
- Controls timing of information flow.
- Performs the computing tasks specified in a programme.

22.2.1.2. Characteristics of microprocessor

In nearly every type of design, with any complexity at all, microprocessors have potential for *drastically reducing component count and shortening design time.* In fact a microprocessor is considered to represent long-awaited next generation of digital building blocks, and that microprocessor will provide the best single approach to the system-level digital integrated circuit.

Some of the characteristics of a microprocessor are listed below :

1. It *handles shorter words than other computers,* usually 4 to as many as 16 bits.
2. It consists of *integrated circuits* from 1 to 30 in number.
3. It contains arithmetic logic unit (ALU), registers, control, random access memory (RAM), data buses and read only memory (ROM) with programmes.

22.2.1.3. Important features

The important features of the microprocessors are :

1. *Low cost*
2. *Small size*
3. *Low power consumption*
4. *Versatile* (The versatility of a microprocessor results from its 'stored programme' mode of operation).
5. *Extremely reliable.*

Note: Probably the term 'micro' in the name of the device can be contributed to its *low cost, small size and low power consumption.* The processing capability of a microprocessor should not, however, be underestimated. Currently available 32-bit microprocessors have a processing power similar to that of the mainframe computer of a few years ago. Even the early 8-bit microprocessors are powerful enough to perform several applications.

22.2.1.4. Functions of Microprocessor

The functions performed by a microprocessor are list below;

1. Averaging the data;
2. Linearization;
3. Alarm or annuciator settings;
4. Totalization;
5. Setting of set points;
6. Amplifier gain control;
7. Control of I/O devices;
8. To provide control actions to control elements;
9. To format output signals;
10. To rest the system;
11. Interface control, etc.

22.2.1.5. Uses of microprocessor

The processing power of the 8-bit microprocessors is more than adequate to satisfy the requirements of most of the *instrumentation applications.* By making an instrument microprocessor-based, it can be made *intelligent by incorporating new features like programmability* which cannot be easily provided in its hard-wired counterpart.

Some important *uses* of microprocessors in instrumentation area are listed below :

1. Frequency meters.
2. Function generators.
3. Frequency synthesizers.
4. Spectrum synthesizers.
5. *Intelligent instruments CRT terminals*
6. Digital millimeters.
7. Oscilloscopes.
8. Counters.
9. **Process control**
   — Instrumentation
   — Monitoring and control
   — Data acquisition
   — Logging and processing.

10. **Medical Electronics**

— Patient-monitoring in intensive care unit

— Pathological analysis

— Measurement of parameters like blood pressure and temperature.

Under this heading the following instruments/machines are included:

(*i*) Microprocessor based medical instrument.

(*ii*) Microprocessor based ECG machines.

(*iii*) Microprocessor based EEG machines etc.

Other Applications of microprocessors :

(*i*) High level language computers.

(*ii*) Replacing hard-wired logic by a microprocessor.

(*iii*) Control of automation and continuous processes.

(*iv*) Computer peripheral controllers.

(*v*) Home entertainment and games.

(*vi*) Inventory control system, pay roll banking etc.

Some Points about Microprocessors worth noting:

- A system designed around a microprocessor, like any other digital computer, needs to be programmed. Thus, a sequence of instructions needs to be formulated and input to the microprocessor-based system for effective operation. *A sequence of instructions designed to perform a particular task is known as* "**program**". A set of programs written for a microprocessor-based system is known as "*software*" for that system. In general both *hardware and software design* are of paramount importance in microprocessor-based system design though, in particular cases, one may be more difficult or significant than other.
- **4–6–8–16–32-bits μPs**—In the beginning there was 4-bit μP followed by 8-bit μPs, 16-bit μPs and 32-bit μPs.

— A μP in which most instructions operate upon *n*-bit data is said to be a "*n*-bit μp".

Example: The 8085 is an 8-bit μP. It can add two 8-bit numbers and can perform several logical operations on two 8-bit patterns. It can perform many other instructions all on 8-bit data.

Since most instructions operate on data stored in on-chip registers, therefore, an n-bit μP will have registers that can hold *n*-bit data. Again, the 8085 has 8-bit registers to hold data on which its instructions operate. It is not neccessary that an *n*-bit μP will *not* have instructions to perform operations on larger length data. For example, though 8085 is an 8-bit μP, it can perform a few operation on 16-bit data also. However, the majority of its opeations are on 8-bit data.

An *n-bit* μP *can* be used to perform operations on larger data, but such operations will generally be *slower* than n-bit operations. For example, the 8085 *can* be used to perform addition of two 32-bit numbers, however, this has to be done under *program control*. However, a 32-bit μP like the Intel 80386 or the Motorola 68030, can add two 32-bit numbers in just one add operation. For this, and a few other reasons, the 32-bit μPs are *much faster* than the 16 or 8-bit μPs.

22.2.2. Evolution of Microprocessors (μP)

The evolution of microprocessors (μP) can be precisely summed up as follows:

- In **1971** the *first microprocessor* was introduced by "*Intel Corporation*".

— This was "Intel 4004", a processor on a *single chip*. It had the capability of performing simple arithmetic and logical operations, e.g. addition, subtraction, comparison, AND, and OR. It also had control unit which could perform various control functions like fetching an instruction from the memory, decoding it and generating control pulses to execute it. It was a 4-bit microprocessor operating on 4 bits of data at a time.

- In the last few years the microprocessor has changed the calculators to video games and computers. Several microprocessors have been manufactured for all sorts of products, some have succeeded and some have not.
- The first microprocessor to make it into a *home computer* was the *Intel 8080*, a complete 8-bit computer on one chip, introduced in *1974*.
- The first microprocessor to make a real splash in the market was the *Intel 8088*, introduced in *1979* and incorporated into the IBMPC which first appeared around *1982*.
- The PC market moved from the 8088 to 80286 to the 80386 to the 80486 to the Pentium to the Pentium II to the Pentium III, to the Pentium 4. All of these microprocessors are made by Intel and all of them are improvements on the basic design of the 8088. The "Pentium 4" can execute any piece of code that ran on the original 8088 but it does it about 5,000 times faster.

The table 22.2 given below indicates the differences among different processors introduced over the years:

Table 22.2: Differences among Different Processors

Name	Date	Transistors	Data width (Bits)	Microns	Clock speed (MHz)	MIPS
4004	1971	2250	4	10	0.1	0.06
8008	1972	3500	8	10	0.2	0.06
8080	1974	6,000	8	6	2	0.64
8085	1974	6,500	8	3	5	0.37
8086	1978	29,000	16	3	5 8 10	0.33 0.66 0.75
8088	1979	29,000	16	3	5	0.33
80286	1982	134,000	16	1.5	16	1
80386	1985	275,000	32	1.5	16	5
80486	1989	1,200,000	32	1	25	20
Pentium	1993	3,100,000	32	0.8	60	100
Pentium II	1997	7,500,000	32	0.035	233	~300
Pentium III	1999	9,500,000	32	0.25	450	~510
Pentium 4	2000	42,000,000	64	0.18	1500	~1,700
Pentium 4 "Prescott"	2004	125,000,000	64	0.09	3.6 Hz	~7,000

Transistors: The number of transistors on the chip.

Data width: The width of ALU.

Microns: The width, in microns, of the smallest wire on the chip.

Clock speed: The maximum rate that chip can be clocked at.

MIPS: stand for "Millions of instructions per second" and is a rough measure of the performance of a CPU.

22.2.3. Microprocessor Systems

Microprocessor systems consist of the following *three* parts :

1. *Central processing unit (CPU) :*
   — This part uses the microprocessor.
   — It recognises and carries out program instructions.
2. *Input and output interfaces :*
   — These interfaces handle communications between the computer and the outside world.
   — For the interface, the term *port* is used.
3. *Memory :*
   — Its function is to hold the program instructions and data.

- *"Microprocessors" which have memory and various input/output arrangements on the same chip are called* ***microcontrollers.***

22.2.3.1. The microprocessor

The microprocessor (generally referred to as CPU) is that part of the processor system which carries out the following functions :

(*i*) Processes the data ;

(*ii*) Fetches instructions from memory;

(*iii*) Decodes and executes the instructions.

Following are the various parts of a microprocessor :

1. *Arithmetic and logic unit (ALU):* This part of the microprocessor *manipulates the data.*

2. *Registers:* Registers are memory locations within the microprocessor and are employed to store information involved in program execution. The various types of registers used are :

(*i*) *Accumulator register :*

— It is a temporary holding register for data to be operated on by the arithmetic and logic unit and also, after the operation, the register for holding the results.

— It deals with all data transfers associated with the execution of arithmetic and logic operations.

(*ii*) *Status register :*

— The status register (also called flag register or condition code register) contains information concerning the result of the latest process carried out in the ALU.

— It contains individual bits with each bit having special significance; the bits are called flags.

— The status of the latest operation is indicated by each flag with each flag being set or reset to indicate a specific status.

(iii) Program counter register (PC) :

— This register, also called instruction pointer (IP), contains the address of the memory location that contains the next program instruction.

— This register is *updated* after the execution of each instruction, so that it contains the memory location where the next instruction to be executed is stored.

(iv) Memory address register (MAR) :

— This register contains the address of data.

(v) Instruction register (IR) :

— This register stores an instruction.

— The control processing unit (CPU), after fetching an instruction from the memory via the data bus, stores it in the instruction register. The microprocessor, after each such fetch, increments the program counter by one with the result that the program counter points to the next instruction waiting to be fetched. The instruction can then be decoded and used to execute an operation. This sequence is known as *fetch-execute cycle.*

(vi) General purpose registers :

— These registers serve as temporary storage for data or addresses and *used in operations involving transfers between other registers.*

(vii) Stack pointer register (SP) :

— The *stack* is a special area of the memory in which program counter values can be stored when a subroutine part of the program is being used.

— The contents of this register form an address which defines the top of the stack in RAM.

The number and form of the registers depends on the microprocessor concerned.

22.2.3.2. Buses

***Buses** are the paths along which digital signals move from one section to another.*

— A bus is just a number of conductors along which electrical signals can be carried. It might be tracks on a printed circuit board or wires in a ribbon cable.

In a microprocessor system there are the following *three forms of bus* :

1. Data bus;
2. Address bus;
3. Control bus.

1. *Data bus :*

— The data bus carries the data associated with the processing function of the CPU.

— Word lengths used may be 4, 8, 16, 32 or 64.

— Each wire in the bus carries a binary signal, *i.e.,* a0 or a1.

— The more wires the data bus has the longer the word length that can be used.

- The earliest microprocessors were 4-bit (word length : $2^4 = 16$) devices and such 4-bit microprocessors are still used in such devices as toys, washing machines etc. They were followed by 8-bit microprocessors (*e.g.,* Motorola 6800, the Intel 8085 A and the Zilog Z80). Now 16-bit, 32-bit and 64-bit microprocessors are available.

2. ***Address bus :***

— It carries signals which indicate where data is to be found and so the selection of certain memory locations or input or output ports.

— Each storage location within a memory device has a unique identification, termed its address, so that system is able to select a particular instruction or data item in the memory.

— Each input/output interface also has an address.

— When a particular address is selected by its address being placed on the address bus, *only that location is open to the communications from the CPU.* The CPU is thus *able to communicate with just one location, at a time.*

— A computer with an 8-bit data has typically a 16-bit wide address bus, *i.e.,* 16 wires. This size of address enables 2^{16} locations to be addressed. 2^{16} is 65 536 locations and is usually written as 64 K, where K is equal to 1024.

3. ***Control bus :***

— This bus *carries the signals relating to control actions.*

— It is also used *to carry the system clock signals;* these are to synchronise all the actions of the microprocessor system.

22.2.3.3. Memory

In a microprocessor, the memory unit stores binary data and takes the form of one or more integrated circuits (ICs).

— The data may be program instruction codes or numbers being operated on.

— *The size of the memory is determined by the number of wires in the address bus.*

Following are the various forms of memory unit :

1. ROM
2. PROM
3. EPROM
4. EEPROM
5. RAM.

1. ***ROM :***

— ROM (*Read Only Memory*) is a memory device in which *data is stored permanently.*

— While manufacturing the integrated circuit ROMs are programmed with the required contents. *No data then can be written into this memory chip in the computer.* The data *can only be read and is used for fixed programs;* memory is not lost when power is removed.

- *Programs stored in ROM are termed as **firmware.***

2. ***PROM :***

— PROM (Programmable ROM) is used for ROM chips that *can be programmed by the user.*

3. ***EPROM :***

— EPROM (Erasable and programmable ROM) is employed for *ROMs that can be programmed and their contents altered.*

— A typical EPROM chip contains a series of small electronic circuits, cells, which can store charge.

— The program is stored by applying voltages to the integrated circuit connecting pins and producing a pattern of charged and uncharged cells. The pattern remains permanently in the chip until *erased by shining ultraviolet light through a quartz window on the top of the device.*

4. *EEPROM :*

— EEPROM (Electrically erasable PROM) is similar to EPROM; erasure, however, is done by the application of a relatively high voltage rather than using ultraviolet light.

5. *RAM :*

— RAM (Random-access memory) is a read/write memory in which data currently being operated on (temporary data) is stored.

— Such a memory can be read or written to.

— When RAM is used for program storage then such programs are referred to as ***software.*** When the system is switched on, software may be loaded into RAM from some other peripheral equipment such as a keyboard or hard disc or floppy disc.

Difference between a software of a computer and a microprocessor :

In computer software is loaded into the computer at the beginning of each computation, software in microprocessor is stored within the computer itself in a ROM chip. The modification of the program is achieved by merely replacing ROM IC with another ROM IC containing a different control program. This is a very notable advantage of software implementation in microprocessor.

22.2.3.4. Input/Output

The transfer of data between the microprocessor and the external world is termed as the ***input/output operation.***

— The pieces of equipment that exchange data with a microprocessor system are called ***peripheral devices.***

— In *input operations* the input device places the data in the data register of the interface chip; this holds the data until it is read by the microprocessor. In *output operations* the microprocessor places the data in the register until it is read by the peripheral.

22.2.4. Intel 8085 Microprocessor

22.2.4.1. Brief history

— Intel Corporation in early seventies introduced the first microprocessor, *Intel 4004.* This microprocessor was a single chip device which was capable of performing simple arithmetic and logic operations such as addition, subtraction, comparison, AND, and OR. Its control unit could perform various functions such as fetching of an instruction from the memory, decoding it and generating control pulses for executing it. It was a 4-bit microprocessor operating upon 4-bits of data at a time.

— Intel introduced *4040* as modified version of microprocessor 4004.

— Intel, later on, introduced 8-bit microprocessors called 8008 and 8080 which could perform arithmetic and logic operations on 8-bit words.

— These days, modified and better version of 8-bit microprocessor is **Intel 8085** which is most widely used and most popular microprocessor.

Now-a-days 12-bit, 16-bit and 32-bit microprocessors are also available.

22.2.4.2. Features of 8085 microprocessor

Features of 8085 microprocessor are given below:

1. It is an 8-bit microprocessor i.e. it can accept, process, or provide 8-bit data simultaneously.
2. It operates on a single +5V power supply connected at V_{CC}; power supply ground is connected to V_{SS}.
3. It operates on clock cycle with 50% duty cycle.
4. It has a chip clock generator. This internal clock generator requires tuned circuit like LC, RC or crystal. The internal clock generator divides oscillator frequency by 2 and generates clock signal, which can be used for synchronizing external devices.
5. It can opeate with a 3 MHz clock frequency. The 8085A-2 version can operate at the maximum frequency of 5 MHz.
6. It has 16 address lines, hence it can access (2^{16}) 64 Kbytes of memory.
7. It provides 8 bit I/O addresses to access (2^8) 256 I/O ports.
8. In 8085, the lower 8-bit address bus (A_0 – A_7) and data bus (D_0 – D_7) are multiplexed to reduce number of external pins. But due to this, external hardware (latch) is required to separate address lines and data lines.
9. It supports 74 instructions with the following addressing modes:
   (*a*) Immediate (*b*) Register (*c*) Direct (*d*) Indirect (*e*) Implied
10. The Arithmetic Logic Unit (ALU) of 8085 performs:
    (*a*) 8 bit binary addition with or without carry
    (*b*) 16 bit binary addition.
    (*c*) 2 digit BCD addition.
    (*d*) 8-bit binary subtraction with or without borrow
    (*e*) 8-bit logical AND, OR, EX-OR, complement (NOT), and bit shift operations.
11. It has 8-bit accumulator, flag register, instruction register, six 8-bit general purpose registers (B, C, D, E, H and L) and two 16-bit registers (SP and PC). Getting the operand from the general purpose registers is more faster than from memory. Hence skilled programmers always prefer general purpose registers to store program variables than memory.
12. It provides five hardware interrupts: TRAP, RST 7.5, RST 6.5, RST 5.5 and INTR.
13. It has serial I/O control which allows serial communication.
14. It provides control signals ($IO/\overline{M}$, $\overline{RD}$, $\overline{WR}$) to control the bus cycles, and hence external bus controller is not required.
15. The external hardware (another microprocessor or equivalent master) can detect which machine cycle microprocessor is executing using status signals ($IO/\overline{M}$, S_0, S_1). This feature is very useful when more than one processors are using common system resources (memory and I/O devices).
16. It has a mechanism by which it is possible to increase its interrupt handling capacity.
17. The 8085 has an ability to share system bus with Direct Memory Access controller. This feature allows to transfer large amount of data from I/O device to memory or from memory to I/O device with high speeds.
18. It can be used to implement three chip microcomputer with supporting I/O devices like IC 8155 and IC 8355.

7.2.4.3. Description of 8085 Microprocessor

Fig. 22.3, shows the block diagram of Intel 8085 microprocessor.

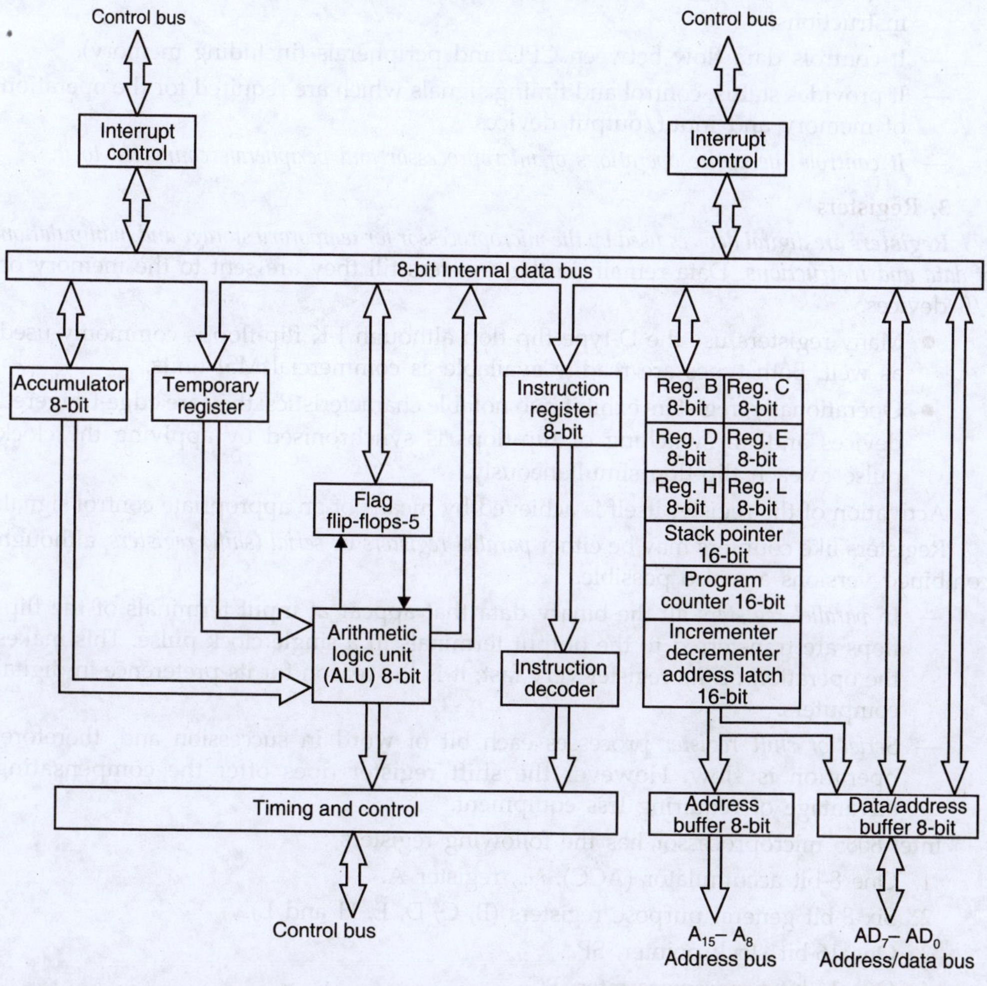

Fig. 22.3. Block diagram of Intel 8085 microprocessor.

Some important functional units/parts are enumerated and discussed as follows:

1. Arithmetic and logic unit (AUL).
2. Timing and control unit.
3. Registers
4. Data and address bus
5. Pin configuration

1. Arithmetic and logic unit (ALU):

ALU performs the following arithmetic and logical operations :

1. Addition
2. Subtraction
3. Logical AND
4. Logical OR
5. Logical EXCLUSIVE OR
6. Complement (logical NOT)
7. Increment (add 1)
8. Decrement (subtract 1)
9. Left shift, Rotate left, Rotate right.
10. Clear etc.

2. Timing and control unit:

— This unit is a section of CPU.

— It *generates timing and control signals* which are necessary for the execution of instructions.

— It controls data flow between CPU and peripherals (including memory).

— It provides status, control and timing signals which are required for the operation of memory and input/output devices.

— *It controls the entire operations of microprocessor and peripherals connected to it.*

3. Registers

***Registers** are digital devices used by the microprocessor for temporary storage and manipulation of data and instructions.* Data remain in the registers till they are sent to the memory or I/0 devices.

- Many registers use the D-type flip-flop although J-K flip-flop is commonly used as well. Both types are readily available as commercial MSI units.
- Operationally, registers exhibit two notable characteristics: they are edge-triggered devices and all switching of flip-flops is synchronised by applying the clock pulse to each flip-flop simultaneously.

Activation of the register itself is achieved by means of an appropriate control signal. Registers like counters may be either *parallel registers* or *serial (shift) registers*, although combined versions are also possible.

— In *parallel registers* all the binary data that appear at input terminals of the flip-flops are transferred to the output terminals in a single clock pulse. This makes the operation of the register very fast; it is the reason for its preference in digital computers.

— *Serial or shift register* processes each bit of word in succession and, therefore, operation is slow. However the shift register does offer the compensating advantage of requiring less equipment.

Intel 8085 microprocessor has the following registers :

1. One 8-bit accumulator (ACC), *i.e.*, register A.
2. Six 8-bit general purpose registers (B, C, D, E, H and L).
3. One 16-bit stack pointer, SP.
4. One 16-bit program counter, PC.
5. Instruction register.
6. Temporary register.

These registers are described below :

1. ***Accumulator (ACC) :***

- It is an 8-bit special purpose register that is a part of the ALU. It is also identified as register A.
- In arithmetic and logical operations the accumulator may store the operand, execute an instruction with the help of other registers, and memory and finally store the result of the operation. In the former case it acts as a source, and in the latter a destination.

2. ***General purpose registers :***

- The 8085 microprocessor contains six 8-bit general purpose registers. These are identified as B, C, D, E, H and L as shown in Fig. 22.3.

- These registers are used in microprocessor for temporary storage of operands or intermediate data in calculations.
- These registers can be used either simply for storage of 8-bit data or in pairs for storage of 16-bit data. When used in pairs, only selected combination can be used for pairing, *i.e.*, B-C, D-E and H-L. When registers are used in pairs the high order byte resides in the first register and low order byte in the second register.

3. ***Stack pointer (SP) :***
   - It is a 16-bit special function register.
   - The *stack* is a sequence of memory locations set aside by a programmer to store/retrieve the contents of accumulator, flags, program counter and general purpose registers during the execution of a program. Any portion of the memory can be used as a stack.
   - In this register, data is stored temporarily on first come and last go basis.
4. ***Program counter (PC) :***
   - It is a 16-bit special-purpose register and is used to hold the memory address of the next instruction to be executed.
   - The contents of the PC are automatically updated by the microprocessor during the execution of an instruction so that at the end of execution it points to the address of the next instruction in the memory.
   - The microprocessor uses the PC for sequencing the execution of instructions.
5. ***Instruction register :***
   - During the execution of a program, microprocessor addresses some memory, which supplies an 8-bit data of instruction code to the data bus which gets stored in the register called the *instruction register*.
   - The instruction register holds the op-code (operation code or instruction code) of the instruction which is being decoded and executed.
6. ***Temporary register :***
   - It is an 8-bit register associated with ALU.
   - It holds data during an arithmetic/logical operation.
   - It is used by the microprocessor and is not accessible to programmer.

Flags. There are five-flops in Intel 8085 microprocessor to serve as status flags; these are : *(i)* Carry Flag (CS), *(ii)* Parity Flag (P), *(iii)* Auxiliary Carry Flag (AC); *(iv)* Zero Flag (Z), and *(v)* Sign Flag (S).

— The flip-flops are set or reset according to the conditions which arise during an arithmetic or logical operation.

— If a flip-flop for a particular flag is set, it indicates 1. When it is reset, it indicates 0.

Instruction decoder. Data from the instruction register is sent to the instruction decoder, where microprocessor decodes it and then translates into specific actions.

4. Data and address bus

The data bus of Intel 8085 microprocessor is 8-bit wide and hence, 8 bits of data can be transmitted in parallel from or to the microprocessor. This microprocessor requires a 16-bit wide address bus as the memory addresses are of 16-bits. The 8 *most significant* bits of the address are transmitted by the address bus, A-bus (pins A_8 to A_{15}). The 8 *least significant* bits of the address are transmitted by address/data bus, AD-bus (pins AD_0–AD_7).

The address/data bus transmits data and address at different movements. At particular moment it transmits either data or address. Thus *AD-bus operates in time shared mode. This technique is called* ***multiplexing***.

5. Pin configuration

Fig. 22.4. shows the schematic diagram of Intel 8085 microprocessor.

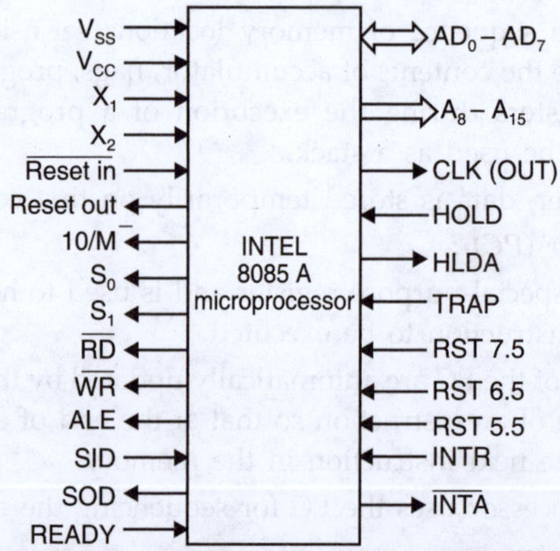

Fig. 22.4. Schematic diagram of Intel 8085 microprocessor.

***ADo–AD_7* (*Input/Output*):** They are used for the least significant 8-bits of the memory address of I/O address during the first clock cycle of a machine cycle. Again they are used for data during second and third clock cycles.

***A_8–A_{15}* (*output*):** These are address bus and are used for the most significant bits of the memory address of 8-bits of I/O address.

***$IO/\overline{M}$* (*Output*):** It is a status signal which distinguishes whether the address is for memory or I/O. When it goes high the address on the address bus is for an I/O device, whereas, when it goes low the address on the address bus is for a memory location.

ALE (Output): It is an *address latch enable signal.* It goes high during first clock cycle of a machine cycle and enables the lower bits of the address to be latched either into the memory or external latch.

***S_0, S_1* (*Output*):** These are status signals sent by the microprocessor to distinguish the various types of operations given in the table below,

S_1	S_0	*Operations/Machine cycle*
0	0	HALT
0	1	WRITE
1	0	READ
1	1	FETCH

$\overline{RD}$ (Output) :

— It is a signal to control READ operation.

— When it goes low the selected memory or I/O device is read.

$\overline{WR}$ (Output) :

— It is a signal to control WRITE operation.

— When it goes low the data on the data bus is written into the selected memory or I/O location.

READY INPUT :

— It is used by the microprocessor to sense whether a peripheral is ready to transfer data or not.

— If READY is high the peripheral is ready, if it is low the microprocessor waits till it goes high.

— A slow peripheral may be connected to the microprocessor through READY line.

HOLD (Input) :

— It indicates that another device is requesting for the use of the address and data bus.

— The microprocessor after having received a HOLD request relinquishes the use of the buses as soon as the current machine cycle is completed. Internal processing may continue.

— The processor regains the bus after the removal of the HOLD signal.

HLDA (Output) :

— This signal indicates that the HOLD request has been received.

— After the removal of a HOLD request the HLDA goes low. The CPU takes over the buses half clock cycle after the HLDA goes low.

INTR (Input); $\overline{INTA}$:

— INTR signal indicates an interrupt request.

— The INTR line is sampled in the last state of the last machine cycle of an instruction.

The microprocessor acknowledges the interrupt signals and issues an $\overline{INTA}$ signal.

$\overline{RESET\ IN}$ (Output) :

— When this signal is applied the program counter is set to zero and resets the Interrupt Enable and HLDA flip flops. Except the instruction register no other register or flag is affected.

— The CPU remains in the reset condition as long as reset is applied.

RESET OUT :

— This is an output signal which shows that CPU is being reset. This can be used as a system RESET.

— The signal is synchronised to the microprocessor clock.

X_1, X_2 (Input) :

— X_1 and X_2 are the terminals to be connected to an external crystal oscillator which drives an internal circuitry of the microprocessor to produce a suitable clock for the operation of microprocessor.

CLK (Output) :

— It is a clock output for user, which can be used for other digital ICs.

— Its frequency is same at which processor operates.

SID and SOD line :

— SID is an input line and it is for serial input data. The serial input data at SID can be identified by an instruction called RIM and serial data can be an instruction called SIM.

— In Intel 8085 microprocessor only serial transmission facility is available. The SID and SOD lines pursuit the input and output serial data. The actual transfer of the data is accomplished by software using the RIM and SIM instructions. Both these instructions are single byte and are also used to read or set/reset interrupt masks.

22.2.5. Microprocessor Programming

The process of developing a program is termed as **Programming**. A microprocessor needs to be programmed before it can perform a useful task.

A "program" is a sequence of instructions that operate on a data to produce desired results. The programs, written to perform a particular task, is stored in the semiconductor memory accessible to the microprocessor. The processor, having been instructed the program, fetches one instruction at a time from the memory and executes it.

- **Instructions and Instruction cycle**

An ***"instruction"*** consists of an *operation code* (abbreviated as *opcode*) and the *address(es) of the data* on which it has to operate. The data on which an instruction operates, is also known as the *operand(s)* for that instruction.

— Most instructions specify directly the address of only one of the operands, the address of the other operand being implicitly known. Some instructions specify the data itself and not the address.

An ***"instruction cycle"*** consists of a *"fetch cycle"* and *"execute cycle"*. The total time required to execute an instruction is the some of time required to fetch an opcode and the time required to execute it.

— The opcode (the Ist byte of an instruction is its opcode; the instruction may be more than one byte long) fetched from the memory goes to the data register, DR (data/address buffer in Intel 8085 microprocessor) and then to instruction register, IR. From the instruction register it goes to the decode circuitry which *decodes* the instruction. The decoded circuitry is within the microprocessor.

— After the instruction is decoded, *execution begins.* If the operand is in the general purpose registers, execution is immediately performed. The time taken in decoding and execution is one clock cycle. If an instruction contains data or operand address which are still in the memory, the CPU has to perform some read operations to get the desired data. After receiving the data it performs execute operation.

A *read cycle* is *similar to a fetch cycle.* In case of a read cycle the quantity received from the memory are data or operand address instead of an opcode. In some instructions write operation is performed.

In *write cycle* data are sent from CPU to the memory or an output device.

In some cases an *execute cycle* may involve one or more read or write cycles or both.

— The necessary steps carried out to perform a fetch, a read or write operation constitute a *"Machine cycle"*. An *instruction cycle consists of several machine cycles.*

- **Machine and Assembly Language (ML and AL) Programming:**

— A *"machine language programe"* is written using binary *opcodes and addresses.*

— In an *"assembly language program"* the instructions are written using *mueonomic codes and symbolic addresses.*

For a microprocessor it is *easier* to write an assembly language programme than to write a machine language programme. However, a *microprocessor can understand only machine language programs.*

- **High Level Language (HLL) Programming:**
  - — For programming on many microprocessors, higher level languages like *fortran,* C, *Pascal, and colbol* are available.
  - — Though programming in *higher level languages is much easier than in machine or assembly languages,* the former *is possible only when a microcomputer with sufficient memory is available.*
  - — Even when a very small system is being designed, it can be programmed by using a *development system* in high level language. The high level language program can be translated into binary and stored in the memory of the small system.
- **The Programming Process:**
  - — The steps involved in a programming process are clearly depicted in Fig. 22.5.

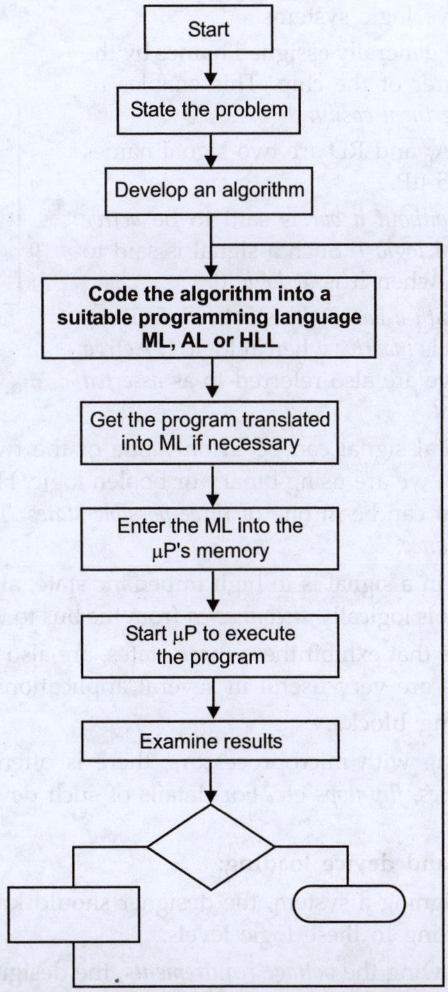

Fig. 22.5. The Programming process.

22.2.6. Digital Logic

Since a microprocessor (μP) is a basically a digital device, therefore, all the signals generated by the μP and data processed, are in digital form.

Analog and digital signals:

- Fig. 22.6 depicts the difference between analog and digital signals.
  - — An *"Analog signal"* is a continuously varying signal. It has *infinitely many levels between a minimum and maximum.*
  - — A *"digital signal"* is characterised by just *two levels.* These levels are denoted by *logical 0 and logical 1.*
  - — In systems that are *positive logic,* the *higher voltage* denotes *logic 1* and the lower voltage denotes *logic 0.* A logic 1 is also referred to as *true,* and logic 0 as *false.* This terminology gets reversed for negative logic systems.

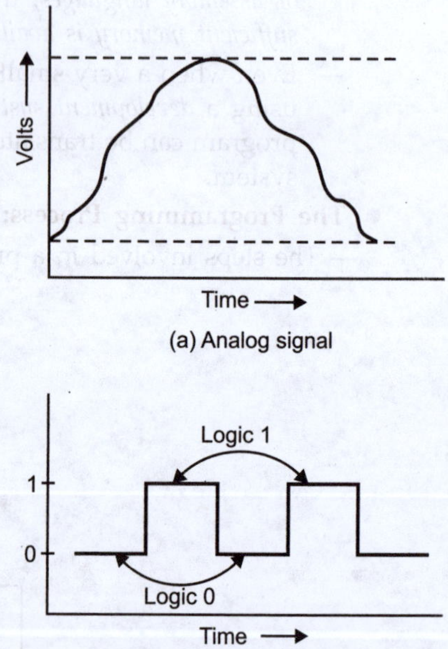

(a) Analog signal

(b) Digital signal

Fig. 22.6. Analog and digital signals

- Signals are generally assigned names by the manufacturer of the chip. This enables us to *reference them easily.*

  Example: A_0 and RD are two signal names in the 8085 μP.

  A signal *without a bar* is said to be *active when it is at logic 1.* Such a signal is said to be *inactive* when it is *at logic 0.*

  A signal *with a bar on it,* is *active when it is at logic 0.* It is *inactive* when at logic 1. Active and inactive are also referred to as *asserted* and *negated.*

- Every digital signal can be in only one of the two stages. This is true in most cases when we are using binary or boolen logic. However some devices generate signals that can be in one of *three possible states.* These states are: *logic 0, logic 1, high impedance.*
  - — When in a signal is in high impedanc state, also known as or *hi_Z* a floating state, it is logically *disconnected* from the bus to which it is physically connected.
  - — Signals that exhibit these three states, are also known as *tri-state signals.* Such signals are very useful in several applications.

Digital building blocks:

When designing with microprocessors, there is often a need to use simple digital devices such as *gates, flip-flops* etc. For details of such devices refer to chapter 6 (Digital electronics).

Signal levels and device loading:

— When designing a system, the designer should know what are the *voltage* levels corresponding to these logic levels.

— Beside knowing the *voltage requirements,* the designer needs to be concerned with the *currents requirements* as well.

Open-collector and Totem-Pole devices:

— The ouputs of several devices which do not have a *pull-up* resistor in their output stage of the internal circuit are specified to be *open-collector* outputs.

Example: The *7405 hex inverter*.

When connecting a device such as an LED, one must use a pull-up resistor when using an open-collector device.

— In totem-pole devices a pull-up resistor is used.

Example: The 74LS04

When using such devices, one can generally connect many devices to the output of one device. The only limitation is as computed by the current-loading considerations.

22.2.7. Types of Microprocessors

The various types of microprocessors are:

1. Vector Processors.
2. Array Processors or SIMD Processors.
3. Scalar and Superscalar Processors.
4. RISC, CISC Processors and EPIC.
5. Digital Signal Processors (DSP)
6. Symbolic Processors
7. I/O Processor
8. Coprocessors
   - Math processors
   - Math coprosessor, Intel 8087
     - Intel 80287, 80C287A
     - Intel 80387 DX
     - Intel 80387 SX
   - Peripheral math processor
   - Graphic processor

Some Other Microprocessors:

1. AMD (Advanced Micro Devices) Microprocessors
2. Sun Microprocessors
   - 1 Ultra SPARC I, II, III, 4+, T1
   - Sun's JVM (Java Visual Machine)
   - Sun Pico Java I, II and Sun MicroJava
   - Sun's MAJC processor
3. MIPS Microprocessors
4. Power PC Microprocessors
5. DEC'S Alpha Microprocessors
6. National Semiconductor Microprocessors
7. ARM Microprocessors
8. Cell Processor
9. i.Max and Cold Fire Processors

10. Motorola Microprocessors
    - Motorola MC6800
    - Motorola MC6809
    - Motorola 68000
    - Motorola 68012
    - Motorola 68020
    - Motorola 68030
    - Motorola 68040
    - Motorola 68060
11. MC 6500
12. Transmeta Corporation's Crusoe Chips
13. Via's C3 Processor
14. Hewlett Packard (HP) Processor.

22.2.8 Processing Speed of a Processor

- *The processing speed of a microprocessor is usually measured in number of instructions executed per second*. The number of instructions executed per second is also known as ***throughout*** of the microprocessor. It is generally expressed in MIPS (Millions of Instructions per Second).
  - — The MIPS rating of a processor *does not* give correct idea of processor's processing ability because instructions differ from processor to processor for high-level language programs. If the same program runs on a RISC as well as CISC processor, the MIPS rating of the RISC processor will be *higher* because the instructions of RISC processor are *simpler* than those of CISC processor.
  - — The floating-point performance of a processor is given MFLOPS (Millions of floating-Point Instructions per Second). MFLOPS rating gives processor's performance for highly mathematical operations.
  - — SPEC (System Performance Evaluation Committee) ratings are used to specify processor's ratings.
- For on-line transaction processing applications, computer performance is measured in TPS (Transactions per second). Such applications demand rapid, interactive processing for a large number of relatively simple transactions. They are supported by very large databases.
- *LINPACK Ratings* are based on a general purpose FORTRON library of mathematical software for *solving dense linear systems of equations of order 100 or higher*.
- *Dhrystone* is a synthetic testing benchmark and measures *integer* performance in KDhrystones/s. Its drawback is its sensitivity to compilers.
- *Whetstone* is a synthetic testing benchmark and measures both *integer and floating-point* performance in KWhetstones/s.
- iCOMP (intel's Comparative Microprocessor Performance Index) consists of a collection of benchmarks to evaluate an intex of relative performance of Intel microprocessors and is based on *integer, floating point, graphics* and *video performance*.

22.3. ORGANISATION OF MICROCOMPUTERS

A "**microcomputer**" *is a small sized and inexpensive computer*. A *'microprocessor' is the heart of a microcomputer*.

- A microcomputer is a device which must be capable of the following *tasks*:

(*i*) *Receiving input* (data and instructions), (*ii*) Performing computations (arithmetic and logical), (*iii*) *Storing data and instructions*, (*iv*) *Displaying the results of any computations*, and (*v*) Controlling all the devices that perform the above mentioned tasks directly or indirectly.

— A device which enables a microcomputer to peform the task (*i*) of the above tasks is known as *"Input device"*:

Typical input devices are: Keyboards, mouse, and toggle switches.

— The device which performs the computations task (*ii*) is known as an "Arithmetic Logic Unit (ALU)".

— *"Memories"* are used to perform task (*iii*) *i.e., storing* and *instructions*. Semiconductor memories are the most commonly used ones. In relatively large computers, Cassettes, Floppy disks and hard disks are used together with semiconductor memories.

— The task of *displaying the results* task (*iv*) computed by the μP, is performed by an *"output device"*.

Some commonly used output devices are: Character printers, cathode ray tubes (CRT) displays, light emitting diodes (LEDs), and laser printers.

— A microprocessor performs tasks (*ii*) and (*v*) in microcomputers.

Fig. 22.7 shows the block diagram of a typical microcomputer. It exhibits various *input/output devices* and *memories* connected to the microprocessor by *busses*.

— On the *"address bus"* the μP transmits the address of the device or a memory location that it needs to address.

— The *"data bus"* is used by the processor to send and receive data to and from different devices. It may be noted that whereas the *address bus is unidirectional*, the *data bus is bidirectional.*

— The *"control bus"* is used for transmitting and receiving control signals between the processor and various devices.

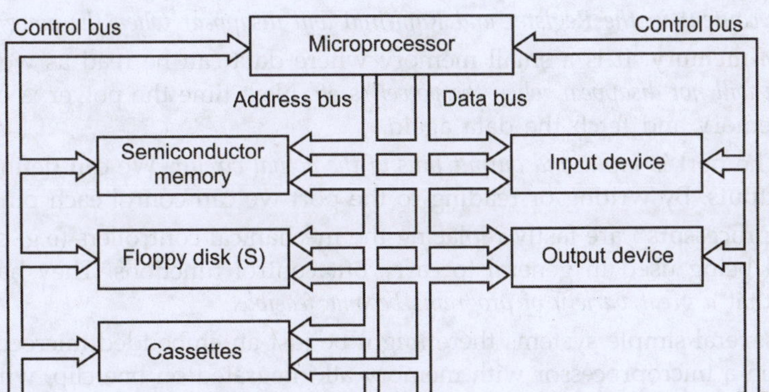

Fig. 22.7. Organisation of a microcomputer.

The microcomputer operates as directed by the user:

— The user writes a sequence of instructions known as a *program* and requests the microcomputer to begin executing each instruction starting from the first one.

— The microprocessor fetches the instructions one by one from the memory and executes them.

22.4. MICROCONTROLLERS

22.4.1. Introduction

The **microcontroller** *is the integration of a microprocessor with memory and input/output interfaces, and other peripherals such as timers, on a single chip.* It is basically *a microcomputer on a single IC.*

Fig. 22.8 shows the simplified block diagram of the microcontroller (microprocessor based controller).

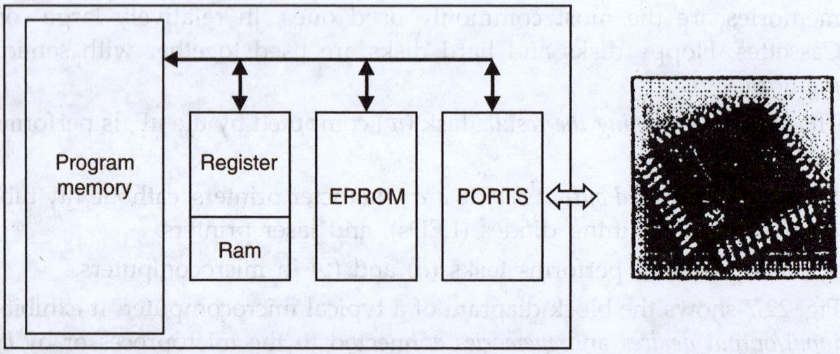

Fig. 22.8. Simplified block diagram of microcontroller.

Program memory. It contains the written program. The program is a set of instructions that the microcontroller performs. The software (instructions) is written in a computer and then programmed burned into the "program memory". This memory is a EPROM memory which can be rewritten thousand times.

Register and Ram box. It contains all the *internal registers and a small Ram* where data can be stored temporarily. *There are several registers with different functions.*

— The Ram memory is not large, *about 64-128 byte.*

— *The content in the Register and Ram-info will disappear when the power is off.*

EPROM-memory. It is a small memory where data can be read as well as written, but the *data will not disappear when the power is off.* Next time the power is on we can go into this memory and fetch the data again.

Ports. The port is *input and output pins of the actual circuit.* We can define the pins as input or outputs. By writing or reading to the port we can conrol each pin as we wish.

"Microprocessors" are fastly replacing the mechanical controllers (e.g. cam-operated switch) and being used in general to carry out control functions. They have the great ***advantage*** that *a great variety of programs become feasible.*

— In several simple systems there might be just an embedded microcontroller, this being a microprocessor with memory all integrated on one clip, which has been specifically programmed for the task concerned.

Examples of Microcontrollers

(*i*) *8-bit microcontrollers (data path 8-bit wide):*

— The Motorola 68 HC11;

— The Intel 8051;

— The PIC16 C6 X/7X.

(*ii*) *16-bit microcontroller:*

— The Motorola 68 HC 16.

(iii) 32-bit microcontroller:

— The Motorola 68300.

- Microcontrollers have limited amounts of ROM and RAM. These are *widely used for embedded control systems.*
- A microprocessor system with separate memory and input/output chips is more suitable for processing information in a computer system.

22.4.2. Characteristics of Microcontrollers

The *characteristics of microcontrollers* are:

1. Low cost
2. Versatility
3. Ease of programming
4. Small size.

- *Microcontrollers are attractive in mechatronic system design since their small size and broad functionality allow them to be physically embedded in a system to perform all of the necessary control functions.*

22.4.3. Applications

Microcontrollers are used in wide array of applications including :

- Entertainment equipment;
- Air planes;
- Home appliances;
- Toys;
- Automobiles;
- Office equipment;
- Trucks.

All these products involve devices that *require some sort of intelligent control*, based on various inputs; ***Examples*** being :

— In a *microwave oven*, the microcontroller monitors the control panel for user input, updates the graphical displays when necessary, and controls the timing and cooking functions.

— In an *automobile*, there are many microcontrollers to control various subsystems, including cruise control, antilock braking, ignition control, keyless entry, environmental control, and air and fuel flow.

— A *toy robot dog* has various sensors to detect inputs from its environment and an on board microcontroller actuates motors to mimic actual dog behaviour based on their input.

— An *office fax machine controls actuators* to feed papers, use photo sensors to scan a page, sends or receives data on a phone line, and provides a user interface complete with menu-driven controls.

All the above mentioned devices are *controlled by 'microcontrollers' and the 'software' running on them.*

- Typically, *'microcontrollers'* have *less than 1 kilobyte to several tens of kilobytes* of program memory, compared with *'microcomputers'* whose ram memory is measured in *megabytes or gigabytes*. Also, microcontroller clock speeds are *slower than those used for microcomputers.*

— A selected microcontroller, for some applications, may not have enough speed or memory to satisfy the needs of the application. The manufacturers of microcontrollers usually provide a wide range of products to accommodate different applications.

— When more memory or Input/Output capability is required, the functionality of the microcontroller can be expanded with *additional components, e.g.*, RAM or EEPROM chips, external A/D and D/A converters, and other microcontrollers.

22.4.4. Classification of Microcontrollers

The classification of microcontrollers is made into *four types* on the following bases:

1. Width of data format.
2. Instruction set.
3. Data types handled.
4. Semiconductor technology adopted in their design.

1. Width of data format:

The *"width of the microcontroller" means the size of the operand upon which the processor can operate.* Based on this we have 4 bit, 8 bit, 16 bit and 32 bit microcontrollers. Of the above *8 bit is most commonly used; 4 bits are used in domestic appliances.*

— *Wider the width of a data format faster will be processor and expensive are the devices using it.*

2. Instruction set:

— On the above basis, microcontrollers may be of two types namely *"Complex Inistruction set computing (CISC)"* or *"Reducing Instruction set computing (RISC)."*

— CISC microcontrollers are *slower* compared to RISC microcontrollers but are *easier to program.*

3. Data types handles:

On the above basis, microcontrollers may be *"Floating point type"* and *"Fixed point type"*. These microcontrollers *differ* in the following *two ways*:

(*i*) Their architecture of ALU is different;

(*ii*) A floating point microcontroller can handle data using floating point format, whereas *additional software is necessary for fixed point-controller* to handle floating point data, which results in *increased processing time.*

4. Semi-conductor technology adopted:

Following are the *most commonly employed semiconductor technologies*:

(*i*) Transistor-Transistor Logic (TTL).

(*ii*) Complimentary Metal Oxide Semiconductor (CMOS).

(*iii*) Emitter Coupled Logic (ECL).

— TTL technology is *commonly used.*

— CMOS consumers low power because of high input impedance of the basic devices built into it.

— ECL technology is used to *design fast processor*.

22.4.5 Description of a Typical Full-featured Microcontroller

Fig. 22.9 shows a block diagram for a *typical full-featured microcontroller*, indicating also the lists of typical external devices that might interface to the microcontroller.

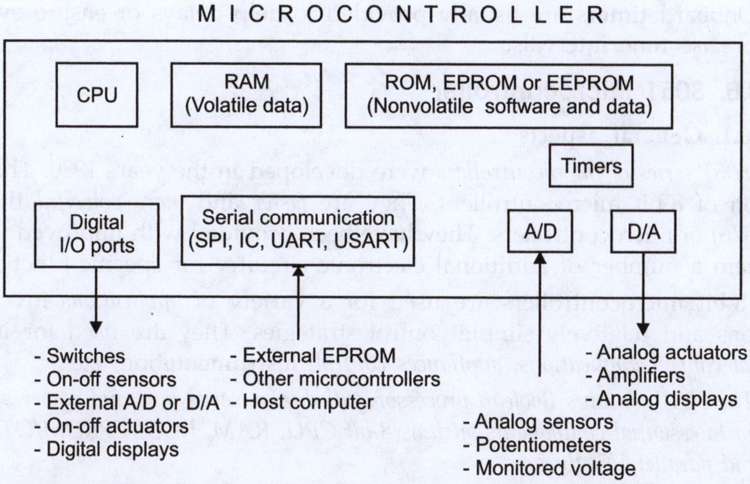

Fig. 22.9. Block diagram for a typical full-featured microcontroller.

The components of a microcontroller include the CPU, RAM, ROM, digital I/O ports, a serial communication interface, timers, A/D converters, and D/A converters.

— The **RAM** is used to *store settings and values used by an executing program.*

— The **ROM** is used to *store the program and any permanent data.* A designer can have a program and data permanently stored in ROM by the chip manufacturer, or the **ROM** can be in the form of *EPROM*, or *EEPROM*, which can be reprogrammed by the user.

Software permanently sotred in ROM is referred to as **firmware.**

Microcontroller manufacturers offer programming devices that can download a compiled machine code file from a PC directly to the EPROM of the microcontroller, usually via the PC serial port and special-purpose pins on the microcontroller. These pins can usually be used for other purposes once the device is programmed. Additional EEPROM may also be available and used by the program to store settings and parameters generated or modified during execution. The data in EEPROM is *nonvolatile*, which means the program can access the data when the microcontroller power is turned off and back on again.

— The **digital I/O ports** allow binary data to be transferred to and from the microcontroller using external pins on the IC. These pins can be used to read the state of switches and on-off sensors, to interface to external analog-to digital and digital-to-analog converters, to control digital displays, and to control on-off actuators.

The I/O ports can also be used to transmit signals to and from other microcontrollers to coordinate various functions.

— The microcontroller can also use a *serial port* to transmit to and from external devices, provided these devices support the same serial communication protocol. There are various standards or protocols for serial communication including SPI (Serial peripheral interface), I^2C (Integrated circuit), UART (Universal asynchronous receiver-trasmitter), and USART (Universal synchronous-asynchronous receiver-transmitter).

— The **A/D converter** allows the microcontroller to convert an external analog voltage (*e.g.*, from a sensor) to a digital value that can be processed or stored by the CPU. The **D/A converter** allows the microcontroller to output an analog voltage to a nondigital device *e.g.*, a motor amplifier.

— Onboard **timers** are usually provided to help delays or ensure events occur at precise time intervals.

22.4.6. 8051 Microcontroller

22.4.6.1. General aspects

The *8051 series of microcontrollers* were developed in the years 1980. They are second generation of 8-bit microcontrollers. They are *faster* and *more powerful* than Intel 8048 series (1976) of microcontrollers. They have been provided with improved instruction set and contain a number of additional electronic circuitry for specific functions.

The 8-bit microcontrollers are used for a variety of *applications* involving limited calculations and relatively simple control strategies. They are used for industrial and *commercial control applications, appliances control*, instrumentation, etc.

— *The 8051 contains Boolean processor, full duplex serial port and power saving circuitry in addition to essential components such as 8-bit CPU, RAM, ROM/EPROM/OTPROM, timer/counter and parallel I/O lines.*

Its family-MCS-51 includes 8031, 8051 and 8751 microcontrollers. The summary of MCS-51 microcontrolles is given in the table 22.3 below:

Table 22.3: Summary of MCS-51

Device	Internal memory		Timer/ Event counters	Interrupts
	Program	Data		
8052AH	8 K × 8 ROM	256 × 8 RAM	3 × 16-Bit	6
8051AH	4 K × 8 ROM	128 × 8 RAM	2 × 16-Bit	5
8051	4 K × 8 ROM	128 × 8 RAM	2 × 16-Bit	5
8032AH	none	256 × 8 RAM	2 × 16-Bit	6
8031AH	none	128 × 8 RAM	2 × 16-Bit	5
8031	none	128 × 8 RAM	2 × 16-Bit	5
8751H	4 K × 8 EPROM	128 × 8 RAM	2 × 16-Bit	5
8751H-12	4 K × 8 EPROM	128 × 8 RAM	2 × 16-Bit	5

- The 8051 microcontrollers have *two versions*: COMMERCIAL and EXPRESS. The operating temperature range for commerical versions is 0°C to 70°C and that for the express version is 40°C to +85°C.

Fig. 22.10 shows the block diagram of Intel 8051.

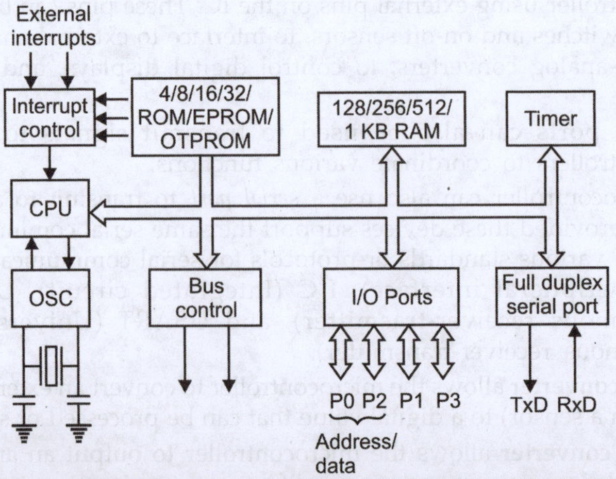

Fig. 22.10. Block diagram of 8051 microcontroller.

8051 microcontroller is a *generic part* number that includes entire family of microcontroller ranging from numbers 8031 to 8751 and is available in N-channel intel oxide silicon (NMOS) and complementary metal oxide silicon (CMOS) in variety of packages.

8051, basically, is a 40-pin DIP integrated circuit that includes the following *features*:

1. Internal RAM and ROM
2. I/O ports with programmable pins
3. Timers and counters
4. Serial data communication.

22.4.6.2. Architecture of 8051 micro controller

The *different parts of the 8051 microcontroller* are as follows:

1. Central processing unit (CPU), which includes program counter, ALU, working registers and block circuit.
2. 8-bit CPU with registers (the accumulator) and B
3. 16-bit program counter (PC) and data pointer (DPTR)
4. 8-bit program status word (PSW)
5. 8-bit stack pointer (SP)
6. Internal ROM of 4 KB
7. Internal RAM of 128 bytes
8. Four register banks each containing 8 registers.
   - Sixteen bytes, which may be addressed at bit level
   - Eighty bytes of general-purpose data memory
9. 32-input/output pins arranged as four 8-bit ports P_0–P_3
10. Two 16-bit timer/counters T_0 and T_1.
11. Full duplex serial data receiver/transmitter SUBF
12. Control registers—TCON, TMOD, SCON, PCON, IP and IE
13. Two external and three internal interrupt sources
14. Oscillator and clock circuits.

22.4.7. Microchip Controllers

- The microchip microcontrollers use a form of architecture termed Harvard architecture (Fig. 22.11). With this architecture, instructions are fetched from program memory using accessible variables.
- Harvard architecture enables faster execution speeds to be achieved for a given clock frequency.

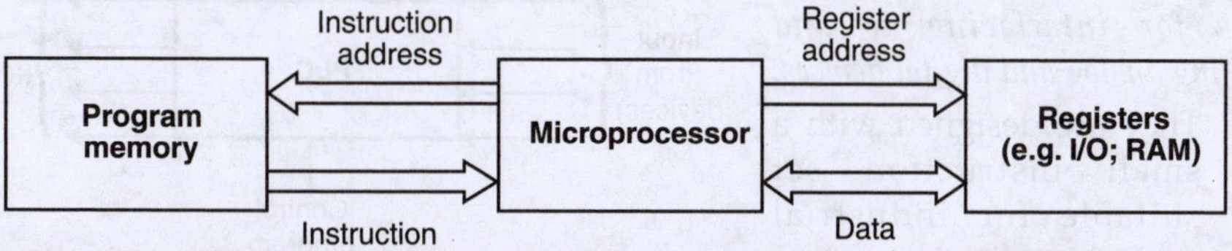

Fig. 22.11. Harvard architecture.

22.4.8. Selection of a Microcontroller

While selecting a microcontroller the following *factors* should be considered :

1. Number of input/output pins.

2. Interfaces required.
3. Memory requirements.
4. The number of interrupts required.
5. Processing speed required.

22.4.9. Comparison of Microprocessors and Microcontrollers

The comparison of microprocessors and microcontrollers is given the table 22.4 below:

Table 22.4: Comparison between Microprocessors and Microcontrollers

S. No.	Aspects	Microprocessors	Microcontrollers
1.	*Components/parts*	CPU, interrupt circuits and memory-addressing circuits.	Besides all parts is of micro-processor, they also contain timers, parallel and serial I/O, internal RAM and ROM.
2.	*Access times for memory and I/O devices*	More	Less
3.	*Number of operational codes (for moving data from external memory to CPU)*	Many	One or two
4.	*Nature of deal with rapid movement of codes and data*	For external address to chip.	Within the chip.
5.	*Operation as a digital computer*	By adding external digital parts.	Without adding external digital parts.
6.	*Memory map for data and code.*	Single	Separate
7.	*Hardware used*	More	Less
8.	*Flexibility from design point of view*	More	Less

22.5. PROGRAMMABLE LOGIC CONTROLLERS (PLCs)

22.5.1. Introduction

PLCs *are specialised industrial devices for interfacing to and controlling analog and digital devices.*

— They are designed with a small instruction set suitable for industrial control applications.

— They are usually programmed with *"ladder logic"*, which is graphical method of laying out the connectivity and logic between system inputs and outputs.

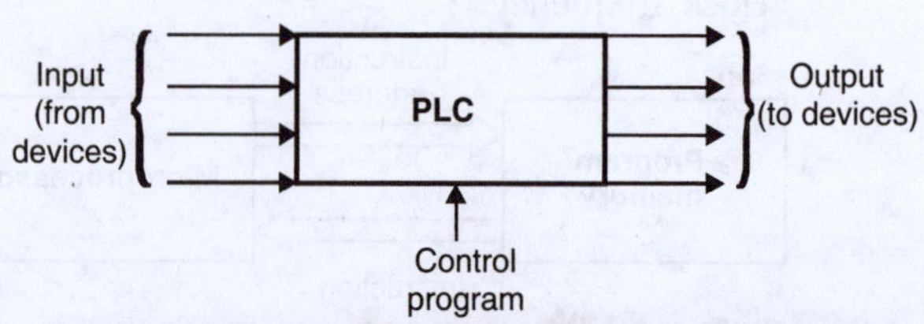

Fig. 22.12. Programmable logic controller.

— They are designed with industrial control and industrial environments specifically in mind. Therefore, in addition to being *flexible and easy to program, they are robust and relatively immune to external interference.*

• A programmable logic controller (first conceived in 1968), is shown in Fig. 22.12. *It is a "digital electronic device" that uses a programmable memory to store instructions and to implement functions such as logic sequencing, timing, counting and arithmetic in order to control machines and processes.*

It has been specifically designed to *make programming easy.*

22.5.2. Advantages of a PLC:

The advantages of a PLC are :

1. *Low cost.*
2. *Easy to install.*
3. Ensures *increased productivity.*
4. *Faster* operational speed.
5. Provides *high reliability and easier maintenance.*
6. Can *withstand* harsh industrial environment/manufacturing environment.
7. Provides *consistency* in manufacturing.
8. Can input/output *both analog and digital signals.*
9. *Reduced cost of scrap and rework* (since the user/designer can change the PLC program without changing any hardware or scraping the present hardware).
10. *Small size.*
11. PLC modules can be *added,* depending upon the input/output requirements.
12. *Easier troubleshooting.*

22.5.3. Selection of a PLC

For selection of a PLC, the following *criteria* need to be considered:

1. *Types of inputs/outputs* required, such as:
   — Isolation;
   — Out-board power supply for inputs/outputs;
   — Signal conditioning.
2. *Input/Output capacity* required.
3. *Size of memory* required. This is linked to the number of inputs/outputs and the complexity of program used.
4. *Speed and power* required for CPU—This is linked to the number of types of instructions that can be handled by a PLC.

22.5.4. Uses of PLCs

PLCs are widely used and *extend from small-contained units for use with perhaps 20 digital inputs/outputs to modular systems which can be used for large numbers of inputs/outputs, handle digital or analog inputs/outputs, and also carry out PID control modes.*

22.5.5. Special features

Although PLCs are similar to computers, yet they have the following *specific features* to their use as controllers:

1. The interfacing for inputs and outputs is *inside* the controller.
2. *Easily programmable.* They have an easily understood programming language.
   — Programming is mainly concerned with *logic* and *switching operation.*
3. *Rugged* and *designed to withstand vibrations, temperatures, humidity and noise.*

22.5.6. Architecture basic internal structure

Fig. 22.13 shows the architecture/internal structure of a programmable logic controller (PLC):

A PLC consists of the following *main components* :

1. Central processing unit (CPU);
2. Memory;
3. Input/Output circuitry.

1. *Central processing unit (CPU) :*

- It controls and processes all the operations within the PLC.

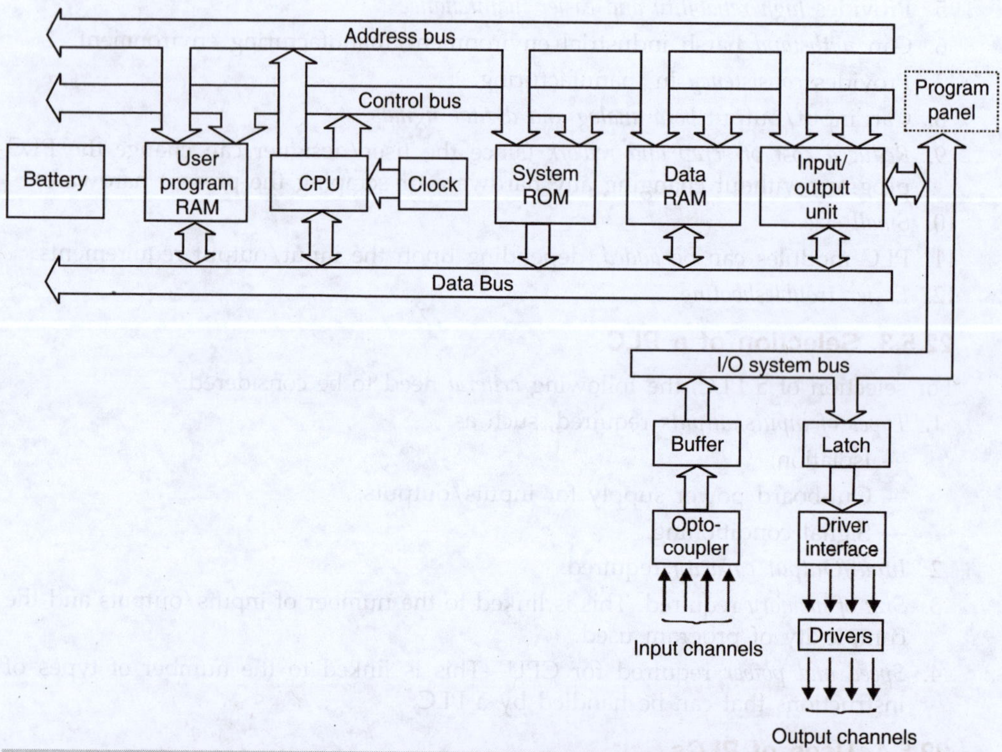

Fig. 22.13. Architecture of a programmable logic controller (PLC).

- It is provided with a *"clock"* with a frequency of typically between 1 and 8 MHz. This frequency determines the operating speed of the PLC and provides the timing and synchronisation for all elements in the system.
- A *"bus system"* carries information and data to and from the CPU, memory and input/output units.

2. *Memory:* The various memory elements available in a PLC are:

(*i*) A system ROM to give permanent storage for the operating system and fixed data.

(ii) RAM for user's program.

(iii) Temporary buffer stores for input/output channels.

- The programs in RAM can be changed by the user. However, to prevent the loss of these programs when the supply is switched off a battery is likely to be used in the PLC to maintain the RAM contents for a period of time.
- Specifications for small PLCs often specify the program memory size in terms of the number of program step (A program step is an instruction for some event to occur) that can be stored. Typically the number of steps that can be handled by a small PLC is of the order of 300 to 1000, which is generally more than adequate for most control situations.

3. ***Input/Output (I/O) circuitry:*** The I/O unit provides the interface between the system and outside world.
   - Programs are entered into the I/O unit from a panel which can vary from small keyboards with liquid crystal displays to those using a visual display unit with keyboard and screen display. The programs, alternatively, can be entered into the system by means of a link to a personal computer which is loaded with an appropriate software package.
   - The I/O channel provides signal conditioning and isolation functions so that sensors and actuators can be generally directly connected to them without the need for other circuitry.
   - The basic form of programming commonly used with PLCs is *ladder programming*. This involves each program task being specified as though a rung of a ladder.

   Following methods can be used for *I/O processing* :

   1. Continuous updating. 2. Mass I/O copying.

Timers: The timers are commonly regarded as relays with coils which, when energised, result in the closing or opening of input contacts after some preset time.

A *timing circuit* is specified by stating the interval to be timed and the conditions or events that are to start and/or stop the timer.

- PLCs are generally provided with only *delay-on timers, i.e.,* a timer which comes on after a time delay.

Internal relays:

- These relays are often used when there are programs with multiple input conditions.
- The internal relays are also used for the starting of multiple outputs.

Counters: The use of counters is restored to when there is *a need to count a specified number of contact operations.*

- Counter circuits are supplied as an internal feature of PLCs.

Shift registers: Several internal relays can be grouped together to form a *register* which can provide a storage area for a series sequence of individual bits. Thus a 4-bit and a 8-bit registers would be formed by using four and eight internal registers respectively.

The term *shift register* is used because the bits can be shifted along by one bit when there is a suitable input to the register.

Shift registes have *three inputs*:

- One to load data into the first element of the register (OUT);
- One as the shift command (SFT);
- One for resetting (RST).

22.5.7. Input/Output Processing and Programming

Input/Output processing:

"Ladder programming" is the *basic form of programming commonly used with PLCs. This involves each program task being specified as through a rung of a ladder*. Thus such a rung could specify that the state of switches *A* and *B*, the inputs, be examined and if A and B are both closed then a solenoid, the output, is energised.

A PLC, when carrying out a program, follows the sequence indicated below:

1. Scanning the inputs associated with a rung of the ladder.
2. Solving the logic operation involving these inputs.
3. Setting/reseting the outputs for that rung.
4. Moving on to the next rung and repeating operations 1, 2, 3.
5. Moving on to the next rung and repeating operations 1, 2, 3.
6. Moving on to the next rung and repeating operations 1, 2, 3; continuing until the end of the program.

For *input/output processing the following two methods* are used:

1. Continuous updating:

In this method the *CPU scans the input channels* as they occur in the program instructions. Each input point is examined individually and its effect on the program determined.

There will be a *built in delay*, typically about 3 ms, when each point is examined in order to ensure that only *valid* input signals are read by the microprocessor. This delay enables the microprocessor to *avoid counting* an input signal twice, or more frequently, if there is contact bounce at a switch. A number of inputs may have to be scanned, each with a 3 ms delay, before the program has the instruction for a logic operation to be executed and an *output to occur*. The *outputs are latched* so that they *retain* their states until the next updating.

2. Mass input/output copying:

In case of continuous updating method since there has to be a 3 ms delay in each point, therefore, the time taken for examining several hundred input/output points can become comparatively *long*.

In order to have a *more rapid execution of program, a specific area of RAM is used as a buffer store between the control logic and input/output unit*. Each input/output has an adress in this memory. At the start of each program cycle the CPU scans all the inputs and copies their status into the input/output addreses in RAM.

As the program is executed logic operations are carried out. The resulting output signals are stored in the reserved input/output section of RAM. *At the end of each program cycle all the inputs are transferred from RAM to the output channels*. The outputs are latched so that they retain their status until the next updating.

Programming:

When a PLC programming is carried out, using ladder diagrams, a program is written in a similar manner to drawing a switching circuit.

The ladder diagram consists of two *vertical lines* representing the *power raits*. *Circuits* are connected as *horizontal lines*, i.e., the rungs of the ladder between these two verticals. While, drawing the circuit line for a rung, *inputs must always precede*, *outputs* and there *must be at least one output on each line. Each rung must start with an input or a series of inputs and end with an output*. The inputs and outputs are *numbered*, the notation used depending upon the manufacturer of PLC.

- *Logic function:* AND, OR, NOR, NAND, EXCLUSIVE – OR(XOR)
- *Latching:* Sometimes there are situations where it is necessary to hold a coil energised, even when the input which energises it ceases. The circuit which is used to carry out such an operation is termed as *latch circuit*. It is a *self-manufacturing circuit in that, after being energised, it maintains that state until another input is received; it remembers its last state.*
- *Sequencing:* Invariably there are control situations where sequences of outputs are required, with the switch from one *output* to another being conrolled by sensors.

22.5.8. Mnemonics

In a *ladder program* each horizontal rung on the ladder represents a line in the program and entire ladder gives the complete program in 'ladder language'. The programmer can enter the program into the PLC using a keyboard with the graphic symbols for the ladder elements, or using a computer screen and a mouse to select symbols, and the program panel or computer than translates these symbols into machine language that can be sotred in the PLC memory.

Alternatively, a *program can also be entered by translating the ladder program into "**mnemonics**"*, each code corresponding to a ladder element, and then enter these into the programming panel or computer. These are then translated into machine language. The mnemonics used by different PLC manufacturers differ.

22.5.9. Master and Jump Controls

A whole block of outputs can be simultaneously turned off or on by using the same internal relay contacts in each output rung so that switching it on or off affects every one of the rungs. An *alternative way of programming* to achieve the same effect is to use a ***master relay***.

- *Jumps.* A function which is often provided with PLCs is the *conditional jump* function. Such a function enables programs to be designed so that if a certain condition exists then a *section of program is jumped.*

22.6. MICROPROCESSOR-BASED INSTRUMENT SYSTEM – INTELLIGENT INSTRUMENT

An instrument which operates on microprocessor system or is microprocessor-based itself is called **Intelligent instrument.**

The *microprocessor-based systems* (in which involvement of human being is completely eliminated or restricted) are employed for wide applications in industries *for monitoring the data as well as controlling the processes.*

Fig. 22.14 shows block diagram of a *microprocessor-based system* :

It consists of *"hardware section"* which includes the following :

— Transducer ;

— Signal conditioning device/element ;

— Analog-to-digital (A/D) converter ;

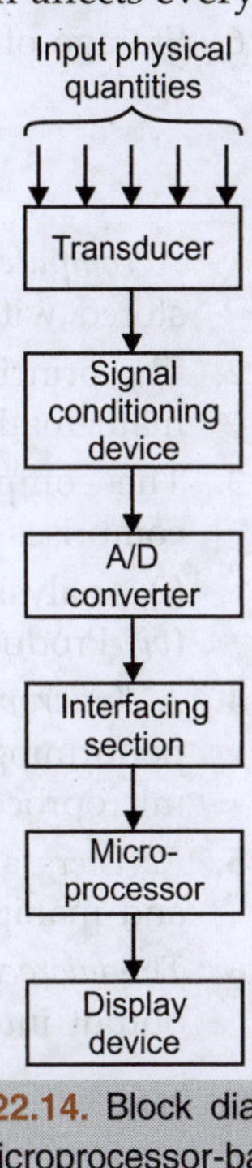

Fig. 22.14. Block diagram of microprocessor-based system/instrument.

— Interfacing section ;
— Microprocessor ;
— Display devices.

The "software section" contains the following :

— Programs ;
— Subroutine for controlling various processes ;
— Calibration ;
— Averaging ;
— Totalization etc.

- The overall designing of an intelligent instrument is employed in the following :

— Linear circuits ;
— Digital electronics ;
— Control elements ;
— Controllers ;
— Software.

Advantages. The microprocessor-based instruments entail the following *advantages* :

1. Higher flexibility.
2. Increased operational speed.
3. Fast and automatic control actions feasible.
4. Easy to change the system's configuration by altering the software section.
5. Functions like averaging, linearization, totalization, calibration, alarming, etc. become automated.
6. Storage of information or data is available through memory interfacing facility.

HIGHLIGHTS

1. A *computer* is a machine that processes data according to set of instructions stored within the machine.
2. The principle of operation of *analog computers* is to create a physical analog of mathematical problems. The *digital* computers *accept digits and alphabets as inputs.*
3. The complete programming process followed by programmer for writing comprises the following *steps:*
   (*i*) Analysis; (*ii*) Flow charting; (*iii*) Coding; (*iv*) Debugging; (*v*) Documentation; (*vi*) Production.
4. A *"microprocessor"* is a large scale integration (LSI) chip that is capable of performing arithmetic and logic functions as defined by a given programme. A microprocessor consists of: (*i*) ALU; (*ii*) Register unit; (*iii*) Control unit.
5. *Registers* are digital devices used by the microprocessor for temporary storage and manipulation of data and instructions.
6. The *microcontroller* is the integration of a microprocessor with memory and input/output interfaces, and other peripherals such as liners, on single chip.

OBJECTIVE TYPE QUESTIONS

Fill in the blanks or Say 'Yes' or 'No'

1. Charle's Babbage is called "Father of computers".
2. A .... is a machine that processes data according to set of instructions stored within the machine.
3. Speedometer is an example of analog computer.
4. .... are popularly known as personal computer (PC).
5. A hybrid computer is a combination of both analog and digital computers.
6. Chips are rated in terms of their .... and ....
7. RAM chip is made with Metal Oxide Semiconductor (MOS).
8. .... are used to multiply a variable quantity by a constant.
9. Servomultipliers are used to multiply two variables.
10. A .... is a large scale integration (LSI) chip that is capable of performing arithmetic and logic functions as defined by a given programme.
11. Microprocessors which have memory and various input/output arrangements on the same chip are called ....
12. .... are memory locations within the microprocessor.
13. MAR (Memory address register) contains the address of data.
14. General purpose registers serve as .... storage for data or addresses and used in operations involving transfers between other registers.
15. .... are the paths along which digital signals move from one section to another.
16. .... bus carries the signals relating to control actions.
17. The size of the memory is determined by the number of wires in the address bus.
18. .... is a device in which data is stored permanently.
19. RAM is a memory that can be read only.
20. Intel 8085 is an 8-bit, NMOS microprocessor.

ANSWERS

1. Yes	2. Computer	3. No	4. Microcomputers
5. Yes	6. Capacity, speed	7. Yes	8. Attenuators
9. Yes	10. microprocessor	11. microcontroller	12. Registers
13. Yes	14. temporary	15. Buses	16. Control
17. Yes	18. ROM	19. No	20. Yes.

THEORETICAL QUESTIONS

1. What is a 'Computer'? Explain.
2. List the characteristics of a computer.
3. What are the limitations of a computer?
4. How are computers classified?
5. How are digital computers classified on the basis of size and capabilities?
6. Explain briefly the following :
   Super computer; Main frame computers; Minicomputer; Microcomputers.
7. What are the differences between analog and digital computers?
8. Draw the block diagram of a computer and explain briefly its various parts.
9. How are chips rated?

10. What do you mean by the term 'Peripheral'? Explain briefly the following devices:
(*i*) Input devices; (*ii*) Output devices.
11. Explain briefly 'storage devices'.
12. List the steps which are required for computer programming process for writing programs.
13. What is a "Microprocessor"?
14. Draw the block diagram of a microcomputer and explain briefly the three segments (ALU, Register and Control unit) of a microprocessor.
15. What are the characteristics of microprocessor?
16. Mention the important features of the microprocessors.
17. What are the uses of microprocessors?
18. Explain briefly the various parts of a microprocessor system.
19. Explain briefly the following registers:
Accumulator register; Status register; Program counter register (PC); memory address register; Instruction register; General purpose registers; Stack pointer register.
20. What are 'buses'? Explain briefly the following buses:
Data bus, Address bus, Control bus.
21. Explain briefly the following forms of memory unit:
ROM; PROM; EPROM; EEPROM; RAM.
22. Explain briefly Intel 8085 microprocessor with the help of a block diagram.
23. Write a short note on 'Microprocessor programming'.
24. What is a 'Microcontroller'? Explain briefly.
25. How are microcontroller classified? Explain briefly.
26. Give the description of a typical full-featured microcontroller.
27. Describe briefly the 8051 microcontroller.
28. What are 'Microchip controllers'?
29. What are programmable logic controllers (PLCs)?
30. What are the advantages of PLCs?
31. State the special features of PLCs.
32. Explain briefly with a neat diagram, the Architecture of a programmable logic controller (PLC).
33. Explain briefly programming of a PLC.

SHORT ANSWER QUESTIONS

Short Answer Questions

Q.1. *What is an instrument? What are the primary requirements of any measuring instrument?*

Ans.
- ***Instrument*** may be defined as a device for measuring the value or magnitude of a quantity or variable.
- The primary requirements for any measuring instrument are :
  (*i*) Power consumption should be as small as possible.
  (*ii*) No alteration in the circuit conditions with the induction of instrument in the circuit.

Q.2. *List the types of instruments that are mainly used as ammeters and voltmeters.*

Ans. The following types of instruments are mainly used as ammeters and voltmeters :

(*i*) PMMC; (*ii*) Moving iron; (*iii*) Electrodynamometer; (*iv*) Hot wire and thermocouple; (*v*) Electrostatic; (*vi*) Induction type; (*vii*) Rectifier.

Q.3. *What are the main advantages of 'Rectifier instruments'?*

Ans. Following are the main advantages of rectifier instruments :

(*i*) They can operate well in the RF (radio-frequency) range with proper use of silicon or germanium rectifiers

(*ii*) Measurements of current and voltage at several hundred megahertz are possible.

(*iii*) This type of meter is more sensitive than any other type of A.C. meter.

Q.4. *What is meant by the term "Measurand"?*

Ans. The quantity to be measured is known as the ***measurand***.

Q.5. *Why absolute instruments are rarely used in practice?*

Ans. Absolute instruments are rarely used in practice because working with these instruments is time-consuming.

Q.6. *What is difference between analog and digital signals?*

Ans. *"Analog signals"* vary in the 'continuous fashion' and assimilate an infinite number of values in any given range, while the 'digital signals' vary in discrete steps and thus comprise only finite different values in a given range.

Q.7. *What is meant by resolution of an instrument?*

Ans. The ***resolution*** of any instrument is the smallest change in the input signal (quantity under measurement) which can be detected by the instrument. It may be expressed as an actual value or as a fraction or percentage of the full-value scale.

Q.8. *What are systematic errors?*

Ans. 'Systematic errors' remain constant or change according to a definite law on repeated measurement of given quantity. These errors can be evaluated and their influence on the results of measurement can be eliminated by the introduction of proper corrections.

Q.9. *What is the difference between fundamental and derived units?*

Ans. The size of the 'fundamental units' is quite arbitrary and can be selected to fit a certain set of circumstances i.e., mass, length, time, etc. whereas the 'derived units' can be expressed in terms of fundamental units with the help of physical equations *i.e.,* area, volume, force etc.

Q.10. *What is the difference between primary and secondary standards?*

Ans. *"Primary standards"* are absolute standards of such high accuracy that they can be used as the ultimate reference standards while the *"secondary standards"* are the basic reference standards used in industrial measurement laboratories. Secondary standards are calibrated and compared against the primary standards in National standard laboratories on periodic basis.

Q.11. *What is Josephon effect?*

Ans. A flow of current between two pieces of a superconducting material separated by an insulator, in known as ***Josephon effect***.

Q.12. *Why is mica used as dielectric in precision capacitors?*

Ans. Mica capacitors are used as precision capacitors because of their 'small tolerances' and 'high stability' under varying temperature and electric stress. Mica is easily available in the form of very thin sheets and have a high resistivity and 'dielectric strength' both of which characteristics are necessary for the purpose for which these are used.

Q.13. *How is gravity control inferior to spring control?*

Ans. The scales of gravity controlled instruments are not uniform but are crowded in the beginning. Secondly, these are to be used in vertical position in order that the control may operate.

Q.14. *On what principle does D'Arsonval galvanometer operate?*

Ans. The following principle is employed in galvanometers.

'When a current carrying conductor is placed in a magnetic field, a mechanical force is experienced on a conductor which tends to move the conductor in a direction given by Flemming's left hand rule'.

Q.15. *What is ballistic sensitivity?*

Ans. ***Ballistic sensitivity*** is defined as the first maximum deflection in mm caused on a scale at a distance of one metre from mirror by a unit charge suddenly passed in the galvanometer.

Q.16. *What is the difference between an ammeter and a voltmeter?*

Ans. An *"ammeter"* is a low resistance indicating instrument while a *"voltmeter"* is high resistance one.

Q.17. *List the advantages of Electrodynamic instruments.*

Ans. *Advantages:*

1. Can be used on both D.C. as well as A.C. systems.
2. Free from hysteresis and eddy current errors.
3. It is possible to construct ammeters upto 10A and voltmeters upto 600 V with precision grade accuracy.

Q.18. *Write different methods of measurement of medium resistance.*

Ans. The methods of measuring medium resistances are :

(*i*) Voltmeter ammeter method; (*ii*) Ohm-meter method;
(*iii*) Bridge method; (*iv*) Substitution method.

Q.19. *What is meant by drift?*

Ans. By *"drift"* we mean variation in the value of quantity under measurement with time. When it is repeatedly measured, if there is no change in the value, it means zero drift.

Q.20. *What are the essentials of indicating instruments?*

Ans. Indicating instruments consist essentially of a pointer moving over a calibrated scale and attached to the moving system pivoted on jewelled bearings. For satisfactory working of indicating instruments the torques required are :

(*i*) Deflecting torque; (*ii*) Controlling torque; (*iii*) Damping torque.

Q.21. *What are prepayment meters?*

Ans. *"Prepayment meter"* is special type of house supply meter, which itself is essentially a standard pattern energy element with the addition of a coin collector box and the prepayment mechanism geared to the recorder train.

Q.22. *Why are the rectifier instruments indispensable for measurements in communication circuits?*

Ans. The *"rectifier instruments"* are indispensable for measurements in communication circuits' since they have very high sensitivity and as such the power required for their operation is small enough to be within the capacities of the circuits.

Q.23. *What materials are used for ammeter shunts?*

Ans. 'Manganin' for D.C. circuits and 'constantan' for A.C. circuits.

Q.24. *How is range of an electrostatic voltmeter increased?*

Ans. The range of an electrostatic voltmeter is increased by employing a 'multiplier' in the form of either a 'resistance' or 'capacitor potential divider'.

Q.25. *How does a current transformer differ from a conventional transformer?*

Ans. The current transformer (CT) is usually a step-up transformer in terms of primary-secondary turn-ratio. It differs slightly in operation from that conventional transformer. In case of a CT, as the load impedance or 'burden' on the secondary is very small so the CT operates on short-circuit conditions. Also the current in secondary is not governed by load impedance on the secondary but depends upon the current flowing in the primary winding (i.e. main circuit current).

Q.26. *What are instrument transformers? How do they differ from power transformers?*

Ans.
- Instrument transformers are devices used with instruments (Ammeter, voltmeter etc.) for A.C. measurement. They *increase* the range of the measuring instruments.
- The power transformers work at 50 Hz and used for stepping up and down the voltage or current. These are connected between generating station and the receiving station.

Q.27. *Define transformation ratio of a current transformer.*

Ans. ***Transformer ratio*** in case of instrument transformers, in general, is defined as the ratio of the primary phasor to the secondary phasor.

Transformation ratio (*Actual*) for a CT,

$$K_{act} = \frac{\text{Primary winding current (Actual)}}{\text{Secondary winding current (Actual)}}$$

Transformation ratio (*Nominal*) for a CT,

$$K_{nom.} = \frac{\text{Rated primary winding current}}{\text{Rated secondary winding current}}$$

Q.28. *What is ratio correction factor (RCF)?*

Ans. The ***ratio correction factor*** is the transformation (actual) ratio divided by nominal ratio.

In other words, $$RCF = \frac{K_{act.}}{K_{nom.}}$$

- The ratio marked on transformers is their nominal ratio.

Q.29. *What do you understand by the term 'Burden of CT'?*

Ans. The product of voltage and current on the secondary side when the CT is supplying the instrument with its maximum rated value of current, is known as ***rated burden*** and is expressed in volt-ampere.

Q.30. *Why the secondary of current transformer is never kept open-circuited?*

Ans. The secondary of current transformer is never kept open-circuited because of the following two *reasons* :

(*i*) There will be no secondary m.m.f. and since the primary current (and m.m.f) is fixed, the core flux will increase enormously. This will cause large eddy current and hysteresis losses and the resulting high temperature may damage the insulation or even the core.

(*ii*) A very high voltage will be induced in the multi-turn secondary and this high voltage may be dangerous both to life and to the insulation.

Q.31. *How can average reading VTVM be converted to a RMS reading voltmeter?*

Ans. The average reading of VTVM can be converted into a RMS reading voltmeter by calibrating its meter scale in terms of the r.m.s. value of a sinusoidal wave.

Q.32. *What is meant by the deflection factor of a CRO?*

Ans. The deflection factor of a CRO is the reciprocal of the deflection sensitivity.

Q.33. *What is a Lissajous pattern?*

Ans. Lissajous is a pattern that results from applying periodic signals to the deflection plates of a CRO.

Q.34. *List the factors that may lead to inaccuracies in measurement by A.C. bridges.*

Ans. The factors that may lead to inaccuracies in measurement by A.C. bridges are :

(*i*) Stray conductance effect due to imperfect insulation;

(*ii*) Mutual inductance effect due to magnetic coupling between various components of the bridge.

Q.35. *Name different types of frequency meters.*

Ans. The frequency meters are of following types :

(*i*) Mechanical resonance type; (*ii*) Electrical resonance type; (*iii*) Weston type; (*iv*) Ratiometer type; (*v*) Saturable core type; (*vi*) Digital frequency meters.

Q.36. *Why is sensitivity of the A.C. section of multimeter less than the D.C. section?*

Ans. Due to complex circuitry, the sensitivity of A.C. section of a multimeter is less than the D.C. section.

Q.37. *For what VTVM stands for?*

Ans. The VTVM stands for vacuum tube volt meter.

Q.38. *What is Cathode Ray Oscilloscope (CRO)?*

Ans. CRO is a device that allows the amplitude of electrical signals, whether they be voltage, current, power, etc. to be displayed primarily as a function of time. It is basically an electron beam voltmeter.

It is very useful and versatile laboratory instrument used for display, measurement and analysis of waveforms, and other phenomena in electrical and electronic circuits.

Q.39. *Name main components of a cathode ray tube (CRT).*

Ans. The main components of a CRT are :

(*i*) Electron gun assembly; (*ii*) Deflection plates assembly;
(*iii*) Fluorescent screen; (*iv*) Glass envelope;
(*v*) Base.

Q.40. *Define a transducer and distinguish between a sensor and a transducer.*

Ans.
- A ***transducer*** is a device which converts the energy from one form to another.
- A ***sensor*** senses the condition, state or value of the process variable and produces an output which reflects this condition, state or value. A *transducer* transforms the energy of the process variable to an output of some other type of energy which is able to operate some control device. Sometimes a *secondary transducer* may be employed to transform the output of *primary sensor* to still another type of energy.

Q.41. *What is 'LVDT'?*

Ans. ***LVDT*** (Linear-variable-differential transformer) is a passive inductive transducer and is commonly employed to measure force (or weight, pressure and acceleration, etc. which depend on force) in terms of the amount and direction of displacement of an object.

Q.42. *What is frequency counter?*

Ans. A ***frequency counter*** is a digital instrument that can measure and display the frequency of any periodic waveform.

Q.43. *What is a load cell and what for is it used?*

Ans. A ***load cell*** is an electromechanical device that converts weight or force into an electrical signal and is widely used for measurement of static and dynamic forces.

Q.44. *What is IEEE 488 bus system?*

Ans. For a set of instruments to be interfaced with a single computer, a general purpose bus called the "IEEE 488 bus system", has been developed to which a number of instruments can be attached so that communication between the computer and the instruments is possible over a common cable.

Q.45. *What is meant by loading effect?*

Ans. The incapability of the system to faithfully measure, record, or control the input signal (measurand) in undistorted form is known as ***loading effect***.

Q.46. *What are the special features of vibration galvanometers?*

Ans. The special features of vibration galvanometers are large sensitivity, small damping, and large restoring constant.

Q.47. *What are the shortcomings of multimeter?*

Ans. It may be noted that A.C. sensitivity of a half-wave A.C. meter is only 45% of the D.C. sensitivity, therefore, the loading effect is greater while measuring A.C. voltages than that while measuring corresponding D.C. voltages.

The rectifying action depends upon the frequency to some extent due to capacitance effect and rectifier instruments show lower readings. Instrument indications may be in error by as much as 0.5% decrease for every 1 kHz rise in frequency.

Q.48. *What is the usual current-rating of CT's secondary?*

Ans. The usual current rating CT's secondary is 5*A*.

Q.49. *How is electrostatic wattmeter superior to other types of wattmeters?*

Ans. An electrostatic wattmeter is free from errors on account of waveforms, frequency, and eddy currents.

Q.50. *What is the difference between an energy meter and a wattmeter?*

Ans. The essential difference between an 'energy meter' and a 'wattmeter' is that the former is fitted with some type of registration mechanism whereby all the instantaneous readings of power are summed over a definite period of time, whereas the latter indicates the value of power at particular instant when it is read.

Q.51. *What is meant by meter constant of an energy meter?*

Ans. The number of revolutions made by the aluminium disc for 1 kWh of energy consumption is called 'meter constant' of an energy meter.

Q.52. *What is trivector meter?*

Ans. *"Trivector meter"* is an instrument which measures kVAh and also kVR of maximum demand. It consists of a kWh meter and a reactive kivolt ampere-hour meter in the common case with a special summation between them.

Q.53. *On what principle does Weston frequency meter operate?*

Ans. *"Weston frequency meter"* is a moving iron instrument which operates on the principle of variation of impedance in an inductive circuit with the variation in supply frequency.

Q.54. *Distinguish between, D.C. and A.C. potentiometers.*

Ans. In 'D.C. potentiometers' only magnitude of unknown e.m.f. is measured whereas in 'A.C. potentiometers' magnitude and phase of the unknown e.m.f. is measured. Accuracy of measurement of D.C. potentiometer is higher as compaerd to a A.C. potentiometer.

Q.55. *Why is screening of bridge elements required?*

Ans. If each component in a bridge has a definite screen connected to one end, a very high accuracy in measurement can be achieved by using 'Wagner's earthing device' in the bridge circuit. This device removes all the earth capacitances from the bridge network.

Q.56. *Why are Ballastic (or D.C) tests conducted?*

Ans. (*i*) To determine flux/flux density in a specimen; (*ii*) To determine B.H. curves; (*iii*) To plot hysteresis loop.

Q.57. *Name the two methods available for the determination of B.H. curve of a specimen.*

Ans. (*i*) Method of reversals; (*ii*) Step-by-step method.

Q.58. *What is the function of a magnetic potentiometer?*

Ans. The 'magnetic potentiometer' measures the magnetic potential drop along any part of a magnetic circuit and hence distribution of m.m.f. round the circuit. It can also be used to determine magnetic potential drop across a joint in the magnetic circuit or between two points between which magnetic leakage is suspected.

Q.59. *What are permeameters?*

Ans. Normally it is not a practical arrangement to have a $\frac{l}{d}$ as large as possible for a bar specimen to be tested. *"Permeameters"* have been developed to overcome the difficulties and inaccuracies encountered in performing measurements on bar specimens of magnetic materials. Thus permeameters test bar specimens under controlled field conditions.

Q.60. *Enumerate the methods which are used for measurement of iron losses.*

Ans. (*i*) Wattmeter method; (*ii*) A.C. bridge methods; (*iii*) A.C. potentiometer method; (*iv*) Oscillographic method.

Q.61. *Why magnetic measurements are not as accurate as other types of measurements in electrical engineering?*

Ans. This is mainly due to the fact that magnetic flux cannot be measured directly by the mere insertion of instrument in a circuit as an electric current; voltage, and power. The other difficulties arise from the fact that flux unlike current does not confine itself to a definite path.

Q.62. *Why underground cables meant for operation on A.C. systems are tested with direct current systems?*

Ans. If alternating current is employed for the test, heavy charging current will be drawn by the cable and so the transformer required for the purpose would be of inconveniently large capacity and cause transportation difficulties of moving the testing transformer to the site of installation. In case, direct current is used for the tests, the plant required would only be of small capacity and can be constructed in portable form.

Q.63. *What are the main components of a 'Universal bridge'?*

Ans. A 'Universal bridge' consists of signal sources (D.C. and 1 kHz) bridges (Wheatstone bridge, Maxwell's and Hay's bridges and De Sauty bridge), imbalance amplifier and metered detector in addition to power supply.

Q.64. *Why is a triggering circuit provided in a CRO?*

Ans. In a CRO, a 'triggering circuit' is provided for synchronising two types of deflections so that horizontal deflection starts at the same point of input vertical signal each time it sweeps.

Q.65. *What do you mean by the deflection factor of a CRO?*

Ans. The deflection factor of a CRO is the reciprocal of the deflection sensitivity.

Q.66. *What is an oscilloscope probe?*

Ans. Probe is a means that connects the test circuit to the oscilloscope without altering loading, or disturbing the test circuit.

Q.67. *What is a digital instrument?*

Ans. A 'digital instrument' indicates the measured value in the form of decimal digits.

Q.68. *What are the essential components of a DVM?*

Ans. A DVM comprises of a clock generator, a circuit that converts the voltage to be measured into pulses, and a digital frequency counter.

Q.69. *Enumerate the various types of high voltage tests.*

Ans. (*i*) Sustained low-frequency tests; (*ii*) Constant direct-current tests; (*iii*) High-frequency tests; (*iv*) Surge or impulse tests.

Q.70. *Name the apparatus which are usually essential for all hight voltage tests.*

Ans. (*i*) High-voltage transformers; (*ii*) Apparatus for voltage regulation; (*iii*) Control gear and high-voltage connections, including safety and protective devices; (*iv*) Apparatus for voltage measurement.

Q. 71. *What is a spectrum analyser? What information is provided by the spectrum analysis of a signal?*

Ans.
- ***Spectrum analyser*** is an instrument which graphically presents an energy distribution of the signal as a function of frequency on its CRT.
- The spectrum analysis of a signal provides the information about the following *particulars/parameters:*

  (*i*) Bandwidth; (*ii*) Effect of different types of modulation and spurious signal generation; (*iii*) Frequency stability; (*iv*) The component levels; (*v*) Spectral purity; (*vi*) Attenuation.

Q. 72. *What are the applications of Linear-variable-differential transformer (LVDT)?*

Ans. Following are the *applications* of LVDT:

(*i*) Measurement of material thickness in hot strip or slab steel mills.

(*ii*) In accelerometers.

(*iii*) Jet engine controls in close proximity to exhaust gases.

Q. 73. *Differentiate between magnetic tape and digital tape recording.*

Ans. The differences between magnetic tape and digital tape recording are given below:

S.No.	Magnetic recording	Digital recording
1.	It is an analog system.	It is a digital system.
2.	More distortion.	Less distortion.
3.	The signal noise ratio is poor.	The signal noise ratio is good.
4.	Vow and flutter are present.	Vow and flutter are totally absent.
5.	Equalisation is required.	Equalisation is not required.
6.	It suffers from accumulation of dirt and dust on its surface.	It is free from such problems due to lacquer coating.

Q. 74. *Nixie tube display is based on what principle?*

Ans. The ***Nixie tube*** works on the *principle* that when a gas breaks down, a glow discharge is produced. There is one common anode and ten cathodes, on which ten (0 to 9) numerals are placed. When a negative voltage is applied to the selected cathode, a gas diode is formed which glows the selected cathode (numeral).

Q. 75. *What are the applications of 'strain gauge'?*

Ans. The ***strain gauge*** is a measurement transducer for measuring strain and associated stress in experimental stress analysis. Secondly, many other detectors and transducers, notably the load cells, torque meters, diaphragm type pressure gauges, temperature sensors, accelerometers and flow meters, employ a strain gauge as a secondary transducer.

Q. 76. *What is 'current transformer (CT)'?*

Ans. A ***current transformer*** is an instrument to measure higher currents of transmission lines with the help of a low range ammeter.

Q. 77. *List the advantages of Electrodynamic instruments.*

Ans. *Advantages:*

1. Can be used on both D.C. as well as A.C. systems.
2. Free from hysteresis and eddy current errors.
3. It is possible to construct ammeters up to 10 A and voltmeters up to 600 V with precision grade accuracy.

Q. 78. *Name different types of 'frequency meters'.*

Ans. The frequency meters are of following types:

(*i*) Mechanical resonance type; (*ii*) Electrical resonance type; (*iii*) Weston type; (*iv*) Ratiometer type; (*v*) Saturable core type; (*vi*) Digital frequency meters.

Q. 79. *Write different methods of measurement of medium resistance.*

Ans. The methods of measuring *"medium resistances"* are:

(*i*) Voltmeter ammeter method (*ii*) Ohm-meter method

(*iii*) Bridge method (*iv*) Substitution method.

Q. 80. *What is meant by drift?*

Ans. By ***drift***, we mean variation in the value of quantity under measurement with time. When it is repeatedly measured, if there is no change in the value, it means zero drift.

Q. 81. *What is ratio correction factor (RCF)?*

Ans. The ***ratio correction factor*** is the transformation (actual) ratio divided by nominal ratio.

In other words, $$RCF = \frac{K_{act.}}{K_{nom.}}$$

- The ratio marked on transformers is their nominal ratio.

Q. 82. *How will you differentiate 7 segment display with 14 segment display?*

Ans.
- The 7 segment display is used for display of numbers only.
- The 14 segment display is used for display of numbers as well as of alphabets.

Q. 83. *What is the role of recorders in measurement system?*

Ans. A ***recorder*** records electrical and non-electrical quantities as a function of time. The recorder serves the following *objectives:*

(*i*) It preserves the details of measurement at a particular time.

(*ii*) It provides at a glance the overall picture of the performance of the unit.

(*iii*) It provides immediate reflection on the action taken by the operator.

Q. 84. *What is the working principle of 'photoelectric transducer'?*

Ans. The ***photoelectric transducer*** operates on the principle that when light strikes special combination of materials then following may result:

(*i*) Electrons may flow.

(*ii*) A voltage may be generated.

(*iii*) A resistance change may take place.

Q. 85. *What are the applications of strip chart recorders?*

Ans. Some of the important applications of ***strip chart recorders*** are:

(*i*) Temperature recording.

(*ii*) Sound level recording.

(*iii*) Recording amplifier drift.

Q. 86. *What VTVM stands for?*

Ans. The VTVM stands for vacuum tube volt meter.

Q. 87. *What is Cathode Ray Oscilloscope (CRO)?*

Ans. ***CRO*** is a device that allows the amplitude of electrical signals, whether they be voltage, current, power etc., to be displayed primarily as a function of time. It is basically an electron beam voltmeter.

It is very useful and versatile laboratory instrument used for display, measurement and analysis of waveforms and other phenomena in electrical and electronic circuits.

Q. 88. *Name main components of a cathode ray cube (CRT).*

Ans. The main components of a CRT are:

(*i*) Electron gun assembly
(*ii*) Deflection plates assembly
(*iii*) Fluorescent screen
(*iv*) Glass envelope
(*v*) Base.

Q. 89. *List the various applications of a 'wave analyser'.*

Ans. The various applications of ***wave analyser*** are:

(*i*) To carry out complete harmonic analysis.

(*ii*) To measure relative amplitudes of single frequency components in a complex waveform.

(*iii*) To measure the signal energy with well defined bandwidth.

(*iv*) To be used to separate and display a large number of harmonics (say 50).

(*v*) The wave analysers are applied industrially in the field of reduction of sound and vibrations generated by rotating electrical machines and apparatus.

Q. 90. *Define a transducer and distinguish between a sensor and a transducer.*

Ans.
- A ***transducer*** is a device which converts the energy from one form to another.
- A ***sensor*** senses the condition, state or value of the process variable and produces an output which reflects this condition, state or value. A *transducer* transforms the energy of the process variable to an output of some other type of energy which is able to operate some control device. Sometimes a *secondary transducer* may be employed to transform the output of primary senser to still another type of energy.

Q. 91. *Enumerate main elements of a digital acquisition system.*

Ans. The main elements of a digital acquisition system are:

(*i*) Transducer
(*ii*) Signal conditioner
(*iii*) Multiplexer
(*iv*) Signal converter
(*v*) Analog to digital converter
(*vi*) Digital recorder.

Q. 92. *Where are 'Digital data acquisition systems' used?*

Ans. ***Digial data acquisition systems*** are used when the physical quantity being monitored has a narrow bandwidth (*i.e.*, when the quantity varies slowly). These systems are also used when high frequency and low cost per channel is required.

These systems are, in general, *more complex* than analog systems, both in terms of instrumentation involved and the volume and complexity of input data they can handle.

Q. 93. *What is 'telemetry'?*

Ans. ***Telemetry*** is defined as a technology which allows users to collect information from inaccessible and inconvenient locations and transmit it to the accessible places to process, record and display the information in presentable form.

It is used for remote measurement of quantities.

Q. 94. *Why is sensitivity of the A.C. section of multimeter less than the D.C. section?*

Ans. Due to complex circuitry, the sensitivity of A.C. section of a multimeter is less than the D.C. section.

Q. 95. *What are the shortcomings of multimeter?*

Ans. It may be noted that A.C. sensitivity of a half-wave A.C. meter is only 45% of the D.C. sensitivity, therefore, the loading effect is greater while measuring A.C. voltages than that while measuring corresponding D.C. voltages.

The rectifying action depends upon the frequency to some extent due to capacitance effect and rectifier instruments show lower readings. Instrument indications may be in error by as much as 0.5% decrease for every 1 kHz rise in frequency.

Q. 96. *Why are the rectifier instruments indispensable for measurements in communication circuits?*

Ans. The ***rectifier instruments*** are indispensable for measurements in communication circuits because they have very high sensitivity and as such the power required for their operation is small enough to be within the capacities of the circuits.

Q. 97. *What are the essentials of indicating instruments?*

Ans. ***Indicating instruments*** consist essentially of a pointer moving over a calibrated scale and attached to the moving system pivoted on jewelled bearings. For satisfactory working of indicating instruments the torques required are:

(*i*) Deflecting torque; (*ii*) Controlling torque; (*iii*) Damping torque.

Q. 98. *What are prepayment meters?*

Ans. ***Prepayment meter*** is special type of house supply meter, which itself is essentially a standard pattern energy element with the addition of a coin collector box and the prepayment mechanism geared to the recorder train.

Q. 99. *List the factors that may lead to inaccuracies in measurement by A.C. bridges.*

Ans. The factors that may lead to inaccuracies in measurement by A.C. bridges are:

(*i*) Stray conductance effect due to imperfect insulation.

(*ii*) Mutual inductance effect due to magnetic coupling between various components of the bridge.

(*iii*) Stray capacitance effect due to electrostatic fields between conductors at different potentials.

(*iv*) Residual-in components, *e.g.*, the existence of small amount of series inductance or shunt capacitance in normally non-reactive resistors.

Q. 100. *How can average reading of VTVM be converted to a RMS reading voltmeter?*

Ans. The average reading of ***VTVM*** can be converted into a RMS reading voltmeter by calibrating its meter scale in terms of the r.m.s. value of a sinusoidal wave.

Q. 101. *What is meant by the deflection factor of a CRO?*

Ans. The deflection factor of a CRO is the reciprocal of the deflection sensitivity.

Q. 102. *What is a Lissajous pattern?*

Ans. ***Lissajous*** is a pattern that results from applying periodic signals to the deflection plates of a CRO.

Q. 103. *What is a function generator?*

Ans. A ***function generator*** is an instrument that has the capability of producing different types of waveform as its output signal. The most common output waveforms are: Sine waves; Triangular waves; Square waves; Sawtooth waves.

Q. 104. *How does a wave analyser functionally differ from a spectrum analyser?*

Ans. A *"wave analyser"* is used to measure relative amplitude of signal frequency components in a complex waveform while a *spectrum analyser* is used to study energy distribution across the frequency spectrum of a given electrical signal.

Q. 105. *What is digital-to-analog (D/A) conversion?*

Ans. ***D/A conversion*** basically is the process of taking a value represented in digital code and converting it into a voltage or current which is proportional to the digital value.

Q. 106. *What is frequency counter?*

Ans. A ***frequency counter*** is a digital instrument that can measure and display the frequency of any periodic waveform.

Q.107. *What is IEEE 488 bus system?*

Ans. For a set of instruments to be interfaced with a single computer, a general purpose bus, called the IEEE 488 bus system, has been developed to which a number of instruments can be attached so that communication between the computer and the instruments is possible over a common cable.

Q. 108. *What is the difference between 3-digit and $3\frac{1}{2}$-digit digital meters?*

Ans. A 3-digit meter has a resolution of $\frac{1}{1000}$th of the range and has a maximum display of 999. In contrast, the $3\frac{1}{2}$-digit digital meter has a resolution of $\frac{1}{2000}$th of the range, and the maximum value displayed is 1999.

Q. 109. *Why is dual slope ADC preferred in a digital voltmeter?*

Ans. Dual slope ADC is preferred in a digital voltmeter (DVM) because it has low cost and high accuracy.

Q. 110. *What are Liquid Crystal Displays (LCDs)?*

Ans. ***LCDs*** are passive displays characterised by very low power consumption and good contrast ratio.

Q. 111. *List the displays commonly used in digital electronic field.*

Ans. Following ***displays*** are commonly used in digital electronic field:

(*i*) Cathode ray tube (CRT)
(*ii*) Light emitting diode (LED)
(*iii*) Liquid crystal display (LCD)
(*iv*) Nixies
(*v*) Electro-luminescent (EL) displays
(*vi*) Incandescent displays
(*vii*) Electrophoretic image displays (EPID)
(*viii*) Liquid vapour display (LVD).

Q. 112. *What is a Data Acquisition System (DAS)?*

Ans. ***DAS*** may be defined as a system used for data processing, data conversion, data transmission and data storage.

A typical data acquisition consists of individual sensors with necessary signal conditioning, multiplexing, data conversion, data processing, data handling and associated transmission, storage and display system.

Q. 113. *What do you understand by oscillographic recorders?*

Ans. The oscillographic recorders are instruments which record information on paper or film with ink, powder, electrostatic charge, liquid, electric arc or liqht.

Although galvanometer recorders, pen recorders, X-Y recorders, strip chart recorders and CRT recorders all may be considered oscillographic recorders, yet here the word 'oscillograph' implies either a galvanometer or a CRT recorder–the recorders having *bandwidths more than 20 kHz*.

Q. 114. *What is a load cell and what for is it used?*

Ans. A ***load cell*** is an electromechanical device that converts weight or force into an electrical signal and is widely used for measurement of static and dynamic forces.

Q. 115. *What are resolvers?*

Ans. ***Resolver*** is a synchro whose rotor is mechanically driven to translate rotor angle into electrical information corresponding to the sine and cosine of rotor angle. It can interchange rectangular and polar coordinates. It is also called a sine-cosine generator and synchro-resolver.

Q. 116. *What is the major difference between the current and voltage telemetering system?*

Ans. The major difference between the two systems is that in the current telemetering system the current in the communication circuit is maintained at a measurable value, irrespective of line conditions, whereas the ideal voltage telemetering system draws from the circuit a current so small that the resistive drop in the line is negligible. Thus, in voltage telemetering system voltage at the measuring end is measured by a null or balance method in order to keep the line current zero at the time of measurement.

Q. 117. *What is multiplexing in telemetering system?*

Ans. Transmission of two or more messages at the same time over a single intercollecting link is called the ***multiplexing*** in ***telemetering system***, *i.e.*, multiplexing is a method of transmitting more than one measurement over a single transmission channel.

Q. 118. *Why is it essential to use large moving coil and strong magnetic field in a galvanometer type strip-chart recorder?*

Ans. As the friction torque in a recorder is high due to large weight of the moving system and friction between the pen and the chart, so the large deflecting torque is required. Obviously, the controlling torque required will also be large. So in such recorders large moving coil and strong magnetic field are required.

Q. 119. *What is meant by instrumentation?*

Ans. ***Instrumentation*** refers to the art and science of collection of several instruments and auxiliary equipment and their utilisation for conducting successfully a test or an experiment on a system, process or plant.

Q. 120. *What is meant by 'loading effect'?*

Ans. The incapability of the system to faithfully measure, record or control the input signal (measurand) in undistorted form is known as ***"loading effect"***.

Q. 121. *What are the special features of vibration galvanometers?*

Ans. The special features of vibration galvanometers are large sensitivity, small damping and large restoring constant.

Q. 122. *Why is microwave channel most commonly used?*

Ans. ***Microwave channel*** is most commonly used because of heavy requirement in industries like power and gas transmission and due to the overcrowding of lower radio-frequency spectrum. The initial cost on transmitter and receiver is high in this system but cost of additional channels is small.

PART – III

GATE and UPSC Examinations' Questions with Answers/Solutions

A. **Conventional Questions with Answers/Solutions**

B. **Multiple-choice Questions with Answers and "*Explanations*"**

GATE and UPSC Examinations' Questions with Answers/Solutions

A. CONVENTIONAL QUESTIONS WITH ANSWERS/SOLUTIONS

Q. 1. ***(Errors in measurements)*** *The power factor of a circuit is determined by* $\cos\phi = P/VI$ *where P is the power in watt, V is the voltage in volt and I is the current in ampere. The relative errors in power, current and voltage are respectively ±0.5%, ±1% and ±1%. Calculate the relative error in power factor. Also calculate the uncertainty in power factor if the errors were specified as uncertainties.* **(GATE)**

Solution. *Given:* $\cos\phi = \dfrac{P}{VI}$; Relative errors in *P*, *V* and *I*: $\dfrac{\delta P}{P} = \pm 0.5\%$; $\dfrac{\delta I}{I} = \pm 1\%$; $\dfrac{\delta V}{V} = \pm 1\%$.

Relative error in power factor (cos ϕ), $\dfrac{\delta \cos\phi}{\cos\phi}$:

$$\frac{\delta\cos\phi}{\cos\phi} = \pm\left(\frac{\delta P}{P}+\frac{\delta V}{V}+\frac{\delta I}{I}\right) = \pm(0.5+1+1)\% = \pm 2.5\% \text{ (Ans.)}$$

Relative uncertainty in power factor:

Relative uncertainties in measurement of *P*, *V* and *I*,

$$\frac{\omega_P}{P} = \pm 5\%,\ \frac{\omega_V}{V} = \pm 1\%;\ \frac{\omega_I}{I} = \pm 1\%$$

Relative uncertainty in measurement of power factor (cos ϕ),

$$\frac{\omega_{\cos\phi}}{\cos\phi} = \sqrt{\left(\frac{\omega_P}{P}\right)^2+\left(\frac{\omega_V}{V}\right)^2+\left(\frac{\omega_I}{I}\right)^2}$$

$$= \sqrt{(\pm 0.5)^2+(\pm 1)^2+(\pm 1)^2} = \mathbf{\pm 1.5\%\ (Ans.)}$$

Q. 2. ***(Transducer)*** *The output relation for a thermistor transducer is given by*

$$R = R_0 e^{\beta\left(\frac{1}{T}-\frac{1}{T_0}\right)}$$

For $T_0 = 300$ *K,* $\beta = 3420$, $R_0 = 1$ *kΩ and* $R = 2$ *kΩ, calculate T.* **(IES)**

Solution. *Given:* $T_0 = 300$ K; $\beta = 3420$; $R_0 = 1$ kΩ; $R = 2$ kΩ.

$\boldsymbol{T} = ?$

Output equation of a thermistor transducer:

$$R = R_0 e^{\beta\left(\frac{1}{T}-\frac{1}{T_0}\right)}$$

Inserting the various values, we get:

$$2 = 1\times e^{3420\left(\frac{1}{T}-\frac{1}{300}\right)}$$

Taking log on both sides, we get:

$$ln\,(2) = 3420\left(\frac{1}{T}-\frac{1}{300}\right) = 3420\left(\frac{300-T}{300\,T}\right)$$

or, $$0.693 = 11.4\left(\frac{300-T}{T}\right)$$

or, $$0.693\,T = 3420 - 11.4\,T$$

or, $$T = \frac{3420}{(0.693+11.4)} = \mathbf{282.8\ K\ (Ans.)}$$

Q. 3. ***(Digital Instruments)*** *(a) Describe the principle of frequency measurement using digital instrument. Draw its block diagram.*

(b) The unknown input signal of 2 V square wave is 3.5 kHz. Determine the display indication if the gate enable time is (i) 0.1 second, (ii) 1 second, and (iii) 10 seconds. **(GATE)**

Solution. (*a*) Refer to Article 8.3.5.

(*b*) *Given*: The unknown input signal of 2V square wave, $f = 3.5$ kHz = 3500 Hz

Display indication:

Frequency of unknown signal, $f = \frac{N}{t}$

where, N = Number of counts displayed by counter, and

t = Time internal between start and stop of gate.

Assuming 5 digit display, we have:

(*i*) $t = 0.1s$: $f = \frac{N}{t}$ or $N = ft = 3500 \times 0.1 = 350$

∴ Display = **00350 (Ans.)**

(*ii*) $t = 1s$: $f = \frac{N}{t}$ or $N = ft = 3500 \times 1 = 3500$

∴ Display = **03500 (Ans.)**

(*iii*) $t = 10s$: $f = \frac{N}{t}$ or $N = ft = 3500 \times 10 = 35000$

∴ Display = **35000 (Ans.)**

Q. 4. ***(Measurement of Power)*** *Three single phase loads can be connected in either star or in delta to form three-phase load. Which of these connections results in higher current when connected to a three-phase supply? During the measurement of power by two-wattmeter method, the total input power to a 3-phase 440-V motor running at power factor of 0.8 was found to be 25 kW. Find the readings of the two wattmeters.* **(IES)**

Solution.

Given: Total input power = 25 kW; Voltage = 440 V;

Power factor, $\cos\phi = 0.8$ or $\tan\phi = \frac{3}{4}$.

We know that:

In *start connection*: Line voltage = $\sqrt{3}$ × phase voltage

Line current = Phase current

In *delta connection*: Line voltage = Phase voltage

Line current = $\sqrt{3}$ × phase current

Hence, delta connection results in higher current when connected to a 3-phase supply.

Now, total power, $25 \text{ kW} = W_1 + W_2$ ...(*i*)

Also, $$\tan \phi = \frac{\sqrt{3}(W_1 - W_2)}{(W_1 + W_2)}$$ ...[Eqn. (4.157)]

or, $$\frac{3}{4} = \frac{\sqrt{3}(W_1 - W_2)}{25}$$

or, $W_1 - W_2 = 10.825 \text{ kW}$ ...(*ii*)

From eqn. (*i*) and (*ii*), we get:

$W_1 =$ **17.01 kW,** and $W_2 =$ **7.09 kW (Ans.)**

Q. 5. ***(Concept of measurements)*** *A circuit was tuned for resonance by eight different students, and the values of resonant frequency in kHz were recorded as 532, 548, 543, 535, 546, 531, 543 and 536.*

Calculate:

(i) The arithmetic mean.
(ii) Deviations from mean.
(iii) The average deviation.
(iv) The standard deviation.
(v) Variance.

Solution. *Given:* Values of resonant frequency: 532, 548, 543, 535, 546, 531, 543, 536.

(*i*) Arithmetic mean:

Let q_1, q_2, q_3 etc., be the set measured values, then:

$$\text{Arithmetic mean, } \bar{q} = \frac{\Sigma q}{n} = \frac{532+548+543+535+546+531+543+536}{8}$$

$$= \mathbf{539.25 \text{ (Ans.)}}$$

(*ii*) Deviations from mean:

The deviations are:

$$d_1 = q_1 - \bar{q} = 532 - 539.25 = -7.25$$
$$d_2 = q_2 - \bar{q} = 548 - 539.25 = 8.75$$
$$d_3 = q_3 - \bar{q} = 543 - 539.25 = 3.75$$
$$d_4 = q_4 - \bar{q} = 535 - 539.25 = -4.25$$
$$d_5 = q_5 - \bar{q} = 546 - 539.25 = 6.75$$
$$d_6 = q_6 - \bar{q} = 531 - 539.25 = -8.25$$
$$d_7 = q_7 - \bar{q} = 543 - 539.25 = 3.75$$
$$d_8 = q_8 - \bar{q} = 536 - 539.25 = -3.25$$

(*iii*) Average deviation:

$$\text{Average deviation} = \frac{7.25+8.75+3.75+4.25+6.75+8.25+3.75+3.25}{8}$$

$$= \mathbf{5.75 \text{ (Ans.)}}$$

(For calculation of average deviation, the *signs of deviations are not considered.*)

(*iv*) Standard deviation, *s*:

$$\text{Standard deviation, } s = \sqrt{\frac{\Sigma d^2}{n-1}}$$

on $s = \sqrt{\dfrac{(-7.25)^2 + (8.25)^2 + (3.25)^2 + (-4.25)^2 + (6.25)^2 + (-8.25)^2 + (3.25)^2 + (-3.25)^2}{(8-1)}}$

$= \mathbf{6.54}$ **(Ans.)**

(v) Variance, *V*:

$$\text{Variance } V = s^2 = (6.54)^2 = \mathbf{42.77(kHz)^2} \textbf{ (Ans.)}$$

Q. 6. ***(Analog electromechanical instruments)*** *It is required to measure the voltage across 10 kΩ resistor of the circuit shown in Fig. 1 below.*

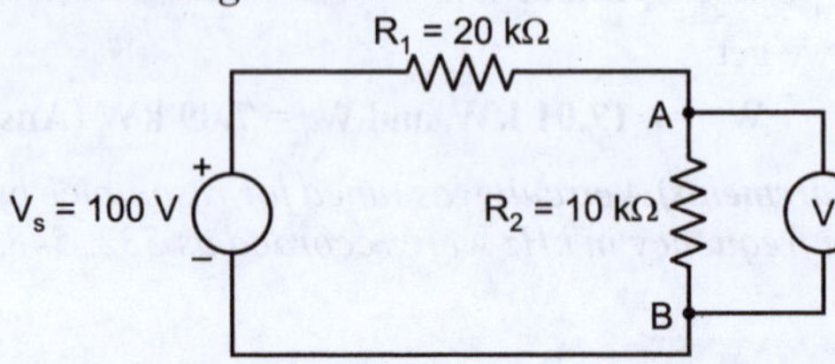

Fig. 1

Two voltmeters are available. Voltmeter A is 1 mA movement and voltmeter B is 50 µA movement. Both use their 50 V scales. Calculate:

(i) The reading of each voltmeter.

(ii) The error from true reading. **(IES)**

Solution. *Given*: $V_s = 100$ V; $R_1 = 20$ kΩ; $R_2 = 10$ kΩ.

Using voltage division rule, actual voltage across 10 kΩ resistor when voltmeter is not connected, is given as:

$$V_{AB} = V_s\left(\frac{R_{AB}}{R_1 + R_{AB}}\right) = 100 \times \left(\frac{10}{20+10}\right) = 33.33 \text{ kV}$$

Case I: Refer to Fig. 2

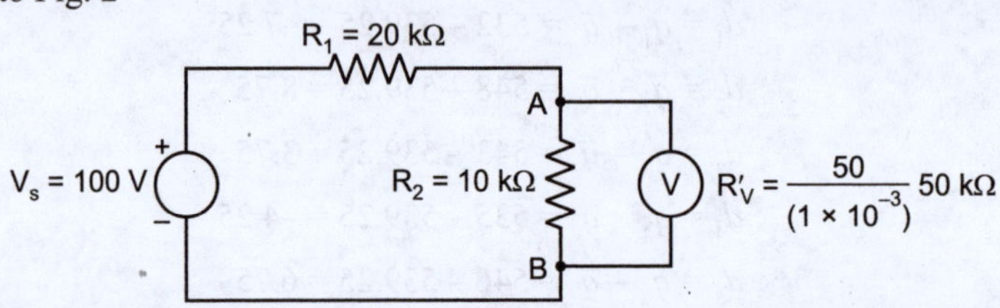

Fig. 2

Resistance across the terminals of voltmeter,

$$R'_{AB} = \frac{10 \times 50}{10 + 50} = 8.33 \text{ kΩ}$$

$$\text{Reading of voltmeter } V'_{AB} = V_s \times \left(\frac{8.33}{20 + 8.33}\right) = 100 \times \left(\frac{8.33}{20 + 8.33}\right) = \mathbf{29.4\ V} \textbf{ (Ans.)}$$

$$\%\ Error = \left(\frac{\text{Measured value} - \text{True value}}{\text{True value}}\right) \times 100$$

$$= \frac{29.4 - 33.33}{33.33} \times 100 = \mathbf{-11.79\%} \textbf{ (Ans.)}$$

Case II: Refer to Fig. 3

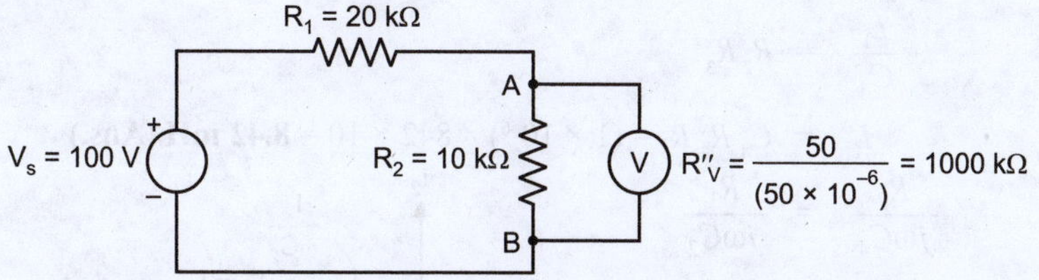

Fig. 3

Resistance across the terminals of voltmeter,

$$R''_{AB} = \frac{10 \times 1000}{10 + 1000} = 9.9\ k\Omega$$

$$\text{Reading of voltmeter, } V''_{AB} = V_s \times \left(\frac{9.9}{20+9.9}\right) = 100 \times \left(\frac{9.9}{20+9.9}\right) = \mathbf{33.11V\ (Ans.)}$$

$$\%\ Error = \frac{33.11 - 33.33}{33.33} \times 100 = \mathbf{-0.66\%\ (Ans.)}$$

Q. 7. ***(A.C. Bridge)*** *A bridge circuit for measurement of effective resistance and inductance of an iron-cored coil consists of*

Arm AB, the unknown impedance

Arm BC, a pure resistance R_3

Arm CD, a lossless capacitor C_4

Arm DA, a capacitor C_2 *in series with a resistance* R_2.

Under balanced condition,

$R_3 = 10\Omega$, $R_2 = 842\Omega$, $C_2 = 0.135\ \mu F$ *and* $C_4 = 1\mu F$.

Calculate the value of effective resistance and self-inductance at a supply frequency of 100 Hz. Derive the equations of balance and draw the phasor diagram under balanced condition.

(IES)

Solution. *Given*: Under balanced conditions: $R_3 = 10\Omega$; $R_2 = 842\Omega$; $C_2 = 0.135\mu F$; $C_4 = 1\mu F$; $f = 100$ Hz.

Equations of balance: Refer to Fig. 4.

Let R_1 and L_1 be the effective resistance and inductance of specimen respectively.

Under balanced conditions:

$$\bar{Z}_1 \bar{Z}_4 = \bar{Z}_2 \bar{Z}_3 \quad ...(i)$$

$$(R_1 + j\omega L_1) \times \frac{1}{j\omega C_4} = \left(R_2 + \frac{1}{j\omega C_2}\right) \times R_3 \quad ...(ii)$$

...Eqns. of balance.

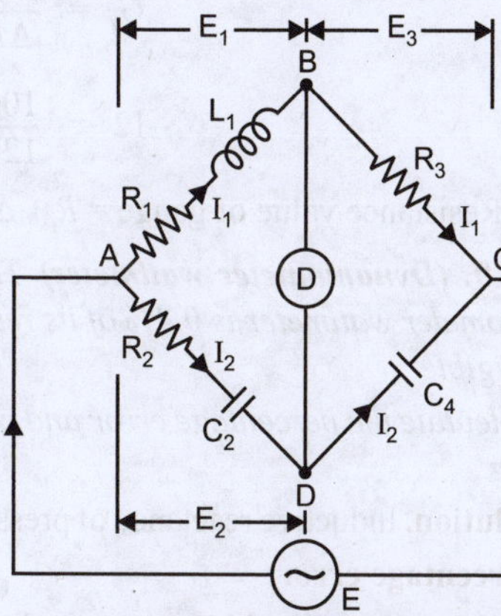

Fig. 4. Bridge circuit.

Value of R_1 and L_1:

Expanding eqn. (*ii*), we get:

$$\frac{R_1}{j\omega C_4} + \frac{L_1}{C_4} = R_2 R_3 + \frac{R_3}{j\omega C_2}$$

Comparing the real and imaginary parts, we have:

$$\frac{L_1}{C_4} = R_2 R_3$$

or, $$L_1 = C_4 R_2 R_3 = (1 \times 10^{-6}) \times 842 \times 10 = \mathbf{8.42\ mH\ (Ans.)}$$

$$\frac{R_1}{j\omega C_4} = \frac{R_3}{j\omega C_2}$$

or, $$R_1 = R_3 \times \frac{C_4}{C_2}$$

$$= 10 \times \frac{(1 \times 10^{-6})}{(0.135 \times 10^{-6})}$$

$$= \mathbf{74.07\Omega\ (Ans.)}$$

Fig. 5 shows the **phasor diagram** under balanced condition. **(Ans.)**

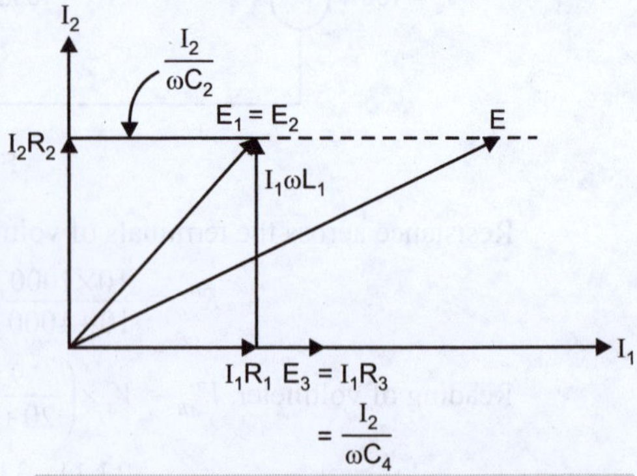

Fig. 5. Phasor diagram.

Q. 8. ***(Strain gauges)*** *Describe the construction of a foil-type strain gauge and explain the advantages over wire-wound strain gauge. A strain gauge has a resistance of unstrained gauge of 120 Ω and the gauge factor of –12. What is the resistance value of the gauge when it is subjected to a strain of 1% ?*

(GATE)

Solution. (*a*) Refer to *Article* **16.13.2.2**

(*b*) *Given*: Resistance of unstrained gauge, $R = 120\Omega$; Gauge factor, $G_f = -12$; Strain = 1% = $\frac{1}{100}$

Resistance value of gauge:

$$\text{Strain} = \frac{\Delta L}{L} = \frac{1}{100}$$

$$G_f = \frac{\Delta R/R}{\Delta L/L} = \frac{\Delta R}{120} \times \frac{1}{(1/100)} = \frac{100}{120}\Delta R$$

or, $$-12 = \frac{100}{120}\Delta R \text{ or } \Delta R = \frac{-12 \times 120}{100} = -14.4$$

∴ Resistance value of gauge = $R + \Delta R = 120 - 14.4 =$ **105.6 Ω (Ans.)**

Q. 9. ***(Dynamometer wattmeter)*** *The inductive reactance of the pressure-coil circuit of a dynamometer wattmeter is 0.4% of its resistance at normal frequency (50 Hz) and the capacitance is negligible.*

Calculate the percentage error and correction factor due to the reactance for load at 0.707 p.f. lagging. **(IES)**

Solution. Inductive reactance of pressure coil = 0.4% resistance; $f = 50$ Hz; $\cos\phi = 0.707$ lagging.

Percentage error:

The pressure coil of dynamometer has an inductance and as such the current lags the voltage by angle θ, given by the relation:

$$\theta = \tan^{-1}\left(\frac{\text{reactance of pressure coil}}{\text{resistance}}\right) = \tan^{-1}\left(\frac{0.4}{100}\right) = 0.229°$$

Now, $\cos\phi = 0.707$ (given), or $\phi = \cos^{-1}(0.707) = 45^\circ$

$\therefore$ The percentage error $= \tan\theta.\ \tan\phi \times 100$

$= \tan(0.229) \times \tan 45^\circ \times 100 =$ **0.399% (Ans.)**

Correction factor:

The term $\dfrac{\cos\phi}{\cos\theta.\cos(\theta+\phi)}$ for logging pf. in a dynamometer wattmeter is known as *correction factor*.

$$\therefore \quad \text{Correction factor} = \frac{\cos\phi}{\cos\theta.\cos(\theta-\phi)}$$

$$= \frac{0.707}{\cos(0.229^\circ).\cos(0.229^\circ - 45^\circ)} = \mathbf{0.9959\ (Ans.)}$$

Q. 10. ***(Strain gauge)*** *Calculate the gauge factor of a strain gauge if a 1.5 mm diameter conductor that is 24 mm long changes length by 1 mm and diameter by 0.02 mm under a compression force.* **(GATE)**

Solution. *Given*: $D = 1.5$ mm; $L = 24$ mm; $\Delta L = 1$ mm, $\Delta D = 0.02$ mm.

Gauge factor, G_f:

The gauge factor can be written as:

$$G_f = 1 \quad + 2\mu \quad + \frac{\Delta\rho/\rho}{e}$$

$$= \begin{bmatrix}\text{Reslitance}\\ \text{change due}\\ \text{to length}\end{bmatrix} \begin{bmatrix}\text{Resistance}\\ \text{change due}\\ \text{to change}\\ \text{in area}\end{bmatrix} \begin{bmatrix}\text{Resistance change}\\ \text{due to piezo-resistive}\\ \text{effect}\end{bmatrix}$$

or, $$G_f = 1 + 2\mu + \frac{\Delta\rho/\rho}{\Delta L/L} \qquad (\mu = \text{Poisson's ratio})$$

If the change in the value of resistivity of a material when strained is *neglected* the gauge factor can be expressed as:

$$G_f = 1 + 2\mu$$

Here, $$\mu = \frac{\text{Lateral strain}}{\text{Longituozinal strain}} = \frac{\Delta D/D}{\Delta L/L} = \frac{(0.02/1.5)}{(1/24)} = 0.32$$

$$\therefore \quad G_f = 1 + 2 \times 0.32 = \mathbf{1.64\ (Ans.)}$$

Q. 11. ***(Moving iron instrument)*** *The inductance of a moving iron instrument is given by the following expression:*

$$L = (30 + 10\theta - 2\theta^2)\ \mu H$$

where θ is the deflection in radions.

The actual spring constant is 25×10^{-6} Nm/rad.

Calculate the value of deflection for a current of 5A. **(IES)**

Solution. *Given:* Inductance, $L = (30 + 10\theta - 2\theta^2)$ μH; Spring constant, $C = 25 \times 10^{-6}$ Nm/rad; Current, I = 5A.

Value of deflection, θ:

Rate of change of inductance with deflection,

$$\frac{dL}{d\theta} = \frac{d}{d\theta}(30 + 10\theta - 2\theta^2) = (10 - 4\theta)\ \mu\text{H/rad}$$

The deflection (θ) is given as:

$$\theta = \frac{1}{2}\cdot\frac{I^2}{C}\cdot\frac{dL}{d\theta} \quad \text{...[Eqn. (4.76)]}$$

Inserting the various values, we get:

$$\theta = \frac{1}{2}\times\frac{5^2}{(25\times10^{-6})}\times(10-4\theta)\times10^{-6}$$

$$\theta = 5 - 2\theta \text{ or } \theta = 1.67 \text{ rad or } \mathbf{95.5° \ (Ans.)}$$

Q. 12. ***(Electrodynamic wattmeter)*** *An electrodynamic wattmeter is employed to measure power in a single phase circuit. The load voltage is 230 V and the load current is 5 amps. at a lagging power factor of 0.1. The wattmeter potential coil has a resistance of 10,000 Ω and an inductive reactance negligible compared to resistance. Determine percentage error in the wattmeter reading. Inductance of coil is 100 mH and pressure coil is connected on load side.* **(IES)**

Solution. *Given*: Load voltage, V = 230 volts; Load current, I = 5 amps; Power factor = 0.1 lagging; Resistance of potential coil, R_P = 10000 Ω; Inductance of coil, L_P = 100 mH.

Percentage error in the wattmeter reading:

When the pressure coil is connected on the load side, the wattmeter measures power loss in the pressure coil circuit in addition to load power. Then,

$$\text{True power} = \text{V I} \cos\phi = 230 \times 5 \times 0.1 = 115 \text{ W}$$

The angle by which the current lags behind the voltage,

$$\theta = \tan^{-1}\left(\frac{\omega L_P}{R_P}\right) = \tan^{-1}\left(\frac{2\pi\times50\times100\times10^{-3}}{10000}\right) = 0.003142 \text{ rad}$$

Actual reading of the wattmeter

$$= \text{True power } (1 + \tan\theta\tan\phi)$$

$$= 115\,[1 + 0.003142 \times \tan(\cos^{-1} 0.1)] = 118.6 \text{ W}$$

Power loss in pressure coil circuit

$$= \frac{V^2}{R_P} = \frac{230^2}{10000} = 5.29 \text{ W}$$

∴ Reading of wattmeter considering the power loss in pressure coil circuit

$$= 118.6 + 5.29 = 123.89 \text{ W}$$

$$\therefore \quad \text{Percentage error} = \frac{123.89 - 118.6}{118.6}\times100 = \mathbf{4.46\% \ (Ans.)}$$

Q. 13. ***(Concepts of measurements)*** *During the measurement of a capacitor, following ten readings were obtained:*

1.002, 0.998, 1.005, 1.009, 0.995, 0.997, 1.004, 1.008, 1.003, 0.994 μF.

Calculate:

(a) The arithmetic mean.

(b) The deviation from the mean.

(c) Average deviation.

(d) Standard deviation. **(IES)**

Solution. *Given*: Readings obtained during the measurement of a capacitor: 1.002, 0.998, 1.005, 1.009, 0.995, 0.997, 1.004, 1.008, 1.003, 0.994 μF.

(*i*) Arithmetic mean:

We know that arithmetic mean,

$$\bar{q} = \frac{1}{n}\sum_{i=1}^{n} q_i$$

$$= \frac{q_1 + q_2 + q_3 + q_4 + q_5 + q_6 + q_7 + q_8 + q_9 + q_{10}}{10}$$

$$= \frac{1.002 + 0.998 + 1.005 + 1.009 + 0.995 + 0.997 + 1.004 + 1.008 + 1.003 + 0.994}{10}$$

$$= \mathbf{1.0015\ \mu F\ (Ans.)}$$

(*ii*) Deviation from mean:

$$d_1 = q_1 - \bar{q} = 1.002 - 1.0015 = 0.0005\ \mu F$$

$$d_2 = q_2 - \bar{q} = 0.998 - 1.0015 = -0.0035\ \mu F$$

$$d_3 = q_3 - \bar{q} = 1.005 - 1.0015 = 0.0035\ \mu F$$

$$d_4 = q_4 - \bar{q} = 1.009 - 1.0015 = 0.0075\ \mu F$$

$$d_5 = q_5 - \bar{q} = 0.995 - 1.0015 = -0.0065\ \mu F$$

$$d_6 = q_6 - \bar{q} = 0.997 - 1.0015 = -0.0045\ \mu F$$

$$d_7 = q_7 - \bar{q} = 1.004 - 1.0015 = 0.0025\ \mu F$$

$$d_8 = q_8 - \bar{q} = 1.008 - 1.0015 = 0.0065\ \mu F$$

$$d_9 = q_9 - \bar{q} = 1.003 - 1.0015 = 0.0015\ \mu F$$

$$d_0 = q_{10} - \bar{q} = 0.994 - 1.0015 = -0.0075\ \mu F\ \textbf{(Ans.)}$$

(*iii*) Average deviation, $\bar{d}$:

$$\bar{d} = \frac{\Sigma |d|}{n}$$

$$= \frac{0.0005 + 0.0035 + 0.0035 + 0.0075 + 0.0065 + 0.0045 + 0.0025 + 0.0065 + 0.0015 + 0.0075}{10}$$

$$= \mathbf{0.0044\ \mu F\ (Ans.)}$$

(*iv*) Standard deviation, *s*:

$$s = \sqrt{\frac{\Sigma d^2}{n-1}} \qquad \text{...[Eqn. (3.68)]}$$

$$= \sqrt{\frac{(0.0005)^2 + (-0.0035)^2 + (0.0035)^2 + (0.0075)^2 + (-0.0065)^2 + (-0.0045)^2 + (0.0025)^2 + (0.0065)^2 + (0.0015)^2 + (-0.0075)^2}{(10-1)}}$$

$$= \mathbf{0.005276\ (Ans.)}$$

Q. 14. *(Potentiometers)* *(a) Discuss the merits and demerits of potentiometers used as transducers.*

(b) A linear resistance potentiometer is 100 mm long and is uniformly wound with a wire of total resistance 10,000 Ω. Under normal conditions the slider is at the centre of the potentiometer. Determine the linear displacement when the resistance of the potentiometer as measured by wheatstone bridge is 3700 Ω. If it is possible to measure a minimum value of 5 Ω resistance with the above arrangement, determine the resolution of potentiometer in mm. **(IES)**

Solution. (*a*) Potentiometers when used as transducers entail the following *merits* and *demerits*:

Merits:

(*i*) Simple to operate.

(*ii*) Inexpensive.

(*iii*) Very high electrical efficiency.

(*iv*) The resolution in 'wire-wound potentiometers' is limited while in 'cermet and metal film potentiometers', it is infinite.

(*v*) Very useful for applications when the requirements are not particularly severe.

Demerits:

(*i*) They require a large force to move their sliding contacts.

(*ii*) Since the sliding contacts can be contaminated, can wear out and generate noise, therefore, the life of the transducer is limited.

(*b*) *Given*: Length of the linear resistance potentiometer = 100 mm; Total resistance of wire = 10000 Ω; Resistance of the potentiometer as measured by wheatstone bridge = 3700 Ω. Minimum resistance that can be measured = 5 Ω.

$$\text{The resistance of potentiometer at its normal position} = \frac{10000}{2} = 5000\ \Omega$$

$$\text{Resistance of potentiometer per unit length} = \frac{10000}{100} = 100\ \Omega/\text{mm}$$

Change of resistance from its normal position = 5000 – 3700 = 1300 Ω

$$\therefore \textit{Displacement of wiper from its normal position} = \frac{1300}{100} = \mathbf{13\ mm\ (Ans.)}$$

Resolution of potentiometer = Minimum measurable resistance × mm / Ω

$$= 5 \times \frac{1}{100} = \mathbf{0.05\ mm\ (Ans.)}$$

Q. 15. *(Energy meters)* *(a) What is meant by phantom or fictitious loading of energy meters? (b) An a.c. energy meter is tested for half hour run at a supply voltage of 230 V with a current of 12 amps at 0.8 p.f. lag. The dial reading at the beginning of test was 58.5 and at the end was 59.5. The meter constant is 1200 revolutions per kWh. The meter revolutions registered during the test were 1150. Find out*

(i) error in registration and

(ii) error in r.p.m. of energy meter?

How error in r.p.m. can be rectified? **(IES)**

Solution. (*a*) Refer to *Article* **4.10.10.4;**

(*b*) *Given*: Period of testing, $t = \frac{1}{2} h$; Supply voltage = 230 V, I = 12 A, p.f. = 0.8,. Dial readings: 58.5, 59.5; Meter constant = 1200 revolutions per kWh; Meter revolutions registered during the test = 1150.

(*i*) Error in registration:

$$\text{Actual kWh consumed} = V\,I\cos\phi \times t \times 10^{-3}$$

$$= 230 \times 12 \times 0.8 \times \frac{1}{2} \times 10^{-3} = 1.104 \text{ kWh}$$

$$\text{Registered kWh} = \frac{\text{Meter revolutions registered during the test}}{\text{Meter constant}}$$

$$= \frac{1150}{1200} = 0.958$$

$$\therefore \quad \text{Error in registration} = \frac{1.104 - 0.958}{1.104} \times 100 = \mathbf{13.22\%\ (Ans.)}$$

(*ii*) Error in r.p.m. of energy meter:

$$\text{Speed of the disc} = \text{Energy supplied/unit} \times \text{meter constant}$$

$$= \frac{1.104}{60} \times 1200 = 22.08 \text{ r.p.m.}$$

$$\text{Revolutions registered/min} = \frac{1150}{30} = 38.33 \text{ r.p.m.}$$

$$\therefore \quad \text{Error in r.p.m.} = \frac{38.33 - 22.08}{22.08} \times 100 = \mathbf{73.6\%\ (fast)\ (Ans.)}$$

- Error in r.p.m. can be rectified by *bringing braking magnet near to the centre of the disc* **(Ans.)**

Q. 16. *(Moving iron instrument)* *The inductance of a moving iron instrument is given by*

$$L = (0.01 + K_1\,\theta)^2\ \mu H$$

where θ is angular deflection in radians from zero position. The angular deflection in radians corresponding to currents of 1.5 and 4.5 amps. are π/6 and π/3 respectively. Find the value of K_1.

(GATE)

Solution. *Given*: Inductance, $L = (0.01 + K_1\theta)^2$ μH; Angular deflections corresponding to 1.5 A and 4.5 A = π/6, π/3 respectively.

Value of K_1:

$$L = (0.01 + K_1\,\theta)^2$$

On differentiation w.r.t.θ. we get:

$$\frac{dL}{d\theta} = 2(0.01 + K_1\,\theta).\,K_1$$

Also, $$\theta = \frac{1}{2}.\frac{I^2}{K}.\frac{dL}{d\theta}, \text{ where } I \text{ represents current.}$$

$$\therefore \quad \frac{\pi}{6} = \frac{1}{2} \times \frac{1.5^2}{K}.\,2\left(0.01 + K_1 \times \frac{\pi}{6}\right).\,K_1 \quad ...(i)$$

and, $$\frac{\pi}{3} = \frac{1}{2} \times \frac{4.5^2}{K}.\,2\left(0.01 + K_1 \times \frac{\pi}{3}\right).\,K_1 \quad ...(ii)$$

Dividing (*i*) by (*ii*), we have:

$$\frac{(\pi/6)}{(\pi/3)} = \frac{(1.5)^2}{(4.5)^2} \times \frac{(0.01 + K_1 \times \pi/6)}{(0.01 + K_1 \times \pi/3)}$$

or, $$(0.5 \times 9) = \frac{(0.01 + K_1 \times \pi/6)}{(0.01 + K_1 \times \pi/3)}$$

or, $4.5(0.01 + K_1 \times \pi/3) = 0.01 + K_1 \times \pi/6$

or, $0.045 + 1.5\,\pi K_1 = 0.01 + \frac{\pi}{6} K_1$

or, $K_1\left(1.5\pi - \frac{\pi}{6}\right) = 0.01 - 0.045 = -0.035$

or, $K_1 = -\frac{0.035}{4.189} = \mathbf{-8.355 \times 10^{-3}}$ **(Ans.)**

Q. 17. ***(Schering bridge)*** *A capacitor bushing forms the arm AB of a high voltage schering bridge, a standard capacitor of 500 μμF and negligible loss forms the arms AD. Arm BC consists of a non-inductive resistance of 300 Ω. The bridge is balanced with the arm CD having a resistance of 72.6 Ω and capacitance of 0.148 μF in parallel. 50 Hz supply is connected between A and C. Calculate from first principles the capacitance and dielectric loss angle of bushing.*

Solution. Refer to Fig. 6.

Given: $C_2 = 500$ μμF; $R_4 = 300\ \Omega$; $R_3 = 72.6\ \Omega$, $C_3 = 0.148$ μF; $f = 50$ Hz.

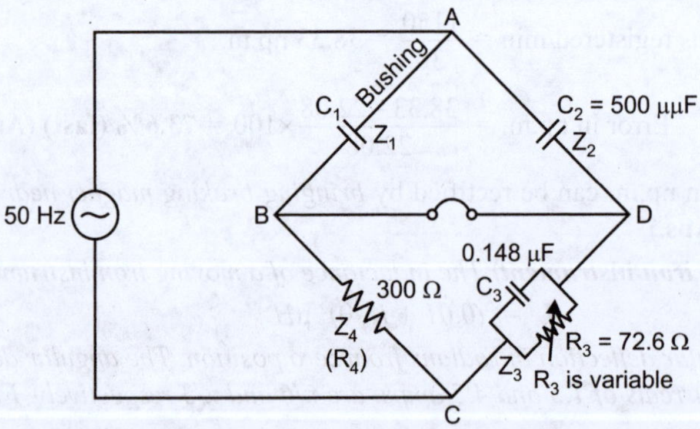

Fig. 6. Schering bridge.

Capacitance of bushing C_1:

When the bridge is balanced, we have:

$$\bar{Z}_1 \bar{Z}_3 = \bar{Z}_2 \bar{Z}_4$$

or, $$\bar{Z}_1 = \bar{Z}_2 \bar{Z}_4 \bar{Y}_3 \quad ...(i)$$

where, $$\bar{Z}_1 = \frac{1}{j\omega C_1};\ \bar{Z}_2 = \frac{1}{j\omega C_2};\ \bar{Z}_4 = R_4;\ \bar{Y}_3 = \left(\frac{1}{R_3} + j\omega C_3\right)$$

Substituting the values in eqn. (*i*), we get:

$$\frac{1}{j\omega C_1} = \frac{1}{j\omega C_2} \times R_4 \times \left(\frac{1}{R_3} + j\omega C_3\right)$$

$$-\frac{j}{\omega C_1} = -\frac{j}{\omega C_2} \times \frac{R_4}{R_3} + \frac{C_3 R_4}{C_2} \quad ...(i)$$

Equating imaginary parts, we have:

$$-\frac{1}{\omega C_1} = -\frac{1}{\omega C_2} \times \frac{R_4}{R_3}$$

or, $$C_1 = \frac{C_2R_3}{R_4} = \frac{500\times10^{-12}\times72.6}{300} = 121 \times 10^{-12}\text{ F} = \textbf{121 pf (Ans.)}$$

Dielectric loss angle of bushing δ:

We known that:

$$\tan\delta = \omega C_1 R_1 = \omega\times\left(\frac{C_2R_3}{R_4}\right)\times R_1 \quad ...(ii)$$

Eqn. (*i*) is modified, due to loss is capacitance as:

$$R_1 = \frac{j}{\omega C_1} = -\frac{j}{\omega C_2}\times\frac{R_4}{R_3}+\frac{C_3R_4}{C_2}$$

Equating real parts, we get:

$$R_1 = \frac{C_3R_4}{C_2}$$

Substituting this value of R_1 in eqn. (*ii*), we have:

$$\tan\delta = \frac{\omega C_2R_3}{R_4}\times\frac{C_3R_4}{C_2}$$

$$= \omega R_3C_3$$

$$= (2\pi\times50)\times72.6\times(0.148\times10^{-6}) = 3375.6\times10^{-6}$$

or, $$\delta = \tan^{-1}(3375.6\times10^{-6}) = \textbf{0.19 (Ans.)}$$

Q. 18. ***(Moving iron voltmeter)*** *A moving iron voltmeter designed to read up to 100 volts has resistance of 2000 Ω and inductance of 0.6 henry. How it can be modified to read up to 300 volts?* **(GATE)**

Solution. *Given*: Resistance of the voltmeter $R_v = 2000\ \Omega$; Inductance, $L = 0.6$ H; Voltage to be read = up to 300 V; $v = 100$ V.

Let R_s = The external series resistance to increase the range.

$$\text{Voltmeter current, } I = \frac{V}{\sqrt{(R_v+R_s)^2+(\omega L)^2}}$$

$$= \frac{300}{\sqrt{(2000+R_s)^2+(2\pi\times50\times0.6)^2}}$$

Voltage drop across instrument,

$$v = I\sqrt{R_v^2+(\omega L)^2}$$

$$= \frac{V}{\sqrt{(R_v+R_s)^2+(\omega L)^2}}$$

or, $$\frac{V}{v} = \frac{\sqrt{(R_v+R_s)^2+(\omega L)^2}}{\sqrt{R_v^2+(\omega L)^2}}$$

or, $$\frac{300}{100} = \frac{\sqrt{(2000+R_s)^2+(2\pi\times50\times0.6)^2}}{\sqrt{(2000)^2+(2\pi\times50\times0.6)^2}}$$

Squaring both sides, we get:

$$9[(2000)^2+(60\pi)^2] = (2000+R_s)^2+(60\pi)^2$$

$$\text{or,} \quad 36 \times 10^6 + 9 \times (60\,\pi)^2 = (2000 + R_s)^2 + (60\,\pi)^2$$

$$\text{or,} \quad 36 \times 10^6 + [8 \times (60\,\pi)^2] = (2000 + R_s)^2$$

$$\text{or,} \quad 36.284 \times 10^6 = (2000 + R_s)^2$$

$$\text{or,} \quad R_s = \mathbf{4023\ \Omega\ (Ans.)}$$

Q. 19. ***(Two-wattmeter method)*** *For measurement in a 3-phase load, two-wattmeter method was used with 3-phase, 440 volt system of supply and the current coils of two wattmeters connected in R and B phases. Determine the line currents, reading of each wattmeter and total input power. The load is resistive with 10 Ω, 15 Ω, 20 Ω (in star) respectively in R, Y and B phases.* **(IES)**

Solution. Refer to Fig. 7.

Assuming V_R as the reference vector:

$$V_R = \frac{440}{\sqrt{3}} \angle 0^\circ = 254 \angle 0^\circ \text{ V}$$

$$V_X = 254 \angle -120^\circ$$

$$V_B = 254 \angle -240^\circ$$

$$I_R = \frac{254 \angle 0^\circ}{10} = \mathbf{25.4 \angle 0^\circ\ A\ (Ans.)}$$

$$I_Y = \frac{254 \angle -120^\circ}{15} = \mathbf{16.9 \angle -120^\circ\ A\ (Ans.)}$$

$$I_B = \frac{254 \angle -240^\circ}{20} = \mathbf{12.7 \angle -240^\circ\ A\ (Ans.)}$$

where,

$$V_{RY} = 440 \angle 30^\circ$$

$$V_{BY} = 440 \angle 90^\circ$$

$$\theta_R = 30^\circ \text{ (angle between } V_{RY} \text{ and } I_R)$$

$$\theta_B = 90^\circ - (-240^\circ) = 330^\circ \text{ (angle between } V_{BY} \text{ and } I_B)$$

Now,

$$W_R = V_{RY} I_R (\cos \phi_R)$$

$$= 440 \times 25.4 \times \cos 30^\circ = 9678.7 \text{ W}$$

and,

$$W_B = V_{BY} I_B \cos \phi_B$$

$$= 440 \times 12.7 \times \cos 330^\circ = 4839.3 \text{ W}$$

$\therefore$ Total input power $= W_R + W_B = 9678.7 + 4839.3 = 14518$ W or **14.518 kW (Ans.)**

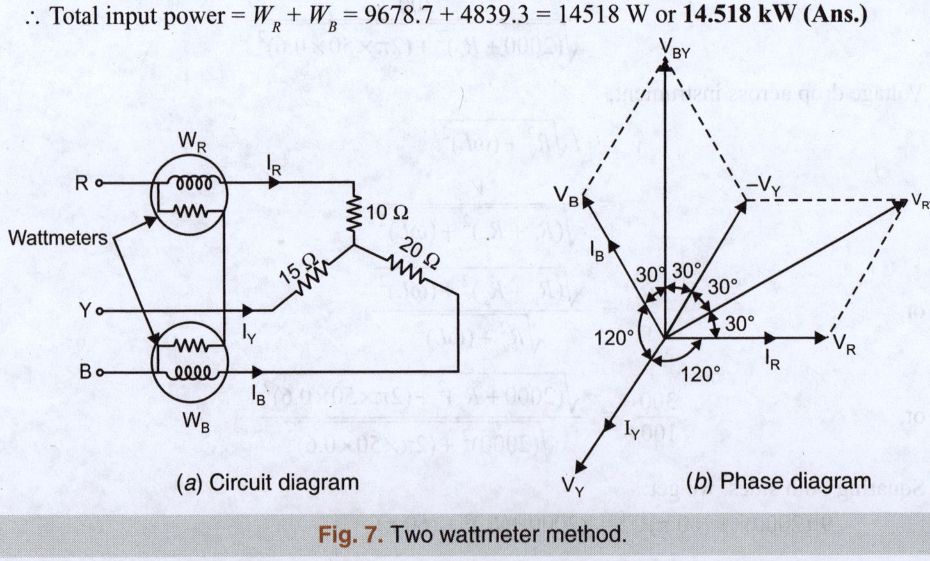

Fig. 7. Two wattmeter method.

Q. 20. (Resistive bridge) *A four arm resistive bridge is used with a temperature sensor (RTD) having resistance temperature coefficient α = 0.042 Ω/°C in one of the arms. Initially the bridge is balanced with all arms having a resistance of 400 Ω. DC supply to the bridge is limited by a maximum permissible current of 30 mA through the temperature sensor. Bridge output is connected to a current responding meter having input resistance of 100 Ω and deflection sensitivity of 2°/μA. Determine the deflection in the output meter if the temperature sensor undergoes a temperature rise of 30°C.* **(IES)**

Solution. *Given:* α = 0.042 Ω/°C; R = 400 Ω (each arm), Max. permissible current. I = 30mA; Input resistance of current responding meter = 100 Ω, deflection sensitivity = 2°/μA; Temperature rise of sensor = 30°C.

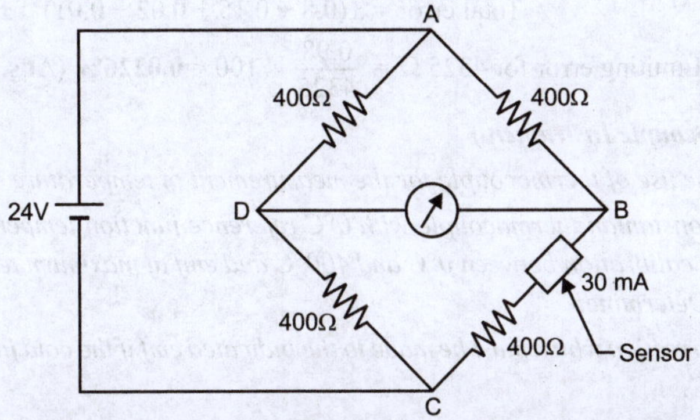

Fig. 8. Four arm resistive bridge.

Deflection in the output meter:

Total resistance of the sensor when it undergoes a temperature rise of 30°C = 400 + 0.042 × 30 = 401.26 Ω.

Current through arm ABC = 30mA (since current through sensor is restricted to 30 mA)

∴ Voltage across ABC = $(400 + 401.26) \times 30 \times 10^{-3} \simeq 24$ V (D.C. supply voltage)

Thevenin resistance = (400) || (400 + 400) || 401.26 = 400.315 Ω

Now, $$V_{th} = V_{AD} - V_{AB} = 24 \times \left(\frac{400}{800} - \frac{400}{801.26}\right) = 0.0188 \text{ V}$$

∴ Current through meter $= \dfrac{0.0188}{(400.315 + 100)} = 37.58\ \mu A$

Hence, deflection of the meter = 37.58 × 2 = **75.16° (Ans.)**

Q. 21. (Errors in measurements)

The limiting errors for a four dial resistance box are:

Units: ±0.2%

Tens: ±0.1%

Hundreds: ±0.05%

Thousands: ±0.02%

If the resistance value is set at 4325 Ω calculate the limiting error for this value. **(GATE)**

Solution. *Given:* Limiting errors for the resistance box: Units: ±0.2%; Tens: ±0.1%; Hundreds: ±0.05%; Thousands: ±0.02%.

Limiting error for 4325Ω:

For thousand (set at 4000 Ω) error $= \pm 4000 \times \frac{0.02}{100} = \pm 0.8\ \Omega$

Similarly, For hundred error $= \pm 300 \times \frac{0.05}{100} = \pm 0.15\ \Omega$

and, For ten error $= \pm 20 \times \frac{0.1}{100} = \pm 0.02\ \Omega$

and, For unit error $= \pm 5 \times \frac{0.2}{100} = \pm 0.01\ \Omega$

Hence, Total error $= \pm(0.8 + 0.15 + 0.02 + 0.01) = \pm 0.98\ \Omega$

$$\%\text{ Limiting error for } 4325\ \Omega = \frac{0.98}{4325} \times 100 = \mathbf{0.0226\%}\ \textbf{(Ans.)}$$

Q. 22. ***(Thermocouple Instrument)***

(a) Describe the use of thermocouple for the measurement of temperature.

(b) A copper-constantan thermocouple with 0°C reference junction temperature was found to have linear calibration between 0°C and 400°C and emf at maximum temperature equal to 20.68 mV. Determine:

(a) The correction which must be made to the indicated emf if the cold junction temperature is 25°C.

(b) The temperature of the hot junction if the indicated emf is 8.92 mV with cold junction temperature of 25°C. **(IES)**

Solution. (*a*) Refer to Article **4.8.10.**

(*b*) *Given:* Emf at maximum temperature = 20.68 mV; Cold junction temperature = 25°C.

(*a*) Correction to be made to the indicated emf:

Assuming linear relationship between emf (e) generated and temperature (t) we have;

$$e = K\Delta t$$

or $$K = \frac{e}{\Delta t} = \frac{20.68}{400} = 0.0517 \text{ mV/°C}$$

Voltage at 25°C = 0.0517 × 25 = 1.2925 mV

Hence, **1.2925 mV correction** must be made to the indicated output **(Ans.)**

(*b*) Temperature of the hot junction:

Indicated emf = 8.92 mV

Corrected emf = Indicated emf + correction

= 8.92 + 1.2925 = 10.2125 mV

Now, since 20.68 mV corresponds to 400°C, therefore, 10.2125 mV corresponds to $\frac{400}{20.68} \times 10.2125 =$ **197.53°C (Ans.)**

Q. 23. ***(Errors in measurements)*** *Resistors R_1 and R_2 have the nominal value 3Ω and 6Ω and tolerance of ±10% and ±5% respectively. What would be the tolerance in the equivalent resistance value when R_1 and R_2 are connected in parallel.* **(GATE)**

Solution. R_1 = 3Ω (tolerance = ±10%), R_2 = 6Ω (tolerance = ±5%)

Tolerance in equivalent circuit:

Equivalent resistance in parallel, when $R_1 = 3\ \Omega$ amd $R_2 = 6\ \Omega$,

$$R_{eq} = \frac{R_1 \times R_2}{R_1 + R_2} = \frac{3\times6}{3+6} = 2\ \Omega$$

Now, value of R_1 will be between 2.7 Ω and 3.3 Ω, so taking maximum value of $R_1 = 3.3\ \Omega$;

Similarly, value of R_2 will be between 5.7 Ω and 6.3 Ω, so taking maximum value of $R_2 = 6.3\ \Omega$, we have:

$$\therefore \quad R_{eq} = \frac{R_1 R_2}{R_1 + R_2} = \frac{3.3\times6.3}{3.3+6.3} = 2.165\ \Omega$$

Hence, tolerance in equivalent circuit = $\pm\frac{2.165-2}{2}\times100$ = **±8.25% (Ans.)**

Q. 24. ***(Dynamometer type wattmeter)*** *The reading of a dynamometer type wattmeter with pressure coil phase angle of 2° is 700 watts, when it is used to measure power of a single-phase inductive load supplied by 240 V single-phase a.c. source. When the wattmeter is replaced by a second wattmeter with a phase angle of 1° for the pressure coil circuit, a reading of 620 watts is obtained. Calculate the actual power and current taken by the load assuming all errors of the wattmeter, except those due to pressure coil inductance, are neglected.* **(IES)**

Solution. *Given:* Pressure coil phase angle, θ = 2°, Reading of wattmeter = 700W, Supply voltage = 240 V … First wattmeter;

θ = 1°, Reading of wattmeter = 620 W … Second wattmeter

True power taken by this load:

$$\text{True power} = \frac{\cos\phi}{\cos\theta.\cos(\phi-\theta)} \times \text{Actual reading of wattmeter} \qquad \text{[Eqn. (4.133a)]}$$

This can be simplified as:

$$\text{Actual wattmeter reading} = \text{True power}\,(1 + \tan\phi\tan\theta)$$

$$\therefore \quad 700 = \text{True power}\,(1 + \tan\phi\tan 2°) \quad \ldots(i)$$

$$\text{and,} \quad 620 = \text{True power}\,(1 + \tan\phi\tan 1°) \quad \ldots(ii)$$

Dividing (*i*) by (*ii*), we get:

$$\frac{700}{620} = \frac{1+\tan\phi.\tan2°}{1+\tan\phi\tan1°} = \frac{1+0.035\tan\phi}{1+0.0175\tan\phi}$$

$$\text{or,} \quad 700(1 + 0.0175\tan\phi) = 620(1 + 0.035\tan\phi)$$

$$\text{or,} \quad 700 + 12.25\tan\phi = 620 + 21.7\tan\phi$$

$$\text{or,} \quad (21.7 - 12.25)\tan\phi = 700 - 620 = 80$$

$$\text{or,} \quad \tan\phi = 8.4656 \text{ or } \phi = 83.26°$$

$$\text{Hence,} \quad \text{True power} = \frac{\text{Actual wattmeter reading}}{1+\tan\phi\tan\theta}$$

$$= \frac{700}{1+(8.4656\times0.035)} = \mathbf{540\ W\ (Ans.)}$$

Current taken by this load, *I*:

$$VI \cos \phi = 540$$

or, $$240 \times I \cos 83.26° = 540$$

$$\therefore \quad I = \frac{540}{240 \times \cos 83.26°} = \mathbf{19.17\ A\ (Ans.)}$$

Q. 25. ***(Error in measurements)*** *A 100.00 V reference d.c. source is used for calibration of a digital multimeter in the 200 V range. The uncertainty quoted by a lab of higher echelon in national calibration network in the calibration certificate of reference standard is ±0.01 volt. The observations made during calibration are 100.2, 100.3, 100.2 and 100.1 volts respectively. Find the assigned value and uncertainity associated with measurements. (Assume a confidence level 95% and corresponding student factor, t = 2.78).* **(IES)**

Solution. *Given:* Voltage of reference d.c. source = 100.00 V; Range of digital multimeter = 200 V; Uncertainty quoted = ±0.01 V; Observations made: 100.2, 100.3, 100.2 and 100.1 V respectively; Confidence level = 95%, corresponding student factor, $t = 2.78$.

Assigned value and uncertainty limits:

The four observations including uncertainty of ±0.01 V are: 100.2 ± 0.01, 100.3 ± 0.01, 100.2 ± 0.01, and 100.1 ± 0.01 V, i.e.

100.21	100.19
100.31	100.29
100.21	100.19
100.11	100.09

$$\text{Mean} = 100 + \frac{0.84}{4} = 100.21 \qquad \text{Mean} = 100 + \frac{0.76}{4} = 100.19$$

Let us assume that mean $\overline{X} = 100.21$. then:

X	$\overline{X}$	d	d^2
100.21	100.21	0	0
100.31	100.21	±0.10	0.01
100.21	100.21	0	0
100.11	100.21	–0.10	0.01

$$\sum_{1}^{4} d^2 = 0 + 0.01 + 0 + 0.01 = 0.02$$

Now, $$\text{Variance, } V = \frac{\Sigma d^2}{n-1} = \frac{0.02}{4-1} = \frac{0.02}{3}$$

and, $$\text{Standard deviation, } s = \sqrt{V} = \sqrt{\frac{0.02}{3}} = 0.082$$

True mean of this voltage being measured lies between the limits

$$= \overline{X} \pm 2.78\, s = 100.21 \pm 2.78 \times 0.082 = 100.21 \pm 0.228 \text{ V}$$

Hence, *assigned value* is **100.21** and *uncertainty limits* lie between **99.982 V** and **100.438 V (Ans.)**

Q. 26. *Discuss the difficulties faced in measuring high resistances. Explain the procedures adopted in the laboratory for measuring high resistances.* **(IES)**

Ans. Refer to *Article* **10.2.4.**

Q. 27. *Explain with a real diagram how Wien's bridge can be used for experimental determination of frequency. Derive the expression for frequency in terms of bridge parameters.* **(GATE)**

Ans. Refer to *Article* **10.3.11.**

Q. 28. *Draw a circuit diagram of De Sauty bridge for the measurement of capacitance and obtain an expression for the unknown capacitance. What are the defects of this bridge?*

Ans. Refer to *Article* **10.3.6.**

Q. 29. *The power in a balanced 3-phase, 3-wire, star connected system is to be measured by two-wattmeters. Draw neat connection and phasor diagrams. Do not derive the expressions for the power measured by each wattmeter.* **(IES)**

Ans. Refer to *Article* **4.9.8.2.**

Q. 30. *Explain the basic principle of operation of a thermocouple.* **(GATE)**

Ans. Refer to *Article* **4.8.10.**

Q. 31. *Explain with a neat diagram the construction and working of a frequency counter.* **(GATE)**

Ans. Refer to *Article* **8.3.5.**

Q. 32. *Define the term telemetry and explain why it is necessary to use it in an instrumentation system. Name different types of telemetry systems. Give salient features of RF telemetry.* **(IES)**

Ans. Refer to *Article* **19.2.**

Q. 33. *What are capacitive transducers? What are their advantages and disadvantages? Explain the use of capacitive transducers for measurement of the liquid level.* **(GATE)**

Ans. Refer to *Articles* **16.8.** and **21.10.1.**

Q. 34. *Draw the circuit of Anderson bridge. Derive the null conditions.*

Ans. Refer to *Article* **10.3.4.**

Q. 35. *What are the advantages of digital voltmeters (DVMs) compared to analog indicating type voltmeters?* **(GATE)**

Ans. Refer to *Article* **8.42.**

B. MULTIPLE-CHOICE QUESTIONS, WITH ANSWERS AND "*EXPLANATIONS*"

Q.1. A reading of 120 is obtained when standard inductor was connected. The circuit of a Q-meter and the variable capacitor is adjusted to a value of 300 pF. A lossless capacitor of unknown value C_x is then connected in parallel with the variable capacitor and the same reading was obtained when the variable capacitor is readjusted to a value of 200 pF. The value of C_x in pF is

(*a*) 100 (*b*) 200 (*c*) 300 (*d*) 500.

Ans. (*a*)

EXPLANATION:

Let C' be the readjusted value and $C = 300$ pF

Then,
$$Q = \frac{1}{\omega RC} = \frac{1}{\omega R(C_x + C')}$$

or,
$$C = C' + C_x$$

∴
$$C_x = 300 - 200 = \mathbf{100\ pF\ (Ans.)}$$

Q.2. The simplified block diagram of a 10-bit A/D converter of dual slope integrator type is shown in Fig. 1. The 10-bit counter at the output is clocked by a 1 MHz clock. Assuming negligible timing overhead for the control logic, the maximum frequency of the analog signal that can be converted using this A/D converter is approximately

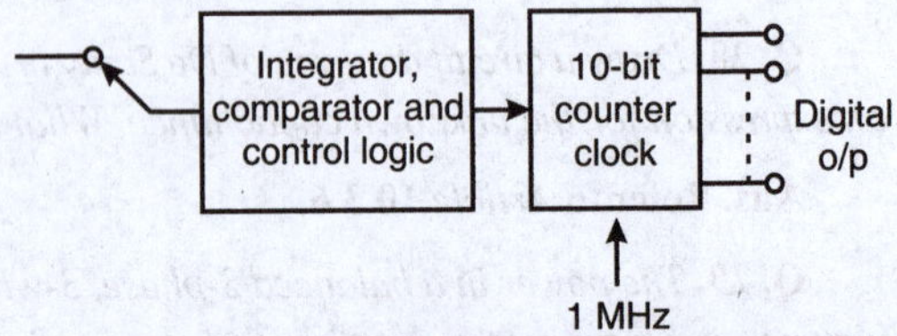

Fig. 1

(*a*) 2 kHz (*b*) 1 kHz (*c*) 500 Hz (*d*) 250 Hz.

Ans. (*b*)

EXPLANATION:

The waveform for dual slope integrater is shown in Fig. 2. The maximum frequency can be attained when $T_2 - T_1 = 0$, and as

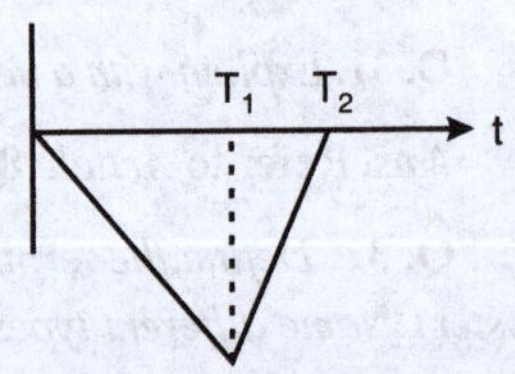

Fig. 2

$$T_1 = 2^n T_C\ (T_C = \text{clock period})$$

∴
$$f_{max} = \frac{1}{2^n \cdot T_C} = \frac{f_C}{2^n}$$

$$= \frac{10^6}{1024} \approx 1\ \text{kHz}\ \textbf{(Ans.)}$$

Q.3. A rectifier type ac voltmeter consists of a series resistance R_s, an ideal full-wave rectifier bridge and PMMC instrument as shown in Fig. 3. The internal resistance of the instrument is 100 Ω and a full scale deflection is produced by a dc current of 1 mA. The value of R_s required to obtain full scale deflection with an ac voltage of 100 V (rms) applied to the input terminals is

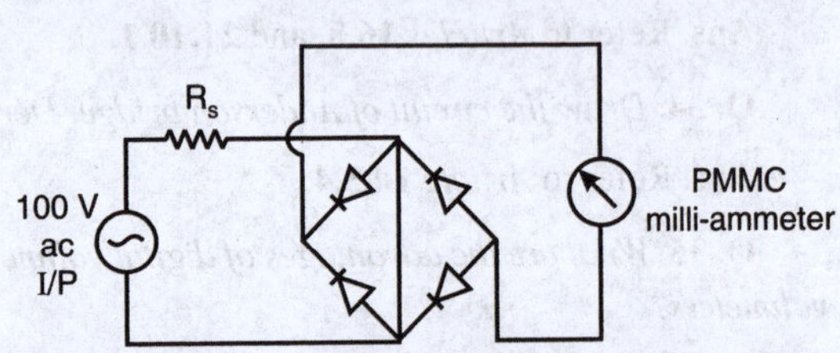

Fig. 3

(*a*) 63.5 Ω (*b*) 69.9 Ω (*c*) 89.9 kΩ (*d*) 141.3 kΩ.

Ans. (*c*)

EXPLANATION:

d.c. sensitivity,
$$S_{dc} = \frac{1}{I_{(FS)}} = \frac{1}{1\text{mA}} = 10^3\ \Omega/\text{V}$$

For full-wave rectifier, a.c. sensitivity,
$$S_{ac} = 0.9\, S_{dc} = 0.9 \times 10^3 = 900\ \Omega/\text{V}$$

Resistance of multiplier, $R_s = S_{ac} V - R_m - 2 R_d$

Since, diodes are ideal, R_d $1 = 0$,

Then, $R_s = 900 \times 100 - 100$

$= 89.9$ kΩ **(Ans.)**

Q.4. A wattmeter reads 400 W when its current coil is connected in the *R* phase and its pressure coil is connected between this phase and the neutral of a symmetrical 3-phase system supplying a balanced star connected 0.8 p.f. inductive load. This phase sequence is RYB. What will be the reading of this wattmeter if its pressure coil alone is reconnected between the B and Y phases, all other connections remaining as before?

(*a*) 400.0 (*b*) 519.5 (*c*) 300.0 (*d*) 692.8.

Ans. (*b*)

Explanation: Refer to Fig. 4, 5, 6.

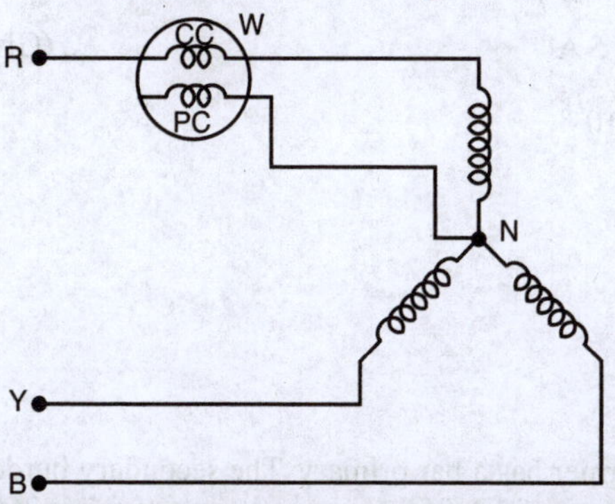

Fig. 4

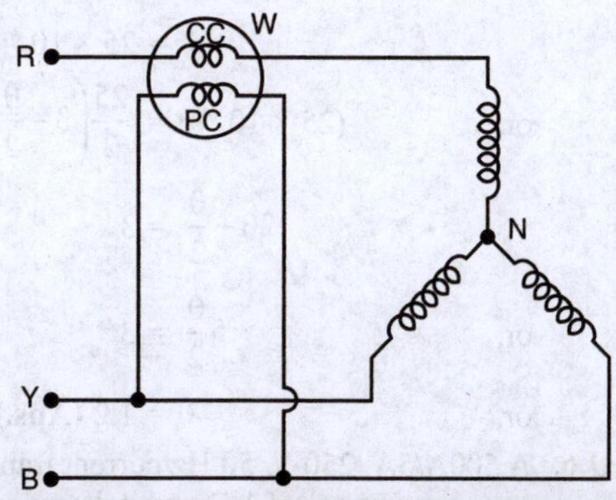

Fig. 5

$$\phi_1 = \cos^{-1} 0.8 = 36.86^\circ$$

In this case, reading of wattmeter,

$$W_1 = \frac{V_L I_L}{\sqrt{3}} \cos\phi_1$$

$$= \frac{V_L I_L}{\sqrt{3}} \times 0.8 = 400 \text{ W}$$

or, $$V_L I_L = \sqrt{3} \times \frac{400}{0.8}$$

In second case, angle ϕ_2 between I_2 and V_{YB},

$$\phi_2 = 30^\circ + 23.14^\circ$$

$$= 53.14^\circ$$

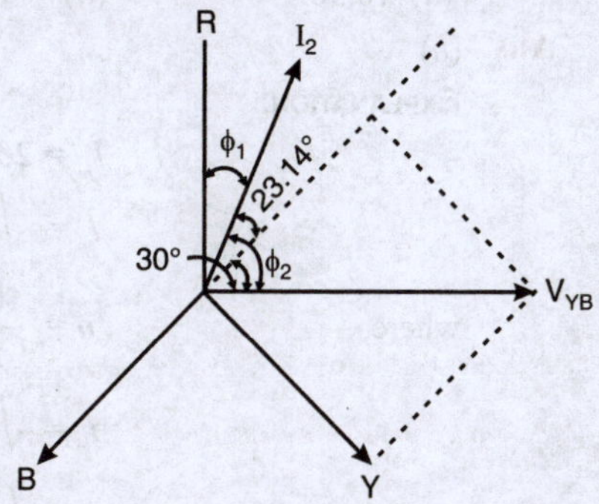

Fig. 6

$\therefore$ $$W_2 = V_{YB} I_L \cos(53.14^\circ)$$

$$= \sqrt{3} \times \frac{400}{0.8} \times \cos(53.14^\circ)$$

$$= 519.5 \text{ W } \textbf{(Ans.)}$$

Q.5. The inductance of a certain moving-iron ammeter is expressed as $L = 10 + 3\theta - (\theta^2/4)$ μH, where θ is the deflection in radians from the zero position. The control spring torque is 25×10^{-6} Nm/radian. The deflection of the pointer in radian when the meter carries a current of 5A, is

(*a*) 2.4 (*b*) 2.0 (*c*) 1.2 (*d*) 1.0.

Ans. (*c*)

EXPLANATION:

$$L = 10 + 3\theta - \left(\frac{\theta^2}{4}\right) \mu\text{H} \quad \text{...(Given)}$$

$$\therefore \quad \frac{dL}{d\theta} = 3 - \frac{\theta}{2}$$

At equilibrium,

$$K\theta = \frac{1}{2} i^2 \frac{dL}{d\theta}$$

$$(25 \times 10^{-6})\theta = \frac{1}{2} \times (5)^2 \, i^2 \left(3 - \frac{\theta}{2}\right) \times 10^{-6}$$

$$(K = 25 \times 10^{-6};\ i = 5 \text{ A} \quad \text{... (Given)}$$

or, $$(25 \times 10^{-6})\theta = \frac{25}{2}\left(3 - \frac{\theta}{2}\right) \times 10^{-6}$$

$$\therefore \quad 2\theta + \frac{\theta}{2} = 3$$

or, $$5\frac{\theta}{2} = 3$$

or, $$\theta = 1.2 \text{ (Ans.)}$$

Q.6. A 500A/5A, 250 V, 50 Hz current transformer has a bar primary. The secondary burden is a pure resistance of 1 Ω and it draws a current of 5 A. If the magnetic core requires 250 AT for magnetisation, the percentage ratio error is

(*a*) 10.56 (*b*) –10.56 (*c*) 11.80 (*d*) –11.80.

Ans. (*b*)

EXPLANATION:

$$I_m = 250/1 = 250 \text{ A}$$

$$I_p = \sqrt{(nI_s)^2 + (I_m)^2}$$

where, $$n = \frac{500}{5} = 100$$

$$\therefore \quad I_p = \sqrt{(5 \times 100)^2 + 250^2} = 559.0169$$

$$\therefore \quad R = \frac{I_p}{I_s} = \frac{559.0169}{5} = 111.8033$$

$$\therefore \quad \% \text{ Ratio error} = \frac{100 - 111.8033}{111.8033} \times 100 = \mathbf{-10.56 \text{ (Ans.)}}$$

Q.7. The voltage-flux adjustment of certain 1-phase 220 V induction watt-hour meter is altered so that the phase angle between the applied voltage and the flux due to it is 85° (instead of 90°). The errors introduced in the reading of this meter when the current is 5 A at power factors of unity and 0.5 lagging are respectively

(*a*) 3.8 mW, 77.4 mW (*b*) –3.8 mW, –77.4 mW

(*c*) –4.2 W, –85.1 W (*d*) 4.2 W, 85.1 W.

Ans. (*c*)

EXPLANATION:

At, $\Delta = 85^\circ, \phi = 60^\circ$ (p.f. = 0.5)

Energy registered = VI $\sin(\Delta - \phi)$

Then, error = VI $\sin(\Delta - \phi)$ – VI $\cos \phi$

$= 220 \times 5\, [\sin (85 - 60)^\circ - \cos 60^\circ)]$

$= -85.1$ W **(Ans.)**

At, $\Delta = 85^\circ, \phi = 0^\circ$ (p.f. = 0)

$= \text{VI}\, [\sin (\Delta - 0)^\circ - \cos 0^\circ]$

$= \text{VI}\, (\sin 85^\circ - \cos 0^\circ)$

$= 220 \times 5 \times (0.9962 - 1)$

$= -4.18$ W **(Ans.)**

Q.8. A dc potentiometer is designed to measure up to about 2 V with a slide wire of 800 mm. A standard cell of emf 1.18 V obtains balance at 600 mm. A test cell is seen to obtain balance at 680 mm. The emf of the test cell is

(*a*) 1.00 V (*b*) 1.34 V (*c*) 1.50 V (*d*) 1.70 V.

Ans. (*b*)

EXPLANATION:

For the d.c. potentiometer, we have:

$$\frac{E_1}{E_2} = \frac{l_1}{l_2}$$

or, Emf of the test cell,

$$E_2 = E_1 \cdot \frac{l_2}{l_1}$$

$$= (1.18) \times \frac{680}{600} = 1.34 \text{ V } \textbf{(Ans.)}$$

Q.9. The circuit shown in Fig. 7 is used to measure the power consumed by the load. The current coil and the voltage coil of the wattmeter have 0.02 Ω and 1000 Ω resistances respectively. The measured power compared to the load power will be

(*a*) 0.4% less (*b*) 0.2% less

(*c*) 0.2% more (*d*) 0.4% more.

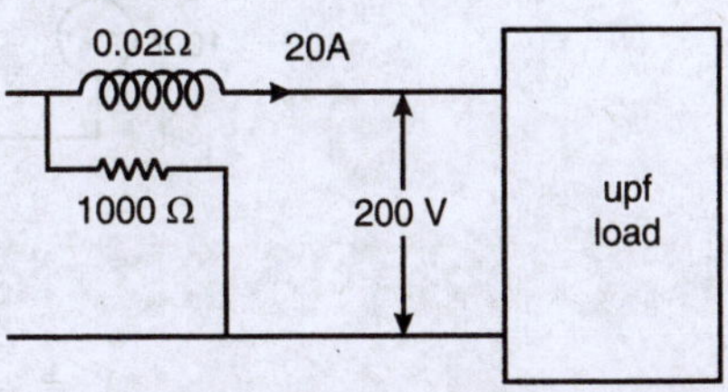

Fig. 7

Ans. (*c*)

EXPLANATION:

Power indicated by wattmeter

= Power consumed by load + power loss in (CC)

$= 200 \times 20 + 20^2 \times 0.02$

$= 4008$ W

Now, Load power $= 200 \times 20 = 4000$ W

$$\therefore \quad \% \text{ Increase} = \frac{4008 - 4000}{4000} \times 100$$

$= 0.2\%$ more **(Ans.)**

Q.10. A galvanometer with a full scale current of 10 mA has a resistance of 1000 Ω. The multiplying power (the ratio of measured current to galvanometer current) of 100 Ω shunt with this galvanometer is

(*a*) 110 (*b*) 100 (*c*) 11 (*d*) 10.

Ans. (*c*)

EXPLANATION: Refer to Fig. 8

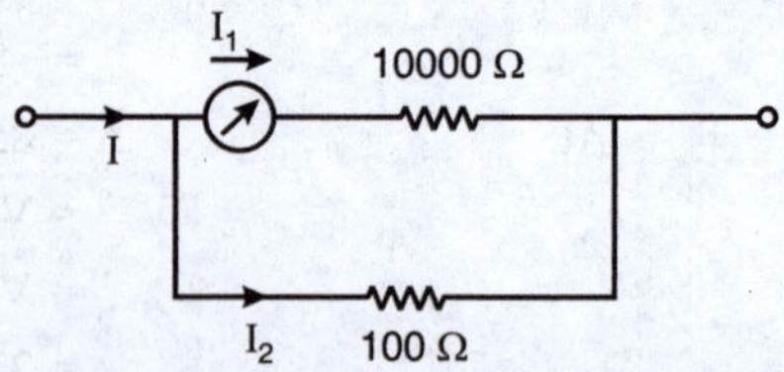

Fig. 8

$$\text{Multiplying factor} = m = \frac{I}{I_1}$$

Now, $$\frac{I_1}{I_2} = \frac{100}{1000}$$

or, $$\frac{I_1}{I_1+I_2} = \frac{100}{100+1000} \text{ or } \frac{1}{11}$$

or, $$\frac{I_1}{I} = \frac{1}{11} \text{ or } \frac{I}{I_1} = 11 \textbf{ (Ans.)}$$

Q.11. A CRO probe has an impedance of 500 kΩ in parallel with a capacitance of 10 pF. The probe is used to measure the voltage between *P* and *Q* as shown in Fig. 9. The measured voltage will be

(*a*) 3.53 V (*b*) 4.37 V

(*c*) 4.54 V (*d*) 5.00 V.

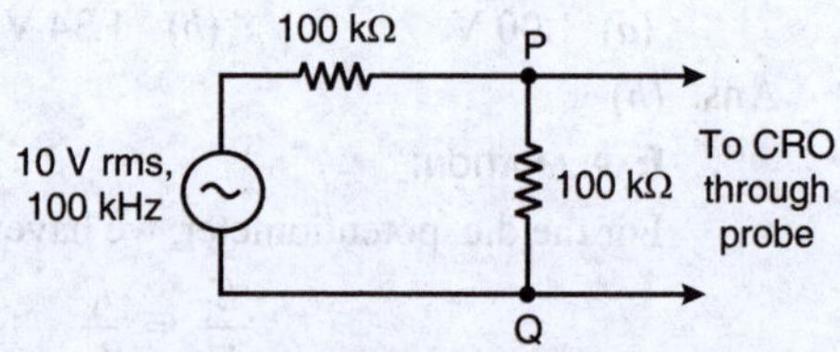

Fig. 9

Ans. (*a*)

EXPLANATION: Refer to Fig. 10.

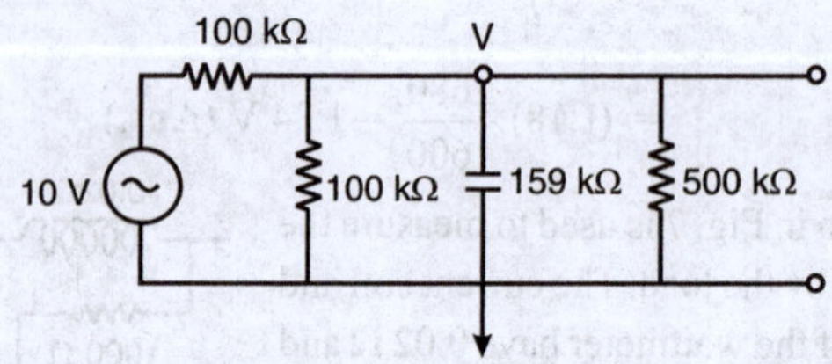

Fig. 10

$$X_c = \frac{1}{2\pi fC} = \frac{1}{2\pi\times(100\times10^3)\times10\times10^{-12}}\times10^{-3}\,\text{k}\Omega$$

$$= 159\ \text{k}\Omega$$

$$\therefore \quad \frac{10-V}{100} = V\left(\frac{1}{100}+\frac{1}{159}+\frac{1}{500}\right)$$

$$= 0.01829\ \text{V}$$

$$10 - V = 1.829\ \text{V}$$

$$\therefore \quad V = \frac{10}{2.829} = 3.53\ \text{V} \textbf{ (Ans.)}$$

Q.12. A moving coil of a metal has 100 turns, and a length and depth of 10 mm and 20 mm respectively. It is positioned in a uniform radial flux density of 200 mT. The coil carries a current of 50 mA. The torque on the coil is

(*a*) 200 μNm (*b*) 100 μNm (*c*) 2 μNm (*d*) 1 μNm.

Ans. (*a*)

EXPLANATION:

Given :

$$B = 200 \times 10^{-3}\ T\ (i.e., \text{Wb/m}^2);$$
$$I = 50 \text{ mA};\ N = 100;$$
$$A = (10 \times 10^{-3})(20 \times 10^{-3}) \text{ m}^2$$

$$\text{Deflecting torque} = T_d = \text{BINA};\ A = l \times d$$

Inserting the values, we get:

$$T_d = (200 \times 10^{-3}) \times (50 \times 10^{-3}) \times 100 \times (10 \times 10^{-3})(20 \times 10^{-3})$$
$$= 2 \times 10^{-4} \text{ Nm or } 200\ \mu\text{Nm}\ \textbf{(Ans.)}$$

Q.13. A dc A-h meter is rated for 15 A, 250 V. The meter constant is 14.4 A-sec/rev. The meter constant at rated voltage may be expressed as

(*a*) 3750 rev/kWh (*b*) 3600 rev/kWh (*c*) 1000 rev/kWh (*d*) 960 rev/kWh.

Ans. (*c*)

EXPLANATION:

Meter constant is 14.4 A-sec per rev.,

$$\text{or,} \quad \frac{14.4}{3600} \text{ A-h} = \frac{14.4 \times 250}{3600} = 1 \text{ whr/rev}$$

Hence, 1 rev/Wh = 1000 rev/k Whr **(Ans.)**

Q.14. A moving iron ammeter produces a full scale torque of 240 μNm with a deflection of 120° at a current of 10 A. The rate of change of self inductance (μH/radian) of the instrument at full scale is

(*a*) 2.0 μH/radian (*b*) 4.8 μH/radian

(*c*) 12.0 μH/radian (*d*) 114.6 μH/radian.

Ans. (*b*)

EXPLANATION:

$$\text{At full scale position,} \quad T_C = \frac{1}{2} I^2 \frac{dL}{d\theta}$$

$$240 \times 10^{-6} = \frac{1}{2} \times (10)^2 \times \frac{dL}{d\theta}$$

$$\therefore \quad \left(\frac{dL}{d\theta}\right) = 4.8\ \mu\text{H/radian}\ \textbf{(Ans.)}$$

Q.15. A single-phase load is connected between R and Y terminals of a 415 V, symmetrical 3-phase, 4 wire system with phase sequence RYB. A wattmeter is connected in the system as shown in Fig. 11. The power factor of the load is 0.8 lagging. The wattmeter will read

(*a*) –795 W (*b*) –597 W

(*c*) +597 W (*d*) +795 W.

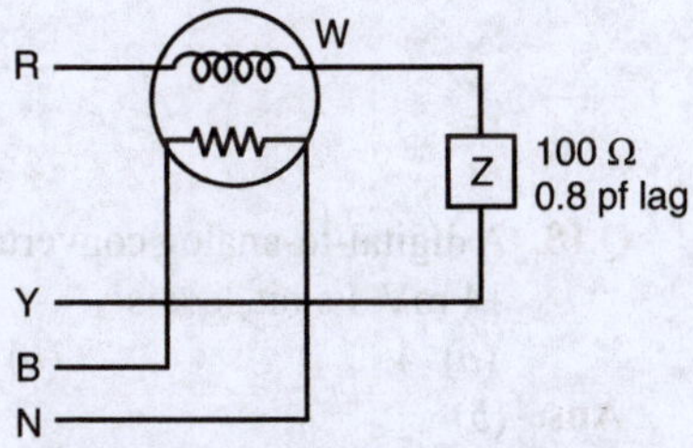

Fig. 11

Ans. (*b*)

EXPLANATION:

From Figs. 11, 12, we have:

$$V_{RY} = 415 \angle 30^\circ$$

$$V_{BN} = \frac{415}{\sqrt{3}} \angle 120^\circ$$

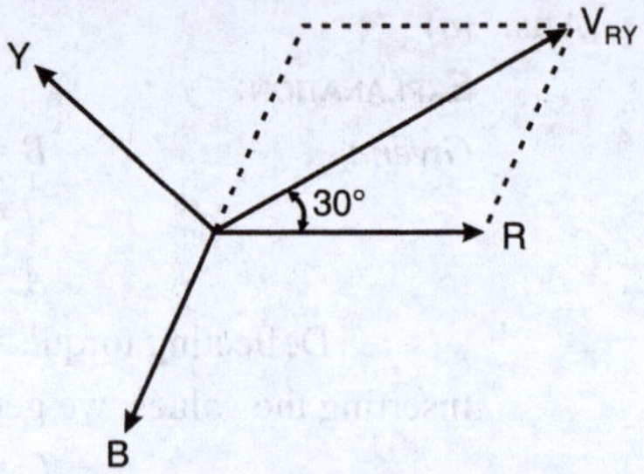

Fig. 12

Now, current through current coil is,

$$I_C = \frac{V_{RY}}{Z} = \frac{415\angle 30^\circ}{100\angle 36.87^\circ}$$

$$= 4.15 \angle -6.87^\circ$$

$$S = VI = \frac{415}{\sqrt{3}} \times 4.15 \angle -126.87^\circ$$

$$= 994.34 \angle -126.87^\circ$$

$$P = -597 \text{ W } \textbf{(Ans.)}$$

Q.16. A 50 Hz, bar primary CT has a secondary with 500 turns. The secondary supplies 5 A current into a purely resistive burden of 1 Ω. The magnetising ampere-turns are 200. The phase angle between the primary and secondary current is

(*a*) 4.6° (*b*) 85.4° (*c*) 94.6° (*d*) 175.4°.

Ans. (*a*)

EXPLANATION:

Phase angle between the (I_s) and (I_p) is $\theta \approx$ degree $\approx \left(\frac{180}{\pi}\right) \times \left(\frac{I_m}{I_p}\right)$

(Here, $I_p = K_t I_s = I_s$)

$$\simeq \left(\frac{180}{\pi}\right) \times \left(\frac{200}{500 \times 5}\right)$$

$$\simeq 4.6^\circ \textbf{ (Ans.)}$$

Q.17. The core flux in the CT of Q. 16, under the given operating conditions is

(*a*) 0 (*b*) 45.0 μWb (*c*) 22.5 mWb (*d*) 100.0 mWb.

Ans. (*b*)

EXPLANATION:

Voltage induced in the secondary,

$$E_s = I_s \times Z_s = 5 \times 1 = 5 \text{ volts}$$

$$E_s = 4.44 f \phi_m N_s$$

$$\therefore \quad \phi_m = \frac{E_s}{4.44 f N_s}$$

$$= \frac{5}{4.44 \times 50 \times 500}$$

$$= 45 \text{ μWb } \textbf{(Ans.)}$$

Q.18. A digital-to-analog converter with a full-scale output voltage of 3.5 V has resolution close to 14 mV. Its bit size is

(*a*) 4 (*b*) 8 (*c*) 16 (*d*) 32.

Ans. (*b*)

EXPLANATION:

$$\text{Resolution} = \frac{V_o}{2^n} = 14 \times 10^{-3} \text{ V (given)}$$

or, $$2^n = \frac{3.5}{14 \times 10^{-3}} = 250$$

or, $$n \ln(2) = \ln(250) \text{ or } n = 8 \textbf{ (Ans.)}$$

Q.19. The simultaneous applications of signals $x(t)$ and $y(t)$ to the horizontal and vertical plates, respectively, of an oscilloscope, produces a vertical figure-of-8 display. If P and Q are constants, and $x(t) = P \sin(4t + 30°)$, then $y(t)$ is equal to

(*a*) Q sin (4t – 30°) (*b*) Q sin (2t + 15°) (*c*) Q sin (8t + 60°) (*d*) Q sin (4t + 30°).

Ans. (*b*)

EXPLANATION:

We know that: $$\frac{f_y}{f_x} = \frac{x-\text{peak}}{y-\text{peak}}$$

Here, x-peak = 1 and y-peak = 2

$$\therefore \quad f_y = \frac{f_x}{2}$$

Here, $$x(t) = P \sin(4t + 30°)$$

$$\therefore \quad y(t) = Q \sin\left(\frac{4t + 30°}{2}\right)$$

$$= Q \sin(2t + 15°) \text{ (Ans.)}$$

Q.20. A DC ammeter has a resistance of 0.1 Ω and its current range is 0–100 A. If the range is to be extended to 0–500 A, then meter requires the following shunt resistance

(*a*) 0.010 Ω (*b*) 0.011 Ω (*c*) 0.025 Ω (*d*) 1.0 Ω.

Ans. (*c*)

EXPLANATION:

$$R_{sh} = \frac{R_m}{m-1}$$

where, m = multiplication factor

$$= \frac{500}{100} = 5$$

$$\therefore \quad R_{sh} = \frac{0.1}{4} = 0.025\ \Omega \text{ (Ans.)}$$

Q.21. The set-up in the Fig. 13. is used to measure resistance R. The ammeter and voltmeter resistances are 0.01 Ω and 2000 Ω, respectively. Their readings are 2 A and 180 V, respectively, giving a measured resistance of 90 Ω. The percentage error in the measurement is

(*a*) 2.25% (*b*) 2.35%

(*c*) 4.5% (*d*) 4.71%.

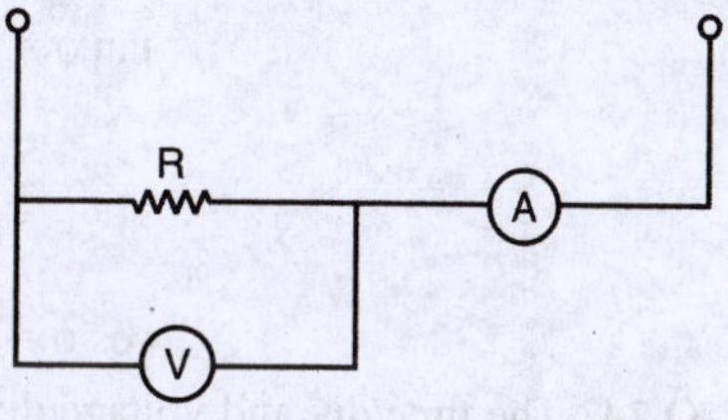

Fig. 13

Ans. (*c*)

EXPLANATION: Refer to Fig. 13.

Current through voltmeter, $$I_V = \frac{180}{2000} = 0.09 \text{ A}$$

Current through (R), $$I_R = 2 - \frac{9}{100} = 1.91 \text{ A}$$

Since, $$1.91\, R = 180$$

$$\therefore \quad R = \frac{180}{1.91} = 94.24\ \Omega$$

Hence,
$$\%\ \text{error} = \frac{94.24-90}{94.24}\times 100 = 4.5\%\ \textbf{(Ans.)}$$

Q.22. A 1000 V DC supply has two 1-core cables as its positive and negative leads; their insulation resistances to earth are 4 MΩ and 6 MΩ, respectively, as shown in the Fig. 14. A voltmeter with resistance 50 kΩ is used to measure the insulation of the cable. When connected between the positive core and earth, then voltmeter reads

(*a*) 8 V (*b*) 16 V
(*c*) 24 V (*d*) 40 V.

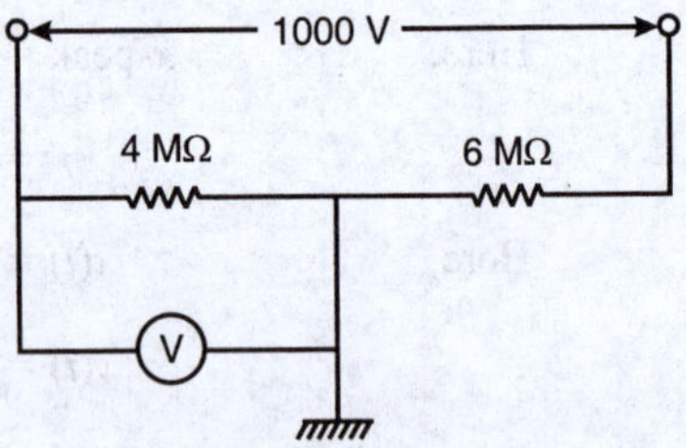

Fig. 14

Ans. (*a*)

EXPLANATION: Refer to Fig. 14

Resistance across the terminal,

$$R_T = \frac{4\times 0.05}{4.05} + 6$$
$$= 6.05\ \text{M}\Omega$$
$$I_T = \frac{1000}{6.05}\ \mu\text{A}$$
$$\therefore \quad V = \frac{1000}{6.05}\times 0.05 = 8\ \text{V}\ \textbf{(Ans.)}$$

Q.23. Two wattmeters, which are connected to measure the total power on a three-phase system supplying a balanced load, read 10.5 kW and –2.5 kW, respectively. The total power and the power factor, respectively, are

(*a*) 13.0 kW, 0.334 (*b*) 13.0 kW, 0.684 (*c*) 8.0 kW, 0.52 (*d*) 8.0kW,0.334.

Ans. (*d*)

EXPLANATION:

Given: $W_1 = 10.5$ kW; $W_2 = -2.5$ kW

$\therefore$ Total power = 10.5 – 2.5 = 8 kW **(Ans.)**

$$\tan\phi = \frac{\sqrt{3}(W_1 - W_2)}{(W_1 + W_2)}$$
$$= \frac{\sqrt{3}[(10.5)-(-2.5)]}{(10.5-2.5)} = 2.815$$
$$\therefore \quad \cos\phi = \cos[\tan(2.815)] = 0.334\ \textbf{(Ans.)}$$

Q.24. The time/div and voltage/div axes of an oscilloscope have been erased. A student connects a 1 kHz, 5 V p-p square wave calibration pulse to channel 1 of the scope and observes the screen to be as shown in the upper trace of the Fig. 15. An unknown signal is connected to channel 2 (lower trace) of the scope. If the time/div and voltage/div on both channels are the same, the amplitude (p-p) and period of the unknown signal are respectively

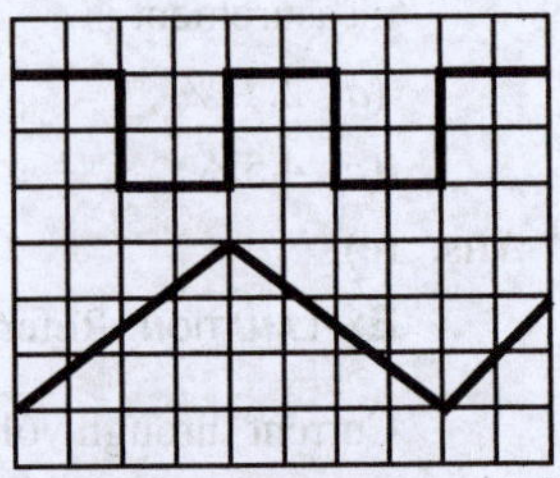

Fig. 15

(*a*) 5 V, 1 ms (*b*) 5 V, 2 ms
(*c*) 7.5 V, 2 ms (*d*) 10 V, 1 ms.

Ans. (*c*)

EXPLANATION:

Peak-peak (p-p) division of upper trace voltage = 2 and value of (p-p) voltage = 5 V

$$\therefore \quad \frac{\text{Voltage}}{\text{Divisions}} = 2.5 \text{ V}$$

Now it will be same for unknown voltage (p-p) division of unknown voltage = 3

$$\therefore \quad \text{p-p voltage} = 3 \times 2.5 = 7.5 \text{ V } \textbf{(Ans.)}$$

$$\text{Frequency of upper trace} = 1 \text{ kHz}$$

$$\therefore \quad \text{Time period} = \frac{1}{10^3} = 1 \text{ ms}$$

$$\text{Division of x-axis (upper)} = 4$$

$$\text{Division of x-axis (lower)} = 8$$

$$\therefore \quad \text{Period of unknown signal} = \frac{8}{4} = 2 \text{ ms } \textbf{(Ans.)}$$

Q.25. A sampling wattmeter (that computes power from simultaneously sampled values of voltage and current) is used to measure the average power of a load. The peak to peak voltage of the square wave is 10 V and the current is a triangular wave of 5 A p-p as shown in Fig. 16. The period is 20 ms. The reading in W will be

(*a*) 0 W (*b*) 25 W

(*c*) 50 W (*d*) 100 W.

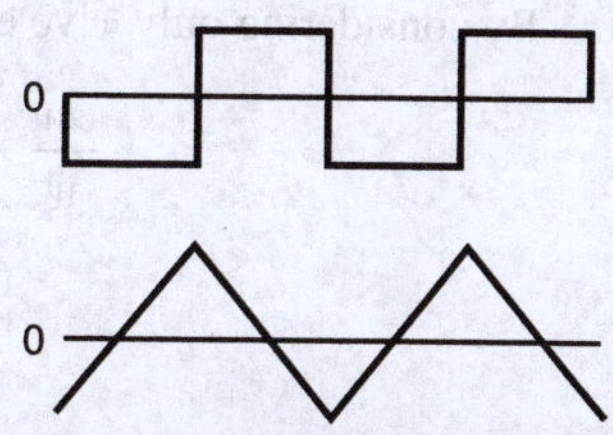

Fig. 16

Ans. (*a*)

EXPLANATION: Refer to Fig. 17.

If we consider both the waves, we can see +ive power and –ive power in each case is equal

$\therefore$ Net power = 0 **(Ans.)**

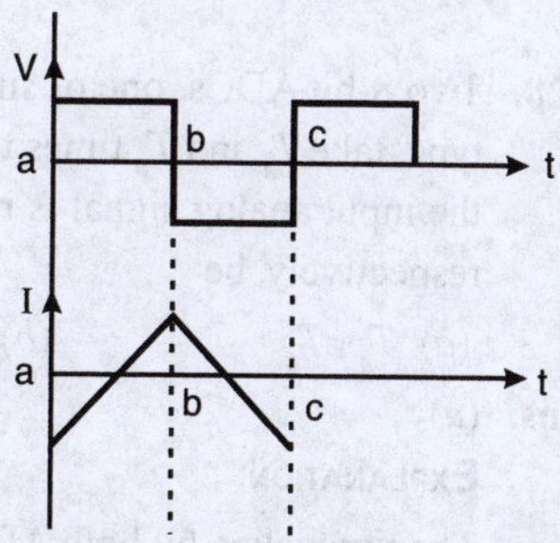

Fig. 17

Q.26. A current of $-8+6\sqrt{2}$ (sin ωt + 30°) A is passed through three meters. They are a centre zero PMMC meter, a true rms meter and a moving iron instrument. The respective readings (in A) will be

(*a*) 8, 6, 10 (*b*) 8, 6, 8 (*c*) –8, 10, 10 (*d*) –8, 2, 2.

Ans. (*c*)

EXPLANATION:

'*PMMC instrument reads*' *only dc value* and since it is centre zero type so it will give –8 value. RMS meter will read *r.m.s value* of the current.

So,
$$\text{RMS} = \sqrt{8^2 + \left(\frac{6\sqrt{2}}{\sqrt{2}}\right)^2} = 10 \text{ A}$$

'Moving iron' also reads *rms value*, so its reading will also be 10 A.

Hence ultimately it will be (–8, 10, 10) **(Ans.)**

Q.27. A variable 'w' is related to three other variables x, y, z as $w = xy/z$. The variables are measured with meters of accuracy ±0.5% reading, ±1% of full scale value and ±1.5% reading. The actual readings of the three meters are 80, 20 and 50 with 100 being the full scale value for the three. The maximum uncertainty in the measurement of 'w' will be

(*a*) ±0.5% rdg (*b*) ±5.5% rdg (*c*) ±6.7% rdg (*d*) ±7.0% rdg.

Ans. (*d*)

EXPLANATION:

Given, $$w = \frac{xy}{z}$$

Taking log, on both sides, we get:

$$\log w = \log x + \log y - \log z$$

By considering only +ive error, differentiating w.r.t. w, we have:

$$\therefore \quad \frac{dw}{w} = \left(\frac{dx}{x} + \frac{dy}{y} + \frac{dz}{z}\right) \times 100$$

$$= \left(\frac{0.005 \times 80}{80} + \frac{0.01 \times 100}{20} + \frac{0.015 \times 50}{50}\right) \times 100$$

$$= \left(\frac{0.4}{80} + \frac{1}{20} + \frac{0.75}{50}\right) \times 100$$

= 7% reading **(Ans.)**

Q.28. Two 8-bit ADCs, one of single slope integrating type and other of successive approximate type, take T_A and T_B times to convert 5 V analog input signal to equivalent digital output, If the input analog signal is reduced to 2.5 V, the approximate time taken by two ADCs will respectively, be

(*a*) T_A, T_B (*b*) $T_A/2$, T_B (*c*) T_A, $T_B/2$ (*d*) $T_A/2$, $T_B/2$.

Ans. (*a*)

EXPLANATION:

The time taken by both ADCs will remain same T_A, T_B. **(Ans.)**

Q.29. Two sinusoidal signals $p(\omega_1, t) = A \sin \omega_1 t$ and $q(\omega_2, t)$ are applied to X and Y inputs of a dual channel CRO. The Lissajous figure displayed on the screen is shown in Fig. 18. The signal $q(\omega_2, t)$ will be represented as

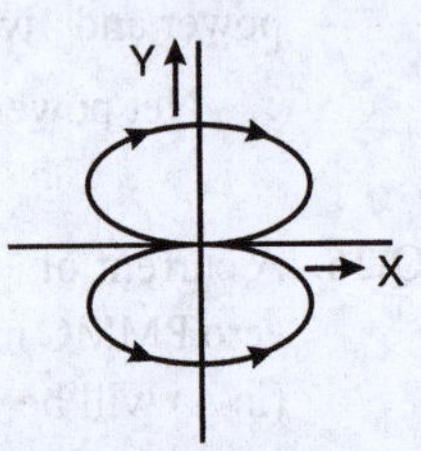

Fig. 18

(*a*) $q(\omega_2\ t) = A \sin \omega_2\ t, \omega_2 = 2\omega_1$

(*b*) $q(\omega_2\ t) = A \sin \omega_2\ t, \omega_2 = \omega_1/2$

(*c*) $q(\omega_2\ t) = A \cos \omega_2\ t, \omega_2 = 2\omega_1$

(*d*) $q(\omega_2\ t) = A \cos \omega_2\ t, \omega_2 = \omega_1/2$.

Ans. (*d*)

EXPLANATION: Refer to Fig. 19

Here, $p(\omega_1\ t) = A \sin \omega_1\ t$,

y_1-line cut '4' times the Lissajous pattern, and

x_1-line cut '2' times the Lissajous pattern.

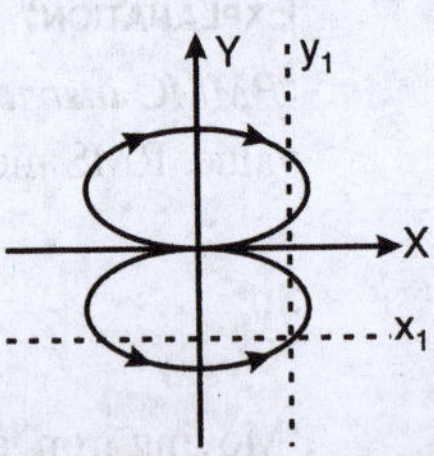

Fig. 19

$$\text{Now,}\quad \frac{\omega_y}{\omega_x} = \frac{\text{Maximum number of intersection of a horizontal line with Lissajous pattern}}{\text{Maximum number of intersection of a vertical line with Lissajous pattern}}$$

$$\text{or,}\quad \frac{\omega_y}{\omega_x} = \frac{2}{4} = \frac{1}{2}$$

$$\text{or,}\quad (\omega_x = \omega_1) \text{ and } (2\omega_y = 2\omega_2)$$

$$\text{or,}\quad \omega_2 = \frac{\omega_1}{2} \textbf{ (Ans.)}$$

and $q(\omega_2, t)$ will lead $p(\omega_1, t)$ by 90°

$$q(\omega_2, t) = A \sin(\omega_2 t + 90°)$$

$$q(\omega_2, t) = A \cos \omega_2 t \textbf{ (Ans.)}$$

Q.30. The ac bridge shown in Fig. 20 is used to measure the impedance Z.

If the bridge is balanced for oscillator frequency $f = 2$ kHz, then the impedance Z will be

(*a*) $(260 + j0)\Omega$ (*b*) $(0 + j200)\Omega$

(*c*) $(260 - j200)\Omega$ (*d*) $(260 + j200)\Omega$.

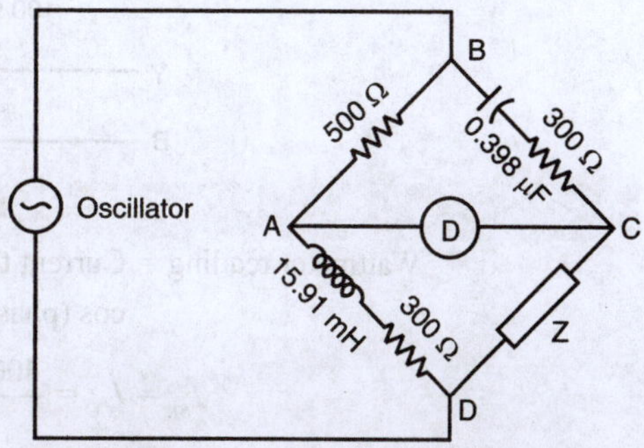

Fig. 20

Ans. (*a*)

Explanation:

$$Z = \frac{Z_{BC} \times Z_{AD}}{Z_{AB}}$$

$$= \frac{(R + j\omega L)(R + 1/j\omega C)}{500} \left(300 - \frac{j}{2 \times 3.14 \times 10^3 \times 0.398 \times 10^{-6}}\right)$$

$$= \frac{(300 + j199.82)(300 - j200)}{500}$$

$$\text{or,}\quad \approx [260 + (0)j]\Omega \textbf{ (Ans.)}$$

$$= \frac{[300 + (j \times 2\pi \times 2 \times 10^3 \times 15.91 \times 10^{-3})]\left[300 - \frac{j}{2 \times \pi \times 2 \times 10^3 \times 0.398 \times 10^{-6}}\right]}{500}$$

$$= \frac{(300 + j199.93)(300 - j200)}{500}$$

$$= \frac{90000 - j60000 + j59979 + 39986}{500}$$

$$\approx (260 + 0j)\Omega \textbf{ (Ans.)}$$

Q.31. The pressure coil of a dynamometer type wattmeter is

(*a*) highly inductive (*b*) highly resistive

(*c*) purely resistive (*d*) purely inductive.

Ans. (*b*)

EXPLANATION:

Since all the voltage appears across it hence (*b*) is the correct choice **(Ans.)**

Q.32. Fig. 21 shows a three-phase delta connected load supplied from a 400V, 50 Hz, 3-phase balanced source. The pressure coil (PC) and current coil (CC) of a wattmeter are connected to the load as shown, with the coil polarities suitably selected to ensure a positive deflection. The wattmeter reading will be

(*a*) 0 (*b*) 1600 W

(*c*) 800 W (*d*) 400 W.

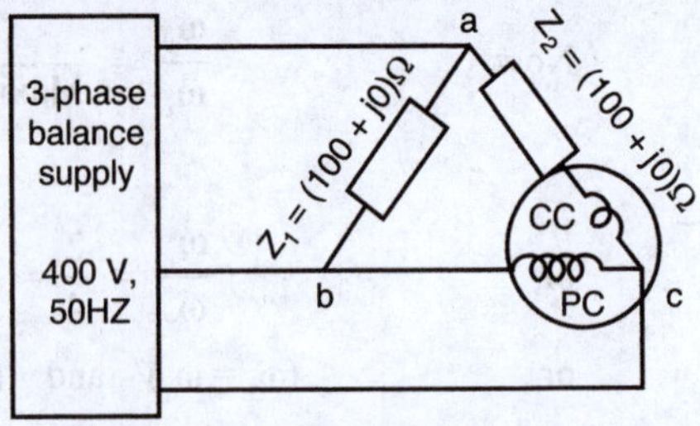

Fig. 21

Ans. (*c*)

EXPLANATION: Refer to Fig. 22

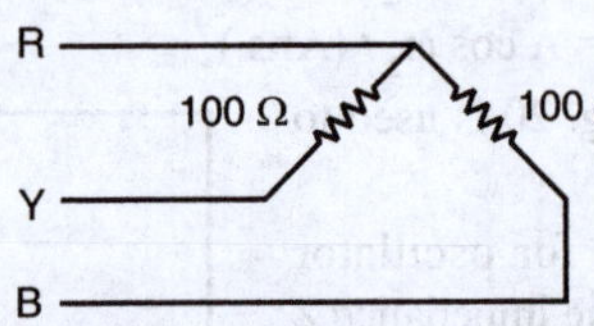

Fig. 22

Wattmeter reading = Current through C.C. × voltage across P.C. × cos (phase angle between *V* and *I*)

$$I_{BR} = I_{CC} = \frac{400\angle 120^{\circ}}{100} = 4\angle 120^{\circ}$$

$$V_{YB} = V_{PC} = 400\angle -120^{\circ}$$

$$W = (400\angle -120^{\circ}) \times (4\angle 120^{\circ}) \times \cos 240^{\circ}$$

$$= 800\text{ W}(-)\ \textbf{(Ans.)}$$

Q.33. An average-reading digital multimeter reads 10V when fed with a triangular wave, symmetric about the time-axis. For the same input an rms-reading meter will read

(*a*) $\frac{20}{\sqrt{3}}$ (*b*) $\frac{10}{\sqrt{3}}$ (*c*) $20\sqrt{3}$ (*d*) $10\sqrt{3}$.

Ans. (*d*)

EXPLANATION:

For triangular wave:

$$\text{Average value} = \frac{V_m}{3};\ \text{RMS value} = \frac{V_m}{\sqrt{3}}$$

$$\therefore \quad \frac{V_m}{3} = 10\text{ V} = V_m = 30\text{V}$$

$$\text{RMS value} = \frac{30}{\sqrt{3}} = 10\sqrt{3}\ \textbf{(Ans.)}$$

Q.34. An ammeter has a current range of 0 – 5A, and its internal resistance is 0.2 Ω. In order to change the range to 0 – 25 A, we need to add a resistance of

(*a*) 0.8 Ω in series with the meter (*b*) 1.0 Ω in series with the meter

(*c*) 0.04 Ω in parallel with the meter (*d*) 0.05 Ω in parallel with the meter.

Ans. (*d*)

EXPLANATION:

To increase the range of ammeter smaller resistance is to be connected in parallel with the ammeter.

Now, $$m = 1+\frac{R_{\text{ammeter}}}{R_{\text{parrallel}}}$$

or, $$\frac{25}{5} = 1+\frac{R_{\text{ammeter}}}{R_{\text{parallel}}}$$

or, $$R_{\text{parallel}} = \frac{0.2}{4} = 0.05\ \Omega \textbf{ (Ans.)}$$

Q.35. The Maxwell's bridge shown in Fig. 23 is at balance. The parameters of the inductive coil are

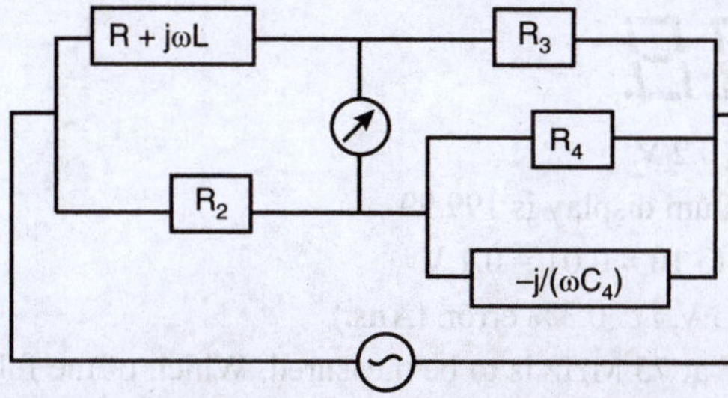

Fig. 23

(*a*) $R = R_2R_3/R_4, L = C_4R_2R_3$ (*b*) $L = R_2R_3/R_4, R = C_4R_2R_3$

(*c*) $R = R_4/R_2R_3, L = 1/(C_4R_2R_3)$ (*d*) $L = R_4/R_2R_3, R = 1/(C_4R_2R_3)$.

Ans. (*a*)

EXPLANATION:

When the bridge is balanced, we have:

$$\frac{R+j\omega L}{R_2} = R_3\left(\frac{1}{R_4} - \frac{\omega C_4}{j}\right)$$

$$= R_3\left(\frac{1}{R_4} + j\omega C_4\right)$$

Comparing real and imaginary parts, we get:

$$\frac{R}{R_2} = \frac{R_3}{R_4} \text{ or } R = \frac{R_3R_2}{R_4} \textbf{ (Ans.)}$$

$$\frac{\omega L}{R_2} = R_3\omega C_4$$

$$L = R_2 R_3 C_4 \textbf{ (Ans.)}$$

Q.36. Consider the following statements:

(*i*) The compensating coil of low power factor wattmeter compensates the effect of the impedance of the current coil.

(*ii*) The compensating coil of a low power factor wattmeter compensates the effect of the impedance of the voltage coil circuit.

(*a*) (*i*) is true (*ii*) is false (*b*) (*i*) is false but (*ii*) is true

(*c*) both (*i*) and (*ii*) are true (*d*) both (*i*) and (*ii*) are false.

Ans. (*b*)

EXPLANATION:

Normally a low power factor wattmeter has compensating windings to compensate impedance of voltage coil, so that the power of voltage coil does not affect wattmeter reading. **(Ans.)**

Q.37. A $4\frac{1}{2}$ digit DMM has the error specification as: 0.2% of reading +10 counts. If a dc voltage of 100 V is read on its 200 V full scale, the maximum error that can be expected in the reading is

(*a*) ±0.1% (*b*) ±0.2% (*c*) ±0.3% (*d*) ±0.4%.

Ans. (*c*)

EXPLANATION:

$4\frac{1}{2}$ digit is shown as:

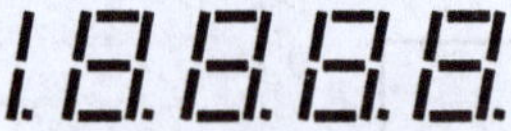

0.2% of reading is 0.2 V.

200 V scale maximum display is 199.99

10 counts amount to 10 × 0.01 = 0.1 V

∴ Total error is 0.3 V, *i.e.* 0.3% error. **(Ans.)**

Q.38. A signal of 10 mV at 75 MHz is to be measured. Which of the following instruments can be used?

(*a*) VTVM (*b*) Cathode ray oscilloscope

(*c*) Moving iron voltmeter (*d*) Digital multimeter.

Ans. (*b*)

EXPLANATION:

At high frequencies, cathode ray oscilloscope is used. **(Ans.)**

Q.39. An average response rectifier type electronic a.c. voltmeter has a d.c. voltage of 10 V applied to it. The meter reading will be

(*a*) 7.1 V (*b*) 10.0 V (*c*) 11.1 V (*d*) 22.2 V.

Ans. (*b*)

EXPLANATION:

Average value = D.C. value = 10 V **(Ans.)**

Q.40. The Wheatstone bridge method of resistance measurement is ideally suitable for the measurement of resistance values in the range of

(*a*) 0.001 Ω to 1 Ω (*b*) 0.1 Ω to 100 Ω

(*c*) 100 Ω to 10 kΩ (*d*) 100 kΩ to 10 MΩ.

Ans. (*c*)

EXPLANATION:

Wheatstone bridge is suitable for *medium value of resistance*. **(Ans.)**

Q.41. To the y-input of a CRO a signal defined by 10 sin 100 *t* is applied. To the x-input, the signal 10 cos 100 *t* is fed. The gain for both x-channel and y-channel is same. The screen shows

(*a*) sinusoidal signal (*b*) a straight line (*c*) an ellipse (*d*) a circle.

Ans. (*d*)

EXPLANATION:

The input signal = 10 cos 100t = 10 sin (100t + 90°)

∴ When phase difference is 90° the screen shows a *circle*. **(Ans.)**

Q.42. The phenomenon of 'creeping' occurs in

(*a*) Ammeters (*b*) Voltmeters

(*c*) Wattmeters (*d*) Watt-hour meters.

Ans. (*d*)

EXPLANATION:

Watt-hour meter is the other name of energy meter, in which phenomenon of creeping occurs. **(Ans.)**

Q.43. In a single-phase power factor meter, the controlling torque is

(*a*) provided by spring control (*b*) provided by gravity control

(*c*) provided by stiffness of suspension (*d*) not required.

Ans. (*d*)

EXPLANATION:

There will be two deflecting torques, one acting on coil A and the other on coil B. The coil windings are so arranged that the torques due to the two coils are *opposite* in direction. So, there is *no* controlling torque. **(Ans.)**

Q.44. In electrodynamometer type wattmeter, the inductance of pressure coil produces error. The error is

(*a*) constant irrespective of the power factor of the load

(*b*) higher at higher power factor loads

(*c*) higher at lower power factor loads

(*d*) highest at unity power factor loads.

Ans. (*c*)

EXPLANATION:

The error caused by pressure coil inductance is VI sin ϕ tan β. With low power factor, the value of ϕ is *large* and, therefore, the error is *correspondingly large.* **(Ans.)**

Q.45. If an induction type energy meter runs fast, it can be slowed down by

(*a*) lag adjustment

(*b*) light load adjustment

(*c*) adjusting the position of braking magnet and moving it closer from the centre of the disc

(*d*) adjusting the position of braking magnet and moving it away from the centre of the disc.

Ans. (*d*)

EXPLANATION:

Braking torque $T_D \propto R^2$, where R is the radial position of braking magnet with respect to the centre of the disc. So, by adjusting the position of braking magnet and *moving it away from the centre of disc* the meter can be slowed down.

Q.46. The reflecting mirror mounted on the moving coil of a vibration galvanometer is replaced by a bigger size mirror. This will result in

(*a*) lower frequency of resonance & lower amplitude of vibration

(*b*) lower frequency of resonance but the amplitude of vibration is unchanged

(*c*) higher frequency of resonance & lower amplitude of vibration (*d*) higher frequency of resonance but the amplitude of vibration is unchanged.

Ans. (*a*)

EXPLANATION:

Bigger size mirror results in increased inertia constant J.

$$f_n = \frac{1}{2\pi}\sqrt{\frac{K}{J}} \quad \& \quad A = \frac{GI_m}{\sqrt{(D\omega)^2 + (K - J\omega^2)^2}},$$

as J increases both f_n & A decrease. **(Ans.)**

Q.47. In calibration of a dynamometer wattmeter by potentiometer, phantom loading arrangement is used because

(*a*) the arrangement gives accurate results

(*b*) the power consumed in calibration work is minimum

(*c*) the method gives quick results

(*d*) the onsite calibration is possible.

Ans. (*b*)

EXPLANATION:

Actual loading arrangements would involve a considerable waste of power. In order to avoid this Phantom or 'Fictitious' loading is done. **(Ans.)**

Q.48. The accuracy of Kelvin's double bridge for the measurement of low resistance is high because the bridge

(*a*) uses two pairs of resistance arms.

(*b*) has medium value resistance in the ratio arms.

(*c*) uses a low resistance link between standard and test resistances.

(*d*) uses a null indicating galvanometer.

Ans. (*a*)

EXPLANATION:

Kelvin bridge overcomes the difficulties that arise in a Wheatstone bridge on account of the resistance of the leads and the contact resistances while measuring low valued resistors. The Kelvin double bridge incorporates the idea of a second set of ratio arms, hence the name double bridge-and the use of four terminal resistors for the low resistance arms. The *second set of ratio arms eliminate the effect of connecting lead resistance.* **(Ans.)**

Q.49. An energy meter having a meter constant of 1200 revolutions per kWh is found to make 5 revolutions in 75 s. The load power is

(*a*) 500 W (*b*) 100 W (*c*) 200 W (*d*) 1000 W.

Ans. (*c*)

EXPLANATION:

Number of revolutions = $K \times P \times t$

$\therefore$ $$5 = 1200 \times P \times \frac{75}{3600}$$

or, $P = 0.2$ kW = 200 W **(Ans.)**

Q.50. In a Q-meter measurement to determine the self-capacitance of a coil, the first resonance occurred at f_1 with $C_1 = 300$ pF. The second resonance occurred at $f_2 = 2f_1$ with $C_2 = 60$ pF. The self-capacitance of coil works out to be

(*a*) 240 pF (*b*) 60 pF (*c*) 360 pF (*d*) 20 pF.

Ans. (*d*)

EXPLANATION:

Self-capacitance,

$$C_d = \frac{C_1 - 4C_2}{3}$$

$$= \frac{300-(4\times 60)}{3} = 20 \text{ pF } \textbf{(Ans.)}$$

Q.51. Which one of the following statements is *correct*?

The deflection of hot wire instrument depends on

(*a*) r.m.s. value of the a.c. current (*b*) r.m.s. value of the a.c. voltage

(*c*) average value of the a.c. current (*d*) average value of the a.c. voltage.

Ans. (*a*)

EXPLANATION:

The expansion (deflection) is proportional to the heating effect of the current and hence to the square of the rms value of the current. Therefore, the meter may be calibrated to read the rms value of the current.

Q.52. Chopper stabilized d.c. amplifier type electronic voltmeter overcomes the effect of

(*a*) Amplifier CMRP (*b*) Amplifier sensitivity

(*c*) Amplifier drift (*d*) Electromagnetic interference.

Ans. (*c*)

EXPLANATION:

Simple a.c. amplifiers may be used to amplify a dc input through the use of an additional circuit component known as *'chopper'*. The opproach is used to built a dc amplifier where in the dc is first converted to an equivalent a.c. signal that is amplifed by a standard a.c. amplifier. The a.c. signal is finally converted back to a dc dignal. Chopper stabilized amplifier eliminates the effects of dc offset currents & the drift or other dc parameters by using an a.c. coupled amplifier for necessary gain. **(Ans.)**

Q.53. Which one of the following defects is responsible for creeping in an induction type energy meter?

(*a*) Imperfect lag compensation (*b*) Over friction compensation

(*c*) Imperfect overload compensation (*d*) Misalignment of brake magnet.

Ans. (*b*)

EXPLANATION:

In some meters a slow but continuous rotation is observed even when there is no current flowing through the current coil and only pressure coil is energized. This is called creeping. The *major cause for creeping is over-compensation for running friction.* **(Ans.)**

Q.54. A rectifier instrument is used to measure an alternating square wave of amplitude 100 V. What is meter reading?

(*a*) 100 V (*b*) 70.7 V (*c*) 111 V (*d*) None of these.

Ans. (*a*)

EXPLANATION:

Rectifier instrument measures *average value*; in this case it is 100 V (for square wave). **(Ans.)**

Q.55. What is range for a $3\frac{1}{2}$ digital meter?

(*a*) 0 to 1999 (*b*) 0 to 1500 (*c*) 0 to 999 (*d*) 0 to 19999.

Ans. (*a*)

EXPLANATION:

A three digit display on a digital voltmeter for 0-1 V range will be able to indicate values from zero to 999 mV. In practice a fourth digit usually capable of indicating either 0 or 1 only is placed to the left of active digits. This permits going above 999 to 1999 to give overlap between ranges for convenience. This is called *over ranging. This type of display is known as a* $3\frac{1}{2}$ *digit display.* **(Ans.)**

Q.56. When reading is taken at half scale in the instrument, the error is

(*a*) exactly equal to half of full-scale error (*b*) equal to full-scale error

(*c*) less than full-scale error (*d*) more than full-scale error.

Ans. (*c*)

EXPLANATION:

Accuracy is defined as 'percentage of true value'. **(Ans.)**

Q.57. T = Ki is the torque equation of a

(*a*) PMMC instrument (*b*) Rectifier instrument

(*c*) Moving iron instrument (*d*) D' arsonval instrument.

Ans. (*d*)

EXPLANATION:

D' arsonval instrument shows *linear deflection.* **(Ans.)**

Q.58. A digit voltmeter has read-out range from 0 to 9999 counts. When full scale reading is 9.999 V, the resolution of the full scale reading is

(*a*) 0.001 (*b*) 1000 (*c*) 2 digit (*d*) 1 mV.

Ans. (*d*)

EXPLANATION:

Resolution $= \frac{9.999}{9999} = 1 \times 10^{-3}$ V = 1 mV. **(Ans.)**

Q.59. The source in the circuit shown (Fig. 24) is a sinusoidal source. The voltage across various elements are marked in the figure. The input voltage is

(*a*) 10 V (*b*) 5 V

(*c*) 27 V (*d*) 24 V.

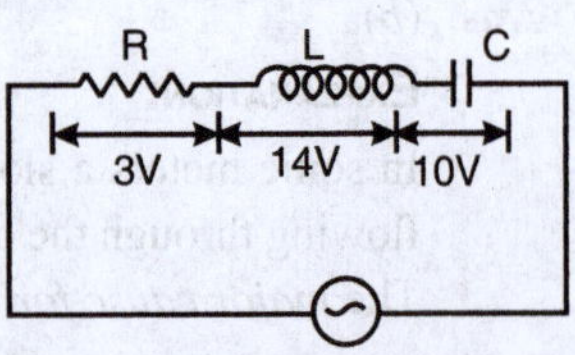

Fig. 24

Ans. (*b*)

EXPLANATION:

$= \sqrt{3^2 + (14-10)^2} = 5$ V **(Ans.)**

Q.60. The gauge factor of strain gauge is 2, stress is 1050 kg/cm², $Y = 2.1 \times 10^6$ kg/cm² and R is 1000 ohms. The value of ΔR will be

(*a*) 2 Ω (*b*) 3 Ω (*c*) 4 Ω (*d*) 1 Ω.

Ans. (*d*)

EXPLANATION:

$$G = \frac{\Delta R/R}{\Delta Y/Y}$$

or, $$2 = \frac{\Delta R/1000}{1050/(2.1\times10^6)}$$

or, $$\Delta R = 1\ \Omega \textbf{ (Ans.)}$$

Q.61. An alternating current source having voltage $E = 110 \sin\left(\omega t + \frac{\pi}{3}\right)$ is connected in an a.c. circuit. If the current drawn from the circuit varies as $I = 5 \sin\left(\omega t - \frac{\pi}{3}\right)$, impedance of the circuit will be

(*a*) 22 ohms (*b*) 16 ohms (*c*) 30.8 ohms (*d*) none of these.

Ans. (*a*)

EXPLANATION:

$$\text{Impedance of the circuit} = \frac{|V|}{|I|} = \frac{110}{5} = 22\ \Omega \textbf{ (Ans.)}$$

Q.62. The resistance of a 125 ohm strain gauge changes by one ohm for 4000 microstrain. The gauge factor of the strain gauge is

(*a*) 1.5 (*b*) 2 (*c*) 2.5 (*d*) 3.

Ans. (*b*)

EXPLANATION:

$$G = \frac{\Delta R/R}{e} = \frac{1/125}{4000\times10^{-6}} = \frac{10^6}{125\times4000} = 2 \textbf{ (Ans.)}$$

Q.63. In a set-up for the determination of self capacitance of a coil, the standard variable capacitor has value of 460 pF at 2 MHz and a value of 100 pF at 4 MHz. The coil self-capacitance is

(*a*) 2 pF (*b*) 20 pF (*c*) 60 pF (*d*) 200 pF.

Ans. (*b*)

EXPLANATION:

$$C_d = \frac{C_1 - 4C_2}{3} = \frac{460 - 4\times100}{3} = 20 \text{ pF} \textbf{ (Ans.)}$$

Q.64. The percentage limiting error, in the case of an instrument reading of 8.3 V with a 0 to 150 voltmeter having a guaranteed accuracy of 1% full-scale reading, is

(*a*) 1.810% (*b*) 0.181% (*c*) 18.07% (*d*) 0.0018%.

Ans. (*c*)

EXPLANATION:

$$\% \text{ Limiting error} = \frac{150\times\frac{1}{100}}{8.3} = \frac{1.5}{8.3} = 0.1807 = 18.07\ \% \textbf{ (Ans.)}$$

Q.65. It is desired to convert a 0–1000 μA meter with an internal resistance of 100 ohms, into a 0–100 mA meter. The required value of shunt resistance is about

(*a*) 1 ohm (*b*) 10 ohms (*c*) 99 ohms (*d*) 100 ohms.

Ans. (*a*)

EXPLANATION: Refer to Fig. 25

We know that $$R_{sh} = \frac{R_m}{m-1}$$

Also, $$m = \frac{100\times10^{-3}}{1000\times10^{-6}} = \frac{100000}{1000} = 100$$

$\therefore$ $$R_{sh} = \frac{100}{100-1} \approx 1\Omega \text{ (Ans.)}$$

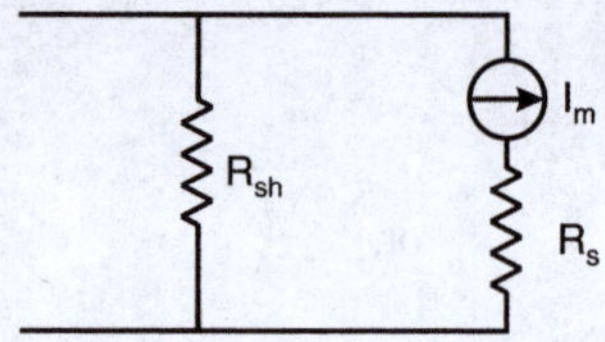

Fig. 25

Q.66. A moving coil ammeter has a fixed shunt of 0.02 ohm resistance. If the coil resistance of the meter is 1000 ohms, a potential difference of 500 mV is required across it for full-scale deflection. Under this condition, the current in the shunt would be

(*a*) 2.5 A (*b*) 25 A (*c*) 0.25 A (*d*) 0.025 A.

Ans. (*b*)

EXPLANATION:

$$\text{Current in the shunt} = \frac{500\times10^{-3}}{0.02} = 25 \text{ A } \textbf{(Ans.)}$$

Q.67. In the Wheatstone bridge shown in Fig. 26, if the resistance in each arm is increased by 0.05%, then the value V_{out} will be

(*a*) 50 mV (*b*) 5 mV

(*c*) 0.1 V (*d*) zero.

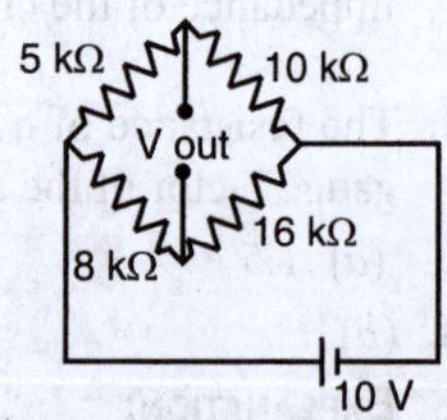

Fig. 26

Ans. (*d*)

EXPLANATION:

Before change, the bridge is in balance. After each resistance change by 0.05%, the current division in each branch changes proportionally, and also the voltage changes proportionally, hence V_{out} remains *zero*. **(Ans.)**

Q.68. The current flowing through the resistor R is as shown in the given Fig. 27. The computed value of power is

(*a*) 400 ± 0.42 W (*b*) 400 ± 4.60 W

(*c*) 400 ± 8.80 W (*d*) 400 ± 10.65 W.

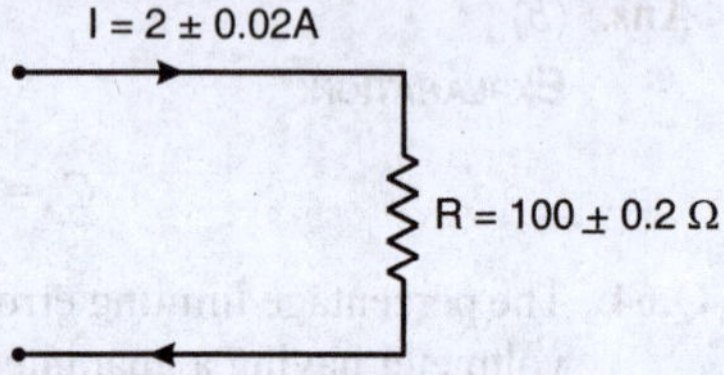

Fig. 27

Ans. (*c*)

EXPLANATION:

$$P = I^2R$$

$$\frac{dP}{P} = \frac{2\times\delta I}{I} + \frac{\delta R}{R}$$

$$= \frac{2\times0.02}{2} + \frac{0.2}{100} = 0.02 + 0.002$$

$$= 0.022$$

Now, $$P = 2^2 \times 100 = 400$$

$$dP = 0.022 \times 400 = 8.8$$

$\therefore$ Measured power = 400 ± 8.8W **(Ans.)**

Q.69. A peak-responding ac voltmeter employing a half-wave precision rectifier has been calibrated to give a reading of 1 V for 1 V rms sinusoidal input. For 1 V dc input, the meter reading will be

(*a*) 0.707 V (*b*) 0.707 V or 0 V (*c*) 1.414 V (*d*) 1.414V or 0V.

Ans. (*b*)

EXPLANATION:

$$\text{Reading with 1 Vdc} = \frac{\sqrt{2}\text{V}}{2\text{V}} = \frac{1}{\sqrt{2}} = 0.707 \text{ V}$$

Now if the dc input polarity is changed then the meter reading becomes *zero*. **(Ans.)**

Q.70. If the secondary burden of a current transformer is 15 VA and the secondary current is 5A, then the impedance of the connected load will be

(*a*) 0.6 ohm (*b*) 5 ohms (*c*) 6 ohms (*d*) 10 ohms.

Ans. (*a*)

EXPLANATION:

$$\text{Impedance} = \frac{15}{5^2} = 0.6\ \Omega \textbf{ (Ans.)}$$

Q.71. The three impedances of an ac bridge shown in Fig. 28 are:

$Z_1 = 200\ \Omega \angle 60°$, $Z_2 = 400\Omega \angle -90°$, $Z_3 = 300\ \Omega \angle 0°$

The value of Z_4 for the bridge to be balanced is

(*a*) $150\ \Omega \angle 30°$ (*b*) $400\ \Omega \angle -90°$ (*c*) $300\ \Omega \angle 90°$ (*d*) $600\Omega \angle -150°$.

Ans. (*d*)

EXPLANATION:

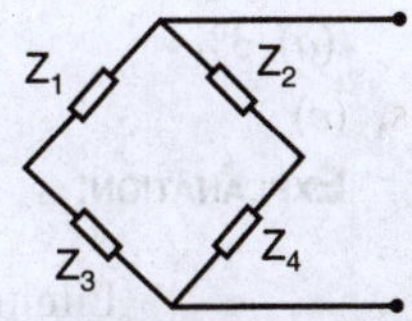

Fig. 28

Under balanced condition:

$$\frac{Z_2}{Z_4} = \frac{Z_1}{Z_3}$$

or,
$$Z_1 Z_4 = Z_2 Z_3$$

$$Z_4 = \frac{Z_2 Z_3}{Z_1}$$

$$= \frac{400\angle -90° \times 300\angle 0°}{200\angle 60°} = 600\ \Omega \angle -150° \textbf{ (Ans.)}$$

Q.72. An A.M. broadcast radio transmitter radiates 10 kW power, when the modulation percentage is 60. The power of the carrier is

(*a*) 5.00 kW (*b*) 7.69 kW (*c*) 8.47 kW (*d*) 9.17 kW.

Ans. (*c*)

EXPLANATION:

$$P_t = P_c\left(1 + \frac{m^2}{2}\right)$$

or,
$$10 = P_c\left(1 + \frac{0.6^2}{2}\right) = P_c\left(1 + \frac{0.36}{2}\right) = 1.18\, P_c$$

$$\therefore \quad P_c = \frac{10}{1.18} = 8.47 \text{ kW } \textbf{(Ans.)}$$

Q.73. The coil of a moving coil ammeter of 100 turns is 40 mm long and 30 mm wide. The control torque is 240×10^{-6} Nm on full scale. If the magnetic flux density in the air-gap is 1 Wb/m², the range of the ammeter will be

(*a*) 1 mA (*b*) 2 mA (*c*) 3 mA (*d*) 4 mA.

Ans. (*b*)

EXPLANATION:

$$T_c = T_d = NBA\,i$$
$$240 \times 10^{-6} = 100 \times 1 \times (40 \times 30 \times 10^{-6}) \times i$$
$$\therefore \quad i = 0.002 \text{ A} = 2 \text{ mA. } \textbf{(Ans.)}$$

Q.74. A 1 mA ammeter has a resistance of 100 ohms. It is to be converted into a 1 A ammeter. The value of shunt resistance required is

(*a*) 0.001 Ω (*b*) 0.1001 Ω (*c*) 100 Ω (*d*) 10000 Ω.

Ans. (*b*)

EXPLANATION:

$$m = \frac{1}{0.001} = 1000$$
$$R_{sh} = \frac{R_m}{m-1} = \frac{100}{1000-1} = 0.1001\ \Omega \textbf{ (Ans.)}$$

Q.75. A 0 to 200 V dc moving coil voltmeter has a guaranteed accuracy of 0.75% of full-scale reading. The voltage measured by instrument is 100 V. The limiting error is

(*a*) 3% (*b*) 2% (*c*) 1.5% (*d*) 0.75%.

Ans. (*c*)

EXPLANATION:

$$\text{Limiting error} = \frac{200 \times (0.75/100)}{100} = \frac{1.5}{100} = 1.5\% \textbf{ (Ans.)}$$

Q.76. For an AM wave, the maximum voltage was found to be 10 V and the minimum voltage was found to be 5 V. The modulation index of the wave would be

(*a*) 0.33 (*b*) 0.52 (*c*) 0.40 (*d*) 0.1.

Ans. (*a*)

EXPLANATION:

$$m = \frac{10-5}{10+5} = \frac{5}{15} = 0.33 \textbf{ (Ans.)}$$

Q.77. A piezoelectric force transducer has a charge sensitivity of 20 pC/N. It is connected to a charge amplifier and the combination of transducer and amplifier has an overall gain of 50 mV/N. The gain of the amplifier is

(*a*) 1 mV/pC (*b*) 1.5 mV/pC (*c*) 2.5 mV/pC (*d*) 4 mV/pC.

Ans. (*c*)

EXPLANATION:

$$\text{Gain} = \frac{50}{20} = 2.5 \text{ mV/PC } \textbf{(Ans.)}$$

Q.78. An average responding rectifier type electronic ac voltmeter has its scale calibrated in terms of the rms value of a sine wave. If a square wave voltage of peak magnitude 100 V is measured using this voltmeter, what will be the reading indicated by the meter?

(*a*) 111 V (*b*) 100 V (*c*) 90.09 V (*d*) 70.7 V.

Ans. (*a*)

EXPLANATION:

Form factor for rectangular wave = 1.11

∴ Reading indicated by the meter = 1.11 × 100 = 111 V **(Ans.)**

Q.79. A wire strain gauge element subjected to a strain has a gauge factor of 2, a resistance of 125 ohms and is of 1 m length. For a change in length of the wire by 0.005 m, the change in the resistance will be

(*a*) 0.25 ohm (*b*) 0.5 ohm (*c*) 1.25 ohm (*d*) 2.5 ohm.

Ans. (*c*)

EXPLANATION:

$$G = \frac{\Delta R/R}{\Delta l/l}$$

or, $$2 = \frac{\Delta R/125}{0.005/1}$$

or, $$\Delta R = 1.25\ \Omega \textbf{ (Ans.)}$$

Q.80. A variable reluctance type tachometer has 60 rotor teeth. The counter records 3600 counts/minute. The device speed is

(*a*) 600 r.p.s. (*b*) 1800 r.p.m. (*c*) 3600 r.p.s (*d*) 7200 r.p.m.

Ans. (*c*)

EXPLANATION:

$$\text{Speed of the tachometer} = \frac{60 \times 3600}{60} = 3600 \text{ rps } \textbf{(Ans.)}$$

Q.81. An accelerometer has a seismic mass of 0.05 kg and a spring constant of 3×10^3 N/m. The maximum permissible displacement of the mass is ±1 mm. The maximum measurable acceleration is

(*a*) 30 m/s² (*b*) 60 m/s² (*c*) 90 m/s² (*d*) 120 m/s².

Ans. (*b*)

EXPLANATION:

Natural frequency, $$\omega_n = \sqrt{\frac{k}{m}} = \sqrt{\frac{3 \times 10^3}{0.05}}$$

$$= 245 \text{ rad/s}$$

Maximum acceleration $= \omega^2 \times \text{displacement}$

$$= (245)^2 \times 0.001 = 60 \text{ m/s}^2 \textbf{ (Ans.)}$$

Q.82. The time base of a high frequency CRO whose screen diameter is 10 cm is set at 10 μ sec/cm. The lowest frequency of the signal that can be fully displayed on this CRO for measurement and analysis with great accuracy is

(*a*) greater than 100 kHz (*b*) equal to 100 kHz

(*c*) equal to 10 kHz (*d*) less than 10 kHz.

Ans. (*c*)

EXPLANATION:

Maximum time period, $T = 10 \times 10 \times 10^{-6}$ sec $= 10^{-4}$ sec

∴ Lowest frequency, $$f = \frac{1}{T} = \frac{1}{10^{-4}} = 10 \text{ kHz. } \textbf{(Ans.)}$$

Q.83. Given, $Z_a = 100\angle 50°$, $Z_b = 300\angle -90°$, $Z_c = 200\angle 0°$, the value of Z_d for the bridge shown in Fig. 29 to be balanced is

(*a*) $600\angle -40°$ (*b*) $600\angle 140°$

(*c*) $600\angle -140°$ (*d*) $150\angle 40°$.

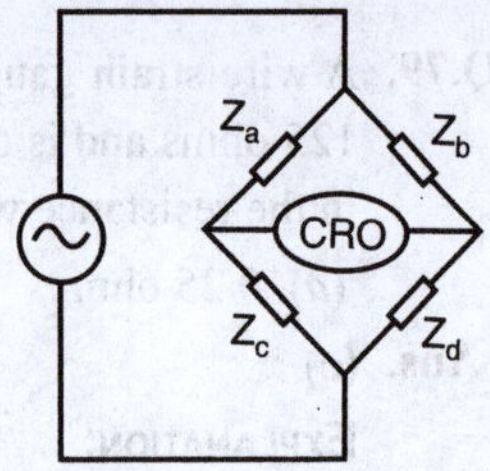

Fig. 29

Ans. (*c*)

EXPLANATION:

Under balanced condition:

$$\frac{Z_a}{Z_c} = \frac{Z_b}{Z_d}$$

or, $$Z_d = \frac{Z_c Z_b}{Z_a}$$

$$= \frac{200\angle 0° \times 300\angle -90°}{100\angle 50°} = 600\angle -140° \textbf{ (Ans.)}$$

Q.84. A variable reluctance type tachometer has 150 teeth on the rotor. The counter records 13,500 pulses per second. The rotational speed is

(*a*) 4800 rpm (*b*) 5400 rpm (*c*) 6000 rpm (*d*) 7200 rpm.

Ans. (*b*)

EXPLANATION:

$$\text{Speed} = \frac{13500}{150} \times 60 = 5400 \text{ rpm } \textbf{(Ans.)}$$

Q.85. An LVDT is used to measure displacement. The output of the LVDT is connected to a voltmeter of range 0 to 5 V through an amplifier having a gain of 250. For a displacement of 0.5 mm, the output of LVDT is 2 mV. The sensitivity of the instrument would be

(*a*) 0.1 V/mm (*b*) 0.5 V/mm (*c*) 1 V/mm (*d*) 5 V/mm.

Ans. (*c*)

EXPLANATION:

$$\text{Sensitivity of the instrument} = \frac{2\times10^{-3}\times250}{0.5\times10^{-3}} = 1000 \text{ V/m} = 1 \text{ V/mm } \textbf{(Ans.)}$$

Q.86. A resistance wire strain gauge with a gauge factor of 2 is bonded to a steel structural member subjected to a tensile stress of 100 MN/m^2. The modulus of elasticity of steel is 200 GN/m^2. The percentage change in gauge resistance due to the applied stress is

(*a*) 0.1 (*b*) 0.2 (*c*) 0.4 (*d*) 0.5.

Ans. (*a*)

EXPLANATION:

$$G_f = \frac{\Delta R/R}{\Delta l/l}$$

or, $$2 = \frac{\Delta R/R}{\frac{100\times10^6}{200\times10^9}}$$ $\left[\because \delta\Delta l/l = \text{strain} = \frac{\text{stress}}{E}\right]$

or, $$\frac{\Delta R}{R} = 2\times\frac{1}{2}\times10^{-3} = 1\times10^{-3}$$

$\therefore$ $$\%\frac{\Delta R}{R} = 0.1\% \textbf{ (Ans.)}$$

Q.87. A 300 V full-scale deflection voltmeter has an accuracy of ±2%, when it reads 222 V. The actual voltage lies between

(*a*) 217.56 V and 226.44 V (*b*) lies between 217.4 V and 226.6 V
(*c*) 216 V and 228 V (*d*) None of these.

Ans. (*c*)

EXPLANATION:

$$\text{Error in reading} = \frac{300\times(\pm 2)}{100} = \pm 6 \text{ V}$$

$$\therefore \quad \text{Actual reading} = 222 \pm 6$$

$$= 216 \text{ V and } 228 \text{ V} \textbf{ (Ans.)}$$

Q.88. A 5-channel dc to 60 Hz telemetry system uses PAM and PCM systems. For a good quality data transmission, the minimum sampling rate must be

(*a*) 300 samples/s (*b*) 500 samples/s
(*c*) 1500 samples/s (*d*) 1250 samples/s.

Ans. (*a*)

EXPLANATION:

Minimum sampling rate $= 5f_n = 5 \times 60 = 300$ sample/s **(Ans.)**

Q.89. A 0–150 V voltmeter has an accuracy of 1% of full-scale reading. The voltage measured by this instrument is 75 V. The limiting error is

(*a*) 1% (*b*) 2% (*c*) 2.5% (*d*) 3%.

Ans. (*b*)

EXPLANATION:

$$\text{Error} = \frac{150\times 1}{100} = 1.5$$

$$\therefore \quad \text{Limiting error} = \frac{1.5}{75}\times 100 = 2\%. \textbf{ (Ans.)}$$

Q.90. Wheatstone bridge requires a change of 6 ohms in the unknown arm of the bridge to produce a change in deflection of 3 mm of the galvanometer. The sensitivity of the instrument is

(*a*) 0.5 per cent (*b*) 2.0 per cent (*c*) 0.5 mm/ohm (*d*) 2.0 ohm/mm.

Ans. (*c*)

EXPLANATION:

$$\text{Sensitivity} = \frac{3}{6} \text{ mm/ohm} = 0.5 \text{ mm/ohm} \textbf{ (Ans.)}$$

Q.91. A digital voltmeter has a read-out range form 0 to 9999 counts. When full scale reading is 9.999V, the resolution of the full scale reading is

(*a*) 0.001 (*b*) 1000 (*c*) 3 digit (*d*) 1 mV.

Ans. (*d*)

EXPLANATION:

$$\text{Resolution} = \frac{9.999}{9999} = 1 \times 10^{-3} \text{ V or } 1 \text{ mV} \textbf{ (Ans.)}$$

Q.92. The resistance of a circuit is found by measuring current flowing and the power fed into the circuit. If the limiting errors in the measurement of power and current are ±1.5% and ±1.0% respectively, the limiting error in the measurement of resistance will be

(*a*) ±1% (*b*) ±1.5% (*c*) ±2.5% (*d*) ±3.5%.

Ans. (*d*)

EXPLANATION:

$$P = I^2R \text{ or } R = \frac{P}{I^2}$$

$$\frac{\delta R}{R} = +\frac{\delta P}{P} \pm \frac{2\delta I}{I} = \pm 1.5 \pm (2 \times 1) = \pm 3.5\% \textbf{ (Ans.)}$$

Q.93. In the statement "the wattmeter commonly used for power measurement at commercial frequencies is of the X-type. This meter consists of two coil systems, the fixed system being the Y-coil and moving system being the Z-coil". X, Y and Z stand respectively for

(*a*) dynamometer, voltage and current (*b*) dynamometer, current and voltage
(*c*) induction, voltage and current (*d*) induction, current and voltage.

Ans. (*b*)

EXPLANATION:

In dynamometer type wattmeter, the fixed coil is current coil and moving coil is voltage coil or pressure coil **(Ans.)**

Q.94. Two-wattmeter method is employed to measure power in a 3-phase balanced system with the current coils connected in the A and C lines. The phase sequence is ABC. If the wattmeter with its current coil in A-phase line reads zero, then the power factor of the 3-phase load will be

(*a*) zero lagging (*b*) zero leading (*c*) 0.5 lagging (*d*) 0.5 leading.

Ans. (*c*)

EXPLANATION:

When power factor cos ϕ = 0.5, then one wattmeter reads zero. The load is lagging in this case **(Ans.)**

Q.95. The meter constant of a single phase 240 V induction watt-hour meter is 400 revolutions per kWh. The speed of the meter disc for a current of 10 amperes at 0.8 p.f. lagging will be

(*a*) 12.8 rpm (*b*) 16.02 rpm (*c*) 18.2 rpm (*d*) 21.1 rpm.

Ans. (*a*)

EXPLANATION:

$$\text{Meter constant} = 400 \text{ revolutions per kWh}$$

$$= 400 \times \frac{240 \times 10 \times 0.8}{1000} \text{ revolutions per hour}$$

$$= 768 \text{ revolution per hour} = \frac{768}{60} \text{ rpm} = 12.8 \text{ rpm } \textbf{(Ans.)}$$

Q.96. A Lissajous pattern on an oscilloscope has 5 horizontal tangencies and 2 vertical tangencies. The frequency of the horizontal input is 1000 Hz. What is the frequency of the vertical input?

(*a*) 400 Hz (*b*) 2500 Hz (*c*) 4000 Hz (*d*) 5000 Hz.

Ans. (*b*)

EXPLANATION:

$$\frac{f_y}{f_x} = \frac{\text{No. of horizontal tangent}}{\text{No. of vertical tangent}}$$

$$\frac{f_y}{1000} = \frac{5}{2} \text{ or } f_y = 2500 \text{ kHz } \textbf{(Ans.)}$$

Q.97. A zero to 300 V voltmeter has an error of ±2% of the full scale deflection. If the true voltage is 30 V then the range of reading on this voltmeter would be

(*a*) 20 V to 40 V (*b*) 24 V to 36 V
(*c*) 29.4 V to 30.6 V (*d*) 29.94 V to 30.06 V.

Ans. (*b*)

EXPLANATION:

$$\frac{\delta V}{V} = 2\% \text{ or } \frac{\delta V}{300} = \frac{2}{100}$$

or, $$\delta V = 300 \times \frac{2}{100} = 6\text{V}$$

So, range of reading = 30 ± 6 = 24 V to 36 V **(Ans.)**

Q.98. A high frequency a.c. signal is applied to a PMMC instrument. If the rms value of the a.c. signal is 2 V, then the reading of the instrument will be

(*a*) zero (*b*) 2 V (*c*) $2\sqrt{2}$ V (*d*) $4\sqrt{2}$ V.

Ans. (*a*)

EXPLANATION:

Because PMMC instrument reads only d.c. value or average value. **(Ans.)**

Q.99. In the circuit shown in Fig. 30, if the ammeter indicates 1A and the voltmeter having an internal resistance of 1 kΩ indicated 100 V, then the value of R would be

(*a*) 111.11 Ω (*b*) 105.2 Ω
(*c*) 100 Ω (*d*) 90.9 Ω.

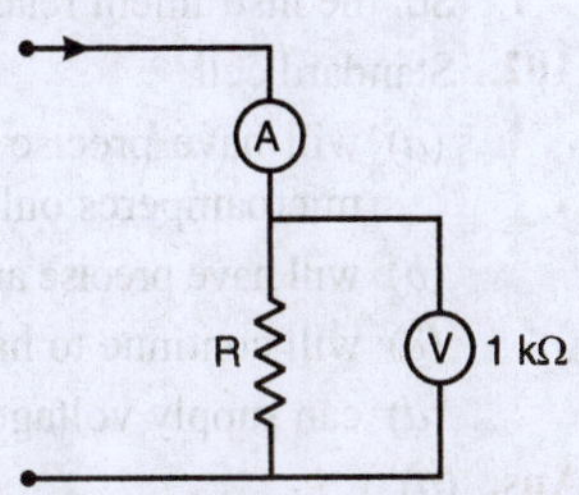

Fig. 30

Ans. (*a*)

EXPLANATION: Refer to Fig. 31

Now, $V = I_1 R$

Also, $$I_1 = \left(\frac{10^3}{10^3 + R}\right) I$$

where, I = 1A and V = 100 volts ...(Given)

$$\therefore \quad 100 = \left(\frac{10^3}{10^3 + R}\right) \cdot R,$$

or, R = 111.11 Ω **(Ans.)**

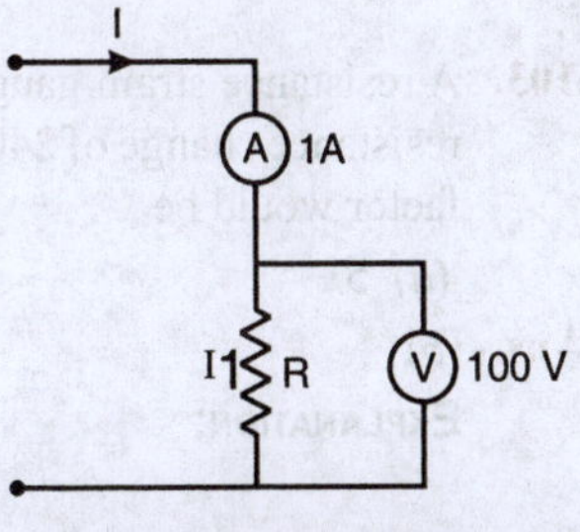

Fig. 31

Q.100. A current $i = (10 + 10 \sin t)$ amperes is passed through an ideal moving iron type ammeter, its reading will be

(*a*) zero (*b*) 10A (*c*) $\sqrt{150}$A (*d*) $10\sqrt{2}$A.

Ans. (*c*)

EXPLANATION:

Moving iron type ammeter reads average value

$$\therefore \quad \text{Reading} = \sqrt{10^2 + \frac{10^2}{2}} = \sqrt{150}\text{A} \ \textbf{(Ans.)}$$

Q.101. In PMMC instrument, the central spring stiffness and the strength of the magnet decrease by 0.04% and 0.02% respectively due to a rise in temperature by 1°C. With a rise in temperature of 10°C, the instrument reading will

(*a*) increase by 0.2% (*b*) decrease by 0.2%

(*c*) increase by 0.6% (*d*) decrease by 0.6%.

Ans. (*a*)

EXPLANATION:

For 10°C spring stiffness K′ = K – 0.4% of K

$$K' = 0.996\text{ K}$$

For 10°C, New B' $\quad B' = B - 0.2\% \text{ of } B = 0.998\, B$

$$\frac{K\theta}{K'\theta'} = \frac{NBAI}{NB'AI}$$

$$\frac{\theta}{\theta'} = \frac{0.996\, K \times E}{K \times 0.998\, B}$$

$$\theta' = 1.002\theta$$

So, the instrument reading will increases by 0.2% **(Ans.)**

Q.102. Standard cell

(*a*) will have precise and accurate constant voltage when current drawn from it is few microamperes only

(*b*) will have precise and accurate constant voltage when few milliamperes are drawn from it

(*c*) will continue to have constant voltage irrespective of loading conditions

(*d*) can supply voltages up to 10 V.

Ans. (*a*)

EXPLANATION:

(*b*) The makers specify the maximum value as 100 μA. This means that the current drawn from the cell should be less than 100 μA and this current should flow momentarily. **(Ans.)**

Q.103. A resistance strain gauge is fastened to a beam subjected to a strain of 1×10^{-6} yielding a resistance change of 240 μΩ. If the original resistance of the strain gauge is 120Ω, the gauge factor would be

(*a*) 5 (*b*) 2 (*c*) 1 (*d*) 0.2.

Ans. (*b*)

EXPLANATION:

$$\text{Gauge factor} = \frac{\frac{\Delta R}{R}}{\frac{\Delta L}{L}} = \frac{\frac{240\times10^{-6}}{120}}{1\times10^{-6}} = 2 \ \textbf{(Ans.)}$$

Q.104. A voltage of $\{200\sqrt{2}\sin 314t + 6\sqrt{2}\sin(942t + 30°) + 8\sqrt{2}\cos(1570t + 30°)\}$ V is given to a harmonic distortion meter. The meter will indicate a total harmonic distortion of approximately

(*a*) 4.99% (*b*) 6.99% (*c*) 7.99% (*d*) 8.99%.

Ans. (*a*)

EXPLANATION:

Total harmonic distortion

$$= \frac{\text{Total harmonics}}{\text{Total rms}} \times 100 = \frac{\sqrt{6^2 + 8^2}}{\sqrt{6^2 + 8^2 + 200^2}} \times 100$$

= 4.99% **(Ans.)**

Q.105. A spring controlled moving iron voltmeter draws a current of 1 mA for full scale value of 100 V. If it draws a current of 0.5 mA, the meter reading is

(*a*) 25 V (*b*) 50 V (*c*) 100 V (*d*) 200 V.

Ans. (*a*)

EXPLANATION:

In moving iron voltmeter, deflection $\alpha\ I^2$

$$\frac{100}{V_1} = \left(\frac{1}{0.5}\right)^2, \text{ or } V_1 = 25 \text{ V } \textbf{(Ans.)}$$

Q.106. In the measurement of power on balanced load by two-wattmeter method in a 3-phase circuit, the readings of the wattmeters are 3kW and 1 kW respectively, the latter being obtained after reversing the connections to the current coil. The power factor of the load is

(*a*) 0.277 (*b*) 0.554 (*c*) 0.625 (*d*) 0.866.

Ans. (*a*)

EXPLANATION:

$$\tan\phi = \sqrt{3}\left(\frac{W_1 - W_2}{W_1 + W_2}\right)$$

$$= \sqrt{3}\left[\frac{3-(-1)}{3-1}\right] = 3.464$$

or, $\phi = 73.9°$

Hence, power factor, $\cos\phi = \cos 73.9° = 0.277$ **(Ans.)**

Q.107. A CRO screen has ten divisions on the horizontal scale. If a voltage signal 5 sin (314 t + 45°) is examined with a line base setting of 5 m sec/div, the number of cycles of signal displayed on the screen will be

(*a*) 0.5 cycles (*b*) 2.5 cycles (*c*) 5 cycles (*d*) 10 cycles.

Ans. (*b*)

EXPLANATION:

Line base setting = 5 m sec./div.

For 10 division line base setting = 10 × 5 = 50 m sec

$\omega = 314$ ...(given)

or, $2\pi f = 314$ or $f = \frac{314}{2\times\pi} = 50$ Hz

∴ No. of cycles = $50 \times 10^{-3} \times 50 = 2.5$ cycles **(Ans.)**

Q.108. A metal strain gauge has gauge factor of 2. Its nominal resistance is 120 ohms. If it undergoes a strain of 10^{-5}, the value of change of resistance in response to the strain is

(*a*) 240 ohms (*b*) 2×10^{-5} ohms

(*c*) 2.4×10^{-3} ohms (*d*) 1.2×10^{-3} ohms.

Ans. (*c*)

Explanation:

$$\text{Gauge factor} = \frac{\frac{\Delta R}{R}}{\frac{\Delta L}{L}}$$

or,
$$2 = \frac{\frac{\Delta R}{120}}{10^{-5}}$$

$\therefore$
$$\Delta R = 2 \times 10^{-5} \times 120 = 2.4 \times 10^{-3}\ \Omega \textbf{ (Ans.)}$$

Q.109. A 0-10 mA PMMC ammeter reads 4 mA in a circuit. Its bottom control spring snaps suddenly. The meter will now read nearly

(*a*) 10 mA (*b*) 8 mA (*c*) 2 mA (*d*) Zero.

Ans. (*a*)

Explanation:

When control spring snaps it gives full scale deflection, because there is no controlling torque present. **(Ans.)**

Q.110. A Lissajous pattern, as shown in Fig. 32 is observed on the screen of a CRO when voltages of frequencies f_x and f_y are applied to the x and y plates respectively. $f_x : f_y$ is then equal to

(*a*) 3 : 2 (*b*) 1 : 2

(*c*) 2 : 3 (*d*) 2 : 1.

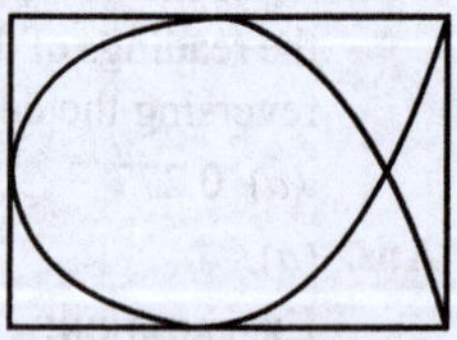

Fig. 32

Ans. (*c*)

Explanation:

$$\frac{f_x}{f_y} = \frac{\text{No. of vertical tangents}}{\text{No. of horizontal tangents}}$$

$$\frac{f_x}{f_y} = \frac{1}{1.5} = \frac{2}{3} \text{ i.e., } f_x : f_y = 2:3 \textbf{ (Ans.)}$$

Q.111. In an energy meter disc makes 10 revolutions in 100 seconds when a load 450 W is connected to it, the meter constant (in rev/kWh) is

(*a*) 1000 (*b*) 500 (*c*) 1600 (*d*) 800.

Ans. (*d*)

Explanation:

$$\text{Revolution per kW} = \frac{10}{450\times10^{-3}} = \frac{10^4}{450}$$

$$\text{Revolution per kW} = \frac{10^4\times3600}{450\times100} = 800 \textbf{ (Ans.)}$$

Q.112. The minimum number of wattmeter (s) required to measure 3-phase, 3-wire balanced and unbalanced power is

(*a*) 1 (*b*) 2 (*c*) 3 (*d*) 4.

Ans. (*b*)

EXPLANATION:

Two wattmeter method will be used to measure 3-phase, 3-wire balanced and unbalanced power **(Ans.)**

Q.113. A 150 mA meter has accuracy of ±2 percent. Its accuracy while reading 75 mA will be

(*a*) ±1% (*b*) ±2% (*c*) ±4% (*d*) ±20%.

Ans. (*c*)

EXPLANATION:

$$\text{Error while reading 150 mA} = \frac{2}{100}\times 150$$

$$= 3 \text{ mA}$$

∴ Accuracy while reading 75 mA

$$= \frac{3}{75}\times 100 = \pm 4\% \textbf{ (Ans.)}$$

Q.114. A resistor of 10 kΩ with 5% tolerance is connected in series with a 5 kΩ resistor of 10% tolerance. What is the tolerance limit of the series network?

(*a*) 5% (*b*) 6.67% (*c*) 10% (*d*) 8.33%.

Ans. (*b*)

EXPLANATION:

$$\text{Tolerance for 10 k}\Omega = \frac{5}{100}\times 10 \text{ k}\Omega = 0.5 \text{ k}\Omega$$

$$\text{Tolerance for 5 k}\Omega = \frac{10}{100}\times 5 \text{ k}\Omega = 0.5 \text{ k}\Omega$$

$$\text{Total tolerance} = 0.5 + 0.5 = 1 \text{ k}\Omega$$

$$\text{Total resistance in series} = 10 + 5 = 15 \text{ k}\Omega$$

$$\therefore \quad \text{Tolerance limit} = \frac{1}{15}\times 100 = 6.67\% \textbf{ (Ans.)}$$

Q.115. Two voltmeters have the same range 400 V. The internal impedances are 30,000 ohms and 20,000 ohms. If they are connected in series and 600 V be applied across them, the readings are

(*a*) 360 V and 240 V (*b*) 300 V and 300 V

(*c*) 400 V and 200 V (*d*) None of these.

Ans. (*a*)

EXPLANATION:

$$\text{Total impedance} = 30000 + 20000$$

$$= 50000 \ \Omega$$

$$\therefore \quad \text{Current through them} = \frac{600}{50000} = 0.012 \text{ A}$$

Voltage across 1st voltmeter = 0.012 × 30000 = 360 V **(Ans.)**

Voltage across IInd voltmeter = 0.012 × 20000 = 240 V **(Ans.)**

Q.116. A meter has a full-scale angle of 90° at a current of 1 A. This meter has perfect square-law response. What is the current when the deflection angle is 45°?

(*a*) 0.5 A (*b*) 0.65 A (*c*) 0.707 A (*d*) 0.87 A.

Ans. (*c*)

EXPLANATION:

$\theta \propto I^2, \quad \therefore \quad \frac{\theta_1}{\theta_2} = \frac{I_1^2}{I_2^2}$

or, $\frac{\frac{\pi}{2}}{\frac{\pi}{4}} = \frac{(1)^2}{I_2^2}$ or $I_2 = 0.707$ A **(Ans.)**

Q.117. Which of the following electronic instruments (or equipment) can be used to measure correctly the fundamental frequency component of a waveform and its higher harmonics?

1. Cathode ray oscilloscope
2. Vacuum tube voltmeter
3. Spectrum analyzer
4. Distortion factor meter.

Select the correct answer using the codes given below

(*a*) 1 and 2 (*b*) 2 and 3 (*c*) 3 and 4 (*d*) 1 and 4.

Ans. (*c*)

EXPLANATION:

Both spectrum analyzer and distortion factor meter are signal analysers. A spectrum analyser sweeps the signal frequency band and displays a plot of amplitude versus frequency. Distortion factor meter tunes out the fundamental signal and gives an indication of the harmonics. **(Ans.)**

Q.118. A dc circuit can be represented by an internal voltage source of 50 V with an output resistance of 100 kΩ. In order to achieve accuracy better than 99% for voltage measurement across its terminals, the voltage measuring device should have a resistance of at least

(*a*) 10 MΩ (*b*) 1 MΩ (*c*) 10 kΩ (*d*) 1 kΩ.

Ans. (*a*)

EXPLANATION:

$$\frac{E_L}{E_0} = \frac{99}{100} = \frac{1}{1+\frac{100\,k\Omega}{Z_L}} \text{ or } \frac{100}{99} = 1 + \frac{100\,k\Omega}{Z_L} \text{ or } Z_L \approx 10\text{ M}\Omega \textbf{ (Ans.)}$$

Q.119. An analogue voltage signal whose highest significant frequency is 1 kHz is to be coded with a resolution of 0.01 per cent for a voltage range of 0–10 V. The minimum sampling frequency and the minimum number of bits should respectively be

(*a*) 1 kHz and 12 (*b*) 1 kHz and 14 (*c*) 2 kHz and 12 (*d*) 2 kHz and 14.

Ans. (*d*)

EXPLANATION:

Minimum sampling frequency = Nyquist sampling rate = 2 kHz, and

$$\text{Resolution, } 0.01 = \frac{1}{2^n - 1} \times 100$$

or, $$2^n - 1 = \frac{100}{0.01} = 10{,}000$$

or, $$2^n = 10001$$

or, $$n\ln(2) = \ln(10001)$$

or, $$n = \frac{\ln(10001)}{\ln(2)} \simeq 14 \textbf{ (Ans.)}$$

Q.120. The total current $I = I_1 + I_2$ in a circuit is measured as $I_1 = 150 \pm 1$ A, $I_2 = 250 \pm 2$ A, where the limits of error are given as standard deviations. I is measured as

(*a*) (400 ± 3) A (*b*) (400 ± 2.24) A (*c*) (400 ± 1/5) A (*d*) (400 ± 1) A.

Ans. (*b*)

EXPLANATION:

$$\sigma_1 = \sqrt{\left(\frac{\partial I}{\partial I_1}\right)^2 \sigma_1^2 + \left(\frac{\partial I}{\partial I_2}\right)^2 \sigma_{I_2}^2}$$

$$\frac{\partial I}{\partial I_1} = \frac{\partial I}{\partial I_2} = 1$$

$$\therefore \quad \sigma_1 = \sqrt{(1)^2(1)^2 + (1)^2(2)^2} = 2.24 \text{ A}$$

$$\therefore \quad I = (400 \pm 2.24) \text{ A. } \textbf{(Ans.)}$$

Q.121. A spring-controlled moving iron voltmeter draws a current of 1 mA for full scale value of 100 V. If it draws a current of 0.5 mA, the meter reading is

(*a*) 25 V (*b*) 50 V (*c*) 100 V (*d*) 200 V.

Ans. (*a*)

EXPLANATION:

$$\theta \;\alpha\; I^2 \quad \text{or} \quad \frac{100}{\theta} = \left(\frac{1}{0.5}\right)^2$$

or, $$\theta = 25 \text{ V } \textbf{(Ans.)}$$

Q.122. A 1 cm piezoelectric transducer having a g-coefficient of 58 V/kg/m² is subjected to a constant pressure of 10^{-3} kg/m² for about 15 minutes. The piezoelectric voltage developed by the transducer will be

(*a*) 116 mV (*b*) 58 mV (*c*) 29 mV (*d*) 0 mV.

Ans. (*b*)

EXPLANATION:

$$V = g.p = 58 \times 10^{-3} = 58 \text{ mV } \textbf{(Ans.)}$$

Q.123. Measurement of flow, thermal conductivity and liquid level using thermistors make use of

(*a*) Resistance decrease with temperature (*b*) Resistance increase with temperature

(*c*) Self-heating phenomenon (*d*) Change of resistivity.

Ans. (*a*)

EXPLANATION:

Thermistors have *negative* resistance temperature coefficient. **(Ans.)**

Q.124. A set of independent current measurements taken by four observers was recorded as: 117.02 mA, 117.11 mA, 117.08 mA and 117.03 mA. What is the range of error?

(*a*) ±0.045 mA (*b*) ±0.054 mA (*c*) ±0.065 mA (*d*) ±0.056 mA.

Ans. (*a*)

EXPLANATION:

$$\text{Average current} = \frac{117.02 + 117.11 + 117.08 + 117.03}{4}$$

$$= 117.06 \text{ mA}$$

As, $$I_{max} = 117.11 \text{ mA} \;\&\; I_{min} = 117.02 \text{ mA}$$

$$\therefore \quad \text{Range of error} = \pm\frac{(I_{max} - I_{av}) + (I_{av} - I_{min})}{2} = \pm\frac{0.05 + 0.04}{2}$$

$$= \pm 0.045 \text{ mA } \textbf{(Ans.)}$$

Q.125. A 12 bit A/D converter has a range 0–10 V. What is the approximate resolution of the converter?

(*a*) 1 mV (*b*) 2.5 mV (*c*) 2.5 μV (*d*) 12 mV.

Ans. (*b*)

EXPLANATION:

$$\text{Approximate resolution} = \frac{10}{(2^{12} - 1)} = 2.5 \text{ mV } \textbf{(Ans.)}$$

Q.126. A moving-coil instrument gives full-scale deflection for 1 mA and has a resistance of 5 Ω. If a resistance of 0.55 Ω is connected in parallel to the instrument, what is the maximum value of current it can measure?

(*a*) 5 mA (*b*) 10 mA (*c*) 50 mA (*d*) 100 mA.

Ans. (*b*)

EXPLANATION:

$$V = 5 \times 1 = 5 \text{ mV}$$

$$\text{New resistance} = \frac{5 \times 0.55}{(5 + 0.55)} = \frac{55}{111}\Omega$$

$$\therefore \quad I_{max} = \frac{5 \times 111}{55} \approx 10 \text{ mA } \textbf{(Ans.)}$$

Q.127. A single slide wire is used for the measurement of current in a circuit. The voltage drop across a standard resistance of 1.0 Ω is balanced at 70 cm. What is the magnitude of the current if the standard cell having an e.m.f. of 1.45 volts is balanced at 50 cm?

(*a*) 3.09 A (*b*) 2.65 A (*c*) 2.03 A (*d*) 1.45 A.

Ans. (*c*)

EXPLANATION:

Voltage drop per unit length = 1.45/50 = 0.029 V/cm

Voltage drop across 70 cm. length = 0.029 × 70 = 2.03 V

$$\therefore \text{ Current through resistor} = \frac{2.03}{1\,\Omega} = 2.03 \text{ A } \textbf{(Ans.)}$$

Q.128. What is the series resistance required to extend the 0–100 V range of a 20000 Ω/V meter to 0–1000 V?

(*a*) 10 MΩ (*b*) 16 MΩ (*c*) 18 MΩ (*d*) 20 MΩ.

Ans. (*c*)

EXPLANATION:

Initial resistance = 20000 × 100 = 2 MΩ

Final resistance = 20000 × 1000 = 20 MΩ

∴ Series resistance = 20 – 2 = 18 MΩ. **(Ans.)**

Q.129. Which of the following can be used/modified for measurement of angular speed?

1. LVDT 2. Magnetic pick-up
3. Tacho-generator 4. Strain gauge

Select the correct answer using the code given below

(*a*) Only 1 and 2 (*b*) Only 2 and 3

(*c*) Only 3 (*d*) Only 2, 3 and 4.

Ans. (*b*)

EXPLANATION:

LVDT and strain gauge measure linear displacement, hence (*b*) is the correct answer **(Ans.)**

Q.130. The meter constant of a signal-phase, 240 V induction wattmeter is 400 revolution per kWh. The speed of the meter disc for a current of 10 amps of 0.8 p.f. lagging will be

(*a*) 12.8 rpm (*b*) 16.02 rpm (*c*) 18.2 rpm (*d*) 21.1 rpm.

Ans. (*a*)

EXPLANATION:

$$\text{Power} = \frac{240 \times 10 \times 0.8}{1000} = 1.92 \text{ kW}$$

Required speed of the meter disc

$$= \frac{400}{60} \times 1.92 = 12.8 \text{ rpm } \textbf{(Ans.)}$$

Q.131. A 10-bit A/D converter is used to digitise an analog signal in the 0 to 5 range. The maximum peak-to-peak ripple voltage that can be allowed in the d.c. voltage is

(*a*) nearly 100 mV (*b*) nearly 50 mV (*c*) nearly 25 mV (*d*) nearly 5 mV.

Ans. (*d*)

EXPLANATION:

The smallest incremental change is

$$= \frac{1}{12^{10}} = \frac{1}{1024}$$

The maximum peak-to-peak voltage that can be allowed in d.c. voltage

$$= 5 \times \frac{1}{1024} \approx 5 \text{ mV } \textbf{(Ans.)}$$

Q.132. The sensitivity of voltmeter using 0 to 5 mA meter movement is

(*a*) 50 Ω/volt (*b*) 100 Ω/volt (*c*) 200 Ω/volt (*d*) 500 Ω/volt.

Ans. (*c*)

EXPLANATION:

$$\text{Sensitivity} = \frac{1}{5 \text{ mA}} = 200 \ \Omega/\text{V } \textbf{(Ans.)}$$

Q.133. Two millimeters with a full scale current of 1 mA and 10 mA are connected in parallel and they read 0.5 mA and 2.5 mA respectively. Their internal resistances are in the ratio of

(*a*) 1 : 10 (*b*) 10 : 1 (*c*) 1 : 5 (*d*) 5 : 1.

Ans. (*d*)

EXPLANATION:

$$I \propto \frac{1}{R}$$

$$\therefore \quad \frac{R_1}{R_2} = \frac{I_2}{I_1} = \frac{2.5}{0.5} = \frac{5}{1} \textbf{ (Ans.)}$$

Q.134. The full scale deflection current of a meter is 1 mA and its internal resistance is 10 Ω. This meter is to have full deflection when 100 V is measured. What is the value of series resistor to be used?

(*a*) 99.99 kΩ (*b*) 100 kΩ (*c*) 99.99 Ω (*d*) 100 Ω.

Ans. (*a*)

EXPLANATION:

$$\text{Series resistor} = \left[\frac{100}{1mA} - \frac{10}{1000}k\Omega\right] = \left(100\ k\Omega - 1\times\frac{1}{100}k\Omega\right) = 99.99\ k\Omega\ \textbf{(Ans.)}$$

Q.135. A bridge type rectifier meter, and a thermocouple meter employ moving coil movement for indication. Both are calibrated on a 100 Hz sinusoidal waveform. If a 100 Hz rectangular waveform is applied to each, the ratio of their readings will be

(*a*) 2 : 1 (*b*) 1.11 : 1 (*c*) 1.41 : 1 (*d*) 1.21 : 1.

Ans. (*b*)

EXPLANATION:

In bridge type rectifier meter, reading is in $V_{rms} = 1.11\ V_{av}$

In thermocouple meter for square wave, reading is in $V_{rms} = V_{av}$

∴ Ratio = 1.11 : 1 **(Ans.)**

Q.136. Pulses from the clock of frequency 100 kHz pass through the counter of digital multimeter during a gate period 5.75 m sec. The number of pulses counted by the counter will be

(*a*) 57500 pulses (*b*) 5750 pulses (*c*) 575 pulses (*d*) 57.5 pulses.

Ans. (*c*)

EXPLANATION:

No of pulses = 5.75 × 100 = 575 pulses **(Ans.)**

Q.137. In a CRT 3×10^{17} electrons are accelerated through a potential difference of 10,000 V over a distance of 40 mm per minute. The average power supplied to the beam of electrons is

(*a*) 2 W (*b*) 4 W (*c*) 6 W (*d*) 8 W.

Ans. (*d*)

EXPLANATION:

Total energy supplied by the source in one minute,

$$W = 3 \times 10^{17} \times 1.6 \times 10^{-19} \times 10000$$
$$= 4.8 \times 10^{2}\ \text{J} = 480\ \text{J}$$

∴ Average power supplied to the beam = 480/60 = 8 W **(Ans.)**

Q.138. Calculate the maximum velocity of the beam of electrons in a CRT having a cathode anode voltage of 1000 V. Assume the electrons to leave the cathode with zero velocity. Charge of electron = 1.6×10^{-19}C and mass of electron = 9.1×10^{-31} kg.

(*a*) 0.1875×10^{6} m/s (*b*) 0.1875×10^{7} m/s

(*c*) 0.1875×10^{8} m/s (*d*) 0.1875×10^{9}. m/s.

Ans. (*c*)

EXPLANATION:

Work done by the electric field = eV

or, Kinetic energy = $\frac{1}{2}mv^2$ = eV

or, $$v = \sqrt{\frac{2eV}{m}} = \sqrt{\frac{2\times1.6\times10^{-19}\times1000}{9.1\times10^{-31}}} = 0.1875 \times 10^{8}\ \text{m/s}\ \textbf{(Ans.)}$$

Q.139. A zero to 300 V voltmeter has a guaranteed accuracy of 1% full scale reading. The voltage measured by the instrument is 83 V. The per cent limiting error is

(*a*) 0.95 (*b*) 1.81 (*c*) 3.61 (*d*) 4.85.

Ans. (*c*)

EXPLANATION:

1% accuracy means that a maximum possible error of $\frac{300\times1}{100} = 3$ V may be present in any reading. Since the deflection is 83V, the present limiting error is $\frac{3}{83}\times 100 = 3.61$ **(Ans.)**

Q.140. A utility type voltmeter with an accuracy of ±3% of full scale (at 25°C) is used on 300 V scale to measure 230 V (*a*) What is the possible percentage error ? (*b*) What range will the actual voltage fall within if the instrument reads 200 V?

(*a*) 3.9%, 200 V (*b*) 3.9%, 191 – 209 V

(*c*) 7.6%, 221–239 V (*d*) 7.6%, 200 V.

Ans. (*b*)

EXPLANATION:

$$\text{Error} = \frac{3}{100}\times 300 = 9\text{ V}$$

Possible % error to measure 230 V = $\frac{9}{230}\times 100 = 3.9\%$ **(Ans.)**

Range = 200 ± 9 = 191 – 209 V **(Ans.)**

Q.141. In a dual slope integrating type digital voltmeter the first integration is carried out for 10 periods of the supply frequency of 50 Hz. If the reference voltage used is 2 V the total conversion time for an input of 1 V is

(*a*) 0.01 sec (*b*) 0.05 sec (*c*) 0.1 sec (*d*) 1 sec.

Ans. (*c*)

EXPLANATION:

In a dual slope integrating type digital voltmeter,

$$V_{in} = V_{ref}\,(t_2/t_1)$$

where, t_1 is the first integration time.

Now, $t = 10 \times 1/50 = 0.2$ sec

$V_{ref} = 2$ volts

$$\therefore \quad t_2 = V_{in}\times\frac{t_1}{V_{ref}}$$

$$= 1\times\frac{0.2}{2} = 0.1 \text{ sec } \textbf{(Ans.)}$$

Q.142. When testing a coil having a resistance of 10 Ω, resonance occurred when the oscillator frequency was 10 MHz and the rotating capacitor was set at 500/2πpF. The effective value of the Q of the coil is

(*a*) 20 (*b*) 254 (*c*) 314 (*d*) 542.

Ans. (*a*)

EXPLANATION:

$$Q = \frac{L\omega_0}{R} = \frac{1}{C\omega_0 R}$$

$$= \frac{1}{(500/2\pi)\times10^{-12}\times2\pi\times10^{7}\times10} = 20 \textbf{ (Ans.)}$$

INDEX

A

B

C

D

F

G

H

I